E. Schiller

Free Radicals and Inhalation Pathology

Springer-Verlag Berlin Heidelberg GmbH

Erich Schiller

Free Radicals and Inhalation Pathology

Respiratory System
Mononuclear Phagocyte System
Hypoxia and Reoxygenation
Pneumoconioses and other Granulomatoses
Cancer

Forewords by Wilhelm Kriz and Helmut Bartsch

With 350 Figures, Some in Colour, and 58 Tables

Springer

Dr. Erich Schiller
Am Hirschgraben 8

63150 Heusenstamm
Germany

Additional material to this book can be downloaded from http://extras.springer.com.

ISBN 978-3-642-62201-4 ISBN 978-3-642-18619-6 (eBook)
DOI 10.1007/978-3-642-18619-6

Library of Congress Cataloging-in-Publication Data
Schiller, Erich, 1922 – Free radicals and inhalation pathology : respiratory system, mononuclear phagocyte system : hypoxia and reoxygenation : pneumoconioses and other granulomatoses : cancer/ Erich Schiller p. ; cm. Includes bibliographical references and index.
ISBN 978-3-642-62201-4
1. Lungs-Pathophysiology. 2. Active oxygen-Pathophysiology. 3. Active nitrogen-Pathophysiology. 4. Free radicals (Chemistry)-Pathophysiology. 5. Granuloma.
I. Title. [DNLM: 1. Lung Diseases-etiology. 2. Free Radicals-adverse effects. 3. Lung-physiology. WF 600 S355f 2003]
RC756.S34 2003 616.2'407-dc21 2003045745

http://www.springer.de

Originally published by Springer-Verlag Berlin Heidelberg New York in 2004
Softcover reprint of the hardcover 1st edition 2004

Cover-Design: design & production GmbH, Heidelberg
Typsetting: Mitterweger & Partner, Plankstadt

Printed on acid free paper 24/3150 hs – 5 4 3 2 1 0

Forewords

This book is unique. It is written by a distinctive author who grew up as the son of an astronomer right in the middle of the observatory in Leipzig. In 1940, he became a student of Max Clara and Wolfgang Bargmann, both of whom evoked an enormous enthusiasm for histobiology in the young student, causing him to pursue this science over the course of his life.

Anatomy perceived as a dynamic science, both in its functional orientation and experimental approach, has molded the author into a multifaceted scientist. In the present work, the author successfully combines morphology, biochemistry, and physiology, providing the basis for a comprehensive understanding of the normal and abnormal structure and function of the lung.

Although many people suffer from pulmonary diseases in modern industrial countries, pneumonology as a whole is a neglected science in Germany. This book hopes to stimulate pulmonary research in Germany and to lead to a new impetus of progress in lung disease treatment. The book centers around oxygen and nitrogen in their reactive species; as superoxide and hydroxide radicals and nitrogen oxide radicals, these compounds are Janus-faced. One the one hand, these compounds are important mediators of pulmonary function, on the other hand, their damaging capabilities decisively contribute to many pulmonary diseases. This book describes their formation, reactivity, suppression, and detoxification and illuminates the ways to avoid their damaging effects in order to maintain health and longevity.

The book is an admirable opus that I am happy to most warmly commend.

Wilhelm Kriz
Professor of Anatomy
and Cell Biology

When about 50 years ago the author together with the internist, Günther Worth published "Die Pneumokoniosen", this book had become a well-known standard for more than twenty years in Germany, until in 1976 the Handbook of internal medicine toock up pulmonary industrial diseases to form an individual volume. Meanwhile, air pollution has grown to a worldwide problem outside the industrial setting where everywhere and everybody may be exposed by inhalation to harmful dusts, vapours, and gases. Basic scientific research has meanwhile revealed that free radicals are not only a matter of chemistry as they were a century ago at Fenton's times, but are recognised now as foe and fried in biochemistry and histobiology. Crushing of mineral dusts produces unpaired electrons on the surface of the particles and phagocytosis of any particle involves the production of reactive oxygen species. As a student of Max Clara in Leipzig E. Schiller became very interested in the histobiology of the cells of the bronchial tree named Clara cells. First under the light microscope and then by transmission electron microscopy, he showed remarkable changes to occur in the number, size and structure of the bronchiolar epithelial cells of the rat after dust inhalation and drug therapy. In cardiovascular pharmacology, the author learned to fight against cardiac and cerebral hypoxia where reoxygenation produced reactive oxygen species too. The introduction of the sydnonimines as nitric oxide donors into therapy of ischaemic heart disease opened to hom the field of the ultrastructural analysis of the hypoxic, cold/immobilization- and isoproterenol-stressed rat myocardium. After he retired from industry the author re-entered basic pneumoconiosis research; together with the late K. D. Friedberg at the Institute of Pharmacology, Faculty of Clinical Medicine, Mannheim, University of Heidelberg he wrote the chapter "Silicon" in the "Handbook on Toxicology of Inorganic Compounds". With J. W. Zeller from the German Cancer Research Center he became engaged in quantitative research on free radical by luminescence analysis of alveolar macrophages *in vitro*.

The comprehensive treatise that E. Schiller has now completed after many years of meticulous work is a concise summary of a complex field and moreover a unique source of about 25,000 relevant literature citations which are brought up-to-date. Also the introduction, describing the historical development of the field, is woth reading as it is superbly illustrated by photographs of leading personalities together with facsimiles of the original cover pages of books and articles from medieval times. This volume would not only be of interest to students in life sciences but is highly informative also for clinicians, toxicologists and scientists working in the field of occupational medicine and related health services. The readability of the text in each of the 21 chapters is greatly facilitated by the high quality of many superb graphical illustrations and micrographs and the clear description of important mechanistic aspects of free radical pathogenesis. This multifaceted field has lately gained an immense importance in understanding chronic degenerative diseases, including aging, to which a chapter in this volume is dedicated.

Helmut Bartsch, Ph.D.
Professor of Toxicology,
University of Heidelberg

Preface

When I started doing experimental pneumoconiosis research more than 50 years ago, Selye's concept of stress was the leading explanation of pathogenesis of connective tissue and of vascular disease. This led me to attempt experimental prophylaxis by administering antistress hormones to rats and mice injected with particulate matter. Once my field of research changed from pulmonary to cardiovascular and cerebral hypoxic states, I aimed my experiments at oxidative stress during reperfusion following hypoxia. In addition to electron microscopy and histochemistry, I also made use of biophysics (a field that my younger daughter is qualified to lecture on) to study photoemission from reactive species arising from crushing quartz particles and engulfing particulate matter by mononuclear phagocytes. The question of carcinogenesis by respirable industrial waste – particles and vapours – led to the topics of oxidative DNA lesions, chromosomal breaks and mutagenicity.

I wish to gratefully acknowledge all those who have helped in experimental work and in the preparation of this book: Prof. H. Brettschneider (Essen), Prof. L. Vollrath (Mainz) and Prof. U. Bleyl (Mannheim), who provided me with access to their electron microscopy facilities and much good advice. I owe thanks to Prof. W.J. Zeller (German Cancer Research Centre, Heidelberg) for enabling me to do biophysical studies on photoemission by reactive species. In particular, I wish to express my appreciation for the kindness, assistance and patience manifested by the staff of the Library of the German Cancer Research Centre while I was gathering the data for this volume. Furthermore, I wish to thank Prof. W. Kriz and Prof. H. Bartsch (University of Heidelberg) for the kind forewords they penned for this monograph. And finally, let me extend my special thanks to Springer-Verlag for the excellent layout of this book and for the presentation of the voluminous bibliography on a compact disk, allowing for a literature search by either author or subject.

Erich Schiller

Contents

PART I

History and Scientific Background

CHAPTER 1

Introduction

In spite of his intelligence man exposes himself to industrial hazards. The understanding of disease and its provocation has not prevented him from endangering himself to some extent in his behaviour and activities (ERICH MÜLLER 1969, p. 92). There is not any gradual evolution from the animals' instinct to man's intellect, but, according to ARNOLD GEHLEN (1950), just a tendency to their mutual exclusion. "The individual, however, who owns nothing but his labor power, has little choice. Since he is obliged to make a living, he must accept work wherever the labor market can absorb him, whether this work is dangerous or not" (SIGERIST 1943). "The incidence of cancer in middle and old age can, in principle, be reduced by 80%–90% and the risk worldwide could be halved, although the methods required are not always socially acceptable" (DOLL 1998).

In industrialised nations, more than 20% of all severe diseases concern the respiratory organs (FERLINZ 1983). Bronchitis (Table 1), emphysema (Table 2) and pulmonary carcinoma (Table 3) are found to be caused in whole or in part by environmental pollution (TOOMEY and PETERSILGE 1944, PEMBERTON and GOLDBERG 1954, PEMBERTON 1961, FOURNIER 1964, UDO SMIDT 1983, TOMATIS 1990, PLESS-MULLOLI et al. 1998).

Cigarette smoking is associated with airway inflammation and the development of chronic obstructive pulmonary disease (COPD). A variety of cells appear to be involved in this inflammatory process including neutrophils (1), eosinophils (2), mononuclear phagocytes (3), and T-lymphocytes (4). Cigarette smoke causes release of interleukin-8 from epithelial cells in dependence on epidermal growth factor activation, and the autocrine production of transforming growth factor-α makes a substantial contribution to this response (RICHTER et al. 2002). In male A/J *mice*, a Bowman-Birk protease inhibitor concentrate, while being effective against 3-methylcholanthrene, failed to modulate tobacco smoke-induced lung-tumour development (WITSCHI and ESPIRITU 2002).

Adult Sprague-Dawley *rats* exposed to concentrated **ambient particle aerosols** for 5 h showed significant oxidative stress, determined as in situ chemiluminescence in the lungs and heart, but not liver (GURGUEIRA et al. 2002). Increases in oxidant levels were also triggered by highly toxic residual oil fly ash particles but not by particle-free air or by inert carbon black aerosols (control particles). Increases in chemiluminescence showed strong association with the content of iron, manganese, copper, and zinc in the lung and with iron, aluminium, silicon, and titanium in the heart.

Residual oil fly ash (ROFA) is an emission-source air pollutant which is generated from the burning of low sulphur residual oil. ROFA is composed of metals, sulphates, acids, and unknown materials, which are complexed to an insoluble, "inert" particulate core (PRITCHARD et al. 1996, DREHER et al. 1997). ROFA is an industrial air pollutant, the tracheal instillation of which produced airways hyperresponsiveness and acute lung injury, consisting of

Table 1. Aetiology of chronic bronchitis

I. Exogenous factors
1. Smoking
2. Occupation (dust, nitrous gases, SO_2)
3. Air pollution/dwelling area (overcrowded areas)
4. Socio-economic status (?)
5. Climate (damp and cold)
II. Past illness
1. Acute bronchitis (viruses, SO_2, formaldehyde)
2. Pulmonary disease in childhood
3. Sinusitis
4. Disturbances of respiratory mechanics (thoracic deformities, hyperalimentosis)
5. Standard of living (alcohol, air conditioning, sprays)

Table 2. Aetiology of pulmonary emphysema

I. Exogenous factors
1. Smoking (chronic bronchitis → centrilobular emphysema)
2. Allergy (asthma bronchiale)
II. Endogenous factors
1. Proteolysis (α_1-antitrypsin deficiency → panlobular emphysema)

epithelial damage, pulmonary oedema, haemorrhage, and the influx of neutrophils, macrophages, and eosinophils into the pulmonary interstitium (Dreher et al. 1997, Dye et al. 1997, Kodavanti et al. 1997). ROFA exposure causes the release of inflammatory cytokines and reactive oxygen species, and apoptosis in cultures of *rodent* macrophages, whereas it produces an oxidative burst followed by apoptosis in *human* alveolar macrophages (Becker et al. 1996). In the BEAS-2B cell line derived from a *human* bronchial epithelial tumour cell, ROFA immediately (<30 s) increased intracellular calcium levels, increased key inflammatory cytokine transcripts within 2-h exposure, and subsequent release of interleukins-6 and -8 cytokine protein after 4-h exposure (Veronesi et al. 1999).

When alveolar macrophages and airway epithelial cells are directly exposed to inhaled atmospheric particles, these small particles are phagocytised by both cell types (Mukae et al. 2001). Several studies have shown that particulate matter with a diameter of $< 10\ \mu m$ stimulates the production of reactive oxygen species and inflammatory mediators by alveolar macrophages (Monn and Becker 1999, Mukae et al. 2000, Van Eeden et al. 2001) and airway epithelial cells (Carter et al. 1997, Mills et al. 1999, Fujii et al. 2001).

The oil fly ash particulates showed a significant dose-dependent increase in 2'-deoxyguanosine hydroxylation to 8-oxo-2'-deoxyguanosine formation over the control 2'-deoxyguanosine (Prahalad et al. 2001). Metal ion chelators and dimethylsulphoxide, a hydroxyl radical scavenger, inhibited this hydroxylation. In contrast, desert dust, coal fly ash, and urban air particles induced 8-oxo-2'-deoxyguanosine with yields ranging from 0.003 to 0.006%, respectively, with levels unaffected by pre-treatment of the particles with metal ion chelators or addition of dimethylsulphoxide to the incubation mixture.

In response to injury caused by various pollutants, toxicants, and carcinogens, proximal and distal airway epithelial cells lose their normal secretory functions and express squamous and keratinising properties (Plopper 1997).

Brochopulmonary carcinomas, thought to be generated by tobacco smoking and various occupation exposures, are the most common malignant tumours and are the major cause of cancer death (Vainio and Boffetta 1994, Steenland et al. 1996, Parkin et al. 1999). Although all or almost all of the current chemoprevention studies in the USA use serial bronchoscopic biopsies to evaluate response, there is no information available on the efficacy of this approach. While histopathological evaluation is the «gold standard», many studies utilise molecular or other biological endpoints. For these reasons, Wistuba and Gazdar (2000) undertook an evaluation of the size and frequency of the molecularly altered (allelic loss and microsatellite alterations) clonal patches in smoking damaged epithelium.

On the basis of chemical analysis of 256 lung specimens with carcinomas resected in Japan in the period 1991–1996, the concentration of 1-nitropyrene was 19.7 ± 10.5 pg/gdry weight, and that of the dinitropyrenes was 3.50 ± 0.12 to 6.26 ± 1.76 pg/g (Tokiwa et al. 1998). Aryl hydrocarbon hydroxylase biotransforming polycyclic **aromatic hydrocarbons** from tobacco smoke to the proper carcinogen (Korsgaard et al. 1983, Trell et al. 1984) was induced in *human* alveolar macrophages and peripheral lymphocytes (McLemore and Martin 1977, McLemore et al. 1977, 1978). Zeller and Schmähl (1985, p. 75) would like to pick out certain risk groups of smokers owing to an increased inducability of the enzyme. Smokers showed a significant decrease in complex IV activity of the mitochondrial respiratory chain, while the rest of the complexes of the mitochondrial respiratory chain were unaffected (Miró et al. 1999). Lipid peroxidation of lymphocyte membranes was increased in smokers compared to the non-smokers and this increase correlated positively with succinate oxidation activation and, to a lesser extent, with complex IV inhibition, although it did not reach statistical significance. This mitochondrial dysfunction could contribute to increased endogenous production of reactive oxygen species and could play a role in tobacco carcinogenicity. Nakachi et al. (1993) found an association of the susceptibility of a Japanese population to lung cancer in relation to cigarette dose and the polymorphism of the CYP1A1 and glutathione transferase genes. Cytochrome P_{450} localised in the bronchiolar Clara cells (Serabjit-Singh et al. 1980) will occupy us in its correlation to the respiratory burst and the activation of xenobiotics. Deilhaug et al. (1985) attributed the pathoklisis of Clara cells for some carcinogens not only with their cytochrome P_{450}-dependent monooxygenase system (Serabjit-Singh et al. 1980): their poor supply with enzymes-repairing DNA as O^6-alkylguanine-DNA alkyltransferase and uracil-DNA glycosylase and their function as stem cells for regenerative processes in the bronchiolar epithelium (Jeffery and Reid 1977) predestine the Clara cell for tumours induced by xenobiotics. *In vitro*, oxygen radicals significantly augmented the genotoxicity of tobacco-specific nitrosamines (Weitberg and Corvese 1990). In a multicentre study on lung cancer *patients* Bartsch et al. (1992) demonstrated a pronounced effect of tobacco smoke on pulmonary metabolism of xenobiotics and prooxidant

Table 3. Aetiology of pulmonary carcinoma

I. Exogenous factors

1. Smoking (Auerbach et al. 1957, 1961, 1979, Ochsner et al. 1960, Oettlé 1963, Ochsner 1973; cf. Eck et al. 1969, Zeller and Schmähl 1985, Vineis 1997)
2. Occupation

   Arsenic compounds in winegrowers (Bauer 1954, Roth 1956, 1957, 1958, Denk et al. 1969, Brain and Valberg 1979), employees formulating and packaging insecticides that contain arsenic (Ott et al. 1974, Chovil 1979), and smelters exposed to arsenic trioxide (Lee and Fraumeni Jr 1969, Dahlgen 1979)

   Asbestos (Gloyne 1935, Nordmann 1938, Linzbach and Wedler 1941, Boemke 1943, 1953, Werber 1952, Bohlig 1955, Chauvet and Feuardent 1955, Jacob and Bohlig 1955, Bohlig and Jacob 1956, 1958, Chauvet 1958, Böhme 1959, Kannerstein and Churg 1972, Whitwell et al. 1974, Brain and Valberg 1979, Chovil 1979, Beck and Schmidt 1984, Johnson et al. 1997, Kraus et al. 1997, McLean and Patel 1997, Hodgson and Darnton 2000)

   Beryllium compounds (Kahlau 1954, p.381, Hardy 1962, Niemöller 1963)

   Cadmium (Lemen et al. 1976, Takenaka et al. 1983, Stayner et al. 1992, Sorahan and Lancashire 1997)

   Chloromethyl ether (Figueroa et al. 1973, Sakabe 1973, Bettendorf 1976, 1977, Weiss and Figueroa 1976, Reznik et al. 1977, Chovil 1979, Weiss et al. 1979)

   Chromates (Lehmann 1932, Teleky 1936, Alwens and Jonas 1938, Alwens 1939, Gross and Koelsch 1943, Letterer 1944, Letterer et al. 1944, Macgle and Gregorius 1948, Baetjer 1950, Bidstrup 1951, Hueper 1951, Mancuso 1951, 1975, Imprescia 1952, Spannagel 1953, Kahlau 1954, Bidstrup and Case 1956, Rinck 1956, Grushko 1961, Enterline 1974, Ohsaki et al. 1974, 1978, Langìrd and Norseth 1975, Michel-Briand and Simonin 1977, Takemoto et al. 1977, Brain and Valberg 1979, Chovil 1979, Zober 1979, Langìrd et al. 1980, 1990, Abe et al. 1982, Sheffet et al. 1982, Tsuneata 1982, Brochard et al. 1983, Frenzel-Beyme 1983, Langìrd and Vigander 1983, Davies 1984, Watanabe and Fukuchi 1984, Becker et al. 1985, Kim et al. 1985, Nishiyama et al. 1985, Norseth 1986, Persson et al. 1986, De Marco et al. 1988, Hayes et al. 1989, Alcedo and Wetterhahn 1990, Langìrd 1990, 1993, Costa 1991, Standeven and Wetterhahn 1991, Popper et al. 1992, Triebig 1992, Sorahan et al. 1998, Sorahan and Harrington 2000)

   Coal and biomass indoor exposure (Kleinerman et al. 2002)

   Cobalt (Brain and Valberg 1979)

   Diesel exhaust (Garshick et al. 1987, Oberdörster et al. 1992, Lipsett et al. 1997)

   Ethylene oxide (Kirchhoff et al. 1999)

   Nickel compounds (Bidstrup 1950, Znamenskii 1963, Doll et al. 1970, 1977, Pedersen et al. 1973, Kreyberg 1978, Brain and Valberg 1979, Chovil 1979, Chovil et al. 1981, Ivankovic et al. 1988)

   Products of coal carbonisation (Lloyd 1971, Redmond et al. 1972)

   Quartz dust (Katabami et al. 2000, Latza and Baur 2000, Latza et al. 2000, Ulm et al. 2000)

   Radon (Pirchan and Šikl 1932, De Villiers and Windish 1964, Perraud et al. 1970, Lundin et al. 1971, Saccomanno et al. 1973, Ševc et al. 1976, Tomášek et al. 1994, Plaček et al. 1997, Brüske-Hohlfeld et al. 1998)

   Vinyl chloride (Waxweiler et al. 1976)

II. Endogenous factors

1. Lack of class μ isoenzymes of glutathione transferase (Seidegìrd et al. 1990)

state and suggested the existence of a metabolic phenotype at higher risk for tobacco-associated lung cancer. Superoxide generated by cigarette smoke damages the respiratory burst and induces physical changes in the membrane order and water organisation of *rat* polymorphonuclear granulocytes (Tsuchiya et al. 1992, 1993) and Jurkat T lymphocytes (Tsuchiya et al. 1993). Asbestos fibres and cigarette tar act in a co-operative or synergistic way in the generation of hydroxyl radical spin adducts (Valavanidis et al. 1996). Asbestos induced apoptosis of *human* and *rabbit* pleural mesothelial cells via reactive oxygen species (Broaddus et al. 1996). Incubated for ≤27 h in phosphate buffer, pH 7.5, chrysotile asbestos induced hydroxyl radical-mediated hydroxylation of deoxyguanosine about ten times more than did man-made rock wool and glass fibres (Leanderson et al. 1988). Exposure to cigarette smoke can increase apoptosis in the *rat* gastric mucosa through a reactive oxygen species-mediated and a p53-independent pathway (Wang et al. 2000). The apoptotic effect could be blocked by pre-treatment with a xanthine oxidase inhibitor (allopurinol, 20 mg/kg i.p.) or a hydroxyl radical scavenger (dimethyl sulphoxide). Neither of these treatments had any effect on p52 levels of the mucosa. Whereas high dietary β-carotene could inhibit the conversion of skin papillomas to carcinomas, such treatment would not be expected to inhibit smoke-induced lung tumours (Wolf 2002).

Cigarette smoking causes 87 % of lung cancer, 60 % of upper aerodigestive cancer, and 8 % of other cancers. In the last group, smoking is a recognised cause of cancer of the pancreas, bladder, kidney, liver and colon (IARC 1986, Doll 1996, Giovannucci 2001). Studies in *humans* demonstrated that tobacco constituents could reach breast tissue (Hecht 2002). The uptake and metabolic activation of mammary carcinogens such as polycyclic aromatic hydrocarbons and 4-aminobiphenyl are frequently higher in smokers than in non-smokers.

The debate on the interaction of asbestos and smoking in lung cancer has been concentrated on two hypotheses: additive (asbestos and cigarette smoke act independently) and multiplicative (asbestos produces an effect proportional to the effect of smoking). Although case-referent studies seemed to support the multiplicative hypothesis, the information from them is essentially unreliable. Thus it cannot weaken the conclusions from the cohort studies, that the multiplicative hypothesis is untenable and that the relative risk of lung cancer from asbestos exposure is about twice as high in non-smokers as in smokers (LIDDELL 2001). The absolute risks are, of course, substantially less in non-smokers than in smokers.

The variations in cancer incidence between different regions of the world give the impression that the majority of *human* cancers (around 80% according to HIGGINS 1969), are due to environmental factors. The relative lack of importance of genetic differences between races and populations can be deduced from the data on cancer incidence in migrants (KMET 1970), as the cancer morbidity of these populations changes over time and becomes similar to that of the country to which they migrate. The epidemiological data on cancer morbidity are vital in the studies on environmental carcinogenesis (GRICIUTE 1978). Our goal can only be the reduction of environmental carcinogens to unavoidable background levels (TOMATIS 1969).

Urban levels of air pollution (downtown São Paulo) modified the progression of urethane (3 g/kg)-induced lung tumours in *mice* (CURY et al. 2000). Urban particles consist of three modes: ultrafine particles, accumulation mode particles (which together form the fine particle mode) and coarse mode particles. Ultrafine particles (those of <100 nm diameter) contribute very little to the overall mass, but are very high in number, which in episodic events can reach several hundred thousand/cm^3 in the urban air. The hypothesis that ultrafine particles are causally involved in adverse responses seen in sensitive *humans* is based on several studies summarised by OBERDÖRSTER (2001). TIMBLIN et al. (2002) demonstrated the development of dose-related proliferation and apoptosis after exposure of an alveolar epithelial cell line (C10) to particulate matter or to ultrafine carbon black, a component of particulate matter. Ribonuclease protection assays demonstrated that increases in mRNA levels of the early response protooncogenes c-*jun*, *jun*B, *fra*-1, and *fra*-2 accompanied cell proliferation at low concentrations of particulate matter whereas apoptotic concentrations of particulate matter caused transient increases in expression of *fos* and *jun* family members and dose-responsive increases in mRNA levels of receptor-interacting protein, Fas-associated death domain, and caspase-8. Significant increases in steady-state mRNA levels of protooncogenes and apoptosis-associated genes, TNFR-associated death domain, and Fas were also observed after exposure of epithelial cells to ultrafine carbon black, but not fine carbon black or glass beads, respectively, suggesting that the ultrafine particulate component of particulate matter is critical to its biological activity.

Low-boiling ($<1600\,°C$) elements Pb; Ba; Y, Sr, Rb, As, and Zn can be concentrated several orders of magnitude higher in fly ash particles collected on a Millipore backup filter following a cascade impactor, than in those collected on the last impactor stage (SPARKS). The high-boiling elements Fe, Cu, and Ga did not show this effect. Scanning electron microscopic analyses (HULETT) of individual fly ash particles that had been etched with a beam of argon ions confirm that at least Ni, Cr, and Zn are considerably more concentrated on particle surfaces than in their interiors.

Using electron paramagnetic resonance, DELLINGER et al. (2001) examined samples of airborne fine particles with a mean aerodynamic diameter of less than 2.5 μm and found large quantities of radicals with characteristics similar to semiquinone radicals. Semiquinone radicals are known to undergo redox cycling and ultimately produce biologically damaging hydroxyl radicals. Aqueous extracts of these samples induced damage to DNA in *human* cells and supercoiled phage DNA. Superoxide dismutase, catalase, and deferoxamine abolished this DNA damage, implicating the superoxide radical, hydrogen peroxide, and the hydroxyl radical in the reactions inducing DNA damage.

Studies of large numbers of lung cancers have demonstrated different patterns of involvement between the two major groups of lung carcinomas, namely small cell lung carcinoma and non-small cell lung carcinoma (VIRMANI et al. 1998) and between the three major histologic types of lung carcinomas, small cell lung carcinoma, squamous cell carcinoma and adenocarcinoma (SHIVAPURKAR et al. 1999, WISTUBA et al. 1999, 2000). Thus, genetic abnormalities in lung cancer can be classified into two groups, those common to all lung cancers and those associated with a specific histologic type of lung cancer.

Mucosal changes in the large airways that may precede or accompany invasive squamous cell carcinoma include hyperplasia of basal and goblet cells, squamous metaplasia, squamous dysplasia and carcinoma in situ (SACCOMANNO et al. 1976). GERSING

(1956, p. 13) attributed the ability of quartz dust to perforate the bronchial mucous film and irritate the ciliated cells underneath to its hydrophilia. In the *rat*, hypertrophy and hyperplasia 6 months after an intracheal instillation of quartz dust were demonstrated by SCHILLER (1962). However, inert particles as elementary silicon also damaged the epithelium. Low concentrations of quartz (7–10 mg/m^3 for 7 h/d, 5 d/week) inhaled up to 60 days induced a prominent increase in bronchiolar Clara cells containing large vacuoles indenting the nucleus (DHOM and SAUER 1967). Short-time inhalations of quartz dust and of other cytotoxic modifications of silica considerably increased the mitosis rate in the alveolar region (STRECKER 1967). On the first days after quartz inhalation, the bronchial epithelium appeared to be irritated and more mitoses were evident than in control *rats* or after the inhalation of innocuous dusts. Pulmonary malignancies (epidermoid tumours and adenocarcinoma) in female specific pathogen-free Sprague-Dawley *rats* exposed to a mixture of coal + quartz dusts (200 mg/m^3) for up to 24 months (5 h/day, 5 days/week, every second week) were detected in 32 of 72 *rats*, while in 485 control *rats* of same age no lung tumour was observed (MARTIN et al. 1977).

Although chronic bronchitis is frequently found among miners, the relatively low frequency of bronchial cancer among Ruhr area coalminers makes it improbable that chronic bronchitis had a dominant effect on the development of bronchial carcinoma, and therefore bronchial carcinoma should not be regarded as an occupational disease among **anthracosilicosis** patients (SCHIMANSKI and ROSMANITH 1974).

The association between exposure to **asbestos** dust and the development of lung carcinoma and diffuse mesotheliomata of the pleura and peritoneum has been well documented (SELIKOFF and LEE 1978). Both crocidolite and amosite were capable of augmenting the oncogenic effect of benzo[a]pyrene. This putative synergistic effect was evident when fibres and chemicals were added to cultures of *murine* fibroblast cultures as simple mixtures and when benzo[a]pyrene was adsorbed to the surface of the fibres (BROWN et al. 1983).

Mesothelioma is expected to increase during the next 20 years (PETO et al. 1995), but 20 % of patients with malignant mesothelioma have no detectable evidence of asbestos exposure (BROCHARD et al. 1993, PAIRON et al. 1994). CICALA et al. (1993) and CARBONE et al. (1994) demonstrated the presence of SV40-like DNA in mesotheliomas and thus suggested a potential implication of SV40 in mesothelioma oncogenesis (CARBONE et al. 1994). At exposure levels seen in occupational cohorts HODGSON and DARNTON (2000) concluded that the exposure specific risk of mesothelioma from the three principal commercial asbestos types is broadly in the ratio 1:100:500 for chrysotile, amosite and crocidolite, respectively.

In keeping with the role of the **cytochromes P_{450}** as one of the body's main lines of defence against toxic foreign chemicals, the expression in normal tissues of the major xenobiotic-metabolising families of cytochromes P_{450} (cytochrome P_{450} families 1, 2, and 3) is characteristically in those tissues which are potentially exposed to environmental chemicals (the liver, lung, small intestine, and kidney). Cytochrome P_{450} expression has been identified in carcinomas developing in several different tissue including the lung (MCLEMORE et al. 1990). However, in many *human* tumours the level of cytochrome P_{450} enzymes, including CYP2B6 4-hydroxylating cyclophosphamide, is reduced (HAYES and WOLF 1990), but many of the activation reactions of anticancer drugs are dependent on the action of members of the cytochrome P_{450} gene superfamily (NEBERT et al. 1991). The important variations in lung cancer include polymorphism at the cytochrome P_{450} gene (CYP) loci and the glutathione *S*-transferases (GST) M1 gene cluster. Individuals with the rare combination CYP1A1*2A/*2A or *2A/*B and GSTM1*0/*0 showed significantly higher benzo[*a*]pyrene diol-epoxide DNA adduct levels (ROJAS et al. 1998). Therefore, combination of homozygous mutated CYP1A1 and GSTM1*0/*0 genotypes lead, at a similar or even lower smoking dose, to a stronger increase of *anti*-benzo[*a*]pyrene diol-epoxide DNA adduct levels than found in individuals with CYP1A1 and GSTM1 wild-type. CYP2D2 (debrisoquine 4-hydroxylase) bioactivates 4-(methylnitrosamino)-1-(3-pyridyl)-1-butanone as well as nicotine. The relationship between lung cancer risk and the CYP2D6 phenotype and/or genotype has been the subject of numerous studies (CAPORASO et al. 1995, CHRISTENSEN et al. 1997). CYP3A5, but not CYP3A4, is present constitutively (RAUNIO et al. 1999). The presence of the mutated MSP fragment in the CYP1A1 gene may be responsible for an increased incidence of lung cancer in the Japanese population (HAYASHI et al. 1991). HAYASHI et al. (1992) analysed a high susceptibility to lung cancer in terms of combined genotypes of CYP1A1 and μ isoenzymes of glutathione transferase. However, as carcinogen metabolism comprises a chain of chemical reactions involving numerous enzymes and enzyme-coding genes, research performed hitherto is able to offer only a very limited explanation of the association between genetic polymorphism and the individual's susceptibility to cancer (INDULSKI and LUTZ 2000).

Epoxides, or oxiranes, are oxygen-containing heterocyclic compounds. Due to the large ring strain associated with the three-membered ring, they are reactive molecules. Because of their reactivity they are important intermediates in chemical industry, especially in polymer production. Wide-ranging industrial applications of epoxides have resulted also in considerable *human* exposure. Compounds such as ethylene, propylene, butadiene, styrene, vinyl chloride and acrylamide are metabolised to 1,2-epoxides by cytochrome P_{450}-dependent (CYP) monooxygenases.

Epoxides are alkylating agents *in vivo* being able to react with different nucleophilic centres of cellular macromolecules including proteins and DNA. DNA adducts in turn have shown considerable association with carcinogenic processes (HEMMINKI 1993, HEMMINKI et al. 1994). The specific DNA adducts induced by some mono-substituted epoxides were recently reviewed by KOSKINEN and PLNÁ (2000). In the *rat*, 600 ppm vaporised styrene inhaled for 12 h per day, 4 days per week for 4 weeks caused a severe outer hair cell loss and an increase of cytoplasmic vesiculation and vacuolisation, and abnormal mitochondria with disrupted cristae and formation of membrane-bound spherical bodies in the outer hair cells in the organ of Corti (MÄKITIE et al. 2002).

GONZALEZ-RECHE et al. (2001) detected dimethyl- and ethyl-phenylmercapturic acids in the urine of workers handling xylenes and ethyl benzene. These minor metabolites were formed via aromatic epoxides.

Benzo[*a*]pyrene-3,6-quinol and other quinols are involved in toxic quinone/quinol redox cycles (LORENTZEN and TS'O 1977, LORENTZEN et al. 1979). Quinols are formed from the corresponding quinones by several reductases. They are rapidly autoxidizes while superoxide anions are formed. Benzo[*a*]pyrene-3,6-quinone has been shown to be mutagenic in the Ames test, using tester strain TA 104 (CHESIS et al. 1984) or TA 102, strains which are particularly sensitive to reactive oxygen species. In male Sprague-Dawley *rats*, a rapid increase of unmetabolised benzo[*a*]pyrene was observed in sera 3 h after benzo[*a*]pyrene treatment followed by a sharp decrease (KIM et al. 2000). The time-dependent pattern of serum lipid peroxidation and the level of erythrocyte antioxidant enzymes were shown to be related to the concentrations of the formation of benzo[*a*]pyrene-quinones, oxidatively altered lipids and antioxidant enzymes in the blood.

After an intratracheal instillation of benzo[a]pyrene (1 mg dissolved in 50 μl tricaprylin), the Clara cell secretory protein dominant-negative mutant form of p53 transgenic *mice* was more susceptible to the development of lung adenocarcinoma than wild-type *mice* (TCHOU-WONG et al. 2002).

For benzo[*a*]pyrene diol-epoxide-DNA adducts in leucocytes of coke-oven workers exposed to polycyclic aromatic hydrocarbons, a sensitive, specific high-performance liquid chromatography-fluorescence assay was developed and validitated to monitor the levels of (+)-*anti*-benzo[*a*]pyrene diol-epoxide bound to DNA by the release of the respective benzo[*a*]pyrene-tetrols from tissue or leucocyte DNA (ALEXANDROV et al. 1992).

Urinary 1-hydroxypyrene could be a suitable biomarker of internal dose of polycyclic aromatic hydrocarbons, as recently shown in carbon black workers, on the condition that both respiratory (including gaseous PAHs and particle-bound PAHs) and dermal exposures have been assessed (TSAI et al. 2002). Urinary 1-hydroxypyrene concentrations were higher among road pavers than among office workers serving as referents (HEIKKILÄ et al. 2002).

There is now considerable evidence supporting the causal role of tobacco-specific and betel-nut specific **nitroso compounds** in lung, oral and upper respiratory tract cancers (HECHT and HOFFMANN 1991). In *rats*, *N*-nitrosodimethylamine produced by oxidative stress induced lipid peroxidation *in vivo and* in isolated hepatocytes (AHOTUPA et al. 1987), and, when produced by long-term inhalation, nasal squamous cell and mucoepidermoid carcinomas (HERMANN 1997).

A chronic exposure to caffeidine-derived *N*-nitroso compounds produced exogenous and endogenously has been strongly implicated in several cancers among high risk populations (BARTSCH and MONTESANO 1984). A high incidence of both oesophageal (SIDDIQI and PREUSSMANN 1989) and stomach cancer (SIDDIQI and PREUSSMANN 1989, KHUROO et al. 1992) occurred due to considerable exposure of the local population in Cashmere to endogenously produced mononitrosocaffeidine and dinitrosocaffeidine formed under the acidic milieu of the stomach in the presence of high dietary nitrate from caffeidine contained in salted tea (SIDDIQI et al. 1992). IVANKOVIC et al. (1998) induced neuroepitheliomas of the olfactory epithelium and squamous cell carcinoma of the nasal cavity of *bd-ix rats* treated with mononitrosocaffeidine dissolved in the drinking water.

Nitrosodimethylamine is considered to be an occupational carcinogen (NIOSH 1997). Peripheral blood lymphocytes of workers from a rubber plant had detectable concentrations of N^7-methylguanine adducts, ranging from 0.1 to 133.2 adducts/10^7 deoxyguanosine nucleosides (REH et al. 2000). Although no association was found between occupational personal breathing zone exposure and either

N^7-methylguanine adduct concentration or O^6-alkylguanine-DNA alkyltransferase activity, the significant positive association between working in and near the rubber vehicle seal department and the presence of O^6-methylguanine adducts, which have mutagenic potentials, provided evidence to link nitrosamine exposure one step closer to *human* cancer by demonstrating an association between external nitrosamine exposure and cancer-related biological effects.

N-Morpholino-*N*-nitrosoaminoacetonitril as a **nitric oxide (•NO) donor** (BOHN and SCHÖNAFINGER 1989, FEELISCH et al. 1989) formed from *N*-ethoxycarbonyl-3-morphlinosydnonimine (molsidomin) via 3-morphlinosydnonimine by $O_2^{\bullet -}$ formation is a safe vasodilator in *man*, but a potential nasal carcinogen in the *rat*, probably due to the vast smooth endoplasmic reticulum with its cytochrome P_{450} (HADLEY and DAHL 1982, ADAMS et al. 1991). The question whether the new series of mesoionic 3-aryl substituted oxatriazole-5-imine derivatives (LÄHTEENMÄKI et al. 1998) have the same side effects is outstanding.

Nitric oxide (•NO) formation is increased in biological tissues during ischaemia, and there is an enzyme-independent pathway of •NO generation due to the reduction of tissue nitrite under the acidic conditions which occur. The generation and accumulation of •NO from typical nitrite concentrations found in biological tissues increased 100-fold when pH decreased from normal values of 7.4 to the acidic values found in ischaemic *rat* tissues, such as the heart, where pH falls to 5.5 (SAMOUILOV et al. 1998).

Air pollution is associated with increased level of exhaled nitric oxide in non-smoking healthy subjects (VAN AMSTERDAM et al. 1999).

Prolonged exposure to arsenic (0.41 ± 011 μg/ml drinking water from tube wells) in an endemic area in Inner Mongolia resulted in an impaired production of endothelial nitric oxide (PI et al. 2000). *In vitro, human* endothelial **nitric oxide synthase** (NOS) activity was significantly suppressed by arsenite (18% inhibition, $P < 0.05$) and arsenate (27% inhibition, $P < 0.01$) at 100 μM. Neither arsenite nor arsenate showed inhibitory action on neuronal NOS activity.

During allograft acute rejection, the production of •NO as measured nitrite/nitrate levels, has been demonstrated in several experimental models and inhibition of inducible nitric oxide synthase (iNOS) activity in a *rat* lung transplant model dramatically improved the survival rate of allografts (SHIRAISHI et al. 1995, WORRALL et al. 1997). Studies carried out on transplant samples from patients with end-stage obliterative bronchiolitis have shown that in addition to being present in inflammatory cells, iNOS immunoreactivity can also be seen in damaged bronchiolar epithelium; the intensity of immunostaining was associated with the extent of epithelial damage (MASON et al. 1998).

Peroxynitrite is a highly reactive oxidant produced by the combination of •NO and $O_2^{\bullet}$–. There are indications that it may contribute to cell death and tissue injury in a number of human diseases, including arthritis (KAUR and HALLIWELL 1994), sepsis (SZABO et al. 1995), inflammatory bowel disease (RACHMILEWITZ et al. 1993) and stroke (OURY et al. 1993). Peroxynitrite decomposition catalysts (SALVEMINI et al. 1998) may represent a unique class of antiinflammatory agents.

Nitrogen dioxide (NO_2) at levels found in healthy subjects indoors caused mild airway inflammation, effects on blood cells, and increased susceptibility of airway epithelial cells to injury from respiratory viruses (FRAMPTON et al. 2002).

As more than 40 different cell types were described in the respiratory organs (SOROKIN 1970), EMURA et al. (1990) developed Syrian *hamster* (foetus on day 15 of gestation) and *human* (foetus of 18 – 22 weeks of gestation) cell systems consisting of a certain type of stem cell capable of differentiating into at least two bronchoalveolar cell types: the Clara cell and the alveolar type II epithelial cell. AUFDERHEIDE et al. (1994) used this M3E3/C3 *hamster* cell line for a new *in vitro* approach to study cytotoxicity and carcinogenic effects of mineral fibres (crocidolite and silicon carbide).

Macrophages gained either by bronchoalveolar lavage (SENIOR et al. 1981) or directly from bone marrow (STEWART 1981) were used for studies of the **oxygen burst** induced by pure (ZELLER et al. 1992, 1993, Brehm et al. 1996) and industrial mixed dusts (MUNDER and MODOLELL 1987, MUNDER et al. 1989). *In vivo*, SCHAPIRA et al. (1995), trapping HO• by sodium salicylate, found significantly more 2,3-dihydroxybenzoic acid in *rat* lungs exposed to 50 mg silica compared with lungs instilled with 50 mg titanium dioxide 7 days before.

The oxygen burst of monocytes induced by ambient air particles (with the exception of oil fly ash particles) was less than the response elicited by polymorphonuclear leucocytes (PRAHALAD et al. 1999). The luminol-enhanced chemiluminescence response of polymorphonuclear leucocytes separated from heparinised *human* blood was generally increased with all washed particles, with oil fly ash and one urban air particle showing statistically significant ($P < 0.05$) differences between dH_2O-washed and unwashed particles. The luminol-enhanced chemiluminescence activity in polymorphonuclear leucocytes induced in post particles and dH_2O-washed particles was significantly correlated with the insoluble Si, Fe, Mn, Ti and Co content of particles ($P < 0.05$). No relationship between luminol-enhanced chemiluminescence activity in polymorphonuclear leucocytes and soluble transition metals such as V, Cr, Ni, and Cu was noted.

Fibroblasts play a major role in the control of tissue remodelling, through their production of growth factors and other regulatory molecules, including •NO (LAVNIKOVA and LASKIN 1995). GANSAUGE et al. (1997) have shown that some inflammatory mediators can induce iNOS expression in *rat* lung fibroblasts, which leads to their proliferation. Up-regulation of iNOS in fibroblasts paralleled the onset and progression of fibrosis in an experimental model of post-transplant obliterative airway disease (ROMANSKA et al. 2000).

Free radical reactions have been observed to influence molecular and biochemical processes and to directly cause some of the changes observed in cells during differentiation, ageing, and transformation (ABATE et al. 1990, SOHAL and ALLEN 1990, ALLEN and TRESINII 2000).

8-Hydroxydeoxyguanosine, the most commonly studied biomarker for oxidative DNA damage, offers a specific and quantitative biomarker for damage caused by reactive oxygen species in *human* sperm (SHEN and ONG 2000).

As to **mineral fibres**, ROSSITER (1994) – in a paper on the research still needed – put forward the question of free radicals and the preventive influence of lung surfactant and endogenous antioxidants in the action of fibres on biomembranes to be evaluated by an *in vitro* test, giving quick and accurate prediction of the carcinogenic potential of fibres. Although asbestos (above all Rhodesian chrysotile) induced 8-hydroxy-2'-deoxyguanosine in *calf* thymus DNA *in vitro* (ADACHI et al. 1992), there was no significant difference in peripheral blood cells of 7 male outpatients (aged 66–75 years) with asbestosis and 6 healthy non-smokers (aged 62–81 years) without exposure (HANAOKA et al. 1993).

Constitutional aspects of clinical pneumoconiosis research were revealed by PARRISIUS and IM BRAHM (1953, 1954). Silicosis was concordant in 25 of 28 monozygous pairs of twins, but only in 14 of 26 heterozygous pairs. The localisation of the silicotic masses in the homozygous twins was striking. In pairs of monocygotic twins tracheobronchial clearance patterns of an aerosol of 6–7 μm monodisperse fluorinated ethylene propylene (Teflon 120) particles tagged with technetium 99m (half life 6.0 h) were highly similar (CAMNER et al. 1972). Within-pair variance of bronchial mucociliary clearance (using ^{99}Tc-labelled albumin aerosols) in concordant smoking monozygotic twins was significantly smaller than among-pair variance (KAMISHIMA et al. 1982).

Workers belonging to different generations exposed to silica dust in a plant producing refractories for about 70 years showed family susceptibility in the development of silicosis (NOWEIR et al. 1978). Members of some families were highly susceptible to silica dust while other families were resistant; and susceptibility varied to a great extent among the different families.

YUCESOY et al. (2001) in a case–control study evaluated the interaction between silica exposure and minor variants in the genes coding for interleukin-1α, interleukin-1 receptor antagonist and tumour necrosis factor-α as risk factors associated with silicosis in 325 ex-miners with moderate and severe silicosis and 164 miners with no lung disease. The odds ratio of disease for carriers of the minor variant, TNF-α (–238), was markedly higher for severe (4.0) and significantly lower for moderate silicosis (0.52). Regardless of disease severity, the odds ratios for carriers of the IL-1RA (+2018) or TNF-α (–308) variants were elevated. In contrast, MCCANLIES et al. (2002) used data generated from these studies to illustrate the utility of genetic information for the purposes of risk assessment and clinical prediction. Their results indicated that genetic information plays a valuable role in effectively characterising risk groups and mechanisms of disease operating in a substantial proportion of the population. However, in the case of fibrotic lung disease caused by silica exposure, information about the presence or absence of the minor variants of interleukin-1α, interleukin-1 receptor antagonist and tumour necrosis factor α is unlikely to be a useful tool for individual classification.

In the *mouse*, BALB/c and DBA/2 strains have been identified as high and low responders, respectively, to DQ12 quartz dust (2.5 mg) injected into a hind footpad (STARK et al. 1988).

Significant interstrain variation in polymorphonuclear leucocyte and protein response was observed between inbred groups of *mice* with 1× and 5× exposures to 1.0 mg ZnO per m^3, which indicated that genetic background has an important role in the development of pulmonary tolerance (WESSELKAMPER et al. 2001). The BALB/cByJ strain and the DBA/2J strain were the most tolerant and intolerant, respectively.

Analyses of macrophage dysfunction phenotypes of segregant and non-segregant populations derived from susceptible C57BL/6J and resistant C3H/He inbred *mice* indicated that two unlinked genes susceptibility to the inhalation of sulphate-associated carbon particles (OHTSUKA et al. 2000). A genome-wide linkage analysis of an intercross (F_2) cohort identifies significant and suggestive quantitative trait loci (QTLs) on chromosomes 17 and 11, respectively. Candidate susceptibility genes were identified for *mice* and *humans* by comparative mapping. Importantly, both QTLs overlap previously identified QTLs for susceptibility to another common pollutant, ozone. Ozone inhalation increases inducible nitric oxide synthase in both macrophages and type II epithelial cells of the lung, associated with the activation of nuclear transcription factor-ϰB (LASKIN et al. 1998).

The inheritance of information based on gene expression levels is known as **epigenetics**, as opposed to genetics, which refers to information transmitted on the basis of gene sequence. The main epigenetic consequence in *humans* is the methylation of cytosine located within the dinucleotide CpG. Epigenetic changes, particularly DNA methylation, are susceptible to change and are excellent candidates to explain how certain environmental factors may increase the risk of cancer (ESTELLER and HERMAN 2002).

Environmental and heritable factors in the causation of cancers were discussed by LICHTENSTEIN et al. (2000), who analysed data on 44,788 pairs of twins from Sweden, Denmark, and Finland. If studies of groups of twins show that concordance for cancer is higher among monozygotic twins (who share all genes) than among heterozygotes (who, on average, share 50 per cent of their segregating genes), genetic effects are likely to be important. If, however, the concordance is similar for both types, the shared environmental effects are probably unimportant. They conclude that inherited genetic factors make a minor contribution to susceptibility to most types of neoplasms. Shared environment, the sum of the common family experiences and habits of the twins, accounted for 0 to 20 per cent of causation, but none of these values were statistically significant, partly because studies of twins have limited power to detect shared environmental effects in dichotomous phenotypes, e.g. the presence or absence of cancer (NEALE et al. 1994).

Aberrant DNA methylation can be used as a biomarker of malignant transformation. The development of the methylation-specific PCR technique (HERMAN et al. 1996) has popularised this approach because it offers a quick, easy, nonradioactive, and sensitive way to detect hypermethylation in CpG islands of tumour suppressor genes. This strategy is further enhanced by three facts: the PCR signal is a positive signal that cannot be masked through contamination by normal cells; promoter hypermethylation occurs early in tumour progression, allowing early diagnosis; and all tumours appear to have one or more hypermethylated loci when panels of these markers are examined together. Aberrant CpG islands methylation has been used as a tool to detect cancer cells in bronchoalveolar lavages (AHRENDT et al. 1999), lymph nodes (SANCHEZ-CESPEDES et al. 1999, ESTELLER and HERMAN 2002), and sputum (PALMISANO et al. 2000). ESTELLER et al. (1999) showed that it was possible to screen for hypermethylated promoter loci in serum DNA from non-small cell lung cancer patients.

Caplan's syndrome occurs in persons who have a genotype disposing to rheumatoid arthritis (CAPLAN 1953, 1965, CLERENS 1953, COLINET 1953, MIALL et al. 1953, MIALL 1954, VAN DER MEER 1954, GOUGH et al. 1955, CAMPBELL 1958, CAPLAN et al. 1958, 1962, KANTOR and MORROW 1958, MAKIOKA 1958, HUZL 1960, POSNER 1960, CHATGIDAKIS and THERON 1961, CHIESURA et al. 1961, ROSMANITH and BRÜCKNER 1962, LAMVIK 1963, BARNI 1965, GOUGH 1965, SNOEK et al. 1965, ROSMANITH and SITAJ 1967, NIEDOBITEK 1969, FRITZE 1970, RUSSO et al. 1977, WAGNER and DARKE 1979). Rheumatoid lung changes may be also observed in asbestosis (RICKARDS and BARRETT 1958, TELLESSON 1961, MORGAN 1964, GREAVES 1979). The radiological severity of the lung changes does not necessarily parallel the degree of arthritis. Pulmonary lesions usually develop contemporaneously with the arthritis but either manifestation may precede the other by several years (MIALL et al. 1953).

Homozygous or heterozygous α_1-antitrypsin deficiency offers a proteolytic mechanism and possible genetic basis for some cases of **panlobular emphysema in miners** (FALK and BRISCOE 1970). The homozygous deficiency, which is known to be associated with emphysema, has been identified in 1–10 % of various groups of patients (BRISCOE et al. 1966, KUEPPERS et al. 1969, LIEBERMAN 1969).

Certain particulate air pollutants may play an important role in the increasing prevalence of **respiratory allergy** by stimulating T helper 2 cell (The)-mediated immune responses to common antigens. VAN ZIJVERDEN et al. (2000) studied the primary response after subcutaneous injection of 1 mg of particle together with 10 µg of receptor antigen TNP-OVA (2,4,6-trinitrophenyl coupled to ovalbumin) into the hind paw of specific pathogen-free BALB/c *mice*. The number of interleukin-4 containing $CD4^+$ T cells increased between day 2 and day 5 in diesel exhaust particle- and carbon black particle-exposed *mice*, in contrast to silica particle-treated animals. All particles acted as adjuvant, but the different particles stimulated different types of immune responses to TNP-OVA. Diesel exhaust particles can be phagocytosed by *human* bronchial epithelial cells, including the release of cytokines (MARANO et al. 2002). MAP kinase pathways (i.e. ERK1/2 and P38) were triggered as well as the activation of the nuclear factor-ϰB. Reactive oxygen species were strongly incriminated in this response because diesel exhaust particles induce the increase of intracellular hydroperoxides and antioxidants inhibit the release of diesel exhaust particle-induced cytokines, the activation of MAP kinases and nuclear factor-ϰB. Organic compounds adsorbed on diesel exhaust particles seemed to be involved in the response and the production of reactive oxygen species.

For many volatile substances inhalation is the most important route of absorption. The relevance of physical activity for the kinetics of inhaled gaseous substances has been recently reviewed by CSANÁDY and FILSER (2001).

Occupational exposure to **hydrocarbon solvents** (aliphatic, aromatic, chlorinated, benzene, other organic solvents) has been found to be associated with an increased risk of exocrine pancreas cancer, the *human* tumour with the highest prevalence of K-*ras* mutations (ALGUACIL et al. 2002). A significantly higher proportion of cases with a mutation from glycine to valine (GGT→GTT) or to aspartic acid (GGT→GAT) were exposed to a hydrocarbon solvent.

Benzene is effectively absorbed via the lungs and about 50 per cent of the inhaled dose is retained (YVES BOIS et al. 1996). The leucaemogenic effect of benzene may be related to its metabolism and to induction of oxidative DNA damage, as shown by the formation of 8-dihydroxy-2'-deoxyguanosine both in HL-60 cells *in vitro* and in bone marrow *in vivo* (KOLACHANA et al. 1993). Chromosomal aberration analysis in workers exposed to concentrations of benzene up to 15 ppm and of toluene up to 50 ppm with a mean occupation exposure of 17 years revealed a significant increase in dicentric incidence in the exposed group compared to the controls ($P = 0.004$) (BOGADI-ŠARE et al. 1997). The increased production of reactive oxygen species/reactive nitrogen species including peroxynitrite during inflammation combined with benzene exposure may increase the genotoxicity of benzene by a mechanism that includes formation of metabolites from the chemical reaction between benzene and reactive oxygen species/reactive nitrogen species (TUO et al. 1998). Reactive oxygen species/reactive nitrogen species mediated hydroxylation and nitration of benzene during immune activation represent a novel mechanism for generation of proximal carcinogens of benzene. Benzene is probably eliciting more current interest as a possible environmental cause of non-Hodgkin's lymphoma than any other substance (O'CONNOR et al. 1999).

The *in vitro* effect of benzene on the ultrastructure of *human* peripheral blood polymorphonuclear cells, lymphocytes and monocytes was examined with a transmission and scanning electron microscope and was found to affect both the internal and external architectures of polymorphonuclear leucocytes and monocytes in a dose-dependent manner (KALIR et al. 1989). No effect was observed on lymphocytes. Pre-incubation with 2-aminoethylthiosulfuric acid, a free radical scavenger, prevented the observed effects of benzene.

Benzene generated DNA strand breaks in the livers of both wild type and inducible nitric oxide synthase (iNOS)-deficient *mice* (VESTERGAARD et al. 2002). In the bone marrow benzene and lipopolysaccharide generated strand breaks only in wild type *mice*. The effects were additive, suggesting that both a redox cycling and an iNOS-dependent pathway may be involved. GASKELL et al. (2002) identified (3"-hydroxy)-1,N^2-benzetheno-2'-deoxyguanosine 3'-monophosphate and a new product, (3",4"-dihydroxy)-1,N^2-benzetheno-2'-deoxyguanosine 3'-monophosphate.

Synaptosomes from *rats* prenatally exposed to **toluene** exhibited an increased level of oxidative stress when incubated with toluene *in vitro* compared to synaptosomes from unexposed offspring (EDELFORS et al. 2002).

It has been suggested, based on epidemiology data, that paternal exposure to **metal dusts** and fumes, including welding fumes, may increase the incidence of cancer in the progeny (BUNIN et al. 1992, BUCKLEY 1994). Nickel is one of the primary suspects, and sperm DNA may be considered as its most likely target. *Human* protamine P2 has an amino acid motif at its N-terminus that can serve as a heavy metal trap, especially for Ni(II) and Cu(II). BAL et al. (1997) synthesised a pentadecapeptide modelling this motif and described its complexes with Ni(II) and Cu(II), including their capacity to mediate oxidative DNA degradation. Using circular plasmid pUC19 DNA as a target, and the single/double strand breaks and production of oxidised DNA bases, as end points, LIANG et al. (1999) found Ni(II) alone to promote oxidative DNA strand scission and base damage, while Cu(II) alone produced the same effects, but to a much greater extent.

Cadmium and nickel can produce delayed effects in *human* cells *in vitro*, which are characteristic if gnomic instability (CONE et al. 2001). The effects even occur at levels where no acute toxic effects can be demonstrated. The consequences of this gnomic instability are not yet known but it is possible that many of the systemic symptoms associated with exposure to low concentrations of these metals could involve delayed expression of cellular damage. It is also clear that these effects cannot be predicted from acute toxicity data.

Lung tissue concentration of chromium was two-fold increased in cancer patients compared to controls, while bronchial nickel content was three-times that of controls (Morkve et al. 1989). Only three cancer patients had a possible occupational exposure to chromium.

Iron overload has been associated with damage of the liver and other organs. In Wistar *rats* injected peritoneally with Fe-dextran (825 and 1650 mg/kg, respectively) the kidney iron was increased (Dimitriou et al. 2000). Iron appeared to accumulate mainly in the lysosomes, inducing distinct changes in the behaviour of the organelles as judged by subcellular fractionation studies. Lysosomes became more fragile and showed increased density.

Iron and oxygen are at a metabolic crossroad where mismanagement leads to inflammation, toxicity, and tissue damage. Many genes encode proteins to provide efficient and safe use of iron and oxygen for crucial reactions in cell division, respiration, regulation and detoxification. A subset of such genes contains non-coding sequences that are transcribed to the mature mRNA to regulate protein synthesis in response to environmental changes of iron, free radicals, and metabolic signals. With the iso-iron response element/iso-iron regulatory protein system has evolved co-ordinated combinatorial control of iron and oxygen metabolism that may exemplify control of mRNAs in other metabolic pathways, viral reproduction and oncogenesis (Theil 2000).

Environmental molecular epidemiology integrates new knowledge in molecular toxicology, DNA mutagenesis, and polymorphism of procarcinogen-metabolising enzyme systems. Biomarkers may represent signals in a continuum of events between causal exposure and resultant disease (Schulte 1989). Borm (1994), taking the framework of mineral dust-induced lung disorders as an example, demonstrated how carefully designed follow-up studies are a prerequisite to test the validity and use of events often put forward as biomarkers. Individual susceptibility to multistep lung carcinogenesis may be measurable at any or all steps along the pathway, from procarcinogen exposure to the development of frank, clinically detectable carcinoma. In a recent review Spivack et al. (1997) focused on procarcinogen bioactivation and DNA damage.

Phenytoin and related xenobiotics can be bioactivated by embryonic prostaglandin H synthase to a teratogenic free radical intermediate. Parman et al. (1998) evaluated the mechanism of free radical formation using photolytic oxidation with sodium persulphate and by electron paramagnetic resonance spectrometry. Incubation of 2'-deoxyguanosine with phenytoin and prostaglandin H synthase-1 resulted in a 5-fold increase in the oxidation to 8-hydroxy-2'-deoxyguanosine.

Based on the updated use of prophylactic and therapeutic strategies in pneumoconiosis Gulumian (1999) thinks it unlikely that a single strategic approach will suffice for adequate treatment of the whole spectrum of mineral particle-induced pulmonary inflammation, fibrosis and carcinogenesis. Yu et al. (1995) presented a combined treatment of experimental silicosis with tetrandrine plus polyvinylpyridine *N*-oxide or tetrandrine plus QOHP in *rats*.

CHAPTER 2

Protagonists of Industrial Medicine

The writings of HIPPOCRATES (about 460 BC - 377 BC) contain numerous allusions to relationships between occupation and disease and the same holds for GALEN (120–199). HIPPOCRATES in his "Epidemics" [in a rather obscure passage of which RAMAZZINI (1700, p.23; 1703, p.23), in his knowledge of occupational diseases, though differing with GALEN and FARR (1780), gives an interpretation closely adherent to that of LITTRÉ (1846)] speaks of a metal digger (ὁ ἐκ μετάλλων) who, we may presume, was exposed to dust in his work like the metalliferous miner of later days, as a man who "has his right hypochondrium bent, a large spleen, and a costive belly; he breathes with difficulty (πνευματώδης), is of pale wan complexion, and is liable to swelling in his left knee".

BERNARDINO RAMAZZINI's "De morbis artificum diatriba" (1700) was the first comprehensive monograph on industrial medicine. LUIGI DEVOTO (1864–1936) edited the first journal of industrial medicine, "Il Lavoro" (Pavia 1901), since 1925 entitled "La Medicina del Lavoro". In Germany, E.W. BAADER (1892–1962) became a well-recognised protagonist of clinical industrial medicine, while in Bavaria FRANZ KOELSCH (1876–1970) as the chief medical inspector of factories was an expert in porcelain pneumoconiosis. As to coalworkers' pneumoconiosis in the Ruhr district, the clinical work of BÖHME (1878–1962) and REICHMANN (1881–1956) was correlated to post-mortem morphology by DI BIASI (1896–1981). Siderosilicosis in the Sieg district was studied by CEELEN (1883–1964) and GERSTEL (1902–1978). The latter (GERSTEL 1934, 1935, 1938, 1941) complemented post-mortem morphology with extended chemical dust analyses. JULIUS WÄTJEN (1883–1968), full professor of pathology at Halle University and member of the German Academy of Scientists «Leopoldina», showed deep-

Fig. 1. VIKTOR REICHMANN (1881–1956) in 1920 came from Jena University to Bochum to become an expert in the clinics and roetgenology of coal workers' pneumoconiosis

Fig. 2. WALTER PARRISIUS (1891–1977) during a research trip to the United Kingdom in June 1950. Taken at Penarth, Glamorgan

black fibrotic foci as a characteristic feature of the copper schist pneumoconiosis from the Mansfield area (Wätjen 1933, 1936, 1952).

The idea of a predisposition was put forward by BROCKMANN (1851, p.137), LINDEMANN (1913), LOCHTKEMPER (1935, 1936) and GEISLER (1937) and made more precise by the results in homozygous twins by PARRISIUS and im BRAHM (1953, 1954). The expression of antioxidative enzymes is controlled by genes, and thus the damage induced by oxidative stress will be determined individually.

In France, BALGAIRIES (1951, 1953), director of the Centre d'Études Médicales Minières, Houillères du Bassin du Nord et du Pas-de-Calais at Douai, distinguished himself in the radiological diagnosis of pneumoconiosis. The Cardiff-Douai Classification was a result of his international efforts.

In Great Britain, field studies by HART and ASLETT (1942) and FLETCHER (1948) and his colleagues from The Pneumoconiosis Research Unit at Cardiff into pneumoconiosis of coal workers have gained international recognition.

In experimental pneumoconiosis research, JULIUS ARNOLD (1885), professor of pathology in Heidelberg from 1866–1907, and KARL WILHELM JÖTTEN, professor of hygiene in Münster (Westphalia) from 1924–1958, were German pioneers. KARL THOMAS (1883–1969) held meetings and colloquia on basic research in silicosis and gained an international reputation through this and the work of his team (WEITZEL 1970).

In England, the biochemist EARL JUDSON KING, who was first confronted with silicosis research when BANTING in Toronto had been asked by a group of gold miners to carry out silicosis research, moved to London in 1934 and worked on experimental silicosis and pneumoconiosis for the rest of his life (on October 31, 1962).

In France, ALBERT POLICARD (1881–1972) was nominated professor of histology at Lyons and became organiser of a laboratory of biology applied to pulmonary silicosis at the Charbonnages de France from 1950–1968.

In Italy, experimental work was done by BIONDI (1906), FELIZIANI (1906), TOMMASI CRUDELI (1907), and others. DEVOTO and CESA-BIANCHI (1911), at the 3rd national congress on industrial diseases in Turin, reported that *guinea pigs* subjected to the inhalation of very fine limestone dust were less susceptible to tuberculosis than those, which had inhaled silica dust. Tedeschi and Abbo (1909) induced serious inflammatory changes by grain silo dust.

Fig. 3. JULIUS ARNOLD (1835–1915) in search of a connection between morphology and physiology performed large-scale dust inhalation experiments in *rabbits* and *dogs* and compared dust metastasis with post-mortem findings in *man*. (Courtesy of Prof. Dr. HERWART F. OTTO, present holder of the professorship)

Fig. 4. KARL WILHELM JÖTTEN (1886–1958) visiting the Silicosis Research Institute "Rheinpreussen" on June 12, 1953. He began pneumoconiosis research under privy councillor WALTHER KRUSE at the Department of Hygiene, University of Leipzig, and was appointed to the chair of hygiene at the University of Münster in Westphalia in 1924. In 1953 he was awarded the Devoto Prize

Fig. 5. EARL JUDSON KING, PhD. of Toronto and honorary M.D. of the Universities of Oslo and Iceland. As HARRISON (1964) stated in his obituary, silicosis was the favourite subject of both KING and KETTLE, who persuaded KING to leave Bart's [St. Bartholomew's Hospital, London] and found the Department of Pathology at the new British Postgraduate Medical School that was to be opened in the spring of 1935. Unfortunately, the chance of a fruitful collaboration was an idle hope, for KETTLE was to survive for only 20 months, and others were to collaborate with KING on his silicosis work. – The photograph, sent to me by KING himself, corresponds to those in HARRISON's obituary in the Journal of Pathology and Bacteriology (88: 601) and ISAAC's (1993) monograph on the first forty years of the British Occupational Hygiene Society (KING was president in 1954/55), but the background has been omitted

Fig. 6. ALBERT POLICARD visiting the Silicosis Research Institute "Rheinpreussen". Born in Paris, he studied medicine at the École de Service de Santé militaire in Lyons. In 1913 he became reader in histology. After World War I he was nominated professor of histology at Lyons and after his superannuation became organiser of a laboratory of biology applied to pulmonary silicosis at the Charbonnages de France from 1950–1968. In 1963 he was elected a non-resident member of the Académie des Science in Paris

In the United States, LEROY UPSON GARDNER (1888–1946) "devoted his life and talents to the cure of the illness of others and the correction and control of industrial hazards to enable his fellowmen to have better and safer conditions under which to perform their daily tasks". In South Africa, W. WATKINS-PITCHFORD, first director of the South African Institute for Medical Research from 1912–1926, was ably supported in pneumoconiosis research by MAVROGORDATO (deceased 1944), SIMSON (1883–1948) and STRACHAN (1891–1949).

2.1 The Neolithic Period

COLLIS (1915) supposed that Neolithic miners digging for flint might have suffered from pneumoconiosis, and, indeed, the man found in the ice at the Hauslabjoch did show anthracosis, and his hair was loaded with arsenic, copper, nickel and manganese (SPINDLER 2001).

2.2 Slaves as Miners in Antiquity

In Egypt, the work in the mines was in the hands of slaves, many of whom were not Egyptians. Semitic inscriptions, dating back to about 1500 B.C. and found by PETRIE at Serabit-el Khadem, show that Syrian workers were among the Sinaitic miners.

The mummy of Ḥar-mosĕ, the Singer of the eighteenth dynasty, who died about 1490 B.C. (cf. LANSING and HAYES 1937) showed pulmonary anthracosis, emphysema, and an old pleural adhesion (SHAW 1938). The pulmonary lymph nodes were loaded with carbon pigment but no silica was detected with crossed Nicol prisms.

The earliest allusion to a mining hazard is to be found in the Bible. HIRAM (about 950 B.C.), the king of Tyre, is supposed to have been cognizant of the poisonous effects of cobalt. When Solomon, his ally, presented him with a gift of 20 villages in the district of Cabul in Galilee, he refused them, because they contained veins of cobalt (LEGGE 1936, p. 302). ROSEN (1943, p. 33) remarked that the biblical verses upon which the above statements are based (Kings 1: 9,11–13) do not furnish any basis for such claims.

The earliest Hellenic medical reference that may be interpreted as relating to a miner is to be found in the fourth book of epidemics in the Corpus Hippocraticum:

> ὁ ἐκ μετάλλων, ὑποχόνδριον δεξιὸν ἐκτεταμένον, σπλὴν μέγας, καί κοιλίη ἐντεταμένη. ὑπόσλκηρα, πνευματώδης, ἄχροος τούτῳ ἐς γόνυ ἀριστερὸν, ὑποστροφὴ, δι' ὅλου ἐκρίθη.
>
> A man from the mines had the right hypochondrium distended. His spleen was enlarged, and his abdomen was distended and hard. His respiration was laboured, and he was both pale and livid. The disease then attacked the left knee. There was a relapse in the entire body and finally a crisis.

Just as doubtful is a reference from Hippocrates' Coacæ prænotiones, quoted in Thomson's (1838, p. 343) paper on black expectoration: "independently of the habitual inspiration of an atmosphere, which can be supposed to be peculiarly loaded with carbonaceous matters" (l.c. p. 343).

> 'Ολέθριοι δὲ καὶ οἱ τὰ μέλανα λιγνυώδεα πτύοντες, ἢ οἷσιν ἀπὸ οἴνου μέλανος γίνεται πτύσματα.
>
> Perniciem etiam denuntiant sputa nigra, fuliginosa, aut quibus qualia ex vino nigro fiunt.

References to Hippocrates and pulmonary disease: Cucchiani Acevedi (1945), Samitier Azparren (1946)

Plautus' (before 251 B.C. to 184 or 183 B.C.) character Tyndarus, consigned there as a punishment, found nowhere so much like Hell as the stonequarry: "nullus ădaequest Ăchĕruns atquĕ ŭbi ĕgŏ fui ĭn lăpĭcīdīnis" (Captivi, versus 999).

Strabo (Στράβων, ca. 63 B.C. to 20 A.D.) referred to a mine employing slaves sold cheaply because of their crimes "where the air is said to be destructive of life and scarcely endurable in consequence of the grievous odour issuing from masses of ore" (Geography XII: 3,40), but he was referring to ore containing realgar (arsenic disulphide, As_2S_2) encountered in a silver mine.

Ramazzini (1700) left it undecided if P. Ovidius Naso (43 B.C. to ca. 17 A.D.) had the avarice of men in mind or the dangers detrimental to their health when writing in his Metamorphoses (I, 138–140):

> "... sed itum est in vescera terrae,
> Quasque recondiderat Stygiis admoverat umbris,
> Effodiuntur opes, inritamenta malorum."

Silius Italicus (ca. 26–101) referred to the avaricious Asturian, who is pale as the gold he tears out of the earth:

> ... Astur avarus
> visceribus lacerae telluris mergitur imis,
> et redit infelix effosso concolor auro.
>
> I, 231–233

T. Lucretius Carus (De rerum natura liber sextus, verses 806–817) wrote:

> nonne uides etiam terra quoque sulpur in ipsa
> gignier et taetro concrescere odore bitumen,
> denique ubi argenti uenas aurique secuntur,
> terrai penitus scrutantes abdita ferro,
> qualis expiret Scaptensula subter odores?
> quidue mali fit ut exalent aurata metalla,
> quas hominum reddunt facies qualisque colores.
> nonne uides audisue, perire in tempore paruo
> quam soleant, et quam uitai copia desit,
> quos opere in tali cohibet uis magna necessus?
> hos igitur tellus omnis exaestuat aestus,
> expiratque foras in apertum promptaque caeli.

C. Plinius Secundus Maior (23–79) recorded in his Historia naturlis (XXXIII, 60: 122):

> Qui minium in officinis poliunt faciem laxis vesicis inligant, ne in respirando pernicialem pulverem trahant et tamen ut per levia spectent.
>
> Those who clean native cinnabar in the factories tie loose bladders over their faces to prevent the inhalation of pernicious dust as they breath.

Plutarch (ca. 50–ca. 125) in his life of Sertorius (chap. 17) described the tactical use of a choking dust in the campaign against the Characitanians, in what is now Portugal. He likened the dust to ἀσβεστινον, but this word was regularly used in Greek to refer to slaked lime; ἀμίαντος was the word used to describe "the stone which could be combed and woven" (i.e., chrysotile asbestos).

Galen (129–199) had personal experience of the occupational hazards of miners. During one of his journeys he visited the island of Cyprus and spent some time inspecting the mines where copper sulphate was obtained. The miners worked in an atmosphere of suffocating fumes, and Galen mentions that he himself was almost overpowered by the stench. Those workers who transported the vitriolic fluid out of the mine did so as rapidly as possible, to avoid suffocation.

2.3 Treatment of Consumptive Miners in the Middle Ages and the Renaissance

Mining during the Middle Ages is marked by the appearance of the free miner and the free mining towns. The miner developed from a feudal serf to a free artisan. The development took several centuries and coincided with a perfection of mining technique, which was due in part to this very process.

The oldest pieces of advice and prescriptions for consumptive miners were directed above all to combatting irritations of the throat. They should promote expectoration.

A Latin prescription, containing Old Magyar words, was written between 1410 and 1430 in an Hungarian mining community (cf. Wilsdorf 1959, Eis 1960). The following substances were recommended ad pectus (not particularly for miners' consumption): gengber (ginger), ffahay (cinnamon), dragorium (tragacanth; the question mark put by Wilsdorf was deleted by Eis), kechgetej (goats' milk), diusmak (nut kernel), fenymak (pine seed? Wilsdorf), hyssop (Hyssopus officinalis), saluia (sage), obruta (Abrotanum = Artemisia abrotanum L, Middle High German: abrute; Eis), chyahaia (after Eis had wrongly taken this for dya; Dyalthaea = an althaea dispensation), ficus (fig) and mel (honey). According to Eis (l.c.) this Old Magyar prescription, by comparing it with other texts, may be easily derived from the international mediaeval tradition, especially the School of Salerno.

The Swabian physician Ulrich Ellenbog (born at Feldkirch in Vorarlberg, became M.A. at Heidelberg on March 17, 1455, MD at Pavia in 1459, and died 1499) in his pamphlet on the poisonous evil vapours and fumes of metals, instructed goldsmiths and other metal workers to protect themselves from the noxious effects of vapours of silver, mercury and lead. Written in 1473 the paper was printed for the first time at Augsburg in 1524 (Fig. 7). "Warmth" should counteract against the "cold" of the metals and their vapours. This idea of the arrangement of elements was found in Aristoteles' medicine and in alchemy up to contemporary times. Therefore in this pamphlet, which Koelsch (1927) termed the "first instruction in industrial hygiene in world literature", Ellenbog recommended treacle taken in tea or vermouth.

Soon after the invention of book printing using moveable type, Wenceslaus Beyer (1523), born at Elbogen in 1488 and with the pen name of Dr. Cu-

Fig. 7. Title page of Ellenbog's pamphlet on the poisonous evil vapours and fumes of metals. The wood engraving first decorated a disputation by Hans Sachs from 1524. The shoes in Sachs' hands were scratched out, and so were the woman and the house depicted in the original engraving

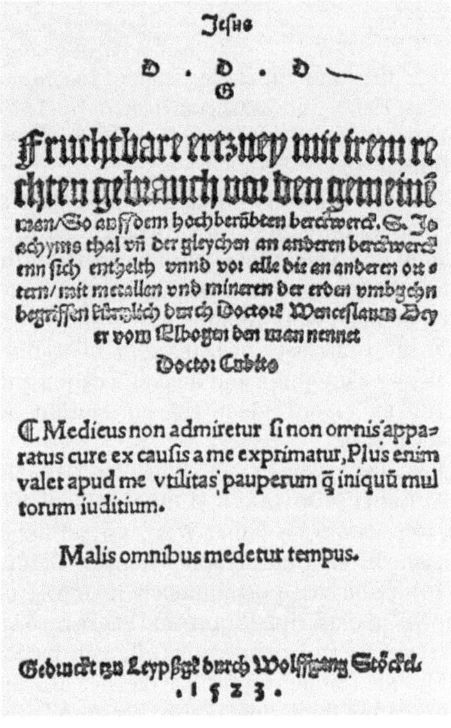

Fig. 8. Title page of Beyer's "Fecund remedy with its signature for the man working in the famous mine of St. Joachimsthal." Printed in 1523 by Wolffgang Stöckel of Leipzig, who had also published a sermon by Martinus Luther delivered on June 29, 1519, during the Leipzig disputation

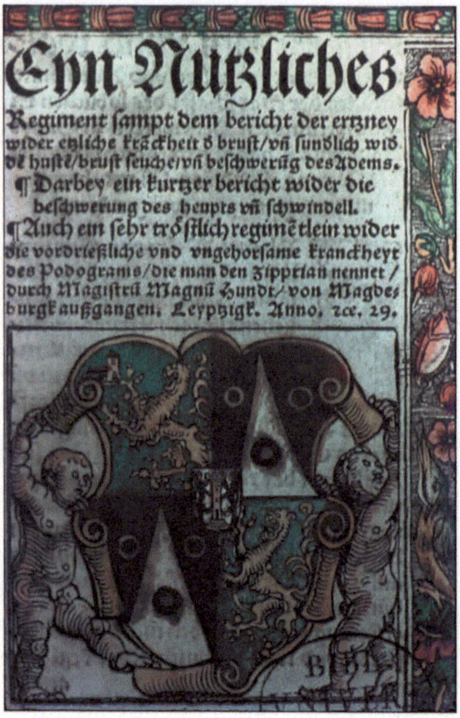

Eyn Nutzliches
Regiment sampt dem bericht der ertzney
wider etliche kranckheit d brust/vñ sündlich wid
dē hustē/brust seuche/vñ beschwerūg des Adems.
¶ Darbey ein kurtzer bericht wider die
beschwerung des heupts vñ schwindell.
¶ Auch ein sehr tröstlich regimētlein wider
die vordrießliche vnd vngehorsame kranckheyt
des Podogramis/die man den zipperlan nennet/
durch Magistrū Magnū Hundt/von Magde-
burgk außgangen. Leyptzigk. Anno. xc. 29.

Fig. 9. Title page of HUNDT's "Useful rule including a report of remedy for some thoracic diseases". Hand-coloured specimen owned by the Leipzig University Library. The coat of arms of the lords of the St. Joachimsthal mines, the Courts Schick, has five compartments, corresponding to their principal estates

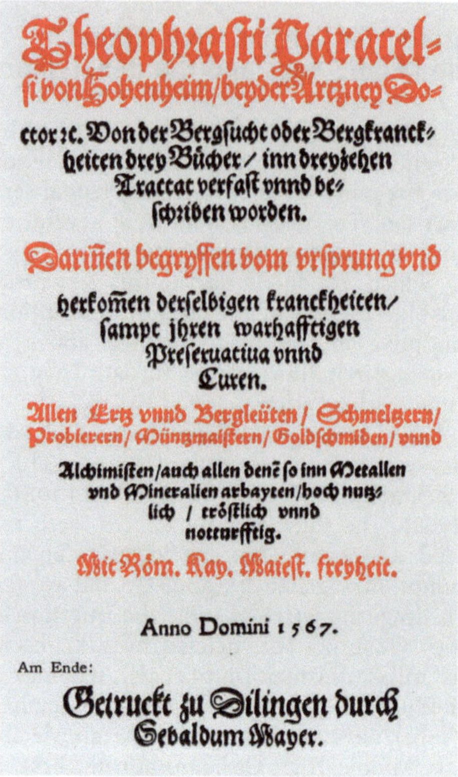

Theophrasti Paracel-
si von Hohenheim/beyder Artzney Do-
ctor ꝛc. Von der Bergsucht oder Bergkranck-
heiten drey Bücher/ inn dreyzehen
Tractat verfaßt vnnd be-
schriben worden.
Darinnen begryffen vom vrsprung vnd
herkomen derselbigen kranckheiten/
sampt jhren warhafftigen
Preseruatiua vnnd
Curen.
Allen Ertz vnnd Bergleüten/ Schmeltzern/
Probierern/ Müntzmaistern/ Goldschmiden/ vnnd
Alchimisten/auch allen denē so inn Metallen
vnd Mineralien arbayten/hoch nutz-
lich / tröstlich vnnd
notturfftig.
Mit Röm. Kay. Maiest. freyheit.
Anno Domini 1567.
Am Ende:
Getruckt zu Dilingen durch
Sebaldum Mayer.

Fig. 10. Title page of PARACELSUS' "Von der Bergsucht" as printed for the first time by Sebaldus Mayer at Dillingen

bito, the "first municipal physician of Joachimsthal" (CLEMEN 1904), and MAGNUS HUNDT JR (1529) edited instructions on medicine for miners' consumption (Figs. 8 and 9).

STURM (1965) emphasised that WERNICH's "Biographisches Lexikon der hervorragenden Ärzte aller Zeiten und Völker" was wrong in attributing the "Useful rule" to the MAGNUS HUNDT born in Magdeburg in 1449, who attained a doctorate in medicine at the University of Leipzig in 1499, but later worked as a clergyman and died as a canon at Meissen. JETTER (1966) identified our author as the younger ("Jr").

PARACELSUS, whose real name was THEOPHRASTUS AUREOLUS BOMBAST (1493–1541) of Hohenheim, was, as was his father WILLIAM, an industrial physician. In his monograph "Von der Bergsucht" (Fig. 10), published posthumously in 1567 from an unknown manuscript (SUDHOFF 1925) by SAMUEL ARCHITECTUS and printed at Dillingen by SEBALDUS MAYER (Bibliographia Paracelsica No 88), he recommended the following prescription:

Rec. Liquoris tartari two ounces
Olii colcotarini one scruple
Laudani purissimi half a dram

mixed and administered with three grains of barley in weight.

PARACELSUS became a symbol of medicine in the eyes of his contemporaries. He carried far greater weight than many important humanists, and his influence never disappeared entirely from medicine. Nevertheless, PAGEL's (1958, p. 202) criticism of defining PARACELSUS as the "founder" of industrial medicine (Strebel 1948) must be appreciated: "To call him the 'founder' of any branch of medicine or pathology is … misleading."

AGRICOLA (1494–1555), born at Glauchau, entered Leipzig University in summer 1514 under his family name GEORGIUS PAWER* (cf. the facsimile from the student directory on the cover of the University of Leipzig catalogue, April 1994) to study Latin and Greek and took the B.A. degree in 1515. It

* AGRICOLA is the latinised form of his name, not a nom de plume as erroneously stated by COLLIS (1915). ACKERMANN (1783, pp. 5, 7, 118) erroneously retranslated AGRICOLA's name as 'Ackermann' and thus made himself a namesake of the famous man. Indeed much of the private life of AGRICOLA has remained unknown, particularly since all of the municipal archives of the little town of Glauchau were destroyed by fire in 1712 (HARTMANN 1953, p. 11).

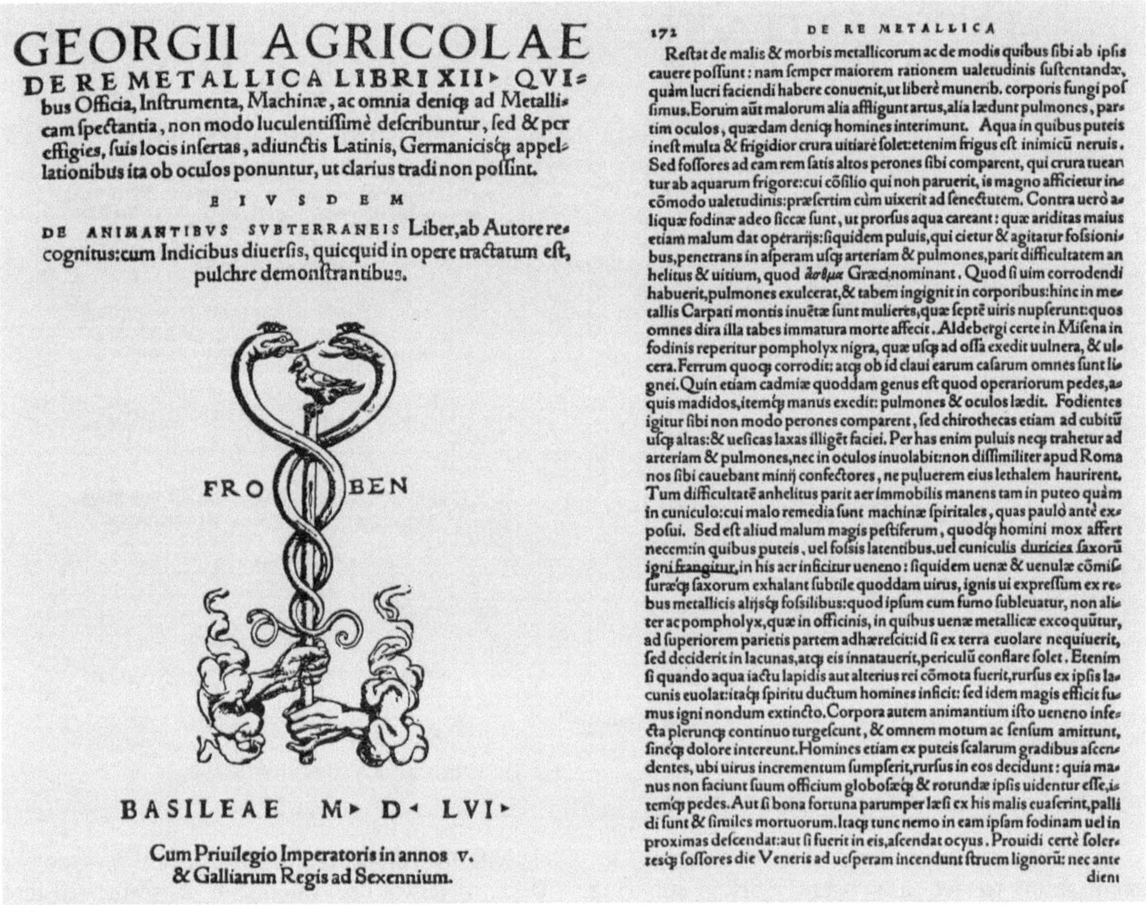

GEORGII AGRICOLAE DE RE METALLICA LIBRI XII. QVIbus Officia, Inſtrumenta, Machinæ, ac omnia deniq; ad Metallicam ſpectantia, non modo luculentiſſimè deſcribuntur, ſed & per effigies, ſuis locis inſertas, adiunctis Latinis, Germanicisq; appellationibus ita ob oculos ponuntur, ut clarius tradi non poſſint.

EIVSDEM

DE ANIMANTIBVS SVBTERRANEIS Liber, ab Autore recognitus: cum Indicibus diuerſis, quicquid in opere tractatum eſt, pulchre demonſtrantibus.

BASILEAE M. D. LVI.

Cum Priuilegio Imperatoris in annos V. & Galliarum Regis ad Sexennium.

172 DE RE METALLICA

Reſtat de malis & morbis metallicorum ac de modis quibus ſibi ab ipſis cauere poſſunt: nam ſemper maiorem rationem ualetudinis ſuſtentandæ, quàm lucri faciendi habere conuenit, ut liberè munerib. corporis fungi poſſimus. Eorum aũt malorum alia affligunt artus, alia lædunt pulmones, partim oculos, quædam deniq; homines interimunt. Aqua in quibus puteis ineſt multa & frigidior crura uitiarè ſolet: etenim frigus eſt inimicũ neruis. Sed foſſores ad eam rem ſatis altos perones ſibi comparent, qui crura tueantur ab aquarum frigore: cui cõſilio qui non paruerit, is magno afficietur incõmodo ualetudinis: præſertim cùm uixerit ad ſenectutem. Contra uerò aliquæ fodinæ adeo ſiccæ ſunt, ut prorſus aqua careant: quæ ariditas maius etiam malum dat operarijs: ſiquidem puluis, qui cietur & agitatur foſsionibus, penetrans in aſperam uſq; arteriam & pulmones, parit difficultatem anhelitus & uitium, quod ἄσθμα Græci nominant. Quod ſi uim corrodendi habuerit, pulmones exulcerat, & tabem ingignit in corporibus: hinc in metallis Carpati montis inuẽtæ ſunt mulieres, quæ ſeptẽ uiris nupſerunt: quos omnes dira illa tabes immatura morte affecit. Aldebergi certe in Miſena in fodinis reperitur pompholyx nigra, quæ uſq; ad oſſa exedit uulnera, & ulcera. Ferrum quoq; corrodit: atq; ob id claui earum caſarum omnes ſunt lignei. Quin etiam cadmiæ quoddam genus eſt quod operariorum pedes, aquis madidos, itemq; manus exedit: pulmones & oculos lædit. Fodientes igitur ſibi non modo perones comparent, ſed chirothecas etiam ad cubitũ uſq; altas: & ueſicas laxas illigẽt faciei. Per has enim puluis neq; trahetur ad arteriam & pulmones, nec in oculos inuolabit: non diſſimiliter apud Romanos ſibi cauebant minij confectores, ne puluerem eius lethalem haurirent. Tum difficultatẽ anhelitus parit aer immobilis manens tam in puteo quàm in cuniculo: cui malo remedia ſunt machinæ ſpiritales, quas pauló antè expoſui. Sed eſt aliud malum magis peſtiferum, quodq; homini mox affert necem: in quibus puteis, uel foſsis latentibus, uel cuniculis duricies ſaxorũ igni frangitur, in his aer inficitur ueneno: ſiquidem uenæ & uenulæ cõmiſſuræq; ſaxorum exhalant ſubtile quoddam uirus, ignis ui expreſſum ex rebus metallicis alijsq; foſsilibus: quod ipſum cum fumo ſubleuatur, non aliter ac pompholyx, quæ in officinis, in quibus uenæ metallicæ excoquũtur, ad ſuperiorem parietis partem adhæreſcit: id ſi ex terra euolare nequiuerit, ſed deciderit in lacunas, atq; eis innatauerit, periculũ conflare ſolet. Etenim ſi quando aqua iactu lapidis aut alterius rei cõmota fuerit, rurſus ex ipſis lacunis euolat: itaq; ſpiritu ductum homines inficit: ſed idem magis efficit fumus igni nondum extincto. Corpora autem animantium iſto ueneno infecta plerunq; continuo turgeſcunt, & omnem motum ac ſenſum amittunt, ſineq; dolore intereunt. Homines etiam ex puteis ſcalarum gradibus aſcendentes, ubi uirus incrementum ſumpſerit, rurſus in eos decidunt: quia manus non faciunt ſuum officium globoſæq; & rotundæ ipſis uidentur eſſe, itemq; pedes. Aut ſi bona fortuna parumper læſi ex his malis euaſerint, pallidi ſunt & ſimiles mortuorum. Itaq; tunc nemo in eam ipſam fodinam uel in proximas deſcendat: aut ſi fuerit in eis, aſcendat ocyus. Prouidi certè ſolertesq; foſſores die Veneris ad ueſperam incendunt ſtruem lignorũ: nec ante dieni

Fig. 11. Title and page 172 of AGRICOLA's "De re metallica"

has not been established (MÜLLER 1994) whether he began his medical studies here or in Bologna, Venice and Padova. As a physician he worked at St. Joachimsthal and Chemnitz.

His most famous monograph "De re metallica" (1556) described miners' diseases of the lungs as well as of the eyes and the ulcerative effects of uraninite down to the bones. Dry pits are more harmful to the miners, since they allow the dust to penetrate much more easily into the trachea and the lungs. Pulmonary ulceration made the affected miners waste away. In some Carpathian mining communities there were women who had married seven times.

Augustus, elector of Saxony (1526–1586) in a letter dated January 18, 1555 urged AGRICOLA to have his yet unpublished book translated into German as quickly as possible (RAEDTS 1955). This translation by PHILIPP BECHIUS was printed by Froben and Bischoff in 1557. SCHIFFNER (1928) edited a recent German translation which was published by the Agricola Society in conjunction with the German Museum. HERBERT CLARK and LOU HENRY HOOVER (1912) translated the text from the first Latin edition of 1556 with a biographical introduction, annotations and appendices upon the development of mining methods, metallurgical processes, geology, mineralogy, and mining law from the earliest times to the 16th century.

Magister JOHANNES MATHESIUS (1504–1565), born at Rochlitz in Saxony, worked, with short interruptions, at St. Joachimsthal (Jáchymov) as a teacher at a grammar school and later as priest. In the first sermon of his "Bergpostille or Sarepta" (folio II), first published at Nuremberg in 1562 and praised by SCHEFFLER (1770, pp. 112 and 183), he recommended the mineral springs and health resorts near the mines such as Gastein and Karlsbad (Karlovy Vary) as a pharmacy for the poor mining people.

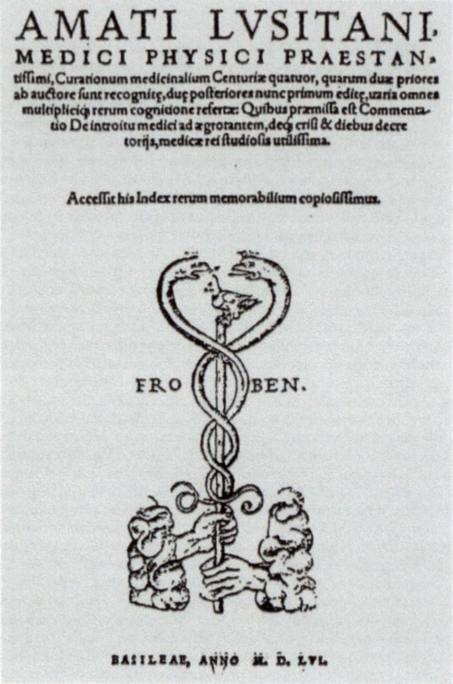

AMATI LVSITANI MEDICI PHYSICI PRAESTANtissimi, Curationum medicinalium Centuriæ quatuor, quarum duæ priores ab auctore sunt recognitę, duę posteriores nunc primum editę, uaria omnes multiplicique rerum cognitione refertæ: Quibus præmissa est Commentatio De introitu medici ad ægrotantem, deque crisi & diebus decretorijs, medicæ rei studiosis utilissima.

Accessit his Index rerum memorabilium copiosissimus.

BASILEAE, ANNO M. D. LVI.

Fig. 12. Title page of Amatus Lusitanus' 400 cures, printed by Froben at Basel in 1556

SAREPTA Darinn von allerley Bergwerck vnnd Metallen / Was jr eygenschafft vnd natur / vnd wie sie zu nutz vnd gut gemacht / guter bericht gegeben. Mit tröstlicher vnd lehrhaffter erklerung aller sprüch / so in Heiliger Schrifft von Metall reden / Vnnd wie der Heilig Geist inn Metallen vnnd Bergarbeit die Artickel vnsers Christlichen glaubens fürgebildet.

Auff ein newes mit fleiß durchsehen / corrigirt / vnd gebessert mit einem Register / Was für sprüch auß altem vnd newem Testament hierinn erkleret / vnd an welchem blat ein jeder zufinden ist.

Sampt der Jochimsthalischen kurtzen Chroniken.

Durch M. Johann Mathesium Pfarrer in S. Jochimsthal / selber für seinem seligen ende verfertigt.

PSALM. CXLVIII.
Berg vnnd Thal lobet den HERREN.

Mit Römischer Keyserlicher Maiestat Freyheit / nicht nachzudrucken auff sechs jar.

Gedruckt zu Nürnberg bey Dietrich Gerlatz.
M. D. LXXI.

Fig. 13. Title page of Mathesius' Sarepta

In his 341st* cure, Amatus Lusitanus (1556) successfully treated a gypsum worker suffering from fever and headache with acetic syrup and a remedy containing bile and slime.

2.4 Professional Lung Diseases in the Baroque Period (ca. 1600 to 1740)

Hippolytus Guarinoni (1571–1654), capitulary and municipal physician of Hall in Tyrol noticed the mountain sickness of the black pitmen caused by continuous working in deep ore mines; in the salt mine at Hall where miners worked underground just as long as this disease was unknown.

The booklet by Martin Pansa (1614), municipal physician at Annaberg in the Ore Mountains, may be a work of plagiarism or at most a compilation of gleanings. Rosner (1957) compared Pansa's text with the corresponding passages of Agricola's "De re metallica libri duodecim" as translated into German by Philipp Bechius (1557) and of Joh. Mathesius' (1563) "Sarepta". However, Pansa's booklet may have satisfied the information needs of the endangered miners. It was reprinted in 1681 (for the title page and three pages of the text, see Schiller 1954, Fig. 4). Jetter (1966) emphasised that it would have been of little importance to quote famous authors as is customary in papers designed for learned readers and defended Pansa against the disreputable suspicion of plagiarism.

Daniel Sennert (1572–1637), professor of medicine at Wittenberg, in chap. 9 of his "Consent and dissent of the chemists and the Aristotelians and the Galenists" (1619) was informed by a mine physician that the bodies of mine workers contained the very metal they had been employed to dig (cf. Ackermann 1783, footnote on p. 52). Sennert had clear insight into the object-specificity of volatile effluvia and their difference from air and water-vapour. Smoke from oak-wood, he said, is substantially different from fumes given off by wood from other trees or from manure. Effluvia emanating from the cadaver of a dog differ in quality from those of any other animal, as is indicated by their smell. Moreover, all these are different in kind from vapour, which is but water volatilised

* Ramazzini (1703, p. 64) and Ackermann [although right on p. 153 of volume 1 (1780)], in vol. 2 (1783, p. 171), erroneously quoted the 5th century. This cure was contained in the Basel edition of 1556 comprising both the first two centuries in a revised printing and the third and fourth centuries edited for the first time (Fig. 12). All seven centuries were first published at Venice in 1563 in duodecimo and in 1566 in octavo. Some later editions were printed at Lyons in 1580 (in duodecimo), at Bordeaux in 1620 (in quarto), and at Frankfort in 1686 (in folio).

CONSILIVM PERIPNEV-
MONIACVM:
Das iſt/
Ein getrewer Rath
in der beſchwerlichen Berg- vnd Lungen-
ſucht/ darinnen verfaſſet/ was die fürnemſten Vrſachen
ſeyn beyderley Beſchwerungen/ beydes der gifftigen / die vom
Bergwerck entſtehet: ſo wol der gemeinen / die von Flüſſen her-
rühret: Zuvor aber/ wie der Menſch mit der kleinen Welt/ vnd mit dem
Bergwerck artlich zu vergleichen / vnd wie beyde Suchten
zu vertreiben ſeyn:
Allen Lung- vnd Bergſüchtigen zum beſten:
Benebens XXX. diſputirlichen Fragen/von dieſer
Materia geſtellet/
Durch
MARTINVM PANSAM,
Schleúſingenſem, Franc. Philoſophum, Medi-
cum, verordneten Stadt-Doctorem auff
S. Annenbergk.
Zuvor niemals in Druck außgangen.
Horatius lib. 1. Epiſtolarum, in fine
Epiſt. 6.
—— —— *Si quid noviſti rectius iſtis*
Candidus imperti: ſi non, his utere mecum
Gedruckt zu Leipzig bey Lorentz Kober/In verlegung
Thomæ Schürers/Buchhändlers.
Im Jahr 1614.

Das neunzehende Capitel. 47
Schwaden in einem ſonderlichen ort entſpringet/ vnd nicht
mag hingenommen oder geringert werden nach vnſerm gefal-
len/ſo bleibt die Grube liegen/ ob ſie gleich von Silber reich
were: Denn es iſt mehr an der geſundheit/ denn an Geldt vnd
Gut gelegen.

Das neunzehende Capitel.
Wie die Bergkſucht im Men-
ſchen entſtehe.
WEil nu der Rauch oder dampff mit ſampt der
Lufft den Athem geben/ſo iſt hinfort zu betrachten die
beſchwerung/die aus ſolchem angezogenem Gifft ent-
ſpringet. Paracellus nennet jhn Mercurium peccantem:
Denn ſo derſelbe in die Lung kömpt / ſo nimpt er jhm eine
Wand für / daran er ſich kleibet/vnd wird bey etlichen pu-
trificirt in ein oliter, oder feſtigkeit bey etlichen in eine ſaltzi-
ſche materien: Oder in eine Tragantiſche reſolution: Oder
in einen kalten Rotz: Oder in ein gantz zehen vnd läderigen
Schleim. Denn weil ſolcher Gifft von der Natur nicht mag
verzehret werden/ſo legt er ſich zum theil in die Lufftadern vnd
Röhrlein der Lungen/zum theil in die kleinen lufftlöchlein der-
ſelben/ zum theil in die lobos oder flügel der Lungen / er durch-
dringet auch die engen poros vmb die gegend der Nieren: Das
man meint es ſey der Lendenſtein: Dannenhero ſolche materia
manchmal in dem Vrin gefunden wird. In Summa / dieſe
Sucht legt ſich in dreyerley wege der Lungen an / Erſtlich auff
Mercurialiſch/ wie ein ſublimirter Rauch / der ſich coagulie-
ret vnd zuſammen ſetzt: Zum andern wie ein Saltzgeiſt / der
H iij ſich

Fig. 14. Title page and p. 47 of PANSA's "Consilium peripneumoniacum" published in 1614. A second edition dating from 1681 showed a view of Freiberg in Saxony

and hence neither identical with air nor object-specific. Metal-fume is atomised metal and as such inhaled and deposited in the miner's lung. SENNERT was a near contemporary of VAN HELMONT. His relevant work appeared before that of the latter, but one may assume that, at a time well after VAN HELMONT, he finished the bulk of his chemical experiments and reaped their intellectual fruit, notably his concept of gas (PAGEL 1982, p. 70).

J. B. VAN HELMONT (1577–1644), an ioatrochemist, wrote "Ortus medicinae" (Fig. 15), which was posthumously edited in 1648 by his son, FRANCISCUS MERCURIUS VAN HELMONT. On p. 370 of his chapter on "Asthma et tussis", VAN HELMONT cited several occupational groups such as miners (fossores) and metal smelters (fusores metallici) soon to be affected by asthma. Masked mercury vapours (fumus Mercurii utut larvatus) obstruct and constrict the larynx. Thus asthma arises. In another asthmatic who suddenly died from suffocation, VAN HELMONT found but very little fluid in the lung; he described a dorsal pleural adhesion of the right lung as the cause of death. For therapy ptisan, cortex Chinae or similar remedies were recommended (p. 380).

M. L. URSINUS (1618–1664) in his "Disputatio medica inauguralis de morbis metallariorum" published in Leipzig in 1652 distinguished two different pulmonary diseases: one not caused by external poisons, but "ab humoribus", the other by minerals within and on the earth, from which poisons escape either spontaneously or on roasting.

SAMUEL STOCKHAUSEN, a mine physician at Goslar, also dealt, in his "Hütten Katze" (1656), with the diseases of mine - and smelter - workers. Most of the book is in Latin but STOCKHAUSEN added an appendix in German on the miners' sickness, indicating that he realised the practical importance of such knowledge to both mine officials and mine workers.

ATHANASIUS KIRCHER (1601–1680) in the second volume of his "Mundus subterraneus" (1664, 1665, 1678), devoted chapter 2 of book 10 to miners' diseases and the following chapter to their cure.

THOMAS WILLIS (1621–1675) made several allusions to persons voiding, with a slight cough, sputa like black ink, often in a day, and especially every

Fig. 17. The allegorical frontispiece of A. Kircher's "Mundus subterraneus" shows Heaven, Sun and Moon. Underneath this, suspended on Homer's Golden Chain, Earth is blown by the Winds. The banner across the lower part of the chain reproduces hexameters from Virgil's Aeneid (VI 726 and 727): "Spiritus intus alit, totamque infusa per artus Mens agitat molem"

«Causa Tragœdiæ fumi metallici sunt, quia de Paralysi non aliorum sed tantum Metallariorum hic sermo est» (l.c. p. 5).

Ramazzini (1700, p. 21; 1703, p. 21) recommended milk for throat complaints:

> Pro faucium, & gingivarum erosione gargarismata ex lacte egregiam præstabunt operam, ut quæ particulas illas corrosivas inibi hærentes demulceant, ac absorbeant.

> For the corrosion and soreness of the throat and gums, gargarisms of milk are extremely serviceable, as being apt to qualify and absorb the corrosive particles lodged in those parts. (Translation quoted from Rosen 1943, p. 119)

In his reflections on "Ramazzinian sources" from the "De morbis artificum diatriba" Carozzi (1937) complained that Ramazzini had not any conception of social solidarity, of the interdependence of the individuals forming a social unit within the city, and had but a very imperfect idea of the social aspect of occupational disease. However, at the begin-

GEORGII WOLFF-
GANGI WEDELII,
MED. DOCTORIS, CONSILIARII ET ARCHIATRI
DUCALIS SAXONICI, ET PROFESSORIS
PUBLICI,
PATHOLOGIA
MEDICA DO-
GMATICA.
JENÆ,
Sumptibus JOHANNIS BIELKII, Bibliopol.
Typis KREBSIANIS.
ANNO M DC XCII.

Fig. 18. Title page of Georg Wolffgang Wedel's "Pathologia medica dogmatica"

Fig. 19. Georg Wolfgang Wedel (1645–1721), engraving published 1718 in Acta eruditorum

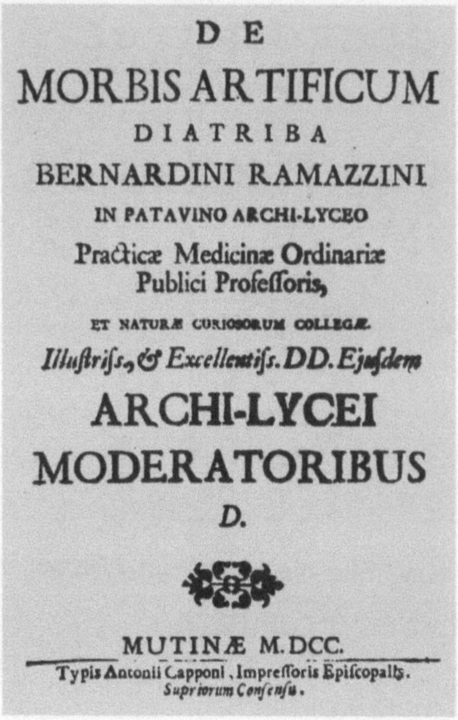
DE
MORBIS ARTIFICUM
DIATRIBA
BERNARDINI RAMAZZINI
IN PATAVINO ARCHI-LYCEO
Practicæ Medicinæ Ordinariæ
Publici Profeſſoris,
ET NATURÆ CURIOSORUM COLLEGÆ.
Illuſtriſs., & Excellentiſs. DD. Ejuſdem
ARCHI-LYCEI
MODERATORIBUS
D.
MUTINÆ M.DCC.
Typis Antonii Capponi, Impreſſoris Epiſcopalis.
Superiorum Conſenſu.

Fig. 20. RAMAZZINI's "De morbis artificum diatriba". Title page of the first edition

ning of his book (p. 12, in the German translation of 1705 on p. 4) RAMAZZINI wrote:

> Extat elegans Epistula D. Cypriani ad complures Episcopos, & Diacones, quos imperatorum, barbaries Metallorum fossioni addixerat, in qua eosdem hortatur, ut jis in fodinis, è quibus aurum, & argentum erruerent se verum Christi aurum probarent.

> Der heilige Cyprianus ermahnet in einer herrlichen Epistel die oeffentlich am Tage liegt / viele Bischoeffe und Pfarrer / die von den tyrannischen Kaeysern in die Bergwercke religiret worden; Sie moechten doch in den Gold- und Silber-Minen das wahre und rechtschaffene Gold Christi bewaehren.

LEGGE (1946, p. 30) stated: "There is a fine humanity, free from any taint of ludicrous, expressed in passages in his book". It is interesting to note the preventive measures advocated for avoiding the sucking in of smoke (mercury fumes) into their mouths by turning their backs to the wind when gilders fired amalgam:

"Cum enim non nisi per amalgamationem id peragi possit, dum postea Mercurium igne propellunt, non tam cauti esse possunt, licet faciem avertant, quin virosus halitus per os excipiant, quare

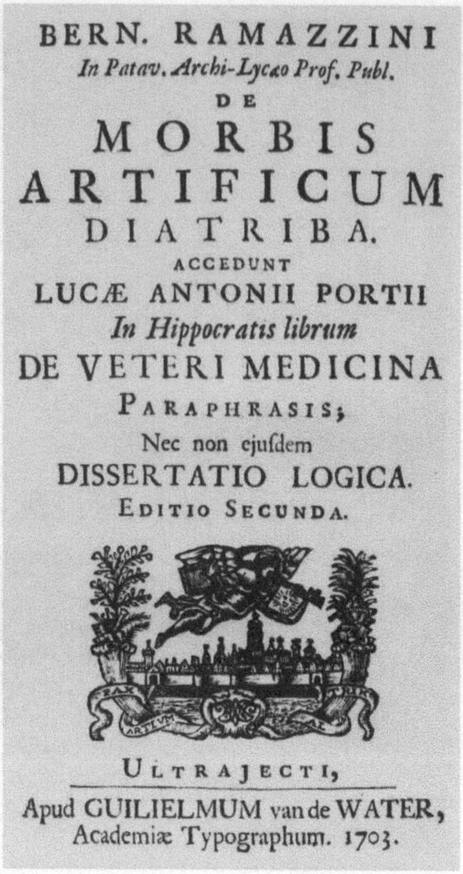
BERN. RAMAZZINI
In Patav. Archi-Lycæo Prof. Publ.
DE
MORBIS
ARTIFICUM
DIATRIBA.
ACCEDUNT
LUCÆ ANTONII PORTII
In Hippocratis librum
DE VETERI MEDICINA
PARAPHRASIS;
Nec non ejuſdem
DISSERTATIO LOGICA.
EDITIO SECUNDA.
ULTRAJECTI,
Apud GUILIELMUM van de WATER,
Academiæ Typographum. 1703.

Fig. 21. RAMAZZINI's "De morbis artificum diatriba". Title page of the second edition

hujusmodi Artifices vertiginosi, asthmatici, paralytici citissime evadunt, habitusque cadaverosos contrahunt" (RAMAZZINI 1703, p. 25).

According to ACKERMANN (1780, folio 4), who cited HALLER's Bibliotheca medicinae practicae, tome III, L.X. p. 483, RAMAZZINI's achievements became apparent in nine reprints of the original Latin edition and seven prints of his subsequent works. "De morbis artificum" was translated into German (Fig. 22 and Fig. 23), English (1705, 1746), Italian and Dutch. DE FORCROY (1777) attended to a French translation from the Latin, which was extended and furnished with footnotes. After ACKERMANN (1780, 1783) some more adaptations and enlargements were published, e.g. by PATISSIER (1822) and SCHLEGEL (1823; Fig. 25). In his preface SCHLEGEL (1823) mentioned 17 reprints and the adaptations by MORGAGNI, DE FORCROY and ACKERMANN. While ADELMANN (1803) referred to 2771 craftsmen who fell ill in Würzburg within 16 years, PATISSIER included nearly 20,000 artisans and workmen from all the Parisian hospitals in 1807 alone.

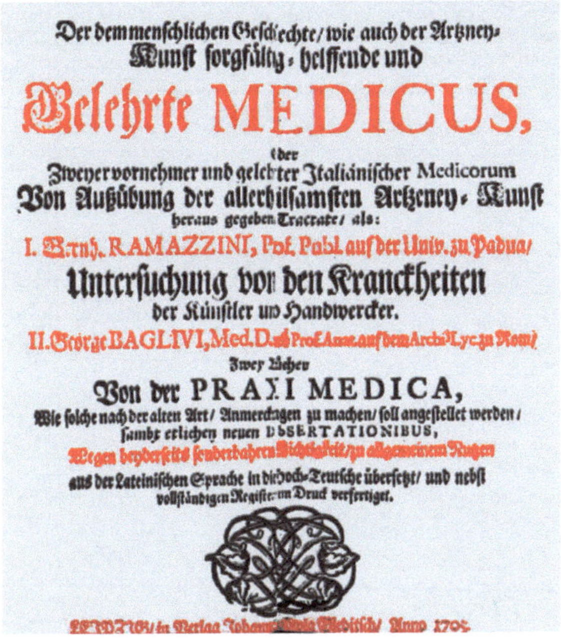
Der dem menschlichen Geschlechte/ wie auch der Artzney-Kunst sorgfältig-helffende und

Gelehrte MEDICUS,

oder

Zweyer vornehmer und gelehrter Italiänischer Medicorum

Von Außübung der allerheilsamsten Artzeney-Kunst

heraus gegeben Tractate/ als:

I. Bernh. RAMAZZINI, Prof. Publ. auf der Univ. zu Padua/

Untersuchung von den Kranckheiten

der Künstler und Handwercker.

II. Georg BAGLIVI, Med. D. und Prof. Anat. auf dem Archi-Lyc. zu Rom/

Zwey Bücher

Von der PRAXI MEDICA,

Wie solche nach der alten Art/ Anmerckungen zu machen/ soll angestellet werden/ sambt etlichen neuen DISSERTATIONIBUS,

Wegen beyderseits sonderbahren Wichtigkeit/ zu allgemeinem Nutzen

aus der Lateinischen Sprache in die Hoch-Teutsche übersetzt/ und nebst vollständigen Register im Druck verfertiget.

LEIPZIG/ im Verlag Johann [illegible] Gleditsch/ Anno 1705.

Fig. 22. Title page of the first German translation of RAMAZZINI's treatise, published together with a contemporaneous medical book. The translator is anonymous

BERNHARDI RAMAZZINI,

Med. Doct. und Professoris Publ. zu Padua,

Untersuchung

Von denen

Kranckheiten

der Künstler und Handwercker,

Worinnen

die Kranckheiten, womit fast alle Künstler und Handwercker gemeiniglich befallen werden, genau beschrieben, wie durch die Kunst oder Handwerck solche zugezogen werden, und wie man solche hernachmals aufs beste und leichteste curiren kan.

LEIPZIG,

bey Moritz George Weidmann.

Anno 1718.

Fig. 23. German translation of RAMAZZINI's treatise. Weidmann's publishing and bookselling house was founded at Frankfort on the Main in 1680 and transferred to Leipzig in 1681. Its most famous authors included WIELAND, GELLERT, LESSING, and LAVATER

Bernhard Ramazzini's

ehemaligen Professor primarius der Arzneywissenschaft zu Padua

Abhandlung

von den

Krankheiten

der

Künstler und Handwerker

neu bearbeitet und vermehret

von

Dr. Johann Christian Gottlieb Ackermann.

Stendal,

bey D. C. Franzen und J. C. Grosse. 1780.

Fig. 24. ACKERMANN's adaptation and enlargement of RAMAZZINI's treatise. Title page of the first of two volumes. The title page of volume 2 identifies him as a member of the Imperial Academy «Leopoldina»

SIEMENS (1705) of Goslar advised metallurgists to protect themselves from the vapours of mercury, antimony, sulphur, arsenic, and lead by avoiding their inhalation and by the swallowing of saliva. It is best to work in the open air. Protective drugs include ginger, cardamom, and cinnamon. For beverages SIEMENS mentioned decocts of veronica, mallow and ivy, and coffee with plenty of warm cow milk.

According to B. ZORN (1714, p. 393) pickmen and miners take pulverised lovage root in wine against the damp and poisonous metallic vapours before descending into the pit or working in the smeltery.

DE PRÉ's (1719) dissertation at Erfurt University is entitled "De phthisi pulmonali samiatorum, vulgo – die Schleiferkrankheit". This disease was mentioned by HIRT (1871, p. 22) as a non-tuberculous phthisis associated with a dusty trade.

JOHANNES BUBBE (1721) in his dissertation at Halle University entitled "De spadone Hippocratico lapicidarum Seebergensium hæmoptysin et phthisin pulmonalem, vulgo: Der Seeberger Steinbrecher Kranckheit" for the first time published some details on a disease caused by the inhalation of stone

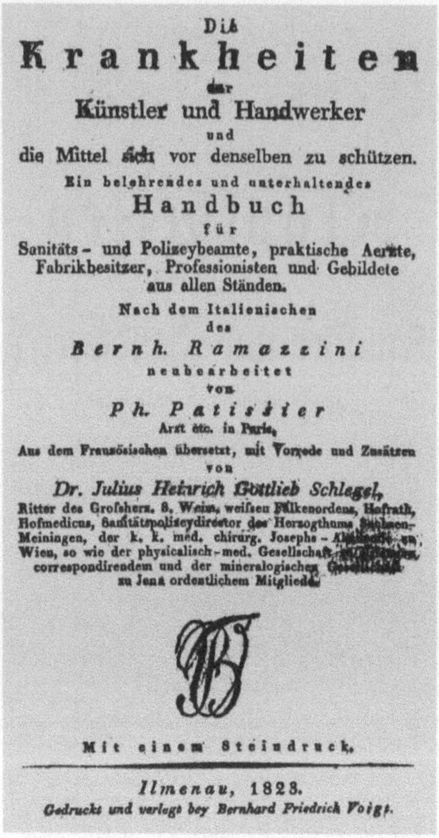

Die
Krankheiten
der
Künstler und Handwerker
und
die Mittel sich vor denselben zu schützen.
Ein belehrendes und unterhaltendes
Handbuch
für
Sanitäts- und Polizeybeamte, praktische Aerzte,
Fabrikbesitzer, Professionisten und Gebildete
aus allen Ständen.
Nach dem Italienischen
des
Bernh. Ramazzini
neubearbeitet
von
Ph. Patissier
Arzt etc. in Paris,
Aus dem Französischen übersetzt, mit Vorrede und Zusätzen
von
Dr. Julius Heinrich Gottlieb Schlegel,
Ritter des Großherz. S. Weim. weißen Falkenordens, Hofrath,
Hofmedicus, Sanitätspolizeydirector des Herzogthums [illegible]-
Meiningen, der k. k. med. chirurg. Josephs-[illegible] zu
Wien, so wie der physicalisch-med. Gesellschaft [illegible]
correspondirendem und der mineralogischen [illegible]
zu Jena ordentlichem Mitgliede.

Mit einem Steindruck.

Ilmenau, 1823.
Gedruckt und verlegt bey Bernhard Friedrich Voigt.

Fig. 25. Schlegel's adaptation of Ramazzini's treatise based on Patissier's adaptation, translated from French into German, and extended with a preface and further additions

▼ **Fig. 26.** Bubbe's dissertation on the quarrymen's disease occurring at Seeberg

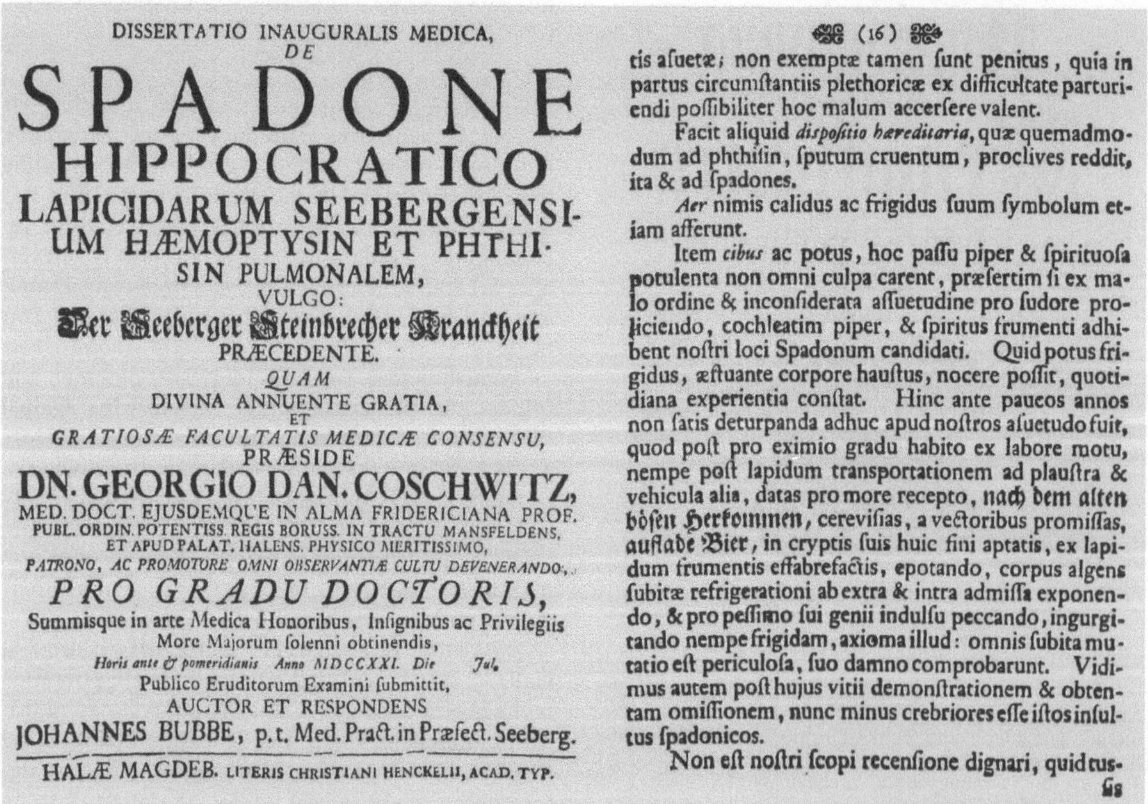

DISSERTATIO INAUGURALIS MEDICA,
DE
SPADONE
HIPPOCRATICO
LAPICIDARUM SEEBERGENSIUM HÆMOPTYSIN ET PHTHISIN PULMONALEM,
VULGO:
Der Seeberger Steinbrecher Kranckheit
PRÆCEDENTE.
QUAM
DIVINA ANNUENTE GRATIA,
ET
GRATIOSÆ FACULTATIS MEDICÆ CONSENSU,
PRÆSIDE
DN. GEORGIO DAN. COSCHWITZ,
MED. DOCT. EJUSDEMQUE IN ALMA FRIDERICIANA PROF.
PUBL. ORDIN. POTENTISS. REGIS BORUSS. IN TRACTU MANSFELDENS,
ET APUD PALAT. HALENS. PHYSICO MERITISSIMO,
PATRONO, AC PROMOTORE OMNI OBSERVANTIÆ CULTU DEVENERANDO,
PRO GRADU DOCTORIS,
Summisque in arte Medica Honoribus, Infignibus ac Privilegiis
More Majorum folenni obtinendis,
Horis ante & pomeridianis Anno MDCCXXI. Die Jul.
Publico Eruditorum Examini fubmittit,
AUCTOR ET RESPONDENS
JOHANNES BUBBE, p.t. Med. Pract. in Præfect. Seeberg.
HALÆ MAGDEB. LITERIS CHRISTIANI HENCKELII, ACAD. TYP.

(16)

tis afuetæ; non exemptæ tamen funt penitus, quia in partus circumftantiis plethoricæ ex difficultate parturiendi poffibiliter hoc malum accerfere valent.

Facit aliquid *difpofitio hæreditaria*, quæ quemadmodum ad phthifin, fputum cruentum, proclives reddit, ita & ad fpadones.

Aer nimis calidus ac frigidus fuum fymbolum etiam afferunt.

Item *cibus* ac potus, hoc paffu piper & fpirituofa potulenta non omni culpa carent, præfertim fi ex malo ordine & inconfiderata affuetudine pro fudore proliciendo, cochleatim piper, & fpiritus frumenti adhibent noftri loci Spadonum candidati. Quid potus frigidus, æftuante corpore hauftus, nocere poffit, quotidiana experientia conftat. Hinc ante paucos annos non fatis deturpanda adhuc apud noftros afuetudo fuit, quod poft pro eximio gradu habito ex labore motu, nempe poft lapidum transportationem ad plauftra & vehicula alia, datas pro more recepto, nach dem alten bösen Herkommen, cerevifias, a vectoribus promiffas, ausflade Bier, in cryptis fuis huic fini aptatis, ex lapidum frumentis effabrefactis, epotando, corpus algens fubitæ refrigerationi ab extra & intra admiffa exponendo, & pro peffimo fui genii indulfu peccando, ingurgitando nempe frigidam, axioma illud: omnis fubita mutatio eft periculofa, fuo damno comprobarunt. Vidimus autem poft hujus vitii demonftrationem & obtentam omiffionem, nunc minus crebriores effe iftos infultus fpadonicos.

Non eft noftri fcopi recenfione dignari, quid tus-
fis

dust. He classified his paper into the chapters: nomen (pp. 8–11), natura (pp. 11–24) and curatio (pp. 24–28). He speculated on some inheritable disposition for consumption and blood-tinged expectoration (p. 16).

KOCHLATSCH (1721) from Neusohl (Banská Bystrica), a formerly Hungarian, now Slovak mining community, in his dissertation at Halle University entitled "De metallicolarum nonnullis morbis" referred to ETTMÜLLER's "De respiratione laesa".

JOHANNES GOTTLIEB NEUMANN (1721) from Freiberg (Saxony) in his dissertation at Halle University entitled "De praeservandis metallicolarum morbis" pointed out that some diseases of miners and metallurgists are incurable. As adolescents are particularly susceptible, prophylaxis is most important for them. On account of the dust hazard arising from picking, this work should be performed in the open air or in well aerated rooms. To protect mouth and nose from penetrating dust the face should be covered with gauze.

THOMAS BENSON, a working painter, in an application made in 1726 for a patent for the wet grinding of flints for use in earthenware manufacture, stated that when the process was carried out by the usual dry method 'persons ever so healthful cannot survive above two years' (quoted from MEIKLEJOHN 1947, 1960).

J.J. WEPFER (1620–1695), famous for his studies in organic and inorganic toxicology, described quarriers' disease from a mountain cavern near Waldshut at the Rhine (1727, p. 444). He regretted being unable to report on any necropsy.

WEPFER (1679) discovered the glandulae duodenales, but they were named after his son-in-law, JOHANN CONRAD BRUNNER (1653–1727), professor at Heidelberg and personal physician of the Elector Palatine. BRUNNER (1686) published on this subject under the expressive title: "Novarum glandularum intestinalium decriptio", while WEPFER's discovery published under the title: "Cicutae aquaticae historia et noxae" might have been overlooked when personal names were introduced in the anatomical nomenclature.

MOLLER (1730) of Neusohl, obtained his doctorate at Halle University. In his "Dissertatio inaugura-

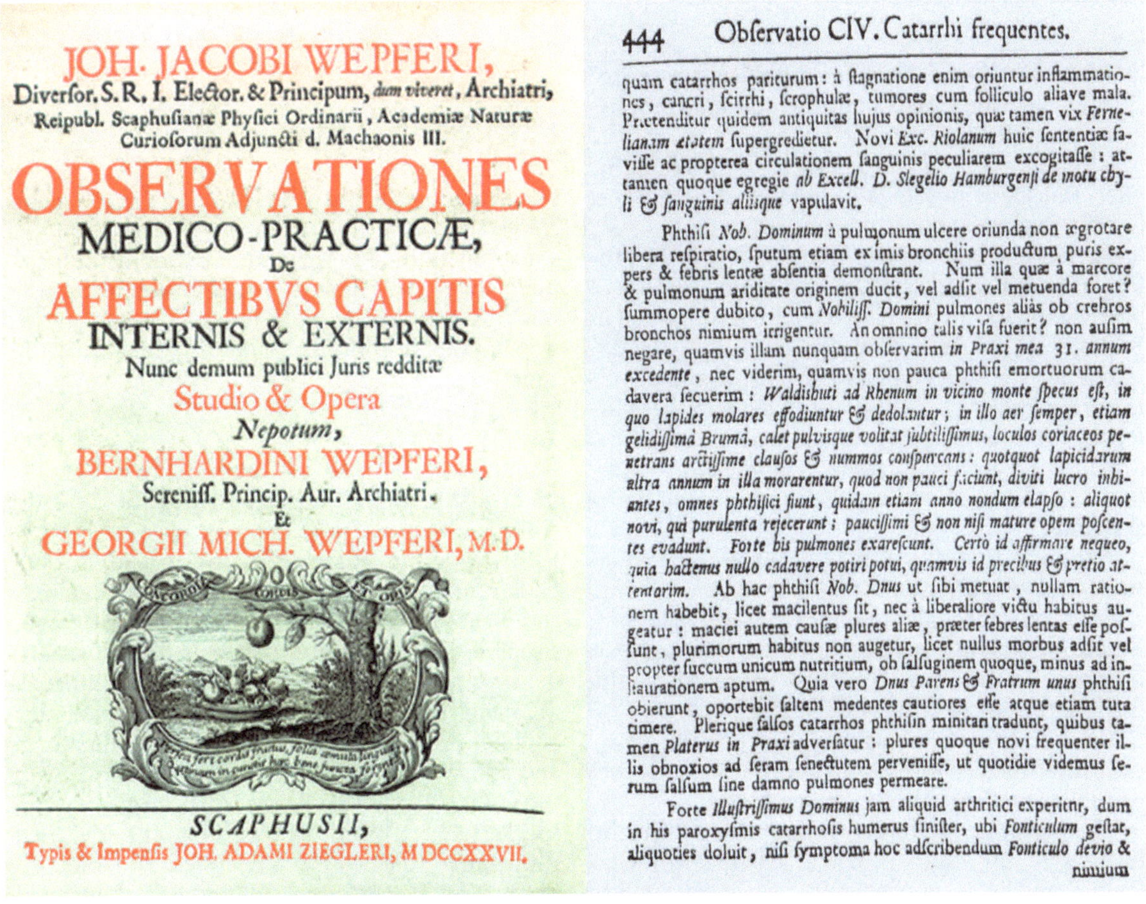
JOH. JACOBI WEPFERI,
Diverſor. S. R. I. Elector. & Principum, dum viveret, Archiatri,
Reipubl. Scaphuſianæ Phyſici Ordinarii, Academiæ Naturæ
Curioſorum Adjuncti d. Machaonis III.

OBSERVATIONES
MEDICO-PRACTICÆ,
De
AFFECTIBVS CAPITIS
INTERNIS & EXTERNIS.
Nunc demum publici Juris redditæ
Studio & Opera
Nepotum,
BERNHARDINI WEPFERI,
Sereniſſ. Princip. Aur. Archiatri,
Et
GEORGII MICH. WEPFERI, M.D.

SCAPHUSII,
Typis & Impenſis JOH. ADAMI ZIEGLERI, M DCCXXVII.

444 Obſervatio CIV. Catarrhi frequentes.

quam catarrhos pariturum: à ſtagnatione enim oriuntur inflammationes, cancri, ſcirrhi, ſcrophulæ, tumores cum folliculo aliave mala. Prætenditur quidem antiquitas hujus opinionis, quæ tamen vix *Fernelianam ætatem* ſupergredietur. Novi *Exc. Riolanum* huic ſententiæ faviſſe ac propterea circulationem ſanguinis peculiarem excogitaſſe: attamen quoque egregie *ab Excell. D. Slegelio Hamburgenſi de motu chyli & ſanguinis aliisque* vapulavit.

Phthiſi *Nob. Dominum* à pulmonum ulcere oriunda non ægrotare libera reſpiratio, ſputum etiam ex imis bronchiis productum puris expers & febris lentæ abſentia demonſtrant. Num illa quæ à marcore & pulmonum ariditate originem ducit, vel adſit vel metuenda foret? ſummopere dubito, cum *Nobiliſſ. Domini* pulmones aliàs ob crebros bronchos nimium irrigentur. An omnino talis viſa fuerit? non auſim negare, quamvis illam nunquam obſervarim *in Praxi mea* 31. *annum excedente*, nec viderim, quamvis non pauca phthiſi emortuorum cadavera ſecuerim: *Waldishuti ad Rhenum in vicino monte ſpecus eſt, in quo lapides molares effodiuntur & dedolantur; in illo aer ſemper, etiam gelidiſſimâ Brumâ, calet pulvisque volitat ſubtiliſſimus, loculos coriaceos penetrans arctiſſime clauſos & nummos conſpurcans: quotquot lapicidarum ultra annum in illa morarentur, quod non pauci faciunt, diviti lucro inhiantes, omnes phthiſici fiunt, quidam etiam anno nondum elapſo: aliquot novi, qui purulenta rejecerunt; pauciſſimi & non niſi mature opem poſcentes evadunt. Forte his pulmones exareſcunt. Certò id affirmare nequeo, quia hactenus nullo cadavere potiri potui, quamvis id precibus & pretio attentarim.* Ab hac phthiſi *Nob. Dnus* ut ſibi metuat, nullam rationem habebit, licet macilentus ſit, nec à liberaliore victu habitus augeatur: maciei autem cauſæ plures aliæ, præter febres lentas eſſe poſſunt, plurimorum habitus non augetur, licet nullus morbus adſit vel propter ſuccum unicum nutritium, ob ſalſuginem quoque, minus ad inſtaurationem aptum. Quia vero *Dnus Parens & Fratrum unus* phthiſi obierunt, oportebit ſaltem medentes cautiores eſſe atque etiam tuta timere. Plerique ſalſos catarrhos phthiſin minitari tradunt, quibus tamen *Platerus in Praxi* adverſatur: plures quoque novi frequenter illis obnoxios ad ſeram ſenectutem perveniſſe, ut quotidie videmus ſerum ſalſum ſine damno pulmones permeare.

Forte *Illuſtriſſimus Dominus* jam aliquid arthritici experitur, dum in his paroxyſmis catarrhoſis humerus ſiniſter, ubi *Fonticulum* geſtat, aliquoties doluit, niſi ſymptoma hoc adſcribendum *Fonticulo devio &* nimium

Fig. 27. WEPFER's posthumous monograph containing his report on Waldshut quarriers' disease

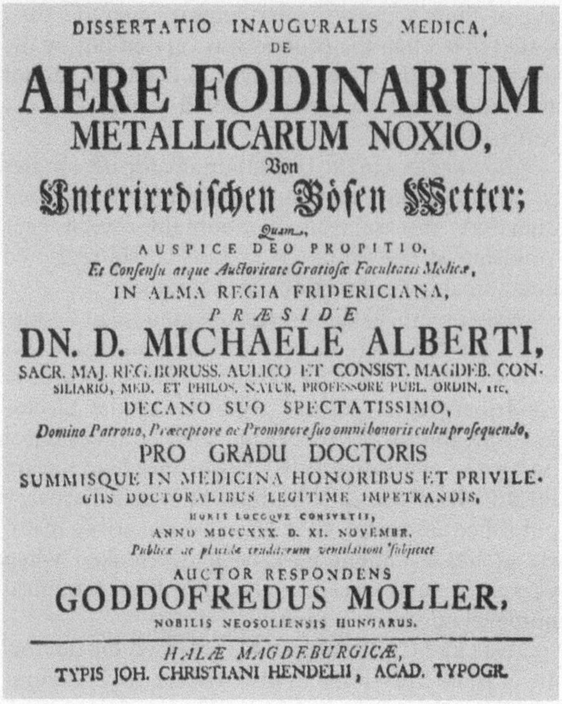
DISSERTATIO INAUGURALIS MEDICA,
DE
AERE FODINARUM
METALLICARUM NOXIO,
Von
Unterirrdischen Bösen Wetter;
Quam,
AUSPICE DEO PROPITIO,
Et Consensu atque Auctoritate Gratiosæ Facultatis Medicæ,
IN ALMA REGIA FRIDERICIANA,
PRÆSIDE
DN. D. MICHAELE ALBERTI,
SACR. MAJ. REG. BORUSS. AULICO ET CONSIST. MAGDEB. CONSILIARIO, MED. ET PHILOS. NATUR. PROFESSORE PUBL. ORDIN. etc.
DECANO SUO SPECTATISSIMO,
Domino Patrono, Præceptore ac Promotore suo omni honoris cultu prosequendo,
PRO GRADU DOCTORIS
SUMMISQUE IN MEDICINA HONORIBUS ET PRIVILEGIIS DOCTORALIBUS LEGITIME IMPETRANDIS,
HORIS LOCOQUE CONSUETIS,
ANNO MDCCXXX. D. XI. NOVEMBR.
Publicæ ac placidæ eruditorum ventilationi subjiciet
AUCTOR RESPONDENS
GODDOFREDUS MOLLER,
NOBILIS NEOSOLIENSIS HUNGARUS.

HALÆ MAGDEBURGICÆ,
TYPIS JOH. CHRISTIANI HENDELII, ACAD. TYPOGR.

Fig. 28. Title page of MOLLER's dissertation "De aere fodinarum noxio" [on noxious mine air]

D. Joh. Friedrich Henkels,
Kön. Pohln. und Churf. Sächs. Berg-Raths.
Medicinischer
Aufstand und Schmelz-Bogen
Von der
Bergsucht
und
Hütten-Katze,
auch einigen andern,
Denen
Bergleuten und Hütten-Arbeitern
zustoßenden Krankheiten,
Vor dieselben und diejenigen
So
in Stein, Erz, Metall und Feuer
arbeiten, ausgestellet.

Dreßden und Leipzig,
bey Friedrich Hekel, 1745.

Fig. 29. Title page Henkel's monograph on the "miners' and smelters' sickness"

lis medica de aere fodinarum metallicarum noxio" (Fig. 28) he considered both therapy and prophylaxis in detail.

CARL LINNAEUS (VON LINNÉ, 1707–1778) in his "Iter Dalekarlicum" (1734) referred to the short lives of sandstone quarrymen of Orsa.

FRIEDRICH HOFFMANN (1660–1742), 1693 professor of medicine at Halle and from 1708–1712 personal physician to Frederick I, king of Prussia in Berlin, in his "De metallurgia morbifera" (1738) stated that the men employed in digging and manufacturing cobalt at Kuttenberg in Bohemia are affected with vomiting, syncope, anxiety, cardialgia, difficulty of breathing, sense of suffocation, tremors, etc.; and that they have the appearance of living skeletons.

Although HOFFMANN was an experienced and successful practitioner, for him medicine became science only through its philosophical treatment. According to HERMAN BOERHAAVE (1668–1738) it is neither useful nor even possible to investigate the final metaphysical causes of disease:

Ultimae quoque metaphysicae et primae physicae causae medico investigatu necessariae, utiles vel possibiles non sunt (Prolegomena § 28).

The physician might best accomplish his task by turning to the causa proxima. HOFFMANN, who nearly equally BOERHAAVE in pedagogic success and international reputation (STEUDEL 1954), was not content with this restriction to the attainable. He established a final principle to derive the vital functions in the healthy and ill organism in a simple manner. He believed he had turned medical art into a teachable science in this way, a thought that was in keeping with the brand of rationalism of his time.

2.5 Occupational Lung Diseases in the Age of Enlightenment (ca. 1740–1830)

For workers in quarries HECQUET and BOUDON (1742, p. 27) recommended a scent bag containing some garlic malaxated with camphor. The protective effect of garlic can be attributed to its antioxidant properties because it has been found that the following compounds of garlic: *S*-allylcysteine, *S*-allylmercaptocysteine, *S*-allylcysteine sulphoxide, and allicin (thio-2-propene-1-sulphinic acid *S*-allyl ester) have the ability to scavenge reactive oxygen species (AUGUSTI and SHEELA 1996, IDE et al. 1996, RABINKOV et al. 1998). *S*-Allylcysteine inhibited NO production through the suppression of inducible nitric oxide synthase mRNA and protein expression in the *murine* macrophage cell line RAW264.7, which had been stimulated with lipopolysaccharide and interferon-γ (Kim et al. 2001).

Stephen Hales (1677–1761), rector and perpetual curate, diocesan proctor, chamber clerk to the Princess of Wales, botanist, physiologist, chemist, and applied scientist, in his later years became a pioneer of public health. In 1743 he described ventilators: whereby… fresh air may… be conveyed into mines, prisons, hospitals, etc. A second part appeared in 1758. The book was translated into French by Demours (1744).

In 1741, two other individuals were investigating ventilation: Sutton, a coffeehouse keeper in Aldergate Street and Martin Triewald, a captain of mechanics to the King of Sweden. Sutton's devices differed from those of Hales but Triewald's seemed identical. Hales did not mention the work of Sutton and denied knowledge of that of Triewald. He stated in the preface on his book on ventilators: "It was a very extraordinary circumstance that two persons at so a great distance from each other, should happen to hit on inventing a like useful Engine."

Henkel (1745) in his "Medicinischer Aufstand und Schmelz-Bogen. Von der Bergsucht und Hütten-Katze, auch einigen andern, denen Bergleuten und Hütten-Arbeitern zustoßenden Krankheiten, vor dieselben und diejenigen so in Stein, Erz, Metall und Feuer arbeiten, ausgestellt" refused blood letting (p. 95).

Richard Mead (1683–1754) in his "Monita et præcepta medica" (1752, p. 9) refers to a miner, who died at the age of 109 years and 3 months as quoted by Ackermann (1783, p. 3, footnote **).

Rousseau (1755) made sociocritical comments on how the unsound professions shortened life or destroyed health:

Qu'on ajoute à tout cela cette quantité de métiers malsains qui abrègent les jours or détruisent le tempérament; tel que sont les travaux de mines, les diverses préparations des métaux, des minéraux, surtout du Plomb, du Cuivre, du Mercure, du Cobalt, de l'Arsenic, du Réalgar; ces autres métiers périlleux qui coûtent tous les jour de la vie à quantité d'ouvriers, les uns Couvreurs, d'autres Charpentiers, d'autres Massons, d'autres travaillant aux carrières; qu'on réunisse, dis-je, tous ces objets, et l'on pourra voir dans l'établissement et la perfection des Sociétés les raisons de la diminution de l'espèce, observée par les plus d'un Philosophe.

DISCOURS
SUR L'ORIGINE ET LES FONDEMENS
DE L'INEGALITÉ PARMI LES HOMMES.
Par JEAN JAQUES ROUSSEAU
CITOYEN DE GENÈVE.
Non in depravatis, ſed in his quæ bene ſecundum naturam ſe habent, conſiderandum eſt quid ſit naturale. Aristot. Politic. L. 2.
A AMSTERDAM,
Chez MARC MICHEL REY.
M D C C L V.

Fig. 30. Title page of Rousseau's «Discour sur l'origine et les fondemens de l'inégalité parmi les hommes»

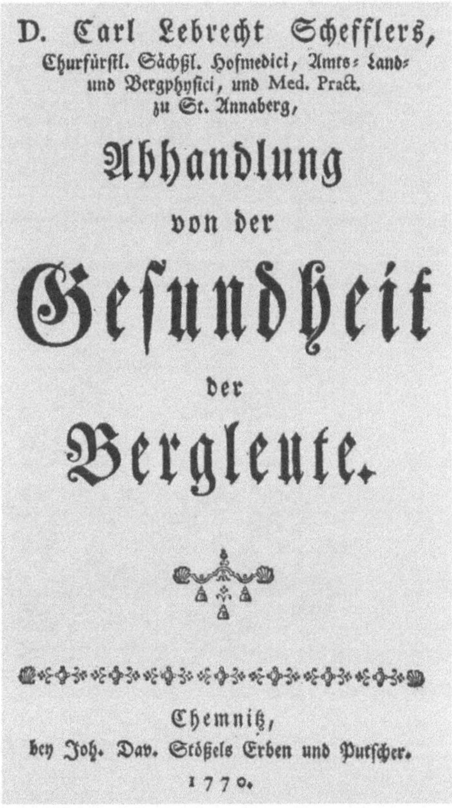
D. Carl Lebrecht Schefflers,
Churfürstl. Sächßl. Hofmedici, Amts- Land- und Bergphyſici, und Med. Pract. zu St. Annaberg,
Abhandlung
von der
Geſundheit
der
Bergleute.
Chemnitz,
bey Joh. Dav. Stößels Erben und Putſcher.
1770.

Fig. 31. Title page of Scheffler's treatise on the health of the miners

Le luxe, impossible à prévenir chez les hommes avides de leur propres commodités et de la considération des autres, achève bientôt le mal que les Sociétés ont commencé, et sous prétexte de faire vivre les pauvre qu'il n'eut pas fallu faire, il appauvrit tout le reste, et dépeuple l'État tôt-ou tard.

Scheffler (1770, p. 199) warned his colleagues about blood letting and purging in ill miners, but proposed to prescribe a true Sal vitale antisepticum of vegetable origin (Herb. Rutae, agrimoniae menthae Crispae, fragariae melissae oreoselini, pulegii, Radix chinae ponderosae Taraxaci, Semen nigelli Napii).

Johann Peter Frank (1745–1821), a great-granduncle of Franz Koelsch, published a six-volume "System of a complete medical police" (1770, 1780, 1783, 1788, 1813, the 6th volume in two parts in 1817 and in 1819). Volume 1, which was out of print after one year, was reprinted in 1784. Volumes 1 to 4 were reedited in an amended and supplemented form in 1786, 1787 and 1790, respectively. Frank emphasised the hygienic importance of mining ventilation (cf. Koelsch 1966, pp. 1813 and 1886).

Thomas Percival (1740–1804) reported on the outbreak of fever in cotton mills at Radcliffe northwest of Manchester in 1784 (cf. Rosen 1958, Meiklejohn 1959, and Buess 1960).

Louis Le Blanc (1775), professor of anatomy and surgery at the École Royale de Chirurgie d'Orléans, wrote a note on the formation and hardening process of sand stone with a description of the singular disease which attacks the workmen hewing and grinding this sort of rock. Under the microscope, the fine, very delicate particles look like small faceted diamonds, but a little oblong in a sharp-cornered form (un petit diamant à facettes, mais un peu plus allongé en forme de coin; p. 562). As the dust particles were able to penetrate the glass of a bottle, they might also invade the human body and no antidote might be found (p. 582).

Hebenstreit (1791, § 179) advised a government paternally concerned for the welfare of its vassals to arrange as far as possible to replace noxious and poisonous materials worked by workmen and artisans with unharmful ones or at least invent easy practicable process engineering.

Lehrsätze
der
medicinischen
Polizeywissenschaft
von
D. Ernst Benjamin Gottlieb Hebenstreit
d. A. W. öffentl. Lehrer zu Leipzig.

Leipzig
im Verlage der Dykischen Buchhandlung.
1791.

Fünfter Abschn. Sorge für die ꝛc. 87

§. 177.

Ohnerachtet es wohl nicht möglich seyn möchte, alle Arten von Gewerben, Künsten und Handwerkern von allen mit ihnen verbundnen Gefahren zu trennen, so wird doch auch schon die Verminderung dieser Gefahren ein verdienstliches Werk von Seiten der Policei seyn, und sich selbst durch Erhaltung vieler dem Staate nützlichen Bürger belohnen.

§. 178.

Am meisten fallen in die Augen die körperlichen Uebel und Krankheiten, welchen diejenigen ausgesetzt sind, die sich mit Gewinnung und Verarbeitung giftiger, durch Einathmung oder Verschluckung der Gesundheit schadender Materien, zumal an engen und eingeschloßnen Orten, wo die Luft nicht genugsam erneuert werden kann, beschäftigen; wohin z. B. die Berg- und Hüttenarbeiter, die Vergolder, die Blei- und Zinngießer, die Glas- und Spiegelfabrikanten, die Farbenreiber, gewissermaßen auch die Färber, Gerber, Salz- und Salpetersieder gehören.

§. 179.

Es ist einer weisen und für das Wohl ihrer Unterthanen väterlich besorgten Regierung würdig, Untersuchungen zu veranlassen, in wie fern es möglich sei, bei einigen dieser Künste und Handwerke, die schädlichen und giftigen Materien, welche sie bearbeiten, durch unschädliche zn ersetzen; wenn sich dieses aber nicht thun läßt, wenigstens leicht anwendbare Mittel und Verfahrungsarten erfinden und

F 4 bekannt

Fig. 32. Hebenstreit's proposition of a medical police science. The title page shows a vignette by [Adam Friedrich] Oeser, director of the academy of art in Leipzig. Section 5 deals with the obligation for safety in the means of making a living

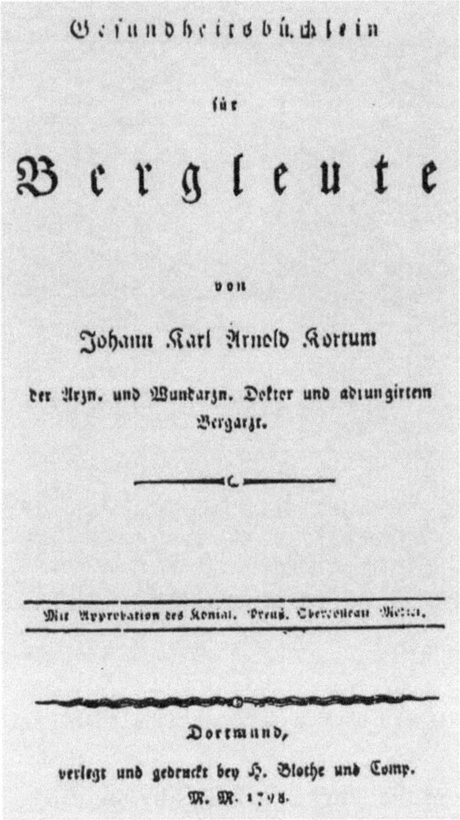
Gesundheitsbüchlein

für

Bergleute

von

Johann Karl Arnold Kortum

der Arzn. und Wundarzn. Dokter und adjungirtem Bergarzt.

Mit Approbation des Königl. Preuß. [illegible]

Dortmund,

verlegt und gedruckt bey H. Blothe und Comp.

1798.

Fig. 33. Title page of KORTUM's health booklet for miners

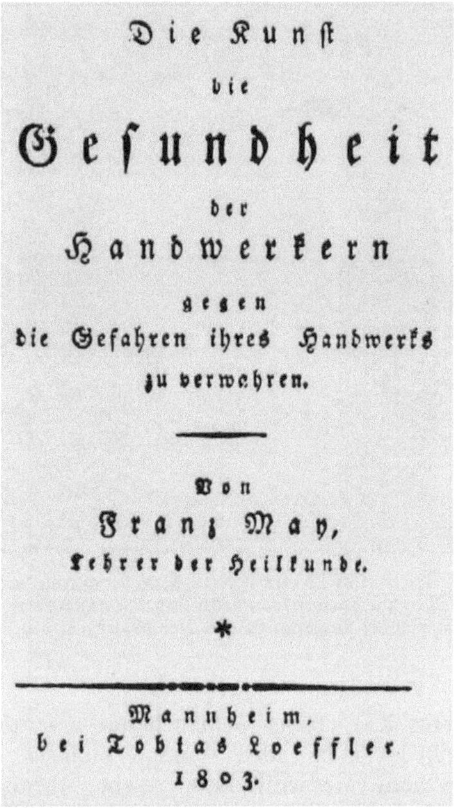
Die Kunst

die

Gesundheit

der

Handwerkern

gegen

die Gefahren ihres Handwerks

zu verwahren.

Von

Franz May,

Lehrer der Heilkunde.

Mannheim,

bei Tobias Loeffler

1803.

Fig. 34. Title page of FRANZ ANTON MAY's "The art of prolonging the health of craftsmen against the dangers of their trade"

The mine physician KARL ARNOLD KORTUM (1745–1824), known for his "Jopsiade" (1784), recommended the following in his health booklet for miners (1798, p. 40): "If paralysis of the limbs, shortness of breath or such illness really exist, one must consult the physician in order to avoid a deeper rooting of the evil, which may be cured by early application of useful remedies. Herb-tea or draughts of melissa, salvia, hyssop, betony, thyme, arnica, and similar herbs one may surely use for an adjuvant." FRANZ ANTON MAY (1803), whose portrait decorates a medal awarded by the Society for Health Education (Fig. 35), recommended to the masons: "Bricks too dry which intensively dust when hewn with the bricklayers' hammer or the clay pick and tear off iron splints, should be sprinkled with the spray brush. When hewing the mason should draw his face back as often as possible to avoid that heated iron splints might damage his eyes or even be inhaled into the lungs."

In Japan, in 1811, a physician of Tatara Goldmine wrote a letter to a professor of a medical school, in which he described precisely the clinical symptoms of silicosis. It was said at that time that plum was effective in preventing silicosis and many plum-trees were planted at Ikuno Silver-mine at the suggestion of the governmental officer. Also in the 19th century a kind of dietetic therapy was given to silicotic patients and some official compensation was made at Omori Silver-mine (quoted from MIURA 1961).

2.6 Professional Lung Diseases in the Age of Industry

THACKRAH's (1832) account of morbidity and mortality among mine workers was, as ROSEN (1943) stated, not very far removed factually from that of RAMAZZINI. THACKRAH conceived that his own observations in practice were avowedly imperfect. When he wrote "The pulmonic diseases, however, chiefly prevail among the miners of coal, which is loaded with pyrites, and which produces consequently considerable dust", we can now say that the dust contained the Fenton catalyst by itself to produce most pathogenic reactive oxygen intermediates.

Fig. 35. Portrait of Franz Anton May decorating a medal awarded by the Society for Health Education (courtesy of the German Cancer Research Centre, Heidelberg)

Fig. 36. Portrait of Charles Turner Thackrah (1795–1833) in the Medical Library of the University of Leeds

James Kay (1831), a Manchester practitioner, described what he called "spinners' phthisis":

"In many cases which have presented themselves at the Ardwick and Ancoats Dispensary, the disease has appeared to me *to differ from ordinary chronic bronchitis*" (emphasis added).

Petrenz (1844) complained of the nearly complete lack of information on quarriers' disease in scientific papers and reference books. Practising medicine at Schandau in the mountain range on the Elbe river in Saxony, he was well-acquainted with this fatal disease. He advised a minimum age of 25 years for admission to the profession, wetted sponges to prevent the dust from entering the respiratory system, and change of occupation in case of respiratory disease.

According to Halfort (1845) expectorants can remove dust only from the bronchi, but not from the alveoli. As Holland (1843) did, he recommended a change in occupation.

Vernois (1858) complained of most physicians' ignorance of details in production engineering. Since the working class comprised the largest number of diseased persons, he urged his colleagues to inspect their working places. Since 1853 the hospitals of Paris recorded deaths by profession, and in the years 1853 to 1856 there were no coal workers(including coal merchants and trimmers). But there were 633 water carriers admitted, 76 of whom died. Thus Vernois concluded that coal dust would not be harmful.

E.H. Greenhow, the first medical inspector of factories in England, in 1864 identified particles of silica in the lungs of grinders by the use of polarised light. He was probably the first to use polarised light for this purpose.

Ludwig Hirt (1844–1907), in order to solve a prize question suggested by professor Heinrich von Bamberger of Würzburg, investigated the insanity of tradesmen using statistical references from larger infirmaries. He distinguished between diseases favoured by dust inhalation and diseases directly caused by dust inhalation. The former cover catarrh of the airways, pulmonary emphysema, bronchiectasis, pneumonia, and phthisis non tuberculosa; the latter pneumoconiosis. Individual trades and factories associated with more or less dust exposure and proposals and measures to compensate the harmful effects of the different types of dust were discussed in detail. However, Hirt's valuable material and personal investigation found little interest, they were ignored, and his results were derided (Koelsch 1967, p.114). Ludwig Hirt found a late appreciation as a pioneer in industrial social hygiene of by Stefanie Hirt (1940).

John Thomas Arlidge (1822–1899), after a short flash of publicity and glory following the publication of his Milroy Lectures in 1889 and his monograph on "The hygiene, diseases and mortality of occupations" in 1892, was almost completely forgotten, but Meiklejohn (1946, 1954, 1963, 1966, 1969) and Posner (1973) reawoke interest in him in several papers, and Isaacson (1956) privately

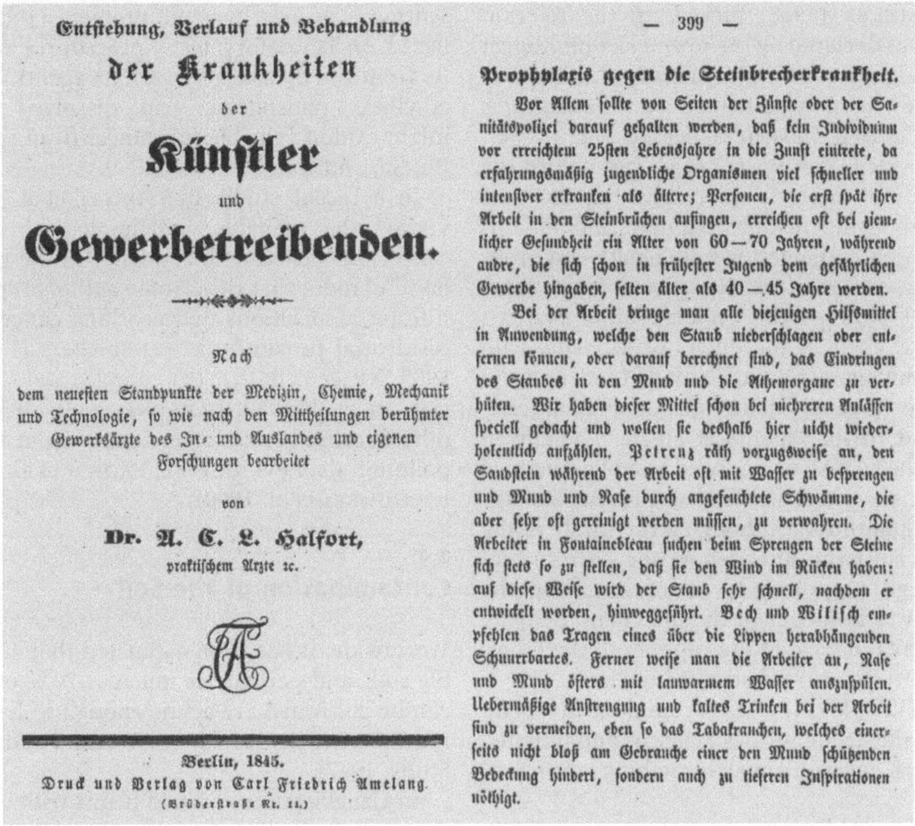

Entstehung, Verlauf und Behandlung

der Krankheiten

der

Künstler

und

Gewerbetreibenden.

Nach

dem neuesten Standpunkte der Medizin, Chemie, Mechanik und Technologie, so wie nach den Mittheilungen berühmter Gewerksärzte des In- und Auslandes und eigenen Forschungen bearbeitet

von

Dr. A. C. L. Halfort,

praktischem Arzte ꝛc.

Berlin, 1845.

Druck und Verlag von Carl Friedrich Amelang.

(Brüderstraße Nr. 11.)

— 399 —

Prophylaxis gegen die Steinbrecherkrankheit.

Vor Allem sollte von Seiten der Zünfte oder der Sanitätspolizei darauf gehalten werden, daß kein Individuum vor erreichtem 25sten Lebensjahre in die Zunft eintrete, da erfahrungsmäßig jugendliche Organismen viel schneller und intensiver erkranken als ältere; Personen, die erst spät ihre Arbeit in den Steinbrüchen anfingen, erreichen oft bei ziemlicher Gesundheit ein Alter von 60—70 Jahren, während andre, die sich schon in frühester Jugend dem gefährlichen Gewerbe hingaben, selten älter als 40—45 Jahre werden.

Bei der Arbeit bringe man alle diejenigen Hilfsmittel in Anwendung, welche den Staub niederschlagen oder entfernen können, oder darauf berechnet sind, das Eindringen des Staubes in den Mund und die Athemorgane zu verhüten. Wir haben dieser Mittel schon bei mehreren Anlässen speciell gedacht und wollen sie deshalb hier nicht wiederholentlich aufzählen. Petrenz räth vorzugsweise an, den Sandstein während der Arbeit oft mit Wasser zu besprengen und Mund und Nase durch angefeuchtete Schwämme, die aber sehr oft gereinigt werden müssen, zu verwahren. Die Arbeiter in Fontainebleau suchen beim Sprengen der Steine sich stets so zu stellen, daß sie den Wind im Rücken haben: auf diese Weise wird der Staub sehr schnell, nachdem er entwickelt worden, hinweggeführt. Bech und Wilisch empfehlen das Tragen eines über die Lippen herabhängenden Schnurrbartes. Ferner weise man die Arbeiter an, Nase und Mund öfters mit lauwarmem Wasser auszuspülen. Uebermäßige Anstrengung und kaltes Trinken bei der Arbeit sind zu vermeiden, eben so das Tabakrauchen, welches einerseits nicht bloß am Gebrauche einer den Mund schützenden Bedeckung hindert, sondern auch zu tieferen Inspirationen nöthigt.

Fig. 37. Halfort's monograph on the origin, course, and treatment of diseases of artists and tradesmen according to the latest developments in medicine, chemistry, mechanics and technology, adapted from the reports of famous industrial physicians at home and abroad and personal research work. Prophylactics against stone masons' disease include prohibition of work done by minors, sprinkling of rock to be wrought, avoidance of dust inhalation and washing of nose and mouth with lukewarm water (p. 399)

printed a short monograph on Arlidge's life and work entitled "The forgotten physician". Arlidge (1892) acknowledged the high incidence in the past, but noted the widespread belief that the serious lesions of the lungs associated with the calling of coal-getters belonged to past history or, at the most, were very uncommon. He attributed this situation to better ventilation, shorter hours of work, and greater attention to hygiene in the mines. With regard to prevention of chest diseases in the pottery industry of Staffordshire Arlidge (1892, p. 312) praised the master potters for the introduction of ventilating fans but regretted that they had not been universally adopted, "the explanation of which fact is to be found in the absence in many factories of steam power to drive the fans."

2.7 Treatment of Hypoxaemia and Hypercapnia

Brockmann (1844, p. 533) recommended combatting hypercapnia with ferrous drugs. However, "chemical decarbonisation" proved less successful and problematic (Brockmann 1851, p. 144). Thus he proposed alkalisation by K_2CO_3 and Na_2CO_3. He warned against pure oxygen for therapy (1844, p. 532; 1851, p. 145), a strikingly modern position confirmed by the results of Zorn (1955), Zorn and Tophof (1956) and Drasche (1958).

2.8 Environmental Protection

Environtology has its origin in Zarathustra (Zoroaster 630–553 BC), who taught that water, air, and earth must not be polluted (Payandeh 1976).

As Evelyn (1661) mentioned, there was a decree in the Dutch city of Harlem prohibiting the use of coal. Biersteker (1968) picked up the relevant manuscript as declared by the town crier on August 27, 1608, and preserved in the municipal archives of Harlem. The original text including a transcript was reproduced in Ned Tijdschr Geneeskd 112: 33–34.

Ackermann (1783, pp. 137–138) pointed to the responsibility of the state in establishing chemical factories producing noxious vapours and inflammable materials endangering both health and property of adjacent residents. He quoted Ramazzini (1703, pp. 39–40) reporting on a lawsuit of a citizen of Finale ‘ a town in the duchy of Modena, who raised a claim against a merchant of Modena, who had a large (ingens) factory producing mercury chloride and vitriol vapours at Finale. The citizen asked that the factory be removed to a place outside the town or elsewhere in order to prevent damage to the neighbourhood. Although a certificate of the local physician (who attended to the citizen) and the death registry spoke in favour of the citizen, the judgement was in favour of the merchant, denying the detrimental nature of the vapours to the lungs. Ramazzini (l.c. p. 40) found this injurious.

Hebenstreit (1791, p. 88) recommended a prize donated by the Académie des Sciences and initiated by the French government as something worth emulating.

A more recent historical example of the problem of how to handle industrialisation with regard to its ecological consequences was given by Wiesing (1989) in the story of the establishment of a glass factory in Bamberg. Anton Dorn (1802) warned of the emission of sulphur dioxide from the combustion of steam coal, its oxidation to sulphuric acid and precipitation from the air damaging the vegetation.

Beryllium compounds contained in some Czechoslovak brown coal undergoing combustion in power stations polluted the environment, representing a serious hygienic problem (Jiřele 1967). Based on laboratory tests, procedures of leaching Be from the fly ashes by various agents with di-(2-ethylhexyl)phosphoric acid dissolved in petrol might exploit valuable substances from coal such as Be, Ge, and Ga.

In a recent study, Benedetti et al. (2001) reviewed the available epidemiological studies on cancer associated with residence in the neighbourhood of industrial sites. Some authors reported significant associations between lung cancer risk and residential proximity to (a) smelters (Pershagen 1985, Xu et al. 1991), (b) complex industrial areas (Shear et al. 1980, Barbone et al. 1995), and (c) other localised emission estimated from air and soil pollution data (Vena 1982, Brown et al. 1984, Jedrychowski et al. 1990).

2.9 Contamination of the Soil

Worldwide, it has been estimated that 40 % of arable soils and perhaps as much as 70 % of land that can be cultivated are acidic enough to have an aluminium toxicity problem (World Food Nutrition Study, 1977).

In Uzbekistan in the arid plains with dusty soils, domestic animals, especially sheep showed pulmonary lesions consisting of perivascular and peribronchial nodular masses and a thickening of the alveolar walls (Матусевич 1955).

Contamination of the soil by fibrous minerals (anthophyllite, tremolite, sepiolite) is widespread, and evidence of pleural plaqueing in agricultural workers has been found in Turkey (Baris et al. 1978, 1979), Greece (Bazas et al. 1981), Bulgaria (Zolov et al. 1967, Burilkov 1968, Burilkov and Babadjov 1970, Burilkov and Michailova 1972), Yugoslavia, Czechoslovakia, and Austria.

CHAPTER 3

Aetiopathogenesis of Pulmonary Fibrosis and Malignancy

JOHANNES BAPTISTA MORGAGNI (1761) in his treatise "De sedibus, et causis morborum per anatomen indagatis" (Fig. 38) gave morphologic pathology a scientific foundation. While Morgagni's thesis of localisation of disease was not an original discovery, his work far surpassed the similarly oriented investigations of preceding centuries. His accurate observations and precise records reveal him as a natural scientist in the tradition of FRANCIS BACON (1561–1626), who in his "Advancement of learning" (1605) had already pleaded for the performance of autopsies on the victims of illness. BONETUS (1629–1689) shared this opinion and the title of his "Sepulchretum" makes reference to the hidden causes revealed by the alterations of organs. WEPFER (1620–1695) and his pupils were more circumspect; their aim was the correlation of clinical symptoms only and anatomic lesions.

In the field of pneumoconiosis research VAN DIEMERBROECK (1672) recognised the penetration of dust into the lungs, and ZENKER (1867) its intracellular deposition.

RUDOLF VIRCHOW (1855, 1858) in his "Cellularpathologie" localised pathogenic processes in the cells as the smallest viable entities*. The "biologic doctrine" established by SCHLEIDEN, SCHWANN, and VIRCHOW according to BARGMANN (1958) increasingly proves itself as a keystone of biology, when biochemistry entered the laboratories of modern cytochemistry.

By means of molecular pathology and pharmacogenetics, biochemistry invaded the core of pathobiology. Molecular anatomy as envisaged by ANDERSON and ANDERSON (1979) on the occasion of the 75th anniversary of the Behringwerke, "must clearly be an interdisciplinary one, borrowing techniques from biochemistry, biophysics, nuclear physics, chemistry, engineering, and mathematics, while retaining the essential viewpoints of anatomy, including the notion of completeness."

* According to BÜCHNER (1955) historic fairness requires reference to the English anatomist JOHN GODSIR who in 1843 had outlined a cellular pathology as VIRCHOW (1852) emphasised.

SAMUEL THOMAS [VON] SOEMMERING (1755–1830), professor of anatomy at the university of Mainz from 1784 to 1797 and Royal Bavarian Privy Aulic Councillor (Fig. 41), agreed with FRANZ DANIEL REISSEISEN of Strasbourg (1808) that the black matter found in the bronchial glands was transported by the lymphatic vessels* from the lungs, but he did not concur in the latter's opinion regarding the origin. On the contrary, he remarked, "all anatomists are sufficiently aware that the bron-

* From SHELDON (1752–1808) at Oxford and MONRO (1733–1817) at Edinburgh SOEMMERING acquired the technique of injecting lymph vessels and became the first to introduce this technique in Germany.

JO. BAPTISTÆ
MORGAGNI
P. P. P. P.
DE SEDIBUS, ET CAUSIS
MORBORUM
PER ANATOMEN INDAGATIS
LIBRI QUINQUE.
DISSECTIONES, ET ANIMADVERSIONES, NUNC PRIMUM EDITAS, COMPLECTUNTUR PROPEMODUM INNUMERAS, MEDICIS, CHIRURGIS, ANATOMICIS PROFUTURAS.
Multiplex præfixus eſt Index rerum, & nominum accuratiſſimus.
TOMUS PRIMUS
DUOS PRIORES CONTINENS LIBROS.
VENETIIS,
MDCCLXI.
Ex Typographia Remondiniana.
SUPERIORUM PERMISSU, AC PRIVILEGIO

Fig. 38. Title page of MORGAGNI's "De sedibus, et causis morborum", Venice, 1761

510 *Anatomes Liber* II.

que per venam pulmonarem ad cordis ſiniſtrum ventriculum promptius deflueret.

Aliena in pulmone reperta. Præter autopſiam, pulmonis ſubſtantiam totam veſiculoſam eſſe, docet etiam ratio: nam multoties ſputa rotunda, craſſa, & fœtida, vomicæ, folliculi, vermes (ſic ante paucos annos vermem vivum, magnitudine & forma bombycem majuſculum referentem, ſed ſubrubrum, cum valida tuſſi à muliere quadam excuſſum, vidimus) calculi, aliæque res præter naturam in pulmonibus gignuntur, ut obſervationum ſcriptores exemplis teſtantur, & nos quoque multoties in praxi vidimus, & cum tuſſi rejectas, & poſt mortem in pulmone repertas: quæ certè non in vaſis ſanguiferis, nec etiam in bronchiis (ſuffocationem enim, aſthma violentum, perpetuamque tuſſim induxiſſent) ſed neceſſario in hiſce veſiculis gigni, ac longo tempore contineri debuerunt & potuerunt.

Obſervatio. Anno 1649. in noſtro noſocomio diſſecui famulum lapicidæ, aſthmate mortuum, in cujus pulmone maximam pulveris lapidum copiam, inſpiratione cum aere attractam, atque veſiculas fere omnes infarcientem, inveni, ita ut in diſſectione pulmonis (qui admodum durus erat) cultro meo quaſi per acervum arenæ, aut per corpus aliquod arenoſum ſcinderem, quo pulvere repletæ veſiculæ aerem inſpiratum admittere non potuerant, atque ſic æger aſthmate extinctus fuerat. Sequenti anno adhuc duo ſimiles caſus lapicidarum ſimili modo interemptorum, eodem in noſocomio à nobis viſi & demonſtrati ſunt. Tunc temporis retulit nobis horum famulorum magiſter lapicida, quod, dum lapides inciduntur, tam ſubtilis pulvis ex iis aſſurgat in aerem, qui poros veſicæ bubulæ inflatæ & ſiccæ, in officina pendentis, penetrare poſſit, ita ut circa finem anni ad minimum hujus pulveris manipulus unus in interiore capacitate veſicæ inveniretur, atque illum eſſe qui tam multos lapicidas, qui ſibi ab iſto pulvere non ſatis diligenter cavent, præmatura morte interimat. Si talis tamque copioſus lapidum pulvis in pulmonis veſiculas inſpiratione penetret, minimè dubitandum eſt aerem in inſpiratione illas omnes permeare. Vidimus etiam alium longo aſthmate tandem extinctum, qui plumas illas, quibus lecti infarciuntur, expurgare, ſeu à detritis ſordibus emundare ſolebat, cui pulmonis veſiculæ à tenui plumarum pulvere penitus erant infarctæ.

Membrana investiens. Dicta veſiculoſa ſubſtantia extrinſecus inveſtitur membrana tenui ac poroſa, quam à pleura dirivatam plerique Medici tradunt: at ego

Fig. 39. Frontispiece of van Diemerbroeck's "Anatome corporis humani". The first necropsy of a stone-mason is on p. 510 of the first edition of 1672

Fig. 40. Portrait of Friedrich Albert [von] Zenker, owned by the Department of Pathology, University of Erlangen. (Courtesy of Prof. Dr. T. Kirchner, present holder of the professorate)

Fig. 41. Portrait of Samuel Thomas [von] Soemmering owned by the Department of Anatomy, University of Mayence

chial glands are filled by a black matter, true pine-soot, particularly among the common people who burn bad tallow coarse oil, which matter can find its way into the bronchial glands only through the air passages."

MORTON (1689) was the first author, as THOMSON (1838, p. 346) stated, who attributed the production of black sputa to the bronchial glands.

ANTOINE PORTAL, who became professor in 1770 and worked at the Muséum National d'Histoire Naturelle in Paris from 1793–1832 (see CORDIER 1955, p. 72), stated in 1780 that he had seen black-streaked sputa in healthy as well as tuberculous persons. He believed that this black matter could be divided into three varieties: one arising from the inhalation of carbonaceous matter; a second derived from extravasated blood; and the third variety being scratched from the bronchial glands.

In town dwellers or in residents of places of great consumption of coal, PEARSON (1813) examined the bronchial lymphatic glands and the black spots of the lungs and found the black pigment resistant to lye of KOH, HCl, HNO_3. He thought it "highly reasonable to suppose, that the particles of charcoal should be retained in the minutest ramifications of the air tubes, or even in the air vesicles, under various circumstances, to produce the coloured appearance on the surface, and in the substances of the lungs" (l.c. p. 165).

BICHAT (1771–1802) believed that the black pigmentation was due to small glands projecting into the bronchi, and appearing under the surface of the pleura. Among the six kinds of pulmonary consumption described by BAYLE (1810) in his «Recherches sur la phtisie pulmonaire» the black matter found in the lungs was regarded as a morbid product produced by the body. CHOMEL (1817) observed that the sputa of persons who had spent a considerable time in an atmosphere loaded with the vapours of oil or of tallow frequently exhibited black streaks. MAURICE (1862), who had been a mine physician in the St.-Étienne district for more than 20 years, reported in a «Note sur la maladie ou l'état noir des poumons des charbonniers» a case of black lungs in a miner. At autopsy the lungs were found intensively black, and on being sectioned an inky fluid was expressed from the tissues. The cut surface presented a grey and black marbled appearance. Almost all the black spots appeared to have a small, indurated point, about the size of a pinhead, at their centres where the pigmentation was deepest. Chemical analysis of the black pigment washed out of the lungs led to the conclusion that it was coal. Since it was not a true melanosis, MAURICE proposed the term melanidie (μέλας = black; εἶδος = appearance).

GABRIEL ANDRAL (1797–1876), professor of pathology and therapy in Paris, took the position in his «Cours de pathologie interne» (1836) that in most cases in which an organ is at the same time indurated and coloured black, the induration is, in fact, independent of the black colour and is the simple result of **chronic inflammation**.

HENLE (1841) was uncertain whether the black spots were organised formations or mere deposits of inhaled coal dust, and preferred the former view. Even less understandable is RUDOLF VIRCHOW's statement (1847, p. 466) that the pulmonary pigment will become black and not remain brown or yellow. It would be wrong to consider the influence of respiration on melanogenesis as the pigments of the costal pleura and the bronchial lymph nodes as well as those of other parts of the body are not related to respiration.

Pathologic pigmentation was induced, as VIRCHOW (l.c. p.379) stated, by stained lipids, bile pigment, and haematochrome. Here humoralism (ROKITANSKY 1846) and cellular pathology (VIRCHOW 1858) clashed. ROKITANSKY's misfortune was, as HOLZNER (1978) in a historic retrospective on pathology in Vienna remarked, his desire to complete a system oriented to pure morphology by means of a chemistry not yet ready for the questions under consideration and ultimately only able to produce unsatisfactory hypotheses.

Fig. 42. Bust of FR. GUST. JACOB HENLE, created in 1882 by F[erd] Hartzer, property of the Department of Anatomy, University of Heidelberg

Fig. 43. Portrait of CHARLES baron ROKITANSKY decorating a medal created by A. Scharff, awarded to the participants of the 62nd meeting of the Deutsche Gesellschaft für Pathologie, Vienna, May 16–20, 1978

In his 11th lecture (March 27, 1858) VIRCHOW started with malaria pigmentation. Citing TIGRI's "milza nera",. VIRCHOW declared that his cellular pathology had been established on the basis of experience, but with the help of speculative thinking and to establish a speculative notion (ALTMANN 1992, p. LXIII). Indeed, he had not any material in pneumoconiosis, neither *human* nor experimental. When TRAUBE* (1860) demonstrated square and hooked black particles in the spit and later in the pulmonary tissue of a 54-year-old charcoal trimmer who had inhaled this type of dust for 12 years, there was no longer any doubt about the penetration of fine carbon particles into the respiratory apparatus. There was not any trace of newly formed connective tissue. Lewin (1862) made *rabbits* inhale powdered charcoal and found the dust particles in their alveoli. Powdered charcoal admixed to the food was resorbed and transferred to the mesenteric lymph nodes.

For KUSSMAUL** (1867) there seemed to be no doubt that the 3 % sand found in the ashes of the bronchial glands of two individuals with 4.2 and 9.5 % sand in lung ashes, respectively, were carried there from the lungs by the lymph channels.

NATALIS GUILLOT (1845) and JEAN CRUVEILHIER (1791–1874), who held the anatomy chair in Paris for 10 years and before assuming the chair for pathology (see CORDIER 1955, p.76), occupied themselves with the problems of pulmonary pigmentation.

While GUILLOT's investigations were not directly concerned with miners, they are important in this connection for they were a significant contribution to the histopathology of black lungs similar to those found in coal miners. CHARLES ROBIN (1821–1885), who in 1862 obtained – 60 years after BICHAT's death – the first professorate of histology (see CORDIER 1955, p. 81), expressed his conviction that the black miners' lungs were due to the inhalation of coal dust. BERGERON (1859) supported this view by demonstrating the infiltration of the lungs of a copper miller with carbonaceous matter, which was certified by chemical analysis. In 1854, Dr. Tardieu of the hospital la Riboisière had appealed to the public heath specialist and to medicine to pay attention to this serious form of chronic bronchitis due to an insidious and progressive deposition of coal dust within the lungs. This is the miners' disease long described in England under the name of anthracosis.

One cellular mechanism involved in the process of the translocation of particles to the regional lymph nodes is the endocytosis and transcellular transport of particles from the alveoli to the interstitial space by type I pneumocytes (GIESEKING 1958, ADAMSON and BOWDEN 1978, 1981). Once the particles reach the pulmonary interstitium they can apparently gain entrance to the peripheral lymphatic channels and subsequently pass to the tracheobronchial lymph nodes as free particles (FERIN and FELDSTEIN 1978). On the other hand, some of the particles that enter the interstitial space may be engulfed by interstitial macrophages (SOROKIN and BRAIN 1975). According to CASARETT (1964) these cells may then migrate into the alveoli or enter lymphatic vessels where they serve as vehicles for transporting the particles to the lymph nodes.

The term "silicosis" was coined by VISCONTI, who first entered this word into the list of necropsies established at the Institute of Pathological Anatomy, Milan (Fig. 44).

Physico-chemical interactions of the different SiO_2 modifications and silicates, respectively, with arginine (STÖBER 1966) other amino acids (DALE and KING 1953), and proteins (DALE and KING 1953, HOLT and BOWCOTT 1954, SCHEEL et al. 1954, WILLY 1954, SCHEEL 1955, RÜTTNER and ISLER 1956, LICHT 1957, 1960, MCFEE and TYE 1964, 1965, STÖBER 1966, ANTWEILER and DJIE 1971, SAKABE et al. 1971, JONES et al. 1972, ISHIYAMA et al. 1974, VALERIO et al. 1986, 1987, DONALDSON et al. 1995) marked the beginning of a **molecular understanding** of pathogenicity.

An increase in protein-bound silica was found in the blood plasma (trichloroacetic acid precipitate)

---

* LUDWIG TRAUBE (1818–1876) in 1853 came to the Charité Hospital in Berlin.

** ADOLF KUSSMAUL (1822–1902), lecturer in Heidelberg, Prof. in Erlangen, Freiburg, and Strasbourg

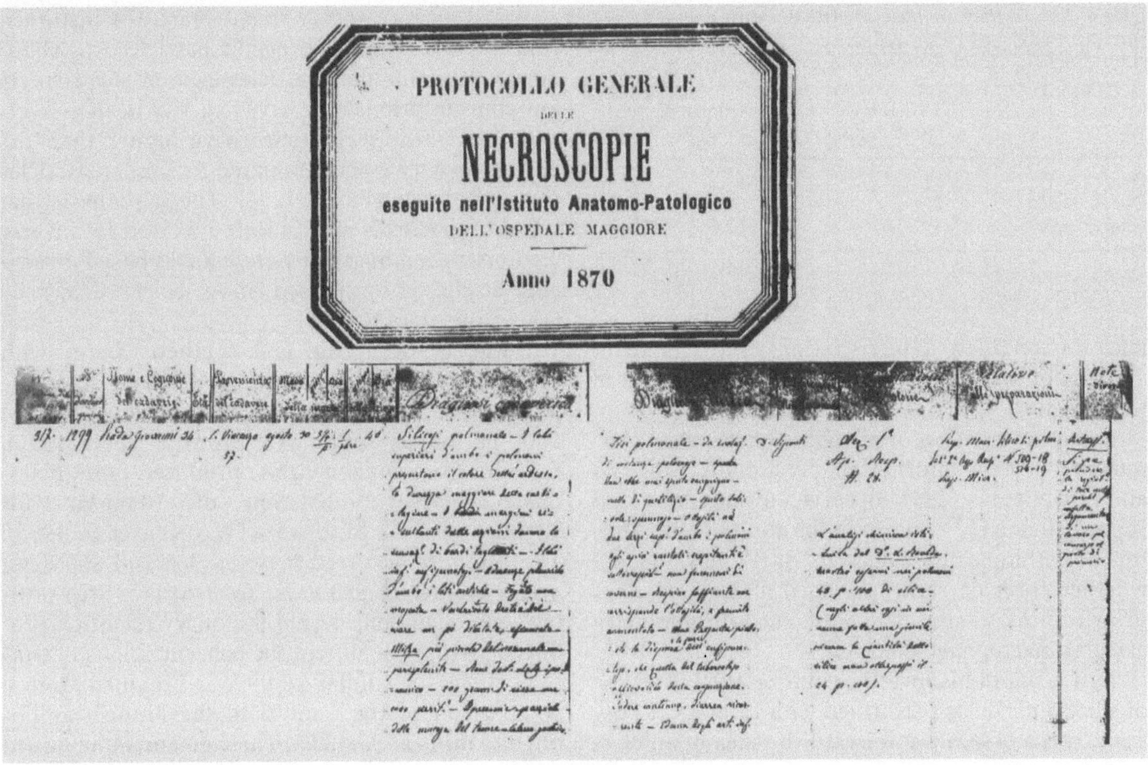

Fig. 44. Two-page record by VISCONTI using the term "silicosis" (*underlined*)

of *rats* instilled intratracheally with 50 mg of slate-dust (SINGH et al. 1985).

Sub-microscopic glass particles (most frequent diameter <0.1 μm) suspended in an electrolyte solution were capable of acting on developing synthetic lipid membranes or pre-formed bilayer lipid membranes by achieving contact with the surface of such membranes and finally entering them (MAJER 1971). *Rat* erythrocytes pre-treated with lipophilic ethyl-3,5,6-tri-*O*-benzyl d-glucofuranoside emulsified in isotonic phosphate buffer were protected from haemolysis by glass particles in a dose-related manner (MAJER 1975).

The adsorption of linoleic acid onto silica gel from petroleum ether solution conformed to a Langmuir isotherm, consistent with the formation of a monolayer (PORTER et al. 1972). Confirming the finding of HONN et al. (1951) with soybean oil, it was found that the most rapid uptake of oxygen occurred at a linoleic acid–silica ratio close to that for the monolayer. Without included antioxidant, oxidation commenced at a nearly linear rate without observable induction period. Time for consumption of one-half mole of oxygen per mole of linoleic acid was ca. 60 min on acid-washed silica. If very small amounts of α-tocopherol were included in the layer, virtually no oxygen uptake measurable in this system occurred during the induction period, the length of which was approximately proportional to tocopherol content. The inflection point at the commencement of rapid oxidation was very sharp; the ensuing oxidation rate approximated that of the unprotected acid. The induction period of linoleic acid with the same tocopherol content was as much as 100 % longer when exposed in monolayer than in a bulk form. However, the rate after commencement of rapid oxidation was 8–10 times greater in the monolayer. Acid washing of the silica reduced its iron content by 75 %. Acid washing also reduced by 60 % the rate of autoxidation without α-tocopherol and increased the length of the induction period four-fold when α-tocopherol was present. The effect of pre-treatment of the silica by adsorption of the acid synergists, ascorbic, phosphoric, citric and ethylenediamine tetraacetic acid was qualitatively similar to the effect of acid washing.

Hydrophilicity or hydrophobicity of the surface determines cell-surface adherence and the propensity for membranolysis. The great influence of surface properties was particularly demonstrated by the unusually high transport rates of polyvinylpyridine-*N*-oxide-coated SiO_2 particles from the peritoneal cavity of *rats* to the lymph nodes compared with the same dusts uncoated (STRECKER 1967). After subcutaneous injection, however, clear-

Table 4. Cytotoxicity of quartz particles with contaminated surfaces established by the reduction of triphenyl tetrazolium chloride (TTC) (from ROBOCK and KLOSTERKÖTTER 1975)

	$TTC\text{-}RA_{120}$ [% of control]	Quartz content [%]
Quartz + Ti	32	89.8
Quartz + Mn	27	88.0
Quartz + Fe	49	87.0

ance was retarded, demonstrating, that cellular reaction, not coating of the particles, was the effect of the polymer.

Using the reaction of nitroblue tetrazolium (formula [18]) to a diformazan precipitate, CILENTO and GEORGELLIS (1991) observed a time delay and suppression of $O_2^{\bullet}-$ release in the ability of dipalmitoyl lecithin-coated quartz dust (250 µg/ml), whether fresh or stale (i.e. aged at least 15 days prior to use), to stimulate adherent *rat* pulmonary alveolar macrophage cultures.

As a result of his investigations on the solid state of silicogenic dusts performed with ROBOCK, KLOSTERKÖTTER (1967) put forward the idea of an **electron transfer**. According to their content of the transition metals, Ti, Mn, and Fe dusts containing nearly the same percentage of quartz showed varying cytotoxicity to guinea pig peritoneal macrophages after incubation for 120 min, as tested with 2,3,5-triphenyl tetrazolium chloride (Table 4).

Chemiluminescence initiated 40 s after the approach of the quartz particle and the cell plainly shows an electron transfer as a primary reaction. MUNDER and MODOLELL (1987) increased the sensitivity of the method by a previous stimulation of the macrophages with zymosan inducing an oxygen burst.

Based on research in recent years, the historical concept of inert and fibrogenic particles was abandoned. It seems that particles even at surprisingly low concentrations may have negative health effects and that ultrafine particles have higher than expected toxicity when compared to similar particles of a larger size (FERIN 1994). The hypothesis that ultrafine particles are causally involved in adverse responses seen in sensitive *humans* is based on several studies summarised in a brief review by OBERDÖRSTER (2001).

Naturally occurring and synthetic fibres with certain physico-chemical characteristics have the potential to induce lung fibrosis and lung, pleural and peritoneal tumours (BIGNON et al. 1995). Erionite from Oregon (USA) produced mesotheliomas following inhalation of 10 mg/m^3 for 12 months in 27 of 28 *rats* (WAGNER et al. 1985). The tumours occurred between 385 and 800 days, on an average of 580 days. Total surface area of fibres achieving long-term pulmonary retention produced virtual equality in the concentrations (TIMBRELL 1984). A similar result was obtained from a comparison of the Cape with the Finnish anthophyllite mine at Paakkila. This agreement, obtained without invoking the differences in fibre type, suggested that the dominant factor in asbestosis is fibre size through its influence on respirability and retention. KAMP et al. (1992) explored the evidence supporting the hypothesis that free radicals and other reactive oxygen species are an important mechanism by which asbestos mediates tissue damage. Unlike the characteristic progressive nature of silica-related lesions, all respirable-sized carbon fibre-induced inflammatory effects were reversible within 10 days after exposure and any significant histopathologic effects were observed at any time post-exposure (WARHEIT et al. 1994).

CHAPTER 4

Formulation of the Problems

4.1 Tumour Defence

The concept of immunological surveillance against cancer is based on the postulate of PAUL EHRLICH (1909) that tumour cells arise with enormous frequency, and that they are normally eliminated by immune mechanisms. BURNET (1970) has restated it in modern terms. One of the arguments of immunologic surveillance is the infiltration of primary tumours by lymphocytes and macrophages. In *human* carcinomas, an impairment of unspecific cellular defence, represented chiefly by the macrophage system, seems to play an important role as shown by KOHOUT (1972) using the skin window test of REBUCK and CROWLEY (1955). In the *mouse*, endotoxin-mediated necrosis and regression of established tumours may be mediated by the tumour necrosis factor (TNF) probably released by activated macrophages (MACPHERSON and NORTH 1986).

BUGELSKI et al. (1983) identified intratumoral macrophages by their capacity to ingest colloidal iron particles from the interstitial fluid. Since colloidal iron is retained in a stable form within these cells for a considerable time, new macrophages that emigrate into the tissue after injection of the colloidal iron are identified by their ability to ingest a second colloid (lanthanum), which can be reliably distinguished from the initial iron label. Preexisting (colloidal iron label) and newly recruited macrophages (lanthanum label) were identified in serial sections by histochemical methods using H_2O_2 oxidation to detect iron (blue reaction product) and cleavage of phosphate esters to demonstrate lanthanum (magenta reaction product). The macrophage content and macrophage recruitment were found substantially in individual metastases within the same host. BUGELSKI et al. (1985) demonstrated an inverse correlation between size of metastases with macrophage content in the B16 *murine* melanoma. The failure of non-specific immunotherapy may be related to the low macrophage density in metastases larger than 1000 cells. Although a strong host response has been associated with a better prognosis for certain *human* tumours (FISHER et al. 1975, WOOD and GALLAHON 1977), the mere presence of large numbers of macrophages in a tumour is not necessarily indicative of a meaningful host response (FIDLER and POSTE 1982).

A major challenge is to determine how immunopotentiators work, and much interest has been focused on the macrophage as the key cell that mediates their effects (MELTZER et al. 1979, 1982; OETTGEN 1979, p. 103; CHIRIGOS et al. 1980; DI LUZIO 1981; MELTZER 1981; SONE et al. 1982; LAVELLE 1983; KOPPER and LAPIS 1985; NAKATA et al. 1985; SOMEYA 1986). Activated macrophages have been found to inhibit and kill cancer cells in tissue culture under conditions when normal cells are not harmed. Macrophages have also been considered as a possible source of the tumour necrosis factor (TNF), found by CARSWELL et al. (1975), GREEN et al. (1977), CLARK et al. (1981)), and HA et al. (1983, 1985) in the serum of *mice* primed with BCG, *Corynebacterium parvum*, *Plasmodium vinckei* subsp. *petteri*, *Mycobacterium lepraemurium*, or *Listeria monocytogenes* and subsequently injected with bacterial endotoxin. After BEUTLER et al. (1985) the TNF is identical with cachetin secreted by macrophages. Binding of calmodulin to the microfilament network correlates with induction of a macrophage tumoricidal response (MECHAM et al. 1985).

The cytotoxicity of activated macrophages depends of the presence of L-arginine biotransformed to L-citrulline and NO_2^- (HIBBS et al. 1987). N^G-monomethyl-L-arginine prevented the synthesis of both these products as well as the expression of cytotoxicity.

$^{\bullet}NO$ was the most likely inorganic nitrogen oxide intermediate in the pathway of NO_2^- to NO_3^- synthesis in macrophages. MARLETTA et al. (1988), HIBBS et al. (1988), and STUEHR et al. (1989) demonstrated that NO synthesised from L-arginine was indeed the precursor of NO_2^- and NO_3^- in these cells. 3-Morpholinosydnonimine spontaneously released $^{\bullet}NO$ (Feelisch and Noack 1987). Fülle et al. (1991) used the substance as a tool in the search for specific cellular functions that may be regulated by $^{\bullet}NO$. $\geq 10\ \mu M$ 3-Morpholinosydnonimine reduced the lipopolysaccharide-induced synthesis of IL-1β, and to a minor degree of TNFα, and markedly elevated cGMP levels in freshly isolated *human* blood mononuclear cells (Fülle et al. 1991). The release of $O_2^{\bullet-}$ (Feelisch et al. 1989) may also contribute to the suppression of monokine production.

4.2 Aetiopathogenesis of Silicosis

According to Giese (1951) toxicity and fibroblastic activity of mineral dusts are contrary: quartz (s.g. 2.6±0.1), the fibroblastic activity of which is well known, is not toxic, whilst extremely fine amorphous silica (s.g. 2.0) demonstrates a high toxicity (King 1947, Jötten and Klosterkötter 1952, Klosterkötter 1952, Schiller 1952, 1955) but produces no fibrosis (King 1947, Schiller 1955), or only a fine network of reticulin more resembling that in Boeck's sarcoid or beryllium lungs (Jötten and Klosterkötter 1952). Coesite (s.g. 3.01), a high-pressure SiO_2, is little fibrogenic (Charbonnier et al. 1965, Strecker 1965). Stöber (1967) therefore substituted the specific gravity for the enthalpy as a parameter of fibrogenicity.

Increased solubility of quartz has been observed particularly in soils rich in organic material; however, no direct link between organic carbon and dissolved silica has been identified (Chesworth and Macias-Vasquez 1985). Bennett and Siegel (1987) presented evidence for an increase in the solubility of quartz in a natural water brought about by dissolved organic compounds. These compounds were produced by the biodegradation of petroleum, and consist largely of a complex mixture of organic acids. Bennett and Siegel proposed that silica is being complexed and mobilised by these organic acids in waters having close to neutral pH.

There is a great discrepancy between the amount of SiO_2 introduced into the body, when quartz of a particle size of 2 μm is introduced, or when a tolerable dose of extremely fine amorphous SiO_2 is given. Ray et al. (1951) showed the variety of the histological appearance of silicotic nodules produced in the lungs of *rats* after intratracheal injection of 2 mg, 5 mg, 10 mg, 30 mg, and 75 mg of quartz, respectively. 2 mg of quartz did not produce any lesions within the lungs, and a dose of 5 mg revealed but a loose network of fine reticulin fibrils without any collagen, resembling the picture described by Jötten and Klosterkötter (1952). The differences regarding fibrosis after intratracheal injection (King 1947) and inhalation (Jötten and Klosterkötter 1952) of the two sorts of silica dust thus seem to be due to the doses used, particularly as a high percentage of very fine dust is exhaled again (van Wijk and Patterson 1940) and the extremely fine silica retained is also removed from the lungs very quickly (King 1947, Klosterkötter 1967).

The dust breathed by miners is partly fresh and partly old (King 1945). Operations such as shotfiring, drilling, shovelling, etc., disturb the deposity of dust from floor, walls, rafters, etc., as well as producing new dust. However, there was no evidence for either a greater solubility or a greater pathogenicity in the freshly formed dusts from South Wales rocks. *In vitro*, freshly fractured quartz caused greater peroxidation of membrane lipids, greater membrane leakage, and a greater decrease in cell viability than aged quartz. Freshly fractured quartz is also a more potent stimulant of reactive species production by alveolar macrophages *in vitro* (Vallyathan et al. 1988, Castranova et al. 1996). In comparison to aged quartz, inhalation of freshly milled quartz resulted in a significantly enhanced cytotoxicity (lipid peroxidation of lung tissue and lavage levels of red blood corpuscles, protein, and lysosomal enzymes) and elevated inflammation (lavageable neutrophils, generation of oxidant species from alveolar macrophages, and histological scoring of lung infiltrates) (Vallyathan et al. 1995, Castranova et al. 1996). Evidence suggests that a direct relationship exists between $HO^{\bullet}$ generation and cytotoxicity, as the production of both $HO^{\bullet}$- and quartz-induced biological reactions decrease in a similar fashion with time after fracturing (Vallyathan et al. 1988, Castranova et al. 1996). *In vivo* exposure of *rats* to aged α-quartz potentiated the production of oxidants by alveolar macrophages in response to the *in vitro* addition of zymosan particles, that is, zymosan-stimulated chemiluminescence was significantly elevated in alveolar macrophages harvested after inhalation of aged silica (Castranova et al. 1996). In contrast, resting chemiluminescence (without zymosan) was not substantially altered by inhalation of quartz. This indicated that pulmonary phagocytes exposed to silica *in vivo* are primed to produce greater quantities of oxidants upon subsequent contact with particles. Inhalation of aged quartz caused significant induction of nitric oxide synthase (i.e., increasing N^{ω}-nitro-L-arginine methyl ester-inhibitable chemiluminescence well above the control level). As with total zymosan-stimulated chemiluminescence, nitric oxide synthase-dependent chemiluminescence was significantly greater after exposure to freshly milled quartz compared with aged quartz.

Homolytic fission of $\equiv$Si–O–Si$\equiv$ bonds is produced by crushing silica in the mining and other industries (FRIPIAT et al. 1971). $SiO^{\bullet}$ radicals may react with water molecules to form $HO^{\bullet}$ radicals. Measurement of surface radicals by electron spin resonance spectroscopy indicated that freshly milled quartz particles exhibited 54% more silicon-based free radicals than milled quartz that was aged for 2 months prior to aerisolization into the exposure chamber (CASTRANOVA et al. 1996).

Radical formation by crashing silica

$$\equiv Si{-}O{-}Si\equiv \xrightarrow{crash} \equiv Si{-}O^{\bullet} + {}^{\bullet}Si\equiv \quad [1]$$

$$\equiv SiO^{\bullet} + H_2O \longrightarrow \equiv SiOH^{\bullet} + HO \quad [2]$$

$$\equiv SiO^{\bullet} + HO^{\bullet} \longrightarrow \equiv SiOOH \quad [3]$$

$$\equiv SiOOH + H_2O \longrightarrow \equiv SiO^{\bullet} + H_2O_2 \quad [4]$$

The relations of the various forms of chemisorbed oxygen on the surface of freshly ground quartz were studied with the emission of singlet oxygen, ${}^1\Delta_gO_2$ (ZAV'YALOV et al. 1985). The emission of singlet oxygen proceeds in a process of irreversible loss of centres of chemisorbed oxygen in the form of $O_2^{\bullet-}$.

Hydroxyl groups bonded to silicon to form either monomeric silicic acid or silica (quartz) surfaces are only weakly acid (THOMAS 1951, HOFMANN 1962). The silanol groups at the surface of quartz can be detected by ionic binding of rhodamine from aqueous solutions (STÖBER 1955). Chlorosilanes, known for their high affinity to silanol groups form siloxane bonds on both quartz and Aerosil® (STÖBER 1956). A direct observation of silanol groups and their different bonds on silica surfaces can be achieved by infrared absorption measurements in the spectral range near 2.7 μm. The results of the studies of KRIEGSEIS et al. (1977) emphasise the significance of silanol groups and adsorption properties in relation to the fibrogenic activity of quartz dusts.

Due to the presence of surface hydroxyl groups, tissue culture polystyrene (TCPS) behaves better in term of its cellular responses than expected on the basis of its wettability (VAN WACHEM et al. 1985, DAVIES 1988). TCPS is a variant of regular polystyrene, assumed to be made by H_2SO_4 oxidation (CURTIS et al. 1983, BENTLEY and KLEBE 1985) or glow discharge treatment (AMSTEIN and HARTMAN 1975, ERTEL et al. 1990) of regular polystyrene to create a high concentration of hydroxyl groups.

On the other hand, trimethylsilyl aerosil, though unwettable by water is well phygocytosed (STRECKER 1956). But it is just a matter of time until the surface trimethylsilyl groups will be removed and pathogenic reactions will start.

As a result of his investigations on the solid state of silicogenic dusts performed with ROBOCK, KLOSTERKÖTTER (1967) put forward the idea of an **electron transfer**. The luminescence induced by activating DQ12 quartz and 39 coal mine dust samples suspended in Tyrode's solution showed varying intensities in the range of 200 nm to 500 nm and 20 °C to 50 °C as a manifestation of their electron structures (ROBOCK and KLOSTERKÖTTER 1971). However, no direct correlation is possible between the low-temperature luminescence and electron spin resonance (KRIEGSEIS et al. 1977). KRIEGSEIS et al. (1979) stated: "The pathogenic nature of SiO_2 dusts is determined exclusively by electron orbitals of the outermost atomic layers. Chemical impurities and structural imperfections within the SiO_2 grain volume do not exert any influence."

Quartz, coal, and asbestos dusts, like ionising radiation, paraquat, and bleomycin induce pulmonary fibroses by radical reactions. Dioxygen reduction in both biotic (e.g. phagocytosis) and abiotic systems (e.g. radiolysis) induces $O_2^{\bullet-}$. $O_2^{\bullet-}$ can exert deleterious effects directly or by engendering more potent oxidants, such as protonation, by reaction with vanadate or manganous salts, or by a metal salt-catalysed interaction with H_2O_2. $O_2^{\bullet-}$ is capable to initiate and propagate free-radical chain reaction and of damaging cell components. Membranes can be destroyed by peroxidation of their lipids, and chromosome strand breaks can be induced. *In vitro*, α-quartz induced hydroxyl radical generation, thymine glycol production and DNA strand breakage (DANIEL et al. 1995).

High **iron** contamination of quartz produced approximately 57% more reactive species in water than quartz with low Fe contamination (CASTRANOVA et al. 1997). Compared to inhalation of quartz with low Fe contamination, high Fe contamination of quartz resulted in increases in the following responses: leucocyte recruitment (537%), laveable red blood corpuscles (157%), macrophage production of oxygen radicals measured by electron spin resonance or chemiluminescence (32 or 90%, respectively), nitric oxide production by macrophages (71%), and lipid peroxidation of lung tissue (38%).

Different electron structure of two quartz dust samples taken from the same locality and indiscernible by mineralogical methods, were attributable to the incorporation of **aluminium** ions into the SiO_4 lattice (BECK et al. 1973). They also differed in their toxicities to macrophages studied with triphenyltetrazolium chloride and in the pulmonary hydroxyproline and phospholipid contents 4 to 12 weeks after intratracheal application to *rats*. Excitation by 2.8×10^4 r revealed significantly different activation energies of their electron traps. Surface aluminosilicate contamination or occlusion of respirable-sized high-silica particles could be detected by the multiple-voltage scanning electron microscopy-energy dispersive X-ray analysis (WALLACE et al. 1994).

While RAY et al. (1951) did not find any inhibitory effect on the action of quartz by anthracite and coal mine dusts mixed with quartz when introduced by intratracheal instillation in the lungs of *rats*, REHN et al. (1995) raised the question whether the formation of reactive oxygen species by macrophages engulfing coal mine dust containing free silica could be depressed by the coal part.

The involvement of reactive oxygen intermediates in nuclear factor-ϰB activation has been demonstrated in studies where antioxidants such as pyrrolidine dithiocarbamate and *N*-acetyl-L-cysteine prevented the degradation of its inhibitor, I-ϰB and NF-ϰB nuclear translocation (SCHRECK et al. 1992, MEYER et al. 1993, SCHENK et al. 1994). I-ϰB kinase is an oxidative stress-activated kinase (FLOHE et al. 1997) and it has been shown that specific regulation of NF-ϰB activity is via modulation of I-ϰB phosphorylation in the cytoplasm, which is in turn governed by the redox pathways (JIN et al. 1997, KRETZ-REMY et al. 1998). Depletion of the intracellular antioxidant GSH often accompanies increased levels of inflammatory mediators in lung diseases (RAHMAN and MACNEE 1998). Activation of NF-ϰB has been demonstrated after quartz (DRISCOLL et al. 1997), asbestos (MOSSMAN et al. 1997, DRISCOLL et al. 1998) and ceramic fibres (GILMOUR et al. 1997). KREJSA and SCHIEVEN (1997) reviewed the contribution of oxidative stress to the effects of phosphotyrosine phosphatase inhibition by vanadium-based compounds in lymphocytes. Although the inactivation of phosphotyrosine phosphatases can lead to nuclear factor-ϰB mobilization in the presence of antioxidants, the other effects noted appear to require a threshold of intracellular oxidation. The combined effects of oxidative stress on signal transduction cascades reflect a synergy between the initiation of signals by phosphotyrosine phosphatases and the loss of control by phosphotyrosine phosphatases. This suggests a mechanism by which environmental agents that cause oxidative stress may alter the course of cellular responses through induction or enhancement of signalling cascades leading to functional changes or cell death.

Both drugs that inhibit oxygen metabolite production by neutrophils and alveolar macrophages (e.g. pentoxifylline) and radical scavengers (e.g. poly-2-vinylpyridine-*N*-oxide) might suppress both acute and chronic phagocyte-dependent inflammatory reactions. Calcium channel blockers (e.g. verapamil, diltiazem, nisoldipin) inhibit $O_2^{\bullet-}$ formation by *human* neutrophil granulocytes stimulated by phorbol-myristate-acetate, and so do tetrandrine and cepharanthine.

The formation of singlet oxygen ($^1O_2$) in systems known to produce $O_2^{\bullet-}$ has been manifested by photoemission upon decay of $^1O_2$ to the ground state. Phagocytosis of DQ12 quartz and inert TiO_2 particles by bone marrow derived macrophages induced different periods and intensities of luminescence demonstrating the formation and degradation of oxygen radicals (MUNDER and MODOLELL 1987).

Electron spin resonance proposed for silicosis research by ARENDS et al. (1963) represents another interesting physical method. It detects configurations of unpaired electrons. The surface of silica ground in air bears the characteristic radicals $\equiv Si^\bullet$, $\equiv Si{-}O^\bullet$, $\equiv Si{-}O{-}O^\bullet$ (peroxyradical) and $O_2^{\bullet-}$ (superoxide anion). With time some 30 % of the total radical population decay (FABINI et al. 1990). After three days the recorded spectrum does not appreciably vary over the period of weeks; aged dusts, as DQ12 and Min-U-sil 5, still exhibited well defined spectra. A new grinding process modified the situation: most of the fine components of the spectra disappear into a broad envelope made up by the major components. Left standing in air, the spectrum slowly evolved into its original shape.

HF (KING et al. 1953, VIGLIANI 1958, MARKS and NAGELSCHMIDT 1960, PERNIS et al. 1960, SAFFIOTTI 1962, DANIEL et al. 1995) and **NaOH** (MARKS and NAGELSCHMIDT 1960, PERNIS et al. 1960, BAUMANN 1965) are the most common chemicals used to attack SiO_2. Treatment with both also erodes surface radicals, but due to the low rate of reaction of crystalline silica, the total elimination of radicals is hard to achieve (FUBINI et al. 1990). As the ozonide features on HF treated samples prevail over the rest of the spectra, HF might create some radicals alongside erosion, or, at least, favour the formation of the specific site giving the "ozonide" form upon contact with O_2. Etching with NaOH increased the acid phosphatase activity of glycogen-elicited macrophages, while haemolysis was decreased (SAKABE et al. 1971). The release of lactate dehydrogenase from 10^6 *guinea pig* alveolar macrophages was increased by 25 µg or 50 µg DQ12 treated with 5 n HCl or 5 n NaOH, but decreased after pretreatment with acetone (TILKES and BECK 1987). Treatment with H_3PO_4 or **HCl** influenced lactate dehydrogenase release from 10^6 cells without or with 5 % serum (KRIEGSEIS et al. 1987). Due to the amount of iron associated with unwashed silica the potential to generate free radicals is greater with unwashed than with HCl-washed material (Miles et al. 1994). Removal of iron eliminates one of the major sources of free radicals and consequent DNA damage: when removed by acids both effects were in fact dramatically decreased (Daniel et al. 1995).

Heat-treating ($\geq$ 800 °C) fully deprived cristobalite of surface radicals and induced hydrophobicity, indicating that hydrophobicity is at least one of the surface properties determining the cytotoxic potential of a dust as tested on proliferating cells of the *mouse* monocyte macrophage cell line J774 (Fubini et al. 1999).

In addition to reactive oxygen species, **nitric oxide** may be a potential mediator of silica-induced toxicity (Castranova et al. 1996). Nitric oxide adsorbed readily to DQ12 quartz (Robertson et al. 1982). At room temperature it remained adsorbed for up to several weeks. Thermal desorption of $^{\bullet}$NO was much lower (43 %) than of NO_2 (67 %). Vallyathan et al. (1997) studied the kinetic clearance of instilled stable nitroxide radicals (2,2,6,6-tetramethyl piperidine *N*-oxyl) and showed an oxidative stress in adult male Sprague-Dawley *rats* exposed to 10 mg Min-U-Sil quartz $<$ 5 μm. The generation of reactive oxygen species increased by silica was associated with enhanced levels of superoxide dismutase ($P<0.05$) and lipid peroxidation.

Radzig (1993) labelled the active sites on **reactive silica** with $H_3C^{\bullet}$ radicals:

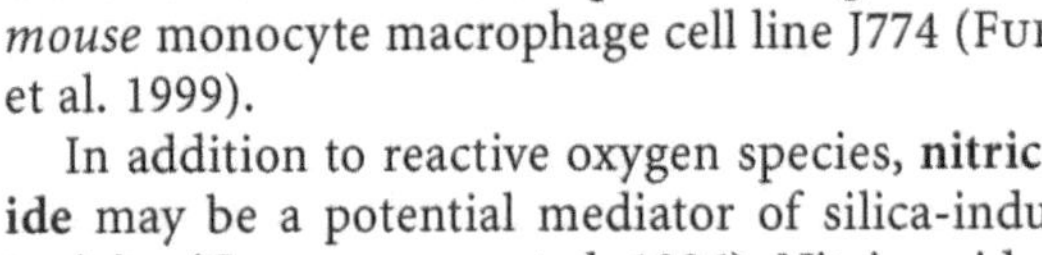

[5]

Fromme et al. (1966) localised tritiated poly-2-vinylpyridine-*N*-oxide of about 55 kDa injected subcutaneously into *rats* (FW 49) on the surface of tridymite dust given into the peritoneal cavity 0, 3, or 5 d before. The free radical scavenging properties of PVPNO may be a mechanism for its action in pneumoconiosis (Gulumian and van Wyk 1987).

Silicates causing pneumoconiosis function as **Fenton catalysts** to generate hydroxyl radicals when incubated with hydrogen peroxide and a reducing substance (Kennedy et al. 1989). In contrast, silicates, which do not cause pneumoconioses demonstrate no Fenton activity. Catalytic activity is decreased by pre-treatment of silicates with the iron chelators, deferoxamine or transferrin. Hæmolysis from silicates is decreased by interventions, which remove superoxide anion or hydrogen peroxide from the medium, or by pre-treatment of dusts with iron chelators.

Serum added to the incubation medium strongly influenced the lag and intensity of cytotoxic manifestations (Modolell et al. 1967). Cells unable to phagocytise but having adsorbed silicogenic particles to their plasmalemmata are damaged after a longer interval.

HL-60 cell derived macrophages were resistant to cytotoxic effects of quartz (Brückner-Nieder et al.1992). Schedle et al. (1995) induced apoptosis by 33–1000 μM metal cations/l. Crocidolite asbestos suppressed the differentiation of HL-60 cells induced by DMSO (Ueki et al. 1992).

P388D$_1$ macrophage-like cells were used in France to test cytotoxicity (viability and lactate dehydrogenase and acid phosphatase levels) at a serum content of 4 % (Davis et al. 1982). Because of uncontrolled variation occurring in animal cells (*rabbit* alveolar macrophages obtained by pulmonary lavage 3–4 weeks after stimulation by i.v. injection of Freund's adjuvant, thioglycolate evoked *rat* peritoneal macrophages) Daniel and Le Bouffant (1980) preferred P388D$_1$ cells, paticularly when investigating dusts with low toxicities. The extent of killing P388D$_1$ macrophages is dependent on both the dose of silica and the concentration of Ca^{2+} ions in the medium (Kane et al. 1980). In the presence of extracellular calcium ions, after 3 h of exposure to 350 μg of silica, 69 % of the cells had lost viability, and ATP content was reduced to 21 % of control level (Kane et al. 1985).

Quartz and other fibrogenic dusts stimulate the macrophages (Schiller 1980, Vigliani 1983), or their death is preceded by a period of stimulation. The fibroblast-stimulating factor released is identical with the lymphocyte-activating interleukin-1 (Schmidt et al. 1982, 1984). The stimulated T-lymphocytes produce lymphokines and, in particular, the "macrophage Ia recruitment factor" (MIRF) which causes macrophages to produce Ia antigens. Ia Antigens are necessary for the macrophages' presentation of exogenous or endogenous antigens to T-lymphocytes. Additional activation by lymphokines increases the immune function of the macrophages, and a vicious circle is triggered. The stimulated T-lymphocytes become hyperactive against all T-dependent antigens. Their functions include the development of a delayed-type hypersensitivity and an action on B-lymphocytes that makes them transform into plasmoblasts and plasma cells. The gammaglobulins produced by the latter precipitate locally on collagen fibres. An increased incidence of lymphoreticular tumours has been described in *rats* inoculated with silica particles (Wagner and Wagner 1972, Wagner et al. 1980).

In Rat2 fibroblasts silica particles stimulated intracellular reactive oxygen species generation, evidenced by 2',7'-dichlorofluorecein oxidation (Cho et al. 1999). The time course of elevation of the intracellular reactive oxygen species was paralleled by the increases of mitogen-activated protein kinase and extracellular signal-regulated protein kinase phosphorylation. Silica-induced extracellular signal-regulated protein kinase phosphorylation was also effectively attenuated by catalase and diphenyleneiodonium chloride. However, superoxide dismutase enhanced the silica-induced extracellular signal-regulated protein kinase phosphorylation, indicating a role for H_2O_2 in extracellular signal-regulated protein kinase activation. Furthermore, extracellular signal-regulated protein kinase and mitogen-activated protein kinase phosphorylation are reproduced by H_2O_2 treatment.

Because proinflammatory cytokines are believed to be central to the development of silica-induced pulmonary toxicities, inhibition of the production of these cytokines should result in protection. The anti-inflammatory, glucocorticoid steroid dexamethasone has been demonstrated to inhibit the production of proinflammatory mediators (GUYRE et al. 1988, KERN et al. 1988, VAN FURTH et al. 1995, OHTSUKA et al. 1996). These inhibitory effects may occur through gene repression by activated glucocorticoid receptor binding to negative glucocorticoid response elements in the 5'-flanking region of the proinflammatory gene. Binding at the glucocorticoid response elements down-regulates the activity of the nuclear transcription factor, NF-ϰB (KLEINERT et al. 1996). NF-ϰB plays an essential role in the production of silica-induced proinflammatory mediators (CHEN et al. 1995). In addition to its inhibition of proinflammatory mediator production, dexamethasone has the ability to enhance the production of anti-inflammatory cytokines, such as IL-10 by leucocytes (VAN FURTH et al. 1995). Tracheal instillation of liposomes containing dexamethasone attenuated silica-induced pulmonary inflammation and fibrosis in *rats* (DIMATTEO and REASOR 1997). One of the parameters used was luminol-dependent chemiluminescence of the lavagable cells.

4.3 Silicosis and Carcinoma

4.3.1 Epidemiology

Hitherto, most authors refused a causal relationship between silicosis and bronchial carcinoma (BERBLINGER 1931, FISCHER 1931, PANCOAST and PENDERGRASS 1933, SAUPE 1933, WÄTJEN 1933, 1936, ALLEN 1934, KOELSCH 1934, KOLLMEIER 1934, FISCHER-WASELS 1936, STAEMMLER 1937, BERGERHOFF 1938, HOLSTEIN 1941, WEDLER 1943, SCHMIDT 1947, EHRHARDT 1949, RÜTTNER 1949, WESTERMANN 1951, FRUHLING and OPPERMANN 1952, SPÖRLEIN 1952/53, SCHOCH 1954, MITTMANN 1959, OTTO and BREINING 1959, OTTO 1963, NICOD 1967, RÜTTNER and HEER 1969, OTTO and VON HINÜBER 1972, ROOKE et al. 1979, AMES et al. 1984). The same was concluded by VORWALD and KARR (1937, 1938) on the basis of extended statistics. Only DIBLE (1934), ANDERSON and DIBLE (1938), KLOTZ (1939), ECK et al. (1969) and GUDBERGSSON et al. (1984) supposed a causal relation between silicosis and pulmonary or bronchial carcinoma. In a review based upon 50 cases of silicosis, KLOTZ (1939) found 9 instances of carcinoma, including all organs (= 18 %). 4 of these were primarily bronchogenic, which is 8 % of the total number of cases of silicosis and 45 % of all the cancers found in this group. These observations were compared with a group of 4,500 autopsies performed on patients dying in the Toronto General Hospital. Of 808 carcinomas of all organs (= 17.7 %) 53 were primarily bronchogenic, which is 1.17 % of all autopsies or 6.5 % of all cancers. In spite of the lack of direct evidence, KLOTZ was led back to the silica as the offending agent (cf. KAHLAU 1954, 1961, GROSSE 1956). In 5 cases of pulmonary carcinoma in preexisting silicosis, ECK et al. (1969) assumed a positive causal relation, although the identity of locality could not be taken as evidence. K.H. BAUER (1963) has claimed: "Again and again one must be on one's guard against the paralogism that a mass statistic rarity would a priori vote against a causal relation. For probability the sum of the individual cases gives a preliminary clue only; de facto et de jure it has to be redetermined for each individual case". SCHAUTZ and KLEIN (1960) refusing quartz to be a direct aetiologic factor for carcinogenesis considered a concomitant bronchitis to be a promoting factor. KÖNN et al. (1976) quoted tobacco smoke as an additional irritant for the bronchial epithelium. CHIYOTANI (1984) thought the recent higher risk of lung cancer to be closely connected with the remarkable longevity of hospitalised pneumoconiotic patients. Information on exposure to silica and smoking habits collected from hospital records showed an elevated risk, supported by a clear dose-response, due to smoking (MASTRANGELO et al. 1988). Exposure to silica also appeared to increase the risk of lung cancer, but only in presence of silicosis. The risk estimated tended to increase both with amount of smoking and duration of exposure to silica, with the magnitude of risk being, however, much smaller for the latter effect.

Nine cohort mortality studies published in 1987–1997, in the opinion of the International Agency for Research on Cancer, provided the least confounded evidence and this time led to the Working Group's final evaluation: 'Crystalline silica in the form of quartz or cristobalite from occupational sources is carcinogenic to humans' (Group 1). It is important to note, McDonald (2000) in a recent editorial on silica and lung cancer stated, that the conclusion was preceded by the following paragraph:

> In making the overall evaluation, the Working Group noted that carcinogenicity in humans was not detected in all industrial circumstances studied. Carcinogenicity may be dependent on inherent characteristics of the crystalline silica or on external factors affecting its biological activity or distribution of its polymorphs.

Ebihara and Kawami (1990) in 140 consecutive necropsies of silicotics found 25 with carcinoma, i.e. 19.9 %. The predominant type was squamous cell carcinoma (54.2 %) followed by small-cell carcinoma (22.9 %) and adenocarcinoma (14.6 %). In cases of progressive massive fibrosis the majority of tumours arose in the segmental bronchi leading to the fibrotic masses. The majority of peripheral lung tumours were closely adjacent to fibrotic areas.

The cancer morbidity and mortality figures of three different Finnish granite areas (Vehmaa: Balmoral red granite; Kuru: grey granite; Viitasaari: black granite), combined with the differences in biological activity (lactate dehydrogenase release from *rat* macrophages and luminol-enhanced chemiluminescence of *human* polymorphonuclear leucocytes) of the granite dusts and a hypothesis that there is a cancer-inducing mechanism for reactice oxygen species, pointed to a direct role of quartz in cancer induction (Koskela et al. 1994).

Thus epidemiological evidence strongly suggests that chronic silicosis predisposes to an increased lung cancer risk, which is due either to high silica exposure or to the fibrotic process or to both factors (Goldsmith 1994). However, the carcinogenic effects of silica are less dramatic than those of asbestos (Allison 1996).

The relative risk of developing lung cancer by inhalation of crystalline silica is about 1.18 (Ulm 1999). This increased risk can also be explained by the different smoking behaviour among the exposed workers. However, workers compensated for silicosis have an about 2 to 2.5 times higher lung cancer risk compared to the general population. Workers without silicosis seem to have no increased risk.

Diffuse interstitial fibrosis-type nonasbestos pneumoconiosis had an exceedingly high concurrence of lung cancers when compared with pneumoconiosis without diffuse interstitial fibrosis (Katabami et al. 2000). Squamous cell carcinomas of the lung from pneumoconiosis with diffuse interstitial fibrosis exclusively comprised peripheral types, as compared with squamous cell carcinomas from pneumoconiosis without diffuse interstitial fibrosis (13 [100 %] of 13 versus 33 [72 %] of 46, $P = 0.03$). Lung cancers arose frequently from the area of diffuse interstitial fibrosis in pneumoconiosis with diffuse interstitial fibrosis.

Soutar et al. (2000) found that parts of the evidence were coherent but there were contradictions. The main scientific uncertainties in the evidence are:

> Smoking habits, socio-economic class differences and inappropriate comparison population
>
> Weakness in the available data on which the exposure data are based
>
> Excess of cancer risk as the result of selection and diagnostic bias
>
> Is there an increased lung cancer risk from exposure to silica also found in subjects without silicosis?
>
> Is it justifiable to assume that quartz and cristobalite have similar health effects?

Reviewing published occupational epidemiological literature directly pertinent to the interrelations among silica exposure, silicosis, and lung cancer, Checkoway and Franzblau (2000) think that until more conclusive epidemiological findings become available, population-based or individually-based assessments should treat silicosis and lung cancer as distinct entities whose cause/effect relations are not necessary linked.

Pilger et al. (2000) did not find any significant differences in the levels of 8-hydroxydeoxyguanosine in leucocyte DNA and the rate of urinary excretion of 8-hydroxydeoxyguanosine between patients with silicosis and quartz-exposed healthy workers. However, in the group of the patients with an increased oxidative DNA damage, urinary excretion of 8-hydroxydeoxyguanosine was lower than in the corresponding group of active workers without silicosis. No association of the formation and/or elimination of 8-hydroxydeoxyguanosine with the period of employment, field of activity, smoking, or age was detected.

In 1517 coal miners' post-mortem examinations from the Mecsek region, Hungary, Kádas (1996) found an incidence of lung cancer of 6.3 % at all and of 6.8 % in the 480 miners with silicosis, but 9.2 % in 6031 non miners aged over 40 years (controls) concluding that the dusty workplaces represented no lung cancer risk.

4.3.2
In Vitro Studies and Bioassay

DONALDSON and BORM (1998) presented a hypothetical sequence of events that lead to production of cancer by quartz (Fig. 45). By utilising suitable chosen endpoints, it is therefore possible to explore the role of the quartz surface in the events leading to carcinogenesis without carrying experiments through to lifetime carcinogenesis studies, which are prohibitively expensive.

FUBINI (1998) modified this version by adding the form of the particle, crystallinity and the distribution of silanol groups (≡SiOH) at the surface. Also dissociated silanols ($\equiv SiO^-$) and features related to mechanical activation (surface radicals and charges) might play a role in phagocytosis. He considered Al^{3+} (CZERNICHOWSKI et al. 1991) and other metal ions (QUINOT and CLAEYS 1957), hydrophobic surface (STRECKER 1956, 1960), and coating with polyvinylpyridine-1-oxide (BECK et al. 1965, FROMME et al. 1966, DOBREVA et al. 1975) or conversion into siloxane bridges (≡Si–O–Si≡) by thermal treatments imparting hydrophobicity to the surface.

In short-term genotoxicity studies silica was not mutagenic to *Salmonella typhimurium* (MORTELMANS and GRIFFIN 1981) nor induced sister-chromatid exchanges in V79-4 Chinese *hamster* lung fibroblasts (PRICE-JONES et al. 1980). Micronuclei were induced in *hamster* embryo cells (HESTERBERG et al. 1986), but not in *mice in vivo* (VANCHUGOVA et al. 1985). At 160 $\mu g/cm^2$ and 320 $\mu g/cm^2$ both Min-U-Sil 5 and Min-U-Sil 10 induced micronucleus formation in V79 and (at a lesser degree) in Hel 299 cells, but no chromosomal aberrations occurred (NAGALAKSHMI et al. 1995). *In vitro*, five preparations of α-quartz (Min-U-Sil 5, Min-U-Sil 5 pre-treated with hyrofluoric acid, Chinese standard α-quartz, DQ-12 and F600), cristobalite and tridymite induced strand breakage, thymine glycol production, and hydroxyl radical generation (DANIEL et al. 1995). By Fourier-transform infrared spectroscopy MAO et al. (1994) proposed that crystalline silica can bind closely to DNA (presumably by hydrogen bonding to phosphate ester groups) and generate oxygen radicals that can damage DNA (SHI et al. 1994), thereby increasing risk of cancer development.

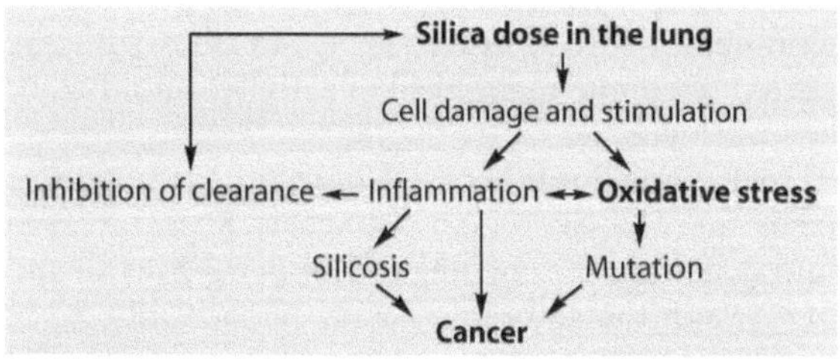

Fig. 45. A hypothetical sequence of events leading to quartz pathology based on studies in animals and cells *in vitro* (DONALDSON and BORM 1998)

A single intratracheal instillation of 2 mg submicronic (300 nm) silica particles in male Swiss Webster *mice* by 24 h induced acute focal necrosis in type I pneumocytes beneath an exudate of polymorphonuclear leucocytes intermingled with a lesser number of alveolar macrophages (BOWDEN and ADAMSON 1984). The presence of silica particles within the cytoplasm of type I cells, indicated the inability of alveolar phagocytes to cope with the acute load. Within 24 h particles were found lying free in the interstitium and within the phagosomes of interstitial macrophages, which by 3 days were aggregated into nodular collections adjacent to terminal bronchioles and perivascular spaces.

From 1 to 5 days after a single intratracheal instillation of 10 mg Min-U-Sil silica (particle size 5 μm) into SPF Wistar *rats* 8-hydroxy-2'-deoxyguanosine levels in lung tissue increased 2.24- to 2.86-fold, suggesting the possible carcinogenicity of silica (YAMANO et al. 1995).

As to mixed-dust exposure in coal miners, BORM and TRAN (2002) required more refined physiologically based pharmacokinetic modelling for a better estimate, also including interindividual difference in lung clearance. They suggested that the difference in pathology between coal and quartz dusts is due mainly to differences in the biopersistence and the intrinsic activity of both substances. In agreement with the increased biosolublity of the MMVF34/HT stone wool fibre, the pathology after 3, 6, 12, 18 and 24 months of inhalation (30 mg/m^3) in *rats* showed minor histopathological changes compared to both MMVF21 high (30 mg/m^3) and medium (16 mg/m^3) dosage groups (KAMSTRUP et a l. 2001).

4.4
Fibrous Dusts and Malignancy

Research on the health effects of naturally occurring and synthetic fibres has shown that fibres with certain physico-chemical characteristics have the potential to induce lung fibrosis and lung, pleural and peritoneal tumours (BIGNON et al. 1995). The potency of inducing tumours is thought to be related to their biopersistence (DAVIS 1986, POTT et al. 1989, VALLYATHAN et al. 1997, HESTERBERG et al. 1998, RÖDELSPERGER et al. 1998, MILLER et al. 1999, MOOLGAVKAR et al. 2000). While measuring the *in vitro* dissolution at neutral pH (7.2–7.8) rep-

Table 5. Surface area, percentage SiO_2, and surface silanol density of fibrous silicates (from Ghio et al. 1994)

Silicate	Surface area (m^2/g)	Percentage SiO_2	Surface silanol density (groups/nm^2)
Amosite	2.3±0.4	49.2±0.5	7.5±1.8
Crocidolite	8.7±1.0	48.5±0.3	4.7±0.6
Chrysotile	28.8±1.4	39.8±0.9	1.8±0.1

resent the dissolution in the extracellular lung fluid, *in vitro* dissolution rates at acidic pH (4.5–5) represent the pH environment within the phagolysosomes of the pulmonary alveolar macrophages (Bauer et al. 1994, Christensen et al. 1994, Knudsen et al. 1996, Guldberg et al. 1998). Collier et al. (1995) found differences in glass fibre biopersistence between the peritoneal cavity and the lung of *rats:* For fibres longer than 20 μm, durability in the peritoneal cavity was greater than in the lung. This may result in higher pathogenicity of such fibres in the peritoneal cavity. Pott (1995) found the intraperitoneal model to be much more specific and sensitive for testing the carcinogenicity of inorganic fibres than the inhalation model. The results with inhaled fibres were broadly consistent with those from intraperitoneal injection studies of the same fibres, in that the responses were dependent on both the durability of the fibres and the numbers of long thin fibres (Miller et al. 1999). Whereas *rat* pleural mesothelial cells *in vitro* showed dose-dependent and significant increases of 8-hydroxydeoxyguanosine formation in response to crocidolite asbestos or iron-chelated crocidolite fibres (but not after exposure to glass beads), a *human* mesothelial line (MET5A), showed decreases in 8-hydroxydeoxyguanosine (Fung et al. 1997). Both cell types exhibited elevations in message levels of manganese superoxide dismutase. In comparison with *human* MET5A cells, *rat* pleural mesothelial cells exhibited increased cytotoxicity and apoptosis in response to asbestos. In human peripheral blood cells, 8-hydroxy-2'deoxyguanosine was not a sensitive biomarker for past asbestos exposure at low levels (Hanaoka et al. 1993).

Surface roughness of fibrous minerals may be evaluated by comparison of Brunauer-Emmet-Teller (1938) surface area (SA_{BET}) and calculated (SA_{cal}) one (Ono-Ogasawara and Kohyama 1999). Natural mineral fibres, as expected from scanning or transmission electron microscopic observations, have complicated cleavage or degraded surfaces due to weathering. As a famous example, the shape of a chrysotile fibrils a hollow cylinder (Yada 1967) with a large surface area because is has both an inside and outside surface (Naumann and Drescher 1966). In addition, chrysotile exists in fibre bundles together with many fibrils that produce irregularly rough surfaces. Because of the results, the large ratio SA_{BET}–SA_{cal} does not express exclusively surface roughness for chrysotile.

According to Liddell (1997), after many electron microscope studies, there is a convincing explanation of the reasons for **differential effects of asbestos in animals and *humans*** (Elmes 1991). Massive doses of chrysotile fibres, once retained in the lung of a *rat*, split up into finer bundles, and eventually into fibrils (the basic crystal units, which are less than 0.1 μm in diameter), and so provided a much increased surface area for biological reaction. Within the animal's life-span, clearance from the lung was not great, and doses were so enormous (at the limit oftolerance, well over 2000 fibres/ml) that substantial proportions of animals developed tumours. On the other hand, however, the same dose of amphibole fibres was retained virtually without increase in surface area, so that the effective dose of amphibole in rodent lung was much smaller than that of chrysotile. In *man*, on the other hand, although the initial build-up of surface area was much greater, exposure for exposure, with chrysotile than with amphibole, the chrysotile was largely cleared from the lung before tumours could develop, whereas amphibole was retained definitively. The **effective dose** of amphibole in the *human* lung was thus very much greater than of chrysotile.

In *man*, an increased asbestos fibre concentration was correlated to galectin-1-binding and the presence of epitopes for natural immunoglobulin G subfractions with selectivity to α-galactosides and α-mannosides (Kayser et al. 2000). The survival of patients with primary and secondary lung tumours was negatively associated with the fibre concentration.

Two interferon-γ-controlled metabolic pathways accounting for some of the cytostatic effects of interferon-γ in *rat* pleural mesothelial cells transformed *in vitro* with benzo[a]pyrene or chrysotile asbestos were not efficient in *human* mesothelioma cells, and suggested that cytokine-induced growth inhibition is mediated by a different pathway in *human* mesothelioma cell lines (Phan-Bitch et al. 1997).

In the current consensus model of the pathogenesis of mesothelioma, it is suggested that neoplastic growth is a result of two concurrent, perhaps mutually reinforcing processes: firstly, genetic changes in the mesothelial cells, whereby their responsiveness to autocrine and paracrine growth signals is altered, and secondly, a chronic inflammatory response. Either or both may by induced by mineral fibres of the appropriate shape and chemical composition or by other carcinogenic/pro-inflammatory agents (GERWIN 1994, JAURAND and BARRETT 1994, FITZPATRICK et al. 1995). BIELEFELDT-OHMANN et al. (1994, 1996) added a third process, immuno-evasion.

While uninflamed appendices elicited mesothelial cells to be positive for the cell adhesion molecule ICAM-1 and negative for VCAM-1 or E-selectin, 6 h after the onset of clinical symptoms of appendicitis a strong reaction on ICAM-1 was found (KLEIN et al. 1995). The expression of VCAM-1 was detectable after 12 to 16 h. E-selectin was not found in acute appendicitis. Identical constellation in the expression of ICAM-1, VCAM-1 and E-selectin were observed on human omental mesothelial cells grown in fibronectin-coated flasks *in vitro*. Activated mesothelial cells from pleural exudates induced the exprimation of $\alpha 3$ and $\alpha 6$ chains, matrix proteins as laminin, tenascin, fibronectin and type IV collagen were increased (BARTH et al. 1995).

MOC31 recognising a transmembrane glycoprotein present on most types of epithelial cells, including adenocarcinoma, but not on normal or malignant mesothelial cells (RUITENBECK et al. 1994, EDWARDS et al. 1995), in a case of an asbestos sprayer played a pivotal role in reaching the diagnosis, and had important implications in claims for compensation for mesothelioma (RYAN et al. 1997).

Asbestos initiates the formation of oxygen radicals through at least two mechanisms: activation of respiratory burst in phagocytic cells and iron-catalysed reactions (KAMP et al. 1992). The dramatic enhancement of release of superoxide anions found by HILL et al. (1995) when long fibre amosite was opsonized with IgG, confirmed the greatly increased biological activity after opsonization (SCHEULE and HOLIAN 1989, NYBERG and KLOCKARS 1990, PERKINS et al. 1991, DONALDSON et al. 1992). Treatment of rat alveolar macrophage cultures with chrysotile dose-dependently stimulated tumour necrosis factor α secretion, which was inhibited by the addition of deferoxamine (SIMEONOVA and LUSTER 1995). When lymphatic clearance of 120 µg short crocidolite fibres was inhibited by obstructing peritoneal stomata 2 days before by intraperitoneal injection of Amicon agarose blue A beads (50–150 µm in diameter) there was no difference to the effects of 200 µg of mixed (native) fibres in the LDH activities in the peritoneal lavage fluids and the trypan blue staining of the diaphragms of *mice* (GOODGLICK and KANE 1990).

Table 6. Definition and materials of man-made fibres (Outline from VDI Instruction 3469, p. 1, 1989)

Inorganic crystalline fibres
Steel
Carbon fibres
Potassium titanate
Silicon carbide whiskers
Other inorganic crystalline fibres
Inorganic amorphous fibres
Textile vitreous fibres
A-Glass
C-Glass
D-Glass
E-Glass
R-Glass
Z-Glass
Silicic acid fibre (SiO_2)
Non-textile glass fibres
Ceramic fibres (alumino-silicate fibres)
Glass wool
Rock wool/slag wool
Organic fibres
Polyester
Polyaramide
Polytetrafluoroethylene (Teflon®)
Polyamide
Viscose (Reyon)
Polypropylene
Polyacrylonitrile
Polyacrylamide

Iron can be mobilised from asbestos in acellular *in vitro* systems, in cultured cells, and *in vivo* (Table 7).

The ability of man-made vitreous fibres, to produce reactive oxygen metabolites in *human* polymorphonuclear leukocytes was studied by luminol-enhanced chemiluminescence, and it was generally less than that produced by quartz (LUOTO et al. 1997). There was a dose-dependent increase in the production of reactive oxygen metabolites by all man-made vitreous fibres, quartz, titanium dioxide and chrysotile. In *rat* alveolar macrophages lucigenin-enhanced chemiluminescence was concentration-dependent for each JM-100 glass microfibre length (BLAKE et al. 1998). Chemiluminescence declined and lactate dehydrogenase rose with increasing fibre concentrations.

Supercoiled ϕX174 RF1 plasmid DNA was used for a target of respirable industrial fibre-induced free radical injury. The results presented by GILMOUR et al. (1995) revealed that the insulation wool MMVF 21 has much less free radical activity than asbestos fibres, but does have a detectable harmful

Table 7. Asbestos and iron

Dust	Model system	*In vivo/ in vitro*	Parameter of toxicity	Result	Reference	Remarks
Crocidolite, amosite, Canadian chrysotile	Phospholipid emulsions	*In vitro*	Lipid peroxidation	All 3 types of asbestos were able to catalyze lipid peroxidation in the native state. This catalytic activity was inhibited by pre-washing of the asbestos with desferroxamine.	WEITZMAN and WEITBERG (1985)	Treatment with Fe chelators diminished lipid peroxidation.
Crocidolite, amosite	Mobilisation of iron	*In vitro*	O_2 consumption	Desferrioxamine B, which forms a redox inactive chalate with Fe(III), inhibited O_2 consumption by 90–100%.	LUND and AUST (1990)	Intracellular mobilisation of Fe → ROS
Crocidolite	Leaching of iron from fibres, acellular	*In vitro*	Reactivity of iron-bound asbestos compared with iron mobilised from asbestos	Mobilisation of iron from crocidolite by chelators such as citrate, nitrilotriacetate or EDTA greatly enhances its redox activity. This may lead to increased production of ROS.	LUND and AUST (1991)	Fe on or in crocidolite may be responsible for redox activity
Asbestos, attapulgite, fiberglass, whiskers of K_2TiO_3, basic $MgSO_4$ and $CaSO_4$, PO_3 polymer	Calf thymus DNA, fibres; additions: 1) none, 2) 5 mM H_2O_2, 3) 0.5 mM H_2O_2 + 0.5 mM $FeSO_4$.	*In vitro*		Asbestos induced 6.6–99.8 of 8-OH-dGuo per 10^5 dGuo. Levels of 8-OH-dGuo in MMMF 3.6–9.4. Amount of 8-OH-dGuo strongly stimulated by addition of H_2O_2 only in asbestos, not in MMMF. Stimulation of 8-OH-dGuo formation by $FeSO_4$ addition in attapulgite, fiberglass, K_2TiO_3 whisker, and metaphosphate polymer. Fiberglass and basic magnesium sulphate whisker higher amounts of 8-OH-dGuo after mannitol addition than these fibres and H_2O_2.	ADACHI et al. 1992	8-OH-2'-deoxyguanosine as a cause of point mutation
Crocidolite		*In vitro*	Surface iron after reduction to Fe^{2+} and chelation by citrate	Corresponding to decrements in surface Fe^{3+}, thiobarbiturate-reactive products and HO. diminished. Deferoxamine provides similar results by chelating Fe.	GHIO et al. 1992	Surface acidic functional groups
Crocidolite, amosite, tremolite, chrysotile	Closed circular superhelical ϕX174 RFI DNA	*In vitro*	DNA single strand breaks (SSB)	Mobilisation of iron by chelators, followed by redox cycling, greatly enhanced crocidolite-dependent formation of DNA SSB.	LUND and AUST (1992)	
Crocidolite	*Calf* thymus DNA	*In vitro*	8-OH-2'-Deoxyguanosin formation	Under simple incubation with 2 mg *calf* thymus DNA 5 mg crocidolite or deferrized crocidolite for 3 h at 37 °C yielded 14.6 and 30.2 OH^8-2'-deoxyguanosine/10^5 2'-deoxyguanosine, respectively. In incubation systems supplemented with 0.5 mM H_2O_2 plus 0.1–1.0 mM Fe_2O_3, 0.5 mM Fe_2O_3 + 0.5 mM ascorbate, 0.5 mM Fe_2O_3 + 1 mM EDTA, or 0.5 mM $FeSO_4$ + 0.5 mM ascorbate, deferrized crocidolite induced higher levels of OH^8-2'-deoxyguanosine.	ADACHI et al. (1994)	Ferric iron promoted DNA oxidation
Crocidolite, amosite	Closed circular superhelical ϕX174 RFI DNA	*In vitro*	DNA single strand breaks	32% DNA damage induced by crocidolite and 52% by amosite when citrate (1 mM) and ascorbate (1 mM) were added. Further addition of 25 μM H_2O_2 increased the figures to 56% and 76%, respectively. During 30 min incubation with 1 mM nitrilotriacetate, 14 μM iron was mobilised from crocidolite compared with 4 μM from amosite.	AUST et al. (1994)	Citrate-chelated iron, reduced by ascorbic acid to Fe^{2+} acts as a Fenton catalyst to generate $HO^{\bullet}$ which induces DNA SSB.
Crocidolite, neutron activated	A549 cells	*In vitro*	^{55}Fe and ^{59}Fe mobilised from crocidolite	3600 pM iron/10^6 cells mobilised of which 54 pM/10^6 cells was in a low-molecular-weight (LMW) fraction. Intracellular concentration: 1.4 mM iron of which 22 μM was in the LMW fraction.	CHAO et al. (1994)	Fe overload in cells after phagocytosis of crocidolite
Asbestos bodies (*human*), amosite	Closed circular superhelical ϕX174 RFI DNA	*In vitro*	DNA single strand breaks	Asbestos bodies + DNA → no SSB; asbestos bodies + ascorbate + DNA → 10% SSB; asbestos bodies + ascorbate + chelator + DNA → 21% (citrate) and 77% (EDTA) SSB	LUND et al. (1994)	Iron deposited on asbestos fibres *in vivo* may increase the damage to DNA

Table 7. (Continued)

Dust	Model system	*In vivo/ in vitro*	Parameter of toxicity	Result	Reference	Remarks
Crocidolite	Supercoiled ϕX174 RF1 plasmid DNA	*In vitro*	DNA single-strand breaks (SSB)	Native crocidolite with additional Fe bound did not change its ablity to cause DNA-SSB. Pre-soaking of crocidolite with deferoxamine and addition of Fe resulted in an increased DNA damage. In Fe-containing medium, more SSB than in regular culture medium. Crocidolite may bind Fe from intracellular sources and this may increase the fibre's lifetime.	HARDY and AUST (1995)	Surface binding of Fe from intracellular sources
Amosite, chrysotile, TiO_2, carbonyl iron	Primary *rat* alveolar type II cells, transformed WI-26 and A549 cells	*In vitro*	Cellular damage, production of ROS, DNA strand breaks	Amosite damaged cultured human epithelial cells (WI-26) and induced DNA strand breaks. All dusts caused release of $HO^{\bullet}$. These effects were attenuated by addition of phytic acid.	KAMP et al. (1995)	Mineral dusts may be directly genotoxic
Chrysotile, crocidolite, TiO_2	*Rat* alveolar macrophages	*In vitro*	ROS	Dose-dependent stimulation of TNF-α secretion by asbestos was inhibited by deferoxamine (iron chelator). Role of ROS confirmed by radical regenerating hypoxanthine-xanthine oxidase system	SIMEONOVA and LUSTER (1995)	ROS in signalling TNF-α stimulation
Amosite UICC	*Horse* spleen ferritin	*In vitro*	Ferritin adsorption	Desorbed protein showed subunits of ~13 and 15 kDa, aside from the 20 kDa subunit present in the native protein	FUBINI et al. (1997)	Exposed iron core of the absorbed protein can trigger Fenton-type reactions
Crocidolite (27 % Fe)	*Rat* lung fibroblasts	*In vitro*	LDH, collagen, DNA, thiobarbituric acid	Collagen content of fibroblasts increased in a dosedependent manner. No cell damage, proliferation or lipid peroxidation	GARDI et al. (1997)	Fe release important in asbestos fibrosis
Long-fibre amosite, SiC fibre, glass fibres, RCF1 ceramic fibres	ϕX174 plasmid DNA	*In vitro*	Hydroxyl radical generation, Fe^{3+} release	Amosite and RCF1 released $HO^{\bullet}$. Soluble iron caused hydroxylation of salicylate.	BROWN et al. (1998)	None of the noncarcino-genic fibres demonstrated free radical activity. $HO^{\bullet}$ by RCF1 was inhibited by an iron chelator (DSF-B)
LFA, C100/475, MMVF10, SiC, RCF1	*Sheep* surfactant	*In vitro*	Fe^{3+} release	Surfactant coated fibres released mor Fe^{3+} than native fibres at both pH 4.5 and 7.2	FISHER et al. (1998)	

effect on plasmid DNA which is consistent, but small enough to not attain statistical significance in their series of experiments. The amelioration by mannitol of the small amount of damage to plasmid DNA caused by MMVF treatment implicated hydroxyl radicals, but chelation of iron with desferrioxamine-B had no effect.

Homologous recombination is one mechanism, which might readily contribute to both the mutagenicity and carcinogenicity of asbestos fibres. All types of asbestos fibres have been shown to mediate cell transfection (APPEL et al. 1988, GAN et al. 1993). Transfection is frequently employed in the laboratory to generate homologous recombination (WALDMAN 1992, SAVRANKY et al. 1994). LEZON-GEYDA et al. (1996) have developed a new assay which determines the extent to which a marker gene present in DNA introduced by asbestos can recombine with homologous genes residing in a transfected cell. Asbestos-induced recombination events may play a significant role in asbestos mutagenesis and carcinogenesis, and promotion of recombination may underlie the well-recognised synergy of asbestos with other carcinogens.

The micronucleus assay is one of the most frequently used short-term assays for the detection of genotoxic agents and potential carcinogens. (MAVOURNIN et al. 1990). In Chinese *hamster* lung fibroblast cell line V79 Manville 100 microfibre and Owens Corning AAA-10 microfibre induced significant numbers of multinucleated and micronucleated cells in a concentration-related manner (ONG et al. 1997). Immunofluorescent staining demonstrated a significant dose-related increase in the proportion of kinetochore-positive micronuclei in cells treated with the two microfibres. Unlike the two microfibres, the larger fibres, Owens Corning building insulation fibre, neither induced micronuclei nor inhibited cytokinesis in V79 cells. Thus, the genotoxic potential of glass fibres in V79 cells may be related to their size.

JENSEN et al. (1996) found that long crocidolite asbestos fibres (10–55 μm), trapped by the cleavage furrow, sterically blocked cytokinesis, sometimes resulting in the formation of a binucleated cell. The ends of the blocking fibres were usually found within invaginations of the newly formed nuclei. Nuclear envelop-fibre attachment was evident when a chromatin strand ran with the fibre into the intercellular bridge. Such strands may break, causing chromosome structural rearrangements.

Relatively minimal information is available on cancers associated with fibre exposure. LECHNER et al. (1997) summarised these carcinogenesis mechanisms with primary emphasis on the alterations described in *human* mesotheliomas. Of the known changes, the most frequent are in the tumour-suppressor genes $p16^{INK4a}$ (KRATZKE et al. 1995) and *NF2* (BIANCHI et al. 1995, SEKIDO et al. 1995) and possibly the SV40 virus (CRISTAUDO et al . 1995) large T-antigen oncogene.

Among 3685 rock-slag wool workers lung cancer incidence was increased (BOFFETTA et al. 1999).

CHAPTER 5

Material and Methods

5.1 Freund's Adjuvant

FREUND's (1947, 1956) complete adjuvant is the most popular adjuvant. It consists of killed and dried mycobacterial cells suspended in an emulsifying oil.

5.2 Implants

The rather simple **cotton-pellet-test** in *rats* was introduced by MEIER et al. (1950). An important source of variability arises from the initial size, i.e. the weight and particularly the surface of the cotton-pellet (BENZI and FRIGO 1964, ROBINSON and ROBSON 1964). The initial phase of the inflammatory reaction to subcutaneous implantation is characterised by increased permeability of the vessels in the connective tissue surrounding the implant, with the pellet undergoing a rapid saturation ('soaking') by the fluid escaping from the vessels (MEYER et al. 1953). Early permeability changes in the implant region are responsible for a transient transudative phase during the first three hours, which is followed by an exudative phase occurring between 3 and 72 h (SWINGLE and SHIDEMAN 1972). The fluid absorbed by the pellet greatly influences the wet weight of the granuloma, so that reproducible assessments are only achieved by determining the dry weight, which correlates well with the amount of granulomatous tissue formed. A continuous increase in the granuloma dry weight has been observed for up to 3 months (DI PASQUALE and MELI 1965). However, the greatest rate of increase occurs during the first few weeks; thus, the majority of assessments have been confined to 7-14 days granulomata.

Cylindrical "bullets" (approximately 3 mm diameter and 5 mm long) of **agar**, **gelatine** and **egg white** were implanted in the peritoneal cavity of *rats* (CURRAN and CLARK 1964). The gelatine implants were rendered insoluble by 24 hours' treatment with formalin, which was then removed by prolonged and repeated soaking in balanced salt solution. We (SCHILLER 1973) used Marbagelan®, an absorbable denatured spongeous gelatine preparation for subcutaneous implantation.

5.3 Dusts

Grinding of Materials. The mechanical disintegration of a solid leads to the formation of a new surface. The formation of this surface begins with the rupture of interatomic bonds and is concluded with an ordering of structure of the layer near the surface. As a rule, the formation of the new surface is accompanied by dissipation of excess energy as evidenced by the evolution of heat, the emission of electrons, ions, or neutral particles, luminescence, and other phenomena.

When milled under an atmosphere of H_2 at 10-12 torr the following reaction occurs (BYSTRIKOV et al. 1980]:

$$\equiv Si-O-Si\equiv + H_2 \rightarrow \equiv Si-H + \equiv Si-O-H \qquad [6]$$

Active centres that react with hydrogen may be generated as a result of either particle splitting or processes of friction (the uptake of gases during milling is sometimes termed tribosorption (τριβή = attrition). The ≡ Si-H bonds are readily hydrolysed by water:

$$\equiv Si-H + H_2O \rightarrow \equiv Si-OH + H_2, \qquad [7]$$

and the yield of H_2 in reaction [7] must be equal to the concentration of the bonds (STRELETSKII and BUTYAGIN 1980).

Quartz. A tertiary quartz sand was broken by jaw crushers and metallic ball mills at Dörentrup Sand and Clay Plant Ltd and sold under the name "Ground Product No. 12" (in short: DQ12), with a size distribution of <60 µm. A <5 µm size fraction was prepared from this material by centrifugal separation in air using the laboratory separator of Messrs. Walther-Staubtechnik of Cologne. This standard quartz is officially designed as DQ12 <5 µm (Robock 1973). The maximum of the particle number distribution is ca. 0.8 µm; that of the surface distribution ca. 1 µm; while the maximum of the volume or mass distribution is ca. 1.3 µm. The upper size of the quartz particles is between 5 and 6 µm. The specific surface found by the BET-method is 7.4 m^2/g.

The characteristic glow curve of DQ12 annealed in air at 520 K for 24 h after X-raying (30 kV, 45 mA, 1 min) at 133 K as depicted by Kriegseis et al. (1977) is comparable to thermally stimulated luminescence curves which have been obtained by other authors from natural crystalline and amorphous SiO_2 samples as well as synthetic quartz single crystals (Medlin 1963, Schlesinger 1965, Halperin et al. 1979, Mattern 1973).

Robock and Klosterkötter (1973) and Robock (1974) have shown that thermally stimulated luminescence of SiO_2 materials is extremely sensitive to surface treatment. The effect of heat treatment at 850 K on the thermally stimulated luminescence was dependent on the gas atmosphere (air, N_2, O_2).

Crystallised Quartz

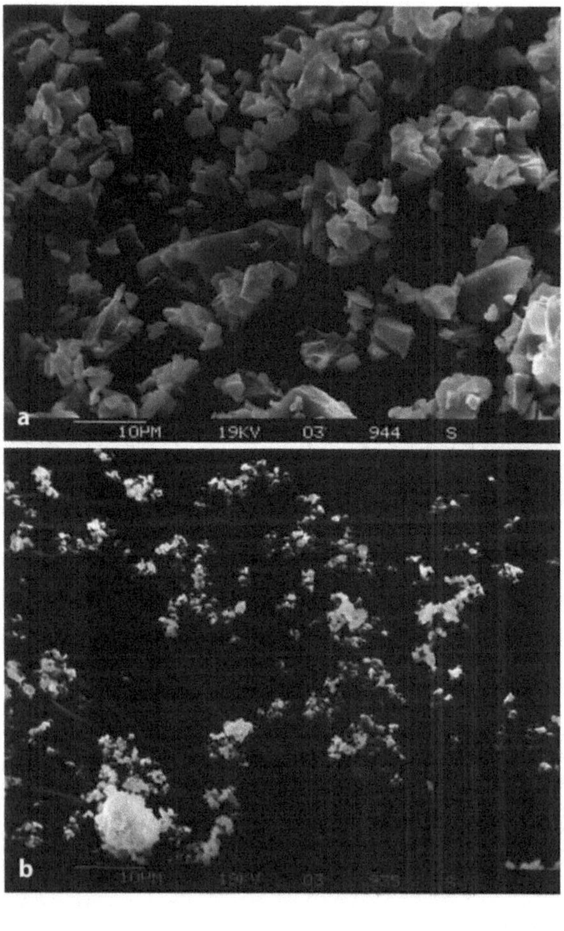

Fig. 46a, b. Brazilian crystallised quartz of two different particle sizes: **a)** specific surface 2 m^2/g; and **b)** specific surface 9.1 m^2/g. Gold coating. Cambridge Stereoscan 150 operated at 19 kV. APh-R. 214/80, negative 03 944, and APh-R. 212/80, negative 03 935, respectively. – Small particles stick to bigger ones, firmly held by the surface charges produced by grinding

Tridymite

Fig. 47. Tridymite (particles 2–4 µm in size). Gold coating. Cambridge Stereoscan 150 operated at 19 kV. APh-R. 216/80, negative 03 953

Dust mixtures of crystalline and amorphous silica cannot be quantified by infrared spectroscopic methods alone, due to a severe overlap of the primary spectral SiO_2 band in the most characteristic infrared region, i.e. 850–750 cm^{-1}. This problem has been solved with the use of multivariate calibration by partial last-squares regression applied to infrared spectroscopy (Bye 1994).

Titanium Dioxide

Anatase (TiO_2 P 25, Degussa, Frankfort on the Main) was produced by flame hydrolysis of $TiCl_4$. Boehm (1966) comprehensively studied the functional groups on the surface of this type of dust.

Fig. 48. Anatase (TiO_2 P 25, Degussa, Frankfort on the Main). Gold coating. Cambridge Stereoscan 150 operated at 19 kV. APh-R. 211/80, negative 03 932

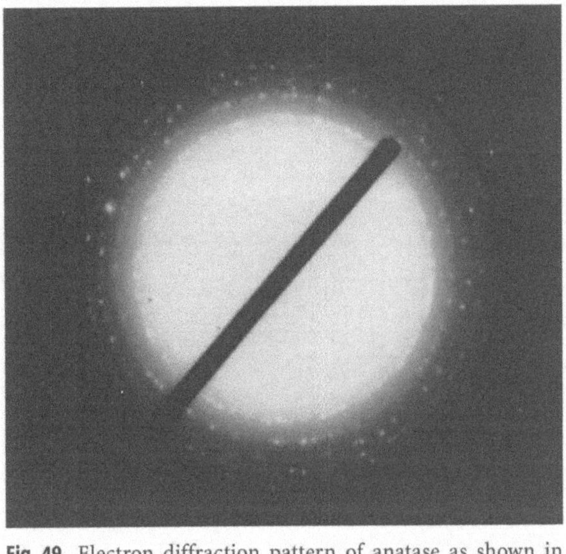

Fig. 49. Electron diffraction pattern of anatase as shown in Fig. 48. 50 mg of the dust were suspended in 2 ml saline and given intraperitoneally to a 194 g Sprague-Dawley *rat* (Charles River, France). 4 days later under pentobarbital anaesthesia (30 mg/kg), the animal was perfused from the abdominal aorta with 2.5% glutaraldehyde in 0.1 M sodium cacodylate buffer (pH 7.4). Postfixation with 1% osmium tetroxide in sodium cacodylate buffer. Embedded in Epon 812 and sectioned at 50 nm. Stained with lead citrate and uranyl acetate. – Particles analysed were within phagolysosomes of Kupffer cells (block 4460). Film 224/79

Aluminium

Table 8. Aluminium powder

Particle Size [μm]	Number [%]
0– 2	46.6
2– 4	45.9
4–10	5.8
10–20	1.4
20–25	0.3

Aluminium Oxides

Besides α-Al_2O_3 (corundum) there are at least eight oxides mostly classed under the term of γ-Al_2O_3 (cf. GINSBERG et al. 1957). Some of these forms may be non-stoechiometric oxide-hydroxides.

Clayey Shale

Quartz-free clayey shale was taken from the surrounding strata of Erda seam, Fürst Leopold-Baldur mine at Harvest-Dorsten. The fraction <5 μm according to ROBOCK and KLOSTERKÖTTER mainly contains crystalline kaolinite. It was pressed to square bars and re-dispersed in POLLEY's (1963, 1965) dust channel 4 h per day, 5 days per week.

Montmorillonite

Montmorillonite clays have an octahedral layer of Al_2O_3-$Al(OH)_3$ sandwiched between two tetrahedral layers of SiO_4 silicate units. The specific surface of particles <2 μm is of the order of 500 m^2/g, sometimes as high as 760 m^2/g, as measured with adsorption of N_2 in the standard Brunauer-Emmell-Teller (BET) method.

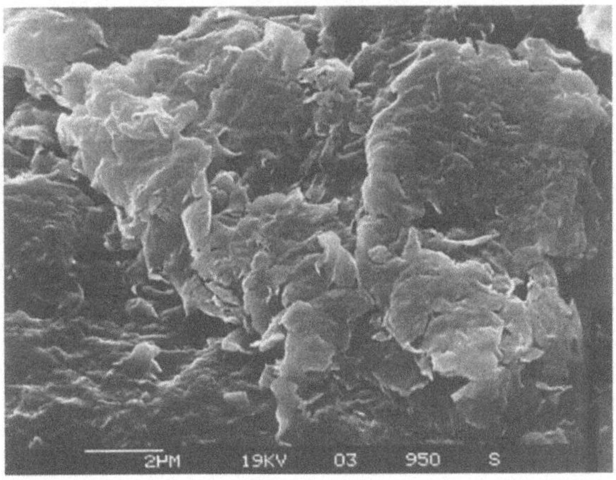

Fig. 50. Montmorillonite API H 26 (particles less than 2 μm in size). Gold coating. Cambridge Stereoscan 150 operated at 19 kV. APh-R. 215/80, negative 03 950

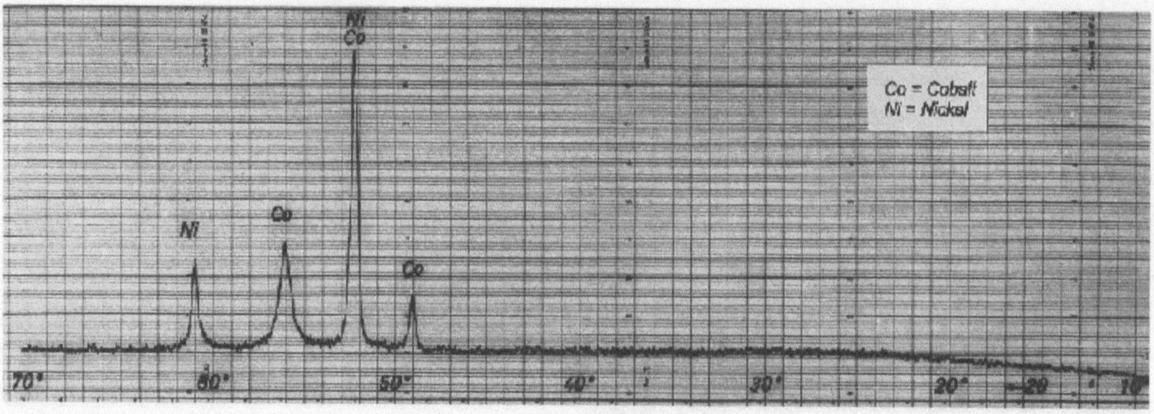

Fig. 51. X-ray diffraction pattern of the powdered cobalt-nickel alloy from the sintered carbide metal production of the Deutsche Edelstahlwerke, Krefeld. No other crystalline material can be detected in the sample, SiO_2 modifications included. Courtesy of Dr. R.-W. SCHLIEPHAKE, Bergbau-Forschung, Essen

Magnetite

Magnetite, $[Fe^{3+}]^{IV}$ $[Fe^{2+}Fe^{3+}]^{VI}$ O_4, is a cubic mineral and member of the spinal structure type, space group Fd3m.

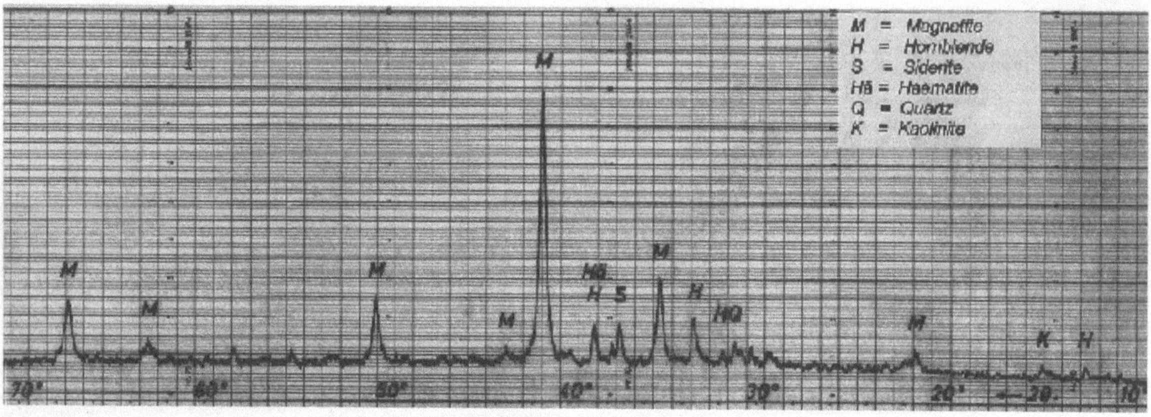

Fig. 52. X-ray diffraction diagram of the powdered magnetite from Grängesberg, Sweden. Besides magnetite (Fe3O4) the sample contains some hæmatite (Fe2O3), siderite (FeCO3), kaolinite [Al2Si2$O_5$$(OH)_4$], an amphibolic mineral, and 3 per cent quartz

Dusts Isolated from Pneumoconiotic Lungs

While CARTWRIGHT and NAGELSCHMIDT (1961) isolated lung dusts by refluxing 5 g dried ground lung with 100 ml of 50 per cent hydrochloric acid (HCl) for 24 h, followed by repeated washing with alcohol, EINBRODT and KLOSTERKÖTTER (1965), EINBRODT et al. (1965) and STÖBER et al. (1967) recovered the dusts by the formamide method described by THOMAS and STEGEMANN (1954).

Fig. 53. a Dust No. 17 gained after formamide treatment at 135 °C of the lungs of an 38-year-old coal miner who had worked underground for 15 years. Gold coating. Cambridge Stereoscan 150 operated at 22 kV. APh-R 664/81, negative 11 613. Sample provided by Prof. H. J. Einbrodt, Aachen

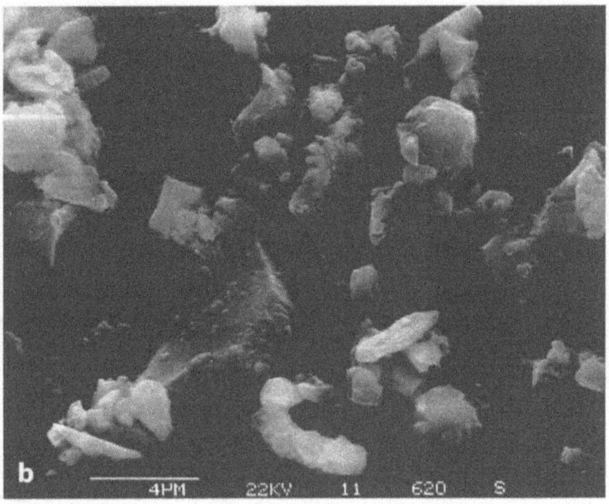

Fig. 53. b Dust No. 17 gained after formamide treatment at 135 °C of the lungs of an 38-year-old coal miner who had worked underground for 15 years. The quartz portion (free SiO_2) of 10 % of the total dust was removed by HF. Gold coating. Cambridge Stereoscan 150 operated at 22 kV. APh-R 665/81, negative 11 620. Sample provided by Prof. H. J. Einbrodt, Aachen

5.3.1 Preparation of Dusts for Animal Experiments

It is clear that the surface properties of substances, which in the form of particulate matter cause lung injuries play an exceedingly important role in the development of the lesions. Thus the preparation of the material to be tested must not alter the physicochemical properties at the interface of the material with its environment before getting in contact with cells and tissues.

Interaction between a surface and solution invariably leads to change in the physical form of the surface, and this may often be detected by reflectance measurements. Nicoll (1942) showed the variation in reflectivity of two surfaces (glass and calcite) with time of exposure to 1 % sodium hydrogen phosphate at 50 °C. Smith and Waddams (1954), in addition, used electron micrographs to demonstrate corrosion of glass. The dissolution of glasswool fibres was greater in medium than in cell culture, whereas the opposite was true for rockwool fibres (Luoto et al. 1998). In glass-based fibre compositions, the replacement of Na^+ by Ca^{2+} increases fibre durability in saline solutions.

It is important to distinguish between true dissolution of **quartz** involving the production of molecularly dispersed silicic acid and the production of mosaic silica (Richardson and Waddams 1955). The latter is not a case of true dissolution but a separation of fragments of crystalline silica which forms a separate phase.

The effect of potassium permanganate on dissolution gave inconclusive results (Waddams 1958). This was thought to result from interaction between hydrated manganese dioxide formed when the anionic vacancies were oxidised and either the centres at which dissolution occurs or the silicic acid formed. Using H_2O_2 as oxidant considerably enhanced 48 h solubility values. When the volume of air in the tubes used for solubility tests was varied from 0 to 50 times the volume of extractant, the effects of 48 h solubility values increased noticeable over the range covered. Welch (1955) has commented on the ability of oxidising and reducing atmospheres to modify the reactivity of solids by changing the concentrations of lattice defects. Thusfar, however, no such simple room-temperature effects as those described here have been reported.

There is thus reason to believe that vacancies in the quartz lattice play an important role in the following processes:

- oxidation-reduction phenomena
- luminescence
- chemisorption
- catalysis
- dissolution of monosilicic acid

It is well known that when quartz particles are extracted with water, two forms of silica are liberated, ammonium molybdate reactive silica or mono(oligo)silicic acid and non- ammonium molybdate reactive silica or polysilicic acid (Ritchie et al. 1952, Richardson and Waddams 1955). A saturated solution of monosilicic acid, probably $Si(OH)_4$, contains 0.01 % silicic acid (Friedberg and Schiller 1988). The $-Si(OH)_2-O$ chain can crosslink by further elimination of water to give a network (silicic acid jelly) and finally colloidal silica with $-OH$ groups on the outer surface.

Polymerisation of monosilicic acid could take place on templates (Hasirci 1976, Erdogdu and Hasirci 1994).

Hydration of silica surfaces by the humidity of the air prevents desorption of water by vacuum evaporation at > 100 °C (Miller and Stöber 1954). The hydroxyl groups at the silica surface – frequently called silanol groups – are assumed to be the active adsorption sites (Nash et al. 1966, Stöber and Brieger 1968). As indicated by the results of infrared measurements, these are responsible for the adsorption of numerous organic compounds which are attached by hydrogen bonding (Stöber 1956, Hair 1967). Boehm (1966) and Peri and Hensley (1968) ascertained an irreversible destruction of silanol groups at temperatures > 720 K favouring the formation of very stable and inactive siloxane groups (Si–O–Si). Corresponding to this behaviour Robock (1973) observed a drastic decrease of the cytotoxic activity of Dörentrup quartz due to heating above 900 K. In aqueous suspensions the silanol groups of annealed silica are regenerated within several months. This process is accelerated by live steam. Thermally stimulated luminescence of quartz peaks at about 171, 238 and 257 K (Kriegseis et al. 1977). Furthermore, a peak in the range between 300 and 350 K as well as a broad peak between 400 and 450 K of lower intensity and varying temperature position from sample to sample, have been detected. Robock and Klosterkötter (1973) and Robock (1974) have shown that the thermally stimulated luminescence of SiO_2 materials is extremely sensitive to surface treatment. Alterations of the thermally stimulated luminescence can be achieved by washing the SiO_2 dusts in hot distilled water or by treatment with acids or bases (HCl, H_3PO_4, NaOH). The effect of heat treatment at 850 K on the thermally stimulated luminescence was dependent on the gas atmosphere (air, N_2, O_2) (Kriegseis et al. 1977).

On the surface of **TiO_2** hydroxyl groups were detected by infrared absorption (Yates 1961, Smith 1964). While after degassing of anatase at 150 °C water molecules persisted, at 350 °C two kinds of hydroxyl groups absorbing at 3715 and 3675 cm^{-1}, respectively, remained (Boehm 1966). Under equal conditions, rutile showed one band at 3680 cm^{-1}. Measurements of water vapour adsorption and adsorption heat (Ganitschenko and Kisselew 1961, Ganitschenko et al. 1961, Hollabough and Chessik 1961), and of wetting heath in water (Wade and Hackerman 1961) indicated that some OH groups persisted at the surface of TiO_2 after degassing at high temperatures. Mays and Bride (1956) measuring nuclear resonance showed water to be adsorbed very stable at 350 °C.

The same considerations apply to the surfaces of the many crystalline modifications of **aluminium oxide** (Boehm 1966). Besides α-Al_2O_3 (corundum) there are at least eight oxides mostly classed under the term of γ-Al_2O_3 (cf. Ginsberg et al. 1957). Some of these forms may be non-stoechiometric oxide-hydroxides.

5.3.2 Dust Feed Mechanisms

Inhalation studies would be the best method for testing a dust under physiological and environmental conditions. Arnold (1885) in Heidelberg published his experiments on inhalation and metastasis of dust in *rabbits* more than 100 years ago. Modern dust feed mechanisms were constructed by Wright (1950), LeBouffant (1958), Polley (1963), and Hemenway et al. (1983, 1986, 1990). Holt and Young (1960) used a Micro Hammer Mill for fibre breakdown adapted for dust formation. The dust generator of Tanaka and Akiyama (1984) consists of a screw feeder supplying a fluidized bed, which has an overflow pipe.

Table 9. Dust inhalation in *rats* (electron microscopy)

Type of dust	Treatment	No. of *rats*	Exposure [h]	Date
Magnetite		5	152, 160	Aug. 24, 1967 to Jan. 15, 1968
Aluminium		5	148, 204	Aug. 16, 1967 to Jan. 15, 1968
Shale		2	320	1967
Shale	Tobacco Smoke	2	320	1967
Shale	P 204	1	320	1967
Co Ni Alloy		2	76	October 2 to 30, 1967
Melamine resin		10	$8^1/_2$, $16^1/_2$, $32^1/_2$, $62^1/_4$, $125^1/_4$	Jan. 26 to Feb. 16, 1976

5.3.3 Intratracheal Application

Since KETTLE (1930) and KETTLE and HILTON (1932) intratracheal instillation has been routinely used for exposure of animals to particles. It has the advantages of accurate application of test materials very rapidly to the lower respiratory tract, bypassing the upper airways defences as the nose and confining the dose to a small volume, which reduces greatly the amount of equipment after exposure to highly toxic, carcinogenic, or radioactive materials. However, instillation is a less physiologic method of exposure than inhalation. Particles administered by instillation may not behave like similar particles in an aerosol. After instillation, particles are increasingly deposited in the basal regions of the lung (BRAIN et al. 1976), and have a significantly less homogeneous particle distribution than at inhalation exposure (PRITCHARD et al. 1985, DORRIES and VALBERG 1992). The use of a vehicle for suspending may both increase the agglomeration of particles and influence the inflammatory response of the lung (HENDERSON et al. 1995).

5.3.3.1 Intratracheal Instillation

COSTA et al. (1986) under enfluorane anaesthesia depressed the tongue and illuminated the pharynx with a fibre-optic laryngoscope. A bevelled Teflon tube was inserted into the trachea to a level just cephalic of the carina. Injection of the material to be delivered was accomplishes rapidly with the use of a No. 19 gauge hypodermic needle inserted directly into the tracheal tube. TROŠIC et al. (1996) used halothan putting *rats* for 30 s into a chamber saturated with the anaesthetic, which has been shown to produce a negligible effect of pulmonary monooxygenases, as well as on cell viability (MORGAN et al. 1980).

BRAIN (1971) increasingly used *hamsters* since they are more resistant to lung diseases than *rats* and have tolerated intratracheal injections better than *rats*.

In Sprague-Dawley *rats*, addition of *pig's* pulmonary surfactant (10 mg) to quartz did not influence the outcome of any parameter after 5 weeks (ZETTERBERG et al. 1998). *In vitro*, respirable fibres coated with *sheep* surfactant released more Fe^{3+} than native fibres at both pH 4.5 and 7.2 (FISHER et al. 1998).

Intratracheal instillation is a highly artificial system and must be interpreted with great caution (SEKHON et al. 1995). The primary mechanism responsible for rapid alveolar liquid clearance depends on the ability of the alveolar epithelium, in particular alveolar epithelial cells, to increase Na transport from the alveolus to interstitium through the stimulation of apical Na channels and basolaterally located Na,K-ATPase (PITTET et al. 1994). CLERICI (1998) reviewed the modulation of the sodium transport in alveolar epithelial cells by O_2 tension.

However, insoluble suspensions tend to localise in the medium-sized bronchi and seldom reach the alveoli (BRAIN et al. 1976).

Table 10. Intratracheal instillation in the *rat*

Volume of the suspension medium
2 ml
BELT et al. (1940)
1.5 ml
RAY et al. (1951, 1952), KING et al. (1953, 1955, 1958), JÖTTEN and KLOSTERKÖTTER (1955)
1 ml
BELT et al. (1940), KING et al. (1947), LLOYD DAVIS and HARDING (1949), KLOSTERKÖTTER (1952, 1959, 1960), WEBER (1952), ATTYGALLE et al. (1956), SCHLIPKÖTER and ROTHES (1956), ZAIDI et al. (1956), GROSS et al. (1958, 1960, 1966, 1967), SCHLIPKÖTER and LINDNER (1959, 1961), BROCKHAUS and SCHLIPKÖTER (1960), GOVERNA et al. (1961), JERRENTRUP (1961), SCHLIPKÖTER and BROCKHAUS (1961), ANTWEILER et al. (1962, 1963), GOLDSTEIN et al. (1962, 1967), GROSS (1963), HAUBENSAK (1965), HÖER and STEURICH (1965), SWENSSON et al. (1968), WENZEL et al. (1969), HOLLENBACH et al. (1971), KAW et al. (1971), KYSELÁ et al. (1973), GUPTA et al. (1974), SINGH et al. (1975)
0.75 ml
CHVAPIL et al. (1979)
0.5 ml
KING et al. (1953), SCHLIPKÖTER and ROTHES (1956), SCHLIPKÖTER et al. (1957), SAFFIOTTI and TOMMASINI DEGNA 1957–1958, 1958, 1959), PERNIS et al. (1958), SCHILLER (1958), SAFFIOTTI (1962), HÖER and STEURICH (1965), REIF et al. (1965), ROSMANITH (1975), ROSMANITH and BREINING (1975), BÜSCHER and PETT (1976), WELLER (1977), MORGAN et al. (1980), BERGER et al. (1981), ROSMANITH et al. (1981, 1983, 1995), WHITE et al. (1983), REISNER et al. (1983), HILSCHER et al. (1991, 1995), LI et al. (1991, 1993), ARDEN and ADAMSON (1992), BRUCH et al. (1993), TORNLING et al. (1993), BRAMMERTZ et al. (1995), FRIEMANN et al. (1995), SEKHON et al. (1995), ADAMIS et al. (1997), DIMATTEO and REASOR (1997), LEIGH et al. (1997), NEHLS et al. (1997), TOYA et al. 1997), ZETTERBERG et al. (1998)
0.4 ml
TROŠIC et al. (1996)
0.3 ml
FRITSCH et al. (1975), CASTRANOVA et al. (1991), MA et al. (1999), NELIN et al. (2002)
0.25 ml
GREEN et al. (1981)
0.2 ml
MORIMOTO et al. (1997), BONNER et al. (1998)
0.5 ml/kg
GROSS ET AL. (1977)

5.3.3.2 Intratracheal Inhalation

A method of particle exposure whereby anaesthetised *rats* intratracheally inhale, at regulated breathing rate and pressure, an aerosolised test material has been developed by OSIER et al. (1997). LEONG et al. (1998) inserted the tip of a nebulization probe through the otoscope speculum to approximately 5 mm beyond the vocal cord into the trachea of an anaesthetised *rat*. During administration of the aerosol, the operator followed the slow respiratory thoracic movements of the anaesthetised *rat* and manually triggered the nebulization system during the inspiratory phase at intervals of three to five breaths. In each activation, 10 µl of Pelikan drawing ink (Diazo dye, 17 Black; Pelikan AG) was aerosolised by a puff of air. Each puff was 5 ml, which approximates to the volume of lungs fully inflated in situ but is larger than the average tidal volume, which was determined to be approximately 1.55 ml for an anaesthetised *rat* of approximately 250 g in body weight (CROSFILL and WIDDICOMBE 1961).

5.4 Pharmaca

N-**Benzhydryl**-*N'*-*p*-**hydroxybenzylpiperazine HCl** (Cas 72–0031, UCB F 241; 5 mg/kg × day) was injected intraperitoneally to *rats* 30 min before starting hypoxic exposure.

Carbocromene (15 mg/kg × day, 30 mg/kg × day) was applied to *rats* either intraperitoneally or by gastric intubation.

Cortisone acetate (10 mg/kg × day) was given to *mice* in order to antagonise the acute toxicity of Aerosil® (SCHILLER 1952) and to delay silicotic fibrosis (SCHILLER 1951, 1954).

Diethylstilbestrol dipropionate (12.5 µg/kg × day) was given to 25 male *mice* for 39 days.

Isotactic **sodium polyacrylate** (8–12 kDa; 150 mg/kg) was given once intraperitoneally to NMRI *mice*. The drug was synthesised by MÜCK et al. (1977) and kindly supplied to me by Dr. Rolly.

Molsidomine (5,000 ppm) was added to the food (powdered Altromin® R) of *rats*.

Pilocarpinum hydrochloricum (1 mg/kg) was given subcutaneously to 3 female *rats* (breeder: Winkelmann, Borchen-Kirchborchen) weighing 228–248 g. The animal were sacrificed 30 min, 60 min and 120 min later, respectively.

Piracetam (2-oxo-1-pyrrolidine acetamide) was given intraperitoneally to *rats* 30 min before starting hypoxic exposure. In old Wistar *rats*, 400 mg piracetam per kg body weight × day were added to the food (powdered Altromin® R) and drinking water.

Poly-2-vinylpyridine ***N*****-oxide** with a molecular weight > 40,000 dissolved (8 %) in physiological saline and injected intraperitoneally into *mice* may be retained (HASIRICI and HOLT 1977) and reduce further cytotoxic effects of silica. After inhalation of micronized solid particles by *rats*, the drug was found in lysosomes of alveolar macrophages (SCHILLER 1971). Transition metal complexes of pyridine *N*-oxide were prepared and characterised by CARLIN (1961).The metal ions tend to attain their maximum coördination number towards pyridine *N*-oxide as a ligand. The complexes were decomposed rapidly when placed in water. Pyridine *N*-oxide is very hygroscopic but only the manganese complex is affected by standing in air. With iron perchlorate, pyridine *N*-oxide yields a yellow complex: $[Fe(C_5H_5NO)_6](ClO_4)_3 \cdot C_5H_5NO$. With copper there are two complexes, a green-blue one, $[Cu(C_5H_5NO)_4](ClO_4)_2$, and a yellow-green one of 739 nm, $[Cu(C_5H_5NO)_6](ClO_4)_2$.

Retinol palmitate (82.5 mg Arovit®/kg) was given to 5 female *rats* (breeder: Winkelmann, Borchen-Kirchborchen) weighing 252±4 g five times a week from April 17 to June 20, 1967, that were 45 applications at all.

Testosterone propionate (2.5 mg Testoviron®/kg × day) was injected subcutaneously into 10 male *mice* for 30 days.

Thyroxine (12,500 ng/ kg × day) was given subcutaneously to 5 female *rats* (breeder: Winkelmann, Borchen-Kirchborchen) weighing 239±4 g) five times a week from April 12 to June 20, 1967.

α-Tocopherol acetate (Ephynal®) was injected intramuscularly to 5 female *rats* (breeder: Winkelmann, Borchen-Kirchborchen) weighing 246±5 g five times a week from April 12 to June 12, 1967.

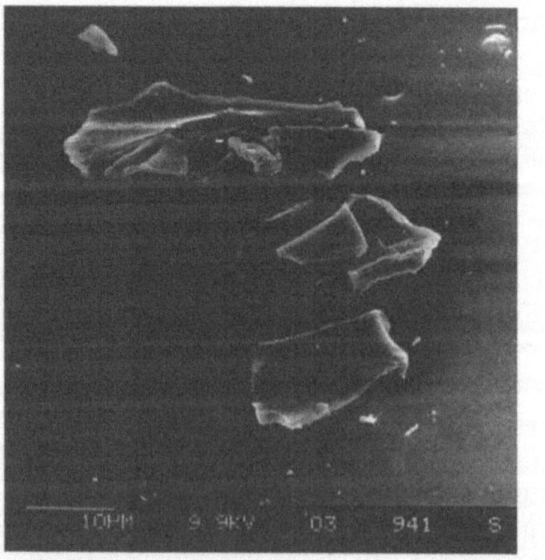

Fig. 54. Polyvinylpyridine *N*-oxide, micronized. Gold coating. Cambridge Stereoscan 150 operated at 9.9 kV. APh-R. 213/80, negative 03 941

5.5 Animals

40 *Dogs* weighing 20–30 kg were given 1 ml Combilen® i.m. 60 min before intravenous thiopental anaesthesia (5–10 mg/kg). After 1 mg succinyl choline/kg intravenously the animals were intubated for nitrous oxide narcosis ($N_2O:O_2 = 3:1$). The anterior descending and circumflex branches of the left coronary artery were exposed and an Ameroid constrictor (LITVAK et al. 1957) with 2.5 mm diameter was set. Epicardiectomy, cardio-pneumopexy, and omentopexy were performed to induce collateral circulation (for details of operation techniques see WERNITSCH 1970).

Table 11. Ameroid constrictors in *dogs*

Experimental group	n	Survival [days]	Extracoronary collateral vessels
Controls	10	19.6± 1.3	–
Epicardiectomy	10	21.2± 1.4	no
Cardiopneumopexy	10	23.6± 1.9	no
Cardioomentopexy	10	126.5±22.0	moderate to numerous

Winkelmann (Borchen-Kirchborchen, Westphalia), Sprague-Dawley (Charles River, France) and Wistar *rats* of either sex were used in the experiments. They were fed Altromin® R ad libitum.

5.6 Tissue Culture

The recent development of culture systems for the growth and transformation of respiratory epithelial cells (MARCHOK et al. 1977, 1978; TERZAGHI and NETTESHEIM 1979; WOODWORTH et al. 1982; WU et al. 1982; LECHNER et al. 1983; PAI et al. 1983; THOMASSEN et al. 1983; NETTESHEIM and BARRETT 1984) will make it possible to study the potential transforming effects of silica particles directly on epithelial cells. It is possible that combined culture systems in which macrophages and/or cell cultures may provide versatile *in vitro* models to investigate cell to cell interactions and the role of silica in multicellular pathways.

BION et al. (2002) prepared slices from *rat* lungs instilled with agarose and cleared of blood elements by vascular perfusion. Slices were positioned on the titanium grid of a Teflon rolling insert (Vitron) and placed into vials with open caps allowing free access to the gaseous phase. One hour after seeding the medium was renewed. Slices were then exposed (typically for 3 h) to a continuous flow of diluted diesel exhaust through the rotating chambers.

For the cultivation of airway epithelial cells, the cells are usually plated onto tissue culture plates coated with collagen or extracellular matrix components. Hormonally defined Ham's F12 medium contained 1 % penicillin-streptomycin, 5 µg/ml insulin, 5 µg/ml transferrin, 25 ng/ml epidermal growth factor, 15 µg/ml epithelial growth factor supplement, 2×10^{-11} M triiodothyronin, and 10^{-7} M hydrocortisone.

To induce and maintain differentiated morphology, in a double chamber culture system described by CLARK et al. (1995) bronchial epithelial cells were plated on the upper chamber of Trans Well culture plates in hormonally supplemented Ham's F12 medium. After 7 days the epithelial cells showed a ciliated differentiation (TAKIZAWA 1999).

To facilitate studies of tracheal epithelium from transgenic *mouse* models of *human* disease, DAVIDSON et al. (2000) have developed a primary culture model of differentiated *mouse* tracheal epithelium. When grown on semipermeable membranes at an air interface, dissociated cells formed confluent polarised epithelia with high transepithelial resistances (~12 kΩ·cm²) that remained viable for up to 80 days.

An exposure system for adherent growing cells to native gaseous compounds was developed by RITTER et al. (2001) using air/liquid culture techniques on the basis of the Cultex system (Patent No. DE 19801763/PCT/EP99/00295). Exposures of *human* lung fibroblasts (Lk004 cells) and *human* lung epithelial cells (HFBE-21 cells) to synthetic air, ozone (202 ppb, 510 ppb) and nitrogen dioxide (75 ppb to 1200 ppb) established that cells could be treated for 120 min without significant loss of cellular viability. At the same time, the experiments confirmed that such exposure times are long enough to detect biological effects of environmentally relevant gas mixtures. The analysis of viability (viable cell number, tetrazolium salt cleavage) and intracellular endpoints (oxidised/reduced glutathione, ATP/ADP), showed that both gases induced relevant cellular changes.

Assays that employ pulmonary alveolar macrophages have demonstrated good correlation between macrophage toxicity and pulmonary fibrogenicity for many inorganic compounds (FISHER and PLACKE 1987). FISHER et al. (1983), VALENTINE and FISHER (1984), ZELLER et al. (1992, 1993), GERCKEN et al. (1993), and BREHM (1996) used *bovine* macrophages, as one typical harvest from a *mouse* lung supplies only 10^5 cells, and from a *rat* lung 10^6 cells, whereas an *in vitro* macrophage assay may well utilise 10^7 to 10^8 pulmonary alveolar macrophages, which can be harvested from a single lobe of *cattle*.

After adherence to plastic dishes in an experimental culture system, an impressive number of genes associated with the inflammatory response are expressed, for example, IL-1β, TNF-α, and IL-8. In the 5'-upstream region of these inducible genes, the transcription factor NFϰB binding site is commonly found (Juliano and Haskill 1993). The messenger RNA stability element, AUUUA, is also commonly found in the 3'-untranslated region of these mRNAs. Adhesion may be sufficient for activation of NFϰB to induce these genes and to suppress the degradation rates of the mRNAs of these genes in monocytes. In *human* mononuclear cells the elastin receptor is linked to a pertussis toxin-sensitive G-protein which activated phospholipase C, resulting in a rapid mobilisation of the phosphoinositol pathway, activation of phosphokinase C, production of diacylglycerol, and a rapid increase in intracellular free calcium (Jacob et al. 1987). Another important function of the elastin receptor in mononuclear cells is the rapid increase in O_2 consumption accompanied by the extrusion of elastase and cathepsin G from polymorphonuclear leucocytes as well as the **liberation of superoxide radical anion** (Labat-Robert and Labat 2000). The repeating hexapeptide, Val-Gly-Val-Ala-Pro-Gly, present in seven copies in the *human* elastin gene, was shown to be chemotactic to monocytes and fibroblasts (Senior et al. 1984). Brehm et al. (1996) showed that the process of cell adherence per se generated reactive oxygen species with a peak about 20 min after transferring the suspended macrophages into the Biolumat test tubes; basal activity was regained approximately 100 min later. Another, although lower peak of lucigenin-enhanced chemiluminescence occurred about 20 min after twice washing by Veronal buffer to eliminate non-adherent cells and other materials.

To detach mononuclear phagocytes adhering to glass or plastic surfaces, trypsinisation was usually ineffective (Nathan 1981). Trypsin removed about one-quarter of the total sialic acid and L-fucose from the BHK_{21}/C_{13} cell and 10 % of the protein, a considerable degree of damage (Buck et al. 1970, Warren and Buck 1974). Repair took place by turnover. Trypsin altered phagocytic receptors (Unkeless 1978). Scraping with a rubber policeman in divalent cation-deficient media at low temperature often gave low yields and poor viability. Stewart (1981) did not find any procedure for removing all monocytes in a viable state.

The optimum of the cell number was 1×10^6 macrophages per vial. Johnston et al. (1978) and Bielefeldt Ohmann and Babiuk (1984) found an inverse relationship between $O_2^{\bullet-}$ release and cell density, i.e. mg cell protein per culture. This effect has to be considered only when cells are assayed after adherence, as found by Badwey et al. (1983).

5.7 Detection of Reactive Species

5.7.1 Electron Spin Resonance (ESR)

Electron spin resonance (ESR) is a technique that can be applied to free radicals, since it detects the presence of unpaired electrons. An unpaired electron has a spin of either $+1/2$ or $-1/2$ and behaves as a small magnet. If it is exposed to an external magnetic field, it can align itself either parallel or antiparallel (in opposition) to that field, and thus can have two possible energy levels. If electromagnetic radiation of the correct energy is applied, it will be absorbed and used to move the electron from the lower energy level to the upper one. Thus an absorption spectrum is obtained, usually in the microwave region of the electromagnetic spectrum.

ESR is a very sensitive method and can detect radicals at concentrations as low as 10^{-10} mol l^{-1}, provided that they stay around long enough to be measured. Highly reactive radicals, by normal ESR, are allowed to react with a compound to produce a long-lived radical. When a radical is spin trapped, what one observes via EPR is not the radical itself but an adduct of the radical. Perkins (1980) in his review of spin trapping, points out four questions which must be addressed when desingning a spin trapping experiment.

1. Will the spin trap participate in reactions other than those with reactive radicals generated in the experiment? Can these alternative reactions yield nitroxides, which will appear as spin adduct impostors?
2. How ready can the spectrum be interpreted and the structure of $R^{\bullet}$ be determined?
3. How fast is the trapping reaction, and how stable are the spin adducts formed?
4. Does the appearance of a spin adduct signify a major reaction pathway, or can it be a minor side reaction?

To these questions Mottley and Mason (1989) have added a fifth.

5. Is there an isotopically labelled (^{13}C, ^{2}H, ^{17}O ^{15}N, etc.) material available for partial proof of structure? If not, is there an independent synthesis for the spin adduct?

The 'ideal' trap should react rapidly and specifically with the radical one wants to study, to produce a product that is stable and has a high characteristic ESR spectrum.

Spin trapping methods have often been used to detect the presence of **superoxide** and **hydroxyl radicals** in biological systems (HARBOUR and BOLTON 1975), and also the formation of organic radicals during lipid peroxidation.

Bronchoalveolar lavage cells (1×10^6) suspended in Dulbecco's phosphate buffered saline (0.5 ml) containing 5,5-dimethyl-1-pyrroline-1-oxide (DMPO; 100 mM) in duplicate samples, one with and one without superoxide dismutase (500 units/ml) by electron paramagnetic resonance measurements recorded **superoxide anion** generation at sites of antigen challenge (SANDERS et al. 1995).

Spin trapping is generally considered the most specific technique for **hydroxyl radical** detection. DMPO reacts with $HO^•$ to yield 2,2-dimethyl-5-hydroxy-pyrrolidinyloxyl (DMPO-OH). However, DMPO-OH is also a decomposition product of DMPO-OOH generated by $O_2^{•-}$, making it unreliable as evidence for $HO^•$ production (BRITIGAN et al. 1986, COHEN et al. 1988). The decomposition products of DMPO-•OOH are ERP-silent except for is reduction product DMPO/•OH. The decomposition of DMPO/•OOH has been widely reported to form $HO^•$ as a minor product, but even this has been questioned (BUETTNER 1993).

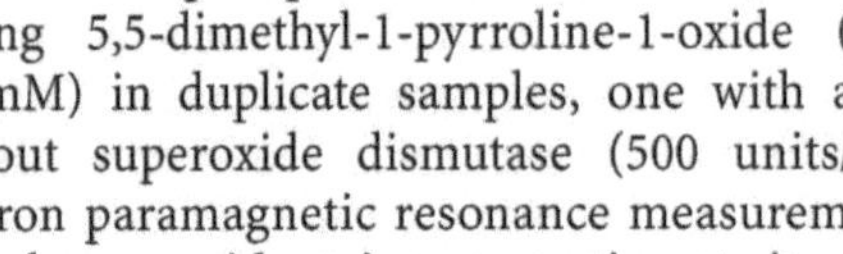

5,5-**Dim**ethyl-1-**pyr**roline-1-**oxide**-HO• adduct [8]

$HO^•$ reacts with dimethyl sulfoxide (DMSO) to form methyl radical (BRITIGAN et al. 1986).

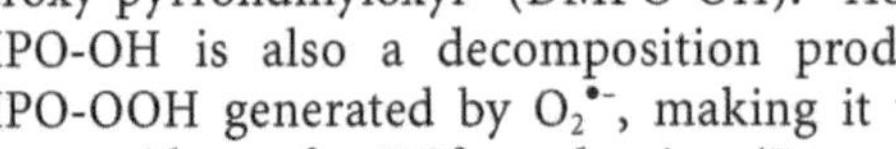

Formation of methyl radical (•CH_3) [9]

When DMSO is present in excess of DMPO as in the study of BRITIGAN et al. (1988), $HO^•$ production is manifested primarily as DMPO-CH_3, providing a more specific detecting system for $HO^•$.

α-(4-Pyridyl-1-oxide)-*N*-*tert*-butyl-nitrone (POBN) was chosen for spin-trap by GONTHIER et al. (1991) because of its solubility in water and buffer and because of its relative stability in the presence of light and oxygen. A phosphate buffer was preferred to the more commonly used Tris-buffer, because it did not give an ESR signal in their mixtures whereas with Tris, a signal was observed with splitting constants (a_N = 15.79 G and a^β_H = 2.84 G) corresponding to a Tris-derived free radical adduct (SAPRIN and PIETTE 1977).

Tertiary **peroxyl radicals** can be detected directly by electron spin resonance (KALYANARAMAN et al. 1983). However, primary and secondary alkylperoxyl radicals disproportionate with a nearly diffusion-controlled, bimolecular rate constant (BENNETT 1987, NIKOLAEV et al. 1992) and, therefore, have extremely short lifetimes and cannot be observed directly in biological samples. However, using ^{17}O-isotope labelling, the methylperoxyl and methoxyl radical adducts should be distinguishable. DIKALOV and MASON (1999) suggested that detection of ^{17}O-alkoxyl radical adduct from ^{17}O-labelled molecular oxygen can be used as indirect evidence for peroxyl radical generation.

For quantification of basal vascular **nitric oxide** (•NO) production from isolated vessels, KLESCHYOV et al. (2000) incubated *rabbit* aortic or venous strips with 250 μM colloid Fe/diethyldithiocarbamate, which resulted in a linear increase in tissue-associated NO-Fe/diethyldithiocarbamate electron paramagnetic resonance signal during 1 h. Removal of endothelium or addition of 3 mM N^G-nitro-L-arginine methyl ester inhibited the signal.

A water-soluble iron complex with *N*-dithiocarboxysarcosine has been developed as an electron spin resonance spin trapping agent for •NO and successfully applied to electron spin resonance imaging for endogenous •NO production in *mice*. KOSAKA and YANEYAMA (2000), however, attempting to measure •NO from purified neuronal NO synthase (EC 1.14.13.39) by this method did not detect •NO. The complex markedly inhibited NOS activity with IC_{50} value of 9.7 ± 0.7 μM in citrulline formation assay. *N*-Dithiocarboxysarcosine alone did not inhibit the activity. An iron complex with *N*-methyl-D-glucamine dithiocarbamate, similar spin trap for •NO, also inhibited the activity with IC_{50} value of 25.1 ± 2.9 μM. Iron-*N*-dithiocarboxysarcosine suppressed cytochrome *c* and ferricyanide reduction activities of nNOS. Iron-*N*-dithiocarboxysarcosine complex markedly increased the nNOS-mediated NADPH oxidation.

Using ESR HOCHSTRASSER and ANTONINI (1972) studied dangling bonds created and maintained under ultrahigh vacuum on **silica surfaces**. The E'_s centres are formed by an unpaired electron trapped on a surface silicon atoms and thus are Si dangling bonds. The signal was stable for months in ultrahigh vacuum or in an inert atmosphere, such as argon. The presence of alkali ions lowered the number of E'_s centres. An ESR study of radicals formed by adsorption of CO_2 was made by LUNSFORD and JAYNE (1965). The $CO_2^{•-}$ signal overlaps the E'_s one, but can be separated from it by graphical difference, or, better, by the fact that, contrary to E'_s, it does not saturate, even at 1 mW microwave power (HOCHSTRASSER and ANTONINI 1972).

5.7.2 Chemiluminescence

Holzapfel et al. (1960) found that a sample of quartz dust excited by X-rays for 1 s would emit 13,000 photons within 2 s.

Robock and Klosterkötter (1971) excited dust samples suspended in Tyrode solution (pH 7.2) by X-rays (2.8×10^4 r) and measured the luminescence as a function of wave length in the range of 200 to 550 nm and as a function of the temperature in the range of 20 °C to 50 °C using a multiplier.

We used a LB 9505 multidetector unit for the simultaneous and continuous recording of luminescence from six samples (Berthold and Maly 1987). This instrument is equipped with six rubidium-bialkali photomultiplier tubes placed in close proximity to the vials inside six independent counting chambers. Reflectors ensure optimum counting efficiency. Each measuring chamber has its own temperature control. Since the results from all six detectors should allow direct comparison, they were standardised by determination of its individual sensitivity constant (λ-factor).

The emission of light as a result of the formation of reactive oxygen metabolites can be amplified substantially (3 to 4 orders of magnitude) in the presence of suitable "chemilumigenic probes". Although both lucigenin and luminol have been widely used, a controversy exists concerning whether they act as both substrate and generator of some reactive oxygen species.

Lucigenin (10,10'-dimethyl-9,9'-biacridinium nitrate)-dependent phagocytic chemiluminescence was described by Williams and Cole (1981) for the assessment of alveolar macrophage metabolic activity in response to stimulation by opsonised particles or soluble agents. The requirement for superoxide anion ($O_2^{\bullet-}$) in the production of chemiluminescence was suggested by inhibition (95 %) using superoxide dismutase. However, lucigenin must be univalently reduced before it reacts with $O_2^{\bullet-}$ yielding an unstable dioxetane, the decompensation of which to the acridone is on the light-yielding pathway; and univalently reduced lucigenin readily autoxidizes, with production of $O_2^{\bullet-}$. Thus lucigenin, like paraquat, or nitroblue tetrazolium, can mediate $O_2^{\bullet-}$ production in systems not producing $O_2^{\bullet-}$ in the absence of these redox active compounds (Liochev and Fridovich 1997). Barber et al. (1997) reported a lucigenin-enhanced chemiluminescence signal emanating from *human* saphenous veins that was not inhibited by superoxide dismutase and lasted for more than 24 h. A larger lucigenin-enhanced chemiluminescence signal with similar properties was produced by saphenous veins that had been dehydrated. A similar, non superoxide dismutase-inhibitable lucigenin-enhanced chemiluminescence was produced with a variety of phospholipids and phosphatidic acid. The absolute magnitudes of modifications in cardiac muscle-derived chemiluminescence during hypoxia and reoxygenation need to the interpreted with caution, since they could be affected by processes that potentially alter the sensitivity of the lucigenin method, such as changes in the redox state, pH, and alternations in lipid metabolism (Xie et al. 1998).

Incubation of isolated mitochondria with lucigenin at non-redox cycling concentration produced lucigenin-derived chemiluminescence (LDCL), which was increased markedly by mitochondrial substrates, pyruvate/malate or succinate (Li et al. 1999). The LDCL was reduced greatly by the membrane permeable superoxide dismutase mimetics, 2,2,6,6-tetramethylpiperidine-*N*-oxyl and Mn(III)-tetrakis(1-methyl-4-pyridyl)porphyrin, but not by Cu,Zn-superoxide dismutase. With an ion-pair HPLC method, a concentration-dependent accumulation of lucigenin was detected within mitochondria. The accumulation of lucigenin by mitochondria was reduced markedly in the presence of carbonyl cyanide *p*-(trifluoromethoxy)phenylhydrazone, an uncoupler known to dissipate the mitochondrial membrane potential. With submitochondrial particles, (Li et al. 1999) observed that both complexes I and III of the mitochondrial electron transport chain appear to be able to catalyse the one electron reduction of lucigenin, a critical step involved in LDCL. After incubation of mitochondria with lucigenin at non-redox cycling concentrations, formation of *N*-methylacridone (formula [10]), the proposed end product of the reaction pathway leading to LCD, within the mitochondria fraction was also detected.

The use of lucigenin chemiluminescence as a measure of $O_2^{\bullet-}$ has been questioned because lucigenin has been shown to be capable of mediating the production of $O_2^{\bullet-}$. Nevertheless, perhaps because of the convenience and sensitivity of luminescence methods, the use of lucigenin^{2+} as a measure of $O_2^{\bullet-}$continues.

Lucigenin enhances the rate of cytochrome *c* reduction with xanthine as substrate, but does not increase the rate of xanthine oxidation (Afanas'ev et al. 1999). When NADH is used as a substrate, lucigenin inhibits the SOD-dependent components of cytochrome *c* reduction and enhances both the SOD-independent cytochrome *c* reduction and NADH oxidation, being a sole acceptor of an electron from the enzyme.

Spasojevic et al. (2000) presented cyclic voltammetry data that established the redox potential of

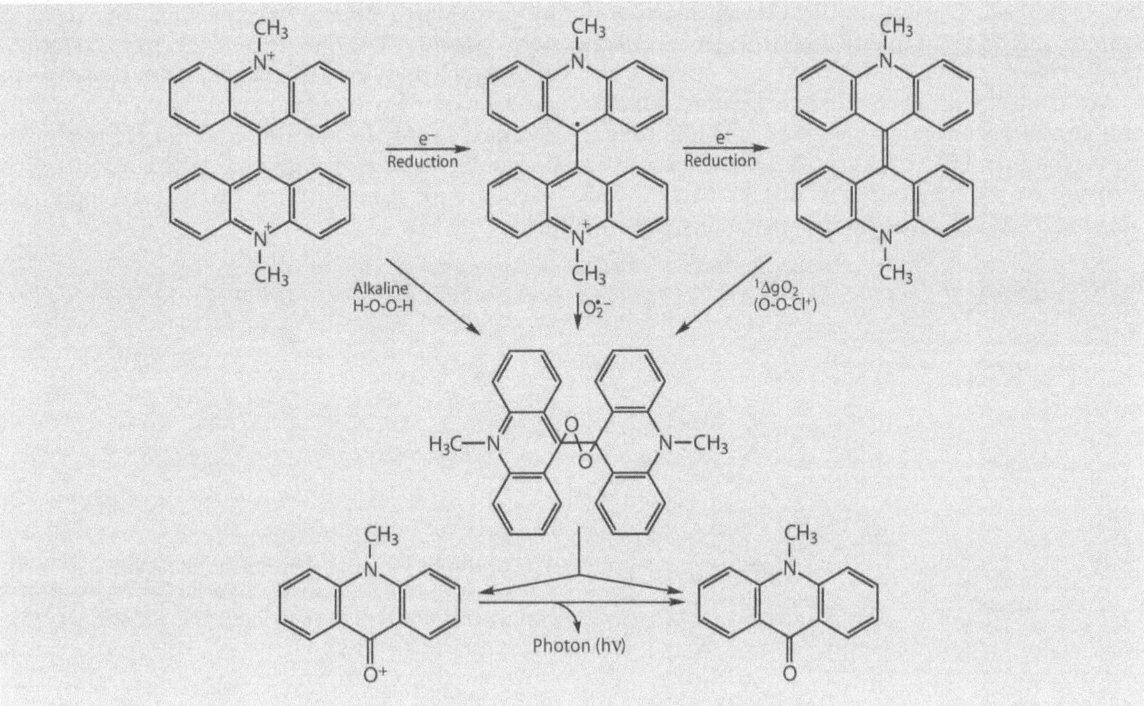

Reaction of lucigenin with reactive oxygen species [10]

lucigenin^{2+} in aqueous and media and which is compatible with its previously demonstrated ability to redox cycle with production of $O_2^{\bullet-}$. The reduction potential for lucigenin is -0.14 ± 0.02 V versus the normal hydrogen electrode. This value applies to both the first and the second electron transfers to lucigenin and it is in accord with the facile mediation of $O_2^{\bullet-}$ production by this compound.

Common quinone radicals $Q^{\bullet-}$ react with O_2 much faster than lucigenin$^{\bullet+}$, raising the possibility that some quinones could act as redox mediators in lucigenin$^{\bullet+}$/O_2 redox cycling. This is a further complication in the use of lucigenin^{2+} (Wardman and Vojnovic 1999).

Chemiluminescence assays with *rabbit* vascular rings or homogenates showed preferential signals with NADPH (vs. NADH) with 5 and 50 μM lucigenin, which were blocked by diphenylene iodonium, superoxide dismutase, or its cell-permeable mimetic MnTBAP (Janiszewski et al. 2002). With 250 μM lucigenin, the relative signal with NADH became larger than with NADPH, and was poorly inhibited by all three antagonists above. All superoxide dismutase/diphenylene iodonium-resistant signals were effectively blocked by the electron acceptor nitrobluetetrazolium. Spin trapping with 5,5-**di**methyl-1-**p**yrroline-1-**o**xide showed an approximate doubling of DMPO-OH radical adduct signal upon addition of 5 μM lucigenin to *rabbit* vascular homogenates incubated with either NADPH or NADH. With 50 or 250 μM lucigenin, much larger increases were observed with NADH, as opposed to NADPH. Oxygen consumption measurements showed analogous results.

In a *rat* liver microsomal system, both **ascorbic acid** and **glutathione** (GSH) as scavengers effectively diminish lucigenin amplified chemiluminescence in a concentration dependent manner to almost zero (Klinger et al. 1996). The spin trap substance *N-t*-butyl-α-phenylnitrone ($C_{11}H_{15}NO$) diminished lucigenin amplified chemiluminescence more effectively than luminol amplified chemiluminescence.

Hepes [4-(2-hydroxyethyl)-1-piperazineethanesulfonic acid], **Tricine** [*N*-[2-hydroxy-1,1-bis(hydroxymethyl)ethyl]-glycine) and **Tris** [2-amino-2-hydroxymethylpropane-1,3-diol] are efficient scavengers of $HO^{\bullet}$ radicals (Hicks and Gebicki 1986). Rate constants for the reactions of these commonly used buffers and hydroxyl radicals as measured using steady-state competition kinetics with thymine were 5.1×10^9, 1.6×10^9 and 1.1×10^9 $l \cdot mol^{-1} \cdot s^{-1}$, respectively. The findings that under some conditions not all the $HO^{\bullet}$ radical damage is inhibited by these buffers (Greenwald and Moy 1980, Burk 1983, Yukawa et al. 1983) suggests either that the secondary radicals derived from the buffers can initiate damage or that the $HO^{\bullet}$ radicals are produced at sites inaccessible to the buffer molecules. There is a growing body of evidence for 'site-specific' formation of $HO^{\bullet}$ which is not affected by scavengers present in the aqueous phase (Samuni et al. 1981, 1983).

Any contact with **phenol red** long before cell culture increased lucigenin-enhanced chemiluminescence (Brehm 1996). A carbon-centred radical is formed via a one-electron transfer (Senne and Marple 1970). This radical is a very bulky and sterically hindered species that cannot be disproportionated quite rapidly.

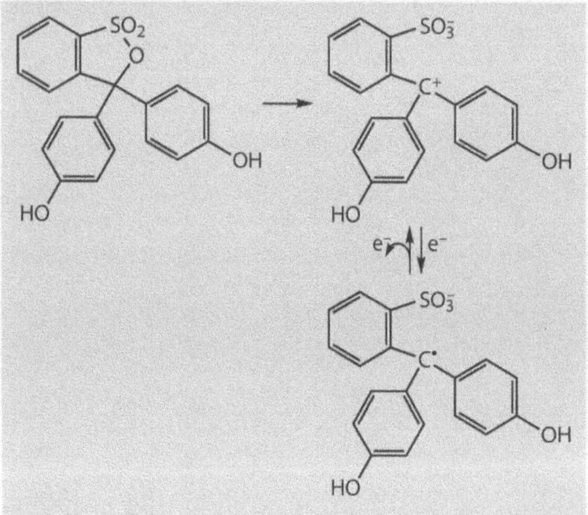

Phenol red radical anion [11]

A more sensitive assay for **peroxidase** using a new chemiluminescent reagent based on 10,10'-dimethyl-9,9'-biacridinium nitrate was obtained by light irradiation of lucigenin in an organic polar solvent as *N,N*-dimethylacetamide (Katsuragi et al. 2000). The detection limit for peroxidase activity using this reagent is 10^{-19} mol/assay. The reagent was stable for more than a year.

Luminol-dependent chemiluminescence shows very little ability to discriminate between individual oxygen or radical species (Vilim and Wilhelm 1989). Furthermore, in biological systems it is extremely prone to interfere with reactive oxygen species metabolising enzymes. As it cannot be totally inhibited by superoxide dismutase, it might be only partially $O_2^{\bullet-}$-dependent. Luminol-peroxidase reaction catalysed by peroxidases is useful for the detection of peroxides and active oxygen species (Seitz 1978, Jones and Hancock 1994). Luminol oxidation by horseradish peroxidase in the presence of H_2O_2 at pH 8.3, has been documented, and it was concluded that the luminol oxidation proceeds by a one-electron transfer similar to oxidation of other electron donors (Cormier and Pritchard 1968). The luminol oxidation at pH 7.0 gave an increase in adsorbance at 400 nm (Nakamura and Nakamura 1998). One- or two-electron oxidations of luminol by peroxidases initiate the reactions that lead to light emission. The one- or two-electron oxidations of luminol may be explained on the basis of the mechanism of peroxidase reactions. The resulting common precursor after the oxidation of luminol is concluded to be an azaquinone, which reacts subsequently with H_2O_2 to form excited aminophthalic acid.

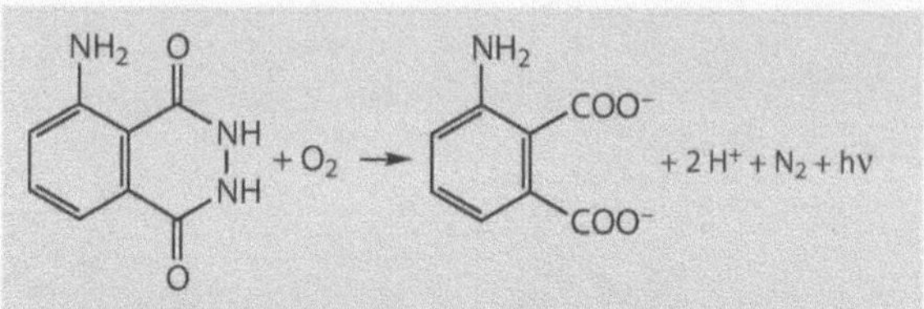

Reaction of luminol with reactive oxygen species [12]

In a *rat* liver microsomal system, both ascorbic acid and glutathione (GSH) as scavengers effectively diminish luminol amplified chemiluminescence in a concentration dependent manner to almost zero (Klinger et al. 1996).

When concentrations of superoxide anion are very low, assay sensitivity may need to be enhanced. Some enhancement reagents, such as indophenol, are phenolic-based compounds. Unfortunately, such enhancers are toxic to live cells and denaturing to some components of subcellular systems; hence, they cannot be used for *in situ* assays. LumiMax^{TM} increases photon output but is nontoxic and does not denature cellular components used in assaying live cell activity (Hoang and Pfefferkorn 1998).

The pool of enzymes which are in charge of the destruction of the free radicals resulting from an oxidative shock is found in the cytoplasmic membrane as well as in the mitochondrial ones. As the molecule of luminol is smaller than that of lucigenin it may enter the cell and react with the reactive oxygen species generated in the mitochondria, while lucigenin is only capable to detect the species discharged into the environment.

Luminol can react intracellularly as well as extracellularly and produces light in the simultaneous presence of H_2O_2 and a variety of co-oxidators (but not $O_2^{\bullet-}$). Bottu (1991) injected a H_2O_2 solution in phosphate buffer (pH 7.2) with luminol and Fe^2+ (from Mohr's salt, $(NH_4)_2\ [Fe^{(II)}(SO_4)_2]$) producing a brief flash of light. The luminescence could be strongly quenched by the typical $HO^{\bullet}$ quenchers ethanol, *t*-butanol, mannitol and sodium benzoate and also by SOD. This suggested that luminol reacted first with $HO^{\bullet}$ and then with $O_2^{\bullet-}$, both being produced in the reaction medium:

$$Fe^{2+} + H_2O_2 \longrightarrow Fe^{3+} + HO^- + HO^\bullet \quad [13]$$

$$H_2O_2 + HO^\bullet \longrightarrow O_2^{\bullet-} + H_2O + H^+ \quad [14]$$

$$Fe^{3+} + H_2O_2 \longrightarrow Fe^{2+} + O_2^{\bullet-} + 2H^+ \quad [15]$$

Cl^- injected into a buffered solution of luminol and H_2O_2 produced a brief flash of light, while a solution of luminol and H_2O_2 injected into a solution of horseradish peroxidase produced luminescence that rose fast and decayed slowly (with a half-life of 85 s).

$$ClO^- + H_2O_2 \longrightarrow Cl^- + H_2O + {}^1O_2 \quad [16]$$

In monocytes, luminol-enhanced chemiluminescence is an entirely extracellular event requiring both the release of myeloperoxidase and the generation of superoxide (Albrecht and Jungi 1992).

Both "spontaneous" and zymosan (0.5 mg/ml final concentration)-stimulated luminol-enhanced chemoluminescence of normal and *Trypanosoma cruzi*-infected *mouse* peritoneal cells were inhibited by 100 μM azide and by 0.8 μM superoxide dismutase, suggesting the involvement of hemoproteins and superoxide anion in the measured responses (Cardoni et al. 1990).

The different sensitivity of lucigenin and luminol to reactive oxygen species is reflected by some opposite results with certain compounds demonstrated by a low correlation coefficient for lucigenin- *versus* luminol-amplified chemiluminescence (Müller-Peddinghaus and Wurl 1987).

The **inactivation of aconitase** (EC 4.2.1.3) by $O_2^{\bullet-}$ (Gardner and Fridovich 1992, Gardner et al. 1995) is one assay for intracellular $O_2^{\bullet-}$ which is reliable, although technically demanding (Liochev and Fridovich 1997).

Pholasin® is the photoprotein of the marine, rock-boring bioluminescent mollusc, *Pholas dactylus*, otherwise known as the common piddock. It emits light in the presence of superoxide anion, singlet oxygen, nitric oxide and possibly hydroxyl radical and/or ferryl radical. A variant of the test for halogenated oxidants is based on the use of chloramine T which when added to Pholasin® will result in the emission of light. The presence in the sample being tested of antioxidants capable of scavenging chloramine T will result in inhibition of emitted light. On its own, Pholasin® does not emit light.

Nitric oxide synthase dependent chemiluminescence (counts per min with zymosan minus counts per min with zymosan + N^{ω}-nitro-L-arginine methylester) was determined as the amount of zymosan-stimulated chemiluminescence that was inhibitable by 1 mM N^{ω}-nitro-L-arginine methylester added during the preincubation period (Jorens et al. 1993).

Firefly **luciferase** is an enzyme isolated from the biolucent organism photinus pyralia and can catalyse its substrate luciferin and coenzyme A to produce luminescence in the presence of adenosine triphosphate and O_2 (Gould and Subramani 1988, Kricka 1988). Zhang et al. (1994) transfected the luciferase gene into two *murine* tumour lines, i.e. c162 melanoma and M109 lung carcinoma, and determined the luciferase activity associated with the cells by a rapid chemiluminescent reaction. Brehm (1996) quantified ATP in *bovine* alveolar macrophages after phagocytosis of micronized quartz under the influence of Fenton catalysts.

5.7.3 Tetrazolium Compounds

Various tetrazolium compounds, introduced into enzyme research by Kuhn and Jerchel (1941) and Lakon (1942), are reduced to scarlet-red (2,3,5-triphenyl-2H-tetrazolium), purplish (neoetrazolium) or blue (ditetrazolium) formazans, insoluble in water and soluble in fats. Formazan generation is practically inhibited by superoxide dismutase as well as by catalase (Brömme et al. 1999). When superoxide radical generation via the xanthine-xanthine oxidase system was measured by chemiluminescence an addition of nitroblue tetrazolium immediately abolished lucigenin-enhanced chemiluminescence. Nitroblue tetrazolium did not inhibit xanthine oxidase.

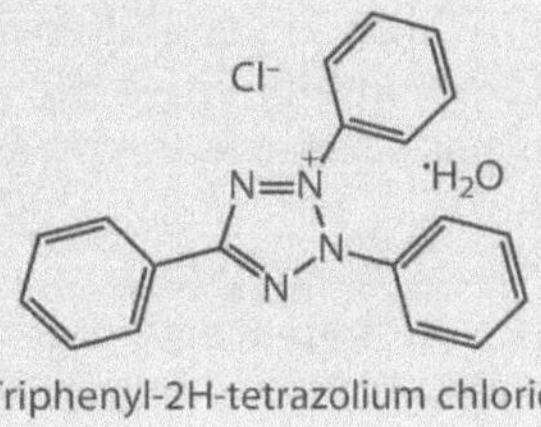

2,3,5-Triphenyl-2H-tetrazolium chloride hydrate [17]

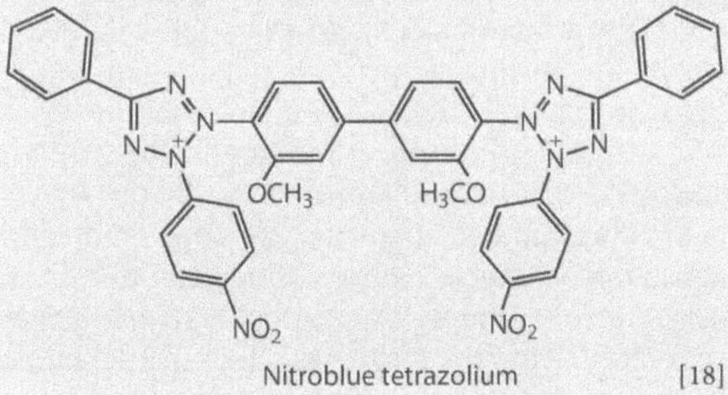

Nitroblue tetrazolium [18]

The nitroblue tetrazolium method has the main advantage of sensitivity. Oberley and Spitz (1985) used diethylenetriamine-pentaacetic acid (DETAPAC) instead of ethylenediamine tetraacetic acid (EDTA), as they found that DETAPAC enhances the sensitivity of the assay, probably because Fe-DETAPAC does not react with $O_2^{\bullet-}$, whereas Fe-EDTA does (BUETTNER et al. 1978). They include catalase in their assay mixture as H_2O_2 inhibits and inactivates CuZnSOD, and for kinetic reasons, so that the equilibrium will not be shifted in favour of $O_2^{\bullet-}$ production.

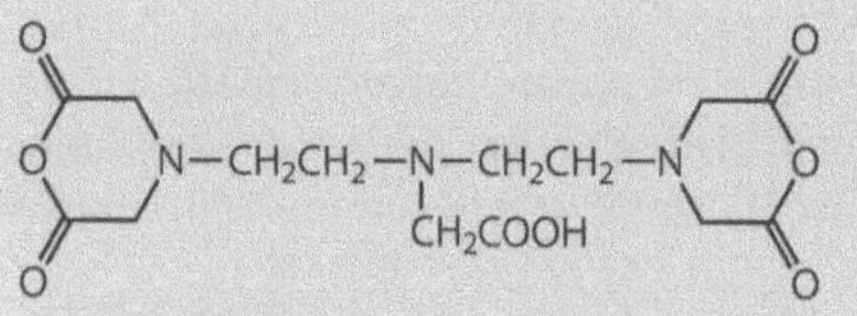

Diethylenetriamine-pentaacetic acid (DETAPAC) [19]

Dioxane proved to be the most suitable solvent for extracting the formazan formed in the nitroblue tetrazolium test (MÜLLER et al. 1981). At 85 °C dioxane completely extracted nitroblue tetrazolium (absorption maximum at $\lambda = 580$ nm, absorption coefficient = 0.0256 ml × $nmol^{-1}$ × cm^{-1}) from the cell pellet within 15 min, and nitroblue tetrazolium diluted in dioxane was sufficiently stable. As a possible aqueous contamination under conditions of an acid pH caused a slower decrease of the optical density of the extract than did pure water, 1 *N* HCl as washing agent to remove remaining nitroblue tetrazolium from the cell pellet is recommended.

NALBANDIAN (1982) and NALBANDIAN et al. (1983) modified the nitroblue tetrazolium immunobead test (O'DONNELL et al. 1979, JABS et al. 1980) to detect cases in which failing and absent phagocytic activity and respiratory burst intensity in aliquots of polymorphonuclear leucocytes have been functionally restored to normal by *in vitro* piracetam treatment of such neutrophils.

The combination of short-length non-denaturing isoelectric focusing and measurement of superoxide dismutase (EC 1.15.1.1) activity by water-soluble tetrazolium salt made it possible to quantify superoxide dismutase activity in a small sample (SHIMAZAKI et al. 2002).

In electron microscopy, SEDAR and ROSA (1961) used nitro-blue tetrazolium to visualise the activity of the succinic dehydrogenase system. Dinitroformazan deposits were found within or on mitochondrial profiles, although occasional extramitochondrial deposits were encountered in some micrographs. Specific intramembranous localisation was impossible since the smallest diameter of the dinitroformazan particles measured ~ 30 nm.

The demonstration of xanthine oxidoreductase (EC 1.1.1.204) activity by tetrazolium salts was optimised by KOOIJ et al. (1991) and was applied validly for quantitative purposes (FREDERIKS et al. 1995). Due to the presence of an intermediate electron carrier in the incubation medium, activity of both xanthine oxidoreductase and xanthine oxidase (EC 1.1.3.22) are detected.

5.7.4 Fluorescent Probes

Two fluorescent probes, 2',7'-dichlorofluorescin diacetate and dihydrorhodamine 123, have been used for detecting intracellular reactive oxygen species production. Properties of an ideal agent for this purpose would include:

- cell membrane permeability such that significant intracellular concentrations are achieved
- minimal cellular toxicity
- conversion from the nonfluorescent to fluorescent form of the probe should be relatively specific for reactive oxygen species reaction with low rates of conversion due to spontaneous and non-reactive oxygen species dependent reactions
- intracellular sequestration of the fluorescent, oxidised form of the probe.

The esterified form of dichlorofluorescin, dichlorofluorescin diacetate, crosses cell membranes and then undergoes deacetylation by intracellular esterases. The resulting compound, dichlorofluorescin, is proposed to be trapped within the cell and susceptible to reactive oxygen species-mediated oxidation to the fluorescent compound, **dichlorofluorescein.**

Recently ROTA et al. (1999) demonstrated that dichlorofluorescein fluorescence cannot be used reliably to measure $O_2^{\bullet-}$ in cells because $O_2^{\bullet-}$ itself is formed during the conversion of dichlorofluorescin to dichlorofluorescein by peroxidases. The disproportionation of superoxide forms H_2O_2 which, in the presence of peroxidase activity, will oxidise more dichlorofluorescin to dichlorofluorescein with self-amplification of the fluorescence. Because the deacetylation of dichlorofluorescin diacetate, even by esterases, can produce H_2O_2, the use of this probe to measure H_2O_2 production in cells is problematic.

The fluorescent product of dihydrorhodamine 123 oxidation, **rhodamine 123**, is a positively charged lipophilic compound which has been used as a vital stain for mitochondria and thus would be sequestered intracellularly (JOHNSON et al. 1980). Oxidation of dihydrorhodamine 123 does not occur with H_2O_2 alone, but is mediated by a variety of secondary H_2O_2-dependent intracellular reactions including H_2O_2-cytochrome *c* and H_2O_2-Fe^{2+} (ROYALL and ISCHIROPOULOS 1993).

2,3,4,5,6-Pentafluorodihydrotramethylrosamine (1–5 μM), a fluorogenic indicator for oxidative activity, is internalised through a nonendocytic pathway, then oxidised in the cytosol, followed by immediate targeting to active mitochondia, resulting in fluorescent staining in this organelle. In both proliferating and quiescent normal *rat* kidney fibroblasts, subcellular distribution of the oxidised dye could be altered by treatment with an oxidant (H_2O_2) or an antioxidant (*N*-acetyl-L-cysteine), indicating a regulatory relationship between cell proliferation and oxidative activity (CHEN and GEE 2000). In solution assay, this probe can be oxidised by a broad spectrum of oxidising species including horseradish peroxidase, H_2O_2 and horseradish peroxidase, cytochrome *c*, cytochrome *c* and H_2O_2, superoxide and H_2O_2, nitric oxide (or nitrite), peroxynitrite, and lipid hydroperoxide.

C11-BODIPY$^{581/591}$ (boron dipyrromethene difluoride) is a fluorescent ratio probe for indexing lipid peroxidation and antioxidant efficacy in model membrane systems and living cells, with excellent characteristics (DRUMMEN et al. 2002):

- emission in the visible range of the electromagnetic spectrum, with good spectral separation of the nonoxidised (595 nm) and oxidised (520 nm) form;
- has a high quantum yield and because of this, low labelling concentrations can be used, ensuring minimal perturbation of the membrane whilst retaining favourable signal to noise ratios;
- has a good photo-stability and displays very few fluorescence artefacts:
- is virtually insensitive to environmental changes, i.e. pH or solvent polarity;
- is lipophilic and as such easily enters membranes;
- once oxidised, C11-BODIPY$^{581/591}$ remains lipophilic and does not spontaneously leave the lipid bilayer;
- C11-BODIPY$^{581/591}$ localizes in two distinct pools within the lipid bilayer, a shallow pool at 1,8 nm and a deep pool at <0.75 nm from the centre of the bilayer;
- is not cytotoxic to *rat*-1 fibroblasts up to 50 μM;
- is sensitive to a variety of oxy-radicals and peroxynitrite, but not to superoxide, nitric oxide, transition metal ions, and hydroperoxides per se;
- its sensitivity to oxidation is comparable to that of endogenous fatty acyl moieties.

5.8 Light Microscope Techniques

In the Mach-Zehnder interference microscope described by GREHN (1959, 1960) and WALTER (1962) the object contrast does not depend on the diffraction by the specimen but is produced by a double-beam interferometer system built within a double-microscope. The object modifies the phase relationship between the two mutually interfering waves by an amount depending on its optical thickness and refracting index.

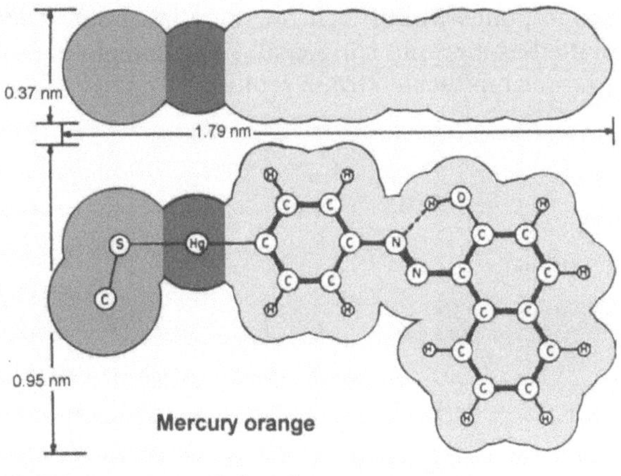

Fig. 55. Structure and size of mercury orange

Crystals and collagen differ in their refractive indices and thus in their phase displacements in the fringe field or in the interference contrast colours (SCHILLER 1974). Coal dust is opaque and thus does not give any change of colour using the wedge compensator. A further advantage of interference microscopy is the study of dust phagocytosis in tissue cultures previously performed by phase contrast microscopy using Heine's condensor (SCHILLER 1954).

Another application of the Mach-Zehnder interference microscope is to detect free sulfhydryl groups as in reduced glutathione, using mercury orange (SCHILLER 1974, 1975).

Optical density in the autoradiographs was analysed with a computerized image processing system (SCHEICH 1983, GONZÁLEZ-LIMA and SCHEICH 1984, MAIER and SCHEICH 1987). The system consisted of a TV camera mounted on a Wild microscope, a custom-made A-D converter, and a Hewlett-Packard 21 MX computer. A given area of an autoradiograph was scanned with the TV camera and was converted into a 256 × 256-point matrix with 8-bit resolution of optical density. The digitalized image was stored on tape and reconstructed on a storage screen. Intensities in each analysed brain section were referred to the average intensity of an area in the corpus callosum of that section in order to allow a comparison between different brains. Arrays of densitometric profiles were generated from each matrix as follows. The 256 horizontal lines of the matrix were divided into 32 horizontal rows with 8 lines by the computer. Relative optical densities of points within each row were integrated and plotted as a profile horizontally. The complete matrix was represented by 32 profiles.

5.9 Electron Microscope Techniques

Rats were anaesthetised by intraperitoneal injection of 30 mg pentobarbital/kg and perfused from the abdominal aorta with 2.5 % glutaraldehyde in 0.1 M sodium cacodylate buffer (pH 7.4), postfixed with 1 % osmium tetroxide in sodium cacodylate buffer and embedded in Epon 812. Sections cut at 50 nm were stained with lead citrate and uranyl acetate.

CHAPTER 6

Reactive Oxygen and Nitrogen Species

The term "radical" is often used loosely in chemistry to refer to various groups of atoms that behave as a unit, such as the carbonate radical (CO_3^{2-}), nitrate radical (NO_3^-), and the methyl radical ($CH3^-$). According to this definition, Lavoisier (1789) has already indicated that the four elements: phosphorus, sulphur, carbon, and hydrogen not only by themselves can take up oxygen, but may also mutually combine to likewise oxidizable compounds. BERZELIUS (1817) formulated: in inorganic nature all oxides contain a simple radical, while all organic compounds are oxides of composed radicals; the radical of vegetable matter mostly consists of carbon, that of animal matter besides carbon contains nitrogen.

HALLIWELL and GUTTERIDGE (1989) avoid this use and define a "free radical" as follows:

> A free radical is any species capable of independent existence that contains one or more impaired electrons.

Classically, radical reactions begin with an initiation process to form a radical species, such as the reaction of H_2O_2 with transition metals (Fenton chemistry) to form hydroxyl radical, $^\bullet$HO (WARDMAN and CANDEIAS 1996).

Table 12. Transition metals

Metal ion	Number of unpaired *d*-electrons
Ti(III)	1
Ti(II), V(III)	2
V(II), Cr(III), Mn(IV)	3
Cr(II), Mn(III)	4
Mn(II), Fe(III)	5
Fe(II)	4
Co(II)	3
Ni(II)	2
Cu(II)	1

Transition Metals in Natural and Man-Made Fibres

1) Iron is a **constant component** of crocidolite and amosite present in rather large percentage. *In vitro*, ferrozine, a strong Fe(II) chelator, mobilised Fe(II) from crocidolite (6.6 nM/mg asbestos × h) and amosite (0.4 nM/mg × h) in 50 mM NaCl, pH 7.5 (LUND and AUST 1990). Inclusion of ascorbate increased the rates to 11.4 and 4.9 nM/mg × h, respectively. *In vivo*, iron mobilisation from crocidolite by low-molecular-weight chelators may lead to the increased production of reactive oxygen species (LUND and AUST 1991).
2) The metal **substitutes for another cation.** In chrysotile Fe^{2+} and Mn^{2+} replace Mg^{2+}.
3) The metal represents an **impurity** caught in manufacturing (CRALLEY et al. 1967).

Two classes of dioxygen-independent enzymatic reaction are now recognized to proceed through radical-based mechanisms, those catalysed by S-adenosylmethionine/iron-sulphur enzymes, and the adenosylcobalamine-dependent enzymes, as recently reviewed by FREY (2001).

6.1 The Respiratory Burst

Free radicals are chemical species with one or more unpaired electrons in their outer orbital. Their production is essential to normal metabolism but they are theoretically destructive unless tightly controlled.

Reactive oxygen species are generated in cells by both enzymatic and nonenzymatic reactions. They should be generated at the cell surface, as a response against the stimulus caused by the presence of a pathogen.

Redox reactions are essential for the function of cell membranes (DEL CASTILLO-OLIVARES et al. 2000). It should be stressed that every bioenergetically competent cell membrane does contain redox systems (SKULACKEV 1988). A plasma membrane electron transport or redox system has been found in every living cell tested. VOEGTLIN et al. (1925) examined a relation between the redox state and cancer. The redox potential of cytochrome b_{558}, a component of NADPH oxidase, is −245 mV. This is atypically low for a cytochrome *b*, but this fact enables the **reduction of oxygen to superoxide** (CROSS et al. 1981). The rate of electron flow has been demonstrated to be matched by the rate of superoxide production (CROSS et al. 1985).

Using *bovine* neutrophils, YAMAMORI et al. (2002) examined the role of p38 mitogen-activated protein kinase in the signalling pathway of the NADPH oxidase activation. Superoxide production was induced by stimulation with serum-promised zymosan and attenuated by p38 mitogen-activated protein kinase inhibitor, SB203580. Serum-promised zymosan stimulation induced the translocation of $p47^{phox}$ and Rac to the plasma membrane and SB203580 completely blocked the translocation of Rac, but only partially blocked that of $p47^{phox}$. Furthermore, SB203580 abolished the serum-promised zymosan-elicited activation of Rac, which was assed by detecting the guanosine triphosphate-bound form of this protein. Phosphatidylinositol 3-kinase inhibitors, wortmannin and LY294002, blocked not only p38 mitogen-activated protein kinase activation but also Rac activation. However, SB203580 showed no effect on the phosphatidylinositol 3-kinase activity.

A growing body of evidence suggests a potential role for oxygen-derived radicals such as superoxide anion ($O_2^{\bullet-}$) and hydrogen peroxide (H_2O_2) as intracellular signalling molecules. Numerous studies have implicated a dynamic change in the intracellular redox state as an important determinant in a host of cellular decisions ranging from growth, to apoptosis, to cellular senescence (FINKEL 1999).

6.1.1 Reactive Oxygen Intermediates

Atmospheric oxygen ($^3O_2$) is relatively unreactive because it is a diradical with parallel spin state. Thus its divalent reduction is kinetically limited by the relatively slow spin inversion process. The spin conservation rule states that spin must be conserved during the time for a chemical reaction to occur. The spin restriction means that when $^3O_2$ is involved in metabolic oxidation it has to be activated, allowing for spin inversion of one electron at a time, in order to have productive collision. This univalent pathway requires the generation of intermediates, among **superoxide** ($O_2^{\bullet-}$) is the first reduced product.

One-electron reduction of oxygen

$$O_2 \xrightarrow{+e^-} O_2^{\bullet-} \xrightarrow{e^- + 2H^+} H_2O_2 \xrightarrow{+e^-} HO^{\bullet} \xrightarrow{+e^- + H^+} H_2O_2$$

Dioxygen → Superoxide radical → Hydrogen peroxide → Hydroxyl radical → Water [20]

The product of the *Krev-1/rap1A* gene was copurified as a component of the superoxide generating system from *human* neutrophils (QUINN et al. 1989).

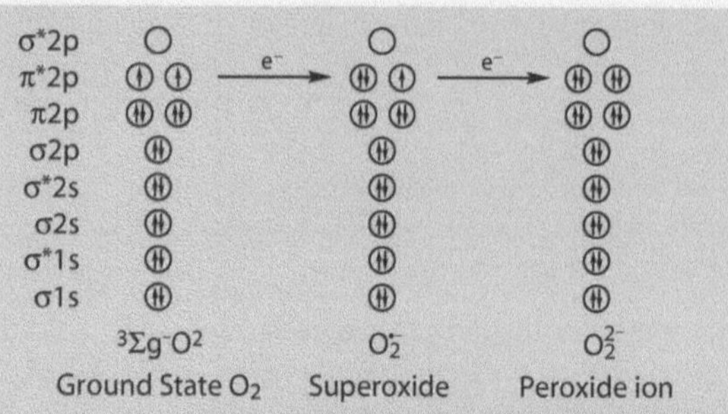

Bonding in the diatomic oxygen molecule [21]

Superoxide under physiological conditions *in vivo* may be a source of electrons for the oxidative phosphorylation of ADP (MAILER 1990). Scavenging of $^{\bullet}NO$ as an important messenger makes $O_2^{\bullet-}$ an antagonist to $^{\bullet}NO$ and already simply by this property $O_2^{\bullet-}$ becomes a messenger molecule itself (WOLIN 1996, OKATANI et al. 1998, MEYER et al. 1999). The existence of defined sources for $O_2^{\bullet-}$ favours the hypothesis of a superoxide-driven redox regulation (ULLRICH and BACHSCHMID 2000). At unphysiological low pH values the protonated $HOO^{\bullet}$ radical is a more reactive species bit at physiological pH its concentration is too low ($pK = 4.8$) and its radical mechanism not suited to cause sulphoxidations of methionine, sulphenic acid formation or the oxidation of vicinal dithiols to disulphides. However, taking into account the reducing capacity of $O_2^{\bullet-}$ it has been confirmed that ferritin bound ferric ions can be reduced and released (FRIDOVICH 1986) and also Cu^{I} and Mn^{II} levels may increase as a consequence of superoxide formation. These metals tend to associate with protein structures and then would be able to react with $O_2^{\bullet-}$ by formation of peroxo or oxo species; e.g.

$$Fe^{3+} + O_2^{-} \xrightarrow{10^6\ M^{-1}\cdot s^{-1}} Fe^{2+} + O_2 \quad [22]$$

$$Fe^{3+} + O_2^{-} \longrightarrow Fe^{III}OO^{2-} \text{ or } Fe^{V} = O \quad [23]$$

Such species are more reactive than H_2O_2 and act locally and hence selectively at metal-binding domains.

Tumour necrosis factor-α (TNF-α) transiently increased intracellular superoxide and other reactive oxygen species production in *human* fibroblasts (MEIER et al. 1989) and *human* oral carcinoma SCC-25 cells (LIU et al. 2000).

Enhanced exhalation of H_2O_2 and thiobarbituric acid reactive substances has been reported in various inflammatory lung diseases (DOHLMAN et al. 1993, ANTCZAK et al. 1997, NOWAK et al. 1999). In healthy volunteers the H_2O_2 exhalation revealed diurnal variation with two-peak values 0.45 ± 0.29 μM and 0.43 ± 0.22 μM at 12:00 and 24:00 h (NOWAK et al. 2001). The lowest concentrations, 0.26 ± 0.13 μM and 0.25 ± 0.26 μM, were found at 20:00 and 8:00 h. Type II pneumocytes, alveolar macrophages, and endothelial cells produce H_2O_2 (KINNULA et al. 1991).

The role of H_2O_2 as a second messenger has been demonstrated in various signal transduction systems stimulated by diverse ligands including cytokines and peptide growth factors through tyrosine kinase and G-protein-coupled receptors (KRIEGER-BRAUER and KATHER 1992, OHBA et al. 1994, SUNDARESAN et al. 1995, BAE et al. 1997). Besides the stimulation of protein phosphorylation and transcription factor activation, H_2O_2 is known to modulate K^+ channels expressed in *Xenopus* oocytes (VEGA-SAENZ DE MIERA and RUDY 1992, SZABO et al. 1997). H_2O_2 is also known to regulate Ca^{2+} signalling (REEVES et al. 1986, ROVERI et al. 1992). On of the ways H_2O_2 is able to modulate the intracellular Ca^{2+} level is to induce Ca^{2+} influx by directly activating Ca^{2+} channels on the plasma membrane (DOAN et al. 1994, LI et al. 1998). The H_2O_2 induced ionic current was not transient but long lasting in a Ca^{2+}-free medium (KIM and HAN 2002). As H_2O_2 was still able to induce current in oocytes loaded with either catalase (EC 1.11.1.6) or *N*-acetyl-L-cysteine, H_2O_2 scavengers, H_2O_2 induces this ionic current possible through the activation of Ca_o^{2+}-inactivated channels by an extracellular mechanism.

Classically, radical reactions begin with an initiation process to form a radical species, such as the reaction of H_2O_2 with transition metals (**Fenton chemistry**) to form hydroxyl radical, $^{\bullet}HO$ (WARDMAN and CANDEIAS 1996).

The reaction requires two steps, where superoxide first reduces ferric iron to ferrous iron:

$$Fe^{3+} + O_2^{\bullet-} \xrightarrow{10^6\ M^{-1}\cdot s^{-1}} Fe^{2+} + O_2 \quad [22]$$

and the ferrous iron attacks hydrogen peroxide to generate hydroxyl radical (KOPPENOL 1994):

$$Fe^{2+} + H_2O_2 \xrightarrow{10^3\text{–}10^5\ M^{-1}\cdot s^{-1}} Fe^{3+} + HO^{\bullet} + HO^- \quad [24]$$

Reactive radical entities can interact with neutral molecules, and the product of these reactions also yield radicals. This is the propagation step of radical reactions, which often proceed through tens to thousands of molecular intermediates. Radicals are eliminated when termination reactions occur in which two radical entities react together or alternatively when the process of propagation results in the reduction in the product radical energy to such an extent that it cannot further react with another molecule, e.g., abstract a hydrogen atom.

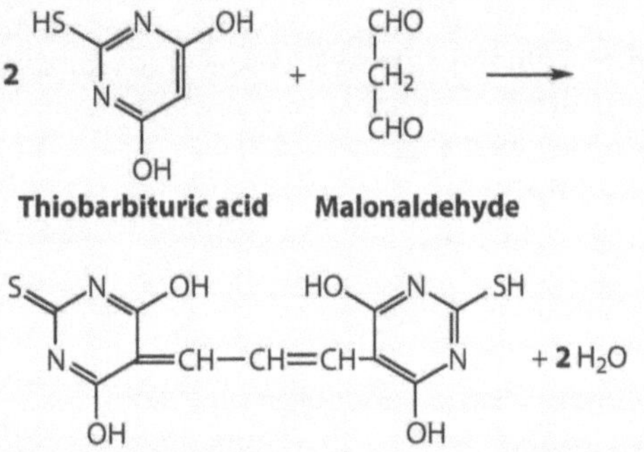

Thiobarbituric acid test [25]

Enhanced exhalation of H_2O_2 and thiobarbituric acid reactive substances has been reported in vari-
ous inflammatory lung diseases (DOHLMAN et al. 1993, ANTCZAK et al. 1997, NOWAK et al. 1999). In healthy volunteers the H_2O_2 exhalation revealed diurnal variation with two-peak values 0.45 ± 0.29 μM and 0.43 ± 0.22 μM at 12:00 and 24:00 h (NOWAK et al. 2001). The lowest concentrations, 0.26 ± 0.13 μM and 0.25 ± 0.26 μM, were found at 20:00 and 8:00 h. Type II pneumocytes, alveolar macrophages, and endothelial cells produce H_2O_2 (KINNULA et al. 1991).

The production of hydroxyl radicals by tetrachlorohydroquinone, a major metabolite of the widely used biocide pentachlorophenol in the presence of H_2O_2 was markedly inhibited by hydroxyl radical scavenging agents dimethyl sulphoxide and ethanol, as well as by tetrachlorohydroquinone radical scavengers desferrioxamine and other hydroxamic acids (ZHU et al. 2000). In contrast, their production was not affected by the nonhydroxymate iron chelators, diethylenetriaminepentaacetic acid, bathophenanthroline disulphonic acid, and phytic acid, as well as the copper-specific chelator bathocuprione disulphonic acid.

One-electron reduction of oxygen can, in principle, occur by oxidation of any substance the redox potential of which is lower than or equal to −0.15 V (the redox potential of the O_2/superoxide pair). Compounds with high kinetic barriers of reaction with O_2 were selected by evolution. Highly reactive coenzymes and prosthetic groups of enzymes operating at the initial and middle steps of the respiratory chain, such as coenzyme Q semiquinone ($CoQH^{\bullet}$) are exception of this rule. $CoQH^{\bullet}$ as an electron reductant can probably commit errors, donating an electron to molecular oxygen rather than to cytochrome b_1, which is its natural oxidant.

One-electron reduction of quinones by ascorbate, followed by the oxidation of the semiquinone by molecular oxygen, results in the catalytic oxidation of ascorbate (with quinone as catalyst) and formation of active forms of oxygen. Along with enzymatic redox cycling of quinone, this process may be related to quinone cytotoxicity and underlie an antitumour activity of some quinones. Roginsky et al. (1999) studied the kinetics of oxygen consumption accompanying the interaction of ascorbate with 55 quinones including substituted 1,4- and 1,2-benzoquinones, naphthoquinones and other quinoid compounds using the Clark electrode technique.

In Jurkat T-cells submicromolar levels of H_2O_2 induced apoptosis through Fenton chemistry independent of the cellular thiol state (Antunes and Cadenas 2001).

6.1.1.1
Singlet Oxygen

Singlet oxygen is a more reactive form of oxygen, which can be generated by an input of energy. The $^1\Delta gO_2$ state has an energy 22.4 kcal above the ground state. The $^1\Sigma^+O_2$ state is even more reactive, 37.7 kcal above the ground state. In both forms of singlet oxygen the spin restriction is removed and so the oxidising ability is greatly increased. Steinbeck et al. (1993) developed a specific and sensitive assay which could be used to measure the amount of intracellular $^1\Delta gO_2$ produced in activated polymorphonuclear leucocytes. Their method was based on the reactivity rates of $^1\Delta gO_2$ with perylene and 9,10-diphenylanthracene, and identifying the product formed in the reaction of $^1\Delta gO_2$ with 9,10-diphenylanthracene (Wasserman et al. 1972, Turro et al. 1981). This reaction is highly specific and forms a stable and quantitative product: 9,10-diphenylanthracene-endoperoxide. Wefers (1987) used 9,10-di(2-ethylene)anthracene to quantify $^1O_2$ through measurement of the endoperoxide as the stable reaction product.

The two states of singlet oxygen

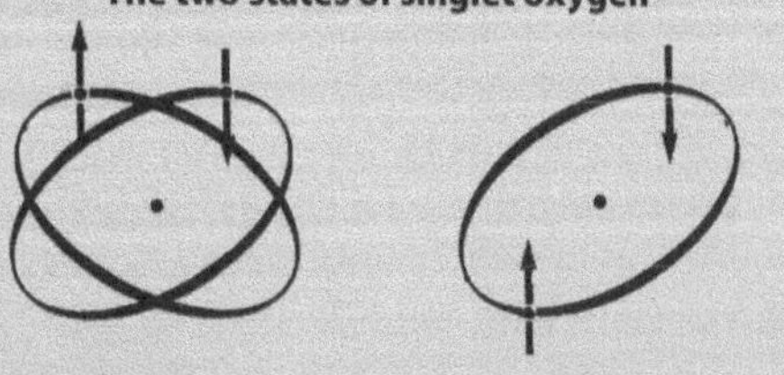

$^1\Sigma g^+O_2$ $\quad$ $^1\Delta gO_2$ [26]

The 570–590 and 630–700 nm bands of the O_2-dependent light emission of macrophages stimulated with concanavalin A correspond with the dimol emission of $^1O_2$ in several electronic excited and vibrational states: $[^1\Delta g]$ $[^1\Delta g]$ (1,0) for 580 nm, and $[^1\Delta g]$ $[^1\Delta g]$ (0,0) and $[^1\Delta g]$ $[^1\Delta g]$ (0,1) for 634 and 703 nm, respectively (Seliger 1960, Khan and Kasha 1970, Inaba et al. 1979). Although the fluorescence spectrum between resting and stimulated macrophages did not differ significantly, it is possible that the blue-green photoemission bands (420–450 and 460–480 nm) are due to decay of fluorescence species (432 and 484 nm) inherent in the macrophages suspension and independent of the activation process (Cadenas et al. 1981).

Erythrocuprein inhibited the luminol-enhanced chemiluminescence caused by singlet oxygen produced in a xanthine-xanthine oxidase system as measured in a Packard scintillation counter (Weser and Paschen 1972).

6.1.1.1.1
Formation of Singlet Oxygen

Electronically exited oxygen in the $^1\Delta_g$ state has been shown to be a product of a number of enzymatic reactions (Piatt et al. 1977, Piatt and O'Brien 1979, Karnofsky 1983, 1984, 1985, 1988, 1991, Karnofsky et al. 1983, 1984, 1985, 1988, Khan et al. 1983).

Myeloperoxidase (EC 1.11.1.7) reaction

$$Cl^- + H_2O_2 \longrightarrow ClO^- + H_2O \qquad [27]$$

$$ClO^- + H_2O_2 \longrightarrow {}^1O_2 + Cl^- + H_2O \qquad [28]$$

Lipid peroxidation in microsomal fraction initiated by hydroperoxides or iron/ascorbate, as well as lipid peroxidation of isolated hepatocytes under oxygenation, has been studied extensively by Cadenas et al. (1981) and Cadenas and Sies (1982). The available evidence points to $^1O_2$ formation by the Russell mechanism (Russell 1957, Howard and Ingold 1968). Secondary lipid-peroxy radicals react via a transient tetroxide to yield an alcohol, a carbonyl and oxygen, the carbonyl or oxygen being in an electronically excited state.

In phorbol myristate acetate-stimulated polymorphonuclear leucocytes it is important to note that the generation of $^1\Delta gO_2$ was dependent on the dose of phorbol myristate acetate added and that stimulation with high-dose phorbol myristate acetate resulted in lower amounts of detectable $^1\Delta gO_2$ (Karnofsky 1994). This finding was not entirely unexpected since higher concentration of phorbol myristate acetate lead to a greater production of

$O_2^{\bullet-}$ and little net accumulation of $^1\Delta gO_2$, as $O_2^{\bullet-}$ at high concentrations quenches $^1\Delta gO_2$ (Rosenthal 1975, Guiraud and Foote 1976, Khan 1977).

Eosinophils oxidise physiological concentrations of bromide ion (about 100 μM) in the presence of the 1000 fold higher physiological concentrations of chloride ion (Weiss et al. 1986). When activated with the physiological agonists C5a and leukotriene B_4, *guinea pig* eosinophils released $^1O_2$ (Teixeira et al. 1999). This release, which occurred at agonist concentrations as low as 10^{-7} M occurred more rapidly than activation with phorbol ester (10^{-6} M), was similar in level, but was more transitory. The release of $^1O_2$ occurred in absence of added bromide ions.

Exposure of *human* red blood cells from healthy adult volunteers of either sex to high concentration of HOCl and HOBr (>40 nmol/10^7 cells) resulted in rapid cell lysis and the detection of an additional nitrogen-centred, protein-derived radical adduct (Hawkins et al. 2001). HOBr induced red blood cell lysis at approximately 10-fold lower concentration than HOCl, whereas with monocyte (HTP1) and macrophage (J774) cells HOCl and HOBr induced lysis at similar concentrations. Erythrocytes exposed to nonlytic doses of HOCl generated novel nitrogen-centred radicals the formation of which is GSH dependent.

Peroxidases can also catalyse the formation of singlet oxygen by a halide-independent mechanism. Oxidation of indole-3-acetic acid by either lactoperoxidase or horseradish peroxidase results in singlet oxygen formation (Straight and Spikes 1985). Singlet oxygen is again a secondary reaction product. The peroxidase-catalyzed oxidation of indole-3-acetic acid is known to produce high concentrations of carbon-centred 3-methylindole (skatole) radicals (Mottley and Mason 1986). These skatole radicals react with oxygen to form peroxyl radicals, which then undergo a bimolecular reaction to form singlet oxygen. Pneumotoxicity of 3-methylindole was described by Huang et al. (1977), Turk et al. (1984) and Becker et al. (1985).

Polymer-bound rose bengal as a singlet oxygen generating system (Perez 1985) is merchandised under the names Sensitox I and Sensitox II. Sensitox II is compatible with aqueous and alcohol-containing media. The lifetime of $^1\Delta gO_2$ in H_2O is 2–5 μs versus 700 μs in CCl_4.

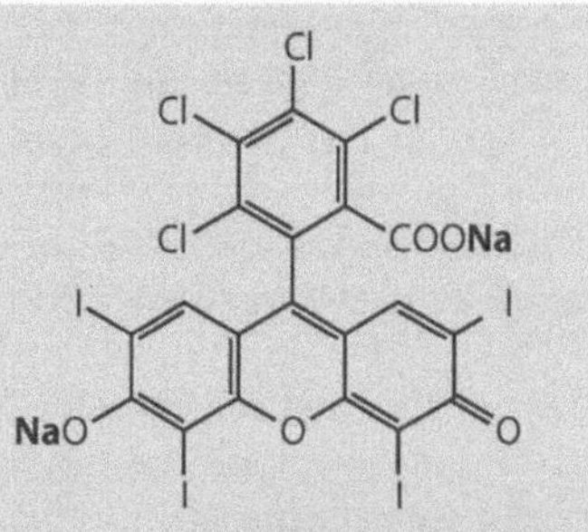

Rose Bengal [29]

Other popular sensitisers of singlet oxygen formation are acridin orange, methylene blue, and toluidin blue (formula [79]); but many biological compounds are also effective *in vitro*, such as the water-soluble vitamin riboflavin and its derivatives flavin mononucleotide and flavin adenine dinucleotide.

Singlet oxygen

$$HO_2^{\bullet} + O_2^{\bullet-} + H^+ \rightarrow {}^1O_2 + H_2O_2 \quad [30]$$

$$O_2^{\bullet-} + H_2O_2 \rightarrow {}^1O_2 + HO^- + HO^{\bullet} \quad [31]$$

$$O_2^{\bullet-} + HO^{\bullet} \rightarrow {}^1O_2 + HO^- \quad [32]$$

$$2\,O_2^{\bullet-} + R{-}\overset{O}{\overset{\|}{C}}{-}O{-}O{-}\overset{O}{\overset{\|}{C}}{-}R \rightarrow 2\,{}^1O_2 + 2\,RCO_2^- \quad [33]$$

The formation of $^1O_2$ in systems known to produce $O_2^{\bullet-}$ has often been postulated, though conclusive evidence in that respect is still missing. The detection of $^1O_2$ in biological systems has been aided by different chemical methods (Foote et al. 1984) as well as the observation of photoemission upon decay of $^1O_2$ to the ground state (Cadenas et al. 1984).

The reaction of potassium superoxide with water is the single most important chemical source of oxygen for breathing purposes in hospitals, mines, submarines (Clarke 1956), and space capsules (Bovard 1960). Even if only a small fraction of singlet oxygen survives quenching, it could prove to be a serious health hazard (Khan 1970).

Reactive oxygen species as $O_2^{\bullet-}$, H_2O_2, and $^1O_2$ can react with lucigenin to form photons. Photoemission was measured using a Berthold Biolumat demonstrating the formation and degradation of these oxygen species. Chemiluminescence generated by granulocytes during the phagocytosis of zymosan particles was markedly impaired by 0.1 mM azide as a scavenger of $^1O_2$ (Sagone et al. 1977). However, azide reacts with $HO^{\bullet}$ to give a reactive $N_3^{\bullet}$ radical.

$$N_3^- + HO^{\bullet} \rightarrow N_3^{\bullet} + OH^- \quad [34]$$

The selective para hydroxylation of phenol or aniline by $^1O_2$ may be used as an indicator for the involvement of singlet oxygen as compared to $HO^\bullet$- or cytochrome P_{450}-mediated reactions (BRIVIBA et al. 1993). The hydroxylation of aniline was suppressed when β-carotene was added to *rat* liver microsomes, while the addition of sodium azide (NaN_3) enhanced the reaction (OSADA et al. 1999). In contrast, the microsomal o-deethylation of 7-ethoxycoumarin was suppressed by the addition of NaN_3. $^1O_2$ was detectable during the reaction of microsomes and NADPH by electron spin resonance spin trapping when 2,2,6,6-tetramethyl-4-piperidone (TMPD) was used as a spin trap, and the $^1O_2$ was quenched by the additions of β-carotene, NaN_3, aniline, and 7-ethoxycoumarin. The enhancement effect of NaN_3 in the hydroxylation of aniline appeared to be due to the conformational change of P450 protein, which in turn enhances the binding of aniline to P450 in terms of the spectral dissociation constant (K_s). In contrast, $^1O_2$ appeared to be active in the o-deethylation of 7-ethoxycoumarin.

6.1.1.1.2
Reactions of Singlet Oxygen

Singlet oxygen can interact with other molecules in essentially two ways: it can either **combine chemically** with them, or else it can transfer its excitation energy to them, returning to the ground state while the molecule enters an excited state. The latter phenomenon is known as **quenching**. Sometimes both can happen. The best studies chemical reactions of singlet oxygen are those involving compounds containing carbon-carbon double covalent bonds.

Fumaric acid first combines with $^1O_2$ by dioxetane formation. Dioxetanes are instable and decompose to form two molecules of glyoxylic acid, one of which is an excited species (equation [35]).

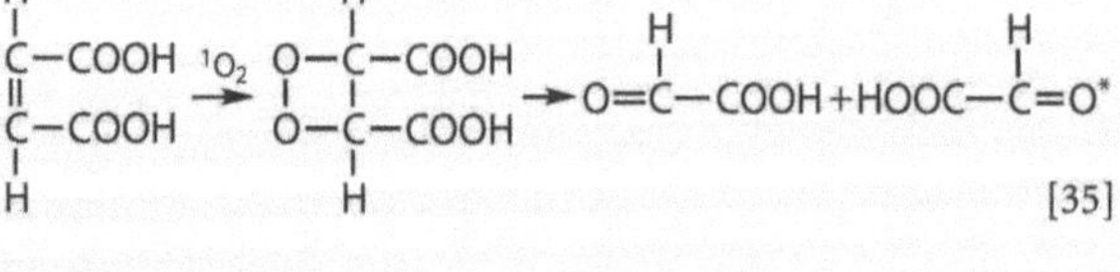

$$\begin{array}{c} H \\ | \\ C-COOH \\ \| \\ C-COOH \\ | \\ H \end{array} \xrightarrow{^1O_2} \begin{array}{c} H \\ | \\ O-C-COOH \\ | \\ O-C-COOH \\ | \\ H \end{array} \longrightarrow O=C-COOH + HOOC-C=O^\bullet \qquad [35]$$

By combining $^1O_2$ with fumaric acid an instable dioxetane is formed. By decomposing this dioxetane a excited carbonyl species is generated.

Singlet oxygen adds to the double bonds of unsaturated fatty acids producing lipid hydroperoxides. SCHAFER and BUETTNER (2000) used electron paramagnetic resonance spin trapping with α-(4-pyridyl-1-oxide) *N-tert*-butylnitrone to detect lipid-derived radical formation from HL-60 *human* leukaemia cells. Free radical formation increased with increasing iron concentration; in the absence of ex-

tracellular iron, radical formation was below the limit of detection and lipid hydroperoxides accumulated in the membrane. In the presence of iron, lipid-derived radical formation in cells was pH dependent; the lower the extracellular pH (.5 – 5.5), the higher the free radical flux; the lower the pH, the greater the membrane permeability induced in K-562 cells, as determined by trypan blue dye exclusion.

Azide reacts with singlet oxygen with high rate constant ($k = 2.2 \times 10^9$; WILKINSON and BRUMMER 1981), producing the strong one-electron oxidant azidyl radical, $N_3^\bullet$ (eq. [36]).

$$N_3^- + {}^1O_2 \longrightarrow N_3^\bullet + O_2^- \qquad [36]$$

α-Tocopherol mostly reacts by quenching. Combining of $^1O_2$ with tocopherol gives α-tocopherylquinone hydroperoxide. Opening of the pyran ring gives various products as shown in equation [37] (NEELY et al. 1988).

FUKUZAWA et al. (1997) investigated the scavenging of singlet oxygen by α-tocopherol in liposomes. The rates of oxidation of α-tocopherol differed depending on the photosensitising dye and the membrane charge: in the methylene blue system, α-tocopherol was oxidised fast in negatively charged dimyristoylphosphatidylcholine liposomes containing dicetylphosphate and slowly in neutrally charged dimyristoylphosphatidylcholine liposomes containing stearylamine, but in the lipid-soluble 12-(1-pyrene)dodecanoic acid-system, the oxidation rate was independent of the membrane charge.

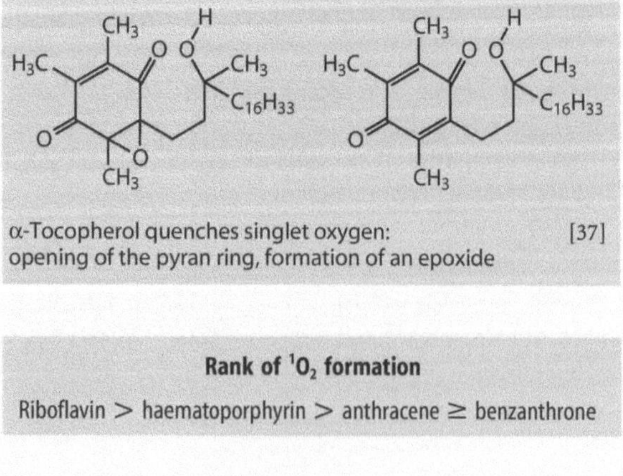

α-Tocopherol quenches singlet oxygen: [37]
opening of the pyran ring, formation of an epoxide

Rank of $^1O_2$ formation

Riboflavin > haematoporphyrin > anthracene ≥ benzanthrone

Riboflavin or vitamin B_2 makes a prosthetic group of flavin enzymes where it can be reversibly reduced by hydrogen atoms (formula [38]). When exposed to light, riboflavin absorbs energy and reacts, via a triplet exited state, with other molecules such as protonated substrates or molecular oxygen, generating reactive species. In the type I photodynamic reaction, energy is transferred from a triplet sensitiser to O_2 with the formation of $^1O_2$.

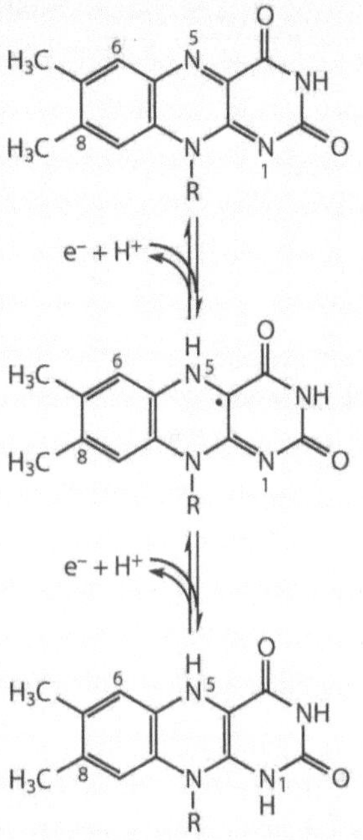

A radical intermediate in the reduction of riboflavin [38]

Certain fourocoumarins may produce potentially damaging singlet oxygen (De Mol and Beijersbergen van Henegouwen 1979, 1981a, 1981b, Anders et al. 1983, Decuyper et al. 1983, Pathak and Joshi 1984). Beaumont et al. (1985), however, did not see any evidence that singlet oxygen is produced from psoralen and 4'-aminomethyl-4,5',8-trimethylpsoralen, when they were intercalated between the strands of *calf* thymus DNA.

Yamazaki et al. (1999) identified singlet oxygen adduct of cholesterol, 3β-hydroxy-5α-cholest-6-ene-5-hydroperoxide, in skin of *rats* pre-treated with oral doses of pheophorbide a and subsequent visible irradiation.

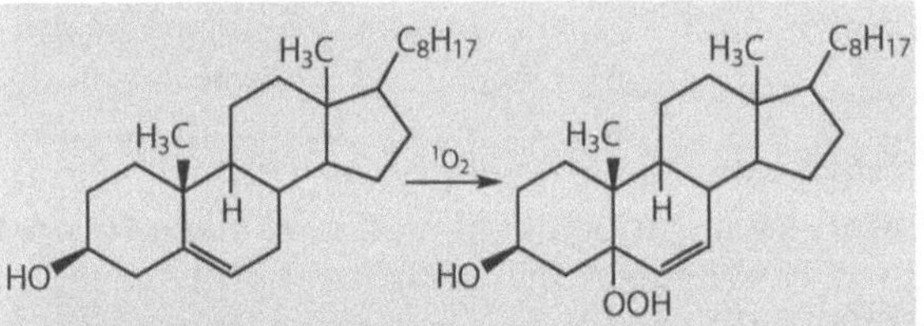

Combining of $^1O_2$ with cholesterol gives cholesterol 5α-hydroperoxide as the major product [39]

Pyrroles are very sensitive to the action of singlet oxygen ($^1O_2$). Wasserman (1970) has observed a novel type of oxidation in the case of aryl-substituted pyrroles exemplified by the photooxidation of 2,3,4,5-tetraphenylpyrrole in methanol. There is special interest in the reactions of imidazoles with $^1O_2$ since it has been shown that photooxidative inactivation of certain enzymes involves destruction of histidine residues, and more specifically oxidation of the imidazole ring (Wasserman and Lenz). These heterocyclic system behave in many respects like furans and pyrroles, but are more prone to cleavage through reactions resembling the oxidation of enamines by $^1O_2$ (Wasserman et al. 1968).

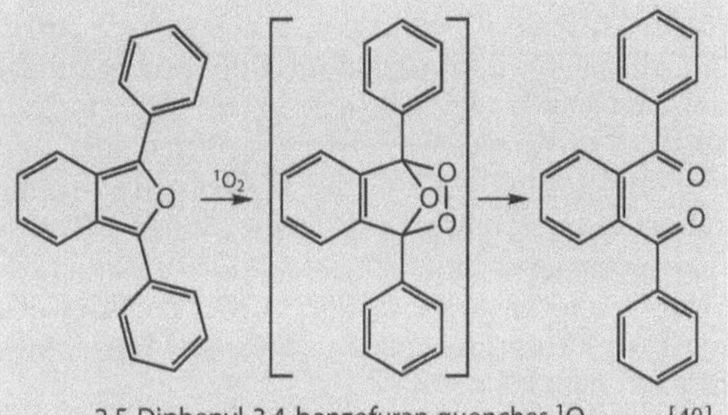

2,5-Diphenyl-3,4-benzofuran quenches $^1O_2$ [40]

It is well established that singlet oxygen is able to oxidise DNA with a much higher specificity than hydroxyl radical. The main stable oxidation products of the reaction of $^1O_2$ with 2-deoxyribose guanine were identified as the 4R* and 4S* diastereoisomers of 4-hydroxy-8-oxo-4,8-dihydro-2'-deoxyguanosine on the basis of extensive NMR and mass spectrometry measurements (Ravanat et al. 1992, Ravanat and Cadet 1995). Similar oxidation products were generated by the type II photooxidation reaction of 2'-deoxyguanosylyl-(3'-5')-thymidine (Buchko et al. 1992). Azide and methionine as $^1O_2$ quenchers decrease the DNA damage strongly, reaching almost control values of the transforming activity (Wefers 1987). Methionine sulphone is used as a further control, the effect of methionine sulphone being much smaller than that of methionine (Wefers et al. 1987).

Lledías et al. (1998, 1999) reported that catalases are oxidized *in vitro* and *in vivo* by singlet oxygen, giving rise to different catalase activity bands in zymograms. Michán et al. (2002) observed mobility changes of Cat-1 and Cat-3 during the asexual life of *Neurospora crassa*. Cat-1 was modified when growing mycelium was subjected to heat shock or paraquat. Paraquat (5 mM) treatment caused Cat-1 to disappear in the course of 6 h and heat shock for more than 2 h produced a smear of Cat-1 activity.

After 3 h in the presence of riboflavin (formula [38]) and light, the electrophoretic mobility of Cat-3 increased similar to other catalases. The presence of singlet oxygen quenchers such as histidine and 5-aminosalicylic acid prevented this change.

Both extracellularly [incubation with disodium 3,3'-(1,4-naphthylidine) dipropionate-1,4-endoperoxide] and intracellularly (rose Bengal and irradiation with white light) generated singlet oxygen induced a concentration-dependent protein oxidation and removal of the singlet-oxygen-damaged proteins by the proteasomal system (GRUNE et al. 2001).

Bacterial killing in *human* polymorphonuclear leucocyte phagosome is principally mediated by singlet oxygen ($^1\Delta_gO_2$) produced by the myeloperoxidase reaction (TATSUZAWA et al. 1999).

Collagen fibril formation was inhibited by singlet oxygen (VENKKATASUBRAMANIAN and JOSEPH 1977). The oxidation of native collagen prepared from-old *rat* tail tendon and its CNBr peptides by singlet oxygen immediately formed blue (430 nm)-fluorescent cross links (FUJIMORI 1989). Singlet oxygen did not degrade collagen CNBr peptides just like UV (300 nm) irradiation, though hydroxyl radicals did degrade them.

Singlet oxygen-challenged unmedicated *rat* cardiomyocytes all hypercontracted as a consequence of Ca^{2+} overload and produced 463.6±143.6 nM malondialdehyde (VER DONCK et al. 1990). Protective Ca^{2+} antagonists reduced the amount of damaged cells, but did generally not affect malondialdehyde production.

6.1.2 Oxygen Activating Enzymes

Two general classes of enzymes are involved in oxygen metabolism: oxidases, transferring electrons from a substrate to oxygen, and oxygenases, transferring oxygen to a substrate after reductive splitting of molecular oxygen. Oxygenases can be divided into dioxygenases and monooxygenases. Mono-oxygenases (mixed function oxygenases) catalyse the incorporation of one atom of $^1\Sigma g^-$ molecular oxygen into a substrate with the concomitant reduction of the other atom to water (MASON 1957, HAYAISHI and NOZAKI 1969).

Piperonyl butoxide is an inhibitor of the mixed-function oxidase system.

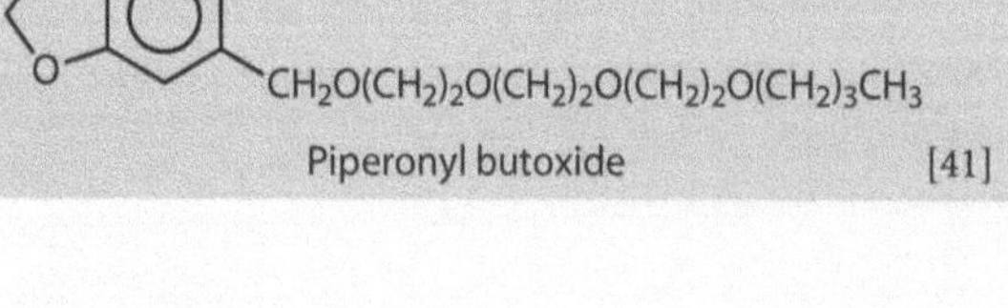

Piperonyl butoxide [41]

6.1.2.1 Aldehyde Oxidase

The cytosolic enzyme, aldehyde oxidase (EC 1.2.3.1) is important in the biotransformation of numerous aldehydes (BEEDHAM 1987). This molybdenum-containing enzyme is generally involved in nucleophilic oxidation although aldehyde oxidase has been shown to catalyse many reduction reactions (BEEDHAM 1985). It is also involved in the metabolism of allopurinol, quinine and methotrexate (BEEDHAM 1985). In the guinea pig, BEEDHAM et al. (1989) observed a marked circadian variation in aldehyde oxidase activity with several substrates (phthalazine, phenathridine, *N*-phenylquinolinium and 3,4-dihydro-4-hydroxy-3-methyl-2-quinazolinone). The main peak occurred at 3 h with minimum activity from 12 to 18 h, the differences between rhythmic extremes being statistically significant ($P < 0.005$). Exogenously administered melatonin caused a significant increase in aldehyde oxidase activity at 9 o'clock and 12 o'clock.

6.1.2.2 Cytochrome Oxidase

NADPH-cytochrome P_{450} oxidoreductase (EC 1.6.2.4) (CYPOR) is a ubiquitous enzyme involved in many aspects of cellular metabolism, most notably, the mixed function oxidase system. This system, which is composed of CYPOR and the cytochromes P_{450}, is pivotal in the metabolism of numerous endogenous substrates (e.g. steroids, fatty acids, and prostaglandins), in the metabolism and detoxification of drugs and environmental pollutants, and in the metabolic activation of chemical carcinogens. CYPOR, a 78,275-Da membrane-bound flavoprotein, functions as an internal electron transport chain, transferring electrons from NADPH to flavin-adenine dinucleotide (FAD) to flavin mononucleotide (FMN) and finally to cytochrome P_{450} (SHEN and KASPER 1993).

O'LEARY and KASPER (2000) delineated the importance of multiple *cis*-acting elements contained within the proximal promoter for basal expression of the CYPOR gene. Transcription factor binding sites within this region included two upstream Sp1 motifs, a SEC element containing overlapping Sp1/Egr-1/CACCC box motifs, and a novel site designated the OxidoReductase Upstream element (ORU). Mutational modification of the ORU element, leading to a loss of protein binding, resulted in an ~90 % decrease in transcriptional activity in H4IIE cells. Similarly, inactivation of the Egr-1/CACCC segment of the SEC element dramatically reduced promoter activity to less than 10 % of wild-type, while mutagenesis of the contiguous Sp1 site did not affect basal transcription.

Cytochrome P_{450} enzymes perform three general types of oxidative reactions:

1. insertion of an oxygen atom into the bond between the hydrogen and a heavier atom to yield the corresponding hydroxyl derivative
2. addition of an oxygen atom across the two carbons of a π-bond to yield an epoxide
3. addition of an oxygen atom to the electron pair of a heteroatom to give a dipolar oxide.

Cytochromes P_{450} belong to external monooxygenases. This implies that they need an external electron donor which transfers the electrons necessary for oxygen activation and the subsequent substrate hydroxylation. Two main classes of cytochromes P_{450} principally different with respect to their electron-supporting system can be defined:

1. microsomal type
2. mitochondrial/bacterial type.

Microsomal cytochromes P_{450} are membrane-bound. They accept electrons from a microsomal NADPH-cytochrome reductase, containing flavin adenine dinucleotide and flavin mononucleotide. All cytochromes P_{450} metabolising drugs and xenobiotica isolated so far belong to this class.

Experiments on **isolated mitochondria** have indicated that a relatively small fraction of each of several components of the electron transport chain is sufficient to sustain a normal respiration rate. These experiments, however, may have not reflected the *in vivo* situation, due to the possible loss of essential metabolites during organelle isolation and the disruption of the normal interactions of mitochondria with the cytoskeleton, which may be important for the channelling of respiratory substrate to the organelles. Therefore, VILLANI and ATTARDI (2000) have developed an approach for measuring cytochrome *c* oxidase activity in **intact cells**, by means of cyanide titration, either as an isolated step or as a respiratory chain-integrated step. The method has been applied to a variety of *human* cell types, including wild-type mtDNA mutation-carrying cells, several tumour-derived semidifferentiated cell lines, as well as specialised cells removed from the organism. The results obtained strongly support the following conclusion:

- the *in vivo* control of respiration by cytochrome *c* oxidase is much tighter than has been generally assumed on the basis of experiments performed on isolated mitochondria:
- cytochrome *c* oxidase threshold depends on the respiratory fluxes under which they are measures:
- measurements of relative enzyme capacities are needed for understanding the role of mitochondrial respiratory complexes in *human* physiopathology.

How is the Oxygen Chemistry Energetically Coupled to the Pumping of Protons?

Some insights have come from mutation studies (HOSLER et al. 1993, 1996, THOMAS et al. 1993, FETTER et al. 1995, GARCÍA-HORSMAN et al. 1995) and the X-ray structures of cytochrome *c* oxidase (IWATA et al. 1995, TSUKIHARA et al. 1995, OSTERMEIER et al. 1997, YOSHIKAWA et al. 1998) revealing at least two pathways that are common to bacterial and mammalian oxidases, designated as the D and K channels due to an aspartate (D132) and a lysine (K362) which are key residues in these proton channels.

Aside from the involvement of different proton paths, one major difference between current models is the extent to which the neutralisation of electron input is evoked as contributing to the energy of proton pumping.

The relationships between the amounts of P_{450}/reductase, the ionic strength of the medium, and the presence of cytochrome b_5 was shown to affect the amount of reactive oxygen species formed in P_{450} systems (ZHUKOV et al. 1989). The interaction between P_{450} and the electron donors (reductase or cytochrome b_5) obviously plays an important role in determining the amount of reactive oxygen species. Thus, the "tightness" of the electron donor coupling to P_{450} will influence the formation of reactive oxygen species in microsomal P_{450} systems at the level of P_{450} and, to a minor extent, at the level of reductase. Since this interaction is affected not only by the relative amounts of P_{450}/electron donor but also by the lipid composition, pathophysiological as well as nutritional aspects will modulate the amount of reactive oxygen species formed (BERNHARDT 1996).

The oxidative burst of phagocytic neutrophil granulocytes is mediated by a multi-component NADPH oxidase regulated by **Rac** (ABO et al. 1991). Both $p67^{phox}$ and p21 activated kinase (PAK) are targets for Cdc42 and Rac1 in neutrophils (Prigmore et al. 1995)., where PAK may mediate stimulus-activated phosphorylation of $p47^{phox}$, an event required for association of cytosolic components with the membrane (KNAUS et al. 1995). Peptides containing residue Ser-328 of $p47^{phox}$ represent good substrates for PAK, this being the site of *in vivo* phosphorylation upon neutrophil activation (BENNA et al. 1994). It would be of interest to determine effects of PAK inhibition on the oxidative burst response of these cells (MANSER and LIM 1999).

SULCINER et al. (1996) proposed a model in which Rac upregulates the production of reactive oxygen species and that an elevated level of this compound subsequently activates NFκB. NFκB is indeed potently activated by reactive oxygen species (BOETTNER and VAN AELST 1999).

The activity of the neutrophil NADPH oxidase is regulated in part by the Rac family of GTPases (BOKOCH 1994). A variety of emerging evidence suggests that a similar role for Rac proteins exists in non-phagocytic cells. Generation of reactive oxygen species in fibroblasts is regulated by Racl (SUNDARESAN et al. 1996), and oxidants mediate mitogenic signalling in Ras-transformed fibroblasts (IRANI et al. 1997). Racl and oxygen radicals play a role in collagenase-1 expression induced by cell shape change (KHERADMAND et al. 1998).

Activation of *human* neutrophils with opsonized particles in the presence of a nontoxic dose of 1-naphthol resulted in inhibition of superoxide anion production but not of the phagocytic activity of the cells ('T HART et al. 1990). The inhibition is not at the level of cellular activation since the N-formyl-met-leu-phe-induced rise of intracellular free calcium was unaffected. The (metabolic) activation of 1-naphthol to 1,4-naphthoquinone by reaction with H_2O_2 from the oxidative burst is a necessary event for the inhibition to occur.

An **aged linked decline** in the cytochrome oxidase activity in *rat* heart mitochondria has been reported by PARADIES et al. (1993, 1994). This decline was attributed to a specific decrease in the cardiolipin content of the inner mitochondrial membrane, due probably to a peroxidative attack of this phospholipid by oxy-radicals which are produced during ageing process (PARADIES et al. 1997). In vitro, *rat* heart mitochondrial membranes exposed to the free radical generating system *tert*-butylhydroperoxide/Cu^{2+} inducing lipid peroxidation established a close correlation between oxidative damage to cardiolipin and alterations in the cytochrome oxidase activity (PARADIES et al. 1998).

Piracetam (2-oxo-pyrrolidine-1-acetamide) stimulated ethanolamine-**plasmalogen** formation, which might be mediated by an increased synthesis or turnover of cytochrome b_5 (WOELK and PEILER-ISHIKAWA 1978). The use of various specific antibodies against NADH cytochrome b_5 reductase indicated the role of microsomal flavoproteins in the supply of reducing equivalents from NADPH or NADP to the cyanide-sensitive factor. cytochrome b_5 seems to be functional as an electron carrier between flavoproteins and the cyanide-sensitive factor. Plasmalogens are quite abundant in certain cells and tissues including the erythrocyte (ENGELMANN et al. 1992), cardiac tissue (GROSS 1984), nervous tissue (MASUZAWA et al. 1989), spermatozoa (LENZI et al. 1996), as well as inflammatory cells (MUELLER et al. 1984). They have been considered to play an important role in diseases such as cancer (F. and C. SNYDER 1975), atherosclerosis (ENGELMANN et al. 1994), and ageing (PÉRICHON et al. 1998) facilitating ion channels in the cardiac sarcolemma (CHEN and GROSS 1994), and as intermediates driving lipid mediator biosynthesis (FONTEH and CHILTON 1992). ENGELMANN et al. (1994) investigated *in vitro* oxidation of low density lipoprotein by 2,2'-azo-bis(2-amidinopropane hydrochloride), which decomposes in aqueous solutions to yield carbon centred radicals, and Cu(II) and found that both α-tocopherol as well as plasmalogen phospholipids were decreased in a parallel manner. Plasmenyl phospholipids were inhibitory on the formation of thiobarbituric acid reactive substances in Cu(II)-induced oxidation (ZOMMARA et al. 1995). When plasmalogens were present in total brain liposomes treated with Fe(III)-ADP and ascorbate, there was a substantial decrease in the total production of thiobarbituric acid reactive substances and decrease in the loss of polyunsaturated fatty acids (SINDELAR et al. 1999).

The biosynthesis of plasmenyl phospholipids is quite complex and entirely separate from the phospholipid biosynthetic pathways of phosphatidyl (1,2-diacyl) glycerophospholipids taking place in the endoplasmic reticulum (LEE 1998). The peroxisome is a central and necessary organelle for ether phospholipid generation (VAN DEN BOSCH et al. 1993).

Nitric oxide (•NO) and its derivative **peroxynitrite** ($ONOO^-$) inhibit mitochondrial respiration by distinct mechanisms. Nanomolar concentrations of •NO specifically inhibit cytochrome oxidase in competition with oxygen, and this inhibition is fully reversible when •NO is removed. The •NO inhibition of cytochrome oxidase may be involved in the cytotoxicity of •NO, and may cause increased oxygen radical production by mitochondria, with may in turn lead to the generation of peroxynitrite. Mitochondrial damage by peroxynitrite may mediate the cytotoxicity of •NO, and may be involved in a variety of pathologies (for review see BROWN 1999). Under turnover conditions, depending on the cytochrome c^{2+} concentration, either the cytochrome a_3^{2+}-NO or the nitrite bound enzyme is formed (SARTI et al. 2000). The predominance of one of the two inhibitory pathways depends on the occupancy of the turnover intermediates. In the dark, the respiration recovers at the rate of NO dissociation ($k' = 0.01\ s^{-1}$ at 37 °C). Illumination of the sample speeds up recovery rate only at high reductant concentrations, indicating that the inhibited species is cytochrome a_3^{2+}-NO. When the reaction occurred with the oxidised binuclear site, light had no effect and NO was oxidised to harmless nitrite eventually released in the bulk, accounting for catalytic NO degradation.

The electron spin resonance, high performance liquid chromatography electron spin resonance, and high performance liquid chromatography electron spin resonance mass spectroscopy analyses showed that cytochrome *c* catalyses formation of pentyl and octanoic acid radicals from 13-hydroperoxide octadecadienoic acid (IWAHASHI et al. 2002). The reaction of 13-hydroperoxide octadecadienoic acid with cytochrome *c* was inhibited by chlorogenic acid, caffeic acid and ferulic acid via two possible mechanisms, i.e. reducing cytochrome *c* (chlorogenic acid and caffeic acid) and scavenging the radical intermediates (chlorogenic acid, caffeic acid and ferulic acid).

6.1.2.2.1 Cytochrome P_{450} in Nasal Nonciliated Columnar Cells

The nasal mucosa, the first tissue of contact for inhaled xenobiotics, possesses substantial xenobiotic-metabolising capacity.

THORNTON-MANNING et al. (1996) confirmed the presence of CYP2A6 in *human* respiratory mucosa. *Human* CYP2A6 metabolises the olfactory-toxic, highly odorous component 3-methylindole to both reactive and nontoxic metabolites (THORNTON-MANNING et al. 1996). The ability of CYP2A6 to accept a number of procarcinogens and toxicants as substrates suggests that this enzyme may play a protective role in nasal mucosa. However, metabolism often results in the formation of reactive metabolites, and CYP2A6 activates a number of compounds, including 6-aminochrysene, *N*-nitrosodiethylamine, *N*-nitrosodimethylamine, and aflatoxin B_1 to toxic or genotoxic metabolites (DAVIES et al. 1989).

In the nasal cavity of the *rat*, the nonciliated columnar cells have an extensive accumulation of smooth endoplasmic reticulum in the apical cytoplasm (MONTEIRO-RIVIERE and POPP 1984, POPP and MONTEIRO-RIVIERE 1985) which in some cases was so dense as to exclude other organelles such as mitochondria. The nonciliated columnar epithelial cell was identified on the surfaces of the conchae and the lateral nasal wall but not identified on the septum.

The accumulation of smooth endoplasmic reticulum suggests that nonciliated columnar cells may be the source of cytochrome P_{450} and P_{450}-associated enzymes that have been previously described in the nasal mucosa (HADLEY and DAHL 1982). Nasal mucosa from *rat* foetuses and neonates metabolises the nasal carcinogen phenacetin (BRITTEBO and ÅHLMAN 1984).

In the *Cynomolgus monkey*, cytochrome P_{450}, NADPH-cytochrome P_{450} reductase and some monooxygenase activities, especially ethoxycoumarin *O*-deethylase activity, were present in respiratory epithelium, although in lower levels than in liver (LONGO et al. 1992).

Comparing *dog*, *rabbit*, *guinea pig*, *Syrian hamster* and *mouse*, the highest nasal microsomal cytochrome P_{450} concentration was found in the *Syrian hamster*, and the lowest in the *dog* (HADLEY and DAHL 1983). As a percentage of liver cytochrome P_{450} concentration, the nasal epithelial membrane P_{450} concentrations ranged from 8 % in the *mouse* to 41 % in the *Syrian hamster*. In general, the nasal epithelial membrane cytochrome P_{450} concentrations were closer to the concentrations found in lung microsomes than in liver microsomes. In *dog* and *rat* the highest microsomal cytochrome P_{450} concentration was found in the ethmoturbinates.

The known nasal carcinogen hexamethylphosphoramide is rapidly *N*-demethylated by the *dog's* nasal microsomes to produce another nasal carcinogen, formaldehyde (DAHL et al. 1982). DAHL and HADLEY (1983) reported 18 substances which are metabolised to formaldehyde by *rat* nasal microsomes and which are frequently deposited in the nose due to their occurrence as air pollutants or their use as drugs, solvents, or essences. With some exceptions, the best substrates for metabolism to formaldehyde by nasal microsomes tend to be those containing *N*-methyl groups and which have some water solubility. The *N*-demethylation of ^{14}C-aminopyrine by the nasal mucosa of C57 Bl *mice* and Sprague-Dawley *rats* was induced by phenobarbital pre-treatment and susceptible to inhibition with metyrapone and SKF 525 A suggesting the presence of a cytochrome P-450-dependent enzyme system in the tissue (BRITTEBO 1982). Immediately after injection of ^{14}C-aminopyrine in rats a uniform distribution of radioactivity in the body was recorded. After 30 min, however, a preferential localisation of radioactivity was found in the nasal mucosa and in the liver. By treatment with metyrapone the uptake of radioactivity in the nasal mucosa and in the liver was blocked suggesting that the observed accumulation of radioactivity was due to metabolites.

Olfactory microsomes from male Fischer 344/N *rats* were 6-fold more active toward hexamethylphosphoramide metabolism than microsomes from *rat* liver and nasal respiratory epithelia and 19-fold more active than microsomes from *rat* lungs (THORNTON-MANNING et al. 1995).

A reiterative administration of chlormethiazole, a specific inhibitor of 2E1 in liver, strongly inhibited many CYPs, including 2E1, 1A2, 2G1, and 2A in the *rat* nasal mucosa, but did not influence expression of 2B or 3A as determined by immunoblotting or catalytic activities (LONGO et al. 2000). The chlormethiazole-mediated inhibition of 1A1 and 2E1 was demonstrated to be at the mRNA level.

6.1.2.2.2
Cytochromes P_{450} in the Lung

In *humans* the CYP3A subfamily consists of at least three members, CYP3A4, CYP3A5, and CYP3A7 (NELSON et al. 1996). These CYP forms are involved in the metabolism of polycyclic aromatic hydrocarbons and other procarcinogens in combustion products, tobacco smoke, and smog (GUENGERICH 1993, NELSON et al. 1996). Using specific antibodies, ANTTILA et al. (1997) in 27 *patients* undergoing surgery to remove tumorous lung lesions localised CYP3A5 in the ciliated and mucous cells of the bronchial wall, bronchial glands, bronchiolar columnar and terminal cuboidal epithelium, type I and type II alveolar epithelium, vascular and capillary endothelium, and alveolar macrophages, whereas CYP3A4 was found in bronchial glands, bronchiolar columnar and terminal epithelium, type II alveolar epithelium, and alveolar macrophages. By reverse-transcriptase-polymerase chain reaction with gene-specific primers, CYP3A7 mRNA was detected in none of the samples.

In the *marmoset* lung CYP1A1 was detected in the connective tissue of interalveolar septa, in the bronchiolar epithelium, in the vascular and bronchiolar smooth muscle cells and in chondrocytes (MÜLLER et al. 1999).

CYP26 expression in normal *human* tracheobronchial epithelial cells was compared with that in *human* lung carcinoma cell lines (KIM et al. 2000). CYP26 mRNA could be induced by the retinoic acid receptor-selective retinoid 4-(5,6,7,8-tetrahydro-5,5,8,8-tetramethyl-2-anthracenyl)-benzoic acid but not by the retinoid X receptor-selective retinoid SR11217 or the anti-activator-protein 1-selective retinoid SR11302. Retinoic acid receptor α-, β, and γ-selective retinoids were able to induce CYP26; this induction was inhibited by the retinoic acid receptorα-selective antagonist Ro41-5253. The induction of CYP26 correlated with increased biotransformation of retinoic acid into 18-hydroxy-, 4-oxo-, and 4-hydroxy-retinoic acid.

In the *rabbit*, microsomes of peripheral lung tissue, airways, small and large vessels, and lysates of alveolar macrophages all expressed proteins of ~50 kDa which cross reacted with a primary antibody raised against *rat* liver cP450 4A1 (ZHU et al. 1998).

In the lungs of male and female *rats* cytochrome P_{450b} was present and its expression in this tissue was independent of phenobarbital treatment (RAMPERSAUD and WALZ Jr (1986). Formation of 1-hexanol in *rat* lung was catalysed by a cytochrome P_{450} isozyme different from the major isozymes induced by either phenobarbital or β-naphthoflavone (TOFTGÌRD et al. 1986). Similarly, formation of 2,5-hexanediol from 2-hexanol was catalysed by a P_{450} isozyme different from cytochrome P_{450}-phenobarbital-B and present in liver but not in lung microsomes. In silica-treated *rats*, pulmonary total activity of cytochromes P_{450} (all isoenzymes) was increased, and so was that of CYP2B1 (MILES et al. 1993).

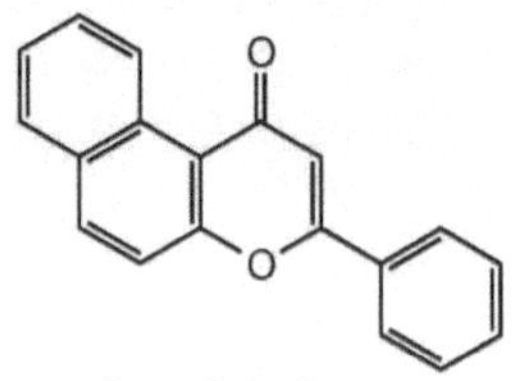

β-Naphthoflavone [42]

Cytochrome P_{450} isozyme 5 is a major component of the *rabbit* pulmonary P_{450} system (SLAUGHTER et al. 1981). CYP2B4 and 4B1 each represent 30–40 % of the total cytochrome P_{450} levels in *rabbit* lung microsomes (DOMIN et al. 1986). Spectral measurements of P_{450}s in microsomes prepared from isolated pulmonary cells of *rabbits* indicated that the highest concentration of cytochrome P_{450} is in Clara cells (AUNE et al. 1985). In male Fischer *rats*, C57BL *mice*, Hartley *guinea pigs* and *monkey*, rates of total and isozyme 5-catalysed metabolism of 2-aminofluorene were grater with pulmonary than with hepatic microsomal preparations from untreated animals (VANDERSLICE et al. 1987).

Cytochrome P_{450} 1A was induced in *hamster* lung by 5 mg 6-nitrochrysene per day for 3 days (CHEN et al. 1998).

6.1.2.2.3
Cytochromes P_{450} in the Liver

A variety of hydrazine derivatives such as the drugs iproniazid, phenelzine (2-phenylethylhydrazine), isoniazid, isocarboxacid, and hydralazine have been associated with liver injury, and a common activation mechanism has been proposed (NELSON et al. 1980, REYNOLDS and MOSLEN 1980). Phenelzine reacted with the prosthetic hem of cytochrome P_{450} and inactivated the enzyme (MUAKKASSAH and

YANG 1981). Support for the theory of alkyl radical transfer was provided by ORTIZ DE MONELLANO et al. (1983) who used the electronic paramagnetic resonance spin-trapping technique to detect the 2-phenylethyl radical formed during microsomal biotransformation of phenelzine. Production of the α-(4-pyridyl 1-oxide)-*N-tert*-butylnitrone/2-phenylethyl radical adduct was dependent on the presence of active microsomes, phenelzine, NADPH (or NADH), and spin trap α-(4-pyridyl 1-oxide)-*N-tert*-butylnitrone (ORTIZ DE MONELLANO et al. 1983). The addition of catalase and superoxide dismutase resulted in a 28.5 and 24 % decrease in radical production, respectively (RUMYANTSEVA et al. 1991). The concentration of the α-(4-pyridyl 1-oxide)-*N-tert*-butylnitrone/2-phenylethyl radical adduct decreased significantly in the presence of metal chelators, i.e. EDTA, diethylenetriaminepentaacetic acid (DTPA), or deferoxamine mesylate.

CYP2E1 (also termed P450j or LM3a) is involved in the bioconversion of ethanol. Incubation of hepatocytes of adult male Sprague-Dawley *rats* with 1 and 10 mM ethanol increased the production of reactive oxygen species by 72 and 151 %, respectively, which was associated with mild decreases in cell viability (BAILEY et al. 1999). Antimycin, a mitochondrial complex III inhibitor, elicited a 17-fold increase in the levels of reactive oxygen species and markedly decreased hepatocyte viability and ATP levels. Ethanol increased reactive oxygen species production and the cytosolic $NADH/NAD^+$ ratio of antimycin-treated cells. Rotenone, a mitochondrial complex I inhibitor that allows electron flow to complex III, significantly increased reactive oxygen species production in antimycin plus ethanol-treated cells. Diphenyliodonium, a mitochondrial complex I inhibitor that inhibits electron flow through flavin mononucleotide, attenuated reactive oxygen species generation in all groups.

A specific hepatic microsomal isoenzyme of cytochrome P_{450} (termed CYP4A1 or P452) with a narrow substrate specificity for ω-hydroxylation of fatty acids was induced by peroxisome proliferators clofibrate (GIBSON et al. 1982, 1990, TAMBURINI et al. 1984, GIBSON 1992) and phthalates.

6.1.2.2.4
Cytochromes in the Testis

The testicular level of xenobiotic metabolising cytochrome P_{450}-dependent monooxygenase activity is regulated by the anterior pituitary gland via the secretion of the gonadotropic hormone, luteinizing hormones (LEE et al. 1980). In the *rat*, a significant decrease in cytochrome P_{450} content occurred after hypophysectomy.

Cytosolic NAD(P)H quinone oxidoreductase is a two-electron reductase that can either bioactivate or detoxify quinones and, thus by the latter function, prevent their participation in redox cycling and oxidative stress (ERNSTER et al. 1962). In hypophysectomized Sprague-Dawley *rat*, a 2.2-fold increase was found in the testicular cytosolic fraction (MEHROTRA et al. 1999).

6.1.2.2.5
Cytochromes in the Nervous System

Cytochromes are responsible for electron transport and oxidative phosphorylation yielding adenosine triphosphate needed for vital processes such as protein synthesis, maintenance of the resting membrane potential, and rapid axoplasmic transport within neurones (OCHS and RANISH 1970, OCHS and HOLLIGSWORTH 1971).

An involvement of CYP2D6 in the pathogenesis of Parkinson's disease has been postulated, although this suggestion has been a matter of controversy (STURMAN and WILLIAMS 1991, TANNER 1991, SMITH et al. 1992).

In the *rabbit* brain, the Nadi (α-naphthol, dimethyl-*p*-phenylenediamine) reaction showed marked activity of cytochrome oxidase in the caudate nucleus, the putamen, the anterior nucleus of the thalamus, the optic tectum, the interpeduncular nucleus, Goll's and Burdach's nuclei, and the inferior olivary nucleus (SHIMIZU et al. 1957). A moderate reaction was seen in the neocortex, hippocampus, dentate gyrus, substantia nigra, cerebellar cortex and the nuclei of the cranial nerves.

Using diaminobenzidine, in the squirrel *monkey* the levels of activity of cytochrome oxidase differed from one region of the brain to another (WONG-RILEY 1976). The reactive neurones belonged to the extrapyramidal motor system (globus pallidus, substantia nigra [pars diffusa]. nucleus ruber, pontine and mesencephalic reticular formation), to certain cranial nerve nuclei (oculomotor, trochlear, abducens, mesencephalic trigeminal, vestibular) and to relay nuclei of the auditory pathway (superior olivary nucleus and nucleus of the trapezoid body). Under the electron microscope, reaction products were localised prominently along the cristae and limiting membranes of the mitochondria and within peroxisome-like bodies of the stained neurones. Non-reactive neurones had little or no reaction products in their organelles.

Cerebral blood vessels are a potentially important site of drug metabolism and P_{450} in the vessels may influence what substances gain access to the brain (WARNER et al. 1998). Number of CNS-active drugs are also good inducers of cytochrome P_{450}.

6.1.2.2.6
Cytochromes P_{450} and P_{448} in the Skin

Cytochrome P_{450} was first shown to be present in *rat* skin microsomes following the skin application of polychlorinated biphenyls (Bickers et al. 1974). Skin application of 1,1,1-trichloro-2,2-bis(*p*-chlorophenyl)ethane (DDT) caused a twofold induction of hepatic cytochrome P_{450} whereas polychlorinated biphenyls evoked a greater than three-fold induction of cytochrome P_{448}. The primary inducibility of cytochrome P_{448} in the skin is of considerable interest since the metabolic activation of polycyclic hydrocarbon carcinogens appears to occur best with cytochrome P_{448}-dependent enzymatic activity (Bickers 1980, Ioannides and Parke 1987). Chemicals which interact with cytochromes P_{448}, such as 7-ethoxyresorufin, 9-hydroxyellipticine and 3-methylcholanthrene, are essentially rigid, planar molecules, whereas compounds that interact with the phenobarbitone-cytochromes P_{450}, but elicit no interaction with cytochromes P_{448}, are non-planar, bulky molecules (Lewis et al. 1986).

6.1.2.2.7
Cytochrome b_{558} in the Carotid Body

Using a monoclonal antibody against the large cytochrome b_{558} subunit, gp91phox, and other antibodies serving as neural (PGP 9.5) and monocyte/macrophage markers (ED1, ED2), Dvorakova et al. (2000) demonstrated at light and electron microscope levels that monocytes/macrophages are abundantly present in the rat carotid body and represent the major source of cytochrome b_{558} in this organ.

6.1.2.2.8
Cytochrome b_{558} in the Neutrophil

In neutrophils, cytochrome b_{558} is located in the plasmalemma and in the membrane of specific granules at a 3/7 or lower ratio (Segal and Jones 1979, Borregaard et al. 1983). Hence, in the resting neutrophil cytochrome b_{558} is mainly located in intracellular membranes. After activation, these granules fuse with the plasma membrane, transferring the cytochrome b_{558} to the cell surface.

Cytochrome b_{558} from *human* neutrophil plasma membranes directly interacted with $p47^{PHOX}$ (Dang et al. 2001). This interaction increased when the proteins were incubated in the presence of Rac1-guanosine triphosphate/guanosine diphosphate.

6.1.2.2.9
Cytochrome *c*-Derived Microperoxidase 8

Microperoxidase 8 is a heme octapeptide, obtained by enzymatic hydrolysis of heart Cytochrome *c*, in which a histidine is axially co-ordinated to the heme iron, and acts as a fifth ligand. It exhibits two kinds of activities: a peroxidase-like activity and a cytochrome P_{450}-like activity. Ricoux et al. (2000) have shown that microperoxidase 8 is not only able to oxidise various aliphatic and aromatic hydroxylamines with the formation of microperoxidase 8-Fe(II)-nitrosoalkane or -arene complexes absorbing around 414 nm, but also that these complexes can be obtained by reduction of nitroalkanes.

6.1.2.3
NADH Dehydrogenase

NADH dehydrogenase, by other name NADH cytochrome *c* reductase (EC 1.6.99.3; formerly EC 1.6.2.1), a flavoprotein containing iron-sulphur centres, was detected in pituicytes of the *rat* (Bock and Goslar 1969) using 2,2',5,5'-tetra-*p*-nitrophenyl-3,3'-(3,3'-dimethoxy-4,4'-biphenylene)-ditetrazolium chloride according to Rosa and Tsou (1961).

In vascular cells, a major source of oxygen radical production is a membrane-bound, flavin-containing NADH/NADPH-dependent oxidase, which is regulated *in vitro* and *in vivo* by angiotensin II. Homogenates of the aorta of *rabbits* treated with nitro-glycerine patches (1.5 μg/kg/min × 3 d) displayed activities of 67 ± 12 vs. 28 ± 2 nmol $O_2^{\bullet -}$/min × mg protein vs. aortic homogenates from untreated animals (Münzel 1997). Acute addition of hydralazine (10 μM) to nitro-glycerine-tolerant vessels immediately inhibited $O_2^{\bullet -}$ production and NADH-oxidase activity in vascular homogenates. The lucigenin-enhanced chemiluminescence signal was inhibited by a recombinant heparin-binding superoxide dismutase demonstrating the specificity of this assay for $O_2^{\bullet -}$.

6.1.2.4
NADPH Dehydrogenase

NADPH dehydrogenase (EC 1.6.99.6) catalyses the univalent reduction of O_2 to $O_2^{\bullet -}$ according to the reaction:

$$2\,O_2 + NADPH \longrightarrow 2\,O_2^{\bullet -} + NADP^+ + H^+ \qquad [43]$$

NADPH dehydrogenase has also been shown to catalyse the divalent reduction of O_2 to H_2O_2 under certain experimental conditions (Green and Wu 1986, Green and Pratt 1987). The K_m value for

NADPH of NADPH dehydrogenase amounts to 30–80 μM, that for NADH to 0.4–0.9 μM, and that for oxygen to about 10 μM (BABIOR et al. 1975, 1976, GABIG and BABIOR 1979, COHEN et al. 1980, LEW et al. 1981, CHAUDHRY et al. 1982, WAKEYAMA et al. 1982, YAMAGUCHI et al. 1983, SUZUKI et al. 1985, TAMURA et al. 1988).

NADPH dehydrogenase is an enzyme system, which apparently consists of multiple components localised in the plasmalemma, in specific granules, and in the cytosol of phagocytes. A number of components has been suggested to be involved in the redox chain of NADPH dehydrogenase, including flavine adenine dinucleotide (FAD), quinones, and a phagocyte-specific cytochrome referred as cytochrome b_{-245}.

There is substantial experimental evidence for the assumption that cytochrome b_{-245} is the terminal redox component of NADPH dehydrogenase. Cyotochrome b_{-245} has been identified in various types of phagocytes including neutrophils, eosinophils, HL-60 cells, and mononuclear cells (SEGAL et al. 1981).

6.1.2.5 Galactose Oxidase

Galactose oxidase (EC 1.1.3.9)

$$R\text{-}CH_2OH + O_2 \longrightarrow R\text{-}CHO + H_2O_2 \quad [44]$$

Galactose oxidase (EC 1.1.3.9) is one of a group of copper-containing oxidative enzymes that includes monoamine oxidase, laccase (EC 1.10.3.2), ascorbate oxidase (EC 1.10.3.3), tyrosinase and dopamine monohydroxylase. The enzyme consists of a single polypeptide chain of 639 amino acid residues (MCPHERSON et al. 1992) with molecular mass of 68 kDa. The reaction catalysed by galactose oxidase is the oxidation of primary alcohols (e.g. the hydroxyl group at the C6 position of D-galactose) to aldehydes, accompanied by the reduction of the molecular oxygen to hydrogen peroxide:

Specificity for the alcohol substrate is very broad, ranging from small molecules (e.g. propanediol) to polysaccharides. Galactose oxidase is strictly stereo specific. It does not oxidise either D-glucose or L-galactose.

A covalent linkage between Tyr272 and Cys228 has been observed, whose functional role may relate to the presence of a tyrosine free radical at Tyr272. The tyrosine free radical could be stabilised by delocalization to Cys228 and stacking interactions with Tryp290 (ITO et al. 1994).

The formally two-electron redox reactions, the alcohol oxidation and the O_2 reduction, are performed at a molecular Cu(II) centre, where the bound tyrosyl radical (Tyr272$^\bullet$) acts as the additional one-electron redox site. Thus, the active species (fully oxidised state) of the enzyme is the Cu(II)-phenoxyl radical complex that is transformed into a Cu(I)-phenol species (fully reduced state) in the alcohol oxidation via the following mechanism:

- deprotonation from the -OH group of the bound substrate,
- inner-sphere electron transfer from the deprotonated substrate to Cu(II),
- α-hydrogen atom abstraction of the resulting ketyl radical by the phenoxyl radical of Tyr272 (the ordering of electron transfer and hydrogen atom abstraction steps could be reversed.

Hydrogen atom abstraction by Cu(II)- and Zn(II)-phenoxyl radical complexes was used as a model for the active form of galactose oxidase (TAKI et al. 2000).

6.1.2.6 Xanthine Oxidase

The molybdoenzyme, xanthine oxidoreductase is a homodimer of 150 kDa subunits and exists in two interconvertible forms, dehydrogenase (EC 1.1.1.204) and xanthine oxidase (EC 1.1.3.22). Reduction of oxygen by either form of the enzyme yields superoxide radical anion and hydrogen peroxide with xanthine or hypoxanthine as substrates. Xanthine dehydrogenase preferentially reduces NAD^+, whereas xanthine oxidase does not reduce NAD^+, preferring molecular oxygen.

The molybdenum in these enzymes is bound by a special organic pterin cofactor, and is not held directly by the sidechains of proteins. The pterin cofactor actually is a dithiolate complex. The molybdenum in the enzymes is not re-oxidised directly by molecular dioxygen and the ancillary flavin and Fe/S centres have to do with the way in which dioxygen is activated; oxygen transfer by molybdenum enzymes is of oxygen atoms from water and not from dioxygen.

Xanthine dehydrogenase from *Eubacterium barkeri* has a mass of 530 kDa and three types of subunits. It contains molybdopterin as the molybdenum-complexing cofactor and 1 mol of selenium in a non-selenocysteine form per mol of native enzyme (SCHRÄDER et al. 1999).

Xanthine oxidase from buttermilk produced $O_2^{\bullet-}$ and H_2O_2 in a ratio of 25:75, while a microbial xanthine oxidase was unable to produce any $O_2^{\bullet-}$ (WIPPICH et al. 2001).

Xanthine oxidoreductase is asymmetrically localised not only in the cytoplasm but also on the outer surface of *human* endothelial (umbilical vein endothelial cells and EA-hy-926 permanent endothelial cell line) and epithelial (HB4a, a mammary epithelial cell line conditionally immortalised by transfection with SV40 virus) cells *in culture* (ROUQUETTE et al. 1998).

Pulse radiolytic investigations of the hypoxanthine-xanthine-uric acid system 0.8 μs after the generation of electron pulses (800 KeV, 4 ns) showed transient species produced by action of $HO^•$ radical on hypoxanthine (SANTAMARIA et al. 1984). The rate of formation of the transient depends on hypoxanthine concentration. The radical decay leads directly to the formation of xanthine. Another radical was produced by oxidation of xanthine by $HO^•$ in less than 1.6 μs. Such a reaction was xanthine concentration dependent. The decay of this radical after ca. 400 μs did not lead directly to the formation of uric acid. SANTAMARIA et al. (1984) suppose that its disproportionation occurs through another transient, which could be a dimer.

Not only $O_2^{•-}$ and $HO^•$, but also alkyl or alkoxyl radicals ($R^•$) were formed when saccharides such as glucose, fructose and sucrose were added into the xanthine oxidase/hypoxanthine system containing iron (LUO et al. 2001). The generated amount of $R^•$ depended on the kind and concentration of saccharides added into this system. In the absence of saccharides no $R^•$ were detected, indication that there is an interaction between the saccharide molecules and the free radicals generated from the iron containing hypoxanthine/xanthine oxidase system.

Xanthine oxidase is an ubiquitous O_2 metabolite-producing enzyme which has been implicated in **ischaemia-reperfusion** (MCCORD 1985), **hypoxia reoxygenation** (TERADA et al. 1992) and **remote organ** (TERADA et al. 1992) **injury**. The oxidase form is derived from posttranslational modification of xanthine dehydrogenase (AMAYA et al. 1990), and accounts for about 10 % of the total enzymatic activity (ENGERSON et al. 1987, SAKSELA and RAIVIO 1996).

Relevant to the proposed role of xanthine oxidase in postischemic tissue injury, hypoxia was found to induce a gradual tissue increase in xanthine oxidase-xanthine dehydrogenase activity over 16–72 h in cultured endothelial cells (TERADA et al. 1992, HASSOUN et al. 1994, POSS et al. 1996). In a more rapid fashion, hyperoxia decreases xanthine oxidase-xanthine dehydrogenase activity in both cultured cells and *rat* lungs (TERADA et al. 1988, PANUS et al. 1992). The control of xanthine oxidase-xanthine dehydrogenase levels by oxygen tension is a complex process involving pre- and posttranslational points of regulation (TERADA et al. 1997).

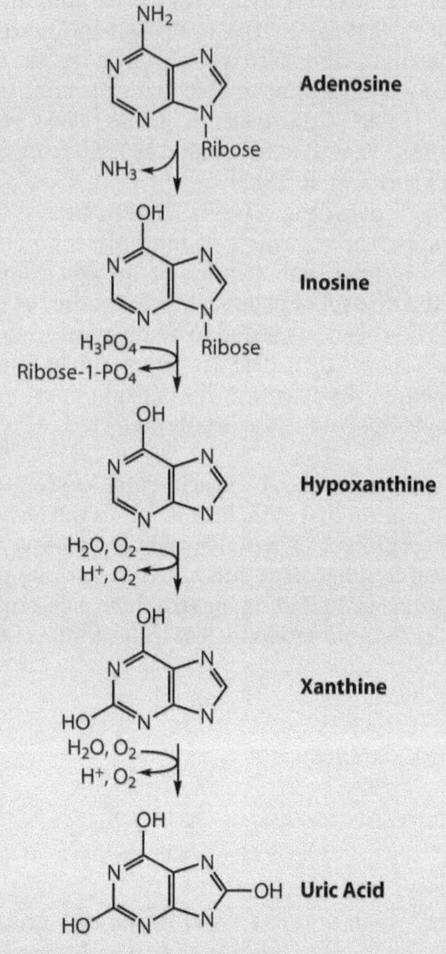

Substrates of xanthine oxidase [45]

Sustained elevation of $[Ca^{2+}]_i$ in ischaemia can promote the conversion of xanthine dehydrogenase to xanthine oxidase (MCCORD 1985). Under low energy conditions, where large parts of ATP are converted to hypoxanthine, this may result in a massive generation of reactive oxygen species. The activation of xanthine oxidase has been implicated for ischaemic neuronal death *in vivo* (COYLE and PUTTFARCKEN 1993) and in kainate toxicity to cerebellar granule cells *in vitro* (DYKENS et al. 1987).

In adult respiratory distress syndrome plasma xanthine oxidase activity (as measured by monitoring the rate of conversion of ^{14}C-hypoxanthine to ^{14}C-uric acid) was ≥75 mIU/l in 8 of 15 patients, but in none of the control patients (RAGSDALE et al. 1986).

Intravenous administration of xanthine (0.225 mg/kg) plus xanthine oxidase (3.0 units/kg) to anaesthetised *rats* resulted in a rapid fall in the arterial pressure and a mortality rate of over 80 % during 120 min observation period (JACINTO and

JANDHYALA 1992). Pre-treatment of the *rats* with superoxide dismutase alone or combined with catalase enhanced survival rate to 60 % confirming that the toxicity after xanthine plus xanthine oxidase administration is due to the generation of oxygen free radicals. Pre-treatment of the rats with either felodipine, a dihydropyridine calcium antagonist or verapamil, a structurally different Ca^{2+}-channel blocker was most effective in promoting survival rat to 90 %; in contrast hydralazine, an arteriolar dilator but not a calcium antagonist, was ineffective in significantly enhancing survival.

In the absence of oxygen, nitro compounds or other bioreductive drugs act as **alternative electron acceptors**. Nitric oxide synthases (NOS) are cytochrome P_{450}-related enzymes that convert arginine to $^{\bullet}NO$ and citrulline. Two constitutive forms are expressed in neurones (nNOS) or endothelial cell (eNOS), respectively, and are activated by Ca^{2+}/calmodulin. Other isoforms (iNOS) in microglia or astroglia are inducible by a variety of stimuli, such as cytokines, and function at basal Ca^{2+} concentrations. Activation of bNOS following Ca^{2+} entry through the *N*-methyl-D-aspartate receptor (NMDA-R) has been implicated in excitotoxicity to cortical neuronal cultures (DAWSON et al. 1991) and in ischaemia due to middle cerebral artery occlusion (HUANG et al. 1994). The possible terminal cytotoxic mediator may be peroxynitrite formed from $^{\bullet}NO$ and $O_2^{\bullet -}$.

L-Arginine–nitric oxide pathway

L-Arginine $\xrightarrow{\text{NOS}}$ **Nitric oxide**($^{\bullet}$NO) $\xrightarrow{+O_2^{\bullet -}}$ Peroxynitrite ($ONOO^-$)

[46]

N-Methyl-D-aspartate receptor activation in perfused, ventilated *rat* lungs triggered acute injury, marked by increased pressures needed to ventilate and perfuse the lung, and by high-permeability oedema (SAID et al. 1996). The injury was prevented by competitive NMDA receptor antagonists or by channel blocker MK-801, and was reduced in the presence of Mg^{2+}. As with NMDA toxicity to central neurones, the lung injury was $^{\bullet}NO$ dependent: it required L-arginine, was associated with the increased production of $^{\bullet}NO$, and was attenuated by either of two NO synthase inhibitors, N^{ω}-nitro-L-arginine methyl ester and L-nitroarginine.

CLARKE et al. (1982) measured the kinetics of reduction of nitroimidazoles by xanthine/xanthine oxidase and estimated the initial rates at a concentration probably rather less than the Michaelis K_m value. The redox dependence of this flavoenzyme was quite similar to that found with free flavins.

Incubation of A 10 cells, a *murine* vascular smooth muscle cell line, with xanthine oxidase and purine resulted in an enhancement of **adenylate cyclase activation** (TAN et al. 1995). The effect of purine and xanthine oxidase was not blocked by coincubation with superoxide dismutase (which catalyses the conversion from superoxide anion to H_2O_2). This suggests that the generation of the superoxide anion is not involved in the mechanism of enhancement of adenylate cyclase activation.

Xanthine oxidase increased the rate of **actin polymerisation** and accelerated the conversion of F(ATP)actin into F(ADP,P_i)actin (LANZARA et al. 1988).

Exposure of cultivated *human* glial cells (U-787CG, U-622CG, U-1508CG) to a hypoxanthine-xanthine oxidase reactive oxygen metabolites (ROM) generating-system showed acute and irreversible injury (THAW et al. 1983): increased formation of endocytotic vacuoles, enhanced membrane ruffling activity, followed by cellular contraction and nuclear as well as cytoplasmic condensation. Subcellular alterations consisted of damage to all organelles (endoplasmic reticulum, mitochondria, nuclei) and a highly increased rate of autophagic activity.

Non-competitive **inhibition** was induced by metal ions (MONDAL et al. 2000). The inhibition constant for Cu^{2+} and Hg^{2+} are in the micromolar and that for Ag^+ is in the nanomolar range. pH dependence studies of the inhibition indicated that at least two ionisable groups of xanthine oxidase are involved in the binding of these metal ions. Xanthine oxidase formed optically observable complexes with Cu^{2+} ion (SAU et al. 2001). The pH dependence studies of the formation of Cu^{2+}-xanthine oxidase by optical spectroscopy and circular dichroism showed that at least one ionizable group may be responsible for the formation of the complex. EPR studies showed that Cu^{2+} ion binds to xanthine oxidase with sulphur and nitrogenous ligands. A transient kinetic study of the interaction of Cu^{2+} with xanthine oxidase showed the existence of two Cu^2 bound xanthine oxidase complexes formed at two different time scales of the interaction, one at ≤ 5 ms and the other one at around 20 s. The complex formed at longer time scale may be responsible for the inhibition of the enzyme activity.

A marked dose-dependent inhibition of xanthine oxidase activity by rutin (200–300 μM) and naringin (200–400 μM) was shown in the xanthin-xanthine oxidase system, while the addition of catechin (200–400 μM) exhibited a lower effect on uric acid formation (RUSSO et al. 2000).

6-Formylpterin, a mixed-type xanthine oxidase inhibitor (OETTL and REIBNEGGER 1999), reacted with NAD(P)H and consumed oxygen catalysing the conversion of NADH to NAD (Arai et al. 2001). Electroparamagnetic resonance spin trapping experiments demonstrated that this reaction is accompanied with the generation of reactive oxygen species, superoxide anion and hydrogen peroxide. When 6-formylpterin was administered to HL-60 cells, intracellular generation of reactive species was observed and apoptosis was induced.

Esculetin (IC_{50} = 20.91 μM), umbelliferone (7-hydroxycoumarin; IC_{50} = 43.65 μM), and 7-hydroxy-4-methylcoumarin (IC_{50} = 96.70 μM) are strong xanthine oxidase inhibitors (Chang and Chiang 1995). The structure of 7-hydroxycoumarin plays a very important role in xanthine oxidase inhibition. The 6-hydroxy group present in the molecule of 7-hydroxycoumarin, e.g. esculetin enhanced the activity, whereas substitution by the 6-methoxy group, e.g. scopoletin, reduced the inhibitory effect.

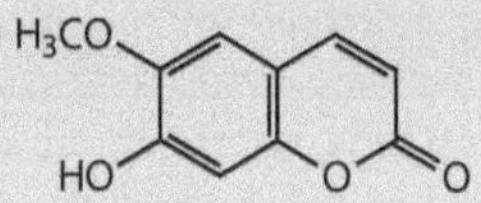

Scopoletin (6-methoxy-7-hydroxycoumarin) [47]

Esculetin (100 μM for 24 h) reduced the level of cyclin-dependent kinase (CDK) 4, up-regulated p27, and down-regulated cyclin D1 in the cell cycle of *human* leukaemia HL-60 cells (Wang et al. 2002).

Protein kinase C ζ was activated by both interleukin-3 and insulin (Mendez et al. 2001). The insulin response required the presence of IRS-1 and was inhibited by the addition of wortmannin.

Quercetin, 3'-methylquercetin, quercetin-4'-glucuronide, and quercetin-3'-glucuronide inhibited xanthine oxidase at low micromolar concentrations whereas quercetin-3-sulphate, quercetin-3-glucuronide, and quercetin-7-glucuronide had 50–800-fold higher inhibitory constants (Day et al. 2000). Quercetin-4'-glucuronide was also as effective at scavenging superoxide radicals generated by xanthine oxidase as quercetin aglycone at the same concentration. Quercetin (5 mg/kg i.p.) protected lung tissues of Swiss albino female *mice* from sulphur mustard (1 LC_{50} = 42.3 mg/m^3 for 1 h duration) intoxication (Kumar et al. 2001).

On the other hand, quercetin in common with (–)-epicatechin (Chrisey et al. 1988) and 4'-(9-acridinylamino)methanesulphon-*m*-anisidine (Wong et al. 1984), can reduce oxygen to produce superoxide without the involvement of a thiol or other reducing agent (Fazal et al. 1990). The hydroxyl radical can be produced from quercetin and Cu(II) by alternative pathways. The strand scission of DNA was shown to occur under conditions in which Cu(II), quercetin and either hydrogen peroxide or oxygen were present and superoxide was not a necessary intermediate. Strand scission involved the hydroxyl radical and a radical DNA intermediate.

6.1.2.7
L-4,5-Dihydroorotate:Oxygen Oxidoreductase

Production of superoxide radical anion ($O_2^{\bullet-}$) during oxidation of dihydroorotate in *rat* liver mitochondria was not affected by antimycin A, thenoyltrifluoroacetone, or added ubiquinone but was inhibited by orotate, a product inhibitor of dihydroorotate dehydrogenase (Forman and Kennedy 1975). It appears likely that superoxide is generated at the primary dehydrogenase. Dihydroorotate dehydrogenase differs from succinate dehydrogenase both in its utilisation of ubiquinone and in the mechanism of cytochrome *b* reduction. Formation of orotate is only partially inhibited by thenoyltrifluoroacetone and the inhibitor does not prevent the reduction of cytochrome *by* dihydroorotate.

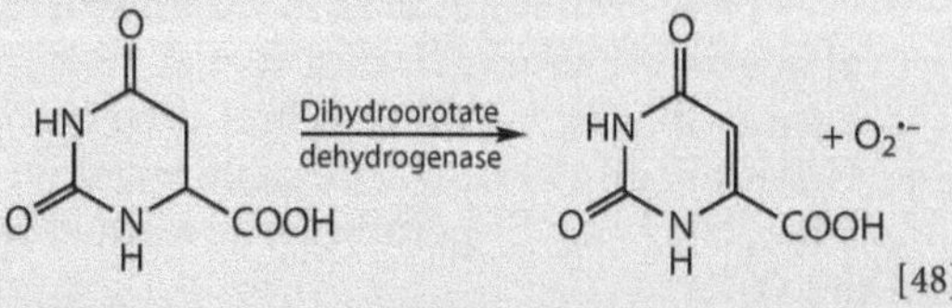

[48]

6.1.2.8
Cucumber Monodehydroascorbate Reductase

Cucumber monodehydroascorbate reductase (EC 1.6.5.4) is capable of reducing phenoxyl radicals which are generated by horseradish peroxidase with H_2O_2 (Sakihama et al. 2000). The addition of monodehydroascorbate reductase plus NADH suppressed the horseradish peroxidase/H_2O_2-dependent oxidation of quercetin, accompanied by the oxidation of NADH. ESR confirmed the quenching of the quercetin radical by monodehydroascorbate reductase plus NADH. Monodehydroascorbate reductase with NADH also suppressed the horseradish peroxidase/H_2O_2-dependent oxidation of hydroxycinnamates, including ferulic acid, coniferyl alcohol, and chlorogenic acid.

6.1.2.9
Prostaglandin H Synthase

Prostaglandin H synthase (EC 1.14.99.1), also called cyclooxygenase, is the rate-limiting enzyme in the conversion of arachidonic acid to prostanoids. To date, two cyclooxygenase isoforms (COX-1 and COX-2) have been cloned; they share over 60% identity at the amino acid level and have similar enzymatic activities (Taketo 1998). COX-1 is constitutively expressed in most tissues and is considered as a housekeeping gene involved in cytoprotection of the stomach, vasodilatation in the kidney, and control of platelet aggregation. COX-2 is an immediate-early gene that can be induced by various stimuli such as hormones, cytokines, growth factors, and tumour promoters. Increased expression of COX-2 has been associated with inflammatory processes and carcinogenesis (Williams et al. 1999, Van Rees et al. 2002).

NO activates COX-1 and COX-2 (Corbett et al. 1993, Salvemini et al. 1993),

Non-steroidal anti-inflammatory drugs inhibit cyclooxygenases and thus the conversion of arachidonic acid to prostanoids.

6.1.3 Modulation of the Respiratory Burst

6.1.3.1 Activation

6.1.3.1.1 Protein Kinase C

Protein kinase C represents a family of more than 11 phospholipid-dependent serine/threonine kinases that are involved in a variety of pathways that regulate cell growth, death, and stress responsiveness.

Protein kinases C transduce the of signals mediated by phospholipid hydrolysis. Activation of G protein coupled receptors, tyrosine kinase receptors, and nonreceptor tyrosine kinases can activate protein kinase C, by stimulation of either phospholipases C to yield diacylglycerol or phospholipase D to yield phosphatidic acid and then diacylglycerol. Phospholipase D has been implicated in the process of generation of reactive oxygen species by neutrophil granulocytes (for review see Exton 2002, p. 53).

The isoenzymes of protein kinase C fall into three subclasses according to their dependence on Ca^{2+} for activation. The Ca^{2+}-dependent group includes PKC subtypes α, β_I, β_{II}, and γ, and the Ca^{2+}-independent group includes PKC subtypes δ, ε, ε', η, and ϑ. The regulation of the atypical subtypes ζ, ι, and λ has not been clearly established, although their activities are stimulated by phosphatidylserine. Protein kinase μ (*human*) and its *murine* homologue, protein kinase D, form a distinct class in that the kinase core is actually most similar to that of the calmodulin-dependent kinases and no pseudosubstrate motif has been identified (Newton 1997). Immunocytochemical analysis has clearly established that different protein kinase C isoenzymes localise to different subcellular compartments (Jaken 1996).

Rat aortic smooth muscle cells cultured with 20 mmol/l glucose showed statistically significant increases in protein kinase C activities, the expression of PKC-β_{II} isoform and platelet derived growth factor-β receptor protein, and proliferation activities, compared with smooth muscle cells cultured with 5.5 mmol/l glucose (Nakamura et al. 2001). Although epalrestat and LY333531 inhibited protein kinase C activation induced by glucose in the same degree, the effect of epalrestat on proliferation activities and expression of the platelet derived growth factor-β receptor were more prominent than those of LY333531. Epalrestat improved the glucose-induced decrease in free cytosolic NAD^+: NADPH ratio and reduced glutathione content, but LY333531 did not. The increased expression of membranous PKC-β_{II} isoform was normalised by epalrestat.

The atypical protein kinase ζ is ubiquitously expressed. It is considered atypical in that it is not activated by diacylglycerol or pre-treatment with phorbol esters. In addition, the protein does not undergo translocation from cytosol to membrane when activated. Phosphatidylinositol 3,4,5-triphosphate has been shown to result in a large stimulation of protein kinase ζ (Nakanishi et al. 1993).

The protein kinase C molecule contains two domains: the amino-terminal regulatory domain interacts with calcium, phosphatidylserine, and diacylglycerol, and, the carboxyl-terminal catalytic domain with ATP and protein substrate-binding sites. The bisindolylmaleimide GF 109203X is a competitive inhibitor of the ATP site (Toulec et al. 1991). The regulatory domain, however, was activated through the concerted action of diacylglycerol, calcium and phosphatidylserine. When cells are activated there is usually an increase in calcium, which may contribute to the activation of protein kinase C perhaps by increasing the binding affinity of the enzyme for diacylglycerol (Kojima et al. 1985, Dougherty and Niedel 1986). In intact cells the enzyme can be activated either by adding a monoacyl-derivative of diacylglycerol such as 1-oleoyl-2-acetylglycerol or by means of phorbol esters (Nishizuka 1984, 1986, Ashendel 1985). PMA showed a positive action that initially activated PKC, then a negative action that initiated the degradation of the enzyme during sustained activation (Nishizuka 1988).

Ischaemic preconditioning of the *rabbit* heart caused selective translocation of the ε and η isoforms without demonstrable changes in total myocardial protein kinase C activity, implying that measurements of total protein kinase C activity are not sufficiently sensitive to detect the involvement of protein kinase C in preconditioning (PING et al. 1997).

The expression of protein kinase C δ gene in the *murine* peritoneal macrophages was inhibited by nitric oxide (JUN et al. 1994).

The involvement of protein kinase C in the enhancement of malignant cell transformation induced by 2,3,7,8-tetrachlorodibenzo-*p*-dioxin was studied by WÖLFLE and MARQUARDT (1996). The protein kinase inhibitor H-7 markedly reduced the *in vitro* promoting activity of 12-*O*-tetradecanoylphorbol-13-acetate but did not affect the promotion by 2,3,7,8-tetrachlorodibenzo-*p*-dioxin. In accord with these results, 12-*O*-tetradecanoylphorbol-13-acetate, but not 2,3,7,8-tetrachlorodibenzo-*p*-dioxin, enhanced the protein kinase C activity in C3H/M2 fibroblasts. Since 12-*O*-tetradecanoylphorbol-13-acetate-mediated activation of protein kinase C was not affected by ascorbic acid plus α-tocopherol, that the antioxidants interfere with tumour promotion at a step beyond protein kinase C activation.

Protein Kinase C Activators

Phorbol Esters

The tumour-promoting phorbol esters, of which 12-*O*-tetradecanoylphorbol-13-acetate is the most potent, are non-physiological activators of protein kinase C. The same holds for Aplysiatoxin and Teleocidin.

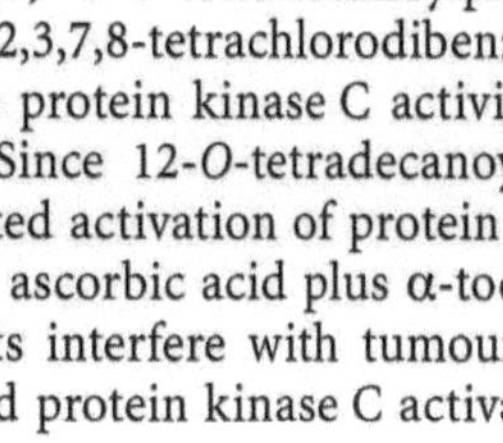

12-*O*-Tetradecanoylphorbol-13-acetate [49]

Phorbol ester-induced ornithine decarboxylase (EC 4.1.1.17) activity associated with tumour promotion (O'BRIEN 1976) was inhibited by deguelin, tephrosin, (-)-13α-hydroxytephrosin, and (-)-13α-hydroxydeguelin, four rotenoids obtained from the African legume *Mundulea sericea* (GERHÄUSER et al. 1995). 17 nmole of retinoic acid, when applied 1 h before treatment of *mouse* skin with 5 nmole of 12-*O*-tetradecanoylphorbol-13-acetate, inhibited by about 90 % the induction of ornithine decarboxylase activity (VERMA 1982).

Treatment of *human* HL-60 cells with 12-*O*-tetradecanoylphorbol-13-acetate appeared to induce the iron regulatory protein-1 activity up to the threefold (EISENSTEIN et al. 1993).

After 12-*O*-tetradecanoylphorbol-13-acetate treatment, a highly metastatic variant of a *human* colorectal carcinoma cell line was three times more invasive in Matrigel than the parental cell line (KOMADA et al. 1993).

BOLES et al. (2000) demonstrated that 12-*O*-tetradecanoylphorbol-13-acetate enhanced the adherence of U-937 cells to fibronectin matrices by increasing the expression of both the α_5- and β_1-subunit mRNAs and the surface expression of the protein. Modulation of $\alpha_5\beta_1$ expression may be important for regulation of monocytic cell function in lung inflammation after injury.

As to the multivesicular bodies of U-937 GTB cells, incubation with 12-*O*-tetradecanoylphorbol-13-acetate (10^{-7} M) caused pronounced changes (FORSBECK et al. 1988, Nilsson et al. 1989). 15 min after stimulation, the relative section area of the multivesicular bodies (23.5 ± 1.4 versus 12.5 ± 1.1 in the controls) as well as the number of inclusion vesicles in multivesicular bodies per cell profile (29.8 ± 2.0 versus 7.0 ± 0.6 in the controls) were significantly ($P < 0.001$) increased. After 60 min the multivesicular bodies appeared to be completely filled with inclusion vesicles.

β-(1→4)-Galactosyltransferase I (EC 2.4.1.38) activity was higher (> ×2) in 12-*O*-tetradecanoylphorbol-13-acetate-treated compared with untreated HL-60 leukaemia cells (PASQUALETTO et al. 2000).

The rate of superoxide generation of *guinea pig* intraperitoneal neutrophils by 12-*O*-tetradecanoylphorbol-13-acetate was increased by 2-bromo-2-chloro-1,1,1-trifluoroethane (halothane) (TSUCHIYA et al. 1988). This increase was inhibited by 1-(5-isoquinolinesulphonyl)methylpiperazine dihydrochloride, a specific inhibitor of Ca^{2+}- and phospholipid-dependent protein kinase C. Halothane was found to activate partially purified protein kinase C significantly.

Iron deprivation induced by desferrioxamine blocked 12-*O*-tetradecanoylphorbol-13-acetate-induced differentiation and induced S-phase arrest and apoptosis in up to 60 % of HL-60 cells (GAZITT et al. 2001). Iron is required for transcription of p21(WAF1/CIP1) in cells induced by 12-*O*-tetradecanoylphorbol-13-acetate.

The binding of [20-^{3}H]phorbol 12,13-dibutyrate to peritoneal macrophages was similar in both normal and BCG-infected *mice* irrespective of their ability to produce H_2O_2 in response to phorbol diesters (WEINBERG and MISUKONIS 1983). Phorbol 12,13-dibutyrate (30 nm) stimulated cytotoxic activity of *human* blood lymphocytes against K562 and Daudi target cells (RAMOS et al. 1983).

In isolated *porcine* leucocytes, 12-*O*-tetradecanoylphorbol-13-acetate produced a concentration-related luminol-enhanced chemiluminescence, that was markedly ($P < 0.05$) suppressed when the nitric oxide donor, pirsidomine was given intravenously 10 min before (WAINWRIGHT and MARTORANA 1993).

The interaction of $^{\bullet}NO$ and $O_2^{\bullet -}$ generated from 12-*O*-tetradecanoylphorbol-13-acetate-stimulated polymorphonuclear leucocytes was studies by a nitroxide spin trap, DMPO (formula [8]). ZHAO et al. (1996) found that addition of L-arginine to the system would significantly decrease the trapped $O_2^{\bullet -}$ by DMPO and addition of N^G-monomethyl-L-arginine would significantly increase the trapped $O_2^{\bullet -}$ by DMPO. It was proven that the formation of $ONOO^-$ by the reaction of $^{\bullet}NO$ and $O_2^{\bullet -}$ was the main reason for the decrease of the trapped $O_2^{\bullet -}$ in the experiment with xanthine/xanthine oxidase and irradiation of riboflavin systems. The generation dynamic of $^{\bullet}NO$ was studied by a luminol-dependent chemiluminescence technique and it was found that after stimulation of polymorphonuclear leucocytes with 12-*O*-tetradecanoylphorbol-13-acetate, there would by an immediate, significant chemiluminescence, which came mainly from the reactive oxygen species generated by polymorphonuclear leucocytes. If L-arginine was added to this system, the chemiluminescence would increase about 100-fold, but N^G-monomethyl-L-arginine inhibited the increase of the chemiluminescence. 10 min after addition of L-arginine, this increase did not change, the chemiluminescence peak decreased gradually, but the half life increased.

In anaesthetised new-born *pigs* (1 to 5 days old) of either sex, injury of moderate severity (1.9 to 2.1 atm) induced by the lateral fluid percussion brain injury technique phorbol 12,13-dibutyrate (1 μmol/l) increased superoxide dismutase-inhibitable nitroblue tetrazolium reduction from 1 ± 1 to 37 ± 5 pmol/mm^2 (ARMSTEAD 1999).

Time (5 min–24 h)-dependent variations in protein kinase C activity after 10^{-7} M 12-*O*-tetradecanoylphorbol-13-acetate stimulation differed significantly between 6-day-old *rats* and adult *rats*; protein kinase C activity decreased in adult alveolar macrophages (50 %) but remained stable in 6-day-old alveolar macrophages (DELACOURT et al. 1997). Leupeptin, used as a calpain inhibitor, inhibited the decrease in protein kinase C activity after exposure of adult alveolar macrophages to 12-*O*-tetradecanoylphorbol-13-acetate and induced a greater than threefold increase in 12-*O*-tetradecanoylphorbol-13-acetate-induced gelatinase secretion.

12-*O*-Tetradecanoylphorbol-13-acetate enhanced L-isoproterenol and prostaglandin E_1 stimulated cyclic AMP formation in clones of *mouse* myeloid leukaemia cells (SIMANTOV and SACHS 1982).

12-*O*-Tetradecanoylphorbol-13-acetate-induced cyclic mononucleotide phosphodiesterase activation in *human* blood mononuclear cells was totally abolished by ethanol, which strongly reduced phosphatidic acid accumulation in response to the phorbol ester (ZAKAROFF-GIRARD et al. 1999).

12-*O*-Tetradecanoylphorbol-13-acetate-primed THP-1 cells (a *human* monocyte cell line), which express a scavenger receptor, were stimulated by mucins through the macrophage scavenger receptor, resulting in enhanced secretion of IL-1β (INOUE et al. 1999). The activity was abolished by treatment of the mucins with sialidase, indicating that sialic acid is involved in the binding.

12-Deoxyphorbol-13-*O*-phenylacetate-20-acetate is not protein kinase C-β isozyme-selective *in vivo* (KILEY et al. 1994). Prolonged treatment (>6 h) of cultures in down-modulation studies is complicated by the metabolism of 12-deoxyphorbol-13-*O*-phenylacetate-20-acetate to 12-deoxyphorbol-13-phenylacetate, a compound which activates all protein kinase C isozymes tested *in vitro* (RYVES et al. 1991).

Expression of tissue thromboplastin and urokinase-type plasminogen activator (EC 3.4.21.73) receptor were stimulated in monocyte-like U-937 cells treated with 5 ng 12-*O*-tetradecanoylphorbol-13-acetate per ml (HAASE et al. 1993).

The expression of many immediate-early genes, such as *c-fos* and *c-jun*, can be induced by 12-*O*-tetradecanoylphorbol-13-acetate. 12-*O*-Tetradecanoylphorbol-13-acetate increased *c-fos* mRNA, cellular cyclic AMP, and protein kinase A activity in the first 30 min with similar inductive time courses (HUANG et al. 1999). Treatment of NIH 3T3 cells with *N*-[2-(*p*-bromocinnamylamino)ethyl]-5-isoquinoline sulphonamide (H-89), a protein kinase A specific inhibitor, suppressed 12-*O*-tetradecanoylphorbol-13-acetate induction of protein kinase A activity and *c-fos* mRNA in a concentration-dependent manner, but did not inhibit serum-induced transcription. H-89 did not inhibit 12-*O*-tetradecanoylphorbol-13-acetate and serum induction of *c-jun* mRNA. H-89 interfered with 12-*O*-tetradecanoylphorbol-13-acetate-stimulated serum-responsive element-binding activity in a concentration-dependent manner, but did not inhibit 12-*O*-tetradecanoylphorbol-13-acetate-induced mitogen-activated protein kinase 1/2 activity or Elk-1 phosphorylation. 12-*O*-Tetradecanoylphorbol-13-acetate stimulation of a *c-fos* promoter reporter construct was inhibited by overexpression of the dominant negative regulatory protein of protein kinase A. In deletion studies, the H-89 inhibitory element was found to be localised between –563 and –379 in the *c-fos* promoter region.

Both troglitazone and 15-deoxy-$\Delta^{12,14}$-prostaglandin J_2 seemed to inhibit phorbol ester-induced TNF-α release from the *human* monocytic cell line THP-1 (Naitoh et al. 2000). On the other hand, neither pioglitazone nor rosiglitazone inhibited phorbol ester-induced TNF-α release. Because the cytotoxicity of troglitazone and 15-deoxy-$\beta^{12,14}$-prostaglandin J_2 was significantly stronger than that of pioglitazone and rosiglitazone, the inhibition of TNF-α release seemed to parallel the lack of cell viability.

Since neoplastic transformation is closely related to cytoskeletal changes, Panagopoulou et al. (2002) developed a culture system of purified populations of Sertoli cells from 20-d-old *rats*. After the addition of 10^{-7} M 12-*O*-tetradecanoylphorbol-13-acetate Sertoli cells began to round up and their cytoplasm retracted towards a central region. Actin bundle organisation was disrupted and vinculin assumed a punctuate distribution throughout the cell.

Bryostatins

Bryostatins are macrocyclic lactones which were isolated from the marine bryozoan *Bugula neritina* (Pettit et al. 1970).

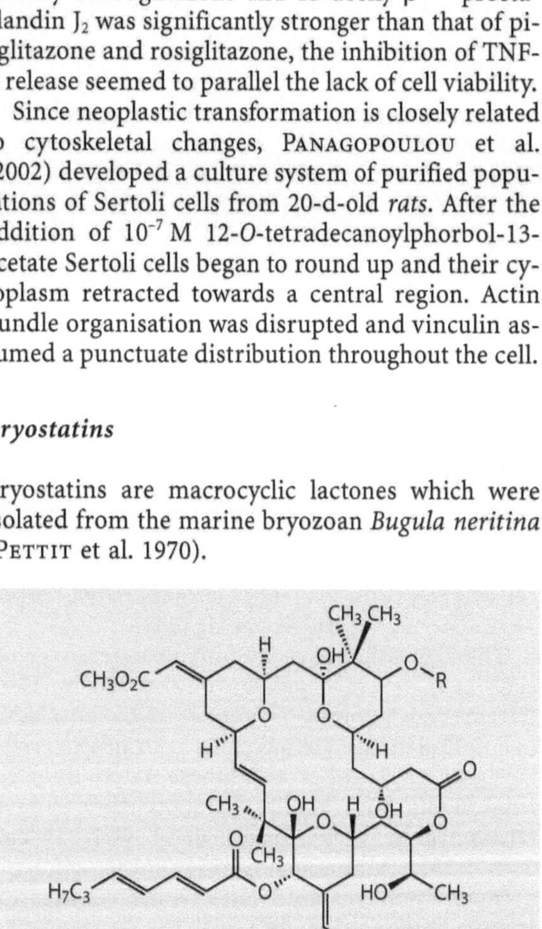

Bryostatins: Bryostatin 1: R = OAc; Bryostatin 2: R = H [50]

Bryostatins, unlike 12-*O*-tetradecanoylphorbol-13-acetate do not include monocytic differentiation of HL-60 cells (Kraft et al. 1986). They bind to the phorbol esterbinding sites in *human* neutrophils and HL-60 cells, activate purified protein kinase C, induce protein phosphorylation patterns similar to 12-*O*-tetradecanoylphorbol-13-acetate, and induce $O_2^{\bullet-}$ formation in *human* neutrophils (Berkow and Kraft 1985, Kraft et al. 1986, Wender et al. 1988). Bryostatin 1 increased the susceptibility of U-937 cells to taxol-induced apoptosis and inhibition of clonogenicity (Wang et al. 1998). Bryostatin induced multiubiquitinylation of protein kinase C-α *in vitro* and in renal epithelial cells (Lee et al. 1996). *In vitro* multiubiquitinylation required ATP (or ATPδS), membranes containing the 76-kDa, nonphosphorylated form of protein kinase C, and a cytsol fraction (Lee et al. 1996). In primary cultures of *human* dermal fibroblasts bryostatin 1 and phorbol myristate acetate down-modulated protein kinase C-α and -ε via the ubiquitin/proteasome pathway (Lee et al. 1997).

Bryostatins stimulated phagocytosis of *human* granulocytes *in vitro* up to 233 ± 21 % in a concentration range from 1 mM to 10 nM (Eisemann et al. 1995). This high phagocytic potential could also be confirmed for bryostatin 1 in the *in vivo* carbon clearance model (conc. of 100 μg/kg NMRI *mouse* injected intravenously, 350 ± 19 %). In the luminol-amplified chemiluminescence test human granulocytes were stimulated by bryostatins 1, 2 and 5 between 1 μg and 10 ng to produce O_2 radicals. The maximal O_2 burst was measured after 8 min at a concentration of 1μM. At 10 μM revealed a reduced burst capacity obviously due to the cytotoxic effect of the bryostatins (37 ± 9 %) at this concentration as estimated by trypan blue staining.

In NHI 3T3 fibroblasts bryostatin 1 showed similar potency to 12-*O*-tetradecanoylphorbol-13-acetate for translocating PKCα to the cell membrane but was a much more potent downregulator of PKCα activity than 12-*O*-tetradecanoylphorbol-13-acetate. It was also a much more potent translocator and downregulator of PKCδ and PKCε than 12-*O*-tetradecanoylphorbol-13-acetate (Szallasi et al. 1994). The compound inhibited the proliferation of *human* A549 lung carcinoma (Dale and Gescher 1989), *human* MCF-7 breast cancer (Kennedy et al. 1992), *murine* renal adenocarcinoma, B16 melanoma, M5076 reticulum cell sarcoma, L10A B-cell lymphoma and exhibited antitumour activity in the *murine* P388 leukaemia screening system (Hornung et al. 1992).

Bryostatin 1 activated T cells that have antitumour activity (Tuttle et al. 1992).

Gnidimacrin

Gnidimacrin, a diaphnane-type diterpene isolated from *Stellera chamaejasme* L. is another protein kinase C activator, which induces downmodulation of the enzyme following long term exposure. The compound exhibits antitumour activity *in vivo* (Yoshida et al. 1996).

Protein Kinase C Inhibitors

Protein kinase C inhibitors specific for either the regulatory domain of the enzyme, such as calphostin (Kobayashi et al. 1989), or its catalytic domain, such as staurosporine (Tamaoki et al. 1986) and CGP 41 251 (Meyer et al. 1989) provided an opportunity to determine more precisely the implication of protein kinase C in macrophage activation. In the *murine* macrophage cell line RAW 264.7 extensively used for the study of LPS stimulation (Virca et al. 1989), particularly at the level of TNF-α (Jue et al. 1991) and •NO (Marletta et al. 1988) biosyntheses, the LPS-triggered IL-6 secretion was strongly enhanced by the inhibitor (Tremblay et al. 1995). TNF-α synthesis and •NO production, however, were strongly inhibited, indicating that their upregulation involves protein kinase C.

Certain chemopreventive and growth-inhibiting agents such as selenite (Gopalakrishna et al. 1997), or polyphenolics such as curcumin (Chen et al. 1996), 4-hydroxy-tamoxyfen (Gundimeda et al. 1995), or ellagic acid can inactivate protein kinase C by oxidising the vicinal thiols present within the catalytic domain.

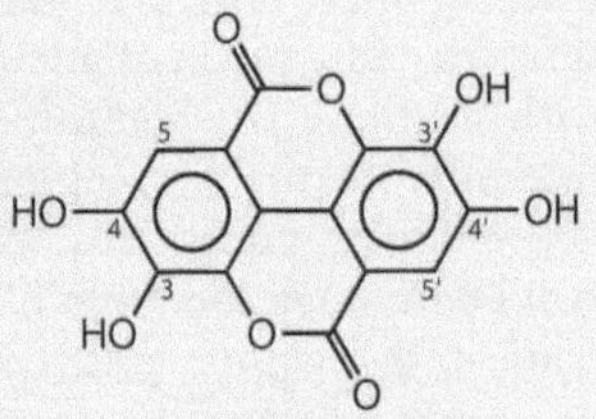

Ellagic acid [51]

Gö-6976, a Ca^{2+}-dependent protein kinase isozyme inhibitor, reduced responses to angiotensin II; however, it did not alter responses to serotonin, norepinephrine, or U-46619, whereas Gö-6976 enhanced BAY K 8644 responses (De Witt et al. 2001). Rottlerin, a protein kinase C-δ isozyme/calmodulin-dependent kinase III inhibitor, reduced responses to angiotensin II and norepinephrine, did not alter responses to serotonin or U-46619, and enhanced responses to BAY K 8644. Immunohistochemistry of *feline* pulmonary arterial smooth muscle cells demonstrated localisation of protein kinase C-α and -δ isozymes in response to 12-*O*-tetradecanophorbol-13-acetate and angiotensin II. Localisation of protein kinase C-α and -δ isozymes decreased with administration of Gö-6976 and rottlerin, respectively.

Calphostin

Calphostin A is a lipophilic, light-sensitive perylenequinone that generates singlet oxygen when illuminated (Wang et al. 1993). It inhibited the activity of protein kinase C (IC_{50} = 250 nM), but only in the presence of light.

Staurosporine

Staurosporine, a microbial alkaloid from *Streptomyces* (Omura et al. 1977), is a potent inhibitor of phospholipid Ca^{2+}-dependent protein kinase A (IC_{50} = 7.0 nM), protein kinase C (IC_{50} = 0.7 nM) and protein kinase G (IC_{50} = 8.5 nM). Platelet aggregation induced by collagen or adenosine diphosphate is also inhibited. Pre-treatment of tracheal smooth muscle cells with staurosporine (10^{-8} M) significantly attenuated sensitisation- and specific antigen challenge-induced changes in membrane potential and isomeric force (Souhrada and Souhrada 1993).

In anaesthetised new-born *pigs* injury of moderate severity (1.9 to 2.1 atm) induced by the lateral fluid percussion brain injury technique staurosporine (10^{-7} M/l) blocked the nitroblue tetrazolium reduction after phorbol 12,13-dibutyrate and blunted the nitroblue tetrazolium reduction observed after lateral fluid percussion brain injury (1±1 to 15±2 versus 1±1 to 5±1 pmol/mm^2 after lateral fluid percussion brain injury in the absence versus presence of staurosporine) (Armstead 1999).

The derivative CGP 41 251 had reduced protein kinase C activity with an IC_{50} of 50 nM bur showed a high degree of selectivity when assayed for inhibition of cAMP-dependent protein kinase (IC_{50} 2.4 μM), S6 kinase (IC_{50} 5.0 μM) and tyrosine kinase specific activity of epidermal growth factor receptor (IC_{50} 3.0 μM) (Meyer et al. 1989). Staurosporine and CGP 41 251 exerted growth inhibition in the *human* bladder carcinoma line T-24, *human* promyelocytic leukaemia lone HL-60 and *bovine* corneal endothelial cells at concentrations which correlated well with *in vitro* protein kinase C inhibition. In addition, both inhibited the release of H_2O_2 from *human* monocytes pre-treated with 12-*O*-tetradecanoylphorbol-13-acetate at non-toxic concentrations.

Melatonin

A melatonin mechanism of action may be through modulation of Ca^{2+}-activated calmodulin. Melatonin binds to calmodulin with a high affinity (Benítez-King et al. 1993), and has been shown to act as a calmodulin antagonist. Among the calmodulin-dependent enzymes, Ca^{2+}/calmodulin-dependent protein kinase II is a particularly abundant enzyme in the nervous system (ERONDU and KENNEDY 1985). *In vitro* inhibition of Ca^{2+}/calmodulin-dependent protein kinase II by 10^{-9} M melatonin was nearly of 30 % (BENÍTEZ-KING et al. 1996). 10^{-5} M Melatonin abolishes autophosphorylation. The effect of melatonin on Ca^{2+}-activated calmodulin- dependent protein kinase II activity was specific, since neither serotonin, *N*-acetylserotonin, or 6-hydroxymelatonin inhibited its activity.

6.1.3.1.3 Guanine Nucleotide Binding Proteins

G-Proteins consist of α-, β-, and γ-subunits (DOHLMAN et al. 1987, NEER and CLAPHAM 1988, CASEY et al. 1988, FREISSMUTH et al. 1989). The α-subunit contains a guanine nucleotide-binding site and in the nonactivated state GDP is bound in the site. Interaction between the G-protein heterodimer and the receptor increases the affinity of the receptor for an agonist. When the receptor is activated by an agonist, a conformational change occurs in the G-protein which facilitates the exchange of GTP for GDP in the guanine nucleotide binding site. On binding GTP, the activated G-protein dissociates, releasing the β- and γ-subunits.

6.1.3.2 Inhibition

6.1.3.2.1 cAMP Increasing Agents

Pentoxifylline reduced the superoxide anion production of *human* polymorphonuclears and monocytes during phagocytosis of latex particles, probably due to the increased intracellular levels of cyclic AMP induced by the drug (BESSLER et al. 1986). In *N*-formyl-methionyl-leucyl-phenylalanine (10^{-7} M)-stimulated *human* polymorphonuclear leucocytes pentoxifylline (10^{-6}–10^{-2} M) and adenosine (5×10^{-8}–1×10^{-3} M) dose-dependently inhibited superoxide anion production (THIEL et al. 1991). The adenosine receptor antagonist, 8-phenyltheophylline only diminished the inhibition mediated by adenosine, but totally failed to affect pentoxifylline. Inhibition by both agents was antagonised by the cyclic AMP antagonist Rp-cAMPS (250 μM; THIEL et al. 1992). Pentoxifylline potentiated adenosine both in inhibiting superoxide anion production and in raising intracellular cyclic AMP.

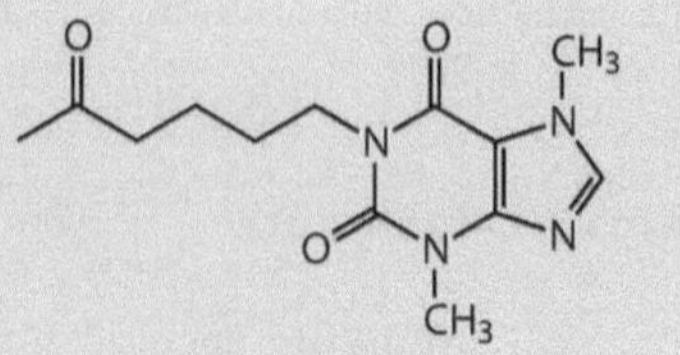

Pentoxifylline [52]

6.1.3.2.2 β-Adrenergic Agonists

Fentolol dose-dependently suppressed of the production of oxygen radicals by zymosan-stimulated *human* polymorphonuclear leucocytes and macrophages as studied *in vitro* by means of luminol enhanced chemiluminescence (SCHOPF and LEMMEL 1983), the ID_{50} being approximately 10^{-6} M both for polymorphonuclear leucocytes and macrophages. When incubated together with the β-adrenergic antagonist propranolol at 10^{-6} and 10^{-7} M the suppressive effect of fentolol could be reversed in dose-dependency.

In purified *rat* cardiac membranes the affinities of both β- and α-adrenergic receptors for [^{3}H]dihydroalprenolol and [^{3}H]prazosin, respectively, were depressed by reactive oxygen species as generated in the xanthine-xanthine oxidase system (KANEKO et al. 1991). The situation with respect to number of binding sites, however, was complex in nature. Treatment of cardiac membranes for 10–30 min with xanthine plus xanthine oxidase increased the number of the β-receptor antagonist, [^{3}H]dihydroalprenolol, binding sites, whereas no effect was seen at 60 min. On the other hand, a significant depression in α-receptor antagonist, [^{3}H]prazosin, was seen only 60 min after the treatment of membranes with xanthine plus xanthine oxidase. High concentration of H_2O_2 produced no effect on the number of adrenergic receptor binding sites, whereas 0.1 mM H_2O_2 plus 0.2 mM $FeSO_4$ significantly increased the number of β-adrenergic receptor sites for [^{3}H]dihydroalprenolol only. Because oxygen free radicals are known to increase membrane permeability (LEVEDEV et al. 1982, MAK et al. 1986), it is possible that the observed increase in the number of [^{3}H]dihydroalprenolol binding sites may be due to the lipophilic nature of this ligand. This view is substantiated by the fact that no effect of oxygen free radical system was observed on the B_{max} values when a hydrophilic ligand, [^{3}H]CGP-12177, was employed for monitoring the state of β-adrenergic receptor activity (STAEHELIN et al. 1983).

6.1.3.2.3
Phosphodiesterase Inhibitors

Phosphodiesterase inhibitors potentiated the inhibitory effects of β-adrenergic agonists on superoxide formation and their stimulatory effects on intracellular cyclic AMP (LAD et al. 1985).

The predominant phosphodiesterase isozyme class in inflammatory cells is composed of the enzymes that belong to the phosphodiesterase IV family (TORPHY and UNDEM 1991). This group of enzymes has a marked preference for cyclic AMP (K_m = 3–10 μM) as a substrate and demonstrates little if any catalytic activity against cyclic GMP. Furthermore, this enzymes are selectively inhibited by compounds such as rolipram and Ro 20-1724 (TORPHY and UNDEM 1991, GIEMBYCZ 1992).

Ro 20-1724 and unspecific inhibitors of phosphodiesterases, e.g., the methylxanthines 3-isobutyl-1-methylxanthine, theophylline, and pentoxiphylline inhibited the respiratory burst (LAD et al. 1985, BESSLER et al. 1986, BURDE et al. 1989). In contrast, rolipram increased cyclic AMP in *human* neutrophils without significantly inhibiting superoxide formation (ELLIOTT and LEONARD 1989). The suppression of superoxide production in *guinea pig* eosinophils is correlated with inhibition of phosphodiesterase IV catalytic activity (BARNETTE et al. 1995). Methylxanthines above 100 μM inhibits phosphodiesterases and inhibit the respiratory burst presumably via an increase in cyclic AMP (SCHMEICHEL and THOMAS 1987, YUKAWA et al. 1989). In contrast, methylxanthines below 100 μM act as competitive antagonists at A_2 receptors and potentiate fMet-Leu-Phe -induced superoxide anion formation in *human* neutrophils and eosinophils (SCHMEICHEL and THOMAS 1987, YUKAWA et al. 1989). Adenosine desaminase mimics the stimulatory effects of methylxanthines, and adenosine counteracts the stimulatory effects of these agents, suggesting that endogenous adenosine plays an inhibitory role in the regulation of NADPH oxidase (SCHMEICHEL and THOMAS 1987). The cell-permeant analogue of cyclic AMP, dibutyryl cAMP, inhibits chemotactic peptide-induced $O_2^{\bullet-}$ formation. (KITAGAWA and TAKAKU 1982, LAD et al. 1985, BESSLER et al. 1986, KRAMER et al. 1988, BURDE et al. 1989).

Zardaverine, a selective inhibitor of cAMP-specific phosphodiesterase isoenzymes III and IV, inhibited superoxide release from zymosan-stimulated *human* polymorphonuclear leucocytes at an IC_{50} of 4×10^{-6} M (BEUME et al. 1989).

Selective cyclic GMP phosphodiesterase inhibitors (M&B 22948 and MY 5445) potentiate inhibition of platelet adhesion by nitric oxide (MONCADA et al. 1988). The phosphodiesterase V inhibitor zaprinast decreased pulmonary arterial pressure in foetal *lambs* (SKIMMING et al. 1996). Recent evidence suggests that the levels of phosphodiesterase V as well as other phosphodiesterase isoforms are increased in pulmonary arteries of hypertensive *rats* (MAC LEAN et al. 1998, HANSON et al. 1998). This is consistent with the use of phosphodiesterase V inhibitors as vasodilators (COHEN et al. 1996).

6.1.3.2.4
Retinoic Acid

The intracisternal injection of either all-*trans*-retinoic acid or [α]-difluoro-methylornithine into the brain of 9-day-old ICR *mice* blocked (>90 %) phorbol ester-induced ornithine decarboxylase (EC 4.1.1.17) activity in a concentration-dependent manner (COPE 1986). This inhibition was not evident with the use of the biologically impotent furyl analogue of retinoic acid. In a similar manner, retinoic acid reduced the soluble protein kinase-C activity by 60 % as well as total [ethylenebis(oxonitrilo)]tetraacetic acid (EGTA)-sensitive kinase activity (66 %) associated with the plasma membrane. Sixty-six percent of the retinoic acid-induced loss of protein kinase-C activity in the soluble fraction could be accounted for by the translocation of protein kinase-C to the plasma membrane as measured by the specific binding of 12-*O*-[^{3}H]tetradecanoylphorbol-13-acetate. [α]-Difluoro-methylornithine and furyl-retinoic acid were not effective in altering protein kinase-C activity or 12-*O*-tetradecanoylphorbol-13-acetate binding to protein kinase-C. In the presence of retinoic acid, however, there was a 2.3-fold increase in specific 12-*O*-[^{3}H]tetradecanoylphorbol-13-acetate binding in the plasma membrane fraction, which was 3.4-fold greater than that lost from the cytosol.

Increase at 843 nm in 5 min at room temperature given by TCNQ (1.84×10^{-4} M) and test compound (2.3×10^{-5} M) dissolved in acetonitrile as a polar organic solvent.

Retinoic acid behaves similarly to retinol in its reactions with the strong electron acceptor 7,7,8,8-

Table 13. Production of 7,7,8,8-tetracyanoquinodimethane radical anion ($TCNQ^{\bullet-}$) from TCNQ and vitamin A derivatives (from LUCY 1969)

β-Carotene	0.48	Retinoic acid	0.06
α-Carotene	0.37	β-Ionylidene acetaldehyde	0.06
All-*trans*-retinol	0.24	9-*cis* Retinal	0.06
Retinyl methyl ether	0.21	β-Ionylidene ethanol	0.03
Retinyl acetate	0.10	13-*cis* Retinal	0.03
3-Dehydroretinol	0.09	All-*trans* retinal	0.03
Retinyl palmitate	0.09	5,6-Monoepoxiretinal	−0.02
13-*cis* Retinol	0.08		

tetracyanoquinodimethane (TCNQ), but it appears to be a weaker **electron donor** than retinol (LICHTI and LUCY 1969).

The effect of β-carotene on *human* serum albumin oxidation by 2,2'-azobis (2-amidinopropane) dihydrochloride under 15, 150, and 760 torr of O_2 to form carbonyls was related to O_2 tension, antioxidant concentrations and interaction between mixtures of antioxidants (ZHANG and OMAYE 2000). High concentration of β-carotene produced more protein oxidation in the presence of high O_2 tension by a prooxidant mechanism. Mixtures of β-carotene, α-tocopherol and ascorbic acid provided better protective effects on protein oxidation than any single compound.

The staurosporine (200 nM)-induced apoptotic damage in *chick* embryonic neurones was reduced by 1 nM–10 μM retinoic acid in a concentration-dependent manner by depressing the production of reactive oxygen species (AHLEMEYER and KRIEGLSTEIN 1998).

6.1.3.2.5 α-Tocopherol

Inhibition by α-tocopherol of protein kinase C has been reviewed in AZZI et al. (1992, 1995, 1996) and NEWTON (1995). Such an inhibition is not caused by a direct binding of α-tocopherol to the enzyme but by preventing its activation via phosphorylation (TASINATO et al. 1995). α-Tocopherol exerts its action independently of its free-radical scavenging capacity and most probably by interacting with a yet not characterised receptor molecule in smooth muscle cells (BOSCOBOINIK et al. 1991, 1994, 1995). α-Tocopherol prevents uniquely protein kinase C-α phosphorylation and its functional activation.

12-*O*-Tetradecanoylphorbol-13-acetate-induced superoxide generation is α-tocopherol-sensitive at a concentration much lower than that for the inhibition of 12-*O*-tetradecanoylphorbol-13-acetate-activated phospholipid-dependent protein kinase (ID_{50} = 30 μM). The α-tocopherol-sensitive superoxide generation is also observed in neutrophils induced by dioctanoylglycerol and calcium ionophore A23187 but not by formyl-methionyl-leucyl-phenylalanine, opsonized zymosan, and sodium dodecylsulphate. The pattern of inhibition by α-tocopherol is quite similar to that of staurospirine, a specific inhibitor of protein kinase C. These results indicate that neutrophil content of α-tocopherol is an important factor in superoxide generation (KANNO et al. 1996).

6.1.3.3 Influences of ATP, ADP, and Cyclic Nucleotides on the Formation of Reactive Oxygen Intermediates

In paraffin-elicited *guinea-pig* peritoneal macrophages, extracellular ATP induced superoxide generation mediated by $[Ca2+]_i$ and by pertussis toxin sensitive G protein (NAKANISHI et al. 1991). ATP, when added to *human* polymorphonuclear neutrophils at concentrations similar to those attained extracellularly at sites of platelet thrombus formation (0.1 to 20 μM), caused an enhancement of *N*-formyl(methionyl)leucylphenylalanine-stimulated superoxide anion generation, but was by itself an ineffective agonist for $O_2^{\bullet-}$ generation (KUHNS et al. 1988). ATP primed *human* neutrophils for enhanced $O_2^{\bullet-}$ generation at low concentrations (1×10^{-6} to 3.2×10^{-4} M) but inhibited $O_2^{\bullet-}$ generation at high ($\geq 6.4 \times 10^{-4}$ M) concentration (NAUM et al. 1991). ADP, when added to stimulatory concentrations of ATP, also caused inhibition of $O_2^{\bullet-}$ production.

The data of TAN et al. (1995) suggest that oxygen-derived species mediating the enhancement of adenylate cyclase is either H_2O_2 itself or the hydroxyl radical. Incubation of A 10 cells with xanthine oxidase and purine resulted in a qualitatively similar enhancement of adenylate cyclase activation. The effect of purine and xanthine oxidase was not blocked by coincubation with superoxide dismutase (which catalyzes the conversion from superoxide anion to H_2O_2). This suggests that the generation of the superoxide anion is not involved in the mechanism of enhancement of adenalate cyclase activation.

6.1.4 Radical Scavengers

6.1.4.1 Ascorbic Acid

Ascorbic acid is the major water-soluble antioxidant present in cells and plasma. It will quench reactive oxygen species as $O_2^{\bullet-}$ (NISHIKIMI 1975), $HO^{\bullet}$ (BIELSKI et al. 1975), and $^1O_2$ (BODANNES and CHAN 1979). On the other hand, it reduces Fe^{3+} to Fe^{2+} and thus will stimulate Fenton catalysis of $H_2O_2 \rightarrow HO^{\bullet}$. Hydroperoxide-dependent lipid peroxidation in *rat* liver microsomes was enhanced by ascorbic acid (LAUDICINA and MARNETT 1990). Ascorbic acid protected cardiac microsomes against lipid peroxidation and oxidative damage (MUKHOPADHYAY et al. 1993). It diminished both luminol- and lucigenin-amplified H_2O_2 derived chemiluminescence in concentrations $>10^{-5}$ (KLINGER et al. 1996).

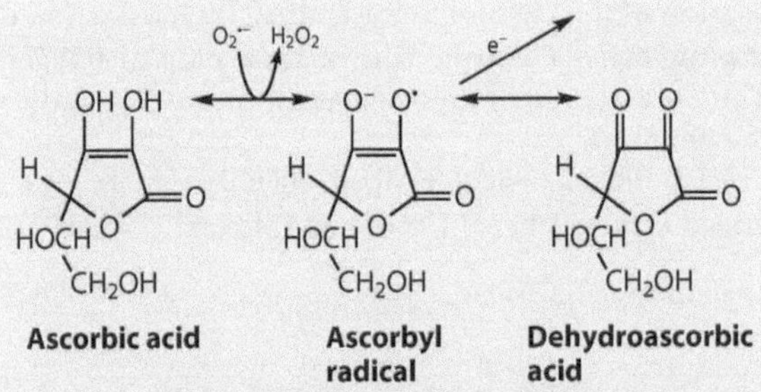

Ascorbyl radical formation [53]

The ascorbyl free radical (formula [53]) as the first product of oxidation can be detected spectrophotometrically at 360 nm (SCHULER 1977). The electron spin resonance spectrum of the ascorbyl radical in solution consists of a doublet first detected during ionising radiation studies (REXROAD and GORDY 1959). The intermediate free radical behaves both as one-electron oxidant and as one-electron reductant (COASSIN et al. 1991). Although ascorbyl free radical is a relatively stable, non-hazardous biological free radical, ascorbate oxidation seems to contribute to the generation of other free radicals and reactive oxygen species, including hydroxyl and superoxide radicals and hydrogen peroxide (MILLER and AUST 1989).

Ascorbic acid forms a complex with disulphides such as oxidised glutathione (GSSG) and cystine (FLEMING et al. 1983). Ascorbic acid is known to accept one electron from thiyl free radicals (FORNI et al. 1983) and to reduce phenoxyl radicals (SCHULER 1977). The effect of ascorbic acid on a series of redox systems, which included 6-hydroxydopamine resulted for the most part in the ascorbyl free radical (BORG et al. 1978).

Recycling of ascorbic acid from its oxidised forms is required to maintain intracellular stores of the vitamin in most cells. Since the ubiquitous selenoenzyme thioredoxin reductase can recycle dehydroascorbic acid to ascorbate, MAY et al. (1998) presented evidence that *mammalian* thioredoxin can catalyse a one-electron reduction of the ascorbyl free radical to ascorbate, using NADPH as the electron donor.

Because of its dithiol/disulphide exchange activity, thioredoxin determines the oxidation state of protein thiols. This small (~12 kDa) protein is evolutionarily conserved between prokaryotes and eukaryotes from yeast to plants and animals. A characteristic feature of most thioredoxins is the presence of a conserved catalytic site Trp–Cys–Gly–Pro–Cys–Lys in a protrusion of the three-dimensional structure of the protein. The two cysteine residues of the site can be reversibly oxidies to form a disulphide bridge and, thereafter, ne reduced by action of the selenoenzyme thioredoxin reductase in the presence of NADPH {$NADPH + H^+ + thioredoxin\text{-}S_2 \rightarrow NADP^+ + thioredoxin\text{-}(SH)_2$}. Thioredoxin reductase activity is decreased by selenium deficiency (HILL et al. 1997). Thioredoxin, secreted by normal and neoplastic cells using a leaderless secretory pathway (RUBARTELLI et al. 1992), stimulates the proliferation of fibroblasts.

In *guinea pigs* (albino, male 350-400 g) fed ascorbate-deficient diets for 20 days, oxidative damage is evidenced by lipid peroxidation (GHOSH et al. 1997). It caused accumulation of conjugated dienes, malondialdehyde and fluorescent pigment in the microsomal membranes of different tissues. Conjugated dienes in nmoles per mg protein increased above that of pair-fed controls (taking extinction coefficient at 234 nm = 25 mM^{-1}) by 74 ± 10 (n = 4) in the liver, 54 ± 6 in the kidney and 77 ± 8 in the lung. The levels of MDA in pmoles per mg microsomal protein were 320 ± 36 (n = 4) in the liver; 265 ± 30 in the kidney: and 290 ± 32 in the lung. Fluorescent pigment per mg microsomal protein was 5.6 ± 0.5 in the kidney and 5.0 ± 0.6 in the lung. Accumulations of MDA, conjugated dienes and fluorescent pigment were reversed after oral ascorbic acid therapy at a dose of 50 mg ascorbate per *guinea pig* per day for two days followed by 15 mg per *guinea pig* per day for the next 8 days.

Flow cytometric analyses demonstrated that pretreatment of Chinese *hamster* ovary cell line AS52 with 50 μM ascorbic acid before exposure to the hypoxanthine-xanthine oxidase radical generating system enhanced cell cycle arrest at the G_2/M DNA damage checkpoint when compared to cells treated with hypoxanthine-xanthine oxidase without premedication (BIJUR et al. 1999). Ascorbic acid had no effect on cell cycle progression in the absence of oxidative stress.

In sickle cell membranes, ascorbic acid proved to be an effective antioxidant (RICE-EVANS et al. 1986). It is possible that the mechanism of action of ascorbate may involve the scavenging of alkoxy or peroxy radicals by a chain-termination reaction (HALLIWELL and GUTTERIDGE 1984) or the decreased initiation of lipid peroxidation.

L-**Ascorbyl-6-palmitate** is a fat-soluble synthetic ester of ascorbic acid. It has been used extensively as an antioxidant in foods, pharmaceuticals and cosmetics, particularly as a preservative for various edible oils and waxes The inhibition of polymerisation of *bovine* serum albumin by $HO^•$ radicals generated by the Fenton reaction indicated ascorbyl radical exerts a considerable protective effect against polymerisation by scavenging $HO^•$ (PERRICONE et al. 1999). L-Ascorbyl-6-palmitate was 1 order of magnitude faster in scavenging these radicals than 5,5-dimthyl-1-pyrroline-*N*-oxide (DMPO; formula [8]). Oxidative modification by ^{60}Co-γ irradiation of 80 krad resulted in a strong increase in protein carbonyl content, which was very efficiently inhibited by L-ascorbyl-6-palmitate.

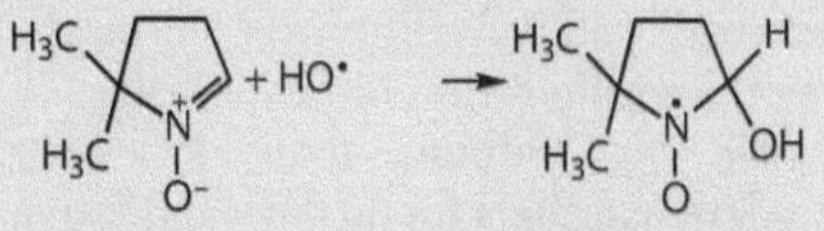

5,5-**Dim**ethyl-1-**p**yrroline-1-**o**xide-HO• adduct [8]

A series of **2-*O*-alkylascorbic acids** combine hydrophilic and lipophilic properties in one molecule. 2-*O*-octadecylascorbic acid markedly inhibited lipid peroxidation ($IC_{50} = 4.3 \times 10^{-6}$ M) and alleviated myocardial lesions induced by ischaemia-reperfusion at an oral dose of 1 mg/kg in *rats* (Kato et al. 1988).

3-*O*-Alkylascorbic acids acted as free radical quenchers and, dependent upon their hydrophobicity, inhibited Fe^{3+}-ADP-induced lipid peroxidation (Nihro et al. 1991). HX-0112 and HX-0113, which are stable lipophilic ascorbic acid derivatives, alleviated myocardial lesions induced by ischæmia-reperfusion treatment in *rats* (Sasamori et al.).

6.1.4.2 Tocopherols

Vitamin E of natural origin consists of a group of compounds, namely α-, β-, γ-, ε-, ζ_1-, ζ_2-, and η-tocopherol. Of these compounds *RRR* α-tocopherol has the highest vitamin potency *in vivo* (Witting 1980). The α-tocopherol molecule is composed of a chroman head and a phytyl side chain. It is generally believed that the phytyl chain intercalated with the fatty acid residues of phospholipids, while the chroman head – responsible for the antioxidant effect – faces the cytosol, although the chroman ring is still located in the hydrophobic zone of the lipid bilayer (Erin et al. 1988). In the antioxidant activity of vitamin E, a radical (R•) abstracts a hydrogen atom from the aromatic hydroxyl group of the chroman head (ArOH) rather than from a polyunsaturated fatty acid, and a chromanoxyl radical is formed (ArO•).

$$R^\bullet + ArOH \longrightarrow RH + ArO^\bullet \qquad [54]$$

The chromanoxyl radical is fairly stable, due to delocalization of the unpaired electron. The oxygen in the heterocyclic chroman ring is fixed in such a position that there is considerable overlap between the 2p-type orbital of the lone electron pair of the oxygen and the aromatic π-system (Burton and Ingold 1981, Burton et al. 1985). This permits stabilisation of the chromanoxyl radical by interaction of the unpaired electron with a lone pair of oxygen. Thus the degree of delocalization of the free radical is enhanced.

The lipid peroxyl radical (LOO•) can be scavenged by tocopherol (TOH) as follows:

$$LOO^\bullet + TOH \longrightarrow LOOH + TO^\bullet \qquad [55]$$

The initial phase is termed lag phase time (t_{inh}) or inhibition period.

$$V_{inh} = \frac{d\,[LOOH]}{dt} = \frac{k_p\,[LH]R_i}{n\,k_{inh}\,[TOH]} \qquad [56]$$

$$t_{inh} = \text{length of lag phase} = \frac{n\,[TOH]}{R_i} \qquad [57]$$

The factor *n* is defined as the number of peroxyl radicals LOO• trapped by each molecule of antioxidants. For vitamin E the value is 2, since both vitamin E and vitamin E radical (tocopheryl radical) trap LOO•. The length of the lag phase is inversely proportional to the rate R_i, by which the initiating radicals are formed.

α-Tocopheryl succinate epitomizes a compound with a shift in biological activity due to provitamin-to-vitamin conversion (Neuzil 2002). The chargeable succinyl moiety as a functional domain is an absolute requirement for vitamin E analogues with phytyl side chain to be pro-apoptotic (Qian et al. 1997, Neuzil et al. 2001). This is supported by the findings that β-tocopheryl succinate (Neuzil et al. 2001) and γ-tocopheryl succinate (unpublished) lacking the protein kinase C inhibitory potential as well as the vitamin E-unrelated cholesteryl hemisuccinate (Fariss et al. 1994, Neuzil et al. 2001, 2002) were capable of triggering apoptosis, albeit to a lower extent than α-tocopheryl succinate.

Deprivation of vitamin E in B6C3 *mice* for 15 weeks resulted in an approximately 5-fold **increase of mitochondrial hydrogen peroxide production** in skeletal muscle and a 1-fold increase in liver when compared with the vitamin E-supplemented (50 IU vitamin E/kg diet) group (Chow et al. 1999).

In BL6 *murine* melanoma cell cultures supplemented with 7 and 10 μg/ml α-tocopherol succinate, respectively, **free radical levels** were **significantly increased** (Ottino and Duncan 1997). The cyclooxygenase activity in BL6 cells supplemented with 1–10 μg/ml α-tocopherol succinate showed a similar trend to that observed for free radicals and lipid peroxidation experiments with a significant increase in enzyme activity occurring at 7 and 10 μg/ml respectively.

Palace et al. (1999) showed that cardiac dysfunction following myocardial infarction is associated with a decrease in vitamin E levels in the myocardium. Commercial preparations of vitamin E, which are used in experimental and clinical studies, contain only a single isoform, α-tocopherol. However, OHRVALL et al. (1996) suggested, that it is the γ-tocopherol isoform, and not α-tocopherol, that is decreased in patients with coronary artery disease. γ-Tocopherol is more effective than α-tocopherol in inhibiting lipid peroxidation and trapping mutagenic electrophiles (CHRISTEN et al. 1997, WOLF 1997). In a rat model of oxidant-induced thrombosis, a mixed-tocopherol preparation rich in γ-tocopherol was shown to be superior to α-tocopherol in inhibiting thrombus formation (SALDEEN et al. 1999). δ-Tocopherol found in natural foods also offers antioxidant protection (CHOPRA and BHAGAVAN 1999). *In vitro*, mixed-tocopherol preparation was much superior to α-tocopherol in terms of *rat* cardiomyocyte protection from the adverse effects of hypoxia-reoxygenation (CHEN et al. 2002).

Raxofelast (IRFI 016; (±)-5-(acetyloxy)-2,3-dihydro-4,6,7-trimethyl-2-benzofuranacetic acid) is a new hydrophilic vitamin E-like antioxidant (CAMPO et al. 1997). Its activity depends on biotransformation to the deacetylated active metabolite IRFI 005, which has, therefore, been used *in vitro* in chemical, subcellular and cellular models.

In a *rat* model of coronary artery occlusion for 1 h followed by 6 h of reperfusion, raxofelast limited myocardial necrosis and reduced lipid peroxidation (CAMPO et al. 1998).

In streptozotocin-induced diabetic *rats* the vitamin E analogue, **Trolox** almost completely normalised the depressed glutathione peroxidase and superoxide dismutase activities as well as the GSH/GSSG ratios in erythrocytes (BATKO and WITMANOWSKI 1999).

Glutathione peroxidase (EC 1.11.1.9)

$$2\,GSH + H_2O_2 \longrightarrow GSSG + 2\,H_2O \qquad [58]$$

The protein contains a selenocysteine residue. Steroid and lipid hydroperoxides, but not the product of reaction of EC 1.13.11.12 on phospholipids, can act as acceptor, but more slowly than H_2O_2 (cf. EC 1.11.1.12)

Trolox (500 μg/kg i.p.) protected lung tissues of Swiss albino female *mice* from sulphur mustard (1 LC_{50} = 42.3 mg/m^3 for 1 h duration) intoxication (KUMAR et al. 2001).

Epigallocatechin gallate suppressed the initiation rate and prolonged the lag phase duration of peroxyl radical-induced oxidation in a phospholipid liposome model to a greater extent ($P < 0.01$) compared to both Trolox and α-tocopherol (HU and KITTS 2001).

6.1.4.3 Dihydrolipoate-Lipoate

Lipoate synthase is an iron-sulphur protein with the cysteine motif CxxxCxxC characteristic of radical S-adenosyl-L-methionine enzymes (SANYAL et al. 1994, MÉJEAN et al. 1995). Lipoate synthase is implicated in the last step of lipoate biosynthesis by genetic evidence (VAN DER BOOM et al. 1991, REED and CRONAN jr. 1993). Cells deficient in *lip*A cannot produce lipoate. Lipoate synthase contains [4Fe–S] centres when purified anaerobically (ÖLLAGNIER-DE CHOUDENS et al. 2000).

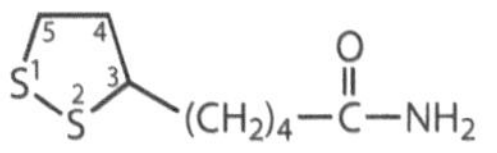

α-Lipoamide [59]

The dihydrolipoic acid/lipoic acid redox couple has been found to exert a synergistic action in the antioxidant recycling mechanisms of natural membranes and low-density lipoproteins *in vitro* and in the protection against oxidative injury *in vivo* (PACKER 1991).

Lipoamide exists in five proteins in eukaryotes, where it is covalently attached to a lysyl residue. Four of these proteins are found in the three α-keto acid dehydrogenase complexes, the pyruvate dehydrogenase complex, the branched chain keto acid dehydrogenase complex, and the α-ketoglutarate dehydrogenase complex.

Combined treatment with D,L-α-lipoic acid and *meso*-2,3-dimercaptosuccinic acid reversed the decreases in the activities of renal γ-glutamyl transferase (EC 2.3.2.2) and *N*-acetyl β-D-glucosaminidase (EC 3.2.1.17) evident in *rats* after the administration of lead citrate in drinking water for 5 weeks (SIVAPRASAD et al. 2002).

Using the xanthine-xanthine oxidase system for producing superoxide radical anions PACKER (1991) showed that dihydrolipoic acid, but not lipoic acid, quenched these radicals.

α-Lipoic acid is a potent **hydroxyl radical scavenger.** 1 mM completely eliminated the Fenton (2 mM H_2O_2 + 0.2 mM $FeSO_4$)-derived 5,5-dimethyl-pyrroline-*N*-oxide-HO$^{\bullet}$ adduct signal (SUZUKI et al.

Table 14. Lipoic acid → hepatocellular dehydrogenases (from GRZYCKI and KRÓLIKOWSKA-PRASAL 1972)

Lactate dehydrogenase	EC 1.1.1.27	↑
Isocitrate dehydrogenase	EC 1.1.1.42	↓
6-Phosphogluconate dehydrogenase	EC 1.1.1.44	↑
Glucose-6-phosphate dehydrogenase	EC 1.1.1.49	↓
Succinate dehydrogenase	EC 1.3.99.1	↓
Glutamate dehydrogenase	EC 1.4.1.3	↑

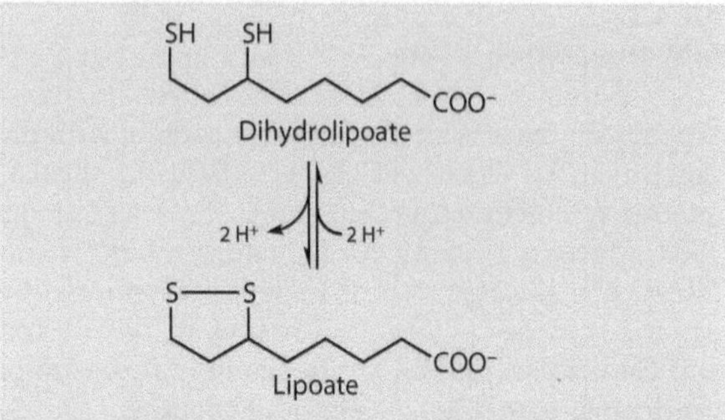

Dihydrolipoic acid/lipoic acid redox couple [60]

1991). Scott et al. (1994) using 2.8 mM H_2O_2, 0.05 mM $FeCl_3$, 0.1 mM EDTA, and 0.1 mM ascorbate as a hydroxyl radical-generating system and deoxyribose degradation for the radical assay, also found α-lipoic acid to be a hydroxyl radical scavenger. Using the hydroxyl radical-producing compound, *N,N'*-bis(2-hydroperoxy-2-methoxyethyl)-1,4,5,8-naphthalenetetracarboxylic diimide (NP-III), Matsugo et al. (1995) confirmed that α-lipoic acid is a hydroxyl radical scavenger and eliminated the possibility that the effect was simply due to metal chelation.

The effect of dihydrolipoic acid on recycling of **chromanoxyl radicals** was strongly dependent on the presence or absence of other redox-cycling compounds (Packer 1991, Kagan et al. 1992). In the presence of dihydrolipoate, Trolox radicals were suppressed until both dihydrolipoate and endogenous levels of ascorbate in *mouse* skin homogenates were consumed (Guo and Packer 2000). Dihydrolipoate regenerated greater amounts of **ascorbate** at a much faster rate than equivalent concentrations of GSH.

6.1.4.4 1,2-Diselenolane-3-pentanoic Acid

1,2-Diselenolane-3-pentanoic acid, in which the sulphur atoms of α-lipoic acid are replaced with selenium, displayed markedly different antioxidant properties when compared to α-lipoic acid (Matsugo et al. 1997). 1,2-Diselenolane-3-pentanoic acid was unable to inhibit protein oxidative modification of *human* low density lipoprotein and *bovine* serum albumin induced by copper ion or hydroxyl radical, whereas α-lipoic acid showed significant protection. However, 1,2-diselenolane-3-pentanoic acid was able to inhibit the formation of lipid peroxidation products in low density lipoprotein after oxidation by copper, while α-lipoic acid did not.

6.1.4.5 Melatonin

Melatonin, the chief secretory product of the pineal gland, is a direct free radical scavenger and indirect antioxidant (Reiter et al. 1998). In terms of scavenging activity, melatonin has been shown to quench the hydroxyl radical, superoxide anion radical, singlet oxygen, peroxyl radical, and the peroxynitrite anion. Additionally, the antioxidant actions of melatonin probably derive from its stimulatory effect of superoxide dismutase, glutathione peroxidase (EC 1.11.1.9), glutathione reductase (EC 1.6.4.2), and glucose-6-phosphate dehydrogenase and its inhibitory action on nitric oxide synthase. By stabilising cell membranes, melatonin makes them more resistant oxidative attack. Due to its lipophilicity (calculated log $P = 0.88$), melatonin can easily reach various biochemical targets in a number of subcellular compartments (Reiter 1993, Reiter et al. 1995). In contrast to other intracellular antioxidants such as ascorbic acid and tocopherol,

Fig. 56. Melatonin (1 mM) decreases lucigenin-enhanced chemiluminescence in the xanthine (0.42 mM)-xanthine oxidase (0.05 U/ml) system as measured in a Berthold Biolumat® 9505

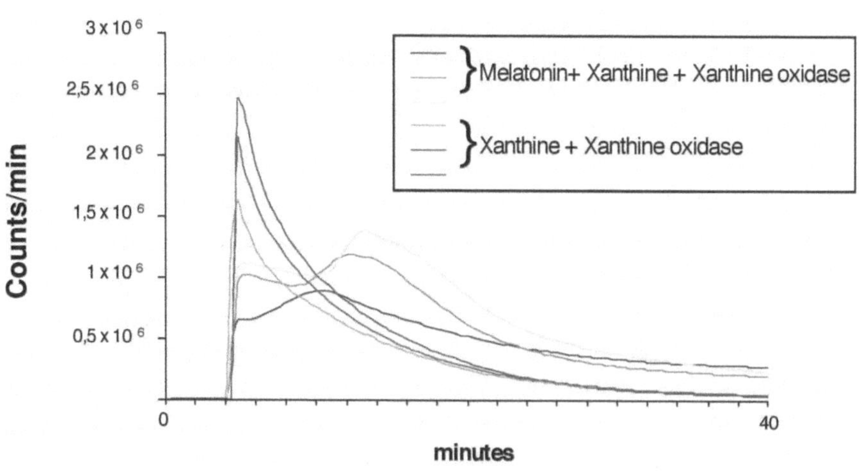

which act primarily in the cytosol and membranes, respectively, melatonin has shown ubiquitous antioxidant effects in membranes, cytosol, and nuclei (Reiter et al. 1995).

The quantum-chemical descriptor ΔH_{ox} (relative adiabatic oxidation potential) and the shape of the singly occupied molecular orbital indicate that the stabilisation of its radical cation partially explain the well-documented antioxidant efficacy of melatonin (Migliavacca et al. 1998). This stabilisation may result from electrostatic interactions and from hyperconjugative effects existing in a family of conformers of the melatonin radical cation having the side chain almost perpendicular to the plain of the aromatic rings. Fowler et al. (2003) have reexamined claims that melatonin directly scavenges H_2O_2 and shown them to be unfounded.

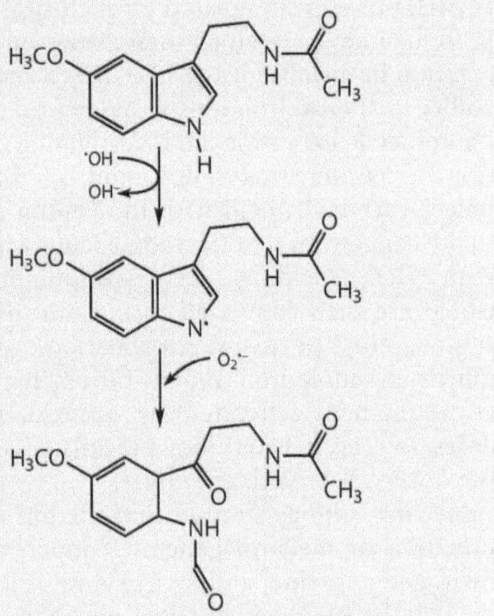

Upon electron donation to the hydroxyl radical (HO˙) to form OH-, melatonin gives rise to the **melatoninyl cation radical**, which is able with the superoxide anion radical ($O_2^{\cdot-}$) to *N*-acetyl-*N*-formyl-5-methoxykynuramine [61]

The evidence that melatonin detoxifies singlet oxygen ($^1O_2$) comes only from indirect evidence. Cagnoli et al. (1995) showed that the neurotoxic effects of $^1O_2$ *in vitro* were counteracted by melatonin. To achieve this, new-born *rat* cerebellar granule cells were treated with rose bengal (formula [29]) and exposed to light, a procedure which generated $^1O_2$ (Agarwal et al. 1991). Both neuronal apoptosis and inhibition of kreatin kinase activity were prevented by the addition of melatonin.

Treatment of *rats* with melatonin and vitamins E plus C significantly ($P < 0.05$) reduced the chlorpyrifos-ethyl-induced increase of thiobarbituric acid-reactive substance in erythrocytes, and overcame the inhibitory effect of chlorpyrifos-ethyl on superoxide dismutase (EC 1.15.1.1) and catalase (EC 1.11.1.6), but not on antioxidant defence potential (Gultekin et al. 2001). Melatonin treatment significantly ($P < 0.05$) increased only glutathione peroxidase (EC 1.1.1.9) activity, irrespective of the effect of chlorpyrifos-ethyl.

Melatonin is an excellent substrate for myeloperoxidase-I, but reacts slowly with myeloperoxidase-II (Allegra et al. 2001). Spectral and kinetic analysis revealed that both compound I and compound II oxidise melatonin via one-electron processes. The second-order rate constant measured for compound I reduction at pH 7 and pH 5 are $(6.1 \pm 0.2) \times 10^6\ M^{-1}\ s^{-1}$ and $(1.0 \pm 0.08) \times 10^6\ M^{-1}\ s^{-1}$, respectively. The rates for the one-electron reduction of compound II back to the ferric enzyme are $(9.6 \pm 0.3) \times 10^2\ M^{-1}\ s^{-1}$ and $(1.0 \pm 0.08) \times 10^3\ M^{-1}\ s^{-1}$, respectively.

A comparative study on the reactivity of five indole derivatives (tryptamine, *N*-acetyltryptamine, tryptophan, melatonin, and serotonin), with the redox intermediates compound I (k_2) and compound II (k_3) of the plant enzyme horseradish peroxidase and the two mammalian enzymes lactoperoxidase and myeloperoxidase, was performed using

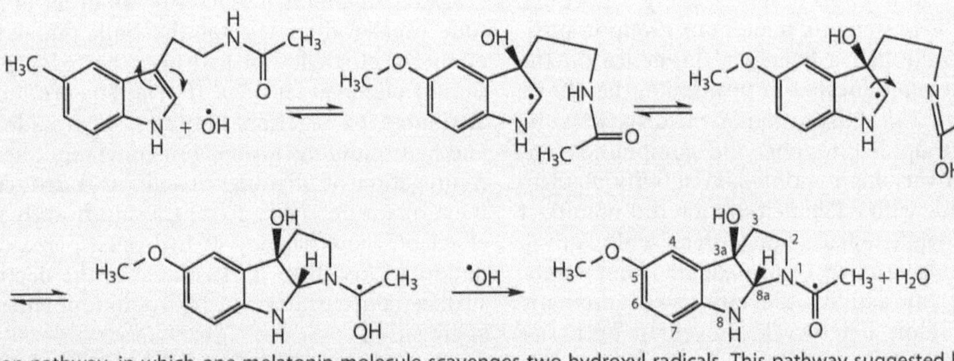

Reaction pathway, in which one melatonin molecule scavenges two hydroxyl radicals. This pathway suggested by Tan et al. (1998) differs significantly from classic free seavenging processes. [62]

the sequential-mixing stopped-flow technique (JANTSCHKO et al. 2002). The calculated biomolecular rate constants (k_2, k_3) revealed substantial differences regarding the oxidizability of the substrates by redox intermediates at pH 7.0 and 25 °C. With horseradish peroxidase it as shown that k_2 and k_3 are mainly determined by the reduction potential (E°') of the substrate with k_2 being 7–45 times higher than k_3. Compound I of the mammalian peroxidases was a much better oxidant than horseradish peroxidase compound I with the consequence that the influence of the indole structure on k_2 of lactoperoxidase and myeloperoxidase was smaller varying of a factor of only 88 and 38, respectively, which is in strong contrast to a factor of 160,000 determined for k_2 of horseradish peroxidase. The k_3 values for all three enzymes were very similar. Oxidation of substrates by mammalian peroxidase compound II is strongly constrained by the nature of the substrate. The k_3 values for the five indoles varied by a factor of 3,570 (lactoperoxidase) and 200,000 (myeloperoxidase), suggesting that the reduction potential of compound II of mammalian peroxidase is less positive than that of compound I, which is in contrast to the plant enzyme.

In addition to its direct role as a free radical scavenger and antioxidant, melatonin stimulates the activity of antioxidant enzymes: glutathione peroxidase (REITER 1995, ANTOLÍN et al. 1996, BLASK et al. 1997, REITER et al. 1997), glutathione reductase (REITER et al. 1997), gluatathione *S*-transferase, an antioxidant enzyme that detoxifies xenobiotics (KOTHARI and SUBRANANIAN 1992), Mn-superoxide dismutase (ANTOLÍN et al. 1996), and Cu,Zn-superoxide dismutase (ANTOLÍN et al. 1996). Melatonin (1–10 μM) induced γ-glutamylcysteine synthetase mediated by activator protein-1 in ECV304 *human* umbilical vascular endothelial cells in a dose-dependent manner (URATA et al. 1999).

Changing the acyl residue generally resulted in more active products (GOZZO et al. 1999). The nonanoyl derivative showed a level of activity comparable to that of phenols despite lacking a phenolic function. The presence of a methoxy group in position 5 generally had a beneficial influence on the activity, but when located in position 6, the effects were various. The substitution of a hydroxyl for the methoxy group led to phenolic compounds endowed with very high antioxidant activity. Replacing the amide with a ketone function did not affect the activity while replacement with an amine group in some cases resulted in prooxidant compounds. Comparing the activity the efficacy of aromatic rings, the indole heterocycle proved to be better than benzofurane and naphthalene rings.

6.1.4.6
5,6-Dihydroxyindoles

Increasing evidence supports the view that diffusible melanin-related metabolites may also act as modulators of the responses of the pigmentary cell melanocyte to external stimuli, especially to inflammation. 5,6-Dihydroxyindole-2-carboxylic acid caused a powerful inhibition of the H_2O_2-Fe(II)/EDTA oxidation under both aerobic and anaerobic conditions, roving to be more efficient than typical hydroxyl radical scavengers even at low concentration with respect to deoxyribose (NOVELLINO et al. 1999). Conversely, 5,6-dihydroxyindole in air was prooxidant at low indole: Fe(II) ratios, but shifted to an antioxidant at higher rations (>6). The magnitude of the prooxidant effect increased by lowering the pH of the medium or by replacing Fe(II) with Fe(III), but was suppressed by exclusion of oxygen. Both the indoles retained their effects on the Fenton reaction in the absence of EDTA, as a result of their ability to chelate iron ions as evidenced by spectrophotometric experiments. Investigation of the reaction of 5,6-dihydroxyindole and 5,6-dihydroxyindole-2-carboxylic acid with the Fenton reagent led to the conclusion that the indoles interact efficiently with $HO^•$, yielding indolesemiquinone species which are then converted to melanin pigments by self-coupling or disproportionation. At low 5,6-dihydroxyindole:iron molar ratios, the ability of semiquinones, generated by autoxidation of indoles, to recycle Fe(II) ions prevails, accounting for the observed prooxidant effect.

In vitro studies showed that the ultraviolet illumination of melanin generates superoxide anion, hydrogen peroxide, and hydroxyl radicals (FELIX et al. 1978,1979). NOFSINGER et al. (2002) examined by studying the UV-B induced oxidation and reduction of cytochrome *c* by reactive oxygen species generated by different aggregation states of eumelanin isolated from the cuttlefish *Sepia officinalis*. The quantum yield for superoxide anion by unaggregated oligomers is 7.4×10^{-3}, an order of magnitude greater than that of the bulk pigment. The quantum efficiency of hydrogen peroxide production by oligomers is 5.7×10^{-3}, and its production is attributed to reaction between superoxide anion and hydroquinone groups on eumelanin oligomers. Aggregation of oligomers results in a reduction of these quantum yields, having a significantly greater effect on the efficiency of hydrogen peroxide production. This effect is attributed to the decrease in surface concentration of hydroquinone sites upon aggregation.

6.1.4.7
4b,5,9b,10-Tetrahydroindeno[1,2*b*]indole

4b,5,9b,10-Tetrahydroindeno[1,2*b*]indole has been shown to inhibit lipid peroxidation and is thought to act as a free radical scavenger (Shertzer and Sainsbury 1991). In Jurkat T cells treated with the cytotoxic agents camptothecin, actinomycin D and ultraviolet irradiation, 4b,5,9b,10-tetrahydroindeno [1,2*b*]indole was found to inhibit the morphological features of apoptosis (Devitt et al. 1999). In UV-irradiated cells, 4b,5,9b,10-tetrahydroindeno[1,2*b*] indole partly inhibited $O_2^{\bullet-}$ production. 4b,5,9b,10-Tetrahydroindeno[1,2*b*]indole was unable to inhibit mitochondrial depolarisation in UV, camptothecin or anti-Fas-treated cells.

6.1.4.8
Pinoline

The methoxylated β-carboline, pinoline (6-methoxy-1,2,3,4-tetrahydro-β-carboline or 5-methoxytryptoline) has been primarily investigated as a free radical scavenger. In the pineal gland pinoline occurs during the metabolism of melatonin with 5-methoxytryptamine as an intermediate (Araksinen et al. 1993). It possesses biological activity in *mammals* in nanomolar concentrations, and recently interest has focused on this molecule as a potential free radical scavenger. According to Pähkla et al. (1997) pinoline possesses free radical scavenging activity similar to that of melatonin.

6.1.4.9
Indolinonic Aminoxyls

Aminoxyls can undergo reversible transformation to the corresponding hydroxylamines in the superoxide dismutase mimic reaction (Samuni et al. 1988, Mitchell et al. 1990) or irreversible transformation to non-paramagnetic alkylated hydroxylamines with carbon-centred radicals (Beckwith et al. 1988, Chateauneuf et al. 1988, 1992, Bowry and Ingold 1992). The antioxidant properties of aminoxyls and their ability to penetrate cell membranes (Hu et al. 1989) makes them attractive compounds for highly sensitive methods in the study of lipid oxidation kinetics as well as membrane accessible antioxidants in biological systems (Nilsson et al. 1989, Mitchell et al. 1991). On thermally and peroxyl radical-induced oxidised linolenic acid micelles different concentrations of aminoxyls malondialdehyde production indicated that indolinonic aminoxyls could be used as effective antioxidants in biological systems (Antosiewicz et al. 1993).

6.1.4.10
Fluvastain and its Metabolites

The inhibitory effects of 5-hydroxyfluvastain (metabolite 2 of fluvastain) and 6-hydroxyfluvastain (metabolite 3 of fluvastain) on the formation of $^1O_2$, $O_2^{\bullet-}$, $HO^{\bullet}$, and OCl^- were stronger than that of pravastatin, simvastatin, probucol and α-tocopherol (Nakashima et al. 2001). Scavenging of $^1O_2$ by the des-isopropyl metabolites 4 and 5 of fluvastain, (+)-fluvastain and (–)-fluvastain was also noted.

6.1.4.11
Benzylisoquinolines

The benzylisoquinoline alkaloids are comprised of a large group of secondary plant metabolites, which display a variety of pharmacological actions. There is a considerable variation in structure between the groups of isoquinoline alkaloids as the result of different stages in the common biogenetical pathway using tyrosine as precursor. Nevertheless, many of them possess phenolic or other reactive groups, which suggest their possible participation in redox reactions.

Some biscoclaurine alkaloids (bisbenzylisoquinoline derivatives) were able to inhibit superoxide production by phorbol ester-stimulated *human* polymorphonuclear leucocytes (Haisong et al. 1990) and lipid peroxidation in biological membranes (Shiraishi et al. 1980).

Bulbocapnine, boldine, glaucine, and stepholidine acted as scavengers of hydroxyl radicals in the deoxyribose degradation by Fe^{3+}-ethylenediamine-tetraacetic acid (EDTA) + H_2O_2 (Ubeda et al. 1993). On the contrary, laudanosoline, apomorphine, protopapaverine, anonaine, and tetrahydroberberine increased deoxyribose degradation by a mechanism related to the generation of superoxide anion. Only apomorphine had a stimulating effect in the system using citrate instead of EDTA as well as in the absence of chelator. Apomorphine also stimulated DNA damage by Cu^{2+}. The iron ion reducing ability of apomorphine and laudanosoline was confirmed using cytochrome *c*. Both compounds scavenged hydroxyl radicals in an aqueous medium, while in Fe^{3+}-induced microsomal lipid peroxidation apomorphine acted as an inhibitor and laudanosoline stimulated the process.

6.1.4.12
Bilirubin

Bilirubin and its metabolic precursor biliverdin at micromolar concentrations inhibited the formation of peroxyl radical-induced oxidation of linoleic acid in homogeneous solution in a concentration-dependent way (Stocker et al. 1987). They can scavenge the chain-carrying peroxyl radical either by donating a hydrogen atom attached to the C-10-bridge of the tetrapyrrole molecule to form a carbon-centred radical with resonance stabilisation extending over the entire bilirubin molecule or by some other path. The antioxidant activity of bilirubin increased as the experimental concentration of oxygen was decreased from 20 % (that of normal air) to 2 % (physiologically relevant concentration). Under 2 % O_2, in liposomes, bilirubin suppressed the oxidation more than α-tocopherol. While bilirubin functioned as a chain breaking antioxidant (peroxyl radical reductant), the biliverdin acted as peroxyl radical trap. Bilirubin in a complex with serum albumin, as found in blood, prevents the peroxidation of albumin bound fatty acids, and protects the bound albumin against $HO^•$ mediated degradation (Stocker et al. 1987, Neuzil and Stocker 1993). Conjugated bilirubin, as found in bile, prevents lipid peroxidation in liposomes (Stocker and Ames 1987). Both aqueous phase biliverdin and conjugated bilirubin synergize with α-tocopherol in preventing lipid peroxidation in liposomes (Stocker and Peterhans 1989). Bilirubin and biliverdin inhibit α-tocopherol consumption, possibly by reducing its chromanoxyl radical. In plasma, exogenous bilirubin inhibits lipid peroxidation after the depletion of endogenous circulating antioxidants, which are consumed in the order: ubiquinol-10, ascorbate, and bilirubin (Neuzil and Stocker 1994). Other antioxidant effects of bilirubin include reactions with $O_2^{•-}$ and HOCl (Stocker and Peterhans 1989), quenching of $^1\Delta gO_2$ (Stevens and Small 1976), inhibition of photooxidative damage to protein (Pedersen et al. 1977), and inhibition of chemiluminescence in active macrophages.

Clark et al. (2000) found that hemin-mediated increase in the inducible isoform of haem oxygenase protein expression and haem oxygenase activity is associated with augmented bilirubin levels. The majority of bilirubin production occurred early after exposure of *bovine* vascular smooth muscle cells to hemin. Hemin pre-treatment also resulted in high resistance to cell injury caused by an oxidant-generating system. Tin protoporphyrin IX, an inhibitor of haem oxygenase activity, significantly reduced bilirubin generation and reversed cellular protection afforded by hemin treatment. Addition of bilirubin to the culture medium markedly reduced the cytotoxicity produced by oxidants.

6.1.4.13 Trimethyluric Acid and its Analogues

Bhat et al. (2001) prepared new water-soluble analogues of 1,3,7-trimethyluric acid with N-1 methyl replaced by various groups and tested their ability to scavenge hydroxyl radicals as well as their protective potential against lipid peroxidation in erythrocyte membranes. The deoxyribose degradation method indicated that all the analogues tested effectively scavenge hydroxyl radicals and some of them show better activity than uric acid and methyluric acids. These effects were shown to be concentration dependent and were more potent at low concentrations (10–50 μM). Among the analogues tested, 1-butyl-, 1-propargyl- and 1-benzyl-3,7-dimethyluric acids showed high hydroxyl radical scavenging properties with a reaction rate constant (Ks) of $3.2–6.7 \times 10^{10}$ $M^{-1}s^{-1}$, $2.3 –3.7 \times 10^{10}$ $M^{-1}s^{-1}$, and $2.4–3.7 \times 10^{10}$ $M^{-1}s^{-1}$, respectively. The effectiveness of these analogues as hydroxyl radical scavengers appeared to be better than mannitol (Ks, $1.9–2.5 \times 10^9$ $M^{-1}s^{-1}$). With the exception of 1-pentyl- and 1-(2'-oxopropyl)-3,7-dimethyluric acids, all other analogues tested were effective inhibitors of *tert*-butylhydroperoxide-induced lipid peroxidation in *human* erythrocyte membranes.

6.1.4.14 Curcumin

While most of the natural antioxidants possess either a phenolic function or a β-diketone group, curcumin, (1,7- *bis* 4-hydroxy-3-methoxyphenyl)-1,6-heptadiene-3,5-one also known as diferuloylmethane, and its analogues are unique, having both phenolic and β-diketone functional groups on the same molecule. Like α-tocopherol, curcumin is a lipid soluble antioxidant and is believed to be localised within the membranous subcellular fraction of cells. Micellized curcumin reacted with haloperoxyl radicals superoxide, and lipid peroxyl radicals with rate constants of 5×10^8, 4.6×10^4, and 5.3×10^5 $M^{-1}s^{-1}$, respectively (Priyadarsini 1997). Curcumin inhibited the $^1O_2$-dependent 2,2,6,6-tetramethylpiperiodine *N*-oxyl (TEMPO) formation in a dose-dependent manner (Das and Das 2002). At 2.75 μM it caused 50 % inhibition of TEMO-$^1O_2$ adduct formation. However, curcumin only marginally inhibited (24 % maximum at 80 μM) reduction of ferricytochrome *c* in a xanthine-xanthineoxidase system demonstrating that it is not an effective superoxide radical scavenger.

In the presence of Cu(II) curcumin caused strand cleavage in DNA through generation of reactive oxygen species, particularly $HO^•$ and H_2O_2

(AHSAN and HADI 1998). Studying the structure-activity relationship between curcumin and its two naturally occurring derivatives, demethoxycurcumin and bisdemethoxycurcumin, Ahsan et al. (1999) found curcumin to be most effective in the DNA cleavage reaction and a reducer of Cu(II) followed by demethoxycurcumin and bisdemethoxycurcumin. The rate of formation of hydroxyl radicals by the three curcuminoids studied also showed a similar pattern. The relative antioxidant activity was examined by studying the effect of these curcuminoids on cleavage of plasmid DNA by Fe(II)-EDTA system (hydroxyl radicals) and the generation of singlet oxygen by riboflavin. The results indicated that curcumin was considerably more active both as an antioxidant as well as an oxidative DNA cleavage agent. The DNA cleavage activity is the consequence of binding of Cu(II) to various sites of the curcumin molecule. AHSAN et al. (1999) proposed three binding sites for Cu(II). Two of these sites are provided by the phenolic and methoxy groups on the two benzene rings and the third site is due to the presence of 1,3-diketone system between the rings.

Translocation of the transcription factor NF-κB in A549 epithelial cells, caused by carcinogenic SiC fibres (60.86 % < 10 μm; 27.6 % < 20 μm) could be significantly reduced by adding 50 μM curcumin to the culture medium (BROWN et al. 1999).

Exposure of *bovine* aortic endothelial cells to curcumin (5–15 μM) resulted in both a concentration- and time-dependent increase in heme oxygenase-1 mRNA, protein expression and heme oxygenase activity (MOTTERLINI et al. 2000). Hypoxia (18 h) also caused a significant ($P < 0.05$) increase in heme oxygenase activity, which was markedly potentiated by the presence of low concentration of curcumin (5 μM). Prolonged incubation (18 h) with curcumin in normoxic or hypoxic conditions resulted in enhanced cellular resistance to oxidative damage; this cytoprotective effects were considerably attenuated by tin protoporphyrin IX, an inhibitor of heme oxygenase activity. In contrast, exposure of cells to curcumin for a period of time insufficient to up-regulate heme oxygenase-1 (1.5 h) did not prevent oxidant-mediated injury.

Generation of $O_2^{\bullet -}$ was inhibited by curcumin, when blood neutrophils from *rhesus monkeys* were stimulated with arachidonic acid, serum treated zymosan and *N*-formyl-methionyl-leucyl alanine (SRIVASTAVA 1989).

In a p53 null *human* oral squamous carcinoma cell line curcumin (5 μM and (10 μM) arrested growth in S/G_2 which was confirmed with an increased bromodeoxyuridine labelling (WEIR and HAGUE 2002). Treated cells showed increased poly(ADP-ribose) polymerase (PARP) cleavage and cleaved caspase 3 activity. The expression of E-cadherin decreased slightly by 72 h, however, there was no change in the protein expression or localisation of β-catenin. However β-catenin activity did slightly increase.

6.1.4.15 Amidothionophosphates

In contrast to ascorbic acid and α-tocopherol amidothionophosphates did not have any prooxidative effects as measured by oxygen consumption from buffer solutions containing the drug and cupric sulphate as a source of redox-active metal ions (TIROSH et al. 1996). Amidothionophosphates reduced significantly and in a dose-dependent manner the oxygen burst in *human* neutrophils as measured by luminol-dependent luminescence, and they also markedly depressed the killing of *human* fibroblasts by mixtures of glucose oxidase and streptolysin S. The toxicity of these molecules was tested by intraperitoneal injection of doses up to 1000 mg/kg to white Sabra *mice*. No mortality was observed 30 d after administration of up to 500 mg/kg.

6.1.4.16 Poly-2-vinylpyridine-*N*-oxide and Other Nitroxides

Nitroxide radicals protect cultured mammalian cells exposed to ionising radiation (DE GRAFF et al. 1992) and microsomal membranes against lipid peroxidation (MIURA et al. 1993). They also protect bacterial cells exposed to H_2O_2, hypoxanthine/xanthine oxidase (SAMUNI et al. 1991), cytotoxic drugs such as streptonigrin (KRISHNA et al. 1994), and the naphthoquinones juglone and menadione (ZHANG et al. 1994). Nitroxides oxidise reduced metals, thus inhibiting their participation in metal-catalyzed, free radical generating reactions (SAMUNI et al. 1991, BOWRY and INGOLD 1992).

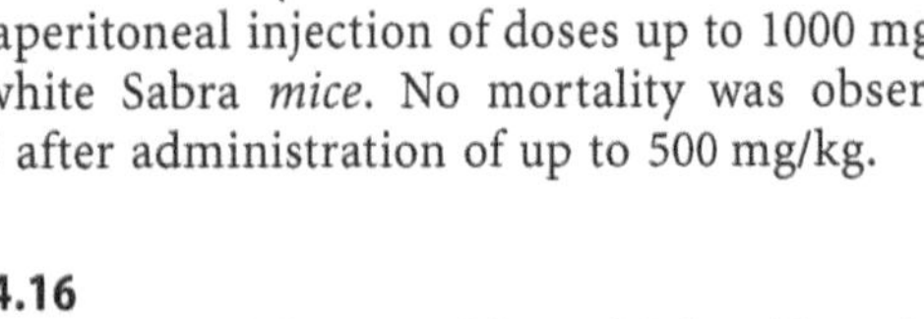

Poly-2-vinylpyridine-N-oxide (PVNO) PVNO radical-adduct [63]

Subcutaneous injections of polyvinylpyridine-*N*-oxide inhibited the development of severe silicosis after intratracheal instillation of quartz DQ 12 in *rats* and prolonged the life span; the total tumour incidence and especially the rate of malignancies increased (POTT et al. 1994). Tumour induction was significantly retarded when 0.3 mg actinolite was injected intraperitoneally suspended in 1 ml of a 2 % solution of polyvinylpyridine-*N*-oxide instead of 0.9 % NaCl (POTT et al. 1987).

Indolinonic and quinolinic aromatic nitroxides have been shown to efficiently scavenge all kinds of radicals. They couple with carbon-centred radicals, giving alkylated hydroxylamines (CARLONI 1991, STIPA et al. 1997), and unlike the aliphatic nitroxides, they react with all oxygen-centred radicals such as hydroxyl (DAMIANI et al. 1999), alkoxyl (GRECI 1982), peroxyl (CARDELLINI et al. 1989), and aroyloxyl (BERTI et al. 1977) to form nonparamagnetic compounds. From a study by DAMINANI et al. 2000) it seems that a structure-activity relationship determined by the type of ring system and its substituents (to which the nitroxide function is attached) could exist.

6.1.4.17 3-Methyl-1-phenyl-2-pyrazolin-5-one

3-Methyl-1-phenyl-2-pyrazolin-5-one (MCI-186) is a potent scavenger of hydroxyl radicals inhibiting not only hydroxyl radicals but iron-induced peroxidative injury (WATANABE et al. 1988, MUROTO et al. 1990). After the reaction with peroxy radicals, MCI-186 changes into 2-oxo-3-(phenylhydrazone)-butanoic acid (YAMAMOTO et al. 1996, KAWAI et al. 1997).

6.1.4.18 Chlorpromazine

Phenothiazines, which are frequently antihistaminic as well as antipsychotic, concentrate in lung tissue after any route of administration. Subcutaneous or intravenous injection of chlorpromazine into *rats*, *rabbits*, and *guinea pigs* resulted in a similar pattern of distribution (BERTI and CIMA 1955, HACKMAN et al. 1970). Studies in the *cat* showed similar concentrations of chlorpromazine in the lung, which remained almost constant from 1 to 48 h after intravenous injection (GOTHELF and KARCZMAR 1963).

Chlorpromazine, which inhibited vasopressin release in the *rat* (MOSES 1964), is an excellent electron donor and thus activated to a cation free radical by the myeloperoxidase system of the *human* neutrophil (VAN ZYL et al. 1990). The relatively high stability of this radical is due to resonance stabilisation and the absence of α hydrogens. The radical ion interacted with deoxyribonucleic acid (OHNISHI and MCCONNELL 1965).

Chlorpromazine inhibited both constitutive nitric oxide synthase and the induction of nitric oxide synthase after LPS challenge (PALACIOS et al. 1993).

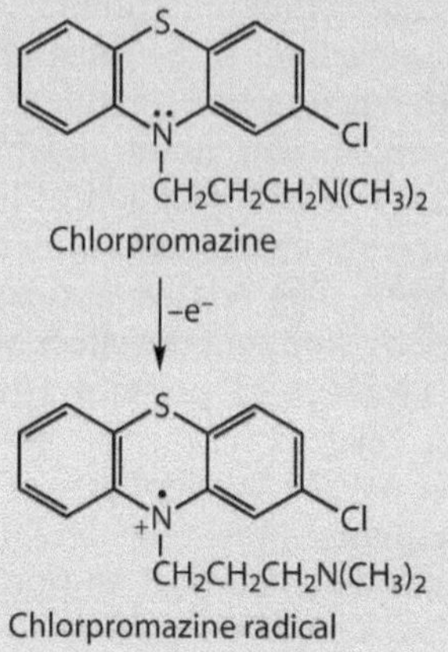

[64]

Both phototoxic and photoallergic reactions occurred in patients receiving low doses of chlorpromazine and several other phenothiazine tranquillizers (ZELICKSON and ZELLER 1964). High dosage and prolonged treatment can produce severe dermatitis that is frequently accompanied by darkening of the skin due to the deposition of melanin in lower layers of the dermis. Such patients may also suffer retinal damage, ocular opacity, and loss of vision.

Chlorpromazine sulfoxide, a chlorpromazine metabolite formed in *man* and several other *mammalian* species, produced when irradiated with near-UV light large amounts of the highly reactive hydroxyl radical (BUETTNER et al. 1986).

6.1.4.19 Ebselen

Description of the glutathione peroxidase-like activity of the biologically active selenoorganic compound ebselen (then called PZ 51) (MÜLLER et al. 1984, WENDEL et al. 1984) led to extended research on this interesting molecule ranging from studies on radical reactivity through its biological properties in cells and organs to clinical settings, notably afflictions of the central nervous system. Basically, the compound is considered capable of contributing to the antioxidant defence in tissues, so that a potential pharmacological application becomes of interest (SIES and MASUMOTO 1996).

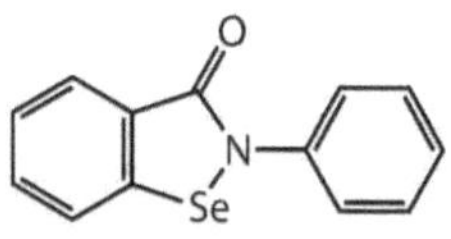

Ebselen, 2-phenyl-1,2-benzisoselenazol-3(2*H*)-one [65]

Ebselen suppressed the oxidation of methyl linoleate emulsions in aqueous dispersion induced by iron (10 μM Fe^{2+}), the spontaneous oxidation of *rat* brain and liver homogenates, but did not suppress the oxidation of these homogenates induced by 10 mM of the free radical initiator, 2,2'-azobis(amidinopropane) dihydrochloride (NOGUCHI et al. 1992).

The reaction of ebselen with **singlet oxygen** is only sluggish (SCURLOCK et al. 1991). The reactivity of ebsenen and related selenoorganic compounds with **1,2-dichloroethane radical** cations and **halogenated peroxyl radicals** was studied by pulse radiolysis (SCHÖNEICH et al. 1990). Ebselen (10 mg/kg) decreased **ozone** (2 ppm for 4 h)-induced pulmonary inflammation in *rats* (ISHII et al. 2000). Although treatment with ebselen did not alter the macrophage expression of inducible nitric oxide synthase after the ozone exposure, it did markedly inhibit the nitration reaction of tyrosine residues, suggesting that ebselen scavenged peroxynitrite during ozone-induced pulmonary inflammation. Treatment with ebselen also enhanced the pulmonary expression of both copper, zinc, and manganous superoxide dismutases at the same time.

Ebselen (<10 mg/kg) significantly inhibited late airway responses to an ovalbumin challenge in *guinea* pigs, but did not inhibit immediate airway response at any dose (ZHANG et al. 2002). Bronchoalveolar lavage examination showed that airway inflammation was significantly suppressed by ebselen at 10 mg/kg. The generation of $O_2^{\bullet -}$ and H_2O_2 occurred on endothelial cells of late airway response bronchi, and was inhibited by 10 mg/kg ebselen.

The **peroxynitrite**-induced luminol chemiluminescence emitted from Kupffer cells (WANG et al. 1991) was inhibited in the presence of low concentrations of ebselen (WANG et al. 1992). Ebselen rapidly reacts with peroxynitrite in a bimolecular fashion, yielding the selenoxide of the parent molecule, ebselen Se-oxide [2-phenyl-1,2-benzisoselenazol-3(2*H*)-one 1-oxide], as the sole selenium-containing product at 1:1 stoichiometry (MASUMOTO and SIES 1996).

A diselenide, 2,2'-diseleno-bis-β-cyclodextrin accepts a variety of hydroperoxides as substrates (LIU et al. 2000). The glutathione peroxidase-like activities, reduction of H_2O_2, *tert*-butyl hydroperoxide and cumenyl hydroperoxide by glutathione were 7.4, 4.5 and 10.2 U/μmol, respectively. In contrast to ebselen, the diselenide displayed high glutathione peroxidase-like activity. The reduction of hydroperoxide by glutathione in the presence of a radical trap showed that the mimic catalyses the reaction via a non-radical mechanism.

6.1.4.20 Thiols

The role of thiols as antioxidants protecting cell against oxidative processes and free radical attack including the antioxidant efficiency of thiols in membranes has been the subject of many studies (SCHÖNEICH et al. 1990, BARCLAY et al. 1995). However, in addition to the chemical repair of radical damage by the reducing thiol group, the role of the thiyl radical formed as a source of a reactive oxidant has to be addressed (STOCKER and FREI 1991). Thiyl radicals attack polyunsaturated fatty acids at the bisallylic methylene groups forming pentadienyl radical by hydrogen abstraction. Thiyl radical additions occur to the double bonds. Thiyl radical-catalysed lipid isomerization takes place within the adduct of the thiyl radical to an olefinic group of unsaturated fatty acids, but not within the pentadienyl radical (SPRINZ et al. 2000).

Carbon-centred radicals formed on the carbohydrate moieties of DNA can be "repaired" through hydrogen transfer from thiols (VON SONNTAG 1987):

$$\mathrm{DNA{-}C^{\bullet} + RSH \longrightarrow DNA{-}CH + RS^{\bullet}} \qquad [66]$$

Rate constants are on the order of 10^4 $M^{-1}s^{-1}$, with the highest average value for 2-deoxy-D-ribose $(2.7 \pm 1.0) \times 10^4$ $M^{-1}s^{-1}$, and the lowest average value for 2-deoxy-D-glucose $(1.6 \pm 0.2) \times 10^4$ $M^{-1}s^{-1}$, based on two ways of kinetic analysis, standard competition kinetics and stochastic simulation of the experimental results, respectively (POGOCKI and SCHÖNEICH 2001). In general, thiyl radicals attack preferentially the C^1-H bond of the carbohydrates, to an extent of ca 72% in 2-deoxy-D-ribose and 90% in 2-deoxy-D-glucose.

Data on the hydrogen transfer to deoxyuridine-1'-yl radicals in model oligonucleotides have shown that this process my occur with remarkable stereoselectivity, restoring predominantly the naturally occurring β-deoxynucleotide (HWANG and GREENBERG 1999).

6.1.4.20.1 Tetradecylthioacetic Acid

Tetradecylthioacetic acid is a synthesised saturated fatty acid where a sulphur atom substitutes the third methylene group from the carboxylic end:

$$\mathrm{CH_3{-}(CH_2)_{13}{-}S{-}CH_2{-}COOH} \qquad [67]$$

Tetradecylthioacetic acid cannot be β-oxidised (LAU et al. 1988) but can be metabolised by sulphur- and ω-oxidation to dicarboxylic acids, which are subsequently excreted via the kidneys (Bergseth and Bremer 1990). Tetradecylthioacetic acid and the more potent tetradecylselenoacetic acid

$$CH_3—(CH_2)_{13}—Se—CH_2—COOH \quad [68]$$

increased the lag time before the onset of copper-induced low-density lipoprotein oxidation in a dose-dependent manner (Muna et al. 2000). Tetradecylthioacetic acid and tetradecylselenoacetic acid were shown to reduce the iron-ascorbate-induced microsomal lipid peroxidation. In the presence of iron, they interacted with the superoxide radical as assessed by direct and indirect testing methods. Both failed to scavenge 1.1-diphenyl-2-picrylhydrazyl radicals. Tetradecylselenoacetic acid bound copper ions as shown by the wavelength spectra measurement.

6.1.4.20.2
Mercaptohistidine Derivatives

2-Mercaptohistidine trimethylbetaine (ergothioneine) is known to be formed in micro-organisms (Melville et al. 1955, Melville 1959, Hartman 1990). It possesses beneficial radioprotective effects (Rougee et al. 1988, Hartman 1990), scavenges singlet oxygen (Dahl et al. 1988), hypochlorous acid, hydroxyl radicals (Akanmu et al. 1991, Asmus et al. 1996), azide ($N_3^•$) radicals (Asmus et al. 1996) and trichloromethylperoxyl ($CCl_3O_2^•$ radicals (Asmus et al. 1996), possesses antimutagenic properties (Hartan and Hartman 1987), and has been linked to the metabolism of iron, copper and zinc (Motohashi et al. 1976) and the inhibition of metallo-enzymes (Hanlon 1971). Ergothioneine inhibited peroxynitrite (1 mM)-induced oxidative damage in isolated *calf* thymus DNA and DNA in the neuronal hybridoma cell line N-18-RE-105 cells (Aruoma et al. 1999). Its concentration in *human* and *mammalian* tissues has been estimated to be up to 1–2 mM (Melville et al. 1955, Melville 1959, Muda et al. 1988, Hartman 1990), which suggests that ergothioneine may serve as a non-toxic thiol buffering antioxidant *in vivo* and may find application in pharmaceutical preparations where oxidative stability is desired.

Of the 4-mercaptohistidine derivatives to which belong the ovothioles from the eggs of marine *echinoderms* and *mollusces*, 1,5-dimethyl-4-mercaptoimidazole due to the unusual stability of its thiyl radical is superior to glutathione as a one-electron donor (Holler and Hopkins 1990). Danen and Newkirk (1976) have similarly rationalised the remarkable stability of thionitroxide radicals.

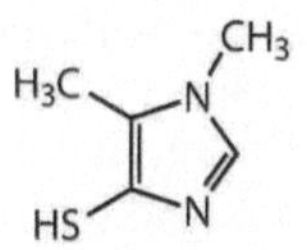

1,5-Dimethyl-4-thioimidazol [69]

Zoete et al. (1997) tested the radical-scavenging properties of fourteen thioimidazols using two stable free radicals, Fremy's salt and the 2,2-diphenyl-1-picrylhydrazyl radical. Seven compounds emerged as the most active molecules far above glutathione and close to ascorbic acid. Each of these active compounds were substituted on position 2 or 5 of the imidazol ring by strongly withdrawing groups like chlorophenyl or trifluoromethylphenyl.

6.1.4.20.3
Thioproline

Thioproline (thiazolidine-4-carboxylic acid) can act as intracellular sulfhydryl antioxidant and free radical scavenger (Weber et al. 1982) protecting cellular membranes from damage due to oxygen-derived reactions. This antioxidant, when administered to old *mice*, had a favourable effect on lymphocyte functions (De la Fuente et al. 1993). Thioproline had no effect on adherence of *murine* (BALB/c *mice* 14 to 20 weeks old) peritoneal macrophages to a smooth plastic surface (Eppendorf tubes) while the phagocytosis of latex beads (1.09 μm) was stimulated (Del Rio et al. 1998). Random migration, chemotaxis, ingestion and superoxide anion production were increased.

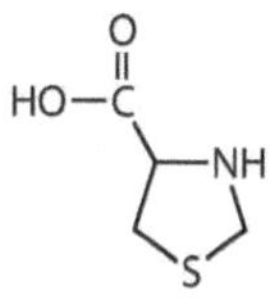

Thioproline [70]

6.1.4.20.4
3-(2-Mercaptoethyl)quinazoline-2,4(1*H*,3*H*)dione

3-(2-Mercaptoethyl)quinazoline-2,4(1*H*,3*H*)dione (2 mM) isomerized 5 mM oleic acid without a significant decay of the thiol concentration (Sprinz et al. 2000). However, in fatty acid chains with bisallylic groups two competing processes occurred: adduct formation (including the isomerization) and the hydrogen abstraction of a bisallylic hydrogen.

$$RS^• + PUFA \rightarrow L^• + RSH \quad [71]$$

6.1.4.20.5 Metallothioneins

Although it is generally accepted that the principal roles of metallothionein lie in the detoxification of heavy metals and regulation of the metabolism of essential trace elements, there is increasing evidence that in can act as a free radical scavenger (SATO and BREMNER 1993). The metallothionein contains two separate metal thiolate clusters, one with four (cluster A) and one with three (cluster B) metal ions, respectively (KÄGI et al. 1990). The affinity of metal ions for the binding sites on metallothionein differs quire markedly, the rank order being Zn < Cd < Cu < Hg < Ag (HOLT et al. 1980). Metallothionein is characterised by a high thiol content and absence of aromatic amino acids, including tyrosine, tryptophan and phenylalanine (KOJIMA and KÄGI 1978). Twenty of the 61 amino acid residues in the molecule are cysteinyl residues, all of which are involved in metal binding.

Oxidative stress induced by several toxicants increased metallothionein levels in *mouse* tissues (BAUMAN et al. 1991).

6.1.4.21 Carvedilol

Carvedilol, 1-(9H-carbazol-4-yloxy)-3-[[2-(2-methoxyphenoxy)ethyl]amino]-2-propanol, competitively inhibits β-receptors (SPONER et al. 1987, NICHOLS et al. 1989, DE MEY et al. 1994, SPONER and FEUERSTEIN 1999), blocks α_1-receptors (DE MEY et al. 1994, SPONER and FEUERSTEIN 1999), and acts as an antioxidant (YUE et al. 1992, 1999, ARUMANAYAGAM et al. 2001). Hydroxylation of the carbazol moiety significantly increases the antioxidative effect. Some metabolites have a ten-times higher oxygen radical-scavenging effect than carvedilol itself (FEUERSTEIN et al. 1993). A hydroxylated analogue of carvedilol affords exceptional antioxidant protection to postischemic *rat* hearts (KRAMER and WEGLICKI 1996).

Four oxidative metabolites: 1-hydroxycarvedilol, 8-hydroxycarvedilol, 4'-hydroxycarvedilol, and *O*-desmethylcarvedilol were formed by incubation of *R*(+)- and *S*(−)-carvedilol with *rat* liver microsomes (FUJIMAKI 1994). As expected from in *vivo* metabolism studies, 1-hydroxycarvedilol and 8-hydroxycarvedilol were the major products for both enantiomers used as a substrate. The *S*/*R* enantiomeric ratios for intrinsic clearance (V_{max}/K_M) of 1-hydroxycarvedilol, 8-hydroxycarvedilol, *O*-desmethylcarvedilol and 4'-hydroxycarvedilol were 0.40, 1.99, 0.77, and 2.71, respectively, showing that stereospecific oxidation occurs in this species. The cause of the difference in intrinsic clearance for 1-hydroxycarvedilol and 8-hydroxycarvedilol between the two enantiomers was based on the difference in affinity in the catalysing enzyme. The main enzyme concerned in the 1- and 8-hydroxylation of both enantiomers is considered to be CYP2D1. In the *O*-demethylation of both enantiomers CYP2C11 is probably the main catalysing enzyme, because sex differences ' but not strain differences, were observed in both Sprague-Dawley and Dark Agouti *rats*, and also anti-CYP2C11 strongly inhibited this demethylation.

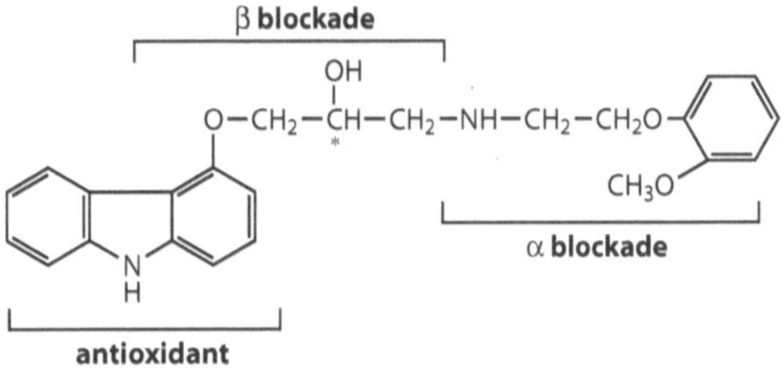

Chemical structure of carvedilol [72]

Since the antioxidant capacity of carvedilol is believed to reside in the carbazole moiety (YUE et al. 1992, 1994, MACKERELL jr. et al. 1995), MIGLIAVACCA et al. (1998) used 4-methoxycarbazole as a model compound for carvedilol, and 1-hydroxy- and 3-hydroxy-4-methoxycarbazole for its 1- and its 3-hydroxylated metabolites. 1-Hydroxy- and 3-hydroxy-4-methoxycarbazole have good antioxidant activities based of their ΔH_{abs} values. These calculations suggest that both compounds scavenge free radicals via direct H-atom transfer, since their oxyl radical is relatively stable. Hydroxylated carbazole derivatives such as carazostatin and carbazomycin show a considerable antioxidant activity, whereas non-hydroxylated carbazoles are inactive (IWATSUKI et al. 1993, KATO et al. 1993).

There was no evidence from cell-free assay systems that carvedilol is a scavenger for $O_2^{\bullet-}$ or $^{\bullet}NO$ (ÅSBRINK et al. 2000). Carvedilol did not affect other reactions dependent on $^{\bullet}NO$, e.g. spontaneous of formyl-methionyl-leucyl-phenylalanine-stimulated polymorphonuclear leucocyte migration or lipoxin A_4-, fMLP-, or A23187-induced neutrophil cytotoxicity for *human* umbilical vein endothelial cells. Thus, these effects point to the possibility that carvedilol modulates the NADPH oxidase of polymorphonuclear leucocytes but leaves the nitric oxide synthase of phagocytes intact. Carvedilol exerted poor reactivity toward phenoxyl, alkoxyl, and peroxyl radicals in acetonitrile solution nor did it show an appreciable antioxidant effect against either the peroxyl radical-induced oxidation of methyl linoleate in acetonitrile or against phosphatidylcholine liposomal membranes in aqueous suspension (NOGUCHI et al. 2000). Carvedilol completely inhibited the ferric ion-induced oxidation of methyl linoleate micelles by sequestering ferric ions, but not by reducing hydroperoxide. It was shown that carvedilol enhanced the oxidation of micelles induced by either methemoglobin or peroxyl radical.

6.1.4.22 Angiotensin Converting Enzyme Inhibitors

[^{14}C]Alacepril is converted to captopril via desacetylalacepril in *rat* liver, kidney and intestine homogenates, but not in lung homogenate and plasma, where only deacetylation occurred (MATSUMOTO et al. 1986). Desacetylalacepril and captopril, and other -SH angiotensin converting enzyme inhibitors are effective free radical scavengers (CHOPRA et al. 1992, NODA et al. 1997). WESTLIN and MULLANE (1988) suggested that captopril scavenges the superoxide radical. However, neither hypoxanthine nor xanthine oxidase had any effect on captopril -SH as measured with 5,5'-dithiobis-(2-nitrobenzoic acid) (CHOPRA et al. 1992). In the *rat*, captopril may protect the lung (WARD et al. 1992), bowel (YOON et al. 1994), and kidney (COHEN 1994, 1996) from the development of radiation injury. The free radical scavenging action of captopril is further substantiated by the observation that captopril, but not lisinopril, inhibited $FeCl_3$/ascorbic acid-induced lipid peroxidation in whole tissue homogenates of *rabbit* aorta to a level comparable to that of superoxide dismutase (MITTRA and SINGH 1998).

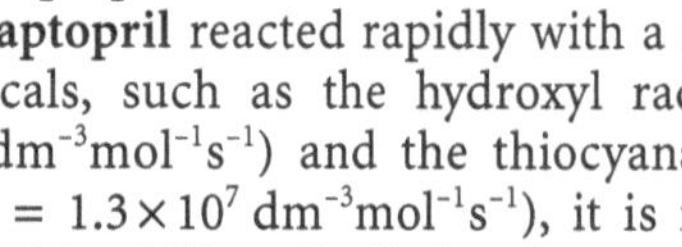

Captopril [73]

Using the pulse radiolysis technique FORNI et al. (1996) obtained absolute rate constants for the reaction of captopril with several free radicals. Although **captopril** reacted rapidly with a number of free radicals, such as the hydroxyl radical (k = 5.1×10^9 $dm^{-3}mol^{-1}s^{-1}$) and the thiocyanate radical anion (k = 1.3×10^7 $dm^{-3}mol^{-1}s^{-1}$), it is not exceptional in this ability. Similarly, the reactions with carbon centred radicals although rapid are an order of magnitude slower than those observed with glutathione. Additional lipid peroxidation studies demonstrated that captopril is a much less effective antioxidant than glutathione. The data go some way to supporting the view that any attenuation of reperfusion injury by captopril is not through a direct free radical scavenging mechanism but may be afforded by other, non radical-mediated mechanisms.

Captopril inhibited (IC_{50} = 7.6 μM) the opsonized zymosan-induced luminol-enhanced chemiluminescence of *human* blood polymorphonuclear neutrophils (NAGY et al. 1997).

Captopril succeeded in suppressing oedema evolution in hind paws of Freund's arthritic Wistar *rats*, during all phases of the disease (AGHA and MANSOUR 2000). During the chronic phase of inflammation, in both Freund's arthritic and mixed-type hypersensitive *rats*, captopril reduced the elevated serum and exudate (local) leukotriene B_4 and IL-6 levels. The effect of leukotriene B_4 was more pronounced in the exudate and tended to be dose-related. The antiarthritic effect of captopril was also accompanied by augmentation of serum level of protein thiols, with reduction or normalisation of elevated systemic and/or local levels of lipid peroxide, superoxide dismutase (EC 1.15.1.1) and glutathione.

Captopril significantly decreased Cu concentration in liver, adrenals, jejunum, urine and hair of male Hartley-Albino *guinea pigs* as measured by flame atomic absorption spectrophotometry (KOTSAKI-KOVATSI et al. 1997). A significant increase was observed in heart, epididymal and faecal Cu.

In order to assess of sulfhydryl–SH group in the effects of captopril, a SH containing drug S8 and a disulphide DG4, both are deficient in angiotensin converting enzyme inhibitory properties *in vitro*, PI and CHEN (1989) found that S8 (180 μmol/l) provided a significant protection while DG4 showed no protective effect.

Ramiprilat, a non-SH-containing angiotensin converting enzyme inhibitor, inhibited free radical-induced damages mainly by stimulation of prostacyclin synthesis and/or release (PI and CHEN 1989). Ramiprilat (10^{-10} M) significantly potentiated the release of nitrite (the hydration product of $^{\bullet}NO$) from isolated coronary microvessels induced by bradykinin (10^{-10} – 10^{-7} M). and kallikrein (0.5–10 U/ml) (ZHANG et al. 2000). The calcium channel blocker, amlodipine (10^{-10} M) markedly enhanced nitrite production by ramiprilat (10^{-7} M) from 122±9 to 168±14 pmol/mg ($P<0.05$ vs. ramiprilat. Nitrite release potentiated by ramiprilat and amlodipine was entirely blocked by N^{ω}-nitro-L-arginine methyl ester, an inhibitor of nitric oxide synthase.

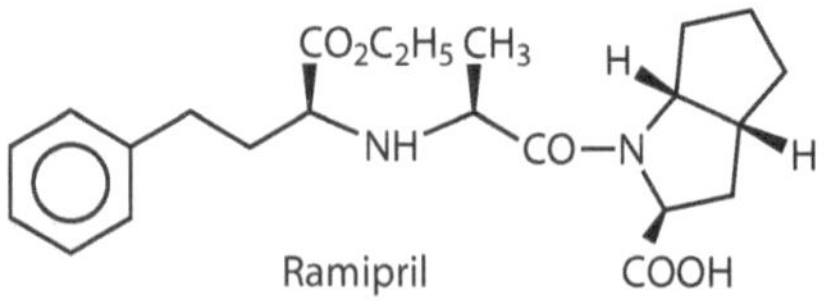

Ramipril [74]

Serum iron levels significantly decreased by 3.3 µmol/l (from 15.5±4.9 to 12.2±3.9 µmol/l) in 11 patients with chronic congestive heart failure medicated with 5 mg ramipril per day for 2 weeks (VERHO et al. 1993).

Cilazapril (10 mg/kg × day) abolished myocardial reactive oxygen species in Dahl salt-sensitive *rats* on high-salt (8% NaCl for 10 weeks) diet (TSUTSUI et al. 2001).

Enalapril maleate modified the activities of NADH oxidase and NADH cytochrome *c* reductase, probably inhibiting electron transfer between complex I and complex III in *rat* kidney mitochondria (BASSO et al. 1991).

Lisinopril, a lysine analogue of enalaprilat, hardly scavenged the superoxide or the hydroxyl radicals *in vitro*, using an ESR method (NODA et al. 1997).

6.1.4.23 Propofol

The anaesthetic propofol (2,6-diisopropylphenol) has, as phenol-based free radical scavengers (R–OH), antioxidant properties (KAHL 1984, MURPHY et al. 1996). By studying the in vitro interaction of propofol and peroxynitrite by chemiluminescence, KAHRAMAM and DERMIRYÜREK (1997) affirmed that propofol lowered the luminol or lucigenin-enhanced chemiluminescence of peroxynitrite. MOUITHYS-MICKALAD et al. (1998) by addition of peroxynitrite to propofol in alkaline solution (pH 12) detected a short lifetime ESR signal corresponding to a phenoxy radical. This finding was confirmed by a UV-visible study, resulting in the appearance of 427 nm peak and the disappearance of the peak located at 239 nm. The 291 nm peak remained unchanged.

6.1.4.24 Marchantin H

Marchantin H is a natural compound isolated from *Marchantia diptera* (WU 1990). Marchantins are naturally phenolic structures isolated from different species of liverwort (TORI et al. 1985, ASAKAWA et al. 1987). Marchantin H could scavenge the stable free radical 1,1-diphenyl-2-picrylhydrazyl and peroxyl radical derived from 2,2'-azobis(2-amidinopropane) dihydrochloride in aqueous phase, but not the peroxyl radical derived from 2,2'-azobis (2,4-dimethylvaleronitrile) in hexane (HSIAO et al. 1996). It was reactive toward superoxide anion generated by the xanthine/xanthine oxidase system. Marchantin H inhibited copper-catalysed oxidation of *human* low-density lipoprotein, as measured by fluorescence intensity, thiobarbituric acid-reactive substance formation, and electrophoretic mobility in a concentration-dependent manner.

6.1.4.25 Salen-Manganese Complexes

EUK-8, a salen-manganese complex may be regarded as a prototype molecule of a class of synthetic catalytic scavengers with combined superoxide dismutase/catalase activity. The superoxide dismutase activity of EUK-8 and other salen-manganese complexes (BAUDRY et al. 1993) has been demonstrated using a coupled "indirect" assay method (MCCORD et al. 1973). In this procedure, xanthine and xanthine oxidase continuously generate superoxide, which is monitored by its ability to reduce the indicator molecule to a spectrophotometrically detectable product. Addition of an agent with superoxide dismutase activity results in suppression of the rate of adsorbance change. EUK-8 suppresses the rate of reduction of the indicator molecule nitro blue tetrazolium in such an assay system (BAUDRY et al. 1993). EUK-8 also exhibited catalase activity, based on the ability to generate oxygen in the presence of H_2O_2.

GONZALEZ et al. (1995) evaluated EUK-8 in a highly stringent *porcine* model for sepsis-induced **adult respiratory distress syndrome** (ARDS). EUK-8 abrogated the increase in lung malondialdehyde, indicating that it prevented tissue lipid peroxidation.

SHARPE et al. (2002) investigated EUK-8 and EUK-134 [manganese 3-methoxy *N,N'*-bis(salicylidene-ethylenediamine chloride] as possible therapeutic agents in **neurological disorders** resulting from oxidative stress, including Alzheimer's disease, Parkinson's disease, stroke, and multiple sclerosis. They found that in the presence of a per-species (H_2O_2, $ONOO^-$, peracetate and persulphate), the Mn-salen complexes are oxidised to the corresponding oxo-species (oxoMn-Salen). oxoMn-Salens are potent oxidants, and they can rapidly oxidise NO to NO_2 and also oxidise nitrite (NO_2^-) to nitrate (NO_3^-). Thus these Mn-salens have the potential to ameliorate cellular damage caused by oxidative and nitrosative stressors, by the catalytic breakdown of $O_2^{\bullet -}$, H_2O_2, $ONOO^-$, and $^{\bullet}NO$ to benign species: O_2, H_2O, NO_2^-, and NO_3^-.

6.1.4.26 Schizandrins

Schizandrins are effective components extracted from the Chinese traditional drug Fructus Schizandrae. Their antioxidant effects may protect *hu-*

man polymorphonuclear leucocytes stimulated with 12-*O*-tetradecanoylphorbol-13-acetate from the reactive oxygen species produced (ZHAO et al. 1990). As judged by electron spin resonance spin trapping on hydroxyl radicals, the scavenging effects of schizandrin B and schizandrol A were greater than those of vitamin E and ascorbic acid. The scavenging effects on $O_2^{\bullet -}$ surpassed those of vitamin E but were less than by ascorbic acid.

6.1.5 Antioxidants

Antioxidants minimise oxidation of the lipid components in foods. It is important to evaluate such natural and/or synthetic compounds fully for both antioxidant and pro-oxidant properties. AESCHBACH et al. (1994) characterised the properties of thymol, carvacrol, 6-gingerol, hydroxytyrosol and zingerone. Thymol, carvacrol, 6-gingerol and hydroxytyrosol decreased peroxidation of phospholipid liposomes in the presence of Fe(III) and ascorbate, but zingerone had only a weak inhibitory effect on the system. The compounds were good scavengers of peroxyl radicals ($CCl_3O_2^{\bullet}$; calculated rate constant $> 10^6\ M^{-1}s^{-1}$) generated by pulse radiolysis. Thymol, carvacrol, 6-gingerol and zingerone were not able to accelerate DNA damage in the bleomycin-Fe(III) system. Hydroxytyrosol promoted deoxyribose damage in the deoxyribose assay and also promoted DNA damage in the bleomycin-Fe(III) system. This promotion was inhibited strongly in the deoxyribose assay by the addition of bovine serum albumin to the reaction mixtures.

Carotenoids act as antioxidants in solution, micelles, and liposomes (HILL et al. 1995, MORTENSEN and SKIBSTED 1997, MORTENSEN et al. 1997, WOODALL et al. 1997). The scavenging ability of the carotenoids: β-carotene, 8'-apo-β-caroten-8'-al, canthaxanthin, 7'-apo-7',7'-dicyano-β-carotene, ethyl 8'-apo-β-caroten-8'-oate, and 7,7'-diapo-7,7'-diphenylcarotene towards radical $HOO^{\bullet}$ correlated with their redox properties (POLYAKOV et al. 2001).

Adhatoda vesica (*Justicia adhatoda*) leaf extract contains a number of different principles: alkaloids (vesicine, vesicinone, vesinol), essential oil (betane), vitamins (ascorbic acid, β-carotene), a non-crystalline steroid (vasakin) and a mixture of fatty acids contributing to the observed medicinal effects of the plant. SINGH et al. (2000) examined the modulatory effect of the extract on the liver, lung, kidney and forestomach of 8 weeks old Swiss albino *mice*. Significant increase in the activities of acid soluble sulphydryl content, cytochrome P450, NADPH-cytochrome P450 reductase, cytochrome b_5, NADPH-cytochrome b_5 reductase, glutathione S-transferase, DT-diaphorase, superoxide dismutase, catalase, glutathione peroxidase and glutathione reductase were observed in the liver. *Adhatoda vesica* acted as bifunctional inducer since it induced both phase I and phase II enzyme systems. Treated groups showed significant decrease in malondialdehyde formation in liver, suggesting its role in protection against prooxidant induced membrane damage.

COE et al. (2002) investigated the role of the antioxidant **glutathione** in the response of embryonic stem cells to oxidative stress. Embryonic stem cells express γ-glutamylcysteine synthetase, a critical enzyme in glutathione (GSH) biosynthesis. Treatment with the pro-oxidant menadione led to elevation of GSH, a strong apoptotic response and reduced clonogenic survival. Addition of D,L-buthionine-[*S*, *R*]-sulphoximine, a specific γ-glutamylcysteine synthetase inhibitor depleted GSH pools and prevented the menadione-induced increase in GSH, sensitising the cells to oxidative insult.

The **peroxiredoxins**, a novel family of antioxidant proteins which catalyse the reduction of peroxides using their conserved cysteine residues, can be divided into two subgroups, one containing a single conserved cysteine residue, and the other containing an additional conserved cysteine residue (RHEE et al. 1999). 1-cys-peroxiredoxin is abundant in the lung (KIM et al. 2002). While there was little change in 1-cys-peroxiredoxin expression during the prenatal period, a marked increase in expression occurred immediately after birth. Enzymatic (peroxidase and phospholipase) activities increased gradually after birth and reached adult level at 7–14 postnatal days. Expression of the protein was induced in the presence of dexamethasone in cultures *human* and *rat* lung epithelial cells and also was upregulated in neonatal *rat* lung *in vivo*.

In COS-1 cell, **flavonoids** (onion extract and quercetin) increased the intracellular glutathione level by transactivation of the γ-glutamylcysteine synthetase catalytical subunit promoter (MYHRSTAD et al. 2002).

Chalcones (1,3-diaryl-2-propen-1-ones) are flavonoids lacking a heterocyclic C ring. Also this category of flavonoids displays a broad spectrum of bioactivities such as anticancer, antifungal, antibacterial, antiviral, and anti-inflammatory properties (CALLISTE et al. 2001). Dihydrochalcones, which do not have α–β double bond, comprise phloretin [β-(4-hydroxyphenyl)-1-(2,4,6-trihydroxypropiophenone) and its glucoside, phloridzin (phloretin 2- β-D-glucose). Comparison with structurally related compounds revealed that the antioxidant pharmacophore of phloretin is 2,6-dihydroxyacetophenone (REZK et al. 2002). The potent activity of 2,6-dihydroxyacetophenone is due to stabilisation of its radical via tautomerisation. The antioxidant pharmacophore of the dihydrochalcone phloretin, i.e. the 2,6-dihydroxyacetophenone group, is different from the antioxidant pharmacophores previously reported in flavonoids.

The anti-oxidant and pro-oxidant properties of **tannic acid** and its structural component gallic acid were compared by KHAN et al. (2000). It was shown that tannic acid is the most efficient generator of the hydroxyl radical in the presence of Cu(II), as compared with gallic acid and its analogues syringic acid and pyrogallol. On the other hand, tannic acid provided the maximum protection against cleavage of plasmid DNA, while gallic acid and its structural analogues were found to be non-inhibitory or partially inhibitory. Restriction analysis of treated phage DNA and thermal melting profiles of *calf* thymus DNA indicated that tannic acid strongly binds to DNA.

An aqueous extract from an infusion of *Ilex paraguariensis* (Aquifoliaceae) inhibited the enzymatic and nonenzymatic lipid peroxidation in *rat* liver microsomes in a concentration-dependent fashion, with IC_{50} values of 18 μg/ml and 28 μg/ml, respectively (SCHINELLA et al. 2000). The extract also inhibited the H_2O_2-induced peroxidation of red blood cell membranes with an IC_{50} of 100 μg/ml and exhibited radical scavenging properties toward $O_2^{\bullet-}$ (IC_{50} = 15 μg/ml) and 2,2-diphenyl-1-picrylhydrazyl radical. In the range of concentrations used the extract was not a scavenger of $HO^{\bullet}$.

Interplay between different protective mechanisms can occur. α-Tocopherol interacts with the activity of other antioxidants like GSH (HAENEN and BAST 1983) and ascorbic acid (MCCAY 1985, WIJESUNDARA and BERGER 1994, HAMILTON et al. 2000). α-Tocopherol inhibited *human* glutathione *S*-transferase π in a concentration-dependent manner, with an IC_{50} value of 0.5 μM (VAN HAAFTEN et al. 2001). At α-tocopherol additions above 3 μM there was no glutathione *S*-transferase π activity left.

Caffeic acid (3,4-dihydroxycinnamic acid) is a widespread phenolic acid with a well-known antioxidant activity (NARDINI et al. 1995, VIEIRA et al. 1998). In *human* monocytic U937 cells it inhibited both ceramide-induced NF-κB binding activity and apoptosis at μmolar concentrations (NARDINI et al. 2001). Other antioxidants were totally ineffective in inhibiting apoptosis, although affecting NF-κB activation. Caffeic acid was found to inhibit protein tyrosine kinase activity, suggesting that this mechanism can be on the basis of the inhibition of apoptosis.

Apigenin, chrysin, and kaempferol strongly enhanced the inhibition of inducible cyclooxygenase and inducible nitric oxide synthase promoter activities in lipopolysaccharide-activated macrophages, which contain the peroxisome proliferator-activated receptor γ expression plasmids (LIANG et al. 2001).

Comparison of the pK_a values to the pH-dependent antioxidant profiles, determined by the Trolox equivalent antioxidant capacity, revealed that for various hydroxyflavones the pH-dependent behaviour is related to hydroxyl moiety deprotonation, resulting in an increase of the antioxidant potential upon formation of the deprotonated form (LEMAŃSKA et al. 2001).

OLLILA et al. (2002) studied the interaction between flavonoids and membranes composed of dipalmitoylphosphatidylcholine by means of noncovalent immobilized artificial membrane chromatography and flavonoid-induced calcein release from fluid egg phosphatidylcholine vesicles. Flavonoids with more hydroxyl groups showed longer retention delays in the immobilized artificial membrane studies, suggesting stronger interactions between the flavonoids, which are rich in hydroxyl groups, and the dipalmitoylphosphatidylcholine membrane interface. Both polar and nonpolar forces were shown to have a significant impact on the flavonoid-biomembrane interactions.

Rosemary (*Rosmarinus officinalis*) contains flavonoids, phenols, volatile oil and terpenoids. Topical application of rosemary extract, carnosol or ursolic acid to *mouse* skin inhibited the covalent binding of benzo[*a*]pyrene to epidermal DNA (HUANG et al. 1994), tumour initiation by 7,12-dimethylbenz [*a*]anthracene (SINGLETARY and NELSHOPPEN 1991), 12-*O*-tetradecanoylphorbol-13-acetate-induced tumour promotion, ornithine decarboxylase (EC 4.1.1.17) activity and inflammation. Carnosol showed potent antioxidative activity in α,α-diphenyl-β-picrylhydrazyl free radicals scavenge and DNA protection from Fenton reaction (LO et al. 2002).

The formation of flavonoid metal complexes increased the capacity of rutin and dihydroquercetin to protect peritoneal macrophages against chrysotile asbestos-induced injury (KOSTYUK et al. 2001). Metal complexes of all flavonoids were found to be considerably more potent than parent flavonoids in protecting red blood corpuscles against asbestos-induced injury.

Hyperforin, a component of *Hypericum perforatum* L. (Saint John's-wort) incubated with *human* coronary endothelial cells reduced the expression of endothelial leucocyte adhesion molecules ICAM-1 (intercellular cell adhesion molecule-1, CD54) and VCAM-1 (vascular cell adhesion molecule-1, CD196), whereas the expression of E-selectin

(CD62E) was unchanged (Fitzl et al. 2002). Electron microscopy showed slight ultrastructural changes of the endothelial cells, e.g., enhanced vesiculation, disturbance of mitochondria, and stronger adhesion of the cells to the substrate.

Dimerumic acid, isolated as the active component with a radical scavenging action from the mould *Monascus anka*, and traditionally used for the fermentation of foods, inhibited NADPH- and iron(II)-dependent lipid peroxidation of rat liver microsomes at 20 and 200 μM, respectively (Taira et al. 2002). The antioxidant action of dimerumic acid is due to one electron donation of the hydroxamic acid group in the dimerumic acid molecule toward oxidants resulting if formation of nitroxide radical.

The inhibition of lipid peroxidation by cinnarizine seems to be independent of the oxidant system used to induce the peroxidation, as was verified when using xanthine oxidase and iron (Janero et al. 1988), as well as simple exposure to air (for *rat* liver homogenates; Fernandes et al. 1991), copper (for *human* plasma and erythrocytes; Fernandes et al. 1991) or H_2O_2 (for *human* erythrocytes; Fernandes et al. 1991).

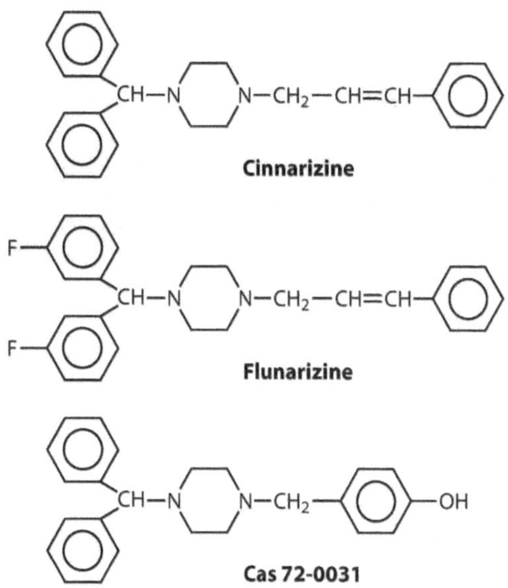

4-Diphenylmethylpiperidine derivatives [75]

6.2 The L-Arginine-Nitric Oxide Pathway

The demonstration of the synthesis of nitric oxide from the amino acid L-arginine by vascular endothelial cells has led to the elucidation of the importance of the L-arginine-nitric oxide pathway as a regulator of cell function in a number of tissues (Moncada et al. 1989). The reaction involves a 5-electron oxidation of one of the chemically equivalent guanidino-nitrogens of L-arginine, leading to the concomitant production of L-citrulline and ${}^{\bullet}NO$. It is accompanied by an NADPH-dependent reduction of molecular oxygen (Mayer et al. 1991), which is incorporated into both reaction products (Kwon et al. 1990, Leone et al. 1991).

L-Arginine–nitric oxide pathway

$$\text{L-Arginine} \xrightarrow{\text{NOS}} \textbf{Nitric oxide}({}^{\bullet}\text{NO}) \xrightarrow{+\mathbf{O_2^{\bullet-}}} \text{Peroxynitrite } (\text{ONOO}^-) \quad [46]$$

$$\text{L-Arginine} + \text{NADPH} + \text{H}^+ + \text{O}_2 \xrightarrow{\text{NOS}} N^{\omega}\text{-hydroxy-L-arginine} + \text{NADP}^+ + \text{H}_2\text{O} \quad [76]$$

$$N^{\omega}\text{-Hydroxy-L-arginine} + {}^1/_2\,\text{NADPH} + {}^1/_2\,\text{H}^+ + \text{O}_2 \xrightarrow{\text{NOS}} \text{Citrulline} + {}^{\bullet}\textbf{NO} + {}^1/_2\,\text{NADPH} + \text{H}_2\text{O} \quad [77]$$

NOS-positive neurones and activated neuroglial cells were the most prominent **citrulline**-positive structures (Keilhoff et al. 2000). Lack of citrulline-immunoreaction in neurones of nNOS knockout *mice* emphasised the dependency of citrulline positivity on NOS activity, and likewise there was no citrulline staining after application of the NOS inhibitors 7-nitroindazole and L-N^5-(1-iminoethyl)lysine. The inhibition of argininosuccinate synthetase by α-methyl-DL-aspartate increased the number of citrulline-positive cells, apparently due to the reduction of the turnover rate of citrulline. Cells positive for NOS but negative for citrulline may indicate that the enzyme is either not activated or inhibited by cellular control mechanisms. The fact that not all citrulline-positive cells were NOS positive may be explained by an insufficient detection sensitivity or by disparate sites of citrulline production and recycling.

Using **chemiluminescence** resulting from the **reaction of ${}^{\bullet}NO$ and O_3**, Maurer and Fung (2000) characterised enzyme activity for purified murine macrophage NOS II. They also estimated the inhibitory parameters for a series of competitive antagonists and mechanism-based inactivators of NOS II. The estimated parameters were in agreement with those reported using other methods.

The most promising of all methods proposed thus far for direct measurement of ${}^{\bullet}NO$ appears to be the use of spin traps from stable paramagnetic adducts detectable by the **electron paramagnetic resonance** (EPR) method (Vanin 1999). Such traps were found to be complexes of bivalent iron with hydrophobic and hydrophilic derivatives of dithiocarbamate, a representative of thiocarbonic acids. The possibility of such trap application for scavenging and detecting •NO in animal organisms was demonstrated for the first time by VANIN et al. (1984). 10 mg Sodium diethyldithiocarbamate (DETC) in 0.2 ml of saline injected intraperitoneally into *mice* entered organ tissues, bound endogenous iron and formed NO traps, hydrophobic Fe^{2+}-DETC which locates themselves in cell membranes.

To a culture of NO-producing macrophages from *murine* bone marrow (5×10^6 cells in 2 ml of cultural medium) superoxide dismutase (10^{-6} M), Na-DETC (1 mg/ml) and $FeSO_4 \cdot 7\,H_2O$ (10^{-5} M) are added successively (VANIN et al. 1991, 1993). In 2 h, the cells harvested with the medium are centrifuged for 10 min at $1{,}500 \times g$, reconstituted in 0.3 ml of supernatant, and frozen in liquid nitrogen for the EPR analysis. The formed hydrophobic mononitrosyl iron complexes with DETC are located in membranous compartments of all cells.

A magnetic resonance imaging (MRI) technique for non-invasive detection of stable NO-iron complexes, such as dinitrosyl-iron complex and mononitrosyl-iron-dithiocarbamate, is based on the electron spin resonance-enhanced nuclear Overhausen effect of the paramagnetic species on proton magnetic relaxation (proton-electron double resonance imaging). FICHTLSCHERER et al. (1997) analysed the contrasting effect of NO-iron complexes in conventional proton NMR imaging buffered solutions of mononitrosyl-iron-dithiocarbamate with proline- and *N*-methyl-D-gluconate-dithiocarbamate as well as dinitrosyl-iron complex with L-cysteine, glutathione and bovine serum albumin for their effect on changes in T_1 (longitudinal) and T_2 (transverse) proton relaxation using a 1.5 Tesla whole body NMR imager. All NO complexes exhibited contrast agent-like properties (decrease in T1 and T2) most strongly the protein bound dinitrosyl-iron complex.

Exposure of *rat* liver to sodium nitroprusside by *ex vivo* and *in situ* perfusion induced a composite X-band electron spin resonance spectrum of isolated liver characteristic of a mononitrosyl-iron complex and dinitrosyl-iron complex (MÜLSCH et al. 1999). On storage of the tissue, the mononitrosyl-iron complex signal disappeared and the dinitrosyl-iron complex signal intensity increased. Correspondingly, in cross-sectional proton-electron-double-resonance imaging images taken at room temperature, the nitroprusside-exposed livers initially exhibited a weak signal that strongly increased with time.

The aqueous-soluble complex of Fe and *N*-methyl-D-glucamine dithiocarbamate (MGD) formed MGD_2-Fe-NO complex with a characteristic triplet EPR signal (a_N 12.5 G and $g_{iso} = 2.04$) at room temperature, in native isolated *rat* hearts following 40 min global ischaemia and 15 min reperfusion (KOMAROV et al. 1997). Diethyldithiocarbamate (DETC) and Fe formed in ischaemic reperfused myocardium the lipophilic $DETC_2$-Fe-NO complex exhibiting an EPR signal ($G_\perp = 2.04$ and $g_{II} = 2.02$ at 77 K) with a triplet hyperfine structure at $g_\perp$. Diethyldithiocarbamate-Fe-NO complexes detected by both trapping agents were abolished by the •NO synthase inhibitor, N^G-nitro-L-arginine methyl ester. Quantitatively, both trapping procedures provided similar values for tissue •NO production, which were observed primarily during ischaemia.

Although •NO rapidly reacts with molecular oxygen under air atmospheric conditions, thereby losing its biological functions, the lifetime of this gaseous radical increases under physiologically low intracellular oxygen tensions (INOUE et al. 1999). Kinetic analysis revealed that •NO enhanced the generation of cyclic GMP and induced vasorelaxation of resistance arteries more potentially under physiologically low oxygen tensions than under hyperbaric conditions. •NO reversibly inhibited the respiration of isolated mitochondria, intact cells and *Escherichia coli*; The inhibitory effect was more marked under hypoxic conditions than under hyperbaric conditions. Kinetic analysis revealed that •NO has pivotal action to increase arterial supply of molecular oxygen for the generation of ATP in peripheral tissues and to suppress energy production in mitochondria and cells in an oxygen dependent manner. These functions of •NO are enhanced by decreasing oxygen tension in situ and suppressed by locally generated superoxide radicals.

6.2.1 Nitric Oxide Synthase (NOS)

Macrophages form nitric oxide by an enzyme inducible by cytokines as interferon-γ (TAYEH and MARLETTA 1991), which is Ca^{2+}-independent and requires NADPH and the monooxygenase cofactor tetrahydrobiopterin. Hepatocytes and macrophages express an identical cytokine-inducible nitric oxide synthase gene (WOOD et al. 1993). Iron regulates nitric oxide synthase activity by controlling nuclear transcription (WEISS et al. 1994). ADAMS et al. (1994) found that endogenous nitric oxide production may be linked to 1,25-dihydroxy vitamin D_3 synthesis in HD-11 cells *in vitro*, indicating that macrophage NO-generating capacity could be functionally linked to endogenous synthesis of the active vitamin D metabolite.

The nitric oxide synthases are single polypeptides that encode a haeme domain, a calmodulin-binding motif, and a flavoprotein domain with sequence similarity to P450 reductase. Despite this simple structural similarity, the three major NOS isoforms differ significantly in their rates of •NO synthesis, cytochrome *c* reduction, and NADPH utilisation and in the Ca^{2+} dependence of the rates. The maximal rate of •NO synthesis is determined by the maximum intrinsic ability of the reductase domain to deliver electrons to the haeme domain (Nishida and Ortiz de Montellano 1998). The Ca^{2+} independence of iNOS requires interactions of calmodulin with both the calmodulin binding motif and the flavoprotein domain. The effects of tetrahydrobiopterin and L-arginine on electron transfer rates are mediated exclusively by haeme domain interactions. Moali et al. (2001) recently illustrated the key role of tetrahydrobiopterin in the reaction mechanism and substrate selectivity. The equilibrium dissociation constant (K_d) for arginine is approximately 0.5 μM for the tetrahydrobiopterin replete neuronal (nNOS) and inducible (iNOS) isoforms of nitric oxide synthase, while the endothelial isoform (eNOS) has a slightly higher K_d (1.5 μM) (Smith et al. 2001). *N*-OH-arginine (an intermediate) binds to nNOS with a K_d of around 0.2 μM, while the inhibitors *N*-methyl-arginine and *N*-nitro-arginine bind more tightly.

All isoenzymes use L-arginine as a substrate and all are inhibited by N^{ω}-monomethyl-L-arginine (L-NMMA) and N^{ω}-nitro-L-arginine (L-NA). Types Ia, II, and III have been cloned by several laboratories, show about 50 to 60% homology, and obviously represent separate gene products. Some isoforms may also present posttranslational modifications. It is also recognised that this classification given by Murad (1994) is probably incomplete, as additional isoforms are expected from future purification, cloning and posttranslational modification experiments.

Eight cDNA sequences have been reported deriving from three known NOS genes in *human*, *cow*, *rat* and *mouse* (Nathan and Xie 1994).

The inhibitory effect of aluminium on *in vitro* tetrahydrobiopterin synthesis in brain preparations may be due to competition with the magnesium required to convert dihydroneopterin triphosphate to 8-tetrahydrobiopterin (Ganrot 1986).

Constitutive nitric oxide synthase (cNOS) has been localised in the hypothalamus, particularly in the supraoptic and paraventricular nuclei, and throughout the neurohypophysis (Bredt et al. 1990, Vincent and Kimura 1992). cNOS-reactive cells are significantly more numerous in the supraoptic nucleus of female than of male Long-Evans *rats* (Wang and Morris 1996). While in normal osmotic conditions, only half of the cells showed NOS reactivity, in Brattleboro *rats* nearly all cells were NOS positive. NO synthase activity, reflected by the intensity of NADPH diaphorase staining, was heterogenously distributed in the supraoptic and paraventricular nuclei and markedly increased and decreased in this areas after water and food deprivation, respectively (Gundlach et al. 1993). In hypothalamic neuronal regeneration NO synthase expression was increased (Wu and Scott 1993).

In the *rat* anterior pituitary gland, nitric oxide synthase is present in gonadotrophs and in folliculo-stellate cells (Ceccatelli et al. 1993). Castration and ovariectomy, respectively, resulted in markedly increased nitric oxide synthase mRNA levels. In the male *rat* a small, but not significant, increase was seen 12 h after surgery, whereas at 3 h levels were still similar to sham-operated rats. After 24 h there was a >3-fold increase in mRNA levels, and this remained elevated for 14 days, the longest period studied. Also in the female *rats*, there was a clear but less strong increase at 4 and 14 d after ovariectomy. Substitution treatment with testosterone or estrogen completely prevented these increases in male and female *rats*, respectively.

In a model of idiopathic pneumonia syndrome after bone marrow transplantation, iNOS deletional mutant *mice* (-/-) given donor bone marrow and spleen T cells (BMS) exhibited improved survival compared with matched BMS controls (Yang et al. 2001). Bronchoalveolar lavage fluids obtained on day 7 post bone marrow transplantation from iNOS(-/-) BMS *mice* contained less tumour necrosis factor-α and interferon-γ, indicating that •NO

Table 15. Isoforms of NO synthase

Type	Cosubstrates, cofactors	Regulated by	M_r [kDa]
Ia (soluble)	NADPH, BH_4, FAD/FMN	Ca^{2+}/calmodulin	155
Ib (soluble)	NADPH	Ca^{2+}/calmodulin	135
Ic (soluble)	NADPH, BH_4, FAD	Ca^{2+} (**not** calmodulin)	150
II (soluble)	NADPH, BH_4, FAD/FMN	?	125
III (particulate)	NADPH, BH_4, FAD/FMN	Ca^{2+}/calmodulin	135
IV (particulate)	NADPH	?	?

stimulated the production of proinflammatory cytokines. However, despite suppressed inflammation and decreased nitrotyrosine staining, iNOS(-/-) *mice* given both donor T cells and cyclophosphamide died earlier than iNOS-sufficient BMS + cyclophosphamide *mice*. Alveolar macrophages from iNOS(-/-) BMS + cyclophosphamide *mice* did not produce •NO but persisted to generate strong oxidants as assessed by the oxidation of the intracellular fluorescent probe 2',7'-dichlorofluorescein.

Endothelial nitric oxide synthase (eNOS, previous alternative abbreviations: Type III NOS and NOS-3) is highly membrane-bound (BUSCONI and MICHEL 1993) and can be found in the Golgi apparatus (SESSA et al. 1995) and in small protein-rich invaginations of the plasmalemma called caveolae (Shaul et al. 1996). Within the caveolae of endothelial cells, eNOS is bound to a protein called caveolin-1 - this interaction involves a conserved 20 amino acid region within caveolin-1 and a proposed caveolin-binding sequence in eNOS in the oxidative domain (FERON et al. 1996, MICHEL et al. 1997). A microenvironment within caveolae might modulate eNOS activity by controlling the local concentration of cofactors and substrate. Additional membrane binding interactions also exist. The presence of glycine as the second amino-acid residue found in eNOS, serves as an acceptor site for tetradecanoic acid ($C_{13}H_{27}COOH$) which is critical for the binding of eNOS to the plasma membrane (BUSCONI and MICHEL 1993, 1994). Artificial site-directed mutagenesis of this residue converts eNOS from a membrane-bound to a cytosolic enzyme (BUSCONI and MICHEL 1993). Palmitoylation ($C_{15}H_{31}COOH$) of eNOS at two cysteine residues near the N-terminus stabilises this membrane binding (ROBINSON and MICHEL 1995, ROBINSON et al. 1995). This dual acylation is unique among the NOS isoforms (MICHEL and FERON 1997).

The presence of eNOS was detected in epithelial cells of mucosa and in endothelium of vascular tissues and myosalpinx during all studied days of the *porcine* oestrous cycle (GAWRONSKA et al. 2000).

One major difference between arteries and veins is the intensity of endogenous •NO production in endothelial cells. Stimulation of venous endothelium results in a low production of •NO as demonstrated by the weak endothelium-dependent vasorelaxation in veins of different species, including *humans* (DE MEY and VANHOUTTE 1982, LÜSCHER et al. 1988, KOJDA et al. 1994). The different intensity of endogenous •NO production by the vascular endothelium reduces the vasodilator potency of organic nitrates such as glyceryl trinitrate (ALHEID et al. 1987, MONCADA et al. 1991, KOJDA et al. 1994). •NO (3 μM, 30 min) significantly impaired bioactivation of glyceryl trinitrate as indicated by a 30–50 % reduction in the accumulation of 1,2-glyceryl dinitrate and 1,3- glyceryl dinitrate, whereas unchanged glyceryl trinitrate was increased in rings of *porcine* coronary arteries (KOJDA et al. 1998).

Polymorphisms of the *human* endothelial nitric oxide synthase genes were demonstrated by HINGORANI (1997) and VALLANCE and HINGORANI 1999). The variable number tandem repeat polymorphism in intron 4 is bi-allelic with individuals having 4 or 5 repeats of a 27-bp sequence element. The multiallelic CA repeat polymorphism in intron 13 is highly polymorphic with allele sizes ranging from 18 to 36 CA repeats. Thus far, only one genetic variant affecting the amino-acid sequence has been identified. A G→T substitution in exon 7 of the gene predicts a glutamic acid (Glu) → aspartic acid (Asp) substitution at residue 298 of the mature protein (HINGORANI et al. 1995, YASUE et al. 1995).

Endothelial nitric oxide synthase gene-deficient *mice* (eNOS -/-) demonstrated marked retardation in postnatal bone formation, reduced bone volume, and defects in osteoblast maturation and activity (AGUIRRE et al. 2001).

In *bovine* aortic endothelial cells 16 h treatment with cyclosporine A induced a transcriptional mediated increase of eNOS gene expression (NAVARRO-ANTOLÍN et al. 2000). A 2 h treatment with cyclosporine A induced a dose-dependent increase in the intracellular formation of •NO (NAVARRO-ANTOLÍN et al. 2001). An elevation of the cyclosporine A-induced relative amounts of formation of superoxide anion and •NO provided data consistent with a role of $O_2^{\bullet-}$, and not •NO, as the limiting factor in the intracellular formation of $ONOO^-$ in the vascular endothelial cells (NAVARRO-ANTOLÍN et al. 2002).

Despite intracellular L-arginine concentration that should saturate eNOS nitric oxide production depends on extracellular L-arginine. HARDY and MAY (2002) addressed this 'arginine paradox' in *bovine* aortic endothelial cells by simultaneously comparing the substrate dependence of L-arginine uptake and intracellular eNOS activity, the latter measured as L-[^{3}H]-arginine conversion to L-[^{3}H]-citrulline. Whereas K_m of eNOS for L-arginine was 2 μM in cell extracts, the L-arginine concentration of half-maximal eNOS stimulation was increased to 29 μM in intact cells. This increase likely reflected limitation by L-arginine uptake, which had a K_m of 108 μM. The effect of inhibitors of endothelial nitric oxide synthesis also suggested that extracellular L-arginine availability limits intracellular eNOS activity. Treatment of intact cells with calcium ionophore A23187 reduced the L-arginine concentration of half-maximal eNOS activity, which is consistent with a measured increase in L-arginine uptake. Increases in eNOS activity induced by several agents were closely correlated with enhanced L-arginine uptake into cells (r = 0.89).

In *rabbit* pial arterioles, topical application of 10 μM L-arginine, but not D-arginine or L-lysine, induced moderate vasodilatation of 4.0 ± 0.9 % (HABERL et al. 1991). When cumulative application of L-arginine (100 nM to 10 μM) was followed by addition of 100 μM angiotensin [3–8] an angiotensin peptide fragment, potentiated dilation of 21.2 ± 2.9 % was seen. Angiotensin [3–8] itself did not induce dilation. Methylene blue, a known inhibitor of endothelium-dependent responses, abolished the dilation to both L-arginine and L-arginine followed by angiotensin [3–8]. N^G-monomethyl-L-arginine acetate (300 μM), an inhibitor of •NO formation from L-arginine, did not block endothelium-dependent responses to acetylcholine and angiotensin II.

CHEN et al. (1996) demonstrated ecNOS in *human* platelets. Tetramethylpyrazine concentration (50 and 200 μM)- and time (15 and 30 min)-dependently triggered ecNOS protein expression in *human* platelets (SHEU et al. 2000).

In numerous inflammatory settings the capacity to express **inducible nitric oxide synthase** (iNOS, previous alternative abbreviations: Type II NOS, NOS-2, macNOS, and hepNOS) appears to signal predominantly deleterious effects.

Human bronchial epithelial cells stimulated with 50 ng/ml interleukin-1β, tumour necrosis factor-α, and interferon-γ express iNOS mRNA, protein and increased nitrite in the cell culture media, which was inhibited by the selective iNOS inhibitor 1400 W (DONNELLY and BARNES 2002). Cells derived from *patients* with asthma produced less nitrite than cells from normal subjects (6.59 ± 0.99 μM nitrite, n = 15 versus 3.89 ± 0.42 μM nitrite, n = 20; $P < 0.05$). This was not attributed to steroid treatment of subjects with asthma because there was no difference in the amount of nitrite released from steroid-naïve and steroid-treated cells (3.51 ± 0.46 versus 4.27 ± 0.7 μM nitrite, n = 10). Neither dexamethasone nor budesonide inhibited iNOS mRNA induction, protein expression, or nitrite accumulation. The cells were not steroid insensitive because steroid inhibited GM-CSF release.

Intratracheal instillation of crocidolite asbestos fibres resulted in an increased iNOS mRNA and protein expression in the lungs from wild-type *mice* (DÖRGER et al. 2002). In contrast, iNOS knockout *mice* displayed an attenuated oxidant-related tissue injury reflected in a decrease in protein leakage and lactate dehydrogenase release into the alveolar space as well as weaker nitrotyrosine staining of lung tissue compared to wild-type *mice*.

Incubation of *rat* mesangial cells with chemically modified tetracyclines resulted in a time- and dose-dependent inhibition of NO production that was maximal at 48 h (< 20 % of control) and at a drug concentration of 5 μg/ml ($P < 0.05$) associated with parallel alterations in steady-state iNOS mRNA abundance and protein expression (TRACHTMAN et al. 1996).

Inflamed mucosa biopsies from *patients* with ulcerative colitis and Crohn's disease showed strong expression of iNOS in the epithelial cells (DIJKSTRA et al. 1998). The distribution was focal, with more intense staining at the apical sites of the crypts. Uninflamed mucosa of two ulcerative colitis and four Crohn's disease *patients* showed no iNOS expression.

Pentoxifylline potentiated •NO production and the expression of iNOS in *porcine* hepatocytes after stimulation with lipopolysaccharide (HOEBE et al. 2001).

12-*O*-Tetradecanoylphorbol-13-acetate synergistically increased interferon regulatory factor-1 and iNOS induction in interferon-γ-treated RAW 264.7 cells (MOMOSE et al. 2000).

In stimulated macrophages, arginase (EC 3.5.3.1) and nitric oxide synthase compete for their common substrate, L-arginine. N^{ω}-Hydroxy-nor-L-arginine as a selective arginase inhibitor was about 40-fold more potent than N^{ω}-hydroxy-L-arginine, an intermediate in the L-arginine/NO pathway, to inhibit the hydrolysis of L-arginine to L-ornithine catalysed by unstimulated *murine* macrophages (IC_{50} values 12 ± 5 and 400 ± 50 μM, respectively; Tenu et al. 1999). Stimulation of *murine* macrophages with interferon-γ and lipopolysaccharide resulted in clear expression of iNOS and an increase in arginase activity. N^{ω}-Hydroxy-nor-L-arginine was also a potent inhibitor of arginase in interferon-γ + lipopolysaccharide-stimulated macrophages (IC_{50} values 10 ± 3 μM]. In contrast to N^{ω}-hydroxy-L-arginine, N^{ω}-hydroxy-nor-L-arginine is neither a substrate nor an inhibitor of iNOS.

In cultures of purified microglial cells and astrocytes from newborn *rats*, the immunocytochemical localisation of iNOS and the release of •NO showed that microglia were primarily responsible for •NO production upon endotoxin stimulation (VINCENT et al. 1996). In a discussion of mechanisms of •NO genotoxicity, LAVAL et al. (1997) concluded that genotoxicity either by direct chemical alterations of DNA or interference with the repair system would be from an iNOS source.

There was significantly increased activity of both total NOS ($P < 0.04$) and iNOS ($P < 0.05$) in chronic venous ulcer tissue compared with normal skin, and significantly increased activity of arginase (P

<0.01) in chronic venous ulcer tissue in comparison with normal skin (ABD-EL-ALEEM et al. 2000).

Isoform-specific effects on NOS-catalysed L-citrullin formation was found by SCHRAMMEL et al. (1998). Salt inhibited iNOS monotonously, whereas nNOS and eNOS were stimulated up to 3-fold at low, and inhibited at high ($\geq 0.1-0.2$ M) salt concentrations. The effectiveness of different ions mostly followed the Hofmeister series, indicating that the effects can for a large part be ascribed to changes in protein solvation. K_m(Arg) increased in the presence of NaCl, demonstrating the importance of charge interactions for substrate binding. The coupling of NADPH oxidation to NO production was not affected by KCl. Salts (≤ 1 M) had no major impact on the tertiary and quaternary structure, or on the state of the heme.

ROMAGNANI et al. (1999) assessed the expression of iNOS mRNA and protein in the kidneys of patents with graft failure due to chronic rejection. In chronic allograft nephropathy, iNOS protein was localised not only in inflammatory cells, but also in vascular, glomerular, and, more rarely, tubular structures.

In Japanese encephalitis virus infection of inbred Swiss *mice*, NOS activity and particularly that of iNOS was significantly enhanced (SAXENA et al. 2001). The response was sensitive to anti-macrophage-derived factor antibody treatment and N^G-monomethyl-L-arginine.

Both iNOS and nitrotyrosine, a marker of peroxynitrite formation, were localised in *rat* skeletal muscle after a period of 2 h warm ischaemia and reperfusion exclusively to mast cells except after 24 h reperfusion when some macrophages and neutrophils also showed positive immunoreactivity (MESSINA et al. 2000).

Neuronal nitric oxide synthase (nNOS, previous alternative abbreviations: Type I NOS, NOS-1, and bNOS) does not produce nitric oxide unless high concentrations of superoxide dismutase are added, suggesting that nitroxyl (NO^-) or a related molecule is the principal reaction product of NOS, which is superoxide dismutase-dependently converted to •NO (MURPHY and SIES 1991). However, KOMAROV et al. (2000) found identical *N*-methyl-D-gucamine dithiocarbamate-Fe-NO complexes both from *S*-nitroso-*N*-acetyl-penicillamine and Angeli's salt but not from nitrite. Moreover, the yield of *N*-methyl-D-gucamine dithiocarbamate-Fe-NO complex from Angeli's salt was stoichiometric even in the absence for superoxide dismutase.

nNOS is anchored to receptor complexes through protein-protein interactions. It must be located in close proximity to the synaptic membrane. So •NO can be efficiently produced, released, and bound to the postsynaptic target cells to elicit a particular response. nNOs is targeted to the synaptic membrane by binding PDZ containing proteins such as synthrophin, PSD-95⁻SAP90, or PSD-93 (BRENMAN et al. 1996). This association, which brings nNOS and *N*-methyl D-aspartate receptors together, is proposed to be the reason why calcium influx following glutamate activation of the *N*-methyl D-aspartate receptor leads to rapid nNOS activation. Additionally, *N*-methyl D-aspartate receptor mediated neurotoxicity, neurotransmitter release, and cAMP elevation all require nNOS activity as these actions are blocked by nNOS-specific inhibitors, demonstrating the importance of nNOS for proper *N*-methyl D-aspartate functioning (DAWSON and DAWSON 1996). JAFFREY et al. (1998) identified a nNOS protein interactor called CAPON (carboxy terminal PDZ ligand of nNOS), which is a soluble cytoplasmic protein competing with PSD-95⁻SAP90, and PSD-93 for nNOS binding indicating this protein may participate in the translocation of nNOS and may impede the activation of the nNOS enzyme, blocking •NO production. CAPON is very selective for nNOS, a contrast from other nNOS binding proteins such as PSD-95⁻SAP90 and calmodulin, which bind to other proteins.

In the *rat* subiculum ultrastructural criteria suggest that both pyramidal and nonpyramidal neurons are immunopositive for nNOS (LIN and TOTTERDELL 1998).

Performing immunoblot analysis and quantification of formazan produced by its specific NADPH diaphorase activity, PLANITZER et al. (2001) found nNOS to be enriched in *rat* skeletal muscles with a high proportion of fast-switch myofibres. The NO donors 1-hydroxy-2-oxo-3-(*N*-methyl-6-aminohexyl)-3-methyl-1-triazene (NOC-9) and *S*-nitroso-*N*-acetyl-D,L-penicillamine both reduced cytochrome oxidase activity in all myofibres.

Protein kinase B has been demonstrated to be involved in the regulation of NOS3 activity (DIMMELER et al. 1999, FULTON et al. 1999, GALLIS et al. 1999). Protein kinase B is a serine/threonine kinase that is itself activated by two distinct phosphorylation events, that probably both involve the PH domain-containing kinase phosphatidylinositol (3,4,5)P_3-dependent protein kinase-1, which is upstream of protein kinase B.

A NOS variant localised in *rat* brain mitochondria (**mtNOS**) was detected in synaptosomes (RIOBÓ et al. 2000). This 144 kDa protein was present only in purified mitochondria. It was not recognised either by antibodies against the N-terminal (1-181) region of NOS I or against the other NOS isoforms. The K_m for L-arginine of the mtNOS variant was higher that that for cytosolic NOS I

(12.7 μM vs. 2.0 μM). The activity was dependent on Ca/CaM, NADPH and BH4 and inhibited by N^G-nitro-L-arginine, coincubation with an anti NOS I antibody or proteinase K treatment.

mtNOS in freshly isolated mitochondria was continuously active (GHAFOURIFAR and RICHTER 1997). Upon exposure to respiratory substrates and Ca^{2+} mitochondria formed peroxynitrite in addition to •NO (GHAFOURIFAR et al. 1999). Intramitochondrially formed $ONOO^-$ stimulated the specific, NAD^+-linked Ca^{2+} release from mitochondria (BRINGOLD et al. 2000).

Liver submitochondrial particles supplemented with 0.25–2 μM $ONOO^-$ showed a $O_2^{•-}$ production that indicated ubiquinone formation and autooxidation (VALDEZ et al. 2000). The nitration of mitochondrial proteins produced after addition of 200 μM $ONOO^-$ was observed by Western blot analysis. Protein nitration was prevented by the addition of 50–200 μM ublquinol-0 or reduced glutathione (GSH). An intramitochondrial steady state concentration of about 2 nM $ONOO^-$ was calculated, taking into account the rate constants and concentrations of $ONOO^-$ coreactants.

The **catalytic efficiency** of purified recombinant neuronal and macrophage nitric oxide synthases markedly decreased in the order

arginine > N^ω-hydroxy-L-arginine > homo-L-arginine > homo-N^ω-hydroxy-L-arginine,

as shown by the 20- and 10-fold decrease of k_{cat}/K_m observed for NOS I and NOS II, respectively, when comparing arginine to homo-*N*-hyrdoxy-L-arginine (MOALI et al. 1998). The greater loss of catalytic efficiency for homo-L-arginine, when compared to that for arginine appears to occur at the first step (*N*-hydroxylation) of the reaction.

The **equilibrium between inactive and active NOS** is differentially regulated at the posttranslational level to give rise to the low and high output pathways (NATHAN and XIE 1994).

Activation of neuronal NOS following Ca^{2+} **entry** through the *N*-methyl-D-aspartate receptor has been implicated in excitotoxicity to cortical neuronal cultures (DAWSON et al. 1991) and in ischaemia due to middle cerebral artery occlusion in *mice* (HUANG et al. 1994). Peroxynitrite may be the possible terminal cytotoxic mediator. Some neurones, especially cortical neurones expressing high levels of neuronal NOS, seem to be resistant to •NO toxicity (KOH et al. 1986, KOH and CHOI 1988), but may kill neighbouring neurones because of their Ca^{2+}-induced •NO production (DAWSON et al. 1993). In cerebellar granule cells, elevated $[Ca^{2+}]_i$ caused both NOS activation and cytotoxicity. However, in cerebellar granule cells, glutamate-triggered, Ca^{2+}-mediated cell death is independent of endogenous •NO production (LAFON-CAZAL et al. 1993). Exposure to •NO donors stimulated the *N*-methyl-D-aspartate receptors, probably because NO-related species stimulate the release of endogenous agonists (LEIST et al. 1997). This sort of autocrine stimulation eventually causes apoptosis (BONFOCO et al. 1996).

Calmodulin plays a critical role in activating NOS, because its bonding triggers electron transfer to the heme (ABU-SOUD et al. 1994). The NOSs display different affinities toward calmodulin with the general order being iNOS >> eNOS > nNOS. According to MAYER (2000) enzymatic formation of superoxide/hydrogen peroxide is a general feature of NOS and reduced oxygen species are certainly not 'non-specific by-products' of the reaction, as suggested by XU (2000). Both 5-ethoxycarbonyl-5-methyl-1-pyrroline *N*-oxide (EMPO) and 5-diethoyphosphoryl-5-methyl-1-pyrroline *N*-oxide (DEPMPO) reacted with superoxide generated from eNOS to yield more persistent superoxide adducts than 5,5-dimethyl-1-pyrroline *N*-oxide (DMPO) as demonstrated by the higher signal-to-noise ratio of the electron paramagnetic resonance spectra at the same rate of superoxide formation (VÁSQUEZ-VIVAR and KALYANARAMAN 2000). Superoxide was only marginally detected with resting enzyme and was abolished by the addition of superoxide dismutase.

Melatonin, a pineal indole hormone inhibited *rat* cerebellar nitric oxide synthase activity (POZO et al. 1994). A significant inhibition of enzyme activity (>22 %) was observed at 1 mM melatonin, which is in the range of the physiological serum concentration of the hormone at night. The inhibitory effect of melatonin was observed exclusively in the presence of Ca^{2+}. The melatonin-induced suppression of NOS activity is believed to be a consequence of the binding of calmodulin by melatonin (POZO et al. 1997, ANTON-TAY et al. 1998). With a drop of •NO synthesis, the formation of $ONOO^-$ is curtailed, and the potential oxidative damage resulting from $ONOO^-$ is averted (PRYOR and SQUADRITO 1995). In the *rat* hypothalamus, melatonin reduced NOS activity (BETTAHI et al. 1996). Whether melatonin reduces NOS activity in all tissues containing this enzyme is unknown.

Muramyldipeptide and granulocyte-macrophage colony-stimulating factor enhanced interferon-γ-induced nitric oxide production by *rat* alveolar macrophages (JORENS et al. 1993).

Relaxin increased the expression of eNOS in *mouse* uterine surface epithelium, glands, endometrial stromal cells, and myometrium, leaving iNOS expression unaffected (BANI et al. 1999). Moreover, relaxin inhibited myometrial contractility, and this effect was blunted by nitro-L-arginine, thus indicating that the L-arginine-NO pathway is involved in the relaxant action of relaxin on the myometrium.

Autoinhibition of neuronal nitric oxide synthase. With the use of purified NOS-1, L-arginine turnover initially operated at V_{max} (0–15 min, phase I), although despite the presence of excess substrate and cofactors, prolonged catalysis (15–90 min, phase II) was associated with a rapid decline in L-arginine turnover (KOTSONIS et al. 1999). TSIKAS et al. (2000) presented evidence for the hypothesis that the endogenous NOS inhibitors methylarginines, asymmetric dimethylarginine being the most powerful (IC_{50} 1.5 µM), are responsible for the L-arginine paradox.

In Sprague-Dawley *rats*, asymmetric dimethylarginine decreased mean arterial blood pressure and heart rate simultaneously. Intracerebroventricular injection of both N^{ω}-nitro-L-arginine methylester and asymmetric dimethylarginine significantly inhibited the baroreflex function, indicating a regulatory role of central nitric oxide in controlling baroreflex function (JIN and D'ALECY 1996). In contrast to the central effect, intravenous injection of asymmetric dimethylarginine caused dose-dependent increases in mean arterial blood pressure that could be blocked by N^{ω}-nitro-L-arginine methylester pretreatment.

6.2.1.1 Nitric Oxide Synthase Inhibitors

Hypoxia (0.2% O_2, 6.2% CO_2) inhibited the production of •NO but did not affect the transcription of iNOS mRNA in *rat* smooth muscle cells treated with IFN-γ, lipopolysaccharide, or both (HONG et al. 2000).

Geldanamycin and 17-allylamino-17-demethoxygeldanamycin both dose-dependently reduced nitrite accumulation, iNOS steady-state mRNA levels, and the cytokine-dependent activation of a rat 2.2-kB iNOS promoter construct stability expressed in *rat* glioma C6 cells (MURPHY et al. 2002).

The critical role played by the L-arginine/•NO pathway and the three major isoforms of NOS in regulating physiological and pathophysiological processes has induced significant efforts to develop isoform-selective NOS inhibitors.

6.2.1.1.1 Substrate Analogues

PARKINSON (2000) reviewed the L-arginine-based substrate analogues, the most commonly used of which are **N^G-methyl-L-arginine** and **N^G-nitro-L-arginine.** Modification of the guanido (but not the N^{α}) terminal of L-arginine appears to be a prerequisite for prevention of endothelium dependent relaxation since N^{α}-monocarbobenzoxy-L-arginine was inactive whilst insertion of either a nitro (as in N^G-nitro-L-arginine) or a methyl group (as in N^G-methyl-L-arginine) into the guanido terminus of L-arginine produces potent inhibitors (MOORE et al. 1990). However, the natue of the chemical substitution in the region of the molecule is important since the introduction of an N^G-tosyl moiety dis not produce an effective inhibitor. Bulky substitutes on the N^{α} region results in weak, non-specific inhibitors of endothelium-dependent relaxation, while addition of an α-methyl group to the carboxyl moiety of L-arginine does not impair the metabolism of L-arginine.

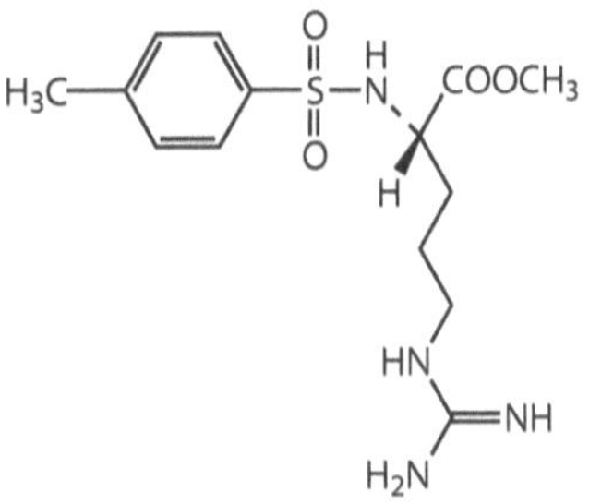

N^{α}-Tosylarginine [78]

A naturally occurring compound, asymmetric **N^{ω}-N^{ω}-dimethyl-L-arginine** (L-ADMA), has been suggested as an inhibitor of the inducible isoform of nitric oxide synthase (VALLANCE et al. 1992) increasing blood pressure after intravenous injection (CALVER et al. 1993). ADMA concentration is increased in certain diseases including renal failure (VALLANCE et al. 1992), muscle dystrophy (INOUE et al. 1979, LOU 1979), hypercholesterolaemia (YU et al. 1994), and pregnancy with preeclampsia (FICKLING et al. 1993). ADMA concentration in urine was increased in premature infants, in whom it is closely related to protein breakdown and was not related to dietary intake of arginine (YUDKOFF et al. 1984). *In vitro* ADMA (10^{-4} M) in phenylephrine-preconditioned *rat* aortic rings significantly increased the concentration of acetylcholine for the threshold response (EC_{15}) and half-maximum response (EC_{50}), indicating that ADMA inhibited the constitutive isoform of nitric oxide synthase in the endothelium (JIN and D'ALECY 1996).

N-Iminoethyl-L-ornithine (100 µM) significantly ($P < 0.05$) decreased the capillary-like tube networks formed by *human* umbilical vein endothelial cells cultured on reconstituted basement membrane matrix Matrigel thus antagonising treatment with the NO donor, *S*-nitroso-*N*-acetylpenicillamine (100 μM) which significantly ($P < 0.05$) increased the capillary network area (Lee et al. 2000).

L-Canavanine, $H_2N-C(NH)-NH-O-(CH_2)_2-CH(NH_2)-COOH$, is an antimetabolite of arginine (Walker 1955).

7-Nitroindazole is a selective inhibitor of nNOS, roughly equipotent to other NOS inhibitors (Moore et al. 1993). A comparison of studies shows that 7-nitroindazole given systemically or via the dialysis probe (Babbedge et al. 1993, Moore et al. 1993, Desvignes 1999) shows a similar relationship to that seen with N^G-nitro-L-arginine (Salter et al. 1996). 7-Nitroindazole increased extracellular dopamine levels when administered alone and reversed the effects of *N*-methyl-D-aspartate in the frontal cortex and raphe nuclei of the freely moving *rat* (Smith and Whitton 2001).

7-Nitroindazole and 3-bromo-7-nitro indazole reduced the EEG power density in all frequency bands in the *rat* (Dzoljic et al. 1997). This effect of 7-nitroindazole was more prominent during the day than during the night, indicating a circadian variation in the NOS response to NOS inhibitors.

6.2.1.1.2
Heme Ligands

The presence of heme in the NOS active site and the critical function of heme in NOS catalysis led to the discovery and characterisation of heme-binding compounds as NOS inhibitors. Imidazole and imidazole-containing compounds are known inhibitors of heme-containing enzymes, particularly P_{450} (Cole and Robinson 1990).

6.2.1.1.3
Pterin Antagonists

On account of the requirement of NOS for tetrahydrobiopterin, two general approaches toward a modulation of NOS activity seem feasible, manipulation of intracellular tetrahydrobiopterin levels and pterin-binding site antagonists. However, recombinant tetrahydrobiopterin-free NOS II catalysed the oxidation of four *N*-hydroxyguanidines tested by NADPH and O_2, with formation of NO_2^- and NO_3^- at rates between 20 and 80 nmol min^{-1} (mg of protein)$^{-1}$ (Moali et al. 2001). In the case of *N*-(4-chlorophenyl)-*N'*-hydroxyguanidine, formation of the corresponding urea and cyanamide was also detected besides that of NO_2^- and NO_3^-. These tetrahydrobiopterin-free NOS II-dependent reactions were inhibited by modulators of electron transfer in NOS such as thiocitrulline or imidazole, but not by L-arginine, and were completely suppressed by superoxide dismutase.

2,4-Diamino-6-hydroxypyrimidine, 6R-5,6,7,8-tetrahydro-L-biopterin and **L-sepiapterin** were potent inhibitors of the recombinant interferon-γ-induced production of nitrogen oxides in intact cultured *rat* alveolar macrophages with I_{50} values for 6*R*-5,6,7,8-tetrahydro-L-biopterin and L-sepiapterin of approximately 10 μM (Jorens et al. 1992).

4-Amino-tetrahydrobiopteridin is a potent pterin antagonist of nitric oxide synthases. The essential pterin requirement for de novo synthesis of functionally intact iNOS renders it possible that this drug exhibits selectivity towads the inducible isoform of NOS *in vivo* (Pitters et al. 1997).

2-Thiouracil is a selective inhibitor of nNOS antagonising tetrahydropteridin-dependent enzyme activation and dimerisation (Palumbo et al. 2000). It caused a 60% inhibition of H_2O_2 production in the absence of L-arginine and tetrahydropteridin, and antagonised tetrahydropteridin-induced dimerisation of nNOS, but did not affect cytochrome *c* reduction.

Cloricromene (2, 20 or 200 μM) inhibited the expression but not the activity of the inducible form of nitric oxide synthase in lipopolysaccharide (100 ng/ml)-stimulated *murine* J774 macrophages in a concentration-dependent manner (Zingarelli et al. 1993). Maximal inhibition (84.0±8.0%) was observed when cloricromene (200 μM) was added to the cells 6 h before lipopolysaccharide, whereas it was ineffective when given 6 h after endotoxin.

Subcutaneous application of **Fe-citrate** (Fe doses of 120 and 240 μM/kg *mouse*) inhibited iNOS expression after a simultaneous intravenous injection of 6 mg *Escherichia coli* lipopolysaccharide per *mouse* (Komarov et al. 1998), while exogenous iron did not affect systemic NO levels when given at 6 h after LPS injection, i.e. after iNOS expression (Komarov et al. 1997). In exercised (after a training period of 30 and 60 min, swimming for 120 min per day for 3, 6, and 12 months, respectively) *rats*, plasma nitric oxide and iron concentrations were negatively correlated (Xiao and Qian 2000).

Chloroquine markedly reduced nitric oxide production from stimulated *murine* (C57BL/6) peritoneal macrophages in a dose-dependent manner and induction of iNOS mRNA was also suppressed by chloroquine pre-treatment (Park et al. 1999).

Nitric oxide production was inhibited by 39 and 54% in brain submitochondrial particles from **haloperidol** (1 mg/kg)-treated *mice* (single- and double-dose treatments, respectively) (Arnaiz et al. 1999). Haloperidol neurotoxicity might be mediated by a decreased mitochondrial •NO production, a decreased intramitochondrial •NO steady-state concentration, and by an inhibition of mitochondrial electron transfer with enhancement of $O_2^{\bullet-}$ and H_2O_2 production. This inhibition does not seem to be caused by increased •NO or $ONOO^-$ formation.

Troglitazone, a thiazolidine derivative, inhibited the expression of iNOS in adipocytes in 3T3-L1 cells and in Otsuka Long-Evans Tokushima Fatty *rats* (DOBASHI et al. 2000).

The anticonvulsant **zonisamide** (50 mg/kg, i.p.) reduced NOS activity, accelerated by *N*-methyl-D-aspartate (30 mg/kg, i.p.) with/without L-buthionine-[*S*,*R*]-sulfoxime (150 mg/kg, i.p.) treatment, to the control levels in the *rat* hippocampus (NODA et al. 1999).

C_3-tris-malonyl-C_{60}-**fullerene** and D_3-tris-malonyl-C_{60}-fullerene derivatives inhibited citrulline and •NO formation by all three NOS isoforms (WOLFF et al. 2000).

Lipoprotein(a) oxidised by hypochlorite decreased iNOS mRNA synthesis in lipopolysaccharide/interferon stimulated *mouse* macrophages (J774A.1) as shown by reverse transcription-polymerase chain reaction (MOESLINGER et al. 2000).

6.2.2 Activation of Guanylate Cyclase by Nitric Oxide

Bovine lung soluble guanylate cyclase (EC 4.6.1.2) contains a pentacoordinated high-spin ferrous haeme with histidine as the proximal ligand (STONE and MARLETTA 1994). Unlike most other haemoproteins (e.g. haemoglobin and myoglobin) it does not bind oxygen to the sixth coordination position of the haeme. This highly unusual feature enables the enzyme to be stimulated by binding NO under aerobic conditions. The formation of a pentacoordinated ferrous-nitrosyl-haeme complex supports the hypothesis that the bound of the proximal histidine ligand to the iron is broken upon binding of NO to the haeme (TRAYLOR and SHARMA 1992).

Nitric oxide production, first associated with vascular endothelium (RADI et al. 1992) and now linked with a multitude of cell types (MONCADA and HIGGS 1991), activates cell cyclic guanosine monophosphate (cGMP) synthesis by inducing a structural change in guanylate cyclase (IGNARRO 1992). This occurs via incompletely understood reactions with enzyme thiol or heme-iron moieties and is catalysed not only by •NO, but also by HO• (MITTAL and MURAD 1977), thiol-reactive agents (WU et al. 1992), and other oxidant states of •NO (FUKUTO et al. 1992). DENNINGER and MARLETTA (1999) reviewed the role of soluble guanylate cyclase in signalling, its relationship to the other nucleotide cyclases, and on what is known about soluble guanylate cyclase genetics, heme environment and catalysis.

•NO has been shown to open potassium channels through a cGMP-dependent protein kinase (ARCHER et al. 1994). Opening of potassium channels seems to be involved in the hyporesponsiveness to vasoconstrictors in isolated *rat* aorta obtained from endotoxaemic animals (CHEN et al. 1999). Incubation with nitric oxide donors (3–300 μM sodium nitroprusside or 70–200 μM *S*-nitroso-acetyl-D,L-penicillamine) reproduces the hyporesponsiveness to phenylephrine (TERLUK et al. 2000). •NO alone accounts for most, if not all, the refractoriness to vasoconstrictors present in septic shock. Soluble guanylate cyclase activation and opening of potassium channels, more specifically the calcium-activated subtype, play a predominant role in this NO-induced hyporesponsiveness to phenylephrine in the *rat* aorta.

•NO, through generation of cyclic GMP and activation of G kinase, modifies profoundly 17β-œstradiol early signalling and contributes to regulate long-term actions of the steroid (FALCONE et al. 2002).

Deactivation is an essential process in the control of local levels of cGMP (WALDMAN and MURAD 1987). STONE and MARLETTA (1995) have suggested that simple dissociation of NO from soluble guanylate cyclase could be the deactivation mechanism. Deactivation occurs within seconds to minutes (PALMER et al. 1987), with half-lives for NO dissociation from five-co-ordinate NO-hemes, as it is found in NO-activated soluble guanylate cyclase, are generally found to be in the range of 12–12,000 min (TRAYLOR and SHARMA 1992). Since the oxidation state of thiols (BRANDWEIN et al. 1981, BRAUGHLER 1982), the Fe atom in the heme (BURSTYN et al. 1995), and NO (DIERKS and BURSTYN 1996) are important in soluble guanylate cyclase function, redox potentiation seems a reasonable path for enzyme deactivation. DIERKS and BURSTYN (1998) found deactivation of soluble guanylate cyclase from *bovine* lung by air occurring slowly, while deactivation by ferricyanide was faster and methylene blue fastest. The mechanism of deactivation of soluble guanylate cyclase by dioxygen in the air id straightforward: the heme is oxidised to Fe(III)heme and nitrate is formed. This reaction is similar to that of dioxygen with NOHb and NOMb as occurs in cured meats. Methylene blue and ferricyanide deactivate soluble guanylate cyclase by a different, as yet undetermined mechanism.

The ability of some non-biological substances and in particular methylene blue (formula [79]) to act as electron carrier in a way similar to that of the coenzymes incited Pullman and Pullman (1959) to calculate energies of molecular orbitals for this compound. The relative small absolute value of the coefficient of the energy of the lowest empty molecular electronic orbital in oxidized methylene blue (MeB^+), which is of the same order of magnitude as that of oxidizes diphosphopyridine nucleotide or oxidized flavin mononucleotide, indicates high electroaffinity and accounts thus for the well-known electron-acceptor properties of this substance.

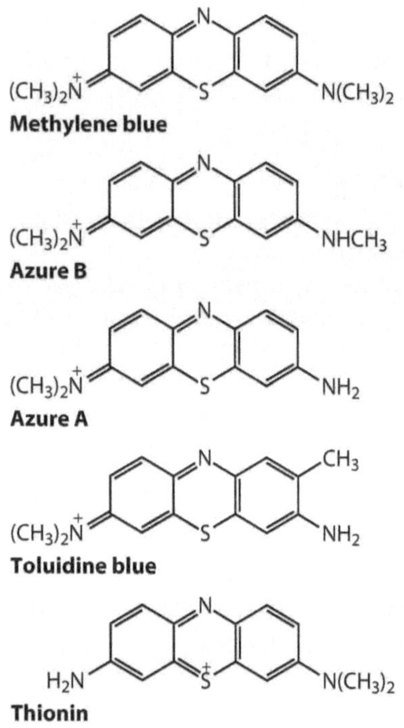

Thiazine dyes [79]

In a *p*-benzoquinone-induced writhing model in the *mouse*, L-arginine displayed antinociceptive effects at the doses of 0.125–1.0 mg/kg (Abacioglu et al. 2000). When the doses of L-arginine were increased gradually to 10–100 mg/kg, a dose-dependent triphasic pattern of nociception-antinociception-nociception was obtained. N^G-nitro-L-arginine methyl ester (18.75–150 mg/kg) possessed antinociceptive activity. Methylene blue (5–160 mg/kg) also produced a dose-dependent triphasic response. When L-arginine (50 mg/kg) was combined with N^G-nitro-L-arginine methyl ester (75 mg/kg), L-arginine-induced antinociception did not change significantly. Cotreatment of L-arginine with 5 mg/kg methylene blue significantly decreased methylene blue-induced antinociception and reversed the nociception induced by 40 mg methylene blue to antinociception.

Cyclic GMP stimulation by the spontaneous NO donor SIN-1, which releases •NO independently of enzymatic catalysis, remained unimpaired in LLC-PK1 kidney epithelial cells pre-treated with glyceryl trinitrate or NO-aspirin (Grosser and Schroeder 2000). Prolonged treatment with NO-aspirin caused down-regulation of the cellular cyclic GMP response, suggesting that tolerance may occur during therapy with NO-aspirin.

2-Hydroxy-benzoic acid 3-nitrooxymethyl-phenyl ester (B-NOD) in vitro increased cGMP production in platelets in a dose-dependent manner (Bing et al. 2000). The effect of sodium nitroprusside donor was greater than that of B-NOD. Inhibition of platelet aggregation was approximately 25 % as compared to the control sodium nitroprusside.

Cyclic GMP possibly represents a factor inhibiting autophagy, as opposed to stimulation by cyclic AMP (Pfeifer 1976).

Isatin, an endogenous indole, can be transported to human platelets, where it inhibited NO-stimulated soluble guanylate cyclase activity (Medvedev et al. 2002). The effect was most pronounced at 10^{-8} M isatin and is the most potent effect of isatin yet observed. The dose response curve was bell-shaped with higher doses becoming less effective. The maximal inhibition observed was 40 %. Isatin had no effect of protoporphyrin IX-stimulated guanylate cyclase.

6.2.3 Indirect Inhibition of Mitochondrial Dihydroorotate Dehydrogenase Activity

Dihydroorotate dehydrogenase (EC 1.3.99.11) catalyses the oxidation of dihydroorotate to orotate in the pyrimidine biosynthesis pathway. Inhibition of cytochrome *c* oxidase indirectly inhibits dihydroorotate dehydrogenase activity. In digitonin-permeabilized cells, sodium 1,1-diethyl-2-hydroxy-2-nitroso-hydrazine, a chemical nitric oxide donor, induced a dramatic decrease in dihydroorotate-dependent O_2 consumption (Beuneu et al. 2000). The inhibition was reversible and more pronounced at low O_2 concentration; it was correlated with a decrease in orotate synthesis.

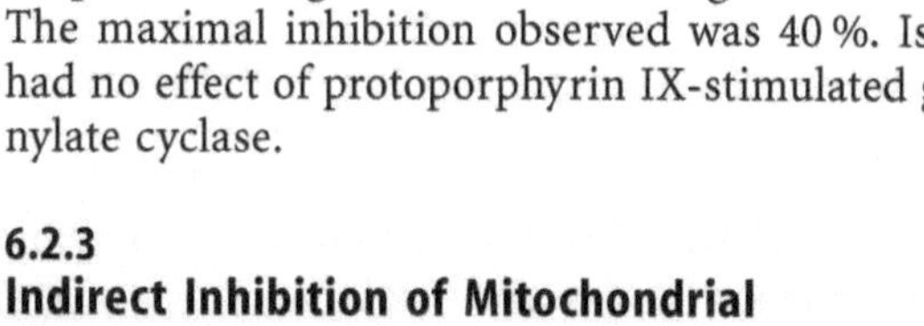

[48]

6.2.4 Nitric Oxide and Iron Proteins

Iron sulphur clusters have long been recognized as molecular targets of •NO generated by endogenous synthases or exogenous donors (for review see CAIRO et al. 2002). DOBRY-DUCLAUX (1960) has shown that black Roussin's salt, $[Fe_4S_2(NO)_7]K$, effectively inhibits the enzyme alcohol dehydrogenase at very low concentrations ($\sim 1 \times 10^{-7}$ M). GORDY and REXROAD (1961) demonstrated that complexes of nitric oxide with haemoglobin and cytochrome *c* exhibited electron spin resonance spectra that could be employed to elucidate the electronic structures of the iron atoms in these biologically important molecules. MCDONALD et al. (1965) investigated complexes formed in aqueous solutions from Fe(II), NO, and a variety of inorganic and organic ligands.

Ferrimyoglobin (*horse* skeletal muscle) at pH 7.4 binds nitric oxide to yield nitric oxide adducts (REICHENBACH et al. 2001). In the presence of glutathione (GSH) nitrosoadducts of metmyoglobin^{III} react with it to give nitrosoglutathione.

Cellular Fe both regulates and is regulated by iNOS (NATHAN and XIE 1994) as well as serving as •NO's most important target (COOPER 1999; for review see BOUTON 1999). The reaction of •NO with iron can be divided depending upon whether they involve mere binding or binding + metabolism. The redox state of the iron (ferrous, ferric or ferryl) also dramatically modulates whether binding is favoured over metabolism, and in the former case the strength of the iron-NO bond.

6.2.4.1 Iron Regulatory Protein-1 RNA-Binding Activity

Iron regulatory protein (IRP)-1 RNA-binding activity was induced by nitric oxide (DRAPIER et al. 1993, WEISS et al. 1993). Activation in cells required a time of treatment similar to desferrioxamine, and while it remains possible that •NO indirectly induces IRP-1 via modulation of intracellular iron levels, the *in vitro* analysis of recombinant IRP-1 suggested that •NO can direct affect the [Fe-S] cluster (DRAPIER et al. 1993).

When prepared in the presence of oxygen, IRP-1 contains both the $[3Fe\text{-}4S]^{1+}$ and the $[4Fe\text{-}4S]^{2+}$ forms (2000 to 5000 nmol of *cis*-aconitate/min × mg protein) (SOUM et al. 2000). When exposed to nitric oxide generated by 3-morpholino-sydnonimine (SIN-1) in the presence of superoxide dismutase for 2 h at 37 °C, IRP-1 lost about 80 % of its aconitase activity and its iron responsive element-binding capacity was increased 3- to 5-fold when further exposed to low concentrations of reductants. 3 to 5 iron atoms were released per molecule of IRP-1. Electron paramagnetic resonance experiments performed at 12 K revealed that IRP-1 treated under the same conditions gave a progressively weaker $[3Fe\text{-}4S]^{1+}$ electron paramagnetic resonance signal.

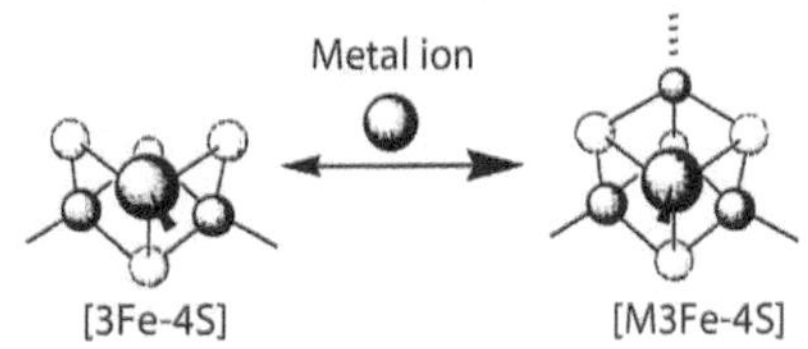

Addition of a metal ion to a [3Fe-4S] cuboidal cluster to yield a [4Fe-4S] cubane [80]

From a pathophysiologic viewpoint, reactive nitrogen species-induced inactivation of IRP-2 may be of grater relevance than a concomitant activation of IRP-1. IRP-2 is highly expressed in macrophages, and exhibits stronger affinity than IRP-1 for ferritin iron responsive elements (THEIL 2000). IRP-2 is highly homologous to IRP-1 (79 % at the amino acid level) but lacks aconitase activity, probably because it inability to assemble a [4Fe-4S] cluster, and is characterized by the presence of a 73 amino acid insertion in the *N*-terminus (HENDERSON 1996, HENTZE and KUHN 1996).

The presence of a mitochondrial ferritin encoded by an intronless gene on chromosome 5q23.1 (LEVI et al. 2001) in mammals is particularly stimulating, because this organelle is tightly involved in iron trafficking, and has a key role in important cellular activities, including respiration, the production of reactive oxygen species and the regulation of apoptotic pathways. Heme is specifically synthesized inside the mitochondrial matrix by the ferrochelatase enzyme, which inserts ferrous iron inside protoporphyrin IX (PONKA 1997). There is an increasing body of evidence showing that Fe-S clusters are synthesized only inside the mitochondria and are locally used by mitochondrial enzymes or exported for insertion in cytosolic and nuclear enzymes (LILL et al. 2000).

Deletion mutants of a plasmid containing the iNOS promoter and a chloramphenicol-acetyltransferase (CAT) indicator gene were transfected into *murine* fibroblasts (NIH3T3) and *murine* J774 macrophages (DLASKA et al. 1997). When using the parent plasmid containing the whole iNOS promoter stimulation of the cells with IFN-γ/LPS and iron resulted in reduced CAT activity as compared to IFN-γ/LPS treated cells, while addition of the iron chelator, desferroxamine enhanced CAT expression. After deleting the promoter region upstream of base 750 (thus deleting "region II" which contains multiple binding sites for various transcription factors) the relative effects of iron perturbation on CAT expression were reduced but still present. This indicated that a factor in this region must be in part responsible for transducing iron-mediated regulation towards iNOS expression. By means of electromobility shift assays DLASKA et al. (1997) identified a transcription factor, induced by stimulation of cells with interferons, the binding affinity of which towards the iNOS promoter was modulated by iron perturbations. Since iron-mediated regulation was still evident after cutting out this region although to a lesser extent they then tried to identify other factors contributing to iron-mediated transcriptional regulation of iNOS: region I (extending from −50 to −300 bp upstream and containing another cluster of transcription factor binding sites) iron also modulated the binding of a yet not identified protein to this iNOS promoter region.

Catalase (EC 1.11.1.6) rapidly formed a reversible complex stoichiometrically with nitric oxide with the Soret band shifting from 406 to 426 nm and two new peaks appeared at 540 and at 575 nm, consistent with the formation of a ferrous-nitrosyl complex (Brunelli et al. 2001). Catalase consumed more nitric oxide upon the addition of hydrogen peroxide. Conversely, micromolar concentrations of nitric oxide slowed the catalase-mediated decomposition of hydrogen peroxide. Catalase pre-treated with nitric oxide and hydrogen peroxide regained full activity after dialysis.

6.2.5 Modulation of •NO Production in the Respiratory Tract

In the lower respiratory tract, nitric oxide is produced by inflammatory cells, alveolar and bronchiolar epithelial cells, and vascular endothelial cells. DEMIRAKÇA et al. (1997) measured expiratory •NO concentration in 20 healthy adult subjects (aged 20–25 y), 20 healthy children (aged 4–16 y) and 7 patients (aged 4–14 y) given L-arginine to trigger growth hormone incretion. Increasing O_2 concentration from 21 % to 50 %, at the end of the expiration •NO concentrations raised from 10 ± 3 ppb to 13 ± 4 ppb. In 10 % O_2 hypoxia (voluntary adults only) at the end of the expiration •NO concentrations declined from 7 ± 2 ppb to 3 ± 1 ppb. Under arginine application, within 30 min •NO concentrations at the end of the expiration raised from 3 ± 2 ppb to 10 ± 5 ppb, thereafter they declined but, after 120 min, failed to reach the starting values.

In the isolated perfused lungs of Sprague-Dawley *rats* given silica (50 mg) intratracheally, nitrate production as an index of •NO production, was significantly greater (53.5 ± 12.0 nmol/90 min) compared with controls (22.5 ± 5.1 nmol/90 min; $P < 0.05$) accompanied by grater ($P < 0.0001$) 90-min [^{3}H]L-arginine uptake (NELIN et al. 2002). The cell type involved in the increase in L-arginine transport found in the silica-treated lungs was not directly determined, but SCHAPIRA et al. (1998) found that L-arginine uptake was increased in alveolar macrophages and neutrophils harvested from silica-exposed *rat* lungs.

Guinea pig tracheal epithelial cells maintained in primary culture produced significantly more reactive nitrogen species when stimulated with either 10 ng TNFα/ml (specific activity 2.86×10^7 U/mg) or 10 ng lipopolysaccharide/ml, the latter having a greater stimulatory effect (ROCHELLE et al. 1998).

6.2.6 Modulation of Endocrine Feedback Control Systems

Nitric oxide modulates the release of corticotropin-releasing hormone from *rat* hypothalamic tissue cultivated *in vitro* (COSTA et al. 1993). L-Arginine is a potent stimulus to growth hormone secretion in *man* (KNOPF et al. 1969, MERIMEE et al. 1969). *Human* growth hormone-releasing hormone may stimulate the synthesis of NO *in vitro*, at least partly through cAMP, thereby partly inhibiting hGHRH-induced GH secretion (KATO 1992). *In vitro* studies of dispersed *rat* anterior pituitary cells suggested that •NO inhibits gonadotropin-releasing-hormone-stimulated luteinizing hormone release (CECCATELLI et al. 1993). Thus there may be a dual mechanism for •NO in the control of anterior pituitary hormone secretion, an autocrine mediation of luteinizing hormone release on gonadotrophs, and a paracrine effect on growth hormone secretion involving folliculo-stellate cells closely related to somatotrophs. CECCATELLI et al. (l.c.) speculated that •NO may participate in producing the pulsatile secretion pattern of the two pituitary hormones. AKI et al. (1994) suggested that •NO may participate in V_2 vasopressin-receptor-mediated renal vasodilatation. The anti-ulcer activity of calcitonin gene-related peptide was inhibited, in a dose-dependent manner, by pre-treatment of *rats* with N^G-nitro-L-arginine methyl ester, a nitric oxide synthase-inhibitor (CLEMENTI et al. 1994). Endogenous nitric oxide induced by interleukin-1β in *rat* islands of Langerhans and HIT-T15 cells caused significant DNA damage (DELANEY et al. 1993).

Deta nonoate [2,2'(hydroxynitrosohydrazono)*bis*-ethanamine], a zwitterion nitric oxide donor, potently inhibited forskolin- and angiotensin II-stimulated aldosterone production in *human* adrenocortical H295R cells in a concentration-dependent manner (KREKLAU et al. 1999).

6.2.7 Modulation of DNA Repair Proteins

Asbestos induced nitric oxide production synergistically with interferon-γ (THOMAS et al. 1994). In the presence of aerobic NO **formamidopyrimidine-DNA glycosilase** (Fpg protein) activity was inhibited (WINK and LAVAL 1994). This enzyme repairs alkylating damage to guanine, such as 2,6-diamino-4-hydroxy-5-N-methylformamidopyrimidine residues (Fapy) and oxidative damage such as 8-oxoguanine residues (reviewed by LAVAL 1996). This protein has a glycosylase activity, incised DNA at abasic sites by a β–δ elimination reaction, has a dRPase activity and contains a zinc finger motif which is mandatory for its various activities (O'CONNOR et al. 1993). It was suggested that N_2O_3 nitrosated the thiol residues, resulting in the ejection of the Zn^{2+}. KRONCKE et al. (1994) showed that the zinc finger protein, LAC9, was degraded in the presence of ˙NO. Using Raman spectroscopy, KRONCKE et al. (1994) showed that S-nitrosothiol adducts were formed.

Exposure of *human* foetal kidney cells transfected with a 4-kilobase fragment of the complementary DNA encoding *human* iNOS to generate ˙NO (EcR293 clone 11) to muristerone A (1 μM and 10 μM) resulted in a 4–5-fold increase in expression of the DNA-dependent protein-kinase catalytic subunit, one of the key enzymes involved in repairing double-stranded DNA-breaks (XU et al. 2000). This NO-mediated increase in enzymatically active DNA-protein kinase not only protected cells from the toxic effects of ˙NO, but also provided crossprotection against clinically important DNA-damaging agents, such as X-ray radiation, adriamycin, bleomycin and cisplatin. Treatment of EcR293 clone 11 cells with SIN-1 (300 μM and 900 μM) resulted in 2-fold and 2.8-fold increases, respectively, in activity of the DNA-dependent protein kinase catalytic subunit promoter.

6.2.8 Nitric Oxide and Apoptosis

Low doses of ˙NO inhibited apoptosis in *human* B lymphocytes (MANNICK et al. 1994, GENARO et al. 1995). In contrast, ˙NO functions as an apoptotic inducer for macrophages, hepatocytes, neurones and glial cells. The proapoptotic activity of ˙NO is mediated via the release of mitochondrial cytochrome *c* into the cytosol, the sequential loss of mitochondrial membrane potential, and the activation of members of the caspase family of proteases (UEHARA et al. 1999, BROOKES et al. 2000, MORIYA et al. 2000). CIBELLI et al. (2002) analysed the NO-induced apoptotic signalling cascade in *human* SH-SY5Y neuroblastoma cells and observed a striking increase in early growth response (Egr)-1 promoter activity as a result of NO-induced cell death. Likewise, they detected an activation of the transcriptional activation potential of the ternary complex factor Elk1, a key transcriptional regulator of serum response element-driven gene transcription. Egr-1 is a zinc finger transcription factor that couples extracellular signals to long-term responses by altering expression of Egr-1 target genes. The Egr-1 5'-flanking region contains five serum response elements (SRE) that function as genetic elements for stimulus-transcription coupling. Moreover, these SREs are binding-sites for Elk1, suggesting that NO activated Egr-1 gene transcription via activation of Elk1.

6.2.9 Nitric Oxide and Lipid Peroxidation

Nitric oxide can both promote and inhibit lipid peroxidation (HOGG and KALYANARAMAN (1999). By itself, ˙NO acts as a potent inhibitor of the lipid peroxidation chain reaction by scavenging propagatory lipid peroxyl radicals (formula [81]). It can also inhibit many potential initiators of lipid peroxidation, such as peroxidase enzymes. In the presence of $O_2^{\bullet-}$, ˙NO forms peroxynitrite (see equation [46]), a powerful oxidant capable of initiating lipid peroxidation and oxidising lipid soluble antioxidants.

Lipid peroxyl radical can be scavenged by nitric oxid

$$LOO^{\bullet} + {}^{\bullet}NO \rightarrow LOONO \quad \text{chain determination} \quad [81]$$

Exposure to ˙NO donors, S-nitrosoglutathione (250 μM) or sodium nitroprusside (500 μM) induced lipid peroxidation in primary cultures of cerebellar granule cells of 7-day-old Wistar *rats* as indicated by the significant increase in thiobarbituric acid reactive substances (WEI et al. 1999). In cells pre-treated with L-ascorbic acid 2-[3,4-dihydro-2,5,7,8-tetramethyl-2-(4,8,12-trimethyltridecyl)-2*H*-1-benzopyran-6-yl-hydrogen phosphate] potassium salt (EPC-K1), superoxide dismutase or NO scavenger haemoglobin, the formation of thiobarbituric acid reactive substances was markedly suppressed.

Brain-derived neurotrophic factor (50 ng/ml) accelerated nitric oxide donor (100 μM sodium nitroprusside)-induced apoptosis of cultured cortical neurones from brains of embryonic day 20 Wistar *rats* (ISHIKAWA et al. 2000).

6.2.10 Peroxynitrite

Superoxide anion and nitrogen monoxide engender peroxynitrite at almost diffusion-controlled rates ($6.7 \times 10^9\ M^{-1} \cdot s^{-1}$; HUIE and PADMAJA 1993; for review see DUCROCQ et al. 1999). Its high rate constant means that $^{\bullet}NO$ competes effectively with superoxide dismutase for reaction with $O_2^{\bullet-}$ (HOGG et al. 1992).

$$O_2^{\bullet-} + {}^{\bullet}NO \longrightarrow ONOO^- \ (pK_a = 6.8) \qquad [82]$$

$$ONOO^- + H^+ \longrightarrow ONOOH \qquad [83]$$

$$ONOOH \longrightarrow HO^{\bullet} + {}^{\bullet}NO_2 \ (20\text{–}30\%\ \text{yield}),\ NO_3^- + H^+ \ (\text{rest}) \qquad [84]$$

Peroxynitrite can be generated when 2- and 4-hydroxyœstrogens are incubated *in vitro* with $^{\bullet}NO$ donor compounds (PAQUETTE et al. 2001). Using dihydrorhodamine 123 as a specific probe for $ONOO^-$ formation, a ratio of 100 μM dipropylenetriamine NONOate to 10 μM 4-hydroxyœstradiol gave an optimal $ONOO^-$ production of 11.9±1.9 μM. Quantification of $ONOO^-$ was not modified by mannitol, supporting the idea that the hydroxyl radical was not involved. This production of $ONOO^-$ required the presence of the catechol structure of œstrogen metabolites since all methoxyœstrogens that were tested were inactive.

Oxidations by peroxynitrite can take place either directly by ground-state peroxynitrous acid, ONOOH, or indirectly by ONOOH* where ONOOH* is an activated form of peroxynitrous acid (GOLDSTEIN et al. 1996). In the direct oxidation pathway the reaction is first order in peroxynitrite and first order in substrate, and the oxidation yield approaches 100%. In the indirect oxidation pathway the reaction is first order in peroxynitrite and zero order in substrate. In the presence of sufficient concentrations of a substrate that reacts by the indirect oxidation pathway, about 50–60% of the ONOOH directly isomerize to nitric acid, and about 40–50% of the ONOOH is converted to ONOOH*. The involvement of hydroxyl radicals in indirect oxidations by peroxynitrite is ruled out on the basis of kinetics and oxidation yields.

Peroxynitrite itself is not a free radical because the two unpaired electrons of superoxide and nitric oxide have combined to form a new bond. Peroxynitrite is an isomer of nitrate, but is 36 kcal × mol-1 higher in energy (KOPPENOL et al. 1992). The half-life of peroxynitrous acid (HOONO), which is 7 s at 0 °C and 1 s at 37 °C, should be long enough to allow the molecule to diffuse inside the cell and react with DNA. The decomposition pathway of HOONO, leading to the formation of nitric acid, remains open to debate (CADET et al. 1997). Peroxynitrite decomposition catalysts (SALVEMINI et al. 1998) may represent a unique class of antiinflammatory agents.

Peroxynitrite can be used as a source of biologically relevant reactive species in a competition reaction method to evaluate antioxidant activity of different compounds (BALAVOINE and GELETII 1999). Pyrogallol red was the most advantageous detecting molecule. It has high reactivity towards radicals generated from peroxynitrite and a strong absorption band. This allows the use of very low concentrations of antioxidants (typically 10 μM to 1 mM) to quantify their activity.

Nitrogen dioxide ($^{\bullet}NO_2$) is a key biological oxidant. It can be derived from peroxynitrite via the interaction of nitric oxide with superoxide, from nitrite with peroxidases, or from autoxidation of nitric oxide. FORD et al. (2002) generated submicromolar concentrations of $^{\bullet}NO_2$ in $<1\ \mu s$ using pulse radiolysis, and the kinetics of scavenging $^{\bullet}NO_2$ by glutathione, cysteine, or uric acid were monitored by spectrophotometry. The formation of the urate radical was observed directly, while the production of the oxidising radical obtained on reaction of $^{\bullet}NO_2$ with glutathione, cysteine, and urate were estimated as $\sim 2\times10^7$, 5×10^7, and $2\times10^7\ M^{-1} \cdot s^{-1}$, respectively. The lifetime of $^{\bullet}NO_2$ in cytosol is $<10\ \mu s$. Reactions between $^{\bullet}NO_2$ and thiols/urate severely limit the likelihood of reaction of $^{\bullet}NO_2$ with $^{\bullet}NO$ to form N_2O_3 in the cytoplasm.

Among the biological molecules in human plasma whose rates of reaction with peroxynitrite have been reported, **carbon dioxide** is one of the fastest with a pseudo-first-order rate constant $k_{CO2/plasma} = 46\ s^{-1}$ (SQUADRITO and PRYOR 1998). Nitrosoperoxycarbonate ($ONOOCO_2^-$) is formed as a reactive intermediate, which then rapidly homolyzes to form a pair of caged radicals $[CO_3^{\bullet-}/{}^{\bullet}NO_2]$, which can then diffuse apart and become free radicals or combine to form the nitrocarbonate ($O_2NOCO_2^-$) which then decomposes to NO_3^- and CO_2. The last reaction is responsible for the catalytic action of CO_2 during the decomposition of peroxynitrite (SQUADRITO and PRYOR 1998). The therapeutic effect of uric acid may be related to the scavenging of the radicals $CO_3^{\bullet-}$ and $^{\bullet}NO_2$ that are formed from the reaction of peroxynitrite with CO_2 (SQUADRITO et al. 2000).

Aldehydes, like CO_2, react rapidly with peroxynitrite and catalyse its decompensation (UPPU et al. 1997). The pH dependence of the reaction is consistent with the addition of $ONOO^-$ (not ONOOH) to the carbonyl carbon atom of the free aldehyde forming a 1-hydroxyalkylperoxynitrite anion adduct, which structurally resembles the nitrosoperoxycarbonate adduct formed from the reaction $ONOO^-$ with CO_2. 1-Hydroxyalkylperoxynitrite or the secondary products derived from it, dacay to give NO_3^- and regenerated aldehyde, with small but significant yields of H_2O_2, organic acids, and organic nitrates. In analogy with the peroxynitrite/CO_2 system, UPPU et al. (1997) suggested that 1-hydroxyalkylperoxynitrite undergoes homolytic or heterolytic cleavage at the O–O bond, giving a caged radical pair [$RCH(OH)O^•/^•NO_2$] or intermediate ion pair [$RCH(OH)O^-/^+NO_2$]. These radical and ions can recombine within the solvent to form 1-hydroxyalkylnitrate [$RCH(OH)ONO_2$], which can then dissociate to give nitrate and regenerate the aldehyde.

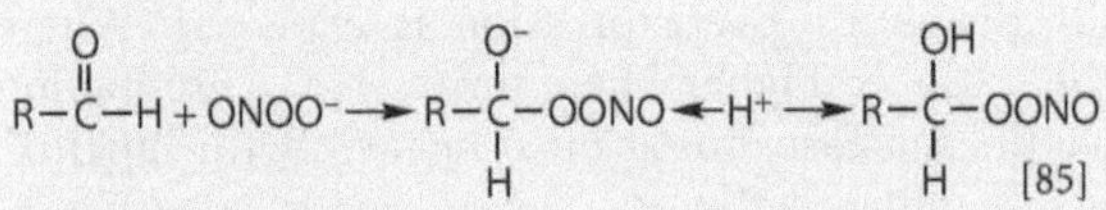
[85]

Aconitase is readily inactivated by peroxynitrite, but not by its precursor, nitric oxide (CASTRO et al. 1994).

In many instances, it is becoming apparent that peroxynitrite serves as a mediator in oxidative actions originally attributed to nitric oxide or other oxygen-derived species, as noted for aconitase inhibition (CASTRO et al. 1994). Peroxynitrite is now being revealed as a key contributing reactive species in pathological events associated with stimulation of tissue production of $^•NO$, e.g. systemic hypertension, inhibition of intermediary metabolism (RADI et al. 1994), ischaemia-reperfusion injury (MATHEIS et al. 1992), ultraviolet B radiation-induced skin injury (DELICONSTANTINOS et al. 1996), immune complex-stimulated pulmonary oedema, cytokine-induced oxidant lung injury, and inflammatory cell-mediated pathogen killing/host injury (ZHU et al. 1992, DENICOLA et al. 1993). Peroxynitrite oxidised α-tocopherol, ascorbate, and many thiols in synaptosomes isolated from *rat* brain (VATASSERY 1996).

Dopamine is oxidised by peroxynitrite (molar ratio 1:1) to 6-hydroxyindole-5-one as characterised by HPLC and photodiode array spectra, akin to the products of the tyrosinase-dopamine reaction, but no evidence of dopamine nitration was obtained (KERRY and RICE-EVANS 1999). Although peroxynitrite did not cause nitration of dopamine in vitro, the catecholamine is capable of inhibiting the formation of 3-nitrotyrosine from peroxynitrite-mediated nitration of tyrosine. The plant-derived phenolic compounds, caffeic acid and catechin, inhibited peroxynitrite-mediated oxidation of dopamine. In contrast, the monohydroxylated hydroxycinnamates, *p*-coumaric acid and ferulic acid, which inhibit tyrosine nitration through a mechanism of competitive nitration, did not inhibit peroxynitrite-induced dopamine oxidation.

Melatonin (4 mM in a phosphate buffer) upon vortex mixing with an equal volume of a solution of peroxynitrite (4 mM) formed 6-hydroxymelatonin and 5-methoxy-2-hydropyrroloindole (ZHANG et al. 1998). The formation of the melatoninyl radical cation in the peroxynitrite/melatonin reaction provides a direct evidence for the one-electron oxidation ability of peroxynitrite. Melatonin is a scavenger of peroxynitrite (GILAD et al. 1997). The overall peroxynitrite/melatonin reaction is first-order in melatonin and first-order in peroxynitrite, but the hydroxylation of melatonin is presumed to be zero-order in melatonin (ZHANG et al. 1999). Melatonin is metabolised in the liver, mainly to 6-hydroxymelatonin, which is a better chain-breaking antioxidant than melatonin. CO_2 modulates the reactions of peroxynitrite. The reaction of peroxynitrite with melatonin in the absence of added bicarbonate produces mainly 6-hydroxymelatonin and 1,2,3,3α,8,8α-hexahydro-1-acetyl-5-methoxy-8α-hydroxypyrrolo[2,3-*b*]indole, with some isomeric 1,2,3,3α,8,8α-hexahydro-1-acetyl-5-methoxy-3α-hydroxypyrrolo[2,3-*b*]indole. In the presence of added bicarbonate, product yields decrease and 6-hydroxymelatonin is not formed.

Peroxynitrite ($ONOO^-$ accompanied by $ONOOCO_2^-$) in aqueous-buffered solutions partially transforms melatonin into 1-nitrosomelatonin and several oxidised derivatives. Kinetic studies by stopped-flow spectrophotometry indicated that these conversions followed the first-order kinetic rate of peroxynitrite decay. The same melatonin transformation occures with $^•NO$ in the presence of oxygen, but at a lower kinetic rate. Under these conditions, nitrosation is the sole transformation of melatonin (BLANCHARD and DUCROCQ 2000).

Reactive nitrogen intermediates such as peroxynitrite also react with **DNA**, forming (among others) the adduct 7,8-dihydro-8-oxo-2'-deoxyguanosine (INOUE and KAWANISHI 1995, LIU and HOTCHKISS 1995, DOUKI and CADET 1996). 7,8-Dihydro-8-oxo-2'-deoxyguanosine is far more susceptible to peroxynitrite than 7,8-dihydro-2'-deoxyguanosine, which emphasises the fact that more stable oxidative end products than 7,8-dihydro-8-oxo-2'-deoxyguanosine exist (DOUKI et al. 1996, UPPU et al. 1996). SPENCER et al. (1996) treating *calf* thymus DNA (0.2 mg/ml) with peroxynitrite (1 mM) found, in addition to 8-nitroguanine, increased levels of both oxidised and deaminated base products, including 5-hydroxyhydantoin, 5-(hydroxymethyl)-uracil, thymine glycol, 4,6-diamino-5-formamidepyrimidine, 2,6-diamino-5-formamidepyrimidine, 8-oxoadenine, 8-oxoguanine, hypoxanthine, and xanthine.

DNA strand breakage in *calf* thymus DNA occurred after exposure to authentic peroxynitrite, or to SIN-1, a sydnonimine compound that generates both $^{\bullet}NO$ and $O_2^{\bullet-}$ and thus can be considered as a peroxynitrite donor (Inoue and Kawanishi 1995). Yermilov et al. (1995), however, suggested that peroxynitrite did not increase C8-oxoguanine levels in DNA. In addition to oxidation products, 8-nitroguanine has also been detected as a product of peroxynitrite reacting with guanine, suggesting that nitration could be an important process (Yermilov et al. 1995). In cell culture, deRojas-Walker et al. (1995) have suggested that oxidative damage to DNA in activated macrophages is due to the formation of peroxynitrite. In J774 macrophages and cultured *rat* aortic smooth muscle cells exposure to authentic peroxynitrite caused a dose-dependent increase in DNA strand breakage (Szabó et al. 1996).

The conditions to induce strand breaks, oxidation and deamination of DNA require high concentrations of reactive nitrogen oxide species or NO which may not be physiologically relevant (Laval et al. 1997). *In vivo*, antioxidants and reactive nitrogen oxide species scavengers, such as ascorbate and reduced glutathione, are abundant decreasing the chance that these chemical species directly modify DNA.

SIN-1 (0.1–15 mM), generating $ONOO^-$ by the simultaneous release of $^{\bullet}NO$ and $O_2^{\bullet-}$, dose- and time-dependently induced single-strand breaks in isolated *human* peripheral blood lymphocytes as assessed by single cell gel electrophoresis (Doulias et al. 2001). Exposure of the cells to SIN-1 (5 mM) in the presence of excess of superoxide dismutase (0.375 mM) increased the formation of single strand-breaks significantly, whereas 1000 U/ml catalase significantly decreased the quantity of single strand-breaks. The simultaneous presence of both superoxide dismutase and catalase before the addition of SIN-1 brought the level of single-strand breaks to that of the untreated cells. Pre-treatment of the cells with the intracellular Ca^{2+}-chelator BAPTA/AM inhibited SIN-1-induced DNA damage, indicating the involvement of intracellular Ca^{2+} changes in this process. Pre-treatment of the same cells with ascorbate or dehydroascorbate did not offer any significant protection in this system.

Toxic effects of superoxide dismutase overexpression are commonly attributed to increased H_2O_2 production. Still, published experiments yield contradictory evidence on whether superoxide dismutase overexpression increases or decreases H_2O_2 production. Gardner et al. (2002) analyzed this issue using a minimal mathematical model. The most relevant mechanisms of superoxide consumption are trated as pseudo first-order processes, and both superoxide production and the activity of enzymes other than superoxide dismutase were considered constant. Even within this simple framework, superoxide dismutase overexpression may increase, hold constant, or decrease H_2O_2 production. At normal superoxide dismutase levels, the outcome depends on the ratio between the rate of processes that consume superoxide without forming H_2O_2 and the rate of processes that consume superoxide with high (≥ 1) H_2O_2 yield. In cells or cellular compartments where this ration is exceptionally low (< 1), a modest decrease in H_2O_2 production upon superoxide dismutase overexpression is expected. Where the ratio is higher than unity, H_2O_2 production should increase, but at most linearly, with superoxide dismutase activity.

In confluent cultures of the *murine* lung epithelial cell line, LA-4 stimulated with cytokines to express iNOS, peroxynitrite caused a concentration-dependent decrease in $^{\bullet}NO$ (Robinson et al. 2001). Similar results were obtained when SIN-1 was added to the flasks.

Treatment of A549 *human* pulmonary epithelial cells with 1 mM peroxynitrite resulted in **ADP-ribosylation**, $NADP^+$ depletion, inhibition of mitochondrial respiration, and increased epithelial paracellular permeability (Szabó et al. 1997). The poly (ADP-ribose) synthetase (PARS) inhibitor 3-aminobenzamide (1 mM) provided a significant, partial protection against the energetic and functional changes. Similarly, inhibition of PARS activity by 3-aminobenzamide reduced the peroxynitrite-induced suppression of mitochondrial respiration in BEAS-2B *human* bronchial epithelial cells.

In *human* lung epithelial cells authentic $ONOO^-$ (100–500 μM) increased Mn-superoxide dismutase mRNA transcripts in a concentration-dependent fashion but did not change HPRT (Jackson et al. 1997). $ONOO^-$ stimulated luciferase gene expression driven by a 2.5 kb fragment of the *rat* Mn-superoxide dismutase gene 5' region. Induction due to $ONOO^-$ was inhibited by 10 mM L-cyteine, 50 mM *N*-acetylcysteine and partially inhibited by 10 mM PDTC. Neither H_2O_2 nor NO_2^-, breakdown products of SIN-1 and $ONOO^-$, had any effect on Mn-superoxide dismutase mRNA expression. $ONOO^-$ has direct, stimulatory effects on Mn-superoxide dismutase gene expression.

A study in the saline-perfused heart of knockout *mice* (with a defective eNOS gene) proved clear evidence that endogenously formed $^{\bullet}NO$ significantly contributed to ischaemia-reperfusion injury, most likely by peroxynitrite formation from $^{\bullet}NO$ (FLÖGEL et al. 1999).

On exposure of cultured *mouse* embryo cardiac myocytes to 0.2 mM $ONOO^-$, $[Ca^{2+}]_i$ increased to beyond the systolic level within 5 min with a concomitant decrease in spontaneous contraction of myocytes followed by complete arrest (ISHIDA et al. 1996). Addition of a L-type Ca^{2+} channel blocker or removal of extracellular Ca^{2+} prevented the $ONOO^-$-induced increase in $[Ca^{2+}]_i$, indicating that the increase in $[Ca^{2+}]_i$ was caused by the enhanced influx of Ca^{2+} through the plasma membrane and not by the enhanced release from sarcoplasmic reticulum. Plasma membrane fluidity and concentration of the thiobarbituric acid-reactive substance in the cells remained unchanged under the $ONOO^-$ treatment.

Peroxynitrite causes irreversible inhibition of mitochondrial respiration and damage to a variety of mitochondrial components via oxidising reactions as reviewed by BROWN (1999). Thus it inhibits or damages mitochondrial complexes I, II, IV and V, aconitase, creatine kinase, the mitochondrial membrane, mitochondrial DNA, superoxide dismutase, and induces mitochondrial swelling, depolarisation, calcium release and permeability transition. Peroxynitrite exposure to *rat* heart mitochondrial preparations resulted in significant inactivation of electron carriers such as succinate dehydrogenase and NADH dehydrogenase as well as the mitochondrial ATPase (RADI et al. 1994).

In *mouse* embryo ventricular heart muscle cells $ONOO^-$ increased $[Ca^{2+}]_i$ through disturbance of Ca^{2+} transport systems in the plasma membrane and impaired contractile protein (ISHIDA et al. 1996). Immunohistology demonstrated that the $ONOO^-$-mediated nitration product nitrotyrosine was formed in postischaemic but not in normally perfused controls (WANG and ZWEIER 1996). Most of the reactive oxygen species-positive cells in inflamed mucosa biopsy specimens from inflammatory bowel disease *patients* and control subjects were also positive for nitrotyrosine (DIJKSTRA et al. 1998). Nitrotyrosine-modified proteins were found on the surface of reactive oxygen species-positive cells.

Reaction of peroxynitrite with **tyrosine** residues gives 3-nitrotyrosine and dityrosine, as well as hydroxylated ring products (ISCHIROPOULOS et al. 1992, VAN DER VLIET et al. 1994, ISCHIROPOULOS 1998). 3-Nirotyrosine appears to be a specific marker of reactive nitrogen species, though it should be noted that this species can undergo further reaction with excess oxidant to give 3,5-dinitrotyrosine (YI et al. 1997). Low levels of peroxynitrite can cause inactivation of the heme-thiolate protein prostacyclin (PGI_2)-synthase by nitration of a tyrosine residue (MEHL et al. 1999). Spectral and kinetic analyses allow to conclude on a ferryl nitrogen dioxide complex as an intermediate which decomposes in the presence of an excess of peroxynitrite under formation of dioxygen, nitrite, and nitrate. This occurs in a catalytic cycle which was more efficient with $P450_{nor}$ than with microperoxidase. If phenol was added to the reaction mixture of peroxynitrite and the ferric complexes the ratio of hydroxylated to nitrated phenols decreased compared to the metal-free system. Phenol competed with the formation of dioxygen indicating that the ferryl intermediate was involved in both pathways. Alternatively, the ferry nitrogen dioxide complex can oxidise a second peroxynitrite molecule to the radical, $ONOO^{\bullet}$, which can decompose to dioxygen and $^{\bullet}NO$. The latter forms N_2O_3 with the remaining $^{\bullet}NO_2$ radical. A third pathway consists in the isomerization to nitrate which also is catalysed by the heme proteins since the ratio of nitrite/nitrate does not change significantly during the catalytic reaction with excess of peroxynitrite.

Immunohistology demonstrated that the $ONOO^-$-mediated nitration product nitrotyrosine was formed in postischaemic but not in normally perfused controls (WANG and ZWEIER 1996). Most of the reactive oxygen species-positive cells in inflamed mucosa biopsy specimens from inflammatory bowel disease *patients* and control subjects were also positive for nitrotyrosine (DIJKSTRA et al. 1998). Nitrotyrosine-modified proteins were found on the surface of reactive oxygen species-positive cells.

Both peroxynitrous acid and peroxynitrite react with **methionine,** $k_{acid} = (1.7 \pm 0.1) \times 10^3\ M^{-1}\ s^{-1}$ and $k_{anion} = 8.6 \pm 0.2\ M^{-1}\ s^{-1}$, respectively, and with *N*-acetylmethionine $k_{acid} = (2.8 \pm 0.1) \times 10^3\ M^{-1}\ s^{-1}$ and $k_{anion} = 10.0 \pm 0.1\ M^{-1}\ s^{-1}$, respectively, to form sulfoxides (PERRIN and KOPPENOL 2000).

D, L-**Selenomethionine** is oxidised by peroxynitrite by two competing mechanisms, a one-electron oxidation that leads to ethylene, and a two-electron oxidation that gives methionine selenoxide (PADMAJA et al. 1997). Kinetic modelling of the experimental data suggests that both peroxynitrous acid and the peroxynitrite anion react with D, L-selenomethionine to form methionine selenoxide with rate constants of $20{,}460 \pm 440\ M^{-1}s^{-1}$ and $200 \pm 170\ M^{-1}s^{-1}$, respectively at 25 °C. In the presence of added bicarbonate, the yield of ethylene obtained from the reaction of 0.4 mM peroxynitrite with 1.0 mM selenomethionine, decreased by 35 % with an increase in the concentration of bicarbonate from 0 to 25 mM. Kinetic simulations showed that the decrease in the yield of methionine selenoxide is due to reaction of the peroxynitrite anion with CO_2.

When peroxynitrite (100–200 nmol $site^{-1}$) was injected into the skin of anaestethized *rats*, the loss of nitrated proteins appeared biphasic with an initial ($t_{1/2}$ = 2 h) and slower loss ($t_{1/2}$ = 22 h), and a major nitrated protein was identified as albumin by Western blot analysis (GREENACRE et al. 1999).

Peroxynitrite at low concentrations (3–10 μM) inhibited agonist-induced platelet aggregation by a mechanism not dependent on the formation of cyclic guanosine monophosphate (LOW et al. 2002). Platelets recovered completely from peroxynitrite-induced inhibition within 30 min. Peroxynitrite induced nitration of cytosolic proteins, but this diminished to near basal levels within 60 min of exposure to the oxidant. During this period there was a reduction in tyrosine phosphorylation of specific proteins such as syk, but this was not due to the direct nitration of the same proteins. The inhibition of phosphorylation was reversible with platelet proteins recovering the ability to be phosphorylated within 15 min of exposure to peroxynitrite. Conversely, peroxynitrite increased phosphorylation of other proteins, but again these events were not directly linked to nitration.

The peroxynitrite generator, 3-morpholinosydnonimine (SIN-1), *in vitro* caused a concentration-dependent inhibition of **eosinophil chemotactic activity** (SATO et al. 1999). Peroxynitrite reduced RANTES (regulated on activation, normal T-cell expressed and secreted) and interleukin-5 binding to eosinophils and resulted in nitrosotyrosine formation. *In vitro* peroxynitrite significantly ($P < 0.05$) attenuated eotaxin-induced chemotactic activity in a dose-dependent manner (SATO et al. 2000). The inhibitory effect was not significant on eotaxin-induced chemotactic activity induced by leukotriene B_4 or complement-activated serum incubated with peroxynitrite. The reducing agents deferoxamine or dithiothreitol reversed eotaxin-induced chemotactic activity inhibition by peroxynitrite, and exogenous L-tyrosine abrogated the inhibition by peroxynitrite. 3-Morpholinosydnonimine (SIN-1) caused a concentration-dependent inhibition of eotaxin-induced chemotactic activity by eotaxin. Consistent with its capacity reduce eotaxin-induced chemotactic activity, peroxynitrite treatment reduced eotaxin binding to eosinophils.

In the central nervous system, $OONO^-$ induced modification of neuronal function. It stimulated the release of neurotransmitters including γ-aminobutyric acid (GABA) and acetylcholine from *mouse* cerebral cortical neurones (OHKUMA et al. 1995). Hydroxyl radical scavengers such as *N,N'*-dimethylthiourea (DMTU), mannitol, and uric acid significantly increased $OONO^-$-evoked [^{3}H]GABA release, whereas urea showed no effects on the release (HIGO et al. 1998). Removal of Ca^{2+} from incubation buffer abolished the enhancement of the release by DMTU, although DMTU showed no effects on the basal release with and without Ca^{2+} in extracellular space.

Uric acid is present in *human* plasma at much higher levels than those encountered in other *primates* because the enzyme urate oxidase is absent from *human* tissues (CUTLER 1984). it has therefor e been proposed that uric acid is an important antioxidant for *humans* (AMES et al. 1981). Urate reacts with peroxynitrite with an apparent second order rate constant of $4.8 \times 10^2\ M^{-1} \cdot s^{-1}$ in a complex process, which is accompanied by oxygen consumption and formation of allantoin, alloxan, and urate derived radicals (SANTOS et al. 1999). The main radical was identified as the aminocarbonyl radical by the electrospray mass spectra of its 5,5-dimethyl-*l*-pyrroline-*N*-oxide adduct. Mechanistic studies suggested that urate reacts with peroxynitrous acid and with the radicals generated from its decomposition to form products that can further react with peroxynitrite anion.

The degradation of the glycosaminoglycan hyaluronan by peroxynitrite was prevented to varying extents by addition of alternate target molecules (LI et al. 1997). Thiourea was extremely effective as a protective agent, dimethyl sulfoxide was moderately effective, while sodium benzoate and mannitol were slightly effective. A similar pattern of protection was observed when hyaluronan was degraded by hydroxyl radical generated by a metal ion/H_2O_2 system.

Phosphorylation activity of five **tyrosine kinases** of the *src* family from both *human* erythrocytes and *bovine* synaptosomes was stimulated by treatment with 30–250 μM peroxynitrite (MALLOZZI et al. 1999). This effect was not observed with *syk*, a non-*src* family tyrosine kinase. Higher concentrations of peroxynitrite inhibited the activity of all kinases, indicating enzyme inactivation.

In order to quantify **peroxynitrite scavenging** by different antioxidants BALAVOINE and GELETII (1999) developed a convenient "tube" assay. Peroxynitrite was employed as a biologically relevant source of radicals with Pyrogallol Red as a detecting molecule. A variety of compounds have been examined, namely polyphenols, uric acid, glutathione, and ascorbic acid. Competition kinetics were observed for the majority of examined compounds, except thymol and ascorbic acid. Pyrogallol Red was fully protected by ascorbic acid against the bleaching by peroxynitrite until its total consumption. The deviation from competition kinetics in the case of thymol was due to the formation of radicals from thymol and their subsequent reaction with Pyrogallol Red. Quercetin was the most efficient scavenger of free radicals. The measurements of relative antioxidant activities using Pyrogallol Red and other detecting molecules, such as gallocyanine and carminic acid, were in fair agreement.

Carotenes (lycopene, α-carotene, and β-carotene) and **oxocarotenoids** (β-cryptoxanthin, zeaxanthin, and lutein) prevented the formation of rhodamine 123 from dihydrorhodamine 123 caused by peroxynitrite, suggesting that the carotenoids react with peroxynitrite (PANASENKO et al. 2000). Oxocarotenoids were as effective as biothiols (glutathione, cysteine, and *bovine* serum albumin), known scavengers of peroxynitrite, whereas lycopene, α-carotene, and β-carotene exhibited a considerably more pronounced effect. Peroxynitrite caused a loss of carotenoids in low-density lipoprotein. Class II congestive heart failure (according to the New York Heart Association) *patients* showed significantly lower malondialdehyde levels and significantly higher levels of vitamin A, vitamin E, lutein, and lycopene than class III *patients* (POLIDORI et al. 2002).

Hydroxytyrosol, one of the *o*-diphenolic compounds in extra virgin olive oil, was highly protective against the peroxynitrite-dependent nitration of tyrosine and DNA damage by peroxynitrite *in vitro* (DEIANA et al. 1999). The lignans (+)-1-acetoxypinoresinol and (+)-pinoresinol are major components of the phenolic fraction of olive oils (OWEN et al. 2000). These lignans, which are potent antioxidants, are absent in seed oils and virtually absent in refined virgin oils but are present in concentrations of up to 100 mg/kg (mean ± SE, 41.53 ± 3.93 mg/kg; range, 0.65–99.97 mg/kg) in extra virgin oils.

Green tea extract and its **polyphenols** markedly inhibited luminol-dependent chemiluminescence activated by peroxynitrite or SIN-1 (VAN DYKE et al. 2000). At high doses of tea or antioxidants, the light generated from the reaction of peroxynitrite and luminol was clearly inhibited. However, at dilute concentrations of antioxidants, one can often observe stimulation of light. WÖRTH et al. (2000) developed an analytical method capable of separating six different catechins and caffeine in tea by micellar electrokinetic chromatography in only 20 min without extensive sample preparation and compared the amount of catechins and caffeine in several teas and different preparation methods. VALCIC et al. (2000) isolated and identified a previously unreported reaction product of (–)-epigallocatechin gallate and three reaction products of (–)-epigallocatechin. In the (–)-epigallocatechin gallate product, the B-ring was transformed into a ring-opened unsaturated dicarboxylic acid moiety. The (–)-epigallocatechin products included a seven-membered B-ring anhydride and a symmetrical (–)-epigallocatechin dimer, both analogues of previously described (–)-epigallocatechin gallate oxidation products. The third (–)-epigallocatechin product was an unsymmetrical dimer.

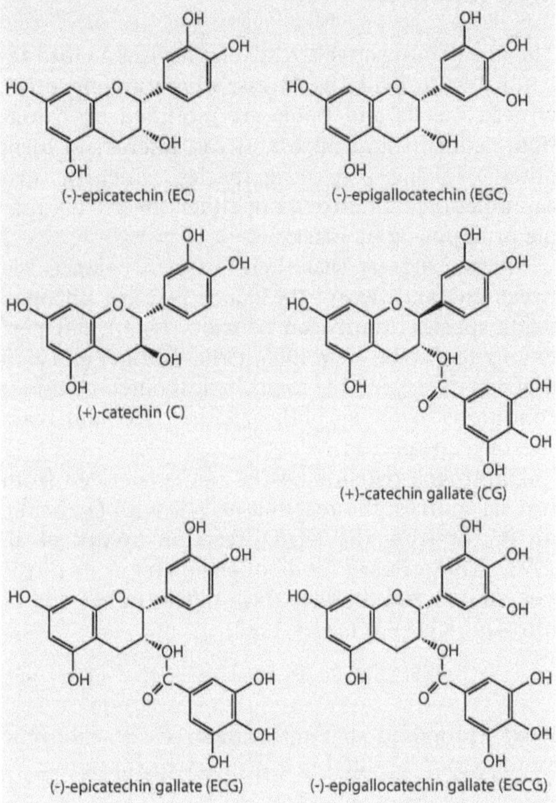

Structures of catechins in tea extracts (from WÖRTH et al. 2000) [86]

Catechins inhibited invasion of *mouse* MO_4 cells into embryonic *chick* heart fragments *in vitro* (BRACKE et al. 1991). The anti-invasive effects could be ranked as follows:

(+)-catechin > (–)-epicatechin > 3-*O*-methyl-(+)-catechin > 3-*O*-palmitoyl-(+)-catechin

Most of the catechins are unstable in cell culture media, and their spontaneous rearrangement products tend to bind to extracellular matrix.

Iron(III) *meso*-tetra(2,4,6-trimethyl-3,5-disulphonato)porphyrine chloride (FeTMPS) is oxidised by peroxynitrite to produce oxoFe(IV)TMPS and NO_2 ($k_1 = 1.3 \times 10^5\ M^{-1}s^{-1}$) (SHIMANOVICH and GROVES 2001). The porphyrin is then reduced back to Fe(III)TMPS by nitrite, but this rate ($k_2 = 1.4 \times 10^4\ M^{-1}s^{-1}$) is not sufficient to maintain the catalytic process at the observed rate. A "cage -return" reaction between the generated oxoFe(IV)TMPS and NO_2 ($k_8 = 5.4 \times 10^4\ M^{-1}s^{-1}$), affording Fe(III)TMPS and nitrate, and a reaction between oxoFe(IV)TMPS and peroxynitrite ($k_7 = 2.4 \times 10^4\ M^{-1}s^{-1}$) that afford oxoFe(IV)TMPS and nitrate are presented.

6.2.11 N_2O_3: Nitrosative Stress

The term "nitrosative stress" emphasises a close relationship to oxidative stress, where among other targets, metals and thiols are modified by oxidation. Oxidants and nitrosants can interact to form other oxidising/nitrosating species, which in turn can potentiate the toxicity of either species is capable of imposing by itself.

Studies suggest that there exists a balance between NO and two of the major reactive nitrogen oxide species, dinitrogen trioxide (N_2O_3) and peroxinitrite (MILES et al. 1995, 1996, WINK et al. 1997) that can determine the toxicological outcome under conditions involving both or either •NO and reactive oxygen species.

Dinitrogen trioxide (N_2O_3) can be derived from several sources: the reaction of •NO with O_2, acidic nitrite, or from the NO/O_2-reaction (WINK et al. 1993). The primary mode of chemistry is to nitrosate amines and thiols to form nitrosamines and S-nitrosothiols (WILLIAMS 1988).

$$NO_2 + NO \rightarrow N_2O_3. \quad [87]$$

It was proposed that nitrosation of the exocyclic amide group resulted in a primary nitrosamine:

$$NH_2\text{–}R + N_2O_3 \rightarrow R\text{–}NHNO + NO_2^- \quad [88]$$

This was then followed by rapid deamination resulting in hydroxy group:

$$R\text{–}NHNO \rightarrow RNNOH \rightarrow R\text{–}OH + N_2. \quad [89]$$

Nitrosylated albumin was detected in activated macrophage supernatants using an anti-NO-acetylated cysteine antibody (GOBERT et al. 1999). It was estimated that 10 % of N_2O_3 produced by activated cells participate in extracellular nitrosylation. N_2O_3 thus appears to be a new effector molecule of the immune system, as an agent for the nitrosylation of albumin, the main nitric oxide carrier *in vivo*.

The •NO-mediated deamination of **nucleobases** depends on the presence of oxygen, indicating that N_2O_3 is involved (NGUYEN et al. 1992). This chemistry would result in the conversion of cytosine to uracil, guanine to xanthine, methylcytosine to thymine, and adenine to hypoxanthine. Single-stranded DNA is far more susceptible to nitrosative chemistry than double-stranded DNA (MERCHANT et al. 1996). It has been proposed that these mechanisms involving NO could contribute to the spontaneous deamination which occurs *in vivo*. Another possible pathway for the mutagenicity of N_2O_3 is the endogenous formation of nitrosamines, which are powerful DNA-alkylating agents (BARTSCH et al. 1990).

$Na_2N_2O_3$, Angelis' salt, as a source of NO^- was found to be cytotoxic to Chinese *hamster* V79 lung fibroblast cells over a concentration range of 2–4 mM (WINK et al. 1998). The presence of equimolar ferricyanide, $Fe(III)(CN)_6^{3-}$, which converts NO^- to •NO, afforded dramatic protection against Angelis' salt-mediated cytotoxicity. Treatment of V79 cells with L-buthionine sulfoximine to reduce intracellular glutathione markedly enhanced the cytotoxicity of Angelis' salt, which suggests that GSH is critical for cellular protection against the toxicity of NO^-. At a molecular level, exposure to Angelis' salt resulted in DNA double strand breaks in whole cells, and this effect was completely prevented by coincubation of cells with ferricyanide or 4-hydroxy-2,2,6,6-tetramethylpiperidine-1-oxyl (TEMPOL).

The beneficial effect of TEMPOL, however, is limited because its short lifetime in tissues (KUPPUSAMY et al. 1996). Polynitroxylated macromolecules such as polynitroxyl-*human* serum albumin enable maintaining a sustained concentration of TEMPOL for longer periods in tissue. In *rats* medicated with polynitroxyl-*human* serum albumin + TEMPOL ischaemic myocardial infarct volume was significantly smaller than that in control *rats* given saline (LI et al. 2002).

Because of their versatility and favourable properties, in addition to their relative nontoxicity in both *in vitro* and *in vivo* systems (SWARTZ et al. 1995, DAMIANI et al. 2000), it is likely that nitroxides will find increasing use in viable biological systems. The results obtained overall by DAMIANI et al. (2000) in an in vitro study of oxidative DNA damage and the effect of nitroxides shed more light on the antioxidant activity of this particular class of compounds and lead to a better understanding of the full exploitation of these powerful tools.

6.2.12 Nitric Oxide/Nucleophile Complexes

Adducts of •NO with nucleophiles (NONOates) have long been known in the chemical literature (DRAGO 1962). Their biological properties, however, have not been explored until recently. The biologically important polyamine, spermine or sulphite ions form adducts showing vasorelaxant properties with

EC_{50} values of 6.2 μM and 62 μM, respectively (Morley and Keefer 1993). NONOates may have the potential to address some of the limitations of currently available nitrovasodilators. They maintain •NO stability during storage in solid form and are highly soluble in aqueous media. The rate at which •NO is generated on dissolution of these compounds in a biologic system can be chemically predicted and adjusted by altering the carrier nucleophile. In acute lung injury induced by intravenous injection of oleic acid (*cis*-9-octadecanoid acid) in the *pig*, aerosolised nonoate following surfactant pre-treatment improved oxygenation and reduced pulmonary hypertension (Jacobs et al. 2000).

Exposure of *rat* skeletal muscle to diethylamine NONOate (5 mM) resulted in a 50% reduction in catalase (EC 1.11.1.6), glutathione peroxidase (EC 1.11.1.9), and a dose-dependent inhibition of Cu,Zn superoxide dismutase (EC 1.15.1.1) (Lawler and Song 2002).

When [^{14}C]acetaminophen was incubated at pH 7.4 with reactive nitric oxygen species, the same new product (3NAP) was produced by at least three separate pathways represented by the following conditions: myeoloperoxidase oxidation of NO_2^-, NO_2Cl, and $ONOO^-$ or 3-morpholinosydnonimine (Lakshmi et al. 2000). *Human* polymorphonucler neutrophils incubated with [^{14}C]acetaminophen and stimulated with 12-*O*-tetradecanoylphorbol-13-acetate produced 3NAP in the presence of NO_2^- Neutrophil 3NAP formation was verified by mass spectrometry and was consistent with myeoloperoxidase oxidation of NO_2^-. Spermine NONO supported 3NAP formation by stimulated cells in the absence of NO_2^-. Results demonstrated that 3NAP is a product of nitrating reactive nitric oxygen species generated by at least three separate pathways and may be a biomarker for nitrating mediators of inflammation.

6.2.13
S-Nitrosothiols as Intermediates in the Metabolism or Bioactivity of •NO

In view of the fact that the plasma and cellular milieu contains reactive species that can rapidly inactivate •NO, it has been postulated that •NO is stabilised by a carrier molecule that preserves its biological activity. Reduced thiol species are candidates for this role. as they readily react in the presence of •NO to yield biologically reactive *S*-nitrosothiols that are more stable, and possibly more potent than •NO itself (Ignarro et al. 1981, Mendelsohn et al. 1990). Sulfhydryl groups in proteins, and free cysteine and GSH represent an abundant source of reduced thiol in biological systems. There is increasing evidence that at least part of the activity of •NO is attributable to *S*-nitrosothiols derived from the reaction of •NO with intracellular thiol compounds like cysteine or GSH (Ignarro 1990, Myers et al. 1990). *S*-Nitrosothiols may play the same role in the mechanism of action of endothelium-derived relaxing factor (EDRF) as •NO; the potent and long-lasting effects of vasodilatation and platelet inhibition that they cause are mediated by guanylate cyclase activation (Ignarro et al. 1981, Mellion et al. 1983). These observations suggest that *S*-nitrosothiol groups in proteins may serve as intermediates in the cellular metabolism or bioactivity of •NO and that their formation may represent an important cellular regulatory mechanism (Stamler et al. 1992). *S*-Nitrosothiols have also been proposed as biologically active intermediates in the metabolism of organic nitrates (Ignarro et al. 1981, Mellion et al. 1983, Loscalzo 1985).

The reactivity of •NO with sulfhydryl (R–SH) groups depends on the electron configuration of its 2p-Π antibonding orbital, as reviewed by Stamler et al. (1992). The presence of one (radical) electron in this orbital does not ordinarily confer reactivity with R–SH groups, though it clearly allows reaction with thiyl radical species. On the other hand, loss of this electron to form NO^+ confers strong electrophilicity and reactivity towards most biological R–SH species (Stamler et al. 1992, Arnelle et al. 1995). A second electron in the 2p-H orbital, forming NO^-, may under certain circumstances confer reactivity with relatively electropositive R–SH species, particularly in the presence of ferrous ion or other transition metals (Stamler et al. 1992, Arnelle et al. 1995, Vanin et al. 1997).

The reductive decomposition of *S*-nitrosoglutathione by L-ascorbic acid yields •NO which was monitored both electrochemically and spectrophotometrically (Smith and Dasgupta 2000). The rate of reaction and •NO release was found to be pH dependent in a manner which dramatically increases with pH demonstrating that the L-ascorbic acid dianion is by far the most reactive species of L-ascorbic acid. *S*-Nitrosoglutathione is relatively stable in the dark, aqueous medium and even in the presence of trace quantities of Cu^{2+}. Induced catalytic decomposition of *S*-nitrosoglutathione only becomes significant above ca. 10 μM Cu^{2+}, but beyond this it shows linear dependency.

Gorren et al. (1996) have suggested that the Cu^+-promoted decomposition of *S*-nitrosoglutathione does not involve reoxydation of Cu^+, since no complex formation occurred when the specific Cu^{2+} chelator cuprizone was added. Noble et al. (1999) found that Cu^{2+} is complexed more strongly to *S*-nitrosoglutathione than it is to cuprizone, since when Cu^{2+} was added to equal concentrations of cuprizone and *S*-nitrosoglutathione, no cuprizone complex was formed. However, they also found that cuprizone stopped the reaction of *S*-nitroso-*N*-acetylpenicillamine, added either at the start or after ~60 % reaction. In this case the disulphide does not have an inhibiting effect on the reaction.

One-electron reduction of *S*-nitrosoglutathione (GSNO) has been found to proceed via a pathway to give glutathione and nitric oxide (Atsipapa et al. 1999):

$$GSNO + e^- \rightarrow GS^- + {}^{\bullet}NO \quad [90]$$

rather than thiyl radicals and nitroxyl anion:

$$GSNO + e^- \rightarrow GS^{\bullet} + NO^- \quad [91]$$

The carbon dioxide radical anion reacted with GSNO at a rate first-order in [GSNO] with $k = 7\times10^8\ dm^3\ mol^{-1}\ s^{-1}$. The differential absorption spectrum (product - reactants) after 100 μs, when reaction was complete, corresponded closely to the bleaching of the ground state of GSNO (λ_{max} = 34 nm). This indicates little (if any) production of $GS^{\bullet}$, which absorbs significantly at 250 - 450 nm. The lack of production of $GS^{\bullet}$ was confirmed by the negligible production of $ABTS^{\bullet-}$ radical from 2,2'-azinobis(3-ethylbenzothiazoline-6-sulphonate), characteristic of oxidation of the chromophore.

Atsipapa et al. (2000) used kinetic competition experiments involving superoxide reacting with GSNO, compared with its dismutation, to measure the reactivity of $O_2^{\bullet-}$ to GSNO. Superoxide reacted with GSNO about seven orders of magnitude slower than superoxide dismutase.

Noble and Williams (2000) speculated what might occur *in vivo* with reference of ${}^{\bullet}NO$ release from GSNO. At micromolar levels, ${}^{\bullet}NO$ will be generated rapidly from GSNO if there is present a sufficient concentration (also around micromolar levels) of a Cu^{2+} source (peptide of protein bound; Dicks and Williams 1996) together with again a low concentration of a thiol or other reducing agent to effect Cu^+ formation. The importance of the transnitrosation process now loses its significance, although it may occur in parallel, but may be too slow at low thiol concentrations to compete. The same conclusion applies to the possible enzymatic transformation, which leads to SNO-CysGly, which behaves much like GSNO at micromolar concentrations in its ability to generate ${}^{\bullet}NO$ by the copper-catalysed reaction.

Adrenaline and its cyclic derivatives (adrenochrome and adrenolutin) may be oxidised by *S*-nitrosoglutathione (Rigobello et al. 2001). Adrenaline was first oxidised to adrenochrome that, after isomerization to adrenolutin, was further oxidised to products monitored as fluorescence decrease. To occur to a significant extent, this oxidation requires copper ions that, in addition to a direct effect on the oxidation of the *o*-diphenol moiety, are also able to compose nitrosothiols, giving rise to nitric oxide. The latter, after interaction with oxygen and superoxide, produces nitrogen oxides and peroxynitrite, respectively, that are important contributors to the oxidative process. In this context, catecholamines might act as regulatory factors toward nitric oxide and its derivatives.

Oxidation mechanism of adrenaline and its cyclic derivatives by the system $GSNO/CuSO_4$ in aerobic conditions. RH_2, $RH^{\bullet}$, and R indicate the *o*-diphenol, semiquinone, and *o*-quinone forms of adrenaline and its cyclic derivatives (leucoadrenochrome and adrenolutin), respectively. (from Rigobello et al. 2001)

$$RH_2 + O_2 \rightarrow RH^{\bullet} + O_2^{\bullet-} + H^+ \quad [92]$$

$$RH_2 + Cu^{2+} \rightarrow Cu^{1+} + RH^{\bullet} + H^+ \quad [93]$$

$$2\ GSNO\text{–}Cu^{1+} \rightarrow GSSG + 2\ {}^{\bullet}NO \quad [94]$$

$$RH_2 + {}^{\bullet}NO \rightarrow RH^{\bullet} + NO^- + H^+ \quad [95]$$

$$2\ {}^{\bullet}NO + O_2 \rightarrow 2\ {}^{\bullet}NO_2 \quad [96]$$

$$RH_2 + {}^{\bullet}NO_2 \rightarrow RH^{\bullet} + HNO_2 \quad [97]$$

$$O_2^{\bullet-} + O_2^{\bullet-} \rightarrow H_2O_2 + O_2 \quad [98]$$

$${}^{\bullet}NO + O_2^{\bullet-} \rightarrow ONOO^- \text{—}H^+ \rightarrow ONOOH \quad [99]$$

$$2RH_2 + ONOOH \rightarrow 2\ RH^{\bullet} + HNO_2 + H_2O \quad [100]$$

$$6\ RH^{\bullet} \rightarrow 3\ RH_2 + 3\ R \quad [101]$$

Vanin et al. (2002) analysed the influence of copper and iron ions on the process of decomposition of *S*-nitrosocysteine, the most labile species among *S*-nitrosothiols. Copper or iron ions, especially in the presence of the reducing agent ascorbate, accelerated the decomposition of *S*-nitrosocysteine and markedly attenuated the amplitude and duration of the relaxant effect of *S*-nitrosocysteine. By contrast, the iron and copper chelators bathophenantroline disulphonic acid and bathocuproine disulphonic acid exerted a stabilizing effect on *S*-nitrosocysteine, prolonged its vasorelaxing effect, and abolished the influence of ascorbate. In the presence of ascorbate, bathophenantroline disulphonic acid displayed a selective inhibitory effect toward the influence of iron ions (but not toward copper ions) on *S*-nitrosocysteine decomposition and vasorelaxant effect, while bathocuproine disulphonic acid prevented the effects of both copper and iron ions.

Although inhibition of NOS, transfection with iNOS, or addition of ${}^{\bullet}NO$ donors did not affect matrix metalloproteinase-9 induction in airway epithelial cells by inflammatory cytokines, addition of

S-nitrosothiols dramatically inhibited matrix metalloproteinase-9 expression, which was potentiated by depletion of cellular GSH (OKAMOTO et al. 2002). Cytokine-induced matrix metalloproteinase-9 expression involves the activation of the transcription factor NF-ϰB, and S-nitrosothiols, in contrast to •NO, were found to inhibit cytokine-induced nuclear translocation and DNA binding of NF-ϰB.

6.2.14 Cell Protection Against NO_x

Both *in vivo* and *in vitro* cell protection against the peroxynitrite anion and the nitrogen dioxide radical was demonstrated for β-carotene in the presence of vitamin E and ascorbic acid (BÖHM et al. 1998).

$$NO_2^{\bullet} + \text{Carotene} \longrightarrow NO_2^{-} + \text{Carotene}^{\bullet+} \quad [102]$$
$$RO_2^{\bullet} + \alpha\text{-Tocopherol} \longrightarrow \alpha\text{-Tocopherol}^{\bullet} + H^{+} \quad [103]$$

In the presence of ascorbic acid (Asc), BÖHM et al. (1998) observed a synergistic rather than an additive effect in their studies in *human* lymphoid and Jurkat cells and this implied an interaction between these individual anti-oxidant components.

$$\text{Carotene}^{\bullet+} = \text{AscH} \longrightarrow \text{Carotene} + \text{Asc}^{\bullet-} + H^{+} \quad [104]$$
$$\text{Carotene}^{\bullet+} = \text{AscH}_2 \longrightarrow \text{Carotene} + \text{AscH}^{\bullet} + H^{+} \quad [105]$$

While there is direct evidence for the regeneration of the parent carotenoid from its radical cation via ascorbic acid, an interaction between α-tocopherol and carotene is more speculative. In polar environment α-tocopherol$^{\bullet+}$ would be rapidly deprotonated (PARKER and BISBY 1993) so that no interaction between α-tocopherol$^{\bullet+}$ and carotene would be possible. Indeed, for the water-soluble vitamin E analogue, Trolox C, they saw the quenching of carotene$^{\bullet+}$ by Trolox C.

TSUCHIHASHI et al. (1995), using 2,2'-azobis(2,4-dimethylvaleronitrile) as a radical initiator and conventional spectroscopic techniques, suggested that any protective effect of β-carotene which they observed was **not** due to carotene$^{\bullet+}$ production, but rather due to the formation of **uncharged** species via addition reaction of the type:

$$LO_2^{\bullet} + \text{Carotene} - \text{LOO} - \text{Carotene}^{\bullet} \longrightarrow {}^{02}\text{LOO} - \text{Carotene-OO}^{\bullet} \quad [106]$$

where $LO_2^{\bullet}$ is a peroxyl radical deep inside the non-polar region of the membrane.

The glutathione redox system may regulate NO_x-mediated protein thiol modification by participating in the repair of protein thiols modified by NO_x (PADGETT and WHORTON 1997). Many protein thiols have pK_a values lower than free cysteine and are therefore more reactive toward NO_x. These protein thiols would be expected to react preferentially with NO_x over glutathione. The observed requirement for glutathione may thus be to reverse the nitrosylation of protein thiols in a manner similar to the role for glutathione in reversing protein thiols oxidized by H_2O_2.

PART II

Respiratory System: Orthology and Pathology

CHAPTER 7

Tracheo-bronchial Epithelium

While the tracheal epithelium of *man* is pseudostratified and columnar, in the *rat* it is of a simple columnar type (RHODIN 1959).

7.1 Cell Types

The tracheal epithelium consists of mainly four kinds of cells: ciliated cells, goblet cells, brush cells, and basal cells (RHODIN and DALHAMN 1956). In the *rat's* bronchial epithelium, JEFFERY and REID (1975) recognised eight epithelial cell types and two mesenchymal ones (which are probably migratory). All epithelial cells were attached to the basement membrane, but not all reached the airway lumen:

1) Ciliated
2) Serous
3) Clara
4) Goblet
5) Intermediate
6) Brush
7) Basal
8) Kulchitsky.

In the *mouse*, TAIRA and SHIBASAKI (1978) distinguished four types of cells: ciliated cells, goblet cells, basal cells, and non-ciliated cells. PACK et al. (1980) found *mouse* tracheal epithelium markedly different from that of other species. However, the near absence of goblet cells and the presence of a large number of secretory Clara-like cells make it similar to the distal airways of other species.

In the trachea of the *guinea pig*, DALEN (1983) recognised four epithelial cells: basal cells, goblet cells, ciliated cells and intermediate cells.

In the *pig*, BASKERVILLE (1970b) described ciliated cells, goblet cells, basal cells, and intermediate cells. Brush cells were not seen, although they have been described in bronchioli of the *pig* (BASKERVILLE 1970a).

In the *sheep*, MARIASSY and PLOPPER (1983) observed 6 distinct, granule-containing secretory cells: 4 mucous cell categories (M1–M4), serous cells, and nonciliated bronchiolar (Clara) cells. Smooth endoplasmic reticulum was observed only in M3 (33.8 %) and Clara cells (49 %). Granular endoplasmic reticulum was most abundant in serous (21 %) and least plentiful in M4 (2.2 %) cells (MARIASSY and PLOPPER 1984).

7.1.1 Ciliated Cells

The elongated ciliated cells stand on the basement membrane and reach the cell surface, this distance being about 15 μm in the tracheal epithelium of the male Wistar *rat* (RHODIN and DALHAMN 1956, p. 348) and 28 μm in the bronchial epithelium of the mongrel *dog* (SCHILLER 1958). In transverse sections they are polygonal.

7.1.1.1 Kinocilia

The kinocilia are evenly distributed over the cell surface with a distance of about 300–400 nm from centre to centre. This makes, with a cell surface area of 33 μm^2, a total number of about 270 cilia per cell (RHODIN and DALHAMN, l.c. p. 357). The cilia have a length of about 5 μm and a mean width of 240 nm. They are covered by a unit membrane (ROBERTSON 1959, 1960) 7 nm thick. In transverse sections they have a circular form and possess a ring of 9 peripheral and 2 central filaments.

Cilia longer than 8 μm have not been found in healthy *mammals* (MENCO 1980). In a patient described by AFZELIUS et al. (1985), the cilia were 10–13 μm long. The clearance rate for radioactively labelled albumin microspheres was 0.30 % in the left lung and 0.42 % in the right lung (normal value 1.50 %). In a patient described by NIGGEMANN et al. (1992), the cilia were up to about 15 μm long. The saccharine transport test showed slightly prolonged transport time. The frequency of the ciliary beat was decreased to 7.5 Hz (normal value for *children* is above 10 Hz).

Contrary to the situation in the usual types of immotile cilia syndrome, the proportion of ciliated cells is increased, not decreased, in the long cilia syndrome. The number of cilia per cell may be normal (AFZELIUS et al. 1985) or increased (NIGGEMANN et al. 1992).

It will be evident from the figures presented above that mucociliary clearance is defective. GHADIALLY (1997) stated two reasons for this: (1) the long cilia beat slowly like flagella (reduced beat frequency); and (2) the long cilia may pile up and block small bronchioles. The long cilia syndrome is classified as a subgroup of immotile cilia syndrome, in which ciliary dyskinesis stems from abnormally lung cilia, and not from defects in the axonome.

In 15 out of 20 apparently healthy domestic *pigs*, RADNER and STOCKINGER (1992) observed extended aggregates of intracellular axonemal derivatives within the apical cytoplasm. They are more likely to be due to a failure of ciliary maturation than to a degradation of incorporated mature cilia.

Microtubules are essential organelles involved in ciliary function. Tectins, which share some structural homology with intermediate filament proteins but are uniquely different, could act to stabilize the pf-ribbon located in the doublet microtubule and thus the microtubule itself (LINCK 1995).

7.1.1.2 Microvilli

In the trachea of the *rat*, the filiform projections have a length of 0.8–1 μm and a width of 100–150 nm (RHODIN and DALHAMN 1956). In a tangential section just above their bases there are 5 projections per μm^2. The projections are bordered by a unit membrane 9 ± 0.2 nm thick.

7.1.1.3 Zonula Occludens

By freeze-fracturing BARTELS (1981) depicted a zonula occludens at the transition from apical to lateral plasmalemmata of the ciliated and Clara cells of the bronchiolar epithelium of the *rat*. The lateral cell membrane of the ciliated cells showed square arrays of particles.

7.1.1.4 Mitochondria

The *human* bronchial mucosa shows mitochondrial aggregates in the apical cytoplasm adjacent to the ciliary border (GHADIALLY 1997).

7.1.1.5 Golgi Apparatus

The Golgi complex is located in the supranuclear region (DALEN 1983).

7.1.1.6 Histochemistry

Ciliated epithelial cells of the *cat* trachea demonstrate diffuse immunoreactivity for **cyclic AMP**, more intense at the apex, sparing nuclei (LAZARUS et al. 1984).

In vitro, the amount of a histochemically detectable 100-kDa **ATPase** subunit in *human* nasal turbinate epithelium correlated with dynein activity and thus with ciliary activity (SCHÜTZ et al. 2002).

Immunohistochemical analysis of the distribution of **P_{450} reductase** in lung tissue demonstrated that in the control *mice* this enzyme was expressed most strongly in the ciliated cells of the proximal and distal bronchi and in the Clara cells of the small or terminal bronchioles (LIM et al. 1998). Cu/Zn superoxide dismutase and Mn superoxide dismutase were detected primarily in the ciliated cells of the bronchial epithelium.

Endogenous **peroxidase** activity was demonstrated in ciliated cells and secretory cells of the laryngeal epithelium and gland of *rats* (KATAOKA 1971). In the septum nasi and trachea of Swiss *mice*, BÖCK (1973) found less numerous peroxisomes than there were in the oviduct ciliated cells.

Both plasma membrane association and non-association of **guanylate cyclase** (EC 4.6.1.2) activity and soluble guanylate cyclase protein were found in ciliated cells of the *rodent* respiratory epithelium (GOSSRAU et al. 2001).

The monoclonal antibody, **RTE3**, specifically reacts with the plasma membrane of ciliated *rat* tracheal epithelial cells cultured on collagen gel-coated membranes at an air-liquid interface in hormone- and growth factor-supplemented medium (CLARK et al. 1995).

ERM (ezrin/moesin/radixin), three actin-binding proteins, which link actin filaments to the plasma membrane (TSUKITA et al. 1997), are primary located in microvilli on the apical surface of epithelial cells, although they have also been observed at intercellular boundaries (BRETSCHER et al. 1997).

Multi-drug resistance associated protein (MRP), member of the ATP binding Cassette transporter proteins, is particularly involved in glutathione-conjugates detoxification. In *human* bronchial ciliated cells MRP mRNA immunostaining was observed at the basolateral plasma membrane (BRÉCHOT et al. 1997).

Using primary passage-1 *human* tracheobronchial epithelial cell cultures and an immortalised *human* bronchial epithelial cell line, HBE1, HARPER et al. (2001) observed that tumour necrosis factor-α enhanced **nuclear factor-κB** (NF-κB) transcriptional activity. TNF-α activation coincided with translocation of NF-κB p65 from the cytoplasm to the nucleus. Pre-treatment with *N*-acetylcysteine (1–10 mM) inhibited TNF-α-induced activation of NF-κB transcriptional activity and IL-8 promoter-mediated reporter gene expression. In contrast, elevated thioredoxin protein levels in cells enhanced TNF-α-dependent NF-κB transcriptional activity and IL-8 promoter activity.

7.1.2 Serous Cells

The serous cell of the *rat* has a relatively electron-lucent cytoplasm but also abundant rough endoplasmic reticulum. Its secretory granules are discrete and electron dense and do not fuse with one another. SPICER et al. (1990) found numerous mitochondria ranging from slightly larger than normal to several micrometers in diameter (giant) in about one-half of the serous secretory cells in the surface epithelium of the normal *gerbil* trachea and proximal bronchi.

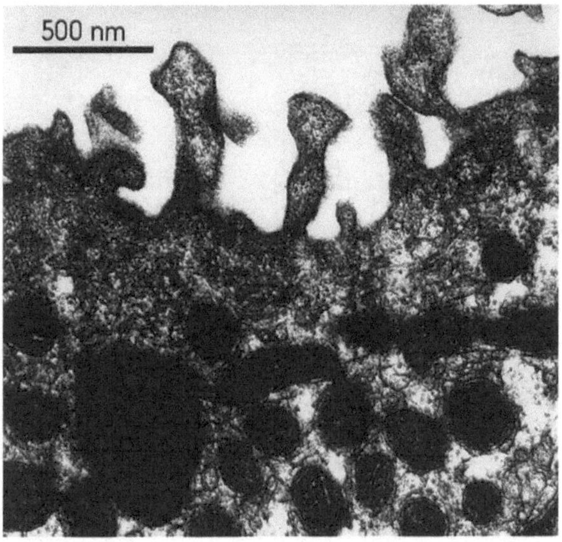

Fig. 58. Apical part of a "serous" bronchiolar epithelial cell of an unmedicated female white *rat* (breeder: Winkelmann, Borchen-Kirchborchen) fixed on January 15, 1968 under methitural anaesthesia by intratracheal instillation of 2.5 % glutaraldehyde in phosphate buffer (pH 7.4) before opening the thorax. Postfixation with 1 % osmium tetroxide in phosphate buffer (pH 7.4). Contrasted en bloc with 0.5 % uranyl acetate in 70 % ethanol. Embedded in a 2:8 mixture of methyl and butyl methacrylate. Sectioned at 50 nm. Lead citrate after REYNOLDS (1963). Film 203/84

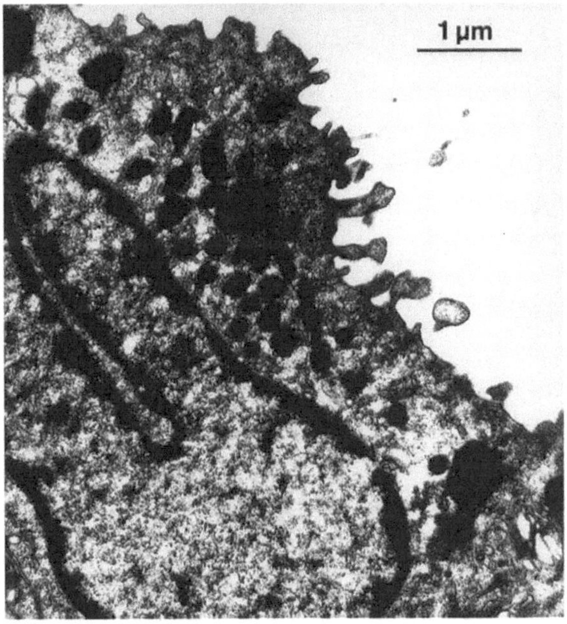

Fig. 57. A non-ciliated cell from a bronchiole of an unmedicated female white *rat* (breeder: Winkelmann, Borchen-Kirchborchen) fixed on January 15, 1968 under methitural anaesthesia by intratracheal instillation of 2.5 % glutaraldehyde in phosphate buffer (pH 7.4) before opening the thorax. Postfixation with 1 % osmium tetroxide in phosphate buffer (pH 7.4). Contrasted en bloc with 0.5 % uranyl acetate in 70 % ethanol. Embedded in a 2:8 mixture of methyl and butyl methacrylate. Sectioned at 50 nm. Lead citrate after REYNOLDS (1963). Film 202/84

In addition to producing the glycoproteins to be added to the bulk phase of the mucous blanket, serous cells produce specific substances, which are probably important in defence against infectious agents. By immunoperoxidase staining, the granules in the serous cells of the seromucous tracheobronchial glands contain lysozyme (BOWES and CORRIN 1977, BALS 1997), an enzyme which is active in the hydrolysis of components of certain gram-positive bacterial cell walls, and lactoferrin, an iron-containing protein which catalyses hydroxyl radical production (AMBRUSO and JOHNSTON JR 1981, WINTERBOURN 1983). BANISTER et al. (1982) demonstrated hydroxyl radicals, measured by PER spin trapping and by ethylene production from α-keto-γ-methiol butyric acid to be produced by a Fenton-type Haber-Weiss reaction catalysed by lactoferrin.

A protease inhibitor with a low molecular weight of about 15 kDa inhibits the activity of the leucocyte-derived proteases cathepsin G (EC 3.4.21.20) and elastase (EC 3.4.21.37) (MOOREN et al. 1982). This protease inhibitor has been called "antileucoprotease". Its importance lies in the fact that during infection the bronchial mucosa becomes purulent and leucocytes degenerate, die, and

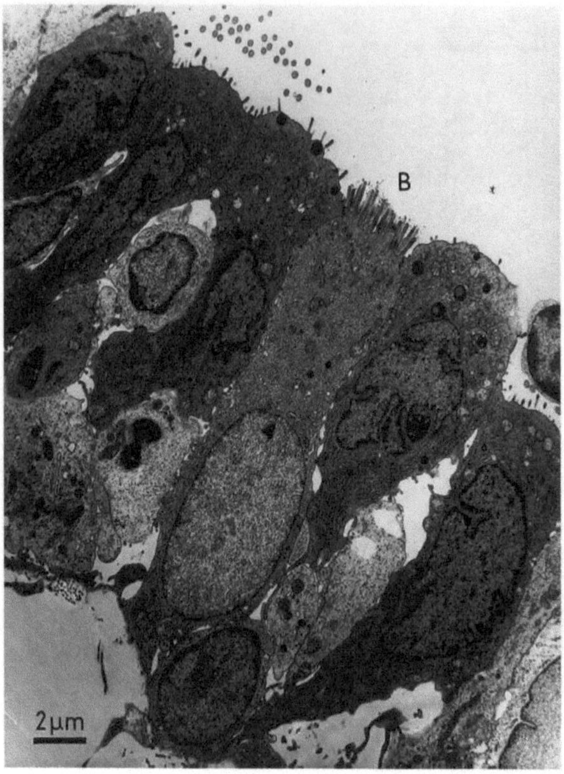

Fig. 59. Five serous cells and a brush cell (B) from the tracheal epithelium (block 675) of an unmedicated 200 g male *rat*. On June 1, 1976 under pentobarbital anaesthesia (30 mg/kg), the animal was perfused from the abdominal aorta with 2.5 % glutaraldehyde in 0.1 M sodium cacodylate buffer (pH 7.4). Postfixation with 1 % osmium tetroxide in sodium cacodylate buffer. Embedded in Epon 812 and sectioned at 50 nm. Lead citrate and uranyl acetate. Plate 3301

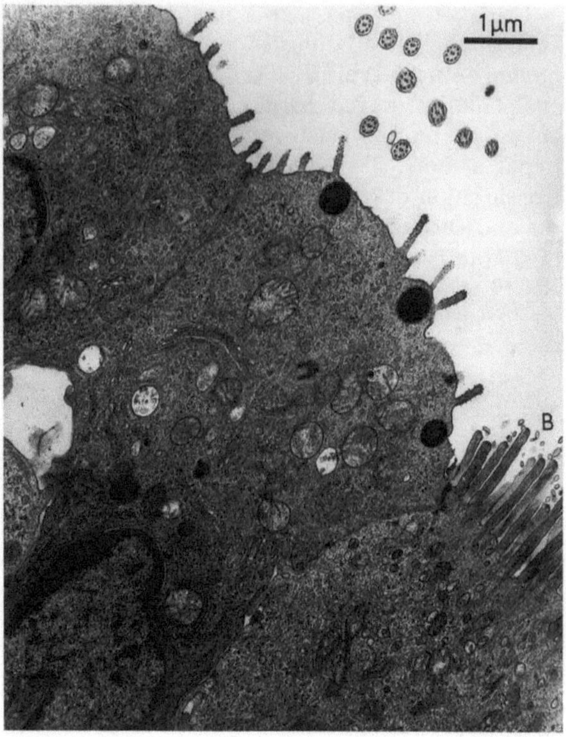

Fig. 60. Supranuclear regions of two serous cells and a brush cell (B) from the tracheal epithelium (block 675) of an unmedicated 200 g male *rat*. On June 1, 1976 under pentobarbital anaesthesia (30 mg/kg), the animal was perfused from the abdominal aorta with 2.5 % glutaraldehyde in 0.1 M sodium cacodylate buffer (pH 7.4). Postfixation with 1 % osmium tetroxide in sodium cacodylate buffer. Embedded in Epon 812 and sectioned at 50 nm. Lead citrate and uranyl acetate. Plate 3302

release potentially injurious proteases. Presumably, the antileucoprotease prevents damage to the cells lining the airways by inactivating the protease (Kuhn III 1985).

7.1.3 Clara Cells

Nonciliated bronchiolar epithelial cells were first described by Koelliker (1881), but the earliest detailed study was that of Clara (1937), after whom they were named by Policard (1955).

Table 16. Scanning electron microscopy of Clara cells

Species	Author(s)
Man	Ebert and Terracio (1975), Kuhn III (1988)
Macaca mulatta	Castleman et al. (1975), Plopper and Dungworth (1987)
Macaca arctoides	Castleman et al. (1975)
Cat	Plopper and Dungworth (1987)
Rat	Andrews (1981), Hung et al. (1982), Doster et al. (1983), Fortoul et al. (1983), Imamura et al. (1983), Dinsdale et al. (1984), Gandy et al. (1984), Konno et al. (1984), Plopper and Dungworth (1987), Pankow et al. (1989), Suzuki et al. (1992), Peão et al. 1993)
Mouse	Mahvi et al. (1977), Obara et al. (1979), Pack et al. (1980), Widdicombe and Pack (1982), ten Have-Opbroek (1986), Forkert and Troughton (1987), Moussa and Forkert (1992)
Guinea pig	Davis et al. (1984)
Rabbit	Morgenroth and Hoerstebrock (1978, Plopper et al. (1983), Scheuermann (1987)
Goat	Breeze and Wheeldon (1977)
Horse	Drommer et al. (1987)

Fig. 61. RUDOLF ALBERT VON KOELLIKER (1817–1905). Oil painting is the property of the Department of Anatomy, Würzburg University (courtesy Prof. Drenckhahn)

Fig. 62. MAX CLARA (1899–1966) in his laboratory at the Department of Anatomy, Leipzig University, mounting stained slices from celloidin-embedded materials (April 3, 1941)

In the light microscope, Clara cells appear columnar, have a deeply invaginated central nucleus, and protrude beyond the ciliated cells. The cell apex is often crowned by a round or fingerlike process, a large vacuolated droplet to which is sometimes attached a second or third droplet with clear, almost transparent contents (CLARA 1937, ETHERTON et al. 1973).

7.1.3.1 Ultrastructure

7.1.3.1.1 Granules

In *primates*, including *man*, the Clara cell has the features of a serous secretory cell that produces a protein-rich secretion.

Several granule types with morphologically distinct contents are bounded by unit membranes. In the *rat*, YOUNG et al. (1986) described dark granules without a crystalline substructure and light granules with an eccentric dense core ultrastructurally similar to the granules in tracheal mucus goblet cells. Rod-shaped granules varying up to 4 μm in length and with a diameter between 60 and 400 nm, showing 9 to 10 nm filaments in a pale matrix and always surrounded by a unit membrane were found in the apical cytoplasm or at the lateral sides of the cell nucleus (BAERT and VANDENBERGHE 1981). Three-dimensional reconstruction showed the rod granule to be a flattened oval, sometimes like a plate (YOUNG et al. 1986). Only rarely did they appear rod-shaped three-dimensionally.

Table 17. Quantification of Clara cells

Parameter	EM	Reference
Cell number per area	SEM	GANDY et al. (1984)
Mean diameter [μm]	SEM	GANDY et al. (1984)
Height [μm]	TEM	PLOPPER et al. (1983), MARIASSY and PLOPPER (1984)
Width [μm]	TEM	PLOPPER et al. (1983), MARIASSY and PLOPPER (1984)
Volume density of cell components (nucleus, smooth ER, glycogen, smooth ER/glycogen, mitochondria, rough ER, secretory granules, Golgi apparatus)	TEM	PLOPPER et al. (1983), MARIASSY and PLOPPER (1984)

7.1.3.1.2 Mitochondria

In the *rabbit*, mitochondria in the Clara cells appear of two types (Hook et al. 1990): One form is normal with prominent cristae and the other form is larger, generally rounded, and contains a few indistinct cristae. An apparent continuity between the large and small forms suggests that the different forms must be closely related.

In the *guinea pig*, Davis et al. (1984) observed two different populations of mitochondria throughout the cytoplasm of the most Clara cells of the terminal bronchioles. A few ellipsoid mitochondria with typical cristae were interspersed among large, oval mitochondria which were characterised by sparse cristae. The large mitochondria, were somewhat more electron dense than the smaller ones and the larger organelles were easily visible with the light microscope.

In the *gerbil*, Clara cells contained abundant giant mitochondria in addition to normal tubular mitochondria and the second population of enlarged spherical mitochondria that have been described in the Clara cells of several genera (Spicer et al. 1990). A survey of a number of cell types in *gerbils* failed to disclose hypertrophied mitochondria outside the tracheobronchial surface epithelium and bronchioles. The mitochondrial enlargement resulted from an increase of matrix but not cristae. The expansion of matrix displaced the relatively sparse cristae into small collections compressed against the outer membrane. The prevalence of giant mitochondria and of granular endoplasmic reticulum was similar among cells, and these two organelles were codistributed within cells. The megamitochondria and granular reticulum occupied a central stratum, whereas normal mitochondria occurred in the apical and basal regions.

7.1.3.1.3 Endoplasmic Reticulum

In the *rabbit*, the largest concentration (20 %) of **granular endoplasmic reticulum** in the Clara cells was found in 27-day old foetuses (Plopper et al. 1983). This was a statistically significant difference from all postnatal age groups 1 week and older, except at 5 weeks. The lowest concentration was found in 25-week-old postnatal animals, and was significantly lower than in all other age groups except the 12- and 17-week-old animals. Granular endoplasmic reticulum constituted about 15 % of the cytoplasm in the other foetal and the neonatal age groups and varied between 7 and 15 % in the other postnatal age groups.

In *mice* and *rats*, but not *humans*, the **smooth endoplasmic reticulum** of the Clara cell tends to present as a meshwork of branching tubules and/or vesicles. It is believed to be involved in the detoxification of drugs. Evidence demonstrating that the Clara cell is the primary site of cytochrome P-450 monooxygenase activity within the lung derives from four different approaches:

- immunocytochemical localisation with antibodies specific for cytochromes;
- autoradiographic demonstration of binding of compounds that are substrates of the system;
- identification of enzyme activity in isolated cells;
- characterisation of the pathogenic response to xenobiotic compounds which are substrates (see section 7.3.3).

Table 18. Types of Clara cells

Ultrastructure of the luminal projections
1. Agranular endoplasmic reticulum
2. Agranular endoplasmic reticulum showing vesicular dilatation
3. Agranular endoplasmic reticulum + mitochondria
4. Agranular endoplasmic reticulum + secretory granules
5. Densely packed secretory granules

The prominent organelle in the cytoplasm of Clara cells from both wild-type strain 129 and Clara cell secretory protein –/– 129 *mice* is the smooth endoplasmic reticulum, which occupies about 20 % of the cell cytoplasm (Stripp et al. 2002). Despite the similarities in abundance there were two aspects of the smooth endoplasmic reticulum that differed markedly between both genotypes of *mice*. First, electron-dense material that was abundant within the lumen of smooth endoplasmic reticulum of Clara cells from wild-type *mice* was not as evident within the lumen of smooth endoplasmic reticulum of Clara cells from Clara cell secretory protein –/– 129 *mice*. Second, Clara cells from Clara cell secretory protein –/– 129 *mice* were unique in that they possessed large concentric whorls of endoplasmic reticulum located within the apical portion of the cell. Membrane-bound inclusions, which appeared to be surrounded by at least one layer of endoplasmic reticulum, were found at the centre of these whorls and contained material of varying electron density and compaction. Some of these inclusions had the characteristics of mitochondria based on the presence of crista-like membrane inclusion.

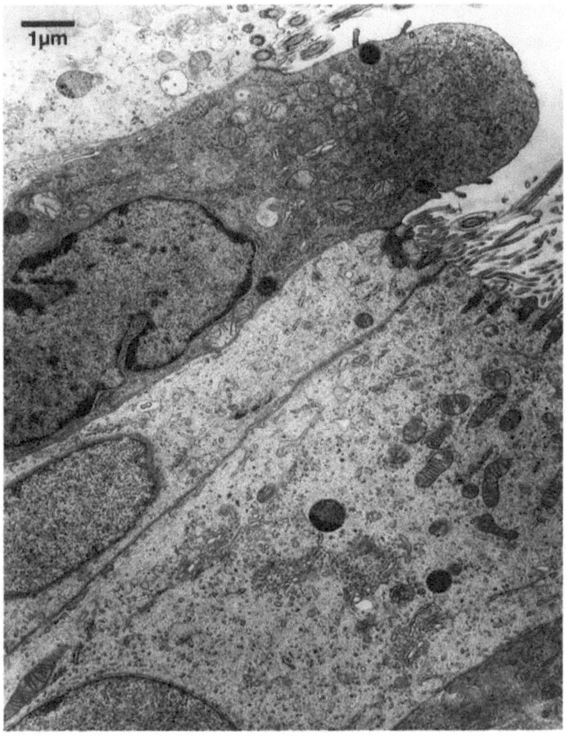

Fig. 63. A Clara cell between ciliated cells from the tracheal epithelium (block 675) of an unmedicated 200 g male *rat*. On June 1, 1976 under pentobarbital anaesthesia (30 mg/kg), the animal was perfused from the abdominal aorta with 2.5 % glutaraldehyde in 0.1 M sodium cacodylate buffer (pH 7.4). Postfixation with 1 % osmium tetroxide in sodium cacodylate buffer. Embedded in Epon 812 and sectioned at 50 nm. Lead citrate and uranyl acetate. Plate 3297

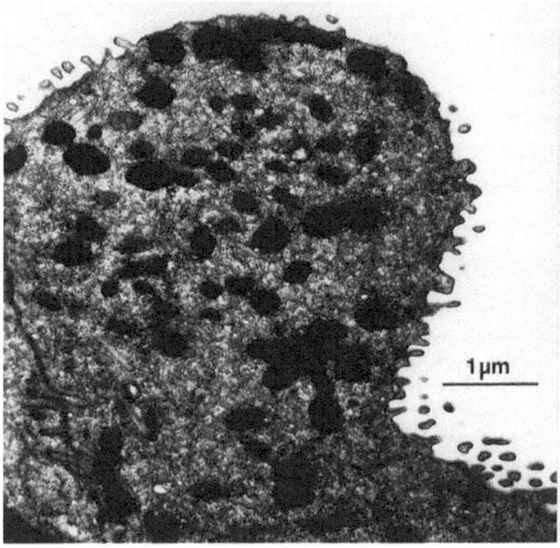

Fig. 64. Apical part of a type 4 Clara cell from a bronchiole of a female white *rat* (No. 5; breeder: Winkelmann, Borchen-Kirchborchen) which inhaled 10 mg powdered Grängesberg magnetite/m^3 4 h per day, 5 days per week from August 24 to October 19, 1967 for a total of 40 days. Fixed on January 15, 1968 under methitural anaesthesia by intratracheal instillation of 2.5 % glutaraldehyde in phosphate buffer (pH 7.4) before opening the thorax. Postfixation with 1 % osmium tetroxide in phosphate buffer (pH 7.4). Contrasted en bloc for 12 h with 0.5 % uranyl acetate in 70 % ethanol. Embedded in a 2:8 mixture of methyl and butyl methacrylate. Sectioned at 50 nm. Lead citrate after Reynolds (1963). Film 239/84

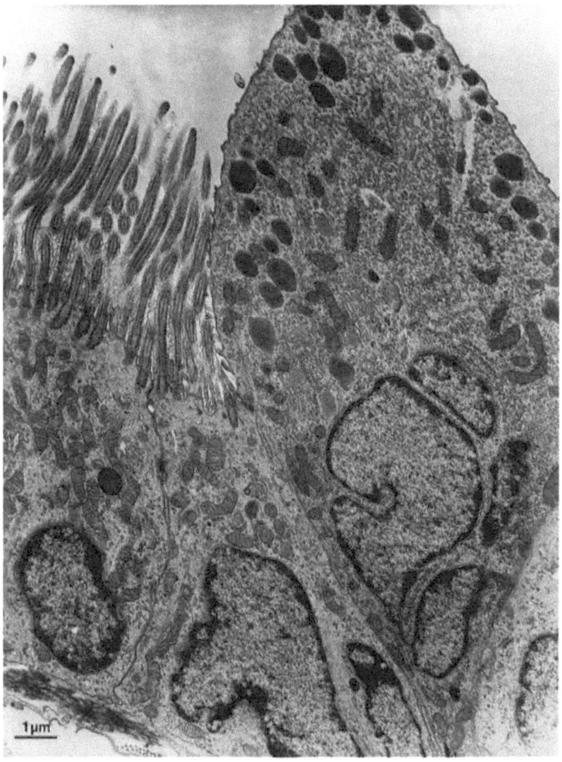

Fig. 65. Prominent endoplasmic reticulum and ovoid electron-dense secretory granules in the apical cap of a type 4 Clara cell from a bronchiole (block 465) of a male *rat* (No. 5/104 E) which inhaled 71.5 mg trimethylol melamine (Madurit) dust/l for a total of 32 h 30 min on 4 consecutive days. Fixed under pentobarbital anaesthesia (30 mg/kg) by intratracheal instillation of 2.5 % glutaraldehyde in 0.1 M sodium cacodylate buffer (pH 7.4) before opening the thorax. Postfixation with 1 % osmium tetroxide in sodium cacodylate buffer. Embedded in Epon 812 and sectioned at 50 nm. Lead citrate and uranyl acetate. Plate 4050

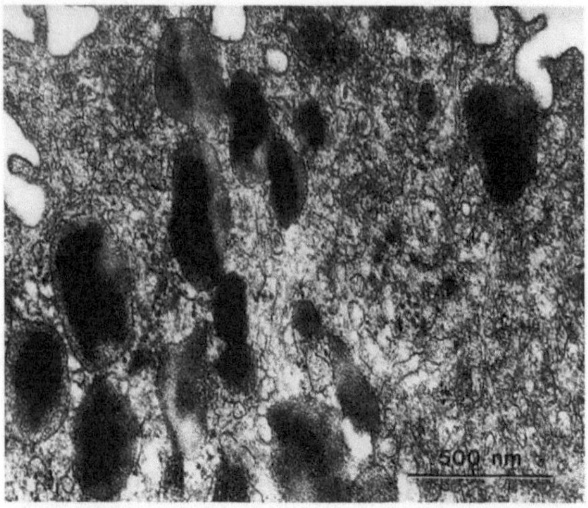

Fig. 66. Electron dense large inclusions in the cytoplasm presenting a stippled appearance, which at high magnification was shown to be due to a regular arrangement of minute hexagonal globules or tubes. Bronchiolar Clara cell (type 4) from a female *rat* (No. 4; breeder: Winkelmann, Borchen-Kirchborchen) which inhaled 10 mg powdered aluminium/m³ 4 h per day, 5 days per week from August 16 to October 27, 1967 for a total of 51 days. Fixed on January 15, 1968 under methitural anaesthesia by intratracheal instillation of 2.5 % glutaraldehyde in phosphate buffer (pH 7.4) before opening the thorax. Postfixation with 1 % osmium tetroxide in phosphate buffer (pH 7.4). Contrasted en bloc for 12 h with 0.5 % uranyl acetate in 70 % ethanol. Embedded in a 2 : 8 mixture of methyl and butyl methacrylate. Sectioned at 50 nm. Lead citrate after Reynolds (1963). Film 601/86

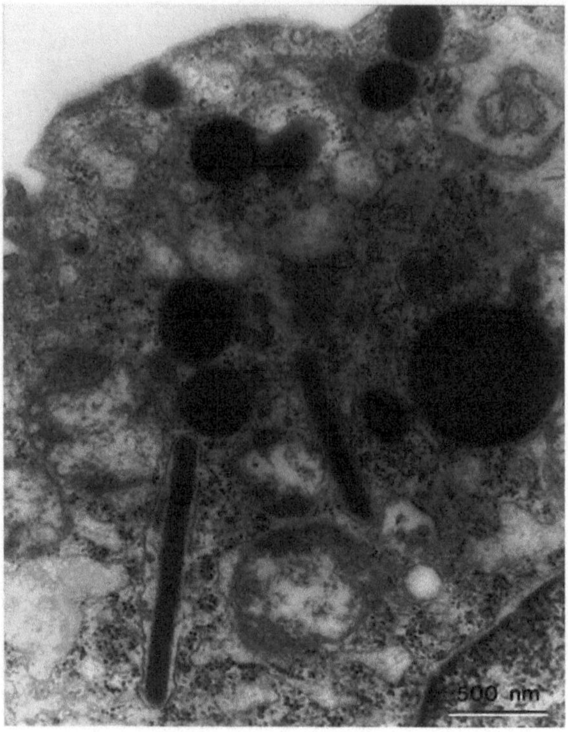

Fig. 67. Rod-shaped granules and a multivesicular body in a Clara cell (block 4797) from an unmedicated male Sprague-Dawley *rat* (Charles River, France). On July 10, 1979 under pentobarbital anaesthesia (30 mg/kg), the animal was perfused from the abdominal aorta with 2.5 % glutaraldehyde in 0.1 M sodium cacodylate buffer (pH 7.4). Postfixation with 1 % osmium tetroxide in sodium cacodylate buffer. Embedded in Epon 812 and sectioned at 50 nm. Stained with lead citrate and uranyl acetate. Film 893/85

Fig. 68. Type 5 Clara cell from a bronchiolus of a female white *rat* (breeder: Winkelmann, Borchen-Kirchborchen), which received intramuscular injection of 150 mg DL-α-tocopherol acetate in a colloidal solution (Ephynal) per kg body weight × day, 5 days per week from April 12 to June 12, 1967 for a total of 40 days. Fixed on June 12, 1967 under methitural anaesthesia by intratracheal instillation of 2.5 % glutaraldehyde in phosphate buffer (pH 7.4) before opening the thorax. Postfixation with 1 % osmium tetroxide in phosphate buffer (pH 7.4). Contrasted en bloc for 12 h with 0.5 % uranyl acetate in 70 % ethanol. Embedded in a 2 : 8 mixture of methyl and butyl methacrylate. Sectioned at 50 nm. Lead citrate after Reynolds (1963). Plate 278/12

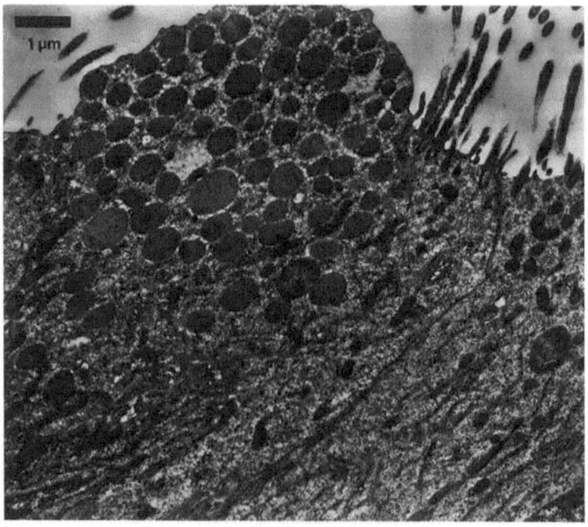

7.1.3.1.4
Lysosomes

In most Clara cells of the *rat* SMITH et al. (1974) found three or four large lysosomes (average diameter 550 nm). These were usually circular or occasionally oval in section and were bounded by a membrane.

7.1.3.1.5
Peroxisomes

Peroxisomes are ubiquitous, spherical, organelles bounded by a single membrane and having a diameter of 100 nm to 1 μm in eukaryotes. They contain at least one H_2O_2-producing oxidase and H_2O_2-decomposing catalase (EC 1.11.1.6).

The theory of the endosymbiotic origin of the peroxisome, while more problematic than that of the mitochondrium, has some support (e.g. de DUVE 1969, CAVALIER-SMITH 1987, BORST 1989, IGUAL et al. 1992). Catalase is the prototypical peroxisomal enzyme, and even cytoplasmic catalase is apparently derived from peroxisomes (e.g., MASTERS and CRANE 1995). This arrangement is advantageous to the host cell because the properties of catalatic and peroxidatic reactions suggest that catalase is a more effective regulator of the concentration of intracellular H_2O_2 than the mitochondrial glutathione (CHANCE 1969, CHANCE et al. 1979). While mitochondria possess a small genome and are thus capable of heritable variation and evolution, peroxisomes lack a genome and therefore cannot be a unit of evolution. Locating catalase, the enzyme best able to regulate the concentration of intracellular H_2O_2, in the peroxisome effectively gives the nucleus greater control over H_2O_2 and thus control over signalling with this molecule as well (BLACKSTONE 1998).

The peroxisomal membrane forms a permeability barrier for a wide variety of metabolites required for and formed during fatty acid β-oxidation (succinate, malate, acetylcarnitine, carnitine, 2-oxoglutarate, isocitrate and oxaloacetate). Continuous β-oxidation depends on the availability of acyl-CoA, NAD^+, NADPH, free coenzyme A and on export of acetyl groups. During β-oxidation of fatty acids, NAD^+ is reduced to NADH. For continuation of β-oxidation, NADH must be re-oxidised to NAD^+. Evidence for a crucial role of intraperoxisomal NAD^+ came from VAN ROERMUND et al. (1995).

7.1.3.1.6
Multivesicular Bodies

The term multivesicular body is employed to describe a vacuole containing vesicles set in a lucent or dense matrix. The data of ROBINEAUX et al. (1980) on phytohaemagglutinin-transformed lymphocytes clearly demonstrated that the pale multivesicular bodies have an endocytic origin. Maturation of 'early' or 'sorting' endosomes involves the removal of residual, recycling surface receptors, delivering of lysosomal hydrolases, and involution of the surrounding membrane to form multivesicular bodies (FUTTER et al. 1996). Multivesicular body fractions prepared from the livers of oestradiol-treated *rats* were positive for acid phosphatase (EC 3.1.3.2) and arylsulfatase (EC 3.1.6.1), consistent with the prelysosomal nature of this department (JOST-VU et al. 1986). Having delivered their cargo to lysosomal hydrolases, mannose-6-phosphate receptors are removed and recycled to the trans-Golgi network before the late endosomes fuse with the lysosomes. Cholesterol is required for the cation-independent mannose 6-phosphate receptor exit from multivesicular late endosomes to the Golgi (MIWAKO et al. 2001).

From three-dimensional reconstruction of *rat* nonciliated bronchiolar (Clara) cells, YOUNG et al. (1986) found a significant polarisation of the multivesicular bodies toward the Golgi. This is consistent with a metabolic or a precursor-product relationship between these structures and secretory granules as has been proposed for the type II pneumocytes (CHEVALIER and COLLET 1972).

7.1.3.1.7
Nucleus

The nucleus has a basal location, and the Golgi apparatus is perinuclear in location. In the *rat*, SMITH et al. (1974) found the nucleus situated in the middle zone. It is elongated and bilobed, the arms of the lobes pointing towards the base of the cell. This is in contrast to the ciliated cells in which the nucleus is oval and not divided into lobes.

7.1.3.2
Histochemistry

Table 19. Secretory Products of Clara Cells

1. Clara cell 10-kDa protein/uteroglobulin
2. Surfactant proteins A, B, and D
3. Leucocyte protease inhibitor
4. 29 kDa β-galactoside binding lectin
5. Trypsinlike protease

7.1.3.2.1 Periodic Acid-Schiff Reaction

Clara cells are negative for saccharides detectable by the periodic acid-Schiff reaction or with methenamine-silver (Schiller 1962).

7.1.3.2.2 Glutathione Biosynthesis

The GSH content of freshly isolated *rabbit* Clara cells was 2.03 ± 0.59 nmol/10^6 cells; it was depleted to 0.75 ± 0.06 nmol/10^6 cells by incubating with 200 μM diethyl maleate for 20 min (Horton et al. 1987). Cysteine was best able to support resynthesis of GSH. There was no evidence for participation of a cystathionine pathway for glutathione synthesis. In steady-state conditions, the GSH measured in isolated *mouse* Clara cells was in the femtomole range, but varied 4-fold between individual cells (West et al. 2000). Clara cells analysed *in situ* and *in vitro* confirmed this heterogeneity. The response of these cells to compounds that modulate GSH was also variable. Dimethylmaleate depleted GSH, whereas GSH monoethylester augmented it. However, both acted nonuniformly in isolated Clara cells. The depletion of intracellular GSH caused a striking decrease in cell viability upon incubation with naphthalene.

7.1.3.2.3 Clara Cell 10-kDa Protein/Uteroglobulin

Clara cell secretory protein, known by a number of other names including urinary protein 1, uteroglobulin (Peri et al. 1995, Mukherjee et al. 1999), PCB-binding protein, and Clara cell 10 kDa protein (Mantile et al. 1993, Dodge et al. 1994, Singh and Katyal 1997) has been described as a "multifunctional protein".

Clara cell 10-kDa protein (CC10), composing 7% of the total protein in normal bronchioloalveolar fluid, is the major protein product of the Clara cells that are precursors for neoplastic and non-neoplastic peripheral airway epithelia (Van Winkle et al. 1996). CC10 protein and mRNA-containing cells have been detected throughout the *human* tracheobronchial tree with the greatest abundance in the distal airways (Broers et al. 1992, Linnoila et al. 1992, Jensen et al. 1994). During epithelial injury, CC10 is coexpressed with an increasing number of pulmonary neuroendocrine cells (Stripp et al. 1995, Reynolds et al. 2000). Khoor et al. (1996) have detected a close spatial relationship between CC10 and neuroendocrine foci insofar as to conclude an interaction between them during lung development. To further assess the *in vivo* role of CC10 during carcinogenesis and neuroendocrine differentiation, Castro et al. (2000) studying the lungs of CC10-deficient *mice* suggested that this protein may not be critical for normal neuroendocrine cell ontogenesis.

7.1.3.2.4 29 kDa β-Galactoside-Binding Lectin

A 29 kDa β-galactoside-binding lectin is secreted by brochiolar Clara cells of the *rat* (Wasano and Yamamoto 1989). Pneumocin (M_r 165 kDa) is a sialoglycoprotein decorating the surface membrane of the Clara cells, but not that of the neighbouring ciliated calls (Lwebuga-Mukasa 1991).

7.1.3.2.5 Lipoproteins

The lipoprotein dye, luxol fast blue, gave a strong reaction in the Clara cells but a very weak reaction in the ciliated bronchiolar epithlial cells and the alveolar cells (Azzopardi and Thurlbeck 1969). Phosphomolybdic acid, that preferentially reacts with lecithin and choline but not with phosphatidyl serine, stained Clara cells intensely only after pretreatment for 5 min with 25% acetic acid.

7.1.3.2.6 Polypeptide Hormones

Both transforming growth factors (TGF)-β_1 and TGF-β_3 mRNA transcripts were present in normal *mouse* lung, and gene expression for both isoforms was predominant in bronchiolar epithelial cells (Coker et al. 1997).

7.1.3.2.7 Enzymes (s. Table 20)

Oxidoreductases

The Embden-Meyerhof and Krebs cycles and the hexose monophosphate shunt are functional in the Clara cells (Azzopardi and Thurlbeck 1969). With strongly reacting enzymes, namely L-lactate dehydrogenase (EC 1.1.1.27), glyceraldehyde-3-phosphate dehydrogenase (EC 1.2.1.12), nicotinamide-adenine dinucleotide diaphorase (EC 1.6.99.3), nicotinamide-adenine dinucleotide phosphate diaphorase (EC 1.6.99.1), and malate dehydrogenase (EC 1.1.1.37), the nonciliated cells gave a stronger reaction than the ciliated ones after 5 min incubation.

Table 20. Clara cells: enzymes

Enzyme	EC	Author(s)
Lactate dehydrogenase	1.1.1.27	AZZOPARDI and THURLBECK (1969)
Malate dehydrogenase	1.1.1.37	AZZOPARDI and THURLBECK (1969)
Glucose-6-phosphate dehydrogenase	1.1.1.49	ROTH (1973)
Carbonyl reductase	1.1.1.184	MATSUURA et al. (1990, 1994)
Formaldehyde dehydrogenase	1.2.1.1	KELLER et al. (1990)
Glyceraldehyde-3-phosphate dehydrogenase	1.2.1.12	AZZOPARDI and THURLBECK (1969)
NADPH cytochrome *c* (P-450)-reductase	1.6.2.4	DEES et al. (1980)
Nicotinamide-adenine dinucleotide phosphate dehydrogenase	1.6.99.1	AZZOPARDI and THURLBECK (1969)
Catalase	1.11.1.6	GOLDENBERG et al. (1978)
Nitric oxide synthase III	1.14.13.39	NIKOLOV et al. (1998, 1999)
Prostaglandin H synthase (cyclooxygenase)	1.14.99.1	SIVARAJAH et al. (1983), YU et al. (1988), LAPORTE et al. (1991)
O^6-Alkylguanine-DNA-alkyltransferase	2.1.1.63	DEILHAUG et al. (1985)
Lipase	3.1.1.3	YONEDA (1978)
Lysophospholipase	3.1.1.5	YONEDA (1978)
Alkaline phosphatase	3.1.3.1	INAYAMA et al. (1995)
Acid phosphatase	3.1.3.2	ACHTERATH and BLÜMCKE (1975)
Guanylate cyclase	4.6.1.2	CETIN et al. (1995)

Carbonyl Reductase (EC 1.1.1.184)

Carbonyl reductase (secondary alcohol: $NADP^+$ oxidoreductase) catalyses the NADPH-linked reduction of various carbonyl compounds to the corresponding alcohols. The lungs of *guinea pigs* and *mice* have been shown to contain large amounts of tetrameric carbonyl reductase that is distinct from the monomeric and dimeric forms of the enzyme in extrapulmonary tissues (NAKAYAMA et al. 1982, 1986), especially in its low substrate specificity for aliphatic and aromatic carbonyl compounds and dual cofactor specificity for NADPH and NADH. Antibodies against *guinea-pig* pulmonary carbonyl reductase react with the *mouse* lung enzyme (NAKAYAMA et al. 1986), and specifically mark the enzyme in the non-ciliated bronchiolar cells (Clara cells), the ciliated calls and the type II pneumocytes of *guinea pig* and *mouse* lungs. as demonstrated at the light microscope level (MATSUURA et al. 1990). By an electronmicroscopical immunogold procedure using monospecific antibodies against the enzyme, MATSUURA et al. (1994) localised tetrameric carbonyl reductase within the mitochondrial matrix of the Clara cells of the *mouse*. No significant labelling was detected in the other cellular components such as nuclei, secretory granules and endoplasmic reticulum.

The **cytochrome P_{450}-type system** not only "inactivates" toxic chemicals in the *mammalian* organism, but is likewise able to "activate" or "toxify" a variety of chemicals which otherwise would be harmless for *mammals*. Cytochrome P_{450} is a major integral membrane protein of both smooth and rough endoplasmic reticulum and is exclusively (>95%) synthesised by polysomes associated with ER membranes (BAR-NUN et al. 1980). The large variety of foreign compounds entering the body by inhalation of polluted air and being deposited on the epithelia of the nasal cavity, the bronchial tree and the lung alveoli call for a local detoxification. In the nasal cavity of the *rat*, nonciliated columnar cells have an extensive accumulation of smooth endoplasmic reticulum (POPP and MONTEIRO-RIVIERE 1985) which may be the source of cytochrome P_{450} and P_{450}-associated enzymes described by HADLEY and DAHL (1982). Bronchiolar Clara cells may be sites of a cytochrome P_{450}-dependent mixed-function oxidase activity (Boyd 1977).

The enzymes controlling the generation or inactivation of carcinogenic and mutagenic chemicals often exhibit markedly differing properties in *man* and the available test systems. While the pharmacologically and toxicologically important cytochrome P_{450}-type drug-metabolising enzyme system develops its activity in *rodents* perinatally or even clearly postnatally, this system is already operative in *primates* during the foetal and even the embryonic periods–in *humans* as early as in the 6th to 7th week of gestation (cf. NAU and NEUBERT 1978).

Cytochrome P_{450} isozyme 2B was detected in dome-shaped, non-ciliated cells (JI et al. 1995). The reaction product was present in the apical portion of some epithelial cells, especially at the youngest stage (1 postnatal day). In 21 and 100 postnatal day *rats*, the protein was distributed throughout the cytoplasm surrounding the nucleus in most cells. The portion of stained cells in the bronchiolar epithelium increased with age.

The CYP2E1 protein comprises a minor component of the total pulmonary cytochrome P_{450} content in *rabbit* (PORTER et al. 1989), *rat* (TINDBERG and INGELMAN-SUNDBERG 1989, SHIMIZU et al. 1990, CARLSON and DAY 1992), *mouse* (FORKERT 1995) and *hamster* (UENG et al. 1991, 1993).

The concentration of P_{450} isozyme 6 was about 20-fold increased in preparations of the Clara and type II cell fractions of *rabbits* treated intraperitoneally with 10 μg 2,3,7,8-tetrachlorodibenzo-*p*-dioxin per kg body weight (DOMIN et al. 1986).

Most experiments with cytochrome P_{450} and other microsomal enzymes were performed on the liver. But Clara cells have even higher levels of cytochrome P_{450} than do the hepatocytes (SERABJIT-SINGH et al. 1980), and are rather selectively injured by number of simple compounds as listed in Table 21.

However, KUHN III (1985, p. 94) questioned whether the results on *rodents* can be applied to other species, since in many species, including *primates*, the smooth endoplasmic reticulum is not well developed.

Transferases

*O*⁶-Alkylguanine-DNA alkyltransferase (EC 2.1.1.63) activity of Clara cells isolated from lungs of male New Zealand white *rabbits* was >0.75 fM/10^6 cells and thus 4 to 20-fold lower than in type II pneumocytes (5.82 fM/10^6 cells) and alveolar macrophages (15.81 fM/10^6 cells) (DEILHAUG et al. 1985). Formation of the promutagenic adduct O^6-methylguanine in the Clara cells of male Fischer 344 *rats* treated with the tobacco-specific nitrosamine, 4-(methylnitrosamino)-1-(3-pyridyl)-1-butanone in doses ranging from 0.03–50 mg/kg correlated with the pulmonary tumour incidence (BELINSKY et al. 1991). However, the hyperplasia detected in this study all arose within the alveolar areas, and ultrastructural examination of these lesions in addition to adenomas and carcinomas revealed morphologic features characteristic of the type II pneumocyte (BELINSKY et al. 1990).

γ-Glutamyl transpeptidase (EC 2.3.2.2) is lacking in the bronchiolar epithelium of the Syrian golden *hamster* (MOORE et al. 1987).

Hydrolases

The secretory granules of the Clara cells of the *rat* showed a strong reaction of lysophospholipase (EC 3.1.1.5) and a weak reaction of lipase (EC 3.1.1.3) (YONEDA 1978).

Clara cells in the bronchioles of the *rabbit* not infrequently showed alkaline phosphatase (EC 3.1.3.1) reactivity by electron microscopy (INAYAMA et al. 1995).

Acid phosphatase (EC 3.1.3.2) was found in the apical secretion vacuoles of Clara cells of the *rat* (ACHTERRATH and BLÜMCKE 1975).

Lyases

Guanylin, a peptide involved in the activation of guanylate cyclase C is exclusively confined to the Clara cells. In *cattle*, *guinea pig* and *rat*, CETIN et al. (1995) found guanylin immunoreactivity in a distinct population of secretory granules mostly localised underneath the apical cell membrane.

7.1.3.3 Comparative Anatomy and Biochemistry

7.1.3.3.1 Non-human Primates

In the stumptail (*Macaca arctoides*) and bonnet (*Macaca radiata*) monkeys, the cytoplasm of the Clara cells contained abundant mitochondria, a medium-sized Golgi complex, and a moderate amount of granular endoplasmic reticulum (CASTLEMAN et al. 1975). In bonnet monkeys, the cells frequently contained round secretory droplets, which were homogeneously electron dense. Secretory droplets were less frequently observed in the non-ciliated bronchiolar cells of the stumptail monkey. The non-ciliated bronchiolar cells of the rhesus monkey (*Macaca mulatta*) were not observed to contain any secretory droplets.

7.1.3.3.2 Rodents

In the *rabbit*, the Clara cell is immature at birth, and differentiation occurs primarily during week 3 and 4 of postnatal life (PLOPPER et al. 1983).

In the neonate *rat*, the apical region of the Clara cell lacks the profusion of smooth endoplasmic reticulum typical of the adult (SMITH et al. 1974); also, in all but a few neonatal Clara cells there is less smooth and rough endoplasmic reticulum in the middle zone than occurs in the adult.

7.1.3.3.3 Sauropsida

The airways of *avian* species do not contain any Clara cells; accordingly, lung microsomes prepared from *chickens* or Japanese *quails* were completely unable to catalyse the NADPH-dependent covalent binding of either 4-ipomeanol or benzo[*a*]pyrene although these reactions occur in liver or kidney microsomes from these species (BUCKPITT and BOYD 1978, 1982). Correspondingly, the administration of 4-ipomeanol to *quail* resulted in hepatic necrosis (and covalent bonding), but not pulmonary lesions.

7.1.4 Goblet Cells

The distribution of goblet cells in *human* tracheal epithelium has been investigated by a whole-mount alcian blue, periodic acid Schiff stain technique (TOS 1970, MOGENSEN and TOS 1976). There is an over-all increase in the number of goblet cells per mm^2 toward the caudal part of the trachea, and a greater density of goblet cells in the cartilaginous than in the membranous wall. Very few goblet cells are found in the bronchioles of normal persons (BUCHER and REID 1961), and this is probably the case in most species.

MUC2, MUC5AC and MUC5B of nine mucin genes have been shown to be subject to regulation by a variety of inflammatory mediator, among them bacterial lipopolysaccharide (LPS) in airway epithelial cells (ROSE and GENDLER 1997). In NCI-H292 *human* airway epithelial cells, MUC2 transcripts were detected after 2 h of exposure to IL-1β and reached maximal level after 8 h (KIM et al. 2000). Actinomycin D experiments indicated that the IL-1β-mediated MUC2 expression was controlled by transcriptional regulation.

In vivo, the constitutive expression of IL-9 in transgenic *mice* resulted in elevated MUC2 and MUC5AC gene expression in airway epithelial cells and periodic acid-Schiif-positive staining (LOUAHED et al. 2000). Similar results were observed in C57BL/6J *mice* after IL-9 intratracheal instillation. In contrast, instillation of the T-helper 1-associated cytokine interferon γ failed to induce mucin production. *In vitro*, IL-9 also induced expression of MUC2 and MUC5AC in *human* primary lung cultures and in the *human* mucoepidermoid NCI-H292 cell line, indicating a direct effect of IL-9 on inducing mucin expression in these cells.

7.1.4.1 Ultrastructure

By electron microscopy the appearance of goblet cells in all species is similar. The cell is distended by electron-lucent secretory granules. The granules are larger than those of the serous cell and sometimes have an electron-dense core. Each granule usually has an incomplete membrane and there is fusion between them. The cytoplasm is electron dense, and like the serous cell, the nucleus irregular in outline.

Secretory granule release was studied in the *guinea pig* tracheal goblet cells (NEWMAN et al. 1996). Arrest of secretion using 1 % tannic acid led primarily to release of the electron lucent material with apparent retention, or retardation of release, of the electron dense core. Secretion continued, however, in the presence of tannic acid, even after secretagogue administration, making quantification of the secretory process possible. In unstimulated cells, after 15 min tannic acid incubation, 13 % of cells (n = 100) contained simple exocytoses involving fusion of single granules with the plasma membrane; 2 % of cells contained more complex compound exocytoses, with granule/granule fusion forming a central cavity to the cell; 1 % of cells had apokrine-like release, involving loss of the apical portion of the cell. After 5 min preincubation in 80 mM KCl, 27 % of cells (n = 100) contained single exocytoses, 13 % had compound exocytoses, and 3 % apocrine-like secretion.

In the *rat* nasal mucosa, goblet cells after conventional fixation with 2.5 % glutaraldehyde in 0.1 M phosphate buffer (pH 7.4) for 1 h and postfixation

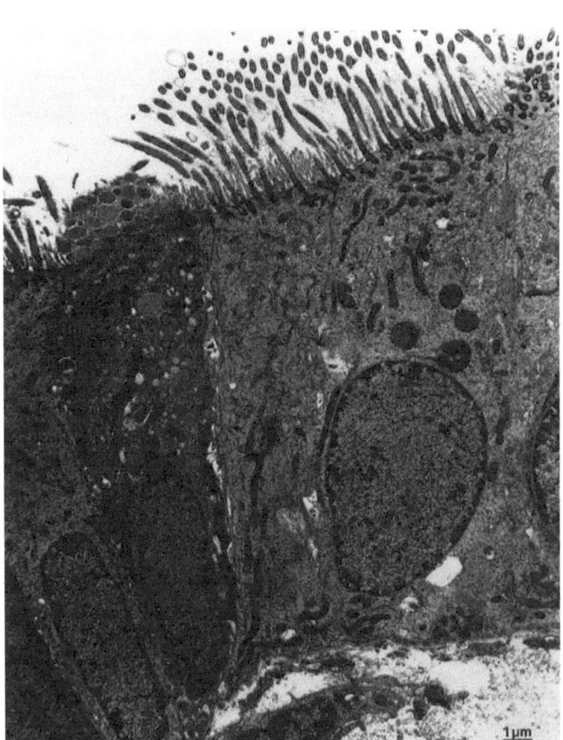

Fig. 69. Discharge of mucus granules from a goblet cell situated between two ciliated cells within the tracheal epithelium (block 675) of an unmedicated 200 g male Wistar *rat*. Under pentobarbital anaesthesia (30 mg/kg), the animal was perfused from the abdominal aorta with 2.5 % glutaraldehyde in 0.1 M sodium cacodylate buffer (pH 7.4). Postfixation with 1 % osmium tetroxide in sodium cacodylate buffer. Embedded in Epon 812 and sectioned at 50 nm. Stained with lead citrate and uranyl acetate. Plate 3295

with 2% OsO_4 in phosphate buffer for 1 h showed that the membrane structure of mucus granules was not clearly evident (SHIMOMURA et al. 1996). Most of the secretory granules had a homogenous matrix structure and some had a double-layered matrix, which occasionally contained a moderately dense core and an electron-lucent halo. In contrast, after quick-freezing in liquid nitrogen and freeze-substitution according to YOSHIHARA et al. (1986) the cells did not have a goblet shape but a columnar shape, and their secretory granules were small in size, mostly round in shape, and had a three-layered unit membrane. Their heterogeneous matrix showed either a moderate dense or a loosely filamentous appearance. Replica electron micrographs of cross-fractured apical cytoplasm displayed the fusion site between the granule membrane and the plasmalemma.

7.1.4.2 Histochemistry

Cat tracheal epithelial goblet cells did not show any immunoreactivity for cyclic AMP, while both serous and mucous cells in submucosal glands immunoreactive cyclic AMP (LAZARUS et al. 1984).

7.1.5 Intermediate Cells

"Intermediate" cells have been described in the respiratory epithelium of a wide variety of species. The cells are called intermediate because they lack either secretory granules or cilia and their apical processes are few and short (Fig. 70). In this respect they are "undifferentiated" and capable of transformation into any of the epithelial cell types. The electron density of the cytoplasm is variable, presumably electron dense if it is to form a secretory or electron-lucent it is to form a ciliated cell. The cell is found both in proximal and distal airways.

7.1.6 Brush Cells

DIMAIO et al. (1988) suggested that brush cells were a response to lung injury.

7.1.6.1 Ultrastructure

While lacking secretory granules the brush cell (Fig. 71) is distinct from the intermediate cell in having a pronounced brush border of fingerlike projections, 1 μm–1.5 μm in length and 50–100 nm in diameter (RHODIN and DALHAMN 1956).

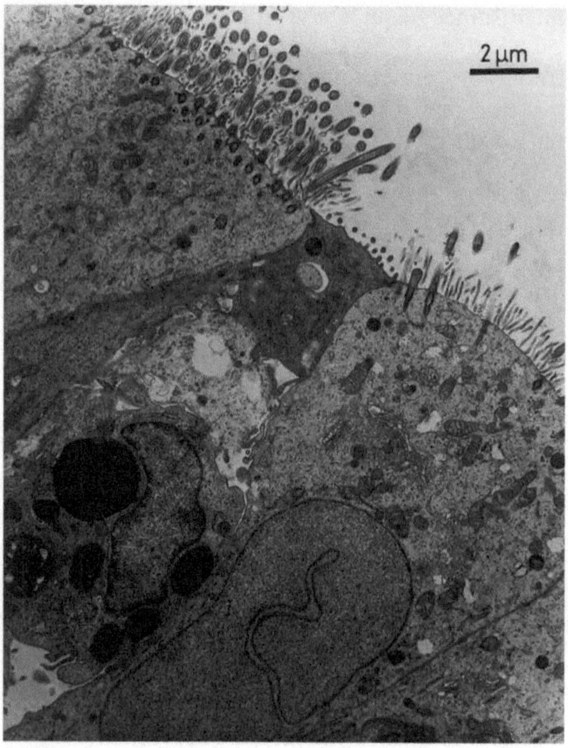

Fig. 70. An "intermediate" cell as an intergrade between ciliated and non-ciliated cells from the trachea (block 675) of an unmedicated 200 g male Wistar *rat*. On June 1, 1976 under pentobarbital anaesthesia (30 mg/kg), the animal was perfused from the abdominal aorta with 2.5% glutaraldehyde in 0.1 M sodium cacodylate buffer (pH 7.4). Postfixation with 1% osmium tetroxide in sodium cacodylate buffer. Embedded in Epon 812 and sectioned at 50 nm. Lead citrate and uranyl acetate. Plate 3299

JEFFERY and REID (1975), however, quoted about 2 μm. Bundles of 10 to 15 fibrils, 3 to 4 nm thick have a very irregular and wavy course. The brush cells of the *rat* trachea are 30–35% larger than that of the terminal bronchioles and alveoli (CHANG et al. 1986).

7.1.6.2 Histochemistry

Villin and fimbrin, two actin filaments-bundling proteins, which were demonstrated in the *rat's* brush cells by HÖFER and DRENCKHAHN (1992), are a hallmark to differentiate this cell type from neighbouring epithelial cells both in the respiratory and the intestinal tracts at the light microscope level.

7.1.7 Basal Cells

The role of the basal cells as stem cells for ciliated, goblet, and brush cells is expressed in their high mitotic capacity. Previously DRASCH (1879) by

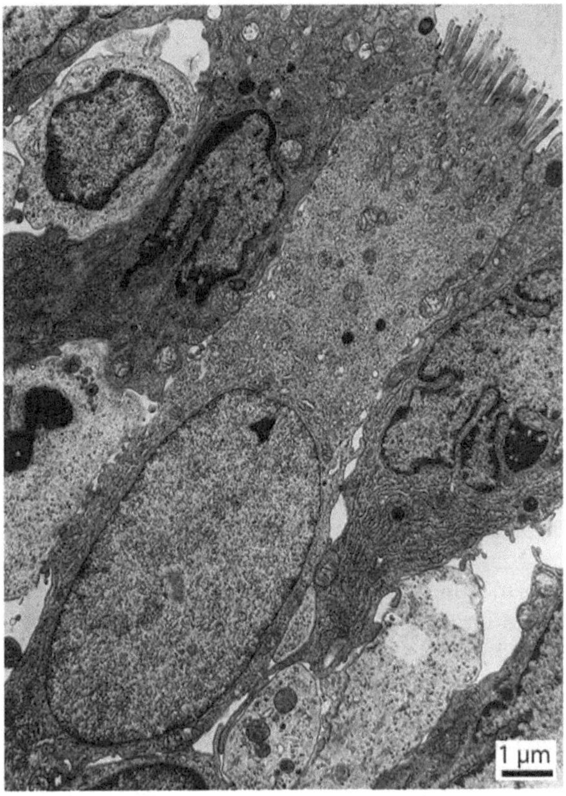

Fig. 71. Brush cell in the tracheal epithelium (block 675) from a 200 g unmedicated male Wistar *rat*. On June 1, 1976 under pentobarbital anaesthesia (30 mg/kg), the animal was perfused from the abdominal aorta with 2.5 % glutaraldehyde in 0.1 M sodium cacodylate buffer (pH 7.4). Postfixation with 1 % osmium tetroxide in sodium cacodylate buffer. Embedded in Epon 812 and sectioned at 50 nm. Lead citrate and uranyl acetate. Plate 3303

maceration in MÜLLER's (1857) fluid and regeneration experiments after burning of the tracheal epithelium of the *rabbit* suggested that the basal cells via an intermediate cell type ("Keilzellen") were the precursors of ciliated and mucous cells. However, he was at a loss for the demonstration of mitoses (l.c. p. 214). BREUER et al. (1990) recorded normal adult *hamster* ^{3}H-thymidine deoxyribose flash labelling indices. In two types of basal cells, B1 and B2, distinguished by height of the nuclei above the basal line, the labelling index was 28 % or 33 %, respectively. This constituted a high percentage of labelled cells (50.6 %) since basal cells only contribute 6.8 % of all the epithelial cells present. Grain count over nuclei only fell in B1 cells, and a cell cycle time of 20.6 days was estimated, with a DNA synthesis time $(T)_s$ of 7.5 h.

BOERS et al. (1998) studied normal *human* lungs (24–84 years) using the proliferation marker MIB-1 (Ki-67) on sections to stain for a labelling index equivalent. Over 101 000 were cells scored between seven patients studies, No age-related differences were found. Basal and parabasal cells were defined by their cytokeratin 5 and 14 positivity. In the upper airways, these contributed 31 % and 7 % of cells, respectively, but also 51 % and 33 % of cells in the proliferative compartment. In the lower airways, only 6 % of the cells were basal, but they contributed 30 % of proliferating cells (there were no parabasal cells in airways below 0.5 mm in diameter). Overall, MIB-1-positive cells were 0.87 % of all cells present, yet 48 % of proliferating cells were in the basal position and 15 % in the parabasal.

7.1.7.1 Ultrastructure

Basal cells are polygonal or triangular cells localised immediately above the basement membrane. Usually, they do not reach the free surface of the epithelium. Their cytoplasm is electron-dense. Tonofilaments are arranged in perinuclear array. Golgi complex and endoplasmic reticulum are poorly developed.

In both smokers and non-smokers typical basal cells were more numerous than atypical basal cells, which were distinguished by their spindle-shaped nucleus, polar cytoplasm, and processes that extended along the basal lamina (BALDWIN 1994). The nucleus of the atypical basal cell was consistently closer to the muco-ciliary surface than was the nucleus of the typical basal cell.

Throughout the bronchial tree both types of basal cells contribute to the maintenance of epithelial cohesion by providing desmosomal attachment for columnar cells.

7.1.7.2 Histochemistry

Basal cells of the *rabbit* trachea and bronchus have fairly high specificity for **alkaline phosphatase** of a non-specific isozyme (92.2 % and 95.6 %, respectively). Therefore, this enzyme is considered to be a useful marker for basal cells (INAYAMA et al. 1995).

α6 and **β4** integrin molecules were expressed strongly at the junction of the basal cells with the underlying basement membrane (MONTEFORT et al. 1992).

Multi-drug resistance associated protein (MRP), member of the ATP binding Cassette transporter proteins, is particularly involved in glutathione-conjugates detoxification. In *human* bronchial basal cells MRP mRNA immunostaining was observed at the whole plasma membrane (BRÉCHOT et al. 1997).

7.1.8
APUD Cells

Endocrine-like cells with biochemical characteristics of **A**mine **P**recursor **U**ptake and **D**ecaboxylase (APUD) were quantified in the tracheal, bronchial, and bronchiolar epithelia of the *guinea pig* by four histochemical stains: Grimelius, en-block silver, lead-haematoxylin-en block silver and periodic acid-Schiff-lead haematoxylin (Marchevsky et al. 1983). There were significant differences between the number of epithelial cells in the various locales of the respiratory tree: 318±5.63 epithelial cells/mm in the trachea, 263±17.11 epithelial cells/mm in the bronchi and 193±4.21 epithelial cells/mm in the bronchioles. Boers et al. (1996) did not find any difference in the number of neuroendocrine cells between large (airway diameter >4.5 mm) and small (airway diameter <1.2 mm) conducting airways in 9 *human* subjects without pulmonary disease out of 250 autopsy cases.

In the Wistar *rat*, there was a decrease in the percentage of bronchi containing Feyrter cells with increasing age (Moosavi et al. 1973). At the age of 7 days, 89 % of the bronchi of one *rat* contained Feyrter cells, but at the age of 31 days only 45 % of the bronchi of another *rat* contained such cells.

Normally there is no proliferative activity in neuroendocrine cells (Hoyt et al. 1990, Montuenga et al. 1992), although this may not be absolute (Stahlman and Gray 1984).

The immunocytochemical localisation of bioactive peptides is now widely applied in the investigation of bioptically obtained *human* lung tissues that contain neoplastic counterparts of neuroepithelial endocrine cells. These APUDomas express a larger spectrum of neuropeptides as compared to the healthy *human* lung, e.g. they may express immunoreactivity for ACTH (Gould et al. 1983, Tsutsumi et al. 1983, Polak and Bloom 1984, Tsutsumi 1990), neurotensin, motilin, glicentin, corticotropin-releasing factor, growth hormone-releasing factor, vasoactive intestinal peptide (Tsutsumi 1990).

Chromogranin A is a recognised 'gold standard' marker of neuroendocrine tumours including the 'silent tumours' which are not associated with the overproduction of any known bioactive peptide or hormone (Hendy et al. 1995). It has been suggested that this protein is a peptide precursor, because of a number of conserved dibasic cleavage sites and the identification of a number of bioactive peptides derived from chromogranin A. These include pancreastatin (Tatemoto et al. 1986), parastatin (Fasiotto et al. 1993), and vasostatin, which inhibits contractile response in segments of endothelium-denuded saphenous vein (Aardal and Helle 1993). Patients with flushing as a result of lung carcinoids were immunonegative for vasostatin (Cunningham et al. 1999).

After exposure to **hypoxia** the APUD cells secrete their amine-rich vesicles by exocytosis (Lauweryns and Cokelaere 1973) establishing the secretory nature of the APUD cells. When *rabbits*, 1 or 2 days old, were exposed to a hypoxia of 10 % oxygen for 20 min, (Lauweryns et al. 1977) found the intensities of the microspectroscopically recorded emission spectra of neuroepithelial bodies significantly decreased. A distinct exocytosis of the dense core vesicles was observed at the basal cell membrane of the corpuscular cells. From studies in which the hypoxic *rabbits* received normoxic blood in the arteria pulmonalis from a donor *rabbit* by means of an arterio-arterial cross circulation with mutual exchange transfusion, Lauweryns et al. (1978) concluded, that neuroepithelial bodies reacted directly to the hypoxia of the inhaled air and not to the hypoxaemia of the pulmonary blood. After short-term (1 h) infranodose as well as long term (3 days) supranodose left vagotomy, the ipsilateral neuroepithelial bodies, though still intact, but no longer connected to the central nervous system, showed decreased fluorescence intensity and increased basal degranulation (Lauweryns and Van Lommel 1986).

At the age of 4 days there was no significant difference between the number of Feyrter cells in hypoxic (380 mm Hg simulating an altitude of 5,488 m) Wistar *rats* and controls (Moosavi et al. 1973).

O_2 regulation of K^+ channels in chemoreceptive neuroepithelial bodies and their immortal counterparts, H146 cells, involves altered reactive oxygen species generation by NADPH oxidase. In contrast, this enzyme complex is not involved in O_2 sensing by the carotid body and pulmonary vasculature. O'Kelly et al. (2001) provided pharmacological evidence to support a role for NADPH oxidase in hypoxic inhibition of K^+ currents in H146 cells. Two structurally unrelated NADPH oxidase inhibitors, diphenylene iodonium and phenylarsine oxide, suppressed hypoxic inhibition of K^+ currents recorded using the parch-clamp technique. Most importantly, however, neither inhibitor fully blocked this response.

7.2
Pharmacology

A non-invasive aerosol inhalation technique using monodisperse aerosol allowed the **estimation of *human* airway diameters** (Schiller-Scotland et al. 1990, Schiller-Scotland 1997). Brochoconstriction was achieved by carbachol (1.8 mg; 2.5 % solution), bronchodilation by oxitropiumbromide (0.2 mg; MDI). Formoterol (24 μg; MDI) has been evaluated as a long acting (>10 h) bronchodilator. Both bronchoconstriction and bronchodilation were most pronounced in proximal airways.

Hyperosmolarity caused a time- and concentration-dependent increase in interleukin-8 expression and secretion in 16HBE14o bronchial epithelial cells (LOITSCH et al. 2000). These effects could be blocked by antioxidants, such as dimethyl sulphoxide (0.5 %), 1,3-dimethyl-2-thiourea (20 mM), dithiothreitol (500 μM), and β-mercaptoethanol (500 μM), suggesting an involvement of reactive oxygen intermediates in the signal transduction of hyperosmolarity-induced IL-8 synthesis.

Table 21. Pharmacological classification of adrenergic receptors

α_1-Adrenergic receptors	
	Agonists: *p*-aminoclonidine > epinephrine > norepinephrine > isoproterenol
	Antagonists: prazosin > phentolamine > yohimbine
α_2-adrenergic receptors	
	Agonists: *p*-aminoclonidine > epinephrine > norepinephrine > isoproterenol
	Antagonists: yohimbine > phentolamine > prazosin
β_1-adrenergic receptors	
	Agonists: isoproterenol > epinephrine > norepinephrine
	Antagonists: betaxolol > ICI 118,551
β_2-adrenergic receptors	
	Agonists: isoproterenol > epinephrine > norepinephrine
	Antagonists: ICI 118,551 > betaxolol
β_3-adrenergic receptors	
	Agonists: pindolol
	Antagonists: BRL 37344

7.2.1 Adrenergic Activators

Adrenergic receptors are located throughout the body on neuronal and non-neuronal cells where they mediate a diverse range of responses to the endogenous catecholamines adrenaline and noradrenaline. To date, nine adrenergic receptors have been cloned. The classification of epinephrine receptors into α-receptors and β-receptors stems from AHLQUIST (1948). Epinephrine and five related substances in a number of bioassays acted in either an excitatory or an inhibitory manner. The results could only be explained by postulating two different receptors, i.e. the α- and the β-receptors.

Three genes encoding α_2-adrenergic receptor subtypes have been identified from several species, termed α_{2A}, α_{2B}, and α_{2C}, respectively (for review see HEIN 2001). The α_{2B}-receptor shows primarily peripheral expression in kidney, liver, lung, and heart.

β-Adrenergic receptors (for review see DANNER and LOHSE 1999) are prototypical members of the family of G-protein-coupled receptors, which comprise a large group of sevenfold membrane-spanning helix cell surface receptors for such diverse stimuli as light (HARGRAVE and MCDOWELL 1992, KHORANA 1992), hormones and neurotransmitters (DOHLMAN et al. 1991, LOHSE 1993) and olfactory stimuli (LANCET 1986, BUCK and AXEL 1991). Upon agonist stimulation β-adrenergic receptors couple to the stimulatory G-protein, G_s, which in turn activates adenylate cyclase leading to an increase in intracellular cyclic AMP concentrations and subsequently to an activation of protein kinase A (HAUSDORFF et al. 1990, COLLINS et al. 1991, DOHLMAN et al. 1991, LOHSE 1993). All three components of this signal transduction cascade are subject to complex regulation on both mRNA and protein levels. To date, three β-adrenergic receptor-subtypes, termed β_1-adrenergic receptor, β_2-adrenergic receptor, and β_3-adrenergic receptor have been cloned and sequenced from different species.

However, molecular comparison of α_1- and α_2-receptors suggests that these proteins are structurally related "isoreceptors" (SHREEVE et al. 1985). STILES et al. (1983) demonstrated structural similarities of cardiac β_1- and β_2-receptor subtypes by photoaffinity labelling. The molecular basis of the striking differences in response to ligands that behave as β_1- or β_2-adrenoceptor antagonists and β_3-adrenoceptor agonists is still unexplained. Although all three β-adrenoceptor subtypes are coupled to the G_s protein and stimulate adenylate cyclase, it is possible that one or each of the receptors actually interacts with different combinations of the α_s-subunit with either of one of the seven or eight gene products encoding the β- and γ-subunits (STROSBERG and PIETRI-ROUXEL 1996).

The β_1-adrenergic receptor is encoded by an intronless and TATA-less gene that is rich in GC sequences in the first 0.5 kb of the 5'-flanking region (MACHIDA et al. 1990, SHIMOMURA and TERADA 1990, COHEN et al. 1993, COLLINS et al. 1993, PADBURY et al. 1995, SEARLES et al. 1995). The transcriptional start site in the *rat* β_1-adrenergic receptor genes occurs 253 bp 5' to the initiator ATG, whereas in the *ovine* and *murine* β_1-adrenergic receptor genes the transcriptional start site occurs 415 and 660 bp 5' to the ATG, respectively (COHEN et al. 1993, PADBURY et al. 1995, SEARLES et al. 1995). To identify the domains involved in retinoic acid-mediated activation of β_1-adrenergic receptor gene transcription, BAHOUTH et al. (1998) ligated three kb of 5'-flanking sequence of the β_1-adrenergic receptor gene to a luciferase reporter gene and transiently transfected them into F9 *mouse* teratocarcinoma stem cells that were pre-exposed to 100 nM retinoic acid for 2 days. By generating deletions in the β_1-adrenergic receptor promoter, a region between −125 and −100 was found to mediate a 3-fold induction in cells exposed to retinoic acid for an additional two days. Through site-directed mutagenesis of the region, it was determined that the retinoic acid responsive element was organised as a direct repeat separated by 5 nucleotides in which the 5'-most AGGTCG half-site was between nucleotides −106 and −101 and the 3'-most AGGTCA half-site was between nucleotides −117 and −112. The retinoic acid receptor α isoform bound to the oligomer representing the sequences between −125 and −100 as a heterodimer complex with the retinoid X receptor α.

The β-adrenergic receptor-adenylyl cyclase system is essential for the homeostatic control of airway calibre, airway reactivity, and normal responses to inflammatory stimuli (Nijkamp et al. 1992). IL-1β exerts complex cell type-specific effects on the function of the β-adrenergic receptor-adenylyl cyclase system in a variety of cell types. In BEAS-2B cells, a transformed normal *human* airway epithelial cell line, 200 pM IL-1β for 18 h significantly increased β-adrenergic receptor density (B_{max}) ~2.5-fold ($P < 0.01$ by one-way analysis of variance) without affecting the receptor binding affinity (Kelsen et al. 1997). IL-1β also significantly increased the percentage of the β_2-receptor subtype from 67±9 to 91±8% in control and IL-1β-treated groups, respectively ($P < 0.001$).

Pharmacodynamic influences on bronchiolar Clara cells (Schiller 1957) are mediated by β-adrenergic receptors visualized by fluorescence microscopy after blocking with 9-amino-acridyl propranolol (Massaro and Davis 1984).

While receptor function is regulated rapidly (over seconds or minutes), regulation of **receptor number** takes much longer, commonly many hours. In isolated cells it is often maximal only 24 h of continuous agonist exposure (Danner and Lohse 1999). Changes in β-adrenergic receptor number can be effected by two classes of mechanisms:

- enhanced proteolytic receptor degradation and
- decreased receptor synthesis, i.e. modulation of gene expression.

Aksoy et al. (2001) examined the role of several nuclear transcription factors in the protein kinase C-activated upregulation of β_2-adrenergic receptor expression.

Prolonged exposure of lung and other tissues to adrenergic agonists results in opposing physiologic effects on β-adrenergic receptors. The process whereby diminished receptor action, or desensitization, takes place may involve both downregulation (decrease in receptor number) and uncoupling (inability of the agonist-bound receptor to initiate the usual physiologic sequelae). This destabilization of the high affinity state of the receptor is characterized by a decreased ratio of the dissociation constants of the agonist for the two forms of the receptor (termed the K_L/K_H ratio) (Kent et al. 1980).

The catechol nucleus represents an important component for **optimum activity** at adrenoreceptors. Elimination of either phenolic group results in a large decrease in activity (α and β). Replacement of the 3-phenolic group by $HOCH_2^-$ or $CH_3SO_2NH^-$ resulted in salbutamol and soterenol, respectively.

7.2.1.1 Superoxide Oxidises Catecholamines

The oxidation of catecholamines by the hydroperoxidase activity of lipoxygenase (EC 1.13.11.12) is documented by Rosei et al. (1994) and Núñez-Delicado et al. (1996). *o*-Diphenols are easily studies spectrophotometrically since, when oxidised, they render coloured compounds, quinones, or their corresponding aminochromes. In the case of isoprenaline the maximum of the oxidation product was developed at 490 nm (Núñez-Delicado et al. 1999), which corresponds to that of the aminochrome product (Jiménez et al. 1985, Núñez-Delicado et al. 1996).

Cohen and Heikkila (1974) have shown that several oxidised catecholamines including 6-hydroxydopamine and 6-aminodopamine could form the cytotoxic species $O_2^{\bullet-}$, H_2O_2 and $HO^{\bullet}$, and this was confirmed by Graham et al. (1978) and Graham (1984). Graham et al. (1978) also showed that the quinone autoxidation products of L-dopa and dopamine exerted toxicity on C1300 neuroblastoma cells *in vitro* via their nucleophilic reactivity. Ambani et al. (1975) observed that the substantia nigra has delectable levels of H_2O_2 and high levels of monoamino oxidase, able to evolve H_2O_2. The polymerisation of the catecholamine-derived quinones to form the neuromelanin for which the substantia nigra is named is a $O_2^{\bullet-}$ evolving autooxidative process (Graham (1984). Ambani et al. (1975) have reported lowered levels of peroxidase and catalase in the substantia nigra of patients with Parkinson's disease.

The macrophage migration inhibitory factor (MIF) converted the oxidised catecholamines 3,4-dihydroxyphenylaminechrome, epinephrinechrome, and norepinephrinechrome to 5,6-dihydroxyindole, 3,5,6-trihydroxy-1-methylindole, and 3,5,6-trihydroxyindole, respectively (Matsunaga et al. 1999). These products are precursors of neuromelanin (Smythies 1996, D'Ischia and Prota 1997), a pig-

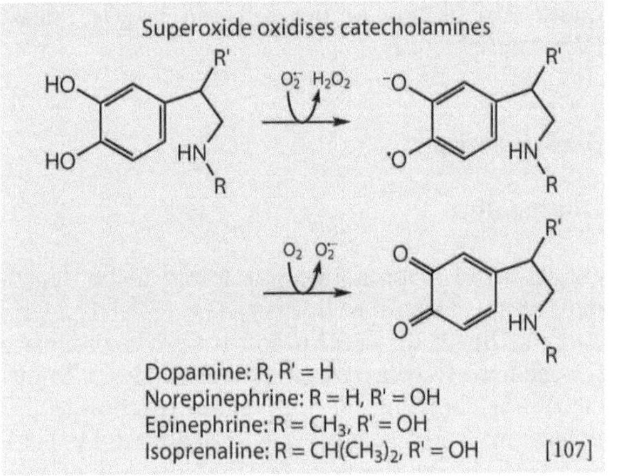

ment found in neurones and glial cells (Zecca et al. 1992, Enochs et al. 1993, Prota and D'Ischia 1993) to function as a sink for toxic metabolites of catecholamine biosynthesis.

7.2.1.2
β_2-Agonists Have Antioxidant Function

In addition to their bronchodilatory action in asthma, β-agonistically active substances of diphenolic structure offer radical scavenging properties toward reactive oxygen species as shown in a model system by electron paramagnetic resonance spectroscopy and photometric approaches (Zwicker et al. 1998). The substances under study showed activity in superoxide radical scavenging under aprotic and protic conditions as well. The efficiency of the reaction decreased in the order fenoterol > salbutamol > reproterol > terbutaline > oxyfedrine when 5,5-dimethyl-1-pyrroline-*N*-oxide (DMSO) was used as an aprotic solvent. In an aqueous system, the rate constants decreased in the order: fenoterol > reproterol > salbutamol.

7.2.1.3
Link Between Catecholamine and Nitric Oxide Metabolism

Dopamine and norepinephrine in aerobic buffer (pH 7.4) were almost completely converted to their 6-nitro-derivatives by nitric oxide at room temperature, while epinephrine was nitrated and above all oxidised (Daveu et al. 1997). Girgin et al. (1999) found insignificant increases in nitrite and nitrate levels in cardiac tissues of aged *rats* compared to young ones. This could be explained by observations of Girgin et al. (1999) that ageing *rat* tissues have higher catecholamine levels, which would lead to higher monoamine oxidase activities, while this catecholamine increase does not seem to have any direct effect on nitrate-nitrite levels. Through the increased activation of monoamine oxidase enhanced production of H_2O_2 and NH_3 occurs, and NH_3 is incorporated into glutamine and L-arginine, the latter of which serves as a substrate for nitric oxide synthase (Zhang et al. 1993). Li et al. (2000) demonstrated that ciliated *rat* tracheal epithelial cells produce •NO, which is correlated with the ciliary beat frequency.

7.2.1.4
α-Adrenergica

Ephedrine

The presence of a β-C substituent limits the pathway of biotransformation of ephedrine to *p*-hydroxylation, *N*-dealkylation, and deamination. Dealkylation and subsequent deamination is the major route of metabolism of ephedrine in most species, excluding the *rat* (Williams et al. 1973).

Phenylephrine

Phenylephrine is an α_1-adrenergic agonist. However, it simultaneously activated β-adrenoceptors and can therefore be used as a selective α_1-adrenoceptor agonist only in the presence of β-blockade (van Meel et al. 1981). Phenylephrine was reported to induce Ca^{2+} efflux, but not influx, in the perfused *rat* liver in the presence of 1.3 mM Ca^{2+} in the medium (Altin and Bygrave 1985). Kleineke and Soling (1987), however, reported that Ca^{2+} influx can occur with phenylephrine under these conditions.

Phenylephrine, norepinephrine and epinephrine (all 5 μmoles/kg), but not isoprenaline caused a loss of about 35 per cent of the ascorbic acid content of lung tissue of anaesthetised male *mice* and an increase of about 20 per cent in lung weight within 15 min (Willis and Kratzing 1974b). The ability of phenoxybenzamine (irreversible in action because it forms a stable covalent bond with the α-receptor; Nickerson 1957) to block both the lung ascorbate loss and the lung oedema caused by phenylephrine, norepinephrine and epinephrine showed that these changes depended on α-receptor activity.

Adrenaline

Effect on Clara Cells

Adrenaline in very large dose (2.5 mg/kg *mouse*) rapidly dilated the cisternae of the smooth endoplasmic reticulum of the Clara cells with subse-

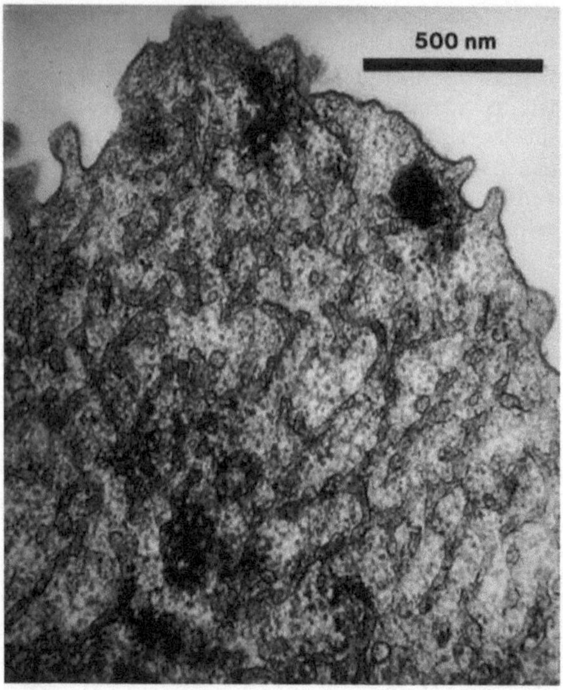

Fig. 72. Smooth endoplasmic reticulum in the apical part of a type 1 Clara cell from a female *rat* (breeder: Winkelmann, Borchen-Kirchborchen) which inhaled 300 μg micronized (mass median diameter 2.45 μm) phenylephrine bitartrate/puff from a suspension type self-propelled aerosol (Medihaler). 12 Puffs/min were dispersed into a 164.5 l box where the animals stayed for 15 min. 30 min later under methitural anaesthesia, the lung was fixed by intratracheal instillation of 2.5 % glutaraldehyde in phosphate buffer (pH 7.4) before opening the thorax. Postfixation with 1 % osmium tetroxide in phosphate buffer (pH 7.4). Contrasted en bloc for 12 h with 0.5 % uranyl acetate in 70 % ethanol. Embedded in a 2 : 8 mixture of methyl and butyl methacrylate. Sectioned at 50 nm. Lead citrate after Reynolds (1963). Plate 53/05

quent vesiculation of the cell and vesicle release (Wang et al. 1971).

7.2.1.5 β-Adrenergica

Isoprenaline

Oxidation of isoprenaline was found to be dependent on the amount of lipoxygenase (EC 1.13.11.12) and was linear up to 0.2 μM of the soybean lipoxygenase used (Núñez-Delicado et al. 1996). To discount any contribution of a Fenton reaction in this oxidation, lipoxygenase was substituted by $FeCl_3$ and $FeCl_2$ in the presence of H_2O_2, the rate of aminochrome production being negligible under these conditions. The inhibitory effect of cyclodextrins on the oxidation of xenobiotics is based on their degree of hydrophobicity and the charge (isoprenaline < 4-methylcatechol < 4-*tert*-butylcatechol < 4-*tert*-octylcatechol) (Núñez-Delicado et al. 1999). This inhibitory effect was due to the complexation of xenobiotics in the hydrophobic cavity of cyclodextrins.

Pre-treatment of *human* bronchial epithelial cells with 10 μM isoprenaline or 100 μM salbutamol augmented the adhesion of fluorescently labelled THP-1 cells, a monocyte/macrophage cell line, to *human* bronchial epithelial cell monolayers by about 40–60 % (Romberger et al. 2000). The increase in THP-1 cell adhesion occurred within 10 min of isoprenaline pre-treatment of *human* bronchial epithelial cells and gradually declined but persisted with up to 24 h of isoprenaline exposure.

Fig. 73. Multivesicular body in a type 3 Clara cell from a female white *rat* (breeder: Winkelmann, Borchen-Kirchborchen) 30 min after an inhalation of 300 μg micronized (mass median diameter 2.45 μm) phenylephrine bitartrate/puff from a suspension type self-propelled aerosol (Medihaler). 12 Puffs/min were dispersed into a 164.5 l box where the animals stayed for 15 min. 30 min later under methitural anaesthesia, the lung was fixed by intratracheal instillation of 2.5 % glutaraldehyde in phosphate buffer (pH 7.4) before opening the thorax. Postfixation with 1 % osmium tetroxide in phosphate buffer (pH 7.4). Contrasted en bloc for 12 h with 0.5 % uranyl acetate in 70 % ethanol. Embedded in a 2 : 8 mixture of methyl and butyl methacrylate. Sectioned at 50 nm. Lead citrate after Reynolds (1963). Plate 44/05

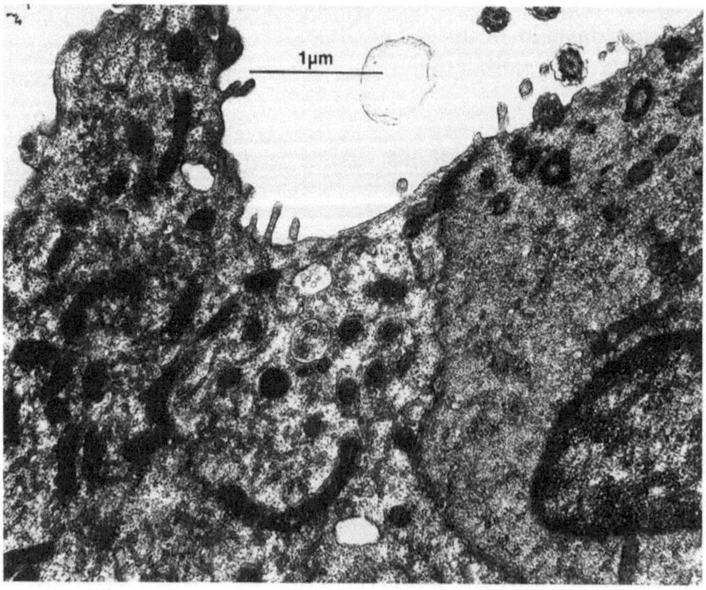

Effect on Clara Cells

0.5 mM Isoprenaline added after two hours of incubation had no effect on the secretion of Clara cell secretory protein by lung explants (HOOK et al. 1990).

Effect on Goblet Cells

Isoprenaline increased mitotic index and goblet cell number at most of the airway levels studied in the Sprague-Dawley *rat* (BOLDUC and REID 1978).

In the *pig*, isoprenaline caused a significant increase in the number of goblet cells in the bronchial epithelium, but there was no change in their glycoprotein type, most containing exclusively sialomucin (BASKERVILLE 1976).

Cardiac Side Effects

It has been known since the work of RONA et al. (1959) that sympathomimetic amines at dose levels, which would be lethal for *humans* produce extensive necrosis in the *rat* myocardium. The lesions and those produced in the same species by hypoxia have morphological features in common (NILES et al. 1968). A proteolipid binding ^{3}H-isoprenaline was localized in the plasma membrane of the *feline* heart ventricular myocytes (OCHOA et al. 1972). Exposing myocytes to catecholamines lead to down-regulation of the number of receptors within minutes (LINDEN et al. 1984).

Isoprenaline decreases coronary artery perfusion and heart muscle oxygenation under circumstances of increased demand.

The exact mechanism of isoprenaline-induced myocardial damage has not been clarified, but a mismatch of oxygen supply versus demand following coronary hypotension and myocardial hyperactivity may offer the best explanation for the complex morphological alterations in the presence of a patent coronary vasculature (YEAGER and IAMS 1981).

In view of the accumulation of H_2O_2 in the myocardium due to ischaemia-reperfusion and changes in β-adrenoceptor mechanisms in the ischaemic-reperfused heart PERSAD et al. (1998) investigated the **effects of H_2O_2 on the β-adrenoceptor**, G-protein and adenylyl cyclase complex. *Rat* hearts were perfused with 1 mM H_2O_2 for 10 min before isolating membranes for measuring the biochemical activities. The stimulation of adenylyl cyclase by different concentration of isoprenaline was depressed upon perfusing hearts with H_2O_2. Both the affinity and density of β_1-adrenoceptors as well as the density of the β_2-adrenoceptors were decreased whereas the affinity of β_2-adrenoceptors was increased by H_2O_2 perfusion. Competition curves did not reveal any effect of H_2O_2 on the proportion of coupled receptors in the high affinity state. The basal as well as forskolin-, NaF- and Gpp(NH)p-stimulated adenylyl cyclase activities were depressed by perfusing the heart with H_2O_2. Catalase alone or in combination with mannitol was able to significantly decrease the magnitude of alterations due to H_2O_2. The positive inotropic effect of 1 μM isoprenaline was markedly attenuated upon perfusing hearts with 200–500 μM H_2O_2 for 10 min.

Isoprenaline Plus Phenylephrine

In brochodilation the greater effect of isoprenaline plus phenylephrine than of isoprenaline alone is due to the spasmolytic effect of isoprenaline plus the vasoconstrictive and decongestant properties of phenylephrine (KALLÓS and KALLÓS-DEFFNER (1964). Oedema of the bronchial mucosa is reduced and mucus secretion is diminished by the action of phenylephrine. The protective effect against dys-

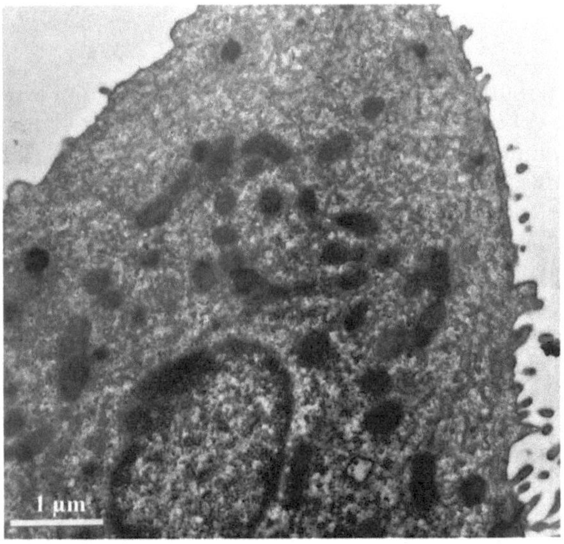

Fig. 74. Smooth endoplasmic reticulum in the apical part of a type 3 Clara cell from a female white *rat* (breeder: Winkelmann, Borchen-Kirchborchen) 2 h after an inhalation of 200 mg micronized isoprenaline HCl + 300 μg micronized (mass median diameter 2.45 μm) phenylephrine bitartrate/puff from a suspension type self-propelled aerosol (Medihaler). 12 Puffs/min were dispersed into a 164.5 l box where the animals stayed for 15 min. 30 min later under methitural anaesthesia, the lung was fixed by intratracheal instillation of 2.5 % glutaraldehyde in phosphate buffer (pH 7.4) before opening the thorax. Postfixation with 1 % osmium tetroxide in phosphate buffer (pH 7.4). Contrasted en bloc for 12 h with 0.5 % uranyl acetate in 70 % ethanol. Embedded in a 2 : 8 mixture of methyl and butyl methacrylate. Sectioned at 50 nm. Lead citrate after REYNOLDS (1963). Film 374/84

pnea on exertion in asthmatic patients is maintained for 30 min by isoprenaline and for more than 120 min by isoprenaline plus phenylephrine.

Deposition and retention of quartz dust inhaled (10 mg/m^3) for 3 h each on 20, 40, and 80 days, respectively, in the lungs of female *rats* premedicated for 5 min with isoprenaline (300 mg/m^3) plus phenylephrine (450 mg/m^3) were reduced by 16 per cent after 20 inhalations and by 24 per cent after 80 inhalations, compared with non-medicated animals (EINBRODT and SCHILLER 1967).

Salbutamol

Salbutamol was synthesised by HARTLEY et al. (1968) with a bronchoselectivity index of 2.7 in the *dog* to 55 in the *guinea pig* (WARDELL et al. 1974).

Terbutaline

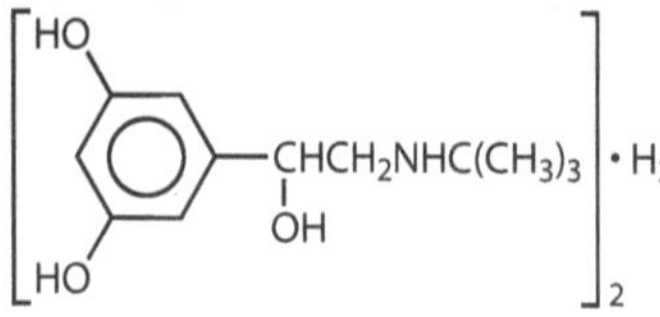

Terbutaline [108]

Terbutaline, (RS)-2-*tert*-butylamino-1-(3,5-dihydroxyphenyl)ethanol, and isoprenaline increased immunoreactive cAMP in ciliated epithelial cells of *dog* and *cat* trachea and in both serous and mucous gland cells of *cat* trachea (LAZARUS et al. 1984). Epithelial goblet cells did not respond to β-adrenergic agonists in either species.

Proper coordination of inhalation and metered dose inhalers canister activation is an important determinant of availability of medication to airways and subsequent clinical response. WILSON et al. (1991) compared coordination of terbutaline and pirbuterol metered dose inhalers by evaluating clinical effects, deposition in a model of the human upper airway, rate of aerosol settling a tank, and particle size distribution. The maximum clinical effects and least deposition in the upper airway occurred when both metered dose inhalers were activated at the onset of inhalation. When metered dose inhalers were activated before inhalation, pirbuterol was less effective than terbutaline. Pirbuterol metered dose inhaler aerosol is slightly smaller than terbutaline metered dose inhaler aerosol.

Salmeterol

While terbutaline (formula [108]) and salbuterol have a short duration of action (3–5 h), several more effective and longer acting β_2-adrenoceptor agonists have been developed. Salmeterol given by inhalation was effective for 6–8 h and, in contrast to its potency on respiratory smooth muscle, >3000-fold weaker than isoproterenol in cardiac tissue, indicating high β_2-adrenoceptor selectivity (JOHNSON 1990). In [^{125}]iodocyanopindolol-labeled bronchial membranes of *guinea pigs*, salmeterol and formoterol induced high-affinity states of the β_2-receptor, the latter inducing a significantly higher percentage (ROUX et al. 1996).

Prolonged exposure to β-agonists such as isoprenaline results in down-regulation of β_2-adrenergic receptor gene expression and function, while long acting β_2-agonists such as salmeterol remain efficacious over many hours. WANG and COLLINS (1996) examined the function of salmeterol, vs. isoproterenol, to regulate β_2-adrenergic receptor expression and stimulation of its effector adenylate cyclase in BEAS-2B cell.

BURGES and BLACKBURN (1972) found that salbutamol selectively activated *rat* lung adenylate cyclase at concentrations that had no effect on the *rat* heart enzyme. Nebulized salbutamol can be as effective as that given intravenously (LAWFORD et al. 1978) but has the added advantage of fewer metabolic and symptomatic side-effects. The S isomer of salbutamol had no activity at extrapulmonary β_2-adrenoceptors in *man* (LIPWORTH et al. 1997). The R isomer demonstrates approximately equivalent activity to racemic salbutamol when compared on a 1:2 microgram ratio. This suggests that the R isomer may offer possible therapeutic advantages over racemic salbutamol in view of less adverse airway effects with comparable systemic adverse effects.

Pirbuterol

The aromatic ring of salbutamol has been replaced by a pyridine ring with no significant loss of potency or β_2-selectivity (STEEN et al. 1974, WILLEY et al. 1976). In 64 healthy subjects pirbuterol inhalation (0.4 mg) all obstruction-related parameters showed significant bronchodilation ($P<0.001$) in both smokers and non-smokers (KRONENBERGER et al. 1991).

Formoterol

NIX et al. (1990) found the protective effect of formoterol against methacholine-induced bronchoconstriction in 12 asthmatic *patients* lasting for at least 5 h. The rapid pulmonary absorption of inhaled formoterol and its slow elimination from plasma are in agreement with the fast onset and the long duration, respectively, of bronchodilation of inhaled formoterol fumarate in clinical trials (LECAILLON et al. 1997). There was no indication that formoterol produced a more pronounced tachyphylaxis to β_2-adrenoceptor stimulation in the bronchial muscle of asthmatics than the β_2-adrenoceptor agonists normally used, e.g. salbutamol (LARSSON et al. 1988, ARVIDSSON et al. 1989). Formoterol powder is as potent as solution (CLAUZEL et al. 1991).

In vitro, formoterol fumarate has been found to be about 50 times more potent than salbutamol on bronchial smooth muscle (IDA 1976, DECKER et al. 1982) and at least as β_2-selective as salbutamol and terbutaline (DECKER et al. 1982).

7.2.2 Adrenolytic Drugs

Although the adrenergic α-blocking agents are useful in the treatment of certain diseases of the respiratory system there is no corresponding therapeutic application for the adrenergic β-blocking agents. The use of nonselective β-blockers (β_1 and β_2) such as propranolol is contraindicated in about 5% of patients suffering from disease of the heart and systemic circulation because the drug provokes an asthmatic attack. The development of β-blockers with cardioselective action (β_1 only) has been an important consideration in the improvement of nonselective β-blockers.

7.2.2.1 Propranolol

Propranolol is a cationic compound, which has a high affinity to acidic phospholipids. As an amphiphilic drug it partially penetrates into the hydrophobic core of the membrane. Propranolol has been suggested to have two different binding sites in phospholipid membranes (KUBO et al. 1986, KODAVANTI and MEHENDALE 1990). Using X-ray diffraction, ALBERTINI et al. (1990) found propranolol to increase water layer thickness on DPPC membrane surface. Propranolol causing the formation of pyrene lipid-enriched microdomains decreased membrane fluidity (JUTILA et al. 1998). The detachment of cytochrome *c* from liposomes by propranolol could de detected by stopped-flow. However, the process was too fast (complete within <3 ms) so as to allow for detailed analysis.

The occurrence of **bronchospasm** which appears with the use of propranolol for the treatment of cardiovascular diseases, can be explained by the predominance of α-mediated bronchoconstriction as well as parasympathetic or vagal induced influences on the airways (AVIADO and MIOZZI 1981).

Although total flow to **ischaemic region** may be reduced, BECKER et al. (1971) suggested that propranolol causes favourable redistribution of flow within the ischaemic area. PARRAT (1980) argued that the results in their Table 2 showed that in the *dogs* subsequently given propranolol there was an abnormally large reduction in the endo/epi flow ratio following coronary artery occlusion (from a mean of 1.15 to 0.62). This was much larger than the reduction in the other two groups (1.13 to 0.85 and 1.21 to 0.84, respectively). In the presence of propranolol, coronary occlusion resulted in an endo/epi ration of 0.84. Studies of BECKER et al. (1975) and KLONER et al. (1976), also using radiolabelled microspheres (15 μm and 8 μm diameter, respectively) have failed to demonstrate a significant increase in subendocardial perfusion or an alteration in the endo/epi flow ratio after propranolol administration in anaesthetised *dogs* subjected to coronary artery occlusion. It seems unlikely (KLONER et al. 1976) that propranolol-induced reductions in "infarct size" are related to increased collateral blood flow or to its distribution within the ischaemic area.

In conscious *dogs*, VATNER et al. (1977) measured regional blood flow in myocardial ischaemia with radio-labelled microspheres of 9 μm diameter before coronary occlusion, 10–15 min after occlusion, and 15–20 min after giving 0.5–2 mg propranolol/kg. There was a gradient of blood flow ranging from 0.31 ml/g × min (severely ischaemic zone) through 0.83 ml/g × min (moderately ischaemic area) to 1.1 ml/g × min (in regions outside the area supplied by the occluded vessel).

7.2.2.2 Nifenalol

Nifenalol (1-*p*-nitrophenyl-2-isopropylaminoethanol; INPEA), a so-called "pure" β-receptor blocker (DOTTI et al. 1968), induced its β-blocking effect without bradycardia or significant changes in blood pressure. PIVA and ONGARI (1968) in 11 patients suffering from coronary artery disease by prophylactic intravenous infusion of 2 mg nifenalol/kg over 5 min prevented increase of both systolic and drop in diastolic blood pressure, tachycardia, shortening of the PQ, depression of the ST segment and flattening or biphasic T-wave induced by 60–90 ng epinephrine/kg × min for 20 min. PHILLIPS (1980), p.19) indicated nifenalol to be a non-selective blocker, while its α-methyl derivative (SOMANI 1969) is β_2-selective.

7.2.2.3 4-(2-Hydroxy-3-isopropylaminopropoxy)-acetanilide

4-(2-Hydroxy-3-isopropylaminopropoxy)-acetanilide in contrast with propranolol even in high doses does not reduce myocardial contractility (SOMANI and LADDU 1969). It principally blocks myocardial β_1-receptors, nut not β_2-receptors in the bronchial mucosa (PALMER 1959). In the heart, ICI 50172 is unable to block the β_1-effects of isoprenaline completely without impairing the β_2-effects on the vessels (PARRAT and WADSWORTH 1969).

7.2.2.4 Pindolol

Compounds with electron-withdrawing groups on the aromatic ring as H87/07 (formula [109]) and pindolol (formula [110]) tend to be partial agonists.

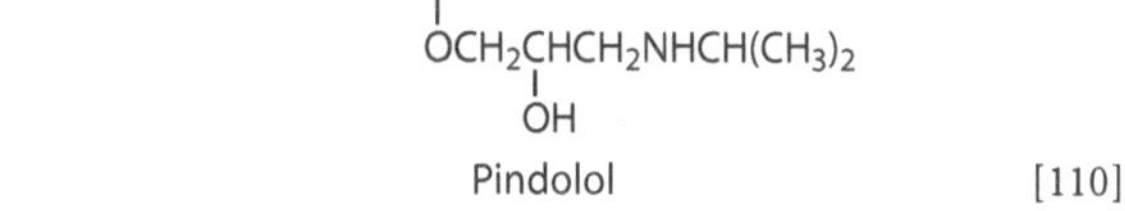

H 87/07 [109]

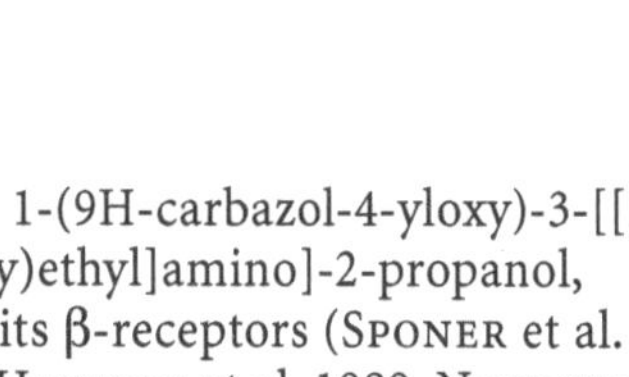

Pindolol [110]

7.2.2.5 Carvedilol

Carvedilol, 1-(9H-carbazol-4-yloxy)-3-[[2-(2-methoxyphenoxy)ethyl]amino]-2-propanol, competitively inhibits β-receptors (SPONER et al. 1987, SEKI et al. 1988, HATTORI et al. 1989, NICHOLS et al. 1989, DE MEY et al. 1994, SPONER and FEUERSTEIN 1999), blocks α1-receptors (DE MEY et al. 1994, SPONER and FEUERSTEIN 1999), and acts as an antioxidant (YUE et al. 1992, 1999).

Carvedilol is extensively metabolised with less than 2% of the dose excreted unchanged in the urine (NEUGEBAUER et al. 1987). The primary metabolic pathways include glucuronidation, aromatic ring oxidation and aliphatic side chain oxidation. The major route of elimination for carvedilol metabolites is via the bile into the faeces.

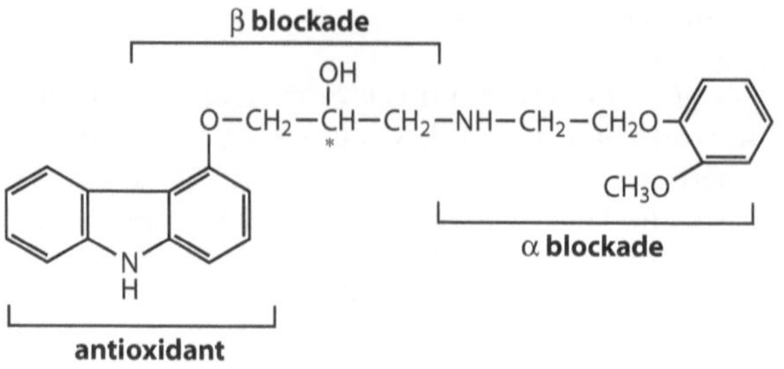

Chemical structure of carvedilol [72]

Since the **antioxidant capacity** of carvedilol is believed to reside in the carbazole moiety (YUE et al. 1992, MACKERELL jr. et al. 1995), MIGLIAVACCA et al. (1998) used 4-methoxycarbazole as a model compound for carvedilol, and 1-hydroxy- and 3-hydroxy-4-methoxycarbazole for its 1- and its 3-hydroxylated metabolites. 1-Hydroxy- and 3-hydroxy-4-methoxycarbazole have good antioxidant activities based of their ΔH_{abs} values. These calculations suggest that both compounds scavenge free radicals via direct H-atom transfer, since their oxyl radical is relatively stable. Hydroxylated carbazole derivatives such as carazostatin and carbazomycin show a considerable antioxidant activity, whereas non-hydroxylated carbazoles are inactive (IWATSUKI et al. 1993, KATO et al. 1993).

7.2.3 Cholinergic Activators

Acetylcholine receptors in the plasma membrane were initially identified by immunoelectronmicroscopy using anti-acetylcholine receptor antibodies (KLYMKOWSKY and STROUD 1979). Transversing the lipid bilayer is a central channel measured by structural analysis as about 0.7 nm diameter (KISTLER et al. 1982), close to the maximum diameter derived from electrophysiological measurements of conductivity of organic cations of various size (FURUKAWA and FURUKAWA 1959, MAENO et al. 1977, HUANG et al. 1978, DWYER et al. 1980). Acetylcholine receptors are sevenfold membrane-spanning receptors (KUBO et al. 1986, HESCH 1991).

Five distinct **muscarinic receptors** have been identified (M_1–M_5) by molecular cloning, and there is evidence to suggest that at least four of them (M_1-M_4) are of functional importance within the airways (BARNES 1993). M_1, M_3, and M_5 receptors stimulate polyphosphoinositide hydrolysis, leading to an increase in inositol monophosphate. This occurs predominantly through the stimulation of pertussis toxin-insensitive G proteins of the G_q and G_{11} class (Bou-Hanna et al. 1994). The M_2 and M_4 receptors, which primarily inhibit adenylyl cyclase, are coupled to pertussis toxin-sensitive G proteins of the G_i/G_o class (Parker et al. 1991). Steel and Hanrahan (1997) investigated the G protein coupling mechanism by incubating golden Syrian *hamster* airway epithelial secretory cells for 16 h with pertussis toxin. Mucus secretion could be stimulated by muscarinic receptor stimulation (100 μM carbachol) coupled to a pertussis toxin-sensitive G protein and appears to require the activation of a phorbol ester-insensitive isoform of protein kinase C. There was evidence to suggest, that muscarinic-induced mucus secretion does not involve NO, as inhibition of NO production by N^G-nitro-L-arginine methyl ester had no significant effect on basal high-molecular-weight glycoconjugate secretion.

Autoradiographs of lungs from adult male Sprague-Dawley *rats* injected with the muscarinic binding probe ^{3}H-quinuclidinyl benzilate showed little labelling despite a 6 month exposure time (Smith and Sidhu 1984).

7.2.3.1 Acetylcholine

Goblet cell in the *rat* nasal mucosa after stimulation with 10 μM acetylcholine had a typical goblet shape after conventional fixation with 2.5 % glutaraldyhyde in 0.1 M phosphate buffer (pH 7.4) for 1 h and postfixation in 2 % osmium tetroxide in phosphate buffer for 1 h. (Shimomura et al. 1996). The disrupted plasma membrane and the granule contents extruded into extracellular space were often observed in the stimulated goblet cells. After quick-freezing in liquid nitrogen and freeze-substitution according to Yoshihara et al. (1986) secretory granules by stimulation with acetylcholine became large in size, and tended to come into contact with each other. A pentalaminal structure was often detected.

7.2.3.2 Carbamylcholine

0.1 mM Carbamylcholine added after two hours of incubation had no effect on the secretion of Clara cell secretory protein by lung explants (Hook et al. 1990).

7.2.3.3 Pilocarpine

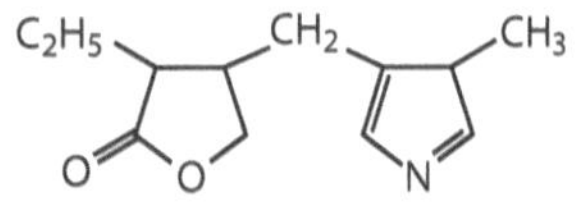

Pilocarpine [111]

Pilocarpine, (3S,4R)-3-ethyl-4,5-dihydro-4-(1-methyl-5-imidazolylmethyl)-2(3H)-furanon, 150 mg/kg *rat* subcutaneously, induced both the release of available and the formation of new secretory granules, which migrated from the basal to the apical part of the Clara cells and were extruded 1 to 2 h later (Yoneda 1977). Quantitatively, Massaro et al. (1979) found 150 mg pilocarpine/kg *rat* given intraperitoneally to decrease the volume density of secretory granules of Clara cells for about 25 % within 30 min and about 50 % within 60 min ($P < 0.01$). Within 2 h after giving pilocarpine, the volume density of the secretory granules had returned to control levels. 14 mg Atropine/kg *rat* blocked the fall as measured 1 h after the injection of the two drugs.

Goblet cell population increased in the *guinea pig* (body weight 400 to 500 g) trachea after intraperitoneal application of 12 daily successive doses of 2.5 mg pilocarpine (Ranga and Kleinerman 1982).

7.2.3.4 Nicotine

The gradual decline of the concentration of pulmonary neuroendocrine cell immunoreactive calcitonin in newborn *hamster* lung cells cultured over 4 weeks was prevented when the medium was supplemented with nicotine for 3 weeks (Nylen et al. 1993).

7.2.4 Anticholinergic Drugs

7.2.4.1 Atropine

Pre-treatment of cultured *hamster* tracheal goblet cells with 100 μM atropine 1 h before addition of 100 μM carbachol significantly inhibited the carbachol-induced mucin secretion (Steel and Hanrahan 1997).

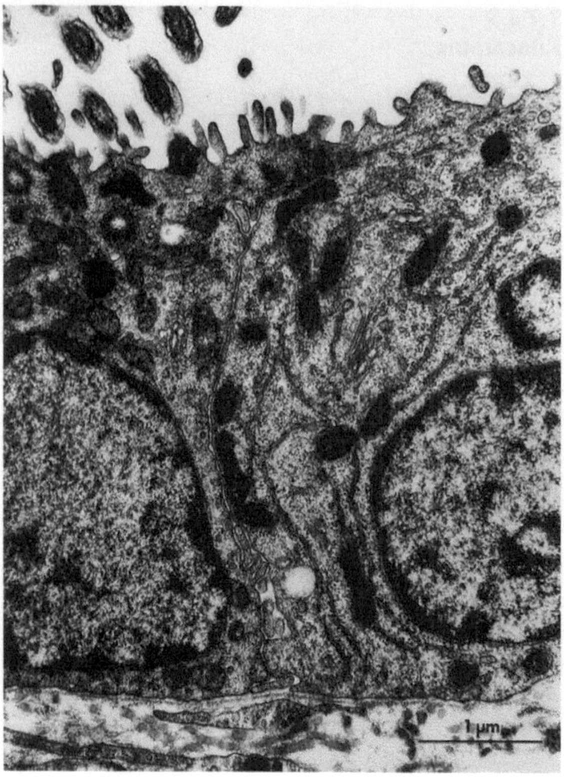

Fig. 75. A ciliated and a serous cell from the bronchiolar epithelium of a female white *rat* (breeder: Winkelmann, Borchen-Kirchborchen) which inhaled micronized deptropine citrate and isoprenaline hydrochloride aa from a suspension type self-propelled aerosol (Medihaler). 12 Puffs/min were dispersed into a 164.5 l box where the animals stayed for 15 min. 1 h later under methitural anaesthesia, the lung was fixed by intratracheal instillation of 2.5% glutaraldehyde in phosphate buffer (pH 7.4) before opening the thorax. Postfixation with 1% osmium tetroxide in phosphate buffer (pH 7.4). Contrasted en bloc for 12 h with 0.5% uranyl acetate in 70% ethanol. Embedded in a 2:8 mixture of methyl and butyl methacrylate. Sectioned at 50 nm. Lead citrate after Reynolds (1963). Plate 52/03

7.2.4.2 Benztropine

3-(10,11-dihydro-5H-dibenzo[a,d]cyclohepten-5-yloxy)-tropane citrate inhibiting salivary secretion in the *rabbit* at an ED_{50} of 110 mg/kg s.c. and 28 mg/kg p.o. (Funcke et al. 1964) was tested in rats inhaling micronized drug in combination with the β-adrenergic isoprenaline (Fig. 75).

7.2.5 Vitamins

7.2.5.1 Retinoids

Retinoids play a major role in lung embryology and organogenesis, deficiencies of dietary retinoic acid or vitamin A are risk factors for the development of lung cancer, and retinoids are being tested in lung cancer chemoprevention trials (Chytil 1996, Minna and Mangelsdorf 1997).

Hind et al. (2002) described the temporal and spatial expression of the retinoid-synthesizing enzymes Aldh-1 and Raldh-2 in the postnatal *mouse* lung. Both enzymes are upregulated during the period of maximal alveolar wall cell proliferation. Aldh-1 is located in the bronchial epithelium and alveolar parenchyma, and Raldh-2 is restricted to the bronchial epithelium and pleural mesothelial cells.

Retinoid function, in general, appears to be supported by at least two **receptors**, cellular retinoic acid-binding protein (Sundelin et al. 1985b) and cellular retinol-binding protein (Sundelin et al. 1985a). These proteins together with cytosolic fatty acid-binding protein constitute a multigene family (Bass 1988, Kaikaus et al. 1990). Abnormalities of retinoic acid receptors (RAR) have been implicated in the pathogenesis of lung cancer with several studies indicating abnormalities of the expression or function of RARβ wich maps to chromosome region 3p24, a frequent site of allele loss in lung cancer (Gebert et al. 1991, Nervi et al. 1991, Geradts et al. 1993, Zhang et al. 1994, Moghal and Neel 1995).

Squamous differentiation of *rabbit* tracheal epithelium in primary culture was associated via expression of cytokeratins CK13/CK4 and transglutaminase I specific markers of metaplasia. Treatment with retinoic acid receptor subtype RARα (CD336) and β (CD2019) agonists or RAR panagonist, but not the RARγ (CD437) agonist or ARX agonist (CD2624), is required for the inhibition of squamous metaplasia, evidenced by inhibition of CK13/CK4 and transglutaminase I expression (Boisvieux-Ulrich et al. 2000). The expression of CK10 cytokeratin of keratinizing epithelia, CK14/CK5 basal cell cytokeratins, and CK6 marker of cell proliferation decreases upon exposure of the RARα/β and RXR agonists. The RARγ agonist CD437, inactive in the decrease in CK13/CK4, CK10 and CK14, reduces CK5/CK6 amounts. CD437 is responsible for the dose-dependent apoptotic response. Nuclear labelling with propidium iodide and electron microscopy revealed chromatin condensation and nuclear fragmentation.

7.2.5.1.1 Retinoic Acid

Retinoic acid modulated secretion of normal *human* tracheobronchial epithelial cells in air-liquid interface cultures (Yoon et al. 1997). 1.5 nM was required to achieve maximum mucin production compared with 50 nM of retinol. Retinoid deficient cultures produced very small, but measurable amounts of mucin. While removal of retinoic acid from the media caused a dramatic reduction of mucin secretion, lysozyme secretion increased >10-fold and secretory leucocyte protease inhibitor secretion increased almost 2-fold.

The antioxidant activity of retinoids against lipid peroxidation has long been observed both *in vitro* (Nicotra et al. 1975, Halevy and Sklan 1987, Vile and Wintebourn 1988, Das 1989, Livrea et al. 1992) and *in vivo* (Kartha and Krishnamurthy 1977, Ciaccio et al. 1993). Samokyszyn and Marnett (1987) showed that 13-*cis*-retinoic acid may react with peroxyl radicals by addition but not H-atom abstraction.

Inhaled aerosolization of all-*trans*-retinoic acid for targeted pulmonary delivery in *rats* showed a significantly longer pulmonary half-life of the drug (5–17 h), lower peak serum concentrations (71±31 ng/ml) and lower liver levels (111±28 ng/g) than the same dose administered intravenously (2 h, 838±0.56 ng/ml, 4258±1006 ng/g, respectively; $P<0.05$ for each comparison) (Brooks et al. 2000). Histologic examination of lungs and trachea showed no focal irritation attributable to the drug after single-dose administration.

[^{3}H]Retinoic acid-treated *human* tracheobronchial epithelial cells contained several polar retinoic acid metabolites that coeluted with 4-oxo-, 4-hydroxy-, and 18-hydroxy-retinoic acid (Kim et al. 2000). 4-Hydroxy-retinoic acid could effectively induce MUC2 and MUC5AC mRNA expression in *human* tracheobronchial epithelial cells, indicating that it is an active retinoid able to induce mucous cell differentiation.

Retinoic acid 5,6-epoxidation was catalysed by *human* blood cells but not by *human* plasma, and purified human haemoglobin also catalysed the epoxidation of retinoic acid to 5,6-epoxyretinoic acid (Iwahashi et al. 1985).

7.2.5.1.2 Retinol

All-*trans*-retinol showed the highest **antioxidant activity** (Livrea et al. 1992). Tesoriere et al. (1993) initiating the oxidation of 315 mM methyl linoleate in methanol by 1.5 mM of the lipid-soluble 2,2'-azobis (2,4-dimethylvaleronitrile) achieved distinct short, distinct inhibition periods, linearly dependent on the concentration (8 and 16 μM) of the added all-*trans*-retinol, after which lipid peroxidation occurred at a rate similar to that observed in its absence. When 2,2'-azobis (2,4-dimethylvaleronitrile) was incorporated into the lipid bilayer, the oxidation of multilamellar phosphatidylcholine liposomes was inhibited by 3.3 μM α-tocopherol for 140 min, and by 3.3 μM all-*trans*-retinol for 220 min. Because of its polyenoic chain, all-*trans*-retinol mobility in the liposome lipid bilayer may be higher than that of α-tocopherol that penetrates the lipid bilayer with its saturated phytil tail. This circumstance would strongly favour interactions of all-*trans*-retinol with lipid radicals and could explain the higher antioxidant potency of all-*trans*-retinol when radicals are generated inside the lipid bilayer, as compared to α-tocopherol (Takahashi et al. 1989).

The role of vitamin A in maintaining **mucus-secreting** epithelia depends upon the ability of the vitamin to facilitate the secretion of mucopolysaccharides or glycoproteins from epithelial cells by an action on the plasma membrane of these cells.

Synthesis of **keratins** 5, 6, 14, 16, and 17 was greatly reduced in tracheobronchial tissue from *patients* undergoing surgery for bronchial obstruction or from autopsy specimen excised within 24 h after death or *rhesus monkeys* aged 2 to 23 y cultured in MEM + 1 μM retinol (Huang et al. 1994). Synthesis of keratins 7, 8, 10, 13, 15, 18, and 19, however, was slightly enhanced. These changes were also reflected at the mRNA level as demonstrated by cell-free translation and by cDNA cloning of *human* keratin genes based on differential hybridisation.

7.2.5.2 Tocopherols

Vitamin E of natural origin consists of a group of compounds, namely α-, β-, γ-, ε-, ζ_1-, ζ_2-, and η-tocopherol. Of these compounds *RRR* α-tocopherol has the highest vitamin potency *in vivo* (Witting 1980). The α-tocopherol molecule is composed of a chroman head and a phytyl side chain. It is generally believed that the phytyl chain intercalated with the fatty acid residues of phospholipids, while the chroman head – responsible for the antioxidant effect – faces the cytosol, although the chroman ring is still located in the hydrophobic zone of the lipid bilayer (Erin et al. 1988). In the antioxidant activity of vitamin E, a radical ($R^\bullet$) abstracts a hydrogen atom from the aromatic hydroxyl group of the chroman heard (ArOH) rather than from a polyunsaturated fatty acid, and a chromanoxyl radical is formed ($ArO^\bullet$).

$$R^\bullet + ArOH \longrightarrow RH + ArO^\bullet \qquad [54]$$

The chromanoxyl radical is fairly stable, due to delocalization of the unpaired electron. The oxygen in

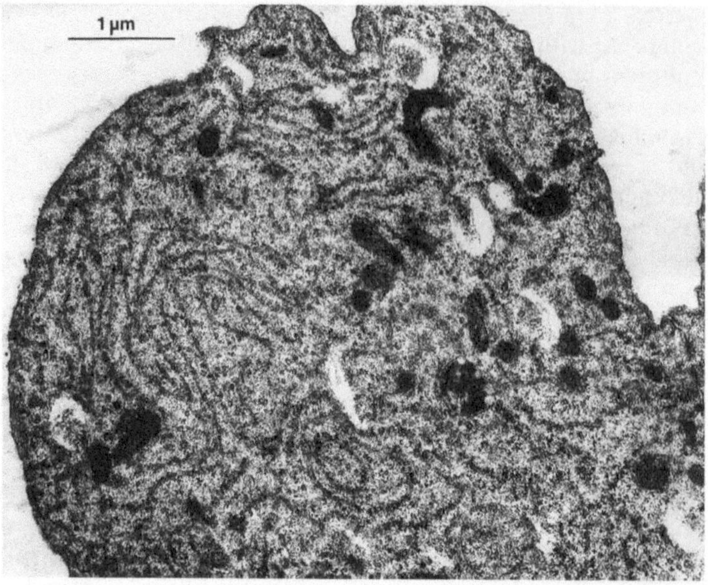

Fig. 76. Plenty of endoplasmic reticulum and polysomes in the apical part of a Clara cell from a female *rat* (breeder: Winkelmann, Borchen-Kirchborchen), which received daily intragastric applications of 82 mg retinol palmitate per kg body weight × day, 5 days per week from April 17 to June 20, 1967 for a total of 42 days. Fixed on June 20, 1967 under methitural anaesthesia by intratracheal instillation of 2.5 % glutaraldehyde in phosphate buffer (pH 7.4) before opening the thorax. After washing in phosphate buffer the tissue was postfixed with 1 % osmium tetroxide in phosphate buffer for 2 h. Contrasted en bloc for 12 h with 0.5 % uranyl acetate in 70 % ethanol. Embedded in a 2 : 8 mixture of methyl and butyl methacrylate. Sectioned at 50 nm. Lead citrate after REYNOLDS (1963). Plate 12/07

the heterocyclic chroman ring is fixed in such a position that there is considerable overlap between the 2p-type orbital of the lone electron pair of the oxygen and the aromatic π-system (BURTON and INGOLD 1981, BURTON et al. 1985). This permits stabilisation of the chromanoxyl radical by interaction of the unpaired electron with a lone pair of oxygen. Thus the degree of delocalization of the free radical is enhanced.

The lipid peroxyl radical (LOO•) can be scavenged by tocopherol (TOH) as follows:

$$LOO^{\bullet} + TOH \longrightarrow LOOH + TO^{\bullet} \qquad [55]$$

The initial phase is termed lag phase time (t_{inh}) or inhibition period.

$$V_{inh} = \frac{d\,[LOOH]}{dt} = \frac{k_p\,[LH]R_i}{n\,k_{inh}\,[TOH]} \qquad [56]$$

$$t_{inh} = \text{length of lag phase} = \frac{n\,[TOH]}{R_i} \qquad [57]$$

The factor n is defined as the number of peroxyl radicals LOO• trapped by each molecule of antioxidants. For vitamin E the value is 2, since both vitamin E and vitamin E radical (tocopheryl radical) trap LOO•. The length of the lag phase is inversely proportional to the rate R_i, by which the initiating radicals are formed.

By confocal laser microscopy and flow cytometry KOLLECK et al. (2002) showed that type II cells accumulate protein-labelled high-density lipoprotein

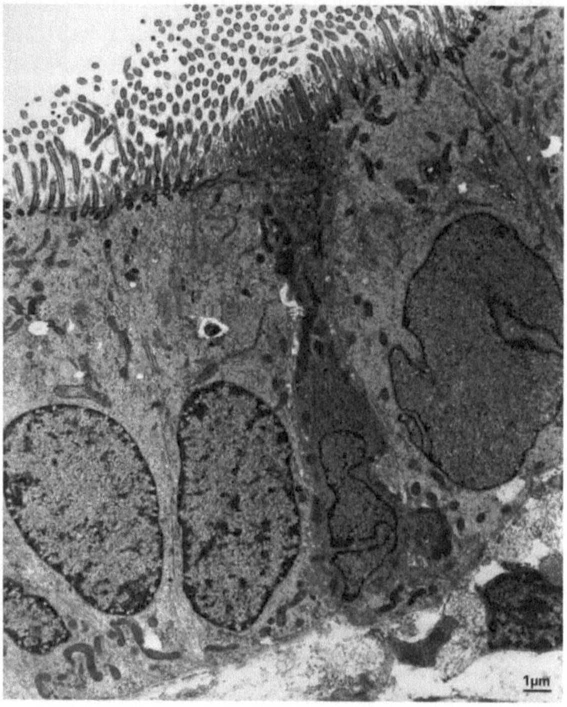

Fig. 77. Brush cell from the bronchiolar epithelium of a female *rat* (breeder: Winkelmann, Borchen-Kirchborchen) which received daily intragastric applications of 82 mg retinol palmitate per kg body weight × day, 5 days per week from April 17 to June 20, 1967 for a total of 42 days. Fixed on June 20, 1967 under methitural anaesthesia by intratracheal instillation of 2.5 % glutaraldehyde in phosphate buffer (pH 7.4) before opening the thorax. After washing in phosphate buffer the tissue was postfixed with 1 % osmium tetroxide in phosphate buffer for 2 h. Contrasted en bloc for 12 h with 0.5 % uranyl acetate in 70 % ethanol. Embedded in a 2 : 8 mixture of methyl and butyl methacrylate. Sectioned at 50 nm. Lead citrate after REYNOLDS (1963). Plate 3601

particles. Vitamin E depletion in *rats* increased high-density lipoprotein particle uptake in alveolar type II cells and the expression of megalin. The expression of cubilin did not change. Refeeding with vitamin E reversed high-density lipoprotein particle uptake and megalin expression. Long-time incubation of type II cells with 12-*O*-tetradecanoylphorbol-13-acetate reduced high-density lipoprotein holoparticle uptake and megalin expression.

Cubilin is a membrane protein that does not contain typical transmembrane sequences and convincing evidence has been presented that cubulin and megalin interact in concert to mediate particulate uptake of high-density lipoprotein as well as albumin absorption. Megalin contains a short transmembrane sequence (Moestrup et al. 1998, Hammad et al. 2000, Zhai et al. 2000). Both cubilin and megalin are expressed in the lung (Kounnas et al. 1994, Zheng et al. 1994)

Pope et al. (2000) extracted the vitamin E metabolites from urine of *rats* fed ^{14}C-labelled α-tocopherol and a healthy *human* volunteer after a single oral dose of 300 mg of d_6-α-tocopheryl acetate. They have identified α-tocopheronolactone and the 2,5,7,8-tetramethyl-2-(2'-carboxyethyl)-6-hydoxychroman metabolites derived from α-, δ-, and γ-tocopherol. In addition they have tentatively identified a novel group of vitamin E metabolites, which are related to the 2,5,7,8-tetramethyl-2-(2'-carboxyethyl)-6-hydoxychromans but have three extra carbons in the side chain.

7.2.5.3 Pantothenic Acid

Pantothenic acid is one of the three substrates needed to synthesise coenzyme A. It is phosphorylated to 4'-phosphopantothenic acid by the action of pantothenate kinase (Abiko et al. 1972). The formation of 4'-phosphopantetheine is a two-step process in which 4'-phosphopantothenic acid and L-cysteine are first converted to 4'-phosphopantothen-

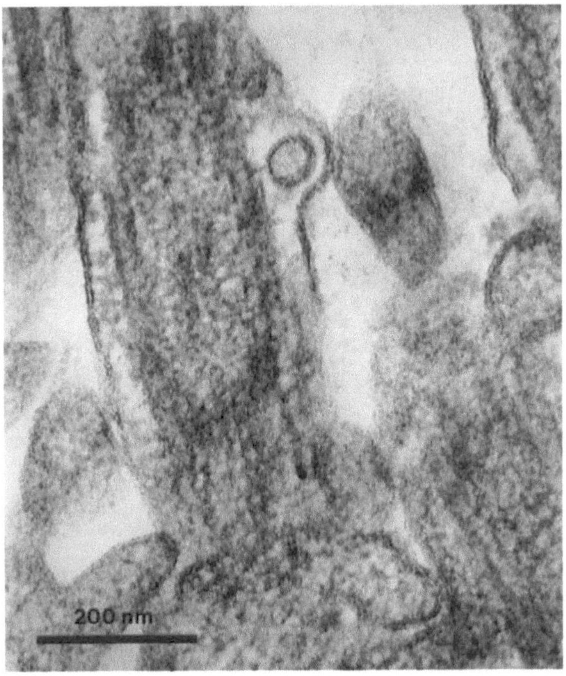

Fig. 78. Kinocilia with a vesicular component in the outer zone alongside the axonema. Bronchiolar epithelial cell of a female *rat* (breeder: Winkelmann, Borchen-Kirchborchen) treated with 150 mg DL-α-tocopherol acetate per kg body weight × day, 5 days per week. The colloidal aqueous solution was injected intamuscularly from April 12 to June 20, 1967 for a total of 45 days. Fixed under methitural anaesthesia by intratracheal instillation of 2.5 % glutaraldehyde in phosphate buffer (pH 7.4) before opening the thorax. After washing in phosphate buffer the tissue was postfixed with 1 % osmium tetroxide in phosphate buffer for 2 h. Contrasted en bloc for 12 h with 0.5 % uranyl acetate in 70 % ethanol. Embedded in a 2 : 8 mixture of methyl and butyl methacrylate. Sectioned at 50 nm. Lead citrate after Reynolds (1963). Plate 31/10

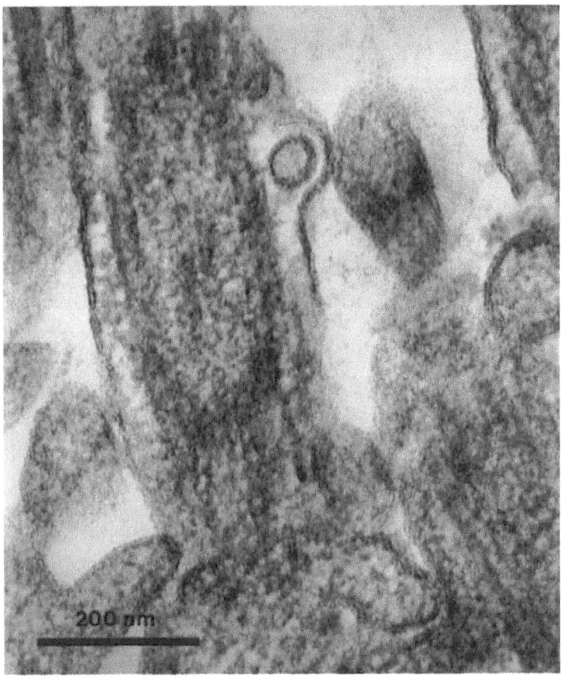

Fig. 79. Bronchiolar kinocilium showing membrane bound vesicles in the outer zone alongside the axonema. Bronchiolar epithelium of a female *rat* (breeder: Winkelmann, Borchen-Kirchborchen) treated for 5 days per week with 150 mg DL-α-tocopherol acetate per kg body weight × day. The colloidal aqueous solution was injected intramuscularly from April 12 to June 20, 1967 for a total of 45 days. Fixed under methitural anaesthesia by intratracheal instillation of 2.5 % glutaraldehyde in phosphate buffer (pH 7.4) before opening the thorax. After washing in phosphate buffer the tissue was postfixed with 1 % osmium tetroxide in phosphate buffer for 2 h. Contrasted en bloc for 12 h with 0.5 % uranyl acetate in 70 % ethanol. Embedded in a 2 : 8 mixture of methyl and butyl methacrylate. Sectioned at 50 nm. Lead citrate after Reynolds (1963). Plate 31/12

oyl-cysteine by formation of a peptide linkage followed by decarboxylation of the cysteine. In the final steps in the pathway, 4'-phosphopantetheine is adenylated to form dephospho-coenzyme A, and dephospho-coenzyme A is phosphorylated at the 3' position of ribose to form coenzyme A.

Pantothenyl alcohol, an alcohol analogue corresponding to pantothenic acid, has a pantothenic acid-like activity in mammals (Pfaltz 1943, Jürgens and Pfaltz 1944, Hegsted 1948, Weiss et al. 1950, Lih et al. 1951, Ludovici and Axelrod 1951, Hirabayashi 1964, Inagawa 1964). The oxidation of the alcohol to pantothenic acid *in vivo* (Burlet 1944, Schmidt 1945, 1947, Drekter et al. 1948, Rubin 1948, Rubin et al. 1948, 1950, Crokaert 1953, Braekkan 1955, Yamawaki and Ishida 1961) and *in vitro* (Abiko et al. 1969) was confirmed by many investigators. Abiko et al. (1969) purified the enzyme responsible for the first step of the oxidative reactions from rat liver and evidenced its identity with alcohol:NAD oxidoreductase (EC 1.1.1.1).

Daily intraperitoneal administration of pantothenic acid (100 mg/kg) for 5 days conferred significant protection against the peroxidative actions of a 0.5 ml/kg intraperitoneal dose of CCl_4 in *rats* (Nagiel-Ostaszewski and Lau-Cam 1990). Lipid peroxidation by incubation of Ehrlich ascites tumour cells with $FeCl_2$ + H_2O_2 was partly prevented by preincubation with D-pantothenate, 4'-phosphopantothenate, D-pantothenol, or pantethine (Slyshenov et al. 1995). *Rats* exposed to γ radiation from a ^{60}Co source, receiving 0.25 Gy at weekly intervals were protected from the deleterious effects by 26 mg pantothenol/kg × day given for 2 d before each irradiation (Slyshenov et al. 1998).

Homopantothenic acid, which is not a coenzyme A precursor, did not exert a protective effect against reactive oxygen species (Slyshenov et al. 1995). Slyshenov et al. (1995, 1995, 1998, 1999) therefore proposed that the observed protective effects of pantothenic acid and some of its derivatives against cell and tissue injury may be due to stimulation of biosynthesis of coenzyme A and glutathione.

Schafer and Buettner (2001) discuss how the redox state of the glutathione disulphide- glutathione couple (GSSG/2GSH) can serve as an important indicator of redox environment. There are many redox couples in a cell that work together to maintain the redox environment; the GSSG/2GSH couple is the most abundant redox couple in the cell. Changes of the half-cell reduction potential (E_{hc}) of the GSSG/2GSH couple appear to correlate with the biological status of the cell: proliferation $E_{hc} \approx$ -240 mV; differentiation $E_{hc} \approx$ -200 mV; or apoptosis $E_{hc} \approx$ -170 mV. These estimates can be used to more fully understand the redox biochemistry that results from oxidative stress.

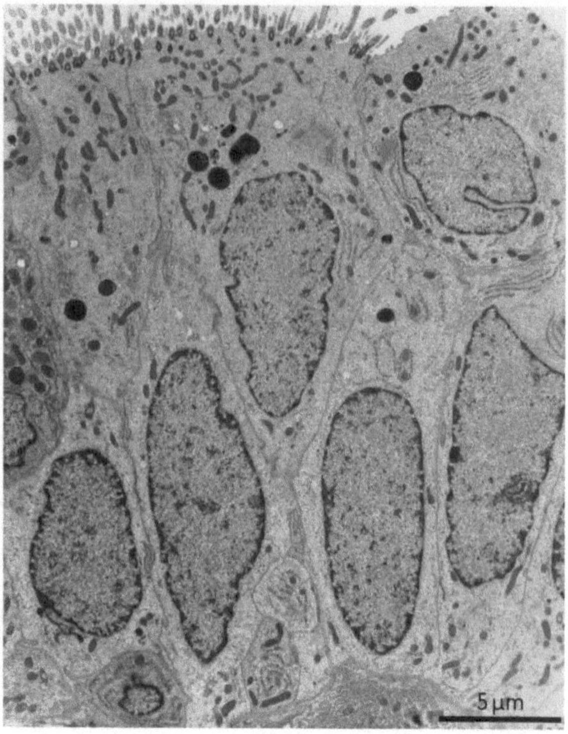

Fig. 80. Two ciliated and a Clara cell from a bronchiole (block B5) of a female *rat* (breeder: Winkelmann, Borchen-Kirchborchen) medicated for 5 days per week with 125 mg dexpanthenol in a 5 % aqueous solution (Bepanthen) per kg body weight × day from April 17 to August 14, 1967 for a total of 82 days. Fixed on August 14, 1967 under methitural anaesthesia by intratracheal instillation of 2.5 % glutaraldehyde in phosphate buffer (pH 7.4) before opening the thorax. After washing in phosphate buffer the tissue was postfixed with 1 % osmium tetroxide in phosphate buffer for 2 h. Contrasted en bloc for 12 h with 0.5 % uranyl acetate in 70 % ethanol. Embedded in a 2:8 mixture of methyl and butyl methacrylate. Sectioned at 50 nm. Lead citrate after Reynolds (1963). Plate 3599

7.2.6 Bradykinin

Bradykinin (H-Arg-Pro-Pro-Gly-Phe-Ser-Pro-Phe-Arg-OH) releases bronchodilator •NO from the airway epithelium (Ricciardolo et al. 2000). Bradykinin-induced •NO release was higher in tracheal than in main bronchial segments of the *guinea pig*. The selective bradykinin B_2 receptor antagonist D-Arg^0-[Hyp^3,Thi^5,D-Tic^7,Oic^8]bradykinin (1 μM) inhibited •NO release induced by a submaximum concentration of bradykinin (1 μM).

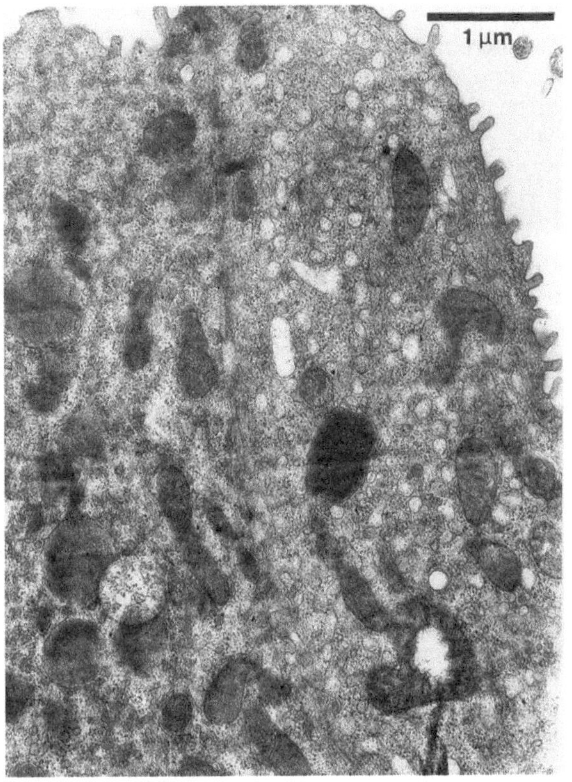

Fig. 81. Vacuolar transformation of the endoplasmic reticulum in the apical region of a type 2 Clara cell from a ca. 235 g female *rat* (breeder: Winkelmann, Borchen-Kirchborchen) treated for 5 days per week with 125 mg dexpanthenol in a 5 % aqueous solution (Bepanthen) per kg body weight × day from April 17 to August 14, 1967 for a total of 82 days. Fixed on August 14, 1967 under methitural anaesthesia by intratracheal instillation of 2.5 % glutaraldehyde in phosphate buffer (pH 7.4) before opening the thorax. After washing in phosphate buffer the tissue was postfixed with 1 % osmium tetroxide in phosphate buffer for 2 h. Contrasted en bloc for 12 h with 0.5 % uranyl acetate in 70 % ethanol. Embedded in a 2 : 8 mixture of methyl and butyl methacrylate. Sectioned at 50 nm. Lead citrate after REYNOLDS (1963). Plate 9/07

7.2.7 Expectorants

7.2.7.1 Bromhexine

Bromhexine (4 mg Bisolvon given 3 times per day) after several days induced both serous and muciparous cells of *human* bronchial glands to increase their secretory activity (GIESEKING and BALDAMUS 1968). The lysosome-like secretory granules of the serous cells, which are suggested to be enzyme complexes, dissolve after their touch with mucus discharges by the mucous cells.

7.2.7.2 Cistinexine

Cistinexine is a compound derived from cystine that has shown an expectorant action similar to bromhexine in preclinical pharmacological trials (RECORDATI 1989).

7.2.7.3 Ozothin

The secretolytic oxidised oleum terebinthinae "Landes" dispensation, Ozothin (7.5 ml intravenously for 2 days) stimulated the serous cells in the sero-mucous acini of the *human* bronchial glands (BAUER 1973).

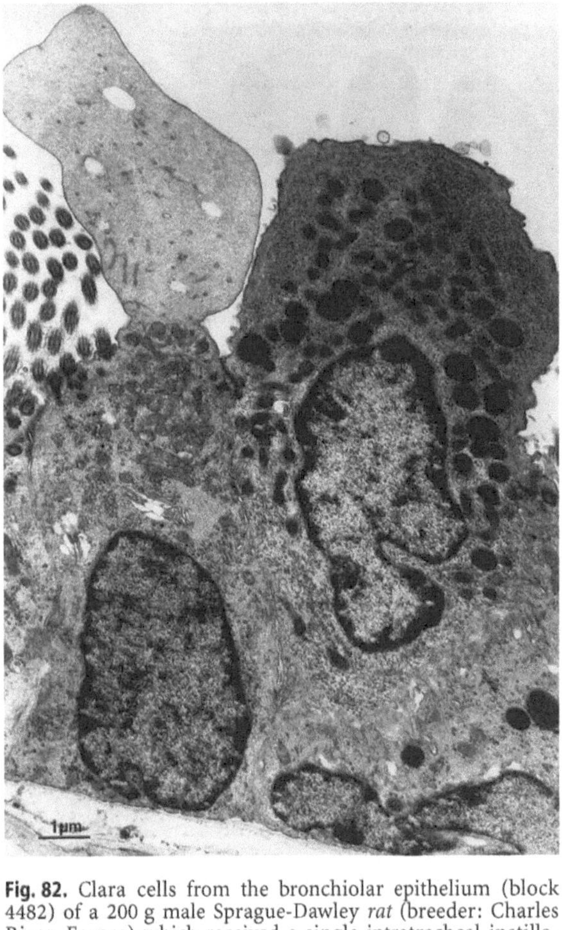

Fig. 82. Clara cells from the bronchiolar epithelium (block 4482) of a 200 g male Sprague-Dawley *rat* (breeder: Charles River, France) which received a single intratracheal instillation of 50 mg anatase (titanium dioxide P 25, Degussa, Frankfort on the Main) suspended in 0.5 ml saline. Animal medicated for 4 days with intraperitoneal injection of 15 mg carbocromene in 10 % methylcellulose per kg body weight × day. On July 25, 1978, under pentobarbital anaesthesia the lung was fixed by intratracheal instillation of 2.5 % glutaraldehyde in 0.1 M sodium cacodylate buffer (pH 7.4) before opening the thorax. Postfixation with 1 % osmium tetroxide in sodium cacodylate buffer. Embedded in Epon 812 and sectioned at 50 nm. Lead citrate and uranyl acetate. Plate 4139

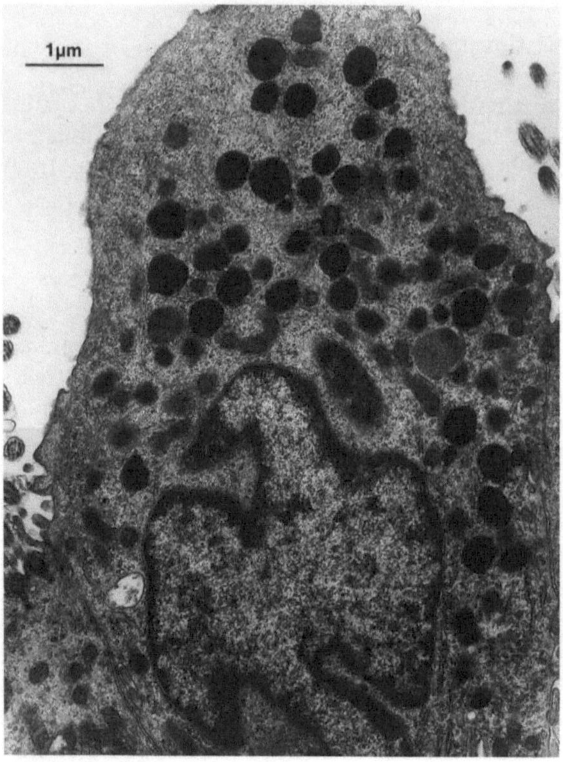

Fig. 83. Type 4 Clara cell from the bronchiolar epithelium (block 4482) of a 200 g male Sprague-Dawley *rat* (breeder: Charles River, France), which received a single intratracheal instillation of 50 mg anatase (titanium dioxide P 25, Degussa, Frankfort on the Main) suspended in 0.5 ml saline. Animal medicated daily with intraperitoneal injection of 15 mg carbocromene in 10 % methylcellulose per kg body weight × day. On July 25, 1978, under pentobarbital anaesthesia the lung was fixed by intratracheal instillation of 2.5 % glutaraldehyde in 0.1 M sodium cacodylate buffer (pH 7.4) before opening the thorax. Postfixation with 1 % osmium tetroxide in sodium cacodylate buffer. Embedded in Epon 812 and sectioned at 50 nm. Lead citrate and uranyl acetate. Plate 4137

7.2.8 Carbocromene

In the *rat*, carbocromene (also known as Cassella 4489, Abbott-27053, and Intensain hydrochloride) was hydrolysed more rapidly by pulmonary or hepatic homogenates than by cardiac or renal ones (Klarwein and Nitz 1965). As to *in vitro* conversion by *human* plasma, the half-life was calculated to be 2.68 ± 0.96 min (Martin and Wiegand 1970).

7.2.9 Isoniazide

Isoniazide (isonicotinic acid hydrazide) has been reported to get oxidised by the peroxidase system forming an excited product (Zinner et al. 1977, Albano and Tomasi 1987). Except for the two hydrazones, all metabolites are ineffective in tuberculostasis. Therefore the portion of isoniazid, which is acetylated should be minimised by medicating *p*-amino salicylic acid or other salicylates simultaneously (Bartmann 1974, Jungbluth and Reimers 1981). The mechanism of acetate transfer involves two consecutive steps: acetylation of the enzyme *N*-acetyltransferase (EC 2.3.1.5), which is found in the cytosol of mammalian cells, by acetyl-CoA giving acetyl-acetyltransferase, which then transfers the acetyl group to an acceptor substrate as was shown by the isolation of [^{14}C]acetyl-acetyltransferase, which donated its labelled acetate group to isoniazid (Steinberg et al. 1971). Besides, continuous administration of 1-acetyl-2-nicotinoyl-hydrazine (0.4 % in the drinking water) also induced lung tumours in Swiss *mice* (Toth and Shimizu 1973). Using 50–100 mM phenyl-*tert*-butyl-

Fig. 84. Bronchiolar Clara cells of a male *mouse* (M 91/61) which was medicated with intraperitoneal injections of 100 mg isoniazide (Neoteben) per kg body weight every week and was sacrificed under methitural anaesthesia after 41 days of experimentation. Fixed in Carnoy's fluid (ethanol – chloroform – glacial acetic acid). Paraffin embedding. Trichrome stain after Goldner. Eyepiece Leitz Periplan 4×; objective Leitz Pl Oil 100/1.32

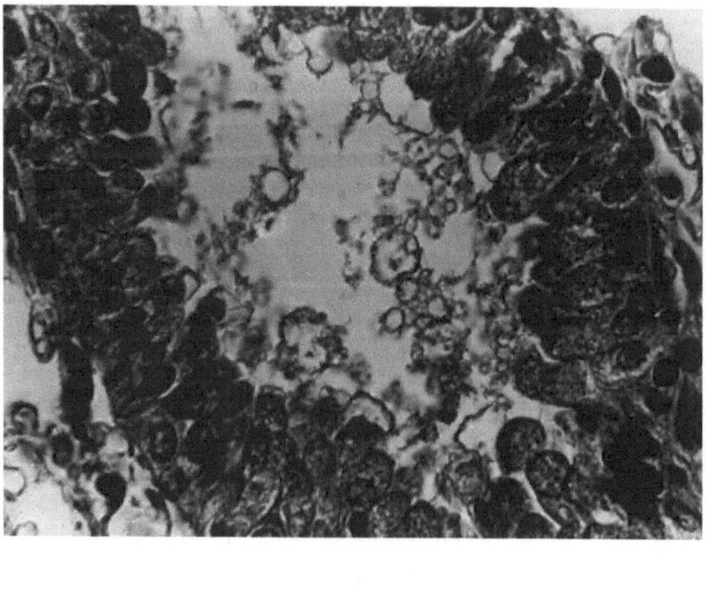

nitrone as the stable spin trap, SINHA (1987) found that perfusion of the male Sprague-Dawley *rat* liver with isoniacid (5–10 mM) resulted in the formation of some carbon-centred radical which was shown to be the acetyl radical. Employing the ESR spin trapping technique *in vivo* to detect the formation of the 5,5-dimethyl-1-pyrroline *N*-oxide/haemoglobin thiyl free radical adduct in the blood of *rats* following administration of isoniazide at its LD_{50} level (650 mg/kg), MAPLES et al. (1988) could not detect any ESR signal.

Aminoguanidine, *N,N'*-diaminoguanidine, isoniazid, hydroxylamine and hydrazine are weak substrates for horseradish peroxidase in the reaction:

$$\text{Enzyme-Fe(III)} + H_2O_2 \longrightarrow \text{Enzyme}^{\bullet}\text{Fe(IV)=O} + H_2O \quad [112]$$

In the reaction with hydrogen peroxide [112] the enzyme itself is oxidised and returns to its ground state by oxidising a suitable substrate (reactions [113] and [114]).

$$\text{Enzyme}^{\bullet}\text{-Fe(IV)=O} + \text{substrate} \longrightarrow \text{H-Enzyme-Fe(IV)=O} + \text{substrate} \quad [113]$$

$$\text{H-Enzyme-Fe(IV)=O} + \text{substrate} \longrightarrow \text{Enzyme-Fe(III)} + H_2O + \text{substrate}^{\bullet} \quad [114]$$

The same compounds inhibited the reaction of *Mycobacterium tuberculosis* peroxidase-catalase with hydrogen peroxide, and hyroxylamine was found to be a weak substrate for this enzyme (BRIMNES et al. 1999).

7.2.10 Streptozotocin-Induced Diabetes

Streptozotocin (75 mg per kg body weight) markedly decreased the number of dense oval granules in the Clara cells of the *rat* (PLOPPER and MORISHIGE 1979). When present, they were usually smaller in size, but generally retained their electron density. The rode-shaped granules present in the apical portion of the Clara cell increased in number: six or more were observed in 28 of the 57 cells examined. Therapy with insulin successfully reversed changes produced by streptozotocin-induced diabetes.

7.3 Toxicology

7.3.1 Ciliated Cells

STOCKINGER et al. (1989) divided the pathological findings on the ciliary apparatus as observed in the respiratory mucosa of *humans* into three groups of arrest levels, referring to the physiological development of the ciliary apparatus:

aplasia of the ciliary apparatus (GÖTZ and STOCKINGER 1983),
disturbances in kinetosome differentiation (LUNGARELLA et al. 1985),
disorders of the peripheral ciliary shaft.

In **primary ciliary dyskinesia** patients showed a range of ultrastructural defects, such as reduced numbers of dynein arms – including inner, outer and both inner and outer dynein arms – as well as excentrically located central microtubules, extra single microtubules, and displaced peripheral doublet microtubules and compound cilia (DE IONGH and RUTLAND 1995). Other microtubule defects seen commonly included absence of one or more peripheral doublets, supernumerary doublets, and absence of the "B" tubule in peripheral doublets. Abnormal orientation of cilia was also commonly seen in patients with primary ciliary dyskinesia.

Focal swelling of the ciliary membrane have been produced in the respiratory mucosa after exposure to heat (VON MECKLENBURG et al. 1974), ionising radiation (BALDETORP et al. 1977), ozone (SATO et al. 1976) and influenza virus (IRIVANI et al. 1978). Local sheet-like outgrowth of the ciliary membrane, named ciliary knob genesis, may be a fixation artefact depending on the osmotic pressure of the buffer systems (BRUNK et al. 1975, ERICSSON et al. 1978), different physiological conditions at the time of fixation (EHLERS and EHLERS 1978) or may indeed by a degenerative phenomenon (DALEN 1981).

Hydrogen peroxide in concentrations from 16 to 64 μM significantly decreased mucociliary transport in frog palate preparations (MACCHIONE et al. 1998). Palates submitted to 64 μM of H_2O_2 returned to their baseline level of mucociliary transport within 30 min of recovery in Ringer's solution.

Exposure of *human* nasal ciliated epithelium to reactive oxidants generated by the enzymatic xanthine – xanthine oxidase superoxide/hydrogen peroxide ($O_2^{\bullet -}/H_2O_2$) and glucose – glucose oxidase H_2O_2-generating systems, or to reagent H_2O_2 or hypochlorous acid (HOCl) resulted in significant alterations in ciliary beating (FELDMAN et al. 1994).

Ni^{2+} in an *in vitro* model of the trachea of the Syrian golden *hamster* significantly decreased ciliary activity in concentrations as low as 0.011 mM (ADALIS et al. 1978). Validation of the *in vitro* results occurred when *hamsters* were exposed to a $NiCl_2$ aerosol at concentrations of 100 to 275 μg of nickel per m^3.

In beagle *dogs* exposed over a period of 290 days, 22.5 h/d and 7 days/week, at an S(IV) concentration of 0.3 mg m^{-3} equivalent to **SO_2** concentration of 0.6 mg m^{-3}, focal lesions of the tracheal epithelium were characterised by the disappearance of cilia and presence of microvilli-like short rods (TAKENAKA et al. 1992). The basal bodies frequently presented abnormal features.

In southwestern metropolitan Mexico City *children* repeatedly exposed to a complex mixture of **air pollutants,** nasal biopsies showed ciliary abnormalities, including absent central microtubules, supernumerary central and peripheral tubules, ciliary microtubular discontinuities, and compound cilia (CALDERÓN-GARCIADUEÑAS et al. 2001).

In *rabbits* exposed to polluted air, the ciliated cells, less numerous than in normal cases, showed an evident decrease in the number and size of the cilia, exposing apical microvilli (GULISANO et al. 1995).

Diesel exhaust (mass median diameter $<10\ \mu$m) at 300 $\mu g/m^3$ associated with concentrations of NO_2 of 1.6 ppm, $^{\bullet}NO$ of 4.5 ppm for 1 h, CO of 7.5 ppm, total hydrocarbons of 4.3 ppm, formaldehyde of 0.26 mg/m^3, and 4.3×10^6 suspended particles/ml in healthy *human* volunteers exposed for 1 h in bronchial epithelium gained by biopsies taken 6 h later had induced a 198 % median relative increase in area staining for interleukin-8 and a 229 % median relative increase in growth-regulated oncogene-α (SALVI et al. 2000).

Diesel exhaust particles were taken up by *human* airway epithelial cells *in vitro* and induced a time- and dose-dependent membrane damage (BOLAND et al. 1999). They induced a time-dependent increase in interleukin-8, granulocyte-macrophage colony-stimulating factor, and interleukin-1β release. This inflammatory response occurred later than phagocytosis, and its extent seems to depend on the content of adsorbed organic compounds because carbon black had no effect on cytokine release.

Acetaldehyde reduced *bovine* bronchial epithelial cell ciliary motility by inhibiting cAMP-mediated phosphorylation (WYATT et al. 1997). When ciliary axonemata were phosphorylated *in vitro*, 1 mM acetaldehyde (a subciliastatic concentration) inhibited the cAMP-dependent phosphorylation of a 38 kDa phosphoprotein.

Acrolein induced a cell death pathway in *human* bronchial epithelial cells, which retain key features of apoptosis, as indicated by phosphatidylserine externalisation and DNA fragmentation (NARDINI et al. 2002). Acrolein-induced apoptosis was associated with depletion of cellular GSH and intracellular generation of oxidants. Supplementation of cells with either α-tocopherol or ascorbic acid was found to strongly inhibit acrolein-induced apoptosis and to prevent the increase in the generation of intracellular oxidants, although GSH depletion was unaffected. Moreover, recovery of cellular GSH levels after acrolein exposure was enhanced following either α-tocopherol or ascorbic acid supplementation.

In *rats* exposed to **tobacco smoke** for 4 hours per day for two weeks, 5 days each week, and for 6 weeks, 4 days each week, respectively, the cytoplasmic ground substance of the ciliated cells was increased from 46 per cent. to about 60 per cent. of the cell area and contributed most of the cell hypertrophy seen in both groups (JEFFERY and REID 1981). There was no evidence of cell oedema. The nucleus maintained its normal size and, consequently, the proportion of the cell occupied by the nucleus fell from 34 per cent. to 25 per cent. There was little change in the proportion of the cell occupied by organelles (normally 10–15 per cent.) but the mitochondria were hypertrophied, being obviously increased in length but not in width. Mitochondria in ciliated cells of controls had a mean maximum length of 0.75 μm; in those animals exposed to smoke alone this was increased to 1.6 μm ($P<0.01$) and in those exposed to smoke with an anti-inflammatory drug, phenylmethyloxadiazole, to the tobacco to 1.2 μm ($P<0.05$). The maximum length found was 3 μm, this in an animal exposed to smoke alone. Some branching of mitochondria was normal, but this was more often found in the ciliated cells of the animals exposed to tobacco smoke without phenylmethyloxadiazole. Microvilli were different from the normal in both groups: they were significantly increased in length ($P<0.01$) but not in width. Cilia were normal for size, structure and concentration.

Exposure to cigarette smoke increased the permeability of *human* bronchial epithelial cells from biopsy material obtained from never-smokers who had normal pulmonary function, smokers with normal pulmonary function, and smokers with chronic obstructive pulmonary disease, but the most marked effect was observed in the latter group (RUSZNAK et al. 2000). Compared with exposure to air, exposure to cigarette smoke led to a significantly increased release of IL-1β and soluble intracellular adhesion molecule-1 (sICAM-1) from cultures of the never-smoker group (mean 250.0 % increase in ILL-1β and mean 175.3 % increase in sI-CAM -1 24 h after exposure) and chronic obstructive pulmonary disease group (mean 383.3 % increase in IL-1β and mean 97.4 % increase in sICAM-1 24 h after exposure). In contrast, cigarette smoke exposure did not influence significantly the release of either mediator from the cells of smokers with normal pulmonary function. Levels of intracellular reduced glutathione (GSH) were significantly higher in cultures of bronchial epithelial cells derived from smokers, both those with normal pulmonary function and those with chronic obstructive pulmonary disease, compared with cultures from healthy never-smokers. Exposure to cigarette smoke significantly decreased the concentration of intracellular GSH in all cultures. However, the fall in intracellular GSH was significantly greater in cells from patients with chronic obstructive pulmonary disease (mean 72.9 % decrease) than in cells from never-smokers (mean 61.4 % decrease; $P < 0.048$) or smokers with normal pulmonary function (mean 43.9 % decrease; $P < 0.02$).

Scanning electron microscopy of the regio olfactoria of *rats* exposed to cigarette smoke for 10 min three times a day during three months showed lesions of the cilia, the microvilli and the sensory cells (Ortuğ 2002).

Exposure of *human* respiratory tract epithelial cell (HBE1) cultures to **nitrite** (1 mM) at pH 7.4, for up to 5 h resulted in a significant increase in levels of hypoxanthine and xanthine over the first 2 h followed by a slow decrease in levels from 3 to 5 h (Spencer et al. 2000). The increase of these products was time dependent with levels of xanthine peaking at ~3.8 nmol/mg DNA over baseline levels after a period of 2 h. In contrast, levels of the oxidatively modified base products measured did not change significantly at any time point. Analysis of DNA strand breakage by alkaline unwinding-fluorescence detection showed that there was a significant amount of strand breakage in cells exposed to 1 mM nitrite for 60 min or grater at pH 6.0 and pH 7.4.

Gossypol acetic acid (10^{-4} M) at 60 min induced extensive swelling of explanted adult *rabbit* tracheal ciliary cell mitochondria (Duckett et al. 1986). Cristae were greatly altered and mitochondrial matrix appeared to be almost entirely washed out. In many cases the mitochondria were distorted, with no evidence of their original filamentous nature remaining. The cytoplasm appeared to contain increased levels of endoplasmic reticulum and the number of Golgi bodies also seemed to be increased. However, cellular membranes were intact and no significant change in ciliary or microvillar structure was evident. Cellular junctions also appeared to be normal.

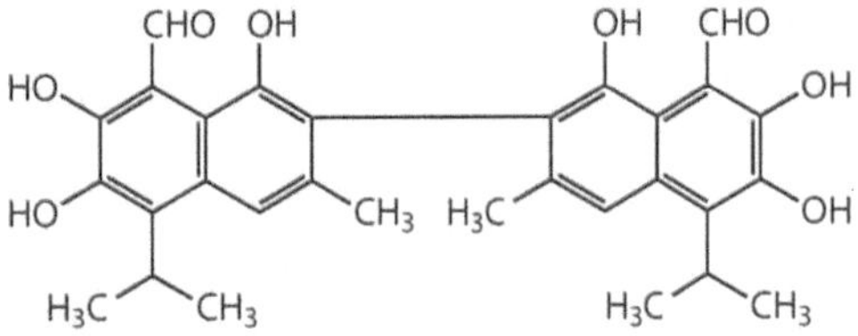

Gossypol ($C_{30}H_{30}O_8$) [115]

In the Syrian golden *hamster*, 10 instillations (3 times per week) of 5 mg **haematite** particles (93 % by weight < 5 μm in diameter) suspended in 0.9 % NaCl solution brought about a loss of ciliated cells and broad areas of abnormal, enlarged nonciliated cells with roughened or wrinkled surfaces (Port et al. 1973). The tracheobronchial epithelium returned to normal 7 weeks after completion of the treatment.

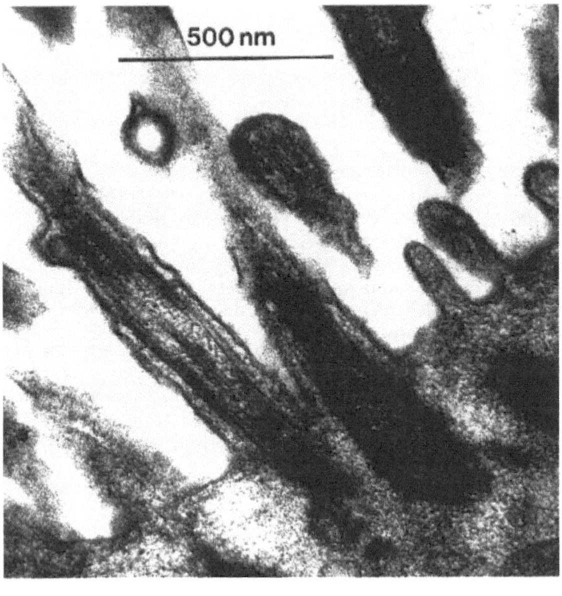

Fig. 85. Bronchiolar ciliated epithelial cell showing membrane bound vesicles between the axonema and the ciliary membrane. Lung of a female *rat* (No. 4 breeder: Winkelmann, Borchen-Kirchborchen) which had inhaled 10 mg powdered Grängesberg magnetite/m3 4 h per day, 5 days per week from August 24 to October 19, 1967 for a total of 40 days. Fixed on January 15, 1968 under methitural anaesthesia by intratracheal instillation of 2.5 % glutaraldehyde in phosphate buffer (pH 7.4) before opening the thorax. Postfixation with 1 % osmium tetroxide in phosphate buffer (pH 7.4). Contrasted en bloc for 12 h with 0.5 % uranyl acetate in 70 % ethanol. Embedded in a 2:8 mixture of methyl and butyl methacrylate. Sectioned at 50 nm. Lead citrate after Reynolds (1963). Film 33131

Fig. 86. Bronchiole (block 498) from a female Wistar *rat* (No. 182/103 E) medicated for 5 months with daily oral applications of 100 mg *N*-benzhydryl-*N*'-*p*-hydroxybenzyl-piperazine HCl per kg body weight × day. Under pentobarbital anaesthesia (30 mg/kg), the animal was perfused from the abdominal aorta with 2.5 % glutaraldehyde in 0.1 M sodium cacodylate buffer (pH 7.4). Postfixation with 1 % osmium tetroxide in sodium cacodylate buffer. Embedded in Epon 812 and sectioned at 50 nm. Lead citrate and uranyl acetate. Plate 30338

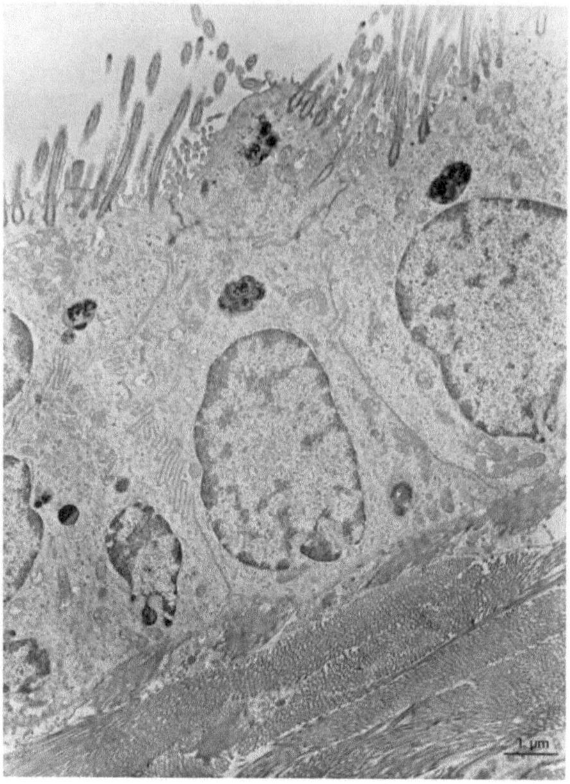

Fig. 87. Bronchiolar cilia showing membrane bound vesicles in the outer zone alongside the axonema. Lung (block S2) of a female white *rat* (breeder: Winkelmann, Borchen-Kirchborchen) which had inhaled 20 mg quartz-free clayey shale dust/m^3 (particles < 5 µm in size origin: Erda seam, Fürst Leopold-Baldur mine at Harvest Dorsten) in Polley's (1963, 1965) dust channel 4 h per day, 5 days per week. Fixed under methitural anaesthesia by intratracheal instillation of 2.5 % glutaraldehyde in phosphate buffer (pH 7.4) before opening the thorax. Postfixation with 1 % osmium tetroxide in phosphate buffer. Contrasted en bloc for 12 h with 0,5 % uranyl acetate in 70 % ethanol. Embedded in Epon 812. Sectioned at 50 nm. Lead citrate after Reynolds (1963). Plate 3593

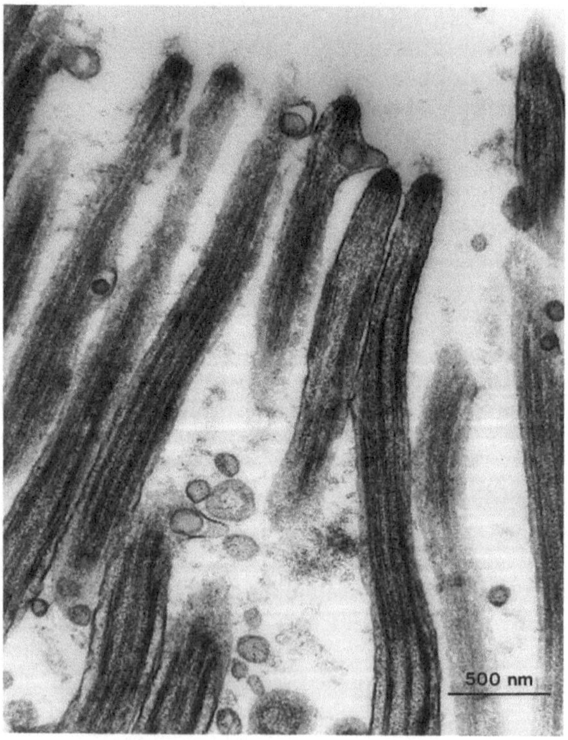

7.3.2
Serous Cells

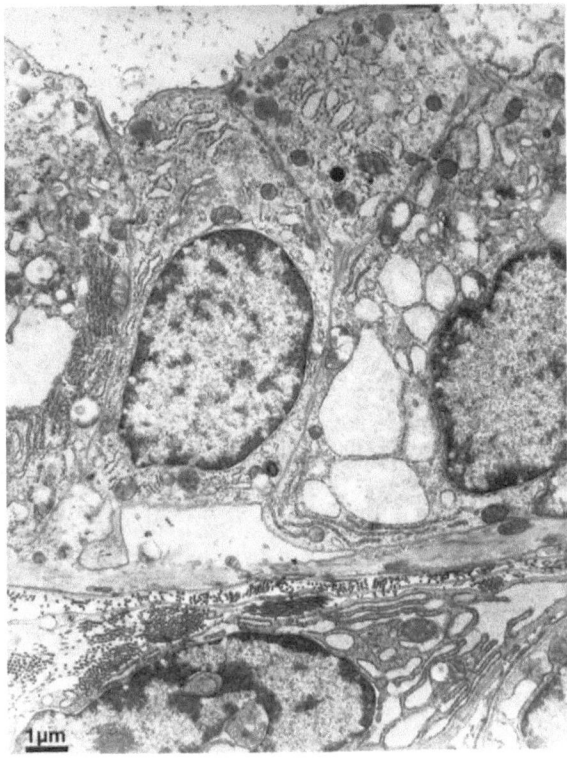

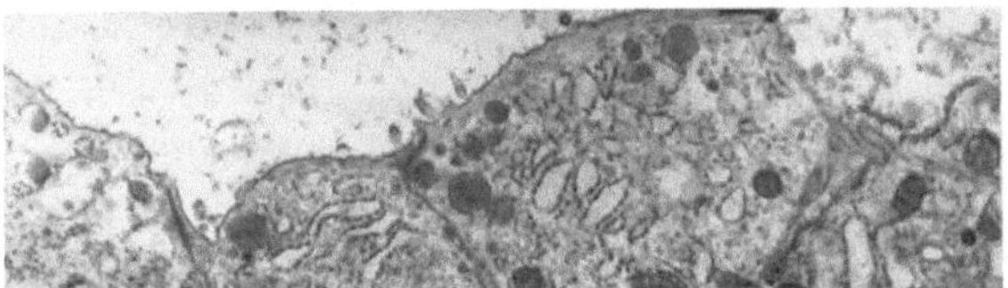

Fig. 88. Dilatation of the rough endoplasmic reticulum in the non-ciliated cells of the trachea (block 467) from a male *rat* (No.6/104 E) which inhaled 71.5 mg trimethylol melamine (Madurit) dust/l for a total of 32 h 30 min on 4 consecutive days. Fixed under pentobarbital anaesthesia (30 mg/kg) by intratracheal instillation of 2.5 % glutaraldehyde in 0.1 M sodium cacodylate buffer. Postfixation with 1 % osmium tetroxide in sodium cacodylate buffer. Embedded in Epon 812 and sectioned at 50 nm. Lead citrate and uranyl acetate. Plate 4052

7.3.3
Clara Cells

The presence of cytochrome P 450 monooxygenases within Clara cells has been correlated with their susceptibility to metabolically activated pulmonary cytotoxicants. Species-specific differences in susceptibility to cytotoxicants exist between Clara cells located in the trachea and bronchi and those in distal bronchioles (PLOPPER et al. 1992).

Analogy for hepatocyte and Clara cell is given by:

1. Both cell types are situated at the entrance of foreign materials into the body, the liver for the substances transported from the intestine via the portal vein, and the Clara cell for xenobiotics inhaled.
2. Both are able to transform xenobiotics by the cytochrome P_{450}-type enzyme system induced in the endoplasmic reticulum.

While liver microsomes from male *rats* metabolise xenobiotics fast than liver microsomes from female *rats*, enzyme preparations from *rat* lung (benzphetamine demethylase, *p*-chloro-*N*-methylaniline demethylase, benzpyrene hydroxylase, UDP-glucuronyltransferase) and gut did not show sex differences in metabolism of several drug substrates (CHHABRA and FOUTS 1974).

Liver and lung NADPH-cytochrome P-450 reductases (EC 1.6.2.4) of well bled male Akkaraman *sheep* appear to have a similar kinetic and spectral properties (ISCAN and ARINÇ 1988). Both reductases supported aniline 4-hydroxylation and ethylmorphine *N*-demethylation reactions to the same extent in the constituted systems. However, *sheep* lung reductase appeared only 36.5 and 14.8 % as effective in catalysing benzo[*a*]pyrene reaction as an equivalent amount of reductase from liver in the presence of liver cytochrome P-450 and 3MC-treated *rat* liver cytochrome P-448, respectively.

In the presence of daunorubicin (200 µM) the NADPH-cytochrome P-450 reductase in *mouse* lung microsomes was approximately 60 % of the liver microsomal activity (MIMNOUGH et al. 1983).

Ozone (1.0 ± 0.1 ppm) transiently increased IL-6 mRNA showing peak levels at 4 h of ozone exposure in both wild-type and Clara cell secretory protein -/- *mice* (MANGO et al. 1998). Differential metallothionein and IL-6 mRNA expression demonstrated increased ozone-induced oxidative stress associated with Clara cell secretory protein deficiency.

Cobalt aerosols inhaled by *rats* at a concentration of 2.12 ± 0.55 mg/m^3 for 5 h per day on 4 consecutive days induced Clara cell hypertrophy with abundant rough endoplasmic reticulum and electron lucent glycogen granules which were normally absent in adult *rats* (KYONO et al. 1992). At 8 days after exposure, the long strands of rough endoplasmic reticulum had disappeared.

Methylene chloride is biotransformed in the microsomal cytochrome P450 mixed function oxidase pathway to carbon monoxide (KUBIC and ANDERS 1975, 1978) and carbon dioxide (GARGAS et al. 1986). The cytosolic glutathione *S*-transferase pathway leads to the formation of formaldehyde and subsequently carbon dioxide (AHMED and ANDERS 1976, 1978). The rate of the methylene chloride metabolism by the glutathione *S*-transferase pathway is significantly greater in the mouse than in the *rat* (GREEN 1989). *In vitro* observations have confirmed this finding, with the cytosolic glutathione *S*-transferase activity in *mouse* liver and lung an

Table 22. Clara cells: toxicity

	Anorganic compounds
O_3	Stephens et al. (1974), Evans et al. (1976), Bassett et al. (1988), Mango et al. (1997)
NO_2	Evans et al. (1976), Foster et al. (1985), Gordon et al. (1986)
$CdCl_2$	Henderson et al. (1979), Fortoul et al. (1983), Martin and Witschi (1985)
Cadmium acetate	Lìg et al. (2002)
Co (20 nm particles)	Kyono et al. (1992)
^{210}Po	Kennedy et al. (1977)
	Chlorinated hydrocarbons
CH_2Cl_2	Foster et al. (1992)
CCl_4	Longo et al. (1978), Boyd et al. (1980)
1,1-Dichloroethylene	Gram et al. (1983), Krijgsheld et al. (1984), Forkert et al. (1985)
Trichloroethylene	Forkert et al. (1985), Forkert and Troughton (1987), Odum et al. (1992)
Hexachlorocyclopentadiene	Rand et al. (1982)
Chlorobenzene	Brittebo and Brandt (1984)
	Brominated hydrocarbons
Bromotrichloromethane	Lungarella et al. (1987)
Bromobenzene	Forkert (1985), Casini et al. (1986), Becher et al. (1989)
	Polycyclic aromatic hydrocarbons
Naphthalene	Mahvi et al. (1977), Tong et al. (1981), Warren et al. (1982), Gram et al. (1983), Buckpitt et al. (1986), Kanekal et al. (1990), Chichester et al. (1994), Fanucchi et al. (1997), Van Winkle et al. (1999), West et al. (2000)
2-Methylnaphthalane	Buckpitt et al. (1986)
4-Nitroquinoline-1-oxide	Terao and Otsu (1973), Ito (1985)
3-Methylindole	Huang et al. (1977), Turk et al. (1984), Becker et al. (1985), Nichols et al. (1989), Thornton-Manning et al. (1993)
	Endogenous polyamines
Spermidine	Foster et al. (1990)
	Nitrosamines
N-Diethylnitrosamine	Reznik-Schüller (1976)
N-Dibutylnitrosamine	Reznik-Schüller (1976)
N-Nitrosomorpholine	Reznik-Schüller (1977, 1978)
Nitrosoheptamethylenimine	Reznik-Schüller and Lijinsky (1979), Reznik-Schüller and Hague jr. (1980)
N-Ethyl-*N*-nitrosourea	Sato and Kauffman (1980)
2,2'-Dihydroxy-di-*n*-propylnitrosamine	Rao and Reddy (1980)
4-(Methylnitrosamino)-1-(3-pyridyl)-1-butanone	Belinsky et al. (1988, 1990, 1991), Alaoui-Jamali et al. (1990)
	Furan derivatives
2-Methylfuran	Franklin and Boyd (1978)
3-Methylfuran	Boyd et al. (1978)
3-Hydroxymethylfuran-n-ethylcarbamate	Cottrell et al. (1984)
4-Ipomeanol (1-[3-furyl]-4-hydroxypentanone; NSC-349438)	Boyd (1977, 1980), Doster et al. (1983), Durham et al. (1985), Newton et al. (1985), Christian et al. (1989), Li and Castleman (1990)
	Diesel exhaust
Diesel-exhaust	Rabovsky et al. (1984)
	Organic phosphates
O,O,S-Trimethyl-phosphorthiolate	Imamura et al. (1983), Gandy et al. (1984, 1987), Durham et al. (1988)
O,O,S-Trimethyl-phosphordithiolate	Dinsdale et al. (1984), Konno et al. (1984)
	Organic metallocarbonyl compounds
Methylcyclopentadienyl-manganese-tricarboxyl	Hakkinen and Haschek (1982), Haschek et al. (1982)
Piperonyl-butoxide pre-treatment + $MnC_6H7(CO)_3$	Haschek et al. (1982)
	Phenobarbital
Phenobarbital	Dees et al. (1980)
	Pyridinium compounds
Paraquat	Etherton and Gresham (1979), Masek and Richards (1990)

order of magnitude greater than that detected in the same organs of the *rat* and *hamster* (Green 1989, Reitz et al. 1989). In the lung the capacity for methylene chloride metabolism decreased in the order *mouse, hamster, rat* (Reitz et al. 1989). In *mouse* Clara cells methylene chloride inhalation (4000 ppm, 6 h/d, 5 d/week for 13 weeks) caused decreases in the levels of microsomal enzymes as assessed by the measurement of ethoxycoumarin *O*-dealkylation (Foster et al. 1992). At the end of each 5-days exposure period large decreases in this enzyme activity could be detected (79 % at day 5, 83 % at day 40, and 40 % at day 89). This response was progressive up to day 89, after which evidence of recovery could be detected. These effects followed the pattern of events established from nitro blue tetrazolium staining of cultured Clara cells in that after nonexposure periods varying degrees of recovery were detected (days 8, 43, and 92). The measurement of aldrin epoxidation in Clara cells from toxicated animals followed a similar trend, although those microsomal enzyme activities reflected by ethoxycoumarin *O*-dealkylation levels appeared more sensitive to methylene chloride. Again the trend towards recovery by the end of the study was illustrated by the activities of these enzymes.

Carbon tetrachloride is suggested to be metabolised by the P_{450} system to give the trichloromethyl radical, a carbon centred radical:

$$CCl_4 \xrightarrow[\text{Cytochrome } P_{450}]{e^-} {}^{\bullet}\mathbf{CCl_3} + Cl^- \qquad [116]$$

ESR and high-performance liquid chromatography experiments with electrochemical detection for analysis of the major redox form of the α-phenyl-*N*-*tert*-butylnitrone (PBN)$^-CCl_3$ adduct of the liver extract of *rats* pre-treated with 4-methylpyrazole (for induction of cytochrome P450 2E1) given CCl_4 intraperitoneally revealed that the *in vivo* spin trapping of ${}^{\bullet}CCl_3$ with PBN leads to a preferential formation of the ESR silent PBN$^-CCl_3$ hydroxylamine (Stoyanovsky and Cederbaum 1999).

The trichloromethyl radical might combine directly with biological molecules, causing covalent modification, as well as abstracting hydrogen from membrane lipids, setting off the chain reaction of lipid peroxidation. The most rapid reaction is with molecular oxygen to form the trichloromethylperoxy radical:

$$ {}^{\bullet}\mathbf{CCl_3} + O_2 \longrightarrow \mathbf{CCl_3O_2^{\bullet}} \qquad [117]$$

$CCl_3O_2^{\bullet}$ reacts much more rapidly with arachidonic acid ($k = 6 \times 10^6$ at pH 7), promethazine, ascorbate, thiol compounds, and the tyrosine and tryptophan residues of proteins than does the trichloromethyl radical (Halliwell and Gutteridge 1989). Formation of the trichloromethylperoxy radical could explain why small amounts of phosgene gas ($COCl_2$) arise by the reactions below:

$$\mathbf{CCl_3O_2^{\bullet}} + \text{lipid-H} \longrightarrow CCl_3O_2H + \mathbf{lipid^{\bullet}} \qquad [118]$$

$$CCl_3O_2H \longrightarrow COCl_2 + HOCl \qquad [119]$$

Mice exposed to the choking agent gas phosgene (8 ppm for 20 min) showed scattered degeneration and necrosis of the terminal airway epithelium, characterized by individual cell necrosis and partial denudation of the bronchiolar and terminal bronchiolar epithelial layers (Duniho et al. 2002).

1,1-Dichloroethylene in *mice* is metabolised by CYP2E1 to its epoxide in a dose-dependent manner (Forkert 1999). Epoxide levels were reduced by pre-treatment with diallyl sulfone, an inhibitor of this P_{450} isozyme. 1,1-Dichloroethylene epoxide is formed *in vivo*, is localised in Clara cells, and correlates with the cytotoxicity manifested in this cell type.

Halogenated aromatic hydrocarbons are metabolised by cytochromes P450 resulting in the formation of an electrophilic intermediate, most commonly depicted as an epoxide, which covalently interacts with tissue macromolecules. The spontaneous isomerization of epoxides to phenols, the conversion of epoxides to dihydrodiols as catalysed by epoxide hydrolases, and the conjugation of epoxides with glutathione, either enzymatically via the glutathione *S*-transferase (EC 2.5.1.18), are alternative pathways that compete with the reaction of epoxides with tissue macromolecules (Daly et al. 1972, Jerina and Daly 1974). As such, these pathways are detoxifying. However, secondary P450-catalysed oxidations of phenols to hydroquinones, followed by their enzymatic and/or uncatalysed oxidation to benzoquinones, represents an additional possibility for bioactivation of benzene derivatives, as quinones may elicit oxidative stress through redox cycling and may also act as electrophiles and arylate macromolecules (Rossi et al. 1986, Powis et al. 1987, Gant et al. 1988, van Ommen et al. 1988, den Besten et al. 1991).

Naphthalene dissolved in corn oil and administered intraperitoneally to young C57BL/6J *mice* selectively damaged Clara cells (Mahvi et al. 1977). At 2 and 3 h after treatment of male, viral antibody-free Swiss Webster *mice* with 200 mg naphthalene/kg, when most Clara cells had early signs of injury, few ethidium homodimer-1 permeable cells were detected (Van Winkle et al. 1999). Many Clara cells had apical membrane blebs that contained abundant, swollen, smooth endoplasmic reticulum and few other organelles. By 6 h after treatment many Clara cells were membrane-permeable.

The toxic effects of naphthalene are mediated by the cytocchrome P450 monooxygenase Cyp2F2 which catalyses the conversion of naphthalene to (1*R*,2*S*)-naphthalene oxide, the principle metabolite responsible for Clara cell injury. In order to determine whether *mice* made tolerant by sublethal naphthalene injection decreased expression of the Cyp2F2 gene, Royce et al. (1996) isolated RNA from minor daughter airways and from liver and measured mRNA for Cyp2F2 and Clara cell secretory protein relative to 18SrNRA by Northern blot and computer digitised photodensitometry. *Mice* were intraperitoneally injected with 200 mg naphthalene per kg body weight × 7 days. Cyp2F2 mRNA expression was significantly decreased at one day (0 vs. 2.1, $P < 0.003$), but did not differ from controls at later times. CCSP mRNA was significantly less than control at 1 (0.17 vs. 4.4, $P < 0.0001$) and 3 (1.1 vs. 3.6, $P < 0.003$) days. These observations suggested that while initial injury by sublethal doses of naphthalene resulted in decreased expression of Clara cell-specific genes, recovery occurred rapidly. By day three, mRNA levels were the same as control values even though previous studies showed that production of 1*R*,2*S* naphthalene oxide remains depressed for at least one week following naphthalene exposure. This suggests that transcription control of Cyp2F2 activity is not likely a major point of regulation for the tolerance that develops after repeated doses of naphthalene. In explants of distal airways of adult *mice* treated with naphthalene, Clara cells exhibited the characteristic cytotoxic responses: cell swelling, formation of cytoplasmic vacuoles, and exfoliation of injured cells into the airway lumen 1–2 days post treatment with naphthalene (Van Winkle et al. 1996).

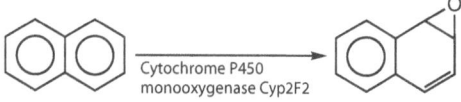

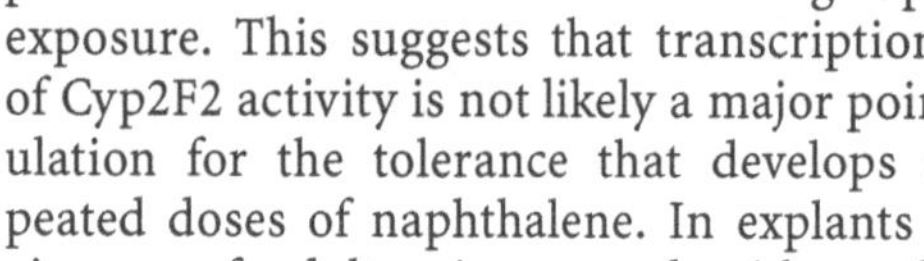

Cytochrome P450 monooxygenase Cyp2F2 catalyses the conversion of naphthalene to (1R,2S)-naphthalene oxide [120]

Wild-type $P450_{cam}$ and all Y96 mutants catalysed the oxidation of naphthalene to give a mixture of 1-naphthol and 2-naphthol, as confirmed by co-elution with authentic samples by gas chromatography (England et al. 1998). There was no evidence of further oxidation of the naphthols to catechols or other compounds.

While bronchiolar epithelium in normal C57BL/6 *mice* contains cells that express high levels Clara cell 10 kDa secretory protein and CYP2F, and exhibit the characteristic pattern of naphthalene injury and repair, Clara cell 10 kDa secretory protein and CYP2F expression was less abundant in transgenic C57BL/g γδ T cell "knock-out" *mice* (Van Winkle et al. 1997). In "knock-out" *mice* 1 to 2 days post injury, injury was not uniform within the terminal bronchiole. Distal ends of the bronchiole contained Clara cell 10 kDa secretory protein positive epithelial cells and were not injured.

Oxidative stress induced by chronic administration of naphthalene (110 mg/kg *rat* × day p.o. in corn oil up to 120 days) resulting in tissue damaging effects as maximum increases in hepatic and brain lipid peroxidation and DNA fragmentation was reported by Bagchi et al. (1998).

No significant difference in *murine* glutathione *S*-transferase (EC 2.5.1.18) α class subunit 3 content of lung homogenates was observed between male Swiss Webster *mice* made naphthalene-tolerant by daily intraperitoneal application of 200 mg naphthalene per kg body weight dissolved in corn oil and control (corn oil only) *mice* (Mitchell et al. 2000).

The **indole** moiety is present in many substances of biological occurrence. Its biotransformation, in most cases, involves an oxidative pathway. In the horseradish peroxidase/H_2O_2 system 2-methylindole and 2,5-dimethylindole showed a two to three orders of magnitude higher chemiluminescence than indole itself (Ximenes et al. 2001). Electronic effects do not seem to be important since neither electron-releasing groups nor electron-withdrawing groups produced effects. On the other hand, 3-methylindole presented low emission.

Indole derivatives autoxidize in the presence of transition metals presumable to nitrogen-centred radicals. 3-Methylindole and serotonin for a substrate in the H_2O_2-peroxidase reaction system incubated with 200 μM GSH took up 48.2 ± 5.6 μM and 63.1 ± 7.1 μM oxygen, respectively (O'Brien 1988). The one-electron oxidation potential of indole derivatives is not known, but indoles slowly reduce cytochrome *c*, which suggests that the potential is less than that for cytochrome *c* (Perez-Reyes and Mason 1981). Gillam et al. (2000) reported the definitive identification of the pigments indigo and indirubin as products of *human* cytochrome P450-catabolized of indole by visible, ^{1}H NMR and mass spectrometry. CYP2A6 from *Escherichia coli* membranes was most active in the formation of these two pigments, followed by CYP2C19 and 2E1. Indoxyl (3-hydroxyindole) was observed as a transient product of CYP2A6-mediated metabolism;

isatin, 6-hydroxyindole, and dioxindole accumulated at low levels. Oxindole was the prominent product formed by CYPs 2A6, 2E1, and 2C19 and was not transformed further. A stable end product was assigned the structure 6*H*-oxazolo[3,2*a*:4,5*b*']diindole by UV, ^1H NMR, and mass spectrometry.

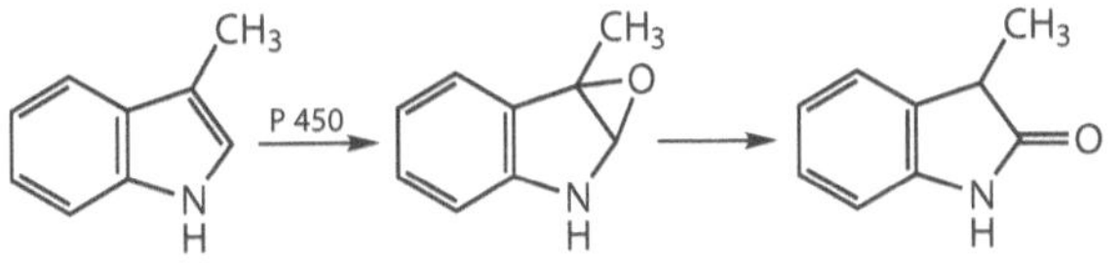

3-Methylindole 2,3-epoxy-3-methylindole 3-methyloxindole [121]

3-Methylindole is species-selective pneumotoxic. *Ruminants* are highly susceptible to 3-methylindole; an intravenous dose of 30–40 mg/kg to a *goat* is usually fatal (CARLSON and BREEZE 1983). The pathways of bioactivation of 3-methylindole to electrophilic reactive intermediates have been partially elucidated by YOST (1989). Using vaccinia-expressed P450 enzymes, the metabolites of [^{14}C]-3-methylindole produced by 14 individual P450s were identified and quantified by high performance liquid chromatography (THORNTON-MANNING et al. 1996). Indole-3-carbinol was produced from incubations of 3-methylindole with only four enzymes. Although 3-methyloxindole was a product of several P450 enzymes, *human* CYP1A2 was most efficient in producing this metabolite. The toxic intermediate of 3-methylindole is believed to be a reactive methylene imine, 3-methyleneindolenine. THORNTON-MANNING et al. (1996) detected this intermediate as its mercapturate adduct, when *N*-acetylcysteine was added to the incubations. 3-Methyleneindolenine was produced by CYP2A6 at a rate of 50.9 ± 8.9 pmol/mg protein $\times$ h and by CYP2F1 at a rate of 205.7 ± 12.5 pmol/mg protein $\times$ h. The *mouse* 1a-2 and *rabbit* 4B1 enzymes produced the reactive intermediate in amounts that exceeded that of the *human* 2F1 enzyme by 1.4-fold and 1.9-fold, respectively. The toxicity of 3-methylindole is believed to be due to covalent binding of a P450 generated intermediate to critical pulmonary proteins.

It is likely that the 2,3-epoxide of 3-methylindole is a precursor of 3-methyloxindole and that the epoxide may bind to DNA (THORNTON-MANNING et al. 1996).

The factors that determine the partitioning between pathways of desaturation and hydroxylation or epoxidation are most likely controlled by the protein environment and not by differences in the reactivity of the porphyrin-$[FeO]^{3+}$ species (JONES et al. 1990).

The addition of *N*-acetylcysteine to incubation of 3-methylindole with *rabbit* Clara cells and macrophages resulted in markedly decreased amounts of 3-methyloxindole formed, as compared to incubations that did not include *N*-acetylcysteine (THORNTON-MANNING et al. 1993).

Involvement of prostaglandin H synthase in the metabolism of 3-methylindole has been demonstrated in *in vitro* systems containing either *ram* seminal vesicle microsomes or *goat* lung microsomes. *Ram* seminal vesicle microsomes, which lack NADPH-dependent mixed function oxidase activity, were used as a rich source of prostaglandin H synthase complex (FORMOSA et al. 1988). Incubation with 3-methylindole resulted in an increase of prostaglandin H synthase activity as indicated by an increase rate of oxygen consumption. This effect was arachidonic acid-dependent and was inhibited by the cyclooxygenase inhibitor indomethacin. Addition of 3-methylindole resulted in a concentration-dependent increase in prostaglandin H synthase-catalysed prostaglandin biosynthesis. Electron spin studies demonstrated the presence of a 3-methylindole free radical generated from the metabolism of 3-methylindole by horseradish peroxidase, a model system for prostaglandin H synthase hydroperoxidase. Testing *goat* lung, FORMOSA and BRAY (1988) after adding 3-methylindole found a pronounced increase in prostaglandin H synthase activity as indicated by both the initial rate and the total oxygen consumption. Incubation of various indolic compounds with *goat* lung microsomes showed that only 3-methylindole was able to generate a free radical in the NADP-dependent microsomal system, as tested by spin trapping (Kubow et al. 1983). Radical formation in the microsomal incubation of 3-methylindole depended on all three components of the system since omission of NADPH, or 3-methylindole, or thermal inactivation of the microsomes in a steam bath for 20 min, prevented the appearance of the nitrogen free radical signal. The splitting constants were the same as those seen with potassium superoxide (KO_2) incubations with 3-methylindole. ACTON et al. (1991) showed that acetyl salicylic acid (two doses of 150 mg/kg body weight per dose) was protective against 3-methylindole (100 mg/kg body weight)-induced lung disease when giver before, but not after 3-methylindole dosing.

4-Ipomeanol is activated in many species by pulmonary cytochrome P-450 mixed function oxidase enzymes to a high reactive metabolite that binds covalently to cellular macromolecules to induce pulmonary toxicity (BOYD et al. 1978, BOYD 1980, BOYD and DUTCHER 1981, STATHAM and BOYD 1982, JONES et al. 1983). There are species and sex variations in target organ alkylation that account for hepatic toxicity in *hamsters* and *birds* (Buckpitt and Boyd 1978) and for renal toxicity in male *mice* (Boyd and Dutcher 1981). Experimental studies on pulmonary toxicity in *rodents* (Doster et al. 1983), *lagomorphs, dogs*, and *calves* (Li and Castleman 1990) indicate that the toxin induces necrosis of the Clara cells and pulmonary oedema.

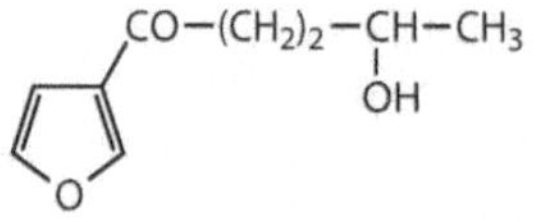

4-Ipomeanol (NSC-349438) [122]

Richards et al. (1990) examined 26 pneumotoxic agents (concentrations ranging from 10^{-7} M to 10^{-3} M) for their ability to reduce the attachment efficiency of functionally competent *mouse* Clara cells isolated by the technique of Oreffo et al. (1990) and calculated TD_{50} values (the amount of toxin required to reduce normal attachment efficiency by 50%). With the possible exception of some halogenated hydrocarbons, the simple toxicity test *in vitro* correlated well with the known effects of the bronchiolar necrotic agents *in vivo*. For 13 compounds studied there was a direct correlation between TD_{50} values *in vitro* and LD_{50} values (mostly oral) *in rodents in vivo*, the correlation coefficient of the regression line being 0.783.

Pulmonary microsomal preparations were much more active than hepatic preparations in mediating the formation of reactive electrophilic products from 4-ipomeanol (Wolf et al. 1982). Both of the *rabbit* pulmonary cytochrome P-450 isozymes, P-450_I and P-450_{II}, were active in the metabolism of 4-ipomeanol. In the assay for covalent binding to protein, P-450_I was slightly more active than P-450_{II} at high substrate concentrations and significantly more active at low substrate concentrations.

Coumarin (*cis-o*-coumarinic acid lactone, 1,2-benzopyrone) oral dosages ≥150 mg/kg caused selective injury to Clara cells in the distal bronchiolar epithelium of the B6C3F1 *mouse* (Born et al. 1998). At 12 h after a single dose of 200 mg/kg, Clara cell swelling was apparent along with the onset of necrosis and bronchiolar epithelial disorganisation. At 24–48 h, necrotic Clara cells were observed sloughed into the lumina of the terminal bronchioles, with concomitant thinning of the epithelium and flattening of the remaining ciliated cells. By 72–96 h, there was epithelial hypertrophy and hyperplasia, and by 7 days after dosing, the Clara cells had regenerated and the bronchiolar epithelial architecture appeared nearly normal. 3,4-Dihydrocoumarin, which is not a *mouse* carcinogen, did not cause Clara cell injury when dosed to *mice* at 800 mg/kg. This finding suggested, because 3,4-dihydrocoumarin lacks a 3,4-double bond, that bioactivation of coumarin to a 3,4-epoxide intermediate may contribute to *mouse* lung Clara cell toxicity.

Coumarin is metabolised by Cyp2A6 (Yamano et al. 1990). Urinary 7-hydroxycoumarin has been the basis for measuring *in vivo* expression of Cyp2A6 (Cholerton et al. 1992).

Transition metal organometallic compounds such as manganese, chromium, and iron carbonyls by the intraperitoneal route, and nickel by inhalation (*mice*) or intravenously (*rats*), induced selective necrosis of the Clara cells (Haschek et al. 1982). The pulmonary toxicity of methylcyclopentadienyl manganese tricarbonyl, representative of this group of compounds, was enhanced by pretreatment with piperonyl butoxide, an inhibitor of the mixed-function oxidase system.

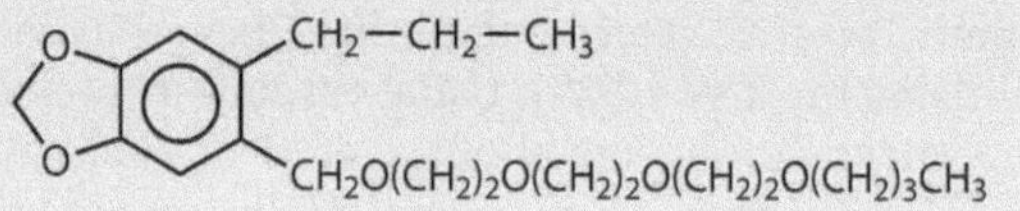

Piperonyl butoxide [41]

Halatek et al. (2000) found that the decrease in serum Clara cell protein of shipyard welders chronically exposed to manganese contained in welding fumes correlated negative with welding fume levels, expressed by ambient air Mn concentration (r^2 = 0.47). Moreover, reduced synthesis and/or removal of Clara cell protein by the Clara cells negatively correlated with Mn level in blood (r^2 = 0.52). On the other hand, in heavily exposed welding fume workers (>1 mg Mn/m^3), increased serum Clara cell protein levels were observed, which correlated positively with Mn air levels (r^2 = 0.85) and Mn concentrations in urine (r^2 = 0.38).

Lipopolysaccharide (10 μg in 0.2 ml saline) introduced into the trachea of the Wistar *rat* within 8 to 12 h in the Clara cells induced well-developed smooth and rough endoplasmic reticula and many free ribosomes (Ooi et al. 1994). Many Clara cells showed a large apocrine-like protrusion filled with an amorphous substance. At 24 h, Clara cells took a stratified arrangement, suggesting enhanced proliferation of progenitor cells and their differentiation into Clara cells.

7.3.4 Goblet Cells

Lipopolysaccharide, a major constituent of the cell walls of gram-negative bacteria, and thus present in a wide variety of occupational and general environmental noxious agents, when inhaled by *guinea pigs* at about 5 mg *Escherichia coli* LPS/kg caused a time-dependent decrease in mucus score, with the maximal response being from 542±49 to 92±20 arbitrary units ($P < 0.001$) after 3 h, which was accompanied by an increase in the number of neutrophils in the tracheal mucosa (TAMAOKI et al. 1997). The LPS-induced mucus discharge was inhibited by oral pre-treatment for 7 days with clarithromycin and erythromycin in a dose-dependent manner (5 mg and 10 mg/kg×day), whereas amoxicillin and cefaclor had no effect. Each dose of clarithromycin and erythromycin, but not of amoxicillin or cefaclor, likewise attenuated the LPS-induced recruitment of neutrophils. The results suggested that LPS stimulates goblet cell secretion (demonstrated with alcian blue and periodic acid-Schiff) and neutrophil accumulation in the airways and that macrolides may be of value in protecting against neutrophil-associated hypersecretion.

SHIMIZU et al. (1996) produced hypertrophic and metaplastic changes of goblet cells in *rat* nasal respiratory epithelium by intranasal instillation of endotoxin. The number of goblet cells increased and that of nongranulated secretory cells decreased time-dependently after endotoxin instillations. Mitotic rates examined after a 6-h colchicine metaphase blockade were very low at any time point studied, and cell division did not play a major role in this process.

Residual oil fly ash induced cytotoxicity and mucin secretion by *guinea pig* tracheal epithelial cells via an oxidant-mediated mechanism (JIANG et al. 2000). Of the soluble transition metals contained therein (Ni, Fe, V), only V individually, or combinations of the metals containing V, provoked secretion. It was suggested that residual oil fly ash enhanced mucin secretion and generated toxicity *in vitro* to airway epithelium via a mechanism(s) involving generation of oxidant stress, perhaps related to the depletion of cellular antioxidant capacity. Deleterious effects of inhalation of residual oil fly ash in the respiratory tract *in vivo* may relate to these cellular responses. Vanadium, a component of residual oil fly ash, may be important in generating these reactions.

7.3.5 Brush Cells

7.3.6 Neuroendocrine (APUD) Cells

Although immunohistochemical staining has been invaluable in defining the abundance of neuroendocrine cells in experimental and disease conditions, quantitative comparison among animals of various ages or experimental groups has been limited in a study by JOAD et al. (1995), significant differences in neuroendocrine cell number per length of airway basal lamina were noted with long-term exposure to environmental tobacco smoke. However, localised changes in neuroepithelial body frequency could not be determined. AVADHANAM et al. (1997) fixed *rat* lungs by intratracheal instillation of ethanol/glacial acetic acid (99:1) at 30 cm pressure for 1 hour. After microdissection of the infracardiac lobe and placing the microdissected airways in 70 % ethanol a polyclonal antibody against calcitonin gene-related peptide as a specific neuroendocrine marker was used to identify and quantify neuroendocrine bodies. Whole-mount airway preparations allow for the analysis and comparison of larger sample sizes per experimental group without labor-intensive approaches.

In intact *rabbits*, airway hypoxia evoked exocytosis of serotonin-containing granules from the basal cytoplasm (LAUWERYNS et al. 1977, 1978). In cell cultures from foetal *rabbit* lungs CUTZ et al. (1993) found that the effects of hypoxia could be stimulated by Ca^{2+} ionophore A-23187. YOUNGSON et al. (1993) have identified an oxygen-binding protein on the endocrine cell surface membrane and demonstrated the existence of an associated oxygen-sensitive K^+ channel. Exposure of foetal *rat* lung organ cultures to HCO_3^- suggested that pulmonary endocrine cells may respond to hypercapnia by releasing bioactive peptides like calcitonin gene-related peptide, thus stimulating afferent nerves and altering patterns of ventilation and perfusion (EBINA et al. 1997).

After naphthalene (300 mg/kg i.p.) treatment of *mice* and epithelial repair, neuroepithelial bodies were significantly increased along the walls of the airways as well as on branch point ridges (PEAKE et al. 2000). REYNOLDS et al. (2000) provided the following evidence that the neuroepithelial body microenvironment serves as a source of airway progenitor cells that contribute for focal regeneration of the airway epithelium:

- nascent Clara cells and neuroepithelial bodies localise to the same spatial domain;
- within neuroepithelial body, both Clara cell secretory protein- and calcitonin gene-related peptide-immunopositive cells are proliferative;
- neuroepithelial body microenvironment of both steady-state and repairing lung includes cells that are dually immunopositive for Clara cell secretory protein and calcitonin gene-related peptide, which were previously identified only within the embryonic lung;

Table 23. Pulmonary APUD cell proliferation induced by toxicants

Toxicant	Species	Author(s)
N-Diethylnitrosamine	*Hamster*	Reznik-Schüller (1976), Linnoila et al. (1984)
	Rat	Kleinerman et al. (1981)
	Rabbit	Huntrakoon et al. (1989), Hung et al. (1991)
N-Dibutylnitrosamine	*Hamster*	Reznik-Schüller (1977)
N-Nitrosomorpholine	*Hamster*	Reznik-Schüller (1977)
Asbestos (UICC Chrysotile B)	*Rat*	Johnson and Wagner (1980)
Asbestos (Chrysotile, Crocidolite)	*Rat*	Johnson et al. (1980)
Asbestos (UICC Crocidolite)	*Rat*	Johnson (1987)
Urban ambient air (Yokohama)	*Rat*	Ito et al. (1989)
Ozone	*Rat*	Ito et al. (1994)

- neuroepithelial bodies harbour variant Clara cells deficient in cytochrome P_{450} 3F2-immunoreactive protein.

Groups of 40–50 neurone specific enolase immunoreactive cells, forming microtumours, were seen in the epithelium of larger airways of *rats* exposed to asbestos fibres up to 3 years (Cole et al. 1982).

Pulmonary neuroendocrine cell 'hyperplasia' may result after carcinogen treatment in hamsters given supranormal oxygen (Sunday et al. 1994).

To Wilson et al. (1992) it seems unlikely, that the products of pulmonary endocrine cells play any role in the pathogenesis of or response to asbestos-induced pulmonary fibrogenesis in *man*.

CD10/neutral endopeptidase inhibition augmented pulmonary neuroendocrine cell hyperplasia in *hamsters* treated with diethylnitrosamine and hyperoxia (Willett et al. 1999).

7.4 Dust Clearance

The mucociliary clearance mechanism is generally recognised as a defence against infectious agents and a means to remove inhaled particulate matter. Inhaled particles reach different locations in the lung depending on their size. Above 5 μm, the conducting airways filter dusts, where ciliated and mucous cells cooperate and waft the particles upwards.

Balásházy and Hofmann (1995) explored the effects of asymmetry in airway diameter, branching angle, and flow division on both inspiratory and expiratory particle deposition patterns for various particle sizes and flow rates.

Besides the intensity and duration of the exposure, the period of life at the beginning of the dusty job may be important in the susceptibility of a person (Worth and Muysers 1967). At high levels of exertion children had a greater percentage of total deposition in the tracheobronchial region than do adults for equivatent levels of activity (Schum et al. 1994). Deposition rate depends on the deposition per breath as well as on tidal volume and breathing frequency (Schiller-Scotland et al. 1994). For inhalation risk assessments, the number of particles deposited per unit time (deposition rate), rather than the deposition per breath, has to be taken (Schiller-Scotland et al. 1996). Total deposition of particles was generally increased in patients suffering from obstructive pulmonary disease as compared with healthy subjects (Schiller-Scotland 1997). With increasing particle size deposition rose, while the differences between the two collectives decreased.

Whereas much is known about clearance of particles from the airways, less is known about the interaction between the surface of the particles and the cellular substrate of the clearing mechanism.

The surface of airway epithelium is covered by **airway surface liquid**, the physical properties of which are determined by active ions and water transport. Baconnais et al. (1997) developed a technique to study the elemental composition of Na, Mg, P, S, Cl, K and Ca of native airway surface liquid collected in germ-free *mice*.

Incubation with TNF-α (100 ng/ml for 18 h) potentiated the increases in short-circuit current responses (as an index of chloride secretion) to cyclic adenosine monophosphate- and Ca^{2+}-dependent agonists in *human* bronchial epithelial cells cultured in Ussing chambers (Bunce and Matthews 1997).

Manipulating cyclic guanosine monophosphate levels by sodium nitroprusside (10^{-4} M), atrial natriuretic peptide (10^{-7} M), brain natriuretic peptide (10^{-7} M), and C-type natriuretic peptide (10^{-7} M) all had no effect on short-circuit current measurements when added both the serosal and mucosal surfaces of *ovine* tracheal epithelia mounted in modified Ussing chambers (Range et al. 1997). 10^{-4} M 3-Isobutyl-1-methylxanthine (a non-specific phosphodiesterase inhibitor) when added to the mucosal surface resulted in a mean reduction in short-circuit current of 4.3 μA cm^{-2} (8.9 % of baseline).

5 min after instillation of charcoal particles or *bovine* liquid extract surfactant, the extracellular airway lining layer in the trachea of *horses* showed a considerably larger amount of unstrctured material pushing the osmiophilic film away from the cilia by about one time their length (GERBER et al. 1997).

Ciliary beat frequency is upregulated by •NO (LI et al. 2000). L-Arginine dose- and time dependently increased ciliary beat frequency of *rat* tracheal epithelial cell explants incubated in Dulbecco's modified Eagle's medium, and so did NO donors [1-hydroxy-2-oxo-3-(*N*-ethyl-2-aminoethyl)-3-ethyl-1-triazene and S-nitroso-L-glutathione], 8-Br-cGMP, and Zaprinast. Inhibitors of NOS, soluble guanylate cyclase, and cGMP-dependent protein kinase G can block the stimulant effect of L-arginine on ciliary beat frequency.

In vivo **mucus transport rates** were studied in *humans* and *donkeys* by external measurement of the rate of clearance of insoluble monodisperse γ-tagged ^{99m}Tc and/or ^{198}Au aerosols (LIPPMANN et al. 1977). The influence of temperature, environmental toxicants and drugs which affect the autonomic nervous system were studies by determining the changes in clearance produced by them in individual subjects. In *donkeys*, increased pre-test ambient temperature accelerates clearance by >1.7 %/°C. Smoking 2–7 cigarettes reduced the duration of bronchial clearance by ~50 % in both *humans* and *donkeys*. In *man*, atropine slowed clearance, while the adrenergic stimulating drugs, isoprenaline and epinephrine, both accelerated it by ≥ 4 times, as did isoprenaline when given subsequent to atropine. The cholinergic stimulating drug methacholine increased mucociliary transport in the *donkey*. Administration of a tap water aerosol for 10–15 min in *humans* increased bronchial clearance rates by ~25 %.

Wheel-running exercise for 60 days did not alter either the rate of clearance of instilled magnetite (0.2 mg in 0.15 ml saline) from *hamster* lungs or macrophage organelle motility (SWEENEY et al. 1994).

For the *human* respiratory tract, the International Commission on Radiological Protection (ICRP) presented a revised compartment model for the time-dependent particle transport from each region (BAIR 1994). This model is based on the premise that the large differences in radiation sensitivity of respiratory tract tissues, and the wide range of dose they receive, argue for calculating specific tissue doses rather than average lung doses for radiation protection purposes. The model describes three clearance pathways. Material deposited in the anterior nasal passages of the extrathoracic region, ET_1, is removed by extrinsic processes, such as nose blowing. For the other regions, clearance of inhaled material is competitive between particle transport processes (such as macrophage uptake and ciliary action) to the gastro-intestinal tract and to lymph nodes and absorption into blood.

In short-time experiments both *in vitro* and *in vivo*, ANTWEILER (1956) examined the transport velocity of soot, coal, quartz, Aerosil and aluminium particles along the tracheal epithelium. Mucus secretion was more stimulated by hygroscopic dust as quartz and Aerosil than by hydrophobe coal dust. Sublethal doses of atropine decreased transport velocity to 20 % of the standard of 12 mm/min. IRAVANI and VAN AS (1972) found pulmonary clearance of inhaled charcoal particles to be more rapid in mature and bronchitic animals than in normal younger SPF *rats*. In the *rabbit*, H_2SO_4, NO_2 and O_3 qualitatively and quantitatively influenced the clearance of ^{85}Sr-labelled 3 μm polystyrene latex microspheres, delivered via an endotracheal tube (SCHLESINGER et al. 1988). Ozone was considered to be about 10–20× as toxic as NO_2. HUANG (1997) exposed *human* tracheal epithelial cells to two coals, one of Utah with high buffering capacity and low acid soluble Fe^{2+}, the other from Pennsylvania with low buffering capacity and high acid soluble Fe^{2+}. The cells treated with the coal of Pennsylvania (10 mg/cm^2) showed a 36 % increase in oxidant formation over control. For the same exposure conditions, the coal from Utah mine had no effect.

Through activation of both capsaicin-pH sensitive irritant **receptors**, particulate matter as residual oil fly ash immediately (<30 s) increased intracellular calcium levels ($[Ca^{2+}]_i$)in a *human* bronchial epithelial cell line (BEAS-2B) and, within 2 h exposure, key inflammatory cytokine transcripts (i.e., IL-6, IL-8, TNF-α) and subsequent release of IL-6 and IL-8 cytokine protein after 4 h exposure (VERONESI et al. 1999).

In the *rat*, ozone consistently delayed short-term (0–50 h post-deposition) clearance and accelerated long-term (50–400 h post-deposition) clearance (PHALEN et al. 1994).

Mucus transport of ^{99m}Tc labelled *human* albumin minimicrospheres (mean aerodynamic diameter 0.75 μm) in chronic bronchitis patients treated with cistinexine dihydrochloride for two weeks was significantly promoted in those patients who showed a greater impairment of transport rate before treatment (SANTOLICANDRO et al. 1995).

Table 24. Bronchial epithelium: damage by dust inhalation

Dust	Species	Author(s)
Quartz	*Rabbit*	GROSS (1927), GERSING (1956)
	Rat	GERSING (1956), KLOSTERKÖTTER (1957), DHOM and SAUER (1967)
Bituminous Coal Mine	*Dog*	SCHILLER (1957, 1961)
Chromite	*Rat*	SCHILLER (1958)
Tantalum oxide	*Rat*	NEMETSCHEK-GANSLER et al. (1975)
Schneeberg Ore Mine	*Mouse*	DÖHNERT (1938)
Asbestos	*Mouse*	NORDMANN and SORGE (1941)
Melamine-Formaldehyde Resine	*Rat*	SCHILLER (1978)
Cotton	*Rat*	GORDON and HARKEMA (1995)

7.4.1 Aluminium

In *rats* and *guinea pigs*, inhaled aluminium (0.25, 2.5, and 25 mg aluminium chlorhydrate/m^3, respectively, for periods up to 24 months) was retained in pulmonary tissue and in the peribronchial lymph nodes but it was largely excluded from other tissues (STEINHAGEN et al. 1978, STONE et al. 1979). This is further suggested by the fact that lung aluminium concentration is greater than that of other tissues, it increases with age, unlike the aluminium levels of other tissues, and pulmonary aluminium levels do not correlate with aluminium levels of other tissues (ALFREY 1980, ALFREY et al. 1980).

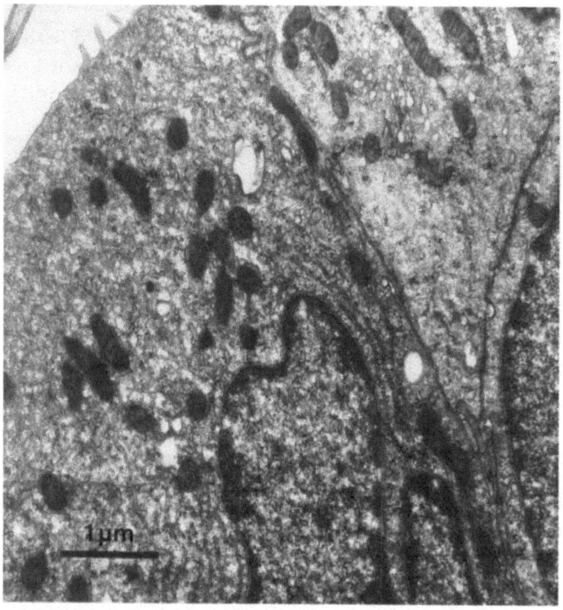

Fig. 89. Type 3 Clara cell (left) from a bronchiole from a female white *rat* (No.4; breeder Winkelmann, Borchen-Kirchborchen) which inhaled 10 mg powdered aluminium/m^3 4 h per day, 5 days per week from August 16 to October 27, 1967 for a total of 51 days. Fixed on January 15, 1968 under methitural anaesthesia by intratracheal instillation of 2.5% glutaraldehyde in phosphate buffer (pH 7.4) before opening the thorax. Postfixation with 1% osmium tetroxide in phosphate buffer (pH 7.4). Contrasted en bloc for 12 h with 0.5% uranyl acetate in 70% ethanol. Embedded in a 2:8 mixture of methyl and butyl methacrylate. Sectioned at 50 nm. Lead citrate after REYNOLDS (1963). Film 206/84

The permeability of *rat* olfactory epithelium for aluminium (20.60±1.96 μg Al/m^3) after nose-only exposure to aluminium chlorohydrate dispersed by a venturi powder disperser for 6 hours/day for 12 days was studied using microbeam particle-induced X-ray emission (DIVINE et al. 1999). The results suggested a direct transport of Al to the olfactory bulb.

Little is known about the temporal pattern of pulmonary clearance of aluminium compounds from the lungs with repeated exposure or the potential subsequent translocation to other organs (SCHLESINGER et al. 2000).

Urinary excretion of aluminium among welders indicated that aluminium is retained and stored in at least two functional compartments of the body and is eliminated from these compartments at different rates (SJÖBERG et al. 1988).

No bronchial hyperreactivity was found in male aluminium potroom workers (LARSSON et al. 1989).

Rats that inhaled aluminium dust for 25 h over a period of five days and were killed on the following day showed dust cells entering the bronchioles (SCHILLER 1957, 1961). In the *rabbit* insufflation of powdered aluminium induced hyperaemia of the bronchial mucosa with an emigration of polymorphonuclear leucocytes invading the aluminium dust deposits, and a desquamation of large areas of the epithelium (DE MARCHI 1947).

7.4.2
Clayey Shale

Fig. 90. Oedema in a Clara cell of a bronchiole (block S2) from a female *rat* (breeder: Winkelmann, Borchen-Kirchborchen) which had inhaled 20 mg quartz-free clayey shale dust/m³ (particles less than 5 μm in size; origin: Erda seam, Fürst Leopold-Baldur mine at Harvest-Dorsten) in POLLEY's (1963, 1965) dust channel 4 h per day, 5 days per week. Fixed under methitural anaesthesia by intratracheal instillation of 2.5 % glutaraldehyde in phosphate buffer (pH 7.4) before opening the thorax. Postfixation with 1 % osmium tetroxide in phosphate puffer (pH 7.4). Contrasted en bloc for 12 h with 0.5 % uranyl acetate in 70 % ethanol. Embedded in Epon 812. Sectioned at 50 nm. Lead citrate after REYNOLDS (1963). Plate 3392

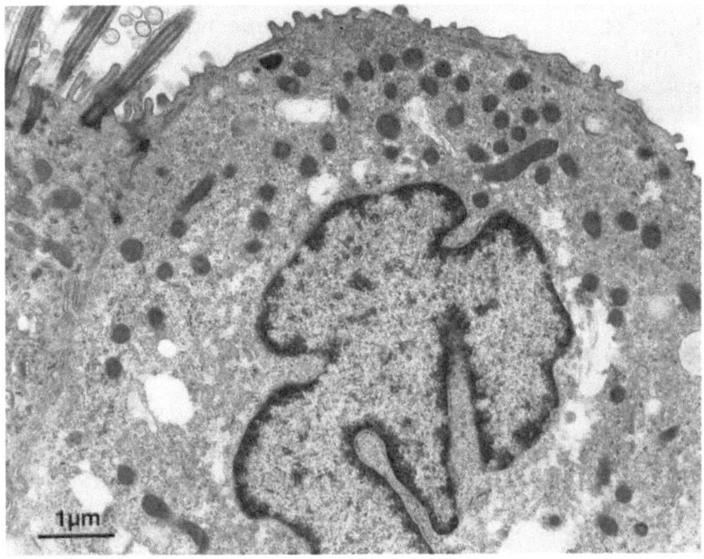

Fig. 91. Prominent oedema in a Clara cell of a bronchiole (block S2) from a female *rat* (breeder: Winkelmann, Borchen-Kirchborchen) which had inhaled 20 mg quartz-free clayey shale dust/m³ (particles less than 5 μm in size; origin: Erda seam, Fürst Leopold-Baldur mine at Harvest-Dorsten) in POLLEY's (1963, 1965) dust channel 4 h per day, 5 days per week. Fixed under methitural anaesthesia by intratracheal instillation of 2.5 % glutaraldehyde in phosphate buffer (pH 7.4) before opening the thorax. Postfixation with 1 % osmium tetroxide in phosphate puffer (pH 7.4). Contrasted en bloc for 12 h with 0.5 % uranyl acetate in 70 % ethanol. Embedded in Epon 812. Sectioned at 50 nm. Lead citrate after REYNOLDS (1963). Plate 3591

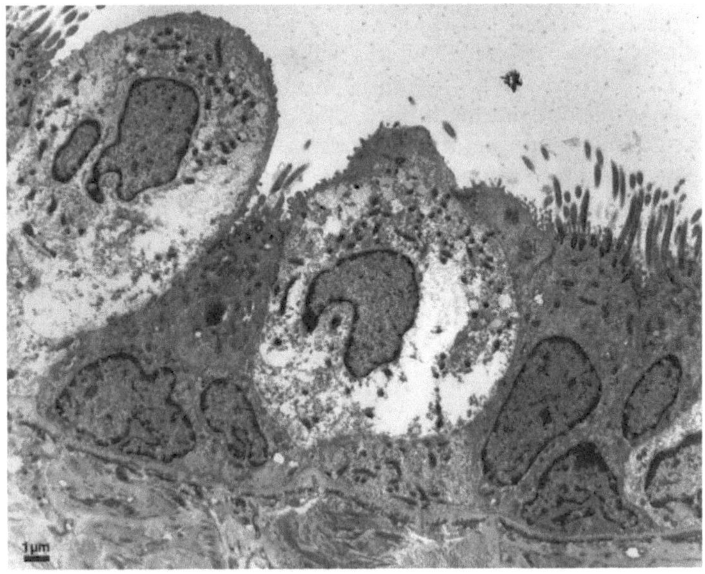

7.4.3
Amosite

In *hamster* tracheal epithelial cells exposed to amosite asbestos for varying periods of time, labelling with rhodamine phalloidin showed a dose responsive progressive aggregation of actin filaments over time (CHURG et al. 1997a). Gel electrophoresis demonstrated increased dose and time dependent increases in the amount of polymerised actin, and flow cytometry revealed an abnormal right angle scatter pattern which could be reduced by treatment with dithiotreitol or catalase. H_2O_2 produced a generally similar pattern.

Cigarette smoke pre-treatment of *rat* tracheal explants increased binding of amosite particles to the epithelial surface (CHURG et al. 1997b). Ozone (0.01 to 1.0 ppm for 10 min) enhanced the uptake of UICC amosite in a dose-response fashion (CHURG et al. 1996).

7.4.4
Crocidolite

Native crocidolite was capable of binding up to 57 nmol Fe^{2+}/mg fibre in 60 min, while desferrioxamine-B-incubated crocidolite was capable of binding only 5.5 nmol Fe^{2+}/mg fibre (HARDY and AUST 1995). The rate of iron binding for the first 5 min of exposure was independent of the concentration of iron in the solution, suggesting that there was a group of rapidly saturable sites, $\approx 1.5\times10^{18}$ binding sites/m^2 crocidolite surface, which was responsible for the immediate binding. Iron mobilization from crocidolite resulted in enhanced iron-catalyzed oxygen consumption and hydroxyl radical generation in the presence of cysteine (LUND and AUST 1991).

BEAS 2 B cells, a *human* bronchial epithelial cell line, exposed to various concentrations of crocidolite showed an increase of DNA strand breaks as follows (% positive cells): after 2 h exposure (2% at 10^{-8} g/cm^2 and 3% at 5×10^{-6} g/cm^2), after 24 h exposure (16% at 10^{-8} g/cm^2 and 20% at 5×10^{-6} g/cm^2) (GILLISSEN et al. 1996). In contrast, even at highest concentrations (24 h exposure) man-made mineral fibre basalt wool did not cause any increase of DNA strand breaks. Scanning electron microscopy confirmed that both fibre types were in part incorporated into the cells. Manganese superoxide dismutase gene expression was induced by 2 µg UICC crocidolite/cm^2 (JAWORSKA et al. 1997).

Hamster tracheal epithelial cells exposed to 5.0 µg crocidolite asbestos/cm^2 showed increases in p65/50 and p50 protein complexes binding to the NFκB-binding consensus DNA sequence (JANSSEN et al. 1995, MOSSMAN et al. 1997). When *N*-acetylcysteine, an agent boosting cellular glutathione levels, was added to cells for 18 h, both crocidolite-induced *c-fos* and *c-jun* mRNA levels and activator protein-1 to DNA-binding activity were diminished, suggesting that oxidative stress was involved. How reactive oxygen species elaborated by asbestos may initiate these signalling pathways is suggested in studies using *rat* pleural mesothelial cells (ZANELLA et al. 1996).

7.4.5
Chrysotile

In *humans* chrysotile (cleared in months) might have less effect in inducing cancer than the amphibole fibres (cleared in years). A detailed elaboration of this argument has been published by BERRY (1999).

SV-40-transformed *human* bronchial epithelial cells co-cultured (transwell system) with *human* blood monocytes and exposed to 100 µg chrysotile B per 10^6 cells showed deoxyribonucleic acid strand lesions induced by the reactive oxygen intermediates released from the mononuclear phagocytes engulfing the fibres (KIENAST et al. 2000). The addition of 200 U catalase (EC 1.11.1.6) per ml or 100 µM desferoxamine to the culture medium blocked almost completely the induction of DNA strand lesions in this system (maximum 85%).

7.4.6
Quartz

Inhalation of quartz dust (F. GROSS 1927, KLOSTERKÖTTER 1957, DHOM and SAUER 1967) or mixed dusts containing quartz (SCHILLER 1958) just as a single intratracheal instillation of quartz (SCHILLER 1961, 1963) induced epithelial lesions resulting in a disability to remove alveolar macrophages containing dust and free particles impinged upon the airway surface.

Experiments to study the **influence of electrostatic charges of dust particles on their retention** in *rat* lungs showed that electric charges do not influence the retention of quartz dust particles in deep

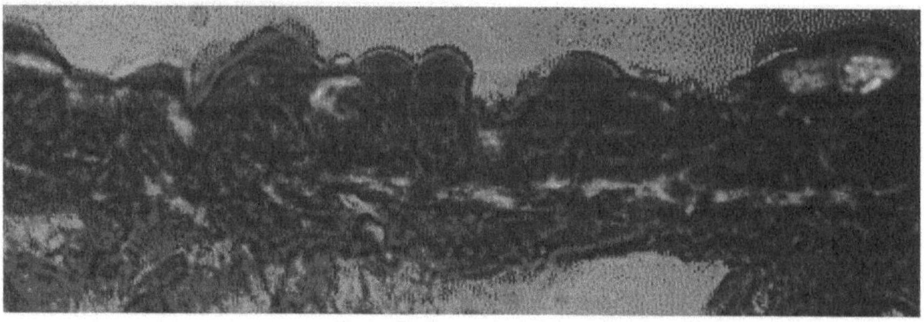

Fig. 92. Bronchiolar epithelium of a 167 g male Wistar *rat* (R 21/63) fed 82 mg retinol palmitate (Arovit)/kg body weight×day from February 18 to 22, 1963 by stomach tube. On the 3rd day of the experiment the animal was exposed to an atmosphere containing 15 mg quartz DQ 12 (particle size <3 µm) per m^3 air for 5 h using POLLEY's dust tunnel. Fixed in alcohol-chloroform-glacial acetic acid after CARNOY. Paraffin section, unstained, embedded in glycerol. Interference contrast Fl 100×/1.36/∞/–. Quartz particles engulfed by an alveolar macrophage are visualised by both interference contrast and polarisation

respiratory tract (SHEVCHENKO 1971). The severity of the silicotic changes, however, depended upon the electric charge on the particles. Exposure to unipolar (positive) and bipolar charged quartz dust for 6 months was followed by more collagen in the lungs of *rats* in the second group; 47.36±1.6 mg and 61.98±3.1 mg ($P < 0.01$) respectively. A study of the functional state of dust phagocytes by vital staining showed that of the animals dusted with electro-charged dust, more dust phagocytes were in the state of partial necrosis. Alkaline phosphatase activity was suppressed in macrophages in *rats* that had inhaled charged dust.

Influence of silica inhalation on the pulmonary clearance of *Listeria monocytogenes*. Pre-exposure of *rats* to silica (15 mg/m^3) for 59 days (6 h/day, 5 days/week) caused substantial increases in the number of lavagable neutrophils and lactate dehydrogenase activity compared with the air control, whereas silica inhalation for both 21 and 59 days significantly enhanced the pulmonary clearance of *Listeria monocytogenes* compared with air controls (ANTONINI et al. 2000).

Influence of pneumoconstriction on dust deposition. Bronchoconstriction in *man* can be caused by exposure to cigarette smoke as shown by LOOMIS (1956), NADEL and COMROE (1961), and GUYATT et al. (1970), or by exposure to inert dusts as shown by DAUTREBANDE et al. (1948), and DUBOIS and DAUTREBANDE (1958). The mechanisms involved have been discussed by NADEL et al. (1965), NADEL (1968) and DUBOIS (1969). Bronchoconstriction, by reducing the cross section for flow in the conductive airways, results in increased air velocities and turbulence. Increased velocity can result in greatly increased deposition by impaction at the airway bifurcations, while increased turbulence can account for an increase in deposition by eddy diffusion (LEHMANN 1938, WORTH and SCHILLER 1951).

7.4.7 Coal Dust

While HALDANE (1918) thought coal dust that was not inhaled in excessive quantities would not do any harm and possible even some good as to help the lungs in eliminating other kinds of harmful particles, such as tubercle bacilli, there is now growing evidence that **any dust is harmful.**

In both the Rhondda (South Wales) and in Leigh (Lancashire) the miners without pneumoconiosis had rather more 'chronic bronchitis' and rather lower mean maximum breathing capacity than those with simple pneumoconiosis (HIGGINS 1960). Higgins found a somewhat higher prevalence of respiratory symptoms in those men who had worked for 10 years and longer, and has confirmed the downward trend in the maximum breathing capacity with increasing dust dosage that was first demonstrated in Czech miners by KADLEC and VYSKOCIL (1950). ROGAN et al. (1973) reported a significant decrease in FEV_1 with increasing dust exposure in all the age/smoking habit subgroups. There was a loss of 150 ml at a cumulative dust exposure of 240 gram hour/m^3. MARINE and GURR (1988) found between 90 and 100 ml FEV_1 decrements for every 100 gram hour/m^3 exposure to coal mine dust in smokers and non-smokers. ATTFIELD and HODOUS (1992) confirmed the exposure-response relationship between various pulmonary function parameters and estimated cumulative dust exposure in a large cohort of U.S. coal miners.

In animal experiments the inhalation of soot increased mucus formation (KNAUFF 1867, GROSS 1927). In *rabbits* and mongrel *dogs* kept underground in a bituminous coal mine (about 50 mg coal dust/m^3) nearly continuously for up to 6 years, SCHILLER (1958) found a hypertrophy of the bronchial epithelium with a prevalence of goblet cells after 3 years exposure, but an atrophy at 6 years exposure. In the bronchioles, Clara cells showed an increase in apocrine secretion.

In the female Wistar *rat*, intratracheal instillation of coal dusts of different rank suspended in saline induced a significant Clara cell hyperplasia as compared with TiO_2 dust ($P < 0.05$) or pure saline ($P < 0.03$) (ALBRECHT et al. 2000). Proliferation and hyperplasia by coal dust were independent of the quartz content of the dust samples.

7.4.8 Magnetite

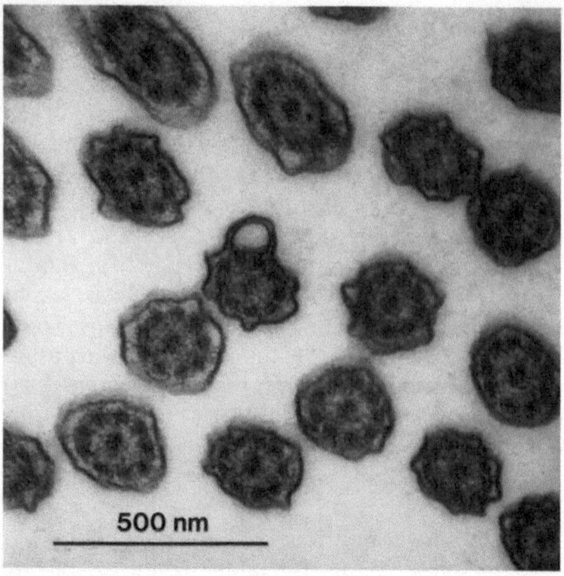

Fig. 93. Cross section of a bronchiolar cilium showing a membrane bound vesicle between the axonema and the ciliary membrane. Lung of a female *rat* (No. 4; breeder: Winkelmann, Borchen-Kirchborchen) which had inhaled 10 mg powdered Grängesberg magnetite/m^3 4 per day, 5 days per from August 24 to October 19, 1967 for a total of 40 days. Fixed on January 15, 1968 under methitural anaesthesia by intratracheal instillation of 2.5 % glutaraldehyde in phosphate buffer (pH 7.4) before opening the thorax. Postfixation with 1 % osmium tetroxide in phosphate puffer (pH 7.4). Contrasted en bloc for 12 h with 0.5 % uranyl acetate in 70 % ethanol. Embedded in a 2 : 8 mixture of methyl and butyl methacrylate. Sectioned at 50 nm. Lead citrate after Reynolds (1963). Film 33130

7.4.9 Manganese Dioxide

Within 15 min of contact with MnO_2 dust (80 % $< 1\ \mu m$) suspended in 0.1 ml normal saline the bronchial epithelial cells discharge their mucus, and their supranuclear cytoplasm shrinks and in some cases appears to be shed (Lloyd Davies and Harding 1949).

7.4.10 Tantalum Oxide

Up to the 46th day after termination of tantalum oxide (Ta_2O_5) dust inhalation (150 mg/m^3) for 10 days (10 h/d), the bronchial and bronchiolar ciliated epithelia of inbred male Sprague-Dawley *rats* showed defects of various sizes (Nemetsckek-Gansler et al. 1975).

7.4.11 Titanium Dioxide

By electron microscopy, Churg et al. (1998) found TiO_2 particles in the epithelial cells of *rat* tracheal explants cultured in Dulbecco's modified Eagle's medium. The volume proportion of both fine (120 nm) and ultrafine (21 nm) particles in the epithelium increased from 3 to 7 days; it was greater for ultrafine particles at 3 days but was greater for fine particles at 7 days. Ultrafine particles appeared to enter the epithelium faster, and once in the epithelium, a greater proportion of them was translocated to the subepithelial space compared with fine particles. However, if it was assumed that the volume proportion was representative of particle number, the number of particles reaching the interstitial space was directly proportional to the number applied, i.e., overall, there was no preferential transport from lumen to interstitium by size.

7.4.12 Vanadium

Vanadium compounds induce oxidative stress in normal *human* bronchial epithelial cells. Both nuclear translocation of nuclear factor-κB and enhanced κB-dependent transcription induce by V^{IV} were inhibited by overexpression of catalase, but not Cu,Zn superoxide dismutase, indicating that peroxides rather than superoxides initiated signalling (Jaspers et al. 2000). Catalase selectively blocked the response to V^{IV} because it inhibited neither NF-κB translocation nor κB-dependent transcription evoked by the proinflammatory cytokine tumour necrosis factor-α. The V^{IV}-induced κB-dependent transcription was dependent upon activation of the p38 mitogen-activated protein kinase because overexpression of dominant-negative mutants of the p38 MAPK pathway inhibited V^{IV}-induced κB-dependent transcription. The inhibition was not due to suppression of NF-κB nuclear translocation because NF-κB DNA binding was unaffected by the inhibition of p38 activity. Overexpression of catalase, but not Cu, Zn superoxide dismutase, inhibited p38 activation, indicating that peroxides activated p38. Catalase failed to block V^{IV}-induced increases in phosphotyrosine levels, suggesting that the catalase-sensitive signalling components were independent of V^{IV}-induced tyrosine phosphorylation.

Heparin-binding epidermal growth factor-like growth factor mRNA expression and protein were seven-fold increased in *human* bronchial epithelial cell cultures exposed to V_2O_5 (Zhang et al. 2001)

7.4.13
Cotton Dust

In *rats* exposed to 15 mg cotton dust/m^3 for 2 h/d for 3 d, the major histologic alteration in the lungs was mucous cell metaplasia with increased amounts of intraepithelial mucosubstances in the surface epithelium lining large-diameter main axial airways (Gordon and Harkema 1995).

Gossypol ($C_{30}H_{30}O_8$) promoted the formation of oxygen radicals when incubated with *rat* liver microsomes (De Peyster et al. 1984) and induced sister chromatid exchanges and chromosome damage in bone marrow cells in *mice* (Nayak and Buttar 1986).

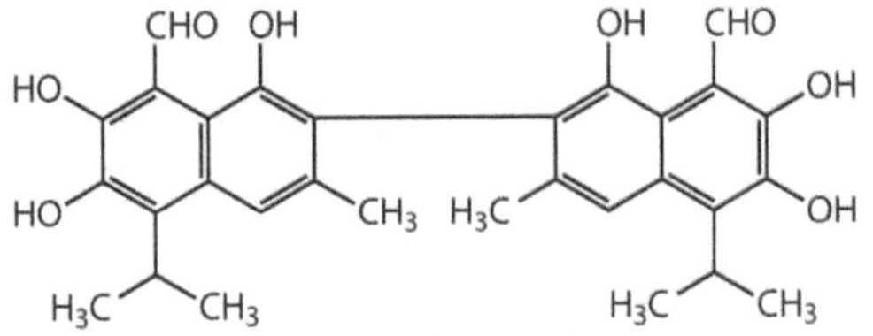

Gossypol ($C_{30}H_{30}O_8$) [115]

As a contribution to the byssinotic effects of cotton seed Duckett and Kennedy (1983) exposed *rabbit* tracheal explants to gossypol and found a dose dependent (10^{-5} M, 5×10^{-4} M, 10^{-4} M) **inhibition of ciliary function**. Mitochondria appeared swollen and cristae were disrupted.

7.4.14
Diesel Exhaust

Diesel exhaust particles forming agglomerates of about 0.1–0.5 μm are composed of a carbon core that adsorbs trace amounts of heavy metals and a great number of organic compounds that could represent up to 40 % of the particle (Westerholm et al. 1991).

Human bronchial epithelial cell subclone 16HBE14o cultured on collagen type I showed an increase in granulocyte-macrophage colony stimulating factor (GM-CSF) mRNA levels after exposure to diesel exhaust particles (Boland et al. 2000). Radical Scavengers inhibited the diesel exhaust particle-induced GM-CSF release, showing involvement of reactive oxygen species in this response. Genistein, a tyrosine kinase inhibitor, abrogated the effects of diesel exhaust particles on GM-CSF release, whereas protein kinase C, cyclooxygenase, or lipoxygenase inhibitors had no effect. Cytochalasin D, which inhibits the phagocytosis of diesel exhaust particles, reduced the increase in GM-CSF release after diesel exhaust particle treatment.

7.4.15
Trimethylolmelamine (Madurit)

Trimethylolmelamine by condensation with formaldehyde forms a synthetic, which is dried by spraying and put on the market in a micronized form. After inhalation of the merchandise (73.6 mg/l) for various periods (see Table 9) *rats* showed a marked dilatation of the rough endoplasmic reticulum of the serous cells in the tracheo-bronchial epithelium (Schiller 1978; Fig. 88).

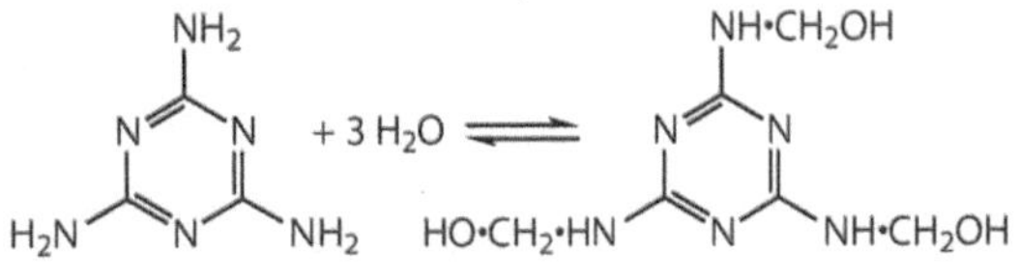

Release of formaldehyde from trimethylolmelamine [123]

The major pathways of metabolism of inhaled formaldehyde are oxidation to formate and incorporation into biological macromolecules via tetrahydrofolate-dependent one-carbon biosynthetic pathways (Huennekens and Osborne 1959, Koivusalo et al. 1982). The most important pathway for oxidation appears to be that catalysed by formaldehyde dehydrogenase (EC 1.2.1.1), an enzyme that requires both glutathione and NAD$^+$ as cofactors. Uotila and Koivusalo (1974) showed that the true substrate is the hemithioacetal adduct of formaldehyde and glutathione and the product formed is the thiol ester of formic acid, *S*-formylglutathione.

arginine > N^{ω}-hydroxy-L-arginine > homo-L-arginine > homo-N^{ω}-hydroxy-L-arginine,

7.5
Submucosal Glands

The submucosal glands are the other primary site of mucin production in the lung. They are present in the cartilaginous airways and are composed of an interconnecting network of tubules and ducts that secrete a proteinaceous fluid into the airways. Two predominant types of tubules in submucosal glands are composed of either serous or mucous acinar and ductular cell types (Meyrick and Reid 1970). In *sheep*, serous cells contain two types of discrete membrane-bound granules (Mariassy et al. 1984). One type is large, ovoid and either biphasic with a homogenous core and fine granular rim, or homogenous without a rim. The other is a slender rod-shaped granule with a dense matrix and is found only in the cell apex. The predominant mucous cell of the glands is M4, containing biphasic granules, with a wide, low-density rim surrounding a dark eccentric homogenous core.

Serous cells secrete a number of bactericidal proteins such as lysozyme (EC 3.2.1.17) (BALS 1997) and lactoferrin (BOWES et al. 1981), whereas mucous cells primarily secrete mucin (STROUS and DEKKER 1992). Serous cells express high levels of cystic fibrosis transmembrane conductance regulator (CFTR) protein, which in part serves to secrete fluid and electrolytes important in glandular secretion and hydration of mucins (ENGELHARDT et al. 1993). SHARMA et al. (1998) used *in situ* hybridisation and immunocytochemical methods to determine the cellular distribution of MUC5B and MUC7 expression in cystic fibrosis and non-cystic fibrosis *human* bronchus to demonstrate that MUC5B expression is limited exclusively to submucosal glands. Specifically, MUC5B expression was confined to all mucous tubules, whereas MUC7 expression was seen in a subset of lysozyme expressing serous tubules of submucosal glands.

Neutrophil elastase is a strong secretagogue of the airway submucosal gland (SOMMERHOFF et al. 1990), and suppressing the pulmonary recruitment of neutrophils by reducing the production of neutrophil chemotactic factors such as IL-8 could expectedly result in a decrement of excessive airway secretion (FAHY et al. 1992, SAKITO et al. 1996). This neutrophil-associated hypersecretion appears to be also involved in goblet cell mucus discharge induced by the inhalation of *Escherichia coli* lipopolysaccharide, which was inhibited by macrolides in *guinea pig in vivo* (SASAKI and GALLACHER 1990). The suppressive effect of erythromycin on respiratory mucus secretion was noticed more directly in an in vitro preparation of *human* airways (GOSWAMI et al. 1990) in which erythromycin reduced both spontaneous (baseline) and stimulated (by either histamine or cholinergic agonist) respiratory glycoconjugate secretion from airways in culture. In airway glands from *feline* tracheae therapeutic concentrations of erythromycin or clarithromycin attenuated the whole cell currents evoked by acetylcholine (IROKAWA et al. 1999).

DUAN et al. (1998) studied a newborn *ferret* tracheal model to probe the contribution of tracheal submucosal gland stem cells to repopulating the surface mucosa. They used this model since it shows a very similar anatomy of submucosal gland distribution to that in *humans*. *Mice* have no submucosal glands in the lower cartilaginous airways.

CHAPTER 8

Pneumocytes

Alveolar epithelium consists of three types of cells. Roughly 98 % of the alveolar surface is covered by squamous (type I) epithelial cells, from which thin sheets of cytoplasm extend to cover large areas of surface of even several alveoli (Haies et al. 1981). Type II epithelial cells, though more numerous but compact, occupy only 2 % of the alveolar surface. Type III cells, also called alveolar brush cells, are rare.

Induction of alveolar epithelial cell phenotypes via mesenchymal signalling regulates pneumocyte differentiation not only during lung development (Hogan and Yingling 1998, Warburton et al. 1999), but also during the response to a myriad of injuries inciting fibrogenesis (McCormack 1998). Direct communication between epithelial and mesenchymal cells is achieved via intercellular contacts between type II pneumocytes and fibroblasts. Type II cells may inhibit fibroblast growth by this direct interaction, and fibroblast proliferation may occur when epithelial injury interferes with cellular contact (Adamson et al. 1990). Terasaki et al. (2000) reported that mesenchymal cell expression of epimorphin, a key regulator of lung epithelial cell morphogenesis (Hirai et al. 1992), coincides with the process of re-epithelialization during the response to bleomycin-induced injury. Epimorphin is a member of a gene family, localised to *human* chromosome 7 (Zha et al. 1996), and also encodes cytoplasmic trafficking proteins known as synthaxins (Hirai et al. 1992).

8.1 Type I Pneumocytes

8.1.1 Ultrastructure

Caveolae, omega-shaped invaginations of the plasmalemma possessing a cytoplasmic membrane protein coat of caveolin are present in the *in vivo* alveolar epithelial type I lung cells, but absent in its progenitor, the type II alveolar epithelial cells. As freshly isolated *rat* type II cells acquire a type I-like phenotype in primary culture caveolin-1 expression increases, with caveolin-1 signal at 192 h postseeding up to 50-folds greater than at 60 h (Campbell et al. 1999). Caveolae were morphologically evident only after 132 h.

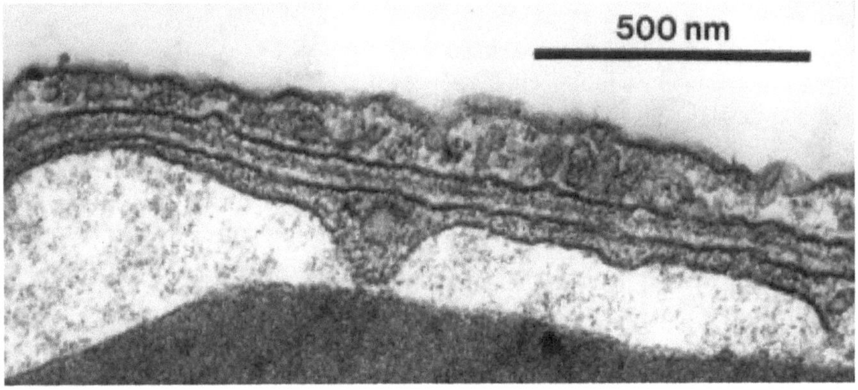

Fig. 94. Air-blood interface: pneumocyte I, basement membrane, and capillary endothelium. Lung of an unmedicated female *rat* (breeder Winkelmann, Borchen-Kirchborchen). Fixed under methitural anaesthesia by intratracheal instillation of 2.5 % glutaraldehyde in phosphate buffer (pH 7.4) before opening the thorax. Postfixation with 1 % osmium tetroxide in phosphate puffer (pH 7.4). Contrasted en bloc for 12 h with 0.5 % uranyl acetate in 70 % ethanol. Embedded in a 2:8 mixture of methyl and butyl methacrylate. Sectioned at 50 nm. Lead citrate after Reynolds (1963). Plate 13/06

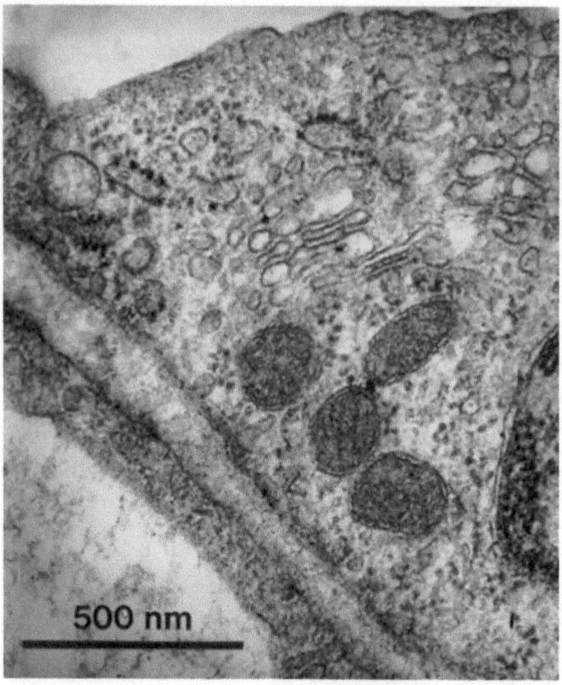

Fig. 95. Perikaryon of a type I alveolar epithelial cell from an unmedicated female *rat* (breeder: Winkelmann, Borchen-Kirchborchen). Fixed under methitural anaesthesia by intratracheal instillation of 2.5 % glutaraldehyde in phosphate buffer (pH 7.4) before opening the thorax. Postfixation with 1 % osmium tetroxide in phosphate buffer (pH 7.4). Contrasted en bloc for 12 h with 0.5 % uranyl acetate in 70 % ethanol. Embedded in a 2:8 mixture of methyl and butyl methacrylate. Sectioned at 50 nm. Lead citrate after Reynolds (1963). Plate 13/07

8.1.2 Histochemistry

A limited number of type I cell markers have been identified (Dobbs et al. 1988, 1999, Brody and Williams 1992, Danto et al. 1992, Christensen et al. 1993, Vanderbilt and Dobbs 1998). Dobbs et al. (1988) described *rat* type I cell 40 kDa-protein, a glycoprotein defined by a monoclonal antibody raised against partially purified type I cells. In the lung it was localised to the apical plasma membrane of type I pneumocytes. The corresponding gene spans 35 kilobase pairs; it contains 6 exons and at least 6 *rat* Identifier repetitive elements (Vanderbilt and Dobbs 1998).

HTI_{56}, a 56-kDa protein, is unique to the *human* lung (Dobbs et al. 1999). By immunoelectron microscopy, it is localised to the type I cell apical plasma membrane. The pI of HTI_{56} is 2.5–3.5. HTI_{56} is glycosylated and has the biochemical characteristics of an integral membrane protein. HTI_{56} is detectable by week 20 of gestation and its expression increases in foetal lung explant culture.

In the *rat*, the monoclonal antibody MEP-1 stained all cells lining the alveolar space, except the type II pneumocytes (Kasper et al. 1996). Double fluorescence staining employing type I cell-specific lectin BPA (Kasper et al. 1994) revealed the type I cell specificity. Immunoelectron microscopy confirmed this selective reaction of the MEP-1 antibody. The polyclonal anti pan-cathedrin antibody selectively decorated type I pneumocytes, alveolar macrophages and endothelial cells of large blood vessels. Caveolin is a selective marker of type I pneumocytes (Kasper and Reimann 1997).

8.1.3 Pharmacology

After subcutaneous injection of 150 mg pilocarpine per kg *rat*, type I alveolar cells showed no detectable changes at any stage of the experiment (Goldenberg et al. 1969).

8.1.4 Toxicology

8.1.4.1 Aluminium

s. Fig. 96.

8.1.4.2 Ammonia

The lungs of two individuals dying acutely of NH_3 inhalation showed marked swelling and inhibitional oedema of the type I alveolar epithelial cells (Burns et al. 1983).

8.1.4.3 Asbestos

Two days after intratracheal instillation, electron microscopy demonstrated acute injury to some type I epithelial cells of the *mouse*, and although most crocidolite fibres were cleared as free particles or were phagocytozed, some small fibres were observed in the type I epithelium (Bowden and Adamson 1985).

After intratracheal instillation of chrysotile asbestos *hamster* alveolar epithelial cells, can transform into large phagocytic cells which ingest asbestos fibres and produce asbestos bodies (Suzuki et al. 1972). The epithelial cells frequently lose their original type I or type II cell structure and change into a form intermediate between the epithelial cell and the alveolar macrophage.

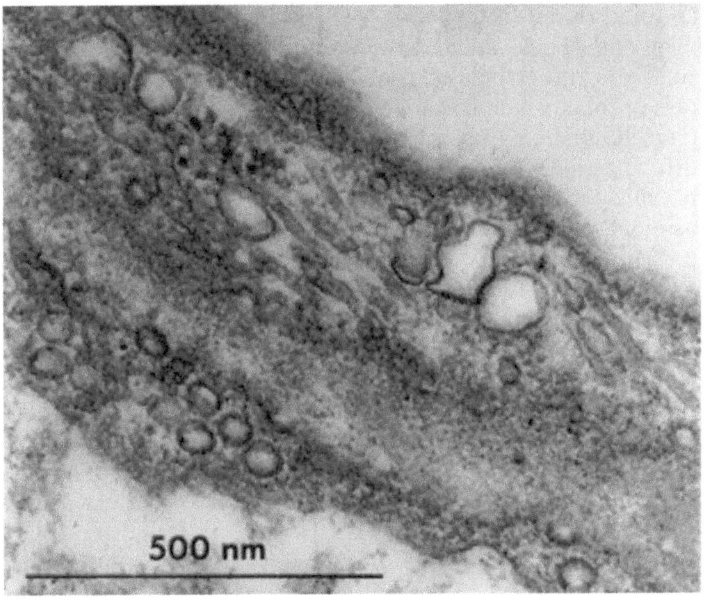

Fig. 96. Prominent microtubules in a cell process of a type I pneumocyte from a female white *rat* (No.4; breeder: Winkelmann, Borchen-Kirchborchen), which inhaled 10 mg powdered aluminium/m^3 4 h per day, 5 days per week from August 16 to October 27, 1967 for a total of 51 days. Fixed on January 15, 1968 under methitural anaesthesia by intratracheal instillation of 2.5 % glutaraldehyde in phosphate buffer (pH 7.4) before opening the thorax. Postfixation with 1 % osmium tetroxide in phosphate buffer (pH 7.4). Contrasted en bloc for 12 h with 0.5 % uranyl acetate in 70 % ethanol. Embedded in a 2:8 mixture of methyl and butyl methacrylate. Sectioned at 50 nm. Lead citrate after REYNOLDS (1963). Plate 38/02

Liposome-encapsulated catalase (20 units) attenuated H_2O_2-induced (50 μM for 30 min), but not amosite-induced (25 μg/cm^2) DNA strand break formation in cultured type I like pulmonary epithelial cells (WI-26), suggesting that the DNA damaging effects of asbestos occur by both H_2O_2-dependent and H_2O_2-independent mechanisms (KAMP et al. 1997).

8.1.4.4 Cadmium

In the Syrian golden *hamster*, type-I pneumocytes were extremely sensitive for alveolar deposition of toxic substances as cadmium compounds (THIEDEMANN et al. 1992).

8.1.4.4.1 Cadmium Oxide

In *rats* exposed to CdO for 3 days (18 h/d, 270 μg Cd/m^3; sacrificed two days after the end of the exposure) necrosis of type-I pneumocytes and extensive proliferation of epithelial cells were most common, especially in the proximal region of the alveolar duct (THIEDEMANN et al. 1989, PAULINI et al. 1992).

8.1.4.4.2 Cadmium Chloride

In the Sprague-Dawley *rat* exposed to an aerosol of 0.1 per cent $CdCl_2$ in physiologic saline (10 mg $CdCl_2$/m^3) for 2 h, cytoplasmic oedema caused type I pneumocytes to bulge out into the alveolar lumen (HAYES et al. 1976).

Intratracheal instillation of 400 μg $CdCl_2$ into the left lung of the *rat* within 1 day damaged type I epithelial cells, and denuded basement membranes were observed (DAMIANO et al. 1988).

8.1.4.5 Cobalt

Cobalt aerosols inhaled by *rats* at a concentration of 2.12 ± 0.55 mg/m^3 for 5 h per day on 4 consecutive days induced ballooning of type I cells (KYONO et al. 1992). Residual bodies or vacuoles with electron dense inclusions were frequently associated with foci of interstitial oedema. on the other hand, round hypertrophic cells showed a large nucleolus, abundant smooth endoplasmic reticulum, a prominent Golgi apparatus' and residual inclusion bodies. Cytoplasmic continuity was always recognised between the cells showing above changes and the thin membranous cytoplasm, were pinocytotic vesicles were observed. These were only temporary changes, which appeared two hours after the end of the exposure and had almost ceased 3 days after the exposure.

8.1.4.6 Nickel Compounds

After an intravenous injection in LD_{50} dosage of nickel carbonyl, i.e. 22 mg Ni/kg of body weight, to Sprague-Dawley *rats*, the most pronounced alterations occurred in the membranous pneumocytes

(Hackett and Sunderman 1968). Great enlargement of the nuclei and nucleoli was observed two to six days. The cytoplasm contained prominent arrays of rough endoplasmic reticulum, as well as increased numbers of free ribosomes and mitochondria. In some of the type I pneumocytes, the endoplasmic reticulum was distended with flocculent osmiophilic material. Abundant Golgi zones were present. Multivesicular bodies were found two and four days after $Ni(CO)_4$ application. They were usually surrounded by a single membrane.

A significant decrease of reduced glutathione was found in lung homogenates of *rats* following daily subcutaneous injections of one fifth of the LD_{50} of $NiSO_4$ for 45 days (Mancas et al. 1996). The lipoperoxide levels in blood and tissues showed that lipoperoxidation involved the antioxidant defence systems.

8.1.4.7 Ozone

Ozone (0.5 ppm, 2-6 h) first induced a mild swelling in the attenuated peripheral cytoplasm of type I rat pneumocytes (Stephens et al. 1974). This process continued to increase in severity until it was quite pronounced. Mitochondria, both in the attenuated portions of the cell as well as those around the nucleus, became progressively swollen until they were scarcely recognisable. This was followed either by a sloughing away of the entire cell from the basement lamina or by rupture of the plasma membrane and disintegration of the cell, leaving the basement lamina devoid of a epithelial covering.

8.1.4.8 Plutonium

Plutonium particles not phagocytized by alveolar macrophages and removed from the lung are found in type I alveolar epithelial cells, which phagozytize plutonium particles within a few hours after deposition (Sanders and Adee 1970). These cells appear relatively radioresistant. Fine structural studies of lung exposed to radiation doses up to 13.000 rads from ^{239}Pu showed no evidence of type I cell detachment or removal (Sanders et al. 1971).

8.1.4.9 Zinc

Zinc salts isolated from the respirable range particulate material EHC-93 instilled intratracheally into Swiss *mice* induced rapid focal necrosis of type I alveolar epithelial cells followed by inflammation and elevation of protein levels in lavage fluid over a 2-week period (Adamson et al. 2000).

8.1.4.10 Butylated Hydroxytoluene

Mice given a single intraperitoneal injection of 400 mg butylated hydroxytoluene per kg body weight dissolved in corn oil showed disintegration of their type I pneumocytes (Hirai et al. 1977). Thereafter, type II epithelium starts to divide and transform themselves as early as 2 days into a cell resembling type I in order to replace the original type I epithelium. This new cell changes its shape to become squamous, eliminates lamellar bodies, and increases the amount of ribosomes and mitochondria. The shape of the peroxisomes becomes indistinct, the density of their matrix is reduced in proportion to the regulation of the peroxidative activity of catalase (Hirai et al. 1983). The presence of pinocytotic vesicles and the extension of cytoplasmic processes covering the denuded basement membrane strongly suggests a type I epithelium profile.

8.2 Type II Pneumocytes

The alveolar type II cells are spherical pneumocytes which comprise only 4 % of the alveolar surface area, yet they constitute 60 % of alveolar epithelial cells and 10–15 % of all lung cells. Four major functions have been attributed to alveolar type II cells:

1) Synthesis and secretion of surfactant
2) Xenobiotic metabolism
3) Transepithelial movement of water
4) Regeneration of the alveolar epithelium following lung injury.

Therefore, alveolar type II cells play important roles in normal pulmonary function and in the response of the lung to toxic compounds, which may cause lung damage.

Several methods have been employed to isolate and purify type II cells, including enzymatic digestion of pulmonary tissue to free lung cells from the pulmonary epithelium and purification of type II pneumocytes by discontinuous density gradients (Kikkawa and Yoneda 1974), primary culture (Dobbs and Mason 1979), centrifugal elutriation (Jones et al. 1982), or unit gravity sedimentation (Brown et al. 1984).

Type II pneumocytes can be identified at the light microscope level by antibodies against cytokeratin 18 (Schlichenmaier et al. 2000). This intermediate filament was thereby predominantly localised in the apical compartment and the intensity of the immunostaining depends on the stage of development. Type I alveolar cells as well as alveolar macrophages were consistently negative to cytokeratin 18 in postnatal *porcine* lung tissue.

The lamellar bodies characteristic for type II pneumocytes can be visualised using light microscopy with the Papanicolaou stain. They are identified as deep blue granules surrounded by a clear halo (Kikkawa and Yoneda 1974). Mason and Williams (1976) have shown that a lipophilic fluorescent dye, phosphine 3R, is concentrated in the lamellar bodies. Therefore, type II pneumocytes can be identified under the fluorescent microscope by their yellow inclusion bodies (Castranova et al. 1988).

8.2.1 Ultrastructure

The **lamellar body** has a distinct three-dimensional architecture (Gil and Reiss 1973, Weibel 1973, Stratton 1976). Overall the lamellae are bell-shaped and organised around a cylindrical core that often includes a protrusion of the limiting membrane. The lamellae insert into or arise from a thin layer of amorphous, moderately electron-dense material, located along the lateral limiting membrane, but also seen in the core. Sections in different angles result in a variety of morphological appearances, from circumferential to longitudinal. Histochemical localisation of enzymes has always been made in the amorphous material of the core or the periphery. Lamellae do not close laterally and there is no continuity or contact between contiguous lamellae. It is known that lamellar bodies derive from multivesicular bodies, which are organelles related to the lysosome system. Therefore, the amorphous material occasionally can be seen containing a few small vesicles, which suggests that it may be related to the original matrix of the multivesicular bodies.

In the *pig*, the granular pneumocytes characteristically contain numerous large partially "filled" vacuoles (Epling 1966). An electron-opaque mass of irregular shape frequently lines the vacuoles. Lamellar forms of vacuolar contents were occasionally observed. The vacuoles were observed in varied states of fusion, and some were observed releasing their contents into the alveolar lumen. In the neonatal and postnatal *porcine* lung, osmiophilic, lamellated bodies (up to 2 μm in diameter) were typical (Winkler and Cheville 1984). Mitochondria, granular endoplasmic reticulum, and a well-developed Golgi field were present.

In the pigtail *macaque*, the number of lamellar bodies per alveolar type II cell of older animals (15, 20 and 20 years, all female, $n = 73$ cells) showed a significant decrease ($P < 0.01$) compared with that of young animals (1.5, 1.5, 4.8 years, all female $n = 66$ cells) (Shimura et al. 1986).

The volume density values for foetal type II cells of *rabbits* fasted (by complete withdrawal of food) for 7 days (days 21–28 of gestation) were, with the exception of the Golgi apparatus, found to be intermediate between those of the controls and a pilocarpine (5 mg/kg administered subcutaneously on days 24 through 27 of gestation)-treated group (Smith et al. 1982). A high percentage of lamellar inclusion bodies within type II cells of the fasted group appeared morphologically abnormal, being characterised by a highly irregular outline and tightly packed lipid membranes.

Intracellular macrophage inflammatory protein-2 was found mainly in the **endoplasmic reticulum** of primary cultured *rat* pneumocytes in response to lipopolysaccharide stimulation (Xavier et al. 1999). The amount of MIP-2 produced in these cells is at a magnitude similar to that from macrophages. Its production in *rat* pneumoctes is regulated at

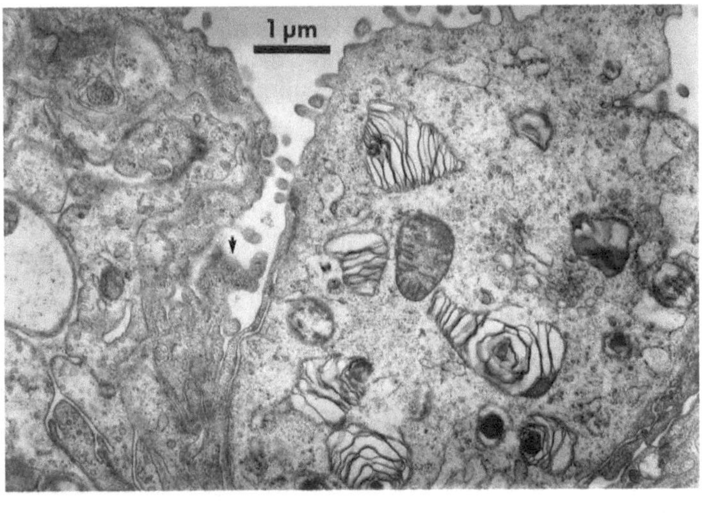

Fig. 97. Membrane vesiculation (arrow) in a type I pneumocyte (left) and lamellar bodies in a type II pneumocyte (right) of an unmedicated male *rat* (No. 2573, block BNh 3767). Under pentobarbital anaesthesia (30 mg/kg), the animal was perfused from the abdominal aorta with 2.5 % glutaraldehyde in 0.1 M sodium cacodylate buffer (pH 7.4). Postfixation with 1 % osmium tetroxide in sodium cacodylate buffer. Embedded in Epon 812 and sectioned at 50 nm. Lead citrate and uranyl acetate. Plate 2515

both the transcriptional and translational levels. Isowa et al. (2000) found that brefeldin A, by blocking anterograde transport from the endoplasmic reticulum to the Golgi apparatus, a process commonly regulated by microtubules, decreased lipopolysaccharide-induced MIP-2 inthe culture medium and increased its storage in cells.Breeldin A at concentrations as high as 10 μg/ml failed to alter the assembly of phosphatidylcholine into lamellar bodies (Osanai et al. 2001). The same concentration of brefeldin A was also ineffective at altering the secretion of newly synthesised phosphatidylcholine from alveolar type II cells isolated from male Sprague-Dawley *rats*. In contrast, concentrations of the drug of 2.5 μg/ml completely arrested newly synthesised lysozyme secretion from the same cells, indicating that brefeldin A readily blocked protein transport processes in pneumocytes II. The disassembly of the Golgi apparatus following brefeldin A treatment was also demonstrated by showing the redistribution of the resident Golgi protein MG-160 to the endoplasmic reticulum.

Type II pneumocytes can be identified at the light microscope level by antibodies against cytokeratin 18 (Schlichenmaier et al. 2000). This intermediate filament was thereby predominantly localised in the apical compartment and the intensity of the immunostaining depends on the stage of development. Type I alveolar cells as well as alveolar macrophages were consistently negative to cytokeratin 18 in postnatal *porcine* lung tissue.

Alveolar type II cells of the *rat* were able to internalise, rapidly and in relatively large amounts, certain types of tracer substances (ferritin, dextran) and, after internalisation, deposit them in **multivesicular bodies** and lamellar bodies (Williams 1984). The studies of Hubbard (1982) on hepatic metabolism of asialoglycoproteins, of McKanna et al. (1984) on endocytosis of ferritin-labelled epidermal growth factor, and of Goldstein et al. (1985) on receptor-mediated uptake of low density lipoproteins and the resultant suppression of cholesterol synthesis, illustrated the general functional roles of multivescular bodies. Male Sprague-Dawley *rat* alveolar type II cells seem to have more vesicular bodies than do other types of cells (0.6 + 0.1 % of their total cell volume or 3±0.6 μm^3 as absolute value for this compartment per cell) and they contain at least three different types of multivesicular bodies, classed loosely as electron-lucent (and large), electron dense (and small) and composite bodies that contain a few phospholipid lamellae (Young et al. 1985). The most striking polarisation observed was the strong association of multivesicular body volume in the quartile closest to the Golgi surface. In *rats* given paraquat orally (680 μmol/kg) the multivesicular bodies were absent (Sykes et al. 1977).

Peroxisomes were observed in differentiating *mouse* epithelial cells at 15 days of gestation (Schneeberger 1972). They were present before the first appearance of cytosomes, a presumed storage site of surfactant. Immediately before birth there was a 3-fold increase in number of peroxisomes per cell, and their number continued to increase during the subsequent 16 postnatal days studied. The peroxisome to mitochondrion ratio was as low as 1/5 during the gestational period and gradually rose to adult levels (1/1) by the 1st week after birth.

Lysosomes isolated from type II cells of male New Zealand White *rabbits* injected intratracheally with [^{3}H]dipalmitoylphosphatidylcholine and [^{14}C]1,2-dihexadecyl-*sn*-glycero-3-phosphocholine are the primary catabolic organelle for alveolar surfactant dipalmitoylphosphatidylcholine following reuptake by type II cells *in vivo* (Rider et al. 2000).

In contrast to alveolar type I cells, the **plasma membrane** of alveolar type II cells appears to lack plasmalemmal invaginations (Campbell et al. 1999, Newman et al. 1999). Other cells lacking caveolae/caveolin are fully functional in responding to growth factors and have microdomains, namely rafts, that are resistant to Triton X-100 solubilization (Langlet et al. 2000). Thus caveolae are not the only type of plasmalemmal subcompartment organised to transduce signalling cascades. Nanjundan and Possmayer (2001) showed lipid phosphate phosphohydrolase activity, phosphatidylcholine plus sphingomyelin, and cholesterol enriched in the detergent-insoluble domains. However, nearly 80 % of the total lipid phosphate phosphohydrolase activity partitioned to the detergent-soluble fractions. The transferrin receptor, a clathrin-coated pit marker, as well as protein kinase Cα, localised predominantly to the detergent-soluble fractions.

8.2.2 Histochemistry

8.2.2.1 Glutathione

The GSH content of freshly isolated *rabbit* type II pneumocytes was 0.43±0.11 nmol/10^6 cells; it was depleted to 0.15±0.03 nmol/10^6 cells by incubating with 200 μM diethyl maleate for 20 min (Horton et al. 1987). Cysteine was best able to support resynthesis of GSH. There was no evidence for participation of a cystathionine pathway for glutathione synthesis.

Type II pneumocytes isolated from pathogen-free Sprague-Dawley *rats* may increase intracellular glutathione levels by reducing cystin to cysteine, which is then rapidly transported into the cell and used as a substrate for intracellular GSH synthesis (DENEKE et al. 1995).

8.2.2.2 Glycoproteins

Pneumocin (M_r 165 kDa) marks the apical membrane surface of type II pneumocytes (LWEBUGA-MUKASA 1991). Expression of glycoprotein is lost when type II cells differentiate into type I cells.

Tenascin is a large, 190–240 kDa, extracellular matrix glycoprotein with a hexameric, multidomain structure composed of disulphide-linked subunits (ERICKSON and BOURDON 1989). It is produced by the *human* type II alveolar epithelial cell line A549 (MARTIN et al. 1995). Basal levels of tenascin production are down regulated by the presence of fibronectin and collagens I and IV, and upregulated by IL-1β, TNF-α and TNF-γ the cytokines having a synergistic effect. Upregulation of tenascin and TGFβ production occurred in A549 cells by antibody against a pulmonary auto-antigen (WALLACE and HOWIE 2001).

Messenger RNA of receptors for advanced glycation end products (AGEs), final products of nonenzymatic glycation and oxidation of proteins or lipids, is abundantly expressed in type II pneumocytes (KATSUOKA et al. 1997).

Two high density lipoprotein-binding proteins purified from *rat* lung tissue were identified as HB2 (LUTTON and FIDGE 1994) and a glycosyl phosphatidylinositol-anchored membrane dipeptidase (WITT et al. 2000). The level in type II pneumocyte membranes of both binding proteins increased when the plasma lipoprotein concentration was reduced by treatment of *rats* with 4-aminopyrazolo[3,4-d]-pyrimidine, consistent with a function to facilitate lipid uptake *in vivo* (WITT et al. 2000). The binding proteins were also dramatically upregulated by feeding *rats* a vitamin E-depleted diet. Vitamin E uptake requires interaction between high density lipoprotein and type II pneumocytes, suggesting a role of HB2 and membrane dipeptidase also in this process.

8.2.2.3 Enzymes

8.2.2.3.1 Oxidoreductases

Oxidoreductases with NAD^+ or $NADP^+$ as Acceptor

Lactate Dehydrogenase (EC 1.1.1.27)

Lactate dehydrogenase (EC 1.1.1.27)

S-Lactate + NAD^+ → Pyruvate + NADH [125]

During culture of *rabbit* type II pneumocytes in normoxia for48 h, lactate dehydrogenase activity increased to 257% of its initial value (PANUS et al. 1988).

Glucose-6-phosphate Dehydrogenase (EC 1.1.1.479)

Glucose-6-phosphate 1-dehydrogenase (EC 1.1.1.49)

D-Glucose-6-phosphate + $NADP^+$ →
D-Glucono-1,5-lactone-6-phosphate+ NADPH [126]

Glucose-6-phosphate dehydrogenase activity in primary cultures of type II *rat* pneumocytes was 442±61 units per mg DNA or 33.1±4.6 units per mg protein while the figures for activity in whole lung trimmed of large bronchi and major blood vessels were 285±8 units per mg DNA or 14.0±1.0 units per mg protein (FORMAN and FISHER 1981, 1985).

Carbonyl Reductase (EC 1.1.1.184)

By an electronmicroscopical immunogold procedure using monospecific antibodies against the enzyme, MATSUURA et al. (1994) localised tetrameric carbonyl reductase within the mitochondria of the type II pneumocytes of the *mouse*. No significant labelling was detected over other components of the epithelial cells.

Oxidoreductases with Oxygen as Acceptor

Xanthine Dehydrogenase (EC 1.1.1.204) and ***Xanthine Oxidase*** (EC 1.1.3.22)

PANUS and FREEMAN (1988) detected xanthine dehydrogenase/xanthine oxidase in isolated type II pneumocytes from several species and the activity was followed in culture. Cells were lysed by sonication in a buffer designed to inhibit artifactual conversion of dehydrogenase to xanthine oxidase by proteolysis and sulfhydryl oxidation. Xanthine dehydrogenase/xanthine oxidase activity was detected by conversion of pterin to isoxanthopterin, a fluorescent product (λ_{ex} = 345 nm and λ_{em} = 390 nm), using methylene blue as the electron acceptor for xanthine dehydrogenase. Isolated *rat* (n = 3) and *rabbit* (n = 5) type II pneumocytes contained 243±197 and 33±24 μU/mg protein, respectively, of which 33±14 % and 25±10 %, respectively, was in the xanthine oxidase form. A 50 % xanthine dehydrogenase to xanthine oxidase conversion would result in intracellular H_2O_2 and $O_2^{\bullet -}$ production of 73 and 49 pmol/min/mg protein respectively, in *rat* type II cells, and 10 and 7 pmol/min/mg protein, respectively, in *rabbit* type II cells. Production of H_2O_2 and $O_2^{\bullet -}$ at these rates would be cytotoxic. Culture of *rat* pneumocytes II resulted in a continuous loss of xanthine dehydrogenase/xanthine oxidase activity becoming 8 % of freshly isolated cell values after 135 h of culture. In *rabbit* type II cells, 39 h of culture resulted in loss of xanthine dehydrogenase/xanthine oxidase activity below detectability.

Oxidoreductases with Reduced NAD or NADP as Donators

Glutathione Reductase (EC 1.6.4.2)

Treating alveolar type II cells isolated from Sprague-Dawley *rats* with *N,N'*-bis-(2-chlorethyl)-*N*-nitrosourea (BCNU) markedly decreased cellular GSSG-reductase activity associated with a time-dependent increase in cellular glutathione (GSH) concentrations (Jenkinson et al. 1994).

Oxidoreductases with Hydrogen Peroxide as Donator

Catalase (EC 1.11.1.6)

Detoxification of hydrogen peroxide by catalase (EC 1.11.1.6)

$$2\,H_2O_2 \rightarrow 2\,H_2O + O_2 \qquad [127]$$

In freshly isolated alveolar type II cells of male, 250- to 300-g, pathogen-free *rats* catalase activity was 2,710±180 U/mg DNA (Walther et al. 1991).

Glutathione Peroxidase (EC 1.11.1.9)

Glutathione peroxidase consists or four protein subunits, each containing one atom of selenium at its active site, probably incorporated into cyteine molecules replacing their sulphur atoms.

Glutathione peroxidase (EC 1.11.1.9)

$$2\,GSH + H_2O_2 \rightarrow GSSG + 2\,H_2O \qquad [58]$$

The protein contains a selenocysteine residue. Steroid and lipid hydroperoxides, but not the product of reaction of EC 1.13.11.12 on phospholipids, can act as acceptor, but more slowly than H_2O_2 (cf. EC 1.11.1.12

The constitutive enzyme activity of glutathione peroxidase in pneumocytes II isolated from SPF Sprague-Dawley *rats* is 2830±220 units/mg DNA or 211±16 units/mg protein (Forman and Fisher 1981, 1985).

During culture of *rabbit* type II pneumocytes for 48 h in normoxia, glutathione peroxidase activity decreased to 13 % of its initial value (Panus et al. 1988).

A549-cells, a *human* airway epithelial carcinoma cell line with type II alveolar epithelial cell differentiation, dislodged before use with a 0.25 per cent trypsin solution, did not contain any detectable glutathione peroxidase (Rietjens et al. 1985).

Oxidoreductases Acting on Paired Donors with Incorporation of Molecular Oxygen

Prolyl 4-hydroxylase (EC 1.14.11.2)

Prolyl 4-hydroxylase (EC 1.14.11.2) is a key enzyme required for the posttranslational hydroxylation of proline residues in collagen. The enzyme consists of a tetramer composed of two pairs of nonidentical subunits ($\alpha_2\beta_2$) of ≈60 kDa each (Berg et al. 1979). Peptide mapping has demonstrated that the α and β subunits are products of different genes (Berg et al. 1979). The β subunit has been shown (Pihlajaniemi et al. 1987) to be virtually identical with the enzyme protein disulphide isomerase (EC 5.3.4.1).

Kasper et al. (1994) reported the selective immunohistological localisation of protein disulphide isomerase in *human* type II alveolar and bronchial epithelial cells. The detection of the hidden antigen with the monoclonal antibody 5B5 usually failed in paraffin sections but succeeded after microwave pre-treatment of tissue slices.

Nitric Oxide Synthase (EC 1.14.23)

L-Arginine–nitric oxide pathway

$$\text{L-Arginine} \xrightarrow{\text{NOS}} \textbf{Nitric oxide}(^{\bullet}NO) \xrightarrow{+O_2^{\bullet -}} \text{Peroxynitrite } (ONOO^-) \qquad [46]$$

In A549-cells, stimulation with a combination of interferon γ, tumour necrosis factor α, interleukin-1β, and LPS for 12, 24 and 48 h induced inducible nitric oxide synthase (WATKINS et al. 1997). TNF-α alone does not induce •NO production directly; however, it does have a stimulatory effect on IL-1β-induced •NO production (KWON and GEORGE 1999). IL-1β and interferon γ both •NO production alone, yet at different concentration thresholds, and act synergistically when present together. In the presence of all three cytokines, the net effect of •NO production exceeded the predicted additive effect of each individual cytokine and the two-way interactions.

Using anti-digoxigenin-alkaline phosphatase and nitroblue tetrazolium PUNJABI et al. (1994) visualised inducible nitric oxide synthase mRNA in *rat* type II pneumocytes. iNOS mRNA was induced in A549-cells by interferon-γ and interleukin-1β (DWEIK et al. 1996). TNFα enhanced IFNγ/IL-1β expression 4-fold, while IL-4 attenuated the IFNγ/IL-1β expression 4-fold. Addition of lipopolysaccharide did not affect IFNγ/IL-1β iNOS expression.

Type 2 foetal *rat* alveolar epithelial cells preincubated with IFN-γ (100 U/ml), TNF-α (500 U/ml), IL-1β (300 pM), and lipopolysaccharide exposed to 15% CO_2 (hypercapnia) revealed a significant increase in •NO production and NOS activity (LANG et al. 2000). Cell 3-nitrotyrosine content as measured by both HPLC and immunofluorescence staining also increased when exposed to the same conditions. Hypercapnia significantly enhanced cell injury as evidenced by impairment of monolayer barrier function and increased induction of apoptosis. The results were attenuated by the NOS inhibitor *N*-monomethyl-L-arginine (1 mM).

Type II pneumocytes can be identified at the light microscope level by antibodies against cytokeratin 18 (SCHLICHENMAIER et al. 2000). This intermediate filament was thereby predominantly localised in the apical compartment and the intensity of the immunostaining depends on the stage of development. Type I alveolar cells as well as alveolar macrophages were consistently negative to cytokeratin 18 in postnatal *porcine* lung tissue.

Oxidoreductases with Superoxide Radical as Acceptor

Superoxide Dismutase (EC 1.15.1.1)

Superoxide dismutase (EC 1.15.1.1)

$$2\,O_2^{\bullet-} + 2\,H^+ \longrightarrow H_2O_2 + {}^3\Sigma_g^- O_2 \qquad [128]$$

Human alveolar type II cells obtained from surgical resections expressed the extracellular form of superoxide dismutase mRNA (SU et al. 1997).

In *human* A549 lung epithelial cells transfected with Mn superoxide dismutase promoter-reporter constructs with or without thymidine kinase-driven *Renilla* luciferase, exposure to hypoxia (<2.5% O_2) for 24 h caused a significant increase in luciferase (OHMAN et al. 1999).

In freshly isolated alveolar type II cells of male, 250- to 300-g, pathogen-free *rats* SOD activity was 213±11 U/mg DNA (WALTHER et al. 1991).

In a *rat* type II pneumocyte analogue, the L2 cell line, exposed for 6 h to a combination of interferon-γ (2,000 U/ml) and tumour necrosis factor-α (500 U/ml), extracellular superoxide dismutase and inducible nitric oxide synthase transcription was upregulated (BRADY et al. 1997). Transcription of both genes was linked by activation of the transcription factor nuclear factor-κB.

8.2.2.3.2 Transferases

O^6-Alkylguanine-DNA alkyltransferase (EC 2.1.1.63)

O^6-Alkylguanine-DNA alkyltransferase (EC 2.1.1.63) of type II pneumocytes isolated from male New Zealand White *rabbits* was 5.82 fM/10^6 cells (DEILHAUG et al. 1985).

Animal **fatty acid synthase** (EC 2.3.1.85) is a homodimer of approx. 570 kDa and in the *rat* and *human* is the product of an 18–19 kb gene (SMITH 1994). LU et al. (2001) transfected A549 cells with various fatty acid synthase gene constructs ligated to the firefly luciferase gene and cultured them with dexamethasone for 24 h after which luciferase activity was measured. Dexamethasone increased luciferase expression in response to a fragment in the promoter and 5'-flanking region of the fatty acid synthase gene, from −1592 to +65 bp. This increase was antagonised by triiodothyronine (T_3) and transforming growth factor-β1. Serial deletions showed that the full response to dexamethasone and T_3 were retained in the 89 bp -33/+56 bp fragment whereas the response to TGF was mediated by the immediately upstream -104/-34 bp sequence.

γ-Glutamyltransferase (EC 2.3.2.2)

γ-Glutamyltransferase (EC 2.3.2.2), a membrane-bound enzyme found on the external surface of the cell (MEISTER et al. 1981), breaks the γ-glutamyl peptide bond of glutathione (GSH), releasing cysteinylglycine. The γ-glutamyl moiety is then transferred to an amino acid forming a γ-glutamyl-amino acid, which is transported into the cell (GRIFFITH and MEISTER 1979): the amino acid is released in the cell by the action of γ-glutamyl-cyclotransferase. The cysteinylglycine moiety is also taken up into the cell as an intact molecule, or as cysteine and glycine after hydrolysis by a dipeptidase. Inside the cell glutamate, cysteine and glycine may be recycled for de novo GSH synthesis (Griffith et al. 1979). Exposure of primary cultures of *rat* type II pneumocytes to the GSH depleting agent L-buthionine-[*SR*]-sulfoximine (BSO) for 3 h led to a dose dependent increase in of γ-glutamyltransferase activity 24 and 48 h later, with almost twice the activity at 125 μM BSO (Van Klaveren et al. 1997). Restoration of intracellular GSH levels by addition of GSH to the culture medium completely prevented the increase in γ-glutamyltransferase activity by BSO, while addition of catalase or dimethylthiourea had no effect.

Glutathione S-transferase (EC 2.5.1.18)

Glutathione S-transferase theta I was localised in type II pneumocytes of the *mouse* (Quondamatteo et al. 1998).

Glutathione S-transferase (EC 2.5.1.18)

RX + Glutathione → HX + R–S–G [129]

8.2.2.3.3 Hydrolases

Phosphoric Monoester Hydrolases

Alkaline Phosphatase (EC 3.1.3.1)

Alkaline phosphatase, a dimeric glycoprotein that catalyses the hydrolysis of orthophosphoric monoesters, is a membrane-bound enzyme comprising four isoenzymes encoded by different structural genes: tissue non-specific, intestinal, placental and germ-cell (Moss 1992).

Meban (1975) localised alkaline phosphatase in the granular pneumocytes of *hamster* lung. Alkaline phosphatase expression by type II cells appears to be regulated in concert with the synthesis of the phospholipid and apoprotein components of pulmonary surfactantin both adult *rat* type II cells in primary culure and in the foetal lung (Edelson et al. 1988).

An increase of a pulmonary isoform of a tissue nonspecific alkaline phosphatase isoenzyme activity has been observed in the bronchoalveolar lavage fluid (separated from cells) of *patients* affected by idiopathic pulmonary fibrosis, sarcoidosis, chronic bronchitis, and silicosis (Capelli et al. 1997).

The alkaline phosphatase isolated from the surfactant from lipopolysaccharide-instilled *rats* was tissue non-specific, but some of the activity was characteristic of intestinal-type alkaline phosphatase (Harada et al. 2002).

Acid phosphatase (EC 3.1.3.2) was chiefly found in the characteristic laminated vacuoles of the type II pneumocytes of the *rat* (Corrin et al. 1969). Not all such vacuoles nor indeed all such cells contained the enzyme; but in many, acid phosphatase reaction product formed a thin rim about the periphery of the vacuoles. The enzyme was not limited to the smaller of these cytosomes and was often present in the larger and presumably older vacuoles. Occasionally reaction product was aligned alongside or within channels of the endoplasmic reticulum. The multivesicular bodies described by Sorokin (1966) and Goldenberg et al. (1967) generally contained acid phosphatase at their periphery and between the individual vesicles.

Serine Proteinases (EC 3.4.21)

Urokinase-type plasminogen activator (EC 3.4.21.73)

Urokinase-type plasminogen activator (EC 3.4.21.73) is synthesised in a single-chain form and is converted to an active, two chain form after it binds to its cell surface receptor, uPAR. Treatment *human* of alveolar type II cells (A549) with 1 to 5 μM chromium(VI) for 4 and 12 h decreased both the specific activity and the amount of uPA protein (Shumilla and Barchowsky 1999). Chromium reduced uPA protein levels by inhibiting protein synthesis and had no effect on uPA mRNA levels or the rate of uPA protein degradation. In contrast, both mRNA and protein levels for the uPA receptor were increased by treatment with concentrations of chromium(VI) that did not completely inhibit protein synthesis.

Cysteine Endopeptidases (EC 3.4.22)

Gelatinase (EC 3.4.24)

Gelatinase A (EC 3.4.24.24)

Cleavage of gelatin type I and collagenes types IV, V, VII, X. Cleaves the collagen-like sequences

Pro–Gln–Gly–|–Ile–Ala–Gly–Gln [130]

Gelatinase A (EC 3.4.24.24), a 70-kDa protein, was produced *in vitro* by type II pneumocytes isolated from pathogen-free Sprague-Dawley *rats* (D'Ortho et al. 1997). A 95-kDa gelatinolytic activity was also present but usually in small amounts compared with the 70-kDa gelatinases. A549 cells under basal conditions expressed only 70-kDa gelatinase activity. Gelatinase activities were inhibited by EDTA but not by phenylmethylsulfonyl fluoride or *N*-ethylmaleimide, demonstrating that they are metalloproteases. Zymography of samples incubated in the presence of 4-aminophenyl-mercuric acetate before electrophoresis showed shifts of 95- to 88-kDa gelatinase and 70- to 62-kDa gelatinase.

Metalloproteinase-2 gelatinolytic activity was expressed in A549 cells, and exposure to ionising radiation (16 Gy) increased this activity (ARAYA et al. 2001). In contrast to treatment with p53 sense oligonucleotide or with lipofectin alone, treatment with p53 antisense oligonucleotide almost perfectly abrogated the irradiation-induced accumulation of p53 protein in A549 cells 3 h after radiation exposure.

Foetal *rat* alveolar epithelial cells (19–20 days) had higher matrix metalloproteinase-2 and metalloproteinase-9 gelatinase expression than adult alveolar epithelial cells, with fivefold higher matrix metalloproteinase-9 activity, and were migratory through gelatin, responding to epithelial growth factor, and fibroblast growth factor-10 (BUCKLEY et al. 2001).

Hydrolases Acting on Acid Anhydrides

Na^+-K^+-ATPase (EC 3.6.1.3)

Na^+-K^+-ATPase (EC 3.6.1.3)

$$ATP + H_2O \longrightarrow ADP + Orthophosphate \quad [131]$$

Exposure of *rat* alveolar type II cells plated on polycarbonate filters in serum-free medium to 10 ng keratinocyte growth factor/ml on day 4 resulted in dose-dependent increases in short-circuit current (I_{sc}) across alveolar epithelial cell monolayers compared with controls by day 5, with further increase occurring through day 8 (BOROK et al. 1998). Relative Na^+-K^+-ATPase α_1-subunit mRNA abundance was increased by 41 % on day 6 and 8 after exposure to keratinocyte growth factor, whereas α_2-subunit mRNA remained only marginally detectable in both the absence and presence of keratinocyte growth factor. Levels of mRNA for the β_1-subunit of Na^+-K^+-ATPase did not increase, whereas cellular α_1- and β_1-subunit protein increased 70 and 31 %, respectively, on day 6. mRNA for α-, β-, and γ-*rat* epithelial Na channel all decreased in abundance after treatment with keratinocyte growth factor.

Na^+-K^+-ATPase glycosylated β-subunit protein was identified in *rat* type II pneumocytes grown in primary cultures as a broad ~50 kDa band when blotted with the polyclonal anti β_1 antibody SpET but could not be detected by blotting with other anti-β antibodies (ZHANG et al. 1997). The lung β-subunit may be glycosilated differently from kidney and other tissues.

Differentiation of *rat* type II pneumocytes to a pneumocyte I-like phenotype resulted in a decrease of α_1- and an increase of α_2-mRNA and protein abundance without changes in the β_1-subunit (RIDGE et al. 1997). The existence of the distinct functional classes of Na^+-K^+-ATPase in type II pneumocytes and pneumocyte type I-like cells was confirmed by ouabain inhibition of Na^+-K^+-ATPase activity. Ouabain inhibition of type II pneumocytes was consistent with expression of the α_1-isozyme (IC_{50} = 4×10^{-5} M); whereas, in pneumocyte I-like cells, it was consistent with the presence of both α_1- and α_2-isozymes (IC_{50} = 9.0×10^{-5} and 1.5×10^{-7} M, respectively); [^{3}H]ouabain binding studies corroborated these findings.

8.2.2.3.4 Lyases

Guanylate Cyclase (EC 4.6.1.2)

Guanylate cyclases exist in both the soluble and membrane fractions of most tissue homogenates, probably across all animals (FOSTER et al. 1999). The conservations of amino acids within "signature' is pronounced, and thus a cyclase homology domain, known to process either adenylyl or guanylyl cyclase activity, and an amphipathic region found in both soluble and membrane forms of guanylate cyclase are easily recognised at the primary amino acid level.

In *rat* lung membranes, atrial natriuretic peptide stimulated the guanylate cyclase 2.1-fold compared to the control at an optimal concentration of 0.1 mM (NASHIDA et al. 2000). The stimulation by atrial natriuretic peptide was smaller in the presence of Mn^{2+} than in the presence of Mg^{2+}. The addition of inorganic phosphate to the reaction mixture altered the guanylate cyclase activities in the presence of Mn^{2+} with or without atrial natriuretic peptide and/or ATP. In the presence of 10 mM phosphate, ATP dose-dependently stimulated the basal guanylate cyclase 5-fold compared to the control at a concentration of 1 mM and augmented the atrial natriuretic peptide-stimulated guanylate cyclase, which was 4.2-fold compared to the basal guanylate cyclase, 5.5-fold compared to the control at a concentration of 0.5 mM. Protein phosphatase inhibitors, okadaic acid (100 nM), H8 (1 μM) and staurosporin (1 μM), did not alter the activity. Orthovanadate (1 mM), an inorganic phosphate analogue, significantly stimulated both the basal guanylate cyclase and the atrial natriuretic peptide-stimulated guanylate cyclase, which were inhibited by ATP.

Ca^{2+} reversibly inhibited both the basal and NO-stimulated forms of *bovine* lung soluble guanylate cyclase (SERFASS et al. 2001). Inhibition was independent of the activator identity and concentration, revealing that Ca^{2+} interacts with a site independent of the heme regulatory site.

8.2.2.3.5
Ligases

Glutamine synthetase (EC 6.3.1.2) was inhibited in A549 *human* lung adenocarcinoma cells during hyperoxia (McGrath-Morrow and Stahl 2002).

8.2.2.4
Endothelin

Endothelin-1, a peptide initially purified from *porcine* aortic endothelial cells, has been shown to be expressed by *human* pneumocytes type II in lung fibrosis (Giaid et al. 1993), by *rat* pneumocytes II (Crestani et al. 1994), and a cell line derived from *rat* pneumocytes II (L2; Markevitz et al. 1995). ET-1 has potent vascular and bronchial smooth muscle cell constrictor properties, acts as a mitogen for different cell types, such as fibroblasts or smooth muscle cells (Battistini et al. 1993), and is involved in the modulation of the inflammatory response through a direct effect on alveolar macrophages (Nagase et al. 1990) or mast cells (Ehrenreich et al. 1992). ET-1 production is inhibited by IL-1β (Odoux et al. 1997). IL-1β effect was dependent on protein synthesis, was partially prevented by indomethacin, and was totally prevented by dexamethasone.

8.2.2.5
Tenascin Immunoreactivity in Cryptogenic Fibrosing Alveolitis

Examination of lung biopsies from patients with cryptogenic fibrosing alveolitis has indicated that prominent type II cell hyperplasia is often associated with deposition of tenascin (Wallace et al. 1995), which is produced at sites of active scarring (Mackie et al. 1988).

8.2.2.6
Fas

Fas is a member of the TNF/NGF receptor family (Itoh et al. 1991, Nagata 1995). Fas, a type I membrane receptor protein, transduces a signal culminating in apoptosis after binding to the Fas ligand. Fine et al. (1997) found Fas gene expression in RNA derived from fresh isolates of primary *rat* type II pneumocytes; restriction of Fas expression to a subset of alveolar type II cells by *in situ* hybridisation and immunohistochemistry of the normal *mouse* lung; induction of apoptosis in a *mouse* lung type II epithelial cell line (MLE) after activation of Fas; and induction of apoptosis in a subpopulation of type II cells after the intratracheal instillation of an activating anti-Fas antibody in *mice*.

8.2.3
Surfactant

Pulmonary surfactant is a lipid-protein complex secreted by type II pneumocytes into the fluid lining the alveoli. Pulmonary surfactant contains specific proteins, categorised as surfactant proteins A–D (Possmayer 1988). Hydrophobic protein B and C together make up about 2 weight per cent of the lipids, although the exact amount of each has not yet been determined with certainty. Protein C, an acylated, hydrophobic, α-helical peptide, enhances the surface activity of pulmonary surfactant lipids (Nag et al. 1996). In *humans*, the amount of saturated phosphatidylcholine does not change from infancy to old age (Rebello et al. 1996).

Expression of interleukin-4 in nonciliated epithelial cells in the conducting airways of transgenic *mice* resulted in a 6.5-fold increase in saturated phosphatidylcholine, a 15-fold in surfactant protein D in the alveolar pool relative to wild type *mice* (Ikegami et al. 2000).

Surfactant protein A (SP-A) belongs to a subgroup of the C-type lectin superfamily, which has collagen-like domains and referred to as collectins. SP-A inhibits lipid secretion from and augments lipid uptake by type II pneumocytes. The mannose-binding protein (MBP) A from *rat* sera that is structurally analogous to SP-A was not able to regulate the uptake and secretion of lipids by type II cells (Kuroki et al. 1997). However, three chimeric molecules constructed by PCR using the overlapping extension method and cDNAs of *rat* SP-A and *rat* MBP-A and consisting of

Asn^{1}-Cys^{218} of SP-A and Gln^{210}-Ala^{221} of MBP,
Asn^{1}-Lys^{203} of Sp-A and Cys^{195}-Ala^{221} of MBP,
Asn^{1}-Gly^{194} of SP-A and Glu^{185}-Ala^{221} of MBP,

retained the ability to augment lipid uptake by and inhibit lipid secretion from type II pneumocytes. SP-A exists as a set of isoforms, which arise by alternate proteolytic processing, are differentially glycosylated, and are important for complete disulphide-dependent oligomeric assembly (Elhalwagi et al. 1997). SP-A is expressed in both alveolar type II cells and in Clara cells. Feinstein et al. (1997) analysed the control of transcription by the *rat* SP-A promoter in two lung cell lines, NCI-H441 (a *human* line thought to derive from a Clara cell) and MLE-15 (a *mouse* line derived from a type II pneumocyte). Reporter gene constructs carrying SP-A promoter deletions joined to the bacterial chloramphenicol acetyl transferase (CAT) gene were transfected into each of the cell lines. CAT activity was assayed 48–72 h after transfection and normalised to the activity of a control plasmid. Maximal activity in both cell lines was obtained with a construct containing 163 nucleotides upstream of the transcriptional start. In MLE-15 cells, deletion of the region between 163 and 133 nucleotides before the transcriptional start (−163 to −133) decreased expression of the reportor gene construct by about 50 %. This region contains several potential binding sites for thyroid transcription factor 1 (TTF-1). A deletion extending from −163 to −77 did not further decrease expression from the levels seen with the shorter deletion. In contrast, in NCI-H441 cells, deletion from −163 to −133 did not affect activity but deletion from −163 to −77 decreased expression by about 50 %. In both cell lines mutation of the nucleotides between −92 and −89, in the context of the −163 SP-A/CAT construct, decreased expression by about 50 %. Both cell lines contain nuclear proteins which bind to an oligonucleotide corresponding to the region between −95 and −77.

There is a significant positive correlation between surfactant protein A and Clara cell 10-kDa protein values in bronchoalveolar lavage fluids of healthy subjects (SHIJUBO et al. 1998).

In adult *human* lungs, only type II pneumocytes could be identified as surfactant protein A-positive cells within the parenchymal region (OCHS et al. 2002). SP-A was localized mainly in small vesicles and multivesicular bodies close to the apical plasma membrane. Only few lamellar bodies were weakly labelled at their outer membranes. Stereologic analysis showed this weak signal to be due to specific labelling. In the alveolar space, lamellar body-like surfactant forms in close proximity to tubular myelin were labelled for SP-A at their periphery. The strongest SP-A labelling was found over tubular myelin figures. Labelling for SP-A was also found in close association with the surface film and unilamellar vesicles.

In midtrimester *human* foetal lung explants treatment with antisense epidermal growth factor receptor oligonucleotide (15–90 μM) decreased both surfactant protein A mRNA and protein compared with controls with no effect in the sense condition (KLEIN et al. 2000). The oligonucleotides did not affect tissue viability as measured by the release of lactate dehydrogenase.

SP-A inhibited production of NO and iNOS in isolated *rat* alveolar macrophages stimulated with smooth lipopolysaccharide, which did not significantly bind SP-A, or rough LPS, which avidly bound SP-A (STAMME et al. 2000). In contrast, SP-A enhanced production of NO and iNOS in cells stimulated with IFN-γ or IFN-γ plus LPS. Neither SP-A nor SP-D affected baseline NO production, and SP-S did not significantly affect production of NO in cells stimulated with either LPS or IFN-γ.

After internalisation into isolated *rat* type II pneumocytes, SP-A and lipid were taken up via the coated-pit pathway and resided in a common compartment, positive for the early endosomal marker EEA1 but negative for the lamellar body marker 3C9 (WISSEL et al. 2001). SP-A then recycled rapidly to the cell surface via Rab4-associated recycling vesicles. Internalised lipid was transported toward a Rab7-, CD63-, 3C9-positive compartment, i.e., lamellar bodies. Inhibition of calmodulin led to inhibition of uptake and transport out of the EEA1-positive endosome and thus of resecretion of both components. Inhibition of intravesicular acidification (bafilomycin A_1) led to decreased uptake of both surfactant components. It inhibited transport out of early endosomes for lipid only, not for SP-A.

Surfactant protein B (SP-B), an 8 kDa hydrophobic protein essential for surfactant function, is produced from the intracellular processing of the 381 amino acid residues, 40 kDa preprotein SP-B within the type II pneumocytes.

Surfactant protein C (SP-C) is synthesised by type II pneumocytes as a 21 kDa propeptide which is proteolytically processed in subcellular compartments distal to the trans-Golgi network to yield the 35 residue mature form.

Surfactant protein D (SP-D), a hydrophilic glycoprotein with a reduced molecular mass of 43 kDa appears in sera of healthy adults as a background leakage from alveolar space into blood vessels and increases in sera from patients with idiopathic pulmonary fibrosis, pulmonary alveolar proteinosis, and interstitial pneumonia with collagen disease (HONDA et al. 1995). Studies in respiratory disease-free children (KANEKO et al. 1997) revealed that change in serum SP-D levels are not age-related and not gender-dependent, and that mechanisms of the SP-D leakage in healthy adults and children might be independent of that of SP-A leakage. SP-D, isolated from the lavage fluid of silica-treated *rats*, significantly enhanced the uptake of three of six strains of *Pseudomonas aeruginosa*, an important pulmonary pathogen, by rat alveolar macrophages as analysed by both fluorescence and electron microscopy (RESTREPO et al. 1999). SP-D had only minimal effects on phagocytosis of *Haemophilus influenzae*.

Surfactant convertase, a diisopropylfluorophosphate (DFP)-binding and DFP inhibitable carboxylesterase (GROSS et al. 1997) of 70 kDa (reduced), is required for conversion of heavy density (tubular myelin) to light density (small vesicle) subtypes of surfactant. Carboxylesterase ES-2 mRNA was localised to type II pneumocytes and alveolar macrophages but not to Clara cells (CLARK et al. 1997). Lamellar bodies contain both glycosylated (70 kDa) and deglycosilated (60 kDa) forms of convertase (DHAND et al. 1997).

Surfactant protein gene expression as influenced by the cytoskeleton was tested in cultured *rat* alveolar type II cells by SHANNON et al. (1998). Cytochalasin D, colchicine, or nocodazole did not result in dramatic cell shape changes, but ultrastructural examination revealed that the cytoplasm of cells treated with cytochalasin D was markedly disorganised; cells treated with colchicine did not exhibit such changes. Treatment with any of the three drugs resulted in a reduction in surfactant protein mRNAs. These decreases were not the result of cell toxicity, since overall protein synthesis was unimpaired by drug treatment. Washing the cells followed by an additional 2 days of culture resulted in an reaccumulation of surfactant protein mRNAs in cytochalasin D-treated but not in colchicine-treated cells. Washing of nocodazole-treated cultured resulted in partly recovery. Surfactant protein mRNA stability was estimated in the presence or absence of cytoskeleton-disrupting drugs. Disruption of either microfilaments or microtubules affected the half-lives of mRNAs for SP-A, SP-B, and SP-C.

It has been suggested that the air/water interface of the lung is formed by a monomolecular layer enriched in the saturated **phospholipid** dipalmitoyl-phosphatidylcholine. This layer is thought to be essential for reducing the surface tension of the alveoli to near zero minimum values during the breathing cycle (CLEMENT 1977). Employing sensitive real time assays, MEYBOOM et al. (1999) have studied the lipid binding characteristics of the surfactant protein A phospholipid liposome interaction. From the final equilibrium level of the resonant mirror binding signal, as apparent dissociation constant of ca. $K_d = 5\ \mu M$ is obtained for the complex between SP-A and dipalmitoyl-phosphatidylcholine liposomes. At nanomolar SP-A concentrations, this complex is formed with a subsecond (0.3 s) reaction time, as measured by light-scattering signals evoked by photolysis of caged Ca^{2+}. With palmitoyloleoyl-phosphatidylcholine, the complex formation proceeds at half the rate, compared to dipalmitoyl-phosphatidylcholine, leading to a lower final equilibrium level of SP-A lipid interaction. Disteoroylphosphatidylcholine showed a stronger interaction than dipalmitoyl-phosphatidylcholine.

Free radical generation by activated leucocytes may promote surfactant dysfunction during acute respiratoy distress syndrome. Surfactant incubated for 2 h with 0.5 mM $FeCl_2$/0.25 mM H_2O_2 to generate oxidised surfactant demonstrated loss of surface activity by bubble surfactometry at both 1 and 5 mg/ml (MARK and INGENITO 1996). Fractions of oxidised surfactant combined with normal surfactant 1:1 and 1:5 inhibited surface activity. H_2O_2 and $HO^{\bullet}$ impaired surfactant compressibility at γ = 20 mN/m by a process that may involve lipid peroxidation (LEE et al. 1997).

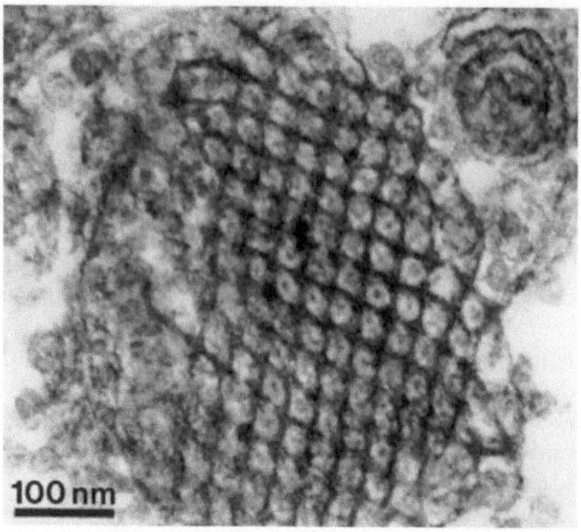

Fig. 98. Tubular myelin in the alveolar space of a 296 g female white *rat* (No. 3; breeder: Winkelmann Borchen-Kirchborchen) which inhaled 10 mg powdered aluminium/m³ 4 h per day, 5 days per week from August 16 to October 27, 1967 for a total of 51 days. Fixed on October 30, 1967 under methitural anaesthesia by intratracheal instillation of 2.5 % glutaraldehyde in phosphate buffer (pH 7.4) before opening the thorax. Postfixation with 1 % osmium tetroxide in phosphate buffer (pH 7.4). Contrasted en bloc for 12 h with 0.5 % uranyl acetate in 70 % ethanol. Embedded in a 2:8 mixture of methyl and butyl methacrylate. Sectioned at 50 nm. Lead citrate after REYNOLDS (1963). Plate 18/08. (from SCHILLER 1971)

$$H_2O_2 + Fe^{2+} \longrightarrow \mathbf{HO^{\bullet}} + HO^- + Fe^{3+} \qquad [132]$$

Caeruloplasmin can remove both hydrogen peroxide and lipid hydroperoxides at physiologically relevant concentrations of reduced glutathione known to be present in lung and lung lining fluid (PARK et al. 1999).

MUDWAY et al. (2001) observed a significant loss of ascorbate from proximal (–45.1 %, $P < 0.01$) and distal (–11.7 %, $P < 0.05$) respiratory tract lining fluid in 15 healthy subjects 6 h after a challenge to 0.2 ppm **ozone** for 2 h. This was associated with increased glutathione disulphide (GSSG) in these compartments.

Ozone (0.8 ppm) affected the surface activity of surfactant in *rats* exposed for 2 or 12 h, whereas distinct morphological changes in bronchoalveolar lavage or in the surfactant subtypes were not observed (PUTMAN et al. 1997). Adsorption experiments indicated that bronchoalveolar lavage from *rats* exposed for 12 h to O_3 remained at lower equilibrium surface pressures than lavage from control *rats*. These observations suggest interference of inflammatory proteins with the surface film. Extracted surfactant, containing only lipids and surfactant proteins B and C, had a decreased adsorption rate after O_3 exposure. These results suggest that the activity of one or both of the hydrophobic surfactant proteins (SP-B and SP-C) was affected by O_3.

Nitric oxide as supplied by the NO-donor *S*-nitroso-*N*-acetyl-D,L-penicillamine, protected *rat* alveolar type II cells against nuclear condensation and DNA fragmentation induced by stretch (30 % at 60 cycles per min) as well as by 0.4 M sorbitol (EDWARDS et al. 2000). *S*-Nitroso-*N*-acetyl-D,L-penicillamine depleted of •NO had no protective effect, and the NO scavenger 2-phenyl-4,4,5,5-tetramethylimidazoline-1-oxyl 3-oxide (0.2 mM) blocked the anti-apoptotic effect of *S*-nitroso-*N*-acetyl-D,L-penicillamine. Alveolar macrophages isolated from *rat* lung lavage fluid actively synthesised and secreted •NO. Using a novel technique in which alveolar macrophages were cocultured with type II pneumocytes while adhered to floating membrane rafts, EDWARDS et al. (2000) found that •NO released from alveolar macrophages was effective in protecting alveolar type II cells from undergoing apoptosis.

L-Arginine–nitric oxide pathway

$$\text{L-Arginine} \xrightarrow{\text{NOS}} \textbf{Nitric oxide}(^{\bullet}\text{NO}) \xrightarrow{+O_2^{\bullet-}} \text{Peroxynitrite } (ONOO^-) \qquad [46]$$

Peroxynitrite inhibited the hydrolysis of surfactant by *C. atrox* and *porcine* pancreatic secretory phospholipases A_2 in a dose dependent manner via nitration of essential tyrosine residues (SMITH et al. 1997).

Peroxynitrite is an isomer of nitrate, but is 36 kcal × mol-1 higher in energy (KOPPENOL et al. 1992). The half-life of peroxynitrous acid (HOONO), which is 7 s at 0 °C and 1 s at 37 °C, should be long enough to allow the molecule to diffuse inside the cell and react with DNA. The decomposition pathway of HOONO, leading to the formation of nitric acid, remains open to debate (CADET et al. 1997).

Formation and degradation of peroxynitrite

$$O_2^{\bullet-} + {}^{\bullet}NO \rightarrow ONOO^- \ (pK_a = 6.8) \qquad [133]$$

$$ONOO^- + H^+ \rightarrow ONOOH \qquad [134]$$

$$ONOOH \rightarrow HO^{\bullet} + NO_2^{\bullet} \ (20\text{–}30\%\ \text{yield}),\quad NO_3^- + H^+ \ (\text{rest}) \qquad [135]$$

The surface tension forces of the alveolar lining film are reduced when *rats* breath air containing ± 70 mg/m^3 of respirable **crystalline silica** 7 h daily for 10 days (MCDERMOTT et al. 1977), 500 mg/m^3 Al_2O_3 for 30 min (ROSENBERG et al. 1962), or are injected intratracheally with saline suspensions of powdered anatase, AlOOH, OX 50 SiO_2 (Degussa), or Dörentrup quartz $<3\ \mu m$ (STALDER and LADANI 1980), respectively. The total amount of lung surfactant factor phospholipids was increased in female Wistar *rats* inhaling either 20 mg DQ12 silica/m^3 for 5 h/day on 5 consecutive days or 20 mg DQ12 silica/m^3 for 5h/day on day 1 and 20 mg SiC/m^3 for 5 h/day on the consecutive 4 days (BRUCH et al. 1993).

The natural process of surfactant coating of inhaled particles in the alveoli may play a role in their phagocytic removal by pulmonary alveolar macrophages. Coating of both freshly ground and stale quartz particles with dipalmitoyl lecithin delayed the production of a diformazan precipitate from nitroblue tetrazolium by $O_2^{\bullet-}$ produced during their ingestion by *rat* alveolar macrophages *in vitro* (CILENTO and GEORGELLIS 1991). Both Min-U-Sil and kaolin abrogated the release of lactate dehydrogenase, β-glucuronidase, and β-*N*-acetyl glucosaminidase from *rat* alveolar macrophages and lysis of *sheep* erythrocytes (WALLACE jr et al. 1985).

The presence of *porcine* surfactant inhibited the quartz-induced down-regulation of complement receptor type 1 expression on activated *human* granulocytes (ZETTERBERG et al. 1997). The preincubation of granulocytes with surfactant before quartz exposure was sufficient to achieve this effect. Complement receptor type 3 expression was unaffected.

Chrysotile, as well as chrysotile after SO_2 sorption, instilled via the *rabbit*'s trachea induced an increase in the unsaturated fatty acid content of lung surfactant, which was similar to that observed in the respiratory distress syndrome of the *new-born*, and an increase in the protein level of pulmonary washings which may be explained by an increase of the permeability of the blood-air barrier (OBLIN et al. 1978).

GOLDENBERG et al. (1969), after stimulating surfactant production by giving pilocarpine to pathogen-free Wistar *rats*, found numerous free alveolar macrophages containing material similar to that lying free in the lumina.

Alveolar macrophages isolated by lung lavage of male Sprague-Dawley *rats* bound and degraded surfactant protein D in a time- and temperature dependent fashion (DONG and WRIGHT 1998). After 100 min of incubation, the formation of trichloroacetic acid-soluble degradation products increased 4-fold in the medium and 30-fold in the cells. The degradation of surfactant protein D was via a cell-associated process because surfactant protein D was not degraded when incubated in medium previ-

Fig. 99. A thin layer of surfactant covers both non-ciliated (Clara) and ciliated bronchiolar epithelial cells. 167 g male Wistar *rat* (R 21/63) fed 82 mg retinol palmitate (Arovit)/kg body weight daily from February 18 to 22, 1963 by stomach tube. On the 3rd day of the experiment the animal was exposed to an atmosphere containing 15 mg quartz DQ 12 (particle size $<3\ \mu m$) per m^3 air for 5 h using POLLEY's dust tunnel. Fixed in alcohol-chloroform-glacial acetic acid after Carnoy. Paraffin section, unstained, embedded in glycerol. Interference contrast Fl 100×/1.36/∞/–. An alveolar macrophage containing (birefringent) quartz particles is being eliminated via the airways

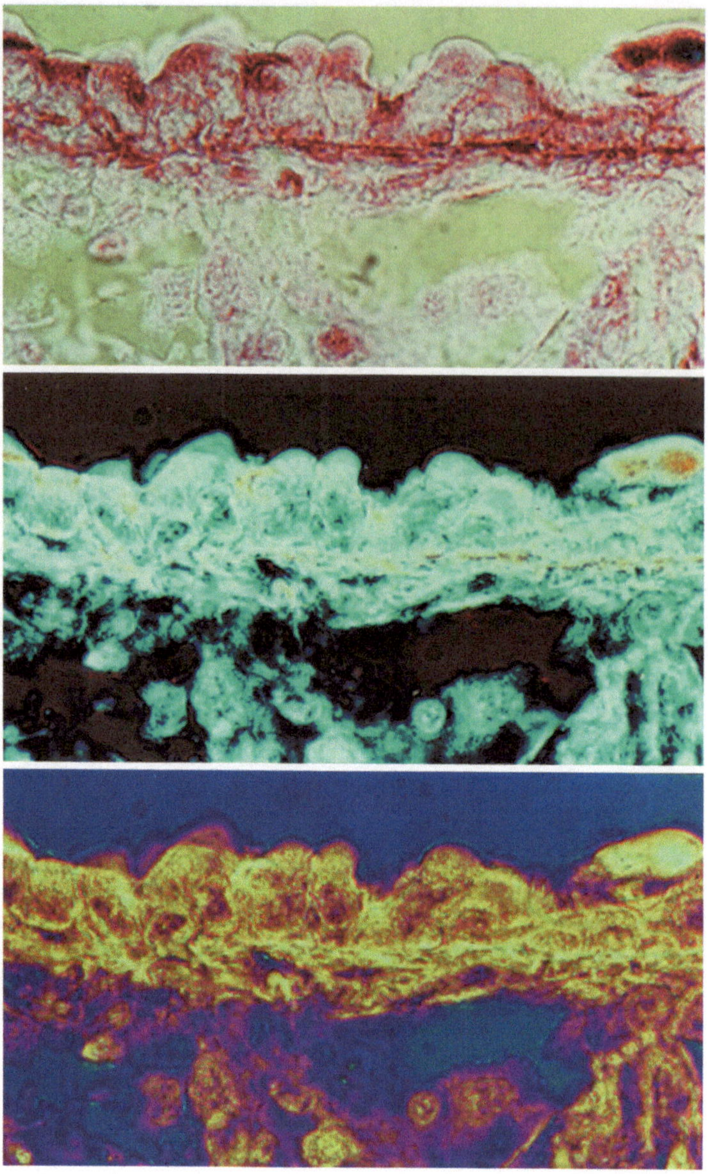

ously conditioned by alveolar macrophages. Gel autoradiography of cell lysate samples after incubation with ^{125}I-labelled surfactant protein D demonstrated an increase in degradation products, further confirming the degradation of surfactant protein D by alveolar macrophages. The degradation of surfactant protein D was not affected by coincubation with surfactant protein A or surfactant-like liposomes containing either phosphatidylglycerol or phosphatidylinsotol.

Various concentrations of total surfactant isolated from healthy adult *rats* suppressed proliferation of stimulated lymphocytes by up to 95% of mitogen-stimulated cells alone (YAO et al. 2001). Large aggregate subfractions of total surfactant had no effect on proliferation, whereas small aggregate subfractions significantly enhanced the lymphocyte proliferation at lower concentrations (7.8 μg/ml) compared to mitogen-stimulated cells alone. Higher concentrations (62.5 μg/ml) inhibited lymphocyte proliferation.

A modified *porcine* surfactant preparation (Curosurf) *in vitro* diminished the nitroblue tetrazolium reduction by both live group B *streptococci* and group B *streptococci*-stimulated polymorphonuclear leucocytes (BOUHAFS et al. 2000). Surfactant was peroxidised by reactive oxygen species from both group B *streptococci* and group B *streptococci*-stimulated polymorphonuclear leucocytes in a time-dependent manner. Vitamin E sig-

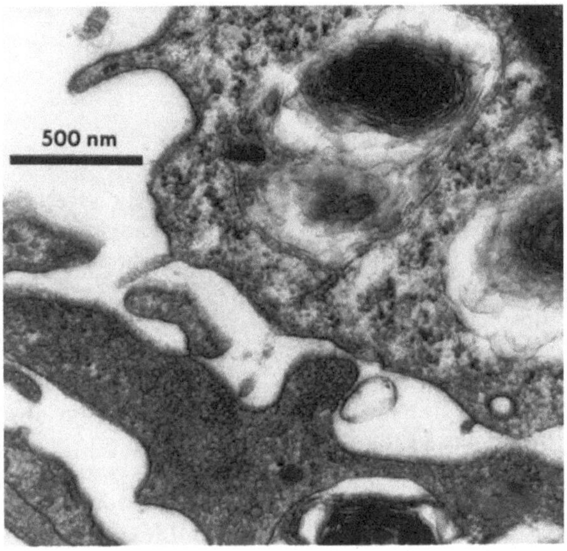

Fig. 100. Portions of two free alveolar macrophages showing mature inclusions of osmiophilic surfactant material. Female white *rat* (No. 4; breeder: Winkelmann, Borchen-Kirchborchen) which inhaled 10 mg powdered aluminium/m³ 4 h per day, 5 days per week from August 16 to October 27, 1967 for a total of 51 days. Fixed on January 15, 1968 under methitural anaesthesia by intratracheal instillation of 2.5% glutaraldehyde in phosphate buffer (pH 7.4) before opening the thorax. Postfixation with 1% osmium tetroxide in phosphate buffer (pH 7.4). Contrasted en bloc with 0.5% uranyl acetate in 70% ethanol. Embedded in a 2:8 mixture of methyl and butyl methacrylate. Sectioned at 50 nm. Lead citrate after REYNOLDS (1963). Film 207/84

nificantly reduced the peroxidation level of surfactant in both cases. Surfactant peroxidation was associated with a reduction in the number of live bacteria.

Opsonization of inhaled particles by surfactant proteins [which are known to have opsonizing properties (Van IWAARDEN et al. 1991, MASON et al. 1998)], antioxidants, and serum-derived and other locally produced proteins may change the surface-charge forces in favour of aggregation. The processed human bronchoalveolar lavage fluid used by KENDALL et al. (2002) may represent the hypophase of lung lining liquid into which the surface surfactant film has been shown to submerge particles <6 μm (SCHÜRCH et al. 1990, 1999). It is suggested that when particles reach the hypophase, they are then more readily cleared by normal clearance mechanisms. This aggregation mechanism is highly significant because macrophages do not readily phagocytize the smaller agglomerates of 35-nm spheres that dominate urban air in developed countries (BERUBE et al. 1999). Epithelial cells have been demonstrated to internalise these ultrafine particles (GRIESE and REINHARDT 1998).

Several reports have suggested the presence of **surfactant outside the bronchoalveolar tree.** HILLS et al. (1982) found surface active phospholipids on the *canine* pleura and speculated on their role of boundary lubricants. However, the demonstration of phospholipids is not a surfactant-specific binding. HAMM et al. (1991) investigated *human* pleural fluids. Mean surfactant protein A was 700 ng/ml (range 280–1750) in transudates and 350 ng/ml (150–500) in exudates. These results suggested that SP-A in pleural effusions does not originate from the alveolar space since increased pleural permeability should lead to higher SP-A values in exudates. Transmission electron microscopy of pleural tissue demonstrated lamellae bodies characteristic of surfactant synthesising cells. Scanning electron microscopy showed surfactant-like material on the mesothelial surface.

8.2.4 Pharmacology

Phosphatidylcholine exocytosis in cultured type II pneumocytes can be stimulated and inhibited by a variety of pharmacological agents. Surfactant secretagogues include agents that act via at least three distinct signal-transduction mechanisms (ROONEY et al. 1994). The signalling mechanisms are activation of the adenylate cyclase with formation of adeonsine 3',5'-cyclic monophosphate, and activation of cAMP-dependent protein kinase, activation of protein kinase C either directly or via generation of diacylglycerol, and elevation of intracellular Ca^{2+} and activation of a Ca^{2+}/calmodulin-dependent protein kinase (ROONEY et al. 1994).

Exposure of type II cells to the secretagogues ATP, 12-*O*-tetradecanoylphorbol-13-acetate, and cAMP resulted in a 1.5- to 2-fold enhancement of binding and uptake of the monoclonal antibody 3C9, which recognises an integral lamellar body-limiting membrane protein of 180 kDa (BATES et al. 2000). Calphostin C inhibited 12-*O*-tetradecanoylphorbol-13-acetate-stimulated phospholipid secretion and also reduced binding and uptake of the monoclonal antibody 3C9 by type II cells.

To follow the retrieval of lamellar body membrane from the cell surface in *rat* type II pneumocytes, SCHALLER-BALS et al. (2000) instilled the monoclonal antibody 3C9 into *rat* lungs. *In vivo* it was endocytosed by type II cells but not by other lung cells. In type II cells that were isolated from *rat* lungs by elastase digestion and cultured on plastic for 24 h, Mab 3C9 first bound to the cell surface, then was found in endosomes, vesicular structures, and multivesicular bodies and, finally, clustered on the luminal face of lamellar body membranes.

Within a few days after an intravenous injection of Freund's complete adjuvant in *rabbits*, hyperplasia of type II alveolar cells is present on the surface of the septa in which an inflammatory reaction is developing (Gard et al. 1977). Mitosis of type II cells is detected 12 h after medication and is common over the next 120 h.

The A549 *human* lung adenocarcinoma cell line may be useful for the studying the metabolic and macromolecular processing contributions of alveolar type II cells to mechanisms of drug delivery at the pulmonary epithelium (Foster et al. 1998). It grew rapidly and consistently in 150-ml flasks; the cells were passaged approximately once a week. The monolayers consisted of cuboidal and polygonal cells, which appeared closely packed. Confluent monolayers were generally formed 3 days postseeding when plated at 10^5 cell/cm^2 in 12-well plates. A549 cells comprising monolayers exhibited a nice and uniform distribution of lamellar bodies revealed through a tannic acid staining procedure and transmission electron microscopy.

Transport measured as ^{86}Rb and ^{22}Na uptake in A549 cells exposed to normoxia, hyperoxia (30 or 40 % O_2) ' or hypoxia (3 % O_2) together with donors of reactive oxygen species and scavengers revealed that $H_2O_2 < 1$ mM did not affect transport, whereas 1 ml H_2O_2 activated ^{22}Na uptake (+200 %) but inhibited ^{86}Rb uptake (–30 %) (Heberlein et al. 2000). Also hyperoxia, aminotriazole plus menadione, and diethyldiathiocarbamate inhibited ^{86}Rb uptake. *N*-acetyl-L-cysteine, diphenyleniodonium, and tetramethylpiperidine-*N*-oxyl, used to reduce reactive oxygen species, inhibited ^{86}Rb uptake, thus mimicking the hypoxic effects, whereas deferoxamine, superoxide dismutase, and catalase were ineffective. Also hypoxic effects on ion transport were not prevented in the presence of H_2O_2, diethyldiathiocarbamate, and *N*-acetyl-L-cysteine.

H_2O_2 (200 or 400 μM for 4 h) inhibited [^{3}H]thymidine incorporation into A549 cells and cell division while inducing a G2/M-predominant growth arrest within 24 h (Shenberger et al. 2002). In addition H_2O_2 increased cell size, [^{3}H]leucine incorporation/cell, and total cell protein. Although time had little effect on eukaryotic translation initiation factor 4E and 4E-binding protein 1 expression and phosphorylation state of control cells, H_2O_2 induced a 2- to 3-fold increase in eukaryotic translation initiation factor 4E and 4E-binding protein 1 expression, a 5-fold increase in eukaryotic translation initiation factor 4E phosphorylation, and a shift in the distribution of 4E-binding protein 1 phosphorylation favouring lesser phosphorylated forms.

8.2.4.1 Hormones

8.2.4.1.1 Neuropeptide Hormones

[Arg^8]-vasopressin stimulated surfactant secretion in primary cultures of *rat* type 2 pneumocytes independently of adenosine 3',5'-cyclic monophosphate (Brown and Wood 1989). A 50 % loss of tritiated phosphatidylinsositol 4,5-biphosphate (PIP_2) occurred from cells prelabeled with myo[^{3}H]inositol within 15 s (Brown and Chen 1990). Consistent with vasopressin-induced PIP_2 hydrolysis the two breakdown products, 1,2-diacylglycerol and inositol 1,4,5-triphosphate, was observed. Vasopressin stimulated protein kinase C activity twofold over the basal activity of 0.74 ± 0.07 nM/min $\times$ mg protein. The [Arg^8]-vasopressin antagonist, 1-deamino-8-D-arginine vasopressin, inhibited [Arg^8]-vasopressin activation of protein kinase C.

8.2.4.1.2 Thyroxine

Exogenous triiodothyronine administration to pregnant *rats* resulted in a stimulation of surfactant system maturation but a paradoxical depression in foetal lung antioxidant enzyme development. Sosenko et al. (1988) adding propylthiouracil (0.01 %) to the drinking water of pregnant *rats* for the final 7–10 days of gestation thus decreasing foetal T3 serum levels, on the 21st day of gestation achieved significant ($P < 0.05$) increase in superoxide dismutase and catalase activities/mg DNA, while glutathione peroxidase (EC 1.11.1.9) activity was significantly ($P < 0.05$) decreased.

In 7 non-surviving AIDS *patients* with *Pneumocystis carinii* pneumonia triiodothyronine levels were markedly ($P < 0.001$) lower compared to 19 survivors (Fried et al. 1988).

8.2.4.1.3 Dexamethasone

Glucocorticoids regulate the major surfactant-associated protein SP-A in a biphasic fashion, i.e., at nanomolar concentrations, SP-A mRNA is increased in the foetal lung, while at higher concentrations and after prolonged exposure, a decrease is observed (Liley et al. 1988, Odom et al. 1988). This was explained on the molecular level by an increase in SP-A gene transcription at low glucocorticoid concentrations that is overcome by a reduction in mRNA stability at higher glucocorticoid concentrations (Boggaram et al. 1989, 1991). Transcriptional

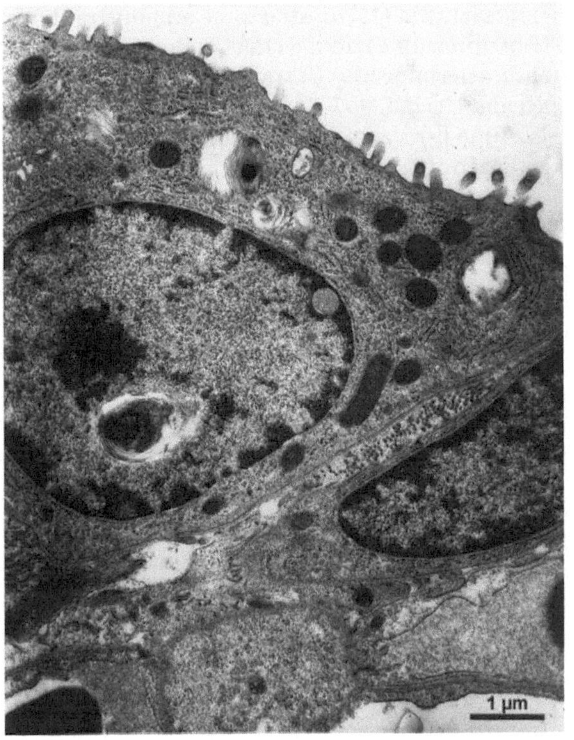

Fig. 101. Pneumocyte type II from a female white *rat* (breeder: Winkelmann, Borchen-Kirchborchen) treated with subcutaneous injection of 12.5 µg thyroxine per kg body weight × day, 5 days per week from April 12 to June 20, 1967. Fixed under methitural anaesthesia by intratracheal injection of 2.5 % glutaraldehyde in phosphate buffer (pH 7.4) before opening the thorax. Postfixation with 1 % osmium tetroxide in phosphate buffer (pH 7.4). Contrasted en bloc with 0.5 % uranyl acetate in 70 % ethanol. Embedded in Epon 812 and sectioned at 50 nm. Lead citrate. Plate 3603

induction of the SP-A gene by glucocorticoids might further be influenced by cAMP, which also stimulates SP-A gene activity (ALCORN et al. 1993). O'REILLY et al. (1991) reported an enhancing effect of glucocorticoids on pulmonary SP-B and SP-C gene transcription. The biosynthesis of the major lipid component of surfactant, phosphatidylcholine, is also stimulated by dexamethasone via activation of choline phosphate cytidyltransferase. Dexamethasone (10^{-5} M) after 48 h of incubation increased Na,K-ATPase α_1 mRNA levels in a *rat* lung foetal day 18 type II cell line 4.6 fold, while β_1 mRNA levels increased 1.6 fold (CHALAKA et al. 1996). In *human* A 549 pneumocytes HART et al. (1996) found a four-fold induction of p50 mRNA after 2 h stimulation with IL-1β (1 ng/ml) and a lesser induction by TNF-α (1 ng/ml) and phorbol-myristate-acetate (0.1 µM). Pre-treatment for 1 h with dexamethasone (1 µM) followed by IL-1β stimulation resulted in a 2 fold reduction in p50 mRNA. IL-1β, TNF-α and phorbol-myristate-acetate produced a lesser induction of p65 mRNA with the addition of dexamethasone having little effect. IL-1β produced a 7 fold increase in IκBα mRNA within 1–2 h and a similar effect was seen with IL-1β plus dexamethasone. Dexamethasone alone produced a 2 fold, more gradual increase in IκBα mRNA. Following 1 h pre-treatment with dexamthasone, a 20 % reduction in IL-1β induced NF-κB DNA binding activity occurred. However, the long term effect of dexamethasone was to increase mRNA expression of the inhibitory protein IκBα as well as decreasing expression of p50 and possibly of p65 and also increasing the ratio of transcriptionally inactive p50 to p65 monomers.

In a *mouse* lung epithelial cell line, dexamethasone decreased retinoic acid receptor-β and surfactant protein C expression to 75 and 70 % of the control values, respectively, with greatest effects at 48 h and at 10^{-7} M (GRUMMER and ZACHMAN 1998). There was no effect of dexamethasone on either retinoic acid receptor-β or surfactant protein C mRNA disappearance with actinomycin D. However, cycloheximide prevented the effect of dexamethasone. Despite dexamethasone, retinoic acid increased both retinoic acid receptor-β and surfactant protein C mRNA.

8.2.4.1.4
1α,25-Dihydroxy Vitamin D_3

1α,25-Dihydroxy vitamin D_3 may be involved in the control of lung maturation by inducing surfactant synthesis and secretion (MARIN et al. 1989). Analysis of semi-thin sections of foetal (18 days of gestation) *rat* lung tissue explanted with 1 nM 1α,25-dihydroxy vitamin D_3 increased the amount of extracellular surfactant by about 170 %, expressed as percent of the surface area of the air spaces. *In vitro* 1α,25-dihydroxy vitamin D_3 modulated the proliferation of adult alveolar type II cells (EDELSON et al. 1994).

8.2.4.2
β-Adrenergic Agonists

Type II pneumocytes isolated from the lungs of male *rats* by treatment with elastase, discontinuous density centrifugation, and adherence in primary culture, after β-adrenergic stimulation (10 µM L-epinephrine, L-isoprenaline, DL-terbutaline, respectively) released significantly increased quantities of [^{14}C]disaturated phosphatidylcholine, the major component of surface-active material (DOBBS and MASON 1979).

In *mice*, targeted transgenic expression of *human* β_2-adrenergic receptors to type II cells increased alveolar fluid clearance (MCGRAW et al. 2001).

8.2.4.2.1 Isoprenaline

The release of [^{14}C]disaturated phosphatidylcholine by *rat* type II pneumocytes cultured in the presence of 10 μM L-isoprenaline was 300±20% as compared with the controls (DOBBS and MASON 1979).

8.2.4.2.2 Salbutamol

Salbutamol (1.5 μM) significantly ($P<0.001$) increased disaturated phospholipids in isolated *rat* lungs perfused for 30 min (BARR et al. 1988).

8.2.4.2.3 Salmeterol

Salmeterol is a highly selective, >5 h acting β_2 adrenergic agonist (JOHNSON 1990) which binds to an exosite domain of the β_2 receptor. Salmeterol stimulated adult type II pneumocytes incubated for 20–22 h with 2 μC [^{3}H]choline/ml and placed in fresh media to increase phosphatidylcholine secretion in a concentration dependent manner (KUMAR et al. 1996). The maximum effective concentration was 30 nM, which resulted in a maximal stimulation of 75.4% above control. The effective concentration which yields 50% of maximal stimulation, EC_{50} was 8.1 nM.

8.2.4.2.4 Orciprenaline

Due to the *meta*-position of the phenolic hydroxyl groups orciprenaline and terbutaline (formula [108]), in contrast to isoprenaline, are not substrates for catechol-*O*-methyltransferase. Comparing the concentrations measured by DENGLER and HENGSTMANN (1976) after oral administration of ^{3}H-orciprenaline and the effect of bronchiolar resistance determined by ULMER et al. (1968) and BAVING and ULMER (1970), the therapeutic plasma levels of the β-receptor stimulating agent might be in the order of 0.5–5 ng/ml.

8.2.4.2.5 Terbutaline

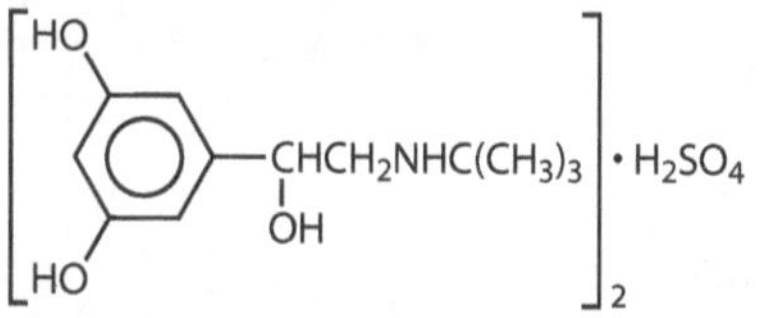

Terbutaline [108]

Terbutaline (10^{-4} M) after 2 days significantly ($P<0.05$) increased the expression of α-rENaC mRNA in *rat* alveolar type II cells isolated by elastase digestion, purified by a differential adherence technique, and cultured in minimum essential medium containing 10% foetal *bovine* serum (MINAKATA et al. 1996). *Rat* type II cells plated on Transwell membrane (but not on plastic dishes) incubated with 10^{-4} M terbutaline showed an increased binding of surfactant protein A (CHEN et al. 1996). 10^{-4} M terbutaline increased Na^+-K^+-ATPase activity in cultured *rat* alveolar type II cells in an exposure time-dependent manner over 7 days in culture (MINAKATA et al. 1998).

8.2.4.3 Cholinergic Agonists

A significant age-related decrease in the density of muscarinic acetylcholine receptors was observed on membranes of *guinea-pig* lungs dissected free of large vessels and bronchi (SUZUKI et al. 1985).

Table 25. Muscarinic acetylcholine receptors in the lungs of male *guinea-pigs* in relation to age (from SUZUKI et al. 1985)

Age	Body weight [g]	[^{3}H]-Quinuclidinyl benzilate bound [fmol/mg protein]	Dissociation constant [nM]	Number of animals
2 weeks	140– 160	67.6±6.0[a,b,c]	0.345±0.029	6
4 weeks	230– 320	68.0±9.9[d]	0.348±0.023	7
2 months	390– 450	59.1±4.7[e]	0.319±0.015	7
3 months	510– 610	57.6±7.0[d]	0.315±0.021	6
6 months	720– 960	48.6±2.3[d]	0.303±0.016	6
1 year	980–1120	49.9±1.4[d]	0.310±0.015	7
2 years	1060–1260	39.9±2.9	0.301±0.039	5

[a] $P<0,02$: different compared with 6 -month-old group.
[b] $P<0,02$: different compared with 1-year-old group.
[c] $P<0,005$: different compared with 2-year-old group.
[d] $P<0,05$: different compared with 2-year-old group.
[e] $P<0,02$: different compared with 2-year-old group.

8.2.4.3.1 Pilocarpine

Preformed osmiophilic inclusions were extruded by type II pneumocytes of adult pathogen-free Wistar *rats*, and myelin figures were present in the alveoli as early as 30 min after giving 150 mg pilocarpine per kg body weight subcutaneously (GOLDENBERG et al. 1969). Simultaneously and thereafter, the cells displayed prominently dilated endoplasmic cisternae and developed many multivesicular bodies in their basal poles. These bodies increased in osmiophilia and size during their migration apically. After 2 to 4 h, newly matured inclusions reached the cell surface and were discharged into alveolar lumina by a merocrine process. After the increased secretory phase, alveolar cell organelles remained quiescent for several hours. Complete cell recovery apparently occurred 18 to 21 h after pilocarpine injection.

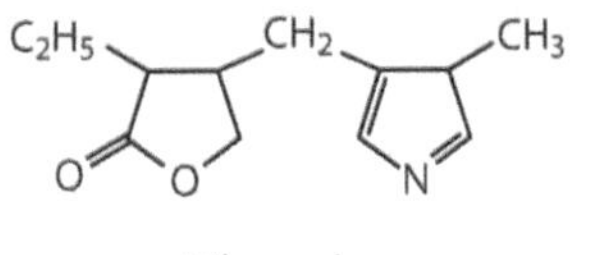

Pilocarpine [111]

In the foetal *rabbit* lung, pilocarpine administered subcutaneously to the mother (5 mg/kg) on days 24 through 27 of gestation, morphometry revealed a significant increase in the number of mature type II cells, both per unit area and per 1,000 lung cells of any kind (SMITH et al. 1979). By electron microscopy, the alveolar epithelium demonstrated morphologic correlates of increased maturation. Type II cells contained (as indicated by morphometric analysis) more and larger lamellar inclusion bodies, as well as more multivesicular bodies than those of the controls. Biochemical determination indicated that the glycogen content of foetal lung, but not liver, was reduced significantly in the pilocarpine-treated group.

8.2.4.4 Serotonin

A family of serotonin receptors have been identified based operational, structural, and transductional data (JULIUS 1991). HUMPHREY et al. (1993) specified seven classes and at least 10 subtypes. WANG et al. (1999) reported expression and localisation of the serotonin receptor 2c subtype in Wistar *rat* alveolar type II cells and the messenger RNA level using reverse transcriptase-polymerase chain reaction and nonisotopic *in situ* hybridisation, and at the protein level using immunohistochemistry. The expression of mRNA signal differed between foetal and adult *rat* type II cells, with weak, predominantly perinuclear localisation in the former and strong cytoplasmic localisation in the latter. Immunocytochemistry, using monoclonal antibody against 5-HT_{2c}-receptor, showed perinuclear localisation in foetal type II cells; whereas in adult type II cells 5-HT_{2c}-receptor immunoreactivity was confined mostly to the plasmalemma, as demonstrated by laser confocal microscopy.

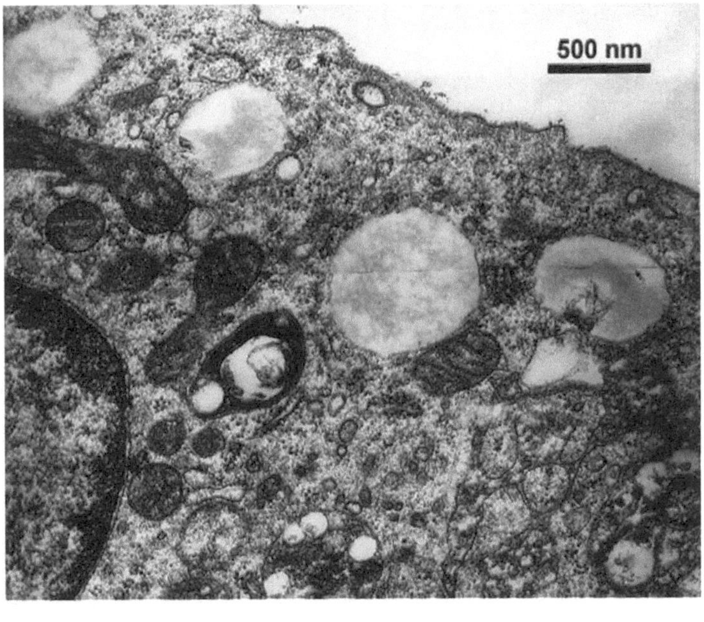

Fig. 102. Conspicuous dilatation of the rough endoplasmic reticulum cisternae in a type II pneumocyte of a 228 g female *rat* (breeder: Winkelmann, Borchen-Kirchborchen) 30 min after a subcutaneous injection of 1 mg pilocarpine HCl per kg body weight. Fixed under methitural anaesthesia by intratracheal instillation of 2.5 % glutaraldehyde in phosphate buffer (pH 7.4) before opening the thorax. Postfixation with 1 % osmium tetroxide in phosphate buffer (pH 7.4). Contrasted en bloc for 12 h with 0.5 % uranyl acetate in 70 % ethanol. Embedded in a 2:8 mixture of methyl and butyl methacrylate. Sectioned at 50 nm. Lead citrate after REYNOLDS (1963). Plate 26/06

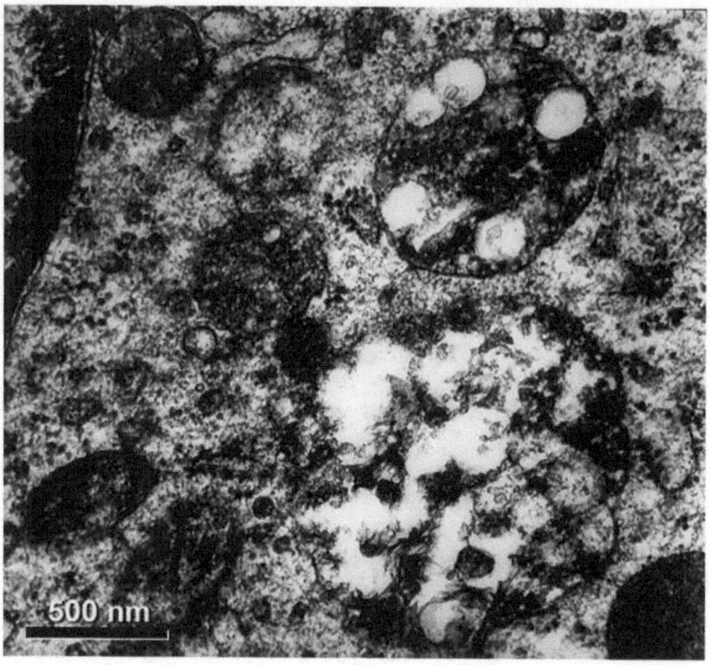

Fig. 103. Multivesicular bodies in various stages of development 30 min after a subcutaneous injection of 1 mg pilocarpine HCl per kg body weight. Lung of a 228 g female *rat* (breeder Winkelmann, Borchen-Kirchborchen) fixed under methitural anaesthesia by intratracheal instillation of 2.5 % glutaraldehyde in phosphate buffer (pH 7.4) before opening the thorax. Postfixation with 1 % osmium tetroxide in phosphate buffer (pH 7.4). Contrasted en bloc for 12 h with 0.5 % uranyl acetate in 70 % ethanol. Embedded in a 2:8 mixture of methyl and butyl methacrylate. Sectioned at 50 nm. Lead citrate after Reynolds (1963). Plate 26/05. Detail from Fig. 102

8.2.4.5 Vitamins

8.2.4.5.1 Retinoic Acid

Using a *mouse* lung epithelial cell line, Grummer and Zachman (1998) explored retinoic acid-dexamethasone interactions through the study of retinoic acid and dexamethasone effects on retinoic acid receptor and surfactant protein C mRNA expression. Retinoic acid increased expression of retinoic acid receptor-β (5.5 times) and surfactant protein C (2 times) mRNA, with maximal effects at 24 h and at 10^{-6} M. The retinoic acid induction was not inhibited by cycloheximide, suggesting retinoic acid affects transcription. With addition of actinomycin D, retinoic acid did not affect the disappearance rate of retinoic acid receptor-β mRNA, but surfactant protein C mRNA degradation was slowed, indicating an effect on surfactant protein C mRNA stability.

8.2.4.5.2 Tocopherols

α-Tocopherol is the major component of the tocopherols, which constitute vitamin E and is the principal lipid-soluble chain-breaking antioxidant of *human* tissues. Freeman and Panus (1988) modulated cellular α-tocopherol content by incorporating up to 42 mol % α-tocopherol into the hydrophobic membrane region of water-soluble dipalmitoylphosphatidylcholine liposomes. A 12-fold increase in α-tocopherol content (0.07 μmol/mg protein to 0.86 μmol/mg protein) could be induced after a 6 h incubation of type II cells isolated from *rabbit* lung with liposomes added to culture medium. Increases of type II cell α-tocopherol content were dependent on incubation time, liposome concentration in media and α-tocopherol concentration in liposomes. Importantly, no significant increases in type II cell α-tocopherol content could be detected in cells similarly incubated with albumin bound α-tocopherol.

The measurement of the uptake kinetics indicated that high-density lipoprotein might be the primary source of the vitamin E uptake by type II pneumocytes (Kolleck et al. 1999). Vitamin E depletion of *rats* caused an increase of vitamin E uptake by isolated type II pneumocytes from high-density lipoprotein but not from low-density lipoprotein or very low-density lipoprotein. Type II pneumocytes express the scavenger receptor class B type 1 (SR-B1), a high-density lipoprotein-specific receptor. Vitamin E depletion caused an increased expression of SR-B1 by a post-transcriptional mechanism. The increased vitamin E uptake from high-density lipoprotein and the increased expression of the SR-B1 were reversed by refeeding the vitamin.

Pre-term neonates and neonates in general exhibit physiological vitamin E deficiency (Kelly et al. 1990) and are at increased risk for the development of acute lung diseases (HAAGSMAN and VAN GOLDE 1985). SINHA et al. (2002) evaluated the hypothesis that vitamin E deficiency predisposes alveolar type II cells for apoptosis. In type II pneumocytes freshly isolated from Wistar *rats* (body mass 80 to 90 g) fed the vitamin E deficient diet Altromin C1018, the pro-apoptotic Bax increased (70±3 versus 64±1 in the controls; $P < 0.05$) which might be responsible for the decrease of the mitochodrial transmembrane potential ($\Delta\psi$ = 203±14 versus 278±28; $P < 0.05$) followed by an increased release of cytochrome *c* into the cytosol (55±6.7 ng/10^6 cells versus 46±4.2 ng/10^6 cells in the controls; $P < 0.05$). Vitamin E depletion did not change the GSH/GSSG ratio and the activities of antioxidant enzymes.

8.2.4.6 Benzodiazepine Receptor Agonists

The binding of [^{3}H]Ro5-4864 to the benzodiazepine receptors of isolated *guinea-pig* alveolar type II cells indicated one population of binding sites with high affinity (K_p = 5.7 nM) and saturability (B_{max} = 4.582 fM/mg membrane protein) (DAS et al. 1987).

8.2.4.7 Protein Synthesis Inhibitors

8.2.4.7.1 Puromycin

Puromycin, though not a direct inhibitor of peptidyl-transferase, inhibits normal peptide chain elongation by participating in the peptidyl transferase reaction.

8.2.4.8 Hyperlipidemic Agents

Male *rats* treated with either clofibrate [2-(*p*-chlorophenoxy)-2-methyl-propionic acid ethyl ester] or nafenopin [2-methyl-2-(*p*-1,2,3,4-tetrahydro-1-naphthyl)phenoxypropionic acid], two peroxisome proliferating compounds with potent hypolipidemic properties, showed identical structural changes in their lungs (FRINGES and REITH 1988).

8.2.4.8.1 Clofibrate

The size of the type II pneumocytes was conspicuously increased by about 30% in Lewis *rats* treated with 300 mg clofibrate/kg body weight × day by intraperitoneal injections for 7 days (FRINGES and REITH 1988). Peroxisomes proliferated and there was an increase in the number of surfactant-rich lamellar bodies (FRINGES et al. 1988).

Table 26. Peroxisome proliferators

Aspirin	acetyl salicylic acid
Bezafibrate[1)]	2-[*p*-[2-(*p*-chlorobenzamido)ethyl]phenoxy]-2-methylpropionic acid
Bifonazole	1-([1,17-biphenyl]-4 ylphenyl-methyl)1H-imidazole
BM 15766	4-[2-[1-(4-chlorocinnamyl)piperazin-4-yl]ethyl]benzoic acid
Cetaben	sodium *p*-hexadecylaminobenzoate
Ciprofibrate	2-[4-(2,2-dichlorocyclopropyl)phenoxy]methylpropionic acid
Clofibrate[2)]	ethyl-α-(*p*-chlorophenoxyisobutyrate)
Clofibric acid	2-(*p*-chlorophenoxy)-2-methylpropanoic acid
CPP	DL-2-(4-chlorophenoxy)-propionic acid
DEHA	diethylhexyl adipic acid ester
DEHP	di(2-ethylhexyl)phthalate
EHA	2-ethyl-hexanoic acid
ETYA	5,8,11,14-eicosatetraynoic acid
Fenofibrate[3)]	isopropyl[4'(*p*-chlorobenzoyl)-2-phenoxy-2-methyl]propionate
FOE 3798	
Gemfibrozil[4)]	5-(2,5-dimethylphenoxy)-2,2-dimethylpentanoic acid
Linolenic acid	Octadecatrienoic acid
Medica 16	$HOOC\text{-}CH_2\text{-}C(CH_3)_2\text{-}(CH_2)_{10}\text{-}C(CH_3)_2\text{-}CH_2\text{-}COOH$
MEHP	mono(2-ethyl-hexyl)phthalate
Nafenopin	2-methyl-2-[*p*-(1,2,3,4-tetrahydro-1-naphthyl)phenoxy]propionic acid
Thyroxine[5)]	
Tiadenol	bis(hydroxyethylthio)-1,10-decane
Tridiphane	2-(3,5-dichlorophenyl)-2-(2,2,2-trichloroethyl)oxirane
Valproic acid	2-propyl-pentanoic acid
Wy-14643	[4-chloro-6(2,3-xylidino)2-pyrimidinylthio]acetic acid

[1)] Befibrat Beza-Lande, Cedur, durabezur, Regadrin Sklerofibrat; [2)] Regelan; [3)] durafenat, Lipanthyl, Lipidil, Normalip; [4)] Gevilon; [5)] Dynothel

0.5% (w/w) clofibrate in the food for 7 days did not have much effect on acyl-CoA oxidase, catalase, dihydroxyacetone phosphate (DHAP) acyltransferase, and alkyl-DHAP synthase levels in *rat* type II pneumocyte peroxisomes (OSSENDORP et al. 1997).

8.2.4.8.2 Nafenopin

The increase in the number of lamellar bodies per type II cells was close to 60% in Wistar *rats* fed a diet containing 0.2% nafenopin (w/w) for 3 weeks (FRINGES and REITH 1988).

8.2.4.9 Expectorants

8.2.4.9.1 Bromhexine

A morphometric examination of type II pneumocytes from *rats* medicated for 3 days with 200 mg bromhexine per kg body weight × day demonstrated that the volumetric density of their lamellar bodies had increased from 18.0 ± 0.9% in the controls to 20.0 ± 0.6% in bromhexine-medicated *rats* (GIL and THURNHEER 1971).

8.2.4.9.2 Ambroxol

Ambroxol (2-amino-3,5-dibromo-*N*-[*trans*-4-hydroxycyclohexyl]benzylamine) is the metabolite VIII of bromhexine. It stimulates the formation and release of surfactant by type II pneumocytes (CURTI 1972). Wistar *rats* treated with 100 or 200 mg Ambroxol per kg body weight × day for 3 or 6 days showed an increase in the volume of their type II pneumocytes (CERUTTI and KAPANCI 1979) and increased phospholipids in this type of cells (ECKERT et al. 1983). Ultrastructural lesion were absent although, both in control and experimental animals, mitochondria occasionally contained some myelin figures (CERUTTI and KAPANCI 1979). ZIEGLER and DISSE (1986) observed a significant ($P < 0.05$) increase in phospholipids in type II pneumocytes after intraperitoneal medication of *rats* with Ambroxol (50 mg/kg × day for 3½ days). Ambroxol (oral ED_{50} = 17.3 mg/kg) increased the surfactant content in the *rat* pulmonary lavage fluid in a dose dependent manner (ENGLER and REINHARDT 1986). EROKHIN et al. (1991) used Ambroxol (15 mg/kg × day i.v.) for the stimulation of surfactant production in destructive pulmonary tuberculosis in *rabbits*.

10^{-4} and 10^{-3} M Ambroxol significantly increased the release of lactate dehydrogenase (EC 1.1.1.27) from *rat* type II pneumocytes *in vitro* (WIRTZ 2000).

By pulse radiolysis experiments Ambroxol and bromhexine were shown to be scavengers of both superoxide and hydroxyl radicals (FELIX et al. 1996). The dismutation of superoxide was accelerated 3-fold by bromhexine and 2.5-fold by Ambroxol over the rate of spontaneous dismutation. The reaction constants of hydroxyl radicals with bromhexine and Ambroxol were determined by competition kinetics to be $1.58 \pm 0.15 \times 10^{10}$ $M^{-1}s^{-1}$ and $1.04 \pm 0.1 \times 10^{10}$ $M^{-1}s^{-1}$, respectively.

In ozone-exposed Sprague-Dawley *rats*, Ambroxol (75 mg/kg i.p.) reduced the significantly increased expression of surfactant protein A (CLOSTERMANN et al. 2000).

In an *in vitro* system, Ambroxol inhibited only HOCl (JAWORSKA et al. 1996).

8.2.4.9.3 YM46A

YM46A, 1-(2-dimethylaminoethyl)-1-(3,4,5-trimethoxphenyl)urea is a novel and potent surfactant secretagogue. KIMOTO et al. (1996) examined the mucociliary transport velocity in *guinea pigs* exposed to SO_2 gas for 5 days by measuring distance of movement of mucus mimic containing dye on tracheal mucosal surface. Normal velocity of 6.3 ± 0.3 mm/min was depressed to 3.4 ± 0.5 mm/min, but at an oral dose of 10 mg YM46A/kg returned to normal levels. YM46A dose-dependently (10^{-5} to 10^{-3} M) increased the release of tritiated phosphatidylcholine from chopped lung of *guinea pigs*.

8.2.4.10 Pentoxifylline

Pentoxifylline antagonised the inhibition of surfactant synthesis induced by TNF-α in type II pneumocytes isolated from *human* cadaveric multiple organ donors and lung-cancer patients (BALIBREA-CANTERO et al. 1994).

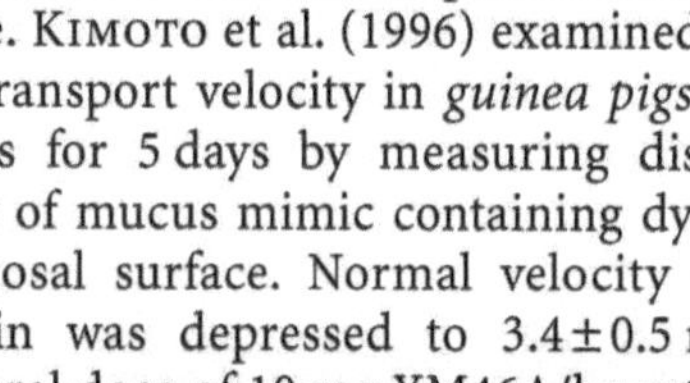

Pentoxifylline [52]

8.2.4.11
Cytokines

Interferon-γ and/or **interleukin-1β** or a combination of **INF-γ** and **TNF-α** induced nitric oxide synthase in *rat* type II pneumocytes *in vitro*; the production of nitric oxide was inhibited by N^G-monomethyl-L-arginine in a dose-dependent manner (Punjabi et al. 1994).

Nacystelyn, a lysine salt of *N*-acetyl-L-cysteine augmenting cellular antioxidant defence *in vitro* (Gillissen et al. 1997), abolished the transcriptional activation of **interleukin-8** in an IL-8-chloramphenicol acetyltransferase reporter system, transfected into A549 cells (Antonicelli et al. 2002).

8.2.4.12
Poly-2-vinylpyridine-1-oxide

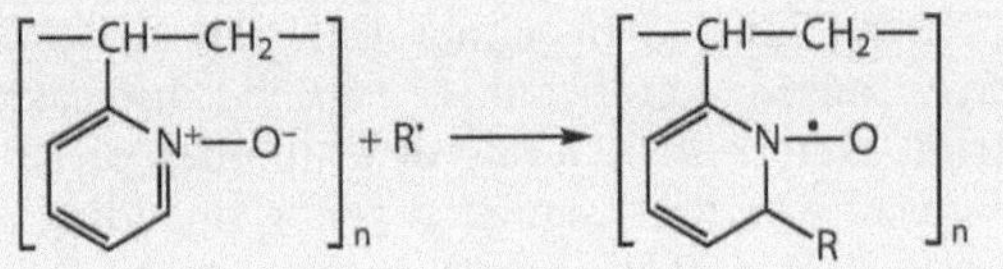

Poly-2-vinylpyridine-1-oxide — Poly-2-vinylpyridine-1-oxide adduct [64]

8.2.4.13
Sydnonimines

1,3-Morpholinosydnonimine (SIN-1) generates $ONOO^-$ by releasing $O_2^{\bullet -}$ and $^{\bullet}NO$ essentially in a simultaneous manner. Within 8 h of exposure SIN-1 type II cells isolated from anaesthetised pathogen-free male Sprague-Dawley *rats* displayed a concentration (0.01–1 μM $^{\bullet}$NO/min and 0.025–2.5 μM peroxynitrite/min, respectively)-dependent loss of cellular viability (Gow et al. 1998). $^{\bullet}$NO was generated by spermine-NONOate and papa-NONOate. $^{\bullet}$NO-mediated cellular injury within type II cells was reduced by preincubation with 2 mM *N*-acetylcysteine for 1 h to increase the intracellular concentration of reduced thiol.

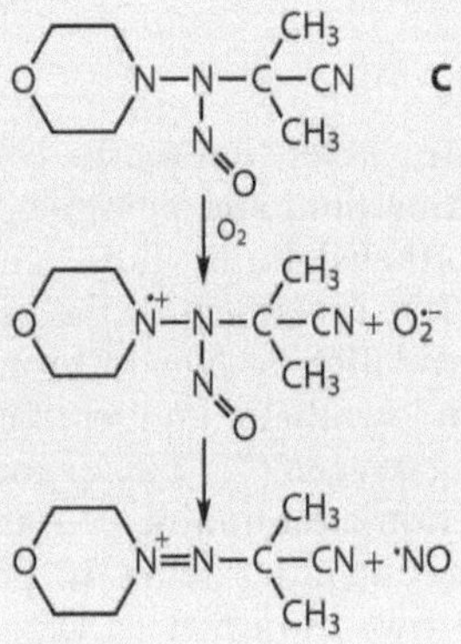

C 78 0652

Formation of both $^{\bullet}$NO and $O_2^{\bullet -}$ from sydnonimine metabolites, e.g. open ring compound C 78-0652 [136]

8.2.4.14
Ferritin

Ferritin is internalised by A549 cells after 30 min in a concentration-dependent manner (Foster et al. 1998). These data agree with observations for pre-

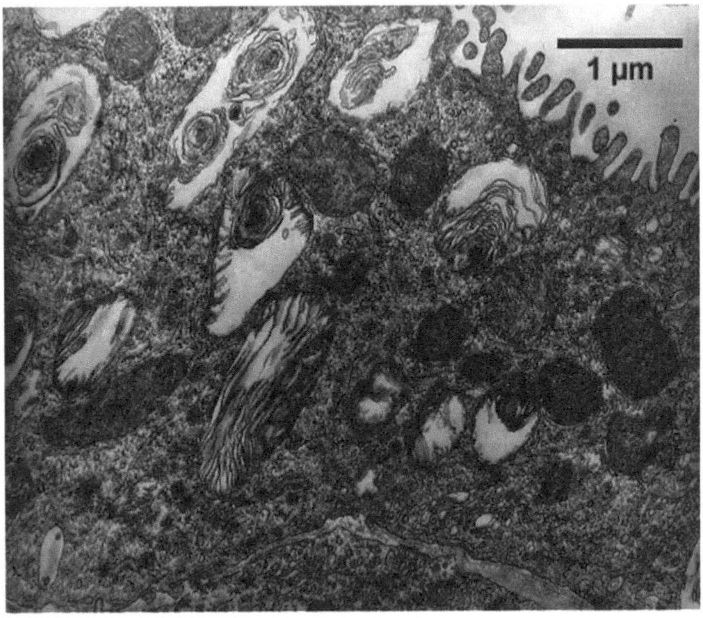

Fig. 104. Type II pneumocyte of a female white *rat* (breeder: Winkelmann, Borchen-Kirchborchen) which inhaled micronized solid poly-2-vinylpyridine 1-oxide passable to the alveoli from July 25, 1967 to January 15, 1968 for 15 min per day, 5 days per week. Fixed on January 15, 1968 under methitural anaesthesia by intratracheal instillation of 2.5 % glutaraldehyde in phosphate buffer (pH 7.4) before opening the thorax. Postfixation with 1 % osmium tetroxide in phosphate buffer (pH 7.4). Contrasted en bloc for 12 h with 0.5 % uranyl acetate in 70 % ethanol. Embedded in a 2:8 mixture of methyl and butyl methacrylate. Sectioned at 50 nm. Lead citrate after Reynolds (1963). Plate 46/11

fixed lung tissues of *rat* type II pneumocytes (Ito et al. 1992). Williams (1984) has shown that type II cells take up cationized ferritin to a greater extent than both neutral and anionic ferritin. This is thought to be due to the presence of negatively charged domains on the surface of the cell. The experimental K_m value for the cationized ferritin is approximately 60 nM, which is consistent with a non-specific absorptive process (Foster et al. 1998).

8.2.5 Toxicology

Inhaled particulate matter first makes contact with lung epithelial lining fluid and possible chemically interacts with epithelial lining fluid components. Sun et al. (2001) investigated the particulate matter-induced oxidation of *human* bronchoalveolar lavage fluid and synthetic lung epithelial lining fluid by measuring oxygen ($^{18}O_2$) incorporation and antioxidant depletion after adding 0–200 μg/ml residual oil fly ash containing about 10 per cent by weight of soluble transition metals. Oxygen incorporation was increased by addition of residual oil fly ash and was enhanced by ascorbic acid and mixtures of ascorbic acid and glutathione (GSH). Ascorbic acid became inhibitory to oxygen incorporation when it was present in high enough concentrations that it was not depleted by residual oil fly ash. Physiological and higher concentrations of catalase, superoxide dismutase, and glutathione peroxidase had no effect of oxygen incorporation. Both protein and lipid were found to be targets for oxygen incorporation; however, lipid appeared to be necessary for protein oxygen incorporation to occur.

Morphological studies have described qualitatively that lung epithelial cells bind and ingest various types of particulates, albeit less avidly than the alveolar macrophages. *In vitro* studies showed that the *human* alveolar epithelial cell line A549 can bind and ingest asbestos fibres (Rosenthal et al. 1994). Stringer et al. (1996) quantified the particle-specific binding and interleukin-8 production by A549 cells. Right-angle light scatter (for technique see Stringer et al. 1994) of A549 cells increased when incubated with 40 μg TiO_2/ml, 100 μg Fe_2O_3/ml, 200 μg α-quartz/ml, or 40 μg concentrated ambient air particulates/ml (fold increase above control: 8.1±0.9, 4.3±0.4, 2±0.1, 1.6±0.1, respectively). Both polyinosinic acid and heparin inhibited particle binding, whereas the control polyanionic ligand chondroitin sulphate did not. The observation that heparin blocks particle binding by A549 cells indicated, that the receptors on epithelial cells are distinct from the heparin-insensitive scavenger-like receptor found on alveolar macrophages (Kobzik 1995).

Interleukin-8 secretion from the *human* alveolar epithelial cell line A549 showed that dust samples from paper- and mail-sorting plants had a significant lower inflammatory potential as compared with dust from plants handling mixed household waste (Allermann and Poulsen 2000). No correlation was observed between the potency factor (IL-8 released per mg dust) and the concentration of endotoxin in the samples.

Compared with fractions from phosphate-buffered saline-treated particulate air pollutants from London and Edinburgh air (1–50 μg/ml) which activated NF-κB in A549 cells, fractions from particulate air pollutants treated with deferoxamine and ferrozine did not stimulate NF-κB activity above background levels (Jiménez et al. 2000). Particulate matter of an aerodynamic diameter of <2.5 μm inhaled by either C3H/HeJ or C57/BL6 *mice* caused significant ($P \leq 0.05$) increases in steady-state mRNA levels of a number of NF-κB-associated and/or -regulated genes, including TNF-α and -β, IL-6, IFN-γ, and transforming growth factor β (Shukla et al. 2000). Lung mRNA levels of lymphotoxin-β and macrophage migration inhibitory factor were unchanged. In *murine* C10 alveolar cells exposure to $PM_{2.5}$ at noncytotoxic concentrations resulted in an increase in transcriptional activation of NF-κB-dependent gene expression, which was inhibited in the presence of catalase. Early and persistent increases in intracellular oxidants, were observed in epithelial cells exposed to $PM_{2.5}$ and ultrafine carbon black particles.

Interactions between alveolar macrophages and epithelial cells may promote inflammatory responses to air pollution particles. For TiO_2 (~1 μm diameter), α-quartz (~1 μm diameter), residual oil fly ash, and urban air particles, a dose-dependent (0–50 μg/ml) increase in tumour necrosis factor-α and macrophage inflammatory protein-2 release was observed in co-cultures of *rat* alveolar macrophages freshly obtained by bronchoalveolar lavage and the *rat* alveolar type II epithelial cell line RLE/6TN, but not in RLE/6TN or alveolar macrophage monocultures (Tao and Kobzik 2002). This contact-dependent cytokine potentiation could not be blocked with anti-CD18/anti-CD54, arginine-glycine-aspartate peptide, or heparin.

8.2.5.1 Inorganic Agents

Takano et al. (2002) determined the cytotoxicity of heavy metals by measuring the lactate dehydrogenase release and ^{51}Cr release from lysed *rat* alveolar type II cells. With respect to the LC_{50} values, drug concentrations causing a 50% loss in cell viability, the mean value for Hg was 110 μM, and that of Cd was 220 to 250 μM. Cytotoxicity was graded high for Hg and Cd, moderate for Pb and Ni, and negligible for Mn. Additional morphological observations of cell membrane integrity by scanning electron microscopy were compatible with the results of biochemical measurements.

8.2.5.1.1 Cadmium

Inhalation of cadmium oxide smoke in *man* has been known to produce progressive emphysema (LANE and CAMPBELL 1954), a condition which, as MACKLIN (1954) has pointed out, would be expected to follow from localised defects in the lung lining.

Exposure of the *human* cell line A549 to 1–10 μM $CdCl_2$ triggered a slight accumulation of glutathione, whereas this compound was depleted after exposure to concentrations of 25–100 μM of cadmium (GAUBIN et al. 2000). treatment of cells with 20 or 40 mM *N*-acetyl-L-cysteine, which traps free radicals, was found to increase by 30% the glutathione level and to suppress the overexpression of stress proteins.

Isolated *rat* type II lung cells were more sensitive than Clara cells to cadmium-induced apoptosis and cell viability (LÌG et al. 2002). On exposure to 10 μmol/l cadmium acetate, the levels of the apoptosis-modulating proteins p53 and Bax were increased at 2 h and 5–12 h, respectively. The expression of p53 preceded the expression of Bax and the apoptotic process. The exposure to 10 μmol/l cadmium acetate did not significantly increase the formation of cellular reactive oxygen species. However, after the exposure to a high concentration of cadmium acetate (100 μmol/l), a 30% increase of the reactive oxygen species level was observed. Catalase, superoxide dismutase, dimethyl sulphoxide, or tetramethylthiourea did not protect against cadmium-induced apoptosis.

Exposure of an immortalised *rat* type II alveolar epithelial cell line to 10 μM $CdCl_2$ for 24 h caused a significant decrease in viability (TIMBLIN et al. 1998). After 48 h of exposure to 1, 5, or 10 μM $CdCl_2$ a dose-dependent decrease in viability was observed at all concentrations as determined by trypan blue exclusion and cell counting. Significant increases in reduced and oxidised glutathione levels were observed at 24 and 48 h in *rat* lung epithelial cells exposed to $CdCl_2$. HART et al. (1999) observed a maximum level of apoptosis (5-fold higher than control) in cultures exposed for 48 h to 20 μM $CdCl_2$. Although the mechanisms by which Cd initiates apoptosis in these cells are presently unknown, reactive oxygen species are likely to play a role.

A well-characterised line of alveolar epithelial cells, resembling untransformed type II pneumcytes (LI et al. 1983) exposed to 10 μmol/l $CdCl_2$ caused time-dependent increases in steady-state mRNA levels of the γ-glutamylcysteine synthetase catalytic (heavy) subunit and of glutathione S-transferase (EC 2.5.1.18) isoforms α and π (SHUKLA et al. 2001). The expression of γ-glutamylcysteine synthetase was significantly increased as early as 2 h after addition of cadmium. Maximal induction of γ-glutamylcysteine synthetase mRNA (~4-fold), at 8 h, was subsequently followed by increases of γ-glutamylcysteine synthetase activity/protein and glutathione (GSH) levels. Cadmium-induced oxidative stress, assessed by alterations in GSH homeostasis and an accelerated rate of intracellular oxidant production, could constitute early events in the signal transduction pathway mediating these responses.

Glutathione S-transferase (EC 2.5.1.18)

$$RX + Glutathione \rightarrow HX + R\text{-}S\text{-}G \qquad [129]$$

8.2.5.1.2 Chromates

The first observation which aroused the suspicion concerning the existence of a cancer hazard to the lung in workers exposed to chromates was made in 1911 and 1912 in two workers engaged in Germany in the production of alizarin which required the use of chromates as oxidising agents (PFEIL 1935). Various concentration of potassium dichromate (0–84.57 μM) added to the Ham's F12 medium were toxic to *rat* type II pneumocytes in a dose-related manner (POPPER et al. 1992). Superoxide dismutase, reduced glutathione (4–8–16 μg/ml), EDTA (50–100 μM), and DMSO (0.1–1–10%) when added to the culture medium had no effect on survival of pneumocyte colonies, whereas dithiothreitol (50–100 μg/ml), desferal (20–100 μM), butylhydroxytoluene (10–50–100 μM) and trolox (50–200 μM) were effective in preventing dichromate cytotoxicity in a dose dependent manner.

Using particle-induced X-ray emission analysis of individual V79 Chinese *hamster* lung cells, DILLON et al. (1998) investigated the permeabilities of Cr(V) complexes. The Cr uptake increased in the order: $[Cr(1,10\text{-phenanthroline})_2\text{-}(H_2O)_2]^{3+}$ < [CrO (2-ethyl-2-hydroxybutanoato)$_2$]$^-$ < [CrO(5,6-(4,5-dichlorobenzo)-3,8,11,13-tetraoxo2,2,9,9-tetramethyl-12,12-diethyl-1,4,7,10-tetraazacyclotridecanate)]$^-$ < $[Cr_2O_7]^{2-}$. Clonal assays showed that Cr(VI) exhibited an expectedly higher cytotoxicity than the Cr(V) complexes. While the genotoxicities of the Cr(V) and Cr(VI) complexes increased according to their permeabilities, the genotoxicities of the Cr(V) complexes were equal to, if not greater than, that of Cr(VI) in terms of the amount of Cr entering the cell.

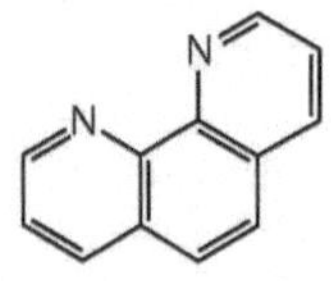

1,10-Phenanthroline [137]

The carcinogen Cr(VI) does not significantly react with and damage DNA unless it is reduced metabolically to lower oxidative states. The intermediates of Cr(VI) reactions with small molecules and/or enzymes include Cr(V), Cr(IV) and Cr(III), as well as oxygen- (STANDEVEN and WETTERHAHN 1991), sulphur- or carbon-centred (KADIISKA et al. 1994, STEARNS and WETTERHAHN 1994) free **radicals.** All these species can damage DNA, though in different ways (WETTERHAHN et al. 1989). The genotoxic effects include formation of DNA-chromium and DNA-radical adducts, DNA interstrand and intrastrand cross-links, DNA stand breaks, and especially DNA-protein cross-links (SHI and DALAL 1989, 1990, SUGIYAMA et al. 1991).

$$2\,Cr^{VI}O_4^{2-} + 9\,H_2O_2 + 2\,OH^- \longrightarrow 2\,Cr^{V}(O_2)_4^{3-} + 10\,H_2O + O_2 \quad [138]$$

$$2\,Cr^{V}(O_2)_4^{3-} \longrightarrow 2\,Cr^{VI}O_4^{2-} + 2\,O_2^{\cdot-} + 2\,{}^1\mathbf{O_2} \quad [139]$$

$$2\,H_2O_2 + 2\,O_2^{\cdot-} \longrightarrow 2\,HO^{\cdot} + 2\,OH^- + 2\,O_2 \quad [140]$$

Equation [138] contains both the reduction of Cr(VI) by hydrogen peroxide and the ligand replacement of O^{2-} with O_2^{2-}. With respect of equation [139] KAWANISHI et al. (1986) presented some evidence for the formation of singlet oxygen and also considered on the basis of the effect of superoxide dismutase that superoxide is generated before the formation of hydroxyl radicals. PETERS et al. (1972) demonstrated that singlet oxygen is generated during the decomposition of potassium tetraperoxychromate(V) in aqueous solution. The decomposition of potassium tetraperoxychromate(V) in aqueous solutions generates a reductant of ferricytochrome *c* as well as an oxidant of ferricytochrome *c* (HODGSON and FRIDOVICH 1974). The reductant was intercepted by superoxide dismutase, and the oxidant was scavenged by azide or histidine. HODGSON and FRIDOVICH (1974) suggested on the basis of these data that one $O_2^{\bullet-}$ and one $^1O_2$ are released during the course of the decompensation of $Cr^V(O_2)_4^{3-}$ to CrO_4^{2-}. This was confirmed by PETERS et al. (1975). The same decompensation is considered likely to occur in the reaction of sodium chromate(VI)with H_2O_2. Equation [140] represents Haber-Weiss reaction that superoxide anion and hydrogen peroxide can combine together directly to generate the hydroxyl radical. The rate constant of this reaction is very small in the absence of matals as iron or copper, however. $HO^\bullet$ may be formed then a chromium(V)-peroxide complex decompses into Cr(VI) complex and hydroxyl radical as in the Fenton reaction:

$$Cr^{V}(O_2)_4^{3-} + n\,O_2^{\cdot-} + n\,H^+ \longrightarrow Cr^{V}(O_2)_{4-n}(O)_n^{3-} + n\,HO^{\cdot} + n\,O_2 \quad [141]$$

$$Cr^{V}(O_2)_4^{3-} + H^+ \longrightarrow Cr^{VI}(O_2)_3(O)^{2-} + HO^{\cdot} \quad [142]$$

The Fenton catalyst, Fe^{2+} reduces chromium(VI) as follows:

$$Fe^{2+} + HCrO_4^- + 2\,H^+ \longleftrightarrow Fe^{3+} + H_3CrO_4 \quad [143]$$

$$Fe^{2+} + H_3CrO_4 + H^+ \longrightarrow Fe^{3+} + Cr(VI) \quad [144]$$

Iron stimulated the rate of reduction of hexavalent chromium by *human* microsomes (MYERS and MYERS 1998). The ratio of *human* microsomal Cr(VI) reduction rates under aerobic versus anaerobic conditions remained fairly constant, regardless of iron concentration. Small increases in intracellular iron could therefore lead to large increases in the rate and extent of microsomal Cr(VI) reduction. Individuals that are simultaneously exposed to Cr(VI) and to agents that increase intracellular iron could therefore be at potentially greater risk for Cr(VI) toxicity and carcinogenicity.

In buffer solutions which were treated to remove Fenton-active metal ions as well as in those not further purified, chromate and reduced glutathione (GSH) induced similar numbers of single-strand breaks in isolated PM2 DNA (KORTENKAMP et al. 1996). Molecular oxygen was found to be essential for the formation of single-strand breaks, but Cr(V) species arising from chromate/GSH, unless activated by oxygen, appeared to be unreactive toward DNA.

The readiness of **thiol groups** to interact with the chromate ion in addition to its strong reducing ability was studied with L-cysteine and its ethylester derivative, L-cysteine-OEt (KAIWAR and RAO (1995). The reductive ability of GSH was about 2.5 times higher to that of L-cysteine while the reductive ability of L-cysteine-OEt was 1.4 times higher than that of GSH. Treatment of *rat* thymocytes with the thiol(SH)-blocking agent diethyl maleate caused a marked inhibition of chromate uptake in the cells

(Debetto et al. 1988), possibly because diethyl maleate can decrease the rapid intracellular reduction of Cr(VI) to Cr(III) by inactivating the SH-groups of the anion carrier or by depleting GSH and other cellular reductants. Dithiothreitol, when added to thymocytes just prior to $K_2{}^{51}Cr_2O_7$, significantly increased Cr(VI) uptake by the cells, probably by maintaining membrane protein thiol groups in their reduced state. The capacity of Cr(VI) anion to interact directly with SH-groups of the cell membrane was documented by Polak (1983).

Models which ascribe the induction of chromium-DNA adducts to chromium(V) and the generation of oxidative DNA damage including single-strand breaks to hydrogen peroxide are oversimplistic (Kortenkamp et al. 1996). A combination of GSH, molecular oxygen, and chromium(VI) can damage DNA via non-Fenton pathways.

8.2.5.1.3 Coal

Primary cultures of *rat* type II pulmonary epithelial cell responded to an exposure to anthracite coal dust PSOC 867 (1.65 µm median particle size; Primrose Mine) freshly ground in a zirconium oxide ball mill with increased synthesis of extracellular matrix components (Lee and Rannels 1998). Elevation of protein synthesis by the dust was further increased by TNF-α and TGF-β_1.

8.2.5.1.4 Cobalt

Cobalt dust inhaled by *rats* at a concentration of 2.12 ± 0.55 mg/m^3 for 5 h per day at 4 consecutive days induced proliferation of type II cells (Kyono et al. 1992). Cobalt dust released lactate dehydrogenase from freshly isolated *rat* and *human* type II pneumocytes cultured *in vitro* for 48 h (Roesems et al. 1997). For *rat* type II cells TD_{50} values were 672 µg (95% C.I. = 264–1706 µg) for pure Co and 101 µg (95% C.I. = 59–172 µg) for Co in Co/WC mixture. No toxicity was found in *human* type II pneumocytes for either Co, WC or Co/WC. Since WC catalyses the oxidation of Co resulting in the production of reactive oxygen species, it appears that *rat* (and possible also *human*) type II pneumocytes are more resistant to particle-induced oxidative stress than *rat* alveolar macrophages.

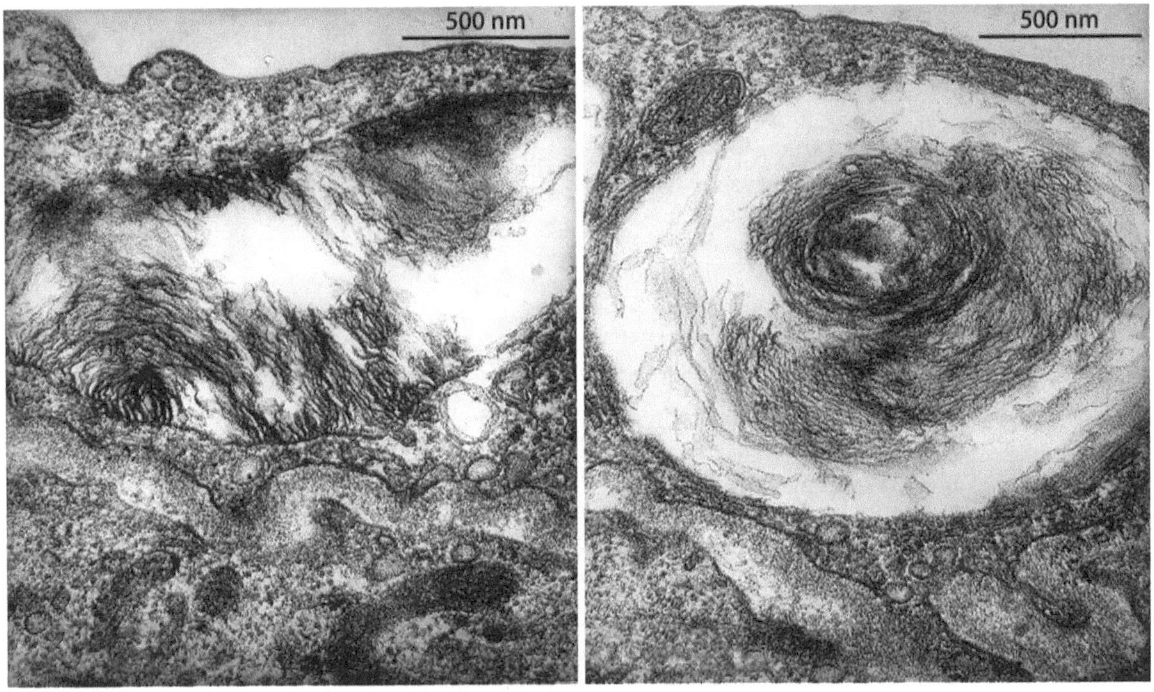

Fig. 105. Lamellar bodies in a type II pneumocyte from a female white *rat* (breeder: Winkelmann, Borchen-Kirchborchen) which inhaled 10 mg of a powdered cobalt-nickel alloy/m^3, used in the production of hard metal by Deutsche Edelstahlwerke in Krefeld, 4 h per day, 5 times per week from October 2-28, 1967 for a total of 19 days. Fixed on October 30, 1967 under methitural anaesthesia by intratracheal instillation of 2.5% glutaraldehyde in phosphate buffer (pH 7.4) before opening the thorax. Postfixation with 1% osmium tetroxide in phosphate buffer (pH 7.4). Contrasted en bloc for 12 h with 0.5% uranyl acetate in 70% ethanol. Embedded in a 2:8 mixture of methyl and butyl methacrylate. Sectioned at 50 nm. Lead citrate after Reynolds (1963). Plates 16/05 and 16/06

One-electron transfer catalysed by cobalt

$O_2 + Co^{2+} \leftrightarrow \mathbf{Co^{3+} - O_2^{\bullet-}} \rightarrow \mathbf{O_2^{\bullet-}} + Co^{3+}$ [145]

$\mathbf{O_2^{\bullet-}} + Co^{2+} + 2\,H^+ \rightarrow H_2O_2 + Co^{3+}$ [146]

$H_2O_2 + Co^{2+} \rightarrow \mathbf{HO^{\bullet}} + HO^- + Co^{3+}$ [147]

Type II cell noduli of two to nine cell profiles were found in all *rabbits* exposed to 0.6 ± 0.5 mg/m^3 Co^{2+} as $CoCl_2$ for four months, 5 days/week, and 6 h/d (Johansson et al. 1992). The separate cells were often enlarged, were filled with lamellar bodies, and had dilated endoplasmic reticulum. Immature cells were also seen.

As an early response to (per)oxidative stress, the **hexose monophosphate shunt** was activated in primary cultures of *rat* type II pneumocytes exposed to freshly prepared cobalt metal-containing particles (Hoet et al. 2002). Maximal stimulation of the hexose monophosphate shunt was reached after 90 min and different concentrations (15, 75, 300 and 1200 μg/well) dose-dependently showed significant increases (2.0 ± 1.2, 2.9 ± 0.4, 3.3 ± 1.6, and 4.0 ± 0.4 fold, respectively). Co^{2+} (1 mM $CoCl_2$, i.e. 120 μg/well), however, initially had no influence on the activity of the hexose monophosphate shunt, but after 4 h of exposure a small, but significant increase in activity was seen.

8.2.5.1.5 Nickel Compounds

The first indication that an occupational inhalation of metallic nickel dust or nickel carbonyl vapours might be associated with a cancer hazard to the nasal cavity, paranasal sinuses and lung was contained in the Annual Report of the Chief Inspector of Factories for the Year 1932, which was published in 1933. The high incidence of pulmonary cancer in Welsh nickel workers was referred to by Baader (1937, p.116). Doll (1958) reported that 35.5 percent of nickel workers in Wales died of cancer of the lung or upper respiratory tissue whereas the incidence among colliery workers was only 1.5 per cent. The relationship between the inhalation of nickel and pulmonary cancer in *rats* induced by heavy concentration of nickel carbonyl has been investigated by Sunderman et al. (1957, 1959), and the implication of nickel as a pulmonary carcinogen in tobacco smoke by Sunderman and Sunderman (1961). Sunderman jr (1963) found Ni in purified preparations of ribonucleic acid obtained from the livers and lungs of normal *rats*. At alkaline pH, under anaerobic conditions, Ni(II) formed complexes with pyrimidines (Weiss and Hein 1959). Huang et al. (1995) found that Ni^{2+} binds to histone H_1 and core histones as determined by ^{63}Ni autoradiography of proteins on nitrocellulose membranes. *In vitro* studies showed that commercially purified histone H_1, and to a considerably lesser extent core histones, enhanced the $NiCl_2$ and H_2O_2 catalysed formation of 7,8-dihydro-8-oxo-2'-deoxyguanosine in a reaction containing free deoxyguanosine base. Since histone H_1 is a lysine- and alanine-rich protein, the levels of 7,8-dihydro-8-oxo-2'-deoxyguanosine induced by $NiCl_2$ and H_2O_2 were studied in the presence of these amino acids and found to be enhanced by them.

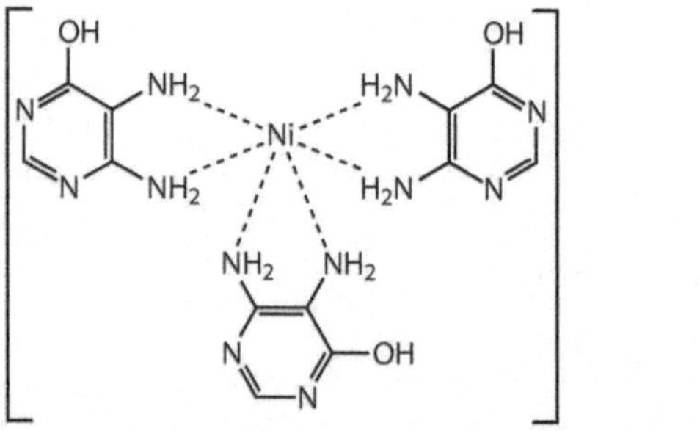

Tris[4,5-diamino-6-hydroxy-pyrimidine]-nickel(II) [148]

Nickel Ions

One-electron transfer catalysed by nickel

$O_2 + Ni^{2+} \leftrightarrow Ni^{3+} - \mathbf{O_2^{\bullet-}} \rightarrow \mathbf{O_2^{\bullet-}} + Ni^{3+}$ [149]

$\mathbf{O_2^{\bullet-}} + Ni^{2+} + 2\,H^+ \rightarrow H_2O_2 + Ni^{3+}$ [150]

$H_2O_2 + Ni^{2+} \rightarrow \mathbf{HO^{\bullet}} + HO^- + Ni^{3+}$ [151]

After intratracheal instillation in *rats*, Ni^{2+} was carcinogenic in a dose-dependent manner.

The *human* alveolar epithelial cell line A549, adult *rat* lung-derived L2 pneumocytes, and embryonic *feline* lung-derived AK-D cells incubated with Ni^{2+} at 10^{-4} M accumulated nickel in a time-dependent manner (Saito and Menzel 1986). Some metabolic inhibitors (NaN_3, NaCN, $NaAsO_2$) reduced the accumulation of Ni by up to 74 %. Nickel was shown to be distributed to three different compartments in the cells, i.e., cell membrane, cytoplasm and intracellular constituents. These results suggested that plasmalemma is readily permeable to nickel and that nickel-binding sites might exist in the pneumocytes examined.

Mixtures of $NiCl_2$ and $CoCl_2$ induced a synergistic (that is, greater than additive) toxic response in cultures of immortalised f£tal *rat* alveolar epithelial type II cells (Cross et al. 2001).

Nickel Sulphides

Hildebrand et al. (1990) studied the cytotoxicity, biological transformation and interaction with plasma membranes of αNi_3S_2 and βNiS on *human* embryonic pulmonary epithelial cells (L132 cell line) in culture. βNiS crystals were preferentially phagocytized in their original form and then probably dissolved in the vacuoles, whereas αNi_3S_2 was transformed in the extracellular space and in the phagocytic vacuoles into minute particles (10 nm) that were recovered bound to the cell membrane, phagocytic vacuoles and lysosomal membranes, respectively. Energy dispersive spectrometry revealed that the particles bound to cell membranes no longer contained sulphur but only phosphorus and nickel as inorganic compounds. This observation suggested the formation of a Ni/P complex with the phosphate groups of membranous phospholipids and/or phosphotransferring proteins.

Nickel Carbonyl

Four days after an intravenous injection in LD_{50} dosage of nickel carbonyl, i.e. 22 mg Ni/kg of body weight, to Sprague-Dawley *rats*, some granular pneumocytes were completely roofed by the cytoplasm of membranous pneumocytes, forming "blisters" which contained accumulations of laminar whorls and lattice-works morphologically identical to the inclusions of the normal granular pneumocytes. During the period from two to six days after nickel carbonyl, the type II pneumocytes appeared to be increased in number. Their nuclei and cytoplasm were enlarged, and the cytoplasm contained prominent arrays of rough endoplasmic reticulum. Many granular pneumocytes were depleted of their lamellar cytoplasmic inclusions. At the height of the pathologic response, prominent cisternal structures developed in the type II pneumocytes, and type II pneumocytes were observed in various stages of detachment from the alveolar septum (Hackett and Sunderman 1968).

Nickel Oxide and Cadmium Oxide

In *rats* instilled intratracheally with 0.5 ml of a suspension of 1 mg NiO + 0.1 mg CdO one week before, the type II pneumocytes showed a fewer number of mitochondria of varying shapes, numerous residual bodies with distinct membranous whorls, and extrusion of membranous structures into the alveolar lumen (Murthy and Holovack 1991).

8.2.5.1.6 Nitric Oxide and Peroxynitrite

A steady-state concentration of ~1.5 μM •NO (generated from 200 μM spermine NONOate) inhibited active transepithelial sodium absorption in *rat* type II pneumocytes cultured on permeable support until they formed confluent monolayers (Guo et al. 1997, 1998). Amiloride (10 μM), applied to the apical side of the monolayers, completely abolished equivalent short-circuit current, suggesting that the entry of sodium occurred via amiloride sensitive ion channels. When added into both sides of the monolayers, it significantly decreased equivalent short-circuit current from $8.2 \pm 1.0\ \mu A/cm^2$ to $1.1 \pm 0.2\ \mu A/cm^2$ and increased transepitheial resistance from $1.3 \pm 0.1\ k\Omega cm^2$ to $2.0 \pm 0.3\ k\Omega cm^2$. Oxyhaemoglobin (50 μM) totally blocked the inhibitory effect of spermine NONOate on the equivalent short-circuit current.

Primary cultured type II pneumocytes isolated from specific pathogen-free Sprague-Dawley *rats* exposed to 1,3-morpholinosydnonimine (SIN-1) endured cell injury due to the autoxidation of the drug and the formation of peroxynitrite (Gow et al. 1998). Cells loaded with rhodamine 123 one hour before exposure reduced the number of cells with R-123 fluorescence relative to control, indicating depolarisation of intracellular membranes, and increased the number of cells with YO-PRO fluorescence, indicating loss of membrane integrity. In addition, there was an increase in background fluorescence indicative of leakage of R-123 to the medium. There was no evidence of loss of cellular viability when type II cells were exposed to the decomposed donor compound, SIN-1C.

8.2.5.1.7 Nitrous Fumes

Chronic exposure to as little as 0.5 ppm NO_2 or acute exposure to higher concentrations has been shown to cause hyperplasia of the type II pneumocytes (Kleinerman 1970, Yuen and Sherwin 1971).

Glutathione *S*-transferase (EC 2.5.1.18) activity in $9000 \times g$ supernatants of lung homogenates derived from female Sprague-Dawley *rats* exposed to 50 ppm NO_x continuously for 2 weeks was significantly ($P < 0.05$) decreased from 28.0 ± 7.0 nmol/mg protein to 13.5 ± 1.0 nmol/mg protein (Pool et al. 1988). NO_x did not markedly affect the enzyme leakage from explanted lung cells.

8.2.5.1.8 Ozone

Exposure of *animals* and *humans* to ozone damages the lungs. Although ozone is not a radical itself, it can react with a wide variety of organic molecules, including membrane lipids, to produce radical species, and it can thus stimulate lipid peroxidation. Sodium oleate cosolubilized with lysozyme in reverse micellar solutions was shown to inhibit the ozone-mediated oxidation of tryptophan residues in the protein (UPPU and PRYOR 1994).

Ozone degrades under physiological conditions yielding the hydroxyl radical

$$O_3 + H_2O \rightarrow 2\,O_2^{\bullet -} + 2\,H^+ \quad [152]$$

$$O_3^{\bullet -} \rightarrow O_2 + O^{\bullet -} \quad [153]$$

$$O_3^{\bullet -} + H^+ \rightarrow O_2 + \mathbf{HO^{\bullet}} \quad [154]$$

$$O_3^{\bullet -} + 2\,H^+ + e^- \rightarrow O_2 + H_2O \quad [155]$$

$$O_3^{\bullet -} + O_3^{\bullet -} + 2\,H^+ \rightarrow O_2 + H_2O_2 \quad [156]$$

The surface activity of surfactant was affected by ozone exposure (0.8 ppm) of *rats* for 2 or 12 h, whereas distinct morphological changes in bronchoalveolar lavage or in surfactant subtypes were not observed (PUTMAN et al. 1997). Adsorption experiments indicated that bronchoalveolar lavage from *rats* exposed for 12 h to ozone remained at lower equilibrium surface pressures than lavage from control *rats*.

Exposure of native *rat* surfactant protein A to 0.75 ppm ozone led to a dose-related decrease in multiple Ca^{2+}-dependent functions including phospholipid binding and aggregation, carbohydrate binding and resistance to trypsin cleavage (CONRADO and MCCORMACK 1997).

Ozone inhalation (2 ppm, 3 h) caused a marked increase in spontaneous **production of nitric oxide** by type II cells (PUNJABI et al. 1994). Cells from ozone-treated *rats* also produced significantly more nitric oxide in response to IFN-γ alone and in combination with IL-1β or TNF-α than did cells from control *rats*. Northern blot analysis demonstrated that both increased spontaneous and IFN-γ-induced production of nitric oxide by cells from ozone-exposed animals was due to an increase in inducible NOS mRNA expression. The promoter region for nitric oxide synthase gene contains consensus sequences for NfϰB (). Treatment of type II pneumocytes from both control and ozone (2 ppm, 2 h) exposed rats with pyrrolidine dithiocarbamate, an inhibitor of NfϰB activity, suppressed the ability of these cells to produce nitric oxide (HECK et al. 1996).

8.2.5.1.9 Plutonium

In *rats*, inhaled plutonium caused proliferation, desquamation, and clearance of type II epithelial cells out to the ciliated bronchioles (SANDERS et al. 1971). In *baboons*, that were 3-times as sensitive as *dogs* to acute internal irradiation, $^{239}PuO_2$ aerosol inhalation caused type II pneumocyte hyperplasia and desquamation (METIVIER et al. 1977).

8.2.5.1.10 Silicon Compounds

After exposure to Dörentrup **quartz** $<3\,\mu m$ type II pneumocytes are increased both in number and secretory activity (HEPPLESTON 1971, HEPPLESTON and YOUNG 1972).

Subchronic exposure (6 h/d, 5 d/week) of male Fischer 344 *rats* to either 0, 2, 10, or 20 mg Min-U-Sil (MMAD 2.5 μm, σg 2.05)/m^3 showed only slight epithelial proliferation after 13 weeks of exposure, while after 26 weeks of exposure frank hyperplasia of type II and end-airway epithelia was apparent in all treatment groups (COSTA and KUTZMAN 1983).

Type II pneumocytes isolated 14 d after intratracheal instillation of 10 mg silica to *rats* and plated at high density ($0.5 \times 10^6/cm^2$) on fibrinogen/collagen matrices in Dulbecco's modified Eagle's medium supplemented with *rat* serum showed an increased DNA synthesis as compared with type II pneumocytes from saline instilled controls (BAKER et al. 1988). These differences had disappeared by 4 d. The rates of [^{14}C]choline incorporation into phospholipids by hypertrophic and by non-hypertrophic cells were similar after 2 and 4 d but were 150 % faster than by cells from saline control *rats* at 2 d; however, no differences in the rates of incorporation of [^{3}H]palmitate were found between the groups.

Type II cell hyperplasia occurred after intratracheal instillation of quartz to F344 *rats* at 2.0 mg and 20 mg, but not at 0.2 mg (DRISCOLL et al. 1996). HPRT mutant frequencies determined 3 months after exposure were significantly increased after quartz with 20 ± 6, 45 ± 5, 56 ± 6, and 106 ± 15 mutants/10^6 epithelial cells detected for saline, 0.2, 2.0 and 20 mg quartz treatment groups, respectively.

Two populations of type II cells can be isolated from the lungs of *rats* instilled intratracheally with 10 mg silica suspended in 0.5 ml saline: type IIA, similar to normal type II cells, and type IIB which are hypertrophic and exhibit increased synthesis of surfactant (ROONEY and VIVIANO 1997). After culturing the cells for 18-20 h, there was a 8- to 10-fold increase in surfactant protein A mRNA in type IIA and IIB cells 1 day after silica instillation. The increase diminished with time but was still ~4-fold after 7 days. SP-B, SP-C, and SP-D mRNA levels in type IIA cells were increased 2- to 4-fold on day 3 but were not significantly increased on days 1 or 7 or in type IIB cells at any time.

A549 pneumocytes cultured in serum free medium with 1 mg silica particles (particle size 0.5 to 10 μm with approximately 80% between 1 to 5 μm, boiled in 3% slurry in 1 N HCl to remove contaminates such as Fe_2O_3)/10^7 cells for 7 d showed no significant release of total LDH (i.e. 1.0%) when compared to A549 pneumocytes cultured in serum free medium alone (1.3%) (ROTHMAN et al. 1994). Increasing the silica concentration to 10 mg silica-particles/10^7 cells had significant toxic effects on A549 cells as reflected by a 70% increase in total LDH release, above that detected in non silica exposed cultures. Ultrastructure demonstrated at 1 mg/10^7 cells that although the cells internalised silica particles, there was no evidence of cellular toxicity: normal lamellar bodies and microvilli. A549 pneumocytes cultured in serum free medium produced C3 at a rate of 78.8±29.4 ng C3/ml×24 h, which is similar to the rate of 88.5±41.5 ng C3/ml×24 h determined for A549 cells cultures in serum free medium plus silica particles. In contrast, A549 pneumocytes maintained in medium containing FBS and silica particles produced 42.0±4.7 ng C3/ml×24 h, that were significantly ($P < 0.01$) lower absolute amounts of antigenic C3 which resulted in depressed total accumulated levels during 11 d in culture, in comparison to A549 cells cultured without silica (94±23.5 ng C3/ml×24 h). A549 cells cultured in serum free medium plus 1 mg silica/10^7 cells produced similar levels of absolute and total accumulated antigenic C5 when compared to A549 pneumocytes cultured in serum free medium without silica. In contrast, A549 pneumocytes cultured in medium supplemented with 15% FBS and silica produced significantly ($P < 0.02$) lower levels of antigenic C5 (129±1.3 ng/ml×24 h) when compared to A549 cell cultured without silica particle insult (273±70.2 ng C5/ml×24 h). Silica particles in WI-38 fibroblast cultures, maintained in serum free medium, did not modulate antigenic C3 or antigenic C5 levels when compared with WI-38 fibroblasts maintained in the absence of silica.

Although both Dörentrup quartz (7.4 m^2/g) and TiO_2 (6.6 m^2/g) were found to cause IκBα degradation in A549 cells, only quartz elicited a mild IκBα depletion, first appearing at 4 h (SCHINS et al. 2000). Up-regulation of IL-8 expression was found to persist with quartz only. Cotreatment with pyrrolidine dithiocarbamate (50 μM) and curcumin (50 μM) reduced particle-elicited IL-8 response, whereas cycloheximide (25 μg/ml) caused enhancement of IL-8 mRNA expression in both the quartz- and the TiO_2-treated cells.

Prostaglandin E_2 secretion was dose-related increased when *rat* type 2 pneumocytes were exposed to 50–200 μg Minusil <5 μm/ml culture medium (Dulbecco's modified Eagle medium containing 10% foetal *calf* serum and 1% antibiotics) as shown by radioimmunoassay (KLIEN and ADAMSON 1989).

Transfection of a reporter construct generated using the *human* growth hormone gene and the 0.6 kb fragment of the *rat* MIP-2 promoter into the *rat* type II cell line RLE-6TN followed by exposure to quartz demonstrated that the 0.6 kb MIP-2 fragment conferred quartz responsiveness (DISCOLL et al. 1997). Deletion mutant analysis of the 0.6 kb fragment demonstrated quartz-induced expression depended on a single NFκB consensus binding site (GGGGATTTCCC) located 144 bases 5' of the ATG site. No sites 5' of the κB site appeared to influence transcription in response to quartz and loss of this site eliminated quartz responsiveness.

The growth-promoting activity for bronchoalveolar lavage fluids (BALF) was further enhanced in silicotic conditions and the difference in growth activity on type II cells between silicotic and control BALF from *human* and *sheep* origin was very clear in the studies presented by LESUR et al. (1992). This contrasted with the data of LESLIE et al. (1989), who also found a stimulatory activity in healthy *rat* BALF, but no additional increase in DNA synthesis with BALF from their *rat* model of silica exposure (PANOS et al. 1990).

In the *murine* alveolar type II cell line, MLE-15, the addition of serum to the culture media reduced the in vitro **cristobalite** (particle diameter ~1.2 μm)-induced chemokine response (i.e., shift in the dose-response curve) as specific serum component, apolipoprotein-A1 selectively binds to silica (BARRETT et al. 1999). This serum protein also binds to other non-fibrous (TiO_2) and fibrous (asbestos) particles.

Extracellular reduced glutathione could completely attenuate the cristobalite (particle diameter ~1.2 μm)-induced expression of monocyte chemotactic protein-1 and macrophage inflammatory protein-2 mRNAs by a *murine* alveolar type II line, whereas TNF-α mRNA levels were unaltered (BARRETT et al. 1999). Using the oxidant sensitive dye, 6-carboxyl-2'7'-dichlorodihydrofluorescein diacetate di(acetoxymethyl ester) treatment of this cell line with cristobalite (18 μg/cm^2 = 100 μg/ml) and TNF-α (1 ng/ml) resulted in the production of reactive oxygen species, which could be inhibited with extracellular GSH treatment. In the case of cristobalite-induced reactive oxygen species, inhibition was also achieved with an anti-TNF-α antibody.

The volume of the epithelial compartment of the *rat* lung doubled after three months of inhalation exposure to either short-range or intermediate-range **chrysotile asbestos** fibres (CRAPO et al.

1980). In the animals exposed to the short-range fibres, the volume of the epithelium did not change significantly between three months and 12 months of exposure. In the animals exposed to the intermediate-range fibres, the volume of the epithelium continued to increase ($P < 0.05$), to reach a three-fold increase after 12 months of exposure. The changes that occurred in the epithelial compartment included increases in the volume of both the alveolar type I and type II epithelium.

After intratracheal instillation of chrysotile asbestos *hamster* alveolar epithelial cells can transform into large phagocytic cells which ingest asbestos fibres and produce asbestos bodies (Suzuki et al. 1972). The epithelial cells frequently lose their original type I or type II cell structure and change into a form intermediate between the epithelial cell and the alveolar macrophage.

Exposure of an immortalised *rat* type II alveolar epithelial cell line to 10 $\mu g/cm^2$ **crocidolite** asbestos fibres for 8 h caused a significant decrease in viability (Timblin et al. 1998). After 24 and 48 h of exposure to 5 or 10 μM $\mu g/cm^2$ crocidolite a dose-dependent decrease in viability was observed at both concentrations as determined by trypan blue exclusion and cell counting. There were no alterations in the levels of heat shock proteins HSP72/73 (constitutive and inducible forms of HSP70, respectively) in *rat* lung epithelial cells exposed to crocidolite fibres.

In living *Taricha granulosa* (newt) lung epithelial cells using high-resolution time lapse video-enhanced light microscopy the keratin cage surrounding the mitotic spindle inhibited fibre migration, resulting in spindles with few fibres (Ault et al. 1995). As in interphase, fibres displayed microtubule-mediated saltatory movements. Fibre position was only slightly affected by the ejection forced of the spindle asters. Physical forced between crocidolite fibres and chromosomes occurred randomly within the spindle and alongside its edge. Crocidolite fibres showed no affinity toward chromatin and most encounters ended with the fibre passively yielding to the chromosome. In a few encounters along the spindle edge the chromosome yielded to the fibre, which remained stationary as if anchored to the keratin cage.

The vimentin system of the foetal Syrian *hamster* lung epithelial cell line M3E3/C3 collapsed completely after **crocidolite** exposure while a concentration of vimentin could be observd at the sites of contact between **silicon carbide** fibres and the cell (Aufderheide et al. 1994).

Of a panel of mineral fibres causing translocation of the transcription factor NF-κB to the nucleus in A549 lung epithelial cells treated with H_2O_2 (1 to 5 mM) for 30 min, 1 h or 2 h, man-made vitreous fibre (MMVF10) was the most potent of the non-pathogenic fibres, causing significant dose-dependent nuclear translocation of NF-κB at high fibre number (Brown et al. 1999). Using three antioxidants (50 μM), curcumin, pyrrolidine dithiocarbamate, and Nacystelin (a lysinated form of *N*-acetyl-L-cysteine), translocation caused by carcinogenic fibres could be significantly reduced. Curcumin was significantly effective only for SiC and did not significantly reduce the number of positively staining cells treated with long fibre amosite or refractory ceramic fibre1. While SiC fibres had no free radical activity, the released Fe^{3+} from long fibre amosite or refractory ceramic fibre1 and hydroxyl radical generation measured as hydroxylated salicylic acid derivative, 2,3-dihydroxybenzoic acid, could be inhibited by the iron chelator, desferrioxamine-B (Brown et al. 1998).

8.2.5.1.11
Tantalum Oxide

Nemetschek-Gansler et al. (1975) occasionally found tantalum oxide dust particles in type II pneumocytes of *rats* after an inhalation of 150 mg/m^3 Ta_2O_5 for 10 days (10 h/d).

8.2.5.1.12
Titanium Dioxide

After 3 h of TiO_2 (50 nm diameter) exposure, A549 cells internalised aggregates of the ultrafine particles which were observed in cytosolic, membrane-bound vacuoles (Stearns et al. 2001). After 24 h of exposure there were considerably more intracellular aggregates of membrane-bound particles, and aggregated particles were also enmeshed in loosely and tightly packed lamellar bodies. Throughout 24 h of exposure a preponderance of particles remained associated with the free surface of the cells and were not internalised.

8.2.5.1.13
Vanadium Oxides

Vanadium compounds have a toxic effect on pneumocytes similar to chromates. Although generally less soluble the vanadium compounds studies by Popper et al. (1992) were ingested or phagocytosed by *rat* pneumocytes II cultured as monolayers in Ham's medium F12. V_2O_5 was the most toxic, while VO_2 was the least toxic oxide. Superoxide dismutase, reduced glutathione, EDTA, and DMSO when added to the culture medium had no influence on the survival of the cultured pneumocytes, while dithiotreitol, desferal, butylhydroxytoluene and trolox prevented the cells from the toxicity of vanadium oxides in a dose dependent manner.

8.2.5.1.14 Zinc Chloride

A dose- and time-dependent decrease in cellular glutathione (GSH) content and glutathione reductase activity were found after incubation of various lung cell lines with zinc chloride (Wilhelm et al. 2001). Changes in glutathione reductase activities were earliest affected and were most marked compared with the other parameters examined. Decrease of enzyme activity was not due to a decrease in the cosubstrate NADPH. In A549 and L2 cells, initial increases in GSH synthesis rates occurred up to about 175 % of control. Later, GSH synthesis decreased to levels below controls. In 16Lu cells, GSH synthesis decreased after 2 h of zinc exposure. Measurement of adenosine triphosphate content did not show any influence of zinc on cellular ATP. Lactate dehydrogenase (EC 1.1.1.27) leakage occurred after 6 h of zinc treatment in the non-malignant cells examined, and after 16 h in malignant A649 cells.

8.2.5.2 Organic Agents

8.2.5.2.1 Amiodarone

Amiodarone (INN) is a iodinated benzofuran derivative, 2-butyl-3-benzofuranyl-[4-(2-diethyl-aminoethoxy)-3,5-diiodophenyl]-ketone, the chemical structure of which resembles that of thyroxin (Heger et al. 1981). It was introduced into Belgium, France, Germany (Cordarex), the Netherlands, and South Africa for use as a coronary vasodilator, but proved to be antiarrhythmic (Rosenbaum et al. 1974, 1976). An attempt to classify amiodarone according to its toxic ability to interfere with the integrated function of electron transport enzymes (Ribeiro et al. 1997) confirmed the effect of the antiarrhythmic drug on complex I and allowed the placement of amiodarone in class A of the classification established by Knobeloch et al. (1990) i.e. together with rotenone, amytal, detergents and solvents. The drug itself and its main metabolite, desethylamiodarone accumulate in lung (Daniels et al. 1989, Reasor et al. 1989), liver (Pirovino et al. 1986) and eye (Bahel et al. 1970) cells and within the ubiquitous monocyte/macrophage system to inhibit the lysosomal phospholipases. Pichler et al. (1988) noted that most patients receiving a cumulative dose of >100 g amiodarone had side effects and that only one patient with an anti-amiodarone titre of >10 density units had no side effect.

In the *rat*, amiodarone (175 mg per kg body weight × day for 3 weeks) increased the number of lamellar bodies in type II pneumocytes (Padmavathy et al. 1993).

After intratracheal administration of amiodarone (1.83 μmol/day of days 0 and 2) or an equivalent volume (0.4 ml) of distilled water to male Fischer 344 *rats*, Northern and immunoblot analyses demonstrated that lung transforming growth factor (TGF)-β1 (mRNA and protein) expression was increased 1.5- to 1.8-fold relative to control during the early inflammation period and 1 day, 1 week, and 2 weeks post amiodarone treatment (Chung et al. 2001). Lung c-*jun* protein was increased 3.3-fold relative to control.

8.2.5.2.2 Bleomycin

In 14-day bleomycin (1.5 mg in sterile saline i.p. daily)-treated *rats* (205 ± 6 g body weight) type 2 alveolar epithelial cells were swollen with enlarged lamellar inclusion bodies (Karam et al. 1998). Biochemical study of freshly isolated cells displayed a significant decrease of lactate dehydrogenase (EC 1.1.1.27) released by these cells when isolated from 14-day-treated *rats* as compared with 7-day. By contrast, bleomycin induced an increase in superoxide dismutase (EC 1.15.1.1) and glutathione peroxidase (EC 1.11.1.9) activities. Cell content of glutathione was decreased and γ-glutamyl transpeptidase activity was markedly increased.

A phase of epithelial cell proliferation, as measured by labelled nuclei in autoradiographs, was coincident with peak basement membrane degradative activity in Sprague-Dawley *rats* 2 weeks after intratracheal bleomycin (Bakowska and Adams 1998).

Double intratracheal instillation of keratinocyte growth (150 μg/kg) factor at 48 h before and 24 h after an instillation of 5 mg bleomycin per kg prevented bleomycin-induced lung fibrosis in *rats* (Sugihara et al. 1998), as keratinocyte growth factor dramatically increased mRNA for SP-A and SP-B at the lowest dose of 1 ng/ml in *rat* alveolar type II cells (Sugihara et al. 1995), which indicates that even a small dose of keratinocyte growth factor proved a potent growth and differentiation factor for type II pneumocytes.

8.2.5.2.3 Ethanol

In two alcoholic *women* aged 64 and 56 years, respectively, post mortem electron microscopy showed type II pneumocytes containing small cytoplasmic vesicles and single large vesicles of neutral lipid and a reduced number of lamellar inclusions (Sridhar and Ryan 1979). In both patients, extreme hypoxia was caused by acute alveolar injury associated with the adult respiratory distress syndrome. The lungs of six patients with severe alcoholic liver injury and steatosis but without acute lung injury revealed no pulmonary lipid increase. On the other hand, the regenerating epithelial cells in the lungs of many non-alcoholics with extreme hypoxia secondary to acute alveolar injury have not shown lipid accumulation (Nash et al. 1974, Bachofen and Weibel 1977).

In *rats* chronically fed ethanol, type II cell mitochondrial GSH was depleted, and tumour necrosis factor-α-induced generation of mitochondrial reactive oxygen species and apoptosis were potentiated (Brown et al. 2001). When added to the ethanol diet, the GSH precursor (–)-2-oxo-4-thiazolidinecarboxylic acid, but not *N*-acetylcysteine normalised the type II cell mitochondrial GSH. Likewise, (–)-2-oxo-4-thiazolidinecarboxylic but not *N*-acetylcysteine normalised TNF-α-induced mitochondrial reactive oxygen species and apoptosis.

8.2.5.2.4 Tobacco Smoke

Cigarette smoke contains not only potent carcinogens, such as benzo[a]pyrene and 4-(methylnitrosamino)-1-(3-pyridyl)-1-butanone, but also a large number of different oxidants (Pryor et al. 1976).

In tissue obtained from the central sites of lungs from 14 current smokers the mean levels of 8-hydroxy-2'-deoxyguanosine were 1.43-fold higher than that of nine non-smokers ($P = 0.0262$; Asami et al. 1997).

In vitro, both whole and vapour smoke condensates induced a recoverable, concentration-dependent increase in *human* type II alveolar epithelial cell (A549 cell line) monolayer permeability to ^{125}iodine-labeled bovine serum albumin associated with a profound fall in intracellular reduced glutathione (Donaldson et al. 1994). Overnight incubation with 0.5 mM GSH did not enhance intracellular GSH content and thus had no protective effect against the oxidant-mediated injury of A549 epithelial cells by cigarette smoke condensate. However, increasing intracellular GSH by 0.5 mM gluathione-monoethyl ester partially protected cells against the effect of smoke.

When cigarette smoke extract was exposed to air before application to A449 cells, the cytotoxic effects were attenuated (Hoshino et al. 2001). Cigarette smoke extract caused cell death without direct contact with the cells. Acrolein and hydrogen peroxide, two major volatile factors in cigarette smoke, extract caused cell death in a similar manner. Aldehyde dehydrogenase, a scavenger of aldehydes, and *N*-acetylcysteine, a scavenger of oxidants and aldehydes, completely inhibited cigarette smoke extract-induced apoptosis.

After internalis tion into isolated *rat* type II pneumocytes, SP-A and lipid were taken up via the coated-pit pathway and resided in a common compartment, positive for the early endosomal marker EEA1 but negative for the lamellar body marker 3C9 (Wissel et al. 2001). SP-A then recycled rapidly to the cell surface via Rab4-associated recycling vesicles. Internalised lipid was transported toward a Rab7-, CD63-, 3C9-positive compartment, i.e.,

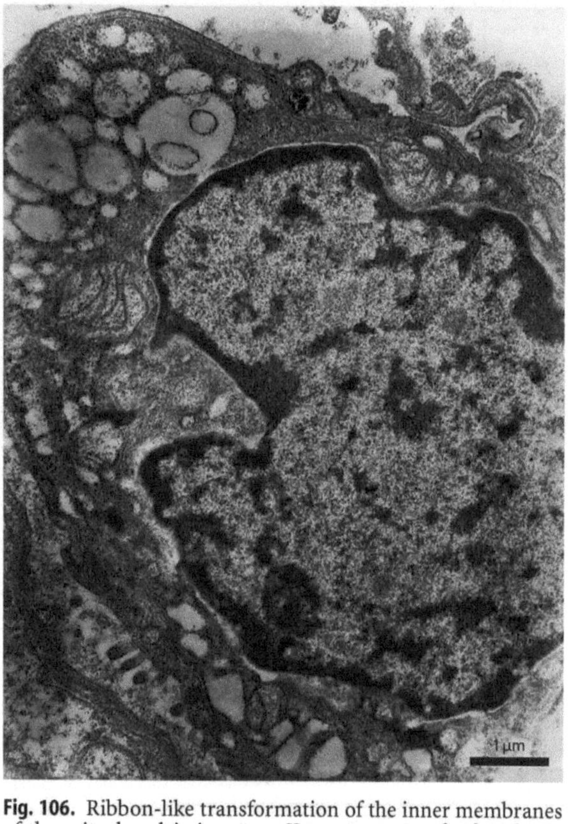

Fig. 106. Ribbon-like transformation of the inner membranes of the mitochondria in a type II pneumocyte of a female *rat* (breeder: Winkelmann, Borchen-Kirchborchen) which inhaled 20 mg shale <5 µm/m³ 4 hours per day, 5 days per week for 4 months. Immediately after each inhalation the animal was exposed to cigarette smoke generated by a smoking machine after Seehofer et al. (1965; cf. Klosterkötter 1967, Klosterkötter and Gono 1969, and Klosterkötter et al. 1969). Fixed under methitural anaesthesia by intratracheal instillation of 2.5 % glutaraldehyde in phosphate buffer (pH 7.4) before opening the thorax. Postfixation with 1 % osmium tetroxide in phosphate buffer (pH 7.4). Contrasted en bloc for 12 h with 0.5 % uranyl acetate in 70 % ethanol. Embedded in a 2:8 mixture of methyl and butyl methacrylate. Sectioned at 50 nm. Lead citrate after Reynolds (1963). Plate 4074

lamellar bodies. Inhibition of calmodulin led to inhibition of uptake and transport out of the EEA1-positive endosome and thus of resecretion of both components. Inhibition of intravesicular acidification (bafilomycin A_1) led to decreased uptake of both surfactant components. It inhibited transport out of early endosomes for lipid only, not for SP-A.

In an *in vivo* model of pulmonary oxidative stress induced by cigarette smoke in *mice*, CAVARRA et al. (2001) showed that cigarette smoke caused a significant decrease in total antioxidant capacity in cell-free bronchoalveolar lavage fluid and significant changes in oxidised glutathione, ascorbic acid, protein thiols, and 8-epi-$PGF_{2\alpha}$. Intratracheal *human* recombinant secretory leukoprotease inhibitor significantly increased bronchoalveolar lavage fluid antitryptic activity. Cigarette smoke induced a 50 % drop in the inhibitory activity of *human* recombinant secretory leukoprotease inhibitor. Pre-treatment with *N*-acetylcysteine prevented the cigarette smoke-induced loss of *human* recombinant secretory leukoprotease inhibitor activity, the decrease in antioxidant defences, and the elevation of 8-epi-$PGF_{2\alpha}$.

8.2.5.2.5 Cotton Dust

Cotton mill dust and field dried bract caused significant dose and time dependent lysis and detachment of A549 pneumocytes (AYARS et al. 1984). Preparations of cotton mill dust, field dried bract, and green bract treated with polyvinylpyrrolidone to remove the tannins, did not cause significant lysis or detachment nor did tannin-free, and tannin and terpenoid aldehyde-free cotton mill dust prepared by organic extraction. In contrast, purified tetrahydroflavindiol, condensed tannin from cotton mill dust, caused lysis and detachment of pneumocyte targets at concentration as low as 1 μg/ml. Also, field dried bract and tetrahydroflavindiol caused dose dependent inhibition of protein synthesis. Similar overall results were obtained when *rat* type II pneumocytes were used as targets.

8.2.5.2.6 Dinitrophenol

Dinitrophenol is an inhibitor of oxidative phosphorylation. Mitochondria isolated from the lungs of *rats* kept at 85±2 % O_2 for 7 days, when treated with 0.5 mM 2,4-dinitrophenol, in the presence of 1 mM CN^- were maximally uncoupled, since state 3U respiration in the presence of 2,4-dinitrophenol was identical to the mitochondria oxygen consumption measured after ADP addition when 2,4-dinitrophenol was absent (FREEMAN and CRAPO 1981).

8.2.5.2.7 *O,O,S*-Trimethyl Phosphothioate

A single sublethal dose of 40 mg *O,O,S*-trimethyl thiophosphoate per kg *rat* dissolved in corn oil and given per gavage resulted in hypertrophy and hyperplasia of type II alveolar epithelial cells (DURHAM and GIJBELS 1989). On day 10 enlarged cells contained a paucity of organelles, except for numerous glycogen rosettes and polyribosomes, and few large osmiophilic lamellar bodies. Numerous cells of this type were markedly enlarged and binucleated. Another, less frequent type II cell variation contained numerous organelles and abundant, variable-size osmiophilic lamellar bodies. On day 30 the hypertrophied type II cells contained increased numbers of osmiophilic lamellar bodies that virtually occupied the entire cytoplasmic volume, and the frequency of binucleated cells war reduced. At 6 months large type II cells contained increased numbers of osmiophilic lamellar bodies were occasionally observed, and at 1 year these changes were less frequent and severe. Mitoses were observed at 90 days.

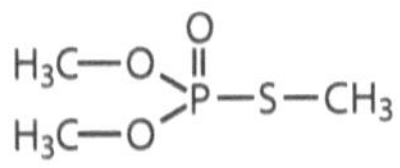

O,O,S-Trimethyl phosphorothioate [157]

8.2.5.2.8 Uracil Mustard

Uracil mustard is a potent inducer of lung tumours in *mice* (ABELL et al. 1965). Its carcinogenic risk in *man* was evaluated in 1975.

The biological efficacies of nitrogen mustard and its derivatives are closely related to the functional 2-chloroethylamine groups, which again depend on the basicity of the central nitrogen atom.

Four hours following an intraperitoneal injection of ca. 3 μM uracil mustard/kg *mouse* in 0.1 ml tricaprylin (glyceryl tricaprylate; 1 ml = approx. 0.95 g) membrane bound intranuclear inclusions containing floccular material of moderate electron density in an electron lucent background were occasionally noted in type II pneumocytes (CURRY et al. 1969). After 4 doses applied three times a week, there were a slight increase in rough endoplasmic reticulum and an increase in the number of free ribosomes in some type II pneumocytes. Six weeks after commencing injections, when the changes mentioned were more pronounced, there was some irregular dilatation of the Golgi zone.

8.2.5.2.9
Nitrosourea Compounds

BCNU long-term therapy (total dose 2760 mg and 3210 mg, respectively) induced marked hyperplasia and hypertrophy of type II pneumocytes (MITSUDO et al. 1984).

8.2.5.2.10
Paraquat

The lung is a major target organ of paraquat toxicity (ROSE et al. 1974, 1976). Mean specific rates of appearance of dipyridylium cation radicals in organ homogenates incubated with paraquat, diquat, or morfamquat were liver > lung > kidney (BALDWIN et al. 1975). BERISHA et al. (1994) have demonstrated that the pulmonary damage due to paraquat is attenuated by N^{ω}-nitro-L-arginine, and that this attenuation is reversed by the NO precursor, L-arginine, but not by its enantiomer D-arginine, suggesting the involvement of •NO in the pulmonary toxicity of paraquat.

BERISHA et al. (1994) have demonstrated that the pulmonary damage due to paraquat is attenuated by N^{ω}-nitro-L-arginine, and that this attenuation is reversed by the NO precursor, L-arginine, but not by its enantiomer D-arginine, suggesting the involvement of •NO in the pulmonary toxicity of paraquat.

24 h after subcutaneous injection of 35 mg paraquat/kg to female Sprague-Dawley *rats* mitochondria of the granular pneumocytes were swollen and their lamellae became disrupted while the cytoplasm appeared vacuolated (MODÉE et al. 1972).

Paraquat [158]

[Me-^{14}C]Choline incorporation into phosphatidylcholines was decreased in isolated *rat* alveolar epithelial type II cells exposed to paraquat (5–10 μM) and hyperoxia (90% O_2) (HAAGSMAN et al. 1987). The incorporation of [1-^{14}C]acetate into phosphatidylcholines, phosphatidylglycerols and neutral lipids appeared to be very sensitive to inactivation by paraquat. The rate of [1-^{14}C]palmitate incorporation into lipids was much less sensitive; it even increased at low paraquat concentrations. At 10 μM paraquat both NADPH and ATP were significantly decreased.

8.2.5.2.11
Streptozotocin

Streptozotocin (75 mg per kg body weight i.v.) dilated the granular endoplasmic reticulum of the type II pneumocytes of the *rat* (PLOPPER and MORISHIGE 1978). This ranged in degree from massive changes, which occupied large portions of the cell cytoplasm to focal dilation associated with normal appearing granular endoplasmic reticulum. The cisterna of this reticulum was filled with a fine granular material, which was lower in electron density than the surrounding cytoplasm. Of the 81 granular pneumocytes examined in diabetic *rat* lungs, 67 had dilated granular endoplasmic reticulum. No alteration in lamellar body number was observed. Therapy with exogenous insulin eliminated the change, indicating that the massive dilation in granular endoplasmic reticulum is a direct result of insulin deprivation.

8.3
Type III Pneumocytes

8.3.1
Ultrastructure

Pneumocytes type III, also called alveolar brush cells, have been found by MEYRICK and REID (1968, 1970), FOLIGUET and ROMANOVA (1980), and DORMANS (1985) under the TEM and by HIJIYA (1978), FOLIGUET and ROMANOVA (1980), and DORMANS (1983, 1985) under the SEM, respectively. Its free edge is restricted by cytoplasmic flanges from adjacent type I pneumocytes. Where the cell abuts against a type II pneumocyte the membranes may be folded. 150–200 Microvilli protruding into the alveolar space are blunt, wide structures approximately 0.8–1.2 μm long and 200 nm wide, containing fine filaments extending deep into the cytoplasm. Moreover, bundles of fine filaments have a perinuclear orientation over the irregularly shaped nucleus located in the basal part of the cell. Numerous pinocytotic vesicles and mitochondria are obvious in the supranuclear part of the cell. Centrioles are often seen. Some lysosomes (showing evidence of acid phosphatase activity), rough endoplasmic reticulum and free ribosomes are present. β-Glycogen is abundant through the cytoplasm, often in clumps.

8.3.2
Histochemistry

Alveolar brush cells are selectively decorated with cytokeratin 18- and villin-specific antibodies (Höfer and Drenckhahn 1992, Kasper et al. 1994).

8.3.3
Pharmacology

8.3.4
Toxicology

8.3.4.1
Pyridinium Compounds

Paraquat administered to *rats* by gavage or intravenously at doses, which were approximately equitoxic (680 μmoles/kg and 65 μmoles/kg respectively) induced a dilatation of the rough endoplasmic reticulum of the brush cells in the *rat's* lung (Sykes et al. 1977).

CHAPTER 9

Alveolar Macrophages

The bone marrow origin of the alveolar macrophages as well as the other cells of the **mononuclear phagocyte system** in the peritoneal and pleural cavities, the liver (Kupffer cells), the kidney, and other organs including the central nervous system (microglial cells) has been shown in a large number of chimera studies (for review see VAN FURTH 1980). The kinetics of mononuclear phagocytes during an acute inflammation have been studied in great detail in animal models, the inflammatory stimulus consisting of an intraperitoneal or an intravenous injection of new-born *calf* serum, polystyrene particles, or silica and/or other dust particles. Compared with experiments on alveolar macrophages gained by bronchopulmonary lavage these methods avoid bacterial or fungal contaminations. Thus MUNDER and MODOLELL (1987) used only bone macrophages to elevate the damaging capacity of reactive oxygen species produced during the interaction of airborne dust particles with phagocytic cells. BRÜCKNER-NIEDER et al. (1992) developed a *human* cell model using HL-60 cells, a promyelocytic line. These cells can be differentiated by phorbol myristate acetate into non-proliferating macrophage-like cells (HL-60-M). Sikron F600 quartz particles were engulfed. At the ultrastructural level, the quartz was seen inside phagolysosomes, but the morphology of the cell remained normal. During the differentiation of HL-60 cells, both an optimal concentration of calcitriol (10^{-8} M to 10^{-7} M) and of ethanol is necessary (ZOLLER et al. 1996, ZOLLER and ZELLER 2000). Differentiation of HL-60 cells in the presence of 0.1 % (v/v) ethanol significantly reduced their ability to produce $O_2^{\bullet -}$; an ethanol concentration of 2 % (v/v), on the other hand, was toxic. HL-60-G cells exposed to a high concentration of calcitriol (400 nM) slowly developed the ability to reduce nitroblue tetrazolium and did not proliferate (STUDZINSKI et al. 1997).

9.1 Electron Microscopy

All 48 alveolar macrophages that were seen in the serially sectioned *human* alveoli fixed in the inflated state were found within or bordering on alveolar junction zones (PARRA et al. 1986). A computer-reconstruction showed the predilection of alveolar macrophages and type II pneumocytes located in septal junction zones. Superposing alveolar macrophages and type II cells in such a reconstruction obliterated almost all gaps in the basement membranes, except for those produced by pores of Kohn when these were free of cells.

The size of the alveolar macrophage varies from 15 μm to 30 μm (BOWDEN 1971). The cytoplasm is faintly basophilic. The nuclei of the smaller cells are round but in the larger cells they may be deeply indented. Their chromatin is much finer than that of the lymphocytes. Nucleoli are inconspicuous.

The basic ultrastructure of macrophages obtained from lung, lymph nodes, liver and spleen is similar (KARRER 1958). The structural heterogeneity of the alveolar macrophages largely depends on the phase of cellular activity and, in particular, on the nature and quantity of material engulfed (Fig. 108). The cytoplasma contains numerous mitochondria of the crista type, plenty of endoplasmic reticulum, many ribosomes, and polyribosomes (Fig. 122). Primary granules, which are generally small and round or very elongated (Fig. 107) are without a halo. The dense bodies, which are bounded by a single unit membrane, do not contain any recognisable structures. Myelin and lattice forms were phagocytosed from the alveolar lumen (KISSLER 1983). According to SCHAFFNER et al. (1967) they are surfactant, but ADAMSON and BOWDEN (1970) objected that the saturated phospholipids of surfactant cannot be demonstrated by conventional staining methods.

Rat alveolar macrophages obtained from unstimulated lungs by endobronchial lavage showed heterogeneity with respect to cell size (88–20 μm diameter), surface morphology and cytochemistry

(HOLT et al. 1982): the largest cells exhibited surface characteristics typical to stimulated macrophages (spreading, prominent peripheral lamellopodia with short, blunt filopodia protrudes to the substratum) while the smallest cells more closely resembled monocytes (poorly spread). Morphometry of pulmonary alveolar macrophages *in situ* and lavaged macrophages revealed significant differences in their volume fractions of nucleus, cytoplasm, ectoplasm, mitochondria, lysosome-like structures, lipid droplets, vacuoles and phagosomes/autophagosomes (LUM et al. 1983).

Microinjection and transfection studies have demonstrated that Cdc42 induces the formation of **filopodia** in several mammalian cell types, including fibroblasts and macrophages (ALLEN et al. 1997, TAPON and HALL 1997).

Lamellipodia are plasma membrane protrusions containing a meshwork of actin filaments, and extend over the substratum to form new adhesive contacts known as focal complexes (WELCH et al. 1997). They are commonly found at the leading edge of migrating cells, driving the forward extension of cells. On adherent cells, membrane ruffles are similar in structure to lamellipodia, but protrude upwards from the dorsal surface of the cell. Both lamellipodium extension and membrane ruffling involve active actin polymerisation occurring adjacent to the plasma membrane.

Bovine alveolar macrophages cultivated for 20 h were heterogeneous, too (WILCZEK 1991). While freshly isolated cells prepared for electron microscopy varied in size from 8.5 µm to 20 µm with a mean of 13 µm, in culture cell diameters were from 12 µm to 40 µm with a mean of 22 µm (FOX 1973). BIELEFELDT OHMAN and BABIUK (1984) first used *bovine* alveolar macrophages for the *in vitro* generation of superoxide anion and hydrogen peroxide. The spontaneous superoxide anion release was very high immediately after lavage and became almost nil after 20 h, as measured by cytochrome *c* reduction; 20 h after lavage, the electrophoretic profiles obtained 3 or 4 subpopulation (POLZER et al. 1991).

9.1.1 Lysosomes

For an organelle to be interpreted as a lysosome, at least one structural requirement must be met: it must have a single limiting membrane with the relative large dimensions of the exoplasmic space membranes (DE DUVE 1969). Furthermore, the limiting membrane is commonly separated from the matrix by a clear halo (DAEMS et al. 1969, NOVIKOFF 1973).

9.1.1.1 Primary Lysosomes

With certain exceptions, primary lysosomes are more difficult to identify than secondary lysosomes. In most cells the primary lysosomes take the form of small vesicles deriving from the Golgi region (NOVIKOFF 1973, WHALEY 1975). The only types of cells in which primary lysosomes are found in a virgin condition are the granulocytes and monocytes. It is of interest to mention here that even the group of primary lysosomes is heterogeneous in the sense that – in monocytes – various types of primary lysosomes can be distinguished,

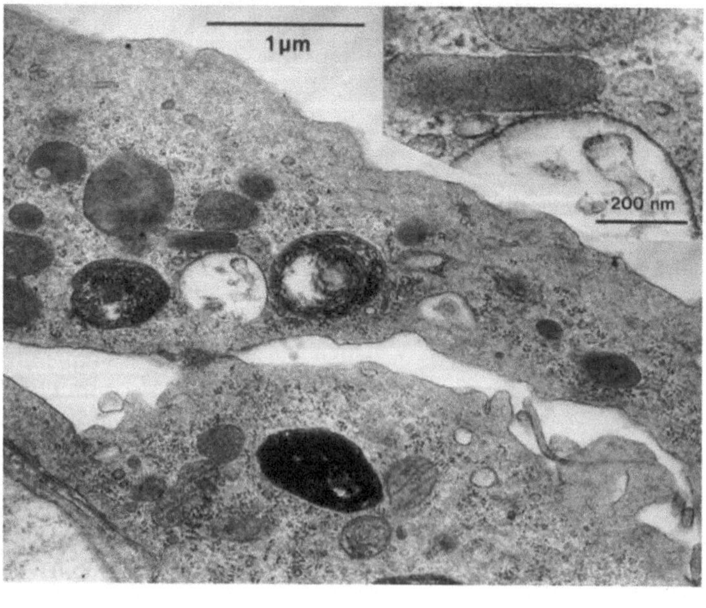

Fig. 107. Elongated primary granule in an alveolar macrophage from a ~250 g female white *rat* (breeder: Winkelmann, Borchen-Kirchborchen). Fixed under methitural anaesthesia by intratracheal instillation of 2.5 % glutaraldehyde in phosphate buffer (pH 7.4) before opening the thorax. Postfixation with 1 % osmium tetroxide in phosphate buffer (pH 7.4). Contrasted en bloc for 12 h with 0.5 % uranyl acetate in 70 % ethanol. Embedded in a 2:8 mixture of methyl and butyl methacrylate. Sectioned at 50 nm. Lead citrate after REYNOLDS (1963). Plates 13/11 and 13/12

not only morphologically and cytochemically, but also on a functional bases (DAEMS and BREDEROO 1973, DAEMS et al. 1973, 1975, NICHOLS and BAINTON 1975).

The fate of most, if not all, of the primary granules is fusion with phagosomes to form a secondary lysosome.

9.1.1.2 Secondary Lysosomes

9.1.1.2.1 Ingestion of Particles

Alveolar macrophages ingest and remove inhaled particulates from the lung (BRAIN 1986). The uptake can be seen easily under the phase contrast microscope and documented by time-lapse cinematography (SCHILLER 1954). HOLM (1972, 1974) and HOLM and HAMMARSTRÖM (1973) devised an *in vitro* system in which haemolysis and phagocytosis by human peripheral blood monocytes were quantitated by the release of radioactivity from red blood corpuscles labelled with ^{51}Cr. GOLDSMITH et al. (1997) found a dose-dependent increase in alveolar macrophage-associated right angle light scatter after uptake of residual oil fly ash or concentrated ambient air particulates.

Actin polymerisation is necessary for phagocytosis (ALLEN and ANDEREM 1996), and Rac and Cdc42, two Rho-related proteins, are required for Fcγ receptor mediated phagocytosis (COX et al. 1997). It is not known precisely which stage of phagocytosis involves Rac and Cdc42, although it is intriguing that phosphoinositide 3-kinase has been shown to play a role in the final membrane fusion step of phagocytosis (ARAKI et al. 1996). Given that phosphoinositide 3-kinase can act upstream of Rac (PARKER 1995), it is possible that Rac also plays a role at this stage. In contrast to Rac and Cdc42, Rho is required during phagocytosis for the initial clustering of Fcγ receptors on macrophages (HACKAM et al. 1997).

Rac has been shown to regulate the activity of a membrane-associated NADPH oxidase complex (SEGAL and ABO 1993; BOKOCH 1995). This multiprotein complex derives electrons from NADPH on its cytosolic face and pumps them into the lumen of phagosomes where they consequently serve to generate superoxide anion ($O_2^{\bullet -}$) and subsequently hydrogen peroxide (H_2O_2) and hydroxyl radicals ($HO^{\bullet}$) as well as hypochlorous acid (HOCl). The latter compounds represent the killing agents of the cell in the fight against phagocytosed material. The basic e^- transport function of the complex resides in the unusual cytochrome b_{558}, consisting of two subunits $gp91^{phox}$ and $p22^{phox}$, which are ingested into the plasma membrane. Proper regulation of the complex requires two additional soluble proteins, $p47^{phox}$ and $p67^{phox}$, which, upon activation, are tethered to the membrane components of the complex.

In search of a specific effector molecule of Rac in the oxidase complex, DIEKMANN et al. (1994) revealed $p67^{phox}$ as a Rac interacting protein. SEASTONE et al. (1998) have shown that a phorbol ester-binding protein is important for RacC-mediated phagocytosis. Phorbol 12-myristate 13-acetate stimulated phagocytosis of recombinant Sindbis viruses encoding Rab5:WT and increased the number and the size of endocytic vesicles, even in the presence of Rab5:S34N, a dominant negative mutant of the rate-limiting regulator of endosome fusion in BHK-21 cell monolayers (ABALLAY et al. 1999). Zinc depletion with *N,N,N',N'*-tetrakis-(2-pyridylmethyl) ethylenediamine and addition of calphostin C, an inhibitor of protein kinase C that interacts with zinc and phorbol ester binding motifs, inhibited both basal and Rab5-stimulated fluid phase endocytosis. These two reagents also inhibited the size and number of endocytotic vesicles promoted by Rab5.

Phagocytosis and Membranes

The best-documented way of endocytosis is receptor-mediated uptake of ligands via clathrin-coated vesicles (reviewed in SCHMID 1997). Receptors are recruited and concentrated into clathrin-coated pits at the plasma membrane. After coated vesicle formation, the clathrin coat is removed by the concerted action of auxilin and heat shock protein 70 (UNGEWICKELL et al. 1995). The uncoated vesicles fuse with early endosomes in a rab5-regulated manner (RUBINO et al. 2000).

In the *Chinese* hamster ovary cell line ts20, containing a thermosensitive ubiquitin-activating enzyme, E1, SACHSE et al. (2002) showed that this coat is predominantly present on early endosomes and has a characteristic bilayered appearance in the electron microscope. The coat contains clathrin heavy as well as light chain, but lacks the adaptor complexes AP1, AP2, and AP3, by which it differs from clathrin coats on endocytic vesicles and recycling endosomes. The coat is insensitive to short incubations with brefeldin A, but disappears in the presence of the phosphatidylinositol 3-kinase inhibitor wortmannin.

A role for scavenger-type receptors in alveolar macrophage uptake of components of residual oil fly ash or concentrated ambient air particulates was identified by marked inhibition of right angle light scatter increases in alveolar macrophages pretreated with the specific scavenger-type receptor inhibitor polyinosinic acid (GOLDSMITH et al. 1997).

Phagocytosis and the Golgi Apparatus

Golgi-specific agents (brefeldin, AlF_4^-, monensin) which induce vesiculation all enhanced phagocytosis of liposomes by *rat* alveolar macrophages (Perry et al. 1996).

Phagocytosis and Cyclic AMP

The level of intracellular cyclic nucleotides is a regulatory factor in a variety of immune processes. Increases in intracellular cyclic AMP (cAMP) and/or cyclic GMP (cGMP) concentration by the inhibition of phosphodieserase have been shown to modulate the inflammatory response.

Phagocytosis of polystyrene latex beads (0.481 µm diameter) by *rabbit* alveolar macrophages was accompanied by an increase of cAMP, which preceded or coincided with the onset of other metabolic events (Schmidt-Gayk et al. 1975). Dibutyryl cAMP, which mimics several action of cAMP, stimulated oxygen consumption and hexose monophosphate activity in phagocytosing cells, whereas in resting cells no significant change was observed.

Adenosine triphosphatase (ATPase) was localised in the plasmalemma of *guinea pig* peritoneal macrophages (North 1966). Enzymatic activity could be removed from the cells by trypsin. While scanning electron microscopy of endothelial cells monolayers harvested using 0.25% trypsin or 0.125% trypsin + 0.01% EDTA failed to reveal any distinctive differences in their surface morphology (Kirkpatrick et al. 1986), *bovine* alveolar macrophages showed blebbing (Brehm 1996).

Intracellular dissolution of a large variety of inhaled inorganic particles not readily soluble in the pulmonary epithelial lining fluid constitutes an important long-term clearance mechanism of the lungs. Fluorescence microscope photometry and dual laser flow cytometry of intraphagolysosomal pH in *canine* and *rat* alveolar macrophages were compared using fluoresceinisothiocyanate labelled, amorphous silica particles (Nyberg et al. 1989, Heilmann et al. 1992). The higher intraphagolysosomal pH in *rat* alveolar macrophages correlated with the smaller intracellular particle dissolution rate in Wistar *rat* alveolar macrophages when compared with the data in Beagle *dogs* (Kreyling et al. 1992).

9.1.1.2.2 Phagolysosomes

The distinction between phagosomes and secondary lysosomes cannot always be made with certainty. Special difficulties are encountered when a distinction of this kind is essential for the understanding of the course of a given process, e.g. an intracellular parasitism (see Draper and D'Arcy Hart 1975).

Destruction of the Lysosomal Membranes by SiO_4 Tetrahedra

Late cytotoxicity of silica is evidently due to the capacity of silica particles after proteins and other protective biological material have been removed by digestion to interact with, and to make permeable, the membranes surrounding secondary lysosomes (Allison et al. 1966).

Like other indigestible polymers, polyvinylpyridine-*N*-oxide accumulates in secondary lysosomes, where it can interact with silanol groups of enclosed particles, preventing them from damaging the lysosomal membrane.

9.1.1.3 Autophagous Vacuoles

Autophagy is a widespread phenomenon in the degradation of cellular components (Dunn 1994). Depending on the mechanism involved, different forms of autophagic degradation have been described: **crinophagy** (Fig. 266) for degradation of secretory proteins after fusion of secretory granules with lysosomes (Smith and Farquhar 1966, Lenk et al. 1991), **carrier-mediated proteolysis** for the degradation of cytosolic protein bearing KFERQ sequence as a signal for a lysosomal receptor/transporter protein (Olson 1989), **microautophagy** where cytosolic domains including ribosomes or glycogen are internalised into lysosomes via an endocytosis-like process (Marzella and Glaumann 1987), and **macroautophagy** for degradation of cytosolic domains and intracellular membrane compartments or specialised regions of the plasma membrane (Kovacs and Reith 1982, Elsässer et al. 1993). Autophagic vacuoles are dynamic structures which can be induced by several experimental regimen. In experimental silicosis research, Bruch (1970) was the first to describe membrane-organelles in pulmonary histiocytes that were phagocytizing DQ12 quartz particles. Schiller (1980, 1993) found such structures both in unmedicated *rats* (Figs. 109, 110) and macrophages stimulated with particulate matter or methylcellulose (Figs. 161, 162) or activated with complete Freund's adjuvant.

Structure, function and turnover of autophagic vacuoles have been studied biochemically after enrichment of the organelles by cell fractionation techniques (MARZELLA et al. 1982, FELLINGER and RÉZ 1990), or by ultrastructural methods (FURUNO et al. 1990, TOOZE et al. 1990). YOKOTA et al. (1995) described the formation of autophagosomes during the degradation of excess peroxisomes induced by di-(2-ethylhexyl)-phthalate, a peroxisome proliferator.

Within 24 h of exposure to cholesterol oxidation products (28 and 56 μM), J774 cells suffered lysosomal destabilisation, release to the cytosol of the lysosomal marker-enzyme cathepsin D (EC 3.4.23.5), apoptosis, and postapoptotic necrosis (YUAN et al. 2000). Enhanced autophagocytosis and chromatin margination was found 12 h after the exposure to cholesterol oxidation products, whereas apoptosis and postapoptotic necrosis were pronounced 24 and 48 h after the exposure. Some lysosomal vacuoles were then filled with degraded cellular organelles, indicating phagocytosis of apoptotic bodies by surviving cells. Because caspase-3 activation was detected in the cells exposed to cholesterol oxidation products, lysosomal destabilisation may associate with the leakage of lysosomal enzymes, and activation of the caspase cascade. Manganese superoxide dismutase mRNA levels were markedly increased after 24 h of exposure to cholesterol oxidation products, suggesting associate induction of mitochondrial protection or turnover.

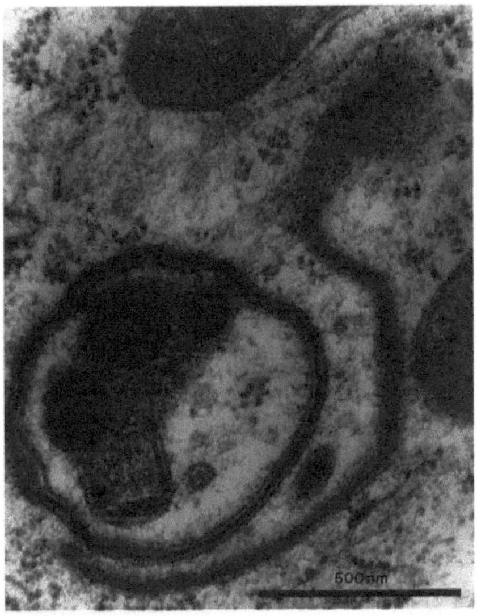

Fig. 109. Phagophore in a mesenteric lymph node macrophage (block 4421) from an unmedicated 240 g Sprague-Dawley *rat* (Charles River, France). Under pentobarbital anaesthesia (30 mg/kg), the animal was perfused from the abdominal aorta with 2.5 % glutaraldehyde in 0.1 M sodium cacodylate buffer (pH 7.4). Postfixation with 1 % osmium tetroxide in sodium cacodylate buffer. Embedded in Epon 812 and sectioned at 50 nm. Stained with lead citrate and uranyl acetate. Plate 4197

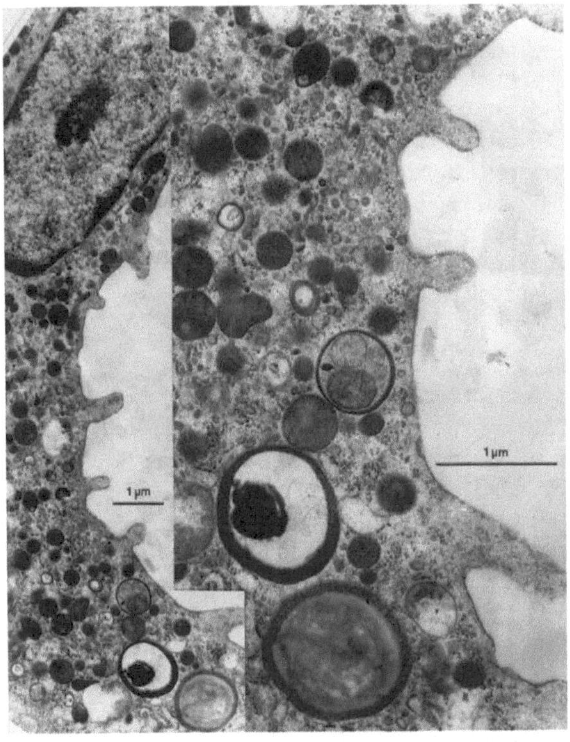

Fig. 108. Cytosegresomes in an alveolar macrophage (block 4424) from an unmedicated 235 g Sprague-Dawley *rat* (Charles River, France). On July 25, 1978 under pentobarbital anaesthesia (30 mg/kg), the animal was perfused from the abdominal aorta with 2.5 % glutaraldehyde in 0.1 M sodium cacodylate buffer (pH 7.4). Postfixation with 1 % osmium tetroxide in sodium cacodylate buffer. Embedded in Epon 812 and sectioned at 50 nm. Lead citrate and uranyl acetate. Plates 4199 and 4200

BARTH et al. (2001) identified *AUT10* as a novel gene required for both the cytoplasm to vacuole targeting of proaminopeptidase I and starvation-induced autophagy.

9.1.2 Peroxisomes

Peroxisomes are ubiquitous, spherical organelles bounded by a single membrane and having a diameter of 100 nm to 1 μm. DRATH et al. (1982) found catalase activity in similar amounts in both supernatants and granule-rich fractions of *rat* pulmonary macrophages isolated by exhaustive pulmonary lavage and postlavage homogenisation techniques.

9.1.3 Cytoskeleton

One of the characteristic features of mononuclear cell activation is the deformation of their cytoskeleton. While PHAIRE-WASHINGTON et al. (1980) determined the spreading of peritoneal macrophages stimulated by phorbol myristate acetate by draw-

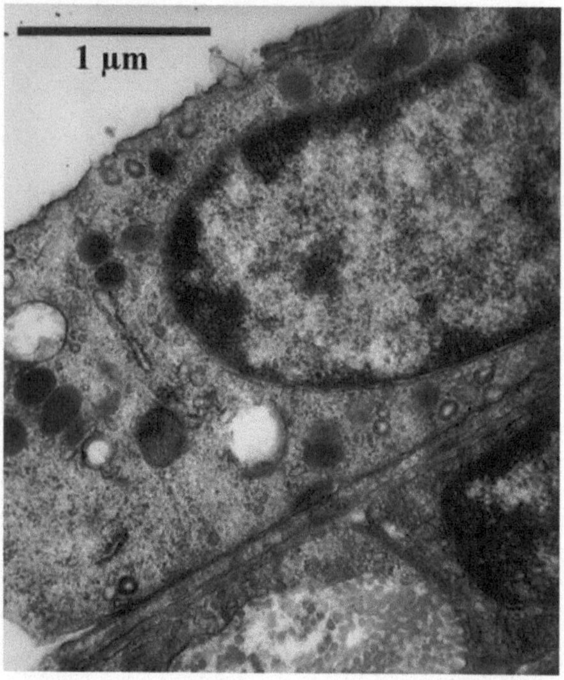

Fig. 110. Peroxisomes in an alveolar macrophage adjacent to a type I pneumocyte of a female *rat* (breeder: Winkelmann, Borchen-Kirchborchen) which inhaled micronized deptropine citrate and isoprenaline hydrochloride āa from a suspension type self-propelled aerosol (Medihaler®) containg a mixture of 28 μg hexadecylpyridinium chloride/puff + sorbitan trioleate for surfactants. 12 Puffs/min were dispersed into a 164.5 l box where the animals stayed for 15 min. 1 h later under methitural anaesthesia, the lung was fixed by intratracheal instillation of 2.5 % glutaraldehyde in phosphate buffer (pH 7.4) before opening the thorax. Postfixation with 1 % osmium tetroxide in phosphate buffer (pH 7.4). Contrasted en bloc for 12 h with 0.5 % uranyl acetate in 70 % ethanol. Embedded in a 2:8 mixture of methyl and butyl methacrylate. Sectioned at 50 nm. Lead citrate after REYNOLDS (1963). Plate 20/02. (from SCHILLER 1971)

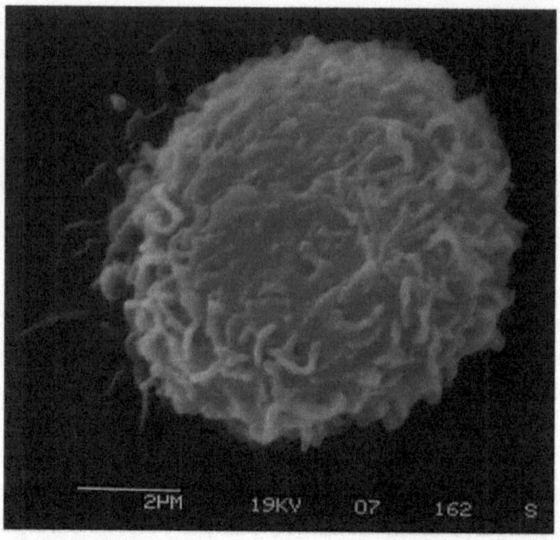

Fig. 112. Ruffled coat of a peritoneal macrophages from a SPF-NMRI *mouse* 48 h after a single intraperitoneal injection of 1 ml NaCl. The cells harvested with the ascites were centrifuged for 10 min at 1000 r.p.m. and incubated in TCM 199 + 20 % fetal calf serum on a plastic cover slip for 2 h, rinsed with phosphate buffered saline after Dulbecco and fixed in 1 % glutaraldehyde in 0.07 M phosphate buffer (pH 7.4). Ethanol, amyl acetate. Critical point drying. Gold coating. Cambridge Stereoscan 150 operated at 19 kV. APh-R. 854/80, negative 07 162. (from SCHILLER 1982)

ing, cutting out, and weighing (cf. ROMEIS 1968, § 916), SCHOEVAERT et al. (1983) measured the increase of the surface area induced by immunomodulators with a 720 lines Plumbicon camera and analysed it using a Quantimet 720 interfaced with a PDP 11/34.

The microtubules, microfilaments, and intermediate filaments of the cytoskeleton influence the shape of free cells such as macrophages, but also

Fig. 111. Peritoneal macrophages from a SPF-NMRI *mouse* 48 h after a single intraperitoneal injection of 1 ml NaCl. The cells harvested with the ascites were centrifuged for 10 min at 1000 r.p.m. and incubated in TCM 199 + 20 % fetal calf serum on a plastic cover slip for 2 h, rinsed with phosphate buffered saline after Dulbecco and fixed in 1 % glutaraldehyde in 0.07 M phosphate buffer (pH 7.4). Ethanol, amyl acetate. Critical point drying. Gold coating. Cambridge Stereoscan 150 operated at 19 kV. APh-R. 102/80, negative 03 798. (from SCHILLER 1982)

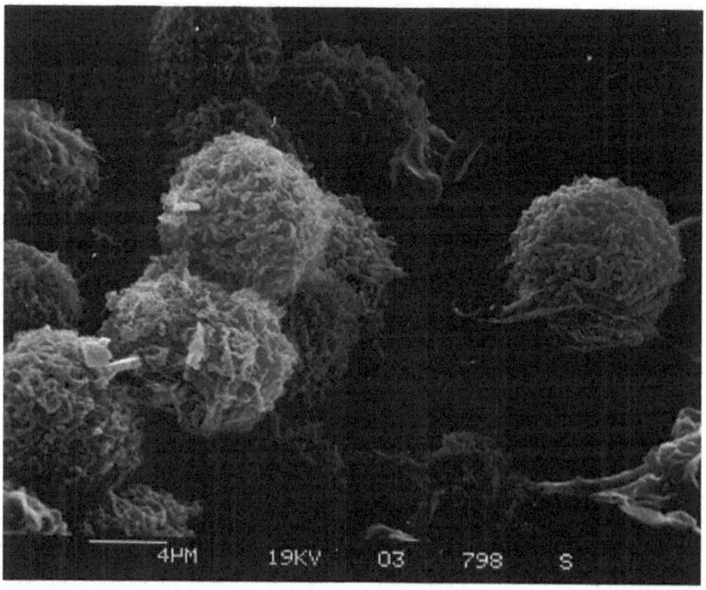

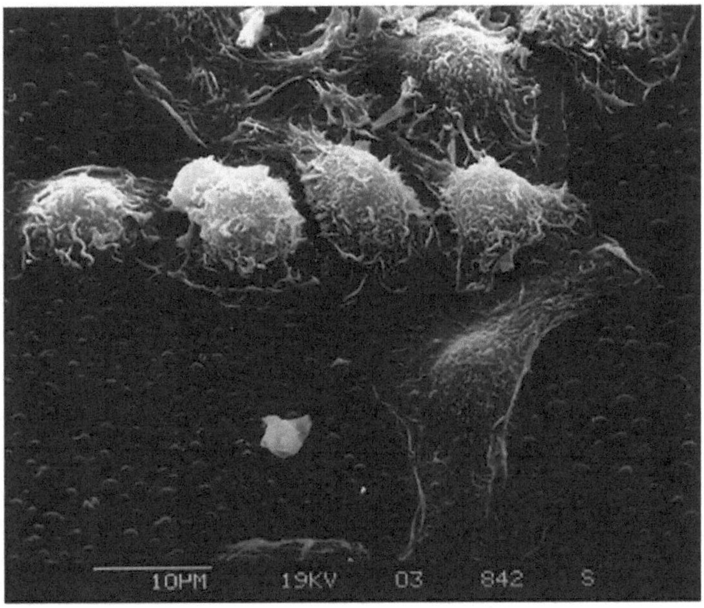

Fig. 113. Both rounded and spread peritoneal macrophages from a SPF-NMRI *mouse*, the former showing surface ridges and ruffles attaching themselves to the underlying substrate by means of thin veils of cytoplasm spreading beneath the dome-shaped nuclear pole. The cells were harvested 48 h after a single intraperitoneal injection of 1 ml saline by centrifugation of the ascites for 10 min at 1000 r.p.m. and incubated in TCM 199 + 20% foetal calf serum on a plastic cover slip for 2 h, rinsed with phosphate buffered saline after Dulbecco and fixed in 1% glutaraldehyde in 0.07 M phosphate buffer (pH 7.4). Ethanol, amyl acetate. Critical point drying. Gold coating. Cambridge Stereoscan 150 operated at 19 kV. APh-R. 102/80, negative 03 842

the locomotion of the cell organelles such as phagosomes, lysosomes, phagolysosomes and secretory granules, and the process of ciliogenesis. The outer surface of macrophages under the influence of saline (Figs. 111–114) or isotactic polyacrylic acid (Figs. 160–165) depends on the cytoskeleton.

Oxidant injury produces dramatic changes in cytoskeletal organisation and cell shape as shown in the P388D$_1$ cell line by Hinshaw et al. (1988).

9.1.3.1 Actin and Microfilaments

At least six different isoforms of actin have been identified in vertebrate tissues, each encoded by a different gene (Vanderkerckhove and Weber 1978, Reddy et al. 1990). Non-muscle cells contain two general cytoplasmic actins, β and γ. They are very similar to each other (only 4 amino acid substitutions) (Vandekerckhove and Weber 1979). Ultrastructural evidence that calponin binds selectively to "cytoskeletal" actin has led some investiga-

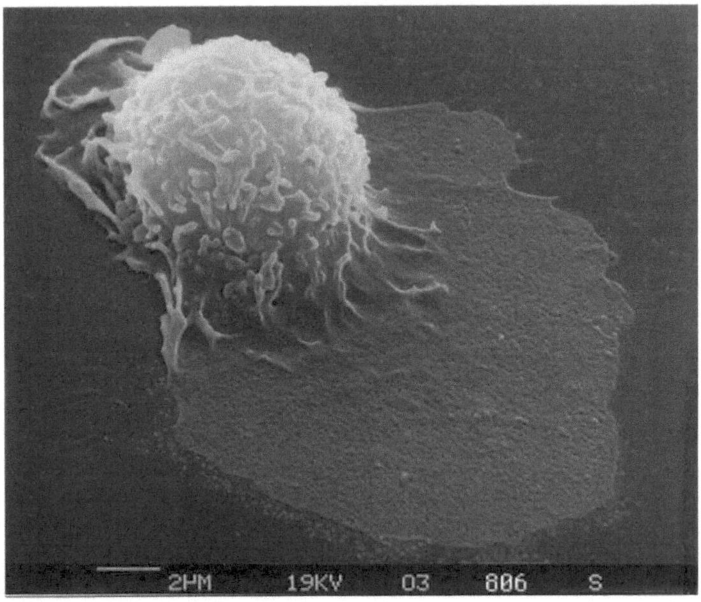

Fig. 114. Rounded peritoneal macrophages from a SPF-NMRI *mouse*, showing surface ridges and ruffles attaching itself to the underlying substrate by means of thin veils of cytoplasm spreading beneath the dome-shaped nuclear pole. The cell was harvested 48 h after a single intraperitoneal injection of 1 ml saline by centrifugation of the ascites for 10 min at 1000 r.p.m. and incubated in TCM 199 + 20% foetal calf serum on a plastic cover slip for 2 h, rinsed with phosphate buffered saline after Dulbecco and fixed in 1% glutaraldehyde in 0.07 M phosphate buffer (pH 7.4). Ethanol, amyl acetate. Critical point drying. Gold coating. Cambridge Stereoscan 150 operated at 19 kV. APh-R. 102/80, negative 03 806. (from Schiller 1982)

tors to propose that the function of calponin in smooth muscle may be structural rather than regulatory. Based on evidence that calponin binds to desmin as well as β actin, MABUCHI et al. (1997) proposed that calponin may function as a bridging protein between actin and intermediate filament networks at dense bodies. Ultrastructural analysis of thin filaments by HODGKINSON et al. (1997) indicated that the location of calponin on F-actin is similar to that of the actin cross-linking protein fimbrin, as well as to that of α-actinin and gelsolin. They suggested that a possible function of calponin may be the competitive inhibition of the binding of these or other actin-binding proteins to the actin filament. This could serve to regulate the building or remodelling of the actin cytoskeleton and thereby affect its mechanical properties.

Monomeric actin is stable in distilled water only.

Actin has been isolated from alveolar macrophages by HARTWIG and STOSSEL (1975, 1982), STOSSEL and HARTWIG (1975, 1976), and HARTWIG et al. (1977). Under the electron microscope, ALLISON et al. (1971) and SENDA et al. (1975) saw arrow heads typical for actin microfilaments with heavy meromyosin.

The reversible assembly of cytoplasmic actin can be regulated at different levels (KORN 1982, STOSSEL et al. 1985, POLLARD 1986). Dissociation of the actin monomer-binding protein, profilin (CARLSSON et al. 1977) of profilin-actin complexes would increase the local concentration of free actin monomers available for nucleation or to polymerise onto nuclei formed through another process. A second class of regulatory proteins, the filament-severing proteins bind to the barbed end of actin filaments and block the binding of actin monomers to that end, such as gelsolin (YIN and STOSSEL 1979). These molecules may function to cap filament ends in resting cells and dissociate as a consequence of agonist binding. Once dissociated, actin monomers could access to the filament ends leading to filament elongation. This mechanism would increase filament length, but not increase the number of filaments in a region of cytoplasm. Net filament growth could also occur by the formation of new nuclei, either by fragmentation of existing filaments or by the activation of proteins expressing sites equivalent to the barbed end of filaments. Nuclei serve as templates for the assembly of new filaments. Nucleation differs from uncapping in that the number of filaments in a region of a cell increases not just the length of pre-existing filaments.

HARTWIG and JANMEY (1989) found that phorbol 12-myristate 13-acetate would increase the nucleation activity of macrophages washed out of the lungs of New Zealand white *rabbits* given complete Freund's adjuvant intravenously. This increment was completely cytochalasin-sensitive, indicating that exposure to PMA leads to formation of free barbed ends.

Xanthine oxidase increased the rate of actin polymerisation and accelerated the conversion of F(ATP)actin into F(ADP,Pi)actin (LANZARA et al. 1988).

HOCl induced a rapidly increasing yield of carbonyl groups in *rabbit* skeletal muscle actin (DALLE-DONNE et al. 2001). However, when carbonylation became evident, some cysteine and methionine residues had been already oxidised. HOCl-mediated oxidation induced the progressive disruption of actin filaments and the inhibition of F-actin formation. The molar ratios of HOCl to actin that lead to inhibition of actin polymerisation seemed to have effect only on cysteines and methionines.

In *human* intestinal (Caco-2) monolayers exposed to reactive oxygen metabolites (H_2O_2; HOCl), preincubation with OPC (Otsuka America Pharmaceutical Company) compounds 12759 or rebamipide {[2-(4-chlorobenzoylamino)-3-[2-(1H)-quinolinon-4-yl] propionic acid} and 6535 {6-[2-(3,4-diethoxyphenyl)thiazol-4-yl]pyridine-2-carboxylic acid} pr stable F-actin (BANAN et al. 2001).

In the P338D_1 *murine* cell line 5 mM H_2O_2 induced side-to-side aggregates or bundles of microfilaments (HINSHAW et al. 1988) suggesting that declining levels of ATP either from metabolic inhibition or H_2O_2 injury are correlated with the fragmentation and shortening of microfilaments into aggregates. No net change in monomeric or polymeric actin was necessary for this to occur. However, at later time points after H_2O_2 exposure some actin assembly did occur.

N-Formyl-methionyl-leucyl-phenylalanine-induced neutrophil activation (actin reorganisation and chemotaxis) was inhibited by the nitric oxide donor, 1,2,3,4-oxatriazolium,5-amino-3-(3,4-dichlorophenyl)-chloride (WARD et al. 2000).

The ability of lipopolysaccharide to cause altered phosphate labelling of β/γ-actin suggests a participation of the microfilament network in lipopolysaccharide-induced monocyte activation (HAUSCHILDT et al. 1997).

Nitric oxide has been reported to be involved in the regulation of pseudopodia formation (JUN et al. 1996). KE et al. (2001) separated the globular and filamentous actins from quiescent or •NO-stimulated macrophage-like cell line RAW 264.7 cells. Predominant G-actin coexisted with Triton X-100-insoluble filamentous (TIF) and Triton X-100-soluble filamentous actin in resting RAW 264.7 cells. The exogenous •NO produced by (±)-(*E*)-2-[(*E*)-hydroxyimino]-6-methoxy-4-methyl-5-nitro-hexenamide (NOR1), the endogenous •NO induced by lipopolysaccharide plus interferon-γ, and dibutyryl-cGMP increased the contents of TIF-actin in dose- and time-dependent manners and altered its morphology. The increase in the TIF-actin contents induced by NOR1 or lipopolysaccharide plus interferon-γ was efficiently blocked by the radical scavenger 2-(4-carboxyphenyl)-4,4,5,5-tetramethylimidazoline-1-oxyl 3-oxide and the soluble guanylate cyclase inhibitor 1*H*-[1,2,4]oxodiazolo[4,3-*a*] quinoxalin-1-one or the arginine analogue N^G-monomethyl-L-arginine acetate, respectively. Preincubation with the calmodulin antagonist W-7 almost completely blocked the •NO-induced TIF-actin increase and morphological change. On the other hand, preincubation with C3 transferase, an inhibitor of Rho protein, efficiently prevented the change in cell morphology, but had no effect on the TIF-actin increase.

In *human* intestinal (Caco-2) monolayers exposed to reactive oxygen metabolites (H_2O_2; HOCl), preincubation with OPC (Otsuka America Pharmaceutical Company) compounds 12759 or rebamipide {[2-(4-chlorobenzoylamino)-3-[2-(1H)-quinolinon-4-yl] propionic acid} and 6535 {6-[2-(3,4-diethoxyphenyl)thiazol-4-yl]pyridine-2-carboxylic acid} prevented actin oxidation, decreased depolymerized G-actin, and enhanced the stable F-actin (Banan et al. 2001).

Nitric oxide has been reported to be involved in the regulation of pseudopodia formation (Jun et al. 1996). Ke et al. (2001) separated the globular and filamentous actins from quiescent or •NO-stimulated macrophage-like cell line RAW 264.7 cells. Predominant G-actin coexisted with Triton X-100-insoluble filamentous (TIF) and Triton X-100-soluble filamentous actin in resting RAW 264.7 cells. The exogenous •NO produced by (±)-(*E*)-2-[(*E*)-hydroxyimino]-6-methoxy-4-methyl-5-nitro-hexenamide (NOR1), the endogenous •NO induced by lipopolysaccharide plus interferon-γ, and dibutyryl-cGMP increased the contents of TIF-actin in dose- and time-dependent manners and altered its morphology. The increase in the TIF-actin contents induced by NOR1 or lipopolysaccharide plus interferon-γ was efficiently blocked by the radical scavenger 2-(4-carboxyphenyl)-4,4,5,5-tetramethylimidazoline-1-oxyl 3-oxide and the soluble guanylate cyclase inhibitor 1*H*-[1,2,4]oxodiazolo[4,3-*a*]quinoxalin-1-one or the arginine analogue N^G-monomethyl-L-arginine acetate, respectively. Preincubation with the calmodulin antagonist W-7 almost completely blocked the •NO-induced TIF-actin increase and morphological change. On the other hand, preincubation with C3 transferase, an inhibitor of Rho protein, efficiently prevented the change in cell morphology, but had no effect on the TIF-actin increase.

Rho family proteins control actin organisation (Ridley and Hall 1992). Like Ras, most members of the Rho family cycle between an inactive GDP-bound form and an active GTP-bound form. Three major regulators control their activity: RhoGDIs interact with the geranyl-geranylated form of the proteins to keep them in a "resting" cytosolic complex (Zalcman et al. 1996, 1999).

Paraquat known to induce production of $O_2^{\bullet -}$ through ist reduction/oxidation (formula [158]) affected actin cytoskeleton by an increase of the filamentous pool of actin and a parallel decrease of the monomeric actin (Cappelletti et al. 1996). Paraquat induced an increase of de novo synthesis of actin, but did not affect the actin degradation rate.

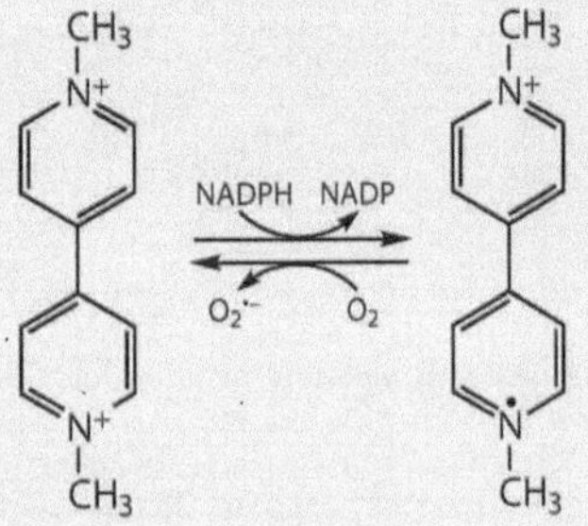

Paraquat [158]

The copper complex tetraanhydroaminobenzaldehyde (16 atoms in a macrocyclic ring, four donor nitrogens associated with Cu^{2+}) as to its functional similarity to superoxide dismutase is able to catalyse superoxide to molecular oxygen and paraquat interacted with the actin cytoskeleton in a competitive-like manner (Urbancíková and Korytár 1999).

The dynamics of the actin cytoskeleton are controlled by a collection of actin-binding proteins.

Severin is a member of a family of actin-binding proteins which consist of a tandem arrangement of homologous domains which have developed slightly modified activities. Severin has three such domains, as does fragmin, while gelsolin and villin have six. Although the first domain of severin can cap the ends of actin filaments the second domain is required to sever them. It is therefore assumed that the second domain binds to side of the actin filament while the first is forced into the filament disrupting it. Calcium has been implicated in the regulation of the activity and calcium-induced changes in the NMR spectrum of severin has identified residues which may bind calcium ions.

Myosin from *vertebrate* macrophages is composed of two heavy chains of M_r 200,000, two light

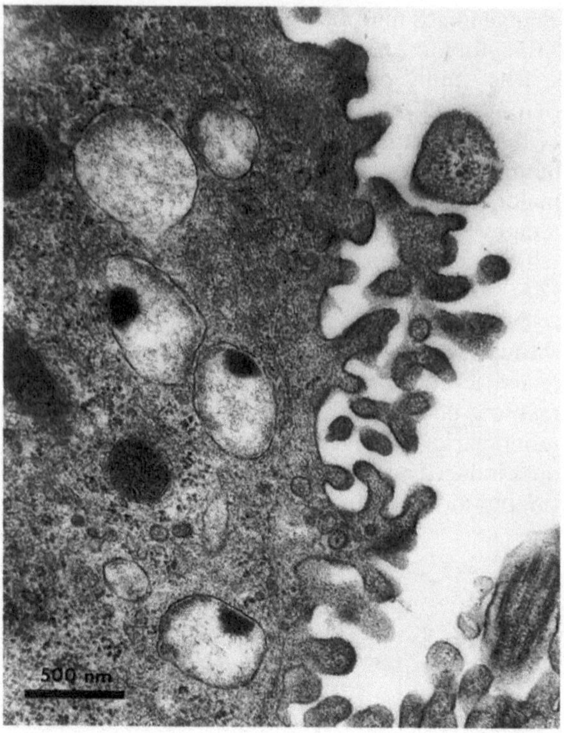

Fig. 115. Numerous subplasmalemmal microfilaments in an alveolar macrophage eliminated via the ciliated epithelium. Lung (block S2) of a female *rat* (breeder Winkelmann, Borchen-Kirchborchen) exposed to 20 mg quartz-free shale dust/m^3 (particles less than 5 μm in size; origin: seem Erda, Fürst Leopold-Baldur mine at Harvest-Dorsten) in POLLEY's (1963, 1965) dust channel 4 h per day, 5 days per week. Fixed under methitural anaesthesia by intratracheal instillation of 2.5 % glutaraldehyde in phosphate buffer (pH 7.4) before opening the thorax. Postfixation with 1 % osmium tetroxide in phosphate buffer (pH 7.4). Contrasted en bloc for 12 h in 0.5 % uranyl acetate in 70 % ethanol. Embedded in methylmethacrylate : butylmethacrylate 2 : 8. Sectioned at 50 nm. Lead citrate after REYNOLDS (1963). Plate 3594 (from SCHILLER 1982)

Table 27. Actin-associated proteins

Myosin
Tropomyosin
Fimbrin
Alpha-Actinin
Gelsolin
Vinculin
Talin
Villin
Synaptopodin
Myopodin

chains of M_r 15,000 and two light chains of M_r 20,000 (HARTWIG and STOSSEL 1975, TROTTER and ADELSTEIN 1979). It structure is, by analogy with other *vertebrate* myosins (WEEDS and LOWEY 1971, BURRIDGE 1979), likely to be characterised by two head regions, each of which is composed of a part of the heavy chain in association with one of each of the two classes of light chain. The 20 kDa light chain plays a regulatory role: it must be phosphorylated in order for the myosin to be activated by actin (TROTTER and ADELSTEIN 1979). Myosin from *rabbit* alveolar macrophages is heterogeneous with respect to its 20 kDa light chain in a ratio of 2 : 1 (TROTTER et al. 1983).

The multiple non-muscle-type myosin heavy chain isoforms can be grouped into two myosin heavy chain A and B types, the latter being predominantly expressed in brain (SUN and CHANTLER 1992, TAKAHASHI et al. 1992) and the former essentially in muscle and in non-muscle-non-brain tissues (MURAKAMI et al. 1993).

Tropomyosin may be involved in the regulatory subunits that mediate cooperativity within thin filament (reviewed by EL-SALEH et al. 1986). Head-to-tail polymerisation of tropomyosin within the thin filament (MAK and SMILLIE 1981) could provide a means for co-operative interaction between functional groups. In a test of this idea WALSH et al. (1984) found that the response of regulated actin-activated myosin ATPase to Ca^{2+} was unaffected by removal of regions of overlap between adjacent tropomyosin molecules. However, PAN et al. (1989) found that removal of head-to-tail overlap of tropomyosin molecules reduced the cooperativity of S1-ADP binding to reconstitutes thin filaments, and they concluded that overlap of adjacent tropmyosin is necessary for near-neighbour interactions. The difference in results between the two studies may be due to the low ionic strength of the assay system (approximately 20 mM) used by WALSH et al., and their low rations (1 : 100) of myosin relative to actin. Thus far, the role of tropomyosin overlap in the apparent cooperativity of tension development has not been studies directly since it as not been feasible to modify the tropomyosin content of skinned fibres.

The expression of nonmuscle tropmyosin was restored to pretransformation levels in ouabain-selected revertants of v-Ki-ras transformed NIH/3T3 cells (BASSIN and NODA 1987).

Alpha-Actinin is a rod-like (3–4 nm × 30–40 nm) cytoskeletal protein belonging to the same family as spectrin, dystrophin and utrophin. α-Actinin is a homodimer with a subunit molecular weight of 94–103 kDa in which the subunits are antiparallel in orientation. The molecule can be devised into three domains, an N-terminal actin binding domain (approximately residues 1–245), four internal 120 residue repeats, and a C-terminal region containing two EF-hand Ca^{2+}-binding motifs (BARON et al. 1987, BLANCHARD et al. 1989). Apart from actin, α-actinin has been reported to bind to the cytoskeletal protein vinculin (Wachsstock et al. 1987), nebulin (Nave et al. 1990), clathrin (Merisko et al. 1988), and to the cytoplasmic domain of the β1-family of integrins, receptors for extracellular matrix proteins (Otey et al. 1990).

Gelsolin is the most potent actin filament severing protein identified to date (Sun et al. 1999). Severing is the weakening of enough non-covalent bonds between actin molecules within a filament to break the filament in two. Membrane ruffling is a functional readout for a co-ordinated series of membrane and cytoskeletal events, and it is activated by the small GTPase, Rac. Gelsolin null fibroblasts have increased Rac expression (Azuma et al. 1998), and Rac-GTP dissociates gelsolin-actin complexes (equivalent to uncapping) in cell extracts but not purified gelsolin-actin complexes (Arcaro 1998).

The rapid depletion of plasma gelsolin following major trauma in patients who subsequently develop respiratory distress suggested that this actin-scavenging protein might protect against delayed pulmonary complications. Christofidou-Solomidou et al. (2002) measured gelsolin levels in three *murine* models of oxidant injury: immunotargeting of pulmonary endothelium with an H_2O_2-generating enzyme; continuous exposure to >95 % O_2; and single high-dose thoracic irradiation. The degree of lung injury was inversely related to gelsolin levels in *mice* treated with glucose oxidase-conjugated antibodies against platelet endothelial cell adhesion molecule-1 ($P < 0.0001$). By 60–72 h of hyperoxic exposure, gelsolin levels had dropped precipitously in all *mice* who sustained major lung damage ($P < 0001$), establishing a quantitative association between gelsolin concentration and hyperoxic lung injury (r = -0.72; 95 % confidence interval: -0.81 to -0.59). Gelsolin levels modestly but progressively fell in irradiated *mice* over the 3 days following treatment ($P = 0.012$) despite the development of only microscopic lung damage during this timeframe.

Gelsolin is expressed at low levels in most cancer cell lines and is up-regulated during in vitro differentiation induced by agents such as phorbol esters and histone acetylase inhibitors in leukaemic (Kwiatkowski 1988) and epithelial cell lines (Jarrard et al. 1998). Gelsolin has also been shown to have decreased expression in lung cancers compared with histologically normal surrounding lung (Dosaka-Akita et al. 1998). Ciglitizone treatment resulted in a prominent induction of gelsolin in multiple cell lines, although 15-deoxy-$\Delta^{12,14}$-prostaglandin J_2 had a lesser effect in the same cell line (Chang and Szabo 2000).

Vinculin and talin are major components of focal contacts that interact with each other and that are thought to be involved in linking actin filaments to integral membrane proteins. Microinjection studies with anti-vinculin antibody (Nuckolls et al. 1992) indicated that vinculin is a key protein in the microfilament-membrane linkage and that talin is essential for the development of focal contacts. In cultured chick embryo fibroblasts, a fraction (2–5 %) of the newly synthesised vinculin and talin reached maximal levels in the cytoskeleton in 30–45 min (Lee and Otto 1997). Both proteins had 2–3 times shorter half-lives in the cytoskeletal pool ($t_{1/2}$ = 6–7 h) than in the cytosolic pool ($t_{1/2}$ = 14–15 h), which suggested that the incorporation of cytosolic vinculin and talin into the cytoskeleton does not involve a simple equilibrium between the two pools.

Talin is a large molecule (270 kDa). Below 0.7 mg/ml it exists as a monomer. Above this concentration it begins to self-associate to form dimers (Molony et al. 1987). In many types of cultured cells, talin in concentrated in focal adhesions (Burridge and Cornell 1983), regions where bundles of actin filaments attach to the cytoplasmic face of the membrane and where the external face of the membrane adheres most tightly to the underlying substratum. Talin is also found in ruffling membranes and subjacent the bundles of extracellular matrix on the cell surface (Collier et al. 1982).

Talin interacts with the β1 integrin subunit on the one hand and vinculin and actin on the other hand (Horwitz et al. 1986). These proteins recognise distinct binding sites on talin (Horwitz et al. 1986) and, as they are colocative to adhesion sites on the plasma membrane (Burridge et al. 1990), it is likely that talin plays a pivotal role in associating matrix recognising integrins with the cytoskeleton.

Bass et al. (1999) defined the three vinculin-binding sites in talin to residues 607–636, 852–876 and 1944–1969; alignment of these sequences shows 59 % similarity, although there are only two identical residues. Predictions of secondary structure indicate that this vinculin-binding motif forms an amphipathic α-helix. The hydrophobic face of helix 607–637 contains three aligned leucines (residues 608, 615 and 622), which show conservative substitution in the other two site.

Phosphorylation of talin is prompted by the exposure of 1,25$(OH)_2$ vitamin D_3 pre-treated CSF-1 dependent macrophage-like BAC 1.25F5 cells to phorbol 12-myristate 13-acetate in place of CSF-1 (Meenakshi et al. 1993).

Binding of calmodulin to the microfilament network correlates with induction of a macrophage tumoricidal response (Mecham et al. 1985).

Calponin (Takahashi et al. 1988, El-Mezgueldi 1996) and caldesmon (Sobue and Sellers

1991) are two thin filament associated proteins that bind to F-actin, tropomyosin and calmodulin. Interaction of 34 kDa calponin with F-actin and tropomyosin occurs in a Ca^{2+}-independent manner, whereas that with calmodulin is regulated in a Ca^{2+}-dependent manner. Despite their apparent functional similarity, sequence analysis indicated that calponin and caldesmon are not related proteins. They act by different mechanisms of inhibition and bind to distinct thin filament populations in smooth muscle cells (SOBUE and SELLERS 1991, TAKAHASHI and NADAL-GINARD 1991, NORTH et al. 1994).

Ultrastructural evidence that calponin binds selectively to "cytoskeletal" actin has led some investigators to propose that the function of calponin in smooth muscle may be structural rather than regulatory. Based on evidence that calponin binds to desmin as well as β actin, MABUCHI et al. 1997) proposed that calponin may function as a bridging protein between actin and intermediate filament networks at dense bodies. Ultrastructural analysis of thin filaments by HODGKINSON et al. (1997) indicated that the location of calponin on F-actin is similar to that of the actin cross-linking protein fimbrin, as well as to that of α-actinin and gelsolin.

Profilins are small, highly abundant cytoplasmic proteins that bind actin, poly-L-proline (TANAKA and SHIBATA 1985), and polyphosphoinositides, raising the possibility that they are second messengers to carry informations between the polyphosphoinositide signalling pathway and the cytoskeleton. From knockout mutants of profilins, such as generated in *mice*, *Dictyostelium* and *yeast*, it became evident that profilins are essential components of the microfilament system (SCHLÜTER et al. 1997). The soluble actin-binding protein profilin binds up to five molecules of phosphatidylinositol-4,5-biphosphate, but not phosphatidylcholine, phosphatidylserine or phosphatidyl ethanolamine, with reasonably high affinity; this binding inhibits both the interaction between profilin and actin and the hydrolysis of phosphatidylinositol-4,5-biphosphate by soluble phosphoinositidase C. GOLDSCHMIDT-CLERMONT et al. (1991) have shown that in the presence of profilin, phospholipase $C\gamma_1$ activity is dependent upon phosphorylation state; i.e. phosphorylation of phospholipase $C\gamma_1$ by the epidermal growth factor receptor relieves the tonic inhibition of phospholipase $C\gamma_1$ activity by profilin. The considerable increases in a highly charged lipid such as phosphatidylinositol-3,4,5-triphosphate (STEPHENS et al. 1991) are likely to exert profound effects upon the inner leaflet of the plasmalemma and might be expected for this reason alone to modify cytoskeleton-membrane interactions.

Ezrin, moesin and **radixin** (ERM), three actin-binding proteins, which link actin filaments to the plasma membrane (TSUKITA et al. 1997), are primary located in microvilli on the apical surface of epithelial cells, although they have also been observed at intercellular boundaries (BRETSCHER et al. 1997). Rho can colocalise to the plasma membrane with ERM proteins, and Rho inactivation prevents the localisation of ERM proteins and vinculin to the plasma membrane (KOTANI et al. 1997). RhoGDI apparently binds in a complex with CD44 and ERM proteins, and could therefore act as an intermediary to target Rho to ERM proteins, or alternatively ERM proteins could increase Rho activity by titrating out RhoGDI (TAKAHASHI et al. 1997). ERM protein interaction with actin is required for Rho and Rac to induce actin reorganisation and focal complex assembly in permeabilized fibroblasts (MACKAY et al. 1997). A Cdc42-interacting protein, CIP4, shows sequence homology to a small region of the ERM proteins and may act as a transducer to the actin cytoskeleton as it induces actin reorganisation when overexpressed in Swiss 3T3 cells (ASPENSTRÖM 1997).

Synaptopodin (formerly called pp44), a member of a family of actin-associated proteins rich in proline, was found in telencephalic dendrites and renal podocytes (MUNDEL et al. 1997). The cytoskeletal rearrangements of mouse podocyte clones are accompanied by the onset of synaptopodin synthesis.

Myopodin, another member of this family, is associated with the Z disc of *human* skeletal muscle (MUNDEL et al. 1999).

9.1.3.2
Tubulin and Microtubuli

Microtubules are proteinaceous organelles, present in nearly all eukaryotic cells, made of subunits of tubulin molecules assembled into long tubular structures, with an average exterior diameter of 24 nm, capable of changes of length by assembly or disassembly of their subunits. Macrotubules with a diameter of 31 to 52 nm, which are related to the paracrystalline tubulin assemblies produced by *Vinca* alkaloids, result from a helical winding of protofilaments (TILNEY and PORTER 1967). Their formation is an indication of changes in the lateral links of tubulin protofilaments. The protein tubulin is acidic, combined with two molecules of guanine nucleotides.

The αβ tubulin heterodimer is the structural subunit of microtubules. Each tubulin monomer binds a guanine nucleotide, which is non-exchangeable when it is bound to the α subunit, or N site, and exchangeable when bound in the β subunit, or

E-site. The α- and β-tubulins share 40% amino-acid sequence identity, both exist in several isotype forms, and both undergo a variety of post-translational modifications (LUDVEÑA 1998). Limited sequence homology has been found with the proteins FtsZ (MUKHERJEE and LUTKENHAUS 1994) and Misato (GABOR MIKLOS et al. 1997), which are involved in cell division in bacteria and *Drosophila*, respectively. NOGALES et al. (1998) presented an atomic model of the αβ tubulin dimer fitted to a 3.7-ø density map obtained by electron crystallography of zinc-induced tubulin sheets. The structures of α- and β-tubulin are basically identical: each monomer is formed by a core of two β-sheets surrounded by α-helices. The monomer structure is very compact, but can be devided into three functional domains: the amino-terminal domain containing the nucleotide-binding region, an intermediate domain containing the taxol-binding site, and the carboxyl-terminal domain, which probably constitutes the binding surface for motor proteins.

Fresh *human* blood mononuclear cells contained an average of 26 microtubules per cell which significantly increased to 31 microtubules per cell following a 30-min exposure to LPS ($P < 0.001$). Using a nocodazole-based assay of microtubule dynamic instability, the half-life of fresh unstimulated monocyte microtubules was approximately 18 s and extended to 26 s following a 30-min exposure to LPS (ALLEN et al. 1997). Endotoxin caused a rapid alteration in monocyte microtubule stability (ALLEN et al. 1997).

The direct interaction of **peroxisomes** with the microtubular network was visualised by SCHRADER et al. (1994, 1996), showing a specific binding of isolated peroxisomes to microtubules. *In vitro*, the binding of peroxisomes to the microtubular network was sensitive to pre-treatment of the organelles with high concentration of KCl or proteases.

Synaptosomal microtubules, contrary to axon microtubules (MATUS et al. 1981), are associated in *rat* brain with high molecular weight microtubule associated proteins which are susceptible to an endogenous Ca^{2+}-dependent protease, proteolysis being essentially complete in 5 min at 37 °C. This could explain the difficulty in demonstrating the microtubule associated proteins by histochemistry (BURGOYNE and CUMMING 1982).

In **tumour cells**, the 3-D reconstruction gives a clear image of the spatial arrangement of tubulin fibres in relation to cell shape and position of other cellular organelles, particularly the nucleus (STROHMAIER et al. 2000). The tubulin forms an intricate network of fibres of variable thickness. The highest tubulin concentrations appear in the cell periphery and particularly in pseudopodia/invadopodia. This is indicative of an enhanced transport of intracellular material facilitating cell movement and lysis of the extracellular matrix.

Melatonin antagonised the action of colchicine on melanocytes (MALAWISTA 1965). It inhibited the regeneration of cilia in *Stentor*, and this effect was antagonised by colchicine (BANERJEE et al. 1972). On the other hand, the two drugs inhibited synergistically the regeneration of the oral band, suggesting that both bound to microtubule protein. In HeLa and KB cells preincubation for 1 h in 10^{-5} M melatonin protected the microtubuli of the spindle against the action 10^{-7} M colchicine (FITZGERALD and VEAL 1976). High concentrations of melatonin prevented ^{3}H-colchicine binding to brain tubulin (WINSTON et al. 1974), a result contradicted by the findings of POFFENBARGER and FULLER (1976) that melatonin does not affect *in vitro* assembly of *bovine* brain tubulin nor the mitosis of Chinese *hamster* ovary cells. In retinal ganglion cells of male New Zealand albino *rabbits* 1.5 µg or 15 µg melatonin injected 90 min before ^{3}H-leucine inhibited fast axoplasmic flow by 71.9±8.4% and 87.2±4.6%, respectively (CARDINALI and FREIRE 1975). Under these conditions melatonin tended to increase ^{3}H-leucine incorporation into retinal proteins *in vitro* (not significant).

Halothane (20 mM) significantly ($P = 0.001$) prevented repolymerisation of microtubules in the sciatic nerve of the *rat*, when depolymerization was induced *in vitro* by incubation at 0 °C for 30 min (LIVINGSTON and VERGARA 1979).

There is increasing evidence that the microtubules are an important component in the interaction between the cytoskeleton and specific mRNAs. Microtubule preparations have been found to contain ribosomes and polysomes (HAMILL et al. 1994) and inhibitor studies showed that mRNA localisation in oocytes is sensitive to colchicine (YISRAELI et al. 1990, POKRYWKA and STEPHENSON 1994). A cytoskeletal fraction from *Drosophila* oocytes is enriched in bicoid mRNA, and this mRNA is released by microtubule-disrupting agent colchicine (POKRYWKA and STEPHENSON 1994).

Chlorpromazine arrests cultured cells in mitosis and disorganises the organised microtubule structure produced by cyclic adenosine monophosphate (POFFENBARGER and FULLER 1977). It causes a reduction in the number of microtubules in spinal ganglion cells (EDSTRÖM et al. 1973, THYBERG et al. 1977) and neuroblastoma cells (EDSTRÖM et al. 1975) *in vitro*. The micellar form of chlorpromazine interacts preferentially with one site on brain tubulin (CANN et al. 1981). Chlorpromazine has been shown to bind reversibly to tubulin prepared from *mouse* brain via two well-resolved processes (HIN-

Table 28. Microtubule-associated proteins

MAP 1 Protein
MAP 2 Proteins
τ-Proteins
Ankyrin
Dyneins

MAP 1 and MAP 2 are basic proteins (Erickson and Voter 1976).

man and Cann 1976). One molecule binds strongly compared with 8–9 molecules that bind weakly.

Aluminium induced nonenzymatic phospho-incorporation into *human* **τ-proteins** (Abdel-Ghany et al. 1993). While 500 μM Fe^3+ induced [γ-^{32}P]ATP-incorporation only weakly, 500 μM Sc^{3+} were twice more effective than 500 μM Al^{3+}.

Dyneins are proteins with adenosine triphosphatase activity (Gibbons and Rowe 1965). These multipeptide complexes have molecular masses between 5.4×10^6 and 6×10^5 Da. They have been fractionated by proteolysis into fragments of 135 kDa to 400 kDa (Warner and Mitchell 1980). Warner and McIlvain (1982) analysed the binding properties of *Tetrahymena* 21S dynein to doublet A and B subfiber microtubules by both a turbidimetric assay (ΔA350 nm) and electron microscopy. KCl-extracted, sucrose-gradient, purified 21S dynein bound to each of the two kinds of axonemal microtubules in both ATP-insensitive and ATP-sensitive modes, even though only a single type of binding occurred to each of the subfibers in situ. Reed et al. (2000) characterised airway epithelial expression of a gene identified by two *human* expressed sequence tags that encoded peptides with sequence similarity to an *invertebrate* ciliary dynein heavy chain. Molecular analyses showed that the gene has a very large RNA transcript that encodes a very high molecular weight polypeptide with biochemical properties that are characteristic of a dynein heavy chain. Expression of the gene transcript correlated with the presence of ciliated cells in tissues, and immunohistochemical localisation of the gene product confirmed its presence in the cilia of mature airway epithelium.

In addition to axonemal dyneins, there are cytoplasmic dyneins, distinct isoforms that transport molecular cargoes along cytoplasmic microtubules and participate in aspects of cell division (Karki and Holzbaur 1999).

9.1.3.3 Intermediate Filaments

Structure, dynamics, function, and disease of intermediate filaments were reviewed by Fuchs and Weber (1994).

Table 29. Molecular types of intermediate filaments

Cytokeratins
Type I (acidic): 17 polypeptides
Type II (alkaline): >14 polypeptides
Vimentin
Desmin
Glial fibrillary acidic protein (GFAP)
Peripherin
Neurofilament proteins H, M, L
α-Internexin
Lamins A, B_1, B_2, C
Nestin
Filensin
Phakinin

Intermediate Filaments of the Vimentin Type

Intermediate filaments of 7–11 nm diameter (Weber and Osborn 1982) of the vimentin type (Franke et al. 1979) are arranged immediately around the cell nucleus, while the remaining cytoplasm reveals only small amounts of 10 nm filaments, which usually do not extend up to the outer membrane of the mononuclear phagocyte (Cain et al. 1982, 1983). With increasing differentiation of monocytes into mature macrophages and epithelioid cell equivalents, a loosening up of the perinuclear vimentin filament network was observed. This development was associated with a straightening of the filaments, which could now be followed into the ectoplasm and into the cytoplasmic processes.

Western blotting analyses of the extract from thioglycollate-stimulated *mouse* peritoneal macrophages incubated with 1 μM showed selective deimination of vimentin without detectable degradation (Asaga et al. 1998). Double immunofluorescence staining of deiminated proteins and vimentin suggested localisation of deiminated vimentin around the periphery of the round-shaped nucleus, which was thought to be an early morphological sign of apoptosis.

The application of both crocidolite asbestos (10 μg/ml) and silicon carbide (50 μg/ml) affected the vimentin system of the Syrian *hamster* epithelial cell line (M3E3/C3) derived from the lung of a foetus on day 15 of gestation in a time-dependent manner (Aufderheide et al. 1994). The vimentin network, which normally appears as a filigree-like pattern throughout the cytoplasm, after exposure to asbestos for 38 h concentrated in bundles. Exposure to silicon carbide induced a concentration of vimentin filaments within the cells at the expense of the normally anastomosing network.

Human KD fibroblasts on per cell volume basis contained 151.6 ng vimentin/μl (Lai et al. 1993). Protein phosphorylation was augmented by treatment of 600 nM okadaic acid for 1 h in these cells.

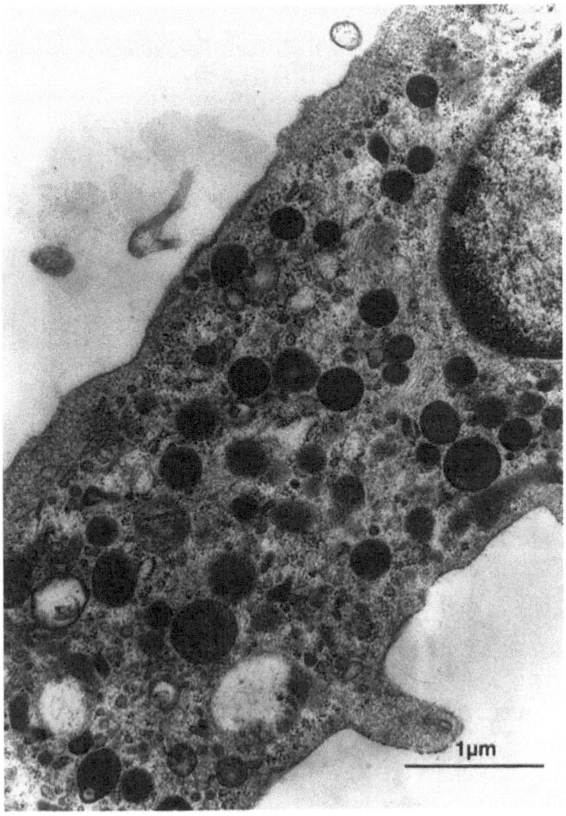

Fig. 116. Numerous intermediate filaments immediately around the cell nucleus in an alveolar macrophage (block 4424) from an unmedicated 235 g Sprague-Dawley *rat* (Charles River, France). On July 25, 1978, under pentobarbital anaesthesia (30 mg/kg), the animal was perfused from the abdominal aorta with 2.5% glutaraldehyde in 0.1 M sodium cacodylate buffer (pH 7.4). Postfixation with 1% osmium tetroxide in sodium cacodylate buffer. Embedded in Epon 812 and sectioned at 50 nm. Stained with lead citrate and uranyl acetate. Plate 4201

During the apparent activation of protein kinases, vimentin became hyperphosphorylated and the phosphorylation level of other nonvimentin phosphoproteins was relatively little affected in KD cells.

Glial fibrillary acidic protein (GFAP) immunostaining as a marker of astrocyte density increased in the oriens layer of the hippocampus of untreated bilateral carotid artery-occluded *rats* as compared to sham occluded animals (De la Torre et al. 1998).

9.2 Histochemistry

9.2.1 Glutathione

Using the GSH-specific enzymatic assay with glyoxalase I, Sibille et al. (1984) observed a GSH content of 5.96±0.61 µM/10^6 *human* alveolar macrophages. The GSH content of freshly isolated *rabbit* lung macrophhages was 3.12± 0.62 nmol/10^6 cells; it was depleted to 1.63± 0.32 nmol/10^6 cells by incubating with 500 µM diethyl maleate for 20 min (Horton et al. 1987). Cysteine was best able to support resynthesis of GSH. There was no evidence for participation of a cystathionine pathway for glutathione synthesis.

9.2.2 Immunophenotypes

The polyclonal anti pan-cathedrin antibody selectively decorated alveolar macrophages, type I pneumocytes, and endothelial cells of large blood vessels (Kasper et al. 1996).

9.2.3 Enzymes

9.2.3.1 Oxidoreductases

While interference with electron transmission and uncoupling of oxidative phosphorylation have no effect on the phagocytic function of polymorphonuclear leucocytes or monocytes they impair the activity of the alveolar macrophage (Karnovsky et al. 1966).

Oxidoreductases with NAD^+ or $NADP^+$ as Acceptor

Lactate Dehydrogenase (EC 1.1.1.27)

Lactate dehydrogenase (EC 1.1.1.27)

S-Lactate + NAD^+ → Pyruvate + NADH [125]

Rat alveolar macrophages contain five lactate dehydrogenase isoenzymes, LDH_5 being the most prominent (Bansal and Kaw 1981). Silica-exposed macrophages liberated LDH_5 in the supernatant culture medium. This LDH isoenzyme also increased in the serum of silicotic *rats* in the early stages of the development of the disease.

9.2.3.1.2 Oxidoreductases with Oxygen as Acceptor

Xanthine Oxidase (EC 1.1.3.22)

Bovine lung macrophages produced 386 pmol uric acid/min × mg protein (Bruder et al. 1983).

Tubaro et al. (1980) reported a marked increase in the xanthine oxidase activity of macrophages obtained from infected animals compared with those obtained from normal *mice*.

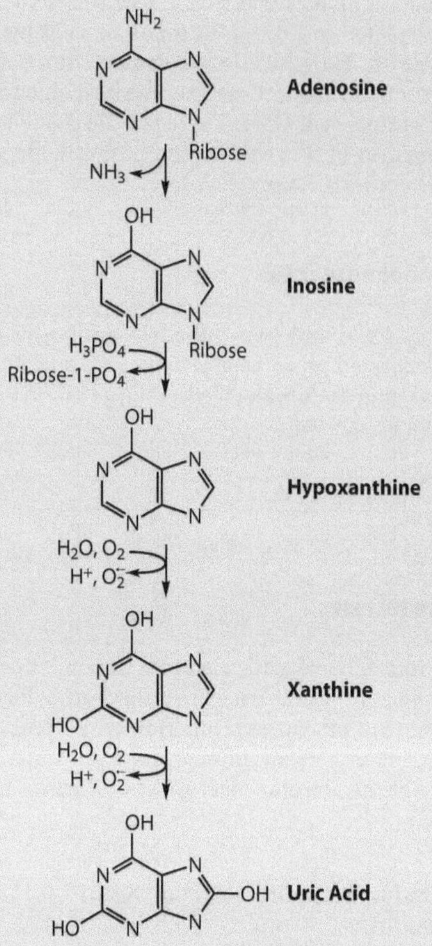

Substrates of xanthine oxidase [45]

9.2.3.1.3
Oxidoreductases with Reduced NAD or NADP as Donators

Increased oxidative stress in the RAW 264.7 macrophage cell line is partially mediated via the *S*-nitrosothiol-induced inhibition of glutathione reductase (EC 1.6.4.2) (Butzer et al. 1999). Fujii et al. (2000) isolated a cDNA for *rat* glutathione reductase and constructed a baculovirus system to produce recombinant glutathione reductase on a large scale. NO donors (*S*-nitrosoglutathione, SIN-1, and *S*-nitroso-*N*-acetyl-D,L-penicillamine) inhibited the enzymatic activities of purified glutathione reductase.

9.2.3.1.4
Oxidoreductases with Hydrogen Peroxide as Donator

Catalase (EC 1.11.1.6)

Detoxification of hydrogen peroxide by catalase (EC 1.11.1.6)

$$2\,H_2O_2 \rightarrow 2\,H_2O + O_2$$ [127]

Salicylic acid may bind to and inactivate catalase, thus affecting H_2O_2 (Sundaresan et al. 1995).

Peroxidase (EC 1.11.1.7)

In HL-60 cells, the peroxidase reaction was observed in the nuclear envelope, primary granules, endoplasmic reticulum and Golgi body (Mikami et al. 1998).

9.2.3.1.5
Oxidoreductases Acting on Single Donors with Incorporation of Molecular Oxygen (Oxygenases)

Lipoxygenase (EC 1.13.11.12) catalyses the stereospecific dioxygenation of polyunsaturated fatty acids containing a 1,4-*cis,cis*-pentadiene system to a **pentadienyl radical** intermediate which reacts with molecular oxygen to yield *cis,trans*-conjugated diene hydroperoxides (Wiseman et al. 1988).

5-Lipoxygenase (EC 1.13.11.12) and 5-lipoxygenase-activating protein are key proteins in leukotriene formation. In both unstimulated normal and ARDS *human* alveolar macrophages the two proteins were found by immonoelectronmicroscopy in the cytoplasm followed by the cell surface and nuclear membranes (Chi et al. 1996). After lipopolysaccharide treatment, approximately twofold increases in the number of immunogold particles/unit area were observed in the cytoplasmic and nuclear membrane compartments. In activated cells, both 5-lipoxygenase and 5-lipoxygenase-activating protein distributed primarily to the cytoplasmic filament network and secondarily to the endoplasmic reticulum.

On stimulation of *rat* alveolar macrophages with the calcium ionophore A-23187 (1 μM for 30 min), synthesis of leukotriene B_4 increased with the degree of maturation, although it was diminished in the oldest subpopulation (Covin et al. 1998). This maturation-dependent upregulation was not explained by increases in arachidonic acid release but was associated with increased expression of 5-lipoxygenase protein as determined by immunoblot analysis. Whereas 5-lipoxygenase is primarily cytosolic in monocytes, it is known to be primarily intranuclear in unfractionated alveolar macrophages. Covin et al. (1998) investigating the localisation of 5-lipoxygenase by immunofluorescence microscopy found it predominantly nuclear in all alveolar macrophage subpopulations. By contrast, the protein was cytosolic in interstitial macrophages isolated by mechanical and enzymatic lung digestion.

9.2.3.1.6 Oxidoreductases Acting on Paired Donors with Incorporation of Molecular Oxygen

Prolyl 4-hydroxylase (EC 1.14.11.2) is a key enzyme required for the posttranslational hydroxylation of proline residues in collagen. The enzyme consists of a tetramer composed of two pairs of nonidentical subunits ($\alpha_2\beta_2$) of ≈60 kDa each (Berg et al. 1979). Peptide mapping has demonstrated that the α and β subunits are products of different genes (Berg et al. 1979). The β subunit has been shown (Pihlajaniemi et al. 1987) to be virtually identical with the enzyme protein disulphide isomerase (EC 5.3.4.1).

Alveolar macrophages obtained by pulmonary lavage from normal male New Zealand white *rabbits* and *humans* evaluated for their lung disease contained significant amounts of prolyl hydroxylase (Kelleher et al. 1977).

When freshly isolated *rat* peritoneal macrophages were incubated in suspension with [^{14}C]proline, they synthesised a small but significant amount of non-diffusible hydroxy[^{14}C]proline (Myllylä and Seppä 1979).

Nitric oxide synthases (EC 1.14.13.39), the inducible form of which has been obtained from the *murine* RAW 264.7 macrophage cell line (Hevel et al. 1991, Stuehr et al. 1991) was found by Liu et al. (1996) in the normal Lewis *rat* alveolar interstitium by immunostaining, whereas alveolar macrophages were iNOS negative. Interferon γ maximally stimulated NO production by alveolar macrophages.

L-Arginine–nitric oxide pathway

$$\text{L-Arginine} \xrightarrow{\text{NOS}} \textbf{Nitric oxide}(^{\bullet}\text{NO}) \xrightarrow{+\mathbf{O_2^{\bullet-}}} \text{Peroxynitrite } (\text{ONOO}^-) \quad [46]$$

iNOS mRNA of *bovine* alveolar macrophages obtained by bronchoalveolar lavage immediately after isolation was below the detection limit (Höckele et al. 1997). Two or four h later increasing mRNA levels could be observed in both control and LPS stimulated cells. After 24 h treatment with LPS high levels of iNOS mRNA were induced while in control macrophages iNOS mRNA had declined to a low but detectable level.

Upon lipopolysaccharide stimulation (1 μg/ml), Prabhu et al. (2002) found significantly higher iNOS transcript and protein expression levels with an increase in NO production in selenium-deficient RAW 264.7 cells than in the Se-complemented cells. Electrophoretic mobility-shift assays, nuclear factor-κB-luciferase reporter assays and Western blot analyses indicated that the increased expression of iNOS in Se deficiency could be due to an increased activation and consequent nuclear localization of the redox-sensitive transcription factor NF-κB.

Co-cultivation of *mouse* RAW264.7 macrophages stimulated with interferon-γ + lipopolysaccharide with *human* lymphoblastoid TK6 and Chinese *hamster* ovary AA8 cells resulted in a significant increase in mutant fraction in the endogenous genes of target cells and in the macrophages themselves, accompanied by a substantial decrease in cell viability (Zhuang et al. 2002). Addition of *N*-methyl-L-arginine abrogated much of the cytotoxicity and genotoxicity in both target and macrophage cells, verifying the role of •NO in the induction of these responses.

The *human* iNOS gene, containing 26 exons, excodes a protein of 131 kDa. Alternative messenger splicing from a single transcript allows for the generation of various forms of mRNA that can be translated into different protein products. These may have distinct functions and regulatory properties. Eissa et al. (1996) identified four sites of alternative splicing which included deletion of: i) exon 5; ii) exons 8 and 9; iii) exons 9, 10 and 11; and iv) exons 15 and 16. Deletion of exon 5 (149 bases) leads to a translational frame shift and a premature stop codon in exon 6 predicting a truncated protein.

iNOS cDNA was cloned and characterised from CD-1 *mice* 4 days after an intraperitoneal injection of thioglycollate broth after incubation of isolated peritoneal macrophages for 16 h with IFN-γ (10 ng/ml) and lipopolysaccharide (1 μg/ml) (Xie et al. 1992).

Activation of the peroxisome proliferator-activated receptor-γ reduced proinflammatory cytokine and iNOS expression in macrophages (Lemberger et al. 1996, Colville-Nash et al. 1998, Ricote et al. 1998), microglial cells (Petrova et al. 1999), and monocytes (Jiang et al. 1998, Combs et al. 2000).

The amino acid sequence of *mouse* macrophage iNOS is only 51 % identical to the deduced sequence of *rat* cerebellar constitutive nitric oxide synthase (cNOS) (Bredt et al. 1991).

Constitutive nitric oxide (cNOS) was produced by naive unstimulated rat alveolar macrophages

(MILES et al. 1998). Using antibodies against two known types of cNOS, i.e., eNOS and bNOS, and an antibody against iNOS, positive results were obtained with the anti-eNOs antibody only. The amount of •NO formed was much less than that produced by eNOS in other cells, i.e., alveolar type II cells and endothelial cells. Some properties of the alveolar macrophage eNOS are similar to and some are different from the eNOS in these other cell types. Alveolar macrophage •NO levels do not seem to be related to cellular metabolism. •NO production was increased approximately threefold in the presence of dipalmitoyl phosphatidylcholine vesicles or pulmonary surfactant.

Type III (endothelial) NOS was detected by immunoperoxidase staining of bronchoalveolar lavage cytospins and normal *human* and *rat* lung tissue cryostat sections (KOBZIK et al. 1996). In ≥90 % of alveolar macrophages positive labelling was found by two separate monoclonal anti-type III NOS antibodies. Using cytofluometry to measure the effects of NOS substrate, L-arginine (1–5 mM), on intracellular basal or phorbol 12-myristate 13-acetate (2×10^{-7} M)-stimulated oxidative metabolism, L-arginine, but not its inactive stereoisomer, D-arginine caused a substantial, dose-dependent decrease in intracellular oxidation of dichlorofluorescin. Since nitric oxide can reversibly inhibit mitochondrial respiration, KOBZIK et al. (1996) hypothesised that type III NOS in alveolar macrophages may function to modulate mitochondrial-derived reactive oxygen species production and redox-based signalling.

NOS enzymes are multidomain proteins consisting of an NH_2-terminal oxygenase domain that contains the active site, a COOH-terminal reductase domain that shuttles electrons from NADPH to the heme iron (SHETA et al. 1994, STUEHR 1997), and a central calmodulin domain that governs electron flow between the two domains (ABU-SOUD and STUEHR 1993). The NOS reductase domain shares sequence homology with the flavoenzyme, NADPH-cytochrome P450 oxidoreductase (EC 1.6.2.4), the activating enzyme of the cytochrome P450 monooxygenase complex (BREDT et al. 1991).

Apigenin and kaempferol were markedly active inhibitors of transcriptional activation of iNOS in the *mouse* macrophage cell line RAW 264.7, with IC_{50} <15 μM (LIANG et al. 1999). In *murine* J774-macrophages cyclosporin A (but not tacrolimus [FK506] at 1 μM was inhibitory when co-incubated with the inducing agent but not when the cells were treated with the immunosuppressant before or after the inducer (DUSTING et al. 1999). Cyclosporin A suppressed expression of mRNA for the inducible NO synthase 2 in a concentration dependent manner when co-incubated with lipopolysaccharide (*E. coli* serotype 0111:B4) as shown in RNA extracted from J774-macrophages and subjected to reverse transcription-polymerase chain reaction.

For L-arginine supply to nitric oxide synthases by cationic amino acid transporters (CATs), CLOSS et al. (2000) studied *mouse* J774A1 macrophages and *human* EA.hy926 endothelial cells. CAT-1 was expressed in both cell types, whereas CAT-2B was only expressed in activated macrophages. Apparent K_M values for transport of L-arginine in both cell types were consistent with the expression of the system y^+ carriers CAT-1 (and CAT-2B in macrophages). L-Arginine transport was Na^+ independent and sensitive to *trans*-stimulation. At 2 h preincubation of activated macrophages in 2 mM L-lysine (which is exchanged for L-arginine by the CATs) reduced the intracellular L-arginine concentration from 2 mM to 160 μM. At the same time, NOS II activity was completely abolished. NOS II activity could be restored with extracellular L-arginine.

Cytochrome P450 (CYP; EC 1.14.14.1)

CYPs (EC 1.14.14.1) are a multiple family of constitutive and inducible enzymes that play a control role in the metabolic activation and detoxification of various xenobiotics, including small molecular weight compounds that cause allergic reactions. BARON et al. 1998) demonstrated that *human* blood monocytes and monocyte-derived macrophages constitutively express P450 isoenzymes. Of particular interest is the strong and constitutive expression of CYP1B1 in monocytes and macrophages demonstrated both on the mRNA and protein levels. CYP1B1 is the only gene of the CYP1B gene family which has been mapped in the *human* chromosome 2p21–22, which contains three exons and two introns, and is present especially in extrahepatic tissues (SUTTER et al. 1994, TANG et al. 1996).

Macrophage subtype 27E10, which has been shown to possess pro-inflammatory activity (ZWADLO et al. 1986, BHARDWAJ et al. 1992), expressed 1B1, 2E1, and 2B6/7 (Baron et al. 1998). on the other hand, in the anti-inflammatory macrophage subtype RM3/1, predominantly 1B1 and to some extent 2B6/7 were found. Treatment with cyclosporin A, phenobarbital, benzanthracene or ethanol resulted in induction of the expression of 3A3/4. Cyp 1B1 was the predominant isoenzyme in all monocytes and macrophages. In monocytes purified by adherence or induced by benzanthracene, lipopolysaccharide or 12-*O*-tetradecanoyl-phorbol-13-acetate, 1A1 was also expressed.

Prostaglandin H Synthase (EC 1.14.99.1)

Prostaglandin H synthases, also called cyclooxygenases, catalyse the two first steps in the biosynthesis of prostanoids such as prostaglandin and thromboxane. These membrane-linked enzymes exhibit heme-dependent cyclooxygenase and peroxidase activities which are responsible for the regio- and stereospecific synthesis of prostaglandin hydroxyendoperoxide PGH_2 from arachidonic acid. As with ribonucleotide reductase and photosystem II, prostaglandin H synthases proceed via radical chemistry. The enzymatic reaction begins with the removal of the 13-*pro* S hydrogen atom of arachidonic acid to create a carbon-centred radical. A free radical chain transfer reaction then allows the addition of a first molecule of dioxygen on C_{11}. A subsequent intramolecular cyclisation leads to a cyclic endoperoxide. Incorporation of a second molecule of dioxygen in C_{15} followed by the transfer of one hydrogen atom back to the substrate results in the formation of the cyclic hydroperoxyendoperoxide PGG_2. The second step of the enzymatic reaction performed by the peroxidase activity involves a two-electron reduction of PGG_2 to the cyclic hydroxyendoperoxide PGH_2.

The tyrosyl radical has been shown to be involved in the initiation of the cyclooxygenase reaction, since it is able to generate an arachidonyl radical from arachidonic acid under anaerobic conditions (Tsai et al. 1995, 1998). Addition of dioxygen to this substrate-radical-bearing enzyme resulted in the regeneration of the tyrosyl radical, completing cyclooxygenase turnover. The crucial role of Tyr_{385} in prostaglandin H synthase-1 (Tyr_{371} in prostaglandin H synthase-2) in the catalytic mechanism was demonstrated using chemical modifications of tyrosine residues with tetranitromethane, peptide mapping and also by site-directed mutagenesis experiments (Shimokava et al. 1990, Tsai et al. 1998).

Vitamin E (30 ppm) inhibited cyclooxygenase activity in macrophages from old *mice* by reducing peroxynitrite production (Beharka et al. 2002). Increasing $^{\bullet}NO$ levels alone using *S*-nitroso-*N*-acetyl-penicillamine or $O_2^{\bullet -}$ levels, using xanthine/xanthine oxidase, had no effect; however, increasing peroxynitrite levels using SIN-1 or xanthine/xanthine oxidase + *S*-nitroso-*N*-acetyl-penicillamine significantly increased cyclooxygenase activity in macrophages from old *mice* fed 500, but not those fed 30 ppm vitamin E.

Desferrioxamine, an iron chelator, upregulated cyclooxygenase-2 expression and prostaglandin production in a *human* macrophage cell line, U937 (Tanji et al. 2001).

Aqueous cigarette tar extracts increased prostaglandin H synthase activity in isolated *rat* pulmonary alveolar macrophages 3-fold above the initial activity within 2 h of incubation and gradually decreased it below the initial activity after 8 h of incubation (Hwang et al. 1999).

Procainamide induced prostaglandin H synthase-2 in *mouse* peritoneal macrophages (Goebel et al. 1999).

In RAW 264.7 macrophages, lipopolysaccharide-induced cyclooxygenase-2-dependent synthesis of prostaglandin E_2 was suppressed by **aspirin** (IC_{50} = 5.35 μM), whereas no significant inhibition was observed in the presence of **sodium salicylate** and the salicylate metabolite **salicyluric acid** at concentrations up to 100 μM (Hinz et al. 2000). However, the salicylate metabolite gentisic acid (2,5-dihydroxybenzoic acid; 10–100 μM) and salicyl-coenzyme A (100 μM), the intermediate product in the formation of salicyluric acid from salicylic acid, significantly suppressed LPS-induced PGE_2 production. In contrast, γ-resorcylic acid (2,6-dihydroxybenzoic acid) as well as unconjugated coenzyme A failed to affect prostanoid synthesis, implying that the *para*-substitution of hydroxyl groups and the activated coenzyme A thioester are important for cyclooxygenase-2 inhibition.

9.2.3.1.7 Oxidoreductases with Superoxide Radical as Acceptor

The **superoxide dismutases** are a family of enzymes responsible for metabolising superoxide free radical ($O_2^{\bullet -}$) to produce hydrogen peroxide and water. Three types of SOD are found in mammalian systems: mitochondrial Mn-SOD, cytosolic CuZn-SOD, and extracellular SOD (Marklund 1984, Zelko et al. 2002).

Superoxide dismutase (EC 1.15.1.1)

$$2\,O_2^{\bullet -} + 2\,H^+ \longrightarrow H_2O_2 + {}^3\Sigma_g^- O_2 \qquad [128]$$

The ability of the **CuZn-SOD** to stimulate oxidation of substrates in the presence of H_2O_2 was first reported by Hodgson and Fridovitch (1975). The formation of carbonate (Goss et al. 1999) and nitro radicals (Zhang et al. 1999) which carry the oxidising power out of the active site, accounts for the "peroxidase' activity of the dismutase. The formation of the copper oxidant at the active site has been examined by Jewett et al. (1999) in terms of the sequence of the events leading to inactivation through 2-oxohistidine formation, copper(I) loss, site-specific and random peptide fragmentation, and inactivation:

$$2\,Cu^+ + H_2O_2 \rightleftarrows 2\,Cu^+ + O_2 + 2\,[H^+] \quad [159]$$

$$Cu^+ + H_2O_2 \longrightarrow (CuOH)^{2+} + OH^- \quad [160]$$

$$(CuOH)^{2+}HCO_3^- \longrightarrow HCO_3^{\bullet} + H_2O \quad [161]$$

CuZn-SOD is an iron-regulated protein in *Aspergillus nidulans* and *A.fumigatus* (OBEREGGER et al. 2000). After 24 h growth under iron deplete conditions, this 16-kDA protein was approximately 5-fold upregulated.

CuZn-SOD IgG_1 immune complexes induced a Fcγ-dependent intracellular delivery of the antioxidant enzyme in IFN-γ-activated *murine* J774 macrophages (VOULDOUKIS et al. 2000). The concomitant stimulation of the Fcγ-receptor and the translocation of the SOD1 in the cytoplasm of IFN-γ-activated macrophages not only reduced the production of superoxide anion but also induced the expression of iNOS and the related •NO production. This inducing effect in the absence of superoxide anion production reduced mitochondrial damage and cell death by apoptosis and promoted the intracellular antioxidant armature.

At constant peroxide, nitrite and azide only partially protect the enzyme against loss of copper(I) and activation up to one anion per copper (JEWETT et al. 2000).

Peroxynitrite reacts with CuZn-SOD to form a complex that selectively nitrates other proteins (ISCHIROPOULOS et al. 1992). The Fe-SOD and Mn-SOD also catalyse nitration, but are slowly inactivated by peroxynitrite. The CuZn-SOD is not appreciably inactivated by the reaction with peroxynitrite. While the copper was necessary for nitration, the active site of superoxide dismutase could by modified to reduce the superoxide scavenging activity, but not significantly affecting the nitrating activity. For example, the essential amino acid arginine 141 which hydrogen bonds to the superoxide in the active site can be covalently modified with phenylglyoxyl with a 90 % loss in superoxide scavenging activity (BEYER et al. 1987).

Mutation in CuZn-SOD causes 25 % of familial amyotrophic lateral sclerosis cases. LIU et al. (2000) examined one such mutant, His46Arg, which has no superoxide dismutase activity yet presumably retains the gain-of-function activity that leads to disease. They demonstrated that Cu^{2+} does not bind to the copper-specific catalytic site of His46Arg CuZn-SOD and that Cu^{2+} competes with other metals for the zinc binding site. Most importantly, Cu^{2+} was found to bind strongly to a surface residue near the dimer interface of His46Arg CuZn-SOD. Cysteine was identified as the new binding site on the basis of multiple criteria including UV-visible spectroscopy, resonance Raman spectroscopy and chemical derivatization. Cysteine 111 was pinpointed as the position of reactive ligand by tryptic digestion of the modified protein and by mutational analysis.

Extracellular superoxide dismutase mRNA expression was demonstrated in the alveolar macrophages of four *patients* undergoing surgical resections for carcinoma and carcinoid tumour, respectively (SU et al. 1997).

Analysis of a series of chimeric and point mutated extracellular SODs showed that the *N*-terminal region contributes to the oligomeric state of the extracellular SODs, and that a single amino acid, a valine (*human* amino acid position 24), is essential for the tetramerization (CARLSSON et al. 1996). This residue is replaced by an aspartate in the *rat*. *Rat* extracellular SOD carrying an Asp → Val mutation is tetrameric and has a high heparin affinity, while *mouse* extracellular SOD with a Val → Asp mutation is dimeric and has lost its high heparin affinity.

9.2.3.2 Transferases

Histamine *O*-methyltransferase (EC 2.1.1.8) is present in *human* monocytes at a mean concentration of 10.08 nM/h/mg protein (ZEIGER et al. 1976).

O^6-Alkylguanine-DNA-alkyltransferase (EC 2.1.1.63) activity of alveolar macrophages isolated from male New Zealand White *rabbits* was 15.81 fM/10^6 cells (DEILHAUG et al. 1985).

γ-Glutamyl transferase (EC 2.3.2.2), a glycoprotein enzyme widely distributed on cell surfaces, catalyses the conversion of leukotriene C_4 to leukotriene D_4. Since LTD_4 is biologically much more potent than LTC_4 (LEWIS and AUSTEN 1984, PIPER 1984), the partial degradation of the glutathione moiety to the cysteinylglycine derivative LTD_4 may be considered a biosynthetic reaction generating the ligand for the LTD_4/LTE_4 receptor.

Radical scavenging effects of leukotrienes

$LTD_4 > LTC_4 > LTB_4$

γ-Glutamyltransferase was found by OHLSSON et al. (1982) in *canine* alveolar macrophages gained by broncho-alveolar lavage and cultured in medium 199. The activity of the enzyme was Ca^{2+}-dependent and did not increase after preincubation with thrombin. The incorporation of [^{14}C]putrescine into casein catalysed by this transferase was partially inhibited by small concentrations of monodansylthiacadaverine.

R-Glutaminyl-peptide : amine γ-glutamyltransferase (EC 2.3.2.13) activity increased during the course of differentiation of immature *human* myeloblastoid $M1^-$ to mature $M1^+$ cells more than 10-times (Kannagi et al. 1982).

Glutathione S-transferase (EC 2.5.1.18)

Glutathione S-transferase (EC 2.5.1.18)

RX + Glutathione → HX + R–S–G [129]

The glutathione S-transferases are a group of enzymes, which catalyse the nucleophilic attack of GSH to many kinds of electrophiles (MANNERVIK and DANIELSON 1988). Most of glutathione S-transferase isoenzymes localize in the cytoplasm and are classified into 3 groups: class α, μ and π. *Human* glutathione S-transferase π can be selectively reduced by hydrogen peroxide (TAMAI et al. 1990). *Rat* class μ enzymes were activated by the xanthine-xanthine oxidase system (MURATA et al. 1990).

Glutathione S-transferase activity in pulmonary alveolar macrophages obtained from 11 *patients* by bronchoalveolar lavage was 0.22 ± 0.12 nmol/min × mg protein (PETRUZZELLI et al. 1988). Glutathione S-transferase activity was 0.31 ± 0.12 nmol/min × mg protein in non-smokers and 0.14 ± 0.07 nmol/min × mg protein in smokers ($P < 0.001$).

Adenosine kinase (ATP: adenosine 5'-phosphotransferase, EC 2.7.1.20) activity in 2-week *human* pulmonary alveolar macrophage cultures obtained from cigarette smokers (one to three packs per day) exhibited significant substrate inhibition with increasing adenosine concentrations (ZUCKERMAN and DOUGLAS 1980).

Phosphofructokinase (EC 2.7.1.30) catalyses the phosphorylation of fructose-6-phosphate to fructose-1,6-biphosphate as the key regulatory enzyme of glycolysis. Inhibition of phosphofructokinase by adenosine triphosphate and its activation by adenosine monophosphate and inorganic phosphate is held responsible for the induction of the Pasteur effect (for review see RAMAIAH 1974).

9.2.3.3 Hydrolases

Phospholipases A_2 (EC 3.1.1.4) hydrolyse the fatty acyl ester bond at the *sn*-2 position of membrane phospholipids to yield free fatty acids such as arachidonic acid. Recognised isoforms include: i) low molecular weight, 14 kDa proteins that are inhibited by reducing agents, require calcium for activation and exhibit no preference for the hydrolysis of arachidonic acid *vs.* other fatty acids; ii) cytosolic, an 85 kDa protein that is resistant to reducing agents, calcium-requiring and preferentially hydrolysed arachidonic acid; and iii) calcium-independent phospholipase A_2, which is non-reducible and exhibits no preference for arachidonic acid hydrolysis, *Rat* alveolar macrophages obtained by bronchoalveolar lavage, disrupted by dounce homogenization and fractioned to yield 4 subcellular fractions (cytosolic, non-nuclear particulate, nuclear particulate and nuclear soluble). The 85 kDa protein was found only in the cytosolic and non-nuclear particulate fractions, whereas the 14 kDa protein was demonstrated in non-nuclear particulate and nuclear particulare fractions and the non-arachidonic acid selective calcium-independent isoenzyme activity was found only in the non-nuclear particulate fraction (OJO and PETERS-GOLDEN 1996).

Freund's adjuvant (1.4–140 μl/25 × 10^6 peritoneal exudate cells of *mice*) induced a significant activation of phospholipase A (MUNDER et al. 1973).

Triacylglyceroprotein acylhydrolase (EC 3.1.1.34) was secreted by *human* blood monocytes and New Zealand White *rabbit* alveolar macrophages cultivated *in vitro* (MAHONEY et al. 1982).

Alkaline phosphatase (EC 3.1.3.1) was visualised (Ca-Co method of GOMORI 1939) in *guinea pig* peritoneal macrophages cultivated in vitro (RILKE and KESSEL 1963). In *human* skin window macrophages, however, LEDER and NICOLAS (1963) and NICOLAS and LEDER 1965) were unable to detect this enzyme using KAPLOW's (1955) technique.

Acid phosphatase (EC 3.1.3.2) is the classic lysosomal enzyme. The lead substitution method has localised acid phosphatase in the phagosomes of macrophages (COHN and WIENER 1963) and can be used for electron microscopy.

Tartrate-resistant acid phosphatase is characterised for activated macrophages (EFSTRATIADIS and MOSS 1985). Tartrate-resistant acid phosphatase contains a binuclear iron centre that has been shown to generate reactive oxygen species. RÄISÄNEN et al. (2001) showed that the *murine* macrophage-like cell line RAW-264 overexpressing tartrate-resistant acid phosphatase produced elevated levels of hydroxyl radicals compared to parental cells.

5'-Nucleotidase (EC 3.1.3.5) hydrolyses the phosphoester linkage in 5'-mononucleotides, liberating a nucleoside and inorganic phosphate. It has been identified as an ecto enzyme (EDELSON 1980) in mouse alveolar macrophages (JAMES and EDELSON).

Alkaline phosphodiesterase I (EC 3.1.4.1) hydrolyses polyribonucleotides or oligodeoxyribonucleotides which have a free 3'-OH group, sequentially liberating 5'-nucleoside monophosphate (KHORANA 1961). This enzyme has been recognised as a component of the plasma membrane of *rabbit* alveolar macrophages (WANG et al. 1976). In *mouse* peritoneal macrophages the specific activity of the enzyme was from two to fourfold higher in thioglycollate-stimulated cells than it was in any of the other macrophage varieties examined. Alkaline phosphodiesterase I levels per square micrometer of cell surface area were increased in certain fractions of line 66 vs. those of lines 168 and 67 of *mouse* mammary tumour-associated macrophages (MAHONEY et al. 1983).

Cyclic 3',5'-Adenosine Monophosphate Phosphodiesterase (EC 3.1.4.17)

Two of the seven cyclic nucleotide phosphodiesterase gene families PDE3 and PDE4, have a high affinity of cAMP and are specifically inhibited by cilostamide and rolipram, respectively. Using these inhibitors in assays with 0.1 μM [^{3}H]cAMP as substrate, GAO et al. (1996) found PDE3 activity > PDE4 in primary blood macrophage cultures and in a *human* monocyte cell line THP-1, whereas in cultured U937 mononuclear cells, PDE4 >> PDE3.

Using reverse transcription and polymerase chain reaction, GAO et al. (1997) identified mRNAs of PDE3 and PDE4 subtypes in *human* inflammatory cells, including T lymphocytes, NK lymphocytes, B lymphocytes, *human* peripheral monocytes and macrophages, Jurkat T cells and cultured *human* mononuclear cells U937 and THP-1. PDE3B, but not PDE3A, mRNA was expressed in all these cells except in B lymphocytes and U937 cells, which also exhibited low PDE3 activity (assays with 0.1 μM [^{3}H]cAMP as substrate. In fresh peripheral elutriated monocytes, PDE4 > PDE3, whereas PDE3 > PDE4 in macrophages. mRNA for all PDE4 isoenzymes were present in U937 and T lymphocytes.

Arylsulphatase (EC 3.1.6.1) from a 96 h *rabbit* peritoneal exudate macrophage was present in segments of the rough endoplasmic reticulum, and perinuclear cisterna as well as within numerous small vesicles in the Golgi region, probably corresponding to secondary lysosomes (NICHOLS et al. 1971, DAVIES and BONNEY 1980).

Glucosaminidase (EC 3.2.1.17): Lysozyme

Lysozyme (EC 3.2.1.17, mucopeptide *N*-acetylmuramoyl hydrolase or muramidase) catalyses hydrolysis of a β(1→4)-glycosidic linkages of polysaccharide component of the peptidoglycan (mucopolymer of the bacterial cell wall), which is composed of alternating *N*-acetylmuramic acid and *N*-acetylglucosamine residues with alternating β(1→4) and β(1→6) linkages, cross-linked with peptide chains. The hydrolytic action of lysozyme gives rise to disaccharide units attached to the peptide chains called muropeptides.

High concentrations of lysozyme were found in monocytes (SCHMALZL and BRAUNSTEINER 1970, GORDON et al. 1974) and alveolar macrophages (MYRVIK et al. 1961, COHN and WIENER 1963, HEISE and MYRVIK 1967, LEAKE and MYRVIK 1968, McCLELLAND and VAN FURTH 1975). Lysozyme is considered to be one of the constitutive enzymes of the macrophages (GORDON et al. 1974, GORDON 1978). It is the major secretory product and forms about 25 % of the extracellularly secreted protein of macrophages.

Hen and *turkey* egg-white lysozymes differ by seven amino acids but have very similar tertiary structures (SARMA and BOTT 1977). Azide radicals ($N_3^•$) react first only with tryptophan residues (PRÜTZ and LAND 1979, PRÜTZ et al. 1980), giving an indolyl type radical. Tyrosine residues are then oxidised by a long-range intramolecular electron transfer to tryptophan, followed by dimerization. Using $N_3^•$ free radicals, the initial yields of dimerization are equal to $(8.6 \pm 0.7) \times 10^{-9}$ mol J^{-1} for both proteins (AUDETTE et al. 2000). Using HO$^•$ free radicals, they become equal to $(1.23 \pm 0.1) \times 10^{-8}$ and $(4.42 \pm 0.1) \times 10^{-8}$ mol J^{-1} for *hen* and *turkey* egg-white lysozymes, respectively (γ radiolysis).

β-Glucuronidase (EC 3.2.1.31)

Epoxide hydrolase (EC 3.3.2.3) activity in pulmonary alveolar macrophages obtained from 11 *patients* by bronchoalveolar lavage was 0.24± 0.10 nmol/min × mg protein (PETRUZZELLI et al. 1988). Epoxide hydroxylase activity was 0.16± 0.02 nmol/min × mg protein in non-smokers and 0.31 ± 0.08 nmol/min × mg protein in smokers ($P < 0.001$).

Leucine aminopeptidase (EC 3.4.1.2) is a plasma membrane-bound enzyme present on macrophages (WACHSMUTH and STOYE 1977).

Angiotensin converting enzyme (EC 3.4.15.1) is a membrane-bound ectoenzyme of *human* blood monocytes augmented by dexamethasone in a biphasic dose-dependent manner with maximum effect after 6 days in culture at 10^{-8} M concentration (VUK-PAVLOVIC et al. 1989). In a pure population of *human* alveolar macrophages obtained by centrifugal elutriation of bronchoalveolar lavage fluid from 7 sarcoid patients the distribution pattern of angiotensin converting enzyme closely resembled that of NADPH-cytochrome-*c*-reductase and sialyltransferase, markers of the endoplasmic reticulum and the Golgi complex, respectively, indicating a common localisation (EKLUND et al. 1987).

Cathepsin B (EC 3.4.22.1), a lysosomal cysteine endopeptidase was localised in *human* alveolar macrophages by the indirect peroxidase-antiperoxidase technique (BURNETT et al. 1983).

Gelatinase B (EC 3.4.24.35)

This 92-kDa gelatinase (= matrix metalloproteinase 9 = type V collagenase = 92-kDa type IV collagenase = macrophage collagenase) is similar to collagenase A, but possesses a further domain. It is expressed in macrophages and eosinophils. Stimulated neonatal alveolar macrophages secreted four to five times more gelatinase than stimulated adult alveolar macrophages (Delacourt et al. 1995). After stimulation by phorbol 12-myristate 13-acetate, Delacourt et al. (1997) observed a dose-dependent increase in gelatinase secretion that was significantly more marked in alveolar macrophages from 6-day-old *rats* than in alveolar macrophages from adult *rats* and that was inhibited by the protein kinase C inhibitor calphostin C. Adenosine 3',5'-cyclo-monophosphate mimetics or concanavalin A failed to induce an increase in gelatinase secretion by alveolar macrophages.

Characterisation of matrix metalloproteinase production by isolated lung fibroblasts, endothelial cells, type II epithelial cells, and alveolar macrophages revealed that only the macrophage has the same spectrum for matrix metalloproteinase activity as seen in the bronchoalveolar lavage fluids in lipopolysaccharide-induced acute lung injury in Long-Evans *rats* (Gibbs et al. 1999).

Concomitant treatment with catalase (EC 1.11.1.6) greatly inhibited metalloproteinase 9 production by *rat* alveolar macrophages in response to immune complexes, but this treatment had little effect on basal production of either metalloproteinase 9 or metalloproteinase 2 by macrophage (Warner et al. 2000).

Arginase (EC 3.5.3.1) catalyses the thermodynamically favoured hydrolysis of L-arginine to L-ornithine and urea, whereas nitric oxide synthase (EC 1.14.13.39) catalyses the oxidation of L-arginine to L-citrulline and •NO with formation of the intermediate N^{ω}-hydroxyl-L-arginine (Kerwin et al. 1995). At least two distinct arginase genes are coding for immunologically distinct isoforms (Gotoh et al. 1996, Vockley et al. 1996, Jenkinson et al. 1997, Perozich et al. 1998). In cells possessing an inducible nitric oxide synthase in addition to an active arginase, N^{ω}-hydroxy-nor-L-arginine appears as an attractive pharmacological tool to study possible interactions between these two pathways.

Arginase was induced in resident *mouse* peritoneal macrophages of male $B_6D_2F_1$ *mice in vitro* by two nondialyzable factors contained in sera from *man*, *rat*, *mouse* and *calf* (Jakway et al. 1980).

The incubation of resident peritoneal macrophages with bacterial lipopolysaccharide induced high arginase activity as judged by the consumption of ^{14}C (U)-L-arginine and the release of labelled ornithine into the cell supernatant (Kriegbaum and Dröge 1985).

The simultaneous presence of iNOS and arginase may play an important regulatory role, through the control of •NO production, in lipopolysaccharide-activated macrophages (Chang et al. 1998). Rouzaut et al. (1999) reported that monocytes, entering a process of activation and differentiation, responded with a differential expression of mRNAs corresponding to iNOS and arginase II and showed different arginase and NOS enzyme activities. An increase in arginase II expression seemed to rescue monocytes from apoptosis, whereas the down-regulation of arginase II, concomitant with an up-regulation of iNOS, led to apoptosis.

N^{ω}-Hydroxy-nor-L-arginine is much more potent than N^{ω}-hydroxy-L-arginine to inhibit *murine* macrophage arginase and neither induces nor inhibits NOS (Tenu et al. 1999).

In cultured DBA/2 *mouse* peritoneal macrophages pre-treated with supernatants of concanavalin A-stimulated spleen cells containing immune interferon (IFN-γ) activity, the acquisition of an intrinsic restriction to herpes simplex virus replication correlated with the generation of appreciably elevated levels of arginase in supernatant (Sethi 1983),

Adenosine deaminase (adenosine aminohydrolase; EC 3.5.4.4) catalyses the deamination of adenosine and deoxyadenosine. Thioglycollate-stimulated C57BL/6 *mouse* peritoneal macrophages contained high levels of adenosine deaminase activity (Chan 1979). In the presence of deoxycoformycin (1 μg/ml), a potent inhibitor of adenosine deaminase (Agarwal et al. 1977), thioglycollate-stimulated *mouse* peritoneal macrophages excreted deoxyadenosine (Chan 1979).

Ca^{2+}-dependent adenosine triphosphatase (EC 3.6.1.3) of a plasma membrane fraction isolated from lysates of Bacillus Calmette-Guérin-induced *rabbit* alveolar macrophages was markedly stimulated by concentrations of Ca^{2+} from 6×10^{-8} to 1×10^{-5} M, with an apparent K_m (Ca^{2+}) of 1×10^{-6} M (Gennaro et al. 1979). ATPase activity and phosphorylation by [γ-^{32}P] ATP of isolated plasma membrane of alveolar macrophages were stimulated in a parallel fashion by physiologic concentrations of Ca^{2+}, with half-maximal activating effect of this ion at $(3-7)\times10^{-7}$ M (Schneider et al. 1979). For various membrane preparations, a direct proportionality existed between Ca^{2+}-dependent ATPase activity and amount of ^{32}P incorporated.

Mg^{2+}-dependent adenosine triphosphatase (EC 3.6.1.3) activity of the *rabbit* alveolar macrophage actomyosin plus cofactor, whether assembled from purified components or studied in a complex collected from crude macrophage extracts, was not influenced by the presence or absence of calcium ions (Stossel and Hartwig 1975).

9.2.3.4
Lyases

Guanylate cyclase (EC 4.6.1.2) obtained from the cytosol of *murine* bone marrow macrophages which were cultured for 7 days in Teflon bags and seeded on culture plates was activated in a concentration-dependent manner and synergistic mode by L-arginine and NADPH (Mülsch et al. 1991). Activation was not observed with D-arginine or NADH, in accordance with the specific requirements for the oxidative L-arginine pathway (Marletta et al. 1988). •NO synthesis in *murine* peritoneal macrophages (Marletta et al. 1988) and bone marrow macrophages (Mülsch et al. 1991) was Ca^{2+}-independent.

Rat growth hormone *in vitro* enhanced soluble guanylate cyclase activity in the lungs of 15-day-old Sprague-Dawley *rats*; in 10 min of incubation the production of [α-^{32}P]cGMP per mg protein significantly ($P < 0.001$) increased from 1,760 ± 12 pmol in the control to 3,470 ± 15 pmol and 3,608 ± 13 under 10 nM and 1 μM growth hormone, respectively (Vesely 1981). Heat-treated growth hormone caused no enhancement of guanylate cyclase activity.

The guanylate cyclase inhibitor LY83583 inhibited *human* alveolar macrophage spreading and staining for cGMP-dependent protein kinase at the cell periphery (Pryzwansky et al. 1995).

9.2.4
Lipids

The total lipid and protein content per cell after the differentiation of *human* blood monocytes to macrophages due to the increase in cell size increased by approximately four-fold (Visioli et al. 2000). The percentage of docosapentaenoic acid (22:5*n*-3) increased from 1.47 ± 0.23 % to 2.93 ± 0.32 % ($P < 0.05$) and that of docosahexaenoic (22:6*n*-3) acid from 2.48 ± 1.02 % to 5.96 ± 1.21 % ($P < 0.05$), while linoleic acid decreased from 7.91 ± 1.54 ± to 3.24 ± 1.23 % ($P < 0.05$).

Rat alveolar macrophages contained only neutral fats but no complex lipids (Caulet et al. 1968).

9.2.5
Ascorbic Acid

Ascorbic acid is able to scavenge reactive oxygen species as $O_2^{\bullet-}$ (Nishikimi 1975), •HO (Bielski et al. 1975) and $^1O_2$ (Bodannes and Chan 1979). On the other hand ascorbic acid can reduce Fe^{3+} to Fe^{2+} and thus give rise to Fenton catalysis of H_2O_2 to form $HO^\bullet$.

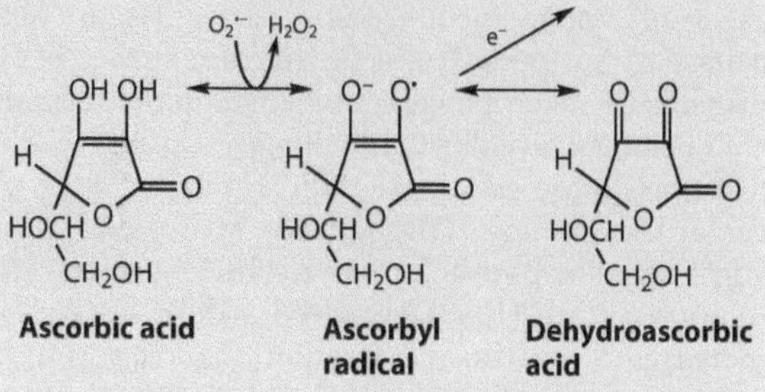

Formation of the ascorbyl free radical [53]

Dehydroascorbic acid, the two-electron oxidised form of the vitamin, is taken up on the glucose transporter and reduced to ascorbate to a much greater extent than ascorbate itself is accumulated in human monocytic U-937 cells (May et al. 1999). In contrast to dehydroascorbic acid, ascorbate enters the cells in a sodium- and energy-dependent transporter.

Matzner (1938) detected ascorbic acid within the "detached alveolar phagocytes", but not within the sessile alveolar wall cells. He supposed the density of vitamin C load to be an indicator of the instantaneous performance of the cell. Occasionally the vitamin was located at the Golgi apparatus. The load with "real" granules varies extremely (Toepfer 1961, p. 92). About 50 % of the ascorbic acid content of sliced *rat* lungs was removed in 2 min by washing with Krebs-phosphate solution suggesting that it was not intracellular (Willis and Kratzing 1974a). Ascorbic acid removed in this way was not present in alveolar macrophages since it remained in solution when these cells were removed by centrifugation. Since 30 % of the ascorbic acid of an intact lung could be removed by washing the air spaces it was concluded that about half the ascorbic acid in lung is present in the fluid lining the respiratory epithelium. Fresh *mouse* lung contained 368 μg/g of total (oxidised plus reduced) ascorbic acid of which 75 μg/g was oxidised (Willis and Kratzing 1974c).

Dehydroascorbic acid, the two-electron oxidised form of the vitamin, was taken up by *human* monocytic U-937 cells on the glucose transporter and reduced to ascorbate to a much greater extent than ascorbate itself was accumulated in the cells (May et al. 1999). In contrast to dehydroascorbic acid, ascorbate entered the cells on a sodium- and energy-dependent transporter. Intracellular ascorbate enhanced the transfer of electrons across the cell membrane to extracellular ferricyanide. Rates of ascorbate-dependent ferricyanide reduction was saturable, fivefold greater than basal rates, and facilitated by intracellular recycling of ascorbate. Whereas reduction of dehydroascorbic acid concentrations above 400 μM consumed reduced glutathione (GSH), even severe GSH depletion by 1-chloro-2,4-dinitrobenzene was without effect on the ability of the cells to reduce concentrations of dehydroascorbic acid likely to be in the physiologic range (<200 μM). Dialysed cytosolic fractions from U-937 cells reduced dehydroascorbic acid to ascorbate in a NADPH-dependent manner that appeared due to thioredoxin reductase. However, thioredoxin reductase did not account for the bulk of dehydroascorbic acid reduction, since the activity was also decreased by treatment of intact cells with 1-chloro-2,4-dinitrobenzene.

9.2.6 Pantothenol

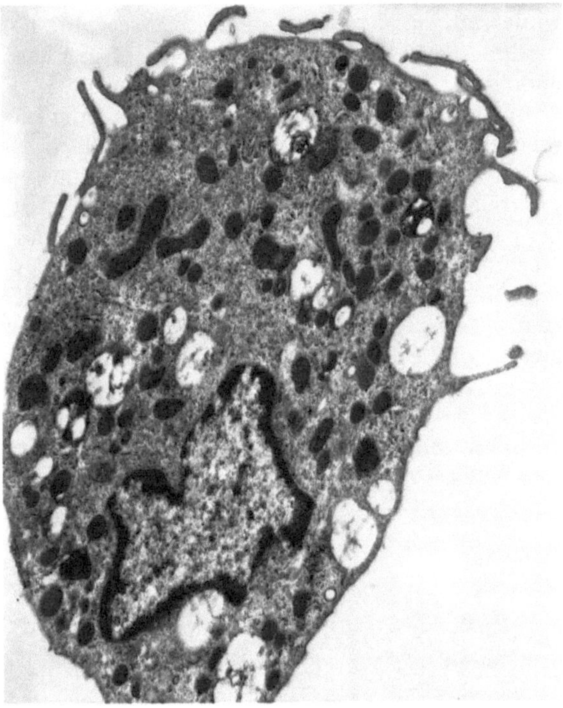

Fig. 117. Plenty of mitochondria and free ribosomes in an alveolar macrophage from a female white *rat* (breeder: Winkelmann, Borchen-Kirchborchen) medicated with 125 mg dexpanthenol per (kg body weight × day) in a 5% aqueous solution during a 5-day work week. Oral application took place from April 17 to August 14, 1967 for a total of 82 days. Fixed on August 14, 1967 under methitural anaesthesia by intratracheal instillation of 2.5% glutaraldehyde in phosphate buffer (pH 7.4) before opening the thorax. Postfixation with 1% osmium tetroxide in phosphate buffer (pH 7.4). Contrasted en bloc for 12 h with 0.5% uranyl acetate in 70% ethanol. Embedded in a 2:8 mixture of methyl and butyl methacrylate. Sectioned at 50 nm. Lead citrate after Reynolds (1963). Plate 6/08

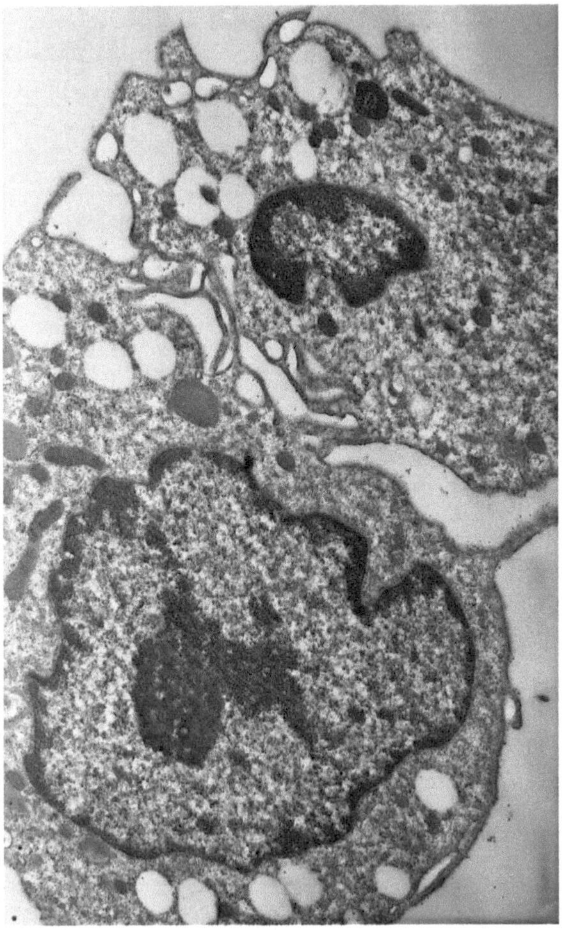

Fig. 118. Beginning of the formation of a multinucleate giant cell by fusion of mononuclear alveolar phagocytes. Female white *rat* (breeder: Winkelmann, Borchen-Kirchborchen) medicated with 125 mg dexpanthenol in a 5% aqueous solution per kg body weight × day during a 5-day work week. Oral application took place from April 17 to August 14, 1967 for a total of 82 days. Fixed on August 14, 1967 under methitural anaesthesia by intratracheal instillation of 2.5% glutaraldehyde in phosphate buffer (pH 7.4) before opening the thorax. Postfixation with 1% osmium tetroxide in phosphate buffer (pH 7.4). Contrasted en bloc for 12 h with 0.5% uranyl acetate in 70% ethanol. Embedded in a 2:8 mixture of methyl and butyl methacrylate. Sectioned at 50 nm. Lead citrate after Reynolds (1963). Plate 5/05

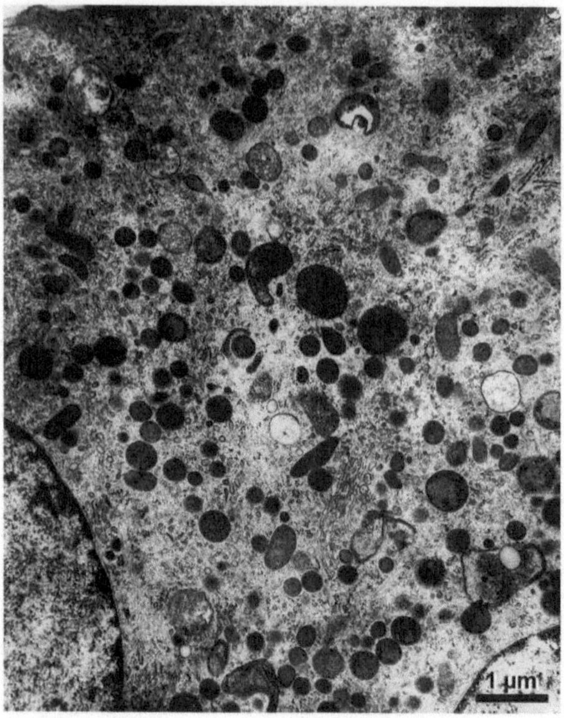

Fig. 119. Numerous autophagosomes and autophagolysosomes, respectively, in an alveolar macrophage polykaryon (block B22) from a ca. 235 g female white *rat* (breeder: Winkelmann, Borchen-Kirchborchen) medicated with 125 mg dexpanthenol in a 5 % aqueous solution per kg body weight × day during a 5-day work week. Oral application took place from April 17 to August 14, 1967 for a total of 82 days. Fixed on August 14, 1967 under methitural anaesthesia by intratracheal instillation of 2.5 % glutaraldehyde in phosphate buffer (pH 7.4) before opening the thorax. Postfixation with 1 % osmium tetroxide in phosphate buffer (pH 7.4). Contrasted en bloc for 12 h with 0.5 % uranyl acetate in 70 % ethanol. Embedded in a 2:8 mixture of methyl and butyl methacrylate. Sectioned at 50 nm. Lead citrate after REYNOLDS (1963). Plate 3606

9.2.7 Retinoids

The *human* promyelocytic leucaemia cell line, HL-60, can be induced to differentiate into granulocyte like cells when cultured in the presence of 10^{-6} M all-*trans* retinoic acid for several days (LADOUX et al. 1986). About 80 % of the viable cells (70 % of the total number of cells) reduced nitroblue tetrazolium.

Retinoids failed to activate NADPH oxidase in dibutyryl cAMP differentiated HL-60 cells (SEIFERT and SCHÄCHTELE 1988). In HL-60 cells retinoids have been found to potentiate formyl-methionyl-leucyl-phenylalanine-induced and phorbol myristate acetate-induced $O_2^{\bullet-}$ formation (SEIFERT and SCHÄCHTELE 1988). Using receptor specific retinoid analogues it was demonstrated that retinoid X receptor-retinoic acid receptor heterodimers mediate retinoid-induced differentiation of HL60 cells, while retinoid X receptor-retinoid X receptor homodimers mediate subsequent retinoid-mediated apoptosis (NAGY et al. 1995, KIZAKI et al. 1996).

In untreated U937 (*human* monoblastic leucaemia cell line) cells retinoid receptor-α mRNA levels increased after induction of differentiation via phorbol myristate acetate (BROWN et al. 1997). Using plasmids containing sense or antisense retinoid receptor-α sequences under the control of an inducible promoter, Brown et al. generated stable transfected cell lines which expressed either increased or decreased levels of retinoid X receptor-α, respectively. The sense cell lines showed an increased sensitivity to 9-*cis* retinoid acid, while the antisense cell lines presented decreased sensitivity. Combined 9-*cis* retinoic acid and 1α,25-dihydroxy

Fig. 120. Alveolar macrophage from a female *rat* (breeder: Winkelmann, Borchen-Kirchborchen) which received daily intragastric applications of 82 mg retinol palmitate per kg body weight × day, 5 days per week from April 17 to June 20, 1967 for a total of 42 days. Fixed on June 20, 1967 under methitural anaesthesia by intratracheal instillation of 2.5 % glutaraldehyde in phosphate buffer (pH 7.4) before opening the thorax. After washing in phosphate buffer the tissue was postfixed with 1 % osmium tetroxide in phosphate buffer for 2 h. Contrasted en bloc for 12 h with 0.5 % uranyl acetate in 70 % ethanol. Embedded in a 2:8 mixture of methyl and butyl methacrylate. Sectioned at 50 nm. Lead citrate after REYNOLDS (1963). Film 92671

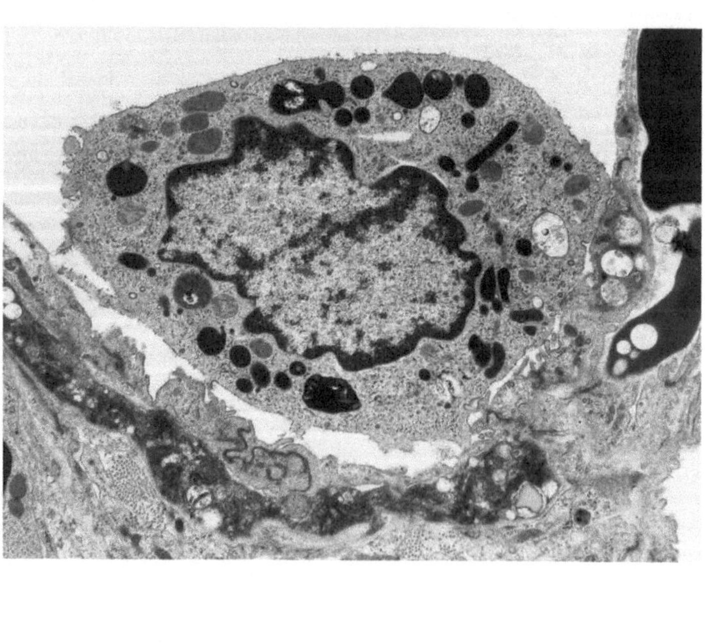

vitamin D_3 exhibited significantly higher levels of phagocytosis of antibody-coated *ovine* erythrocytes (61±7 cpm) compared to MEP cells (U937 cells harbouring the pMEP4 expression plasmid without an insert; Hewison et al. 1994) (35±4 cpm).

9.2.8 Tocopherol

As opposed to an initial *human* blood monocytic vitamin E content of 4.75 pmol/10^6 cells, macrophagic vitamin E levels were undetectable (Visioli et al. 2000).

Six hours after an intravenous application of 30 mg of [^{3}H]tocopheryl nicotinate radioactivity was detected in *rat* alveolar macrophages (Gallo-Torres 1980).

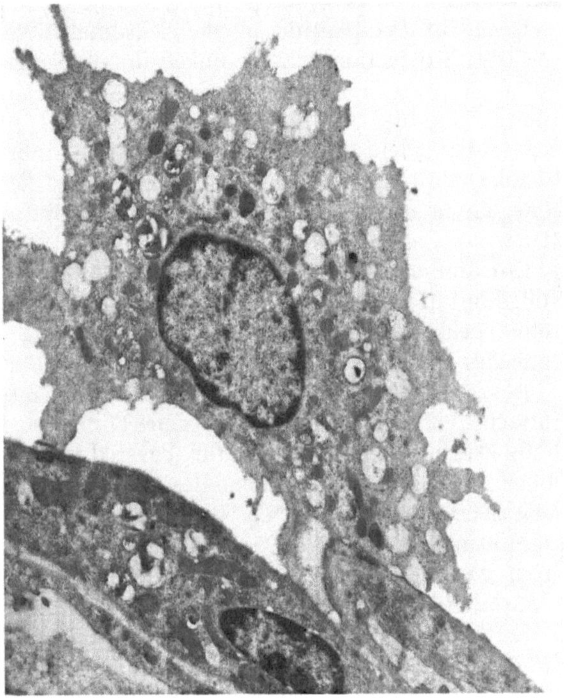

Fig. 121. Alveolar macrophage from a female *rat* (breeder: Winkelmann, Borchen-Kirchborchen), which received intramuscular injection of 150 mg DL-α-tocopherol acetate in a colloidal solution (Ephynal®) per kg body weight×day, 5 days per week from April 12 to June 12, 1967 for a total of 40 days. Fixed on June 12, 1967 under methitural anaesthesia by intratracheal instillation of 2.5% glutaraldehyde in phosphate buffer (pH 7.4) before opening the thorax. Postfixation with 1% osmium tetroxide in phosphate buffer (pH 7.4). Contrasted en bloc for 12 h with 0.5% uranyl acetate in 70% ethanol. Embedded in a 2:8 mixture of methyl and butyl methacrylate. Sectioned at 50 nm. Lead citrate after Reynolds (1963). Plate 275/03

9.3 Origin of Alveolar Macrophages

In the preterm *Macaca nemestrina* (130–156 days of gestation), the number of alveolar macrophages is low and increases rapidly after birth (Jacobs et al. 1985). In foetal *rabbits*, macrophage appearance coincides with maturation of the lung near term and in the first days after birth (Zeligs et al. 1977, Sieger 1978).

While monoblasts pinocytose very infrequently (Goud et al. 1975) promonoxytes (van Furth 1978) and monocytes/macrophages (Cohn and Benson 1965, Cohn and Parks 1967, Ehrenreich and Cohn 1968, Bowden 1971, Edelson et al. 1975, Kaplan and Nielsen 1978, van Furth 1978, 1981, Von Melchner and Hilgard 1979, Kaplan and Keogh 1981, Papadimitriou and Van Bruggen 1982, Burgert and Thilo 1983, Pratten and Lloyd 1983, Abrass 1988) do pinocytose avidly.

9.4 Secretion

Macrophages, originally considered as mere scavenger cells, are now respectfully regarded as one of the most productive cell population. First Carrel (1922) and Carrel and Ebeling (1922) in the *chicken* found that extracts of leukocytes secrete substances that determine the multiplication of fibroblasts.

From about 15 known activities in 1976 (Unanue 1976) the list comprised approximately fifty substances in 1984 (Takemura and Webb 1984) and about one hundred products in 1987 (Nathan 1987). Many substances formerly defined according to their biological activities, could now be presented with accurate molecular definitions.

In an attempt to identify novel molecules selectively expressed by alternatively activated macrophages, Kodelja et al. (1998) applied subtractive cloning methods and differential hybridisation and, finally, cloned a novel *human* CC-chemokine, alternative macrophage activation-associated CC-chemokine-1 (AMAC-1). The AMAC-1 gene consists of three exons (Goerdt et al. 1999). Several putative regulatory sequences for interleukin 4- and interferon-γ-dependent transcriptional pathways were found including STAT6 and STAT1 binding sites as well as several AP-1 and C/EBP elements. Interestingly, a combined STAT6/STAT1 binding element is located in the direct vicinity of the first putative transcription start point. Competitive binding of IL-4-induced STAT6 versus interferon-γ-induced STAT1 to this site may explain the antagonistic effects these cytokines exert on AMAC-1 expression.

9.4.1 Polypeptide Hormones

9.4.1.1 Interleukin 1

The biological basis for interleukin-1 was review by DINARELLO (1996) giving 586 references.

Three members of the interleukin-1 family have been cloned (DINARELLO 1996). All bind both types of receptors but only IL-1α and IL-1β (encoded on chromosome 2) are agonists. The third is the interleukin 1 receptor antagonist protein (IRAP). All share 20–25 % amino acid homology. Mature IL-1α and IL-1β have similar three-dimensional open barrel structures of β sheets.

Interleukin-1β is a potent immunomodulatory cytokine that is produced in large quantities by mononuclear phagocytes, and by other cells present in essentially all tissues. It affects cell targets by binding with high affinity to well-characterised cell-surface receptors. *Murine* IL-1β is initially synthesised as a **33-kDa precursor protein** that has no measurable affinity for the IL-1 receptor and that carries no biological signal. This precursor protein is activated by proteolytic cleavage between Asp^{117} and Val^{118}. The C-terminal 17-kDa fragment represents the mature IL-1β protein with high receptor affinity.

Although several different proteinases can cleave the IL-1β precursor *in vitro*, the natural cleavage of the IL-1β precursor in macrophages and monocytes occurs via the action of a novel cysteine proteinase designated the **IL-1β-converting enzyme.**

There are two **IL-1 receptors**, the type I receptor transduces a signal, whereas the type II receptor binds IL-1 but does not transduce a signal. In fact, IL-1 receptor II acts as a sink for IL-1β and has been termed a "decoy" receptor, which is somewhat unique to cytokine biology (COLOTTA et al. 1994). When IL-1 binds to IL-1 receptor I, a complex is formed that then binds to the IL-1 receptor accessory protein, resulting in high-affinity binding (GREENFEDER et al. 1995).

An **IL-1 receptor antagonist** (IRAP) has been isolated and purified from activated monocytes. IRAP inhibits IL-1 bioactivity by binding competitively at the IL-1 receptor (EISENBERG et al. 1990, HANNUM et al. 1990, MAZZEI et al. 1990). *In vitro* IRAP can attenuate IL-1-dependent fibroblast proliferation, collagenase synthesis, and prostaglandin E_2 release (HANNUM et al. 1990). IRAP suppressed IL-1-induced monocyte production of IL-1, tumour necrosis factor-α, or interleukin 6. Multiple lipid oxidation products in low density lipoproteins induce IL-1β release from human blood mononuclear cells (THOMAS et al. 1994).

Glucocorticoids control IL-1 via inhibition of mitogen-induced transcription of the IL-1β gene and decreased stability of its mRNA in monocytes of several species (KNUDSEN et al. 1987, LEE et al. 1988, AMANO et al. 1993, HUETHER et al. 1993). In contrast, the IL-1 receptor is induced by glucocorticoids in a protein and mRNA neosynthesis-dependent fashion (AKAHOSHI et al. 1988, GOTTSCHALL et al. 1991), as is the type II receptor, a potential decoy target for IL-1 (RE et al. 1994).

The **nitric oxide donor,** S-nitroso-*N*-acetyl-D,L-penicillamine induced dose-dependent inhibition of IL-1 production in LPS-stimulated *rat* alveolar macrophages in which endogenous nitric oxide production was blocked (PERSOONS et al. 1996).

Silica (Quartex 265, <3μm) had a dramatic and selective effect on mature IL-1 release from peritoneal macrophages obtained from BDF1 *mice* harvested without elicitation in a dose-dependent fashion (SARIH et al. 1993).

Levels of cytoplasmic oligonucleosomal DNA (cell death detection enzyme-linked immunosorbent assay) were significantly enhanced for **chrysotile** (3–25 μg/ml) and crocidolite (25–75 μg/ml) in a dose-dependent manner, a process that was inhibitable by 10 μM Z-Val-Ala-Asp fluoromethyl ketone, an interleukin-converting enzyme inhibitor (HAMILTON et al. 1996).

Crocidolite or **potassium octatitanate** whisker instilled intratracheally into male Wistar *rats* (2 mg in saline) peaked levels of IL-1α mRNA in alveolar macrophages at 1 month thereafter (TSUDA et al. 1997).

Rat alveolar macrophages lavaged 1 day after intratracheal instillation of **welding fumes** (1 mg/100 g body weight) collected during flux-covered manual metal arc welding (using a stainless steel consumable electrode containing Cr, Ni, and Mn), released significantly more ($P < 0.05$) IL-1β than did fumes from gas metal arc welding using a mild steel electrode (ANTONINI et al. 1996).

A 30-min exposure to $\mathbf{SO_2}$ induced a significant ($P < 0.05$) decrease in lipopolysaccharide-stimulated IL-1β release from *human* alveolar macrophages obtained by bronchoalveolar lavage and cultured *in vitro* (KNORST et al. 1996).

When expressed by transfected tumour cells, IL-1 decreased the **tumorigenicity** of oncogene cell lines (DOUVDEVANI et al. 1992) and oncogene transformed NIH_3T_3 fibroblasts (SAITO et al. 1994).

9.4.1.2 Interleukin 4

Interleukin-4 (IL-4) can prime *murine* macrophages (PHILLIPS et al. (1990). Pre-treatment of murine bone-marrow derived macrophages with 10 U/ml murine IL-4 for 48 h enhanced the respiratory burst following subsequent stimulation with 10^{-6} M phorbol myristate acetate or 1 mg/ml zymosan.

IL-4 increased the motility of human blood monocyte-derived macrophages *in vitro* (Dugast et al. 1997). The cells spread in thin cytoplasmic lamellas, regrouped in clusters, and within 1–3 weeks, differentiated into giant cells.

IL-4 in addition to TGFβ and TGF-α was shown to induce collagen synthesis by cultured fibroblasts (Elias et al. 1990, Sempowski et al. 1994). IL-4 was produced by macrophages during development of post-irradiation lung fibrosis in *rats* (Büttner et al. 1997).

9.4.1.3 Interleukin 6

Interleukin-6 is a pluripotent cytokine with immunomodulatory effects (Akira et al. 1990). It is secreted mainly by cells of the monocyte or macrophage lineage and by endothelial cells (Hirano et al. 1990). Both cyclooxygenase and 5-lipoxygenase pathways are involved in platelet-activating factor-induced IL-6 production by alveolar macrophages (Thivierge and Rola-Pleszczynski 1994). *Human* alveolar macrophage response to endotoxin (LPS) includes the production of IL-6. LPS increased the relative abundance of NFϰB and NF-IL6 (Hopkins et al. 1996). This increase in NFϰB and NF-IL6 corresponded temporally with the rise in IL-6 mRNA. In RAW 264.7 macrophages upregulation of IL-6 production appeared to be protein kinase C independent, while its down-regulation is mediated by a protein kinase C-dependent feedback mechanism (Tremblay et al. 1995).

In *human* venous blood IL-6 mean values were higher during the light period of the day without, however, any significant differences with respect to the nocturnal mean levels (Lissoni et al. 1998).

Hypoxia (3 % O_2) in primary cultures of *human* pulmonary fibroblasts and pulmonary vascular smooth muscle cells induced transcription and translation of IL-6 (4- to 5-fold) in both cell types (Tamm et al. 1998). Hyperoxia-induced expression of IL-6 was suppressed by 50 % to 60 % in the presence of platelet-activating factor antagonist WEB2170, or neutralising anti-platelet-derived growth factor antibodies.

IL-6 has been involved in the pathogenesis of inflammatoy and/or autoimmune diseases, including rheumatoid arthritis and systemic lupus erythematosus (Hirano et al. 1990). It could be implicated in the pathogenesis of systemic sclerosis, as IL-6 (Needleman et al. 1992) or anti-IL-6 auto antibodies (Takemura et al. 1992) are more frequently detected in sera from systemic sclerosis patients than in sera from control patients. Crestani et al. (1994) demonstrated that during systemic sclerosis lung disease spontaneous and stimulated IL-6 secretion by blood monocytes is increased, compared with secretion by healthy control subjects. By contrast, IL-6 secretion by alveolar macrophages recovered by bronchoalveolar lavage is normal despite of mild alveolitis.

*r*IL-6 by itself did not exhibit any chemotactic activity and it could not activate human mononuclear leucocytes for an oxidative burst response (Kharazmi et al. 1989). Preincubation with *r*IL-6 at concentrations of 5 and 50 ng/ml primed them for enhanced generation of reactive oxygen species as documented by the lucigenin-enhanced chemiluminescence following stimulation with the chemotactic peptide f-Met-Leu-Phe or phorbol myristate acetate.

Human alveolar epithelial cells exposed to residual oil fly ash release inflammatory cytokines including IL-6, IL-8, and tumour necrosis factor (Carter et al. 1997). The IL-6 response was inhibited by the metal chelator deferoxamine and the free radical scavenger *N*-acetyl-L-cysteine, suggesting that the activation of NFϰB may be mediated through reactive oxygen intermediates generated by transition metals found in residual oil fly ash (Quay et al. 1998).

5 ppm **SO_2** exposure of *human* alveolar macrophages for 30 min yielded a small but not significant ($P > 0.05$) reduction from 611 ± 76 pg spontaneous IL-6 production/ml to 585 ± 85 pg/ml (Knorst et al. 1996).

Phorbol 12-myristate 13-acetate (PMA; 10 nM) strongly stimulated IL-6 production by *human* blood monocytes cultured in RPMI1640 supplemented with 2 mM glutamine, 50 mg/ml gentamicin and 10 % heat-inactivated foetal *calf* serum (Costanzo et al. 1999).

Enhanced secretion of IL-6 by cultured **sarcoid** alveolar macrophages has been demonstrated, though previous studies show no difference in bronchoalveolar lavage fluid IL-6 levels between sarcoid patients and healthy individuals (Jones et al. 1991, Homalka and Müller-Quernheim 1993, Steffen et al. 1993, Girgis et al. 1995).

When expressed by transfected tumour cells, IL-6 decreased the **tumorigenicity** of 3LL lung carcinoma cells by delaying their doubling time (Porgador et al. 1992), but there was no influence on B16-F10 melanoma cells (Dranoff et al. 1993) and TS/A adenocarcinoma cells (Allione et al. 1994).

9.4.1.4
Interleukin 8

Interleukin-8, part of a supergene family of novel chemotactic cytokines (CXC and CC chemokines), is secreted by macrophages/monocytes, neutrophils, fibroblasts, and endothelial and epithelial cells in response to a variety of stimuli. A potent chemoattractant and activator of neutrophils, IL-8 may also be chemotactic for T lymphocytes (LARSEN et al. 1989, BAGGIOLONI et al. 1993). IL-8 induces the respiratory burst in monocytes, and the structurally related chemokine monocyte chemotactic protein-1 (MCP-1) is a potent chemotactin for monocytes (BAGGIOLINI et al. 1993).

Oxidised LDL induces monocytic cell expression of IL-8 (TERKELTAUB et al. 1994).

Hypoxia (3% O_2) in primary cultures of *human* pulmonary fibroblasts and pulmonary vascular smooth muscle cells induced transcription and translation of IL-8 (5- to 6-fold) in both cell types (TAMM et al. 1998).

Eosinophil infiltration in the bronchoalveolar lavage fluid in *guinea pigs* was induced by recombinant *human* IL-8 (LAGENTE et al. 1995). Administration of the phosphodiesterase IV isoenzyme inhibitor, rolipram (5 mg/kg) or betamethasone (10 mg/kg) significantly ($P < 0.05$) reduced the IL-8-induced eosinophil infiltration in bronchoalveolar lavage fluid. Betamethasone may act directly on eosinophils to inhibit their infiltration by IL-8.

In patients with fibrosing alveolitis, occurring alone or in association with rheumatological diseases such as systemic sclerosis, more monocytes secreted IL-8 than autologous macrophages and there was heterogeneity in the *in vitro* IL-8 secretion by alveolar macrophages and peripheral blood monocytes (PANTELIDIS et al. 1997). IL-8 secretion by alveolar macrophages was significantly higher in subjects with fibrosing alveolitis compared with subjects without fibrosing alveolitis, due to a higher percentage of IL-8-secreting alveolar macrophages in the fibrotic group both in the absence ($P < 0.002$) and presence of lipopolysaccharide ($P < 0.04$) and correlated with bronchoalveolar lavage neutrophil percentage. Using the MoAbs RFD1, RFD7 and RFD9, that distinguish subsets of alveolar macrophages, the authors have been able to identify associations between secretion of IL-8 and smaller cells and the cells identified by the MoAb RFD7. *In situ* hybridisation of the bronchoalveolar lavage cell population revealed that alveolar macrophages are the predominant source of IL-8 in the lung.

Haemolytic phospholipase C from *Pseudomonas aeruginosa* up to 1 unit/4×10^5 *human* monocytes induced a dose-dependent increase in IL-8 release and IL-8-specific mRNA expression (KÖNIG et al. 1997).

The protein kinase C inhibitor, staurosporine (100 nM) significantly inhibited the production of IL-8 by a *human* monocytic cell line, MonoMac6 stimulated by LPS, and so did the tyrosine kinase inhibitor, genistein (25 μM) (FUJISHIMA et al. 1997).

9.4.1.5
Interleukin 10

IL-10 is a potent antiinflammatory cytokine that shares with adenosine the ability to control TNF release after LPS challenge (DE WAAL MALEFYT et al. 1991, GÉRARD et al. 1993, MARCHANT et al. 1994).

Macrophages from chronically infected lungs of cystic fibrosis patients appeared to be one source of IL-10, but BONFIELD et al. (1995) found little or no intracellular IL-10 in bronchoalveolar lavage macrophages from healthy control subjects. MARTINEZ et al. (1997) quantified IL-10 and TNF-α mRNA levels in alveolar macrophages obtained by bronchoalveolar lavage from patients with idiopathic pulmonary fibrosis (IPF) and bronchiolitis obliterans with organising pneumonia (BOOP) and in normal healthy volunteers. While the levels of TNF-α mRNA in macrophages obtained from IPF and BOOP patients were not significantly different from normal healthy subjects, macrophages from patients with IPF and BOOP expressed increased levels of IL-10 mRNA compared with healthy controls.

O'FARRELL et al. (1998) characterised IL-10 responses of the J774 *mouse* macrophage cell line, and of J774 cells expressing wild-type hIL-10R, mutant hIL-10R lacking two membrane-distal tyrosines involved in recruitment of Stat-3 (hIL-10R-Tyr^{FF}), a truncated Stat3 (ΔStat3) which acts as a dominant negative, or an inducibly active Stat3-gyraseB chimera (Stat3-GyrB). A neutralising anti-mIL-10R monoclonal antibody was generated to block the function of endogenous mIL-10R. IL-10 inhibited proliferation of J774 cells and of normal bone marrow-derived macrophages, but not J774 cells expressing hIL-10RTyr^{FF}. Dimerization of Stat3-GyrB by coumermycin mimicked the effect of IL-10, and expression of ΔStat3 blocked the anti-proliferative activity of IL-10. For macrophage de-activation responses, hIL-10R-Tyr^{FF} could not mediate inhibition of lipopolysaccharide-induced TNFα, IL-1β or CD86 expression, while ΔStat3 did not interfere detectably with these IL-10 responses.

Pre-incubation with **adenosine** dose-dependently enhanced IL-10 release by TNF stimulated *human* monocytes (+29, +58, +116% at 1, 10, and 100 μM, respectively). Adenosine also significantly enhanced

IL-10 production after hydrogen peroxide and LPS stimulation and dose-dependently inhibited TNF secretion. The enhanced IL-10 production was not observed when cells were preincubated with adenosine A_1 or A_2 receptor agonists (*R*-phenylisopropyladenosine, 5'-*N*-ethylcarboxamide-adenosine, and 2-chloroadenosine) and was not affected by pre-treatment with theophylline, an antagonist of both A_1 and A_2 receptors, or with dipyridamole, an inhibitor of adenosine cellular uptake (LE MOINE et al. 1996).

In a *mouse* model of inflammatory lung reaction induced by intratracheal instillation of **silica**, the levels of IL-10 protein both in cells obtained after bronchoalveolar lavage and in lung tissue homogenates were significantly increased when compared with controls (HUAUX et al. 1998). After in vitro lipopolysaccharide stimulation (1 μg/ml), bronchoalveolar lavage cells obtained from silica-treated animals produced significantly more IL-10 protein and mRNA than cells obtained from control animals.

When expressed by transfected tumour cells, IL-10 decreased the **tumorigenicity** of Chinese hamster ovarian carcinoma cells (RICHTER et al. 1993), CL8-1 cells (SUZUKI et al. 1995), TS/A cells (GIOVARELLI et al. 1995) and B16-F1 melanoma cells (GÉRARD et al. 1996).

9.4.1.6 Interleukin 12

Interleukin 12 was originally discovered as a product of B cells in the spent medium of Epstein-Barr virus-transformed *human* lymphoblastoid cell lines (KOBAYASHI et al. 1989). However, this cytokine is now known to be produced predominantly by macrophages and dendritic cells but minimally by B cells, especially by resting B cells (D'ANDREA et al. 1992, TRINCHIERI 1994). Interleukin 12 production by macrophages/dendritic cells is induced by two different pathways of stimulation; one that occurs when the cells are exposed to various pathogens or their products (D'ANDREA et al. 1992, MA et al. 1995), and the other, when macrophages/dendritic cells as antigen-presenting cells interact with activated T cells (MACATONIA et al. 1995, SHU et al. 1995, MARUO et al. 1996). TAKENAKA et al. (1997) showed that T cell-dependent and -independent IL-12 protein production is regulated by IL-10 and IL-4 corresponding to the levels of mRNA accumulation for the p40 and p35 IL-12 genes, whereas the presence of IL-6 during stimulation decreases IL-12 protein production without affecting steady-state mRNA levels. ISLER et al. (1999) studied the autocrine regulation of the heterodimeric, biologically active form of IL-12 (IL-12 p70), as well as the expression of its subunits of 35 kDa (p35) an 40 kDa (p40). Alveolar macrophages cultured in medium alone expressed only p35 mRNA. Both p35 and p40 mRNA levels were induced by lipopolysaccharide and were further increased by interferon-γ. Lipopolysaccharide alone induced IL-12 p40 but not IL-12 p70 production in monocytes and in alveolar macrophages. However, IL-12 p70 was released when the autocrine production of IL-10 was neutralised by IL-10 blocking antibody, and IL-12 p40 production increased. In CD-1 *mice*, *Brucella abortus* 2308 was a potent inducer of IL-12p40 (FERNÁNDEZ-LAGO et al. 1999).Secretion of IL-12p70 was also demonstrated *in vivo*, although at much lower levels. Production of IL-12 over the first 7 days after infection was accompanied by active multiplication of *B. abortus* in the spleens of infected mice. CD-1 cultured macrophages secreted only IL-12p40 in response to *B. abortus* infection and no production of IL-12p70 was observed. In contrast, CD-1 peritoneal macrophages secreted detectable amounts of IL-12p70 in response to purified lipopolysaccharide from *B. abortus* 2308.

In *human* macrophages, interferon-β, a naturally occurring protein produced by virus-infected cells – unlike interferon-γ that can only prime the production of IL-12 – is able to produce this cytokine in response to antigenic stimuli (JIANG and DHIB-JALBUT 1998).

In BALB/c *mice* treated with one dose of alcohol (5 mg/kg) or buffered saline and challenged 30 min later with lipopolysaccharide (LPS; 0.1 mg/kg) or mycobacterial lipoarabinomannan (araLAM; 4 mg/kg) intratracheally under anaesthesia, and sacrificed 6 h later lung homogenates analysed for IL-12 (p75) showed a weak suppression of this cytokine by alcohol (MASON et al. 1997).

When expressed by transfected tumour cells, IL-12 decreased the **tumorigenicity** of MCA-207 and MCA-102 fibrosarcoma cell lines (TAHARA et al. 1995), but had no influence on C26 colon carcinoma cells (MARTINOTTI et al. 1995).

9.4.1.7 Interleukin 13

Human interleukin 13 (IL-13) is a cytokine that has a profound effect on primary immune cells by inducing immunoglobulin production, proliferation of B cells, and the differentiation of cells of the monocytic lineage. Alveolar macrophages obtained by bronchoalveolar lavage expressed more IL-13 mRNA in patients with fibrotic lung disease, in comparison with those from healthy volunteers (HANCOCK et al. 1998). IL-13 protein was detectable in the bronchoalveolar lavage fluid of 8 of 13 patients with pulmonary fibrosis, but in none of the control subjects.

9.4.1.8 Interleukin 18

In 6 day granulocyte-macrophage colony-stimulating factor-treated *human* macrophages, an "inactive" IL-18-recognizing monoclonal antibody detected the IL-18 proform (24 kDa) and a 48-kDa protein, which were gradually increased concomitant with maturation stage (KIKKAWA et al. 2001).

9.4.1.9 Tumour Necrosis Factor-α

Tumour necrosis factor is a pleiotropic cytokine, initially isolated on the basis of its capacity to induce tumour necrosis (BEUTLER and CERAMI 1988). It exists in two forms, TNF-α and TNF-β, both having molecular masses around 17 kDa. TNF has a wide range of biological activities, including fibrogenic activity. *In vitro*, it enhances the proliferation of some fibroblast lines but inhibits others (SUGARMAN et al. 1985), and it inhibits collagen secretion (DAYER et al. 1985). *In vivo*, it is strongly fibrogenic (PIGUET et al. 1990). In patients with adult respiratory distress syndrome, significant amounts of TNF were observed in the bronchopulmonary secretions, although no TNF was detected in samples of control patients (MILLAR et al. 1989). Using a *human* bronchial epithelial cell line (BEAS-2B), Clara cell secretory protein mRNA levels increased in response to TNF-α (20 ng/ml) stimulation after 8 to 36 h with the peak increase at 18 h (YAO et al. 1998). Immunoblotting of CCSP protein released into the culture media demonstrated that TNF-α induced the synthesis and secretion of CCSP protein released in a time dependent manner over 8 to 18 h.

TNF has been shown to be present in the blood of patients with rheumatoid arthritis (COPE et al. 1992).

The gene of the TNF protein is located on the short arm of chromosome 6, in close proximity to the major histocompatibility complex. It is closely related to another cytokine discovered by RUDDLE and WAKSMAN (1968) and GRANGER and WILLIAMS (1968). The former group called the protein cytotoxic factor while the latter group named it lymphotoxin (LT). TNF and LT are 35 % homologous (LI et al. 1987). They lie closely spaced within the class III region of the major histocompatibility complex, and they bind to the same receptors and share many activities, such as killing of L929 cells, induction of necrosis in Meth A sarcomas and activation of polymorphonuclear leucocytes (GRAY et al. 1984; SHALABY et al. 1985). These observations prompted them to change the name LT to TNF-β, and that of TNF to TNF-α.

ProTNF-α is initially produced as a membrane-bound form that is processes by TNF-α converting enzyme (TACE), a type I membrane-bound metalloproteinase (BLACK et al. 1997, MOSS et al. 1997). TACE represents a potential target for the development of a therapeutic agent for inflammation and related diseases (PATEL et al. 1998). TACE is inhibited by TIMP-3 but not by TIMP-1, -2 and -4 (AMOUR et al. 1998).

TNF is a macrophage/monocyte-derived anticancer cytokine (CARSWELL et al. 1975, MATTHEWS 1978, NIITSU et al. 1985) with strong cytotoxicity to tumour cells *in vitro* (HELSON et al. 1975, NIITSU et al. 1988, WATANABE et al. 1985, 1988). In the transformed *murine* macrophage cell line RAW 264.7, TNF-α up-regulation 4 h after lipopolysaccharide addition seems to involve protein kinase C, as staurosporine strongly inhibited (over 70 %) the expression of TNF-α mRNA while calphostin had no significant influence (TREMBLAY et al. 1995).

Recombinant *human* TNF induced increased **hydroxyl radical production** in TNF-sensitive cells (YAMAUCHI et al. 1989). LIOCHEV and FRIDOVICH (1997) explained the $O_2^{\bullet-}$-dependent toxicity of TNF on the basis of a variant of the Fenton reaction:

$$Fe(II) + LOOH \longrightarrow Fe(III) + OH^- + LO^\bullet \qquad [162]$$

where L denotes a polyunsaturated fatty acid such as arachidonate. The alkoxy radical ($LO^\bullet$) so produced can initiate the oxidation of polyunsaturated lipids in membranes by a free radical chain reaction (TAPPEL 1972, BARCLAY and INGOLD 1981). The initiation and propagation reactions of lipid peroxidation can be blocked by antioxidants such as butylated hydroxy anisole (BREKKE et al. 1994), *N*-acetyl cyteine, pyrrolidine dithiocarbamate and nordihydroguaiaretic acid. Each of these antioxidants has been seen to diminish the cytotoxicity of TNF.

Using the stable free radical, 1,1-diphenyl-2-picrylhydrazyl (DPPH) and the electron spin resonance technique, MATSUBARA et al. (1991) showed TNF-α to have scavenging properties *in vitro*.

TNF-α synthesis was induced in resting *human* blood monocyte cultures by actinomycin D (VOITENOK et al. 1989). Cycloheximide inhibited the production of TNF-α. The data imply that TNF-α synthesis by *human* blood monocytes can be induced by posttranscriptional regulation of TNF-α mRNA presynthesized *in vivo*.

Serum TNF-α levels were increased, although nonsignificantly, in male adult older C57BL/6J mice that received either **melatonin** (10 μg/ml drinking water, only during the night hours from 18.00 and 9.00) or **Zn** (22 μg/ml drinking water) supplementation for 3 months (CHEN et al. 1999).

$PbCl_2$ increased TNF-α secretion from *human* peripheral blood mononuclear cells in a concentration- and time-dependent manner, suggesting that $PbCl_2$ increased TNF-α expression by post-transcriptional mechanisms and enhanced the reactivity and uptake of TNF-α by increasing the surface expression of TNF-R p55 (GUO et al. 1996).

A 30-min exposure to **SO_2** induced a significant ($P < 0.001$) dose-dependent decrease in spontaneous and lipopolysaccharide-stimulated TNF-α release from *human* alveolar macrophages cultured *in vitro* (KNORST et al. 1996).

NO production by *murine* peritoneal macrophages co-cultured with $CD4^+$ T cells prepared from BCG-infected *mice* was induced by TNF-α together with IFN-γ (SAITO and NAKANO 1996).

Rat alveolar macrophages lavaged 1 day after intratracheal instillation of **welding fumes** (1 mg/100 g body weight) collected during flux-covered manual metal arc welding (using a stainless steel consumable electrode containing Cr, Ni, and Mn), released significantly more ($P < 0.05$) TNF-α than did fumes from gas metal arc welding using a mild steel electrode (ANTONINI et al. 1996).

In NMRI *mice* bronchoalveolar lavage fluid gained 3 days after an intratracheal instillation of 2 mg **Dörentrup quartz** DQ12 suspended in 100 μl saline levels of TNF-α protein from fibrotic, inflammatory, and non-inflammatory models were significantly lower than in controls (HUAUX et al. 1999). Although this downregulation of TNF-α was maintained for up to 120 days in the fibrotic model, TNF-α levels returned to control values after 15 and 30 d in the non-inflammatory and resolving inflammatory models, respectively. The pattern of TNF-α mRNA expression in the lung was opposite to that of the protein. In the resolving and inflammatory models, mRNA levels were upregulated as compared with those of controls. This effect was present for up to 30 d in the resolving alveolitis model, and at all time points examined in the fibrotic model. No effect on TNF-α mRNA levels was observed in the non-inflammatory model. The downregulation of TNF-α protein was concurrent with the accumulation of recruited polymorphonuclear neutrophils in alveoli, and coculture experiments showed that polymorphonuclear leucocytes explanted from the lungs of *mice* treated with silica particles were able to downregulate the expression of TNF-α protein by naive alveolar macrophages. In addition, polymorphonuclear leucocyte depletion prevented the downregulation of TNF-α induced by silica.

Significantly ($P < 0.05$ vs. control) increased amounts of TNF-α mRNA were found in Wistar *rat* alveolar macrophages exposed to **crocidolite, chrysotile A, chrysotile B, MMVF 21, CRF 1,** or **SiC** whiskers for 90 min, and significantly ($P < 0.05$ vs. control) increased activities of TNF-α were found in the medium of macrophages exposed to crocidolite, chrysotile A, chrysotile B or MMVF 21 for 4 h (LJUNGMAN et al. 1994).

Long **amosite** stimulated more TNF release from HAN Wistar *rat* alveolar macrophages than short amosite ($P < 0.05$) (DONALDSON 1992). Short **chrysotile** fibres augmented the production of TNF (481±122 U/ml) by male Wistar *rat* alveolar macrophages to a significantly greater extent than did the long fibres (260±65 U/ml; $P < 0.01$) (MORIMOTO et al. 1993). While the short **ceramic fibres** appeared to stimulate TNF production less than did the long fibres (112±21 U/ml vs. 161±76 U/ml), the difference was not statistically significant.

Both **wool** and **grain dust** samples were capable of stimulating TNF release from *rat* alveolar macrophages isolated by bronchoalveolar lavage, in a dose-dependent manner (BROWN and DONALDSON 1996). Leaches prepared from the dusts contained LPS and also caused TNF release but leachable LPS could not account for the TNF release and it was clear that non-LPS leachable activity was present in the grain dust and that wool dust particles themselves were capable of causing release of TNF.

The antibody for recombinant *human* TNF-α completely inhibited the formation of multinucleated giant cells from *rat* bone marrow-derived macrophages treated with acetyl lignin (SORIMACHI et al. 1995). By an autocrine and/or paracrine mechanism TNF-α could interact immediately with its receptor.

TNF is inactivated rapidly in and cleared from the circulation. The clearance mechanism seems to consist of two phases. First, after the release of TNF in circulation, inactivation of the biologic activity of TNF is observed. Thereafter, clearance of the TNF protein takes place, mainly by the kidney and to a much lesser extent by the liver (BEUTLER et al. 1985, PESSINA et al. 1987).

When expressed by transfected tumour cells, TNF decreased **tumorigenicity** of Chinese *hamster* ovary carcinoma cells (OLIFF et al. 1987, QUIN et al. 1995), J558L plasmacytoma cells (BLANKENSTEIN et al. 1991, HOCH et al. 1993), 1591-RE skin tumour cells (TENG et al. 1991), MCA-205 fibrosarcoma cells (ASHER et al. 1991), and TS/A *murine* mammary adenocarcinoma (ALLIONE et al. 1994), but had no influence on MCA-102 fibrosarcoma cells (KARP et al. 1993), and B16-F10 melanoma cells (DRANOFF et al. 1993). TNA-α gene suppressed

Friend leukaemia cells (Ferrantini et al. 1993, 1994), and TNF-γ gene C1300 neuroblastoma cells (Watanabe et al. 1989), CMS-5 fibrosarcoma cells (Gansbacher et al. 1990), SP1 and CT-26 colon carcinoma cells (Esumi et al. 1991).

9.4.1.10 Interferons

Interferons are a family of proteins secreted by vertebrate cells in response to various stimuli. They were identified by Isaac and Lindeman (1957) as antiviral agents and they have long been known for their antitumoral and antiproliferative activities (Paucker et al. 1962, Sen and Lengyel 1992).

Initially classified according to their cellular origin, the interferons are now typed according to their sequences and their activities. Interferon type I regroups IFN-α (leukocyte IFN) and IFN-β (fibroblast IFN); IFN type II: IFN-γ (immune IFN).

After induction by 10-carboxymethy-9-acridanone, *murine* bone marrow-derived macrophages produced two activities of interferon β differing in the degree of glycosilation (Brehm et al. 1986).

Using the stable free radical, 1,1-diphenyl-2-picrylhydrazyl (DPPH) and the electron spin resonance technique, Matsubara et al. (1991) showed both interferon α and interferon γ to have scavenging properties *in vitro*.

IFN-γ alone or combined with lipopolysaccharide induced iNOS expression and increased nitrite production in $iNOS^{+/+}$ macrophages, but not in $iNOS^{-/-}$ macrophages (Niu et al. 2000). Formation of thiobarbituric acid reactive substances from low-density lipoprotein was suppressed in IFN-γ- and IFN-γ/lipopolysaccharide-treated $iNOS^{+/+}$ macrophages but was increased in IFN-γ-treated $iNOS^{-/-}$ macrophages. In the presence of N^G-monomethyl-L-arginine the suppressive effect of IFN-γ and IFN-γ/lipopolysaccharide was abolished and the formation of thiobarbituric acid reactive substances was even increased to a level above that of untreated $iNOS^{+/+}$ macrophages. NOC 18, an NO donor, dose-dependently inhibited macrophage-mediated low-density lipoprotein oxidation. IFN-γ increased superoxide and thiol production in both types of macrophages.

Collagen synthesis in fibroblast cultures is suppressed by interferons (Rosenbloom et al. 1984, Rockey et al. 1992). Interferon-α is less potent than interferon-γ.

9.4.1.11 Fibroblast Growth Factors (FGFs)

Fibroblast growth factors (FGFs) comprises a family of peptides with a potent fibrogenic activity *in vitro* and *in vivo* (Sprugel et al. 1987) but whose biological role is poorly understood. They are present in a wide variety of cell types but it is not established whether their synthesis and secretion are regulated and/or if they are released only by damaged cells (D'Amore 1990).

9.4.1.12 Colony Stimulating Factor (CSF)

The macrophage colony stimulating factor, CSF-1, is responsible for survival, proliferation, and differentiation of mononuclear phagocytes from bone marrow progenitor cells to mature macrophages (Tushinski et al. 1982). It is a homodimeric glycoprotein growth factor, and its pleiotropic effects are mediated via a high affinity cell surface CSF-1 receptor (Guilbert and Stanley 1980, Byrne et al. 1981, Bartelmez and Stanley 1985) identical to the c-fms protooncogene product (Sherr et al. 1985). $1,25(OH)_2$ vitamin D_3 prompted appearance of CSF-1 receptor in murine bone marrow precursors (Perkins and Teitelbaum 1991). Specific ^{125}I-CSF-1 binding by BAC 1,2F5 cells is enhanced ($P < 0.001$) after 24 h exposure to 10 nM $1,25(OH)_2$ vitamin D_3 (Meenakshi et al. 1993).

Macrophage activation with yeast **glucan** induced a pronounced elevation of colony stimulating activity in serum (Satoh et al. 1982, Chihara 1983). The increase of colony stimulating activity was dose dependent. While doses of 62.5 mg krestin (β-1,4; β-1,3; β-1,6, protein complex; PSK)/kg and 125 mg PSK/kg produced no increase in colony stimulating activity, a dose of 250–1000 mg/kg was associated with a rapid increase in colony stimulating activity followed by a rapid decline. Yeast glucan has been demonstrated to be very effective in enhancing CSF at doses where PSK is ineffective (Patchen and McVittie 1983).

Overexpression of granulocyte-macrophage colony-stimulating factor induces pulmonary granulation tissue formation and fibrosis by induction of transforming growth factor-β1 and myofibroblast accumulation (Xing et al. 1997).

9.4.2 Complement Components

Most of the serum complement proteins are produced primarily in the liver, but monocytes and macrophages constitute an important source of extrahepatic production of many complement components. Local production of complement at tissue sites of inflammation may be of considerable importance in host defences and immunopathological reactions.

Maturation of mononuclear phagocytes into tissue macrophages may be signalled or enhanced by changes in their capacity to synthesise and secrete biologically active complement proteins. *Human* breast milk macrophages secrete haemolytically active C2 and factor B immediately *in vitro*, while a lag of 3 to 6 days in complement production is noted in peripheral blood monocytes from the same *woman* (Cole et al. 1980). The rate of secretion of both C2 and C4 by macrophages from different tissues may be compared by simultaneous plaque assay and tissue culture (Cole et al. 1980). While alveolar macrophages from *guinea pig* secrete less C2 and C4 than peritoneal macrophages *in culture*, the rates of C2 and C4 secretion by complement producing alveolar macrophage are greater than the rates per complement producing peritoneal macrophage (Cole et al. 1980).

Posttranslational modification of precursor proteins appears to be an important regulatory step in the synthesis and secretion of complement proteins, as it is with other proteins (Ooi et al. 1980). Hall and Colten (1978) showed in the *guinea pig*, that deficiency of C3 resulted from a translational defect in the synthesis of C4 protein. Under cell-free conditions, hepatic polysomes from C4-deficient *guinea pigs* synthesised only nascent C4 polypeptides that remain polysome-bound. Thus, the mRNA appears to be present but not completely translated.

The complement peptides C5a and C5a-desArg have been found to elicit the respiratory burst: *human* C5a has a threshold of 10^{-9} M for $O_2^{\bullet-}$ generation; maximum is still not reached at $>10^{-7}$ (Dahinden et al. 1983, McPhail and Snyderman 1983, Gerard et al. 1986). C5a-desArg from the same species is approximately 50 times less potent (Webster et al. 1980). The activity of *human* C5a is greatly enhanced in the presence of a yet unidentified serum factor (protein?) (Maly et al. 1983).

9.4.3 Platelet-Activating Factor

The synthesis of the platelet-activating factor (PAF; 1-alkyl-2-acetyl-*sn*-glycero-3-phosphocholine) by the action of phospholipase A_2 on phosphatidylcholine, resulting in the generation of lysophosphatidylcholine. Alkyl species of phosphatidylcholine are a major source of PAF. They are converted to lysoplatelet-activating factor (lysoPAF) and then by PAF acetyltransferase to PAF (Sturk et al. 1989, Snyder 1990).

9.4.4 Proteolytic Enzymes

9.4.4.1 Metalloproteinases

Metalloproteinases are usually synthesised as proenzymes in balance with activating and inhibiting proteins. By blocking the synthesis of the major inhibitor TIMP (tissue inhibitor of metalloprotease by the antisense technique), the importance of the fine regulation of this balance was deduced. Upon removal of TIMP, the cells adopt a partially transformed phenotype, namely, anchorage-independent growth and soft agar colony formation (Khokha et al. 1989). Another feature of metalloprotease activity has been revealed by overexpressing collagenase type I or stromelysin in *rat* 2 cells that had been transformed by the activated *ras* gene.

Human matrix metalloproteinases may be processed from their proenzyme forms to their active forms by two new and unique mechanisms (Maeda et al. 1998): Firstly, by bacterial proteases such as *Pseudomonas* elastase and *Vibrio cholerae* protease, which cleave off the *N*-terminal autoinhibitory domain (so-called cysteine switch) from pro-matrix metalloproteinases. The second mechanism depends on free radical generation by activated polymorphonuclear leucocytes. In this case, peroxynitrite ($ONOO^-$) or nitrogen dioxide radical ($^{\bullet}NO_2$) are the key reagents.

$$O_2^{\bullet-} + {}^{\bullet}NO \longrightarrow ONOO^- \ (pK_a = 6.8) \quad [82]$$

$$ONOO^- + H^+ \longrightarrow ONOOH \quad [83]$$

$$ONOOH \longrightarrow HO^{\bullet} + {}^{\bullet}NO_2 \ (20\text{–}30\%\ \text{yield}),\ NO_3^- + H^+ \ (\text{rest}) \quad [84]$$

Both $O_2^{\bullet-}$ and $^{\bullet}NO$ are generated by activated macrophages and polymorphonuclear leucocytes as a result of immunologic responses involving various proinflammatory cytokines. $^{\bullet}NO_2$ or $ONOO^-$ seems to interact with a single cysteine residue in the propeptide autoinhibitory domain, or so called cysteine-switch of pro-matrix metalloproteinases, thus transforming pro-matrix metalloproteinases into their active conformation.

TIMP-3 inhibited TNF-α converting enzyme (Amour et al. 1998).

9.4.4.1.1
Elastase

The detection and quantification of macrophage-derived elastolytic enzymes are important considerations when studying model systems designed to evaluate the impact of macrophage secretory proteinases on lung and vascular elastin. The *mouse* peritoneal macrophage elastase is a neutral metalloproteinase (BANDA and WERB 1981) catalytically and immunochemically distinct from *mouse* pancreatic and *mouse* granulocyte elestases, both of which are serine proteinases. Macrophage elastase was inhibited by α_2-macroglobulin, but not by α_1-proteinase inhibitor. *Mouse* macrophage elastase catalysed the limited proteolysis of selected subclasses of *mouse* immunoglobulins, including monomeric IgG_{2a}, IgG_3, and some forms of IgG_{2b} (BANDA et al. 1983). *Mouse* IgG_1 was resistant to elastase degradation; however, *human* IgG_1 was degraded. IgG_3 in immune complexes was cleaved in a manner similar to that of monomeric IgG_3. Degradation by macrophage elastase was limited to the heavy chain, resulting in products that did not compete for binding to the macrophage Fc receptor. Macrophage elastase usually produced a pepsin-like rather than a papain-like pattern of proteolysis, resulting in the release of $F(ab')_2$ and Fc' subfragments. This degradation of IgG differed from the papain-like cleavage of IgG by granulocyte elastase. Macrophage elastase degraded papain-generated Fc fragments of IgG_{2a} into multiple fragments. Therefore, macrophage elastase at concentrations found in culture medium has a potential to regulate some aspects of cellular events associated with immunoglobulins.

The C-terminal domain of *human* macrophage elastase inhibited elastin degradation (GRONSKI jr. et al. 1996).

9.4.4.1.2
Collagenase

Reactive oxygen species and serine proteases activated latent 70–75 kDa *human* neutrophil collagenase (SAARI et al. 1990). The hydroxyl radical may cause the oxidative activation of collagenase by cleaving the bond between Cys^{73} and the forth coordination site of active site Zn atom and oxidise cysteine to cysteic acid that cannot be liganded by the Zn atom (SPRINGMAN et al. 1990). Hydrogen peroxide (H_2O_2) increased the steady-state mRNA levels of collagenase/MMP-1 in *human* dermal fibroblasts (BRENNEISEN et al. 1997).

9.4.4.1.3
Stromelysin

Stromelysin 1 (= matrix metalloproteinase 3) has been implicated as playing a pivotal role in joint-degrading diseases like arthritis (HASTY et al. 1990, HEMBRY et al. 1995). Its synthesis by chondrocytes and synoviocytes can be induced by inflammatory mediators like interleukin-1 (HASTY et al. 1990, MACNAUL et al. 1990).

Stromelysin 2 (EC 3.4.24.22 = matrix metalloproteinase 10 = transin 2)

Stromelysin 3 (= matrix metalloproteinase 11) from *rat* skin has 491 amino acids, and shows 83, 95 and 58 % homology with *human*, *mouse* and *Xenopus* ST3, respectively (OKADA et al. 1997). In contrast to other matrix metalloproteinases, ST3 is secreted into the extracellular space as a potentially active molecule (PEI and WEISS 1995, SANTAVICCA et al. 1996). By in situ hybridisation, WOLF et al. (1993) detected ST3 transcripts specifically in fibroblastic cells in primary breast carcinomas and so did OKADA et al. (1997) in fibroblastic cells localised in the superficial dermis of healing *rat* skin wounds.

9.4.4.1.4
Gelatinase B (EC 3.4.24.35)

This 92-kDa gelatinase (= matrix metalloproteinase 9 = type V collagenase = 92-kDa type IV collagenase = macrophage collagenase) is similar to collagenase A, but possesses a further domain. It is expressed in macrophages and eosinophils, and is stored in the tertiary granules of neutrophils as well. To assess the role of the 92 kDa gelatinase in neutrophil, macrophage, and eosinophil function, SHIPLEY et al. (1996) generated *mice* deficient in this enzyme by targeted mutagenesis. The *murine* 92 kDa gelatinase gene was cloned from a 129Sv genomic P1 library. In the targeting construct, most of exon 2 is replaced with a cassette consisting of the neomycin phosphotransferase cDNA driven by the phosphoglycerate kinase promoter. The construct contains 5 kb of 5' homolgy and 3 kb of 3' homology. The targeting construct was introduced into 129Sv embryonic stem cells by electroporation, and eight independent correctly-targeted clones were obtained from 120 G418-resistant colonies. Chimeric *mice* were bred with C57BL/6J *mice* to generate *mice* heterozygous for the targeted allele.

FINLEY et al. (1997) found that expression of gelatinase B mRNA was enhanced in macrophages from patients with chronic obstructive pulmonary disease.

The functions of gelatinase B as regulator and effector in leucocyte biology were reviewed by OPDENAKKER et al. (2001).

Matrix metalloproteinases-2 and -9 facilitated tumour invasion via basement membrane degradation. Metastatic colorectal cancer cells (CRC line W620) induced matrix metalloproteinase release by *human* THP-1 monocytes (SWALLOW et al. 1996).

9.4.4.1.5
68 kDa Gelatinase

9.4.4.1.6
Angiotensin Convertase (EC 3.4.15.1)

Angiotensin converting enzyme (EC 3.4.15.1) is a membrane-bound ectoenzyme of *human* blood monocytes augmented by dexamethasone in a biphasic dose-dependent manner with maximum effect after 6 days in culture at 10^{-8} M concentration (VUK-PAVLOVIC et al. 1989).

9.4.4.2
Serine Proteinases

9.4.4.2.1
Urokinase-type Plasminogen Activator (EC 3.4.21.73)

Plasminogen activator, a neutral serine protease, is secreted by many different cell types and is thought to be important in cell migration, malignant cell transformation, tissue remodelling, and cytolysis of tumour cells. Alveolar macrophage plasminogen activator could be important in resolution of fibrin clots, promotion of inflammation, or defence against malignant cells in the lung. Secretion of plasminogen activator by *rabbit* alveolar macrophages derived from normal animals and rabbits pretreated with **bacillus Calmette-Guérin (BCG)** to activate these macrophages indicated by an increase in the phagocytic index was detected by SCHUYLER and FORMAN (1984) in the conditioned medium of day 2–3 alveolar macrophage cultures from 17 of 22 animals treated with BCG, opposed to 2 of 14 normal animals ($P < 0.001$, χ^2 test).

Endotoxin (5 mg per kg body weight intraperitoneally) rapidly increased plasminogen activator release by *rat* alveolar macrophages (ETOH et al. 1984). There was significant positive correlation between plasminogen activator release by alveolar macrophages and fibrinolytic activity in bronchoalveolar lavage fluid.

In *murine* C57BL/6 peritoneal macrophages elicited with either proteose-peptone or fresh thioglycollate, enhancement of the secretion of plasminogen activator by lymphokine did not require bacterial lipopolysaccharide in addition to lyphokine (JONES et al. 1984).

Concanavalin A and **phorbol myristate acetate** greatly increased secretion of *rabbit* alveolar macrophage plasminogen activator, although the higher concentration of 10.0 μg Con A/ml had an adverse effect on viability (SCHUYLER and FORMAN 1984).

Urokinase produced by alveolar macrophages is operative not only at the alveolitis stage but also later in the fibrotic process, produced by **silica** particles, supporting the role of urokinase-type plasminogen activator in fibrogenesis (LARDOT et al. 1998).

Asbestos increased urokinase-type plasminogen activator receptor at the surface of *rabbit* and *human* mesothelial cells, suggesting that altered expression of this receptor could be involved in asbestos-induced remodelling of the pleural mesothelium (PERKINS et al. 1999).

Athersclerotic vessels studied by immunohistochemistry showed that urokinase plasminogen activator, plasminogen activator inhibitor type 2, and macrophages were mainly expressed within plaques while tissue plasminogen activator and plasminogen activator inhibitor type 1 were also expressed outside plaque lesions (FALKENBERG et al. 1998).

9.4.4.2.2
Cytolytic Proteinase

9.4.4.3
Aspartyl Proteinases

9.4.4.3.1
Cathepsin D (EC 3.4.23.5)

Cathepsin D is a lysosomal protease present in normal cells at a very low concentration. It is first produced in a precursor form, pro-cathepsin D (52 kDa), and the processed in the cell to an intermediate form of 48 kDa, then finally to the mature forms of 34 kDa and 14 kDa. Phagocytosis of intravenously injected IgG-coated *sheep* red blood cells by *rat* liver Kupffer cells did not change the activity of cathepsin D (BROUWER et al. 1981).

9.4.4.4
Cysteine Proteinases (EC 3.4.22)

Cysteine proteinases are most implicated in matrix degradation and thus in tumour growth and spontaneous metastasis. A comprehensive knowledge of their biochemistry, pharmacology and therapeutic influencabilty is useful as to tumour growth and metastasis prevention.

L-*trans*-Epoxysuccinyl-leucylamido(guanidino) butane (E-64) and its analogues inhibited cysteine proteinases including cathepsins B, H, and L (Barrett et al. 1982). Conway et al. (1996) tested some sulphamide derivatives by s.c. implantation of mini-pumps, which delivered compound throughout the period of primary tumour growth and spontaneous metastasis to the lung at steady state drug concentration orders of greater magnitude than the concentrations needed to either inhibit collagenase, gelatinase or stromelysin *in vitro*. Inhibitor treatment showed the growth of primary s.c. Mat Ly Lu *rat* prostate tumour, LOX *human* melanoma tumours by 40–60 % but had no significant effect on the growth of primary M27 *murine* Lewis lung tumours.

9.4.4.4.1
Cathepsin B (EC 3.4.22.1)

Cathepsin B is a lysosomal cysteine endopeptidase with broad specificity for peptide bonds. It preferentially cleaves -Arg-Arg- bonds in small molecule substrates (thus differing from cathepsin L).

Cathepsin B/L activity in **microglia** compared to both elicited or resident *rat* peritoneal macrophages was significantly ($P < 0.001$) higher (Banati et al. 1993). The significanty higher ($P = 0.029$) fluorescence of thioglycolate-elicited versus resident *rat* peritoneal macrophages demonstrated that the activation of macrophages is accompanied by an increase of cathepsin B/L activity expression.

The extracellular release of cathepsin B was demonstrated in the amyloid deposits of **Alzheimer** brain (Cataldo and Nixon 1990), where microglial activation is a prominent neuropathological feature (McGeer et al. 1988, Styren et al. 1990).

Cigarette smoke (twice daily 10 puffs) stimulated cathepsin B activity in alveolar macrophages of Sprague-Dawley *rats* by 43 % at both 4- and 10-week exposure points (Gairola et al. 1989).

9.4.4.4.2
Cathepsin L (EC 3.4.22.15)

Human alveolar macrophages synthesise and express an active form of cathepsin L (Mason et al. 1986, Chapman et al. 1987).

9.4.5
Other Enzymes

9.4.5.1
Lipases

Human blood mononuclear leucocyte lysosomal acid lipase (triacylglycerol acylhydrolase, EC 3.1.1.3) in two individuals hydrolysed 9.5 and 9.0 nmol substrate per min × mg protein (Coates et al. 1979). The enzyme is under thyroid hormone regulation (Coates et al. 1982).

9.4.5.2
Glucosaminidase: Lysozyme

Lysozyme (EC 3.2.1.17, mucopeptide *N*-acetylmuramoyl hydrolase or muramidase) catalyses hydrolysis of a β(1→4)-glycosidic linkages of polysaccharide component of the peptidoglycan (mucopolymer of the bacterial cell wall), which is composed of alternating *N*-acetylmuramic acid and *N*-acetylglucosamine residues with alternating β(1→4) and β(1→6) linkages, cross-linked with peptide chains. The hydrolytic action of lysozyme gives rise to disaccharide units attached to the peptide chains called muropeptides.

High concentrations of lysozyme were found in monocytes (Schmalzl and Braunsteiner 1970, Gordon et al. 1974) and alveolar macrophages (Myrvik et al. 1961, Cohn and Wiener 1963, Heise and Myrvik 1967, Leake and Myrvik 1968, McClelland and van Furth 1975). Lysozyme is considered to be one of the constitutive enzymes of the macrophages (Gordon et al. 1974, Gordon 1978). It is the major secretory product and forms about 25 % of the extracellularly secreted protein of macrophages.

In 6 from 11 patients with **sarcoidosis**, initial 24-h secretion of lysozyme by monocytes exceeded the normal range of values for cells from 11 age- and sex-matched control individuals (Bodel et al. 1979). Cells with initially augmented secretion rates continued to secrete increased amounts of lysozyme for 3 days. A correlation was noted between *in vitro* secretion of lysozyme by monocytes and serum levels of lysozyme in the same *patient*.

9.4.5.3
Lysosomal Acid Hydrolases

9.4.5.3.1
Proteases

9.4.5.3.2
Lysosomal Acid Lipase

Lysosomal acid lipase (EC 3.1.1.13) is a lipolytic enzyme involved in the intracellular metabolism of cholesteryl esters and triacylglycerols derived from plasma lipoproterins (Goldstein et al. 1975). The enzyme is synthesised in a variety of cells, including fibroblasts (Sando and Henke 1982, Sando and Rosenbaum 1985), mononuclear leucocytes

(Coates et al. 1982), lymphocytes (Coates et al. 1975) and liver cells (Ameis et al. 1994). After synthesis in the endoplasmic reticulum, the enzyme is targeted to the lysosomal compartment (Sando and Henke 1982). Ameis et al. (1994) give a complete nucleotide sequence and deduced protein sequence for *human* LAL cDNA. There are some nucleotide differences of liver and fibroblast LAL cDNA.

9.4.5.3.3 (Deoxy)ribonuclease

9.4.5.3.4 Phosphatases

Acid phosphatase (EC 3.1.3.2) is the classic lysosomal enzyme. The lead substitution method has localised acid phosphatase in the phagosomes of macrophages (Cohn and Wiener 1963) and can be used for electron microscopy.

9.4.5.3.5 Glycosidases

Glycosidase (*N*-acetyl-β-D-glucosaminidase, *N*-acetyl-β-D-galactosaminidase, β-D-galactosidase, α-L-galactosidase, α-D-mannosidase, α-L-fucosidase and β-D-glucuronidase) activities were higher in the alveolar macrophages obtained by bronchoalveolar lavage from smokers than in those from nonsmokers (Scharfman et al. 1980).

Swainsonine reversibly inhibited macrophage lysosomal α-mannosidase (EC 3.2.1.24) *in vitro* (Greenaway et al. 1983).

9.4.5.3.6 Sulphatases

Arylsulphatase (EC 3.1.6.1) from a 96 h *rabbit* peritoneal exudate macrophage was present in segments of the rough endoplasmic reticulum, and perinuclear cisterna as well as within numerous small vesicles in the Golgi region, probably corresponding to secondary lysosomes (Nichols et al. 1971, Davies and Bonney 1980).

9.4.5.4 Deaminase: Arginase

Rat peritoneal macrophages elicited with λ-carrageenan had 10 times greater arginase activity than resident cells at harvest (Albina et al. 1988).

9.4.6 Inhibitors of Enzymes

Minactivin is a *human* monocyte product which specifically inactivates urokinase-type plasminogen activators (Golder and Stephens 1983). This inhibitory activity of monocyte culture supernatant was enhanced after culture with muramyl dipeptide. Inhibition was specific for M_r 52,000 and 36,000.

9.4.7 Proteins of Extracellular Matrix or Cell Adhesion

9.4.7.1 Fibronectin

Alitalo et al. (1980) showed that cultured *human* blood monocyte-derived macrophages secreted fibronectin, but did not deposit a pericellular fibronectin matrix as do various connective tissue cell in culture.

Ozone exposure (0.1 ppm for 10, 20, or 30 min) to alveolar macrophages obtained by bronchoalveolar lavage from healthy, nonsmoking volunteers between the ages of 18 to 25 years, did not produce increased amounts of fibronectin (Devlin et al. 1994).

In **idiopathic pulmonary fibrosis**, alveolar macrophages have glucocorticoid receptors, but glucocorticoid therapy does not suppress alveolar macrophage release of fibronectin (Lacronique et al. 1984).

Rom et al. (1984) studying alveolar macrophages obtained by bronchoalveolar lavage from 7 individuals exposed to asbestos, 4 coal miners and 2 patients with silicosis, suggested that similar to the pathogenesis of fibrosis in the interstitial lung disorders of unknown aetiology, activation of alveolar macrophages to release increased amounts of fibronectin and alveolar macrophage derived growth factor might play a role in the development of fibrosis in the **pneumoconioses**. In coalworkers' pneumoconiosis, fibronectin was detected by immunofluorescence microscopy on the surface of macrophages laden with dust particles (Friemann et al. 1985). Fibronectin provides a mechanism for attaching the fibroblast to the connective tissue matrix and plays a role as a competence signal to move fibroblasts into the early portion of G_1 of the replication cycle (Erdogdu and Hasirci 1998).

H_2O_2 alone induced no changes of fibronectin purified from *human* plasma, even at an 800-fold molar excess (Vissers and Winterbourn 1991). Radiolytic $HO^{\bullet}$ caused a rapid loss of tryprophan fluorescence, an increase in bityrosine fluorescence, and extensive crosslinking.

9.4.7.2
Gelatin-Binding Protein/95 kDa Gelatinase

Vartio (1985) isolated a 95 kDa protein from *human* blood monocytes cultured in a 1:1 mixture of medium 199 and RPMI 1640 supplemented with 5% newborn *calf* serum by preparative polyacrylamide gel electrophoresis. The ability of this fraction to degrade gelatine was calcium-dependent and was inhibited by serum, sulfhydryl and metal-chelating agents, but not with serine proteinase inhibitors. Gelatine was degraded optimally at pH 7–9 and at 41 °C and 37 °C and less effectively at 22 °C. Native type I collagen was degraded at 41 °C but not at 37 °C or 22 °C. The results showed that cultured *human* macrophages secreted highly specific gelatine-degrading metal-proteinase activity associated with the 95 kDa gelatine-binding protein.

9.4.7.3
Thrombospondin

9.4.7.4
Chondroitin Sulphate Proteoglycans

Resident *mouse* peritoneal macrophages synthesise and secrete equivalent amounts of chondroitin sulphate and heparan sulphate proteoglycans (Kolset 1987).

9.4.7.5
Heparin Sulphate Proteoglycans

Resident *mouse* peritoneal macrophages synthesise and secrete equivalent amounts of heparan sulphate and chondroitin sulphate proteoglycans (Kolset 1987).

9.4.8
Bioactive Oligopeptides

9.4.8.1
Glutathione

Glutathione (L-γ-glutamyl-L-cysteinyl-glycine, GSH) was released by *rat* pulmonary alveolar macrophages cultured in RPMI 1640 supplemented with 10% foetal *calf* serum exposed for 2 h to 0–100 μg quartz particles (Min-U-Sil, 4.5 ± 1.0 μm) or crocidolite asbestos (IUAC, fibre length 2.1 ± 0.31 μm) in a concentration-dependent manner (Boehme et al. 1992).

In plasma from 24 healthy individuals aged 25–35 years, Jones et al. (2000) found the concentration of GSH (2.8 ± 0.9 μM) much higher than that of GSSG (0.14 ± 0.04 μM). The redox potential of the GSSG/2GSH pool (−137 ± 9 mV) was considerably more oxidised than values for tissues and cultured cells (−185 to −258 mV). This indicates that a rapid oxidation of GSH occurs upon release into plasma. The difference in values between individuals was remarkably small, suggesting that the rates of reduction and oxidation in the plasma are closely balanced to maintain this redox potential.

Glutathione thiyl radicals (GS•) can be produced e.g. when radicals from oxidable drugs redox cycle (Schreiber et al. 1989), or in the repair of oxidative stress (O'Brien 1988) and radiation damage (Quintiliani et al. 1977).

$$GSH + HO^{\bullet} \longrightarrow GS^{\bullet} + H_2O \quad [163]$$

The fate of GS• depends critically on four key, rapidly established equilibria:

$$GSH \longleftrightarrow GS^- + H^+ \quad [164]$$

$$GS^{\bullet} + GS^- \longleftrightarrow GSSG^{\bullet -} \quad [165]$$

$$GS^{\bullet} + O_2 \longleftrightarrow GSSO^{\bullet} \quad [166]$$

$$GS^{\bullet} + HCR^1R^2R^3 \longleftrightarrow GSH + {}^{\bullet}CR^1R^2R^3 \quad [167]$$

and on a fast electron transfer reaction:

$$GSSG^{\bullet -} + O_2 \longrightarrow GSSG + O_2^{\bullet -} \quad [168]$$

The kinetics of [164]–[167] have been investigated, mostly by pulse radiolysis, although some rate constants are known only for analogous reactions involving other thiols.

Glutathione esters have been shown to increase the levels of GSH in the liver and kidney of *mice*. Administration of γ-glutamylcysteinylglycine monomethyl or monoethyl esters at doses of 10 mmol/kg intraperitoneally raised GSH levels 4-fold in *mice* with GSH levels depleted by prior treatment with L-buthionine sulphoxime (Puri and Meister 1983). The improved uptake of the esters, and their subsequent intracellular hydrolysis to GSH may be explained by improved permeability through lysosomal membranes, and by specific hydrolysis by lysosomal esterases (Goldman 1973, Goldman and Naider 1974).

Glutathione has long been suspected to be the primary source of oxidative power for protein folding. It has now been shown to be just the opposite, namely a source of reductants (Bader et al. 1999). The ultimate origin of oxidants has become even more of a mystery.

Detoxification by glutathione is involved in the cellular resistance of cancer cells to anticancer drugs (Morrow and Cowan 1990).

While the IGROV ovarian cancer cell line was very sensitive to cisplatin (IC_{50} = 0.2 μg/ml), the IC_{50} of the drug-resistant subline $IGROV_{CDDP}$ was 4.0 μg/ml ($P < 0.01$) (SCHELLENS et al. 1995). Total GSH in the resistant cell line was not increased: 7.65±0.84 in the parental cell line and 4.11±1.15 μg GSH/mg protein in $IGROV_{CDDP}$ ($P < 001$).

9.4.9 Arachidonic Acid and Prostaglandins

Elaboration of eicosanoids, oxygenated derivatives of arachidonic acid, is one means by which macrophages can regulate the functions of other effector cells, and thereby modulate inflammation and immune processes as well as tissue injury and repair (NATHAN 1987). Inasmuch as eicosanoids have been implicated in physiologic and pathologic processes in the lung (HENDERSON jr 1987), an understanding of the alterations in arachidonic acid metabolism that accompany macrophage differentiation in the lung is of substantial interest.

9.4.9.1 Arachidonic Acid Release from Membrane Phospholipids

Phospholipid breakdown is one of the earliest cellular events elicited by many extracellular signalling substances. Phospholipases A_2, C and D are involved in these signal transduction pathways. Phospholipase A_2 (EC 3.1.1.4) or phosphatidylcholine 2-acylhydrolase is thought to be the key enzyme of the arachidonic acid cascade inducing the liberation of arachidonic acid from phospholipids and consequently the formation of eicosanoids, which are important biological mediators in many inflammatory processes. Phospholipase C (EC 3.1.4.3) hydrolysing inositol phospholipids produces diacylglycerol, which both serves as the major physiological regulator of protein kinase C and as substrate for diacylglycerol lipase (EC 3.1.1.34) leading to the liberation of arachidonic acid and thus to the formation of eicosanoids.

When *rat* pineal glands were incubated in culture, time-dependent release of arachidonic acid was significantly inhibited by a known 85-kDa cytosolic phospholipase A_2 inhibitor, methyl arachidonyl fluorophosphonate (LI et al. 2000). Co-incubation with melatonin inhibited the arachidonic acid release in a concentration dependent manner, and this decrease was accompanied by a reduction of cytosolic phospholipase A_2 protein and mRNA expression. Melatonin-receptor agonists, 2-iodo-*N*-butanoyl-5-methoxytryptamine and 5-methoxycarbonylamino-*N*-acetyltryptamine, also decreased arachidonic acid release and cytosolic phospholipase A_2 protein and mRNA levels, while pre-incubation with the melatonin-receptor agonists luzindole and 2-phenylmelatonin abolished the melatonin effects.

Mineral oil-elicited *guinea pig* peritoneal macrophages released arachidonic acid in excess of that of unstimulated cells when stimulated with the ionophore, A23187, followed in decreasing potency by phorbol myristate acetate, a group of agents of similar activities (opsonized zymosan, phospholipase C, NaF, and fMet-Leu-Phe), and by concanavalin A and wheat germ agglutinin (BROMBERG and PICK 1983).

In *rat* alveolar macrophages incubated with ≤20 μM eicosapentaenoic acid and further incubated with 1 mg SiO_2 for 90 min, the production of LTB_4 (eicosatrienoic acid) was inhibited dose-dependently. While production of LTB_5, a metabolite of eicosapentaenoic acid, was increased at ≤10 μM eicosapentaenoic acid and decreased at >10 μM (SAKU et al. 1996).

Rabbit alveolar macrophages exhibited an apparent discrepancy between the abundance of immunospecific cP450 4A protein and no 20-hydroxyeicosatetraenoic acid production (ZHU et al. 1998). •NO generated by alveolar macrophages may autoinhibit the P450 4A enzymes and production of 20-HETE in these cells.

Human bronchoalveolar macrophages produce leukotriene B_4 (FELS et al. 1982). A rapid conversion of LTB_4 into dihydro-LTB_4 (5,12-dihydroxyeicosatrienoic acid) is mediated by reduction of one of the conjugated double bonds, while no ω-oxidised products were detected in pure macrophage suspensions (SCHÖNFELD et al. 1988).

LY293111 is a potent leukotriene B_4 receptor antagonist with exceptional oral activity (SOFIA et al. 1997). It inhibited binding of [^{3}H]LTB_4 to *guinea pig* lung membrane receptors and to receptors on *human* neutrophils with a K_i = 7.1±1.1 nM and an IC_{50} = 17.6±4.8 nM, respectively. *In vivo* the agent blocked LTB_4-induced airway obstruction when administered both intravenously (ED_{50} = 0.014 mg/kg) and orally (ED_{50} = 0.40 mg/kg). Histological evaluation of total granulocyte infiltration into lung tissue induced by LTB_4 showed when given orally, LY29311 sodium salt inhibited cell migration with an ED_{50} ≅ 3.0 mg/kg. Analysis of the lung lavage fluid from these animals demonstrated that neutrophil migration into the lumen was also inhibited (ED_{50} = 0.2 mg/kg).

9.4.9.2
Inhibition of Arachidonic Acid Metabolism by Radical Scavengers

Epidemiological and clinical evidence indicates a strong chemopreventive effect of prostaglandin synthesis by nonsteroidal antiinflammatory drugs on *human* colorectal cancer (Marks et al. 1996). The initiation-promotion approach of *mouse* skin carcinogenesis provides a model for mechanistic evaluation and further improvement of such chemopreventive measures which are aiming at an interruption of tumour development at a premalignant stage.

Lipoxygenase (EC 1.13.11.12) catalyses the stereospecific dioxygenation of polyunsaturated fatty acids containing a 1,4-*cis,cis*-pentadiene system to a **pentadienyl radical** intermediate which reacts with molecular oxygen to yield *cis,trans*-conjugated diene hydroperoxides (Wiseman et al. 1988).

The phospholipase A_2 inhibitor **dibromoacetophenone**, the anti-inflammatory steroid **fluocinolone acetonide** or the lipoxygenase inhibitor **nordihydroguiaretic acid** just prior to intraperitoneal injection of 100 ng 12-*O*-tetradecanoylphorbol-13-acetate into unmanipulated CD-1 female *mice* resulted in a dose-dependent decrease in the number of peritoneal exudate cells producing superoxide anion radical as assessed by the reduction of nitroblue tetrazolium, while the cycloxygenase inhibitor indomethacin had no effect on the number of formazan-positive peritoneal exudate cells caused by PMA treatment (Czerniecki and Witz 1989).

Bryostatin 1 could inhibit the effect of 12-*O*-tetradecanoylphorbol-13-acetate both in stimulation of [^{3}H]arachodonic acid release and the induction of prostaglandin H synthase in Madin Darby *canine* kidney cells cultivated in Dulbecco's modified Eagle's medium (Parker et al. 1988).

γ-Glutamyl transferase (EC 2.3.2.2), a glycoprotein enzyme widely distributed on cell surfaces, catalyses the conversion of leukotriene C_4 to leukotriene D_4. Since LTD_4 is biologically much more potent than LTC_4 (Lewis and Austen 1984, Piper 1984), the partial degradation of the glutathione moiety to the cysteinylglycine derivative LTD_4 may be considered a biosynthetic reaction generating the ligand for the LTD_4/LTE_4 receptor.

Radical scavenging effects of leukotrienes

$LTD_4 > LTC_4 > LTB_4$

9.4.10
Cyclic Nucleotides

In RAW264 *murine* macrophages, Takahashi et al. (2000) identified three cAMP inducible mRNAs, named cI-1, cI-2, and cI-3 (for cAMP inducible genes 1–3. The cI-3 probe was identical to a previously known gene, gly96.

PGE_2 within the range of 1.4×10^{-9} to 1.2×10^{-8} M inhibited PGI_2 (2.8×10^{-6} M) stimulation of the adenylate cyclase of starch elicited *rat* peritoneal macrophages at PGE_2 concentrations that have no effect by themselves on the level of cyclic andenosine 3',5'-monophosphate (Adolfs and Bonta 1982). With higher concentrations of PGE_2 the inhibition was either non-existent of masked by the effect of PGE_2 per se on cyclic AMP levels. Cyclic AMP formation is stimulated in resident *rat* peritoneal cells by exogenous arachidonic acid (Elliott et al. 1983).

9.4.11
Cytotoxic Substances

Normal *mouse* peritoneal macrophages cultivated for <8 h exerted a potent cytolytic activity against extracellular ^{51}Cr-labelled syngeneic erythrocytes, as demonstrated by isotope release to the medium (Melsom et al. 1975). This lytic reaction was due to the liberation of a labile macrophage cytolytic factor <1 kDa which required 10^{-4} M cystine for its detection. Experiments with 10^{-4} M 5,5'-dithio-bis-(nitrobenzoic acid) demonstrated that the factor depended on free sulphydryl groups. 10 μg α-Tocopherol/ml abolished both the sulphydryl and the ascorbic acid mediated lytic reaction supporting the theory of lipid peroxidation as the molecular mechanism of target cell lysis. While reduced glutathione and dihydrolipoic acid were effective, oxidised glutathone and lipoic acid were not.

Rat macrophages activated by endotoxin *in vitro* released a factor cytotoxic for sarcoma cells but not for normal cells (Currie and Basham 1975).

9.5
Pharmacology

9.5.1
Receptors

A receptor is an integral membrane protein which transduces biological information, conferred in signals.

9.5.1.1
Cell Surface Receptors

The macrophage cell surface as a crucial interface studded with receptors links antibody dependent molecular immune recognition triggering a broad spectrum of cellular responses including phagocytosis, cytolysis, release of lysosomal enzymes and mediators such as prostaglandin E_2, and release of reactive oxygen metabolites.

Membrane receptors contain single, fourfold, sevenfold, and 24-fold membrane-spanning domains.

9.5.1.1.1
Single Membrane-Spanning Receptors

Characteristic for all partners of the single membrane-spanning receptors, which are stimulated by transformation growth factors, is the fact that endocytosis usually takes place after the ligand binds to the receptor.

Growth Hormone Receptor

The growth hormone receptor is especially interesting because a high degree of homology exists between the growth hormone receptor membrane protein and the growth hormone serum-binding protein (Hesch 1991). To determine if growth hormone would stimulate *human* alveolar macrophages, Keane et al. (1996) incubated cells gained by bronchoalveolar lavage with 10 nM HGH and 100 nM HGH, respectively, for 4 h and stimulated them with fMet-Leu-Phe and *E. coli*, respectively. There were no significant differences between control and HGH-incubated macrophages as to superoxide anion release.

9.5.1.1.2
Fourfold Membrane-Spanning Receptors

The main function of the fourfold membrane-spanning receptors is at synapses, where they rapidly modulate informational intensity. GABA and glycine receptors contain mainly a positively charged area composed of positively charged amino acids located on the extracellular and cytosolic surfaces of the receptor, whereas the nicotinic acetylcholine receptor is thought to contain mainly negatively charged areas.

Phorbol esters were observed to reduce the rate of acetylcholine receptor synthesis in cultured *chick* myotubes (Miskin et al. 1978). These findings were confirmed by Bursztajn et al. (1988) and by Klarsfeld et al. (1989), who, in addition, demonstrated that phorbol esters enhance receptor expression upon chronic exposure, thought to lead to a depletion of protein kinase C from the treated cells.

9.5.1.1.3
Sevenfold Membrane-Spanning Receptors

β-Adrenergic Receptor

Computer analysis of [^{125}I]iodocyanopindolol competition studies using the relatively selective β1-adrenoceptor antagonist, ICI 89406, and the β_2-selective antagonist, ICI 118551, on *rabbit* arterial blood mononuclear leucocyte plasmalemmal preparations favoured a two-site model indicating that both β_1- and β_2-adrenoceptor subtypes were present in approximately equal numbers (Tenner jr. et al. 1989).

Exposure of THP-1 cells, a monocyte/macrophage cell line, to 10 μM isoprenaline or 100 μM salbutamol for 30 min before an adhesion assay to *human* bronchial epithelial cells did not result in an increase in adhesion (Romberger et al. 2000).

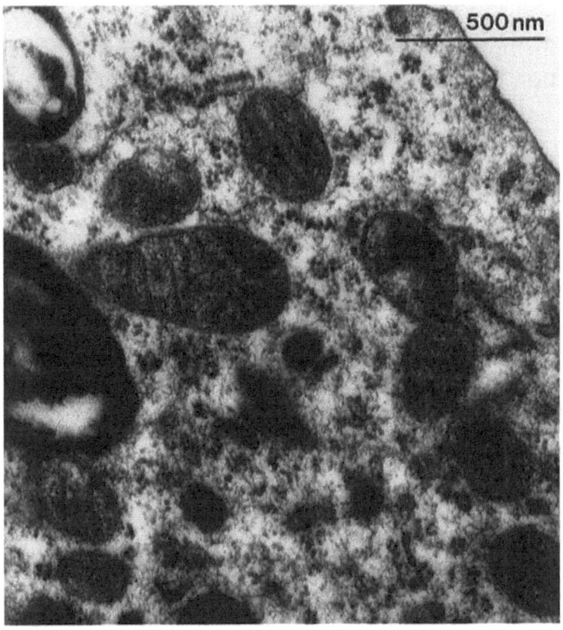

Fig. 122. Polysomes in an alveolar macrophage of a female *rat* (breeder: Winkelmann, Borchen-Kirchborchen) 30 min after inhaling 200 μg micronized isoprenaline HCl + 300 μg micronized phenylephrine bitartrate/puff from a Medihaler®. 12 Puffs/min were dispersed into a 164.5 l box where the animals stayed for 15 min. 30 min later under methitural anaesthesia, the lung was fixed by intratracheal instillation of 2.5 % glutaraldehyde in phosphate buffer (pH 7.4) before opening the thorax. Postfixation with 1 % osmium tetroxide in phosphate buffer (pH 7.4). Contrasted en bloc for 12 h with 0.5 % uranyl acetate in 70 % ethanol. Embedded in a 2:8 mixture of methyl and butyl methacrylate. Sectioned at 50 nm. Lead citrate after Reynolds (1963). Film 345/84

In *human* monocytes isoprenaline accumulates cAMP (GRIESE et al. 1990).

Formoterol induced a significant ($P < 0.01$) increase of basal $O_2^{\bullet -}$ production from *human* alveolar macrophages recovered by bronchoalveolar lavage from 11 patients with chronic obstructive pulmonary disease (CAPELLI et al. 1989). The effect was independent from dose and comparable to control data obtained with zymosan.

9.5.1.2 Thyrotropin Receptors

Binding of [^{125}I]TSH to monocytes is reversible and saturable (CHABAUD and LISSITZKY 1977). The cells exhibit high-affinity, low-capacity and low-affinity, high-capacity sites.

9.5.1.2.1 Cytosolic/Nuclear Receptors

Cytosolic/nuclear receptors are primarily localised in the cytosol. They bind small lipophilic molecules, such as steroid or thyroid hormones, which pass the cell membrane by passive diffusion and, after activation, move towards the nucleus where the ligand-receptor complex exerts its specific function by directly or indirectly modifying DNA transcription.

Vitamin D Receptor

The active form of vitamin D_3, 1α,25-dihydroxyvitamin D_3, is an important regulator of calcium metabolism and elicits most of its biological effects by binding to a high-affinity receptor in target tissues. In the formation of 1α,25-dihydroxyvitamin D_3, vitamin D_3 is hydroxylated in two sequential steps. An initial 25-hydroxylation in the liver is followed by an 1α-hydroxylation in the kidney (DELUCA and SCHNOES 1983). The 25-hydroxyvitamin D_3 1α-hydroxylase (1α-hydroxylase) is the key enzyme in the determination of the level of 1α,25-dihydroxyvitamin D_3 and plays a vital role in calcium homeostasis. It is present in the inner mitochondrial membrane of renal proximal tubular cells (KAWASHIMA et al. 1981, PAULSON and DELUCA 1985). While under physiological conditions, the kidney is the only site of 1α,25-dihydroxyvitamin D_3 (DELUCA 1988), in sarcoidosis or lymphoma, 1α-hydroxylase may be expressed at other sites (ARMBRECHT et al. 1992).

The *human* vitamin D receptor is a member of the ligand-inducible nuclear transcription factors superfamily that regulate transcription through binding to hormone response elements in the target genes (FULLER 1991, WAHLI and MARTINEZ 1991, GRONEMEYER 1992, DARWISH and DELUCA 1993, ROSS et al. 1994). These hormone response elements are basically divided into two groups: the first consists of palindromic sequences which include the glucocorticoid response elements subfamily (MADER et al. 1989, UMESONO and EVANS 1989). The second group which includes vitamin D response elements consists of two direct repeats (UMESONO et al. 1991, CARLBERG 1995). In the latter group, the spacing between the two repeat elements plays an important role in determining hormone receptor specificity (UMESONO et al. 1991). The data of JIN et al. (1996) clearly demonstrated the existence of an activation domain in the *human* vitamin D receptor that is separable from the domain involved in dimerization. Factors that couple the *human* vitamin D receptor to the general transcription apparatus in *yeast* through the activation domain in the *human* vitamin D receptor, however, appear to be unrelated or dissimilar to those in COS-1 cells.

The vitamin D receptor mediates the signal of 1α,25-dihydroxyvitamin D_3 by binding to vitamin D responsive elements in DNA as a homodimer or as a heterodimer composed of one vitamin D receptor subunit and one retinoid X receptor subunit (NISHIKAWA et al. 1995). Exogenous $1,25(OH)_2D_3$ stimulated $24,25(OH)_2D_3$ synthesis in freshly isolated *human* monocytes, whereas matured macrophages showed no $24,25(OH)_2D_3$ synthesis (KREUTZ et al. 1992). Alveolar macrophages gained by bronchoalveolar lavage from 8 healthy volunteers and 6 patients with sarcoidosis II and incubated with various concentrations (10^{-6}–10^{-12} mol/l) of vitamin D_3 and 24,25-dihydroxyvitamin D_3 with and without priming with interferon-γ expressed the intercellular adhesion molecule (ICAM-1) in a dose-dependent manner and released TNF-α and reactive oxygen species in 6 of 8 healthy volunteers with highest values using a concentration of 10^{-6} mol of vitamin D_3/l, while in sarcoidosis these effects seemed to be less pronounced (BRAUN et al. 1996). $1,25(OH)_2D_3$ induces *human* promyelocytic leukaemia cell line, HL-60, to differentiate into monocytes/macrophages (TANAKA et al. 1983, MANGELSDORF et al. 1984, INABA et al. 1986, 1987). While a high-salt extract of LG HL-60 cells displays specific $1,25(OH)_2[^3H]D$ binding activity, ATCC HL-60 cells after sonication showed little or no specific binding explainable by the action of a serine protease in these cells (INABA and DELUCA 1989).

New analogues of vitamin D have been developed in an attempt to separate the calcaemic actions from the effects on cell replication and differentiation. The most effective approach appeared to be modification of the side chain of the vitamin D molecule. EB1089 and KH1060 were more effective in the differentiation of U937 and HL-60 cells than the native hormone (James et al. 1997). KH1060 is characterised by altered stereochemistry at the carbon 20 position.

The conversion of vitamin D-binding protein to a very potent macrophage activating factor (Yamamoto and Homma 1991) involves stepwise deglycosilation of the glycosilated vitamin D-binding protein molecules (natural vitamin D-binding protein is only 0%–5% glycosilated) by membranous β-galactosidase and sialidase of inflammation-primed B and T cells. While vitamin D-binding protein-macrophage activating factors obtained from natural vitamin D-binding protein significantly enhanced macrophage activity over the control, recombinant vitamin D-binding protein-macrophage activating factor had no effect (Ray 1996).

The absence of specific 1,25-$(OH)_2[^3H]D_3$ binding activity in HL-60 cell preparations is due to proteolysis of the 1,25-$(OH)_2[^3H]D_3$ receptor (Inaba and DeLuca 1989). An endogenous proteinase may act on the steroid-binding domain of the 1,25-$(OH)_2D_3$ receptor or in its immediate vicinity, because it destroyed the 1,25-$(OH)_2D_3$-binding ability of unoccupied receptor, but did not digest occupied receptor. The proteinase appeared to belong to the serine class of proteases, which is specifically inhibited with diisopropylfluorophosphate and not hydrolysed diisopropylfluorophosphate.

Contact with UMR106 *rat* osteoblast-like cells and the presence of 1α25-dihydroxyvitamin D_3 and *human* macrophage colony stimulating factor were absolute requirements for differentiation of *human* breast carcinoma tumour-associated macrophages into mature functional osteoclasts (Quinn et al. 1998).

9.5.2 Serotonin

In *murine* macrophages serotonin antagonists (250 μM methysergide, cyproheptadine or spiperone) inhibited the binding of serotonin and muramyl peptides and the respiratory burst induced by these stimuli (Silverman et al. 1985). These data suggest that muramyl peptides and serotonin act via the same receptor.

During the respiratory burst of *human* blood monocytes serotonin acts as a **radical scavenger** and is oxidised to a dimer, probably 5,5'-dihydroxy-4,4'-bitryptamine (Schuff-Werner et al. 1995).

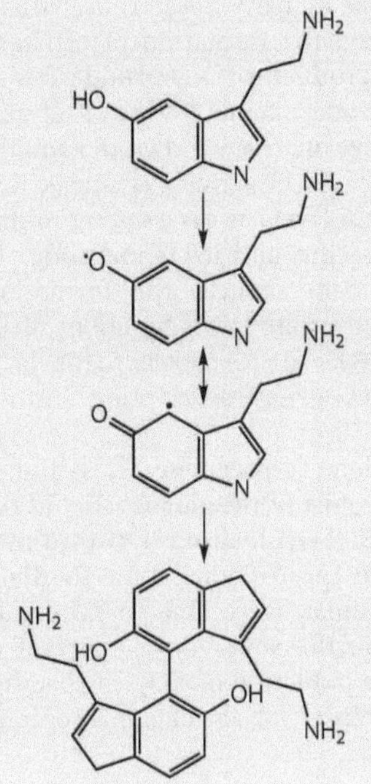

Autoxidation of serotonin [167]

9.5.3 Histamine

Bovine alveolar macrophages in response to histamine in a dose- ($P < 0.001$) and time- ($P < 0.001$) dependent manner released eosinophilic chemotactic activity (Nomura et al. 1996). The activity was predominantly ethyl acetate extractable. The release of eosinophilic chemotactic activity was inhibited by lipoxygenase inhibitors ($P < 0.01$). Leukotriene B_4 receptor inhibitor inhibited the eosinophilic chemotactic activity ($P < 0.01$).

9.5.4 Dexamethasone

Glucocorticoid (dexamethasone or prednisolone)-induced reduction of **prostanoid synthesis** in PMA (5 nM)-differentiated U937 cells is mainly due to a reduced cyclooxygenase activity (Koehler et al. 1990). Cyclooxygenase-2 as a target of glucocorticoid regulation was confirmed by *ex vivo* studies (Masferrer et al. 1994): cyclooxygenase-2 levels were undetected in macrophages from normal *rats*, but were elevated following adrenalectomy, and not detectable in adrenalectomized animals receiving dexamethasone replacement therapy. These data show that also under normal physiological conditions, cyclooxygenase-2 levels are regulated by endogenous glucocorticoids.

Treatment of RAW 264.7 cells with dexamethasone reduced the formation of nitrite, one of the stable end products of **•NO production** as measured in culture supernatants (WALKER et al. 1997). The IC_{50} is approximately 9 nM of dexamethasone. The reduction of iNOS activity is caused by decreased iNOS protein levels as assessed by immunoblotting using a specific anti-iNOS-antibody. Dexamethasone treatment reduces the formation of iNOS mRNA steady state levels by about 50 % of IFN-γ-induced iNOS mRNA levels. This is due to decreased iNOS gene transcription and iNOS mRNA stability.

Cortexolone (4-pregnene-17,21-diol-3,20-dione), a partial agonist of the glucocorticoid receptor, and RU 38 486 [17β-hydroxy-11β-(4-dimethylaminophenol) 17α-(prop-1-ynyl)estra-4,9-diene-3-one], a pure antagonist, were able to modulate and partially inhibit the suppressive effect of dexamethasone on the induction of NO_2^- in specific pathogen-free male Wistar *rat* alveolar macrophages (JORENS et al. 1992).

9.5.5 Dehydroepiandrosterone

Dehydroepiandrosterone added *in vitro* to alveolar macrophages lavaged from 11 non-smoking asbestos workers significantly reduced superoxide anion release (ROM and HARKIN 1991).

9.5.6 Phosphodiesterase Inhibitors

Theophylline, pentoxifylline, and amrinone [5-amino-3,4'-bipyridin-6(1H)-on, a specific PDE 3 inhibitor] by therapeutic-range levels down-regulated the release of H_2O_2, nitrite and TNF-α in a *rat* alveolar macrophage cell line, NR8383, which responds vigorously and predictably to various soluble and particulate stimuli by producing these inflammatory mediators (HELMKE et al. 1996). The release of IL-6, however, while inhibited by ≥20 μg/ml amirone, was enhanced by theophylline or pentoxifylline at concentrations of ≥5 mM. This effect was seen only after exposure to particulate zymosan and was absent with LPS, a soluble stimulus.

Pentoxifylline (1 mM) stimulated the rate of *rabbit* alveolar macrophage spreading on glass coverslips more than twofold (WANG et al. 1996). It reduced superoxide generation induced by phorbol-myristate-acetate by >50 %. Production of TNF-α induced by lipopolysaccharide was suppressed >85 %.

9.5.7 H⁺-Transport Inhibitors

A plasmalemmal V-type H^+ pump (V-ATPase) and a Na^+/H^+ exchanger interact to control the cytosolic pH (pH_i) of alveolar macrophages. The NADPH oxidase system that generates superoxide anions in activated phagocytes is sensitive to pH. In *rabbit* alveolar macrophages superoxide production induced by phorbol myristate acetate was reduced (~50 %) by 10 μM bafilomycin A_1, a specific inhibitor of ATPases, and marginally reduced (~15 %) by 1 mM amiloride, an inhibitor of Na^+/H^+ exchange (HEMING et al. 1996). Neither amiloride nor Na^+-free media altered macrophage pH_i. MOLSKI et al. (1986) showed that most of the fMet-Leu-Phe-stimulated Na^+ influx is not coupled to H^+ efflux. H-7 [1-(5-isoquinolinesulfonyl)-2-methylpiperazine], a potent inhibitor of both protein kinase C and cyclic nucleotide-dependent protein kinases, blocks fMet-Leu-Phe-induced cell alkalinization without inhibiting superoxide anion formation, indicating that an increase of the intracellular pH is not obligatory required for activation of superoxide anion radical formation.

Bafilomycin (10 μM, pH 7.4) caused shortening of microvilli, focal loss of surface ruffles, and marked cytoplasmic vacuolation in resident *rabbit* macrophages by 1 h incubation, with significant macrophage fragmentation and apoptosis at 48 h incubation (HAQUE et al. 1997). Amiloride (100 μM, pH 7.4) caused elongation of microvilli and significant apoptosis by 1 h incubation.

In *human* alveolar macrophages gained by bronchoalveolar lavage from healthy subjects bafilomycin A_1 inhibited phagolysosomal sealing and Fc-mediated phagocytosis (COAKLEY et al. 1996).

9.5.8 Phenylbutazone

Phenylbutazone (100 μM) significantly ($P \doteq 0.01$) inhibited H_2O_2 production by *human* monocytes/macrophages stimulated by 1 mg non-opsonized zymosan/10^6 cells or 0.1 μM 12-*O*-tetradecanoyl-phorbol-13-acetate (BAGGIOLINI et al. 1985).

A phenylbutazone radical has been identified spectrophotometrically using pulse radiolysis during which a solution of phenylbutazone was bombarded with hydroxyl radicals (Evans and Aruoma, unpublished work). It is therefore likely that ferryl systems oxidise the drug to free radical forms which cause oxidative damage to arachidonic acid and α_1-antiproteinase.

9.5.9 Phenothiazines

9.5.9.1 Promethazine

Promethazine or chlorpromazine but not thiazinamium chloride reduced the respiratory burst during phagocytosis of zymosan by *rat* alveolar macrophages gained by lung lavage in a dose-dependent manner (CHANG et al. 1983).

9.5.9.2 Thiazinamium Chloride

Thiazinamium chloride did not reduce the respiratory burst during phagocytosis of zymosan by *rat* alveolar macrophages gained by lung lavage (CHANG et al. 1983).

9.5.9.3 Chlorpromazine

The univalent oxidation of chlorpromazine *in vitro* to a free-radical intermediate has been clearly demonstrated in a variety of EPR experiments (BORG and COTZIAS 1962, PIETTE and FORREST 1962). In each case oxidation was produced by either a metal ion (Fe^{3+}, Ce^{4+}), concentrated acid, or controlled electrolysis. The univalent oxidation produced a coloured intermediate with an absorption maximum at 255 nm and 530 nm. CAVANAUGH (1957) reported chlorpromazine to be a substrate for both peroxidase (donor:H_2O_2 oxidoreductase, EC 1.11.1.7) and catalase (H_2O_2:H_2O_2 oxidoreductase, EC 1.11.1.6) system. He reported the formation of a coloured product (530 nm) in the reaction of chlorpromazine with both peroxidase-H_2O_2 and catalase over a pH rase, EC 1.11.1.6) se, EC 1.11.1.6) nge of 3 to 6.3. GILLETTE and KAMM (1960) have shown that chlorpromazine is completely oxidised to both the sulphoxide and sulphone by the enzymatic action of liver microsomes.

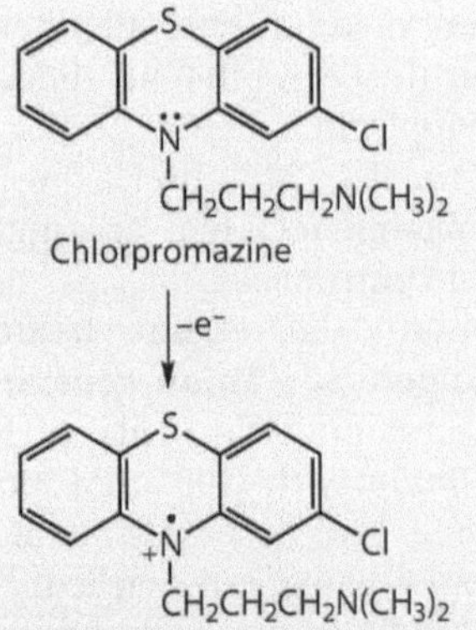

Chlorpromazine radical [64]

PIETTE et al. (1964) measured the dismutation reaction constant, k_d, at pH 4.8 to be 15.0 $M^{-1} \times s^{-1}$. This extreme stability of the free radical allowed further oxidation of the free radical by horse radish peroxidase in the substrate-limited case; the rate constant for this reaction PIETTE et al. (1964) found to be 3.9×10^4 $M^{-1} \times s^{-1}$, approximately 10 times slower than k_4, the first oxidation.

Chlorpromazine sulphoxide, a chlorpromazine metabolite formed in *man* and several other *mammalian* species, produced when irradiated with near-UV light large amounts of the highly reactive hydroxyl radical (BUETTNER et al. 1986).

9.5.10 Thiosemicarbazones

The primary lesion created in cells by the heterocyclic carboxaldehyde thiosemicarbazones is interference with the biosynthesis of DNA, and this action is primarily due to the potent inhibition of ribonucleoside diphosphate reductase activity. 5-Hydroxypicolinaldehyde thiosemicarbazone (NSC-107392) inhibits ribonucleoside diphosphate reductase by the chelation of iron. KRAKOFF et al. (1974) tried the drug by intravenous injection to 30 *patients* with leukaemia or various solid tumours. Marked haemolysis, iron chelation, and urinary excretion of iron occurred.

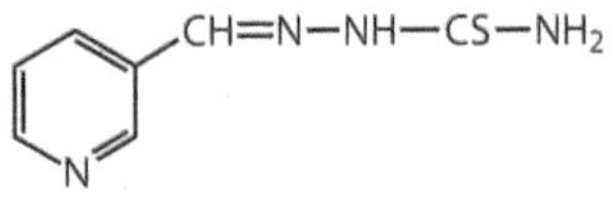

Pyridine-3-aldehyde thiosemicarbazone [168]

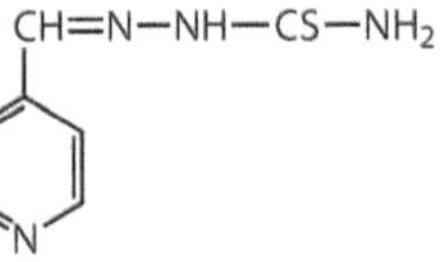

Pyridine-4-aldehyde thiosemicarbazone [169]

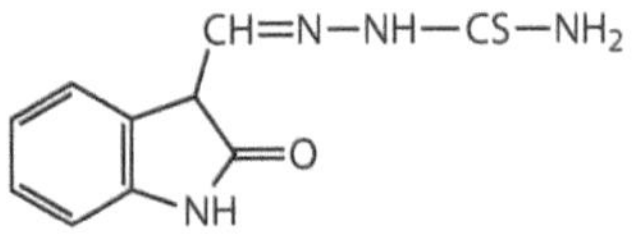

2-Oxo-indole-3-carbaldehyde-thiosemicarbazone [170]

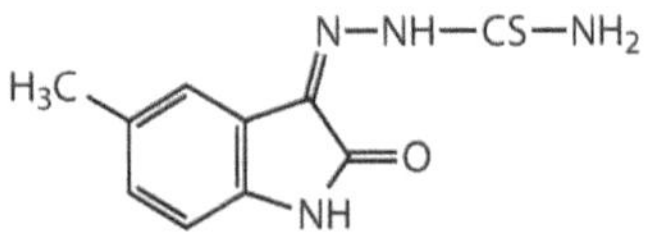

5-Methylisatin-3-thiosemicarbazone [171]

1-Methylisatin-3-thiosemicarbazones and Cu^{2+} ions form a stable complex with which any type of nucleic acid of at least 25 kDa would firmly associate (Mikelens et al. 1976). Among the copper(II)-bisthiosemicarbazone chelates, Cu(II)[2,3-butanedionebis(N^4-methylthiosemicarbazone)] is a most effective superoxide dismutase-like compound (Wada et al. 1994). Copper(II)bisthiosemicarbazones rapidly distributed to all tissues and readily crossed cell membranes as well as the blood-brain barrier with high distributions to the brain and the heart. The superoxide dismutase-like activities of Cu(II)[2,3 - butanedionebis(N^4 - methylthiosemicarbazone)], as measured by several well-known methods, were compared to that of $Cu(II)_2Zn(II)_2$ superoxide dismutase. The Cu(II)[2,3-butanedionebis (N^4-methylthiosemicarbazone)] showed high IC_{50} values in these experiments. However these IC_{50} values were approximately two orders of magnitude higher than the corresponding IC_{50} values of $Cu(II)_2Zn(II)_2$ superoxide dismutase.

9.5.11 Isoniazide

Isoniazide, an antituberculous drug, was determined in alveolar macrophages by high-performance liquid chromatography (Guillaumont et al. 1982).

9.5.12 Histamine Liberator 48/80

After an intravenous injection of Compound 48/80 (1 mg/kg body weight) the phagocytic function of the reticuloendothelial system (India ink clearance) of female albino *rats* remained unchanged (Lázár et al. 1968). Carbon granules could not be seen in the lungs or kidneys.

9.5.13 Dinitrophenol

Rats instilled intratracheally with 50 mg Dörentrup quartz and medicated with 2,4-dinitrophenol aerosol inhalations for 19 months initially showed some retardation in the development of pulmonary fibrosis, but after 9 and 12 months an increased proliferation of connective tissue occurred which was only insignificantly different from that in unmedicated quartz dust-instilled control *rats* (Beck and Schlipköter 1960).

9.5.14 Chlorophyllin

Chlorophyllin, the sodium-copper salt of chlorophyll is used in the treatment of several *human* ailments, particularly in geriatric patients (Young and Bergei 1980). Under *in vitro* conditions chlorophyllin is known to be a highly efficient antimutagenic agent which can significantly reduce or eliminate the mutagenic activity of an impressive range of chemical mutagens/carcinogens like aflatoxin B_1, 2-aminoanthracene, benzo[*a*]pyrene, *N*-methyl-*N*-nitro-*N*-nitrosoguanidine, diemethylnitrosamine and nitrosamines related to tobacco, and complex mixtures such as cigarette smoke.

Treatment with chlorophyllin inhibited nitric oxide production in the lipopolysaccharide-stimulated *murine* RAW 264.7 cells in a dose-related manner (Cho et al. 2000). Competitive RT-PCR analysis, using a DNA competitor as an internal standard, demonstrated that the treatment with 1, 10, and 50 μM chlorophyllin decreased lipopolysaccharide-induced inducible nitric oxide synthase mRNA expression in a concentration-dependent manner. Chlorophyllin down-regulated the NF-κB DNA binding on its cognate recognition site.

9.5.15 Chloroquine

Chloroquine is rapidly taken up by *mouse* peritoneal macrophages cultured on cover slips in Leighton tubes as observed by phase-contrast microscopy and recorded by motion pictures (Fedorko et al. 1968). The drug is localised in small foci in the Golgi region.

Autophagic vacuoles containing various cytoplasmic components and acid phosphatase were produced in *mouse* peritoneal macrophages by addition of 30 μg chloroquine/ml culture medium (Fedorko et al. 1968). The early toxic vacuoles appeared in the perinuclear region within 15 min; on electron microscopy, they showed irregular shape, amorphous moderately dense content, apparent double membranes, and in some instances curved thin tubular extensions with a central, dark linear element. When chloroquine was removed by changing the culture medium after 4 h, the cells survived and 24 h later they exhibited no abnormality except for large cytoplasmic dense bodies packed with membrane lamellae. 24 h after exposure to chloroquine the macrophages had accumulated less hydrolases than control cells.

Inhibition of the serum treated zymosan-stimulated **superoxide anion generation** from *human* blood mononuclear cells preincubated for 30 min or 60 min with 0.001 mM–0.1 mM chloroquine was time- and dose-dependent (Hurst et al. 1986). The drug could not markedly influence the oxygen burst induced by phorbol-myristate-acetate, A23187 or fluoride ion, excluding a significant effect on protein kinase C, calmodulin-dependent kinase(s) or the membrane-bound superoxide-generating NADP(H)-oxidase.

Chloroquine (0.1 mM) caused a decrease in *mouse* peritoneal macrophage **antigen catabolism** associated with inhibition of antigen presentation (^{125}I-labeled *Listeria monocytogenes*) to T cells (ZIEGLER and UNANUE 1982).

9.5.16 Azaspiranes

The immunomodulatory azaspiranes are novel cationic amphiphilic drugs with beneficial effects in a number of animal models of autoimmune disease and transplantation. WAITES et al. (1995) compared *N,N*-dimethyl-8,8-dipropyl-2-azaspiro[4,5]decane-2-propanamine HCl (SK&F 105685) and two analogues, SK&F 106615 and SK&F 103811 with chlorphentermine and chloroquine for their ability to induce phospholipid accumulation and suppressor cell activity. Oral administration of SK&F 105685 and SK&F 106615 caused phospholipid accumulation in bronchoalveolar lavaged *rat* macrophages but to a far lesser extent (three- to fivefold) than chlorphentermine. Neither the immunologically unreactive azaspirane SK&F 103811 nor chloroquine affected phospholipid levels. Alveolar macrophages from *rats* treated with SK&F 105685 or SK&F 106615 expressed more potent suppressor cell activity than chlorphentermine. Neither SK&F 103811 nor chloroquine induced suppressor cell activity.

9.5.17 Poly-2-vinylpyridine-1-oxide

The uptake of poly-2-vinylpyridine-1-oxide into lysosomes of alveolar macrophages of *rats* (GRUNDMANN 1967) and *guinea pigs* (BECK and BOJE 1967) and hepatic Kupffer cells of *mice* (BAIRATI and CASTANO 1968) has been documented.

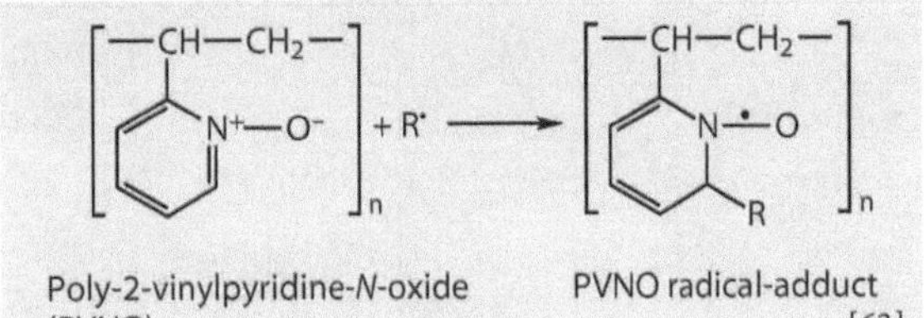

[63]

NASH et al. (1966) synthesised another compound, polyvinylpyridinioacetic acid, which has even greater hydrogen bonding capacity than poly-2-vinylpyridine-1-oxide and very efficiently protects macrophages from silica toxicity. This protection was confirmed by SAKABE and KOSHI (1967), who also showed that it prevents the increase in acid phosphatase in macrophages following exposure to quartz particles <2 μm in diameter. However, polybetaine was slightly more active than poly-2-vinylpyridine-1-oxide.

The protective effects of syndiotactic and isotactic poly-2-vinylpyridine-1-oxide remarkably differ in the spatial arrangement of the structural units (HOLT et al. 1970). The isotactic polymer has probably a helical conformation. When the cultures

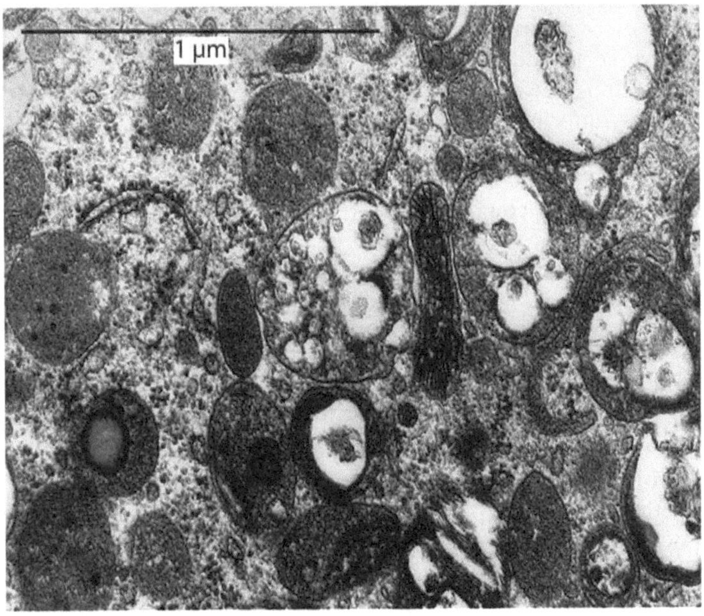

Fig. 123. Alveolar macrophage from a female *rat* (breeder: Winkelmann, Borchen-Kirchborchen) which inhaled micronized poly-2-vinylpyridine-1-oxide for 15 min per day, 5 times per week from July 25, 1967 to January 15, 1968. Fixed on January 15, 1968 under methitural anaesthesia by intratracheal instillation of 2.5 % glutaraldehyde in phosphate buffer (pH 7.4) before opening the thorax. Postfixation with 1 % osmium tetroxide in phosphate buffer (pH 7.4). Contrasted en bloc for 12 h with 0.5 % uranyl acetate in 70 % ethanol. Embedded in a 2:8 mixture of methyl and butyl methacrylate. Sectioned at 50 nm. Lead citrate after REYNOLDS (1963). Plate 47/07

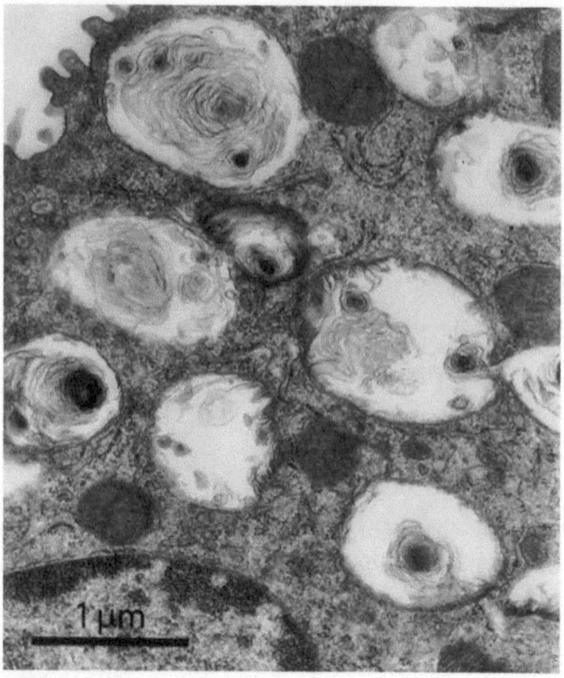

Fig. 124. Alveolar macrophage from a female *rat* (breeder: Winkelmann, Borchen-Kirchborchen) which inhaled micronized poly-2-vinylpyridine-1-oxide 15 min per day, 5 times per week from July 25, 1967 to January 15, 1968. Fixed on January 15, 1968 under methitural anaesthesia by intratracheal instillation of 2.5% glutaraldehyde in phosphate buffer (pH 7.4) before opening the thorax. Postfixation with 1% osmium tetroxide in phosphate buffer (pH 7.4). Contrasted en bloc for 12 h with 0.5% uranyl acetate in 70% ethanol. Embedded in a 2:8 mixture of methyl and butyl methacrylate. Sectioned at 50 nm. Lead citrate after Reynolds (1963). Plate 40/08

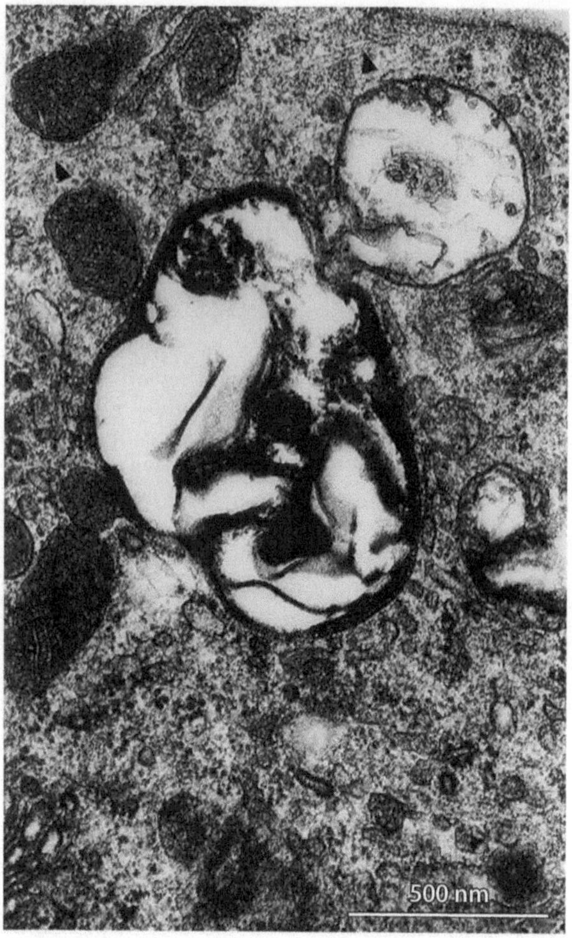

Fig. 125. Microtubules (arrowheads) and microfilaments (arrows) in an alveolar macrophage from a female *rat* (No. 1; breeder: Winkelmann, Borchen-Kirchborchen) which inhaled micronized poly-2-vinylpyridine-1-oxide for 15 min per day, 5 times per week from July 25, 1967 to January 15, 1968. Fixed on January 15, 1968 under methitural anaesthesia by intratracheal injection of 2.5% glutaraldehyde in phosphate buffer (pH 7.4) before opening the thorax. Postfixation with 1% osmium tetroxide in phosphate buffer (pH 7.4). Contrasted en bloc for 12 h with 0.5% uranyl acetate in 70% ethanol. Embedded in a 2:8 mixture of methyl and butyl methacrylate. Sectioned at 50 nm. Lead citrate after Reynolds (1963). Plate 47/05

of guinea pig peritoneal macrophages were pre-treated it was as active against the pathogenic effects of Dörentrup quartz as the atactic polymer, but when the quartz was pre-treated it gave little protection. The syndiotactic polymer has a planar zigzag conformation. When the cell cultures were pre-treated, no protection was given to the cells against silica: when the quartz was pre-treated, the protection was comparable with that given by the atactic polymer previously used.

Natta et al. (1966) found that polymers inhibiting of the lysis of macrophages that had engulfed silica particles belong to two chemical classes characterised by the presence in the monomeric unit either of the N→O function or the N-methylene or N-ethylene groups. Their activity decreased and vanished when molecular weight decreased.

Poly-*N*-(4-morpholino) ethyl acrylamide, poly-*N*-ethyl-*N*-β-(4-morpholino) ethyl acrylamide, and poly-*N*-isopropyl-*N*-β-(4-morpholino) ethyl acrylamide, though showing a good protection *in vitro* were completely inactive in protecting Swiss *mice* from hepatic silicosis after intravenous injection of 5 mg tridymite dust (Chiappino et al. 1967).

Nash et al. (1966) synthesised another compound, polyvinylpyridinioacetic acid, which has even greater hydrogen bonding capacity than poly-2-vinylpyridine-1-oxide and very efficiently protects macrophages from silica toxicity. This protection was confirmed by Sakabe and Koshi (1967), who also showed that it prevents the increase in acid phosphatase in macrophages following exposure to silica. Holt et al. (1970) reported that some poly-2-vinylpyridinium salts prevented quartz toxicity in cultures. Thus the presence of hydrogen accepting groups (either N-oxides of pyridinium groups) is frequently associated with protection. A difference between the protective effect of syndiotactic and isotactic polyvinylpyridine *N*-oxide polymers differ only in the spatial arrangement of the structural units.

9.5.18 Coumarins

Coumarins used in postischaemic cardiomyopathy, have a reputation as anticoagulant, antimutagenic, tumoristatic (Von Angerer et al. 1994, Weber et al. 1998), antimetastatic, antiinflammatory (Chaturvedi et al. 1974), immunostimulatory (Berkarda et al. 1983), anticonvulsant and hypotensive agents. They are able to scavenge (quench) reactive oxygen species, stimulate respiration ionophoretically, inhibit 5- and 12-lipoxygenases and inhibit xanthine oxidase and phenylalanine hydroxylase.

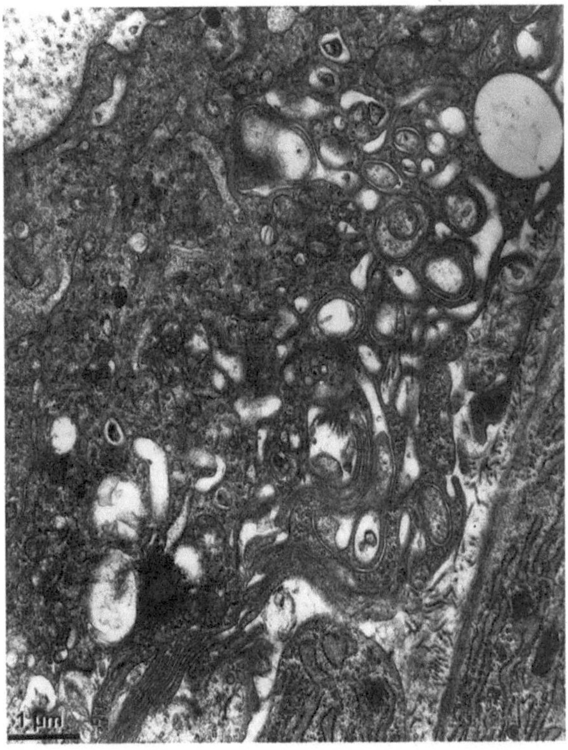

Fig. 126. Macrophage polykaryon with many microvilli, numerous cisternae of the endoplasmatic reticulum, polysomes, mitochondria, bundles of filaments and lipid droplets. Marbagelan®-induced (7 days) resorption granuloma (block 171) of a male Sprague-Dawley *rat* (No. 10) treated for 14 days with intragastric application of 15 mg carbocromene per kg body weight × day. Perfused under pentobarbital anaesthesia (30 mg/kg) from the abdominal aorta with 2.5 % glutaraldehyde in 0.1 M sodium cacodylate buffer (pH 7.4). Postfixation with 1 % osmium tetroxide in cacodylate buffer. Embedded in Epon 812 and sectioned at 50 nm. Lead citrate and uranyl acetate. Plate 2846. Detail from Fig. 155

9.5.18.1 Carbocromene

In Marbagelan®-induced resorption granulomas, carbocromene enhanced the formation of polycaria, but prevented the translocation of nuclei to form typical Langhans cells (Schiller 1982).

9.5.18.2 Cloricromene

Cloricromene or 8-monochloro-3*β*-diethylaminoethyl-4-methyl-7-ethoxycarbonyl-methoxy-coumarine (Aporti et al. 1978) inhibited the induction of the inducible isoform of nitric oxide synthase in *murine* J774 macrophages (Zingarelli et al. 1993).

9.5.19 Benzhydryl Piperazines

Analogues containing the benzhydryl piperazine structure are mainly *N*-dealkylated (King et al. 1965, Kuntzman et al. 1965, Narrod et al. 1965, King and Howell (1966). A striking sex difference was observed for the *in vitro* metabolism of flunarizine in *rat*. In male *rats*, oxidative *N*-dealkylation at one of the piperazine nitrogens, resulting in bis(4-fluorophenyl) methanol, was a major metabolic pathway, whereas aromatic hydroxylation at the phenyl of the cinnamyl moiety, resulting in hydroxyl-flunarizine, was a major metabolic pathway in female *rats* (Lavrijsen et al. 1992).

Epoxidation by the introduction of an oxygen atom into flunarizine produced 1-[bis(4-fluorophenyl)methyl]-4-[(3-phenyloxiran-2-yl)methyl] piperazine (metabolite 2) and epoxide hydration to a diol, 3-[4-[bis(4-fluorophenyl)methyl]-1-piperazinyl]-1-phenyl-1,2-propanediol (metabolite 10). Lavrijsen et al. (1992) found metabolites formed by epoxidation at the double bond (metabolite 2) and epoxide hydration (metabolite 10) in incubates with subcellular hepatocyte fractions of male and female *rats*. Metabolites formed by epoxidation and epoxide hydration were not detected *in vivo* (Meuldermans et al. 1983), probably because the resulting metabolites were metabolised *in vivo*, much more quickly than *in vitro*, into secondary metabolites. A diol metabolite, however, was described for the metabolism of 1-butyl-4-cinnamylpiperazine in *guinea pigs* (Morishita et al. 1978). With supernatant fractions a rapid disappearance of the epoxide intermediate from incubate was observed. This seems to indicate that, for the epoxide hydrolysis, besides microsomal epoxide hydrolase, cytosolic epoxide hydrolase might also be involved.

Cinnarizine, 1-(diphenylmethyl)-4-(3-phenyl-2-propenyl)-piperazine is a selective calcium entry blocker, and is extensively used in the treatment of cerebral and peripheral insufficiency (Godfraind et al. 1982, Singh 1986). Its oxidative metabolism to 1-(diphenylmethyl)piperazine (M-1), 1-(diphenylmethyl)-4-[3-(4'-hydroxyphenyl)-2-propenyl]-piperazine (M-2), benzophenone (M-3) and 1-[4'-hydroxyphenyl)-phenylmethyl]-4-(3-phenyl-2-propenyl) piperazine (M-4) in *rat* liver microsomes required NADPH, and was inhibited by carbon monoxide and SKF 525-A, typical inhibitors of P_{450} (Kariya et al. 1992). Only M-2 formation was suppressed by sparteine or metoprolol, and was significantly lower in female Dark Agouti *rats* than in Wistar *rats* of both sexes.

Cinnarizine was found to inhibit spontaneous lipid peroxidation in *rat* liver homogenates, copper-induced lipid peroxidation in *human* plasma and copper-induced and hydrogen peroxide-induced lipid peroxidation in *human* erythrocytes (Fernandes et al. 1991).

Cyclicine, cyproheptadine and some other structurally related antihistaminics, when administered to *rats*, provoked a marked degranulation of the pancreatic beta cells together with a dramatic dilatation of the rough endoplasmic reticulum (Hruban et al. 1972, Longnecker et al. 1972, Fischer et al. 1973, 1975). In fed, fasting and adrenalectomized *rats* cyproheptadine (10–40 mg/kg) caused a hyperglycaemia and a fall in plasma insulin (Poser et al. 1975).

9.5.19.1 *N*-Benzhydryl-*N'*-*p*-hydroxybenzylpiperazine

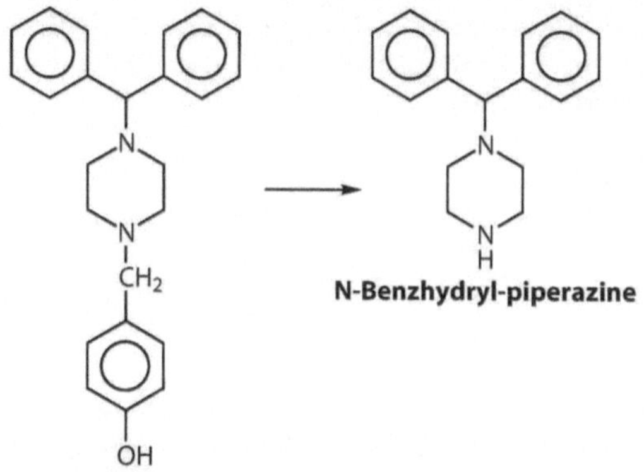

Biotransformation of *N*-Benzhydryl-*N'*-*p*-hydroxybenzylpiperazine [172]

9.5.20 Lazaroids

The lazaroids, or 21-aminosteroids, are potent inhibitors of iron-dependent lipid peroxidation. U-83836E is a derivative of vitamin E which contains a trolox ring and the aminosteroid portion of the lazaroids. U-83836E protected a *rat* macrophage cell line (NR8383) against a dose-dependent lipid peroxidation (13.8± 2.4–423.7 ±76 μM malondialdehyde/10^6 cells) induced by an exposure to 0.5–5 mM H_2O_2 for 1 h (Alessandrini et al. 1996). U-83836E pre-treatment reduced both lactate dehydrogenase activity (46.58±13.5 U/l versus 124.91±13.5 U/l) and proinflammatory cytokine mRNA induction (Alessandrini et al. 1997).

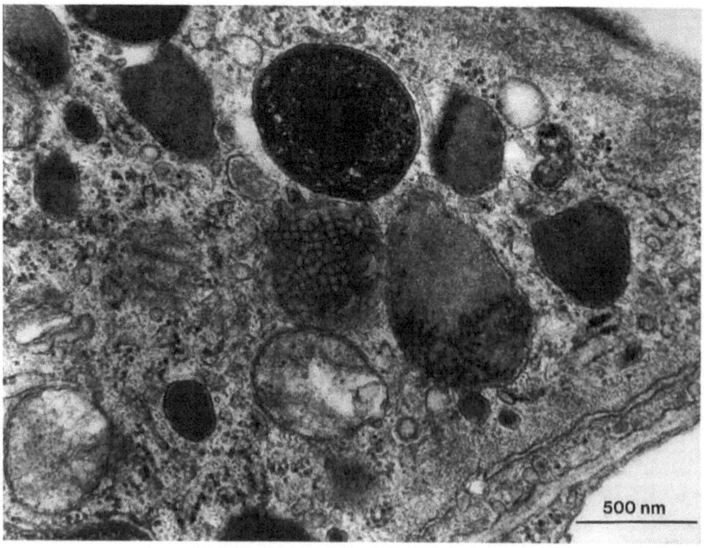

Fig. 127. Alveolar macrophage (block BNh 3406) of a male Wistar *rat* (No. 2569) 90 min after a single intraperitoneal injection of 5 mg *N*-benzhydryl-*N'*-*p*-hydroxybenzyl-piperazine HCl (UCB F 241) per kg body weight. Under pentobarbital anaesthesia (30 mg/kg), the animal was perfused from the abdominal aorta with 2.5 % glutaraldehyde in 0.1 M sodium cacodylate buffer (pH 7.4). Postfixation with 1 % osmium tetroxide in sodium cacodylate buffer. Embedded in Epon and sectioned at 50 nm. Lead citrate and uranyl acetate. Plate 2397

9.5.21
Biphosphonates

The effect of biphosponates on macrophages might explain the acute phase response occurring in some patients who received an amino-biposphonate intravenously for the first time showing a transient pyrexia of 1–2 °C, sometimes more, accompanied by flu-like symptoms (ADAMI et al. 1987). The pyrexia has been shown to be accompanied by an increase in circulating IL-6 and TNF-α (SCHWEITZER et al. 1995, SAUTY et al. 1996).

9.5.22
Ginkgo biloba Extract

Oxidative burst triggered in *murine* J774-macrophages by zymosan was significantly ($P < 0.05$) attenuated when the cells were preincubated for 1 h with ≥50 μg *Ginkgo biloba* extract/ml (RONG et al. 1996).

9.6
Toxicology

9.6.1
Physical Hazards

After a single thoracic x-irradiation of 1,800 r, GROSS and BALIS (1978) divided the *murine* alveolar macrophages lavaged 2, 4, and 8 weeks later, respectively, into two populations: the larger cell population consisted of cells that were either on the alveolar surface at the time of irradiation, or still in the interstitium but at a postmitotic radioresistant stage. As this population aged (4 weeks) the cell size increased. It was finally replaced by a small cell population which was absent 2 weeks after irradiation and appeared in the alveolar lumina in the following two weeks. Small cells reappeared in small numbers at 4 weeks. By 8 weeks they had increased in size and number, and had replaced the older large celled population.

9.6.2
Chemical Agents

9.6.2.1
Inorganic Agents

Intracellular dissolution of a large variety of inhaled inorganic particles not readily soluble in the pulmonary epithelial lining fluid constitutes an important long-term clearance mechanism of the lungs. Fluorescence microscope photometry and dual laser flow cytometry of intraphagolysosomal pH in *canine* and *rat* alveolar macrophages were compared using fluoresceinisothiocyanate labelled, amorphous silica particles (NYBERG et al. 1989, HEILMANN et al. 1992). The higher intraphagolysosomal pH in *rat* alveolar macrophages correlated with the smaller intracellular particle dissolution rate in Wistar *rat* alveolar macrophages when compared with the data in Beagle *dogs* (KREYLING et al. 1992).

9.6.2.1.1
Aluminium

Aluminium has the atomic number 13 and the valence electrons are those on the $3s^2 3p^1$ shell. Aluminium has a fixed oxidation number of +3 and therefor cannot participate in the redox reactions that lead to radical formation. However, aluminium salts accelerate peroxidation of membrane lipids stimulated by iron salts (GUTTERIDGE et al. 1985).

The soluble hexahydrates species of aluminium $[Al(H_2O)_6]^{3+}$, as those of iron $[Fe(H_2O)_6]^{3+}$, and zinc $[Zn(H_2O)_6]^{2+}$, according to the law of mass action. is balanced with the pentahydrate, $[Al(OH)(H_2O)_5]^{2+} + H^+$.

Aluminium is complexed by two carboxylates and the alkoxy group of citrate, leaving a free carboxylate (GREGOR and POWELL 1986). The availability of a free carboxylate may be of importance to the ability of Al citrate to cross membranes.

Aluminium can bind to phosphates and other oxygen donating ligands and form stable complexes. By doing so, Al may disrupt the enzyme activity in the mitochondria and thus effect the electron transport chain. This may lead to an increase in reactive oxygen species formation which in turn can lead to oxidative injury (HALLIWELL 1992). Mitochondrial respiratory activity in *rat* glioma (C-6) cells grown in RPMI medium with 10% foetal *bovine* serum was found to be significantly higher in aluminium (500 μM)-treated cells (CAMPBELL et al. 1999). At 500 μM, the oxidant effects of Al were maximal (BONDY and KIRSTEIN 1996).

Aluminium salts stimulated luminol-enhanced chemiluminescence production by *human* neutrophils (STANKOVIC and MITROVIC 1991). KONG et al. (1992) described an Al(III) complex with $O_2^{\bullet-}$ which was a stronger oxidant than $O_2^{\bullet-}$ itself and which may contribute to the adverse biological effects of Al(III). Aluminium can enhance hydroxyl radical production by iron, but it is not in itself capable of catalysing the generation of hydroxyl radical from the Fenton or Haber-Weiss reactions (GUTTERIDGE et al. 1985). Aluminium facilitation of iron-mediated lipid peroxidation is dependent on substrate, pH (greater at pH 5.5 than 7.4), and aluminium and iron concentrations (XIE and YOKEL 1996). In male *rhesus monkeys*, 25 mg Al^{3+}/kg body weight every alternate day for 52 weeks resulted in a significant ($P < 0.001$) enhancement in the levels of lipid peroxidation in hippocampus (82 %), cerebral cortex (55 %), and corpus striatum (32 %) (SARIN et al. 1997). Regional alterations in calcium homeostasis might enhance the production of free radicals as a cause of augmented lipid peroxidation.

Aluminium and iron may share important biological pathways (CANNATA et al. 1984, 1991, CANNATA and DIAZ LÓPEZ 1991). Both are transported by transferrin (TRAPP 1983, MARTIN 1986). *Human* serum transferrin released Al^{3+} when the pH varied from pH 7.4 (FATEMI et al. 1991). The ubiquitous iron-storage protein, ferritin formed complexes with aluminium in the livers and brains of *rats* fed $AlCl_3$ for 1 year (FLEMING and JOSHI 1987). Ferritin isolated from the brains of patients who died of Alzheimer's disease contained more aluminium and more iron than that from age-matched controls. In neuroblastoma cells, aluminium enhanced iron uptake and expression of neurofibrillary tangle protein (ABREO et al. 1999).

The iron-binding agent, 54 inducing a negative iron balance (BREITHAUPT et al. 1986) also chelated Al^{3+}. It relieved the inhibitory action of aluminium on parathyroid cells and osteoblasts (RAPOPORT et al. 1987). However, COURNOT-WITMER and PLACHOT (1990) showed that in the parathyroid glands of aluminium-intoxicated *patients* the presence of aluminium deposits neither induced cellular damage or chief cell necrosis nor interfered with the production of parathyroid hormone. In normal *rats*, Fe and Al after systemic administration chelated as ferroxamine and aluminoxamine, respectively (ALLAIN et al. 1988).

1,2-Dimethyl-3-hydroxypyrid-4-one (deferiprone, L1) has been demonstrated to be a clinical effective iron and aluminium chelator, generally safe, and less expensive than desferrioxamine, being to date the most promising alternative to desferrioxamine for longer use in humans (AL-REFAIE et al. 1995, KONTOGHIORGHES 1995, OLIVIERI et al. 1998). However, in some cases the long therapy with L1 has also been associated with complications such as neutropenia, agranulocytosis, hepatic fibrosis, arthralgias, and gastrointestinal disturbances (DOMINGO 1996, OLIVIERI et al. 1998, TÖNDURY et al. 1998). In addition to the potential adverse effects, L1 has also a lower therapeutic ratio than desferrioxamine. Oral administration of L1 was the most effective treatment in enhancing urinary Al excretion in both young (21 days old) and old (18 months) Al-loaded *rats* (GÓMEZ et al. 1999). Concurrent administration of desferrioxamine and L1 had no advantages over the use of either single agent, while 1-(*p*-methylbenzyl)-2-ethyl-3-hydroxypyrid-4-one was not effective in mobilising Al from Al-loaded *rats*.

Aluminium preferentially binds within the nuclear compartment (DEBONI et al. 1974, CRAPPER et al. 1978, WEN and WISNIEWSKI 1985, WEDRYCHOWSKI et al. 1986), particularly to heterochromatin (CRAPPER et al. 1980) and DNA phosphates and bases (KARLIK et al. 1980).

Preincubation of DNA and RNA from *mouse* and *rat* lung and liver with $AlCl_3$ markedly inhibited the

Table 30. Effects of aluminium on enzyme activities

Enzyme	EC	Authors
NAD^+-dependent isocitrate dehydrogenase	1.1.1.41	YOSHINO et al. (1992)
$NADP^+$-dependent isocitrate dehydrogenase	1.1.1.42	YOSHINO et al. (1992)
Glucose-6-phosphate dehydrogenase	1.1.1.49	CHO and JOSHI (1989)
Cytochrome *c* oxidase	1.9.3.1	ENGELBRECHT and JORDAN (1972)
Dihydropteridine reductase	1.14.16	ALTMANN et al. (1987)
Ferroxidase	1.16.3.1	HUBER and FRIEDEN (1970)
Choline acetyltransferase	2.3.1.6	YATES et al. (1980), HOFSTETTER et al. (1987), BILKEI-GORZÓ (1994), CHERRORET et al. (1994)
Hexokinase	2.7.1	HARRISON et al. (1972), WOMACK and COLOWICK (1979), TRAPP (1980), CHO and JOSHI (1988)
Acetylcholinesterase	3.1.1.7	MARQUIS and LERRICK (1882), MARQUIS (1983)
Acid phosphatase (potato)	3.1.3.2	HELFERICH and SCHMITZ (1953)
Ca^{2+}-calmodulin-dependent phosphodiesterase	3.1.4	RICHARDT et al. (1985)
Phospholipase C	3.1.4.3	SHI et al. (1993)
Amylase	3.2.1.1	MCGEACHIN et al. (1962)
α-Chymotrypsin	3.4.21.1	CLAUBERG and JOSHI (1993)
Calpains I and II	3.4.22.17	ZHANG and JOHNSON (1992)
ATPase	3.6.1.3	MISSIAEN et al. (1989), HYPPÖNEN et al. (1993)
Na^+-K^+-ATPase	3.6.1.3	LAI et al. (1980)
Mg^{2+}-ATPase	3.6.1.3	LAI et al. (1980), HYPPÖNEN et al.(1993)
Glutamate decarboxylase	4.1.1.15	HOFSTETTER et al. (1987)
δ-Aminolaevulinic acid dehydratase	4.2.1.2.24	MEREDITH et al. (1974), ZAMAN et al. (1993)

binding of 4-hydroxyaminoquinoline 1-oxide to either nucleic acid (YAMANE and OHTAWA 1978). Al exists in a variety of species in aqueous solution, depending upon the degree of hydroxylation (JÄGER and JÄGER 1941, JÄGER 1949). This complexity affects aluminium binding to DNA, which is pH-depending and produces a shift from phosphate to base binding with decreasing pH (KARLIK et al. 1980).

Human Pathology

An open lung biopsy from a 44-year-old male employed for the previous 6 years as an aluminium rail grinder working in an extremely dusty environment without wearing a protective mask showed numerous areas where the alveolar spaces were filled with granular, hypocellular eosinophilic material, which was strongly and uniformly periodic-acid-Schiff-positive material, diastase resistant, and metachromatic with toluidine blue (MILLER et al. 1984). The electron microscopic findings were typical for alveolar proteinosis. Only rare alveolar macrophages were found, and their cytoplasm was distended with granules similar to those free in alveolar spaces. The cytoplasmic borders of the macrophages were abnormally smooth and rounded and only occasional primary lysosomes were present.

Experimental Pathology

Inhalation experiments in *rats* demonstrated an eager phagocytosis of aluminium dust by alveolar macrophages and their elimination via the bronchi (SCHILLER 1957). Although there is no disintegration of the alveolar macrophages produced by aluminium, characteristic phagolysosomal changes at the electron microscope level were described by SCHILLER (1970, 1971). *Rats* inhaling aluminium dust up to 750 h accumulated the particulate matter in interstitial macrophages (SCHILLER 1959). After

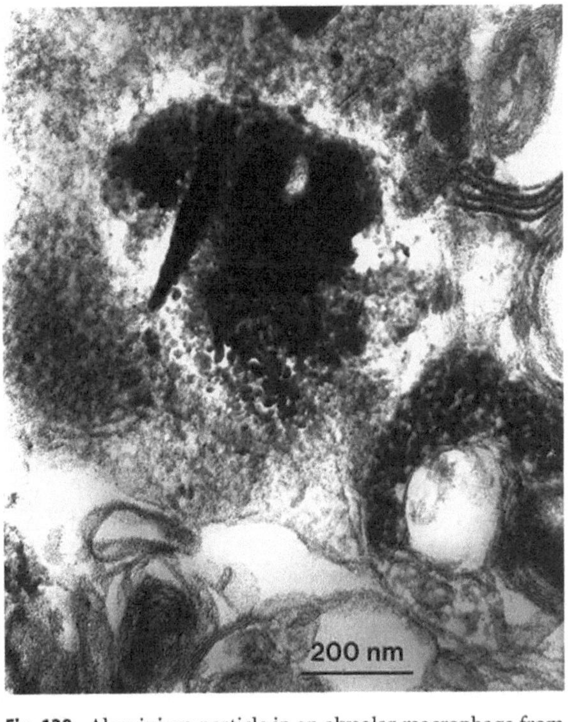

Fig. 128. Aluminium particle in an alveolar macrophage from a 296 g female white *rat* (No. 3; breeder: Winkelmann Borchen-Kirchborchen) which inhaled 10 mg powdered aluminium/m^3 4 h per day, 5 days per week from August 16 to October 27, 1967 for a total of 51 days. Fixed on October 30, 1967 under methitural anaesthesia by intratracheal instillation of 2.5 % glutaraldehyde in phosphate buffer (pH 7.4) before opening the thorax. Postfixation with 1 % osmium tetroxide in phosphate buffer (pH 7.4). Contrasted en bloc for 12 h with 0.5 % uranyl acetate in 70 % ethanol. Embedded in a 2:8 mixture of methyl and butyl methacrylate. Sectioned at 50 nm. Lead citrate after REYNOLDS (1963). Plate 19/09

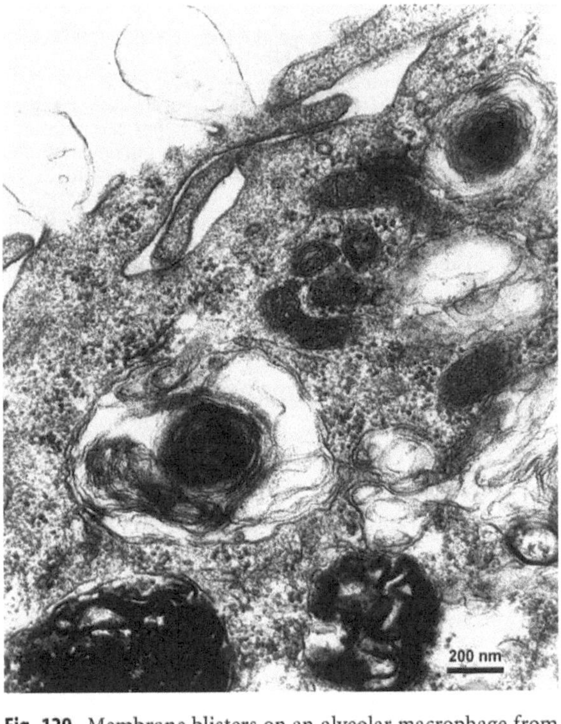

Fig. 129. Membrane blisters on an alveolar macrophage from a female *rat* (No. 4; breeder: Winkelmann, Borchen-Kirchborchen) which inhaled 10 mg powdered aluminium/m^3 4 h per day, 5 days per week from August 16 to October 27, 1967 for a total of 51 days. Fixed on January 15, 1968 under methitural anaesthesia by intratracheal instillation of 2.5 % glutaraldehyde in phosphate buffer (pH 7.4) before opening the thorax. Postfixation with 1 % osmium tetroxide in phosphate buffer (pH 7.4). Contrasted en bloc for 12 h with 0.5 % uranyl acetate in 70 % ethanol. Embedded in a 2:8 mixture of methyl and butyl methacrylate. Sectioned at 50 nm. Lead citrate after REYNOLDS (1963). Plate 45/05

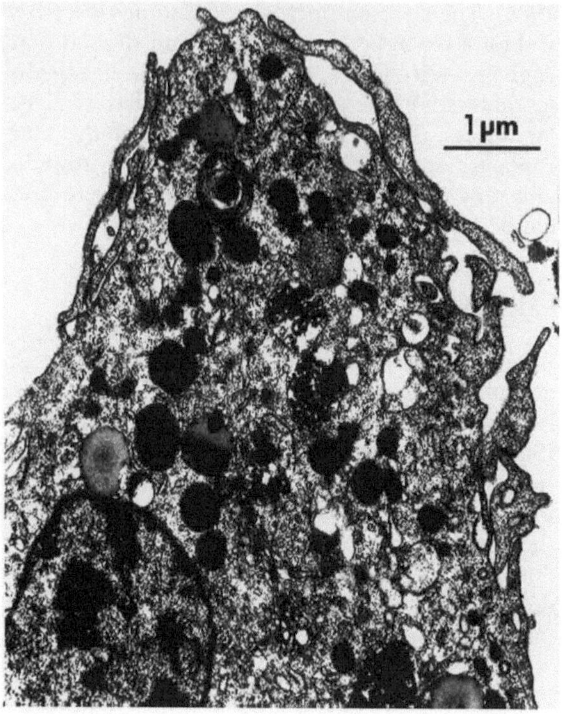

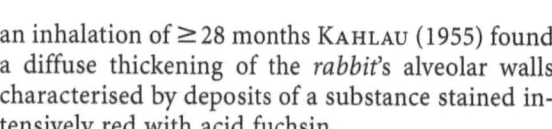

Fig. 130. Alveolar macrophage of a female *rat* (No. 4; breeder Winkelmann, Borchen-Kirchborchen) which inhaled 10 mg powdered aluminium/m^3 4 h per day, 5 days per week from August 16 to October 27, 1967 for a total of 51 days. Fixed on January 15, 1968 under methitural anaesthesia by intratracheal instillation of 2.5 % glutaraldehyde in phosphate buffer (pH 7.4) before opening the thorax. Postfixation with 1 % osmium tetroxide in phosphate buffer (pH 7.4). Contrasted en bloc for 12 h with 0.5 % uranyl acetate in 70 % ethanol. Embedded in a 2:8 mixture of methyl and butyl methacrylate. Sectioned at 50 nm. Lead citrate after REYNOLDS (1963). Plate 38/10

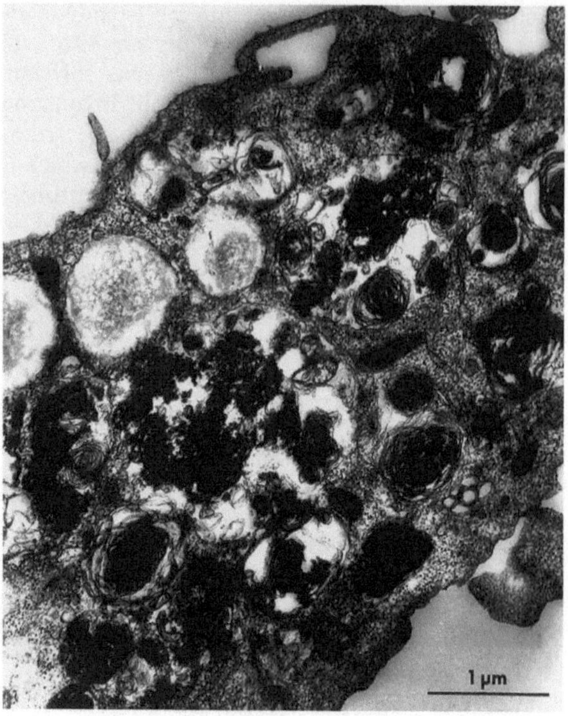

Fig. 131. Alveolar macrophage with myelin-figure inclusions and complex phagolysosomes from a 296 g female *rat* (No. 3; breeder Winkelmann, Borchen-Kirchborchen) which inhaled 10 mg powdered aluminium/m^3 4 h per day, 5 days per week from August 16 to October 27, 1967 for a total of 51 days. Fixed on October 30, 1967 under methitural anaesthesia by intratracheal instillation of 2.5 % glutaraldehyde in phosphate buffer (pH 7.4) before opening the thorax. Postfixation with 1 % osmium tetroxide in phosphate buffer (pH 7.4). Contrasted en bloc for 12 h with 0.5 % uranyl acetate in 70 % ethanol. Embedded in a 2:8 mixture of methyl and butyl methacrylate. Sectioned at 50 nm. Lead citrate after REYNOLDS (1963). Plate 19/07

an inhalation of ≥28 months KAHLAU (1955) found a diffuse thickening of the *rabbit*'s alveolar walls characterised by deposits of a substance stained intensively red with acid fuchsin.

10 mg powdered aluminium/m^3 (for particle size see Table 8) inhaled by female *rats* over a period of 10 weeks and a half (4 h per day for a total of 51 days) was phagocytosed by alveolar macrophages and became included within phagosomes (Fig. 128). Secondary lysosomes contain electron dense reaction products of Al^{3+} with cellular constituents. They are unchanged for several months, but their membranes will be disrupted finally. The number of free ribosomes is increased and so is the activity of the Golgi apparatus. Microtubules are well preserved, but there are also some macrotubules (Fig. 133). Microfilaments are preferably seen in the ruffled border. Some nuclei show elliptical vacuoles or fibrillary degeneration (Fig. 134).

9.6.2.1.2 Americium

Americium has the atomic number 95 and the valence electrons are on the 5f^7 shell. Half life of ^{241}Am is 462 years.

At concentrations of <0.125 µCi $^{241}AmO_2$/ml (4.63 kBq/ml), Beagle *dog* alveolar macrophages obtained by bronchoalveolar lavage did not suffer significant changes of their viability and their ability to phagocytose opsoninized *sheep* erythrocytes (TAYA and MEWHINNEY 1987). Cell killing started at 0.5 µCi ^{241}Am/ml (18.5 kBq/ml), whereas the reduction of phagocytic ability started at 0.125 µCi ^{241}Am/ml at 20 h after exposure. After 72 h, 10 % of the $^{241}AmO_2$ particles (physical diameter 0.1 to 1 µm) were dissolved *in vitro*.

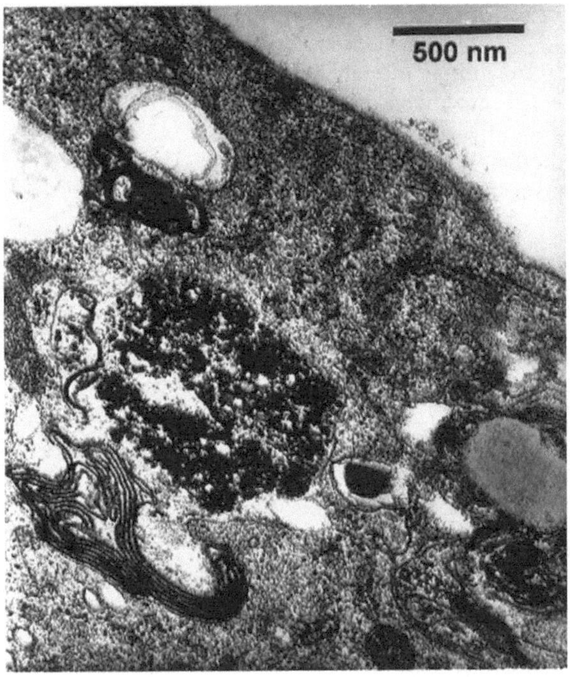

Fig. 132. Complex phagolysosomes and myelin figures in an alveolar macrophage of a female *rat* (No. 4; breeder Winkelmann, Borchen-Kirchborchen) which inhaled 10 mg powdered aluminium/m^3 4 h per day, 5 days per week from August 16 to October 27, 1967 for a total of 51 days. Fixed on January 15, 1968 under methitural anaesthesia by intratracheal instillation of 2.5 % glutaraldehyde in phosphate buffer (pH 7.4) before opening the thorax. Postfixation with 1 % osmium tetroxide in phosphate buffer (pH 7.4). Contrasted en bloc for 12 h with 0.5 % uranyl acetate in 70 % ethanol. Embedded in a 2:8 mixture of methyl and butyl methacrylate. Sectioned at 50 nm. Lead citrate after Reynolds (1963). Plate 43/10

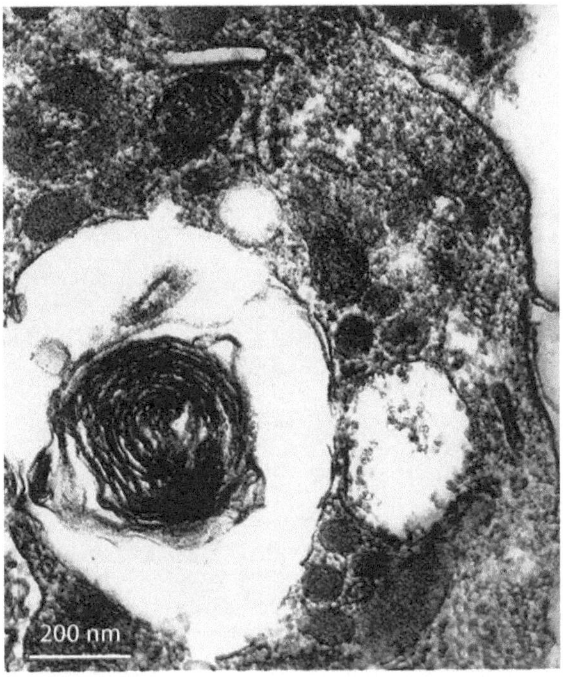

Fig. 133. Macrotubule in an alveolar macrophage from a 296 g female *rat* (No. 3; breeder Winkelmann, Borchen-Kirchborchen) which inhaled 10 mg powdered aluminium/m^3 4 h per day, 5 days per week from August 16 to October 27, 1967 for a total of 51 days. Fixed on October 30, 1967 under methitural anaesthesia by intratracheal instillation of 2.5 % glutaraldehyde in phosphate buffer (pH 7.4) before opening the thorax. Postfixation with 1 % osmium tetroxide in phosphate buffer (pH 7.4). Contrasted en bloc for 12 h with 0.5 % uranyl acetate in 70 % ethanol. Embedded in a 2:8 mixture of methyl and butyl methacrylate. Sectioned at 50 nm. Lead citrate after Reynolds (1963). Plate 18/07

9.6.2.1.3 Antimony

Antimony has the atomic number 51 and the valence electrons are on the $5s^2 5p^4$ shells. Its oxidation states are −3, +3 and +5.

Inhalation of 1200 mg Sb_2O_3 dust/m^3 (average particle size 1.1 µm) in *rats* and 900 mg Sb_2O_3 dust/m^3 in *hamsters*, respectively, was associated with a significant increase in the mean macrophage count; whereas the inhalation of Fe_2O_3 or of SiO_2 dusts did not result in a such significant increase in either species (Gross et al. 1969).

Zymosan-stimulated alveolar macrophages isolated from *rabbits* treated with Freund's adjuvant *in vitro* released less $O_2^{\bullet-}$ when challenged with 200 µg/10^6 cells of a waste incinerator dust sonicated five times for 15 min (Gulyas et al. 1990). Decrease correlated best with logarithms of antimony, lead and arsenic contents and with logarithms of particle numbers and dust surfaces.

9.6.2.1.4 Arsenic

Arsenic has the atomic number 33 and the valence electrons are on the $4p^3$ shell. Its oxidation states are −3, +3 and +5.

Pulmonary alveolar macrophages lavaged from As(V)-exposed male Sprague-Dawley *rats* showed significant increases in $O_2^{\bullet-}$ production and in basal release of TNF-α, while As(III) did not induce significant alterations (Lantz et al. 1995). Neither arsenical inhibited prostaglandin E_2 production.

9.6.2.1.5 Beryllium

Beryllium has the atomic number 4 and the valence electrons are those of the $2s^2$ shell. +2 is its only oxidation state. Beryllium tends to form polynuclear complexes.

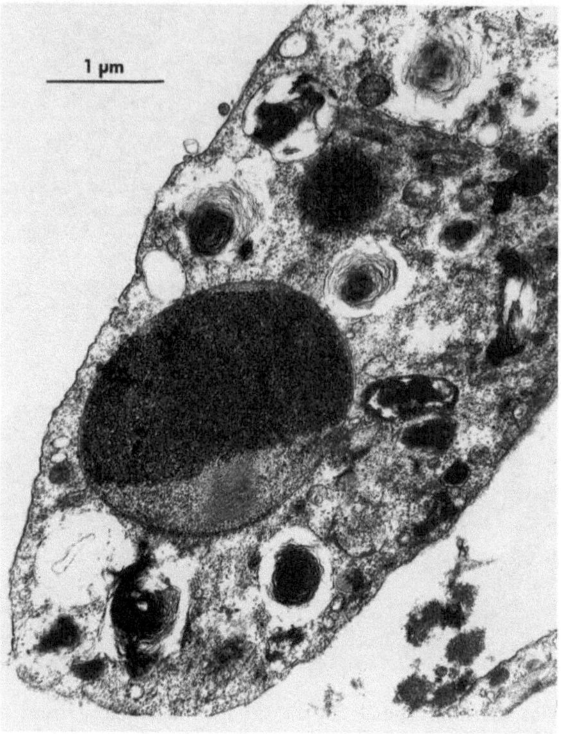

Fig. 134. Fibrillary nuclear degeneration in an alveolar macrophage of a female *rat* (No. 4; breeder Winkelmann, Borchen-Kirchborchen) which inhaled 10 mg powdered aluminium/m^3 4 h per day, 5 days per week from August 16 to October 27, 1967 for a total of 51 days. Fixed on January 15, 1968 under methitural anaesthesia by intratracheal instillation of 2.5 % glutaraldehyde in phosphate buffer (pH 7.4) before opening the thorax. Postfixation with 1 % osmium tetroxide in phosphate buffer (pH 7.4). Contrasted en bloc for 12 h with 0.5 % uranyl acetate in 70 % ethanol. Embedded in a 2:8 mixture of methyl and butyl methacrylate. Sectioned at 50 nm. Lead citrate after REYNOLDS (1963). Plate 45/09

Table 31. Effects of beryllium on enzyme activities

Enzyme	EC	Authors
Tryptophan pyrrolase	1.13.11.11	WITSCHI and MARCHAND (1971)
Glutamic-oxalacetic aminotransferase	2.6.1.1	VACHER et al. (1975)
Glutamic-pyruvate aminotransferase	2.6.1.2	VACHER et al. (1975)
Alkaline phosphatase	3.1.3.1	KLEMPERER et al. (1949)
β-Glucuronidase	3.2.1.31	VACHER et al. (1975)

Beryllium has been shown to induce an enhanced pathological and antibody response similar to Freund's adjuvant (SALVAGGIO et al. 1965).

Beryllium Metal

In the *guinea pig*, intratracheal instillation of 75 mg beryllium metal after 3 months induced focal lymphocytic and large mononuclear cell reaction immediately about visible deposits of the substance (VORWALD 1950).

Beryllium Oxide

In *Macaca irus* monkeys exposed to beryllium oxide (3.30 to 4.38 mg Be/m^3) three times, for 30 min each, at monthly intervals CONRADI et al. (1971) after 2 years found histiocytes in the pulmonary interstitial space containing heterogeneous dark granular material and lipid droplets within membrane-bounded space.

After intratracheal instillation of beryllium oxide to *guinea-pigs*, POLICARD (1950) found hypertrophied histiocytes, sometimes 30 to 40 μm in diameter with voluminous nuclei (20–25 μm), irregular wrinkled, abundant chromatin and several large nucleoli.

Beryllium Hydroxide

In the *guinea pig*, intratracheal instillation of 150 mg beryllium hydroxide after 4 months induced focal lymphocytic and large mononuclear cell reaction with proliferation of local tissue cells and mild fibrosis (VORWALD 1950).

Beryllium Sulphate

Beryllium sulphate (84.3 mg $BeSO_4 \cdot 4\ H_2O/m^3$) inhalation in *rats* induced rather marked pulmonary oedema accompanied by a moderate infiltration of neutrophils and monocytes (SPRAGUE et al. 1950).

H36.12j clonally derived, C57BL/6N-*mouse* hybrid precursor macrophages were found useful to evaluate the mechanisms by which Be stimulates macrophage cytokine production, and by which T cell derived IFN-γ amplifies TNF-α production in granulomatous disease (SAWYER et al. 2000). The response was maximal at 100 μM $BeSO_4$ and did not occur when 12j cells were stimulated with either aluminium sulphate of cobalt sulphate. Beryllium stimulated the production of 725±25 pg/ml TNF-α protein by 12j cells as measured by ELISA of culture supernatants after 24 h. As measured by RT-PCR, Be-stimulated 12j cell TNF-α protein production was accompanied by an increased intracellular TNF-α mRNA at 3 and 24 h. The addition of 10 U or 100 U of recombinant-Mu-IFN-γ to Be-stimulated 12j cells further increased TNF-α production 1.5–4 fold (1.6±0.1 ng/ml) respectively. Bacterial lipopolysaccharide (LPS, 1 μg/ml) stimulated production of TNF-α in 12j culture supernatants after 6 h (515±151 pg/ml).

Recombinant-Mu-IFN-γ (10 U) + beryllium (100 μM $BeSO_4$)-stimulated H36.12j macrophages produced 1195±225 pg TNF-α/ml, while H36.12j cells stimulated with rMu-IFN-γ by itself produced only 284±31 pg TNF-α/ml (HAMADA et al. 2000).

Beryllium Stearate

12 Months after intratracheal instillation of 150 mg beryllium stearate, the alveolar spaces of the *guinea pig* contained scattered dust laden macrophages in various stages of degeneration (VORWALD 1950).

Zinc Beryllium Manganese Silicate

In the *guinea pig*, zinc beryllium manganese silicate inhalation (8 h/d) for 40 months induced mild lymphocytic and large mononuclear cell reaction indicative of slight irritation about dust particles (VORWALD 1950).

9.6.2.1.6 Cadmium

Cadmium has the atomic number 48 and the valence electrons are those of the $4d^{10}5s^2$ shells. +2 is its only oxidation state. The co-ordination number in its complexes is 4.

Table 32. Effects of cadmium on enzyme activities

Enzyme	EC	Authors
Lactate dehydrogenase	1.1.1.27	HART (1986)
Catalase	1.11.1.6	ZIKIC et al. (1998)
Glutathione peroxidase	1.11.1.9	ZIKIC et al. (1998)
Lysine oxidase	1.13.12.2	CHICHESTER et al. (1981)
Proline hydroxylase	1.14.11.2	CHICHESTER et al. (1981)
Superoxide dismutase	1.15.1.1	ZIKIC et al. (1998)
Alkaline phosphatase	3.1.3.1	HART (1986)
Acid phosphatase	3.1.3.2	HART (1986)
Na^+-K^+-ATPase	3.6.1.3	LAI et al. (1980)
Mg^{2+}-ATPase	3.6.1.3	LAI et al. (1980), HYPPÖNEN et al. (1993)

Cadmium Oxide

In specific-pathogen-free male Wistar *rats*, focal areas of epithelial hyperplasia, a mononuclear interstitial infiltrate, and increased numbers of alveolar macrophages were observed after a single 3-h exposure to an aerosol of insoluble CdO (BUCKLEY and BASSETT 1987).

In male Sprague-Dawley *rats* instilled intratracheally with 5 mg CdO suspended in 0.5 ml saline one week before, MURTHY et al. (1982) noted two types of alveolar macrophages. The first showed a significant increase in size when compared to those seen in control animals. The cytoplasm of these cells was filled with numerous small vacuoles which contained bizarre-shaped electron-dense aggregations. Some vacuoles seemed to coalesce with others which resulted in larger vacuoles. The mitochondria were smaller in size and fewer in number when compared with controls. Some mitochondria were phagocytized in membrane-bound autophagosomes. A distinct nucleolus was lacking and chromatin material seemed to be fragmented and scattered. The second type of alveolar macrophages was smaller in size and their nucleus contained a prominent nucleolus. The cytoplasm contained fewer vacuoles, more profiles of rough endoplasmic reticulum, and mitochondria which were larger than those seen in the first type.

In alveolar macrophages from male Sprague-Dawley *rats* intratracheally intubuated with a suspension (0.5 ml) of 0.1 mg CdO plus 0.2 mg CuO particles, MURTHY and HOLOVACK (1991) found large aggregates of multivesicular bodies.

Cadmium Sulphide

In conventionally kept Wistar *rats* into which 15 mg cadmium sulphide (Merck, Darmstadt) had been injected intratracheally six times at weekly intervals, KISSLER (1983) observed pulmonary fibrosis of varying intensity from 2½ weeks after the last administration. Many alveolar septa were thickened by connective tissue. They did not show any inflammatory infiltrates. They were coated with continuous rows of isoprismatic type II pneumocytes with foamy cytoplasm. Numerous alveolar macrophages with foamy cytoplasm were seen from the beginning; the majority of them had phagocytosed CdS particles well recognised by their bright yellow colour. Abundant fine-granular but otherwise unstructured eosinophilic material was found in varying amounts between the phagocytes. Acid phosphatase reaction is intensified in intraalveolar macrophages only (KISSLER and SCHERBECK 1981, KISSLER 1983).

Cadmium Telluride

Cadmium telluride (CdTe) particles (mean diameter 1.70 μm) suspended in sterile, phosphate-buffered saline, pH 7.2, and instilled intratracheally into specific pathogen-free female Sprague-Dawley *rats* at doses of 12 mg, 24 mg, 48 mg, or 96 mg/kg body weight after 3 days had induced an intaalveolar inflammatory exudate consisting of neutrophils, macrophages, fibrin, and protein (MORGAN et al. 1995). A minimal amount of black cadmium telluride particles was present.

Cd^{2+}

Goat alveolar macrophages gained by bronchoalveolar lavage of the isolated lung and exposed to ≥0.04 mM $CdCl_2$ for 20 h showed strongly reduced viability ($P < 0.01$). The lactate dehydrogenase activity of the supernatant culture medium was elevated ($P < 0.01$). Phagocytic potential had decreased ($P < 0.01$). In cultures exposed to 0.20 mM $CdCl_2$ macrophages had withdrawn their pseudopodia, as seen by the scanning electron microscope (WASEEM et al. 1993).

In Long-Evans *rats*, cadmium ions depressed the oxidative burst of alveolar macrophages during phagocytosis of zymosan (CASTRANOVA et al. 1980): 10^{-3} M Cd^{2+} given *in vitro* for 15 min significantly inhibited oxygen consumption (by 76±1%), glucose metabolism (by 83 ± 7%), and release of active oxygen species (by 73 ± 1%). This cadmium-induced inhibition of oxidative processes of macrophages was postulated to diminish their antibacterial activity.

The oxidative metabolism of NMRI *mouse* peritoneal macrophages induced by phorbol myristate acetate was enhanced by $CdCl_2$, the zymosan-induced oxidative metabolism was markedly reduced as measured by means of the lucigenin-enhanced chemiluminescence in a Berthold Biolumat 9500 up to 120 min (HILBERTZ et al. 1986).

After a single incubation of *mouse* resident peritoneal macrophages for 18 h in medium containing 10 μM Cd^{2+}, a five-fold increase in **•NO production**, with respect to incubation in medium alone, was observed (RAMIREZ and GIMENEZ 2000). After 18 h of culture, NO inductor agents induced significant expression of inducible nitric oxide synthase (iNOS) and •NO production in peritoneal macrophages. When cells were simultaneously incubated for 18 h in medium containing 10 μM Cd^{2+} and NO inductor agents, an increase in response to lipopolysaccharide (1 μg/ml), okadaic acid (50 nM) and 12-*O*-tetradecanoylphorbol-13-acetate (200 nM), and a decrease in the synergistic effect of lipopolysaccharide/okadaic acid, in relation to culture in medium without Cd^{2+} were observed.

9.6.2.1.7
Carbon

While the surface of **diamonds** from volcanic chimneys is hydrophobic, that from secondary deposits is hydrophilic due to the formation of oxides (BOEHM et al. 1964). Powdered diamonds with oxides on their surface show a higher wetting heat than those without. Treating diamonds oxidised on their surface with water forms carboxyl groups. Estimation of active hydrogen after ZEREWITINOFF or by deuterium exchange revealed that nearly all hydrogen atoms were carboxylic (SAPPOK and BOEHM, quoted from BOEHM 1966). Diamonds cleared of oxides at 10^{-5} to 10^{-6} torr in vacuo at 800 °C do not show an essential resonance absorption (11 μval free radicals/g diamond).

In the layers of the **graphite** lattice the atoms are combined with three neighbours by σ-bindings

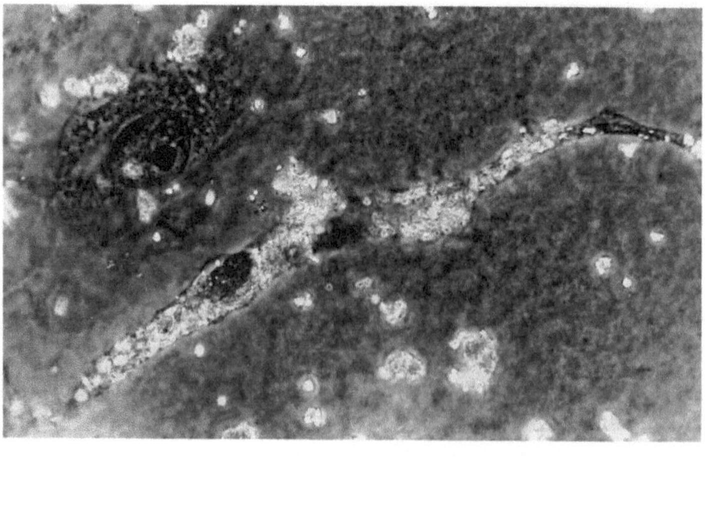

Fig. 135. Alveolar macrophages having engulfed powdered diamond. 3-Day old culture of a 3-day old *mouse*. The preparation was fixed in Carnoy's fluid, stained with iron hematoxylum after Weigert, and mounted in toto in Rhenohistol®. Eye-piece Leitz Periplan 4×; objective Leitz Pv Fl oil 70/1.15. Positive phase contrast (from SCHILLER 1954)

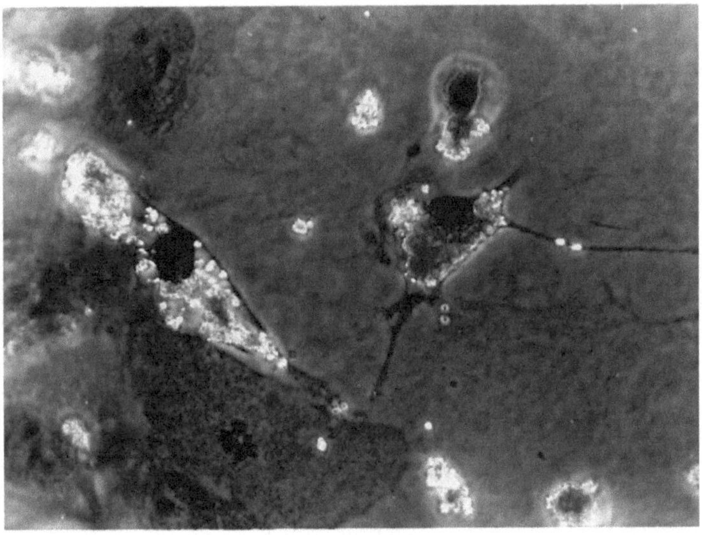

Fig. 136. Alveolar macrophages having engulfed powdered diamond. 4-Day old culture of a 3-day old *mouse*. The preparation was fixed in Carnoy's fluid, stained with iron hematoxylum after Weigert, and mounted in toto in Rhenohistol®. Eye-piece Leitz Periplan 4×; objective Leitz Pv Fl oil 70/1.15. Positive phase contrast (from SCHILLER 1954)

(sp^2-hybridisation). The fourth valence electrons are bound as π-electrons which can easily change from one atom in the layer to another. The layers are held together by Van der Waals' forces (distance 0.335 nm).

Twenty and 30 s following intravenous injection, BURKE and SIMON (1970) localised **carbon** predominantly in tiny clumps in the sinuses of the marginal zone and the red pulp of the *rabbit*'s spleen. At this stage some carbon was already being phagocytosed by macrophages arising from the Billroth cords.

In vitro, *human* alveolar macrophages collected by bronchoalveolar lavage and loaded with small masses (0.03–3 μg/10^6 cells) of ultrafine particle aggregates induced a dose-related impairment of both the attachment and the ingestion processes of fluorescein-labelled silica particles (3 μm) with a marked impairment down to a carbon particle dose around 0,2 μg/10^6 cells (LUNDBORG et al. 2001). Incubation with interferon-γ (12.5 U/ml) also induced significant impairments in both the attachment and the ingestion processes. Loading with carbon further aggravated the effects of interferon-γ. In contrast to earlier studies in *rat* alveolar macrophages (LUNDBORG et al. 1999), interferon-γ did not impair the oxidative metabolism (nitroblue tetrazolium reduction) at rest in these *human* alveolar macrophages; instead the oxidative metabolism was increased. This difference was due to a difference between *rat* and *human* alveolar macrophages and not between *rat* and *human* interferon-γ.

Unlike the characteristic progressive nature of silica-related lesions, all of respirable-sized **carbon fibre**-induced inflammatory effects were reversible within 10 days after exposure and any significant histopathologic effects were observed at any time post-exposure (WARHEIT et al. 1994).

9.6.2.1.8 Cerium

Cerium has the atomic number 58 and the valence electrons are those of the $4f^2 6s^2$ shells. The oxidation states of Ce are +3 and +4.

Cerium forms complexes with riboflavin (SEKHON and CHOPRA 1974). Cerium chloride (2 mg/kg *rat* intravenously) impaired the activities of cytochrome P-450 and NADPH-cytochrome *c* reductase of *rat* liver, which could be prevented by α-tocopherol acetate (100 mg/kg×day i.p.) pre-treatment (ARVELA 1974). Ce^{3+} promoted the binding of creatine kinase to Cibacron blue F_3 GA, the substrate analogue of the enzyme, even in the absence of Mg^{2+}, its native cofactor (SHIVAKUMAR et al. 1989).

In a lung biopsy of a female aged 31 years, who had polished spectacle lenses with cerium oxide for three years, SINICO et al. (1982) found multiple macrophagic granulomas containing dust particles identified by analytical electron microscopy as cerium oxide.

Table 33. Effects of cerium on enzyme activities

Enzyme	EC	Author(s)
Glucose-6-phosphate dehydrogenase	1.1.1.49	ARVELA (1974)
NADPH-cytochrome *c* reductase	1.6.99.3	ARVELA (1974)
Creatine kinase	2.7.3.3	SHIVAKUMAR et al. (1989).

9.6.2.1.9
Chromium

Chromium has the atomic number 24 and the valence electrons are those on the $3d^5 4s^1$ shells. The oxidation states of chromium include +2, +3 and +6.

$$[CrO_4]^{2-} + 4\,H_2O + 3\,e^- \rightarrow Cr(OH)_3(s) + 5\,OH^- \quad [175]$$

$$2\,[CrO_4]^{2-} + H_3O^+ \rightarrow [Cr_2O_7]^{2-} + 3\,H_2O \quad [176]$$

$$Cr(VI) + 3\,V(VI) \rightarrow Cr(III) + 3\,V(V) \quad [177]$$

$$Cr(VI) + V(IV) \rightarrow Cr(V) + V(V) \quad [178]$$

$$Cr(V) + NADH \rightarrow Cr(III) + NAD \quad [179]$$

Cr^{VI} is formed from Cr^{III} and from metallic Cr by oxidation, primarily by ozone. In gas shielded arc welding, ozone itself is formed by the interaction of oxygen and ultra-violet radiation emitted from the arc.

Hydrogen peroxide or superoxide form a paramagnetic complex in the reaction with chromium(VI) oxide in an alkaline water solution at room temperature (LAGERCRANTZ 1999). The complex $[Cr(OH)_5O_2]^{5-}$ with the g-value equal to 1.9734 is believed to contain hydroxyl groups derived from the alkaline solution and dioxygen derived from H_2O_2 or $O_2^{\bullet-}$.

Cr^{III} supplementation prevented the increase in TNF-α levels and oxidative stress caused by high levels of glucose (30 mM) in cultured U937 monocytic cells (JAIN and KANNAN 2001). Similarly, chromium supplementation prevented elevated TNF-α secretion and lipid peroxidation levels in H_2O_2-treated U937 cells.

Alveolar macrophages from both *guinea pig* and *man* tolerated Cr(VI) salts up to 4.97 μg/ml (RATSCHEK et al. 1988). An increase in the number of macrophages in the bronchoalveolar lavage fluid was seen in *rats* inhaling $Na_2Cr_2O_7$ for 22 h per day on 30 or 90 consecutive days (GLASER et al. 1990).

Chromite

Chromium ore (chromite, $FeO \cdot Cr_2O_3$) has generally been absolved of playing in causal role in chromium-induced malignancies as chromite miners allegedly were free from lung cancers and because chromium ore was thought to be harmless on account of its insolubility in the body (KOELSCH 1938). A similar type of reasoning was advanced in placing the responsibility on zinc chromate as the cause of lung cancer among chrome pigment workers because it is more soluble than barium or lead chromate (GROSS and KOELSCH 1943). MANCUSO and HUEPER (1951) added powdered chromium ore to citrated *human* blood for 16 h at room temperature and found 0.000 and 0.001 mg Cr/ml plasma, respectively, or 0.10 and 0.12 mg Cr/ml washed cells, respectively. In genotoxic studies Cr(III) was absolutely inactive unless a direct interaction with purifies DNA was permitted by the test conditions (BIANCHI et al. 1983).

Macrophages outgrowing from *mouse* lung explants eagerly engulfed chromite dust particles (2 μm) without suffering any damage detectable by phase contrast microscopy (WORTH and SCHILLER 1955). In the *rabbit*, inhaled Cr^{3+} reduced the capacity of alveolar macrophages to catabolize surfactant (JOHANSSON et al. 1992).

9.6.2.1.10
Coal Dust

Coal is characterised by a non-uniform molecular structure within which aromatic structural units can be distinguished. They are composed of one to five condensed or directly joined aromatic rings substituted by heterocyles and aliphatic (mainly methyl) groups. Nitrogen is involved in pyridine and pyrrole rings and a part of sulphur in thiophene rings. C- and O-alkylation may occur by the reductive alkylation reactions of coal. LAZAROV et al. (1984) presented a structural model which fits the values of the structural parameters obtained by ^1H and ^{13}C NMR spectroscpy of alkylated coal.

H_2O_2 was formed when H_2SO_4 was adsorbed on **charcoal** in the presence of freshly adsorbed O_2, the yield increasing with the concentration of H_2SO_4 (BURSHTEIN and FRUMKIN 1941). The quantity of H_2O_2 reached a maximum in about 10 min, and then decreased because of the catalytic decompensation on charcoal. The yield was very low when inert gases instead of O_2 were bubbled through the solution. The formation of H_2O_2 stopped then the adsorption of the acid was complete. At a polarized cathode in the presence of freshly adsorbed O_2, the yield of H_2O_2 as the concentration increased from 6.5×10^{-5} to 130×10^{-5} A per 0.1 g of charcoal with further increase of the concentration, the yield decreased. The yield was little dependent on the composition of the solution.

$$C_xO_2 + 2\,H^+ + 2\,X^- \rightarrow C_x^{2+}(X^-)_2 + H_2O_2 \quad [180]$$

Phagocytosis by *mouse* macrophages of dust particles from a **coal** miner's lung (Fig. 53) is not only a local reaction between the particle and the cell membrane, but a reaction in which both size and shape of the macrophage is changed (Figs. 137, 138).

Coals have an appreciable concentration of free radicals in their native form, presumably formed in

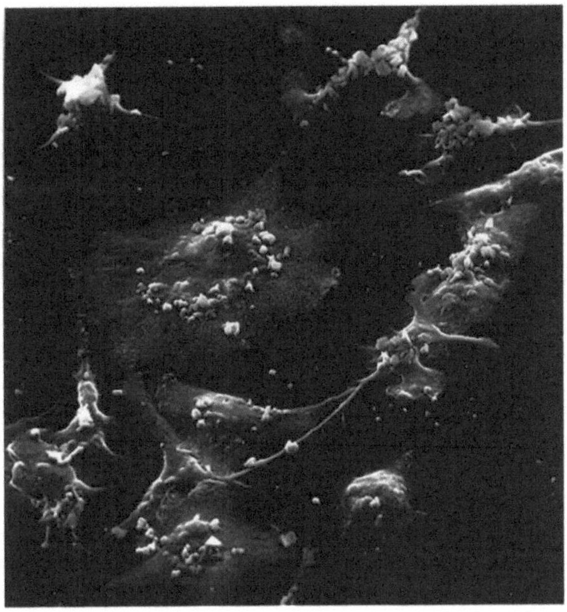

Fig. 137. Peritoneal macrophages from a SPF-NMRI *mouse* cultured in a Leighton tube after CHANG (1964) 4 h after addition of dust particles from a coal miner's lung (No. 17 of EINBRODT 1965, aged 38 years, underground for 15 years), digested by formamide. Fixed with 1 % glutaraldehyde in 0.07 M phosphate buffer (pH 7.4). Ethanol, amyl acetate. Critical point drying. Gold coating. Cambridge Stereoscan Mark 2 A operated at 20 kV, negative 78-168/1. ×600 (from SCHILLER 1982)

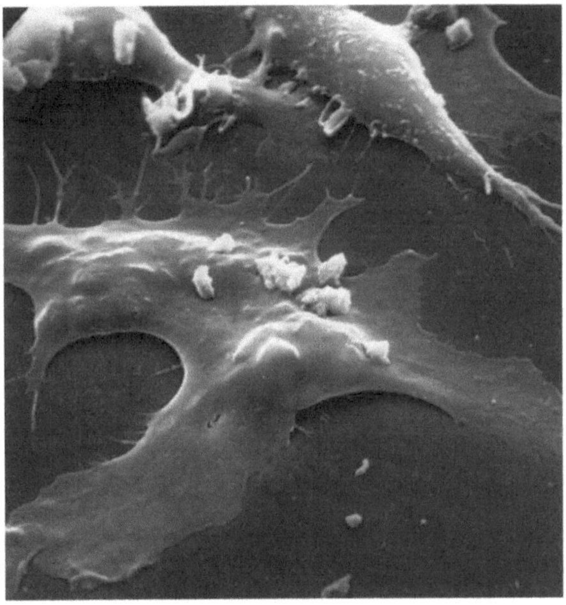

Fig. 138. Peritoneal macrophages from a SPF-NMRI *mouse* cultured in a Leighton tube after CHANG (1964) 4 h after addition of dust particles from a coal miner's lung (No. 17 of EINBRODT 1965, aged 38 years, underground for 15 years), digested by formamide. Fixed with 1 % glutaraldehyde in 0.07 M phosphate buffer (pH 7.4). Ethanol, amyl acetate. Critical point drying. Gold coating. Cambridge Stereoscan Mark 2 A operated at 20 kV, negative 78-166/15. ×2,400 (from SCHILLER 1982)

the "coalification" process. In addition, during crushing of coking coal paramagnetic centres are formed and their concentration increases with the milling time (LEBEDEV et al. 1978). These centres are **free radicals** produced by breaking chemical bonds with an unpaired electron on the primary and secondary carbon atoms.

The nature and concentration of free radicals present in a given coal are determined by rank, thermal history, and particle size (PETRAKIS and GRANDY 1978).

All the coal mine dust samples from Utah Blind Canyon and Western Virginia Pittsburgh seam examined by DALAL et al. (1995) generated varying levels of $HO^•$ radicals from H_2O_2 in the presence of a $HO^•$ spin trap 5,5-dimethyl-1-pyrroline-*N*-oxide. $HO^•$ radical generation by the coal from H_2O_2 was effectively inhibited by deferoxamine and catalase, but only partly inhibited by superoxide dismutase. Metal chelators DETAPAC and EDTA enhanced the radical generation. The $HO^•$-generating potential of all coal dusts showed a positive correlation with the surface iron content of coal mine dust.

Instillation of 500 mg generic anthracite dust No. 867 (3.6 % silica; 7 µm maximum diameter) and generic bituminous dust No. 1361 (0.7 % silica; 7 µm maximum diameter), respectively, suspended in 50 ml PBS into the right caudal lung lobe of 3 female pigtail macaque monkeys (*Macaca nemestrina*) each using a flexible fibreoptic bronchoscope increased the numbers of macrophages in the dust exposed lobes and significantly elevated the *N*-acetyl-β-D-glucosaminidase activity within the bronchoalveolar lavages taken at 2-weeks intervals for 12 weeks thereafter (MACK et al. 1995).

In the *hamster*, a single intratracheal instillation of 4 mg bituminous coal dust increased both alveolar macrophages and neutrophils in broncho-alveolar lavage fluid (KLEINERMAN and IP 1990). The elastolytic activity of the culture fluid in which the macrophages were sustained was increased. The total concentration of the anti-proteases in the alveolar fluid, both α_1 protease inhibitor and α_2 macroglobulin were not significantly different from control values. Multiple coal dust instillations performed at 5–7 day intervals over 4–5 weeks showed an increase of neutrophils and macrophages in the broncho-alveolar lavage fluid obtained 3 and 90 days after the last instillation as compared with controls. The total elastolytic activity of the leucocytes was greater at day 3 than at day 90. However, the α_1 protease inhibitor and α_2 macroglobulin concentrations in the alveolar fluid were also increased as compared to controls at both 3 and 90 days following the last dust instillation. The data indicate a simultaneous increase in the elastolytic burden and

Table 34. Petrologic components (macerals) in coal and their groupings

Maceral groupings	Macerals or components
Vitrinite	telinite
	collinite
Exinite	resinite
	cerinite
	sporinite (exinite)
	cutinite
	suberinite
	alginite
Inertinite	massive micrinite
	granular micrinite
	sclerotinite
	semifusinite
	fusinite

in the protease inhibitor activity of the alveolar fluid. Emphysema was not present by histologic study suggesting that no significant imbalance between proteolytic and anti-proteolytic forces occurred as a result of the coal dust treatments.

Fenton's method of very fine grinding of the coals, followed by fractionation in liquids of differing specific gravity, resulted in **exinite** fractions rich in sporinite. BENT and BROWN (1961) noted that an absorption band at about 1,700 cm^{-1}, attributed to a carbonyl group, and an increase in non-aromatic C–H absorption, were the only significant differences between the spectra of lower rank sporinites and the dominant maceral of normal coals, vitrinite. The occurrence of margins and larger areas of higher reflectance than the main mass of resinite globules has frequently been observed on polished surfaces of coal examined under oil immersion in incident light (MURCHISON 1963). These more highly reflecting regions have been attributed by STACH (1962) to the effects of oxidation on the resin before it was finally incorporated in the peat in a reducing environment.

In dusts isolated by EINBRODT from the fibrotic lungs of coal miners SCHULZ et al. (1995) found a positive correlation of the amount of **inertinite** to the stage of fibrosis.

Fusinite, a coal component discernible from semifusinite and other opaque ingredients by its strong reflecting power in reflected light was used as a electron paramagnetic resonance (EPR) oxygen probe. It is highly inert in biological systems (VAHIDI et al. 1994, SANTINI et al. 1998). This stability is very important since fusinite particles are randomly internalised in the cytoplasm and may remain there for long periods of time (PETRAKIS and GRANDY 1978). From the physical point of view, fusinite has an exchange-narrowed EPR spectrum which can be broadened by interaction with oxygen principally through the adsorption of molecules of this gas to the fusinite surface. The feasibility of using this EPR probe in measuring the intracellular molecular oxygen concentration in cultured cells has been demonstrated by VAHIDI et al. (1994) and SANTINI et al. (1995, 1996).

The question of the biological importance of substances leached from coals of different rank (gasflame coal to anthracite) was first put forward by THAER (1952, 1953, 1955). SORENSON et al. (1974) analysed the concentrations of Cd, Cu, Fe, Ni, Pb and Zn in bituminous coals from mines with differing incidences of coal workers' pneumoconiosis. ROSMANITH et al. (1977) thought that Cd, Pb and other trace elements might be engaged in the fibrogenicity of mineral coal. KOBER et al. (1976) found that humic substances leached from Pennsylvania and Utah coals with high and low pathogenicities, respectively, differed in the quantities of material removed and in their metal-binding and galactose oxidase (EC 1.1.3.9) inhibiting activities.

Polycyclic aromatic hydrocarbons contained in coal dust were not eluted by *pig* lung homogenates or *human* gastric juices, nor were they capable of crossing the skin barrier *in vitro* (FOÀ et al. 1998).

In U937 cells and *human* macrophages aryl hydrocarbon receptor, with its partner cofactor aryl hydrocarbon receptor nuclear translocator, was expressed and CYP1A1 mRNA expression was induced in the presence of aryl hydrocarbon receptor ligand 3-methylcholanthrene (KOMURA et al. 2001).

Coal workers' pneumoconiosis is associated with increased basal and PMA-stimulated $O_2^{\bullet-}$ formation in their alveolar macrophages in comparison to control subjects (WALLAERT et al. 1990). In 96 coke oven workers NADIF et al. (1997) found erythrocyte catalase activity significantly higher and erythrocyte glutathione peroxidase activity significantly lower than in 105 randomly recruited non-exposed workers from a power plant.

Differences in the cytotoxicity (triphenyltetrazolium chloride method) of coal dusts to *guinea-pig* macrophages depend on coal rank expressed as the percentage of volatile matter (REISNER and ROBOCK 1977). The Horst strata, with a mineral content of nearly 50 % (by weight), caused a 35 % de-

Table 35. Classification of coal by rank

	Volatile matter [%]	Strata
Flame coal	>40	Dorsten
Gasflame coal	40–35	Horst
Gas coals	35–28	Essen
Bituminous coals	28–19	Bochum
Steam coals	19–12	Witten
Semi-anthracites	12–10	Sprockhövel
Anthracites	<10	

pression of the biological activity (TTC-RA = 65 %). Similar depressions were caused by upper Essen strata dusts with 33 % mineral; by lower Essen and Bochum strata dusts (mineral ≅ 20 %) and by Witten strata dusts (mineral ≅ 15 %).

As measured by electron spin resonance spectroscopy, a freshly ground anthracite coal (95 % carbon) produced greater concentration of free radicals than a bituminous coal (72 % carbon), and the radical reactivity was also greater for the anthracite (Dalal et al. 1989). The reactivity of the newly produced radicals in the anthracite dust correlated with the dust's toxicity. Furthermore, similar coal-based free radicals were detected in the lung tissue of autopsied coal miners, suggestive of persistent reactivity by the embedded coal dust leading to the progressive disease process.

In the *rat*, long-term clearance of low rank coal dust (Lea Hall, Staffordshire) tended to be greater than that of high rank (Ffeldau, South Wales), although a peculiar macrophage reaction to low rank dust may have affected its removal from the lung (Heppleston et al. 1971).

The particulate matter in the lungs of *rats* exposed to aerosols (6.6 ± 1.9 mg/m^3 and 14.9 ± 6.2 mg/m^3) of high-potential coal workers' pneumoconiosis bituminous coal dust from a mine in Appalachia for up to 20 months was contained almost entirely within the cytoplasm of alveolar macrophages so burdened with the material that cellular detail was obscured (Busch et al. 1981).

Interferon induction by influenza virus in *monkey* kidney (LLC-MK_2) cell monolayers pre-treated with coal dust was inhibited in relation to coal rank (Hahon 1983). Maximal inhibition of viral interferon induction was noted with high rank coal and the degression of this activity was related to coal's position in the carboniferous series: anthracite, bituminous, subbituminous, lignite, and peat. Adsorption of poly(4-vinylpyridine-*N*-oxide) to bituminous coal dust from the Pittsburgh seam, Cambria County, Pa., not only occurred at a more rapid rate than to cell monolayers, but also less polymer was required to pretreat coal dust than cell monolayers to achieve comparable amelioration of interferon production (Hahon 1976).

The exact consequences of the cellular mechanisms that occur in the lungs of subjects chronically exposed to coal dust are still much of a puzzle (Schins and Borm 1999).

9.6.2.1.11 Cobalt

Cobalt has the atomic number 27 and the valence electrons are those of the $3d^7 4s^2$ shells. The oxidation states of cobalt are +2 and +3.

Most **complex compounds** of cobalt are of oxidation state +3 and six-co-ordinate, or of oxidation state +2 and four- or six-co-ordinate. Five-co-ordinate complexes are the anions $[Co(CN)_5]^{3-}$ and $[Co(CN)_5]^{2-}$, and the oxygen-carrying Schiff's base compounds described by Earnshaw et al. (1963). The readiness of cobalt to form five-co-ordinate compounds can be related to the tendency of the metal to achieve (cobalt (I), $3d^8$) or approach (cobalt (II), $3d^7$) the effective atomic number of the next inert gas (Nyholm 1961). The correlation between the co-ordination number, stereochemistry, spectra, and magnetism of bivalent cobalt complexes has been widely investigated. The crystal-field stabilisation energy of the $d_\gamma^4 d_\varepsilon^{3-}$ configuration is favourable for a tetrahedral arrangement and in many compounds of the type $MX_2(Hal)_2$ where X can be, for example, pyridine (Gill and Nyholm 1961). In general, the ligands of high electronegativity, such as water, favour a co-ordination number of six, whereas those which are more easily polarised, such as the halide ions, give rise to four-co-ordinate tetrahedral complexes.

As models for natural oxygen carriers, many cobalt(II) complexes coordinated with nitrogen-bases have been shown to reversibly interact with molecular oxygen to form CoO_2 or Co_2O_2 complexes depending upon the nature of the ligands.

Hearon (1947) showed that a co-ordination complex formed between **histidine** and cobaltous ion possesses the ability to combine reversibly with molecular oxygen. In contrast with this, histidine complexes with other transition metals tested (Cu^{2+}, Cu^+, Ni^{2+}, Fe^{2+}, Mn^{2+}) failed to exhibit the property of combining reversibly with oxygen gas. The Co^{2+}-histidine complex does not readily, if ever, become oxidised to Co^{3+} from the presence of O_2 (Hearon 1948). Cobalt complexes of various histidine derivatives: histamine, 1-methyl histidine, 3-benzyl histidine, dibenzyl histidine, and the peptides anserine, carnosine, L-alanyl-L-histidine, and α-L-aspartyl-D-histidine combined with Co^{2+} and the resulting complex was capable of carrying oxygen reversibly, although in certain cases (most notably carnosine) the irreversible reaction sets in exceedingly rapidly even at room temperature (Hearon et al. 1949).

Co(Salen) catalyses the oxygenation of 3-substituted indoles and flavonols via a proton transfer process, representing a model for tryptophan 2,3-dioxygenase (EC 1.13.11.11) or indoleamine 2,3-dioxygenase (heme enzyme) and quercetinase (EC 1.13.11.24; Cu enzyme). The 3-substituted indoles give formylkynurenine-type products (Nishinaga 1975, Nishinaga et al. 1981).

Phosphate buffer solutions of carnosine and anserine produced active oxygen species as measured by bleaching of *N,N*-dimethyl-4-nitrosoaniline (HARTMAN and HARTMAN 1992). The formation of active oxygen species appeared to depend predominantly on the presence of metal ions contaminating the phosphate buffer since prior Chelex treatment of buffer substantially reduced bleaching activity. Activity was restored by addition of micromolecular levels of Cu(I), Cu(II) and to a lesser extent, by Co(II) but not by addition of ferrous or ferric ions. Catalase eliminated most activity but superoxide dismutase had little effect.

One-electron transfer catalysed by cobalt

$$O_2 + Co^{2+} \leftrightarrow \mathbf{Co^{3+} - O_2^{\bullet-}} \rightarrow \mathbf{O_2^{\bullet-}} + Co^{3+} \quad [145]$$

$$\mathbf{O_2^{\bullet-}} + Co^{2+} + 2\,H^+ \rightarrow H_2O_2 + Co^{3+} \quad [146]$$

$$H_2O_2 + Co^{2+} \rightarrow \mathbf{HO^{\bullet}} + HO^- + Co^{3+} \quad [147]$$

Cobalt(II) reacts with $\mathbf{H_2O_2}$, to produce $\mathbf{HO^{\bullet}}$, $^1O_2$, and metal-oxygen complexes, which cause site-specific DNA damage (YAMAMOTO et al. 1989, KAWANISHI et al. 1994).

Though essential in forming cobalamin (vitamin B_{12}), cobalt is an **industrial poison** taken up from the catalyst in the Fischer-Tropsch synthesis (FISCHER and PICHLER 1936) of benzene and in the sintered carbide metal industry (SCHILLER 1958, 1961, BECH et al. 1962, KÜHNE 1962, 1965, COATES and WATSON 1971, HARTUNG et al. 1982, HILLERDAL and HARTUNG 1983, FIGUEROA et al. 1992, LISON 1996). When both cobalt and tungsten carbide particles are associated, electrons provided by cobalt metal are easily transferred to the surface of the carbide particles where reduction of oxygen can occur at a rate greatly increased (LISON et al. 1995). Further, cobalt compounds are used as dryers in paints (e.g. cobalt linoleates), and as pigments in glass and pottering industries.

Co^2 is effective in stimulating cyclic **GMP dependent protein kinase** from a variety of sources, with optimal concentration of 0.5 mM, equivalent to a K_m of about 0.1 mM (KUO and SHOJI 1982). The maximal activity stimulated by Co^{2+} is about 40%–80% of that by Mg^{2+} for the *mammalian* cGMP dependent protein kinase (KUO et al. 1976, SHOJI et al. 1977a, b). Co^{2+} stimulates *mammalian* cAMP dependent protein kinase to a greater extent than does Mg^{2+} (KUO et al. 1970). 5 mM Co^{2+} almost completely inhibits the activity of *bovine* lung cGMP stimulated protein kinase stimulated by 10 mM Mg^{2+} (GILL et al. 1977).

Alveolar macrophages of *rabbits* exposed for 4 months (5 days/week, 6 h/day) to 0.6 ± 0.5 mg/m^3 Co^{2+} as $CoCl_2$, had significantly more laminated, surfactant-like inclusion, compared with the controls (JOHANSSON et al. 1992).

The **carbonyls** of cobalt, like other metal carbonyls [see $Ni(CO)_4$], are the most toxic cobalt compounds. $Co_2(CO)_8$, $Co_4(CO)_{12}$ and $Co_6(CO)_{16}$ are solid substances which are prepared from cobalt and carbon monoxide using elevated temperature and pressure.

Inhalation of cobalt hydrocarbonyl, $HOCo(CO)_4$, for 30 min established a LD_{50} of 165 mg Co/m^3 (PALMES et al. 1959).

Cobalt-Nickel Alloy

The toxicity of cobalt and cobalt-nickel alloys in experimental hard metal pneumoconiosis is well documented. Metallic cobalt powder instilled intratracheally in the lungs of *rats* had an acute irritant action and lead to severe alterations in capillaries (HARDING 1950, SCHILLER 1958, 1961). The solubility of cobalt in plasma is some 500 times greater than in saline, but a fairly large (10 ml) intraperitoneal dose of plasma saturated with cobalt was without evident effects on the *rat* (HARDING 1950).

In *rabbits* exposed to soluble cobalt 0.4–2 mg/m^3 $CoCl_2$ for 1–4 months, the number of alveolar macrophages was increased and some of them, close to the type II cell nodules, were enlarged and filled with laminated, surfactant like inclusions (JOHANSSON et al. 1983, 1986). In lavage fluids the number of macrophages and the percentage of these cells with smooth surface and intracellular surfactant-like inclusions were more increased in group Co^{2+} + Cr^{3+} than in group Co^{2+} as were oxidative metabolic and phagocytic activities of the macrophages (JOHANSSON et al. 1992).

In hard-metal workers, giant cells have been found in the lung lavage (DAVIDSON et al. 1983, TABATOWSKI et al. 1988).

9.6.2.1.12 Copper

Copper has the atomic number 29 and the valence electrons are those on the $3d^{10}4s^1$ shells. Its oxidation states are +1 and +2. The difference of 1 e^- in the 3*d* shells of cuprous and cupric ions allow a participation in radical reactions. Copper does not really fit the definition of a transition element since its 3*d*-orbitals are full, but it readily forms the Cu^{2+} ion by loss of two electrons, one from the 4*s*- and one from the 3*d*-orbital. This leaves an unpaired electron.

One-electron transfer catalysed by copper

$$O_2 + Cu^+ \rightarrow \mathbf{O_2^{\bullet-}} + Cu^{2+} \quad [181]$$

$$\mathbf{O_2^{\bullet-}} + Cu^+ + 2\,H^+ \rightarrow H_2O_2 + Cu^{2+} \quad [182]$$

$$H_2O_2 + Cu^+ \rightarrow \mathbf{HO^{\bullet}} + HO^- + Co^{2+} \quad [183]$$

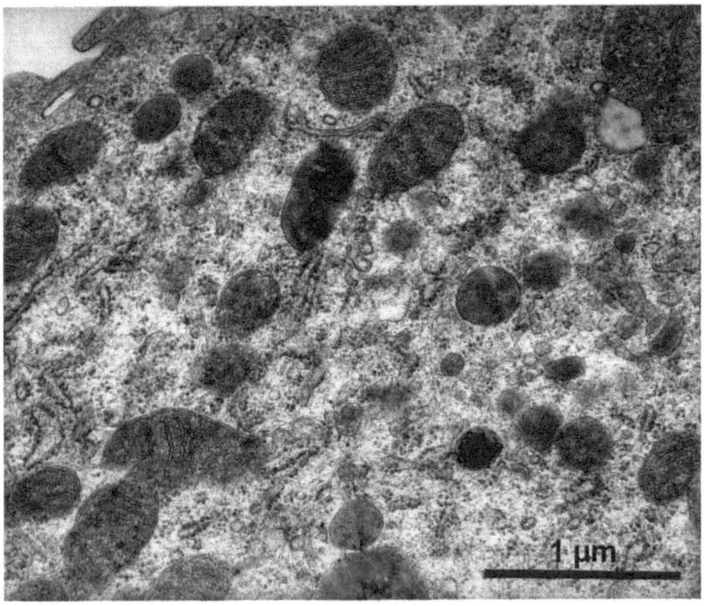

Fig. 139. Alveolar macrophage with a tubular invagination from a 220 g female white *rat* (No. 1 breeder: Winkelmann, Borchen-Kirchborchen) which inhaled 10 mg of a powdered cobalt-nickel alloy/m³, used in the production of hard metal by Deutsche Edelstahlwerke in Krefeld, 4 h per day, 5 times per week from October 2–28, 1967 for a total of 19 days. Fixed on October 30, 1967 under methitural anaesthesia by intratracheal instillation of 2.5 % glutaraldehyde in phosphate buffer (pH 7.4) before opening the thorax. Postfixation with 1 % osmium tetroxide in phosphate buffer (pH 7.4). Contrasted en bloc for 12 h with 0.5 % uranyl acetate in 70 % ethanol. Embedded in a 2:8 mixture of methyl and butyl methacrylate. Sectioned at 50 nm. Lead citrate after REYNOLDS (1963). Plate 16/01

As with iron, copper serves to convert superoxide ($O_2^{\bullet -}$) and hydrogen peroxide, formed during the metabolism of oxygen (particularly in the presence of redox-cycling xenobiotics; RUMYANTSEVA and WEINER 1988, MASON 1990) to the highly reactive hydroxyl radical (equations [181]–[183]), which is considered to be the oxidising species responsible for the induction of biomolecular damage.

Copper-dependent hydroxyl radical formation has been demonstrated in many model systems,

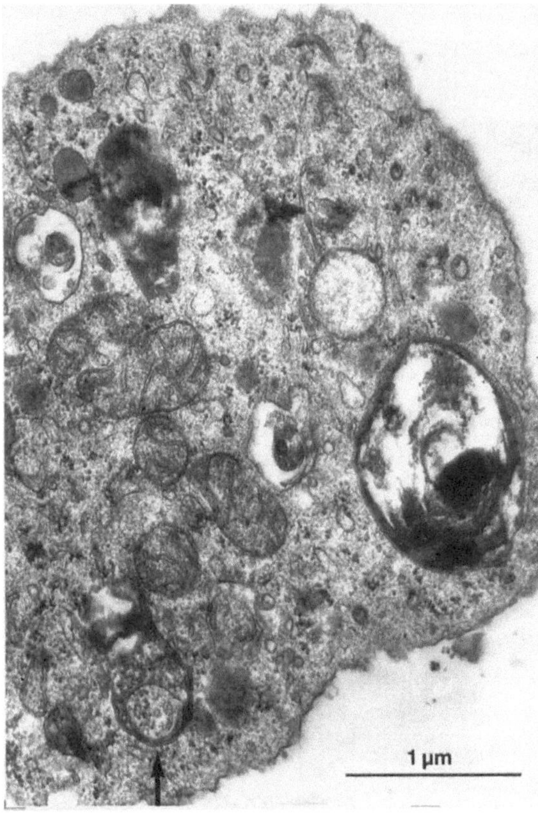

Fig. 140. Phagophore (*arrow*) in an alveolar macrophage from a 225 g female white *rat* (breeder: Winkelmann, Borchen-Kirchborchen) which inhaled 10 mg of a powdered cobalt-nickel alloy/m³, used in the production of hard metal by Deutsche Edelstahlwerke in Krefeld, 4 h per day, 5 times per week from October 2–28, 1967 for a total of 19 days. Fixed on October 30, 1967 under methitural anaesthesia by intratracheal instillation of 2.5 % glutaraldehyde in phosphate buffer (pH 7.4) before opening the thorax. Postfixation with 1 % osmium tetroxide in phosphate buffer (pH 7.4). Contrasted en bloc for 12 h with 0.5 % uranyl acetate in 70 % ethanol. Embedded in a 2:8 mixture of methyl and butyl methacrylate. Sectioned at 50 nm. Lead citrate after REYNOLDS (1963). Plate 35/11

generally via the application of standard methods of $HO^{\bullet}$ determination. VAN STEVENINCK et al. (1985) have used salicylate hydroxylation in systems containing H_2O_2, Cu^{II} and reducing agents (equation [184]). However, the most direct method of hydroxyl radical detection is electron spin resonance (ESR) spectroscopy in conjunction with a spin trapping reagent (BUETTNER and MASON 1990, MASON et al. 1994). The $HO^{\bullet}$ radical was detected as its adduct to the spin trap 5,5-dimethyl-1-pyrroline *N*-oxide (DMPO) (equation [8]).

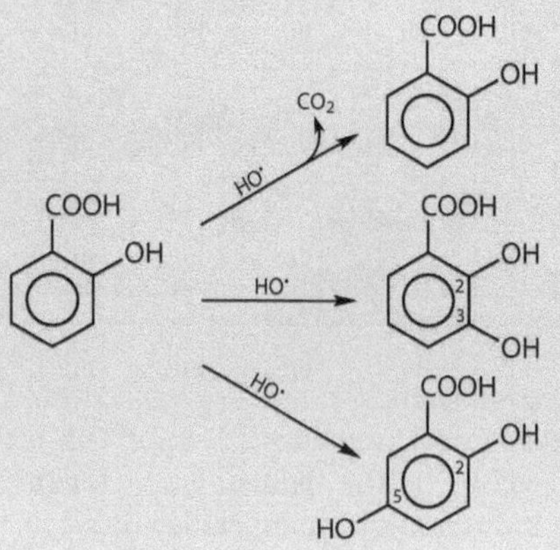

Salicylate hydroxylation by HO· [184]

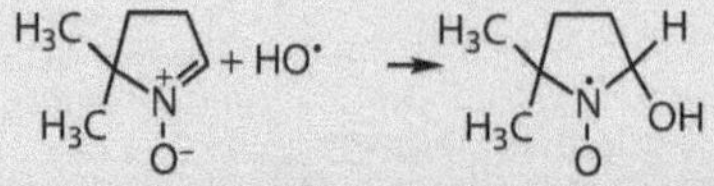

5,5-**Dim**ethyl-1-**p**yrroline-1-**o**xide-HO· adduct [8]

The possible involvement of Cu(III), Fe(IV), or Mn(III) in the mechanism of the toxicity of $O_2^{\bullet-}$ is not surprising as the redox potential of the couple $O_2^{\bullet-}/H_2O_2$ at pH 7.4 is 0.94 V (KOPPENOL and BUTLER 1985), which implies a strong oxidising capability for $O_2^{\bullet-}$ (GOLDSTEIN and CZAPSKI 1990). However, there is no direct evidence for the formation of a Cu^{III} species obtainable from electron spin resonance spin trapping studies (BURKITT et al. 1995).

Tetrakis-μ-3,5-diisopropylsalicylatodiaquodicopper(II) is known to disproportionate $O_2^{\bullet-}$ at the same rate as superoxide dismutase (SORENSEN 1989).

Copper bound to 1,10-phenanthroline was found to be a good catalyst of hydroxyl radical formation: the Cu^{II} (1,10-phenanthroline)$_2$ complex can be reduced by H_2O_2 and the Cu^{I} (1,10-phenanthroline)$_2$ generated reacts with the peroxide to form $HO^{\bullet}$.

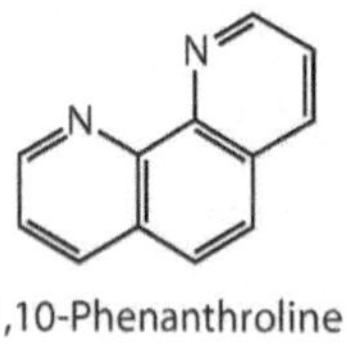

1,10-Phenanthroline [137]

The rate constant for the Fenton reaction is higher for copper than for iron (HALLIWELL and GUTTERIDGE 1989).

2,9-Dimethyl-1,10-phenanthroline (10 μM) significantly increased the inhibitory effects of both $CuSO_4$ and Cu(2,9-dimethyl-1,10-phenanthroline)$_2$ NO_3 on NADH oxidase activity (SMIT et al. 1982).

A transition complex with triplet molecular oxygen may also be formed:

$$Cu^+ + {}^3\Sigma g^- O_2 \rightarrow (Cu^{2+} - O_2^{\bullet-}) \quad [185]$$

Copper and its co-ordination compounds are known to produce severe cytotoxicity (FREEDMAN et al. 1986, BYRNES et al. 1990).

Tetrahedral co-ordination dominates with two and three co-ordination much less common. The co-ordination number rarely exceeds four although five co-ordination is known (GAGNÉ 1976).

In solutions of $CuCl_2$ and adenine copper can be bound to adenine. Two Cu(adenine)$_2$ complexes $[Cu(C_5H_4N_5)_2]^{2+}$/Cu $(C_5H_4N_2)_2)$ are in equilibrium with free adenine (BRUSTON et al. 1999). Copper-adenine complexes present a catalytic activity (e.g., H_2O_2 disproportionation into O_2 and water) but depending on complex concentration H_2O_2 also strongly oxidises the adenine within the complexes. Raman spectroscopy quantifies copper-adenine complex formation and H_2O_2 consumption.; polarography quantifies O_2 production. As for C_{40} catalase, optimal catalytic capacities depend on physiological conditions, such as pH and temperature. The comparative analysis of kinetic parameters shows that the affinity for H_2O_2 of Cu(adenine)$_2$ is 37-fold lower than that of C_{40} catalase and that the molar activity for O_2 production is 200-fold weaker for Cu(adenine)$_2$ than for the enzyme. In the 10^{-6} to 10^{-3} M range, the strong decrease of activity with raising complex concentration is explained by aggregation or stacking, which protects Cu(adenine)$_2$ entities from H_2O_2 oxidation, but also decreases O_2 production.

Glutathione and Cu^{2+} form several complexes (PETIT et al. 1975). Reduction by glutathione triggered a *N,N'*-bis(2-pyridylmethylene)-1,4-butanediamine (*N,N',N",N'"*)-Cu(II) diperchlorate-supported redox cycle with oxygen yielding H_2O_2 (STEINKÜHLER et al. 1991). Whereas reduction by ascorbate was reversible.

Under physiological conditions of pH, peptide (class A) chelates are only formed by those bidentate amide ligands with X being either an imidazole (sp^2) nitrogen or a terminal (sp^3) amino nitrogen. Mercaptide sulphur must also be considered to belong to this group of strong copper(II) binding sites.

The apolipoprotein B of low-density lipoprotein has two sulphydryl groups exposed to its surface and they could play a role in the reductive activation of the transition metal ions, e.g. Cu^{2+}. Once bound, Cu^{2+} must be reductively activated by a net transfer of one electron:

$$Cu^{2+} + e^- \rightarrow Cu^+ \quad [186]$$

The formation of hydroxyl radicals ($HO^•$) by the reaction of the Cu(II) complex of ethylenediamine tetraacetic acid (EDTA) with H_2O_2 in the presence of biological reductants, such as L-ascorbic acid and L-cysteine, has been demonstrated by ESR spectroscopy using water-soluble spin-traps (Ozawa et al. 1992).

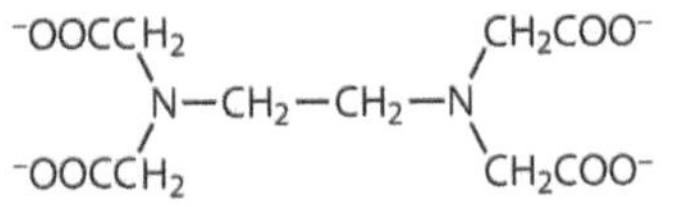

Ethylenediamine tetraacetic acid (EDTA) [187]

Thiourea (1–10 mM) provided marked and dose-dependent protection against protein oxidation in three copper-containing systems: Cu(II)/ascorbate, Cu(II)/H_2O_2, and Cu(II)/H_2O_2/ascorbate [Cu(II), 0.1 mM; ascorbate, 1 mM; H_2O_2; 1mM] (Zhu et al. 2002). In contrast, only minor protection was observed with dimethyl sulphoxide and mannitol, even at concentrations as high as 100 mM. Strong protection was also observed with dimethylthiourea, but not with urea or dimethylurea. Thiourea also significantly inhibited copper-catalysed oxidation of ascorbate, and competed effectively with histidine and 1,10-phenanthroline for binding of cuprous, but not cupric, copper, as demonstrated by both UV-visible and low temperature electron spin resonance measurements.

Both phospholipid-containing delipidated *human* low-density lipoprotein ghosts and trilinolein-reconstituted low-density lipoprotein were devoid of antioxidants and were extremely susceptible to 2,2'-azobis-(2-amidino propane) hydrochloride-induced oxidation but, paradoxically, were rather resistant to copper-mediated oxidation (Visioli et al. 2000). The dynamic reduction of Cu(II) to Cu(I) was quantitatively decreased in low-density lipoprotein ghosts and in trilinolein-reconstituted low-density lipoprotein, also lacking the initial rapid reduction and the subsequent inhibition phases, due to the absence of endogenous antioxidants. Conversely, the rate of copper reduction was linear and likely due to lipid peroxides, either already present of formed during copper-induced oxidation.

Ternary copper(II) complexes constituted by a 1,10-phenanthroline and an amino acid as ligands accelerated both the decomposition of peroxynitrite and its nitration of 4-hydroxyphenylacetic acid at pH > 7 (Ferrer-Sueta et al. 1997). The enhancing effect of Cu^{2+} on the decomposition of *N*-ethyl-*N*-nitrosourea (Preussmann et al. 1975) can be inhibited by complex formation with ethylenediamine tetraacetic acid (Zeller and Ivancovic 1972).

The formation of oxidising species from bolus H_2O_2 and cupric ions was found to be profoundly dependent on the choice of buffer. In phosphate and cacodylate buffers, as well as unbuffered water, Sandström et al. (1994) found barely detectable amounts of methane sulphinic acid. However, in Hepes buffer they could detect a strong formation of methane sulfinic acid. The addition of 2-[(2-bis[carboxymethyl]amino-5-methylphenoxy)-methyl]-6-methoxy-8-bis[carboxymethyl]-aminoquinolin (quin2) inhibited 97 % of the methane sulfinic acid formation. With the deoxyribose assay, a significant 2-thiobarbituric acid reaction was found in all media tested, the strongest in cacodylate buffer and water, and weaker in Hepes and phosphate. In this assay quin2 inhibited the formation of oxidising species by 65 % in phosphate buffer, 85 % in water and more than 95 % in both Hepes and cacodylate buffers.

Cu^{2+} (2–10 µM) catalysed the conversion of xanthine dehydrogenase (EC 1.1.1.204) to xanthine oxidase (EC 1.1.3.22) which was prevented by oleic acid, arachidonic acid, eicosapentaenoic acid and docosahexaenoic acid (50–200 µM) as shown in the perfused *rabbit* liver (Fujita et al. 1995).

The contribution of Esterbauer and Ramos (1996) on the chemistry and pathophysiology of oxidation of low-density lipoprotein focuses mainly on *in vitro* oxidation of LDL by Cu^{2+}.

LDL was able to reduce Cu^{2+} to Cu^+ in a time and concentration-dependent manner (Proudfoot et al. 1997a). Blocking of free –SH groups on LDL apoprotein B by preincubation with dithionitrobenzoic acid had no significant effect on the rate and extent of Cu^{2+} reduction. Consumption of tocopherol in LDL undergoing oxidation with Cu was very rapid (rate = 6×10^{-10} M s^{-1}). When Cu^+ formed during incubation with LDL was complexed with neocuproine, there was significant inhibition of LDL oxidation, as indicated by lipid peroxide formation and mobility on agarose gel electrophoresis. Tocopherol consumption was even more rapid in the presence of neocuproine, consistent with a shift in Cu^{2+}/Cu^+ equilibrium and faster reduction of Cu^{2+} by α-tocopherol.

$$LOOH + Cu^{2+} \rightarrow Cu^+ + LOO^• + H^+ \quad [188]$$

$$Cu^{2+} + \alpha\text{-Toc}H \rightarrow Cu^+ + \alpha\text{-Toc}^• + H^+ \quad [189]$$

$$\alpha\text{-Toc}^• + LH \rightarrow \alpha\text{-Toc}H + L^• \quad [190]$$

When oxidising isolated LDL, there was a decrease in lag time with increasing concentration of Cu^{2+} until a minimum "lag time" was reached at a Cu:LDL ratio of about 50:1 (Proudfoot et al. 1997b). In serum, an initial decrease in "lag time" occurred with increasing Cu concentrations up to 12.5 μM. However, with higher Cu concentrations "lag time" to oxidation increased, contrary to expectation, until a maximum was reached at about 50 μM Cu. This dose response observed for Cu oxidation of diluted serum was highly reproducible in a number of individual subjects. When serum was gel-filtered to remove low molecular weight compounds, the resulting filtrate behaved the same as isolated LDL. Uric acid was found to be an important component of the low molecular weight fraction for the paradoxical effect of Cu concentration on serum oxidation. The same paradoxical effect was found when isolated LDL was incubated with uric acid in the presence of *human* serum albumin and Cu. The incubation of *human* serum albumin with reducing agents such as uric acid or bilirubin in the presence of high Cu concentration, produced a "peroxidase-like" activity, capable of breaking down hydrogen peroxide as well as lipid hydroperoxides.

The oxidation of low-density lipoprotein by *mouse* resident peritoneal macrophages or that initiated by Cu^{2+} ions was inhibited by SB209995, a metabolite of carvedilol in *human*, with IC_{50}s of 59 nmol/l and 1.7 μmol/l, respectively (Feuerstein and Yue 1994). Under the same conditions, the IC_{50}s of carvedilol were 3.8 and 17.1 μmol/l, respectively.

There are at least two pathways by which copper is transported across the cell membrane (DiDonato and Sarkar 1997). Ceruloplasmin, the most abundant copper-protein in plasma, can contribute copper to cells (Hsie and Frieden 1975, Campbell et al. 1981, Dameron and Harris 1987, Mas and Sakar 1992, Saenko et al. 1994). Studies of ceruloplasmin-mediated copper transport in cells have shown that copper derived from ceruloplasmin enters the cell but the protein does not (Percival and Harris 1990). Stimulation of copper uptake by ascorbate and the inhibition of the process by cuprous chelators suggest that copper takes place in the form of Cu(I) rather than Cu(II) (Percival and Harris 1989, Harris 1991).

Ceruloplasmin is a plasma glyoprotein containing about seven Cu atoms per molecule and ≈95% of the total plasma Cu (Rydén 1984, Fox et al. 1995). It is a 132-kDa monomer comprised almost entirely of three major domains that have 35–40% homology to each other (Ortel et al. 1984) and to three homologous domains in factors Va and VIIIa (Church et al. 1984). The prooxidant site was localised to a region containing His_{426} because ceruloplasmin$_{His426A}$ almost completely lacked prooxidant activity whereas the other mutants expressed normal activity (Mukhopadhyay et al. 1997).

Copper is distributed in all cellular organelles including the nucleus, mitochondria, lysosomes, endoplasmic reticulum and cytosol (Linder 1991). Lindquist (1968) presented biochemical and electron microscopic evidence that $CuCl_2$ injected intraperitoneally into rats was sequestered by liver lysosomes and initiated lipid peroxidation of lysosomal membranes with resulting lysosomal rupture and release of hydrolytic enzymes. Glutathione has been implicated as the main chelator of copper in the cytoplasm. In copper-loaded hepatoma cells, over 60% of cytosolic copper exists as a Cu(I)-GSH complex (Freedman et al. 1989). ^{1}H NMR of copper(II) complexes of adenosine (2') (3') (5')-monophosphates have been compared with the spectra of adenosine 2',3'- and 3',5'-cyclic phosphates. A comparison of the chemical shifts and broadening effects with varying copper(II) concentrations suggested that greater base stacking and no phosphate-base deshielding occurred in the cyclic nucleotides relative to the nucleoside monophosphates (Berger and Eichhorn 1971). Activation of guanylate cyclase in a post-mitochondrial supernatant fraction of *rat* lung homogenates was found to be dependent on temperature, Cu^{2+}, and the presence of oxygen (White et al. 1976). The activation was inhibited by thiols, bovine serum albumin, KCN, and sodium diethyldithiocarbamate, a Cu^{2+} chelator.

Viral **interferon** induction was inhibited in *rhesus monkey* kidney (LLC-MK_2) cells treated with 0.1 mg copper at 49.4% (Hahon et al. 1980). Interferon yields were maximally inhibited by the smallest particles (1 μm). Particles of increased size (10 μm) were progressively ineffectual in depressing interferon production.

Alveolar macrophages from male Sprague-Dawley *rats* instilled intratracheally with 5 mg CuO suspended in 0.5 ml saline one week before showed significant increase in size and contained numerous vacuoles (Murthy et al. 1982). Their polymorphic nuclei with margination of the chromatin and well developed nucleoli were eccentrically placed. Lipid droplets in the cytoplasm were numerous, and often appeared to fuse to form large vacuoles. Other cytoplasmic constituents included numerous bizarre-shaped membranous bodies, myelin figures, and crystalloid-like inclusions. The mitochondria appeared smaller when compared with those of the control animals. The plasmalemma displayed considerable undulations and numerous cytoplasmic processes.

Copper as a DNA-Cleaving Agent

By the change of valences, copper salts can act as a Fenton reagent in the presence of hyrogen peroxide (von Sonntag 1987). Cu^{2+} and H_2O_2 induced strong cleavage of both *calf* thymus DNA and *mammalian* chromatin at quite low concentrations (10^{-6} and 10^{-5} M) (Sagripanti and Kraemer 1989, Yamamoto and Kawanishi 1989) in a reaction that appears to involve Cu^{2+}. In contrast to studies with iron ions, complexation of copper with either nitrilotriacetate or with *N,N,N',N'*-ethylenediaminetetraacetic acid inhibits the damage to DNA bases. Cu^{2+} can bind to the DNA (Yamamoto and Kawanishi 1989), but whether the mechanism goes via the hydroxyl radical (Sagripanti and Kraemer 1989) or some sort of metal bound complex (Yamamoto and Kawanishi 1989) is uncertain. Studies directly on Cu^+/H_2O_2 suggested that free hydroxyl radicals are not formed but may remain bound to the copper (Johnson et al. 1985). Drouin et al. (1996) induced base modifications by Cu(II)/ascorbate/H_2O_2 distinguishing base damage from frank strand break. Modified base production predicted by computer simulation at an initial Cu(II) concentration of 50–70 μM was experimentally validated. The copper ion binding sites on DNA were saturated at 50 μM bound copper ion, or when ≈40% of the DNA phosphates were occupied. The k_{app} indicated that DNA base damage occurred slowly in relation to the rate of DNA-Cu(I) oxidation (Stoewe and Prutz 1987).

$$\text{DNA—Cu}^{2+} + \text{Asc}^{2-} \rightarrow \text{DNA—Cu}^{+} + \text{Asc}^{\bullet-} \qquad [191]$$

$$\text{DNA—Cu}^{+} + \text{H}_2\text{O}_2 \rightarrow \text{DNA—Cu}^{2+} + \text{OH}^- + \text{HO}^\bullet \qquad [192]$$

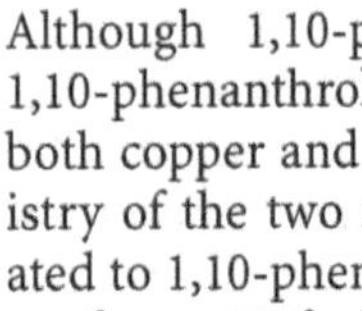

1,10-Phenanthroline [137]

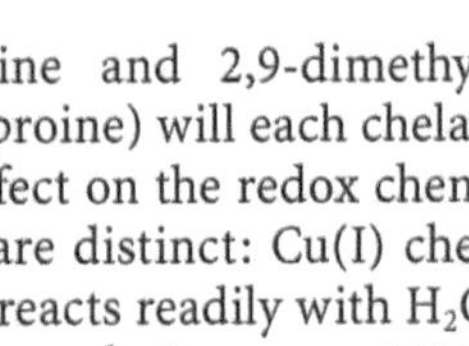

2,9-Dimethyl-1,10-phenanthroline (neocuproine) [193]

Although 1,10-phenanthroline and 2,9-dimethyl-1,10-phenanthroline (neocuproine) will each chelate both copper and iron, the effect on the redox chemistry of the two metal ions are distinct: Cu(I) chelated to 1,10-phenanthroline reacts readily with H_2O_2 to form $HO^\bullet$ (Goldstein and Czapski 1986), whereas reaction of the Fe(II) complex is inhibited (Mello-Filho and Meneghini 1991). Neocuproine inhibits the reaction of Cu(I) with H_2O_2 (Czapski and Goldstein 1986) but appears to have little effect of Fe(II) (Mello-Filho and Meneghini 1991).

Copper-thiosemicarbazide complexes interact with both the bases and the phosphate groups of native *calf* thymus DNA (Pillai et al. 1977).

Malignancies in Copper Miners and Smelters

In the Mansfield copper-bearing shales (Kupferschiefer) miners' pneumoconiosis Wätjen (1933) did not find any malignancy in 54 necropsies. In 1936, under additional 65 cases Wätjen detected three primary bronchial carcinomas: two miners had severe pneumoconioses after 41 and 43 years of work underground, respectively, the third miner, who had worked underground for 43 years had a slight degree of pneumoconiosis.

Epidermoid or squamous cell carcinoma of the lung has been found to be predominant among copper smelter workers (Newman et al. 1976, Axelson et al. 1978). Of 71 consecutive occupational lung cancers among male Japanese copper smelter workers Kreyberg group I (squamous, large cell, and small cell carcinomas) was prevalent, which could not be explained by age, sex, ethnicity, air pollution, or smoking per se (Tokudome et al. 1988).

Copper fume is likely to provide the principal potential health hazard in the arc-air gouging process (Sanderson 1968).

Copper Gallium Diselenide and Copper Indium Diselenide

Copper gallium diselenide ($CuGaSe_2$) and copper indium diselenide ($CuInSe_2$) particles (mean diameters 2.47 μm and 2.55 μm, respectively) suspended in sterile, phosphate-buffered saline, pH 7.2, and instilled intratracheally into specific pathogen-free female Sprague-Dawley *rats* at doses at doses of 12 mg, 24 mg, 48 mg, or 96 mg/kg body weight after 3 days were found within the cytoplasma of macrophages (Morgan et al. 1995).

9.6.2.1.13 Gallium

Gallium has the atomic number 31 and the valence electrons are those on the $4p^1$ shell. Its only oxidation state is +3.

In two *patients* inhaling an aerosol containing ^{67}Ga citrate 24 h before resection surgery Kennedy et al. (1985) found the nuclide only in alveolar macrophages but not in epithelial cells and interstitial mononuclear phagocytes.

Pulmonary lesion of *rats* exposed for 4 weeks (2 h/d, 5 d/week) to 27 mg Ga_2O_3/m^3 (particle size

220 nm) 1 day thereafter consisted primarily of accumulations of proteinaceous debris in the alveoli (alveolar proteinosis) and a slight increase in the number of alveolar macrophages and neutrophils in alveolar air spaces within or adjacent to alveolar ducts. In the *rats* sacrificed at 6 months after exposure, several large foci of alveolar macrophages were scattered throughout the parenchyma, but most often centred within centriacinal regions or subpleural locations. In these areas, the macrophages were large and markedly vacuolated (WOLFF et al. 1989).

After intratracheal instillation into *rats*, Ga_2O_3 particles were being actively phagocytized while nodules of macrophages and Ga_2O_3 were found adhered to the alveolar walls (WEBB et al. 1986).

9.6.2.1.14 Germanium

Germanium has the atomic number 32 and the valence electrons are those on the $4p^2$ shell. Its only oxidation state is +4.

GeO_2 occurs in **three modifications**: one is isostructural to α-quartz, the other to cristobalite (BÖHM 1968), and the third isostructural to rutile. The rutile form was prepared at 5×10^9 Pa and 1000 °C from the hexagonal modification in a belt-type apparatus (SCLAR et al. 1964). KLOSTERKÖTTER (1956) in 3 of 55 animals found fibrohyalin nodules after intraperitoneal application of hexagonal GeO_2. Phagocytes containing dust particles showed a honeycomb-like eosinophilic protoplasm.

9.6.2.1.15 Indium

Indium has the atomic number 49 and the valence electrons are those of the $5s^25p^1$ shells. The usual oxidation states of indium are +1 and +3.

Indium trichloride ($InCl_3$) inhaled by female Fischer 344 *rats* initiated an inflammatory response (BLAZKA et al. 1994). Seven days following inhalation of 20 mg $InCl_3/m^3$ for 1 h the total cell number, fibronectin, and TNF-α levels in the bronchoalveolar lavage fluid were 8, 40, and 5 times higher than the control, respectively. While the increase was due primarily to an influx of polymorphonuclear leucocytes, a portion could be attributed to alveolar macrophages, which increased from 10.4 to 24.0×10^6.

Indium phosphide (InP) is a II–V compound of semiconductor materials widely employed for optoelectronic devices. KABE et al. (1996) showed that InP particles were hardly soluble in the synthetic lung fluid in vitro and that the LDL_0 of InP was >5,000 mg/kg, which was considered relative low toxicity. UEMURA et al. (1997) showed phagocytosis of InP by macrophages in bronchoalveolar lavage fluid of male Fischer 344 *rats*. After intratracheal instillation of 0, 1.2, 6.0 and 62.0 μg InP/kg *rat*, respectively, ODA (1997) found a dose-related mild elevation of superoxide dismutase activity and lactate dehydrogenase activity in bronchoalveolar fluid on day 1 without increase of inflammatory cells and total protein in BAL, which suggested the response of neutrophils and alveolar macrophages to instilled InP, and/or the manifestation of a very early stage of inflammation.

Autoradiography of *human* alveolar macrophages after phagocytosis of ^{111}In-oxine showed a highly significant ($P<0.001$) concentration of grains over the nucleus (DAVIS et al. 1980). All cytoplasmic structures except the lysosomes showed less than predicted labelling. If nuclear labelling was not considered, and only cytoplasmic labelling compared, there was a significant increase in labelling of lysosomes ($P<0.0001$). The cell bound activity at 1 h was 93 % and at 24 h it was 83 %. Radiolabelled macrophages were similar to unlabeled cells by dye exclusion, adherence and phagocytosis both at 1 and 24 h after labelling.

9.6.2.1.16 Iron

Iron has the atomic number 26 and the valence electrons are those of the $3d^64s^2$ shells. The usual oxidation states of iron are +2 and +3, less frequently +4 and +6.

As all the metals in the first row of the *d*-block in the Periodic Table with the sole exception of zinc, iron contains **impaired electrons** and can thus qualify as a **radical** (HALLIWELL and GUTTERIDGE 1989, p. 15), and ferrous ions can participate in electron transfer reactions with molecular oxygen.

In his review on potential use of iron chelators against oxidative damage, GALEY (1997) recalled that although triplet dioxygen cannot directly react with biomolecules in the ground state, iron, as well as other transition metals, can relieve the spin restriction of oxygen and dramatically enhances the rates of oxidation (MILLER et al. 1990).

Heme, an essential **iron chelate**, serves in respiration, oxygen transport, detoxification, and signal transduction processes. The potential toxicity of heme and hemoproteins points to a critical role for heme degradation in cellular metabolism. The heme oxygenases provide this function and participate in cellular defence.

The autoxidation of ferrohemoglobin to methemoglobin generates $O_2^{\bullet-}$ and its dismutation product H_2O_2 that could react with reduced iron to form

HO$^{\bullet}$ (KALYANARAMAN et al. 1983). However, it remains unclear if such a process would be catalysed in vivo by the heme iron itself, or by iron released from heme (CADENAS 1989). Indeed, whether any form of free heme or hemoprotein can act as a true Fenton catalyst to generate HO$^{\bullet}$ remains controversial (RYTER and TYRRELL 2000).

Various biological iron chelates were assayed for efficiency of Fenton catalysis (percentage conversion of $O_2^{\bullet-}$ to HO$^{\bullet}$) and ranked:

citrate >> pyrophosphate > lactate > adenosine triphosphate > hematin, transferrin, hemin, haemoglobin.

Polynuclear **ferric citrate complex** (metal-to-ligand molar ration 1:1)-preincubated *mural* peritoneal macrophages gave a positive Prussian blue reaction, while mononuclear iron complex (metal-to-ligand molar ration 1:10)-preincubated cells failed to do so (GEBRAN et al. 1993).

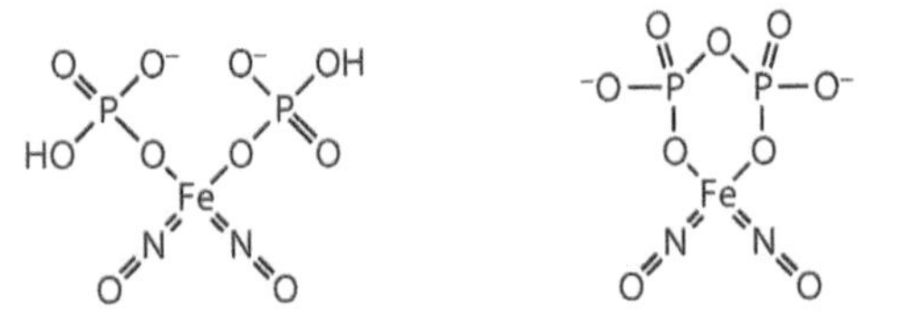

Fe(II)-NO-phosphate and pyrophosphate complexes [194]

Fe^{2+} is **complexed by dihydrolipoic acid** (PAGANI et al. 1989).

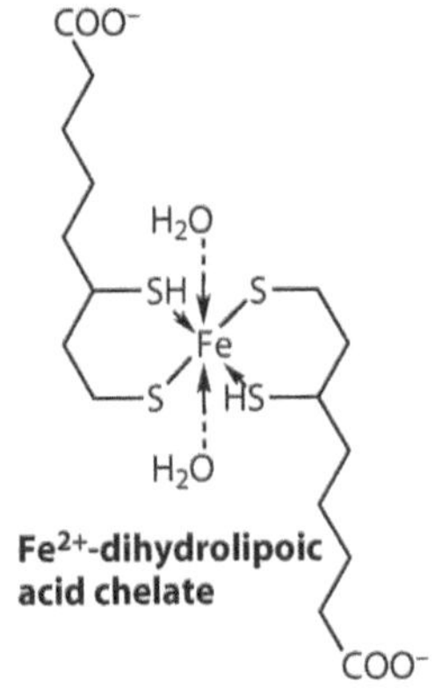

Fe^{2+}-Dihydrolipoic acid chelate [195]

Complexation of ferric iron with D-tagatose, a cell-permeable zero-energy producing ketohexose, protected against oxidative cell injury (VALERI et al. 1997). Lethal liver cell injury induced *in vitro* by 300 μM nitrofurantoin was completely prevented by high concentrations (20 mM) of D-tagatose, whereas equimolar concentrations of glucose, mannitol, or xylose were ineffective (PATERNA et al. 1998).

Surface-complexed iron [Fe^{3+}] observed after intrapleural injection of 30 mg of either amosite, crocidolite, and chrysotile both genuine and saturated with Fe^{3+} in Sprague-Dawley *rats* corresponded to oxidant generation, measured as barbituric acid reactive products of deoxyribose, and more covalently closed, circular DNA strand scission induced by these asbestos fibres (GHIO et al. 1994).

Iron chelates as Fe(III) nitrilotriacetate were genotoxic in HeLa cells and V79 cells (HARTWIG et al. 1992, 1993). Ferric nitrilotriacetate promoted *N*-diethylnitrosamine-induced renal tumorigenesis in the *rat* (ATHAR and IQBAL 1998). In renal cell carcinoma induced by ferric nitrilotriacetate, glutathione *S*-transferase Yp isozyme was over expressed and Ya isozyme concomitantly down-regulated (TANAKA et al. 1998).

N-(2-hydroxybenzyl)-L-serine is a tridentate iron chelator (KITAZAWA and IWASAKI 1999) that does not induce hypoxia inducible factor-1 (CREIGHTON-GUTTERIDGE and TYRRELL 2002).

Lactobionic acid, a major constituent of a solution used to preserve organs prior to transplantation, can chelate ferric iron (ISAACSON et al. 1989). Relative to iron(III) chelated to ethylenediamine tetraacetic acid (EDTA), the lactobionic acid-iron complex is less able to participate in the Fenton reaction as measured by formaldehyde generation from dimethylsulphoxide and bleaching of *p-N,N*-dimetylnitrosoaniline.

N,N'-bis (2-hydroxybenzyl) ethylenediamine-*N,N'*-diacetic acid (HBED) facilitated Fe(II) oxidation but blocked $O_2^{\bullet-}$-induced reduction of Fe(III) and consequently pre-empted production of HO$^{\bullet}$ or hypervalent iron through the Haber-Weiss reaction cycle (SAMUNI et al. 2001). The efficacy of HBED as a 1-electron donor was demonstrated by reduction of the 2,2'-azino-bis(3-ethylbenzothiazoline-6-sulphonate)-derived nitrogen-centred radical cation (ABTS$^{\bullet+}$), accompanied with a short-lived phenoxy radical.

Ferrocene, di-2,4-cyclopentadien-1-yliron, $Fe(C_5H_5)_2$, which is used in the manufacture of integrated circuits (FOX and RADER 1988) and acetylferrocene react with cyanomethyl and ethoxycarbonylmethyl carbon-centred radicals leading to homo- and heteroannular disubstituted products (BACIOCCHI et al. 1993). Histochemists used 1-chloromercuriferrocene as an electron-opaque stain for aldehydes and thiol groups (ALLEN and PERRIN 1974, SWIFT 1975).

The findings from several animal model studies of the pathology of iron indicate that **oxidative damage** to the membranes of cell organelles may be a crucial event in toxicity. There also exists a large body of indirect evidence from *in vitro* studies involving isolated organelles (BURKITT and GILBERT 1989, RICE-EVANS et al. 1989, BACON and BRITTON 1990, SEVANIAN et al. 1990), cells (POLI et al. 1987), and tissue homogenates (ARTHUR et al. 1988) to suggest that oxidative damage due to hydroxyl radical formation is responsible for the toxic effects of iron. AUST et al. (1993) have employed a secondary radical-trapping technique in which the hydroxyl radical reacts with dimethyl sulfoxide to form the methyl radical $^{\bullet}CH_3$, which is then detected as its adduct of the spin trap PBN in the bile of animals given an intragastric dose of ferrous sulphate.

One-electron transfer catalysed by iron

$$O_2 + Fe^{2+} \longleftrightarrow Fe^{3+}\text{—}\mathbf{O_2^{\bullet -}} \longrightarrow \mathbf{O_2^{\bullet -}} + Fe^{3+} \quad [196]$$

$$\mathbf{O_2^{\bullet -}} + Fe^{2+} + 2\,H^+ \longrightarrow H_2O_2 + Fe^{3+} \quad [197]$$

$$H_2O_2 + Fe^{2+} \longrightarrow HO^{\bullet} + HO^- + Fe^{3+} \quad [198]$$

The intermediate formed in equation [196] is the **perferryl ion,** a resonance hybrid of Fe^{3+}–$O_2^{\bullet -}$ and Fe^{2+}–O_2. Perferryl has been widely implicated as an initiating species (PEDERSON and AUST 1975, SVINGEN et al. 1978, TIEN et al. 1981, SUGIOKA et al. 1983) since the concept was first introduced by HOCHSTEIN and ERNSTER (1963) and HOCHSTEIN et al. 1964). However, iron in the **ferryl ion** (FeO^{2+}) is considerably more reactive than in the perferryl ion, and the ferryl ion has been considered as alternative to the $HO^{\bullet}$ in Fenton chemistry (KOPPENOL 1985).

Ferric ions readily precipitate out of neutral aerobic solution to form insoluble ferric hydroxides, and it was recognised that complexing with ligands such as adenosine diphosphate (HOCHSTEIN and ERNSTER 1963, HOCHSTEIN et al. 1964) or ethylenediaminetetraacetate (EDTA) overcame this problem. Complexing with EDTA helps to keep iron in a reactive form in solution and may favourably alter the redox potential of the iron (AUST et al. 1985; GROOTVELD and HALLIWELL 1986). Depending on its molecular ratio to iron, EDTA can either stimulate of inhibit lipid peroxidation (GUTTERIDGE et al. 1979). EDTA promotes the autoxidation of ferrous ions (Fe^{2+}) and formation of $HO^{\bullet}$ (COHEN and SINET 1980), but EDTA can inhibit lipid peroxide decomposition (GUTTERIDGE 1984).

LAMB and ELDER (1931) found solutions of $FeSO_4$ containing sodium pyrophosphates to oxidise 1000 times faster than $FeSO_4$ in 0.1 to 3.0 M H_2SO_4. While the rate expression for Fe^{2+} autoxidation in the presence of sulphate shows as second-order dependence of Fe^2 concentration, that for phosphate is first order (CHER and DAVIDSON 1954). The rate of Fe^{2+} autoxidation is also dependent upon the nature of its chelators. Anions with low affinity for Fe^{2+} do not appear to affect the rate of autoxidation (LAMBETH et al. 1982). However, chelation of Fe^{2+} by compounds with oxygen donor atoms tends to enhance Fe^{2+} autoxidation, perhaps because of their affinity for Fe^{3+} (CHER and DAVIDSON 1954, HARRIS and AISEN 1973).

Table 36. Spin state of oxidised iron (Fe^{3+})

1	1	1	1	1	S = 5/2 (high-spin state)
11	11	1			S = 1/2 (low-spin state)
11	1	1	1		S = 3/2 (intermediate state).

$$[Fe(H_2O)_6]^{3+} + e^- \longrightarrow [Fe(H_2O)_6]^{2+} \quad [197]$$

Octahedral d^5 complexes have been shown to exist in two different magnetic and thermodynamic stable forms only: the low- or the high-spin state. The temperature dependence of the spin equilibrium shows a very low energy barrier between the two spin states.

WHITE and COON (1980) have postulated a highly resonance-stabilised perferryl species:

$$S^-Fe(5+)O^{2-}\ S^-(4+)O^- \longleftrightarrow S^-Fe(3+)O \longleftrightarrow S^{\bullet}Fe(3+)O^- \quad [200]$$

Under physiological conditions, iron is only slightly soluble, tending to form a precipitate with anions such as OH^- and inorganic phosphate. However, a variety of chelating agents greatly increase the solubility of iron. For example, the addition of EDTA to an $O_2^{\bullet -}$ generating system in the presence of Fe^{3+} markedly potentiated cytotoxicity (ROSEN et al. 1981), $HO^{\bullet}$ formation (MCCORD and DAY 1978, ROSEN et al. 1981), and lipid peroxidation (GUTTERIDGE et al. 1979). The powerful iron chelators DTPA and Desferal (desferrioxamine B methanesulfonate) totally inactivate Fe^{3+}. Desferrioxamine is a hexadentate ligand, i.e. it occupies all six coordination sites of an iron ion. The desferrioxamine molecule 'wraps around' the iron ion, encasing it in an envelope of organic material.

Hydroxypyridones are bidentate ligands, i.e. they can occupy only two coordination sites. Hence three molecules of hydroxypyridones are needed to complex one Fe(III) completely. Upon dilution, 3 : 1 hydroxypyridone : iron ion complexes might dissociate into 2 : 1 complexes, leaving available coordination sites on the iron that could allow catalysis on free radical reactions.

Quinolinic acid, α-picolinic acid, fusaric acid, and 2,6-pyridinedicarboxylic acid enhanced the Fenton reaction in phosphate buffer, respectively (IWAHASHI et al. 1999). The ultraviolet-visible absorption spectrum of the mixture of α-picolinic acid with ferrous ion showed a characteristic band with a λ_{max} at 443 nm, suggesting that α-picolinic acid chelate of Fe^{2+} ion forms in the solution. Similar characteristic absorbance band was also observed for the mixture of Fe^{2+} ion with quinolinic acid (or fusaric acid, or 2,6-pyridinedicarboxylic acid). The chelation seems to be related to the enhancement by quinolinic acid, α-picolinic acid, fusaric acid, and 2,6-pyridinedicarboxylic acid of the Fenton reaction.

The inhibitory effect of **diethyldithiocarbamate** on Fenton reagent-generated hydroxyl radicals may be due either to its capacity to scavenge hydroxyl radicals or to **chelate and then oxidise ferrous ions** (Liu et al. 1996).

Peroxidation of lysosomal membranes can be induced by ascorbate/Fe(III)complex (Mak and Weglicki 1985) or by the autoxidizing, superoxide producing, compound dihydroxyfumarate in combination with a Fe(III)complex (Mak et al. 1983). These processes probably involve the initiation of lipid peroxidation by either iron-centred radicals or by hydroxyl radicals. Zdolsek and Svensson (1993) found that the damaging effects of hydrogen peroxide in the presence of Fe(III)-complexes depended on the additional presence of a reducing agent, such as cysteine. The reducing agents would reduce Fe(III) to Fe(II) making it prone to catalyse Fenton reactions generating hydroxyl radicals in the presence of hydrogen peroxide.

The mechanism of Fe^{2+}-initiated lipid peroxidation in a liposomal system was studied by Tang et al. (2000). They found that a second addition of ferrous ions within the latent period lengthened the time lag before lipid peroxidation started. The apparent time lag depended on the total dose of Fe^{2+} whenever the second dose of Fe^{2+} was added, which indicated that Fe^{2+} has a dual function: to initiate lipid peroxidation on one hand and suppress the species responsible for the initiation of the peroxidation on the other. When the pre-existing peroxides (LOOH) were removed by incorporating triphenylphosphine into liposomes, Fe^{2+} could no longer initiate lipid peroxidation and the acceleration of the Fe^{2+} oxidation by the liposomes disappeared. However, when extra LOOH were introduced into liposomes, both enhancement of the lipid peroxidation and shortening of the latent period were observed. When the scavenger of lipid peroxyl radicals (LOO•), *N,N'*-diphenyl-*p*-phenylene-diamine, was incorporated into liposomes, neither initiation of the lipid peroxidation nor acceleration of the Fe^{2+} oxidation could be detected.

Lipid peroxidation, measured by Fontecave et al. (1990) by malonaldehyde formation is induced in *rat* liver microsomes by the insoluble iron-containing minerals, pyrite, magnetite, nemalite (a fibrous brucite where about 8 % of Mg^{2+} is substituted by Fe^{2+}) and minette de Lorraine.

Fe^{3+} reacts with the **silanol groups** at the surface of amorphous silica forming

$$Fe^{3+} + m(SiOH) \rightleftarrows Fe(OSi)_m^{3m+} + mH^+ \quad [201]$$

Fe^{3+} reacts with **prostaglandin** G_2 as follows:

$$Fe^{3+} + PGG_2 \rightarrow [FeO]^{3+} + PGH_2 \quad [202]$$

$$[FeO]^{3+} + PGG_2 \rightarrow Fe^{3+} + PGH_2 + {}^1O_2 \quad [203]$$

Rötig et al. (2001) examined the expression levels of various **genes involved in iron uptake, reduction, and storage,** in Fe–S protein biogenesis, in mitochondrial electron transport chain, plus the two superoxide dismutase genes, in *human* adult tissues by Northern blot analysis. Most of these genes were ubiquitously expressed, but their transcription showed strongly different levels in the various tissues investigated denoting different mechanisms for iron utilisation in various organs.

Normally the **entry of iron(III) into cells** and its intracellular distribution is carefully controlled. However, if iron(III) is complexed by nitrilotriacetate (NTA), it circumvents the controlled transferrin path, and may even penetrate into the cell nucleus where it catalyses detrimental oxidative reactions with DNA. The Fe(III) NTA-complex has been shown to cause cancer in *rats* (Ebina et al. 1986) and DNA breaks and increased frequency of sister chromatid exchanges in Chinese *hamster* V79 cells (Hartwig et al. 1993).

Sequestration of iron by macrophages diminishes the potential of the metal to catalyse oxidant generation (Olakanmi et al. 1993). The storage of iron within intracellular **ferritin** confers an antioxidant function to this protein and, in certain cells, provides cytoprotection in vitro against oxidants (Cozzi et al. 1990, Balla et al. 1995). Ferritin expression is regulated by a posttranscriptional mechanism. A specific sequence at the 5'-untranslated end of ferritin mRNA called the iron responsive element (IRE) binds a cubane iron-sulphur cluster, the iron regulatory protein (IRP), when the IRP exists in the apoprotein form (Leibold and Guo 1992, Beinert and Kennedy 1993). Iron included in atmospheric particles could react with IRP, which alters its conformation. This decreases the affinity of the protein to mRNA, and it is displaced, allowing translation of ferritin to proceed.

The binding activity of IRP-1 was increased and the ferritin synthesis was suppressed when the macrophages were cultures under hyperoxia, and the reverse occurred under hypoxia (KURIYAMA-MATSUMURA et al. 2001). Iron diminished the IRP-1-binding activity and the enhanced synthesis of ferritin. However, this effect was arrested under hyperoxia. Consistently, hypoxia-induced loss of binding activity of IRP-1 and the enhanced synthesis of ferritin was blocked in the presence of an iron chelator deferoxamine. These alteration in the binding activity of IRP-1 in response to oxygen were not reproduced in the cell-free extract.

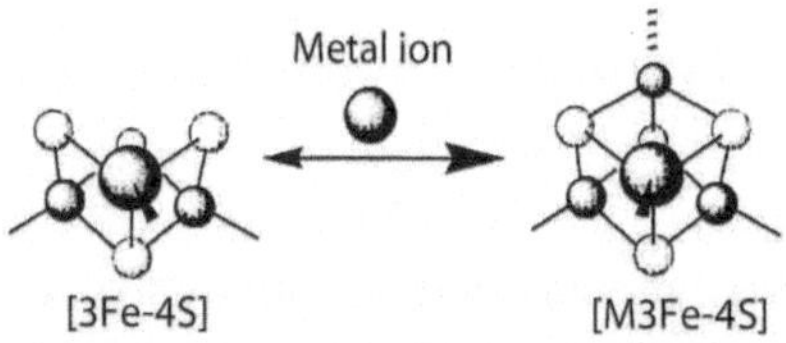

Addition of a metal ion to a [3Fe-4S] cuboidal cluster to yield a [4Fe-4S] cubane [80]

The loading of iron into ferritin needs to be initiated with Fe(II), with oxygen as an oxidant, to deposit iron in the core of ferritin as ferric oxyhydroxides.

The iron stored in ferritin can be released by $O_2^{\bullet-}$ (THOMAS et al. 1985) and other reducing radicals, e.g. dipyridyls, nitroaromatics, and semiquinones (THOMAS and AUST 1986). Reducing agents like flavins (SIRIVECH et al. 1974) and ascorbate (BOYER and MCCLEARY 1987) also release iron slowly from ferritin. Polyhydroxypyrimidines: dialuric acid, isouramil, divicine, and acid-hydrolysed vicine in aerated phosphate-buffered saline, pH 7.3, at 37 °C, all caused concentration-dependent release of iron from ferritin, which was measured as the progressive formation of Fe^{2+}-ferrozine complex (MONTERO and WINTERBOURN 1989). When superoxide was anaerobically generated by dissolving KO_2, no metal complex with bathophenantroline built up from *horse* spleen ferritin (CASSANELLI and MOULIS 2001). Similar to superoxide, initial reports (REIF and SIMMONS 1990) of iron release by nitric oxide have been questioned (LAULHÈRE and FONTECAVE 1996).

The addition of NifS to *horse* spleen ferritin does not sustain iron release at a high rate after an initial burst (CASSANELLI and MOULIS 2001). This increase in soluble iron can be attributed to endogenous compounds able to expel iron from ferritin. Addition of the NifS substrate, L-cysteine, increases the rate.

Organically bound iron is one of the most potent initiators of phagocytic activity in macrophages (RHODES et al. 1969, BAKER and MORGAN 1970, RHODES 1970, WADA et al. 1970).

Intramacrophagic iron was 0.33 ± 0.21 μg × 10^{-6} cells in healthy non-smoking subjects without occupational exposure (CORHAY et al. 1992). Intramacrophagic iron was increased in smokers, iron-steelworkers, and in patients with chronic obstructive pulmonary disease or lung cancer even in the absence of pulmonary haemorrhage. About 80 % of the whole bronchoalveolar lavage fluid iron content was in the cells.

To study the fate of internalised Fe, MCGOWAN and PARRISH (1985) and MCGOWAN et al. (1986) incubated *human* alveolar macrophages with ^{59}Fe saturated transferrin at 37 °C, lysed the cells in a French pressure cell at 8000 psi and 4°, and fractionated the lysate by differential centrifugation (at 900 × g for 5 min and then at 11,500 × g for 20 min). ^{59}Fe internalised initially bound to transferrin was distributed to a cytoplasmic, largely ferritin-associated, pool more slowly in smokers than in non-smokers, during a 24-hour incubation *in vitro*. Significantly less newly internalised iron was returned to the culture medium by alveolar macrophages from smokers, which by 24 h had released 11.0 ± 3.7 % of the initially internalised ^{59}Fe compared with $36 \pm 2{,}3$ % for non-smokers ($P < 0.01$). According to MCGOWAN and HENLEY (1988) the majority of the iron contained in alveolar macrophages is insoluble and is most likely incorporated into haemosiderin. However, approximately 25 % of the intracellular iron is associated with ferritin and could potentially be released to catalyse the formation of hydroxyl radicals through the Haber-Weiss reaction.

The iron burden in alveolar macrophages of smokers considerably exceeds that of non-smokers. Of 440–1,150 μg iron contained in 1 g cigarette 0.1 % gets into the mainstream smoke (MUSSALO-RAUHAMMAA et al. 1986). Furthermore, bronchitis increases the iron content of alveolar macrophages from 3.2times to 5.4times of that of a healthy non-smoker.

A study on monocytic THP-1 cells induced to differentiate with phorbol myristate acetate showed that further stimulation with interferon-γ strongly increased H-ferritin expression and also induced a greater release of TNF-α (SCACCABAROZZI et al. 2000). This effect is possibly mediated by a stabilization of H-ferritin transcript caused by phorbol myristate acetate and the relevant mRNA region was localised within a pyrimidine rich region in the 3' untranslated region (AI et al. 1999). Most of the stimuli related to inflammation and directed to ferritin synthesis seem to upregulate H-ferritin preferentially over the L-ferritin, thus determining an increase of the catalytic sites and a reduction of cell iron availability. These events may be accompanied by indirect effects via the induction of nitric oxide that modify the activity of iron regulatory proteins and regulate equally both subunits.

When stimulated with either phorbol myristate acetate (10 ng/ml) of opsonized zymosan (3 mg/ml) neither *human* monocytes nor monocyte-derived macrophages exhibited spin trap evidence of **hydroxyl radical** formation unless provided with an exogenous Fe^{3+} catalyst (BRITIGAN et al. 1988). Using a spin adduct resistant to superoxide-mediated destruction, BRITIGAN et al. (1990) confirmed the lack for an endogenous capacity of significant $HO^{\bullet}$ production during the respiratory burst of *human* phagocytes.

In HL-60 cells treated with H_2O_2, BYRNES (1996) measured diffusion distances, i.e. the distances between the sites of generation of presumed hydroxyl radicals by low molecular weight forms of Fe and the site of their reaction with substrate. In HL-60 cells made deficient of Fe by treatment with bathophenanthroline disulphonic acid and ascorbic acid, and subsequently allowed to incubate in fresh media during short periods, an additional diffusion distance of 1.9 nm was measured. This was assumed to derive from Fe acquired from extracellular sources and was associated with ATP, as depletion of cellular ATP led to decreased single-strand break generation (LI and BYRNES 1999). LI et al. (1999) measured two diffusion distances for Fe complexes each of ethylene diamminetetraacetic acid (FeEDTA) and nitrilotriacetic acid for generation of malondialdehyde-type products from deoxyribose and of single-strand breaks in the plasmid pBR322. The closer diffusion distances for pBR322 single-strand break generation (5–6 nm) were considerably greater than the diffusion distances for malondialdehyde generation in the deoxyribose assay (2–3 nm). This is consistent with charge-charge interactions playing an important role in defining diffusion distance. The diffusion distances for FeNTA, FeEDTA, and other Fe species generating single-strand breaks in isolated Ehrlich ascites tumour cell nuclei ranged from 2.1 to 14 nm.

Nitric oxide ($^{\bullet}NO$) in a concentration-dependent (0.4–1.8 μM) manner inhibited lipid peroxidation induced in HL-60 cells by an oxidative stress of 20 μM Fe^{2+} (KELLEY et al. 1999).

Lipid peroxyl radical can be scavenged by nitric oxide

$LOO^{\bullet} + {}^{\bullet}NO \longrightarrow LOONO$ chain determination [204]

The **phagocytic capacity** of ferric citrate-treated peritoneal macrophages of Balb/c *mice* was inhibited in a dose-related manner, 50 μM being the lowest concentration achieving a significant inhibition of phagocytosis *in vitro* (GEBRAN et al. 1993). However, such inhibition was only observed in the presence of ferric citrate with a metal-to-ligand molar ratio of 1:1, but not with ferric citrate with a metal-to-ligand molar ratio of 1:10 in which the hydrolyzation and polymerization of iron in physiological solutions is prevented. Lipid peroxidation at concentrations of 125 μM ferric citrate showed an absorbance of thiobarbituric acid reactive substances of 0.403±0.012 and 0.154±0.005, respectively, while the untreated control macrophages gave figures of 0.029±0.003, read at 532 nm.

Iron Oxides

The three iron oxides correspond to the idealised compositions FeO, Fe_2O_3 and Fe_3O_4. However, often they were not composed stoichiometrically.

Three distinct submicronic aerosols of iron oxide can be reproducibly generated by combustion of iron pentacarbonyl vapours under varying conditions (VALBERG and BRAIN 1979). The "feathers" haematite has a mass median aerodynamic diameter of 0.17 μm and a large surface area because it is an agglomerate of units 0.005 μm in diameter. The "birdshot" haematite has a median aerodynamic diameter of 0.31 μm, but has a smaller surface area because the subunits are 0.03 μm in diameter. The γ-oxide has a median aerodynamic diameter of 0.73 μm and has crystalline subunits of 0.2 μm in diameter; it is a magnetic form of haematite, γ-Fe_2O_3. The magnetic properties of γ-Fe_2O_3 permit enhancement of deposition and non-invasive detection.

Taconite, a low-grade siliceous formation in Minnesota, from which high grade iron ore is derived, consists of fine-grained silica with variable ratios of hæmatite and magnetite totalling less than 30 per cent iron.

Table 37. Iron Ores

Mineral	Formula	Fe (%)	Deposit
Magnetite	Fe_3O_4	72	Norway, Sweden (Kiruna)
Hæmatite	α-Fe_2O_3	70	Sieg area, Cumberland, Spain, Morocco
Goethite	α-$Fe^{3+}O(OH)$	63	Central Europa
Lepidocrocite	γ-FeO(OH)	63	Sieg area, Hesse
Linonite	FeO(OH)	63	Salzgitter, Westerwald
Siderite	$FeCO_3$	48	Sieg area, Styria, Bergamo
Chamosite		36–42	Lorraine, Luxembourg
Thuringite		31–35	Lorraine, Luxembourg

Hæmatite

Hæmatite (α-Fe_2O_3) is hexagonal. It is most widely encountered as the henna-red pigment of many diverse rocks. Some sedimentary rocks and their metamorphosed products contain such high percentages of hæmatite that they have been recovered as iron ore.

Following intrapulmonary instillation of 2.6-μm-diameter Fe_2O_3 particles in *human* subjects, particles were cleared from the lavagable alveolar macrophage compartment in a biphasic pattern, with a rapid-phase clearance half-time of 0.5 d and long-term clearance half-time of 110 d (LAY et al. 1998). The intracellular distribution of particles within lavaged alveolar macrophages was similar in bronchial and alveolar bronchoalveolar lavage fractions. Alveolar macrophages with high intracellular particle burdens disappeared from the lavagable phagocytic alveolar macrophage population disproportionately more rapidly (shorter clearance half-time) than did alveolar macrophages with lower particle burdens.

The zymosan-stimulated hydrogen peroxide and superoxide anion release of alveolar macrophages isolated from *rabbits* treated with Freund's adjuvant were not significantly decreased by ≤50 μg Fe_2O_3/10^6 cells prior to zymosan stimulation (GULYAS et al. 1990).

Magnetite

Magnetite is a cubic mineral and member of the spinel structure type, space group Fd3m, with composition $[Fe^{3+}]^{IV}[Fe^{2+}Fe^{3+}]^{VI}O_4$. The most spectacular ore body occurs at Kiruna in northern Sweden, but magnetite can also be prepared by heating hæmatite (α-Fe_2O_3) in a reducing atmosphere. Oxidation of magnetite leads to maghemite (γ-Fe_2O_3), which gradually inverts to hæmatite.

Magnetite particles of an aerodynamic diameter of 2.8 μm allowed to deposit in the alveolar region of the *human* lung by the non-invasive method of magnetopneumography characterised two important properties of the alveolar macrophage: the cell-energy and the viscosity (STAHLHOFEN and MÖLLER 1992). Cell energy was much larger than thermal energy at 37 °C: $kT = 4.3 \times 10^{-21}$ J. The Newton-viscous properties of the cytoskeleton could be determined as $\eta = 87.1 \pm 34.6$ Pa × s.

Alveolar macrophages of *rats* dusted with magnetite from Grängesberg (Central Sweden) for 8 weeks contained large autophagous vacuoles (SCHILLER 1971, 1989; Fig. 141). X-ray microanalysis of the telolysosomes contained in the alveolar macrophages of *rats* instilled intrabronchially with iron dextran showed plenty of iron and phosphorus (FASSKE and MORGENROTH 1985) suggesting an involvement of surfactant. Even 3 months after the last inhalation of Fe_3O_4 alveolar macrophages manifested eager phagocytosis of surfactant (Fig. 143).

J774 *murine* macrophage cell line was used to verify the dynamic processes of spherical magnetite particles of different sizes within the cells *in vitro* (STAHLHOFEN and MÖLLER 1994). Relaxation, that is the decay of a remanent magnetic field after primary magnetisation, was particle-size-dependent. The results might indicate that relaxation is not a volume effect, but depends more on the surface area of the phagolysosomes.

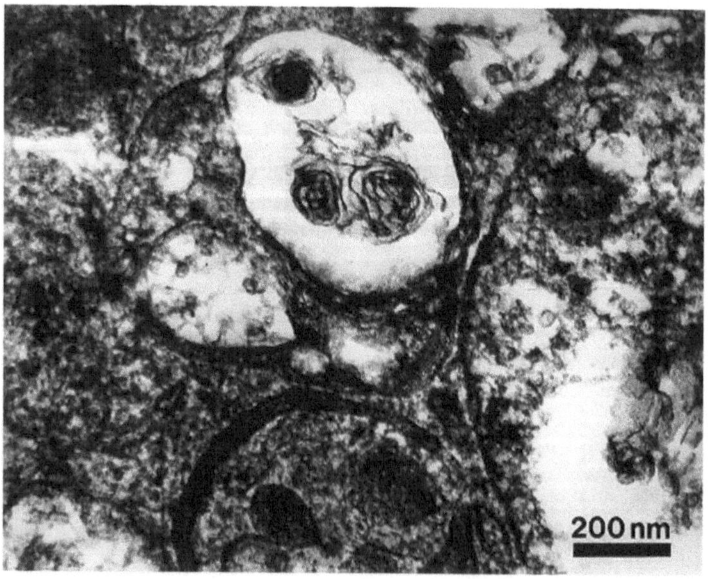

Fig. 141. Alveolar macrophage from a 286 g female white *rat* (No. 3; breeder: Winkelmann, Borchen-Kirchborchen) which inhaled 10 mg powdered Grängesberg magnetite/m³ 4 h per day, 5 days per week from August 24 to October 19, 1967 for a total of 40 days. Fixed on October 30, 1967 under methitural anaesthesia by intratracheal instillation of 2.5 % glutaraldehyde in phosphate buffer (pH 7.4) before opening the thorax. Postfixation with 1 % osmium tetroxide in phosphate buffer. Contrasted en bloc for 12 h with 0,5 % uranyl acetate in 70 % ethanol. Embedded in a 2 : 8 mixture of methyl and butyl methacrylate. Sectioned at 50 nm. Lead citrate after REYNOLDS (1963). Plate 34/12

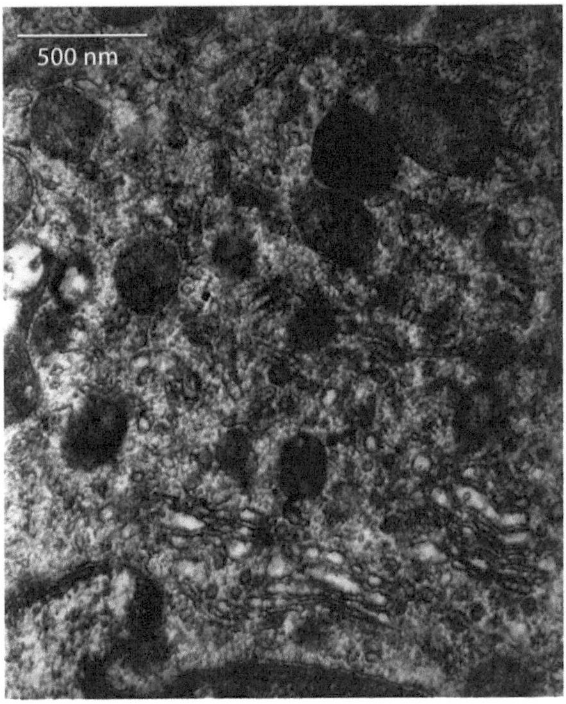

Fig. 142. Rough endoplasmic reticulum and dilated Golgi cisternae in an alveolar macrophage from a 286 g female white *rat* (No. 3, breeder: Winkelmann, Borchen-Kirchborchen) which inhaled 10 mg powdered Grängesberg magnetite/m^3 4 hours per day, 5 days per week from August 24 to October 19, 1967 for a total of 40 days. Fixed on October 30, 1967 under methitural anaesthesia by intratracheal instillation of 2.5% glutaraldehyde in phosphate buffer (pH 7.4) before opening the thorax. Postfixation with 1% osmium tetroxide in phosphate buffer. Contrasted en bloc for 12 h with 0.5% uranyl acetate in 70% ethanol. Embedded in a 2:8 mixture of methyl and butyl methacrylate. Sectioned at 50 nm. Lead citrate after REYNOLDS (1963). Plate 46/06

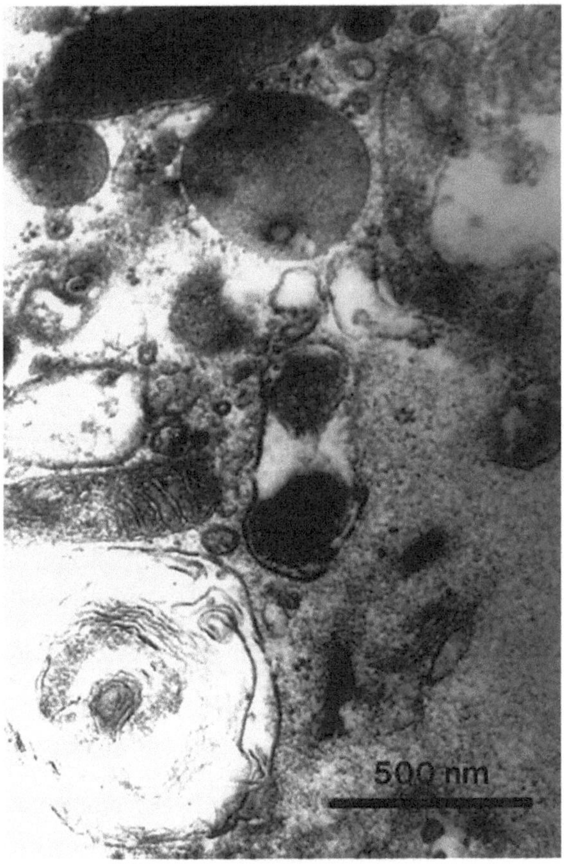

Fig. 143. Tubular myelin engulfed by an alveolar macrophage from a female white *rat* (breeder: Winkelmann, Borchen-Kirchborchen) which inhaled 10 mg powdered Grängesberg magnetite/m^3 4 hours per day during a 5-day work week. Inhalation took place from August 24 to October 19, 1967 for a total of 40 days. Fixed on January 15, 1968 under methitural anaesthesia by intratracheal instillation of 2.5% glutaraldehyde in phosphate buffer (pH 7.4) before opening the thorax. Postfixation with 1% osmium tetroxide in phosphate buffer. Contrasted en bloc for 12 h with 0,5% uranyl acetate in 70% ethanol. Embedded in a 2:8 mixture of methyl and butyl methacrylate. Sectioned at 50 nm. Lead citrate after REYNOLDS (1963). Film 33188

Carbonyl Iron

Pentacarbonyliron, $Fe(CO)_5$, is able to form stable π-complexes with hydrocarbons. RHEILEN et al. (1930) reported the first example of such a complex, tricarbonyl[η^4-1,3-butadiene]iron.

Carbonyl iron inhaled by male Crl:CD BR *rats* for 6 h at 100 mg/m^3 did not produce cellular or biochemical indices for pulmonary inflammation, either in normal or complement-depleted animals (WARHEIT et al. 1991).

Carbonyl iron instilled intratracheally in SPF male Fischer 344 *rats* did not significantly elevate iNOS mRNA in alveolar macrophages lavaged 24 h after treatment (BLACKFORD et al. 1997).

Ferric Ions

2-[(2-Bis[carboxymethyl]amino-5-methylphenoxy)-methyl]-6-methoxy-8-bis[carboxymethyl]-aminoquinolin (quin2) is a powerful transition metal ion chelator. It can efficiently catalyse the formation of oxidising species from the generally unreactive combination of Fe^3 and H_2O_2 (SANDSTRÖM et al. 1994). Using the Fe^{2+} indicator ferrozine, Sandström et al. (1997) found evidence for direct reduction of Fe^{3+}-quin2 by H_2O_2. Superoxide anion radical appeared to be less efficient than H_2O_2 as reductant of Fe^{3+}-quin2 as addition of superoxide dismutase in the ferrozine experiments only decreased the amount of Fe^{2+} available for Fenton reaction by 10–20%.

Nitric oxide synthase activity was decreased by about 50% in homogenates obtained from J744A.1 macrophages treated with interferon γ plus lipopolysaccharide in the presence of 50 μM Fe^{3+} as compared with extracts of cells treated with IFN-γ plus LPS alone (WEISS et al. 1994).

Gaucher Cells

LEE et al. (1977) in a *child* aged $2^1/_2$ years observed a remarkable number of Gaucher cells in alveolar space and in terminal bronchioles heavily laden with iron.

9.6.2.1.17
Lead

Lead has the atomic number 82 and the valence electrons are those of the $6s^2 6p^2$ shells. The oxidation states of lead include +2 and +4.

Lead, an immunomodulator and potential *human* carcinogen, is a major airborne pollutant in industrial environment which possesses a serious threat to *human* health.

In 66 secondary lead smelter workers lead-exposed for 8.09 ± 4.03 years there was a positive correlation between the presence of Pb in blood and significant increases in malondialdehyde levels and superoxide dismutase activity (YE et al. 1999). A positive correlation was found between malondialdehyde and DNA damage.

Inhalation for 3 h/d for 4 d of 30 μg/m³ PbO by male New Zealand white *rabbits* enhanced H_2O_2 and $O_2^{\bullet-}$ production by their pulmonary macrophages (ZELIKOFF et al. 1993). Phagocytic uptake of latex particles was reduced with increasing post-exposure time reaching a maximum inhibition at 72 h. Effects of TNF-α release/activity appeared earliest and were persistent up to 72 h. Immediately and 24 h after exposure, lipopolysaccharide-stimulated activity of TNF-α was depressed by 62% and 50%, respectively. After 72 h, TNF-α release was significantly enhanced compared to control levels. GUO et al. (1996) reported that $PbCl_2$ increased TNF-α secretion from *human* peripheral blood mononuclear cells in a concentration- and time-dependent manner. They suggested that $PbCl_2$ increased TNF-α expression by posttranscriptional mechanisms and enhanced the reactivity and uptake of TNF-α by increasing the surface expression of TNF-R p55.

The oxidative metabolism of NMRI *mouse* peritoneal macrophages induced by phorbol myristate acetate was enhanced by $PbCl_2$, the zymosan-induced oxidative metabolism was weakly reduced as measured by means of the lucigenin-enhanced chemiluminescence in a Berthold Biolumat 9500 up to 120 min. After 20 h of exposure, $PbCl_2$ concentration-dependently (10^{-4}–10^{-8} M) suppressed the oxidative metabolism induced by 100 μg zymosan/ml and 0.5 μg PMA/ml as well as phagocytosis of latex particles (HILBERTZ et al. 1986).

QUINLAN et al. (1988) observed that lead (0.01–0.4 mM) stimulated Fe^{2+}-induced lipid peroxidation, determined by quantification of thiobarbituric acid-reactive species.

5-Aminolaevulinic acid accumulates in all tissues in lead poisoning and could be a source of oxygen radical formation and cellular damage (MONTEIRO et al. 1989). The intermediateness of reactive oxygen species in these processes was evidenced in the inhibitory effects of superoxide dismutase, catalase, and mannitol (MONTEIRO et al. 1986). Melatonin, mannitol and trolox, all of which are free radical scavengers, inhibited the formation of 8-hydroxydeoxyguanosine in a concentration-dependent manner (QI et al. 2001). The concentration of each to reduce DNA damage by 50%, i.e., the IC_{50}, was 0.52, 0.84 and 0.90 mM, respectively.

$$H_2N-CH_2-CO-CH_2-CH_2-CO_2^- \rightleftharpoons H_2N-CH=C(OH)-CH_2-CH_2-CO_2^-$$

$$\downarrow O_2$$

$$[H_2N-CH=C(O^{\bullet})-CH_2-CH_2-CO_2^-]$$

$$\downarrow O_2^{\bullet-}, H_2O_2, HO^{\bullet}$$

$$OHC-CO-CH_2-CH_2-CO_2^- + NH_4^+$$

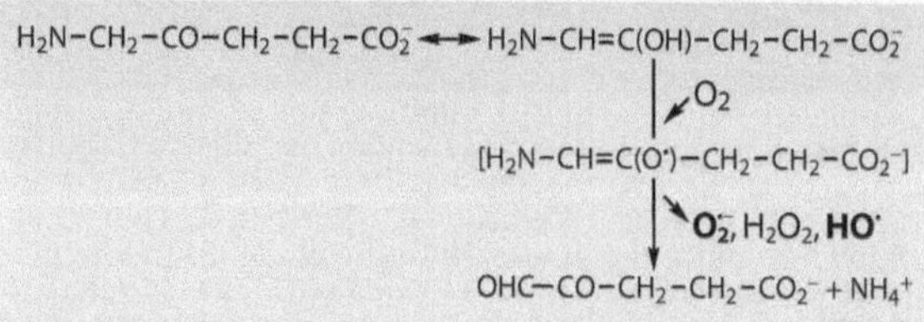

$$[205]$$

Iron is released from *horse* spleen ferritin by both 5-aminolaevulinic acid-generated $O_2^{\bullet-}$ and enoyl radical, which amplifies the chain of 5-aminolaevulinic acid oxidation (ROCHA et al. 2000). Iron chelators such as ethylenediaminetetraacetic acid (EDTA), ATP, but not citrate, and phosphate accelerate this process and 5-aminolaevulinic acid-promoted iron release from *horse* spleen ferritin is faster in *horse* spleen isoferritins containing larger amounts of phosphate in the core.

The intensity of spontaneous chemiluminescence from liver, soleus muscle, and parietal brain region of 5-aminolaevulinic acid-treated *rats* was markedly increased after two doses 40 mg/kg body weight injected intraperitoneally on alternate days (DEMASI et al. 1997). The plasma trapping capacity evaluated by the luminol/2-aminopropane system, gave a parallel response: maximum values after two doses and decreased values after prolonged treatment. After eight doses, the 5-aminolaevulinic acid concentration was found to be 3-fold above the normal value in plasma, 48% higher in liver and 38% higher in total brain. These data indicated that the plasma antioxidant system responded to 5-aminolaevulinic acid treatment and was correlated with tissue chemiluminescence.

Free radicals and α-ketoglutaraldehyde, a source of Schiff adducts with proteins (ACHARYA and MANNING 1980), might be formed *in vivo* when 5-aminolaevulinic acid accumulates in lead poisoning and might constitute a causative factor of tissue damage (HERMES-LIMA et al. 1991).

CARNEIRO and REITER (1998) have reported that the incubation of *rat* cerebral, hippocampal and cerebellar homogenates with δ-aminolaevulinic acid increases the formation of lipid peroxidation products presumably as a result of the induction of free radicals by δ-aminolaevulinic acid. Melatonin co-incubation, in both a concentration-dependent and time-dependent manner, prevented the rises in the damaged lipid products malondialdehyde and 4-hydroxyalkenals. *In vivo* as well, acute melatonin administration reduced damaged lipid products in the brain of *rats* treated with δ-aminolaevulinic acid. PRINC et al. (1997) in similar studies like CARNEIRO and REITER (1998) felt that the ability of melatonin to protect against δ-aminolaevulinic acid toxicity relates to the free radical scavenging activity of the indolamine. Furthermore, PRINC et al. (1998) showed that melatonin administered *in vivo* not only reduced lipid peroxidation in the brain of *rats* treated with δ-aminolaevulinic acid, but also increased enzyme activities of the heme pathway.

The dithiol ***meso*-2,3-dimercaptosuccinic acid** was the first orally administered metal chelating agent to receive U.S. Food and Drug Administration (FDA) approval for the treatment of childhood plumbism. It reduced oxidative stress in lead exposed C57BL/6 *mice in vivo* (ERCAL et al. 1996).

Lead sulphide, 6 times 15 mg given intratracheally to *rats*, was phagocytosed by alveolar macrophages inducing foamy cytoplasm within two months (KISSLER 1983). Alveolar lipoproteinosis occurred 6 months after intratracheal instillation of 50 mg PbS (ROSMANITH et al. 1989).

Lead citrate inhibited oxidative metabolism of *mouse* peritoneal macrophages exposed to macrophage-activating factor (BUCHMÜLLER-ROUILLER et al. 1989).

Particulate PbO, Pb_3O_4, and PbO-coated coal fly ash of respirable size were ingested by New Zealand white *rabbits*' lavaged bronchoalveolar macrophages cultured in vitro (DE VRIES et al. 1983). Swelling of the mitochondria, nuclear membrane, and endoplasmic reticulum was common, as well as were characteristic **reprecipitation complexes** of lead, phosphorus, and calcium within the nuclear heterochromatin and cytoplasm of the cell. It was suggested that solubilization of lead from the ingested particles in phagosomes of macrophages results in the liberation of intracellular lead with the resultant formation of reprecipitation complexes.

Triethyllead $[C_2H_5)PbCl]$ in high (≥50 μM), but not in low concentrations (≤1 μM) inhibited the incorporation of exogenous [^{14}C]arachidonic acid into HL-60 cells (KRUG et al. 1994). Calcium ionophore A 23187-induced stimulation of acetyltransferase, the key enzyme of PAF synthesis, was inhibited only by 40 μM of triethyllead, while 50 μM blocked the enzyme below basal activity.

Lead caused cell rounding in **epithelioid cells** by altering the fluidity of the membrane bilayer, making it more vulnerable to osmotic shock (FILERMAN and BERLINER (1980).

9.6.2.1.18 Manganese

Manganese has the atomic number 25 and the valence electrons are those of the $3d^5 4s^2$ shells. The oxidation states of manganese include +1, +2, +3, +4, +5, +6, or +7.

Unlike Fe and Cu, **inorganic complexes** of Mn are known to exist in high concentrations in certain cells as the chlorplast stroma (KONO et al. 1976) and in many lactic acid *bacteria* (ARCHIBALD and FRIDOVICH 1981, ARCHIBALD 1986). CHETON et al. (1988) used three different free radical generating systems and four $HO^•$ detecting methods to investigate the activity of biologically relevant inorganic Mn complexes. When using $HO^•$ generator systems (hypoxanthine-xanthine oxidase system; formula [45]) in which O_2 derived $O_2^{•-}$ and H_2O_2 are apparently precursors of $HO^•$, simple Mn complexes were effective in reducing oxidative effects generally believed to be $HO^•$ dependent, while in systems not requiring $O_2^{•-}$ and H_2O_2 as precursors for $HO^•$ generation (^{137}Cs γ-irradiation), Mn complexes were ineffective.

Magnetic resonance techniques have been applied to study the stability of the complexes formed between Mn(II) ions and **NADP** in aqueous solutions at a pH of 7.5 and 20° (GREEN and KOTOWYCZ 1979).

A purely chemical system for NAD(P)H oxidation to biologically active $NAD(P)^+$ has been developed and characterised by PAOLETTI et al. (1990). Suitable amounts of EDTA, manganous ions and mercaptoethanol, combined at physiological pH, induce nucleotide oxidation through a chain length also involving molecular oxygen, which eventually undergoes quantitative reduction to hydrogen peroxide. Mn^{2+} is specifically required for activity, while both EDTA and mercaptoethanol can be replaced by analogues. Two steps appear of special importance in nucleotide oxidation: (a) the supposed transient formation of **NAD(P)$^•$** from the reaction between NAD(P)H and thiyl radicals; (b) the oxidation of the reduced complex by superoxide to keep thiol oxidation cycling.

The manganese in resting **MnSOD** is trivalent, and the catalytic cycle of this enzyme involves reduction and then reoxidation of the metal centre during successive encounters with $O_2^{\bullet-}$. In contrast, the stable form of manganese in aqueous solutions is Mn(II), while Mn(III), a strong oxidant, tends to equilibrate with Mn(II) and Mn(IV).

Replacement of haeme in prostaglandin H synthase-1 by **mangano protoporphyrin IX** gives a holoenzyme that retains most cyclooxygenase activity but only ~4% of the peroxidase activity (ODENWALLER et al. 1992, STRIEDER et al. 1992, KULMACZ et al. 1994). LASSMANN et al. (1991) and KULMACZ et al. (1994) demonstrated that a 35–38 G radical was produced in mangano prostaglandin H synthase-1 during aerobic reaction with either fatty acid peroxide, 15-hydroperoxyeicosatetraenoic acid or arachidonic acid. The mangano prostaglandin H synthase-1 radical structure was sensitive to the tyrosine modifying agent, tetranitromethane, and the kinetics of its formation appeared to correlate with production of prostaglandin G_2/prostaglandin H_2 (KULMACZ et al. 1994). The EPR data of TSAI et al. (1998) for the arachidonic acid-derived radical formed by prostaglandin H synthase-2 and mangano prostaglandin H synthase-1 could be accounted for by a planar pentadienyl radical with two strongly interacting β-protons at C10 of arachidonic acid.

Mn^{II} forms an adduct with $O_2^{\bullet-}$, much as vanadate does (BIELSKI and CHAN 1978):

$$Mn^{II} + O_2^{\bullet-} \leftrightarrow Mn^{I}OO^{\bullet} \quad [206]$$

$Mn^{I}OO^{\bullet}$ can either act as an oxidant directly or, in the presence of chelating agents capable of stabilising Mn^{III}, can yield Mn^{III} which, in turn, acts as an oxidant:

$$Mn^{I}OO^{\bullet} + NADPH \leftrightarrow Mn^{I}OOH + NADP^{\bullet} \quad [207]$$

$$Mn^{I}OOH + H^+ \leftrightarrow Mn^{II} + H_2O_2 \quad [208]$$

$$Ligand\text{-}Mn^{I}OO^{\bullet} + 2\,H^+ \leftrightarrow Ligand\text{-}Mn^{III} + H_2O_2 \quad [209]$$
$$Ligand\text{-}Mn^{III} + NADPH \leftrightarrow Ligand\text{-}Mn^{II} + NADP^{\bullet} + H^+ \quad [210]$$

H_2O_2 is a powerful agent reducing Mn^{III}. ARCHIBALD and FRIDOVICH (1982) quote the following reactions:

$$Mn^{III} + H_2O_2 \rightarrow MnO_2^{+} + {}^2H^+ \quad [211]$$
$$MnO_2^{+} + Mn^{III} \rightarrow (Mn\text{-}O\text{-}O\text{-}Mn)^{+4} \quad [212]$$
$$(Mn\text{-}O\text{-}O\text{-}Mn)^{+4} \rightarrow 2\,Mn(II) + O_2 \quad [213]$$

The hydroxyl radical, $HO^{\bullet}$, reacts with Mn^{2+}:

$$Mn^{2+} + HO^{\bullet} \rightarrow Mn(OH)^{2+} \quad [214]$$

A purely chemical system for NAD(P)H oxidation to biologically active $NAD(P)^+$ has been developed by PAOLETTI et al. (1990). Mn^2 is specifically required for activity, while both EDTA and thioethanol can be replaced by analogues. Optimal molar ratios of chelator/metal ion (2:1) yield an active coordination compound which catalyses thiol autoxidation to thiyl radical. The latter is further oxidised to disulphide by molecular oxygen whose one-electron reduction generates superoxide radical.

$$(EDTA)_2\text{-}Mn^{(n)} + RS^- \rightarrow (EDTA)_2\text{-}Mn^{(n-1)} + RS^{\bullet} \quad [215]$$
$$RS^{\bullet} + RS^- \rightarrow RSSR^{\bullet-} \quad [216]$$
$$RSSR^{\bullet-} + O_2 \rightarrow RSSR + O_2^{\bullet-} \quad [217]$$
$$(EDTA)_2\text{-}Mn^{(n-1)} + O_2 + 2\,H^+ \rightarrow (EDTA)_2\text{-}Mn^{(n)} + H_2O_2 \quad [218]$$
$$\text{Sum: } 2\,RS^- + O_2^{\bullet-} + {}^2H^+ \rightarrow RSSR + H_2O_2 \quad [219]$$

Reaction [215] represents the initial event in thiyl radical formation as proposed for the so-called spontaneous autoxidation of thiols catalysed by metal ion complexes (MISRA 1974, SAEZ et al. 1982, HARMAN et al. 1984). The propagation (reactions [216] and [217]) leads to thiol oxidation and contemporary superoxide radical formation. The latter might be reduced to hydrogen peroxide by reacting with $(EDTA)_2\text{-}Mn^{(n-1)}$ to yield back the oxidised complex and close the catalytic cycle.

The importance of manganese and its compounds for industrial medicine and the uncertainty of the MPC values valuable now were emphasised by PFLAUMENBAUM et al. (1990).

Manganese is extracted from pyrolusite (MnO_2), braunite (Mn_2O_3), hausmannite (Mn_3O_4), and rhodochromite ($MnCO_3$, manganese spar).

CAMNER et al. (1985) found an increase in the size of alveolar macrophages lavaged from *rabbits* exposed to an aerosol of 3.9±1.0 mg Mn^{2+}/m^3 for 4–6 weeks (5 days/week, 6 h/day).

The toxicity of insoluble manganese dioxide dusts depends on the particle surface area as demonstrated *in vitro* by lactate dehydrogenase release from NMRI *mouse* peritoneal macrophages elicited by casein hydrolysate (LISON et al. 1997). *In vivo*, the lung inflammatory response was assessed by analysis of bronchoalveolar lavage after intratracheal instillation in *mice* (LDH activity and protein concentration in the cell-free fraction, recruitment of polymorphonuclear leucocytes). Freshly ground particles with a specific surface area of 5 m^2/g *in* vitro exhibited an enhanced cytotoxic activity, which was almost equivalent to that of 62 m^2/g particles, indicating that undefined reactive sites produced at the particle surface by mechanical cleavage may also contribute to the toxicity of insoluble particles.

Respiratory burst activity in lung macrophages increased at 61 % and 68 % after exposure to 10 μM and 50 μM manganese, respectively (GRABOWSKI et al. 1996). No increase in the degree of tyrosine phosphorylation in lysate from manganese exposed lung macrophages was evident. Despite the lack of increased tyrosine phosphorylation, the respiratory burst by manganese was reduced approximately 50 % by the NADPH oxidase inhibitor, 2-deoxy-D-glucose.

9.6.2.1.19 Mercury

Mercury has the atomic number 80 and the valence electrons are those of the $6s^1$ shell. The oxidation states of mercury include +1 and +2.

Nitrosothiols are very readily decomposed by the addition of mercuric ion (well-known for its coordinating ability to sulphur). This reaction results in NO^+ expulsion leading to nitrous acid in acid solution and nitrite anion formation at pH > ca 3 (equation [220]).

$$RSNO + Hg^{2+} \rightleftarrows \underset{\substack{| \\ Hg}}{[RSNO]^{2+}} \rightleftarrows RSHg^{+} + NO^{+} \qquad [220]$$

Even when carried out anaerobically there is no •NO formation and this reactions shows very little of the structural dependence so pronounced in the copper reactions (SWIFT and WILLIAMS).

Primary exposure to mercury occurs through environmental contamination as the result of mining, smelting, and industrial discharge, and include ingestion via inhalation and the food chain (GOYER 1991). The decrease in free sulfhydryl groups may lead to the formation of an oxidative stress, resulting in tissue-damaging effects.

Alveolar macrophages gained by lung lavage of freshly slaughtered, healthy *calves*, incubated with Hg^{2+} in the range of 0.25, 0.5, 1.0, 1.5, 2.0, and 5.0 μmol/l showed almost no rosetting with opsonized erythrocytes in the range of 1 to 5 μmol/l after 1 h incubation (BUSS et al. 1991).

9.6.2.1.20 Montmorillonite

The K10 montmorillonite clay catalysed the clean backbone rearrangement of cholest-5-ene into cholest-13(17)-enes; the transformation shown is quantitative and can be run on the gram scale (SIESKIND and ALBRECHT 1985).

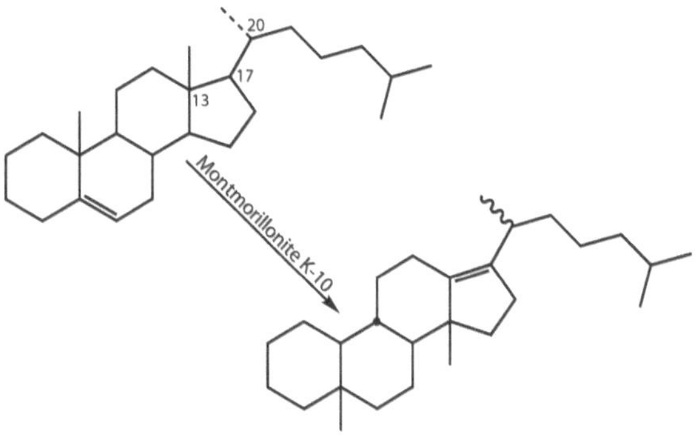

Rearrangement of cholest-5-ene into cholest-13(17)-enes [221]

The term "fuller's earth" is applied to a number of clay minerals capable to remove impurities from animal, plant and mineral oils. The soft, flocky, mica like minerals are open-pit mined and purified by calcination and washing.

Pneumoconioses after the inhalation of particulate fuller's earth from the Nutfield district were reported by MIDDLETON (1940) and CAMPBELL and GLOYNE (1942), also from Surrey by TONNING (1949), and from Olmstead, Illinois, USA by MCNALLY and TROSTLER (1941). After an exposure of 38 years, CAMPBELL and GLOYNE (1942) found the alveoli in the more severely affected areas of the lung involved in a reticular fibrosis surrounding the dust particles; macrophages, mononuclear cells, fibroblasts and leucocytes were present, but they were not very numerous and no foreign body-giant cells were observed. In TONNING's case, the exposure was some 50 years ago, and the dust was contained in large macrophages held in a mesh of fine reticulin fibres.

In *rats*, illite and muscovite, but not kaolin produced lesions of alveolar proteinosis (MARTIN et al. 1977). REISNER et al. (1982) and ROSMANITH et al. (1982) emphasised that the relation between the kaolinite content of the lymph nodes and that of the lungs six months after an intratracheal injection of different West German coal mine dusts showed an exponential regression function by the method of least squares.

In vitro illite and bentonite decreased intracellular lactate dehydrogenase activity of glycogen elicited peritoneal macrophages (ADAMIS and TIMÁR 1976, 1978). Groups which adsorbed paraquat cations may have a role in the cytotoxic effect of bentonite (ADAMIS and TIMÁR 1976).

The complex multilayered structure of mica with its surface exposed negatively charged honeycomb arrangements of $Si(Al)O_4$ tetrahedra has been used as a substrate for monolayer formation of amphiphilic organic molecules, such as alkylphosphonic acids (WOODWARD and SCHWARTZ 1996) and organosilanes (SCHWARTZ et al. 1992, OKUSA et al. 1994, HU et al. 1996, XIAO et al. 1996). As an alternative to the attachment of a two-dimensional siloxane network onto the mica surface, efforts have been made to alter the surface chemistry by exchanging the surface cations (mostly potassium) at the basal {001} cleavage plane with other inorganic and organic ions (SHELDEN et al. 1993, HÈHNER et al. 1996). Surfactant adsorption of long-chain alkylammonium salts, such as hexadecyl-trimethylammonium bromide (ERIKSSON et al. 1996, SHARMA et al. 1996) and *N*-dodecyl-pyridinium chloride (SHELDEN et al. 1993) are known to hydrophobize negatively charged minerals. Exchange with bivalent cations has been used to mediate binding of DNA for SPM studies (SHELDEN et al. 1994). Muscovite surfaces reversibly, site-specifically immobilised polyarginine-tagged fusion proteins (NOCK et al. 1997).

Alveolar macrophages were stimulated by montmorillonite not only after inhalation or intratracheal instillation (Fig. 144) of the dust, but also by intraperitoneal application (Figs. 145, 146).

A circulus vitiosus (VON ZGLINICKI and BRUNK 1993) comes into play because several damaged mitochondria have to be degraded in lysosomes. If lipofuscin accumulation in lysosomes decreases their degradative efficiency, damaged mitochondria producing high amounts of hydrogen peroxide will stay undegraded for longer times both outside and within lysomes.

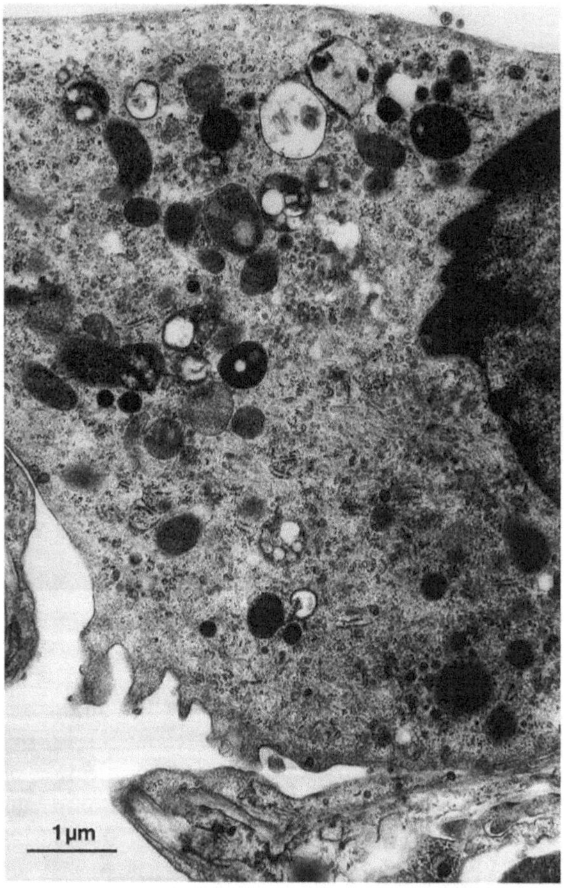

Fig. 144. Alveolar macrophage (block 4480) of a 201 g Sprague-Dawley *rat* (Charles River, France) instilled intratracheally with 50 mg montmorillonite API H 26 suspended in 0.5 ml saline. The animal was treated daily for 4 consecutive days with intraperitoneal injection of 15 mg carbochromene per kg body weight. Under pentobarbital anaesthesia (30 mg/kg), the lung was fixed by intratracheal instillation of 2.5 % glutaraldehyde in 0.1 M sodium cacodylate buffer (pH 7.4). Postfixation with 1 % osmium tetroxide in sodium cacodylate buffer. Embedded in Epon 812 and sectioned at 50 nm. Lead citrate and uranyl acetate. Plate 4132

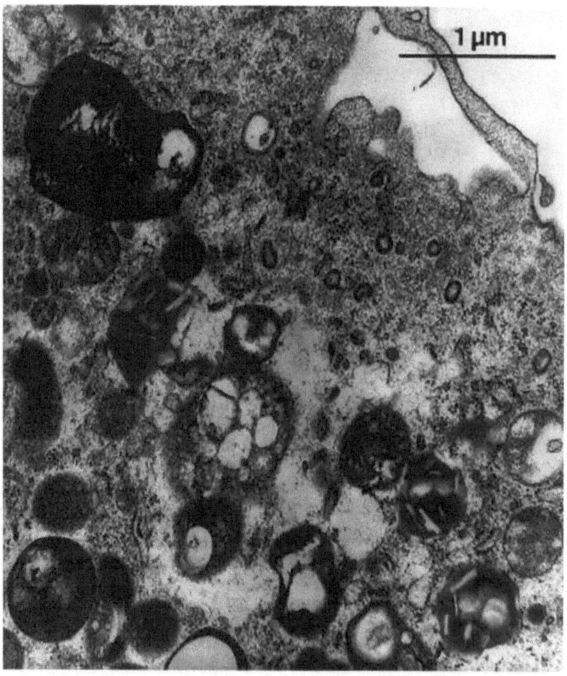

Fig. 145. Alveolar macrophage (block 4473) from a 215 g Sprague-Dawley *rat* (Charles River, France) injected intraperitoneally with 50 mg montmorillonite API H 26 suspended in 2 ml saline. Four days later, on July 25, 1978 under pentobarbital anaesthesia (30 mg/kg), the animal was perfused from the abdominal aorta with 2.5 % glutaraldehyde in 0.1 M sodium cacodylate buffer (pH 7.4). Postfixation with 1 % osmium tetroxide in sodium cacodylate buffer. Embedded in Epon 812 and sectioned at 50 nm. Lead citrate and uranyl acetate. Film 288/79

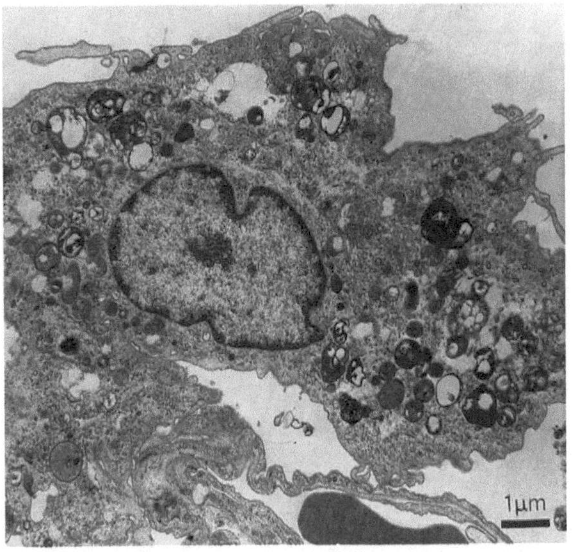

Fig. 146. Alveolar macrophage (block 4473) of a 215 g Sprague-Dawley *rat* (Charles River, France) injected intraperitoneally with 50 mg montmorillonite suspended in 2 ml saline. 4 days later under pentobarbital anaesthesia (30 mg/kg), the animal was perfused from the abdominal aorta with 2.5 % glutaraldehyde in 0.1 M sodium cacodylate buffer (pH 7.4). Postfixation with 1 % osmium tetroxide in sodium cacodylate buffer. Embedded in Epon 812 and sectioned at 50 nm. Lead citrate and uranyl acetate. Film 286/79

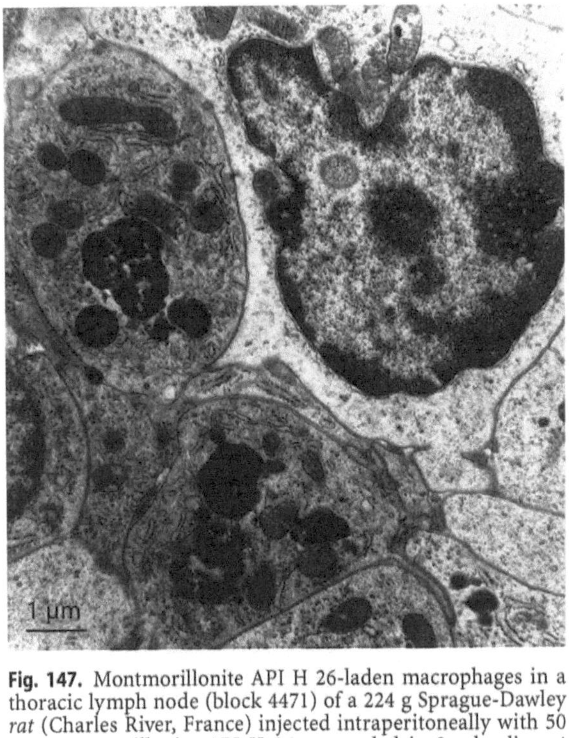

Fig. 147. Montmorillonite API H 26-laden macrophages in a thoracic lymph node (block 4471) of a 224 g Sprague-Dawley *rat* (Charles River, France) injected intraperitoneally with 50 mg montmorillonite API H 26 suspended in 2 ml saline. 4 days later under pentobarbital anaesthesia (30 mg/kg), the animal was perfused from the abdominal aorta with 2.5 % glutaraldehyde in 0.1 ml sodium cacodylate buffer (pH 7.4). Postfixation with 1 % osmium tetroxide in sodium cacodylate buffer. Embedded in Epon 812 and sectioned at 50 nm. Lead citrate and uranyl acetate. Film 295/79

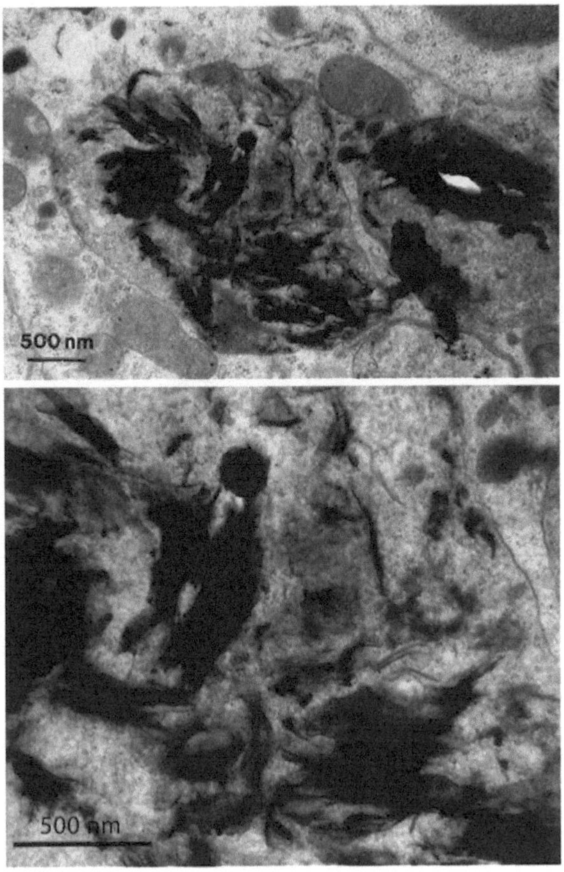

Fig. 148. Details of the montmorillonite API H 26 particles in a dust-laden macrophage in a thoracic lymph node (block 4471) of a 224 g Sprague-Dawley *rat* (Charles River, France) injected intraperitoneally with 50 mg montmorillonite suspended in 2 ml saline. 4 days later under pentobarbital anaesthesia (30 mg/kg), the animal was perfused from the abdominal aorta with 2.5 % glutaraldehyde in 0.1 ml sodium cacodylate buffer (pH 7.4). Postfixation with 1 % osmium tetroxide in sodium cacodylate buffer. Embedded in Epon 812 and sectioned at 50 nm. Lead citrate and uranyl acetate. Films 33121 and 33122

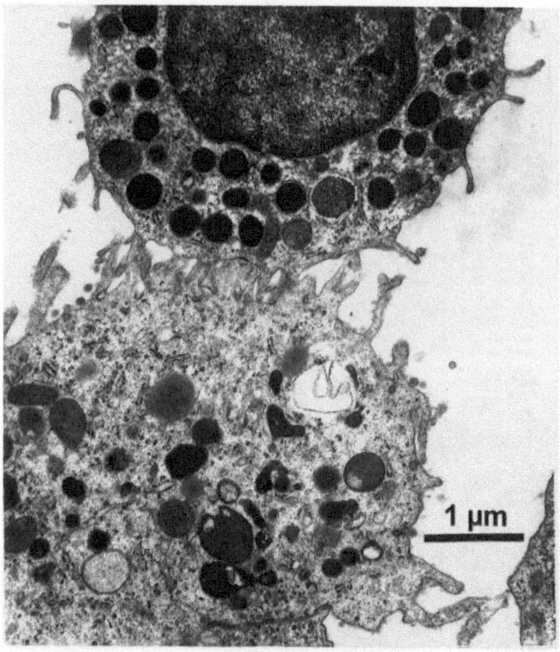

Fig. 149. Intimate contact of the plasmalemmata of a mononuclear phagocyte (bottom) and a mast cell (top) in a mesenteric lymph node (block 4477) of a 215 g Sprague-Dawley *rat* (Charles River, France) injected intraperitoneally with 50 mg montmorillonite API H 26 suspended in 2 ml saline. 4 days later under pentobarbital anaesthesia (30 mg/kg), the animal was perfused from the abdominal aorta with 2.5 % glutaraldehyde in 0.1 ml sodium cacodylate buffer (pH 7.4). Postfixation with 1 % osmium tetroxide in sodium cacodylate buffer. Embedded in Epon 812 and sectioned at 50 nm. Lead citrate and uranyl acetate. Film 271/79

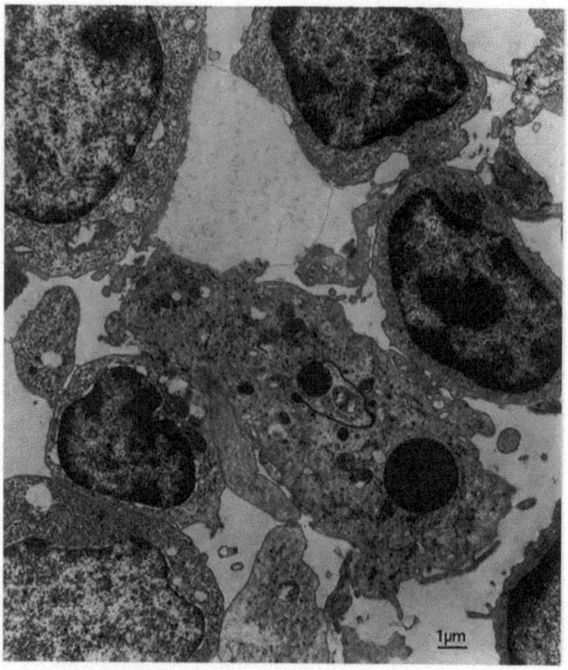

Fig. 150. Segregation of a damaged mitochondrium, cisternae of the rough endoplasmic reticulum, and free ribosomes by a limiting membrane within a macrophage from a mesenteric lymph node (block 4470) of a 224 g Sprague-Dawley *rat* (Charles River, France) injected intraperitoneally with 50 mg montmorillonite suspended in 2 ml saline. 4 days later under pentobarbital anaesthesia (30 mg/kg), the animal was perfused from the abdominal aorta with 2.5 % glutaraldehyde in 0.1 M sodium cacodylate buffer (pH 7.4). Postfixation with 1 % osmium tetroxide in sodium cacodylate buffer. Embedded in Epon 812 and sectioned at 50 nm. Lead citrate and uranyl acetate. Film 298/79

Fig. 151. Segregation of a damaged mitochondrium, cisternae of the rough endoplasmic reticulum, and free ribosomes by a phagophore within a macrophage from a mesenteric lymph node (block 4470) of a 224 g Sprague-Dawley *rat* (Charles River, France) injected intraperitoneally with 50 mg montmorillonite suspended in 2 ml saline. 4 days later under pentobarbital anaesthesia (30 mg/kg), the animal was perfused from the abdominal aorta with 2.5 % glutaraldehyde in 0.1 M sodium cacodylate buffer (pH 7.4). Postfixation with 1 % osmium tetroxide in sodium cacodylate buffer. Embedded in Epon 812 and sectioned at 50 nm. Lead citrate and uranyl acetate. Film 299/79

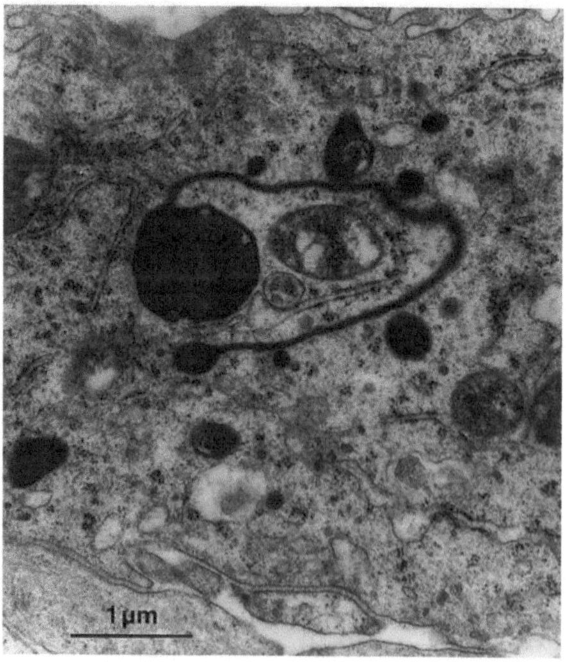

9.6.2.1.21
Nickel

Nickel has the atomic number 28 and the valence electrons are those of the $3d^8 4s^2$ shells. The oxidation states of nickel include −1, 0,+1, +2, +3, or +4, but the most prevalent oxidation state is +2.

The chemical and biological properties of nickel are highly dependent on its surrounding ligands. Both the coordination geometry and oxidation-reduction potential are significantly altered when simple salts forming $Ni\text{-}(H_2O)_6^{2+}$ in aqueous solution are chelated by strong donor ligands. Under physiological conditions, simple nickel salts exhibit little intrinsic oxidation or reduction chemistry particularly in contrast to iron, copper and cobalt (Kasprzak 1991, Hausinger 1993). However, nickel binds to the backbone of peptides and proteins, it becomes quite easily oxidised by O_2, $O_2^{\bullet -}$, and H_2O_2 (Bossu et al. 1977, 1978). GGH-Ni(II), glutathione-Ni(II), and other oligopeptide complexes of Ni(II) have been shown to promote lipid and protein oxidation in the presence of peroxides to a much greater extent than either free Ni^{2+} or ligand alone (Shi et al. 1992). Similarly, GGGG-Ni(II) induced polymerisation of histones (Kasprzak and Bare 1989), and histidine-Ni facilitated oxidation of deoxyguanosine (Datta et al. 1992).

Mononuclear thiolate complexes containing nickel sites in different oxidation state as $[Ni(SC_6H_4O)_2]^{2-}$ and $[Ni(SC_6H_4O)_2]^-$ were described by Köckerling and Henkel (1993). With respect to the chemical as well as electrochemical properties of the nickel sites of various dehydrogenases, $[Ni(SC_6H_4O)_2]^-$ is a relevant model complex of coordination number four.

Ni^{2+} forms a lipophilic complex with pyridinethione by bonding through the oxygenated sulphur atoms. Lipophilic complex formation may promote metal absorption and tissue penetration. Oral administration of sodium pyridinethione together with $^{63}Ni^{2+}$ to *rats, guinea-pigs* and *ferrets* was found to induce highly increased tissue levels of the metal in comparison with animals given the Ni^{2+} alone (Borg-Neczak and Tjälve 1994).

Nickel exists in many forms as an inhalable pollutant which include metallic nickel, nickel subsulphide, nickel chloride, nickel oxide, and nickel carbonyl. Of these compounds, Ni_3S_2 is recognised as the most carcinogenic (Sunderman 1983). The majority of nickel-associated lesions described in *man* have been caused by inhalation exposure (Wehner et al. 1975). Immune mechanisms appear to be particularly vulnerable to the effects of nickel compounds.

In an aqueous medium **reactive oxygen species** were generated by powdered Ni_2O_3 when Ni(II) were converted to Ni(III) ions (Sawatari 1988).

One-electron transfer catalysed by nickel

$$O_2 + Ni^{2+} \leftrightarrow Ni^{3+} - \mathbf{O_2^{\bullet -}} \rightarrow \mathbf{O_2^{\bullet -}} + Ni^{3+} \quad [149]$$

$$\mathbf{O_2^{\bullet -}} + Ni^{2+} + 2\,H^+ \rightarrow H_2O_2 + Ni^{3+} \quad [150]$$

$$H_2O_2 + Ni^{2+} \rightarrow \mathbf{HO^\bullet} + HO^- + Ni^{3+} \quad [151]$$

Solubilities in distilled water and saline were in the order of nickel fumes (97 % NiO, 3 % Ni_2O_3; particle size 5–10 nm) > Ni_2O_3 powder (particle size 2.0±1.69 μm) >> NiO powder (particle size 2.2±1.68 μm) (Toya et al. 1997).

Phagocytic indices of particulate nickel compounds by *rat* peritoneal macrophages *in vitro* were ranked by Kuehn et al. (1982) as follows: NiO > Ni_4FeS_4 > $NiTiO_3$ > NiSe > αNi_3S_2 > Ni > Ni_5As_2 > NiS_2 > NiFe alloy > NiSb > $Ni_{11}As_8$ > Ni_3Se_2 > βNiS > NiTe > NiAs > NiAsS > amorphous NiS. Rank-correlation ($P < 0.03$) was observed between the phagocytic indices of the nickel compounds and their dissolution half-times in *rat* serum. Nickel subsulphide, αNi_3S_2 was a notable exception to the general concordance between phagocytic indices and dissolution half-times: αNi_3S_2 was avidly phagocytized by macrophages, yet it had one of the shortest dissolution half-times. Preliminary results of the carcinogenesis tests of 14 of the nickel compounds did not indicate significant rank-correlation between the phagocytic indices of the nickel compounds and the sarcoma indices at 1 year after intramuscular administration of the compounds to *rats*.

Metallic Nickel

In manual metal arc and shielding gas welding fumes nickel is dissolved in the iron oxide fume particles and in small amounts it exists as separate nickel oxide particles.

Alveolar macrophages from *rabbits* exposed to 0.5 mg Ni dust/m³ 6 h per day during a 5-day work week after 4 weeks showed highly undulating membranes, numerous slender microvilli and long protrusions from the cell surface, excentrically placed polymorphic nuclei with chromatin marginations, and well-developed nucleoli (Camner et al. 1978). There were also membrane-bound bodies containing laminated structures of different sizes, similar to the laminated structures of alveolar type II cells. Lysosomal structures were absent in nickel-exposed cells found to contain large amounts of laminated structures, which was in contrast to the control macrophages. X-ray microanalysis revealed no intracellular Ni in the *in vivo* experiment, while *in vitro* macrophages internalized large amounts of Ni particles.

Instillation of 0.1 g Ni dust suspended in 1 ml isotonic saline into the *rabbit's* trachea induced macrophages with a surface rich in long protrusions and microvilli (WIERNIK et al. 1981). The cytoplasm contained many lamellar bodies and lipid droplets. The nuclei of the cells were polymorphic and had well developed nucleoli. The Golgi areas were prominent and so was the granular endoplasmic reticulum. The nitroblue tetrazolium reduction in macrophages from nickel-exposed *rabbits* showed, without any incubation, values of the same magnitude as from the corresponding cells lavaged from the lungs of unexposed *rabbits* after incubation with the surfactant from nickel-treated animals.

Nickel Subsulphide

In vitro, phagocytosis of particulate Ni_3S_2 resulted in a rapid dissolution of the compound once it had entered the macrophage with subsequent release of Ni^{2+} ion (COSTA et al. 1981). Scanning electron micrographs of bovine macrophages exposed to Ni_3S_2 showed loss of surface features, membrane disruption, and bleb formation (FISHER and PLACKE 1987).

$^{\bullet}$NO generation from RAW 264.7 Abelson virus-transformed *murine* macrophages stimulated with lipopolysaccharide (0.5 or 1 μg/ml) was enhanced by Ni_3S_2 in a dose-dependent manner (KAWANISHI et al. 2001). The addition of 100 μg Ni_3S_2/ml caused strong cell toxicity. Nickel compounds did not induce $^{\bullet}$NO production in the absence of lipopolysaccharide. When RAW 264.7 cells were stimulated with 100 U/ml interferon-γ, Ni_3S_2 enhanced the NO_2^- accumulation. A nonselective NOS inhibitor, N^G-monomethyl-L-arginine, inhibited enhancement of $^{\bullet}$NO production, suggesting that the enhanced accumulation of NO_2^- was due to induction of NOS.

Repeated inhalation of 0.6 or 2.5 mg β-Ni_3S_2/m^3, 6 h/day for up to 22 days by male and female F344/N *rats* aged 7 weeks at the start of the exposure resulted in an exposure-concentration-dependent increase of lactate dehydrogenase activity in the bronchoalveolar lavage fluid (BENSON et al. 1995). β-Glucuronidase activity increased more pronounced in *rats* exposed to 2.5 mg than to 0.6 mg Ni_3S_2/m^3.The viability of the alveolar macrophages was significantly reduced after 2 days of exposure, but more slowly thereafter.

Male *Cynomolgus monkeys* instilled intratracheally with 800 nM Ni_3S_2/g lung showed mild focal accumulation of macrophages and lymphocytes within the pulmonary interstitium and alveoli, perivenular lymphoid infiltration and follicles with a central core of large monocytes, macrophages, and lymphoblasts surrounded by a mantle of small, darkly basophil lymphocytes (HALEY et al. 1987).

Nickel Chloride

Acute and chronic inhalation exposure of *rabbits* and *mice* to $NiCl_2$ consistently resulted in deficient macrophage function as indicated by decreased phagocytic and bactericidal capacity of pulmonary alveolar macrophages (ADKINS et al. 1979, WIERNIK et al. 1983). Viability of macrophages in these studies was unchanged, reflecting a subtile functional abnormality that has been reproduced *in vitro* (GRAHAM et al. 1975).

Washed alveolar macrophages from *rats* exposed (8 h per day, 5 days per week, for 18 days) to $NiCl_2$ (109 μg/m^3) aerosols were found to contain reduced quantities of the various hydrolytic enzymes (except for acetylesterase) when compared with those from control *rats* (MURTHY et al. 1983). On the other hand, a significant increase in enzymatic activity was noted in lung washout fluid from exposed animals.

Nickel Sulphate

The numbers of macrophages and polymorphonuclear leucocytes were increased in bronchoalveolar lavage fluid following the intratracheal instillation of 0.4 ml $NiSO_4$ solution (50 μg Ni/*rat*) or inhalation (36.5 mg Ni/m^3 for 2 h) of nickel sulphate (HIRANO et al. 1994).

Mice deficient in the cytoplasmic domain of Ron (Ron tk-/-) when exposed to aerosolised $NiSO_4$ (115 ± 7 μg/m^3) succumbed to nickel-induced acute lung injury earlier, expressed larger, early increases in interleukin-6, monocyte chemoattractant protein-1, and macrophage inflammatory protein-2, displayed greater serum nitrite levels, and exhibited earlier onset of pulmonary pathology and augmented pulmonary tyrosine nitrosylation (MCDOWELL et al. 2002).

Nickel Oxides

In F344/N *rats* and B6C3F_1 *mice*, repeated inhalation of **NiO** at levels resulting in alveolar macrophage hyperplasia and alveolitis impaired clearance of subsequently inhaled 7.5–16.2 mg ^{63}NiO/m^3 of a mass median aerodynamic diameter of 1.1–2.2 μm (BENSON et al. 1995).

In the *rat*, single (3.8 mg/kg) or repeated (14.3 mg/kg) intratracheal instillations of nickel fumes induced a marked mobilisation of macrophages into the alveoli, the formation of foam cells, and their destruction (Toya et al. 1997).

Alveolar macrophages of *rats* injected intratracheally with NiO contained invaginated bizarre-shaped nuclei, which lacked prominent nucleoli, numerous primary and secondary lysosomes, pigment aggregations, and several membranous whirls (Migally et al. 1982). *In vitro*, >500 μM $\mathbf{Ni_2O_3}$ inhibited H_2O_2 and $O_2^{\bullet -}$ release from alveolar macrophages of *rabbits* injected intravenously with complete Freund's adjuvant by 50 % (Labedzka et al. 1989). 100 μM Ni_2O_3 increased extracellular lactate dehydrogenase activity to 140.0 ± 68.8 % of the controls.

Washed alveolar macrophages from *rats* exposed (8 h per day, 5 days per week, for 18 days) to NiO (120 μg/m^3) aerosols were found to contain reduced quantities of the various hydrolytic enzymes (except for acetylesterase) when compared with those from control *rats* (Murthy et al. 1983). On the other hand, a significant increase in enzymatic activity was noted in lung washout fluid from exposed animals.

In the lungs of *rats* instilled with 0.5 ml of a suspension of 1 mg NiO + 0.1 mg CdO one week before, the alveolar macrophages appeared smaller than those of control, with numerous lamellar bodies of different sizes. The lamellar bodies consisted of membranous whorls, disintegrating structures, as well as accumulation of surfactant-like material (Murthy and Holovack 1991). In addition, the cytoplasm was filled with multivesicular bodies, degenerating mitochondria, as well as primary and secondary lysosomes.

Nickel Carbonyl

Owing to its volatility (boiling point at 43 °C), lack of strong odour, and propensity for inadvertent formation, nickel carbonyl, $Ni(CO)_4$, is generally considered the most hazardous compound encountered in the workplace. 38 per cent of the administered $^{63}Ni(CO)_4$ was detected in the expired air for 6 h after intravenous injection into male Sprague-Dawley *rats*, demonstration that the lung is a major route for excretion of $Ni(CO)_4$ (Sunderman and Selin 1968). Barnes and Denz (1951) and Ghiringhelli and Agamennone (1957) demonstrated that nickel was rapidly mobilized from lung, liver, kidney, brain and blood of *rats* following inhalation of nickel carbonyl.

Chromatographic fractionations performed on the ultracentrifugal supernatants of homogenates of lung and liver of *rats* exposed to $Ni(CO)_4$ demonstrated nickel firmly bound to macromolecular constituents (Sunderman 1964). A Ni^{2+}-binding protein, *pNiXa*, was identified in *Xenopus* oocytes and embryos (45 kDa, isoelectric point ~8.5) with a strong homology to human α_1-antitrypsin, α_1-antichymotrypsin, and other serine protease inhibitors (Sunderman 1993).

9.6.2.1.22 Nitrogen Oxides

Nitrogen has the atomic number 7 and the molecular orbitals of N_2 are $KK\sigma^2 2s\sigma^{*2} 2s(\pi 2p_y \pi 2p_z)^4 \sigma^2 2p_x$. The oxidation states of nitrogen include –3, –2, –1, +1, +2, +3, +4, and +5.

Nitrogen dioxide and nitric oxide have been associated with industrial pulmonary disease for a long time (e.g. Haldane 1902, Lowry and Schuman 1956). In coal mines these gases are present both in diesel exhaust and shotfiring fume. Particular concern has been expressed about possible adverse effects on the health of coal miners from high levels of oxides of nitrogen present after shotfiring (Nicholas and Wall 1971) and Kennedy (1970, 1972, 1974) has suggested that exposure to oxides of nitrogen in shotfiring fume may cause emphysema in coal miners. Robertson et al. (1982) found both nitric oxide and nitrogen dioxide readily absorbed on coal, DQ12 quartz and kaolinite dusts.

Nitroxides have been used to study the membranes of macrophages and also to investigate some of the oxidative intermediates that are produced by macrophages as part of their mechanism of cell killing. As in other cell lines, the reduction of lipophilic nitroxides, which spontaneously localise in membranes, is not preceded by active internalisation and cytoplasmic reduction and does not occur by penetration of external ascorbic acid into the lipid bilayer. The reduction is enzymatic and can be described by first-order kinetics; it can be reversed by potassium ferricyanide, decreased by disulphides, and increased by NO_2 (Rowlands et al. 1978).

Nitrogen Dioxide

Similar to ozone, NO_2 is relatively insoluble and penetrates deep into the respiratory tract.

Concanavalin A agglutinated alveolar macrophages of female Sprague-Dawley *rats* that had inhaled 12.1 ppm nitrogen dioxide for 2 h (Goldstein et al. 1977). Agglutinability was almost completely inhibited by α-methylmannose. Exposure of male Long-Evans *rats* to 40 ppm NO_2 for 5 h resulted in a significant increase in NADPH cytochrome *c* reductase activity (46 %) in macrophages, but no change in any biosynthetic enzymes (WRIGHT et al. 1982).

Adsorption of nitrogen dioxide onto kaolinites produced a small but significant decrease in toxicity to P388D$_1$ cells (ROBERTSON et al. 1982).

Nitrogen Oxide and •NO-forming Drugs

A simple molecular orbital diagram for •NO shows that the unpaired electron resides in an antibondal orbital. The fact that •NO contains one electron in an antibonding orbital explains the bond order of 2.5 (i.e., the total bonding is described as three net bonds gained from the filled σ_z, π_x, and π_y molecular orbitals minus half a bond from the partially filled π^* antibonding orbital). Since the highest occupied molecular orbital is antibonding in nature, it may be expected that this electron is loosely held and should be easily lost to generate $^+$NO. This is indeed the case, as the ionisation potential of NO is only 9.25 eV (compared to, e.g. 15.56 eV for N_2 or 14.1 eV for CO).

NO fluxes could be detected at distances from RAW 264.7 macrophages of 100–500 µm (PORTERFIELD et al. 2001). The initial flux and the distance from the cells at which NO could be detected were directly related to the number of cells in the immediate vicinity of the probe releasing NO. Thus, whereas NO fluxes of ~1 pmol × $cm^{-2} \times s^{-1}$ were measured from individual macrophages, aggregates composed of groups of cells varying in number from 18 to 48 cells produces NO fluxes between ~4 and 10 pmol × $cm^{-2} \times s^{-1}$. NO fluxes required the presence of L-arginine. Signals were significantly reduced by the addition of haemoglobin and by *N*-nitro-L-arginine methyl ester. NO fluxes were greatest when the sensor was placed immediately adjacent to cell membranes and declined as the distance from the cell increased. The NO signal was markedly reduced in the presence of the protein albumin but not by either oxidised or reduced glutathione. A reduction in the NO signal was also noted after the edition of lipid micelles to the culture medium.

One of the most biologically significant aspects of •NO chemistry is its ability to bind to and/or react with metals and metal containing proteins. The activation of the iron heme-containing enzyme guanylate cyclase occurs through a ligation of •NO to the iron heme (see, e.g. IGNARRO 1989, 1992, and the references therein). Pulmonary soluble guanylate cyclase is a hemoprotein containing 1 mol of ferroprotoporphyrin-IX per mol of heterodimer (GERZER et al. 1981). Enzyme stimulation is mediated by binding of •NO to the heme (IGNARRO 1991). STONE and MARLETTA (1994) showed that soluble guanylate cyclase contains a pentacoordinated high-spin ferrous heme with histidine as the proximal ligand.

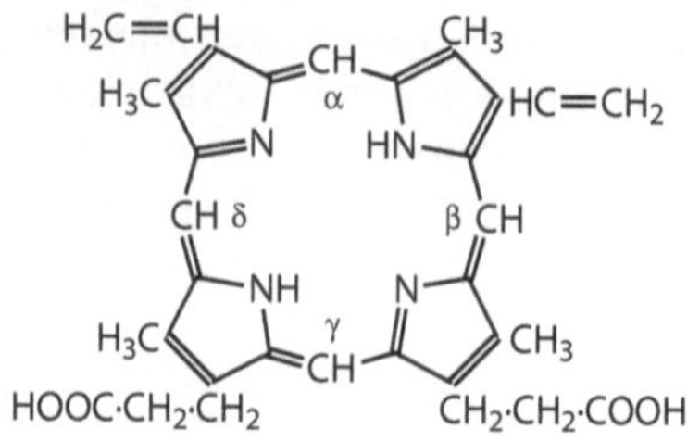

Protoporphyrin IX [222]

Protoporphyrin IX [222], the iron-free precursor of heme, stimulates soluble guanylate cyclase (EC 4.6.1.2) independently of •NO (IGNARRO et al. 1982). As protoporphyrin IX does not contain Fe, its structure resembles that of NO-heme complex in which the iron is moved out of the plane of the porphyrin ring. In both cases, i.e. in the protoporphyrin IX- and the •NO-stimulated enzyme, the axial histidine is unbound.

Phantom solutions of synthetic nitrosyl-iron complexes altered the signal intensity of proton-electron-double-resonance imaging (MÜLSCH et al. 1999). The dinitrosyl-iron complex with serum albumin induced a significantly larger signal alteration than the mononitrosyl-iron complex with dithiocarbamate. Exposure of *rat* liver to sodium nitroprusside by ex vivo and in situ perfusion induced a composite X-band electron spin resonance spectrum of the isolated liver characteristic of a mononitrosyl-iron complex and dinitrosyl-iron complex. On storage of the tissue, the mononitrosyl-iron complex signal disappeared and the dinitrosyl-iron complex signal intensity increased. Correspondingly, in cross-sectional proton-electron-double-resonance images taken at room temperature, the sodium nitroprusside-exposed livers initially exhibited a weak signal that strongly increased with time.

Upon activation with monocyte chemoattractant protein-1, *murine* peritoneal macrophages showed a dose- and time-dependent production of •NO together with increased tumoricidal activity against P815 mastocytoma cells (BISWAS et al. 2001). *N*-Monomethyl-L-arginine inhibited the monocyte chemoattractant protein-1-induced •NO secretion and generation of macrophage-mediated tumoricidal activity against P815 (NO-sensitive, TNF-resistant) cells but not the L929 (TNF-sensitive, NO-resistant) cells.

•NO produced by the cells upon cytokine induction, or added as a gas to the recombinant protein, converts the iron-regulatory protein (also called iron-regulatory factor [IRF] or iron-responsive element-binding protein [IRE-BP]) to the same condition as low iron, causing a binding of the protein to mRNA (DRAPIER et al. 1993, WEISS et al. 1993).

Macrophages isolated 0–48 h after an exposure of male Balb/c *mice* to 80–100 ppm •NO for 5 h produced low levels of •NO, but stimulation with lipopolysaccharide (**LPS**) and interferon-γ resulted in a time-dependent induction of •NO (Laskin et al. 1997). Alveolar macrophages from •NO-exposed *mice* produced significantly more •NO than cells from control animals which was maximum in cells isolated 24–48 h after exposure. This was correlated with increased inducible nitric oxide synthase (iNOS) expression by the cells. Alveolar macrophages from •NO-exposed *mice* were also found to produce significantly more superoxide anion than cells from control *mice*. This was evident *in vitro* in isolated cells and *in situ* in histological slides. •NO reacts rapidly with superoxide anion to form peroxynitrite, a potent oxidising agent. Following inhalation of •NO, alveolar macrophages were also found to produce significant quantities of peroxynitrite.

Nitroxide radicals without charged groups were reduced significantly in the *murine* lung, while radicals with charged groups were only slightly reduced (Takeshita et al. 1999). Permeation rates across lung plasma membrane were not limiting of the stage of reduction of noncharged nitroxides.

L-*Arginine*

Arginine can be synthesised from citrulline in the body, the kidneys being the major site (Dhanakoti et al. 1990). L-Citrulline produced by nitric oxide synthase can be recycled to L-arginine in endothelial cells (Hecker et al. 1990). Sessa et al. (1990) reported that the synthesis of arginine from citrulline in endothelial cells was markedly inhibited by L-glutamine or L-arginine, pointing to a possible role for amino acids in controlling the availability of arginine and the production of •NO. The synthesis of [^{14}C]-arginine from 0.1 mM [^{14}C]-citrulline by *rat* peritoneal macrophages *in vitro* was about 300 pmol/h per 10^6 cells (Wu and Brosnan 1992). Both arginine synthesis from citrulline and nitrate production as an indicator of •NO generation were increased about 3-fold in the cells from lipopolysaccharide-treated animals. The rate of arginine synthesis from citrulline was inhibited by about 20 % by 0.5 mM L-glutamine in both control and lipopolysaccharide-treated *rat* cells, but was inhibited by 0.5 mM L-arginine only in control cells.

L-Arginine–nitric oxide pathway

NOS $+ O_2^{\bullet -}$
L-Arginine → **Nitric oxide**(•NO) → Peroxynitrite ($ONOO^-$)
[46]

When cultured *rat* alveolar macrophages were stimulated with LPS, substantial amounts of nitric oxide were produced (Persoons et al. 1996). Inhibition of the nitric oxide production by the L-arginine analogue, N^G-monomethyl-L-arginine resulted in an increase of IL-1β and IL-6, whereas TNF-α concentrations remained unchanged. Conversely, the nitric oxide donor, S-nitroso-*N*-acetyl-D,L-penicillamine induced dose dependent inhibition of IL-1 production in LPS-stimulated alveolar macrophages in which endogenous nitric oxide production was blocked.

For macrophage cell death, activation of endogenous •NO generation or exogenously applied •NO causes apoptosis. Morphological criteria as well as biochemical analysis revealed characteristic apoptotic features (Messmer and Brüne, unpublished, quoted from Brüne et al. 1995). A chemically heterogeneous group of NO-releasing compounds like sodium nitroprusside, SIN-1, S-nitroso-*N*-acetyl-D,L-penicillamine, spermine-NO, and the diethylamine-nitric oxide complex produced a time- and concentration-dependent effect.

While increasing concentrations of free **calcium** (EC_{50} 300 and 30 nM, respectively) increased •NO synthesis in *porcine* endothelial and cerebellar cytosol, it was unaffected in *murine* bone marrow macrophage cytosol (Mülsch et al. 1991).

Endogenous nitric oxide production may be linked to **1, 25-dihydroxyvitamin D_3** synthesis in *chick* myeolomonocytic HD-11 cells *in vitro*, indicating that macrophage NO-generating capacity could be functionally linked to endogenous synthesis of the active vitamin D metabolite (Adams et al. 1994).

There has been considerable controversy about whether *human* macrophages make nitric oxide. Treatment *in culture* with the same sorts of regimens that stimulate nitric oxide production in *rodent* cells generally failed to elicit significant amounts of nitric oxide production (Padgett and Pruett 1992), though Munoz-Fernandez et al. (1992) described activation for the killing of intracellular *Trypanosoma cruzi* by TNF-α and TNF-γ through a nitric oxide dependent mechanism, and Sherman et al. (1991) cytokine- and *Pneumocystis carinii*-induced L-arginine oxidation by *human* alveolar macrophages.

Sydnonimines

The sydnonimines are one of the most interesting classes of organic nitro compounds as they are highly active in treating coronary artery disease and are nearly devoid of tolerance (i.e. in contrast with organic nitrates long-term therapy does not become ineffective).

Sydnonimines are unstable in aqueous solutions. The sydnonimine ring opens, by a base-catalysed mechanism to give SIN-1A. SIN-1A reduces oxygen, in a one-electron transfer reaction, to give superoxide anion radical ($O_2^{\bullet-}$) and SIN-$1^{\bullet+}$, a cation radical. SIN-$1^{\bullet+}$ decomposes to $^{\bullet}$NO and biologically inactive metabolites like SIN-1C (FEELISCH et al. 1989). By high-pressure liquid chromatography, NOACK and FEELISCH (1989) found that 500 μM SIN-1 were converted to SIN-1A with a half-time of 126 min, while SIN-1C was formed three to four times more quickly with a half-time of 38 min. The thiomorpholinyl analogue of SIN-1, compound C 78-0698, showed that the A form was more rapidly formed than with SIN-1, but was considerably more stable.

30 min after an oral application of 6 mg ^{14}C-*N*-ethoxycarbonyl-3-morpholino-sydnonimine (molsidomine) per kg *rat*, pulmonary radioactivity (2.98±0.892 μg/g wet weight) was increased in relation to blood (1.74±0.714 μg/g wet weight) (TANAYAMA et al. 1970).

In totally hepatectomized *rats*, the *N*-ethoxycarbonyl-3-morpholinosydnonimine level did not decline during the first hour after intravenous injection of the drug (TANAYAMA et al. 1974). Furthermore, the amounts of the metabolites in the blood did not increase with time.

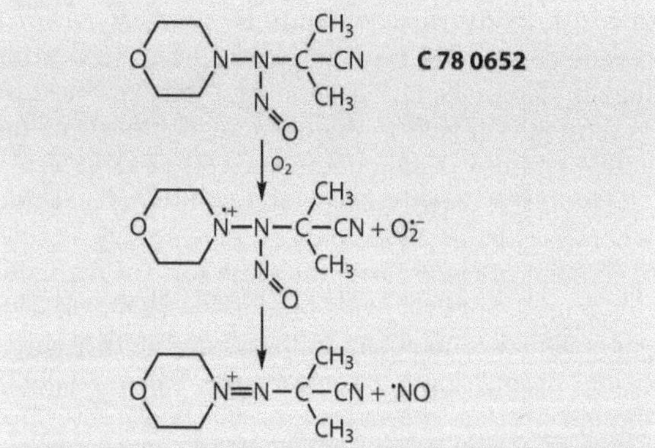

Formation of both $^{\bullet}$NO and $O_2^{\bullet-}$ from sydnonimine metabolites, e.g. open ring compound C 78-0652 [136]

The very different chemical reactivity towards α-keto-γ-methiolbutyric acid by the SIN-1 and iron-ascorbate-generated oxidants indicates that hydroxyl radical is not a major oxidant produced by the SIN-1 system (REGOLI and WINSTON 1999).

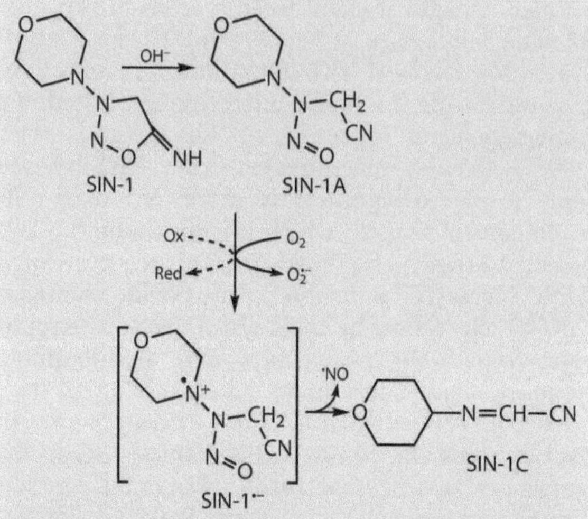

Mechanism of $^{\bullet}$NO production and formation of an imino nitroxide during the reaction of SIN-1 and nitronyl nitroxide (SINGH et al. 1999) [223]

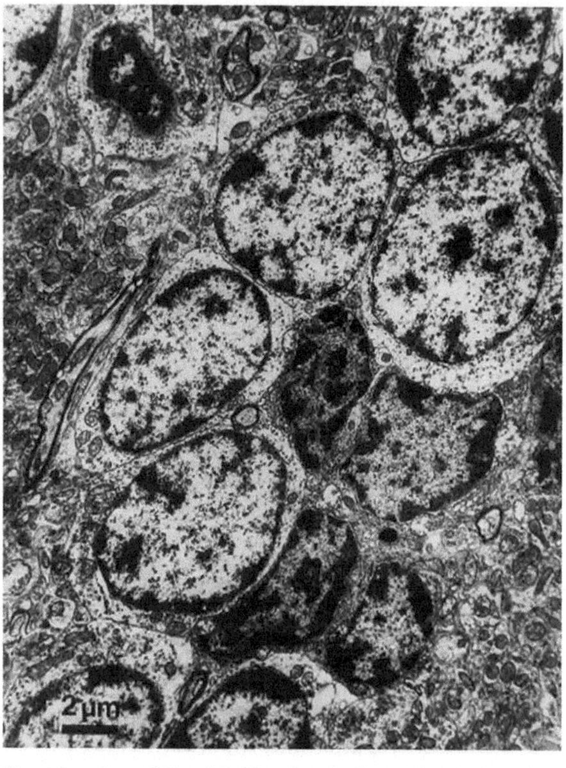

Fig. 152. Microglial cell infiltration in the granule layer of the cerebellum (block 4766) of a 221 g female *rat* (No. 668) medicated with 5,000 ppm molsidomine added to the food (powdered Altromin® R) from July 17/18, 1978 to November 30, 1978. After discontinuation of the medication for 7 weeks, under pentobarbital anaesthesia (30 mg/kg), the animal was perfused from the abdominal aorta with 2.5 % glutaraldehyde in 0.1 M sodium cacodylate buffer (pH 7.4). Postfixation with 1 % osmium tetroxide in sodium cacodylate buffer. Embedded in Epon 812 and sectioned at 50 nm. Lead citrate and uranyl acetate. Film 182/79

Nitronyl nitroxide has been widely used for detecting and antagonising the effect of •NO (AKAIKE et al. 1993, HOGG et al. 1995, KONOREV et al. 1995). Nitronyl nitroxide is converted to an imino nitroxide during its reaction with •NO with a stoichiometry of approximately 0.5:1 (HOGG et al. 1995). PFEIFFER et al. (1997) demonstrated that nitronyl nitroxide enhanced the formation of •NO from SIN-1, as determined by an increase in cyclic GMP levels. SINGH et al. (1999) reported that one-electron oxidising agents, in addition to oxygen, can oxidise SIN-1A, resulting in the release of •NO without the concomitant formation of $O_2^{\bullet -}$. Easily reducible nitroxides, such as the nitronyl and imino nitroxides, are able to oxidise SIN-1. Biological oxidising agents such as ferricytochrome *c* also stimulate •NO production from SIN-1. Decomposition of SIN-1 by *human* plasma or by the homogenate of *rat* liver, kidney and heart tissues resulted in the formation of •NO.

Nitric oxide Synthase Induction as Influenced by Sydnonimines

A single dose of the **molsidomine** metabolite SIN-1 given in the postinduction phase did not inhibit •NO release from *mouse* macrophages (line RAW 264.7) induced by LPS/IFN-γ, but •NO formation was reduced by about 60% when 500 μM SIN-1 were applied for 4 h in the induction phase (OSTROWSKI et al. 1992). Superoxide dismutase antagonised this effect. The pharmacologically inactive molsidomine metabolite SIN-1C and sydnonimine C 3754, a metabolite of **pirsidomine**, did not influence the induction of NO synthase activity in this cell line, while LPS/IFN-dependent induction was lost by 40 and 60%, respectively, when LPS/IFN or *mouse* macrophages were preincubated with SIN-1 for 1 h. [^{3}H] L-citrulline formation mediated by cytosolic *mouse* macrophage NO synthase was reduced by SIN-1, which could not be compensated by superoxide dismutase. Pirsidomine metabolite C 3786 was unable to change nitric oxide synthase activity. In *bovine* alveolar macrophages SIN-1 (100 μM) increased iNOS mRNA after isolation procedure comparable with untreated cells (HÖCKELE et al. 1997).

Oxatriazole-type NO-donors

The novel oxatriazole-type •NO-donors (GEA compounds) are mesoionic 3-aryl substituted oxatriazole-5-imine derivatives releasing •NO. They have vasodilator, antiplatelet and fibrinolytic activities (CORELL et al. 1994, KARUP et al. 1994, KANKAANRANTA et al. 1996). They are also effective in reducing blood pressure (NURMINEN and VAPAATALO 1996) and inhibiting the growth of tumour and haematopoietic cells (VILPO et al. 1994, 1997).

After 60 min incubation of *rat* vascular smooth muscle cells the order of the compounds in their capability to release •NO was GEA 3162 = SIN 1 > *S*-nitroso-*N*-acetylpenicillamine > GEA 5624 (LÄHTEENMÄKI et al. 1998).

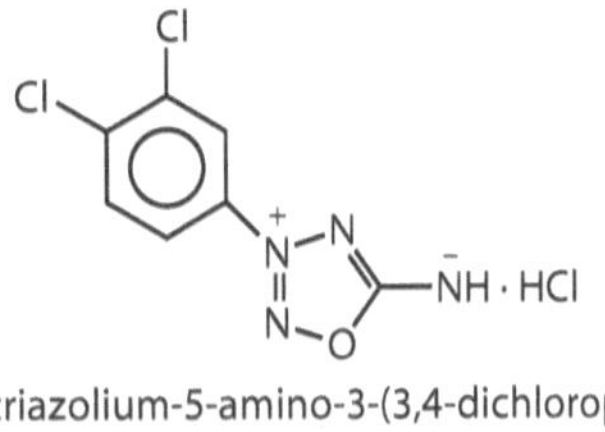

(1,2,3,4-oxatriazolium-5-amino-3-(3,4-dichlorophenyl) [224]

Co-culture of *human* neutrophils with the NO donors GEA 3162 (1,2,3,4-oxatriazolium,5-amino-3-(3,4-dichlorophenyl)-chloride) (10–100 μM) and 3-morpholino-sydnonimine (SIN-1) (0.3–3 mM) caused a dramatic and concentration-dependent induction of apoptosis (WARD et al. 2000).

4-Ethyl-2[(Z)-hydroxyiminol]-5-nitro-3(E)-hexeneamide

4-Ethyl-2[(*Z*)-hydroxyiminol]-5-nitro-3(*E*)-hexeneamide (FK409) is a semi-artificial fermentation product of *Streptomyces griseosporeus* No. 16917 that spontaneously generates NO and possesses potent vaso-relaxant and anti-platelet activities (FUKUYAMA et al. 1995). It is relative unstable, with a half-life of approximately 45 min at 37 °C in phosphate buffer. FK409 has been reported to attenuate experimental liver (OHMORI et al. 1998), cardiac (ISONO et al. 1993), pulmonary (TAKEYOSHI et al. 2000) and renal (MATSUMURA et al. 1998) ischaemia/reperfusion injury. FK409 improved survival and reduced microcirculatory disturbances after intestinal ischaemia/reperfusion injury (KALIA et al. 2001).

Inhibitors of •NO Synthesis

Taurine chloramine inhibited the synthesis of nitric oxide and the release of tumour necrosis factor in activated RAW 264.7 cells (PARK et al. 1993). Inhibition of •NO production was dependent on taurine chloramine concentration and was accounted for by reduced expression of inducible nitric oxide synthase mRNA, regardless of activator combinations (PARK et al. 1997).

Cloricromene (2, 20 or 200 μM) inhibited the expression but not the activity of the inducible form of nitric oxide synthase in lipopolysaccharide (100 ng/ml)-stimulated *murine* J774 macrophages in a concentration-dependent manner (ZINGARELLI et al. 1993). Maximal inhibition (84.0 ± 8.0 %) was observed when cloricromene (200 μM) was added to the cells 6 h before lipopolysaccharide, whereas it was ineffective when given 6 h after endotoxin.

Δ^9-Tetrahydrocannabinol inhibited lipopolysaccharide- and IFN-γ-induced nitric oxide production by thioglycollate-elicited peritoneal macrophages of female BALB/c *mice* (COFFEY et al. 1996).

Nitrous Oxide

Nitrous oxide (N_2O) is a colourless gas (boiling point -88.5 °C) with a slightly sweet taste and odour at high concentration.

$$N_2O + e_{aq}^- \longrightarrow N_2 + HO^\bullet + OH^- \qquad [225]$$

It has been known for many years that the inhalation of "nitrous fumes' may cause acute pulmonary oedema and death in *man*. In the United Kingdom, nitrous fume poisoning is listed under occupational diseases prescribed under the National Insurance Act, 1965 (P.D.17). KENNEDY (1972) summarised some of the many situations where nitrous fumes may occur. In another paper (KENNEDY 1972, 1974) he listed *animal* studies on the toxicology of nitrous fumes. In *rabbits*, KLEINERMAN and WRIGHT (1961) found macrophage infiltration 4 days after an exposure of 25 ppm for 2 h.

Interactions of nitrous oxide with *bovine* heart dioxygen utilisation (SOWA et al. 1987) and cytochrome *c* oxidase (EINARSDÓTTIR and CAUGHEY 1988) were reported.

9.6.2.1.23 Ozone

Ozone is formed in both the upper (stratosphere) and lower (troposphere) sections of the earth atmosphere. Stratospheric O_3 is generated by intensive ultraviolet light. The high energy of this radiation splits molecular O_2 into oxygen atom radicals, $O^\bullet$, which then react with oxygen molecules:

$$O_2 + h\nu \longrightarrow 2\,O^\bullet \qquad [226]$$
$$O_2 + O^\bullet \longrightarrow O_3 \qquad [227]$$

Cyclic generation of ozone in the troposphere occurs as follows:

$$O_3 + H_2O \longrightarrow {}^\bullet NO + O^\bullet \qquad [228]$$
$$O_2 + O^\bullet \longrightarrow O_3 \qquad [229]$$
$$O_3 + {}^\bullet NO \longrightarrow NO_2 + O_2 \qquad [230]$$

Ozone is a major component of photochemical smog. High levels of this pollutant, sufficient to affect *human* health, are found in many urban areas world-wide (LIPMANN 1991).

$$O_3 + H_2O \longrightarrow 2\,O_3^{\bullet -} + 2\,H^+ \qquad [231]$$
$$O_3^{\bullet -} \longleftrightarrow O_2 + O^{\bullet -} \qquad [232]$$
$$O_3^{\bullet -} + H^+ \longrightarrow O_2 + HO^\bullet \qquad [233]$$
$$O_3^{\bullet -} + 2\,H^+ + e^- \longrightarrow O_2 + H_2O \qquad [234]$$
$$O_3^{\bullet -} + O_3^{\bullet -} + 2\,H^+ \longrightarrow O_2 + H_2O_2 \qquad [235]$$

In *Escherichia coli* K-12, glyceraldehyde-3-phosphate dehydrogenase (EC 1.2.1.12) showed the greatest susceptibility to ozone (KOMANAPALLI et al. 1997). When the active site sulphydryl (cysteine) is reversibly blocked by tetrathionate during ozone treatment, enzyme activity is completely retained despite the oxidation of tryptophan, methionine, and histidine residues (KNIGHT and MUDD 1984).

Accumulation of Macrophages *In Vivo*

After an exposure of male Wistar *rats* to 0.5 ppm of O_3, 6 h a day 6 days a week for 2 months, alveolar macrophages accumulated in the centroacinar alveoli (HIROSHIMA et al. 1989). These macrophages were false positive for periodic acid-Schiff staining and partially positive for Berlin blue staining. The intraluminar accumulation of alveolar macrophages lasted until 12 months after exposure.

In C57BL/6 *mice*, steady-state mRNA levels for monocyte chemoattractant protein-1 (MCP-1) in the lung increased at 0.5 ppm ozone and were maximal at 2.0 ppm ozone (ZHAO et al. 1998). After exposure to 2 ppm ozone, macrophage inflammatory protein -2 (MIP-2) mRNA levels peaked at 4 h postexposure, whereas MCP-1 mRNA levels peaked at 24 h postexposure. Neutrophils and monocytes recovered in bronchoalveolar lavage fluid peaked at 24 and 72 h, respectively.

Influence on the Number and Viability of BAL Macrophages

After a 15 min inhalation of 0.5 ppm, 1.0 ppm, and 2.0 ppm ozone, respectively, the number of cells in bronchoalveolar lavage of female Wistar *rats* was dose-dependent reduced (STALDER et al. 1987). *In vitro*, peritoneal macrophages elicited by glycogen showed a reduced phagocytosis of yeast cells (as shown by bioluminescence) and at ≥2 ppm a permeability for nigrosine.

In male Wistar *rats* immediately following an exposure to 1.77±0.03 ppm O_3 for 2 and 4 h, respectively, the number of lavaged macrophages significantly ($P < 0.05$) decreased from $1.44 \pm 0.14 \times 10^6$ cells in the controls to 0.82±0.09 and $0.94 \pm 0.14 \times 10^6$ cells per lung (BASSETT et al. 1988).

2 days after the initiation of exposure of male Fischer 344 *rats* to 0.25 and 0.50 ppm ozone for 20 h/d, alveolar macrophages showed a transient wave of proliferation, as demonstrated by labelling with [methyl-^{3}H]-thymidine (WRIGHT et al. 1987).

Phagocytosis *In Vitro*

Rat alveolar macrophages exposed to ozone for 1 h in concentrations of 0.74, 1.35, 2.48 or 4.48 ppm showed a concentration-related inhibition of phagocytosis of latex particles (WENZEL and MORGAN 1983). The O_3 inhibition was not influenced by the presence of either bovine serum in the medium, or by alternately removing and replacing the cultured medium (rotated exposure) or both.

30 min after an intratracheal injection of approximately 10^8 organisms of Lancefield C *Streptococci*, bronchoalveolar lavage from albino *rabbits* of either sex exposed to ozone concentrations from 0.67 to 9.50 ppm showed a significant ($P < 0.001$) decrease in the numbers of both phagocytic cells and bacteria ingested (COFFIN et al. 1968).

Cells lavaged from male Wistar *rats* exposed to 0.5 ppm ozone for 1 week phagocytised 0.9 μm polystyrene beads at the same levels as that from the control animals, both to the percentage of cells (index I) and the number of particles per cell (index II) (CREUTZENBERG et al. 1995). After an exposure for 2 months, however, the number of phagocytized particles per macrophage increased from 4.8±2.7 to 11.1±4.5.

Lysosomal Enzymes *In Vivo*

In *rats* infected with aerosols of *Staphylococcus aureus* and then exposed for 5 h to 2.5 ppm of ozone, bacterial ingestion and clearance by alveolar macrophages were impaired due to the absence of lysosomal acid phosphatase and β-glucuronidase activities in those cells subjected to the dual insults (GOLDSTEIN et al. 1978).

Formation of Superoxide Anion Radicals

The formation of superoxide anion radicals by 0.2×10^6 lavaged cells was not changed after a 7-day exposure of *rats* to 0.5 ppm ozone compared to the control group, but was increased after a two months ozone exposure (CREUTZENBERG et al. 1995).

In *murine* macrophages, 3-(4,5-dimethylthiazol-2-yl)-diphenyltetrazolium bromide metabolism (monitored by optical density) 24 h after treatment with ozone showed a slight but significant increase ($P = 0.01$) compared to controls for the 20 μg/ml dose and a decrease ($P = 0.02$) for 100 μg/ml dose (CARDILE et al. 1995).

4-Hydroxynonenal Protein Adducts After Ozone Exposure

To determine whether 4-hydroxynonenal could account for the acute effects of ozone on *human* alveolar macrophages, HAMILTON jr. et al. (1998) exposed healthy, non-smoking volunteers to 0.4 ppm ozone or air for 1 h with exercise (each subject served as his/her own control). Six hours after ozone exposure, cells obtained by airway lavage were examined for apoptotic cell injury, presence of 4-hydroxynonenal adducts, and expression of stress proteins. Significant apoptosis was evident in airway lung cells after ozone exposure. Western analysis demonstrated an increase in a 32-kDa 4-hydroxynonenal protein adduct and a number of stress proteins, viz., 72-kDa heat shock protein and ferritin, in alveolar macrophages after ozone exposure. All these effects could be replicated by *in vitro* exposure of alveolar macrophages to 4-hydroxynonenal.

Synthesis of Inflammatory Cytokines

Exposure of *guinea pig* alveolar macrophages to 0.4 ppm O_3 for 60 min increased the IL-6 activity by 252±60% and TNF activity by 202±35% (ARSALANE et al. 1995). The increase of monokine production by *human* alveolar macrophages was 443±208% for TNFα 484±171% for IL-1β, 383±147% for IL-6, and 226±45% for IL-8 after a 60 min exposure to 0.4 ppm O_3. Lowest O_3 concentrations (0.1 and 0.2 ppm) only increased TNFα secretion. Shorter (30 min) or longer (120 min) exposure duration to 0.4 ppm O_3 were also associated with a significant increase in monokine production, suggesting that the effect of ozone on cytokine production did not depend on the exposure duration. The mRNA expression of TNFα, IL-1β, IL-6, and IL-8 was increased in *human* alveolar macrophages, whereas similar amounts of mRNA were detected in each sample.

A statistically significant 29-fold increase in bronchoalveolar lavage fluid IL-6 levels was observed in *rats* exposed to 0.5 ppm O_3 during nighttime hours then compared with daytime hours even though similar kinetics of inflammation were induced by each exposure (McKinney et al. 1998). Animals receiving an initial night-time exposure showed a lesser degree of inflammation following a subsequent O_3 exposure when compared with animals which received an initial daytime exposure.

Ozone and Serum Ceruloplasmin

Decreased serum ceruloplasmin concentration and the cumulated worktime along the week showed a linear relationship in aluminium welders exposed to ozone, which is the major pollutant in arc welding (Pierre et al. 1988). Ceruloplasmin may be a protective molecule against direct oxidant injury to the *human* lung in cigarette smoke and air pollution (Galdston et al. 1984).

9.6.2.1.24 Platinum

Platinum has the atomic number 78 and the valence electrons are those of the $5d^9 6s^1$ shells. The oxidation states of platinum are +2 and +4.

Platinum(IV) complexes of halogenopentammine type are reduced irreversibly to platinum(II) complexes in one step (Hall and Plowman 1955) The polarographic stability of the pentammineplatinum(IV) complexes, $[Pt(NH_3)_3X]$, increases in the following order of ligand X:

$$Br^- < Cl^- < NH_3 < OH^-$$

Cisplatin

The subcutaneous injection of *cis*-diamminedichloroplatinum(II) into ABD2F1 *mice* (8 mg/kg) resulted in alterations of size and surface structure of peritoneal macrophages and in a reduction of the mean number of their concanavalin A binding sites (Peschke et al. 1992).

Murine peritoneal macrophages treated with cisplatin showed an enhanced production of nitric oxide and increased tumoricidal activity against P815 mastocytoma cells (Sodhi and Kumar 1994). The •NO secretion and generation of macrophage-mediated tumoricidal activity were significantly inhibited by L-*N*-monomethyl arginine, a specific inhibitor of L-arginine pathway. Induction of •NO production in macrophages on activation with cisplatin or lipopolysaccharide is inhibited by EGTA, nifedipine, TMB-8 and W-7, suggesting the probable involvement of Ca^{2+} and calmodulin in the induction of •NO release. Protein kinase C and tyrosine kinase inhibitors significantly inhibited the cisplatin/lipopolysaccharide induced •NO production, suggesting that phosphorylation via these kinases may up-regulate the •NO synthase activity in macrophages.

Cisplatin-treated macrophages from pathogen free BALB/c *mice* synthesised and secreted oncostatin M (Singh and Sodhi 1998). The protein kinase C and protein tyrosine kinase inhibitors significantly inhibited oncostatin M production of cisplatin-treated macrophages. The oncostatin M production of cisplatin-treated macrophages was also inhibited in the presence of Ca^{2+} chelators, Ca^{2+} channel blockers and calmodulin/calmodulin-dependent kinase inhibitors.

Cisplatin treatment of pathogen free BALB/c *mouse* macrophages increased the expression and activation of lyn, a protein tyrosine kinase of src family, within 5 min (Singh and Sodhi 1998). Cisplatin-induced expression and activation of lyn involved serine/threonine phosphatases 1/2A, protein tyrosine phosphatases, protein tyrosine kinase and protein kinase C. It was also observed that Ca^{2+}/calmodulin and calmodulin-dependent kinases are involved in the regulation of cisplatin-induced lyn expression and activation in macrophages.

9.6.2.1.25 Plutonium Oxide ($^{239}PuO_2$)

Plutonium particles were located and identified within alveolar macrophages by autoradiography. The paths of the α particles emitted from the plutonium within the cell formed a "star" of tracks of reduced silver halide grains (Morrow and Casarett 1961, Sanders 1970).

Migration of alveolar macrophages lavaged from *rabbits* dusted with PuO_2 was inhibited (Nolibe 1972).

9.6.2.1.26 Selenium

Selenium has the atomic number 34 and the valence electrons are those of the $4s^2 4p^4$ shells. Se is classified in group VIa of the periodic table below sulphur and its chemical and physical properties are very similar to those of the latter. Like sulphur, selenium occurs in the oxidation states −2, 0, +2, +4 and +6 and is present in compounds analogues to those of sulphur such as the inorganic selenides, selenites and selenates and the various organic compounds in which the element is present mainly in the –2 oxidation state such as dimethylselenide, trimethyl selenonium, selenocysteine, selenomethionine and Se-methylselenocysteine.

The two elements differ, as KYRIAKOPOULOS and BEHNE (2002) stated in a review on selenium-containing proteins, with regard to the reduction of their oxyanions. In biological systems the selenium compounds tend to be more easily reduced than sulphur, which is thermodynamically stable in the +6 state. Most important for the significance of selenium in biological systems, however, is the lower pK_a value of the selenohydryl group of selenocysteine (pK_a 5.24) as compared with that of the sulphydryl group of cysteine (pK_a 8.25) (HUBER and CRIDDLE 1967). Accordingly, the selenols of the selenocysteine-containing proteins are anionic at the physiological pH, while the thiols in the cysteine-containing proteins are mainly protonated under these conditions. Replacement of selenocysteine in the active centre of a selenoenzyme with cysteine therefore results in a drastic decrease in its catalytic activity as was shown, for instance, in the case of formate dehydrogenase H from *Escherichia coli* (AXLEY et al. 1991), type I iodothyronine deiodinase (BERRY et al. 1991) and thioredoxin reductase (GASDASKA et al. 1999, LEE et al. 2000).

Selenium is an essential trace element that constitutes the active centre of glutathione peroxidase (FORSTROM et al. 1978, LANDESTEIN et al. 1979, WINGLER and BRIGELIUS-FLOHÉ 1999).

Selenium compounds are more active electron donors compared to analogous sulphur compounds and play a wide role in biochemical systems. They can decompose peroxides (CALDWELL and TAPPEL 1964, 1965) and can act as antioxidants (DILLARD et al. 1978) and free radical scavengers (URSINI and BINDOLI 1987). MASUMOTO and SIESS (1996) and MASUMOTO et al. (1996) have shown that peroxynitrite rapidly oxidises ebselen and its main metabolite to the corresponding selenoxides. The results of PADMAJA et al. (1997) suggest that CO_2 partially protects methionine selenoxide from peroxynitrite-mediated oxidation and that $O{=}N\text{-}OO\text{-}CO_2^-$ or its derivatives do not mediate the oxidation of D,L-selenomethionine or methionine selenoxide.

SeO_3^{2-} and SeO_4^{2-} are weakly mutagenic in the Ames test (TAKANO and SAKURAI 1979).

In vitro, guinea-pig peritoneal macrophages showed nontoxic responses to variable concentrations of $Na_2SeO_3 \times 5\ H_2O$: cell viability was unaffected at 10^{-6} M, migration at 10^{-6} and 10^{-5} M selenite, while for 10^{-4} M selenite no excessive lipid peroxidation occurred (GABOR et al. 1985).

In vivo, rats' lung and lymph node weights were not influenced by selenium supplementation in animals given quartz (GABOR et al. 1986). However, a decrease of lung lipids, phospholipids, and hydroxyproline was observed in *rats* with experimental silicosis.

9.6.2.1.27 Silicon

Silicon has the atomic number 14 and the valence electrons are those of the $3s^2 3p^2$ shells. There are 3 stable isotopes naturally occurring, ^{28}Si, ^{29}Si, and ^{30}Si. Their relative frequencies are 92.18 %, 4.71 % and 3.12 %, respectively. ^{31}Si and ^{32}Si are artificial radionuclides with half-lives of 2.62 h and 101 years, respectively. The annual limit of intake of $^{32}SiO_2$ by inhalation is 2×10^5 Bq/m^3 (BURKART 1987). The oxidation state of Si is +4.

Silicon and oxygen are the two most important elements in the crust and the form is a fundamental SiO_4 tetrahedral unitconsisting of a central silicon ion with oxygen ions attached three-dimensionally at the four "corners" of a tetrahedron. All pathogenic forms of "silica", i.e. silicon dioxide (SiO_2), are composed of these tetrahedra joined by oxygen atoms so that each crystal consists of a giant molecule with an average stoichiometric formula of SiO_2. Being uncombined they are referred to as "free silica". The tetrahedra are linked in various ways by $\equiv$Si-O-Si$\equiv$ chains, and the manner in which metallic cations are included in this linkage decides their form and characteristics.

Electron spin resonance studies of a series of air-annealed samples of glassy SiO_2 having various degrees of enrichment (or depletion) in the ^{29}Si isotope have confirmed that a γ-ray-induced doublet of 420-G splitting is the ^{29}Si hyperfine structure of the well-known *E'* centre (GRISCOM 1979). This finding validated the widely accepted model of the *E'* centre as an unpaired electron spin in a dangling sp^3 hybrid orbital of a silicon bonded to three oxygens in the glass structure and eliminates an alternative model proposed by SHENDRIK and YUDIN (1978).

An experimental study by WU et al. (1998) on the competitive adsorption resulted in the magnitude order of metal ions (Ag^+, Ni^{2+}, Zn^{2+}, Cu^{2+}, Cd^{2+}, Pb^{2+}, Cr^{3+}) adsorbed onto oxide and silicate minerals in near-neutral solution with low ionic strength in mol/nm^2 as follows:

$CaCO_3$ > quartz > hydromuscovite > kaolinite > Ca-montmorillonite > goethite > gibbsite.

Cu^{2+} and Cr^{3+} were adsorbed onto quartz at a higher rate than on calcite.

The distinction between free and combined silica is important: combined silica is SiO_2 in combination with various cations as silicates.

Free silica is the most widespread substance in nature with a fibrogenic potential for living tissues. Free silica (silicon dioxide) occurs in three forms: polymorphic crystalline, cryptocristalline (i.e., minute crystals), and amorphous (i.e., noncrystalline).

The tetrahedral – and thus pathogenic – crystalline phases of silica are as follows:

1. **Quartz** which is stable up to 867 °C but is capable of metastable existence at higher temperatures.
2. **Tridymite** which is stable from 867 °C to 1470 °C and is capable of metastable both above 1470 °C and below 867 °C.
3. **Cristobalite** which is stable from 1470 °C up to its melting point of 1723 °C but is capable of unstable existence at any temperature below 1470 °C. Opaline silica is a variety of cristobalite (Sosman 1965). Pure quartz when heated to temperatures between temperatures between 867 °C and 1470 °C is nearly always converted to cristobalite and not to tridymite unless a catalyst is present.
4. **Coesite** was first synthesised at a pressure of about 3.4 GPa, in the temperature range of 500-800 °C in the laboratory by Coes (1953) and later discovered and identified from shocked Coconino sandstone of the Meteor Crater in Arizona (Chao et al. 1960, Bohn and Stöber 1965), and from the Ries Crater in Nördlingen, Bavaria. It has a specific gravity of 2.915±0.015 and a hardness of about 8. It is biaxial positive with 2V about 64°. Its indices of refraction are α 1.5940, β 1.5955, and γ 1.5970±0.0005. It is nearly insoluble in 5 % HF at room temperature.

Octahedral **stishovite** is 46 % denser than coesite. It has a specific gravity of 4.35 for the synthestic material. In stishovite each oxygen is co-ordinated to three ^{VI}Si, as opposed to two ^{IV}Si in lower-pressure silica minerals. The $^{IV}Si \rightarrow {}^{VI}Si$ transitions increase the Si-O separation as well as the packing efficiency. The $^{IV}Si \rightarrow {}^{VI}Si$ transformations occur over a limited range of pressure.

Stishovite is biologically inert (Strecker 1965, Brieger and Gross 1967).

Different investigators have suggested a number of possible **post-stishovite phases** of SiO_2. Such phases have been obtained from theoretical simulations or observed experimentally in related systems; these are $CaCl_2$, α-PbO_2, and modified α-PbO_2 (space group *I2/a*), α-$PbCl_2$, fluorite and modified fluorite (space group *Pa*3) structures (Tsuchida and Yagi 1989, Tse et al. 1992, Lacks and Gordon 1993, Kingma et al. 1995, 1996). A new high-pressure phase of silica (space group *Pnc*2) was synthesised by heating silica gel or quartz to 2000±50 K at ≈80 GPa (Dubrovinsky et al. 1997). The phase transitions: stishovite → $CaCl_2$-like phase → *Pnc*2 structure → *Pa*3 were found at 45 GPa, 80 GPa, and 220 GPa, respectively.

Moganite, a very rare microcrystalline silica mineral, consists of systemic twinning of one R- and one L-quartz {1011} lattice slice per unit cell.

"Colloidal silica", also defined acid (Lühning 1954), is a poisonous substance (Gye and Purdy 1922). It reacts with rat macrophages without affecting cell integrity (Cantrell and Ellis 1983).

The toxicity of silica has been attributed to hydrogen donation by polymeric silicic acid, forming hydrogen-bonded polymeric complexes notably with phospholipids of cell membranes (Allison and Davies 1974). Breaking a hydrogen bond can result in a significant structural change in biomolecules. Specifically, hydrogen bonding between silica and phosphate groups of membrane phospholipids could be responsible for silica toxicity (Hobza et al. 1981).

The ability to destroy the cell membrane depends upon the particle size: the denaturing effect was found to increase with the increase in size of colloidal silica particles. The proposed mechanism was that large particles would stretch the protein molecule of the membrane by adsorption forces, whereas small particles (2–3 nm) would be too small to separate the protein coils (Harley and Margolis 1961).

Using *human* erythrocytes as a model of interaction, Diociaiuti et al. (1999) due to the high surface/volume of aerosil particles obtained considerable membrane damage with small weight concentrations. Silica induced a considerable reduction of the intramembranous density in the endoplasmic fracture face, while the protoplasmic fracture face was practically unchanged.

The adsorption of water on the silica surfaces provided a useful tool for surface analysis (Stöber 1955, 1956, Bolis et al. 1983). Water is in fact a constituent of a silica surface exposed to the atmosphere, for when equilibrated with water vapour the surface is mostly hydroxylated (surface silanols) and molecular water is bound to the silanols. Water can be released by heating under vacuum, and the process is reversible or irreversible depending on the outgassing temperature. The silanol groups interact with the protein or phospholipid constituents of the membrane and thus damage it (Nash et al. 1966, Summerton et al. 1977).

After intratracheal injection of a fresh suspension of 20 mg of **condensed silica** in saline, the spherical particles of <30 nm diameter formed agglomerates. After 4 h numerous polynuclear phagocytes have emigrated from the alveolar capillaries, followed by macrophages, and adhered to the masses of particles (POLICARD et al. 1955). After 7 days, POLICARD and COLLET (1958) distinguished histiocytes capable to phagocytize and non-phagocytic reticulo-histiocytes with plenty of ergastoplasm. Aerosil® particles (20–50 nm) given intraperitoneally to *rats* after 8 days were found conglutinated by a biologic substance and thus detoxified (KLOSTERKÖTTER and THEMANN 1958).

Min-U-Sil 5 **quartz**, together with Sikron F-600 the most likely candidate for an international α-quartz calibration standard (VERMA and SHAW 2001), shifted the phenotypic ratio of *human* alveolar macrophages obtained by bronchoalveolar lavage of normal, non-smoking adult volunteers of either gender to a more inflammatory condition (HOLIAN et al. 1997). The macrophage phenotypes were characterised by flow cytometry targeting the RFD1 and RFD7 epitopes. Results demonstrated that crystalline silica, as well as chrysotile and crocidolite, but not titanium dioxide or wollastonite, increased the $RFD1^+$ phenotype (inducer or immune activator macrophages) and decreased the $RFD1^+RFD7^+$ phenotype (suppressor macrophages).

Alveolar macrophages obtained by bronchoalveolar lavage from 25 silicotic *patients* under the electron microscope showed silica particles, activated nuclei, autophagic vacuoles and destroyed cell organelles (REISZ et al. 1988). Lysosomal acid phosphatase activity was elevated as compared with healthy volunteers.

Instillation of 500 mg generic quartz dust (7 μm maximum diameter) suspended in 50 ml PBS into the right caudal lung lobe of 3 female pigtail macaque monkeys (*Macaca nemestrina*) using a flexible fibreoptic bronchoscope increased the numbers of macrophages in the dust exposed lobes and significantly elevated the *N*-acetyl-β-D-glucosaminidase activity within the bronchoalveolar lavages taken at 2-weeks intervals for 12 weeks thereafter (MACK et al. 1995).

In *Macaca fascicularis*, DQ12 inhalation caused a shift from large, active (with respect to phagocytosis) alveolar macrophages to small, monocyte-like, less active alveolar macrophages (HILDEMANN et al. 1992).

Silica directly injured *rat* alveolar macrophages as evidenced by a cytotoxic index of 32.9±2.5, whereas the addition of surfactant protein A (5 μg/ml) significantly ($P<0.001$) reduced the cytotoxic index to 16.6±1.2 (SPECH et al. 2000). The effect was reversed when surfactant protein A was incubated with either polyclonal rabbit anti-rat surfactant protein A antibody or D-mannose.

Table 38. Macrophages ($\times 10^6$) in bronchoalveolar lavage fluid. 1 day after intratracheal instillation in *hamsters* (BECK et al. 1987)

α-Quartz	CDM = 1.3 μm	3.1±0.3
Iron oxide	1.3 μm	6.3±0.3
Talc	0.8 μm	5.5±0.3
Granite	0.55 μm	6.7±1.0
Saline	-	6.6±0.5

CDM = Count Median Diameter

The role of functionalities on the surface of α-quartz may play specific roles in the initiating of the cascade of events resulting in the fibrotic response (FUBINI et al. 1990). Sianols and surface radicals that both decreased upon thermal heating, leaving isolated sianols, siloxanes and distorted bridges, and siloxane groups appear to be linked to the phagocytosis and transport of particles, whereas the surface radicals and distorted bridges might be more closely associated with the fibrogenic response (HEMENWAY et al. 1994).

After stimulation with lipopolysaccharide lavaged Sprague-Dawley *rat* alveolar macrophages (10^6 cells/cuvette) in a dose-dependent manner generated **oxygen radicals** when 0.05 to 5 mg silica were added (LIM et al. 1997). At higher concentrations silica-induced chemiluminescence abruptly decreased. Cytosolic $[Ca^{2+}]_i$ was suppressed only in part, perhaps due to a release of calcium from intracellular calcium stores. Staurosporine, a protein kinase C inhibitor to bind to the catalytic domain of protein kinase C (TAMAOKI et al. 1986), inhibited silica-induced chemiluminescence by 66 % at 0.2 μM and at 99 % at 2 μM. Sphingosine, an agent that competes with diacylglycerol/phorbol ester for binding protein kinase C (HANNUN et al. 1991) inhibited silica-induced chemiluminescence by 53 % at 2.5 μM and by 99 % at 50 μM. Of the phospholipase C inhibitors, neomycin inhibited silica-induced chemiluminescence by 29 % at 0.2 mM and at 72 % at 2 mM, and U-73122 by 25 % at 0.1 μM and by 71 % at 1 μM. Genistein which is known to inhibit protein tyrosine kinase activity by binding to ATP binding sites in a competitive manner (AKIYAMA and OGAWARA 1991), inhibited silica-induced chemiluminescence by 83 % at 20 μM, and erbstatin by 99 % at 5 μM.

To re-evaluate macrophages as accessory cells by treating various cell preparations with either silica (O'ROURKE et al. 1978, THIELE and LIPSKY 1982) or L-**leucine** methyl ester (THIELE et al. 1983), which have been reported to be cytotoxic to macrophages, BOWERS et al. (1988) showed that L-leucine methyl ester failed to kill *rat* macrophages or dendritic cells, whereas silica was specifically toxic for *rat* macrophages.

There is an increase in **L-arginine uptake and metabolism** by *rat* bronchoalveolar lavage fluid inflammatory cells following *in vivo* exposure to silica, suggesting an increase in L-arginine utilisation by both arginase and nitric oxide synthase (SCHAPIRA et al. 1996). From 1 day after intratracheal instillation of silica iNOS showed positive reaction, peaked at 3 days and then decreased (SETOGUCHI et al. 1996). NOS activity, as measured by conversion of [^{14}C]arginine to [^{14}C]citrulline, took the same time profile. In situ hybridisation showed that iNOS mRNA was detected in the inflammatory cells phagocytizing silica. iNOS was expressed on macrophages, neutrophils and bronchial epithelial cells by immunohistochemical double staining. The administration of N^{ω}-nitro-L-arginine methyl ester resulted in significant reducing the number of inflammatory cells such as monocyte-derived exudate macrophages and neutrophils in the granulomatous lesions.

The amount of macrophage-derived nitric oxide release varies remarkably between animal species, i.e. *rat* and *hamster* (DÖRGER et al. 1997).

HL-60 cell derived macrophages were recommended as an easily manageable model system for toxicity studies of particulate bronchopulmonary noxious agents (ZOLLER et al. 1996, ZOLLER and ZELLER 2000). During the differentiation of HL-60 cells, both an optimal concentration of calcitriol (10^{-8} M to 10^{-7} M) and of ethanol is necessary. Differentiation of HL-60 cells in the presence of 0.1 % (v/v) ethanol significantly reduced their ability to produce $O_2^{\bullet -}$; an ethanol concentration of 2 % (v/v), on the other hand, was toxic. HL-60-G cells exposed to a high concentration of calcitriol (400 nM) slowly developed the ability to reduce nitroblue tetrazolium and did not proliferate (STUDZINSKI et al. 1997). HL-60 cells differentiated with phorbol myristate acetate were resistant to cytotoxic effects of quartz (BRÜCKNER-NIEDER et al. 1992). HL-60-M cells produced tumour necrosis factor after stimulation with lipopolysaccharide but not with quartz. SCHEDLE et al. (1995) induced apoptosis by 33–1000 μM metal cations/l. Crocidolite asbestos suppressed the differentiation of HL-60 cells induced by DMSO (UEKI et al. 1992).

Some bisbenzimides, Hoechst 33342, but not Hoechst 33258, induced apoptosis in the HL-60 cell in a time- and dose-dependent manner (ZHANG et al. 1999). Endogenous nuclear topisomerase 1 activity in HL-60 cells was inhibited by Hoechst 33342, but not Hoechst 33258.

U-937 cells, a myelomonocytic cell line derived from a *patient* with histiocytic lymphoma, when differentiated by phorbol myristate acetate and exposed to DQ12 quartz (specific area 3 m^2/g), MnO_2 (specific area 59 m^2/g) and TiO_2 (specific area 7.5 m^2/g), particles, respectively, showed specific patterns of changes in the carbohydrate moiety of glycoproteins (TRABELSI et al. 1997). Given the biological implications of carbohydrates, these changes may be critical in modulating macrophage functions. Melatonin reduced H_2O_2-induced DNA damage in U-937 cells (ROMERO et al. 1999).

P388D$_1$ macrophage-like cells were used in France to test cytotoxicity (viability and lactate dehydrogenase and acid phosphatase levels) at a serum content of 4 % (DAVIS et al. 1982) Because uncontrolled variation occurring in animal cells (*rabbit* alveolar macrophages obtained by pulmonary lavage 3–4 weeks after stimulation by i.v. injection of Freund's adjuvant, thioglycolate evoked *rat* peritoneal macrophages) DANIEL and LE BOUFFANT (1980) preferred P388D$_1$ cells, particularly when investigating dusts with low toxicities. The extent of killing of P388D$_1$ macrophages is dependent on both the dose of silica and the concentration of Ca^{2+} ions in the medium (KANE et al. 1980). In the presence of extracellular calcium ions, after 3 h of exposure to 350 μg of silica, 69 % of the cells had lost viability, and ATP content was reduced to 21 % of the control level (KANE et al. 1985). P388D$_1$ cells injured by silica showed a similar dose and time relationship between loss of viability and increased chlorotetracycline fluorescence (GOODGLICK et al. 1986): after 6 h of silica exposure 58.1 ± 4.7 % of the cells were killed and 45.4 ± 5.4 % showed increased chlorotetracycline fluorescence. To test whether disruption of mitochondrial Ca^{2+} sequestration was sufficient to cause cell death, two types of metabolic inhibitors were used. First, the mitochondrial membrane potential was disrupted by *p*-trifluoromethoxy-phenylhydrazone or carbonyl cyanide *m*-chlorophenylhydrazone which caused leakage of rhodamine 123 and release of mitochondrial Ca^{2+} as detected by fura2. After 6 h of *p*-trifluoro-methoxyphenylhydrazone or carbonyl cyanide *m*-chlorophenylhydrazone exposure, 98.3 ± 5.6 % and 96.7 ± 7.7 % of the cells were viable and 2.5 ± 1.0 % showed increased chlorotetracycline fluorescence after 6 h. LUOTO et al. (1998) investigated the dissolution of short and long rockwool and glasswool fibres by P388D$_1$ cells by assessing the dissolution of Si, Fe, and Al from the fibres.

The **contact of dust particles with cell membranes** is the first event during the process of phagocytosis, whatever the way of particle uptake. Membrane fluidity of alveolar macrophages gained from *guinea pigs* by bronchoalveolar lavage was changed more by quartz than by TiO_2 as shown by 1,6-diphenyl-1,3,5-hexatriene fluorescence polarization probe (CAO 1990). Compared with quartz controls, fluidity was decreased when quartz plus aluminium citrate were added to the cell cultures simultaneously, although aluminium citrate did not influence membrane fluidity by itself.

Experiments with autoxidation of linolate, haemolysis of red blood cells, lipid peroxidation in macrophages, and *in vivo* effects of antioxidants on collagen produced by quartz injected subcutaneously, have shown that peroxidation of membrane lipids is not the primary mechanism whereby quartz attacks membranes (Kilroe-Smith 1974).

If silica or chrysotile particles are added to macrophages in serum-free medium, early cytotoxic effects are seen within minutes of exposure (Allison and Davies 1974). These effects are analogous to the haemolysis produced by silica and chrysotile and are thought to be due to the interaction with the plasmalemma. Nash et al. (1966) have pointed out that silica particles unlike inert dusts, have on their surface in aqueous media silicic acid with multiple, rigidly placed hydroxyl groups which can act as hydrogen donors in hydrogen-bonding reactions with membrane phospholipids. Release by silica particles from liposomes composed of phospholipid and cholesterol has been reported.

Phagocytosis of DQ 12 quartz (Munder and Modolell 1987) and Brazilian crystallised quartz (Brehm 1996, Brehm et al. 1996) by bone marrow macrophages and *bovine* alveolar macrophages, respectively, induced different periods and intensities of luminescence demonstrating the formation and degradation of oxygen radicals. RAW 264.7 *mouse* macrophages treated with α-quartz for 48 h indicated many regulatory mechanisms of gene expression were simultaneously triggered (Segade et al. 1995). The silica-induced genes *SIG-12*, *-14*, and *-20* corresponded to the genes for ribosomal proteins L13a, L32, and L26, respectively. *SIG-61* is the *mouse* homologue of p21 RhoC. *SIG-91* is identical to the 67-kDa high-affinity laminin receptor. Four genes (*SIG-41*, *SIG-81*, *SIG-92*, and *SIG-111*) were not identified and are novel. In *rat* alveolar macrophages standard crystalline silica obtained from the Institute of Occupational Medicine, Chinese Academy of Preventive Medicine (Beijing) with a particle size of <95% less than 5 μm in diameter enhanced $O_2^{\bullet-}$ and H_2O_2 formation (Zhang et al. 2000). There were clear dose- and time-dependent relationships in silica-induced cytotoxicity and genotoxicity.

By phase-contrast microcinematography, Schiller (1953, 1954) showed a swift decline in the phagocytosis rate of quartz particles by *muine* alveolar macrophages as compared with that of powdered diamonds of the same size. Quartz dust poisoned the macrophages very quickly. Vacuolisation occurred only in cells that had taken up some quartz particles, while dust-free cells in the neighbourhood entered mitosis, and so did cells that had taken up powdered diamond (Schiller 1961). Interspecies comparisons gave no principal variation of the response between different animal species (Seidel et al. 1990). Brückner-Nieder et al. (1992) developed a *human* cell model using HL-60 cells, a promyelocytic line. These cells can be differentiated by phorbol myristate acetate into non-proliferating macrophage-like cells (HL-60-M). Sikron F600 quartz particles were engulfed. At the ultrastructural level, the quartz was seen inside phagolysosomes, but the morphology of the cell remained normal.

Treatment of *human* alveolar macrophages with crystalline silica resulted in significant **apoptosis** within 6 h. Macrophage scavenger receptors are trimeric integral membrane proteins which exhibit unique ligand-binding characteristics. The scavenger receptor may play an important role in the uptake of charged particulates (e.g. negatively charged silica) by alveolar macrophages. Silica treatment of *human* alveolar macrophages resulted in increased interleukin-converting enzyme (detection of p20 fragments) and cpp32β (enhanced degradation of cpp32) activity (Iyer and Holian 1997). Pretreatment of cells with the interleukin-converting enzyme and cpp32β inhibitors, Z-Val-Ala-Asp-fluoromethyl ketone (10 μM) and DEVD, respectively, significantly inhibited apoptosis. In specific pathogen-free Wistar *rats*, 2.5 mg, 7.5 mg, and 22.5 mg Min-U-Sil 5 silica suspended in 0.5 ml saline, respectively, 10 days after intratracheal instillation induced apoptosis in a dose-related quantity of bronchoalveolar lavage cells (Leigh et al. 1997). Engulfment of apoptotic cells by macrophages was also noted. 56 days after instillation, morphologically apoptotic cells, the majority of which were considered to be differentiated macrophages, could be identified in granulomas.

The RAW 264.7 macrophage cell line was more sensitive, and the IC-21 cell line more tolerant to silica particle (<1 μm) exposure (0.2 or 1 mg/ml for 6 h) as evidenced by significantly higher apoptotic responses in RAW 264.7 ($P < 0.05$) (Gozal et al. 2002). RAW 264.7 macrophages exhibited enhanced TNF-α production and NF-κB activation in response to silica, whereas IC-21 macrophages did not produce TNF-α in response to silica and did not induce NF-κB nuclear binding. Inhibition of NF-κB in RAW 264.7 cells with 50 μM BAY 11-7082, an inhibitor of IκB phosphorylation, significantly increased apoptosis while inhibiting TNF-α release.

In *rat* alveolar macrophages, Shen et al. (2001) showed a temporal pattern of apoptotic events starting with the formation of reactive oxygen species and followed by caspase-9 and caspase-3 activation, poly(ADP-ribose)polymerase clevage and DNA fragmentation. Silica-induced apoptosis was significantly attenuated by a caspase-3 inhibitor, *N*-acetyl-Asp-Glu-Val-Asp aldehyde, and ebselen (formula [65]).

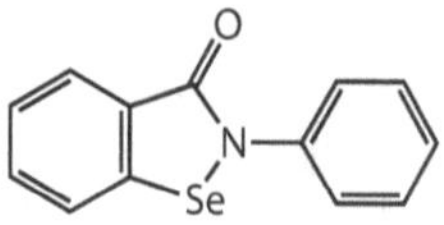

Ebselen, 2-phenyl-1,2-benzisoselenazol-3(2*H*)-one [65]

KIM et al. (2001) showed that pan-caspase inhibitor benzyloxycarbonyl-Val-Ala-Asp-fluoromethyketone or *t*-butyloxycarbonyl-Asp-fluoromethylketone caused the death of lipopolysaccharide-activated RAW 264.7 cells with apoptotic features.

Exposure of 1×10^6 *human* alveolar macrophages to 100 μg silica/ml for 4 h increased ferritin protein concentrations by approximately 50 % of the baseline value in both supernatants and lysates (GHIO et al. 1997). Inclusion of 1.0 mM deferoxamine, an iron chelator, in the reaction mixtures inhibited increases after silica. There were no increases in ferritin after incubation with acid-washed particles or silica with complexed zinc cation. There were no significant differences in levels of ferritin cDNA between any of the exposures suggesting a post-transcriptional control of ferritin expression.

Cortisone (5 mg given to *rats* 5 days a week for 3 weeks and then 3 days a week for the rest of the experiment) produced relative immobilisation of the dust cells, so that they were unable to migrate, and therefore unable to concentrate the quartz (50 mg instilled intratracheally) into foci (HARRISON et al. 1952, KING and HARRISON 1960).

Lazaroids (21-aminosteroid U-75412E) at a concentration of 5 μM protected *rat* alveolar macrophages against crystalline silica (Min-U-Sil <5 μm)-induced cytotoxicity (HUANG et al. 1998). Maximal inhibition of lactate dehydrogenase (EC 1.1.1.27) release from macrophages was achieved at a lazaroid concentration of 20 μM. H_2O_2 secretion from alveolar macrophages was also inhibited at 5 μM lazaroid concentration. *N*-Acetyl-β-D-glucosaminidase (EC 3.2.1.17) showed minimal inhibition (17 %) at all concentration of lazaroid used. Superoxide dismutase (EC 1.15.1.1) and glutathione peroxidase (EC 1.11.1.9) showed concentration-dependent decreases at all the concentration of lazaroids.

In vitro, **disodium cromoglycate**, the disodium salt of 1,3-bis 2-carboxycromon-5-yloxy-hydroxypropan, an electron scavenger (CARMICHAEL et al. 1988), was effective in protecting Chinchilla male *rabbit* alveolar macrophages from silica-induced necrosis only when the quartz dust was pre-treated with the drug dissolved in saline for 60 min at 37 °C, while pre-treatment of the macrophages with cromoglycate for 30 min before phagocytosis of the Dörentrup quartz particles had no or a negligible protective effect on the cells (VLCKOVÁ et al. 1976).

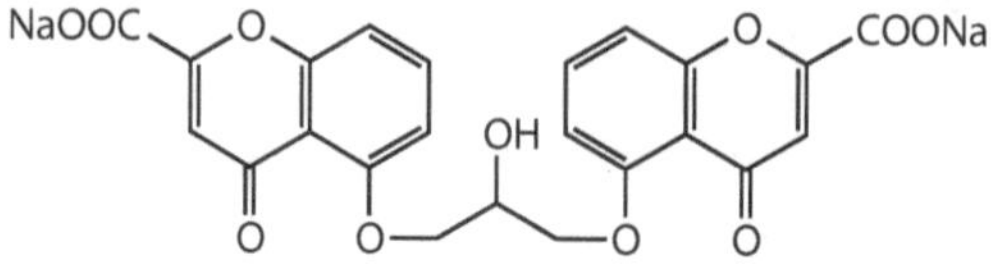

Disodium cromoglycate [236]

Macrophage inflammatory protein (MIP) 1α mRNA was upregulated in human peripheral blood monocytes after a 4–6 h exposure to 100 μg Min-U-Sil/ml (ERKAN et al. 1997).

Treatment of RAW 264.7 *murine* macrophage cultures with dimethyl sulphoxide, extracellular glutathione, or *N*-acetyl-L-cyteine decreased **cristobalite** (particle size range 0.08–1.5 μm; particle dose 35 μg/cm^2 or 54–78 particles/cell)-induced tumour necrosis factor-α mRNA levels by 40 %, 20 %, and 42 %, respectively (BARRETT et al. 1999). Both MIP-1α and MIP-1β mRNA levels were reduced at a magnitude similar to the reduction in TNF-α mRNA levels, whereas monocyte chemotactic protein (MCP)-1 mRNA levels were reduced at a magnitude similar to the reduction in MIP-2 mRNA levels following antioxidant treatment.

Pure cristobalite was considerably less cleared from the lungs of male Fischer 344 *rats* exposed for 6 h per day to 13 ± 2.4 mg/m^3, 40 ± 12.4 mg/m^3, and 81 ± 15.3 mg/m^3, respectively, on twice 4 days with an interval of 2 days compared with two quartz materials (HEMENWAY et al. 1990). There was little or no clearance after the initial 30 days post exposure. Cristobalite showed an early and sustained response with an elevated macrophage, neutrophil, and lymphocyte count through 180 days post exposure.

Human alveolar macrophages produced more reactive oxygen species in response to particles than *rat* alveolar macrophages (RAHMAN et al. 1997).

Granite dust containing quartz crystal (10 to 30 % silicon dioxide) and other aluminium, magnesium, and calcium silicates was present within the majority of alveolar macrophages lavaged from Vermont granite workers with 4 to 36 yr of employment in the industry compared with those from control subjects, as determined by polarising light microscopy and confirmed by scanning electron microscopy with X-ray energy spectroscopy (CHRISTMAN et al. 1985). There were no differences in the phagocytosis of zymosan particles (labelled with ^{99m}Tc) or the viability of macrophages (defined by trypan blue dye exclusion) from granite workers compared with those from nonexposed volunteers.

Kaolin is a nonfibrous hydrated aluminium silicate ($Al_2O_3 \cdot SiO_2 \cdot 2\ H_2O$) that can cause pneumoconiosis in workers processing the clay commercially (THOMAS 1952, LYNCH and MCIVER 1954, HALE et al. 1956, SHEER 1964, KENNEDY et al. 1983, SEPULVEDA et al. 1983, ALTEKRUSE et al. 1984, LAPENAS et al. 1984, MORGAN et al. 1988). Its basic unit of structure consists of tetrahedral and octahedral sheets in which the anions at the exposed surface of the octahedral sheet are hydroxyls. The catalytic activity of kaolin for generating hydroxyl radicals from hydrogen peroxide was studied in a chemical system that measured $HO^•$ as evolution of methane from dimethylsulfoxide. In the presence of 1 mM ascorbate as a reducing agent, and 10 mM H_2O_2, hydrous and calcined kaolin generated CH_4 concentrations of 1634±328 ppm and 1395±29 ppm, respectively (BASER et al. 1990). Surface modification with dipalmitoyl lecithin, the lipid of pulmonary surfactant, blocked generation of $HO^•$ in hydrous kaolin (38±38 ppm CH_4) but not in calcined kaolin (875±262 ppm). Generation was inhibited by the $HO^•$ scavenger dimethylthiourea (15 mM) or by preincubating kaolin for 48 h with the iron chelator deferoxamine (0.5 mg kaolin/ml in aqueous solution of 18 mM deferoxamine).

Decomposition of kaolin on heating

$$\underset{\text{kaolin}}{3\,(Al_2O_3 \cdot 2\,SiO_2)} \longrightarrow \underset{\text{mullite}}{3\,Al_2O_3 \cdot 2\,SiO_2} + \underset{\text{free silica}}{4\,SiO_2} \quad [237]$$

In *human* pulmonary tissue obtained at thoracotomy or autopsy from 5 kaolin workers from Georgia particulate-filled macrophages distended the interstitium, both in septa and in peribronchial-perivascular areas (LAPENAS et al. 1984). The particles were predominantly intracellular in macules. Only in one patient were large numbers of particle-laden macrophages found in the alveolar spaces, presumably because of ongoing kaolinite exposure.

Rats instilled intratracheally with untreated Cornish kaolin showed foci of alveoli filled with large macrophages which contained some dust (KING et al. 1948). Ignited Cornish kaolin in 14 to 73 days produced a simple phagocytic reaction without reticulinosis. In a *rat* which surviving for 140 days there was a phagocytic reaction in the perivascular connective tissue, associated with a strictly local reticulinosis.

Riebeckite, a non-asbestiform polymorph of crocidolite *in vitro* was non-toxic to alveolar macrophages at doses on a mass basis where crocidolite exhibited substantial activity (MOSSMAN and SESKO 1990). CASTRANOVA et al. (1994) starting from surface areas found that riebeckite exhibited cytotoxicity to *rat* alveolar macrophages *in culture* at doses equal to or less than those for crocidolite.

Asbestos dust was readily phagocytosed by macrophages (DAVIS 1963, 1967). Each cell could have several dozen dust containing phagosomes at any one time. Although it was obvious from these experiments that asbestos was not rapidly toxic in the same way as silica, SMITH and DAVIS (1971) investigated the occurrence of acid phosphatase in *guinea-pig* granulomas produced by the intrapleural injection of chrysotile asbestos dust by histochemical staining combined with electron microscopy. In dust-containing macrophages and giant cells, often fewer than 50 per cent of primary lysosomes and as few as 10 per cent of dust-containing phagosomes showed evidence of acid phosphatase. There was no increase in enzyme activity between 2 and 4 weeks.

Phagocytosis of asbestos particles up to 0.5 μm in length occurs by alveolar macrophages (MORGENROTH 1973) attracted to the alveolar duct bifurcations where inhaled asbestos fibres are deposited (WARHEIT et al. 1984). Inhaled asbestos activates a complement-dependent chemoattractant for macrophages (WARHEIT et al. 1985). Phagocytosis is followed by the appearance of hæmosiderin in the cytoplasm, then by intracellular transport of iron micelles from hæmosiderin granules into the phagosomes with the asbestos fibres. Iron micelles progressively concentrate in the vicinity if the fibres. The central fibre, with its coating of hæmosiderin, and the investing membranes of the phago-

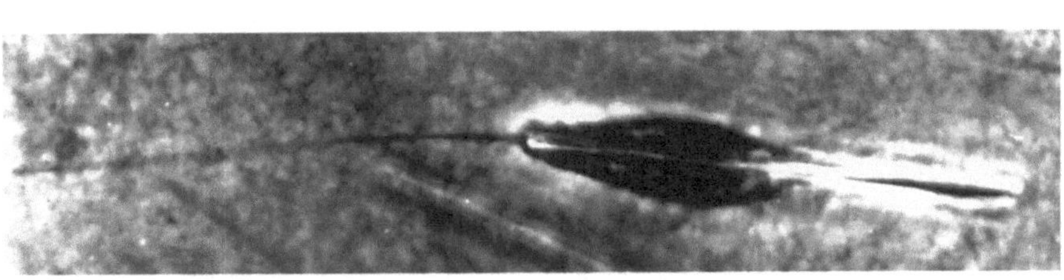

Fig. 153. Frustrated phagocytosis of powdered crocidolite from Griqualand. Organ culture of the lung of a 3-day old *mouse*. The preparation was fixed in Carnoy's fluid, stained with haematoxylin after Böhmer, and mounted in toto in Rhenohistol®. Eyepiece Leitz Periplan 4×; objective Leitz Pv Fl oil 70/1.15. Positive phase contrast (from SCHILLER 1954)

some are regarded as essential elements of the ferruginous body.

Scanning electron micrographs of alveolar macrophages incubated with fibres show the cell membrane placing itself around the fibre like a sleeve (Robock and Klosterkötter 1977). In *mouse* L-fibroblasts, the incomplete incorporation increased permeability of the plasmalemma in the area of invagination (Beck et al. 1971). The early cytotoxic effects of different types of asbestos are correlated with their magnesium content (Harington et al. 1974) and it seems likely that the positively charged magnesium groups on the surface of the fibres interact with negatively charged membrane sialoglycoproteins. Clustering of the membrane proteins follows, and this is associated with increased passive ion flux through the membranes in excess of active cation transport capacity. As a result of the osmotic pressure exerted by entrapped protein, water accumulates and the cell swells until it bursts. Protection against lysis by asbestos but not by silica can be achieved if a non-penetrating solute, such as sucrose, is present in the extracellular medium, counterbalancing the osmotic pressure of intracellular protein (Allison 1974).

With either successful or frustrated (Fig. 153) phagocytosis, a direct contact between lysosomal content (acidic pH, strongly oxidising medium) and fibre will occur, which may yield different redox reactions. This on one hand modifies the surface of the mineral itself, and on the other hand provokes a prolonged generation of reactive oxygen species, the release of which into the surrounding medium can be demonstrated by lucigenin-enhanced chemiluminescence (Zeller et al. 1993, Brehm 1996).

30-min exposure of 10^6 *human* alveolar macrophages to 100 μg chrysotile B caused a significant (up to 2.5-fold) increase in reactive oxygen intermediates release compared with control experiments (Oettinger et al. 1999). 100 μg Chrysotile B induced a significant maximum 4.0-fold up-regulation of NK-κB gene expression.

From *rabbit* alveolar macrophages, UICC chrysotile A and leached (0.1 N oxalic acid) amphiboles selectively released β-galactosidase (Jaurand et al. 1980).

Macrophages obtained from male BD-IX *rats* dusted with UICC crocidolite were able to stimulate both dusted and control lymphocytes to proliferate, demonstrating that crocidolite inhalation resulted in alveolar macrophages with different properties from those of control macrophages (Miller et al. 1980).

Alveolar macrophages are activated and release growth factors that stimulate mesenchymal cell proliferation and enhanced formation of extracellular matrix. Both insulin-like growth factor-I (IGF-I), and transforming growth factor β (TGF-β) regulate cellular growth and promote matrix accumulation and are hypothesised to play important roles in asbestosis. Lee et al. (1997) performed immunohistochemistry using polyclonal antibodies to specific synthetic peptides of the three mammalian isoforms of TGF-β (TGF-β1, -β2, -β3) and to TGF-I on lungs of *sheep* treated intratracheally with UICC Canadian chrysotile asbestos fibres. All three TGF-β isoforms were found in alveolar macrophages. While asbestos had minimal acute effects on cytokine production by the *human* alveolar macrophage, intense chronic exposure to asbestos lead to the enhanced basal release of significant amounts of several cytokines that have activity for the fibroblast, even in the absence of overt fibrosis (Perkins et al. 1993).

Quantification of IGF-I also manifested elevated IGF-I levels in silicotic *rat* bronchoalveolar lavage fluids, which tended to increase with silica exposure *in vivo*, but no alteration in IGF-I level could be found in sera (Chen et al. 1994).

Chrysotile, but not amphiboles (crocidolite, anthophyllite, or amosite), stimulated a rapid (<1 min) and dose-dependent (2.5–50 μg/ml) production of superoxide anion at noncytotoxic dose (2.5–25 μg/ml) by guinea pig alveolar macrophages as monitored by measuring the rate of ferricytochrome *c* reduction at 550 nm (Roney and Holian (1989). The stimulation of superoxide anion production by chrysotile could be blocked by putative protein kinase C inhibitors (staurosporine, sphingosine, and fluphenazine). Chrysotile also stimulated phosphatidylinositol turnover as measured using $^{32}P_i$ incorporation into phospholipids, [^{3}H]diacylglycerol levels, and intracellular calcium mobilisation as measured using fura-2 and ^{45}Ca. Pertussis toxin partially blocked chrysotile stimulated superoxide anion production.

Asbestos initiates the formation of oxygen radicals through at least two mechanisms: activation of respiratory burst in phagocytic cells and iron-catalysed reactions (Kamp et al. 1992). The dramatic enhancement of release of superoxide anions found by Donaldson et al. (1995) and Hill et al. (1995) when long fibre amosite was opsonized with IgG, confirmed the greatly increased biological activity after opsonization (Scheule and Holian 1989, Nyberg and Klockars 1990, Perkins et al. 1991, Donaldson et al. 1992). All asbestos fibres were able to induce chemiluminescence but chrysotile induced maximal chemiluminescence at higher concentrations than amosite and crocidolite (Lim et al. 1997). Protein kinase C inhibitors (sphingosine and staurosporine) suppressed the ability of asbestos to induce oxygen radical generation. Phospholipase C inhibitors (U73122 and neomycin) and protein tyrosine kinase inhibitors (erbstatin and genistein) decreased oxygen radical generation of asbestos stimulated *rat* alveolar macrophages. Oxygen radical formation was not suppressed by an adenylate cyclase activator (forskolin), a protein kinase A inhibitor (H-8), and a protein serine-threonin phosphatase inhibitor (okadaic acid). Phospholipase C and protein tyrosine kinase inhibitors suppressed the increment of phosphoinositide turnover by amosite.

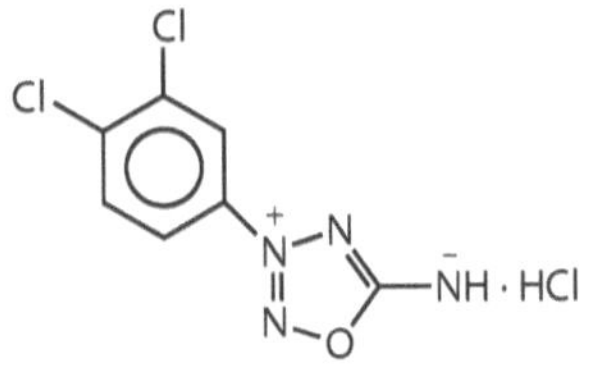

Forskolin [238]

Alveolar macrophages from Fischer 344 *rats* inhaling crocidolite or chrysotile asbestos showed significant elevation ($P < 0.05$) in nitrite/nitrate levels which were ameliorated by N^G-monomethyl-L-arginine (QUINLAN et al. 1998). Temporal patterns of •NO generation from alveolar macrophages correlated with neutrophil influx in bronchoalveolar lavage samples after asbestos inhalation. To determine the molecular mechanisms and specificity of iNOS promoter activation by asbestos, RAW 264.7 cells and alveolar macrophages isolated from control rats were exposed to crocidolite *in vitro*. These cells showed increases in steady-state levels of iNOS mRNA in response to asbestos and more dramatic increases in both iNOS mRNA and immunoreactive protein after addition of lipopolysaccharide (LPS). After transfection of an iNOS promoter/luciferase

Table 39. Toxicity of silicates containing iron

Nesosilicates: isolated SiO_4 tetrahedrons bound to each other only by ionic bonds of the interstitial cations		
Olivines	$(Mg, Fe)_2SiO_4$	KING et al. (1945), GLÖMME and SWENSSON (1963)
Fayalite	Fe_2SiO_4	
Kirschsteinite	$CaFeSiO_4$	
Knebelite	$FeMnSiO_4$	
Garnets		
Almandine	$Fe_3Al_2(SiO_4)_3$	
Andradite	$Ca_3Fe^{3+}{}_2(SiO_4)_3$	
Staurolite	$FeAl_9O_6(SiO_4)_4(O, OH)_2$	
Sorosilicates: two SiO_4 tetrahedrons linked into Si_2O_7 groups		
Epidote	$Ca_2(Al, Fe)Al_2O(SiO_4)(Si_2O_7)(OH)$	
Pistacite = epidote rich in iron		
Cyclosilicates: Si_4O_{12} rings, along with BO_3 triangles and OH groups		
Axinite	$(Ca, Fe, Mn)_3Al_2(BO_3)(Si_4O_{12})(OH)$	
Inosilicates: one-dimensional infinite single (silicon to oxygen ratio of 1:3) or double (silicon to oxygen ratio of 4:11) chains created by the linkage of SiO_4 tetrahedrons		
(SiO_3) or (Si_2O_6) chains		
Pyroxenes	$(Mg, Fe, Ca) SiO_3$	
Hypersthene	$(MgFe)SiO_3$	
Hedenbergite	$CaFe^{2+} [Si_2O_6]$	
Augite	$(Ca, Mg, Fe)(Mg, Fe) Si_2O_6$	
Acmite	$NaFe^{3+}Si_2O_6$	
Ferrosilite	$Fe^{2+}Si_2O_6$	
(Si_4O_{11}) bands, 0.9 nm		
Amphibole asbestos	$(Na, Ca)_{2-3}(Mg, Fe^{2+}, Fe^{3+}, Al)_5$ $(Si, Al)_8O_{22}(OH, O, F)_2$	
Actinolite	$(Ca, Na_2)(Fe, Mg, Al)_5$ $[(OH, F)_2$ $(Si, Al)_8O_{22}]$	POTT et al. (1988)
Anthophyllite	$(Mg, Fe)_6(Al, Fe)[(OH)(Si, Al)_4O_{11}]$	BECK et al. (1967), ROBOCK and KLOSTERKÖTTER (1971)
Amosite	Fe-Anthophyllite	BECK et al. (1967), SZENTEI (1967), ROBOCK and KLOSTERKÖTTER 1971), AUST et al. (1994), BROWN et al. (1994), GHIO et al. (1994)
Crocidolite	$Na_2(Fe^{3+}{}_2, Fe^{2+}{}_3) Si_8O_{22}(OH)_2$	BECK et al. (1971), AUFDERHEIDE et al. (1994), AUST et al. (1994), CASTRANOVA et al. (1994), GHIO et al. (1994)
Cummingtonite	$(Mg, Fe)_7Si_8O_{22}(OH)_2$	
Grunerite	$Fe_7[(OH)_2/Si_8O_{22}]$	
Phyllosilicates: Two-dimensional infinite sheets of composition Si_2O_5 with SiO_4 tetrahedrons linked by the sharing of three corners of each tetrahedron to form a hexagonal mash pattern		
Micas		
Biotite (black mica)	$K(Mg, Fe^{2+})_3(Al, Fe^{3+})Si_3O_{10}(OH, F)_2$	
Leptomelan	rich in Fe^{3+}, therefore called micaceous iron ore	
Zinnwaldite	$K_2(Li, Fe, Al)_6(Si, Al)_8O_{20}(OH, F)_4$	
Glauconite	$(K, Na)(Fe^{3+}, Al, Mg)_2(Al, Si)_4O_{10}(OH)_2$	

reporter construct, RAW 264.7 cells exposed to LPS, crocidolite asbestos and its nonfibrous analogue, riebeckite, revealed increases in luciferase activity whereas cristobalite had no effects.

Silicate complexes with Fe^{3+} were observed by HAZEL et al. (1949). WEBER and STUMM (1965) investigating the equilibria,

$$Fe^{3+} + Si(OH)_4 \leftrightarrow FeSiO(OH)_3^{2+} + H^+ \quad [239]$$

found a formation constant, $\beta_{111} = 0.57$. For the reaction

$$Fe^{3+} + SiO(OH)_3^- \leftrightarrow FeSiO(OH)_3^{2+} \quad [240]$$

they deduced, by using a value of 9.5 for the pk_1 value of silicic acid, a formation constant of $\beta_1 = 1.8 \times 10^9$ ($\log \beta_1 = 9.26$).

Comparing the preference of $SiO(OH)_3^-$ to form complexes with different cations one obtains the series: $Ca^{2+} < Mg^{2+} << Fe^{3+} < H^+$.

Fe^{3+} forms considerable stronger complexes which may, in natural water, be present in significant amounts.

Surface-complexed iron [Fe^{3+}] observed after intrapleural injection of 30 mg of either amosite, crocidolite, and chrysotile both genuine and saturated with Fe^{3+} in Sprague-Dawley *rats* corresponded to oxidant generation, measured as barbituric acid reactive products of deoxyribose, and more covalently closed, circular DNA strand scission induced by these asbestos fibres (GHIO et al. 1994).

Metal complexes of flavonoids were adsorbed by chrysotile fibres considerably better than uncomplexed compounds and probably for this reason flavonoid metal complexes have better protective properties against asbestos-induced haemolysis (KOSTYUK et al. 2001).

The endocytosis of the fibre is associated with oxidant generation and therefor should increase with the concentration of complexed iron and surface area (HOBSON et al. 1990, GHIO et al. 1992). Such endocytosis positions the silicate surface with the complexed transition metal in the proximity of DNA, which also can co-ordinate this metal (ANDRONIKASHVILI et al. 1974, BARRY et al. 1983). Subsequent oxidant generation could then result in strand breaks, genotoxicity, and cancer (LIBBUS et al. 1989).

Table 40. Surface area, percentage SiO_2, and surface silanol density of fibrous silicates (from GHIO et al. 1994)

Silicate	Surface area (m^2/g)	Percentage SiO_2	Surface silanol density ($groups/nm^2$)
Amosite	2.3 ± 0.4	49.2 ± 0.5	7.5 ± 1.8
Crocidolite	8.7 ± 1.0	48.5 ± 0.3	4.7 ± 0.6
Chrysotile	28.8 ± 1.4	39.8 ± 0.9	1.8 ± 0.1

Nitric oxide may interact with the surface of mineral fibres. Cigarette smoke which increases the risk of asbestos-induced lung cancer contains up to 600 µg •NO per cigarette (IARC 1985) raised the fibre-bound •NO from 34 to 85 µg NO/g fibre (LEANDERSON et al. 1997). •NO was found in different amounts on chrysotile B, crocidolite, amosite and silicon carbide whiskers. There was a strong correlation between the amount of •NO and the specific surface area of these fibres ($r = 0.98$). •NO could not be demonstrate on rockwool fibres, man-made vitreous fibres MMV21 and MMV22 or silicon nitride whiskers.

The haemolytic and early cytotoxic effects of different types of asbestos fibres are correlated with their magnesium content (HARINGTON et al. 1974). It seem likely that the positively charged magnesium groups on the surface of the fibres interact with the negatively charged membrane sialoglycoproteins. Clustering of the membrane proteins follows, and this is associated with increased passive ion flux through the membranes in excess of the active cation transport capacity. As a result of the osmotic pressure exerted by entrapped protein, water accumulates and the cells swells until it bursts. Protection against lysis by asbestos but not by silica can be achieved if a non-penetrating solute, such as sucrose, is present in the extracellular medium, counterbalancing the osmotic pressure of intracellular protein (ALLISON 1974). When asbestos fibres are incubated with cells in the presence of fresh serum, haemolysis can occur by activation of the complement system (HARINGTON et al. 1974).

The surface of chrysotile asbestos modified with a chemically bound coating of an alkyl hydrocarbon altered phagocytic activity, cytotoxicity and mutagenicity in *human* alveolar macrophages obtained by fibre optic bronchial lavage (MACE jr et al. 1981). Scanning electron microscopy of cell particle interactions and phagocytosis showed no impairment of the processes by the coated fibres. Cytotoxicity studies by trypan blue exclusion revealed a significant reduction in alveolar macrophage mortality over 24 and 48 h incubation periods, compared to uncoated fibres.

Erionite particles (sizes from colloidal to about 10–15 µm in length) upon phagocytosis by the *rat* lung alveolar macrophage cell line, NR8383 increased generation of reactive oxygen metabolites as shown by fluorescence imaging with 5-(and 6) carboxy-2',7'-dichlorodihydrofluorescein diacetate (LONG et al. 1997). Sharply delineated fluorescence characteristically evolved within fibre-exposed cells during the period up to 90 min following the cell-fibre contact. For the 20–40 min period, the erionite-exposed cells had a mean cellular fluorescence more than three times that of controls. During the 40–60 min period, the erionite-exposed cells had a mean cellular fluorescence still more than twice that of controls. Only during the 60–80 min period did the mean erionite cellular fluorescence decline to a level relatively indistinguishable compared to controls.

Wollastonite, an aciculate or fibrous silicate mineral ($CaSiO_3$), in a dose of 25 μg/ml *in vitro* reduced the viability of *rat* alveolar macrophages from 77.0±11.0% to 66.4±8.9% ($P < 0.05$) and increased lactate dehydrogenase release from 37.2±10.2% to 46.5%±7.3% ($P < 0.05$) within 24 h (PASANEN et al. 1984). Wollastonite (<10 μm) was moderately cytotoxic in both haemolysis and alveolar macrophage enzyme release (lactate dehydrogenase, β-glucuronidase, β-*N*-acetylglucosaminidase) studies whereas shorter fibres (<5 μm) were only mildly cytotoxic (VALLYATHAN et al. 1984).

Wollastonite (up to 400 μg/ml) produced no significant DNA fragmentation in a 24-h culture of *human* alveolar macrophages (HAMILTON et al. 1996).

Glass fibres with diameters ≤3 μm and lengths ≤10 μm are cleared quite efficiently from the *rat* lung, presumably by macrophage-mediated processes (MORGAN et al. 1982). For MMVF34, the inhalation test at different duration showed a similar persistence patter, while the intratracheal test longer elimination half-times especially for long fibres (KAMSTRUP et al. 1998). The MMVF34 fibre is considerably less biopersistent than the traditional MMVF21 fibre when comparing the calculated elimination half-times after short-term inhalation. Throughout the 18 month period, only slight macrophage reaction was seen in the lungs of *rats* exposed to MMVF34 (Wagner grade 2) although in some animals occasional microgranulomas were seen (Wagner grade 3).

Pulmonary lavage washings from glass-fibre-dusted inbred male BD-IX *rats* consisted of 5–10% lymphocytes and a heterogeneous macrophage population, varying in size from 10–40 μm (MILLER 1980). Many cells contained a great number of glass fibres, and multinucleate giant cells comprised between 10–20% of the population. Many macrophages also contained protruding fibres. Such cells still adhered to the glass substrate after 22 h of incubation and retained their characteristic surface ruffles.

Dust- and fibre-induced release of lactate dehydrogenase from *rat* alveolar macrophages at a dose of 10 mg/ml by man-made vitreous fibres was less than by quartz, that by RCF 1 the least of all the fibres, even smaller than that induced by titanium dioxide (LUOTO et al. 1997). MMVF 10 glasswool fibre caused less LDH release than MMVF 21 rockwool fibre, the MMVF 22 slagwool fibre or the ceramic fibre RCF 3 ($P < 0.005$). Refractory ceramic fibres (RF2) and rock wool (RW1) increased the release of lactate dehydrogenase with increasing fibre concentration (KIM et al. 2001). From these parameters, RF2 was shown to exhibit greater cytotoxicity than did RW1.

For each of 5 lengths (generated by a dielectrophoretic classifier) of Manville Code 100 fibres to *rat* macrophages cultured on 96-well plates for 18 h, toxicity depended on concentration ranging from 0–500 μg fibres/ml (BLAKE et al. 1997, 1998). Lucigenin-enhanced chemiluminescence declined and lactate dehydrogenase release increased with increasing fibre concentration. On a μg/ml basis, an intermediate-length fibre (mean length = 17 μm) showed the greatest toxicity, while both shorter and longer fibres caused less damage.

A significantly higher percentage of *rat* alveolar macrophages than peritoneal macrophages endured frustrated phagocytosis of MMVF10 and MMVF21 fibres (DÖRGER et al. 2001). In line with these findings, alveolar macrophages generated higher levels of oxygen radicals than peritoneal macrophages upon exposure to MMVF21 fibres. In contrast, MMVF10 fibres failed to induce the generation of reactive oxygen species by both alveolar and peritoneal macrophages.

P388D$_1$ macrophages grown in RPMI 1640 medium flowthrough (2–3 ml/h) culture dissolved man-made vitreous fibres leading to subsequent morphological changes of fibres as shown by scanning electron microscopy, while the dissolution of Si, Fe, and Al was measured on days 4, 8, 16, and 28 using an atomic absorption spectrometer (LUOTO et al. 1998).

HL-60-M cells challenged by glass wool code A or stone wool code G fibres preincubated in unbuffered saline for about 4 weeks, produced less intense luminol-enhanced chemiluminescence, as compared with freshly suspended fibres (ZOLLER and ZELLER 2000). MMVF 21 and HT-N fibres as well as crocidolite and erionite did not show any decrease in chemiluminescence intensity after incubation in aqueous solutions.

Intracellular adenosine triphosphate content of *rat* alveolar macrophages obtained by bronchoalveolar lavage was decreased by refractory ceramic fibres (RF2) and rock wool (RW1) in a concentration-dependent manner (KIM et al. 2001). This fibres suppressed succinate-triggered oxygen consumption. Polyinsosinic acid, a ligand of the scavenger receptor, inhibited the man-made vitreous fibre-induced decrease in ATP concentration.

Silicon nitride (Si_3N_4), a fibrous material, increased the release of lactate dehydrogenase, but not of acid phosphatase, from lavaged *bovine* pulmonary macrophages *in vitro* (Fisher et al. 1989). Scanning electron micrographs indicated that the fibrous Si_3N_4 is cytotoxic and poorly tolerated by macrophages.

Complexes with **catechol** and other *o*-diphenols (Rosenheim et al. 1931, Weiss et al. 1956, 1959, Baumann 1963) form as follows:

$$3\,C_6H_4(OH)_2 + Si(OH)_4 \longleftrightarrow [Si(C_6H_4O_2)_3]^{2-} + 4\,H_2O + 2\,H^+ \quad [241]$$

In these complexes silicon is octahedrally surrounded by six oxygens (three dihydroxybenzenes, didentately bound to Si). According to Weiss et al. (1961), besides these mononuclear silicon complexes, also binuclear complexes (dimers) seem to be formed.

The relationship between **silicate minerals and the interferon system** was studied by Hahon and Booth (1987) in *monkey* kidney cell monolayers. Minerals within the classes nesosilicate, sorosilicate, cyclosilicate, and inosilicate exhibited either little or marked (50 % or greater) inhibition of interferon induction. Within the inosilicate class, however, minerals of the pyroxenoid group (wollastonite, pectolite, and rhodonite) all significantly showed a two- to threefold increase in interferon production. Silicate materials in the phyllosilicate and tectosilicate classes all showed inhibitory activity for the induction process. When silicate minerals were coated with poly(4-vinylpyridine-*N*-oxide), the inhibitory activity of silicates on viral interferon induction was counteracted.

In the lung-associates lymph nodes of *rats* instilled intratracheally with DQ12 quartz (surface area 9.4 m^2/g) or Aerosil® 150 (surface area $150 \pm 15\ m^2/g$), poly-2-vinylpyridine-*N*-oxide treatment significantly reduced the incidence and severity of inflammation as evidenced by the presence of well circumscribed aggregates of intact particle-laden macrophages without signs of degeneration and accompanying granulocyte infiltrations and fibrosis (Ernst et al. 2002).

Granuloma Formation After Intravenous Injections of Quartz

In an extended study of experimental tissue reactions following intravenous injections of silica and other dusts Simson and Strachan (1940) stated that the distribution of the microscopically visible dust in the organs and tissues of the rabbit in the early stage is the same irrespective of whether the dust later exhibits toxic or non-toxic properties. The chief situations in which it can be found are the organs provided with sinusoids and an active reticulo-endothelial tissue in relation to the blood vascular system, e.g. liver, spleen, bone-marrow and lymph-nodes.

Liver

Gardner et al. (1944) produced massive fibrosis in the liver by 20 intravenous injections of powdered quartz (1–3 μm; total dose 500 mg) suspended in saline into the ear veins of rabbits. When the quartz powder was suspended in 0.33 per cent gelatinous colloidal alumina this silicosis failed to develop, only a marked foreign body reaction being produced in the liver despite the accumulation there of a large amount of quartz mixed with aluminium.

Spleen

Simson and Strachan (1940) noted great variation in the size of the *rabbit* spleen, quartz dust

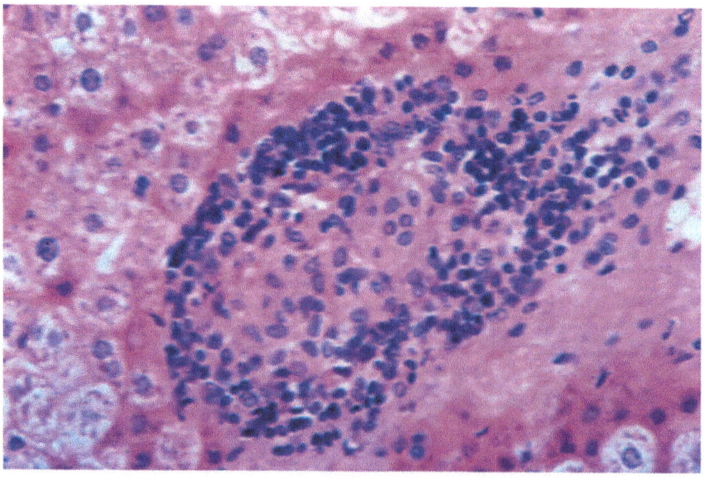

Fig. 154. Silicotic granuloma in the liver (20 g) of a 410 g male white *rat* (No. 36) 12 weeks after an intravenous application of 25 mg Brazilian crystallised quartz (specific surface 2 m^2/g). Fixed by immersion in Bouin's fluid. Paraplast. Haematoxylin and eosin. Objective Leitz Pl 40/0.65. Leitz-Orthomat® (×3.2). Agfachrome 50 L professional

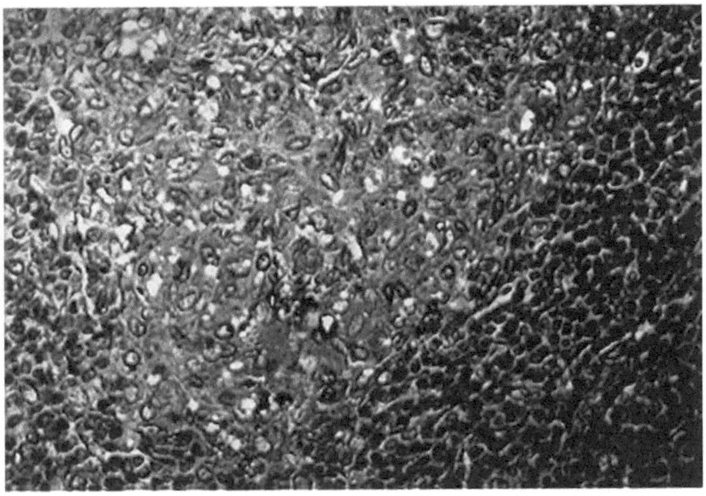

Fig. 155. Macrophages in the marginal zone of the splenic white pulp of a male white *rat* (No. 30) 8 weeks after an intravenous application of 25 mg Brazilian crystallised quartz (specific surface 2 m^2/g). Fixed by immersion in Bouin's fluid. Paraplast. Azan modification using rubine fast red and aniline blue (SPECHT 1973). Objective Leitz Pl 40/0.65. Leitz-Orthomat® (×3.2). Elliptic polarised light. Agfachrome 50 L professional

being responsible for the greatest enlargement. The spleen in texture, colour and consistence was indistinguishable from that of Banti's splenomegaly.

Colloidal, molybdate-inactive silicic acid after prolonged intravenous injection up to 11 months, accumulated in the *rabbit* spleen (up to 4.580 mg/g) inducing splenomegaly (AMMON and MOHN 1958). In the *rat*, colloidal silicic acid was accumulated in histiocytes (MOHN 1963). Lymphocytes did not contain any SiO_2, and there were no lymphocytes in mature silicotic granulomas. Necrosis occurred at an increased rate of SiO_2 administration.

Polyvinylpyridine-*N*-oxide retarded splenic fibrosis of *mice* given 3 mg quartz intravenously as judged from the hydroxyproline content of the organ (DOLGNER and POTT 1967).

9.6.2.1.28 Sulphur

Sulphur has the atomic number 16 and the valence electrons are those of the $3s^2 3p^4$ shells. The oxidation states of sulphur are −2, +4 and +6. These many valence states make sulphur an ideal redox regulator.

Sulphur Dioxide

In the nasal passages and lung, SO_2 is hydrated rapidly according to the equation

$$SO_2 + H_2O \longleftrightarrow HSO_3^- + H^+ \quad [242]$$

with the equilibrium constant 1.7×10^2 M/l (PETERING 1977). Bisulfite, which predominates at physiological pH, is a weak acid which dissociates according to the reaction

$$HSO_3^- + H_2O \longleftrightarrow SO_3^{-2} + H_3O^+ \quad [243]$$

with an equilibrium constant 1.02×10^7 M/l.

Prostaglandin synthase (EC 1.14.99.1), which is immunoreactive in *human* and other *mammalian* alveolar macrophages (YU et al. 1988),

Human alveolar macrophages gained by bronchoalveolar lavage were placed on a polycarbonate membrane through which they were supplied with RPMI 1640 for nutrient. Exposed for 10 min, 20 min and 30 min to 2.5 ppm, 7.5 ppm and 12.5 ppm SO_2, respectively, they showed a significant increase in luminol-enhanced chemiluminescence due to the production of reactive oxygen species (KIENAST et al. 1993). The 12.5 ppm concentration of SO_2 killed 62±9% of the alveolar macrophages and thus reduced chemiluminescence by 63% compared with the 2.5 ppm exposure. A 30-min exposure to SO_2 induced a significant decrease in spontaneous and lipopolysaccharide-stimulated tumour necrosis factor-α ($P < 0.001$) and lipopolysaccharide-stimulated interleukin-1β release ($P < 0.059$), while the release of interleukin-6 and tansforming growth factor-β was not significantly affected (KNORST et al. 1996).

9.6.2.1.29 Tantalum

Tantalum has the atomic number 73 and the valence electrons are those of the $4f^{14}5d^3 6s^2$ shells. The oxidation state of tantalum is +5.

Tantalum oxide (150 mg Ta_2O_5/m^3) dust inhaled for 10 days (10 h/d) was phagocytized by *rat* alveolar macrophages. The cells contained agglomerations of particles of different size, all <1 μm within phagolysosomes, unchanged mitochondria and numerous free and membrane-attached ribosomes (NEMETSCHEK-GANSLER et al. 1975).

In vitro, at two hours more than 90 % of alveolar macrophages obtained from New Zealand white *rabbits* by pulmonary lavage had ingested tantalum oxide particles (1 μm mean particle diameter); at 12, 24 and 30 h lactate dehydrogenase release was significantly higher than in control and latex (0.81 μm mean particle diameter) preparations (MATTHAY et al. 1978). Loss of cell viability and cell sloughing was observed first at 6 h, and by 24 h when LDH levels were high (29.16 ± 5.58 μg) few cells remained adherent and viable. At 12 h lysozyme release in tantalum oxide preparations (26.00 ± 3.12 μg) was equal to that in quartz (0.28 μm mean particle diameter) preparations (25.19 ± 2.20 μg), and both tantalum oxide and silica media lysozyme levels were significantly ($P < 0.01$) higher than those in control (10.53 ± 1.37 μg) and latex preparations (9.75 ± 0.32 μg).

9.6.2.1.30 Thallium

Thallium has the atomic number 81 and the valence electrons are those of the $6s^2 6p^1$ shells. The oxidation states of thallium are +1 and +3. Tl^+ tends to form stable complexes with soft ligand donors, such as sulphur-containing compounds. The well-known mechanism of thallium toxicity is related to the interference with the vital potassium-dependent processes, substitution of potassium in the (Na^+/K^+)-ATPase, as well as a high affinity for sulfhydryl groups from proteins and other biomolecules (AOYAMA et al. 1988). Lipid peroxidation – particularly in the kidney – was increased with a marked decrease in non-protein sulfhydryls as an indicator of glutathion levels, as well as a depletion of glutathione peroxidase (EC 1.11.1.9) activity, suggesting that the tissue damage induced by thallium can be also associated to peroxidative processes.

The effect of thallium on lipid peroxidation was first tested by HASAN and ALI (1981) and compared to nickel and cobalt administration.

In *patients* intoxicated with thallium CAVANAGH et al. (1974) described oedema, moderate congestion, patchy early bronchopneumonia and inflammation.

9.6.2.1.31 Titanium

Titanium has the atomic number 22 and the valence electrons are those of the $3d^2 4s^2$ shells. The oxidation states of titanium include +2, +3 and +4. The tetravalent form is the is the most commonly encountered.

The violet cation Ti^{3+} was introduced by DIXON and NORMAN (1962) to produce $HO^{\bullet}$ from H_2O_2 in the laboratory:

$$Ti^{3+} + H_2O_2 \longrightarrow Ti(IV) + HO^{\bullet} + OH^- \qquad [244]$$

The generation of singlet oxygen ($^1O_2$) by irradiated TiO_2 (anatase) in ethanol was measured by ESR spectroscopy using 2,2,6,6-tetramethyl-4-piperidone as a $^1O_2$-sensitive trapping agent (KONAKA et al. 1999).

Titanium Dioxides

There are three modification of TiO_2: rutile, anatase, and brookite.

Alveolar lumina completely filled with phagocytes some of which have been destroyed were found in a 38 years old patient, who had worked in a titanium dioxide factory for about nine years, all this time as a repair worker in the dusty departments (ELO et al. 1972). Specimens from this and another patient showed alveolar phagocytes with large lysosomes containing electron-dense TiO_2 particles.

Macrophages which contained carbon-like, but birefringent pigment were found in the sputum as well as in extracellular areas of biopsy samples from former workers of a TiO_2 pigment factory, who had been put on pension because of chronic bronchitis and partial pulmonary dysfunction (MÄÄTTA and ARSTILA 1975). Dispersive X-ray analyses from the extremely electron-dense, round or oval-shaped particles inside the lysosomes gave distinct peaks for titanium K_α and K_β in every case. The electron-dense, more or less rectangular particles inside the lysosomes gave typical peaks for silicon K_α and weak signal for aluminium K_α and potassium K_α, as one of the coating materials in the industrial process was aluminium silicate.

Autopsy findings in a 55-year-old *man* known to have been occupationally heavily exposed to titanium dioxide dust showed extensive pulmonary deposition of white pigment identified by energy dispersive X-ray analysis and electron X-ray diffraction as rutile (RODE et al. 1981). The pigment was mainly distributed in the perivascular tissue, but smaller amounts were found in the alveolar walls and also in alveolar macrophages.

The biological activity of ultrafine TiO_2 maybe a consequence of its size and DONALDSON and GILMOUR (1995) hypothesised that there could be free radical activity associated with the surface of ultrafine TiO_2, which caused strand breakage electron-dense detectable as 100 % depletion of supercoiled DNA whilst normal sized TiO_2 had no effect; the injury could be ameliorated by including mannitol, implicating hydroxyl radical.

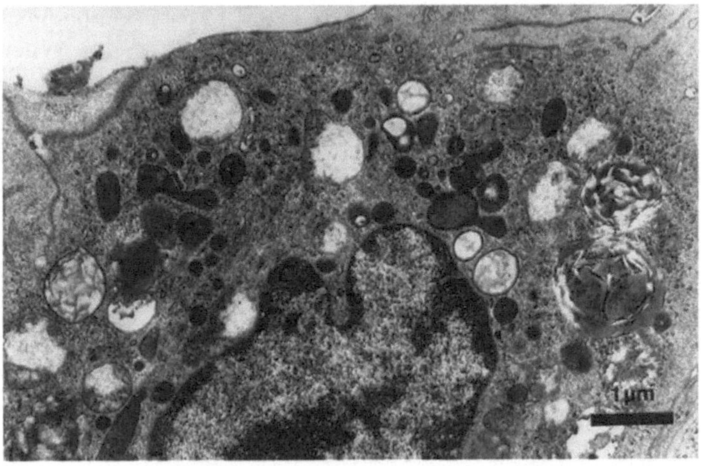

Fig. 156. Alveolar macrophage (block 4459) of a 194 g male Sprague-Dawley *rat* (Charles River, France), injected intraperitoneally with 50 mg anatase (titanium dioxide P 25, Degussa, Frankfort on the Main) suspended in 2 ml saline. 4 days later, on July 25, 1978 under pentobarbital anaesthesia (30 mg/kg), the animal was perfused from the abdominal aorta with 2.5 % glutaraldehyde in 0.1 M sodium cacodylate buffer (pH 7.4). Postfixation with 1 % osmium tetroxide in sodium cacodylate buffer. Embedded in Epon 812 and sectioned at 50 nm. Lead citrate and uranyl acetate. Film 362/79

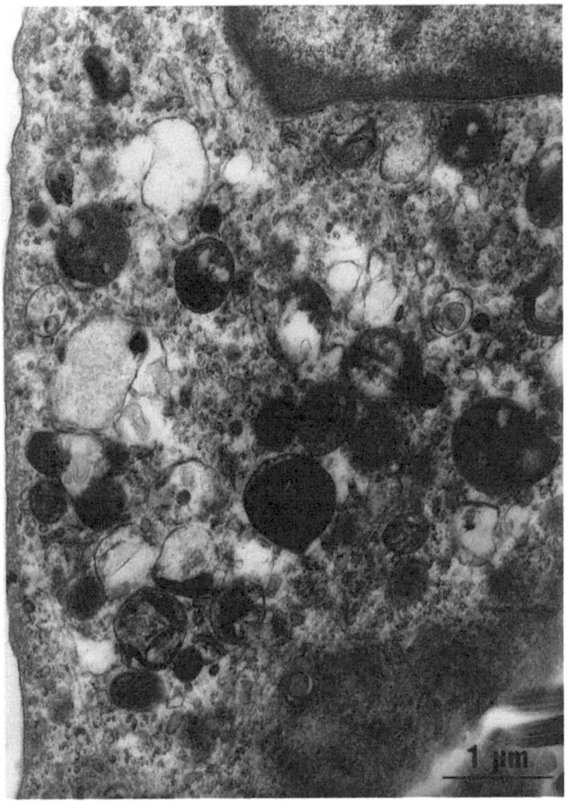

Fig. 157. Macrophage on the bronchiolar epithelium (block 4482) from a 200 g Sprague-Dawley *rat* (Charles River, France) instilled intratracheally with 50 mg anatase (titanium dioxide P 25, Degussa, Frankfort on the Main) suspended in 0.5 ml saline. Animal treated for 4 consecutive days with intraperitoneal injection of 15 mg carbocromen in 10 % methylcellulose per kg body weight × day. On July 25, 1978 under pentobarbital anaesthesia (30 mg/kg), the lung was fixed by intratracheal instillation of 2.5 % glutaraldehyde in 0.1 M sodium cacodylate buffer (pH 7.4) before opening the thorax. Postfixation with 1 % osmium tetroxide in sodium cacodylate buffer. Embedded in Epon 812 and sectioned at 50 nm. Lead citrate and uranyl acetate. Plate 4138

Fig. 158. Microfilaments (MF) and microtubules (MT) in an alveolar macrophage (block 4481) from a 199 g Sprague-Dawley *rat* (Charles River, France) instilled intratracheally with 50 mg anatase (titanium dioxide P 25, Degussa, Frankfort on the Main) suspended in 0.5 ml saline. Animal treated for 4 consecutive days with intraperitoneal injection of 15 mg carbocromen in 10 % methylcellulose per kg body weight × day. On July 25, 1978 under pentobarbital anaesthesia (30 mg/kg), the lung was fixed by intratracheal instillation of 2.5 % glutaraldehyde in 0.1 M sodium cacodylate buffer (pH 7.4) before opening the thorax. Postfixation with 1 % osmium tetroxide in sodium cacodylate buffer. Embedded in Epon 812 and sectioned at 50 nm. Lead citrate and uranyl acetate. Plate 4134

Guinea pigs' alveolar macrophages after 14 days' exposure to rutile (23±7.3 mg/m^3 for 20 h/d) contained TiO_2 particles in phagolysosomes (Baskerville et al. 1988). The macrophage blockade by TiO_2 did not alter the animals' susceptibility to Legionnaires' disease nor increase mortality.

Human alveolar macrophages produced more reactive oxygen species in response to ultrafine titanium dioxide particles than *rat* alveolar macrophages (Rahman et al. 1997). In case of *human* alveolar macrophages, the peak level of chemiluminescence was reached in less than 1 min after the addition of ultrafine titanium dioxide particles.

Instillation of 500 mg TiO_2 dust (7 μm maximum diameter) suspended in 50 ml PBS into the right caudal lung lobe of 3 female pigtail macaque monkeys (*Macaca nemestrina*) using a flexible fibreoptic bronchoscope increased the numbers of macrophages in the dust exposed lobes and significantly elevated the *N*-acetyl-β-D-glucosaminidase (EC 3.2.1.17) activity within the bronchoalveolar lavages taken at 2-weeks intervals for 12 weeks thereafter (Mack et al. 1995).

Titanium Tetrachloride

Rats exposed to titanium tetrachloride ($TiCl_4$) hydrolysis products by inhalation exposure at aerosol concentrations of 0.1, 1.0 and 10 mg/m^3 for 6 h/day, 5 days/week, for 2 years showed alveoli filled with two types of alveolar macrophages: dust-laden macrophages were filled with densely aggregated particles and foamy alveolar macrophages that only contained a small amount of dust particles (foamy dust cells). Many foamy dust cells showed degenerative changes and were disintegrated in the alveolar air spaces releasing dust particles and granular or fibrinous cellular debris. Cholesterol crystal clefts were formed within the proteinaceous material or foamy alveolar macrophages. Subsequently, cholesterol granulomas were developed with accumulation of proteinaceous material, foamy dust cells, and cellular debris from disintegrated foamy dust cells. At the electron microscopic level, the cytoplasm was packed with numerous myelin inclusions, scanty lysosomes, and phagosomes containing a few dust particles (Lee et al. 1986).

Titanium Phosphate

Titanium phosphate, a manmade fibre of 0.2 μm to 0.3 μm in diameter and from 10 μm to 20 μm in length, has squared-off ends and forms tight bundles.

Intratracheal injection of suspension in physiologic saline into *rats'* lungs induced accumulation of mononucleated cells including the fibres and forming syncytia and giant cells and a slight, dose-related fibrogenic response (Gross et al. 1977). After 17 months macrophages had undergone lipidic degeneration.

Potassium Titanate

Four weeks after intratracheal instillation of 2 mg of respirable potassium titanate whiskers into male Wistar *rats* RNA extracted from LPS stimulated alveolar macrophages expression of IL-1α was increased (Tsuda et al. 1996, 1997). The expression of IL-1α mRNA was by fibres was greatest in potassium octatitanate whisker > UICC crocidolite > UICC chrysotile > refractory ceramic fibre-instilled *rat* alveolar macrophages. The levels of IL-1α and TNF-α mRNA in alveolar macrophages (expressed in their ratio to β-actin) peaked at 1 month and 3 days after exposure to octatitanate, respectively.

9.6.2.1.32 Tungsten

Tungsten has the atomic number 74 and the valence electrons are those of the $(4f^{14})5d^4 6s^2$ shells. The oxidation states of tungsten are +2, +3, +4, +5 and +6.

Calcium tungstate (250 μg) instilled intratracheally in CD-1 *mice* was ingested by alveolar macrophages which showed a high density of surface microvilli (Peão et al. 1993).

9.6.2.1.33 Uranium

Uranium has the atomic number 92 and the valence electrons are those of the $5f^3 6d7s^2$ shells. The oxidation states of uranium are +3, +4, +5 and +6.

α-Tracks emanating from $^{234}UO_2$ particles within macrophages of *rats* exposed for 100 min to 45±1.3 mg/m^3 and 87.0±2.6 mg/m^3 uranium dioxide, respectively, were shown in autoradiographs of the airways at times between 2 and 35 days post-inhalation (Gore and Thorne 1977).

9.6.2.1.34 Vanadium

Vanadium has the atomic number 23 and the valence electrons are those of the $3d^3 4s^2$ shells. The oxidation states of vanadium are +2, +3, +4 and +5.

Vanadium, like molybdenum, has two very different catalytic activities associated with it and two very different biological functions. Both elements have a rich oxygen chemistry associated with the oxyanion and oxycation species. Vanadium compounds of a low oxidation state are reductants, while vanadium(V) compounds are oxidants. In aqueous solutions V^{2+}aq, V^{3+}aq, VO^{2+}, and VO_3^- ions exist, but also numerous complexes and polynuclear complexes.

The second class of chemistry of these two metals belongs to sulphur-rich reducing conditions, where vanadium like molybdenum, can form sulphur-containing anionic centres, e.g. VS_4^{3-} (MoS_4^{2-}); sulphur-containing cationic centres, e.g. VS^{2+} (compare $MoOS^{2+}$); and iron-sulphur clusters Fe_6MS_8, associated in biology with nitrogen fixation.

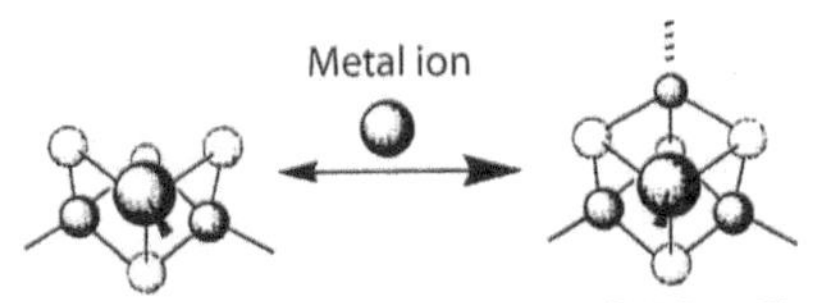

[3Fe-4S] [M3Fe-4S]

Addition of a metal ion to a [3Fe-4S] cuboidal cluster to yield a [4Fe-4S] cubane [80]

In toadstools biology has used another curious device, an apparently simple ligand – *N*-hydroxyimino-di-α-propionate – which forms an extraordinary stable 2:1 complex with V(IV), and is therefore able to extract the metal from soils in which it is scarce relative to its potential competitors, e.g. Fe^{3+} and Cu^{2+} (FRAÚSTO da SILVA 1989).

Vanadium compounds of a low oxidation state are reductants, while vanadium(V) compounds are oxidants. In aqueous solutions V^{2+}aq, V^{3+}aq, VO^{2+}, and VO_3^- ions exist, but also numerous complexes and polynuclear complexes.

Vanadium Pentoxide (V_2O_5) and Vanadates

$$V^{V} + O_2^{\bullet-} \leftrightarrow V^{IV}\text{-}OO^{\bullet} \quad [245]$$
$$V^{IV} - OO^{\bullet} + NADPH \leftrightarrow V^{IV} - OOH + NADP^{\bullet} \quad [246]$$
$$NADP^{\bullet} + O_2 \leftrightarrow NADP^{+} + O_2^{\bullet-} \quad [247]$$
$$V^{IV} - OOH + H^{+} \leftrightarrow V^{V} + H_2O_2 \quad [248]$$

This process has been shown to account for the vanadate stimulation of NADPH oxidation by biological membranes (LIOCHEV and FRIDOVICH 1986). A number of sugars and sugar phosphates can reduce V^V to V^{IV}, and V^{IV} can reduce dioxygen to yield $O_2^{\bullet-}$. It follows that V^V plus sugar will, under aurobic conditions, generate $O_2^{\bullet-}$, which in the presence of V^V will then initiate the oxidation of NADPH (LIOCHEV and FRIDOVICH 1987).

Rabbit alveolar macrophages incubated for 20 h in supplemented Medium 199 with particulate forms of $\mathbf{V_2O_5}$ (13 μg V/ml), $\mathbf{V_2O_3}$ (21 μg/ml) or $\mathbf{VO_2}$ (33 μg V/ml) showed a reduction of their viability by 50 %, while only 9 μg V/ml were required when V_2O_5 was dissolved in media prior to exposure of cells (WATERS et al. 1974).

100 μM Desferal was able to suppress the stimulatory effect of ≤200 μM $\mathbf{V_2O_5}$ on the oxidation of NADH by $O_2^{\bullet-}$ (DARR and FRIDOVICH 1985). This chelating agent could form 1:1 complexes with metavanadate ($V_4O_{12}^{-4}$) or orthovanadate (VO_4^{-3}). The latter was the catalytically active species.

Supernatants from *rat* alveolar macrophages treated with V_2O_5 *in vitro* released upregulatory activity for platelet-derived growth factor receptor-α on cultured lung myofibroblasts, and this activity was blocked by the interleukin-1-receptor antagonist (BONNER et al. 1998).

Ammonium **metavanadate** exposure affected *mouse* WEHI-3 macrophage interferon-γ-binding and -inducible responses (COHEN et al. 1996). It interfered with both the uptake and ultimate intralysosomal killing of *Listeria monocytogenes* by *mouse* peritoneal macrophages (COHEN et al. 1989).

Respiratory burst in lung macrophages significantly increased within 5 min of exposure to metavanadate (10–100 μM), with a maximum increase of 173 % above baseline after 15 min, measured using a DCF-DCFHDA fluorescence assay (GRABOWSKI et al. 1996). Metavanadate (10–100 μM) inhibition of phosphotyrosine phosphatase activity in membrane and cytoplasmic fractions of lung macrophages was dose-dependent with a range of 28–89 %. Suppression of the metavanadate-induced burst by 37–47 % was accomplished using 10 μM of Tyrphostin B50, implicating epidermal growth factor receptor kinase activity in the respiratory burst.

The inducible production of interleukin-6 and interferon-γ by alveolar macrophages lavaged from Fischer 344 *rats* exposed to ammonium metavanadate (2 mg V/m^3) for 8 h/d for 4 d, followed, 24 h later by intratracheal instillation of 0.15 mg polyinosinic:polycytidylic acid in 100 μl saline, within 24 h increased 11-fold and 20-fold, respectively, vs. ≈30-fold and 7-fold, respectively, in air control *rats* (COHEN et al. 1997). At 24 h following poly I:C instillation, there were no significant host exposure dependent differences in either spontaneous of opsonized zymosan-stimulated $O_2^{\bullet-}$ formation. At 48 h a time post poly I:C-dependent effect upon opsonized zymosan-stimulated pulmonary alveolar macrophage $O_2^{\bullet-}$ production was evident; this effect was greater in macrophages from *rats* exposed to V prior to poly I:C instillation than in similarly instilled air control animals.

In HL-60 cells, vanadate stimulated O_2 consumption for the synthesis of superoxide, which triggered the formation of peroxovanadyl [$V^{(4+)}$-OO] and vanadyl hydroperoxide [$V^{(4+)}$-OOH), one or both of these species, rather than vanadate itself being responsible for the respiratory burst (TRUDEL et al. 1991).

Rat alveolar macrophages exposed *in vitro* to 100 μM vanadyl chloride/1 μCi ^{48}V incorporated 8.3 % of the metal after 30 min (GRABOWSKI et al. 1999). Exposure of alveolar macrophages to increasing concentrations of sodium metavanadate resulted in a dose-dependent increase in production of reactive oxygen intermediates as measured by dichlorofluorescin oxidation. The lowest dose yielding a measurable response was 50 μM, whereas 1000 μM increased respiratory burst activity by 173 %. NADPH oxidase inhibitors deoxy-D-glucose (100 mM) and diphenylene iodonium (25 μM) reduced the metavanadate-induced respiratory burst by 62 and 71 %, respectively, implicating NADPH oxidase as the primary cellular source of reactive oxygen intermediates. Enhanced cerium chloride oxidation in response to metavanadate localised to the plasma membrane consistent with increased NADPH oxidase activity. Pre-treatment of alveolar macrophages with the epidermal growth factor receptor inhibitor, tyrphostin B50 (10 μM), reduced the metavanadate-induced respiratory burst, but did not influence overall tyrosine phosphorylation. Metavanadate and H_2O_2 exposure greatly increased overall tyrosine phosphorylation, yielding a similar but distinguishable pattern of phosphorylation in these cells.

Acid phosphatase activity in 10^6 $B_6C_3F_1$ *mouse* peritoneal macrophages was depressed by 22.8 and 44.7 % when ammonium metavanadate (2.5 mg and 10 mg V/kg) was given i.p. every 3 d for 6 weeks, while the activities of β-glucuronidase, N-acetyl-β-D-glucosaminidase, and lysozyme were not significantly influenced (VADDI and WEI 1991).

9.6.2.1.35 Yttrium

Yttrium has the atomic number 39 and the valence electrons are those of the $4d^1 5s^2$ shells. Its only oxidation state is +3.

Yttrium Chloride

MARUBASHI et al. (1998) investigated pulmonary clearance of yttrium and acute lung injury following intratracheal instillation of YCl_3 in saline- or YCl_3-pretrated male Wistar *rats* (30 days before the second challenge). About 67 % of the initial dose of Y remained in the lung even 31 days after intratracheal treatment. The pre-treatment with YCl_3 significantly reduced YCl_3 -induced increases in biochemical inflammatory indicators in bronchoalveolar lavage fluid, such as lactate dehydrogenase, β-glucuronidase, and alkaline phosphatase activities and protein concentration, while the pre-treatment increased the number of polymorphonuclear leucocyte in bronchoalveolar lavage fluid.

Yttrium-Barium-Copper Oxide ($YBa_2Cu_3O_7$) Superconducting Material

Bovine alveolar macrophages incubated for 20 h with 100 μg $YBa_2Cu_3O_7$/ml PPMI 1640 exhibited a sponge-like structure. After 60 min the dust was detected in vacuoles increasing with time. Viability was reduced in a dose-dependent manner (WILCZEK et al. 1990).

9.6.2.1.36 Zinc

Zinc has the atomic number 30 and the valence electrons are those of the $3d^{10}4s^2$ shells. Its only oxidation state is +2. Zinc appears in the active site of a variety of enzymes and, unlike other well-known constituents of metalloenzymes like iron and copper, does not function as a redox site but rather acts as a Lewis acid that can accept an electron pair (PRASAD 1977).

Zinc forms a lipophilic complex with pyridinethione by bonding through the oxygenated sulphur atoms. Lipophilic complex formation may promote metal absorption and tissue penetration. Zinc reacts with several peptide hormones. Adrenocorticotropic hormone (ACTH) binds zinc and copper(II) ions in acid solution (CARR et al. 1952). Zinc glucagon is somewhat less soluble than glucagon, and exhibits prolonged biological effects (WEINGES 1959, SOKAL 1960, BROMER and CHANCE 1969, TARDING et al. 1969, ASSAN and DELAUNAY 1972). If protamine is added to the zinc glucagon suspension, a greater prolongation of action is observed, even though the protamine is not bound to the zinc glucagon (TARDING et al. 1969).

Working with purified plasma membranes of alveolar macrophages from *rabbits*, CHVAPIL et al. (1976) found a definite but small amount of zinc present in the membrane, and most of this metal being linked to a certain fraction.

The viability of thioglycolate-elicited peritoneal macrophages isolated from male CD 1 *mice* pretreated for 4 days with 0.25 mg $ZnCl_2$ per day in Hepes buffer i.p. and exposed to 1 μm particles of silica (0.5–0.7 mg crystalline SiO_2/ml) *in vitro* was decreased from 55.9±2.8% to 37.6±13.7% (KARL et al. 1973). 0.05 mg $ZnCl_2$, however, increased the viability of silica-treated macrophages to 79.1±2.8%.

One week after an intratracheal instillation of 0.5 ml suspension of 10 mg ZnO/ml saline the size of *rat* pulmonary alveolar macrophages was considerably reduced (MIGALLY et al. 1982). The cells contained a prominent nucleolus within an invaginated nucleus. Several types of lysosomes were considered to be primary, secondary, and residual bodies. Short profiles of rough endoplasmic reticulum, small dense mitochondria and numerous profiles of Golgi membranes were also noted.

9.6.2.1.37 Zirconium

Zirconium has the atomic number 40 and the valence electrons are those of the $4d^2 5s^2$ shells. The oxidation states of zirconium are +3 and +4.

In the *hamster*, multiple short-term intratracheal instillations of zirconium lactate produced lesions beginning with exudative pneumonia followed by pneumonitis (interstitial pneumonia) and foreign body granulomas. Electron microscope microprobe analysis demonstrated the metallic component of the instilled compound in membrane-bound cytoplasmic inclusions of macrophages (LEININGER et al. 1977).

In the *rat*, zircon ($ZrSiO_4$) dust (180 mesh) injected intratracheally was found 7 to 9 months later in swollen alveolar histiocytes with pyknotic nuclei and a faintly staining hydropic cytoplasm in which were a few tiny granules of pigment (HARDING 1948).

In the *guinea-pig* immunised with 0.2 mg sodium zirconium lactate in Freund's complete adjuvant 14 days before, weekly intradermal injections of 25 μg of the metal salt in 0.1 ml saline produced nodular granulomas (TURK et al. 1978). Lesions taken at the peak of granuloma development 7 days after skin testing contained epithelioid cells which could be recognised particularly by their content of rough endoplasmic reticulum, fimbriated cell membrane and typical nuclear appearance. Multinucleate giant cells were also seen. They were characterised by their large number of lysosome-like dense bodies and phagocytic structures. A constant feature of the *guinea-pig* lesions was the presence of large numbers of apparently active fibroblasts.

9.6.2.2 Organic Agents

9.6.2.2.1 Acrolein

Acrolein is a pollutant found in both indoor and outdoor environments. It is manufactured as an intermediate in chemical synthesis, used as an aquatic biocide, and is also formed as a combustion product of organic matter including tobacco.

Human alveolar macrophages developed a dose dependent (1–50 μM) induction of stress response, inhibition of cytokine release, apoptosis, and necrosis (LI et al. 1997).

9.6.2.2.2 Amiodarone

Amiodarone (INN) is a iodinated benzofuran derivative, the chemical structure of which resembles that of thyroxin (HEGER et al. 1981). It was introduced into Belgium, France, Germany (Cordarex®), the Netherlands (Sotalol®), and South Africa for use as a coronary vasodilator, but proved to be antiarrhythmic (ROSENBAUM et al. 1974, 1976). An attempt to classify amiodarone according to its toxic ability to interfere with the integrated function of electron transport enzymes (RIBEIRO et al. 1997) confirmed the effect of the antiarrhythmic drug on complex I and allowed the placement of amiodarone in class A of the classification established by KNOBELOCH et al. (1990) i.e. together with rotenone, amytal, detergents and solvents. The drug itself and its main metabolite, desethylamiodarone accumulate in lung (DANIELS et al. 1989, REASOR et al. 1989), liver (PIROVINO et al. 1986) and eye (BAHEL et al. 1970) cells and within the ubiquitous monocyte/macrophage system to inhibit the lysosomal phospholipases. PICHLER et al. (1988) noted that most patients receiving a cumulative dose of >100 g amiodarone had side effects and that only one patient with an anti-amiodarone titre of >10 density units had no side effect.

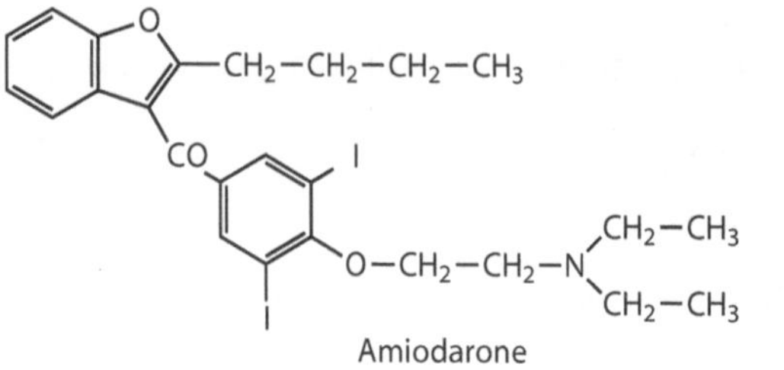

Amiodarone

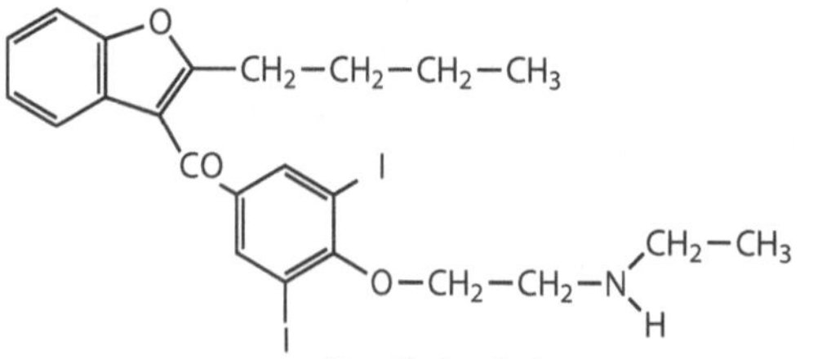

Desethylamiodarone

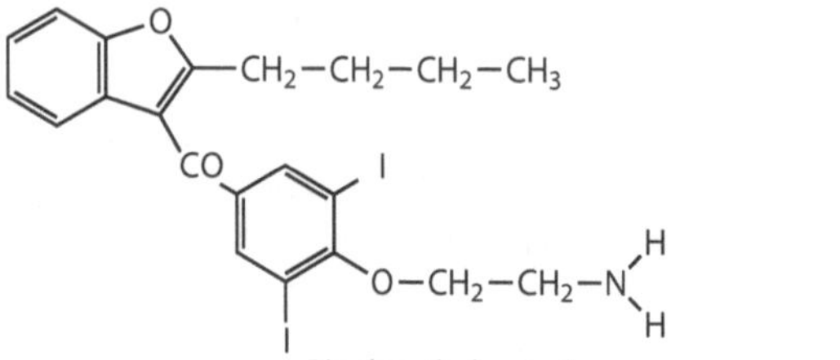

Bis-desethylamiodarone

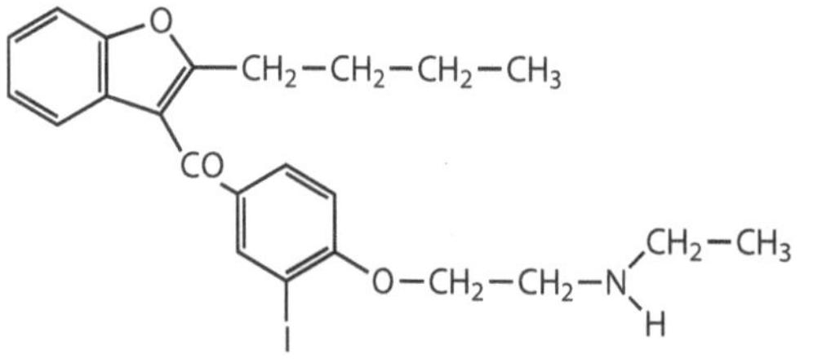

Monodeiodinated-desethylamiodarone

Amiodarone and its derivatives [249–252]

The most serious side effect is the development of pulmonary toxicity (Akoun et al. 1987, Martin and Rosenow 1988), characterised by phospholipidosis (Reasor et al. 1989) and the presence of interstitial and/or alveolar infiltrates (Martin and Rosenow 1988), which may progress to pulmonary fibrosis.

In a 55-year-old male patient Colgan et al. (1984) described alveolar spaces containing large macrophages with abundant pale foamy cytoplasm. Numerous heterogenous, dense, lamellar inclusions within their cytoplasm, which showed a decreased number of normal cytoplasmic organelles.

Transbronchial and open lung biopsies in a 29-year-old *man* treated with amiodarone (800 mg/day) for 10 months showed filling of the alveoli by foamy macrophages (Costa-Jussà et al. 1984).

In isolated single-pass perfused *rat* lungs amiodarone (30 μM + 1 μCi ^{14}C-amiodarone) uptake at 20 min averaged 1312 ± 225 nmoles/g, corresponding to a tissue to medium ration of 44 ± 7.5 (Camus et al. 1989). Efflux of amiodarone was incomplete, only 34 ± 10 % of the amount present at 20 min had effluxed from the lung at 40 min. Chlorphentermine significantly accelerated the efflux of amiodarone. When amiodarone was raised in the inflowing perfusate, accumulation in lung increased proportionally, except at 120 μM where the tissue to medium ration decreased. This concentration also induced pulmonary oedema.

Alveolar macrophages from Fischer-344 *rats* treated with amiodarone showed a dose (50 mg/kg–200 mg/kg) and time (1 day to 9 weeks)-dependent increase of all individual classes of phospholipids (Reasor et al. 1988). Using an oesophageal cannula male *mice* dosed with 200 mg amiodarone/kg × day suspended in 5 per cent methylcellulose up to three months and young adult Sprague-Dawley *rats* given 400 mg amiodarone/kg × day for up to one month showed lamellated bodies in their pulmonary alveolar macrophages after glutaraldehyde/paraformaldehyde perfusion (Costa-Jussà et al. 1984).

The adhesion molecules, CD18 and CD54 (ICAM-1) were significantly more expressed by bronchoalveolar cells lavaged from *rats* treated for 6 weeks with 175 mg amiodarone/kg × day by gavage (Castro Silva et al. 1997). After 12 weeks of treatment, however, the expression of these molecules significantly decreased. Throughout the experiment the cellularity of the BAL was elevated in experimental and control groups with great predominance of alveolar macrophages; the number of neutrophils was higher in experimental groups as well as the percentage of macrophage spreading. Release of H_2O_2, both spontaneously or stimulated by PMA, was variable being equal or higher than the amount released by control alveolar macrophages.

After intratracheal administration of amiodarone (1.83 μmol/day of days 0 and 2) or an equivalent volume (0.4 ml) of distilled water to male Fischer 344 *rats*, Northern and immunoblot analyses demonstrated that lung transforming growth factor (TGF)-β1 (mRNA and protein) expression was increased 1.5- to 1.8-fold relative to control during the early inflammation period and 1 day, 1 week, and 2 weeks post amiodarone treatment (Chung et al. 2001). Lung c-*jun* protein was increased 3.3-fold relative to control.

Tumour necrosis factor-α release by lavaged alveolar macrophages of female Fischer 344 *rats* significantly ($P < 0.05$) increased 3 and 6 weeks after an intratracheal instillation of 6.25 mg amiodarone/kg dissolved in ~200 μl of sterile deionized water (Reinhart and Gairola 1997). 5 mg Desethylamiodarone/kg had no effect.

Pulmonary phospholipidosis induced by 150 mg amiodarone/kg *rat*×day attenuated the acute toxicity of 2.5 mg and 10 mg α-quartz (median area equivalent diameter 3.5 μm) per 100 g body weight (ANTONINI et al. 1994).

In vitro, amiodarone decreased the surface density of *rabbit* alveolar macrophage mitochondria and lysosomes while increasing the surface density of inclusion bodies, increased the incorporation of choline into dipalmitoylphosphatidylcholine, modified the distribution of lysosomal enzymes, and did not affect the uptake and processing of diphtheria toxin (10 nM) but inhibited the degradation of surfactant protein A (BARITUSSIO et al. 2001).

9.6.2.2.3 Benzo[a]pyrene

Benzo[a]pyrene is biotransformed by *guinea pig* alveolar macrophages to give 3-OH-benzo[a]pyrene, 7,8-dihydro-7,8-dihydroxybenzo[a]pyrene, 9,10-dihydro-9,10-dihydroxybenzo[a]pyrene, several chinones and three other unidentified substances (DEHNEN 1975). BENSON et al. (1987) incubated resting *canine* alveolar macrophages *in vitro* with ^{14}C-benzo[a]pyrene and determined the effect of addition of SKF-525A and metyrapone (cytochrome P_{450} inhibitors), indomethacin (cyclooxygenase inhibitor), and nordihydroguiaretic acid (lipoxygenase inhibitor) on the extent of benzo[a]pyrene metabolism and the pattern of metabolites formed. Indomethacin and nordihydroguiaretic acid inhibited benzo[a]pyrene metabolism in resting alveolar macrophages by 35% and 30%, respectively, while SKF-525A and metyrapone had no effect. Activation of alveolar macrophages by zymosan resulted in an approximately 60% reduction in benzo[a]pyrene metabolism.

Due to the presence of polycyclic aromatic hydrocarbons in asphalt, most studies of the health hazards for asphalt exposure have been concentrated on carcinogenicity, while very few studies have been devoted to characterisation of the effects of asphalt fumes on non-cancerous responses in the lung. *In vitro* exposure to paving asphalt fume condensate (<200 μg/ml) to *rat* alveolar macrophage functions did not induce cytotoxicity, oxidant generation, or IL-1 production by alveolar macrophages, but it did a small but significant increase in TNF-α release from alveolar macrophages (MA et al. 2000). *In vitro* exposure of alveolar macrophages resulted in a significant decline of chemiluminescence in response to zymosan or 12-*O*-tetradecanoylphorbol-13-acetate stimulation. *In vivo* studies in *rats* instilled intratracheally with 0.1 or 0.5 mg paving asphalt fume condensate showed that this toxicant did not induce significant neutrophil infiltration or alter lactate dehydrogenase or protein content in acellular lavage samples. Macrophages obtained from these *rats* did not show significant differences in oxidant production or cytokine secretion at rest or in response to lipopolysaccharide in comparison with control macrophages.

9.6.2.2.4 Bleomycin

Bleomycin has been shown to exert its cytotoxic effects through the intracellular generation of oxygen radicals by forming complexes with iron and entering redox cycles (SAUSVILLE et al. 1978, SUGIURA and KIKUCHI 1978, MAHMUTOGLU et al. 1987).

The number of alveolar macrophages recovered by bronchoalveolar lavage from bleomycin-injected *rat* lungs was more than double control values at 10 days and remained above normal at 6 weeks (YOUNG and ADAMSON 1993). Both hyaluronan content in bronchoalveolar lavage and staining for hyaluronan in macrophages accumulating in injured areas of the *rat* lung were maximal at 4 d (SAVANI et al. 2000). Increased hyaluronan in bronchoalveolar lavage correlated with increased locomotion of isolated alveolar macrophages. Hyaluronan-binding peptide was able to specifically block macrophage motility *in vitro*. Importantly, systemic administration of hyaluronan-binding peptide to *rats* before injury not only decreased alveolar macrophage motility and accumulation in the lung, but also reduced lung collagen α (I) mRNA and hydroxyproline contents.

As compared to control *rats*, production of $O_2^{\bullet -}$ by alveolar macrophages from bleomycin-medicated (20 U/kg body weight×day for 5 days) animals was increased upon stimulation with either phorbol myristate acetate (21.04±21.78 vs. 11.45±2.26 nM/10^6 cells×20 min; $P<0.05$) or opsonized zymosan (9.35±0.87 vs. 7.03±0.66 nM/10^6 cells×20 min; $P<0.05$) (SLOSMAN et al. 1990).

Intratracheal instillation of bleomycin (0.1 U/*mouse*) stimulated an early influx of neutrophils followed by an increase in lymphocytes and macrophages in the bleomycin + control diet group (GURUJEYALAKSHMI et al. 2000). Taurine (1% in water) and niacin (2.5% in diet) treatment significantly reduced the numbers of neutrophils, lymphocytes and macrophages in the bleomycin + taurine/niacin group and caused significant reductions in bleomycin-induced increases in the lung hydroxyproline content at 14 and 21 days in the bleomycin + taurine/niacin group. The *mice* in the saline-instilled + control diet and the saline-instilled + taurine/niacin control groups had lower levels of

•NO in bronchoalveolar lavage fluid, whereas *mice* in the bleomycin-instilled + control diet group as compared to the saline-instilled + control diet group had elevated levels of •NO from day 3 through day 21. The increases in •NO levels in bronchoalveolar lavage fluid from bleomycin + control diet group were associated with elevated levels of iNOS gene expression and protein in the lungs. Taurine and niacin suppressed the bleomycin-induced increases in iNOS massage and iNOS protein.

Alveolar macrophages from *rats* or *hamsters* with bleomycin-induced fibrosis secrete cytokines (JORDANA et al. 1988, SUWABE et al. 1988), fibroblast growth (KOVACS and KELLEY 1985, DENHOLM and PHAN 1989, PHAN 1989) and inhibiting (CLARK et al. 1983) factors, and leucocyte chemotactic factors (KAELIN et al. 1983, DENHOLM et al. 1989), all of which are thought to play important roles in mediating the fibrotic reaction of the lungs to this drug. Alveolar macrophages from normal *rats* (JORDANA et al. 1988) or *hamsters* (SUWABE et al. 1988) stimulated with bleomycin *in vitro* induced similar secretory activities.

Activin A immunoreactive and bioactive protein belonging to the transforming growth factor β (TGF-β) supergene family was demonstrated in lavaged *murine* alveolar macrophages 14 days after bleomycin (10 mg/kg×day for 10 days) treatment (MATSUSE et al. 1995).

9.6.2.2.5 Carrageenan

Carrageenan was first described as a mononuclear phagocyte toxin by ALLISON et al. (1966), who ascribed its toxic effect to destabilisation of the membranes of the secondary lysosomes, resulting in release of hydrolytic enzymes into the cytoplasm of treated cells. The λ-form of carrageenan appears in several reports to be the most potent macrophage toxin, compared to other forms of this material (BECKER and RUDBACH 1979). However, carrageenan has a broad spectrum of effects apart from its suppression of mononuclear phagocytes (LEBLANC and RUSSELL 1981), as it can have either a depressive of adjuvant effect on antibody production to either cellular or soluble antigens (ASCHHEIM and RAFFEL 1972, TURNER and HIGGINBOTHAM 1977, BECKER and RUDBACH 1979). In light of these facts, LEBLANC and RUSSELL (1981) consider carrageenan as one of the least specific macrophage inhibitors. REHM (1983) injected type II and type V ι-carrageenan intraperitoneally into BALB/c and outbred ICR *mice:* viable recoverable macrophages progressively declined, reaching a nadir of 15–45 % of control values. After administration of the chemically purer type V, more rapid macrophage depletion was achieved (nadir at days 2–5). Thus REHM (1983) recommended ι-carrageenan as an investigational tool for manipulating alveolar macrophages.

Carrageenan-activated macrophages showed tumoricidal activity against a tissue culture line of BN472 cells (malignant mammary adenocarcinoma) (BIJMA et al. 1988).

9.6.2.2.6 Cellulose Dust

One month after a single intratracheal instillation of 15 mg cellulose powder MN 300 as used for thin layer chromatography (Nagel, Düren), male CFY *rats* showed irregularly shaped, periodic acid-Schiff positive foreign bodies inside the macrophages (TÁTRAI and UNGVÁRY 1992). Three months after the istillation several multinucleated foreign body giant cells were seen in the granulation tissue filling the lumina of brochioli and alveoli.

9.6.2.2.7 Chlorphentermine

p-Chlorphentermine, *p*-chloro-α,α'-dimethylphenethylamine HCl applied intraperitoneally to male Sprague-Dawley *rats* at doses of 50 mg/kg for one to 30 days or at doses of 20 mg/kg for 60 and 72 days, respectively, induced large intraalveolar foam cells (LÜLLMANN-RAUCH et al. 1972). Their tightly packed concentric lamellar bodies (periodicity 4–5 nm) proved to be phospholipids. Foam cells of chlorphentermine-treated (80 mg/kg×day) female Siv 50 *rats* contained mainly phospholipids, i.e. lecithin and only small amounts of neutral lipids, i.e. cholesterol (KARABELNIK and ZBINDEN 1975). Chlorphentermine (60 mg/kg×day) inhibited the incorporation of 1-^{14}C-palmitic acid into lung phospholipids in a time-dependent manner (KARABELNIK and ZBINDEN 1976). The levels of thiobarbituric acid-reactive material, an index of lipid peroxidation, was increased in the lung tissue above saline vehicle controls in all vitamin E dietary *rat* groups (0, 60, 300 ppm) with the vitamin E-deficient group being the highest (GAIROLA et al. 1983).

9.6.2.2.8 Cork Dust

A *man* aged 54 years, who was exposed to cork dust for about 14 years, showed intraalveolar giant cells loaded with coal and cork particles. Some had asteroid bodies in their cytoplasm (REMMELE and

Einbrodt 1962). The formamide method (Thomas and Stegemann 1954) discovered some pseudo-asbestos bodies.

Guinea pigs and *rabbits* inhaling cork dust in different factory workshops from 56 to 1,278 and from 276 to 1,100 days, respectively, showed suberophages and multinucleated giant cells containing larger particles (Lopo de Carvalho Cancella 1959). Pneumoconiotic focal lesions were observed after an exposure for more than a year. In some of these nodules fibroblasts had proliferated scarcely accompanied by the formation of pre-collagenous fibres.

9.6.2.2.9 Diesel Exhaust

Free radical generation from diesel particulate was characterized by Ross et al. (1982) using electron paramagnetic resonance spectroscopy. The free radical signals were sensitive to oxygen, nitric oxide (•NO), nitrogen dioxide, and ultraviolet radiation. Vogl and Elstner (1989) showed that diesel soot particulate could catalyse the release of ethylene from α-keto-γ-methylthiobutyrate, which suggests the release of strong oxidants. They showed that the photodynamic catalysis reaction could be inhibited by radical scavengers (α-tocopherol, 100 %; catalase, 83 %, azide, 40 %; 1,4-diazabicyclo[2,2,2] octane, 83 %).

Clinical Investigations: Diesel exhaust (mass median diameter <10 μm) at 300 μg/m^3 associated with concentrations of NO_2 of 1.6 ppm, •NO of 4.5 ppm for 1 h, CO of 7.5 ppm, total hydrocarbons of 4.3 ppm, formaldehyde of 0.26 mg/m^3, and 4.3×10^6 suspended particles/ml in healthy *human* volunteers exposed for 1 h induced a 240 % median relative increase in interleukin-8 mRNA gene transcripts in the cells contained in bronchial lavage performed 6 h later (Salvi et al. 2000). There were no changes in the mRNA transcript levels for ILL-1β, ILL-4, TNF-α, interferon-γ, and granulocyte-macrophage colony-stimulating factor after exposure to diesel exhaust compared with air.

Animal Experiments: Diesel exhaust particles were inhaled by Fischer-344 *rats* for 6 month and 1 year, respectively, at concentrations of 250 μg/m^3, 750 μg/m^3, 1500 μg/m^3 and 6000μg/m^3. Bronchoalveolar lavage showed nearly all the macrophages containing particle-filled phagosomes, their amount varying with the concentration of the exposure, with some macrophages appearing several times the size of the control macrophages (Strom 1984). The number of alveolar macrophages increased primarily with the inhaled concentration of particles, but was not different from controls in the group exposed to 250 μg/m^3. The macrophages did not increase significantly with the duration of the exposure.

Female ddY *mice* after intratracheal instillation of 0.05 mg diesel exhaust particles suspended in 50 mM phosphate-buffered saline (pH 7.4) containing 0.05 % Tween 80 and sonicated for 3 min showed HO• generation in their lungs as demonstrated *in vivo* by non-invasive L-band ESR spectroscopy and a membrane-impermeable nitroxyl probe (Han et al. 2001). HO• generation was confirmed with the enhancement of *in vivo* ESR signal decay rate of the probe. The decay rate at mid-thorax was significantly enhanced in diesel exhaust particle-treated *mice* compared to that in vehicle treated animals. The enhancement was completely suppressed by the administration of either HO• scavengers, catalase, or desferrioxamine, while the administration of superoxide dismutase further increased the rate. The administration of Fenton's reagents into the lung also enhanced the decay of the probe at mid-thorax of *mice*.

Male ICR *mice* instilled intratracheally with suspensions containing 1 or 2 mg of diesel exhaust particles in 1 ml of 50 mM phosphate-buffered saline (pH 7.4) and 0.05 % Tween 80 prepared by sonication once a week for 10 weeks showed increased levels of inducible NO synthase in the macrophages (Lim et al. 1998).

Rat lung inflammation 24 h after intratracheal instillation of diesel exhaust particles (Standard Reference Material 2975) initially exposed to 0.1 ppm $[^{18}O]O_3$ for 48 h was more potent in increasing neutrophilia, lavage total protein, and lactate dehydrogenase (EC 1.1.1.1.27) activity compared to unexposed diesel exhaust particles (Madden et al. 2000). Exposure of diesel exhaust particles to 1 ppm O_3 led to a decreased bioactivity of the particles.

9.6.2.2.10 Diphenyl-methane 4,4'-diisocyanate

The early acute pulmonary response of Wistar *rats* exposed nose-only to respirable polymeric diphenyl-methane 4,4'-diisocyanate aerosol for 6 h were bronchoalveolar lavage cells markedly loaded with phosphatidylcholine (Pauluhn 2000). Bronchoalveolar lavage cells from *rats* repeatedly exposed to 12.9 mg polymeric diphenyl-methane 4,4'-diisocyanate m^{-3} air (6 h/day, 5 days/week for 14 days; exposure was from days 0–17 followed by a post-exposure period to day 35) increased in number and phospholipid content significantly in a time-dependent manner and returned to almost normal levels within 10 post-exposure days (Pauluhn 2002).

9.6.2.2.11
Ethanol

The detrimental influence of alcoholic beverages on host susceptibility to infection is, historically, a well known clinical observation (Rush 1784, Koch 1884). The incidence of tuberculosis and its severity is profoundly higher for alcoholics (Feingold 1976). The proportion of persons isolating *M. tuberculosis* among the new cases of tuberculosis complicated by chronic alcoholism was 1.3 times higher than that among patients with tuberculosis not complicated by alcoholism (Rudoi and Chubakov 1984). Drew et al. (1984) found both unstimulated and pokeweed mitogen stimulated immunoglobulin synthesis by peripheral blood mononuclear cells *in vitro* and plasma immunoglobulin concentrations abnormal in alcoholics. Glucan significantly modified the susceptibility of chronic ethanol-treated mice to *Staphyloccocus aureus* infection by possibly enhancing macrophage-mediated host defence mechanisms (Di Luzio and Williams 1980). THE ANIMALS MAINTAINED ON A CHRONIC ETHANOL DIET WHICH WERE PRE-TREATED WITH GLUCAN EXHIBITED THE WELL-DOCUMENTED GLUCAN-INDUCED GRANULOMA BUT NO HEPATIC INJURY AS DID THE CONTROL *mice*. ^{131}I labeled microaggregated albumin clearance is significantly depressed in rats given ethanol (Ali and Nolan 1967, Nolan et al. 1980), but two similar studies in *humans* conflicted on whether the delayed clearance of the colloid is due to decreased perfusion (Cooksley et al. 1973) or a direct depression of the R.E.S. function (Liu 1975).

The role of short-chain aliphatic alcohols in the regulation of the **respirarory burst** is controversial, and several mechanisms may be involved in their effects (Seifert and Schultz 1991). Alcohol metabolism has at least four pathways, the most important one being the oxidation of ethanol by alcohol dehydrogenase. After chronic consumption of ethanol or in the presence of high concentration of ethanol, minor pathways such as the microsomal ethanol oxidising system

$$H_3C\text{-}CH_2OH + NAD^+ \longrightarrow NADH + H^+ + H_3C\text{-}CHO \qquad [253]$$

become more important. At least two ethanol microsomal oxidation pathways have been characterised by electron spin resonance (Krikun et al. 1984, Albano et al. 1988). The α-hydroxyethyl radical formed by the one-electron oxidation of ethanol has been detected by spin trapping with α-(4-pyridyl 1-oxide)-*N-tert*-butylnitrone (POBN) in microsomal incubations (Gonthier et al. 1991). Radical production resulted in part from Fenton-type chemistry due to microsomal hydrogen peroxide and trace iron present in the microsomal preparations (McCay et al. 1992). However, with a Desferal-treated system, radical adducts from ethanol and other alcohols could still be detected and were attributed to enzymatic activity by cytochrome P450 IIE1 (Albano et al. (1988, 1991).

The expression of the ethanol-metabolising cytochrome P-450 (CYP2E1) in *human* monocyte-derived macrophages cultured for 3 and 6 days in RPMI 1640, 15 % *human* serum, and 80 µg/ml gentamicin was studied at the mRNA and protein levels. Hutson and Wickramasinghe (1999) showed that CYP2E1 is expressed at a level similar to that demonstrated in other extrahepatic tissues.

Ethanol may scavenge the highly reactive hydroxyl radical to form a much less reactive hydroxyethyl radical:

$$H_3CCH_2OH + HO^{\bullet} \longrightarrow H_3CC^{\bullet}HOH + H_2O \qquad [254]$$

This reaction is made use of in spin-trapping experiments with 5,5-dimethylpyrroline-*N*-oxide.

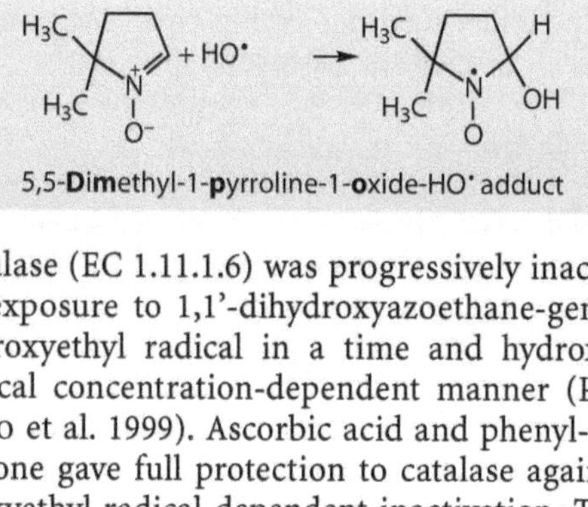

5,5-**Dim**ethyl-1-**p**yrroline-1-**o**xide-HO• adduct [8]

Catalase (EC 1.11.1.6) was progressively inactivated by exposure to 1,1'-dihydroxyazoethane-generated hydroxyethyl radical in a time and hydroxyethyl radical concentration-dependent manner (Puntarulo et al. 1999). Ascorbic acid and phenyl-*t*-butyl nitrone gave full protection to catalase against hydroxyethyl radical-dependent inactivation. The antioxidants 2-*tert*-butyl-4-methylphenol, pyrogallate, and α-tocopherol protected catalase against inactivation by 84, 88, and 39 %, respectively. Other antioxidant enzymes were also sensitive to exposure to hydroxyethyl radical. Glutathione reductase (EC 1.6.4.2), glutathione peroxidase (EC 1.11.1.9), and superoxide dismutase (EC 1.15.1.1) were inactivated by 46, 36, and 39 %, respectively, by hydroxyethyl radical.

In the *rat*, Takahashi et al. (1992), French et al. (1993), Morimoto et al. 1994, 1995) and Albani et al. (1996) correlated the level of CYP2E1 with liver pathology, *in vitro* lipid peroxidation, and the formation of α-hydroxyethyl radical adduct.

Mitochondrial DNA isolated from the livers of *rats* fed ethanol for 42–76 days showed a 21 % increase in the level of 8-hydroxydeoxyguanosine compared with the controls (4.8 ± 0.9 versus 3.98 ± 0.5 8-OH-dG/10^5 dG) (Cahill et al. 1997). This difference increased to 43% in animals that had been fed ethanol for 105–164 days (control, 6.9 ± 1.0; ethanol, 9.9 ± 1.1 8-OH-dG/10^5 dG).

Ethanol inhibited •NO production by thioglycollate elicited *mouse* macrophages as measured by nitrite production (Goldin et al. 1993). Some significant differences between strains suggested that genetic factors may be important in alcoholic liver disease.

9.6.2.2.12 Formaldehyde

The major pathways of metabolism of inhaled formaldehyde are oxidation to formate and incorporation into biological macromolecules via tetrahydrofolate-dependent one-carbon biosynthetic pathways (Huennekens and Osborne 1959, Koivusalo et al. 1982). The most important pathway for oxidation appears to be that catalysed by formaldehyde dehydrogenase (EC 1.2.1.1), an enzyme that requires both glutathione and NAD^+ as cofactors. Uotila and Koivusalo (1974) showed that the true substrate is the hemithioacetal adduct of formaldehyde and glutathione and the product formed is the thiol ester of formic acid, S-formylglutathione.

$$H_2CO + GSH + NAD(P)^+ \longleftrightarrow GSCHO + NAD(P)H + H^+ \quad [255]$$

Rat alveolar macrophages obtained by bronchoalveolar lavage showed a concentration- and time-dependent depression of zymosan induced and lucigenin enhanced chemiluminescence when exposed to formaldehyde (Schroers and Tilkes 1990).

In the male 200 g Sprague-Dawley *rat*, 20 ppm formaldehyde exposure for 4 h at rest delayed short-term clearance (0–50 h) of monodisperse polystyrene latex microspheres (about 1.7 μm in diameter) labelled with ^{51}Cr and inhaled for 20 min just prior to pollutant exposure. Exposure to 10 ppm formaldehyde did not produce significant effects; however, this same concentration during treadmill exercise (8 m/min) caused a delay in short-term clearance (Phalen et al. 1994).

9.6.2.2.13 Glutaraldehyde

Rat alveolar macrophages obtained by bronchoalveolar lavage showed a concentration- and time-dependent depression of zymosan induced and lucigenin enhanced chemiluminescence when exposed to glutaraldehyde (Schroers and Tilkes 1990).

9.6.2.2.14 Glyoxal

The α-oxoaldehydes glyoxal, methylglyoxal and 3-deoxyglucosone are physiological metabolites. Glyoxal is formed by the slow, spontaneous oxidative degradation of glucose (Wells-Knecht et al. 1995), the degradation of glycated proteins (Namiki and Hayashi 1983), and lipid peroxidation (Mlakar and Spiteller 1996).

Rat alveolar macrophages obtained by bronchoalveolar lavage showed a concentration- and time-dependent depression of zymosan induced and lucigenin enhanced chemiluminescence when exposed to glyoxal (Schroers and Tilkes 1990).

Incubation of *murine* P388D$_1$ macrophages with 30 mM H_2O_2 increased the cellular concentration of glyoxal after 3 h ($P < 0.05$) and decreased the concentration of glyoxal in the medium ($P < 0.01$) with respect to control values (Abordo et al. 1999).

9.6.2.2.15 4-Hydroxynonenal

4-Hydroxynonenal is one of the major aldehyde products of lipid peroxidation and has been reported to be the most toxic aldehyde formed (Benedetti et al. 1979, 1980, Benedetti and Comporti 1987). Li et al. (1996) found a formation of protein adducts in alveolar macrophages from C3H/HeJ and C57BL/6J *mice* in a dose-dependent manner. Alveolar macrophages from both strains had extensive apoptosis at 100 μM 4-hydroxyonenal.

9.6.2.2.16 Lindane

1α,2α,3β,4α,5α,6β-Hexachlorocyclohexane, lindane at a concentration of 125 μM *in vitro* stimulated $O_2^{\bullet -}$- production by *guinea pig* alveolar macrophages (Holian et al. 1984). 1.5×10^6 *Human* neutrophil granulocytes exposed to 400 μM lindan formed >11 nM $O_2^{\bullet -}$/min (English et al. 1986),

9.6.2.2.17 Local Anaesthetics

Lidocaine solution (12 mM), as proposed by Stewart (1981) to remove adherent *murine* monocytes, activated superoxide dismutase (Gulyaewa et al. 1987). Lidocaine and related local anaesthetics disrupt ion transport across cell membranes through changes in membrane fluidity. Lidocaine

(2.5 mM, 1-h pre-treatment) caused a small but significant decrease in baseline cytosolic pH (pH_i) of 0.14±0.02 units (BIDANI et al. 1996). Ammonia-prepulse studies with bafilomycin A_1 (a specific inhibitor of V-type H^+ pumps, i.e., V-ATPase) and amiloride (an inhibitor of the Na^+/H^+ exchanger, i.e., NHE) showed that 2.5 mM lidocaine suppressed plasmalemmal V-APTase activity by ~50 % and NHE activity by ~90 %. The median effective concentration (EC_{50}) of lidocaine (1-h pre-treatment) against pH_i recovery from ammonia-prepulse acid-loads was 0.97±0.04 mM. Lidocaine (2.5 mM) diminished the generation of superoxide anions induced using PMA by >60 % and reduced the release of TNF-α induced using lipopolysaccharide by >50 %.

9.6.2.2.18
Melamine Resin Semifinished Products

Powdered Trimethylolmelamine (Madurit®)

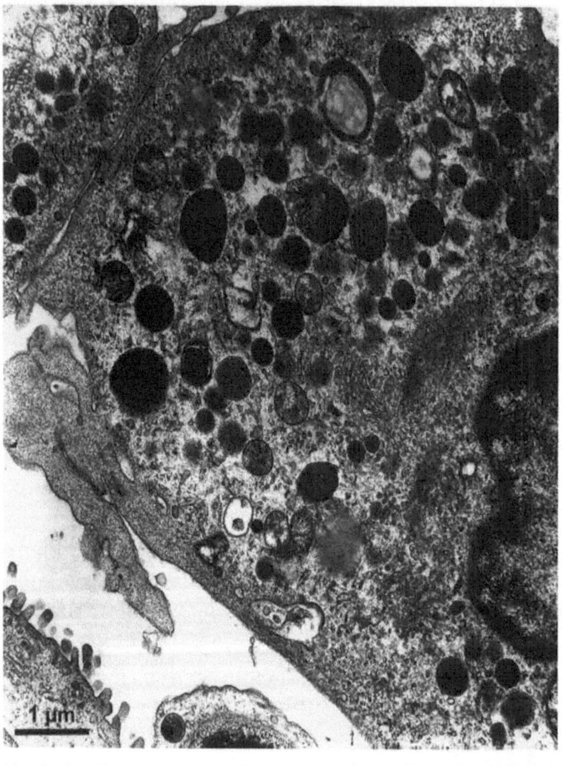

Fig. 159. Alveolar macrophage (block 490) from a male *rat* which inhaled 73.6 mg trimethylol melamine (Madurit®) dust/l for a total of 125 h 15 min on 5 days/week for 16 days. Fixed under pentobarbital anaesthesia (30 mg/kg) by intratracheal instillaTion of 2.5 % glutaraldehyde in 0.1 M sodium cacodylate buffer. Postfixation with 1 % osmium tetroxide in sodium cacodylate buffer. Embedded in Epon 812 and sectioned at 50 nm. Lead citrate and uranyl acetate. Plate 4068

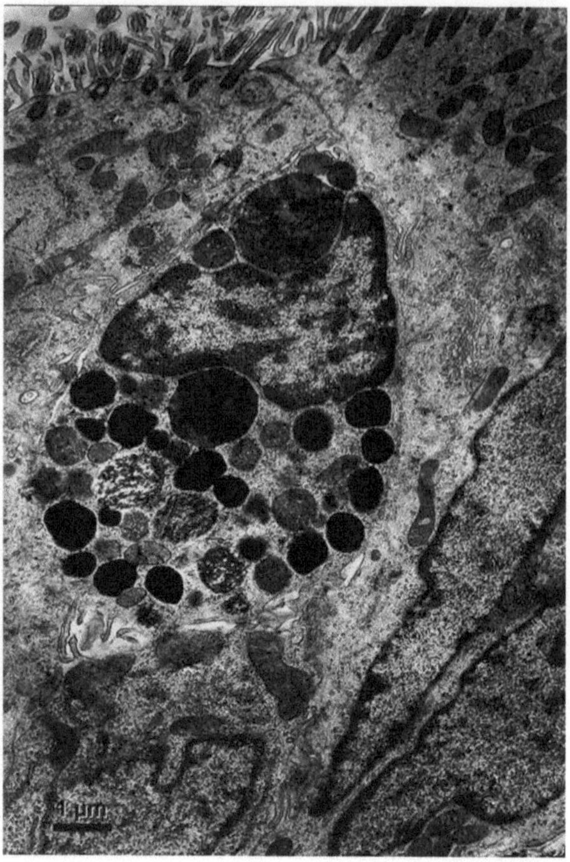

Fig. 160. Mononuclear phagocyte in the bronchial epithelium (block 490) of a male *rat* which inhaled 73.6 mg trimethylol melamine (Madurit®) dust/l for a total of 125 h 15 min on 5 days/week for 16 days. Fixed under pentobarbital anaesthesia (30 mg/kg) by intratracheal instillation of 2.5 % glutaraldehyde in 0.1 M sodium cacodylate buffer. Postfixation with 1 % osmium tetroxide in sodium cacodylate buffer. Embedded in Epon 812 and sectioned at 50 nm. Lead citrate and uranyl acetate. Plate 4108

9.6.2.2.19
Methyl Cellulose

The anæmia caused by administration of methyl cellulose is due to excessive destruction of normal erythrocytes by the cells of the reticulo-endothelial system, especially those in the spleen, which have undergone hypertrophy and hyperplasia in response to the injected inert colloidal material (PALMER et al. 1953, GIBLETT et al. 1956).

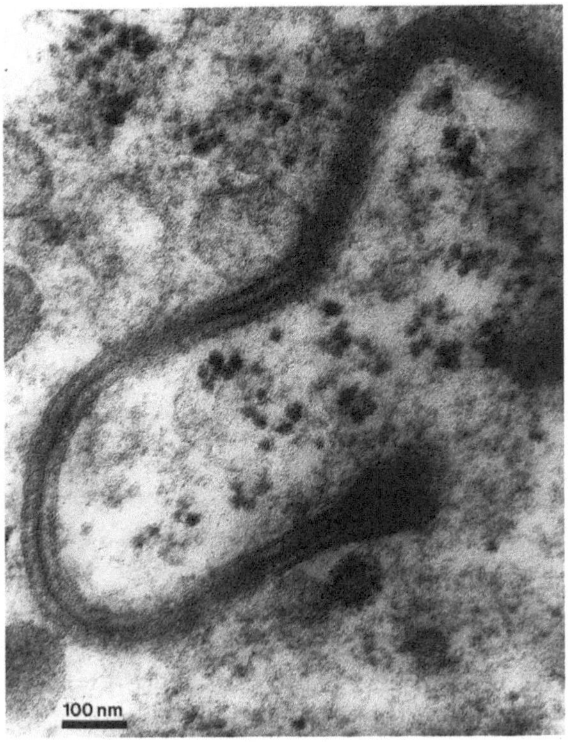

Fig. 161. Phagophore in a lymph node macrophage (block 4435) from a 225 g Sprague-Dawley *rat* (Charles River, France) four days after intraperitoneal injection of 10 mg methylcellulose/kg body weight. Under pentobarbital anaesthesia (30 mg/kg), the animal was perfused from the abdominal aorta with 2.5 % glutaraldehyde in 0.1 M sodium cacodylate buffer (pH 7.4). Postfixation with 1 % osmium tetroxide in sodium cacodylate buffer. Embedded in Epon 812 and sectioned at 50 nm. Stained with lead citrate and uranyl acetate. Plate 4144

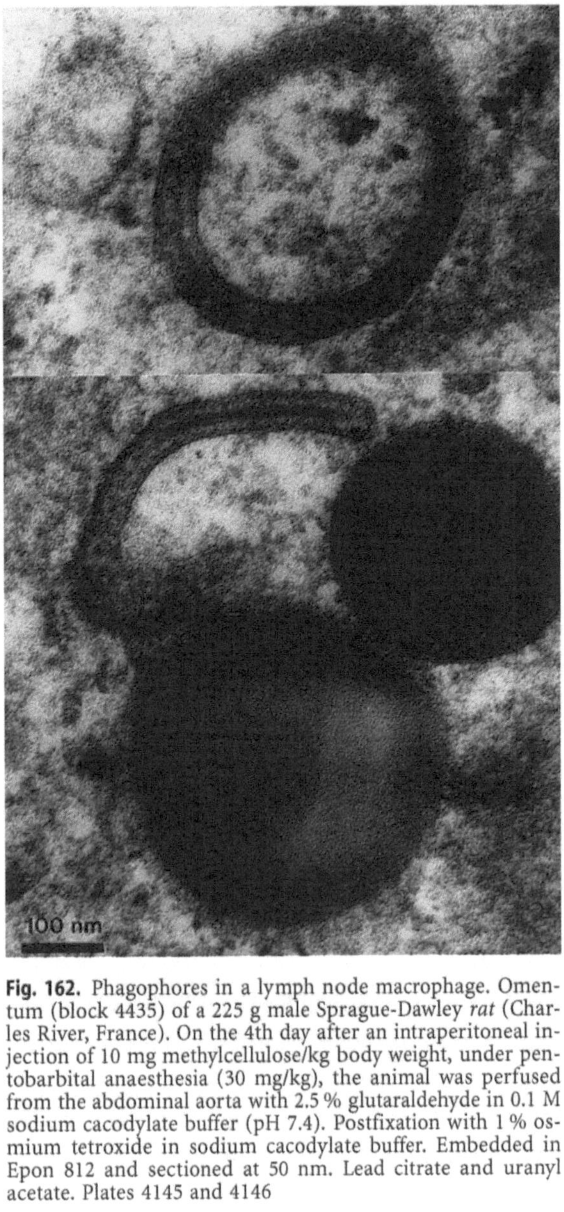

Fig. 162. Phagophores in a lymph node macrophage. Omentum (block 4435) of a 225 g male Sprague-Dawley *rat* (Charles River, France). On the 4th day after an intraperitoneal injection of 10 mg methylcellulose/kg body weight, under pentobarbital anaesthesia (30 mg/kg), the animal was perfused from the abdominal aorta with 2.5 % glutaraldehyde in 0.1 M sodium cacodylate buffer (pH 7.4). Postfixation with 1 % osmium tetroxide in sodium cacodylate buffer. Embedded in Epon 812 and sectioned at 50 nm. Lead citrate and uranyl acetate. Plates 4145 and 4146

9.6.2.2.20 Paprika Dust

One month after a single intratracheal instillation of 15 mg paprika dust suspended in 1 ml saline, male CFY *rats* showed some phagocytosed, irregularly shaped periodic acid-Schiff positive foreign bodies in their alveolar macrophages (Tátrai and Ungváry 1992). 3 Months after instillation intraalveolar foci of foreign body type giant cells had formed.

9.6.2.2.21 Pyridinium Compounds

The pyridinyl radical may be oxidised or reduced in one-electron reactions. Each oxidation state is capable of complexation, with the pyridinium ion acting as acceptor, the 1,4-dihydropyridine ring acting as a donor, and the pyridinyl radical functioning as both a donor and an acceptor towards itself.

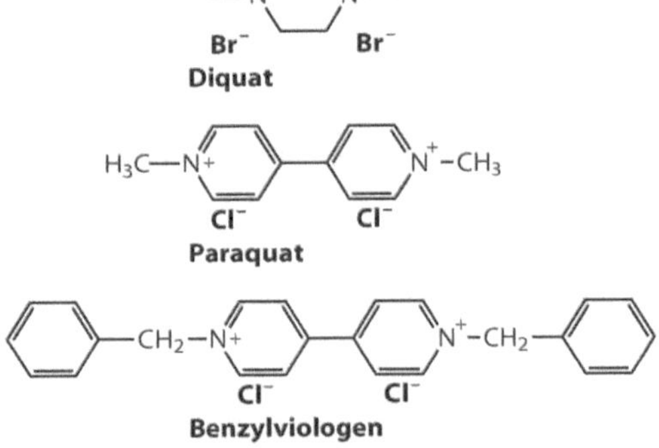

Pyridinium compounds [256]

Pyridinium compounds, e.g. diquat, paraquat and benzylviologen, by enzymatic one-electron transfer become radicals. These radicals react with O_2 forming $O_2^{\bullet-}$ and the starting materials. By redox cycling $O_2^{\bullet-}$ is produced again and again (Bus et al. 1974). Hydroxyl radicals generated via H_2O_2 induce lipid peroxidation and denaturation of proteins and nucleic acids, which are segregated and degraded by lysosomal enzymes. 4 h after an inhalation of hexadecylpyridinium chloride, phagophores were detected in alveolar macrophages (Fig. 162). The Fenton catalyst, Fe^{2+} is segregated from cytoplasmic areas containing ferritin. The autophagous vacuoles included other organelles, membrane structures and fat droplets.

Hexadecylpyridinium Chloride

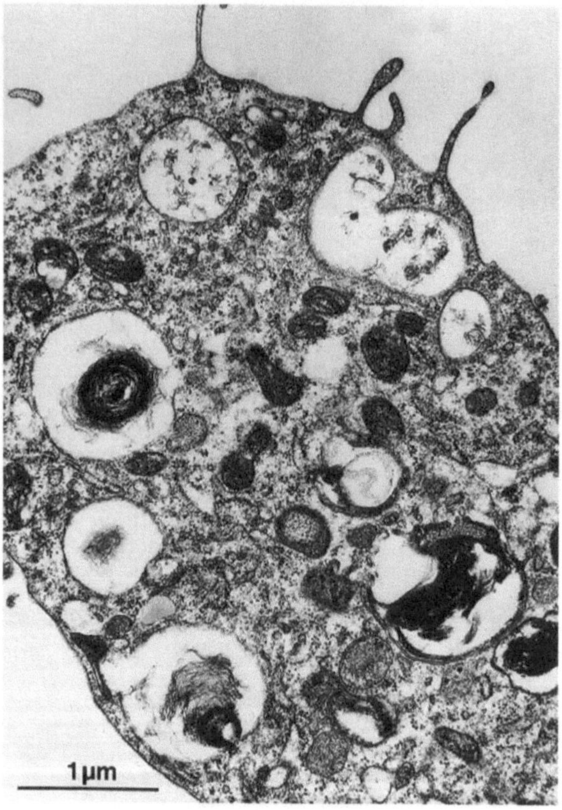

Fig. 163. Phagosomes in a stimulated alveolar macrophages from a female *rat* (breeder: Winkelmann, Borchen-Kirchborchen) 4 hours after an inhalation of a mixture of 28 µg hexadecylpyridinium chloride/puff + sorbitan trioleate (surfactants for metered aerosols) dispersed from a Medihaler®;. 12 puffs/min were dispersed into a 164.5 l box where the animals stayed for 15 min. Fixed under methitural anaesthesia by intratracheal instillation of 2.5 % glutaraldehyde in phosphate buffer (pH 7.4) before opening the thorax. Postfixation with 1 % osmium tetroxide in phosphate buffer (pH 7.4). Contrasted en bloc for 12 h with 0.5 % uranyl acetate in 70 % ethanol. Embedded in a 2:8 mixture of methyl and butyl methacrylate. Sectioned at 50 nm. Lead citrate after Reynolds (1963). Plate 48/05

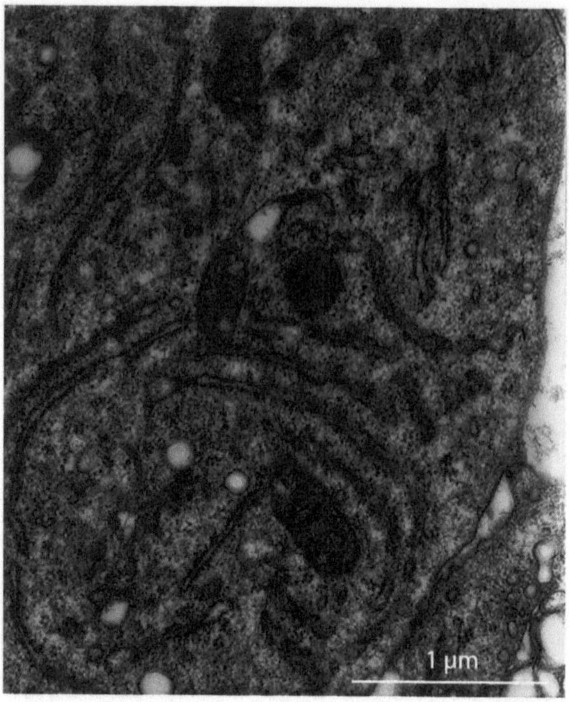

Fig. 164. Rough endoplasmic reticulum in a stimulated alveolar macrophages from a female *rat* (breeder: Winkelmann, Borchen-Kirchborchen) 4 hours after an inhalation of a mixture of 28 µg hexadecylpyridinium chloride/puff + sorbitan trioleate (surfactants for metered aerosols) dispersed from a Medihaler®. 12 puffs/min were dispersed into a 164.5 l box where the animals stayed for 15 min. Fixed under methitural anaesthesia by intratracheal instillation of 2.5 % glutaraldehyde in phosphate buffer (pH 7.4) before opening the thorax. Postfixation with 1 % osmium tetroxide in phosphate buffer (pH 7.4). Contrasted en bloc for 12 h with 0.5 % uranyl acetate in 70 % ethanol. Embedded in a 2:8 mixture of methyl and butyl methacrylate. Sectioned at 50 nm. Lead citrate after Reynolds (1963). Plate 49/08

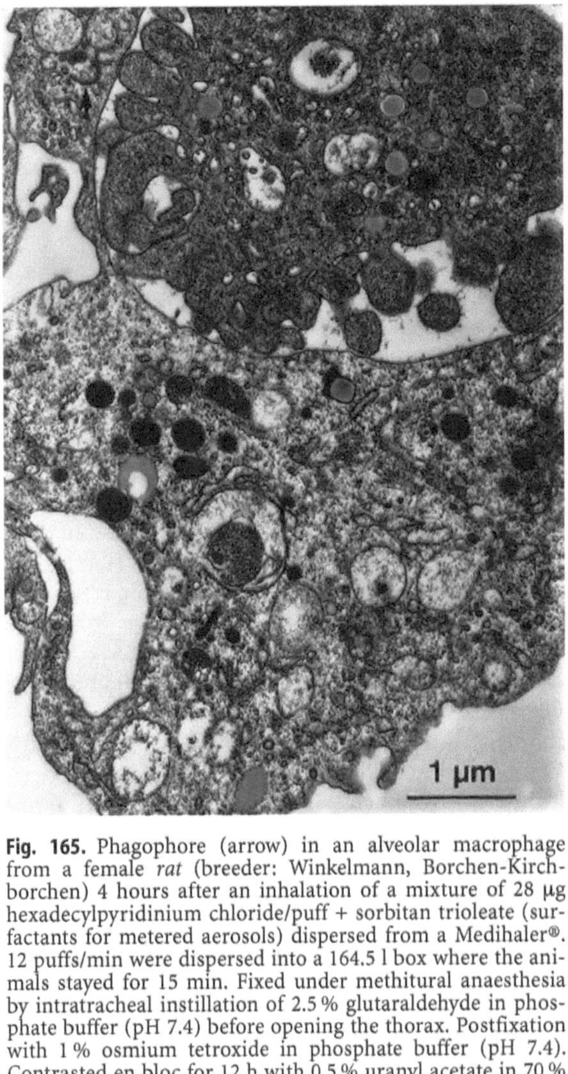

Fig. 165. Phagophore (arrow) in an alveolar macrophage from a female *rat* (breeder: Winkelmann, Borchen-Kirchborchen) 4 hours after an inhalation of a mixture of 28 µg hexadecylpyridinium chloride/puff + sorbitan trioleate (surfactants for metered aerosols) dispersed from a Medihaler®. 12 puffs/min were dispersed into a 164.5 l box where the animals stayed for 15 min. Fixed under methitural anaesthesia by intratracheal instillation of 2.5 % glutaraldehyde in phosphate buffer (pH 7.4) before opening the thorax. Postfixation with 1 % osmium tetroxide in phosphate buffer (pH 7.4). Contrasted en bloc for 12 h with 0.5 % uranyl acetate in 70 % ethanol. Embedded in a 2 : 8 mixture of methyl and butyl methacrylate. Sectioned at 50 nm. Lead citrate after REYNOLDS (1963). Plate 49/11

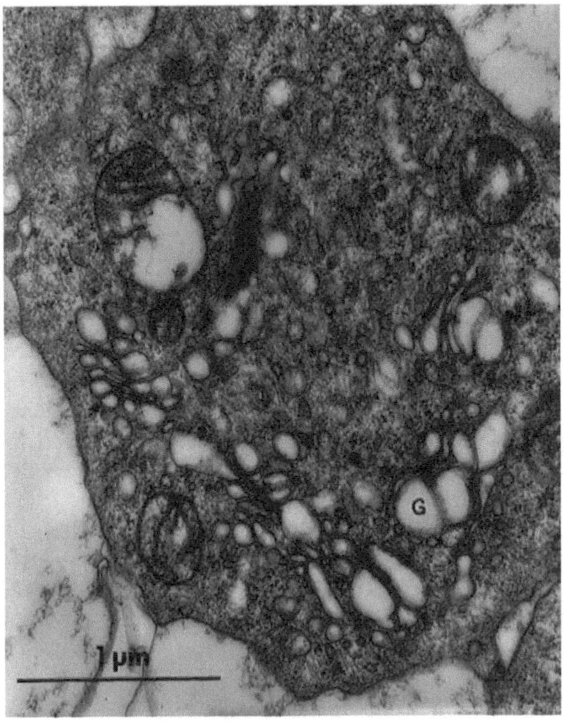

Fig. 166. Dilated Golgi cisternae (G) in a stimulated alveolar macrophage from a female *rat* (breeder: Winkelmann, Borchen-Kirchborchen) 4 hours after an inhalation of a mixture of 28 µg hexadecylpyridinium chloride/puff + sorbitan trioleate (surfactants for metered aerosols) dispersed from a Medihaler®. 12 puffs/min were dispersed into a 164.5 l box where the animals stayed for 15 min. Fixed under methitural anaesthesia by intratracheal instillation of 2.5 % glutaraldehyde in phosphate buffer (pH 7.4) before opening the thorax. Postfixation with 1 % osmium tetroxide in phosphate buffer (pH 7.4). Contrasted en bloc for 12 h with 0.5 % uranyl acetate in 70 % ethanol. Embedded in a 2 : 8 mixture of methyl and butyl methacrylate. Sectioned at 50 nm. Lead citrate after REYNOLDS (1963). Plate 49/07

Paraquat

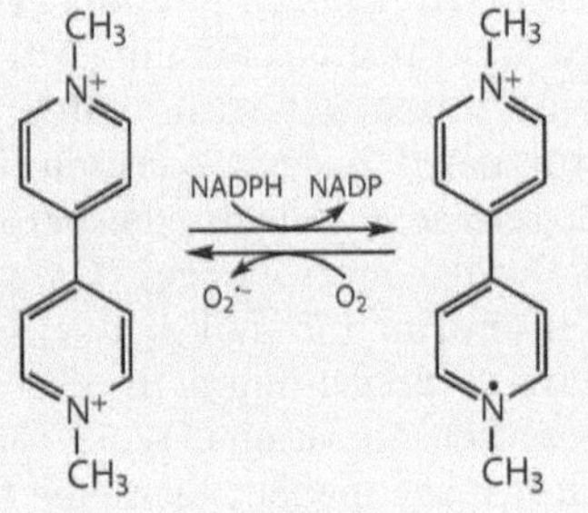

Paraquat [158]

Paraquat (1,1'-dimethyl-4,4'-bipyrdilium dichloride) is a widely used herbicide causing severe injury to the lungs and other organs mediated by reactive oxygen intermediates, NADP cytochrome *c* reductase in microsomal drug metabolising enzyme systems forming radicals (GAGE 1968, BUS et al. 1974, 1976). SHIMADA et al. (1998), however, demonstrated a mitochondrial NADH-quinone oxidoreductase of the outer mitochondrial membrane responsible for paraquat cytotoxicity in *rat* livers.

Paraquat significantly reduced concavalin A-stimulated $O_2^{\bullet -}$ production when incubated at 1 mM with *rat* alveolar macrophages in the absence of glucose (FORMAN et al. 1980). The effect of paraquat was reversed by glucose, but fructose, lactate, and pyruvate could not reverse paraquat inhibition.

Paraquat enhanced oxidation of NADPH (but not NADH) by cell supernatant and increased pentose phosphate shunt activity in resting macrophages, but did not affect mitochondrial respiration or ATP content of alveolar macrophages.

Paraquat (respective LD_{10} dose given intraperitoneally) induced different biochemical responses including different protective responses in Wistar *rats* and Swiss *mice* (ALI et al. 2000). As a protective response, NADPH-specific quinone reductase is induced in *rats*, while catalase is induced in *mice*. It is implied that an early induction of catalase in mice as opposed to *rats* may account for the resistance of Swiss *mice* to paraquat toxicity. Xanthine oxidase, which was induced in *rats*, remains unaffected in *mice* indicating that the enzyme contributed to paraquat toxicity only in Wistar *rats*.

The "lipid peroxide stimulation' mechanism of paraquat acute toxicity has been questioned by SHU et al. (1979), who have shown that it is possible by pre-treatment with *N,N'*-diphenyl-*p*-phenylenediamine, an antioxidant, to prevent the paraquat stimulation of lipid peroxidation without protecting the animals against its lethal effects.

9.6.2.2.22 Tannin from Cotton Bract

Condensed tannin, a polymer of monoflavonoid subunits, is one of the major water-soluble components present in the airborne dust generated from the processing of a wide variety of woody plants (HEMINGWAY 1989). *In vitro* studies have shown that condensed tannin can profoundly change the functional capacity of alveolar macrophages (ROHRBACH et al. 1989, 1992, KREOFSKY et al. 1990, 1992, VUK PAVLOVIC and ROHRBACH 1990).

Arachidonic acid release from *rabbit* alveolar macrophages gained by bronchoalveolar lavage was increased by tannin in a dose-dependent manner (RALSTON and ROHRBACH 1994).

Using the luciferin luciferase reaction, PRÉVOST et al. (1996) showed that tannin markedly depleted ATP content of *rabbit* alveolar macrophages *in vitro*. In inositol-labeled cells, tannin increased inositol-phosphate release in a dose-dependent manner. In lyso-PAF-labeled cells, tannin induced synthesis of phosphatidic acid and diglycerides. In the presence of ethanol, the level of tannin-induced phosphatidic acid was slightly reduced, and phosphatidylethanol was synthesised. No phosphatidylethanol was found in alveolar macrophages stimulated by zymosan in the presence of ethanol. GF 109203X, a specific inhibotor of protein kinase C decreased only tannin-induced phosphatidylethanol synthesis.

9.6.2.2.23 Tetrandrine

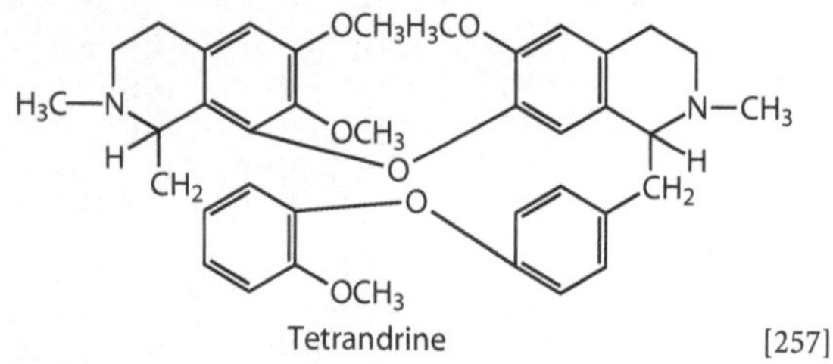

Tetrandrine [257]

Tetrandrine, an antiinflammatory immunosuppressive bisbenzylisoquinoline alkaloid isolated from the Chinese herb *Stephania tetranda*, at 10^{-7} M to 10^{-4} M caused dose- and time-dependent loss of cell viability of *guinea pig* alveolar macrophages, *mouse* peritoneal macrophages, and *mouse* macrophage-like J774 cells (PANG and HOULT 1997). Loss of macrophage viability after tetrandrine treatment was accompanied by the generation of large amounts of prostaglandin E_2, to levels 285–877 % of control. Coincubation with indomethacin abolished PGE_2 generation, but did not prevent cell death. Tetrandrine did not cause generation of nitric oxide. Verapamil also reduced the viability of *mouse* peritoneal macrophages and J774 cells, but did not cause PGE_2 overproduction, except at 10^{-4} M in *mouse* peritoneal macrophages. In macrophages cultured with lipopolysaccharide and interferon-γ to induce the generation of large amounts of both PGE_2 and nitric oxide, tetrandrine reduced mediator release and their forming enzymes (cyclooxygenase-2 and inducible nitric oxide synthase), secondary to cytotoxicity.

Tetrandrine interacted with slowly inactivating calcium channels by completely blocking *d*-cis-[^{3}H]diltiazem binding, partially inhibiting (±)-[^{3}H]D-600 binding, and markedly stimulating [^{3}H]nitrendipine binding in porcine cardiac sarcolemmal membrane vesicles (KING et al. 1988). Tetrandrine reversibly blocked inward Ca^{2+} currents through L-type Ca^{2+} channels in GH_3 anterior pituitary cells grown in a humidified atmosphere containing 5 % CO_2.

Tetrandrine was a potent inhibitor *in vitro* of zymosan-stimulated oxygen consumption, superoxide anion release, and hydrogen peroxide secretion by *rat* alveolar macrophages (CASTRANOVA et al. 1991). Tetrandrine (33 μg/kg × day) was also effective *in vivo* in preventing activation of alveolar macrophages after inhalation (117.5 ± 12.5 mg/m^3 of 1.54 ± 0.01 μm mass median diameter for 6 h per day) or intratracheal instillation (40 mg in 0.3 ml of saline, specific surface 3.97 m^2/g) silica. Tetrandrine also inhibited stimulant-induced chemiluminescence by polymorphonuclear leucocytes.

Lipid peroxidation of liposomes induced by Fe^{2+} was inhibited by tetrandrine at 88 %, while cepharanthine, another biscoclaurine head-to-head diether, inhibited lipid peroxidation at 97 % (SHIRAISHI et al. 1980).

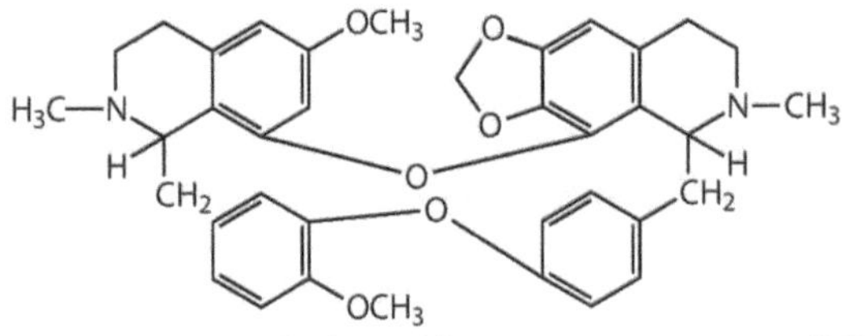

Cepharanthine [258]

In alveolar *rat* macrophages (NR8383), tetrandrine inhibited the activation of NF-κB and NF-κB-dependent reporter gene expression by lipopolysaccharide, 12-*O*-tetradecanoylphorbol-13-acetate and silica in a dose dependent manner (CHEN et al. 1997).

9.6.2.2.24 Tobacco Smoke

By using the electron spin resonance (ESR) spin trapping technique with the *N*-methyl-D-glucamine dithiocarbamate-Fe(II) complex, the generation of •NO was observed in NO spin trapping solution bubbled with the filtered mainstream of cigarette smoke (TOKIMOTO and SHINAGAWA 2002). The ESR signal, with a three-line spectrum characteristic of a •NO radical, which was not observed immediately after bubbling of smoke, started rapidly increasing with time up to around 25 min after the addition of Fe^{2+}, and the slowly approached a peak value dependent on the burned cigarette mass and on the smoking speed. The production of •NO was, however, much affected by air oxidation and enhanced by the addition of ascorbic acid. A certain concentration of $NaNO_2$ solution, in which NO_2^- is assumed as the main origin of the •NO mimicked closely the time course of •NO generation resulting from the smoke of one cigarette. The cigarette smoke was passed through alkaline pyrogallol solution as a deoxidiser; however, it exhibited an unchanged intensity of •NO signal throughout the measurement.

Cigarette smoke serves as a relevant model of alveolar macrophage toxicology, since it represents one of the most common and universal insults to lung defences, and its effects on *human* alveolar macrophages have been extensively studied. In response of chronic cigarette exposure the macrophage population increases in number and individual cell size (DAVIS et al. 1980), and multiple changes occur in microbicidal function, substrate transport (LOW and BULMAN (1977), protein biosynthesis (LOW 1974), and immunologic responsiveness, as reviewed by MARTIN and WARR (1977). The absolute increase in CD11/CD18-positive alveolar macrophages found by SCHABERG et al. (1995) was mainly due to the increased absolute number of positive alveolar macrophages with high density from fractions 3 (>1.40 to ≤1.050 g/ml) and 4 (>1.050 to ≤1.070 g/ml) compared with non-smokers ($P < 0.05$ to < 0.01). These high-density alveolar macrophages of smokers are mainly responsible for the spontaneous $O_2^{\bullet -}$ production of these cells.

ZWILLING (1981) proposed the following sequence of events underlying the formation of emphysematous lesions:

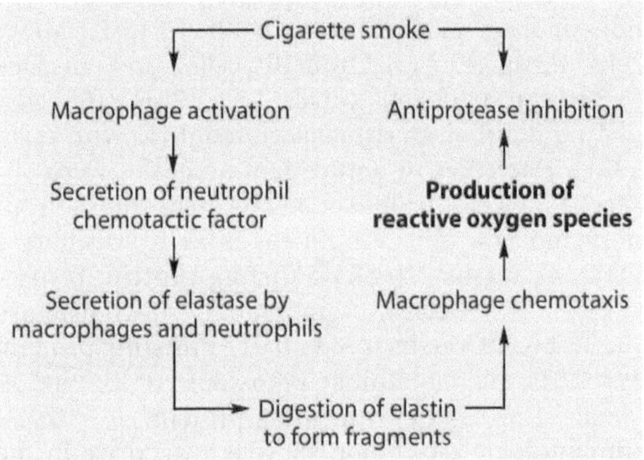

Lavaged pulmonary alveolar macrophages obtained from chronic cigarette smokers contain various quantities of **brownish-yellow pigment** (WARR and MARTIN 1978). In the presence of serum, alveolar macrophages from smokers cultured on glass in TCM 199 were significantly larger in area than those from non-smokers.

REITER (1983) identified sudanophilic chromolipids in paraffin sections of cigarette smokers' alveolar macrophages, even 11 months after burial. Electron micrographs of *human* transbronchial lung biopsies showed typical heterogeneous osmiophilic intralysosomal smokers' inclusions (SÉBASTIEN et al. 1994). Only the five heavy smokers out of 15 patients exhibited similar inclusions on their interstitial macrophages.

Glycosidase (*N*-acetyl-β-D-glucosaminidase, *N*-acetyl-β-D-galactosaminidase, β-D-galactosidase, α-L-galactosidase, α-D-mannosidase, α-L-fucosidase and β-D-glucuronidase) activities were higher in the alveolar macrophages obtained by bronchoalveolar lavage from smokers than in those from non-smokers (SCHARFMAN et al. 1980).

In the alveolar macrophages of *rats* exposed to tobacco smoke for 4 hours per day for two weeks,

5 days each week, and for 6 weeks, 4 days each week, respectively, JEFFERY and REID (1981) found macrophages containing "tar bodies' showing that smoke did reach this distal site.

Bronchoalveolar lavage fluid recovered from upper lobes of smokers contained higher concentrations of **iron** ($P < 0.01$) and ferritin ($P < 0.006$) and lower concentrations of transferrin ($P < 0.003$) compared with the lower lobes (NELSON et al. 1996). In contrast, bronchoalveolar lavage fluid recovered from upper and lower lobes of non-smokers contained much lower concentrations of iron and ferritin, and concentrations were similar in both sites.

Production of **oxygen radical species** was significantly ($P < 0.01$) elevated in alveolar macrophages from current smokers compared with those from non-smokers in the presence ($29.8 \pm 15{,}5$ µM/10^6 cells versus $13.7 \pm 8{,}2$ µM/10^6 cells) and absence (3.8 ± 1.6 µM/10^6 cells versus 2.2 ± 1.2 µM/10^6 cells) of 1 ng phorbol myristate acetate/ml (KONDO et al. 1994). Decreases in antioxidant activities were observed in cells from smokers versus those from non-smokers for Cu,Zn-superoxide dismutase (114 ± 41 versus 210 ± 73 units/mg protein, respectively, $P < 0.01$), glutathione S-transferase (0.217 ± 0.091 versus 0.368 ± 0.017 units/mg protein, $P < 0.01$), and glutathione peroxidase (0.736 ± 0.779 versus 1.590 ± 0.879 units/mg protein, $P < 0.05$). Immunologic estimation showed a decrease in the levels of Cu,Zn-superoxide dismutase in cells from smokers (104.3 ± 46.6 versus 184.1 ± 64.4 ng enzyme/mg protein, respectively, $P < 0.01$). Northern blot analysis of Cu,Zn-superoxide dismutase mRNA showed no apparent difference between the two groups, suggesting not the inactivation of this enzyme but a reduction in the translational step or increased proteolysis.

J774A.1 macrophage cells incubated with 250 µg/ml of an aqueous smokeless tobacco extract released lactate dehydrogenase into the media as an indicator of cellular membrane damage. The amount released was both concentration- and time dependent (BAGCHI et al. 1995). Superoxide dismutase, catalase, mannitol and allopurinol by themselves had no significant effect, while a combination of the four free radical scavengers resulted in a 59 % decrease in the release of lactate dehydrogenase.

Arachidonic acid metabolism: *In vitro*, a significant decrease in prostaglandin E_2 and thromboxane B_2 synthesis but not in prostaglandin $F_{2\alpha}$ synthesis by the smoker pulmonary alveolar macrophages compared with non-smoker PAM was observed by LAVIOLETTE et al. (1981).

9.6.2.2.25 Trifluoperazine

Trifluoperazine, 10-[3-(4-methyl-1-piperazinyl)propyl]-2-trifluoromethylthiophenazine, a drug that binds to Ca^{2+}-calmodulin and inhibits its interaction with other proteins, reversibly inhibited growth and phagocytosis of polyvinyltoluene latex beads (2.97 µm diameter) in a macrophagelike cell line, J774.16 (HORWITZ et al. 1981).

9.6.2.2.26 Wood Smoke Particles

RAW 264.7 *mouse* macrophages exposed to wood smoke from western bark thermolysis showed significant DNA damage, which was intensified by the addition of H_2O_2 (LEONARD et al. 2000). Both sodium formate, an $HO^{\bullet}$ radical scavenger, and deferoxamine, a metal chelator, inhibited DNA damage caused by wood smoke plus H_2O_2 and completely blocked DNA damage from wood smoke in the absence of H_2O_2. Elemental analysis performed on the filter samples presented Fe as the only transition metal present in measurable amounts.

9.6.2.2.27 Wool and Grain Dusts

Wool and grain (wheat and barley) dusts stimulated TNF secretion by *rat* alveolar macrophages *in vitro* in a dose-dependent manner as measured in supernatants with the *mouse* fibroblast cell line L929 bioassay (BROWN and DONALDSON 1996).

9.7 Transformation of Macrophages to Epithelioid Cells

Macrophages turn into epithelioid cells when they become immobilised at the site of inflammation without being called upon to undertake phagocytosis or when phagocytosis or pinocytosis leads to complete elimination of the particle within a few days. On the other hand, uptake of indigestible, non-excretable material prevents macrophages from becoming epithelioid cells (PAPADIMITRIOU and SPECTOR 1971).

9.7.1 Epithelioid Cells in Resorption Granulomas

The mature epithelioid cell has a large nucleus fairly regular in outline with finely marginated nuclear chromatin (Fig. 167). The presence of one or more large, spherical, reticulated nucleoli is typical.

In type A epithelioid cells both smooth- and rough-surfaced endoplasmic reticulum was predominant (Fig. 165). The cytoplasm of type B epithelioid cells contained a well developed Golgi apparatus and a variety of vesicles (Figs. 167, 168).

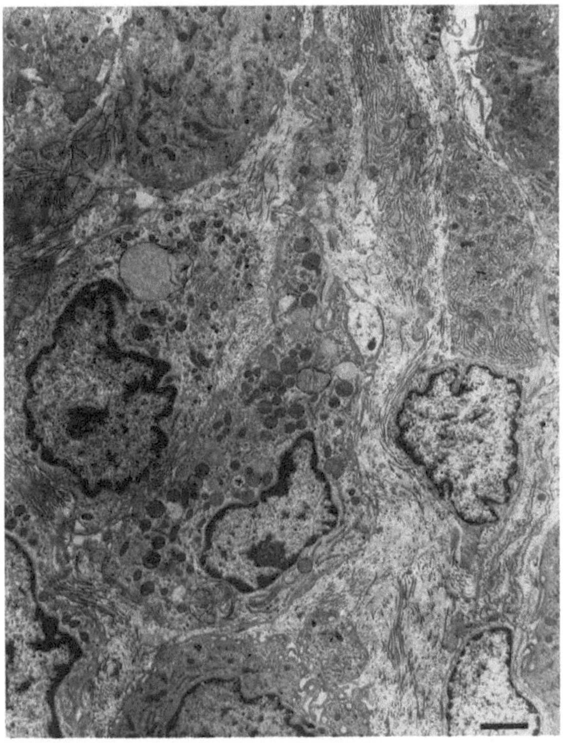

Fig. 167. Type A epithelioid cells with much rough surfaced endoplamic reticulum and type B epithelioid cells showing vesicles of different sizes from a Marbagelan®-induced (7 days) resorption granuloma (block 171) of a male Sprague-Dawley *rat* (No. 10). Animal treated for 14 days with intragastric application of 15 mg carbocromene per kg body weight × day. Perfused under pentobarbital anaesthesia (30 mg/kg) from the abdominal aorta with 2.5 % glutaraldehyde in 0.1 M sodium cacodylate buffer (pH 7.4). Postfixation with 1 % osmium tetroxide in cacodylate buffer. Embedded in Epon 812 and sectioned at 50 nm. Lead citrate and uranyl acetate. Plate 2850

Fig. 168. Type A epithelioid cells with much rough surfaced endoplamic reticulum and type B epithelioid cells showing vesicles of different sizes from a Marbagelan®-induced (7 days) resorption granuloma (block 171) of a male Sprague-Dawley *rat* (No. 10). Animal treated for 14 days with intragastric application of 15 mg carbocromene per kg body weight × day. Perfused under pentobarbital anaesthesia (30 mg/kg) from the abdominal aorta with 2.5 % glutaraldehyde in 0.1 M sodium cacodylate buffer (pH 7.4). Postfixation with 1 % osmium tetroxide in cacodylate buffer. Embedded in Epon 812 and sectioned at 50 nm. Lead citrate and uranyl acetate. Plate 2845

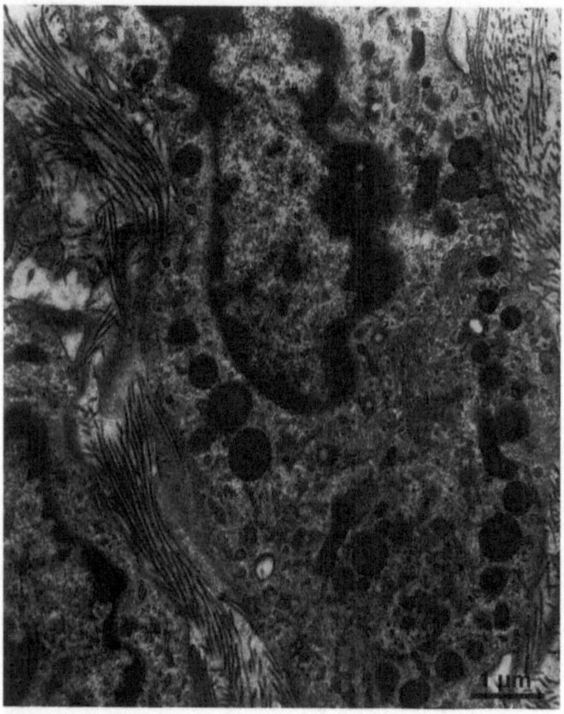

Fig. 169. Type B epithelioid cell showing vesicles of different sizes, few mitochondria and indistinct plasmalemma from a Marbagelan®-induced (7 days) resorption granuloma (block 171) of a male Sprague-Dawley *rat* (No. 10). Animal medicated for 14 days with intragastric application of 15 mg carbocromene per kg body weight × day. Perfused under pentobarbital anaesthesia (30 mg/kg) from the abdominal aorta with 2.5 % glutaraldehyde in 0.1 M sodium cacodylate buffer (pH 7.4). Postfixation with 1 % osmium tetroxide in cacodylate buffer. Embedded in Epon 812 and sectioned at 50 nm. Lead citrate and uranyl acetate. Plate 2849

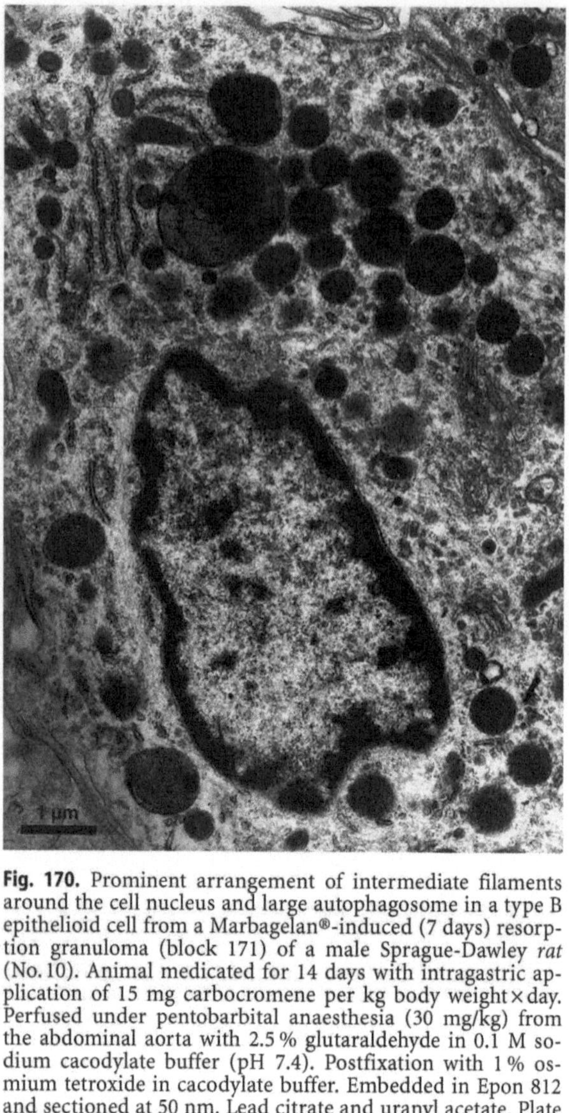

Fig. 170. Prominent arrangement of intermediate filaments around the cell nucleus and large autophagosome in a type B epithelioid cell from a Marbagelan®-induced (7 days) resorption granuloma (block 171) of a male Sprague-Dawley *rat* (No. 10). Animal medicated for 14 days with intragastric application of 15 mg carbocromene per kg body weight × day. Perfused under pentobarbital anaesthesia (30 mg/kg) from the abdominal aorta with 2.5 % glutaraldehyde in 0.1 M sodium cacodylate buffer (pH 7.4). Postfixation with 1 % osmium tetroxide in cacodylate buffer. Embedded in Epon 812 and sectioned at 50 nm. Lead citrate and uranyl acetate. Plate 2848

9.8 Stimulation, Priming and Activation of Macrophages

The word stimulus can be used as a noun, as in inflammatory stimulus, but in this context not in the adjectival form, as in stimulated macrophages (Van Furth 1989). The term stimulated macrophages is imprecise, because stimulation means that a stimulus has been applied that may result in elicitation and/or activation of cells.

The term "immunomodulators" has been employed to define agents or approaches that will modify the relationship between tumour and host by modifying a host's biologic response to tumour cells, with an enhanced therapeutic benefit. The cascade of the cellular immune system is initiated by the uptake and processing of antigen by the macrophage.

Several agents which increase macrophage and lymphocyte tumour cell cytotoxicity appear to do so by interferon production. Poly I:C and maleic anhydride divenyl ether copolymer incubated *in vitro* with resting macrophages and tumour cells caused macrophage killing of the tumour cells. Adding a specific antibody neutralising interferon will abandon tumoricidy.

9.8.1 Non-immunological Stimulation

Di Luzio (1976) established the criteria for an agent which would function as a macrophage stimulant as follows:

1. Non-viable entity
2. Defined chemical structure
3. Non-toxic at employed dose
4. Non-antigenic

5. Activate all phases of macrophage mediated events
   a. Chemotaxis
   b. Phagocytosis
   c. Intracellular killing
   d. Extracellular killing
   e. Expression of humoral and cell mediated immunity
   f. Secretory aspects
6. Induced macrophage response, i.e., hyperfunction-hyperplasia and hypertrophy of RES – must be reversible in nature
7. Readily available and relatively inexpensive.

9.8.1.1 Immunostimulators

The term "immunomodulators" has been applied to define agents or approaches that will modify the relationship between tumour and host by modifying a host biologic response to tumour cells, with an enhanced therapeutic benefit. Several terms have appeared in the literature describing the action of these agents: immuno-augmentators, immunostimulators, immunoenhancers, immunoregulators, immunomodifiers, and immunorestorators. As most agents produce a direct or indirect effect on a cellular component of the immune system, the term "biologic response modifiers" includes the many agents and approaches that have a mechanism of action which involves the individual's biologic responses.

The macrophage appears to play a major role in initiating the cellular immune system with the production and secretion of soluble factors (monokines) which regulate various cellular components. These monokines, along with lymphokines (lymphocyte secreted cell regulators) and specific chemicals comprise the bulk of the list of biologic response modifiers.

Hormones

As reviewed by YAMAMOTO (1985) and BEATO (1989) steroid hormones influence gene expression at essentially all known levels of regulation.

In certain cases, steroid hormones were converted into corresponding radical intermediates in either enzymatic or non-enzymatic systems (KODADMA et al. 1997). Although detection of the free radicals was limited to oestrogens, evidence suggests that glucocorticoids as well as androgens may also share the physiological formation of free radicals.

In general, the concentrations of hormones needed to inhibit peroxidation *in vitro* seem much larger than physiological levels (picomolar range). The phenoxyl radicals produced during antioxidant activity are reactive and capable of damaging proteins and DNA *in vitro* (LIEHR 1996).

Glucocorticoids

Phagocytosis of *Staphylococcus aureus* by macrophages elicited in *rabbits* by intrapleural application of Liebig's meat extract plus carboxymethylcellulose was ≤50% enhanced when the animals had been injected subcutaneously with 0.2 mg cortisone acetate per kg body weight×day (CRABBÉ 1956).

On glucocorticoid stimulation, macrophages elaborate anti-phospholipase proteins such as macrocortin (FLOWER and BLACKWELL 1979) and lipomodulin (HIRATA et al. 1980). Glucocorticoids reduced the mRNA levels of the rate-limiting enzyme in cholesterol synthesis, 3-hydroxy-3-methylglutaryl CoA reductase, in a *mouse* macrophage-like cell line, $P388D_1$ (HELMBERG et al. 1990). The defensive role of neutrophils and mononuclear phagocytes are impaired. Fauve and PIERCE-CHASE (1967), however, have suggested that some synthetic steroids can retain the beneficial anti-inflammatory activities without drastically compromising the defensive role of the macrophage.

While glucocorticoids dampen most immune responses, they stimulate polyclonal immunoglobulin production, e.g., by *human* peripheral blood lymphocytes (GOODWIN and ATLURU 1986), increase the synthesis of IgE by IL-4-stimulated *human* lymphocytes (WU et al. 1991) and induce IgA in serum (but not at mucosal surfaces) and secretory components in serum, saliva, and bile of *rats* (WIRA et al. 1990, WIRA and ROSSOLL 1991).

Oestrogens

Female sex hormones may contribute to the gender-related differences in the immune response (ROBERT and SPITZER 1997). Nitric oxide production by *rat* alveolar macrophages cultured in the presence or in the absence of lipopolysaccharide was higher from female than from male *rats*, but without statistical significance. However, ovariectomy induced significant inhibition in spontaneous production of •NO by alveolar macrophages. In orichidectomized *rats*, the •NO response by alveolar macrophages to lipopolysaccharide stimulation relative to spontaneous •NO production was significantly downregulated.

Oestrogen is a potent initiator of multiplication, mobilisation and phagocytic activities of macrophages and the infiltration of the uterus is presumably a result of the oestrogen surge during oestrus (Eufinger 1932, Nicol and Vernon-Roberts 1965, Vernon-Roberts 1969).

Radicophilic moieties (Kitagawa et al. 1992, Cynshi et al. 1995) in oestrogens are important for efficient antioxidant and radical scavenging activities. Römer et al. (1998) designed and synthesised a novel series of 17β-œstradiol and its homologue $\Delta^{9(11)}$-dyhydro-17β-œstradiol linked with aromatic moieties at the C-17 position, with an aim to improve scavenging of lipid peroxyl and/or superoxide anion radicals ant to enhance Fe(II) chelation as well as stimulation of Fe(III) reduction. While the classical oestrogen, 17β-œstradiol as well as selected phenolic compounds only moderately inhibited iron-dependent lipid peroxidation and stimulated total antioxidant activity, $\Delta^{9(11)}$-dyhydro-17β-œstradiol and its 17α-substituted analogues directly altered the iron redox chemistry and diminished superoxide anion formation in the xanthine xanthine oxidase system.

17β-Oestradiol is a good inhibitor of lipid peroxidation in microsomal and pre-formed liposomal systems (Wiseman 1995). The order of effectiveness in the microsomal system was 4-hydroxytamoxifen > 17β-œstradiol > tamoxifen > ICI 164 384 and in the liposomal system 4-hydroxytamoxifen > 17β-œstradiol > ICI 164 384 > tamoxifen. However, tamoxifen also functions as a genotoxic carcinogen (Hard et al. 1993), a therapeutic regimen including melatonin plus tamoxifen may allow for a lower dose of the latter agent thereby reducing its collateral toxicity (Reiter 1998). Metabolism of tamoxifen by *rat* liver microsomes gave equal amounts of the *R* and *S* enantiomers (Osborne et al. 2001). They have the same chemical properties but, on treatment of *rat* hepatocytes in culture, *R*-(+)-α-hydroxytamoxifen gave at least eight times as many DNA adducts as the *S*-(−)-isomer.

Yen et al. (2001) utilized the fluorescent dye 2',7'-dichlorofluorescin diacetate to measure the generation of reactive oxygen species in *human* umbilical vein endothelial cells. 17β-Oestradiol (54 μM) pretreatment for 18 h or direct co-incubation significantly suppressed both *tert*-butylhydroperoxide- and oxidised LDL-induced stimulation of the generation of reactive oxygen species.

17β-Oestradiol afforded protection against oxidative nerve cell apoptosis. 17β-Oestradiol is indeed a potent antioxidant comparable to other phenolic compound such as α-tocopherol (Behl 1999, Moosmann and Behl 1999).

Increasing evidence indicates that the effector molecule crucial for œstrogen signalling is the short-lived messenger •NO. The observation that estrogens not only increases expression of NO synthesizing enzymes (NOS; Weiner et al. 1994), but also triggers direct activation of eNOS in the first minutes after the addition to cells (Chen et al. 1999, Chambliss et al. 2000, Haynes et al. 2000, Hisamoto et al. 2001), opens the possibility that part of the •NO effects on œstrogen biological actions are also mediated through regulation of the signal transduction cascades initiated by the hormone. Exposure of the MCF-7 *human* breast cancer cell line to 17β-œstradiol in the presence of •NO gave rise to activation of signalling events additional to those triggered by 17β-œstradiol alone, namely tyrosine phosphorylation of specific proteins, including the insulin receptor substrate-1, with recruitment to this adapter of the phosphatidylinositol 3'-kinase and persistent activation of Akt (protein kinase B; Falcone et al. 2002). Active Akt, in turn, prevented 17β-œstradiol from activating p42/44 extracellular signal-regulated kinases (ERK 1/2; Kelly and Levin 2001).

Older references: Nicol and Helmy (1951), Abou-Zirky and Nicol (1952), Heller et al. (1957), Nicol and Snell (1957), Bilbey and Nicol (1958), Nicol et al. (1964)

Diethylstilbestrol

It is widely believed that metabolic activation by peroxidases plays a role in the biological effects of diethylstilbestrol, a synthetic oestrogen which is also a tumour-inducing agent (Metzler 1987). When diethylstilbestrol is exposed to horseradish peroxidase/H_2O_2 in the presence of the cationic surfactant hexadecyltrimethylammonium bromide, the rapid dioxygen consumption is accompanied by a burst of light emission (Knudson et al. 1992). Lipoxygenase (EC 1.13.11.12) catalyses the oxidation of diethylstilbestrol to its corresponding quinone to yield free radical species intermediates (diethylstilbestrol semiquinone and diethylstilbestrol quinone), which are associated with the adverse effects of this synthetic oestrogen (Núñez-Delicado et al. 1997). Due to the low degree of water solubility of diethylstilbestrol, the enzyme works in a range of diethylstilbestrol concentrations below K_m.

The E-diethylstilbestrol-4',4'-semiquinone oxygen-centred radical is converted to a carbon centred radical which reacts with O_2 forming a dioxetane. Unless the dioxetane is rapidly protonated, it will almost instantly cleave into excited carbonyl products because of the intramolecular chemically initiated electron exchange luminescence. In the latter, electron transfer from the phenoxide moiety to the peroxide ring promotes the fluorescence observed from the singlet-excited 4-hydroxy propiophenone product. If any triplet species are formed, they derive from the protonated dioxetane. The latter is expected to go fast intersystem crossing to the lowest triplet state, which is of the π,π* type.

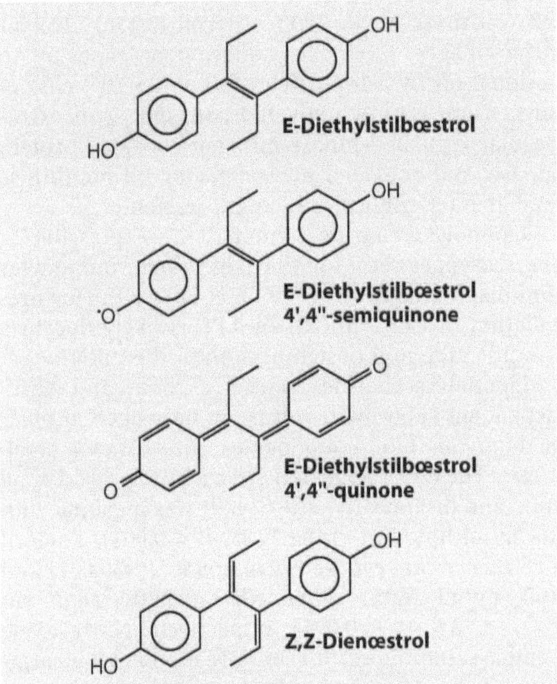

Diethylstilbœstrol biotransformation [259]

When diethylstilbœstrol was administered to female and male CD-1 *mice* as four subcutaneous injections for 1 week at 0, 5, 15, and 30 μg/kg body weight doses, female thymus glands were significantly larger than their male counterparts (CALEMINE et al. 2002). Short-term administration of diethylstilbœstrol to female of male *mice* neither induced thymic atrophy nor altered the relative percentages of thymic subsets. Nevertheless, diethylstilbœstrol treatment of female of male *mice* induced a dose-related apoptosis of $CD4^+8^+$, $CD4^+8^-$ and $CD4^-8^+$ subsets as analysed by 7-amino-actinomycin. Immature $CD4^-8^-$ subsets of thymocytes from females was also affected by high dose diethylstilbœstrol. The pattern of mitogen-induced proliferation of splenic lymphocytes varied with the dose of hormone and the gender. In females, splenic lymphocytes from low dose diethylstilbœstrol (5 μg/kg body weight)-treated *mice* exhibited an increased proliferative response to Con-A, lipopolysaccharide or PMA/ionomycin compared with controls.

Experimental dose dependent palatoschisis due to diethylstilbœstrol treatment of pregnant Swiss *mice* has been reported by GABRIEL-ROBEZ et al. (1972).

Further references: NICOL and HELMY (1951), NICOL et al. (1958, 1960, 1961, 1964), NICOL and WARE (1960), TREJO et al. (1972), STEVEN and SNOOK (1975), BOORMAN et al. (1980), DEAN et al. (1980, 1986)

***trans*-Resveratrol** (3,5,4'-trihydroxy-*trans*-stilbene) is a naturally occurring stilbene found in grapes and certain medicinal plants, where it is believed to provide protection from fungal infections and other stresses (SCHWEKENDIEK et al. 1992, SOLEAS et al. 1997). It has been reported to have a diverse range of pharmacological properties, including antiinflammatory, antiplatelet (PACE-ASCIAC et al. 1995), and antioxidant activities (BELGUENDOUZ et al. 1997, CHANVITAYAPONGS et al. 1997).

Progesterone

In the male *guinea pig*, progesterone (3 mg per day for 3 weeks) had little or no effect on the phagocytic activity of reticulo-endothelial system (NICOL and SNELL 1957).

Progesterone at 10^{-4} M inhibited basal and *in vitro* lipopolysaccharide-stimulated nitric oxide generation by alveolar macrophages of both saline and lipopolysaccharide-treated male *rats* (ROBERT and SPITZER 1997).

Lipids

Human monocytes cultured with a mixture of serum lipids and lipoproteins (cardiolipids) changed morphologically after overnight culture (CHU et al. 1999). Most of them detached from the bottom of the plate and assumed a rounded morphology. IL-1β, IL-6, platelet-derived growth factor, and transforming growth factor β1 all exhibited a significant reduction ($P < 0.05$) in lipid-treated cultures compared to control cultures.

Glyceryl trioleate

Peritoneal macrophages of male *mice* 8 h to 14 d after intraperitoneal injection of 10 mg glyceryl trioleate emulsified in Hanks' solution had longer processes, more acid phosphatase and more lysosomes than controls (CARR 1967).

Further references: COOPER and STUART (1961, 1962), COOPER and WEST (1962), STUART and COOPER (1962), STUART and DAVIDSON (1964), CARR (1966), MOUTON et al. (1975)

Polyunsaturated Fatty Acid (PUFAs)

Six all *cis* polyunsaturated fatty acids (PUFAs) in HL-60 cells induced apoptosis in correlation with the number of double bonds (HAWKINS et al. 1998). Cell death was preceded by a progressively increasing lipid peroxidation. PUFA-induced apoptosis was oxidative, being blocked by both vitamin E acetate (5 or 50 μM) and sodium selenite (10 or 100 nM), the latter in a critically time-dependent manner, as the cytotoxic effect of *cis*-parinaric acid ($\Delta^{9,11,13,15}$, *cis, trans, trans, cis*; 5 μM) on Mia-Pa-Ca-2 cells was prevented only when the selenite was added simultaneously or within 2 h. In combination, addition of PUFA and γ-irradiation induced a significantly greater cell kill than either agent alone.

Lipoarabanomannan

The mycobacterial cell wall glycolipid lipoarabanomannan (LAM) is capable of regulating several macrophage functions. AraLAM, lacking terminal mannosyl units and derived from nonpathogenic organisms, is a potent inducer of proinflammatory cytokines and chemokines. Terminally-mannosylated LAM (ManLAM)from virulent *M. hominis*, only weakly induces these proteins. BERNARDO et al. (1996) studies LAM-induced cell migration, intracellular Ca^{2+}-fluxes and surface receptor expression and function in fresh *human* monocytes and in 48-h monocyte-derived macrophages. Using modified Boyden chambers, they found that AraLAM but not ManLAM (0.01–10 μg/ml) was chemotactic for both monocytes and macrophages. Chemotaxis was blocked by anti-CD14 mAb or by the lipopolysaccharide antagonist *Rhodobacter sphaeroides* lipid A. ManLAM ' but not AraLAM, induced a transient rise in cytosolic Ca^{2+} in a subpopulation of macrophages, but not in monocytes.

Lipopolysaccharides

Lipopolysaccharide initiates a variety of events which ADAMS and HAMILTON (1987, 1989), HAMILTON and ADAMS (1987) and ADAMS et al. (1988) collectively have termed cascades II and III. Included in **cascade II** are enhanced levels of massage for the protooncogenes c-*fos* and c-*myc* (which can be attributed to enhances transcription of these genes), and enhanced mRNA for the competence gene JE (which can be attributed to accumulation or stabilisation of massage). Many of the events in cascade II can be mimicked by platelet activating factor (PAF), a potent bioactive ether lipid for which a receptor has been demonstrated on macrophages (PRPIC et al. 1988). **Cascade III** is typified by synthesis of a distinct set of new polypeptides as well as transcription of the competence gene KC.

Human alveolar macrophages stimulated with lipopolysaccharide (1 μg/ml) activated extracellular regulated kinase/mitogen-activated protein kinase, in addition to JUN amino-terminal kinase (JNK) and p38, and showed translocation of nuclear factor κB (MONDAL et al. 2000). In contrast to alveolar macrophages, monocyte adhesion or exposure to residual oil fly ash particles (10 μg/2×10^5 cells) in suspension rapidly activated p38, JNK, and extracellular regulated kinase/mitogen-activated protein kinase, and activated nuclear factor κB binding as well as interleukin-8 mRNA expression.

Lipopolysaccharide inhibited the expression of the scavenger receptor Cla-1 in *human* monocytes and macrophages (BUECHLER et al. 1999). Downregulation of Cla-1 mRNA by LPS is likely due to a modification and destabilisation of the mRNA.

Lipopolysaccharides caused a time- and dose-dependent **superoxide release** in nonadherent purified *human* blood monocytes (LANDMANN et al. 1995). The effect appeared after 5 min, peaked at 30 min, and disappeared after 2 h. It was maximal with 10 ng/ml lipid A (+148±22 %, $P<0.001$), 1 ng/ml LPS *Escherichia coli* Re (+226±68 %, $P<0.001$) and 100 ng/ml LPS *Salmonella abortus equi* sm (+272±52 %, $P<0.001$), respectively. Nitric oxide synthase inhibitors, 0.1 mM N-monomethyl arginine or 20 mM spermine-HCl lowered the enhanced levels of superoxide released by LPS treated *rat* peritoneal macrophages (JOE and LOKESH 1997). The superoxide dismutase activity in LPS treated macrophages was 51 % lower than that observed in resident cells. NO synthase inhibitors prevented the loss of SOD activity in LPS treated cells. Exogenously added SOD during sensitisation of cells with LPS also inactivated the enzyme. This inactivation of SOD is inhibited by NO synthase inhibitors. While *human* alveolar macrophages incubated for 18 h in media supplemented with 50 μg, 100 μg or 250 μg Fe_2O_3 or TiO_2 dust/ml did not increase release of LDH, TNF-α, or IL-1β, supplementation by 0.1 μg lipopolysaccharide/ml induced alveolar macrophage release of both cytokines with decreased release of **IL-1β** by alveolar macrophages from smokers compared with non-smokers (BING and WESSELIUS 1996). Both *murine* peritoneal exudate cells and a *murine* macrophage cell line, RAW 264.7 produced surface and secreted **TNF-α** in response to lipopolysaccharide in a dose-dependent manner (CHAUDHRI 1997). However, much lower concentrations of LPS (100 ng/ml) were needed for optimal expression of surface TNF-α than for secreted TNF-α (1 μg/ml). Furthermore, concentrations of actinomycin D that inhibit the synthesis of new mRNA and the production of secreted TNF-α did not block the expression of surface TNF-α on LPS-stimulated cells.

Lipopolysaccharide priming amplifies lung macrophage tumour necrosis factor production in response to air particles (IMRICH et al. 1999).

Activation of **nuclear factor-κB** is necessary for the induction of both TNF-α and Mn superoxide dismutase mRNAs by LPS (WHITE et al. 2000). Neither hypoxia (1 % O_2, 5 % CO_2, and 94 % N_2) nor diphenylene iodonium (0.5 to 5 μM) had any effect on LPS activation of nuclear factor-κB in *human* monocytes (WHITE and TSAN 2001). Treatment of *human* alveolar macrophages with LPS, polymeric- and secretory-IgA, but not 12-*O*-tetradecanoylphorbol-13-acetate, induced NF-κB activation through IκBα phosphorylation and subsequent proteolysis (OUADRHIRI et al. 2002).

While low levels of **IL-6** were detected in medium from uninduced *human* blood monocytes cultured in RPMI1640 supplemented with 2 mM glutamine, 50 mg/ml gentamicin and 10 % heat-inactivated foetal *calf* serum, incubation of these cells with LPS and IFN-γ led to a very high IL-6 protein secretion (COSTANZO et al. 1999).

The intracellular oxidation of fluorescent probes dichlorofluorescin, dihydrorhodamine, and hydroethidine by *hamster* alveolar macrophages in response to lipopolysaccharide did not shoe any significant increase in intracellular reactive oxygen species (IMRICH et al. 1999).

In gene transfection studies performed with cytomegalovirus-luciferase as a reporter plasmid and cationic liposome as a transfecting agent, the presence of endotoxin in plasmid DNA preparations severely limited transgene expression in macrophages but had little or no effect in other cell types tested (DOKKA et al. 2000). This decreased transfection was dependent on ROS-mediated cellular toxicity induced by endotoxin. Neutralising the endotoxin by the addition of polymyxin B effectively increased transfection efficiency and reduced toxicity. Electron spin resonance studies confirmed the formation of reactive oxygen species in endotoxin-treated cells and their inhibition by free radical scavengers The ROS scavenger *N-tert*-butyl-α-phenylnitrone, the H_2O_2 scavenger catalase (EC 1.11.1.6), and the $HO^•$ scavenger sodium formate effectively inhibited endotoxin-induced effects, whereas the $O_2^{•-}$ scavenger superoxide dismutase (EC 1.15.1.1) had lesser effects.

The initial events of the cellular response to LPS have now been characterised; endotoxin responses are initiated by the binding of complexes of LPS and LPS-binding protein to **CD14**, which is a phosphatidylinositol-linked surface receptor present mainly in myelomonocytic cells (SCHUMAN et al. 1990, WRIGHT et al. 1990, ZIEGLER-HEITBROCK et al. 1994). Subsequently, activation of these cells occurs, leading to the release of biologically active products. This system is very sensitive, and triggered by picogram levels of LPS corresponding to concentrations that induce endotoxin shock. The levels of *rat* CD14 mRNA expression in resident peritoneal macrophages, alveolar macrophages, and peripheral blood monocytes are constitutively high, whereas that in Kupffer cells is low (TAKAI et al. 1997). IL-1β and iNOS mRNA expressions in blood monocytes, and both peritoneal and alveolar macrophages after stimulation with LPS *in vivo* and *in vitro* are high, whereas those in Kupffer cells are low. Scavenger receptors may regulate nitric oxide production from *mouse* peritoneal macrophages stimulated by LPS (MATSUNO et al. 1997).

Sensitivity to LPS and monophosphoryl lipid A can be enhanced by concurrently loading *mice* with D-galactosamine (ELLIOTT et al. 1991). Significant diurnal variation in susceptibility to lethal toxicity was observed in D-galactosamine loaded in *mice* upon LPS or monophosphoryl lipid A immunostimulant challenge. In *mice* treated with either monophosphoryl lipid A or monophosphoryl lipid A plus D-galactosamine, at the time of greatest toxic sensitivity, serum TNF levels were significantly higher than was seen in *mice* treated at the time of low sensitivity. Peritoneal exudates cells harvested from *mice* treated with either D-galactosamine or monophosphoryl lipid A displayed enhanced *in vitro* superoxide ($O_2^{•-}$) production. Simultaneous treatment with D-galactosamine and monophosphoryl lipid A led to a synergistic enhancement of $O_2^{•-}$ production above that induced by either xenobiotic alone. Pre-treatment with superoxide dismutase mimic Cu(II) (diisopropylsalicylate)$_2$ significantly protected *mice* from the lethal toxicity of D-galactosamine–monophosphoryl lipid A challenge.

Twenty hours after LPS and IFN-γ stimulation, alveolar macrophages recovered by bronchoalveolar lavage from 3-day-old specific pathogen-free, virus-free *rats* produced >2 times more NO_2^- and NO_3^- than did macrophages more mature animals (SHERMAN et al. 1996). Basal and stimulated $O_2^{•-}$ was similar among 3-day-old, 10-day-old and adult alveolar macrophages. EU and STAMLER (1997) investigated the intracellular fate of $^•NO$ and its relationship to the redox state of the RAW 264.7 cell line immunostimulated by LPS and interferon-γ. Following induction of iNOS, the rates of NO_2^-/NO_3^- accumulation in the media remained steady of 24 h. The redox state within the cell, as measured by the intracellular concentrations of reduced glutathione and the ratios of GSH/GSSG, remained stable for 12 h following immunostimulation and declined thereafter, i.e. the cell became progressively oxidised. Three to 4 % of the intracellular protein thiols were

S-nitrosylated during the steady state production of NO_2^-/NO_3^-. The depletion of GSH by buthionine sulfoxime led to further increases in protein S-nitrosothiols, but did not alter the rate of NO_2^-/NO_3^- production.

The combination of aerosolised LPS and IFN-γ to induce enhanced pulmonary alveolar macrophage activation in C57BL/6 *mice* was significantly better in killing of B16-F10 melanoma cells than either treatment modality alone (Eisenberg et al. 1991). Activated pulmonary alveolar macrophages selectively killed tumour cells, but did not kill the 3T3 fibroblast cell line. Peritoneal macrophages from mice treated with LPS + IFN-γ by inhalation were enhanced (indicating a systemic effect), but not to the same extent as pulmonary alveolar macrophages.

The incubation of resident peritoneal macrophages with bacterial lipopolysaccharide induced high **arginase** (EC 3.5.3.1) activity as judged by the consumption of ^{14}C (U)-L-arginine and the release of labelled ornithine into the cell supernatant (Kriegbaum and Dröge 1985).

•NO production in LPS-stimulated RAW 264.7 cells was inhibited by flavonoids such as apigenin, wogonin, luteolin, tectorigenin, and quercetin as measured by nitrite formation at 10–100 μM (Kim et al. 1999). Apigenin, chrysin, and kaempferol strongly enhanced the inhibition of inducible cyclooxygenase and inducible nitric oxide synthase promoter activities in lipopolysaccharide-stimulated RAW 264.7 macrophages, which contain the peroxisome proliferator-activated receptor γ expression plasmids (Liang et al. 2001). A synthetic carbazole compound, 9-(2-chlorobenzyl)-9*H*-carbazole-3-carbaldehyde, was found to have an inhibitory effect ($IC_{50} = 1.3 \pm 0.4\ \mu M$) on lipopolysaccharide-stimulated •NO generation in RAW 264.7 cells (Tsao et al. 2002). This carbazole decreased the transcription of iNOS mRNA through a signal pathway that did not involve NF-κB activation.

Murine peritoneal exudate cells pre-exposed to bacterial lipopolysaccharide showed augmented •NO production by LPS restimulation, in contrast to LPS tolerance with reduced production of TNF-α and IL-6 (Tominaga et al. 1999). Significant amounts of IFN-γ were detected in the peritoneal exudate cell cultures on LPS stimulation, and anti-IFN-γ antibody suppressed the LPS-induced •NO, but not TNF-α and IL-6, production. Addition of anti-IFN-γ antibody to the cultures in the LPS pre-exposure step strongly suppressed the augmented •NO production on LPS restimulation. Anti-IL-12 antibody, which suppressed the LPS-induced IFN-γ production, also suppressed the augmented •NO production, as did anti-IFN-γ antibody. *Mouse* peritoneal macrophages fed with synthetic β-hematin, structurally identical to native hemozoin, no longer produced tumour necrosis factor-α and •NO in response to LPS (Taramelli et al. 2000). β-Hematin-mediated inhibition of macrophage function cannot be ascribed to iron release from β-hematin because neither prevention by iron chelators nor down-regulation of iron-regulatory protein activity was detected. Inhibition appeared to be related to pigment-induced oxidative stress because (a) thiol compounds partially restored macrophage functions, (b) heme oxygenase and catalase mRNA levels were up-regulated, and (c) free radicals production increased in β-hematin-treated cells.

In a mixed cell culture system of astrocytes and microglial cells from the neocortex of new-born Wistar *rats*, nitrite levels, used as an indicator of nitric oxide production, were elevated after the addition of LPS and cytokines (Possel et al. 2000). Immunohistochemistry and the NADPH diaphorase technique demonstrated selective localisation of the iNOS protein in microglial cells, whereas no iNOS protein or NADPH diaphorase activity was detected in astrocytes.

The NO donor DETA NONOate (200 μM) inhibited *mouse* bone marrow-derived macrophage proliferation by approx. 80% (Vadiveloo et al. 2001). However, despite •NO being an antimitogen, LPS was as potent at inhibiting proliferation in marrow-derived macrophages from NOSII-/- *mice* as from wild-type *mice*. Consistent with these findings, LPS-induced cell cycle arrest in normal bone marrow-derived macrophages was not reversed by the addition of NOSII inhibitor *S*-methylisothiourea. Moreover, in both normal and NOSII-/- bone marrow-derived macrophages, LPS inhibited the expression of cyclin D1, a protein that is essential for proliferation of many cell types. Despite inhibiting proliferation DETA NONOate had no effect on cyclin D1 expression.

Blocking of NO synthase with aminoguanidine (4 mg/*mouse* intraperitoneally 1 h before an intravenous application of 300 μg LPS) decreased the levels of NO derivatives in the serum and bronchoalveolar lavage fluid by 45–50% (Heremans et al. 2000). TNF production in the serum was also reduced, but there was no effect on the development of lung oedema unless the dose of aminoguanidine was increased to 8 or 10 mg. Inhibition of NOS synthase by 2-amino-5,6-dihydro-6-methyl-4*H*-1,3-tiazine abrogated LPS-induced increase of L-arginine uptake in *rat* alveolar macrophages (Hammermann et al. 2000).

Cytokine-induced neutrophil chemoattractant production in lipopolysaccharide-stimulated *rat* alveolar macrophages is regulated by the polyamine, putrescine (Kamoi et al. 1997).

Arachidonic acid metabolism of THP-1 monocytic cells grown in RPMI supplemented with 10% foetal bovine serum, 100 U penicillin, 0.1 mg/ml streptomycin, and 2 mM L-glutamine was activated by purified lipopolysaccharide R595 from *S. minnesota* (PFAU et al. 1998).

Peptides

Glycyl-leucyl-phenylalanine

GATTEGNO et al. (1988)

Heptanoyl-γ-D-Glu-(L)meso-α,∈-A₂pm-(L)-AlaOH

Heptanoyl-γ-D-Glu-(L)meso-α,∈-A_2pm-(L)-AlaOH (FK-565) activated peritoneal macrophages harvested from C57BL/6 *mice* to become cytotoxic to syngeneic B16 melanoma cells (INAMURA et al. 1985).

Lauroyltripeptide (RP 56 142)

Lauroyltripeptide (RP 56 142) (N^2-[*N*-(N-lauroyl-L-alanyl)-γ-D-glutamyl]-L, L-2,6-diaminopimelamic acid) induced macrophage activation and enhanced cytotoxicity of natural killer cells in spleen, blood, and liver (FIZAMES et al. 1989).

Tuftsin (Threonyl-lysyl-prolyl-arginine)

The tetrapeptide tuftsin represents the active moiety of leukokinin, the leukophilic fraction of immunoglobulin G. It is derived from leukokinin in two steps by enzymatic cleavage, one by tuftsin endocarboxypeptidase occurring in the spleen, the other by leukokinase on the surface of phagocytic cells. Neutrophils and monocytes possess high-affinity binding sites for tuftsin which cross-react with substance P, suggesting that tuftsin receptors may be considered a subtype of substance P receptors (STABINSKY et al. 1978, BAR-SHAVIT et al. 1980, FRIDKIN and GOTTLIEB 1981, WATSON 1984). Tuftsin has been shown to activate superoxide radical ($O_2^{\bullet-}$) formation in *murine* macrophages with a biphasic concentration-response function (TRITSCH and NISWANDER 1982). The time course of thromboxane B_2 production by albumin-elicited *guinea pig* peritoneal macrophages differed from that of $O_2^{\bullet-}$ and H_2O_2 release by *Corynebacterum parvum*-induced cells in that considerable amounts were measurable at 2 h and plateau values were reached at 12 h (HARTUNG and TOYKA 1984).

Macrophage monolayers from *mice* injected intraperitoneally with thioglycollate broth pulsed *in vitro* with keyhole *limpet* hemocyanin-tuftsin conjugates exerted a stronger immunogenic response effect than keyhole *limpet* hemocyanin alone (DAGAN et al. 1987). *Bovine* serum albumin, which by itself was not immunogenic, when applied to macrophages as tuftsin conjugate evoked a high lymphoproliferative immune response. *In vivo*, *bovine* serum albumin conjugated to tuftsin, when injected in aqueous solutions, augmented significantly antibody production, whereas administration of *bovine* serum albumin alone or *bovine* serum albumin admixed with tuftsin had no immunogenic effect. Studies conducted to elucidate the mechanisms underlying the activation of the immunogenic function of macrophages by the peptide revealed that treatment of cells with antigen and tuftsin increases secretion of interleukin-1 and expression of cell surface Ia encoded antigens.

Antigen-tuftsin covalent conjugates, injected in aqueous solution intramuscularly or intravenously into *mice*, significantly augmented antibody production (DAGAN et al. 1988).

Tuftsin deficiency in *humans* has been observed by CONSTANTOPOULOS et al. (1972) and NAJJAR and CONSTANTOPOULOS (1973).

Further references: NAJJAR and NISHIOKA (1970), NAJJAR and CONSTANTOPOULOS (1972), NAJJAR (1975, 1985), SPIRER et al. (1977), TZEHOVAL (1978), TZEHOVAL et al. (1978), STABINSKY et al. (1980), BRULEY-ROSSET et al. (1981), MARTINEZ and WINTERNITZ (1981), NAJJAR et al. (1981), NISHIOKA et al. (1981), BAR-SHAVIT and GOLDMAN (1982), GOLDMAN and BAR-SHAVIT (1983), TRITSCH and NISWANDER (1983), CHIRIGOS and TALMADGE (1985), FRIDKIN and NAJJAR (1989), SOROKIN et al. (1989), KACHEL et al. (1996)

Dalargin (D-Ala-Gly-Phe-Leu-Arg)

Dalargin (10^{-3} to 10^{-9} M) stimulated the luminol-dependent chemiluminescence of *mouse* whole blood during phagocytosis of polystyrene latex beads (0.8 μm in diameter) recorded over 10 s (ROGOVINE and MUSHTAKOVA 1995).

Valyl-glutamyl-prolyl-isoleucyl-prolyl-tyrosine

GATTEGNO et al. (1988)

Synthetic Hexapeptide $C3a_{72-77}$

The synthetic hexapeptide $C3a_{72-77}$ induced generation of thromboxane B_2 from *guinea pig* peritoneal macrophages (HARTUNG et al. 1984). This effect is specific as it can be blocked by an appropriate antibody. It is not dependent on T cells. Phorbol myristate acetate given as a second stimulus excited release of large amounts of thromboxane B_2.

Substance P

The tachykinin substance P is an undecapeptide

H-Arg-Pro-Lys-Pro-Gln-Gln-Phe-Phe-Gly-Leu-Met-NH_2

structurally related to the formyl peptides in its C-terminal portion. Substance P binds to specific plasma membrane receptors which can be divided into subtypes (WATSON 1984, 1987, REGOLI et al. 1987). High-affinity binding sides have been identified on *guinea pig* macrophages (MARASCO et al. 1981). Substance P has been reported to activate the respiratory burst in *guinea pig* macrophages (HARTUNG and TOYKA 1983). Activation of NADPH oxidase (EC 1.6.99.6) by substance P is accompanied by phospholipase C activation and Ca^{2+} mobilisation. Cytochalasin B enhances substance P-induced phosphoinositide turnover and Ca^{2+} mobilisation, but somewhat surprisingly, cytochalasin B inhibits the respiratory burst (SERRA et al. 1988). Unlike the respiratory burst induced by fMet-Leu-Phe, that induced by substance P is only partially pertussis toxin-sensitive (SERRA et al. 1988).

Neurotensin (H-pyroGlu-Leu-Tyr-Glu-Asn-Lys-Pro-Arg-Arg-Pro-Tyr-Ile-Leu-OH)

BAR-SHAVIT et al. (1982) found the specific binding of [^{3}H]neurotensin to thioglycollate-elicited *mouse* peritoneal macrophages cultivated *in vitro* to be concentration dependent. Neurotensin was competitively displaced by tuftsin, substance P (SP) and SP 81–84.

Gramicidin (HCO-Val-Gly-Ala-D-Leu-Ala-D-Val-Val-D-Val-Trp-D-Leu-Trp-D-Leu-Trp-D-Leu-Trp-$NHCH_2CH_2OH$)

Gramicidins are linear pentadecapeptide ethanolamide antibiotics with a formyl group at the N-terminus (BAMBERG et al. 1976). Gramicidin is similarly potent but less effective than fMet-Leu-Phe. A competitive antagonist at formyl peptide receptors prevents activation by gramicidin, suggesting that this peptide is partial agonist at formyl peptide receptors (JACOB 1988).

Gliadin

Gliadin is an ethanol soluble fraction of gluten. Synthesized gliadin dodecapeptide sequence FQQPQQQYPSSQ elicited the highest TNF-α, IL-10, and RANTES secretion and increased IFN-γ-primed NO production by female Balb/c/Ph *mouse* macrophages (TUČKOVÁ et al. 2002).

Cyclomunine

Cyclomunine is a hexacyclodepsipeptide extracted from the fungus *Fusarium equiseti*. *In vitro* incubation of resident peritoneal macrophages from *rats* with 10^{-1} to 10^{-6} μg cyclomunine/ml for 3–24 h significantly increased the intracellular levels of proteins and lysosomal β-D-glucuronidase, as well as the neutral protease release and glucosamine incorporation (JOSEPH et al. 1981). Superoxide anion generation during zymosan phagocytosis was also increased by preincubation of these cells with cyclomunine.

Cyclosporin

Cyclosporin is a cyclic hendecapeptide extracted from the fungus *Trichoderma polysporum*. There is some evidence that this immunosuppressant acts by interfering with the process by which antigens raise intracellular Ca^{2+} levels (BUTTON and PALACIOS 1982). Interleukin 2 producing T cells do not express receptors for interleukin 1 in the presence of cyclosporin A. Thus, this subset of T cells is rendered unable to produce and/or secrete interleukin 2, which is a requirement for T cell proliferation and growth. When T cells are rendered unresponsive to interleukin 2, progression of T cell help for development of T cytotoxic cells, which are instrumental in rejecting foreign tissue is effectively stopped. It does not appear to inhibit T suppressor cells; rather through its selective toxicity for T helper cells, it permits a greater expression of the suppressive reactivity. In addition, the lymphokine, macrophage activating factor thought to be synonymous with interferon γ, may not be released in the presence of cyclosporin A.

Pulmonary eosinophilia in allergic *mice* (challenged with nebulized ovalbumin) was inhibited by cyclosporin A when given orally at 24 h and 1 h before challenge (UMLAND et al. 1999). There was a complete, dose-dependent inhibition of eosinophils in the bronchoalveolar lavage fluid of challenged animals, with only a partial reduction in interleukin-5 mRNA levels in the lung. For each concentration of cyclosporin A tested there was a greater reduction in eosinophils than of IL-5.

Cyclosporin A, an inhibitor of cyclophilin (TAKAHASHI et al. 1989), does not inhibit but slows the folding of the monomeric protein transferrrin in the endoplasmic reticulum (LODISH and KONG 1991), since it is active in concentrations which only block a minor percentage of the cell's peptidyl-prolyl isomerases. Its complex with peptidyl-prolyl isomerase interferes with signal transduction pathways of eukaryotic cells, and the inhibitor mediates pharmacological and toxic effects primarily by interfering with regulatory functions in the cell (KUNZ and HALL 1993). The same cyclophilin can act in a late step of protein folding by acting as a prolyl isomerase, but also in an early step, by preventing the aggregation of a denatured substrate protein. Both activities are inhibited in the presence of cyclosporin A (FRESGARD et al. 1992). Cyclosporin A slows collagen triple-helix formation in vivo (STEINMANN et al. 1991). This was accompanied by a remarkable post-translational overmodification and increased degradation.

Reactive oxygen species generated during its metabolism and inducing microsomal lipid peroxidation were shown to be a possible explanation for the toxicity of this immunosuppressive drug (SERINO et al. 1993).

All seven subsets of epithelial cells were markedly changed in the *rat* thymus after oral application of 30 mg cyclosporin A per kg body weight for 21 consecutive days (MILICEVIC and MILICEVIC 1997). The abundance of cytokeratin bundles was registered in type 1-‚subcapsular/paraseptal/perivascular' - epithelial cells.

Diabetes mellitus spontaneously occurring in 40–60 % of colony of BioBreeding/Worcester *rats* was significantly reduced in its frequency and delayed in its onset by pre-treatment of susceptible animals for 10-day intervals prior to 70 days of age with cyclosporin A (LIKE et al. 1984).

Cyclosporin A exhibits very little myelotoxicity and does not interfere with phagocytosis by macrophages. Cardiotoxicity of the drug was seen in swine (TOPALIDIS et al. 1989). The ready reversibility of both the nephrotoxicity (SIBLEY et al. 1983, MIHATSCH et al. 1988, 1989) and hepatotoxicity (ANDRÉS et al. 2000) of cyclosporin A has aided its acceptance as a drug of choice.

Poly-L-lysine

Poly-*L*-lysine treatment of lipopolysaccharide-coated particles greatly increased the attachment rate of thioglycolate-elicited peritoneal macrophages of 7-week-old female *mice* (strain ddY) closely correlated with the difference in surface charge density of the particles caused by such treatments, increasing with the decrease in negative charge density (MATSUI et al. 1983).

Poly-*L*-lysine appeared as a potent dissolver of preformed β-amyloid fibrils *in vitro* (NGUYEN et al. 2002). Its efficiency is instantaneous. Poly-*L*-lysine can be used as a universal dissolver of all types of oligomeric β-sheet conformation, precursor of the fibrils.

Poly-L-glutamic Acid

AXLINE et al. (1973)

Glycopeptides

Muramyl dipeptide (*N*-acetyl-muramyl-L-alanyl-D-isoglutamine), which has been synthesized along with hundreds of analogues, has been shown to be effective at both enhancing and suppressing the immune response.

Macrophages were stimulated by muramyl dipeptide to induce polymorphonuclear leucocyte accumulation in the peritoneal cavities of *guinea pigs* (NAGAO et al. 1990).

Muramyl dipeptide-induced immunosuppression appears to be mediated through the generation of suppressor T cells (LECLERC et al. 1984).

Mouse alveolar macrophages were rendered tumoricidal after the intravenous administration of liposomes containing muramyl tripeptide phosphatidyl-ethanolamine, a lipophilic derivative of muramyl dipeptide (FIDLER et al. 1989).

Nucleotides

Isoprinosine

Isoprenosine, a complex of *p*-acetamidobenzoic acid, *N,N*-dimethylamino-2-propanol and inosine at a molar ratio of 3:1, is a white crystalline powder, slightly bitter in taste, soluble in water at room temperature to an extent of 25 % (250 mg/ml) and stable in neutral solution. The dosage commonly administered to *man* is 500 mg tablets (HADDEN and GINER-SOROLLA 1981). Of particular importance is the metabolic lability of the inosine moiety of isoprinosine, the half life of which is 3 min and 50 min, respectively, following intravenous and oral administrations in the *rhesus monkey*. More than 90 % of the labelled inosine is excreted as allantoin and uric acid.

Further references: HADDEN (1978), HADDEN et al. (1979), HADDEN and GINER-SOROLLA (1981)

NPT 15392 [erythro-9(2-hydroxy, 3-nonyl) hypoxynthine]

NPT 15392 augmented lymphokine-induced *guinea pig* macrophage proliferation and phagocytosis-induced chemiluminescence (HADDEN and GINER-SOROLLA 1981).

5-Iodo-2'-deoxyuridine

ANACLERIO et al. (1976)

Synthetic Oligonucleotides

Pre-treatment of *murine* bone marrow macrophages, J774 and RAW264.7 macrophages with synthetic 30-mer phosphorothionate oligonucleotides inhibited NO production induced by CpG oligonucleotides, *E. coli* DNA, or lipopolysaccharide (ZHU et al. 2002).

Synthetic Polynucleotide Complexes

Synthetic polynucleotide complexes have been shown to be effective immune response modulators in animals and *man* (BRAUN et al. 1971, JOHNSON 1979). The polynucleotides are formed following the action of an enzyme, polynucleotide phosphorylase on the synthetic mononucleotide diphosphates. Complexing takes place following the mixing of polymers composed of opposite base pairs. Two have been utilised, polyinosinic acid complexed with polycytidylic acid (poly I · poly C) and polyadenylic acid complexed with polyuridylic acid (poly A · poly U). The single strands mononucleotides are ineffective.

Polyinosine-Polycytidylic Acid

As polyinosine-polycytidylic acid failed to enhance killing of *Escherichia coli* by *mouse* peritoneal macrophages *in vitro*, THALINGER and MANDELL (1972) suggested that it acted not directly on mononuclear phagocytes.

Polyriboinosinic-polyribocytidylic acid augmented resistance of thioglycolate-elicited inflammatory macrophages to infection with herpes simplex virus type 1 (PYO et al. 1991). Polyriboinosinic-polyribocytidylic acid antiviral activity was completely abrogated by antibodies to interferon-β whereas antibodies to other interferons or to other cytokines had no effect.

LEE et al. (1990) used a murine mammary tumour model to test the efficacy of a combination of heparin and the interferon inducer, polyriboinosinic-polyribocytidylic acid on spontaneous metastasis from a subcutaneous primary tumour and on experimental metastasis following intravenous injection of tumour cells. This treatment had no effect on the growth of primary tumours, but lung metastases arising from these tumours were reduced.

Further references: EVANS and ALEXANDER (1976), HAMBURG et al. (1978), PUGH-HUMPHREYS and THOMSON (1979), MARTINEZ et al. (1980), TARAMELLI and VARESIO (1981), REIDARSON et al. (1982), CHIRIGOS and TALMADGE (1985), CHIRIGOS et al. (1985), SOUVANNAVONG and ADAM (1990), COHEN et al. (1997)

Polyadenyl-Polyuridylic Acid

The double stranded polynucleotide, poly A:U, when administered intraperitoneally at the same time as intravenous infection with *Brucella abortus*, suppressed the growth of that organism in the spleen and liver of *mice* (MADRASO and CHEERS 1978). The antiviral function of spleen cells induced by poly A:U was evident in the supernatant fluid when cultured for 48 h at 37 °C (LEE et al. 1992). *Murine* cytomegalovirus-induced plaques were reduced to 52 and 5 % of controls in the plaque assays performed at 37° and 40°, respectively.

Further references: EVANS and ALEXANDER (1976), PUGH-HUMPHREYS and THOMSON (1979), CHIRIGOS and TALMADGE (1985)

Compounds Containing a Sulphur Moiety

Both levamisole and isoprinosine have received extensive *in vivo* immunopharmacologic testing in animals and *man*. Many of the immune functions that are affected by thymic hormones are shown also to be affected by levamisole, and isoprinosine *in vitro* and have been confirmed *in vivo* following administration to experimental animals or *human* subjects (SPECTER and HADDEN 1985). Both have additional actions on macrophages which have not been tested extensively for the thymus hormones. Regarding their therapeutic usefulness, both have limitations as they are relatively mild in their actions. Levamisole treatment often takes weeks to months to achieve effects, and agranulocytosis as a side effect is significant.

Some of the other sulphur-containing compounds are less toxic and may be more potent.

Levamisole (*L*-2,3,5,6-Tetrahydro-6-phenyl-imidazo-[2,1-*b*]thiazole-HCl)

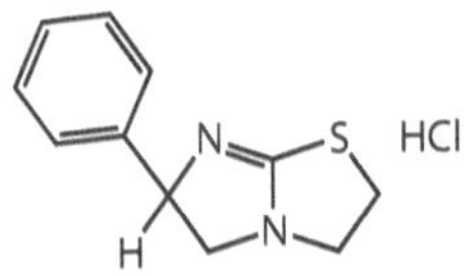

Levamisole [260]

Levamisole is rapidly absorbed by both oral and parenteral routes, with extensive metabolism and rapid excretion of both unchanged drug and metabolites in urine and faeces of *rats*. Two major metabolites and six minor ones have been identified in experimental animals. In *man*, most of the urinary activity consists of hydrophilic compounds: *p*-hydroxy-levamisole {(+)-2,3,5,6-tetrahydro-6-(4-hydroxyphenyl)imidazo [2,1-*b*]thiazole} and its glucuronide conjugate have been identified (ADAMS 1978). A possible mechanism for the formation of (–)-2-oxo-3-(mercaptoethyl) -5-phenylimidazolidine, one of the major metabolites of levamisole, has been proposed by VAN BELLE and JANSSEN (1979). α-Ketoaldehydes, particularly glyoxal and methylglyoxal, are specific catalysts for the hydrolysis of levamisole to (–)-2-oxo-3-(mercaptoethyl)-5-phenylimidazolidine. The proposed mechanism involves a charge-transfer complex followed by a concerted mechanism in which water is carried from the hydrated aldehyde to levamisole with concomitant ring opening.

Further references: RÈNOUX and RÈNOUX (1971), SCHUERMANS (1975), VAN GINCKEL and HOEBEKE (1975), HUSKISSON et al. (1976), RENOUX et al. (1976), SCHMIDT and DOUGLAS (1976), DAUGHADAY et al. (1977), SYMOENS and ROSENTHAL (1977), HADDEN (1978), KELLY (1978), NATHANSON et al. (1978), PLEWIG and LUDERSCHMIDT (1978), HADDEN et al. (1979), BOURA et al. (1984), JOHNSON and REGAL (1985), J.H. SCHILLER et al. (1991), KIMBALL et al. (1992); against: ZIPRIN et al. (1977)

Diethyldithiocarbamate

Sodium diethyldithiocarbamate inhibits prostaglandin synthetase at ~10 μg/ml. The site of action is not at the fatty acid substrate site. It is a copper chelating agent and probably prevents the interaction of oxygen with the enzyme.

Thiabendazole [2-(4-thiazolyl)-benzimidazole]

Thiabendazole primarily used as an anthelmintic in male ICR *mice* depleted of glutathione by treatment with DL-buthionine sulphoximine was much more nephrotoxic than in females (MIZUTANI et al. 1992).

Thioglycollate

KARL et al. (1973), RABINOVITCH and DE STEFANO (1973), GORDON et al. 1974), UNKELESS et al. (1974), BIANCO et al. (1975), EDELSON et al. (1975), WERB and GORDON (1975), WELLEK et al. (1977), Dy and ASTOIN (1978), DY et al. (1978, 1979), SCHNYDER and BAGGIOLINI (1978), SWENSON and KOZEL (1978), TANSEY et al. (1978), VAN DER ZEIJST et al. (1978), WERB et al. (1978), AHO et al. (1979), ERICKSON and HU (1979), JONES and SCOTT-BURDEN (1979), VETVICKA et al. (1979), DAHLGREN et al. (1980), HAGMANN and FISHMAN (1980), JOHNSON and BALISH (1980), MIYASAKA et al. (1980), PHAIRE-WASHINGTON et al. (1980), SAINT-GUILLAIN et al. 1980), WEIDEMANN et al. (1980), BELLER and UNANUE (1981), FORNUSEK et al. (1981), IMAI and TANAKA (1981), KHOO et al. (1981), VRAY et al. (1981), WALDREP and REESE (1981), ROUBIN et al. (1982), THYBERG et al. (1982), BAGGIOLINI et al. (1983), BEELEN and WALKER (1983), BERTON and GORDON (1983), DAHA and VAN ES (1983), GOTTLIEB et al. (1983), WERB and CHIN (1983), HUME and GORDON (1984), DE BAETSELIER et al. (1985), MELLMAN and UKKONEN (1985), METZGER et al. (1986), VEERHUIS et al. (1986), MORSE and MOORE (1988), NEWBURG et al. (1988), ABE et al. (1990), GOLDIN et al. (1993), TRITSCH and EVANS (1993), ZDOLSEK et al. (1993), COSTA ROSA et al. (1994), BRISSEAU et al. (1995), WATANABE et al. (1998), HAN et al. (2001)

DL-2-Oxo-3-(2-mercaptoethyl)-5-phenylimidazolidone

VAN GINCKEL and DE BRABANDER (1979)

Carbohydrates

Glycogen

HEIDELBERGER et al. (1954), COMOLLI and PERIN (1963), VAUGHAN (1965), BOROWSKI et al. (1972), KURISU et al. (1978), STALDER and SCHRÄDER (1980), CIANCIOLA and SNYDERMAN (1981), MANTOVANI (1981), GLASS et al. (1983), YAMAZAKI et al. (1983), GLASS et al. (1984), BRADLEY et al. (1989)

Zymosan

α-Mannans and β-glucans are the major constituents of zymosan, which is prepared from yeast cell walls (PILLEMER and ECKER 1941, PHAFF 1963, BACON et al. 1969).

The mannosyl-fucosyl receptor has broad specificity and is able to bind various glycoconjugates in the order L-fucose = D-mannose > *N*-acetyl glucosamine ≈ D-glucose > D-xylose >> D-galactose = L-arabinose = D-fucose (STAHL et al. 1980, SHEPHERD et al. 1981). The ligand binding site of type 3 complement receptors is involved in macrophage-zymosan binding and uptake with Fab fragments of anti-C3 antibodies and monoclonal antireceptor antibodies M01 and OKM10 (EZEKOWITZ et al. 1985). Unopsonized zymosan is a poor trigger of respiratory burst activity in 7-d adherent *human* blood monocyte-derived macrophages, but induced cell aggregation and secretion of large amounts of superoxide anion, when these cells were co-cultivated with neutrophils in serum-free medium and challenged with zymosan. *Mouse* serum-treated zymosan activated the respiratory burst in primed *murine* macrophages (BERTON and GORDON 1983).

Measuring the production of oxygen free radicals in response to zymosan, KREIPE et al. (1988) found a 20-30 fold increase in multinucleated giant cells when compared with unfused cultured *human* blood monocyte-derived macrophages. This augmentation, however, could not be noted when chemiluminescence was related to the nuclei counted in the syncytia. Opsonization of zymosan is not necessary for stimulation of the *murine* bronchoalveolar macrophage oxidative burst (SUGAR and FIELD 1988). In *rat* cerebral microglia cell cultures, opsonized zymosan produced superoxide radical anions (COLTON and GILBERT 1987).

By injecting zymosan A (50 mg) into an air pouch on male CD (Sprague-Dawley) and Fisher 344 *rats*, SAMS et al. (2000) found significant increases ($P < 0.01$) in 8-hydroxy-2'-deoxyguanosine after 1 day in the DNA from cells lining the air pouch from zymosan-treated versus control rats. By 28 days, 8-hydroxy-2'-deoxyguanosine levels had returned to background in Sprague-Dawley *rats*, but remained elevated in F-344 *rats*.

Pre-treatment of murine bone-marrow derived macrophages with 10 U/ml murine IL-4 for 48 h enhanced the respiratory burst following subsequent stimulation with 10^{-6} M phorbol myristate acetate or 1 mg/ml zymosan (PHILLIPS et al. (1990).

Zymosan-bound histamine activated $O_2^{\bullet-}$ formation in *guinea pig* alveolar macrophages via H1 receptors (DIAZ et al. 1979).

After intravenous injection of quartz dust (particle size 1–2 μm), a serum factor appeared in 8 out of 10 *rabbits* which agglutinated zymosan to a dilution of 1:256 (PERNIS et al. 1959).

Further references: RIGGI and DI LUZIO (1961), DICARLO et al. (1963), WISSE (1974), RINGROSE et al. (1975), DEAN et al. (1979), BITTER-SUERMANN (1980), CASTRANOVA et al. (1980, 1981, 1994), CZOP et al. (1981), ROUBIN and BENVENISTE (1981), ROUZER et al. (1981, 1982), BERRY (1982), FELS et al. (1982), SCHADE and RIETSCHEL (1982), BAGGIOLINI et al. (1983), BAXTER et al. (1983), BERTON and GORDON (1983), BODMER and DEAN (1983), CHANG et al. (1983), SCHOPF and LEMMEL (1983), D'ONOFRIO and LOHMANN-MATTHES (1984), EZEKOWITZ et al. (1984, 1985), PETERS-GOLDEN et al. (1984), PATERSON et al. (1985), DE MAROUSSEM et al. (1986), GAIROLA and TAI (1986), KADISH et al. (1986), COLTON and GILBERT (1987), PETERS-GOLDEN and THEBERT (1987), ANDRE et al. (1988), KREIPE et al. (1988), KUROIWA et al. (1988), SUGAR and FIELD (1988), SCHROERS and TILKES (1990), JORENS et al. (1991), POLZER et al. (1992, 1993), HUWILER and PFEILSCHIFTER (1993), COSTA ROSA et al. (1994), TAPPER and SUNDLER (1995), HELMKE et al. (1996), SHERMAN et al. (1996), SPORN et al. (1996), COHEN et al. (1997), USHIJIMA et al. (1997), BLAKE et al. (1998), FERNÁNDEZ et al. (1999), NAGAISHI et al. (1999), AKIBA et al. (2002)

Glucan (β-1,3-poly-glucopyranose)

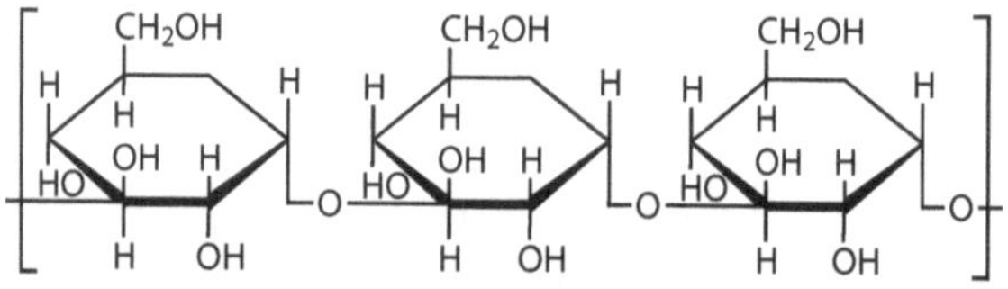

Glucan (β-1,3-poly-glucopyranose) [261]

Macrophage activation with yeast glucan induced a pronounced elevation of colony stimulating activity in serum (SATOH et al. 1982, CHIHARA 1983). The increase of colony stimulating activity was dose dependent. While doses of 62.5 mg krestin (β-1,4; β-1,3; β-1,6, protein complex; PSK)/kg and 125 mg PSK/kg produced no increase in colony stimulating activity, a dose of 250–1000 mg/kg was associated with a rapid increase in colony stimulating activity followed by a rapid decline. Yeast glucan has been demonstrated to be very effective in enhancing CSF at doses where PSK is ineffective (PATCHEN and MCVITTIE 1983).

Pulmonary granuloma formation induced by infusion of 5 mg particulate glucan/*rat* was markedly reduced when neutrophils were depleted by *rabbit* antiserum directed against *rat* peripheral blood neutrophils or by catalase (15,000 U/*rat*) at the time of injection and 24 h after glucan infusion (KILGORE et al. 1997). Neutrophils and reactive oxygen intermediates (H_2O_2) are required for the local induction of monocyte attractant protein-1 to be secreted by endothelial cells.

Further references: Di Luzio and Bierman (1964), Mansell et al. (1975), Browder et al. (1976, 1977, 1978), Di Luzio (1976, 1977, 1983, 1985), Di Luzio et al. (1976, 1978, 1984), Burgaleta and Golde (1977), Burgaleta et al. (1978), Cook et al. (1978, 1980, 1982), Di Luzio and Williams (1978, 1979, 1980), Kokoshis et al. (1978), Niskanen et al. (1978), Stiteler et al. (1978), Williams et al. (1978), Deimann and Fahimi (1980), Way et al. (1980), Bärlin et al. (1981), Seljelid et al. (1981), Agostini et al. (1982), Benach et al. (1982), Chihara (1983), Patchen and MacVittie (1983), Janiak et al. (1984), Kimura et al. (1984), Patchen et al. (1984), Lew et al. (1986), Sherwood et al. (1986), Ralston and Rohrbach (1994), Kilgore et al. (1997)

Lentinan

Lentinan is a β-1,3-glucan having two β-1,6-glucopyranoside branchings for every five β-1,3-glucopyranoside linear linkage (Sasaki et al. 1976). According to Talmadge and Chirigos (1985) there is no influence on the macrophage-mediated cytotoxicity *in vitro* and *in vivo*.

Alginate

Alginates are polysaccharides wih gel-forming properties composed of 1,4-linked β-D-mannuronic acid, α-L-guluronic acid and alternating β-D-mannuronic acid-α-L-guluronic acid blocks. Alginates stimulated *human* monocytes to produce high levels of tumour necrosis factor-α, interleukin-6 and interleukin-1 (Otterlei et al. 1991).

Scholler Lignin

The substance has an unusually high adsorbing power; cell activation could possibly by achieved through physico-chemical influences on the cell and/or intracellular membranes (Flemming 1974).

Further references: Graack and Flemming (1966), Flemming (1967), Flemming and Graack (1967)

Chitin Derivatives

Chitin-based tissues are absent in the *human* body, but *N*-acetylglucosamine, the repeating unit of chitin, and chitobiose are present in glycosaminoglycans and in glycoproteins (Muzzarelli et al. 1986). About 1 % of the secretory proteins of macrophages is chitinase (Boot et al. 1995). Escott and Adams (1995) showed experimentally that on differentiation of the monocytes into macrophages *in vitro*, chitinase is released into the growth medium. Chitinase activity was found to be associated with the monocyte cell line, THP-1 following culture with colloidal chitin (Escott et al. 1996).

The induction of cytotoxic macrophages was enhanced by an increase of negative charge at O-6 and decreased by further modification at O-3 of the glucopyranose-*N*-acetyl residue of *O*-(carboxymethyl)chitins (Nishimura et al. 1986). *O*-(Carboxymethyl)chitins had a minor effect on mitogenic activity that was independent of the site of modification; partially *N*-deacetylated chitins had little activity. Although there was remarkable enhancement of accessibility to lysozyme upon modification at O-6 of the glucopyranose-*N*-acetyl residue, the accessibility was decreased by further substitution of O-3.

The secretion of enzymes, interleukins, tumour necrosis factor, nitric oxide, peroxide and other compounds by activated macrophages provides macroscopic evidence of the biochemical significance of chitosans (Muzzarelli 1996). Various chitosans were found to induce production of tumour necrosis factor-α in human monocytes (Otterlei et al. 1994). Lipopolysaccharides and water-soluble chitosans recognise a binding site on monocytes which involves CD14, a receptor of lipopolysaccharides also present in macrophages. Gorbach et al. (1994) found immunostimulating and antitumour activities in chitooligosaccharide derivatives containing *N*-linked side chains which mimic the lipopolysaccharides.

Intravenous administration of phagocytable chitin particles (1–10 μm) in C57BL/6 *mice* and SCID *mice* primed alveolar macrophages within 3 days up to a 50-fold increase in their oxidative burst when elicited *in vitro* with phorbol myristate acetate (Shibata et al. 1997). C57BL/6 *mice* pre-treated with monoclonal antibodies against mouse γ-interferon showed a markedly decreased level of alveolar macrophage priming following injection of chitin particles.

The consequences of chitosan application to living tissues could possibly go far beyond the enhanced production of nitric oxide by macrophages; additional NO could be produced chemically, but some could be consumed as nitrite by the chitosan amino groups. Thus, extra amounts of NO could be beneficial and help resorb chitosan. The presence of chitosan could presumably have some as yet undocumented consequence on the iron-enzymes (Muzzarelli 1997).

Polymers

Pyran (Divinylether-maleic Anhydride) *Copolymer*

MERIGAN and REGELSON (1967), REGELSON (1967), BRAUN et al. (1970), MUNSON et al. (1970), KAPILA et al. (1971), HIRSCH et al. (1972), KAPLAN et al. (1974), REGELSON et al. (1974), BAIRD and KAPLAN (1975), HARMEL and ZBAR (1975), HERLING (1975), ELZAY and REGELSON (1976), MOHR et al. (1976), SCHULTZ et al. (1976, 1977), LEVINE et al. (1977), MAJESKI and STINNETT (1977), PUCCETTI et al. (1979), STINNETT et al. (1979), KAPLAN et al. (1980), MILLER et al. (1980), BÄRLIN et al. (1981), DEAN et al. (1981), LOVELESS and MUNSON (1981), MÄNNEL (1982), ADAMS et al. (1983), WERB and CHIN (1983), JANIAK et al. (1984), KATAOKA and OH-HASHI (1985), KUUS et al. (1985)

Cyclohexyl-1,3-dioxepin Maleic Anhydride Copolymer

KUUS et al. (1985)

4-Methyl-2-pentenoyl Maleic Anhydride Copolymer

KUUS et al. (1985)

Poly-2-vinylpyridine-N-oxide

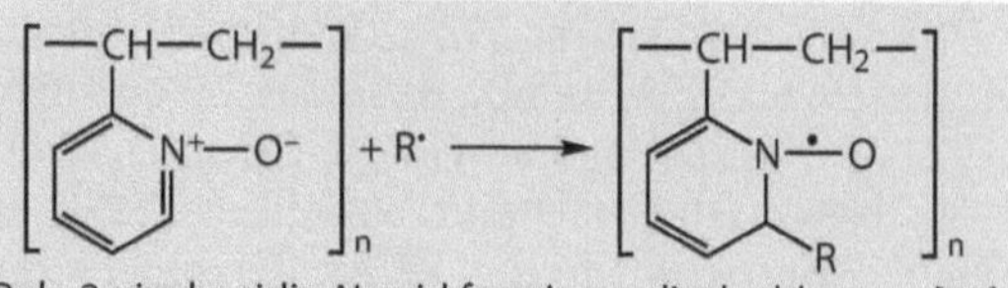

Poly-2-vinylpyridin *N*-oxid forming radical adducts [63]

FLEMMING and NOTHDURFT (1968), NOTHDURFT and FLEMMING (1970), FLEMMING (1974)

Polyacrylic Acid

Synthetic polyanions derived from polyacrylic acid (REGELSON et al. 1960, 1974, SAKAMOTO et al. 1982) having molecular mass $< 10^4$ are potent stimuli to macrophage phagocytic function and exhibit anti-tumour activity. Polymers with higher molecular mass inhibit macrophage function, and only those with molecular mass $> 15{,}000$ have antiviral and immunological stimulating capacities (REGELSON 1979). KIEF et al. (1974) 24 h after a single subcutaneous injection of 10 mg polyacrylic acid (~15 kDa) per animal saw many vacuoles in the Kupffer cells of *mice* in connection with a histochemically demonstrable activation of acid phosphatase.

In *mouse* liver microsomes, cytochrome P-450_{47} increased after monomeric methylmetacrylate at a dose of 45 mg/kg × day administered intraperitoneally for 4 days but was totally repressed at a dose of 600 mg/kg × day, while the total amount of cytochrome P-450 was unaltered (NILSEN et al. 1978).

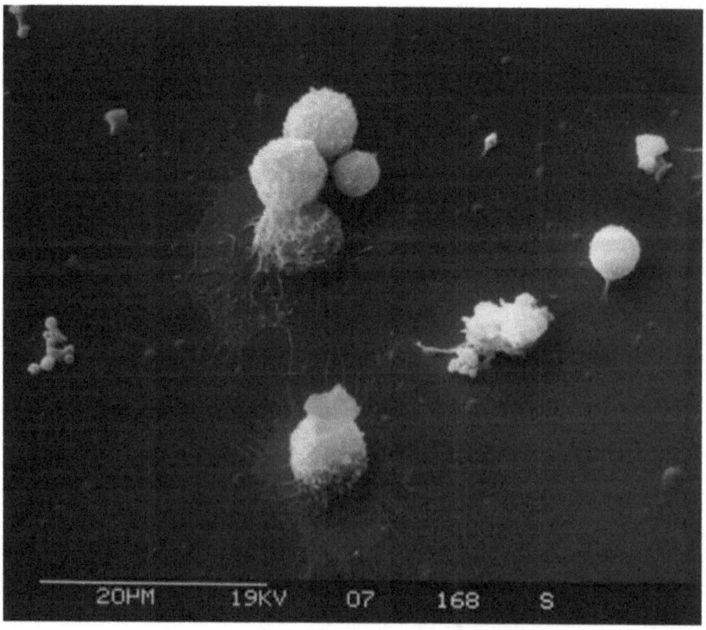

Fig. 171. Two non-immunologically stimulated peritoneal macrophages from a SPF-NMRI *mouse* 2 h after a single intraperitoneal injection of 150 mg isotactic polyacrylic acid (Vi 2329 molecular mass 8–12 kDa, MÜCK et al. 1977) as sodium salt per kg body weight. The produced ascites were centrifuged for 10 min at 1000 r.p.m. and incubated in TCM 199 + 20 % foetal calf serum on a plastic cover slip for 2 h, rinsed with phosphate buffered saline after Dulbecco and fixed in 1 % glutaraldehyde in 0.07 M phosphate buffer (pH 7.4). Ethanol, amyl acetate. Critical point drying. Gold coating. Cambridge Stereoscan 150 operated at 19 kV. APh-R. 856/80, negative 07 168. – The cells are rounded showing surface ruffles and attach themselves to the underlying substrate by means of thin veils of cytoplasm spreading beneath the dome-shaped nuclear pole.

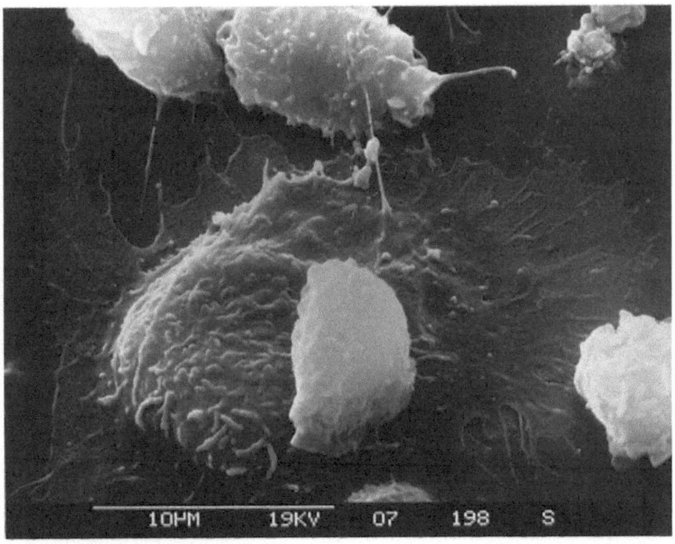

Fig. 172. Spreading of a non-immunologically stimulated peritoneal macrophages from a SPF-NMRI *mouse* 8 h after a single intraperitoneal injection of 150 mg isotactic polyacrylic acid (Vi 2329 molecular mass 8–12 kDa, MÜCK et al. 1977) as sodium salt per kg body weight. The produced ascites were centrifuged for 10 min at 1000 r.p.m. and incubated in TCM 199 + 20 % foetal calf serum on a plastic cover slip for 2 h, rinsed with phosphate buffered saline after Dulbecco and fixed in 1 % glutaraldehyde in 0.07 M phosphate buffer (pH 7.4). Ethanol, amyl acetate. Critical point drying. Gold coating. Cambridge Stereoscan 150 operated at 19 kV. APh-R. 859/80, negative 07 198. (from SCHILLER 1982)

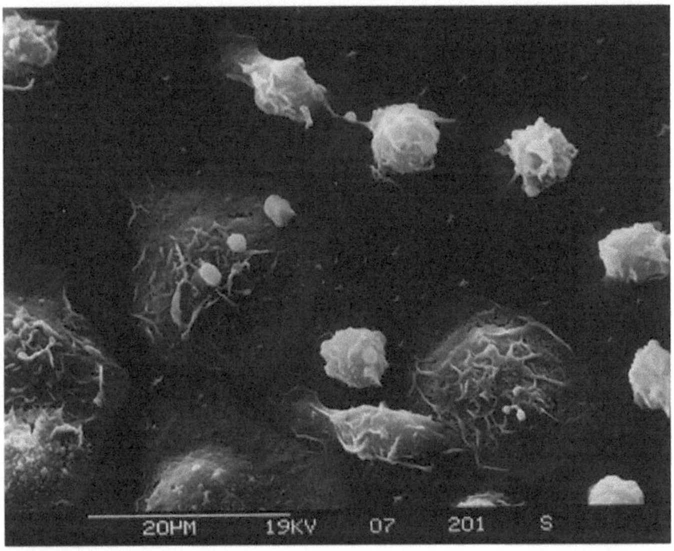

Fig. 173. Spreading of some non-immunologically stimulated peritoneal macrophages from a SPF-NMRI *mouse* 24 h after a single intraperitoneal injection of 150 mg isotactic polyacrylic acid (Vi 2329; molecular mass 8–12 kDa; cf. MÜCK et al. 1977) as sodium salt per kg body weight. The produced ascites were centrifuged for 10 min at 1000 r.p.m. and incubated in TCM 199 + 20 % foetal calf serum on a plastic cover slip for 2 h, rinsed with phosphate buffered saline after Dulbecco and fixed in 1 % glutaraldehyde in 0.07 M phosphate buffer (pH 7.4). Ethanol, amyl acetate. Critical point drying. Gold coating. Cambridge Stereoscan 150 operated at 19 kV. APh-R. 861/80, negative 07 201

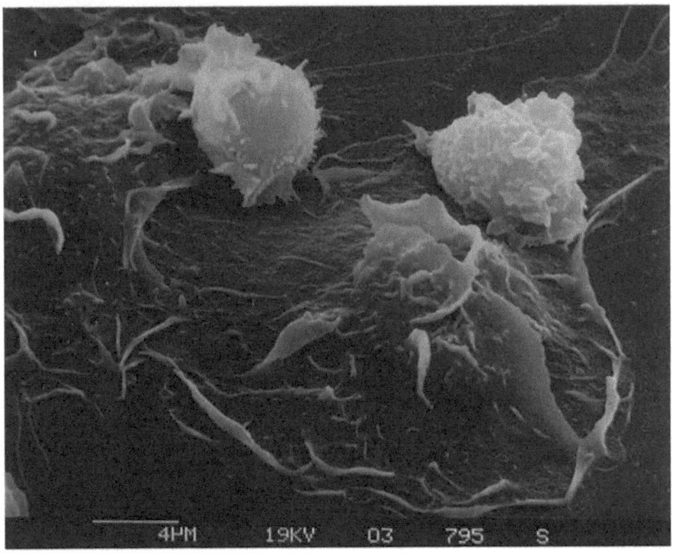

Fig. 174. Spreading of a non-immunologically stimulated peritoneal macrophages from a SPF-NMRI *mouse* 48 h after a single intraperitoneal injection of 150 mg isotactic polyacrylic acid (Vi 2329; molecular mass 8–12 kDa; cf. MÜCK et al. 1977) as sodium salt per kg body weight. The produced ascites were centrifuged for 10 min at 1000 r.p.m. and incubated in TCM 199 + 20 % foetal calf serum on a plastic cover slip for 2 h, rinsed with phosphate buffered saline after Dulbecco and fixed in 1 % glutaraldehyde in 0.07 M phosphate buffer (pH 7.4). Ethanol, amyl acetate. Critical point drying. Gold coating. Cambridge Stereoscan 150 operated at 19 kV. APh-R. 101/80, negative 03 795. (from SCHILLER 1982)

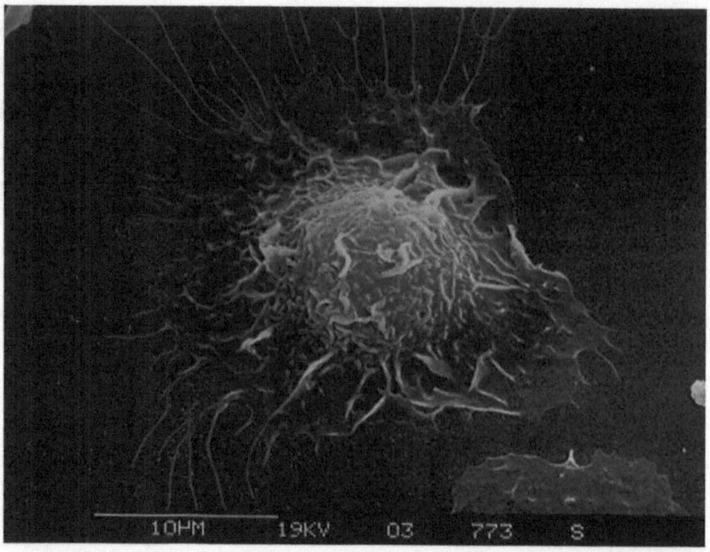

Fig. 175. Non-immunologically stimulated peritoneal macrophages from a SPF-NMRI *mouse* 48 h after a single intraperitoneal injection of 150 mg isotactic polyacrylic acid (Vi 2329; molecular mass 8–12 kDa; cf. MÜCK et al. 1977) as sodium salt per kg body weight. The produced ascites were centrifuged for 10 min at 1000 r.p.m. and incubated in TCM 199 + 20 % foetal calf serum on a plastic cover slip for 2 h, rinsed with phosphate buffered saline after Dulbecco and fixed in 1 % glutaraldehyde in 0.07 M phosphate buffer (pH 7.4). Ethanol, amyl acetate. Critical point drying. Gold coating. Cambridge Stereoscan 150 operated at 19 kV. APh-R. 101/80, negative 03 773. (from SCHILLER 1982)

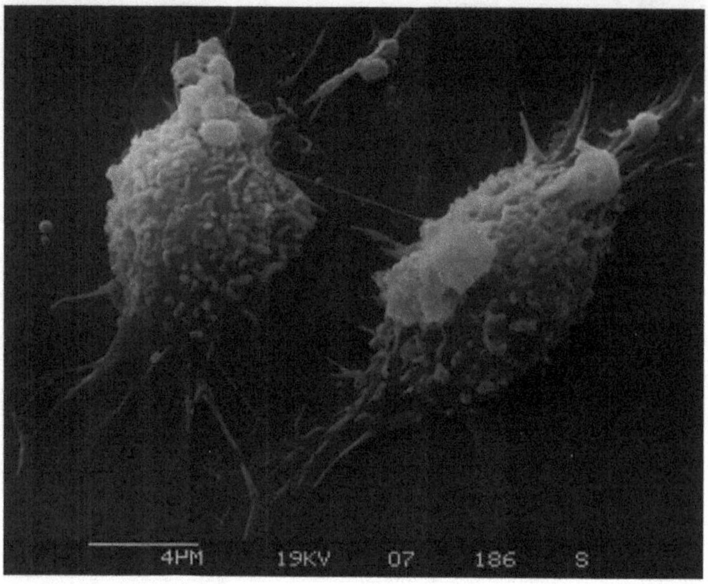

Fig. 176. Spreading of two non-immunologically stimulated peritoneal macrophages from a SPF-NMRI *mouse* 48 h after a single intraperitoneal injection of 150 mg isotactic polyacrylic acid (Vi 2329; molecular mass 8–12 kDa; cf. MÜCK et al. 1977) as sodium salt per kg body weight. The produced ascites were centrifuged for 10 min at 1000 r.p.m. and incubated in TCM 199 + 20 % foetal calf serum on a plastic cover slip for 2 h, rinsed with phosphate buffered saline after Dulbecco and fixed in 1 % glutaraldehyde in 0.07 M phosphate buffer (pH 7.4). Ethanol, amyl acetate. Critical point drying. Gold coating. Cambridge Stereoscan 150 operated at 19 kV. APh-R. 856/80, negative 07 186

Further references: DE CLERCQ and DE SOMERS (1968), REMINGTON and MERIGAN (1969), VANDEPUTTE et al. (1970), VAN DIJCK (1970), RICHMOND (1971), DE CLERCQ (1973), EYCKMANS et al. (1973), SCHILLER (1980, 1982)

Methylmethacrylate

FRIEDMANMOR et al. (1977)

Drugs Activated to Free Radicals

Adriamycin

STOYCHKOV et al. (1979), HASKILL (1981), MARTIN et al. (1982)

Mitomycin C

OGURA et al. (1980), SHINDO et al. (1985)

9.8.2 Immunologically Specific Activation

The term "macrophage activation" was introduced by MACKANESS (1969, 1970) on account of morphological changes of mononuclear phagocytes after immunisation of the host against *Listeria monocytogenes*. Activated macrophages adhere to the glass wall of the culture vessel and spread quickly (NATHAN et al. 1971) showing increased phagocytosis.

ALLISON (1978) preferred the term "activation" for the marked metabolic changes that occur in macrophages over several hours or days, by analogy with those induced in lymphocytes by antigens or mitogens. "Stimulation" can be used for short-term responses to stimuli such as phagocytosis.

Theoretically, activation in one or an other form is necessary to be cured from infective diseases, in which the antigen survives or multiplies in normal macrophages.

FIDLER and RAZ (1981) reserved the term "macrophage activation" for the process whereby noncytotoxic macrophages acquire tumour cytotoxic or tumoricidal properties.

The target specificity of activated macrophages has been controversial:

1. Specifically immune macrophages whose target specificity has been associated with the presence of cytophilic antibody that enhances macrophage-tumour interaction (PELS and DEN OTTER 1974).
2. Armed macrophages that demonstrate specific tumour cell killing after becoming activated by specific macrophage-activating factor (MAF) release by T lymphocytes (EVANS and ALEXANDER 1971, 1972; GRANT et al. 1972; LOHMANN-MATTHES et al. 1973, 1974; KRIPKE et al. 1977).
3. Macrophages that are activated by non-specific agents, including MAF, are rendered tumoricidal.

ADAMS and MARINO (1984) defined "activation" as a competence to produce a given output (i.e. to complete a given function). They pose several critical questions:

1. What functions of macrophages can be activated?
2. What capacities of macrophages are necessary for completion of each function?
3. What signal(s) regulates each capacity?
4. How specific is the relationship between each signal and each capacity? (i.e., How many capacities are induced by each signal? How many signals induce each capacity?)
5. What is the interplay between the various signals? (i.e., Does receipt of one signal alter the responsiveness or the response(s) of the macrophages to a second signal?) Does induction of one preclude development of another?
6. What governs the path of activation followed? Is the path simply a matter of the first signal received or are more complicated regulatory circuits involved?

ad 1) Examples of functions might represent kill tumour cells or microbes, phagocytosis of a given type of particle, processing and presentation of antigen, or chemotaxis.

ad 2) Capacities represent either the expression of distinct gene products or covalent modifications thereof, while functions represent the combined interaction of several capacities.

ad 3) Macrophage activation is known to be regulated by a large number of signals (for review see ADAMS and HAMILTON 1987, HAMILTON and ADAMS 1987). Inductive signals include IFNγ, interferons α/β, ganulocyte-macrophage colony-stimulating factor (GM-CSF 1), B cell stimulating factor (BSF 1), vitamin D_3, retinoic acid, bacterial lipopolysaccharide (LPS), maleylated or acetylated proteins, tumour necrosis factor (TNF), heat-killed gram-positive microorganisms, such as *L. monocytogenes*, and liposome encapsulated muramyl dipeptide. *N*-Acetylmuramyl-L-alanyl-D-isoglutamine, a synthetic substance of a minimal structure required for the adjuvant activity of bacterial cell walls was found to activate macrophages in *mice*, whereas its diastereomer *N*-acetylmuramyl-L-alanyl-L-isoglutamine, which is inactive as adjuvant, did not activate them (TANAKA et al. 1977). Muramyl dipeptide bound to glycosylated serum albumin or bound to gluconoylated and glycosilated poly-L-lysine were more efficient than free muramyl dipeptide in rendering mouse peritoneal macrophages and rat alveolar macrophages cytostatic against various tumour cells (PETIT et al. 1990).

ad 4) Many of these activating signals act through defined surface receptors (ADAMS and HAMILTON 1987, HAMILTON and ADAMS 1987).

9.8.2.1 Activation of the Mononuclear Phagocyte System by Complete Freund's Adjuvant

Experimental animals given a course of injections of homologous antigen and Freund's complete adjuvant develop delayed hypersensitivity to the antigen. These immunised animals may also develop granulomatous lesions at the injection sites (in muscles if the primary injection was made intramuscular, and internally if the injection was made either intavenous or intraperitoneal).

Freund's adjuvant (1.4–140 $\mu l/25 \times 10^6$ peritoneal exudate cells of *mice*) induced a significant activation of phospholipase A (EC 3.1.1.4) (MUNDER et al. 1973). The relative concentrations of lysolecithin (1-acyl-*sn*-glycero-3-phosphorylcholine), neutral lipids and cephalin changed. Cephalin (1,2-acyl-glycero-3-phosphorylethanolamine) was attacked by phospholipase A and degraded at the same rate as lecithin (1,2-acyl-glycero-3-phosphorylcholine).

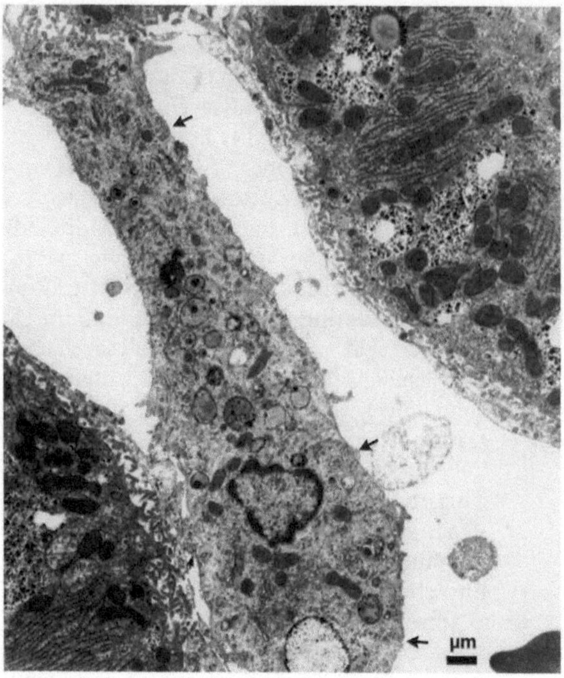

Fig. 177. Kupffer cell (block 559) from an unmedicated female *rat*. Under pentobarbital anaesthesia (30 mg/kg), the animal was perfused from the abdominal aorta with 2.5 % glutaraldehyde in 0.1 M sodium cacodylate buffer (pH 7.4). Postfixation with 1 % osmium tetroxide in sodium cacodylate buffer. Embedded in Epon 812 and sectioned at 50 nm. Lead citrate and uranyl acetate. Film 376/79. – Several tubular invaginations are marked by arrows

In the presence of 1.4 μl Freund's complete adjuvant ^{14}C-oleic acid (10 nM) was transesterified from phospholipid to triglyceride.

While bentonite and bentonite mixed with wax fractions were unable to induce enhancement of carbon clearance on antigen recall in the *chicken*, neutral delipidated *M. avium* cell walls made good potentiators of antigen recall in carbon clearance (AIYEDUN 1972).

Adjuvant fibrosis due to complete and incomplete Freund's adjuvant was used for a model of experimental pulmonary fibrosis by KISSLER (1983), while ANTWEILER et al. (1963), SCHILLER (1980) and KISSLER (1983) studied the influence of adjuvant in pneumoconiosis.

Intravenous injection of complete Freund's adjuvant induced infiltrative and follicular lesion in the *rat* lung, resembling *human* noncaseating granulomatoses of specific type in their localisation and cellular patter comparable in particular to sarcoidal and hypersensitivity granulomatous lesions (BASSET et al. 1974). The *rat* appeared particularly suitable because its pulmonary lesions, in spite of an excellent tolerance, regress more slowly than those in the *dog* (STRAUSS et al. 1970) and *rabbit* (RUPP et al. 1960, MOORE and SCHOENBERG 1963, MOORE 1973).

MASIH et al. (1979) have found that in *guinea pigs* immunised i.v. with 0.15 ml of complete Freund's adjuvant and challenged five days later with 1 mg of BCG i.v. (2 mg/ml) most of the lung parenchyma was replaced by multiple granulomas consisting of epithelioid cells and a few giant cells. Mitogenic and macrophage migration inhibition (MIF) activity were found in aqueous extracts of the lungs of these animals.

In the *rabbit*, subpleural pulmonary lesions were the preferential reactions after an intravenous injection of Freund's adjuvant (AMEMORI and ALTSCHUL 1963). Initial granulamas resemble normal solitary follicles. Later they grow and exceed the lymph follicles by size. Epithelioid cells appear. Advanced granulomas gain a considerable size and undergo central necrosis.

Saccharated iron oxide premedication modified the response of the reticulo-endothelial cells in the liver and spleen of the *rabbit*. There was a decrease in the proliferative response (MOORE and SCHOENBERG 1963).

Histamine release from *human* basophils and mast cells is mediated by nondialyzable factor(s) (molecular mass >2 kDa) secreted by lung macrophages when cocultured for 24 h (SCHULMAN et al. (1985).

N-Acetylmuramyl-L-alanyl-D-isoglutamine (muramyl dipeptide) first isolated from mycobacteria (ELOUZ et al. 1974, KOTANI 1976) and later synthesized (LEDERER 1980), was found to be the minimal structural unit that can replace mycobacterial organisms in Freund's complete adjuvant for activation of macrophages (CHEDID et al. 1979). However, although synthetic muramyl dipeptide affects several macrophage functions *in vitro* (CHEDID et al. 1979), it did not have these effects *in vivo*. This might be, as PARENT et al. (1979) showed, due to the clearance of soluble muramyl dipeptide from the body within 60 min. Multilamellar vesicle liposomes containing muramyl dipeptide were more effective than free muramyl dipeptide in potentiating the tumoridical activity of *human* alveolar macrophages during culture for 4 h (SONE et al. 1984).

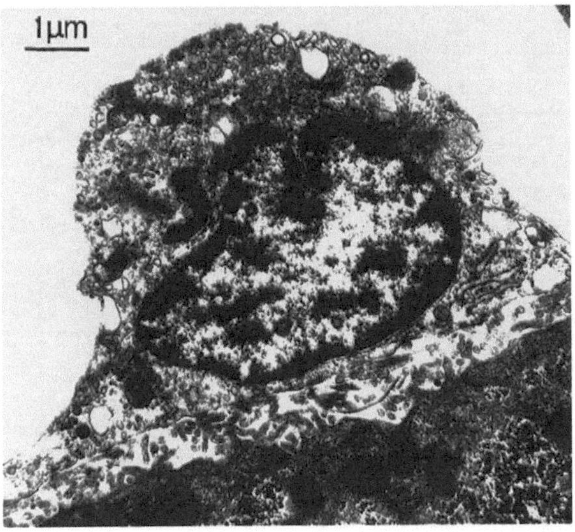

Fig. 178. Kupffer cells (block 4790) from a male Sprague-Dawley *rat* (Charles River, France) 2 weeks after an activation of the mononuclear phagocyte system by intravenous injection of 0.05 ml complete Freund's adjuvant on two consecutive days. Under pentobarbital anaesthesia (30 mg/kg), the animal was perfused from the abdominal aorta with 2.5 % glutaraldehyde in 0.1 M sodium cacodylate buffer (pH 7.4). Postfixation with 1 % osmium tetroxide in sodium cacodylate buffer. Embedded in Epon 812 and sectioned at 50 nm. Lead citrate and uranyl acetate. Film 983/79 and Film 979/79

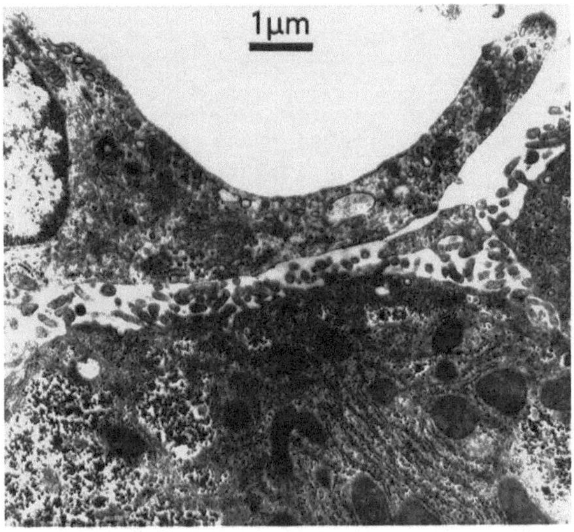

Fig. 179. Lymph node (block 4800) from a male Sprague-Dawley *rat* (Charles River, France) 2 weeks after an activation of the mononuclear phagocyte system by intravenous injection of 0.05 ml complete Freund's adjuvant on two consecutive days. Under pentobarbital anaesthesia (30 mg/kg), the animal was perfused from the abdominal aorta with 2.5 % glutaraldehyde in 0.1 M sodium cacodylate buffer (pH 7.4). Postfixation with 1 % osmium tetroxide in sodium cacodylate buffer. Embedded in Epon 812 and sectioned at 50 nm. Lead citrate and uranyl acetate. Film 1003/79

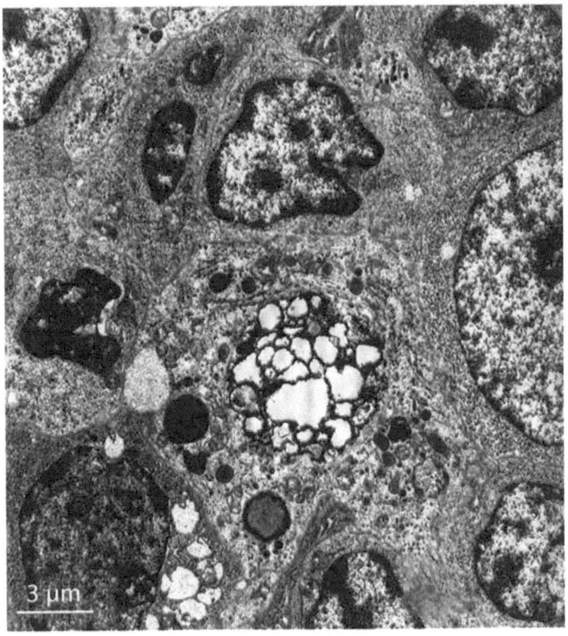

Fig. 180. Lymph node (block 4800) from a male Sprague-Dawley *rat* (Charles River, France) 2 weeks after an activation of the mononuclear phagocyte system by intravenous injection of 0.05 ml complete Freund's adjuvant on two consecutive days. Under pentobarbital anaesthesia (30 mg/kg), the animal was perfused from the abdominal aorta with 2.5 % glutaraldehyde in 0.1 M sodium cacodylate buffer (pH 7.4). Postfixation with 1 % osmium tetroxide in sodium cacodylate buffer. Embedded in Epon 812 and sectioned at 50 nm. Lead citrate and uranyl acetate. Film 1006/79

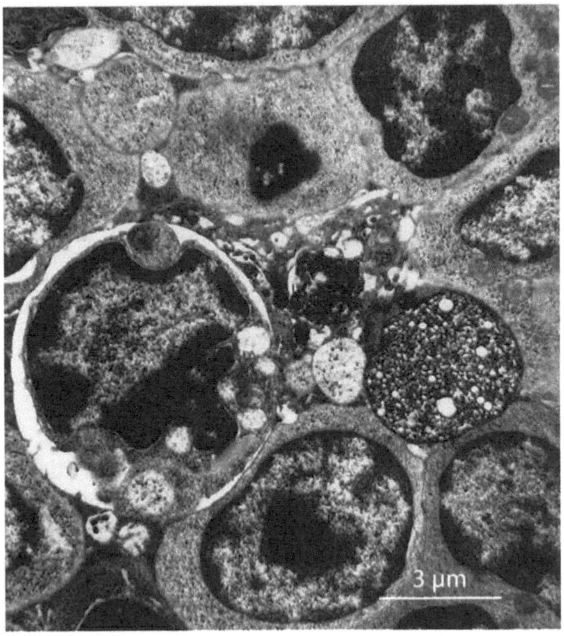

Fig. 181. Splenic white pulp (block 4791) from a male Sprague-Dawley *rat* (Charles River, France) 2 weeks after an activation of the mononuclear phagocyte system by intravenous injection of 0.05 ml complete Freund's adjuvant on two consecutive days. Under pentobarbital anaesthesia (30 mg/kg), the animal was perfused from the abdominal aorta with 2.5 % glutaraldehyde in 0.1 M sodium cacodylate buffer (pH 7.4). Postfixation with 1 % osmium tetroxide in sodium cacodylate buffer. Embedded in Epon 812 and sectioned at 50 nm. Lead citrate and uranyl acetate. Film 991/79

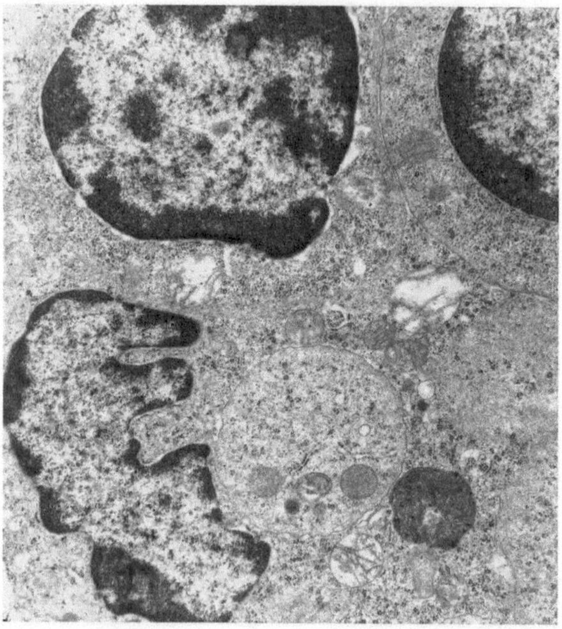

Fig. 182. Lymph node (block 4812) from a 258 g male Sprague-Dawley *rat* (Charles River, France) 4 weeks after an activation of the mononuclear phagocyte system by intravenous injection of 0.05 ml complete Freund's adjuvant on two consecutive days. Under pentobarbital anaesthesia (30 mg/kg), the animal was perfused from the abdominal aorta with 2.5 % glutaraldehyde in 0.1 M sodium cacodylate buffer (pH 7.4). Postfixation with 1 % osmium tetroxide in sodium cacodylate buffer. Embedded in Epon 812 and sectioned at 50 nm. Lead citrate and uranyl acetate. Film 1010/79

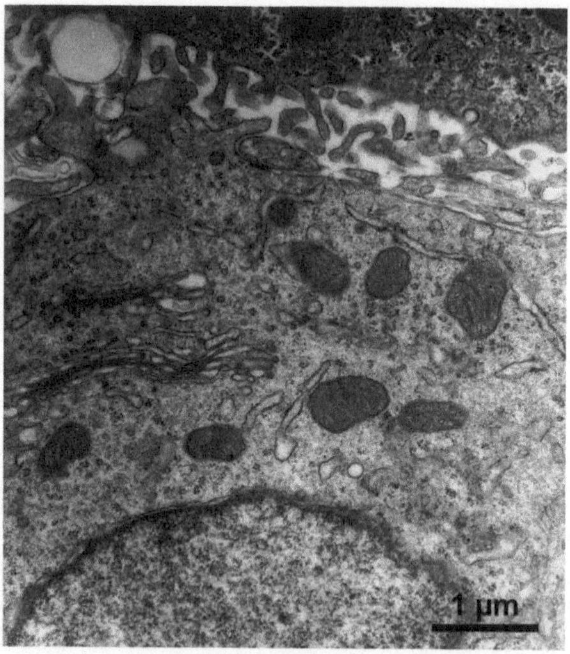

Fig. 183. Prominent Golgi area in a Kupffer cell (block 4838) from a male Sprague-Dawley *rat* (Charles River, France) 16 weeks after an activation of the mononuclear phagocyte system by intravenous injction of 0.05 ml complete Freund's adjuvant on two consecutive days. Under pentobarbital anaesthesia (30 mg/kg), the animal was perfused from the abdominal aorta with 2.5 % glutaraldehyde in 0.1 M sodium cacodylate buffer (pH 7.4). Postfixation with 1 % osmium tetroxide in sodium cacodylate buffer. Embedded in Epon 812 and sectioned at 50 nm. Lead citrate and uranyl acetate. Film 226/84

Fig. 184. Dilated Golgi cisternae (G) in a Kupffer cell (block 4838) from a male Sprague-Dawley *rat* (Charles River, France) 16 weeks after an activation of the mononuclear phagocyte system by intravenous injection of 0.05 ml complete Freund's adjuvant on two consecutive days. Under pentobarbital anaesthesia (30 mg/kg), the animal was perfused from the abdominal aorta with 2.5 % glutaraldehyde in 0.1 M sodium cacodylate buffer (pH 7.4). Postfixation with 1 % osmium tetroxide in sodium cacodylate buffer. Embedded in Epon 812 and sectioned at 50 nm. Lead citrate and uranyl acetate. Film 33059/88

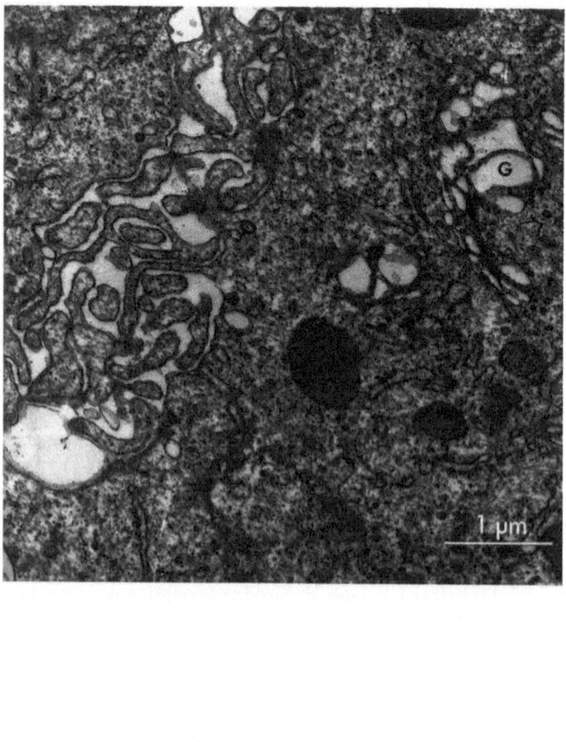

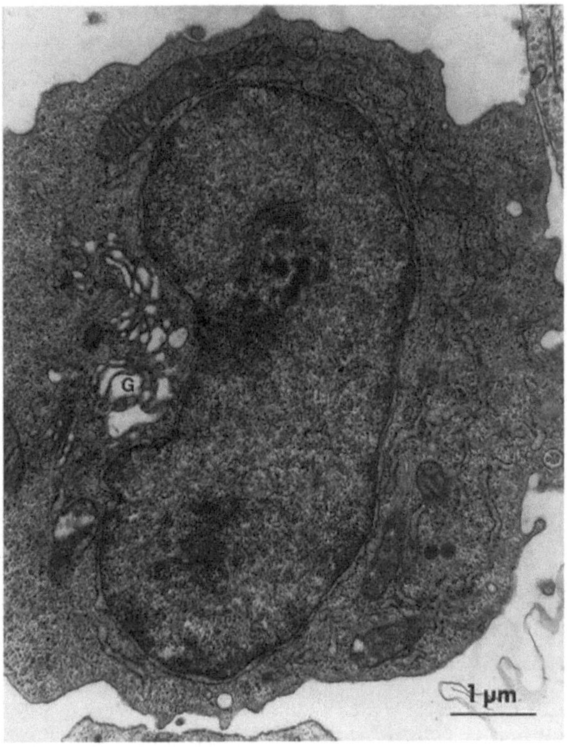

Fig. 185. Dilated Golgi cisternae (G) in a lymph node macrophage (block 4841) from a male Sprague-Dawley *rat* (Charles River, France) 16 weeks after an activation of the mononuclear phagocyte system by intravenous injection of 0.05 ml complete Freund's adjuvant on two consecutive days. Under pentobarbital anaesthesia (30 mg/kg), the animal was perfused from the abdominal aorta with 2.5 % glutaraldehyde in 0.1 M sodium cacodylate buffer (pH 7.4). Postfixation with 1 % osmium tetroxide in sodium cacodylate buffer. Embedded in Epon 812 and sectioned at 50 nm. Lead citrate and uranyl acetate. Film 1289/79

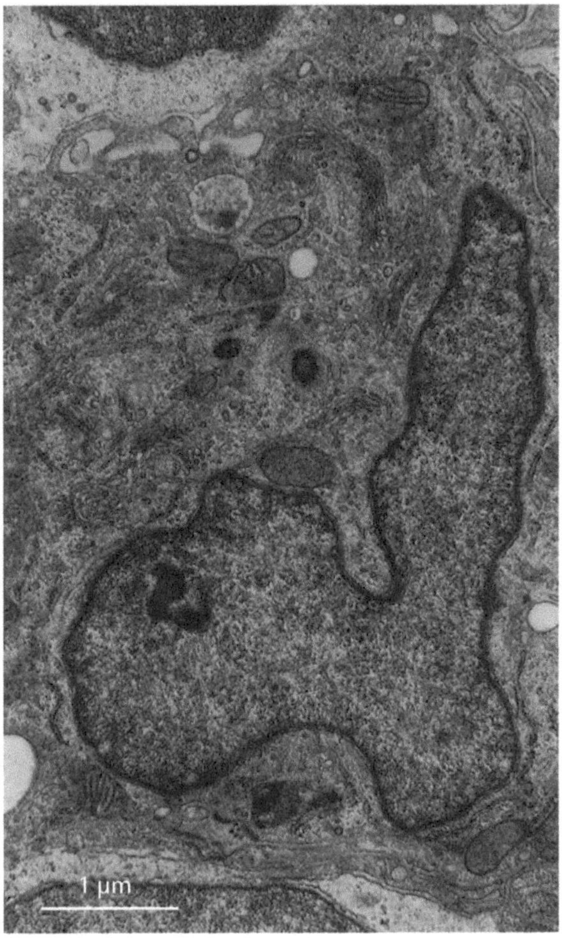

Fig. 186. An epithelioid-like mononuclear cell from a lymph node (block 4841) from a male Sprague-Dawley *rat* (Charles River, France) 16 weeks after an activation of the mononuclear phagocyte system by intravenous injection of 0.05 ml complete Freund's adjuvant on two consecutive days. Under pentobarbital anaesthesia (30 mg/kg), the animal was perfused from the abdominal aorta with 2.5 % glutaraldehyde in 0.1 M sodium cacodylate buffer (pH 7.4). Postfixation with 1 % osmium tetroxide in sodium cacodylate buffer. Embedded in Epon 812 and sectioned at 50 nm. Lead citrate and uranyl acetate. Film 1290/79

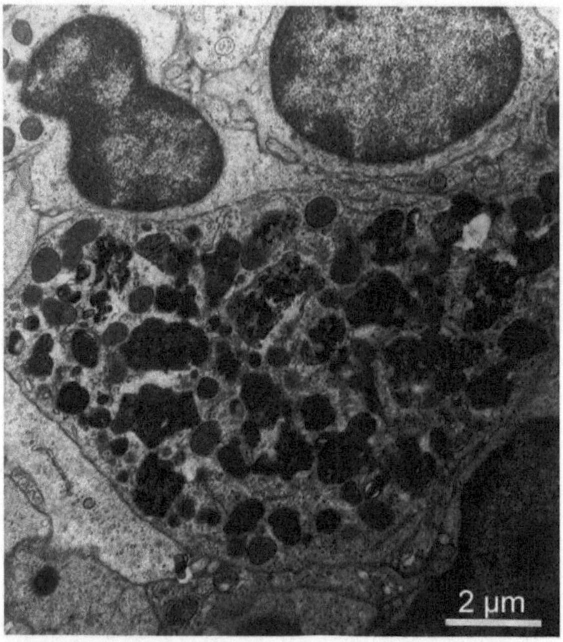

Fig. 187. Pleomorphic inclusions in a lymph node macrophage (block 4837) activated 16 weeks before by two portions of 0.05 ml complete Freund's adjuvant given intravenously at an interval of 24 h. Male Sprague-Dawley *rat* (Charles River, France) perfused under pentobarbital anaesthesia (30 mg/kg) from the abdominal aorta with 2.5 % glutaraldehyde in 0.1 M sodium cacodylate buffer (pH 7.4). Postfixation with 1 % osmium tetroxide in cacodylate buffer. Embedded in Epon 812 and sectioned at 50 nm. Lead citrate and uranyl acetate. Film 219/84

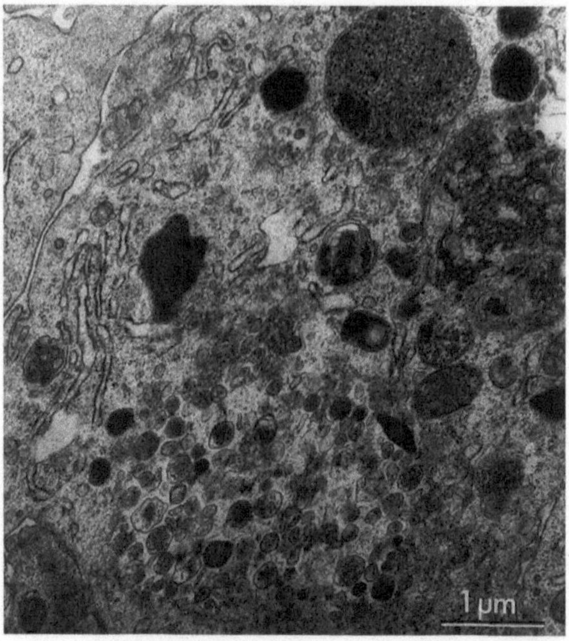

Fig. 188. Tubular invaginations in a splenic macrophage (block 4836) activated 16 weeks before by two portions of 0.05 ml complete Freund's adjuvant given intravenously at an interval of 24 h. Male Sprague-Dawley *rat* (Charles River, France) perfused under pentobarbital anaesthesia (30 mg/kg) from the abdominal aorta with 2.5 % glutaraldehyde in 0.1 M sodium cacodylate buffer (pH 7.4). Postfixation with 1 % osmium tetroxide in cacodylate buffer. Embedded in Epon 812 and sectioned at 50 nm. Lead citrate and uranyl acetate. Film 214/84

Fig. 189. Granules of different sizes and micropinocytotic coated pits in the wall of a tubular invagination. Lymph node macrophage (block 4841) 16 weeks after activation of the mononuclear phagocyte system by intravenous injection of 0.05 ml complete Freund's adjuvant on two consecutive days. Male Sprague-Dawley *rat* (Charles River, France) perfused under pentobarbital anaesthesia (30 mg/kg) from the abdominal aorta with 2.5 % glutaraldehyde in 0.1 M sodium cacodylate buffer (pH 7.4). Postfixation with 1 % osmium tetroxide in sodium cacodylate buffer. Embedded in Epon 812 and sectioned at 50 nm. Lead citrate and uranyl acetate. Film 231/84

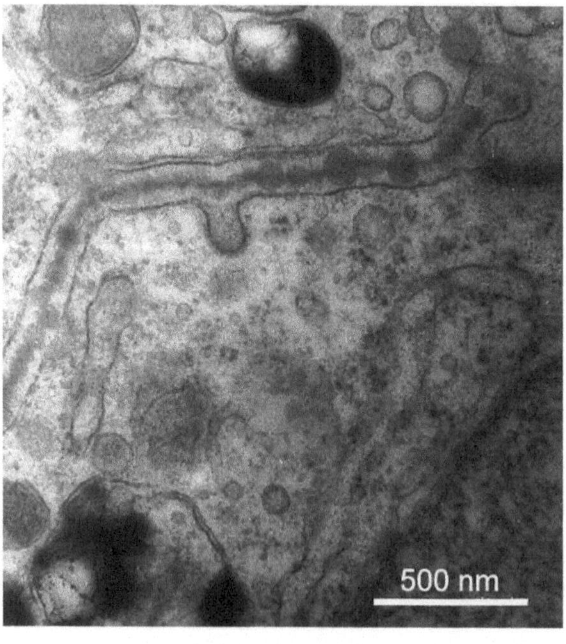

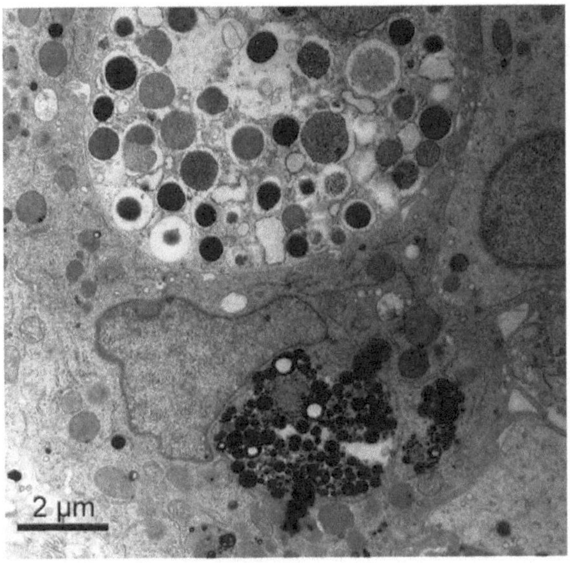

Fig. 190. Intimate contact between a mononuclear phagocyte (bottom) activated by complete Freund's adjuvant and a mast cell (top) in a lymph node (block 4841) from a male Sprague-Dawley *rat* (Charles River, France) 16 weeks after activation of the mononuclear phagocyte system by intravenous injection of 0.05 ml complete Freund's adjuvant on two consecutive days. Under pentobarbital anaesthesia (30 mg/kg) the animal was perfused from the abdominal aorta with 2.5 % glutaraldehyde in 0.1 M sodium cacodylate buffer (pH 7.4). Postfixation with 1 % osmium tetroxide in sodium cacodylate buffer. Embedded in Epon 812 and sectioned at 50 nm. Lead citrate and uranyl acetate. Film 229/84

9.8.3 Prospective Pulmonary Fibrosis-Inducing Potency of Macrophages

The prospective pulmonary fibrosis-inducing potency of the mononuclear phagocyte system of active coal workers and retired coal miners has been studied and discussed as to the usefulness of biological markers as representative signals in a continuum of events between causal exposure and resultant disease (SCHULTE 1989). FOWLE and SEXTON (1992) defined a biomarker as "a measurement of environmental pollutants or their biological consequences after the contaminants have crossed one of the body's boundaries and entered human tissues or fluids, and which serves as an indicator of exposure, effect, and/or susceptibility.

Asbestos or silica inhalation studies in *rats* showed that the antioxidant response and, more specifically, mRNA levels of Mn superoxide dismutase and heme oxygenase, correlated well with the inflammatory response in broncho-alveolar lavage fluid (JANSSEN et al. 1992). The upregulation of ornithine decarboxylase (MARSH and MOSSMAN 1991) and *c-fos/c-jun* protooncogenes (HEINTZ et al. 1993), need further *human* studies.

Tumour necrosis factor-α levels were significantly different between active coal miners from three French mining regions (Nord-Pas de Calais, Lorraine and Provence). However, after correlation for age and region, TNF was found not to be related to dust exposure (PORCHER et al. 1994). In 66 Belgian coal miners exposed to dust at the coal face at least 12 yrs blood mononuclear phagocytes spontaneously and stimulated with coal mine dust, silica or lipopolysaccharide released TNF at a higher rate than blood monocytes from 12 non-exposed controls (BORM et al. 1988). The greatest discriminator between controls and cases with coal workers' pneumoconiosis was coal mine dust-induced TNF release.

In the *rat*, asbestos fibres and silica particles stimulated alveolar macrophages to release TNF (DUBOIS et al. 1989).

CHAPTER 10

Interstice

10.1 Alveolar Septa

The walls of the distal airspaces have a similar structure whether these are alveoli or alveolar ducts (KUHN 1985).

10.1.1 Matrix

The collagen-elastin-proteoglycan matrix is the key constituent of lung parenchyma and plays a major role in the mechanical behaviour of lung tissue. However, the exact composition of the proteoglycan matrix in lungs has not yet been fully determined. Immunohistochemistry on peripheral lung tissue from patients undergoing therapeutic lung resections showed that lumican, a keratan sulphate- proteoglycan belonging to the family of relatively small, leucine-rich repeat- proteoglycans, is a major component of the proteoglycan matrix in adult *human* lungs (DOLHNIKOFF et al. 1998). Details on the organisation and chromosomal location of the lumican gene have been published by GROVER et al. (1995). All members of the small proteoglycan family, including lumican, interact with fibrillar collagen and may influence the interaction of the collagen fibrils with other components of the extracellular matrix, thus participating in the maintenance of the extracellular milieu (SCOTT 1988, KJELLÉN and LINDAHL 1991, RUOSLAHTI and YAMAGUCHI 1991, KRESSE et al. 1993, SCHÖNHERR et al. 1995).

Alterations of the connective tissue composition of the extracellular matrix are an important feature of various inherited and acquired disorders of soft tissue (MILLER 1976). This phenomenon is particularly important in the lung in which elasticity is a vital determinant of function. Interstitial collagen is composed of at least two distinct trimeric polypeptide molecules designated as type I and type III collagens. Type I contains two α_1(I) chains and one α_2 chain, whereas type III contains 3 identical α_1(III) chains. Pulmonary fibroblasts produce more type I than type III collagen (KELLEY et al. 1979). Cultured *human* lung fibroblasts produced 26 to 68% more type III collagen at confluence than at low cell density (TROMBLEY et al. 1981).

Extracellular matrix components had changed in 12 elderly subjects aged 67–88 years (D'ERRICO et al. 1989). The elastic fibres showed a notable decrease along the alveolar walls while type III collagen had increased when compared with that of the controls aged between 28 and 57 years. Using antibodies against type IV collagen and laminin the thickness of the alveolar basement membranes appeared increased in some of the subjects, while antibodies to fibronectin and type V collagen did not reveal any age-related modifications.

Type VI collagen is an extracellular matrix protein with structural and biosynthetic properties strikingly different from those of type I and type III collagens (TIMPL and ENGEL 1987). Collagen VI is resistant to degradation by metalloproteiunases, which usually degrade other collagens (OKADA et al. 1990). It has a relatively short collagenous domain flanked by two large globular domains at the N- and C-terminal ends. The monomers are composed of three genetically distinct protein chains (TIMPL and ENGEL 1987). The α1 and α2 chains (*Mr* 140,000) are encoded on chromosome 21, whereas the distinctly larger α3 chain (*Mr* 330,000) is encoded on chromosome 2 (WEIL et al. 1988). Collagen VI lacks the proteolytic processing with the removal of large N- or C-terminal propeptides (COLOMBATTI and BONALDO 1987). In a unique intracellular polymerisation process disulphide-bonded dimers and tetramers are formed that constitute the building blocks of microfibrils with 100 nm-periodicity (BRUNS et al. 1986, ENGVALL et al. 1986) found on the surface of cells, between collagen fibres, and adjacent to basement membranes (KEENE et al. 1988). Collagen VI has been detected in most foetal and adult tissues (TIMPL and ENGEL 1987) including normal animal (GIBSON and CLEARY 1983, AMENTA et al. 1988) and *human* lungs (VON DER MARK et al. 1984). Collagen VI expression is increased in lung fibrosis, and its degree appears independent of the aetiology of the fibrosis (Specks et al. 1995). There was no evidence for differential regulation of gene expression for the α_1(VI) and α_3(VI) constitutive peptide chains of collagen VI. Collagen VI mRNA is expressed by fibroblasts, mostly with myofibroblast characteristics.

Glycyl-histidyl-lysine is a matrix-derived tripeptide that modulates new connective tissue formation. It was originally isolated from *human* plasma (Pickart and Thaler 1973) and described as a growth factor for differentiated cells (Pickart 1981). Glycyl-histidyl-lysine spontaneously forms a high affinity complex with Cu^{2+}-ions. Cu is bound to the *N*-terminal end of the peptide – the amide bond between glycine and histidine – and the imidazole ring of histidine (Lau and Sarkar 1981). In biological solutions, however, it is likely that the complex contains two peptides for one copper ion. In monolayer cultures of fibroblasts, Maquart et al. (1988) demonstrated that glycyl-histidyl-lysine-Cu induced a dose-dependent stimulation of collagen synthesis, with a maximal effect at the 10^{-9} concentration.

Glutathione and metallothionein protected *mouse* lung fibroblasts against copper toxicity, as shown in buthionine sulphoxime/Zn-pretreated metallothionein MT–/– *mice* (Jiang et al. 2002).

Collagen as Altered by Free Radicals

The generation of free radicals and other reactive oxygen species such as singlet oxygen and hypochlorous acid may play a role in the alterations that are observed in collagen either during natural ageing or certain diseases. Hawkins and Davies (1997) examined the attack of hydroxyl radicals, generated using the Fe(II)/H_2O_2 couple on the major amino acid types present in collagen. All of the amino acids tested, with the exception of glycine, gave clear signals consisting of triplets of doublets with the former splitting arising from the nitroxide nitrogen, and the latter from a single hydrogen atom on the substrate. These spectra are believed to arise from the trapping of carbon-centred radicals of the type •CHR'R" produced by hydrogen abstraction by hydroxyl radicals from the side chains. In no case were any signals detected which could be assigned to a species arising from hydrogen abstraction at the α-carbon. With Cu(I)/H_2O_2 or Cu(II)/H_2O_2 instead of Fe(II)/H_2O_2, evidence has been obtained for: i) altered sites of attack and fragmentation, ii) C-terminal decarboxylation, and iii) hydrogen abstraction at *N*-terminal α-carbon sites. In type I collagen from *calf*skin, the side chain radicals were consistent with attack at lysine, but not proline or hydroxyproline, side-chain sites. This suggested that there is some selectivity within the collagen molecule in the site(s) of hydroxyl radical attack.

Following a 60-hour exposure of *rats* to 98 % O_2 lung collagen content was reduced by almost 20 % (Riley et al. 1983). Brief periods of hyperbaric hyperoxia had no influence (Richmond and D'Aoust 1976).

Oxygen metabolites produced by intratracheal instillation of 38 mg xanthine in 0.25 ml sterile phosphate-buffered saline (pH 7.4) plus 100 µg xanthine oxidase in 0.25 ml ice-cold phosphate-buffered saline on day 0 and 38 mg xanthine plus 250 µg xanthine oxidase on day 5 into *hamsters* increased hydroxyproline from 903 µg/lung in controls to 1080, 1301, 1195 and 1148 µg/lung at days 12, 19, 26, and 33, respectively (Giri et al. 1988).

Isolated heparan sulphate chains were degraded by chemical reactions which generate HO• (Fe^{2+}/EDTA + $H_2O_2 \rightarrow Fe^{3+}$/EDTA + HO• + OH^-) directly proportional to the amount of HO• produced and was inhibited by specific scavengers of HO• (DMSO and thiourea), but not by chemically-related non-HO• scavengers as urea (Hoidal et al. 1985).

10.1.2 Alveolar Myofibroblasts (Contractile Interstitial Cells)

The connective tissue cells have a very irregular outline with complex projections of cytoplasm containing discrete bundles of contractile 3–8 nm filaments. They contact one another through nexus junctions, so that the contraction of several cells is probably functionally co-ordinated (Bartels 1979). Sims and Westfall (1982) recommended the term "6 nm F-maculae adherentes". Kapanci et al. (1974) proposed that these cells may have a function in the fine matching of ventilation to perfusion. Richards et al. (1977) regarded the interstitial fibroblast as the architect cell of lung tissue, being responsible for the formation and degradation of collagen (fibrous, pro and tropocollagen) and protein-mucopolysaccharide complexes (proteoglycans, proteoglycosaminoglycans), which form the matrix necessary for the correct conformation of collagen fibres (Chvapil 1974, Richards and Wusteman 1974). Thus, in lung fibrosis and probably emphysema the ultimate response to toxins or particulate matter must involve the fibroblast. *Human* macrophages derived from blood monocytes by cultivation *in vitro* after incubation with Dörentrup quartz dust and coal mine dusts released a soluble mediator, which significantly increased DNA synthesis by WI-38 *human* lung fibroblasts (Hübner and Seemayer 1992). While *human* foetal lung fibroblasts cultured in the absence of growth factors did not increase in number, fibroblasts cultured with fibronectin + alveolar macrophage derived growth factor did (Bitterman et al. 1983). Transforming growth factor (TGF)-β may potentiate wound healing and fibrosis by stimulating fibroblast collagen deposition. Coker et al. (1997) examined the effects of TGF-β_1, -β_2, and -β_3 on lung fibroblast collagen metabolism *in vitro* and localised their gene expression during bleomycin-induced lung fibrosis *in situ* hybridisation with digoxigenin-labeled riboprobes. All three isoforms stimulated fibroblast procollagen production. TGF-β_3 was the most potent and also reduced procollagen degradation.

There are two distinct **fibroblast subtypes** in the neonatal *rat* lung. The lipid interstitial cell (LIC) accumulates lipid, contains lipoprotein lipase, and myofilaments. The non-lipid interstitial (NLIC) cell does neither contain lipoprotein lipase nor abundant myofibrils. Villanueva et al. (1988) treated both subtypes with TGF-β, an effector molecule that stimulates embryonic fibroblasts to produce collagen and fibronectin. TGF-β caused a significant increase in collagen and total protein production in LIC cultures, as measured by the amount of nondialyzable [^{3}H]hydroxyproline and [^{3}H]proline incorporated into proteins over 24 h. Parallel changes in the amounts of total nanomoles of nondialyzable hydroxyproline indicated that these increases were not the result of alteration in proline pool size. TGF-β did not stimulate an increase in collagen formation by NLIC cultures.

A significant amount of newly synthesized collagen is degraded intracellularly rather than secreted, but there is controversy about whether this process occurs in the lysosomes. Ripley and Bienkowski (1997) studied the distribution of procollagen I in the Golgi and the lysosome/endosome system of cultured *human* foetal lung fibroblasts. When cells were incubated with the proline analogue *cis*-hydroxyproline, which inhibits correct triple helix formation and increases intracellular degradation, the amount of procollagen codistributing with the lysosome/endosome markers cathepsin B (EC 3.4.22.1, a lysosomal cysteine endopeptidase) or LAMP-2 increased greatly. Similar results were obtained in I-cells, which do not have functioning lysosomal hydrolases.

Photoprotection of parenteral multivitamin solutions or total parenteral nutrition prevented light induction of procollagen mRNA in the lungs of 3-day-old *guinea pig* pups (Lavoie et al. 2002). The effect of parenteral multivitamin solutions + light was associated with a peroxide load coupled with a low glutathione level. This was also observed with a 500 μM H_2O_2 group. The addition of GSSG prevented the increase of procollagen mRNA caused by H_2O_2.

Prostaglandin F synthase (EC 1.1.1.188) reduced retinal to retinol in a dose- and time-dependent manner using NADPH as a cofactor (Endo et al. 2001). The conversion of retinal to retinol was observed in cultured *bovine* pulmonary contractile interstitial cells, as demonstrated by the greenish fluorescence characteristic of retinol. Thus, retinal might be one of the natural substrates for prostaglandin F synthase *in vivo*, and retinol synthesised from retinal in contractile interstitial cells may play physiological and pathological roles in the lungs.

In cultures of foetal *rat* lung fibroblasts, differential display technique showed that the mRNA level of **prolyl 4-hydroxylase α(I)** (EC 1.14.11.2), an active subunit that catalyses the oxygen-dependent hydroxylation of proline residues in procollagen, increased 2-3-fold after an 8-h exposure to hypoxia (Takahashi et al. 2000). In *rats* exposed for 11 weeks to hypoxia, simulating an altitude corresponding to 7000 m, in subcutaneously implanted polyurethane sponges the synthesis of newly formed collagen was significantly stimulated as shown by the increase of the specific activity of collagenous ^{14}C-hydroxyproline (Chvapil et al. 1970).

The hydroxylation system consists of bivalent iron complex bound by ethylenediamine tetraacetic acid (formula [187]), ascorbic acid (formula [53]), and hydrogen peroxide (Brodie et al. 1954, Udenfriend et al. 1954). *In vitro*, the major amount of hydroxyproline was formed during the first 30 s (Hurych 1967). If the proline in the incubation mixture was replaced by hydroxyproline a decomposition of about 30 % of the hydroxyproline occurred. Accordingly, the newly formed hydroxyproline was the result of synthetic and degradation processes, which are caused, as is generally assumed (Haber and Weiss 1934, Kolthoff and Medalia 1949), by **free radicals**.

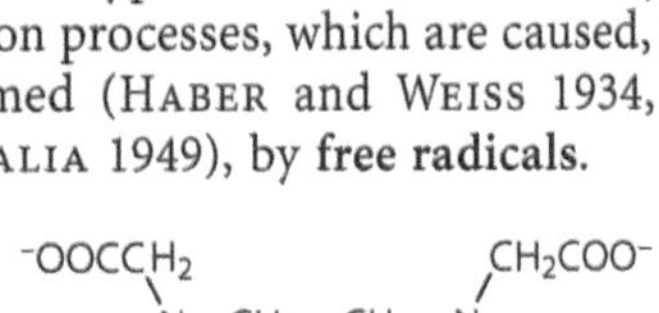

Ethylenediamine tetraacetic acid (EDTA) [187]

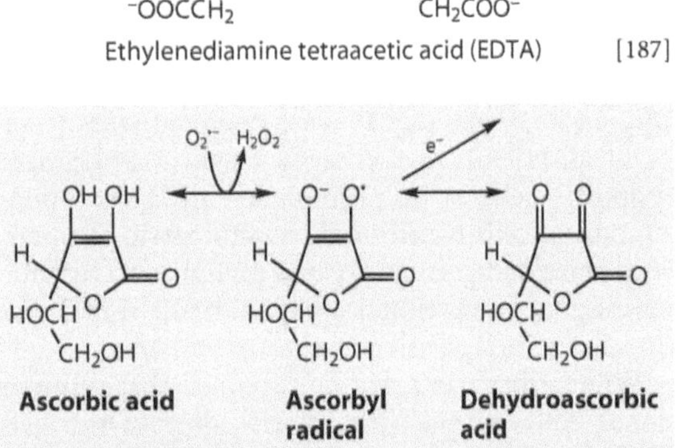

Formation of the ascorbyl free radical [53]

$$H_2O_2 + Fe^{2+} \longrightarrow HO^{\cdot} + HO^- + Fe^{3+} \quad [198]$$

Chvapil et al. (1974) studied the effect of both Fe^{2+} and Fe^{3+} chelating agents in tissues with a relatively high rate of collagen synthesis such as foetal skin and skin of newborn *rats*, 17β-oestradiol-stimulated uterus of immature *rats*, and carrageenan granuloma tissue. Neither systemic nor local injections of 1,10-phenanthroline or desferrioxamine alone inhibited prolyl hydroxylase in any of the models. But simultaneous administration of both agents inhibited prolyl hydroxylase and hydroxylation of collagen in some models. Although local injections of 1,10-phenanthroline into granuloma tissue did not inhibit prolyl hydroxylase activity, hydroxylation of collagen synthesized in the granuloma from 4 to 16 h after injection was reduced significantly. This seeming discrepancy is explained by the finding that prolyl hydroxylase is active seen even without the addition of Fe^{2+} in the assay medium; a strong reducing environment (α-ketoglutarate, ascorbic acid) reduced Fe^{3+}, thus providing ferrous ions essential for prolyl hydroxylase activity. Both forms of iron, therefore, must be chelated simultaneously to affect prolyl hydroxylase activity in the assay system.

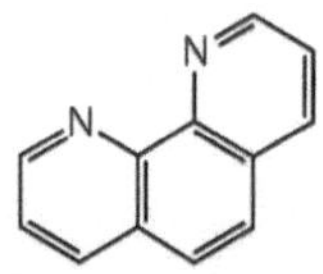

1,10-Phenanthroline [137]

Lysyl oxidase is the enzyme that synthesizes through oxidative deamination of some εNH_2 groups of lysyl or hydroxylysyl residues the aldehyde necessary to form either stable covalent cross links of an aldol-condensate type of Schiff base with reaction with the other free εNH_2 group in collagen polypeptides. The highest activity of the enzyme in young polyvinylalcohol sponge granuloma is related to a high proportion of fibroblasts (Chvapil et al. 1974).

Apoptotic bodies of interstitial fibroblasts in *rat* lung on postnatal day 19 were depicted by Schittny et al. (1998). Awonusonu et al. (1999) judged apoptotic cells to be prominently lipid-containing interstitial cells based on flow cytometric estimates of cell size and granularity and on light-microscopic staining of intracellular lipid with oil red O and Hoechst 33342-positive apoptotic bodies.

When fibroblasts are cultured in three-dimensional collagen gels, fibroblasts contract the gels. This phenomenon has been considered to be an in vitro model of wound contraction and connective tissue morphogenesis (Bell et al. 1979, Ehrlich and Wyler 1983). Several biological factors are known that can modulate fibroblast-mediated collagen gel contraction. β_1-Integrin-mediated gel contraction was stimulated by the platelet-derived growth factor (PDGF) (Gullberg et al. 1990) and the tranforming growth factor-β (Montesano and Orci 1988).

Blood monocytes from health donors (>95% pure) cast into type I collagen gels that contained lung fibroblasts inhibited the fibroblast-mediated gel contractility in a time-and concentration-dependent manner (Sköld et al. 2000). The concentration of PGE_2, a well-known inhibitor of gel contraction, was higher ($P < 0.01$) in media from co-culture; these media attenuated fibroblast gel contraction, whereas conditioned media from either cell type cultures alone did not. Three-dimensional cultured monocytes responded to conditioned media from cocultures by producing interleukin-1β and tumour necrosis factor-α, whereas fibroblasts increased synthesis of PGE_2. Antibodies to interleukin-1β and tumour necrosis factor-α blocked the monocyte inhibitory effect and reduced the amount of PGE_2 produced.

Fibroblast growth factors (FGFs) are potent regulators of cell growth, differentiation and function of a wide variety of cells derived from the mesoderm and neuroectoderm. These proteins play crucial roles in normal development, in maintenance of tissues and in wound healing and repair. They may also contribute to pathological conditions, including tumour growth, rheumatoid arthritis, arteriosclerosis and metastases.

The FGFs vary in size from 155 to 268 amino acids, and share 33–65% amino acid sequence identity. FGFs mediate cellular responses by binding and activating of specific cell surface tyrosine kinase receptors (Jaye et al. 1992).

Fibroblast growth factor (FGF)-2 enhanced cell growth of *human* lung fibroblasts at 1–100 ng/ml while FGF-13 stimulated at 1,000-2,500 ng/ml (Leung et al. 1998). FGF-13 and FGF-2 had little effect on the production of interleukin-6 by lung fibroblasts.

Platelet-derived growth factor receptor-α mRNA expression was induced 24 h after an tracheal instillation of V_2O_5 (2 mg/kg suspended in 200 μl saline) in Sprague-Dawley *rats* (Bonner et al. 1998). Platelet-derived growth factor receptor-β mRNA was constitutively expressed and did not increase. Western blotting showed upregulation of platelet-derived growth factor receptor-α protein by 48 h, and immunohistochemical analysis localised platelet-derived growth factor receptor-α primary in mesenchymal cells residing within the fibrotic tissue. Upregulation of platelet-derived growth factor receptor-α *in vivo* preceded mesenchymal cell hyperplasia (3–7 days) and collagen deposition by day 15. Supernatants from *rat* alveolar macrophages treated with V_2O_5 *in vitro* released upregulatory activity for platelet-derived growth factor receptor-α on cultured lung myofibroblasts, and this activity was blocked by the interleukin-1-receptor antagonist.

Bleomycin sulphate (1.2 mg/ml saline/*rat* intratracheally) induced proliferation of fibroblasts and their "differentiation" into cell types with increasing contractile capability during the development of fibrosis in the animal model (ADLER et al. 1986). In *mice* intratracheally instilled with 0.04–0.08 U bleomycin and *rats* instilled with 1 U of the drug the interstitial cells had the phenotype of an actin (α-actin in the *rat*) and lipid containing interstitial cell with a poorly developed endoplasmic reticulum; in silica (DQ12 <5 μm, 2 mg in *mice* and 6 mg in *rats*) injected animals, in contrast, the interstitial cells were without cytoplasmic actin or lipid but with a markedly developed endoplasmic reticulum (PIGUET et al. 1996). Thus bleomycin and silica induced the growth of two different types of interstitial cells, the myofibroblast and the regular fibroblast.

Bleomycin (0.15 U/*mouse*) given intratracheally induced significantly more pulmonary fibrosis in *mice* deficient in SPARC (secreted protein, acidic and rich in cysteine) compared with that in wild-type control *mice*, with the mutant *mice* demonstrating greater neutrophil accumulation in the lungs (SAVANI et al. 2000). However, in wild-type and SPARC-deficient *mice* given intraperitoneal bleomycin (0.8 U/injection × 5 injections over 14 days), the pattern and severity of pulmonary fibrosis, as well as the levels of leucocyte recruitment, were similar in both strains of *mice*.

Connective tissue growth factor mRNA expression is upregulated in bleomycin-induced lung fibrosis in C57BL/6 and BALB/c *mice* (LASKY et al. 1998).

In idiopathic pulmonary fibrosis, during the inflammatory stage, the alveolar myofibroblasts expressed α-smooth muscle actin, and in very fibrotic and cystic alveolar tissue, i.e. at end stage fibrosis, the number of α-smooth muscle actin positive myofibroblasts diminished (KAPANCI et al. 1995).

Human lung fibroblasts inhibited the production of tumour necrosis factor-α by lipopolysaccharide activated peripheral blood monocytes though the release of soluble factors such as **PGE_2** (VANCHERI et al. 1996, CONTE et al. 1997). Fibroblasts derived from diseased tissue, such as fibrotic lung fibroblasts, exhibit different functional features compared with normal cells, with particular regard to their modulatory role (VANCHERI et al. 2000). Fibrotic fibroblasts spontaneously produced less PGE_2 (3,300 ± 410 pg/ml) compared with normal fibroblasts (7,500 ± 270 pg/ml) and, as a consequence, they showed a reduced ability to down-regulate the production of TNF-α by lipopolysaccharide-activated macrophages.

Upon stimulation with cytokines such as interleukin-1α or tumour necrosis factor-α *human* skin fibroblasts in primary culture released **reactive oxygen species** (MEIER et al. 1989). The primary radical produced was $O_2^{\bullet -}$ as determined by ESR spin trapping and cytochrome *c* reduction. In contrast to the oxidative burst in granulocytes and monocytes, radicals were formed continuously for at least 4 h.

In a *human* foetal fibroblast cell line, MRC-5, cadmium induced cellular damage by producing reactive oxygen species (YANG et al. 1997). A time- and dose-dependent increase in both lactate dehydrogenase leakage and malondialdehyde formation was observed in Cd-treated cells. A close correlation between these two events suggested that lipid peroxidation may be one of the main pathways causing cytotoxicity. Cd-induced cell injury and lipid peroxidation could be inhibited by catalase and superoxide dismutase.

Inosine deriving from the metabolism of adenosine or inosine monophosphate in the fibroblast provides the substrate for xanthine oxidase and is, therefore, an important source of toxic reactive oxygen species. With well-oxygenated medium, adenosine release appeared to be greater for aged than young *human* lung fibroblasts of the IMR-90 cell line (REISERT et al. 2002). In that the adenosine release by young cells was enhanced by reduced oxygenation, the effect of anoxic stress on the release of the purine nucleosides adenosine and inosine by low-passage (PDL 23–26; young) vs. high-passage (PDL 43–51; aged) *human* lung fibroblasts was studied. Immediately following anoxia, adenosine release by young anoxic fibroblasts was 29% greater than for young normoxic fibroblasts, whereas both the young and aged anoxic fibroblasts displayed a 34- and 21-fold greater inosine release, respectively, as compared to normoxic fibroblasts. Deamination of the released adenosine was not the primary source of inosine.

The role of reactive oxygen species generated by hypoxanthine/xanthine oxidase in the induction of apoptosis was studied by GANSAUGE et al. (1997). In *human* fibroblast cell line WL38 after 48 h incubation with 1 mM hypoxanthine/xanthine and 0.05 U xanthine oxidase/ml apoptosis but not necrosis was observed in proliferating fibroblasts. Catalase hindered induction of apoptosis. Cell-cycle analysis revealed a reduction of cells in the S/G2 phase 24 and 48 h after stimulation, suggesting that reactive oxygen species induce a G1 arrest in proliferating fibroblasts. This was supported by an accumulation of p53 and the cdk inhibitor p21^{WAF1CIP1}. Since apoptosis was not inducible in senescent fibroblasts the data indicated that reactive oxygen species mainly induce apoptosis in proliferating cells.

Chinese *hamster* HA-1 fibroblasts exposed to H_2O_2 induced an RNA species, which was found to be degraded when evaluated by Northern blot hybridisation (Crawford et al. 1997). Cloning and subsequent sequencing identified the partially degraded RNA as 16S ribosomal RNA, a major component of mitochondrial ribosomes. Degradation, and associated decreases in the levels of the mature- and precursor species of 16S rRNA, appeared to be dependent upon calcium, but not cytoplasmic protein synthesis nor nuclear transcription. Other decreased mitochondrial RNAs were also identified, including 12S rRNA, NADH dehydrogenase subunit 6, ATPase subunit 6, and cytochrome oxidase subunits I and III. A significant part of many, if not all, of these RNA decreases was due to degradation. As compared with 16S rRNA, significantly less degradation was observed for cytoplasmic 28S/18S rRNAs, even at very high peroxide concentration. Analysis of 21 cytoplasmic mRNAs revealed little or no decrease in mature band response to peroxide, and several cytoplasmic mRNAs were actually up-regulated.

The *human* fibroblast GM00637/MT-III cell line (in which the brain-specific metallothionein MT-III coding region was permanently transfected) displayed significantly more resistance against H_2O_2 than GM00637 cells, as determined by cytotoxicity, lactate dehydrogenase leakage, and lipid peroxidation (You et al. 2002). In addition, the GM00637/MT-III cells were highly protected from the H_2O_2-induced production of reactive oxygen species and DNA damage.

Exposure of L929 fibroblasts to **ozone** resulted in K^+ leakage and inhibition of several enzymes (Van der Zee et al. 1987). Most sensitive to ozone exposure were glyceraldehyde-3-phosphate dehydrogenase (EC 1.2.1.12) and pyruvate kinase (EC 2.7.1.40). The activity of another hydroslic enzyme, lactate dehydrogenase (EC 1.1.1.27), the mitochondrial enzymes glutamate dehydrogenase (EC 1.4.1.3), succinate dehydrogenase (EC 1.3.99.1), cytochrome *c* oxidase (EC 1.9.3.1) and the activity of the lysosomal enzymes acid phosphatase (EC 3.1.3.2) and β-glucuronidase (EC 3.2.1.31) were, initially, not or only slightly affected. After prolonged exposure complete deterioration of the cells were observed and all enzyme activities declined.

Neither H_2O_2 treatment (0.5 mM) nor short-term (1 mM **S-nitroscysteine**; half-life in the range of several minutes at 37 °C) or long term (1 mM **(Z)-1-[N-(2-aminoethyl)-N-(2-ammonioethyl)amino] diazen-1-ium-1,2-diolate**; half-life of about 7–8 h at 37 °C) •NO-exposure affected intracellular GSH levels of L929 fibroblasts, whereas concomitant γ-glutamylcysteine synthetase inhibition, but not glutathione reductase inhibition, completely decreased GSH concentrations (Berendji et al. 1999).

Several cellular signal transduction cascades are affected by oxidative stress. Exposure of HER14 cells with H_2O_2 resulted in a concentration-dependent inhibition of epidermal growth factor (EGF) receptor internalisation (De Wit et al. 2000).

As judged by NADPH-diaphorase staining, nitric oxide synthase activity is increased in TrkA (the nerve growth factor-activated receptor tyrosine kinases)-expressing NIH3T3 (TRK1) cells upon exposure to nerve growth factor (Bulseco et al. 2001). Immunocytochemistry showed that levels of the brain NOS isoform were increased in TRK1, but not E25 cells exposed to nerve growth factor.

Using an oxidation-sensitive compound, dihydrorhodamine, Dugan et al. (1997) measured the formation of reactive oxygen species in a central nervous system-derived cell line, (GT1-1 trk) and in superior cervical ganglion neurones, both of which express the transmembrane nerve growth factor receptor tyrosine kinase, trkA. There was enhanced production of reactive oxygen species in both cell types in the absence of nerve growth factor that was rapidly inhibited by application of nerve growth factor; complete inhibition of reactive oxygen species generation in GT1-1 trk cells occurred within 10 min. Nerve growth factor suppression of ROS formation was prevented by PD 098059, a specific inhibitor of MEK (mitogen/extracellular receptor kinase, which phosphorylates mitogen-activated protein kinase).

Cortexolone (4-pregnene-17,21-diol-3,20-dione), a partial agonist of the glucocorticoid receptor, and RU 38486 [17β-hydroxy-11β-(4-dimethylamino-phenol) 17α-(prop-1-ynyl)estra-4,9-diene-3-one], a pure antagonist, were able to modulate and partially inhibit the suppressive effect of dexamethasone on the induction of NO_2^- in specific pathogen-free male Wistar *rat* lung fibroblast cultures (Jorens et al. 1992).

Peroxyl radicals produced strand breaks and base modification in DNA isolated from *human* male fibroblasts grown in 150 mm dishes to confluent monolayers (Rodriguez et al. 1999). Oxidative base modifications were observed to occur at a greater extent than strand breaks at every concentration measured. A total of 87 % of all guanine positions in the examines sequences was found to be significantly oxidised. The order of reactivity of DNA bases toward oxidation by peroxyl radicals was found to be G>>C>T. Adenine is essentially unreactive. The yield of oxidative base modifications at guanines and cytosines by peroxyl radicals depends on the exact specification of 5' and 3' flanking bases in a polarity dependent manner. Every guanine in the 5'XGC3' motif was found to be oxidised, where X is any 5' neighbour. In contrast, 5' and 3' purine flanks drastically reduced the extent of peroxyl radical G oxidation.

iNOS gene upregulation was associated with the early proliferative response of *human* foetal lung fibroblasts (HFL-1) to cytokine (IFN-γ, IL-1β, TNF-α, cytomix) stimulation (ROMANSKA et al. 2002). Administration of the **NO donor** *S*-nitroso-*N*-acetylpenicillamine (5–100 μM) had also a concentration-dependent effect on HFL-1 cells.

The cytotoxicity of the superoxide anion and nitric oxide radicals releasing compound **SIN-1** to L929 cells was significantly higher in the presence of Hepes than in its absence (LOMONOSOVA et al. 1998). The available amount of peroxynitrite formed from SIN-1, however, was significantly decreased by Hepes as indicated by decreased oxidation of dihydrorhodamine 123. On the other hand, 20 mM Hepes largely increased the formation of H_2O_2 from 1.5 mM SIN-1. Catalase protected the L929 cells from SIN-1 cytotoxicity in the buffer with Hepes. In the buffer without Hepes catalase did not have any protective effect. In contrast, tyrosine (1 mM) and tryptophan (1 mM) provided significant protection against SIN-1 cytotoxicity independent of the presence of Hepes. Tryptophan can be oxidised very rapidly, due to the low oxidation potential of this amino acid, and results in the formation of a wide range of materials including ring hydroxylated compounds (at the 2-, 4-, 5-, 6-, or 7-positions) and ring opened materials such as *N*-formylkynurenine, 3-hydroxykynurenine, kynurenine and further oxidation products (JOVANOVIC et al. 1991, MASCOS et al. 1992, JOSIMOVIC et al. 1993).

Sodium nitroprusside as a nitric oxide donor, stimulated tyrosine phosphorylation in cytoplasmic proteins of *murine* fibroblasts stably transfected with the *human* epidermal growth factor EGF receptor (HER14 cells) (MONTEIRO et al. 2000).

In lysates of the *mouse* fibroblast cell line, LMTK⁻, the effects on •NO at increasing RNA-binding activity were only observed when cells were made Fe-replete (WARDROP et al. 2000). Under these circumstances, iron regulatory protein 1 contains an [4Fe⁻4S] cluster that was susceptible to NO. In contrast, when lysates were prepared from cells treated with the Fe chelator desferrioxamine, NO had no effect on the RNA-binding activity of iron regulatory protein 1. In contrast to the NO generators, *S*-nitroso-*N*-acetylpenicillamine, spermine-NONOate, and *S*-nitroso-glutathione, sodium nitroprusside decreased iron regulatory protein 1 RNA binding when cells were incubated with this compound. However, sodium nitroprusside had no effect on iron regulatory protein 1 RNA-binding activity in lysates, suggesting that the decrease after incubation of cells with sodium nitroprusside was not due to S-nitrosation of critical sulphydryl groups.

Collagen synthesis of normal dermal fibroblasts from male Lewis *rats* cultured in high glucose Dulbecco's minimal essential medium containing 10 % foetal *bovine* serum was enhanced by 74.3 ± 18.2 and 87.5 ± 28.2 % in the presence of 100 and 400 μM *S*-nitroso-*N*-acetylpenicillamine, respectively (WITTE et al. 2000). This effect was not due to increased collagen type I or type III gene transcription. Cellular proliferation measured by thymidine incorporation was significantly decreased in the presence of *S*-nitroso-*N*-acetylpenicillamine, indicating that the increased collagen production was due to a net increase of collagen synthesis by the cells.

Nitric oxide synthase inhibitors (300 μM aminoguanidine) block the neoplastic transformation of C3H 10T1/2 *mouse* fibroblasts induced with the carcinogen 3-methylcholanthrene (10 μM for 1 day) during log phase cell growth (BURNETT et al. 2000). In contrast, treatment initiated after formation of a confluent monolayer was associated with diminished protection, while treatment commencing late in the promotional phase had no protective effect and appeared to enhance the number and stage of foci observed. Although induction of iNOS by treatment with lipopolysaccharide (10 μg/ml) + interferon-γ (30 ng/ml) during the last 2 weeks of the assay was associated with enhanced transformation, the efficacy of aminoguanidine in protecting against transformation was not clearly associated with substantial reduction of •NO synthesis.

Treatment of *murine* fibroblast L929 cells with interferon-γ resulted in excess •NO synthesis and iNOS gene expression (OH et al. 2001). All-*trans*-retinoic acid significantly inhibited •NO synthesis and iNOS gene expression in a dose-dependent manner. Similarly 9-*cis*-retinoic acid also inhibited •NO synthesis, but retinol did not show any inhibitory effect on •NO synthesis.

Peroxynitrite (generated from 0.6 M $NaNO_2$ + 1.2 M H_2O_2/0.6 M HCl or from 5 mM SIN-1 in Earle's balanced salt solution) administered to *human* primary fibroblasts from neonatal skin resulted in a dose- and time-dependent activation of the anti-apoptotic kinase Akt (KLOTZ et al. 2000).

Akt activation was rapid and followed by phosphorylation of glycogen synthase kinase-3, an established substrate of Akt. Akt activation was inhibited in the presence of the phosphoinositide 3-kinase inhibitors wortmannin and LY294002, and by treatment with the platelet-derived growth factor (PDGF) receptor (PDGFR) inhibitor AG1295, indicating a requirement for PDGFR and phosphoinositide 3-kinase in mediating peroxynitrite-induced Akt activation.

The peroxynitrite generator SIN-1 (0.5 mM) induced apoptosis both in *src* oncogene-transformed and non-transformed *rat* fibroblasts 208 F, indicating that peroxynitrite is no selective apoptosis inducer per se, but that selective apoptosis induction in transformed cells by •NO is achieved through selective peroxynitrite generation (HEIGOLD et al. 2002). The interaction of •NO with target cell derived superoxide anions represents a novel concept for selective apoptosis induction in transformed cells. Apoptosis induction mediated by •NO involves mitochondrial depolarisation and is blocked by Bcl-2 overexpression.

Ni^{2+} specifically inhibited the binding, internalisation and degradation of α_2-macroglobulin by *murine* embryonic fibroblasts at 37° (KANCHA and HUSSAIN 1997).

V_2O_5, a cause of occupational asthma and chronic bronchitis, activated the extracellular signal-regulated kinases 1 and 2 in *rat* pulmonary myo-fibroblasts (WANG and BONNER 2000). Activation of these enzymes and tyrosine phosphorylation of the 115-kDa Src homology 2 protein tyrosine phosphatase-binding protein by V_2O_5 is oxidant-dependent. Pre-treatment of cells with the antioxidant *N*-acetyl-L-cysteine (50 mM) blocked V_2O_5-induced mitogen-activated protein kinase activation and 115-kDa protein phosphorylation $>90\,\%$.

Polymeric silicic acid (3 mM) interacted directly with neonatal *rat* lung fibroblasts, which had a dramatic effect on the surface membrane, its subsequent internalisation, and cytoplasmic processing (LINTHICUM 2001). Polymeric silicic acid was the only treatment that caused the formation and appearance of numerous osmiophilic vesicles, lipoid bodies and multivesicular bodies during the entire 72-h time course.

For the pharmacology of polysilicic acid see FRIEDBERG and SCHILLER (1988).

The lung fibroblast as influenced by both granular and fibrous **dust** has been studies *in vitro* by morphological, biochemical, and biophysical methods. In *rabbit* lung fibroblast cultures, Rhodesian chrysotile, anthophyllite, amosite and crocidolite UICC reference dusts at a lower dose favoured a fibrogenic response (HEXT and RICHARDS 1976) as judged from the hydroxyproline determination after STEGEMANN (1958). Brucite, a $Mg(OH)_2$ polymer, which is a major component of the outer layer of chrysotile, did not significantly affect the level of cell mat hydroxyproline in the cultures. Persistent long-term exposure to UICC Rhodesian chrysotile over 37 passages lead to enhancement in cell mat collagen deposition and was accompanied by considerably higher both protein and RNA levels between passages 20 and 27 as compared with the normal strain (HEXT et al. 1977). *Human* macrophages derived from blood monocytes by cultivation *in vitro* after incubation with Dörentrup quartz dust and coal mine dusts released a soluble mediator which significantly increased DNA synthesis by WI-38 *human* lung fibroblasts (HÜBNER and SEEMAYER 1992).

Direct exposure of *rat* lung myofibroblasts to urban ambient particles from Mexico City $\leq 10\ \mu m$ in size elicited upregulation of their platelet-derived growth factor receptor α, and this effect was blocked by recombinant endotoxin neutralising protein and mimicked by lipopolysaccharide, but not vanadium, both constituents present within these samples (BONNER et al. 1998).

Rat lungs treated with a single unilateral intratracheal instillation of 400 μg **$CdCl_2$** showed an intense inflammatory reaction followed by infiltration of fibroblast-like cells, until, by day 7, these cells became surrounded by thick bundles of collagen (STEEGER et al. 1988). Between days 1 and 7 after administration of $CdCl_2$, α_1(I) procollagen mRNA increased 16-fold in the treated lung and 34-fold in the contralateral lung. The α_1 (III) procollagen mRNA increased 4-fold in the treated lung and 12-fold in the contralateral lung. However, in the contralateral lung, a similar increase in collagen mRNAs was not associated with an increased deposition of collagen.

Incorporation of [^{3}H]-tyrosine into proteins of cultured *human* foetal lung fibroblasts was impaired by adding 10 μM **Ce** especially in a Mg^{2+}-deficient (35 μM instead of 790 μM) medium (SHIVAKUMAR and NAIR 1991).

Pre-treatment with desferrioxamine protected *human* fibroblast (cell line GM 05757) monolayer cultures from cytotoxicity and genotoxicity (single-strand breaks) induced by the pentachlorophenol (PCP) metabolite, **tetrachlorohydroquinone** (WITTE et al. 2000). Similar pattern of protection were also observed for three other hydroxamic acids: aceto-, benzo-, and salicylhydroxamic acid. Dimethylsulphoxide, an efficient hydroxyl radical scavenger, provided only partial protection even at high concentrations. *In vitro* studies showed that the hydroxamic acids effectively scavenged the reactive tetrachlorosemiquinone radical and enhanced the formation of the less reactive and less toxic 2,5-dichloro-3,6-dihydoxy-1,4-benzoquinone (chloranilic acid).

Type I **collagen** gels prepared from *rat* tail tendons were markedly contracted when *human* bronchial epithelial cells were plated on their top and incubated for 48 h to allow the cells to attach (LIU et al. 1998). Within 24 h the area was reduced by 88±4% ($P < 0.01$). The degree of gel contraction was dependent on cell density; 12,500 cells/cm^2 resulted in maximal contraction, and half-maximal contraction occurred at 7,500 cell/cm^2. Contraction varied inversely with the collagen concentration (91±1% with 0.5 mg/ml collagen vs. 43+5% with 1.5 mg/ml collagen). MIO et al. (1998) demonstrated that, through the release of factors including transforming growth factor-β_2 which can augment and prostaglandin E which can inhibit, *human* bronchial epithelial cells can modulate fibroblast-mediated collagen gel contraction.

A549 cells, as well as *human* bronchial epithelial cells and *rat* alveolar epithelial cells contracted collagen gels more when they were plated on top of the gel than when they were embedded inside, in contrast to *human* lung fibroblast, which contracted more when cast inside (UMINO et al. 2000).

When *human* foetal lung fibroblasts (HFL-1) were cultured in type I collagen gels and floated in medium containing TNF-α, IL-1β, or IFN-γ alone or in combination (cytomix), all cytokines inhibited the contraction of the gel significantly (ZHU et al. 2001). The potency order was IL-1β, TNF-α, IFN-γ. The cytomix was no more potent than was IL-1β alone. PGE_2 production was increased by TNF-α (5.0 versus 0.16 ng/ml, $P < 0.01$), IL-1β (5.3 versus 0.16 ng/ml, $P < 0.01$), and cytomix (5.9 versus 0.16 ng/ml, $P < 0.01$), and was completely inhibited by indomethacin. Indomethacin ($P < 0.05$) and N^G-monomethyl-L-arginine citrate ($P < 0.05$) alone both partially attenuated the inhibition of contraction caused by cytokines alone or by cytomix. Indomethacin and N^G-monomethyl-L-arginine citrate together attenuated inhibition more than either alone ($P < 0.05$). Exogenous PGE_2 and exogenous NO donors inhibited the contraction significantly. The protein kinase A inhibitor KT5270 and the protein kinase G inhibitor Rp-8-pCPT-cGMPS attenuated the inhibition induced by PGE_2 and •NO, respectively.

Within focal regions where fenestration of the alveolar wall and some confluence of air spaces were identified, BELTON et al. (1977) found randomised electron dense collagen fibrils, whereas the normal alveolar septa contained uniformly parallel collagen fibrils. Many of the randomised fibrils had sufficient swelling to display prominent internal spiralling. When the thickened fibrils were cut transversely, the appeared as a stack of flattened, electron dense lamellae.

Tropoelastin, the soluble precursor of the interstitial lung matrix component elastin, is expressed by lung fibroblasts. Phorbol ester-stimulated alveolar epithelial cells secrete a soluble factor what causes a time- and dose-dependent repression of lung fibroblast tropoelastin mRNA (MARIANI et al. 1998). This alveolar epithelial cell-mediated repressive activity is specific for tropoelastin, is effective on lung fibroblasts from multiple stages of development, and acts at the level of transcription. Partial characterisation of the repressive activity indicated it is an acid-stable, pepsin-labile protein. Gel fractionation of alveolar epithelial cell conditioned medium revealed two peaks of activity with relative molecular masses of ~25 and 50 kDa.

Elastin gene expression in *rat* lung fibroblasts was selectively inhibited by okadaic acid (BERK et al. 1996). *In vitro*, 5 nM had minimal effects (91% control values) on elastin mRNA levels; okadaic acid at 25 nm and 50 nm decreased elastin mRNA to 23 and 6% of control levels, respectively. Inhibition of protein synthesis with cycloheximide did not block okadaic acid-induced suppression of elastin mRNA levels. Okadaic acid had minimal effect of *rat* GTP-binding protein mRNA levels. Sodium orthovanadate, a tyrosine phosphatase inhibitor, induced minor decreases in elastin mRNA levels at micromolar concentration. Protein kinase C desensitisation by prolonged exposure to phorbol 12-myristate 13-acetate did not alter the effect of okadaic acid on elestin mRNA levels.

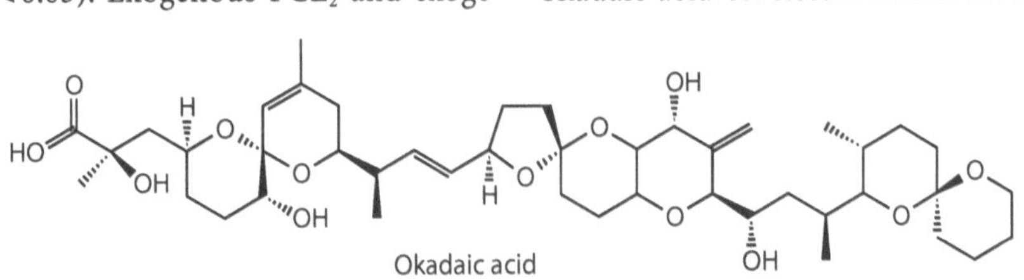

Okadaic acid [262]

The elevated expression of calpastatin, a specific inhibitor of calpain, induced in *human* UV^r-1 fibroblasts by transfection of its cDNA resulted in decreased survival in the presence of okadaic acid but in no apparent alteration in the sensitivity to other drugs such as 5-fluorouracil, mitomycin C and metothrexate (Chi et al. 1999).

In an "emphysematous" variant from C57BL/6J *mice*, the tight skin *mouse* (C57BL6J. Tsk+/+pa), O'Donnell et al. (1997) observed fragmentation of the elastin ultrastructure with fragmentation and disorganisation of alveolar plates.

Elastin shows a distinct increase with age (Blumenthal et al. 1964). Amino acid composition of elastins isolated from *human* pulmonary connective tissue by alkaline digestion demonstrated an increase in glutamic and aspartic acids (Fitzpatrick and Hospelhorn 1962). Within skin elastic fibres, starting of the fourth decade of life, electron dense materials accumulate in an age-dependent manner (Pasquali-Ronchetti and Baccarani-Contri 1997). In very old subjects, these materials seem to have disappeared, leaving behind holes, which give to the fibre a cribriform appearance.

Lovastin induced apoptosis in acute lung injury and idiopathic pulmonary fibrosis fibroblasts as determined by TUNEL, acridine orange staining, and flow cytometry (Tan et al. 1997).

Thiol depletion by culturing normal *human* foetal lung fibroblasts in cystine-free medium or with thiol-depleting agents induced oxidant accumulation and cell death by apoptosis (Aoshiba et al. 1999). The cell death was prevented by the antioxidants ascorbic acid and catalase.

10.1.3 Histiocytes

As are alveolar macrophages, interstitial macrophages have been suggested to be heterogeneous. Chandler et al. (1986) showed marked heterogeneity in cellular volume of interstitial macrophages isolated from male Fischer 344 *rats* by a modification of the method of Hunninghake and Fauci (1976). Macrophages of density 1.046 to 1.075 g/ml exhibited higher receptor activity capability of attaching and phagocytizing *sheep* red blood cells opsonized with small amounts of IgG and towards zymosan. All density-defined interstitial macrophages exhibited similar abilities to attach complement coated *sheep* red blood cells.

Using monoclonal antibody techniques, in the bronchiolar submucosa $M241^+/T6^-$ **dendritic cells** were the predominant population (75 % of the labelled cells), whereas $M241^-/T6^+$ Langerhans cells represented less than 10 % of the labelled cells (Soler et al. 1988). In asthmatics, there were higher numbers of $CD1a^+$ dendritic cells ($P = 0.025$), $L25^+$ dendritic cells ($P = 0.008$), HLA-DR expression ($P = 0.057$) and IgE^+ cells ($P = 0.0003$) in the lamina propria compared with non-asthmatic controls (Möller et al. 1996). *Human* lung dendritic cells obtained from surgical specimens distant from a primary lung carcinoma had an immature phenotype with a rather good endocytic capacity, through their mannose receptor, while maintaining a low CD64 expression (Cochand et al. 1999).

An inducible isoform of **nitric oxide synthase** (iNOS) is expressed by macrophages in the normal *rat* lung interstitium (Liu et al. 1997). When costimulated with granulocyte-macrophage colony-stimulating factor and interferon-γ, interstitial macrophages expressed a marked increase in nitric oxide ($^{\bullet}NO$) production. Intratracheal challenge with heat-killed *Listeria monocytogenes* yielded decreased NO production by interstitial macrophages.

In the *rabbit*, Schwarz (1963) described histiocytes commonly rounded with long projections. Their relatively ample cytoplasm shows a well developed Golgi apparatus, and various numbers of electron-dense inclusions. Nuclear chromatin is aggregated to irregular clumped granules, while electron-lucent cords traverse the nucleus.

10.1.4 Lymphoid Cells

Probably due to the narrow interstitium of the lung, lymphocytes had been overlooked in this area for a long time. In the *rabbit*, Schwarz (1963) described oval to spindle-shaped lymphoid cells with plumpy projections. Holt et al. (1986) carefully isolated, quantified and characterised lymphocytes in the interstitium of the *human* lung. For the whole lung interstitium a total of about 10^{10} lymphocytes were calculated, which is equivalent to the lymphocyte number in the circulating blood pool in *humans*. In the interstitium of the *human* lung natural killer cells (NK cells) have also been identified (Weissler et al. 1987). The lymphocytes in the interstitial pool of the *rat* revealed a subset composition different from that in the peripheral blood (Fliegert et al. 1995).

Both bryostatin 1 and 12-*O*-tetradecanoylphorbol-13-acetate inhibited lymphocyte antibody-dependent cell-mediated cytotoxicity such that, at a 20:1 effector-to-target ratio, control lymphocytes had a mean 49 ± 12 % specific ^{51}Cr release of antibody-coated target cells while bryostatin 1 and 12-*O*-tetradecanoylphorbol-13-acetate cultured lymphocytes had 12 ± 6 and 8 ± 12 %, respectively, in four experiments (Tilden and Kraft 1991).

When B-lymphocytes from healthy *volunteers* were transformed by Epstein-Barr virus and maintained in RPMI 1640 medium with 10 % foetal *bovine* serum, herbimycin A, a specific inhibitor of tyrosine kinase, significantly decreased 12-*O*-tetradecanoylphorbol-13-acetate-stimulated superoxide production in a dose-dependent manner (Yang et al. 2000). The amount of p91, the catalytic subunit of NADPH oxidase, was decreased in the cellular membrane of herbimycin A treated cells compared to untreated controls. Similar results were obtained for the movement of a regulatory subunit of the NADPH oxidase complex, p47.

Several types of lymphocytes, not a single phenotype alone, produced interferon-γ in the normal *mouse* lung (Davis et al. 2000). $CD4^+$ T cells were the most numerous producers of IFN-γ, with lesser numbers of natural killer cells and of $CD4^-CD8^-$ $\gamma\delta TCR^+$ cells. In the normal lung, only 2 to 5 % of the cells within each of these surface antigen phenotypes produced IFN-γ. Both the absolute number and the percentage of IFN-γ-producer cells within each of these three phenotypes increased substantially in fully developed silicosis induced by exposure to cristobalite (70 mg/m^3) for 6 h/d for 12 d. The absolute increase was greatest for $CD4^+$ T-cells, but the relative 3- to 5-fold increase in producer cell number was similar for all free phenotypes.

A 15 kDa selenoprotein was isolated from ^{75}Se-labelled *human* T cells and was shown to contain a selenocysteine residue encoded by TGA (Gladyshev et al. 1998).

Surfactant protein $D^{-/-}$ *mice* demonstrated increased numbers of airway- and vessel-associated lymphocytes without increases in interstitial lymphocytes (Fisher et al. 2002). There was increased proliferative activity of lymphocytes isolated by enzymatic dissociation of minced lung. There was marked T-cell activation in the lungs of surfactant protein $D^{-/-}$ *mice*, as reflected by an increased percentage of both $CD4^+$ and $CD8^+$ T cells expressing CD69 and CD25.

Intratracheal instillation of silica (20 mg in 50 μl sterile saline) into BALB/c *mice* reduced mitogenic responses to T cell receptor stimulation, and markedly increased activation-induced cell death, compared with control lymphocytes from saline-instilled *mice* (Borges et al. 2002). $CD4^+$ T cell death was mediated by Fas ligand, because $CD4^+$ T cells from Fas ligand-deficient *gld mice* did not suffer activation-induced apoptosis.

10.1.5 Plasma Cells

Plasma cells, together with polymorphonuclear leucocytes and smooth muscle cells compose only 2 % of the interstitial cell population of the normal *human* lung (Pinkerton et al. 1988).

Plasma cell iron was found in patients without other morphological changes of alcoholism such as megaloblastosis, erythroid vacuolisation, and ringed sideroblasts (McCurley et al. 1984). Plasma cell iron could be demonstrates in biopsy and postmortem material from extra-marrow sites. Ultrastructural studies showed iron always was located in membrane-bound lysosomal vesicles of plasma cells. It seems that Fe is introduced into the plasma cell by micropinocytosis. In aspirated marrow from a 65-year-old man Shanmugathasa et al. (1979) saw reticular macrophages and plasma cells containing many pleomorphic unit membrane delimited structures with large amounts of iron and heterogeneous deposits of other dense materials (residual bodies). Connections (cytoplasmic bridges) were noted between plasma cells and reticular cells as well as between plasma cells. In these areas, the membranes were fused proving that a true bridge existed. Exchange of materials across these connections was not demonstrated.

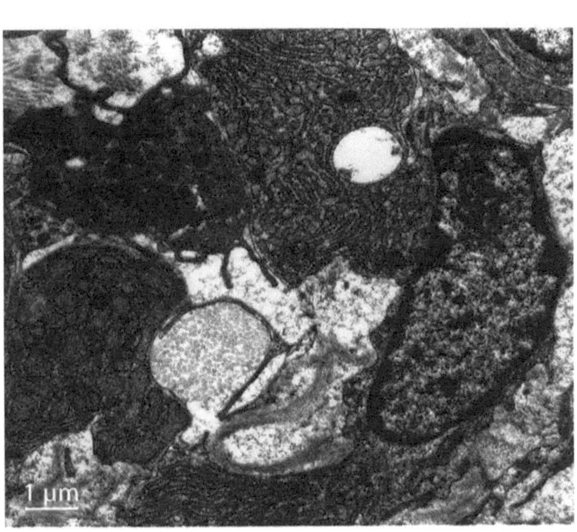

Fig. 191. Plasma cells in the pulmonary interstice of a female *rat* (breeder: Winkelmann, Borchen-Kirchborchen) which inhaled 200 μg micronized deptropine citrate/puff and 200 μg micronized isoprenaline hydrochloride/puff from a suspension type self-propelled aerosol (Medihaler®). 12 Puffs/min were dispersed into a 164.5 l box where the animals stayed for 15 min. 1 h later under methitural anaesthesia, the lung was fixed by intratracheal instillation of 2.5 % glutaraldehyde in phosphate buffer (pH 7.4) before opening the thorax. Postfixation with 1 % osmium tetroxide in phosphate buffer (pH 7.4). Contrasted en bloc for 12 h with 0.5 % uranyl acetate in 70 % ethanol. Embedded in a 2:8 mixture of methyl and butyl methacrylate. Sectioned at 50 nm. Lead citrate after Reynolds (1963). Plate 52/06

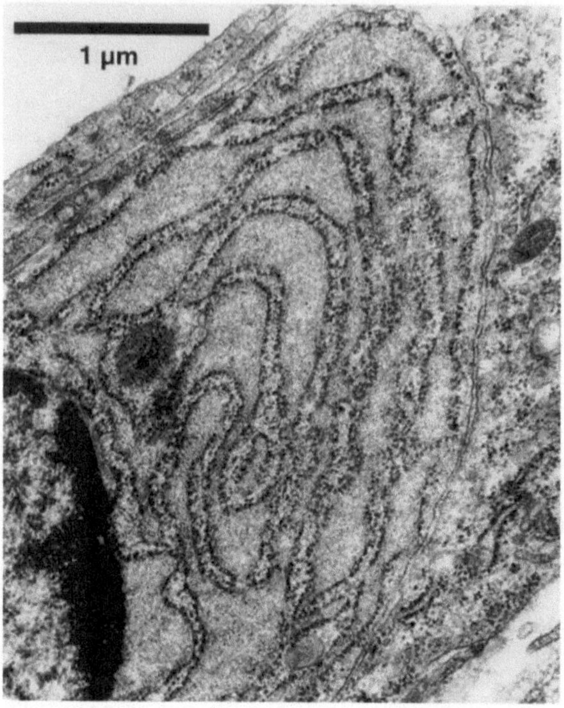

Fig. 192. Plasma cell from the pulmonary interstice of a female *rat* (breeder: Winkelmann, Borchen-Kirchborchen) which inhaled 10 mg powdered Grängesberg magnetite/m^3 4 hours per day, 5 days per week from August 24 to October 16, 1967 for a total of 38 days. Fixed on October 17, 1967 under methitural anaesthesia by intratracheal instillation of 2.5% glutaraldehyde in phosphate buffer (pH 7.4) before opening the thorax. Postfixation with 1% osmium tetroxide in phosphate buffer (pH 7.4). Contrasted en bloc for 12 h with 0.5% uranyl acetate in 70% ethanol. Embedded in a mixture of methyl and butyl methacrylate. Sectioned at 50 nm. Lead citrate after REYNOLDS (1963). Plate 15/09

10.1.6 Mast Cells

Post-mortem lung tissue from two patients dying of non-pulmonary diseases showed mean values of 26.1 and 50.6 mast cells per mm^2 (HEARD et al. 1989).

Mucosal mast cells and connective tissue mast cells differ in **staining properties**. Mucosal mast cells stained blue with copper phthalocyanin dyes, such as astra blue or alcian blue, in a staining sequence with safranin, while connective tissue mast cells stained red (ENERBÄCK 1966). Mucosal mast cell granules, unlike those of connective tissue mast cells do not exhibit a fluorescent binding with the dye berberine (WINGREN and ENERBÄCK 1983). This dye forms a strongly fluorescent complex with heparin in connective tissue mast cells (ENERBÄCK 1974). A staining sequence of berberine followed by toluidine blue can be used to distinguish connective tissue from mucosal mast cells in the same specimen (WINGREN and ENERBÄCK 1983).

Table 41. Non-allergic increase of pulmonary mast cells (from Bienenstock et al. 1986, 1987)

Hypoxia/hypertension
Immunization
Interstitial lung disease
Fibrosis
Asbestosis, silicosis

From *human* lung tissue obtained at thoracotomy $2.1 \pm 0.4\%$ mast cells were found in the dispersed cell population of $15.0 \pm 1.5 \times 10^6$ cells/g tissue gained after digestion with 1 mg collagenase/ml for 4×30 min at 37 °C (RAAIJMAKERS et al. 1988). Due to their seizes and densities the formalin-insensitive alcian blue positive cells with densities < 1.060g/ml could be separated from formalin-sensitive alcian blue positive ones at densities > 1.082 g/ml.

A relationship has often been observed between the increase in numbers of mast cells, signs of mast cell degranulation, and the synthesis and accumulation of collagen leading to fibrosis. This applies to a number of *human* and experimental conditions, notably chronic inflammation, scarring and diffuse pulmonary fibrosis (interstitial fibrosing alveolitis). The pulmonary lesions are basically similar whether they occur in *man* (KAWANAMI et al. 1979, LYKKE et al. 1979) or are induced in *rats* by asbestos (WAGNER et al. 1984), irradiation (WATANABE et al. 1974, TRAVIS et al. 1977, VERGARA et al. 1987) or bleomycin (GOTO et al. 1984, COX et al. 1988). In *human* pulmonary sarcoidosis the bronchoalveolar lavage contained an increased number of basophil and metachromatic cells of uncertain lineage (RANKIN et al. 1987) and cells resembling mucosal mast cell-type (FLINT et al. 1986).

In *rat* mast cells there are two distinct **serine proteases** (WOODBURY and NEURATH 1980, LAGUNOFF 1981) which differ in solubility, structure and antigenicity. One of these, referred to as Rat Mast Cell Protease II (RMCP II), is found in the intestinal mucosa and localised in mucosal mast cells (WOODBURY et al. 1978); the other enzyme is referred to as Rat Mast Cell Protease I and is mainly located in mast cells of the connective tissue mast cell type. Mast cells in the lung parenchyma contained RMCP I exclusively, while cells in the bronchial epithelium contained RMCP II (GIBSON and MILLER 1986). PEARCE et al. (1982) showed that disodium cromoglycate and theophylline were effective against connective tissue mast cells, while they did not inhibit the antigen-induced histamine release from mucosal mast cells.

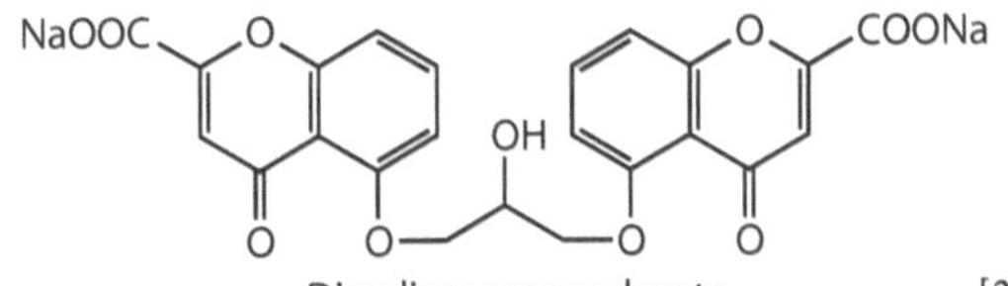

[236]

Mast cells with their **metachromatic granules** (0.3 to 1.0 μm in diameter) as their hallmarks produce histamine, which has marked vasoactive effects and also seems to enhance oxygen radical formation by phagocytes (FRIEDL et al. 1989).

As the injection of $CaCl_2$ (0.1 M) into individual *rat* peritoneal mast cells elicited secretory granule extrusion (KANNO et al. 1973), calcium might be involved in some stage of the secretion process (GOTH 1978). *N*-benzhydryl-*N*'-*p*-hydroxybenzylpiperazine, a potent anti-histaminic and anti-serotoninergic calcium blocker, prevented mast cells from degranulation (Fig. 193), while the introduction of particulate matter into the body induced mast cell degranulation (Fig. 196).

Human mast cell line HMC-1 stimulated by phorbol 12-myristate 13-acetate and ionomycin or interleukin-2 significantly increased interleukin 13 gene expression, which was significantly suppressed by dexamethasone at 1 μM (FUSHIMI et al. 1996).

Lipopolysaccharide (20 μg/ml) stimulation of *rat* mast cells collected from the abdominal and thoracic cavities caused a significant ($P < 0.005$) decrease in the amounts of $O_2^{\bullet -}$ as measured by the increase in adsorbance due to the reduction of ferricytochrome *c* at 550 nm (SALVEMINI et al. 1991). This effect of lipopolysaccharide was in turn attenuated by treating the cells for 1 h with 300 μM N^G-monomethyl-L-arginine ($P < 0.0005$). N^G-Monomethyl-L-arginine alone promoted a small but significant increase in the levels of $O_2^{\bullet -}$ ($P < 0.01$). The release of histamine by $O_2^{\bullet -}$ generated by xanthine (100 μM)-xanthine oxidase (20 mU/ml) was not altered by sodium nitroprusside (80 μM).

De novo expression of eNOS was noted in mast cells located in the perivascular and peribronchial connective tissue of the *rat* lung in response to lipopolysaccharide (ERMERT et al. 2002).

In unstimulated *rat* peritoneal mast cells, GILCHRIST et al. (2002) basally detected small amounts of eNOS. Following stimulation by antigen, interferon-γ, or anti-CD8 antibody, peritoneal mast cells upregulated iNOS mRNA expression. In situ reverse transcriptase-polymerase chain reaction confirmed that iNOS mRNA originated from peritoneal mast cells. Production of iNOS protein was confirmed in stimulated mast cells by immunohistochemistry. Upon stimulation with antigen, interferon-γ, or anti-CD8, nitrite production was increased significantly (8.4±0.6, 7.6±0.9, and 6.6±0.9 μM/2×10^5 cells/48 h NO_2^-, respectively; $P < 0.01$), whereas unstimulated peritoneal mast cells released 2.1±0.3 μM/2×10^5 cells/48 h NO_2^-.

In *rat* gastrocnemius muscle both nitric oxide synthase II and nitrotyrosine were localised exclusively to mast cells except after 24 h reperfusion when some macrophages and neutrophils also showed positive immunoreactivity (MESSINA et al. 2000).

In permeabilized mast cells, recombinant Rac and Rho proteins enhance secretion, whereas C3 tansferase and dominant negative Rac inhibit GTPγS-induced secretion (PRICE et al. 1995). Rac has also been purified from mast cells as a factor that can enhance secretion (O'SULLIVAN et al. 1996). Secretion is accompanied by actin reorganisation and these changes are also mediated by Rac and Rho, but by using inhibitors of actin reorganisation it has been shown that the two responses can be regulated independently (NORMAN et al. 1996).

In chronic fibrotic reactions mast cells appear to interact with other cells as fibroblasts, macrophages (see Figs. 280, 297), and T cells (CLAMAN 1985). The outcome may depend on the balance between inhibitors and stimulators of collagen production. In the lung it may be that all the various cells present are constantly interacting during fibrogenesis (REISER and LAST 1986), but the collagen is primarily produced by fibroblasts (TRELSTAD and BIRK 1985). Evidence of a link between the mast cell, the activation of fibroblasts, and the progression of connective tissue changes in fibrotic sarcoid lung has been presented by BJERMER et al. (1987).

10.1.7 Neutrophil Granulocytes

Neutrophil trafficking in lung involves transendothelial migration, migration in tissue interstitium, and transepithelial migration. In a *rat* model of IgG immune complex-induced lung injury, JOHNSON and WARD (1974) neutrophils accumulated in alveolar compartment and lung interstitium. Production of reactive oxygen species and release of proteinases from infiltrated neutrophils and activated macrophages results in severe tissue damage. Mediators and regulation of neutrophil accumulation in inflammatory responses in lung were reviewed by GUO and WARD (2002).

10.1.8 Eosinophils

The lung is one of the principle sites of the accumulation of eosinophils. That tissue eosinophils are difficult to find in pathogen-free animals strongly suggests that the retention of eosinophils in tissues is disease-related (Spry 1993). *Guinea pig, bovine* and *human* eosinophils, when challenged with platelet-activating factor (PAF), display a marked increase in oxygen consumption and liberate **superoxide anions** extracellularly as a consequence of the activation of the NADPH oxidase. In *guinea pig* cells, this effect occurs at concentrations of PAF >100-fold higher than are necessary to promote chemotaxis, thromboxane production, degranulation, inositol (1,4,5)triphosphate accumulation and Ca^{2+} mobilisation (Kroegel et al. 1989, 1991). However, the finding that oxidant production was antagonised by apafant in those studies indicates that this response also is PAF receptor-mediated. In addition to increasing directly oxidative metabolism, low concentrations of PAF that produce little, if any, superoxide anions per se, prime the eosinophil NADPH oxidase to activation of N-formyl-methionyl-leucyl-phenylalanine (Zoratti et al. 1992) and serum-opsonized zymosan (Coffer et al. 1998).

Budesonide inhibited eosinophil activation primarily through effects on lung fibroblasts, presumably by inhibiting production of granulocyte-macrophage colony-stimulating factor (Spoelstra et al. 2000). After longer incubation periods, budesonide also directly inhibited eosinophil activation. In contrast, formoterol can inhibit eosinophil activation only via inhibitory effects on lung fibroblasts.

10.1.9 Pericytes

Pericytes, contractile cells contained within the sheath of the capillary basement membrane, represent 17 % of the total interstitial cell population in the normal *human* lung (Pinkerton et al. 1988). They demonstrated intensive membrane contacts with endothelial cells and commonly bridged endothelial cell lining adjacent capillary lumina. The close apposition of pericytes with the abluminal surfaces of endothelial cells may serve a potential function of regulating local blood flow through the pulmonary capillary bed.

Whereas pericytes appear to be quite frequent in *human* and *dog* lungs, as well as in the *guinea pig*, they are rarer in *rat* lungs, and Weibel (1974, p. 233) has not been able to find a pericyte process in the lung of the *Etruscan shrew* (body weight 2 g), in which the pulmonary air-blood barrier is extraordinary thin. Haworth (1983) represented a pericyte in the alveolar wall of a *pig* aged 4 hours.

On the side adjacent to the endothelium the cytoplasm contains a dense meshwork of fine filaments, whereas on the outer surface numerous pinocytotic vesicles are observed (Weibel 1974). Pulmonary pericytes influenced the actin distribution of pulmonary microvascular endothelial cells (Shepro and Morel 1993). Pericytes enhanced the formation of a distinct dense peripheral band at microvascular endothelial cell periphery and promoted close apposition of adjacent endothelial cells. Transforming growth factor-β appears to mediate this pericyte-endothelial interaction. The role played by TGF-β is supported by the finding that this agonist modulates extracellular matrix organisation and the formation of tubelike structures with apparent tight junctions in three-dimensional cultures of *rat* epididymal fat pad microvascular endothelial cells (Merwin et al. 1990).

Growth of *rat* lung pericytes is differentially encouraged by culture in DMEM + 10 % homologous serum; that of endothelial cells, by supplementation with 5 % of platelet poor plasma (Davies et al. 1984). The resulting lung pericyte monolayer comprises cells that adhere strongly and exhibit prominent stress fibres. They can be further distinguished from endothelial cells by their lack of contact inhibition and negative staining with anti-Factor VIII antibody, and from lung fibroblasts by positive staining with antibodies against both smooth muscle myosin and platelet myosin.

10.2 Interalveolar Pores

In all *mammalian* species so far examined, there are openings called pores of Kohn penetrating the interalveolar septa, the number varying among species (Macklem 1971). Pores are generally absent at birth, but they appear early in life and are established in *human* lungs by 1 year of age. They are round to oval, varying in diameter with the degree of lung inflation. In humans at total lung volume, they range from 2.5 μm to an upper limit of normal taken somewhat arbitrarily at 12 to 15 μm. Takaro et al. (1979) found 5.3 μm to be the average diameter. Since electron microscopy has invariably shown them to be lined by alveolar epithelium, they are probably not an artefact (Weibel 1971, Cordingly 1972, Takaro et al. 1979). Macklin (1948) described silvered dust cells inserted into alveolar pores of the *mouse*.

Wright (2001) found that *guinea pigs* exposed to cigarette smoke for 12 months had a larger mean number of pores per alveolus ($P < 0.001$), and the distributions of pore size and shape were significantly shifted to indicate a larger and more irregular pore configuration ($P < 0.001$, $P < 0.01$, respectively). These alterations appeared to precede any increase in airspace size.

10.3 Peribronchium

The peribrochial tissue, appreciated for the first time by POLICARD (1938), represents a sheath keeping open the bronchial lumen and preventing its deformation. In the region of the bronchioles it is rudimentary and increases in size from the small brochi across the medium to the large ones (POLICARD 1955, p.223).

In specific-pathogen-free, female BALB/c *mice* sensitised with picryl chloride, cellular infiltrates appeared around the bronchiole and its accompanying blood vessel at 12 h after an intratracheal instillation of picryl sulfonic acid and progressively expanded by 48 h (NISHIDA et al. 1999). As quantitated by computer-assisted morphometry, I-A^+ dendritic cells and CD^+ Th cells significantly increased in number around the bronchiole to a maximum at 24 h, whereas F4/80^+ macrophages were predominantly accumulated around the accompanying vessels with a peak at 48 h. Serial section analysis revealed that dendritic cells were colocalizes with Th cells in the inflamed peribronchiolar tissue. Immunoelectron microscopy demonstrated that dendrtic cells found inside and around the capillaries and venules of peribrochiolar interstitium displayed round forms, indicating their emigration from here, while those situated far from the microvessels were elongated, often in close apposition to the lymphocytes.

Giant cells in the peribronchiolar interstitium were seen one week after instillation of long fibres obtained by sedimentation of UICC crocidolite (BOWDEN and ADAMSON 1987). By two weeks many granulomas had formed at these locations. Electron microscopy detected occasional fibres within giant cells. By six weeks the cellularity of the lesions had decreased and the granulomas became more fibrotic.

In the lungs of *rats* exposed to cigarette smoke, CHANG et al. (2001) found the terminal bronchioles infiltrated predominantly with lymphocytes in the peribronchiolar region and a mild moderate degree of emphysema in the alveolar spaces. The terminal bronchioles also showed marked lipid peroxidation, dilatation, and peribronchiolar fibrosis.

CHAPTER 11

Pulmonary Blood Vessels

The lungs receive blood through both the bronchial and pulmonary arteries. The former carries blood at systemic pressure and has thicker, more muscular walls than the pulmonary arteries which carry blood under one-sixth as much pressure.

Pulmonary artery pressure and size may be affected by the hypoxia that occurs with increasing altitude. According to GIO et al. (1988), however, the radiologically measured pulmonary artery size at median altitudes (Salt Lake City; 1400 m) is the same as that seen at sea level.

For mapping the microvascular pattern of normal and diseased tissue, NETTUM (1995) described a combined vascular-bronchoalveolar casting method using formalin-fixed *canine* lungs and a low viscosity silicone rubber.

11.1 Arteria Pulmonalis

11.1.1 Truncus

The pulmonary trunk and pulmonary arteries larger than 500 μm or 1 mm in diameter are designated elastic pulmonary arteries. Their tunica media consists of multiple concentric elastic laminae separated by smooth muscle, collagen and ground substance containing proteoglycan.

In tissue sections of the *porcine* pulmonary trunk, endothelial cells were specifically immunolabelled with both the monoclonal and polyclonal antibodies raised against **vimentin** (SCHNITTLER et al. 1998). A moderate degree of staining intensity was observed in all endothelial cells of the pulmonary trunk. The pulmonary trunk contained a 2- to 2.5-fold higher amount of vimentin than the endocardial endothelium of the right ventricle. Cultured endothelial cells of the *porcine* pulmonary trunk displayed considerably higher amounts of vimentin than did freshly isolated cells.

11.1.2 Muscular Branches

The morphologic changes of the *human* pulmonary arteries with age showed that almost immediately after birth there was a rapid thinning of the media, which progressed more gradually until it reached the values of adult pulmonary arteries at an age between 6 and 18 months (NAEYE 1961, WAGENVOORT et al. 1961). The density of muscularized pulmonary arteries decreased rapidly during the first few years of life, than more gradually during the first two decades of life (TAKASHI et al. 1983). Concomitantly, the diameter of the smallest muscularized arterial branches increased with age.

11.2 Capillaries

In the lung of *Macacus sinicus* Geoffr. completely injected with India ink PFEIFER (1934) found varying congestion related to the ventilation. In hyperaerated areas, capillary nets were rather dilated, and the capillaries stretched and very thin. Other areas showed a narrow mesh of short, stout capillaries with a slit-like lumen (sheet flow, FUNG and SOBIN 1969). PFEIFER (l.c.) emphasised that congestion not only depended on blood pressure, but also on the variable tightness and relax of the pulmonary tissue.

In the *cat*, FUNG and SOBIN (1969) showed the pulmonary interalveolar capillary bed to be consistent with a sheet-like endothelium-lined space bridged by avascular endothelium-covered posts. Electron microscopy demonstrated that the capillary posts have a highly organised internal structure with abundant collagen and an elastin or elastinlike core (SOBIN et al. 1972). Collagen fibres originate from the alveolar-capillary basement membrane, emerge in a herringbone pattern, and sweep toward the centre of the post in a helical array around the elastinlike amorphous and fibrillar core. The unusual compliance of the microvascular blood vessels in the lung can be correlated with the architectural organisation within the posts.

Substantial reversible morphological changes in the configuration and dimensions of capillaries on *rabbit* lungs fixed by vascular perfusion involving differences in capillary diameter, number of capillaries open and homogeneity of capillary size were found by CIUREA and GIL (1996). In zone 1 (apex), where the air pressure (PA) > the arterial pressure (Pa) > the venous pressure (Pv), and in zone 2 (middle), where Pa > PA > Pv, which showed similar morphology, septal capillaries outside the corners were partly closed and when open, frequently had slit-like configuration. In zone 3 (bottom), where Pa > Pv > PA, all septal capillaries were wide open, of circular outline and of similar dimension. Morphometry showed great dissimilarities of capillary size in zones 1 and 2, and homogeneity in zone 3, as proven by analysis of the size distribution curve of capillary sizes.

The capillary endothelial cells are nonfenestrated and form a continuous lining of the alveolar capillaries. Their surface is flat (SMITH et al. 1971). The rows of tight junction particles between the large gap junctions decrease from two to six rows at the arterial end to one or two rows at the venous one. Microfilaments are few, mainly associated with the junctions, and are not organised into distinct bundles (BENSCH et al. 1964). Weibel-Palade granules are lacking.

While the alveolar capillaries in normal *human* lungs and in nonthickened alveolar walls of patients with various fibrotic lung disorders gave a negative reaction for **Factor VIII** this reaction became positive in thickened alveolar septa (KOMATSU et al. 1989, KAWANAMI et al. 1995).

The endothelium of the pulmonary circulation has functions other than serving as a conduit for respiratory gas exchange and fluid filtration. It also acts as a sort of biochemical filter because of its efficient and highly selective metabolic machinery.

Both normoxic and hypoxic *bovine* pulmonary microvascular endothelial cells constitutively released **xanthine oxidase** (EC 1.1.3.22) activity into their culture media (PARTRIDGE et al. 1992). Incubation of hypoxic or normoxic *bovine* pulmonary microvascular endothelial cells with oxygenated medium (95% O_2) stimulated the release of xanthine oxidase activity into the extracellular medium within 5 min. The xanthine oxidase activity could not be detected in the oxygenated medium after 60 min incubation with 95% O_2. Oestradiol (10 μM) almost completely inhibited xanthine oxidase and dehydrogenase activities in both normoxic and hypoxic *rat* pulmonary microvascular endothelial cells (KAYYALI et al. 1999). Dexamethasone increased the activity of xanthine oxidase, and also induced xanthine oxidase promoter activation in endothelial cells transiently transfected with a *rat* xanthine oxidase promoter-firefly luciferase construct.

Glyceraldehyde-3-phosphate dehydrogenase (EC 1.2.1.12) was induced by hypoxia in *bovine* pulmonary artery endothelial cell monolayers (GRAVEN et al. 1999). Upregulation occurred primarily at the level of transcription. Transient transfection studies using portions of the glyceraldehyde-3-phosphate dehydrogenase promoter linked to a CAT receptor gene identified an endothelial cell specific responsive region that was further characterised (using SV40-promoter-CAT reporter constructs) as a 19-nucleotide sequence (–130 to –112) containing both an hypoxia inducible factor-1 (HIF-1)-binding site and a novel flanking sequence. Electrophoretic mobility shift assays confirmed inducible endothelial cell protein binding to this fragment. Mutation of either the HIF-1-binding site of the flanking sequence resulted in complete loss of function and loss of inducible protein binding. Thus, a single HIF-1-binding site is necessary, but not sufficient, for hypoxic regulation of glyceraldehyde-3-phosphate dehydrogenase in endothelial cells.

Nitric oxide synthase (EC 1.14.13.39) in endothelial cells is constitutively expressed. Its activity is dependent on the intracellular Ca^{2+} concentration. The half-saturating concentration (K_M) of the substrate L-arginine measured *in vitro* is 3 μM. Accordingly, endothelial nitric oxide synthase (eNOS) has been shown to be largely independent of the extracellular arginine supply (PALMER et al. 1988). The constantly high L-arginine concentrations might, at least in part, be due to the ability of endothelial cells to recycle L-arginine from L-citrulline (HECKER et al. 1990). However, under certain pathophysiological conditions such as diabetes (PIEPER and PELTIER 1995), hypertension (LAURENT et al. 1995), or hypercholesterolaemia (COOKE et al. 1991, CREAGER et al. 1992), increased plasma L-arginine levels can induce improved vasodilatation presumably due to enhanced eNOS activity in endothelial cells. Incubation of *human* EA.hy926 endothelial cells in 2 mM L-lysine for up to 24 h decreased the intracellular L-arginine concentration from 3.5 mM to about 600 μM but did not reduce the eNOS activity (CLOSS et al. 2000).

Cultured *rat* pulmonary microvascular endothelial cells did not express NOS II mRNA or protein when exposed to normoxia or hypoxia unless they were pre-treated with interleukin-1β and/or tumour necrosis factor-α for 24 h (ZULETA et al. 2002). Induction of NOS II by interleukin-1β + tumour necrosis factor-α was significantly attenuated by concomitant exposure of endothelial cells to hypoxia or treatment of endothelial cells with antioxidants such as tiron, diphenyliodonium, and catalase (EC 1.11.1.6), suggesting that NOS II expression is dependent on the production of reactive oxygen species.

Lungs from common bile duct ligation *rats* exhibited markedly blunted hypoxic pressor response, increased eNOS, protein expression, and decreased endothelin-1 mRNA and peptide expression (CARTER et al. 2000). The blunted hypoxic pressor response was not reversed by sequential NOS and soluble guanylyl cyclase inhibition by nitro-L-arginine and 1*H*-[1,2.4]oxadiazolo[4,3-*a*]quinoxaline-1-one, respectively, or by NOS inhibition combined with endothelin-1 addition. The blunted hypoxic pressor response was not due to a generalised inability to vasoconstrict because perfusion pressure was equally elevated by increased perfusate KCl in common bile duct ligation and sham lungs. After KCl vasoconstriction, hypoxic pressor response was potentiated and did not differ between common bile duct ligation and sham lungs. Blunted hypoxic pressor response was also completely restored in common bile duct ligation lungs treated with nitro-L-arginine, 1*H*-[1,2.4]oxadiazolo[4,3-*a*]quinoxaline-1-one, and Ca^{2+}-activated K^+ channel blockers apamin and charybdotoxin.

Phenylarsine oxide, an inhibitor of tyrosine phosphatase, caused a dose-dependent decrease in eNOS activity in total membrane and in purified eNOS fractions from *porcine* pulmonary artery endothelial cells, even though the latter had no detectable tyrosine phosphatase activity (SU and BLOCK 2000). Phenylarsine oxide also caused a decrease in sulphydryl content and eNOS activity in purified *bovine* eNOS. The reduction in eNOS sulphydryl content and the inhibitory effect of phenylarsine oxide on eNOS activity were prevented by dithiothreitol, a disulphide-reducing agent.

The activity of **extracellular superoxide dismutase** (EC 1.15.1.1) in *mouse* lungs is 3- to 10-fold higher than that found in most *mammals*, and is 30-fold higher than that found in *rat* lungs (MARKLUND 1984). Extracellular superoxide dismutase labelling is strongest in the matrix of vessels, airways, and alveolar septa, especially in the septal tips (FATTMAN et al. 2000).

Superoxide dismutase (EC 1.15.1.1)

$$2\,O_2^{\cdot-} + 2\,H^+ \longrightarrow H_2O_2 + {}^3\Sigma_g^- O_2 \qquad [128]$$

Choline acetyltransferase mRNA was detected by reverse transcription-polymerase chain reaction at *porcine* pulmonary arterial endothelial cells predominantly of large vessels but extended also to the endothelium of arterioles (HABERBERGER and KUMMER 1996). No labelling was observed at the capillary endothelium. Intense choline acetyltransferase immunoreactivity was observed on freshly cultured endothelial cells of *porcine* pulmonary arteries.

While endothelial cells in *human* control lungs showed very weak histochemical reactions for **metalloproteinases** (MMPs), a family of proteolytic enzymes involved in remodelling of the extracellular matrix, in emphysematous lungs there were stronger reactions for MMP-1, MMP-2 (ED 3.4.24.24) and MMP-9 (ED 3.4.24.35) (HORIBA et al. 1997).

Quercetin significantly inhibited basal and oxidised low-density lipoprotein-stimulated MMP-1 expression in *human* umbilical vein endothelial cells cultured in a 5% CO_2 atmosphere (SONG et al. 2001).

Sprouting capillaries (WAKUI 1988) are characterised by small, slitlike lumina and thick-walled endothelial cells that contain numerous cytoplasmic organelles. In densely fibrotic lesions, regenerating endothelial cells contained numerous free ribosomes, cistern of the rough endoplasmic reticulum, numerous microfilaments, and Weibel-Palade bodies (KAWANAMI et al. 1995). The microfilaments often showed dense insertion sites along the abluminal plasmalemma.

11.3 Pulmonary Veins

11.3.1 Intrapulmonary Veins

The smallest tributaries of pulmonary veins are indistinguishable histologically from the pulmonary arterioles except by tracing their origin and drainage. Thus, the venules have a wall consisting of a single elastic lamina, gradually acquiring a muscular media downstream. Formed near bronchioles, they pass through successive generations to drain into muscular veins in the connective tissue septa between secondary lobules. In young people, the intima of the veins is thin, composed mainly of collagen and a few myofibroblasts, but it gradually thickens with age. The tunica media is slightly irregular in thickness consisting of bundles of obliquely and circularly arranged smooth muscle cells and collagen. Irregular elastic fibrils occur in both the media and the adventitia, and the boundary between these two coats is frequently ill defined. There is usually a distinct internal elastic lamina. The adventitia includes mainly longitudinally oriented elastic fibres and bundles of muscle.

The musculature in small pulmonary veins was found to form small ridges in injection studies in *humans* (Takino and Miyake 1936). The entrance of postcapillary venules was characterised by a muscle ring (Von Hayek 1952). Scanning electron microscopy of casts of *rats'* pulmonary veins showed narrow (1–3 μm) circumferential constrictions about every 30 μm in length (range 20 to 50 μm) and before and after accepting tributaries (Schraufnagel and Patel 1990). These constrictions are caused by muscular sphincters (Aharinejad et al. 1992). The smooth muscle cells are regularly positioned between endothelial layer and elastic lamina. Stained with anti-α-smooth muscle actin they showed discontinuous, periodical thickenings of circular bundles in the walls of the venules, but they became thin and continuous in the larger vessels (or veins) that had a cardiac muscle layer on the outside (Hashizume et al. 1998). Under scanning electron microscopy, the smooth muscle cells formed circular oriented bundles at constant intervals along the venules less than 100 μm in diameter. These bundles had circumferential constrictions in the lumen. The cardiac muscle cells, which appeared in large pulmonary veins of more than 100 μm, ran in a circular or oblique direction and completely surrounded the vessel wall outside of the thin continuous layer of smooth muscle cells. In *cattle*, constrictions occurred in series along the course of veins (9.6/500 μm), giving the cast veins a string-of-pear look, with narrowing of 33–81 % of the outer diameter (Aharinejad et al. 1996). Spanner (1939) described a string-of-beads figure in the pulmonary veins of *horses*.

11.3.2 Extrapulmonary Veins

The extension of cardiac muscle from the left atrium along the walls of the major pulmonary veins at the hilus is most prominent in *rodents* as in the *rat*, the *mouse*, and exceptionally in the *guinea-pig* (Guieysse-Pelissier 1937).

11.4 Bronchial Arteries

The bronchial tree is supplied by arterial blood via the bronchial arteries. On leaving the aorta from its ventral side and branching to both lungs, the bronchial arteries have a coat of circulatory oriented smooth muscle and a thick internal elastic lamina. Once they enter the lungs, they lie in the walls of the bronchi where they are subjected to the stimulus of repeated longitudinal stress. Each bronchus usually has two branches of the bronchial artery coursing along its length. These arteries may be straight in some areas or coiled and have several connections with other branches of the bronchial artery (Tobin 1952).

In response to the longitudinal stress, the bronchial arteries sometimes develop a characteristic layer of longitudinally oriented smooth muscle (Preto Parvis 1954). The smooth muscle cells in this layer were often separated by elastic fibrils. The lamina elastica interna is thin or even lacking. The bronchial arterioles are typical of the systemic vasculature with a thick media of circular myocytes.

Bronchopulmonary anastomoses were described in the *human* lung (Verloop 1948) and in the lungs of *rabbits*, *guinea-pigs*, *rats* and *mice* (Verloop 1949).

Constriction of the bronchial veins by norepinephrine leads to a shift of venous blood to the bronchopulmonary anastomoses, thus increasing venous admixture in the systemic arterial blood. With parenteral injection of epinephrine or isoprenaline in asthmatics, a reduction in oxygen content of the blood occurs (Aviado and Micozzi 1981).

The role of the tracheobronchial circulation in aerosol clearance has been discussed by Wagner (1995). Its supportive role for ciliated cells and mucus glands has not been studied. For this reason it is unclear whether this circulation plays a significant part in maintaining mucociliary clearance. Consequently, the capacity of the airway circulation to influence the clearance of soluble and insoluble particles remains ill-defined and requires investigation at a fundamental level.

11.5 Experimental Pharmacology of Pulmonary Endothelial Cells

Pulmonary endothelial cells are important in the regulation of circulating hormones (Ryan 1982), and consequently it is not surprising that certain xenobiotics, which have physicochemical properties similar to those of the endogenous substrates, also serve as substrates or ligands for the specialised enzymes, receptors, binding sites, and transport mechanisms localised on or in endothelial cells.

Multiple neurohormonal inflammatory/vasoactive factors and ischæmia/reperfusion- and/or leucocyte-induced oxygen radicals promote endothelial cell disruption, likely through elevation of intracellular free Ca^{2+}. Elevated $[Ca^{2+}]_i$ induces endothelial cell permeability by promoting contraction (Garcia and Schaphorst 1995).

A Ca^{2+}-inhibitable adenylyl cyclase is expressed in *rat* pulmonary microvascular endothelial cells (Chetham et al. 1997).

11.5.1 Vasoactive Amines

Certain biogenic amines, including 5-hydroxytryptamine (serotonin), L-norepinephrine, and β-phenylethylamine, are removed from pulmonary circulation whereas others, such as histamine, dopamine, and epinephrine generally are not (GILLIS and GREENE 1977).

11.5.1.1 Serotonin

In perfused lungs, the removal of serotonin occurs via carrier-mediated transport processes that are saturable and sodium- (JUNOD 1972), energy-, and temperature-dependent (IWASAWA et al. 1973). Rapid intracellular metabolism by monoamine oxidase follows (JUNOD 1972). 1 μg Serotonin was inactivated by the isolated perfused lung from male *rats* aged 4 weeks at 52.3±6.3 %, from *rats* aged 7 weeks at 77.1 ± 13.7 %, from *rats* aged 36 weeks at 78.7 ± 13.3 %, from *rats* aged 48 weeks at 61.9 ± 14.6 %, and from *rats* aged 72 weeks at 57.4 ± 12.7 % only (KITAMURA et al. 1983).

Tetrachlorodecaoxygen (20 μM), a wound healing agent, relaxed isolated *calf* pulmonary arteries precontracted with 0.5 μM serotonin, which was rapidly accelerated by addition of 1 μM reduced haemoglobin with or without 20 μg/ml catalase + 10 μg/ml superoxide dismutase (WOLIN et al. 1994).

Indole derivatives autoxidize in the presence of transition metals presumable to nitrogen-centred radicals. Serotonin for a substrate in the H_2O_2-peroxidase reaction system incubated with 200 μM GSH took up 63.1±7.1 μM oxygen (O'BRIEN 1988). The one-electron oxidation potential of indole derivatives is not known, but indoles slowly reduce cytochrome *c*, which suggests that the potential is less than that for cytochrome *c* (PEREZ-REYES and MASON 1981).

During the respiratory burst of *human* blood monocytes serotonin acts as a **radical scavenger** and is oxidised to a dimer, probably 5,5'-dihydroxy-4,4'-bitryptamine (SCHUFF-WERNER et al. 1995). When incubated with a hydroxyl radical ($HO^•$)-generating system (ascorbic acid/Fe^{2+}-EDTA/O_2/H_2O_2), serotonin in rapidly oxidised initially to a mixture of 2,5-, 4,5-, and 5,6-dihydroxytryptamine (WRONA et al. 1995). The major reaction product is 2,5-dihydroxytryptamine, which at physiological pH exists as its keto tautomer, 5-hydroxy-3-ethylamino-2-oxindole. Rapid autoxidation of 4,5-dihydroxytryptamine gives tryptamine-4,5-dione, which reacts with the C(3)-centred carbanion of 5-hydroxy-3-ethylamino-2-oxindole to give 3,3'-bis (2-aminoethyl)-5-hydroxy-[3,7'bi-1*H*-indole]-2,4',5'-3H-trione. The latter slowly cyclises to 3'-(2-aminoethyl)-1',5',7',8'-tetrahydro-5-hydroxy-spiro-[3*H*-indole-3,9'-[9*H*]pyrrolo[2,3-*f*]quinoline]-2,4',5' (1*H*)-trione. A minor amount of tryptamine-4,5-dione dimerizes to give 7,7'-bi-(5-hydroxytryptamine-4-one).

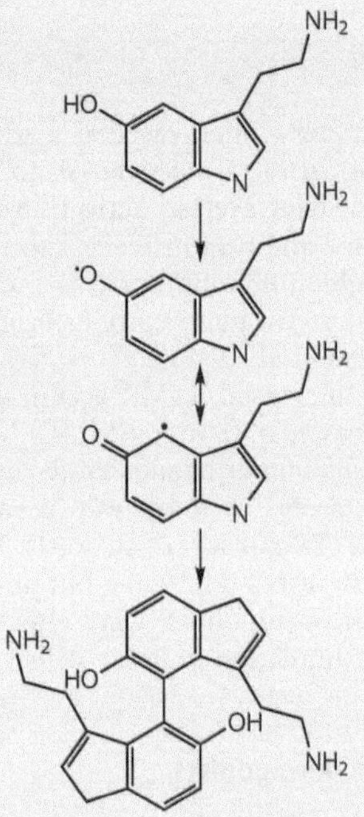

[168]

Serotonin —Ferrocytochrome *c*→ **Serotonin semiquinone-imine radical**

Because $^•$**NO** is not very reactive in an oxygen-free buffer, a significant part of serotonin is transformed by $^•$NO in nondeaerated phosphate buffer, at pH 7.4, into (4-serotonyl)-serotonin, 4-nitroso-serotonin, and 4-nitroserotonin (BLANCHARD et al. 1997). Dimerization and above all nitrosation occur through the HNO_2 reaction in the pH 4–6 range possible via radical mechanism involving N_2O_3. Serotonin is readily a substrate for nitrosation by HNO_2 or N_2O_3, whereas tyrosine was described as not very reactive under the same conditions. Peroxynitrite converts serotonin to the (4-serotonyl)-4-serotonin and to the 4-nitro derivatives.

11.5.1.2 Norepinephrine

In isolated male Sprague-Dawley *rat* lungs perfused for 6 min at 10 ml/min by a peristaltic pump, ^{3}H-L-norepinephrine (10 μM) was rapidly deaminated and O-methylated, and the metabolic products were subsequently returned to the perfusate (NICHOLAS et al. 1974). Electron microscope autoradiographs indicated endothelial cells labelled in the arteries at 12 %, in the small vessels at 90 %, in the capillaries at 31 %, and in the veins at 70 %.

Endothelin inhibited the evoked ^{3}H-norepinephrine overflow in *guinea-pig* pulmonary artery (WIKLUND et al. 1989).

11.5.1.3 Isoprenaline

In *rat* pulmonary microvascular endothelial cells cultivated *in vitro*, responsiveness to isoprenaline or direct adenylyl cyclase activation by forskalin was attenuated and responsiveness to the phosphodiesterase inhibitor, rolipram was increased compared with those in pulmonary arterial endothelial cells (STEVENS et al. 1999).

Rolipram increased cAMP accumulation in *rat* pulmonary vascular endothelial cells than did forskolin, isoprenaline, or adenosine derivatives alone, although extensive synergy was seen with combined agents (THOMPSON et al. 2002). High-affinity phosphodiesterase-4 inhibitors, but not low-affinity or non-selective inhibitors, were effective inducers of cAMP accumulation in intact cells.

11.5.1.4 β-Adrenergic Antagonists

3,4-Dihydro-8(2-hydroxy-3-isopropylamino)-propoxy-3-nitroxy-2*H*-1-benzopyran, nipradilol, is designed as a dual-action drug to produce two complementary pharmacological effects: β-blockade and nitric oxide releasing action (HAYASHI and IGUCHI 1998). It is a non-selective β-blocker, which has approximately two times higher activity than that of propranolol, and has a nitroglycerine-like vasodilating activity (UCHIDA et al. 1993). It improves cGMP production and dilates vessels by releasing nitric oxide (SASAGE et al. 1995, HAYASHI et al. 1997). In cultured *bovine* aortic endothelial cells, this eNOS up-regulatory action was abolished by a β_2-receptor antagonist, erythro-DL-1-(7-methylindan-4-yloxy)-3-isopropylaminobutan-2-ol, ICI-118551 (JAYACHANDRAN et al. 2001).

11.5.2 Polypeptides

11.5.2.1 Angiotensin I

The caveolae of the plasma membrane of endothelial cells contain a carboxypeptidase, angiotensin-converting enzyme (ACE) that converts the decapeptide angiotensin I to the potent vasopressor octapeptide angiotensin II by cleaving off the two C-terminal amino acids, histidine-leucine, in the presence of Cl^- ions (RYAN et al. 1972, RYAN and RYAN 1977).

Angiotensin I (500 ng) was age-related activated by the isolated perfused lung from male *rats* aged 1 week at 57.2 ± 16.5 % (maximum value), from *rats* aged 4 weeks at 26.4 ± 4.5 %, from *rats* aged 7 weeks at 24.2 ± 5.2 %, from *rats* aged 36 weeks at 29.8 ± 4.9 %, from *rats* aged 48 weeks at 43.1 ± 6.4 %, and from *rats* aged 72 weeks at 43.7 ± 14.1 % (KITAMURA et al. 1983).

Dose-response curves to angiotensin I (1 ng–μg) in normoxia showed greatly enhanced pressor responses in chronically hypoxic compared with normal *rats*, probably attributable to increased sensitivity to angiotensin II rather than enhanced conversion of angiotensin I to angiotensin II (RUSSELL et al. 1990). Captopril caused a proportionate reduction in responses in both groups of *rats*.

11.5.2.2 Angiotensin II

Angiotensin II, a hypertrophic/anti-apoptotic hormone, utilises reactive oxygen species as growth-related signalling molecules in vascular smooth muscle cells. Angiotensin II causes rapid phosphorylation of Akt/protein kinase B (USHIO-FUKAI et al. 1999). Exogenous H_2O_2 (50–200 μM) also stimulated Akt/protein kinase B phosphorylation suggesting that Akt/protein kinase B activation is redox-sensitive. Diphenylene iodonium, an inhibitor of flavin-containing oxidases, or overexpression of catalase to block angiotensin II-induced intracellular H_2O_2 production significantly inhibited Akt/protein kinase B phosphorylation, indicating a role for reactive oxygen species in agonist-induced Akt/protein kinase B activation.

As the effect of angiotensin II on apoptosis has not yet be determined with primary cultures of well-differentiated pulmonary vascular endothelial cells the possibility exists that lung endothelial cells *in vivo* respond differently to angiotensin II than the *human* umbilical vein endothelial cell or *human* coronary artery endothelial cell lines (FILIPPATOS et al. 2001). On the other hand, the EC_{50} for induction of apoptosis in these cell lines is far above the plasma angiotensin II concentration, at least under normal conditions. Whether or not the significantly elevated angiotensin II levels attained in acute respiratory distress syndrome (WENZ et al. 1997) or other lung injury settings are within a range capable of inducing pulmonary endothelial cell death is an interesting topics for further inquiry.

11.5.2.3
Angiotensin IV

Angiotensin IV activates the constitutively expressed lung endothelial cell isoform of nitric oxide synthase (ecNOS) by a receptor-mediated pathway, leading to increases in nitric oxide (•NO) release, production of cGMP, and NO-cGMP-mediated *porcine* pulmonary arterial vasodilatation (PATEL et al. 1998, HILL-KAPTURCZAK et al. 1999). Angiotensin IV-induced activation of ecNOS is mediated through mobilisation of Ca^{2+} concentration and by increased expression and release of the Ca^{2+} binding protein calreticulin in *porcine* pulmonary artery endothelial cell monolayers (PATEL et al. 1999).

11.5.2.4
Bradykinin

Bradykinin (10^{-6} to 10^{-5} M) within 5 min released O_2^- from *human* umbilical vein endothelial cells *in vitro* (HOLLAND et al. 1990). O_2^- release was partially inhibited by indomethacin (63±6%).

11.5.2.5
Endothelin

Endothelin-1 is a survival factor for endothelial cells (SHICHIRI et al. 1997). It also antagonised apoptosis of vascular smooth muscle cells induced by serum deprivation and nitric oxide via mitogen-activated protein kinase pathway (SHICHIRI et al. 2000). Endothelin-1 has a protective effect in ω-3 fatty acid-induced apoptosis of vascular smooth muscle cells as evidenced by the inhibition of caspase-3 activation and subsequent DNA fragmentation in late-stage apoptosis (DIEP et al. 2000).

Exposure of foetal *sheep* pulmonary artery smooth muscle cells to endothelin-1 resulted in increases in superoxide production and viable foetal pulmonary artery smooth muscle cells after 72 h (WEDGWOOD et al. 2001). These increases were prevented by pre-treatment with the ET_A receptor agonist PD-156707 (1 μM). Treatment with pertussis toxin blocked the effects of endothelin-1. Wortmannin, LY-294002, diphenyleneiodonium, 4-(2-aminoethyl)benzenesulphonyl fluoride, and apocynin also prevented the endothelin-1-mediated increases in superoxide production and viable cell numbers. Exposure to H_2O_2 or diethyldithiocarbamate increased viable cell number by 37% and 50%, respectively.

Synthesis and release of endothelin depend not only on the transcription and translation, but also on the conversion of inactive to active forms of endothelin, followed by secretion (SCHIFFRIN and TOUYZ 1998). The conversion of the inactive precursor "big endothelin" to the biologically active peptide is catalysed by **endothelin converting enzymes** (OHNAKA et al. 1990, XU et al. 1994, KIDO et al. 1997). LÓPEZ-ONGIL et al. (2002) found that glucose oxidase (EC 1.1.3.4) increases endothelin converting enzyme-1 mRNA, protein and activity in *bovine* aortic endothelial cells. Catalase (EC 1.11.1.6) abolished this effect. Glucose oxidase treatment of endothelial cells transactivated the endothelin converting enzyme-1 promoter.

11.5.2.6
Tumour Necrosis Factor-α

In *calf* pulmonary microvessel endothelial monolayers TNF treatment (1,000 U/ml) for 4 h induced a significant increase in DNA binding of activating protein-1 (GERTZBERG et al. 2000). The effects of TNF were prevented by the superoxide radical scavenger superoxide dismutase (SOD) (100 U/ml), the •NO synthase inhibitor aminoguanidine (100 μM), the guanylate cyclase inhibitor ODQ (100 μM), and the protein kinase G inhibitors KT5823 (1 μM) and 8-bromo-cyclic guanosine monophosphate-thiolate (100 μM). Spermine-NO (1 μM) and L-arginine (400 μM) prevented the aminoguanidine-induced ablation of activating protein-1 activation in response to TNF. Phosphorylation of H-Arg-Lys-Ile-Ser-Ala-Ser-Glu-Phe-Asp-Arg-Pro-Leu-Arg-OH (BPDEtide), a specific substrate for protein kinase G, measured the activity of cGMP-dependent protein kinase. TNF for 0.5 h induced an increase in protein kinase G activity that was prevented by aminoguanidine, ODQ, KT5823, and 8-bromo-cGMP-thioate; however SOD had no effect. The protein kinase G agonist 8-bromo-cGMP (100 μM), when given alone, increased protein kinase G activity but induced significant DNA-binding activity of AP-1 only when given in the ODQ + TNF group. SIN-1 (1 μM) increased DNA-binding activity of AP-1. SOD prevented SIN-1-induced AP-1 activation, a response similar to that of the SOD + TNF group. TNF (50 ng/ml) induced an acute (30 min) increase followed by a protracted decrease (4–24 h) in ecNOS protein levels in *calf* pulmonary microvessel endothelial monolayers (BOVE et al. 2001). The other NOS isotypes, inducible and brain NOS, could not be detected in the endothelial monolayers using RT-PCR and Western blot assay. ecNOS antisense oligonucleotide decreased ecNOS protein, which prevented the increase in •NO and albumin permeability at TNF-4h. Spermine-NONOATE, the NO agonist, ablated the protective effect of ecNOS antisense oligonucleotide on albumin permeability in response to TNF-4h. However, ecNOS antisense oligonucleotide had no effect on the TNF-induced increase in albumin permeability at 24 h despite prevention of the increase in •NO.

11.5.3 Fatty Peroxides and Lung Prostaglandins

Highly purified hydroperoxides prepared from auto-oxidised methyl linoleate when injected intravenously into Sprague-Dawley *rats* (200 mg per kg body weight) induced large changes in the arachidonic (20:4). oleic (18:1), and linoleic (18:2) acid concentrations in lung lipid, and the biosynthetic activity of the tissue of this organ to convert arachidonic acid to prostaglandin E was impaired (TAN et al. 1974). The ratio of arachidonic acid in the lung was doubled by the effects of the hydroperoxide, while those of oleic and linoleic acids decreased.

Ingestion of conjugated linoleic acid (4.2 g per day) for 3 months significantly ($P < 0.0001$) increased urinary 8-iso-prostaglandin$_{2\alpha}$ levels in healthy subjects (BASU et al. 2000). Conjugated linoleic acid had no effect on the serum α-tocopherol levels. However, γ-tocopherol levels in the serum increased significantly ($P < 0.015$) in the conjugated linoleic acid-treated group.

11.5.3.1 Arachidonic Acid Metabolites

The actions of prostaglandins on isolated pulmonary blood vessels were reviewed by PIPER and VANE (1979). Usually, E-type PG relax pulmonary vascular smooth muscle but there is some species variation in the reaction of intrapulmonary blood vessels (PALMER et al. 1973, KADOWITZ et al. 1975). The intrapulmonary vessels will be exposed to the prostaglandins released in the lung by agents such as bradykinin, slow-reacting substance of anaphylaxis or histamine.

11.5.3.1.1 Prostaglandin E_2

Immunoregulatory actions of PGE_2 produced by macrophages might have implications in *human* pathology. Immunosuppression in certain malignancies, e.g., Hodgkin's disease, and chronic inflammatory diseases such as sarcoidosis seems to be a consequence of enhanced PGE_2 production by adherent mononuclear cells.

PGE_2 levels in the circulation have been detected by radioimmunoassay in amounts that could have a biological importance. However, on the basis of measurements of the major urinary and circulating metabolites, it was shown that the true levels are much smaller and in the range of a few picograms per millilitre (HAMBERG and SAMUELSSON 1971, 1973). Under conditions of shock, however, PGE_2 levels were found by bioassay to be elevated to levels of several nanograms per millilitre (JAKSCHIK et al. 1974), and this was subsequently confirmed by gas chromatography-mass spectrometry (FRÖLICH 1977).

11.5.3.1.2 Prostaglandin $F_{2\alpha}$

An infusion of bradykinin into isolated lungs of the *guinea pig* caused release of $PGF_{2\alpha}$ (PIPER and VANE 1971).

8-Iso-prostaglandin $F_{2\alpha}$ is produced from arachidonic acid by free radicals independent of the cyclooxygenase pathway. Its vasoconstrictive effect on the pulmonary vascular bed does not depend on an influx of extracellular Ca^{2+} via voltage-gated Ca^{2+} channels, but is protein kinase C-dependent (WAGNER et al. 1996). SQ29548, a TXA_2 receptor antagonist, blocked the effects of 8-iso-prostaglandin $F_{2\alpha}$ in the pulmonary artery.

11.5.3.1.3 Prostaglandin I_2

Endothelial cells are the most active producers of prostaglandin I_2 (WEKSLER et al. 1977, MACINTYRE et al. 1978), the production persisting even after numerous subcultures *in vitro* (CHRISTOFINIS et al. 1979). The biosynthesis of PGI_2 is inhibited by 15-HPETE and other fatty acid hydroperoxides (GRYGLEWSKI et al. 1976).

11.5.3.1.4 20-Hydroxyeicosatetraenoic Acid (20-HETE)

20-Hydroxyeicosatetraenoic acid (20-HETE) is produced in *human* lungs and is a potent cyclooxygenase-dependent dilator of isolated pulmonary arteries (BIRKS et al. 1997). Small pulmonary arteries of New Zealand White *rabbits* express cP450 4A proteins and vascular smooth muscle cells derived from these arteries synthesise 20-HETE (ZHU et al. 1998).

11.5.3.1.5 Leukotoxin (9,10-Epoxy-12-octadecenoate)

Leukotoxin, 9,10-epoxy-12-octadecenoate dilated *rat* pulmonary arteries by means of nitric oxide synthase activation (ISHIZAKI et al. 1995). The nitric oxide synthase inhibitors, N^G-monomethyl-L-arginine and aminoguanidine and endothelium denudation significantly attenuated leukotoxin-induced vasodilatation (Nakanishi et al. 2000). Aminoguanidine also significantly attenuated leukotoxin-induced vasodilatation in lipopolysaccharide-treated *rat* denuded pulmonary arteries, and attenuated leukotoxin-induced cGMP content increase in denuded pulmonary arterial rings from lipopolysaccharide-treated *rats* and in lipopolysaccharide-treated *human* pulmonary artery smooth muscle cells.

11.5.4 Xanthine Oxidase

Xanthine oxidase is a well known source of $O_2^{\bullet-}$. *Rat* pulmonary artery endothelial cells were found to have 53±8.57 units/10^6 cells of total xanthine oxidase + xanthine dehydrogenase activity, one unit defined as the conversion of 1% of the substrate to product in 30 min of incubation (Phan et al. 1989). Xanthine oxidase comprised 31.6±3.1% of this total activity. Addition of *human* neutrophils stimulated with 12-*O*-tetradecanoylphorbol-13-acetate caused a rapid and dose-dependent increase in *rat* pulmonary artery endothelial cell xanthine oxidase activity from 31.6±3.1% to 71.7±4.8% of total without altering total (xanthine oxidase + xanthine dehydrogenase) activity. Allopurinol, oxypurinol, and lodoxamide inhibited both enzyme conversion and cytotoxicity, catalase, superoxide dismutase, or deferoxamine failed to do so.

Xanthine oxidase can reduce nitrate to nitrite (Westerfield et al. 1959, Fridovich and Handler 1962). Xanthine oxidase and dissimilatory nitrate reductase share structural similarities. Both are molybdoenzymes and contain flavin adenine dinucleotide and Fe/S clusters (McCord 1985, Mitchell 1986, Payne et al. 1997). Zhang et al. (1998) reported that both purified *bovine* buttermilk xanthine oxidase and xanthine oxidase-containing inflamed *human* synovial tissue can generate $^{\bullet}$NO by reducing nitrite in the presence of NADH. This nitrite reductase activity of xanthine oxidase may act as a supplement to the activity of nitric oxide synthase (NOS) to redistribute blood flow to ischaemic tissues when NOS activity is absent.

In male, weanling Sprague-Dawley *rats* iron deficiency caused a loss in intestinal xanthine oxidase activity, but also caused an increase in hepatic xanthine oxidase activity (Kelley and Amy 1984).

Synthetic benzophenones as 2,2',4,4'-tetrahydrobenzophenone, 3,4,5,2',3',4'-hexahydrobenzophenone and 4,4'-dihydrobenzophenone displayed their inhibitory effects on xanthine oxidase with an order of activity of IC_{50} = 47.59, 69.40 and 82.94 μM, respectively (Sheu et al. 2000). The apparent inhibition constant of 3,4,5,2',3',4'-hexahydrobenzophenone and 4,4'-dihydrobenzophenone were 15.61 and 64.86 μM, respectively.

11.5.5 Reactive Oxygen Species

Hypoxic pulmonary vasoconstriction matches lung perfusion with ventilation. Controversy exists whether decreased or increased reactive oxygen species may elicit hypoxic pulmonary vasoconstriction and from which source such oxygen metabolites are derived in *rabbit* lungs, Weissmann et al. (2000) detected transcripts of a nonphagocytic NADPH oxidase subunit homologous to mitogenic oxidase-1 or NADPH oxidase homologue 1. 4-(2-Aminoethyl)benzenesulfonyl fluoride (100–600 μM), a NADPH oxidase inhibitor, induced a transient increase in pulmonary arterial pressure with increased strength of hypoxic pulmonary vasoconstriction. Subsequent to this initial response, normoxic pulmonary arterial pressure was not affected and hypoxic pulmonary vasoconstriction was specifically suppressed. The superoxide dismutase inhibitor, diethyldithiocarbamate (100 μM to 10 mM) turned out to act in a non-specific fashion. However, a second superoxide dismutase inhibitor, triethylenetetramine (1–25 mM) dose-dependently inhibited hypoxic pulmonary vasoconstriction without affecting the pressure response elicited by the stable thromboxane analogue, U-46619. Virtually the same inhibition profile was seen when the experiments were performed after blockade of $^{\bullet}$NO generation with N^G-monomethyl-L-arginine (400 μM).

Whole *rat* lung and cultured *mouse* endothelial cell permeability both increased significantly in response to chemical redox imbalance (Zhao et al. 2001). Thiol depletion also resulted in decreased endothelial cadhedrin content and disruption of the endothelial barrier. These deleterious effects of intracellular redox imbalance were blocked by pretreatment with exogenous glutathione.

Bovine pulmonary artery endothelial cells cultured in RPMI 1640 medium and exposed to varying concentrations of H_2O_2 (0 to 5 mM) at concentrations of $H_2O_2 \leq 1$ mM demonstrated the nuclear features following staining with acridine orange and ethidium bromide characteristic for apoptotic cell death as early as 1 h, and demonstrated significant increases in apoptotic cells by 2 h after H_2O_2 exposure (Lelli et al. 1998). At concentrations of $H_2O_2 \leq 1$ mM, there was no significant increase in the percentage of the cells dying by necrosis as compared to the control. The percentage of the cells undergoing apoptosis after exposure to concentrations of $H_2O_2 \leq 1$ mM was significantly higher than that of the control cells ($P < 0.05$). At the highest concentration of H_2O_2 (5 mM), the pattern of cell death was predominantly necrotic. A freshold level of adenosine triphosphate 25 % of basal levels was required for apioptosis tp proceed after oxidant stress, otherwise necrosis occurred. Agents like glutamine that enhance ATP levels in oxidant stressed cells may be potent means of shifting cell death during inflammation to the noninflammatory form of death-apoptosis.

Brief exposure of *bovine* pulmonary artery endothelial cells to relatively high concentrations of H_2O_2 (1 mM) resulted in a time- and dose-dependent tyrosine phosphorylation of focal adhesion kinase, which reached maximum levels within 10 min (290 % of basal levels) (Vepa et al. 1999). Cytoskeletal reorganisation as evidenced by the appearance of actin stress fibres preceded H_2O_2-induced tyrosine phosphorylation of focal adhesion kinase, and the microfilament disrupter cytochalasin D also attenuated the tyrosine phosphorylation of focal adhesion kinase. Treatment of *bovine* pulmonary artery endothelial cells with 1,2-bis(2-aminophenoxy)ethane-*N,N,N',N'*-tetraacetic acid-AM attenuated H_2O_2-induced increases in intracellular Ca^{2+} but did not show any consistent effect on H_2O_2-induced tyrosine phosphorylation of focal adhesion kinase. Several tyrosine kinase inhibitors, including genistein, herbimycin, and tyrphostin, had no detectable effect on tyrosine phosphorylation of focal adhesion but attenuated the H_2O_2-induction of mitogen-activated protein kinase activity.

The H_2O_2-induced pressor responses in blood-free, perfused, isolated *rabbit* lungs were blocked by indomethacin (cyclooxygenase inhibitor), imidazole (inhibitor of thromboxane synthetase), mepacrine (phospholipase inhibitor), and W7 and trifluoperazine (agents that interfere with calcium-calmodulin function) (Seeger et al. 1986). Treatment with 1,3-bis(2-chloroethyl)-1-nitrosourea (BCNU) dose-dependently inhibited the lung glutathione reductase activity and augmented the metabolic (prostanoid release) and functional (vasoconstriction) responsiveness of the pulmonary vascular bed to H_2O_2. Application of 1-(2-chloroethyl)-1-nitrosourea (CCNU), a control to BCNU, and inhibition of catalase activity by aminotriazole did not increase the sensitivity to externally applied H_2O_2.

SB209995, a metabolite of carvedilol in *human*, protected cultured *bovine* pulmonary artery endothelial cells against hydroxyl radical or superoxide-mediated lipid peroxidation and cytotoxicity, assessed as lactate dehydrogenase release and cell death (Feuerstein and Yue 1994).

After transient exposure of *human* umbilical vein endothelial cells ($1-5 \times 10^6$) to hydroperoxides, such as H_2O_2, the synthesis of 40 polypeptides in mitochondria was found using comparative two-dimensional polyacrylamide gel electrophoresis (Mitsumoto et al. 2002). Eleven proteins were identified: these include 60 kDa heat shock protein (HSP60), a mitochondrial type of 70 kDa HSP (mtHSP70), manganese-dependent superoxide dismutase, three metabolic enzymes in citric acid cycle, two components for respiratory chain complexes, a ribosomal protein for translation in mitochondria (RM12), and an unnamed protein.

11.5.6 Nitric Oxide and Peroxynitrite

Nitric oxide is a potent pulmonary vasodilator that can be delivered via inhalation (Pepke-Zaba et al. 1991, Pison et al. 1993, Rimar and Gillis 1995, Day et al. 1996, Nelin and Hoffman 1998). In the neonatal *pig*, inhaled nitric oxide dilated the smaller arteries more than the larger ones (Guarín et al. 2001).

Endothelial cell-derived products of the L-arginine-nitric oxide pathway have been recognised for their important biological activities (Moncada and Higgs 1993). Endothelial constitutive nitric oxide synthase (ecNOS) requires tetrahydrobiopterin, calcium, calmodulin, and NADPH as cofactors. These enzymes generate •NO via continuous synthesis and at lower concentrations than via the cytokine-inducible NOS (iNOS) pathway (Moncada et al. 1991). The iNOS is calcium-independent but requires NADPH and tetrahydrobiopterin as cofactors. The cofactor is tightly bound to NOS but its exact function remains unclear. It may be important in maintaining NOS in an active configuration (Tzeng et al. 1995). It may be required to limit inactivation of NOS by •NO (Griscavage et al. 1994) and as a result of allosteric interactions, the affinity of NOS for arginine (Klatt et al. 1994, Brand et al. 1995). iNOS is found in macrophages (Lyons et al. 1992), vascular smooth muscle cells (Nakayama et al. 1992), neutrophils (McCall et al. 1989), hepatocytes (Billiar et al. 1990), and endothelial cells (Rosenkranz-Weiss et al. 1994). Increased iNOs protein occurred in astrocytes of the *hph*-1 (tetrahydropteridin deficient) *mouse*, while the formation of •NO was limited (Barker et al. 1998).

L-Arginine–nitric oxide pathway

$$\text{L-Arginine} \xrightarrow{\text{NOS}} \textbf{Nitric oxide}(NO^{\bullet}) \xrightarrow{+O_2^{\bullet-}} \text{Peroxynitrite } (ONOO^-) \qquad [46]$$

Phenotypic heterogeneity among endothelial cell populations may account for important organ-specific behaviours. *Rat* pulmonary microvascular endothelial cells and macrovacular endothelial cells incubated with interferon-γ, tumour necrosis factor-α, and *Salmonella typhimurium* lipopolysaccharide alone or in combination, showed that aortic endothelial cells produced significantly more nitrite than lung microvacular endothelial cells, while the nitrite generation from pulmonary artery endothelial cells was intermediate (GEIGER et al. 1997). In normal *rat* lung 39.8% (680 of 1724) endothelial cells of large arteries and 13% (92 of 533) of alveolar duct and 16% (192 of 1222) of alveolar wall vessels expressed ecNOS (NIKOLOV et al. 1999).

Lipopolysaccharide (0.5 μg/ml) and tumour necrosis factor-α (0.5 ng/ml) treatment increased L-arginine metabolism to both •NO and urea in *bovine* pulmonary arterial endothelial cell cultures and resulted in increased levels of cationic amino acid transporter-2 mRNA (NELIN et al. 2001).

•NO-induced posttranscriptional regulation of the catalytic activity of ecNOS is associated with the formation of intramolecular disulphide bonds within the ecNOS protein because exogenous addition of the disulphide-reducing enzyme thioredoxin reductase was able to restore ecNOS activity in *porcine* pulmonary artery endothelial cells (PATEL et al. 1996). ZHANG et al. (1998) demonstrated that exposure of *porcine* pulmonary artery endothelial cells to 8.5 ppm NO gas for 24 h reduced the expression of thioredoxin and thioredoxin reductase mRNA and protein as well as the catalytic activity of thioredoxin reductase without altering GSH and GSSG contents or the catalytic activity of glutaredoxin. Western blot analysis revealed that IκB-α content was reduced by 20 and 60% in *porcine* pulmonary artery endothelial cells exposed to 8.5 ppm •NO for 2 and 24 h, respectively (ZHANG et al. 1999). Exposure of *mice* to 10 ppm •NO for 24 h resulted in a significant reduction of lung thioredoxin and IκB-α mRNA and protein expression and in the oliginucleotide encoding thioredoxin and NF-κB/DNA binding.

To determine the potential role of nitric oxide in the vascular remodelling induced by hypoxia, QUINLAN et al. (2000) exposed wild-type [WT(+/+)] and endothelial NOS-deficient [(-/-)] *mice* to normoxia or hypoxia (10% O_2) for 2, 4, and 6 days or for 3 weeks. Smooth muscle α-actin and von Willebrand factor immunohistochemistry revealed significantly less muscularization of small vessels in hypoxic eNOS(-/-) *mouse* lungs than in WT(+/+) *mouse* lungs at early time points, a finding that correlated with decreases in proliferating vascular cells (5-bromo-2'-deoxyuridine positive) at 4 and 6 days of hypoxia in the eNOS(-/-) *mice*. After 3 weeks of hypoxia, both *mouse* types exhibited similar percentages of muscularized small vessels; however, only the WT(+/+) *mice* exhibited an increase in the percentage of fully muscularized vessels and increased vessel wall thickness. eNOS protein expression was increased in hypoxic WT(+/+) *mouse* lung homogenates at all time points examined, with significantly increased percentages of small vessels expressing eNOS protein after 3 weeks.

Exposing *human* endothelial cells to hypoxia (pO_2 = 5 mm Hg), GOERGE et al. (2002) found an acute (within minutes) release of von Willebrand factor. Despite acute von Willebrand factor release, potential cellular modulators of secretion, such as intracellular pH and cell volume, remained unchanged. They only detected a small instantaneous increase of cytosolic Ca^{2+} concentration. Although overall cell morphology remained virtually unchanged, high-resolution atomic force microscopy images of hypoxic endothelial cells disclosed secretion pores, most likely the loci of Weibel-Palade bodies exocytosis on luminal plasma membrane.

N^G-Nitro-L-arginine methylester (30 μM), methylene blue (10 μM), and removal of endothelium significantly reduced the electrical field (50 V, 0.2 ms, 0.1–10 Hz for 5 s) stimulation-induced relaxations of *rat* main pulmonary arterial rings pretreated with 3 μM phenylephrine (GÜMÜSEL et al. 2001). The inhibitory action of N^G-nitro-L-arginine methylester was completely reversed by L-arginine (1 mM) but not by D-arginine (1 mM). L-Arginine alone potentiated the magnitude of the relaxations elicited by electrical field stimulation. On the other hand, immunohistochemical work clearly demonstrated the existence of nNOS in the pulmonary artery vessel wall.

1,3-Morpholinosydnonimine (SIN-1), which generates ONOO- by releasing $O_2^{\bullet}$- and •NO essentially in a simultaneous manner, significantly inhibited Ca^{2+} signalling (ELLIOTT 1996). Initially, the inhibitory effect of 1 mM SIN-1 was selective toward agonist-stimulated influx of external Ca^{2+}. At later time points, SIN-1 additionally depleted internal stores of releasable Ca^{2+}.

Bovine pulmonary artery endothelial cells exposed to •NO generated by spermine-NONOate and papa-NONOate and to the same fluxes of $ONOO^-$ generated by SIN-1 resulted in cellular injury and death (GOW et al. 1998). Loss of cellular viability was evident after 18 h postexposure. Events preceding cell death included depolarisation of the mitochondrial membrane, evident as early as 6 h postexposure, loss of cellular redox activity at 16 h, and DNA fragmentation detected by in situ staining at 18 h after exposure. •NO did not affect the cellular viability.

Nitric oxide donors and atrial natriuretic peptide have been shown to block H_2O_2 mediated increase in the permeability of *porcine* pulmonary artery endothelial cells. Erythro-9-(2-hydroxy-3-nonyl)-adenine inhibition of cyclic GMP-stimulated phosphodiesterase reduced this effect (SUTTORP et al. 1996). In *pig* aortic endothelial cells, treatment with H_2O_2 (0.5 mM) for 20 h reduced the number of viable cells to 44 % of control (OBERLE and SCHRÖDER 1996). A 6-h preincubation with SIN-1 (0.5 mM) protected endothelial cells from H_2O_2-mediated cytotoxicity and increased viability to 81 % of control. However, SIN-1 had no protective effect when the preincubation time was reduced to 3 h or when SIN-1 and H_2O_2 were added simultaneously to the cells.

Exogenous •NO in the form of a nitric oxide donor (*S*-nitroso-L-glutathione or diethylamine NO) significantly reduced IL-8 mRNA in cytokine-activated *human* endothelial cell line ECV304 (FOWLER et al. 1999).

•NO synthesised and released by endothelial cells can reduce the adhesion of cancer cells to the endothelium (KONG et al. 1996) and exert a cytotoxic effect on them (GENG et al. 1996). eNOS activity and •NO release are increased in response to stimulation with tumour necrosis factor-α, interferon-γ, interleukin-1β, lipopolysaccharide, or forskolin (MURATA et al. 1994, CHANG et al. 1996).

The extract of *Ginkgo biloba* leaves (50 µg/ml) caused a 30 % reduction of NO metabolites released by a *human* endothelial cell line (ECV304); cellular inducible NO synthase (iNOS) activity was reduced by 28 % with a concomitant reduction in the levels of iNOS protein mass and mRNA (CHEUNG et al. 1999).

11.5.7
S-Nitroso-*N*-acetyl-D-penicillamine

The nitric oxide donor, *S*-Nitroso-*N*-acetyl-D-penicillamine, in *rat* pulmonary microvascular endothelial cells exposed *in vitro* to H_2O_2 through its enzymatic generation by glucose and glucose oxidase or by its direct application prevented the concentration- and time-dependent release of ^{51}chromium (CHANG et al. 1996). It decreased the net oxidation of ferrous to ferric iron by H_2O_2, the iron-catalysed consumption of H_2O_2 in Fenton's reaction, the iron-mediated generation of hydroxyl radicals and the Fe^2+-H_2O_2-catalysed peroxidation of lipid membranes. *S*-Nitroso-*N*-acetyl-D-penicillamine substantially reduced H_2O_2-mediated barrier dysfunction in the absence of 1*H*-[1,2,4]oxadiazole [4,4-*a*]quinoxalin-1-one (GUPTA et al. 2001). •NO increased cGMP and cAMP levels in *porcine* pulmonary microvascular endothelial cells, but treatment with inhibitors of soluble guanylate cyclase or protein kinase G did not abrogate •NO-mediated barrier protection. In contrast, H_2O_2 decreased protein kinase A activity, and inhibiting protein kinase A abrogated the protective effect of •NO. H_2O_2-induced barrier dysfunction was not associated with decreased levels of cGMP and cAMP.

11.6
Experimental Pathology of Pulmonary Endothelial Cells

11.6.1
Arsenite

Environmentally relevant concentrations of arsenite caused oxidant-dependent increases in nuclear transcription factor levels in cultured *porcine* vascular endothelial cells (BARCHOWSKY et al. 1996). Cells exposed to levels of arsenite that initiate cell signalling did not deplete 5'-triphosphate, not did they affect basal or bradykinin-stimulated intracellular free Ca^{2+} levels, indicating that they were not lethal (BARCHOWSKY et al. 1999). Electron paramagnetic resonance (EPR) spectroscopy, including spin trapping with 2-(4-carboxyphenyl)-4,5-dihydro-4,4,5,5-tetramethyl-1*H*-imidazolyl-1-oxy-3-oxide (potassium salt), demonstrated that 5 µM or less of arsenite did not increase •NO levels over a 30-min period relative to •NO release stimulated by bradykinin. However, the same levels of arsenite rapidly increased both oxygen consumption and superoxide formation, as measured by EPR oxymetry and spin trapping with 5,5-dimethyl-1-pyrroline-*N*-oxide (DMPO). respectively. Pretreatment of the cells with diphenyleneiodonium, apocynin, or superoxide dismutase abolished arsenite-stimulated DMPO-OH adduct formation. Arsenite increased extracellular accumulation of H_2O_2, measured as oxidation of homovanillic acid, with the same time and dose dependence, as seen for superoxide formation.

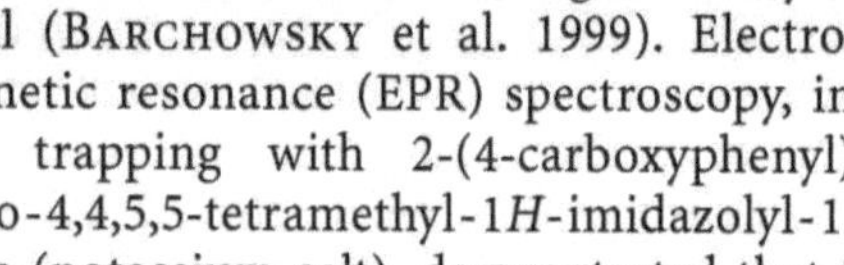

[8]

5,5-**Dim**ethyl-1-**pyr**roline-1-**o**xide-HO• adduct

11.6.2
Carbon Monoxide

Lung capillary leakage was significantly increased 18 h after *rats* had been exposed to CO at concentrations of 50 ppm or more for 1 h (THOM et al.

1999). An elevation of •NO during CO exposure was demonstrated by electron paramagnetic resonance spectroscopy. There was a 2.6-fold increase of •NO over control in the lungs of *rats* exposed to 100 ppm CO. A quantitative increase in the concentration of H_2O_2 was also detected in the lungs during CO exposure, and this change was caused by •NO as it was inhibited in *rats* pre-treated with the nitric oxide synthase inhibitor, N^{ω}-nitro-L-arginine methylester. Production of •NO-derived oxidants during CO-exposure was indicated by an elevated concentration of nitrotyrosine in lung homogenates. The CO-associated elevations in lung capillary leakage and nitrotyrosine concentration did not occur when *rats* were pre-treated with N^{ω}-nitro-L-arginine methyl ester. CO exposure did not change the concentrations of endothelial or inducible nitric oxide synthase in lung and leucocyte sequestration was not detected as a consequence of CO exposure. CO-mediated lung leak and nitrotyrosine elevation were not affected by neutropenia.

11.6.3 Silica

Two weeks after intratracheal instillation of a suspension of 50 mg of silica particles (< 5 μm in diameter) in 1 ml of normal saline into Sprague-Dawley *rats*, some capillary endothelial cells were thin-walled but showed extensive overlapping with the cytoplasm of adjacent cells and had cytoplasmic projections that protruded into the lumina (Kawanami et al. 1995). These changes were often localised to regions adjacent to small bronchioles. In such capillaries, the cellular areas facing the lumen often contained numerous ribosomes and Weibel-Palade bodies. The cytoplasm of these endothelial cells occasionally contained a large number of actinlike microfilaments, which were closely adjacent to the plasma membrane. The microfilaments frequently had insertion sites along the abluminal plasma membrane. The cytoplasm of newly formed endothelial cells became widely adherent to the basement membrane. At 1 month and thereafter, the capillary lumina of sprouting endothelial cells became larger and the widened spaces between adjacent endothelial cells disappeared. Fenestrations were evident in some of the attenuated portions of these endothelial cells.

11.6.4 Diperoxovanadate

Diperoxovanadate, a synthetic diperoxovanadium compound and cell-permeable oxidant that acts as a protein tyrosine phosphatase inhibitor and insulinomimetic, increased phospholipase D activation in endothelial cells (Natarajan et al. 1998). Diperoxovanadate activated extracellular signal-regulated kinase, c-Jun NH_2-terminal kinase, and p38 mitogen-activated protein kinase in a dose- and time-dependent fashion (Natarajan et al. 2001). Treatment of endothelial cells with p38 mitogen-activated protein kinase inhibitors SB-203580 and SB-202190 or transient transfection with a p38 dominant negative mutant mitigated the phospholipase D activation by diperoxovanadate but not by phorbol ester. SB-202190 blocked diperoxovanadate-mediated p38 mitogen-activated protein kinase activity as determined by activated transciption factor-2 phosphorylation.

11.6.5 Bleomycin

In the *mouse*, endothelial blebbing and subendothelial oedema were evident within two days after a single intravenous injection of 120 mg bleomycin/kg (Bowden 1985).

11.6.6 Paraquat

Within 12–18 h after subcutaneous application of 35 mg paraquat/kg *rat* Modée et al. (1972) found capillary engorgement and interstitial oedema in the alveolar walls. After 48 h endothelial cells displayed pronounced cytoplasmic swelling.

Superoxide production from paraquat in a *pig* pulmonary microvascular endothelial cell suspension was demonstrated by Tampo et al. (1999) using 2-methyl-6-(*p*-methoxyphenyl)-3,7-dihydroimidazol[1,2-α]pyrazin-3-one, a chemiluminescence probe, to detect superoxide anions. Increased rates of superoxide production from paraquat, which were sensitive to superoxide dismutase, required the presence of reduced nicotinamide adenine dinucleotide phosphate (NADPH) in the reaction medium, and occurred instantaneously after the addition of NADPH, which is impermeable to cell membranes. NADH as an electron donor was not as effective, and xanthine or succinate had no influence. Paraquat was anaerobically reduced in the presence of NADPH and 2-methyl-6-(*p*-methoxyphenyl)-3,7-dihydroimidazol[1,2-α]pyrazin-3-one to yield a one-electron reduced radical, and the reduction was inhibited by $NADP^+$. Diphenyleneiodonium, an inhibitor of flavoprotein reductase, also markedly inhibited both paraquat reduction and superoxide production.

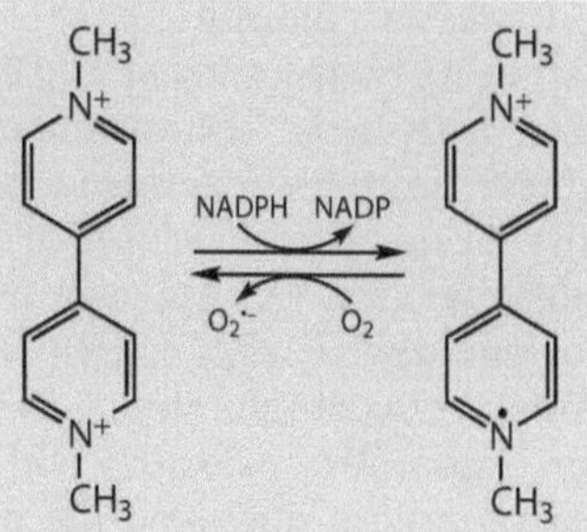

Paraquat [158]

11.6.7 Pyrrolizidine Alkaloids

Besides the liver, the lung is an important target organ for pyrrolizidines. There is strong evidence, that these compounds require metabolic activation to exert their toxic actions. Chemically, the parent alkaloids are rather inert. However, toxic pyrrolizidine alkaloids are metabolised both *in vivo* and *in vitro* to pyrrole derivatives which avidly bind to macromolecular tissue constituents. The pyrrole derivatives are potent alkylating agents, and their reactivity is greatly enhanced by the presence of ester groups at or adjacent to ring positions 1 and 7. As envisaged by MATTOCKS (1972), the ester groups are the most reactive sites, due to cunjugation with the ring nitrogen atom. The ester group(s) therefore could be lost readily, leaving a positively charged dihydropyrrolizine moiety which could react with nucleophilic groups such as amines or thiols to form relatively stable alkylation products.

At 1 week after an intravenous injection of pyrrole derivatives of pyrrolizidine alkaloids (5 mg pyrrole per kg) male albino *rats*' endothelial nuclei were considerably enlarged and bizarre in shape, and the cytoplasm was abundant with an increase in rough endoplasmic reticulum, free ribosomes and mitochondria (BUTLER 1970).

11.6.8 Tobacco Smoke

Cigarette smoke extracts cause an irreversible inhibition of nitric oxide synthase activity in pulmonary artery endothelial cells (SU et al. 1998) and a reduction in constitutive nitric oxide synthase activity in the *rat* in a dose-dependent manner (MA et al. 1999). Conversely, other studies in the *rat* have shown that cigarette smoking results in an increase in nitric oxide synthase gene expression and protein production (WRIGHT et al. 1999). However, lower respiratory tract nitric oxide concentrations were increased following cigarette smoking (CHAMBERS et al. 1998) although, in *humans*, plasma and urinary levels of nitrate, a metabolite of inhaled nitric oxide, have been shown to be unchanged suggesting that nitric oxide is not absorbed from the inhaled smoke.

11.6.9 Diesel Exhaust

Diesel exhaust particle extracts damaged *human* pulmonary artery endothelial cells under both subconfluent and confluent conditions (BAI et al. 2001). Cytotoxicity was markedly reduced by treatment with superoxide dismutase, catalase, *N*-(2-mercaptopropionyl)-glycine, or ebselen. Thus $O_2^{\bullet-}$, H_2O_2, and other reactive oxygen metabolites are likely to be implicated in diesel exhaust particle extracts-induced endothelial cell damage. Moreover, N^G-nitro-L-arginine methyl ester and N^G-monomethyl-L-arginine, inhibitors of nitric oxide synthase, also attenuated diesel exhaust particle extracts-induced cytotoxicity, while sepiapterin, the precursor of tetrahydrobiopterin enhanced attenuated diesel exhaust particle extracts-induced cell damage.

CHAPTER 12

Pulmonary Lymphatic Vessels

12.1 Lymph Capillaries

Pulmonary lymphatics are critical to clearing dust particles which have reached the lung fluid.

Using scanning electron microscopy of casts by a resin cast into the pulmonary blood vasculature of œdematous lungs, Schraufnagel (1992) categorised four forms of pulmonary lymphatics:

1) Prelymphatics, tissue planes with a characteristic structure that connect to reservoir and conduit lymphatics;
2) reservoir lymphatics, flat structures with textured surfaces that can have ribbon-like components with small lateral buds;
3) sacculo-tubular lymphatics, saccular structures that wrap around blood vessels, bronchi, and bronchioles, and also have tubular components;
4) conduit lymphatics, which are tubular structures that can be flat or round, are usually grooved, often twist in a helical structure, may contain valves, and vary in diameter.

From studies in spontaneous hypertensive *rats*, Aharinejad et al. (1999) suggested that pulmonary lymphatic filling is associated with pulmonary venous sphincters, and perivascular muscle action might be a component of the pulmonary lymphatic system.

12.2 Pulmonary Lymph Nodes

12.2.1 Topography

The hilar lymph nodes, which are encountered within the lung at the bifurcations of the large bronchi, are complete nodes with a subcapsular sinus and follicles. From the hilar nodes, the lymph drains via extensively anastomosing channels through tracheobronchial lymph nodes clustered alongside the main stem bronchi, beneath the carina, and along the course of the trachea. In the mediastinum, the pulmonary lymph is joined by lymph from the trachea, heart, esophagus, diaphragm, and chest wall before emptying into the venous system. On the right side, the thoracic lymphatics enter the subclavian vein at is confluence with the jugular vein, either via a separate duct or after joining the lymphatic trunks from the arm, head and neck to form a common duct. On the left, the mediastinal lymphatics may join the ductus thoracicus or to empty into the vena subclavia independently.

The American Joint Committee for Cancer Staging and End Results Reporting (AJC) classified and mapped the regional pulmonary lymph nodes in 1979 (see Tisi et al. 1983).

12.2.2 Free Cells

The cells primarily engaged in phagocytosis are the freely movable **macrophages** (true wandering cells) and **reticulum cells** (fixed macrophages, resting wandering cells). Although the two types of cells are different in shape, they exert their function according to the same principle: dust particles and fibres are taken up by an invagination of the plasmalemma and so engulfed into the cytoplasm in the form of smooth-walled vacuoles, called phagosomes.

Mast cells with their metachromatic granules (0.3 to 1.0 μm in diameter) as their hallmarks produce histamine, which has marked vasoactive effects and also seems to enhance oxygen radical formation by phagocytes (Friedl et al. 1989).

As the injection of $CaCl_2$ (0.1 M) into individual *rat* peritoneal mast cells elicited secretory granule extrusion (Kanno et al. 1973), calcium might be involved in some stage of the secretion process (Goth 1978). *N*-benzhydryl-*N*'-*p*-hydroxybenzyl-piperazine, a potent anti-histaminic and anti-serotoninergic calcium blocker, prevented mast cells from degranulation (Fig. 195), while the introduction of particulate matter into the body induced mast cell degranulation (Fig. 196).

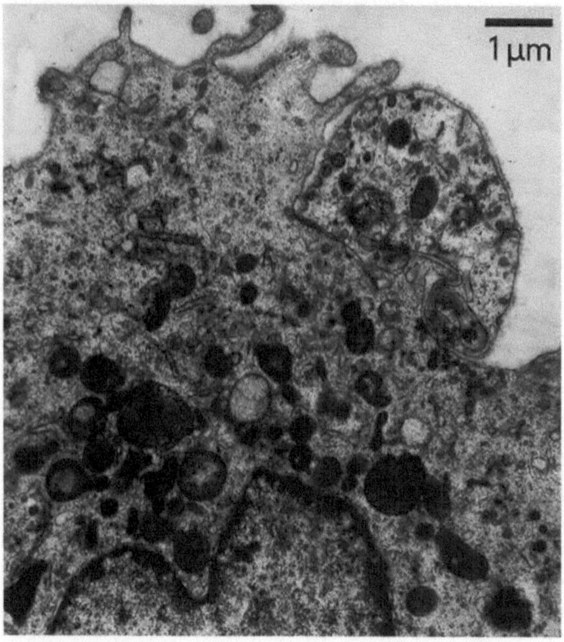

Fig. 193. Numerous secondary lysosomes in a macrophage from a pulmonary lymph node (block 1973) of an unmedicated 345 g *rat*. Under pentobarbital anaesthesia (30 mg/kg), the animal was perfused from the abdominal aorta with 2.5 % glutaraldehyde in 0.1 M sodium cacodylate buffer (pH 7.4). Postfixation with 1 % osmium tetroxide in sodium cacodylate buffer. Embedded in Epon 812 and sectioned at 50 nm. Lead citrate and uranyl acetate. Film 1075/79

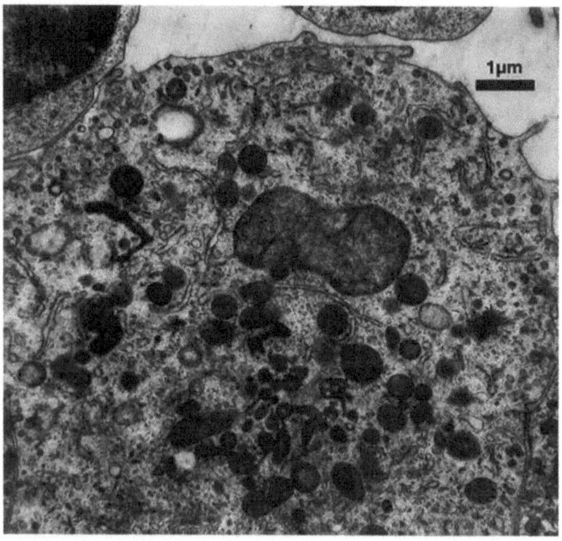

Fig. 194. Pleomorphic lysosomes in a pulmonary lymph node macrophage (block 1973) of an unmedicated 345 g *rat*. Under pentobarbital anaesthesia (30 mg/kg), the animal was perfused from the abdominal aorta with 2.5 % glutaraldehyde in 0.1 M sodium cacodylate buffer (pH 7.4). Postfixation with 1 % osmium tetroxide in sodium cacodylate buffer. Embedded in Epon 812 and sectioned at 50 nm. Lead citrate and uranyl acetate. Film 1076/79

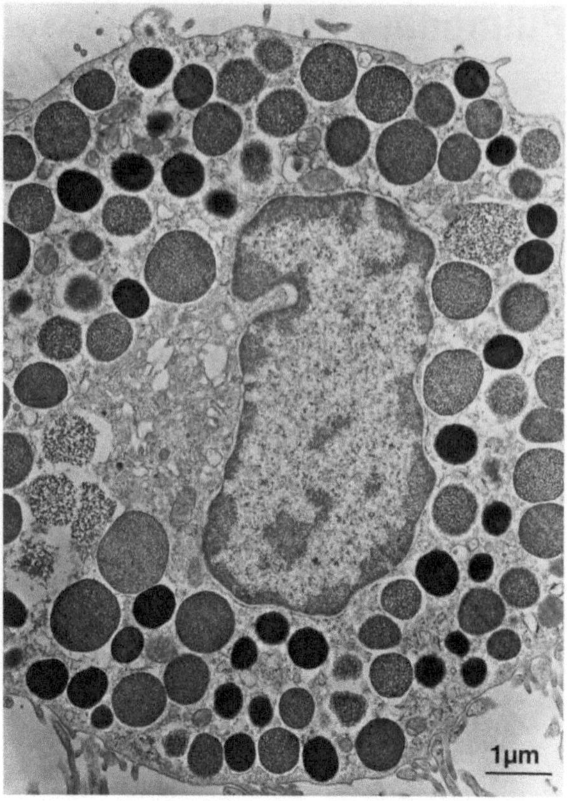

Fig. 195. Mast cell from a lymph node (block 501) of a 218 g female Wistar *rat* (No. 182/103 E) medicated for 5 months with oral application of 100 mg *N*-benzhydryl-*N*'-*p*-hydroxybenzylpiperazine HCl (UCB F 241) per kg body weight × day. Under pentobarbital anaesthesia (30 mg/kg), the animal was perfused from the abdominal aorta with 2.5 % glutaraldehyde in 0.1 M sodium cacodylate buffer (pH 7.4). Postfixation with 1 % osmium tetroxide in sodium cacodylate buffer. Embedded in Epon and sectioned at 50 nm. Lead citrate and uranyl acetate. Plate 3034

12.2.3 Postcapillary Venules

The endothelium of the postcapillary venules (Fig. 197), commonly called **high endothelial venules** (HEVs) was first described by THOMÉ (1898) and VON SCHUMACHER (1899) and reviewed by MAXIMOW (1927, pp. 351–352) and HELLMAN (1930, p. 345; 1943, p. 222). It is composed of cuboidal and squamous cells the surface of which strongly bulges into the lumen. The cytoplasm is provided with a well-developed granular endoplasmic reticulum, a large Golgi zone, and a great number of cytoplasmic filaments. This layer is especially enlarged when an active process to antigens occurs. The differentiated endothelial cells, obviously by virtue of receptor abilities, recognise both T and B lymphocytes and force them to adhere at the luminal surface and subsequently mi-

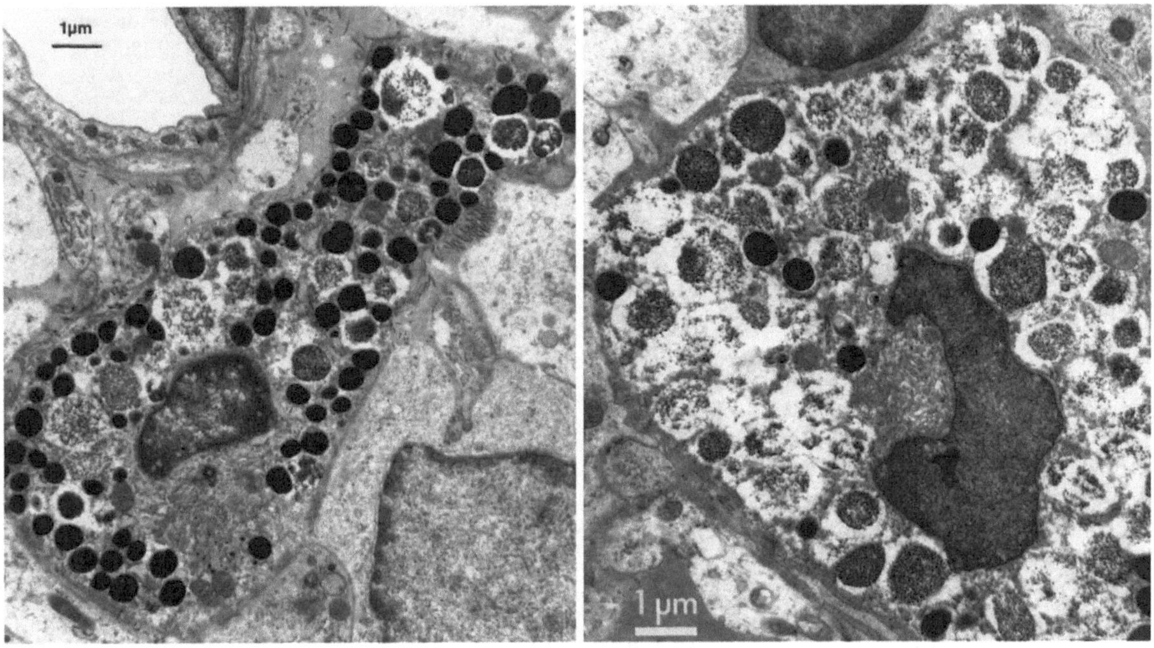

Fig. 196. Mast cells from a thoracic lymph node (block 4471) of a 224 g Sprague-Dawley *rat* (Charles River, France) injected intraperitoneally with 50 mg montmorillonite API H26 suspended in 2 ml saline. 4 days later under pentobarbital anaesthesia (30 mg/kg), the animal was perfused from the abdominal aorta with 2.5 % glutaraldehyde in 0.1 M sodium cacodylate buffer (pH 7.4). Postfixation with 1 % osmium tetroxide in sodium cacodylate buffer. Embedded in Epon 812 and sectioned at 50 nm. Lead citrate and uranyl acetate. Films 294/79 and 296/79

grate through the endothelium. While the T cells stay in the paracortal zone to be available for a new immune response, the B cells proceed to the outer cortical area, where they form compact conglomerates within primary follicles and corresponding structures in the perifollicular zone of the secondary follicles. Like germinal centre B cells, the germinal centre T cells lack homing receptors for high endothelial venules (Jalkanen et al. 1986). Homing receptors and vascular addressins were reviewed by Lakey Berg et al. (1989).

Mebius et al. (1991) reported that the morphological and functional aspects of HEV can be studied by organ culture of *mouse* isolated axillary, brachial, or inguinal lymph nodes. At 24 h of culture, the appearance of the node was still quite normal, whereas the HEV became flat-walled, with a 45–50 % reduction in the capacity to bind lymphocytes. This decrease in function of HEV could be reduced when lymph nodes were cultured in the presence of lipopolysaccharides. The effect of lipopolysaccharides on the function of HEV was presumably mediated by macrophages in the subcapsular sinus, because HEV in lymph nodes, which were depleted of subcapsular sinus and medullary macrophages previous to culture, could not be stimulated by addition of lipopolysaccharides to the culture.

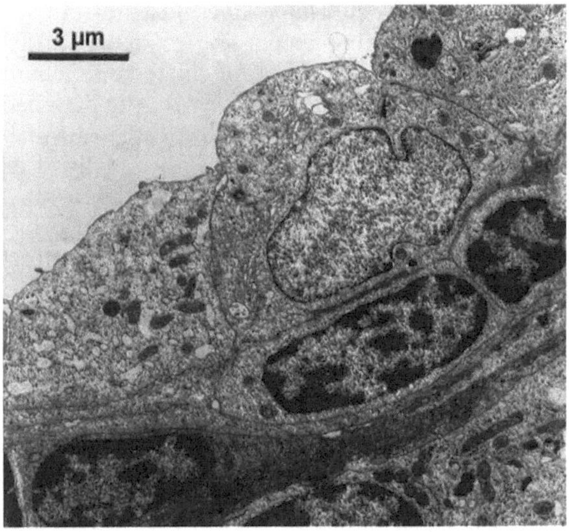

Fig. 197. High endothelial venule (HEV) with a convex adluminal surfaces of the endothelial cells in a thoracic lymph node (block 4478) of a 215 g Sprague-Dawley *rat* (Charles River, France) injected intraperitoneally with 50 mg montmorillonite API H 26 suspended in 2 ml saline. 4 days later under pentobarbital anaesthesia (30 mg/kg), the animal was perfused from the abdominal aorta with 2.5 % glutaraldehyde in 0.1 M sodium cacodylate buffer (pH 7.4). Postfixation with 1 % osmium tetroxide in sodium cacodylate buffer. Embedded in Epon 812 and sectioned at 50 nm. Lead citrate and uranyl acetate. Film 264/79

12.3 Lymphoid Tissue: BALT

From the lungs of 34 *patients* (aged 59±23 years) who had died of non-lung associated diseases and with no history of respiratory tract diseases, no BALT could be identified (Pabst and Gehrke 1990). Hiller et al. (1998) in a study of 27 further lungs of *patients* over 20 years of age in only 3 cases found small BALT structures. Lun'kova et al. (1998) reported similar findings.

BALT is significantly more frequent in smokers and recent ex-smokers (<2 years) as compared to never-smokers and longer term ex-smokers (>2 years) ($P < 0.01$) in age and sample-matched groups (Richmond et al. 1992). Endobronchial biopsies of second- and third-generation bronchi taken from 44 non-smoking elite cross-country skiers showed small BALT in 64% of skiers but in only 25% of their control subjects (Sue-Chu et al. 1998).

In the Wistar *rat*, lymphoid aggregates of the bronchi and bronchioles varied from small collections of mononuclear cells limited to the lamina propria of the airway to large nodules that extended through the muscularis and merged into the peribronchial connective tissue or the walls of adjacent alveoli (Chamberlain et al. 1973).

Scanning electron microscopic photographs of *rabbit* bronchial epithelium showed islands of lympoepithelium varying from a few to a few hundred cells surrounded by a carpet of ciliated epithelium (Bienenstock and Johnston 1976). The flattened cells devoid of cilia consistently possessed small membrane projections, subsequently identified as microvilli by transmission electron microscopy. The lymphoepithelial cells generally contained mitochondria, but endoplasmic reticulum, pinocytotic vesicles, and phagosomes were not a feature.

BALT macrophages can be distinguished from pulmonary alveolar macrophages and pulmonary tissue macrophages on the basis of their enzyme reaction products (Van der Brugge-Gamelkoorn et al. 1985). With three monoclonal antibodies, ED1, ED2, and ED3, that recognize different antigen determinants of *rat* macrophages, the three subpopulations could be clearly distinguished. Pulmonary alveolar macrophages are ED1-positive and pulmonary tissue macrophages ED2-positive. The macrophage subpopulation scattered throughout BALT is ED1-positive; macrophages situated at the peripheral site of BALT, near artery and bronchus, are ED2-positive. No ED3-positive macrophages were observed either in the lung or in BALT.

In the BALT of perfused male Sprague-Dawley *rats*, expression of eNOS (NOS-3) was markedly suppressed under entotoxin challenge, while cellular expression of bNOS (NOS-1) did not change in response to endotoxin (Ermert et al. 2002).

12.4 Lymphatic Clearance

Transfer of intratracheally administered substances to pulmonary lymph nodes was investigated by Takeda et al. (1993) using fluorescein isothiocyanate labelled dextrans. The concentrations of this drug in parapulmonary lymph nodes increased with increasing molecular weight. For selective transfer of anticancer drugs to the parapulmonary lymphatic system, a macromolecular cisplatin derivative synthesised by chemical modification with polyethyleneglycol. The concentrations of this derivative in parapulmonary lymph nodes were much higher than that for the parent drug.

A six compartments model representing the pharmacokinetics of dust movement within the alveolar area of the lungs and lymph nodes was proposed by Smith (1985) including free particles and two macrophage departments on the alveolar surfaces, temporary and encapsulation particles in the interstitial area, and particles in lymph nodes. Seven processes control the quantities of particles in each of the three areas:

1. Deposition of inhaled particles on the alveolar surface;
2. Phagocytosis of free particles by macrophages;
3. Removal of dust laden macrophages from the lungs;
4. Penetration of free particles through the alveolar membrane into the cells and interstitial area;
5. Sequestration of free particles or particles in macrophages in the interstitial area by fibrosis or other processes;
6. Transportation of free or phagocytosed interstitial particles to the lymph nodes; and
7. Dissolution of particles by dissolving soluble materials into body fluids.

A multicompartmental model of the kinetics of dust retention in the pulmonary region and in the tracheobronchial lymph nodes has been developed by Katnelson et al. (1992). Penetration into the interstitium and translocation to the lymph nodes are possible for nonphagocytosed particles only.

Brown and Donaldson (1994), however, suggested that the alveolar macrophages of HAN *rats* after intratracheal instillation of 1 mg of wool dust collected from two wool mills in the North of England, with increased secretion of immunostimulatory cytokines, could be migrating to the lymph nodes, stimulating the local lymphocytes in the node.

Penetration and translocation depend on the degree of damage to macrophages by dust and on the extent of compensatory enhancement in the recruitment of neutrophils taking part in the clearance of the pulmonary region free surface. A certain proportion of initially penetrating particles is continuously returned to this surface together with recruited pulmonary macrophages (KATNELSON et al. 1992).

12.4.1 Lymphotropism of Quartz Dust

In the *rabbit*, inhalation of quartz dust for 1 h per day, in 33 months induced silicotic nodules arranged like a string of pearls (JÖTTEN 1936).

In coal workers' lungs accumulations of coal dust are seen to form a sleeve around the respiratory bronchioles (CUMMING and SEMPLE 1973, p. 422). According to DUGUID and LAMBERT (1964), however, there is not any evidence of the choking of the lymphatics by dust that some authors have described. "It is true that occasional dust cells are seen in lymph channels, and no doubt some may be carried by way of them to the hilar lymph glands, but this does not seem to be a necessary part of the disease process. The essential change is simply an accretion of dust cells in the marginal alveoli with subsequent incorporation of them into the fibrous tissue" (l.c. p.394).

In mongrel *dogs* kept underground in a German bituminous coal mine up to six years continuously (i.e. 24 h a day, 7 days a week), SCHILLER (1956) found the dust particles in the peribronchial and perivascular lymph sheaths, while the alveoli were bare of dust.

12.4.2 Conveyance of Quartz Dust to Other Organs

12.4.2.1 Liver

Dörentrup quartz powder causing fatal acute silicosis a 25-years-old woman packing scouring powder for $2^1/_2$ years induced miliary hepatic granulomas composed of macrophages containing crystalline silica (GIESE 1931).

12.4.2.2 Spleen

In a miner aged 51 years exposed to silica dust for 5 years and 10 months, who died 20 years later, NICOD (1950) found small silicotic nodules rich in macrophages in the white pulp.

In a case of asbestosilicosis of the lung GOLDBERG (1961) did not find any asbestos bodies in the spleen, but the lymphatics of trabelular arteries and veins contained brown particles which were birefringent, being likewise probably silica particles.

12.4.2.3 Bone Marrow

In the bone marrow of Valais miners all stages of silicotic lesions from a simple accumulation of dust-laden macrophages to fibrotic nodules were described (NICOD and GARDIOL 1960).

Phagocytosis of particulate air pollutants by *human* alveolar macrophages stimulated the bone marrow (MUKAE et al. 2000).

CHAPTER 13

Pulmonary Nerves

The lung receives its major sensory and motor innervation from the tenth cranial (vagus) nerve. Postaganglionic fibres from the thoracic sympathetic plexus mingle with the vagus fibres as they enter the lung at the hilus. The combined nerves break up into plexuses accompanying the ramifications of the arteries, bronchi, and veins.

Signal transduction pathway(s) linked to the α-amino-3-hydroxy-5-methyl-4-isoxazolepropionate (AMPA)/kainate receptor subtype plays a major role in transmission of sensory information from the airways to the nucleus tractus solitarii and from nucleus tractus solitarii neurones to the airway vagal preganglionic cells, mediating reflexly increased tracheal blood flow, submucosal gland secretion, and airway tone (Haxhiu et al. 2000).

13.1 Plexus Bronchiales

The nerves to the bronchi form two plexuses (Larsell 1922), a large plexus external to the cartilages and a smaller plexus deep to the cartilage. Beyond the termination of the cartilages, the two plexuses merge and continue distally to the respiratory bronchioles. Ganglia are present within the larger nerves of the extrachondral plexus along the proximal three bronchial generations becoming rare further distally.

The periarterial plexus lies in the adventitia of the arteries, and its bundles anastomose with those of the extrachondral plexus. Fibres of the arterial plexus enter the media with the vasa vasorum and supply its outermost portion, The plexus continues distally as far as the arterioles.

The perivenous plexus accompanies the veins into the perilobular septa and even reaches the visceral pleura to supply the subpleural alveolar walls. Spencer and Leof (1964) have described twigs of the venous plexus reaching and ramifying in the subendothelial space. A few ganglia are present in the perivenous plexus near the hilus.

Anti-constitutive-nitric oxide synthase (c-NOS) labelling of *human* lung specimens showed strong immunostaining of submucosal nerves (Kobzik et al. 1993).

The innervation density of *human* airway smooth muscle by NOS-containing nerve fibres decreases significantly from trachea to large-diameter bronchi to small-diameter bronchi, whereas nitric oxide synthase (n-NOS, type I NOS)-containing nerve fibres are completely absent from bronchioli (Fischer and Hoffmann 1996). Colocalisation of NOS with vasoactive intestinal peptide (VIP) but not with substance P was frequent in these nerve fibres.

The airways are supplied by three basically different types of primary efferent nerve endings:

Fast conducting slow adapting stretch receptors
Fast conducting rapidly adapting stretch or "irritant" receptors
C-fibres.

There is a clear interganglionic segregation of these functionally different types of afferent neurones conveying information from the airways. The jugular ganglion contributes slowly adapting receptors and C-fibres to the airways, and the nodose ganglion provides mainly rapidly adapting ("irritant") receptors located in the mucosa (Kummer et al. 1996).

Neuroepithelial bodies (NEBs) were visualised in adult *rats* with the fluorescent anterograde neuronal tracer DiI injected into the vagal sensory nodose ganglion (Adriaensen et al. 1998). In the lung of the lizard, *Gekko gecko*, NEBs showed serotonin-, substance P- and neurokinin 3 receptor-like immunoreactivity, although not always with an identical intracellular distribution (Adriaensen et al. 1997). NEBs appeared to be supplied with a complex branching varicose substance P-like immunoreactive innervation that was clearly seen to contact NEB cells using dual channel confocal laser scanning microscopy. Quinacrine (6-chloro-9-[4-(diethylamino)-1-methylbutyl]amino-2-methoxy-acridine) fluorescence appears to be a marker for NEBs in the *rat* lung, suggesting the involvement of ATP or related substances in neurotransmission (Brouns et al.

1999). Quantitative analysis of immunolabelled cryostat sections of the lungs revealed that almost 50 % of the NEBs in control adult Wistar *rats* received a calcitonin gene-related peptide immunoreactive innervation, in contrast to only 3 % after capsaicin treatment (Brouns et al. 2001). Double labelling with antibodies against calcitonin gene-related peptide and protein gene-product 9.5, as a general marker for NEBs, clearly demonstrated that all protein gene-product 9.5- immunoreactive NEBs invariably expressed calcitonin gene-related peptide in control and capsaicin-treated animals. Calbindin D28k immunoreactivity of the vagal sensory innervation of NEBs was not altered in capsaicin-treated *rats*.

Fu et al. (2001) reported expression and functional characterisation of serotonin type 3 receptor mRNA in NEB cells in the lungs of different mammals (*hamster*, *rabbit*, *mouse*, and *human*).

13.2 Peribronchial Microparaganglia

Peribronchial microparaganglia may consist of a few solitary chromaffin cells which are related to a common capillary network or small clusters of 10–15 chromaffin cells per section (Böck 1982). The capillaries have been shown to be supplied from bronchial arteries, not from the pulmonary system (Böck 1970). Blessing and Hora (1968) found more than 50 microparaganglia in a serially sectioned lung from a *human* newborn. The paraganglia are situated along the main ramifications of the bronchial tree, connected either to the branches of the pulmonary artery or to bronchi. No paraganglia are found in the periphery of the lung.

In *human* airway intrinsic ganglia, the number of NOS-containing neuronal cell bodies increased from 57 % in the trachea up to 83 % in small bronchi (Fischer and Hoffmann 1996). Around these perikarya, many nerve fibres displaying VIP-immunoreactivity or substance P-immunoreactivity were found.

Electron microscopically, paraganglionic cells, enveloping sustentacular cells, and synaptic connections with unmyelinated axons are seen in peribronchial microparaganglia (Böck 1970).

In the lungs of the salamanders, *Cynops pyrrhogaster* and *Ambystoma mexicanum*, Adriaensen et al. (1993) described NOS-like immunoreactivity in dispersed nerve cells in a plexus composed of thick nerve bundles and in a delicate nerve netwerk made up of fine varicose nerve fibres. There was evidence that at least part of these varicose fibres originated from NOS-containing nerve cells in intrapulmonary ganglia.

13.2.1 Neuronal Nitric Oxide Synthase in Peribronchial Microparaganglia

In *human* airway intrinsic ganglia, the number of NOS-containing neuronal cell bodies increased from 57 % in the trachea up to 83 % in small bronchi (Fischer and Hoffmann 1996). Around these perikarya, many nerve fibres displaying VIP-immunoreactivity or substance P-immunoreactivity were found.

In the *ferret* trachea, anterograde tracers (either rhodamine- or biotin-labelled dextran amines) were found in nerve fibres that formed basketlike complexes associated with neurones of the longitudinal trunk and the superficial muscular plexus ganglia and were observed in the nerve fibres of smooth muscle and tracheal plexus (Zhu and Dey 2001). Some VIP or nNOS positive neurones in the superficial muscular plexus ganglia contained retrogradely transported tracer.

In the lungs of the salamanders, *Cynops pyrrhogaster* and *Ambystoma mexicanum*, Adriaensen et al. (1993) described NOS-like immunoreactivity in dispersed nerve cells in a plexus composed of thick nerve bundles and in a delicate nerve netwerk made up of fine varicose nerve fibres. There was evidence that at least part of these varicose fibres originated from NOS-containing nerve cells in intrapulmonary ganglia.

13.3 Pulmonary Neuropharmacology

Contractile and relaxant responses of the *guinea pig* tracheal smooth muscle have been shown to be mediated by distinct neural pathways. Fischer et al. (1996) localised and characterised the parasympathetic relaxant innervation using electrophysiological stimulation, neuronal tracing and immunohistochemistry of organotypic cultures of trachea-esophagus preparations. Disruption of the connection between the esophagus and the trachea completely abolished parasympathetic relaxant responses and pharmacological blockers of ganglionic tranmission only affected relaxant responses when applied to the esophagus. Most neurones of the esophageal wall retrogradely labelled by microinjection of a fluorescent tracer into the tracheal smooth muscle were immunoreactive for the •NO-generating enzyme NO-synthase (NOS), or vasoactive intestinal peptide (VIP).

Several kinds of bronchoconstrictors, such as **acetylcholine** (Belvisi et al. 1991, Gupta and Prasad 1992, Matera et al. 1995), **histamine** (Nijkamp et al. 1993, Persson et al. 1995, Matsumoto et al. 1997), **bradykinin** (RICCIARDOLO et al. 1994, SCHLEMPER and CALIXTO 1994, FIGINI et al. 1996), and **leukotriene** C_4 (PERSSON et al. 1995) produce •NO during bronchoconstriction, and •NO counteracts this bronchoconstriction.

300 ppm NO inhalation significantly decreased the bronchocontrictor response to acetylcholine administered intravenously in *rabbits* (MENSING et al. 1997). Following the inhalation of an aerosolised solution of *N*-nitro-L-arginine methyl ester (1.2 mM) for 40 min, the response of the dynamic elastance to acetylcholine increased significantly.

In anaesthetized *guinea pigs* both exogenous as well as capsaicin-induced **neurokinin A** increased airway opening pressure (P_{ao}) and the exhaled •NO level, and both were inhibited by an antagonist selective for NK_2 receptor (a receptor for neurokinin A), SR48968 (IMASAKI et al. 2001). The exhaled •NO level became negligible with an inhibitor of NO synthase type 1–3 (N^G-nitro-L-arginine methyl ester, L-NAME) with increased P_{ao}, but not with a NOS type 2 inhibitor. In an *in vitro* study, neurokinin A increased the nitrite/nitrate level in superfused fluid of tracheal segments. Removing smooth muscle reduced nitrite/nitrate in the fluid to negligible levels, while the level was unchanged with removal of the epithelia. Pre-treatment with L-NAME enhanced the tension of epithelia-removed tracheal segments.

Vasoactive intestinal peptide, a hormone which occurs widely in the nervous system and gastrointestinal tract (POLAK et al. 1974, BUFFA et al. 1977, LARSSON 1978, SAID 1980) has a predominantly neural distribution in the lung. It is localised in nerve cell bodies and fibres in the walls of the bronchi, in relation to the secretory glands and smooth muscle, in pulmonary and bronchial vessels as well as in ganglion-like structures in the peribronchial connective tissue (UDDMAN et al. 1978, DEY and SAID 1980). SHIMOSEGAWA and SAID (1988) observed VIP-immunoreactive nerve fibres on smooth muscle bundles of the *guinea-pig* trachea. DEY et al. (1988) measured the distribution of VIP binding sites in bronchial epithelium and smooth muscle by videoimage analysis in autoradiograms of *guinea-pig* lung. Both binding sites appeared to be specific, but the epithelium had a higher capacity to bind VIP. Mediator inhibition studies using ovalbumin-challenged chopped lung showed that preincubation with VIP (10^{-5} M) or calcitonin gene-related peptide (10^{-6} M) inhibited the release of immunoreactive 5-hydroxy-6(*R*)-*S*-glutathionyl-7,9-*trans*-11,14-*cis*-eicosatetraenoic acid (leucotriene C_4, LTC_4) and histamine, but the effect was weak at the concentration of 10^{-7} M in both peptides (NOMURA et al. 1988).

Contractions of bronchial smooth muscle of the *guinea pig* in response to **tachykinin** release from sensory nerves induced by electrical field stimulation were inhibited by **nociceptin** at a threshold concentration of 10^{-9} M (FISCHER et al. 1997). At concentration of 10^{-7} M, nociceptin caused a 64 ± 6 % inhibition of tachykinergic contractions. These inhibitory effects were not blocked by antagonists for the classical μ, κ, and δ-opioid receptors (naloxone, naltrindole, norbinaltorphimine), indicating that ORL-1 is involved. The presence of a ORL-1-mRNA in the jugular ganglion, where the cell bodies of the tachykinin-containing C-fibres are located, was demonstrated by RT-PCR using ORL-1 specific primers.

CHAPTER 14

Radical-Mediated Pulmonary Lesions

Free radical mechanisms (partly) mediate pulmonary damage by ozone, nitrogen oxide, nitrogen dioxide, high doses of pure oxygen or dust particles as quartz and other types of tetrahedral silica (tridymite, cristobalite, coesite) and fibrous dusts (asbestos and man-made mineral fibres). Inorganic dust inhalation may prime lung inflammatory cells and markedly enhance their capacity to release toxic oxygen radicals (CANTIN et al. 1988). CRUZEL et al. (1987) and DAMON et al. (1989) observed a linear correlation between the maximal luminol-enhanced chemoluminescence of zymosan-opsonized *human* alveolar macrophages and the severity of asthma.

Chronic hypoxia causes pulmonary arterial smooth muscle cell depolarisation, elevated endothelin-1, and vasoconstriction. Resting $[Ca^{2+}]_i$ in smooth muscle cells from intrapulmonary arteries of *rats* exposed to 10% O_2 for 21 days was 293.9±25.2 nM (vs. 153.6±28.7 nM in normoxia) (SHIMODA et al. 2000). Resting $[Ca^{2+}]_i$ was decreased after extracellular Ca^{2+} removal but not with nifedipine (10^{-6} M), an L-type Ca^{2+} channel antagonist. After chronic hypoxia, the endothelin-1-induced increase in $[Ca^{2+}]_i$ was reduced and was abolished after extracellular Ca^{2+} removal or nifedipine. Removal of extracellular Ca^{2+} reduced endothelin-1-induced tension; however, nifedipine had only a slight effect.

There is substantial evidence that K^+ channel activity and vascular tone can be modulated by reducing and oxidising agents. The importance of reduction-oxidation (redox) potential of key sulfhydryl groups in the gating mechanism of certain K_v channels appear related to the presence of cysteine groups in the channel (DEMIRA and RUDY 1992). In pulmonary artery smooth muscle cells, oxidants such as *t*-butyl hydroperoxide, reduced/oxidised glutathione (GSH/GSSG), and diamide (WEIR and ARCHER 1995) increase opening of K_v channels causing membrane hyperpolarization and relaxation. ARCHER et al. (1999) hypothesised that the K^+ channels are not themselves "O_2 sensors" but rather respond to the reduced redox state created by hypoxic inhibition of candidate O_2 sensors (NADPH oxidase or the mitochondrial electron transport chain). Both pathways shuttle electrons from donors, down a redox gradient, to O_2. Hypoxia inhibits these pathways, decreasing radical production and causing cytosolic accumulation of unused, reduced, free diffusible electron donors. Pulmonary artery smooth muscle cell K^+ channels are redox responsive, opening when oxidised and closing then reduced. Inhibitors of NADPH oxidase (diphenyleneiodonium) and mitochondrial complex 1 (rotone) both inhibit pulmonary artery smooth muscle cell whole-cell K^+ current but lack the specificity to identify the O_2-sensor pathway. ARCHER et al. (1999) used *mice* lacking the gp91 subunit of the NADPH oxidase [chronic granulomatous disease (GCD) *mice*] to assess the hypothesis that NADPH oxidase is a pulmonary artery O_2-sensor. NADPH oxidase, though a major source of lung radical production, is not the pulmonary vascular O_2 sensor in *mice*.

Overactivation of glutamate receptors can contribute to a wide variety of lesions, including acute pulmonary oedema (SAID et al. 1996), airway hyperesponsiveness (SAID 1999) and inflammation. Blocking of the *N*-methyl-D-aspartate subtype of glutamate receptors by dizocilpine maleate (MK-801; 10 μM) attenuated oxidant injury induced by paraquat (100 mg/kg to the pulmonary circulation) or by xanthine oxidase (SAID et al. 2000).

In primary cell cultures from the tunica media of *human* pulmonary arteries, hypoxia inhibited platelet-derived growth factor-induced $[^3H]$glucosamine incorporation in secreted glycosaminoglycans, especially hyaluronic acid, in vascular smooth muscle cells (PAPAKONSTANTINOU et al. 2000). In contrast, it stimulated glycosaminoglycan secretion, specially heparan sulphate, by fibroblasts.

14.1 Reperfusion Injury

The lung may be susceptible to oxygen radical-mediated injury during ischaemia, as O_2 tension will be high when ventilation is maintained and blood flow impaired. Peroxydation of membrane phospholipids may directly alter vascular permeability. Reperfusion injury in pulmonary tissue has been observed after occlusion of pulmonary arteries due to surgery, or after expansion of a collapsed lung (Bishop et al. 1988, Kennedy et al. 1989). After reperfusion, vasoconstriction, pulmonary œdema and leucocyte infiltration have been observed (Levinson et al. 1986, Bishop et al. 1988, Jackson et al. 1988, Kennedy et al. 1989). In *rabbits*, interleukin-8 was locally produced by alveolar epithelium and macrophages in lung reperfusion injury and this event was accompanied by a massive neutrophil infiltration (Sekido et al. 1993). Neutrophils are not necessary for induction of ischæmia-reperfusion lung injury (Deeb et al. 1990). The late phase of reperfusion injury, however, is known to be neutrophil-dependent. Lung content of myeloperoxidase (MPO) in the ischaemic lung rose during the period of reperfusion when compared with control *rats* (Eppinger et al. 1997). There was a statistically significant rise in MPO activity at 30 min of reperfusion. Levels were significantly elevated by 1 h of reperfusion ($P < 0.01$). At 2 and 4 h of reperfusion, MPO activity peaked and reached a plateau ($P < 0.01$ versus control for each point). These findings pointed to a significant sequestration of neutrophils in the ischaemic-reperfused lung.

In female SPF Wistar *rats*, ultrastructural changes after temporary ischaemia were extraordinarily multiform (Amthor 1978). Type I pneumocytes showed vesiculation of the cytoplasm. From the second day on type II pneumocytes increased in number, but they contained only little osmiophilic material.

Neutrophil apoptosis may be important in regulating the inflammatory process by controlling neutrophil numbers and thus activity. Exogenous inhaled nitric oxide is now a widely used therapy in *patients* with acute lung injury, and its effects on apoptosis may be important. In a model of nitric oxide-treated lung injury, Blaylock et al. (1998) incubated polymorphonuclear leucocytes isolated from venous blood for up to 16 h with and without 1.7 μg/ml lipopolysaccharide and the nitric oxide donor GEA-3162 or the peroxynitrite donor SIN-1. Apoptosis was attenuated when cells were exposed to lipopolysaccharide and both nitric oxide and peroxynitrite dose-dependently inhibited this suppression at all time points and was most apparent at 16 h ($P = 0.004$ and 0.001, respectively).

In the lung, ischaemia-reperfusion does not necessarily imply hypoxia-reoxygenation, if ventilation is maintained during the period when blood flow is impaired. Pulmonary ischaemia-reperfusion injury is clinically manifest as pulmonary œdema (Bishop et al. 1986, 1987, Corris et al. 1987) or local pulmonary œdema (Ward and Pearse 1988), following ventilated and nonventilated ischaemia. To determine whether reactive oxygen species generated during ischaemia mediated ischaemic injury, Becker et al. (1998) measured tissue levels of F_2-isoprostanes as an index of lipid peroxidation, 30 min after administration of glucose (5 mM)-glucose oxidase (GOX, 0.1 U/ml), or after short (45 min) or long (180 min) ventilated ischaemia, in isolated *ferret* lungs. Osmotic reflection coefficient for albumin (σ_{alb}), an estimate of vascular protein permeability, was measured in the same lungs. Tissue F_2-isoprostanes increased 375 % after exposure to glucose-GOX in association with a 42 % decrease in σ_{alb}, and administration of catalase (10^5 U) and superoxide dismutase (25,000 U) completely attenuated this lipid peroxidation. In contrast, tissue F_2-isoprostanes increased only 60 % following 45 min of ischaemia, then did not increase additionally. σ_{alb} was not altered by 45 min of ischaemia, but decreased 72 % following 180 min of ischaemia. Catalase + superoxide dismutase did not alter F_2-isoprostane formation during ischaemia, but partially attenuated vascular injury. These results suggested that tissue levels of F_2-isoprostanes reflect lung lipid peroxidation, but that F_2-isoprostane generation does not directly increase vascular permeability following ventilated pulmonary ischaemia.

In the *rabbit*, prostaglandin E_1 may ameliorate post-ischæmic lung reperfusion injury (Matsuzaki et al. 1993). In the isolated perfused *rat* lung, Reignier et al. (1994) reduced ischæmia-reperfusion injury by pentoxifylline probably mainly due to a decrease in pulmonary neutrophil sequestration during reperfusion. After ischæmia-reperfusion, lung myeloperoxidase and blood neutrophil count decrease were lower with pentoxifylline than in controls (Chapelier et al. 1995).

Erythro-9-(2-hydroxy-3-nonyl)-adenine (EHNA), a specific inhibitor of cyclic GMP-stimulated phosphodiesterase (PDE2) reversed the pulmonary vasoconstriction that follows a hypoxic challenge in the perfused *rat* lung (Haynes et al. 1996). An EHNA-inhibitable PDE2 was shown to be present in the smooth muscle of *rat* lung and was shown to be at least partly responsible for this vasodilatory response. This suggests PDE2 might be involved in the modulation of smooth muscle tone in the *rat* lung.

In a *canine* model of left lung transplantation Günther et al. (1997) analyzed the surfactant properties in fresh explants (control), explants kept ischaemic for 24 h (Euro-Collins/prostacyclin flush, 4 °C) and grafts and native lungs of recipients after 6 and 12 h of reperfusion under anaesthesia and mechanical ventilation. Surfactant treatment (Alveofact®) resulted in a less pronounced decrease in phospholipid-protein ratio values and normalized minimum surface tension values, but still slightly elevated minimum surface tension values in the presence of bronchoalveolar lavage fluid proteins.

In a *rat* model, FEHRENBACH et al. (1997) showed that intraalveolar oedema resulted in a wide separation of epithelium and surface lining surfactant monolayer. The regular lattice-like array of tubular myelin, a surfactant subtype that in controls was seen to occupy extended regions just beneath the surface lining layer, was altered in experimental lungs. Large whorls of disturbed tubular myelin forming zigzag patterns and loops were seen within intraalveolar oedema accumulations. Remnants of lattice-like tubular myelin showed less particles to be associated with its membranes.

Intraalveolar oedema seen in Euro-Collins solution-perfused isolated *rat* lungs after ischaemia (2 h at 4 °C) and reperfusion (40 min) occupied 36±6 % of the gas exchange region as compared to control lungs fixed by vascular perfusion immediately after excision (1±1 %) (OCHS et al. 1999). Relative intraalveolar surfactant composition showed a decrease in surface active tubular myelin (3±1 % vs. 12±0 %; $P = 0.008$) and an increase in inactive unilamellar forms (90±7 % vs. 64±5 %; $P = 0.008$) in the Euro-Collin solution-perfused lungs. These changes occurred both in oedematous (tubular myelin 3±1 %, unilamellar forms 88±6 %) as well as in non-oedematous regions (tubular myelin 4±3 %, unilamellar forms 77 + 5 %). Alveolar atelectasis, predominating in the Euro-Collin-protected group, was reduced during preservation with Celsior solution (FEHRENBACH et al. 1999).

In hypoxia (2 % O_2), [^{14}C]spermidine was prominently localised in explanted conduit, muscularised, and partially muscularised *rat* pulmonary arteries, which was not evident in normoxic lung tissue (BABAL et al. 2000). Hypoxic main pulmonary arterial explants also exhibited substantial increase in [^{14}C]spermidine uptake relative to control explants, and autoradiography revealed that enhanced uptake was most evident in the medial layer. Main pulmonary arterial explants denuded of endothelium failed to increase polyamine transport in hypoxia. Conversely, medium conditioned by endothelial cells cultured in hypoxic, but not in normoxic, environments enabled hypoxic transport induction in denuded arterial explants.

Nitric oxide donors and atrial natriuretic peptide (ANP) have been shown to block H_2O_2 mediated increases in the permeability of *porcine* pulmonary artery endothelial cells. EHNA inhibition of PDE2 reduced this effect (SUTTORP et al. 1996). However, PDE2 is thought to be regulating permeability through the hydrolysis of cGMP rather than cAMP in this system suggesting that PDE2 in these cells could be functioning in a negative feedback response to cGMP synthesis rather than in the mediation of a cAMP regulatory pathway.

In vitro, bovine pulmonary artery endothelial cells stimulated by 10 mM H_2O_2 synthesized platelet-activating factor (PAF), which was reduced by >90 % when catalase (500 U/ml) was included in the incubation buffer (LEWIS et al. 1988).

Pre-infusion of FR139317 {(R)2-[(R)-2-[(S)-2-[[1-(hexahydro-1H-azepinyl)]carbonyl]amino-4-methylpentanoyl]amino-3-[3-(1-methyl-1H-indoyl)]propionyl]amino-3-(2-pyridyl)propionic acid}, a selective endothelin-1 receptor antagonist (SOGABE et al. 1993), prevented post-reperfusion injury of the *rat* lung (OKADA et al. 1995).

14.2 Hyperoxia

Hyperbaric oxygen treatment (i.e. exposure to 100 % O_2 at a pressure of 2.5 ATA for a total of three 20 min periods) of *human* subjects) caused clear and reproducible DNA effects in the comet assay with leucocytes (DENNOG et al. 1996). DNA damage was detected only after the first treatment and not after further treatments under the same conditions, indicating an increase in antioxidant defences. Blood taken 24 h after hyperbaric oxygen treatment was well protected against the *in vitro* induction of DNA damage by hydrogen peroxide (ROTHFUSS et al. 1998).

TANIGUCHI et al. (1986) exposed female Wistar *rats* to hyperoxia (>97 % O_2) for 60 h and found leukotriene B_4 (LTB_4) significantly increased in lung lavages compared with that in normoxic control *rats*. At the same time, the marked increase in the number of polymorphonuclear leucocytes in lung lavages and the decrease in the activity of NADPH-cytochrome *c* reductase in lung microsomes were also observed. The administration of AA861, a 5-lipoxygenase inhibitor, reduced not only the increase in LTB_4, but also the increase in the number of polymorphonuclear leucocytes in lung lavages of *rats* exposed to hyperoxia for 60 h. Furthermore, treatment with AA681 (1 mg/kg or 5 mg/kg intraperitoneally every 12 h from the onset of exposure to the end of exposure) also protected the decrease in the activity of NADPH-cytochrome *c* reductase. The effects of AA681 on these parameters occurred in a dose-dependent manner.

Mice exposed to 100 % O_2 died after 3 or 4 d with diffuse alveolar damage and alveolar oedema (BARAZZONE et al. 1998). Extensive cell death was evident by electron microscopy in the alveolar septa, affecting both endothelial and epithelial cells. The damaged cells showed features of both apoptosis (condensation and margination of chromatin) and necrosis (disruption of the plasma membrane).

Lipopolysaccharide inhalation increased neutrophil number in the bronchoalveolar lavage fluid of *mice*, which was significantly inhibited by hyperbaric oxygen but not hyperbaric air pre-exposure (KANG et al. 2002). However, myeloperoxidase content in the lung was prominently increased by pre-exposure to hyperbaric oxygen, which correlated with the infiltration of polymorphonuclear leucocytes in lung tissue. Further, hyperbaric oxygen + lipopolysaccharide, but not saline inhalation caused a significant increase in the bronchoalveolar lavage fluid protein level and lactate dehydrogenase activity compared with that of lipopolysaccharide inhalation alone. Lipopolysaccharide exposure induced significant increase in plasma NO metabolites, which was not potentiated by hyperbaric oxygen pre-exposure. The iNOS inhibitor, aminoguanidine, significantly attenuated the increases in plasma NO metabolites and tissue myeloperoxidase content as well as lung injuries.

Myeloperoxidase (EC 1.11.1.7) reaction

$$Cl^- + H_2O_2 \rightarrow ClO^- + H_2O \quad [27]$$

$$ClO^- + H_2O_2 \rightarrow {}^1O_2 + Cl^- + H_2O \quad [28]$$

Hyperoxia (incubation of lung slices in 85 % O_2) increased oxygen radical production in *rat* lungs (FREEMAN and CRAPO 1981). Mitochondria account for 15 ± 3 % of the CN resistant respiration in *rat* lungs under hyperoxic conditions and will release H_2O_2 extramitochondrially at a rate of 50 nmol/min × *rat* lung. Pure oxygen breathing most severely damaged the capillary endothelium of the lungs of *mice* (CEDERGREN et al. 1959) and *monkeys* (KAPANCI et al. 1969). Ganglion cells of *rats* exposed to 100 % oxygen at 8 ata showed a hyperchromatic cell shrinkage and cytoplasmic vacuolation (VON SCHNAKENBURG 1971). In *mice* hyperbaric oxygenation significantly decreased brain γ-aminobutyric acid and glutamic acid decarboxylase, but this was not prevented by 200 mg/kg disulfiram, an effective oxygen protectant (FAIMAN et al. 1977).

Mitochondria with longitudinally orientated cristae developed in the type II of the alveolar cells of rats exposed to high concentrations of oxygen (ROSENBAUM et al. 1969, YAMAMOTO et al. 1970), and it is known from biochemical studies that high oxygen concentrations can act as an inhibitor of several kinds of enzymatic activity, including such exclusively mitochondrial enzymes as succinic dehydrogenase and cytochrome oxidase. However, as KARNOVSKY (1962, 1963) has pointed out, low cytochrome oxidase activity is not invariably associated with horizontally orientated cristae.

Genome-wide linkage analyses of intercross (F_2) and recombinant inbred cohorts identified significant and suggestive quantitative trait loci on chromosome 2 (hyperoxia susceptible locus 1 (*Hsl1*) and 3 (*Hsl3*), respectively (CHO et al. 2002). Comparative mapping of *Hsl1* identified a strong candidate gene, *Nfe2l2* (nuclear factor, erythroid derived 2, like 2 or *Nrf2*) that encodes a transcription factor NRF2 that regulates antioxidant and phase 2 gene expression. Strain-specific variation in *mouse* lung *Nrf2* messenger RNA expression and a T → C substitution in B6 *Nrf2* promoter that cosegregated with susceptibility phenotypes in F_2 animals supported *Nrf2* as a candidate gene.

Hyperoxia transcriptionally activated lung cytochrome P4501A1 (CYP1A1) gene expression *in vivo* and in vascular endothelial cells (1995). Cimetidine pre-treatment and treatment every 24 h reduced lung oxidant stress and cytochrome P_{450}-derived oxidation products of arachidonic acid in new-born *lambs* exposed for 72 h to 95 % O_2 (HAZINSKI et al. 1996).

Steady-state levels of the Na^+,K^+-ATPase α_1 and β_1 subunit mRNAs increased in whole lung tissue of adult *rats* during exposure to 85 % O_2 (sublethal hyperoxia) for 7 days, followed by 2, 3, or 4 days in 100 % O_2 (JOHNSON et al. 1998). Stability of the Na^+,K^+-ATPase α_1 and β_1 subunit mRNA messages in whole lung RNA did not change significantly. Thus, lung Na^+,K^+-ATPase gene expression in sublethal hyperoxia appears to be regulated in part at the transcriptional level. Pneumocyte type II cell Na^+,K^+-ATPase α_1 and β_1 subunit proteins, measured by quantitative immunofluorescence, increased significantly after sublethal hyperoxia and 100 % O_2 exposures. Increases in lung fluid clearance after sublethal hyperoxia were associated with increased type II pneumocyte Na^+,K^+-ATPase protein and whole lung Na^+,K^+-ATPase mRNA expression.

Numerous cytokines may contribute to the inflammation seen in hyperoxia-induced lung injury. Hyperoxia increases the production of monocyte chemotactic protein (MCP), a cytokine that attracts monocyte lineage cells which are important in chronic inflammation. U937 cells treated with actinomycin to halt RNA transcription and then either exposed to normoxia or 95 % O_2 hyperoxia showed that the $t_{1/2}$ of MCP mRNA from 4.8 ± 0.44 h in the presence of normoxia significantly ($P < 0.05$) increased to 6.8 ± 0.7 h in the presence of hyperoxia (FULLER et al. 1996).

The epidermal growth factor receptor (EGFR) network known to be a potent modulator of epithelial cell growth was examined on isolated *rat* type 2 cells and SV40T-T2, a type 2 cell line, under normoxic conditions, after 24 h and 48 h of *in vitro* hyperoxia, and after 24 h of normoxic recovery (NICI et al. 1996). EGF induced tyrosine phosphorylation of EGFR in type 2 cells and SV40T-T2, which decreased with hyperoxia and increased above normoxic levels in recovering cells, suggesting biphasic changes in receptor number or function with injury. The EGFR appeared to be stimulated in an autocrine fashion in these cells. There was decreased DNA synthesis and proliferation in SV40T-T2 and isolated type 2 cells treated with tyrphostin B56, a specific EGFR inhibitor. Pre-treatment with suramin, which binds to growth factor, resulted in increased EGFR tyrosine phosphorylation after stimulation, suggesting disruption of normal autocrine receptor downregulation. Transforming growth factor-α was identified in conditioned media from normoxic and hyperoxic SV40T-T2 and type 2 cells.

Anti-neutrophil chemokine preserved alveolar development in hyperoxia-exposed new-born *rats* (AUTEN JR. et al. 2001).

Leptin, a cytokine product of the *ob* gene that is secreted by adipocytes and is involved in the regulation of food intake, does not play an essential role in the direct and short-term effects of oxygen-induced injury. Using leptin-deficient *mice* and wild-type *mice* treated with anti-leptin antibody, BARAZZONE-ARGIROFFO et al. (2001) demonstrated that weight loss was leptin independent. Lung damage by 100% O_2 was moderately attenuated in leptin-deficient *mice* but was not modified by anti-leptin antibody or leptin administration.

Surfactant protein B deficiency worsened hyperoxic injury (95% O_2 at 1 atmosphere for 72 h) to the alveolar epithelium of SP-$B^{+/-}$ *mice* (TOKIEDA et al. 1999).

Glucose depletion was a principal determinant of hyperoxic (95% O_2) death of *human* lung-epithelial-like cells (A549), as lactate dehydrogenase activity increased only after glucose was depleted in the medium (ALLEN and WHITE 1998).

Hyperoxia (100% O_2 for 12 h) increased NOS II mRNA in airway epithelial cells of health individuals by 2.5-fold but did not increase extracellular glutathione peroxidase mRNA (COMHAIR et al. 2000). *In vitro* exposure of epithelial cells brushed from second- or third-order bronchi to reactive oxygen species or reactive nitrogen species also increased extracellular glutathione peroxidase expression.

NOS III (ecNOS; EC 1.14.13.39) protein was highly expressed in Clara cells of *rats* breathing 87% O_2 for 28 days increasing distally, while in ciliated cells the intensity of the signal was lower (NIKOLOV et al. 1998, 1999). Types I and II pneumocytes showed the greatest signal increase in hyperoxia.

Human alveolar epithelial cells (A549) and *human* lung microvascular endothelial cells cultured under hyperoxia (95% O_2) + •NO (derived from chemical donors) died after 4 to 5 days, whereas cells neither died nor divides in •NO alone (NARULA et al. 1998). The toxic effect was clearly synergistic, and cell death did not occur via apoptosis. As an indicator of peroxynitrite formation, nitrotyrosine-containing proteins at molecular masses of 25 and 35 kDa were found to be increased in A549 cells exposed to •NO or •NO + hyperoxia.

Although the morphologic data suggest that pulmonary cells die by necrosis in hyperoxia, recent studies suggest that **apoptotic programmed cell death** is also induced in lungs exposed to high levels of oxygen. The Bcl-2 family of proteins plays a predominant role in the induction and prevention of apoptosis (ADAMS and CORY 1998). Adult *mice* exposed to >95% oxygen concentrations for 48 to 88 h had increased whole-lung mRNA levels of Bax and Bcl-X_L, no change in Bak, Bad, or Bcl-2, and decreased levels of Bcl-w and Bfl-1 (O'REILLY et al. 2000). In situ hybridisation revealed that hyperoxia induced Bax and Bcl-X_L mRNA in uniform and overlapping patterns of expression throughout terminal bronchioles and parenchyma, coinciding with terminal transferase dUTP end-labelling (TUNEL) staining. Electron microscopy and DNA electrophoresis, however, suggested relatively little classical apoptosis. Unexpectedly, Western analysis demonstrated increased Bcl-X_L, but not Bax, protein in response to hyperoxia. Bax and Bfl-1 were not altered by hyperoxia in *p53* null *mice*; however, oxygen toxicity was not lessened by *p53* deficiency.

The induction of the **nitric oxide pathway in** *rat* **alveolar macrophages** *in vitro* stimulated with lipopolysaccharide (100 ng/ml) and *rat* recombinant interferon-γ (100 U/ml) was further upregulated by hyperoxia of 85% O_2 (PEPPERL et al. 2001). The binding capacity of NF-κB, in contrast to that of AP-1, was activated on stimulation with lipopolysaccharide and interferon-γ, and both were further increased under hyperoxia. The antioxidants pyrrolidine dithiocarbamate and *N*-acetyl-L-cysteine inhibited intracellular production of reactive oxygen species and the NO pathway under both normoxic and hyperoxic conditions but had diverse effects on the transcription factors.

Phylogenetic differences to hyperoxia were reported by SOMAYAJULU et al. (1978) who exposed *chickens* to 100% oxygen at 1 atm pressure for prolonged periods and did not find any pathological changes in their lungs, although H_2O_2 did accumulate in their airways. *Rabbits*, however, exposed to hyperoxia all died within 5 days. Their lungs showed generalised oedema and patchy haemorrhage. Because H_2O_2 occurred in considerable amounts in the airways of both rabbit and chickens, it appeared that H_2O_2 might not be the source of oxygen damage in rabbits unless a large species difference to H_2O_2 was evident. GRAM (1997) put forward the question if the deficiency of pulmonary cytochrome P_{450} in *bird* lungs might somehow protect them from oxygen pulmonary toxicity or are other factors involved.

Non-muscular segments of *rat* lung microvessels in hyperoxic pulmonary hypertension become muscularized using locally recruited fibroblasts or cells with morphological properties between smooth muscle cells and pericytes (JONES 1992).

Vasoconstriction induced by hyperbaric oxygen decreases total and regional **cerebral blood flow** in *human* subjects (LAMBERTSEN et al. 1953, VISSER et al. 1996, OMAE et al. 1997) and in *rats* (TURBATI et al. 1978, BERGO and TYSSEBOTN 1992). DEMCHENKO et al. (2000) have implicated lack of $^{\bullet}$NO in hyperoxic vasoconstriction because inhibitors of NOS types I and III did not enhance the vasoconstriction during hyperbaric oxygen and the effects of $^{\bullet}$NO donors that increase regional cerebral blood flow in air-breathing *rats* were attenuated during hyperbaric oxygen exposure. After 30 min exposure at 5 ATA, regional cerebral blood flow had decreased in the substantia nigra, caudate putamen, hippocampus and parietal cortex by 23 to 37 %. These reductions in regional cerebral blood flow were not augmented by exposure to hyperbaric oxygen in animals pre-treated with *N*ω-nitro-L-arginine methyl ester (30 mg/kg i.p.). After 30 min at 5 atmospheres, brain NO_x levels had decreased by 31 ± 9 % and correlated with the decrease in regional cerebral blood flow, while estimated $HO^{\bullet}$ production increased by 56 ± 8 %. The decrease in regional cerebral blood flow at 5 ATA was completely abolished by MnSOD administration into the circulation before hyperbaric oxygen exposure. During exposures of *rats* to hyperbaric oxygen regional cerebral blood flow decreased at 4 atmospheres, decreased for the initial 30 min at 5 atmospheres then gradually increased, and increased within 30 min at 6 atmospheres (DEMCHENKO et al. 2001). Changes in regional cerebral blood flow correlated positively with NO_x production. In *rats* pre-treated with L-N^{ω}-nitro-L-arginine methyl ester, regional cerebral blood remained maximally decreased throughout 75 min of the hyperbaric oxygen at 4, 5, and 6 atmospheres.

Cytoprotection

Hyperoxia (95 % O_2) inhibited growth of H441 cells, a transformed *human* lung cell line with characteristics of Clara cells, and the reduction of 3-(4,5-dimethylthiazol-2-yl)-2,5-diphenyl tetrazoliumbromide (O'DONOVAN et al. 2000). However, cells transfected with **glutathione reductase** cDNA with a mitochondrial leader sequence had greater mitochondrial glutathione reductase activities, sustained more normal growth pattern, and had less inhibition of 3-(4,5-dimethylthiazol-2-yl)-2,5-diphenyl tetrazoliumbromide reduction after exposure to hyperoxia for 48 h than was observed in cells transfected with superoxide dismutase in reverse orientation or in control cells not infected with adenovirus containing glutathione reductase cDNA with a mitochondrial leader sequence. Resistant cells had higher mitochondrial GSH levels and maintained mitochondrial GSH/GSSG ratios in hyperoxia, suggesting that maintaining mitochondrial GSH homeostasis determined critical aspects of cell division in these studies.

In A549 cells pre-treated with hyperoxia (95 % O_2) **glutathione** (GSH) and glutathione monoethylester (2 mM) significantly ($P < 0.001$) increased GSH levels from 171 ± 13 nmol/mg protein to 383 ± 26 nmol/mg protein (RAHMAN et al. 2001). H_2O_2 (0.01 mM) induced NF-ϰB activation, whereas hyperoxia exposure did not affect NF-ϰB activation. Pre-treatment with DL-buthionine (*S*,*R*)-sulfoximine, which decreased intracellular glutathione, increased NF-ϰB binding induced by H_2O_2 and increased lactate dehydrogenase release ($P < 0.001$).

While the number of macrophages recoverable from cultures of *rabbit* alveolar macrophages exposed to hyperoxia for 72 h was decreased compared with that in normal control cultures exposed to normoxia for similar durations, addition of **dimethylthiourea** or **catalase** (EC 1.11.1.6), but not superoxide dismutase (EC 1.15.1.1), increased the number of macrophages recoverable from cultures exposed to hyperoxia (HARADA et al. 1983). Alveolar macrophages exposed to hyperoxia in the presence of dimethylthiourea or catalase, but not superoxide dismutase, did not develop ultrastructural abnormalities as vacuolisation and cytoplasmic and nuclear degeneration.

MUZYKANTOV et al. (1987) postulated that an antioxidant enzyme conjugated with a specific endothelial ligand will bind to endothelial cells and protect them against oxidative injury. SAKHAROV et al. (1987) documented that indirect targeting of catalase mediated by polyclonal anti-endothelial antibody protects cultured endothelial cells against the cytotoxic effects of H_2O_2. ATOCHINA et al. (1998)

utilised two monoclonal antibodies that recognise the constitutive endothelial surface antigens ACE (anti-ACE MAb 9B9) and ICAM-1 (anti-ICAM-1 MAb 1A29). As the lungs contain 20–30 % of the total number of endothelial cells in the body, pulmonary tissue is enriched in these antigens. These antibody-catalase conjugates augmented the antioxidant defence in isolated *rat* lungs perfused for 1 h with catalase conjugated with either MAb 9B9, MAb 1A29, or control *mouse* IgG. Approximately 20 % of the injected dose of Ab-^{125}I-catalase accumulated in the perfused *rat* lungs (vs. 5 % for IgG-^{125}I-catalase). After elimination on nonbound material, the lungs were perfused for 1 h with 5 mM H_2O_2. H_2O_2 induced an elevation in tracheal and pulmonary arterial pressure, lung wet-to dry weight ratio, and ACE release into the perfusate. Both MAb 9B9-catalase and MAb 1A29-catalase significantly attenuated the H_2O_2-induced elevation in *1)* angiotensin-converting enzyme release, *2)* lung wet-to dry weight ratio, *3)* tracheal pressure, and *4)* pulmonary arterial pressure. Nonconjugated catalase, nonconjugated antibodies, non-specific IgG, and IgG-catalase conjugate had no protective effect, thus confirming the specificity of the Mab-catalase.

Blood thiobarbituric acid-reactive substances were significantly decreased in jaundiced neonatal Gunn *rats* compared to nonjaundiced pups on day 3 of hyperoxia (>95 % O_2), and blood thiobarbituric acid-reactive substances were inversely correlated to serum bilirubin on day 3 of hyperoxia (Dennery et al. 1995). Relative lung weight was lower in jaundiced pubs exposed to hyperoxia compared to similarly exposed nonjaundiced pups, suggesting a reduction in hyperoxia-induced lung oedema.

Nuclear factor erythroid 2-related factor (NRF2) has been described in *murine* liver, intestine, lung, and kidney, where detoxification reactions occur routinely (Chan et al. 1996, Itoh et al. 1997). Cho et al. (2002) determined the role of NRF2 in the pathogenesis of hyperoxic lung injury by comparing pulmonary responses to 95–98 % oxygen between mice with site-directed mutation of the gene for NRF2 (*Nrf2*$^{-/-}$) and wild-type *mice* (*Nrf2*$^{+/+}$). Pulmonary hyperpermeability, macrophage inflammation, and epithelial injury in *Nrf2*$^{-/-}$ *mice* were 7.6-fold, 47 %, and 43 % greater, respectively, compared with *Nrf2*$^{+/+}$ *mice* after 72 h hyperoxia exposure. Hyperoxia markedly elevated the expression of NRF2 mRNA and DNA-binding activity of NRF2 in the lungs of *Nrf2*$^{+/+}$ *mice*. MRNA expression for antioxidant response element-responsive lung antioxidant and phase 2 enzymes was evaluated in both genotypes of *mice* to identify potential downstream molecular mechanisms of NRF2 in hyperoxic lung responses. Hyperoxia-induced mRNA levels of NAD(P)H:quinone oxidoreductase 1, glutathione-S-transferase-Ya and -Yc subunits, UDP glycosyl transferase, glutathione peroxidase-2, and heme oxygenase-1 were significantly lower in *Nrf2*$^{-/-}$ *mice* compared with *Nrf2*$^{+/+}$ *mice*. Consistent with differential mRNA expression, NAD(P)H:quinone oxidoreductase 1 and total glutathione-*S*-transferase activities were significantly lower in *Nrf2*$^{-/-}$ *mice* compared with *Nrf2*$^{+/+}$ *mice* after hyperoxia.

14.3 Haemorrhagic Shock

A neutrophil-dependent oxidant injury to the alveolar epithelium prevents the normal upregulation of alveolar fluid clearance by catecholamines after haemorrhagic shock. As haemorrhage increases proinflammatory cytokine expression in the lung partly through the activation of α-adrenergic receptors, Laffon et al. (1999) restored the normal fluid transport of the *rat* alveolar epithelium after haemorrhagic shock with phentolamine, an α-adrenergic receptor antagonist.

There are several mechanisms that could explain the protective effect of α-adrenergic blockade against the shock-mediated cytokine-associated injury to the alveolar epithelium:

- Haemorrhage-induced α-adrenergic stimulation could enhance NF-κB activation and proinflammatory cytokine expression in the lung by direct increasing the production of reactive oxygen species. Catecholamines cause the formation of oxygen radical species through the degradation of quinones (Powis and Appel 1980) or through the release of iron from ferritin under anaerobic conditions (Allen et al. 1994). Activation of NF-κB in the pulmonary mononuclear cell population after the onset of the haemorrhagic shock has been shown to be dependent on the presence of reactive oxygen species generated by the xanthine-xanthine oxidase system (Shenkar and Abraham 1996). Inhibition of the conversion of xanthine dehydrogenase (EC 1.1.1.204) to xanthine oxidase (EC 1.1.3.22) either by a tungsten-enriched diet or by pre-treatment with allopurinol (Modelska et al. 1999) or inhibition of the activation of NF-κB by sulfasalazine (Morris et al. 1999), restored the normal fluid transport capacity of the alveolar epithelium after haemorrhagic shock.
- α-Adrenergic stimulation causes the activation of protein kinase C and thus increases intracellular calcium (Exton 1985). NF-κB can be activated either by protein kinase C-mediated phosphorylation of IκB (Ghos and Baltimore 1990) or by

proteolytic degradation of the carboxyl terminal protein sequence of IκB by calcium-dependent intracellular proteases (FINCO et al. 1994).

- Phentolamine might have affected alveolar liquid clearance in haemorrhaged *rats* by mechanisms that are unrelated to its inhibition of the release of reactive oxygen species. For instance, phentolamine has been shown to block ATP-sensitive potassium channels in cardiac ventricular cells by a mechanism unrelated to its α-adrenergic blockade (WILDE et al. 1994).

14.4 Granulomatous Epithelioid Cell Reactions

Granuloma is a specialised form of inflammatory reaction featuring focal macrophage and T cell accumulation and multinucleated giant cell formation. Most important chronic inflammatory diseases, including rheumatoid arthritis can be regarded as examples of granulomas, i.e., relatively dense collections of macrophages, lymphocytes, plasma cells and fibroblasts with some granulocytes. The predominant cell type is usually the macrophage, and this is seen most obviously in tuberculosis, sarcoidosis or leprosy (SPECTOR 1974). In rheumatoid arthritis, the synovia may be infiltrated with macrophages, but even if it is not, we have to remember that the proliferating synovial cells which are such a feature of the disease, consist in part of cells of the macrophage line. The kinetics of chronic inflammation are therefore essentially the kinetics of the inflammatory macrophage (SPECTOR 1977). LAN's and his coworkers' (1995) study of macrophage proliferation within immunologically induced granulomas in *rat* experimental Goodpasture's syndrome, however, challenges the conventional view. In this disease, granulomatous lesions in the kidney and lung contained 60 to 70 % macrophages of an $ED1^+ED2^-ED3^-$ blood monocyte phenotype. However, double immunohistochemistry showed that up to 75 % of $ED1^+$ macrophages within granulomatous lesions were proliferating on the basis of proliferating cell nuclear antigen expression or bromodeoxyuridine incorporation was detected in blood monocytes, indicating that proliferation of $ED1^+ED2^-ED3^-$ cells was a localised event within granulomatous lesions. >95 % of nuclei within multinucleated giant cells were positive for proliferating cell nuclear antigen, but these nuclei lacked bromodeoxyuridine incorporation. This suggested a novel mechanism of multinucleated giant cell formation involving fusion of macrophages in the G1 phase, which then halted progression into S phase of the cell cycle.

Considerable experimental evidence has been accumulated to support the concept that **polymorphonuclear leucocytes** and their contents are critical to the production *in vivo* of inflammation and tissue injury. Soon after injury, leucocytes begin to leave the axial column of blood cells, to appear in the marginal plasma stream and to impinge from time to time on the venular wall. Leucocyte sticking in inflamed vessels depends upon some alteration in the vascular wall and not in the adhering leukocytes. Diapedesis of neutrophils and monocytes through gaps between the endothelial cells of small veins 6 h, venules and capillaries 12–24 h after an injection of old tuberculin was demonstrated in *guinea-pigs* sensitised 4–5 weeks before by heat-killed tubercle bacilli in mineral oil (WIENER et al. 1967).

The enhanced production of superoxide ion ($O_2^{\bullet-}$) and peroxynitrite ($ONOO^-$) by bloodstream neutrophils and superoxide ion by monocytes from rheumatoid arthritis patents was registered by luminol- or lucigenin-enhanced chemiluminescent measurement (OSTRAKHOVITCH and AFANAS'EV 2001). Superoxide dismutase (EC 1.15.1.1) and rutin were the most efficient suppressors of oxygen radical overproduction by rheumatoid arthritis neutrophils, while mannitol and desferrioxamine were ineffective.

The removal of neutrophils from the inflammatory sites is essential for the resolution of inflammation. Surface changes, including phosphatidylserine exposure, label neutrophils for phagocytosis by macrophages. HAMPTON et al. (2002) demonstrated that externalisation of phosphatidylserine and uptake by monocyte-derived macrophages occurred in *human* neutrophils ingesting *Staphylococcus aureus*. Both processes were dependent on oxidant production from the neutrophil NADPH oxidase. There was no requirement for myeloperoxidase (EC 1.11.1.7), and H_2O_2 was identified as the most likely trigger for phosphatidylserine exposure.

Myeloperoxidase (EC 1.11.1.7) reaction

$$Cl^- + H_2O_2 \rightarrow ClO^- + H_2O \qquad [27]$$

$$ClO^- + H_2O_2 \rightarrow {}^1O_2 + Cl^- + H_2O \qquad [28]$$

The various types of **mononuclear cells** formed large interstitial aggregates (WIENER 1970, pp. 153–154). Frequently, the surface membranes of lymphocytes, monocytes and macrophages were in close proximity, being separated by distances measuring less than 500 nm. In occasional areas the surface membranes between lymphocytes and monocytes of macrophages were either in contact with one another or fused, thereby providing cytoplasmic continuity between lymphocytes and monocytes of macrophages. Cytoplasmic bridges between macrophages and lymphoid and plasma cells have been described in the lymph nodes of animals with immediate hypersensitivity (SCHOENBERG et al. 1964).

The **epithelioid cells** in tuberculosis, sarcoidosis, Kveim tests, Crohn's disease, beryllium disease and swimming pool granuloma, irrespective of their cause, showed identical enzyme patterns, strong pentose cycle activity - biosynthesis, and strong hydrolase activity - phagocytosis (WILLIAMS et al. 1969), which was in agreement with those of electron microscope studies of epithelioid cells in sarcoidosis (HIRSCH et al. 1967, WANSTRUP 1967) or developed *in vitro* by culture of *cockerels*' leucocytes (SUTTON 1967). CAMARERO et al. (1990) described that epithelioid macrophages do not release $O_2^{\bullet-}$ even after stimulation with phorbol myristate acetate. Instead, they release into the medium of short-term cultures an activity that deactivates activated macrophages. This protein of 11 kDa is thermostable, trypsin and pronase sensitive and inhibited $O_2^{\bullet-}$ release from activated *mouse* peritoneal macrophages (CAMARERO et al. 1993). This activity was termed macrophage deactivating factor (MDF). Polyclonal antibodies raised by intrasplenic injection of MDF-positive culture supernatants of epithelioid cells inhibited MDF activity and immunostain cells of 14-day implanted coverslips as well as epithelioid-like cells in BCG-induced granulomas in the *mouse* (MARIANO et al. 1993). Epithelioid cells forming *human* tuberculous granuloma selectively express the calcium-binding protein MRP-14 but not MRP-8 (ODINK et al. 1987, HESSIAN et al. 1993), and so did epithelioid cells from foreign-body granuloma induced by the implantation of coverslips into the subcutaneous tissue of Swiss *mice* (AGUIAR-PASSETI et al. 1997).

In the *human* spleen, germinal centres with an epithelioid appearance may represent the early part of a normal primary immune reaction (MILLIKIN 1970).

Epithelioid cells showed intense immunostaining for CD68, and cathepsin B, cathepsin K, and cathepsin L, irrespective of whether they were associated with non-caseating or caseating granulomas of sarcoid-like lesions, sarcoidosis or tuberculosis (BÜHLING et al. 2001). In granulomas related to sarcoidosis or tuberculosis, some cathepsin K was detected in nearly all epithelioid cells. Cathepsin K expression appeared to differentiate between non-stimulated tissue macrophages and epithelioid cells within sarcoid-like lesions or granulomas.

Multinucleated giant cells are a hallmark of granulomatous reactions. LEMAIRE et al. (1996) cultured resident alveolar macrophages gained from *rat* lung by bronchoalveolar lavage and examined the effects of defined cytokines on alveolar macrophage differentiation and multinucleated giant cell formation. The presence of multinucleated giant cells was found after 3 d in culture with maximal numbers obtained at 7 d and thereafter (up to 21 d). Macrophage colony-stimulating factor and granulocyte-macrophage colony-stimulating factor (25–75 U/ml) stimulated the formation of multinucleated giant cells up to 4-fold, whereas IL-3, IL-10, and IFN-γ had no stimulatory effect. Multinucleated giant cells with distinct phenotypes were: (1) spherical ones with 3–16 nuclei, dense cytoplasm, and lower expression of β_3 integrin (Type 1) and (2) irregular multinucleated giant cells with 3–30 nuclei, thin and vacuolated cytoplasm, and higher expression of β_3 integrin (Type 2). Granulocyte-macrophage colony-stimulating factor promoted, in alveolar macrophage cultures, the appearance of an elongated fibroblastoid phenotype and stimulated mostly the formation of Type 2 multinucleated giant cells. In contrast, macrophage colony-stimulating factor did not cause significant change in the general morphology of regular alveolar macrophages but stimulated the appearance of both Type 1 and Type 2 multinucleated giant cells. Reverse transciptase-polymerase chain reaction analysis demonstrated that, under these conditions, macrophage colony-stimulating factor induced granulocyte-macrophage colony-stimulating factor gene expression in alveolar macrophages. In addition, neutralising antibodies against macrophage colony-stimulating factor selectively decreased the formation of Type 1 multinucleated giant cells, whereas neutralising anti-granulocyte-macrophage colony-stimulating factor inhibited Type 2 formation.

$CD68^+$ multinucleated giant cells of the Langhans type, with nuclei on the cell periphery, and of the Touton type, with nuclei in the centre of the cell, expressed cathepsin K and the immunostaining was even more intense than that of the surrounding epithelioid cells (BÜHLING et al. 2001).

In rheumatoid arthritis, histological analysis demonstrated that the scores for infiltration by **lymphocytes** and **plasma cells**, and the scores for inflammation, were higher than in reactive arthritis (SMEETS et al. 1998). Immunolabelling studies showed that in particular, the scores for infiltration by $CD38^+$ plasma cells, granzyme B^+ cells, and interferon-γ^+ cells were significantly higher in rheumatoid arthritis than in reactive arthritis. Collagen-immunised *rats* treated with different doses of apocynin (4-hydroxy-3-methoxyacetophenone), a strong inhibitor of neutrophil superoxide anion ($O_2^{\bullet-}$) release *in vitro*, in the drinking water starting at the onset of joint-swelling and terminating 14 days later, at the time when joint swelling in the control group was maximal had a normal plasma level of collagen-specific antibodies, but showed a significant reduction of the joint swelling ('T HART et al. 1990).

NK-cells (DUNN and NORTH 1991) and T cells (MORRIS et al. 1992) produce interferon-γ. ASANO et al. (1995) demonstrated that biphasic IFN-γ production, primarily by NK cells and secondarily by T cells, might contribute to granuloma formation in *Rhodococcus aurantiacus*-infected *mice*. Granulomas were not induced in IFN-$\gamma^{-/-}$ *mice* infected with *Rhodococcus aurantiacus* because of the lack of non-T cells producing IFN-γ (YIMIN et al. 2001).

The capacity of the oxidase of Epstein-Barr virus immortalised B lymphocytes was decreased or abolishes in chronic granulomatous disease (MOREL et al. 2000). Cytochrome b_{-245}, the major membrane redox component of the $O_2^{\bullet-}$ generating oxidase, was only slightly expressed in the membrane of Epstein-Barr virus immortalised B lymphocytes.

The role of **interferon-γ** in specific granuloma formation was shown in IFN-γ gene-deficient *mice* (BALB/c and C57BL/6) inoculated with 10^3–10^7 bacilli of various strains of *Mycobacterium tuberculosis* (Kurono, H37Rv, H37Ra and BCG Pasteur) through their tail veins and examined 7 weeks later for granuloma formation (SUGAWARA et al. 1998). The avirulent BCG Pasteur and H37Ra strains (10^3–10^4 bacilli/ml) induced granulomas in the spleen, liver and lungs of IFN-γ-deficient *mice*. The granulomas consisted of epithelioid macrophages and Langhans multinucleate giant cells, but lacked caseous necrosis. The virulent Kurono and H37Rv strains induced disseminated abscesses but not granulomas in various organs of INN-γ-deficient *mice* and Mac-3-positive macrophages were not dectected in the abscess lesions. These results suggest that IFN-γ may be primarily responsible for macrophage activation and that other factor(s) may be involved in the granuloma formation mechanism.

A number of granulomatous diseases are associated with an increase in **angiotensin-converting enzyme** activity (Table 42). It has been proposed that the amount of angiotensin-converting enzyme activity reflects the magnitude of the total body burden of granulomas (SILVERSTEIN et al. 1976, ROHATAGI et al. 1981, REA et al. 1983). The experiments of GILBERT et al. (1993) injecting killed *Mycobacterium butyricum* intravenously into *mice* after a primary intracutaneous injection with Freund's complete adjuvant provided direct evidence that the amount of angiotensin-converting enzyme activity in granulomatous disease reflects the total burden of granulomas. Angiotensin-converting enzyme may not only be a marker of granuloma formation, but it may play also a role in modulating the granulomatous process. In this respect, angiotensin-converting enzyme is a nonspecific protease with the ability to degrade bradykinin, substance P, neurokinin A, and angiotensin I. WEINSTOCK et al. (1981) in granulomatous response to *Schistosoma mansoni* eggs in *mice* and SCHRIER et al. (1982) in *BCG*-induced granulomatous inflammation have shown that there is suppression of granuloma formation after the administration of an angiotensin-converting enzyme inhibitor.

Table 42. Angiotensin-I-converting enzyme in granulomatous disease

BCG	SCHRIER et al. (1982), REA et al. (1983)
M. leprae	REA et al. (1983)
M. butyricum	GILBERT et al. (1993)
Miliary tuberculosis	BRICE et al. (1995)
Silicosis	D'ANDREA et al. (1979), BRICE et al. (1995)
Coalworkers' pneumocosiosis	THOMPSON et al. (1984), WALLAERT et al. (1984)
Sarcoidosis	LIEBERMAN (1975), SILVERSTEIN et al. (1976, 1977), STUDDY et al. (1978), ROHRBACH and DEREMEE (1979, 1982), BAUR et al. (1980), DEREMEE et al. (1980), UEDA et al. (1980), ROHATAGI et al. (1981), STUDDY and JAMES (1983), BAUGHMAN et al. (1984), RØMER et al. (1984), TAKADA et al. (1984), DIETEMANN-MOLARD et al. (1989), SPECKS et al. (1989), STUDDY and BIRD (1989), BRICE et al. (1995), ARBUSTINI et al. (1996), PÉREZ-ARELLANO et al. (1996), SAINANI et al. (1996), TOMITA et al. (1996, 1997), BONE et al. (1997)
Histoplasmosis	LIEBERMAN and REA (1977)
Schistosomiasis	WEINSTOCK et al. (1981), WEINSTOCK and BOROS (1982)

Three types of agents are now known to cause epithelioid cell granulomas: infectious organisms (bacteria, fungi, and parasites), products of plants and animals (pollen, sporangia, proteins), and metallic compounds (Be, Zr). In addition, there is still a group of epithelioid cell granulomatoses with unknown aetiology. Sarcoidosis, one of these granulomatoses, has recently elicited a controversy: by molecular techniques *Mycobacteria* (POPPER et al. 1997) and *Corynebacterium acnes* (NAKATA et al. 1994) have been identifies in sarcoid granulomas and a link to the aetiology of sarcoidosis has been proposed. If these bacteria induce some cases of sarcoidosis by an allergic mechanism, has still to be proven (POPPER 2000).

14.4.1 The Tuberculous Granuloma

Emile et al. (1997) found two types of granuloma in idiopathic disseminated BCG infection in French *children*. The first type (type I, tuberculoid) consisted of well-circumscribed and well-differentiated granulomas, with epithelioid and multinucleated giant cells containing very few acid-fast rods, surrounded by lymphocytes and fibrosis and occasionally with central caseous necrosis. The second type (type II, lepromatous-like) consisted of ill-defined and poorly differentiated granulomas, with few if any giant cells and lymphocytes but widespread macrophages loaded with acid-fast bacilli. Most *children* displayed a single type of granuloma. One half displayed type I lesions and the other half displayed type II lesions. There was a strong correlation between the type of granuloma and the clinical outcome. Tuberculoid lesions were associated with survival, whilst lepromatous-like lesions correlated with death.

When heat-treated BCG Pasteur bacilli were introduced into the lungs of *guinea pigs* by an inhalation exposure apparatus, pulmonary granulomas without necrosis developed (Sugawara et al. 2002). When four kinds of mycolates from *M. tuberculosis* Aoyama B strain were introduced into the lungs by the same method, only trehalose 6,6'-dimycolate and methyl ketomycolate induced pulmonary granulomas without central necrosis. The pulmonary granulomas consisted of epithelioid macrophages and lymphocytes. When a mixture of trehalose 6,6'-dimycolateand anti-trehalose 6,6'-dimycolate antibody was introduced into the lungs, development of granulomatous lesions was reduced.

14.4.1.1 Polymorphonuclear Initial Stage

During the earliest stages, the first cells to react against the tubercle bacilli are the polymorphonuclear leucocytes, which encompass and phagocytose the organisms (Vorwald 1931). After intravenous injection into *rabbits*, this reaction was confined at first to small capillaries, and then extended to adjoining alveolar walls and spaces. Within 14 h, the polymorphonuclear leucocytes with and without tubercle bacilli were phagocytosed by the mononuclear exudate cells, which increased in number to the exclusion of the leucocytes. The tubercle bacilli thus were liberated in the cytoplasm of the exudate cell.

In the *rabbit's* ear transparent chamber Ebert et al. (1948), Ebert and Barclay (1950), Sanders et al. (1951) and Dodson et al. (1954) have described some of the reactions observable microscopically that follow the introduction of tubercle bacilli into the tissue. The earliest stage of tubercle formation was seen as a collection of about 15 to 20 polymorphonuclear leucocytes in the interstitial tissue between capillaries. About 12 h later the number of polymorphs had doubles or trebled, and some leucocytes were sticking to the walls of the adjacent capillaries. The lesion continued to increase in size and cellularity and by 48 h a few macrophages were present.

14.4.1.2 The Monocyte in Tuberculosis

The interaction between *Mycobacterium tuberculosis* and monocytes/macrophages in the early phase of infection is crucial for the pathogenesis of the disease. Pathogens have developed specific mechanisms that allow them to establish a close interaction with host cells and to escape immune response (Ameisen et al. 1994).

Employing immunocytochemical methods, Buchwalow et al. (1997) provided evidence that α-subunits of stimulatory and inhibitory G-proteins and protein kinase C-β as well as two major cytoskeletal components, microfilaments and microtubules, participated in uptake of *Mycobacterium bovis* BCG by *human* macrophages and co-localized in phagosomes.

Macrophages that are unable to kill intracellularly replicating mycobacteria allow their multiplication until the cell bursts. In the first stage of tuberculosis necrosis is the cellular event responsible for macrophage death (Dannenberg jr. and Tomashefski 1988).

Apoptosis was increased three-fold in bronchoalveolar lavage-cells obtained from 7 male patients (mean age 40 years) with pulmonary tuberculosis and even more markedly in alveolar macrophages of 7 male *Mycobacterium tuberculosis*-infected AIDS patients (mean age of 35 years), compared with 7 controls with non-tuberculous pulmonary diseases (Placido et al. 1997). Apoptosis was analysed and characterised by propidium iodide incorporation, terminal deoxytansferase-mediated dUTP-biotin nick end labelling, and tissue transglutaminase expression. The *Mycobacterium tuberculosis*-macrophage interaction was also investigated *in vitro* by infecting monocyte-derived macrophages with the virulent H37Rv strain. The induction of apoptosis by *Mycobacterium tuberculosis* required viable bacteria, was dose-dependent, and was restricted to H37Rv. Infection with either *Mycobacterium avium* complex or HIV-1 and treatment with heat-killed *Mycobacterium tuberculosis* failed to induce apoptosis.

In tissue samples apoptotic cells were not detected in regressive granulomas (FAYYAZI et al. 2000). Whereas productive granulomas without histologically recognisable caseous necrosis revealed only single apoptotic cells, large numbers of apoptotic CD68+ macrophages and apoptotic CD3+, CD45RO+ T cells were observed within caseous foci.

The a-subunit of the **clotting factor XIII** (FXIIIa) has been shown in tuberculous infections as well as in sarcoidosis to be synthesised by macrophages identified by immunohistochemistry predominantly in the periphery of granulomas, whereas the centres were generally devoid of these cells (PROBST-COUSIN et al. 1997).

14.4.1.3 The Epithelioid Cell in Tuberculosis

In fixed and stained tuberculous tissue the term "epithelioid" is applied to the cells constituting roughly circular areas slightly more opaque than the surrounding tissue. In fixed tissues these cells can be distinguished from polymorphs, lymphocytes and fibroblasts. In the infected ear chamber the cells in position corresponding to those seen in fixed tissue have been called epithelioid, and their identity with them was presumed (DODSON et al. 1954). Individual epithelioid cells varied from about 25 to 50 μm in diameter; their most characteristic feature was the presence of refractile granules a little larger than the granules of eosinophil leucocytes. The nucleus could not be seen but usually a clear central area, outlined by granules, was present. With the oil immersion objective DODSON et al. (1954) occasionally saw a waving surface membrane.

14.4.1.4 The Giant Cell in Tuberculosis

In *human* lingual tuberculosis, MILLER et al. (1978) found giant cells containing up to as many as 50 irregular nuclei when observed within a single thin section. Nuclear membranes were usually highly indented, and the chromatin was concentrated peripherally. The nuclei were arranged toward the cell peripherally, and the central cytoplasm appeared quite electron dense. The cytoplasm was usually composed of concentrated cytofilaments, many free ribosomes, peripheral concentration of mitochondria, multiple well-formed Golgi complexes, and an abundance of dense, dumbbell-shaped, lysosomal-like structures. The rough endoplasmic reticulum was well formed and scattered throughout the cell. Many lateral projections or microvilli appeared to emanate from the cytoplasmic membrane into the adjacent intercellular spaces and interdigitate with the cell processes of the adjacent cells. Infrequent, dense, phagocytic bodies, empty membrane-bound vacuoles, and lipid like inclusions were noted within the giant cell cytoplasm.

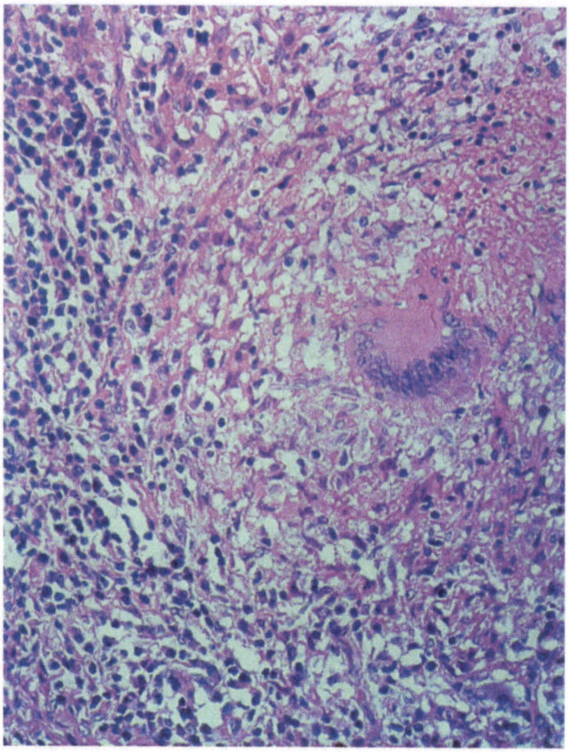

Fig. 198. Langhans cell from a tuberculous submandibular lymph node excised from a *girl* aged 16. Department of Surgery, University of Leipzig (868/43). Formalin, paraffin, hæmatoxylin and eosin. Objective Leitz 40/0.65. Orthomat® (×2)

14.4.2 Morphological Manifestations of Leprosy

Leprosy is probably the best example of a disease that has a spectrum ranging from the anergic to the hypersensitive forms. In tuberculoid form the organised epithelioid cell granulomas are due mainly to the specific cell-mediated immunity.

Patients with **tuberculoid leprosy** have localised lesions with scanty organisms and a strong delayed type hypersensitivity against *M. leprae* which leads to granuloma formation.

In **lepromatous leprosy** disseminated skin lesions contain profusely profusely distributed bacilli and show unresponsiveness to *M. leprae*. This is the field where MODLIN et al. (1988) have demonstrated the important concept of investigating the relevant T-lymphocyte subsets of $CD45R^+$ $CD45R^-$,$4B4^+$

(UCHL1$^+$) phenotype in combination with the isolation of cells from tissue lesions of different types in order to document specific reactivity to *M. leprae* antigen. in lepromatous lesions both CD4$^+$ and CD8$^+$ cells are distributed throughout the lesion without a clear organisation, with an overall dominance of CD8$^+$ cells (T4/T8: 0.6; MODLIN et al. 1983).

Large numbers of bacilli may invade cutaneous and subcutaneous vessel walls (CORUH and MCDOUGALL 1979). If large veins are involved there may be a granulomatous phlebitis with eventual occlusion of the lumen (MUKHERJEE et al. 1983).

RIDLEY (1988) suggested that it is possible to distinguish newly arrived monocyte-macrophages from epithelioid cells or activated macrophages, but not from lymphocytes, by the size and staining characteristics of their nucleus. CAMPBELL and CREE (1992) showed that epithelioid cells and lepromatous macrophages share similar nuclear characteristics. Local pressure effects may influence epithelioid cell and macrophage nuclear morphology in leprosy lesions.

Neutrophils isolated from leprosy patients release TNF-α and exhibit accelerated apoptosis *in vitro* (OLIVEIRA et al. 1999). Thalidomide, a drug known to inhibit TNF-α synthesis on monocytes, exerted an inhibitory effect on TNF-α secretion.

The bone marrow may act as a reservoir for viable lepra bacilli in the absence of a host response in treated and untreated patients with lepromatous leprosy (SUSTER et al. 1989).

STORRS et al. (1978) recommended the nine-banded armadillo (*Daspys novemcinctus* L) as an animal model for lepromatous leprosy. In both *humans* and *armadillos*, clusters or masses of the infected macrophages cause lesions in the skin, dermal nerves, large nerves of the extremities, nose and eyes. Probably due to the animal's low body temperature (32 to 35 °C) lepromatous disease disseminates to tissues as lung, heart, stomach, urinary bladder, spinal cord, and occasionally, brain. The involvement of liver, spleen, and bone marrow is much greater in *armadillos* than in *humans*.

Phagocytosis of M.leprae by BALB7/c mural peritoneal macrophages was inhibited by erbstatin (516 μM) and staurosporine (25 μM), but not by genistein (306 μM); all the protein kinase inhibitors prevented uptake of M.leprae by C57 beige bg/bg mouse macrophages (PRABHAKARAN et al. 2000).

As *M. leprae* proliferate inside macrophages, it has been speculated that catalase (EC 1.11.1.6) encoded by *katG* may protect the bacilli from deleterious effects of peroxide generated from the macrophage and may also play a crucial role in the survival of *M. leprae in vivo*. However, unlike that of M. tuberculosis, the *katG* of *M. leprae* has been reported to be a pseudogene, implication that isoniazid, which is activated to a potent tuberculocide agent by catalase, is unlikely to be of therapeutic benefit to leprosy patients. KANG et al. (2001) found catalase-like activity in *M. leprae* cell lysates by the diaminobenzidine staining method with non-denaturing polyacrylamide gel electrophoresis.

14.4.3 Sarcoidosis

Sarcoidosis is a multi-organ system disease with a marked propensity for involvement of the pulmonary parenchyma and thoracic lymphatic system. It is characterised by non-caseating granulomas, consisting of mononuclear epithelioid cells surrounded by and interspersed with lymphocytes and a lesser number of plasma cells. Multinucleate giant cells, occasionally with asteroid bodies or calcific concretions, are often present. Special stains fail to demonstrate organisms and characteristically little or no necrosis is associated with the granulomas (MORALES et al. 1974, MITCHELL et al. 1977, HAFERKAMP 1980, VAN MAARSSEVEEN et al. 1983). Regressive or older lesions are accompanied by a variable number of fibroblasts, inflammatory cell infiltrates, and fibrosis, which may be dense and hyalinized. GUSEK (1968) found no periodicity of 64 nm characteristic for collagen fibrils and, therefor, used the term "paramyloid" which AZAR et al. (1973, p. 63) think misleading since the various amyloid deposits were all shown to have the same histochemical and ultrastructural characteristics (AZAR 1968).

In the lung, the granulomas are located in the interstitium of the alveolar structures (MITCHELL et al. 1974, 1977, JAMES et al. 1975).

14.4.3.1 The Monocyte in Sarcoidosis

Lymphocytes, monocytes, and platelets were frequently seen in the lumina of alveolar capillaries in specimens with alveolitis compared with nonalveolitis specimens ($P < 0.005$). Adhesion between lymphocytes and monocytes and endothelium was frequently observed. Adhesion between infiltrating cells and endothelium was seen in 80 % of specimens with alveolitis and in 43 % of specimens without alveolitis (TAKEMURA et al. 1995).

In vitro, blood monocytes of patients suffering from sarcoidosis incorporate more 3H-cytidine into their nuclei than do monocytes from healthy persons with a positive tuberculin test (M. SCHMIDT et al. 1970). In pulmonary sarcoidosis, the patent capillaries contain monocytes in increased numbers (Judd et al. 1975). These monocytes are presumably the source of the macrophages in the interstitial tissue and in the less mature granulomas.

14.4.3.2 The Macrophage in Sarcoidosis

Lavaged alveolar macrophages in sarcoidosis were larger than those of control subjects (Iijima et al. 1989). They showed well developed pseudopodia, marked polarity, less nuclear heterochromatin and cytoplasmic granules which were larger and more numerous but less electron dense than normal (Danel et al. 1983). While both lysosomes and phagolysosomes were prominent in macrophages gained by bronchial lavage from normal non-smokers, they were comparatively rare in non-smokers with sarcoidosis (Hawley et al. 1979). Mature 25F9+ macrophages are characteristic for active sarcoidosis (Neuchrist et al. 1988). Strong expression of the 25F9 antigen on the bronchoalveolar lavage cells may be a consequence of prolonged cell life and proliferation in response to enhanced T cell activities on the site of inflammation. 25F9+ macrophages could be responsible for augmented proliferation of fibroblasts and production of collagen.

In biopsies from skin and lymph node lesions, Carr and Norris (1977) described numerous electron-dense inclusions ranging from 50 nm to 250 nm in minimum diameter and elongated up to 2 µm in length. The core of these inclusions is composed of granules about 4 nm in diameter arranged to form lines with an approximate periodicity of 8 nm.

Lucigenin-enhanced **chemiluminescence** of yeast cell wall-stimulated alveolar macrophages gained by brochoalveolar lavage from patients suffering from sarcoidosis was significantly increased as compared with patients with other pulmonary disease or healthy volunteers (Winsel et al. 1987).

Spontaneous release of IL-10 from alveolar macrophages gained by bronchoalveolar lavage and cultured for 24 h increased from 10±4 pg/ml/106 cells in normal subjects to 101±37 pg/ml/106 cells (P <0.0001) in patients with active sarcoidosis (Marques et al. 1997). After stimulation with 1.0 µg lipopolysaccharide/ml the figures were 200±42 pg/ml/106 cells and 643±300 pg/ml/106 cells, respectively.

As the clinical course of pulmonary sarcoidosis has been correlated with patients' individual capacities for spontaneous tumour necrosis factor production by alveolar macrophages (Müller-Quernheim et al. 1992, Zheng et al. 1995), modulation of TNF in a more specific way than corticoids seems to be the most promising therapeutic alternative. Pentoxifylline, which inhibits TNF expression at the TNF mRNA level (Strieter et al. 1988) may improve therapeutic regimens either by sparing or replacing corticosteroids (Zabel et al. 1997).

Increased angiotensin-converting enzyme activity was localised on the surface of plasma membrane and cytoplasmic processes of alveolar macrophages, as demonstrated by immunoelectron microscopy (Takada et al. 1984). An insertion/deletion polymorphism in angiotensin-converting enzyme gene in DNA extracted from mononuclear peripheral blood cells of 156 (47 male and 109 female) sarcoidosis patients showed insertion homozygote type in 62 patients (= 39.7 %), heterozygote type in 69 patients (= 44.2 %), and deletion homozygote type in 25 patients (= 16.0 %) (Tomita et al. 1996).

Osteopontin, a negatively charged, glycosylated sialoprotein, absent in normal lung or granulation tissue and all isotype controls, was extensively expressed in the macrophages in biopsies from patients with pulmonary sarcoidosis (Chupp et al. 1996).

The a-subunit of the clotting factor XIII (FXIIIa) has been shown in sarcoidosis or mycobacterial infections to be synthesised by macrophages identified by immunohistochemistry predominantly in the periphery of granulomas, whereas the centres were generally devoid of these cells (Probst-Cousin et al. 1997).

Allele frequencies in natural resistance associated macrophage protein (NRAMP) gene may play a role in susceptibility of African American versus Caucasian populations to sarcoidosis (Maliarik et al. 1996). In a case control study of 157 African American patients with sarcoidosis and 111 African American control subjects, Maliarik et al. (2000) found – in contrast to those in tuberculosis patients – the less common genotypes of NRAMP gene more often in control subjects than in case patients. In particular, one polymorphism, a $(CA)_n$ repeat in the intermediate 5' region of the gene, was found to have a protective effect ($P = 0.014$).

Alveolar macrophages and T cells from sarcoid, but not normal lung, were permissive to adenovirus infection (Conron et al. 2001).

Sarcoid alveolar macrophages produced less $\mathbf{C_3b}$ and phorbol myristate acetate stimulated chemiluminescence than control alveolar macrophages (Williams et al. 1981).

Oxygen radical formation of alveolar macrophages from the bronchoalveolar lavage fluid is a sign of early macrophage activation (Dalhoff et al. 1992). The samples of 10 patients with sarcoidosis stage I (aged 35±7 years) showed significantly higher photon emission than those of 6 healthy male control persons (non-smokers, aged 26±3 years).

Nitric oxide in exhaled gases and nitrite/nitrate, the stable end products of •NO metabolism, in bronchoalveolar lavage fluid were not significantly changed in sarcoidosis (O'DONNELL et al. 1997).

Melatonin, a pineal indole hormone in two cases of chronic sarcoidosis unresponsive to long-term steroid therapy, resolved symptoms and radiological findings (CAGNONI et al. 1995).

14.4.3.3 Epithelioid and Giant Cells in Sarcoidosis

Fully developed epithelioid cells were not identified in lavage specimens (DANEL et al. 1983).

Schaumann bodies are residual bodies of heterophagic mycobacterial derivation (ANG and MOSCOVIC 1996). They immunostained intensely for the lysosomal proteins muramidase and CD68, variably for some cytoskeletal proteins (tubulin, desmin, vimentin) and not at all for cytokeratin, muscle actin, α1-antichymotrypsin and ferritin. Both cross-reactive and species specific antigenic determinants of Mycobacterium tuberculosis complex were shown to be present. The bodies were clearly labelled with the monoclonal antibodies TB68 and TB71, known to recognise species specific epitopes of Mycobacterium tuberculosis complex.

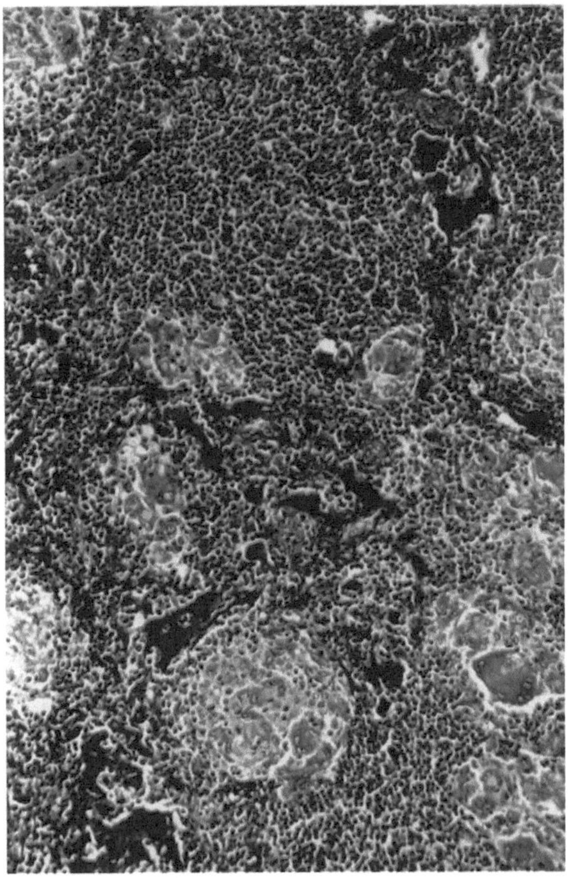

Fig. 199. Epithelioid-cell granulomatosis interpreted as Morbus Boeck. Mediastinal lymph node of a male patient born in 1948 (Nordwestkrankenhaus Frankfort on the Main, autopsy 2109/79). – The slide shows Langhans giant cells but also some foreign body type giant cells. Fixed by immersion into formalin solution. Paraffin embedding. Haematoxylin and eosin. Objective Reichert Pl 25×/0.45/∞/0.17. Green filter. (Material provided by Prof. Höer†)

Tenascin expression in sarcoid appeared much less prominent than in cryptogenic fibrosing alveolitis (WALLACE et al. 1995).

14.4.4 Sarcoidosis-like Granulomas

14.4.4.1 Granulomatous Dermal Reactions

14.4.4.1.1 Cheilitis Granulomatosa

Cheilitis granulomatosa shows small non-caseating and non confluent epithelioid cell granulomas with lymphocytes, plasma cells, and histiocytes mainly located peri- and paravascularly in the oedematous connective tissue of the lip (HORNSTEIN 1961, BUTENSCHÖN 1976, HORSTEIN and HANEKE 1980, WORSAAE et al. 1982). NOZICKA (1985) argues endolymphatic epithelioid granulomas obstructing lymphatic lumina and thus causing lymphostasis, lymphatectasia, and lymphatic oedema.

14.4.4.1.2 Necrobiosis Lipoidica and Granulomatosis Disciformis Chronica et Progressiva

Necrobiotic xanthogranuloma is a condition characterised by necrobiosis with xanthomatous granulomas occurring in association with paraproteinaemia (ANSTEY et al. 1993).

Immunohistochemistry demonstrated strong labelling of epithelioid cells and giant cells with vimentin, KP1 and HAM 56 (ORCHARD et al. 1995).

14.4.4.2 Selenium Granulomas

A 71-year old man employed in selenium refining for 50 years, working until the day of admission, showed numerous perivascular noncaseating pulmonary granulomas (DISKIN et al. 1979). Some were completely cellular with histiocytes, multinucleated giant cells, plasma cells, and lymphocytes, but others were completely replaced by fibrosis.

The selenium content of the lung was 109 ppm (normal 0.15 to 0.21 ppm), that of the peribronchial nodes 26.0 ppm (normal 0.10 ppm).

14.4.4.3 Zirconium Granulomas

In a female aged 50 years working in a plant manufacturing nuclear fuel rods for 17 years, sarcoid-like pulmonary granulomas showed epithelioid cells and multinucleated foreign body giant cells interpreted as a zirconium pneumoconiosis (KOTTER and ZIEGER 1992).

In nonsensitive subjects, mononuclear phagocytes deposed zirconium lactate injected intradermally in membrane-bound phagosomes of strikingly variable size and shape (ELIAS and EPSTEIN 1968). An inner, dense core was surrounded by a less dense, granular matrix. In cells with the largest amount of zirconium, inclusions tended to fuse, producing giant phagosomes consisting of numerous dense, amorphous deposits within the granular matrix.

In contrast to the foreign body response, the hypersensitivity reaction consisted of the organisation of epithelioid cells into tubercles. Multinucleate giant cells occurred both in developing granulomas (4 weeks) and adjacent to necrotic areas (BLACK and EPSTEIN 1974).

14.4.4.4 Hodgkin's Disease Type VI

In Hodgkin's disease, JACKSON jr. and PARKER jr. (1947), LENNERT and MESTDAGH (1968), KADIN et al. (1970, 1971), SACKS et al. (1978) and NOEL et al. (1979) described non-caseating epithelioid granulomas. In 55 records of 608 consecutive patients with biopsy-proven Hodgkin's disease SACKS et al. found groups of epithelioid histiocytes, containing few if any lipid droplets, occasionally associated with giant cells of the Langhans or foreign body types and sometimes accompanied by lymphocytes and eosinophils, without atypical mononuclear cells or Sternberg-Reed cells. Central fibrinoid necrosis was observed occasionally within these granulomas. Schaumann's and asteroid bodies were not seen. POPPENA and LENNERT (1980) presenting biopsies from 278 children <15 years of age found a high epithelioid cell content in 9 boys and 1 girl. There were several or numerous typical Sternberg-Reed cells.

14.4.4.5 Granulomas in Brucellosis

Rather uniform manifestations of brucellosis prevent the histopathologic differential diagnosis of granulomas induced by Brucella abortus, Brucella melitensis, and Brucella suis. Characteristic "brucellomas" (LÖFFLER and MORONI 1952), magnocellular nodules resembling miliary tubercles, were found above all in the spleen, liver and bone marrow. In the spleen NICOD (1935) observed minute, blurred necroses surrounded by few neutrophil granulocytes. Other focal necroses presented lymphocytes, plasma cells and epithelioid cells. The nodular endophlebitis resembles the subintimal typhoid nodules preventing differential diagnosis. AJELLO (1950, 1951) considered the polymorphy of the giant cells to be pathognomonic. Diffuse proliferation of reticulum cells (HASLHOFER 1933, RÖSSLE 1933, VON ALBERTINI and LIEBERHERR 1938) reminds of a reticulosis. NICOD (1935) emphasised a heavy erythrophagocytosis in the spleen. In 5 of 20 patients LÜBCKE (1971) found hepatic granulomas. SHAH and KIST (1983) described non-caseating epithelioid cell granulomas in biopsies of pelvic crista, liver and spleen in a patient aged 50 years, whose diagnosis was made after discharge by serology.

In the field hare, Brucella suis granulomas are characterised by proliferation of epithelioid cells and histiocytes with central

14.4.4.6 Granulomas in Tularaemia

Granulomatous pleuriris caused by *Francisella tularensis* in a 55-yr-old sheep shearer confirmed by elevated serum agglutination titres against *F. tularensis* and detection of the organism in a pleural biopsy showed a moderate to marked infiltrate of mononuclear phagocytic cells and lymphocytes, and contained multiple noncaseating granulomas (SCHMID et al. 1983). Langhans' type giant cells were present in some granulomas.

So-called necrotizing granulomatous lymphadenitis is seen in tularaemia, but also in *yersinial* mesenteric lymphadenitis. The central necrosis mostly contains a scattered number of neutrophils and monocytes and is closely surrounded by macrophages. In contrast to epithelioid cells, these macrophages are not negative but strongly positive to Ki-M8. Like epithelioid cells, they are positive to Mono 1. The ractivity of Ki-M8 is a useful differential diagnostic aid in separating epithelioid cell type from granulomatous necrotizing forms of lymphadenitis (BÖDEWADT-RADZUN et al. 1990).

14.4.4.7 Lymphogranuloma Venereum = Lymphopathia Venerea

A biovar and serovar of Chlamydia trachomatis is often found infecting mononuclear cells, specifically macrophages, with a predilection toward lymph-node involvement (SCHACHTER and OSOBA 1983). Initially microabscesses form in the glands, the central collection of polymorph leucocytes and nuclear debris being surrounded by a wall of histiocytes which are arranged in a pallisaded fashion (SPENCER and HUTT 1973). Outside the histiocytic layer are collections of plasma cells, lymphocytes, eosinophils and occasional giant-cells. The abscesses enlarge and the polymorph leucocytes have a striking stellate shape which is a helpful distinguishing feature from a tuberculous lesion which they very closely resemble. Eventually healing occurs with the formation of much scar and granulation tissue causing destruction of the lymph node and its supplying lymphatic vessels.

Proctitis in lymphopathia venerea shows follicular lymphohistiocytic-plasma cell infiltrates in the submucosa, muscularis propria and serosa (DE LA MONTE and HUTCHINS 1985).

14.4.4.8 Syphilitic Granulomas

14.4.4.8.1 Macrophages in the Initial Sclerosis

AZAR et al. (1970) studied a chancre measuring 3 cm at its base and 2 cm in height. Electron microscopy revealed that Treponema pallidum organisms were principally located in intercellular spaces in the direct vicinity of small blood vessels. These organisms were, furthermore, observed within the cytoplasm of neutrophils, macrophages, endothelial and perivascular connective tissue cells, and within plasma cells.

14.4.4.8.2 Granulomas in Secondary Syphilis

Haematogenous dissemination of Treponema pallidum induces foci of severe angiitis with a mobilisation of cells and an infiltration of round-cells characteristic for secondary stage syphilis. A skin biopsy from the upper back showed well-defined granulomas formed by aggregation of epithelioid cells and a few Langhans giant cells in the upper and midportions of the dermis. Very few lymphocytes and plasma cells were found at the periphery of the granuloma and in the perivascular areas throughout the upper portion of the dermis. A lesion of the face taken one week later revealed a diffuse and dense cellular reaction composed largely of plasma cells including lymphocytes and histiocytes (LANTIS et al. 1969). Five patients with clinically obvious secondary syphilis subjected to biopsy all showed sarcoid-like granulomas in the dermis with epithelioid cells, some with Langhans' giant cells (KAHN and GORDON 1971). In a case reported by KURUMAJI et al. (1987) histological examination showed epithelioid cell granulomas in both a penile ulcer and a solitary lung lesion. Kupffer cell hyperplasia with filopodia and multiple vacuoles was noted in 11 of 19 liver biopsies from patients with secondary syphilis (BROOKS et al. 1979).

14.4.4.8.3 Gummata in Tertiary Syphilis

Granulomatous inflammation is a characteristic feature of the tertiary stage of syphilis. The similarity in the formation of epithelioid and giant cells and the caseation of the lesions in both gummata and tuberculosis can develop analogously (LETTERER 1959, p. 721). Syphilitic gummas of the central nervous system are now rare, although this situation may change, as the incidence of syphilis increases in some Eastern European (TICHONOVA 1997, SMACCHIA et al. 1998), South East Asian and sub-Saharan African countries, especially in high-risk groups such as prostitutes and their partners (LOVE 2001).

14.4.4.9 Granulomas in Benign Lymphoreticulosis (Cat-Scratch Disease)

Cat-scratch disease is an abscess-forming reticular lymphadenitis. Electron microscopy showed rod-shaped, elliptical or round bacteria in the extracellular spaced between degenerated cells of the lymph nodes (OHTANI et al. 1992).

A gram-negative bacterium or its cell wall-defective variants were isolated from lymph nodes of ten patients with cat-scratch disease (ENGLISH et al. 1988). Ultrastructurally, the organism showed a chain-like arrangement, septal formation, branching and clubbing ends (KUDO et al. 1988). The bacterium was enveloped by a tri-lamellar layer that consisted of an outer membrane, a central peptidoglycan-like layer, and an inner plasma membrane (OHTANI et al. 1992).

14.4.4.10 Granuloma Inguinale

Granuloma inguinale, also known as Donovanosis, is a chronic granulomatous disease involving the genitalia and surrounding sites. It is found in specific geographical foci, namely New Guinea, north western Australia, south-east India, the Caribbean, parts of South America, parts on central Africa (RICHENS 1991), and the KwaZulu/Natal region of South Africa (BASSA et al. 1993). Calymmatobacterium granulomatis, the ætiological agent of the disease, as grown in 'Donovan body'-positive biopsy specimens co-cultured with peripheral blood mononuclear cells in vitro was characterised by KHARSANY et al. (1997) as a bacterium with a typical gram-negative cell wall consisting of an outer membrane, middle electron opaque layer and an inner plasma membrane. The capsule was thick end electron dense. Numerous electron dense granules were present within the cytoplasm.

14.4.4.11 Toxoplasmosis

Human toxoplasmosis can be divided into two broad categories, acquired and congenital. Toxoplasma gondii oocysts from cats can infect human beings (MILLER et al. 1972), and epidemic toxoplasmosis in 37 out of 86 patrons in a riding stable in Atlanta, Georgia, became evident by indirect fluorescent-antibody test (TEUTSCH et al. 1979). Ingestion of oocysts by the patrons of the stable might have been occurred in two ways: stirred up in the dust by the horses in the enclosed area might have been inhaled, and the oocysts subsequently swallowed, or hand-to-mouth transmission may have occurred directly or indirectly by oocyst-contaminated food or beverages.

Pulmonary toxoplasmosis, once considered a rare complication of human immunity virus infection, now has been reported with increasing frequency in patients with acquired immunity deficiency syndrome (AIDS). NASH et al. (1994) reported four cases. Structures suggestive of tachyzoites must be distinguishes from potential look-alikes, including intracellular Histoplasma, Pneumocystis, the cytoplasmic inclusions of cytomegalovirus, the amastigote forms of Leishmania sp. and Trypanosoma sp., and foreign bodies ingested by macrophages (SUN 1988, TSCHIRHART and KLATT 1988).

In the dog, COHRS (1956) described pulmonary histiocytic granulomas with central necrosis.

Lymph node toxoplasmosis is characterised by strong hyperplasia but preserved general structure of the lymph node with small groups of epithelioid cells both in the paracortical area and in the germinal centres (MIETTINEN et al. 1980).

14.4.5 Granulomas Without Epithelioid Cell Formation

14.4.5.1 Granulomas in Salmonelloses

Within 48 h of experimental infection of Balb/c mice with a relatively severe challenge dose of Salmonella typhimurium the infiltration of host liver with neutrophilic polymorphonuclear leucocytes, histiocytes and smaller numbers of lymphocytes was clearly recognisable (STUART et al. 1969). Section through a granuloma showed extensive resorption of liver tissue. A noteworthy feature was the invasive nature of the host response whereby lymphocytes and histiocytes actively pushed their way into the hepatic cells. Pseudopodia of macrophages were very close to the nucleus. Phagocytosis of mitochondria was confirmed in vitro by guinea-pig peritoneal macrophages and preparations of guinea-pig liver mitochondria greatly enhanced by the presence of an anti-mitochondrial serum.

The interaction of Salmonella enteritidis with peritoneal macrophages of ICR mice immunised with i.p. injections of 5×10^3 avirulent S. enteritidis, followed in 21 d by 5×10^2 virulent organisms was influenced by the viability of immunogen, optimal time for harvesting macrophages with respect to immune state, physical state of macrophages whether attached or suspended, components of medium, incorporation of antibiotics, and time of maintenance of macrophages after parasitization (SOLOTOROVSKY and TEWARI (1974).

14.4.5.2 Granulomas in Listeriosis

In perinatal listeriosis, KLATT et al. (1986) located microabscesses and granuloma-like lesion peribronchally, although scattered parenchymal lesion were present.

Listeria monocytogenes was killed by *human* peripheral blood monocytes purified by counterflow centrifugal elutriation and treated with recombinant interferons γ, α, or β in a dose-dependent fashion (PECK 1989). While the killing mechanism following treatment with IFN-γ appeared to depend on the activity of superoxide, inhibition of superoxide radicals by superoxide dismutase did not substantially affect killing by IFN-α or IFN-β.

14.4.6 Granulomas in Mycoses

14.4.6.1 Blastomycosis

14.4.6.1.1 North American Blastomycosis

North American blastomycosis is caused by the dimorphic fungus Blastomyces dermatitidis, first described by GILCHRIST and STOKES in 1896. The primary infection is, as a rule, pulmonary with frequent secondary foci in skin, bone, male genital system (INOSHITA et al. 1983), and eventually, spares no organ in widely disseminated cases (SCHWARZ and SALFELDER 1977) except stomach and hair (CHICK 1971). Infection by inoculation is reported by LARSON et al. (1983) and BUTKA et al. (1984). In the lung, a truly "specific granuloma" that could be considered pathognomonic for blastomycosis does not exist. Thus, the demonstration of Blastomyces dermatitidis by simple (e.g. haematoxylin and eosin stained sections; CHICK et al. 1969, p.91; CHICK 1971, p.478) or sophisticated methods (e.g. a combination of double immunodiffusion test and complement fixation test; TANG et al. 1984) is absolutely necessary for diagnosis. Organisms may be found contained within the cytoplasm of multinucleated giant cells or within the inner zone of the tubercles (CHICK et al. 1971, p.478).

Leucocytes are often present in the central portions of the tubercle, a finding which is rare in tuberculosis. Macrophages, epithelioid cells, lymphocytes, and plasma cells appear in varying degrees depending on the type of tissue reaction. Multinucleated giant cells can be found in almost every case, but they are sometimes very rare, at other times quite numerous. When the type of tissue reaction is granulomatous, numerous mononuclear inflammatory cells may be found forming the lesion. This may occur as large aggregates of cells, or may be more diffusely scattered through the remaining parenchymal structures.

In secondary or haematogenous skin lesions, there is a granulomatous reaction of the dermis or subcutaneous layer. Numerous polymorphonuclear leucocytes, lymphocytes, and mononuclear inflammatory cells are seen. In pseudo-epitheliomatous hyperplasia and minute microabscesses there is usually a rather extensive mononuclear cell infiltration (CHICK 1971, Fig. 16).

Prostatic involvement was reported by INOSHITA et al. (1983). Biopsies revealed a widespread granulomatous reaction with a large number of giant cells and neutrophils within and between the gland spaces.

In the thyroid gland, Blastomyces dermatitidis causes small irregular nodules. There is preponderance of macrophages and numerous multinucleated giant cells (CHICK 1971, Fig. 30).

A cerebellar biopsy revealed a granulomatous lesion characterised by multinucleated giant cells, admixed with plasma cells, lymphocytes, macrophages, and fibroblasts (MIRRA et al. 1980). Numerous yeast forms were noted within the cytoplasm of giant cells and macrophages in membrane-bound compartments. The surface of the macrophages bearing the yeasts showed subplasmalemmal linear densities which ranged in length from 0.1–1.6 μm and were often associated with extracellular amorphous material or collagen. In addition, adjoining macrophages formed desmosome-like junctions. Although numerous coated vesicles were seen within such macrophages, a clear relationship of the desmosomes and vesicles was not discerned.

14.4.6.1.2 South American Blastomycosis (Paracoccidioidomycosis)

Paracoccidioidomycosis is a chronic, progressive and granulomatous disease that mainly attacks the lungs, mucosa of the mouth and nose, and neighbouring teguments with frequent spread to the lymph nodes, adrenal glands and other viscera. In the absence of secondary infection, the nodule is formed by epithelioid cells, Langhans cells or foreign body giant cells, plasmocytes, and lymphocytes. Paracoccioides brasiliensis is generally observed inside the giant cells or mixed with the inflammatory cells. The nodules may be necrotic in their centre thus resembling a tubercle unless a special stain of the fungi is performed (cf. ANGULO O. and POLLAK 1971).

14.4.6.3 Coccidioidomycosis

Coccidioidomycosis is a systemic fungal disease endemic in the south-western USA, Mexico, Central and South America. In Europe, due to the climatic conditions unfavourable to the life cycle of the organism, Coccidioides immitis it is seen very rarely. Sporadic cases have been reported from England (REAMS 1960), Belgium (WILLIOT 1966), Norway (NATVIG and FRAAS 1967), the former USSR (STEPANISHTCHEVA et al. 1972), Finland (ALANKO et al. 1975), Switzerland (FERRARI 1976, CAVIN et al. 1984), France (DROUHET and DUPONT 1989), Czechia (TOMSIKOVÁ 1993), Romania (BUIUC et al. 1995), Sweden (PAPADOPOULOS et al. 1996), and Hungary (CSILLAG 1958, ZALATNAI et al. 1998).

Individuals in certain occupations are more frequently exposed to the contaminated soil (dust) and therefore are more likely to be infected with Coccidioides immitis. Among them agricultural and construction workers or military personnel, geologists, archaeologists, etc. are most frequently affected. It is well known that the fungus poses a health hazard to hospital and medical laboratory workers even outside the endemic regions (JOHNSON 1981).

14.4.6.4
Torulosis (Cryptococcosis)

19 of 36 patients with so called granulomatous pneumonia showed varying degrees of inflammatory response, which ranged from acute inflammation to diffuse intra-alveolar granulomas with giant cells (MCDONNELL and HUTCHINS 1985). As the clinical diagnosis of cryptococcosis is usually difficult, GUPTA (1985) described his experience in the sputum cytology diagnosis of pulmonary cryptococcosis using deep cough samples.

When viable *Cryptococcus neoformans* were administered directly into bronchi of the right lung, immunologically intact *rats* developed hard nodular lesions in the ipsilateral lung (GRAYBILL et al. 1983). These contained caseous centres with numerous *cryptococci* seen on histologic examination. Over the course of 9 months, the lesions shrunk and disappeared. In contrast, congenitally athymic (nude) *rats* developed progressive cryptococcosis with widespread dissemination. The *rat* seems to be a useful model for study of pulmonary cryptococcosis. *Rats*, like *humans*, have extremely effective immune mechanisms for controlling pulmonary *Cryptococcus neoformans* infection. The mechanism(s) responsible for efficient immunity in *rat* experimental infection is unknown. Induction of inducible nitric oxide synthase and nitric oxide have been implicated as an important microbicidal mechanism by which activated macrophages effect cytotoxicity against microbes. GOLDMAN et al. (1996) found the most prominent iNOS immunoreactivity localised to epithelioid macrophages ($CD11b/c^+$) within granulomas; $CD4^+$ and $CD8^+$ T cells were numerous around granulomas but did not express iNOS. iNOS immunoreactivity was detected in a selective population of epithelioid macrophages within some granulomas but not others. $iNOS^-$ granulomas were identical to $iNOS^+$ granulomas with respect to morphology and immunohistochemical profiles. Airway iNOS immunoreactivity was limited to the luminal border of rare bronchiolar epithelial cells. iNOS immunoreactivity was missing in uninfected *rats*.

14.4.6.5
Histoplasmosis

Histoplasma capsulatum, a thermally dimorphic fungal pathogen, causes histoplasmosis in *man*. Infection is established by inhalation of conidia from the soil-inhabiting saprophytic mycelial stage. In the lungs, conidia transform into the parasitic yeast form, which causes disease ranging from a benign infection to chronic cavitation or disseminated infection. Reactivation of previously controlled histoplasmosis has recently become a serious problem in patients with the acquired immune deficiency syndrome (AIDS) (GRAYBILL 1988, SALZMAN et al. 1988).

Scar carcinomas of the lung of the nonmucinous (Clara cell) type arising around an old scar due to histolasmosis granuloma were found in a histoplasmosis endemic area of eastern Kentucky (YONEDA 1990).

In wild *rats*, EMMONS and ASHBURN (1948) found granulomas usually situated in the liver and spleen, but also in lungs and adrenals. From Brazil SILVA and PAULA (1956) described natural infections of *rat*.

14.4.7
Granulomas in Helminthiasis

14.4.7.1
Nematode Infections

14.4.7.1.1
Ascaroidea

Ascaridia

Respiration in the parasite nematode worm *Ascaridia galli* was inhibited at O_2 concentrations in excess of 255 μM (PAGET et al. 1998). Mitochondria-enriched fractions isolated from the tissues of *Ascaridia galli* produced H_2O_2 in the energised state; higher rates of H_2O_2 production were observed in the presence of the uncoupler carbonyl cyanide *m*-chlorophenylhydrazone. *o*-Hydroxydiphenyl, an inhibitor of alternative electron-transport pathways, inhibited respiration by 98 % and completely inhibited the production of H_2O_2 in gut-plus-reproductive-tissue mitochondria; respiration and H_2O_2 production in muscle tissue mitochondria were inhibited by 90 and 86 %, respectively.

Toxocara canis

Human Pathology

Microscopically, *toxocara* granulomas consist of a central mass of karyorrhectic nuclear debris, and eosinophilic structureless and often fibrinoid-like material (SPENCER 1973. In less severe tissue reactions the centres of the lesions are occupied by a mass of eosinophils without cellular necrosis. In a minority of lesions a portion of the causative larva may be identified near the centre of the lesion and cross sections may show the characteristic alar ridges projecting from the exterior of the cuticle. Around the central necrotic area the lesions consist of masses of eosinophil cells, plasma cells, lymphocytes and histiocytic cells, the latter often arranged in a characteristic pallisaded fashion.

In a hepatomegalic *child* aged 2 years with an eosinophilia of 11,000/μl (= 33 %) BOGOMOLETZ (1977) described a hepatic granuloma with histiocytes and giant cells and numerous eosinophilic granulocytes. There was not any necrosis nor a parasite or microorganism in the serial sections of the biopsy, but anamnesis revealed to pat *dogs* infected by *Toxocara canis*.

Escape of the larva into the posterior eye chamber results in an endophthalmitis and the formation of a retinal granuloma (ASHTON 1960).

Experimental Pathology

In *mice* 11 days after stomach intubation of embryonated eggs of *Toxocara canis*, the initial granuloma consisted primarily of eosinophils and appeared to develop from the acute inflammatory infiltrate in the musculature (KAYES and OAKS 1978). During the ensuing 48 h, most of the eosinophils appeared to loose their granules and disintegrate. The resulting cellular debris was then taken up by newly arrived macrophages which become the predominant mononuclear cell in the lesion by 28 days of infection. By 11 weeks, the granuloma had become a fibrotically encapsulated epithelioid granuloma surrounding the inciting larva.

14.4.7.1.2 Filarioidea

Brugia

Brugia malayi

In the *mouse*, a monoclonal antibody (IgM isotype) against *Brugia malayi* microfilariae induces resistance against the lymph-dwelling nematode parasite. *In vitro*, it recognises antigenic determinants of 110 kD present on the surface of *B. malayi* but not of *Dipetalonema viteae* (AGGARWAL et al. 1985).

Brugia pahangi

Specific antibodies to the cuticule of microfilariae of *Brugia pahangi* were detected in infected, amicrofilaraemic *cats* (PONNUDURAI et al. 1974).

Dipetalonema viteae

In vitro, macrophage adherence to infective larvae of *Dipetalonema viteae* is complement-mediated (HAQUE et al. 1982). *D. viteae* microfilariae incubated for 8 h, 16 h and 24 h with either normal *rat* peritoneal macrophages and normal serum or normal *rat* peritoneal macrophages and heat-inactivated amicrofilaraemic immune serum were not damaged (OUAISSI et al. 1981). However, microfilariae were killed by eosinophil-rich cell populations (HAQUE et al. 1981) and IgE antibodies (QUAISSI 1981). *In vivo*, WEISS and TANNER (1979) demonstrated the antibody dependent cell-mediated destruction of microfilariae in the *golden hamster*. The contribution of the monocyte to the adhesion reaction increased with the time of implantation and was pronounced when trapped microfilariae were partly disintegrated. This suggests that the monocyte primarily removes debris of microfilariae. In *Mastomys natalensis*, subcutaneous injections of dexamethasone increased the number of microfilariae in the circulating blood (SÄNGER and LÄMMLER 1979).

Onchocerca volvulus

The bulk of cells in the centre of the dermal *Onchocerca* nodules consists of lymphocytes and macrophages in all stages of activation (BURCHARD et al. 1979). A major part of the macrophages shows degenerative changes: fat vacuoles increased while lysosomes decreased. Near the worm epithelioid (KOZEK and FIGUEROA MARROQUÍN 1976, BURCHARD et al. 1979) and giant cells (BURCHARD et al. 1979) may be found. There are also some polymorphonuclear neutrophils and eosinophils, and varying numbers of mature plasma cells and mast cells.

Wuchereria bancrofti

A 34-year-old woman resident in the city of Colombo, Sri Lanka, with tropical pulmonary eosinophilia a few days after starting diethylcarbamazine therapy noticed a small, tender lump in the right breast. Excision biopsy revealed dilated lymphatics and an inflammatory reaction with eosinophils predominating. A 1.1 mm × 0.75 mm granuloma contained 14 separate microfilariae of *Wuchereria bancrofti* (Chandrasoma et al. 1977).

14.4.7.1.3 Oxyuroidea

Enterobius vermicularis

A nodule of the liver was incidentally noted during left colectomy for a large villous adenoma in a 74-year-old *man* (Mondou and Gnepp 1989). Microscopic examination disclosed a granuloma of the liver containing numerous *Enterobius vermicularis* ova and a cross-section of the nematode.

14.4.7.1.4 Trichostrongyloidea

Nippostrongylus brasiliensis

Seitzer et al. (1997) developed a new method for the *in vitro* generation of granulomas in which L3 larvae of *Nippostrongylus brasiliensis* were used as target for the cellular response of *human* mononuclear blood cells. Epithelioid cells and multinucleated giant cells developed. The presence of tumour necrosis factor-α, interleukin-1β, interleukin-6, and inducible nitric oxide synthase transcripts were demonstrated in multinucleated giant cells.

Paget et al. (1987a) have shown that respiration is inhibited, both in whole *worms* and in mitochondria-enriched fractions, by oxygen above a threshold of about 60 μM. Respiration in these mitochondria is predominantly via a mammalian-like phosphorylating electron-transport chain inhibited by antimycin A and terminating in cytochrome aa_3, as well as an alternative salicylhydroxamic acid-sensitive pathway (Fry et al. 1983); some two-third of the respiration may occur through the aa_3 oxidase, although this proportion depends ultimately on the O_2 concentration prevailing (Paget et al. 1987a). Unlike in *mammalian* mitochondria, the addition of antimycin A does not stimulate H_2O_2 production in mitochondria from *Nippostrongylus brasiliensis* (Paget et al. 1987b). Another inhibitor of alternative electron transport, salicylhydroxamic acid, showed a complex mode of action; low concentrations (>0.5 mM) stimulated respiration and H_2O_2 production, whereas 2 mM-salicylhydroxamic acid inhibited respiration by 35 % and stopped H_2O_2 production completely. In the presence of antimycin A the O_2-inhibition threshold and apparent K_m values for O_2 of respiration and H_2O_2 production matched closely, suggesting that the alternative oxidase is a likely site of H_2O_2 production.

14.4.7.1.5 Trichuroidea

Trichinella spiralis

Resistance to the intestinal stage of *Trichinella spiralis* has been shown to be mediated by cellular immunity (Larsh et al. 1966, Love et al. 1976). Macrophages from *mice* infected >11 d earlier inhibited DNA synthesis of syngeneic and allogeneic tumour cells (Wing et al. 1979), a property attributed to activated macrophages. However, macrophages from infected *mice* did not develop the ability to inhibit multiplication of the intracellular pathogen *Toxoplasma gondii*. Resistance against *Trichinella spiralis* is increased by activation of macrophages by either chronic *Toxoplasma gondii* or acute *Listeria monocytogenes* infections (Wing and Remington 1978). When *mice* were irradiated immediately before infection with *Trichinella spiralis* there was a profound and long-lasting interference with the ability to expel adult worms from the intestine (Wakelin and Wilson 1977). Administration of 8 μg diethylstilbestrol/g *mouse* for 5 consecutive days beginning on days –5, 0, +3 or +8 of infection with *Trichinella spiralis* inhibited adult worm expulsion and tissue reactions in the small intestine (Luebke et al. 1984).

14.4.7.2 Platyhelminth Infections

14.4.7.2.1 Cestode Infections

Echinococcus

Human alveolar echinococcosis is endemic in the eastern part of France, in Switzerland, Russia, Japan, Canada and Alaska. Cell-mediated immunity may be responsible for a granulomatous proliferation with epithelioid cells, giant cells and numerous lymphocytes surrounding parasitic cysts.

Lance et al. (1984) developed an experimental model using intrahepatic injection of *Echinococcus multilocularis* larvae in four inbred strains of *mice* differing by their sensitivity to the parasite. Their observation in C57BL6 and C57BL10 *mice* respectively "sensitive" and "resistant" to infection by *Echinococcus multilocularis* although sharing the same H2 (b) determinants argue against a direct link between H2 and resistance to the parasite.

Cysticercus

In a 54 year-old *man*, who had worked underground for 28 years as a stoneworker and for 6 years as a face worker, ZORN and WORTH (1952, p. 290) found silicosis stage III and cysticercosis representing "occasional, round, very dense and well defined shadows irregularly distributed throughout the lung fields, with a loculated appearance".

14.4.7.2.2 Trematode Infections

Schistosomiasis

The blood flukes of the genus *Schistosoma* have a wide geographic distribution. *Schistosoma japonicum* is found in the Far East while *Schistosoma mansoni* is caught in Africa, South America, the Middle East and the Carribean. *Schistosoma haematobium* is common in Africa and the Middle East.

Immunosuppression in the definitive and intermediate hosts of *Schistosoma mansoni* is mediated by the release of immunoreactive peptides as β-endorphin (DUVAUX-MIRET et al. 1992). The release of the glycosyl-phosphatidyl-inositol (GPI)-anchored glycoproteins may contribute to immune evasion of *Schistosoma mansoni* (SAUMA and STRAND 1990).

Rat C-reactive protein can induce platelet-mediated cytotoxic activity against *Schistosoma* sp. (BOUT et al. 1986).

Tissue damage in schistosomiasis is attributable more to a host response than to toxic effects of the **ova**. Thus schistosomal eggs deposited experimentally in the *mouse* in liver sinusoids induced a cell-mediated granulomatous reaction (WARREN et al. 1967), and this granulomatous reaction can be elicited also by a soluble antigen of the eggs (BORUS and WARREN 1970).

Around eggs as well as dying or dead *Schistosoma mansoni* parasites GÖNNERT (1955, p. 315) found macrophages that can develop into epithelioid cells or confluence to form giant cells.

Knockout *mice* with a disrupted interleukin-12 p40 gene exposed to *Schistosoma mansoni* had abundant and very large multinucleated giant cells (> 50 μm) in their lungs concurrent with extensive eosinophilia and a population of large macrophages (ANDERSON et al. 1999).

Inhibition of protein tyrosine kinase by genistin (OGAWARA et al. 1989) and protein kinase C by 1-(5-isoquinolinylsulfonyl)-2-methylpiperazine dihydrochloride (HIDAKA et al. 1984) prevented lymphocyte activation by *Schistosoma mansoni* antigens and reduced *in vivo* granuloma reaction (ALMEIDA et al. 1998).

Nitric oxide production could upregulate the *in vitro* granuloma reaction on *human* schistosomiasis through changes in the cytokine/chemokine profile released by peripheral blood mononuclear cells (OLIVEIRA et al. 1999). The mechanisms involved may lead to a MIP-1α-increased and IL-10-decreased secretion. Reactive nitrogen species influenced the enteric neurones of infected ileum in *mice* (VAN NASSAUW et al. 2001). Quantitative analysis of whole-mounts showed that the percentage of 3-nitrotyrosine-immunoreactive neurones significantly increased with time in both the submucous and myenteric plexus. Caspase-3 immunoreactivity was predominantly found in parasite eggs in infectes *mice*.

14.4.7.2.3 Fascioliasis

In *rabbits* experimentally infected with cercariae of *Fasciola hepatica* URQUHART (1956) found granulomatous lesions in response to the presence of fluke eggs in the tissues. In *sheep* single infestations with ≥750 metacercariae led to delayed maturation of some parasites (THORPE and FORD 1969). The hepatic lesions of focal parenchymal destruction and fibrous scarring were qualitatively similar in *sheep* subjected to large single-dose infestations and multiple-dose infestations. Enzyme histochemical studies showed that during the migratory phase of *F. hepatica* the host liver showed abnormal patterns of enzyme activity in liver lobules free from fluke migration; alkaline phosphatase activity was increased, canalicular adenosine triphosphatase was reduced, and there was a loss of zonal differences in activity, with the oxidative enzyme methods used. Significant increases in the serum levels of sorbitol dehydrogenase, glutamate dehydrogenase, and glutamate-oxaloacetate transaminase were recorded for *sheep* given single doses of 2000 metacercariae, and sheep giver 3 successive doses of 200 metacercariae.

14.4.8 Granulomas Induced by Products of Plants and Animals

Splinters of wood and prickles of cactuses may induce changes similar to granuloma anulare (HORNSTEIN and HANEKE 1980). Both splinters of wood and prickles of cactuses are composed of hexagonal, strongly birefringent cells with a dark dot in their centres. Prickles of cactuses have a single-layered cuticle.

14.5 Pneumoconioses

Chronic inhalation of inorganic dusts induces an inflammation of the lower respiratory tract (Rom et al. 1987). This inflammatory process is dominated by alveolar macrophages releasing excessive amounts of mediators. Despite the fact that silica, coal dust and asbestos are very different agents, alveolar macrophages release similar mediators. In the *rat*, however, the pattern and magnitude of the response to inhalation of silica and chrysotile asbestos are different (Donaldson et al. 1988).

Table 43. Particulate matter free radical toxicology

Dust	Model system	Parameter of toxicity	Result	Reference	Remarks
Bituminous, subbituminous, and lignite coal dusts	Solvent extracts in *Salmonella typhimurium*		High mutagenicity when reacted with $NaNO_2$ at pH ~3.5	Whong et al. (1983)	Elevated incidence of gastric cancer in coal miners
Quartz	Emission of singlet oxygen (Δ^1O_2)		Emission of Δ^1O_2 proceeds in a process of irreversible loss of centers of chemisorbed oxygen in the form of superoxide radicals	Zav'yalov et al. (1985)	
Short crocidolite asbestos fibres, TiO_2 particles	Thioglycolate-elicited *mouse* peritoneal macrophages	H_2O_2 production	Reactive oxygen metabolite scavenging enzymes, superoxide dismutase or catalase, prevented the toxicity of long and short crocidolite fibres. Macrophages were not killed when either long of short fibres were soaked in the iron chelator, desferroxamine.	Goodglick and Kane (1990)	
Quartz, chrysotile, amosie, crocidolite, anthophyllite (PT 311), wollastonites (FW 200, FW 325)	*Human* polymorphonuclear blood leucocytes	Luminol-enhanced chemiluminescence	On an equal weight basis, the particulates induced chemiluminescence in the following order of magnitude: chrysotile, quarth > amosite, crocidolite > anthophyllite, wollastonite. The intensity of CL correlated positively with the Alcian blue binding capacity of the particles. Polyvinylpyridine-*N*-oxide (0.5 μg/ml) inhibited the CL completely the quartz-induced CL, but had little effect on asbestos fibre-induced CL. Carboxymethylcellulose (1.0 μg/ml), however, reduced CL caused by chrysotile but had no effect on CL induved by the other particles.	Klockars et al. (1990)	Surface characteristics, including charge of particles might influence the reaction of inflammatory cells to dust particles.
Crocidolite			Intracellular mobilization of iron from asbestos may increase reactions with O_2 leadung to $HO^•$ formation, DNA damage and cancer.	Aust and Lund (1991)	
Crocidolite, chrysotile	*Hamster* tracheal epithelial cell (HTE) cultures	^{51}Cr release, [^{3}H]thymidine incorporation, ornithine decarboxylase activity	In confluent HTE cells, significant blockage of chrysotile or crocidolite asbestos-stimulated ornithine decarboxylase activity occurred with simultaneous of catalase, but not of superoxide dismutase, in medium. The addition of xanthine plus xanthine oxidase caused a dose-dependent increase in ornithine decarboxylase activity, which was significantly inhibited after addition of catalase or mannitol, indicating that H_2O_2 was the principal oxidant produced in that reaction. Addition of phenazine methosulfate (redox reagent for H_2O_2 production) resulted in a significant elevation of ornithine decarboxylase, which was inhibited by the addition of superoxide dismutase but not catalase. H_2O_2 added to culture medium also caused a	Marsh and Mossman (1991)	H_2O_2 plays a major role in asbestos-stimulated ornithine decarboxylase induction.

Table 43. (Continued)

Dust	Model system	Parameter of toxicity	Result	Reference	Remarks
			potent increase in ornithine decarboxylase activity inhibitable by catalase. $HOCl^-$ caused increase in ornithine decarboxylase activity, although the magnitude of this response was less than that observed with other oxidants. $O_2^{\bullet-}$ and H_2O_2 were more proficitent than $HO^{\bullet}$. All oxidants, except $HOCl^-$, caused a significant increase in [^{3}H]thymidine incorporation at 24 or 48 h after their addition to HTE cells.		
Erionite, crocidolite, amosite, anthophyllite, chrysotile, JM code 100 glass fibres, glass wool	Incubation with H_2O_2	Salycilate adduct formation	Erionite, JM code 100 and glass wool were the most effective initiators of $HO^{\bullet}$ formation, followed, in order, by crocidolite, amosite and chrysotile	Maples and Johnson (1992)	$HO^{\bullet}$ formation figures are compared to *human* and exper. tumour induction
Quartz dust, metal-containing dusts	*Bovine* alveolar macrophages	Superoxide dismutase (120 U/ml)-inhibitable reduction of 80 μM ferricytochrome *c* in the supernatant of 4×10^6 *bovine* alveolar macrophages/well, determined spectrophotometrically at 550 nm	Incubation of BAM with the dusts (12.5–1000 μg/ml medium) showed a concentration dependent increase in the release of reactive oxygen species which was already observed after 15 min and continued up to 90 min. Only in testing the CWIA 1 containing the highest percentage in heavy metals, the release of ROI ceased after 60 min.	Berg et al. (1993)	The release of H_2O_2 correlated best, in descending order, with the content of Fe, Mn, Cr, V, and As in the dusts.
Quartz, alumina	Rabbit alveolar macrophages and monocytes, *human* blood monocytes and granulocytes	Nitroblue tetrazolium reduction, superoxide anion formation, H_2O_2 generation	Phagocytosis of quartz by *rabbit* AM and monocytes and *human* granulocytes was accompanied by NTB reduction to formazan. Low fibrogenic alumina had no effect on H_2O_2 generation	Gusev et al. (1993)	The uncontrolled generation of superoxide and H_2O_2 might immediately cause damage to pulmonary parenchyma.
Crocidolite, amosite	Closed-circular, superhelical φX174 RFI DNA single-strand breaks	Single-strand breaks	32 % DNA damage induced by crocidolite and 52 % by amosite when citrate (1 mM) and ascorbate (1 mM) were added. Further addition of 25 μM H_2O_2 increased the figures to 56 % and 76 %, respectively. During 30 min incubation with 1 mM nitrilotriacetate, 14 μM iron was mobilized from crocidolite compared with 4 μM from amosite.	Aust et al. (1994)	Citrate-chelated iron, reduced by ascorbic acid to Fe^{2+} acts as a Fenton catalyst to generate $HO^{\bullet}$ which induces DNA single-strand breaks.
Amosite	Cultured human mesothelial cells (MET 5A), hypoxanthine-xanthine oxidase system	DNA single strand breaks, extracellular release of nucleotides and their catabolites, LDH release	Superoxide radical and H_2O_2 exposures resulted in the depletion of adenine nucleotides, accumulation of the products of nucleotide catabolism, induction of DNA single strand breaks.	Kinnula et al. (1994)	Amosite did not induce acute oxidant-type injury to mesothelial cells *in vitro*.
Asbestos		Oxygen radicals		Quinlan et al. (1994)	
Crocidolite, chrysotile	*Rat* alveolar macrophages	$^{\bullet}$NO production (measured as NO_2^-)	Time-dependent production of NO was maximal in 48 h cultures.	Thomas et al. (1994)	Nitric oxide pathway also engaged in asbestos toxicity

Table 43. (Continued)

Dust	Model system	Parameter of toxicity	Result	Reference	Remarks
Bituminous coals from Utah and WVa	ESR of H_2O_2/ coal reactions *in vitro*	Hydroxyl radical generation from H_2O_2	All coal dusts generated varying levels of $HO^•$ from H_2O_2 in the presence of a $HO^•$ spin trap. The $HO^•$-generation potential of all coal dusts showed a positive correlation with the surface iron content of coal mine dust.	DALAL et al. (1995)	
Amosite	Chinese *hamster* ovary cells	Inhibition of γ-glutamyl cysteinyl synthetase by buthionine sulphoximine	Decreasing intracellular glutathione may impose an oxidant stress on the cells, which contributes to chromosome damage.	DONALDSON and GOLYASNYA (1995)	
Amosite, crocidolite, RCF 1, 2, 3, and 4, MMVs 10, 11, 21, and 22	Suoercoiled ϕX174 RFl plasmid DNA	Fe^{2+} release, Fe^{3+} relese, oxidative DNA damage	Amosite and crocidolite caused substantial damage to DNA (dose related). MMVF 21 released signif. more Fe(III) than any other fibre. Short amosie and MMVF 21 released large quantities of Fe without causing free radical damage.	GILMOUR et al. (1995)	Fibre-bound Fe not released from the surface of asbestos could be important.
Amosite, chrysotile, TiO_2, carbonyl iron	SV40-transformed WI-26 cells; A549 cells	^{51}Cr release; production of $HO^•$-like species, DNA strand breaks	Amosite asbestos damaged cultured *human* pulmonary epithelial-like cells (^{51}Cr release). Phytic acid (Fe chelator) attenuated this effect. Amosie induced dose-dependentDNA stand breaks in WI-26, A549, and primarily isolated *rat* AEII cells.	KAMP et al. (1995)	Mineral dusts may be directly genotoxic.
Quartz, titanium dioxide	Intratracheal instillation in Sprague-Dawley *rats*	Hydroxyl radical production	7 d after instillation signif more 2,3- and 2,5-dihydroxybenzoic acid formed from salicylic acid in the silica exposed lungs compared with lungs instilled with TiO_2.	SCHAPIRA et al. (1995)	
Chrysotile (Tuva, Russia)	Peritoneal macrophages (Wistar *rat*)	Lucigenin-enhanced chemiluminescence, formation of thiobarbiturate-reactive substances, LDH release, cytolysis	Rutin (IC_{50} = 90 μM) and quercetin (IC_{50} = 290 μM) reduced the LDH release from 2×10^6 peritoneal macraphages/ml caused by 2 mg asbestos/ml; oxidation was superoxide dismutase-sensitive. IC_{50} for the formation of thiobarbiturate-reactive substances were 8 and 80 μM, respectively, and for lucigenin-enhanced chemiluminescence 3 and 30 μM, respectively.	KOSTYUK et al. (1996)	
Asbestos	*Human* alveolar macrophages	Superoxide anion production	Superoxide anion production was augmented over fibres alone by the incubation of fibres with C5, C5 + serum, and serum, but the supernatants with no fibres did not increase superoxide production over control.	PERKINS and HAMILTON (1996)	It is suggested that asbestos alters C5 to produce fragments that stimulate macrophages adherent to fibres or in solution.
Crocidolite, chrysotile	Aqueous buffer suspensions containing asbestos fibres alone and in combination with aqueous cigarette tar extracts	Hydroxyl radical production	UICC crocidolyte generated hydroxyl radicals in aqueous buffer solutions via an iron-catalyzed Fenton reaction. Grinding of asbestos fibres and addition of EDTA (iron chelator) enhanced the intensity of the electron spin signal.	VALAVANIDIS et al. (1996)	Strong evidence for lung cancer in smoking asbestos workers
Quartz	HL-60 cells (1,25-Dihydroxyvitamin D_3-primed), *bovine* alveolar macrophages	Chemiluminescence	Nearly identical results obtained using either system	ZOLLER et al. (1996)	

Table 43. (Continued)

Dust	Model system	Parameter of toxicity	Result	Reference	Remarks
Crocidolite	*Rat* parietal pleural mesothelial (RPM) cells, *human* mesothelial cell line (MET5A)	8-hydroxydeoxyguanosine formation, mRNA levels of manganese containing superoxide dismutase (MnSOD)	RPM cells schowed dose-dependent and signif. increases in 8-OHdG formation in response to crocidolite or iron-chelated crocidolite fibres. MET5A showed decreses in 8-OHdG. Both cell types exhibited elevations in message levels of MnSOD. In comparison with *human* MET5A cells, RPM cells exhibited increased cytotoxicity and apoptosis in response to asbestos.	Fung et al. (1997)	
Man-made vitreous fibres	*Rat* alveolar macrophages	Reduction of cytochrome *c*	Suppression of respiratory burst by surfactant	Brown et al. (1998)	
Crocidolite	JB6 *mouse* epithelial cell line, TRE-luciferase reporter transgenic *mice*	Activator protein-1 activity, protein kinase phosphorylation	Activator protein-1 transactivation	Ding et al. (1999)	
Crocidolite	Transformed *human* mesothelial cells	DNA single strand breaks	Pre-treatment with inhibitors of antioxidant enzymes did not cause any further increases in the mean tail moments.	Ollikainen et al. (1999)	Other mechanisms than free radicals involved
Chrysotile, crocidolite	*Rabbit* mesothelial cells	^{125}I-Urokinase-type plasminogen activator receptor	Asbestos increases urokinase-type plasminogen activator receptor	Perkins et al. (1999)	
Crocidolite; wollastonite and glass beads for control	*Rabbit* pleural mesothelial cells	DNA strand breakage	Asbestos increased intracellular oxidation, DNA strand breakage, and apoptosis, cell-cycle arrest in G2/M	Liu et al. (2000)	Phagocytosis important and perhaps necessary prerequisite
Crocidolite	Nose-only inhalation in transgenic *mice*	Mutagenicity	Mutant frequency significantly increased during the 4th week after exposure	Rihn et al. (2000)	Possible involvement of a DNA repair decrease in crocidolite-treated animals
Crocidolite	Intraperitoneal injection in *rats*	mRNA expression pattern	Upregulation of *c-myc*, *fra-1* and *egfr*	Sandhu et al. (2000)	Fibres may affect integrin-linked signal transduction and extracellular matrix proteins
Chrysotile (NIEHS and UICC), crocidolite (NIEHS and UICC)	*Rabbit* pleural mesothelial cells	Coating of fibres, phagocytosis and dichlorofluorescein assay	Vitronectin adsorption to fibres increases fibre phagocytosis and intracellular oxidation. Apoptosis blocked by integrin-ligand blockade with arginine-glycine-aspartic acid peptides	Wu et al. (2000)	
Crocidolite (NAIMA Fiber Repository, Alexandria, VA	Intratracheal instillation in C57BL6 *mice*	iNOS	iNOS knockout *mice* exhibited an exceeded pulmonary expression and production of TNF-α as well as high influx of neutrophils into the alveolar space than wild-type *mice*.	Dörger et al. (2002)	
Insulation wool HT, stone wool D6 (less biopersistent than MMVF21)	Intraperitoneal injection in *rats*	Survival, carcinogenicity	D6 caused a statistically significant increase of mesotheliomas compared to the negative control, but HT fibres did not cause any mesotheliomas or any increase in other tumour types (pituitary, mamma).	Kamstrup et al. (2002)	

Scheme

Pathophysiology of pneumoconioses and other chronic lung disease

Pneumoconioses and other chronic lung disease

↓

Alveolar hypoxia

↓

Increased expression of HIF-1 in the lung

↓

Pulmonary hypertension

↓

Right ventricular hypertrophy

↓

Decreased pulmonary blood flow

↓

Systemic hypoxia

↓

Death

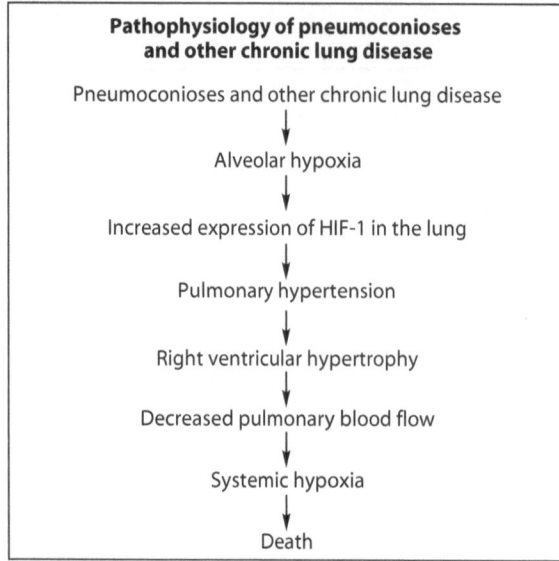

14.5.1 Berylliosis

Occupational exposure to inhaled beryllium (Be) has been associated with two syndromes: an acute chemical pneumonitis (Royston 1949) and an immunologically mediated chronic granulomatous lung disease (Deodhar et al. 1973, Jones Williams and Williams 1983, Kreiss et al. 1994, Meyer 1994, Stange et al. 1996). The immunologic role of beryllium in the pathogenesis of this disease was even more firmly established when the technique of bronchoalveolar lavage was applied to patients with chronic beryllium disease (Epstein et al. 1982, Rossman et al. 1988). For the first time, cells from an area of active inflammation could be harvested and studied. Epstein et al. (1982) and Rossman et al. (1988) demonstrated beryllium-sensitive cells in the lungs of most patients with chronic beryllium disease; an increased percentage of these cells was present in the lung compared to that in peripheral blood. Similar to sarcoidosis, the bronchoalveolar lymphocytosis was largely of T lymphocytes, predominantly $CD4^+$ or T helper cells. *In vitro*, Be stimulated the $CD4^+$ T lymphocytes. Cloned T cells specific for Be did not react with other antigens (Saltini et al. 1989). The proliferative response of Be-sensitive T cells could be inhibited by antibodies to MHC Class 2 but not Class I molecules. Thus Be induces an antigen-driven immunologic response, presumably by acting as a hapten through binding to a self-molecule that is then perceived as foreign. Using anti-T-cell receptor monoclonal antibodies, Fontenot et al. (1998) investigated the T-cell receptor β and α variable ($V\beta$ and $V\alpha$, respectively) in the bronchoalveolar lavage and blood of both chronic beryllium disease patients and healthy controls. There was marked heterogeneity within the bronchoalveolar lavage $CD4^+$ T-cell repertoire in both patients and controls. However, 11 of the 28 patients with chronic beryllium disease demonstrated 16 different T-cell subset expansions within the bronchoalveolar lavage as compared with only one expansion in 10 healthy controls. Five of the 16 expansions in chronic Be disease patients expressed $V\beta3$. Altered T-cell receptor expression within the bronchoalveolar lavage T-cell repertoire appeared to persist over time in patients who underwent repeat evaluation. After *in vitro* stimulation of bronchoalveolar lavage T-cells with beryllium sulphate and interleukin-2, Fontenot et al. (1998) noted further alteration of the bronchoalveolar lavage T-cell receptor repertoire in some individuals.

The granulomas and frequently found inclusion bodies due to beryllium are indistinguishable from those in sarcoidosis (Jones Williams 1977). Non specific interstitial inflammation is more pronounced in beryllium disease than in sarcoidosis and correlates better with prognosis than the number of granulomas (Freiman and Hardy 1970). The diagnosis of beryllium disease thus requires the following additional informations (Jones Williams 1977):

History of exposure;
Consistent clinico-radiological features;
Demonstration of beryllium in tissue and/or urine;
Positive beryllium patch test;
Positive *in vitro* lymphocyte transformation test.

Tinkle and Newman (1997) demonstrated that tumour necrosis factor-α and interleukin-6 participate in the initiation of the chronic beryllium disease-derived bronchoalveolar lavage cells-mediated response to beryllium salts *in vitro*. Tumour necrosis factor-α and interleukin-6 have a central role in initiating the cell-mediated response to antigen and in other granulomatous lung diseases (Kunkel et al. 1989). IL-6 is involved in T-lymphocyte activation, probably controlling early steps in T-lymphocyte activation and, in combination with IL-1, induces T-lymphocyte production of IL-2 and IL-2R (Hirano 1994). Tumour necrosis factor-α is a key mediator of inflammation and triggers the release of cytokines that amplify and extend the inflammatory response (Tracey 1994). The effects of IL-6 are synergistic with tumour necrosis factor-α (Hirano 1994).

In the *mouse*, nose-only inhalation of 1,030 mg Be/m^3 28 weeks after 6 exposures for 90 min each over 3 days induced multifocal granulomatous pneumonia characterised by hypertrophic, vacuolated macrophages in alveoli, variable numbers of neutrophils in alveoli and alveolar septa, and septal and perivascular infiltrates of lymphocytes, plasma cells, monocytes and macrophages (Nikula et al. 1997). Cholesterol clefts and degenerating macrophages were present. Foreign-body giant cells were frequently observed, whereas giant cells of the Langhans type were less commonly seen.

14.5.2 Silicosis

Silicosis, due to the inhalation of crystalline tetrahedral (quartz, tridymite, cristobalite) – but not octahedral (stishovite) – silicon dioxide, is characterised by granulomatous inflammation of the lung. As in various granulomatous diseases, such as other pneumoconioses, sarcoidosis, and tuberculosis, silicosis is characterised by an early mononuclear cell infiltration of the lung parenchyma followed by resident cell activation and proliferation (Becklake 1994).

Four weeks after intratracheal instillation of 50 mg silica in 0.3 ml normal saline into *rats*, concentrations of ionizable iron adsorbed to the surface of the accumulated silica increased significantly from 12.7 ± 1.4 µmol [Fe^{3+}]/g dust to 38.5 ± 1.5 µmol [Fe^{3+}]/g dust (Ghio et al. 1994). Corresponding to this elevation of surface-adsorbed material concentrations in bronchoalveolar lavage fluid, lung tissue, plasma, and liver tissue all increased. Antioxidant molecules in lung tissue, including ascorbate, urate, and glutathione, all decreased, whereas superoxide dismutase increased. Oxidized products in the lung tissue, measured as thiobarbituric acid-reactive products, similarly increased, reflecting an **oxidant stress**. Dietary depletion of iron stores before instillation of silica dust resulted in low iron stores (hæmatocrit values of 21.8 ± 1.9) and low iron concentrations in lavage fluid, lung and liver tissues. *Rats* on iron-depleted diets demonstrated a diminished fibrotic injury after dust instillation.

Hyperbaric conditions (pressure of 2.5×10^5 Pa corresponding to 50% O_2 at normal pressure) in *monkeys* inhaling 5 mg DQ12 quartz/m^3 (8 h/day, 5 days/week) over 20 months suggested that quartz effectively inhibited mixed function oxidation of proteins in the lung (Lenz et al. 1989).

Classical silicotic nodules are rounded, whirled, well demarcated very fibrotic lesions clearly demarcated from the background lung. Microscopically they have a narrow rim of dust containing macrophages admixed with randomly oriented collagen fibres, an intermediate zone of concentrically arranged collagen fibres and a central collagenous core which may be variably hyalinised and calcified.

So-called classical silicosis was not only found in South African gold miners (Strachan and Simson 1930) and in stonemen of the Ruhr district (Husten 1931), but also in grinders (Staub-Oetiker 1916, Koppenhöfer 1935, Kühne 1965), and sand blasters of the era before using artificial abrasives, and steel casting cleaners (Slenský et al. 1966, Slenský 1975). Silicotic type nodules were more

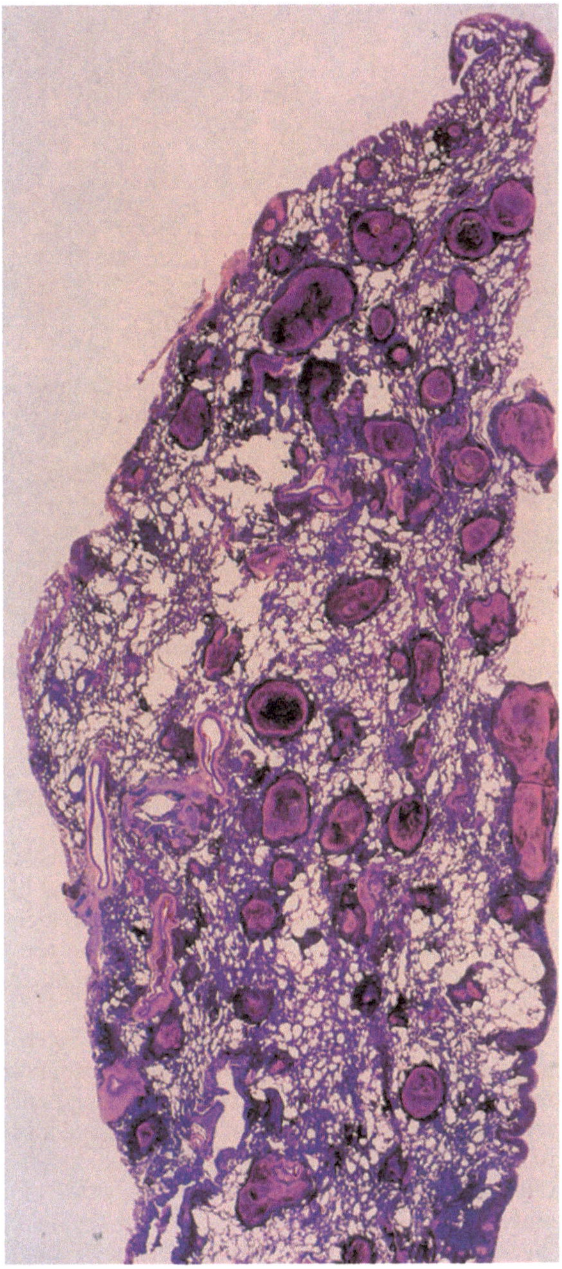

Fig. 200. Stoneman's silicosis from the Ruhr district. Haematoxylin and eosin. (Material kindly provided by Prof. Husten †)

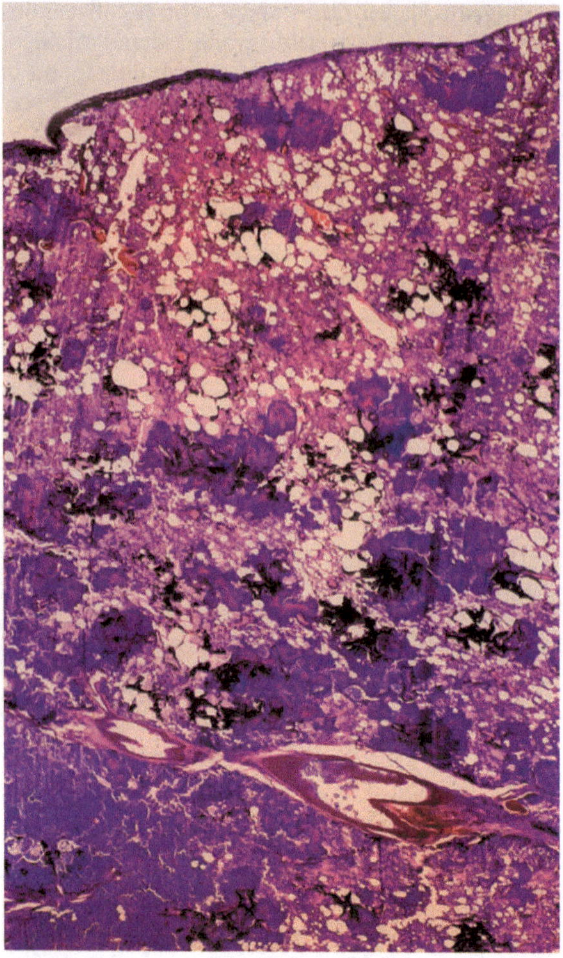

Fig. 201. Coalworker's lung from the Ruhr district. Haematoxylin and eosin. (Material kindly provided by Prof. Husten †)

frequent in the North Wales cases of slate workers' pneumoconiosis than in cases from Vermont (USA), as compared by GIBBS et al. (1988).

According to KALBFLEISCH (1949) the particular localisation of silicotic lesions and their typical bilateral symmetry are due to the functional segmental innervation. MOSINGER (1963) using the technique of CHAMPY et al. (1946) in silicosis found severe lesions of the intrapulmonary nerves.

Mixed dust fibrotic nodules are stellate. Microscopically they are composed of a central zone of collagen with a periphery of linearly and radially arranged collagen admixed with dust laden macrophages. In a series of 13,296 deceased South African mine workers from gold, coal, asbestos and other mines, approximately 1% showed lesions of the mixed dust pneumoconiosis type in their lungs, but there was no obvious relationship to the type of mine (GOLDSTEIN and WEBSTER 1971). In 234 North Wales and 11 Vermont slate workers the pathological findings showed some differences from classical silicosis, in particular considerable interstitial fibrosis (GIBBS et al. 1988). In 62 subjects who were predominantly exposed to either nonfibrous silicates without free silica or with small quantities of free silica, or to nonfibrous silicates and free silica, the pathological changes were similar in kaolin, talc and mica exposed lungs and consisted of interstitial collections of macrophages together with varying numbers of ferruginous bodies and foreign body granulomas (GIBBS et al. 1994). Kaolin plus quartz and mica plus quartz induced silicotic nodules, mixed dust fibrotic lesions and varying degrees of interstitial fibrosis.

The primary lesion in coalworkers' pneumoconiosis is the coal dust macula; it evolves by the incorporation of dust-filled macrophages into the walls of respiratory bronchioles and alveolar ducts (HEPPLESTON 1953, NAEYE and DELLINGER 1979). Aggregations of dust maculae are responsible for the soft, black nodules and irregular streaks seen on cut section of lungs of workers with the anthracotic pneumoconioses. Differences in the volume of dust maculae and their crystals and collagen content may explain most of the differences in coalworkers' pneumoconiosis encountered among miners of different ranked coals in Pennsylvania and West Virginia (NAEYE et al. 1971). The miners of higher ranked coal have a larger volume of dust maculae and silicotic nodules. The maculae in turn have a larger concentration of silica crystals and collagen than do comparable maculae in miners of lower ranked coal.

Partly contradictory results from the epidemiological, animal and cytotoxicological studies from West German coal mines may depend on both the rank of the coal mined and its mineral and quartz content (REISNER and ROBOCK 1977). Both are positively correlated to the pathogenicity of the respirable dust, but negatively correlated to each other. The iron content as analysed by LAser Microprobe Mass Spectrometry (LAMMS) may also play an important role in the toxicity of coal dust (TOURMANN and KAUFMANN 1994). DALAL et al. (1995) measured the total concentrations of silica, iron, cobalt, copper and chromium in American coal dust samples and obtained a positive correlation ($r = 0.93$) at 95% confidence between the dusts' surface iron content and its $HO^{\bullet}$ generating potential. The amount of iron in simple pneumoconiosis lungs from different coalfields in Great Britain is related to their mineral and coal contents and to the factor 'years underground' (BERGMAN and CASSWELL 1972). The strongest relationship is with coal and mineral for coalface workers from England. For pit accident cases (all simple pneumoconiosis) mineral

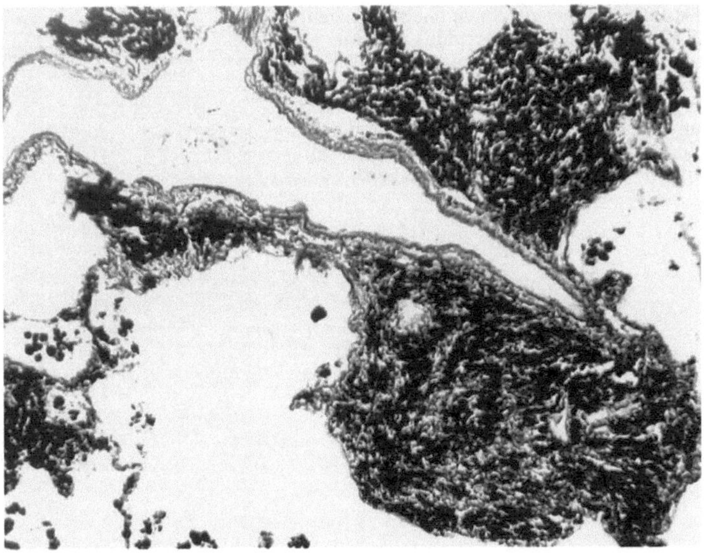

Fig. 202. Simple pneumoconiosis: dust focus with central vessel cut obliquely. Soft coal miner from the Rhondda valley aged 36 years who had worked underground for 16 years. Cause of death: acute myelitis. Post-mortem (Prof Gough, Cardiff) additionally showed pneumoconiosis and enlarged tuberculous hilar lymph nodes. Fixed in formalin. Paraffin section impregnated with silver. Eyepiece Leitz Periplan 4×; objective Leitz 12/0.30, 16 mm apochromat. (from SCHILLER 1954)

is important as well as years underground, while for simple pneumoconiosis cases from Scotland, years underground is the most important factor.

Using data from U.S. coal miners and assumptions about the overloading of alveolar clearance from studies in rats, KUEMPEL (2000) developed a biologically based dosimetric lung model to describe the fate of respirable particles in the *human* lung. This model includes alveolar, interstitial, and hilar lymph node compartments. The form of the model that provides the best fit to the lung dust burden data in these coal miners includes a first-order interstitialisation process and either no dose-dependent decline in alveolar clearance or much less decline than expected from *rodent* studies. These findings are consistent with the particle retention patterns observed in the lungs of *primates* (NIKULA et al. 1997). This *human* lung dosimetry model is useful for investigating the factors that may influence the relationship between the airborne particle exposure, lung dust burden, and fibrotic lung disease.

In 460 autopsied coal miners from West Virginia on pathologic examination 96 % had macules, 70 % micronodules, 45 % macronodules, 15 % silicosis, and 28 % progressive massive fibrosis (VALLYATHAN et al. 1996).

In 20 *women* and 8 *men* from Seoul, Korea, ranging in age from 42 to 86 years (median, 64 years) not occupationally exposed to coal dust or other soot particles complaining of cough and dyspnoea on exertion, pathologic study of the lesion obtained by bronchoscopic biopsy or thoracotomy showed dense bronchial and/or peribronchial fibrosis with interspersed black pigments (CHUNG et al. 1998). Marked perivascular fibrosis and several foci of peribronchial perivascular granulomatous inflammation were noted on microscopic examination.

Biological hazards associated with Candiota coal field in Rio Grande do Sul, the southern state of Brazil, demonstrated an increase in DNA damage in *Ctenomys torquatus*, a subterranean rodent endemic in southern South America (FREITAS and LESSA 1984), detectable by the Comet assay (DA SILVA et al. 2000).

Highly significant positive correlation was found between the severity of coalworkers' pneumoconiosis and the black pigment score in both liver and spleen (LEFEVRE et al. 1982). Significant positive correlations were found between years spent underground, years of retirement, and age at death versus pigment scores. Significant negative correlation was found between smoking and pigment.

Whereas in some British pits there is practically no silica associated with the coal seam, those in Somerset and Dean Forest (Gloucestershire) include Pennant rock, a highly siliceous sandstone (MEIKLEJOHN 1960). HART and ASLETT (1942) concluded that the incidence of pneumoconiosis was somehow related to the hardness or rank of coal. Thus the disease was more present in the anthracite (hard--high rank) than in the bituminous or steam coal (soft–low rank) pits.

The Cu, Fe, Ni, Pb, and Zn content of coal from a Pennsylvania mine having a high incidence of coal workers' pneumoconiosis was found to be higher than a sample of coal from a Utah mine with low incidence (SORENSON et al. 1974). The cadmium content was found to be the same in both coals.

"Labrador lung", an unusual mixed (iron, silica and anthophyllite) dust pneumoconiosis from mining and milling iron-containing rock in the Carol

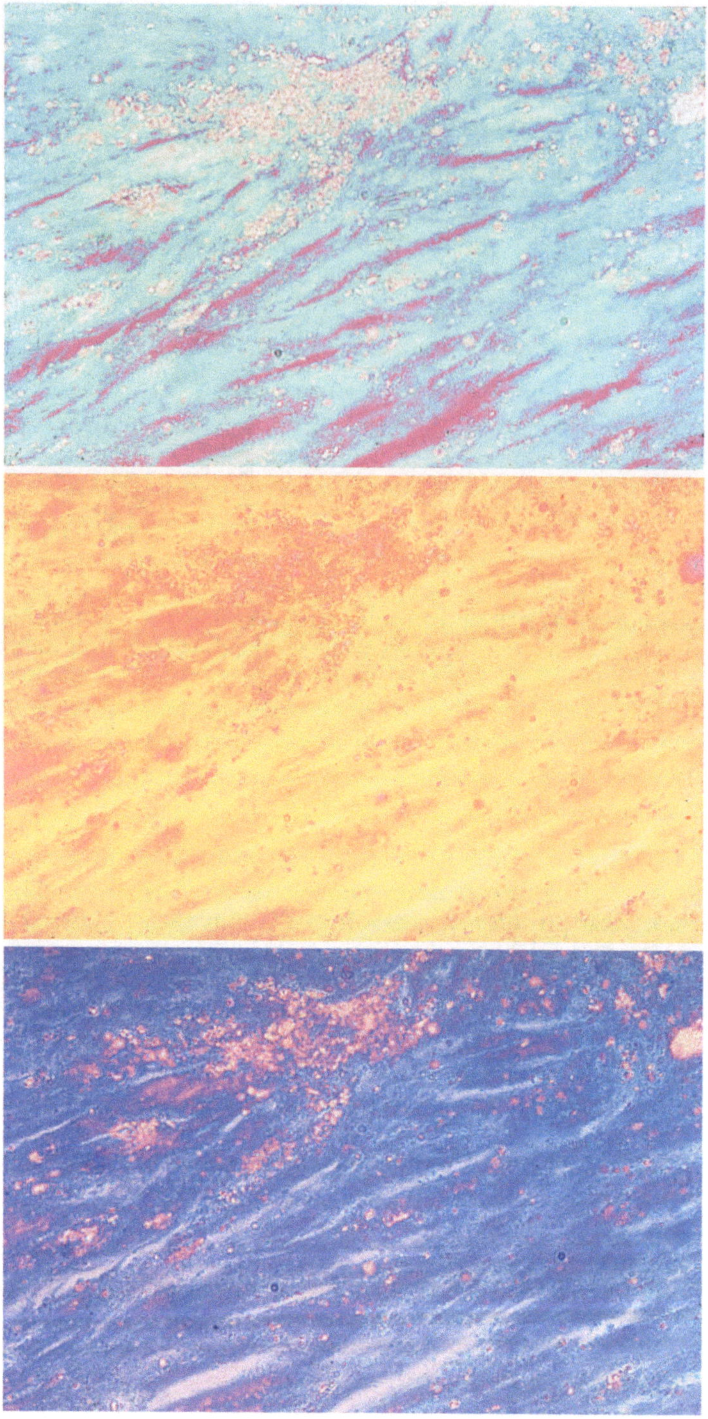

Fig. 203. Autopsy specimen (Gelsenkirchen S. 58/53) of a silico-tuberculous ex-miner who had worked underground from 1907–1924 and from 1935–1949 as a fire-brick layer. Large quantities of stone-dust in an area of massive fibrosis. Fixed in formalin, paraffin section, embedded in glycerol. Interference contrast Fl 100×/1.35/∞/–

Lake area of western Labrador in 13 lung biopsy specimens showed widespread focal fibrosis, large amounts of haemosiderin and silica and many ferruginous bodies, one containing a core of anthophyllite (Edstrom and Rice 1982). A granulomatous reaction was seen in two biopsy specimens and typical silicotic nodules were found in two others. The relatively brief exposure, especially in one case (11 months), and the short latent interval before the development of the pneumoconiosis are reasons for concern for the health of iron ore workers.

In Cornish China clay workers there is no doubt that there is a risk of disabling pneumoconiosis, but it is limited to a quite small proportion of the workers (HALE 1960). However, cases with occupational histories free from any complicating occupation were found, with nodular mottling (comparable to silicosis in appearance) and with respiratory disability. Others showed similar changes with the addition of progressive massive fibrosis comparable to that seen in coal-miners. On of the former type developed complicating tuberculosis, died a year after the first examination, and post-mortem showed emphysematous lungs, and the presence of nodulation, the nodules being up to 6 mm in diameter.

In a South African pottery quartz concentrations in excess of 0.1 mg/m^3 were found in all sections of the manufacturing process from slip production to biscuit firing and sorting (REES et al. 1992a). The proportion of quartz in the respirable dust of these sections was 24% to 33%. This is higher than is usually reported in English potteries. A firm diagnosis of potters' pneumoconiosis was made in 11 of 358 workers radiographed (REES et al. 1992b).

Dust particles are the most important morphological detail to be demonstrated in pneumoconiotic specimens by interference contrast (SCHILLER 1974). The surprising quantity of crystalline matter may be understood if one imagines that the polarising microscope is unable to show all the anisotropic material at one but only successively by rotating the slide. There are large quantities of dust in the cytoplasm of alveolar macrophages, which may get the same dry mass concentrations as the erythrocytes.

Crystals and collagen differ in their refractive indices and thus in their phase displacements in the fringe field or in the interference contrast colours (SCHILLER 1974). Coal dust is opaque and thus does not give any change of colour using the wedge compensator.

Inducible nitric oxide synthase (iNOS; EC 1.14.13.39) mRNA in bronchoalvelar lavage cells (using reverse transcriptase-polymerase chain reaction) and the magnitude of $^{\bullet}$NO-dependent chemiluminescence from alveolar macrophages correlated with the pathology of silica-exposed coal miners' chest radiographs (CASTRANOVA et al. 1998).

Erythrocyte Cu^{2+}/Zn^{2+} superoxide dismutase (EC 1.15.1.1) activity was significantly higher in a group of 34 underground coal miners from Lorraine than in a group of 30 surface workers (PERRIN-NADIF et al. 1996).

Superoxide dismutase (EC 1.15.1.1)

$$2\,O_2^{\bullet -} + 2\,H^+ \longrightarrow H_2O_2 + {}^3\Sigma_g^- O_2 \qquad [128]$$

Glutathione peroxidase (EC 1.11.1.9) activity increased in the plasma of 33 retired miners with pneumoconiosis, compared with 21 retired miners without pneumoconiosis (PERRIN-NADIF et al. 1996).

Decreased glutathione S-transferase (EC 2.5.1.18) was found in erythrocytes of subjects with early stages of coal workers' pneumoconiosis (International Labour Office classification 0/1–1/2) when compared with control miners (EVELO et al. 1993). At further progression of coal workers' pneumoconiosis (≥2/1), the activity of glutathione S-transferase was not different from controls. In the same group with moderate coal workers' pneumoconiosis a decrease in reduced glutathione in red blood corpuscles occurred. Decreases in glutathione S-transferase activity in early stages of coal workers' pneumoconiosis, as well as the decreases in glutathione peroxidase (EC 1.11.1.9) activity and in reduced glutathione concentrations (ENGELEN et al. 1990), may originate from damage caused by reactive oxygen species.

Glutathione S-transferase (EC 2.5.1.18)

$$RX + Glutathione \longrightarrow HX + R\text{-}S\text{-}G \qquad [129]$$

Glutathione S-transferases are a superfamily of enzymes that function to catalyse the nucleophilic attack of glutathione (GSH) on electrophilic groups of a second substrate. Glutathione S-transferases are present in many organs and have been implicated in the detoxification of endogenous α,β unsaturated aldehydes, including 4-hydroxynonenal.

T-Helper lymphocytes in pottery ($P < 0.05$) and B cells in foundry ($P < 0.01$) workers were significantly lower when compared with their controls (BAŞARAN et al. 2002). In addition, silica-exposed foundry workers had a significant reduction in the IgG, IgA and IgM levels.

Acute silicosis is an entity induced both by high concentration of freshly broken micronized silica in tunnelling (UEHLINGER 1950, GRABER 1952) and alkaline-reactivated silica in the manufacture of scouring powder (MACDONALD et al. 1930, GIESE 1931, CHAPMAN 1932, GERLACH and GANDER 1932, GÖRNHARDT 1933, MIDDLETON 1936, RITTERHOFF 1941). Acute silicosis from sand blasting showed plenty of opaque components linked with the inhalation of iron (TERBRÜGGEN and MOHNKE 1953). In another sandblaster MICHEL and MORRIS (1964) found considerable irregular brownish pigmentation both positive and negative for stainable iron. A man aged 39 years, who handled silica, barium sulphate, talc, and dolomite in a dusty environment showed intraalveolar accumulation of a granular proteinaceous substance, presumable in macrophages (ROESLIN et al. 1980). The alveolar walls were infiltrated by lymphocytes as plasma cells and contained many silicotic micronodules with hyaline centres. In the hilar lymph nodes there was some fibrosis more in the medullar than in the marginal sinus.

On account of 19 case-records on sand blasters' silicosis from the archives of the Institute of Industrial Medicine in Lyons GALY et al. (1956) felt compelled to emphasize the extreme severity in the development of silicosis among those workers who despite of disposal had been working without adequate protection against the hazard of this trade.

The **prophylactic action of aluminium** on the development of silicotic fibrosis was first claimed by DENNY et al. (1937, 1939a, b). Using the intratracheal injection technique of KETTLE and HILTON (1932), BELT and KING (1943) and KING et al. (1945) were unable to substantiate this. Inhalation experiments by KING et al. (1950) showed a retardation of about 200 days of the development of silicotic lesions in the lungs of *rats* by the inclusion of 2 % metallic aluminium powder in the quartz. ULMER (1964) and ULMER et al. (1964), in a pilot study with $AlCl_3$ demonstrated the virtually complete protective action of aluminium against the fibrogenic and alveolar damaging effects of inhaled silica dust. McIntyre aluminium powder, however, did not inhibit pulmonary fibrosis (WELLER et al. 1966), while powdered aluminium by itself produced collagen as measured by total hydroxyproline of the lung.

Aluminium chlorohydroxy-allantoinate reduced experimental silicosis and anthracosilicosis in *rats* (LE BOUFFANT et al. 1977) and *monkeys* (LE BOUFFANT et al. 1984). In coalworkers' pneumoconiosis, the compound delayed progression during a 5-years therapy (LE BOUFFANT et al. 1984).

Aluminium lactate instillation into the *ovine* tracheal lobe suppressed the increase in phospholipids induced by silica dust (BÉGIN et al. 1989).

The type of pneumoconiosis in gold miners from the Kolar Gold Field (FFRENCH 1955) and tracheal instillation experiments in *rats* (RAY et al. 1952) of dust suspensions from these mines showed that the presence of a high Al_2O_3 content in the dust may be of great significance as exerting an antidotal effect on the quartz.

In persons professionally exposed to aluminium dust, KAHLAU (1941, 1942, 1947/48, KOELSCH 1941, 1942) and KIRCH (1943, 1942/43) described several cases of fatal lung fibrosis, and EDWARDS (1947), therefore, firmly demanded that "its use should be allowed only under the supervision of expert observers."

The relationship between silica and the interferon system was studied by HAHON and BOOTH (1987) in *monkey* kidney cell monolayers. Minerals within the classes nesosilicate, sorosilicate, cyclosilicate, and inosilicate exhibited either little or marked (50 % or greater) inhibition of interferon induction. Within the inosilicate class, however, minerals of the pyroxenoid group (wollastonite, pectolite, and rhodonite) all significantly showed a two- to threefold increase in interferon production. Silicate materials in the phyllosilicate and tectosilicate classes all showed inhibitory activity for the induction process. When silicate minerals were coated with poly(4-vinylpyridine-*N*-oxide), the inhibitory activity of silicates on viral interferon induction was counteracted.

14.5.3 Asbestosis

Asbestosis must be defined as pulmonary parenchymal fibrosis resulting from inhalation of considerable amounts of, and usually after prolonged ex-

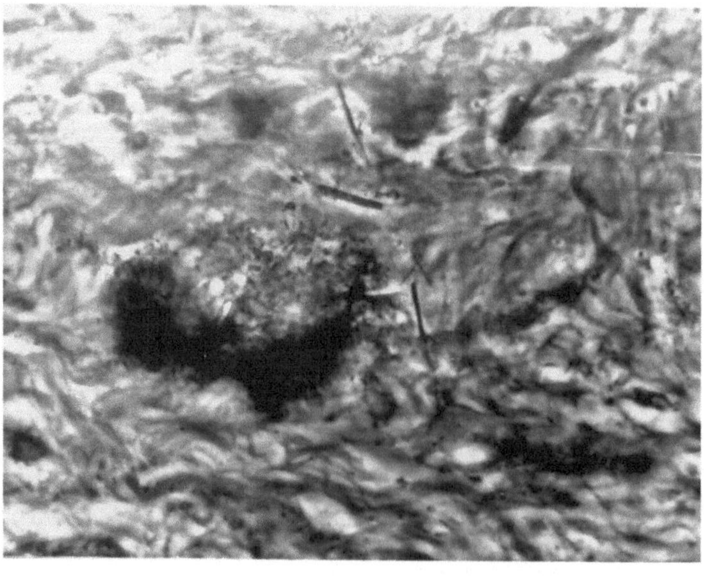

Fig. 204. Asbestos fibres in the lung of a *man* aged 46 years, who has worked in the preparation plant of an asbestos factory from 1931 to 1934. In 1934 he was compensated for asbestosis. In 1948 he grew dyspnoeic. The post-mortem in March 1949 besides an asbestosis showed pneumonia. Material kindly provided by Prof. J. Gough †, Cardiff. Paraffin, hæmatoxylin and eosin. Eyepiece Leitz Periplan 4×; objective Leitz PvFl Oil 70/1.15. Phase contrast. Positive Zernike. (from SCHILLER 1954)

posure to, asbestos fibres, and in which asbestos bodies and fibres are present in the tissues and can usually be seen easily in 5 μm haematoxylin- and eosin-stained sections of lung.

Monoclonal antibody analysis with antibodies that detect surface antigens on the majority of circulating blood monocytes but only on a minority of mature alveolar macrophages demonstrated that an increased proportion of alveolar macrophages of asbestos workers expressed monocyte lineage antigens, suggesting the presence of "young" newly recruited macrophages and thus enhanced recruitment (SPURZEM et al. 1987). Culture of alveolar macrophages of these subjects with [^{3}H]thymidine followed by autoradiography demonstrated an increased proportion of alveolar macrophages synthesising DNA, suggesting the macrophages are replicating at an increased rate in situ.

Bronchoalveolar lavage fluid from patients exposed to crocidolite in Western Australia contained mitogens for *human* lung fibroblasts, but platelet-derived growth factor (PDGF)-BB, tumour necrosis factor-α, insulin-like growth factor-1, or interleukin-1β did not contribute to this activity (MUTSAERS et al. 1998). Some stimulation of proliferation was observed for the PDGF-AB antibody ($P < 0.05$).

The most carcinogenic forms of asbestos, crocidolite and amosite, contain up to 27% **iron** by weight as part of their crystal structure. These minerals can acquire more iron after being inhaled, thereby forming asbestos bodies. SHEN et al. (2000) reported a method for depositing iron on asbestos fibres *in vitro* which produced iron deposits of the same form as observed on asbestos bodies removed from *human* lungs. To assess the effect of long-term binding, crocidolite was incubated in $FeCl_2$ or $FeCl_3$ and amosite in $FeCl_3$ for 14 days. X-ray diffraction suggested that ferrihydrite, a poorly crystallised ferric oxide, had formed. X-ray diffraction also showed that ferrihydrite was present in amosite-core asbestos bodies taken from *human* lung. Auger electron spectroscopy confirmed that Fe and O were the only constituent elements present on the surface of the asbestos bodies, although H cannot be detected by Auger electron spectroscopy and is presumably also present.

Table 44. Types of asbestos bodies (SANO 1961)

A:	Club or rod shape
B:	Dumb-bell shape
C:	Segmented form
D:	Beads or necklace like form
E:	More complicated form du to the process of its destruction

Nickel, which is contained in serpentine soils, was hyperaccumulated by the crucifer plant *Thlaspi montanum* into its tissues. BOYD and MARTENS (1994) have shown that such hyperaccumulation is defensive against *herbivores*. Leaves differing 167-fold in Ni content (3000 vs. 18 ppm) were obtained from plants growing on high- and low-Ni soils. *Steptanthus polygaloides*, another serpentine plant accumulating Ni was protected from pathogen infection (BOYD et al. 1994). It is suggested that Ni is simply coordinated with histidine, since supplying histidine to a non-accumulating plant greatly increased its nickel tolerance and capacity for nickel transport to the soot (KRÄMER et al. 1996).

Blood **copper** levels were significantly ($P < 0.01$) elevated in 11 out of 13 patients with asbestosis (Avolio et al. 1988).

In *man*, it is commonly believed that the differences in the pathogenicity of chrysotile compared to amosite or crocidolite asbestos reflect low **pulmonary retention** of chrysotile compared to amphiboles. MCCONNOCHIE et al. (1987) and CHURG et al. (1993) suggested that the associated tremolite contaminant rather than the chrysotile fibres themselves might be the actual agent of disease induction, particularly for mesothelioma. This theory is consistent with the large amount of tremolite retained in the lungs of chrysolite miners and millers, and with the known greater propensity of amphiboles to induce disease, particularly mesothelioma (MCDONALD and MCDONALD 1986). The alternative hypotheses that transient chrysotile or sequestered retained chrysotile are responsible, cannot be ruled out. If one or both of the latter are true then tremolite may be regarded as an internal dose marker for chrysotile exposure.

Penetration and migration of asbestos in *humans* were detected by COOK and OLSON (1979) after ingestion of drinking water contaminated with 5×10^7 fibres of amphibole asbestos per litre. BOATMAN et al. (1984) did not find significant levels of asbestos in the urine of humans exposed to 2×10^8 chrysotile fibres per litre. CARTER and TAYLOR (1980) identified amphibole asbestos in lung > liver > jejunum of persons exposed to a high oral intake of the mineral.

NAYEBZADEH et al. (2001) used a transmission electron microscope, equipped with an x-ray energy-dispersive spectrometer, to analyse lung mineral fibres of 86 subjects from Thetford-Mines and Asbestos regions and to classify fibre sizes into three categories. The most consistent difference was the higher concentration of tremolite in lung tissues of workers from Thetford-Mines, compared with workers from the Asbestos region. Amosite and crocidolite were also detected in lung tissues of several workers from the Asbestos region. No consistent and biologically important difference was found for fibre dimension; therefore, fibre dimension did not seem to be a factor that accounts for difference in incidence of respiratory diseases between the two groups. Among the mineral fibres studied, retention of tremolite fibres was most apparent.

Actin, the contractile protein within cells, may be responsible for movement of asbestos particles through the epithelium to the lung interstitium where the fibres react with macrophages and fibroblasts (Brody et al. 1983).

Patients with asbestosis seem to have a high protection against neutrophil elastase at the alveolar level, not totally due to α_1-proteinase inhibitor. Electrophoretic studies by Scharfman et al. (1989) demonstrated a functional activity of only a part of immunoreactive α_1-proteinase inhibitor. Therefore, there is no doubt that other inhibitors of neutrophil elastase are present at the alveolar level, one of them being the *human* mucus proteinase inhibitor (or "bronchial inhibitor") but it represents only 14 % of α_1-proteinase inhibitor molar concentration (Boudier et al. 1987).

It has been suggested that sharp fibres in the lungs penetrate the pulmonary **pleura** during respiration and pass directly into the parietal layer, possibly causing traumatic microhemarrhages and fibrin deposition (Heard and Williams 1961, Thomson 1970). Taskinen et al. (1973) discussed lymphogenic metastasis of dusts to the intercostal lymphatic vessels.

Acute asbestosis reported by Jacobson (1936) in a 25-years old machine operator who had been welding with asbestos-covered electrodes seemed to be doubtful as there were no asbestos fibres or asbestos bodies found in the sputum. Three cases seen by Bohlig et al. (1960, p. 147) were more likely due to old age, as the jobs were begun at 55, 60, and 63 years, respectively.

Treatment of asbestosis and idiopathic pulmonary fibrosis with colchicine (0.6 mg orally for 12 weeks) resulted in declines in dyspnoea index, selective improvement in several high-resolution computed tomography scans, but no statistically significant changes in bronchoalveolar lavage cells, cytokines, fibronectin, or hydroxyproline (Addrizzo-Harris et al. 2002). However, there was a decline in hydroxyproline in the bronchoalveolar lavage fluid in 8 of 10 patients.

Carcinoma of the lung was correlated with a markedly elevated asbestos content of lung tissue (Roggli and Sanders 2000). Asbestos content was recorded as total asbestos fibres, commercial amphibole fibres, noncommercial amphibole fibres, and chrysotile fibres 5 µm or greater in length per gram of wet lung tissue.

Wagner (1997) stressed that the majority of **mesotheliomas** occurred after prolonged exposure to large quantities of fibre, a situation rarely existing today, and that at least 10 % of these diffuse mesotheliomas occurred without exposure to asbestos dust. Some were reported before the widespread use of asbestos. Peterson et al. (1984) and Pelnar (1988) reviewed mesotheliomas of other causes than asbestos exposure. However, it has not been possible to make a realistic estimate of the frequency of mesotheliomas which are not due to fibre exposure because, in Western countries, everyone is exposed to some fibres and these fibres will be found in the lung at autopsy whether they are the cause of the mesothelioma or not (Elmes 1994).

Mesothelial tumours are pleomorphic. There is much diversity from tumour to tumour, and in different areas of individual tumours and their metastases, because both epithelial and connective tissue (mesenchymal) elements are usually present (Suzuki et al. 1976). Four distinct histological patterns are recognised (Table 45).

Transmission electron microscopy has been found to be of greater value than scanning electron microscopy in the differential diagnosis of mesothelial reactions, malignant mesothelioma and metastatic carcinoma (Butler and Johnson 1980).

The ultrastructure of neoplastic cells of the epithelial forms varies from well differentiated (marked polarity, microvilli, glycogen granules, junctional structures, tonofilaments, intracellular vacuoles) to poorly differentiated (which lack some of these epithelial characteristics) (Dardick et al. 1984). Mitochondria-rough endoplasmic reticulum complexes consisting of alternated elongated cisternae of rough endoplasmic reticulum and strandlike mitochondria have been seen in one case of malignant pleural mesothelioma (Coleman et al. 1989). Perhaps the most characteristic indicator of mesothelial differentiation is the presence of myriad slender sinuous microvilli, sometimes secondary and tertiary branching (Kobzik et al. 1985, Dardick et al. 1987). Vimentin does not appear to be a simple discriminatory marker of malignant mesothelioma (Jasani et al. 1985). The prekeratin antibody stains the epithelial component; the vimentin antibody decorates the sarcomatous mesenchymal one (Altmannsberger et al. 1982). Calretinin

Table 45. Histological patterns of mesothelial tumours (Whitwell and Rawcliffe 1971)

1. Tubulopapillary type (predominantly epithelial)
2. Sarcomatous type (predominantly mesenchymal)
3. Undifferentiated cell type (predominantly epithelial)
4. Mixed type

proved as a relatively sensitive marker for 97% of the epithelioid and 19% of the sarcomatoid mesotheliomas, while in 90% of pulmonary adenocarcinomas immunohistochemistry failed to decorate calretinin antigen in the tumour cells (WIETHEGE et al. 1997). In a study by CARELLA et al. (2001) 40 of 46 (87%) mesotheliomas were positive with calretinin, 29 (63%) with thrombomodulin, 40 (87%) with cytokeratins 5/6, and 41 (89%) with high-weight cytokeratins; five (11%) were focally reactive with MOC 31, 4 (9%) with Ber-EP4, ad 2 (4%) with carcinoembryonic antigen (CEA; indicators of lung adenocarcinomas).

S-100 protein immunoperoxidase staining may be of use when epithelial and biphasic malignant mesothelioma vs. poorly differentiated carcinoma is a diagnostic problem (RASMUSSEN and LARSEN 1985). *Human* milk fat globulin 2 positivity in a tumour negative for both carcinoembryonic antigen and pregnancy-specific B_1 glycoprotein wood be good confirmatory evidence of malignant mesothelioma (GIBBS et al. 1985).

The **neurofibromatosis 2 (NF2) tumour suppressor gene** was implicated in the genesis of *human* mesothelioma. Transfection of the NF2 gene into NIH/3T3 cells inhibited their growth (LUTCHMAN and ROULEAU 1995) and reversed the v-Ha-*ras*-induced malignant phenotype (TIKOO et al. 1994). KLEYMENOVA et al. (1997) examined *rat* chrysotile-induced primary mesothelioma for alterations in this gene. Their data suggested that the role of NF2 in the development of rodent asbestos-induced mesothelioma may differ significantly from the role in *human* disease.

Crocidolite from the Cape asbestos mines, which contains much iron, was implicated for the mesotheliomas of pleura, and occasionally of peritoneum, observed by WAGNER (1963, 1986) in the workers and those living in the vicinity of the mines. The exposure could be as short as 3 months in some cases. There was a long interval between the first exposure and tumour development, and the majority of the cases had paraoccupational exposures. Cases of pure crocidolite exposure have occurred among women who inserted the crocidolite pads into the canisters of the respirators prepared for the Ministry of Defence during the last World War; in contrast, so far no cases have been recorded from the people involved in packing the civilian respirators, which contained chrysotile (WAGNER 1986). A 46-year-old woman who had worked for 5 years in a windowless decorator's studio polluted with crocidolite sprayed to steel ceilings died from a pleural mesothelioma (SCHNEIDER et al. 2001). A fibre analysis by light microscopy showed 3162 ferruginous bodies per gram of wet lung tissue.

From the data presented by BERRY (2002) it is clear that the amount of crocidolite in the lungs of mesothelioma cases in the United Kingdom has decreased between 1976+1977 and 1990–1996, but the amount of amosite has changed much less. These findings are consistent with the use of amphibole asbestos in the UK. Importation of crocidolite was discontinued by 1979 whilst the importation of amosite continued with little decrease throughout the 1970s (PETO et al. 1995). McDONALD et al. (2001) noted that >90% of the mesothelioma cases started work before 1970 and mainly before 1965 and so many have been exposed to crocidolite in that period.

Surface thermodynamic properties of mesothelioma cells exhibiting either sarcomatous or epitheliomatous growth were studied *in vitro* by GREEN et al. (1990). These properties were correlated with changes in **cell adhesion to substrates.** A two-phase aqueous polymer system (4% dextran T2000/4% PEG, 20000) was used to measure the contact angle of droplets on the dextran-rich phase made on a cell monolayer. This technique is sensitive to changes in the hydration of the cell surface, and also appears to correlate with the ability of a cell to adhere to a substrate. The contact angle of the epitheliomatous mesothelioma was 28.6°±3.6 S.D., n = 10 which was significantly lower than that obtained on normal mesothelium 59.6°±4.34 S.D., n = 8, or sarcomatous mesothelioma cells (61.6°±3.0, S.D. = 10). Interference reflection microscopy was used to study the close adhesion and focal contact pattern of the two phenotypes of the mesothelioma. The sarcomatous cells demonstrated significantly greater adhesion to glass cover at 48 h incubation period than the epitheliomatous cells. Scanning electron microscopy demonstrated marked differences in morphology between the two cell types. The sarcomatous cells tended to become long and spindle shaped and exhibited strong substrate adhesion in culture. By contrast the epitheliomatous cells formed free floating clusters in culture, and showed numerous microvilli. The two cell lines also exhibited differences in metastasis assays indicating that the surface changes may have biological significance.

While **manganese superoxide dismutase**, a superoxide radical-scavenging mitochondrial enzyme, showed low reactivity in human pleural mesothelial cells, malignant pleural mesothelioma cells expressed high levels of MnSOD *in situ* (KAHLOS et al. 1998). The results of Northern blotting, Western blotting, immunochemistry, and the specific activity of MnSOD were consistent with findings of KINNULA et al. (1996). The M38K mesothelioma cell line was the most resistant cell line to both oxidant and epirubicin exposure, and this same cell line had not only the highest MnSOD activity but also the highest catalase activity. *Human* pleural mesothelioma (M14K) cells contained higher basal MnSOD activity than A549 lung adenocarcinoma cells (28.3±3.4 vs. 1.8±0.3 U/mg protein), and MnSOD activity was significantly induced by tumour necrosis factor-α only in A549 cells, but the induction did not offer any protection during subsequent oxidant or drug exposure (Järvinen et al. 2000).

Both **cyclooxygenase-2** and **inducible nitric oxide synthase** were expressed in 30 *human* mesothelioma tissues but were not detectable in nonreactive mesothelial tissues from the same individuals (Marrogi et al. 2000). *In vitro* exposure of *human* mesothelioma cell lines to the COX2 inhibitor, NS398, revealed dose- and time-dependent antiproliferative activity, whereas the NOS2 inhibitor, 1400 W, had no detectable inhibitory effect.

Distinguishing malignant mesothelioma from reactive mesothelial hyperplasia and reactive fibrosis can be a diagnostic problem in small pleural biopsies, made more difficult by the recent recognition of mesothelioma-*in situ* (Whitaker et al. 1992, Henderson et al. 1998). Diffuse linear staining for epitheilal membrane antigen is more specific for malignancy than p53 in differentiating reactive mesothelium from mesothelioma, and more so in epithelioid than in sarcomatoid elements (Cury et al. 1999). However, neither is completely specific.

Simian virus (SV) 40 and SV40-like DNA sequences have recently been detected in several types of *human* tumours, including malignant mesothelioma. However, the presence of SV40 DNA sequences is not sufficient to account for its possible role in tumour development because the viral proteins must be expressed and ultimately impair the function of relevant cell proteins, such as p53 and pRb. Pilatte et al. (2000) investigated SV40 large T antigen (SV40 Tag) protein expression in mesothelioma cell lines, established in their laboratory by Western blotting, immunoprecipitation, and immunocytochemistry using Tag-specific *mouse* monoclonal antibodies (mAbs) Ab-1 (or Pab 419). By Western blotting of cell extracts, none of the mesothelioma cell lines expressed detectable amount of SV40 Tag. However, They found that Ab-1 as well as Pab-101, another SV40 Tag-specific mAB, may generate false-positive signals due to the fact that both antibody preparations were contaminated by a protein of similar size (90 kDa) as SV40 Tag and reacted with the various secondary horseradish peroxidase-conjugated anti*mouse* immunoglobilin Gs tested.

While healthy pleural mesothelium from nonsmokers was negative or very weakly positive for all **peroxiredoxins**, in mesothelioma the most prominent reactivity was observed with peroxiredoxins I, II, V, and VI (Kinnula et al. 2002). Peroxiredoxin I was highly or moderately expressed in 25/36 cases, the corresponding figured for peroxiredoxins II–VI being 27/36 (Prx II), 13/36 (Prx III), 2/36 (Prx IV), 24/36 (Prx V), and 30/36 (Prx VI). Positive staining was observed both in the cytosolic and the nuclear compartment, with the exception of Prx III, which showed no nuclear reactivity.

Prognosis is poor but is somewhat better for patients with epithelioid tumours than for patients with fibrous mesotheliomas and biphasic tumours (Griffiths et al. 1980). There was no significant correlation between prognosis and immunoexpression of VEGF ($P = 0.07$), FGF-1 ($P = 0.3$), or TGFβ ($P = 0.1$), or between intra-tumoural microvascular density and any of the cytokines studied individually (Kumar-Singh et al. 1999).

Mortality trends from mesothelioma for French men aged 50 to 79 years continue to increase, reaching a peak averaging between 1140 (optimistic scenario) and 1300 deaths (pessimistic scenario) annually around the years 2030 and 2040, respectively (Banaei et al. 2000). In Japan, from 1958 to 1996, a total of 1,846 (0.17 %) malignant mesothelioma cases (1,287 male, 558 female, 1 unknown) were registered among 1,056,259 autopsy cases (Murai 2001). The frequency of mesothelioma (number of cases/total autopsy cases) was 0.10 % (461/440,334) for the term 1958–1979, 0.18 % (716/390,124) for 1980–1989, and 0.30 % (669/225,801) for 1990–1996; the frequency of cases increased significantly over the time periods ($P > 0.0001$). Among 1,785 cases for which tumour sites were ascertained, there were 1,213 pleural mesothelioma (68 %), 431 peritoneal (24.1 %), 108 pericardial (6.1 %), 6 tunica vaginalis testis (0.3 %), and 28 "others" (1.6 %).

14.5.4 Pneumoconiosis from Glass Fibres

In the *human* lung, fiber glass after periods which ranged from 16 to 32 years caused no demonstrable gross or microscopic pulmonary damage (Gross et al. 1971).

From the *rat* lung, glass fibres with diameters ≤3 μm and lengths ≤10 μm are cleared quite efficiently, presumably by macrophage-mediated processes (Morgan et al. 1982). The fact that glass fibres are leached *in vivo* suggests that they are unlikely to produce in *man* the pathological effects associated with amphibole fibres. Otherwise with much larger doses of the same fibres, BERNSTEIN et al. (1980) noted granulomatous response. KAHLAU (1947/48) described a case of fatal pneumonia after insulating an emergency dwelling by glass wool. In the Syrian hamster, very fine and durable glass fibres can induce lung carcinomas and not only mesotheliomas (POTT et al. 1984). The prevalence of malignant tumours in *rats*, induced by intraperitoneally injected dusts generated by friction products using a man-made mineral fibre base, did not differ from that obtained from the injection of asbestos-resin and asbestorubber dusts (KOGAN et al. 1994).

14.5.5 Silicon Carbide Pneumoconiosis

MASSÉ et al. (1988) in 3 patient aged 60 to 72 years, who had worked in a silicon carbide plant for 30 to 40 years, found a mixed pneumoconiosis:

- abundance of intraalveolar macrophages associated with a mixture of inhaled particles including carbon, silicon, pleomorphic crystals, silicon carbide, and ferruginous bodies showing a thin black central core;
- nodular fibrosis, generally profuse, containing silica and ferruginous bodies and associated with large amount of carbon pigment;
- interstitial fibrosis, less prominent than the nodular form;
- carcinoma in two cases.

14.5.6 Siderosis

Inhalation of iron and iron oxide dusts free of silica was considered as innocent (cf. SCHILLER 1961, WORTH 1969, KÖNN et al. 1976) until ALGRANTI et al. (1985) observed interstitial fibroses in workers engaged in milling and filling plants for iron oxide. HUEPER (1961, p.338) stated pulmonary siderosis as a stigma of exposure and thinks the production of sarcomas in *rats* and *mice* by injecting an iron-dextran complex (RICHMOND 1959, HADDOW and HORNIG 1960) to be a sufficient indication to pay attention to this problem. HUEPER and PAYNE (1962) obtained granulomas with large amounts of extra- and intracellularly located iron oxide (sideromas), when they implanted *rats* with pellets containing 25 mg of finely powdered metallic iron obtained by degradation of iron carbonyl and suspended in 50 mg of wool fat, but there was no tendency to a malignant change.

14.5.7 Stannosis

There are two dusty recovery products in tin smelting, one called 'fume' from smelting furnace, and another called 'after-stuff' from the refinery, which are saved and used again in the calcine furnace.

In 19 workers exposed to tin dust (1.0% SiO_2) and fumes for <3 to >12 years all 10 men exposed for more than 3 years showed X-ray changes but were practically asymptomatic and their vital capacities, maximal breathing capacities, resting minute volumes and respiratory reserves were within normal ranges (OYANGUREN et al. 1957).

The lungs of a *man* who had worked on the furnaces for over 40 years, had retired for the last 20 years, and was virtually normal died at an age of over 80 years showed dust with no fibrosis (ROBERTSON 1960).

14.5.8 Hard Metal Pneumoconiosis

Granulomatous pulmonary foci with lymphoid follicles, round follicular groups of histiocytes, and a patchy infiltration of eosinophils and much well advanced fibrous tissue had developed in a 33-year-old male who had worked as a precision grinder, grinding a hard metal compound (stellite) with a silicone carbide wheel and without any exhaust ventilation (JOSEPH 1968). The chemical analysis of the lung biopsy showed only titanium, but no tungsten or cobalt.

Electron microscopy of a lung biopsy of a 49-years old white *man* who had spent 12 years in the manufacturing area revealed markedly altered alveolar lining cells and basement membranes (COATES and WATSON 1971). Normal appearing type I pneumocytes were absent.

Unusual cannibalistic multinucleated giant cells in air spaces as seen in giant cell interstitial pneumonia occurred in four of five cases exposed to cemented tungsten carbides and cobalt for 5 to 19 years (AUCHINCLOSS et al. 1992). By electron microscopy, the multinucleate giant cells in three patients who had restrictive ventilatory defects comprised both type II pneumocytes and multinucleate macrophages (DAVISON et al. 1983).

Urinary cobalt concentrations were exceeded in workers exposed to hard metal dust, and the coefficient of correlation between the cobalt concentrations in the air and in the workers' urine was 0.753 (LINNAINMAA and KIILUNEN 1997).

14.6 Experimental Granulomatous Inflammation

14.6.1 Scanning Electron Microscopy in Peritoneal Exudate Macrophages

After adherence to plastic dishes in an experimental culture system, an impressive number of genes associated with the inflammatory response are expressed, for example, IL-1β, TNF-α, and IL-8. In the 5'-upstream region of these inducible genes, the transcription factor NFκB binding site is commonly found (Juliano and Haskill 1993). The messenger RNA stability element, AUUUA, is also commonly found in the 3'-untranslated region of these mRNAs. Adhesion may be sufficient for activation of NFκB to induce these genes and to suppress the degradation rates of the mRNAs of these genes in monocytes. In *human* mononuclear cells the elastin receptor is linked to a pertussis toxin-sensitive G-protein which activated phospholipase C, resulting in a rapid mobilisation of the phosphoinositol pathway, activation of phosphokinase C, production of diacylglycerol, and a rapid increase in intracellular free calcium (Jacob et al. 1987). Another important function of the elastin receptor in mononuclear cells is the rapid increase in O_2 consumption accompanied by the extrusion of elastase and cathepsin G from polymorphonuclear leucocytes as well as the **liberation of superoxide radical anion** (Labat-Robert and Labat 2000). The repeating hexapeptide, Val-Gly-Val-Ala-Pro-Gly, present in seven copies in the *human* elastin gene, was shown to be chemotactic to monocytes and fibroblasts (Senior et al. 1984). Brehm et al. (1996) showed that the process of cell adherence per se generated reactive oxygen species with a peak about 20 min after transferring the suspended macrophages into the Biolumat test tubes; basal activity was regained approximately 100 min later. Another, although lower peak of lucigenin-enhanced chemiluminescence occurred about 20 min after twice washing by Veronal buffer to eliminate non-adherent cells and other materials.

14.6.2 Cotton Pellet Test

This rather simple test for the formation of granuloma tissue after a subcutaneous implantation of a cotton-wool-pellet (originally 40 mg; Meier et al. 1950) has been widely employed for testing anti-inflammatory drugs. This and similar methods have some disadvantages as lack of clear demarcation of the regenerating tissue from the adjacent structures, production of insufficient amounts of tissue for study, presence of foreign material in the tissue, and the development of foreign body reactions (Kwaan and Astrup 1964).

The initial phase of the inflammatory reaction to subcutaneous implantation of cotton pellets in *rats* is characterised by an increased permeability of the vessels in the connective tissue surrounding the implant, with the pellet undergoing a rapid saturation ('soaking') by the fluid escaping from the vessels (Meyer et al. 1953). Early permeability changes in the implant region are responsible for a transient transudative phase during the first three hours, followed by an exudative phase occurring between 3 and 72 h (Swingle and Shideman 1972). The growth of the granulation tissue depends on the proliferation of fibroblasts and the new connective tissue synthesis, the latter occurring approximately 4 h–4 days after the implantation of the cotton pellet (Swingle and Shideman 1972).

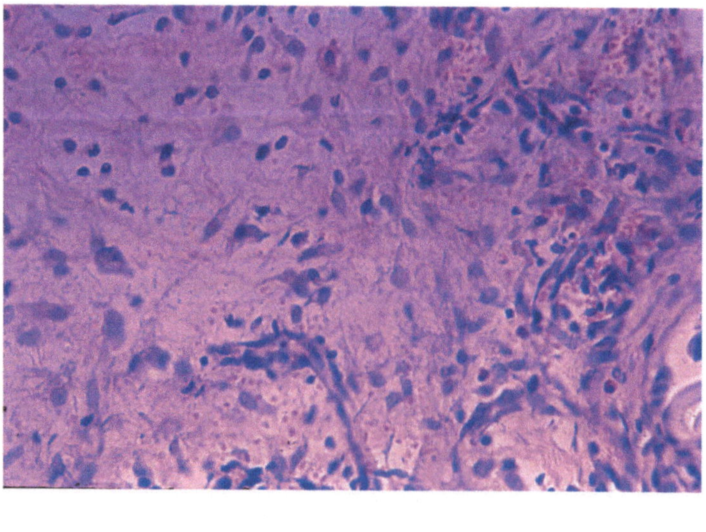

Fig. 205. Cotton pellet granuloma 7 days after subcutaneous implantation in a 120 g male Sprague-Dawley *rat* (No. 3; February 4, 1969). Fixed by immersion in Bouin's fluid. Paraplast 4 μm. Haematoxylin and eosin. Objective Leitz Pl 40/0.65. Leitz-Orthomat®. Agfachrome 50 L professional

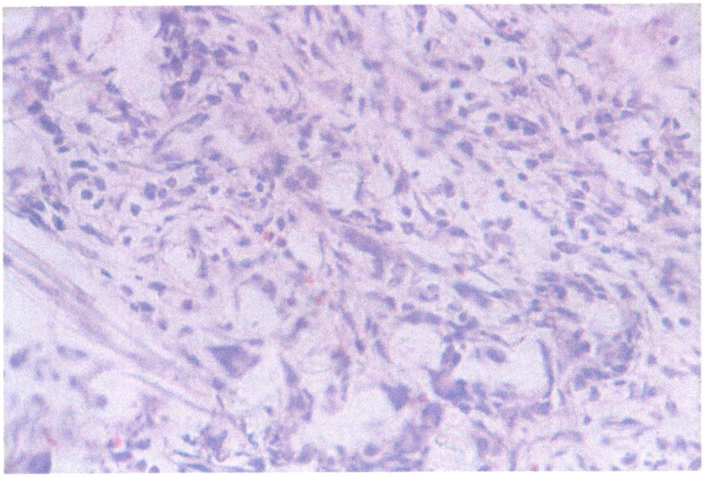

Fig. 206. Cotton pellet granuloma 7 days after subcutaneous implantation in a 120 g male Sprague-Dawley *rat* medicated with 15 mg carbocromene/kg × day for 14 days. The mesenchymal reaction is increased compared with an unmedicated control (Fig. 205). Giant cells are going to wrap the individual cotton fibres. Fixed by immersion in Bouin's fluid. Paraplast 4 µm. Haematoxylin and eosin. Objective Leitz Pl 40/0.65. Leitz-Orthomat®. Agfachrome 50 L professional

Drugs to be tested should always be administered by the oral route in order to avoid local irritation which may produce non-specific granuloma inhibition (Cygielman and Robson 1963).

14.6.3 Resorptive Inflammation of Implanted Gelatine Sponges

Curran and Clark (1964) implanted cylindrical "bullets", approximately 3 mm diameter and 5 mm long, of gelatine rendered insoluble by 24 hours' treatment with formalin, which was then removed by prolonged and repeated soaking in balanced salt solution.

A much larger surface of the implants can be gained by using a foamed preparation as applied for tampon in surgery (Marbagelan®). We used this model of inflammation in testing drugs for resorption of infarcted myocadial tissue (Schiller 1972, 1973, 1975). The accumulation of mononuclear phagocytes at the inner surface of the sponge 7 days after subcutaneous implantation is shown in Fig. 207.

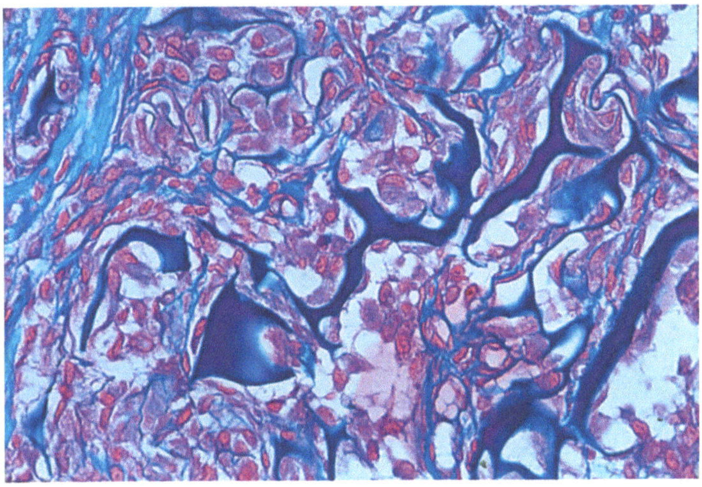

Fig. 207. Resorptive granuloma of a 18 mg Marbagelan® sponge 7 days after subcutaneous implantation into a male Sprague-Dawley *rat* of 220 grams. Fixation after Bouin. Paraplast section 4 µm. Azan stain using ruby fast red and aniline blue after Specht. Rhenohistol®. Objective Leitz Pl 40/0.65. Orthomat®. Agfachrome 50 L professional. – Mononuclear phagocytes join the trabecules of the inner surface of the Marbagelan® sponge

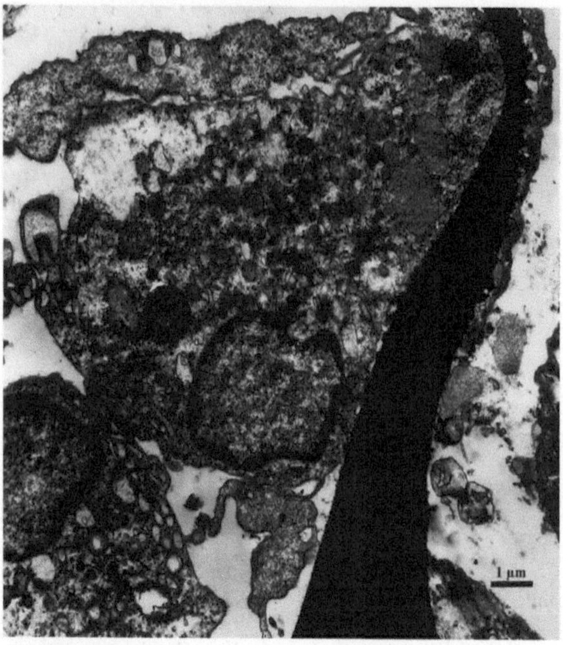

Fig. 208. Two mononuclear phagocytes from a Marbagelan®-induced (7 days) resorption granuloma (block 107) of a male Sprague-Dawley *rat* (No. 6). Under pentobarbital anaesthesia (30 mg/kg) the animal was perfused from the abdominal aorta with 2.5 % glutaraldehyde in 0.1 M sodium cacodylate buffer (pH 7.4). Postfixation with 1 % osmium tetroxide in sodium cacodylate buffer. Embedded in Epon 812 and sectioned at 50 nm. Lead citrate and uranyl acetate. Plate 2836

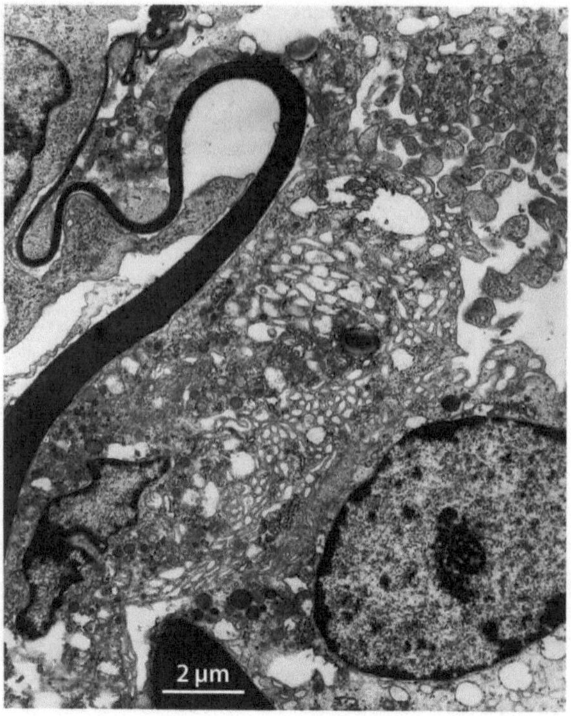

Fig. 209. Evagination of blebs of the nuclear envelope. Macrophage from a Marbagelan®-induced (7 days) resorption granuloma (block 107) of a male Sprague-Dawley *rat* (No. 6). Under pentobarbital anaesthesia (30 mg/kg) the animal was perfused from the abdominal aorta with 2.5 % glutaraldehyde in 0.1 M sodium cacodylate buffer (pH 7.4). Postfixation with 1 % osmium tetroxide in sodium cacodylate buffer. Embedded in Epon 812 and sectioned at 50 nm. Lead citrate and uranyl acetate. Plate 2835

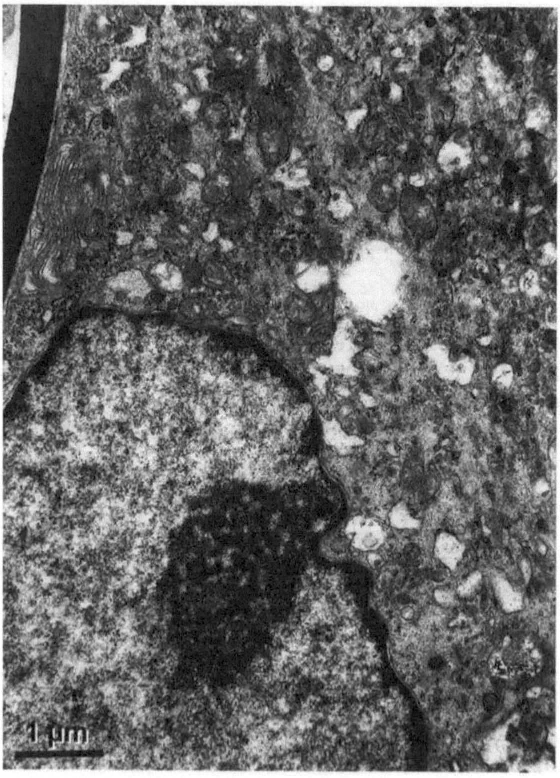

14.6.3.1 Giant Cells in Resorption Granulomas

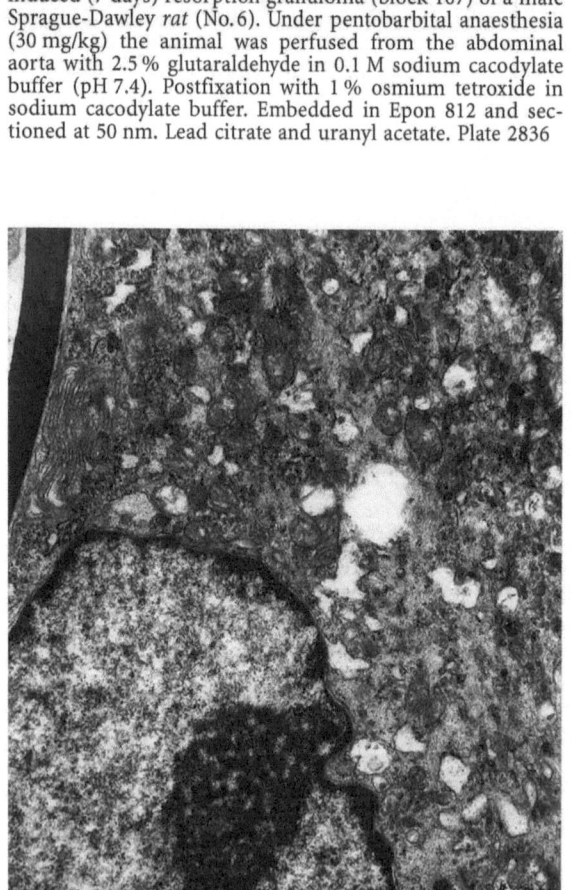

Fig. 210. Macrophage polykaryon from a Marbagelan®-induced (7 days) resorptive granuloma (block 107) of a male Sprague-Dawley *rat* (No. 6). Fixed by perfusion from the abdominal aorta under pentobarbital anaesthesia (30 mg/kg) with 2.5 % glutaraldehyde in 0.1 M sodium cacodylate buffer (pH 7.4). Postfixation with 1 % osmium tetroxide in cacodylate buffer. Embedded in Epon 812 and sectioned at 50 nm. Lead citrate and uranyl acetate. Plate 2829

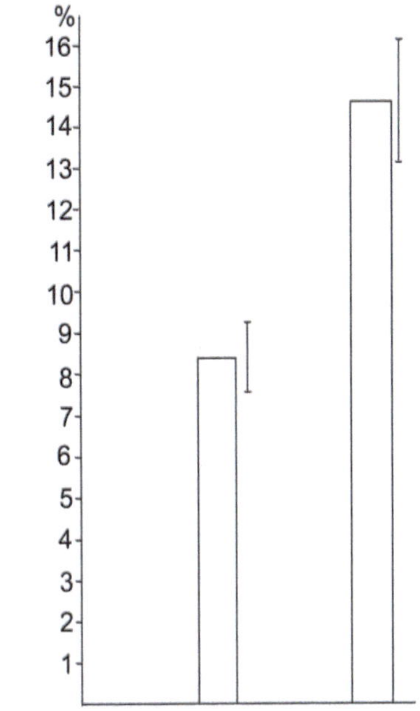

Fig. 211. Pre-treatment with 15 mg carbocromene/kg body weight × day for 7 days and further medication reduced the percentage of area of the Marbagelan® sponge 7 days after subcutaneous implantation in *rats*. Analyses were performed using the Leitz Classimat®

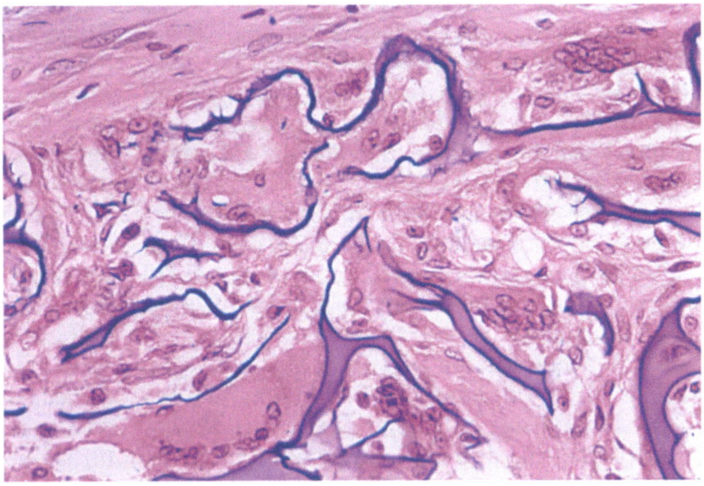

Fig. 212. Resorptive granuloma of a 18 mg Marbagelan® sponge 7 days after subcutaneous implantation into a male Sprague-Dawley *rat* of 220 grams treated for 14 days with intragastric injection of 15 mg carbocromene per kg body weight × day. Fixation after Bouin. Paraplast section 4 µm. Hæmatoxylin and eosin. Objective Leitz Pl 40/0.65. Orthomat®. Agfachrome 50 L professional. – The immigrated mononuclear phagocytes have combined to form multinucleate giant cells that resorb the gelatine

14.7 Ionising Radiation

Ionising radiation interacts with water giving rise to $HO^•$, H^+, and an electron, e^-. FRICKE and MORSE (1927) showed that it makes a great difference if the water is irradiated in the presence or in the absence of oxygen. In the absence of oxygen, small amounts of hydrogen peroxide are formed, together with small amounts of hydrogen; but in the presence of oxygen, much hydrogen peroxide is formed in the irradiated water. WEISS (1944) explained this fact by the assumption that in the absence of oxygen H and O–H react back to water, whereas in the presence of oxygen, the H-atoms are intercepted by the oxygen to form hydrogen peroxide, according to the equation:

$$\begin{matrix} H + OH \\ \qquad + O_2 = H_2O_2 + 2\,HO^• \\ H + OH \end{matrix} \qquad [263]$$

FRICKE showed furthermore that the hydroxyl radicals produced by the irradiation of water can be intercepted by Fe^{2+} ions, according to the equation:

$$Fe^{2+} + HO^• = Fe^{3+} + OH^- \qquad [264]$$

When the two kinds of interceptions were united and water was irradiated in the presence of both, of oxygen + ferro-ions, the following reactions occurred:

$$2\,H_2O = 2\,H + 2\,HO^• \qquad [265]$$
$$O_2 + 2\,H = H_2O_2 \qquad [266]$$
$$2\,Fe^{2+} + 2\,HO^• = 2\,Fe^{3+} + 2\,OH^- \qquad [267]$$
$$2\,Fe^{2+} + H_2O_2 = 2\,Fe^{3+} + 2\,OH^- \qquad [268]$$
$$2\,H_2O + O_2 + 4\,Fe^{2+} = 4\,Fe^{3+} + 4\,OH^- \qquad [269]$$

the net result being that 4 atoms of iron are oxidised, if two molecules of water are split.

Effects of $HO^•$ radicals on biological membranes include alterations in membrane proteins, peroxidation of unsaturated lipids accompanied by perturbation of the lipid bilayer polarity. BERROUD et al. (1996) have measured radiation-induced membrane modifications using two fluorescent lipophilic membrane probes (TMA-DPH and DPH) by the technique of fluorescence polarization of Chinese *hamster* ovary K1 and lymphoblastic RPMI 1788 cell lines. γ-Irradiation from a ^{60}Co source with dose rates of 0.1 and 1 Gy/min for a final dose of 4 and 8 Gy induced a dose-dependent decrease of fluorescence intensity and anisotropy of DPH and TMA-DPH in both cell lines but varied inversely with the dose rate. Moreover, the fluorescence anisotropy measured in lymphoblastic cells using TMA-DPH was found to decrease as early as 1 h after irradiation, and remained significantly lower 24 h after irradiation. The results of NAVARRO et al. (1997) in *human* breast cancer and nonsmall cell lung cancer indicated that glutathione redox ratio in blood can be used as an index of radiation-induced oxidative stress.

Low-dose irradiation with 50 cGy of X-ray induced *in vivo* production of superoxide dismutase-like substances and accelerated antioxidant activity in liver, brain and bone marrow of male C57BL/6 *mice* (YAMAOKA et al. 1999).

NFκB, a transcription factor inducible by oxygen radicals may be an important link in lung fibroses induced by ionising radiation (HAASE et al. 1997).

The higher the degree of unsaturation in the lipid acyl chain, the greater is the degradation caused by ^{60}Co γ-irradiation (SAMUNI and BARENHOLZ 1997). The fully saturated fatty acids, palmitic acid (C16), and stearic acid (C18) showed no significant changes in their levels. Both Tempo and Tempol provided similar protection to acyl chain residues. The nitroxides appear to protect via a catalytic mode. Unlike common antioxidants, such as α-tocopherol, which are consumed under irradiation and are, therefore, less effective against high radiation dose, nitroxide radicals are restored and terminate radical chain reactions in a catalytic manner. Furthermore, nitroxides neither yield secondary radicals upon their reaction with radicals nor act as prooxidants. MIURA and OZAWA (2000) examined the effect of X-irradiation on the signal decay rate of nitroxyl in the upper abdomen of *mice* using *in vivo* electron spin resonance. The signal decay rates increased 1 h after 15 Gy irradiation, and the enhancement was suppressed by preadministration of cysteamine, a radioprotector.

ALEXANDER et al. (1955) attributed the radioprotection by cysteamine (2-aminoethanethiol) to a radical scavenger effect, and SHEWELL et al. (1964) tried the combination of hypoxia and cysteamine in *mice* to protect them from 8 MeV electrons whole-body irradiation.

Cysteamine is oxidized by cysteamine dioxygenase (EC 1.13.11.19) to hypotaurine, which is then oxidized to taurine by an as yet unidentified enzyme.

The cysteinesulphinic acid pathway starts with the oxidation of cysteine by cysteine dioxygenase (EC 1.13.11.20). In the *rat*, cysteine dioxygenase is found only in the liver and consists of two sub-units containing Cu and Fe, one catalytic and the other one stabilizing. Cysteinesulphinic acid is converted into hypotaurine by a specific hepatic decarboxylase (EC 4.1.1.29).

Radiotherapy damages both mitochondria and rough endoplasmic reticulum of endothelial cells (FAJARDO and STEWART 1973). Additional application of doxorubicin (adriamycin) increases the risk (FAJARDO et al. 1976).

VERGARA et al. (1984) characterised the effect of radiation on different *rat* lung cells and tissue compartments by morphometry to assess the effects of radioprotectants as the aminothiol, amifostine (WR-2721). 3000 r radiation caused marked decreases in all cell types except interstitial cells which were increased relative to other cells. The surface areas and volumes of the epithelium and capillaries were also decreased. The interstitium was thickened almost fourfold and the volume was high relative to the overall decrease in volumes. Amifostine alone gave no significant protection from radiation although the capillary endothelium seemed relatively more preserved. The combination of amifostine plus corticosteroids confirmed the trend and also resulted in significantly less interstitial volume and thickness.

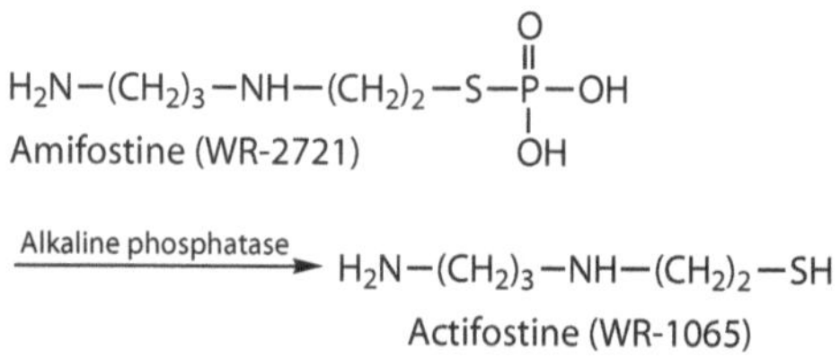

[270]

The mechanism of cytoprotection for amifostine is similar to that of other thiols. Once dephosphorylated to actifostine [equation 270], the active thiol can scavenge oxygen free radicals (OHNISHI et al. 1992) and bind to a variety of electrophilic agents.

α-Emitting Radionuclides

Uranium dioxide particles 7, 10 or 17 days after an inhalation for 100 min by *rats* were either associated with the lumen of the trachea at a distance between 4 and 8 μm from the epithelial cell surface (i.e. in the mucous layer, either free or within cells) or at a depth between 8 and 10 μm from the surface (GORE and PATRICK 1982). GORE and THORNE (1977) depicted UO_2 particles within a macrophage on an airway.

In the *mouse*, 75 min after intravenous injection of 0.5 ml **thorotrast** diluted 1 in 10 in saline small amounts of thorium dioxide lay extracellularly close to the cell membranes, which were separated to accomodate the extracellular material, and a small amount was present in membrane bounded vesicles (FISHER et al. 1968). The splenic macrophages of *mice* killed 24 h after injection of ThO_2 were heavily loaded and showed marked cytoplasmic vacuolation. The mitochondria were swollen and the pattern of cristae was distorted or lost. In one mitochondrium the outer double membrane was disrupted. The normal Y-shaped branching of endoplasmic reticulum was completely lost.

Inhalation of **β-emitting radionuclides** (^{90}Y, ^{91}Y, ^{144}Ce or ^{90}Sr) in fused aluminosilicate particles by Beagle *dogs* induced tumours of pulmonary epithelium, connective tissue and blood vessels (HAHN et al. 1988). The tumours originate in the terminal airways and alveolar portions of the lung, not in the larger airways. Tracheobronchial lymph nodes and heart (the middle tracheobronchial lymph node rests on the base of the heart over the left atrium) developed tumours through direct irradiation of the tissues. Haeangiosarcomas of the heart metastasised widely.

In single brief (<76 min) nose-only inhalation exposures to a polydisperse aerosol of ^{90}Y, ^{91}Y, ^{144}Ce-^{144}Pr or ^{90}Sr-^{90}Y in fused aluminosilicate particles *dogs* showed a wide range of specific activities of particles deposited in their lungs, from about 0.1 100 Bq per particle, three orders of magnitude, with ^{90}Y having the highest specific activity and ^{90}Sr the lowest (HAHN et al. 1995). If specific activity of the particles was an important factor, one would expect the ^{90}Y to result in more risk for lung cancer than ^{91}Y and so on.

$$^{90}Y >> {}^{91}Y = {}^{144}Ce\text{-}{}^{144}Pr \geq {}^{90}Sr\text{-}{}^{90}Y$$

In fact, the lung cancer risks were essentially equal, at doses <50 Gy to the lung. In conclusion, 0.10–30 Bq/particle did not present an extraordinary risk for lung cancer, at least in the 0.5–50 Gy range of doses to the lung. ^{90}Sr (half-life 28.5 years) decays to ^{90}Y, having a half-life of only 64.1 h and a decay energy of 2.3 MeV, that is several times greater than ^{90}Sr (0.5 MeV).

In comparison of BALB/c *mice* (30 % spontaneous incidence of lung tumours) and C57BL/6J *mice* (1.5 % spontaneous incidence of lung tumours) exposed to aerosols containing ^{144}Ce HAHN and GRIFFITH (1995) concluded that absolute risk is more accurate than relative risk for predicting lung tumour risk from high to low lung-tumour incidence strains of *mice*.

Exposure of *mouse* epidermis to high-dose β-radiation showed that before the appearance of a visible or histologically detectable tumour, overexpression of p53 was common at an exposed site and was accompanied by an overexpression of $p62^{c\text{-}fos}$ and $p21^{N\text{-}ras}$ (LESZCZYNSKI et al. 1994).

γ-Emitting Radionuclides

^{137}Cs has a half live of 30 a. The energy of its γ-radiation is about 0.7 MeV. Brazilian crystallised quartz particles (specific surface 2 m^2/g) irradiated with 280 Gy showed an increased lucigenin-derived chemiluminescence as compared with not irradiated particles (BREHM 1996). This type of chemiluminescence was not inhibited by superoxide dismutase (EC 1.15.1.1) of catalase (EC 1.11.1.6). *Bovine* alveolar macrophages engulfing quartz particles activated by γ-irradiation emitted more photons than cells taking up untreated particles. The reactive oxygen intermediates produced by the cells were scavenged by superoxide dismutase, while the chemiluminescence induced by the γ-irradiated quartz particles themselves persisted.

Under the influence of ionising radiation, electrons of the valence band of the solid were raised to the conduction band from where they are able to occupy empty traps in the energy band gap. The simultaneously generated electron holes localise at the so-called activator levels. At the end of the excitation the solid contains negatively (traps) and positively (activators) charged defects being in a state of energy non-equilibrium. As a consequence of addition of thermal energy, electrons are able to leave the solid causing thermally stimulated exoelectron emission, or to recombine with activators under emission of light, thus producing thermally stimulated chemiluminescence.

Deuteron beams (10.000 rad) derived from the 60-inch Brookhaven cyclotron directed to a 9×5 mm area of the dorsal surface of the C57BL/10 *mouse* brain induced greatly elongated mitochondria with uniformly spaced and parallel arrays of cristae in some of the irradiated neurones (SAMORAJSKI et al. 1970, SAMORAJSKI 1975). Granular cells of the cerebellum contained the most unusual mitochondria, which often ranged from 6 to 8 μm in length. The mitochondria were often seen in close apposition to the nucleus and contained from four to ten uniformly spaced cristae which terminated either by fusion at the edge of the mitochondrium or ended as dilations, seemingly within the mitochondrial matrix.

Mice that harboured the **x-ray**-induced low efficiency allele of the major X-linked isozyme of glucose-6-phosphate dehydrogenase (EC 1.1.1.49), $Gpdx^{a\text{-}m2Neu}$, and, in addition, harboured the transgenic shuttle vector for the determination of mutagenesis *in vivo*, pUR288, were employed by FELIX et al. (2002) to further their understanding of the interdependence of general metabolism, oxidative stress control, and somatic mutagenesis. In spite of the mild phenotype, the brains of glucose-6-phosphate dehydrogenase-deficient males exhibited a significant distortion of redox control (~3-fold decrease in the ration of reduced glutathione to oxidised glutathione), a considerable accumulation of promutagenic etheno DNA adducts (~13-fold increase in ethenodeoxyadenosine and ~5-fold increase in ethenodeoxycytidine), and a substantial elevation of somatic mutation rates (~3-fold increase in mutant frequencies in *lacZ*, the target and reporter gene of mutagenesis in the shuttle vector, pUR288). The mutation pattern in the brain was dominated by illegitimate genetic recombinations, a presumed hallmark of oxidative mutagenesis.

Synchrotron x-ray sources have proven to be a boon for structural biological research. CHENG and CAFFREY (1996) showed free radical-mediated x-ray damage of model membranes. The effect of degree of lipid hydration, phospholipid chemical structure, mesophase identity, aqueous medium composition, and incident flux on the severity and progress of damage was quantified using time-resolved x-ray diffraction and chromatographic analysis of damage products. Electron spin resonance measurements of spin-trapped intermediates generated during irradiation suggested a free radical-mediated process.

Ultraviolet A light exposed cells can induce the production of reactive oxygen species which can damage the cellular elements (NISHI et al. 1991). The effects mainly include local inflammation, epidermal hyperplasia, DNA damage, the activation of protein kinase C, and the stimulation of arachidonic acid metabolism (WANG et al. 1991). Quercetin (50 mg/kg body weight) protected female Sprague-Dawley *rats* from UV radiation (18 W/cm^2)-induced damage as demonstrated by a decrease in liver malondialdehyde formation compared with that in unmedicated UV-irradiated control *rats* (ERDEN INAL and KAHRAMAN 2000).

Radicals formed in DNA bases by irradiation. At low temperature, only thymine anion radicals ($T^{\bullet -}$) and guanine cation radicals ($G^{\bullet +}$) can be detected (HERAK et al. 1985, CULLIS and SEYMONS 1985). $G^{\bullet +}$ radicals react fast with thiols.

$$\text{Guanosine}^{\bullet} + \text{Melatonin-H} \rightarrow \text{Guanosine-H} + \text{Melatonin}^{\bullet} \quad [271]$$

Radioprotective effects of melatonin given intraperitoneally to male adult albino *mice* (10 mg/kg) became apparent in a significant reduction in micronuclei polychromatic erythrocytes when animals were treated with melatonin 1 h before and not 30 min after exposure to 1.4 Gy of γ-irradiation (BADR et al. 1999). Aberration in spermatogonal chromosomes decreased from 46 to 32 % in *mice* pre-treated with melatonin. Similarly, the frequency of chromosomal aberration in meiotic metaphases decreased from 43.5 % in the irradiated group to 31.5 % in the irradiated group treated with melatonin 1 h before irradiation.

Nitric oxide synthase (EC 1.14.13.39) induction by interferon-γ or interferon-γ plus lipopolysaccharide in BNL CL.2 *murine* embryonic liver cells was potentiated by ionising radiation (Yoo et al. 2000). H_2O_2 formed by γ-radiation might make the cells responsive for interferon-γ or interferon-γ plus lipopolysaccharide, as catalase (EC 1.11.1.6) blocked H_2O_2 formation and $^{\bullet}NO$ production.

14.7.1 Does Ionising Radiation Promote Pneumoconiotic Fibrosis?

DANTIN GALLEGO (1950,1951) introduced electronics into the pathogenesis of silicosis. After X-raying of natural quartz crystals BRUNNER et al. (1961) in infrared spectrographs observed an intensive band near 3600 cm^{-1}. Absorption at this wave-number as well as near 3685 cm^{-1} may also be caused by some organic compounds adsorbed on the SiO_2 surface (CLARK-MONKS and ELLIS 1973, CURTHOYS et al. 1974). While mine dust samples collected from collieries of the Ruhr and Saar districts in their activation energies of the electron traps (ROBOCK 1967,1968) did not show any relation with their cytotoxicity, unmixed SiO_2 modifications did (ROBOCK and KLOSTERKÖTTER 1971). Both constituents absorbing chemiluminescence and dust components with no cytotoxicity were held responsible for this lacking correlation in mixed dust samples. KRIEGSEIS et al. (1977) used the methods of thermally stimulated luminescence, thermally stimulated exoelectron emission, electron spin resonance, and infrared spectroscopy to investigate whether the specific cytotoxic activities of SiO_2 dusts of various origins can be attributed to physical parameters. Quartz particles (specific surface 2 m^2/g) stored for over 40 years and reactivated by γ-rays from a ^{137}Cs source (280 Gy) immediately before starting the experiment induced more photons when engulfed by *bovine* alveolar macrophages than did untreated quartz dust (ZELLER et al. 1993, BREHM 1996). Lucigenin-enhanced chemiluminescence could not be inhibited by enzymes scavenging $O_2^{\bullet -}$ and H_2O_2, superoxide dismutase (EC 1.15.1.1) and catalase (EC 1.11.1.6), respectively.

PART III

Extrapulmonary Targets

CHAPTER 15

Radical Damage in Nonrespiratory Organ Systems

Cellular O_2 concentrations in the *human* body are precisely regulated to maintain adequate substrate for oxidative phosphorylation and other essential metabolic reactions while minimising the production of reactive oxygen species capable of damaging cellular DNA, lipids, and proteins.

Increased oxidant formation is a significant contribution to disease and often an indispensable requirement, but not just an epiphenomenon. In carcinogenesis, chemoprevention of exocyclic DNA adducts by antioxidants was demonstrated by NAIR et al. (1996). Primary prevention of cancer is not only to avoid hazardous exposures but also involves the intake of protective agents and modulation of the host's defence mechanisms before the onset of malignancy.

Adaptive systems allowing to survive to moderate or even severe hypoxia involve an increase in the expression of genes coding for proteins responsible for anaerobic production of adenosine triphosphate, namely aldolase A (EC 4.1.2.13), enolase-α (EC 4.2.1.11), lactate dehydrogenase (EC 1.1.1.27), pyruvate kinase (EC 2.7.1.40), and glucose transporter-1 (SEMENZA et al. 1996).

Several transcription factors have been reported to be involved in the response to hypoxic stress (for review see MINET et al. 2001).

Hypoxia-inducible factor 1 (HIF-1) is a basic-helix-loop-helix transcription factor that plays essential roles in mammalian development and physiology. HIF-1 is a heterodimer composed of HIF-1α and HIF-1β subunits (WANG and SEMENZA 1995, WANG et al. 1995). The expression and activity of the HIF-1α subunit are tightly regulated by cellular O_2 concentration. Under hypoxic conditions, HIF-1 activates the transcription of genes encoding erythropoietin, glucose transporters, glycolytic enzymes, vascular endothelial growth factor, and other genes the protein products of which increase O_2 delivery or facilitate metabolic adaptation to hypoxia. Pharmacological manipulation of HIF-1 levels may provide a novel therapeutic approach to diseases that represent the most common cases of mortality in Western society, including cancer, chronic lung disease, and myocardial ischaemia (SEMENZA 2000).

In cultured proximal tubular LLC-PK_1 cells, *S*-nitrosoglutathione induced HIF-1α accumulation and concomitant DNA binding (SANDAU et al. 2000). The response was attenuated by the kinase inhibitor genistein and blockers of phosphatidylinositol 3-kinase such as Ly 294002 or wortmannin. •NO appears to regulate HIF-1α via the PI 3K/Akt pathway under normoxic conditions.

Hypoxic stress induced by exposing adult *rabbits* to a simulated altitude of 7,000 m for 6 h increased glucocorticoid levels in plasma by 11 % whereas in erythrocytes was 55 % after hypoxia (SRIDHARAN et al. 1991). Plasma reduced glutathione (GSH), lactate dehydrogenase (EC 1.1.1.27) and isocitric dehydrogenase (EC 1.1.1.41) activities increased and aspartate aminotransferase activity was depressed after hypoxic stress.

Acute hypoxia induced **NO overproduction** in brain (101.2 ± 16.1 vs. 54.1 ± 8.2 ng/g tissue in the controls), but not in liver (48.1 ± 10.2 vs. 45.5 ± 16.4 ng/g tissue in the controls) of male Wistar *rats* decompressed to 170 mm Hg (MALYSHEV et al. 1999). Adaptation to hypoxia for 8 days decreased NO production in liver (19.1 ± 14.8 ng/g tissue) and brain (23.1 ± 11.3 ng/g tissue), prevented NO overproduction in brain, and potentiated NO synthesis in liver (63.7 ± 15.5 ng/g tissue) in subsequent acute hypoxia. Adaptation to hypoxia developed against a background of accumulation of heat shock protein HSP70 in liver and brain. A course of dinitrosyl iron complex reproduced the antihypoxic effect of adaptation. A course of N^{ω}-nitro-L-arginine during adaptation hampered both accumulation of HSP70 and development of the antihypoxic effect. MANUKHINA et al. (1999) showed that both the nitrite/nitrate level and the NO store increased as adaptation to hypoxia developed. The NO store volume significantly correlated with plasma nitrite/nitrate.

Hyperglycaemia induced prior to ischaemia resulted in earlier infarction that correlated with increased immunoreactivity for iNOS, Mn superoxide dismutase and nitrotyrosine (STE-MARIE et al. 2001).

The findings that reperfusion after ischaemia is necessary for NO formation (Lepore et al. 1999) suggests that an inflammatory pathway is responsible for **NOS-independent NO formation** in ischaemia-reperfusion injury to skeletal muscle. In transient (2 h) focal cerebral ischaemia in *rats*, infusion of sodium nitroprusside (0.11 mg/kg) and spermine/NO (0.36, 3.6 mg/kg) reduced the infarct size (Salom et al. 2000).

Data accumulated in the last decade indicate that •NO participates in the regulation of neurotransmission in the central nervous system (reviewed by Kiss 2000). Since •NO is a potential nonsynaptic modulator, it may have an important role of monoaminergic systems and their impairment in a number of severe neuropsychiatric diseases (e.g. depression, Parkinson's disease).

Coumarins used in postischaemic cardiomyopathy, have a reputation as anticoagulant, antimutagenic, tumoristatic, antimetastatic, antiinflammatory (Chaturvedi et al. 1974), immunostimulatory (Zlabinger et al. 1994), anticonvulsant and hypotensive agents. Some are able to scavenge (quench) reactive oxygen species, stimulate respiration iono-phoretically, inhibit 5- (**esculetin** = 6,7-dihydroxycoumarin; Neichi et al. 1983) and 12-lipoxygenases and inhibit xanthine oxidase (Chang and Chiang 1995) and phenylalanine hydroxylase. However, coumarins lacking dihydroxy substitution, did not scavenge superoxide, while **fraxetin** (7,8-dihydroxy-6-methoxycoumarin) was the most potent (IC_{50} = 2.3 μM in the cytochrome assay and 5.8 μM using NTB) scavenger (Paya et al. 1994). In the xanthine-xanthine oxidase system, reactive oxygen species-induced chemiluminescence was inhibited by **7,8-dihydroxycoumarin** and **7,8-dihydroxy-4-methylcoumarin** by 98.9 and 99.6 % at 100 μM (de las Heras and Hoult, unpublished, quoted by Paya et al. 1994).

Esculetin, umbelliferone (7-hydroxycoumarin) and **7-hydroxy-4-methyl coumarin** are strong xanthine oxidase inhibitors (Chang and Chiang 1995). The structure of 7-hydroxy coumarin plays a very important role in xanthine oxidase inhibition. the 6-hydroxy group present in the molecule of 7-hydroxy coumarin, e.g. esculetin enhanced the activity, whereas substitution by the 6-methoxy group, e.g. **scopoletin** (formula [47]), reduced the inhibitory effect. (Chang and Chiang 1995).

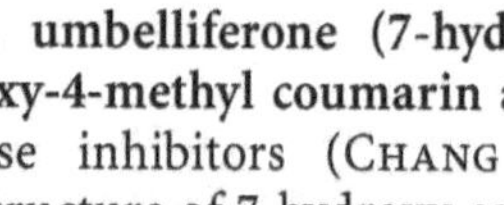

Scopoletin (6-methoxy-7-hydroxycoumarin) [47]

Coumarin is essentially a prodrug, since it is rapidly and totally metabolised to 7-hydroxycoumarin in the *human* liver. The first step in coumarin metabolism is catalysed by the phase I cytochrome P_{450} system. The activity of CYP2A6 was estimated in *humans* by measuring the urinary excretion of 7-hydroxycoumarin (Pelkonen et al. 1994). After 7-hydroxylation coumarin undergoes phase II conjugation, mainly to glucuronide. The pharmacokinetics of 3-[^{14}C]coumarin after a single intraperitoneal injection in *rats* showed a half life of about 43 h and a rapid decline of the brain levels of the drug by 0.66 μg/h during the first 10 h and then plateauing at 1.22 μg/g until 24 h (Piller 1977). 7-Hydroxycoumarin inhibited oncogene-induced transformation of murine fibroblasts (Seliger and Pettersson 1994).

Salicylic acid was identified as a metabolite of warfarin (Deckert 1973).

Carbocromene [3-(2-diethylamino-ethyl)-4-methyl-7-(cabethoxy-methoxy)-2-oxo-1,2-chromene hydrochloride] suppressed compensatory cardiac hypertrophy following hypoxia in *rats* exposed to 300 r X-irradiation (Nitz 1972, 1974). Electron spin resonance studies on the influence of carbocromene on *bovine* heart mitochondria and oligomycin-sensitive mitochondrial ATPase showed that h_0/h_{-1} ratios in repeated scans increased in the presence of 20-40 μmol/l carbocromene, while carbocromene under similar conditions had no effect on *rat* liver mitochondria (Zimmer et al. 1982).

In the central nervous system of the *rat*, ^{14}C-labelled carbocromene showed a strong and long-lasting activity in neocortex, thalamus, hippocampus, nucleus septi, putamen, corpora amygdaloidea, and colliculi inferiores (Möllmann et al. 1972). The local concentration of radioactivity in these areas corresponds to some extent the distribution pattern of the psychoactive agent, 9-(*N*-methylpiperiliden-4')-thioxanthen (Eckert and Hopf 1970).

Cloricromene (8-monochloro-3*β*-diethylamino-ethyl-4-methyl-7-ethoxycarbonyl-methoxy-coumarine) inhibited the induction of nitric oxide synthase (Zingarelli et al. 1993). Rings of thoracic aortas from lipopolysaccharide (4 mg/kg, i.v.)-shocked *rats*, contracted with phenylephrine, showed a progressive decrease in tone, that was of a greater magnitude than that of aortas from naive *rats*. Moreover, a decreased response to the constrictor effect of phenylephrine was observed in aortas from shocked *rats*. *In vivo* treatment with cloricromene (2 mg/kg, i.v.) 30 min before lipopolysaccharide administration partially prevented the loss in tone of aortic rings and improved their reactivity to phenylephrine.

Human monocyte adhesion to *human* cultured umbilical vein endothelial cells was significantly inhibited by 15 to 30 μM cloricromene (TRANCHINA et al. 1994). When monocyte adhesion was mediated by a large set of adhesive receptors, as obtained after treatment of umbilical vein endothelial cells with either interleukin 1β (50 ng/ml) or tumour necrosis factor-α (100 u/ml), the inhibitory effect of cloricromene was considerably reduced. Cloricromene (50–500 μM) inhibited in a concentration-dependent manner the release of the platelet-activating factor (a phospholipid mediator of inflammation) from *human* polymorphonuclear leucocytes stimulated with 10 μM A23187 (RIBALDI et al. 1996).

15.1 Reperfusion Damage

The complications of reoxygenation after hypoxaemia or ischaemia are still the most frequent cause of death in the western hemisphere. Reoxygenation is often more damaging than anoxia (SCHILLER 1993). As an indication of cerebral lipid peroxidation, jugular venous-arterial malondialdehyde differences in 25 *patients* undergoing carotid endarterectomy were significantly enhanced before reperfusion, and an additional rise was observed 15 min after reperfusion (WEIGAND et al. 1999).

Plasma malondialdehyde levels were significantly increased in 25 *patient* with schizophrenia and 23 *patient* with bipolar disorder compared with 20 healthy subjects (KULOĞLU et al. 2002). Superoxide dismutase (EC 1.15.1.1) and glutathione peroxidase (EC 1.11.1.9) activity levels were significantly higher in the schizophrenic group compared with controls. Superoxide dismutase activity levels in the bipolar group were significantly higher than controls whereas there were no significant changes in glutathione peroxidase activity levels in the bipolar group and controls.

Vascular endothelial cells represent a heterogeneous cell group with characteristic morphological and biochemical properties. Their junctional organisation extends from the very leaky junctions in mesenteric and muscular capillaries, to continuous junctions in brain capillaries (CERVOS-NAVARRO et al. 1988).

Ischaemia causes sustained elevation of free intracellular **calcium** ion concentration ($[Ca^{2+}]_i$), that can promote the conversion of xanthine dehydrogenase (EC 1.1.1.204) to xanthine oxidase (EC 1.1.3.22), i.e. the enzyme transfers electrons during the catalytic cycle of molecular oxygen instead of adenin-nicotin dinucleotides (MCCORD 1985). Under low energy conditions, where large parts of adenosine triphosphate are converted to hypoxanthine, this may result in a massive generation of reactive oxygen species. The activation of xanthine oxidase has been implicated for ischaemic neuronal death *in vivo* (COYLE and PUTTFARCKEN 1993) and in kainate toxicity to cerebellar granular cells *in vitro* (DYKENS et al. 1987). The control of xanthine oxidase-xanthine dehydrogenase levels by oxygen tension is a complex process involving pre- and posttranslational points of regulation (TERADA et al. 1997).

In *patients* with transient ischaemic attacks, peripheral blood plasma **adenosine** levels were increased upon admission, peaked on day 2, and steadily decreased to the normal range, reached by day 5 (LAGHI PASINI et al. 2000).

Ischaemic preconditioning by brief periods of ischaemia protected tissues from the consequences of prolonged ischaemia, as first seen in the myocardium (MURRY et al. 1986). Acute or classical preconditioning is effective as long as the interval between the initial, brief preconditioning ischaemia and the later, prolonged insult is less than approximately 2 h. However, a second window of protection, termed delayed preconditioning or ischaemic tolerance, becomes apparent in tissues subjected to the prolonged ischaemia 24 h after the initial insult (KORTHUIS et al. 1998). Although there is evidence suggesting that the factors initiating acute and delayed preconditioning are similar, more recent evidence suggests a role for nitric oxide ($^{\bullet}NO$) as a major trigger of delayed preconditioning. Diminished $^{\bullet}NO$ availability following ischaemia/reperfusion may contribute to the pathogenesis of the multiple organ dysfunction syndrome.

Evidence that hypoxia or ischaemia/reperfusion elicits oxidant-mediated tissue injury suggests that the adaptational response to brief periods of hypoxia or ischaemia could involve augmented production of antioxidant enzymes in affected tissues.

15.1.1 Nervous System

The brain is prone to suffer oxidant damage due to its relatively **low content of antioxidant enzymes** (CARRILLO et al. 1992) and its **high content of iron,** as shown in neuromelanin by Laser microprobe mass analysis (LAMMA) (GOOD et al. 1992). Immunoreactive catalase protein decreased in the hypothalamus and the prefrontal cortex of the ageing *rat* brain (3 vs. 12 months) by 63 and 55 %, respectively (CIRIOLO et al. 1997). Iron will be easily released when the cells are injured and cannot be safely bound as cerebrospinal fluid has not any significant iron binding capacity. The toxicity of $O_2^{\bullet -}$ has been largely explained on the basis of increased "free" iron released from the [4Fe-4S] clusters of dehydratases as a consequence of the oxidative attack of $O_2^{\bullet-}$ on those clusters (Liochev 1996); followed by the reduction of H_2O_2 by Fe(II), with the production of the powerful hydroxyl radical i.e. the Fenton reaction. The rate of lipid peroxidation in brain homogenate from various regions of the *rat* brain correlated with the total iron content of that tissue (Zaleska and Floyd 1985). The only region that did not follow this trend was the striatum. The less than expected amount of lipid peroxidation is most likely due to the high content of dopamine in this brain region. Dopamine is very effective in preventing lipid peroxidation as Zaleska and Floyd (1985) demonstrated by its addition to cerebellum, a brain region which has the same iron content as striatum but unlike striatum has very little dopamine. Information regarding quantitative analyses of iron and the iron regulatory proteins in animals, and normal and diseased states in *humans*, has been reviewed by Connor (1992), Connor and Benkovic (1992), and Connor et al. (1992). The data on cellular distribution of iron and the iron regulatory proteins indicate that oligodendrocytes play a significant role in maintaining iron homeostasis in the brain (Connor 1994). The distribution of transferrin positive oligodendrocytes in grey matter is similar to the iron containing cells; these cells are in perneuronal and pervascular positions (Connor and Fine 1986, Connor et al. 1990, Connor and Benkovic 1992, Connor 1994). In both *monkey* and *human* brain tissue microglial cells immunostained robustly for L-chain ferritin, but did not immunostain for H-chain ferritin (Connor 1994). Oligodendrocytes have been observed which contain only H-chain ferritin, but the majority of the cells contained both H and L ferritin. Using monoclonal antibodies, Connor et al. (1992) found ferritin in the pyramidal neurones of the *monkey* cerebral cortex and so did Connor (1994) in pyramidal cells of the *human* hippocampus.

Iron overload within the mitochondria confers cellular sensitivity to oxidant stress and may play an important role in the pathogenesis of Friedreich's ataxia, which might be treated by iron chelators to mobilise mitochondrial iron (Smith et al. 1999, Richardson et al. 2001). However, desferrioxamine cannot efficiently mobilise iron from cells (Bottomley et al. 1985, Richardson et al. 1994), and it is not effective at mobilising Fe from Fe-loaded mitochondria (Ponka et al. 1979, 1984). Pyridoxal isonicotinoyl hydrazone showed high affinity at mobilising Fe from an experimental model of mitochondrial Fe overload in reticulocytes (Ponka et al. 1979), and so did several of its analogues (Richardson et al. 2001).

From thermodynamic reasons, there seems to be a physical impossibility in designing a clinically efficient iron chelator, with a large safety margin for long-term use (Galey 1997). As a matter of fact, the affinity constant for iron of such a compound should be relatively low in order to avoid side effects related to iron mobilization from iron proteins. At the same time, to compete for iron traces released during oxidative stress, the association constant must be high enough compared to those of endogenous, non-specific metal-binding molecules. Indeed, intracellular medium is rich in various small molecules able to non-specifically bind iron, e.g. amino acids, nucleotides, or citrate. In pure *in vitro* systems, *N,N'*-bis-(3,4,5-trimethoxybenzyl) ethylenediamine *N,N'*-diacetic acid was able to protect biological molecules, proteins, lipids, and DNA against iron-catalysed oxidative damage (Galey et al. 1996).

Copper toxicoses may be of genetic origin as in Wilson's disease in *man* (Underwood 1974) and copper toxicosis in the Bedlington *terrier* (Johnson et al. 1980).

The phospholipid mixture isolated from brain was employed as a model system to study the free-radical oxidation chemistry using Cu(II) and H_2O_2 (Hall and Murphy 1998) to generate a hydroxyl radical by the Fenton reaction. The distribution of plasmenyl phospholipids present in brain is somewhat different from that observed in erythrocyte membranes with the former having a high abundance of plasmenyl species with 18:1, 20:1, and 22:1 fatty acyl groups at the sn-2 position, which is comparable in abundance to those molecular species with 20:4, 22:4, and 22:6 fatty acyl groups at sn-2 (Khaselev and Murphy 1999).

Copper deficiency in Menke's disease, an X-linked genetic disorder characterised by rapidly progressive cerebral degeneration was reported by Danks et al. (1972).

Both **chromium**(VI) and chromium(III) are biologically active oxidation states of this transition metal considered as an essential trace element influencing glucose, protein, and fat metabolism. Chromium(VI) enters many types of cells and under physiological conditions can be reduced intracellularly by H_2O_2, glutathione reductase, carbohydrates, ascorbic acid and reduced glutathione to produce active intermediates and, ultimately, chromium(III).

$$2\,Cr^{VI}O_4^{2-} + 9\,H_2O_2 + 2\,OH^- \longrightarrow 2\,Cr^{V}(O_2)_4^{3-} + 10\,H_2O + O_2 \qquad [138]$$

$$2\,Cr^{V}(O_2)_4^{3-} \longrightarrow 2\,Cr^{VI}O_4^{2-} + 2\,O_2^{\bullet-} + 2\,{}^1\mathbf{O_2} \qquad [139]$$

$$2\,H_2O_2 + 2\,O_2^{\bullet-} \longrightarrow 2\,HO^{\bullet} + 2\,OH^- + 2\,O_2 \qquad [140]$$

The oral administration of potassium dichromate (25 mg/kg×day) for 3 days increased *mouse* brain homogenate chemiluminescence and thiobarbituric acid-reactive substances by 34 and 29%, respectively (TRAVACIO et al. 2000).

Parenteral administration of **nickel** chloride ($NiCl_2$) to *rats* enhanced lipid peroxidation in liver, kidney and lung, but not in brain, heart, spleen or testis, as measured by the thiobarbituric acid reaction for malondialdehyde and related chromogens in fresh tissue homogenates (SUNDERMAN et al. 1985).

The apparent ability of **manganese** in its divalent form to promote formation of reactive oxygen species within a *mouse* cortical mitochondrial-synaptosomal fraction was completely abolished by the addition of one five hundredth of its molarity of desferroxamine, a trivalent metal chelator (HAMAI et al. 2001). Ferric ion was able to dampen the reactive oxygen species-generating capacity of manganous chloride, whereas manganic ion markedly promoted this property attributed to manganous ion.

$$Mn^{2+} + H_2O_2 \leftrightarrow [Mn^{2+}\text{-}OH \leftrightarrow Mn^{3+}\text{-}{}^{\cdot}\mathbf{OH}] \leftrightarrow Mn^{3+}\mathbf{HO}^{\cdot} + OH^- \quad [272]$$

The production of H_2O_2 by *guinea-pig* cerebral cortex synaptosomes, triggered by an increase of intrasynaptosomal Ca^{2+} has been described by ZOCCARATO et al. (1989, 1990). The H_2O_2 output was inhibited by the Fe chelator deferoxamine (ZOCCARATO et al. 1993). After treatment with digitonin, the H_2O_2 production increased by 40–50%. It appears that synaptosomes contain two distinct pools of Ca^{2+}-Fe-dependent oxidase, one intrasynaptosomal, activated by ionomycin and detectable after depletion of glutathione (ZOCCARATO et al. 1989), the other localised superficially and dependent on added NADH, acetylated ferrocytochrome *c*, iron and Ca^{2+}.

Oxygen reactive intermediates have been repeatedly demonstrated to be important mediators of hypoxic and anoxic cell death in the brain (DEMOPOULOS et al. 1980, WATSON et al. 1984, WATSON and GINSBERG 1988, BRAUGHLER and HALL 1989, KONTOS 1989, LAZZARINO et al. 1992, HALL 1993, PEREZ VELAZQUEZ et al. 1997). DANDEKAR (1997) commented the importance of free radicals in memory loss. Cerebral reperfusion in *rats* ischaemic by bilateral clamping of the common carotid arteries for 5 min induced oxidation of hypoxanthine to xanthine and uric acid (LAZZARINO et al. 1992). With respect to preischaemic values, after 10 min reperfusion caused a tenfold increase of inosine and a 1.7 fold increase of adenosine. Oxygen radical-mediated injury was reflected by significant increase in cerebral malondialdehyde. In isolated *rat* brain microvessels a 10 to 20 min anoxic period followed by a 40 min reoxygenation period resulted in an enhanced formation of 2,3- or 2,5-dihydroxybenzoic acid, when salicylate was used as a hydroxyl radical trap (GRAMMAS et al. 1993).

Following hypoxia/reoxygenation (6 h/96 h), cultured neurones from embryonic *rat* forebrain undergo delayed apoptosis (LIÈVRE et al. 2000). Temporal evolution of intraneuronal free radical generation was monitored by flow cytometry using dihydrorhodamine 123. Two distinct peaks of radical generation were depicted, at the time of reoxygenation (+27%) and 48 h later (+25%), respectively. Transcript and protein levels of both Cu/Zn-SOD and Mn-SOD were reduced 1 h after the onset of hypoxia, but activities were transiently stimulated. Reoxygenation was associated with an increased expression (139%), but a decreased activity (21%) of the inducible Mn-SOD, whereas Cu/Zn-SOD protein and activity were low and progressively increased until 48 h post-hypoxia, when the second rise in radical occurred.

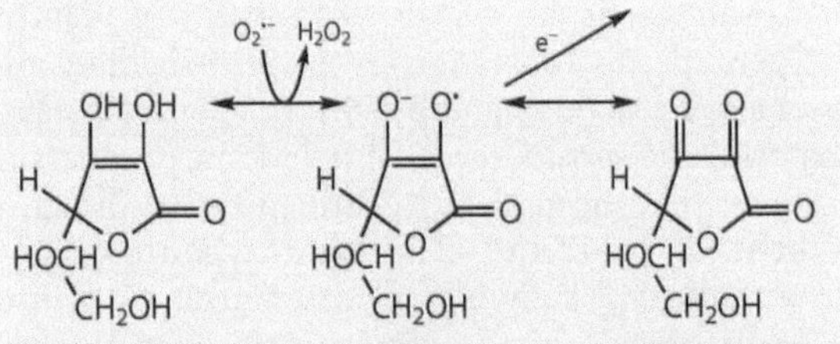

Ascorbic acid **Ascorbyl radical** **Dehydroascorbic acid**

Formation of the ascorbyl free radical [53]

Ascorbate, the ionised form of ascorbic acid, is released during hypoxia, including stroke, and is subsequently oxidised in plasma. The oxidised product (dehydroascorbate) is transported into neurones via a glucose transporter during a reperfusion period. The dehydroascorbate taken up by cells is reduced to ascorbate by both enzymatic and non- enzymatic processes, and the ascorbate is stored in cells. This reduction process causes an oxidative stress, due to the coupling of redox reactions, which can induce cellular damage and trigger apoptosis. Incubating *rat* cerebral cortical slices, SONG et al. (2001) induced a decrease in cellular glutathione content and an increase in lipid peroxide production. Wortmannin, a specific inhibitor of phosphatidylinositol-3-kinase, prevented the ascorbate-induced decrease of GSH, and suppressed ascorbate-induced lipid peroxide production.

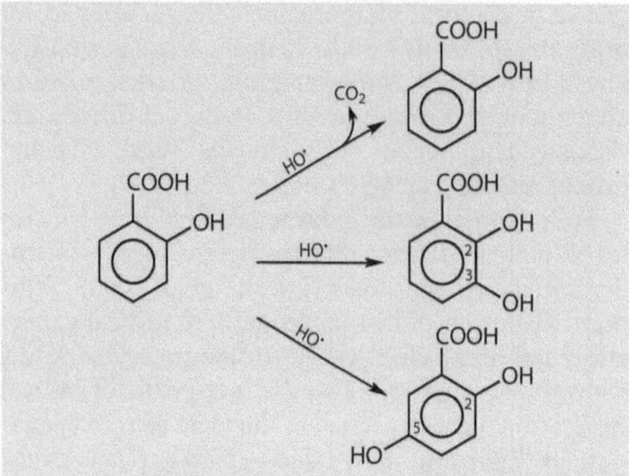

Salicylic acid as a hydroxyl radical trap [184]

A comparison of salicylate and the amino acids tryptophan, phenylalanine, and tyrosine clearly showed that salicylate is the best indicator of hydroxyl radical production (MASKOS et al. 1992). RAUCA et al. (1999) trapped free hydroxyl radicals generated after intraperitoneal application of 48 mg pentylenetetrazol per kg body weight by salicylate 30 min after the application of the convulsant.

5-Aminosalicylic acid in assessment of reactive oxygen species formation by *in vitro* Fenton and ozonation reactions and by *in vivo* ozone-exposure experiments in *rats* revealed oxidation products as follows: salicylic acid, by deamination; 2,3-dihydroxybenzoic acid and 2,5-dihydroxybenzoic acid, from radical or enzymatic hydroxylation; 5-amino-2-hydroxy-*N*,*N*'-bis(3-carboxy-4-hydroxyphenyl)-1,4-benzoquinonediimine, a condensation product of oxidised 5-aminosalicylic acid; and 5-amino-2,3,4,6-tetrahydroxybenzoic acid, attributed to hydroxyl radical attack without deamination, identified by high-pressure liquid chromatography electrochemical detector system analysis and by gas chromatography-mass spectrometry analysis of trimethyl silyl derivatives (KUMARATHASAN et al. 2001).

Mandelic acid, a major metabolite of cyclandelate [3,3,5-trimethylcyclohexyl)-mandelate], is a very effective hydroxyl radical scavenger (HAENEN 1989).

The *N*-methyl-D-aspartate receptor-like complex in synaptic membranes might be among the targets of reactive oxygen species formed by the reaction of xanthine and xanthine oxidase (see formula [45]). Therefore, AGBAS et al. (2002) isolated the receptor-like complex from *rat* brain synaptic membranes and treated it with xanthine (5 μM) plus xanthine oxidase (0.02 U/ml) for 30 min. Xanthine plus xanthine oxidase inhibited more than 80 % of the glutamate-binding activity of the complex. A nearly identical magnitude of inhibition of the *N*-methyl-D-aspartate-sensitive L-[^{3}H]glutamate binding was achieved if Cu,Zn-superoxide dismutase was added at the end of the preincubation period and just prior to the addition of L-[^{3}H]glutamate. The ligand-binding activity of the complex was not affected by either xanthine alone, superoxide dismutase-1 alone, or xanthine plus Cu,Zn-superoxide dismutase. Inhibition of glutamate binding to the complex and glutamate-binding protein by O_2^- was greater than that produced by H_2O_2, another product of the xanthine/xanthine oxidase reaction. Mutation of two cysteine residues in recombinant glutamate-binding protein (Cys^{190,191}) eliminated the effect of O_2^- on L-[^{3}H]glutamate binding. Both S-thiolation reaction of glutamate-binding protein in synaptic membranes with [^{35}S]cysteine and reaction of Cys residues in glutamate-binding protein with [^{3}H]*N*-ethylmaleimide were significantly decreased after exposure of membranes to O_2^-. Inhibition of cysteylation of membrane glutamate-binding protein by O_2^- was still observed after iron chelation by desferrioxamine, albeit diminished, and was not altered by the presence of catalase.

Measurement of effects of direct and indirect acting radical scavengers and/or neuroprotective drugs such as *N*-methyl-D-aspartate receptor antagonists, calcium channel blockers, inhibitors of nitric oxide synthase, dopamine agonists and drugs, which induce antioxidative enzyme activity, and others can be studied in freely moving animals using intracerebral microdialysis in combination with the salicylate hydroxylation assay. TEISMANN and FERGER (2000) presented an *in vivo* approach using local application of glutamate into the striatum of the *rat* and an *in vitro* screening using the Fenton reaction to induce hydroxyl radical formation.

The spin-trapping compound, *N*-tert-butyl-α-phenylnitrone reversed memory loss in aged gerbils (CARNEY et al. 1991) and preserved memory retention in 24-month old *rats* (SOCCI et al. 1995). The generation of HO$^\bullet$ from a reaction between $O_2^{\bullet-}$ and H_2O_2 in the presence of iron or copper may play a role in the regulation of γ-aminobutyric acid/barbiturate receptor-gated chloride ion flux in *rat* synaptoneurosomes by phospholipase A_2 (SCHWARTZ et al. 1988).

8-iso-Prostaglandin $F_{2\alpha}$ and 15-keto-dihydro-prostaglandin $F_{2\alpha}$ measurements in jugular bulb plasma may be used as biomarkers for quantification of free radical catalysed oxidative brain injury and inflammatory response in reperfusion injury (BASU et al. 2000). In *piglets* subjected to 5 min cardiac ventricular fibrillation followed by close-chest cardiopulmonary resuscitation, 8-iso-prostaglandin $F_{2\alpha}$ in the jugular bulb plasma (draining the brain) increased four-fold compared with the control group without cardiac arrest, in which jugular bulb 8-iso-prostaglandin $F_{2\alpha}$ remained unchanged. The 15-keto-dihydro-prostaglandin $F_{2\alpha}$ also increased four-fold in the experimental group.

F_2-Isoprostanes are a reliable indicator of oxidative stress *in vivo* (Roberts and Morrow 2000):

- F_2-Isoprostanes are specific products of lipid peroxidation;
- they are stable compounds;
- levels are present in detectable quantities in all normal biological fluids and tissues, allowing the definition of a normal range;
- their formation increases dramatically *in vivo* in a number of animal models of oxidant injury;
- their formation is modulated by antioxidant status;
- their levels are not affected by lipid content of the diet.

Superoxide anion radicals ($O_2^{\bullet-}$) generated in alloxan diabetic *rat* brain tissue extracts incubated for 10 min at 25 °C with varying concentrations of xanthine (0.05, 0.1, 0.15, and 0.2 mM) plus xanthine oxidase (0.1. 0.2, 0.3, and 0.6 U/ml), caused a decrease in the cytosolic **creatine kinase** (EC 2.7.3.2) activities by 29, 50, 72, and 79 %, respectively (Genet et al. 2000). The addition of 80 μg/ml of superoxide dismutase (EC 1.15.1.1) reversed the depressed creatine kinase activities almost to control values. Hydrogen peroxide (0.001, 0.01, 0.1, and 1 mM) decreased the cytosolic creatine kinase activities by 5, 11, 20, and 40 %, respectively. This decrease in creatine kinase activities was reversed significantly to control values when 10 μg/ml of catalase (EC 1.11.1.6) was added to the reaction mixture during the incubation.

While oxygen free radicals are important mediators of brain injury, questions remain regarding which cell types and enzyme pathways trigger this radical generation.

Microglial cells have been hypothesised to be an important source or radical generation. Owing to their distribution throughout all the regions of the central nervous system, migroglial cells have the potential to modify signalling or to promote oxidative damage in neurones focally and globally (Meda et al. 1995). Microglia, thus, have been implicated as potential effectors of neuronal injury in a variety of chronic neurodegenerative diseases including the AIDS dementia complex (Dickson et al. 1993, Lipton et al. 1994, Lipton and Gendelmen 1995), Alzheimer's disease (Dickson et al. 1993, Yan et al. 1996), and Parkinson's disease (Youdim and Lavie 1994).

In vitro, the oxidative capability of *rat* microglia is double that of thioglycollate-elicited macrophages, and both granulocyte-macrophage colony stimulating factor (GM-CSF) and IL-1β significantly increased the thiobarbituric acid-reactive substances (TBARS) level by 36 % ($P < 0.003$) and 34 % ($P < 0.05$), respectively (Smith et al. 1998).

Primary mixed cultures prepared from the brains of newborn Wistar *rats* were able to remove added oxidized laminin and myelin basic protein from the extracellular environment (Stolzing et al. 2002). Moderately oxidized proteins were degraded most efficiently, whereas strongly oxidized proteins were taken up by the microglial cells without an efficient degradation. Stimulation of microglial cells by lipopolysaccharide from *Escherichia coli* (10 μg/ml) or 12-*O*-tetradecanoylphorbol-13-acetate (10 μM) enhanced the selective recognition and degradation of moderately oxidized protein substrates by proteases. Inhibitor studies also revealed an involvement of the lysosomal and the proteasomal systems in the degradation of extracellular proteins.

In *human* glial cells, cyclooxygenase-2 was localised in the perinuclear region even in control brains (Tomimoto et al. 2000). This immunolabelling was more intense and occurred also in the glial cytoplasm with chronic cerebral ischaemia such as Binswanger's disease. Double-labelling immunohistochemistry confirmed that cycooxygenase-2-immunoreactive glia were mostly microglia. These results indicated that prostanoid synthesis is upregulated in microglia during chronic cerebral ischaemia, and that these cells may be involved in tissue repair or inflammation-mediated responses. Iadecola et al. (2001) demonstrated that cyclooxygenase-2-deficient mice have a significant reduction in the brain injury produced by occlusion of the middle cerebral artery.

In microglial N9 cells grown in RPMI 1640 + 15 % heat-inactivated serum + 5 μM dimethyl sulfoxide, oxidised low density lipoprotein caused a dose-dependent increase in 2,7-dichlorofluorescin fluorescence, which was inhibited by pre-treatment with antioxidants, consistent with the formation of reactive oxygen species (Keller et al. 1999).

Microglia, when stimulated by phorbol ester or opsonified zymosan, gave rise to electron paramagnetic resonance spectra characteristic of superoxide (Sankarapandi et al. 1998). Experiments performed in the presence of superoxide dismutase, catalase, deferoxamine, and dimethyl sulfoxide excluded generation of hydroxyl radicals in significant amounts. Microglial superoxide generation was blocked by the NADPH oxidase inhibitor diphenylene iodonium in a manner similar to that seen in neutrophils, suggesting that a neutrophil like NADPH oxidase was the source of the superoxide production. Western blots of microglia lysates demonstrated that both large (gp91-*phox*) and small (p22-*phox*) NADPH oxidase subunits are expressed in both unstimulated and stimulated microglia.

Ullrich et al. (2001) demonstrated an intracellular signalling pathway, which apparently enables microglial cells to convert from the resting to the activated state, to express migration-relevant adhesion molecules, and to cope with the oxidative challenges during their activated state. Chemical inhibition of the nuclear stress-responsive enzyme poly-ADP-ribose-polymerase or depletion by antisense-experiments contributed to a reduced activation response of BV-2 microglial cells after TNF-α or lipopolysaccharide administration. After supercultivation onto *N*-methyl-D-aspartic acid-damaged organotypic hippocampal slice cultures with a low level of primary neuronal damage contributed to a severe secondary damage by the activated microglial cells. In contrast, supercultivation of antisense poly-ADP-ribose-polymerase-BV-2 cells or chemical inhibition of the poly-ADP-ribose-polymerase reduced distinctly this secondary microglial-induced neuronal damage.

Addition of •NO-generating compounds to *murine* BV-2 microglial cells impaired phagocytosis of coated latex micropheres as compared to untreated microglia (Kopec and Carroll 2000). The addition of nitric oxide synthase inhibitors (1.0 μM sodium 7-nitroindazole and 0.25 mM N^G-monomethyl-L-arginine) to microglial cells resulted in potentiation of phagocytosis, suggesting that constitutive NOS was participating in the regulation of phagocytosis. Cell extracts prepared from untreated microglia were found to contain both neuronal and endothelial NOS isoforms, but not the inducible form.

In BV-2 *mouse* microglial cells and *rat* primary microglial cell cultures exposed to lipopolysaccharide and interferon-γ, •NO was produced in a manner dependent on time and dose of activating agents (Lee et al. 2001). Inhibition of •NO synthesis by iNOS inhibitor *N*-monomethyl-L-arginine blocked the apoptosis of activated microglial cells.

Using *human* ramified microglial cells (which express no constitutive NOS isoforms), it has been conformed that a pre-treatment of cells with nearly physiological levels of exogenous •NO suppressed NOS-II mRNA expression as elicited by lipopolysaccharide plus TNF-α, this effect being ascribed to an inhibitory action of low levels of •NO on NF-κB activation (Colosanti et al. 1995; for review see Colosanti and Persichini 2000).

Microglia are an obvious target for the infective agent in spongiform encephalopathies (Brown et al. 1998) and they have the potential to become activated to produce cytotoxic and inflammatory mediators in response to disease-causing PrP^{Sc} (Peyrin et al. 1999, Silei et al. 1999). PrP mRNA has been detected in isolated microglial cells (Brown et al. 1990), and microglia were immunoreactive for PrP^{Sc} in a *mouse* model for scrapie (Bruce et al. 1989), but this has not been confirmed (Williams et al. 1994, Mühleisen 1995). The accumulation of microglia around prion plaques differs from microglial 'rosettes' observed around amyloid plaques in Alzheimer's disease (Rezaie and Lantos 2001).

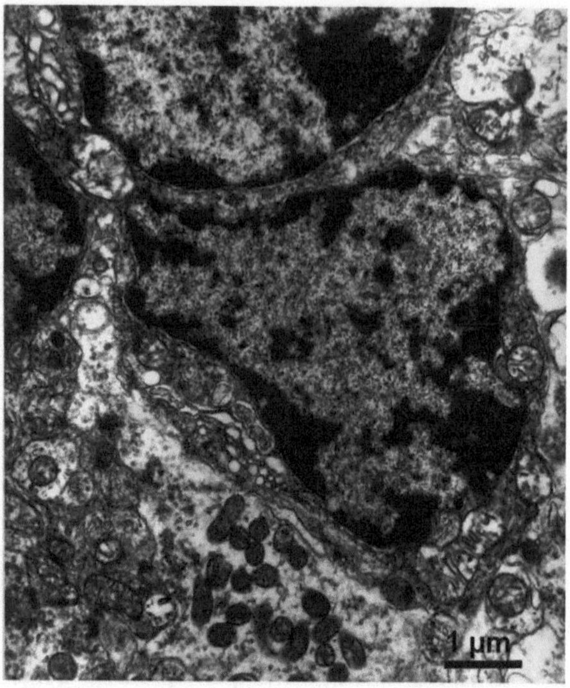

Fig. 213. Prominent Golgi areas in two microglial cells of the cerebellar cortex (block 4766) of a 221 g female *rat* (No. 668) medicated with 5,000 ppm molsidomine added to the food (powdered Altromin® R) from July 17/18, 1978 to November 30, 1978. After discontinuation of the medication for 7 weeks, under pentobarbital anaesthesia (30 mg/kg), the animal was perfused from the abdominal aorta with 2.5 % glutaraldehyde in 0.1 M sodium cacodylate buffer (pH 7.4). Postfixation with 1 % osmium tetroxide in sodium cacodylate buffer. Embedded in Epon 812 and sectioned at 50 nm. Lead citrate and uranyl acetate. Film 76/79

Astrocytes of both the grey and white matter showed a moderate dilatation of the cisternae of the endoplasmic reticulum in the first six days, when *rats* were exposed continuously to 10 % oxygen (Yu et al. 1972). From the 7th day on, in the white matter many astrocytes showed a profuse accumulation of gliofilaments within their cytoplasm pushing aside other organelles. In the grey matter, many astrocytes underwent marked distension with increased translucency of their cytoplasm, greatly dilated endoplasmic reticulum cisternae and bundles of haphazardly arranged gliofilaments in the cytoplasm. The astrocytic processes were often arranged in lamellar whorls sometimes encircling part of the cytoplasm with organelles or wrapping presynaptic terminals.

The mannose receptor in astrocytes was identified and characterised by BURUDI et al. (1999). In the BALD/c *mouse*, its expression is at its highest in the first week of life and dramatically decreases thereafter, being maintained at a low level throughout adulthood. The receptor is present in most brain regions at an early postnatal age (BURUDI and RÉGNIER-VIGOUROUX 2001).

Upon exposure to glucose deprivation or H_2O_2, uninfected and Lac-Z-expressing postnatal (days 1–3) Swiss-Webster *mouse* astrocytes showed an immediate, rapid increase in the accumulation of reactive oxygen species that was slowed or reduced by Bcl-x_L (OUYANG et al. 2002). Changes in the mitochondrial membrane potential ($\Delta\psi_m$) in response to the two insults differed. H_2O_2 induced a decrease in $\Delta\psi_m$ that was initially greater in Bcl-x_L cells, but then held stable. $\Delta\psi_m$ in control and Lac-Z-expressing cells initially declined more slowly, but after about 20 min showed rapid deterioration. Five hours of glucose deprivation caused mitochondrial membrane hyperpolarization followed by a decrease in $\Delta\psi_m$, which was not observed with Bcl-x_L overexpression. Bcl-x_L failed to inhibit the calcium dysregulation seen in control cells exposed to 400 μM H_2O_2, but still improved cell survival. There was no increase in $[Ca^{2+}]_i$ with 5 h glucose deprivation.

In *rat* cortical astrocytes, the interplay between mitochondrial reactive oxygen species and endoplasmic reticulum sequestered Ca^{2+} increased the frequency of transient mitochondrial depolarisations and caused mitochondrial Ca^{2+} loading from ER stores (JACOBSON and DUCHEN 2002). The depolarisations were attributable to opening of the mitochondrial permeability transition pore. Initially, transient events were seen in individual mitochondria, but ultimately, the mitochondrial potential ($\Delta\psi_m$) collapsed completely and irreversibly in the whole population. Both reactive oxygen species and endoplasmic reticulum Ca^{2+} were requires to initiate these events, but neither alone was sufficient.

Imidazoline-2 receptors are highly distributed on astrocyte mitochondrial membranes (TESSON et al. 1992, REGUNATHAN et al. 1993). In primary cultures of neonatal *rat* astrocytes, imidazoline drugs, such as idazoxan, guanabenz, guanfacine, BU224, and RS-45041-190, showed protective effects against naphthazarin-induced oxidative cytotoxicity, as evidenced by lactate dehydrogenase (EC 1.1.1.27) release and Hoechst 33341/propidium iodide staining (CHOI et al. 2002). The imidazoline drugs stabilized lysosomes and inhibited naphthazarin-induced lysosomal destabilization, as evidenced by acridin orange relocation. Guanabenz inhibited the leakage of lysosomal cathepsin D (EC 3.4.23.5) to cytosol, the decreased mitochondrial potential, and the release of mitochondrial cytochrome *c*, which were induced by naphthazarin. The lysosomal destabilization by oxidative stress and other apoptotic signals and subsequent cathepsin D leakage to the cytosol can induce apoptotic changes of mitochondria and eventually cell death.

In TNC1 *rat* astrocytes grown in Dulbecco's modified Eagle's medium + 10% fetal *bovine* serum, oxidised low density lipoprotein caused a dose-dependent increase in 2,7-dichlorofluorescin fluorescence, which was inhibited by pre-treatment with antioxidants, consistent with the formation of reactive oxygen species (KELLER et al. 1999).

Linear regression analysis showed that control mesencephalic astrocytes from the new-born C57BL/6 *mouse* brain (slope coefficient = 0.01) had a three-fold (*F*-test, $P < 0.05$) greater rate of change in reactive oxygen species production when compared to cortical (0.003) or striatal (0.003) astrocytes (WONG et al. 1999). However, when treated with 500 µM 1-methyl-4-phenyl-1,2,3,6-tetrahydropyridine for 120 min, mesencephalic and striatal astrocytes demonstrated a decreased and increased rate of change in reactive oxygen species production, respectively. On the other hand, when exposed to 10 µM 1-methyl-4-phenylpyridinium, a significant increase in the rate of change in reactive oxygen species formation occurred in both mesencephalic and striatal astrocytes, with mesencephalic astrocytes producing a four-fold greater increase when compared to striatal astrocytes. Cortical astrocytes did not show any significant changes in reactive oxygen species production when treated with 1-methyl-4-phenyl-1,2,3,6-tetrahydropyridine or 1-methyl-4-phenylpyridinium. When striatal astrocytes were challenged to 1-methyl-4-phenyl-1,2,3,6-tetrahydropyridine over a 24 h period, their superoxide dismutase activity increased significantly reaching a maximum by 8 h, followed by a return to control levels at 24 h, while mesencephalic astrocytes did not show any change in SOD activity up to 8 h, then a decrease was noted reaching approximately 32% of the control levels by 24 h.

Using concentrations of menadione (100 µM for 1 h) that resulted in comparable initial mtDNA damage, more efficient repair was observed in *rat* astrocytes compared to either oligodendrocytes or microglia (HOLLENSWORTH et al. 2000). The differential susceptibility of glial cell types to oxidative damage and oligodendrocytic and microglial apoptosis did not appear related to cellular antioxidant capacity, because under culture conditions astrocytes had lower glutathione content and superoxide dismutase activity than oligodendrocytes and microglia.

Vasorelaxing nitrogen oxides were released from astrocytes derived from the neonatal *rat* cerebral cortex and maintained in culture when exposed to bradykinin ($EC_{50} \cong 20$ nM), norepinephrine ($EC_{50} \cong 100$ nM), and quisqualate ($EC_{50} \cong 1$ μM) (MURPHY et al. 1991). Norepinephrine and quisqualate activate nitric oxide synthase via the α_1-adrenergic and the metabotropic quisqualate receptors, respectively. Although all free receptors are linked via inositol phospholipid hydrolysis to calcium mobilisation in astroglial cells, the observation that thapsigargin (100 nM) did not cause the release of nitrogen oxides suggests that calcium influx is required.

Astrocytes activated by incubation with lipopolysaccharide/interferon γ for 18 h, induced nitric oxide synthase and, hence, continuously released nitric oxide (BOLAÑOS et al. 1996). coincubation for 24 h of activated astrocytes with neurones caused a limited loss of complex IV activity and had no effect on the activity of complexes I or II–III. However, neurones exposed to astrocytes had a 1.7-fold increase in glutathione concentration compared to neurones cultured alone. Under these coculture conditions, the neuronal ATP concentration was modestly reduced (14 %). This loss of ATP was prevented by the nitric oxide synthase inhibitor, N^G-monomethyl-L-arginine.

Astrocytic $^{\bullet}NO/ONOO^-$ caused significant damage to the activities of complex II/III and IV of neighbouring neurones after a 24-h coculture (STEWART et al. 1998). Using polytetrafluoroethane filters, which are permeable to gases such as $^{\bullet}NO$ but impermeable to NO derivatives, STEWART et al. (2000) demonstrated that astrocyte-derived $^{\bullet}NO$ is responsible for the damage observed in their coculture system. 24 h after removing of $^{\bullet}NO$-producing astrocytes, neurones exhibited complete recovery of complex II/III and IV activities.

L-Buthionine-[*S*,*R*]-sulfoximine-induced glutathione depletion did not alter $^{\bullet}NO$, H_2O_2, or glutamate levels in cultured astrocytes from 1 2 days old Wistar *rats* (MCNAUGHT and JENNER 2000). Inhibition of complex I activity by up to 43 % had no effect on extracellular $^{\bullet}NO$ accumulation, but increased H_2O_2 and glutamate levels. Lipopolysaccharide-induced activation of cultured astrocytes increased extracellular levels of $^{\bullet}NO$, H_2O_2, or glutamate. Extracellular accumulation of $^{\bullet}NO$ and H_2O_2 caused by lipopolysaccharide was markedly less in glutathione-depleted or complex I-inhibited astrocytic cultures compared to normal astrocytic cultures.

SHARPE et al. (2002) using primary *rat* cortical astrocyte cultures investigated EUK-8 and EUK-134 [manganese 3-methoxy *N,N'*-bis(salicylidene-ethylenediamine chloride] as possible therapeutic agents in neurological disorders resulting from oxidative stress, including Alzheimer's disease, Parkinson's disease, stroke, and multiple sclerosis. They found that in the presence of a per-species (H_2O_2, $ONOO^-$, peracetate and persulphate), the Mn-salen complexes are oxidised to the corresponding oxo-species (oxoMn-Salen). oxoMn-Salens are potent oxidants, and they can rapidly oxidise NO to NO_2 and also oxidise nitrite (NO_2^-) to nitrate (NO_3^-). Thus these Mn-salens have the potential to ameliorate cellular damage caused by oxidative and nitrosative stressors, by the catalytic breakdown of O_2^-, H_2O_2, $ONOO^-$, and $^{\bullet}NO$ to benign species: O_2, H_2O, NO_2^-, and NO_3^-.

Although glutamate transporters are expressed by all CNS cell types, astrocytes are the cell type primarily responsible for glutamate uptake. Factors that influence glutamate transporter expression have been reviewed extensively by GEGELASHVILI and SCHOUSBOE (1997). ANDERSON et al. (2000) reviewed the mechanisms of astrocyte glutamate uptake and release, with particular focus on high-affinity Na^+-dependent transporters. FIGIEL et al. (2001) reported that the expression of both GLT-1 and GLAST were robustly promoted by TGFα, EGF, and GDNB as revealed by immunoblot analysis of cultured cortical astroglia. FGF-2 and PDGF-BB can exert similar stimulatory effects on GLT-1 and GLAST expression when the ERK activity is inhibited with PD98059. The stimulatory effects of the various factors on glial glutamate transporter expression are not additive and all involve nuclear factor ϰB as a downstream signal.

Filopodia retraction can be mediated by glutamate or several subtype specific metabotropic glutamate receptor ligands (DEROUICHE et al. 2001). As shown by immuno-electron microscopy in *rat* hippocampus, these metabotropic glutamate receptor subtypes are also preferentially localised in the peripheral glial processes. Their topic relation to the synapse may thus be regulated by glutamate.

Dopamine sensitivity of cortical and striatal astrocytes was altered by long-term application of either dopamine or epinephrine, but not serotonin (REUSS and UNSICKER 2001). Preincubation of cortical and striatal cultures with dopamine (>1 μmol/l) or epinephrine (>0.1 μmol/l) significantly reduced astrocyte dopamine sensitivity. While the dopamine-mediated down-regulation of astroglial dopamine sensitivity was antagonised by SCH23390 (>1 μmol/l) and sulpiride (>10 μmol/l), epinephrine-mediated effects were not. Haloperidol and clozapine both failed to block down-regulation of astroglial dopamine sensitivity by either dopamine or epinephrine.

Exposure of a primary culture of *rat* glial cells to the classical neurotoxicant trimethyltin resulted in the release of prostaglandin E_2 and tumour necrosis factor-α (VIVIANI et al. 2001). Prior treatment of glial cells with either the non-specific inhibitor of cyclooxygenase and lipoxygenase eicosatetraenoic acid or the cyclooxygenase inhibitor indomethacin completely prevented trimethyltin-induced prostaglandin E_2 production and tumour necrosis factor-α release, suggesting a role for cyclooxygenase metabolites in trimethyltin-induced tumour necrosis factor-α release. Exposure of glial cells to increasing concentrations of prostaglandin E_2 or other prostanoids did not increase TNF-α synthesis, while the presence of exogenous prostaglandin E_2 during treatment of glial cells with trimethyltin actually suppressed TNF-α release. The activation of arachidonic acid metabolism produces reactive oxygen species. Scavenging reactive oxygen species by the antioxidant trolox prevented the trimethyltin-induced release of TNF-α from glial cells, while indomethacin was found to suppress the formation of reactive oxygen species induced by 1 μM trimethyltin in glial cells.

Oligodendrocytes do not show anoxic damage as rapidly as the astrocytes (HILLS 1964). A few hours after the anoxic episode, the endoplasm reticulum cisternae and mitochondria may swell slightly. In the later periods of hypoxia oligodendrocytes showed marked dilatation of the cisternae of the endoplasmic reticulum, partly in the peripheral zone of the perikarya (YU et al. 1972).

Oligodendrocyte progenitors were more vulnerable to Cd^{2+} toxicity mediated by reactive oxygen species than were mature oligodendrocytes (ALMAZAN et al. 2000). Pre-treatment with *N*-acetylcysteine, a thiocompound with antioxidant activity and precursor of glutathione, prevented Cd^{2+}-induced reduction in glutathione levels and induction of 72 kDa stress protein and diminished Cd^{2+} uptake and Cd^{2+}-evoked cell death.

Cultures enriched for microglia, astrocytes, or oligodendrocytes treated with *S*-nitroso-*N*-acetyl D,L-penicillamine (•NO releasing chemical) showed a significant decrease in the function of the ferrosulphur-containing mitochondrial enzyme, succinate dehydrogenase, in oligodendrocytes and astrocytes, whereas microglia were unaffected (MITROVIC et al. 1994). A subpopulation of oligodendrocytes were killed by •NO via a necrotic, non-apoptotic mechanism (MITROVIC et al. 1995).

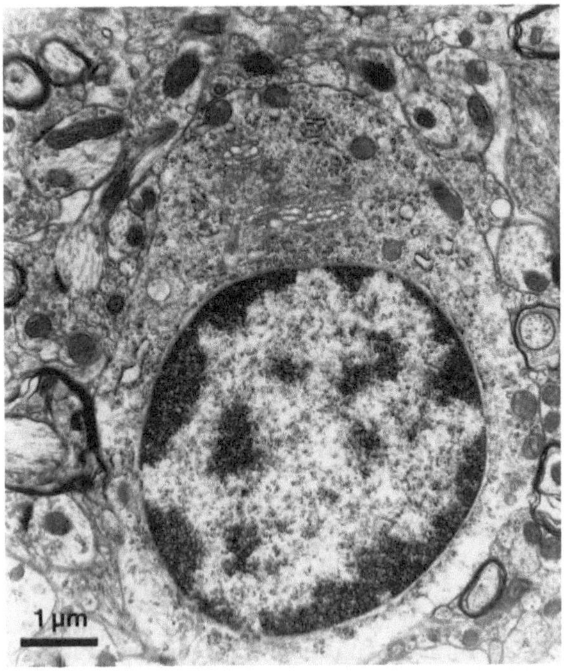

Fig. 214. Oligodendrocyte in the corpus striatum (block 4578) of a *rat* medicated with 5,000 ppm molsidomine added to the food (powdered Altromin® R) from July 17, 1978 to November 30, 1978. After discontinuation of the medication for 7 weeks, under pentobarbital anaesthesia (30 mg/kg), the animal was perfused from the abdominal aorta with 2.5% glutaraldehyde in 0.1 M sodium cacodylate buffer (pH 7.4). Postfixation with 1% osmium tetroxide in sodium cacodylate buffer. Embedded in Epon 812. Lead citrate and uranyl acetate. Film 558/79

Adenosine formation is accelerated under hypoxic conditions (RUDOLPHI et al. 1992), but it is metabolised only after reperfusion. Now superoxide dismutase and catalase are insufficient to serve detoxification of the reactive oxygen species resulting from the biotransformation of hypoxanthine to xanthine and of xanthine to uric acid. While xanthine oxidase inhibitors as oxypurinol do protect against ischaemic damage (HELFMAN and PHILLIS 1989, LIN and PHILLIS 1992), adenosine deaminase inhibitors as trazodone (SHEID 1985) are expected to reduce the formation of hypoxanthine and xanthine, the substrates for xanthine oxidase.

The adenosine precursor **5-aminoimidazole-4-carboxamide riboside**, reduced hippocampal cell death caused by 5-min bilateral carotid occlusion in *gerbils* (CLOUGH-HELFMAN and PHILLIS 1990). R-N^6-phenylisopropyladenosine could prevent kainate-induced damage resembling some forms of ischaemia of the hippocampus at doses as low as 10μg/kg *rat* i.p. (MACGREGOR et al. 1993), and at a dose of 25 mg/kg i.p. the protection could be completely abolished by the simultaneous administration of 1,3-dipropyl-8-cyclopentylxanthine, indicating the involvement of adenosine A_1 receptors in the protection (MacGregor et al. 1996).

Making transgenic *mice* of strain Tg HS/sF-218 carrying SOD-1 genes and thus overexpressing *human* CuZn-**superoxide dismutase** in brain cells allowed Chan (1994) to disprove the specific role of superoxide radicals in reperfusion injury after focal ischaemia. Mn-superoxide dismutase became transiently upregulated in the entire brain after focal cortical ischaemia (Bidmon et al. 1998). Mn-superoxide dismutase and glutathione peroxidase (EC 1.11.1.9) were increased in the surviving tissues and entire brain of preconditioned *rats*, whereas Mn-superoxide dismutase was only increased within or in the vicinity of the lesions induced by transient (1 h) occlusion of the right cerebral artery in non-preconditioned animals indicating that upregulation of MnSOD has the potential to be one of the factors contributing to the protective effect in preconditioning (Bidmon et al. 1999). When a bolus of 6×10^3 IU/2 ml *human* recombinant CuZn-superoxide dismutase per kg *rat* was administered intravenously 72 h before transient occlusion of the middle cerebral artery, the infarct area and volume assessed with the 2,3,5-triphenyltetrazolium stain showed that the administration of *human* recombinant CuZn-superoxide dismutase suppressed the development of neuronal tolerance induced by preconditioning (Mori et al. 2000). Expression of 72-kDa heat-shock protein at 72 h after preconditioning was considerably reduced in *rats* treated with *human* recombinant CuZn-superoxide dismutase compared with those treated with vehicle.

In CuZn-superoxide dismutase-transfected BV-2 *murine* microglial cells, the expression and activity increased after lipopolysaccharide stimulation (Chang et al. 2001). On the other hand, upon activation by LPS, these cells produced less •NO, IL-1β, and TNF-α than the parental microglial cells.

Hypoxia (9 % O_2 for 1 h) preconditioning afforded neuroprotection against a second harmful event as pentylenetetrazol (55 mg/kg i.p.)-induced seizures (Rauca et al. 2000). *Rats* pre-treated with NaCl showed a significantly higher extent of free hydroxyl radicals in the whole brain (without cerebellum) compared with *N-tert*-butyl-α-phenyl-nitrone-injected preconditioned animals and with naive and sham-exposed controls.

The quite rapid reaction of $O_2^{\bullet-}$ with •NO constantly being formed in the brain by neuronal (Schilling et al. 1993, Lüth and Winkelmann 1994), endothelial (Morin and Stanboli 1993) and glial (Murphy et al. 1993) nitric oxide synthase forms peroxynitrite, which at neutral pH instantly degrades to **hydroxyl radicals** and $^{\bullet}NO_2$ radicals.

Cobbs et al. (2001) showed that p53, a key tumour suppressor protein, has evidence of peroxynitrite-mediated modifications in gliomas *in vitro*.

At least three forms of **nitric oxide synthase** have been characterised in brain cells. Neurones produce •NO mainly by Ca^{2+}-dependent activation of neuronal nitric oxide synthase (nNOS or NOS1), which is constitutively expressed in these cells (Knowles and Moncada 1994). In the brain regions of *rat*, *mouse*, and *bovine*, the highest expression of nNOS mRNA has been observed in cerebellum (Sessa et al. 1993). In *human* brain, however, although the level of nNOS mRNA constitutively expressed in cerebellum was high, the highest expression was observed in putamen and caudate nucleus (Park et al. 2000).

Glial cells (astrocytes, oligodendrocytes and microglia cells) synthesise •NO mainly after Ca^{2+}-independent inducible NOS expression (iNOS or NOS 2) by treatment with the endotoxin lipopolysaccharide (LPS) and/or certain cytokines, such as interferon-γ, tumour necrosis factor-α or interleukin-1β (for reviews see Murphy et al. 1993, Murphy and Guzybicki 1996).

Endothelial cells produce ⁻NO by the constitutive, Ca^{2+}-dependent activity of endothelial NOS (eNOS or NOS3) (Knowles and Moncada 1994). Inhibition of NOS activity by L-N^{ω}-nitro-L-arginine (L-NNA) or L-Nω-nitro-L-arginine methyl ester (L-NAME) enhanced the infarcted area in experimental transient ischaemia, suggesting that increased •NO production during ischaemia would be protective because it modulates the cerebral blood flow (Shapira et al. 1994).

•NO mediates certain aspects of synaptic plasticity and neurotoxicity associated with *N*-methyl-D-aspartate receptors, but does not play a major role in other pathways. The coupling between *N*-methyl-D-aspartate receptor activation and nNOS activity is mediated by a scaffolding protein, PSD-95, that employs PDZ protein domains to physically bridge nNOS to the *N*-methyl-D-aspartate receptor (Bredt 2000).

Cerebral global transient ischaemia in *rats* induced a rise in cNOS activity from 62.0 ± 6.1 to 133.3 ± 13.3 pmol/min × mg protein and cGMP levels from 459.3 ± 49.6 to 1974.1 ± 132.1 fmol/mg protein (Fernández et al. 1997). Pre-treatment with 10 mg desmethyl tirilazad/kg intraperitoneally abolished these increases.

The relative amount of **nitric oxide** derived from eNOS and nNOS was accessed using eNOS(–/–) or nNOS(–/–) *mice* and matched wild control *mice* (Wei et al. 1999). •NO was trapped using Fe(II)-diethykldithiocarbamate. In wild-type *mice*, only small •NO signals were seen prior to ischaemia, but after 10 to 20 min of ischaemia the signals increased more than 4-fold. This •NO generation was inhibited more than 70 % by NOS inhibition. In either nNOS(–/–) or eNOS(–/–) *mice* before ischaemia, •NO generation was decreased about 50 % compared to that of wild-type *mice*. Following the onset of ischaemia a rapid increase in •NO occurred in nNOS(–/–) *mice* peaking after only 10 min. The production of •NO in the eNOS(–/–) *mice* paralleled that in the wild type with a progressive increase over 20 min, suggesting progressive accumulation of •NO from nNOS following the onset of ischaemia. NOS activity measurements demonstrated that eNOS (–/–) and nNOS(–/–) brains had 90 % and < 10 %, respectively, of the activity measured in wild type.

Nitric oxide levels changed during global *rat* forebrain ischaemia and reperfusion as measured *ex vivo* by electron paramagnetic resonance spectroscopy (Shutenko et al. 1999). The spin trap used was diethyldithiocarbamate sodium salt (DETC) associated with ferrous citrate. The complex Fe(DETC)$_2$NO was detected at 77 K as a triplet signal at $g = 2.035$. In intact anaesthetised *rats*, the signal was about three times grater in the cortex than in the cerebellum. During ischaemia, the signal rose to 110 % in cortex (not significant) and to 283 % in cerebellum ($P < 0.05$). In reperfusion, it fell again to 91 % of control in cerebellum (not significant) and 35 % in cortex ($P < 0.05$).

In a *rat* endovacular middle cerebral artery occlusion stroke model the relative brain •NO concentration was increased to 131.94 ± 7.99 % at 15 min of artery occlusion (Chen et al. 2002). Neuropeptide Y treatment further increased the relative brain •NO concentration to 250.94 ± 50.48 %, whereas the Y1 receptor antagonist BIBP3226 significantly reduced •NO concentration to 69.63 ± 8.84 %. [Leu^{31},Pro^{34}]-neuropeptide Y or neuropeptide Y3-36 did not affect the brain •NO concentration at 15 min of artery occlusion.

In *murine* primary neurone-glia co-cultures the production of •NO was first detectable 9 h after the exposure to lipopolysaccharide/interferon γ and increased for up to 48 h (Jeohn et al. 2000). A significant neuronal cell death was observed 36–48 h after treatment with LPS/IFN-γ. The •NO generated at the initial stage of •NO synthesis (about 12 h) following exposure to LPS/IFN-γ was found to be critical for LPS/IFN-γ-induced neurotoxicity.

Ni^{2+} inhibited nNOS in a competitive, reversible manner with respect to the substrate L-arginine (K_i = 30 ± 4 μM; Palumbo et al. 2001). The IC_{50} values were dependent on calmodulin concentration, but proved independent of Ca^{2+}, tetrahydrobiopterin and other essential cofactors. Ni^{2+} also inhibited calmodulin-dependent cytochrome *c* reduction, NADPH oxidation, and H_2O_2 production by nNOS.

Pre-treatment with the NOS inhibitor N^G-monomethyl-L-arginine impaired responding for cocaine self-administration when the drug was available and the increase of drug-seeking behaviour upon abrupt cessation of cocaine availability observed in control *rats* significantly reduced after treatment with N^G-monomethyl-L-arginine (Orsini et al. 2002).

Discovered in the normal *mammalian* brain (Shintani et al. 1996) and in the *rat* spinal cord (Chiari et al. 2000), **6-nitrocatecholamines**, e.g. 6-nitrodopamine and 6-nitronorepinephrine, attract increasing interest as potential indicators of biochemical interactions between catecholamine neurotransmitters and the nitric oxide signalling pathway. Oxidation of 6-nitrodopamine and 6-nitronorepinephrine, as well as of the model compounds 4-nitrocatechol and 4-methyl-5-nitrocatechol, with horseradish peroxidase/H_2O_2, lactoperoxidase/H_2O_2, Fe^{2+}/H_2O_2, Fe^{2+}-EDTA/H_2O_2 (Fenton reagent), horseradish peroxidase or Fe^{2+}/EDTA in combination with D-glucose oxidase, or Fe^{2+}/O_2, resulted in the smooth formation of yellowish-brown pigments positive to the Griess assay (Palumbo et al. 2001). In the pH range 3–6, Griess positive pigment induced concentration- and pH-dependent nitrosation of 2,3-diaminonaphthalene, but very poor (up to 2 %) nitration of 600 μM tyrosine.

•NO reacts with the deprotonated form of 6-hydroxydopamine at pH 7 and 37 °C with a second-order rate constant of 1.5×10^3 M^{-1} s^{-1} as calculated by the rate of •NO decay measured with an amperometric sensor (Riobó et al. 2002). Accordingly, the rates of formation of 6-hydroxydopamine quinone were dependent on •NO concentration. The coincubation of •NO and 6-hydroxydopamine with either *bovine* serum or α-synuclein led to tyrosine nitration of the protein, in a concentration-dependent manner and sensitive to superoxide dismutase.

Nitrotyrosine generated via inducible nitric oxide synthase (iNOS) in vascular wall in focal ischaemic-reperfused C57BL/6J *mice* was detected only in wild-type *mice*, not in iNOS knockout or sham-operated *mice* (Hirabayashi et al. 2000). Immunohistochemical reaction for nitrotyrosine was predominantly in the vascular wall in the peri-infarct region of the cerebral cortex in wild-type *mice* after 15 h reperfusion, but not in corresponding knockout *mice*.

Increased levels of both iNOS mRNA and protein nitrotyrosine expression were observed in *rats* with experimental allergic encephalomyelitis, which also showed lower levels of thermolabile NOS inhibitor in whole encephalic mass homogenates and sera than controls (Teixeira et al. 2002).

Inactivation of •NO in both cerebellar cell suspensions and brain homogenates required O_2 and, from measurement of homogenates, the principal end product was NO_3^-, which is also the main product of endogenously formed •NO *in vivo* (Griffiths et al. 2002). Direct chemical reaction with O_2, superoxide anions or haemoglobin was not responsible. The capacity of the •NO sink in cells was limited.

Peroxynitite ($OONO^-$) stimulated the release of neurotransmitters including γ-aminobutyric acid (GABA) and acetylcholine from *mouse* cerebral cortical neurones (Ohkuma et al. 1995). Hydroxyl radical scavengers such as *N,N'*-dimethylthiourea (DMTU), mannitol, and uric acid significantly increased $OONO^-$-evoked [^{3}H]GABA release, whereas urea showed no effects on the release (Higo et al. 1998). Removal of Ca^{2+} from incubation buffer abolished the enhancement of the release by DMTU, although DMTU showed no effects on the basal release with and without Ca^{2+} in extracellular space.

Gutiérrez-Martín et al. (2002) simulated chronic exposure to $OONO^-$ by treatment of *rat* brain synaptosomes or plasma membrane vesicles with repetitive pulses of $OONO^-$ during at most 50 min, which efficiently produced nitrotyrosine formation in several membrane proteins including the Ca^{2+}-ATPase. The plasma membrane Ca^{2+}-ATPase activity at near-physiological conditions (pH 7, submicromolar Ca^{2+}, and millimolar Mg^{2+}-ATP concentrations), which plays a major role in the control of synaptic $[Ca^{2+}]_i$, could be more than 75 % inhibited by a sustained exposure to micromolar $OONO^-$ (e.g., to 100 pulses of 10 μM $OONO^-$). This inhibition was irreversible and mostly due to decreased V_{max}, and to a 2-fold increase of the $K_{0.5}$ for Ca^{2+} stimulation and about 5-fold increase of the K_M for Mg^{2+}-ATP. $[Ca^{2+}]_I$ increased to >400 nM when synaptosomes were subjected to this treatment. Reduced glutathione could afford only partial protection against the inhibition produced by micromolar $OONO^-$ pulses.

Histochemical activation of **guanylate cyclase** (EC 4.6.1.2) in the *rat* hippocampus could be reliably demonstrated only with nitroprusside and nitro-glycerine (Poeggel et al. 1992). The enzyme was localised in small and medium-sized neurones and astroglial cells. Intensely stained were postsynaptic densities and glial cell processes. Surrounding capillaries the reaction was restricted to microvascular glial cell processes and basement membrane. Kainate lesions of the cerebellum resulted in a nearly complete loss of basal guanylate cyclase activity (Biggio et al. 1978).

Incubated tissue slices from different regions of the *rat* brain contained **cGMP** in the following descending order of content: cerebellum > hypothalamus > striatum > thalamus-midbrain > brain stem > hippocampus > cerebral cortex (Palmer and Duszynski 1975). cGMP levels were increased in incubated tissue slices from *rat* cerebral cortex in response to added cholinomimetic agents (carbachol and choline chloride) and neuroleptic compounds (chlorpromazine, 8-hydroxychlorpromazine, 7-hydroxychlorpromazine methiodide, haloperidol, thioridazine, chlorpromazine sulphoxide and promethazine) (Palmer et al. 1976). Calcium ions were required for this effect.

The translocation of extracellular **calcium** into the cell during ischaemia, the precipitous rise in the free cytosolic calcium concentration, and the role of calcium in activating lipases, proteases, kinases, phosphatases, and endonucleases in potentially harmful metabolic cascades supported the idea of calcium to be a mediator of ischaemic brain damage. *In vitro* and *in vivo* experiments suggest that the main route of calcium entry is through channels gated by **glutamate receptors** (Siesjö et al. 1995). Piani et al. (1991) have shown that microglia *in vitro* produce high amounts of glutamate. In *cats*, prolonged transient ischemia induced by middle cerebral artery occlusion resulted in secondary glutamate accumulation (Taguchi et al. 1996). The release of excitatory amino acids from ischaemic brain slices is increased in the presence of oxygen-derived free radicals (Pellegrini-Giampietro et al. 1990). The phospholipase inhibitors 4-bromophenacyl bromide (non-selective inhibitor), 7,7-dimethyleicosadienoic acid (inhibitor of secretory type phospholipase A_2), $AACOCF_3$ (a trifluoromethyl ketone analogue of arachidonic acid which inhibits the Ca^{2+}-dependent cytoplasmic form of phospholipase A_2), HELSS (inhibitor of a Ca^{2+}-independent cytoplasmic phospholipase A_2), and U-73122 {1-[6-((17β-3-methoxyestra-1,3.5(10)-trien-17-yl)amino)hexyl]-1H-pyrrole-2,5-dione}, a selective inhibitor of phospholipase A_2, all significantly attenuated glutamate and aspartate release into the extracellular milieu (Phillis and O'Regan 1996). The protein kinase C inhibitor, chelerythrine chloride, also reduced excitatory amino acid efflux, whereas the protein kinase C activator phorbol 12-myristate 13-acetate enhanced their release. The non-selective kinase inhibitor, staurosporine, and H-89 (N-[2-((*p*-bromocinnamyl)amino)ethyl]-5-isoquinolinesulfonamide, HCl), which selectively inhibits protein kinase A, did not reduce ischaemia-evoked amino acid efflux. In a neuronal cell line, glutamate toxicity was inhibited by monoamine oxidase (monoamine:O_2 oxidoreductase; EC 1.4.3.4) type-A-specific inhibitors, but only at concentrations much higher than those required to inhibit classical type-A monoamine oxidase (MAHER and DAVIS 1996). Toxicity was not inhibited by MAO type-B-specific inhibitors at any concentration. Furthermore, treatment of cells with agents that block monoamine uptake inhibited glutamate toxicity.

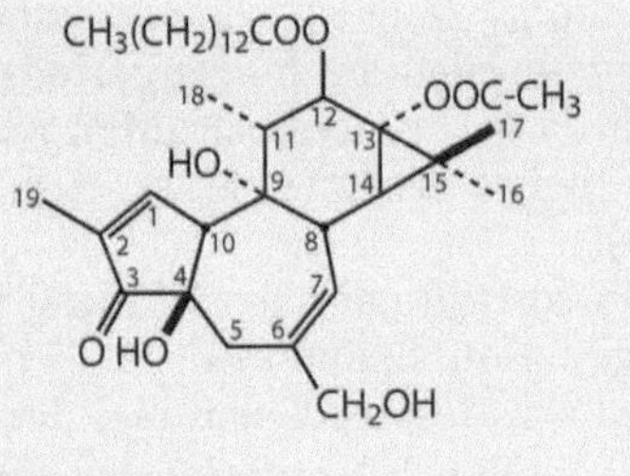

12-*O*-Tetradecanoylphorbol-13-acetate [49]

Glutamate-dependent Ca^{2+} influx into neurones resulted in delayed neuronal cell death and this appears likely to be important in ischaemia. In the studies of cultured neurones, a 30-min treatment with the glutamate receptor agonist, *N*-methyl-D-aspartic acid, induced a transient phosphorylation of eEF2, which preceded neuronal death by several hours (MARIN et al. 1997). Pharmacological inhibition of protein translation at the elongation stage using either cycloheximide or diphtheria toxin protected neurones against the toxicity evoked by low concentrations of *N*-methyl-D-aspartic acid. It has been proposed that glutamate induces either apoptotic or necrotic cell death in neurones, depending on the intensity and duration of the stimulation of *N*-methyl-D-aspartic acid receptors (TOESCU 1998, GWAG et al. 1999).

$HOOC—CH_2—CH_2—CH(NH_2)—COOH$ L-Glutamic acid [273]

The ***N*-methyl-D-aspartic acid receptor complex** consists of a membrane-spanning channel, which is highly permeable to both $Na^+ K^+$ and Ca^{2+} in a voltage-dependent manner and possesses several regulatory sites, including glycine, Zn^{2+}, polyamine, and phenylcyclidine binding sites, all of which allosterically affect glutamate-mediated channel opening (WATKINS et al. 1990). *N*-Methyl-D-aspartate receptor density and binding may be very important in pharmacology of learning abilities (STECHER et al. 1997).

$HOOC—CH_2—CH(NHCH_3)—COOH$ *N*-Methyl-D-aspartic acid [274]

Hypoxia (7 %gen for 1 h)-induced modification of the glycine binding site of the *N*-methyl-D-aspartate receptors might be a potential mechanism of neurotoxicity as shown by RAZDAN et al. (1996) in the *guinea pig*. Blockade of *N*-methyl-D-aspartate receptors by 2-amino-7-phosphonoheptanoic acid may protect against ischaemic damage in the brain (SIMON et al. 1984). Memantine binds non-competitively to the *N*-methyl-D-aspartate receptor and so protects rats from ischaemia of 10 min in a dose-dependent manner (SEIF EL NASR et al. 1990). The non-competitive antagonist dizocilpine, (5*R*, 10*S*)-(+)-5-methyl-10,11-dihydro-5*H*-dibenzo[*a*,*d*]-cyclohepten-5,10-imine (MK-801) in Mongolian *gerbils* subjected to 5 min of bilateral carotid artery occlusion 1 h prior to ischaemia significantly reduced damage to CA1 hippocampal neurones (BUCHAN and PULSINELLI 1996). However, the effect proved to be due to postischaemic hypothermia. After photothrombotic occlusion of the distal middle cerebral artery in *rats*, MK-801 (0.5 mg/kg i.p., 45 min after ischaemia) blocked spontaneous spreading depression waves, but did not shorten penumbra focal ischaemic depolarisation, the decay of which was slowed down to the rate found in the ischaemic core (KOROLEVA et al. 1998). Morphological examination showed that the volume of total and partial necrosis was increased in the MK-801 (medicated for 3 days) group and marginally reduced in the cerebrolysin (2.5 mg/kg medicated at 24-h intervals for 8 days) group. MK-801 administered 15 min before a 90-min decapitation ischaemia (30 °C) in the *rat* induced dose-dependent recovery of population spike with ED_{50} values of 0.3 mg/kg (ARTEMENKO et al. 2000).

In specific regions of the optic lobes of *Sepia officinalis*, stimulation of the *N*-methyl-D-aspartic acid receptors resulted in a selective decrease in α-tubulin levels within 30 min with partial recovery after 4 h (PALUMBO et al. 2002). The effect was suppressed by the NOS inhibitor N^{ω}-nitro-L-arginine. Incubation of optic lobes with 3-nitrotyrosine resulted likewise in a selective loss of α-tubulin, due apparently to incorporation of the amino acid into the C-terminus of detyrosinated α-tubulin to give the nitrated protein purportedly more susceptible to degradation.

There is not any critical hypoxic threshold for modification of the 3-(2-carboxypiperazin-4-yl) propyl-1-phosphonic acid high-affinity binding site of the *N*-methyl-D-aspartate receptor, but modification is coupled to a gradual decrease in brain cell energy metabolism and increase in lipid peroxidation (FRITZ et al. 2001).

In a preparation of rat brain synaptosomes, the efflux of D-[^{3}H]aspartate was found to be enhanced by micromolar concentrations of externally added D-and L-aspartate, L-cysteate and L-cysteine sulphinate (ERECINSKA and TROEGER 1986). The stimulation of release by external amino acids followed Michaelis-Menten kinetics; the apparent K_m values (in μM) were: 14.65±0.98 for D-aspartate; 8.00±1.5 for L-aspartate; 22.31±1.62 for L-glutamate ; 6.76±0.3 for L-cysteate and 7.89±1.23 for L-cysteine sulphinate.

GILMAN et al. (1994) used isolated presynaptic nerve terminals from the *guinea pig* cerebral cortex to examine the actions and interactions of peroxide, iron, and desferrioxamine on the nonvesicular release of excitatory amino acid neurotransmitters. Pre-treatment with peroxide, iron alone, or peroxide with iron significantly increased the calcium-independent basal release of D-[2,3-^{3}H]aspartate. Pre-treatment with desferrioxamine had little effect on its own but significantly limited the enhancement by peroxide. High K^+-evoked release in the presence of Ca^{2+} was enhanced by peroxide but not by iron.

Aquopentacyanoferrate(II), $[Fe^{II}H_2O(CN)_5]^{3-}$, one of the photodegradation products of vasodilator and nitric oxide donor nitroprusside, is a highly potent, competitive, and selective *N*-methyl-D-aspartate receptor antagonist (NEIJT et al. 2001). It blocked *N*-methyl-D-aspartate-induced depolarisation in *rat* cortical slices at submicromolar concentrations, whereas responses to α-amino-3-hydroxy-5-methyl-4-isoxazolepropionic acid (AMPA) and kainate were not affected.

In an *in vitro* experimental model using *rat* brain slices incubated with [^{18}F]2-fluoro-2-deoxy-D-glucose in oxygenated Krebs-Ringer solution at 36 °C, MURATA et al. (2000) found the highly potent, selective, and non-competitive *N*-methyl-D-aspartate antagonist dizocilpine, (5*R*,10*S*)-(+)-5-methyl-10,11-dihydro-5*H*-dibenzo[*a*,*d*]-cyclohepten-5,10-imine (MK-801) and hypothermia, but not the free radical scavenger, α-phenyl-*N*-*tert*-butyl nitrone, to have a protective effect against O_2 and glucose deprivation when administered during deprivation. Excitatory amino acids during O_2 and glucose deprivation loading were found to be the main factor in the neuronal damage, while in hypoxia, excitatory amino acids working in tandem with free radicals immediately after reoxygenation were implicated.

In the *rat*, dizocilpine (75 μg/kg) had no effect on the acquisition of a spatial discrimination task in a Y-maze, but disrupted reversal learning (CROSS et al. 1995). Both the acquisition and reversal of a visual discrimination task were impaired following dizocilpine (75 μg/kg). Dizocilpine (40 μg/kg) also disrupted performance of a five-choice visual reaction time task.

It has been suggested that γ-aminobutyric acid (GABA) transitorily substitutes for the triggering action of α-amino-3-hydroxy-5-methyl-4-isoxazolepropionic acid (AMPA) receptors at immature hippocampal synapses, which only display *N*-methyl-DN-aspartate receptor-based activity and lack functional AMPA receptors (BENKE et al. 1993). This mechanism may be relevant in precociously inducing a functional state in synapses otherwise destined to remain silent for longer periods of time because of the lack of functional AMPA receptors able to trigger their activity (RUMPEL et al. 1998, LIAO et al. 1999).

Cultured cortical neurones of foetal ICR *mice* at 14–15 d gestation maintained in 25 mM glucose succumbed a widespread neuronal death after exposure to *N*-methyl-D-aspartic acid, AMPA, and kainate (SEO et al. 1999). Among these, *N*-methyl-D-aspartic acid toxicity was substantially reduced in neurones maintained in 100 mM glucose. *N*-Methyl-D-aspartic acid induced increase in $[Ca^{2+}]_i$ and reactive oxygen species was attenuated in neurones maintained in high glucose that revealed increased mitochondrial membrane and redox potentials as determined using rhodamine 123 and 3-(4,5-dimethylthiazol-2-yl)-2,5-diphenyl tetrazolium bromide, *p*-trifluoromethoxy-phenylhydrazone, KCN, and rotenone, the selective inhibitors of mitochondrial potential, abrogated neuroprotective effect of high glucose against *N*-methyl-D-aspartic acid.

Another mechanisms that may contribute to cerebral injury after hypoxia/ischaemia is the **generation** within the tissue **of excitatory amino-acid neurotransmitters**. These excitatory neurotransmitters cause neurones to "fire off' continuously until they are damaged. Synthetic neurotoxins of this type include the glutamate analogue **kainic acid** (2-carboxyl-3-carboxymethyl-4-isopropenylpyrrolidine), **ibotenic acid** (α-amino-3-hydroxy-5-isoxazole acetic acid) and **quinolinic acid**. The kainate receptor is selectively blocked by low concentrations of kynurenic acid (4-hydroxyquinoline-2-carboxylic acid) and 6,7-dinitroquinoxaline-2,3-dione. In the male Wistar *rat*, the hippocampal half-life of 2,2,5,5-tetramethylpyrrolidine-1-oxid, a blood-brain barrier-permeable nitroxide radical, after kainate-induced seizures was significantly prolonged ($P < 0.01$), whereas the prolongation of the cortical half-life was not significant (YOKOYAMA et al. 1999). After intracerebroventricular injection of nmol concentrations of kainic acid on postnatal day 7, neuronal loss was observed in the CA3 subfield of the rat hippocampal formation at postnatal day 45 and postnatal day 75 (HUMPHREY et al. (2001).

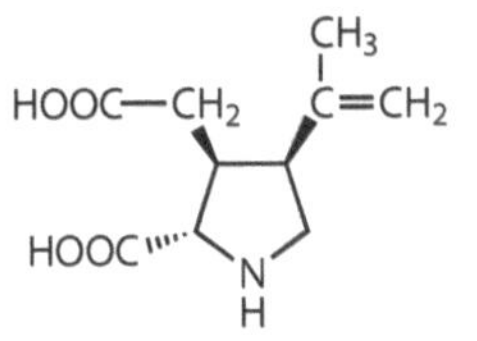

Kainic acid (2-carboxyl-3-carboxymethyl-4-isopropenylpyrrolidine) [275]

An increase in cytochrome *c* oxidase activity and cytochrome *c* oxidase subunit IV mRNA occurred within 1 h after subcutaneous application of kainic acid (15 mg/kg) in all *rat* brain areas tested between 120 % and 130 % of control activity, followed by a significant reduction from control, in amygdala and hippocampus on day three and seven, respectively (MILATOVIC et al. 2001). In amygdala, ATP and phosphocreatine levels were reduced to 44 % and 49 % of control 1 h after seizures. No significant recovery was seen on day tree of seven. Pre-treatment of *rats* with the spin-trapping agent *N-tert*-butyl-α-phenylnitrone (200 mg/kg i.p.) 30 min before kainic acid administration had no effect on status epilepticus, but protected cytochrome *c* oxidase activity and attenuated changes in energy metabolites. Pre-treatment for three days with the antioxidant vitamin E (100 mg/kg i.p.) had an even greater protective effect than *N-tert*-butyl-α-phenylnitrone. Both pre-treatment regimens attenuated kainic acid-induced neurodegenerative changes, as assessed by histology and prevention of the decrease of cytochrome *c* oxidase subunit IV mRNA and cytochrome *c* oxidase activity in hippocampus and amygdala, otherwise seen following kainic acid treatment alone.

Specific inhibitors of Ca^{2+}-dependent phospholipase A_2 (EC 3.1.1.4) prevented the decrease of a neuronal marker, GluR1, and increase in Ca^{2+}-dependent phospholipase A_2 and 4-hydroxynonenal immunoreactivities in *rat* hippocampal slices treated with 100 μM kainate (LU et al. 2001).

Inhibition of nitric oxide synthase by intraperitoneal pre-medication of N^{ω}-nitro-L-arginine for four days dramatically potentiated seizures induced in adult male Sprague-Dawley *rats* by 10 mg kainic acid per kg body weight (MAGGIO et al. 1995).

Other than glutamate, ibotenate was able to stimulate the generation of inositol phosphates in both normal and low extracellular Ca^{2+} (SCHOLZ 1994). The maximal response to ibotenate was approximately equal to that of glutamate, when cultured embryonic *rat* hippocampal pyramidal neurones were stimulated in 50 mM extracellular Ca^{2+}. Ibotenic acid infusions (10 μg/μl) at a rate of 0.2 μl/min bilaterally into the ventral hippocampus of *rats* induced degenerative changes throughout the hippocampus, and in hippocampal efferent projections to forebrain structures, including the septal nucleus and nucleus accumbens, and within the olfactory tubercle and orbital cortex (HALIM and SWERDLOW 2000). Sections stained for tyrosine hydroxylase (EC 1.14.16.2) immunoreactivity released reduced tyrosine hydroxylase labelling through the forebrain dopamine terminal fields 28 days, but not 14 days after ventral hippocampal lesions.

Quinolinic acid-induced degradation of IκB-α, the cytoplasmic binding protein of the nuclear factor κB is almost totally mediated by a caspase-3-dependent mechanism, while kainic acid-induced IκB-α degradation is only partially dependent on caspase-3 (NAKAI et al. 2000). A free radical scavenger (OPC-14117; 600 mg/kg *rat*, orally) attenuated the effects of quinolinic acid but not kainic acid on IκB-α degradation, suggesting that oxidative stress contributes to the quinolinic acid- but not to the kainic acid-induced degradation of IκB-α.

Melatonin (0.1–4 mM) in a dose-response manner reduced kainate (11.7 mM)-induced lipid peroxidation in homogenates of the cerebellum, hippocampus, hypothalamus and striatum of both Wistar and Sprague-Dawley *rats* (MELCHIORRI et al. 1995). *In vivo* melatonin crosses the blood brain barrier (REITER 1991), is rapidly taken up by the brain (MENENDEZ-PELAEZ et al. 1993), and may be retained in at least two brain regions; hippocampal neurones and specific nuclei in the hypothalamus such as the suprachiasmatic nuclei (MARANI and RIETVELD 1987, NARANJO-RODRIGUEZ et al. 1991).

Receptor autoradiography in the somatosensory cortex of the *rat* after occlusion of the arteria cerebri media revealed an increase of excitatory receptors and a decrease in $GABA_A$ receptors (ZILLES et al. 1995). Using 14 nM [^{3}H]corticosterone as radioligand, mineralocorticoid receptor binding was reduced by approximately 50 % in the hippocampus and hypothalamus of adult *rats* that had been born by the Caesarean procedure either with or without an added period of anoxia, in comparison to vaginally born controls (BOKSA et al. 1996). Saturation analysis revealed that these reductions resulted from decreases in affinity of the mineralocorticoid receptor for [^{3}H]corticosterone, with no change in numbers of receptors.

Tigrolysis and cell oedema following ischaemia were first observed and reported by MARINESCO (1896). COLMANT (1965) emphasised posthypoxic tigrolysis in certain circumstances would be ascertainable for several days thereafter. HOCHBERG and HYDÉN (1949) presented the results of systematical measurements of the ribonucleoprotein content of nerve cell cytoplasm after temporary (15 min max.)

ischaemia of the abdominal aorta by UV microspectrography.

Silver impregnation of **degenerating axons** (NAUTA and GYGAX 1954) was used by VANICKY et al. (1991) for selective visualisation of "ischaemic neurons" after global cerebral ischaemia using a highly reproducible model of 15 min cardiac arrest in *dogs* and survival periods of 1 h to 7 d. NADLER and EVENSON (1983) used the silver impregnation method of GALLYAS et al. (1980) to demonstrate degenerated neuronal somata induced by pressure injection (SHIPLEY 1982) of kainic acid or *N*-methyl-DL-aspartic acid into the anterior hypothalamic nucleus.

The intracellular pool of **ascorbate**, which is highly concentrated in neuropils (1–3 mM; GRÜNEWALD 1993) can be released under various conditions including dopaminergic receptor stimulation (CLEMENS and PHEBUS 1984), γ-aminobutyric acid receptor stimulation (BIGELOW et al. 1984), increased extracellular glutamate (GRÜNEWALD and FILLENZ 1984), and depolarisation (MILBY et al. 1981). In male Wistar *rats*, reperfusion after forebrain ischaemia (induced by bilateral carotid artery occlusion with haemorrhagic hypotension for 10 and 15 min, respectively), significantly increased ascorbate release to 504 % and 334 %, respectively (YUSA 2001). The extended time of ischaemia significantly inhibited glutamate re-uptake and ascorbate release during reperfusion.

The role of **iron** in catalysing oxygen-derived radical production is well known, and there is evidence that reactive oxygen species may be a primary cause of cerebral damage during ischaemia and postischaemic reperfusion (KOGURE et al. 1985, BRAUGHLER and HALL 1989, BROMONT et al. 1989). Acidosis (BRALET et al. 1991) or superoxide anion production (SAMOKYSZYN et al. 1988) might delocalize intracellular iron (KRAUSE et al. 1985, HEALING et al. 1990, BRALET et al. 1992, ROTHMAN et al. 1992), providing a source of iron in a form capable of catalysing one-electron reduction of oxygen species. OUBIDAR et al. (1996) assessed the effect of artificially elevated cell iron on oxygen-derived free radical production in brain slices by using an iron ligand, 8-hydroxyquinoline. This Fe^{3+}-complex exhibited a high lipid solubility evidenced by *n*-octanol/water partition coefficient. It was avidly taken up by *rat* brain slices and was significantly more effective in lipid peroxidation than the hydrophilic iron citrate complex.

Oligodendrocytes due to their high intracellular iron concentrations, the high rate of their oxidative metabolism, and rich lipid environment show an increased vulnerability to oxidative stress and injury (GRIOT et al. 1990) which can be prevented by catalase (KIM and KIM 1991, NOBLE et al. 1994).

Lactate as a sole energy substrate is able to support neuronal function in the *rat* hippocampal slice preparation (SCHURR et al. 1988). Lactate-supplied *rat* hippocampal slices (400 μm thick) showed a significantly higher degree of recovery of synaptic function after a 5-min hypoxic (95 % N_2, 5 % CO_2) period than slices supplied with an equicaloric amount of glucose (SCHURR et al. 1997). All slices in which anaerobic lactate production was enhanced by pre-hypoxia glucose overload exhibited functional recovery after a 23-min hypoxia. An 80 % recovery of synaptic function was observed even when glucose utilisation was blocked with 2-deoxy-D-glucose during the later part of the hypoxic period and during reoxygenation. In contrast, slices in which anaerobic lactate production was blocked during the initial stages of hypoxia did not recover their synaptic function upon reoxygenation despite the abundance of glucose and the removal of 2-deoxy-D-glucose.

Brain tissue levels of **interleukin-8**, a potent neutrophil chemotactic cytokine (chemokine), increased significantly at 6 h after reperfusion, but without a noticeable elevation of plasma IL-8 levels (MATSUMOTO et al. 1997). IL-8 protein was detected immunohistochemically in the vascular wall and, to a lesser degree, in infiltrated neutrophils, suggesting a local production of IL-8 in reperfused brain tissues. A neutralising anti-IL-8 antibody significantly reduced brain oedema and infarct size in comparison to *rabbits* receiving a control antibody.

Insulin-like growth factor binding protein (IGFBP) **4** mRNA expression was reduced in regions of neuronal loss following hypoxic-ischaemic injury (ligation of the right carotid artery, followed by a 15 min or 60 min exposure to 8 % O_2) in the 21-day-old *rat* brain (BEILHARZ et al. 1993).

Mice deficient in **Mac-1** (CD11b/CD18) are less susceptible to cerebral ischaemia/reperfusion injury (SORIANO et al. 1999). Mac-1 deficiency reduced neutrophil infiltration ($P = 0.19$) and reduced the infarct volume ($P < 0.05$).

For hypoxic brain damage, the pathologist must at least examine the parasagittal cortex, the Ammon's horn, the thalamus, and the cerebellum (ADAMS 1988).

15.1.1.1 Hippocampus

Cytochrome P_{450} has been identified in multiple forms belonging to subfamilies 1A, 2B, and 2E in mitochondria isolated in eight regions of *human* brain samples obtained at autopsy (BHAGWAT et al. 2000). P_{450}-associated monooxygenase activities including aminopyrine *N*-demethylase (APD), 7-

ethoxycoumarin *O*-deethylase (ECD), *p*-nitrophenol hydroxylase (PNPH), and *N*-nitrosodimethylamine *N*-demethylase (NDMAD) were detectable in the mitochondria from *human* brain regions. The addition of antiserum to microsomal NADPH cytochrome P_{450} reductase did not affect the mitochondrial P_{450}-associated monooxygenase activities, although it completely inhibited the corresponding activities in brain microsomes.

Kageyama and Wong-Riley (1982) showed that **cytochrome oxidase** staining in the somata of CA3 pyramidal cells and various interneurons was more intense than in CA1 pyramidal and dentate granule cells, while very low levels of cytochrome oxidase activity was observed in the stratum lucidum of CA3.

Haloperidol, a dopamine receptor antagonist, was cytotoxic to *mouse* clonal hippocampal HT22 cells in a concentration-dependent manner and caused death by oxidative stress as assessed by 3-(4,5-dimethylthiazol-2-yl)-2,5-diphenyl tetrazolium bromide (Post et al. 1998). The addition of haloperidol to HT22 cells led to an increase in intracellular peroxides and a rime-dependent drop in the intracellular glutathione levels. Haloperidol-induced oxidative cell death was prevented by melatonin, its precursor N-acetyl serotonin, and most effectively by α-tocopherol.

Nitric oxide synthase III immunoreactivity was observed in the dendrites of hippocampal pyramidal cells of knockout and wild type animals indicating that even in functional NOS-III knockouts the non-deleted N-terminal trunk of the protein is still expressed (Zanger et al. 1999). NOS-III immunoreactivity for the C-terminal portion of the protein was found in wild types only. Electron microscopically NOS-III immunoreactivity was not located at synapses but was present in and associated with mitochondria which were mainly located in dendrites compared to the soma and were never seen in axons.

Perforant pathway stimulation resulted in an increase in the number of nNOS-immunoreactive neurones in the stratum radiatum of the CA1 and CA3 subfields of the *rat* hippocampus proper, and the hilus and the dentate gyrus (Lumme et al. 2000). The morphology and distribution of the nNOS immunoreactive neurones resembled that of interneurones.

LiCl (12 mEq/kg i.p.) and tacrine (5 mg/kg i.p.) enhanced the expression of nNOS, but not eNOS, enzyme protein in the *rat* hippocampus during the preconvulsive period and this triggered seizures and hippocampal damage (Bagetta et al. 2002). Systemic administration of 7-nitro indazole (50 mg/kg given i.p. 30 min before tacrine), a selective inhibitor of nNOS, prevented the expression of motor and electrocortical seizures and abolished neuronal cell death in the hippocampus.

From several **nitric oxide-containing ruthenium complexes** tested on their influence on the evoked potentials recorded from the CA1 region of the *mouse*, only *trans*-$[(NO)(P(OEt)_3)(NH_3)_4Ru](PF_6)_3$ (1–2.5 mM) exerted a strong facilitatory action on the population spike, the EPSP, and the spontaneous activity (Wieraszko et al. 2001). Its activity probably depends upon its ability to release NO following reduction. The phosphito ligand is important both in terms of adjusting the reduction potential of the complex to be biologically accessible and in labilising the co-ordinated NO. The effects of this compound could not be reversed by perfusion.

Systemic administration of the specific **NOS inhibitor** 7-nitroindazole (50 mg/kg i.p.) and the non-selective inhibitor of guanylate cyclase and NOS, methylene blue (30 mg/kg s.c.) increased the extracellular levels of 5-hydroxytryptamine and dopamine in the *rat* ventral hippocampus (Volke et al. 2000). Local administration (via the microdialysis probe) of citalopram (1 μM), paroxetine (2 μM), tianeptine (2 μM) and imipramine (20 μM) produced a significant decrease in the levels of $[^3H]$L-citrulline compared to the control activity (Wegener et al. 2000). In contrast, 5-hydroxytryptamine (20 nM) failed to influence the levels of $[^3H]$L-citrulline. Reverse dialysis with the NOS inhibitor, N^{ω}-nitro-L-arginine (2 mM), also decreased the levels of $[^3H]$L-citrulline ($P < 0.05$). The NOS activity *in vitro* was unaffected by citalopram, paroxetine, tianeptine, imipramine and 5-hydroxytryptamine in the concentrations used *in vivo*, but showed a significant decrease after administration of N^{ω}-nitro-L-arginine ($P < 0.05$).

Neuronal loss was initiated by inhibition of the antioxidant enzyme, **superoxide dismutase** (EC 1.15.1.1.) type 1, using the copper chelator diethyldithiocarbamate (Moskowitz et al. 2001). Continuous diethyldithiocarbamate treatment of Sprague-Dawley *rat* or C57/Bl6 *mouse* hippocampal slice cultures induced delayed neuronal loss beginning at 9 days of treatment that lasted for over 4 weeks. Neuronal loss was significantly attenuated in slice cultures that overexpress superoxide dismutase type 1, suggesting that superoxide dismutase inhibition was responsible. Inhibitors of nitric oxide synthase also attenuated diethyldithiocarbamate-induced neuronal loss.

Glutaminergic transmission has been shown to be of crucial importance for synaptic plasticity in the hippocampus. Glutaminergic signal transduction is a key determinant in the endangerment of hippocampal cells viability by several insults such as hypoxia/ischaemia (SAPOLSKI and PULSINELLI 1985).

HT22 cells derived from the *mouse* hippocampus by SV40 transformation probably phenotypically most similar to neuronal precursor cells lack ionotropic glutamate receptors, but are killed by exogenous glutamic acid in the millimolar range (MAHER and SCHUBERT 2000). The initiating event in this form of glutamate-mediated cell death is the blockade of the cystine uptake into the cell via the inhibition of the glutamate/cystine antiporter by exogenous glutamate (BANNAI and KITAMURA 1981). This antiporter, which has been cloned by SATO et al. (1999), normally transports the oxidised form of the essential amino acid cysteine down its concentration gradient into cells, coupled with the export of intracellular glutamate. High extracellular glutamate blocks this process, depriving the cell of cystine. Inside the cell, cystine is reduced to cysteine, which is a substrate for glutathione synthesis. In the absence of GSH, cells become oxidatively 'stressed', produce large amounts of reactive oxygen species, and die – a process termed **oxidative glutamate toxicity** (MURPHY et al. 1989).

Kainate receptors are a subtype of ionotropic glutamate receptors, permeable to cations and thus expected to have an excitatory depolarising on neurones. However, kainate receptor activation inhibits γ-aminobutyric acid release in the hippocampus through activation of protein kinase C in a pertussis toxin-dependent manner, suggesting a coupling of kainate receptors to G proteins. CUNHA et al. (1999) directly investigated the G protein coupling of kainate receptors in the *rat* hippocampus by using a selective kainate receptor agonist, $[^3H](2S,4R)$-4-methylglutamate.

4-Hydroxynonenal was absent from normal hippocampal neurones, but dense staining to 4-hydroxynonenal was observed after kainate injections into the right lateral ventricle of *rats* (ONG et al. 2000). The increase in 4-hydroxynonenal staining occurred as early as 1 day postinjection, at a time when there was no histological evidence of cell death. 4-Hydroxynonenal immunoreactivity was observed in the degenerating CA1 and CA3 fields at 3 days and 1 week postinjection, but was confined to a cluster of neurones at the edge of the degenerating CA fields, at 2 and 3 weeks postinjection.

A 50–60 % decrease in membrane **protein kinase** C activity was observed in CA1 neurones 6 h post-reperfusion in Wistar *rats* subjected to 15-min normothermic forebrain ischaemia (CHAKRAVARTHY et al. 1998).

Cyclic GMP-stimulated phosphodiesterase (PDE2) is present in many of the brain structures that belong to the limbic system including the olfactory cortex, amygdala, hippocampal formation and the anterior thalamic nucleus (JUILFS et al. 1999).

Cyclic AMP-stimulated phosphodiesterase (PDE4) enzymes are characterised by high affinity and specificity for cyclic AMP, low K_m, insensitivity to Ca^{2+}, and selective inhibition by several drugs, including rolipram and Ro 20-1724 (BEAVO 1995). PDE4B was significantly more sensitive to inhibition by both rolipram and diazepam than the PDE4A subtype (CHERRY et al. 2001).

Hypoxia by breathing an atmosphere of 5 % oxygen for 30 min and reoxygenation for 3 h to 21 d preferably damaged the CA3 and CA4 regions, and less severe the granule cell layer of the gyrus dentatus, while the neurones in the CA1 region of the *rat* were relatively resistant (YAMAOKA et al. 1993).

Prolonged anoxic depolarisation exacerbated **NADH hyperoxidation** and promoted poor electrical recovery after anoxia in *rat* hippocampal slices (PÉREZ-PINZÓN et al. 1998).

Cholecystokinin-positive cells have been observed in all parts of the hippocampal formation, i.e. the fascia dentata, the cornu Ammonis, and the subiculum (LORÉN et al. 1979, VANDERHAEGHEN et al. 1980, GREENWOOD et al. 1981). There is evidence indicating that cholecystokinin is present in both intrinsic (HANDELMANN et al. 1981) and extrinsic hippocampal fibres (GREENWOOD et al. 1981). The intrinsic fibres include mossy fibre axons of the dentate gyrus granule cells (STENGAARD-PEDERSEN et al. 1983, GALL 1984, GALL et al. 1986). CCK-8 has an excitatory action on hippocampal pyramidal cells similar to that of the response evoked by glutamate (DODD and KELLY 1981). In the dentate gyrus of the *rat*, cholecystokinin exhibited a biphasic responsiveness to inescapable electric foot-shock (SIEGEL et al. 1984): at 2 min, peptide concentrations were 50 % greater than in controls. Immediately thereafter prestress values were regained, until the second response occurred, leading to significantly elevated cholecystokinin levels at 30 min and at 60 min.

In schizophrenic *patients* ^{125}I-BH cholecystokinin$_{33}$ specific binding was reduced by 40 % ($P < 0.02$) in the hippocampus (FARMERY et al. 1985).

Somatostatin-containing perikarya have been reported in the fascia dentata, the cornu Ammonis, and the subiculum (FINLEY et al. 1981, SHIOSAKA et al. 1982). The lower cortical layers, the CA1 region of the hippocampus, and the habenula are highly enriched in [^{125}I-Tyr^{11}]SS receptor sites, the dentate gyrus has an intermediate density, whereas the hypothalamic region is almost devoid of receptor sites (MAURER and REUBI 1985).

Somatostatin applied to the cell bodies of neurones in the pyramidal layer of the CA1 and CA2 regions of the *rat* hippocampus resulted in a strong excitation which was fast in onset and thus resembled that evoked by glutamate (Dodd and Kelly 1978).

A slight reduction of somatostatin immunoreactive cells was observed in the hilus of the dorsal and ventral hippocampus of male Wistar *rats* seven days after moderate hypoxia (9 % O_2 in N_2 for two times 8 h) (Schwarzer et al. 1996). At the same time, the total number of neuropeptide Y immunoreactive neurones was increased in this area due to a pronounced increase in staining of presumable basket cells. There was also increased staining of neuropeptide Y positive fibres in the outer molecular layer.

VIP-immunoreactivity is located in bipolar nonpyramidal neurones in stratum radiatum and stratum oriens and multipolar neurones in stratum lacunosum-moleculare of the *rat* hippocampus (Léránth and Frotscher 1983). Axon terminals of VIP-like immunoreactive non-pyramidal neurones terminate on cell bodies of the pyramidal cells (Léránth et al. 1984).

The hippocampus of the *rat* contains 357±84 fmol **angiotensin II**/g, that is 7.8±4.6 times that of the cortex (Sirett et al. 1981). No cell body staining was observed in any part of the cerebral cortex, although fibres were widely, if quite sparsely, distributed (Lind et al. 1985). At least a few angiotensin II-immunoreactive fibres were found in all major fields of the hippocampal formation, with no obvious preferential innervation of any region. However, binding studies have provided little, or no evidence for angiotensin II receptors in the hippocampal formation (Bennett and Synyder 1976, Sirett et al. 1977, Harding et al. 1981).

Neutron activation analysis of **iron** in the *human* hippocampus (10 samples from patients aged 23 to 66 y) resulted in $2.38 \pm 0.19 \times 10^{-4}$ terms of element weight per unit dried tissue (Höck et al. 1975).

Ferric ions (demonstrated histochemically by the Prussian blue method) in the dentate gyrus of Swiss-Webster *mice* chronically implanted with either stainless steel or Pt-Ir wire probes correlated with significant ($P < 0.05$) performance deficits tested in a one-trial inhibitory avoidance task (Boast et al. 1975).

Zinc uptake into the hippocampus of the *rat* was increased by phenothiazine derivatives (Czerniak an Ben Haim 1971).

15.1.1.1.1 Pyramidal Cells

The pyramidal cells of the hippocampal end-blade are among the most sensitive neurones in the brain to a variety of deleterious conditions, including anoxia, senile dementia and severe epilepsy.

In the Mongolian gerbil (*Meriones unguiculatus*), Kirino (1982) after bilateral carotid occlusion for 5 min observed three different types of changes in the CA4, CA2 and CA1 subfields. In CA4, the change was rapid and corresponded to ischaemic cell change. The alteration in CA2 was relatively slow, and identical to what has been called reactive change. On the contrary, the change in the CA1 pyramidal cells was very slow, only becoming apparent by light microscopy 2 days following ischaemia. By electron microscopy, the lamellar alignment of proliferated cisterns of the endoplasmic reticulum was the most conspicuous finding in these cells. Four days following ischaemia, almost all the pyramidal cells in CA1 were destroyed.

In the *gerbil* 4 days after 5 min ischaemia, Ishimaru et al. (1996) by light microscopy observed atrophy in most CA1 pyramidal cells, and a reduction in their number was noticeable on day 7. By day 14, almost all pyramidal cells had disappeared from the stratum pyramidale in the hippocampal CA1 subfield. In contrast, no cell damage was observed in the CA3 or dentate gyrus in all experimental animals.

While two-minute bilateral common carotid artery occlusion in the *gerbil* produced no histological neuronal damage in the brain, 3 such occlusions at 1-h intervals consistently lead to severe destruction of hippocampal CA1 pyramidal neurones (Kato and Kugure 1990, Kato et al. 1990). Occluding these arteries for 15 min, Onodera et al. (1987) found a marked decrease in [^{3}H]quinuclidinyl benzilate and [^{3}H]cyclohexaladenosine binding activities in the CA1 subfield 3–27 days after recirculation. Three to 27 days after ischaemia, the adenosine A_1 binding activities in the CA3 subfield of the hippocampus and in the dentate gyrus were reduced despite the normal appearance of these areas throughout the reperfusion period. Muscarinic binding sites in the CA3 subfield were also reduced 27 days after ischaemia. Despite minimal neuronal damage in the lateral septal nucleus and in the substantia nigra, the A_1 binding activity in these regions was reduced by 70 % and 50 %, respectively, providing evidence that the muscarinic receptors in the dorsolateral region of the caudate-putamen are localised postsynaptically on small and medium-sized neurones and those in the CA1 subfield of the hippocampus are localised on the CA1 pyramidal cells.

Forebrain ischaemia (5 min) increased NADH and decreased flavoprotein signals in all hippocampal areas of the *gerbil*, but reduction in mitochondrial redox ratio was greater in CA1 than in other areas of the hippocampus (Shiino et al. 1999). Immediately after recirculation, mitochondrial redox ratio recovery was delayed in the CA1 and the dentate gyrus, and the reduction in mitochondrial redox ratio persisted in CA1.

Primary neuronal cultures of the CA1 region obtained from Wistar *rat* embryos on gestational days 17–19 and exposed to 100 μM iodoacetate for 5 min showed histotoxic hypoxia, 15 min later showed progressive ribosomal desegregation becoming near complete after 1 h (Dux et al. 1996). Changes of the fine structure of the cytoplasmic organelles developed more slowly. After 15-min recovery the only change was vacuolisation of a few mitochondria. Between 1 h and 6 h, however, notable swelling of mitochondria, massive vacuolisation of Golgi apparatus and endoplasmic reticulum, fragmentation of neurofilaments and condensation of cytoplasm became visible.

A high extracellular sodium concentration ($[Na^+]_o$) attenuated the hypoxia-induced response in the CA1 pyramidal cell layer of *rat* hippocampal slices, its onset latency was longer and the time constant of its decay phase was shorter than in controls (Yamaguchi et al. 1997). In contrast, hypoxia in low $[Na^+]_o$ elicited a significantly enhanced response with a short onset-latency and delayed decay phase. This exaggerated response to hypoxia in low $[Na^+]_o$ was reversed by pre-incubation of the slice in low $[Na^+]_o$ prior to the hypoxic insult.

Preconditioning motor activity for 15 min significantly enhanced the resistance of CA1 pyramidal neurones to ischaemia at a temperature of 30 °C (Gerasimov et al. 2001). The period of protection lasted for up to 40 min after the end of motor activity, When ischaemia was started within 5–10 min after the preconditioning, complete restoration of the field potentials to preischaemic control levels could be achieved.

The Schaffer collateral-commissural pathway conveying inputs to pyramidal neurones in the CA1 region exhibits a form of long-term potentiation which critically depends on a primary *N*-methyl-D-aspartate receptor-mediated Ca^{2+} influx into the postsynapse and is, at least partly, due to an increase of presynaptic transmitter release (Bolshakov and Siegelbaum 1995). Freely diffusible nitric oxide (•NO), generated postsynaptically by Ca^{2+}-calmodulin-dependent nitric oxide synthase (NOS), has been identified as the messenger of the underlying retrograde signalling mechanism (Schuman and Madison 1991, Zhuo et al. 1994, Arancio et al. 1995).

Free radical generation in *rat* CA1 pyramidal neurones of organotypic slices subjected to a hypoxic-hypoglycaemic insult was temporally correlated with intracellular calcium elevation, as measured by injection of fluo-3 in individual pyramidal cells, using patch electrodes (Perez Velazquez et al. 1997).

During reoxygenation after anoxia, concentrations of ATP and phosphocreatine rapidly recovered in *guinea pig* hippocampal slices preincubated with 20 mM creatine for 2 h (Yoneda et al. 1983).

Recent findings implicate apoptosis in neuronal degeneration after ischaemic brain injury in animal models of stroke (Mattson et al. 2000). Apoptotic cascades involve: increased levels of intracellular oxyradicals and calcium; induction of expression of proteins such as Par-4 (prostate apoptosis response-4), which act by promoting mitochondrial dysfunction and suppressing antiapoptotic mechanisms; mitochondrial membrane depolarisation, calcium uptake, and release of factors (e.g. cytochrome *c*) that ultimately induce nuclear DNA condensation and fragmentation; activation of cysteine proteases of the caspase family; activation of transcription factors such as AP-1 that may induce expression of "killer genes".

Transgenic *mice* overexpressing *human* Bcl-2 in their neurones showed resistance to hypoxic ischæmia-induced neuronal death (Wang et al. 1999). In the hippocampus, granule cells in the dentate gyrus of NSE-*hbcl*-2 transgeric *mice* were apparently resistant to neuronal degeneration by hypoxic ischæmia, whereas those of wild-type *mice* were susceptible to apoptosis following the same treatment. However, the anti-apoptotic activity of Bcl-2 manifested during high levels of constituent protein expression was not observed in the CA1 neurones of the transgenic *mice* by this treatment, although the onset of neuronal death was delayed as compared with the wild-type *mice*.

Kainic acid (12 mg/kg *rat* administered intraperitoneally) affected the distal basilar dendritic segments of CA1 and sometimes CA2, 3 and 4 pyramidal neurones, as well as the distal apical dendritic segments of CA1, CA3a and CA3b pyramidal neurones and dentate granule neurones (Olney et al. 1979). The dendritic spines that invaginate the mossy fibre terminals – and occasionally portions of the dendritic segments giving rise to them, but not to their cell bodies – were markedly dilated despite absence of pathological alterations in the mossy axon terminals embracing them. The oedematous neuronal elements, either in the mossy fibre zone or elsewhere, could be identified with confidence as dendritic because of the axon terminals

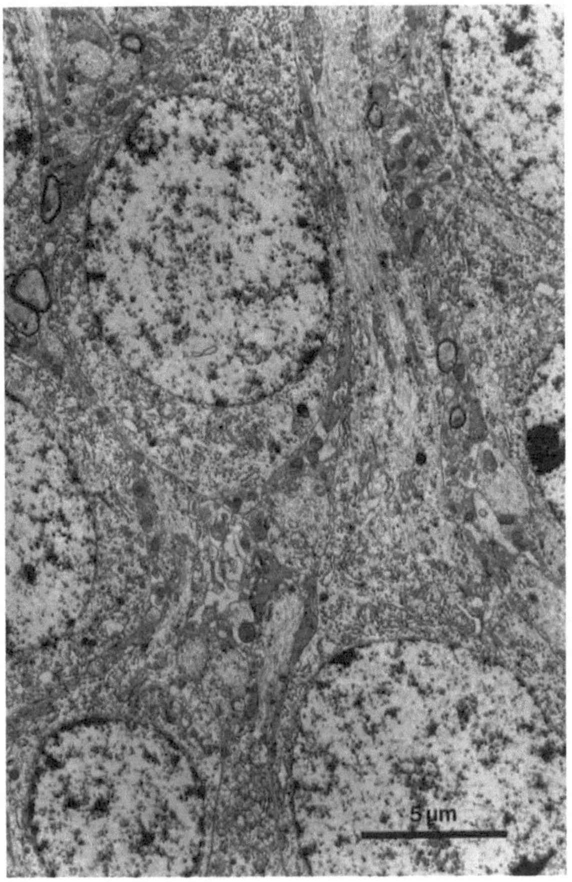

Fig. 215. Reoxygenation damage to hippocampal pyramidal cells (block 1144) of a *rat* (No. 5) treated for 7 consecutive days with intraperitoneal injection of 1.5 ml Tyrode's solution per kg body weight × day. On the last 4 days of experimentation the animal was exposed to an atmosphere containing only 5 % oxygen for 30 min. On August 4, 1976, half an hour after the last exposure under pentobarbital anaesthesia (30 mg/kg), the animal was perfused from the abdominal aorta with 2.5 % glutaraldehyde in 0.1 M sodium cacodylate buffer (pH 7.4). Postfixation with 1 % osmium tetroxide in sodium cacodylate buffer. Embedded in Epon 812 and sectioned at 50 nm. Lead citrate and uranyl acetate. Plate 3491

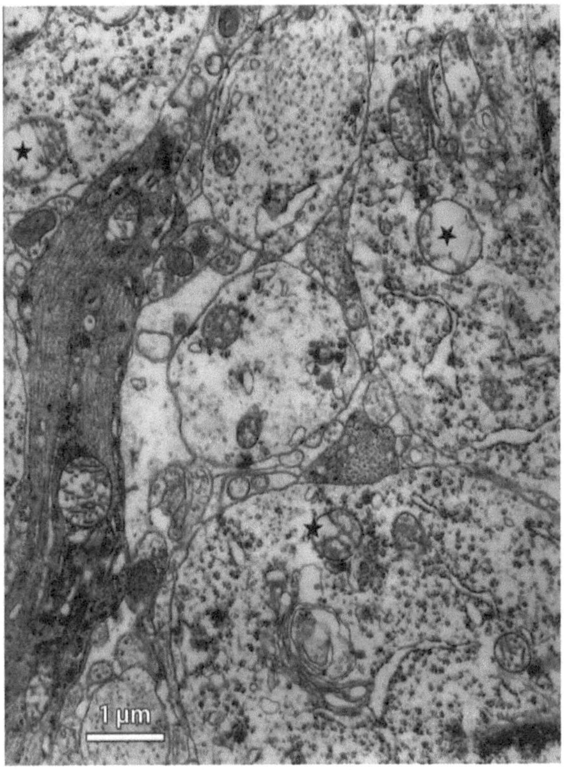

Fig. 216. Severe reoxygenation mitochondrial damage (*) to hippocampal pyramidal cells (block 1144) of a control *rat* (No. 5) treated for 7 consecutive days with intraperitoneal injection of 1.5 ml Tyrode's solution per kg body weight × day. On the last 4 days of experimentation the animal was exposed to an atmosphere containing only 5 % oxygen for 30 min. On August 4, 1976, under pentobarbital anaesthesia (30 mg/kg), the animal was perfused from the abdominal aorta with 2.5 % glutaraldehyde in 0.1 M sodium cacodylate buffer (pH 7.4). Postfixation with 1 % osmium tetroxide in sodium cacodylate buffer. Embedded in Epon 812 and sectioned at 50 nm. Lead citrate and uranyl acetate. Plate 3493

frequently found to be in presynaptic contact with them. These reactive dendritic elements had the same appearance as has been described repeatedly in lesions induced by glutamate or its various excitatory analogues (Olney 1971, Olney et al. 1971, 1974).

Kainate has repeatedly been shown to depress γ-aminobutyric acid-mediated inhibition in CA1 pyramidal cells, but both the mechanisms and the physiological relevance of this effect are unclear (Ben-Ari and Cossart 2000).

CA3 pyramidal neurones are amongst the most responsive neurones to kainate in the brain, because they readily degenerate following local or distal injections of kainate. However, studies using the kainate model have also shown that CA3 pyramidal neurones are highly vulnerable to network hyperactivity per se and readily degenerate following recurrent seizures probably because of a sustained release of glutamate leading to an activation of kainate receptors. Thus, injections of kainate in structures that are distal from the hippocampus, at concentrations that do not diffuse to the hippocampus, are sufficient to generate a seizure and brain damage syndrome that includes CA3 damage (Sloviter 1996).

Interleukin-6 receptors were mainly expressed in hippocampal pyramidal cells of the Mongolian *gerbil* (Vollenweider et al. 2001). Their immunoreactivity remained unaltered 5 and 12 h after transient (7 min) global ischaemia. The disappearance of their expression was correlated to the ischaemic insult and neuronal cell death, which began in CA1 two days after recirculation. After 3 and 4 days, IL-6 receptors were absent in CA1 and CA3.

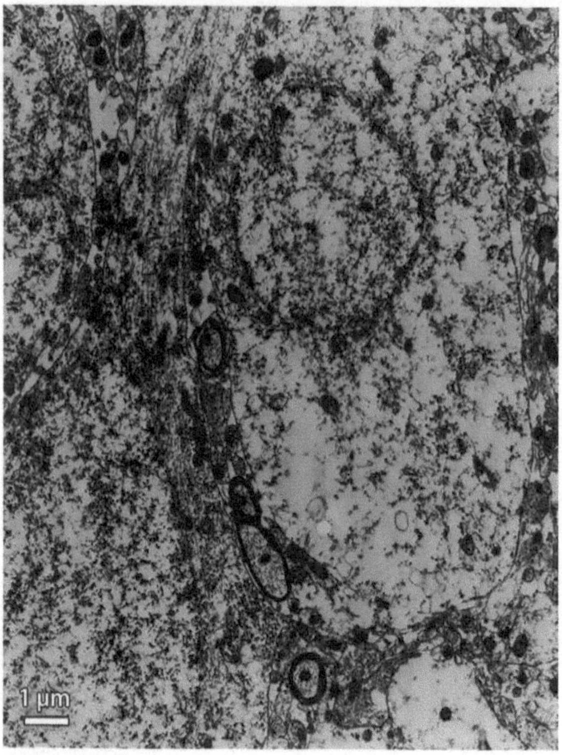

Fig. 217. Severe reoxygenation mitochondrial damage to hippocampal pyramidal cells of a female *rat* (No. 5/14) treated for 4 consecutive days with intraperitoneal injection of 1.5 ml Ringer's solution per kg body weight × day. 30 min after each medication, the animal was exposed to an atmosphere containing only 5 % oxygen for 30 min. Half an hour after the last exposure, under pentobarbital anaesthesia (30 mg/kg), the animal was perfused from the abdominal aorta with 2.5 % glutaraldehyde in 0.1 M sodium cacodylate buffer (pH 7.4). Postfixation with 1 % osmium tetroxide in sodium cacodylate buffer. Embedded in Epon 812 and sectioned at 50 nm. Lead citrate and uranyl acetate. Film 242–32

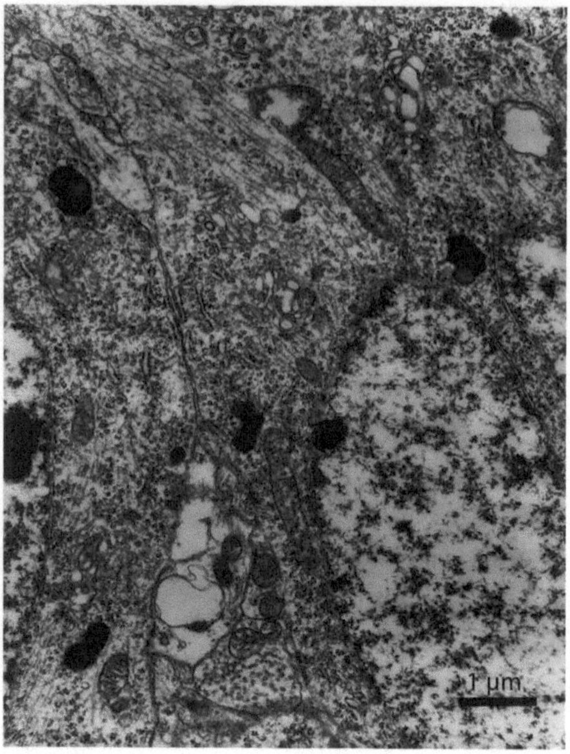

Fig. 218. Reoxygenation damage to the hippocampal pyridamidal cells (block 1234) of a *rat* (No. 8) treated for 7 consecutive days with intraperitoneal injection of 15 mg carbocromene per kg body weight × day. On the last 4 days of experimentation the animal was exposed to an atmosphere containing only 5 % oxygen for 30 min. On August 4, 1976 under pentobarbital anaesthesia (30 mg/kg), the animal was perfused from the abdominal aorta with 2.5 % glutaraldehyde in 0.1 M sodium cacodylate buffer (pH 7.4). Postfixation with 1 % osmium tetroxide in sodium cacodylate buffer. Embedded in Epon 812 and sectioned at 50 nm. Lead citrate and uranyl acetate. Plate 3486

15.1.1.1.2 Microglial Cells

Upon brain damage, microglial cells are rapidly activated and function as tissue macrophages. The first steps in this activation still remain unclear. Several mediators from activated microglia have been proposed as being involved. Both cytokines as the tumour necrosis factor TNF-α (Chao et al. 1993) and transforming growth factor TGF-β (Chao et al. 1992) and reactive oxygen (Tanaka et al. 1994) and nitrogen (Boje and Arora 1992, Chao et al. 1992) intermediates have been incriminated. Priming of *human* foetal microglial cell cultures with interferon-γ or TNF-α enhanced the superoxide production (Chao et al. 1995).

Thrombin induced a transient Ca^{2+} increase in cultured *rat* microglial cells, which persisted in Ca^{2+}-free media (Möller et al. 2000). It was blocked by thapsigargin, indicating that thrombin caused a Ca^{2+} release from internal stores. Preincubation with pertussis toxin did not alter the thrombin-induced $[Ca^{2+}]_i$ signal, whereas it was blocked by hirudin, a blocker of the proteolytic activity of thrombin. Incubation with thrombin led to the production of nitric oxide and the release of the cytokines TNF-α, IL-6, IL-12, the chemokine KC, and the soluble tumour necrosis factor-α receptor II and had a significant proliferative effect.

The mobilisation of transferrin and ferritin-positive phagocytes is linked with the degradation of neurones induced by 5 min ischaemia in the *gerbil's* hippocampus by bilateral common carotid artery clamping (Ishimaru et al. 1996).

The findings of Banati et al. (1992) that propentofylline significantly depressed the respiratory

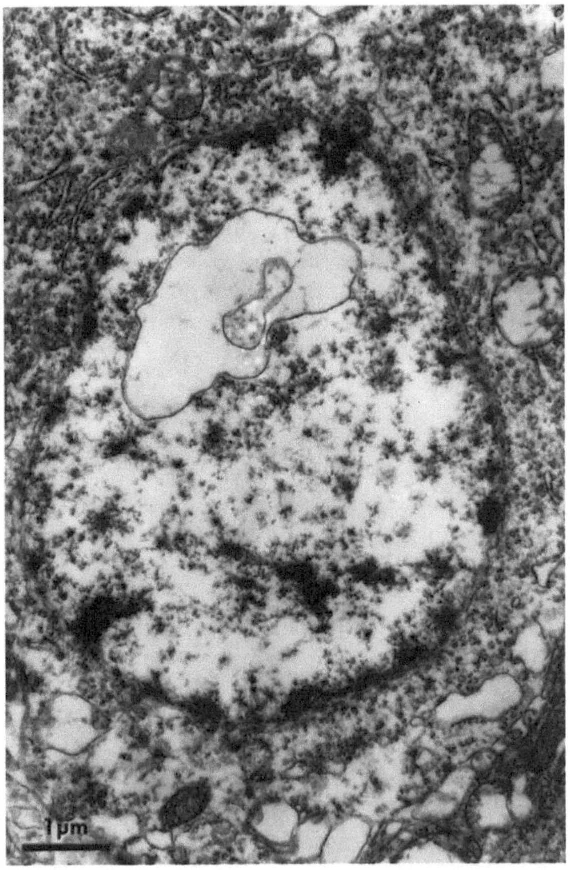

Fig. 219. Severe reoxygenation damage to nucleus, mitochondria and endoplasmic reticulum of a hippocampal pyramidal cell (block 1325) of a *rat* (No. 11) treated for 8 consecutive days with intraperitoneal injection of 15 mg carbocromene per kg body weight × day. On the last 5 days of experimentation the animal was exposed to an atmosphere containing only 5 % oxygen for 30 min. On August 5, 1976, half an hour after the last exposure under pentobarbital anaesthesia (30 mg/kg), the animal was perfused from the abdominal aorta with 2.5 % glutaraldehyde in 0.1 M sodium cacodylate buffer (pH 7.4). Postfixation with 1 % osmium tetroxide in sodium cacodylate buffer. Embedded in Epon 812 and sectioned at 50 nm. Lead citrate and uranyl acetate. Plate 3497

burst activity of microglial cells and Con A-stimulated peritoneal macrophages supports the assumption that the neuroprotection by propentophylline against ischaemia-induced brain damage might result, to some extent, from an inhibition of the formation of reactive oxygen intermediates in microglial cells (Banati et al. 1993). Treatment of cultured microglial cells with 50 μM propentophylline increased the release of nitric oxide by approximately 30 %, which is proportional to the decrease in the production of reactive oxygen intermediates. It seemed, therefore possible to Banati et al. (1993) that the reactive nitrogen and oxygen intermediates interacted with each other. $^{\bullet}NO$ can readily react with $O_2^{\bullet -}$ to form peroxynitrite ions (Beckman et al. 1990), which then decompose to highly reactive hydroxyl radicals and nitrogen dioxides. Further, superoxide dismutase (EC 1.15.1.1) catalyses the conversion of superoxide radicals to hydrogen peroxide, leading to a potentiation of the nitric oxide-dependent cGMP increase. This appears to be due to the dismutation of superoxide anion radicals, which can then no longer interact with endogenously formed nitric oxide.

Immunocytochemical staining with different polyclonal anti-macrophage colony-stimulating factor receptor antibodies revealed differential receptor expression on microglia throughout the central nervous system, with very low levels in the dentate gyrus (Haas et al. 1996).

15.1.1.1.3 Capillaries

Microvessel density in the pyramidal cell layer of the *mouse* hippocampus was significantly higher in sector CA3a than in sector CA1 (Shimada et al. 1992).

Endothelial Cells

Endothelial cells of brain arteries, veins and capillaries of the Wistar *rat* showed a positive NADPH-diaphorase (EC 1.6.99.1) reaction in form of 2-(2'-benzothiazolyl)-5-styryl-3-(4'-phthal-hydrazidyl) tetrazolium chloride-formazan contrast at the nuclear envelope, membranes of mitochondria and endoplasmic reticulum (Stanarius et al. 1997). Endothelial nitric oxide synthase (EC 1.14.23) immunostaining without tyramide signal amplification yielded electron dense 3,3'-diaminobenzidine precipitates that were seen in the endothelium if arteries, veins and some capillaries. Its pattern corresponded roughly with the membrane-bound 2-(2'-benzothiazolyl)-5-styryl-3-(4'-phthal-hydrazidyl) tetrazolium chloride-formazan deposits generated by NADPH-diaphorase.

Endothelial swelling 4 h and 10 days after up to 12 periods of hypoxia (2 % O_2) for 3 to 5 min each and alternate normoxic periods of 10 min each was argued for the severity of neuronal damage in the neocortex of the *guinea-pig* (Winkelmann et al. 1972).

In those *rats* where endothelial oedema did nor occur the endothelium remained unchanged, or showed some degree of hypertrophy within 24 h of anoxia (Hills 1964). In the latter case there was an increase in the amount of free ribosomes, and many pseudopodial processes appeared which projected into the capillary lumen (Fig. 220).

Endothelial dysfunction is characterised by a loss of ability to release endothelium-derived relaxing factor (now known to be nitric oxide; Moncada et al. 1990), occurring within 2.5 min in *rats* (Tsao and Lefer 1990) and *cats* (Tsao et al. 1990) subjected to myocardial ischaemia and reperfusion. In the cerebral microcirculation of the parietal cortex of the *rabbit*, topically applied L-arginine (10 μM) induced moderate vasodilatation of 4±0.9% (Haberl et al. 1991). In *canine* basilar arteries, nitro L-arginine (an inhibitor of NOS) contracted rings with endothelium, but only minimally contracted those without (Schini et al. 1993). The contractions are likely to reflect an increase in the myogenic tone after the impairment of the basal production of •NO in endothelial cells.

Using a combined ribonuclease protection/semi-quantitative reverse transcriptase-polymerase chain reaction approach, Mandriota et al. (2000) demonstrated that hypoxia up-regulates angiopoietin-2 mRNA levels by up to 3.3-fold in two *human* endothelial cell lines. In *bovine* microvascular endothelial cells, the flavoprotein oxidoreductase inhibitor diphenylene iodonium and the related compound iodonium diphenyl mimicked induction of angiopoietin-2 but not vascular endothelial growth factor by hypoxia; in combination with hypoxia, diphenylene iodonium further increased angiopoietin-2 expression but had no effect on the induction of vascular endothelial growth factor by hypoxia. Neither angiopoietin-2 nor vascular endothelial growth factor were increased by cyanide or rotenone, suggesting that failure in mitochondrial electron transport is not involved in the oxygen-sensing system that controls their expression. In ischaemic *rat* dorsal skin flaps or in the brains of *rats* maintained for 12 h under conditions of hypoxia, angiopoietin-2 mRNA was up-regulated 7.7- or 17.6-fold, respectively. Vascular endothelial growth factor was concomitantly increased, whereas expression of angiopoietin-1, endothelial cell tyrosine kinase receptor Tie-2, and the related receptor Tie-1 was unaltered.

Porcine brain endothelial cells exposed to 120 min of hypoxia (95% N_2/5% CO_2) showed a decline in adenosine triphosphate content, enhanced lactate dehydrogenase liberation, and lactate release tended to increase in comparison to the normoxic controls (Blasig et al. 1994). Within the first 30 min of reoxygenation, the indicator of radical-induced lipid peroxidation, thiobarbituric acid-reactive substances, exhibited a slight increase during hypoxia and a more dramatic rise to about 3.5-fold greater than the control. Electron microscopy showed intact glycocalix, few pinocytotic vesicles, Weibel-Palade bodies, and characteristically tight junctions in the normal brain endothelial cells. During hypoxia most of the cells appeared to remain intact; however, a relatively small number of cells was damaged by autolysosomal processes. Yet 15 min after the onset of reoxygenation, the destruction in the entire monolayer proceeded markedly: vacuoles of different shape appeared, the number of lysosomes and autolysosomal cells increased, and blebs were formed. Only after 30 min of reoxygenation, lipofuscin granula representing stable end products of lipid peroxidation were detected.

In an anoxia-reoxygenation model using pure cultures of cerebral endothelial cells isolated from *piglet* cortex to measure cerebral endothelial cell oxygen free radical production and determine its

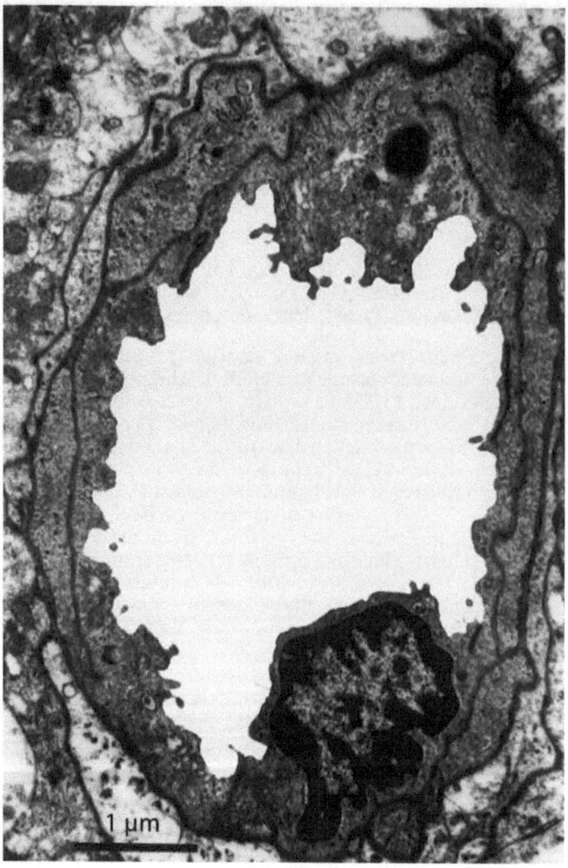

Fig. 220. Unspecific endothelial surface enlargement after intermittent hypoxia. Hippocampus of a female Wistar *rat* (No. 5/5) treated for 4 consecutive days with intraperitoneal injection of 1.5 ml Ringer's solution per kg body weight × day. 30 min after each medication the animal was exposed to an atmosphere containing only 5% oxygen for 30 min. Half an hour after the last exposure, under pentobarbital anaesthesia (30 mg/kg), the animal was perfused from the abdominal aorta with 2.5% glutaraldehyde in 0.1 M sodium cacodylate buffer (pH 7.4). Postfixation with 1% osmium tetroxide in sodium cacodylate buffer. Embedded in Epon 812 and sectioned at 50 nm. Lead citrate and uranyl acetate. Film 242-39

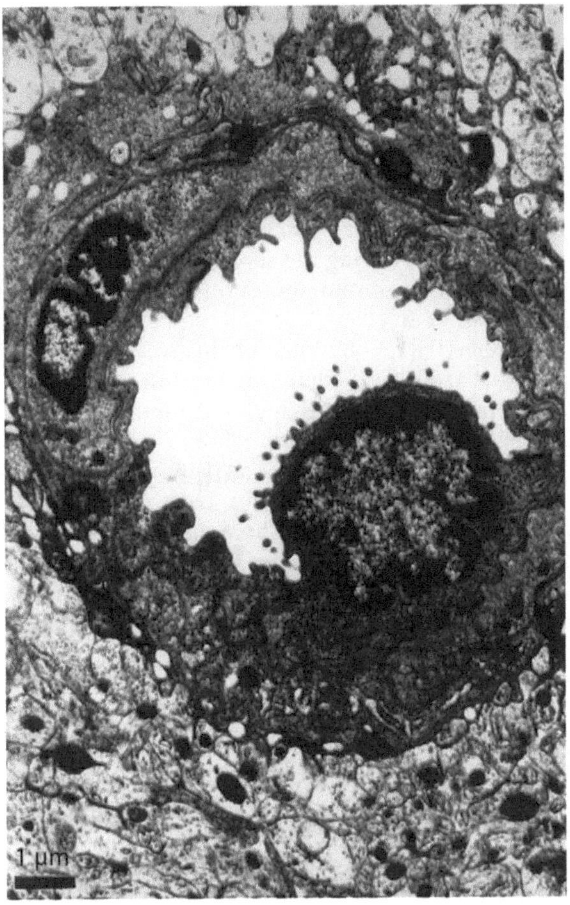

Fig. 221. Unspecific endothelial surface enlargement after intermittent hypoxia. Hippocampus of a female Wistar *rat* (No. 15/8/2) treated for 4 consecutive days with intraperitoneal injection of 300 mg piracetam per kg body weight × day. 30 min after each medication the animal was exposed to an atmosphere containing only 5 % oxygen for 30 min. Half an hour after the last exposure, under pentobarbital anaesthesia (30 mg/kg), the animal was perfused from the abdominal aorta with 2.5 % glutaraldehyde in 0.1 M sodium cacodylate buffer (pH 7.4). Postfixation with 1 % osmium tetroxide in sodium cacodylate buffer. Embedded in Epon 812 and sectioned at 50 nm. Lead citrate and uranyl acetate. Film 234-38

role in anoxia-reoxygenation-induced injury, BEETSCH et al. (1998) reported a progressively increased efflux of lactate dehydrogenase in the culture medium with the duration of the anoxic exposure, becoming significant after 10 h. Reoxygenation significantly increased cerebral endothelial cell anoxic injury in a time-dependent manner. A 55 % increase in oxygen free radical production, determined by fluorescence detection of dihydroethidium oxidation, was measured at the end of a 4-h reoxygenation in cerebral endothelial cells subjected to anoxia-reoxygenation conditions that killed 40 % of the cells. Blockade of oxygen free radical production with superoxide dismutase (250 and 1000 U/ml) or oxypurinol (50 and 200 µM) reduced this injury by 32–36 % and 30–39 %, respectively.

Susceptibility to neutrophil-mediated killing of endothelial cells shows a reciprocal relationship between the generation of $O_2^{\bullet -}$ (more killing) and the production of •NO (less killing).

The endothelial cells of the microvasculature have a 6-10 fold higher concentration of single membrane-spanning transferrin receptors than the brain parenchyma (KALARIA et al. 1992).

Pericytes

Pericytes have been viewed as housekeeping scavenger cells (ANDREWS et al. 1971, SCHEITHAUER et al. 1984, PALMER et al. 1985), and a second line of defence in the blood-brain barrier (BROADWELL and SALCMAN 1981).

FREY et al. (1991) have reported the presence of γ-glutamyl transpeptidase (EC 2.3.2.2) in brain capillary endothelial cells and pericytes, both *in vitro* and *in vivo*. γ-Glutamyl transpeptidase is a heterodimer glycoprotein distributed on the external surface of the cell. It catalysed the transfer of a γ-glutamyl residue from a donor peptide, in particular glutathione, to acceptor peptides. Functionally γ-glutamyl transpeptidase appears to be concerned with the transport of large, neutral amino acids across the blood-brain barrier. The relative roles of endothelial cells and pericytes have not been establishes (ALLT and LAWRENSON 2001). No other microvessels show detectable amounts of γ-glutamyl transpeptidase, which is also absent from endothelial cells of brain regions that lack a blood-brain barrier such as the median eminence.

Ecto-5'-nucleotidase (EC 3.1.3.5) has been identified on the outer surface of the cell membrane and so its reaction product adenosine is released into the extracellular space.

Aminopeptidase N (EC 3.4.11) was shown to occur selectively in pericytes of brain capillaries (KRAUSE et al. 1992). Functionally this enzyme is involved in the degradation of polypeptides including bioactive peptides (KENNEY and TURNER 1987), such as those with vasoactive properties. The enzyme may therefore be a component of the blood-brain barrier modifying neuropeptides which themselves alter vascular permeability. In experimental allergic encephalomyelitis there is a down-regulation of pericyte aminopeptidase N accompanying the well-known transient and focal perturbation of the blood-brain barrier (KUNZ et al. 1995).

The **phagocytic capacity** of pericytes has been well documented in ischaemia of the central nervous system of the *rabbit* (JEYNES 1985). *In vitro*, pericytes showed phagocytic activity when exposed to polystyrene beads and zymosan (BALABANOV et al. 1996).

Pericytes in cortical biopsies from 17 *patients* with brain oedema associated to brain trauma, tumours and congenital malformations exhibited remarkable **oedematous changes** (CASTEJÓN 1984). They were enclosed by a thickened basement membrane, represented increased hypolemmal micropinocytotic transport, slightly dilated rough endoplasmic reticulum and moderate hydropic changes of cytoplasmic matrix. The mitochondria also showed oedematous changes of both matrix and cristae. Besides, lipid droplets, primary and secondary lysosomes, small protein-containing vacuoles, coated vesicles and clear and dark microtubules were found.

In **anoxic-ischaemic lesions** of the *rat* cerebral cortex, pericytes which surround many of the larger capillaries did not swell, but became compressed between the endothelium and the neuropil (HILLS 1964).

15.1.1.1.4 Arteriolar Smooth Muscle Cells

Chronic nicotine (4.5 mg/kg per day of nicotine free base, 15 to 22 days via osmotic minipump) upregulated NO signalling of Ca^{2+} channels and downregulated Ca^{2+}-activated K^+ channels in smooth muscle cells isolated from cerebral lenticulostrate arterioles of *rats* (GERZANICH et al. 2001). In pial window preparations, chronic nicotine blunted NO-induced vasodilatation of pial vessels and the increase in cortical blood flow measured by laser-Doppler flowmetry, demonstrating the importance of Ca^{2+} channel downregulation in NO-induced vasorelaxation.

15.1.1.1.5 Experimental Therapy

The pathophysiological significance of the acute burst in $HO^\bullet$ and lipid peroxidation in the hippocampus is underscored by the pre-ischaemic or prehypoxic treatment with free radical scavengers or antioxidants in attenuating the selective CA1 damage associated with these models. The current state of antioxidant therapy in acute central nervous system injury was recently reviewed by GILGUN-SHERKI et al. (2002).

The extremely short life of CuZn-superoxide dismutase (6 min) in circulating blood and its failure to pass the blood-brain barrier make it difficult to employ enzyme therapy in cerebral ischaemia. However, a modified enzyme therapy with an increased half-life, such as polyethylene glycol-conjugated superoxide dismutase, has been successfully used to reduce infarct volume in *rats* that have been subjected to focal cerebral ischaemia (LIU et al. 1989, HE et al. 1993). Liposome-entrapped superoxide dismutase has an increased half-life (4.2 h), blood-brain barrier permeability, and cellular uptake, and it has also proven to be an effective treatment in reducing the severity of traumatic and focal ischaemic brain injuries (CHAN et al. 1987, IMAIZUMI et al. 1990).

Copper(II)[2,3 - butanedionebis(N^4 - methylthiosemicarbazone)], a stable superoxide dismutase-like copper complex with high membrane penetrability was stable in *mouse* brain homogenates (WADA et al. 1994). In contrast to Cu(II)pyruvaldehydebis (N^4-methylthiosemicarbazone), most of which was metabolised in both brain homogenates and blood by the end of the incubation period, about 90 % of Cu(II)[2,3 - butanedionebis(N^4 - methylthiosemicarbazone)] remained intact after 30 min of incubation in blood and about 60 % remained after incubation in brain homogenate.

Oxypurinol (40 mg/kg), given to *gerbils* intraperitoneally 30 min after a 5 min bilateral occlusion of the carotid arteries protected against CA1 damage (PHILLIS and CLOUGH-HELFMAN 1990). **Deoxycoformycin** (500 μg/kg), which inhibits adenosine deaminase, did not confer protection when administered 30 min after the ischaemic period, even though it is effective when given prior to carotid artery occlusion.

Pre-treatment with the arachidonic acid **cyclooxygenase inhibitor** ibuprofen (20 mg/kg *gerbil* s.c.) decreased (- 68 %) the level of salicylate hydroxylation in the hippocampus, but not the isocortex (HALL et al. 1993).

The highly selective cyclooxygenase inhibitor DFU [5,5-dimethyl-3-(3-fluorophenyl)-4-(4-methylsulphonyl) phenyl-2(5H)-furanone; 10 mg per kg body weight p.o.] reduced neuronal damage when administered several hours after 5 min of transient forebrain ischaemia in *gerbils* (CANDELARIO-JALIL et al. 2002). The extent of ischaemic injury was assessed behaviourally by measuring the increases in locomotor activity and by histopathological evaluation (pyknosis, eosinophilia, karyorhexis and chromatin condensation) of the extent of CA1 hippocampal pyramidal cell injury 7 days after ischaemia.

Dihydrolipoic acid (50–100 mg/kg *rat*) given 30 min before a 10 min clamping of both carotid arteries protected hippocampal neurones against ischaemic/hypoxic damage as observed by light microscopy after 7 d recovery (PREHN et al. 1990). However, *chick* embryo (7 d) cerebral hemispheres cultivated *in vitro* for 4 days and then made hypoxic for 30 min by 1 mM NaCN showed a dose-dependent neurotoxic effect of dihydrolipoic acid characterised by the protein content of the cultures measured 3 days after hypoxia.

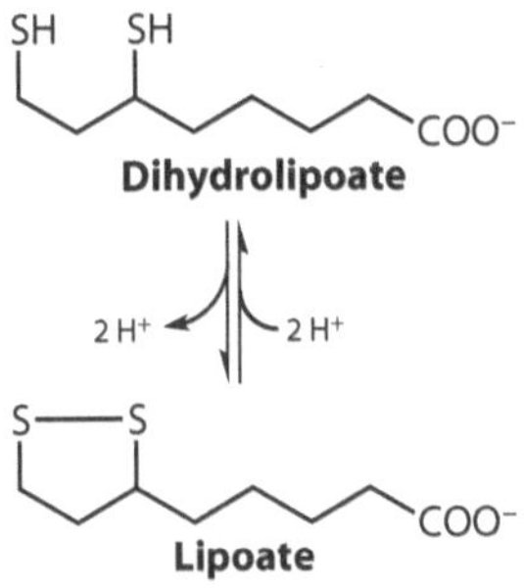

Dihydrolipoate – lipoate redox couple [60]

Pre-treatment with 30 mg dihydrolipoic acid /kg *rat* showed a selective vulnerability from hypoxia induced by breathing only 11 % oxygen for 60 min and a recovery period of 16 h (Oppermann and Derouiche 1996). The number of necrotic acidophilic cells was significantly reduced by dihydrolipoic acid.

α-Lipoic acid (100 mg/kg s.c.) pre-treatment of *mice* and *rats* 2 h, but not 1 h, 4 h or 6 h before middle cerebral artery occlusion by microbipolar electrocoagulation significantly reduced the infarcted areas in the cerebral cortex, but not in the corpus striatum (Wolz and Krieglstein 1996).

Using a new assay for dihydrolipoic acid, Jones et al. (2002) found that EA.hy926 cells rapidly took up and reduced α-lipoic acid to dihydrolipoic acid, most of which was released into the incubation medium. Nevertheless, the cells maintained dihydrolipoic acid following overnight culture, probably by recycling it from α-lipoic acid.

Using the *rat* hippocampal slice technique, Dimpfel (1996) examined the effects of the D- and L-enantiomers of α-lipoic acid in antagonising apamin neurotixicity. Apamin, a polypeptide from bee venom, acts as a most specific blocker of the small conductance Ca^{2+}-dependent K^{+} channel (for review see Dryer 1990). Thioctic acid firstly reversed the effect of 6 mM calcium in the superfusion fluid and secondly reversed the effect of apamin.

One hour of pre-incubation with dihydrolipoic acid (1 μM), but not with α-lipoic acid (1 μM), reduced damage of neurones from *chick* embryo telencephalon caused by 1 mM sodium cyanide or iron ions (Müller and Krieglstein 1995). α-Lipoic acid (1 μM) reduced cyanide-induced neuronal damage when added 24 h before hypoxia, and pre-treatment with α-lipoic acid for >24 h enhanced this neuroprotective effect. Both the *R*- and the *S*-enantiomer of α-lipoic acid exerted a similar neuroprotective effect. Pre-treatment with α-lipoic acid (1 μM) from the day of plating onward prevented the degeneration of chick embryo telencephalic neurones that had been exposed to Fe^{2+}/Fe^{3+}. α-Lipoic acid (1 μM) added to the culture medium the day of plating also reduced neuronal injury induced by 1 mM L-glutamate in rat hippocampal cultures, whereas 30 min of preincubation with α-lipoic acid failed to attenuate glutamate-induced neuronal damage.

Superior cervical ganglion blood flow, 52 % reduced by streptozotocin-induced diabetes, was dose-dependently restored by α-lipoic acid (ED_{50} = 44 mg/kg × day) in the *rat* (Cameron et al. 2001).

A new positively charged water soluble lipoic acid amide analogue, **2**-(*N*,*N*-dimethylamine)ethylamido lipoate-HCl was – independent of its stereochemistry – more effective than lipoic acid in (1) protecting *mouse* hippocampal HT4 cells against glutamate-induced cytotoxicity, (2) preventing glutamate-induced loss of intracellular reduced glutathione (GSH), and (3) disallowing increase of intracellular peroxide level following the glutamate challenge (Tirosh et al. 1999). The protective function of this antioxidant was synergistically enhanced by 1μM of sodium selenite.

The water and lipid soluble vitamin E analogue, **trolox**, protected cultured foetal *murine* cortical neurones against damage induced by exposure to either iron ions (50 μM Fe^{2+} and 50 μM Fe^{3+}) or ultraviolet light of primary wave length (> 60 %) 253.7 nm, consistent with an ability to inhibit free radical-mediated cytotoxicity (Chow et al. 1994). Trolox also reduced neuronal death induced by 24 h exposure to α-amino-3-hydroxy-5-methyl-4-isoxazole propionic acid (10 μM), but not that induced by *N*-methyl-D-aspartate (15 μM). When combined with the *N*-methyl-D-aspartate receptor antagonist dextrorphan (100 μM), trolox also reduced the neuronal injury induced by glucose deprivation.

17β-Oestradiol afforded protection against oxidative nerve cell apoptosis. 17β-Oestradiol is indeed a potent antioxidant comparable to other phenolic compound such as α-tocopherol (Behl 1999, Moosmann and Behl 1999).

Œstrogens can influence brain function in at least three ways:

- direct regulation of different neuronal functions, such as neurotransmitter receptor expression, the availability of neurotransmitters and the excitability of neuronal membranes;
- influence of synapse formation during development and regeneration;

- protection of neurones against exogenous insults by acting as a 'chemical shield' (Behl et al. 1995, Honjo et al. 1995, Green et al. 1997).

Employing various cell-free *in vitro* paradigms of peroxidation reactions, it has been found that œstrogen has a high intrinsic antioxidant activity (Nakano et al. 1987, Sugioka et al. 1987, Liu et al. 1992, Rifici and Khachadurian 1992, Subbiah et al. 1993) which enables this molecule to protect neurones against oxidative stress-induced cell death and excitotoxicity (Behl et al. 1995, Moosmann et al. 1997, Weaver et al. 1997).

In organotypic *rat* hippocampal slice cultures Moosmann et al. (2001) found that 4-dodecylaniline or iminostilbene completely protected these neurones from the toxic effects of H_2O_2 overload. All aromatic amines and imines with at least one free nitrogen–hydrogen bond showed some protective properties, with the less effective compounds exhibiting EC_{50} values between 1 an 10 μM. Substances containing two aromatic ring systems, whether separated or linked, showed higher efficacies, with the bridged compounds iminostilbene, phenoxazine, and phenothiazine having the highest potencies and the lowest EC_{50} values of averaged 75, 45, and 20 nM, respectively. Abrogation of the NH-bonds lead to a loss of cytoprotective activity, as exemplified by the relative inefficiency of *N,N*-dimethyl-1-naphthylamine an *N*-methyl-phenothiazine.

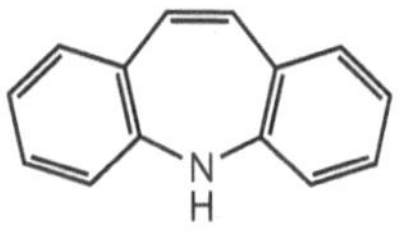

Iminostilbene [276]

This remarkable efficacy could be directly correlated to calculated properties of the compounds by means of a novel, quantitative structure–activity relationship model.

Lutinising hormone releasing hormone (LHRH), enkephalin, angiotensin, oxytocin, and vasopressin are biochemical antioxidants in aqueous medium.

Carvedilol {1-[carbazolyl-(4)-oxyl]-3-[2-methoxyphenoxyethyl) amino]-propanol-(2)} rapidly inhibited Fe^{2+} (0.25 mM $FeCl_2$ + 1 mM ascorbic acid)-initiated lipid peroxidation, measured as thiobarbituric acid reactive substance, in *rat* brain homogenate with an IC_{50} of 8.1 µM (Yue et al. 1992). Carvedilol protected against Fe^{2+}-induced α-tocopherol depletion with an IC_{50} of 17.6 µM. The antioxidant effect of carvedilol mainly resides in the carbazole moiety, and the substitution of a hydroxyl group at certain positions on the phenyl ring on either carbazole or the *ortho*-substituted phenoxylethylamine part of carvedilol resulted in an increase of antioxidant activity. The protective effect of carvedilol analogues against $HO^•$-mediated neuronal death positively correlated to their antioxidant effect. Lysko et al. (2000) showed that in cultured cerebellar neurones, and in brain and heart membranes, carvedilol had far greater antioxidant activities than metoprolol, which was essentially inactive as an antioxidant in these model systems.

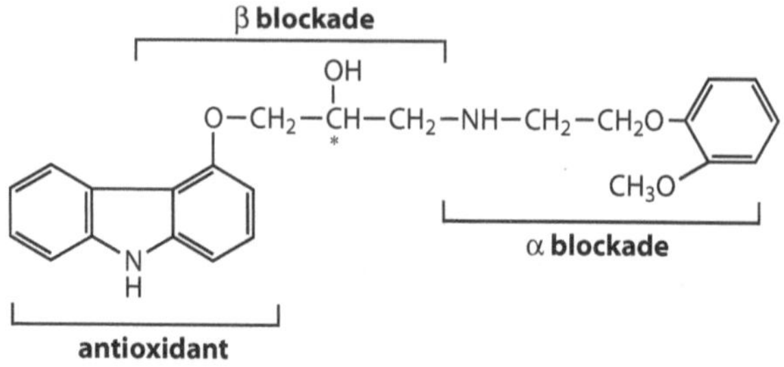

Chemical structure of carvedilol [72]

21-Aminosteroids or "**lazaroids**" (reviewed by Villa and Gorini 1997), have been shown to be potent inhibitors of lipid peroxidation *in vitro*. Using brain homogenates or purified brain synaptosomes as the lipid source, U74006F and U74500A potently inhibited iron-dependent lipid peroxidation, surpassing that of the glucocorticosteroid methylprednisolone (Braughler et al. 1987, 1989, Hall et al. 1988). U-74389G, 21-[4-(2,6-di-1-pyrrolidinyl-4-pyrimidinyl)-1-piperazinyl]-pregna-1,4,9-(11)-triene-3,20-dione,2-butenedioate, protects the antioxidant enzymes in the ischaemia/reperfusion-induced *rat* brain damage (Farbiszewski et al. 1994). **Tirilazad mesylate** (U74006F) posses multiple antioxidant mechanisms. These include the following: (1) scavenging of lipid peroxyl radicals similar to the action of vitamin E (Braughler and Pregenzer 1989), (2) facilitation of the antioxidant properties of the endogenous lipid peroxidation inhibitor vitamin E (Braughler and Pregenzer 1989), (3) scavenging of hydroxyl radicals (Althaus et al. 1991) and membrane stabilisation, i.e. decreased membrane fluidity (Audus et al. 1991). U74006F spared α-tocopherol and exerted a synergistic effect against the oxidation of liposomal membranes (Noguchi et al. 1998). Domenici et al. (1993) analysed the effect of the highly potent, selective, and non-competitive *N*-methyl-D-aspartate receptor antagonist, dizocilpine = (5*R*,10*S*)-(+)-5-methyl-10,11-dihydro-5*H*-dibenzo[*a*,*d*]-cyclohepten-5,10-imine (MK-801), and the 21-aminosteroids U-74500A and U-78517F on the hypoxia-induced electrophysiological changes in *rat* hippocampal slices. U-78517F, but not U-74500A, was able to ameliorate the recovery of the electrical response during reoxygenation significantly. Low concentrations of MK-801 improved the protective activity of ineffective low concentrations of U-78517F.

Dizocilpine or dantrolene, but not FK506, suppressed convulsive seizures in EL *mice* (NAGATOMO et al. 2001). Only MK-801 reduced NO_x in the brain.

3-Methyl-1-phenyl-2-pyrazolin-5-one (MCI-186) is a potent scavenger of hydroxyl radicals inhibiting not only hydroxyl radicals but iron-induced peroxidative injury (WATANABE et al. 1988, MUROTO et al. 1990). After the reaction with peroxy radicals, MCI-186 changes into 2-oxo-3-(phenylhydrazone)-butanoic acid (YAMAMOTO et al. 1996, KAWAI et al. 1997). In a thrombotic *rat* distal middle cerebral artery occlusion model, 3-methyl-1-phenyl-2-pyrazolin-5-one (3 mg/kg) significantly ($P < 0.05$) decreased the size of the cerebral infarct 1 day after artery occlusion (KAWAI et al. 1997).

S-Adenosyl-L-methionine (50 mg/kg given i.p. ever day for 3 days 1 h before ischaemia/reperfusion) in a combined model of permanent focal ischaemia and global reperfusion in the *rat* brain reduced the production of thiobarbituric acid-reactive substances after induction with ferrous salt as an indicator of brain lipid peroxidation (VILLALOBOS et al. 2000). Total glutathione production was increased. These changes were accompanied by an increase in mitochondrial capacity to reduce tetraphenyl tetrazolium.

Taurine (2-aminoethane sulfonic acid; 43 mg/kg i.p.) by itself did not have any effect on ozone (0.7–0.8 ppm for 4 h)-induced memory alterations in young *rats* but in old *rats* it improved memory, supporting the idea that taurine has an antioxidant effect when there is a pre-existing oxidative stress state, in this case old age (RIVAS-ARANCIBIA et al. 2000). However, there are other pharmacological effects of taurine that may be acting on memory improvement seen in old *rats*; taurine has been characterised as an neurotransmitter in the striatal system (BIANCHI et al. 1996) and has a modulatory interaction with dopamine (RUOTSALAINEN et al. 1996). With age there is a decrease in taurine levels in some brain structures, such as corpus striatum, nucleus accumbens, cerebellum and cortex (BENEDETTI et al. 1991), that correlated with a decrease in dopamine striatal content (DAWSON et al. 1999). In the *rat*, increasing the plasma levels of taurine 100-fold does not significantly increase brain levels (LEVI 1968). The rate of entry of taurine into the brain from the blood is the same in immature and mature *rats* (BAÑOS et al. 1971). MARTIN and SHAIN (1979) described the high-affinity, concentrative transport of taurine into LRM55 glial cells. Stimulation by β-adrenergic agonists released taurine in a dose-dependent manner (SHAIN et al. 1983). Propranolol inhibited the agonist-stimulated release of taurine.

Hypotaurine and taurine were found to reside within the cytosolic compartment of *human* neutrophils isolated from venous blood by density centrifugation on ficoll-diatrizoate gradients (GREEN et al. 1991). The ratio of taurine to hypotaurine is approx. 50:1. The concentration of hypotaurine decreased by 80% when resting neutrophils were converted into actively respiring cells by exposure to promised zymosan. Hypotaurine competed with 5,5'-dimethyl-1-pyrroline for hydroxyl radicals.

Taurine is a specific scavenger of HOCl (ARUOMA et al. 1988). In oncogene-transformed *rat* fibroblasts 208 F src3 pre-treated with 20 ng/ml TGF-β for 2 days, taurine (20 mM) inhibited the formation of HOCl and subsequent superoxide anion-dependent hydroxyl radical formation (HERDENER et al. 2000).

Hypotaurine (2-aminoethane sulfinic acid) may quench singlet oxygen to form taurine (PECCI et al. 1999).

Quenching of singlet oxygen by hypotaurine

$$NH_2\text{-}CH_2\text{-}CH_2\text{-}SO_2H + {}^1\mathbf{O_2} \rightarrow NH_2\text{-}CH_2\text{-}CH_2\text{-}SO_2OOH \qquad [277]$$

$$NH_2\text{-}CH_2\text{-}CH_2\text{-}SO_2OOH \rightarrow NH_2\text{-}CH_2\text{-}CH_2\text{-}SO_3H + 1/2\ O_2 \qquad [278]$$

Carnosine (β-alanyl-L-histidine, 10 μM) was demonstrated in the glomerular layer of the *rat* olfactory bulb by immunoelectron microscopy (SAKAI et al. 1988). It inhibited lipid peroxidation induced by 2,2-azobis-(2-amidinopropane) dihydrochloride by about 50% (COHEN et al. 1988).

The *rat* brain carnosine synthetase required DPN, which was used in the overall reaction at a 5.5/1 ratio carnosine/DPN (SKAPER et al. 1973).

Anserine (β-alanyl-L-methylhistidine)

Dimethyl sulfoxide, powerful free radical scavenger and cell membrane protectant, combined with a glycolytic intermediate, fructose 1,6-diphosphate, reduced mortality and neuronal death and improved sensory-motor function following severe traumatic brain injury in *mice* (DE LA TORRE 1995). While *rats* subjected to bilateral carotid artery occlusion after 14 weeks showed severe visuo-spatial memory impairment, dimethyl sulfoxide plus fructose 1,6-diphosphate (250:130 mg/kg) given i.p. for seven days and tested on the Morris water maze showed a 54% improvement in their visuo-spatial memory (DE LA TORRE et al. 1998). When treatment was discontinued improvement was lost and visuo-spatial memory function regressed to pre-treatment levels. Immunohistochemical examination showed minimal neuronal damage in all bilateral carotid artery-occluded *rats* and slight loss of microtubule associated protein-2. Glial fibrillary acidic protein immunostaining increase was observed only in the oriens layer of the hippocampus of untreated bilateral carotid artery-occluded *rats*.

The free radical scavenger ***N-tert*-butyl-α-phenylnitrone** (PBN) has been shown to reduce oedema and hippocampal CA1 neuronal loss in the *gerbil* following global ischaemia by 5 min of transient bilateral carotid artery occlusion (Yue et al. 1992). In focal ischaemic *rats*, PBN treatment significantly reduced infarct volume, oedema formation and neurological deficits (Cao and Phillis 1994). Pazos et al. (1999) determined whether delayed PBN (100 mg/kg *rat*) on days 3, 5 and 7 post insult is protective at 2 months following transient global forebrain ischaemia, and whether additive effects can be observed when PBN is administered in combination with moderate brain hypothermia (30 °C). A significant attenuation of cognitive deficits was observed in the animal group receiving the combination postischaemic hypothermia and delayed PBN treatment. Quantitative CA1 hippocampal cell counts indicated that each of the ischaemic groups exhibited significantly fewer viable CA1 neurones compared to sham controls. However, in *rats* receiving either delayed PBN treatment of 3 h of postischaemic hypothermia, significant sparing of CA1 neurones relative to the normothermic ischaemia group was observed.

Ethyl docosahexaenoate-treated foetal *rat* brain preparations exhibited an almost 70 % decrease in the amount of 5,5'-dimethyl-1.pyrroline-*N*-oxide-OH adducts compared to those from ethyl oleate-treated animals (Green et al. 2001). The decreased lipid peroxide production, as well as increased production of prostaglandin E_2 and nitric oxide by the foetal brain following ethyl docosahexaenoate administration could be mimicked by a synthetic quinone possessing both hydroxyl radical producing and lipid peroxide propagation inhibiting properties.

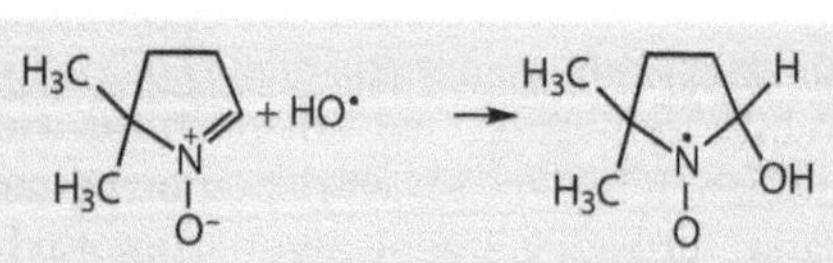

5,5-**Dim**ethyl-1-**p**yrroline-1-**o**xide-HO• adduct [8]

Many, but not all, **flavonoids** protected *mouse* hippocampus-derived HT-22 cells and *rat* primary neurones from glutamate toxicity as from five other oxidative injuries (Ishige et al. 2001). Three structural requirements of flavonoids for protection from glutamate are the hydroxylated C3, an unsaturated C ring, and hydrophobicity. Ishige et al. (2001) found free distinct mechanisms of protection:

- increasing intracellular GSH (quercetin, fisetin),
- directly lowering levels of reactive oxygen species (galangin, beicelein, kaempferol, luteolin, quercetin, fisetin), and
- preventing the influx of Ca^{2+} despite high levels of reactive oxygen species (flavonol, 6-hydroxyflavonol, 7-hydroxyflavonol).

Neurotoxicity induced by oxidised low-density-lipoprotein in cultures of embryonic *mouse* striatal neurones was neither reduced nor enhanced by inhibiting extracellular signal-regulated kinases 1/2 activation with mitogen-activated protein kinase kinase inhibitors, suggesting that this cascade is unlikely to be involved in either oxidised low-density-lipoprotein toxicity or the protective effects of flavonoids (Schroeter et al. 2001).

New neuroprotective strategies have been proposed with **selective nitric oxide synthase inhibitors** for the neuronal (ARL 17477; Zhang et al. 1996) or the inducible (1400W) isoforms or with compounds combining in one molecule selective nNOS inhibition and antioxidant properties (BN 80933), in experimental ischaemia-induced acute neuronal damage.

Desmethyl tirilazad reduced brain nitric oxide synthase activity and cyclic guanosine monophosphate during cerebral global ischaemia in *rats* (Fernandez et al. 1997).

Activation of central β-adrenergic receptors by the lipophilic β_2-adrenergic receptor agonist **clenbuterol** (100 μM) provided protection against glutamate-induced damage in *rat* hippocampal cultures (Semkova et al. 1996). In a *rat* model of transient forebrain ischaemia, clenbuterol (4×1 mg/kg) administered intraperitoneally increased the number of viable neurones in CA1 subfield.

Stimulation of $5\text{-}HT_{1A}$ receptors can cause neuroprotection *in vitro* and *in vivo* (Piera et al. 1995). $5\text{-}HT_{1A}$ receptor agonists can reduce glutamate release (Rupalla et al. 1994), which is most likely mediated through activation of presynaptic $5\text{-}HT_{1A}$ receptors located on glutaminergic terminals (Raiteri et al. 1991). The $5\text{-}HT_{1A}$ receptor agonist **Bay x 3702** ((–)-*R*-2-[4-[[(3,4-dihydro-2*H*-1-benzopyran-2-yl)methyl]amino]-butyl]-1,2-benzisothiazol-3(2*H*)-one 1,1-dioxide monohydrochloride; 4 μg/kg i.v.) caused a 10 % reduction of neuronal damage in the *rat* hippocampal CA1 subfield (Schaper et al. 2000).

There is evidence that GABAergic neurones are resistant to ischaemia-induced damage both *in vitro* (TECOMA and CHOI 1989) and in focal (JOHANSEN et al. 1991) and global (NITSCH et al. 1991) models of ischaemia, so that GABA-mimetic agents could assist in protecting vulnerable cells until the endogenous system starts functioning again. Therefore, compounds that enhance GABA function should be neuroprotective, providing that they enhance function in conditions in which there is little endogenous GABA function or tone.

L-Carnitine, [(−)-β-hydroxy-γ-trimethylammonio]butyrate, vitamin B_γ, accelerated the growth and differentiation of neurones, astrocytes, oligodendrocytes and ependymal cells from neurospheres in long-term cultures (ATHANASSAKIS et al. 2002). In addition to its pivotal role in the transport of activated fatty acids for β-oxidation, L-carnitine can increase membrane stability as tested on erythrocytes (ARDUINI et al. 1990), brain microsomes and liposomes (ARIENTI et al. 1992).

Piracetam, or 2-oxo-pyrrolidine-1-acetamide, was introduced as the first nootropic substance. Chronic treatment of young and aged *rats* with piracetam significantly increased membrane fluidity in some brain regions of the aged animals, but had no measurable influence on membrane fluidity in young ones (MÜLLER et al. 1997). Its relative *in vitro* specificity for the L-glutamate receptor (BERING and MÜLLER 1985) and the elevation of the *N*-methyl-D-aspartate receptor density by about 20 % (COHEN and MÜLLER 1993) could contribute to the therapeutic effects of this drug in *man*.

In normobaric hypoxic hypoxidosis (9.8 % O_2 for a period of 23 min) in healthy volunteers, aged between 20 and 38 years, two piracetam preparations attenuated the vigilance decrement (SALETU et al. 1995). Intravenous infusion of 12 g piracetam showed its encephalotropic peak effects in the earlier hours, oral dosing of 12 g piracetam syrup in the later hours.

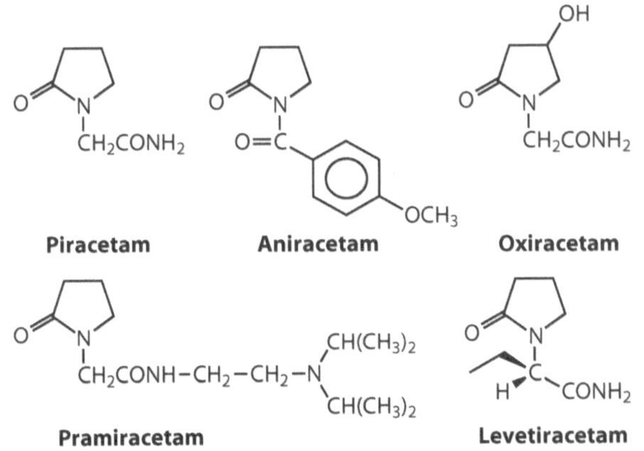

Nootropic drugs [279]

Pentylenetetrazol (45 mg/kg i.p. every 48 h) kindling-induced neuronal loss in the dentate area, in the hilus as well as in the pyramidal layer of CA3 could be prevented by pre-treatment of *rats* with piracetam (100 mg/kg i.p.) (POHLE et al. 1997).

Piracetam influenced the sulphydryl groups of the hypothalamo-hypophyseal system (SCHILLER 1974, 1975). It raised ATP formation in liver mitochondria (PEDE et al. 1971), but there was no specific binding of [^{3}H]piracetam to any of the various subcellular fraction (TACCONI and WURTMAN 1986). The cyclic 3',5'-adenosine monophosphate (cAMP) content of both brain and intestine was elevated (WETH 1982, WETH and GROSS 1982). The discovery of the cAMP-responsive element modulator (CREM) gene opened a new dimension in the study of transcriptional response to cAMP (FOULKES et al. 1991). Expression of CREM antagonist isoforms in several hypothalamic nuclei associated with homeostatic regulation as the magnocellular neurones in the supraoptic hypothalamic nuclei respond to osmotic stimulation by the differential temporal induction of two of the antagonistic isoforms, CREMα and CREMβ (MELLSTRÖM et al. 1993). From the hippocampus of *rats* treated with a single intraperitoneal injection of 300 mg piracetam/kg VILLA et al. (1989) isolated three types of mitochondria: both NADH-cytochrome *c* reductase and cytochrome oxidase activities were lowest in the synaptic heavy mitochondrial subfraction. In addition, other enzyme activities were different in the non-synaptic as compared to both the light and heavy synaptic mitochondria. Acute treatment decreased citrat synthase, glutamate dehydrogenase, NADH-cytochrome *c* reductase, and cytochrome oxidase activities only in the heavy mitochondria obtained from synaptosomes. Piracetam was ineffective to mitochondrial malondialdehyde generation (BLASCHKE et al. 1985). While pyritinol and taminitinol protected *rat* brain cytosol protein from insolubilization by hydroxyl radicals generated by a Fenton-type reaction *in vitro*, piracetam and oxiracetam were without any effect (PAVLÍK and PILAR 1989). Piracetam was protectively effective on the decrease of synaptosomal Ca^{2+}-dependent adenosine triphosphatase (EC 3.6.1.3) activity induced by hypobaric (8.7 kPa) hypoxia of for 18 h (BLASCHKE et al. 1988).

Potentiation by **oxiracetam** was specific for α-amino-3-hydroxy-5-methyl-4-isoxazolepropionic acid (AMPA) receptor-mediated signal transduction, as the drug changed neither the stimulation of $^{45}Ca^{2+}$ influx by kainate or *N*-methyl-D-aspartate nor the activation of inositol phospholipid hydrolysis elicited by quisqualate or (±)-1-aminocyclopentane-*trans*-1,3-dicarboxylic acid. Piracetam, aniracetam, and oxiracetam increased the maximal density of the specific binding sites for [^{3}H]AMPA in synaptic membranes from the *rat* cerebral cortex (Copani et al. 1992).

Aniracetam reversibly potentiated the ionotropic quisqualate responses recorded intracellularly from the pyramidal cells in the CA1 region of *rat* hippocampal slices (Ito et al. 1990). The excitatory postsynaptic potentials in Schaffer collateral-commissural-CA1 synapses was also potentiated by aniracetam.

Levetiracetam (ucb L059; Keppra®), the *S* enantiomer of the α-ethyl analogue of piracetam, in the CA3 subfield of the *rat* hippocampal slice preparation inhibited epileptiform bursting induced by 10 μM bicuculline, a γ-aminobutyric acid $(GABA)_A$-receptor antagonist (Birnstiel et al. 1997). In experiments in which the glutamate receptor agonist, *N*-methyl-D-aspartate was used to generate spontaneous bursting, levetiracetam had no effect on the size of the bursts but decreased bursting frequency. *In vivo*, inhibition of (+)-bicuculline-induced neuronal excitability by levetiracetam may involve calcium-dependent processes not associated with blockade of $GABA_A$ receptors (Margineanu and Wülfert 1997). Levetiracetam dose dependently reduced kainic acid (13.2 mg per kg *rat* subcutaneously)-induced seizures (Klitgaard et al. 1998).

Contrary to piracetam, a novel **piracetam peptide analogue**, ethyl ester of *N*-phenyl-acetyl-L-prolyl-glycine (GSV-111; 1 μM) in isolated neurones of the snail *Helix pomatia* suppressed the high-threshold K^+ current independent of Ca^{2+} $(I_{K(V)})$ by 49±18% (Solntseva et al. 1997).

It has been postulated that the brain neuronal degeneration associated with ischaemia may result from excessive release of excitatory amino acids such as glutamate (Hagberg et al. 1985). This hypothesis has been supported by reports that *N*-methyl-D-aspartic acid antagonists protect the brain from ischaemic damage (Simon et al. 1984, Boast et al. 1987, 1988). (+)-1-Methyl-1-phenyl-1,2,3,4-tetrahydro-isoquinoline HCl, a non-competitive ***N*-methyl-D-aspartic acid receptor antagonist** (Hodgkiss et al. 1993), prevented hippocampal CA1 cell damage at a dose of 10 mg/kg and reduced spontaneous locomotor hyperreactivity in *gerbils* after the development of ischaemia at a dose of 32 mg/kg (Nakanishi et al. 1994).

The non-selective *N*-methyl-D-aspartic acid and α-amino-3-hyroxy-5-methyl-4-isoxazolepropionic acid receptor antagonists were neuroprotective in animal models but clinical development has been hampered due to their side-effect profile. However, subtype selective antagonists will be neuroprotective with reduced side-effects (Gill 1996). Blocking the glycine-binding site of *N*-methyl-D-aspartic acid receptors (Maruoka et al. 1998, Ohtani et al. 1998) by (*S*)-9-chloro-5-[*p*-aminomethyl-*o*-(carboxymethoxy)phenylcarbamoylmethyl]-6,7-dihydro-1*H*,5*H*-pyrido[1,2,3-*de*]quinoxyline-2,3-dione hydrochloride trihydrate (15 or 30 mg/kg intravenously immediately and 2 h after bilateral carotid artery occlusion) prevented the progression of ischaemic pathology induced by bilateral carotid artery occlusion in spontaneously hypertensive *rats* (Ohtani et al. 2000).

Spermine (1 mM) is neuroprotective against either anoxia (10 min) or *N*-methyl-D-aspartate (0.5 mM) in *rat* hippocampal slices (Ferchmin et al. 2000).

ACEA 1021 is a potent and selective competitive glycine/N-methyl-D-aspatate antagonist which is bioavailable, formulated for intravenous administration, is effective as a neuroprotectant in multiple models of cerebral ischaemia. It has a favourable nonclinical central nervous system and cerebrovascular safety profile and an acceptable toxicity profile (Whitehouse 1996).

CERESTAT® (aptiganel HCl, CNS 1102), a selective, non-competitive antagostist of *N*-methyl-D-aspartiate ion channel complex, has robust neuroprotective effects in animal stroke models (Gamzu 1996).

Compounds that increase the resistance to oxidative stress were identified by their ability to increase the IC_{50} of the *human* neuronal cell line IMR-32 for hydrogen peroxide (Fiander and Schneider 1999). All of the compounds identified that increased the IC_{50} (238±53% by 2 μM 1-nitro-1-cyclohexene; 172±51% by 2 μM 3-methylene-2-norbornanone) also increased the specific activity of glutathione *S*-transferase. In addition, compound-caused increases in the specific activity of glutathione *S*-transferase correlated with compound-caused increases in the IC_{50}, the expected behaviour if glutathione *S*-transferase was a critical enzyme.

Hexasulfobutylated C_{60}, a free radical remover, at dosages of 10 and 100 μg/kg, reduced the total volume of infarction in both pre-treatment and treatment groups of male Long-Evans *rats* with focal cerebral ischaemia induced by middle cerebral artery occlusion (Huang et al. 2001). After administration of the drug, the •NO content in plasma was increased and the lactate dehydrogenase (EC 1.1.1.27) levels were decreased.

Vascular endothelial growth factor (VEGF) in an immortalised *mouse* hippocampal neuronal cell line (HN33) reduced cell death associated with an *in vitro* model of cerebral ischaemia (Jin et al. 2000). At a maximally effective concentration of 50 ng/ml, VEGF approximately doubled the number of cells surviving after 24 h of hypoxia and glucose deprivation.

Dipyridamole treatment in patients with prior stroke or transient ischaemic attack substantially reduced stroke recurrence, with a beneficial effect comparable to and additive with that induced by acetylsalicylic acid (DINER 1996). *In vitro*, dipyridamole behaved. as an antioxidant twice as effective as α-tocopherol in inhibiting lipid peroxidation of methyl linoleate (IULIANO et al. 1995), and in the oxidation of low-density lipoprotein induced chemically and by endothelial cells (IULIANO et al. 1996). KUSMIC et al. (2000) during cerebral hypoperfusion with human carotid endarterectomy could attenuate cerebral oxidative stress by pre-treatment with oral dipyridamole.

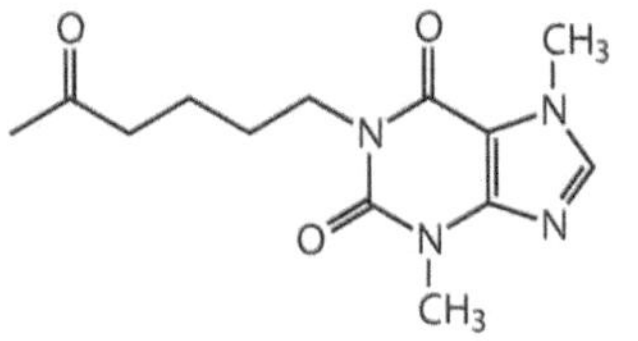

Pentoxifylline [52]

Pentoxifylline (formula [52]) significantly increased the average neuronal mitochondrial size in the hippocampus and cerebral cortex compared with unmedicated ischaemic *gerbils* and normal controls; the treated ischaemic group showed the greatest increase (HARTMANN et al. 1977). These large unvacuolated mitochondria appeared structurally normal in all respect except for their excessive size. Mitochondrial crests were orderly and closely packed, with little or no rarefaction of the mitochondrial matrix. Thus they differed essentially from swollen or vacuolated mitochondria seen in ultrastructural studies of a variety of experimental or pathologic conditions, including untreated anoxia and ischaemia. Furthermore, they showed none of the bizarre internal structure, such as crystalline or granular inclusions or concentric arrays of cristae (Fig. 289).

There are many possible reasons for the discrepancies between pre-clinical studies that usually show very promising results with neuroprotective drugs and the clinical practice that is very disappointing (LEKER and SHOHAMI 2002). The most important reasons for failure of early trials may be related to the fact that all these studies concentrated on blocking just one of the individual death related processes in ischaemia leaving the door open for the other processes to produce cellular death. CHOI (1996) and LEE et al. (1999) elegantly showed that after blocking all excitotoxic activity cells would die of apoptosis instead of excitotoxicity.

Contrary to the previously accepted dogma of an ischaemic cascade wherein one pathological process leads to the sequential activation of other processes it is currently suggested that all the above mentioned mechanisms of cell death are induced in parallel very early after ischaemic onset. Use of combination therapy of drugs capable of neutralizing all individual death-causing pathways, drugs active against all damaging mechanisms or hypothermia is more likely to succeed in reducing damage.

15.1.1.2
Rhinencephalon

The mammalian olfactory bulb is considered to be one of favourable regions for analysing information processing and understanding some general principles of neuronal organisation.

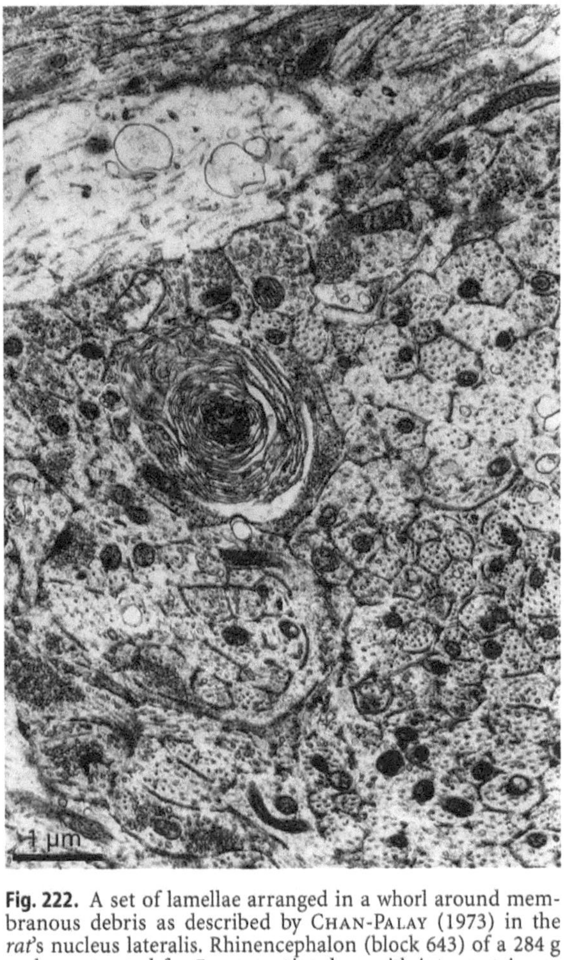

Fig. 222. A set of lamellae arranged in a whorl around membranous debris as described by CHAN-PALAY (1973) in the *rat's* nucleus lateralis. Rhinencephalon (block 643) of a 284 g male *rat* treated for 7 consecutive days with intragastric application of 1.5 ml Tyrode's solution per kg body weight × day. On the last 4 days of experimentation the animal was exposed to an atmosphere containing only 5 % oxygen for 30 min. On May 10, 1976 half an hour after the last exposure under pentobarbital anaesthesia (30 mg/kg), the animal was perfused from the abdominal aorta with 2.5 % glutaraldehyde in 0.1 M sodium cacodylate buffer (pH 7.4). Postfixation with 1 % osmium tetroxide in sodium cacodylate buffer. Embedded in Epon 812 and sectioned at 50 nm. Lead citrate and uranyl acetate. Plate 3148

The coexistence of immunoreactivities for glutamate decarboxylase (EC 4.1.1.15) and tyrosine hydroxylase (EC 1.14.16.2) was revealed in some neurones in the periglomerular region and in the superficial part of the external plexiform layer of the *rat* main olfactory bulb (KOSAKA et al. 1985).

Using slices of the olfactory cortex of the *guinea pig*, adenosine and adenine nucleotides have been shown to depress the amplitude of synaptic responses (KURODA and KOBAYASHI 1975, OKADA and SAITO 1979).

Cyclic GMP-stimulated phosphodiesterase (PDE2) is present in the olfactory cortex (JUILFS et al. 1999).

15.1.1.3 Nucleus Accumbens

Besides those in the corpus striatum, there are large accumulations of dopamine terminals in some nuclei of the limbic system such as the nucleus accumbens and the olfactory tubercle. The L-DOPA-induced increase in *rat* motor activity might be evoked, at least partly, by a stimulation of postsynaptic dopamine receptors in the nucleus accumbens since local application of dopamine bilaterally to this region markedly enhanced the *rat* motor activity (PIJNENBURG and VAN ROSSUM 1973).

Nitric oxide (•NO) may be very much involved in the expression of associative increase in extracellular dopamine in the nucleus accumbens (AFANAS'EV et al. 2000).

15.1.1.4 Corpus Striatum

15.1.1.4.1 Normoxic Cytobiology

The adult *rat* neostriatum has been reported to contain 4 (MORI 1966) or 5 (DIMOVA et al. 1980) ultrastructurally different types of neurones and 4 types are identified in primary cultures of newborn *rat* neostriatum (PANULA et al. 1979). Using the combined Golgi-electron microscope method of FAIRÉN et al. (1977), DIMOVA et al. (1980) identified three kinds of cells, according to cell body size and process morphology, i.e. giant, small (dwarf or neurogliform cells lacking an axon and synapses) and medium-sized cells, the latter consisting of four different types.

Small nerve cells (10–15 μm in diameter) are very numerous (MORI 1966). Groups of round dense bodies, 0.1–0.7 μm in diameter, are usually found in the Golgi region. They are filled with numerous fine granules of 8 nm diameter, spacing a thin light zone of about 12.5 nm width inside the limiting membrane. Another group of dense bodies are extremely electron dense and irregular in shape and contain various droplets or granules in addition to a few fine granules. DIMOVA et al. (1980) termed oval or polygonal cells (10–18 μm in diameter) with 4 to 7 dendrites emerging from the perikaryon medium-sized type I cells. Their type II medium-sized cells are oval (17–20 μm to 10–12 μm) with 4 to 5 dendrites (150–200 μm), and poor in spines. Oval (15–20 μm long and over 10 μm wide) medium-sized type III nerve cells showed a prominent nucleolus (up to 2.3 μm in diameter) and intranuclear inclusions. The cytoplasm displays a well-developed granular endoplasmic reticulum, comprising either single or parallel-arrayed (up to 7) cisternae and numerous free ribosomes. The Golgi apparatus is also well-developed and up to 9 profiles per section were observed. The round or elongated neuronal bodies of type IV medium-sized nerve cells presents nuclei of an irregular, sometimes polylobulated form due to numerous deep invaginations of the nuclear envelope. The nucleus occupies practically the entire cell body and only a scarce quantity of cytoplasm was found around the nucleus. The cytoplasm contains few and most frequently branched granular cisternae, never forming Nissl bodies, a moderate number of ribosomes and a few mitochondria. The Golgi zones are very well developed and some of them are considerably long.

Giant neurones are elongated reaching 20 μm in width and 30–60 μm in length (DIMOVA et al. 1980). Perinuclear cytoplasm is rich in both lysosomes and Golgi zones of considerable size and with narrow lamellae.

Neutron activation analysis of iron in the *human* nucleus caudatus and putamen from patients aged 23 to 66 y resulted in 8.30 ± 0.46 and $8.78 \pm 0.55 \times 10^{-4}$ terms of element weight per unit dried tissue, respectively (HÖCK et al. 1975).

Ascorbic acid having the unique property of reducing an acidified solution of silver nitrate almost instantly (GOMORI 1952, p. 90), can be detected both in nerve and above all in glial cells of the *human* striatum (CLARA 1942, 1953).

Neurones in *rat* neostriatum seemed more intensely immunopositive for glutathione (GSH) than in neocortex, though their overall appearance was similar to that in cortex (MAYBODI et al. 1999). L-Buthionine-[*S*,*R*]-sulphoxime pretreatment, which depletes cellular GSH levels, enhanced zinc-induced oxidative injuries in the nigrostriatal system of SD *rats* (LIN 2001).

Cytochrome oxidase (EC 1.9.3.1) activity was shown histochemically in the caudate nucleus and putamen (SHIMIZU et al. 1959, YAMADA 1961).

Tyrosine hydroxylase (EC 1.14.16.2) is the rate-limiting enzyme which synthesises dopamine. Immunoreactive tyrosine hydroxylase has been demonstrated by fluorescence microscopy in the caudate nucleus (Chiueh et al. 1984). Vacca-Galloway et al. (1988) found a selective decrease of immunoreactive tyrosine hydroxylase in nigrostriatum of adult male *rats* after *N*-methyl-4-phenyl-1,2,3,6-tetrahydropyrdine treatment.

The •NO generator, glutathione-*N*-oxide, enhanced the inhibitory action of 1-methyl-4-phenylpyridinium (MPP^+) on complex I activity in brain submitochondrial particles (Cleeter et al. 2001).

Nitric oxide synthase (EC 1.14.23) mRNA investigated by in situ hybridisation histochemistry was confined to a subpopulation of scattered medium-sized neurones comprising 1.5 to 2 % of *human* striatal neurone population (Nisbet et al. 1994), a finding in agreement with animal studies (Dawson et al. 1991).

The ability of dopamine to increase [^{3}H]cAMP formation in cultured striatal neurones (16-day embryos of Swiss *mice*) was blocked by its nitration by •NO or its nitrogen oxide derivatives (Daveu et al. 1997).

Monoamine oxidase (EC 1.4.3.4)

$$RCHNH_2 + H_2O + O_2 \longrightarrow RCHO + NH_3 + H_2O_2 \qquad [280]$$

Deprenyl is a monoamine oxidase B inhibitor but also possesses many other pharmacological activities. Superoxide dismutase (EC 1.15.1.1) activity was significantly increased in the striatum of *rats* treated for three weeks with 0.50 up to 2.00 mg (–)deprenyl per kg body weight × day subcutaneously (Knoll 1988). This has been confirmed especially for Mn-SOD (Carrillo et al. 1991, 1992, 1993). In contrast to SOD activities, an increase in catalase (EC 1.11.1.6) activity became significant only after two weeks of subcutaneous infusion of 2 mg (–)deprenyl per kg × day, and only in the brain regions where SOD activities were increased earlier. The delay in the increase in catalase activity following deprenyl infusion suggested that this increased catalase activity is an adaptive response to the earlier increase in deprenyl-induced SOD activities rather than a direct effect of deprenyl on catalase activity, although the latter possibility could not be excluded (Carrillo et al. 1992).

In *rats* and *mice* injected intravenously with [^{14}C]tyrosine, chlorpromazine significantly augmented the accumulation of [^{14}C]dopamine in their brains, whereas the accumulation of [^{14}C]noradrenaline was almost unchanged (Nybäck and Sedvall 1968).

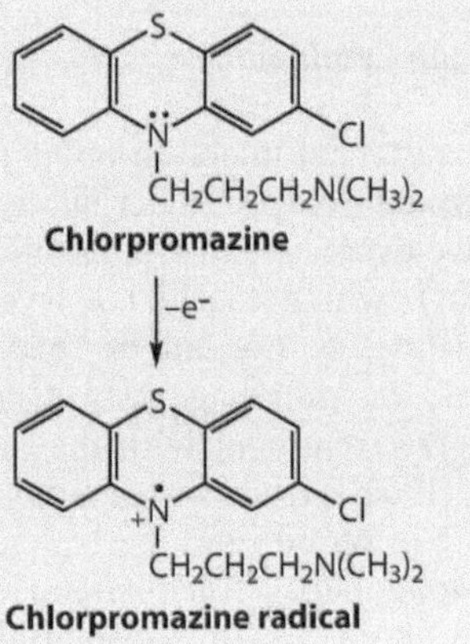

[64]

Cyclic GMP and cyclic AMP were localised at the ultrastructural level using horseradish peroxidase immunocytochemistry (Ariano and Matus 1981). In the *rat* caudate-putamen both of the cyclic nucleotides were detected within postsynaptic terminal boutons and within astroglial processes. cGMP postsynaptic staining was stronger than glial staining, whereas the localisation pattern was reversed for cAMP.

Cyclic GMP-stimulated phosphodiesterase (EC 3.1.4.17) appeared to be present in axons or axon terminals from cell bodies located in the caudate putamen and running to the substantia nigra (Juilfs et al. 1999).

In primary cultures of *rat* neostriatum, glutamate decarboxylase (EC 4.1.1.15) immunoreactivity was restricted to a small nerve cell type with an indented nucleus and rich amounts of free ribosomes in the cytoplasm (Panula et al. 1981). Glial cells were non-reactive.

Soluble guanylate cyclase (EC 4.6.1.2) was highly concentrated in both the caudatoputamen and the nucleus accumbens of the male Wistar *rat* (Walaas 1981). Local injections of kainic acid, which destroyed cholinergic and GABA neurones in the caudatoputamen and in the nucleus accumbens, caused a rapid (70–90 %) decrease in the soluble guanylate cyclase and a slower 50–50 % fall in the particulate guanylate cyclase in these nuclei.

The 5-hydroxytryptamine$_6$ receptor was immunolocalized in olfactory tubercle, piriform cortex, nucleus accumbens, island of Calleja, striatum, hippocampus (CA1 and dentate gyrus) and the molecular layer of the cerebellum (Gérard et al. 1997). Specific binding of [^{3}H]Ro 63-0563 [4-amino-*N*-2,6 bis-methylamino-pyridin-4-yl)-benzene sulphonamide] to recombinant *rat* and *human* 5-hydroxytryptamine$_6$ receptors was saturable, rapid, and reversible with equilibrium dissociation constants (K_d) of 6.8 nM and 4.96 nM, respectively (Boess et al. 1998).

15.1.1.4.2
Reoxygenation Cytobiology

In *cats* intracisternal injection of 0.5 ml/kg of fresh unheparinized autologous arterial blood resulted in a progressive decrease in cerebral blood flow (caudate nucleus), which reached a level of approximately 50 % below normal at three hours after subarachnoid haemorrhage (HALL and BRAUGHLER 1988). Pre-treatment with the antioxidant, α-tocopherol (1000 IU) almost completely prevented this decrease in blood flow.

In the Mongolian *gerbil* bilateral carotid occlusion model there was a substantial cell loss of 40–50 %, measured in haematoxylin and eosin-stained section, 5 days post-insult (STAMFORD et al. 1999). Ascorbic acid (500 mg per kg × day intraperitoneally for 3½ days, first dose 1 h before occlusion) caused significant striatal protection (cell loss reduced from 49 % to 20 %). After occlusion of the right carotid artery for 20 min, HARTMANN et al. (1973) found enlarged mitochondria in the cytoplasm of the neurones of ipsilateral caudate nucleus.

Application of ischaemic challenge to a *rat* striatal brain slice evoked a consistent profile of dopamine release in the absence of drug tratment (STAMFORD et al. 1999). an initial quiescent period lasting 2–3 min was followed by a precipitous monophasic dopamine release event.

Major changes in phosphorylation induced by ischaemia and subsequent reperfusion in *rat* striatum were observed for a 130-kDa protein, tentatively identified as the Ca^{2+} transport ATPase, and calcium/calmodulin-dependent protein kinase II (SANKARAN et al. 1997). A 200–300 % increase in $[^{32}P]8N_3ATP$ photoinsertions was observed in the striatum and hippocampus region of a 43-kDa protein with an isoelectric point of 6.8. This protein was identified as glutamine synthetase and the increase in binding was found to be due to both increased copy number and activation by Mn^{2+}. An increase in $[^{32}P]8N_3ATP$ photoinsertions into a 55-kDa protein, identified as the β-subunit of tubulin, was found only in the striatum and hippocampus indicating the depolymerization of microtubulin in these tissues.

After transient (6, 8, or 10 min) hypoxic ischaemia followed by ventilating with 100 % oxygen, the caudoputamen of male Sprague-Dawley *rats* showed shrunken electron-dense neurones (RADOVSKY et al. 1997).

A massive loss of immunoreactivity to tachykinins and enkephalins was described in striatal neurones of adult *gerbil* after transient forebrain ischaemia (CHESSELET et al. 1990). A massive reduction of the number of preproenkephalin-mRNA containing neurones concomitant with the emergence of apoptotic cells was demonstrated by POPOVICI et al. (1997).

During ischaemia and reperfusion phases, LEI et al. (1997) observed a significant increase in the *rat* striatal H_2O_2 level.

Using sodium salicylate as a **hydroxyl radical** trap, tissue sampled from caudatoputamen of normoglycaemic transient ischaemic male Wistar *rats* failed to show an increase in 2,3- or 2,5-dihydroxybenzoate after 5 and 15 min of recirculation (LI et al. 1999). Hyperglycaemia (infusion with a 25 % glucose solution, 2.3–3.0 ml/h for 30 min before ischaemia) enhanced the hydroxylation of salicylic acid to 2,5-dihydroxybenzoic acid at 15 min of recirculation. Elevated extracellular glutamate levels (15 mM, 1.5 mM, and 150 μM in perfusing solutions) increased the formation of $HO^•$ in the striatum of anaesthetised *rats* (YANG et al. 1995). Dihydrokinate (1 mmol/l in perfusate) and an anion channel blocker, 4,4'-dinitrostilben-2,2'-disulphonic acid (1 mmol/l), suppressed glutamate release in *rat* striatum (SEKI et al. 1999).

Malondialdehyde formation in *rat* striatum slices incubated with a mixture of ascorbic acid (25 μmol/l) and Fe^{3+} (5 μmol/l) for 30 min at 37 °C was inhibited by piracetam (4 mmol/l) by ~40 % (BLASCKE et al. 1985).

In an *in vitro* system that generated $HO^•$ from $FeSO_4$–H_2O_2, bromocriptine dose-dependently ($IC_{50} = 11.25 \pm 0.89 \times 10^{-5}$ M) quenched $HO^•$ radicals, but did not inhibit their formation (OGAWA et al. 1994). Pre-treatment with bromocriptine (5 mg/kg, i.p., 7 days) completely protected against the decrease in *mouse* striatal dopamine and its metabolites induced by intraventricular injection of 6-hydroxydopamine (40 μg) after intraperitoneal administration of desipramine (25 mg/kg, administered i.p. 30 min after the final injection to block noradrenaline re-uptake sites), but similar pre-treatment with L-DOPA/carbidopa (75/7.5 mg/kg, i.p, 7 days) showed only partial protective effect.

6-Hydroxydopamine-induced **release of iron** from ferric-transferrin (BORISENKO et al. 2000) was demonstrated by: (1) low-temperature spectroscopic evidence for decay of the characteristic ferric-transferrin signal ($g = 4.3$) and appearance of the high-spin signal from iron chelated by 6-hydroxydopamine oxidation products; (2) spectrophotometric detection of complexing of iron with the Fe^{2+} chelator ferrozine; (3) redox-cycling of ascorbate yielding EPR-detectable ascorbate radicals; and (4) generation of hydroxyl radicals as evidenced by EPR spectroscopy of their adduct with a spin trap, 5,5'-dimethylpyrroline oxide (DMPO) (DMPO-OH).

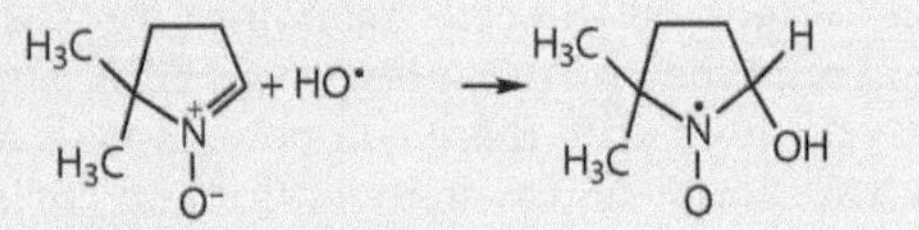
5,5-**Dim**ethyl-1-**p**yrroline-1-**o**xide-HO• adduct [8]

In vitro, the presence of 800 mM iron increased (>100%) the production of HO• by 5 μM 6-hydroxydopamine while Mn^{2+} caused a significant reduction (72%) (MÉNDEZ-ÁLVAREZ et al. 2001). The presence of ascorbate (100 μM) induced a continuous generation of HO• while the presence of hydroxyl reductants (100 μM) limited this production to the first minutes of the reaction. *In vivo*, tyrosine hydroxylase immunohistochemistry revealed that intrastriatal injections of *rats* with 6-hydroxydopamine (30 nmol) + (600 nmol), 6-hydroxydopamine + ascorbate + Fe^{2+} (5 nmol), and 6-hydroxydopamine + ascorbate + Mn^{2+} (5 nmol) caused large striatal lesions, which were markedly reduced (60%) by the substitution of ascorbate by L-cysteine. Injections of Fe^{2+} or Mn^{2+} alone showed no significant difference to those of saline.

Hypoxic (5–16% O_2) exposures to *rats* evoked **dopamine release**, the restitution of which was significantly accelerated by pre-treatment for 7 or 14 days with 100 mg piracetam/kg body weight given two times a day (WUSTMANN et al. 1982). Oxidation of the excess dopamine released during ischaemia could be partially responsible for the production of **free radicals** during reperfusion. *In vitro*, SLIVKA and COHEN (1985) generated hydroxyl radicals in the presence of 1 mM dopamine by the following two mechanisms: 1) a classic Fenton type reaction between hydrogen peroxide and a ferrous chelate (ferrous diethylenetriaminepentaacetate) and 2) the cyclic redox reactions of iron-EDTA/ascorbate. Scavengers of hyroxyl radicals as dimethylsulfoxide, mannitol, and ethanol suppressed the yields of ring-monohydroxylated products in a concentration-dependent manner.

To elucidate the direct effect of reactive oxygen species on **dopamine transport**, FLECKENSTEIN et al. (1997) used xanthine oxidase to generate such species and found that oxygen radicals, like methamphetamine administration (FLECKENSTEIN et al. 1997) diminish dopamine transporter V_{max}. Since dopamine tranporters are the primary means whereby dopamine is cleared from the synaptic cleft, disruption could lead to increased extracellular dopamine and ultimately the production of highly destructive reactive oxygen species. Dopamine clearance in *rat* striatum was decreased 7 days after an infusion of 1 μl of 4.2 nM ferrous citrate into the substantia nigra (LIN et al. 1998). Iron-induced oxidative stress attenuated the effect of nomifensine, a high-affinity dopamine uptake blocker, on dopamine clearance.

In the striatum of the *rat* extracellular glutamate levels increased the formation of hydroxyl radicals (YANG et al. 1995). Dihydrokinate (1 mmol/l in perfusate) and an anion channel blocker, 4,4'-dinitrostilben-2,2'-disulfonic acid (1 mmol/l), suppressed glutamate release in *rat* striatum (SEKI et al. 1999).

Catalase-positive **microperoxisomes** are relatively abundant in catecholaminic areas of the *rat* brain, as MCKENNA et al. (1976) showed in the substantia nigra, the locus coeruleus and in nucleus A_1 of the medulla.

Carbon monoxide (CO) has been proposed as a neurotransmitter, taking part in long-term transmission (VERMA et al. 1993, ZHUO et al. 1993). CO was found to form a hexacoordinated complex without inducing cleavage of the histidine-iron bond in heme (STONE and MARLETTA 1994, YU et al. 1994). CO is enzymatically produced within the body by the heme oxygenases (EC 1.14.99.3). These enzymes catalyse the degradation of heme to biliverdin and CO (MAINES 1988, MARKS et al. 1991).

Endogenous production of CO catalysed by heme oxygenases (EC 1.14.99.3)

$$\text{Haem} + 3\,AH_2 + 3\,O_2 \rightarrow \text{Biliverdin} + Fe^{2+} + CO + 3\,A + 3\,H_2O$$

Requires NAD(P)H + NADPH-cytochrome *P*-450 reductase (EC 1.6.2.4) [281]

Soluble guanylate cyclase was not (YU et al. 1994) or only marginally (STONE and MARLETTA 1994) stimulated by 100% CO. MAYER et al. (unpublished results), however, using the enzyme purified from *bovine* lung (HUMBERT et al. 1990) observed an about tenfold increased activity in the presence of 100% CO.

When pregnant *rats* on day 15 of gestation were exposed to approximately 1000 ppm CO for either 2 or 3 h, the offspring showed ectopic swellings of caudate tissue into the lateral ventricles (DAUGHTREY and NORTON 1983). In the neonatal *rats*' body of the caudate the number of dendritic branches was reduced in Golgi type II bipolar spiny neurones, while in later postnatal ages the branching in control and CO-exposed neurones was not significantly different.

PIANTADOSI et al. (1995) investigated the effects of hyperbaric oxygen on generation of highly reactive hydroxyl radical in the brain after CO poisoning in Sprague-Dawly *rats* using nonenzymatic hydroxylation of salicylic acid to 2,3 dihydroxybenzoic acid (formula [178]) as a probe. In control studies, the concentrations of 2,3-dihydroxybenzoic acid after hyperbaric oxygen in brain mitochondria and postmitochondrial supernatant (cytosol) were similar to air-exposed animals. After CO poisoning, 2,3-dihydroxybenzoic acid concentration increased in brain mitochondria but not in the cytosol. After CO exposure and hyperbaric oxygen administration at 1.5 atmospheres absolute (ATA), a decrease in 2,3-dihydroxybenzoic acid production was detected in brain mitochondria. After CO and hyperbaric oxygen administration at 2.5 ATA, 2,3-dihydroxybenzoic acid concentration increased in both mitochondria and cytosol. The oxidant scavenger dimethylthiourea and the monoamine oxidase inhibitor pargyline, administered to CO poisoned *rats* after hyperbaric oxygen at 2.5 ATA, diminished 2,3-dihydroxybenzoic acid production in both subcellular compartments.

15.1.1.4.3 Manganese Accumulation

Manganese toxicity in *humans* due to chronic inhalation of high concentration or airborne Mn from Mn mines, steel mills, and chemical industries not only causes Mn pneumonia (Baader 1937, Heine 1944, Dervillée et al. 1966, van Beukering 1966), but also affects the brain (Canavan et al. 1934, Peñalver 1957, Cotzias 1974, Donaldson et al. 1982, Donaldson and Barbeau 1985). Chronic Mn intoxication can cause permanent degenerative damage in the nigrostriatal system. Autopsy findings of low dopamine levels, morphologic lesions and the extrapyramidal signs suggest that this type of abnormality is similar to Parkinson's disease (Donaldson et al. 1982). Studies in monkeys and rodents chronically intoxicated with $MnCl_2$ have also demonstrated the depletion of dopamine in the striatum. PC-12 cells widely used as a model for catecholaminergic neurones, grown in the presence of 0.1 mM $MnCl_2$ and 0.1 mM tyrosine with time of exposure showed increased lipid peroxidation and release of lactate dehydrogenase (Sun et al. 1993).

Archibald and Tyree (1987) have demonstrated that Mn^{3+} in a pyrophosphate complex (probably Mn^{3+}[pyrophosphate]$_3$) has the capacity to rapidly destroy the principal brain catecholamines and their precursor, dopa, oxidising them to a best useless and at worst toxic products. In the absence of Mn^{3+}, oxidation of the neurotoxic 6-hydroxydopamine to a stable coloured compound in the O_2-containing reaction mixture took several minutes, while Mn^{3+} accelerated this to completion in <5 s. The observation thar phenylalanine and tyrosine were not oxidised by Mn^{3+}, although their dihydroxy derivatives were, suggested that the site of Mn^{3+} attack on the catecholamines. Indeed, the placement of the hydroxyl groups was all-important in the reactivity with these compounds with Mn^{3+}. Since Mn^{3+} has a similar ionic radius and chelating behaviour to Fe^{3+}, and Fe^{3}_{+} chelates very strongly to catechols (Neilands 1980), it is likely that Mn^{3+} by the dihydroxy moiety is the first step in the oxidative attack of Mn^{3+} upon catecholamines.

Graham et al. (1978) conclude that 6-hydroxydopamine and 2,4,5-trihydroxyphenylalanine kill cells through the production of H_2O_2, $O_2^{\bullet-}$ and $HO^{\bullet}$, while for dopamine and dopa the reaction of quinone oxidation products with nucleophiles probably also contributes to their cytotoxicity. *O*-methylated catecholamines are less susceptible to autoxidation than their nonmethylated precursors (Miller et al. 1996). Melatonin, which has been shown to be a powerful, endogenous hydroxyl radical scavenger (Tan et al. 1993), is capable of scavenging free radicals produced during catecholamine autoxidation (Miller et al. 1996).

Superoxide oxidises catecholamines

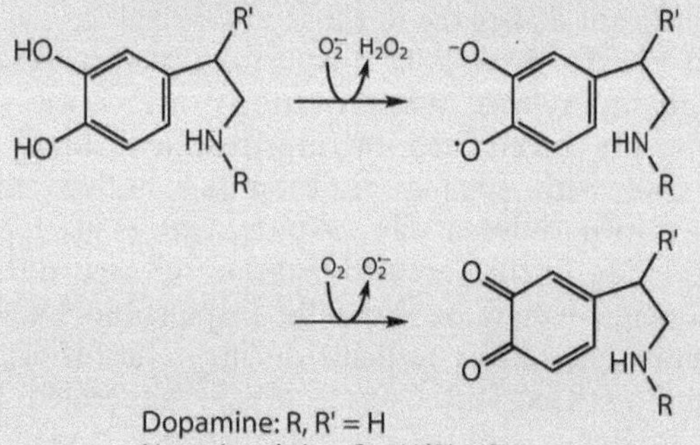

Dopamine: R, R' = H
Norepinephrine: R = H, R' = OH
Epinephrine: R = CH_3, R' = OH
Isoprenaline: R = $CH(CH_3)_2$, R' = OH [107]

O-methylated catecholamines are not only less susceptible to autoxidation than their corresponding non-methylated precursors (Miller et al. 1996), and thus less likely to generate reactive oxygen intermediates, but they are also effective *in vitro* scavengers of $HO^{\bullet}$ (Napolitano et al. 1993, Tan et al. 1994) and are poorly or not reactive toward $^{\bullet}NO$ (D'Ischia and Constantini 1995). The production of $HO^{\bullet}$ by the Fenton reaction was diminished significantly by *O*-methylated catecholamines (*O*-methyldopa, *O*-methyldopamine, *O*-methyltyrosine, and *N*-acetyl-*O*-methyldopamine), whereas radical production was augmented by dihydroxyphenols (DOPA, dopamine, and *N*-acetyldopamine), including those with methylated side chains (N-methyldopamine and α-methyldopamine), while monohydroxyphenyls such as octopamine, tyramine, tyrosine, and α-methyltyrosine had little or no effect on radical production (Nappi and Vass 1998).

o-Quinones are physiological oxidation products of catecholamines that contribute to redox cycling, toxicity and apoptosis, i.e. the neurodegenerative processes underlying Parkinson's disease and schizophrenia. The cyclized *o*-quinones aminochrome, dopachrome, adrenochrome and noradrenochrome, derived from dopamine, dopa, adrenaline and noradrenaline, respectively, are efficiently conjugated with glutathione, in the presence of *human* glutathione transferase M2-2 (Baez et al. 1997).

Studies in workers occupationally exposed to manganese have shown positive correlation between homovanillic acid and manganese urinary elimination, which suggests increased catecholamine turnover elicited by Mn (Buchet et al. 1993). In male *Rhesus monkeys* given 20 mg $MgCl_2$ per kg body weight × day for 18 months, striatal Mn was elevated from 1.39 ± 0.02 µg/g in the controls to 4.79 ± 0.33 µg/g in the Mn^{2+} treated animals (Chandra et al. 1979).

The ability of Mn to enhance the formation of reactive oxygen species, e.g. $O_2^{\bullet -}$, $HO^{\bullet}$, H_2O_2 and oxidation by-products of catecholamines (quinones) has been suggested as the underlying mechanism for Mn neurotoxicity (Graham 1984).

Dopamine in the presence of Mn for up to 60 min produced 6-hydroxydopamine, while the incubation mixture without Mn did not show any dopamine oxidation (Garner and Nachtman 1989).

Manganese alters dopaminergic transmission (Chandra and Shukla 1981, Eriksson et al. 1987).

Subchronic exposure to $MnCl_2$ for 7 days (total dose 1750 µmol/kg) by means of mini-osmotic pumps inplanted s.c. produced significant reduction in GSH-peroxidase (EC 1.11.1.9) activity in the cytosol and mitochondrial fractions of the whole *rat* brain and the striatum (Liccione and Maines 1988). The decrease in GSH-peroxidase was most pronounced in the mitochondrial fraction of the striatum where the activity was reduced to 35 % of the control. Catalase (EC 1.11.1.6) activity was also decreased in the striatum of *rats* treated with Mn but not in the whole brain. GSH content was markedly depleted (20 % of the control) in the striatum, although only modestly decreased in the whole brain (80 % of the control). The treatment of *rats* with Mn also decreased the activity of oxidised glutathione reductase (EC 1.6.4.2); the same treatment increased the activity of γ-glutamyltranspeptidase (EC 2.3.2.2). The activity of γ-glutamylcysteine synthetase was not altered by Mn.

Marked increases ($P < 0.05$) of striatal manganese content were observed in bile-duct obstructed cirrhotic male Wistar *rats* treated (0.5 mg Mn^{2+}/ml) and untreated with manganese in their drinking water (Montes et al. 2001).

15.1.1.4.4
Nitric Oxide Donors: Chronic Oral Toxicity

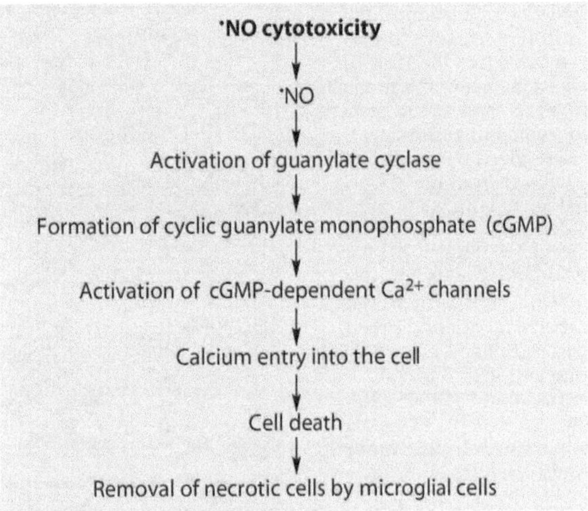

15.1.1.4.5
Superoxide ($O_2^{\bullet -}$) and •NO in Methamphetamine Cytotoxicity

Methamphetamine, a world-wide abused drug, produces neurotoxicity via generation of reactive oxygen and nitrogen (Di Monte et al. 1996) species. Exposure of PC12 cells to methamphetamine (200 µM) for 24 h resulted in a significant depletion of dopamine, and its metabolites DOPAC and homovanillic acid, as well as the significant formation of 3-nitrotyrosine, a marker of peroxynitrite generation (Imam and Ali 2000). Selenium (10 and 20 µM) pre-treatment for 30 min attenuated the depletion of dopamine and its metabolites, 3,4-dihydroxyphenylacetic acid and homovanillic acid and the formation of 3-nitrotyrosine. Adult male C57BL/6N *mice* supplemented with sodium selenite in their drinking water, 1 week before and 1 week after multiple injections of methamphetamine (4 × 10 mg/kg, i.p. at 2 h interval) formed less 3-nitrotyrosine in their striata than *mice* injected with methamphetamine but not supplemented with Na_2SeO_3.

Exposure of PC12 cells to 500 µM peroxynitrite activated extracellular-regulated kinases 1 and 2 and p38 within 5 min and this was followed by gradual decreases in activation over the next 25 min (Jope et al. 2000). Activation of extracellular-regulated kinases 1 and 2 by peroxynitrite was mediated by activation of the epidermal growth factor receptor in a calcium/calmodulin-dependent kinase II- and *sre* family tyrosine kinase-dependent manner, as it was blocked by the selective epidermal growth factor receptor inhibitor AG1478, by

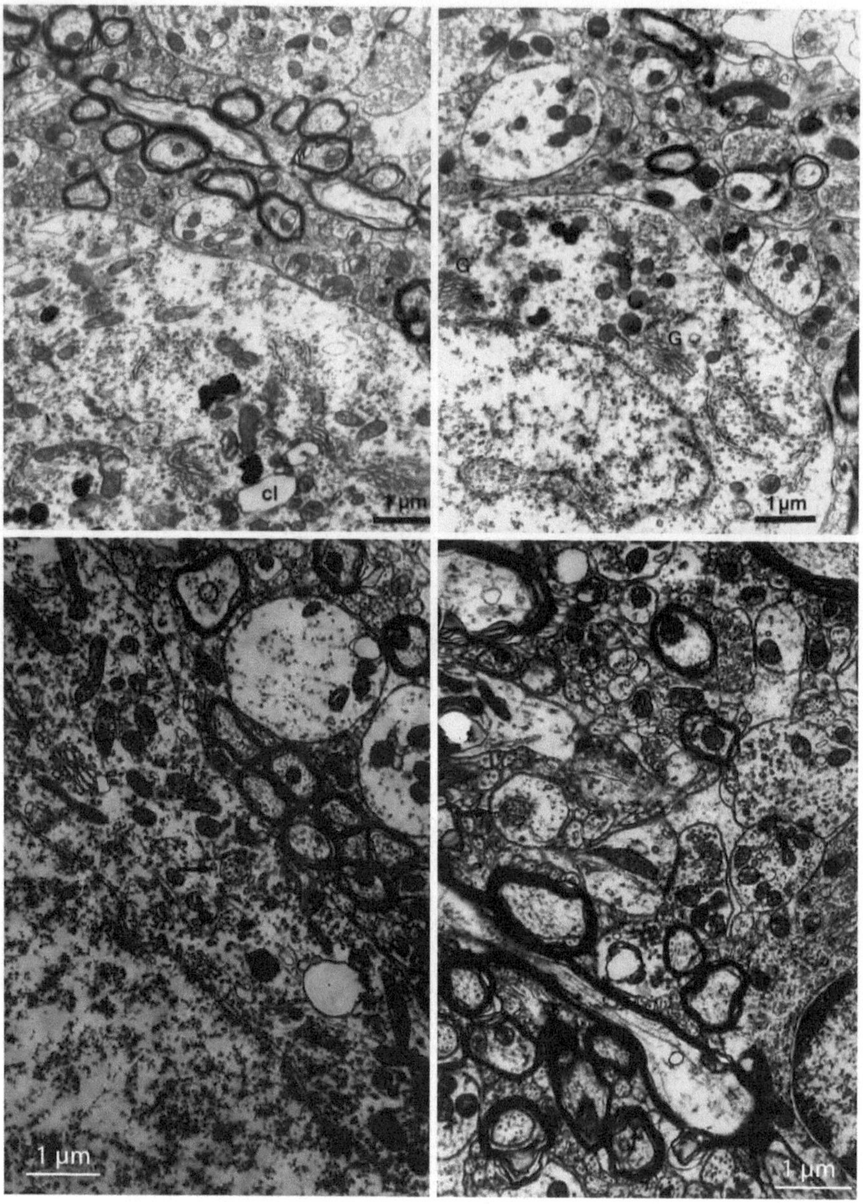

Fig. 223. Corpus striatum (block 4578) of a *rat* medicated with 5,000 ppm molsidomine added to the food (powdered Altromin® R) from July 17, 1978 to November 30, 1978. After discontinuation of the medication for 7 weeks, under pentobarbital anaesthesia (30 mg/kg), the animal was perfused from the abdominal aorta with 2.5% glutaraldehyde in 0.1 M sodium cacodylate buffer (pH 7.4). Postfixation with 1% osmium tetroxide in sodium cacodylate buffer. Embedded in Epon 812. Lead citrate and uranyl acetate. Films 204/79, 591/79, 197/79, and 198/79. – G, Golgi apparatus; cl, clear vacuole. Multivesicular bodies shown in the lower two frames are marked by arrows

KN62, an inhibitor of calcium/calmodulin-dependent kinase II, and by PP1, a *sre* family tyrosine kinase inhibitor. Activation of p38 by peroxynitrite was independent of the epidermal growth factor receptor, required activation of calcium/calmodulin-dependent kinase II and *sre* family tyrosine kinases, and was modulated by nerve growth factor in a time-dependent manner.

15.1.1.4.6
Excitotoxic Lesions Produced by *N*-methyl-D-aspartate

In a primary cell culture model from embryonic day 18 *rat* brains in which striatal neurones undergo a gradual and delayed neurodegeneration after a brief (5 min) with the glutamate receptor agonist *N*-methyl-D-aspartate •NO was generated continuously for up to 16 h after the *N*-methyl-D-aspartate exposure (Strijbos et al. 1996). Neuronal death followed the same general time course except that its start was delayed by ~4 h. Application of the NOS inhibitor nitroarginine after, but nor during, the *N*-methyl-D-aspartate exposure inhibited •NO formation and protected against delayed neuronal death. Blockade of *N*-methyl-D-aspartate receptors or of voltage-sensitive Na^+ channels with tetrodotoxin during the postexposure period also inhibited both •NO formation and cell death.

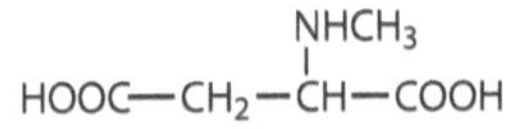

N-Methyl-D-aspartic acid [274]

In anaesthetised 7-days-old Wistar *rats* bearing a microdialysis cannula implanted in the striatum and perfused with a solution containing salicylate as an HO• trap, hydroxyl radical formation was evaluated, after a 3 h postoperative delay, by measuring the 2,3-dihydroxybenzoic acid levels before, during and over 3 h after the administration of glutamatergic agonists or antagonists (CAMBONIE et al. 2000). Administration of *N*-methyl-D-aspartate and of ibotenate dramatically increased the efflux of HO•, 17-fold and sixfold, respectively. Glutamate, used at the same concentration did not produce any significant increase in the HO• release and may even decrease this efflux when given at larger concentrations. The *N*-methyl-D-aspartate-induced HO• response was partially but progressively reduced by glutamate coinjection and completely blunted by (*RS*)-3,5-dihydroxyphenylglycine, a group I metabotropic glutamate receptor agonist. Conversely, (*RS*)-1-aminoindan-1,5-dicarboxylic acid, an antagonist of the same receptors, unmasked an HO• response to glutamate.

Oral administration of 1 % creatine to male Sprague-Dawley *rats* significantly attenuated striatal excitotoxic lesions produced by *N*-methyl-D-aspartate, but had no effect on lesions produced by α-amino-3-hydroxy-5-methyl-4-isoxazolepropionic acid or kainic acid (MALCON et al. 2000).

15.1.1.4.7 Excitotoxic Lesions Produced by Kainic Acid

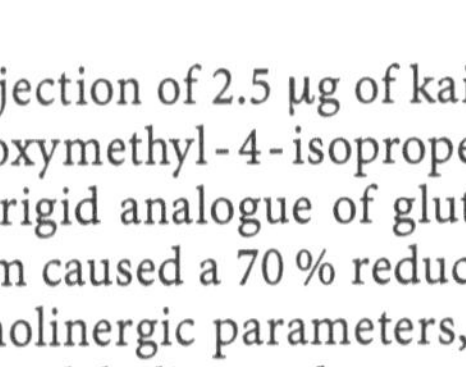

Kainic acid (2-carboxyl-3-carboxymethyl-4-isopropenylpyrrolidine) [275]

Stereotactic injection of 2.5 μg of kainic acid (2-carboxyl-3-carboxymethyl-4-isopropenylpyrrolidine, $C_{10}H_{15}NO_4$), a rigid analogue of glutamic acid, into the *rat* striatum caused a 70 % reduction in the striatum of the cholinergic parameters, choline acetyltransferase, acetylcholine and synaptosomal uptake of choline and a similar reduction in the GABAergic parameters, glutamic acid decarboxylase, γ-aminobutyric acid (GABA) and synaptosomal uptake of GABA (Schwarcz and COYLE 1977). The kainate lesions caused a 85 % decrement in the activity of dopamine-sensitive adenylate cyclase, a 40 % reduction in the specific binding of [^{3}H]quinuclidinyl benzilate and a 195 % increase in the specific binding of [^{3}H]GABA in the striatum. The morphology of the kainate injected striatum was markedly altered with nearly a complete loss of intrinsic neurones, increased number of glial cells but intact internal capsule fibres.

15.1.1.4.8 Hypoxic Modulation of Lesions Induced by Endothelin-1

Under normoxia, intrastriatal stereotaxic injection of exogenous endothelin-1 (40 pmol) induced a significant ($P < 0.05$) reduction ($\leq 29 \pm 12$ %) in the regional (striatal) blood flow measured by Laser Doppler flluometry for up to 40 min in halothane-anaesthetized male Long-Evans *rats* (PARK and THORNHILL 2000). Intrastriatal injection of endothelin-1 10 min after the onset of hypoxia (12 % O_2, balance N_2) tended to blunt but not significantly, the striatal blood flow, which was similar to the effect of endothelin-1 during normoxia. Endothelin-1-induced infarction when administered prior to hypoxia, but not during post-hypoxia, was significantly ($P < 0.05$) exacerbated compared to infarction of endothelin-1 without hypoxia.

15.1.1.4.9 Experimental Therapy

Piracetam (200 mg per kg × day for 14 days) pretreatment of male Wistar *rats* was effective in preventing the post-hypoxic dopamine release inhibition from K^+-stimulated striatum slices (GRÄSSLER et al. 1987).

15.1.1.5 Globus Pallidus

In the globus pallidus of the *monkey* the cell bodies are relatively far apart and dispersed throughout the nucleus (Fox and RAFOLS 1976). The predominate type of synaptic endings has large, egg-shaped, synaptic vesicles (Fox et al. 1974).

Neutron activation analysis of iron in the *human* pallidum (9 samples from patients aged 23 to 66 y) resulted in $10.56 \pm 0.91 \times 10^{-4}$ terms of element weight per unit dried tissue (HÖCK et al. 1975).

Ascorbic acid having the unique property of reducing an acidified solution of silver nitrate almost instantly (GOMORI 1952, p. 90), can be detected only in a few isolated nerve cells of the *human* pallidum (CLARA 1942, 1953).

In the dual circuit scheme, the external pallidum in *primates* (the globus pallidus in *rats*) is believed to be a simple relay with spiking activity strongly linked to changes in the discharge rate of inhibitory striatal neurones projecting to the pallidum. Chesselet and Delfs (1996) have clearly summarised the various objections to this assumption. A thalamic input to the pallidal neurones provides further evidence of the integrative function of the globus pallidus (Féger 1997). It is likely that the combination of increased inhibitory (γ-aminobutyric acid) inputs from the striatum and excitatory inputs from the subthalamic nucleus is responsible for an increased firing activity observed in the subthalamic nucleus after dopamine depletion (Chesselet and Delfs 1997).

In *rats* microinjected unilaterally with kainic acid (0.75 μg) in the globus pallidus, loss of neuronal perikarya and reactive gliosis occurred (Di Chiara et al. 1980).

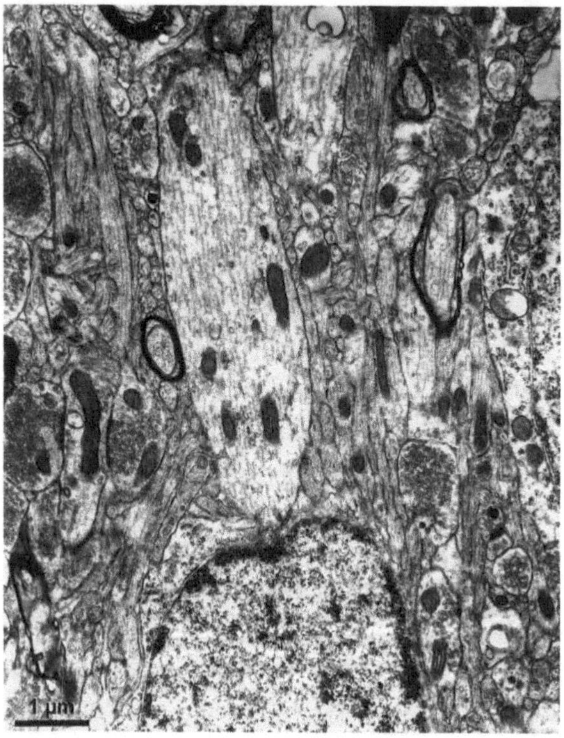

Fig. 224. Almost longitudinally sectioned dendrite covered with synaptic endings in the globus pallidus (block BNh 3433) from a male Wistar *rat* (No. 2570) 90 min after a single intraperitoneal injection of 5 mg *N*-benzhydryl-*N'*-*p*-hydroxybenzylpiperazine HCl per kg body weight. Under pentobarbital anaesthesia (30 mg/kg), the animal was perfused from the abdominal aorta with 2.5 % glutaraldehyde in 0.1 M sodium cacodylate buffer (pH 7.4). Postfixation with 1 % osmium tetroxide in sodium cacodylate buffer. Embedded in Epon 812 and sectioned at 50 nm. Lead citrate and uranyl acetate. Plate 2396

In bicuculline-induced seizures the **blood-brain barrier** in pallidum of the *rabbit* opened to horseradish peroxidase mainly at the level of arterioles (Nitsch et al. 1986). Traversing the endothelial cells via pinocytosis with the tight junctions remaining intact, the tracer accumulated in the extracellular spaces and basement membranes of the arterioles and penetrated from there through the narrow interstitial spaces between the perivascular glial cells into the neuropil. Some of the boutons covering dendrites exhibited reaction product in vesicles opening to the synaptic cleft.

Nitric oxide synthase (EC 1.14.23) mRNA investigated by in situ hybridisation histochemistry was localised in scattered neurones in the medial segment of the *human* globus pallidus and adjacent medial medullary lamina (Nisbet et al. 1994). A number of these cells was spindle-shaped although others were large round cells. The level of NOS mRNA expression per segment of the globus pallidus and not the lateral segment which relays the basal ganglia output to the motor nuclei of the thalamus. In the pallidum, NO is probably co-localised with the inhibitory neurotransmitter γ-aminobutyric acid, which has been shown to be the neurotransmitter of virtually all pallidal neurones (Mugnaini and Oertel 1985).

Bilateral necrosis of the globus pallidus was observed in a 43-year-old *man* who had received treatment for malignant hypertension with sodium nitroprusside (Kim et al. 1982).

15.1.1.6
Corpus Geniculatum Laterale

Monocular deprivation of visual input for 2, 7, or 14 consecutive days in new-born Long Evens *rats* caused p53 accumulation, cell death, and a progressive loss of neurones in the dorsal lateral geniculate nucleus (Nucci et al. 2000). N^{ω}-Nitro-L-arginine, an inhibitor of •NO synthesis, and *N*-methyl-D-aspartate and other glutamate receptor antagonists prevented these lesions. Finally, poly-(ADP-ribose) polymerase knock-out *mice* appeared to be protected from monocular deprivation-induced cell death.

15.1.1.7
Hypothalamus

In addition to the usual structural features of neurones, the neurosecretory neurones in the hypothalamus contain prominent laminal arrays of rough endoplasmic reticulum, large Golgi apparatuses, and numerous membrane-bounded neurosecretory granules in the perikarya and axonal processes.

The neurosecretory neurones concerned with hormone synthesis are segregated into anatomically defined regions, termed nuclei, in the hypothalamus. The supraoptic nucleus is concerned primarily with the synthesis of antidiuretic hormone, whereas oxytocin is produced predominantly by neurones in the paraventricular nucleus.

Histochemical evidence indicated that the neurosecretory material is a protein rich in cystine. SLOPER (1955) incubated sections at 54 °C in alkaline solutions of blue tetrazolium chloride and 2,5-diphenyl-3,4-styrylphenyl tetrazolium chloride. The formazan produced by the former was sufficiently alcohol-insoluble to be mounted in balsam: its production in held by PEARSE (1953) to indicate the presence of cystine or cysteine in the absence of material reacting with the performic acid-Schiff and periodic acid-Schiff techniques. Since these latter reactions were largely negative in the peripheral distributions of neurosecretory material, i.e. in the infundibular stem and process save in some Herring bodies, it seemed possible that formazans produced in these areas at the site of neurosecretory material were due to the presence of cystine or cysteine.

Using ^{35}S-cystine and ^{35}S-methionine GOSLAR and SCHULTZE (1958) found the proportion of incorporation of these thioamino acids into the neurosecretory nuclei was equal to that into other areas with an increased metabolic activity, e.g. nucleus originis nervi oculomotorii and formatio reticularis rhombencephali.

Oxidation of methionine

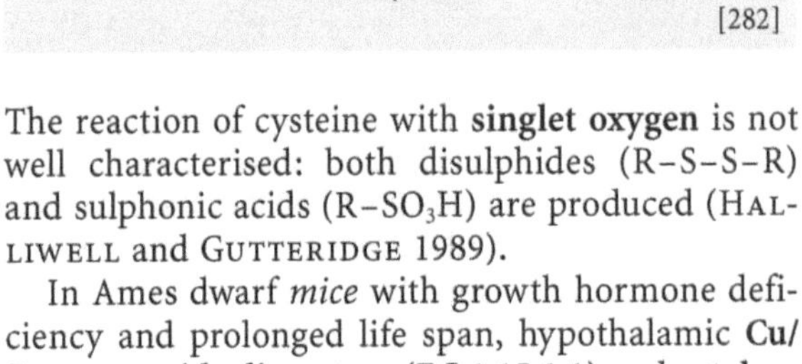

[282]

The reaction of cysteine with **singlet oxygen** is not well characterised: both disulphides (R–S–S–R) and sulphonic acids (R–SO_3H) are produced (HALLIWELL and GUTTERIDGE 1989).

In Ames dwarf *mice* with growth hormone deficiency and prolonged life span, hypothalamic **Cu/Zn superoxide dismutase** (EC 1.15.1.1) and **catalase** (EC 1.11.1.6) activities declined with age, and were higher than the corresponding normal values in young and middle-aged groups (HAUCK and BARTKE 2000). In growth hormone transgenic *mice* with overexpression of growth hormone and reduced lifespan, age-associated decline of both catalase and Cu/Zn superoxide dismutase occurred earlier than in normal animals.

Nitric oxide synthase (EC 1.14.13.39) is most notably in the paraventricular nucleus and the supraoptic nucleus (BREDT et al. 1990). Immunostaining and NADPH-diaphorase staining of hypothalamic neurones were comparable in all hypothalamic nuclei of *rat* and *mouse* except in the arcuate nucleus that stained positive for nitric oxide synthase immunoreactivity but negative for NADPH-diaphorase immunoreactivity (NG et al. 1999). Two dense clusters of NOS-containing neurones were found in the paraventricular and supraoptic nuclei of the *rat* in contrast to their scarcity in the same nuclei of the *mouse* hypothalamus. At the median eminence, strong nNOS immunoreactivity was detected in cell processes in the internal zone (CECCATELLI et al. 1992, HERBISON et al. 1996, YAMADA et al. 1996). Double labelling studies showed that there was no overlapping distribution of nNOS and glial fibrillary acidic protein (GFAP) immunoreactivities within the median eminence, suggesting that nNOS is not present in glial and ependymal cell processes, but only in neuronal processes (HERBISON et al. 1996). The use of a specific anti-eNOS monoclonal antibody revealed numerous immunoreactive cells in the vascular tissue and at the immediate proximity of the neuroendocrine nerve terminals of the external zone of the median eminence (PREVOT et al. 2000). When the anatomical distribution of nNOS immunoreactivity is compared to the distribution of gonadotropin releasing hormone immunolabelling in the median eminence, it strikingly appears that nNOS fibres and gonadotropin releasing hormone fibres are distributed separately in the internal and external zones, respectively (HERBISON et al. 1996).

L-Arginine reduced KCl-evoked vasopressin release in the male Wistar *rat*; this effect was reversed by N^{ω}-monomethyl-L-arginine and reduced by the addition of ferrous *human* haemoglobin (YASIN et al. 1993). Similarly, SIN-1 and sodium nitroprusside attenuated KCl-evoked vasopressin release. L-Arginine also reduced IL-1β-stimulated vasopressin release. •NO appears to directly and specifically inhibit the stimulated release of vasopressin from *rat* hypothalamus explants *in vitro*, similar to its effects on corticotrophin releasing hormone.

Chronic salt loading up-regulated the expression of neuronal NOS mRNA with a concomitant increase in NOS activity in the male Wistar *rat* neuropituitary (KADOWAKI et al. 1994). Intraperitoneal injection of N^{ω}-nitroarginine (10 mg/kg × day) had no effect on the levels of both arginine vasopressin and oxytocin transcripts in the supraoptic nucleus and paraventricular nucleus.

Subcutaneous capsaicin (8-methyl-*N*-vanillyl-6-nonenamide) injection in male *rats*, compared with vehilcle, caused a significant increase in Fos expression in the paraventricular nucleus (PVN), supraoptic nucleus (SON), and medial and cortical amygdala (OKERE et al. 2000). The expression of nicotinamide adenine dinucleotide phosphate diaphorase, a histochemical marker for nitric oxide synthase (NOS), was also increased in these brain areas in addition to the periventricular and lateral hypothalamic area and central amygdaloid nucleus. Also, capsaicin significantly increased the expression of neuronal NOS mRNA and protein in PVN, SON, and medial amygdala as demonstrated by in situ hybridisation and immunohistochemistry, respectively. The capsaicin-induced activation of the PVN and SON neurones and the medial amygdaloid nucleus attenuated in the NOS inhibitor N^{ω}-nitro-L-arginine methyl ester pre-treated animals in comparison with the inactive D-enantiomer.

Glutamate and γ-aminobutyric acid are the most abundant excitatory and inhibitory transmitters in the *mammalian* hypothalamus.

Both glutamate and •NO are important regulators of gonadotropin releasing hormone secretion from the neuroendocrine system (for review see BRANN 1995). Numerous studies have demonstrated the ability of •NO to induce gonadotropin releasing hormone secretion from the hypothalamus *in vivo* and *in vitro*, leading to a surge in luteinizing hormone and subsequently ovulation (RETTORI et al. 1994). *In vitro* studies by RETTORI et al. (1994b) showed that coadministration of haemoglobin, a •NO scavenger, and glutamate to *rat* medial basal hypothalamus fragments completely blocked the stimulation normally induced by glutamate alone, further demonstration •NO is important for gonadotropin releasing hormone stimulation induced by glutamate. MORETTO et al. (1993) and MAHACHOKLERTWATTANA et al. (1994) have shown •NO is capable of stimulating gonadotropin releasing hormone release from immortalised GnRH (GT-7) neurones.

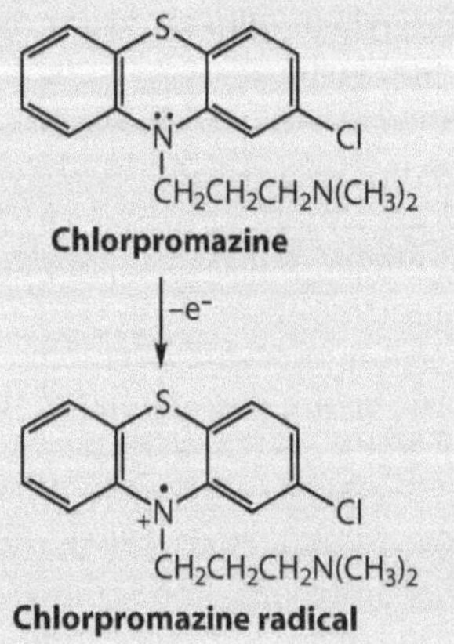

[64]

A single injection of **chlorpromazine** in infant male *rats* interfered with the gonadotrophin secretion (LADOSKY et al. 1970). In the Indian wall lizard, *Hemidactylus flaviviridis* (Ruppell), after a daily injection of 15 µg chlorpromazine for two days the neurones of the paraventricular nucleus showed a slight depletion of aldehyde fuchsin-stained neurosecretory material (HAIDER 1974). In the supraoptic nucleus, neurones were enlarged and many of them showed vacuolisation of their cytoplasm, while the quantity of their neurosecretory material was not reduced as compared with the controls.

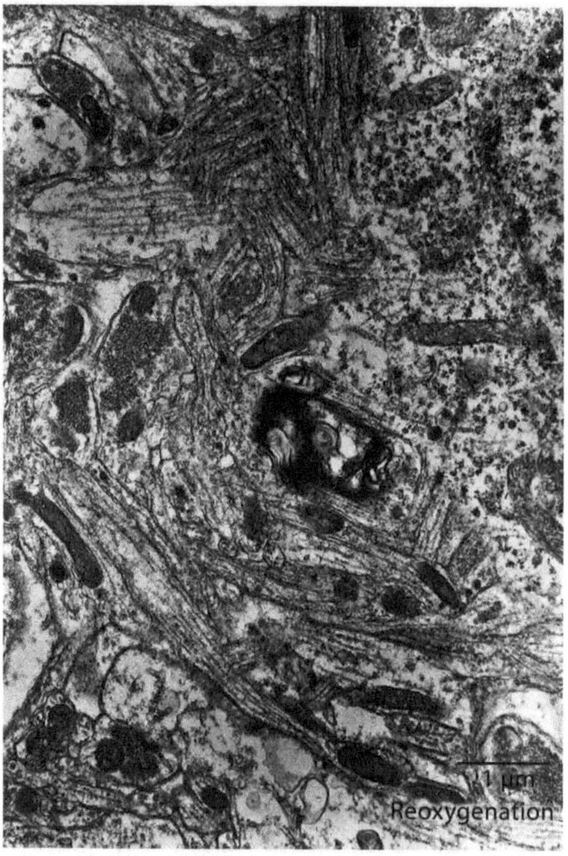

Fig. 225. Paucity of membrane-bounded secretory granules in the hypothalamic neurosecretory neurones (block 404) after reoxygenation of a male Wistar *rat* (No. 36/103 E) medicated for 7 consecutive days with intraperitoneal injection of 1.5 ml Tyrode's solution per kg body weight × day. On the last 4 days of experimentation the animal was exposed to an atmosphere containing only 5 % oxygen for 30 min. On November 26, 1975, half an hour after the last exposure under pentobarbital anaesthesia (30 mg/kg), the animal was perfused from the abdominal aorta with 2.5 % glutaraldehyde in 0.1 M sodium cacodylate buffer (pH 7.4). Postfixation with 1 % osmium tetroxide in cacodylate buffer. Embedded in Epon 812 and sectioned at 50 nm. Lead citrate and uranyl acetate. Plate 3046

15.1.1.7.1
Suprachiasmatic Nucleus

The *mammalian* suprachiasmatic nuclei contain an endogenous pacemaker that generates daily rhythms in behaviour and secretion of hormones. They stimulate evening ACTH secretion in the *rat* (CASCIO et al. 1987). KALSBEEK et al. (1996) used microdialysis-mediated intracerebral administration of the vasopressin V_1-receptor antagonist to study the mechanisms underlying the circadian control of basic corticosterone release in the *rat*. By timed administration of the vasopressin antagonist divided equally over the day/night cycle, they were able to uncover the existence of an additional stimulatory input from the suprachiasmatic nucleus to the hypothalamo-pituitary-adrenal axis.

Pituitary adenylate cyclase-activating polypeptide (PACAP) showed a remarkable modulatory influence on glutamatergic neurotransmission (KOPP et al. 2001). The neuropeptide can reduce or inhibit calcium rises in suprachiasmatic nucleus neurones induced by glutamate acting on metabotropic group I glutamate receptors ($mGlu_1$ and $mGlu_5$). These effects could be mimicked by cAMP analogues. PACAP can also amplify or even induce glutamatergic signalling elicited by an activation of ionotropic non-*N*-methyl-D-aspartate receptors of the α-amino-3-hydroxy-5-methyl-4-isoxazolepropionic acid (AMPA) and kainate types. These activating influences were independent of the second messengers cAMP and calcium. These data from individual suprachiasmatic nucleus neurones indicate that on the subcellular level the photic signal "light' can be modulated by the information "darkness', both directly transmitted from the eyes via the retinohypothalamic tract.

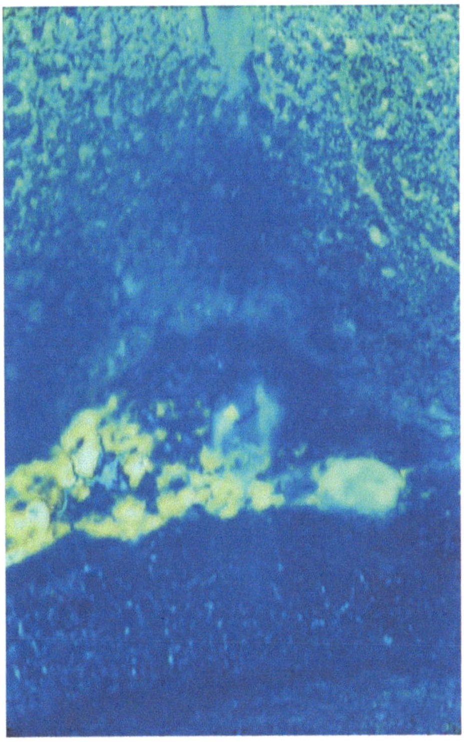

Fig. 226. Free sulfhydryl groups in the neurosecretory ganglion cells of the nucleus suprachiasmaticus of an unmedicated *rat*. The unfixed cryostat section was incubated with mercury orange, mounted in glycerol and photographed under a Mach-Zehnder interference microscope. Interference contrast

The suprachiasmatic nucleus controls synthesis and release of **melatonin**, the hormone of the pineal gland. Melatonin itself feeds back to the suprachiasmatic nucleus (CASSONE et al. 1987, STEHLE et al. 1989, MCARTHUR et al. 1991). Using brain slice technique and immunocytochemistry KOPP et al. (1997) demonstrated that pituitary adenylate cyclase-activating polypeptide induces the phosphorylation of the transcription factor cyclic AMP response element binding protein in the suprachiasmatic nucleus of the Wistar *rat* during the late subjective day and that melatonin inhibits this pituitary adenylate cyclase-activating polypeptide-induced phosphorylation.

15.1.1.8
Pineal Body

In the adult pineal, besides blood vessels and connective tissue, at least two definitive cell types are recognisable: the pinealocytes (parenchymal cells) and the glial cells (interstitial cells or gliocytes).

15.1.1.8.1
Pinealocytes

Normoxic Cytobiolgy

Unlike other neuroendocrine cells, the ultrastructure of the pinealocyte does not give an immediate impression that this cell type is specialised for the storage of large quantities of secretory products (VOLLRATH 1981, 1984). While in the Mongolian *gerbil* pinealocytes contain intriguingly high amounts of synaptic-like microvesicles of variable size ranging from 30–40 nm to sometimes more than 100 nm in diameter (REDECKER et al. 1990, REDECKER and BARGSTEN 1993, REDECKER 1999), in the *rat* microvesicles are not very conspicuous. Results from several independent lines of research indicated that signal molecules of synaptic-like microvesicles interfere with the endocrine activity of pinealocytes. Currently, most of the available data refer to the impact of glutamate on melatonin production. Earlier studies have revealed that glutamate (formula [273]) inhibits arylalkylamine *N*-acetyltransferase activity and melatonin synthesis (GOVITRAPONG and EBADI 1988, KUS et al. 1991). These observations have been confirmed by *in vitro* analyses showing that, at least in the *rat*, L-glutamate (100 μM–1 mM) in a dose-dependent manner suppresses the adrenergic-stimulated secretion of melatonin (VAN WYK and DAYA 1994, KUS et al. 1994, YAMADA et al. 1996). Similar inhibitory effects on melatonin synthesis have been ascribed to L-aspartate (YAMADA et al. 1997).

$$HOOC{-}CH_2{-}CH_2{-}CH(NH_2){-}COOH$$ L-Glutamic acid [273]

The intracellular signal transduction pathway mediating the inhibitory effects of glutamate on melatonin secretion seems to be connected to metabotropic receptors (PETRALIA et al. 1996) since in *rat* pinealocytes activation of the glutamate metabotropic receptor 3 has been reported to inhibit melatonin synthesis by decreasing cyclic AMP levels and arylalkylamine *N*-acetyltransferase activity (YAMADA et al. 1998).

Pinealocytes can express different subunits of all major classes of **glutamate receptors**, i.e. the ionotropic receptors for *N*-methyl-L-aspartate, α-amino-3-hydroxy-5-methyl-4-isoxazole propionate, and kainate type (WISDEN and SEEBURG 1993, SATO et al. 1993, TÖLLE et al. 1993, PETRALIA et al. 1994), as well as the metabotropic receptors (PETRALIA et al. 1996).

The pineal gland also contains reuptake systems for the sequestration of extracellular glutamate. One of these **glutamate transporters**, denominated GLT-1, has been localised in *rat* pinealocytes in which it operates with similar kinetics as in the nervous system (YAMADA et al. 1997). This findings has led to the suggestion that the efficient uptake of glutamate from the extracellular space is necessary for the precise control of melatonin synthesis (YAMADA et al. 1997). REDECKER (1999) pointed out that interstitial glial cells seem to contribute to the pineal expression of glutamate transporters such as GLT-1. PIANI et al. (1991) have shown that microglia *in vitro* produce high amounts of glutamate.

Pituitary adenylate cyclase-activating **polypeptide** (PACAP)-immunoreactive nerve fibres were demonstrated in the *rat* pineal gland (MØLLER et al. 1999). In the superficial pineal grand, the concentration of PACAP-27 was only about 3 % of the concentration of PACAP-38. PACAP is able to upregulate neuropeptide Y in neurones of the superior cervical ganglia (MAY and BRAAS 1995, BRAAS and MAY 1996). Because neuropeptide Y stimulated melatonin synthesis (SIMMONEAUX et al. 1994), this might be an indirect way for PACAP to influence hormone synthesis/secretion in the pineal gland. In cultured *rat* pineal glands peptide N-terminal histidine and C-terminal isoleucine stimulated *N*-acetyltransferase activity and melatonin production in both a time- and dose-dependent manner (MOUJIR et al. 1992). Vasoactive intestinal polypeptide (VIP)- and PACAP-induced cyclic AMP response element-binding protein (CREB) phosphorylation is restricted to subpopulation of neonatal cells and thus displays an adult pattern (SCHOMERUS et al. 1999).

Lactate dehydrogenase (EC 1.1.1.27) activity did not show any sexual differences in the isoenzyme pattern in CFY strain *rats* kept under standard laboratory conditions on 14 h light and 10 h dark periods (ANDA et al. 1976).

Lactate dehydrogenase (EC 1.1.1.27)

S-Lactate + NAD^+ → Pyruvate + NADH [125]

Glucose-6-phosphate dehydrogenase (EC 1.1.1.49) was present in large amounts within oval or rounded type B and sharply angled type C cells, but in only small amounts within elongated or spindle shaped type A cells (resembling fibroblasts), and lacking in the very small rounded type D cells with their darkly staining nuclei and relatively little cytoplasm (probably macrophages) outgrowing in tissue cultures of the *rat* pineal gland (HUXLEY and TAPP 1972).

Glucose-6-phosphate 1-dehydrogenase (EC 1.1.1.49)

D-Glucose-6-phosphate + ADP^+
→ D-Glucono-1,5-lactone-6-phosphate + NADPH [126]

NADH diaphorase activity has been shown to be strong in *human* pinealocytes (BAYEROVÁ and BAYER 1967) and strong to moderate in *rat* pinealocytes *in vivo* (BOSTELMANN 1963, 1968, TAPP et al. 1973) and *in vitro* (TASCĂ et al. 1968, S. MILCU et al. 1968), in *rabbit* pinealocytes (BOSTELMANN 1963) and in *guinea pig* pinealocytes (VOLLRATH and SCHMIDT 1969).

Monoamine oxidase (EC 1.4.3.4) activity of *rat* pinealocytes showed striking regional differences (RICHTER et al. 1992). In tissue culture, HUXLEY and TAPP (1972) found it present in moderate amounts within types B and C cells, while enzyme activity was lacking in types A and D cell.

Monoamine oxidase (EC 1.4.3.4)

$$RCHNH_2 + H_2O + O_2 \rightarrow RCHO + NH_3 + H_2O_2$$ [283]

Cytochrome *c* oxidase (EC 1.9.3.1) has been found to be evenly distributed in *rat* and *rabbit* pinealocytes (BOSTELMANN 1963, 1968). In the *rat*, the enzyme activity has been claimed to be weak (KRSTIC and TARSOLY 1972) to moderate (TAPP et al. 1973). In the *guinea pig* VOLLRATH and SCHMIDT (1969) could not detect any activity.

Catalase (EC 1.11.1.6) activity has been demonstrated in the *rat* pineal (RICHTER et al. 1992).

Detoxification of hydrogen peroxide by catalase (EC 1.11.1.6)

$$2\,H_2O_2 \rightarrow 2\,H_2O + O_2 \qquad [127]$$

Peroxidase (EC 1.11.1.7) activity has been reported to be weak in the *rat* pineal (GUSEK and BUSS 1966).

Nitric oxide synthase (EC 1.14.13.39) expression determined the intensity of the adrenergic cGMP response in *rat* pinealocytes (SPESSERT et al. 1997). In male Wistar *rats*, pineal nitric oxide synthase mRNA was suppressed by continuous exposure to light for 9 days (JACOBS et al. 1997, 1999).

L-Arginine–nitric oxide pathway

$$\text{L-Arginine} \xrightarrow{\text{NOS}} \textbf{Nitric oxide}(NO^{\cdot}) \xrightarrow{+O_2^{\cdot-}} \text{Peroxynitrite } (ONOO^-) \qquad [46]$$

Tryptophan hydroxylase (EC 1.14.16.4) is the rate-limiting enzyme that catalyses the pterin-dependent hydroxylation of tryptophan to form 5-hydroxytryptophan. Northern analysis of *human* pineal gland revealed the presence of two mRNA species (AUSTIN and O'DONNELL 1999). The cellular concentration of tryptophan hydroxylase in pinealocytes was extremely high throughout the pineal gland.

Serotonin *N*-acetyltransferase (EC 2.3.1.5) activity in male Wistar *rats* showed a dark-induced increase closely paralleled by changes in pineal and serum melatonin concentrations (WILKINSON et al. 1977). Activity of *N*-acetyltransferase fell abruptly between 02.00 and 03.00 h. In *rat* pineal organ cultures incubated with cholergan (50 mg/ml) and subsequently in the presence of 10 μM norepinephrine serotonin *N*-acetyltransferase activity was larger than that caused by either treatment alone (MINNEMAN 1977). While significant induction of pineal *N*-acetyltransferase was not observed when *rat* pineals were cultured with phenoxybenzamine (an irreversible α-adrenergic blocking agent) alone, phenoxybenzamine (10^{-5} M) shifted the dose-response curve for *N*-acetyltransferase induction by norepinephrine to the left (ALPHS and LOVERBERG 1984). Modulation of *N*-acetyltransferase induction by phenoxybenzamine in organ culture was blocked by DL-propranolol HCl, suggesting that such modulation is mediated through a β-adrenergic mechanism. Phenoxybenzamine did not modulate norepinephrine-mediated *N*-acetyltransferase induction in pineal extracted from superior cervical ganglionectomized *rats*. The a-agonist phenylephrine HCl also produced a leftward shift of the dose-response curve to norepinephrine.

Vasoactive intestinal polypeptide (VIP) stimulated basal serotonin *N*-acetyltransferase activity in *rat* pineals in organ culture and enhanced the effects of catecholamines in inducing the enzyme (YUWILER 1983).

The locus controlling pineal serotonin *N*-acetyltransferase activity is located on *mouse* chromosome 11 (GOTO et al. 1994).

Molecular cloning of the arylalkylamine-*N*-acetyltransferase showed daily variations of its mRNA expression in the Syrian *hamster* pineal gland, with undetectable levels in the second half of the light period and a dramatic increase at night (GAUER et al. 1999).

N-Bromoacetyltryptamine was shown to be a potent inhibitor of *N*-acetyltransferase *in vitro* and in a *rat* pineal cell culture assay (KHALIL et al. 1999). The mechanism of inhibition is suggested to involve a serotonin *N*-acetyltransferase-catalysed alkylation reaction between *N*-bromoacetyltryptamine and reduced CoA, resulting in the production of a tight-binding bisubstrate analogue inhibitor. This alkyltransferase activity is apparently catalysed at a functionally distinct site compared with the acetyltransferase activity active site on serotonin *N*-acetyltransferase. Such active site plasticity is suggested to result from a subtle conformational alteration in the protein. This plasticity allows for an unusual form of mechanism-based inhibition with multiple turnovers, resulting in "molecular fratricide".

Cytosolic **phospholipase A_2** (EC 3.1.1.4) mRNA levels in the *rat* pineal gland showed an off-phase diurnal pattern in relation to melatonin levels (LI et al. 2000). Intravenous administration of isoprenaline, which has been shown to elevate melatonin production, also decreased the level of cytosolic phospholipase A_2 mRNA significantly. Direct administration of melatonin to *rats* by intravenous injection decreased the levels of cytosolic phospholipase A_2 protein and mRNA in *rat* pineal glands.

Alkaline phosphatase (EC 3.1.3.1) is absent in any of the four cell types sprouting in tissue culture of the *rat* pineal gland (HUXLEY and TAPP 1972).

Acid phosphatase (EC 3.1.3.2) was present in large amounts within type D cells in tissue cultures of the *rat* pineal, while type A, type B, and type C cells showed little activity (Huxley and Tapp 1972).

Cyclic 3',5'-adenosine monophosphate phosphodiesterase (EC 3.1.4.17) activity, assayed in *rat* pineal organ cultures at 1 μM substrate concentration, was significantly increased 4 h after choleragen exposure and was maximally activated (170 percent of control) after 6 h, thereafter declining to basal levels (Minneman and Iversen 1976).

Na^+,K^+-ATPase (EC 3.6.1.3). Expression of the α1, α3, and β2 subunit isoforms was identified in the *rat* pineal gland (Shyjan et al. 1990). No α2 or β1 subunits were detectable in this tissue. The apparent absence of β1 subunit expression in pineal glands suggests that in this tissue, Na^+,K^+-ATPase isoenzymes are likely to consist of either α1/β2 or α3/β2 subunit complexes.

Cytosolic guanylate cyclase (EC 4.6.1.2) in *rat* pinealocytes required NO-dependent activation for cyclic GMP formation after adrenergic stimulation (Spessert et al. 1993).

Pineal **lipid histochemistry** has been studied most frequently and in most detail in laboratory *rats.* Quay (1962) attempted to distinguish and classify kinds of pinealocytes within adult *rat* pineal glands. Type I cells contain cytoplasmic lipid droplets easily seen with the light microscope, both in normal or untreated animals and others that have been operated upon or given hormones or drugs. The type II cells are less uniform and are more open to question as to the nature of their lipid content. They are seen more fully of in larger numbers after massive injection of norepinephrine or amphetamine three to 6 h before removal and fixation of the pineal tissue. Similarity occurs in the specificity of the response to norepinephrine by pineal type II cells in vivo and brain ^{32}P incorporation after treatment with serotonin or histamine (Quay 1958, 1962). Statistically significant increases in the number of type II cells positive to Baker's (1947) acid haematein test per pineal cross section were obtained following subcutaneous injections of DL-norepinephrine or of DL-amphetamine ($P < 0.01 - < 0.02$), and possibly significant increases were obtained following either a vitamin E-deficient diet for 72 days or a rachitogenic diet supplemented with 10 μg calciferol per day for 21 days ($P < 0{,}05$) (Quay 1958, 1962). Gonzalez and Blazquez (1975) described lipid droplets migrating through a pinealocyte terminal pole to the pericapillary space.

Bovine and *hog* pineals contain relatively large quantities of phospholipids as demonstrated by the method of Klüver and Barrera (1953) using luxol fast blue (Gutte and Grütze 1977).

Lipofuscins. In the light of the findings of Pieri et al. (1994, 1995) and Marshall et al. (1996) that melatonin is a potent ROO• scavenger, it would be predicted that the indole may afford protection against free radical-induced lipid peroxidation.

The high permeability of **cyclic nucleotide-gated channels** for Ca^{2+} suggests that cyclic nucleotide-gated channels function as cyclic nucleotide-operated calcium channels which transduce a hormonally induced rise in the intracellular cGMP/cAMP-level into an influx of Ca^{2+}. Such a mechanism has been identified in *chick* (Dryer and Henderson 1991) and *rat* (Schaad et al. 1995) pinealocytes where cyclic nucleotide-gated channels may be involved in the regulation of the synthesis and secretion of melatonin. Cyclic nucleotide-gated channels mediate synaptic feedback by nitric oxide (Savchenko et al. 1997).

Norepinephrine activating α_1- and β-adrenergic receptors rises the intracellular concentration of calcium ions ($[Ca^{2+}]_i$) and cyclic adenosine monophosphate, respectively. The norepinephrine-induced rise in $[Ca^{2+}]_i$ potentiates the increase of cAMP evoked by β-adrenergic stimulation (Vanecek et al. 1985). cAMP finally activates serotonin *N*-acetyltransferase, the key enzyme in melatonin production.

Norepinephrine-regulated calcium signalling and phosphorylation of cyclic AMP response element-binding protein at amino acid serine 133 are already fully developed at birth, i.e. prior to ingrowth of the sympathetic innervation into the pineal parenchyma (Schomerus et al. 1999).

Acetylcholine acting upon nicotinic acetylcholine receptors triggers the release of L-glutamate from pineal microvesicles and thereby inhibits melatonin secretion (Yamada et al. 1998). Activation of muscarinic acetylcholine receptors found at a presynaptic site in the adult *rat* pineal gland apparently inhibits the release of norepinephrine from these sympathetic nerve fibres (Drijfhout et al. 1996). Cholinergic calcium signalling exhibits a developmental switch within the first three postnatal weeks (Schomerus et al. 1999). In postnatal pinealocytes, acetylcholine elevates $[Ca^{2+}]_i$ via muscarinic rather than not affected. Between 1 and 4 days after an exposure of adult Wistar rats to an altitude of 8,000 m for two h in an altitude chamber, the mitochondrial number und lipid droplets in the pinealocytes appeared to be reduced compared with those in control *rats* (Kaur et al. 2002). At 7 days, however, the mitochondrial numbers and lipid droplets were noticeably increased. At the same time interval, the expression of complement type 3 receptors and major histocompatibility class II nicotinic acetylcholine receptors. In the second postnatal week, pinealocytes gain responsiveness to nicotine and gradually lose responsiveness to muscarinic cholinergic stimuli.

γ-**Aminobutyric acid** levels in the *rat* pineal gland were increased by amino-oxyacetic acid, a GABA transaminase inhibitor (WANIESWSKI and SURIA 1977). Isoprenaline did not change the GABA content of the pineal gland. The ability of the pineal gland to accumulate [^{3}H]GABA was significantly reduced in *rats* that had been deprived of the superior cervical ganglion. There was a diurnal fluctuation of the endogenous GABA levels of the pineal.

Reoxygenation Cytobiology

Contrarily to the pathological findings in other organs of the same animals as hippocampus and myocardium, the pinealocytes of hypoxic-reoxygenated *rats* (Figs 129 and 130) commonly showed only slight **mitochondrial** cristolysis and clearing of the matrix (SCHILLER 2000). Osmiophilic granules described by PERRELET et al. (1968) in hypoxic-reoxygenated control *rats* were antigens as detected with the antibodies OX-42 and OX-6, respectively, in macrophages/microglia was up-regulated compared with that in the control *rats* and those killed at earlier times. This was attributed to the increased serum melatonin after the altutude exposure. By 14 and 21 days, the ultrastructure of pinealocytes and immunoreactivity of macrophages/microglia were comparable with that in the control *rats*.

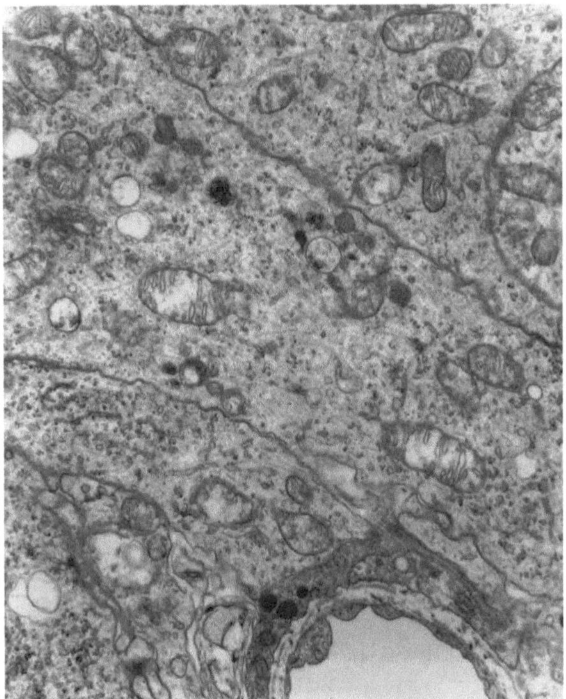

Fig. 227. Mild mitochondrial reoxygenation damage showing moderate cristolysis and clearing of the matrix in the repetitively reoxygenated pineal cells (block 1211) from a *rat* (No.7) treated for 7 consecutive days with intraperitoneal injection of 1.5 ml Tyrode's solution per kg body weight × day. On the last 4 days of experimentation the animal was exposed to an atmosphere containing only 5 % oxygen for 30 min. Half an hour after the last exposure under pentobarbital anaesthesia (30 mg/kg), the animal was perfused from the abdominal aorta with 2.5 % glutaraldehyde in 0.1 M sodium cacodylate buffer (pH 7.4). Postfixation with 1 % osmium tetroxide in sodium cacodylate buffer. Embedded in Epon 812 and sectioned at 50 nm. Lead citrate and uranyl acetate. Plate 3396

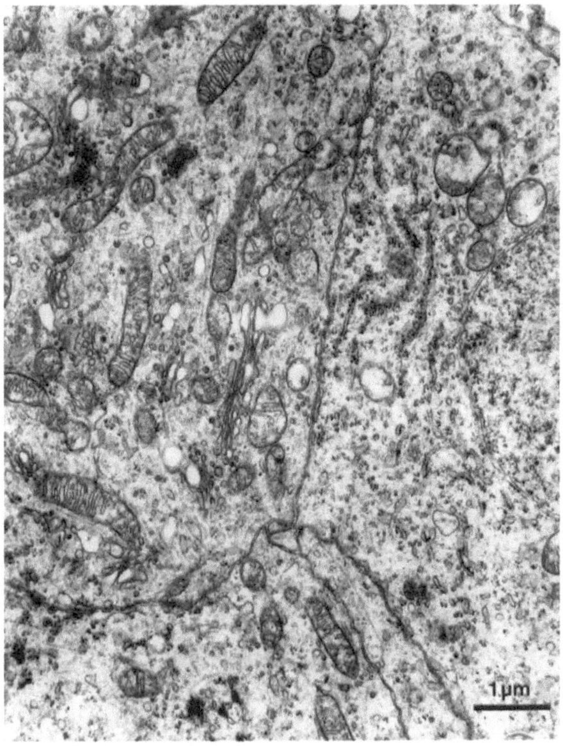

Fig. 228. Several prominent Golgi areas and microtubules, but little damage to the mitochondria in the left pinealocyte. The right cell, however, shows some damaged mitochondria and prominent cisternae of the endoplasmic reticulum. Repetitively reoxygenated pineal (block 1362) of a *rat* (No. 12) treated for 7 consecutive days with intraperitoneal injection of 1.5 ml Tyrode's solution per kg body weight × day. On the last 4 days of experimentation the animal was exposed to an atmosphere containing only 5 % oxygen for 30 min. Half an hour after the last exposure under pentobarbital anaesthesia (30 mg/kg), the animal was perfused from the abdominal aorta with 2.5 % glutaraldehyde in 0.1 M sodium cacodylate buffer (pH 7.4). Postfixation with 1 % osmium tetroxide in sodium cacodylate buffer. Embedded in Epon 812 and sectioned at 50 nm. Lead citrate and uranyl acetate. Plate 3429

Annulate lamellae (Fig. 229) are intracytoplasmic membrane systems composed of parallel arrays of cisternae bearing at regular intervals small annuli or fenestrae. As shown by electron microscopy, the structure of the annulate lamellae pore complex is similar to that of the pore complexes of the nuclear envelope. When melatonin was administered to *rats* or when *rats* were kept in darkness FREIRE and CARDINALI (1975) observed changes in the ultra-

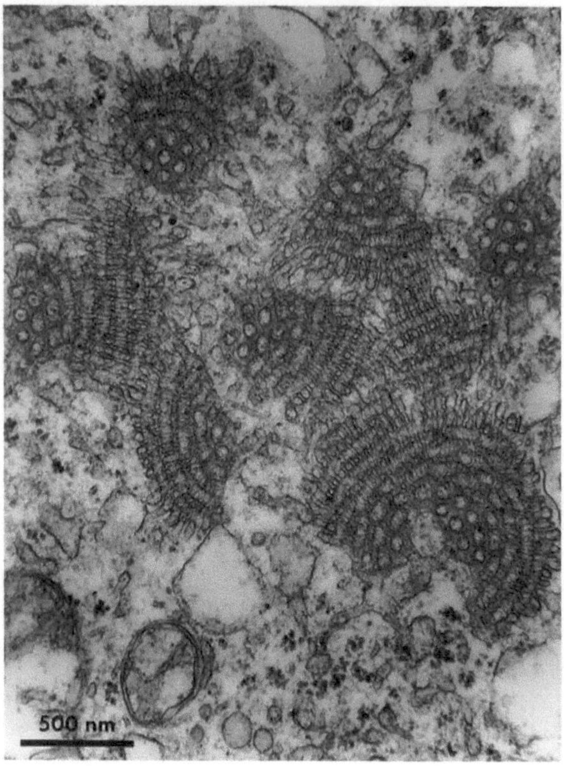

Fig. 229. Annulate lamellae and damaged mitochondria (swelling and cristolysis) in a repetitively reoxygenated pineal cell (block 1211) from a *rat* (No. 7) treated for 7 consecutive days with intraperitoneal injection of 1.5 ml Tyrode's solution per kg body weight × day. On the last 4 days of experimentation the animal was exposed to an atmosphere containing only 5 % oxygen for 30 min. Half an hour after the last exposure under pentobarbital anaesthesia (30 mg/kg), the animal was perfused from the abdominal aorta with 2.5 % glutaraldehyde in 0.1 M sodium cacodylate buffer (pH 7.4). Postfixation with 1 % osmium tetroxide in sodium cacodylate buffer. Embedded in Epon 812 and sectioned at 50 nm. Lead citrate and uranyl acetate. Plate 3395

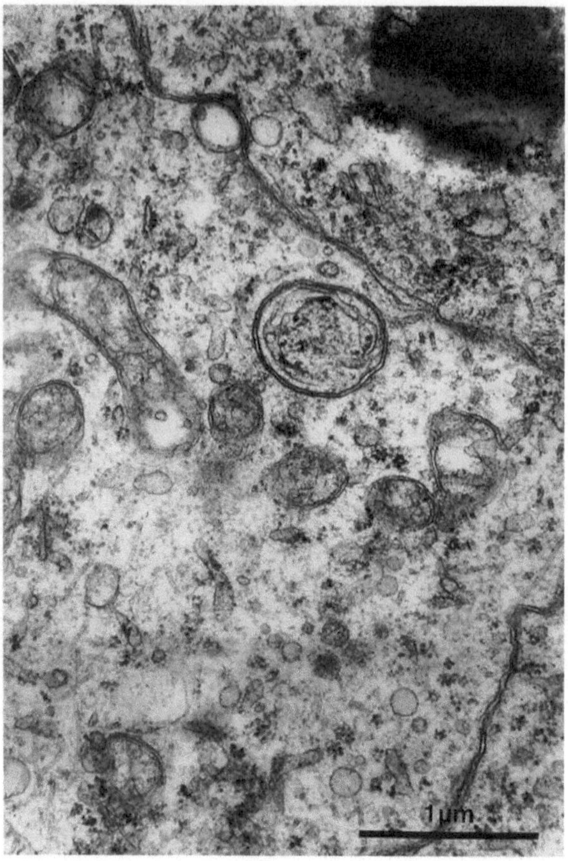

Fig. 230. Advanced intralysosomal degradation of unidentified structures and mitochondrial reoxygenation damage in a pinealocyte (block 1211) from a *rat* (No. 7) treated for 7 consecutive days with intraperitoneal injection of 1.5 ml Tyrode's solution per kg body weight × day. On the last 4 days of experimentation the animal was exposed to an atmosphere containing only 5 % oxygen for 30 min. Half an hour after the last exposure under pentobarbital anaesthesia (30 mg/kg), the animal was perfused from the abdominal aorta with 2.5 % glutaraldehyde in 0.1 M sodium cacodylate buffer (pH 7.4). Postfixation with 1 % osmium tetroxide in sodium cacodylate buffer. Embedded in Epon 812 and sectioned at 50 nm. Lead citrate and uranyl acetate. Plate 3401

structure of the pineal gland compatible with a generalised activation of the organ including annulate lamellae. In the African green *monkey* kidney cell line, RC37 and *bovine* mammary gland epithelial cells, BMGE, CORDES et al. (1996) observed that annulate lamellae and single cisternae containing pore complexes were occasionally in continuity with the ribosome-studded rough endoplasmic reticulum. In contrast, no such spatial relationship was observed between annulate lamellae and the Golgi apparatus. Likewise, no obvious interconnections were observed between annulate lamellae and peroxisomes, which were immunostained with rabbit antibodies against catalase, or between annulate lamellae and lysosomes, which were immunostained with monoclonal antibodies against the lysosomal membrane glycoproteins LAMP-1 and LAMP-2.

Tryptophan is readily oxidised by various oxidising free radicals (FLETCHER and ROSENFELD 1983, 1985, 1988). Except for $HO^{\bullet}$ radical which predominantly adds to the indole ring of tryprophan (OUDERKIRK et al. 1983), free-radical oxidants such as $SO_4^{\bullet-}$ (FLETCHER and ROSENFELD 1983), $Br^{\bullet-}$ (FLETCHER and ROSENFELD 1985), $(SCN)_2^{\bullet-}$ (FLETCHER and ROSENFELD 1985), and $N_3^{\bullet-}$ (FLETCHER and ROSENFELD 1988) react by one-electron transfer. The result is the tryptophan radical cation, which is in equilibrium with the neutral tryptophan radical. The reactivity of the tryptophan radical cation was found to be 1–2 orders of magnitude higher than that of the neutral radical (JOVANOVIC et al. 1991).

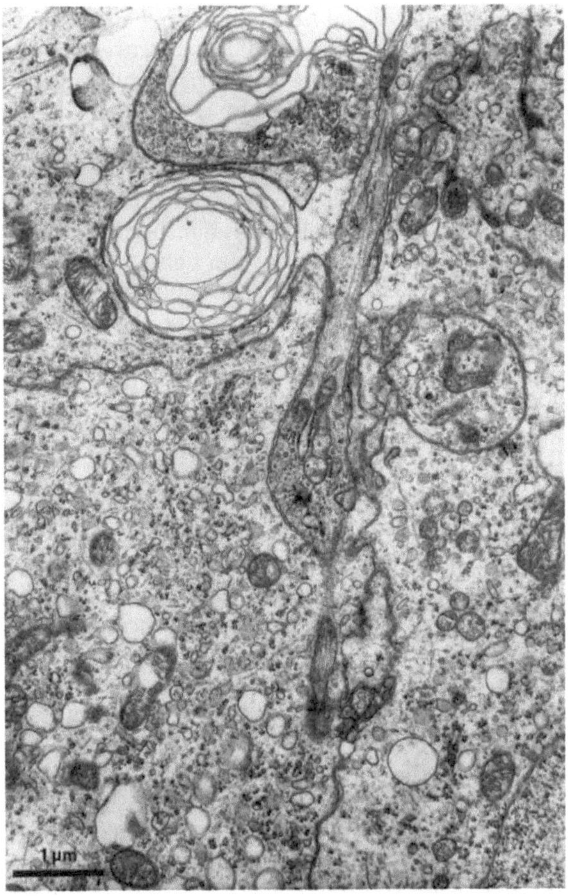

Fig. 231. Reoxygenation damage to the mitochondria and smooth endoplasmic reticulum of the pinealocytes (block 335) from a male Wistar *rat* (No. 26/103E) treated for 11 consecutive days with intragastric injection of 15 mg carbocromene per kg body weight × day. On the last 4 days of experimentation the animal was exposed to an atmosphere containing only 5 % oxygen for 30 min. Half an hour after the last exposure under pentobarbital anaesthesia (30 mg/kg), the animal was perfused from the abdominal aorta with 2.5 % glutaraldehyde in 0.1 M sodium cacodylate buffer. Postfixation with 1 % osmium tetroxide in sodium cacodylate buffer. Embedded in Epon 812 and sectioned at 50 nm. Lead citrate and uranyl acetate. Plate 2902

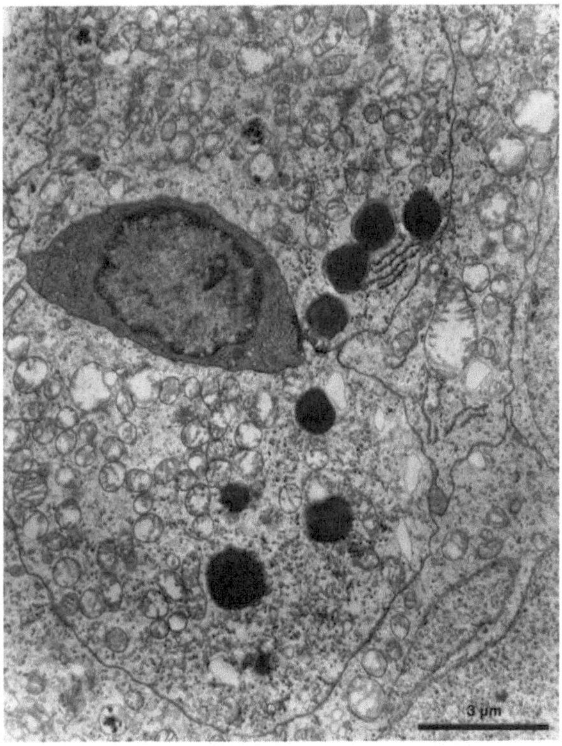

Fig. 232. Mitochondrial reoxygenation damage of the pinealocytes (block 1211) adjacent to a microglial cell from a *rat* (No. 7) treated daily for 7 consecutive days with intraperitonal injection of 1.5 ml Tyrode's solution per kg body weight × day. On the last 4 days of experimentation the animal was exposed to an atmosphere containing only 5 % oxygen for 30 min. Half an hour after the last exposure under pentobarbital anaesthesia (30 mg/kg), the animal was perfused from the abdominal aorta with 2.5 % glutaraldehyde in 0.1 M sodium cacodylate buffer. Postfixation with 1 % osmium tetroxide in sodium cacodylate buffer. Embedded in Epon 812 and sectioned at 50 nm. Lead citrate and uranyl acetate. Plate 3397

The reaction of a tryptophan derivative, *N*-(*tert*-butoxycarbonyl)-L-tryptophan, with hypoxanthine/xanthine oxidase/Fe(III)-EDTA mainly resulted in the oxygenation of the pyrrole ring of the indole nucleus (ITAKURA et al. 1994). 2-[(*tert*-Butoxycarbonyl)-amino]-3-(3-indolyl)propionic acid and *N*-(*tert*-butoxycarbonyl)-*N*'-formylkynurenine were identified as the major products.

Antioxidant Activity of Melatonin

Melatonin, *N*-acetyl-5-methoxytryptamine, is a hormone, synthesised mainly, but not exclusively (RALPH 1980, TAN et al. 1999) by the pineal gland of all *mammals* including *man* (REITER 1991). In tracing indoleamine synthesis, the best autoradiographical results have been obtained with tritiated 5-hydroxytryptophan. Using this compound, a diffuse labelling of pinealocytes was observed in *rat* (ARSTILA et al. 1971), *mouse* (GERSHON and ROSS 1966), *rabbit* (ROMIJN et al. 1977), and *monkey* (LOUIS et al. 1970). In the *rat* pineal organ stimulation of melatonin biosynthesis is tightly linked to activation of β_1-adrenergic receptors by norepinephrine which elevates intracellular cAMP levels and finally causes phosphorylation of the transcriptional factor cAMP responsive element binding (CREB) protein. In neonatal Wistar *rats* norepinephrine (10^{-6} M) induced an increase of $[Ca^{2+}]_i$ in most pinealocytes (SCHOMERUS and KORF 1997). Norepinephrine evoked a distinct phosphorylated CREB immunoreactivity in most pinealocytes. NO synthase type I expression is an essential part of the adrenergic cGMP response of the *rat* pineal (Spessert et al. 1997). Generally, in tissues containing neuronal or endothelial NO synthase, any increase in the intracellular calcium concentration will lead to stimulation of the calcium sensitive soluble guanylate cyclase and eventually to an increased •NO production. Due to its high membrane permeability, •NO not only stimulates soluble guanylate cyclase within the •NO producing cell, but also in adjacent cells.

Serotonin is the key intermediate in the synthesis of melatonin. This compound is formed from tryptophan by the successive actions of tryptophan hydroxylase and L-aromatic amino acid decarboxylase (Snyder and Axelrod 1964). Subcellular fractionation of pineal tissue has shown that the enzyme 5-hyroxylase, which converts tryptophan into 5-hydroxytryptophan, is restricted to the mitochondrial fraction (Hori et al. 1976) and that *p*-chlorophenylalanine, an inhibitor of this enzyme, did not cause any alteration in *rabbit* pinealocyte ultrastructure except for an enlargement of the mitochondria and an increase in number of their cristae (Romijn et al. 1977). These observations suggest that hydroxylation of tryptophan to 5-hydroxytryptophan happens at the mitochondrial membranes, while the conversion of 5-hydroxytryptophan to 5-hydroxytryptamine by aromatic L-amino acid decarboxylase may occur free in the cytosol. In the pineal, serotonin can be *N*-acetylated by serotonin *N*-acetyltransferase using acetyl CoA (Weissbach et al. 1960). The *N*-acetylserotonin that is formed is then *O*-methylated by hydroxyindole-*O*-methyltransferase (Axelrod and Weissbach 1961) to form *N*-acetyl-5-methoxytryptamine, which is melatonin. The regulation of the enzyme serotonin *N*-acetyltransferase appears to be the pivotal step in the regulation of melatonin synthesis, and serotonin *N*-acetyltransferase, in turn, is regulated by cAMP (Klein and Weller 1973). A peak in adenylate cyclase activity in the pineal gland was observed in untreated rats with intact ovaries at 4 weeks (Okatani et al. 1998). Ovariectomy at week 6 resulted in a significant increase in the activity of the enzyme at week 8. At week 10, adenylate cyclase activity resembled that of control animals. Subcutaneous injection of oestradiol benzoate (1.0 μg per day) suppressed the increase in adenylate cyclase activity induced by ovariectomy, similar to the level in control *rats* with intact ovaries. The decline in melatonin synthesis during puberty may be related to an increase in oestrogen level. For the photic inhibition of pineal melatonin neither functional rod receptors nor rod or cone outer segments are required (Lucas and Foster 1999, Lucas et al. 1999). Diurnal profiles of pineal melatonin content were similar in both *rd* and *rds* mutant types and in wild-type C3H *mice*; melatonin peaked between 3–5 h before light on. All three genotypes exhibited irradiances dependent inhibition of pineal melatonin content: 2.6×10^{-2} microwatt/cm^2 509 nm light induced complete suppression in all three genotypes, whereas lower irradiances were ineffective in all cases. Bilateral enucleation abolished responses even to 6 microwatt/cm^2 509 nm light.

Immediately after an intravenous infusion of 500 ng [^{3}H]-melatonin/h for 2 h, 1,17 ± 132 pg of the hormone and its metabolites per g wet weight – that is 0.11 % of the dose infused – were found in the *rat* lung in the following subcellular distribution: plasma membrane ≥ nuclei ≥ mitochondria ≥ supernatant ≥ microsomes (Messner et al. 1998). Six h after the end of infusion the figures were 1,633 ± 283 pg and 0.16 %, respectively.

There have been multiple proposals that melatonin can protect against damage caused by oxygen radicals *in vivo*. Thus, melatonin decreased DNA damage in *rats* treated with the carcinogen safrole (Tan et al. 1993, 1994), limited paraquat damage to the *rat* lung (Melchiorri et al. 1995), suppressed the development of cataracts in newborn *rats* treated with the glutathione depleting agent buthionine sulfoximine (Abe et al. 1994), decreased lipid peroxidation in the liver of CCl_4-exposed *rats* (Daniels et al. 1995), and limited alloxan toxicity in islet cells of *mice* (Pierrefiche et al. 1993). Kainic acid (40 mg/kg s.c.) damage to CD2-F1 *mice* concentrated in the CA3 pyramidal neurones was significantly reduced by melatonin injected intraperitoneally at a single dose of 5 mg/kg 10 min before kainic acid administration (Tan et al. 1998). *In vitro*, membrane fluidity in microsomes of a *rat* liver model in which lipid peroxidation was induced by the addition of $FeCl_3$, ADP and NADPH was reduced by melatonin in a concentration-dependent manner, and so were malonaldehyde and 4-hydroxyalkenal generation (García et al. 1997). In a dispersed linoleic acid system, melatonin did not reduce the rate of peroxidation induced by 2,2'-azobis (2-amidinopropane) dihydrochloride, but did reduce in a concentration-dependent manner, the rate of reaction activated by 0.15–3 mM Fe^{2+}-EDTA (Longoni et al. 1998). However, melatonin was only about one one-hundredth as effective an antioxidant as vitamin E in the micelles system.

The levels of 8-hydroxy-deoxyguanosine which were found to be elevated in the hippocampus and frontal cortex of *rats* treated with kainic acid (10 mg/kg) were significantly ($P < 0.05$) reduced in animals co-treated with 10 mg melatonin/kg (Tang et al. 1998).

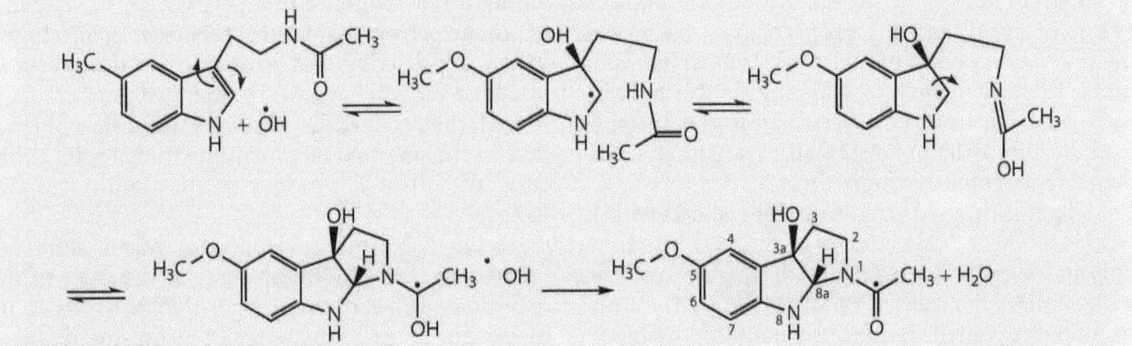

Reaction pathway, in which one melatonin molecule scavenges two hydroxyl radicals. This pathway suggested by TAN et al. (1998) differs significantly from classic free seavenging processes. [61]

Using mass spectrometry, proton nuclear magnetic resonance (^{1}H NMR), COSY ^{1}H NMR analysis, and calculation on the relative thermodynamic stability, TAN et al. (1998) identified a novel melatonin metabolite as **cyclic 3-hydroxymelatonin**. This is the product of reaction of melatonin with $HO^•$ generated in two different cell-free *in vitro* systems. Cyclic 3-hydroxymelatonin also existed in urine of both *human* and *rat*. When *rats* were challenged with ionising radiation (800 cGy = LD_{50} for a one month period), as expected, urinary cyclic 3-hydroxymelatonin was doubled over that of the controls ($P = 0.002$).

The loss of an electron by melatonin causes melatonin itself to become a nitrogen centred radical, the **indolyl cation radical**, which then is believed to scavenge a superoxide anion radical, $O_2^{•-}$ (HARDELAND et al. 1993, POEGGELER et al. 1994). As several chemically related molecules as serotonin, *N*-acetylserotonin, or 5-methoxytryptamine did not compare favourably to melatonin as a neutraliser of $HO^•$ generated by the photolysis of H_2O_2, these structure-activity studies suggested that the methoxy group at position 5 of the indole nucleus and the *N*-acetyl group on the side chain are both necessary for the efficient scavenging activity of melatonin (TAN et al. 1993, SCAIANO 1995).

In a system wherein $FeSO_4$ and H_2O_2 were used to generate $HO^•$, melatonin was quickly oxidised (POEGGELER et al. 1994). By comparison, Fe^{2+} was incapable of efficiently oxidising melatonin and likewise the indoleamine was not oxidised by its exposure to H_2O_2 alone. Systemic infusion of melatonin via osmotic pumps had no effect on iron-induced neurodegeneration (LIN and HO 2000). However, repetitive intraperitoneal injection of melatonin (10 mg/kg *rat*) prevented iron-induced oxidative injuries.

In vitro, melatonin did not scavenge superoxide radical and weakly protected DNA against damage

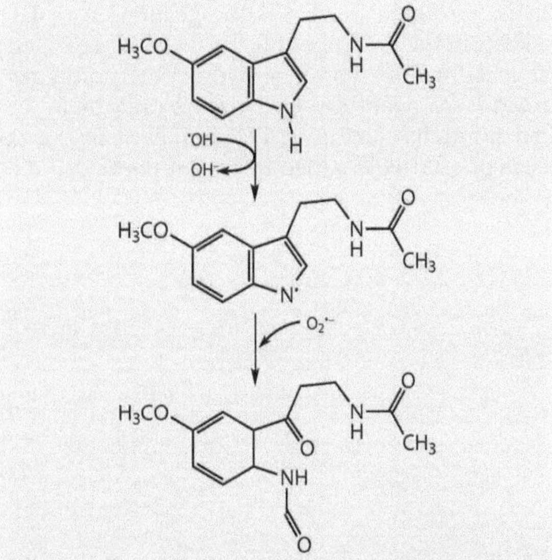

Upon electron donation to the hydroxyl radical (HO·) to form OH-, melatonin gives rise to the **melatoninyl cation radical**, which is able with the superoxide anion radical ($O_2^{•-}$) to *N*-acetyl-*N*-formyl-5-methoxykynuramine [60]

by the ferric bleomycin system (MARSHALL et al. 1996). Melatonin in a dose-dependent manner protected a multilamellar vesicle system composed of dilinoleoylphospatidylcholine from peroxidation by Fe^{2+}-EDTA and was slightly less protective against linoleic acid peroxidation initiated by different free radical-generating systems (LONGONI et al. 1998). However, melatonin was only about one-hundredth as effective an antioxodant as vitamin E in the micelles system.

Photooxidation of melatonin represents one of several possibilities of a more general, biologically highly important property of this indoleamine to act as an extremely efficient, high-affinity radical scavenger (TAN et al. 1993; review: HARDELAND et al. 1993), which through a two-step mechanism, terminates radical reaction chains (cf. HARDELAND 1995). In evolution, this property may have been the primary function of melatonin, which has been utilised by organisms for trapping dangerous radicals resulting from both irradiation and metabolism, including the nonenzymatic, transition metal-dependent Fenton reaction.

Melatonin reacted with hydroxyl radicals with $k_r = 2.7 \times 10^{10}\ M^{-1}s^{-1}$ (MATUSZAK et al. 1997). Other indols and kynurenine reacted with hydroxyl radicals with similar high rates ($k_r > 10^{10}\ M^{-1}s^{-1}$). In contrast to unhydroxylated indoles (melatonin, 6-chloromelatonin, and 5-methoxytryptamine), hydroxylated indoles (5-hydroxytryptamine = serotonin and 6-hydroxymelatonin) may function both as $HO^\bullet$ promoters and $HO^\bullet$ scavengers. The melatonin precursor serotonin promoted the generation of $HO^\bullet$ radicals in the presence of ferric iron and H_2O_2, and the melatonin metabolite 6-hydroxymelatonin generated large quantities of hydroxyl radicals in aerated solutions containing Fe^{3+} ion, even in the absence of externally added hydrogen peroxide.

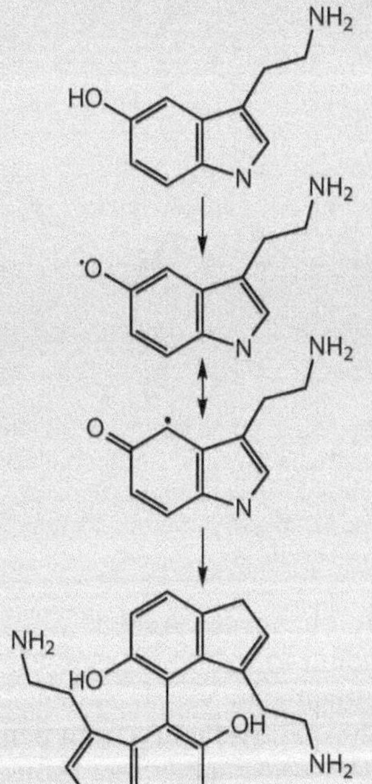

Serotinin ⟶ Serotonin semiquinone-imine radical [169]

Various one-electron oxidants ($HO^\bullet$, *tert*-Bu $O^\bullet$, $CCl_3OO^\bullet$, $Br_2^{\bullet-}$ and $N_3^\bullet$), generated pulse radiolytically in aqueous solutions at pH 7, were scavenged by melatonin to form two main absorption bands with λ_{max} = 335 nm and 500 nm (MAHAL et al. 1999).

The reaction of peroxynitrite with melatonin gives the melatoninyl radical cation ($MLT^{\bullet+}$), providing direct evidence that peroxynitrite is capable of one-electron oxidation (ZHANG et al. 1998). When the reaction was performed in the presence of added bicarbonate, $MLT^{\bullet+}$ was still produced, although the product distributions was different from that in the absence of added bicarbonate. In either case, no nitrated product of melatonin was detected.

Serum NO_x (nitrite + nitrate) values were decreased in male C57BL/6J *mice* (at the start of the experiment 15 months of age) that received 10 μg melatonin/ml drinking water (only during the night hours from 18.00 and 9.00) supplementation for 3 months (CHEN et al. 1999).

Melatonin (2.5 mg/kg i.p., administered four times) partially prevented all decreases in reduced glutathione of kainic acid-treated *rats* (FLOREANI et al. 1997). These neuroprotective effects of melatonin may result from a sparing of glutathione reductase, which decreased in kainic acid-treated but not in kainic acid/melatonin-treated animals. Kainic acid caused a rapid decrease in the reduced glutathione content of cultured cerebellar granule neurones but not astrocytes. FLOREANI et al. (1997) suggested that melatonin prevented the neurotoxic effects of reactive oxygen species linked to kainic acid receptor activation by maintaining cellular reduced glutathione homeostasis. In MCF7 cells, melatonin acted directly as an antioxidant and die not stimulate antioxidant defences (BALDWIN and BARRETT 1998). Glutathione levels were not altered by melatonin treatment.

Homocysteine-induced lipid peroxidation in *rat* brain homogenates was inhibited by melatonin (OSUNA et al. 2002).

Phosphine-induced changes in *rats* were significantly or completely blocked by melatonin (10 mg/kg) given 30 min before PH_3 (2 mg/g i.p.) while ascorbic acid (30 mg/kg) or β-carotene (6 mg/kg) were less effective or inactive (HSU et al. 2000).

Chlorpyrofos-ethyl (diethyl 3,5,6-trichloro-2-pyridyl phosphorothionate)-induced *rat* pulmonary thiobarbituric acid reactive substance was considerably reduced by 10 mg melatonin per kg body weight × day for 6 consecutive days (KARAOZ et al. 2002).

Adriamycin toxicity in CBA *mice* bearing TLX5 lymphoma was reduced by a pharmacological dose of melatonin (20–40 mg/kg) without decreasing the antitumour action of adriamycin (RAPOZZI et al. 1998).

Inhibition of the multifunctional oxidase cytochrome P450 by both endogenous and exogenous melatonin has been suggested by TAN et al. (1994). REITER (1996), however, has shown that melatonin potentiated free radical scavenging by a non-enzymatic process of electron donation. These observations reflect the ability of melatonin itself to act as a free radical (equation [56]), known as the indolyl or melatonyl cation radical. This radical is far less potent than the free radicals it neutralises *in vivo*.

The formation of DNA adducts represent a critical step in the initiation of carcinogenesis (TAN et al. 1993). The ability of exogenous and partly endogenous (TAN et al. 1994) melatonin to suppress the development of carcinogenic processes suggests that melatonin, as a strong component in the antioxidant system, helps to prevent carcinogenesis (BEYER et al. 1998).

In addition to its direct role as a free radical scavenger and antioxidant, melatonin stimulates the activity of antioxidant enzymes. It stimulates the activity of glutathione peroxidase (REITER 1995, ANTOLÍN et al. 1996, BLASK et al. 1997, REITER et al. 1997), glutathione reductase (REITER et al. 1997), glutathione *S*-transferase, an antioxidant enzyme that detoxifies xenobiotics (KOTHARI and SUBRAMANIAN 1992), Mn-superoxide dismutase (ANTOLÍN et al. 1996), and Cu,Zn-superoxide dismutase (ANTOLÍN et al. 1996). Melatonin (1–10 μM) induced γ-glutamylcysteine synthetase mediated by activator protein-1 in ECV304 *human* umbilical vascular endothelial cells in a dose-dependent manner (URATA et al. 1999).

SAINZ et al. (1995) have found melatonin to be anti-apoptotic. In both *in vivo* and *in vitro* studies, apoptosis was induced in the thymus by exposing the cells to the glucocorticoid dexamethasone. DNA fragmentation and increased numbers of apoptotic cells both were in part inhibited when melatonin was also given. SAINZ et al. (1999) demonstrated an effect of melatonin on the mRNA for antioxidant enzymes in thymocytes, also showing an unexpected regulation by dexamethasone of these mRNA. Both an effect of melatonin on the general machinery of apoptosis and a possible regulation on the expression of the cell death-related genes *bcl-2* and *p53* were shown not to be involved.

Daily intraperitoneal injection of melatonin significantly reduced diarrhoea and occult blood in the faeces as well as morphological lesions in *mice* provided with dextran sulphate in the drinking water (PENTNEY and BUBENIK 1995).

At the molecular level, melatonin has been shown to inhibit the activation of the transcriptional regulator nuclear factor (NF)-κB, the activation of which involves free radicals as second messengers. Using HeLa S3 cells, MOHAN et al. (1995) stimulated NF-κB binding to DNA by exposing the cells to TNF-α, phorbol 12-myristate 13-acetate, or ionising radiation, all of which generate intracellular free radicals. 10 μM Melatonin, exogenously added, inhibited the activation of NF-κB by each of these agents, with the mechanism of the inhibitory effect of melatonin probably involving the ability of indole to neutralise free radicals.

Serum TNF-α levels were increased, although nonsignificantly, in *mice* that received 10 μg melatonin/ml drinking water (only during the night hours from 18.00 and 9.00) supplementation (CHEN et al. 1999).

Nitric Oxide Scavenging Activity of Melatonin or its Precursors

In vitro melatonin and its precursors exhibited nitric oxide scavenging activity in the following order:

Nitric Oxide Scavenging Activity
Melatonin > serotonin > 5-hydroxytryptophan ≈ *N*-acetyl-5-hydroxytryptamine > L-tryptophan

(NODA et al. 1999). All compounds exhibited a dose-dependent scavenging activity with different dose thresholds. Melatonin resulted in a significant ($P < 0.05$) increase in •NO scavenging at concentrations of 0.125 mM or greater. Threshold concentration for the onset of scavenging activity occurred at the following concentrations:

melatonin	0.125 mM
serotonin	0.250 mM
5-hydroxytryptophan	0.250 mM
N-acetyl-5-hydroxytryptamine	1.0 mM
L-tryptophan	16 mM

The melatonin levels required to scavenge •NO are much higher than physiological concentrations, but under or near physiological concentrations melatonin modulates NOS activity and thereby influences •NO production (POZO et al. 1997).

15.1.1.8.2 Microglial Cells

Microglial cells were described by DEL RIO HORTEGA (1929), ORLANDI (1928), ORLANDI and GUARDINI 1929), FARINA (1941), and BARGMANN (1943).

Human pineal glands of adult or older age sometimes have iron-rich interlobular pigmented cells. These are probably macrophages and their pigment is probably haemosiderin, resulting from the phagocytosis and destruction of extravasated erythrocytes. Increased iron pigment deposits occur in the *human* pineal gland as well as in a wide variety of other organs in haemochromatosis.

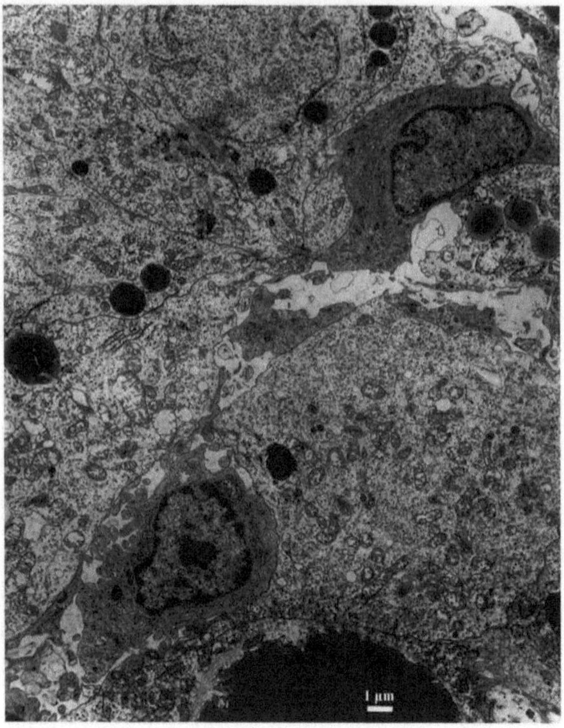

Fig. 233. Microglial cells, the lower with numerous pseudopodia, from the pineal body (block 1453) of a *rat* (No. 15) treated for 8 consecutive days with intraperitoneal injection of 15 mg carbocromene per kg body weight × day. On the last 5 days of experimentation the animal was exposed to an atmosphere containing only 5 % oxygen for 30 min. Half an hour after the last exposure under pentobarbital anaesthesia (30 mg/kg), the animal was perfused from the abdominal aorta with 2.5 % glutaraldehyde in 0.1 M sodium cacodylate buffer (pH 7.4). Postfixation with 1 % osmium tetroxide in sodium cacodylate buffer. Embedded in Epon 812 and sectioned at 50 nm. Lead citrate and uranyl acetate. Plate 3440

15.1.1.9 Substantia Nigra

Melanins are complex bio-polymers, the chemical structure of which has not been satisfactorily determined and whose physiological roles are not fully understood. The ability of melanin to scavenge reactive oxygen species, such as singlet oxygen (SEALY et al. 1984, SARNA et al. 1985), hydroxyl radical (SARNA et al. 1986) and superoxide anion (GEREMIA et al. 1984, KORYROWSKI et al. 1986, SARNA et al. 1986), has been firmly established in model systems, suggesting that melanin could protect pigment cells against oxidative stress. In a study of the interaction of dopa-melanin and cysteinyldopa-melanin with a series of reducing and oxidising free radicals RÓZANOWSKA et al. (1999) showed that melanin interacted via simple one-electron transfer processes, the reaction of both melanins with the strongly oxidising peroxyl radical ($CCl_3O_2^•$) from carbon tetrachloride, involved radical addition.

The binding of iron by melanin is a potentially important phenomenon as detailed knowledge of this binding is essential for understanding the role of melanin and iron in the pathogenesis of oxidative damage in the substantia nigra. Studies by SHIMA et al. (1997) indicated that the spectra at g = 4.3, attributable to Fe^{3+}, provide a useful parameter for determining the amount of paramagnetic iron bound to melanin in intact *human* substantia nigra.

Parkinson's disease is characterised by the selective and progressive destruction of the nigrostriatal dopaminergic neurones. In Parkinson's disease a high susceptibility for oxidative stress exists due to a high concentration of iron (DEXTER et al. 1989, ADAMS and ODUNZE 1991) and low GSH and GSH peroxidase levels in the substantia nigra (SIAN et al. 1994). Quinones of dopamine and 6-hydroxydopamine may play an important role in the formation of oxygen radicals due to the fact that they can undergo redox cycling (SIAN et al. 1994).

Catalase-positive microperoxisomes are relatively abundant in catecholaminic areas of the *rat* brain, as MCKENNA et al. (1976) showed in the substantia nigra, the locus coeruleus and in nucleus A_1 of the medulla. Approximately half of the neurones in the zona compacta substantiae nigrae (nucleus A_9) contain moderate numbers of catalase-positive bodies, while the remaining neurones contain few or none.

1-Methyl-4-phenyl-1,2,3,6-tetrahydropyridine (MPTP) is a frequently used animal model for Parkinson's disease. When injected into animals, MPTP is taken up by the cells in the substantia nigra where monoamino oxidase B (EC 1.4.3.4) converts it to 1-methyl-4-phenylpyridinium ion (MPP^+). This ion induces free radicals and duplicates many of the signs of Parkinson's disease in animals. In mice melatonin attenuated the resulting damage in the central nervous system (ACUÑA-CASTROVIEJO et al. 1997). MPTP-induced neural lipid peroxidation was reduced to control levels when melatonin was co-administered with the herbicide. Neuronal loss and MPTP-induced reduction of tyrosine hydroxylase (EC 1.14.16.2) activity in neurones of the striatum were attenuated by melatonin.

JHA et al. (2002) demonstrated that increasing reduction in total glutathione in dopaminergic PC 12 cells results in corresponding decreases in ubiquitin-protein conjugate levels suggesting that ubiquitination of proteins is inhibited in a glutathione dependent fashion. Decreased ubiquinated protein levels appear to be due to inhibition of E1 activity as demonstrated by reductions in endogenous E1-ubiquitin conjugate levels as well as decreases in the production of de novo E1-ubiquitin conjugates when glutathione is depleted.

Substantia nigra was the brain area of the C57BL/5 *mouse* exhibiting the highest level of heme oxygenase-2, constitutive and inducible heat shock protein (HSP) 70, GSSG, peroxides, iron, and calcium, in contrast with the lowest content in GSH, GSH/GSSG ratio and glutathione reductase activity, compared to the cortex, hippocampus, septal area, striatum and cerebellum (Calabrese et al. 2002).

When glutathione synthesis was inhibited by buthionine sulfoximine so that were was a 50 % depletion of glutathione, the immortalised *rat* mesencephalic cell line CSM14.1.4 showed an enhanced synergistic toxicity of sulphite and peroxynitrite (Marshall et al. 1999). Because sulphite is present normally in the brain as a product of cysteine metabolism, and because increased peroxynitrite formation has been reported in Parkinson's disease, these events might contribute to neuronal death.

Coinfusion of iron and melatonin, 60 µg/µl, but not 20 µg/µl, prevented iron-induced elevation of lipid peroxidation in substantia nigra and iron-induced depletion of dopamine in *rat* striatum (Lin and Ho 2000). Intranigral infusion of melatonin (60 µg/µl) or ethanol (20 %) alone altered neither basal lipid peroxidation in substantia nigra or dopamine content in the ipsilateral striatum. Coinfusion of melatonin (60 µg/µl) prevented the iron-induced reduction in tyrosine hydroxylase immunoreactive fibres in the striatum ipsilateral to iron-infused substantia nigra compared with that of the intact side. By contrast, the density of tyrosine hydroxylase positive axons in the melatonin + iron group was similar to that of the intact side of the same *rat*.

Incubation of *rat* embryonic mesencephalic tissue, rich in dopaminergic neurones, with 0.2 mM *tert*-butyl hydroperoxide in the presence of the spin trap 5,5-dimethyl-1-pyrroline *N*-oxide for 20 min resulted in the trapping of radicals (Karlsson et al. 2000). The main radicals detected in cell suspensions were the *tert*-butoxyl radical and the methyl radical, indicating the one-electron reduction of the peroxide followed by a β-scission reaction. The appearance of electron paramagnetic resonance signals from the trapped radicals preceded the onset of cytotoxicity, which was almost exclusively necrotic in nature. The inclusion of resveratrol (3,5,4'-trihydroxy-*trans*-stilbene) in incubations resulted in the marked protection of cells from *tert*-butyl hydroperoxide.

15.1.1.10 Cerebellum

The **granule cell** bodies of the *rat* cerebellar cortex lie in groups and are in close contacts with each other (Gray 1961). Mature granule cells contain 1.46 ± 0.02 nucleoli (Lafarga et al. 1989). The morphometric measurement of the nucleolar area showed 0.359 ± 0.07 μm^2. The fine texture does not differ substantially from the 'ring-shaped' configuration usually found in cells with low levels of protein synthesis. The Golgi apparatus is generally restricted to a single stack of four to six curved, flat cisternae and some associated vesicles (Palay and Chan-Palay 1977, p. 76). This complex is wedged into one of the thicker parts of the perikaryon that gives rise to dendrites.

Because cerebellar granule cells differentiate after birth, they can be readily cultured from newborn *rodents* and such primary cultures provide the most highly enriched *in vitro* system for the study of a single neuronal cell type. The neuronal preparations fully differentiate in culture and mature into glutamatergic neurones capable of synthesising and releasing glutamate by 8–10 days *in vitro*. These processes strictly require the presence of depolarising concentrations of KCl or an initial activation of *N*-methyl-D-aspartate receptors (Balazs and Leon 1994). Apoptosis can be triggered by a loss of Ca^{2+} homeostasis and elevated free radical production, resulting in the loss of cell membrane integrity (Orrenius et al. 1989, Markesbery 1992, Coyle and Puttfarcken 1993, Trump and Berezesky 1995). Spontaneous neuronal apoptosis was inhibited in dose- and time-dependent manners by antioxidants (U-78439G, α-tocopherol, and melatonin), NO synthase inhibitors (*N*-nitro-L-arginine and *N*-nitro-D-arginine), and a NO chelator (haemoglobin) in the micromolar range (Mason et al. 1999).

Melatonin was more effective to protect primary *rat* cerebellar granular neurones against the toxicity of H_2O_2, glutamate and *N*-methyl-D-aspartate when compared to normelatonin (Lezoualc'h et al. 1998).

In cultured *rat* granular cells, four-hour exposure of the neurones to glutamate receptor agonists, quisqualate, kainate, α-amino-3-hydroxy-5-methyl-4-isoxazolepropionate and *N*-methyl-D-aspartate, increased levels of brain-derived neurotropic factor mRNA (Bessho et al. 1993). Glutamate in combination with agonists of the ionotropic glutamate receptors, 6-cyano-7-nitroquinoxaline-2,3-dione, D-2-amino-5-phosphonovalerate and/or (+)-5-methyl-10,11-dihydro-5*H*-dibenzocyclohepten-5,10-imine hydrogen maleate (MK-801), also increased levels of brain-derived neurotropic factor mRNA. However, the addition of glutamate itself to the cultures produced severe neuronal death and failed to increase the mRNA level.

When cultured rat cerebellar granule cell neurones are transferred from 25 mM KCl to 5 mM KCl caspase-3 and caspase-8, but not caspase-1 or caspase-9, activities were induced and cells died apoptotically (VALENCIA and MORÁN 2001). Granule cell death was triggered by a $[Ca^{2+}]_i$ modification when $[Ca^{2+}]_i$ was reduced from 300 nM to 50 nM in a 5 mM KCl medium. The $[Ca^{2+}]_i$ changes were followed by and increase in reactive oxygen species levels. The generation of both cytosolic and mitochondrial reactive oxygen species occurred at three different times, 10 min, 30 min and 3–4 h but only those reactive oxygen species produced after 3–4 h were involved in the process of cell death. When granule cells cultured in a 5 mM KCl medium were treated with different antioxidants like scavengers of reactive oxygen species (mannitol, dimethyl sulphoxide) or antioxidant enzymes (superoxide dismutase and catalase) phosphatidylserine translocation, caspase activity, chromatin condensation and cell death were markedly diminished. The protective effect of antioxidants is not mediated through a modification in $[Ca^{2+}]_i$. Caspase activation, phosphatidylserine translocation and chromatin condensation were downstream of reactive oxygen species production.

The **Purkinje cell** (Fig. 235) is certainly one of the most spectacular nerve cells known. In the *rat* the cell bodies are 21 μm in diameter and 25 μm on the average. There are 350,000 (ARMSTRONG and SCHILD 1970), or 450,000 (SMOLYANIOV 1971). ARMSTRONG and SCHILD (1970) that the cells are farther apart in the depths of the cerebellar fissures than on the sides and crests of the folia.

The nuclear membrane is not smooth all the way around the nucleus but is wrinkled or puckered on the side facing the origin of the dendritic tree. the wrinkles vary from a shallow dimple in the surface of the nucleus to a deep and complicated crease with many tributary folds. The irregular depression in the nucleus is stuffed with Nissl substance, forming a prominent basophilic nuclear cap, which both RAMÓN Y CAJAL (1911) and JAKOB (1928) especially noted.

In sections stained with basic dyes, some Purkinje cells display a diffuse basophilia with only small, more intensely basophilic areas representing the Nissl bodies (Fig. 235). Other Purkinje cells contain a few large Nissl bodies and a multitude of small ones, distinctly separated from one another by clear spaces. It is also readily apparent in electron micrographs of thin sections that the Nissl substance in all of these cells is highly diffuse, forming what amount to either a close- or loose-mashed network throughout the cytoplasm. The content in ribosomes and polysomes is markedly less in lobule X (priscocerebellum) than in cells from lobule VIa (novocerebellum) (MÜLLER and HEINSEN 1984).

The mitochondria are among the most pleomorphic of neurones.

Lysosomes vary from small, dark vesicles, 100 nm in diameter, to large irregular masses, several micrometers in diameter. With advancing age of the animal the lysosomes become larger and more complex, as lipofuscin pigment accumulates in them (SAMORAJSKI et al. 1965).

The output of Purkinje cells is regulated by two distinct excitatory inputs, the parallel fibres and the climbing fibres. Conjunctive stimulation of parallel fibres and climbing fibres input at low frequencies results in a persistent attenuation of the parallel fibre- Purkinje neurone synapse, a process termed long-term depression (ITO et al. 1982).

Basket cell somata are occupied almost entirely by a large oval nucleus lying with its major axis parallel to the Purkinje cell layer. The nucleus can be deeply creased. Sometimes blocks of condensed chromatin are located near the nuclear envelope. Often a thin knife of cytoplasm slices more than halfway through the nucleus, carrying with it a few mitochondria, the endoplasmic reticulum, and hor-

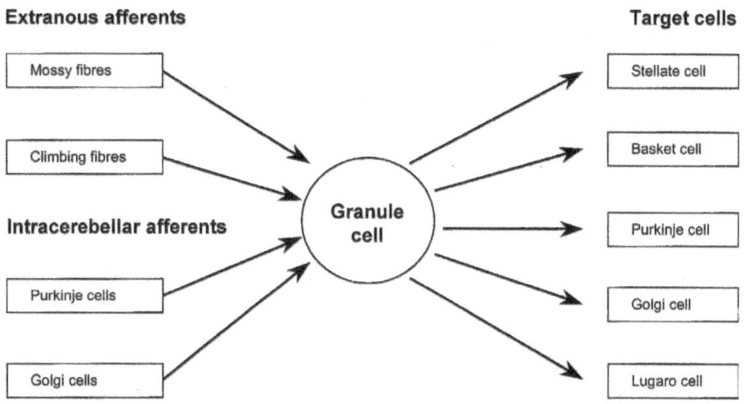

Fig. 234. Synaptic input and output of the cerebellar granule cell

des of ribosomes in clusters. Aside from its irregular contour the nuclear envelop is not remarkable. Except for the region immediately occupied by the Golgi apparatus, the whole perikaryon is pervaded by small clusters of free ribosomes.

Axons are relatively thick and contain many neurofilaments (CHAN-PALAY and PALAY 1970). They give off attenuated collaterals, the rounded viscosities of which are densely packed with vesicles and which form *en passant* synapses with clusters of thorns projecting from the major Purkinje dendrites.

Stellate cells in the superficial half of the molecular layer can be securely identified simply by their location. They have small elliptical perikaryal profiles, 5–9 μm in diameter, that are occupied almost entirely by their rounded nuclei having one or several shallow indentations. The deeper cell bodies can be larger, up to 12 μm in diameter, and they have a more voluminous cytoplasm while their nuclei are more likely to have deep and complicated indentations (CHAN-PALAY and PALAY 1972). The chromatin is usually homogeneously dispersed throughout the karyolymph except for a thin, irregular marginal condensation. The nuclear envelope often throws out irregular streamers that join with the granular endoplasmic reticulum. The sparse, rod-shaped or globular mitochondria range from 200 to 400 nm in diameter and 800 nm to 1.6 μm in length.

In the *human* cerebellum, the **horizontal cell of** LUGARO (1894) containing vast amounts of lipofuscin granules unequivocally distinguish themselves from granule cells containing only a small pigment granule (BRAAK 1974). In parasagittal sections of *rat* cerebellum, the LUGARO cells can be recognised as an ellipsoidal perikaryon (7–8 μm across) lying beneath the Purkinje cell layer. The elongated nucleus contains a large, prominent nucleolus (1.5–2 μm in diameter). The nuclear envelop is deeply indented and creased. The invaginations are filled with cytoplasm bearing free ribosomes in rosettes, helical rows, and other polysomal arrays (PALAY and CHAN-PALAY 1974). The mitochondria are generally short and plump with either transverse or oblique cristae. Both the cell body and the dendrites are partially ensheathed by a highly perforate neuroglial envelope, which the preterminal axons must penetrate in order to reach the postsynaptic surface.

Biochemical Pharmacology

CYP2D in *rat* cerebellum, the most intensely immunoreactive region on Western blots, is strong in both Purkinje cells and stellate and basket cells of the molecular layer (MIKSYS et al. 2000). There was some staining of the Bergmann glial cells, but very little immunoreactivity in the granule cell layer, confined to the granule cells and absent in the glomeruli.

As the unspecific (HALLIWELL and GUTTERIDGE (1989, p.441) xanthine oxidase inhibitor allopurinol protected *rat* cerebellar granule cells from kainate toxicity *in vitro* analogously to reperfusion tissue injury (DYKENS et al. 1987), xanthine oxidase activity was incriminated for neurodegeneration (CHOI 1988, MONYER et al. 1990, COYLE and PUTTFARKEN 1993, MCNAMARA and FRIDOVICH 1993). In addition to their inhibitory effects on xanthine oxidase and other pyrrolases, both allopurinol and its metabolite oxypurinol can serve as effective HO• sinks (MOORHOUSE et al. 1987), suggesting that the ability of allopurinol to protect neurones against kainate toxicity could be due to non-specific HO• scavenging independent of xanthine oxidase inhibition. *Rats'* cerebellar mitochondria produced free radicals when exposed to elevated Ca^{2+} and Na^{+} (DYKENS 1994).

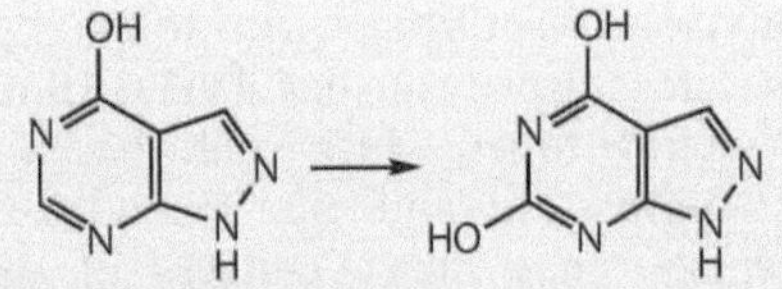

Allopurinol and its major metabolite, oxipurinol. Both are hydroxyl radial scavengers (k_2 approx. 10^9 $M^{-1} \cdot s^{-1}$ and 4×10^9 $M^{-1} \cdot s^{-1}$, respectively) [284]

Endogenous peroxidatic activity was particularly high in the processus around the Purkinje cells as demonstrated in the squirrel monkey, *Saimiri sciureus* (WONG-RILEY 1976).

12-*O*-Tetradecanoylphorbol-13-acetate (150 nM) dramatically reduced the size of the dendritic tree of Purkinje cells from slice cultures of postnatal *rat* cerebellum, while the axonal projections of the Purkinje cells to the deep cerebellar nuclei appeared unaffected (KAPFHAMMER and EGLE 1999). The protein kinase C inhibitor, GF 109203X (5 μM) increased both length and arborization of the dendrites.

Excitatory Amino Acids

Purkinje cell responses to the excitatory amino acids have been studies *in vitro* in *rat* cerebellar slices (CREPEL et al. 1982, 1983, LAMB and MILLER 1984). Quisqualate, glutamate, and aspartate excited these cells, whereas *N*-methyl-D-aspartate only weakly excited or inhibited them indirectly by exciting the small inhibitory interneurones (basket

Table 46. Selective neurotoxicity of different excitatory amino acid receptor agonists in immature *rat* cerebellar slices

Excitatory amino acid		Target authors
N-Methyl-D-aspartic acid	→	Granule cells
GARTHWAITE and GARTHWAITE (1985)		
Kainic acid	→	Golgi cells
MELDRUM and GARTHWAITE (1990)		
α-Amino-3-hydroxy-5-methyl-4-isoxazolepropionic acid	→	Purkinje cells
MELDRUM and GARTHWAITE (1990)		

and/or stellate cells; LAMB and MILLER 1984). *In vivo*, responses of male Wistar *rats* to *N*-methyl-D-aspartate were either biphasic (excitation followed by inhibition) or purely inhibitory and were antagonised by the selective *N*-methyl-D-aspartate-receptor antagonist, 2-amino-5-phosphonovalerate (QUINLAN and DAVIES 1985). Quisqualate and kainate either excited or induced biphasic responses, in these neurones, which were only reduced by amino acid antagonists that acted at non-*N*-methyl-D-aspartate-receptors. The excitatory amino acid-induced inhibitions were also antagonised by the selective γ-aminobutyric acid antagonist, picrotoxin, suggesting that they were indirectly mediated via GABAergic inhibitory interneurones, which could be excited via *N*-methyl-D-aspartate and non-*N*-methyl-D-aspartate receptors. In primary cultures of cerebellar granule cells from 8-day old *rats*, piracetam, aniracetam, and oxiracetam enhanced α-amino-3-hydroxy-5-methyl-4-isoxazolepropionic acid (AMPA)-stimulated $^{45}Ca^{2+}$ influx (COPANI et al. 1992).

N-Methyl-D-aspartate R1 receptor immunoreactivity was detected in *human* Purkinje cells, where the dye not only precipitated on the cell membrane and cell processes but also in the cytoplasm of the somata and along their dendrites (BÖCKERS et al. 1994).

Kainate-induced death of *murine* cerebellar neurones *in culture* was prevented by inhibiting the enzyme xanthine oxidase, a cellular source of cytotoxic superoxide radicals (DYKENS et al. 1987). Moreover, neurones were also protected from exitotoxin-induced death by the addition to the culture medium of either superoxide dismutase or mannitol, which scavenged superoxide and hydroxyl radicals, respectively, or serine protease inhibitor, which forestalled formation of xanthine oxidase. In *rat* cerebellar granule cell cultures melatonin was found to be beneficial in preventing kainate-induced excitotoxicity (GIUSTI et al. 1995).

In slice cultures of postnatal *rat* cerebellum, the glutamate receptor agonist, *trans*-ACPD (50 μM) induced a nearly complete loss of the dendrites of the Purkinje cells (KAPFHAMMER and EGLE 1999).

Coculture of astrocytes with cerebellar neurones from 6-days old *mice* enhanced the uptake of glutamate by astrocytes (BROWN 1999). Inhibition of glutamate uptake in a coculture system led to death of cerebellar cells. This toxicity could be inhibited by the *N*-methyl-D-aspartate receptor antagonist, MK801. However, in the presence of the glutamate uptake inhibitor, there was no increase in glutamate in the cultures compared to when the neurones were not cocultured.

•NO/cGMP/cGMP Kinase Signalling Cascade

Although the participation of •NO/cGMP in long-term depression was already proposed by ITO and KARACHOT (1990), its role remains controversial (PFEIFER et al. 1999). All necessary parts of the •NO/cGMP/cGMP kinase signalling cascade are available in the Purkinje cells: NO synthase is found in the vicinity of Purkinje cells (BREDT et al. 1990), guanylate cyclase is highly expressed in these cells (ARIANO et al. 1982), and they contain high concentrations of cGMP kinase I (KEILBACH et al. 1992). •NO is necessary for long-term depression in cerebellar slice preparations. Block of endogenous •NO production by NOS inhibitors prevents long-term depression and application of •NO can substitute for the climbing fibre stimulation in *rat* slice preparations (SHIBUKI and OKADA 1991). Release of caged •NO inside the Purkinje cell completely replaces parallel fibre activity and synergizes with depolarisation to cause long-term depression (LEV-RAM et al. 1995). Similar experiments with caged cGMP revealed that cGMP can induce long-term depression (LEV-RAM et al. 1997). The effect of cGMP was prevented by inhibition of cGMP kinase activity and was dependent on a subsequent or coincident increase in $[Ca^{2+}]_i$ (LEV-RAM et al. 1997). In contrast to these experiments, cerebellar long-term depression is not attenuated in *mice* deficient either in cGMP kinases I or II. In agreement with this finding are experiments on *murine* Purkinje neurones showing, that the •NO/cGMP pathway is not required for cerebellar long-term depression (LINDEN et al. 1995).

In cerebellar granule cells, glutamate-triggered, Ca^{2+}-mediated cell death is independent of endogenous •NO production (LAFON-CAZAL et al. 1993). A rather inverted mechanism has been demonstrated in these neurones: exposure to •NO donors leads to stimulation of *N*-methyl-D-aspartate receptors, probably because •NO-related species stimulate the release of endogenous agonists (LEIST et al. 1997). This sort of stimulation eventually causes apoptosis (BONFOCO et al. 1996).

cGMP Levels in the Rat Cerebellar Cortex

Measurement of cGMP in the cerebellar cortex allows an estimation of the excitatory and inhibitory input into the cerebellum (GUIDOTTI et al. 1975). Treatment of *rats* with piracetam (400 mg/kg×day) for 15 days significantly decreased cGMP levels (APUD et al. 1983). This effect is not mediated by a γ-aminobutyric acid-ergic mechanism as:

- co-administration of threshold doses of piracetam and flurazepam were unable to reduce cGMP in the *rat* cerebellum;
- piracetam (400 mg/kg) could not antagonise the increase of cGMP induced by isoniazide in the *rat* cerebellar cortex.

cGMP-specific Phosphodiesterase (Phosphodiesterase V)

In cultured *rat* cerebellar granule cells and astroglia, cGMP appeared to be predominantly hydrolysed by a calmodulin-dependent phosphodiesterase (AGULLÓ and GARCÍA 1997). It is known that elevation of calcium in these cultures in response to *N*-methyl-D-aspartate results in cGMP increases by activation of nitric oxide synthase I. Calmodulin-dependent phosphodiesterase activity is also increased by the elevated calcium levels and reduces cGMP levels (BALTRONS et al. 1997). Therefore, the calmodulin-dependent phosphodiesterase activation may be important for cGMP turnover after NOS-I stimulation (JUILFS et al. 1999).

The expression of cGMP-specific phosphodiesterase (phosphodiesterase V) protein and mRNA in both the cell bodies and the dendritic tree of the Purkinje cells of the *mouse* was shown by JUILFS et al. (1999).

Cyclic GMP-stimulated Phosphodiesterase (Phosphodiesterase II)

Golgi cells in the granule cell layer are interneurones involved in modulating input to the Purkinje cells from the mossy fibre pathway. They express phosphodiesterase II (JUILFS et al. 1999). In relation to the number of cells in the cerebellum these represent a minor cell population, however, signal intensity is high suggesting that there is a significant level of phosphodiesterase II protein expressed in these cells.

cGMP Kinase I

cGMP kinase I is found in high concentrations (LOHMANN et al. 1981). Immunoblotting of *rat* cerebellum with isozyme specific antibodies showed that cGMP kinase Iβ was not highly expressed, but relatively high concentrations of cGMP kinase Iα were contained (KEILBACH et al. 1992).

After a 2-h pulse with [*C3*-^{3}H]**sphingosine** at different doses (0.1–200 nmol/mg of cell protein), both cerebellar granule cells and astrocytes efficiently incorporated the long chain base; the percentage of cellular [^{3}H]sphingosine over total label incorporation was extremely low at sphingosine doses of <10 nmol/mg of cell protein and increased at higher dose (RIBONI et al. 2000). Most of the [^{3}H]sphingosine taken up underwent metabolic processing by *N*-acetylation, 1-phosphorylation, and degradation (assessed as $^{3}H_2O_2$ released in the medium).

Glial Cells

The cerebellar cortex is not rich in glial cells (TESTUT and LATARJET 1948, MARINI-ABREU 1973) but all three neuroglial cell types are nevertheless present.

Ascorbic acid having the unique property of reducing an acidified solution of silver nitrate almost instantly (GOMORI 1952, p. 90), in the granule layer of the *human* cerebellum can only be detected in a few glial cells (CLARA 1942, 1953).

The Bergmann fibres belong to the **astroglia** (PETERSEN 1969). They exhibit bundles of glial filaments and are rich in glycogen particles. Most of the cell bodies of the Bergmann fibres lie in the granule cell layer, some in the Purkinje cell layer or the molecular layer.

All **oligodendrocytes** of the *rat* cerebellar cortex showed polymorphic contours (MONTEIRO 1983). The nuclei had a smooth outline and were rounded, elliptical or irregular in shape. They had many chromatin granules, clumped either through the karyolymph or against the nuclear membrane. The cytoplasm was, as a rule, electron-dense chiefly owing to the presence of large numbers of microtubules, rather than to an abundance of ribosomes. The cells lacked filaments, and dense bodies were frequent. Precautions must be taken not to confuse type I cells (paler cytoplasm and nuclei) with granule cells and type II cells (nuclei much more irregular in shape than type I cells) with interstitial form of microglial cells.

Microglial cells exhibited perinuclear cisternae which were not so evident as those of the oligodendrocytes, and dense bodies were more seen (MONTEIRO 1983). The presence in microglial cells of a few long narrow units of rough endoplasmic reticulum, filled with dark contents, allowed distinction between them and oligodendrocytes type II.

Capillaries

The capillaries in the cerebellar cortex of the *cat* have a diameter between 3.5 µm and 12 µm (LANGE and HALATA (1972). In the granule cell layer, the capillaries have a smaller lumen than those in the Pukinje cell and molecular layers. The endothelium forming slender, partially overlapping lamellae is separated from the pericapillary compartment by a basement membrane which often splits into two layers; between these layers processes of the percytes are localised, mostly in the granular layer. Capillaries with more than 10 µm diameter are often surrounded by a perivascular space containing pericytes, fibroblasts, and circularly arranged collagen fibres.

Blood-brain barrier breakdown in the cerebellum is consistently obtained with pentylenetetrazole-induced seizures (NITSCH et al. 1986). With horseradish peroxidase as a tracer, numerous small- to medium-sized vessels were stained, mainly in the granule cell layer. Granule cells did not incorporate horseradish peroxidase.

Intermittent hypoxia in *guinea pigs* induced 'paired cisternae' in the rough endoplasmic reticulum of the Purkinje cells (PALLADINI et al. 1976). A moderate degree of mitochondrial swelling was apparent in a few neurones.

•NO Cytotoxicity

•NO
↓
Activation of guanylate cyclase
↓
Formation of cyclic guanylate monophosphate (cGMP)
↓
Activation of cGMP-dependent Ca^{2+} channels
↓
Calcium entry into the cell
↓
Cell death
↓
Removal of necrotic cells by microglial cells

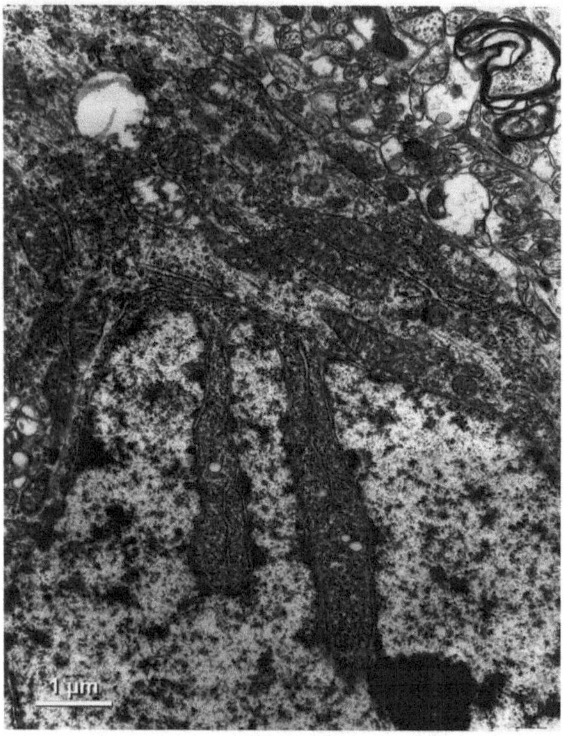

Fig. 235. Deeply invaginated dendritic pole of the nucleus of a Purkinje cell in the reoxygenated cerebellar cortex (block 1333) of a *rat* (No. 11) treated for 7 consecutive days with intraperitoneal injection of 15 mg carbocromene per kg body weight × day. On the last 4 days of the experimentation the animal was exposed to an atmosphere containing only 5 % oxygen for 30 min. On August 5, 1976 under pentobarbital anaesthesia (30 mg/kg), the animal was perfused from the abdominal aorta with 2.5 % glutaraldehyde in 0.1 M sodium cacodylate buffer (pH 7.4). Postfixation with 1 % osmium tetroxide in sodium cacodylate buffer. Embedded in Epon 812 and sectioned at 50 nm. Lead citrate and uranyl acetate. Plate 3421

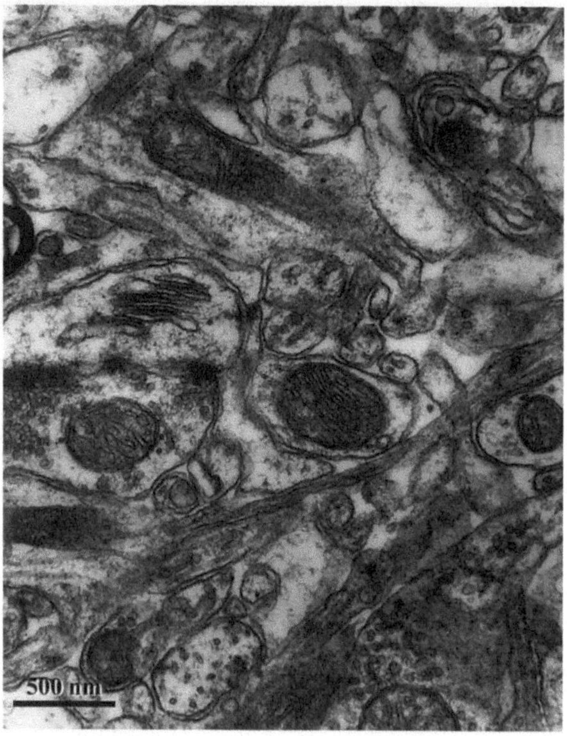

Fig. 236. Purkinje spine synapse in the molecular layer of the reoxygenated cerebellar cortex (block 1333) of a *rat* (No. 11) treated for 7 consecutive days with intraperitoneal injection of 15 mg carbocromene per kg body weight × day. On the last 4 days of experimentation the animal was exposed to an atmosphere containing only 5 % oxygen for 30 min. On August 5, 1976 under pentobarbital anaesthesia (30 mg/kg), the animal was perfused from the abdominal aorta with 2.5 % glutaraldehyde in 0.1 M sodium cacodylate buffer (pH 7.4). Postfixation with 1 % osmium tetroxide in sodium cacodylate buffer. Embedded in Epon 812 and sectioned at 50 nm. Lead citrate and uranyl acetate. Plate 3424

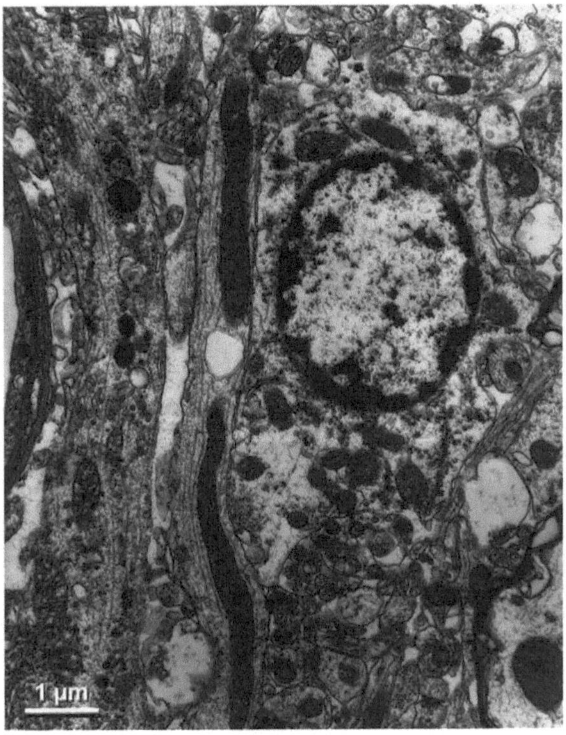

Fig. 237. Climbing fibre with long, slender mitochondria, sparse tubular endoplasmatic reticulum, numerous microtubules, and few neurofilaments. Molecular layer of the reoxygenated cerebellar cortex (block 1333) of a *rat* (No. 11) treated for 7 consecutive days with intraperitoneal injection of 15 mg carbocromene per kg body weight × day. On the last 4 days of the experimentation the animal was exposed to an atmosphere containing only 5 % oxygen for 30 min. On August 5, 1976 under pentobarbital anaesthesia (30 mg/kg), the animal was perfused from the abdominal aorta with 2.5 % glutaraldehyde in 0.1 M sodium cacodylate buffer (pH 7.4). Postfixation with 1 % osmium tetroxide in sodium cacodylate buffer. Embedded in Epon 812 and sectioned at 50 nm. Lead citrate and uranyl acetate. Plate 3419

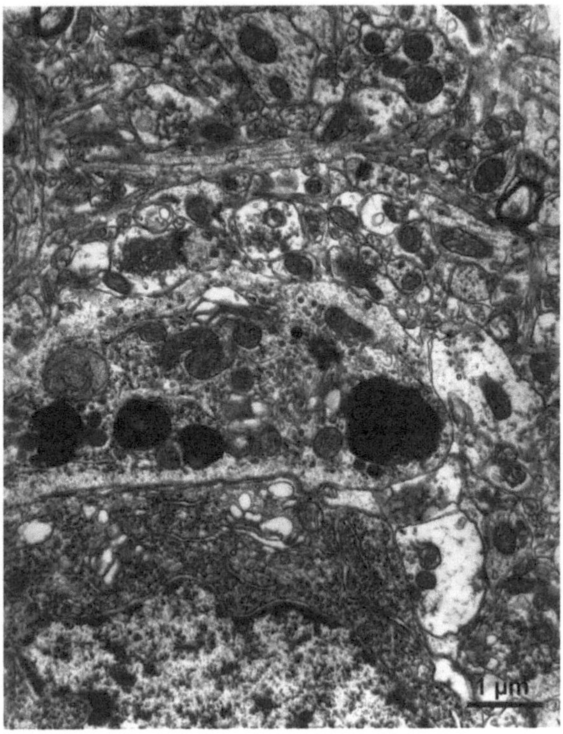

Fig. 238. Molecular layer of the intermittent hypoxic and reoxygenated cerebellar cortex (block 1333) of a *rat* (No. 11) medicated for 8 consecutive days with intraperitoneal injection of 15 mg carbocromene per kg body weight × day. On the last 5 days of experimentation the animal was exposed to an atmosphere containing only 5 % oxygen for 30 min. On August 5, 1976 under pentobarbital anaesthesia (30 mg/kg), the animal was perfused from the abdominal aorta with 2.5 % glutaraldehyde in 0.1 M sodium cacodylate buffer (pH 7.4). Postfixation with 1 % osmium tetroxide in sodium cacodylate buffer. Embedded in Epon 812 and sectioned at 50 nm. Lead citrate and uranyl acetate. Plate 3422

Nitric oxide synthase is present in granule cells as well as in basket cells (Koscheck et al. 1998), but is absent from Purkinje cells. Cyclic nucleotide levels increase sevenfold in response to activation of the *N*-methyl-D-aspartate subtype of the glutamate receptor that is present on granule cells, Purkinje cells and cells of Bergmann (Böckers et al. 1994) and which conceivably triggers the formation of •NO that diffuses to the adjacent Purkinje cells to activate soluble guanylate cyclase (EC 4.6.1.2). In cultured *rat* cerebellar glial cells *S*-nitroso-*N*-acetylpenicillamine increased intracellular cyclic guanosine monophosphate levels (Garg et al. 1992). •NO-treated cells showed decreased conversion of tetrazolium to blue formazan. The •NO donor SIN-1 (3-morpholinosydnonimine *N*-ethylcarbamide), at concentrations which produced a much higher guanylate cyclase activation (i.e. •NO concentration) than *N*-methyl-D-aspartate, was not neurotoxic and did not increase *N*-methyl-D-aspartate-neuronal death (Lafon-Cazal et al. 1993). nNOS expression was tightly correlated with the afferent innervation of granule cells *in vivo*, and primary cell cultures demonstrated that the developmental expression of nNOS is subject to regulation by glutamatergic input (Baader et al. 1996). This transsynaptic regulation of nNOS expression depends on the developmental stage of granule cells. It correlated with intracellular levels of free calcium, and several molecularly defined pathways regulating levels of intracellular calcium appear impinge on nNOS expression. Cultured granule cells from embryonic day 17 donors could be identified by their expression of NOS, a complement of functionally coupled receptors for glutamate and γ-aminobutyrate, and their typical morphology which included the location of their afferent synapses exclusively on their processes, but not on their perikarya (Mertz et al.

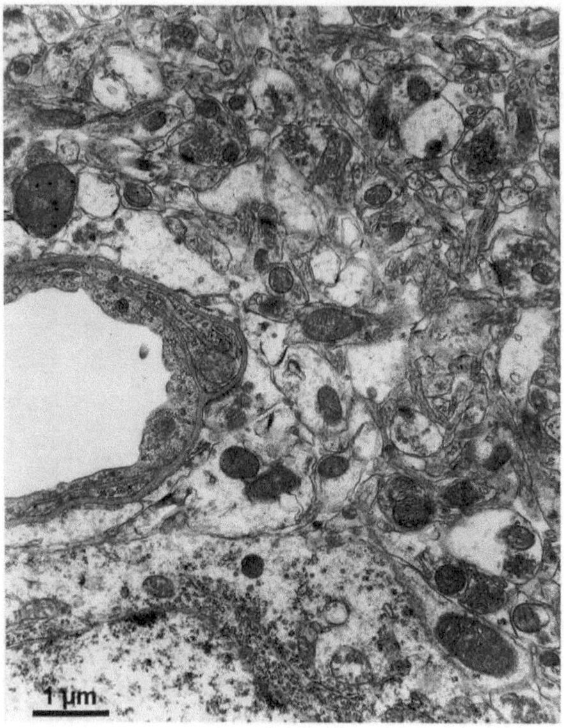

Fig. 239. Capillary showing pericyte processes generally surrounded by a basement membrane. Molecular layer of the intermittently hypoxic and reoxygenated cerebellar cortex (block 1333) of a *rat* (No. 11) medicated for 8 consecutive days with intraperitoneal injection of 15 mg carbochromene per kg body weight × day. On the last 5 days of experimentation the animal was exposed to an atmosphere containing only 5 % oxygen for 30 min. On August 5, 1976 under pentobarbital anaesthesia (30 mg/kg), the animal was perfused from the abdominal aorta with 2.5 % glutaraldehyde in 0.1 M sodium cacodylate buffer (pH 7.4). Postfixation with 1 % osmium tetroxide in sodium cacodylate buffer. Embedded in Epon 812 and sectioned at 50 nm. Lead citrate and uranyl acetate. Plate 3417

1998). Basket/stellate cells could be identified by their expression of the calcium-binding protein, parvalbumin.

In the frog, *Rana esculenta*, nitric oxide synthase immunoreactivity in a high percentage of Purkinje cells (mainly on the ispilateral side) was still present 60 days after an excision of the right eighth cranial nerve between Scarpa's ganglion and the brain stem (Pisu et al. 2002). On the basis of cell density evaluations, it was proposed that the early induction of NOS after neurectomy was linked to the degeneration of a part of the Purkinje neurones, while the permanence of NOS labelling might be due to a neuroprotective role of NO in the restoration phase of the vestibular compensation process.

Rats medicated with 5,000 ppm molsidomine added to the food from July 17/18, 1978 to November 30, 1978, after discontinuation of the medication for 7 week showed a focal increase of microglial cells in the granule cell layer seen under the light microscope (Fig. 241). This was corroborated under the electron microscope (Figs. 142 and 199). The Golgi areas of the microglial cells were enlarged (Fig. 199). Some microglial cells showed telolysosomes containing lipofuscins (Fig. 243).

Exposure of conscious *rats* to **10 % O_2** for 12 or 48 h resulted in a time-dependent increase in cerebellar nitric oxide synthase activity and a significant rise in mRNA and protein expressions of nNOS isoform (Guo et al. 1997). However, endothelial NOS protein expression declined significantly.

Microinjection of bacterial lipopolysaccharide (10 µg) and interferon-γ (20 U) into *rat* cerebellum induced iNOS expression in granule cells and subsequent cell death assessed by staining for DNA

Fig. 240. Cerebellar cortex (block 4748) of an unmedicated 238 g female control *rat* (No. 5). Under pentobarbital anaesthesia (30 mg/kg), the animal was perfused from the abdominal aorta with 2.5 % glutaraldehyde in 0.1 M sodium cacodylate buffer (pH 7.4). Postfixation with 1 % osmium tetroxide in sodium cacodylate buffer. Embedded in Epon 812 and sectioned at 1 µm. Toluidine blue

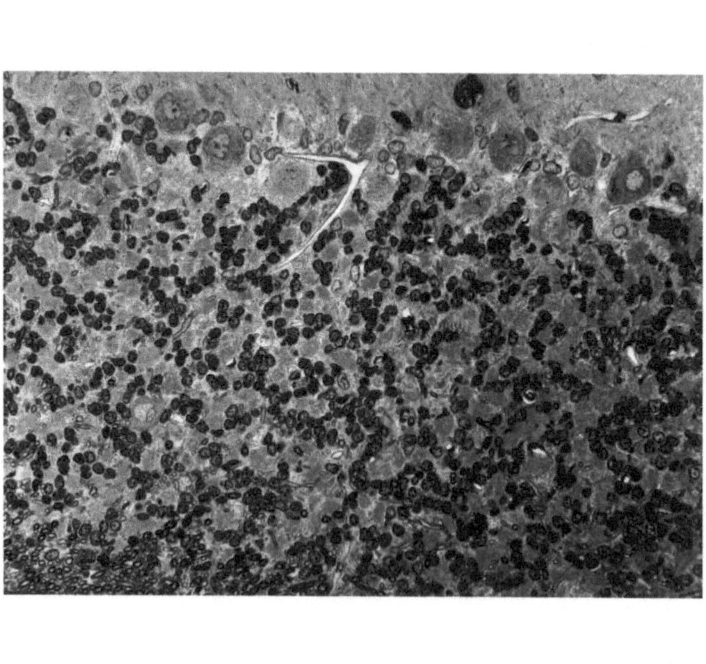

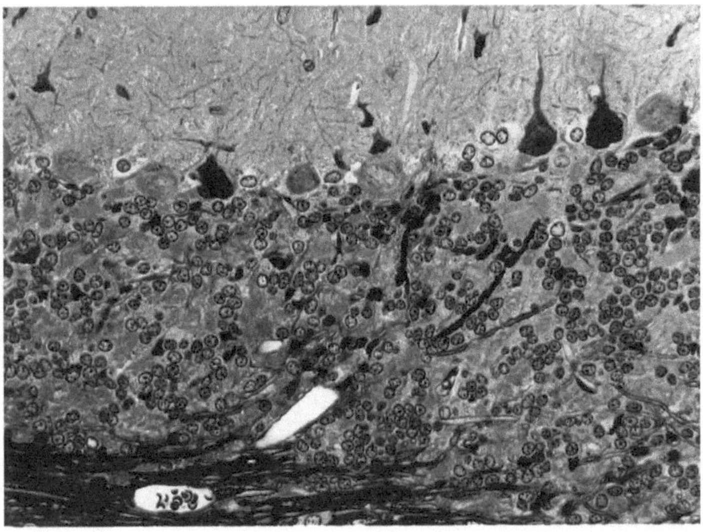

Fig. 241. Cerebellar cortex (block 4775) of a 216 g female *rat* (No. 667) medicated with 5,000 ppm molsidomine added to the food (powdered Altromin®) from July 17/18, 1978 to November 30, 1978. After discontinuation of the medication for 7 weeks, under pentobarbital anaesthesia (30 mg/kg), the animal was perfused from the abdominal aorta with 2.5 % glutaraldehyde in 0.1 M sodium cacodylate buffer (pH 7.4). Postfixation with 1 % osmium tetroxide in sodium cacodylate buffer. Embedded in Epon 812 and sectioned at 1 μm. Toluidine blue. Prominent accumulation of microglial cells in the granule cell layer

fragmentation (HENEKA et al. 2000). Co-injection of three structurally distinct agonists of the peroxisome proliferator-activated receptor-γ, including the antidiabetic thiazolidinedione troglitazone, the non-steroidal anti-inflammatory drug ibuprofen, and the prostanoid 15-deoxy-$\Delta^{12,14}$ prostaglandin J_2, reduced both iNOS expression and cell death, whereas co-injection of the selective cyclooxygenase inhibitor NS-398 had no effect.

Endogenous •NO is released after stimulation of climbing fibres (SHIBUKI and OKADA 1991). Long-term depression evoked by conjunctive stimulation of parallel and climbing fibres is blocked by haemoglobin (which strongly binds NO) or N^{ω}-monomethyl-L-arginine (a NOS synthase inhibitor). Exogenous •NO (from 3 mM sodium nitro-

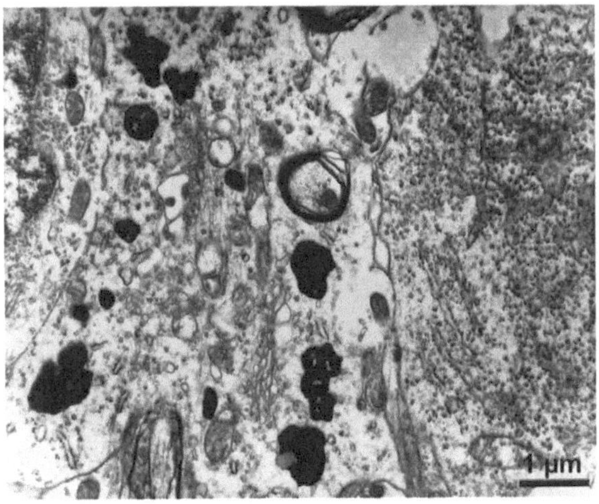

Fig. 242. Purkinje cell layer (block 4784) of a 216 g female *rat* (No. 696) medicated with 5,000 ppm molsidomine added to the food (powdered Altromin® R) from July 17, 1978 to November 30, 1978. After discontinuation of the medication for 7 weeks, under pentobarbital anaesthesia (30 mg/kg), the animal was perfused from the abdominal aorta with 2.5 % glutaraldehyde in 0.1 M sodium cacodylate buffer (pH 7.4). Postfixation with 1 % osmium tetroxide in sodium cacodylate buffer. Embedded in Epon 812. Lead citrate and uranyl acetate. Film 150/79

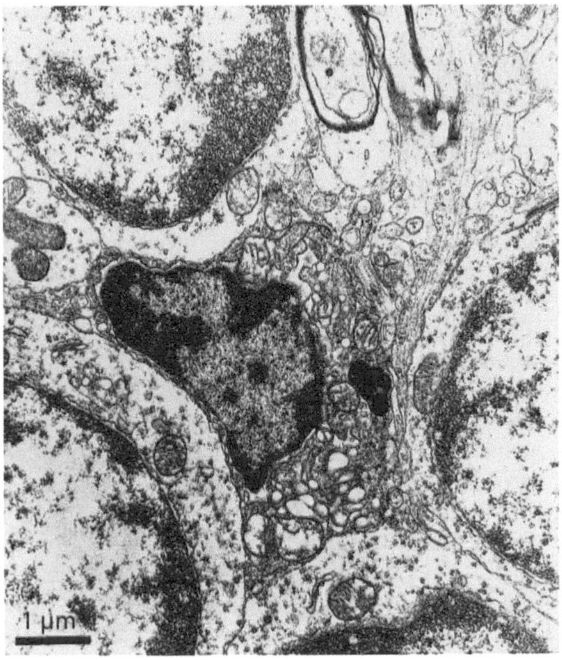

Fig. 243. Telolysosomal lipofuscins in progressive microglial cells in the granule cell layer of the cerebellar cortex (block 4766) of a 221 g female *rat* (No. 668) medicated with 5,000 ppm molsidomine added to the food (powdered Altromin® R) from July 17, 1978 to November 30, 1978. After discontinuation of the medication for 7 weeks, under pentobarbital anaesthesia (30 mg/kg), the animal was perfused from the abdominal aorta with 2.5 % glutaraldehyde in 0.1 M sodium cacodylate buffer (pH 7.4). Postfixation with 1 % osmium tetroxide in sodium cacodylate buffer. Embedded in Epon 812. Lead citrate and uranyl acetate. Films 171/79 and 173/79

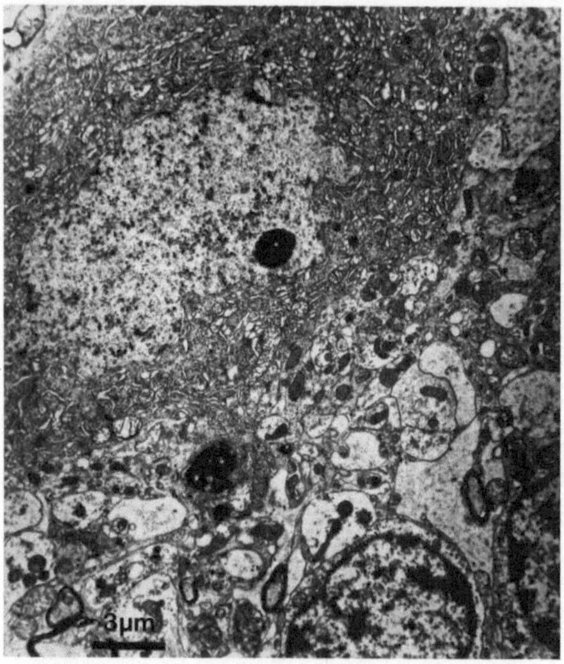

Fig. 244. Purkinje cell perikaryon (block 4766) of a 221 g female *rat* (No. 668) fed an Altromin® diet containing 5000 ppm molsidomine from July 17 to November 30, 1978. 7 Weeks after the discontinuance of the molsidomine treatment, under pentobarbital anaesthesia (30 mg/kg), the animal was perfused from the abdominal aorta with 2.5 % glutaraldehyde in 0.1 M sodium cacodylate buffer (pH 7.4). Postfixation with 1 % osmium tetroxide in sodium cacodylate buffer. Embedded in Epon 812 and sectioned at 50 nm. Lead citrate and uranyl acetate. Film 177/79

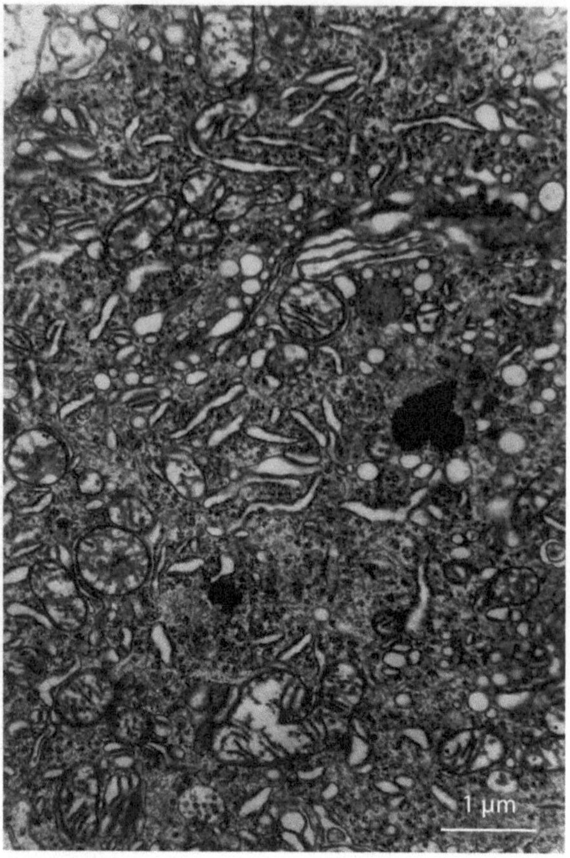

Fig. 245. Mitochondrial changes (swelling, cristolysis and clustered granula osmiophila) and 'paired cisternae' of the endoplasmic reticulum in a Purkinje cell perikaryon (block 4766) of a 221 g female *rat* (No. 668) fed an Altromin® diet containing 5000 ppm molsidomine from July 17 to November 30, 1978. 7 Weeks after the discontinuance of the molsidomine treatment, under pentobarbital anaesthesia (30 mg/kg), the animal was perfused from the abdominal aorta with 2.5 % glutaraldehyde in 0.1 M sodium cacodylate buffer (pH 7.4). Postfixation with 1 % osmium tetroxide in sodium cacodylate buffer. Embedded in Epon 812 and sectioned at 50 nm. Lead citrate and uranyl acetate. Film 548/79

prusside) or 8-bromo-cyclic GMP (1 mM) can substitute for the stimulation of climbing fibres to cause long-term depression in *rat* cerebellar slices. Exposure to cerebellar granule cells to the •NO donor *S*-nitroso-*N*-acetylpenicillamine induced oxidative stress, which was characterised by the accumulation of cytosolic and mitochondrial reactive oxygen intermediates, the increase in the extracellular hydrogen peroxide level, and the formation of lipid peroxidation products (Wei et al. 2000).

In primary cerebellar cultures, suppression of glutamatergic neurotransmission resulted in a drastic up-regulation of NOS expression by granule cells as shown by histochemistry and semi-quantitative immunoblotting (Baader et al. 1994). In contrast, interference with the GABA-ergic neurotransmission had only a minor effect on NOS expression.

N-Methyl-D-aspartate (100 µM), noradrenaline (100 µM), and the Ca^{2+} ionophore A23187 (10 µM) decreased cGMP generated in response to direct stimulation of soluble guanylyl cyclase by *S*-nitroso-*N*-acetylpenicillamine in both granular neurones and astrocytes in primary cultures from cerebella of 7-day-old Sprague-Dawley *rats* (Baltrons et al. 1997). This effect required extracellular Ca^{2+} and was prevented by the calmodulin inhibitor W7 (100 µM). Membrane depolarisation, manipulation of the Na^+ gradient, and intracellular Ca^{2+} mobilisation also decreased NO donor-induced cGMP formation in granule cells. In astrocytes Ca^{2+} entry additionally down-regulated cGMP generated by stimulation of the particulate guanylyl cyclase by atrial natriuretic peptide.

L-Type calcium channel blockers, verapamil (5 mg/kg i.p.) and nifedipine (5 mg/kg i.p.) enhanced L-2-propionic acid (750 mg/kg per os)-induced *rat* granule cell necrosis, assessed by measuring the degree of L-2-propionic acid-induced reductions aspartate concentrations, increases in cerebellar

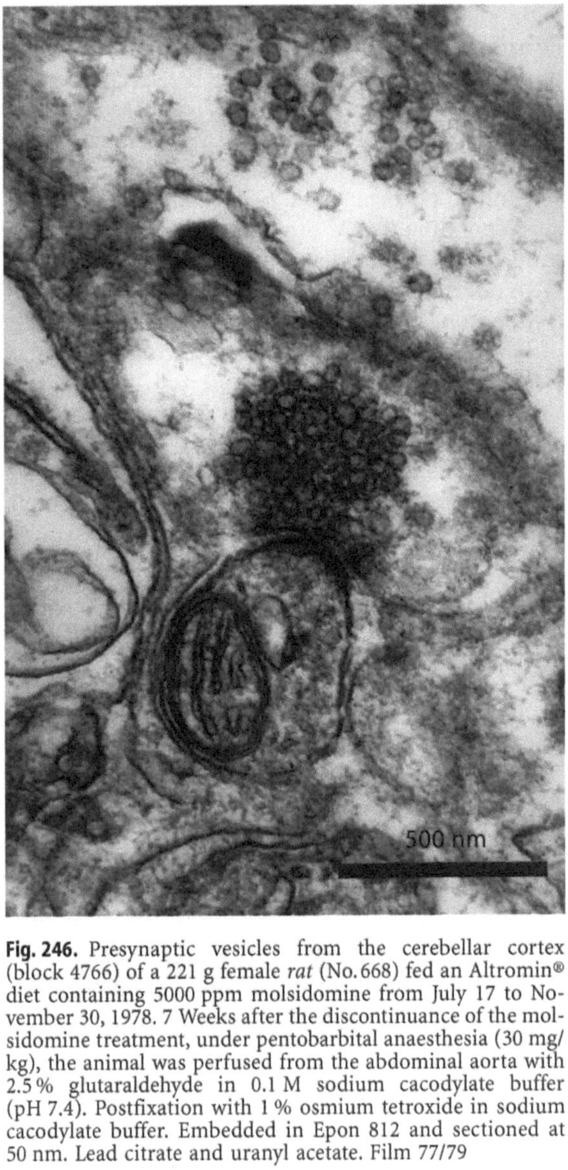

Fig. 246. Presynaptic vesicles from the cerebellar cortex (block 4766) of a 221 g female *rat* (No. 668) fed an Altromin® diet containing 5000 ppm molsidomine from July 17 to November 30, 1978. 7 Weeks after the discontinuance of the molsidomine treatment, under pentobarbital anaesthesia (30 mg/kg), the animal was perfused from the abdominal aorta with 2.5 % glutaraldehyde in 0.1 M sodium cacodylate buffer (pH 7.4). Postfixation with 1 % osmium tetroxide in sodium cacodylate buffer. Embedded in Epon 812 and sectioned at 50 nm. Lead citrate and uranyl acetate. Film 77/79

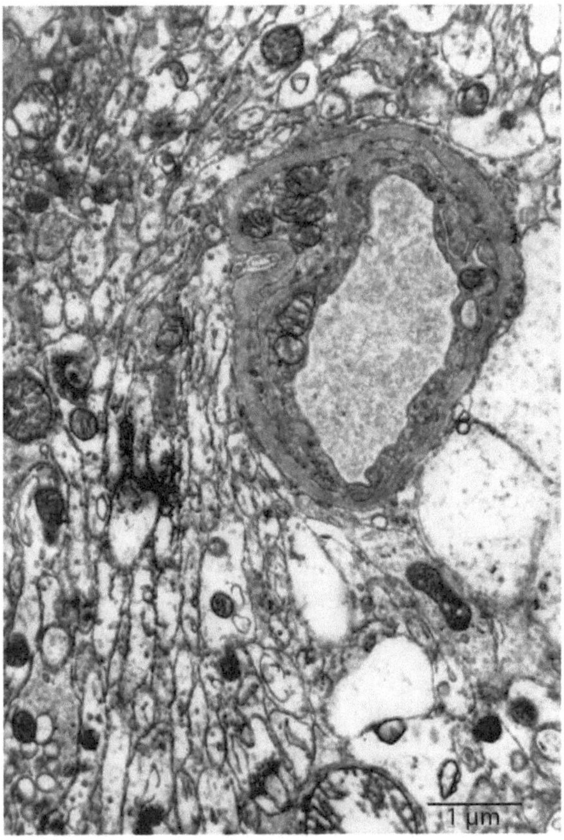

Fig. 247. Capillary with a process of a pericyte generally surrounded by a prominent basement membrane. Molecular layer of the cerebellar cortex (block 4766) of a 221 g female *rat* (No. 668) fed an Altromin® diet containing 5000 ppm molsidomine from July 17 to November 30, 1978. 7 Weeks after the discontinuance of the molsidomine treatment, under pentobarbital anaesthesia (30 mg/kg), the animal was perfused from the abdominal aorta with 2.5 % glutaraldehyde in 0.1 M sodium cacodylate buffer (pH 7.4). Postfixation with 1 % osmium tetroxide in sodium cacodylate buffer. Embedded in Epon 812 and sectioned at 50 nm. Lead citrate and uranyl acetate. Film 72/79

glycine concentrations and the development of cerebellar oedema (WIDDOWSON et al. 1997).

Addition of either insulin-like growth factor I or insulin to the serum-free medium prevented *rat* cerebellar granule cells from developing sensitivity to kainate neurotoxicity whereas brain-derived neurotrophic factor, neurotrophin-3, neurotrophin-4, and nerve growth factor did not (LESKI et al. 2000). The phosphatidylinositol 3-kinase inhibitors wortmannin (10–100 nM) and LY 294002 (0.3–1 µM) abolished the protection afforded by IGF-I.

In *rat* cerebellar neurones treated *in vitro* with 0.8 mM L-buthionine-[*S*,*R*]-sulphoxinine for 48 h, the resulting collapse of the neurites and the lysis of the cell bodies the oxidative phenotype could be reverted nearly to completion with 1 µM dodecylaniline or 100 nM iminostilbene (MOOSMANN et al. 2001).

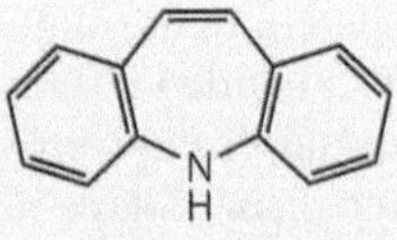

Iminostilbene [276]

15.1.1.11 Vestibular Nucleus

In the *guinea pig*, pre-treatment with the NOS inhibitor, N^{ω}-nitro-L-arginine methyl ester (100 mM) administered by s.c. osmotic minipump for 50 h resulted in a significant decrease of spontaneous nystagmus frequency after unilateral surgical destruction of the labyrinthine receptors (Paterson et al. 2000). Analysis of NOS activity at 50 h after unilateral surgical destruction of the labyrinth + L-NAME-pre-treatment showed that 100 mM L-NAME resulted in a significant decrease in NOS activity in the contralateral medial vestibular nucleus/praepositus hypoglossi ($P < 0.05$) and that NOS activity in the ipsilateral medial vestibular nucleus/praepositus hypoglossi was not significantly affected.

15.1.1.12 Nucleus Tractus Solitarii

Increased extracellular concentrations of substance P were measured during hypoxia (9 % O_2, at three 5-min intervals in one 30-min period) induced in artificially ventilated *cats* (Lindefors et al. 1986).

15.1.1.13 Spinal Cord

In the *cat*, one hour after 50 min of asphyxiation induced by increasing the pressure in the dural cavity above the blood pressure, the cytoplasm of the motoneurones started to disintegrate (Khattab 1967, Van Harreveld and Khattab 1967). The lysosomes were greatly enlarged, often clumped together, and had lost their membranes. In the close vicinity of these bodies there were stained granules which might be glycogen. The enlarged lysosomes seemed to pour their contents into the surrounding cytoplasm due to disruption of their membranes. Degenerating mitochondria frequently lay close to or were attached to these enlarged lysosomes.

Abdominal aorta occlusion for 20 min and recirculation after 6 h to 4 days induced changes in the *rabbit*'s lumbar spinal cord (Fercáková et al. 1991). Neuronal changes ranged from chromatolysis to disintegration of motoneurons and neuropil vacuolation. After 6 h recirculation, electron microscopy showed a marked decrease in endoplasmatic reticulum in motoneurons. Within 12 h necrobiotic changes included vacuolation of cytoplasm, plasmalemmal disruption and degeneration of synapses. After 1 or 2 days survival extensive spongy lesions reaching central part of the ventral horns and intermediate zone were found in the light microscope. Using the silver method of Nauta and Gygax (1954) to impregnate degenerating axons, 1 h, 6 h, and 12 h after reoxygenation, respectively, the majority of Nauta positive neurones and their dendrites were localised in the deep part of the dorsal horn (Saganová and Maršala 1991). No sign of ischaemic damage was seen within laminae I-III. After 40 min and 90 min aortal ligation, respectively, disturbances of calcium homeostasis were observed only after recirculation for 60 min and 24h, respectively, as localised by Ca^{2+}-precipitation with potassium pyroantimonate (Jalc et al. 1991).

Graded postischaemic reoxygenation (10 % O_2 corresponding a pO_2 of 48 ± 12 mm Hg) evidently protected the spinal cord even after 2 days survival (Fercáková et al. 1992). In ultrathin sections, only few neurones were detected displaying high density of cytoplasm and containing a large number of neurofilaments and lysosomes scattered among Nissl bodies. Oedema of dendrites and astrocytes occurring in close vicinity of these neurones was quite evident.

Protein carbonyl content – a marker of protein oxidation – significantly increased at 3–9 h after impact spinal cord injury in male Sprague-Dawley *rats* and the ratio 8-hydroxy-2-deoxyguanosine/deoxyguanosine – an indicator of DNA oxidation – was significantly higher at 3–6 h postinjury in the injured cords than in the sham controls (Leski et al. 2001). Neurofilament protein light chain (NFP-68) was gradually degraded in nerve fibres, neurone bodies, and large dendrites following spinal cord injury. A mixture of Mn(III) tetrakis (4-benzoic acid) porphyrin (10 mg/kg) – a novel superoxide dismutase mimetic – and nitro-L-arginine (1 mg/kg) injected intraperitoneally, increased NFP-68 immunoreactivity and the numbers of NFP-positive nerve fibres after spinal cord injury, correlating NFP degradation in spinal cord injury to free radical-triggered oxidative damage.

15.1.1.14 Ependyma

Numerous reactive microperoxisomes having an average diameter of 150 nm are found in the ependymal cells bordering the third ventricle of the ventral hypothalamus of the *rat* (McKenna et al. 1976). Ependymal cells of the Mongolian *gerbil* which line the third ventricle displayed nitric oxide synthase-immunoreactivity, NADPH-diaphorase activity as well as calmodulin immunostaining (Krstic and Nicolas 1995).

The subcommissural organ, an ependymal differentiation of the Mongolian *gerbil*, *Meriones unguiculatus*, exhibited a strong NADPH-diaphorase reaction in all its ependymocytes (Krstic and Nicolas 1995). The nuclei were clearly negative, while some hypendymal cells and the basal ependymal processes, forming a palisade-like structure, linking the subcommissural organ to leptomeninges, displayed a notable NADPH-diaphorase activity. The

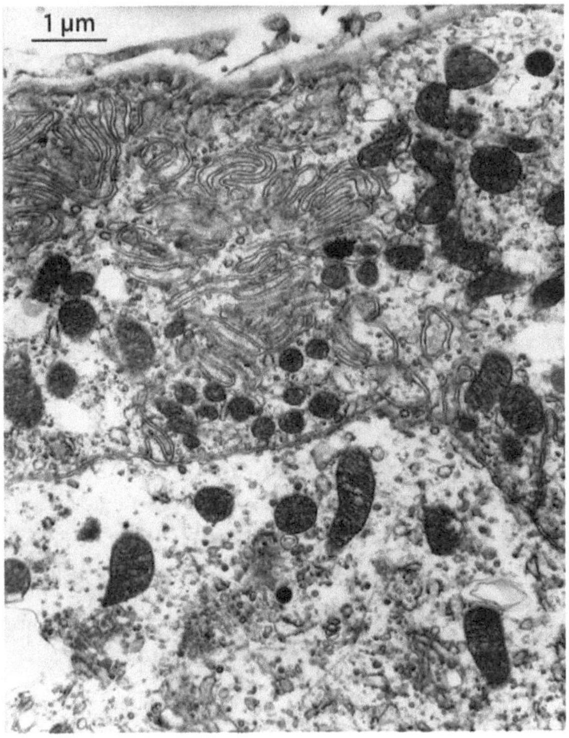

Fig. 248. Reoxygenated ependyma (block 641) from a 284 g male Wistar *rat* exposed to an atmosphere containing only 5 % oxygen for 30 min. Half an hour thereafter, under pentobarbital anaesthesia (30 mg/kg), the animal was perfused from the abdominal aorta with 2.5 % glutaraldehyde in 0.1 M sodium cacodylate buffer (pH 7.4). Postfixation with 1 % osmium tetroxide in sodium cacodylate buffer (pH 7.4). Embedded in Epon 812 and sectioned at 50 nm. Lead citrate and uranyl acetate. Plate 3142

subcommissural organ of the Mongolian *gerbil* did not exhibit nitric oxide synthase-immunoreactivity, which, nevertheless, was clearly expressed in neighbouring verve cells.

15.1.1.15 Choroid Plexus

The epithelium of the choroid plexus with its basal labyrinth (Fig. 249) is similar in many of its ultrastructural features to the epithelium of the renal tubule, and to other cellular structures that participate in transport.

Macrophage-like cells which morphologically resemble epiplexus cells and which are very abundant within the connective tissue stroma and also between the epithelial cells of all choroid plexuses of *Macaca mulatta*, are thought to be monocytes (MERKER 1972).

Very strong immunoreactivity for CYP2D was observed in the endothelial cells of the choroid plexus of the Wistar *rat* (MIKSYS et al. 2000).

LAGRANGE et al. (1994) found significant amounts of superoxide anion production (measured by spectrometric recording of reduction of acetylates ferricytochrome *c* at 550 nm) in homogenates of *rats'* choroid plexus incubated with 2,5-dimethyl benzoquinone (109.3 nM/min/NADPH-cytochrome P450 reductase activity) or menadione (100 nM/min/NADPH-cytochrome P450 reductase activity).

Melatonin can have stimulatory actions *in vivo* on the choroidal ependymal cells in golden *hamsters:* increase in cytoplasmic volume, increase in nuclear volume, increase in mitochondrial area or volume per cell, and increase in length of apical (ventricular) microvilli (DECKER and QUAY 1982). All of these ultrastructural effects were statistically

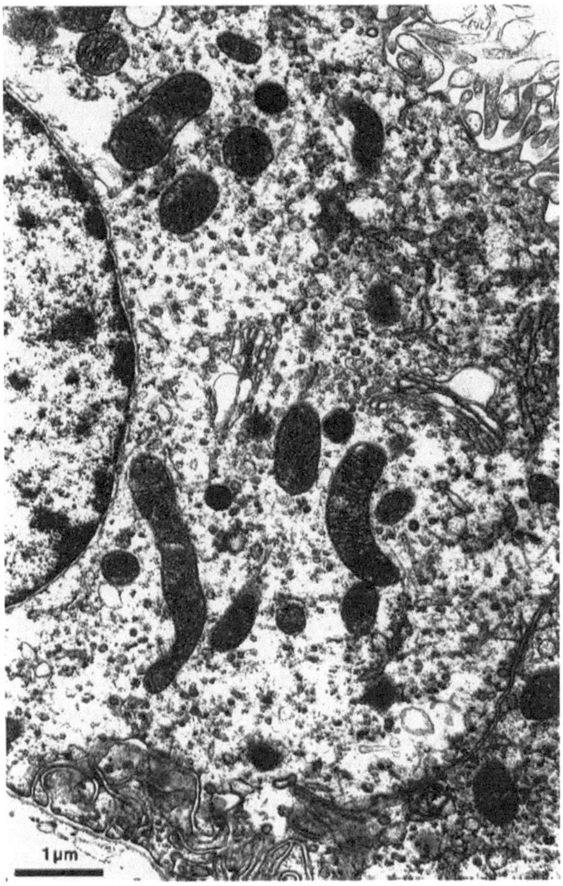

Fig. 249. Reoxygenation damage to the choroid plexus epithelium (block 641) from a 284 g male Wistar *rat* exposed to an atmosphere containing only 5 % oxygen for 30 min. Half an hour thereafter, under pentobarbital anaesthesia (30 mg/kg), the animal was perfused from the abdominal aorta with 2.5 % glutaraldehyde in 0.1 M sodium cacodylate buffer (pH 7.4). Postfixation with 1 % osmium tetroxide in sodium cacodylate buffer (pH 7.4). Embedded in Epon 812 and sectioned at 50 nm. Lead citrate and uranyl acetate. Plate 3141. – The apical surface of the epithelium is composed of tightly packed, irregular microvilli. Basal epithelial infoldings

significant, but only in the choroidal ependymal cells of the lateral ventricles as compared with those of ventricles III and IV.

While the enzyme activities of the γ-glutamyl cycle were widely distributed in the brain, they are present in much higher concentrations in the choroid plexus than in other parts of the brain. The activities observed are of about the same order of magnitude as found in the kidney (TATE et al. 1973). SCHILLER (1974, 1975) demonstrated free sulphydryl groups as in reduced glutathione by 1-(4-chloromercury-phenylazo)-2-naphthol using a Mach-Zehnder microinterferometer.

15.1.1.16 Retina

The high level of polyunsaturated fatty acids of the rod outer segment membranes probably insures a fluid membrane at all times, but makes the membranes particularly vulnerable to oxidative processes. However, bovine rod outer segments possess unusually high concentrations of α-tocopherol (DILLEY and MCCONNELL 1970, FARNSWOTH and DRATZ 1976). In addition, there is a high level of superoxide dismutase (HALL and HALL 1975).

Electron spin resonance spin trapping analysis of the signals obtained from the perfused *rabbit* retina showed that hydroxyl radical adducts were generated during the ischaemic period with no further increase at reperfusion (MULLER et al. 1997). Electron spin resonance detection of the ascorbyl free radical/dimethyl sulfoxide complex in the retina confirmed that an oxidative stress occurred under ischaemia and was not increased under reperfusion.

The flux of hydroxyl radicals, as measured directly by conversion of salicylate to 2,3- and 2,5-dihydroxybenzoic acid (formula [177]), was significantly lower in gallium-desferrioxamine-treated retinas of *cat* eyes subjected to 90 min retinal ischaemia followed by 5 min of reperfusion (BANIN et al. 2000). Gallium-desferrioxamine caused a significant reduction, by 2.56-fold, in lipid peroxidation, as reflected by levels of malondialdehyde. Acorbic acid, a natural antioxidant present in the retina, was severely depleted in untreated eyes. In contrast, in gallium-desferrioxamine-treated eyes, levels were 10 times higher than the control.

In organ culture *rabbit* retinas deprived of oxygen and glucose for only 3 min showed generalised swelling of mitochondria and alteration in the structure of the synapses with loss of synaptic vesicles (WEBSTER and AMES 1965). Extending the combined deprivation caused further mitochondrial swelling and synaptic changes and also led to progressive swelling of the Golgi membranes and the granular endoplasmic reticulum. All these changes were almost completely reversible for up to 20 min but were irreversible after 30 min, at which time multiple discontinuities had appeared in cell and organelle membranes. Anoxia alone produced alterations similar to those found after somawhat shorter periods of the combined deprivation, whereas glucose withdrawal produced only minor changes.

Neural retinal cells in culture were protected from experimental ischaemia/reoxygenation (8 h ischaemia followed by 16 h reoxygenation) when melatonin was in the growth medium (CAZEVIEILLE and OSBOURNE 1997). Since glutamate (formula [273]) was released in excess from these cells during ischaemia and the resulting damage was believed to be related to free radical production after the interaction of the neurotransmitter with its receptor, CAZEVIEILLE and OSBOURNE (1997) presumed that melatonin preserved retinal neuronal integrity due to its free radical scavenging properties.

A model of *in vitro* cell oxidative stress in *bovine* retinal pigment epithelium exposed to ischaemia-like condition obtained by interference with glucose utilisation through both oxidative phosphorylation and glycolysis resulted in a statistically significant decrease of the intracellular ATP levels, which reflects a bioenergetic decline to that associated with mitochondrial damage or loss in normal post-mitotic cells ageing *in vitro* (PALMERO et al. 2000). 300 μM ascorbic acid or 10 mM *N*-acetylcysteine protected cells from this type of oxygen stress.

In the turtle, *Pseudemys scripta elegans*, two morphologically different types of amacrine cells were nNOS/glycine-immunoreactive and three types are nNOS/γ-aminobutyric acid-immunoreactive (HAVERKAMP et al. 2000).

Nitric oxide ($^{\bullet}$NO) stabilises whereas nitrosonium (NO^+) enhances filopodial outgrowth by *rat* retinal ganglion cells *in vitro* (CHEUNG et al. 2000).

To investigate the effect of two doses of acetylsalicylic acid in prevention of retinal ischaemia in stretozotocin-diabetic *rats*, DE LA CRUZ et al. (2002) compared nondiabetic *rats* and diabetic *rats* after 1, 2, and 3 months of diabetes, and in diabetic *rats* treated with 2 mg or 10 mg acetylsalicylic acid per kg body weight per day p.o. from the first day of diabetes. In diabetic rats $^{\bullet}$NO production increased after 2 and 3 months of treatment to levels seen in nondiabetic *rats*.

Lipofuscin harbours two unusual retinoids, the lipophilic cations *N*-retinyl-*N*-retinylidene ethanolamine and its isoform, iso-*N*-retinyl-*N*-retinylidene ethanolamine, first isolated from the eyes of old individuals (ELDRED and LASKY 1993, PARISH et al. 1998). SUTER et al. (2000) showed that *N*-retinyl-*N*-retinylidene ethanolamine induced apoptosis in retinal pigment epithelial and other cells at concentrations found in *human* retina. Apoptosis was accompanied by the appearance of the proapoptotic proteins cytochrome *c* and apoptosis-inducing factor in the cytoplasm and the nucleus. Biochemical examinations showed that *N*-retinyl-*N*-retinylidene ethanolamine specifically targets cytochrome oxidase (SHABAN et al. 2001). In the dark, inhibition is overcome by cardiolipin or other acidic phospholipids. With illumination, inhibition is stronger, becomes complete with prolonged exposure, and is than no longer abrogated by cardiolipin. Cardiolipin effectively displaces *N*-retinyl-*N*-retinylidene ethanolamine from cytochrome *c* oxidase, suggesting noncovalent binding of *N*-retinyl-*N*-retinylidene to the enzyme.

15.1.1.17 Olfactory Mucosa

In the olfactory mucosa of Wistar *rats* exposed four times a day to 175 ml cigarette smoke for 5, 10, and 15 min, respectively, ORTUG and ÖZBEK (2001) found intraepithelial inflammatory cells and especially deep invaginations at the nuclear membrane of supporting cells. Extension between extracellular space, cytoplasmic protrusions in the apical surface of the supporting cells, atrophy of the microvilli and olfactory neurone cilia and numerous electron-dense granular structures, lysosome-like structures were seen in relation to the times of exposure.

15.1.1.18 Trigeminal Ganglion

In streptozotocin-induced diabetic *rats* the number of NADPH-diaphorase-positive neurones significantly decreased compared with controls (RODELLA et al. 2000). Insulin treatment prevented the decreased nociceptive threshold and reduction of the number of NADPH-diaphorase-positive neurones. In the trigeminal ganglion of the *cat*, a large number of ganglion cells, dispersed without any regional preference, were glutamate immunoreactive (STOYANOVA and LAZAROV 2001). Most of them belonged to the medium- or large-sized neuronal subpopulations.

15.1.1.19 Spiral Ganglion

After a global brain ischemia of 30 min followed by reperfusion for 3 and 6 days, a progressive accumulation of lipofuscin granules in type I neurones and in satellite cells of the *rat* spiral ganglion occurred (LUKÁN et al. 1996). 6 Days after ischaemia numerous neurones with peripheral cytoplasmic vacuolisation were seen. The authors supposed the structural changes to be due to free radical generation.

15.1.1.20 Spinal Ganglia

In *rat* spinal ganglia both neuronal and endothelial (the latter predominantly in the smooth endoplasmic reticulum associated with the mitochondria) nitric oxide synthases were localised by their immunoreactivities (HENRICH et al. 2001). In vibratome slices of *rat* spinal ganglia hypoxia (1 % O_2) stimulated nitric oxide formation while the production of reactive oxygen species was not increased.

5.1.1.21 Neuromuscular Junction

After a two-hour ischaemia applied to one of the *rat* hind limbs reperfusion injury lesions were confined to the nerve terminals on the presynaptic side of the neuromuscular junction affecting the mitochondria primarily, the synaptic vesicles, the presynaptic membrane, and finally a large number of terminals degenerated (TÖMBÖL et al. 2000). The Schwann cells were activated, as well as the macrophages. Recovery started 1 day after reperfusion and free synaptic surfaces were found 4 weeks following reperfusion.

15.1.2 Pituitary

15.1.2.1 Pars Nervosa

The posterior lobe is composed of nerve fibres and endings of hypothalamic neurones (Fig. 250), capillaries (Figs. 251, 254, and 255) and characteristic astroglial cells, the pituicytes. In rapidly frozen, freeze-substituted neural lobes of male albino *rats*, pituicytes occupied close to 60 % of the basal lamina at the neurohaemal contact zone, while axons occupied about 20 % (TIAN et al. 1991). TAKEI et al. (1980) classified the *human* pituicytes into five classes based on their ultrastructural features:

1) major pituicytes
2) dark pituicytes
3) ependymal pituicytes
4) oncocytic pituicytes
5) granular pituicytes.

In the *rat*, HARTMANN (1958) distinguished three types of pituicytes.

In tissue culture, the glial cells derived from posterior lobes of *dog* and *rat* had the same morphological characteristics as protoplasmic astrocytes derived from various parts of the brain (HILD 1954). Oligodendroglia cells were sometimes observed in cultures of posterior lobe; macrophages, which according to the investigations of COSTERO (1930) are probably identical with microglial, are a very common observation.

Oxidative enzymes (α-glycerophosphate dehydrogenase, succinic dehydrogenase, and reduced diphospho- and triphosphopyridine nucleotide diaphorases) gave a very intense nitroblue tetrazolium (formula [18]) reaction in the matrix of the posterior lobe from mature female Sprague-Dawley *rats* (FAND and WATTENBERG 1963).

Nitric oxide synthase (EC 1.14.13.39) immunostaining was positive, especially within dilated axon terminals (Herring bodies) of *man* (LLOYD et al. 1995).

Activity increased in the posterior pituitary of salt-loaded *rats* (KADOWAKI et al. 1994). N^{ω}0-Nitroarginine significantly inhibited NOS activity in the posterior pituitary in a dose-dependent manner, but did not influence NOS mRNA levels. Two percent salt loading for 3 or 4 days significantly depleted the contents of both arginine vasopressin and oxytocin in the posterior pituitary, and simultaneous treatment with daily injections of N^{ω}-nitroarginine at a dose of 10 mg/kg significantly enhanced the depletion of both hormones.

Particulate **guanylate cyclase** (EC 4.6.1.2) activity could not be demonstrated in samples incubated in basal medium without *rat* atrial natriuretic factor or *porcine* brain natriuretic peptide or samples incubated in basal medium containing sodium nitroprusside (RAMBOTTI et al. 1994).

Muscarinic cholinergic receptors were characterised by atropine-sensitive binding of L-[^{3}H]-quinuclidinyl benzilate to membrane resuspensions of *sheep* posterior pituitary (TOLLIVER et al. 1981). In *rat* neurointermediate lobes, superior cervical ganglionectomy had no demonstrable effect on [^{3}H]-quinuclidinyl benzilate binding.

The demonstration of **GABA$_A$ receptors** coupled to Cl^- channels in the posterior pituitary (ZHANG and JACKSON 1993) prevailed ZHANG and JACKSON (1995) by activation of these receptors and gating the Cl^- channels to alter membrane potential, action potentials, and the status of voltage-gated channels. Their results supported a depolarisation block mechanism in the inhibition of secretion by GABA.

After repetitive **hypoxia** (5 % O_2 for 30 min each) and reoxygenation, the neurosecretory pathway of the *rat* showed a significant decrease in pituitary neurophysin as visualised by Gömöri's chromium-haematoxylin-phloxine (SCHILLER 1974). **Whole body X-ray exposure** (1000 r) – in contrast to an isolated irradiation of the head – induced depletion of Gömöri-positive material in female Wistar *rats* (KRATZSCH et al. 1962).

Chlorpromazine (20 mg/kg × day) inhibited the depletion of pituitary vasopressin in water deprived male Holtzman *rats* (MOSES 1964).

15.1.2.1.1
Mitochondria

Reoxygenation damage showing cristolysis, clearing of the matrix and depletion of intramitochondrial granules could not be prevented by the drugs tested, neither by the nootropic piracetam, nor by carbocromene, a coumarin devoid of the two hyroxyl groups necessary for scavenging reactive oxygen species, nor by the antihistaminergic and antiserotoninergic *N*-benzhydryl-*N'*-*p*-hydroxybenzylpiperazine.

15.1.2.1.2
Herring Bodies

Masses of hyaline or granular character described by HERRING (1908) in the *dog* and *cat* neurohypophysis and subsequently named after him were investigated in a variety of *mammalian* and other species (ROMEIS 1940, BARGMANN 1950, SCHARRER and SCHARRER 1954, HANSTRÖM 1954, DELLMANN 1962, CHRIST 1966). SLOPER (1955) mentioned a faint sudanophilia of some Herring bodies. With the periodic acid-Schiff technique, which demonstrates unsaturated lipids as well as carbohydrate, there was some reaction in occasional Herring bodies. With one modification of the periodic acid-Schiff technique (SCHIEBLER 1952), Herring bodies were more evident. In some Herring bodies formazan was produced from blue tetrazolium chloride and 2,5-diphenyl-3-4-styrylphenyl tetrazolium-chloride.

Under the electron microscope, DELLMANN and RODRÍGUEZ (1970) distinguished three types of Herring bodies: normal, degenerating, and regenerating. Type II is shown in reoxygenation of *rats* pre-treated with carbocromene (Fig. 252) and *N*-benzhydryl-*N'*-*p*-hydroxybenzylpiperazine; respectively, and type III in reoxygenation after piracetam pre-treatment.

Multilamellar bodies, as seen in the neurohypophysis of the *hedgehog* (HOLMES and KIERNAN 1964), the *rabbit* and the *rainbow trout* (LEDERIS 1964), were also observed in the *human* neurohypophysis (LEDERIS 1965). The multilayered, often concentric lamellae seem to develop in some larger nerve swellings. In such swellings only few mitochondria were seen but numerous small vesicles were usually present. LEDERIS (1964) showed in the *rainbow trout* that the small vesicles apparently arise as a result of disintegration of the osmiophilic lamellae.

15.1.2.1.3 Autophagy of Neurosecretory Granules

Autophagy of neurosecretory granules (Fig. 253) occurs in many instances, including normal condition (PILGRIM 1970, WITTKOWSKI 1970, DELLMANN 1973). Autophagy appeared to be the predominant process for granulolysis and might be considered as an aspect of general turnover of cell constituents, related to the sudden regression of hyperactivity-induced hypertrophy (BOUDIER and PICARD 1976)

15.1.2.1.4 Endothelium

One of the earliest important events after reoxygenation of the ischaemic or anoxic capillaries is a significant degree of endothelial dysfunction manifested by vacuolation (Figs. 254, 255) as described by POCHE et al. (1967) and POCHE (1977) in the anoxic *rat* myocardium.

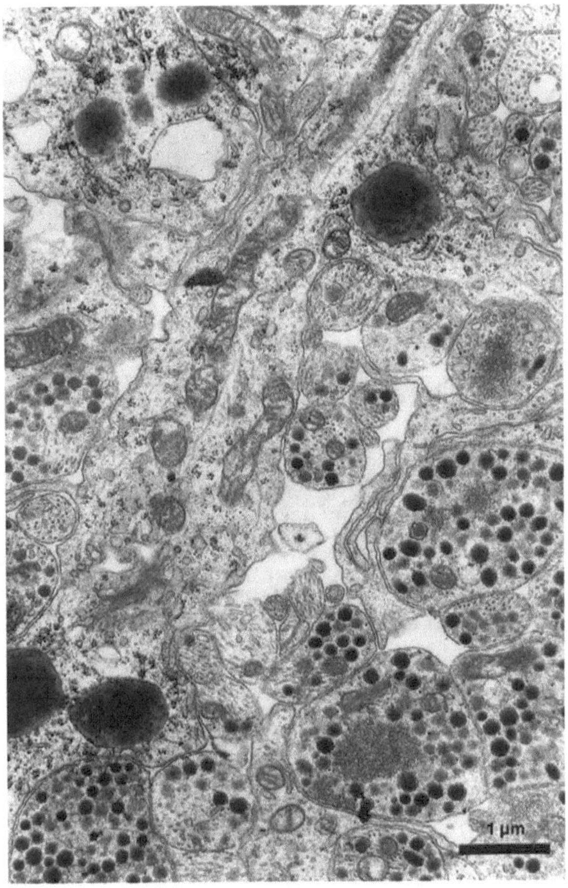

Fig. 250. Reoxygenation-induced mitochondrial damage in the neuropituitary secretory neurones (block 1421) of a *rat* (No. 14) treated for 8 consecutive days with intraperitoneal injection of 1.5 ml Tyrode's solution per kg body weight × day. On the last 5 days of experimentation the animal was exposed to an atmosphere containing only 5 % oxygen for 30 min. On August 6, 1976 half an hour after the last exposure, under pentobarbital anaesthesia (30 mg/kg), the animal was perfused from the abdominal aorta with 2.5 % glutaraldehyde in 0.1 M sodium cacodylate buffer (pH 7.4). Postfixation with 1 % osmium tetroxide in sodium cacodylate buffer. Embedded in Epon 812 and sectioned at 50 nm. Lead citrate and uranyl acetate. Plate 3477

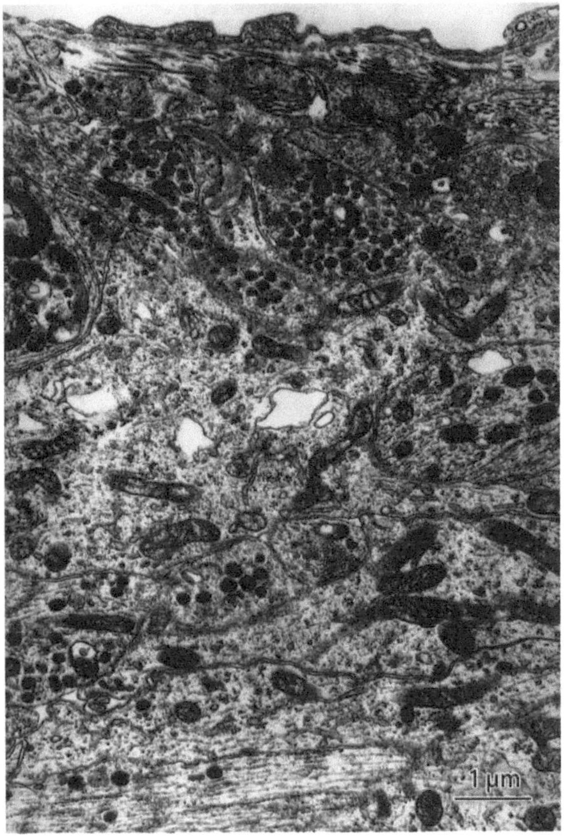

Fig. 251. Scantiness of neurosecretory granules in some neurones (bottom) in the pituitary pars nervosa (block 1330) of a *rat* (No. 11) treated for 8 consecutive days with intraperitoneal injection of 15 mg carbocromene per kg body weight × day. On the last 5 days of experimentation the animal was exposed to an atmosphere containing only 5 % oxygen for 30 min. On August 5, 1976, half an hour after the last exposure under pentobarbital anaesthesia (30 mg/kg), the animal was perfused from the abdominal aorta with 2.5 % glutaraldehyde in 0.1 M sodium cacodylate buffer (pH 7.4). Postfixation with 1 % osmium tetroxide in sodium cacodylate buffer. Embedded in Epon 812 and sectioned at 50 nm. Lead citrate and uranyl acetate. Plate 3462

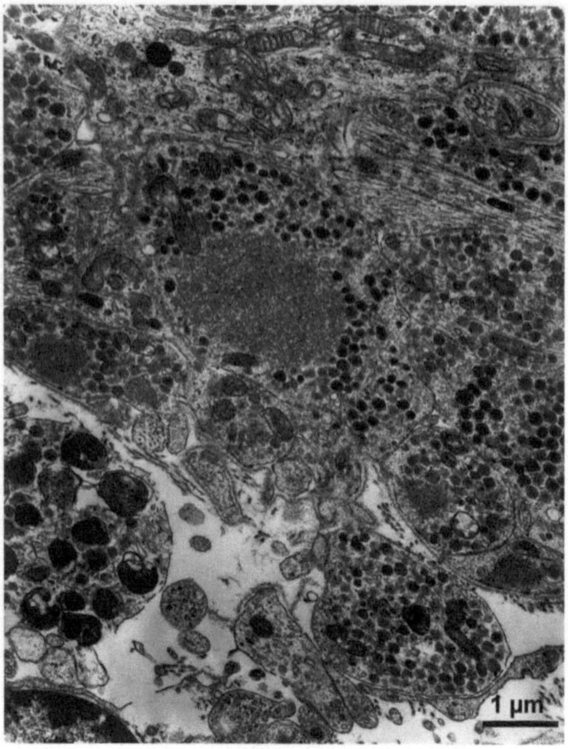

Fig. 252. A type II Herring body showing an accumulation of dense bodies and myelinated bodies. Pituitary pars nervosa (block 1330) of a *rat* (No. 11) treated for 8 consecutive days with intraperitoneal injection of 15 mg carbocromene per kg body weight × day. On the last 5 days of experimentation the animal was exposed to an atmosphere containing only 5 % oxygen for 30 min. On August 5, 1976, half an hour after the last exposure under pentobarbital anaesthesia (30 mg/kg), the animal was perfused from the abdominal aorta with 2.5 % glutaraldehyde in 0.1 M sodium cacodylate buffer (pH 7.4). Postfixation with 1 % osmium tetroxide in sodium cacodylate buffer. Embedded in Epon 812 and sectioned at 50 nm. Lead citrate and uranyl acetate. Plate 3463

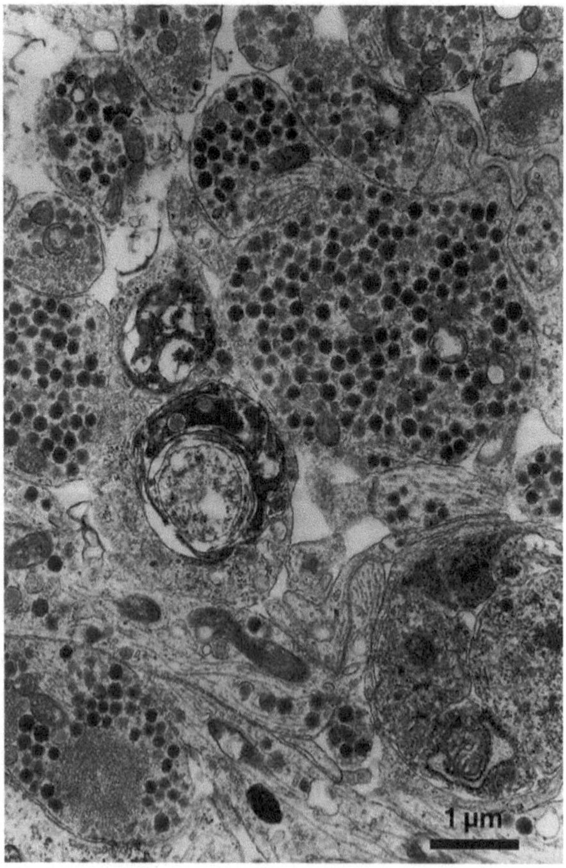

Fig. 253. Autophagic vacuoles in the intermittent hypoxic neurohypophyseal secretory neurones (block 1330) of a *rat* (No. 11) treated for 8 consecutive days with intraperitoneal injection of 15 mg carbocromene per kg body weight × day. On the last 5 days of experimentation the animal was exposed to an atmosphere containing only 5 % oxygen for 30 min. On August 5, 1976 half an hour after the last exposure, under pentobarbital anaesthesia (30 mg/kg), the animal was perfused from the abdominal aorta with 2.5 % glutaraldehyde in 0.1 M sodium cacodylate buffer (pH 7.4). Postfixation with 1 % osmium tetroxide in sodium cacodylate buffer. Embedded in Epon 812 and sectioned at 50 nm. Lead citrate and uranyl acetate. Plate 3461

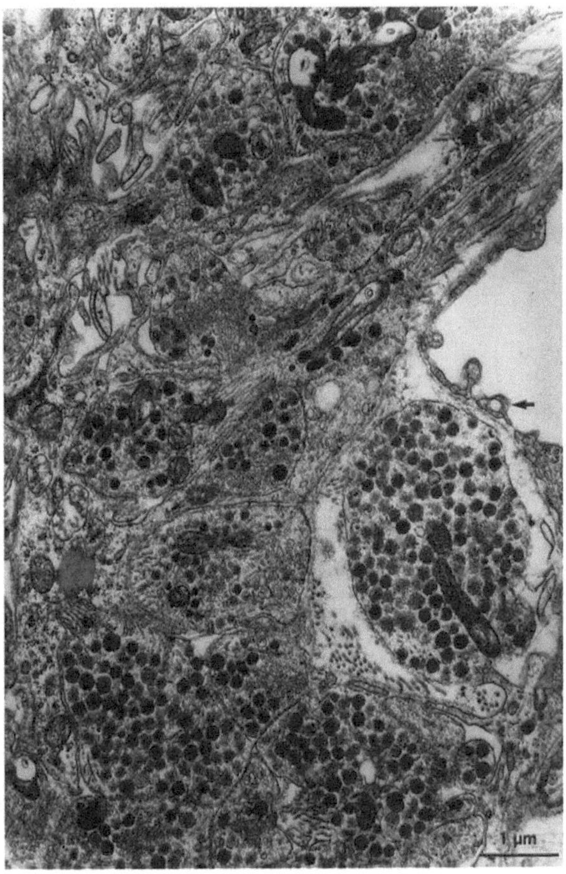

Fig. 254. Vacuoles (arrow) in the vascular endothelium in the intermittent hypoxic and reoxygenated pituitary pars nervosa (block 1330) of a *rat* (No. 11) treated for 8 consecutive days with intraperitoneal injection of 15 mg carbocromene per kg body weight × day. On the 5 last days of experimentation the animal was exposed to an atmosphere containing only 5 % oxygen for 30 min. On August 5, 1976 half an hour after the last exposure, under pentobarbital anaesthesia (30 mg/kg), the animal was perfused from the abdominal aorta with 2.5 % glutaraldehyde in 0.1 M sodium cacodylate buffer (pH 7.4). Postfixation with 1 % osmium tetroxide in sodium cacodylate buffer. Embedded in Epon 812 and sectioned at 50 nm. Lead citrate and uranyl acetate. Plate 3460

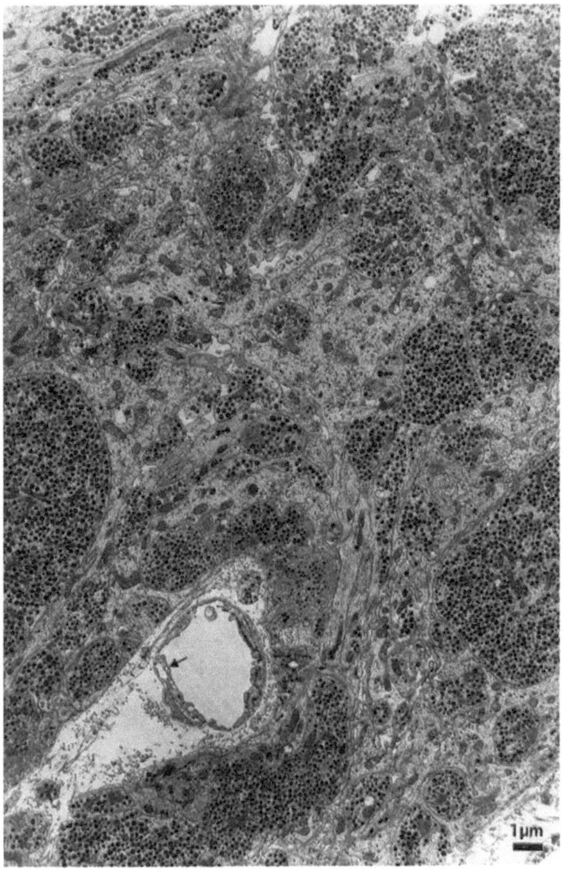

Fig. 255. Vacuoles (arrow) in the vascular endothelium in the intermittent hypoxic and reoxygenated pituitary pars nervosa (block 1451) of a *rat* (No. 15) treated for 8 consecutive days with intraperitoneal injection of 15 mg carbocromene per kg body weight × day. On the 5 last days of experimentation the animal was exposed to an atmosphere containing only 5 % oxygen for 30 min. On August 6, 1976 half an hour after the last exposure under pentobarbital anaesthesia (30 mg/kg), the animal was perfused from the abdominal aorta with 2.5 % glutaraldehyde in 0.1 M sodium cacodylate buffer (pH 7.4). Postfixation with 1 % osmium tetroxide in sodium cacodylate buffer. Embedded in Epon 812 and sectioned at 50 nm. Lead citrate and uranyl acetate. Plate 3479

15.1.2.2 Pars Intermedia

The cells of the pars intermedia are round to ovoid and contain many electron-dense to more lucent secretory granules with an average diameter of 200 nm (RODIN 1974). Rough endoplasmic reticulum is not prominent. Sometimes electron-dense secretory granules are adjacent to a conspicuous Golgi apparatus (Fig. 257). Using the peroxidase-labeled antibody method of NAKANE (1968, 1970, 1975) the cells react mainly with antibody against ACTH. Acid phosphatase (EC 3.1.3.2) activity was restricted to the lamellae of the Golgi apparatus, to granules associated with them, and to lysomomes (HOWE and MAXWELL 1968).

Ciliated cells are occasionally found in the junctional zone between pars intermedia and pars nervosa of the *rat* pituitary particularly in association with the ramifications of the pituitary cleft (HOWE and MAXWELL 1968).

The cells in the pars intermedia of the *rat* pituitary produce a series of biologically active peptides, i.e., α-melanocyte-stimulating hormone, CLIP, β-endorphin, β-lipotropin, and γ-MSH-related peptides. The pro-opiomelanocortin-derived peptides, especially α-MSH and β-endorphin, are suggested to be involved in learning and memory (O'DONO-

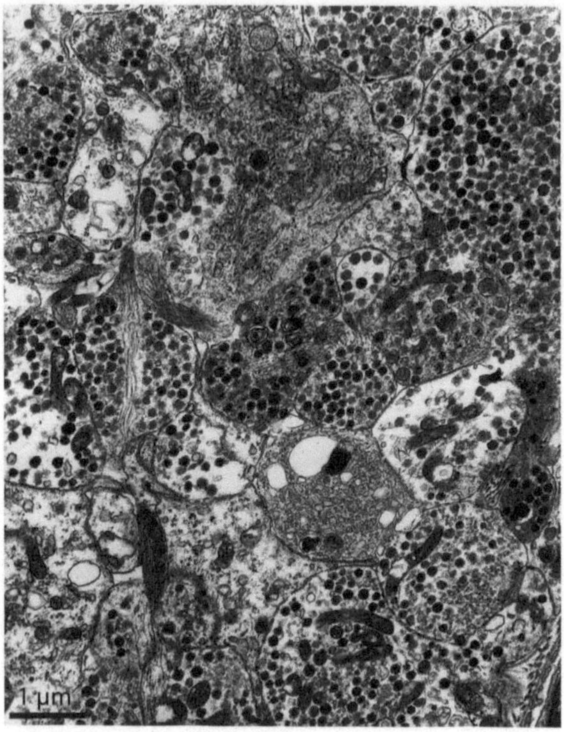

Fig. 256. Neuropituitary (block 370) from a male Wistar *rat* (No. 27/103 E) treated for 7 consecutive days with intraperitoneal injection of 5 mg *N*-benzhydryl-*N'*-p-hydroxybenzyl-piperazine HCl per kg body weight × day. After an exposure to an atmosphere containing only 5 % oxygen for 30 min, under pentobarbital anaesthesia (30 mg/kg), the animal was perfused from the abdominal aorta with 2.5 % glutaraldehyde in 0.1 M sodium cacodylate buffer (pH 7.4). Postfixation with 1 % osmium tetroxide in sodium cacodylate buffer. Embedded in Epon 812 and sectioned at 50 nm. Lead citrate and uranyl acetate. Plate 2926

Scheme. Pituitary pars intermedia: Cell types

- Granular cells
  - Melanotrophs
  - Corticotrophs
  - PAS-positive cells
- Agranular cells
  - Cleft cells
  - Folliculo-stellate cells
  - Bordering or marginal cells
- Miscellaneous cell types
  - Fibroblasts
  - Eosinophil leucocytes
  - Mast cells

HUE and DORSA 1982). RECHARDT and BÄCK (1977) and BÄCK and RECHARDT (1985) have shown degranulation of the cells in the pars intermedia of reserpine-treated *rats*.

Bromocriptine did not induce any change in the volume fraction, number or location of electron-dense secretory granules (BÄCK 1989). Instead, there was a shift toward a more homogeneous cell population containing smaller granules, the mean granule volume being reduced by ~30 %. The volume fraction of electron-lucent granules was markedly reduced, indicating a functional significance of these organelles. The volume of the Golgi apparatus was not significantly altered, but the number of condensing granules within the Golgi area was reduced. In an *in vitro* system that generated $HO^•$ from $FeSO_4$–H_2O_2, bromocriptine dose-dependently ($IC_{50} = 11.25 \pm 0.89 \times 10^{-5}$ M) quenched $HO^•$ radicals, but did not inhibit their formation (OGAWA et al. 1994).

In *rat*, *gerbil*, *hamster* and *guinea pig*, cells of the intermediate lobe were barely or at most moderately stained by the rab3A-specific antibody Cl 42.2 (REDECKER et al. 1995). In addition, this antibody stained numerous punctuate nerve terminals which were distributed throughout the lobe. In contrast, intermediate lobe cells were strongly immunoreactive when sections were incubated with rab3 antibody Cl 42.1. The rab3-positive cells could be unequivocally identified as endocrine intermediate lobe cells by virtue of their expression of proopiomelanocortin-derived peptides, i.e. α-MSH, and β-endorphin. At variance, endocrine cells of the intermediate lobe displayed no detectable immunoreactivity for synaptotagmin I.

A variety of oxidative enzymes in the pars intermedia of the *pig* have been demonstrated by HOWE and THODY (1967). The epithelial cells of the pars intermedia of the *ox* showed a strong positive reaction for cytochrome oxidase, but it was not possible to differentiate between type 1 and type II cells (RAFTERY 1969). Sections treated with 0.05 M sodium azide as inhibitor were negative. The strong reaction for the mitochondria-associated enzymes, succinic dehydrogenase and cytochrome oxidase, indicates high oxidative activity within the cells of the intermedia.

In *Xenopus laevis*, ALLAERTS et al. (2000) showed that peptide release from melanotropes is stimulated by administration of superoxide dismutase added to the superfusion medium to prevent scavenging of $^•NO$ by $O_2^{•-}$. Pretreating the cells with the general NOS inhibitor, N^{ω}-nitro-L-arginine methyl ester for 24 h attenuated the superoxide dismutase-induced stimulation, but did not affect the stimulation by sodium nitroprusside or SIN-1, wheras haemoglobin blocked the combined effect of superoxide dismutase plus $^•NO$ donors. The soluble guanylate cyclase inhibitor 1*H*-[1,2.4]oxadiazolo [4,3α]-quinoxaline-1-one did nor inhibit but even significantly potentiated the effect of NO donors on peptide release without affecting the superoxide dismutase-induced stimulation of peptide release.

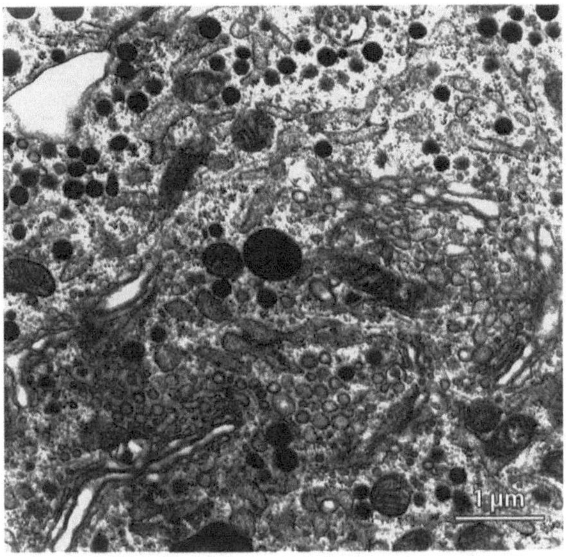

Fig. 257. Pars intermedia of the pituitary gland (block 1029) of a *rat* (No. 1) treated for 7 days with intraperitoneal injection of 1.5 ml Tyrode's solution per kg body weight × day. On August 4, 1976 under pentobarbital anaesthesia (30 mg/kg), the animal was perfused from the abdominal aorta with 2.5 % glutaraldehyde in 0.1 M sodium cacodylate buffer (pH 7.4). Postfixation with 1 % osmium tetroxide in sodium cacodylate buffer. Embedded in Epon 812 and sectioned at 50 nm. Lead citrate and uranyl acetate. Film 632-11

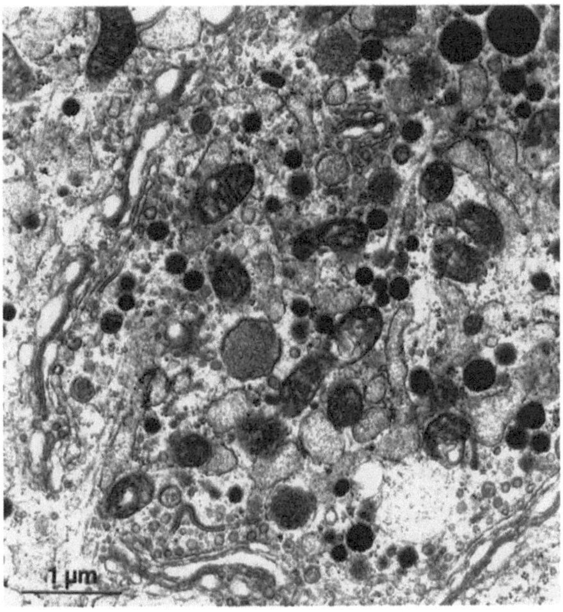

Fig. 258. Reoxygenation mitochondrial damage in the pars intermedia of the pituitary gland (block 1149) of an intermittent hypoxic and reoxygenated *rat* (No. 5) treated for 7 days with intraperitoneal injection of 1.5 ml Tyrode's solution per kg body weight × day. On the last 4 days of experimentation the animal was exposed to an atmosphere containing only 5 % oxygen for 30 min. On August 4, 1976 under pentobarbital anaesthesia (30 mg/kg), the animal was perfused from the abdominal aorta with 2.5 % glutaraldehyde in 0.1 M sodium cacodylate buffer (pH 7.4). Postfixation with 1 % osmium tetroxide in sodium cacodylate buffer. Embedded in Epon 812 and sectioned at 50 nm. Lead citrate and uranyl acetate. Film 633-10

Fig. 259. Pars intermedia of the intermittent hypoxic and reoxygenated pituitary gland (block 1240) of a *rat* (No. 8) treated for 7 days with intraperitoneal injection of 15 mg carbocromene per kg body weight × day. On the last 4 days of experimentation the animal was exposed to an atmosphere containing only 5 % oxygen for 30 min. On August 4, 1976 under pentobarbital anaesthesia (30 mg/kg), the animal was perfused from the abdominal aorta with 2.5 % glutaraldehyde in 0.1 M sodium cacodylate buffer (pH 7.4). Postfixation with 1 % osmium tetroxide in sodium cacodylate buffer. Embedded in Epon 812 and sectioned at 50 nm. Lead citrate and uranyl acetate. Film 633-39

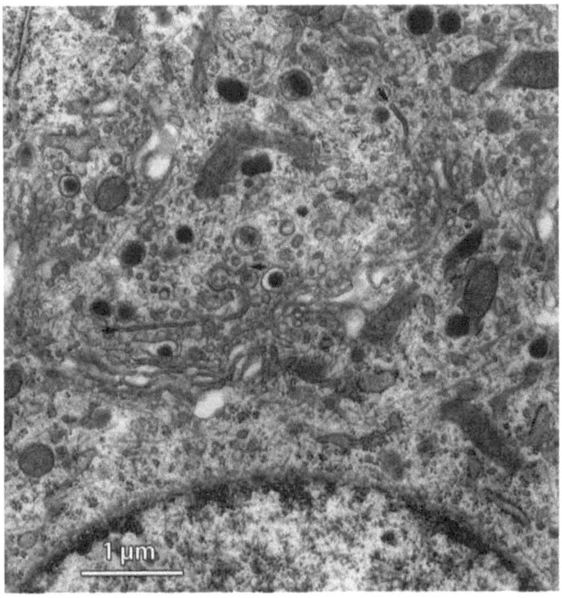

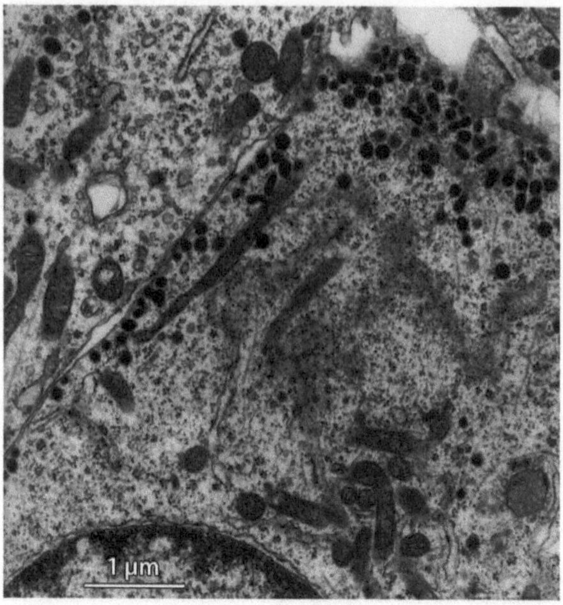

Fig. 260. Slender, twisted mitochondria with longitudinal cristae, and dense, round to oval secretory granules with delicate investing membranes in a pituitary pars intermedia cell (block 1240) of an intermittent hypoxic and reoxygenated *rat* (No. 8) medicated for 7 consecutive days with intraperitoneal injection of 15 mg carbocromene per kg body weight × day. On the last 4 days of experimentation the animal was exposed to an atmosphere containing only 5 % oxygen for 30 min. On August 4, 1976 half an hour after the last exposure, under pentobarbital anaesthesia (30 mg/kg), the animal was perfused from the abdominal aorta with 2.5 % glutaraldehyde in 0.1 M sodium cacodylate buffer (pH 7.4). Postfixation with 1 % osmium tetroxide in sodium cacodylate buffer. Embedded in Epon 812 and sectioned at 50 nm. Lead citrate and uranyl acetate. Film 633-41

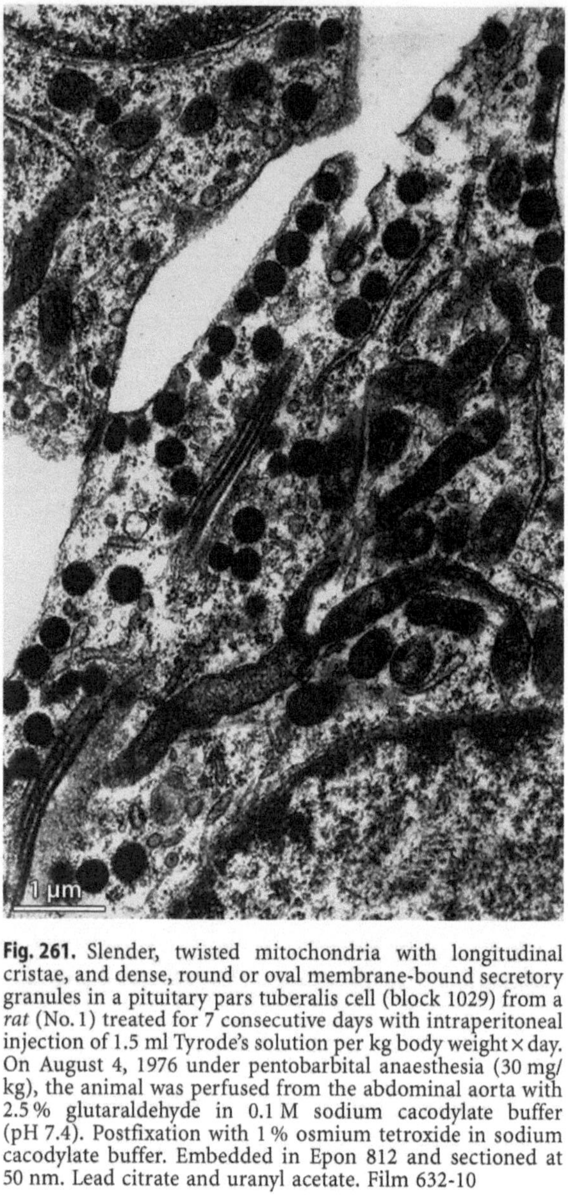

Fig. 261. Slender, twisted mitochondria with longitudinal cristae, and dense, round or oval membrane-bound secretory granules in a pituitary pars tuberalis cell (block 1029) from a *rat* (No. 1) treated for 7 consecutive days with intraperitoneal injection of 1.5 ml Tyrode's solution per kg body weight × day. On August 4, 1976 under pentobarbital anaesthesia (30 mg/kg), the animal was perfused from the abdominal aorta with 2.5 % glutaraldehyde in 0.1 M sodium cacodylate buffer (pH 7.4). Postfixation with 1 % osmium tetroxide in sodium cacodylate buffer. Embedded in Epon 812 and sectioned at 50 nm. Lead citrate and uranyl acetate. Film 632-10

15.1.2.3
Pars Tuberalis

15.1.2.3.1
Normal Cytobiology

Human pars tuberalis consists mainly of gonadotophs interspersed with few corticotrophs and thyreotrophs. Somatotrophs and lactotrophs were not identified by Asa et al. (1983). The subpopulation of gonadotophs possesses all organelles required for synthesis and storage of hormones but showed ultrastuctural features of functional inactivity.

In the *monkey*, many cells of the pars tuberalis produce β-luteinizing hormone and some β-follicle-stimulating hormone (Bock et al. 1999)

In the *rat*, the pars tuberalis consists of several layers of relatively small epithelial cells along the ventral surface of the median eminence (Oota and Kurosumi 1966). The epithelial cells may be classified into two types: specific secretory cells and follicular cells (Dellmann et al. 1974). The follicular cells are devoid of secretory granules, they do not only line the numerous follicular cavities of the pars tuberalis but may also be found in the periphery of the cell cords (border cells). Most of them contain relative large nuclei, a few mitochondria, rough and smooth endoplasmic reticula, and a Golgi apparatus. Sometimes, a colloid-like substance fills the intercellular space, into which a great number of microvilli and cilia may project. The other type of cells is granulated, larger in size, and contains a large number of mitochondria, well-developed rough and smooth endoplasmic reticula and a Golgi apparatus. The most prominent characteristic of this cell type are many electron dense granules, about 150 nm in diameter, resembling basophils of the pars distalis. The characteristic appearance of the lysosomes with a dense body and a cuplike expansion was depicted by DELLMANN et al. (1974).

In *sheep* only the common α-chain of glycoprotein hormones could be detected in pars tuberalis-specific cells by immunocytochemistry while antibodies directed against the β-chain of luteinizing hormone (LH), follicle-stimulating hormone (FSH), thyroid-stimulating hormone (TSH) and β-lipotropin labelled only single "migrated" pars distalis cells in the pars tuberalis (BOCKMANN et al. 1997). In situ hybridisation and northern blot analysis, however, revealed the expression of the common α-chain, β-TSH and, in a for lower extent, prolactin and POMC throughout the entire pars tuberalis.

15.1.2.3.2 Reoxygenation Cytobiology

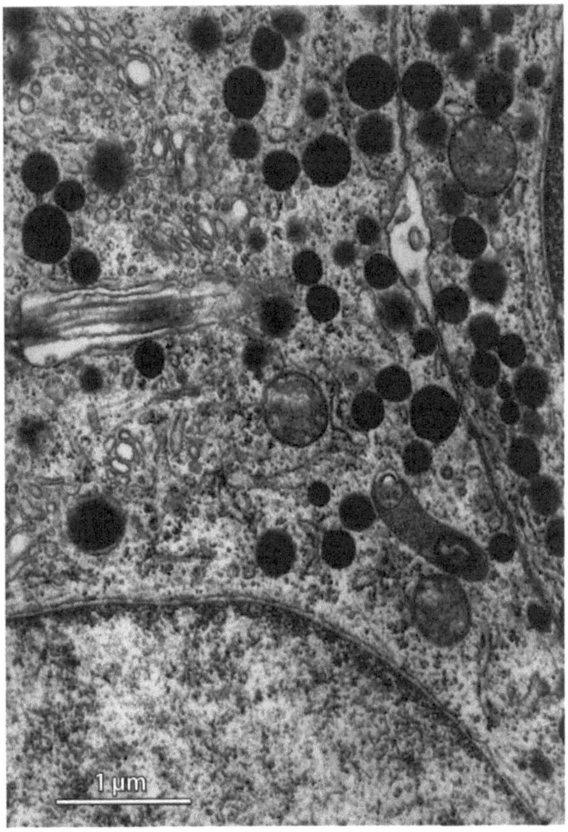

Fig. 262. Secretory granules and lysosomes in the pituitary pars tuberalis (block 1209) after intermittent hypoxemia and reoxygenation of a *rat* (No.7) treated for 7 consecutive days with intraperitoneal injection of 1.5 ml Tyrode's solution per kg body weight × day. On the 4 last days of experimentation the animal was exposed to an atmosphere containing only 5 % oxygen for 30 min. On August 4, 1976 under pentobarbital anaesthesia (30 mg/kg), the animal was perfused from the abdominal aorta with 2.5 % glutaraldehyde in 0.1 M sodium cacodylate buffer (pH 7.4). Postfixation with 1 % osmium tetroxide in sodium cacodylate buffer. Embedded in Epon 812 and sectioned at 50 nm. Lead citrate and uranyl acetate. Film 633-28

15.1.2.3.3 Effects of Melatonin on Pars Tuberalis Cells

During hibernation (mid February), the number of secretory vesicles in the pars tuberalis cells of the *garden door mouse* was considerably reduced (DELLMANN et al. 1974). The profiles of the rough endoplasmic reticulum, after having lost their ribosomes, tended to become closely apposed.

Long and short photoperiods induced conspicious cytological differences between specific secretory cells of the pars tuberalis of the Djungarian hamster, *Phodopus sungorus* (WITTKOWSKI et al. 1984). The cells differ with respect to the shapes of perikarya and nuclei and show diverse amounts of secretory granules, lysosome-like bodies and glycogen.

Functional receptors for melatonin have been localised and characterised on the pars tuberalis of a number of mammalian species (MORGAN et al. 1991, BÖCKERS et al. 1995, BOCKMANN et al. 1996). Using forskolin (1 μM), as a non-specific stimulant of adenylate cyclase, melatonin (10 nM) was shown to inhibit the formation of cyclic AMP by 80–90 % in *ovine* pars tuberalis cells both before and after Percoll centrifugation (MORGAN et al. 1991).

Although melatonin (10 nM) did not affect the expression of fibroblast growth factor (bFGF) mRNA in *ovine* pars tuberalis cells independently, it attenuated the activation of bFGF gene expression by 10 μM forskolin (GRAHAM et al. 1999). The induction (two fold) observed with forskalin (10 μM) was almost completely inhibited by melatonin (10 nM) ($P < 0.01$). Melatonin also significantly inhibited (30 %) 12-*O*-tetradecanoylphorbol-13-acetate-(100 nM) induced basis fibroblast growth factor (bFGF) mRNA expression ($P < 0.01$) in *ovine* pars tuberalis cells.

15.1.2.4 Pars Distalis

Recent evidence of plurihormonality in various pituitary cell types indicates that the once axiomatic one cell-one hormone theory is untenable and that the present perception of pituitary cell types and their function requires modification (HORVATH and KOVACS 1988). Transdifferentiation of somatotrophs to thyrotrophs occurred in the pituitary of female patients with protracted primary hypothyroidism (VIDAL et al. 2000). In early phase of transformation thyrosomatotrophs still showed large somatotroph-like granules, but the majority of secretory granules were small, similar to those of thyrotrophs. Double immunogold labelling revealed thyrosomatotrophs containing bihormonal granules as well as monohormonal granules labelled either for growth hormones or β-thyroid-stimulating hormone.

Sequential application of reverse haemolytic plaque assays for growth hormone and prolactin revealed the presence of individual pituitary cells that release both hormones (FRAWLEY et al. 1985). These dual cells accounted for approximately one third of all growth hormone and/or prolactin secretors in 24-h pituitary cultures from male *rats*.

KIKUTA et al. (1993) showed that 34 % of cells expressing oestrogen receptors were somatotropes.

There is an abundant immunoreactivity of rab3 protein in all endocrine cell types of the anterior pituitary of *rat*, *gerbil*, *hamster* and *guinea pig*, contrasting with the heterogeneous distribution of synaptotagmin I among endocrine adenohypophyseal cells (REDECKER et al. 1995). The subplasmalemmal concentration of rab3 coincided with the intracellular distribution of secretory granules, as shown by the staining pattern elicited by antibodies against the respective pituitary hormones. Of all endocrine cell types, lactotrophs most consistently were unreactive towards antibody Cl 42.2.

Macrophage migration inhibitory factor (MIF) protein was found to be secreted from a pituitary cell line upon stimulation with lipopolysaccharide (BERNHAGEN et al. 1993). Analysis of intact pituitary revealed that MIF protein is pre-formed and comprised ~0.05 % of total pituitary protein content. In comparison, ACTH and prolactin comprise 0.2 % and 0.08 %, respectively, of the total pituitary protein. MIF is localised within three subtypes of secretory granules in both corticotropic and thyrotropic cells: granules with ACTH and MIF, TSH and MIF, or MIF alone (NISHINO et al. 1995). The release of MIF from corticotrophs is stimulated by the hypothalamic hormone corticotrophin-releasing factor (CRF) in a dose-dependent manner, resulting in a concomitant increase in serum MIF above basal levels (MIF normally circulates at 2–4 ng/ml in *human* serum) (NISHINO et al. 1995).

In vivo studies in *rodents* confirmed that MIF is secreted from the pituitary during stress of lipopolysaccharide stimulation and resulted in increased serum levels with an accompanying decrease in pituitary MIF (CALANDRA et al. 1995). Hypophysectomized *mice* injected with lipopolysaccharide have no detectable serum MIF at a time that MIF levels are highest in control *mice*, indicating that the pituitary is the major source of serum MIF during systemic inflammatory response.

Lipopolysaccharide directly stimulated the intrapituitary interleukin-6 production by folliculo-stellate cells via specific receptors and the p38α mitogen-activated protein kinase/nuclear factor-κB pathway (LOHRER et al. 2000).

In the *rat*, after a stimulation with lipopolysaccharide, a small subpopulation of resident macrophages synthesised interleukin-1β and tumour necrosis factor-α (SCHÄFER and ULKE 1999). Their expression by folliculo-stellate cells could be excluded.

In the male *dd mouse* agranular cells are subdivisible into two types (ISHIHARA 1969). Cell of the first type are small in size and rarely encountered in small groups. They are oval or polygonal and show abundant zonula adhaerens at the cellular margin. The small rod-shaped mitochondria are few; they show a small number of cristae arranged transversely in the electron-lucent matrix. The granular reticulum is poorly developed and found as small numbers of vesicles or flatted sacs studded with few ribosomes. The Golgi apparatus lies around the nucleus and shows a sparse membrane system. Cells of the second type are relative large, electron-lucent and oval or irregularly rectangular in shape. The mitochondria are rod-shaped or branched. The granular reticulum is sparsely scattered in the form of round or flatted sacs. The Golgi apparatus appears as a fine small membrane system.

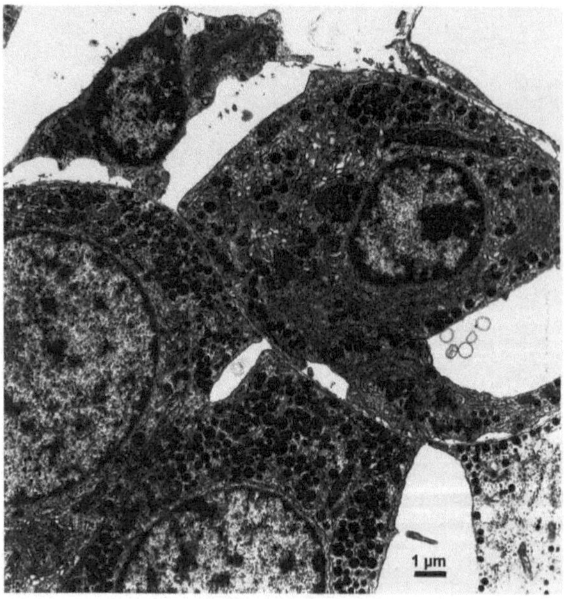

Fig. 263. An agranular cell in the pars distalis of the pituitary gland (block 1240) of a *rat* (No. 8) medicated for 7 consecutive days with intraperitoneal injection of 15 mg carbocromene per kg body weight × day. On the last 4 days of experimentation the animal was exposed to an atmosphere containing only 5 % oxygen for 30 min. On August 4, 1976, half an hour after the last exposure under pentobarbital anaesthesia (30 mg/kg), the animal was perfused from the abdominal aorta with 2.5 % glutaraldehyde in 0.1 M sodium cacodylate buffer (pH 7.4). Postfixation with 1 % osmium tetroxide in sodium cacodylate buffer. Embedded in Epon 812 and sectioned at 50 nm. Lead citrate and uranyl acetate. Film 633-34

In the mink (*mustela vison*) immunoperoxidase labelling of tissue sections demonstrated the presence of two types of S-100 positive cells (Cardin et al. 2000). Type 1 cells were stellate-shaped cells the nuclei of which were localised near the centre of pituitary follicles. In this type, S-100 labelling was strong in anterior pituitary sections obtained during spring, a period characterised by high prolactin pituitary content and low gonadotropin pituitary content. Type 2 cells were rounded cells occupying the periphery of the follicles. During periods of low prolactin pituitary content and high gonadotropin pituitary content the type 2 S-100 positive cells formed aggregates of several cells. Cultured S-100 cells were elongated, polygonal, or rounded. The presence of pseudopodia suggested that cultured folliculo-stellate cells could migrate. The vacuoles may be related to the phagocytic activity ascribed to these cells.

15.1.2.4.1 Peptide Hormones Producing Cells

Adrenocoricotropic Hormone (ACTH) Producing Cells

Many ACTH cells are widely distributed in the pars distalis, and the pars intermedia is composed exclusively of them. The ACTH cells in the pars distalis are characterised by elongated cytoplasm which is in contact with the capillaries. These elongated cells are usually in close relationship with other types of hormone-secreting cells, including growth hormone cells. The size of the dense-cored secretory granules located in the periphery of the cytoplasm averages 200 nm in diameter. The perinuclear area contains rough endoplasmic reticulum and Golgi apparatus. These characteristics do not permit precise identification of cells which secrete ACTH. Thus differential diagnosis of ACTH and TSH cells relies upon immunohistochemical staining (Moriarty 1973, Osamura et al. 1978, Kawarai 1981, Osamura and Watanabe 1985 [*man*]).

Extracts of the intermediate-posterior lobe of the *mouse* pituitary contain approximately a tenth as much ACTH activity as extracts the anterior lobe (Mains and Eipper 1975). In extracts of both the anterior and the intermediate-posterior lobes, about half of the immunological ACTH activity is similar in size to *porcine* ACTH (M_r = 4000–5500). Two higher M_r forms of ACTH account for the remainder of the ACTH activity. About 40 % of the immunological ACTH activity in anterior lobe extracts has a M_r of 6500–9000. Extracts of both the anterior lobe and the intermediate-posterior lobe contain ACTH activity with a M_r of 20,00–30,000. While this 20,00–30,000 M_r ACTH accounts for only 5 % of the immunological ACTH activity in the anterior lobe extracts, it accounts for half the immunological ACTH activity in extracts of the intermediate-posterior lobe.

-tyr-ser-**met**-glu-his-phe-arg

Methionine in ACTH is oxidised by HO$^{\bullet}$, $^{1}O_2$ or H_2O_2 (Dedman et al. 1961) to the sulphoxide and further to the sulphone.

Oxidation of methionine

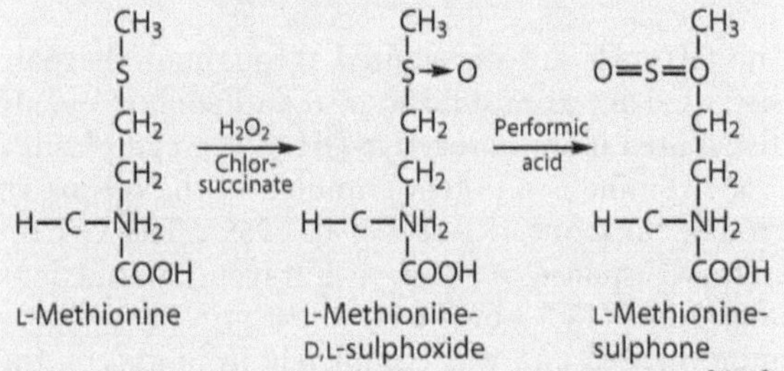

[282]

Subcutaneous injections of both melatonin (1 mg per kg × day) and its vehicle (1 % ethanol) into intact Long-Evans *rats* for 10 days equally decreased pituitary ACTH content by 50 % (Barchas et al. 1969).

Blockade of nitric oxide ($^{\bullet}$NO) formation with 30 mg N^{ω}nitro-L-arginine methylester per kg *rat* produced a dramatic increase in ACTH released by the intravenous injection of 100 ng *human* recombinant interleukin-1β/kg *rat* (Rivier 1995). Blockade of adrenergic receptors with 0.5 mg prazosin/kg *rat* and 2.5 mg propranolol /kg *rat* did not alter the stimulatory effect of interleukin-1β. This treatment did not significantly interfere with the potentiation influence of N^{ω}nitro-L-arginine methylester 30 min after IL-1β injection, but blunted this effect at 60 min. Immunoneutralization of vasopressin did not consistently decrease the ACTH response to IL-1β regardless of whether $^{\bullet}$NO was present. Blockade of prostaglandin synthesis with 10 mg ibuprofen per kg *rat* totally abolished IL-1β-induced ACTH secretion; in addition, it prevented the interaction between N^{ω}nitro-L-arginine and the pituitary response.

AtT-20 anterior pituitary cells, an established line of *mouse* corticotrophic cells, when stimulated with lipopolysaccharide secreted a 12.5 kDa protein that shared very high homology with *human* macrophage migration inhibitory factor MIF (Bernhagen et al. 1993).

In vitro, N^{ω}nitro-L-arginine was able to modify ACTH basal secretion and to block interleukin-1β-induced CRH and ACTH release from *rat* hypothalamic and anterior pituitary cell cultures (BRUNETTI et al. 1993).

Cyproheptadine and reserpine (10^{-9}–10^{-7} M of each) suppressed immunoreactive ACTH and β-endorphin secretion from adenomatous and non-adenomatous tissues of patients with Cushing's disease (SUDA et al. 1983). Inhibition of depolarisation-dependent Ca^{2+} entry into cells as suggested by DONATSCH et al. (1980) in insulin secretion or a nonserotonergic mechanism may be involved.

Growth Hormone (GH) Producing Cells

The GH cells are distributed throughout the pars distalis. They are round to oval and contain evenly distributed immunoreactive GH in the cytoplasm.

L-Arginine is a potent stimulus to the release of growth hormone (KNOPF et al. 1969). The GH secretory response of *sheep* to intravenous infusions of arginine (12.5 g over a 20- to 30-min period) was quite marked and was completely inhibited by the concomitant infusion of 1.0 μg epinephrine/kg × min (KIPNIS et al. 1969). KATO (1992) has demonstrated *in vitro* that N^{ω}-methyl-L-arginine potentiates the growth hormone releasing hormone-stimulated growth hormone secretion without affecting basal growth hormone secretion, findings supported by experiments with the •NO donor sodium nitroprusside. Erythrityl tetranitrate strikingly potentiated Hexarelin-stimulated growth hormone secretion (RIGAMONTI et al. 1999).

Combined immunohistochemical, in situ hybridisation, and biochemical studies by CECCATELLI et al. (1993) provided evidence for nitric oxide synthase synthesis in the **folliculo-stellate cells** that do not produce any known hormones (HORVATH and KOVACS 1988) but may indirectly influence growth-hormone secretion. By interferon-γ VANKELECOM et al. (1997) induced nitric oxide synthase (iNOS) in subpopulation of folliculo-stellate cells and in an unidentifiable population in monolayer cultures of the *rat* anterior pituitary.

The involvement of a **guanosine triphosphate (GTP)-binding protein** coupling occupancy of receptors by somatostatin and acetylcholine (or carbachol) to activation of K^+ channels was demonstrated by three types of experiments (YATANI et al. 1987): 1) K^+ channels activated in the cell-attached mode did not remain active if, after excision, they were placed into solutions without GTP; or if in the presence of GTP, excess of the inhibitor guanosine [β-thio]diphosphate was added; 2) addition of guanosine[β-thio]diphosphate (100 μM) to patches excised without prior activation by ligands led, after lag times that varied between 2 to 10 min, to activation of K^+ channels with the same unit conductance as seen with receptor ligands plus GTP, and 3) addition to excised patches of purified *human* red blood cell pertussis toxin-substrate G_k, within seconds activated K^+ channels, which again had the same unit conductance as those seen after activation by ligands (plus GTP) or by guanosine[β-thio]diphosphate.

A decrease in the number of **microtubuli** has been observed in *bovine* pituitary slices incubated with two non-physiological stimulants, extracellular barium and 3-isobutyl-1-methylxanthine of growth hormone secretion (SHETERLINE et al. 1977). Both growth hormone secretion and depolymerization were prevented by 2-methylpentan-2,4-diol, a microtubule stabiliser.

In normal *rat* anterior pituitary cells, reserpine blocked both basal and K^+-stimulated **calcium** uptake (LOGIN et al. 1985). Reserpine selectively blocked maitoxin (a potent calcium channel activator) but not A23187 (a calcium ionophore)-induced calcium uptake.

15.1.2.4.2 Glycoprotein Hormones Producing Cells

The glycoprotein hormones are composed of two subunits, α and β; they share common α subunits and possess a specific β subunit that controls their specific structure and function.

Gonadotrophs

The gonadotropic (LH and FSH) cells of the *rat*'s pituitary pars distalis contain dilated and vesicular rough endoplasmic reticulum and scattered round, small secretory granules, 100–300 nm in diameter (OSAMURA 1983). Four different subtypes can be determined according to their cytology, cytochemistry, and partly, to their activity (JEZIOROWSKI et al. 1995). About 90 % of the cells contained LH and FSH, and chromogranin A and secretogranin II, the other types preferentially FSH/CgA or LH/Sg II. FSH was restricted to the larger secretory granules, LH was predominantly and Sg II exclusively localised in small ones.

The pituitary gonadotropin-releasing hormone receptor interacts with the decapeptide gonadotropin-releasing hormone which is produced in the hypothalamus. The GnRH receptor structurally differs from all other known G-protein-coupled receptors by the complete lack of a cytoplasmic C-terminus (STOJILKOVIC et al. 1994). The arginine at position 8 of *mammalian* GnRH is crucial for high-affinity binding to *mammalian* GnRH receptors (FLANAGAN et al. 1994).

Stimulation with GnRH caused exocytosis of secretory granules only from those cells with predominantly small dense granules, and virtually all the exocytoses occurred at the perivascular border of the cells (Morris and Durnin 1993). Higher concentrations of GnRH caused the appearance of many multiple exocytoses, and also recruited a larger proportion of cells into secreting pool. All the exocytosed granule cores were dense and about 200 nm in diameter. Gonadotrophs with the larger, variably electron-dense granules did not show signs of exocytosis. However, when 56mMK$^+$ was used to stimulate the pituitary tissue exocytosis occurred from both typed of gonadotrophs, but the exocytoses were still highly polarised to the vascular border of the cells, and some cells showed no release despite this non-specific depolarising stimulus.

Combined immunohistochemical, in situ hybridisation, and biochemical studies provide evidence for **nitric oxide synthase** synthesis in two cell populations of the anterior pituitary gland, gonadotrophs and folliculo-stellate cells, and for an inhibitory effect of •NO on stimulated release of LH (Ceccatelli et al. 1993). The folliculo-stellate cells do not produce any known hormones (Horvath and Kovacs 1988) but may indirectly influence growth-hormone secretion. Thioredoxin was prominently localised in the folliculo-stellate cells of the *mammalian* adenohypophysis while only a minor proportion of the glandular cells were positive (Padilla et al. 1992). Glutaredoxin localisation in the adenohypophysis resembled that of thioredoxin.

Treatment of P11 cells, derived from the transplantable *rat* pituitary tumour 7315a, with interferon-γ (50 U/ml) and lipopolysaccharide (10 μg/ml) resulted in a 23-fold increase in nitrite production and induced expression of iNOS protein (Xu and Miller 2000). The increase in nitrite levels was attenuated by the non-selective nitric oxide synthase (NOS) inhibitor N^{ω}-nitro-L-arginine methylester, but not the neuronal NOS inhibitor 7-nitroimidazole. Incubation of *rat* anterior pituitaries with either N^{ω}-monomethyl-L-arginine or N^{ω}-nitro-L-arginine methylester significantly increased prolactin release (Duvilanski et al. 1995).

***N*-Methyl-DL-aspartate**-treated *rats* exhibited increased serum LH and GH concentrations while kainic acid-treated *rats* showed increases only in serum GH concentrations (Mason et al. 1983). Neither ibotenic acid nor quinolinic acid altered adenohypophyseal hormone levels. The stimulatory action of kainic acid on LH and FSH secretion was age-dependent, since the agonist was completely ineffective on the anterior pituitary of 75-day and 18-month-old male *rats* (Zanisi et al. 1994). 6,7-Dinitroquinoxaline-2,3-dione, a specific antagonist of the kainic acid receptor subtype, was able to block the kainic acid-induced gonadotropin secretion; similarly, 2-amino-5-phosphonovalerate, a competitive *N*-methyl-D-aspartate receptor antagonist, prevented the stimulatory effect of kainic acid on LH and FSH release. An interaction between the opiatergic and the excitatory aminoacid systems emerged from the observation that pulses of kainic acid applied to anterior pituitaries of 50-day-old *rats* during a continuous perfusion with a medium containing morphine (5 μM) failed to increase gonadotropin secretion.

Chlorpromazine (10 mg/kg *rat*×day for 1 week) led to a rise in pituitary gonadotrophic activity (Jarrett 1963). A single injection of 40 mg chlorpromazine per kg body weight in infant male Wistar *rats* induced some alterations in the brain which promoted higher secretion of luteinizing hormone as demonstrated by accelerated spermatogenesis (Ladosky et al. 1970).

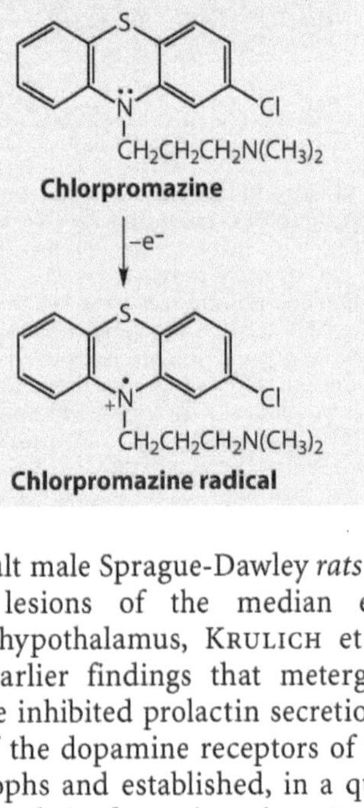

[64]

In young adult male Sprague-Dawley *rats* with large electrolytic lesions of the median eminence–mediobasal hypothalamus, Krulich et al. (1981) confirmed earlier findings that metergoline and methysergide inhibited prolactin secretion through activation of the dopamine receptors of the pituitary lactothrophs and established, in a quantitative manner, that their dopaminergic potencies were comparable to the potency of the dopamine receptor agonist, piribedil, with ED_{50} in the order of 0.35 to 0.22 mg/kg. Cyproheptadine acting by an unknown mechanism had only a weak inhibiting effect (ED_{50} > 20.0 mg/kg).

The volume of secondary crinophagic lysosomes per prolactin cell increased during late oestrus and remained elevated throughout early dioestrum of the *rat* (Poole et al. 1981).

Melatonin implanted into the median eminence of male *rats* reduced plasma levels of luteinizing hormone (Fraschini 1969). This indicated that the

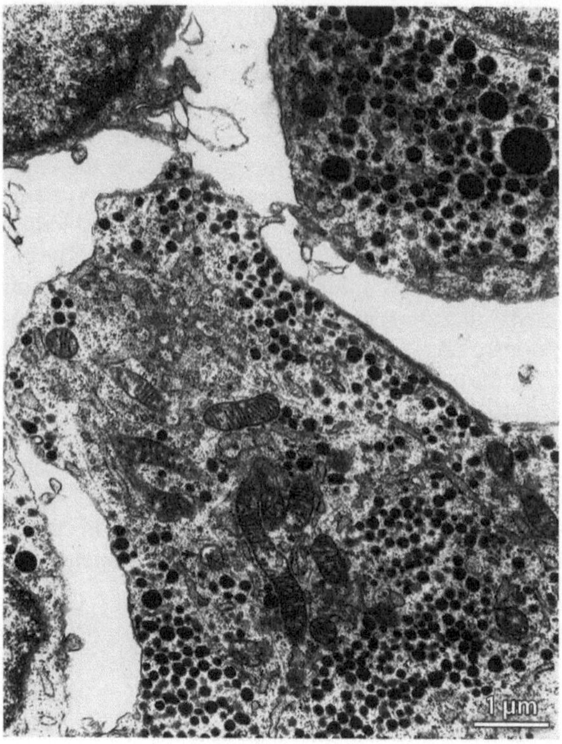

Fig. 264. Reoxygenated gonadotroph with large areas of free ribosomes and with multivesicular bodies (arrows) from the anterior pituitary (block 1390) of a *rat* (No. 13) treated for 8 consecutive days with intraperitoneal injection of 15 mg carbochromene per kg body weight × day. On the last 5 days of experimentation the animal was exposed to an atmosphere containing only 5 % oxygen. On August 5, 1976, half an hour after the last exposure under pentobarbital anaesthesia (30 mg/kg), the animal was perfused from the abdominal aorta with 2.5 % glutaraldehyde in 0.1 M sodium cacodylate buffer (pH 7.4). Postfixation with 1 % osmium tetroxide in sodium cacodylate buffer. Embedded in Epon 812 and sectioned at 50 nm. Lead citrate and uranyl acetate. Plate 3470

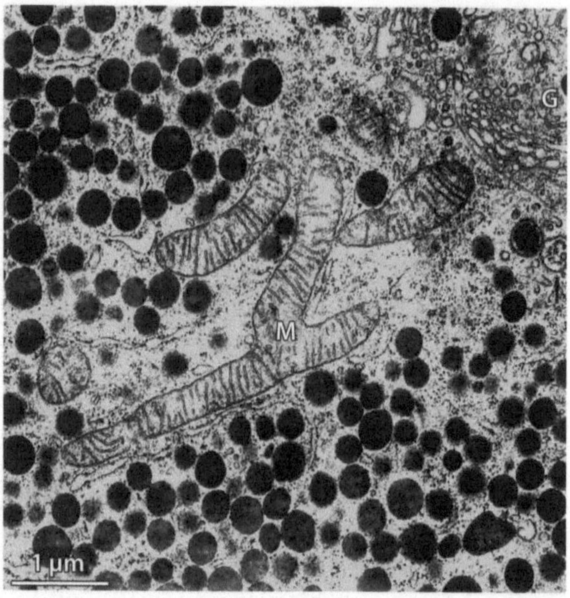

Fig. 265. A multivesicular body (arrow), an extended Golgi field (G), and a branched mitochondrium (M) in a prolactin cell (block 1209) from an intermittent hypoxic and reoxygenated *rat* (No. 7) medicated for 7 consecutive days with intraperitoneal injection of 1.5 ml Tyrode's solution per kg body weight × day. On the last 4 days of experimentation the animal was exposed to an atmosphere containing only 5 % oxygen for 30 min. On August 4, 1976, half an hour after the last exposure under pentobarbital anaesthesia (30 mg/kg), the animal was perfused from the abdominal aorta with 2.5 % glutaraldehyde in 0.1 M sodium cacodylate buffer (pH 7.4). Postfixation with 1 % osmium tetroxide in sodium cacodylate buffer. Embedded in Epon 812 and sectioned at 50 nm. Lead citrate and uranyl acetate. Film 633-27

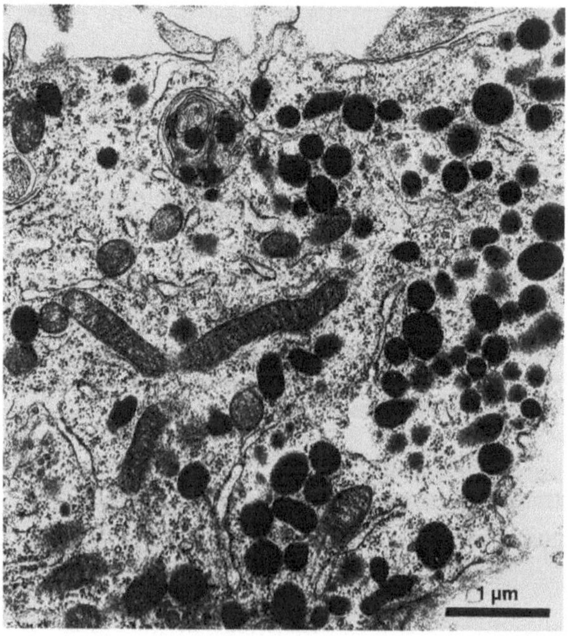

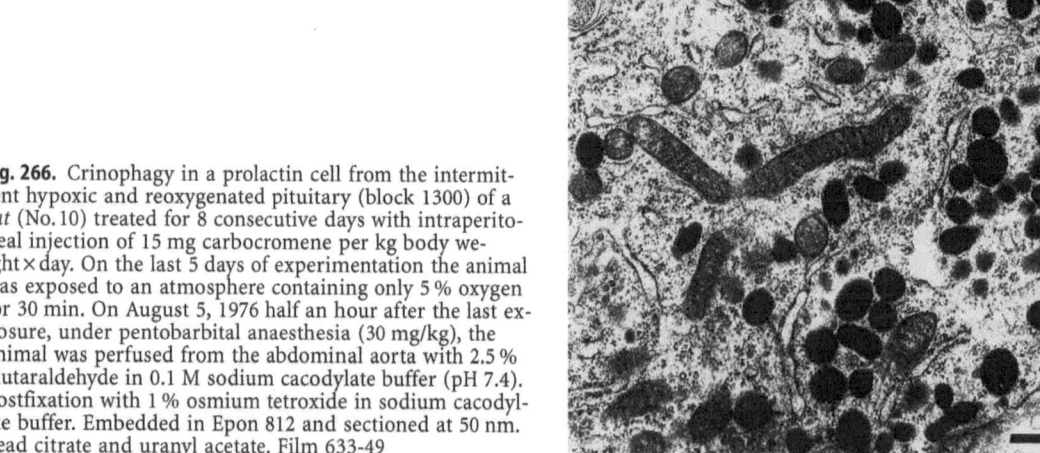

Fig. 266. Crinophagy in a prolactin cell from the intermittent hypoxic and reoxygenated pituitary (block 1300) of a *rat* (No. 10) treated for 8 consecutive days with intraperitoneal injection of 15 mg carbocromene per kg body weight × day. On the last 5 days of experimentation the animal was exposed to an atmosphere containing only 5 % oxygen for 30 min. On August 5, 1976 half an hour after the last exposure, under pentobarbital anaesthesia (30 mg/kg), the animal was perfused from the abdominal aorta with 2.5 % glutaraldehyde in 0.1 M sodium cacodylate buffer (pH 7.4). Postfixation with 1 % osmium tetroxide in sodium cacodylate buffer. Embedded in Epon 812 and sectioned at 50 nm. Lead citrate and uranyl acetate. Film 633-49

release of luteinizing hormone was not activated and suggested that brain implants of melatonin (and probably also those of 5-hydroxytryptophol) inhibited the synthesis of luteinizing hormone and that receptors sensitive to indole compounds (FRASCHINI et al. 1968) may play a major role in the control of luteinizing hormone secretion. During the light period, when the two methoxyindole derivatives, melatonin and methoxytryptophol, cannot be formed due to the inhibition of hydroxyindole-O-methyltransferase, the pituitary gland is relieved from pineal inhibition. In male *dogs*, carotid pretreatment with 10 or 100 µg melatonin per kg body weight significantly inhibited the pituitary to release luteinizing hormone induced by a carotid injection of 5 µg of luteinizing hormone releasing hormone per kg body weight 3 h later (YAMASHITA et al. 1978).

The finding that ^{63}**Ni** following intravenous administration of $^{63}NiCl_2$ (240 µg/kg *rabbit*) was particularly localized in the pituitary (PARKER and SUNDERMAN jr 1974) may have physiological significance, in view of the reports by LABELLA et al. (1973) that Ni(II) depresses *in vitro* release of prolactin from *bovine* pituitary. LaBella et al. have found that intravenous administration of $NiCl_2$ (300 to 600 gg/kg) to chlorpromazine-treated male *rats* resulted in 40 % decrease of serum prolactin levels at 30 min after the injection. ^{63}Ni distribution experiments in *guinea pigs* also showed that following five daily subcutaneous injections of 1 mg per kg body weight the pituitary was especially high in Ni (CLARY 1975).

Treating confluent HC11 mammary epithelial cells with prolactin (10 µg/ml) and cortisol (1 µM) produced a progressive, four- to fivefold, increase in xanthine oxidoreductase activity, while xanthine oxidoreductase activity in control cells remained constant (McMANAMAN et al. 2000).

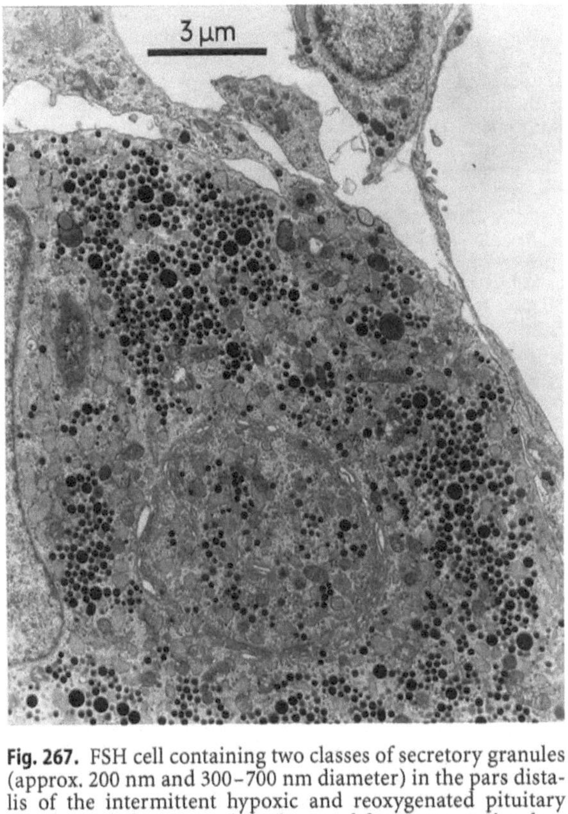

Fig. 267. FSH cell containing two classes of secretory granules (approx. 200 nm and 300–700 nm diameter) in the pars distalis of the intermittent hypoxic and reoxygenated pituitary (block 1209) from a *rat* (No. 7) treated for 7 consecutive days with intraperitoneal injection of 1.5 ml Tyrode's solution per kg body weight × day. On the last 4 days of experimentation the animal was exposed to an atmosphere containing only 5 % oxygen for 30 min. On August 4, 1976 half an hour after the last exposure, under pentobarbital anaesthesia (30 mg/kg), the animal was perfused from the abdominal aorta with 2.5 % glutaraldehyde in 0.1 M sodium cacodylate buffer (pH 7.4). Postfixation with 1 % osmium tetroxide in sodium cacodylate buffer. Embedded in Epon 812 and sectioned at 50 nm. Lead citrate and uranyl acetate. Film 633-23

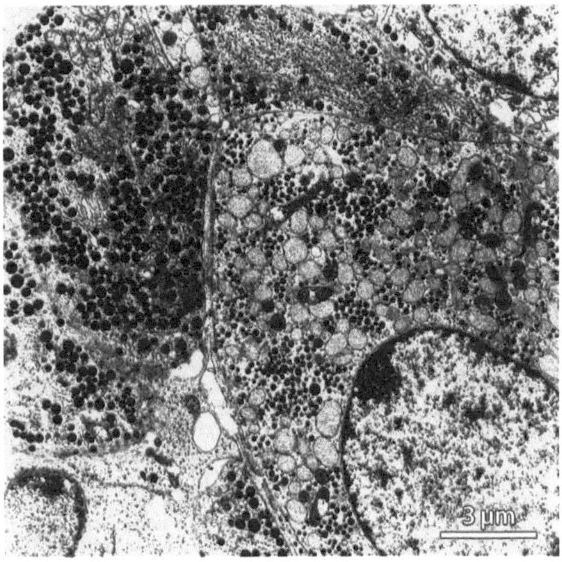

Fig. 268. Saccular cytoplasmic dilatation, small and large secretory granules, damaged mitochondria with intramitochondrial granules (~30 nm diameter), and microtubules in a FSH cell from the intermittent hypoxic and reoxygenated adenohypophysis (block 1149) of a *rat* (No. 5) treated for 7 consecutive days with intraperitoneal injection of 1.5 ml Tyrode's solution per kg body weight × day. On the last 4 days of experimentation the animal was exposed to an atmosphere containing only 5 % oxygen for 30 min. On August 4, 1976 half an hour after the last exposure, under pentobarbital anaesthesia (30 mg/kg), the animal was perfused from the abdominal aorta with 2.5 % glutaraldehyde in 0.1 M sodium cacodylate buffer (pH 7.4). Postfixation with 1 % osmium tetroxide in sodium cacodylate buffer. Embedded in Epon 812 and sectioned at 50 nm. Lead citrate and uranyl acetate. Film 633-11

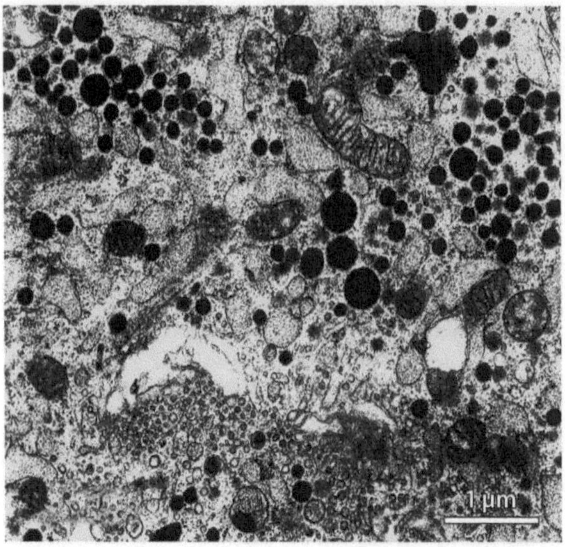

Fig. 269. Saccular cytoplasmic dilatation, small (~200 nm diameter) and large (300–700 nm diameter) secretory granules and damaged mitochondria with imtramitochondrial granules (aprox. 30 nm diameter) in the pars distalis of the intermittent hypoxic and reoxygenated pituitary (block 1209) from a *rat* (No. 7) treated for 7 consecutive days with intraperitoneal injection of 1.5 ml Tyrode's solution per kg body weight × day. On the last 4 days of experimentation the animal was exposed to an atmosphere containing only 5 % oxygen for 30 min. On August 4, 1976 half an hour after the last exposure, under pentobarbital anaesthesia (30 mg/kg), the animal was perfused from the abdominal aorta with 2.5 % glutaraldehyde in 0.1 M sodium cacodylate buffer (pH 7.4). Postfixation with 1 % osmium tetroxide in sodium cacodylate buffer. Embedded in Epon 812 and sectioned at 50 nm. Lead citrate and uranyl acetate. Film 633-31

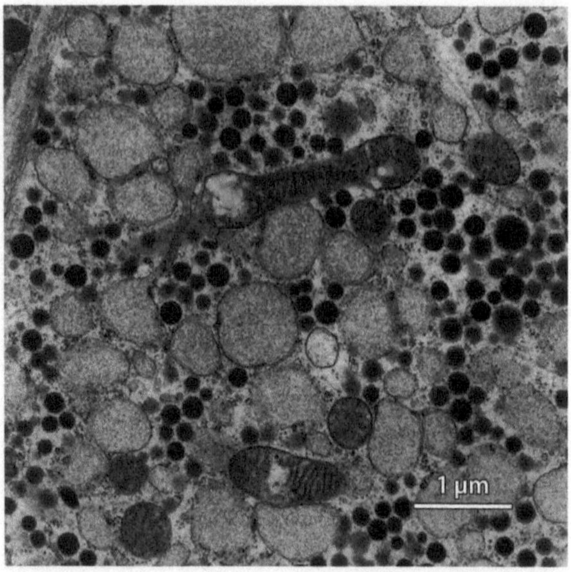

Fig. 270. Saccular cytoplasmic dilatation, small (~200 nm diameter) and large (300–700 nm diameter) secretory granules, damaged mitochondria with imtramitochondrial granules (aprox. 30 nm diameter), and some microtubules in the pars distalis of the intermittent hypoxic and reoxygenated pituitary (block 1149) of a *rat* (No. 5) treated for 7 consecutive days with intraperitoneal injection of 1.5 ml Tyrode's solution per kg body weigh × dayt. On the last 4 days of experimentation the animal was exposed to an atmosphere containing only 5 % oxygen for 30 min. On August 4, 1976 half an hour after the last exposure, under pentobarbital anaesthesia (30 mg/kg), the animal was perfused from the abdominal aorta with 2.5 % glutaraldehyde in 0.1 M sodium cacodylate buffer (pH 7.4). Postfixation with 1 % osmium tetroxide in sodium cacodylate buffer. Embedded in Epon 812 and sectioned at 50 nm. Lead citrate and uranyl acetate. Film 633-09. Detail from Fig. 268

Fig. 271. FSH cell containing two classes of secretory granules (~200 nm and 300–700 nm diameter), damaged mitochondria with intramitochondrial granules (~30 nm diameter), and some microtubules from the reoxygenated adenohypophysis (block 1209) of a *rat* (No. 7) treated for 7 consecutive days with intraperitoneal injection of 1.5 ml Tyrode's solution per kg body weight × day. On the last 4 days of experimentation the animal was exposed to an atmosphere containing only 5 % oxygen for 30 min. On August 4, 1976 half an hour after the last exposure, under pentobarbital anaesthesia (30 mg/kg), the animal was perfused from the abdominal aorta with 2.5 % glutaraldehyde in 0.1 M sodium cacodylate buffer (pH 7.4). Postfixation with 1 % osmium tetroxide in sodium cacodylate buffer. Embedded in Epon 812 and sectioned at 50 nm. Lead citrate and uranyl acetate. Film 633-30

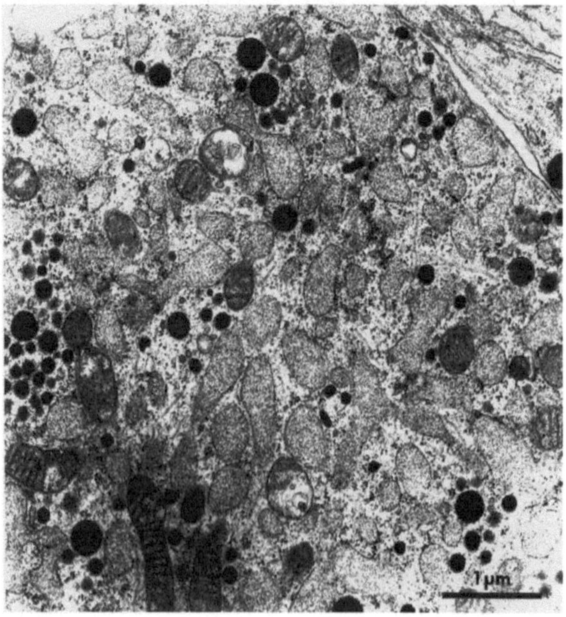

Follicle-Stimulating Hormone Cells

The FSH cells are recognised by the presence of ample cytoplasm containing abundant dilated, rough endoplasmic reticulum, and scattered rounded secretory granules, 200 nm and 300–700 nm in diameter.

Thyroid-Stimulating Hormone Cells

The "typical" thyrotroph is a fairly large, elongated cell that is stellate or angular in shape. The amount of rough endoplasmic reticulum – lamellar and slightly dilated – varies greatly, influencing the overall appearance of the cell. The Golgi apparatus is usually prominent, containing developing secretory granules (Fig. 272). Cytoplasmic microtubules may be conspicuous in the Golgi region. In sparsely granulated cells, the secretory granules often form a single layer under the plasmalemma, whereas in well-granulated thyrotrophs, the granules are distributed throughout the cell, often with heavier accumulation in the basal portion of the cytoplasm. The mostly spherical secretory granules are usually in the range of 100–200 nm, but granules measuring up to 250 nm are also noted.

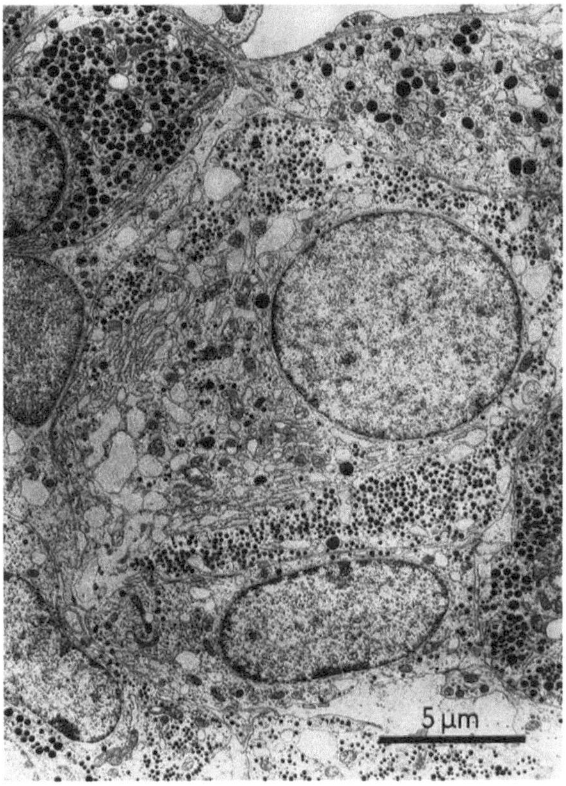

Fig. 272. Reoxygenation damage to a TSH cell from the anterior pituitary gland (block BNh 2417) of a Wistar *rat* (No. 1912) which was injected intraperitoneally with 300 mg piracetam/kg body weight x day for 4 consecutive days. 30 min thereafter the animal was exposed to an atmosphere containing only 5 % oxygen for 30 min. Half an hour after the last exposure, under pentobarbital anaesthesia (30 mg/kg), the animal was perfused from the abdominal aorta with 2.5 % glutaraldehyde in 0.1 M sodium cacodylate buffer (pH 7.4). Postfixation with 1 % osmium tetroxide in sodium cacodylate buffer. Embedded in Epon 812 and sectioned at 50 nm. Lead citrate and uranyl acetate. Plate 2212

Melatonin was undetectable in individual anterior pituitary glands of male Wistar *rats* killed by decapitation during darkness (WILKINSON et al. 1977). Pituitary tissue pooled from the time of maximal pineal and near-maximal serum content (23.30 h) contained <3.5 pg melatonin/pituitary gland. These results suggested that in the male *rat* endogenous melatonin may not be concentrated in pituitary tissue.

15.1.3 Thyroid

A pivotal biochemical event in thyroid physiology is a **superoxide anion radical-mediated activation of iodine** into active $I^{\bullet-}$ giving fairly stable spin adducts with *N-tert*-α-phenyl nitrone (VERMA et al. 1990).

Under the influence of exogenous thyroid stimulating hormone (TSH), the burst of $O_2^{\bullet-}$ radicals after 20 min and its complete scavenging after 60 min, as shown by PREM PRAKASH et al. (1997) in the pigeon, *Columba livia intermedia*, throw light on the thyrotropin influenced changes in thyroidal oxygen status and an adaptive induction of superoxide dismutase. An inverse correlation between superoxide dismutase and $O_2^{\bullet-}$ radical was clearly evident.

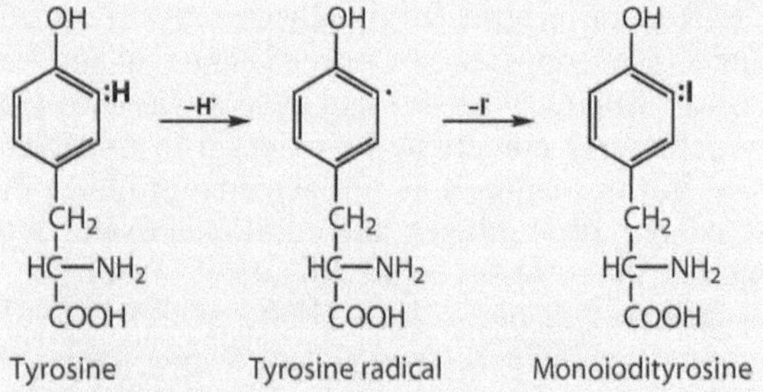

Tyrosine radical as an intermediate in tyrosine monoiodination [285]

The synthesis of thyroid hormones by thyroid peroxidase requires iodination of thyrosine residues in thyreoglobulin, and this reaction depends on H_2O_2 (DEME et al. 1985, NAKAMURA et al. 1987, 1989). Plasma membrane fractions of thyroid cells posess an H_2O_2-generating and NADPH-oxidising enzyme system which shows some properties similar to NADPH oxidase of phagocytes. NAKAMURA et al. (1989) suggested the primary product of the enzyme reaction to be $O_2^{\bullet-}$, and H_2O_2 may be provided by dismutation of $O_2^{\bullet-}$.

Superoxide dismutase (EC 1.15.1.1)

$$2\,O_2^{\bullet-} + 2\,H^+ \longrightarrow H_2O_2 + {}^3\Sigma_g^- O_2 \qquad [128]$$

Even before the subcutaneous injection of 20 μg TSH to immature female Swiss *mice*, thyroid gland homogenates exhibited a fairly good SOD activity (VERMA et al. 1991). But after the administration of 20 μg TSH, there was an immediate surge in SOD activity in the thyroid gland noticeable at 15 min after medication. The activity of this enzyme reached its crescendo ($P < 0.01$) at 30 min after injection, which dropped down significantly ($P < 0.01$) from 45 min onward.

H_2O_2 itself did not modulate basal $[Ca^{2+}]_i$ in *rat* thyroid FRTL-5 cells (TÖRNQUIST et al. 2000). However, H_2O_2 attenuated store-operated calcium entry evoked by thapsigargin, both in a sodium-containing buffer and in sodium-free buffer. The effect of H_2O_2 was abrogated by the reducing agent β-mercaptoethanol. H_2O_2 also attenuated the thapsigargin-evoked entry of barium and manganese. The effect of H_2O_2 was, at least in part, mediated by activation of protein kinase C, as H_2O_2 enhanced the binding of [^{3}H]phorbol 12,13-dibutyrate. H_2O_2 also stimulated the translocation of the isoenzyme PKC_ε from the cytosolic fraction to the particulate fraction. Furthermore, H_2O_2 did not attenuate store-operated calcium entry in cells treated with staurosporine or calphostin C, or in cells with down-regulated PKC.

Tyrosine radicals will combine with nitric oxide ($^\bullet$NO) to form both O- and C-centred nitroso adducts. These appear to be unstable and will add water to regenerate tyrosine and yield nitrite. Nitrosotyrosine may potentially be oxidised to nitrotyrosine, but this requires an initial oxidant to form the tyrosine radical initially and additional oxidants to modify the nitrosotyrosine (BECKMAN 1996).

Horseradish peroxidase- (USHIJIMA et al. 1985) or myeloperoxidase-catalysed tyrosine oxidation (USHIJIMA et al. 1997) accompanies light emission in the visible region. Similar chemiluminescence phenomena have been obtained during the fertilisation of *sea urchin* eggs, which forms bityrosine crosslinks in the egg membrane (TAKAHASHI et al. 1989), during incubation of tyrosine-rich bamboo shoot extract with H_2O_2 (TOTSUNE et al. 1993), and during activation of human phagocytic leucocytes by opsonized zymosan in the presence of absence of added tyrosine (USHIJIMA et al. 1997). Tyrosine phenoxyl neutral radical and/or bityrosine phenoxyl neutral radical generated in the horseradish peroxidase molecule may attack adjacent tyrosine and tryptophan residues in the enzyme molecule to abstract electrons from these aromatic amino acids, yielding the corresponding cation radical, tyrosine phenoxyl cation radical ($Tyr^{\bullet+}$) and the nitrogen-centred cation radical with a pK_a of 4.2 (ADAMS and WARDMAN 1977), respectively. Electrolysis of tyrosine-emitted light which peaked at 490 nm was almost completely quenched by superoxide dismutase, while emission by bityrosine peaked at 530 nm (TOTSUNE et al. 1999). In the horseradish-H_2O_2-tyrosine system the oxidation-reduction of tyrosine-emitted light with two prominent peaks, 490 and 530 nm, and was not quenched by superoxide dismutase. The phenoxyl neutral radical of the tyrosine in horseradish peroxidase-H_2O_2-tyrosine system was detected by electron spin resonance spectrometry using *tert*-nitrosobutane as a spin trap; the spin adduct was found to adhere to the horseradish peroxidase molecule during the enzymatic reaction. further, bityrosine was detected in the horseradish peroxidase-H_2O_2-tyrosine reaction system. Changes in absorption spectra of horseradish peroxidase (HRP)-catalysed oxidation of tyrosine suggests that the photon emission compound III is a candidate superoxide donor to the phenoxyl cation radical of tyrosine on the enzyme molecule. the luminescence observed by TOTSUNE et al. (1999) might be originated from at least two exciplexes involved with the tyrosine cation radical ($Tyr^{\bullet+}$) and the bityrosine cation radical ($Btyr^{\bullet+}$) as

$$\text{Compound III} + Tyr^{\bullet+}) \longrightarrow [Tyr^{\bullet+}\,O_2^{\bullet-}] + HRP \qquad [286]$$

$$\text{Compound III} + Btyr^{\bullet+} \longrightarrow [Btyr^{\bullet+}\,O_2^{\bullet-}] + HRP \qquad [287]$$

and

$$[Tyr^{\bullet+}\,O_2^{\bullet-}] \longrightarrow h\nu\ (490\ nm), \qquad [288]$$

$$[Btyr^{\bullet+}\,O_2^{\bullet-}] \longrightarrow h\nu\ (530\ nm), \qquad [289]$$

where $[Tyr^{\bullet+}\ O_2^{\bullet-}]$ and $[Btyr^{\bullet+}\ O_2^{\bullet-}]$ are exciplexes.

The myxoedematous form of endemic cretinism, which is still prevalent in many parts of the world (CONTEMPRE et al. 1991), is not only due to the lack of iodine, but also of selenium. The selenoprotein type I 5'-deiodinase catalyses the deiodination of the prohormone L-thyroxine (3,3',5,5'-tetraiodo-L-thyronine, T_4) to the biologically active form 3,3',5-triiodo-L-thyronine (L-T_3). During fetal development the maternal organism contributes at least minimal amounts of thyroid hormone for the fetus. After birth, the baby then slowly reaches the condition of thyroid hormone deficiency, because continuous stimulation of the thyroid by thyrotropic hormone under these conditions generates massive amounts of H_2O_2 and oxygen radicals which cannot be used or degraded in the newborn thyroid for two reasons: firstly, the lack of iodine and a substrate for the iodination and coupling reaction of thyroglobulin catalyzed by thyroid peroxidase does not consume H_2O_2 and oxygen radicals; secondly, the normal thyroid gland is known to contain selenium concentrations higher than the kidney and some other endocrine organs (for review see KÖHRLE 1994, 1995). The high concentration of Se in the thyroid is partially due to high concentrations of glutathione peroxidase (EC 1.11.1.9), which is involved in destruction of toxic H_2O_2 and oxygen radicals. As expression of glutathione peroxidase depends strictly of Se supply, fuelling into the fatal circle of enhanced H_2O_2 and oxygen radical production, concomitant to decreased consumption and degradation, finally results in a complete destruction of thyroid tissue.

In the *rat*, even during servere experimental selenium depletion, thyroid 5'-deiodinase activity is still maintained at roughly normal levels, while 5'-deiodinase of liver and kidney are severely decreased or lost (CHANOINE et al. 1993, MEINHOLD et al. 1993).

Plasma membrane association of guanylate cyclase (EC 4.6.1.2) activity and α_1, β_1 and/or β_2 soluble guanylate cyclase immunoreactivity were found in *rodent* thyreocytes (GOSSRAU et al. 2001).

Transcytosis in thyroid follicle cells is of particular interest, because it has been recognized as the cellular prerequisite for the appearance of thyroglobulin in the circulation (HERZOG 1983, 1984). Morphometric analyses showed that the rates of endocytosis and transcytosis are TSH-dependent (ROMAGNOLI and HERZOG 1991). Transcytosis is affected by temperature in a similar way as transfer to lysosomes, suggesting the existence of a common gating step for both pathways. The transfer of endocytic vesicles to multivesicular bodies continued in thyreocytes at 15 °C (ROMAGNOLI and HERZOG 1991), while in Madin-Darby *canine* kidney cells it was completely blocked at 18 °C (VAN DEURS et al. 1990).

For the *ascidian* endostyle FUJITA (1975) wished to emphasize that organic iodide could be reabsorbed into the cytoplasm, to be contained in the multivesicular bodies and lysosomes.

In the Indian rock pigeon, *Columba livia* subcutaneous implantation of melatonin-beeswax pellets (4 mg melatonin + 30 mg wax) after 2 weeks resulted in augmentation in the levels of superoxide radical in the thyroid gland (spin-trapped by EPR spectroscopy) with a concomitant decrease in the levels of the total superoxide dismutase activity (PRAKASH et al. 1998). Significantly lower plasma levels of T_4 and T_3 in treated birds indicated that subcutaneous implantation of melatonin has an inhibitory effect on thyroid and extrathyroid iodine metabolism.

Polychlorinated biphenyls (PCB, 5, 50, and 500 ppm mixed into powdered Purina rat chow using corn oil as a vehicle) given to male Osborne-Mendel *rats* for 4 weeks resulted in a dose-dependent hypertrophy and hyperplasia of follicular cells with abnormal accumulation of large colloid droplets and irregular lysosomes (COLLINS and CAPEN 1980). There was limited evidence of colloid droplet-lysosome interaction necessary for the secretion of thyroid hormones. Microvilli on luminal surfaces were decreased, abnormally shaped and short. Serum thyroxin and triiodothyronine were decreased significantly after feeding PCB.

Incubation of *bovine* Cu,Zn superoxide dismutase (EC 1.15.1.1) with Aroclor 1242 or 2,2',4,4'-

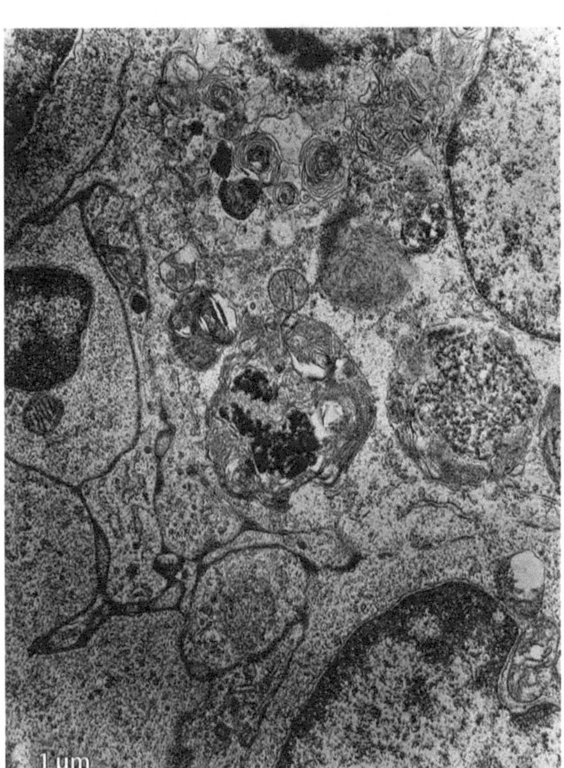

Fig. 273. Thyroid gland (block 381) from a male Wistar *rat* (No. 29/103 E) treated for 7 consecutive days with intraperitoneal injection of 5 mg *N*-benzhydryl-*N'*-*p*-hydroxybenzyl-piperazine HCl per kg body weight × day. After an hypoxia (5 % oxygen) of 30 min, under pentobarbital anaesthesia (30 mg/kg) the animal was perfused from the abdominal aorta with 2.5 % glutaraldehyde in 0.1 M sodium cacodylate buffer (pH 7.4). Postfixation with 1 % osmium tetroxide in sodium cacodylate buffer. Embedded in Epon 812 and sectioned at 50 nm. Lead citrate and uranyl acetate. Plate 3036

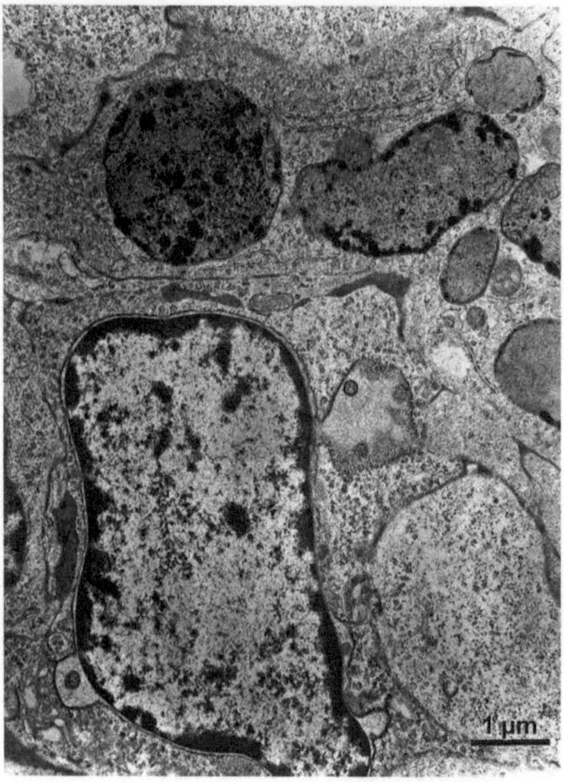

Fig. 274. Evagination of blebs of the nuclear envelope. Thyroid gland (block 381) from a male Wistar *rat* (No. 29/103 E) treated for 7 consecutive days with intraperitoneal injection of 5 mg *N*-benzhydryl-*N'*-*p*-hydroxybenzyl-piperazine HCl per kg body weight × day. After an hypoxia (5 % oxygen) of 30 min, under pentobarbital anaesthesia (30 mg/kg) the animal was perfused from the abdominal aorta with 2.5 % glutaraldehyde in 0.1 M sodium cacodylate buffer (pH 7.4). Postfixation with 1 % osmium tetroxide in sodium cacodylate buffer. Embedded in Epon 812 and sectioned at 50 nm. Lead citrate and uranyl acetate. Plate 3018

tetrachlorobiphenyl (PCB47), a constituent of Aroclor 1242, in a cell-free system reversed the enzyme-mediated inhibition of 6-hydroxydopamine autoxidation, indicating that polychlorinated biphenyls inhibited superoxide dismutase activity (NARAYANAN et al. 1998).

The **cyanide** ion is readily absorbed from the lungs and reacts readily with the trivalent iron of cytochrome oxidase in the mitochondria. Cellular respiration is thus inhibited and cytotoxic hypoxia results (KLAASSEN 1985). In the thyroid gland CN^- competes with iodine. At 4 h ^{131}I thyroid uptake studies showed a much higher concentration of iodine in the glands of cyanide-exposed workers than in those of the control group (EL GHAWABI et al. 1975).

Hyperthyroidism increased the risk of ozone-induced lung toxicity in Sprague-Dawley *rats* (HUFFMAN et al. 2001).

Comparative Physiology

Freshly isolated *pig* thyrocytes and primary cultures of *pig* thyroid exhibit several peculiarities compared to model systems using thyrocytes from other species (KÖHRLE 1997).

- High saturation by iodide;
- Different response to TSH and other growth factors (GÄRTNER et al. 1990), GORETZKI et al. 1990, DUMONT et al. 1992, VASSART and DUMONT 1992);
- Expression of functional 5'-deiodinase is surprisingly low in the thyroid tissue and cells of the "omnivorous" *pig* (similar to the *herbivores*) compared to other omnivores such as *man*, *rodents*, and *guinea pigs* (BEECH et al. 1993). KÖHRLE could not even detect steady-state mRNA levels by Northern blotting analysis or reverse transcriptase polymerase chain reaction using primer pairs complementary to the *rat* or *human* 5'-deiodinase sequence.
- Freshly isolated *pig* thyrocytes contain unusually high (free?) thyroid hormone concentrations (NISHIDA et al. 1991).

15.1.4 Adrenal

Several different fine structural characteristics were found in the **mitochondria** of the zona fasciculata of the of the *rat* adrenal cortex in both sexes regardless of sex and body weight (KADIOGLU and HARRISON 1972). The diameters varied from 1.3 × 0.8 µm to 8.4 × 6.2 µm. Although the shape changed from cell to cell they were generally oval or nearly round. Following complete hypophysectomy, in the outer zona fasciculata of Sprague-Dawly *rats* mitochondrial volume increased from $1.1 \pm 1.23\ \mu m^3$ to $36.4 \pm 27.45\ \mu m^3$ in giant mitochondria, while adjacent mltochondria measured $0.5 + 0.32\ \mu m^3$ (VOLK and SCARPELLI (1966). In *rats* treated with 1 mg progesterone per kg body weight × day for six days the number of mitochondria and mitochondrial cristae increased (VOLK 1971). Aminoglutethimide, which inhibits 11β-hydroxylation in corticoid biosynthesis (LIDDLE 1960), induced mitochondrial alterations consisting of interruption of the external limiting membranes and cavitation of the matrix, whereas mitochondrial hypertrophy predominated at later intervals (RACELA et al. 1969).

A haemoprotein spectroscopically fairly well agreeing with hepatic **cytochrome** b_5 was isolated from *hog* adrenals by KRISCH and STAUDINGER (1958). At the light microscopic level, parenchymal cells of the *bovine* zona fasciculata and zona reticularis stained heavily for cytochrome P-450$_{11\beta}$ and cytochrome P-450$_{scc}$, while the parenchymal cells of the zona glomerulosa stained lightly for both (MITANI et al. 1982).

Cytochrome P-450$_{11\beta}$ is an intrinsic membrane protein embedded in the inner mitochondrial membrane (LOMBARDO et al. 1986). Inclusion of the into artificial liposomes resulted in a marked stabilisation of the cytochrome activity. After treatment of the liposome-integrated cytochrome the native protein moiety (47 kDa) rapidly disappeared, while a major 34 kDa peptide component was formed. This peptide core retained the heme moiety and part of the cytochrome steroid-11β hydroxylase activity. Cytochrome P-450$_{11\beta}$ displayed maximal activity in lipid vesicles composed of saturated lipids, such as dipalmitoyl and dimyristoyl phosphatidylcholines, with turnover numbers ranging from 35 to 60 min^{-1} (SEYBERT 1990). Incremental increases of phospholipids resulted in a progressive inhibition of 11β hydroxylase activity; most of this kinetic effect is attributable to a significant decrease in V_{max} accompanied by modest changes in K_m for the steroid substrate deoxycorticosterone. Diphosphatidyl glycerol (cardiolipin), which has been shown to active cytochrome P450$_{scc}$, is a potent inhibitor of the 11β hydroxylase activity of cytochrome P-450$_{11\beta}$, with half maximum inhibition observed vesicles containing 4–5 mol% diphosphatidyl glycerol. Kinetic analysis demonstrated that this inhibition by diphosphatidyl glycerol is reflected in both a decrease in V_{max} and relative large increases (up to sevenfold) in K_m for the steroid substrate.

Metyrapone, 2-methyl-1,2-di-3-pyridyl-1-propanone, is a potent reversible inhibitor (DOMINGUEZ and SAMUELS 1963, SANZARI and PERON 1966, SATRE and VIGNAIS 1974) of the and cytochrome P-450$_{11\beta}$-hydroxylase enzyme system of the adrenal cortex. A quantitative structure activity relationship has been established between 5-hydroxytryptamine receptor agonist activity of 8-hydroxy-2-alkylamino)tetralins and adrenal cortical 11β-hydroxylase inhibition potency of metyrapone derivatives and the van der Waals volume of the compound (SINGH 1986).

The thickness of the zona glomerulosa decreased in 6-month-old Wistar *rats* exposed to chronic normobaric hypoxia (LORENTE et al. 2002). The main ultrastructural changes were: 1) a decrease in, or complete elimination of lipid droplets content; 2) a marked increase in the lysosome number; and 3) giant mitochondria.

A stratum of cells that did not contain both aldosterone synthase cytochrome P450 (cytochrome P-450$_{ald}$) and cytochrome P-450$_{11\beta}$ was found immunohistochemically between the zona glomerulosa and zona fasciculata of *the* rat adrenal cortex (MITANI et al. 1994). It was 5–10 cells thick under normal feeding conditions, but diminishes to 4–5 cells when animals were maintained under Na restriction, which is known to stimulate the secretion of angiotensin II.

While CYP11A1, catalysing the initial step of steroid hormone synthesis, i.e., the conversion of cholesterol to pregnenolone, is expressed in all steroidogenic tissues (adrenal, testes, ovary), CYP11B1, CYP11B2, and CYP21 are present only in the adrenal (WATERMAN and SIMPSON 1990). CYP11B1 catalyses the 11β-hydroxylation of 11-deoxycortisol and 11-deoxycorticosterone to give cortisol and corticosterone, respectively, and CYP11B2, being 93% identical in its nucleotide and protein sequence to CYP11B1 (MORNET et al. 1989, KAWAMOTO et al. 1990), is involved in the conversion of corticosterone to aldosterone. CYP21 participates in the formation of 11-deoxycorticosterone from progesterone and 11-deoxycortisol from 17-hydroxyprogesterone. Besides tissue specificity, differences in zonal distribution of some cytochromes P450 can also be observed. CYP11A1 was shown to be located in the zona fasciculata as well as in the zona glomerulosa. CYP11B1 has been traced to the zone fasciculata only (HO and VINSON 1993, MITANI et al. 1994, ERDMANN et al. 1995). In contrast, CYP11B2 was found only in the zona glomerulosa (HO and VINSON 1993, MITANI et al. 1994, SANDER et al. 1994).

In Cushing's disease, the P-450$_{11\beta}$ immunoreactivity was intensive in cortical micronodules and inner zona fasciculata and zona reticularis, and relatively intensive in the zona glomerulosa and outer zona fasciculata in idiopathic hyperaldosteronism, corresponding to the sites of active steroidogenesis (SASANO et al. 1988).

Resonance Raman scattering experiments on CO-complexed cytochrome P-450$_{scc}$ from *bovine* adrenocortical mitochondria demonstrated the simultaneous enhancement of ν(Fe–CO) stretching and bound ν(C–O) stretching frequencies at 477 and 1953 cm^{-1}, respectively (TSUBAKI and ICHIKAWA 1985).

In cultured adrenocortical cells CYP11B1 undergoes rapid inactivation in the presence of so-called pseudosubstrates, whereas no such drastic inactivation has been observed with CYP11A1. The inactivation was suggested to result from generation of reactive oxygen species (HORNSBY 1986).

CYP1B1 protein displays a triphasic pattern of expression in adrenocortical microsomes from foetal and neonatal *rats* (BRAKE et al. 1999). It paralleled changes in microsomal metabolism of 7,12-dimethylbenz[a]anthracene, a marker of CYP1B1 activity.

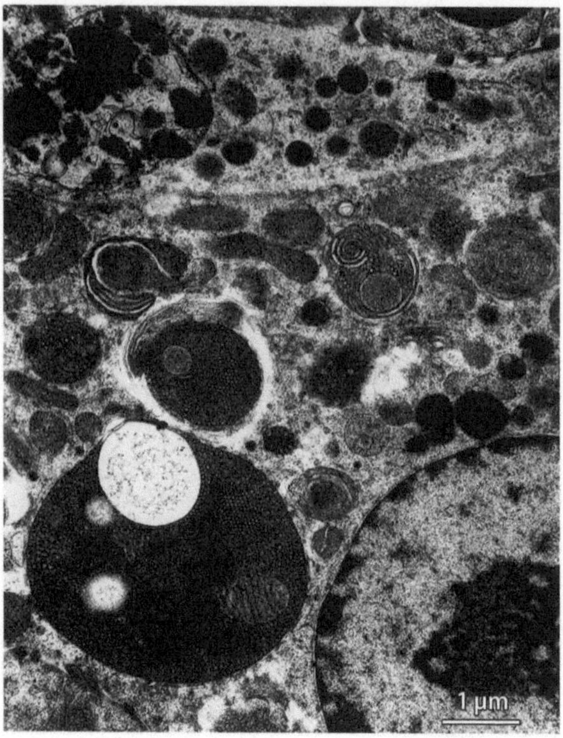

Fig. 275. Giant mitochondria with a very dense arrangement and orientation of the vesicular cristae to a hexagonal pattern. Zona fasciculata (block BNh 2111) of the adrenal cortex of a Wistar *rat* (No. 1889) medicated for 4 consecutive days with intraperitoneal injection of 300 mg piracetam per kg body weight × day. On October 31, 1974 under pentobarbital anaesthesia (30 mg/kg) the animal was perfused from the abdominal aorta with 2.5 % glutaraldehyde in 0.1 M sodium cacodylate buffer (pH 7.4). Postfixation with 1 % osmium tetroxide in sodium cacodylate buffer. Embedded in Epon 812 and sectioned at 50 nm. Lead citrate and uranyl acetate. Plate 2133

Superoxide dismutase (EC 1.15.1.1) activity was found to be very high in young immature Swiss *mice* (VERMA et al. 1991). After a subcutaneous injection of 20 µg TSH, there was a dramatic loss ($P < 0.01$) of SOD in the adrenal. The activity reached its lowest level at 30 min after the injection. But at 45 min after the injection, there was a significant ($P < 0.01$) reversal of the TSH-induced SOD depletion. No significant reversal could be noted thereafter.

In *dogs* subjected to repetitive coronary artery underperfusion (duration of the experiments 1 to 8 h), but not in normal animals or in *dogs* made continuously hypoxic by surgically induced aortic stenosis, BACHMANN (1952, 1953) found polymorphonuclear leucocytes infiltrating the anoxic parenchyma of the adrenal cortex, presumably the zona fasciculata and zona reticularis.

Temporary clamping of the circulation to the *rat's* adrenal for a period of 1 h gave rise to wedges of infarction in the zona fasciculata and zona reticularis not involving either the zona glomerulose and medulla (KOVÁCS et al. 1966). Five days pretreatment with compound 48/80 almost completely prevented the infarction.

Deta nonoate, a zwitterion nitric oxide donor, potently inhibited forskolin- and angiotensin II-stimulated aldosterone production in *human* adrenocortical H295R cells in a concentration-dependent manner (KREKLAU et al. 1999). The half-maximal and maximal inhibition of forskolin-evoked aldosteronogenesis occurred at 0.6 and 100 µM deta nonoate, respectively. The respective half-maximal and maximal deta nonoate inhibition of angiotensin II-stimulated aldosterone generation occurred at 150 µM and 1 mM. In H295R cells, deta nonoate and sodium nitroprusside did not stimulate cGMP production, and the soluble guanylate cyclase inhibitor oxadiazoloquinoxalinone (10 µM) did not block deta nonoate-mediated attenuation of aldosteronogenesis. 25-Hydroxycholesterol (10 µM)-facilitated aldosterone synthesis was also diminished with half-maximal and maximal inhibition occurring at 120 µM and 1 mM deta nonoate, respectively.

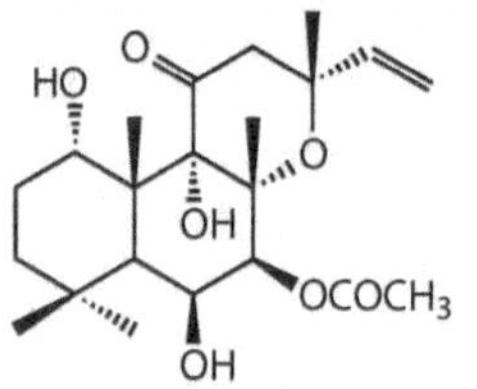

Forskolin [238]

The adrenal catecholamine content of *guinea pigs* and *rats* exposed to 7.5 per cent oxygen in nitrogen for periods of 4, 8, and 12 h decreased significantly (STEINSLAND et al. 1970). The rate of adrenal catecholamine depletion in hypoxic *rats* treated with α-methyl-L-tyrosine (100 mg/k i.p. at 5-h interval) followed a patters which reflects an alteration in catecholamine release after 12 h.

A catecholaminergic cell line, *rat* pheochromocytoma (PC 12) cells, that has been proven to be a useful system to study hypoxia-regulated gene expression, at 5 % O_2 rapidly induced a persistent phosphorylation of cyclic AMP response element binding protein (CREB) on Ser[133] (BEITNER-JOHNSON and MILLHORN 1998), an element that is required for CREB-mediated transciptional activation (GONZALEZ and MONTMINY 1989, BONNI et al. 1995). PC 12 cell cultures exhibited a loss of cells and increase in intracellular oxidative stress when exposed to ethanol for 24 h (LI et al. 2001). Catalase (EC 1.11.1.6) can attenuate this ethanol-induced cell loss and oxidative stress.

H_2O_2 stimulation of PC12 cells resulted in weak activation of both the ERK and JNK signal transduction pathways (Hassan et al. 2002). α-Phenyl-*tert*-butyl nitrone pre-treatment of PC 12 cells followed by H_2O_2 stimulation, resulted in strong and selective activation of the pro-survival ERK pathway. H_2O_2 induction of ERK activity in α-phenyl-*tert*-butyl nitrone pre-treated cells was shown to be dependent on extracellular Ca^{2+} influx. Further analysis of the ERK pathway showed that in α-phenyl-*tert*-butyl nitrone pre-treated cells, epidermal growth factor receptor and the adapter protein SHC were phosphorylated in a Ca^{2+}-dependent, ligand-independent manner following H_2O_2 stimulation.

Utilizing *rat* PC12 cells neuronally differentiated with nerve growth factor, Facchinetti et al. (2002) observed that concentrations of H_2O_2 inducing apoptotic cell death rapidly triggered the expression of Fas mRNA and protein as well as FasL mRNA. Although nerve growth factor addition to naïve PC12 downregulated constitutive Fas and FasL transcription, the H_2O_2-induced Fas and FasL mRNA upregulation invariably occurred either in the presence or in the absence of nerve growth factor. Similarly, 12-*O*- tetradecanoylphorbol-13-acetate, a potent protein kinase C activator, did not modify Fas and FasL mRNA upregulation subsequent to H_2O_2 exposure. On the contrary, forskolin and dibutyryl cAMP, which elevate intracellular cAMP by independent mechanisms, both counteracted H_2O_2-induced Fas, but not FasL, mRNA upregulation and increased constitutive expression of FasL mRNA.

Exposure of PC12 cells to reagent peroxynitrite promoted the release of arachidonic acid mediated by activation of phospholipase A_2 (Guidarelli et al. 2000). Guidarelli and Cantoni (2002) presented experimental evidence consistent with the notion that this response is not directly triggered by peroxynitrite but, rather, by reactive oxygen species generated at the level of complex III of the mitochondrial respiratory chain. In particular, $O_2^{\bullet-}$ (and not H_2O_2) has a pivotal role in peroxynitrite-dependent activation of phospholipase A_2. This observation was confirmed by results showing that superoxide, or peroxynitrite, promotes release of arachidonic acid in isolated mitochondria. Consistently, the release of arachidonic acid elicited by either peroxynitrite or A23187 in intact cells was shown to be calcium-dependent and differentially affected by phospholipase A_2 inhibitors with different levels of specificity. In particular, the effects of peroxynitrite, unlike those of A23187, were both sensitive to low concentrations of two general phospholipase A_2 inhibitors and insensitive to arachidonyltrifluoromethyl ketone, which shows some selectivity towards cytosolic phospholipase A_2.

Exposure of PC12 cells to 4-hydroxynonenal resulted in a decrease in levels of 3-(4,5-dimethylthiazol-2-yl)-2,5-diphenyltetrazolium bromide reduction, which was due to necrotic and apoptotic cell death (Nakajima et al. 2002). Addition of interleukin-6 24 h before 4-hydroxynonenal treatment provided a concentration-dependent protection against 4-hydroxynonenal toxicity, whereas neither IL-1β nor IL-2 had any effect. Addition of glutathione (GSH)-ethyl ester, but not superoxide dismutase or catalase, before 4-hydroxynonenal treatment to the culture medium protected PC12 cells from 4-hydroxynonenal toxicity. Nakajima et al. found that IL-6 increases intracellular GSH levels and the activity of γ-glutamylcysteine synthetase in PC12 cells. Buthionine sulfoximine, an inhibitor of γ-glutamylcysteine synthetase, reversed the protective effect of IL-6 against 4-hydroxynonenal toxicity.

Programmed PC12 cell death induced by glutathione depletion due to buthionine sulphoximine (100 μM) treatment was blocked by inhibitors of 12-lipoxygenase (nordihydroguairetic acid or baicalein), but did not appear to mediated through the formation of 12-hydroxyeikosatetraenoic acid derivatives (Le Foll and Duval 2001).

L-Arginine–nitric oxide pathway

$$\text{L-Arginine} \xrightarrow{\text{NOS}} \textbf{Nitric oxide}(\text{NO}^{\bullet}) \xrightarrow{+O_2^{\bullet-}} \text{Peroxynitrite } (\text{ONOO}^-)$$

[46]

•NO produced a large stimulation of basal catecholamine secretion in chromaffin cells isolated from *bovine* adrenal glands (Oset-Gasque et al. 1994). This effect was calcium- and concentration-dependent (EC_{50} = 64±8 μM) and was not due to unspecific damage of the tissue by •NO. •NO also modulated the catecholamine secretion evoked by nicotine in a dose-dependent manner.

A NO synthase in the adrenal gland has been identified and partially characterised (Palacios et al. 1989). Cyclic GMP levels in the soluble fraction of homogenates of *rat* adrenal glands were stimulated by sodium nitroprusside and by *S*-nitroso-*N*-acetylpenicillamine, known activators of soluble guanylate cyclase, demonstrating the presence of this enzyme in the preparation. The soluble guanylate cyclase from both the cortex and the medulla was also stimulated in the presence of L-arginine and this stimulation was inhibited by haemoglobin. The L-arginine:NO synthase present in both the cortex and the medulla is NADPH- and Ca^{2+}-dependent, like that in the vascular endothelium, the platelet and the brain.

Addition of NO donors, 1mM *S*-nitroso-*N*-acetyl-DL-penicillamine or 1 mM diethylenetriamine-NO adduct (NOC-18), to PC12 cells resulted in a steady-state level of 1–3 μM •NO, rapid and almost complete inhibition of cellular respiration (within 1 min), and a rapid decrease in mitochondrial membrane potential within the cells (BAL-PRICE and BROWN 2000). A 24-h incubation of PC12 cells with NO donors or specific inhibitors of mitochondrial respiration (myxothiazol, rotenone, or azide), in the absence of glucose, caused total ATP depletion and resulted in 80–100% necrosis. The presence of glucose almost completely prevented the decrease in ATP level and the increase in necrosis induced by NO donors or mitochondrial inhibitors.

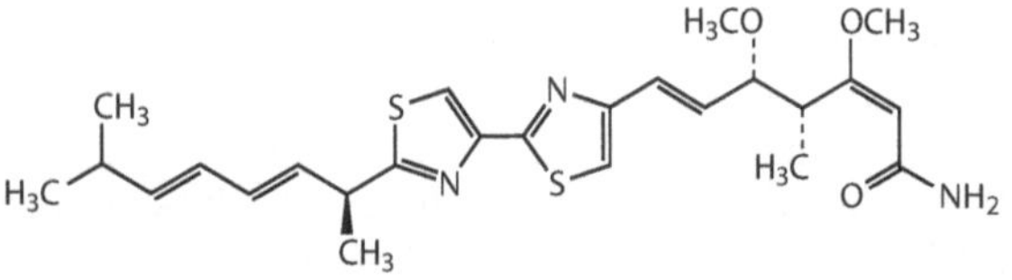

Myxothiazol [290]

Soluble guanylate cyclase (EC 4.6.1.2) in the 100,000 × g supernatant fraction from *bovine* adrenal cortex homogenates were activated by the NO-generating compounds sodium nitroprusside (optimal concentration 30 μM) and *N*-methyl-*N'*-nitro-*N*-nitrosoguanidine (optimal concentration 1 μM), a highly potent carcinogen (STRUCK and GLOSSMANN 1978).

The functional importance of this pathway in regulating adrenal cortical and medullary functions is not clear. Cyclic GMP has been implicated in both steroidogenesis (SHARMA et al. 1974, 1976, PERCHELLET et al. 1978, PERCHELLET and SHARMA 1979) and catecholamine secretion (DEROME et al. 1981, DOHI et al. 1983, O'SULLIVAN and BURGOYNE 1990).

Preincubation of *bovine* chromaffin cells with C-type natriuretic peptide, which increased cGMP levels and activated cGMP-dependent protein kinase, or with cGPM-permeant analogue (which also activates cGMP-dependent protein kinase), in the presence of a broad-spectrum phosphodiesterase inhibitor, resulted in a decrease in subsequent sodium nitroprusside -dependent cGMP elevations (FERRERO et al. 2000).

ZHENG and ZHAO (2001) reported a significant increase in the expression of transferrin receptor mRNA and a corresponding increase in cellular ^{59}Fe net uptake by PC12 phaeochromocytoma cells, but not by *murine* astrocytes, after Mn (200 μM) exposure for 3 days.

15.1.5 Carotid Body

The carotid body is known to function as a chemoreceptor sensitive to hypoxia, hypercapnia and increasing acidity in the arterial blood supply (BISCOE et al. 1970).

The *rat* carotid body possesses an NADPH oxidase, which shows certain properties similar to the ones of NADPH oxidase in neutrophils (ACKER et al. 1989). NADPH oxidase of the carotid body has been suggested to play a role as sensor for the oxygen concentration in the arterial blood (ACKER et al. 1989). The *rat* carotid body shows a typical spectrum of cytochrome *b* and diphenylene iodonium inhibits H_2O_2 formation (CROSS et al. 1990).

In response to intermittent hypoxia (4% O_2; 5% CO_2), incubated *cat* carotid bodies released catecholamines (WANG and FITZGERALD 2002). The M2 muscarinic receptor agonist, methoctramine, enhanced the hypoxia-induced release of catecholamines during each exposure.

In *rats* kept in a hypobaric chamber at a pressure of 460 mm Hg, LAIDLER and KAY (1975) found a linear relation between the combined total volume of the carotid bodies and the combined volume of glomic cells (r = 0.92, $P < 0.001$). There was a linear relation between the volume of glomic cells and the duration of exposure (27, 28, and 35 days, respectively) to chronic hypoxia (r = 0.63, $P = 0.05$). The type I cells were enlarged due to an increase in the volume of their cytoplasm (LAIDLER and KAY (1978). Many of their dense-core vesicles were vacuolated and the core was displaced eccentrically to become adherent to the limiting membrane of the vesicle. The concentration and distribution of dense core vesicles remained unaltered and there were no obvious changes in the mitochondria, ribosomes or Golgi apparatus. There were no structural changes in the type II cells.

Consequently, hypertrophy of the carotid bodies has to be expected under all pathological conditions that are accompanied by hypoxia, such as cyanotic congenital heart disease, emphysema and chronic bronchitis, but the ultrastructure of carotid bodies from *human* beings has been found to be normal when compared to the fine structure of carotid bodies from experimental animals (BÖCK et al. 1970, SIMÁRSZKY and LAPIS 1970).

15.1.6 Testis

The testicular production of free radicals and the function of the antioxidative defence system have a role in infertility caused by defective sperm function and in testicular damage in cryptorchidism (Ahotupa and Huhtaniemi 1992) or on exposure to toxic chemicals as cigarette smoking (Peltola et al. 1994), ethanol (Rosenblum et al. 1985), and 7,12-dimethylbenz(*a*)anthracene (Georgellis et al. 1987).

In *mouse* Leydig cell culture, testosterone inhibited cAMP-induced de novo synthesis of Leydig cell P-$450_{17\alpha}$ by an androgen receptor-mediated mechanism (Hales et al. 1987). The negative effect of testosterone could be mimicked by the addition of the androgen agonist, mibolerone, and prevented by the addition of the antiandrogen, hydroxyflutamide. Neither oestradiol nor dexamethasone had any effect on the synthesis of P-$450_{17\alpha}$. Studies on the degradation of newly synthesized P-$450_{17\alpha}$ demonstrated that testosterone had no effect on the decay of P-$450_{17\alpha}$ during the first 24 h but caused a significant increase in the rate of decay between 24 and 48 h.

Testosterone treatment of male Sprague-Dawley *rats* for 8 days significantly ($P < 0.01$) decreased the levels of the peroxide-metabolising enzymes, catalase, glutathione peroxidase, and glutathione transferase by 44 %, 24 %, and 31 %, respectively (Peltola et al. 1996). These changes predominantly reflect the interstitial tissue, in which catalase and glutathione peroxidase activities were much higher than in the seminiferous tubules. Testicular ZnCu or Mn superoxide dismutase activities, which were high in the seminiferous tubules, were not affected by gonadotropin suppression. The total peroxyl radical-trapping capacity of the testis, or its components, vitamin E and ubiquinol 9, were not affected either. Lipid peroxidation was decreased after 8-day testosterone treatment, as detected by diminished formation of conjugated dienes and fluorescent chromolipids (-30 % and -19 %, respectively; $P < 0.05$ for both).

Cadmium-induced acute testicular toxicity and testicular interstitial cell tumours in *rats* could be prevented by low-dose Cd (3.0 μM/kg subcutaneously) pre-treatment (Wahba et al. 1990).

Parenteral administration of **nickel** chloride ($NiCl_2$) to *rats* enhanced lipid peroxidation in liver, kidney and lung, but not in brain, heart, spleen or testis, as measured by the thiobarbituric acid reaction for malondialdehyde and related chromogens in fresh tissue homogenates (Sunderman et al. 1985).

15.1.7 Islands of Langerhans

Pancreatic B cells are sensitive to reactive oxygen species and this may play an important role in type 1 diabetes and during transplantation. B cells contain low levels of enzyme systems that protect against reactive oxygen species. The weakest link in their protection system is a deficiency in the ability to detoxify hydrogen peroxide by catalase (EC 1.11.1.6) and glutathione peroxidase (EC 1.11.1.9). Targeted overexpression of Cu/Zn superoxide dismutase protected B cells against oxidative stress (Kubisch et al. 1997). In a B cell tumour line, however, superoxide dismutase overexpression without a corresponding increase in catalase was detrimental (Tiedge et al. 1998). Xu et al. (1999) produced transgenic *mice* that have an increase in B cell catalase activity. With respect to islet function, insulin content, and morphology, these islets were completely normal. The increase in catalase activity was sufficient to protect islets against some toxins; H_2O_2 was much less potent in inhibiting insulin secretion and streptozocin was significantly less diabetogenic.

The relationship between **•NO**, endothelins, and prostaglandins have been explored in isolated pancreatic tissue from streptozocin-diabetic *rats* (González et al. 1999). The addition of nitric oxide synthase inhibitors (1 mM N^G-nitro-L-arginine methyl ester or 600 μM N^G-monomethyl-L-arginine) in the incubating medium reduced and NO donors (300 μM SIN-1 or 100 μM spermine NONOate) increased endothelin levels in pancreatic slices from control and diabetic animals.

The receptor tyrosine phosphatase-like protein islet cell autoantigen 512 binds the protein domain named for PSD-95, disc large and ZO-1 domains of β2-synthrophin and **nNOS** in pancreatic B cells (Ort et al. 2000).

The neuropeptide pituitary adenylate cyclase-activating polypeptide (1×10^{-8} M) inhibited the reduction of βTC cell viability, NO production, expression of iNOS mRNA, and iNOS promoter activity caused by the combination of three proinflammatory cytokines (Sekiya et al. 2000). Selective iNOS inhibitors also showed the cytoprotective effect in βTC cells.

Reactive oxygen species induced within the pancreatic B cells by **alloxan** are involved in the pathogenesis of type I diabetes mellitus, while extracellular generation of reactive oxygen species by the xanthin-xanthin oxidase system only retarded insulin release from the B cells (Ebelt et al. 1999). Melatonin protected B cells from alloxan-induced lesion *in vitro*.

Treatment of alloxan diabetic *rats* with *S*-allyl cysteine sulphoxide isolated from garlic (*Allium sativum* L.), ameliorated the diabetic condition almost to the same extent as did glibenclamide and insulin (Augusti and Sheela 1996). *S*-Allyl cysteine sulphoxide significantly stimulated in vitro insulin secretion from B cells isolate from normal *rats*.

In the absence of reducing agents (e.g. reduced glutathione, GSH) alloxan may act as a scavenger for $O_2^{\bullet -}$ (Brömme et al. 1999). In the presence of electron donors, however, alloxan acts as a prooxidant. The production of $O_2^{\bullet -}$ was shown by lucigenin enhanced chemiluminescence as well as by the formation of formazan from nitroblue tetrazolium. Both reactions were inhibited by superoxide dismutase and partially be catalase. Melatonin inhibited alloxan-mediated chemiluminescence.

Pre-treatment with α-phenyl-*tert*-butylnitrone (150 mg/kg i.p.) significantly reduced the severity of hyperglycaemia, the activation of nuclear transcription factor ϰB, and nitric oxide production in both alloxan- and streptozotocin-induced diabetes (Ho et al. 2000). Electron paramagnetic resonance studies showed that α-phenyl-*tert*-butylnitrone could effectively trap alloxan-induced free radicals. Activation of the nuclear transcription factor ϰB may be a key signal leading to the B cell death and insulin-dependent diabetes mellitus. In *rat* insulinoma RIN-5F cells phenyl *N-tert*-butylnitrone prevented the generation of nitrite induced by treatment with TNF-α, IL-1β, and INF-γ in a dose-dependent fashion (Tabatabaie et al. 2000). The generation of $^{\bullet}$NO as a result of cytokine treatment and the inhibitory effect of phenyl *N-tert*-butylnitrone were further confirmed by electron paramagnetic resonance spectroscopy. Aminoguanidine, a selective inhibitor of iNOS, abolished the cytokine-induced nitrite generation whereas N^{ω}-nitro-L-arginine, an inhibitor more selective for other NOS forms, was significantly less effective. Cytokine-induced nitrite formation was also inhibited by the two antioxidant agents α-lipoic acid and *N*-acetylcysteine.

Desmethylcyproheptadine 10,11-epoxide, a biotransformation product of the antihistaminic and antiserotoninergic drug, cyproheptadine inhibited proinsulin and insulin biosynthesis and insulin release in isolated *rat* pancreatic islets (Chow et al. 1989). Measurement of (pro)insulin (proinsulin and insulin) synthesis using incorporation of ^{3}H-leucine showed that desmethylcyproheptadine epoxide, desmethylcyproheptadine and cyproheptadine epoxide were 22, 10, and 4 times, respectively, more potent than cyproheptadine in inhibiting hormone synthesis. In *man*, there was no evidence for metabolic alteration at the tricyclic ethylene bridge (C-10, C-11), whereas *dog*, *cat*, and *rat* all metabolise the drug, at least in part, at this site (Porter et al. 1974). A minor *N*-oxide conjugate as a metabolite of cyproheptadine has been found in *man* (Johnson et al. 1962).

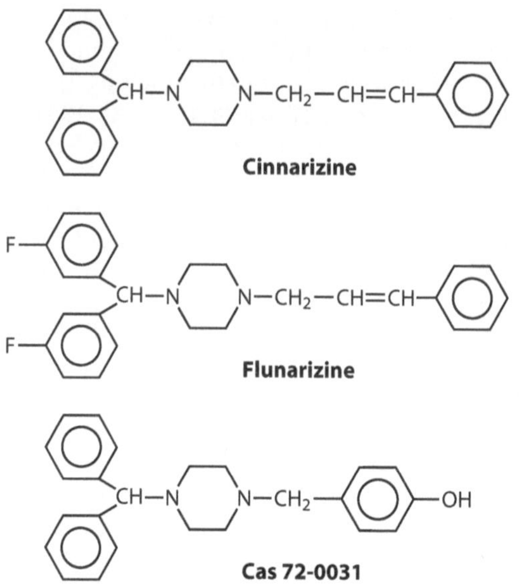

4-Diphenylmethylpiperidine derivatives [75]

Fischer et al. (1973) showed that a structural requisite for this type of islet cell toxicity was a piperidine or piperazine ring coupled to two aromatic groups through a single carbon. The aromatic rings could be free rotating as they are in 4-diphenylmethylpiperidine, or they could be part of a rigid tricyclic ring system as in cyproheptadine. Hintze et al. (1977) showed that on the piperidine ring the position of substitution with an aromatic ring-containing system is also important for the production of islet cell toxicity. Since 2- or 3-diphenylmethylpiperidine did not produce cytoplasmic vacuoles nor B-cell insulin depletion, the structural characteristics necessary for nesotoxicity include an aromatic ring system coupled to a piperidine ring only at the 4-position.

Mice of the nonobese diabetic strain develop a progressive insulinitis resulting in β-cell destruction and autoimmune-type diabetes mellitus with features mimicking the disease in *humans* (Makino et al. 1980, Atkinson and Leiter 1999). Superoxide radicals are abundantly formed by leucocytes and other mechanisms in inflammatory reactions. The overexpression of extracellular–superoxide dismutase presumably by the β-cells in the pancreata of transgenic non-obese diabetic *mice* (Sandström et al. 2002) should reduce formation of peroxynitrite from $O_2^{\bullet -}$ and $^{\bullet}$NO present in the islet interstitium.

Nitric oxide is extremely toxic for islet cells. Lysis of cultured Wistar *rat* islet cells by syngeneic macrophages activated by a heat-inactivated *Corynebacterium parvum* suspension can be inhibited by N^{G}-methyl-L-arginine as a false substrate for nitric oxide synthase in a concentration dependent man-

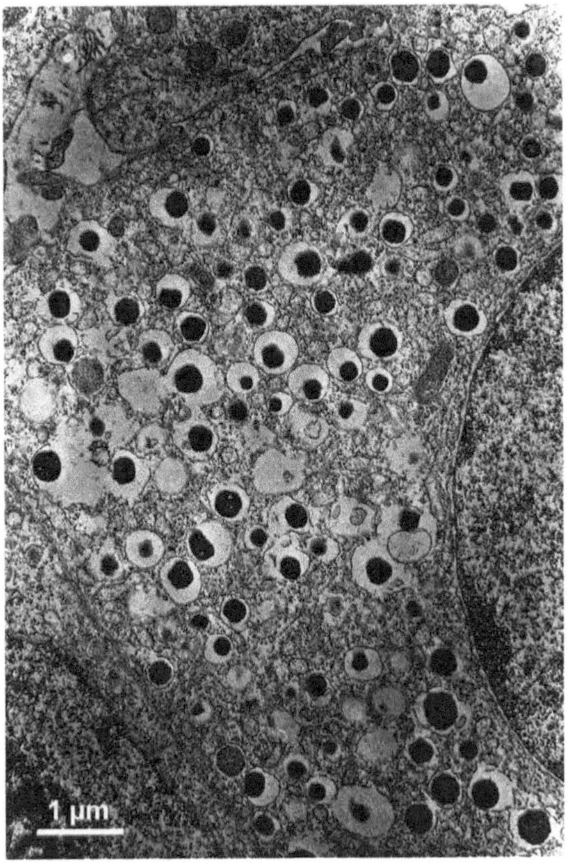

Fig. 276. Scantiness of β-granules in a B-cell of an islet of Langerhans from the pancreas (block 383) of a male Wistar *rat* (No. 29/103E) medicated for 7 consecutive days with intraperitoneal injection of 5 mg *N*-benzhydryl-*N'*-*p*-hydroxybenzyl-piperazine HCl per kg body weight × day. After an hypoxia (5% oxygen) of 30 min, under pentobarbital anaesthesia (30 mg/kg) the animal was perfused from the abdominal aorta with 2.5% glutaraldehyde in 0.1 M sodium cacodylate buffer (pH 7.4). Postfixation with 1% osmium tetroxide in sodium cacodylate buffer. Embedded in Epon 812 and sectioned at 50 nm. Lead citrate and uranyl acetate. Plate 3019

ner (Kröncke et al. 1991, 1993). This competitive inhibition can be reversed by an excess of L-arginine. Islet cells cultivated with the •NO donor, sodium nitroprusside were also lysed. Electron micrographs showed that IL-1β-induced islet cell lysis usually occurred in a chequerboard-like pattern and that islets cultured in the presence of IL-1β (50 U/ml) plus N^{ω}-monomethyl-L-arginine (0.5 mM) exhibited normal morphology comparable to untreated controls (Bergmann et al. 1992). As measured by the 'comet' assay, Delaney et al. (1993) found significant DNA damage in *rat* islets and insulin-containing *hamster* transformed B cells, HIT-T15 by treatment with 3-morpholinosydnonimine (SIN-1), which also releases superoxide radical anion (formula [136]). As damage was not reduced by superoxide dismutase (200 U/ml), it was suggested that •NO itself, rather than superoxide or peroxynitrite may be the active species. Following a 30 min exposure to *S*-nitrosoglutathione (300 µM), a nitric oxide donor thought not to generate superoxide, DNA damage was observed, with an increase in mean comet length from 50.5 ± 2.2 to 77.8 ± 6.8 and a corresponding increase in nitrite from 1.1 ± 0.5 to 95.7 ± 2.9. Damage also resulted following induction of NOS by *human* recombinant interleukin-1β (0.1 nM) in both islets and HIT-T15 cells and was prevented by replacing the substrate, arginine, with N^{ω}-methyl-L-arginine (1 mM).

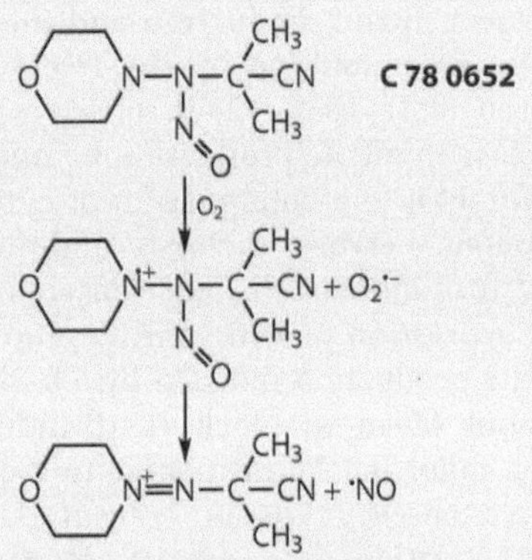

Formation of both •NO and $O_2^{\bullet-}$ from decomposed sydnonimines, e.g. C 78-0652 [136]

In search for marker molecules specifically expressed in nitric oxide (1 mM nitroprusside or 0.1–2 mM *S*-nitroso-*N*-acetyl-penicillamine as •NO donors) treated islet cells to recognise early events in islet destruction, Fehsel et al. (1997) immunocytochemically established the presence of neo-C-reactive protein in *rat* islet cells as early as 2 h after treatment.

•NO mediates zinc release from the zinc-storage protein metallothionein via *S*-nitrosation and subsequent formation of disulphides (Kröncke et al. 1994). Culture of *rat* islet cells for 24 h in the presence of nontoxic concentrations (0.5 mM) of the slow-releasing NO donor diethylenetriamine/NO resulted in a significantly reduced Zn^{2+}-dependent Zinquin (25 µM) fluorescence (Tartler et al. 2000).

The •NO mediated destruction of both *rat* and *mouse* islets of Langerhans and its effects on insulin secretion provide strong evidence for the involvement of •NO in *human* diabetes (Kröncke et al. 1998). Administration of a natural IL-12 antagonist, which suppresses the progression of islet inflammation and concomitant upregulation of iNOS (Rothe and Kolb 1999), and overexpression of the anti-apoptotic gene A20, which abrogated cytokine-induced NO production and protected both *human* and *rat* islet cells against apoptosis (GREY et al. 1999), suggest possible strategies for therapeutic intervention against NO-mediated toxicity in islet inflammation. However, it is not known whether the inhibition of *human* iNOS will reduce the destruction of >90 % of the pancreatic islets found in type-1 diabetes (ZAMORA et al. 2000).

Ebselen (20 μM) prevented the increase in nitrite production by *rat* islet cells exposed to IL-1β for 6 h and induced significant protection against the acute inhibitory effects of alloxan or H_2O_2 exposure, as judged by the preserved glucose oxidation rates (DE MELLO et al. 1996). However, ebselen failed to prevent the increase in nitrite production and the decrease in glucose oxidation and insulin release by *rat* islets exposed to IL-1β for 24 h. Ebselen prevented the increase in nitrite production by *human* islets exposed for 14 h to a combination of cytokines (IL-1β, tumour necrosis factor-α and interferon-γ). In *rat* insulinoma cells, ebselen counteracted both the expression of iNOS mRNA and the increase in nitrite production induced by 6 h exposure to IL-1β but failed to block IL-1β-induced iNOS expression following 24 h exposure to the cytokine. Moreover, ebselen did not prevent IL-1β-induced NF-κB activation.

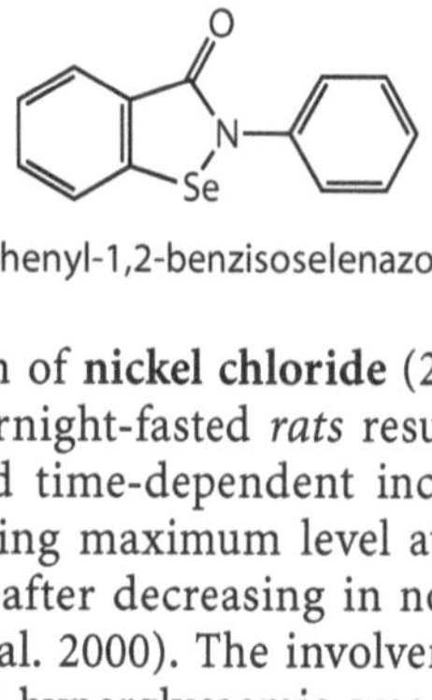

Ebselen, 2-phenyl-1,2-benzisoselenazol-3(2*H*)-one [65]

Administration of **nickel chloride** (25 to 100 μmol/kg i.p.) to overnight-fasted *rats* resulted in significant dose- and time-dependent incrate in plasma glucose, attaining maximum level at 1 h posttreatmentand thereafter decreasing in normal levels by 4 h (GUPTA et al. 2000). The involvement of •NO in nickel-induced hyperglycaemia was evident by the observation that treatment of rats with N^{ω}-monomethyl-L-arginine (10 to 50 μmol/kg i.p.) significantly attenuated the nickel-mediated increase in the plasma glucose levels in a dose-dependent fashion. The activity of cNOS was significantly decreased (2.8-fold) with a concomitant increase (11.6-fold) in inducible NOS at the same time interval. A significant increase in iNOS protein expression in the pancreas was associated with a significant elevation of cyclic GMP levels, possibly via the stimulation of cytosolic guanylate cyclase.

15.1.8 Heart

The heart is the most susceptible of all the organs to premature ageing and free radical oxidative stress. Clinical research has clearly documented the role of free radical damage and the progression of numerous degenerative diseases, particularly cardiovascular disease. This may be the result of acute ischaemia-reperfusion injury, endothelial damage of hyperhomocysteinaemia, as well as chronic oxidative damage secondary to lipid peroxidation. Heavy metals in the body, especially the transitional metals iron and copper, are capable of initiating adverse free radical reactions (SINATRA and DEMARCO 1995).

Treatment with hydrogen peroxide for 10 min following cell-permeable iron augmented the effect of iron with an increase of lactate dehydrogenase (EC 1.1.1.27) and creatine kinase (EC 2.7.3.3) in the coronary effluent of the isolated perfused *rat* heart and myocardial malondialdehyde content (CHEN et al. 2002).

The undamaged myocardium contains **dehydrogenase** and substrate which are capable of reducing the faintly yellowish tetrazolium salt nitroblue tetrazolium (formula [18]) to dark blue formazan. SANDRITTER and JESTÄDT (1957) and JESTÄDT and SANDRITTER (1959) used 2,3,5-triphenyltetrazolium chloride (formula [17]) in *human* material and *rabbits* after ligation of the Ramus descendens of the left coronary artery for microscopic identification of fresh myocardial infarction, while NACHLAS and SHNITKA (1963), ANDERSEN and HANSEN (1973) and NITZ et al. (1982) utilised nitroblue tetrazolium. 3-(4,5-Dimethylthiazolyl-2)-2,5-diphenyltetrazolium bromide also gives a blue formazan (JESTÄDT and SANDRITTER 1959).

Before percutaneous transluminal coronary angioplasty, coronary sinus concentrations of **adenosine** and **hypoxanthine** were 176±34 nMol and 723±73 nMol, respectively (BARDENHEUER et al. 1994). 30 s after balloon deflation adenosine concentrations were enhanced in close proportion to the duration of coronary occlusion: 326±47 nM [30 s], 531±80 nM [60 s], and 793±150 nM [90 s], respectively. During reperfusion similar results were also obtained in the case of hypoxanthine and uric acid.

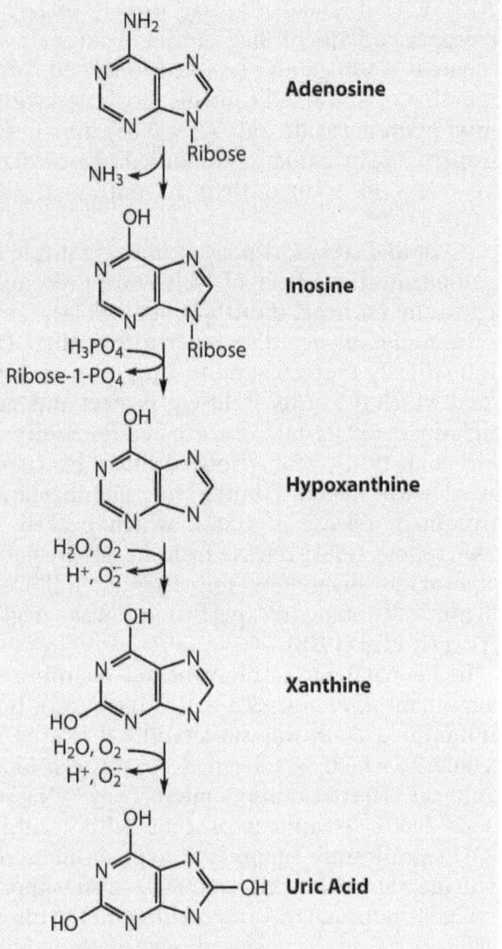

Substrates of xanthine oxidase [45]

Electron paramagnetic resonance spectroscopy has been applied to measure radical generation in the postischaemic heart. Zweier et al. (1989) found $O_2^{\bullet-}$ derived $HO^{\bullet}$, $R^{\bullet}$ and $ROO^{\bullet}$ radicals. In reperfused tissue, 3 prominent radical signals were observed: A, isotropic $g = 2{,}004$ suggestive of a semiquinone; B, anisotropic $g_{||} = 2{,}033$ and $g_{\perp} = 2{,}005$ suggestive of $ROO^{\bullet}$; and C, a triplet $g = 2{,}00$ and $a_N = 24$ G suggestive of a nitrogen centered radical. Peak signal intensities occurred after 15 s of reflow following 30 min of ischaemia.

Electron paramagnetic resonance spectroscopy showed that xanthine/xanthine oxidase in cultured *rat* cardiomyocytes produced superoxide and hydroxyl radicals during 10 min (Durot et al. 2000). The xanthine/xanthine oxidase system altered sharply and irreversibly the spontaneous electrical and mechanical activities of the cardiomyocytes. However, the gas chromatographic analysis showed that these drastic functional damages were associated with comparatively moderate degradation of membrane polyunsaturated fatty acids. Moreover the electron paramagnetic resonance analysis did not reveal the production of lipid-derived free radicals.

Using various free radical scavenging systems, Jolly et al. (1984), Burton (1985), Ambrosio et al. (1986), Bolli et al. (1987) and Bhatnagar (1995) have shown a link between myocardial ischaemia and reperfusion injury and the generation of various toxic free radical species. Sharma et al. (1994) in an electron paramagnetic resonance study found ascorbyl free radical as a real-time marker of free radical generation in briefly ischaemic and reperfused *dog* hearts. Only Uraize et al. (1987) were unable to limit the size of myocardial necrosis after 40 min of ischaemia by superoxide dismutase.

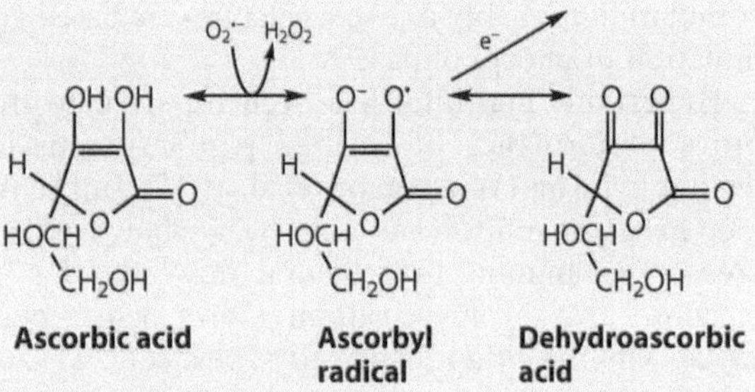

Formation of the ascorbyl free radical [53]

It is not easy to demonstrate the direct effect of free radicals on myocardial tissue. McCord and Fridovich (1969) designed an experiment that involved the electrolytic generation of the $O_2^{\bullet-}$ radical. Jackson et al. (1986) used electrolysis of the perfusion medium to study free radicals mediated tissue injury. Stewart et al. (1988) and Chahine et al. (1991) observed myocardial dysfunction in isolated perfused *rabbit* and *rat* hearts subjected to electrolysis. Lecour et al. (1998) suggested that both hydroxyl and superoxide radicals were formed during electrolysis.

Increased Ca^{2+} influx into cardiomyocytes via L-type channels precedes $O_2^{\bullet-}$ formation. In addition, ATPases have been postulated to be activated by increased calcium entry into the cells, thus further jeopardizing myocyte viability by depleting energy stores. Regulation of this calcium current is important in controlling cardiac contractility. Some of the earliest studies were performed using *frog* ventricular myocytes where cAMP and cGMP were shown to have opposite effects of calcium current (Hartzell and Fischmeister 1986). Intracellular perfusion of *frog* ventricular myocytes with cGMP or application of NO donors were shown to have no effect on basal calcium current. However, cGMP does inhibit the current when it has been pre-stimulated with isoprenaline or cAMP (MERY et al. 1995). Inhibition of cyclic GMP-stimulated phosphodiesterase by erythro-9-(2-hydroxy-3-nonyl)-adenine reverses the cGMP effect on the stimulated current with no effect on basal current indicating that cyclic GMP-stimulated phosphodiesterase is responsible for the cGMP regulation of the cAMP-stimulated calcium current.

Isolated *rat* hearts subjected to 15 min ischaemia followed by 30 min reperfusion showed concomitant accumulation of free calcium (Indo-1 fluorescence technique) and degradation of membrane phospholipids as indicated by an increase of tissue arachidonic acid content (IVANICS et al. 2001). This observation is suggestive for a relationship the Ca^{2+}-related fluorescence and arachidonic acid accumulation probably due to a calcium-mediated stimulation of phospholipase A_2.

Hypertonic **mannitol** has been reported to improve performance of isolated papillary muscles during hypoxia (WILLERSON et al. 1974). In the intact heart, mannitol has also been shown to increase the amount of collateral flow into an ischaemic area of myocardium (WILLERSON et al. 1972). On the other hand, APSTEIN et al. (1976), substituting equiosmolar mannitol for an increase in glucose concentration, could not improve mechanical performance during hypoxia and recovery of developed tension with reoxygenation was slightly but significantly worse.

The ability of **ischaemic preconditioning** to protect the myocardium against prolonged ischaemia may derive from improved energy balance. KOBARA et al. (1996) examined myocardial energy metabolism and mitochondrial oxidative phosphorylation in isolated perfused *rat* hearts which were either subjected or not subjected to preconditioning (5 min ischaemia followed by 5 min reperfusion) prior to 30 min sustained ischemia and 30 min reperfusion. The data suggested that mitochondrial ATPase contributes only slight to ATP depletion during sustained ischaemia, and both creatine phosphate overshoot phenomenon and the decrease in anaerobic glycolysis can be attributed to cardioprotection during sustained ischaemia. The preservation of ATPases and adenine nucleotide translocase activities may be a possible explanation for the restoration of high energy phosphates in the preconditioned hearts.

The activity of DT-diaphorase significantly increased in young male Wistar *rats* injected intraperitoneally with *tert*-butyl hydroperoxide (1 mmol/kg every other day for 3 weeks) and forced to swim 60 min five times a week for 6 weeks and thereafter 90 min for 3 weeks, indicating that an increase in H_2O_2 levels stimulated the activity of this enzyme (RADÁK et al. 2000). Cardiac muscle of *tert*-butyl hydroperoxide-medicated trained animals accumulated significantly less carbonylated proteins than that of untrained controls medicated with *tert*-butyl hydroperoxide only ($P < 0.05$). *tert*-Butyl hydroperoxide by oxidation modified myocardial proteins less in trained than in untrained *rats* ($P < 0.05$).

δ_1-Opioid receptors play an important role in the cardioprotective effect of ischaemic preconditioning in the *rat* heart (SCHULTZ et al. 1998).

In *rabbit* myocardial reperfusion **MnSOD** was dramatically protective up to 5 μg/ml perfusate beyond which it lost its ability to protect and, at very high doses (50 μg/ml), exacerbated the injury (NELSON et al. 1994). While isolated *rabbit* hearts reperfused with standard buffer for 45 min showed a prominent release of GSSG, which peaked 5 min after reflow, GSSH release from hearts treated with superoxide dismutase for 15 min, followed by 30 min of standard perfusion, was negligible (TRITTO et al. 1998).

In neonatal *rat* cardiomyocytes lecithinized recombinant *human* **Cu,Zn SOD** (100 U/ml), but not unmodified SOD, was successfully delivered intracellularly, which was verified by Western blot and confocal laser-scanning microscopy (NAKAJIMA et al. 2000). Treatment of cells with lecithinized SOD significantly suppressed hypoxia-induced cell damage. Since lecithinized SOD also suppressed hypoxia-induced DNA fragmentation, the improved cell survival by lecithinized SOD is thought to be mediated by its antiapoptotic effect.

15.1.8.1
Never Hypoxic Myocardial Cells

Nitric oxide may be produced within the heart by either constitutive or inducible **nitric oxide synthase**. Inducible nitric oxide synthase was found by immunogold labelling in the perinuclear space, Golgi complex, contractile fibres, and mitochondria of cultured neonatal *rat* cardiomyocytes (BUCHWALOW et al. 1997). Incubation of the cells with 10^{-3} M dibutyryl-cAMP for 24 h dramatically increased production of nitrite indicating the upregulation of iNOS. PINSKY et al. (1997) demonstrated that increasing or decreasing ventricular preload *in vivo* is followed by parallel changes in [NO], which may represent a novel autoregulatory mechanism to adjust cardiac performance or perfusion on a beat-to-beat basis. Experiments in *rabbits* and *rats* in which denuding cardiac endothelial and endocardial cells abrogated the NO signal indicated that these cells transduce mechanical stimulation into $^{\bullet}NO$ production in the heart.

In streptozotocin diabetic male Wistar *rats* the activity of total nitric oxide synthase and the amount of mRNAs encoding ecNOS and iNOS depended on the duration of diabetes (STOCKKLAUSER-FÄRBER et al. 2000). When diabetes had lasted for 4 to 6 weeks, both the total activity as well as the mRNA encoding ecNOS and iNOS were elevated. A decrease of NOS activity and the amounts of mRNA of ecNOS and iNOS was only seen after more than 20 weeks of diabetes.

There is evidence that •NO is involved in the chronotropic, the inotropic, and the vasodilator response to β-adrenoceptor agonists (SCHMETTERER et al. 1999). N^G-Monomethyl-L-arginine significantly blunted the *human* heart rate response to β-adrenoceptor stimulation by isoprenaline (0.1–0.8 μg/min) in a dose-dependent manner.

Abundance of caveolin-3 (0.59±0.08 versus 0.20±0.08 arbitrary units, $P = 0.01$) but not caveolin-1 was increased in *canine* pacing-induced heart failure compared with control conditions, assessed by Western blot (HARE et al. 2000). Transmission electron microscopy revealed increased caveolae (2.7±0.4 versus 1.3±0.3) per *μm* myocyte membrane ($P < 0.005$). The association between caveolin-3 and NOS3 at the sarcolemma and T tubules was unchanged in heart failure compared with control myocytes. The impact of NOS inhibition with L-N^G-methylarginine HCl on β-adrenergic inotropy was assessed in conscious *dogs* before and after heart failure. In control *dogs*, dobutamine (5 *μg*×kg^{-1}× min^{-1}) increased +dP/dt by 36±7%, and this was augmented to 66±24% by 20 mg/kg L-N^G-methylarginine but not affected by 10 mg/kg L-N^G-methylarginine.

The β_3 adrenoceptor antagonist, BRL 37344 induced time-dependent changes of eNOS phosphorylation at Ser^{1177} in isolated *human* cardiomyocytes of the right atrium and the left ventricle (BLOCH et al. 2002).

N^ω-Nitro-L-arginine methylester (50 mg/kg × day) raised the blood pressure of Wistar *rats* by 71% (MANDARIM-DE-LACERDA and MEIRELLES PEREIRA 2001). With respect to cardiomyocyte nuclear size, the angiotensin-converting enzyme inhibitor enalapril maleate (15 mg/kg×day) and the calcium channel blocker verapamil hydrochloride (15 mg/kg×day) showed beneficial effects when NO synthesis was blocked.

Much of the damaging action of nitric oxide in the heart may be due to its diffusion-limited reaction with superoxide to form **peroxynitrite**. DIGERNESS et al. (1999) exposed intact and skinned *rat* papillary muscles to a steady state concentration of 4 μM peroxynitrite for 5 min, followed by a 30-min recovery period to monitor irreversible effects. In intact muscles developed force fell immediately to 26% of initial force, recovering to 43% by 30 min. Resting tension increased by 600% immediately, and was still elevated 500% by 30 min. Nitrotyrosine immunochemistry showed that peroxynitrite can induce tyrosine nitration at low concentrations and is capable of penetrating 200–380 μm into the papillary muscle after a 5-min infusion. Decomposed peroxynitrite had no effect on either intact or skinned muscle developed force or resting tension.

•NO → activation of guanylate cyclase
→ guanosine 3',5'-monophosphate (cyclic GMP)↑

Using a novel antibody raised against a C-terminal peptide of the *rat* β_1 subunit of soluble guanylate cyclase, the enzyme is primarily localised to the vasculature rather than to cardiomyocytes with a switch of soluble guanylate cyclase-specific staining from arterial smooth muscle cells to endothelial cells during development (MIETENS et al. 2001). In *rat* adult cardiomyocytes, soluble guanylate cyclase expression was clearly detectable, but low in comparison to whole heart.

Treatment of isolated *rat* cardiomyocytes (for 10 min) with the membrane permeant cyclic GMP analogue 8-(4-chlorophenylthio)-cGMP (200 μM) caused a decrease in glucose transport in non-stimulated (basal) myocytes, as well as in cells stimulated with insulin or with the mitochondrial inhibitor oligomycin B by up to 40% (BERGEMANN et al. 2001). An inhibitory effect was also observed with another cGMP analogue (8-bromo-cGMP), and in cells stimulated by H_2O_2 or anoxia.

15.1.8.1.1 Mitochondria

The mitochondria in the conducting fibres are distinctly smaller than those in the working fibres (KAWAMURA 1961).

The mitochondrial **outer membrane** is a physical barrier separating the metabolic processes of the cytoplasm from those of the mitochondrion. Integrity of the outer membrane of mitochondria can be determined using a cytochrome *c* accessibility assay (DOUCE et al. 1973). Since holocytochrome *c* cannot diffuse across intact outer membranes, the ratio of the initial rate of succinate:cytochrome c oxidoreductase activity by the mitochondria to that of the same mitochondria after deliberate lysis of their outer membranes is proportional to the extent of damage to the outer membranes incurred during mitochondrial isolation.

Cell respiration in mitochondria is catalysed by cytochrome *c* oxidase, which reduces O_2 to water,

Fig. 277. An interfibrillar mitochondrium towering the length of 4 sarcomeres in a cardiomyocyte (between the anterior and posteror papillary muscles of the left ventricles; block 1127) of a never hypoxic *rat* (No. 4) medicated for 7 consecutive days with peritoneal injection of 15 mg carbocromene per kg body weight × day. On August 6, 1976 under pentobarbital anaesthesia (30 mg/kg), the animal was perfused from the abdominal aorta with 2.5 % glutaraldehyde in 0.1 M sodium cacodylate buffer (pH 7.4). Postfixation with 1 % osmium tetroxide in sodium cacodylate buffer. Embedded in Epon 812 and sectioned at 50 nm. Lead citrate and uranyl acetate. Film 627-16

coupled with translocation of four protons across the mitochondrial membrane (Wikström 1977). The catalytic cycle of the enzyme consists of a reductive phase, in which the oxidised enzyme receives electron from cytochrome *c*, and an oxidative phase, in which the reduced enzyme is oxidised by O_2. With the purified enzyme inlaid in liposomes, Verkhovsky et al. (1999) reported time-resolved measurements of membrane potential, which showed that half of the electrical charges due to proton-pumping actually cross the membrane during reduction after a preceding oxidative phase.

Contact sites were first described by Hackenbrock (1968) in thin sections of liver mitochondria as places where the inner and outer mitochondrial membranes were in very close apposition. Van Venetië and Verkleij (1982) and Knoll and Brdiczka (1983) characterized them in freeze-fractured mitochondria. Knoll and Brdiczka (1983) and Brdiczka et al. (1986) postulated that contact sites play an important role in the regulation of the mirochondrial metabolism. Under normoxic conditions, the ATP formed in the mitochondria is converted into creatine phosphate by the activity of the translocase, and the mitochondrial isoenzyme of creatine kinase (Wallimann et al. 1992). So, if the cardiac metabolism is stimulated the mitochondrial ATP formation increases, as does the mitochondrial creatine kinase. Since mitochondrial creatine kinase is active in mitochondrial contact sites (Biermans et al. 1990, Nicolay et al. 1990, Jacob et al. 1992), and can even induce contact site formation (Rojo et al. 1991), the surface density of mitochondrial contact sites in this situation will be high. Mitochondria lose the ability to form contact sites after more than 15 min of ischaemia and this might be a first indication of irreversible injury (Bakker et al. 1995).

The mitochondrial outer membrane contains numerous copies (10^3–10^4/µm^2) of a 31-kDa **porin** protein referred to as voltage-dependent, anion-

Fig. 278. Curved mitochondrium (block 134) in a cardiomyocyte from the posterior wall of the left ventricle of a never hypoxic male *rat* (No. 8) medicated for 14 consecutive days with peritoneal injection of 15 mg carbocromene per kg body weight × day and implanted subcutaneously with a Marbagelan® pellet. Under pentobarbital anaesthesia (30 mg/kg), the animal was perfused from the abdominal aorta with 2.5 % glutaraldehyde in 0.1 M sodium cacodylate buffer (pH 7.4). Postfixation with 1 % osmium tetroxide in sodium cacodylate buffer. Embedded in Epon 812 and sectioned at 50 nm. Lead citrate and uranyl acetate. Plate 2855

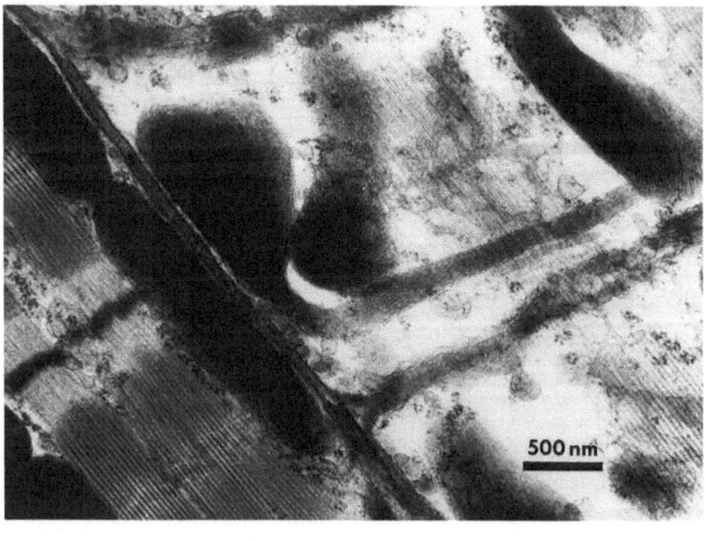

selective channel (VDAC; for review see MANNELLA et al. 1992). It is widely presumed that VDAC represents the main permeability pathway for metabolites through the outer mitochondrial membrane. There are multiple human VDAC genes which may be differentially expressed in different cells (BLACHLY-DYSON et al. 1993, 1994), suggesting heterogeneity in function and/or targeting of the channel proteins. KONSTANTINOVA et al. (1995) Studied the distribution of porin on the outer membranes of *rat* heat mitochondria by means of immunogold labelling with antibodies to the *N*-terminal part of the *human* protein. They found that only a minority of isolated, unfixed mitochondria were labelled by these antibodies, with the gold particles frequently organised in threads or bands. Extensive immunogold labelling was frequently observed on regions of outer membranes stripped away from mitochondria and on regions separating two mitochondrial compartments whose cristae display different configurations (possibly representing two mitoplasts covered by a common outer membrane). Also, pairs of connected mitochondria were sometimes heavily labelled in the "neck" regions, which may represent the junctions in electrical communication between mitochondria in cardiac tissue.

In the **inner membrane** of mitochondria, the ion channel highly selective for K^+ is blocked by adenosine triphosphate and by the antidiabetic sulfonyl urea derivative, glibenclamide (INOUE et al. 1991, PAUCEK et al. 1992). In *beef* heart mitochondria K_d for glibenclamide binding is by one order of magnitude lower, e.g. 300 nM, than in *rat* liver mitochondria (SZEWCZYK et al. 1997). Guanine nucleotides reverse the inhibition of mitoK_{ATP} channel by ATP and ADP (PAUCEK et al. 1996). The mitoK_{ATP} channel may have a dual physiological function. Firstly, a concerted action of the electrophoretic K^+ uniport and the electroneutral K^+/H^+ exchange is believed to be the main mechanism responsible for maintaining potassium homeostasis within the mitochondrium, and thus for controlling intramitochondrial osmolarity and volume of mitochondria.

Fig. 279. Prominent intramitochondrial granules in a myocardial cell between the anterior and posterior papillary muscles of the left ventricle (block 1067) of a never hypoxic *rat* (No. 2) treated for 7 consecutive days with intraperitoneal injection of 1.5 ml Tyrode's solution per kg body weight × day. On August 4, 1976, under pentobarbital anaesthesia (30 mg/kg), the animal was perfused from the abdominal aorta with 2.5 % glutaraldehyde in 0.1 M sodium cacodylate buffer (pH 7.4). Postfixation with 1 % osmium tetroxide in sodium cacodylate buffer. Embedded in Epon 812 and sectioned at 50 nm. Lead citrate and uranyl acetate. Plate 3320

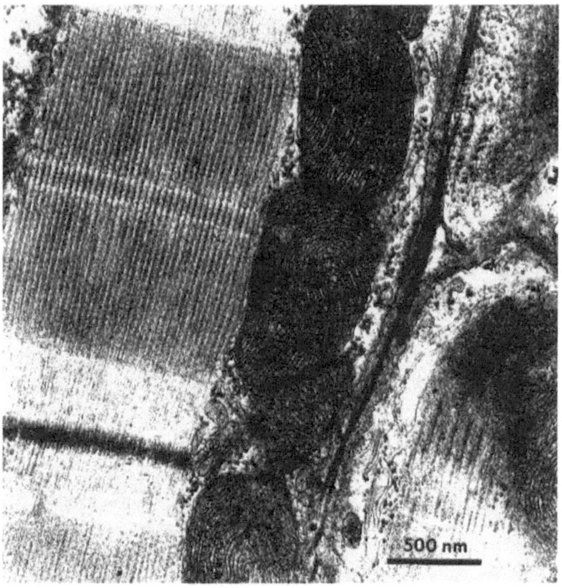

Fig. 280. Prominent intramitochondrial granules in a myocardial cell from the right ventricle (block 1070) of a never hypoxic *rat* (No. 2) treated for 7 consecutive days with intraperitoneal injection of 1.5 ml Tyrode's solution per kg body weight × day. On August 4, 1976, under pentobarbital anaesthesia (30 mg/kg), the animal was perfused from the abdominal aorta with 2.5 % glutaraldehyde in 0.1 M sodium cacodylate buffer (pH 7.4). Postfixation with 1 % osmium tetroxide in sodium cacodylate buffer. Embedded in Epon 812 and sectioned at 50 nm. Lead citrate and uranyl acetate. Film 625-01

Regulatory changes of this volume are regarded as one of the important mechanisms of metabolic control at the mitochondrial level (for review see HALESTRAP 1994).

In the mitochondria of most cells, the **matrix** contains some electron-dense granules 20 to 50 nm in diameter. These are referred to by various terms such as "dense granules", "intramitochondrial granules", "matricial or matrix granules", and "native granules" containing divalent cations such as calcium.

The palmitoylcarnitine-activated oxygen consumption in state 3 of the coupled and uncoupled *rat* heart mitochondria was markedly reduced by carbocromen, while the P/O-ratio remained unchanged (OSTROWSKI 1977). L-Carnitine developed its cardioprotective effects mainly during the reperfusion of ischaemic hearts. Addition of carnitine (5 mmol/l) to the perfusate of Langendorff-perfused *rat* hearts increased the intracellular carnitine concentrations and decreased the formation of malondialdehyde and thus obviously reduced ischaemic damage (LÖSTER and BÖHM 2001).

Generally, mitochondria are able to reduce the cytosolic **Ca^{2+} overload** by sequestering large quantities of Ca^{2+} during stimulation of the cell. However, due to their uptake of Ca^{2+} that would otherwise act as a feed-back inhibitor of influx pathways, functioning mitochondria may, under certain circumstances, increase the total amount of Ca^{2+} influx following an excitotoxic stimulation (BUDD and NICHOLLS 1996). Mitochondrial Ca^{2+} extrusion is an energy requiring process (33 kJ/mol) linked to H^+ exchange. The net effect is the import of two H^+ in exchange of one Ca^{2+} exported.

Ca^{2+} release from mitochondria is stimulated during **oxidative stress**. Oxidation of nicotinamide adenine dinucleotide phosphate (NADH), with subsequent ADP-monoribosylation of mitochondrial proteins or formation of cyclic ADP-ribose, has been suggested as a regulatory mechanism. Such enhanced Ca^{2+} extrusion may be the basis of "Ca^{2+} cycling", i.e. continuous uptake and release of Ca^{2+} into mitochondria, which leads ultimately to the dissipation of the membrane potential and to mitochondrial failure. As a consequence of the breakdown of the membrane potential ($\Delta\Psi$) or other forms of damage, mitochondria may release Ca^{2+} into the cytosol and generate reactive oxygen species. Their generation may play a role in cell death due to ischaemia/reperfusion and in normal ageing of tissues, with one such scenario calling for the

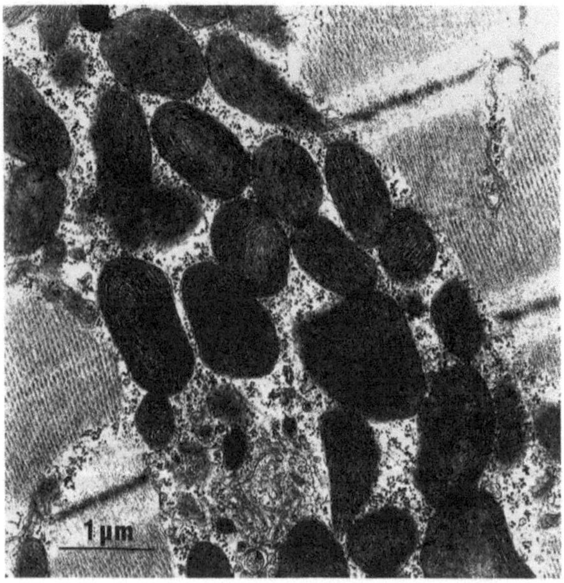

Fig. 281. Prominent intramitochondrial granules in a myocardial cell from the interventricular septum (block 1129) of a never hypoxic *rat* (No. 4) medicated for 7 consecutive days with intraperitoneal injection of 15 mg carbocromen per kg body weight × day. Under pentobarbital anaesthesia (30 mg/kg), the animal was perfused from the abdominal aorta with 2.5% glutaraldehyde in 0.1 M sodium cacodylate buffer (pH 7.4). Postfixation with 1% osmium tetroxide in sodium cacodylate buffer. Embedded in Epon 812 and sectioned at 50 nm. Lead citrate and uranyl acetate. Film 627-33

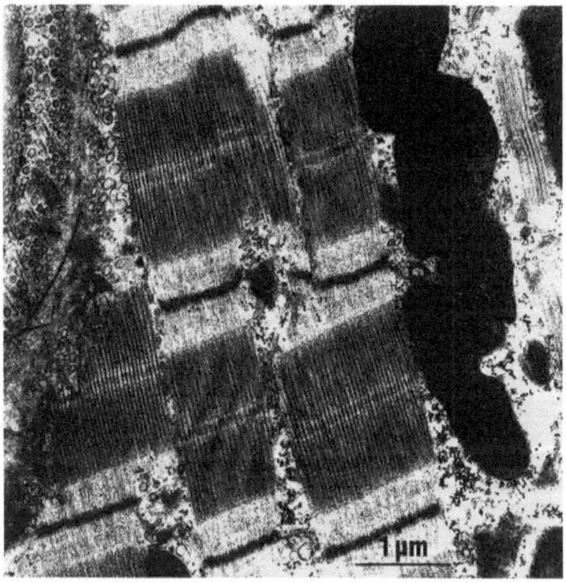

Fig. 282. Numerous intramitochondrial granules in a myocardial cell between the anterior and posterior papillary muscles of the left ventricle (block 1127) of a never hypoxic *rat* (No. 4) medicated for 7 consecutive days with intraperitoneal injection of 15 mg carbocromen per kg body weight × day. Under pentobarbital anaesthesia (30 mg/kg), the animal was perfused from the abdominal aorta with 2.5% glutaraldehyde in 0.1 M sodium cacodylate buffer (pH 7.4). Postfixation with 1% osmium tetroxide in sodium cacodylate buffer. Embedded in Epon 812 and sectioned at 50 nm. Lead citrate and uranyl acetate. Film 627-22

fixation of oxidative damage to mt-DNA as mitochondria undergo many generations of replication in nondividing cells. There have been reports of an increase of reactive oxygen species generation with age in myocardial mitochondria (NOHL and HEGNER 1978, SOHAL et al. 1994, LUCAS and SZWEDA 1998). However, HANSFORD et al. (1997) found an unchanged rate of reactive oxygen species formation in heart mitochondria from *rats* aged 24 months vs. *rats* aged 6 months, but did not rule out an accumulation of oxidative damage with ageing. To the extent that this may be true of mt-DNA (AMES et al. 1993, MECOCCI et al. 1993, AGARWAL and SOHAL 1994), this could be particularly significant.

The main **free radical generation sites** have been localised at Complexes I (TAKESHIGE and MINAKAMI 1979, TURRENS and BOVERIS 1980, HANSFORD et al. 1997, HERRERO and BARJA 1998) and III (TURRENS et al. 1985, NOHL et al. 1996. HERRERO and BARJA 1998) of the respiratory chain. The strong increase in H_2O_2 production observed in *mouse*, *parakeet* and *canary* mitochondria after one-line addition of rotenone, which inhibits electron flow from Complex I to ubiquinone, clearly indicated that Complex I contains a free radical generator site in the free species (HERRERO and BARJA 1998). The rotenone-stimulated increase was totally abolished by ethoxiformic anhydride, which inhibits NADH-ubiquinone but not NADH-ferricyanide reduction (VIK and HATEFI 1984. WYATT et al. 1995). This suggested that the Complex I generator ‘ as in the case of *rats* and *pigeons* (HERRERO and BARJA 1997), can be the Complex I iron-sulphur centres. Complex III also generated free radicals in the three species studied by HERRERO and BARJA (1998).

Incubation of *rat* heart mitochondria with ATP resulted in the phosphorylation of two mitochondrial membrane proteins, one with a M_r of 6 kDA consistent with the NDUFA1 (MWFE), and one at 18 kDa consistent with either NDUSF4 (AQDQ) or NDUFB7 (B18) (RAHA et al. 2002). Phosphorylation of both sununits was enhanced by cAMP derivatives and protein kinase A and was inhibited by protein kinase A inhibitors. When mitochondrial membranes were incubated with pyruvate dehydrogenase kinase, phosphorylation of an 18 kDa protein but not a 6 kDa protein was observed, NADH cytochrome c reductase activity was decreased and superoxide production rates with NADH as substrate were increased. On the other hand, with protein kinase A-driven phosphorylation, NADH cytochrome *c* reductase was increased and superoxide production decreased. Overall there was a 4-fold variation in electron transport rates observable at the extremes of these phosphorylation events. This suggests that electron flow through complex I and the production of oxygen free radicals can be regulated by phosphorylation events.

Knockout *mice* lacking the gene encoding the mitochondrial **manganese superoxide dismutase** develop metabolic acidosis and die a few days after birth of severe heart failure due to dilated cardiomyopathy (LI et al. 1995). Several lines of evidence demonstrate a decline of mitochobdrial enzyme activities with age, particularly complex I (TROUNCE et al. 1989). Lack of an efficient repair mechanism and absence of histones make mtDNA more susceptible to the toxic effects of free radicals. Accumulation of mtDNA mutations are proposed to cause progressive impairment of mitochondrial respiration. Deletion of mtDNA accumulate with age in the *human* heart (HATTORI et al. 1991).

Mitochondria are often located near a supply of substrate, or at sites in the cell known to require the ATP generated by the mitochondrion (LEHNINGER 1965). One of the frequently observed **associations is between mitochondria and lipid droplets**. Instances of continuity or apparent continuity have been reported to occur between the mitochondrial membranes and sarcoplasmic reticulum (BOWMAN 1967). One of the most striking associations is that seen between mitochondria and myofilaments. In longitudinal sections of cardiac muscle, rows of mitochondria are seen lying closely apposed to such filaments, while in transverse section through the I-band region, rings or girdles of mitochondria may be seen surrounding groups of myofilaments.

The enlargement of the mitochondrial compartment in male Wistar *rats* placed on a **copper-deficient** diet at 10 days of age (DALLMAN and GOODMAN 1970) is related to the depletion of cytochromes and in activity of cytochrome oxidase (DALLMAN 1967). By feeding weanling *mice* the copper-chelating agent, cuprizone, TANDLER and HOPPEL (1972) considerably enhanced the number of cardiac mitochondria and the frequency of occurrence of partitioned mitochondria.

15.1.8.1.2
Golgi Complex

The Golgi apparatus is a complex structure that can be schematically viewed as composed of two basic elements: flat disc-shaped cisternae and tubular-reticular networks. Groups of three to eight cisternae piled in stacks are in continuity with cisternae of adjacent stacks through tubular-reticular elements. The Golgi apparatus, despite its complexity, is a very dynamic organelle, as observed most drastically by the rapid and reversible effects of brefeldin A, a fungal macrocyclic lactone.

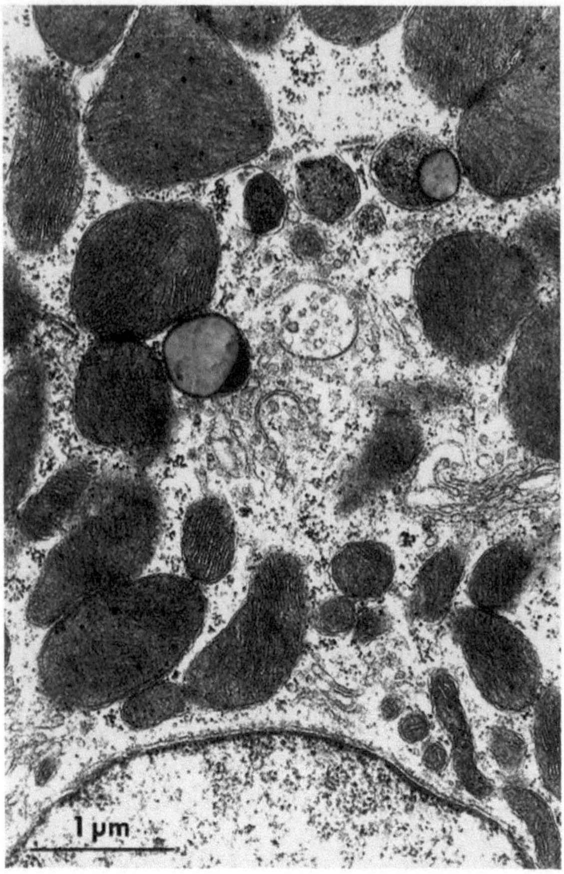

Fig. 283. Large Golgi areas, secondary lysosomes, and a multivesicular body in a cardiomyocyte between the anterior and posterior papillary muscles of the left ventricle (block 1097) of a never hypoxic *rat* (No. 3) medicated for 7 consecutive days with intraperitoneal injection of 15 mg carbocromene per kg body weight×day. On August 4, 1976, under pentobarbital anaesthesia (30 mg/kg), the animal was perfused from the abdominal aorta with 2.5% glutaraldehyde in 0.1 M sodium cacodylate buffer (pH 7.4). Postfixation with 1% osmium tetroxide in sodium cacodylate buffer. Embedded in Epon 812 and sectioned at 50 nm. Lead citrate and uranyl acetate. Film 625-33

15.1.8.1.3
Sarcoplasmic Reticulum

Ryanodine receptors are a family of intracellular Ca^{2+} release channels that were originally identified in the sarcoplasmic reticulum of skeletal muscle cells. Three members of the family were distinguished, RyR2 ryanodine receptors in the cardiac muscle. RyR1 (in the skeletal muscle) and RyR2 function as Ca^{2+} release channels from the sarcoplasmic reticulum intracellular calcium store and play a crucial role in the excitation-contraction cycle. They bind Ca^{2+} and calmodulin and become phosphorylated by various protein kinases including Ca^{2+}/calmodulin- and cAMP-dependent kinases (Lokuta et al. 1995, Mayrleitner et al. 1995). RyR2 is also activated by adenosine and its analogues (McGarry et al. 1994).

An nNOS type NOS is present in cardiac sarcoplasmic reticulum membrane (Xu et al. 1999) and both nNOS and eNOS are associated with cardiac sarcolemmal membrane (Zhou et al. 2000). Cardiac sarcolemmal (Na^+,K^+)-ATPase is an important integral membrane protein that catalyses Na^+ and K^+ active transport across the plasma membrane (Skou 1998). Lack of NOS isoforms in cardiac muscle significantly altered both (Na^+,K^+)-ATPase activity and sarcoplasmic reticulum Ca^{2+}-ATPase function (Zhou et al. 2002).

The possible role of ATP in regulation of L-type channels was investigated in *guinea pig* ventricular cells (Yazawa et al. 1997) and a stimulatory action of MgATP in inside-out patches was observed.

15.1.8.1.4
Sarcomeres

The striking regularity of thin and thick filaments within the sarcomere is not simply the result of a self-assembly properties of their major constituent proteins alone, but rather involves specific interactions with the cytoskeletal lattice (Small et al. 1992). The most obvious structures in this context are the M bands and Z-disks, which are involved in the packing thick and thin filaments, respectively. The sole component that integrates both compartments is obviously the giant protein titin, also called connectin.

The morphology of the M band in cardiac muscle can be correlated roughly with heartbeat frequency. In general, cardiac M bands give a five line pattern, in which the M1-line is relatively much stronger than the other M-lines (Pask et al. 1994). Five proteins have been described to be localised specifically in the M-band: the muscle isoform of creatine kinase, M-protein, myomesin, skelemin and titin.

Superoxide anion radicals ($O_2^{\bullet -}$) generated in alloxan diabetic *rat* heart tissue extracts incubated for 10 min at 25 °C with varying concentrations of xanthine (0.05, 0.1, 0.15, and 0.2 mM) plus xanthine oxidase (0.1. 0.2, 0.3, and 0.6 U/ml), caused a decrease in the cytosolic **creatine kinase** (EC 2.7.3.2) activities by 20, 49, 76, and 86%, respectively (Genet et al. 2000). The addition of 80 µg/ml of superoxide dismutase (EC 1.15.1.1) reversed the depressed creatine kinase activities almost to control values. Hydrogen peroxide (0.001, 0.01, 0.1, and 1 mM) decreased the cytosolic creatine kinase activities by 4, 10, 19, and 35%, respectively. This decrease in creatine kinase activities was reversed significantly to control values when 10 µg/ml of catalase (EC

1.11.1.6) was added to the reaction mixture during the incubation.

Creatine kinase in *rat* heart homogenate was inactivated by indomethacin with horseradish peroxidase-H_2O_2 (MIURA et al. 2001). When indomethacin was incubated with horseradish peroxidase-H_2O_2, the maximum absorption of indomethacin at 280 nm rapidly decreased and a new peak at 410 nm occurred with isosbestic points at 260 and 312 nm. In contrast, under anaerobic conditions, the spectral change of indomethacin was almost absent, indicating indomethacin was oxidised to a yellow substance by horseradish peroxidase-H_2O_2. Adding catalase (EC 1.11.1.6) strongly inhibited the production of yellow substance. Sodium azide also blocked the formation of yellow substance and the inactivation of creatine kinase. Electron spin resonance signals of indomethacin carbon-centred radical were detected using 2-methyl-2-nitrosopropane during the interaction of indomethacin with horseradish peroxidase-H_2O_2 under anaerobic conditions. Oxygen was consumed during the interaction of indomethacin with horseradish peroxidase-H_2O_2. Sulphydryl groups and tryptophan residues of creatine kinase decreased during the interaction of indomethacin with horseradish peroxidase-H_2O_2. Other sulphydryl enzymes, including alcohol dehydrogenase (EC 1.1.1.1) and glyceraldehyde-3-phosphate dehydrogenase (EC 1.2.1.12), were also readily inactivated during the interaction with horseradish peroxidase-H_2O_2.

Titin, the largest protein known to date, is the third most abundant myofibrillar protein in vertebrate skeletal and cardiac muscles (WANG et al. 1979). In solution, native full-length titin (T_1, α-connectin) or its large fragments (T_2, β-connectin) is a long (up to 1 μm), slender, and flexible strand of beaded domains (3–4 nm in diameter), frequently with a larger globule at one end (MARUYAMA et al. 1984, TRINICK et al. 1984, WANG et al. 1984, NAVE et al. 1989). Titin seem to be the major component responsible for the interaction of the thick and thin filament systems of the sarcomere, since individual titin molecules span an entire half sarcomere. This was most firmly established by a panel of monoclonal antibodies that recognised different epitopes along the titin polypeptide and that in immunoelectron microscopy decorated sarcomeres ar distinct positions along the whole half sarcomere (FÜRST et al. 1988, 1989). Biochemistry showed under non-denaturing extraction conditions a fraction of myomesin and M-protein remained firmly bound to the conspicuous globular head of titin (NAVE et al. 1989). Sequencing of titin cDNA portions put the amino-terminus towards the Z-disc (LABEIT and KOLMERER 1995, GAUTEL et al. 1996, YAJIMA et al. 1996), while the carboxy-termini were shown to reside in the M band (GAUTEL et al. 1993, OBERMANN et al. 1996).

Cardiac muscles that lack nebulin display a high degree of variability in the lengths of the thin filaments (ROBINSON and WINEGRAD 1979).

The heart myoglobin radical formed by hydrogen peroxide was identified immunologically (DETWEILER et al. 2002). The reaction of metmyoglobin with H_2O_2 in the presence of 5,5-dimethyl-1-pyrroline *N*-oxide (DMPO) produces an adduct with an electron spin resonance (ESR) signal that decays within minutes (DAVIES 1990). The trapping of this myoglobin radical to form $Mb^{\bullet}$-DMPO is the result of hydrogen peroxide-driven self-peroxidation, which forms a phenoxyl radical at tyrosine-103 as determined by side-specific substitution of tyrosine with phenylalanine (GUNTHER et al. 1998). Mass spectrometric analysis by GUNTHER et al. (1998) demonstrated that the ESR-silent species exhibited an increase in mass consistent with the addition of a single DMPO to myoglobin.

15.1.8.1.5 Lysosomes

The internalisation of lipopolysaccharide into *rat* cardiomyocytes was dependent on endosomal trafficking, because an inhibitor of microfilament reorganization prevented uptake in both cardiomyocytes and whole hearts (COWAN et al. 2001).

Secondary lysosomes of cardiac myocytes and interstitial cells house the bulk of the immunocytochemically identifiable cathepsin D (EC 3.4.23.5) (DECKER et al. 1980).

15.1.8.1.6 Peroxisomes

Peroxisomes (microbodies) are cytoplasmic organelles bounded by a single limiting membrane. The fine granular matrix contains catalase (EC 1.11.1.6) and various H_2O_2-producing oxidases. HAND (1974) and HERZOG and FAHIMI (1974, 1975) described the presence of peroxisomes in the ventricular myocardium of the adult *rat* and *mouse*, respectively. Microbodies were also detected in the left ventricular myocardium of the *guinea pig*, *Mongolian gerbil*, *tree shrew*, and *Macaca java* (HICKS and FAHIMI 1977). In all these species the electron-dense reaction product of catalase was localised in 0.2–0.5 µm oval particles, surrounded by a single limiting membrane and located usually at the junction of I and A bands. The peroxisomes in the hearts of *gerbil* and *Macaca java* were especially long and tortuous. A close spatial association was found between the myocardial peroxisomes and mitochondria, lipid droplets, and the membranes of the sarcoplasmic reticulum, especially the so-called junctional sarcoplasmic reticulum. ZHOU and KANG (2000) observed a few peroxisomes in the cardiomyocytes of non-transgenic *mice*, and the gold particles that identify catalase were found only in peroxisomes. In transgenic *mice*, no obvious ultrastructural changes were found, except that the number of peroxisomes markedly increased from $3.2 \pm 1.8/100\ \mu m^2$ in non-transgenic myocardium to $9.0 \pm 3.4/100\ \mu m^2$ in the transgenic myocardium that contained catalase activities about 60-fold higher than normal. The size of the peroxisomes was also significantly enlarged to $0.43 \pm 0.12\ \mu m$ ($P < 0.01$) in the transgenic myocardium in comparison to $0.29 \pm 0.09\ \mu m$ in the non-transgenic control.

15.1.8.1.7
Hyperpolarization-Activated, Cyclic Nucleotide-Gated Channels

Two **h**yperpolarization-activated, **c**yclic **n**ucleotide-gated channels genes are expressed in *mouse* and *human* heart, HCN2 and HCN4 (BIEL et al. 1999). In situ hybridisation (LUDWIG et al. 1998) and PRC analysis (SANTORO et al. 1998) revealed that HCN2 is expressed throughout the heart including the sinoatrial node (SANTORO et al. 1998). The exact distribution of HCN4 within the heart has not yet been investigated, however the relatively low abundance of HCN4 clones in *human* heart cDNA libraries (Ludwig et al. unpublished data) indicates that this channel may exist in much lower concentrations in total heart than HCN2.

The ionic conductance underlying the cardiac pacemaker depolarisation was identified by BROWN et al. (1977). YANAGIHARA and IRISAWA (1980) and DIFRANCESCO (1981) and called I_f (f for "funny") or I_h (h for hyperpolarization activated). Cyclic nucleotides regulate HCN channel activity by directly binding to a cyclic nucleotide-binding domain (CNBD) which is situated in the C-terminus of the protein. The native I_h channel (DIFRANCESCO and TORTORA 1991) and also the expressed mHCN2 channel reveal an about tenfold higher apparent affinity for cAMP than for cyclic GMP whereas cyclic nucleotide-gated channels are about 40-100 fold more sensitive for cGMP than for cAMP (ZAGOTTA and SIEGELBAUM 1996).

15.1.8.2
Hypoxic Myocardium

In a *murine* model of myocardial infarction and remodelling created by the left anterior descending coronary artery ligation for 4 weeks, hydroxyl radicals, which originated from the superoxide anion, and lipid peroxide formation in the mitochondria were both increased in the noninfarcted left ventricle (IDE et al. 2001). The mtDNA copy number relative to the nuclear gene (18S rRNA) preferentially decreased by 44 % in myocardial infarction by a Southern blot analysis, associated with a parallel decrease (30 % to 50 % of sham) in the mtDNA-encoded gene transcripts, including the subunits of complex I (ND1, 2, 3, 4, 4L, and 5), complex III (cytochrome *b*), complex IV (cytochrome *c* oxidase), and rRNA 12 S and 16 S).

Hypoxia (10 % O_2 for 3 weeks) per se induced a rise in hexokinase (EC 2.7.1.1) activity in left and right ventricles and septum of the *rat* heart and a fall of hydroxyacyl Co-A dehydrogenase (EC 1.1.1.35) activity in both myocardial ventricles (DANESHRAD et al. 2000). The respiratory rate and the citrate synthase (EC 4.1.3.7) activities were unaffected by hypoxia.

There is growing evidence that multiple mitogen-activated protein kinases are activated during ischaemia and/or reperfusion and may contribute to the structural and functional changes after myocardial ischaemia (ABE et al. 2000). Exposing *rat* neonatal cardiomyocytes to ischaemia resulted in a rapid ant transient activation of extracellular signal-regulated kinase, p38, and c-Jun NH_2-terminal protein kinase (YUE et al. 2000). On reoxygenation, further activation of all 3 mitogen-activated protein kinases was noted; peak activities increased (fold) by 5.5, 5.2, and 6.2, respectively. Visual inspection of myocytes exposed to ischaemia/reoxygenation identified 18.6 % of the cells as showing morphological features of apoptosis, which was further confirmed by DNA ladder and terminal deoxyribonucleotide transferase–mediated dUTP nick end labelling (TUNEL). Myocytes treated with PD98059, a mitogen-activated protein kinase/extracellular signal-regulated kinase (MEK1/MEK2) inhibitor, displayed a suppression of ischaemia/reoxygenation-induced extracellular signal-regulated kinase activation, whereas p38 and c-Jun NH_2-terminal protein kinase activities were increased by 70.3 % and 55.0 %, respectively. In addition, the number of apoptotic cells was increased to 33.4 %. With pretreatment of cells with SB242719, a selective p38 inhibitor, or SB203580, a p38 and c-Jun NH_2-terminal protein kinase 2 inhibitor, ischaemia/reoxygenation + PD98059-induced apoptotic cells were reduced by 42.8 % 63.3 %, respectively.

Heart muscle cell cultures from 3-days old *rats* at 40 mm Hg pO_2 during the first 24 h showed some slightly swollen mitochondrial cristae (AUCLAIR et al. 1976). At 48 h, however, all mitochondria were swollen with distorted cristae, and the outer membrane was often interrupted. The matrix was clear and contained diffuse electron-dense zones. Mitochondrial fragmentation was often observed. Widespread changes, such as margination of the nuclear chromatin, nuclear disintegration, and clearing of the sarcoplasm with frequent loss of the cell membrane were also seen. Myofilaments were damaged; Z material was diffuse and some myofibrils were disrupted and randomly oriented.

After breathing 5 % O_2 for 20 min, reoxygenation for 5 h caused an increase in volume densities of sarcoplasmic reticulum, T-tubuli and mitochondria more intensely in young *rats*, and of lipid, vacuoles, mitochondrial destruction and average volume prevailing in senescent animals (WELT et al. 2000).

Post-ischaemic reperfusion highlights the relevance of permeability transition pores in NAD^+ metabolism (DI LISA et al. 2001). Mitochondrial NAD^+ content, which is hardly affected during ischaemia, becomes almost depleted when coronary flow is restored after a prolonged period of ischaemia. The inhibition of mitochondrial NAD^+ depletion exerted by cyclosporin A suggests that upon reperfusion the rise in intracellular Ca^{2+}, along with the recovery of neutral pH and the boosting of oxyradical generation (BOLLI and MARBAN 1999), promoted permeability transition pore opening causing the release of intramitochondrial NAD^+ and its subsequent hydrolysis. The inhibition of permeability transition pores not only prevented the mitochondrial NAD^+ decrease, but also significantly protected cell viability, as documented in different experimental models (NAZARETH et al. 1991, GRIFFITHS and HALESTRAP 1993, MINNERS et al. 2000, DI LISA et al. 2001). The protection afforded by cyclosporin A was mimicked by *N*-methylvaline-4-cyclosporin, a cyclosporin derivative that binds cyclophilins and inhibits the permeability transition pores but does not affect calcineurin activity (ZENKE et al. 1993, DI LISA et al. 2001).

Myocardial 'stunning', a reversible contractile failure often found in ischaemic cardiomyopathy is associated with selective degradation of troponin I, resulting in a truncated molecule that shows considerably reduced calcium sensitivity. It was postulated that intracellular calcium overload due to severe ischaemia could activate the calcium-dependent protease calpain I, which in turn would cleave troponin I (GAO et al. 1996). KLAINGUTI et al. (2000) have shown that myocardial endothelin-1 is increased after 20 min of ischaemia and have suggested that endothelin-1 might play a role as an inducer of apoptosis.

Nonsedimentable cathepsin D (EC 3.4.23.5) activity was significantly elevated in ischaemic tissue by 15 min after occlusion of the circumflex artery of *rabbit* hearts and increased after the next 105 min (Wildenthal et al. 1978). Increases in nonsedimentable glucosaminidase (EC 3.2.1.17) and acid phosphatase (EC 3.1.3.2) activities were somewhat less marked than that of cathepsin D; statistically significant changes in the distribution of these enzymes appeared at 30 min. Unlike cathepsin D and glucosaminidase, changes in acid phosphatase were maximal by 45 min, after which further increases of nonsedimentable activity were not apparent. Immunohistochemical staining of cathepsin D in ischaemic tissue sections revealed apparent diffusion of the enzyme from discrete organelles into the surrounding cytosol by 30 min after occlusion, but not at 15 min.

Preconditioning with one cycle of 5-min ischaemia/5-min reperfusion before 20-min ischaemia significantly reduced the percentage of area at risk from the control value of 49.9 ± 2.0 to 35.4 ± 2.8, and repetitive preconditioning with 2 cycles of 5-min ischaemia/5-min reperfusion further limited the percentage of area at risk to 3.2 ± 0.9 (TANNO et al. 2000). Infarct-size limitation by single-cycle preconditioning was completely abolished by a protein kinase C inhibitor, staurosporine (100 µg/kg). In contrast, the cardioprotection by repetitive preconditioning was only partially blocked by staurosporine, another protein kinase C inhibitor, polymyxin B (5 mg/kg), or a tyrosine kinase inhibitor, genistein (5 mg/kg).

CGP 41251 is a staurosporine derivative with a high selectivity for protein kinase C entered recently in phase 1 clinical trial as an anticancer drug (AKIYAMA et al. 1999). Treatment of *rat* papillary muscle by this drug at 3 µM reduced decreasing of contractility by simulated hypoxia and reperfusion (KOCIC et al. 2001).

A large and time-dependent increase of the electron spin resonance signal corresponding to the NO-Fe(II) *N*-methyl-D-glucamine dithiocarbamate complex was observed in the ischaemic area 8 h and 24 h after myocardial ischaemia induced by occlusion of the left main coronary artery (LECOUR et al. 2001). In contrast, no electron spin resonance triplet was observed in the non-ischaemic region of the heart and in sham-operated *rats*.

In vitro, cardiac myocytes of foetal *mice* resupplied with oxygen and glucose after 1 h of deprivation showed degenerating mitochondria, myofibrils, and glycogen granules surrounded by a single or double layered membrane which at times was incomplete (SYBERS et al. 1979). The vacuoles were most commonly located in the perinuclear region and the nuclear envelope frequently had a concavity adjacent to the vacuoles as if indented by pressure from the vacuole.

JNK-1 and c-*fos* proteins were significantly induced in the ischaemic/reperfused Langendorff *rat* heart, which was inhibited by a proanthocyanidin extract (Sato et al. 2001). In concert, red grape seed proanthocyanidin water-alcohol extract significantly reduced the appearance of reactive oxygen species and apoptotic cardiomyocytes in the ischaemic/reperfused hearts.

15.1.8.3 Intermittently Hypoxic Myocardium

Angiographic studies have shown intermittent coronary occlusion in the acute phase of myocardial infarction and during thrombolytic therapy (Hackett et al. 1987), which may lead to repetitive ischaemic episodes and induction of reperfusion injury on several occasions. Ravn et al. (2000) by external application of two serrated clamps, followed by twisting of the left anterior descending artery by moving one clamp clockwise and the other anticlockwise, created a medial injury resulting in exposure of adventitial tissue to the lumen. Platelet accumulation in the left ventricle of *pigs* was significantly higher in the area of risk compared to the right ventricle.

Adult male Wistar *rat* hearts subjected to 60-min ischaemia followed by reperfusion using a Langendorff perfusion apparatus showed a moderate (30.2 ± 8 %) increase in the production of reactive oxygen species determined by the dihydrorhodamine 123 method and a significant increase in the production of malondialdehyde (from <1 to 23 ± 2.7 nmol/ml) (Szabados et al. 1999).

Adult *rat* cardiac myocytes isolated with 0.1 % collagenase in a modified Langendorff perfusion apparatus and plated in laminin-coated dishes were subjected to anoxia and substrate deprivation for 13, 30, 60, 90, and 120 min and reoxygenated for 2 min (Khalid and Ashraf 1993). Maximum formation of $HO^{\bullet}$ was observed in myocytes subjected to 15 min of anoxia/reoxygenation (2.83 ± 0.27 nmol/mg protein), at which time no injury was observed at light and ultramicroscopic levels. On the other hand, there was no correlation between the amount of $HO^{\bullet}$ production and different parameters of cell injury in myocytes subjected to anoxia/reoxygenation longer than 15 min. Myocytes developed extensive blebbing, loss of cell membrane permeability, and ultrastructural damage. Malondialdehyde increased from 0.78 ± 0.14 nmol/mg protein at 15 min to 1.65 ± 0.35 nmol/mg protein at 120 min. Incubation with 1 mM deferoxamine reduced the $HO^{\bullet}$ production at all anoxic intervals, most significantly at 15 min, but did not decrease lactate dehydrogenase and malondialdehyde release or provide ultrastructural preservation. It was likely that the damage to myocytes in this system was still mediated by free radicals other than $HO^{\bullet}$, as indicated by the protection by diphenylenediamine against cellular injury.

Embryonic *chick* cardiac myocytes grown to confluence on glass cover slips and studied using an epifluorescent imaging system showed that the greatest oxidation of dihydroethidium occurred during ischaemia, increasing from a baseline of 0.30 ± 0.02 to 0.80 ± 0.02 at 30 min, 1.73 ± 0.10 at 1 h, and 3.61 ± 0.27 at 2 h ischaemia (Vanden Hoek et al. 1997). Despite these increases in reactive oxygen species generation, cell death as measured by the exclusion dye propidium iodide remained less than 5 %. Antioxidant treatment significantly reduced dihydroethidium oxidation (1.10 ± 0.13 vs. 1.73) and 2',7'-dichlorofluoresceine oxidation (0.56 ± 0.10 vs. 1.12) during 1 h ischaemia, and improved subsequent viability at 3 h reperfusion (68.4 ± 3.3 % cell death in untreated cells vs. 22.8 ± 5.51 % in treated).

Ischaemic hearts of 7 *patients* aged 48 to 63 years had increased mtDNA damage and OXPHOS gene expression, suggesting that mtDNA damage is associated with OXPHOS deficiency (Corral-Debrinski et al. 1991). Oxidative phosphorylation defects may also play a role in some other forms of cardiac disease as idiopathic dilated cardiomyopathy, hypertrophic cardiomyopathy, myocarditis, brown atrophy and coronary atherosclerosis.

Grabellus et al. (2002) semiquantitatively analysed the immunoreactivity for heme oxygenase-1 in left ventricular tissue of 23 *patients* with end-stage heart failure before and after left ventricular assist devices support, while two unused donor hearts served as controls. Control hearts stained almost negative for heme oxygenase-1, while failing hearts showed immunoreactivity mainly in myocardiocytes, but also in endothelial cells, some smooth muscle cells and fibroblasts. Hearts with ischaemic heart disease (6 *patients*) showed significantly higher heme oxygenase-1 immunorectivity th!n hearts with dilated cardiomyopathy (14 *patients*) or myocarditis/congenital heart disease 3 *patients*). After left ventricular assist devices support, the heme oxygenase-1 content decreased significantly in the dilated cardiomyopathy and ischaemic heart disease groups and was significantly higher in the subendocardium than in the subepicardium. *In vitro*, under hypoxic conditions, neonatal *rat* cardiomyocytes showed an increase of heme oxygenase-1 protein content up to six fold above the normal level, which returned to normal values after normoxic cultivation.

mRNA levels for constitutive nitric oxide synthase III were increased in ischaemic cardiomyopathy compared to non-failing hearts as determined by quantitative RNase protection assay (STEIN et al. 1997). In Western blot, NOS III protein was detected in both failing and non-failing hearts. Immunohistochemical studies with a selective antibody to NOS III showed no obvious differences in the staining of the endothelium of blood vessels from non-failing hearts. However, NOS III-immunoreactivity was significantly more intense in cardiomyocytes from failing compared with non-failing hearts.

15.1.8.3.1 Sarcolemma

The sarcolemma is the first subcellular organelle of the cardiomyocyte to be the target of reactive oxygen species predominantly generated in the endothelial cells (ZWEIER et al. 1988) during reperfusion of the ischaemic myocardium. In preparations of *canine* cardiac sarcolemmal vesicle H_2O_2 significantly depressed [^{3}H]-quinuclidinyl benzilate binding to muscarinic receptors (ARORA and HESS 1985). 10 mM H_2O_2 inhibited 80 % of sarcolemmal Na^+-K^+-ATPase activity at 90 min (KUKREJA et al. 1990). $HO^•$ generated by Fenton's reagent (200 μM Fe^{2+} + 5 mM H_2O_2) significantly decreased maximum binding of ouabain (43.06 ± 1.45 to 31.96 ± 2.37 pmol/mg) and was significantly protected by 5 mM mannitol ($P < 0.05$).

The lipids that constitute the cell membrane, especially those lipids containing unsaturated double bonds, are susceptible to free radical attack, leading to the formation of lipid peroxides and aldehydes (KAKO 1985). A number of short chain fragments produced from peroxidation of polyunsaturated fatty acids as 4-hydroxyperoxy nonenal and 4-hydroxy 2-alkenals react with sulphydryl groups of various enzymes modifying their activities.

A significant amount of lipid peroxidation associated with depressed Na-K-ATPase activity was observed after 15 min of H_2O_2 (300 μM) Langendorff perfusion of the isolated *rat* heart, and consequently the cell membrane permeability was greatly increased (OGURO et al. 1992). When *N'*-diphenyl-1,4-phenylenediamine (2.5 mM) a potent antioxidant, was added to the perfusate, the lipid peroxidation was totally inhibited.

Cell adhesion molecule gene expression is upregulated in ischaemia/reperfusion. KACIMI et al. (1998) studied intercellular adhesion molecule (ICAM) response to cytokines and acute hypoxia in cultured myocardial cells. Northern blot analysis and immunoassay showed that the proinflammatory cytokines interleukin-1β and tumour necrosis factor-α stimulated concentration dependent increase in ICAM mRNA and protein. Pre-treatment with a specific inhibitor of nuclear transcription factor-κB prevented cytokine induction. Inhibition of tyrosine kinase and p38/RK (stress-activates protein kinase) pathways prevented IL-1β-induced ICAM protein synthesis, whereas extracellular signal-regulated protein kinase (ERK1/ERK2) inhibition did not. Neither hypoxia (0 % O_2 for 6 hours) alone nor hypoxia/reoxygenation had any effect on ICAM mRNA. However, hypoxia did enhance IL-1β-induced ICAM mRNA expression in myocytes.

An excessive myocardial accumulation of both Na^+ (KARMAZYN 1988, KARMAZYN and MOFFAT 1993) and Ca^{2+} (TANI 1990) is closely associated with ischaemia/reperfusion injury.

Tetrodotoxin blocking the Na^+ channel developed tension to 56 ± 3 % of the pre-ischaemic level and reduced the resting tension to 126 ± 12 % of the pre-ischaemic level (ENG et al. 1998).

HOE-642, a selective sodium-hydrogen exchange subtype 1 inhibitor (SCHOLZ et al. 1995), when given at 1 μM together with tetrodotoxin (44 μM) protected isolated *rat* hearts from ischaemic/reperfusion injury and lowered creatine phosphokinase activity in the coronary effluent in comparison to that observed in the non-treated hearts (ENG et al. 1998).

Activation of the ATP-sensitive potassium channel has been shown to exert protective effects on the ischaemic and reperfused myocardium (GROVER et al. 1990, 1991, 1992, 1995, BEHLING and MALONE 1995). Pinacidil but not nicorandil opens ATP-sensitive K^+ channels and protects against simulated ischaemia in *rabbit* myocytes (CRITZ et al. 1997). Induction of cromakalim (100 μM), a K_{ATP} channel opener, through a microdialysis probe implanted from the epicardial surface into the left ventricular *rat* myocardium significantly increased the level of 2,3-dihydroxybenzoic acid formed from salicylic acid by $HO^•$ (OBATA and YAMANAKA 2000). Another K_{ATP} channel opener, nicorandil, also increased the level of 2,3-dihydroxybenzoic acid. When iron(II) was administered to cromakalim-pretreated animals, a marked elevation of 2,3-dihydroxybenzoic acid was observed, compared with iron(II)-only-treated animals. A positive linear correlation between iron(II) and formation of $HO^•$, trapped as 2,3-dihydroxybenzoic acid in the dialysate, was shown (r^2 = 0.988). When corresponding experiments were performed with nicorandil-treated animals, a positive linear correlation between iron(II) and 2,3-dihydroxybenzoic acid in the dialysate was shown (r^2 = 0.988). However, the presence of glibenclamide (1–50 μM) decreased the cromakalim-induced 2,3-dihydroxybenzoic acid formation in a concentration-dependent manner

(IC_{50} = 9.1 µM). 5-Hydroxydecanoate (100 µM), another K_{ATP} channel antagonist, also decreased cromakalim-induced HO• formation (IC_{50} = 107.2 µM). In the presence of glibenclamide (10 µM), the heart was subjected to myocardial ischaemia for 15 min by occlusion of the left anterior descending coronary artery. when the heart was reperfused, the normal elevation of 2,3-dihydroxybenzoic acid in the heart dialysate was not observed in animals pretreated with glibenclamide (10 µM). In corresponding experiments using 5-hydroxydecanoate (100 µM), the same results occurred.

Coronary heart disease *patients* on acetylsalicylic acid, diuretics or beta blockers, but not the ones on calcium channel blockers, had significantly higher redox balance than non-users (MATTILA et al. 2001).

Using isolated perfused *rat* heart, GAN et al. (1998) studied whether K_{ATP} activation exerted any effect on the direct deleterious effects of either 200 µM hydrogen peroxide or a free radical generating system consisting of purine plus xanthine oxidase in terms of function and energy metabolite status. On their own, H_2O_2 or the combination of purine plus xanthine oxidase treatment resulted in a time-dependent depression of myocardial contractility, which reached over 90 % after 30 min perfusion, an effect which was associated with approximately 1000 % elevation in left ventricular end-diastolic pressure (LVEDP). The K_{ATP} channel opener cromakalin (500 nM) significantly attenuated the hydrogen peroxide-induced loss in systolic function throughout the treatment period, and reduced the elevation in LVEDP with significant attenuation 10, 15 and 20 min after H_2O_2 addition. Contractile dysfunction induced by H_2O_2 was associated with significantly reduced tissue ATP, creatine phosphate and glycogen content to approximately 70, 60 and 70 % of control, respectively. The depletion of these metabolites was significantly attenuated to 35, 23 and 23 %, respectively, in the presence of cromakalim. The K_{ATP} channel antagonist, glibenclamide (1 µM) abolished the protective effect of cromakalim against both contractile function and depletion in intermediary energy metabolites. Glibenclamide on its own failed to alter the cardiac response to H_2O_2 with respect to any parameter. The response to the free radical generating system consisting to purine plus xanthine oxidase was unaffected by cromakalim.

The effect of HOCl on methionine oxidation in isolated, perfused rat heart was similar to that observed with either tissue slices 500 µm in thickness or isolated myocytes, showing comparable oxidation of this amino acid with increasing concentrations of HOCl (FLISS 1988). Cellular sulfhydryls in the perfused hearts were also susceptible to HOCl. The concentrations of both protein sulfhydryl and reduced glutathione (GSH) decreased significantly (28 % and 39 %, respectively) after perfusion with 100 µM HOCl for 1 h.

Ultrastructural changes induced in isolated *rat* heart by enzymatically generated oxygen radicals included swelling and cristolysis of mitochondria, disruption of filaments, development of intracellular oedema and focal disruption of the sarcolemma (YTREHUS et al. 1987). Stereological examination revealed few alterations after 5 min perfusion with oxygen radicals. After 10 min perfusion with oxygen radicals, however, the V_v (myocyte/myocardium) increased from 0.542±0.042 (mean±s.d.) to 0.663±0.144, and this paralleled the development of V_v (cellular oedema/myocyte) being 0.047 ± 0.028. V_v (capillary wall/capillary) increased from 0.215±0.046 to 0.411±0.123 indicating endothelial swelling. Although the mitochondria appeared swollen, V_v (mitochondria/myocyte) remained constant.

15.1.8.3.2 Mitochondria

Hypoxia exerts a reciprocal control on transcription of glycolytic (increase) and mitochondrial (decrease) enzymes in myotubes (WEBSTER et al. 1990). The effects on glycolytic enzymes may be mediated by the oxygen sensitive transcription factor HIF (hypoxia-inducible factor). In the immortalised *mouse* skeletal muscle cell line (C2C12), when used at low concentrations, azide inhibited more than 70 % of cytochrome oxidase activity without changes in bioenergetics (either lactate production or creatine phosphorylation) or mRNA for mitochondrial enzymes (LEARY et al. 1998).

Lipoxygenase inhibitor FLM 5011, a congener of the 2-hydroxyarylketoximes, which is able to form chelate complexes with metals in a hydrophobic environment and, thus, to interact with the active site of lipoxygenases, protected mitochondria of ischaemic and reperfused *canine* cardiomyocytes (FITZL et al. 1999).

Enalapril, an **angiotensin-converting enzyme inhibitor**, increased the mitochondrial number in ageing *mouse* myocardiocytes (FERDER et al. 1993). Inhibition of angiotensin-converting enzyme possibly increases the calcium content of the mitochondria, as angiotensin II decreased mitochondrial calcium in the *bovine* adrenocortical glomerulosa cell *in vitro* (KRAMER 1990). The free radical scavenging action of captopril – due to its -SH group – was substantiated by the observation that captopril, but not lisinopril (nonsufhydryl angio-

Table 47. Giant mitochondria in myocardium

Hypoxic myocardium of *dogs*	Lozada and Laguens (1966)
Ischaemic/reperfused myocardium of *dogs*	Fitzl et al. (1999)
Dogs with aortic stenosis	Wollenberger and Schulze (1961)
Hypoxic myocardium of *rats* in experim. aortic stenosis	Mölbert (1968)
Hypoxic isolated perfused *rat* hearts	Sun et al. (1969)
Bleeding/reoxygenation in *rats*	Büchner and Onishi (1968)
Rats maintained for long periods on small doses of thyroxin	Zaimis et al. (1969)
Dogs treated with quinidin	Hiott and Howell (1971)
Mice, bats, and *dogs* treated with reserpine	Wilcken et al. (1967), Sun et al. (1968), Hagopian et al. (1972)
Rats treated with tetradecylthioacetic acid	Hexeberg et al. (1995)
Left ventricle in idiopathic hypertrophic non-obstructive cardiomyopathy	Sekiguchi et al. (1970), Sekiguchi and Konno (1971)
Patient on digitalis therapy assumed to have a primary cardiomyopathy	Kraus and Cain (1980)
Patient on Adriamycin therapy for osteosarcoma	Kaduk and Seiler (1977)

tensin-converting enzyme inhibitor), inhibited $FeCl_3$/ascorbic acid-induced lipid peroxidation in whole tissue homogenates of *rabbit* aorta to a level comparable to that of superoxide dismutase (Mittra and Singh 1998). Experiments in *rat* hearts suggested that local inhibition of converting enzyme by ramipril exerted protective effects after ischaemia and reperfusion by reducing arrhythmias and improving cardiac function and metabolism (Linz et al. 1986). Myocardial tissue levels of glycogen, ATP, and creatine phosphate were distinctly increased and lactate levels decreased in isolated working hearts from ramipril-pretreated *rats* (1 mg/kg p.o.) following coronary occlusion and reperfusion in comparison to vehicle-pretreated ischaemic hearts from control animals. Ramipril can reduce reperfusion arrhythmias by inhibition of local angiotensin II formation and bradykinin degradation (Linz and Schölkens 1987).

Megamitochondria

Mitochondria towering the length of a sarcomere were found in reoxygenated cardiomyocytes from any area of both the left (Figs. 284–287) and right ventricles pre-treated with either Tyrode's solution (Figs. 284–287) or carbocromene.

Megamitochondria were induced in cultured *rat* hepatocytes by hydrazine and hydrogen peroxide, inducers of free radicals (Karbowski et al. 1997). Free radicals generated by different mechanisms or via different sources seem to be involved in their

Fig. 284. Long end-to-end fused interfibrillar mitochondrium in a cardiomyocyte from the region between the papillary muscles of the left ventricle (block 1278) from a *rat* (No. 9) treated for 7 consecutive days with intraperitoneal injection of 1.5 ml Tyrode's solution per kg body weight × day. On the last 4 days of experimentation the animal was exposed to an atmosphere containing only 5% oxygen for 30 min. On August 4, 1976 half an hour after the last exposure under pentobarbital anaesthesia (30 mg/kg), the animal was perfused from the abdominal aorta with 2.5% glutaraldehyde in 0.1 M sodium cacodylate buffer (pH 7.4). Postfixation with 1% osmium tetroxide in sodium cacodylate buffer. Embedded in Epon 812 and sectioned at 50 nm. Lead citrate and uranyl acetate. Plate 3362

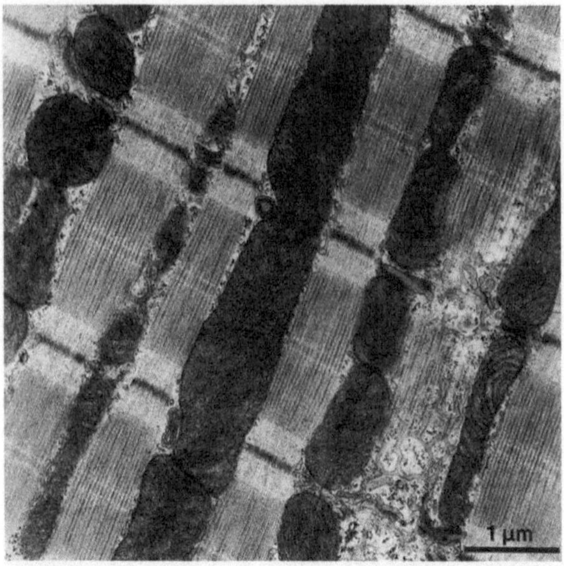

Fig. 285. Interfibrillar mitochondrium towering the length of a sarcomere and L system in a cardiomyocyte (between the anterior and posterior papillary muscles of the left ventricle; block 1429) of a *rat* (No. 14) treated for 8 days with intraperitoneal injection of 1.5 ml Tyrode's solution per kg body weight×day. On the last 5 days of experimentation the animal was exposed to an atmosphere containing only 5 % oxygen for 30 min. Half an hour after the last exposure under pentobarbital anaesthesia (30 mg/kg), the animal was perfused from the abdominal aorta with 2.5 % glutaraldehyde in 0.1 M sodium cacodylate buffer (pH 7.4). Postfixation with 1 % osmium tetroxide in sodium cacodylate buffer. Embedded in Epon 812 and sectioned at 50 nm. Lead citrate and uranyl acetate. Film 619-43

Fig. 286. Interfibrillar mitochondria towering the length of a sarcomere from region of the anterior papillary muscle (block 1277) of a *rat* (No. 9) treated for 7 consecutive days with injection of 1.5 ml Tyrode's solution per kg body weight×day. On the last 4 days of experimentation the animal was exposed to an atmosphere containing only 5 % oxygen for 30 min. On August 4, 1976, half an hour after the last exposure under pentobarbital anaesthesia (30 mg/kg), the animal was perfused from the abdominal aorta with 2.5 % glutaraldehyde in 0.1 M sodium cacodylate buffer (pH 7.4). Postfixation with 1 % osmium tetroxide in sodium cacodylate buffer. Embedded in Epon 812 and sectioned at 50 nm. Lead citrate and uranyl acetate. Film 631-46

induction. H_2O_2 known to be one of the strongest stress inducers resulting in irreversible cell injury by lipid peroxidation (RUBIN and FARBER 1984, STARKE and FARBER 1985) can induce oxidative stress in cardiomyocytes reacting with mitochondrial cytochrome *c* (RADI et al. 1993). Chloramphenicol or ethidium bromide induce megamitochondria by suppression of the dividing process of mitochondria due to their inhibitory action of protein in mitochondria (KING et al. 1972, WAGNER and RAFAEL 1979).

Concentric Cristae

Mitochondria containing cristae that in ultrathin sections present crescent, annular, spiral, or concentric profiles (Fig. 289) have been noted to occur in certain normal and pathological tissues (BÜCHNER and ONISHI 1968, GRADUALLY 1997). They are often found in company with mitochondria with stacked parallel cristae, and it is thought that in such instances the stacked appearance may represent a longitudinal section through the cristae that would appear concentric on transverse section (LEESON and LEESON 1969, HUG and SCHUBERT 1970). In normal mammalian myocardium they were seen on rare occasions (MOORE and RUSKA 1956, STENGER and SPIRO 1961). HUG and SCHUBERT's (1970) observations in the mitochondria in cardiomyopathy support the idea, that such mitochondria are metabolically more active than others, or represent the sequence of change in the life-span of the mitochondrion, heralding degenerative changes leading ultimately to lysosomal degradation and myelin figure formation.

Fig. 287. Irregularly sized subsarcolemmal mitochondrium towering the length of a sarcomere in a cardiomyocyte between the anterior and posterior papillary muscles of the left ventricle (block 1429) of a *rat* (No. 14) treated for 8 consecutive days with intraperitoneal injection of 1.5 ml Tyrode's solution per kg body weight × day. On the last 5 days of experimentation the animal was exposed to an atmosphere containing only 5 % oxygen for 30 min. On August 6, 1976 under pentobarbital anaesthesia, the animal was perfused from the abdominal aorta with 2.5 % glutaraldehyde in 0.1 M sodium cacodylate buffer. Postfixation with 1 % osmium tetroxide in sodium cacodylate buffer. Embedded in Epon 812 and sectioned at 50 nm. Lead citrate and uranyl acetate. Film 619-48

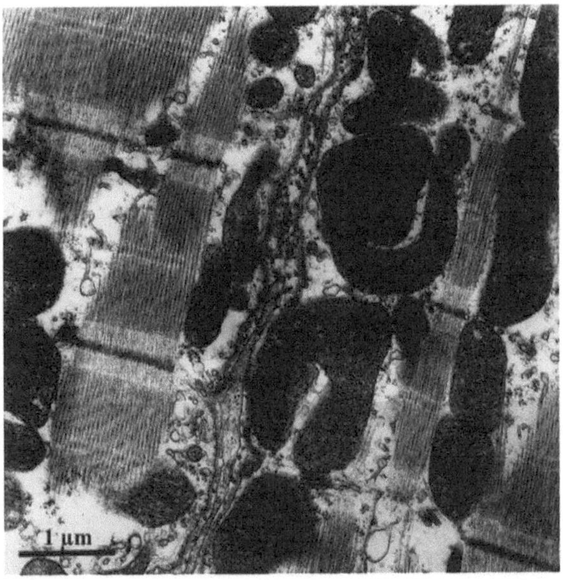

Fig. 288. Variation in mitochondrial size and shape in a myocardial cell from the region of the posterior papillary muscle (block 367) of a male Wistar *rat* (No. 27/103 E) treated for 11 consecutive days with intraperitoneal injection of 5 ml *N*-benzhydryl-*N'*-*p*-hydroxybenzylpiperazine HCl per kg body weight × day. On the last 4 days of experimentation the animal was exposed to an atmosphere containing only 5 % oxygen for 30 min. Half an hour after the last exposure, under pentobarbital anaesthesia (30 mg/kg), the animal was perfused from the abdominal aorta with 2.5 % glutaraldehyde in 0.1 M sodium cacodylate buffer (pH 7.4). Postfixation with 1 % osmium tetroxide in sodium cacodylate buffer. Embedded in Epon 812 and sectioned at 50 nm. Lead citrate and uranyl acetate. Film 506-41

Mitochondrial Compartments

A non selective pore that is triggered by Ca^{2+} and particular metabolic derangements associated with ischaemia/reperfusion injury, namely falling ATP, raised P_i and oxidative stress (Crompton and Andreeva 1993). Once activated, the pore flickers open and closed states and disrupts mitochondrial energy transduction, allowing ATP hydrolysis by the F_1F_0 ATPase. Pore activation was prevented by cyclosporin A, which also retarded the onset of necrosis in heart cells subjected to substrate-free anoxia and allowed partial regeneration of ATP on reoxygenation.

An increase in number and size of **intramitochondrial native dense granules** has been reported in ischaemic *dog* myocardium (Herdson et al. 1965). After low-flow perfusion for 45 min and reperfusion in anaesthetised *rats in vivo* Siedenburg et al. (1999) found that mitochondria contained remarkably more calcium precipitates and were more severely damaged than during low-flow perfusion.

Various alterations of mitochondrial morphology are seen, depending upon the stage reached by the process, and also upon which of the two mitochondrial chambers is primarily involved.

Scheme. Mitochondrial compartments

Outer or intermenbraneous chamber: lying between the two membranes of the mitochondrial envelope and extending either as a potential space or a cleft-like or tubular space within the cristae
Inner chamber: containing the mitochondrial matrix.

Since the permeability of the inner and outer mitochondrial membranes is known to be different, it follows that flooding of one or the other, or of both, chambers may occur.

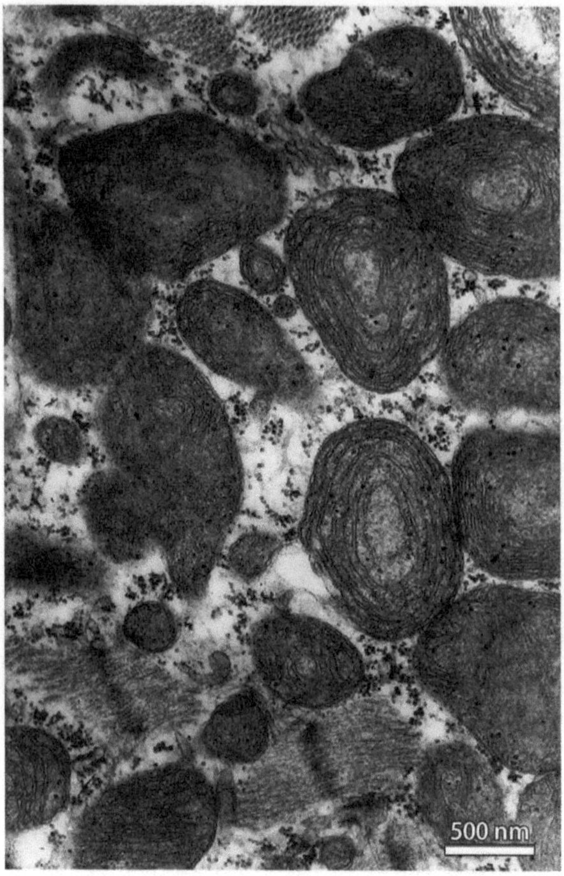

Fig. 289. Circular arrangement of the mitochondrial cristae parallel to the membrana mitochondrialis and plenty of dense matrix granules in a myocardiocyte from the region of the posterior papillary muscle (block 1218) of a *rat* (No. 7) treated for 7 consecutive days with intraperitoneal injection of 1.5 ml Tyrode' solution per kg body weight × day. On the last 4 days of experimentation the animal was exposed to an atmosphere containing only 5 % oxygen for 30 min. On August 4, 1976, half an hour after the last exposure under pentobarbital anaesthesia (30 mg/kg), the animal was perfused from the abdominal aorta with 2.5 % glutaraldehyde in 0.1 M sodium cacodylate buffer (pH 7.4). Postfixation with 1 % osmium tetroxide in cacodylate buffer. Embedded in Epon 812 and sectioned at 50 nm. Lead citrate and uranyl acetate. Plate 3359

Physiologically, permeability of the **inner mitochondrial membrane** is low, and a transmembrane electrochemical gradient of protons ($\Delta\mu_{H+}$), required for oxidative phosphorylation, is formed. Volume changes, accompanying energy-dependent ion accumulation in mitochondria, however, occur (HARRIS et al. 1967). Adenosine diphosphate (ADP) induced an increase of mitochondrial membrane -SH groups, which is followed by decreased swelling (PACKER 1960, ZIMMER 1970) as well as decreased probability of membrane permeability transition (ZORATTI and SZABO 1995). Conversely, oxidative stress and the ensuing of cytosolic Ca^{2+} have been described independently by MACEDO et al. (1988) and ZIMMER and FREISLEBEN (1988) to be correlated with increased disulphide formation from inner membrane thiol (dithiol) groups. The ischaemia/reperfusion situation has been advocated to be connected with the mitochondrial permeability transition (ZORATTI and SZABO 1995). The inner mitochondrial membrane of artificially arrested *human* heart muscle cells gained by needle biopsy during open-heart surgery suffered a typical condensation in ischaemia periods of >40 min (SCHAPER and THIEDEMANN).

By far the commonest variety of swelling seen is that due to the involvement of matrix or inner chamber. With advancing hydration it shows a patchy appearance due to the development of multiple electron-lucent foci, and in time frank cavitation of the mitochondrial matrix occurs (Figs. 275–277). In the perfused heart and isolated heart mitochondria, changes in [Ca^{2+}] within the physiological range increase neither the matrix [PP_i] nor the mitochondrial volume (Griffiths and Halestrap 1993). Respiration and fatty oxidation in heart mitochondria are sensitive to changes in matrix volume in a similar manner to that observed in liver mitochondria (HALESTRAP 1987).

Application of 3 % halothane-air mixture for 15 min induced a swelling of *guinea pig* heart mitochondria and a loss of 30 % of their inner membrane surface area relative to unit volume cytoplasm, a loss confined exclusively the cristae mitochondriales (PFALLER et al. 1974).

Cristal adhesions were recognised ultrastructurally as electron dense, rod-shaped, pentalaminar condensed membranous profiles located in the intracristal compartment (HEGSTAD et al. 1997). In the isolated *rat* heart they occurred predominantly during reperfusion and were associated with moderately and severely altered myocytic mitochondrial alterations (HEGSTAD et al. 1999). Qualitative similar structures have been observed in conditions associated with the generation of reactive oxygen species (LOESSER et al. 1991, HORIKAWA et al. 1995), and have furthermore been observed in energy-deprived myocardium (FEUVRAY 1981).

An excessive myocardial accumulation of both Na^{+} (KARMAZYN 1988, KARMAZYN and MOFFAT 1993) and Ca^{2+} (TANI 1990a) is closely associated with ischaemia/reperfusion injury. Attempts to enhance the intrinsic scavenging systems, superoxide dismutase and catalase, were not successful in improving Na^{+} imbalance and reducing Ca^{2+} overload in the *rat* heart perfused by the Langendorff technique (TANI 1990b).

Ca^{2+} release from mitochondria is stimulated during **oxidative stress**. Oxidation of nicotinamide adenine dinucleotide phosphate (NADH), with sub-

Table 48. Release of calcium from mitochondria

Uncouplers
Quinones
Peroxides
Divicine (2,6-diamino-4,5-dihydroxypyrimidine)
Alloxan
1-Methyl-4-phenyl-1,2,3,6-tetrahydropyridine
Fe^{2+}
Cd^{2+}

sequent ADP-monoribosylation of mitochondrial proteins or formation of cyclic ADP-ribose, has been suggested as a regulatory mechanism. Such enhanced Ca^{2+} extrusion may be the basis of "Ca^{2+} cycling', i.e. continuous uptake and release of Ca^{2+} into mitochondria, which leads ultimately to the dissipation of the membrane potential and to mitochondrial failure. As a consequence of the breakdown of the membrane potential ($\Delta\Psi$) or other forms of damage, mitochondria may release Ca^{2+} into the cytosol and generate reactive oxygen species. Their generation may play a role in cell death due to ischaemia/reperfusion and in normal ageing of tissues, with one such scenario calling for the fixation of oxidative damage to mt-DNA as mitochondria undergo many generations of replication in nondividing cells. There have been reports of an increase of reactive oxygen species generation with age in myocardial mitochondria (NOHL and HEGNER 1978, SOHAL et al. 1994, LUCAS and SZWEDA 1998). However, HANSFORD et al. (1997) found an unchanged rate of reactive oxygen species formation in heart mitochondria from *rats* aged 24 months vs. *rats* aged 6 months, but did not rule out an accumulation of oxidative damage with ageing. To the extent that this may be true of mt-DNA (AMES et al. 1993, MECOCCI et al. 1993, AGARWAL and SOHAL 1994), this could be particularly significant.

Diltiazem has been shown to improve post-ischaemic function when administered prior to the onset of ischaemia (GROVER et al. 1990). This effect is more likely related to its negative inotropic effects and subsequent energy-sparing effect than its direct inhibitory effect on the mitochondrial Ca^{2+} efflux pathway. Hearts pre-treated with cyclosporin A and subjected to 30 min of ischaemia showed better functional recovery and preservation of ATP levels than those studied in its absence (GRIFFITHS and HALESTRAP 1992).

The main **free radical generation sites** have been localised at Complexes I (TAKESHIGE and MINAKAMI 1979, TURRENS and BOVERIS 1980, HANSFORD et al. 1997, HERRERO and BARJA 1998) and III (TURRENS et al. 1985, NOHL et al. 1996, HERRERO and BARJA 1998) of the respiratory chain. The strong increase in H_2O_2 production observed in *mouse*, *parakeet* and *canary* mitochondria after one-line addition of rotenone, which inhibits electron flow from Complex I to ubiquinone, clearly indicated that Complex I contains a free radical generator site in the free species (HERRERO and BARJA 1998). The rotenone-stimulated increase was totally abolished by ethoxiformic anhydride, which inhibits NADH-ubiquinone but not NADH-ferricyanide reduction (VIK and HATEFI 1984, WYATT et al. 1995). This suggested that the Complex I generator ‘ as in the case of *rats* and *pigeons* (HERRERO and BARJA 1997), can be the Complex I iron-sulphur centres. Complex III also generated free radicals in the three species studied by HERRERO and BARJA (1998).

Dibucaine promoted an inhibition of the Ca^{2+}-induced increase in mitochondrial H_2O_2 generation measured on isolated *rat* liver mitochondria by the oxidation of scopoletin in the presence of horseradish peroxidase (KOWALTOWSKI et al. 1998). This decrease increase in mitochondrial H_2O_2 generation may be attributed to the reduction of Ca^{2+} binding to the membrane induced by dibucaine, as assessed by measuring $^{45}Ca^{2+}$ binding to the mitochondrial membrane. Mg^{2+} also inhibited Ca^{2+} binding to the mitochondrial membrane, mitochondrial swelling, membrane protein thiol oxidation, and H_2O_2 generation induced by Ca^{2+}.

Pretreating isolated *rat* heart mitochondria with 100 pg/ml, 1 ng/ml, and 10 ng/ml, respectively, of prostaglandins E_2 $F_{2\alpha}$, and I_2 (prostacyclin) stimulated calcium (30–420 μM)-linked changes in mitochondrial respiration (KARMAZYN 1986). For the most part the effects of prostaglandins was concentration-dependent with the exception of PGE_2, which showed no effect at the highest dose of 10 ng/ml.

Tetrodotoxin blocking the Na^+ channel developed tension to 56±3% of the pre-ischaemic level and reduced the resting tension to 126±12% of the pre-ischaemic level (ENG et al. 1998).

HOE-642, a selective sodium-hydrogen exchange subtype 1 inhibitor (SCHOLZ et al. 1995), when given at 1 μM together with tetrodotoxin (44 μM) protected isolated *rat* hearts from ischaemic/reperfusion injury and lowered creatine phosphokinase activity in the coronary effluent in comparison to that observed in the non-treated hearts (ENG et al. 1998).

Mitochondrial permeability transition as proposed by HALESTRAP et al. (1998) occurs as a result of the binding of mitochondrial cyclophilin to the adenine nucleotide translocase in the inner mitochondrial membrane. This binding is enhanced by thiol modification of the adenine nucleotide translocase caused by oxidative stress or other thiol reagents. Cyclophilin binding enhances the ability of the adenine nucleotide translocase to undergo a conformational change triggered by Ca^{2+}.

An interaction of $O_2^{\bullet-}$ with $^{\bullet}NO$, associated with the formation of $ONOO^-$, appears to be an important process that contributes to cardiac muscle mitochondrial respiration caused by hypoxia and reoxygenation (Xie et al. 1998).

L-Arginine–nitric oxide pathway

$$\text{L-Arginine} \xrightarrow{\text{NOS}} \textbf{Nitric oxide}(\text{NO}^{\bullet}) \xrightarrow{+O_2^{\bullet-}} \text{Peroxynitrite } (\text{ONOO}^-) \quad [46]$$

Peroxynitrite exposure to *rat* heart mitochondrial preparations resulted in significant inactivation of electron carriers such as succinate dehydrogenase and NADH dehydrogenase as well as the mitochondrial ATPase (Radi et al. 1994). Immunohistology demonstrated that the $ONOO^-$-mediated nitration product nitrotyrosine was formed in postischaemic but not in normally perfused controls (Wang and Zweier 1996). Tetrodotoxin blocking the Na^+ channel developed tension to 56±3% of the pre-ischaemic level and reduced the resting tension to 126±12% of the pre-ischaemic level (Eng et al. 1998).

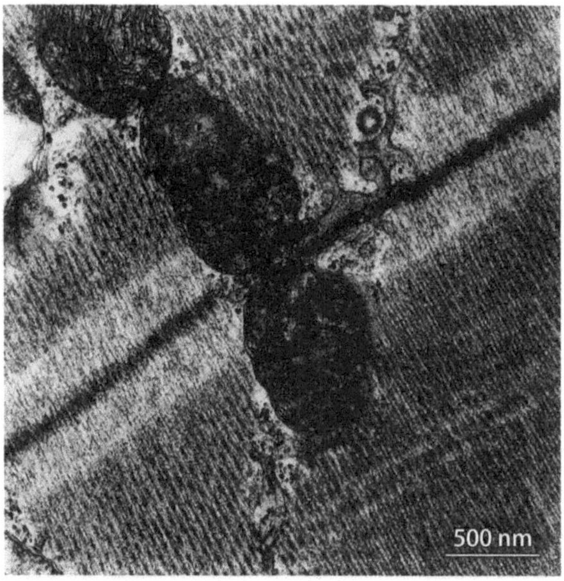

Fig. 290. Varying, occasionally prominent (upper left) mitochondrial reoxygenation damage showing cristolysis, clearing of the matrix and depletion of intramitochondrial granules in a myocardiocyte from the right ventricle (block 1371) of a *rat* (No. 12) treated for 8 consecutive days with intraperitoneal injection of 1.5 ml Tyrode's solution per kg body weight × day. On the last 5 days of experimentation the animal was exposed to an atmosphere containing only 5% oxygen for 30 min. On August 5, 1976, half an hour after the last exposure under pentobarbital anaesthesia (30 mg/kg), the animal was perfused from the abdominal aorta with 2.5% glutaraldehyde in 0.1 M sodium cacodylate buffer (pH 7.4). Postfixation with 1% osmium tetroxide in sodium cacodylate buffer. Embedded in Epon 812 and sectioned at 50 nm. Lead citrate and uranyl acetate. Film 618-25

Fig. 291. Reoxygenation damage to the interfibrillar mitochondria in a cardiomyocyte of the right ventricle (block 1371) of a *rat* (No. 12) treated for 8 consecutive days with intraperitoneal injection of 1.5 ml Tyrode' solution per kg body weight × day. On the last 5 days of experimentation the animal was exposed to an atmosphere containing only 5% oxygen for 30 min. On November 26, 1975, half an hour after the last exposure under pentobarbital anaesthesia (30 mg/kg), the animal was perfused from the abdominal aorta with 2.5% glutaraldehyde in 0.1 M sodium cacodylate buffer (pH 7.4). Postfixation with 1% osmium tetroxide in cacodylate buffer. Embedded in Epon 812 and sectioned at 50 nm. Lead citrate and uranyl acetate. Film 618-27

Biochemical Pharmacology

An excessive myocardial accumulation of both Na^+ (Karmazyn 1988) and Ca^{2+} (Tani 1990a) is closely associated with ischaemia/reperfusion injury. Attempts to enhance the intrinsic scavenging systems, superoxide dismutase and catalase, were not successful in improving Na^+ imbalance and reducing Ca^{2+} overload in the *rat* heart perfused by the Langendorff technique (Tani 1990b).

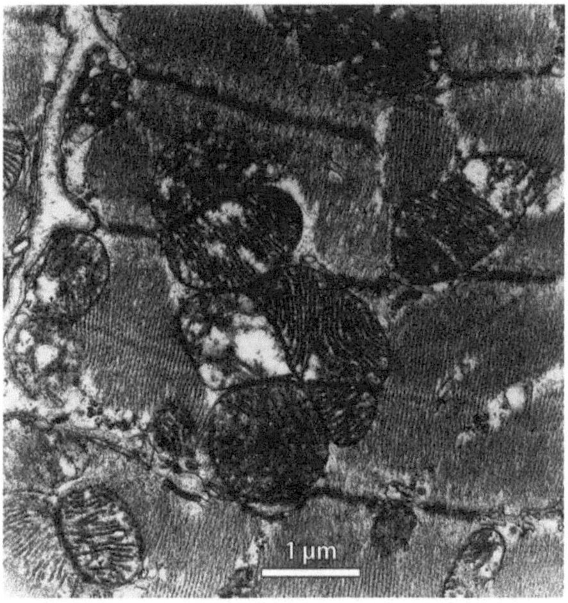

Fig. 292. Varying, but mostly prominent mitochondrial reoxygenation damage showing cristolysis, clearing of the matrix and depletion of intramitochondrial granules in a myocardiocyte from the right ventricle (block 1160) from a *rat* (No. 5) treated for 7 consecutive days with intraperitoneal injection of 1.5 ml Tyrode's solution per kg body weight × day. On the last 4 days of experimentation the animal was exposed to an atmosphere containing only 5 % oxygen for 30 min. On August 4, 1976 under pentobarbital anaesthesia (30 mg/kg), the animal was perfused from the abdominal aorta with 2.5 % glutaraldehyde in 0.1 M cacodylate buffer (pH 7.4). Postfixation with 1 % osmium tetroxide in cacodylate buffer. Embedded in Epon 812 and sectioned at 50 nm. Lead citrate and uranyl acetate. Film 624-11

Fig. 293. Multivesicular body (arrow) in a cardiomyocyte from the right ventricle (block 1371) or a *rat* (No. 12) treated for 7 consecutive days with intraperitoneal injection of 1.5 ml Tyrode's solution per kg body weight × day. On the last 4 days of experimentation the animal was exposed to an atmosphere containing only 5 % oxygen for 30 min. On August 5, 1976 under pentobarbital anaesthesia (30 mg/kg), the animal was perfused from the abdominal aorta with 2.5 % glutaraldehyde in 0.1 M cacodylate buffer (pH 7.4). Postfixation with 1 % osmium tetroxide in cacodylate buffer. Embedded in Epon 812 and sectioned at 50 nm. Lead citrate and uranyl acetate. Film 618-29

Dihyrolipoic acid prevented hypoxic/reoxygenation and peroxidative damage induced by 20 mM H_2O_2 in *rat* heart mitochondria (Scheer and Zimmer 1993).

15.1.8.3.3 Lysosomes

After coronary artery occlusion in the *rabbit*, an aggregation of lysosomal profiles occurred in the paranuclear regions of injured myocytes (Decker and Wildenthal 1978). While no evidence of rupture of lysosomal membranes was observed over the first hour, after 2 h lysosomal profiles were rarely present. Methylprednisolone (30 mg/kg, intravenously) protected lysosomes from the ischaemic damage (Decker and Wildenthal 1978).

15.1.8.3.4 Myofibrils

The contractile material of the *human* heart is less affected by ischaemia than the mitochondria or the nuclei (Schaper et al. 1977). Sarcomeres exhibited moderate contraction and contracture bands were infrequently observed.

If reperfusion is established after short periods of ischaemia, there is usually a prolonged period of impaired contraction, even though eventual recovery can be almost complete. This is referred to as 'stunned myocardium'. The mechanism of this impairment of contractile function appears to be decreased calcium sensitivity of the myofilaments (Hofmann et al. 1993). Reperfusion followed by recovery is associated with moderately elevated Ca_i (Allen et al. 1989, Lee and Allen 1992), and it is thought that this switches on calcium-activated proteases (Barry 1991).

Reperfusion after prolonged ischaemia has been associated with a particular type of microscopical

Fig. 294. An autophagosome in a cardiomyocyte (between the anterior and posterior papillary muscles of the left ventricle (block 1338) of a *rat* (No. 11) treated for 8 consecutive days with intraperitoneal injection of 15 mg carbocromene per kg bodey weight × day. On August 5, 1976, half an hour after the last exposure under pentobarbital anaesthesia (30 mg/kg), the animal was perfused from the abdominal aorta with 2.5 % glutaraldehyde in 0.1 M sodium cacodylate buffer (pH 7.4). Postfixation with 1 % osmium tetroxide in sodium cacodylate buffer. Embedded in Epon 812 and sectioned at 50 nm. Lead citrate and uranyl acetate. Film 630-18

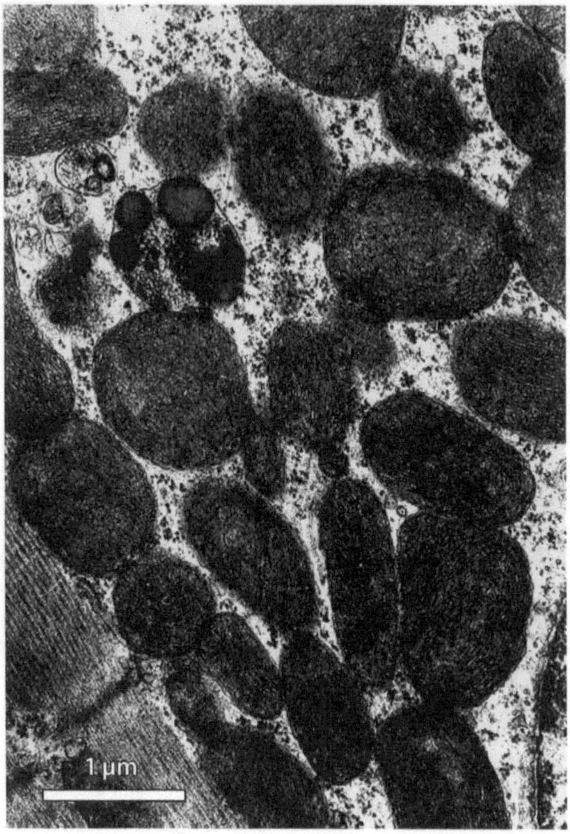

Fig. 295. Autophagosome from the reoxygenated ventricular septum (block 1461) of a *rat* (No. 15) treated for 8 consecutive days with intraperitoneal injection of 15 mg carbocromene per kg body weight × day. On the last 5 days of experimentation the animal was exposed to an atmosphere containing only 5 % oxygen for 30 min. On August 6, 1976, half an hour after the last exposure under pentobarbital anaesthesia (30 mg/kg), the animal was perfused from the abdominal aorta with 2.5 % glutaraldehyde in 0.1 M sodium cacodylate buffer (pH 7.4). Postfixation with 1 % osmium tetroxide in sodium cacodylate buffer. Embedded in Epon 812 and sectioned at 50 nm. Lead citrate and uranyl acetate. Film 622-46

appearance known as **contraction band necrosis.** Contraction bands are aggregates of acidiohilic, hypercontracted sarcomeres that occur singly or in several groups in a cardiomyocyte. Although they are typically associated with reperfusion, they may also be found in a variety of other settings including catecholamine cardiotoxicity (Schiller 1972, 1973), cardioversion, trauma and myocarditis.

Deamination of catecholamine by monoamine oxidase (Spina and Cohen 1989) may contribute to the formation of cytotoxic free radicals in the presence of transition metals such as iron, copper and manganese (Donaldson et al. 1981, Riederer et al. 1989, Obata and Yamanaka 1997). Obata and Yamanaka (2002) examined the effect of iron(III) (50 μM) on the generation of hydroxyl radicals in the extracellular fluid of *rat* myocardium. The generation of $HO^•$ was assessed by infusing sodium salicylate in Ringer's solution (0.5 nmol/μl per min) directly into the myocardium of the anaesthetised *rat* through a microdialysis probe and measuring the non-enzymatic reaction product 2,3-dihydroxybenzoic acid trapped in the dialysate. Tyramine concentration-dependently increased the level of 2,3-dihydroxybenzoic acid. However, in the presence of iron(III), the effect of tyramine was abolished. When iron(III) was administered to tyramine (1 mM)-pre-treated animals, the tyramine-induced stimulation of noradrenaline did not change, but the level of 2,3-dihydroxybenzoic acid decreased significantly ($n = 6$, $P < 0.05$). When desferrioxamine, a strong iron(III) chelator, was administered to tyramine (1 mM)-pre-treated animals, a marked increase in 2,3-dihydroxybenzoic acid was

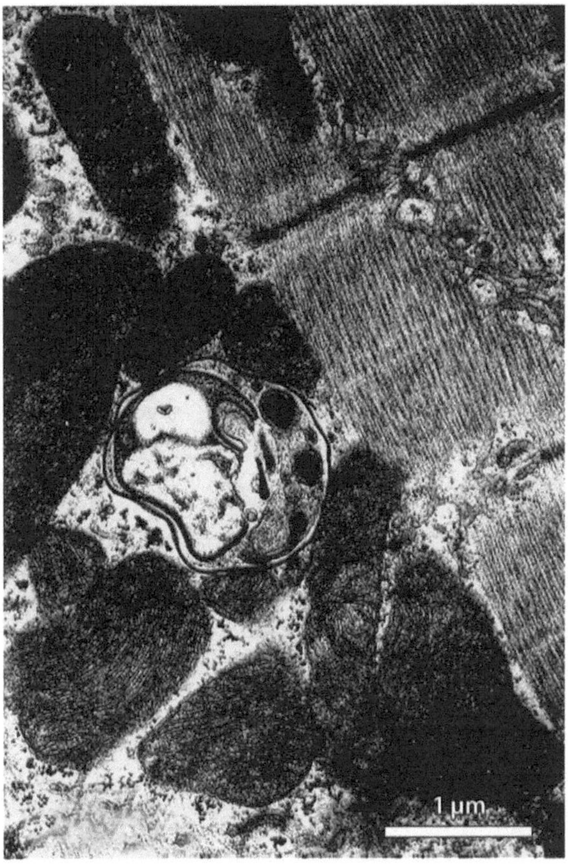

Fig. 296. Autophagosome from the reoxygenated ventricular septum (block 1461) of a *rat* (No. 15) treated for 8 consecutive days with intraperitoneal injection of 15 mg carbocromene per kg body weight × day. On the last 5 days of experimentation the animal was exposed to an atmosphere containing only 5 % oxygen for 30 min. On August 6, 1976 ' half an hour after the last exposure under pentobarbital anaesthesia (30 mg/kg), the animal was perfused from the abdominal aorta with 2.5 % glutaraldehyde in 0.1 M sodium cacodylate buffer (pH 7.4). Postfixation with 1 % osmium tetroxide in sodium cacodylate buffer. Embedded in Epon 812 and sectioned at 50 nm. Lead citrate and uranyl acetate. Film 622-44

seen. Administration of iron(II) to the desferrioxamine-pretreated animals increased 2,3-dihydroxybenzoic acid markedly compared with the iron(II)-only treated group, with a positive linear correlation between iron(II) concentration and HO$^•$ trapped as 2,3-dihydroxybenzoic acid ($R^2 = 0.987$). As to the effect of iron(III) on reperfusion of the *rat* heart after ischaemia induced by clamping the left anterior descending coronary artery branch for 15 min, perfused noradrenaline and 2,3-dihydroxybenzoic acid rose markedly in the heart dialysate. The presence of iron(III) (50 μM) abolished the elevation of 2,3-dihydroxybenzoic acid. Iron(III) also significantly blunted the rise of serum creatine phosphokinase, an index of myocardial damage.

The appearance of contraction bands requires the presence of adenosine triphosphate (VANDERHEIDE et al. 1986), since that is necessary for the actin-myosin interaction which generates hypercontracture in the presence of elevated Ca_i. The total current through the calcium channel was increased by isoprenaline, by the adenylate cyclase activator forskalin, and by injection of cAMP.

Troponin T is part of the troponin-tropomyosin molecular switch of the thin filament of myofibrils in the sarcomere. The globular COOH-terminal domain mediates interactions with troponin I and troponin C, and the NH_2-terminal tail binds to the tropomyosin strand (FARAH and REINACH 1995). For immunoassays, specific antibodies against the single cardiac form (cardiac troponin T) have been developed (KATUS et al. 1993, BAUM et al. 1997). The cardiac troponin T concentration in normal *human* cardiomyocytes is around 12 mg per g of protein (Voss et al. 1995), with no significant difference between left and right ventricles. Of the cardiac troponin T, 6–8 % is cytosolic, the remainder being filament-bound (KATUS et al. 1991, VOSS et al. 1995). Following cell damage, cardiac troponin T concentrations in plasma show a biphasic time pattern. Cytosolic cardiac troponin T is released first into the circulation, causing an initial peak in plasma (KATUS et al. 1991). Persisting ischaemia leads to depolymerization of the fin filament (HEIN et al. 1995) and a protracted release of filament-bound cardiac troponin T (GERHARDT et al. 1991, KATUS et al. 1991). In a study using the maximal values from serial sampling in 502 infarction-suspected patients, GERHARDT and LJUNGDAHL (1998) found a diagnostic sensitivity for non-Q- and Q-wave infarction of 100 %, with a specificity of 99 %.

15.1.8.3.5 Sarcoplasmic Reticulum

A plexus of tubules, diameter 60 nm, embraces the myofibrils to form the sarcoplasmic reticulum. The plexiform arrangement of these tubules is continuous from one sarcomere to another. Flattened saccular extensions from the sarcoplasmic reticulum are applied as footplates to the T-tubules (MUIR 1967).

Since the work of PORTER and PALADE (1957), the numerical relation of this junctional sarcoplasmic reticulum to transverse tubule has been expressed in the terms triad and dyad, respectively. Cardiac muscle can have couplings singly at a transverse tubule (two interior couplings = triad), or at opposite sides of a transverse tubule (two interior couplings = triad); instead, or in addition, it may

Fig. 297. Surface view of the sarcoplasmic reticulum in a myocardiocyte from the anterior papillary muscle (block 1277) of a *rat* (No. 9) treated for 7 consecutive days with intraperitoneal injection of 1.5 ml Tyrode's solution per kg body weight × day. On the last 4 days of experimentation the animal was exposed to an atmosphere containing only 5 % oxygen for 30 min. On August 4, 1976 half an hour after the last exposure under pentobarbital anaesthesia (30 mg/kg), the animal was perfused from the abdominal aorta with 2.5 % glutaraldehyde in 0.1 M sodium cacodylate buffer (pH 7.4). Postfixation with 1 % osmium tetroxide in sodium cacodylate buffer. Embedded in Epon 812 and sectioned at 50 nm. Lead citrate and uranyl acetate. Film 631-47

Fig. 298. Peripheral coupling with junctional sarcoplasmic reticulum (arrow) in a cardiomyocyte from the right ventricle (block 1250) of a *rat* (No. 8) treated for 7 consecutive days with intraperitoneal injection of 15 mg carbocromene/kg body weight × day. On the last 4 days of experimentation the animal was exposed to an atmosphere containing only 5% oxygen for 30 min. On August 4, 1976 under pentobarbital anaesthesia (30 mg/kg), the animal was perfused from the abdominal aorta with 2.5 % glutaraldehyde in 0.1 M sodium cacodylate buffer (pH 7.4). Postfixation with 1 % osmium tetroxide in sodium cacodylate buffer. Embedded in Epon 812 and sectioned at 50 nm. Lead citrate and uranyl acetate. Film 631-41

have couplings at the peripheral sarcolemma (peripheral couplings), that is, unassociated with transverse tubules (Fig. 298).

Glycogen phosphorylase (EC 2.4.1.1) isoenzyme BB is a key enzyme of glycogenolysis. Its degree of association with the sarcoplasmic reticulum glycogenolysis complex depends essentially on the metabolic state of the myocardium (Entman et al. 1977). With the onset of tissue hypoxia, when glycogen is broken down, glycogen phosphorylase isoenzyme BB is converted from the structurally bound into a cytoplasmic form. Plasma concentrations of glycogen phosphorylase isoenzyme BB and cardiac troponin T could be the future scenario for the laboratory testing for myocardial injury (Mair 1998).

15.1.8.3.6 Microtubules

Microtubules are a consistent feature of the cardiac muscle cell. Ranging in size from 24 to 28 nm (Goldstein and Entman 1979, Forbes and Sperelakis 1983) they encircle the nucleus, are associated with around the myofibrils in a helical arrangement ‘ and form a network that runs transversely at the level of the I band and axially between the myofibrils (Goldstein and Entman 1979). They are also closely associated with mitochondria, sarcoplasmic reticulum and T tubules near the Z disc (Goldstein and Entman 1979, Forbes and Sperelakis 1980).

15.1.8.3.7 Nucleus

Neonatal *rat* cardiomyocytes subjected to 24-h hypoxia (95 % N_2, 5 % CO_2) caused an increase of apoptotic rates and the release of lactate dehydrogenase (EC 1.1.1.27), while subsequent 4-h reoxygenation not only further increased the apoptotic

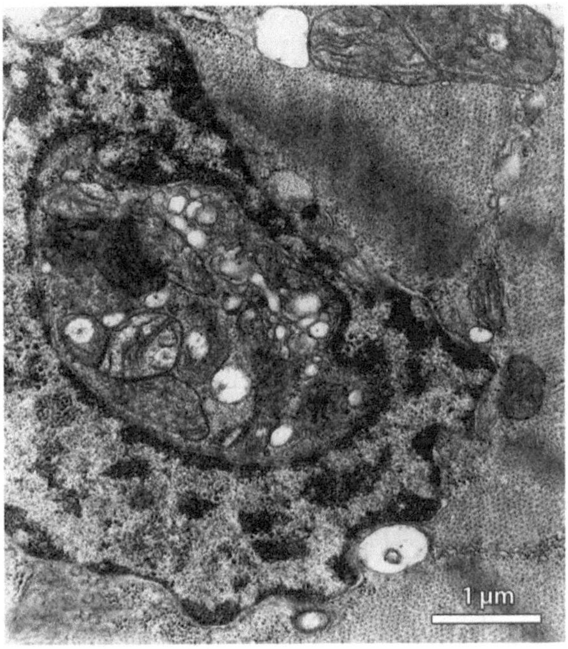

Fig. 299. Numerous microtubules running parallel to the nuclear envelope within a large invagination of the cytoplasm into the nucleus (false nuclear inclusion). Cardiomyocyte from the anterior papillary muscle (block 1367) of a *rat* (No. 12) treated for 8 consecutive days with intraperitoneal injection of 1.5 ml Tyrode's solution per kg body weight × day. On the last 5 days of experimentation the animal was exposed to an atmosphere containing only 5 % oxygen for 30 min. On August 5, 1976, half an hour after the last exposure under pentobarbital anaesthesia (30 mg/mg), the animal was perfused from the abdominal aorta with 2.5 % glutaraldehyde in 0.1 M sodium cacodylate buffer (pH 7.4). Postfixation with 1 % osmium tetroxide in sodium cacodylate buffer. Embedded in Epon 812 and sectioned at 50 nm. Lead citrate and uranyl acetate. Film 617-26

rates and leakage of lactate dehydrogenase, but also induced necrosis of cardiomyocytes (Shen et al. 2000). Chinonin, an effective component isolated from Chinese herb *Rhizoma anemarhenea*, significantly decreased the rates of apoptotic and necrotic cardiomyocytes, and inhibited the leakage of lactate dehydrogenase. It also diminishes NO_2^-/NO_3^- and thiobarbituric acid reactive substances, down-regulated the expression level of p53 protein, and up-regulated bcl-2 protein, respectively.

In adult cardiomyocytes isolated by enzymatic dissociation (0.3 % collagenase) from the hearts of female Sprague-Dawley *rats* reoxygenation (18 h of reoxygenation after 6 h of hypoxia) and prolonged hypoxia (24 h of hypoxia) resulted in a 59 % and 51 % decrease in cellular viability, respectively (Kang et al. 2000). During reoxygenation, cell death occurred predominantly via apoptosis associated with appearance of cytosolic cytochrome *c* and activation of caspase-3 and -9. However, nonapoptotic cell death predominated during prolonged hypoxia. Both caspase inhibition and Bcl-2 overexpression during reoxygenation significantly improved cellular viability through inhibition of apoptosis but had minimal effect on hypoxia-induced cell death. Bcl-2 overexpression blocked reoxygenation-induced cytochrome *c* release and activation of caspase-3 and -9, but caspase inhibition alone did not block cytochrome *c* release.

In the isolated *rat* heart perfused according to Langendorff's technique, the 8-hydroxyguanine content in the DNA significantly increased after an ischaemia of 30 or 60 min followed by reperfusion with oxygenated Krebs-Henseleit buffer (You et al. 2000). The levels of 8-hydroxyguanine did not increase either in the ischaemic hearts reperfused with a nitrogenated solution or in the ischaemic-reperfused hearts treated with superoxide dismutase, mannitol or allopurinol.

15.1.8.3.8 Conduction System

Cardiac arrhythmias are believed to be related to free radicals generated in the heart especially during the period of reperfusion (Tosaki and Das 1994). Melatonin reduced the incidence and severity of arrhythmias induced by ischaemia/reperfusion due to ligation of the anterior descending coronary artery in the isolated *rat* heart for 10 min (Tan et al. 1998). Melatonin was more potent than ascorbic acid in protecting against arrhythmias induced by ischaemia/reperfusion. Besides the function of melatonin as a broad spectrum free radical scavenger, this hormone may also have reduced cardiac arrhythmias due to its regulation of intracellular calcium levels, i.e. by preventing calcium overloading, or due to its ability sympathetic nerve function and reduce adrenergic receptor function in the myocardium.

15.1.8.4 Endothelium of the Myocardial Capillaries

Accumulating evidence suggests that oxygen stress alters many functions of the endothelium, including modulation of vasomotor tone (Cai and Harrison 2000). Inactivation of nitric oxide by superoxide and other reactive oxygen species seems to occur in conditions such as hypertension, hypercholesterolaemia, diabetes and cigarette smoking. Loss of •NO associated with these traditional risk factors may in part explain why they dispose to atherosclerosis.

One of the earliest important events after reperfusion of an ischaemic vascular bed is a significant degree of endothelial dysfunction characterized by the loss of endothelium-derived relaxing factor

(TSAO et al. 1990), now known to be nitric oxide (MONCADA et al. 1990). This may contribute significantly to the reperfusion injury, because **•NO** is known to induce vasorelaxation (FURCHGOTT and ZAWADZKI 1980), inhibit platelet aggregation (RADOMSKI et al. 1991), quench superoxide radicals (GRYGLEWSKI et al. 1986, RUBANYI and VANHOUTTE 1986), and attenuate adherence and activation of polymorphonuclear leucocytes (MCCALL et al. 1988, KUBES et al. 1991). In the ischaemic reperfused *cat*, the •NO donor C87-3754 prevented most of the coronary contraction induced by activated polymorphonuclear leucocytes as well as attenuated most of the ensuing endothelial dysfunction (i.e., a reduced vasorelaxation to endothelium-dependent acetylcholine vis-à-vis a normal vasorelaxation to $NaNO_2$) (LEFER et al. 1993).

The physiological relevance of the inhibitory interaction of caveolar targeting on basal •NO production was provided in a study on intact endothelial cells exposed to high levels of LDL cholesterol (FERON et al. 1999). As originally identified by FIELDING and FIELDING (2000), caveolae also participate in reverse cholesterol transport by increasing caveolin abundance to promote cholesterol trafficking and efflux. The consequence for eNOS function of this cholesterol-induced increase in calveolin abundance is a marked decline in basal •NO release, suggesting that the equilibrium between eNOS bound to caveolin and caveolin-free eNOS determines the basal component of the eNOS-dependent •NO release in endothelial cells.

Superoxide generation was demonstrated near the luminal surface of arterial, capillary, and venular endothelial cells during the first 2 min of reoxygenation of the isolated *rat* heart after 60 min of warm ischaemia (BABBS et al. 1991). The histochemical reaction (cf. BABBS 1994) using a modification of Karnovsky's manganese/diaminobenzidine technique (BRIGGS et al. 1986) was absent or markedly reduced in non-manganese-treated or nonischaemic hearts, as well as in hearts perfused with calcium-free or oxygen-free buffers.

Isolated *pig* coronary arteries with intact endothelium generated $O_2^{\bullet-}$ at a rate of 9.0 ± 0.8 pM per min and mg dry weight (BRANDES et al. 1997); this rate was diminished by about 24 % when the endothelium was removed. The nitroblue tetrazolium staining of arterial ring preparations showed formazan precipitation mainly in the intima.

In Langendorff perfused *rat* hearts $O_2^{\bullet-}$ increased the release of vasoconstrictive (TxA_2 and $PGF_{2\alpha}$) and decreased vasodilating (PGI_2 and PGE_2) prostanoids (GUPTE et al. 2000). Although indomethacin (10 μM), a cyclooxygenase inhibitor, attenuated the rise in coronary perfusion pressure during $O_2^{\bullet-}$ perfusion, the increase was not completely blocked. OKY 046Na (10 μM), a thromoboxan synthase inhibitor, had no effect on $O_2^{\bullet-}$ induced increases in coronary perfusion pressure, whereas ONO 3708 (10 μM), a TxA_2/PGH_2 receptor antagonist, suppressed this effect. Protein kinase C activity was also elevated by > 50 % by $O_2^{\bullet-}$ perfusion. Coronary perfusion pressure typically increased throughout the $O_2^{\bullet-}$ wash-out. This post-$O_2^{\bullet-}$ vasoconstriction was not inhibited by indomethacin, nitro-glycerine or nitrendipine. In contrast, ONO 3708 (10 μM) and two protein kinase C inhibitors, staurosporine (10 nM) and calphostin C (100 nM), completely blocked the rise in coronary perfusion pressure, and even elicited vasodilatation. Phorbol 12,13-dibutyrate enhanced the post-$O_2^{\bullet-}$ vasoconstriction.

Myocardial injury of Langendorff perfused hearts from transgenic *mice* overexpressing **CuZn superoxide dismutase** in coronary vascular cells was attenuated after a 45 min reperfusion after 35 min of global ischaemia as compared with non-transgenic hearts (CHEN et al. 2000). There was a significant reduction in lactate dehydrogenase release from the transgenic hearts.

Human coronary artery endothelial cells were incubated with epinephrine (10^{-9} to 10^{-5} M) alone or with the water-soluble analogue of vitamin E (trolox) (10^{-5} M), the lipid-soluble vitamin E (5×10^{-5} M), or the β_1-adrenergic blocker atenolol (10^{-5} M). At 1 and 24 h of incubation with epinephrine, superoxide anion generation increased by 102 and 81 % (MEHTA and LI 2001). There was a marked increase in both MnSOD and Cu/ZnSOD mRNA and protein. Both MnSOD and Cu/ZnSOD activities were also increased. Pre-treatment of the cells with trolox and vitamin E decreased superoxide anion generation ($P < 0.05$ vs. epinephrine alone) and blocked the subsequent upregulation of SOD mRNA and protein. Treatment of cells with the β-blocker atenolol also blocked the upregulation of SOD ($P < 0.05$ vs. epinephrine alone).

DNA damage was assessed in *human* umbilical vein endothelial cells exposed to superoxide, hydrogen peroxide, nitric oxide, and peroxynitrite (BALLINGER et al. 2000). In both vascular endothelial and *human* aortic smooth muscle cells, the mitochondrial DNA was preferentially damaged relative to the transcriptionally inactive nuclear β-globulin gene. Similarly, a dose-dependent decrease in mtDNA-encoded mRNA transcripts was associated with reactive species treatment. Mitochondrial protein synthesis was also inhibited in a dose-dependent manner by peroxynitrite, resulting in decreased cellular ATP levels and mitochondrial redox function.

Carvedilol conveyed significant protection to endothelial cells subjected to all reactive oxygen intermediates. At concentrations within the clinically relevant plasma concentration observed in patients medicated with 50 mg of carvedilol per day (Louis et al. 1987) cultured endothelial cells were protected from oxidative stress (Yue et al. 1993, 1995, 1999).The two hydroxy-carbazole metabolites of carvedilol, SB209995 and SB211475 also provided potent protection. Carvedilol acts as an antioxidant by directly inhibiting electron adduction. In addition, it protects and replenishes the endogenous antioxidant defence mechanisms. In particular, carvedilol increases levels of glutathione and protein SH as well as those of vitamin E. Carvedilol (10^{-5} M) reduced the rate of apoptosis in *human* umbilical vein endothelial cells incubated for 24 h with serum from patients with class IV congestive heart failure (Ferrari et al. 1999). The beneficial effects of carvedilol were accompanied by maintenance of near normal levels of protein and non-protein SH-groups in the endothelial cells, suggesting that the reduction of the oxidative stress does result in the rate of apoptosis.

Sheridan et al. (1996) used intravital microscopy to record and quantified **leucocyte attachment** to coronary microvascular endothelium during ischaemia-reperfusion in the beating *dog* heart and think it likely that activation and attachment of leucocytes promote endothelial injury and represent the initial step to diapedesis and invasion of the myocardium.

Neutrophil adherence to *human* saphenous vein endothelium incubated for 2 h under hypoxic conditions increased five- to sixfold compared with adherence in normoxic cultures (Arnould et al. 1998).

Maxwell and Gavin (1992) observed that a combination of superoxide dismutase (SOD) and catalase visibly lessened both luminal membrane blebbing and endothelial cell swelling in ischaemic/reperfused *rat* hearts. Trolox, a hydrophilic antioxidant analogue of α-tocopherol as superoxide dismutase reduced the blebbing (Ward and Scoote 1997).

Pre-treatment with the reactive oxygen scavenger *N*-(2-mercaptopropionyl)-glycine completely inhibited **monocyte chemotactic protein-1 induction** by *dog* cardiac venules in reperfused infarcts (Lakshminarayanan et al. 2001). *In situ* hybridisation localised monocyte chemotactic protein-1 message to small venular endothelium in ischaemic areas without myocyte necrosis. Immunohistochemical staining demonstrated reperfusion-dependent nuclear translocation of c-Jun and NF-κB (p65) in small venular endothelium, only in the ischaemic regions of the myocardium, that was inhibited by *N*-(2-mercaptopropionyl)-glycine. *In vitro*, treatment of cultured *canine* jugular vein endothelial cells with H_2O_2 induced a concentration-dependent increase in monocyte chemotactic protein-1 mRNA levels, which was inhibited by the antioxidant *N*-acetyl-L-cysteine, a precursor of glutathione, but not pyrrolidine dithiocarbamate, an inhibitor of NF-κB and activator of AP-1.

Under some circumstances, **•NO** has been demonstrated to protect endothelial cells from the effects of endothelium-derived oxidants produced by ischaemia-reperfusion (Gaboury et al. 1993, Kubes et al. 1991, 1993, Lefer 1993, Kurose et al. 1994, Niu et al. 1994). Okayama et al. (1998) have attempted to examine polymorphonuclear-endothelial adhesion as an important index of oxidant-mediated endothelial stress in *human* umbilical vein endothelial cell oxidant injury using concentrations of •NO and hydrogen peroxide found in several similar studies (Gasic et al. 1991, Suzuki et al. 1991, Johnston 1996), but observed that neither 0–0.1 mM H_2O_2 nor 0–0.5 mM spermine-NONOate (an NO donor) induced any change in neutrophil adhesion; however, in combination (0.01 or 0.1 mM peroxide plus 0.5 mM spermine-NONOate and 0.1 mM spermine-NONOate plus 0.1 mM peroxide), both agents significantly increased the adhesion of neutrophils in cultured endothelium. These data indicated that •NO and peroxide are in fact the important chemical mediators in this model. The increased adhesion promoted by •NO plus peroxide was significantly attenuated by 0.1 mM desferrioxamine as well as 1 mM methionine, suggesting that hydroxyl radicals are formed by the interaction of H_2O_2 with surface-bound iron through the Fenton-type chemical reactions.

The **endothelial type NOS** (NOS III, ecNOS) is constitutively expressed in the cardiovascular system (Bredt et al. 1991) and is tightly regulated by Ca^{2+} (100–500 nmol/litre) (Förstermann et al. 1994) and calmodulin (Palmer et al. 1988, Yoshizumi et al. 1993). An ecNOS sequence contig of 5138 nucleotides length was established containing an open reading frame of 3618 nucleotides (1206 amino acids predicting a 133-kDa protein) and 253 bp 3'-UTR (distal to TGA codon)/1267 bp proximal to ATG codon (containing 5'-UTR and 5'-flanking sequences = putative promoter region) (Schwemmer and Bassenge 1999). Induction of the activity of the 1600 bp NOS III gene promoter by unidirectional and oscillatory share stress is modulated by similar mechanisms that involve NF-κB activation, but do not involve Ras-dependent MAP kinase activation (Silacci et al. 2000). The lack of induction of NOS III gene regulation by oscillatory shear stress can be attributed to the activation of a yet unidentified negative *cis*-acting element present in the NOS III gene.

Ischaemic and receptor-mediated eNOS activation increased NADPH-diaphorase reactivity and eNOS immunoreaction of the *rat* myocardium as measured by antibodies against either amino acids of a central *bovine* eNOS domain or the *human* eNOS N-terminal end (Bloch et al. 2001). In contrast, the antibody against the *human* eNOS C-terminal end exhibited no alteration of eNOS immunoreaction. The transient eNOS activation was associated with increased cGMP content. In human myocardium subjected to ischaemia during cardiac surgery early reperfusion increased eNOS activity.

Inhibition of •NO synthesis by various **NOS inhibitors** increased leucocyte oxidant release and stimulated rolling and adhesion of neutrophils to endothelial cells via mast cell-/oxidant-dependent mechanisms (Davenpeck et al. 1994, Niu et al. 1994, 1996). NO donors and 8-Br-cGMP attenuated some of the NOS inhibitor effects. Adhesion of neutrophils could further be blocked by anti-CD18 or anti-ICAM antibodies. Adhesion between leucocytes and platelets in also influenced by •NO/cGMP pathways. NOS inhibitors (e.g. NAME) can induce aggregation between platelets and leucocytes, an effect that is attenuated by 8-Br-cGMP, as well as by antibodies to P-selectin (Kurose et al. 1993). The upregulation of P-selectin by the mobilisation of Weibel-Palade bodies (Inauen et al. 1990, Zimmerman et al. 1990) may be an important mechanism for the adhesive response.

Adhering leucocytes demonstrated a transient increase in cGMP levels that was coincident with co-localisation of **cGMP-dependent protein kinase** and vimentin, as well as increased phosphorylation of vimentin (Wyatt et al. 1991).

Short-term adhesion of neutrophils to collagen I can be reduced by the NO-donor *S*-nitrosopenicillamine and 8-Br-cGMP (Sundqvist et al. 1994). After prolonged incubation there was no difference between *S*-nitrosopenicillamine and unmedicated cells suggesting that the effect was mediated by •NO which declines as the short-lived *S*-nitrosopenicillamine deteriorates.

Tumour necrosis factor-α-induced **vascular cell adhesion molecule-1** (VCAM-1) expression on endothelial cells is inhibited by NO-donors, 8-Br-cGMP and flow-induced endothelial •NO release (Tsao et al. 1996). •NO and nitrovasodilators also inhibited the IL-1-induced increase in intercellular adhesion molecule-1 (ICAM-1) and VCAM-1, an effect blocked by haemoglobin inhibition of •NO (De Caterina et al. 1995, Takahashi et al. 1996). Kokura et al. (2000) exposed *human* umbilical vein endothelial cell monolayers to 60 min of anoxia, followed by 24 h of reoxygenation, wherein freshly isolated *human* T lymphocytes were added at 6 h during reoxygenation. After an additional 18 h of incubation (i.e. total of 24 h of reoxygenation), the T-cell/endothelial cell coculture media were collected and added to naive endothelial cell monolayers incubated with neutrophils. Although the anoxia/reoxygenation-conditioned media per se had no effect on neutrophil adhesion, the media from T-cell/endothelial cell cocultures significantly increased the adhesion response. This enhanced adhesive interaction was associated with significant increases in TNF-α and interleukin-8 levels in the T-cell/endothelial cell coculture media and was accompanied by a pronounced increase in endothelial E-selectin expression. Treatment of T-cell/endothelial cell coculture media with anti-TNF-α or anti-IL-8 antibodies reduced the media-induced neutrophil adhesion response. The enhanced neutrophil adhesion and the elevated medium levels of TNF-α, but not IL-8, were markedly reduced by inserts that prevented direct T-cell/endothelial cell contact and by monoclonal antibodies directed against vascular cell adhesion molecule-1 or very late antigen-4.

Pentoxifylline decreased tumour necrosis factor production and inhibited its activity on neutrophils (Mandell 1995).

•NO released from the endothelium does not only diffuse into smooth muscle cells but also into the lumen of the blood vessel. There, through the stimulation of soluble guanylate cyclase in platelets, •NO leads to **inhibition of platelet aggregation**. The antiaggregatory effect of •NO is likely to be locally restricted within the lumen of blood vessels as •NO is rapidly inactivated by binding to haemoglobin. •NO stimulates soluble guanylate cyclase binding to the prosthetic heme of the enzyme and leads to an up to 400-fold activation of the purified enzyme (Humbert et al. 1990, Stone and Marletta 1995). Among the three redox forms of NO (NO^-, •NO, NO^+), only the uncharged NO radical (•NO) was shown to significantly activate the enzyme (Dierks and Burstyn 1996).

Bradykinin significantly increased membrane invagination and vesiculation in the endothelium of the isolated perfused *rat* heart (Bloch et al. 1996). Cytopempsis of ferritin was increased by simultaneous suppression of the •NO synthesis by N^G-nitro-L-arginine. Suppression of the basal •NO synthesis by N^G-nitro-L-arginine did not significantly change membrane vesiculation. Sodium nitroprusside decreased the number of vesicles.

Tibolone, a synthetic steroid with mixed oestrogenic and progestogenic/androgenic activity, and its two oestrogenic 3α-OH and 3-β metabolites, but

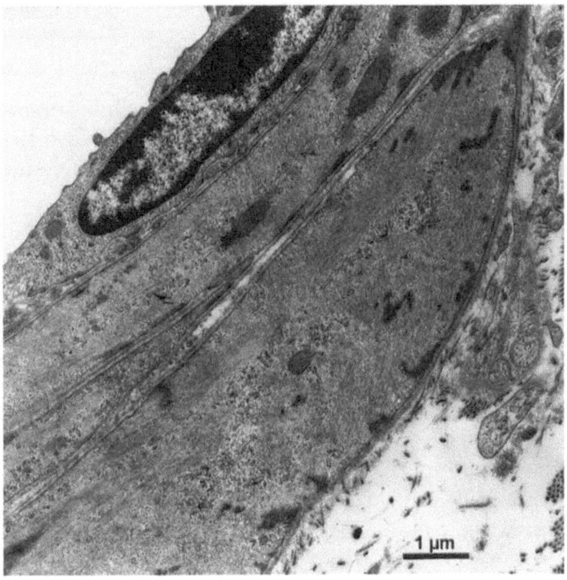

Fig. 300. Endothelium and smooth muscle cells from a coronary arteriole from the right ventricle (block 1130) of a never hypoxic *rat* (No. 4) treated for 7 consecutive days with peritoneal injection of 15 mg carbocromene per kg body weight × day. On August 4, 1976 under pentobarbital anaesthesia (30 mg/kg), the animal was perfused from the abdominal aorta with 2.5 % glutaraldehyde in 0.1 M sodium cacodylate buffer (pH 7.4). Postfixation with 1 % osmium tetroxide in sodium cacodylate buffer. Embedded in Epon 812 and sectioned at 50 nm. Lead citrate and uranyl acetate. Film 627-50

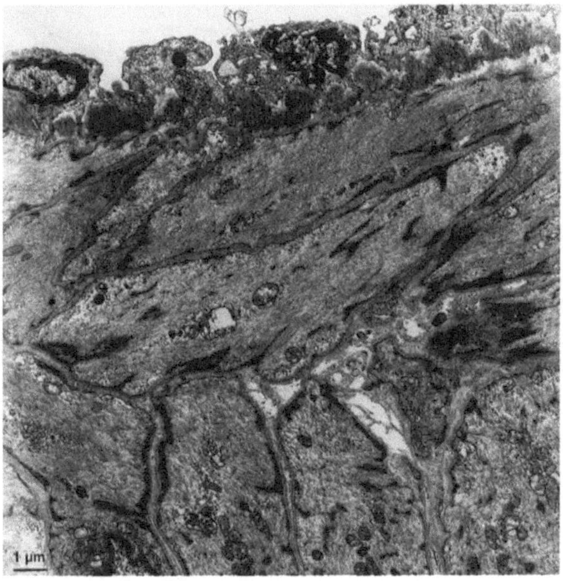

Fig. 301. Reoxygenation damage to both endothelium and smooth muscle cells from a coronary arteriole. Right ventricle (block 1160) from a *rat* (No. 5) treated for 7 consecutive days with intraperitoneal injection of 1.5 ml Tyrode's solution per kg body weight × day. On the last 4 days of experimentation the animal was exposed to an atmosphere containing only 5 % oxygen for 30 min. On August 5, 1976, half an hour after the last exposure under pentobarbital anaesthesia (30 mg/kg), the animal was perfused from the abdominal aorta with 2.5 % glutaraldehyde in 0.1 M sodium cacodylate buffer (pH 7.4). Postfixation with 1 % osmium tetroxide in sodium cacodylate buffer. Embedded in Epon 812 and sectioned at 50 nm. Lead citrate and uranyl acetate. Film 624-05

not the progestogenic/androgenic Δ^4-isomer, concentration-dependently decreased lipopolysaccharide-induced vascular wall adhesion molecule-1 protein expression in human saphenous vein endothelial cells cultured in phenol red-free Medium 199 (Simoncini and Genazzini 2000).

The endothelium can interact with circulating **platelets** in at least two ways. First of all, products of the platelet release reaction (i.e. ADP and serotonin) have been shown to be potent stimulators of endothelium-derived relaxing factor (i.e. •NO). The second mechanism involves a direct and potent

Fig. 302. An activated endothelial cell with pinocytotic vesicles and dilated cisternae of the endoplasmic reticulum from the region between the papillary muscles of the left ventricle (block 1278) from a *rat* (No. 9) treated for 7 consecutive days with intraperitoneal injection of 1.5 ml Tyrode's solution per kg body weight × day. On the last 4 days of experimentation the animal was exposed to an atmosphere containing only 5 % oxygen for 30 min. On August 4, 1976 half an hour after the last exposure under pentobarbital anaesthesia (30 mg/kg), the animal was perfused from the abdominal aorta with 2.5 % glutaraldehyde in 0.1 M sodium cacodylate buffer (pH 7.4). Postfixation with 1 % osmium tetroxide in sodium cacodylate buffer. Embedded in Epon 812 and sectioned at 50 nm. Lead citrate and uranyl acetate. Plate 3366

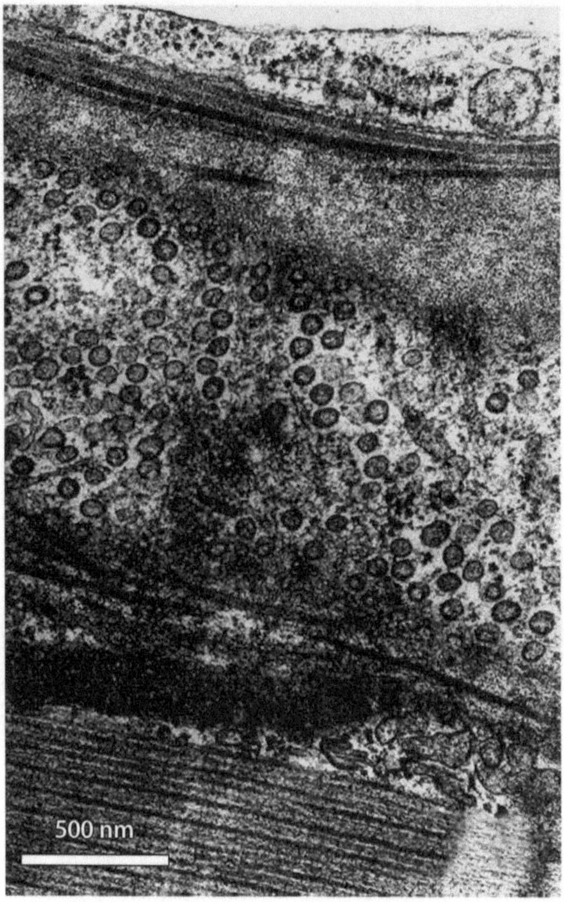

Fig. 303. Dilated endoplasmatic reticulum in an endothelial cell from the region of the posterior papillary muscle (block 1401) of a *rat* (No. 13) treated for 8 consecutive days with intraperitoneal injection of 15 mg carbocromene per kg body weight × day. On the last 5 days of experimentation the animal was exposed to an atmosphere containing only 5 % oxygen for 30 min. On August 5, 1976, under pentobarbital anaesthesia (30 mg/kg), the animal was perfused from the abdominal aorta with 2.5 % glutaraldehyde in 0.1 M sodium cacodylate buffer (pH 7.4). Postfixation with 1 % osmium tetroxide in cacodylate buffer. Embedded in Epon 812 and sectioned at 50 nm. Lead citrate and uranyl acetate. Film 619-11

antiaggregant effect of endothelium-derived relaxing factor on platelets themselves (Azuma et al. 1986, Busse et al. 1987, Furlong et al. 1987).

Prostacyclin (PGI_2) synthesised by endothelial cells and released predominantly towards the luminal side (Bassenge and Heusch 1990), acts via a receptor-mediated adenylate cyclase. In platelets, the increase in cAMP inhibits adhesion, aggregation and the release of pro-aggregatory and vasoconstrictor compounds such as serotonin, thromboxane A_2, and ADP.

15.1.8.5 Immigration of Inflammatory Cells

Gap junctions are aggregates of intercellular channels, formed by head-to-head alignment of two hemichannels (connections), each of which is made up of six connexin protein molecules (Harold et al. 1997). Zhang et al. (1999) demonstrated that hypoxia-reoxygenation induced a temporal reduction of gap junctional intercellular communication in *human* umbilical vein endothelial cells and that the protein tyrosine kinase pathway was primarily responsible for this gap junctional intercellular communication abnormality.

It is well known that neutrophils infiltrate into ischaemic, reperfused myocardial tissue and play a critical role in the development of reperfusion injury (Engler and Schmidt-Schönbein 1983, Engler et al. 1986, Engler 1987, Schmidt-Schönbein 1987, Granger 1988, Mehta et al. 1988, Reynolds and McDonagh 1989, Seko et al. 1996). The role of leucocytes in free radical production during myocardial revascularisation was shown by De Vecchi et al. (1997) when they used leucocyte depleted blood in cardioplegic reperfusion. Polymorphonuclear neutrophils and infarct sizes were significantly reduced when BW755C was applied before and after ischaemia in experimental *porcine* myocardial infarction (Bohle et al. 1991).

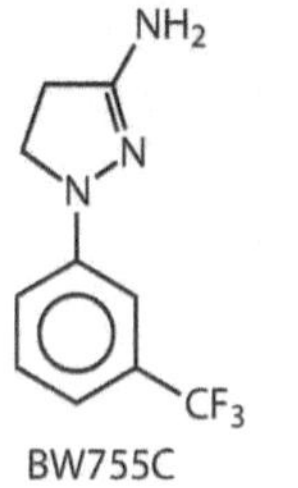

BW755C [291]

Use of C6-deficient and C6-sufficient *rabbits* has shown a considerable evidence that the **cytolytic membrane attack complex** (MAC) participates in mediating the development of myocardial reperfusion injury (Kilgore et al. 1998). When present in high concentration, the MAC promotes target cell lysis. MAC regulated IL-8 expression and subsequent neutrophil recruitment in the setting of myocardial ischaemia/reperfusion injury. In addition to a significant reduction in myocardial infarct size in C6-deficient animals, analysis of myocardial tissue demonstrated a decrease in neutrophil influx into the infarcted region. The reduction in neutrophil influx correlated with the decreased expression of the neutrophil chemotactic cytokine IL-8, as determined by ELISA and immunohistochemical analysis. There was no significant difference in myeloperoxidase activity in the non-infarct related region of the left ventricle of C6-sufficient to that of C6-deficient animals for the first 2 h of reperfusion. However, after this period, the myeloperoxidase levels found in the area at risk from C6-deficient animals demonstrated decreased neutrophil influx as compared to C6-sufficient animals.

On stimulation, endothelial cells express membrane glycoproteins, that recognize ligands on polymorphonuclear leucocytes (CARLOS and HARLAN 1990, GENG et al. 1990, PATEL et al. 1991, PIGOTT et al. 1992, FRANZINI et al. 1995). This phenomenon is responsible for increased polymorphonuclear leucocyte penetration into myocardial tissue and also sustains inflammation.

GENG et al. (1992) investigated whether or not **selectin oligopeptides** corresponding to residues 23–30, 54–63, and 70–79 of the N-terminal lectin domain of *human* P-, E-, and L-selectin, which may represent contact sites for carbohydrate structures on target cells, can inhibit neutrophil adhesion to P-selectin *in vitro*. The 23–30 peptide of E-selectin and the 54–63 peptide of E- and L-selectin were shown effectively to inhibit neutrophil adhesion to immobilized P-selectin. They were also shown effectively to inhibit adhesion of HL-60 promyelocytic cells to E-selectin-transfected COS-7 cells *in vitro* (GENG et al. 1992). SEKO et al. (1996) found almost undetectible expression of P-selectin on the luminal surface of vascular endothelium in the left ventricle of sham-opertated *rats*. Reperfusion for 2 h after 30 min of ischaemia, however, significantly increased the expression of P-selectin on the luminal surface of the vascular endothelium, resulting in the attachment of several neutrophils. An intravenous injection of 10 mg/kg of the 23–30 peptide (YTHLVAIQ) 5 min before coronary artery occlusion reduced the size of the infarcted area from 38 per cent to 16 per cent of the myocardium as investigated after 48 h of reperfusion.

From oxidant-stimulated (2×10^{-4} M hypoxanthine/4.5 mU xanthine oxidase for 15 min) *human* umbilical vein endothelial cells cultured *in vitro*, FRANZINI et al. (1995) concluded that **cAMP** is probably involved in the adherence of polymorphonuclear neutrophils. Preincubation with pentoxifylline inhibited adherence in a concentration-dependent manner. Isobutylmethylxanthine and isoprenaline, which increase intracellular cAMP content, were also inhibitory.

It is still unclear whether neutrophil binding to the endothelial cell surface affects VE-cadherin functions (VESTWEBER 2000). It was proposed that adhesion of polymorphonuclear leucocytes to human umbilical endothelial cells leads to the disorganisation of the VE-cadherin-dependent endothelial adherens junctions. Combined immunofluorescence and biochemical data suggested that following adhesion of polymorphonuclear leucocytes to the endothelial cell surface, β-catenin as well as plakoglobin was lost from the cadherin/catenin complex and from total cell lysates (DEL MASCHIO et al. 1996, ALLPORT et al. 1997). However, MOLL et al. (1998) demonstrated that the adhesion-dependent disappearance of endothelial catenins in these experiments was not mediated by a leucocyte to endothelium signalling event, but was due to the activity of a neutrophil protease which had been experimentally released upon lysis and handling of the co-incubated neutrophils and endothelial cells.

Catecholamine-induced focal disseminated myocardial necroses showed some mesenchymal reac-

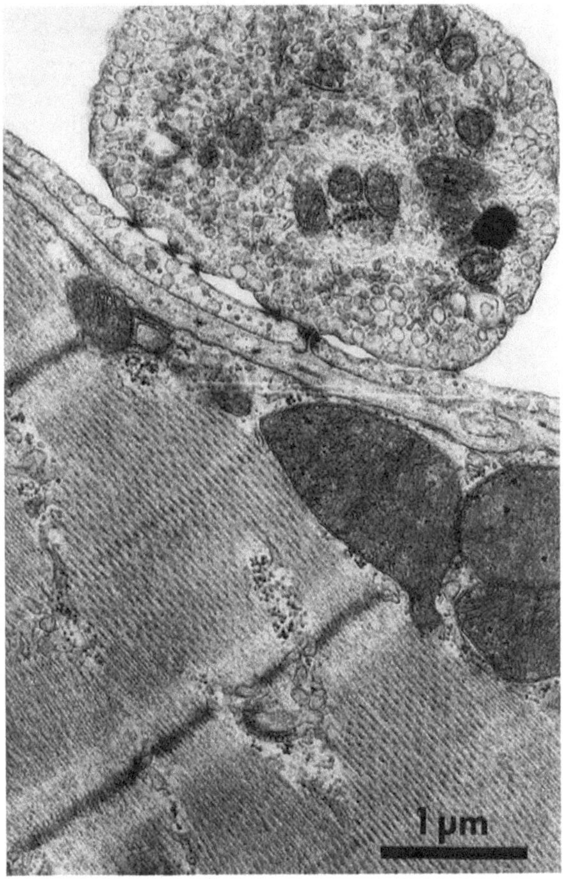

Fig. 304. Attachment of a leucocyte to an interendothelial junction from the interventricular septum (block 1461) of a *rat* (No. 15) treated for 8 consecutive days with intraperitoneal injection of 15 mg carbocromene per kg body weight × day. On the last 5 days of experimentation the animal was exposed to an atmosphere containing only 5 % oxygen for 30 min. On August 6, 1976 under pentobarbital anaesthesia, the animal was perfused from the abdominal aorta with 2.5 % glutaraldehyde in 0.1 M sodium cacodylate buffer. Postfixation with 1 % osmium tetroxide in sodium cacodylate buffer. Embedded in Epon 812 and sectioned at 50 nm. Lead citrate and uranyl acetate. Film 623-01

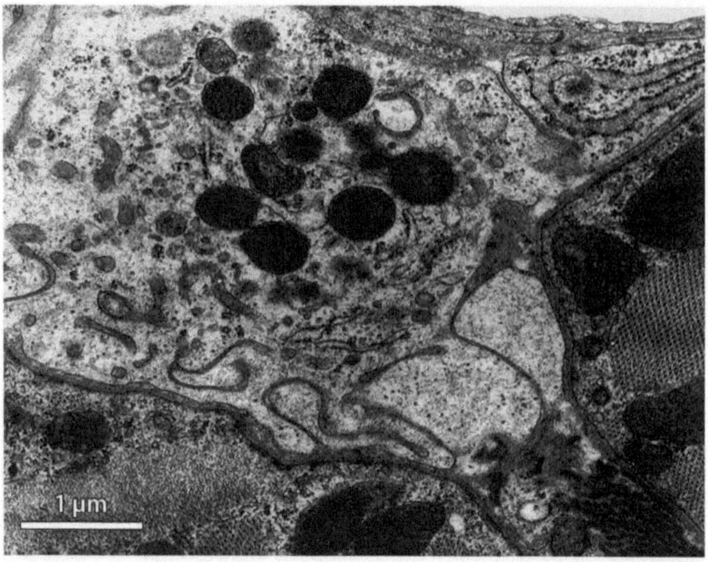

Fig. 305. Myocardial macrophage with a prominent ruffled border from the posterior papillary muscle (block 1309) of a *rat* (No. 10) treated for 8 consecutive days with peritoneal injection of 15 mg carbocromene per kg body weight × day. On the last 5 days of experimentation the animal was exposed to an atmosphere containing only 5 % oxygen for 30 min. On August 6, 1976 under pentobarbital anaesthesia, the animal was perfused from the abdominal aorta with 2.5 % glutaraldehyde in 0.1 M sodium cacodylate buffer. Postfixation with 1 % osmium tetroxide in sodium cacodylate buffer. Embedded in Epon 812 and sectioned at 50 nm. Lead citrate and uranyl acetate. Film 629-30

tions, which lead Szakács and Cannon (1958) to use the term "epinephrine myocarditis". Following isoprenaline (10 mg per kg body weight s.c.)-induced myocardial injury in *rats*, α-smooth muscle actin expressing myofibroblasts were seen in the border of the affected area and appeared in the greatest number on days 5–7 post injectionem, followed by a gradual decrease by day 35 (Nakatsuji et al. 1997). The number of ED1-positive macrophages began to increase as early as day 1, reaching a peak on day 3 within the injured myocardium. The expansion of ED1-positive macrophages preceded in increased number of α-smooth muscle actin-positive myofibroblasts suggesting that myofibroblast proliferation and activation may be mediated by factors released by ED1-positive macrophages in response to myocardial injury. The number of ED2-positive tissue-fixed, resident macrophages gradually increased from day 3 post injectionem, and peaked on day 14, but the number of ED2-positive macrophages was consistently less than that of ED1-positive macrophages during the 35 days of the experiment.

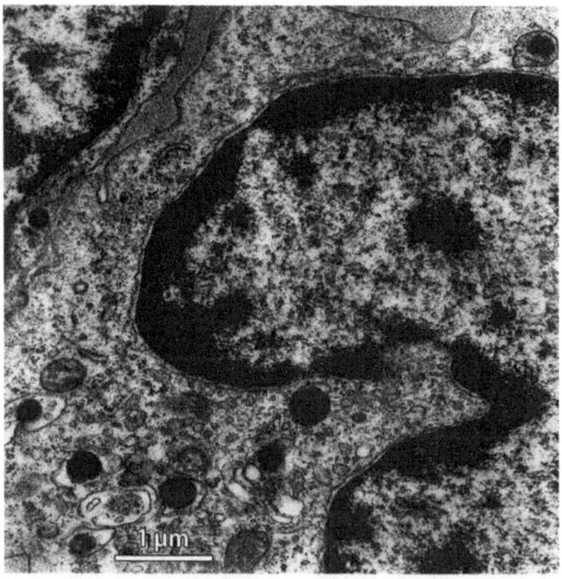

Fig. 306. Numerous microtubules and microfilaments in a myocardial macrophage (right ventricle, block 1462) from a white *rat* (No. 15) medicated for 8 consecutive days with intraperitoneal injection of 15 mg carbocromene per kg body weight × day. On the last 5 days of experimentation the animal was exposed to an atmosphere containing only 5 % oxygen for 30 min. On August 6, 1976 under pentobarbital anaesthesia, the animal was perfused from the abdominal aorta with 2.5 % glutaraldehyde in 0.1 M sodium cacodylate buffer. Postfixation with 1 % osmium tetroxide in sodium cacodylate buffer. Embedded in Epon 812 and sectioned at 50 nm. Lead citrate and uranyl acetate. Film 629-21

It has been shown that the dendritic cells functionally influence not only T lymphocytes but ED2-positive resident macrophages in infarcted (Zhang et al. 1993b) and hypertrophied *rat* heart (Zhang et al. 1993a). ED2-positive resident macrophages provide processed antigen fragments for the antigen-presenting dendritic cells by their phagocytic activity (Zhang et al. 1993b). This implies that ED2-positive cells have a potential to act as immune effector cells in the chronic inflammatory state of the myocardial healing process.

15.1.8.6 Fibrocytes

Elsässer et al. (1998) proposed the hypothesis that a vicious cycle exists in *human* hibernating myocardium between the progression of myocyte degeneration and the development of fibrosis. Elsässer

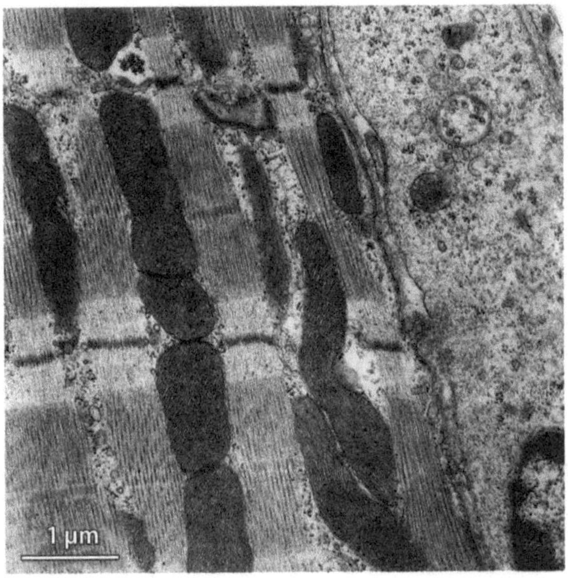

Fig. 307. Two multivesicular bodies in a fibrocyte from the right ventricle (block 1462) of a *rat* (No. 15) medicated for 8 days with intraperitoneal injection of 15 mg carbocromene per kg body weight × day. On the last 5 days of experimentation the animal was exposed to an atmosphere containing only 5 % oxygen for 30 min. On August 6, 1976, half an hour after the last exposure under pentobarbital anaesthesia (30 mg/kg), the animal was perfused from the abdominal aorta with 2.5 % glutaraldehyde in 0.1 M sodium cacodylate buffer (pH 7.4). Postfixation with 1 % osmium tetroxide in sodium cacodylate buffer. Embedded in Epon 812 and sectioned at 50 nm. Lead citrate and uranyl acetate. Film 623-13

et al. (2000) reported an upregulation of TGF-β_1 evident by a 5-fold increase of fibroblasts and macrophages exhibiting a TGF-β_1 content 3-fold larger than in control, and a >3-fold increase in TGF-β_1 mRNA by rt-PCR.

Probing the mechanism of cardiac fibrogenesis in magnesium deficiency, Kumaran and Shivakumar (2001) furnished evidence that serum from magnesium-deficient *rats* has a more marked effect than serum from magnesium-sufficient *rats* on mitogenesis, net collagen production, and superoxide generation in cardiac fibroblasts from young adult *rats*. The enhanced mitogenic response was abolished by superoxide dismutase and N-acetyl cysteine, showing that it is mediated by superoxide anion.

15.1.8.7 Mast Cells

Deleterious effects of resident mast cells on the course of myocardial ischaemia originate from experimental models in which acute myocardial ischaemia is followed by reperfusion (Keller et al. 1988, Mannaioni et al. 1988). In these models, mast cells are thought to be involved in the precipitation of reperfusion arrhythmias, the production of reactive oxygen species, and the impairment of coronary perfusion by the release of preformed mediators such as histamine (Levi et al. 1985, Triggiani et al. 1985, Mannaioni et al. 1988, Pearce 1991) and serotonin. Engels et al. (1995) in Wistar *rats* bred by Winkelmann after ligature of the left anterior descending coronary artery found mast cells accumulated in the subepicardial layer of the infarcted region at 3 weeks after surgery, possibly originating from an influx of mast cells from the pericardium to the myocardium.

15.1.9 Kidney

Renal tissue is often a prime target for oxidative stress, and releases $O_2^{\bullet-}$ and $^{\bullet}NO$, resulting in $ONOO^-$ formation (Narita et al. 1995, Raij and Baylis 1995).

Increased oxygen consumption by the kidney may lead to the generation of reactive oxygen species. Rehan et al. (1984) and Shah (1988, 1989) showed that treatment of animals with scavengers of superoxide anion, hydrogen peroxide, and hydroxyl radical ameliorated the proteinuria and glomerular injury associated with experimental nephritis. While Rehan et al. (1984) assumed neutrophils infiltrating the interstitium were the source of reactive oxygen species, it has now been shown that certain cells intrinsic to the kidney are an important source of reactive oxygen species. *Rat* glomerular mesangial cells in culture produce hydrogen peroxide and superoxide anion in response to opsonized zymosan and immune complexes (Baud et al. 1983), and in the *rabbit* resident glomerular macrophages are also capable of generating reactive oxygen species (Boyce et al. 1989). Primary cultures of *rabbit* proximal tubule, cortical collecting duct, and papillary collecting duct produced augmented levels of reactive oxygen species, specifically $O_2^{\bullet-}$ and H_2O_2, in response to heat-aggregated immunoglobulin and zymosan (Rovin et al. 1990). Although the magnitude of reactive oxygen species generation by renal tubular cells appears to be significantly less than that produced by other renal cell types, it appears that reactive oxygen species secretion by tubules during immune injury of the kidney participates in the development of interstitial disease which accompanies glomerulonephritis.

In subtotally nephrectomised *rats*, the imposition of increased protein intake was sufficient to cause renal damage arising from oxidant stress (Nath et al. 1990). It appears that in the hyperfunctioning surviving nephron, there is increased sodium reabsorption, and hence oxygen consumption, which generates a variety of reactive oxygen species at several subcellular loci. Increased rates of oxygen consumption require increased transport of reducing equivalents through the mitochondrial electron transport chain. These augmented rates of transport may generate free electrons that overwhelm the resident mitochondrial antioxidant mechanisms, resulting in a peroxidation of mitochondrial membrane lipid. A dietary deficiency of antioxidants as vitamin E and selenium, imposed on weaning *rats* caused renal growth and tubulointerstitial damage in the intact, nonmanipulated *rat* kidney with concomitant reduction of overall somatic growth (Nath and Salahudeen 1990), The authors suggested that augmented rats of ammonia production which antedated the phase of renal growth were in part responsible for the renal damage due to these dietary deficiencies. It has been suggested that the accumulation of reactive oxygen species in the kidney may directly stimulate renal ammoniagenic pathways (Nath and Salahudeen 1990), or that H_2O_2 deaminates amino acids via a metal ion-catalysed reaction with the generation of ammonium bicarbonate (Stadtman and Berlett 1988, Berlett et al. 1990). In the same way that reactive oxygen species up-regulate collagen gene expression in fibroblasts, so they may in tubular cells (Houglum et al. 1991). The resultant peritubular fibrosis leads to nephron destruction and disease progression.

Large sex differences have been reported in the cytochrome P_{450} concentration of *mouse* kidney microsomes and in the hydroxylation of testosterone by kidney microsomes (Hawke et al. 1983, Hawke and Welch 1985). When assayed at 500 μM, male renal 7-ethoxycoumarin-*O*-deethylase activity was 3-fold greater than female 7-ethoxycoumarin-*O*-deethylase activity, although this difference was less than that observed in cytochrome P_{450} concentration as indicated by an approximately 2-fold greater turnover value for female renal microsomes.

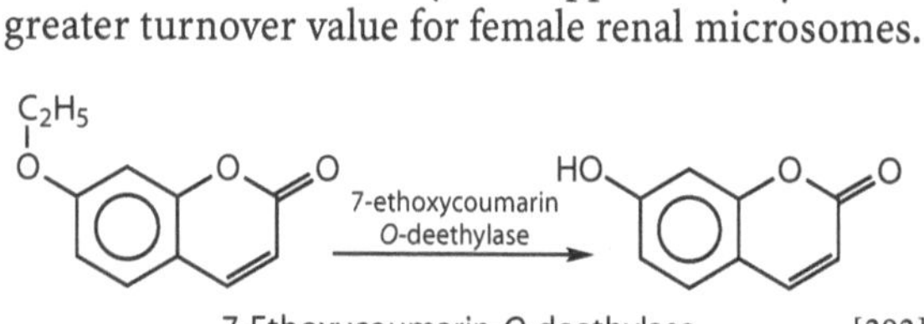

7-Ethoxycoumarin-*O*-deethylase [292]

Styrene inhalation was followed by the induction of renal cytochrome P_{450} (Vainio and Elovaara 1979).

In the presence of daunorubicin (200 μM) the NADPH-cytochrome P-450 reductase in *mouse* kidney microsomes was approximately 30 % of the liver microsomal activity (Mimnough et al. 1983).

Inhaled nitric oxide (107 ± 13 ppm NO for 4 h) increased NOS III protein expression, nitrotyrosine, and phosphotyrosine in the liver and kidney, but not in the lung or spleen of Sprague-Dawley *rats* (Kielbasa and Fung 2001).

Nitric oxide donors [sodium nitroprusside, *S*-nitroso-*N*-acetyl-D,L-penicillamine or 1-hydroxy-2-oxo-3-(*N*-methyl-6-aminohexyl)-3-methyl-1-triazene = NOC-9] did not affect the activities of non-specific alkaline phosphatase and non-specific esterase in *rat* and *mouse* kidney proximal tubulocytes which do not contain nitric oxide synthase I (Dahrmann et al. 1997). However, the activities of succinate dehydrogenase (EC 1.3.99.1), cytochrome *c* oxidase (EC 1.9.3.1), catalase (EC 1.11.1.6), peroxidase (EC 1.11.1.7), and cholinesterase (EC 3.1.1.7) and different types of adenosine triphosphatase (EC 3.6.1.3) were found reduced.

In homogenates of *mouse* kidney perfused with saline before homogenisation, haemoglobin (30 μmol/ml), a •NO scavenger, provided protection from tissue injury caused by sodium nitroprusside (1 mmol/ml) (Yokozawa et al. 1999). The lipid peroxidation produced by 3-morpholinosydnonimine (SIN-1; 1 mmol/ml) as a simultaneous $O_2^{\bullet-}$ and •NO generator, was blocked by superoxide dismutase/catalase (250 units/ml, respectively) or haemoglobin ($P < 0.001$).

Iron has been implicated as a key factor in tubular cell damage in a variety of acute and chronic renal insults (Nankivell et al. 1992), and the tubulointerstitial injury, an invariant finding in the chronically diseased kidney, may be the major determinator of subsequent nephron loss and progressive renal failure (Alfrey 1994).

After incubation of *rat* kidney microsomes and mitochondria in an ascorbate-Fe^{2+} system, at 37 °C during 60 min, Piergiacomi et al. (1996) observed that the total cpm/mg protein originated prom light emission (chemiluminescence) was lower in those organelles obtained from the control group when compared with the vitamin A-supplemented group. The fatty acid composition of microsomes and mitochondria from control group was profoundly modified when subjected to non-enzymatic lipoperoxidation with a considerable decrease of arachidonic acid, C20:4 and docosapentaenoic acid, C22:5 in mitochondria and docosahexaenoic acid, C22:6 in microsomes.

In Fe-loaded *rats* both plasma and renal tissue malondialdehyde and renal tissue nitrotyrosine were increased significantly compared with control *rats* (Zhou et al. 2000). Fe-depleted animals showed a marked reduction in plasma and renal tissue malondialdehyde and nitrotyrosine together with significant elevation of urinary NO_x excretion. In addition, iron-overload was associated with up-regulation of renal eNOS and iNOS expressions when compared with the control and Fe-depleted *rats* that showed comparable values.

Gold (Au), when administered to the *rat* as either $NaAuCl_4$, or sodium aurothiomalate, accumulates in the particulate component of the kidney and in low molecular weight metalloprotein fraction of the cytosol. Induction of metallothionein synthesis by Au not only seems to be less efficient than that in response to mercury (Hg), cadmium (Cd) and bismuth (Bi) (PIOTROWSKI et al. 1979), but also to be determined by changes in copper distribution (MOGILNICKA and WEBB 1981). Pre-treatment with Cd increased the contents of metallothionein-bound Au, Cu and Zn in the *hamster* kidney (MOGILNICKA and WEBB 1982).

The **podocyte** is the major culprit in the progression of glomerular diseases (KRIZ et al. 1998). Glomerular overproduction of oxygen radicals in Mpv17 gene-inactivated *mice* caused podocyte foot process flattening and proteinuria (BINDER et al. 1999). When exposed to increased challenge of any kind, podocytes are unable to maintain their normal differentiated phenotype, but change in appearance in a fairly stereotyped manner (KRIZ et al. 1994, 1995). These changes comprise cell hypertrophy, foot-process effacement, cell body attenuation, pseudocyst formation, cytoplasmic overload with reabsorption droplets, and finally, detachment from the glomerular basement membrane. Cell body attenuation and pseudocyst formation have been shown to result directly from mechanical overextension (NAGATA and KRIZ 1992).

Cultivation of podocytes from isolated glomeruli dates back to KREISBERG et al. (1978). Podocytes in culture grow as so-called "cobblestones", that is, simple polygonal cells corresponding in many respects, such as mitotic activity, and in appearance, to the podocyte precursor cells in ontogeny. Considerable doubts as to whether these cultured cobblestone cells really are derived from podocytes have ruled out by YAOITA et al. (1995). The positive staining of these cells for WT-1 protein clearly identifies them as podocyte derived, and establishes the validity of these cultures (MUNDEL et al. 1997). Podocytes from the transgenic immorto-*mouse* (JAT et al. 1991) cultured at 33 °C vividly proliferate, but stop proliferation at 37 °C and begin to differentiate into what is called arborized cells.

In *mouse* glomeruli, ENDLICH et al. (2002) observed the 80-kDa CD2-associated protein mainly in podocytes. It co-localised with F-actin in foot processes, but was also present in podocyte cell bodies. Immunoelectron microscopy confirmed the presence of CD2AP in podocyte foot processes, where it was found close to the slit diaphragm, at the sole plate, and at other cytoplasmic areas. Time-lapse microscopy of living podocyted revealed that ring-like actin clouds were motile structures emerging and vanishing spontaneously, moving within the cytoplasm. They shrank or expanded with a mean velocity of 1.2 ± 0.4 μm/s.

In the **mesangium** nitric oxide is generated through L-arginine metabolism by constitutive and inducible NOS isoenzymes. Mesangial cells are competing in expressing iNOS in response to various cytokines. Mesangial cells cultured from *rat* glomeruli stained heterogeneously for iNOS, depending on cell passage and iNOS stimulating pathway (NITSCH et al. 1997). Mesangial cells expressing iNOS did not display signs of apoptosis and, vice versa, cells showing characteristics of apoptosis did not stain for iNOS.

Mesangial cells start to change their phenotype as soon as they face •NO (SANDAU and BRÜNE 2000). When they cope with •NO, most stress fibres of their cytoskeleton vanish and only a few contacts to neighbouring cells persist. This phenomenon is fully reversible and does nor result in a detachment of the cells. The mechanism leading to F-actin dissolution is very sensitive to •NO as 500 nM or 1 μM of *S*-nitrosoglutathione was sufficient to achieve these alterations.

Using various •NO donors such as *S*-nitrosoglutathione or spermine-NO MÜHL et al. (1996) established a proapoptotic action in cultured *rat* mesangial cells. Proapoptotic action appears to be transmitted in part by activation of the cJun N-terminal kinases 1/2 (SANDAU et al. 1999). •NO itself promoted strong JNK1/2 activation and apoptosis, in some analogy to Ro 318220, a characterised JNK1/2 activator. In contrast, activation of p42/p44 mitogen activated protein kinases (ERK1/2) afforded some protection toward •NO-evoked apoptosis that was antagonised by the upstream ERK1/2 inhibitor PD 98059.

Peroxynitrite formation by the simultaneous formation of $O_2^{\bullet-}$ antagonised •NO-mediated apoptotic cell death (BRÜNE et al. 1997, SANDAU et al. 1997, SANDAU and BRÜNE 2000).

15.1.9.1 Reperfusion Injury

Isolated *rat* kidneys perfused with a gassed (95% O_2 and 5% CO_2) albumin-Krebs-Henseleit solution, after 2 h displayed a characteristic pattern of cell necrosis, which was confined to the interbundle region of the outer medulla and was not evident in either the cortex or the inner medulla (SCHUREK and KRIZ 1985). In the outer stripe only those proximal

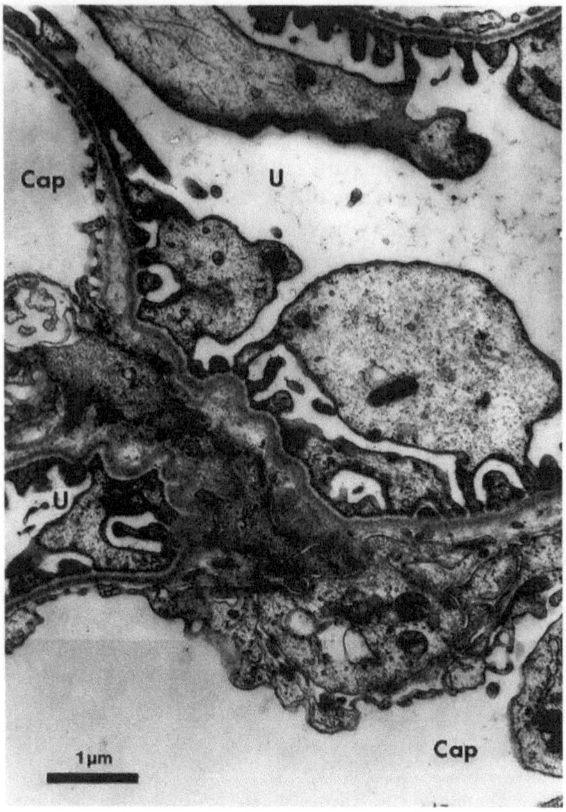

Fig. 308. Reoxygenation damage to podocytes and endothelial cells from a renal glomerulus (block BNh 2145) of a Wistar *rat* (No. 1900) treated for 4 consecutive days with intraperitoneal injection of 1.5 ml Tyrode's solution per kg body weight × day and exposed to an atmosphere containing only 5% oxygen for 30 min. On October 31, 1974, half an hour after the last exposure under pentobarbital anaesthesia (30 mg/kg), the animal was perfused from the abdominal aorta with 2.5% glutaraldehyde in 0.1 M sodium cacodylate buffer (pH 7.4). Postfixation with 1% osmium tetroxide in cacodylate buffer. Embedded in Epon 812 and sectioned at 50 nm. Lead citrate and uranyl acetate. Plate 2172

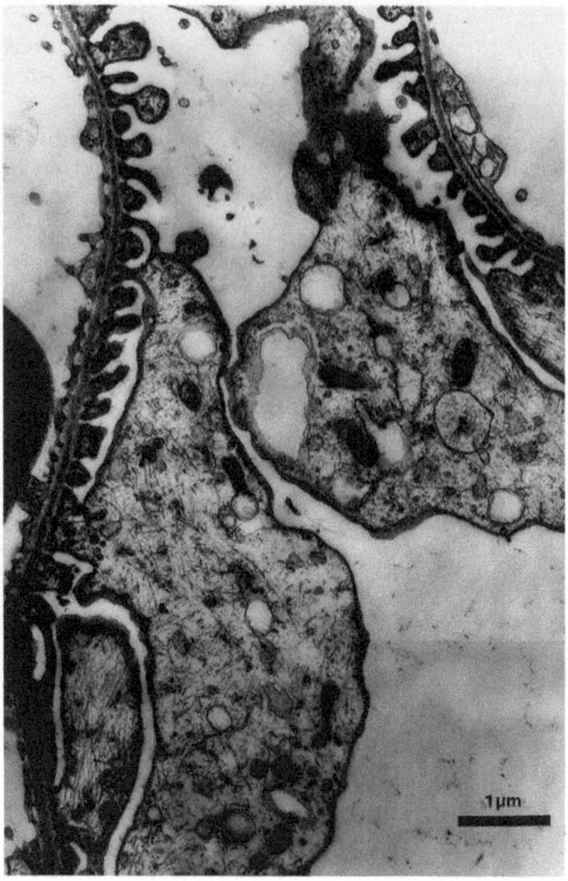

Fig. 309. Reoxygenation damage to podocytes and endothelial cells from a renal glomerulus (block BNh 2145) of a Wistar *rat* (No. 1900) treated for 4 consecutive days with intraperitoneal injection of 1.5 ml Tyrode's solution per kg body weight × day and exposed to an atmosphere containing only 5% oxygen for 30 min. On October 31, 1974, half an hour after the last exposure under pentobarbital anaesthesia (30 mg/kg), the animal was perfused from the abdominal aorta with 2.5% glutaraldehyde in 0.1 M sodium cacodylate buffer (pH 7.4). Postfixation with 1% osmium tetroxide in cacodylate buffer. Embedded in Epon 812 and sectioned at 50 nm. Lead citrate and uranyl acetate. Plate 2174

straight tubules (P_3 segment) farthest from the vascular bundles were damaged. In the inner stripe only those thick ascending loops of Henle at the periphery of the vascular bundles escaped damage; all thick ascending loops of Henle lying farthest from the bundles were severely damaged.

Paller and Jacob (1994) have shown that cytochrome P_{450} mediates hydroxyl radical formation during reoxygenation of the kidney leading to lethal cell injury.

Prior induction of heme oxygenase-1 with L-buthionine-(*S*,*R*)-sulphoximine (2 mmol/kg body weight) 5 h before the occlusion of the left renal artery and vein for 45 min ameliorated the ischaemic injury induced in male Wistar *rats* 3 weeks after the removal of their right kidney (Horikawa et al. 2002).

Gröne et al. (2002) demonstrated HOCl-modified proteins in glomeruli of *patients* with membranous glomerulonephritis using monoclonal antibodies that do not cross-react with other oxidative modifications. Immunostaining was detected in intracapillary cells and immune complex deposits within the glomerular basement membrane. In *human* membranous glomerulonephritis, staining for HOCl-modified proteins was localised at the basement membrane and podocytes.

The change in peroxisomal functions in ischaemic-reperfused *rat* kidney is reflected by a significant decrease in peroxisomal catalase activity (35%) and β-oxidation of lignoceric acid (43%) observed following 90 min of ischaemia (Gulati

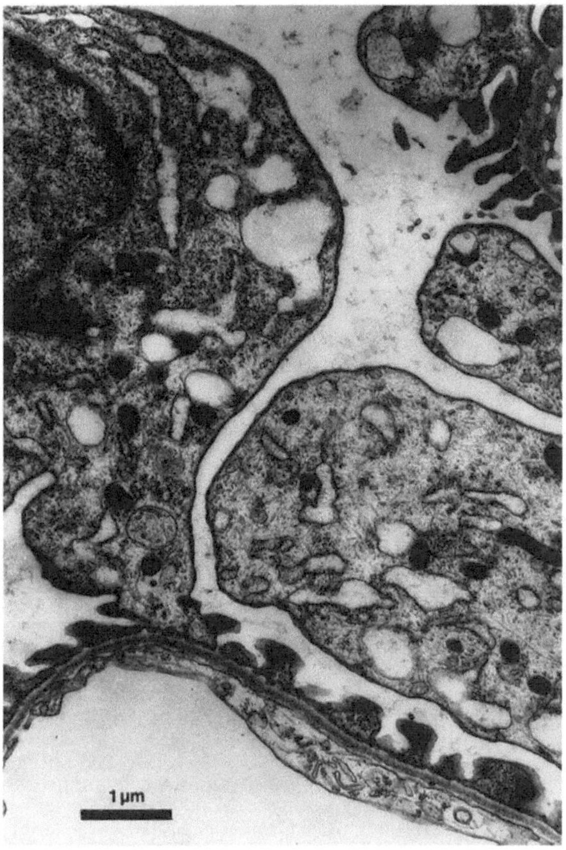

Fig. 310. Reoxygenation damage to podocytes and endothelial cells from a renal glomerulus (block BNh 2152) of a Wistar *rat* (No. 1901) treated for 4 consecutive days with intraperitoneal injection of 1.5 ml Tyrode's solution per kg body weight × day and exposed to an atmosphere containing only 5% oxygen for 30 min. On October 31, 1974, half an hour after the last exposure under pentobarbital anaesthesia (30 mg/kg), the animal was perfused from the abdominal aorta with 2.5% glutaraldehyde in 0.1 M sodium cacodylate buffer (pH 7.4). Postfixation with 1% osmium tetroxide in cacodylate buffer. Embedded in Epon 812 and sectioned at 50 nm. Lead citrate and uranyl acetate. Plate 2175

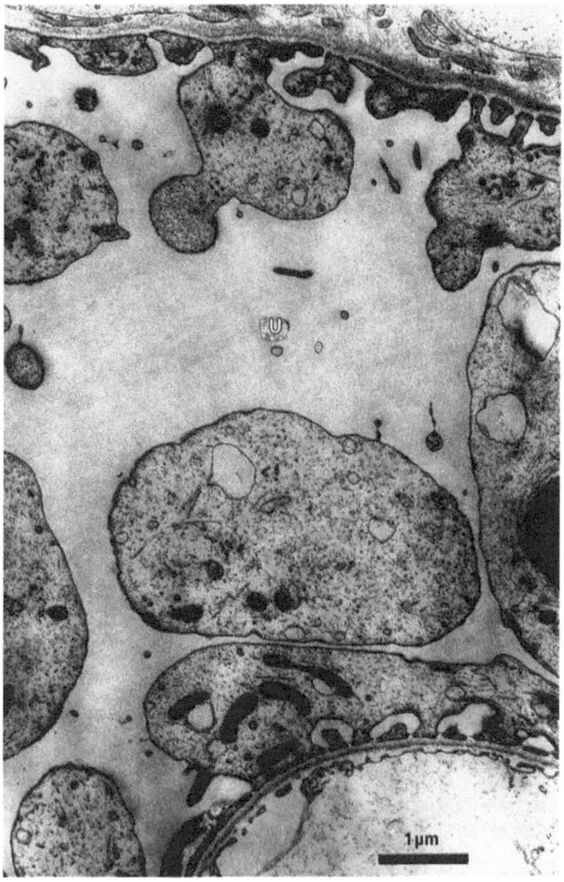

Fig. 311. Reoxygenation damage to podocytes and endothelial cells from a renal glomerulus (block BNh 2159) of a Wistar *rat* (No. 1902) treated for 4 consecutive days with intraperitoneal injection of 1.5 ml Tyrode's solution per kg body weight × day and exposed to an atmosphere containing only 5% oxygen for 30 min. On October 31, 1974, half an hour after the last exposure under pentobarbital anaesthesia (30 mg/kg), the animal was perfused from the abdominal aorta with 2.5% glutaraldehyde in 0.1 M sodium cacodylate buffer (pH 7.4). Postfixation with 1% osmium tetroxide in cacodylate buffer. Embedded in Epon 812 and sectioned at 50 nm. Lead citrate and uranyl acetate. Plate 2176

et al. 1992). The decrease in catalase activity was more pronounced in reperfused kidneys even after a shorter term of ischaemic injury. Peroxisomes from ischaemic kidneys have two populations: one with normal density (1.21 g/cm^3) and the other with a lighter density (1.14 g/cm^3).

An increased urinary excretion of the brush border antigens BB50, BBA, HF5, and IAP (specific for S3 segment) is in all likelihood a reflection of damage to proximal tubular cells (Mutti 1989, Mutti et al. 1985, 1988). Belgian workers exposed to Hg vapour in a chloralkali plant for et least one year presented a significantly higher urinary excretion of BB50, BBA, HF5, and IAP than the controls (Cárdenas et al. 1993).

Apoptosis was observed in the *rat* renal tubular cells after clamping the renal artery for 30 min and two hours of reperfusion (Toronyi et al. 1999). Calcium channel blockers given into the carotid artery at the beginning of reperfusion decreased the number of apoptotic cells.

Arteries of isolated erythrocyte-perfused *rat* kidneys ischaemic for 25 min and recovered for 20 min revealed the presence of occasional focal areas of necrosis of the media. The endothelium appeared normal except for areas of disruption in regions overlying muscle necrosis (Lieberthal et al. 1989).

Polymorphonuclear neutrophil granulocytes seem to participate in the pathogenesis of renal ischaemic reperfusion injury. The ratio of intraglo-

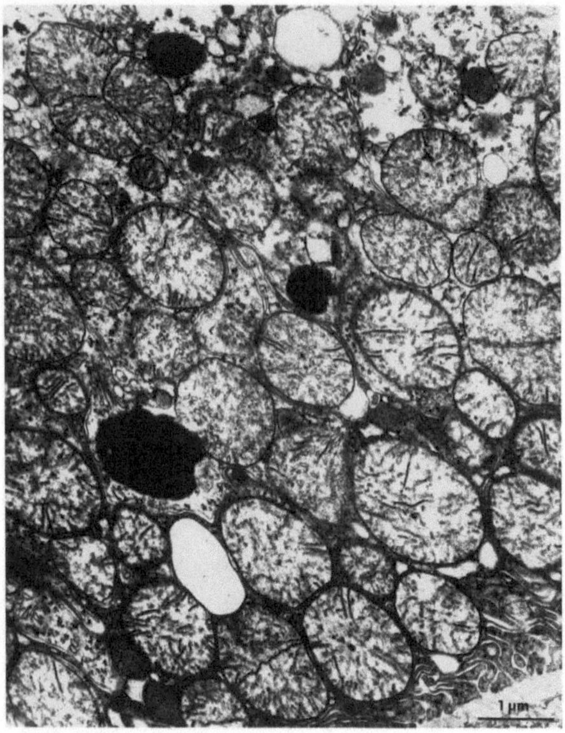

Fig. 312. Partitioned mitochondria in renal proximal tubule cell (block BNh 3372) of a male Wistar *rat* (No. 2567) 90 min after a single intraperitoneal injection of 5 mg *N*-benzhydryl-*N'*-p-hydroxybenzylpiperazine HCl per kg body weight. Under pentobarbital anaesthesia (30 mg/kg), the animal was perfused from the abdominal aorta with 2.5 % glutaraldehyde in 0.1 M sodium cacodylate buffer (pH 7.4). Postfixation with 1 % osmium tetroxide in sodium cacodylate buffer. Embedded in Epon 812 and sectioned at 50 nm. Lead citrate and uranyl acetate. Plate 2449

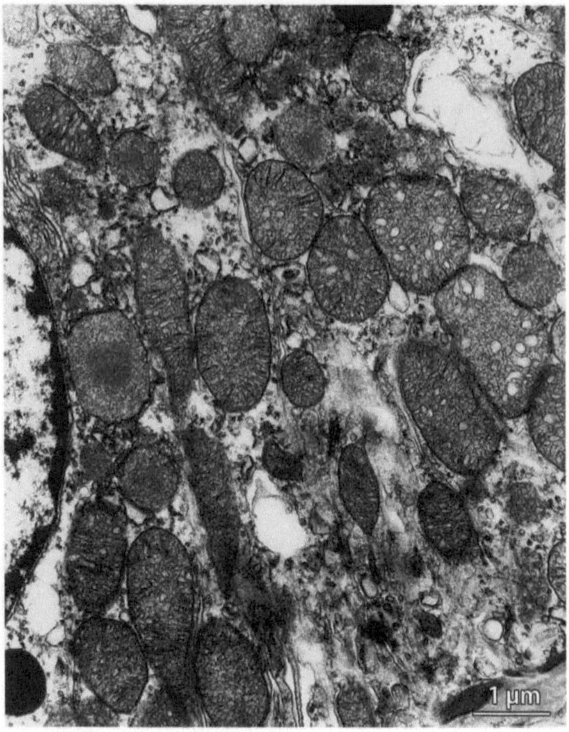

Fig. 313. Renal proximal tubule cell (block BNh 3405) from a male Wistar *rat* (No. 2569) 90 min after a single intraperitoneal injection of 5 mg *N*-benzyl-*N'*-p-hydroxybenzylpiperazine HCl per kg body weight. Under pentobarbital anaesthesia (30 mg/kg), the animal was perfused from the abdominal aorta with 2.5 % glutaraldehyde in 0.1 M sodium cacodylate buffer (pH 7.4). Postfixation with 1 % osmium tetroxide in sodium cacodylate buffer. Embedded in Epon 812 and sectioned at 50 nm. Lead citrate and uranyl acetate. Plate 2478. – Nucleus: most of the chromatin is condensed into large, confluent heterochromatin masses forming a perforated shell immediately under the nuclear envelope. Several microbodies, some with a prominent nucleoid

merular against peritubular neutrophil granulocytes in Sprague-Dawley *rats* was approximately 2 in controls, but 0.5 after 120-min reperfusion interval (Willinger et al. 1992). The outer stripe of the outer medulla contained only a small number of neutrophil granulocytes whereas neutrophil granulocyte counts of 923 ± 197 (n = 4) per cm^2 were found in the inner stripe after 120 min reperfusion. Proximal tubular cells are particularly sensitive for oxygen radical-induced injury (Andreoli et al. 1990).

Pre-treatment with dexamethasone significantly attenuated neutrophil infiltration and expression of intercellular adhesion molecule-1 induced by renal ischaemia/reperfusion (Takahira et al. 2001). Treatment with nitroxyl anion releaser known as Angeli's salt abolished the beneficial effect of dexamethasone. Renal dysfunction and tubular damage induced by renal ischaemia/reperfusion were not ameliorated by pre-treatment with dexamethasone.

Yoshioka et al. (1990) demonstrated enhancement of intrinsic antioxidant enzyme activity after renal ischaemia-reperfusion injury in the *rat* and protection of renal function against injuries induced by reactive oxygen species.

Preincubation of proximal and distal tubular cells isolated from *rat* kidney cortical cell by Percoll density-gradient centrifugation with 5 mM glutathione (GSH) or 5 mM dithiothreitol delayed *tert*-butyl hydroperoxide-induced cytotoxicity, indicating a protective role of glutathione (Lash and Tokarz 1990). Addition of buthionine sulphoximine and acivicin with GSH, to inhibit GSH synthesis and degradation, eliminated the protective effect of GSH, indicating that protection by GSH in distal tubular cells is not dependent on uptake of the intact tripeptide. Incubation of both proximal and distal tubular cells with *tert*-butyl hydroperoxide resulted in oxidation of GSH to glutathione disulphide (GSSG). Activities of five detoxification enzymes

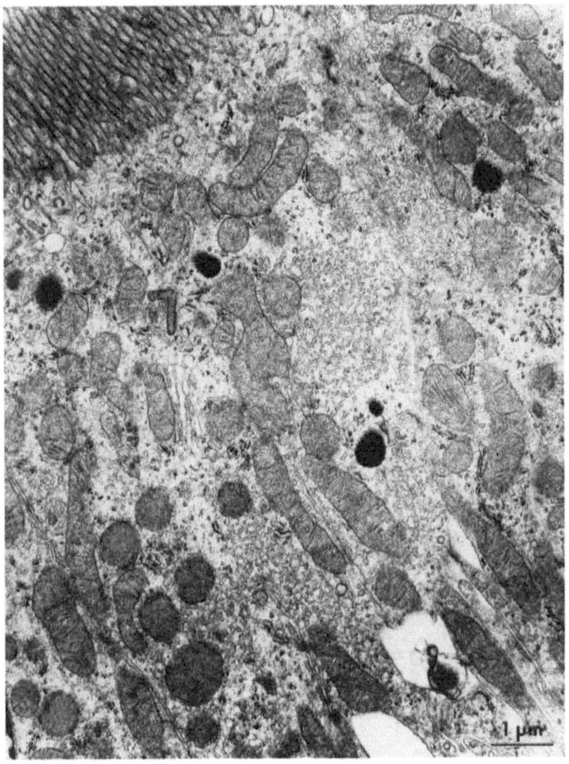

Fig. 314. Smooth endoplasmic reticulum in a proximal tubule cell. Kidney (block BNh 3384) from a male Wistar *rat* (No. 2568) 2 h after a single intraperitoneal injection of 5 mg *N*-benzhydryl-*N'*-p-hydroxybenzylpiperazine HCl per kg body weight. Under pentobarbital anaesthesia (30 mg/kg), the animal was perfused from the abdominal aorta with 2.5 % glutaraldehyde in 0.1 M sodium cacodylate buffer (pH 7.4). Postfixation with 1 % osmium tetroxide in sodium cacodylate buffer. Embedded in Epon 812 and sectioned at 50 nm. Lead citrate and uranyl acetate. Plate 2970

were significantly higher in proximal tubular cells, indicating that a diminished ability to detoxify reactive metabolites may contribute to the higher intrinsic susceptibility of distal tubular cells to oxidative injury.

Six hours following dimethylthiourea administration to male Sprague-Dawley *rats*, renal GSH content was significantly ($P < 0.05$) increased (10 %), and was increased further after 24 h (28 %; $P < 0.001$) (MILNER et al. 1993). Seven days of daily dimethylthiourea medication significantly ($P < 0.001$) increased renal and hepatic GSH content by 36 and 54 %, respectively, which was associated with a significant ($P < 0.001$) increase in the renal activities of glutathione peroxidase (EC 1.11.1.9) by 38 %, glutathione transferase (EC 2.5.1.18) by 92 %, and glutathione reductase (EC 1.6.4.2) by 19 % ($P < 0.05$).

Glutathione peroxidase (EC 1.11.1.9)

$$2\,GSH + H_2O_2 \rightarrow GSSG + 2\,H_2O \qquad [58]$$

A protein containing a selenocysteine residue. Steroid and lipid hydroperoxides, but not the product of reaction of EC 1.13.11.12 on phospholipids, can act as acceptor, but more slowly than H_2O_2 (cf. EC 1.11.1.12)

A novel theta class glutathione transferase isoenzyme from *mouse* termed mGSTT3 has been identified by analysis of the expressed sequence tag database (COGGAN et al. 2002). The gene encoding *mGSTT3* is clustered with the *mGSTT1* and *mGSTT2* genes on chromosome 10 and has an exon/intron structure that is similar to that of the other ϕ class genes. mGSTT3 is expressed strongly in the liver and to a decreasing extent in the kidney and testis.

Apoptosis occurring in the *rat* kidney in the first 24 h after 30 min ischaemia was influenced by the administration of Ca channel blockers verapamil, bepridil, nifedipin and Sensit® (TORONYI et al. 1999).

15.1.9.2 Diabetic Glomerulopathy

Conventional electron microscopy revealed conspicuous lysosomes in the podocytes of streptozotocin (60 mg/kg) diabetic *rats* (INA et al. 2002). Immunoelectron microscopy showed that endogenous *rat* immunoglobulin G and exogenous *goat* immunoglobulin G were present in the lysosomes of podocytes from diabetic *rats*.

Chronic administration of melatonin (0.02 % in drinking water) and taurine (1 % in drinking water) reduced lipid peroxidation in streptozotocin (50 mg/kg i.v.) diabetic *rats* by nearly 50 % (HA et al. 1999).

15.1.9.3 Steroid-Sensitive Nephrotic Syndrome

The 12-*O*-tetradecanoylphorbol-13-acetate-induced release of H_2O_2 from blood monocytes in 19 patients with steroid-sensitive nephrotic syndrome with proteinuria was significantly higher than in remission or normal controls (TANAKA et al. 1996).

15.1.9.4 Cadmium-Provoked Oxidative Stress

Cadmium is one of the most toxic industrial and environmental contaminants. Toxicity of the metal to kidney and liver may result from:

- Cd-induced synthesis of metallothionein and Cd accumulation (NORDBERG 1978),

- metabolic disorders due to cytochrome P_{450} destruction (Klimczak et al. 1984),
- prooxidant effect owing to induction of heme oxygenase and enhanced heme catabolism with release of Fe(II)/Fe(III) as a Fenton catalyst (Aust et al. 1985, Chevion 1988).
- alteration of sulphydryl homeostasis and antioxidative reserves thus contributing to propagation of lipid peroxidation and derangement of biological membranes (Manca et al. 1991, Sugiyama 1994, Wisniewska-Knypl and Wronska-Nofer 1995).

Iron overload potentiated the Cd-provoked lipid peroxidation and organ swelling leading to further compensative activation of glutathione peroxidase (EC 1.11.1.9) in the *rat* kidney in response to aggravated oxidative stress (Wrosnka-Nofer et al. 2000). Iron chelation with desferrioxamine abolished Cd-provoked lipid peroxidation and did not influence the relative organ weights as well as the antioxidative reserve status, the level of GSH and the compensating activation of renal glutathione peroxidase, thus attenuating oxidative stress.

15.1.9.5 Nickel-Induced Crystalline Inclusions in Mitochondria of the Ascending Limb

Nickel(II) exposure induced immortalisation of normal *human* kidney epithelial cells (Tveito et al. 1989). Growth inhibition by transforming growth factor β_1 was abrogated in both the immortalised and transformed *human* kidney epithelial cells (Mollerup et al. 1996).

In *rats* Ni_3S_2 (5 mg in 0.05 ml glycerol injected into each pole of the exteriorised right kidney) induced helicoids, linear, and flexuous crystalline inclusions predominantly in mitochondria of the basal infoldings of the epithelial cells of the ascending limb of the nephron (Jasmin 1978). The cytochrome oxidase staining reaction was apparent in longitudinally disposed cristae but literally absent in the crystalline inclusions. In a similar unilateral intrarenal injection experiment McCully et al. (1982) found a marked glomerulomegaly and hyperplasia of mesangial cells in both kidneys.

It is noteworthy that other divalent metallic salts exert a similar effect on mitochondria of the basal infolding of epithelial cells of the distal nephron (Jasmin and Riopelle 1976). Piracetam that induced paracrystalline structures in the thyreocytes of old *rats* (Fig. 340) induced such an inclusion in a collecting duct epithelial mitochondrion (Fig. 315).

Unilateral intrarenal injection of 2.5 or 5 mg Ni_3S_2 per *rat* induced marked glomerulomegaly and hyperplastic mesangial cells containing prominent ribosomes in both kidneys (McCully et al. 1982).

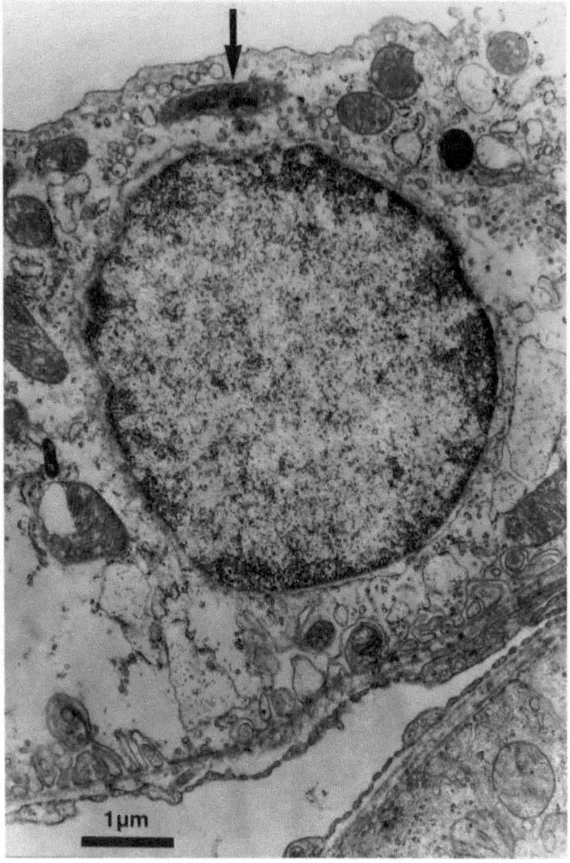

Fig. 315. A paracrystalline mitochondrial inclusion in a cortical collecting duct cell (block BNh 3313) from a Wistar rat (No. 2005) 60 min after a single intraperitoneal injection of 300 mg piracetam per kg body weight. Under pentobarbital anaesthesia (30 mg/kg), the animal was perfused from the abdominal aorta with 2.5 % glutaraldehyde in 0.1 M sodium cacodylate buffer (pH 7.4). Postfixation with 1 % osmium tetroxide in sodium cacodylate buffer. Embedded in Epon 812 and sectioned at 50 nm. Lead citrate and uranyl acetate. Plate 2282

15.1.9.6 Halogenated Hydrocarbon-Induced Nephrotoxicity and Carcinogenicity

15.1.9.6.1 Chloroform

Male Osborne-Mendel *rats* given a 10 per cent solution of $CHCl_3$ in corn oil (gavaged five days each week) had a significant increase in renal epithelial neoplasms after 78 weeks, as compared with controls (Weisburger 1977). No increase in renal neoplasms was noted in C57BL, CBA, or CF1 male *mice* (Roe et al. 1979).

15.1.9.6.2
Carbon Tetrachloride

Although CCl_4 exposure has produced a severe nephropathy in *humans* (Smetana 1939, Sirota 1949, Moon 1950, Guild et al. 1958, Stewart et al. 1963), there have been relative few studies on its nephrotoxicity in experimental animals (Kluwe 1981). In *humans*, nephrotoxicity is more pronounced if exposure occurs by inhalation rather than by ingestion (Guild et al. 1958, Moon 1950). As with $CHCl_3$ in *rats* and *mice*, CCl_4 toxicity in *human* kidneys selectively appears in epithelial cells of the proximal tubule, with few changes in other segments of the nephron (Smetana 1939, Moon 1950).

$$CCl_4 \xrightarrow[\text{Cytochrome } P_{450}]{e^-} {}^{\bullet}CCl_3 + Cl^- \qquad [116]$$

$${}^{\bullet}CCl_3 + O_2 \longrightarrow CCl_3OO^{\bullet} \qquad [117]$$

In male Sprague-Dawley *rats*, CCl_4 (0.25 ml per 100 g body weight in an equal volume of mineral oil given by gastric tube) produced an early, reversible renal lesion limited to the proximal tubule (Striker et al. 1968). The first change occurred in the mitochondria, followed by cell swelling and proliferation of the smooth endoplasmic reticulum.

Pre-treatment of *rats* with MFO inducers, such as polybrominated biphenyl, enhanced the nephrotoxicity of CCl_4 (Kluwe et al. 1982).

15.1.9.6.3
Trichloroethylene

Trichloroethylene is a major environmental contaminant and its industrial use as an incombustible solvent. Most trichloroethylene toxicicty depends on its biotransformation, which follows two pathways: cytochrome P_{450} (substrate, reduced-flavoprotein : oxygen oxidoreductase [RH-hydroxylating or epoxidizing]; EC 1.14.14.1)-dependent oxidation and GSH conjugation. The nephrotoxicity and nephrocarcinogenicity of trichloroethylene have been attributed to formation of reactive sulphur-containing metabolites generated by GSH conjugation and subsequent metabolism by γ-glutamyltransferase (EC 2.3.2.2), dipeptidase (aminopeptidase M; EC 3.4.11.2) and β-lyase (L-cysteine-S-conjugate thiol-lyase [deaminating]; EC 4.4.1.13) (Anders et al. 1988, Goeptar et al. 1995). The initial step of this pathway is catalysed by GSH-S-transferase (EC 2.5.1.8) isofoms, which catalyse the formation of *S*-(1,2-dichlorovinyl)glutathione. This is followed by hydrolysis reactions catalysed by γ-glutamyltransferase and dipeptidase, which cleave the glutamyl and glycyl residues to form *S*-(1,2-dichlorovinyl)-L-cysteine. Then *S*-(1,2-dichlorovinyl)-L-cysteine can undergo either *N*-acetylation to form the mercapturate *N*-acetyl-*S*-(1,2-dichlorovinyl)-L-cysteine or a β-elimination reaction catalysed by the β-lyase to form a reactive thiol. This thiol, in turn, can rearrange to form potent acylating species. Subsequent acylation of proteins and DNA may lead to cytotoxicity and mutagenesis (Anders et al. 1988, Goeptar et al. 1995). Trichloroethylene has been reported to cause a low incidence (3 %) of kidney adenocarcinomas in male Sprague-Dawley *rats* upon inhalation exposure to 600 ppm amine-stabilised, epoxide-free trichloroethylene for 7 h/day, 5 days/week, during 104 weeks (Maltoni et al. 1986). The nephrocarcinogenic effects of trichloroethylene were reported neither for the *mouse* nor for female *rats*, either upon inhalation exposure or upon oral dosing.

However, the kidney receives a high blood flow and possesses transport systems for *S*-conjugates, properties that also contribute to an accumulation of *S*-conjugates in this organ (Lash and Jones 1985).

Trichloroethylene was **acutely cytotoxic** to proximal and distal tubular cells isolated by collagenase perfusion from the kidneys of male Fischer 344 *rats* only at high concentrations, but, under conditions in which cytochrome P_{450} was not inhibited, *S*-(1,2-dichlorovinyl)glutathione formation was detected only in proximal tubular cells (Cummings et al. 2000). Inhibition of cytochrome P_{450} resulted in increases in both cytotoxicity and biotransformation of trichloroethylene.

S-(*trans*-1,2-Dichlorovinyl)-L-cysteine added to the drinking water of mole Swiss-Webster *mice* by 4 weeks dose-related caused cytomegaly, nuclear hyperchromatism and multiple nucleoli in the cells of the pars recta region of the kidney (Jaffe et al. 1984).

15.1.9.6.4
Hexachloro-1,3-butadiene

The formation of the corresponding monoglutathione *S*-conjugate from hexachloro-1,3-butadiene and subsequent cleavage of this conjugate by γ-glutamyltranspeptidase and β-lyase may be responsible for the nephrocarcinogenicity of the parent compound *in vivo*, whereas formation of the bis-glutathione *S*-conjugate probably plays no role in the organ specific effects of hexachloro-1,3-butadiene (Vamvakas et al. 1988).

15.1.9.6.5
Bromobenzene

A severe tubular necrosis was visible in male NMRI albino *mice* intoxicated intragastrically with bromobenzene mixed with two volumes of mineral oil (Casini et al. 1986). The histopathologic effects were particularly marked in the corticomedullary region, in which the tubular cells appeared in various stages of degeneration. Many proximal and distal convoluted tubules were markedly dilated and filled with acidophilic casts, sometimes containing cellular debris. A complete denudation of the tubular epithelium was observed in some areas. No appreciable changes were observed in the glomeruli.

15.1.9.6.6
1,1,2-Trichloro-3,3,3-trifluoro-1-propene

1,1,2-Trichloro-3,3,3-trifluoro-1-propene is structurally closely related to the stable and non-toxic tetrachloroethylene. However, in 1,1,2-trichloro-3,3,3-trifluoro-1-propene, the trifluoromethyl group enhances chemical reactivity with nucleophiles. Oral administration of 25 and 50 mg /kg dissolved in corn oil (1 ml) to female *rats* resulted in a large, dose-dependent increase in urinary excretion of γ-glutamyl transpeptidase indicative of proximal tubular damage (Vamvakas et al. 1989).

15.1.9.7
Doxorubicin-Induced Glomerular Toxicity

Although doxorubicin and menadione (≥25 μM) generate extensive reactive oxygen species and decrease protein synthesis, there was no correlation between the extent of oxidative stress and cytotoxicity in isolated *rat* glomeruli exposed to <100 μM doxorubicin (Morgan et al. 1998).

15.1.9.8
Di-2-ethylhexyl Phthalate-Induced Renal Microsomal P_{450} and b_5 Induction

Phthalate esters are widely used as industrial solvents and plasticizers in the manufacture of different plastics. Di-2-ethylhexyl phthalate, the most commonly used phthalate ester in polyvinylchloride formulations, can constitute up to 40 % of the end product. (Autian 1973). Phthalate esters have been recognised as environmental pollutants (Overturf et al. 1979). Di-2-ethylhexyl phthalate is a representative of peroxisome proliferators in rodents, and also induces enzymes specific for other organelles, DNA synthesis and organ growth (Huber et al. 1996).

In subtotally nephrectomized male Wistar *rats* di-2-ethylhexyl phthalate (2 ml per kg body weight × day) induced an increase in microsomal total cytochrome P_{450} and cytochrome b_5 in the remnant kidneys, suggesting an induction of enzymes (Kertai et al. 2000). The elevation of P_{450} levels preceded those of b_5 and reached a higher proportion at all moments (7, 14, and 21 days of treatment). While the renal cortex became hypertrophied, the medulla was little affected. Glomeruli were hypertrophied and proximal tubular epithelial cells showed degenerative changes and cytoplasmic vacuolisation.

15.1.9.9
Paraquat-Induced Lesions of Tubular Brush Border in *Man*

In a patient aged 60 years, who had accidentally ingested a mouthful of a mixture of paraquat and diquat, a renal biopsy taken on the eighth day of intoxication showed lesions of the brush border at the proximal tubular cells (Bescol-Liversac et al. 1975).

15.1.9.10
Styrene-Induced Tubular Brush Border Villin Urinary Excretion

Bolm-Audorff et al. (2001) examined the prevalence of increased urinary excretion of brush border villin in 182 workers exposed to styrol in 13 Hessian plastics processing factories.

15.1.9.11
Nitric Oxide and Tubular Brush Border

In renal proximal tubules, three isoenzymes of nitric oxide synthase (EC 1.14.23) have been identified. Inducible NOS can be expressed at high levels under specific combinations of lipopolysaccharide and/or cytokines. Once induced, iNOS remains active for sustained periods and produces large amounts of •NO, which is cytotoxic to pathogens and to several types of cells. In primary cultures of male Sprague-Dawley *rat* proximal tubules, the level of iNOS mRNA induced by lipopolysaccharide (10 μg/ml) was enhanced by 500 μM ferric nitrilotriacetate ($P < 0.01$); lower or higher concentrations had no effect (Wu and Qiu 2001). However, the supernatant NO_2^- level did not change significantly ($P > 0.05$), although tubular injury was aggravates ($P < 0.001$). The addition of arginine increased lactate dehydrogenase release from 25.05 ± 8.36 % in the iron group to 38.67 ± 7.67 % in the iron plus LPS group ($P < 0.05$); concomitantly, N^G-nitro-L-argi-

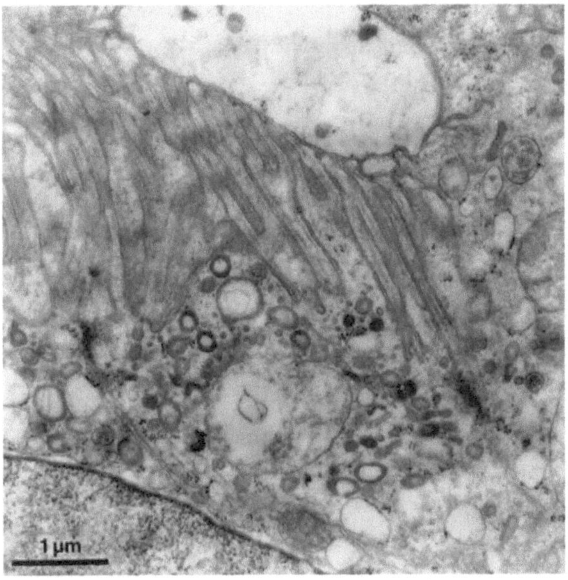

Fig. 316. Brush border of a renal tubulus (block 932) of a male beagle dog (No. 126) medicated for 12 months with 20 mg molsidomine per kg body weight × day contained in the food (Altromin® H). On June 11, 1976, under intravenous pentobarbital anaesthesia, the animal was exsanguinated. small pieces of tissue were fixed by immersion in a 2.5 % solution of glutaraldehyde in 0.1 M sodium cacodylate buffer (pH 7.4). Postfixation with 1 % osmium tetroxide in sodium cacodylate buffer. Embedded in Epon 812 and sectioned at 50 nm. Lead citrate and uranyl acetate. Film 635-50

nine mitigated iron toxicity in lipopolysaccharide-treated proximal tubules ($P < 0.05$). Hydroxyl radical scavengers provided complete protection against iron-mediated cytotoxicity ($P < 0.001$), but the decrease of NO_2^- production was only significant in the lipopolysaccharide-treated groups. In contrast, superoxide dismutase (EC 1.15.1.1) was partially effective in the LPS group ($P < 0.05$) whereas the NO_2^- level in the supernatant was inversely raised ($P < 0.05$). Reduced glutathione had no effect on either iron toxicity or NO_2^- production.

15.1.9.12 Nitric Oxide and the Juxtaglomerular Apparatus

Structural differences of macula densa cells among species concern essentially cell height and amount of mitochondria. In *rabbits*, the macula densa cells are densely stuffed with small mitochondria and are a prominent plaque of distinctly taller cells than the thick ascending limb cells (KAISSLING and KRIZ 1979). In *rat* (KAISSLING and KRIZ 1982), mice, and *Psammomyys*, species with a greater capacity for urinary concentration, the macula densa cells are less or not all prominent and possess only a very few mitochondria.

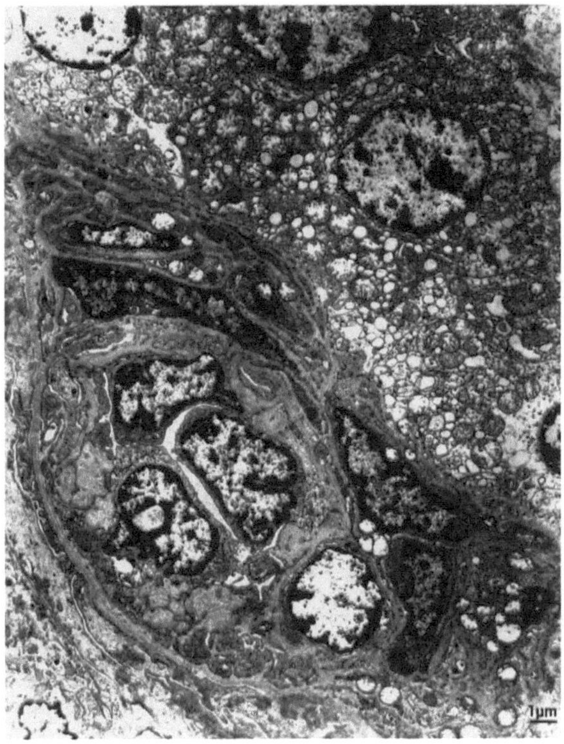

Fig. 317. Juxtaglomerular apparatus with Goormaghtigh cells (partly split up into their processes) and macula densa from the kidney (block BNh 3372) of a male Wistar *rat* (No. 2567) 90 min after a single intraperitoneal injection of 5 mg *N*-benzhydryl-*N'*-p-hydroxybenzylpiperazine HCl per kg body weight. Under pentobarbital anaesthesia (30 mg/kg), the animal was perfused from the abdominal aorta with 2.5 % glutaraldehyde in 0.1 M sodium cacodylate buffer (pH 7.4). Postfixation with 1 % osmium tetroxide in sodium cacodylate buffer. Embedded in Epon 812 and sectioned at 50 nm. Lead citrate and uranyl acetate. Plate 2368

The macula densa takes a special position in its enzyme pattern (BARGMANN 1978, p. 194).

NOS I is predominantly localized in macula densa cells, while NOS III is found endothelial cells of arteries, arterioles and glomerular capillaries (BACHMANN et al. 1995). In order to characterize the promoters active in the macula densa and to study the regulation of NOS I expression in these cells, OBERBÄUMER et al. (1997) cloned the 5' ends of NOS I from *rat* kidney by RT-PCR. They found that the two alternative first exons of brain NOS I were not contained in the main NOS I mRNA from kidney.

Several *in vitro* studies in isolated juxtaglomerular cells showed that •NO either inhibited (KURTZ et al. 1986, HENRICH et al. 1988, GREENBERG et al. 1995) or stimulated (GARDES et al. 1992, SCHOLZ and KURTZ 1993) renin secretion. Studies on the isolated perfused juxtaglomerular apparatus indicated that macula densa-generated •NO is impor-

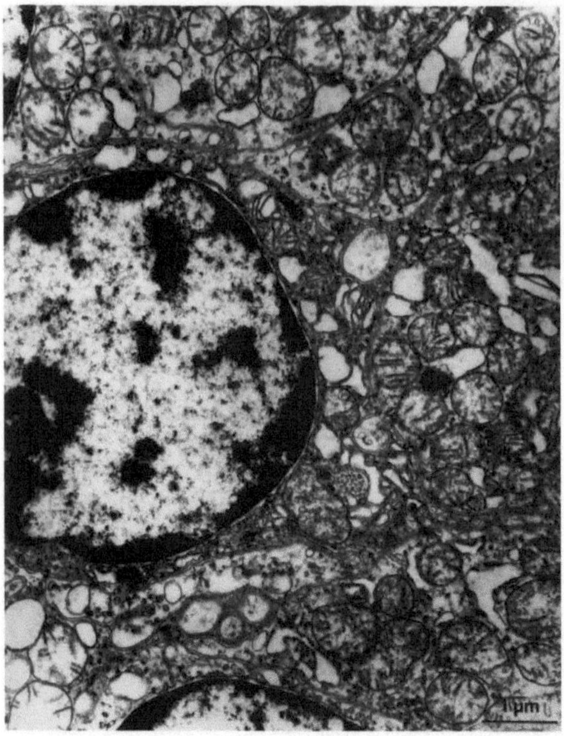

Fig. 318. Macula densa cells from the kidney (block BNh 3372) of a male Wistar *rat* (No. 2567) 90 min after a single intraperitoneal injection of 5 mg *N*-benzhydryl-*N*'-p-hydroxybenzylpiperazine HCl per kg body weight. Under pentobarbital anaesthesia (30 mg/kg), the animal was perfused from the abdominal aorta with 2.5 % glutaraldehyde in 0.1 M sodium cacodylate buffer (pH 7.4). Postfixation with 1 % osmium tetroxide in sodium cacodylate buffer. Embedded in Epon 812 and sectioned at 50 nm. Lead citrate and uranyl acetate. Plate 2453

tant for the stimulation of renin secretion whereas endothelial-derived •NO inhibits renin secretion (He et al. 1995).

15.1.9.13
Nitric Oxide and Reactive Oxygen Species in Wegener's Granulomatosis

Wegener's granulomatosis is a systemic inflammatory disease. In most patients a rapidly progressive form occurs, characterised by fibrinoid necrosis of the glomerular capillaries, the formation of glomerular crescents, and marked glomerular and interstitial infiltration of neutrophils and mononuclear leucocytes, sometimes accompanied by granuloma formation (Stilmant et al. 1979, Couser 1988). In a *rat* model of anti-myeloperoxidase-associated crescentic glomerulonephritis Heeringa et al. (1998) investigated the temporal expression of NO synthases in conjunction with platelet aggregation, inflammatory cell influx, and the generation of reactive oxygen species and nitrotyrosine formation. They observed a marked transient induction of iNOS in polymorphonuclear leucocytes and macrophages, coinciding with the generation of reactive oxygen species and the formation of nitrotyrosine. A significant reduction in eNOS expression was found in glomerular and interstitial tubular capillaries and cortical vessels in Wegener's granulomatosis (Heeringa et al. 2001). Infiltrating inflammatory cells, mainly located in the interstitium, expressed iNOS. H_2O_2-producing cells were detected in glomeruli and were abundantly present in the interstitium. Nitrosotyrosine-positive cells, however, were almost exclusively found in the interstitium.

Small concentrations of •NO are cytoprotective for endothelial cells (Gröne 2001). In a *rat* model of renal angiitis, inhibition of eNOS was detrimental and enhancement of activity of eNOS was beneficial.

Table 49. Enzymology of macula densa epithelial cells

Enzyme	Authors
α-Glycerophosphate dehydrogenase (EC 1.1.1.8)	Krompecher-Kiss and Bucher (1977)
Lactate dehydrogenase (EC 1.1.1.27)	Krompecher-Kiss and Bucher (1977)
Malate dehydrogenase (EC 1.1.1.38)	Krompecher-Kiss and Bucher (1977)
Isocitrate dehydrogenase (EC 1.1.1.42)	Krompecher-Kiss and Bucher (1977)
Glucose-6-phosphate dehydrogenase (EC 1.1.1.49)	Krompecher-Kiss and Bucher (1977)
Succinate dehydrogenase (EC 1.3.99.1)	Krompecher-Kiss and Bucher (1977)
Cytochrome oxidase (EC 1.9.3.1)	Krompecher-Kiss and Bucher (1977)
Nitric oxide synthase (EC 1.14.23)	Bachmann et al. (1993, 1995), Bosse and Bachmann (1994), Oberbäumer et al. (1997), Weichert et al. (1998)
NAD-tetrazolium reductase	Krompecher-Kiss and Bucher (1977)
NADP-tetrazolium reductase	Krompecher-Kiss and Bucher (1977)

15.1.9.14
Experimental Hyperlipaemia-Hyperglycaemia in the *Hamster*

Administration of L-arginine had no effect on the thickened, nodular basal lamina of the renal capillaries in experimentally hyperlipaemic-hyperglycaemic male Syrian golden *hamsters* (Popov et al. 2002).

15.1.9.15
Oxalate-Mediated Oxidative Stress

Calcium oxalate monohydrate (25–250 μg/ml) treatment of proximal (LLC-PK1) and distal (MDCK) tubular epithelial cell cultures increased superoxide production 3–6-fold as measured by both lucigenin chemiluminescence in permeabilized cells and dihydrorhodamine fluorescence in intact cells (Khand et al. 2002). The use of mitochondrial probes, substrates, and inhibitors indicated that increased $O_2^{\bullet-}$ production originated from mitochondria. Treatment with calcium oxalate monohydrate decreased glutathione (total and redox state), indicating a sustained oxidative insult. An increase in NADH in calcium oxalate monohydrate-treated cells suggested this cofactor could be responsible for elevating $O_2^{\bullet-}$ generation.

15.1.10
Intestine

Intestinal cytochrome P-450, in common with its hepatic isozymes, is induced by a wide variety of xenobiotics. Not only intraluminal iron but also selenium is critically required for basal maintenance of intestinal cytochrome P-450 content and its dependent mixed-function oxidase. Accordingly, acute deprivation of either element for as short a period as a single day reduced basal intestinal cytochrome P-450 content by limiting the formation and/or availability of its prosthetic heme (Pascoe et al. 1983). Whereas both intraluminal iron and selenium were required for maintenance of the prosthetic apocytochrome moiety of the constitutive intestinal isozyme, only intraluminal selenium was required for the viability of apocytochrome of the β-naphthoflavone-inducible P-448 (Pascoe and Correia 1985).

In the *human* jejunal mucosa semisynthetic (elemental) diets did not change 1-naphthol glucuronyltransferase and reductase activities, but significantly depressed 7-ethoxycoumarin O-deethylase activity (Hoensch et al. 1984). Male subjects had significantly higher ethoxycoumarin O-deethylase activities than female subjects on semisynthetic diet (6.6±2.3 vs. 3.2±0.9) as well as on home diet (16.3±9.0 vs. 6.4±3.0). Semisynthetic diet also reduced the jejunal villous height significantly when compared with home diet (408±49 vs. 373±44 μm).

A b_{558} cytochrome was isolated from brush border membranes of *rabbit* enterocytes (Knöpfel and Solioz 2002). The purified haemoprotein exhibited ascorbate-stimulated reduction of iron(III) and copper(II). The rate constants, k_1, for these reactions were 1.38±0.12 and 0.64±0.16 min^{-1}, respectively.

In intestinal mucosa of fed adult *rats*, cytochrome P-450 content and activity of benzpyren hydroxylase and *p*-nitroanisole O-demethylase were highest in the upper small intestine and progressively decreased toward the terminal ileum (Hoensch et al. 1975). Among the mucosal cell populations, mature villous tip cells contained 6 to 10 times more cytochrome P-450 and drug-metabolising activity per mg microsomal protein than epithelial crypt cells. On restriction of dietary iron intake for 48 h, cytochrome P-450 content and drug metabolising activity of villous tip cells decreased to 42 % and 13 % of control values, but were restored within 24 h by oral iron supplementation.

Nafenopin, 2-methyl-2-[*p*-(1,2,3,4-tetrahydro-1-naphthyl)phenoxy]propionic acid (1.25 g per kg body weight × day) added to the food for 17 days induced peroxisomal proliferation in the epithelial cells of the small intestine of male Albany strain *mice* (Pšeničnik and Pipan 1977).

Oxygen free radical induced damage during intestinal ischaemia/reperfusion in normal and xanthine oxidase deficient *rats* (Nalini et al. 1993). Following reperfusion of small bowel grafts stored in saline or modified University of Wisconsin solution (Kurzawinski et al. 1994) at 4 °C for 12 h and reperfused for 6 h in syngeneic *rats*, there was crypt and villous epithelial apoptosis, loss of crypt and villous structures, and an increase in mucosal inflammatory cell infiltration (Shah et al. 1997). Ongoing apoptosis was maximum at 1 h, its degree decreasing with increasing reperfusion intervals. Large numbers of apoptotic bodies dominated the picture from 3 h of reperfusion.

Both ischaemia (clamping of the superior mesenteric artery for 30, 60, 90 min) and ischaemia/reperfusion affected respiratory function of isolated *rat* enterocyte mitochondria as compared to control (Madesh et al. 2000). Preconditioning with nitric oxide donor, sodium nitroprusside (1 mM, given into the proximal jejunal lumen at a rate of 1 ml/min), significantly enhanced the recovery of the respiratory control rate. Mitochondrial lipid changes suggestive of activation of phospholipase

A_2 (EC 3.1.1.4) and phospholipase D (EC 3.1.4.4) seen after ischaemia/reperfusion were prevented by the simultaneous presence of a nitric oxide donor in the intestinal lumen.

MORIWAKI et al. (1996) observed xanthine oxidase staining in infiltrating lymphocytes (probably T-lymphocytes but not B-lymphocytes) in inflammatory lesions of the *human* small and large intestine. Its ubiquitous localization suggests that xanthine oxidase is involved in the pathogenesis of reperfusion tissue injury.

Brush-border membrane alkaline phosphatase (EC 3.1.3.1) of the *rat* small intestine was inactivated in an *in vitro* Fe^{2+}/ascorbate oxygen-radical generating system (DUDEJA and BRASITUS 1993). The activities of sucrase, maltase, leucine aminopeptidase, and γ-glutamyl transpeptidase were unchanged.

In anaesthetised male Sprague-Dawley *rats* subjected to 45 min of intestinal ischaemia followed by 3 h of reperfusion BÖRJESSON et al. (2000) found a 76 % increase in alveolar fluid clearance compared with the control values ($P < 0.05$). The stimulated alveolar liquid clearance seen after intestinal ischaemia-reperfusion was not inhibited by propranolol, indicating stimulation through a noncatecholamine-dependent pathway. Administration of a neutralising polyclonal anti-tumour necrosis-α antibody before induction of intestinal ischaemia completely inhibited the increased alveolar liquid clearance observed after intestinal ischaemia-reperfusion.

In monolayers of a *human* colonic cell line (Caco-2) a 30 min exposure to H_2O_2 or HOCl induced a significant increase in the oxidation of tubulin, decrease in the stable S2 polymerised tubulin, and increase in the unstable S1 monomeric tubulin (BANAN et al. 2000). In concert, each reactive oxygen species in a dose-dependent manner damaged the microtubule cytoskeleton and disrupted barrier function. Growth factors (10 nM) were protective. Antibody against the growth factor-receptor (1 μg/ml) and inhibitors of growth factor-receptor tyrosine kinase (25 μM tyrophostin or 150 nM AG 1478) abolished growth factor protection, indicating the involvement of epidermal growth factor-receptor signalling pathway.

Nitric oxide has been designated as (one of) the principal mediator(s) of nonadrenergic, noncholinergic transmission of inhibitors neurones in the enteric nervous system of *rat*, *guinea pig*, and *dog* (BULT et al. 1990, LI and RAND 1990, SHUTTLEWORTH et al. 1991). Antibodies against nitric oxide synthase have revealed immunoreactivity in the *rat* and *guinea pig* enteric nervous system (BREDT et al. 1990, COSTA et al. 1992, FURNESS et al. 1992). In the *pig*, NOS-immunoreactive neurones were mainly encountered in the outer submucous and myenteric plexus (TIMMERMANS et al. 1993). In the inner submucous plexus they were very rare. Both the circular muscle and tertiary plexus and, to a lesser extent, the longitudinal muscle were densely innervated by NOS-immunoreactive nerve fibres.

Both nonselective (N^G-monomethyl-L-arginine) and specific (*N*-iminomethyl-L-lysine) inhibitors of iNOS significantly increased leucocyte binding by normal *human* intestinal microvascular endothelial cells activated with cytokines and lipopolysaccharide, but had no effect on leucocyte adhesion by similarly activated inflammatory bowel disease *human* intestinal microvascular endothelial cells (BINION et al. 2000).

The NO donor 4-ethyl-2[(*Z*)-hydroxyiminol]-5-nitro-3(*E*)-hexeneamide (10 mg FK409/kg *rat*) given intravenously both 30 min prior to ischaemia (induced by occlusion of the superior mesenteric artery with a non-traumatic vessel clamp for 30 min) and 30 min post-reperfusion significantly ($P < 0.001$) decreased villous leucocyte adhesion (KALIA et al. 2002). While collapsed alveoli, thickened interstitial walls, and a dense neutrophilic infiltrate were apparent in the lungs of unmedicated ischaemia/reperfused animals, lung histology was normal in FK409-treated *rats*.

The role of reactive metabolites of oxygen and nitrogen in inflammatory bowel disease (Crohn's disease, ulcerative colitis) was recently reviewed (PAVLICK et al. 2002).

15.1.11 Liver

The electron paramagnetic resonance probe technique described by VALGLIMIGLI et al. (2001) provides a versatile way of gaining rapid and reproducible quantitative measurements of reactive oxygen species in biological systems, ranging from subcellular fractions to whole *animals* and *human* liver.

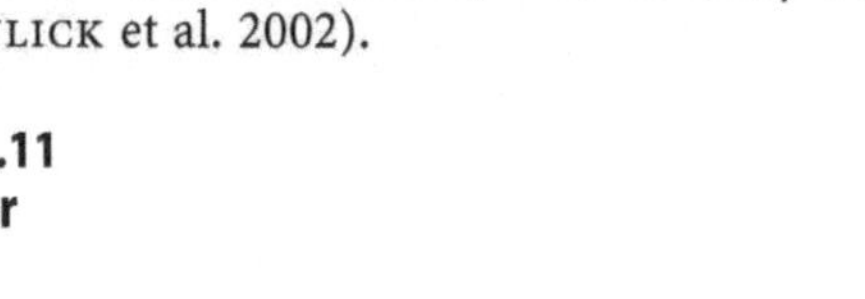

5,5-**Di**methyl-1-**p**yrroline-1-**o**xide-HO• adduct [8]

Spin traps incorporated into cultured foetal *mouse* liver cells after irradiation for 60 s under an UV hand lamp gave reproducible electron spin resonance signals as an evidence of spin trap incorporation (MORGAN et al. 1985). The spin trap 5,5-dimethyl-1-pyrroline-1-oxide was found to be incorporated, while α-phenyl-β-*tert*-butylnitrone and α-(4-pyridyl-1-oxide)-*N*-*tert*-butylnitrone were not. 2-Methyl-2-nitrosopropane was toxic to the cells.

15.1.11.1 Hepatocytes

15.1.11.1.1 Cytochromes

In the mature *mammalian* liver, most genes appear to be expressed in an ascending or descending gradient from the portal to the central vein within the acinus, the microcirculatory unit of the liver (GEBHARDT 1992). This zonation is particularly prominent for the **cytochrome P_{450}** genes (OINONEN and LINDROS 1998). OINONEN et al. (1993) and OINONEN and LINDROS (1995) observed that in hypophysectomized *rats* there is a high expression of CYP2B and 3A in the normally silent periportal region, overturning the constitutive zonation of these forms. Although hypophysectomy suppressed CYP2C7 mRNA and *human* recombinant growth hormone (0.01 IU/h subcutaneously) counteracted it, regulation at this level did not appear to occur in a zone-specific fashion (OINONEN et al. 2000). This indicated that growth hormone-mediated zonal regulation of CYP2C7 protein has additional translational or posttranslational components. Ethanol treatment, which has been shown to affect growth hormone levels, significantly induced CYP2C7 mRNA, but not zone-specifically. In the male Wistar *rat*, after an acute dose (5 mg ethanol per kg body weight) free radical production increased up to 1 h and then plateaued of the next 30 min (NAVASUMRIT et al. 2000). During chronic exposure to 5% ethanol in a liquid diet to provide 36% of the caloric requirement, free radical generation increased significantly after 1 week and then declined again to remain at a low level over the next 2 weeks. This transient increase corresponded closely with the induction of CYP2E1 in response to ethanol feeding. CYP2E1 is expressed predominantly in the liver and, at lower levels, other organs such as kidney, nasal mucosa, lung, intestine, colon, and lymphocytes (HUKKANEN et al. 1997). CYP2E1 was classified into 7 genotypes in the Japanese population and 4 in the Caucasian population (INOUE et al. 2000).

In Sprague-Dawley *rats*, the mRNA transcript for hepatic CYP1A1, 1B1, and 2B1/2 and mammary Cyp1A1 were up-regulated after treatment with indole-3-carbinol at 250 mg/kg (HORN et al. 2002). However, the level of expression of CYP1B1 in the liver was lower than that of other CYPs. Hepatic P_{450} probe activities indicative of induction of CYP1A1, 1A2, and 2B1/2 were increased by indole-3-carbinol in a dose-dependent manner. Treatment with indole-3-carbinol at 250 mg/kg increased the capacity of liver microsomes to metabolise 17β-œstradiol to 2-OH-17β-œstradiol, 2-OH-œstrone, 6α-OH-17β-œstradiol, œstriol, and 15α-OH-17β-œstradiol and œstrone to 2-OH-œstrone, 2-OH-17β- œstradiol, 6(α+β)-OH-œstrone, and 6α-OH-17β-œstradiol. The magnitudes of increases of CYP-dependent activities and rates of œstrogen metabolite formation achieved with indole-3-carbinol at 250 mg/kg were smaller after ten than four treatments. The increased rates of formation of 6α-OH-17β-œstradiol, 6β-OH-17β-œstradiol, and 15α-OH-17β-œstradiol from 17β-œstradiol were also detected after treatment with indole-3-carbinol at 25 mg/kg, and, except for increased 6β-OH-17β-œstradiol from 17β-œstradiol, no other changes in 17β-œstradiol or œstrone metabolism occurred after treatment with indole-3-carbinol at 5 mg/kg.

CYP2E1 is important in **alcoholic** liver pathogenesis because during oxidation of ethanol by CYP2E1, superoxide is generated which enhances lipid peroxidation and other toxic by-products (FRENCH et al. 1993). Experiments using intragastric alcohol-fed *rats* have shown that accumulation of malondialdehyde within the liver is associated with the development of antibodies recognizing protein-malondialdehyde adducts (ALBANO et al. 1996).

Chlormethiazole specifically inhibited the elevation of CYP2E1 mRNA and protein, but did not prevent CYP2B2 and CYP3A1 or CYP1A1 induction caused by treatment of male Sprague-Dawley *rats* with phenobarbital or β-naphthoflavone, respectively (HU et al. 2002).

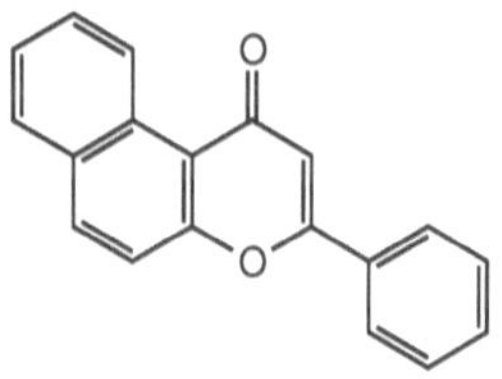

β-Naphthoflavone [42]

A cytochrome P_{450} isozyme belonging to the CYP2D subfamily is involved in the ring hydroxylation of flunarizine, 1-[bis(4-fluorophenyl)-methyl] piperazine (KARIYA et al. 1992).

Using electron paramagnetic resonance (EPR) spectroscopy to detect nitrosyl protein complexes indicating nitric oxide production, CHAMULITRAT and SPITZER (1996) showed that the concentrations of nitrosyl complexes in whole blood and in liver tissues of alcohol-fed Sprague-Dawley *rats* treated with lipopolysaccharide increased 3-fold, compared with those from *rats* on control diet treated with lipopolysaccharide. Inhibition of •NO production by

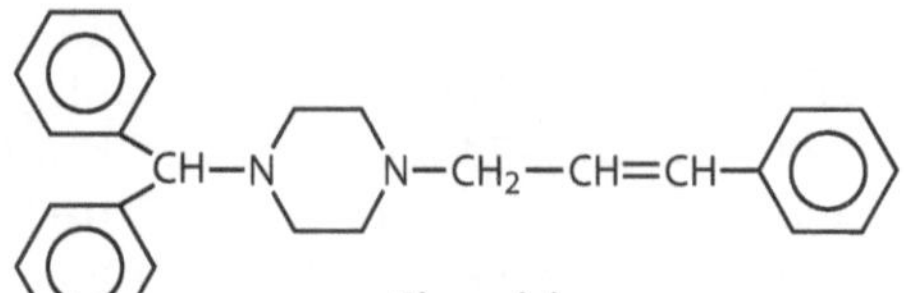

Cinnarizine

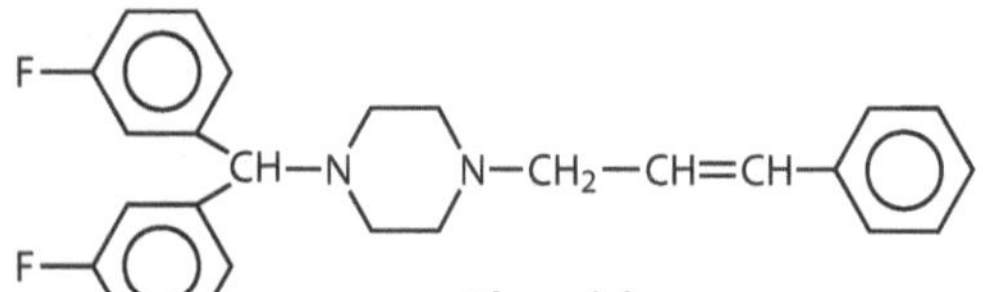

Flunarizine

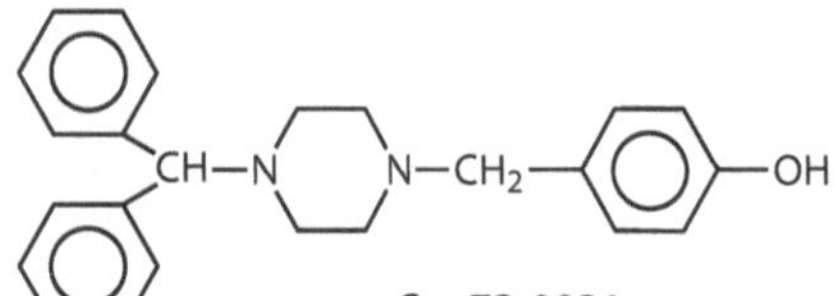

Cas 72-0031

4-Diphenylmethylpiperidine derivatives [75]

aminoguanidine (100 mg/kg, s.c.) treatment attenuated plasma hepatic aspartate aminotransferase (EC 2.6.1.1), and alanine aminotransferase (EC 2.6.1.2) levels in the alcohol + lipopolysaccharide-treated group.

CYP2E1 degradation involves the ubiquitin-proteasome pathway (WANG et al. 1999, GOASDUFF and CEDERBAUM 2000). The CYP2E1 protein is degraded by the 26S proteasome after ubiquitination (KORSMEYER and DAVOLL 1999, WANG et al. 1999). After giving 13 mg ethanol per kg body weight × day intragastrically to male Wistar *rats* for 30 days there was a 90 per cent inhibition of the chymotrypsin-like activity by PS-341, a potent proteasome inhibitor (BARDAG-GORCE et al. 2002). After ethanol withdrawal, the proteasomal chymotrypsin-like activity returned to control levels. In ethanol-withdrawals *rats* injected with PS-341, the chymotrypsin-like activity was significantly inhibited before withdrawal ($P < 0.001$). Ethanol treatment induced a 3-fold increase in CYP2E1 levels determined by Western blot. When ethanol was withdrawn, CYP2E1 decreased to control levels. In ethanol-withdrawn rats injected with PS-341, CYP2E1 remained at the induced level.

Dehydroepiandrosterone was extensively metabolised by *human* and *rat* liver microsomal fractions which was potently inhibited in both species by miconazole, demonstrating a principle role for cytochrome P_{450} (FITZPATRICK et al. 2001).

Arachidonic acid caused toxicity and induced lipid peroxidation and mitochondrial membrane damage in cells overexpressing CYP2E1 but had little or no effect in control cells not expressing CYP2E1 (PÉREZ and CEDERBAUM 2001). The toxicity appeared to be both apoptotic and necrotic in nature. 4-Hydroxy-[2,2,6,6-tetramethylpiperidine-1-oxyl] (Tempol) and α-(4-pyridyl-1-oxide)-N-*tert*-butyl nitrone (POBN) protected against the decrease in cell viability and the apoptosis and necrosis. These spin traps prevented the enhanced lipid peroxidation and the loss of mitochondrial membrane potential. Tempol and POBN had little or no effect on cellular viability or on CYP2E1 activity at concentrations which were protective. *tert*-Butyl hydroperoxide mobilizes arachidonic acid from membrane phospholipids in *rat* hepatocytes under cytotoxic conditions, thus leading to an increase in intracellular arachidonic acid, which precedes cell death. Cells trated with *tert*-butyl hydroperoxide maintained viability and energy status at 10 min (MARTÍN et al. 2001). *tert*-Butyl hydroperoxide depleted GSH, as well as inducing lipid peroxidation and formation of reactive oxygen species, detected by dichlorofluorescein fluorescence. *tert*-Butyl hydroperoxide also significantly increased (32.5 %) the intracellular [^{14}C]- arachidonic acid from [^{14}C]arachidonic acid-labelled hepatocytes. The phospholipase A_2 inhibitor, mepacrine, completely inhibited the [^{14}C]arachidonic acid response. The addition of antioxidants to the cell suspension affected the *tert*-butyl hydroperoxide-induced lipid response differently. The [^{14}C]arachidonic acid accumulation correlated directly with reactive oxygen species and negatively with endogenous GSH. Promethazine prevented lipid peroxidation and did not affect the [^{14}C]arachidonic acid increase.

While isolated *rat* hepatocytes do not liberate appreciable amounts of O_2^-, simple quinones, such as 2,5-dimethyl-*p*-benzoquinone stimulate the formation of O_2^- up to 15 nmoles per min and 10^6 cells (POWIS et al. 1981). Hepatocyte O_2^- formation stimulated by a variety of simple quinones and more complex antitumor quinones is maximal at a quinone one-electron reduction potential (E_7^1) of –70 mV and qualitatively similar to the pattern of O_2^- formation seen with mitochondrial NADH:ubiquinone oxidoreductase and microsomal NADH-cytochrome b_5 reductase. O_2^- production, by microsomal NADPH-cytochrome P-450 reductase is maximal at a quinone E_7^1 of –200 mV. Phenobarbital induction, which increases NADPH-cytochrome P-450 reductase, had no effect on O_2^- formation by hepatocytes.

Acetaminophen-induced hepatotoxicity is mediated by cytochrome P_{450} metabolism to *N*-acetyl-*p*-benzoquinone imine. Following low doses of acetaminophen the metabolite is efficiently detoxified by glutathione (GSH); however, following large doses hepatic GSH is depleted and to *N*-acetyl-*p*-benzoquinone imine covalently binds to proteins as acetaminophen-cysteine adducts (Cohen et al. 1997). Immunohistochemical studies have revealed that the histological site of covalent binding, as well as the relative amount of binding, correlated with the development of toxicity. The large majority of cells which develop acetaminophen-protein adducts subsequently become necrotic (Roberts et al. 1991, Hart et al. 1995). In *mice* lacking iNOS activity exposed to acetaminophen (300 mg/kg) the serum alanine aminotransferase levels were approximately 50 % of the wildtype *mice* (Michael et al. 2001). However, histological examination of liver sections indicated similar levels of centrilobular hepatic necrosis in both wild-type and NOS2 null *mice*. In wild-type *mice* hepatic tyrosine nitration was greatly increased relative to saline treated controls. Tyrosine nitration increased in NOS2 null *mice* also, but the increase was much less. Acetaminophen increased hepatic malonaldehyde levels (lipid peroxidation) in NOS2 null *mice* only.

The acetaminophen free radical (*N*-acetyl-4-aminophenoxyl) is very reactive and forms melanin-like polymeric products (West et al. 1984). A model system, leading to more stable metabolites, can be obtained by introduction of methyl groups next to the oxygen, 3',5'-dimethyl-acetaminophen (3',5'-dimethyl-4'-hydroxyacetanilide). Fischer et al. (2002) analysed the ESR spectrum of the free radical formed.

In transfected *human* hepatoma G2 cells, carbaryl and thiobendazole were found to activate CYP1A1 at the level of transcription, as demonstrated by the dose-dependent increase in reporter chloramphenicol acetyl transferase and CYP1A1 mRNAs (Delescluse et al. 2001).

Soils from several industrial sites in France, Germany, and Portugal contaminated with **polycyclic aromatic hydrocarbons** (60–4700 mg/kg soil) led to induction of CYP1A1 in *rat* duodenal mucosa cells, regardless of their extent of contamination, showing that relevant doses were mobilized in the gastrointestinal tract and absorbed (Roos 2002). Subsequent distribution of non-metabolised compounds was indicated by induction of CYP1A1 in the liver.

72 h after a single dose of the LD_{50} of 2,3,7,8-tetrachlorodibenzo-*p*-**dioxin** (25 μg/kg), 2,4-dichlorophenoxyacetic acid (375 mg/kg and dieldrin (38 mg/kg) by gavage in female Sprague-Dawley *rats*, increased expression of CYP1A1, CYP1A2 and CYP1B1 was observed in the liver, kidney and mammary tissue (Badawi et al. 2000). Since CYP1A1, CYP1A2 and CYP1B1 are the major enzymes catalysing 2- and 4-hydroxylation of 17β-oestradiol, respectively, the effect of these chlorinated hydrocarbons on the metabolism of 17β-oestradiol to 2- and 4-catechol oestrogens was increased in a tissue-specific manner in response to treatment.

The protein levels of normally expressed CYPs 1A2, 2B1/2, and 2E1 increased significantly in liver microsomes from *rats* treated with the **pyrrolizidine alkaloid** retrorsine (12,18-dihydroxysenecionan-11,16-dione; β-longilobine; 30 mg/kg i.p.) compared to untreated control *rats* ($P < 0.05$), but protein levels of CYP 4A3, CYP 3A1, and CYP reductase were unchanged after retrorsine treatment (Gordon et al. 2000). In addition, CYP 1A1 mRNA and protein, which are not detectable in the livers of control *rats*, were induced after retrorsine exposure.

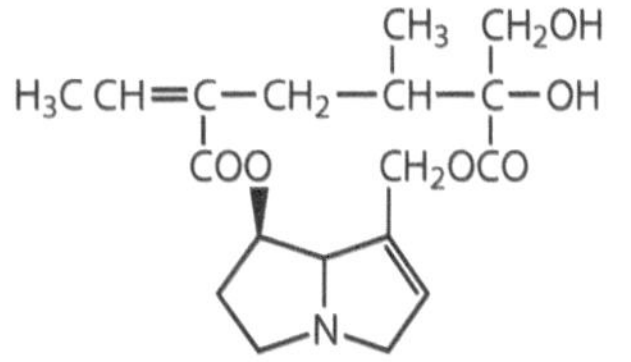

Retrorsine [293]

Glyceryl trinitrate was denitrated in *rat* hepatic subcellular fractions, with formation of glyceryl dinitrates and glyceryl mononitrates (Delaforge et al. 1993). Among differently treated *rat* liver microsomes, the highest microsomal activity was obtained under anaerobic conditions with microsomal preparations from dexamethasone-treated *rats* and NADPH. The reaction was inhibited by O_2, CO, miconazole, dihydroergotamine and troleoandomycin showing that it was catalysed by **CYP3A** isoforms.

2,6-Di-*tert*-butyl-4-methylphenyl *N*-methylcarbamate (**terbutol**, a mitotic disrupting herbicide; Hoffman and Vaughn 1994) increased P_{450} and cytochrome b_5 contents in F344 *rat* liver (Suzuki et al. 2001). In male *rats*, P_{450} and cytochrome b_5 contents, and NADPH cytochrome *c* reductase activity in liver microsomes were increased about 2-fold by 1 % terbutol administered in the diet for 7 to 28 days. Among the P_{450}-dependent monooxygenase activities in liver microsomes, 7-benzyloxyresorufin-*O*-debenzylase activity was greatly increased by 100-fold, and 7-ethoxyresorufin-*O*-deethylase, 7-ethoxycoumarin-*O*-deethylase, and aminopyrine-*N*-demethylase activities were elevated 2- to 3-fold.

The azole antifungal drug ketoconazole was found to inhibit Fe(III)-ascorbate dependent lipid peroxidation using either *rat* liver microsomes or *ox*-brain phospholipid liposomes as the substrates (Wiseman et al. 1991). It also inhibited microsomal peroxidation induced by the Fe(III)-ADP/NADPH system. The related azoles, miconazole and clotrimazole, were much weaker inhibitors than ketoconazole. Ketoconazole was approximately equipotent with the triphenylethylene anticancer drug tamoxifen in the microsomal system and was almost as effective as 4-hydroxytamoxifen in the liposomal system. Ketoconazole introduced into phospholipid liposomes during their preparation inhibited Fe(III)-ascorbate induced lipid peroxidation to a greater extent than similarly introduced cholesterol, ergosterol or tamoxifen. Miconazole and clotrimazole were again poor inhibitors of lipid peroxidation in this system.

Metabolism of tamoxifen by *rat* liver microsomes gave equal amounts of the *R* and *S* enantiomers (Osborne et al. 2001). They have the same chemical properties but, on treatment of *rat* hepatocytes in culture, *R*-(+)-α-hydroxytamoxifen gave at least eight times as many DNA adducts as the *S*-(−)-isomer.

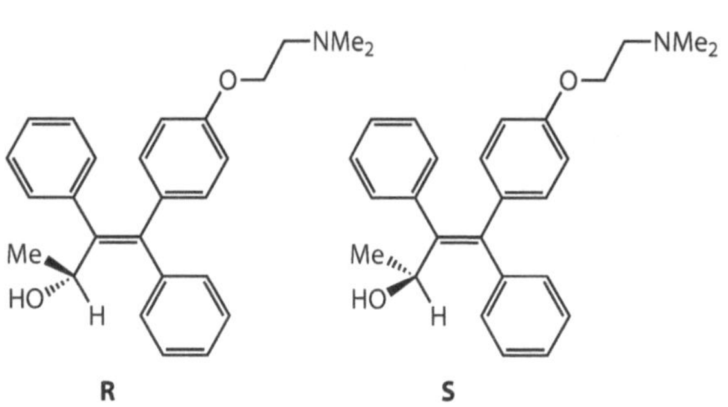

Structure of –(+)-α-hydroxytamoxifen, R and S stereoisomers [294]

Tamoxifen, 4-hydroxytamoxifen, nafoxidine, 17β-oestradiol and ICI 164,384 were all found to protect *rat* liver nuclei against Fe(III)-ascorbate-dependent lipid peroxidation (Wiseman and Halliwell 1994). The order of effectiveness of these compounds was 4-hydroxytamoxifen > 17β-oestradiol > nafoxidine > tamoxifen > ICI 164,384. The idea of a protection by tamoxifen against the formation of the genotoxic reactive intermediates and products of lipid peroxidation in the nuclear membrane and thus of an anticarcinogenic benefit was later questioned: Carthew et al. (2001) found a clear dose-response relationship of tamoxifen-induced DNA adducts in the *rat* liver and the subsequent increase in the development of liver cancer, with and without phenobarbital promotion. In the absence of phenobarbital promotion there was a threshold value for of tamoxifen-induced DNA adducts (180 adducts/10^8 nucleotides) and the subsequent induction of liver cancer.

Notley et al. (2002) investigated the P_{450} forms responsible for covalent drug-protein adduct formation and the possibility that covalent adduct formation might occur via alternative pathways to catechol formation. Recombinant P_{450} 3A4 catalysed adduct formation, and this correlated with the level of uncoupling in the P_{450} incubation, consistent with a role of oxygen species in potentiating adduct formation after enzymatic formation of the catechol metabolite. Whereas P_{450}s 1A1, 2D6, and 3A5 generated catechol metabolite, no covalent adduct formation was observed with these forms. By contrast, P_{450} 2B6, 2C19, and *rat* liver microsomes catalysed drug-protein adduct formation but not catechol formation. Drug protein adducts formed specifically with P_{450} 3A4 in incubations using membranes isolated from bacteria expressing P_{450} 3A4 and reductase, as well as in reconstitutions of purified 3A4, suggesting that the electrophilic species reacted preferentially with the P_{450} enzymes concerned.

Tamoxifen induced signs of autophagy in the FM3 *murine* breast tumour cell line, which was enhanced when it was combined with medroxyprogesterone (Bilir et al. 2001).

In different *in vitro* biological systems pure **P_{450} reductase** was able to generate 1-hydroxyethyl radicals from ethanol under anaerobic conditions (Díaz Gómez et al. 2000). Its formation was shown to be strongly dependent on the presence of NADPH. No 1-hydroxyethyl formation was observed when pure isocitric dehydrogenase from *porcine* heart was incubated with ethanol in the presence of isocitrate and $NADP^+$ or NAD^+. 1-Hydroxyethyl radicals were generated by nuclear biotransformation of ethanol to a lower extent than in the case of microsomes but still dependent on the presence of NADPH. The reaction was completely inhibited by adding P_{450} antibody to the incubation mixture.

Hepatocyte injury caused by **hypoxia** occurs following a period of marked reductive stress as reflected by an increase in the hepatocyte lactate/pyruvate ratio and β-hydroxybutyrate/acetoacetate ratio. Increasing the reductive stress with NADH generators results in increased cytotoxicity whereas offsetting the reductive stress with NADH oxidisers averts cytotoxicity. Cytotoxicity was also prevented by desferoxamine, a ferric ion chelator and scavenger of reactive oxygen species. This reductive stress resulted in the intracellular release of iron and formation of reactive oxygen species (Khan and O'Brien 1995, Niknahad et al. 1995).

In an endoplasmic reticulum fraction from *pig* liver enriched in transitional endoplasmic reticulum vesicles capable of forming 50–60 nm buds in the presence of ATP and retinol, the **NADPH oxidase** activity was inhibited by micromolar and submicromolar concentrations of retinol (Sun et al. 2000). Retinol at 1 mM stimulated the activity. The inhibition was confined to two activity maxima separated in time by about 5 min. In contrast, with a dithiodipyridine substrate, the activity was stimulated by retinol and the stimulations were in the part of the oscillatory pattern where retinol inhibition of NADH oxidation was observed.

Cytochrome *c* release from intact *rat* liver mitochondria depended strictly on pore opening and not on membrane potential, and a calcium-dependent permeability transition-enhancing oxidative stress also augmented cytochrome *c* release (Kantrow et al. 2000). Compared to mitochondria from normal *mouse* livers, fatty liver mitochondria had a 50% reduction in cytochrome *c* content and produced superoxide anion at a grater rate (Yang et al. 2000).

15.1.11.1.2 Glutathione

They also contained 25% more GSH and demonstrated 70% greater manganese superoxide dismutase activity and a 35% reduction in glutathione peroxidase activity. Mitochondrial generation of H_2O_2 was increased by 200% and the activities of enzymes that detoxify H_2O_2 in other cellular compartments were abnormal.

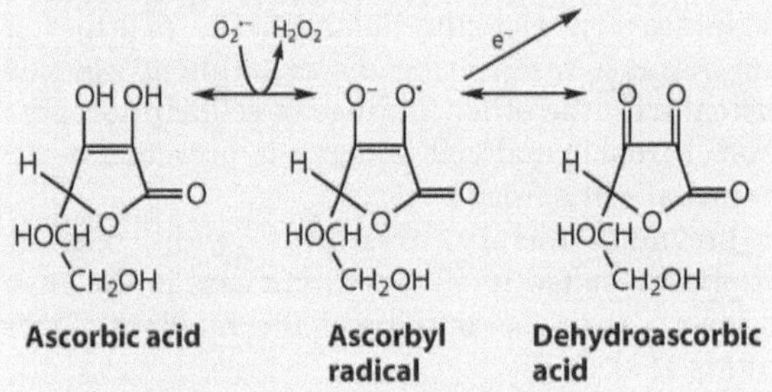

Formation of the ascorbyl free radical [53]

Freshly prepared mitochondria contain ascorbate, as do mitoplasts, that lack the outer mitochondrial membrane (Li et al. 2001). Both mitochondria and mitoplasts rapidly take up oxidised ascorbate as dehydroascorbic acid and reduce it to ascorbate. Ascorbate concentrations in mitochondria and mitoplasts rise into the low micromolar range during dehydroascorbic acid uptake, although uptake and reduction are opposed by ascorbate efflux. Mitochondrial dehydroascorbic acid reduction depends mainly on GSH, but mitochondrial thioredoxin reductase may also contribute. Reactive oxygen species generated within mitochondria oxidise ascorbate more readily than they do GSH and α-tocopherol.

Glutathione peroxidase (EC 1.11.1.9)

$$2\,GSH + H_2O_2 \rightarrow GSSG + 2\,H_2O \qquad [58]$$

A protein containing a selenocysteine residue. Steroid and lipid hydroperoxides, but not the product of reaction of EC 1.13.11.12 on phospholipids, can act as acceptor, but more slowly than H_2O_2 (cf. EC 1.11.1.12)

A protein isolated from *pig* liver was found to exhibit glutathione peroxidase activity with phosphatidylcholine hydroperoxide, a substrate that did not react with cytosolic glutathione peroxidase (Ursini et al. 1982).

After *in vivo* labelling with [^{75}Se]selenite, the intracellular distribution of selenoproteins in the liver was investigated in selenium-adequate and selenium-deficient Wistar *rats*. Behne et al. (1990) identified 12 selenoproteins. Glutathione peroxidase was concentrated in the cytosol and in the mitochondria. With the newly detected selenoproteins, some were enriched in the cytosol, one was mainly found in the nuclear fraction and some, which were present mainly in the mitochondrial and microsomal fractions, are most probable membrane-bound. In the livers of selenium-depleted *rats* the selenium administered was used predominantly to restore the levels of some of the newly found selenoproteins, while in the livers of selenium-adequate animals most of the selenium retained was incorporated into the glutathione peroxidase.

Selenite induced the oxidation and cross-linking of protein thiol groups, mitochondrial permeability transition, a decrease in the mitochondrial membrane potential ($\Delta\Psi_m$), and the release of cytochrome *c* in mitochondria isolated from *rat* liver (Kim et al. 2002). Induction of mitochondrial permeability transition by selenite (1 mM) was prevented by cyclosporin A (2 μM), ethylene glycol-bis(2-aminoethyl)-*N,N,N',N'*-tetraacetic acid (5 or 10 μM), or *N*-ethylmaleimide (50 or 100 μM).

The decrease in 3,3',5-triiodothyronine (T_3) production from thyroxine (T_4) found in the livers of selenium-deficient *rats* (Arthur et al. 1990, Behne et al. 1990) shows that an adequate selenium supply is important for thyroid hormone metabolism.

During reperfusion (1 h) of an ischaemic (2 h) male *rat* liver Jaeschke et al. (1987) found no evidence for the intracellular generation of reactive oxygen species (evaluated as increased biliary or sinusoidal **glutathione** (GSSG) **efflux** or elevated hepatic GSSG content). Intravital microscopy revealed accumulation of leukocytes within the postischaemic hepatic microvasculature with stasis in sinusoids (75.9 ± 8.9 per liver lobule) and adherence to the endothelial lining of postsinusoidal venules (534.7 ± 125.3 per mm^2 endothelial surface) (Müller et al. 1996). Concomitantly, compromised microvascular reperfusion was charcterized by perfusion deficits of individual sinusoids (25.6±4.0 %). The xanthine oxidase inhibitor allopurinol (50 mg/kg body weight orally) and the radical scavenger superoxide dismutase (60000/IU/kg body weight intravenously) effectively ($P < 0.01$) inhibited both sinusoidal leukostasis (16.1±2.6 and 32.1±3.1 cells/lobule) and venular leucocyte adherence (247.6±7.9 and 205.0±38.0 cells/mm^2), and, hence, reduced microcirculatory deterioration, indicated by the attenuation of sinusoidal perfusion failure (2.8±0.8 and 9.0±3.1 %). Using a Saticon ultrasensitive videocamera with image intensifier, device able to record the organ live image and to measure the tissue **photon emission** at a single photon level, Gasbarrini et al. (1998) detected photoemission starting after a few minutes of reperfusion of ischaemic male Wistar *rat* liver, reaching its maximum after 15–20 min and disappearing within 50–60 min. Chemiluminescence emitted by livers from younger (4 months) *rats* however, was significantly higher when compared to chemiluminescence emitted by organs isolated from 30 months old *rats* (0.8±0.1 vs 0.44±0.08 photons×10^5/s, respectively, after 15 min; $P < 0.01$).

Rat hepatocytes kept at 4 °C for 20 h after reworming to 37 °C for up to 3 h showed time-dependent production of reactive oxygen species, lipid peroxidation, chromatin condensation and membrane blebbing, decrease in GSH concentration, and protein sulphydryl groups (Vairetti et al. 2001). Cold preservation and rewarming in the presence of *N*-acetyl-L-cysteine induced a significant improvement in the morphology, less oxidative stress and apoptosis. Conversely, GSH depleting agents, diethylmaleate and buthionine sulphoximine, caused a marked increase of oxygen stress, deterioration of the morphology, and apoptosis.

The early post-ischaemic cell death was more extensive in *rats* with low initial GSH content (diethylmaleate-pre-treated or fasted for 48 h) than in *rats* with high GSH content (Jennische 1984).

Lindane (60 mg/kg p.o.) diminished the content of hepatic GSH 4 h after treatment (Junqueira et al. 1993). The activities of glutathione peroxidase, glutathione reductase, glutathione-*S*-transferases and γ-glutamyltransferases in the livers of lindane-treated *rats* and control animals were comparable. Liver GSH turnover, measured after a pulse of [35S]cysteine, was enhanced by 69 % ($P < 0.05$) in lindane-treated rats 24 h after intoxication compared with controls, with a 63 % ($P < 0.05$) increase in the estimated rate of GSH synthesis.

2-Mercaptoethanesulphonate (mesna) gave a slight stimulation of lipid peroxidation by *rat* hepatic microsomes incubated with 0.2 mmol/l ascorbate + 10 μmol/l Fe^{2+} (Bast et al. 1987). GSH (1 mmol/l) delayed lipid peroxidation, GSSG was ineffective. Direct measurement of GSH showed that excessive reduction of GSSG to GSH by mesna indeed occurred.

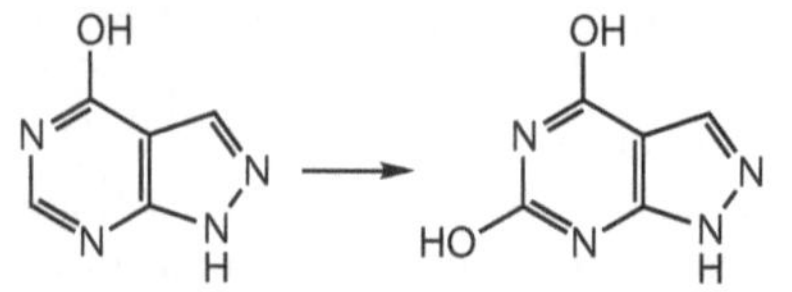

Allopurinol and oxipurinol. Both are hydroxyl radical scavengers (k_2 approx. $10^9\ M^{-1} \cdot s^{-1}$ and $4 \times 10^9\ M^{-1} \cdot s^{-1}$, respectively) [284]

In order to correlate morphological and functional findings, ^{31}P nuclear magnetic resonance spectroscopy and electron microscopy were used to investigate metabolic and ultrastructural changes during 6 h of reperfusion with an oxygenated erythrocyte/buffer solution after Ross (1972) of removed *rat* livers immediately transferred into a basin filled with lactated Ringer's preservation solution for 4 h at 4 °C and slowly rewarmed for 30 min. Ellermann et al. (1992) noted the simultaneous presence of vacuolary degenerated mitochondria and mitochondria of increased activity. ^{31}P nuclear magnetic resonance spectra demonstrated initially a partial ATP-recovery. Hepatocellular release of aspartate aminotransferase during the experiment was consistent with the other findings of cellular degeneration. Parenchymal cell injury was preceded by endothelial cell damage.

Previous dietary treatment with selenium (1 ppm) resulted in a marked increase in the glutathione peroxidase activity of the rat liver (Benedetti et al. 1974).

Rat liver microsomes exhibit selenium-independent glutathione peroxidase (EC 1.11.1.9) activity which is associated with glutathione *S*-transferase (EC 2.5.1.18) activity (Reddy et al. 1981).

Glutathione S-transferase (EC 2.5.1.18)

RX + Glutathione ⟶ HX + R–S–G [129]

Many distinct gene families, including the α, μ, π, κ, σ, ϑ, ζ and ω families, as well as over 20 distinct soluble glutathione *S*-transferases have been identified in mammals, including *humans* (Hayes and Pulford 1995, Board et al. 2000). In *mouse* liver mitochondria Raza et al. (2002) identified three distinct forms of glutathione *S*-transferase: GSTA1-1 and GSTA4-4 of the α family, and GSTM1-1 belonging to the μ family.

Glutathione *S*-transferases A1, A4, and M1 mRNA and protein levels were increased in the livers of iron-overloaded male Balb/c *mice* (Desmots et al. 2002). An acute exposure of primary cultures of *mouse* hepatocytes to iron-citrate (10 μM) strongly induced oxidative stress and cellular injury and resulted in an increase in glutathione *S*-transferase A4 expression, while cotreatment with iron-citrate and either desferrioxamine (10 μM) or vitamin E (250 μM) prevented both toxicity and glutathione *S*-transferase A4 induction.

A novel theta class glutathione transferase isoenzyme from *mouse* termed mGSTT3 has been identified by analysis of the expressed sequence tag database (Coggan et al. 2002). The gene encoding *mGSTT3* is clustered with the *mGSTT1* and *mGSTT2* genes on chromosome 10 and has an exon/intron structure that is similar to that of the other ϑ class genes. mGSTT3 is expressed strongly in the liver and to a decreasing extent in the kidney and testis.

The reaction of glutathione with haloalkenes is catalysed by both microsomal and cytosolic glutathione *S*-transferases (Vamvakas et al. 1989, Koob and DeKant 1990, Jin et al. 1996) or, in some cases, exclusively by cytosolic glutathione *S*-transferase (DeKant et al. 1998, Lash et al. 1998). The energy of the lowest uncoupled molecular orbital (E_{LUMO}) values for haloalkenes were inversely related to the specific activity of the microsomal glutathione *S*-transferase 1-catalyzed reaction but not the cytosolic glutathione *S*-transferase-catalyzed reaction (Jolivette and Anders 2002).

Glutathione *S*-transferases are susceptible to inactivation by reactive nitrogen species. Treatment of isolated glutathione *S*-transferases or *rat* liver homogenates with either peroxynitrite, the myeloperoxidase/hydrogen peroxide/nitrite system, or tetranitromethane, resulted in loss of glutathione *S*-transferase activity with a concomitant increase in the formation of protein-associated 3-nitrotyrosine (Wong et al. 2001).

In vitamin A-deficient *rats*, the mitochondrial GSH/GSSG ratio was significantly lower and the levels of malondialdehyde and 8-oxo-7,8-dihydro-2'-deoxyguanosine were higher when compared to control *rats* (Barber et al. 2000). These values were partially restored in re-fed *rats*. The mitochondrial membrane potential of vitamin A-deficient *rats* was significantly lower than in control *rats* and returned to normal levels in restored vitamin A *rats*. Two populations of mitochondria were found in vitamin A-deficient *rats* according to the composition of membrane lipids. One population showed a similar pattern to the control mitochondria and the second population had a higher membrane lipid content.

Captopril, an angiotensin-converting enzyme inhibitor, at high doses depleted endogenous thiols because of the requirement of cysteine and glutathione for the metabolism of captopril and captopril-protein conjugates (Yeung et al. 1983).

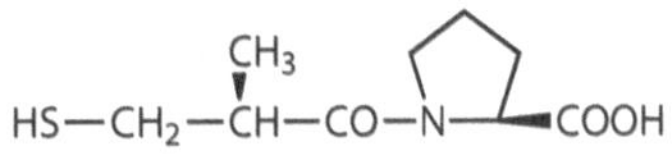

Captopril [73]

Helliwell et al. (1985) observed a dose-dependent depletion of hepatic glutathione in *mice* following intraperitoneal administration of 50–300 mg/kg. Serum glutamate pyruvate transaminase levels were increased and hepatic necrosis observed.

In male ddY strain *mice*, oral medication of a standardized extract of *Ginkgo biloba* leaves (100–1000 mg/kg) and bilobalide (10–30 mg/kg) once a day for 4 days caused a dose-dependent elevation of glutathione-*S*-transferase activity (Sasaki et al. 2002). Gingkolide A (20 mg/kg, for 4 days) also significantly elevated glutathione-*S*-transferase activity, whereas gingkolide B and gingkolide C at the same dose had no effects. Extract of *Ginkgo biloba* leaves significantly increased the protein level of glutathione-*S*-transferase π, and bilobalide significantly increased those of glutathione-*S*-transferase α and glutathione-*S*-transferase μ. Moreover, the medication of the standardized extract of *Ginkgo biloba* leaves and bilobalide treatment caused significant elevation in DT-diaphorase activity and in hepatic glutathione content.

15.1.11.1.3 α-Tocopherol

Liver secretion of very low density lipoproteins contributed to the daily 'turnover' of plasma *RRR*-α-tocopherol. Patients with the autosomal recessive neurodegenerative disease called ataxia with isolated vitamin E deficiency have an impaired ability to incorporate α-tocopherol into very low density lipoproteins secreted by the liver, because of mutations in the gene encoding the tocopherol transfer protein.

Pre-treatment of *rat* hepatocytes with α-tocopheryl succinate dramatically enriched cells and mitochondria with α-tocopherol and provided these membranes with complete protection against ethyl methanesulphonate-induced oxidative damage (Zhang et al. 2001). α-Tocopherol pre-treatment suppressed ethyl methanesulphonate-induced cellular production of reactive oxygen species, generated from mitochondrial complex I and III sites.

Dietary deficiency of α-tocopherol for 8 weeks decreased hepatic mitochondrial and microsomal concentrations of α-tocopherol by 80–90 % (Burc-

ZYNSKI ET AL. 2001). INCUBATION OF HEPATIC MITOCHONDRIA AND MICROSOMES FROM CONTROL *rats* with $FeSO_4$ (1.0 mM) caused a time-dependent stimulation of lipid peroxidation as indicated by the formation of thiobarbituric acid reactive substances (TBARS); the rate of TBARS production increased in preparations from α-tocopherol-deficient animals.

In order to estimate the interference of nitrates (0.02 or 0.04 % $NaNO_3$ in drinking water) in the presence of vitamin E (10 mg per kg body mass × day) with the metabolism of the *rat*'s liver, PAWŁOWSKA-GÓRAL et al. (2002) determined the composition and the amount of glycosaminoglycans. It was stated that the total amount of glycosaminoglycans increased in livers in all *rats*. Basing on all fractions of the examined glycosaminoglycans it was determined that the most significant differences between the individual groups appeared in the amount of heparane sulphate. There was a normalising influence of vitamin E on the quantitative composition of glycosaminoglycans of rats, which ingested nitrates.

Ginkgo biloba extract 761 (50–200 μg/ml) had no effect on *rat* liver mitochondrial function before anoxia, but had a specific dose-dependent protective effect after anoxia/reoxygenation *in vitro* (DU et al. 1999).

15.1.11.1.4 Transition Metals

When *rat* liver microsomes were incubated with NADPH, the major products were hydroperoxides which increased with time indicating that endogenous **iron** content is able to promote lipid peroxidation (MORINI et al. 1990). The addition of either 5 μM Fe^{2+} or Fe^{3+} ions strongly enhanced the hydroperoxide formation rate. However, due to the hydroperoxide breakdown, hydroperoxide concentration decreased with time in this case. Higher ferrous or ferric iron concentration did not change the situation much, in that both hydroxide breakdown and formation were similar to those when NADPH only was present in the incubation medium. After lipid peroxidation, analysis of fatty acids indicated that the highest amount of peroxidized polyunsaturated fatty acids occurred in the presence of 5 μM of either Fe^{2+} or Fe^{3+}. As far as the optimum Fe^{2+}/Fe^{3+} ratio required promoting the initiation of microsomal lipid peroxidation in *rat* liver is concerned, the highest hydroperoxide formation was observed with a ratio ranging from 0.5 to 2.

Lipid peroxidation of liver and kidney microsomes induced a highly characteristic sequence of morphological changes by detachment of ribosomes and formation of large aggregates of vesicles bound together by dense amorphous material and myelin figure-like debris (ARSTILA et al. 1972). The trilamellar structure of the membrane is, however, retained even after complete peroxidation, though its spacing may be increased. The aggregates resemble lipofuscin pigment as well as the membranous aggregates of endoplasmic reticulum seen in the liver after carbon tetrachloride poisoning.

The effects of sulphur amino acid deprivation on the changes in cellular calcium and free iron pool, prooxidant production and the ferritin light chain expression were comparatively evaluated in Hepa1c1c7 and Raw264.7 cell lines (KIM and KIM 2002). $[Ca^{2+}]_i$ was rapidly increased by sulphur amino acid deprivation. Sulfhydryl-containing compounds prevented the increase in $[Ca^{2+}]_i$. Inhibition of Ca^{2+} mobilization decreased the fluorescence of Phen Green SK inside cells, representing the inhibition of free iron release. Both inhibition of Ca^{2+} mobilization and iron chelation decreased the dichlorofluorescein oxidation, indicating the possibility that the increase in $[Ca^{2+}]_i$ affected that in cellular free iron and peroxidant production.

Administration of a tungsten-supplemented diet to young *rats* for 10 weeks resulted in an 80 %–90 % reduction in the liver xanthine dehydrogenase activity and a 100 % increase in the liver non-heme iron content ad compared to control animals (TOPHAM et al. 1982). A time course study of the effect of dietary tungsten in adult *rats* demonstrated that an increase in the non-heme iron content of liver was not observed until the liver xanthine dehydrogenase activity was reduced by 50 % of that of the control animals. Reduction of the liver xanthine dehydrogenase activity beyond 50 % of the control value was accompanied by a progressive and continual increase in the non-heme iron content of liver. Removal of the tungsten from the diet of the remaining *rats* resulted in a rapid increase of the liver xanthine dehydrogenase activity, which was accompanied by a decrease in the liver non-heme iron content.

Iron overload following dietary administration of dicyclopentadienyl iron (ferrocene) becomes primarily evident in hepatocytes with only a small amount present in Kupffer cells (VALERIO and PETERSEN 2000). A striking pattern of iron deposition occurred in ferrocene-fed *mice* as evidenced by large circular deposits of haemosiderin in the centrilobular hepatocytes. Masson's trichrome stain revealed a mild increase in liver collagen associated with iron-loaded centrilobular hepatocytes. Hydroxyproline was significantly ($P < 0.05$) increased. Malondialdehyde concentrations as a marker of lipid peroxidation were increased fourfold ($P < 0.05$).

Because current iron overload therapy used only Fe^{3+} chelators, such as desferrioxamine, HUANG et al. (2002) tested a hypothesis that addition of a Fe^{2+} chelator, 2,2'-dipyriyl, may be more efficient and effective in preventing iron-induced oxidative damage in *human* liver HepG2 cells than desferrioxamine alone. 2,2'-Dipyriyl or desferrioxamine alone decreased levels of iron and lipid peroxidation in cells treated with iron. Desferrioxamine + 2,2'-dipyriyl together had the most significant effect in preventing cells from lipid peroxidation but not as effective in decreasing overall iron levels in the cells. Using electron spin resonance spin trapping technique, HUNG et al. (l.c.) tested factors that can influence the production of reactive oxygen species by Fe^{2+} with dissolved O_2 in a cell-free system. Oxidant formation increased with increasing Fe^{2+} concentrations and reached a maximum at 5 mM of Fe^{2+}. When the concentration of Fe^{2+} was increased to 50 mM, the oxidant producing activity of Fe^{2+} sharply decreased to zero.

One-electron transfer catalysed by iron

$$O_2 + Fe^{2+} \longleftrightarrow Fe^{3+}\text{—}O_2^{\bullet-} \longrightarrow O_2^{\bullet-} + Fe^{3+} \quad [196]$$

$$O_2^{\bullet-} + Fe^{2+} + 2\,H^+ \longrightarrow H_2O_2 + Fe^{3+} \quad [197]$$

$$H_2O_2 + Fe^{2+} \longrightarrow HO^{\bullet} + HO^- + Fe^{3+} \quad [198]$$

Isolated Wistar *rat* liver mitochondria accumulated $^{59}Fe(III)$ partly by an energy-dependent and partly by an energy independent mechanism (ROMSLO and FLATMARK 1973). When the iron-loaded mitochondria were disrupted mechanically and the mitochondrial subfractions isolated by density gradient centrifugation, the iron accumulated by the energy-dependent mechanism was recovered mainly in the soluble matrix and intermembrane space (approx. 50 % of the total activity) and the inner membrane (approx. 30 %; ROMSLO and FLATMARK 1974). On the other hand, most of the energy-independent iron accumulation was confined to the outer and inner membranes (approx. 35 % of the total activity in each).

For ethyl methanesulphonate- and Fe^{2+}-treated *rat* hepatocytes, the *in vivo* pretreatment with D-α-tocopherol offered no protection against cell death and lipid peroxidation over a 6 h incubation period (FARISS et al. 2001). In contrast, hepatocytes isolated from *rats* treated with D-α-tocopheryl succinate were completely protected against ethyl methanesulphonate-induced lipid peroxidation for the entire incubation period, which also resulted in dramatic protection against ethyl methanesulphonate-induced cell death. Similarly, D-α-tocopheryl succinate administration *in vivo* dramatically protected isolated hepatocytes against Fe-induced cell death and lipid peroxidation even at Fe^{2+} concentrations up to 5 mM.

Ferrous ion-induced lipid peroxidation of *rat* liver mitochondria was accelerated by phosphate (YAMAMOTO et al. 1974). Preincubation of *rat* liver microsomes with iron (Fe)/ascorbate (50 μM/200 μM), known to induce peroxidation, resulted in a significant inhibition of (i) the rate-limiting enzyme in cholesterol biosynthesis, HMG-CoA reductase (46 %, $P < 0.01$, (ii) the crucial enzyme controlling the conversion of cholesterol in bile acids, cholesterol 7α-hydroxylase (48 %, $P < 0.001$), and (iii) the central enzyme for cholesterol esterification, acyl-CoA:cholesterol acyltransferase (ACAT, 80 %, $P < 0.0001$) (BRUNET et al. 2000). The disturbances of these key enzymes coincided with a high rate of malondialdehyde production (350 %, $P < 0.007$) and the loss of polyunsaturated fatty acids (36.19 ± 1.06 % vs. 44.24 ± 0.41 % in controls, $P < 0.0008$). While α-tocopherol simultaneously neutralised lipid peroxidation, preserved microsomal fatty acid status, and restored ACAT activity, it was not effective in preventing Fe/ascorbate-induced inactivation of both HMG-CoA reductase (44 %, $P < 0.01$) and cholesterol 7α-hydroxylase (71 %, $P < 0.0001$).

The pentyl and ethyl radicals can be formed by peroxidation of linoleic acid at C-13, of linolenic acid at C-16, and of arachidonic acid at C-15 when *rat* liver microsomes were incubated with ADP, NADPH and ferric chloride (IWAHASHI et al. 1992).

Incubation with Fe-citrate with either H_2O_2, L-ascorbate, or L-cysteine induced single- and double-strand breaks in supercoiled plasmid pZ189 in a concentration- and time-dependent fashion (TOYOKUNI and SAGRIPANTI 1993). DNA strand breaks produced by Fe-citrate plus H_2O_2 increased at reduced pH (≤6.9). Catalase and free radical scavengers inhibited the DNA breakage produced by Fe-citrate in combination with each reductant, suggesting that H_2O_2 and finally $HO^{\bullet}$ are responsible DNA damaging species.

When indomethacin was incubated with Wistar *rat* liver microsomes in the presence of horseradish peroxidase, H_2O_2 and ADP-Fe^{3+}, lipid peroxidation was time-dependent (MIURA et al. 2002). Catalase and desferrioxamine almost completely inhibited lipid peroxidation, indicating that H_2O_2 and iron are necessary for lipid peroxidation. ESR signals of indomethacin radicals were detected during the interaction of indomethacin with horseradish peroxidase-H_2O_2. However, the indomethacin radical by itself did not reduce the ferric ion.

Ferric iron used under its complexed form with nitrilotriacetic acid exhibited a prooxidant activity corresponding to an increase in free malondialdehyde in adult *rat* hepatocytes and in the culture medium (Morel et al. 1990). The toxic effect of Fe-nitrilotriacetic acid on hepatocyte cultures was a function of the incubation time (from 0 to 48 h) and of the concentration of ferric iron loading (i.e. 5, 20 and 100 μM). Superoxide dismutase (250, 500, 1000 and 1500 I.U./ml), mannitol (10, 25 and 50 mM), α-tocopherol (250, 500 and 1000 μM), and β-carotene (1.8, 4.5 and 9 mM) presented an intracellular activity, as they reduced mostly lipid peroxidation. Catalase (100, 1000 and 10,000 I.U./ml), dimethylpyrroline *N*-oxide (3.2, 16 and 32 mM), and thiourea (50, 100 and 500 mM) seemed essentially efficient in protecting the plasmalemmata, as shown by an important decrease in lactate dehydrogenase (EC 1.1.1.27) leakage.

Diquat dibromide-treated *rats* (20 mg/kg body weight) had increased levels of hepatic low molecular weight chelatable iron and decreased levels of hepatic ferritin iron when compared to saline-treated animals (Reif et al. 1988).

Copper is stored by the liver in the storage protein metallothionein and excreted by the transport protein ceruloplasmin into the bile. When 10^6 *rat* hepatocytes were incubated with 50 μM Cu^{2+} for 2 h, the formation of reactive oxygen species as determined by oxidation of dichlorofluorescin diacetate to dichlorofluorescein increased from 90 ± 5 intensity units in the controls to 412 ± 9 intensity units in the metal-treated cells (Pourahmad and O'Brien 2000). Malondialdehyde UV absorption increased from 0.048 ± 0.006 to 0.662 ± 0.012 units obtained at λ_{max} = 532 nm ($P < 0.001$). The ED_{50} concentrations found for Cu^{2+} and Cd^{2+} (i.e. 50 % membrane lysis in 2 h) were 50 and 20 μM, respectively (Pourahmad and O'Brien 2000). However, reactive oxygen species formation, GSH oxidation and lipid peroxidation were induced by Cu^{2+} at these concentrations more rapidly than by Cd^{2+}. The decline of mitochondrial membrane potential though occurred at the same time and to the same extent for both metals.

Wistar *rats* intraperitoneally injected with a daily dose of cupric nitrilotriacetate, which contained 4 to 7 mg of copper per kg body weight, showed massive liver necrosis, haemolytic anaemia, and acute renal tubular necrosis at the beginning of the experiment and intermittently after 4 weeks of injections (Toyokuni et al. 1989). Electron microscopic X-ray analysis at day 93 revealed that copper (> 250 μg/g dry weight) stored in secondary lysosomes was always accompanied by a proportional amount of sulphur (correlation coefficient, 0.98, $P < 0.005$).

In vitro additions of 0.4 mM $CuCl_2$ to *rat* liver lysosome preparations induced a release of acid phosphatase (EC 3.1.3.2) activity from lysosomes to 20,000 × *g* supernatants which was directly related to the molarity of copper(II) and length of incubation (Lindquist 1968). Addition of $CuCl_2$ to emulsions of unsaturated fatty acids initiated peroxidation of these lipids as indicated by the thiobarbituric acid reagent for lipid peroxidation products. The lipids of lysosomes isolated from copper-loaded *rat* livers absorbed in the 230- to 236 nm region, which is characteristic of conjugated dienes.

Supercoiled plasmid DNA pZ189 treated with Cu(II) or Fe(III) in the presence of different reducing agents produces single-strand breaks linearly related to 8-hydroxy-2'-deoxyguanosine measured by high performance liquid chromatography (Toyokuni and Sagripanti 1996).

Long-Evans *rats* with a cinnamon-like coat colour (LEC *rats*) lack in ability of the efflux of Cu in the liver and accumulate Cu in a form bound to metallothionein (Li et al. 1991, Sugawara et al. 1991, Okayasu et al. 1992, Sakurai et al. 1992, Suzuki 1995, 1996). However, when Cu accumulates to more than the capacity of the liver to synthesise metallothionein, Cu not bound to metallothionein is assumed to be a toxic form of Cu (Suzuki 1995). Intense ascorbate radicals were produced together with weal hydroxyl radicals at first in the liver of LEC *rats* before the onset of jaundice (Suzuki 1997). The former ascorbate radicals disappeared within 5 min, while hydroxyl radicals trapped by 5,5-dimethyl-1-pyrroline oxide increased after the disappearance of the ascorbate radicals. In LEC *rats* treated with tetrathiomolybdate, the hepatic Cu concentration was 60 μg/g wet weight, compared to 170 μg/g in untreated *rats* (Sugawara et al. 1999). Yamamoto et al. (2001) quantified $HO^{\bullet}$ production in plasma and liver of hepatitic LEC *rats* by trapping $HO^{\bullet}$ with salicylic acid (formula [184]). The ratios of 2,3-dihydroxybenzoic acid /salicylic acid were significantly higher in plasma and livers of hepatitic LEC *rats* than those of Wistar *rats* and LEC *rats* showing no signs of hepatitis. Furthermore, the ratios of 2,3-dihydroxybenzoic acid/salicylic acid in plasma and livers of hepatitic LEC *rats* were almost the same as those of Wistar *rats* treated orally with $CuSO_4$ (0.5 mmol/kg) 2 h before acetylsalicylic acid injection. D-Mannitol (500 mg/kg) treatment significantly reduced hepatic mitochondrial lipid peroxidation.

Yamamoto et al. (2001) quantified $HO^{\bullet}$ production in plasma and liver of hepatitic LEC *rats* by trapping $HO^{\bullet}$ with salicylic acid (formula [184]). The ratios of 2,3-dihydroxybenzoic acid /salicylic acid were significantly higher in plasma and livers of hepatitic LEC *rats* than those of Wistar *rats* and LEC *rats* showing no signs of hepatitis. Furthermore, the ratios of 2,3-dihydroxybenzoic acid/salicylic acid in plasma and livers of hepatitic LEC *rats* were almost the same as those of Wistar *rats* treated orally with $CuSO_4$ (0.5 mmol/kg) 2 h before acetylsalicylic acid injection. D-Mannitol (500 mg/kg) treatment significantly reduced hepatic mitochondrial lipid peroxidation.

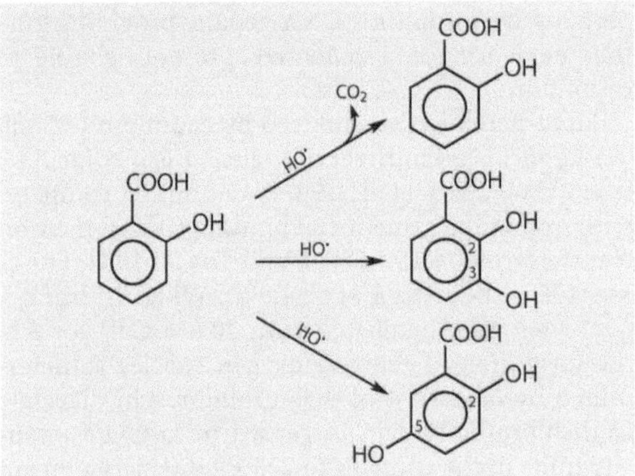

Salicylic acid as a hydroxyl radical trap [184]

When a *rat* liver microsomal suspension (1 mg protein per ml) was incubated at 37 °C with 5 mM salicylic acid and 0.2 mM NADPH, the amounts of thiobarbituric acid reactive substances and 2,5-dihydroxybenzoic acid increased with the incubation time (Doi et al. 2002). Simultaneously, spontaneous chemiluminescence was found to be generated here. The addition of SKF-525A, an inhibitor of cytochrome P_{450}, to the reaction mixture inhibited the chemiluminescence generation together with the inhibition of the oxidative metabolism. The anti-oxidants and singlet oxygen scavengers like *N,N*-diphenylphenylenediamine and histidine suppressed the chemiluminescence generation. The addition of 1,4-diazabicyclo [2.2.2] octane, a singlet oxygen quencher, to the reaction mixture generating chemiluminescence enhanced chemiluminescence transiently and then chemiluminescence decreased markedly. Thus chemiluminescence observed here may possibly originate from the singlet oxygen.

Cinnarizine was found to inhibit spontaneous lipid peroxidation in *rat* liver homogenates (Fernandes et al. 1991).

In 15 of 15 cases of primary biliary cirrhosis including early cases with minimal pathology, and in 5 of 5 cases of Wilson's disease, Elmes et al. (1989) found copper-associated protein. Metallothionein distribution was abnormal in most copper-associated protein-positive livers. Necrotic hepatocytes were intensely metallothionein-positive and in Wilson's disease had a characteristic appearance.

Wilson's disease is an autosomal recessive inherited disorder of copper metabolism marked by neuronal degeneration and hepatic cirrhosis. The Wilson protein is a copper-transporting P-type ATPase localized in the TGN, and retention of mutant misfolded versions of this protein in the endoplasmic reticulum is the molecular basis of the disease (Payne et al. 1998).

Parenteral administration of **nickel** chloride ($NiCl_2$) to *rats* enhanced lipid peroxidation in liver, kidney and lung, but not in brain, heart, spleen or testis, as measured by the thiobarbituric acid reaction for malondialdehyde and related chromogens in fresh tissue homogenates (Sunderman et al. 1985).

Nickel carbonyl inhibited ^{14}C-orotic acid incorporation into *rat* liver RNA (Beach and Sunderman 1969).

Calcium complexes significantly decreased the biliary excretion of ^{52}**manganese** as well as its concentration in the liver of the *guinea pig* (Koutenský et al. 1967). $Na_3CaEDTA$ was found the most effective. After administration of $^{52}MnCl_2$ and the Ca complexes, certain forms of chelates were excreted in the urine. The manganese and calcium complexes of the respective aminopolycarbonic acid may be detected as well as a bi-calcium complex. Analogous results were achieved after administration of $Na_2MnEDTA$ and $Na_3MnDTPA$.

In welders working with electrodes containing Mn compounds in the coating intravenous medication of 15 ml of 20 % CaEDTA significantly raised the daily excretion of Mn (Barbořík and Sehnalová 1967).

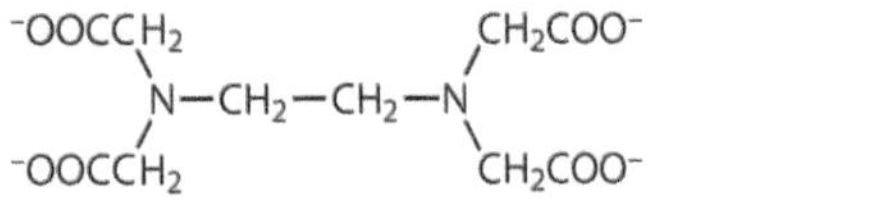

Ethylenediamine tetraacetic acid (EDTA) [187]

Manganese (200 ppm manganese chloride during 10 weeks in drinking water) in *rats* induced an increase in the amount of rough endoplasmic reticulum, groups of polyribosomes, proliferation of smooth endoplasmic reticulum in the centrilobular area, prominent Golgi field in the biliary poles of the cells, and polymorphous mitochondria with electron dense matrices (Wassermann and Wassermann 1977). Lipid droplets were scarce although were slightly more common than in controls. Cells almost entirely packed with lipid droplets were seen occasionally. A fatty degeneration, however, as observed by Kobert (1883) and Browning (1969) was not noted under these experimental conditions.

Zinc inhibited *in vitro*-induced lipid peroxidation in liver microsomes (Chvapil et al. 1972, 1974) and protected the *rat* liver against the toxic effects of carbon tetrachloride (CHVAPIL et al. 1973). Zinc acetate (50 mg per kg×day) administered intragastrically to *rats* receiving 2 ml CCl_4, per kg body weight twice a week intraperitoneally, significantly lowered the content of malonaldehyde in the microsomal, as well as mitochondrial fraction of the liver.

There was no effect of dietary **zinc deficiency** on free radical production in the *rat* liver homogenate and whole liver under acute oxidative stress *in vitro* and *in vivo* as indicated by electron spin resonance spin trapping (XU and BRAY 1994).

Cadmium (2 mg $CdCl_2$ i.p. per kg body weight) changed distribution of Cu in different cellular protein fractions in the livers of adult male ICR *mice* (YANG et al. 2000). Cu in the high molecular weight protein pool reduced over time. Beginning on the second day, an increasing amount of Cu was seen in the metallothionein fraction.

Toxicity of cadmium to liver and kidneys may result from:

- Cd-induced synthesis of metallothionein and Cd accumulation (NORDBERG 1978),
- metabolic disorders due to cytochrome P_{450} destruction (KLIMCZAK et al. 1984),
- prooxidant effect owing to induction of heme oxygenase and enhanced heme catabolism with release of Fe(II)/Fe(III) as a Fenton catalyst (AUST et al. 1985, CHEVION 1988).
- alteration of sulphydryl homeostasis and antioxidative reserves thus contributing to propagation of lipid peroxidation and derangement of biological membranes (MANCA et al. 1991, SUGIYAMA 1994, WISNIEWSKA-KNYPL and WRONSKA-NOFER 1995).

When male *rats* were given either a single dose of cadmium (3.58 mg $CdCl_2 \cdot 6\ H_2O$/kg, i.p.) 72 h prior to sacrifice or a single dose of nickel (59.5 mg $NiCl_2 \cdot 6\ H_2O$/kg, s.c.) 16 h prior to sacrifice, the activities of ethylmorphine *N*-demethylase, aminopyrine *N*-demethylase and aniline 4-hydroxylase, and the levels of cytochrome P_{450} and microsomal heme were significantly decreased (İŞCAN et al. 1992). Cadmium decreased the cytochrome b_5 level significantly, whereas it did not alter the NADPH-cytochrome *c* reductase activity significantly.

An interactive effect of cadmium (0.18 mg portions of particles <60 μm in diameter suspended in 0.4 ml sterile Tyrode's solution instilled intratracheally) with UICC standard reference chrysotile B (1.82 mg) or with *N*-nitrosoheptamethyleneimine in the induction of pulmonary hyperplasias or tumours in *rats* could not be evaluated (HARRISON and HEATH 1988). However, previous work by HARRISON and HEATH (1986) demonstrated and apparent synergy between cadmium and *N*-nitrosoheptamethyleneimine in the presence of crocidolite in the induction of lung tumours, and MANDEL and RYSER (1987) have shown that cadmium can enhance the mutagenic effect of another *N*-nitroso compound (*N*-methyl-*N*-nitrosourea), perhaps by inhibiting DNA repair, providing further evidence for a potential 'promoting' role of cadmium.

Lipid peroxidation induced by cadmium (2.5 mg per kg *rat*) is an indirect one since it is mediated by iron (CASALINO et al. 1997). Cadmium competes with iron at the same metal binding sites in the iron transfer proteins (SCHÄFER and FORTH 1984, FOULKES 1988, SUGAWARA et al. 1988). When 10^6 *rat* hepatocytes were incubated with 20 μM Cd^{2+} for 2 h, the formation of reactive oxygen species as determined by oxidation of dichlorofluorescin diacetate to dichlorofluorescein increased from 90±5 intensity units in the controls to 225±9 intensity units in the metal-treated cells (POURAHMAD and O'BRIEN 2000). Malondialdehyde UV absorption increased from 0.048±0.006 to 0.248±0.008 units obtained at λ_{max} = 532 nm (P <0.001).

Cadmium exposure (1.23 mg Cd/kg = 2.0 mg $CdCl_2$/kg) markedly decreased the activity of acetohexamide reductase catalysing the ketone-reduction of acetohexamide, an oral antidiabetic drug, in liver microsomes of male *rats* (SHIMADA et al. 2002).

Selenium treatments (10 μM SeO_2/kg given subcutaneously at −24, 0 and +24 h) enhanced cadmium (45 μM $CdCl_2$/kg given subcutaneously) accumulation at 24 h in the liver (23 %) of male Wistar *rats* (WAHBA et al. 1993). The synthesis of metallothionein in liver was unaffected.

15.1.11.1.5 Other Metals

Mercury toxicity is related to the induction of oxidative stress, as revealed by the decrease in antioxidant enzymes as glutathione *S*-transferase (REDDY et al. 1981). Examining interactions of purified Mn-superoxide dismutase (1 μM) with $HgCl_2$ indicated that mercury ions suppressed Mn-superoxide dismutase activity by reduction of the native form (SHIMOJO et al. 2002). Due to the minimal hepatic accumulation of inorganic mercury after the subcutaneous application of $HgCl_2$ (0.25–3 mg/kg) to *mice* the hepatic Mn-superoxide dismutase might be unaffected while the renal enzyme due to a 34–75 times higher accumulation of mercury in the kidney was decreased in a dose-dependent manner.

Glutathione S-transferase (EC 2.5.1.18)

RX + Glutathione → HX + R–S–G [129]

Lead exposure (1 mg Pb^{2+} per kg body weight i.p.) of *rats* for a period of 4 weeks followed by a period of 1 week to recover, resulted in significantly ($P < 0.05$) higher accumulation of lead, associated with significant ($P < 0.05$) increases in lipid peroxide level in the liver and brain (PATRA et al. 2001). Treatment with antioxidants alone resulted in reversal of oxidative stress without significant decline in tissue lead burden.

Beryllium poisoning in *mice* was antagonised by an intravenous or intraperitoneal injection of 600 mg sodium salicylate per kg body weight up to 8 h after an intravenous injection of $BeSO_4$ in LD_{95} amounts (FINKEL and WHITE 1952). Pre-treatment before the Be was administered was ineffective.

In all $BeSO_4$-injected *rats* GOLDBLATT et al. (1973) found masses of smooth endoplasmic reticulum in a large proportion of the cells. Rough-surfaced cisternae appeared to be reduced in number, but were still evident. Swelling of the endoplasmic reticulum, swelling of mitochondria with extraction of matrix density, and swelling of Golgi complexes were frequently encountered, but were difficult to evaluate. The most consistently altered organelle was the lysosome although a variety of morphologic forms was evident. Scattered large autophagic vacuoles were occasionally seen.

Deoxythymidine kinase, partially purified from 24 h regenerating *rat* liver, was markedly inhibited by $BeSO_4$ (MAINIGI and BRESNICK 1969). The inhibition was competitive with the cofactor, magnesium.

Cerium chloride injected intravenously into male *mice* (0.5, 1.0 and 2.0 mg/kg body weight) was more toxic in DBA/2N *mice* than in C57BL/6N *mice* (ARVELA et al. 1991). At 24 h, coumarin 7-hydroxylase activity was increased in a dose-dependent manner in DBA/2N animals, whereas no change was seen in C57BL/6N animals. A significant increase in all other enzymes studies, cytochrome P_{450}, ethoxycoumarin *O*-deethylase and ethoxyresorufin *O*-deethylase, was seen in DNA/2 *mice* injected with the highest dose of Ce. At 72 h Ce increased coumarin 7-hydroxylase activity, as well as other enzymes, in C57BL/6N *mice* in a dose-dependent manner, whereas in DBA/2 *mice* the increase was only seen after the two lower doses, the highest dose causing severe morphological changes in the liver structure and a decrease in coumarin 7-hydroxylase and other activities. The distribution studies with ^{141}Ce showed that C57BL/6N livers contained more Ce than DBA/2 livers after the highest dose.

Selenocompounds (phenyl methyl selenide and 1-hexynyl methyl selenide) did not inhibit hepatic δ-aminolevulinic acid dehydratase (FARINA et al. 2001). However, the products of reaction of phenyl methyl selenide or 1-hexynyl methyl selenide with H_2O_2 inhibited δ-aminolevulinic acid dehydratase from rat liver with IC_{50} values in the micromolar range. Although the selenides did not affect dithiothreitol oxidation, the products of reaction between phenyl methyl selenide and 1-hexynyl methyl selenide with H_2O_2 decreased the total amount of free -SH groups from dithiothreitol. Ethanol-induced severe fatty acid accumulation, mild inflammation, and necrosis in the *rat* liver were blunted significantly by ebselen (KONO et al. 2001). While there were no significant effects of either ethanol or ebselen on glutathione peroxidase (EC 1.11.1.9) activity in serum or liver tissue, ebselen blocked the increase in serum nitrate/nitrite caused by ethanol.

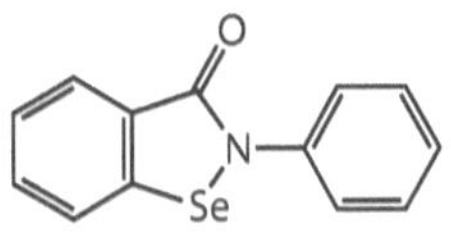

Ebselen, 2-phenyl-1,2-benzisoselenazol-3(2*H*)-one [65]

15.1.11.1.6 Endoplasmic Reticulum

The induction of formation of endoplasmic reticulum membranes by phenobarbital leads to a considerable change in lipid composition of the membranes. The molar ratio of cholesterol/total phospholipid in the membranes of the endoplasmic reticulum falls to 73 % of the control and the proportion of linoleic acid in phosphatidylcholine and phosphatidylethanolamine rises to 120–125 % of the control, whereas the proportion of oleic acid, arachidonic acid docosahexaenoic acid decreases (DAVISON and WILLS 1976). The capacity of reducing added hydroperoxides is increased in isolated microsomal fractions from phenobarbital-pretreated *rats* as compared to controls (SIES and SUMMER 1975).

Since many cytochromes P_{450} are inducible, the amount of reactive oxygen species can increase by the effect of an inducer; e.g. halogenated aromatic compounds, including 2,3,7,8-tetrachlorodibenzo-*p*-dioxin (AL-BAYATI and STOHS 1987), lindane (JUNQUEIRA et al. 1986, VIDELA et al. 1990), or phenobarbital (SCHOLZ et al. 1990). Lindane elicited a dose-dependent enhancement of total oxygen uptake by the isolated perfused *rat* liver, which was largely inhibited by 0.55 mM desferrioxamine (VIDELA et al. 1989).

Chloroform (0.75 ml/kg p.o.)-induced hepatotoxicity was prevented by dimethyl sulphoxide (2 ml/kg i.p) by mechanisms other than by inhibition of bioactivation by cytochrome P_{450} (LIND et al. 2000). However, the ability of the two CYP2E1 inhibitors, diallyl sulphide (100 mg/kg i.p.) and aminobenzotriazole (30 mg/kg i.p.) were as effective as DMSO at interrupting the development of the $CHCl_3$-induced hepatic lesions, and aminobenzotriazole was shown to inhibit the bioactivation of residual $CHCl_3$. DMSO is known to be an antioxidant and scavenge free radicals (BRAYTON 1986). Although $CHCl_3$ can produce free radicals during its bioactivation (TOMASI et al. 1985), LIND et al. (2000) found no elevated levels of malondialdyhyde.

$$CCl_4 \xrightarrow[\text{Cytochrome } P_{450}]{e^-} {}^{\bullet}CCl_3 + Cl^- \quad [116]$$

$$^{\bullet}CCl_3 + O_2 \longrightarrow CCl_3OO \quad [117]$$

$CCl_3OO^{\bullet}$ reacts with phenols forming phenoxyl radicals (NETA et al. 1989, ALFASSI et al. 1993). The reactions of $CCl_3OO^{\bullet}$ with *trans*-3,5,4'-trihydroxystilbene (*trans*-resveratrol) and its analogues showed that the *para*-hydroxyl group of *trans*-resveratrol scavenges free radicals more effectively than its *meta*-hydroxyl groups (STOJANOVIC et al. 2001).

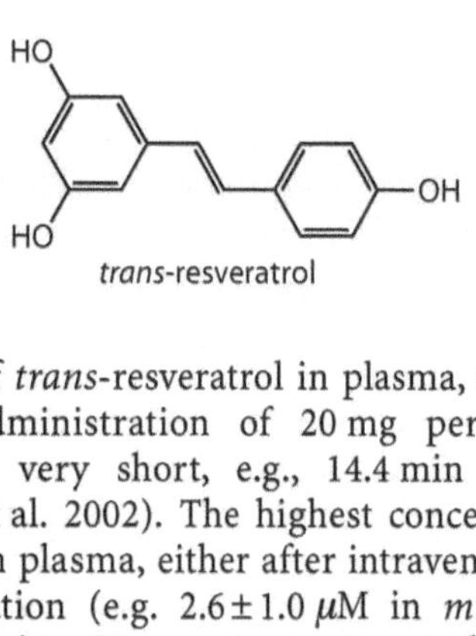

trans-resveratrol [295]

Half-life of *trans*-resveratrol in plasma, after intravenous administration of 20 mg per kg body weight, is very short, e.g., 14.4 min in *rabbits* (ASENSI et al. 2002). The highest concentration of the drug in plasma, either after intravenous or oral administration (e.g. $2.6 \pm 1.0\ \mu M$ in *mice* 2.5 min after receiving 20 mg *trans*-resveratrol/kg orally) was reached within the first 5 min in all animals studied. Extravascular levels (brain, lung, liver, and kidney) of *trans*-resveratrol, which paralleled those in plasma, were always < 1 nmol/g fresh tissue.

The trichloromethylperoxyl radical ($CCl_3OO^{\bullet}$), involved in **carbon tetrachloride** hepatotoxicity (SLATER 1987), has a reduction potential of ≥ 1.1 V, but < 1.3 V from its lack of reactivity with deoxyadenosine 5'-monophosphate (JOVANOVIC and SIMIC 1988). The rate constant for the reaction of $CCl_3OO^{\bullet}$ + ergothioneine was determined to be $1.2 \times 10^9\ M^{-1} \cdot s^{-1}$ (ASMUS et al. 1996), surprisingly double the estimate obtained by JOVANOVIC and SIMIC (1991).

CCl_4 in male Wistar *rats* induce oxidative stress mainly in the aqueous phase of the liver cell as ascorbic acid concentration was significantly decreased, while hydrophobic vitamin E was not influenced significantly (SUN et al. 2001).

Ethanol-treated *rats* had higher CYP2E1-dependent microsomal activities and CCl_4-induced lipid peroxidation than controls. Simultaneous chlormetriazole treatment inhibited CYP2E1 expression in *rat* liver and abolished and CCl_4-induced lipid peroxidation (HU et al. 1994). The protective effect of cimetidine in $^{\bullet}CCl_3$-induced lipid peroxidation and liver injury can be attributed to a reduction in cytochrome P_{450} (MERA et al. 1994).

Methanol inhalation (0, 1000, 2500, or 10000 ppm for 6 h) potentiated CCl_4 hepatotoxicity in Fischer 344 male *rats* (ALLIS et al. 1996). Hepatic microsomes showed increased *p*-nitrophenol hydroxylase activity but no increase in pentoxyresorufin-*O*-dealkylase or ethoxyresorufin-*O*-deethylase activities. Hepatic antioxidant levels, glutathione levels and glutathione-*S*-transferase activity in methanol-treated animals were not different from controls. Pre-treatment with allyl sulphone, a specific chemical inhibitor of CYP2E1, abolished the difference in microsomal metabolism between exposed and control animals.

The daily intraperitoneal medication with D-pantethine (500 mg/kg), D-(+)-pantothenic acid (100 mg/kg) or cystamine (50 mg/kg) for 5 days conferred significant protection against the hepatotoxic and peroxidative actions of a 0.5 ml CCl_4 intraperitoneal dose in male Sprague-Dawley *rats* (NAGIEL-OSTASZEWSKI and LAU-CAM 1990). All three treatments lessened the increases in serum ALT and liver thiobarbituric acid reactive substances, and the reductions in serum triglyceride levels, and prevented the development of hepatic steatosis caused by the halocarbon. Pantethine was found to offer the greatest protection.

Intravenous administration of trypan blue (3 ml of a 2.5 % suspension in physiol. saline) afforded a marked protection from the hepatotoxicity of CCl_4, when *rats* were challenged 24 h after the dye injection (PETRELLI and STENGER 1969). There was no evidence, either by light or electron microscopy, that the dye had even entered the hepatic parenchymal cells.

Nitric oxide was produced in liver tissues of *rats* that had been treated with a nonlethal dose of CCl_4 (1.3 g/kg) followed by a low dose of lipopolysaccharide (100 μg/kg) (CHAMULITRAT et al. 1994). EPR spectroscopy was used to detect nitrosyl-protein complexes. Haemoglobin-nitrosyl complexes were detected in both whole blood and liver. By performing analyses of EPR spectra obtained from hepatocytes exposed to •NO, Chamulitrat et al. (l.c.) were able to identify EPR signals attributable to nitrosyl-cytochrome P_{420} in *rat* liver. Nitrosyl complex formation was inhibited by treatment with N^G-monomethyl-L-arginine, suggesting enzymatic biosynthesis of •NO. A small but significant inhibition of nitrosyl complex formation by gadolinium trichloride pre-treatment was found in the liver, suggesting that Kupffer cells were also involved in •NO biosynthesis, because this treatment decreased Kupffer cells. There was a synergistic effect of CCl_4 and lipopolysaccharide on the serum levels of the hepatic enzymes aspartate aminotransferase (EC 2.6.1.1), alanine aminotransferase (EC 2.6.1.2), lactate dehydrogenase (EC 1.1.1.27), and sorbitol dehydrogenase (EC 1.1.1.14), which are indices of parenchymal cell damage. N^G-Monomethyl-L-arginine treatment increased these hepatic enzyme activities, suggesting a protective role for •NO.

The •NO levels in *mouse* liver tissues and NO_2^-/NO_3^- concentration in serum were found to decrease significantly both in a dose-dependent manner and in time course after CCl_4 treatment (Zhu and Fung 2000). The nitric oxide synthase (NOS) II activity in the liver, in contrast, was found to increase significantly. The •NO donor sodium nitroprusside (50 or 100 µg/kg) treatment resulted in decreases of lactate dehydrogenase (EC 1.1.1.1.27), glutamic pyruvic transaminase, and thiobarbituric acid reactive substances levels, leading to a protective effect in CCl_4-treated *mice*. On the other hand, N^G-nitro-L-arginine methyl ester (100 or 300 mg/kg) caused more severe liver damage.

Oleanolic acid, 3β-hydroxyolea-12-en-28-oic acid, is a pentacyclic triterpenoid present in many kinds of medical plants, protected against the acute hepatotoxicity produced by CCl_4, acetaminophen and cadmium (Liu et al. 1994). In *mouse* peritoneal macrophages oleanolic acid upregulated inducible nitric oxide synthase expression to produce nitric oxide (Choi et al. 2001).

Co-administration of colchicine (100 µg per day) with sub-lethal doses of CCl_4 in *rats* over 10 weeks did not prevent progression of cirrhosis (Das et al. 2000). However, *rats* made cirrhotic with repeated CCl_4 challenge and subsequently treated with colchicine (20–30 µg per day) for 12 months, all showed histologic regression of cirrhosis. The antioxidant effect of colchicine *in vitro* was evident only at very high concentrations compared to other plasma antioxidants.

Although covalent binding of $^{14}CCl_4$ to hepatic Cd,Zn-metallothionein-II was detected following incubation in the presence of a *rat* metallothionein hepatic microsomal bioactivation system (Suntres and Lui 1990), it did not account for the CCl_4-induced loss of metallothionein thiol groups for the following reasons:

- prior oxidation of sulphydryl groups of metallothionein by hydrogen peroxide did not alter the binding;
- anaerobiosis did not alter the extent of covalent binding but obliterated the inhibitory effect of CCl_4 on metallothionein thiol content.

Measurement of the thiol content of CCl_4-treated metallothionein after treatment with 1,4-dithiothreitol revealed that all the thiol groups that were lost subsequent to CCl_4 treatment could be regenerated.

Sulphaguanidine (2 % in the diet) completely protected centrilubular hepatocytes of *mice* from CCl_4-incuced necrosis (Leduc 1973). Transient changes occurred which resembled those in unprotected *mice*. Distention of cisternae of the rough endoplasmic reticulum began similarly but persisted 12 h or more; then normal configuration of RER membranes was restored beginning at 18 to 24 h. Associated ribosomes were partially lost. Glucose-6-phosphatase activity was wiped out through 12 h with some restoration evident at 18 h. [^{3}H]-Leucine incorporation was inhibited through 6 h with gradual restoration beginning at 12 h and sometimes complete at 24 h. Peripheral hepatocytes remained normal in structure and glucose-6-phosphatase activity but [^{3}H]-leucine incorporation was heightened throughout this period.

Previous dietary treatment with selenium (1 ppm) decreased the hepatic content of malonic dialdehyde of the liver of *rats* fed CCl_4 (2.5 ml/kg body weight) to nearly the normal values (Benedetti et al. 1974).

A significant reduction ($P < 0.05$) by about 34 % of lipid peroxidation products was observed in erythrocytes of *rats* treated simultaneously with CCl_4 and kolaviron, a biflavonoid fraction of the defatted alcoholic extract of *Garcinia kola* Heckel (family *Guttiferae*) seed, when compared to CCl_4-treated rats (Adaramoye and Akinloye 2000).

Vinyl fluoride, **vinylidene fluoride**, and **fluroxene** were found to cause major losses of cytochrome P450 when incubated with microsomes from phenobarbital-pretreated *rats* in the presence of NADPH (Ortiz de Montellano et al. 1982). A high yield of green pigment was isolated from the livers of phenobarbital-pretreated *rats* exposed to vinyl fluoride, vinylidene fluoride, and fluroxene.

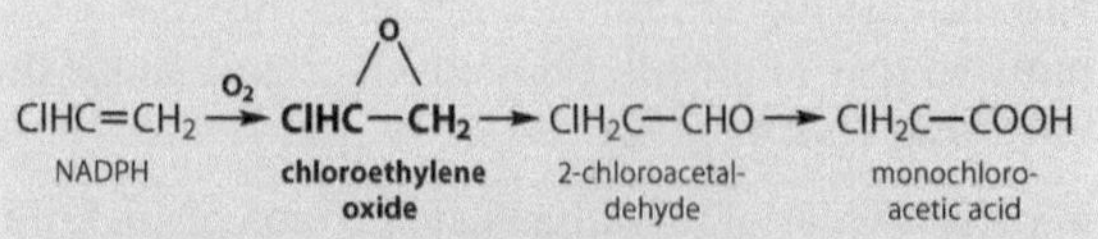

Chloroethylene oxide as the reactive toxic intermediate in the biotransformation of vinyl chloride [296]

Liver biopsies taken from 15 workers at a polyvinylchloride producing plant – dependent on the duration of exposure – showed focal hydropic swelling of the hepatocytes, disseminated toxic steatosis, ample lipofuscin deposits, paracrystalline inclusions in enlarged mitochondria, focal cytoplasmic degradation, and occasionally necrotic cells (SCHATTENBERG et al. 1977).

Tetrachloroethylene, or perchloroethylene, C_2Cl_4, has considerable industrial use (about 10^8 kg annually in the United States alone) and is of toxicological interest because of a variety of effects. Most of the existing literature presents tetrachloroethylene oxide as a critical intermediate in the oxidative metabolism of tetrachloroethylene to Cl_3COOH, oxalic acid, and products covalently bound to proteins, including trichloroacetyl derivatives of lysine (YOSHIOKA et al. 2002). A tetrahedral Fe^{III}-O-tetrachloroethylene oxide P_{450} intermediate can collapse via Cl^- migration to yield trichloroacetylchloride or via formation of tetrachloroethylene oxide, with the former being favoured.

Bromodichloromethane was metabolised by recombinant *rat* and *human* cytochrome P_{450} (ALLIS and ZHAO 2002). In *rat*, CYP2E1, CYP2B1/2 and CYP1A2 were the principal metabolising enzymes and CYP3A1 may also have some activity. In *human*, CYP2E1, CYP1A2 and CYP3A4 showed substantial activity and CYP2A6 also measurably metabolised bromodichloromethane. In both species CYP2E1 is the low K_m isoenzyme, with K_m approximately 27-fold lower than those for the isoenzymes with the next lowest K_m.

Hypoxia was found to promote hepatotoxicity of **halothane** (EL-BASSIOUNI et al. 1998). When oxygen concentration is low, the reduced P_{450} is able to donate its electron to an alternate electron acceptor producing a carbon radical that may be released to initiate lipid peroxidation (VAN DYKE et al. 1988, WHITE 1994). The C–Br bond is the weakest carbon-halogen bond in halothane. Formation of chlorodifluoroethane from the radical may be catalysed by cytochrome P_{450} donating an electron to remove fluorine as F^-. The $F_3CC^{\bullet}HCl$ radical might also combine with oxygen to form a reactive peroxyl radical, $F_3CCO_2^{\bullet}HCl$.

A time-dependent depletion of GSH and vitamin E in liver and plasma and ascorbic acid in the liver of the *dog* accompanied by a simultaneous increase in the levels of malondialdehyde were in the following order: halothane + hypoxia > hypoxia > halothane + oxygen > oxygen (EL-BASSIOUNI et al. 1998). The greatest depletion was observed for vitamin E and the least for ascorbic acid. Hypoxia resulted in inhibition of liver superoxide dismutase activity.

Ornithine decarboxylase (EC 4.1.1.17) activity was elevated in the *rat* liver following exposure to halothane, enflurane and isoflurane (VAN DYKE et al. 1982).

1,1-Dichloro-1-fluoroethane (HCFC-141b) has been developed as a substitute for ozone-depleting chlorofluorocarbons. Anaerobic incubations of liver microsomes from pyridine-induced rats with HCFC-141b in the presence of the spin-trapping agent *N*-*t*-butyl-α-phenylnitrone resulted in the formation of a typical ESR radical signal (ZANOVELLO et al. 2001). In the presence of HCFC-141b, a dose-dependent formation of conjugated dienes was observed that was partially inhibited by *N*-*t*-butyl-α-phenylnitrone, glutathione (GSH) and ascorbic acid. HCFC-141b increased the release of lactate dehydrogenase (EC 1.1.1.27) and the depletion of cellular glutathione in isolated *rat* hepatocytes under both normoxic and hypoxic conditions. HCFC-141b-dependent cytotoxicity was completely prevented by *N*-*t*-butyl-α-phenylnitrone under both conditions and it was partially prevented under normoxic conditions by the broad-spectrum P450 inhibitor metyrapone, the P4502E1 specific inhibitor 4-methylpyrazole and the P4503A-specific inhibitor troleandomycin.

1,1-Dichloro-2,2,2-trifluoroethane (HCFC-123) has also been developed as a substitute for ozone-depleting chlorofluorocarbons. Besides trifluoroacetic acid, chlorodifluoroacetic acid and inorganic fluoride were identified as products of the enzymatic oxidation of HCFC-123 in *rat* and *human* liver microsomes by ^{19}F-NMR and mass spectrometry (URBAN et al. 1994). These metabolites were not detected in incubations with halothane. HCFC-123 and halothane were transformed by liver microsomes from untreated *rats* at low rates. Microsomes from ethanol- and pyridine-treated *rats* metabolised both HCFC-123 and halothane at much higher rates. In *human* liver microsomes, rabbit anti-rat P450 2E1 IgG recognised a single protein band corresponding in apparent molecular weight to *human* P450 2E1.

Incubation of **hexafluoropropene** (1 mM) with microsomes and cytosol in the presence of glutathione (GSH) yielded *S*-(1,2,3,3,3-pentafluoropropenyl) glutathione and *S*-(1,1,2,3,3,3-hexafluoropropenyl) glutathione as identified by thermospray mass spectrometry and ^{1}H-NMR (KOOB and DEKANT 1990).

Pentachlorobenzene is part of a small class of chlorobenzenes consisting of 12 different isomers. These isomers display a range of physicochemical properties that are specifically related to the degree of chlorination and account for the broad range of application for chlorinated benzenes in both home and industry (PEATTIE and LINDSAY 1984). THOMAS et al. (1998) reported that pentachlorobenzene significantly promotes the growth of glutathione S-transferase π foci in the medium-term liver foci bioassay.

Glutathione S-transferase (EC 2.5.1.18)

RX + Glutathione ⟶ HX + R–S–G [129]

Using pentachlorobenzene as a promoter and diethylnitrosamine as an initiator, THOMAS et al. (2000) applied two types of models for describing the time-course models in glutathione S-transferase π foci. The two cell model, which was parameterized to describe a negative selection mechanism, produced adequate simulations of both the site and number of foci. This model-based analysis suggested that the differences between pentachlorobenzene-treated and untreated *rats* were primarily in parameters involving the rates of cell death.

Lipid peroxidation in the liver occurred in male NMRI albino *mice* intoxicated intragastrically with **bromobenzene** (15 mmol /kg body weight) mixed with two volumes of mineral oil (CASINI et al. 1986). Serum glutamate pyruvate transaminase activity was elevated when the hepatic GSH depletion reached a threshold value. Elevation of both serum glutamate pyruvate transaminase and malondialdehyde contents of the liver was observed after 9 h, even though at 9 h the GSH content was already decreased by 84 %, as compared with that of glucose-fed nonintoxicated animals.

2,3,7,8-Tetrachlorodibenzo-*p*-dioxin (30 nmol/kg) altered both the immunofluorescence staining intensity and distribution of cytochrome P_{450} isozymes of the *rabbit* liver (DEES et al. 1982). $P450_{LM4}$ and $P450_{LM6}$ stained intensely throughout the liver lobules, in contrast to their uneven distribution in the normal livers, while $P450_{LM2}$ was markedly reduced.

Tetrachloro-1,4-benzoquinone, a carcinogenic metabolite of the wood-preservative pentachlorophenol, when incubated with the spin-trapping agent 5,5-dimethyl-1-pyrroline *N*-oxide (DMPO) and H_2O_2 produced the DMPO/HO• adduct (ZHU et al. 2002). The formation of DMPO/HO• was markedly inhibited by the HO• scavenging agents dimethyl sulphoxide (DMSO), ethanol, formate, and azide, with the concomitant formation of the characteristic DMPO spin trapping adducts with $^{\bullet}CH_3$, $^{\bullet}CH(CH_3)OH$, $^{\bullet}COO^-$, and $^{\bullet}N_3$, respectively. The formation of DMPO/HO• and DMPO/$^{\bullet}CH_3$ from tetrachloro-1,4-benzoquinone and H_2O_2 in the absence and presence, respectively, of DMSO was inhibited by the trihydroxamate compound desferrioxamine, accompanied by the formation of the desferrioxamine-nitroxide radical. In contrast, DMPO/HO• and DMPO/$^{\bullet}CH_3$ formation from tetrachloro-1,4-benzoquinone and H_2O_2 was not affected by the nonhydroxamate iron chelators bathophenanthroline disulphonate, ferrozine, and ferene, as well as the copper-specific chelator bathocuproine disulphonate.

The photoreduction of **benzophenone** triplets in micellar solution leads to the generation of isolated radical pairs, the behaviour of which resembles that of biradicals (SCAIANO et al. 1982). Radical pair decay is controlled by intersystem crossing and by radical exit from the micelle. 2-Hydroxy-4-methoxybenzophenone, a commonly used sunscreen, was neither genotoxic in the *Drosophila* somatic mutation and recombination test nor clastogenic in the cytogenetic assay in *rat* bone marrow cells (ROBISON et al. 1994).

The incubation of freshly isolated *rat* hepatocytes with benzophenone (0.25–1 mM) elicited a concentration- and time-dependent cell death, accompanied by loss of intracellular ATP and depletion of adenine nucleotide pools (NAKAGAWA et al. 2000). Benzophenone at a low-toxic level (0.25 mM) in the hepatocyte suspension was converted to benzhydrol, *p*-hydroxybenzophenone and its sulphate conjugate, without marked loss of cell viability. The amounts of benzhydrol and sulphate conjugate increased with time. Since benzophenone was essentially stable in Krebs-Henseleit buffer under carbogen flow during a 3-h incubation and did not elicit an increase in oxygen consumption which indicates that autoxidation through the formation of superoxide anion does not occur readily (NAKAGAWA and MOLDÉUS 1992), these results support that the production of metabolites is due to enzymatic reactions in *rat* hepatocytes.

2,2',4,4'-Tetrahydroxybenzophenone, 3,4,5,2',3',4'-hexahydroxybenzophenone, and 4,4'-dihydroxybenzophenone displayed an inhibitory effect on xanthine oxidase with an order of activity of IC_{50} = 47.59, 69.40 and 82.94 µM, respectively (SHEU et al. 1999). The apparent inhibition constants (Ki) of 3,4,5,2',3',4'-hexahydroxybenzophenone and 4,4'-dihydroxybenzophenone were 15.61 and 64.86 µM, respectively, and both of the induced mixed-type (non-competitive-uncompetitive) inhibitions of the substrate xanthine.

***t*-Butylhydroperoxide** induced swelling of isolated *rat* liver mitochondria only in the presence of exogenous Ca^{2+} (>2 µM), whereas diamide was

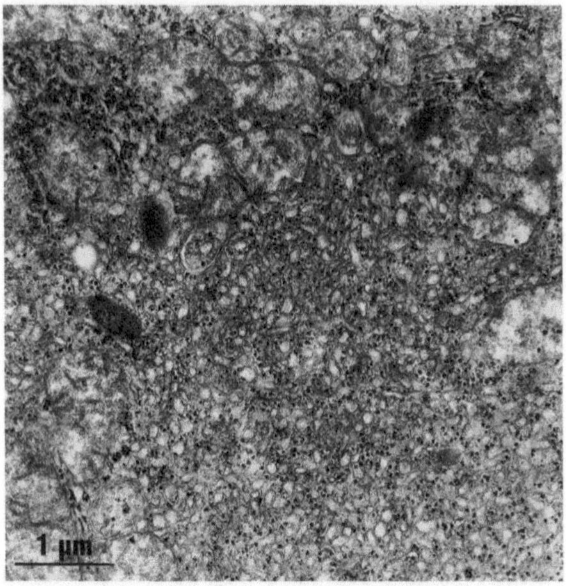

Fig. 319. Smooth endoplasmic reticulum induced by daily oral application of 60 mg 2-(5-cyano-N,3,4,-trimethyl-3-aza-pentane-amido)-5-chloro-2'-fluoro-benzophenone HCl per kg body weight in the liver (block 2336_2) of a male *dog* (No. 119). Fixed by immersion with 2.5 % glutaraldehyde in 0.1 M sodium cacodylate buffer (pH 7.4). Postfixation with 1 % osmium tetroxide in sodium cacodylate buffer. Embedded in Epon 812 and sectioned at 50 nm. Lead citrate and uranyl acetate. Film 635-31

effective in its absence (KUSHNAREVA and SOKOLOVE 2000). In the absence of exogenous inorganic phosphate (P_i), both prooxidants caused a collapse of the membrane potential ($\Delta\Psi$) that preceded the onset of mitochondrial swelling.

Diethyl maleate (5 mM) and **ethyl methanesulfonate** (35 mM) treatments rapidly depleted cellular reduced glutathione below detectable levels (1 nM/10^6 cells), and induced lipid peroxidation and necrotic cell death in freshly isolated *rat* hepatocytes (TIRMENSTEIN et al. 2000). In hepatocytes incubated with 2.5 mM diethyl maleate and 10 mM ethyl methanesulfonate, however, the complete depletion of cellular GSH observed was not sufficient to induce lipid peroxidation or cell death. Instead, diethyl maleate- and ethyl methanesulfonate-induced lipid peroxidation and cell death were dependent on increased reactive oxygen species production as measured by increases in dichlorofluorescein fluorescence. The addition of antioxidants (vitamin E succinate and deferoxamine) prevented lipid peroxidation and cell death suggesting that lipid peroxidation is involved in the sequence of events leading to necrotic cell death induced by diethyl maleate and ethyl methanesulfonate.

Thioacetamide (500 mg/kg body weight i.p.) administration to Wistar *rats* significantly increased the activity of caspase-3-like protease in the liver compared to that in the control group (SUN et al. 2000). Thioacetamide caused apoptosis. At 24 h, the concentration of liver lipid hydroperoxides, a mediator of radical reactions, was 2.2 times as high as that of control *rats*. After 12 and 24 h of thioacetamide administration, the liver concentration of vitamins C and E decreased significantly.

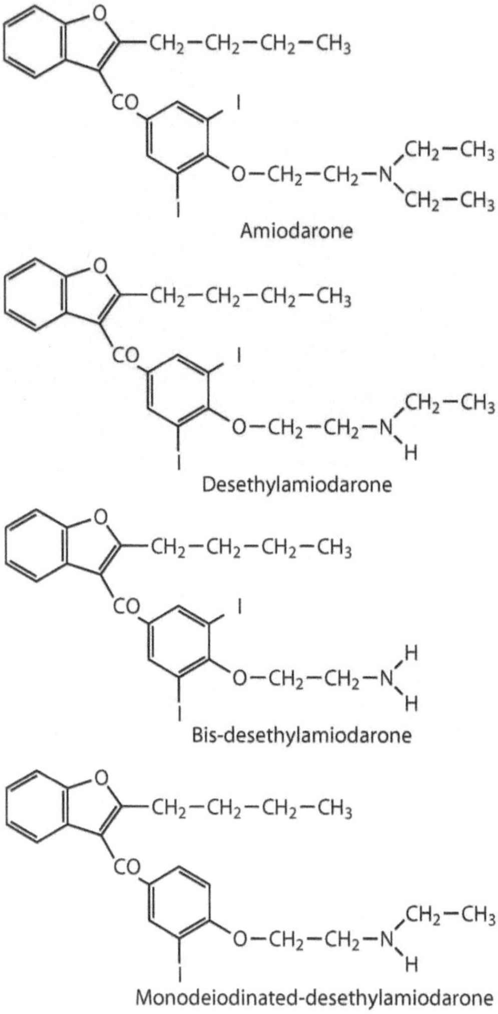

Amiodarone and its derivatives [249–252]

Amiodarone (formula [249]) and its desethyl metabolite desethylamiodarone (formula [250]) induced striking lamellated electron dense cytoplasmic inclusion bodies in *human* hepatocytes and Kupffer cells (ADAMS et al. 1983).

Amiodarone- or desethylamiodarone-induced hepatotoxicity was prevented by DL-α-tocopherol acetate as shown in primary culture of *rat* hepatocytes (RUCH et al. 1991). However, several other antioxidants of diverse nature (Cu,Zn-superoxide dismutase, catalase, *N,N'*-diphenylphenylenediamine, butylated hydroxytoluene, and *N*-acetylcysteine) were unable to prevent amiodarone or desethylamiodorane toxicity. The onset of hepatocyte death by amiodarone or desethylamiodorane was not related to hepatocyte lipid peroxidation. Both drugs inhibited NADPH-dependent *rat* liver microsomal superoxide production in a dose-dependent manner, while **paraquat** increased $O_2^{\bullet-}$ generation in relation to its concentration (0.05, 0.5, and 5 mM). The activation of guanylate cyclase (EC 4.6.1.2) by paraquat was not blocked by KCN, an inhibitor of superoxide dismutase (Vesely et al. 1979). Catalase (EC 1.11.1.6) did not block the paraquat activation of guanylate cyclase. Shimada et al. (2002) examined the paraquat detoxicative system in *mouse* liver. The survival rate of *mice* receiving 50 mg paraquat per kg body weight was 41 % at 7 days and significantly rose to 88, 64, 69 % with pre-treatment with phenytoin, Phenobarbital, and rifampicin, respectively. Phenytoin induced activity in NADPH-cytochrome P450 reductase, CYP3A, CYP2B, and CYP2C that was 3 to 4 times higher than that of the controls. Phenobarbital induced CYP2B and rifampicin induced CYP3A, respectively, in addition to NADPH-cytochrome P450 reductase. Repeated intravenous injections of α-tocopherol to paraquat-loaded *mice* significantly reduced the paraquat mortality and when these *mice* were pre-treated with rifampicin, 100 % of them survived.

2,2'-Azobis(2-amidinopropane) dihydrochloride as a hydrophilic radical-generating azo compound by perfusion (30 min) induced mitochandrial swelling in *rat* hepatocytes in a dose-dependent manner (Yasuda et al. 1995). Both endogenous vitamin E in the membranes and a water-soluble vitamin E analogue, 2-carboxy-2,5,7,8-tetramethyl-6-chromanol added simultaneously with the radical initiator suppressed the hepatic damage. In BALB/c *mice*, 2,2'-azobis(2-amidinopropane) dihydrochloride (100 mg/kg, i.p.) induced a transient decrease in hepatic energy charge, liver mitochondrial respiratory activity, and succinate dehydrogenase (EC 1.3.99.1) activity and a transient increase in blood ammonia level (Saibara et al. 1994).

L-Arginine (540 mg/kg i.v. 1 h before clamping the hepatic hilum, at clamping, at reperfusion, and at 1h and 2 h after reperfusion) reduced damage to liver tissue after ischaemia/reperfusion in a *pig* model (Calabrese et al. 1997). The administration of L-arginine may protect cells against apoptosis. $^{\bullet}$NO formed may inactivate $O_2^{\bullet-}$ radicals (see formula [46]).

L-Arginine–nitric oxide pathway

$$\text{L-Arginine} \xrightarrow{\text{NOS}} \textbf{Nitric oxide}(\text{NO}^{\bullet}) \xrightarrow{+O_2^{\bullet-}} \text{Peroxynitrite } (\text{ONOO}^-) \qquad [46]$$

The incorporation of L-[guanido-^{14}C]arginine into membranes of the rough and smooth endoplasmic reticulum in post-ischaemic *rat* liver showed a lag phase of about 2 h, and was then normally resumed in reversibly injured liver cells (Ferrero et al. 1980).

In vitro, $^{\bullet}$NO showed a biphasic action on oxidative killing of isolated *rat* hepatocytes: low concentrations protected from oxidative killing, while higher doses enhanced killing, and these two effects occurred by distinct mechanisms (Joshi et al. 1999). While low doses of $^{\bullet}$NO from (Z)-1-[*N*-(3-ammonio propyl)-*N*-(n-propyl)-amino]-diazen-1-ium-1,2_2 diolate or *S*-nitroso-*N*-acetyl-L-penicillamine prevented killing of *rat* hepatocytes by *tert*-butylhydroperoxide, further increasing doses resulted in increased killing. Hepatocytes, when stimulated to produce $^{\bullet}$NO endogenously, became resistant to *tert*-butylhydroperoxide killing, indicative of the presence of an NO-triggered antioxidant defence mechanism.

After stimulation with lipopolysaccharide, pentoxifylline potentiated $^{\bullet}$NO production and the expression of inducible nitric oxide synthase (iNOS) in primary *porcine* liver cell cultures (Hoebe et al. 2001). The increased expression of iNOS and concurrent production of $^{\bullet}$NO was also observed when liver cell cultures were incubated with dibutyryl cyclic adenosine monophosphate.

Basolateral membranes isolated from male Wistar *rat* hepatocytes by ultracentrifugation in sucrose gradients incubates with *S*-nitroso-*N*-acetyl-penicillamine or SIN-1 by polarisation of fluorescence showed increased membrane fluidity induced by $^{\bullet}$NO and decreased membrane fluidity induced by peroxynitrite, respectively, in a concentration-dependent manner (Muriel and Sandoval 2000). Na^+/K^+-ATPase (EC 3.6.1.37) activity was reduced by $^{\bullet}$NO or peroxynitrite.

N^{ω}-Hydroxy-nor-L-arginine is a potent competitive inhibitor of *rat* liver arginase (Custot et al. 1997) but neither a substrate nor an inhibitor of purified recombinant iNOS (Moali et al. 1998).

The microsomal P_{450}-catalysed dehydrogenation of valproic acid (2-*n*-propyl-pentanoic acid) and the major serum metabolite 2-*n*-propyl-2-pentenoic acid give 2-*n*-propyl-4-pentenoic acid (Rettie et al. 1987) and 2-*n*-propyl-2,4-pentadienoic acid (Kasahun and Baillie 1993), respectively. Tabatabaei and Abbott (1999) incubating *rabbit* liver microsomes with *human* blood lymphocytes supported the hypothesis that the metabolism-dependent valproic acid-induced *in vitro* cytotoxicity is the result of generation of hydrogen peroxide in the medium that can readily cross cell membranes and subsequently interact intracellularly with iron to produce highly reactive hydroxyl radicals.

Repeated inhalation exposure to **octamethylcyclotetrasiloxane** vapours (≤700 ppm for 6 h per day, 5 days per week) by female Fischer 344 *rats* increased liver-to-body weight ratios (McKim jr. et al. 2001). Hepatic incorporation of 5'-bromo-2-deoxyuridine following exposure was highest on day 6 (labelling index = 15–22 %) and was at or below control (0.05 % phenobarbial in drinking water over a 4-week period) values by day 27.

15.1.11.1.7 Mitochondria

Under normal physiological conditions the mitochondria are the major source of reactive oxygen species in the hepatocyte (Boveris and Cadenas 1997). Approximately 80–90 per cent of the oxygen utilised by hepatocytes is metabolised by mitochondria (Thayer and Rubin 1982) and it is estimated that 2 per cent of the oxygen consumed is converted to superoxide anions (Boveris and Chance 1973).

A recent review by Bailey and Cunningham (2002) is focused on observations, which indicate that the ability of ethanol to increase mitochondrial reactive oxygen production is linked to its metabolism via oxidative processes and/or ethanol-related alterations to the mitochondrial electron transport chain. One of the earliest effect of chronic ethanol consumption on liver is an alteration in mitochondrial structure in which the organelle is often enlarged (see Table 50) and misshapen and contains disrupted cristae.

Kurose et al. (1997) have examined the alteration of mitochondrial membrane potential ($\Delta\Psi m$) in *rat* hepatocytes exposed to ethanol using Rh123 as an indicator of mitochondrial membrane potential. Acute ethanol administration significantly decreased $\Delta\Psi m$ in hepatocytes within 30 min, suggesting that mitochondrial depolarisation is acutely observed as an early event of ethanol-induced hepatocyte injury. This phenomenon is known as the mitochondrial membrane permeability transition induced by the opening of the mitochondrial megachannel, also known as PT pore (Bernardi 1996). In a recent review, Adachi and Ishii (2002) addressed the mechanism of mitochondrial alterations and liver injury induced by ethanol.

Patel and Cunningham (2002) showed in male Sprague-Dawley *rats* that an alteration in the shape of ethanol mitoribosomes is responsible for the reduced sedimentation rate and that the decrease in the translation activity of ethanol mitoribosomes is associated with the dissociated particles. Both the shape change and increased dissociation may arise from posttranscriptional modification in ribosomal proteins.

Mitochondrial reduced glutathione (GSH) plays an important role in the maintenance of cell functions and viability by metabolism of oxygen free radicals generated by the respiratory chain. Marí et al. (2002) showed that the ability of pyruvate to protect against a loss of GSH in HepG2 cells treated with buthionine sulphoximine (BSO), a specific inhibitor of the rate-limiting enzyme in GSH synthe-

Table 50. Giant mitochondria in pancreas, liver, adrenal, and kidney

Pancreas in kwashiokor	Blackburn and Vinijchaikul (1969)
Liver and adrenals of protein-deficient *rats*	Svoboda and Higginson (1964), Svoboda et al. (1966)
Liver of riboflavin-deficient *mice*	Tandler et al. (1968)
Liver in a *child* with sialuria	Dupont et al. (1967)
Liver in alcoholics	Albot et al. (1968), Albot and Parturier-Albot (1969, 1971), Bianchi et al. (1973), Stewart and Dincsoy (1982)
Liver of morbidly obese *patients*	Friedman et al. (1977)
Human liver in leptospirosis icterohaemorrhagica	Sandborn et al. (1966)
Human liver surrounding a large but well-differentiated hepatoma	Ghadially and Parry (1966), Ghadially (1997)
Biopsy specimens of livers containing metastatic carcinoma	Albukerk and Duffy (1976)
Human and *rat* livers in hypoxia-hypercapnia	Dupont et al. (1967)
Reperfused isolated *rat* liver	Ellermann et al. (1992)
Rat hepatocytes cultured in the presence of 2 mM hydrazine	Karbowski et al. (1997)
Liver in hydrazine-fed *rats*	Wakabayashi et al. (1987)
Liver in hydrazine-fed *mice*	Wakabayashi et al. (1983)
Liver in 2,4-dinitrophenylhydrazine-fed *mice*	Wakabayashi et al. (1984)
Liver in cuprizone-fed *mice*	Suzuki and Kikkawa (1969), Albring et al. (1973, 1975)
Liver in ethidium bromide (50 μg/g) injected *mice*	Albring et al. (1973)
Liver of chloramphenicol-fed *mice*	Matsuhashi et al. (1996)
Kidney in nephrotic syndrome	Thoenes (1966)
Proximal nephron in glomerulonephritis	Schuurmans Stekoven and v. Haelst (1970)
Proximal nephron in cyclosporin nephropathy	Waldherr et al. (1982), Mihatsch et al. (1989)

sis, required a functional GSH pool, especially in the mitochondrial compartment, and that in the absence of GSH, pyruvate increased cell injury by damaging the mitochondria.

Endogenous formation of reactive oxygen species was markedly increased in catalase-inhibited or GSH-depleted isolated *rat* hepatocytes (SIRAKI et al. 2002). Respiratory chain inhibitors or hypoxia markedly increased the formation of reactive oxygen species before cytotoxicity ensued.

Cup-shaped and elongated mitochondria with **longitudinal cristae** appeared in hepatocytes of male Sprague-Dawley *rats* fed chow with cadmium (MILLER et al. 1974). These were not seen with semipurified diets and cadmium.

Hepatic **megamitochondria** were reversibly induced by feeding *rats* and *mice* diets containing a wide spectrum of ammonia derivatives (WAKABAYASHI et al. 1984). Ammonia derivatives with electron releasing groups, such as hydrazine, phenylhydrazine, hydroxylamine and aniline were effective, while derivatives with electron withdrawing groups, such as formamide, sulfamic acid and acetamide were ineffective.

Electron-releasing groups in the chemical structure of ammonia derivatives were linked with the formation of megamitochondria (WAKABAYASHI et al. 1984, 1987).

$$N_2H_5^+ + 2\,H_2O \longrightarrow 2\,{}^{\cdot}NO + 9\,H^+ + 8\,e^- \quad [297]$$

Compositions of fatty acids in hydrazine-treated mitochondrial membranes showed decreases in relative amounts of palmitic (16:0) and stearic (18:0) acids and increases in those of oleic (18:1) and linoleic (18:2) acids (WAKABAYASHI et al. 1987). Among the relative amounts of phospholipid species, the increases in amounts of phosphatidylinositol, phosphatidylserine, and phosphatidylethanolamine were observed. A ratio of phosphatidylethanolamine vs. phosphatidylcholine was also increased.

Hydrazine markedly inhibited *rat* liver extramitochondrial glutamic-oxaloacetic transaminase activity, but caused delayed inhibition of intramitochondrial glutamic-oxaloacetic transaminase activity (STEIN et al. 1971). Both extra- and intramitochondrial glutamic-oxaloacetic transaminases were inhibited by hydrazine *in vitro*. Pyridoxal phosphate and pyridoxamine phosphate partially reversed this inhibition when added directly to the assay mixture. However, the *in vivo* administration of pyridoxal phosphate, pyridoxine, pyridoxamine, and pyridoxal failed to reverse the hydrazine inhibition of either extra- and intramitochondrial glutamic-oxaloacetic transaminase activity.

15.1.11.1.8 Peroxisomes

Catalase (EC 1.11.1.6), a peroxisomal marker enzyme, was found by immuno-electron microscopy in *guinea pig* hepatocytes not only in peroxisomes, but also in the cytoplasm (BEIER et al. 1988). BULITTA et al. (1996) have been able to distinguish in *guinea pig* liver homogenates between the cytosolic catalase and that part of the enzyme activity which is due to leakage of the enzyme from peroxisomes by adding 4 % polyethylene glycol to the homogenisation medium. This approach revealed that approximately 40 % of the catalase activity and almost all of the α-hydroxy-acid oxidases are peroxisomal, while 60 % of catalase is of genuine cytosolic origin. The cytosolic catalase exhibited a slightly higher M_r (≈1000) and a less acidic p*I* than the peroxisomal enzyme. Immunoprecipitation with an antibody against *guinea pig* catalase followed by high-resolution polyacrylamide gel electrophoresis revealed two polypeptide bands differing slightly in M_r.

Very-long-chain fatty acids as erucic acid (Δ^{13}-docosenoic acid, C22:1), polyunsaturated fatty acids, methyl-branched fatty acids, dicarboxylic fatty acids, prostaglandins, and the cholesterol side chain in bile acid synthesis are preferentially or exclusively oxidised in peroxisomes. Peroxisomal β-oxidation starts with introduction of a Δ2,3-double bond catalysed by acyl-CoA oxidase, which consumes O_2 and produces H_2O_2 (FOERSTER et al. 1981).

YOKOTA et al. (1995) described the formation of autophagosomes in *rat* hepatocytes during the degradation of excess peroxisomes induced by di-(2-ethylhexyl)-phthalate.

Rat liver peroxisomal *trans*-prenyltransferase activity increased linearly when the peroxisomal protein concentration was increased up to 0.1 mg protein in the incubation medium (TEKLE et al. 2002). The enzyme has an absolute requirement for dival-

Table 51. Hepatic peroxisomal H_2O_2-generating systems

Glycolate oxidase	EC 1.1.3.1	MCGROARTY et al. (1974)
L-α-Hydroxy acid oxidase	EC 1.1.3.α	MCGROARTY et al. (1974)
D-Amino acid oxidase	EC 1.4.3.3	DABHOLKAR (1992)
Urate oxidase	EC 1.7.3.3	YAMAMOTO and FAHIMI (1987), PILL et al. (1992)

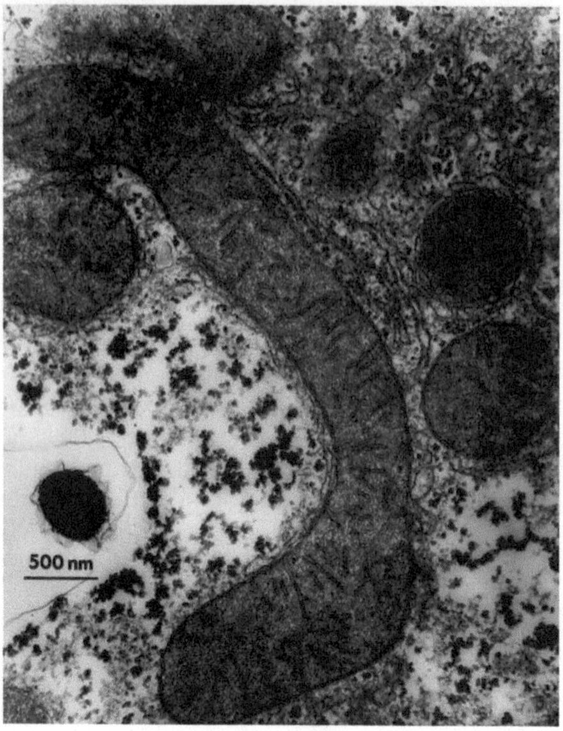

Fig. 320. Hepatocellular (block 668) peroxisome containing eccentrically placed nucleoid material. Unmedicated male 200 g *rat*. Under pentobarbital anaesthesia (30 mg/kg), the animal was perfused from the abdominal aorta with 2.5 % glutaraldehyde in 0.1 M sodium cacodylate buffer (pH 7.4). Postfixation with 1 % osmium tetroxide in sodium cacodylate buffer. Embedded in Epon 812 and sectioned at 50 nm. Lead citrate and uranyl acetate. Plate 3158

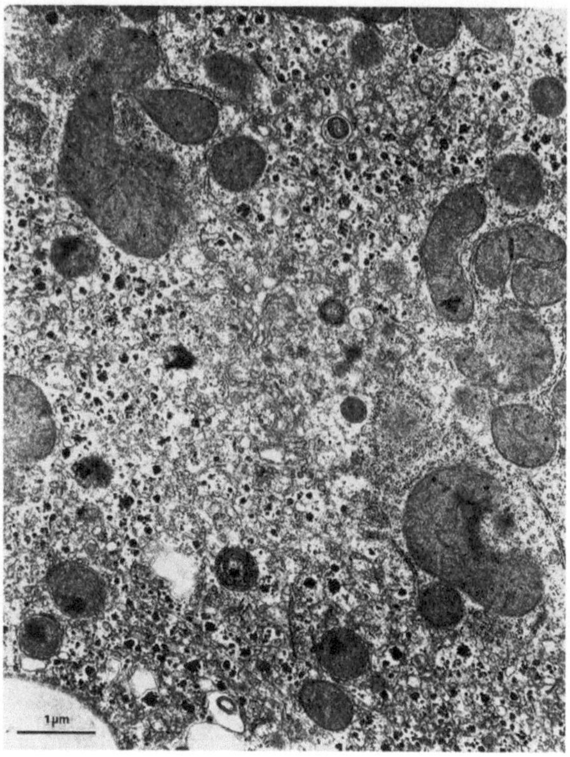

Fig. 321. Several peroxisomes containing eccentrically placed nucleoid material in a hepatocyte (block 492) from a female Wistar *rat* (No. 182/103 E) medicated daily for 5 months with oral application of 100 mg *N*-benzhydryl-*N'*-p-hydroxybenzyl-piperazine HCl per kg body weight × day. Under pentobarbital anaesthesia (30 mg/kg), the animal was perfused from the abdominal aorta with 2.5 % glutaraldehyde in 0.1 M sodium cacodylate buffer (pH 7.4). Postfixation with 1 % osmium tetroxide in sodium cacodylate buffer. Embedded in Epon 812 and sectioned at 50 nm. Lead citrate and uranyl acetate. Plate 3025

ent cations, and in the absence of cations no activity was observed. The peroxisomal transferase activity was maximally activated by 0.1 mM Mn^{2+}, while 0.5 mM completely inhibited the enzyme.

Kovacs et al. (2002) reviewed some of the recent findings related to the localization of cholesterol biosynthetic enzymes in peroxisomes and discussed the impairment of cholesterol biosynthesis in peroxisomal deficiency disease.

A specific hepatic microsomal isoenzyme of cytochrome P_{450} (termed CYP4A1 or P452) with a narrow substrate specificity for ω-hydroxylation of fatty acids was induced by peroxisome proliferators clofibrate (Gibson et al. 1982, 1990, Tamburini et al. 1984, Gibson 1992) and phthalates.

Linear dose-response relationships were observed for induction of cyanide-insensitive palmitoyl CoA oxidation used as an enzyme marker of peroxisome proliferation, by di(2-ethylhexyl) adipate, 2-ethylhexanol and 2-ethylhexanoic acid in *rats* and *mice* (Keith et al. 1992). On a molar basis, di(2-ethylhexyl) adipate was twice as potent as 2-ethylhexanol or 2-ethylhexanoic acid which were equipotent.

In liver tissue from *rat* and *guinea pig* specific immunolabelling of alkyl-dihydroxyacetonephosphate synthase (EC 2.5.1.26) involved in the biosynthesis of ether phospholipids was detected predominantly on the luminal side of the peroxisomal membranes (Biermann et al. 1999).

Peroxysome proliferator-activated receptors are ligand-activated transcription factors of the nuclear hormone receptor superfamily. Their α isoform regulates the transcription of target genes involved in lipid metabolism (Schoonjans et al. 1996). Long-chain fatty acids and eicosanoids were even more potent ligands for *human* peroxysome proliferator-activated receptors than the hitherto most potent peroxysome proliferator-activated receptor-α ligand WY-14643, [4-chloro-6(2,3-xylidino)2-pyridinylthio]acetic acid (Murakami et al. 1999).

BR-931 [4-chloro-6(2,3-xylidino)2-pyridinylthio-(*N*-β-hydroxyethyl)-acetamide], a potent inducer of liver peroxisomes and of mitochondrial carnitine acetyltransferase, appeared to be a hypolipidemic agent of high efficacy and low toxicity for the clinical treatment of hyperlipidaemias and atherosclerosis (SIRTORI et al. 1978).

Classical responses to peroxisome proliferators, cyanide-insensitive acyl-CoA oxidase activity and increased 12-hydroxylation of lauric acid, were elevated in a dose-related manner in male B6C3F1 *mice* maintained on trichloroacetate and clofibric acid (positive control), but not with dichloroacetate, dibromoacetate or bromochloroacetate (PARRISH et al. 1996).

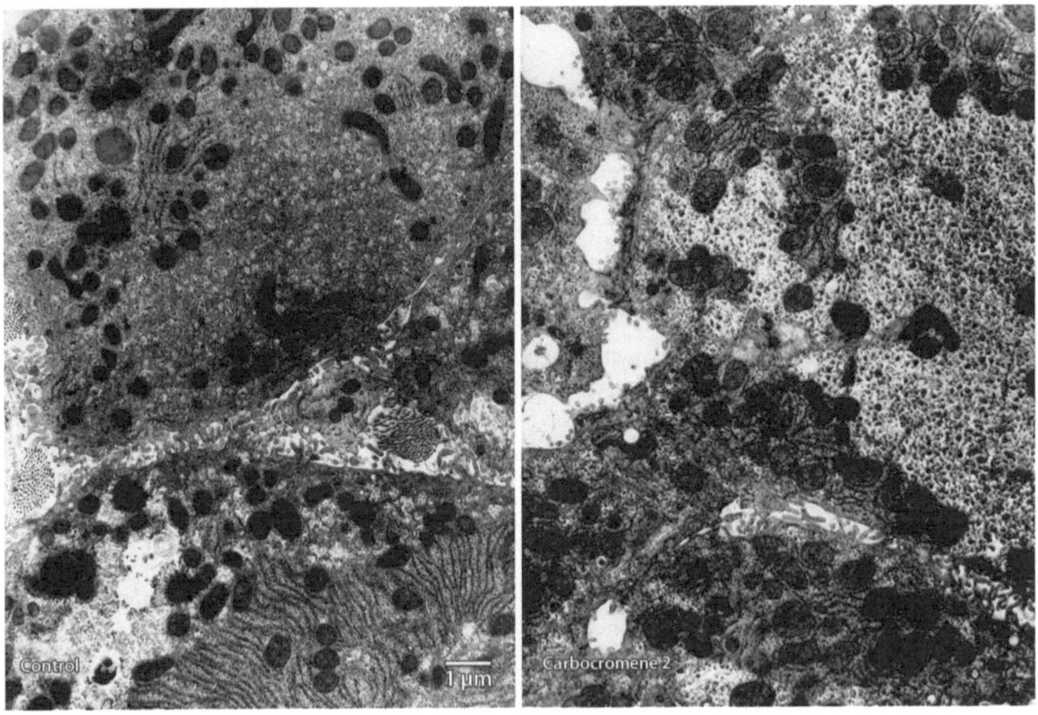

Fig. 322. Glyconeogenesis 2 h after an oral application of 30 mg carbocromene per kg (*rat* No. 6, block 2750; right panel) and aq. dest. control showing smooth endoplasmic reticulum (*rat* No. 19, block 2815; left panel). Under pentobarbital anaesthesia (30 mg/kg), the *rats* were perfused from the abdominal aorta with 2.5 % glutaraldehyde in 0.1 M sodium cacodylate buffer (pH 7.4). Postfixation with 1 % osmium tetroxide in sodium cacodylate buffer. Embedded in Epon 812 and sectioned at 50 nm. Lead citrate and uranyl acetate. Films 667-17 (control) and 659-45 (carbocromene)

Fig. 323. Circadian variations of hepatic glycogen in per cent of the total area after an oral application of 30 mg carbocromene per kg body weight and aq. dest. in the control *rats*. Analyses were performed using a Leitz Classimat®

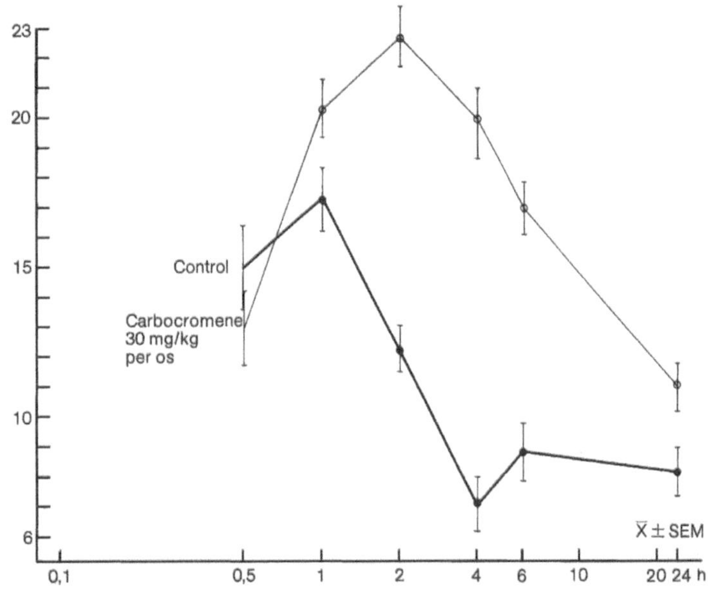

Prolonged treatment of obese and diabetic *mice*, but not of lean control *mice*, with the selective peroxisome proliferator-activated receptor γ ligands and activators, thiazolidinediones, including troglitazone, rosiglitazone, or pioglitazone, resulted in the development of severe hepatic centrilobular steatosis (Boelsterli and Bedoucha 2002).

15.1.11.1.9 Coumarin Derivatives

Centrilobular hepatic necrosis by single doses of coumarin (1,2-benzopyrone, *cis-o*-coumarinic acid lactone) have been reported in the *rat* (Lake 1984, Lake et al. 1989, Fentem et al. 1992), whereas chronic administration resulted in bile duct lesions (Hagan et al. 1967, Cohen 1979, Evans et al. 1989). The mechanism of acute coumarin-induced hepatotoxicity in the *rat* has been investigated by comparing the effects of coumarin with those of a number of methyl-substituted coumarin derivatives (Lake et al. 1994). Coumarin administration produced dose-related hepatic necrosis and a marked elevation of plasma alanine aminotransferase and aspartate aminotransferase activities. In contrast, non of the coumarin derivatives examined produced either hepatic necrosis or elevated plasma transaminase activities. Coumarin reduced hepatic microsomal ethylmorphine *N*-demethylase and 7-ethoxycoumarin *O*-deethylase activities, whereas one or both mixed function oxidases appeared to be induced by treatment with 3,4-dimethylcoumarin, 4-methylcoumarin, 3-methyloctahydrocoumarin and 4-methyloctahydrocoumarin. These results provides an evidence that acute coumarin-induced hepatotoxicity in the *rat* is due to the formation of a coumarin 3,4-epoxide intermediate.

3-(β-Diethylamino-ethyl)-4-methyl-7-carbethoxy-methoxy-2-oxo-(1,2-chromen) is accumulated in the perinuclear cisterna and the endoplasmatic reticulum (Schiller et al. 1977). Glycogen deposition is augmented (Figs. 317, 318). Glyconeogenesis increased by an activation of glycogen-α-4-glycosyltransferase (EC 2.4.1.11) was demonstrated by the method of Takeuchi and Glenner (1960) as improved by Sasse (1966). Glucose-6-phosphate dehydrogenase (EC 1.1.1.49) activity was significantly increased as demonstrated with thiazolyl blue tetrazolium bromide.

15.1.11.1.10 Reactive Nitrogen Species

When *rat* liver was subjected to ischaemia followed by reperfusion, but not ischaemia alone, an NF-κB complex composed of p50/p65 heterodimer and p50 homodimer was rapidly activated within 1 h and remained elevated for up to 3 h, and then tended to decline after 5 h of reperfusion (Hur et al. 1999). Also, the expression of iNOS mRNA was initiated after 1 h and continued to increase after 5 h of reperfusion during the time course studied. This upregulated **iNOS** mRNA expression coincides with increased iNOS enzyme activity and NF-κB binding activity after hepatic ischaemia/reperfusion. administration of *N*-acetylcysteine (20 mg/kg intravenously 10 min before reperfusion) not significantly inhibited the expression of iNOS mRNA but also blocked upregulated NF-κB binding activity after reperfusion.

3,3',5-Triiodothyronine (0.1 mg T_3/kg *rat* × day) for three consecutive days increased the rate of •NO generation, with significant increases after 2 days (47 %) and 3 days (70 %) of T_3 treatment, and a net 45 % ($P < 0.05$) enhancement in the N^G-monomethyl-L-arginine sensitive •NO production, compared to control values (Fernández et al. 1997). These enhancement effects were reversed to control levels after 3 days of hormone withdrawal, concomitantly with the normalization of hepatic respiration. Enhancement of liver NOS activity in hyperthyroid animals was diminished by 27 % ($P < 0.05$) by the selective *in vivo* inactivation of Kupffer cells by gadolinium chloride ($GdCl_3$), without direct action of $GdCl_3$ on the enzyme.

Whereas nitric oxide produced by constitutive **eNOS** is protective to the liver, •NO produced by iNOS can be either toxic or protective depending on the conditions (Li and Billiar 1999). The availability of selective iNOS inhibitors and *mice* lacking various NOS isoforms made it possible to elucidate the precise role of •NO in the liver.

In carbon tetrachloride-induced liver damage, endogenous •NO protected the *rat* liver from lipid peroxidation, collagen accumulation, and damage (Muriel 1998). One mechanism for the protection against oxidative damage is through the upregulation of heme oxygenase-1 (Kim et al. 1995). This results in the production of the potent antioxidant biliverdin. Direct scavenging of oxygen radicals by •NO is another possibility; however, the ration between •NO and $O_2^{\bullet-}$ is likely to be one important factor.

3-Morpholinosydnonimine-*N*-ethylcarbamide (**molsidomine**) is enzymatically transformed to 3-morpholinosydnonimine (SIN-1), which hydrolyses in a pH-dependent manner to form open-ring chemical compounds such as *N*-morpholino-*N*-nitrosoaminoacetonitrile (SIN-1A), which then decomposes to form nitric oxide and biologically inactive metabolites like SIN-1C (Noack and Feelisch 1989). In an aqueous solution 500 μM SIN-1 were degraded to SIN-1A with a half-time of

126 min, while SIN-1C was formed three to four times more quickly with a half-time of 38 min. Experiments with the thiomorpholinyl analogue of SIN-1, compound C 78-0698, showed that the A form was more rapidly formed than with SIN-1, but it was considerably more stable. The structure of the open ring compound C 78-0652 (formula [129]) is analogue to that of SIN-1A.

In comparison with control groups, significantly higher concentrations (C_{max}, area under the plasma concentration-time curve) of molsidomine were observed after oral administration in 3 studies on a total of 21 patients with impaired liver function (Rosenkranz et al. 1996). The absolute bioavailability was higher in patients with impaired liver function (93 %) than in healthy individuals (44 to 59 %), while clearance was also reduced to about 25 %. In a study by Huber et al. (1992) in which molsidomine was administered intravenously to patient with impaired liver function, the plasma concentration and $t^1/_2$ (3.36 ± 1.74 h) of molsidomine were both markedly increased in comparison with the respective values after intravenous administration of molsidomine to healthy individuals.

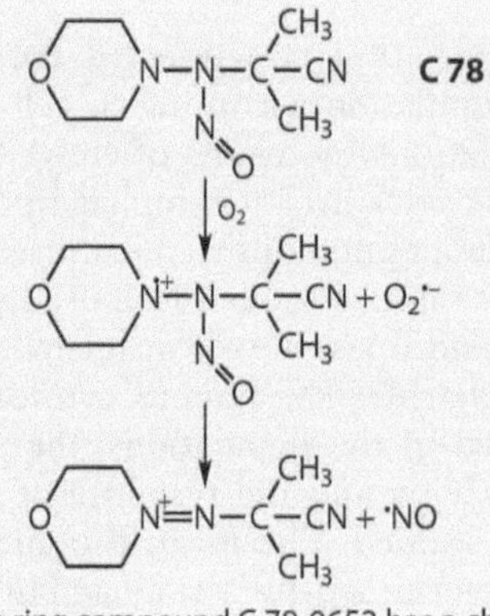

C 78 0652

The open ring compound C 78-0652 has a chemical structure analogue to SIN-1A [136]

The pharmacokinetics and metabolism of molsidomine were profoundly altered In patients with **alcoholic cirrhosis**: after an oral dose of 2 mg, elimination half-life was prolonged from 1.2 ± 0.2 h in 6 healthy male volunteers to 13.1 ± 10.0 h in the 7 patients ($P < 0.01$) because of a decrease in its apparent plasma clearance from 590 ± 73 ml/h × kg in the volunteers to 39.8 ± 31.9 ml/h × kg in the cirrhotic patients (Spreux-Varoquaux et al. 1991). The elimination half-life of SIN-1 was also prolonged in cirrhotic patients 7.5 ± 5.4 h *vs* 1.0 ± 0.19 h, $P < 0.05$).

Isolated *rat* hepatocytes exposed to IFN-γ, TNF-α, IL-1β, and lipopolysaccharide demonstrated a g = 2.04 axial electron paramagnetic resonance signal

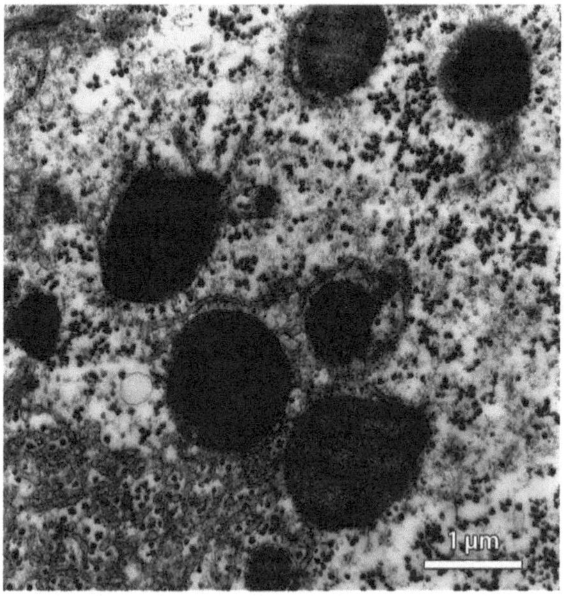

Fig. 324. Smooth endoplasmatic reticulum (bottom left) in a hepatocyte (block 933) of a male beagle *dog* (No. 126) which received 20 mg molsidomine per kg body weight × day per os for 12 months. On June 11, 1976, under intravenous pentobarbital anaesthesia, the animal was exsanguinated. Small pieces of tissue were fixed by immersion in a 2.5 % solution of glutaraldehyde in 0.1 M sodium cacodylate buffer (pH 7.4). Postfixation with 1 % osmium tetroxide in sodium cacodylate buffer. Embedded in Epon 812 and sectioned at 50 nm. Lead citrate and uranyl acetate. Film 636-07

indicative of the formation of nonheme-iron **nitrosyl complexes** (Stadler et al. 1993). Concurrent incubation with L-N^G-monomethylarginine, a competitive inhibition of NOS, prevented the appearance of the signal. The g = 2.04 signal was localized in the cytosolic fraction of hepatocyte extracts. Treatment of hepatocytes with exogenous •NO or nitroprusside generated an identical g = 2.04 signal of much greater intensity than with cytokines plus lipopolysaccharide. Treatment with nitroprusside also caused the appearance of a signal from pentacyanonitrosylferrate ion.

Following biliary obstruction the formation of **nitrotyrosine** increased from 2.00 ± 0.5 ng/mg to 6.3 ± 0.9 ng/mg (Frost et al. 2000). Lipopolysaccharide caused a modest but insignificant increase in protein nitrotyrosine levels in the *rats* with biliary cirrhosis (8.4 ± 2.2 ng/mg) or control animals (3.6 ± 1.3 ng/mg). Chronic treatment with lipoic acid prevented the increase of liver nitrotyrosine following biliary cirrhosis (3.1 ± 0.5 vs 6.3 ± 0.5 ng/mg, $P < 0.05$).

Ischaemic liver may be toxic to weakly metastatic *human* colorectal cancer (CRC) cells on reoxygenation (Jessup et al. 1999). This toxicity involves reactive oxygen and nitrogen species.

15.1.11.2 Ductular Progenitor Cells

When hepatocyte proliferation for restoration of centrolobular injury induced by carbon tetrachloride is inhibited by treatment with *N*-2-acetylaminofluorene, ductular progenitor cells proliferate into the liver lobule (Yin et al. 2002). The neoductular proliferation is accompanied by fibronectin-positive Kupffer cells and desmin-positive stellate cells, which may play critical roles not only in controlling proliferation and differentiation of ductular progenitor cells, but also in re-establishing hepatic cord structure.

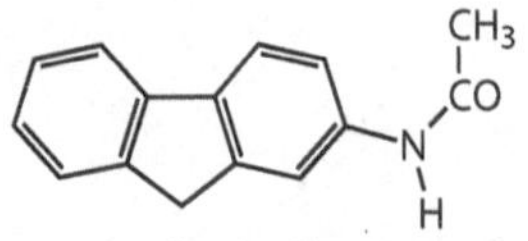

2-Acetylaminofluorene [298]

15.1.11.3 Endothelial Cells

Endothelial cells have been shown to regulate vascular contraction. In circular muscle preparations of the *rat* portal vein with intact endothelium inhibition of •NO synthesis by *N*ω-nitro-L-arginine induced a tonic contraction (Shimamura et al. 2000). The contraction was inhibited by L-arginine, sodium nitroprusside or nifedipine. *N*ω-Nitro-L-arginine did not induce contraction in endothelium-damaged preparations.

Endothelial cells containing large foamy vacuoles were described in a case of Morquio's disease (Wisse et al. 1974). Neoplastic lesions of endothelial cells, angiosarcomas, have been noted in *human* liver and in experimental animals. Several toxic environmental agents, such as vinyl chloride, arsenic, and thorotrast, have been implicated in the pathogenesis of these lesions (Popper et al. 1978).

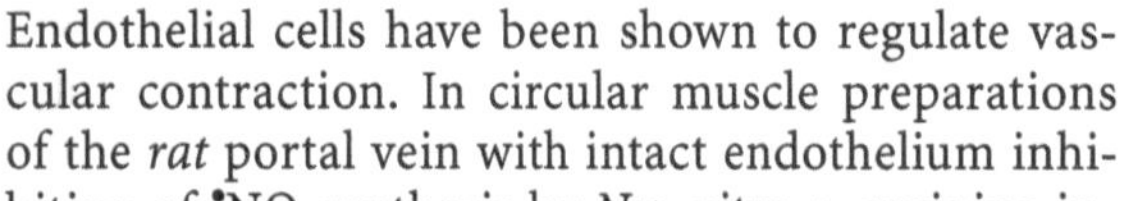

Chloroethylene oxide as the reactive toxic intermediate in the biotransformation of vinyl chloride [296]

15.1.11.4 Kupffer Cells

Kupffer cells, resident macrophages in the liver, constitute the largest population of macrophages in the body and are critical effector cells for the defence against invading micro-organisms and foreign molecules. They are activated on perfusion in a nonrecirculating system with oxygen-saturated Krebs-Henseleit-bicarbonate buffer (pH 7.4) pumped through the *rat* liver via a cannula inserted into the portal vein (Lindert et al. 1992). Upon reflow, colloidal carbon-containing perfusate revealed carbon uptake to be elevated approximately threefold to 234 mg/g × h. Electron microscopy showed numerous pseudopodia and lamellopodia. These changes were absent in controls or in livers under low-flow conditions. Prevention of Kupffer cell activation with methylpalmitate improved survival slightly after orthotopic liver transplantation in the *rat* (Marzi et al. 1991). When Kupffer cells were inactivated by intravenous pre-treatment of *rats* with $GdCl_3$, release of enzymes and malondialdehyde from the isolated liver perfused from the portal vein was reduced significantly by >50 %, and hepatic cell death was almost completely absent (Bremer et al. 1994). A significant difference in spontaneous chemiluminescence between $GdCl_3$-pretreated (10 mg/kg *rat* given intravenously in physiological saline) and unmedicated ischaemia-reperfused liver surface was shown by Curtín et al. (1998).

Daemen et al. (1991) fractionated the *rat* liver macrophage population according to cell size into three subpopulations by means of elutriation centrifugation. The total liver macrophage population and the three subpopulations were cultured and exposed to the immunostimulators muramyl dipeptide, in a free and liposome-encapsulated for, and/ or lipopolysaccharide. The tumour cytotoxic activity thus induced in the populations, the preservation of this activity, and the response of a second stimulus were studied. The *in vitro* induced cytotoxicity was determined by a radioactivity release assay, using C26 colon adenocarcinoma cells, labelled with [*methyl*-^{3}H]thymidine, as target cells. Muramyl dipeptide or lipopolysaccharide readily activated the total macrophage population in maintenance culture to a tumour cytotoxic state during the first 2 days after isolation. Four days after isolation, the activation induced with both muramyl dipeptide and lipopolysaccharide was strongly reduced. The small to intermediate-size macrophages could be activated to tumour cytotoxic activity with muramyl dipeptide for up to 3 days and with lipopolysaccharide for up to 4 days in culture. The large-size macrophages could only by activated up to 2 days in culture with muramyl dipeptide or lipopolysaccharide or both.

Xanthine oxidoreductase (EC 1.1.1.204) activity was detected in vesicles and occasionally on granular endoplasmic reticulum of *rat* Kupffer cells (Fre-

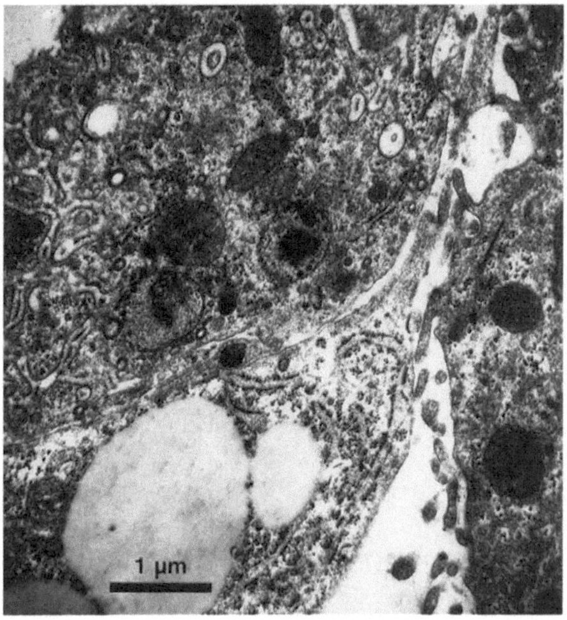

Fig. 325. A Kupffer cell (top) showing many tubular invaginations and a fat storing Ito cell (bottom) in the liver (block 559) of an unmedicated 235 g female *rat*. Under pentobarbital anaesthesia (30 mg/kg), the animal was perfused from the abdominal aorta with 2.5% glutaraldehyde in 0.1 M sodium cacodylate buffer (pH 7.4). Postfixation with 1% osmium tetroxide in sodium cacodylate buffer. Embedded in Epon 812 and sectioned at 50 nm. Lead citrate and uranyl acetate. Film 258/84

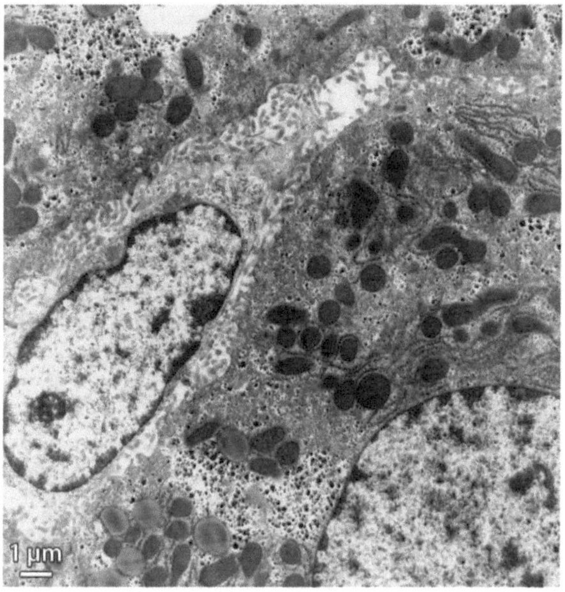

Fig. 326. Kupffer cell (block 1869) from a male *rat* (No. 4) medicated for 12 consecutive days with 600 mg molsidomine per kg body weight × day. 18 days after discontinuance of the medication, on September 21, 1976, under pentobarbital anaesthesia (30 mg/kg), the animal was perfused from the abdominal aorta with 2.5% glutaraldehyde in 0.1 M sodium cacodylate buffer (pH 7.4). Postfixation with 1% osmium tetroxide in sodium cacodylate buffer. Embedded in Epon 812 and sectioned at 50 nm. Lead citrate and uranyl acetate. Film 637-34

DERIKS and VREELING-SINDELÁROVÁ 2002). Xanthine oxidase (EC 1.1.3.22) activity was not found in Kupffer cells and sinusoidal endothelial cells.

Treatment of animals with large doses of retinol stimulated Kupffer cell functions including phagocytic activity and release of tumour necrosis factor and superoxide anion ($O_2^{\bullet-}$) (MOBLEY et al. 1991). OHATA et al. (2000) reported mRNA by *rat* Kupffer cells of retinoic acid receptor and retinoid X receptor subtypes and their binding activities to the retinoic acid responsive element or retinoid X responsive element.

In C57BL/10 *mice* depleted from macrophages by intravenous injection of liposomes containing dichloromethylene diphosphonate (DMDP), acetaminophen (paracetamol) caused less severe early liver damage than in control mice given empty liposomes (GOLDIN et al. 1996). Cytochrome P_{450} converts acetaminophen to *N*-acetyl-*p*-benzoquinone imine by a pathway utilising oxygen and NADPH. This product then undergoes redox cycling to give $O_2^{\bullet-}$ (VAN DE STRAAT 1987, VAN DE STRAAT et al. 1988).

Kupffer cells are involved in the generation of protein adducts with both acetaldehyde and ethanol-induced lipid peroxidation products in alcoholic liver disease (NIEMELÄ et al. 2002).

Intravenously administered particles of beryllium phosphate (50 µM/kg) were endocytosed by *rat* Kupffer cells within 15 min of administration but they were never found in the endothelial or parenchymal cells of the liver (DINSDALE 1982). Endocytosed particles were rapidly degraded within vesicles which were identified as secondary lysosomes.

Kupffer cells are major targets for **lipopolysaccharides**. Hepatic macrophages engulf lipopolysaccharide by phagocytosis (MATHISON and ULEVITCH 1979). Lipopolysaccharide increased $O_2^{\bullet-}$ production in Kupffer cells (BAUTISTA et al. 1990). Additionally, lipopolysaccharide induces migration of activated polymorphonuclear neutrophils into the liver (LEVY et al. 1967). Since these cells are the other source of free radicals, the intensity of polymorphonuclear neutrophil infiltration determines the degree of hepatic endotoxemia (HEWETT et al. 1992).

$^{\bullet}$NO generation by Kupffer cells isolated from male Wistar *rats*, cultured for 48 h *in vitro* and then stimulated with lipopolysaccharide increased 5- to 10 fold over the basal level (GAILLARD et al. 1991). This increase could be further enhanced by prostaglandin E_2 (1 µM) and dibutyryl cAMP (20 µM), especially when added 1 h after lipopolysaccharide. •NO generation after stimulation with lipopolysaccharide or lipopolysaccharide + PGE_2 depended on the simultaneous production of PGE_2 by the stimulated Kupffer cells. It could be partly inhibited by anti-PGE_2 antibody or acetylsalicylic acid. While murine TNF-α did not stimulate •NO synthesis significantly, added PGE_2 raised •NO synthesis about 5-fold. The addition of dibutyryl cAMP to TNF-α in the same concentration as with lipopolysaccharide, however, had no effect.

Ricin in the liver first targets the Kupffer cells which are heavily damaged as early as 4 h after the intravenous inoculation of 6 LD_{100} into *mice* (BINGEN et al. 1987). Four hours after the inoculation of ricin every Kupffer cell bore several worm-like structures, which sometimes took a spindle-like aspect. About 14 % of the completely lysed cells (29 of 208) lay on the endothelial lining within the sinusoid. *In vivo*, Kupffer cells accumulated I-labelled ricin to a much greater extent than parenchymal cells (SKILLETER et al. 1981).

Ricin, a glycoprotein present in the seeds of *Ricinus communis* consists of two subunits joined by a disulphide bond. The A chain subunit or "effectomer" enzymatically inactivates the 60S ribosomal subunit, thus inhibiting protein synthesis (OLSNES and PIHL 1972) and causing cell death. The B chain subunit or "haptomer" serves to bind the ricin molecules to the galactose containing receptors that are present on most eucaryotic cell membranes. Carbohydratebinding specificity of ricin is as follows (VIERBUCHEN 1991):

[Galβ(1–3)GalNAcα1-Ser/Thr]$_4$glycopeptide
> Triantennary oligosaccharides containing N-acetyllactosamine at non-reducing end
> [Galβ(1–3)GalNAcα1-Ser/Thr]glycopeptide
> Galβ(1–3)GalNAcα1-Ser/Thr > GalNAc α1-Ser/Thr.

KAMINSKI et al. (1986) observed that ricin was considerably more toxic to macrophages when conjugated to monoclonal antibody B-6 than unconjugated ricin.

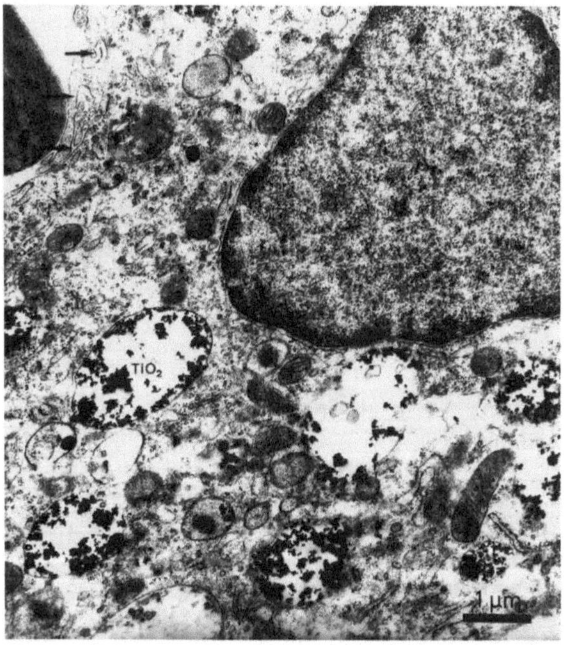

Fig. 327. A Kupffer cell (block 4460) showing several tubular invaginations (arrows) and large phagolysosomes containing anatase from a 194 g Sprague-Dawley *rat* (Charles River, France) injected intraperitoneally with 50 mg anatase (titanium dioxide P 25, Degussa, Frankfort on the Main) suspended in 2 ml saline. 4 days later, on July 25, 1978 under pentobarbital anaesthesia (30 mg/kg), the animal was perfused from the abdominal aorta with 2.5 % glutaraldehyde in 0.1 M sodium cacodylate buffer (pH 7.4). Postfixation with 1 % osmium tetroxide in sodium cacodylate buffer. Embedded in Epon 812 and sectioned at 50 nm. Lead citrate and uranyl acetate. Film 208/80

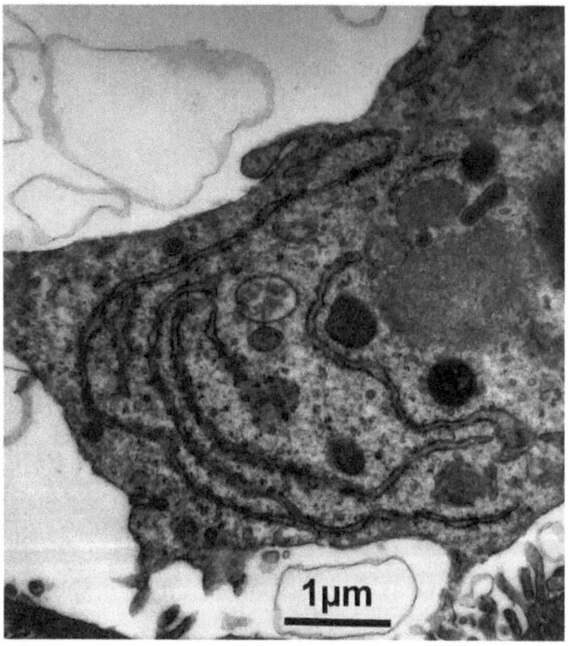

Fig. 328. Multivesicular body in a Kupffer cell (block 4838)) from a male Sprague-Dawley *rat* (Charles River, France) 16 weeks after an activation of the mononuclear phagocyte system by intravenous injection of 0.05 ml complete Freund's adjuvant on two consecutive days. Under pentobarbital anaesthesia (30 mg/kg), the animal was perfused from the abdominal aorta with 2.5 % glutaraldehyde in 0.1 M sodium cacodylate buffer (pH 7.4). Postfixation with 1 % osmium tetroxide in sodium cacodylate buffer. Embedded in Epon 812 and sectioned at 50 nm. Lead citrate and uranyl acetate. Film 220/84

Like other indigestible polymers, **polyvinylpyridine-N-oxide** accumulated in secondary lysosomes, where it can interact with silanol groups of enclosed particles, preventing them from damaging the lysosomal membrane.

Liver biopsies taken from 15 workers at a **polyvinylchloride** producing plant showed an increased number of enlarged and elongated Kupffer cells with long, occasionally lobed condensed nuclei (SCHATTENBERG et al. 1977).

Large numbers of Kupffer cells were found in liver parenchyma surrounding colon **carcinomas** when compared with levels of control livers, but these cells were not activated (GRIFFINI et al. 1996). The low activation status of Kupffer cells both in terms of production of reactive oxygen species and Ia-antigen expression and the absence of significant numbers of pit cells at tumour sites suggested that Kupffer cells and pit cells do not play a significant role in advanced stages of tumour growth.

15.1.11.5 Perisinusoidal Ito Cells

Perisinusoidal Ito cells (fat-storing cells, hepatic lipocytes, vitamin A-storing cells) are the primary matrix-producing cells in liver fibrosis. They are difficult to visualize in standard histological preparations but can be identified in sections of *rat* liver by immunohistochemical detection of the intermediate filament protein desmin; this protein is not expressed by other sinusoidal cells (YOKOI et al. 1984, BURT et al. 1986).

Both *in vivo* and culture studies indicated that lipocytes can undergo activation, a process characterised by increased proliferation and fibrogenesis (FRIEDMAN et al. 1989, GEERTS et al. 1990, MAHER and MCGUIRE 1990). In culture-activated Sprague-Dawley *rat* lipocytes, interferon-γ (1,000 U/ml) significantly inhibited lipocyte proliferation as assessed by [^{3}H]thymidine incorporation assay and nuclear autoradiography (ROCKEY et al. 1992). FAHIMI et al. (1976) demonstrated the association of peroxisomes with lipid droplets in fat-storing cells. A marked activity of γ-glutamyl transpeptidase (EC 2.3.2.2) was reported by TANAKA et al. (1976). A specific lesion of fat-storing cells In *rat* liver induced by the carcinogen N-nitrosomorpholine was noted by BANNASCH et al. (1981). A significant increase in number and size of fat-storing cells has been reported in patients with psoriasis receiving methotrexate (4-amino-10-methyl-pteroylglutamic acid) therapy (HOPWOOD and NYFORS 1976).

Carbon tetrachloride (a single intragastric bolus of 1 ml of 40 per cent CCl_4 in liquid paraffin) induced an expansion of the desmin-positive perisinusoidal cell population, predominantly within the damaged perivenular zones, which reached a peak on days 3 and 4 following administration of CCl_4 (JOHNSON et al. 1992).

FERLUGA and ALLISON (1978) have shown that splenic macrophages play a major role in the pathological process of endotoxaemia-associated hepatic injury. Migrating into endotoxin-damaged liver they cause further hepatic injury. Ligation of the vena lienalis has been found to effectively prevent accumulation of splenic macrophages in the liver and to attenuate the liver dysfunction induced by lipopolysaccharide administration (SHIRATORI et al. 1990). Inhibition of adhesive interactions between splenic macrophages and sinusoidal endothelial cells with specific monoclonal antibodies directed against adhesion molecules also diminished lipopolysaccharide-associated liver injury (LUSTER et al. 1994). WATANABE et al. (2001) incubated splenic macrophages isolated from Wistar *rats* with either lipopolysaccharide or interferon-γ and cocultured them with hepatocytes. NO release and nitrotyrosine expression in hepatocytes increased after 8 h of coculture with activated macrophages, and this coculture also induced increases in the 8-hydroxy-deoxyguanosine/deoxyguanosine ratio and single-stranded DNA in the hepatocytes. The alterations were attenuated by superoxide dismutase (EC 1.15.1.1) and •NO synthesis inhibitors. A similar pattern of alterations was observed in hepatocytes incubated with SIN-1 (1 and 10 mM), and these changes were also prevented by superoxide dismutase.

During culture on uncoated plastic surfaces, fat storing cells loose their normal phenotype and within 1–2 weeks transform into highly productive myofibroblasts, which have the capacity to produce soluble, most profibrogenic mediators stimulating proliferation, transformation and enhanced proteoglycan synthesis of quiescent fat storing cells in a paracrine mode (BACHEM et al. 1992).

Vimentin, desmin, and smooth muscle cell α actin were expressed in all cultures of GRX *murine* hepatic stellate cells (GUMA et al. 2001). Glial fibrillary acidic protein showed a heterogeneous intensity of expression and did not form a filamentous cytoskeletal network, showing a distinct punctuate cytoplasmic distribution. When activated by inflammatory mediators, GRX cells increased expression of desmin and glial fibrillary acidic protein. Retinol-mediated induction of the lipocyte phenotype elicited a strong decrease of intermediate filament protein expression and the collapse of the filamentous structure of the cytoskeleton.

In *human* liver, fat-storing cells were increased after anabolic steroid and corticosteroid therapies, in some stages of alcoholic liver disease, and in extrahepatic biliary obstruction (Bronfenmajer et al. 1966). Rozycka et al. (1978) noted that in 60 per cent of cases of chronic alcoholism secondary magnesium deficiency has been observed as a result of the decrease in the magnesium absorption from the alimentary tract with concurrent increase in its excretion in the urine (McCollister et al. 10957, Pounier 1965).

Attempting to integrate conceptionally present knowledge on the molecular and cellular interaction of fat storing cells in a hypothetic model of cell activation, Gressner and Chunfang (1995) proposed a three-step cascade mechanism of fat storing cell activation, which implies the sequential cross-talk between fat storing cells cells, hepatocytes, Kupffer cells, thrombocytes, endothelial cells and myofibroblasts (transformed fat storing cells). Reactive oxygen species released by Kupffer cells may damage the membranes of both parenchymal and fat storing cells.

15.2 Manganese-Induced Neuronal Cell Injury

Manganese toxicity in *humans* due to chronic inhalation of high concentration or airborne Mn from Mn mines, steel mills, and chemical industries not only causes Mn pneumonia (Baader 1937, Heine 1944, Dervillée et al. 1966, van Beukering 1966), but also affects the brain (Canavan et al. 1934, Peñalver 1957, Cotzias 1974, Donaldson et al. 1982, Donaldson and Barbeau 1985). Chronic Mn intoxication can cause permanent degenerative damage in the nigrostriatal system. Autopsy findings of low dopamine levels, morphologic lesions and the extrapyramidal signs suggest that this type of abnormality is similar to Parkinson's disease (Donaldson et al. 1982). Studies in monkeys and rodents chronically intoxicated with $MnCl_2$ have also demonstrated the depletion of dopamine in the striatum. PC-12 cells widely used as a model for catecholaminergic neurones, grown in the presence of 0.1 mM $MnCl_2$ and 0.1 mM tyrosine with time of exposure showed increased lipid peroxidation and release of lactate dehydrogenase (Sun et al. 1993).

Archibald and Tyree (1987) have demonstrated that Mn^{3+} in a pyrophosphate complex (probably Mn^{3+}[pyrophosphate]$_3$) has the capacity to rapidly destroy the principal brain catecholamines and their precursor, dopa, oxidising them to a best useless and at worst toxic products. In the absence of Mn^{3+}, oxidation of the neurotoxic 6-hydroxydopamine to a stable coloured compound in the O_2-containing reaction mixture took several minutes, while Mn^{3+} accelerated this to completion in <5 s. The observation thar phenylalanine and tyrosine were not oxidised by Mn^{3+}, although their dihydroxy derivatives were, suggested that the site of Mn^{3+} attack on the catecholamines. Indeed, the placement of the hydroxyl groups was all-important in the reactivity with these compounds with Mn^{3+}. Since Mn^{3+} has a similar ionic radius and chelating behaviour to Fe^{3+}, and Fe^{3}_{+} chelates very strongly to catechols (Neilands 1980), it is likely that Mn^{3+} by the dihydroxy moiety is the first step in the oxidative attack of Mn^{3+} upon catecholamines.

Graham et al. (1978) conclude that 6-hydroxydopamine and 2,4,5-trihydroxyphenylalanine kill cells through the production of H_2O_2, $O_2^{\bullet -}$ and $HO^{\bullet}$, while for dopamine and dopa the reaction of quinone oxidation products with nucleophiles probably also contributes to their cytotoxicity. *O*-methylated catecholamines are less susceptible to autoxidation than their nonmethylated precursors (Miller et al. 1996). Melatonin, which has been shown to be a powerful, endogenous hydroxyl radical scavenger (Tan et al. 1993), is capable of scavenging free radicals produced during catecholamine autoxidation (Miller et al. 1996).

Superoxide oxidises catecholamines

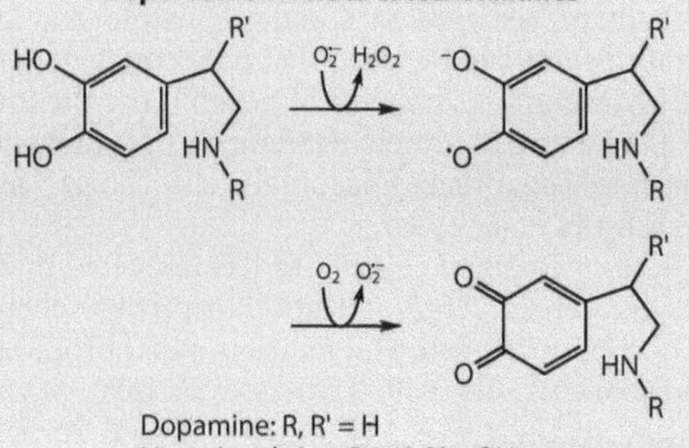

Dopamine: R, R' = H
Norepinephrine: R = H, R' = OH
Epinephrine: R = CH_3, R' = OH
Isoprenaline: R = $CH(CH_3)_2$, R' = OH [107]

O-methylated catecholamines are not only less susceptible to autoxidation than their corresponding non-methylated precursors (Miller et al. 1996), and thus less likely to generate reactive oxygen intermediates, but they are also effective *in vitro* scavengers of $HO^{\bullet}$ (Napolitano et al. 1993, Tan et al. 1994) and are poorly or not reactive toward $^{\bullet}NO$ (D'Ischia and Constantini 1995). The production of $HO^{\bullet}$ by the Fenton reaction was diminished significantly by *O*-methylated catecholamines (*O*-methyldopa, *O*-methyldopamine, *O*-methyltyrosine, and *N*-acetyl-*O*-methyldopamine), whereas radical production was augmented by dihydroxyphenols (DOPA, dopamine, and *N*-acetyldopamine), including those with methylated side chains (*N*-methyl-dopamine and α-methyldopamine), while monohydroxyphenyls such as octopamine, tyramine, tyrosine, and α-methyltyrosine had little or no effect on radical production (Nappi and Vass 1998).

o-Quinones are physiological oxidation products of catecholamines that contribute to redox cycling, toxicity and apoptosis, i.e. the neurodegenerative processes underlying Parkinson's disease and schizophrenia. The cyclized *o*-quinones aminochrome, dopachrome, adrenochrome and noradrenochrome, derived from dopamine, dopa, adrenaline and noradrenaline, respectively, are efficiently conjugated with glutathione, in the presence of *human* glutathione transferase M2-2 (Baez et al. 1997).

15.3 Immobilization- and Hypothermia-Induced Myocardial Damage

Pilný et al. (1969, 1971) and Schiller (1974) used polarizing microscopy to evaluate experimental myocardial lesions in rats induced by coronary artery constriction, isoproterenol, or immobilization/hypothermia.

The concept of "myocardial distress" includes the following changes in the contractile apparatus of the cells:

1. Dilatation of the isotropic strip
2. Dilatation of the anisotropic strip
3. Dilatation of the isotropic and anisotropic strips
4. An increase in the birefringence values
5. Contraction of the cell and a decrease of the visibility of its transverse striation
6. Cell contraction associated with dilatation of the intercellular spaces.

Using the interference microscope the object contrast does not depend on diffraction by the specimen but is produced by a double-beam interferometer system built within a double-microscope. The object modifies the phase relationship between the two mutually interfering waves by an amount depending on its optical thickness and refracting index.

In the interference contrast transverse sections of normal myocardial cells produce a homogeneous interference colour on a background of a different colour. Erythrocytes due to their high dry matter concentration show another colour.

Fig. 329. Myocardium of a 300 g female *rat* immobilized and undercooled for twice two hours at 12 °C on three successive days. Sacrificed on the following day. Fixed by immersion in Bouin's fluid. Paraplast. Haematoxylin and eosin. Objective Leitz Pl 40/0.65. Circular polarised light. Leitz-Orthomat® (×3.2). Agfachrome 50 L professional

Fig. 330. Myocardium of a *rat* (No. 321) 6 h after a subcutaneous injection of 15 mg isoproterenol per kg body weight. Fixed in Bouin's fluid, paraffin section 4 μm in thickness, unstained, embedded in glycerol. Interference contrast Fl 50×/0.85/∞/0.17

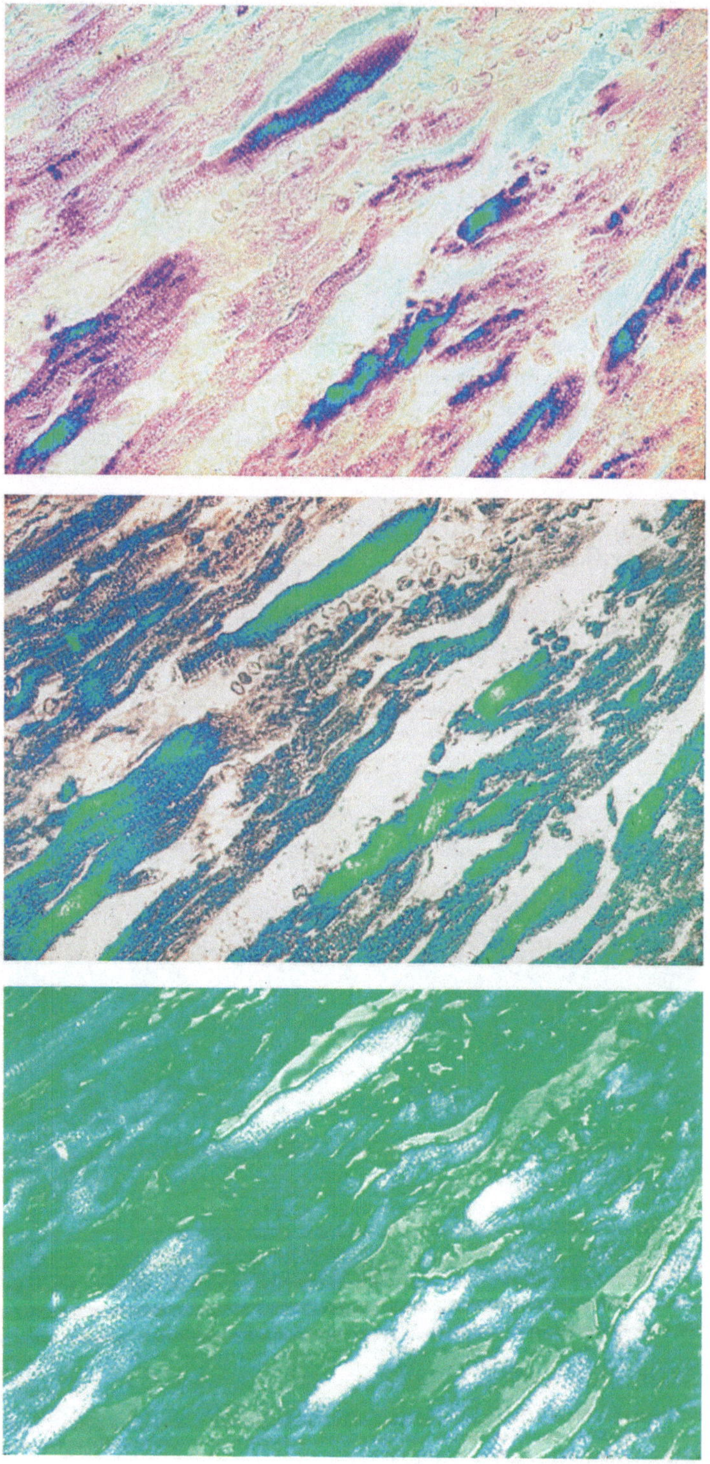

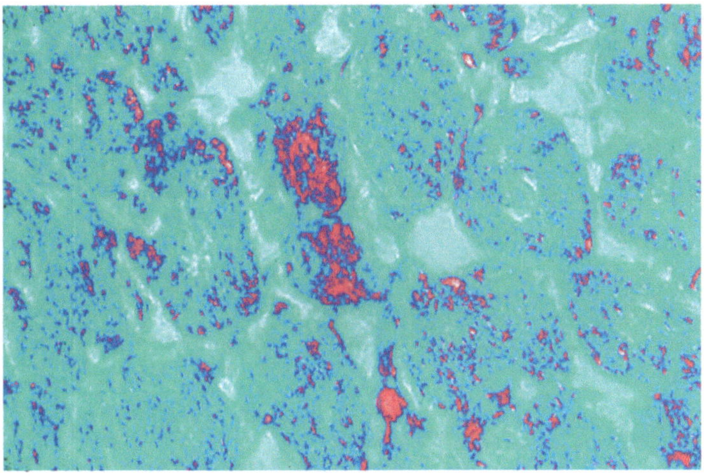

Fig. 331. Different interference colours of the myofilaments within the myocardiocytes of a *rat* (No. 299) 24 h after a subcutaneous injection of 30 mg isoproterenol per kg body weight. Fixed in Bouin's fluid, paraffin section 4 μm in thickness, unstained, embedded in glycerol. Interference contrast Fl 100×/ 1.36/∞/–

15.4 Norcocaine Nitroxide-Induced Hepatotoxicity

The hepatotoxicity of cocaine is known to be associated with its *N*-oxidative pathway. Cocaine is first *N*-demethylated to norcocaine, followed by oxidation to *N*-hydroxynorcocaine and then to norcocaine nitroxide. Norcocaine nitroxide is the active metabolite. Rauckman et al. (1984) presented a scheme of nonooxygenase-catalyzed reduction of nitroxide and superoxide formation. In DBA/2Ha *mice* a single dose of 60 mg per kg body weight intraperitoneally within 1 h induced a dilatation of rough endoplasmic reticulum in centrilubular hepatocytes, coincident with an onset of increased conjugated diene absorption in microsomal lipids (Gottfried et al. 1986). Progression of ultrastructural changes included focal mitochondrial membrane disruption followed by more extensive mitochondrial swelling and disruption with increasing swelling of rough endoplasmic reticulum.

PART IV

Ageing

CHAPTER 16

Molecular Ageing

Among the numerous theories raised to explain ageing, the free radical theory of ageing has become especially relevant (HARMAN 1956, SOHAL 1984, 1993, 2002, SOHAL and SOHAL 1991, AGARWAL and SOHAL 1993, ORR and SOHAL 1994, SOHAL et al. 1994, 1995, MARTIN et al. 1996, CADENAS and DAVIES 2000, SZWEDA et al. 2002). According to this theory, random tissue damages caused by oxygen radicals, produced by normal aerobic metabolism, accumulate during life and lead to various breakdown events at cellular or molecular levels.

The "Disposable soma theory of ageing" (KIRKWOOD 1987) included cellular maintenance systems. The amount of oxidatively damaged proteins, which results from the balance between the oxidation and elimination rate of these proteins, can be explained either by increased damage production, decreased elimination, or the combination of both phenomena (see FRIGUET 2002).

Oxidized amino acids can be incorporated into proteins by protein synthesis (RODGERS et al. 2002). The level of incorporation into proteins was dependent on the concentration of oxidized amino acid supplied to the cells. At low level of incorporation, the oxidized amino acids examined increased the degradation rate of the cell proteins. Degradation of certain proteins containing high levels of DOPA (but not *ortho* or *meta* tyrosine) was decreased to below the basal degradation rates suggesting that DOPA may contribute to proteins becoming resistant to proteolysis. Changes in the degradation rates of the oxidized amino acid-containing proteins was shown to have no impact on the degradation rates of native proteins, indicating that the activity of the degradative machinery was not affected. Oxidized proteins were selectively degraded by the proteasomes. KELLER et al. (2002) reviewed what is currently known about the proteasome in the central nervous system, and described established age-related alterations in proteasome biology.

Non-enzymatic glycation of reactive amino groups in model proteins increased the rate of free radical production at physiologic pH nearly fifty-fold over non-glycated protein (MULLARKEY et al. 1990). Superoxide generation was confirmed by electron paramagnetic resonance measurements with the spin trap α-phenyl-β-*tert*-butylnitrone. Both Schiff base and Amadori glycation products were found to generate free radicals in a ratio of 1:1.5.

The transcriptional regulation of the expression of vascular endothelial growth factor by advanced glycation end products and possible involvement of reactive oxygen species in the induction was recently investigated by URATA et al. (2002) who employed an advanced glycation end product of *bovine* serum albumin prepared by an incubation of serum albumin with D-glucose for 40 weeks and N^{ε}-carboxymethyl-lysine, a major advanced glycation end product. In RAW 264.7 *mouse* macrophage, advanced glycation end product-*bovine* serum albumin-stimulated activator protein-1 activity showed a peak at 5 h, which paralleled the formation of reactive oxygen species. Reduction of advanced glycation end product-*bovine* serum albumin with $NaBH_4$ or addition of vitamin E attenuated the advanced glycation end product-*bovine*

Table 52. Amino acids most sensitive to oxidation

Amino acid	Oxidation products
Cysteine	Disulphide bridges
Methionine	Methionine sulphoxide, methionine sulphone
Tryptophan	Hydroxytryptophan, kynorenine, hydroxy- and formyl-kynorenine
Phenylalanine	Hydroxyphenylalanine, dihydroxyphenylalanine
Tyrosine	Dihydroxyphenylalanine, tyrosine-tyrosine bridges, nitrotyrosine
Histidine	Oxohistidine, asparagine, aspartic acid
Arginine	Glutamic semialdehyde
Lysine	α-Aminoadipic semialdehyde
Proline	Glutamic semialdehyde, hydroxyproline, pyrrolidone
Threonine	2-Amino-3-keto-butyric acid
Glutamic acid	Oxalic acid, pyruvic acid

serum albumin-stimulated signalling pathway leading to the same pattern as for N^{ε}-carboxymethyllysine–*bovine* serum albumin-stimulated signal.

A low, yet significant, spontaneous superoxide anion production, matching with enhanced levels of basal adherence, was detected in fibronectin-plated neutrophils of blood donors over 65 years old (TORTORELLA et al. 2000). In contrast, although neutrophil stimulation with tumour necrosis factor-α, granulocyte macrophage-colony stimulating factor, formyl-methionyl-leucyl-phenylalanine or 12-*O*-tetradecanoylphorbol-13-acetate gave rise to a massive and prolonged fibronectin-primed $O_2^{\bullet -}$ release, a significant impairment of oxidative response occurred in the aged group as a result of granulocyte macrophage-colony stimulating factor or formyl-methionyl-leucyl-phenylalanine cell challenge.

It has been suggested that protein modifications by malondialdehyde, a major product of lipid peroxidation, contribute to the fluorescence formation of lipofuscin (KIKUGAWA and BEPPU 1987, TSUCHIDA et al. 1987, YIN 1996). However, ITAKURA and UCHIDA (2001) isolated an aminoenimine, *N,N'*-bis[5-(*tert*-butoxy-carboxamido)-5-carboxypent]-1-amino-3-iminopropene, formed from the reaction of malondialdehyde with a lysine derivative, N^{α}-*tert*-butoxycarbonyl-L-lysine, at neutral pH, and confirmed that the purified *N,N'*-bis[5-(*tert*-butoxy-carboxyamido)-5-carboxypent]-1-amino-3-iminopropene, exhibited no fluorescence.

A malondialdehyde-deoxyguanosine adduct (9.0 ± 1.6 pM/100 μg DNA) was isolated from *rat* liver DNA (AGARWAL and DRAPER 1992). SETO et al. (1983, 1985) demonstrated that both carbonyl groups of malondialdehyde react with guanine, guanosine, and deoxyguanosine to form pyramidinopurines.

Hyperinsulinaemia is an important ageing factor via enhanced nonenzymatic glycosilation of proteins (MONNIER and CERAMI 1981). XU and BADR (1999) have demonstrated in Sprague-Dawley *rats* that a sixfold increase in serum insulin concentrations, maintained for 1 week, inhibited both peroxisomal oxidation of fatty acids and catalase activity. Insulin causes •NO-mediated vasodilatation in skeletal muscle by stimulating nitric oxide synthase (NOS), as such effect is blunted by NOS inhibitors (SHERRER et al. 1994, STEINBERG et al. 1994). Both •NO and peroxynitrite inhibit multiple enzymes of the mitochondrial respiratory chain, such as complex I, II, and ARP-synthase (BROWN 1999). FACCHINI et al. (2000) suggested that an overall increase of insulin signalling pathways is a pro-ageing factor, probably of greater relevance than hyperinsulinaemia itself.

γ-Glutamylcysteine synthetase activity was significantly decreased with increased age in liver, kidney lung, and erythrocytes of Fisher 344 *rats* (LIU and CHOI 2000). Parallel with the decreased enzyme activity, the protein and mRNA content of both γ-glutamylcysteine synthetase subunits also changed inversely with age in liver, kidney and lung, implying a decreased γ-glutamylcysteine synthetase gene expression during ageing.

Chronic stimulation of GSH synthesis increased proliferative live span of *human* diploid fibroblasts by about 40 %, while chronic inhibition of GSH synthesis decreased proliferative life span (HONDA and MATSUO 1988). Treatment of early passage cells with oxidants such as H_2O_2 caused them to assume many characteristics of a senescent phenotype (CHEN et al. 1998). Senescent cells typically exhibit growth arrest in an aberrant state more similar to G_1 than to G_0 (CRISTOFALO and KRITCHEVSKY 1969, CRISTOFALO and PIGNOLO 1993, CRISTOFALO 1996). This differs from the effects of some forms of oxidative stress such as hyperoxia, which block cells in G_2 (BALIN et al. 1978). Increases in •NO, however, tend to block mitosis in G_1 (SARKAR et al. 1997).

Considering 75 "noncentanarians" not supplemented by antioxidants (vitamins or minerals), MECOCCI et al. (2000) observed a consistent behaviour in the plasma antioxidant pattern, with a decrease of the nonenzymic antioxidants and an increase in the enzymic antioxidants relative to age. Remarkably, 32 healthy **centenarians** were characterised as having the highest levels of retinol and α-tocopherol, whereas the activities of both plasma and erythrocyte superoxide dismutase, which increase with age, decreased in centenarians.

16.1 Oxidatively Damaged DNA Accumulates in Senescence

Both spontaneous DNA hydrolysis and reactive oxygen intermediates caused by metabolism represent the sources of endogeneously caused damage to DNA. Modern chromatography methods revealed a broad spectrum of oxidative base modifications (DIZDAROGLU 1992). POULSEN et al. (1996) found a 33 % increased rate of oxidatively modified DNA in 20 men subjected to extensive exercise up to 10 h per day over a period of one month. They measured 8-oxo-7,8-dihydro-2'-deoxyguanosine excreted in urine as metabolic parameter indicating the excision-repair of the oxidised base, thereby documenting the risk to damaged DNA by excessive oxygen consumption. In healthy young organisms DNA repair (review: DEMPLE and HARRISON 1994) and antioxidative cellular defence mechanisms counteract the deleterious effects of reactive oxygen intermediates.

MIQUEL and FLEMING (1986) suggested that mitochondria may play a key role in cellular ageing and that mitochondrial DNA (mtDNA) is the major target of free radical attack. mtDNA from brain and liver of old *rats* exhibited oxidative damage that was significantly higher than that from young *rats* (SASTRE et al. 1998). LUCAS an SZWEDA (1998) have shown that modification of mitochondrial proteins by the lipid peroxidation product 4-hydroxy-2-nonenal occurs solely upon reperfusion of senescent *rat* heart. This observation suggested that ageing may predispose cardiac mitochondria to reperfusion-induced oxidative damage, possibly due to a number of factors including age-related

increase in the production of free radicals by mitochondria,
decrease in the ability to protect oxidative damage,
changes in mitochondrial composition and structure.

Mitochondria are the main source of energy in the cell. They contain own DNA (mtDNA), a small 16.5 kb circulatory molecule which is exclusively concerned with coding for 13 polypeptides, 22 tRNAs and 2 rRNAs, all components of the respiratory-chain/oxidative phosphorylation system (ANDERSON et al. 1981). Mutations in mtDNA could accumulate because of the intrinsic instability of the mitochondrial genome, which is not protected by proteins and by efficient DNA repair mechanisms, and is situated in close proximity to oxygen radical sources (MIQUEL and FLEMING 1984, ZEVIANI et al. 1998). The overall activity of Complex I was decreased in liver, heart and gastrocnemius muscle from 24 month aged *rats*, nevertheless direct assay of Complex I using artificial quinone acceptors may underevaluate the enzyme activity (LENAZ et al. 1997). A decrease of NADH oxidation and its rotenone sensitivity was observed in nonsynaptic mitochondria, but not in synaptic 'light' and 'heavy' mitochondria of cerebral cortex from aged *rats*.

Endothelial cell cultures from *human* umbilical vein after 11 passages showed a significant increase in senescent cells as detected by acidic β-galactosidase expression and a reduction of telomere length (VASA et al. 2000). Telomerase activity was reduced after the seventh passage, thereby preceding the development of endothelial cell senescence. The repeated (every 48 h) addition of the NO-donor *S*-nitrosopenicillamine (20 μmol/l) starting at the eighth passage significantly reduced endothelial cell senescence and delayed age-dependent inhibition of telomerase activity, whereas inhibition of endogenous •NO synthesis by N^G-monomethyl-L-arginine (1 mmol/l) every 12 h starting at the sixth passage had an adverse effect.

Senescence-associated β-galactosidase activity seems to reflect the expression of lysosomal β-galactosidase rather than a specific enzyme induced only during senescence, and even in tissue culture model systems "senescence-associated" β-galactosidase can hardly be considered a specific marker of senescence (COATES 2002). In the upper gastrointestinal tract, activity is low in mucosal proliferation compartments and increased with cellular differentiation, especially in native or metaplastic intestinal mucosae (GOING et al. 2002). Senescence-associated β-galactosidase activity persists in dysplastic mucosae but may show some reduction or loss in adenocarcinomas ($P = 0.0012$). Loss of senescence-associated β-galactosidase activity is not, therefore, an early event in glandular dysplasia-neoplasia of the upper gastrointestinal tract.

Vascular smooth muscle cells isolated from aortas of corpulent JCR:LA-cp *rats* (a strain characterised by insulin resistance, hyperinsulinaemia, and hyperlipidaemia, factors strongly associated with both ageing and cardiovascular disease) exposed to glucose (5–100 mM) or reactive oxygen species (generated in the xanthine-xanthine oxidase system) with growing age of the donors (6, 12, and 17 months, respectively) accumulated mtDNA mutations (FUKAGAWA et al. 1999).

16.2 Decline of Mitochondrial Oxidative Phosphorylation

The genetic information located in the mitochondria has been identified as a major genetic component. Instabilities of the mitochondrial DNA (mtDNA) lead to mitochondrial dysfunction and increased oxidative stress (OSIEWACZ 1997).

Differences in mitochondrial function can be very important for attaining successful or unsuccessful ageing. A mtDNA germline inherited variant (haplogroup J) is associated with successful ageing and longevity in an Italian population (DE BENEDICTIS et al. 1999, 2000). In Japanese people, three associated mtDNA germline mutations have been found at higher frequency in centenarians than in controls (TANAKA et al. 1998).

Studies in *human* skeletal muscle (COOPER et al. 1992, BOFFOLI et al. 1994), liver (YEN et al. 1989), and brain (MÜLLER-HÖCKER et al. 1993) have revealed that mitochondrial oxidative phosphorylation declines during ageing.

Cell cultures of hybrids between *human* cells lacking mtDNA (ϱ^0 cells) (KING and ATTARDI 1989) and mitochondria from young and old subjects demonstrated respiratory chain deficiency in hybrids containing mitochondria from older subjects (LA-

DERMAN et al. 1996). In contrast, other hybrid experiments showed that nuclear rather than mtDNA mutations were responsible for the observed age-related reduction of oxidative phosphorylation (ISOBE et al. 1998).

KWONG and SOHAL (2000) measured the activities of respiratory complexes I, II, III, and IV in mitochondria isolated from brain, heart, skeletal muscle, liver, and kidney of young (3.5 months), adult (12–14 months), and old (28–30 months) C57BL/6 *mice*. Activities of some individual complexes were decreased in old animals, but no common pattern could be discerned among various tissues. In general, activities of the complexes were more adversely affected in tissues such as brain, heart, and skeletal muscle, the parenchyma of which is composed of postmitotic cells, than those in the liver and kidney, which are composed of slowly dividing cells. the main feature of age-related potentially dysfunctional alterations in tissues was the development of a shift in activity ratios among different complexes, such that it would tend to hinder the ability of mitochondria to effectively transfer electrons down the respiratory chain and thus adversely affect oxidative phosphorylation and/or autooxidizability of the respiratory components.

MARTINEZ et al. (1996) observed a linear inverse relationship between protein carbonyl content and complex IV/complex I ration (which was used as an index of imbalance between mitochondrial respiratory complexes) in synaptic mitochondria in five age groups of female OF-1 *mice* ($r = -0.99$, $P < 0.001$).

In the male *Drosophila melanogaster*, no statistically significant influences of age on activities of Complexes I and II or citrate synthase were observed (SCHWARZE et al. 1998). In contrast, from 2 to 45 days post-eclosion, declines were found in Complex IV cytochrome *c* oxidase activity (40 % decline) and ATP abundance (15 %), while lipid peroxidation increased 71 %. In *flies* that were either genetically or chemically oxidatively stressed to determine the effect on levels of mitochondrial-encoded oxidase I RNA and cytochrome oxidase activity. A catalase null mutant line had 48 % of cytochrome *c* oxidase I RNA compared to the wild type. In Cu/Zn superoxide dismutase (SOD) null *flies*, the rat of cytochrome oxidase I RNA was greater than in controls. Cytochrome oxidase I RNA also declined with increasing hydrogen peroxide treatment, which was reflected in reduced cytochrome *c* oxidase activity. These results showed that oxidative stress is closely associated with reductions in mitochondrial transcript levels and supported the hypothesis that oxidative stress may contribute to mitochondrial dysfunction and ageing in *Drosophila melanogaster*.

A mutant of the nematode *Caenonabditis elegans* (called *age*-1), which lives about 65 % longer than the wild type has been identified by JOHNSON (1990). The gene for *age*-1 was mapped to chromosome II, and is closely linked to the *sod*-1 gene. Although the *age*-1 and *sod*-1 genes are not identical, the *age*-1 mutation leads to an increase in SOD-1 activity starting at about 10 days and continuing until death at about 25–30 days (LARSON 1993). Catalase activity increased with age in parallel with SOD-1 in the *age*-1 mutant.

Oxidative damage of mitochondrial proteins with formation of reactive carbonyl groups has been found to increase with age (AMES et al. 1991). These oxidised proteins could undergo crosslinking, proteolysis and loss of functional activity. In rat heart and brain, ageing has been found to be associated with decrease in mitochondria of the content of the β-F_1 subunit and depression of the catalytic activity of the F_0F_1 ATP synthase (GUERRIERI et al. 1992). The same decrease in content and activity of the ATP synthase was observed when the formation of the $HO^{\bullet}$ radical was promoted by exposing the isolated mitochondrial membrane to Co irradiation (GUERRIERI et al. 1993).

16.3 Biorheusis of Transitional Metals

Serum **iron** levels increase about two-fold with age, and this is accompanied by similar increase in serum ferritin (CHOI and YU 1994). Caloric restriction reduced iron content roughly by half at all ages.

Copper and **zinc** have pro-oxidant and antioxidant properties, respectively, so that their imbalance may be expected to condition oxidative stress status. As oxidative stress is relevant in ageing and age-related degenerative diseases, MEZZETTI et al. (1998) investigated blood content of copper, zinc and ceruloplasmin as well as of lipid peroxides in 81 health and 62 disabled octo-nonagenarians affected by chronic diseases, and in 81 healthy adults. Serum copper/zinc ratio and ceruloplasmin were significantly higher in the elderly than in the health adults. Moreover, all these parameters were significantly higher in the disabled than in the health elderly. Notably, the increased copper/zinc ratio found in health elderly was due to high copper values, whereas in the disabled, both high copper and low serum zinc concentrations were present. The copper/zinc ratio was significantly and positively related to systemic oxidative stress status in all groups. The higher the serum copper/zinc ratio the higher the lipid peroxides plasma content.

Serum alkaline phosphatase (EC 3.1.3.1), an enzyme containing zinc, significantly ($P < 0.001$)

decreased in *men* ages 61–80 years but increased in *women* of this age group (STRUCK and HILLESHEIM 1990). Leucyl aminopeptidase (EC 3.4.11.1) only in *men* showed a significant decrease with age.

The glyco- and lipoxidation product **N^{ϵ}-(carboxymethyl)-lysine** in proteins has metal binding properties owing to their EDTA/glycine-like configuration (DWYER and MELLOR 1964). N^{ϵ}-(Carboxymethyl)-lysine is a glycoxidation product of the Maillard reaction which accumulates in ageing *human* skin collagen (DUNN et al. 1991) and lens crystallins (DUNN et al. 1989). N^{ϵ}-(Carboxymethyl)-lysine-rich poly-L-lysine and *bovine* serum albumin were found to bind non dialyzable Cu^{2+}, Zn^{2*} and Ca^{2+} (SAXENA et al. 1999). N^{ϵ}-(Carboxymethyl)-lysine-*bovine* serum albumin-copper complexes oxidised ascorbate and depolymerized protein in the presence of H_2O_2. N^{ϵ}-(Carboxymethyl)-lysine-rich tail tendons implanted for 25 days into the peritoneal cavity of diabetic *rats* had a 150 % increase in copper content and oxidised ascorbate three times faster than controls. N^{ϵ}-(Carboxymethyl)-lysine-rich proteins immunoprecipitated from serum of uraemic patients oxidised four times more ascorbate than control and generated spin adducts of 5,5-dimethyl-1-pyrroline-1-oxide (DPMO) in the presence of H_2O_2. The chelator diethylene diamine tetraacetic acid (DTPA) suppressed ascorbate oxidation thereby implicating transition metals in the process. In ageing and disease; N^{ϵ}-(carboxymethyl)-lysine accumulation may result in a deleterious vicious cycle since N^{ϵ}-(carboxymethyl)-lysine formation itself is catalysed by lipoxidation and glycoxidation.

CHAPTER 17

Ageing of Organs

17.1 Brain

In aged *rats*, the rates of transient redox responses of cytochrome oxidase (i.e. initial oxidation followed by re-reduction) are slowed by about 50 % in comparison to young *rats* (SYLVIA et al. 1983). Cortical norepinephrine was similar in both age groups. However, while depletion of cortical norepinephrine causes slowing of the rate of re-reduction in young *rats* by about 50 %, such depletion had no effect on the already slow kinetics of the redox shift of aged *rats*.

Cognitive deficits such as learning impairment and delayed amnesia are the debilitating consequences of ageing. In the brain free radical may be responsible for neurodegeneration occurring in age-related diseases, such as Alzheimer's (EVANS et al. 1989, VOLICER and CRINO 1990, HENSLEY et al. 1993, BUTTERFIELD et al. 1994, ZHOU et al. 1995, MARKESBERY 1997, MULTHAUP et al. 1997, 2002, PERRY et al. 2000, SMITH et al. 2000, GIASSON et al. 2002, PRATICò 2002) and Parkinson's (ADAMS and ODUNZE 1991, POIRIER and THIFFAULT 1993, YOSHIKAWA 1993, FISHER and GAGE 1995, GIASSON et al. 2002) diseases.

While these diseases undoubtedly involve high complex pathophysiological processes which have not rapidly yielded to intensive investigative effort, it seems likely free radical-induced cell loss is a process related to these conditions. In Alzheimer's disease this seems to be the consequence of amyloid β protein while in Parkinson's disease it is the autooxidation of dopamine (cf. formula [107]) which induced cell death. PAPPOLLA et al. (1997) scrutinised the potential utility of melatonin in preventing the cytotoxic actions of amyloid β protein in *murine* neuroblastoma cells grown in the presence of 25–35 amino acid residues (50 mM) of the amyloid β protein. At the end of 24 h incubation period, roughly 80 % of the neurones were killed by apoptosis, while the scrambled peptide used for control killed only about 10 % of the cells. 10 mM melatonin reduced the toxicity of the amyloid β protein fragment to the level of the controls.

While KISH et al. (1986) did not find any increase in the activity of glutathione peroxidase, LOVELL et al. (1995) reported increased activities of glutathione peroxidase (EC 1.11.1.9), glutathione reductase (EC 1.6.4.2) and catalase (EC 1.11.1.6). GSELL et al. (1995) rather found decreased catalase activity in brains of patients with Alzheimer's disease. The findings on the expression and activity of superoxide dismutase in brains of patients with Alzheimer's disease are also highly controversial (MARKLUND et al. 1985, BALASZ and LEON 1994, GSELL et al. 1995, LOVELL et al. 1995) that may be due to the post mortem delay.

Statistically significant elevations ($P < 0.05$) of total oxidised proteins were observed both patients suffering from Alzheimer's disease and their relatives when compared with non-Alzheimer controls (CONRAD et al. 2000). Moreover, a protein band (e.g., MW = 78-kDa) was uniquely oxidised in the plasma of Alzheimer's disease patients. This protein from Alzheimer's disease patients was more susceptible to *in vitro* oxidation.

CASTEGNA et al. (2002) presented the first proteomics approach to identify specifically oxidised proteins in Alzheimer's disease, by coupling 2D fingerprinting with immunological detection of carbonyls and identification of proteins by mass spectrometry. The powerful techniques, emerging from application of proteomics to neurodegenerative disease, revealed the presence of specific targets of protein oxidation in Alzheimer's disease brain: creatine kinase BB, glutamine synthase, and ubiquitin carboxy-terminal hydrolase L-1.

Although reactive oxygen species in cells from patients with Alzheimer's disease were only slightly less than in cells from controls under basal conditions (−10 %) or after exposure to H_2O_2 (−16 %), treatment with antioxidants revealed clear differences (GIBSON et al. 2000). Pre-treatment with dimethylsulphoxide, a hydroxyl radical scavenger, reduced basal H_2O_2-induced levels of reactive oxygen species significantly more in cells from controls (−22 %, −22 %) than in those from patients with Alzheimer's disease (−4 %, +14 %). On the other hand, pre-treatment with Trolox diminished H_2O_2-induced reactive oxygen species significantly more in cells from patients with Alzheimer's disease (−60 %) than control subjects (−39 %).

The metal ion homeostasis is severely deregulated in Alzheimer's disease (Ehmann et al. 1986, Thompson et al. 1988, Samudralwar et al. 1995, Deibel et al. 1996, Cornett et al. 1998, Lovell et al. 1998). Increased concentrations of copper, iron, and zinc were detected in the neuropil of the brain where they were highly concentrated within amyloid plaques and reached concentrations up to 0.4 μM (Cu) and 1 μM (Fe and Zn) (Smith et al. 1997, Lovell 1998). A likely reason is that Aβ binds equimolar amounts of Cu(II) and Zn(II) at pH 7.4 (Bush et al. 1993, 1994, Huang et al. 1997, Atwood et al. 1998). The amyloid precursor protein is a transmembrane glycoprotein that undergoes extensive alternative splicing (Sandbrink et al. 1994) and has been shown to bind Zn(II) and Cu(II) at two distinc sites (Bush et al. 1993, Hesse et al. 1994). Experimentally induced disturbances of the homeostasis of Zn(II) and Cu(II) affect the metabolism of the amyloid precursor protein (Borchardt et al. 1999, 2000).

Eto et al. (2002) showed that the levels of H_2S were severely decreased in the brains of Alzheimer's disease patients (76.4±2.3 years) compared with the brains of the age matched normal individuals (71.5±7.2 years). In addition to H_2S production cystathionine β-synthase also catalyses another metabolic pathway in which cystathionine is produced from the substrate homocysteine. *S*-adenyl-L-methionine, a cystathionine β-synthase activator, is much reduced in Alzheimer's disease brains (Morrison et al. 1996, Eto et al. 2002) and homocysteine accumulates in the serum of Alzheimer's disease patients (Clarke et al. 1998, Eto et al. 2002).

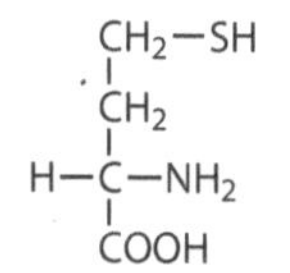

Homocysteine [299]

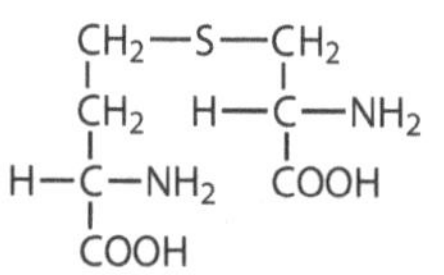

Cystathionine [300]

The Tg2576 transgenic *mouse* model utilized by Van Ess et al. (2002) expresses a double mutant *human* amyloid precursor protein identified in a Swedish family with exceptionally early familial Alzheimer's disease. Pedersen et al. (1999) documented an altered stress response in these tg(hAPP) *mice* characterised by severe hypoglycaemia and death following every-other-day feeding, as well as hypoglycaemia hypercortisolinaemia following moderate restraint stress. Van Ess et al. (2002) found elevated hepatic and depressed renal cytochrome P450 activity in aged tg(hAPP) *mice* compared with wild-type.

The discovery of the isoprostanes, recent studies performed in living patients (Praticò et al. 2000), and the development of transgenic animal models of Alzheimer's disease amyloidosis are three important factors helping us to understand and define the role that reactive oxygen species might play in the pathogenesis of this disease (for review see Praticò 2002).

Isoprostanes $F_{2\alpha}$ III and $F_{2\alpha}$-VI were markedly elevated in both frontal and temporal poles of brains affected with Alzheimer's disease but not in the other groups (Praticò et al. 1998).

In a transgenic *mouse* model, which expressed a gene encoding 18 residues of signal peptide and 99 residues of carboxyl-terminal fragment of the amyloid β protein precursor, this protein in the pancreas accumulated in an age-dependent manner (Shoji et al. 2000). Amyloid β protein fibril deposits closely correlated with degeneration of pancreatic acinar cells and macrophage activation.

α-Synuclein is a major component of the abnormal protein aggregation in Lewy bodies of Parkinson's disease and senile plaques of Alzheimer's disease. The aggregation of α-synuclein was induced by Cu(II) and H_2O_2 system. When α-synuclein was incubated with both Cu,Zn-superoxide dismutase and H_2O_2, α-synuclein was induced to be aggregated (Kim et al. 2002). This process was inhibited by radical scavengers and spin trapping agents such as 5,5'-dimethyl-1-pyrroline *N*-oxide and *tert*-butyl-α-phenylnitrone. Copper chelators diethyldithiocarbamate and penicillamine, also inhibited the Cu,Zn-superoxide dismutase/H_2O_2 system-induced α-synuclein aggregation.

Poly-*L*-lysine appeared as a potent dissolver of preformed β-amyloid fibrils *in vitro* (Nguyen et al. 2002). Its efficiency is instantaneous. Poly-*L*-lysine can be used as a universal dissolver of all types of oligomeric β-sheet conformation, precursor of the fibrils.

Thees et al. (2002) have used the *baboon* as a potential non-human primate model for age-related pathology which afflicts the *human* brain. While the neurones from young *baboons* had a bright, punctuate cytoplasmic distribution of cytochrome *c*, clearly excluded from the nuclear space in pyramidal cells, in aged animals cytochrome *c* containing structures clustered in the vicinity of the nucleus, probably in the region of the Golgi apparatus. In the dentate gyrus of most examined aged animals, the clustered, punctuate pattern of cytochrome *c* was also present in the nuclear space as well as in the vicinity of the nucleus.

In *rat* brain cortex labelled with the lipid-specific spin probe, 5-nitroxyl stearate, stimulation of the mitochondrial electron transport chain was accomplished using 20 mM succinate at 25 °C for 3 h. Mitochondrially derived free radicals, when reacted with the paramagnetic centre of the spin probe, resulted in a loss of paramagnetism resulting in loss of intensity. GABBITA et al. (1998) found a significant lowering (23 %, $P < 0.0001$) in the signal amplitude (B_0) of 5-nitroxyl stearate, indicative of generation of oxyradicals. The order of parameter, an inverse EPR-measure of membrane fluidity of the 5-nitroxyl stearate spin labelled mitochondrial and synaptosomal membranes, also decreased following mitochondrial respiratory stimulation ($P < 0.005$). Changes in the physical state of cytoskeletal and transmembrane proteins due to succinate oxidation were measured using MAL-6 (2,2,6,6-tetramethyl-4-maleimidopiperidin-1-oxyl), a thiol-specific nitroxide spin label.

In *rats* aged 31–38 months, the ultrastructure of hippocampal pyramidal neurones and interneurones was increasingly transformed (BRICHOVÁ et al. 2000). Lesions in CA1 resembled those found in hypoxic tissue, but only a small number of neurones were necrotised. Astrocyte processes situated between these neurones were hypertrophied and they were swollen around heavily damaged neurones. In the cytoplasm of the astrocytes phagocytosed particles were observed. Swollen processes were also present in the proximity of degenerating synapses. In the degenerating axons, the distances between myelin sheaths were increased. Activated microglial cells were present in the aged tissue. Both endothelial cells and basement membrane of the capillaries often were thickened, and there were pericytes with a large amount of engulfed material in the cytoplasm surrounded by a duplicature of the basement membrane. The capillary wall was often covered by the swollen astrocyte footpads.

Quantitative electron microscopic analysis of the supragranular zone of the dentate gyrus molecular layer has shown that the number, volume fraction and surface area of dendritc shaft profiles were significantly decreased in senescent Fischer-344 *rats*, relative to young adults (GEINISMAN et al. 1978a). The number and volume fraction of profiles of astroglial processes were significantly increased (GEINISMAN et al. 1978b). Memory-impaired aged *rats* showed a loss of perforated axospinous synapses in the dentate gyrus in comparison with either young adults or aged rats with goof memory (GEINISMAN et al. 1986).

In *gerbil* brain, levels of hydroxyl radicals and neurotransmitters such as glutamate, aspartate and γ-aminobutyric acid are low at birth, reach a plateau and decrease with age (DELBARRE et al. 1992). On the other hand, when *gerbils* were exposed to an ischaemia/reperfusion insult, the older animals had a higher stroke index and $HO^{\bullet}$ as well as glutamate and other neuromediators were concomitantly increased. Oxidative damage to mitochondrial DNA is inversely related to maximum life span in the brain of *mammals* (BARJA and HERRERO 2000)

The prominent postnatal decrease of kainate receptors in immunodeficient NZB *mice* as compared with CFW *mice* express an early beginning decrepitude (CLAUDIUS et al. 2001).

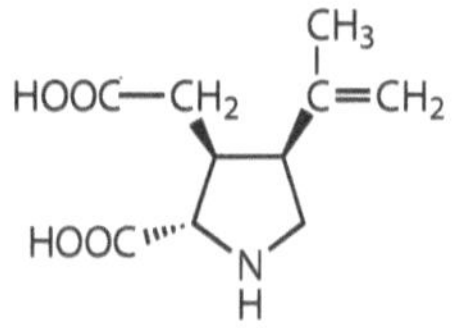

Kainic acid (2-carboxyl-3-carboxymethyl-4-isopropenylpyrrolidine) [275]

In homogenised cerebrum, basal ganglia, cerebellum, medulla, and cervical cord of *rats* aged 3, 12 and 22 months, the formation of thiobarbituric acid-reactive substances declined with age (ANSARI et al. 1989). Between the ages of 12 and 22 months, thiobarbituric acid-reactive substances in homogenates of spinal cord and cerebellum were increased. The rostro-caudal gradient for glutathione peroxidase (EC 1.11.1.9) activity was found to be opposite in direction to that for thiobarbituric acid-reactive substances. Age was not found to have a significant effect on catalase (EC 1.11.1.6) activity.

In the brains of longeval *mammals*, ONO and OKADA (1984) found a positive correlation between the specific activity of superoxide dismutase (EC 1.15.1.1) and lifespan. The mitochondrial form of superoxide dismutase (SOD-2) may be the most critical form of SOD in reducing chronic low level oxidative damage and retarding ageing. LI et al. (1995) have reported that new-born *mice* lacking SOD-2 activity are hypothermic, and many die soon after birth with enlarged and anatomically altered hearts. The new-borns appear to be normal at birth, suggesting that the mitochondrial SOD-2 activity is essential for normal muscle function, presumably because of its role in protecting mitochondria from self-inflicted oxidative damage.

CuZn-superoxide dismutase in 26-month-old *rats* compared with. 8-month-old animals had decreased from $2.05 \pm 0.09 \times 10^2$ units per g cerebral tissue to $1.85 \pm 0.10 \times 10^2$ units per g cerebral tissue (GOMI and MATSUO 1995).

Melanotransferrin which is able to bind and internalise iron into cells (KENNARD et al. 1995, JEFFE-

RIES et al. 1996) is selectively expressed on reactive microglial cells in patients with Alzheimer's disease (JEFFERIES et al. 1996). The reactive microglia in senile plaques express high levels of ferritin.

In the *human* cortex age-related regression involves the intraneuronal "nucleus-ribosome system" (SPOERRI et al. 1981). Such regressive signs are characterised by the polymorphous outline, invagination, reduced size of the nucleus, and in the nuclear texture. Recognisable changes in the cytoplasm are alterations in the density and distribution of free ribosomes and in the structure and organisation of the rough endoplasmic reticulum cisternae. Furthermore light cytoplasmic areas, and increased number of microtubules and the gradual congestion of the perikaryon by age pigment are reliable criteria for identifying the sequence of morphologic events that occur during neuronal ageing.

The activity of **catalase** (EC 1.11.1.6), the main enzyme responsible for detoxification against hydrogen peroxide, significantly decreases in prefrontal cortex and hypothalamus of aged *rats* (CIRIOLO et al. 1997). The reductions of the enzyme activity appears to be due to a decreased protein expression rather than impaired function of the native enzyme.

Detoxification of hydrogen peroxide by catalase (EC 1.11.1.6)

$$2\,H_2O_2 \longrightarrow 2\,H_2O + {}^3\Sigma g^- O_2 \qquad [127]$$

Age-associated changes in hypothalamic catalase activity and level, and Cu/Zn superoxide dismutase activity were examined in Ames dwarf *mice* with growth hormone deficiency and prolonged lifespan, in PEPCK-hGH transgenic *mice* with overexpression of growth hormone and reduced lifespan, and compared to values measured in normal controls (HAUG and BARTKE 2000). Hypothalami from young (3–4 months), middle-aged (9–10 months), and old (19–23 months) male *mice* were examined using spectrophotometric assay and Western blot. In dwarf *mice*, Cu/Zn superoxide dismutase and catalase activities declined with age, and were higher than the corresponding values in young and middle-aged groups. Catalase levels also declined with age, but were similar to values in normal controls. In growth hormone transgenic *mice* age-associated decline of both calalase and Cu/Zn superoxide dismutase occurred earlier than in normal animals. Catalase levels and activities in transgenic animals were similar to controls, whereas Cu/Zn superoxide dismutase activity was higher in trangenic than in normal *mice*.

Table 53. Twisted tubules

Alzheimer's presenile dementia
Old age
Down's syndrome
Subacute sclerosing encephalitis
"Dementia pugulistica"

A significant decrease (55 %) in **tyrosine hydroxylase** (EC 1.14.16.2) – the step-limiting enzyme in the biosynthesis of dopamine – was discovered in the substantia nigra of aged *rats* (DE LA CRUZ et al. 1996). When *rat* striatal homogenate was incubated with H_2O_2, there was a time-dependent decrease in tyrosine hydroxylase activity, which highly correlated with measurements of carbonyl groups content of tyrosine hydroxylase enzyme.

Basal expression of **glutamic acid decarboxylase** (EC 4.1.1.15) 65 mRNA was increased in the medial preoptic area and posterior bed nucleus of the stria terminalis in aged Fischer 344/Brown Norway F1 hybrid *rats* in relation to both middle-aged and young groups (HERMAN and LARSON 2001).

Significant correlation was found between plasma and cerebrospinal fluid **ascorbic acid** levels in demented *patients* (BARABÁS et al. 1995). As a consequence, intravenous infusion of ascorbic acid (2 g), a slow, but marked increase of the concentration in the cerebrospinal fluid was measured by high performance liquid chromatography with electrochemical detection. Ascorbic acid level might be an important factor representing the protection of the central nervous system against free radicals.

The primary process of ageing is not a generalised or diffuse phenomenon, but appears localised and sometimes restricted to some neuronal complexes (HASSLER 1965). The following areas are more susceptible: substantia nigra, locus caeruleus, striatum, centre median of the thalamus, the dorsal nucleus of the vagus, oliva inferior, nucleus dentatus and nucleus basalis. Many of these areas are pigmented (substantia nigra, locus caeruleus, nucleus dorsalis nervi vagi) and involved in the synthesis and regulation of monoaminergic fibre systems.

Monoamine oxidase (EC 1.4.3.4)

$$RCHNH_2 + H_2O + O_2 \longrightarrow RCHO + NH_3 + \mathbf{H_2O_2} \qquad [268]$$

Mitochondrial genesis of **lipofuscin** became evident from electron microscope studies of brain, neuronal tissue culture and heart (GLEES et al. 1974). GLEES and HASAN (1976) considered mitochondria to be most sensitive organelles, which due to ageing, disease or stress will give rise to lipofuscin formation. The biochemical basis of lipofuscin, ceroid, and age pigment-like fluorophores was reviewed by YIN (1996).

In the perikarya of the pyramidal cells of layers III and V of area 10 (Brodmann) of the frontal brains of

6 *men* aged 32 to 77 years, lipofuscin contents considerably varied from 2.94±1.26% to 9.03±3.84% and from 2.18±1.59% to 6.27±3.86% of the areas, respectively (Rösler and Kemnitz 1983).

The increase in intraneuronal lipofuscin in the hippocampus CA1 zone of the Fischer 344 *rat* from 11 to 17 months was 100%, from 17 to 29 months it was 35%, and from 11 to 29 months it was 169% (Brizee and Ordy 1979). In visual area 17, the increase in intraneuronal lipofuscin from 11 to 17 months was 33%, from 17 to 29 months it was 25%, and from 11 to 29 months it was 67%. In the Purkinje cells of the cerebellum of ageing C57BL/10 mice, Samorajski et al. (1968) observed numerous large pigment granules at 20 and 30 months concentrated between the nucleus and the apical dendrite, while at four and eight months pigment granules were rarely detected. In the nucleus olivaris caudalis of the musquash (*Ondatra zibethica*), the great fruit bat (*Pteropus giganteus*), and the guinea-pig (*Cavia cobaya*) Wünscher et al. (1967) found a significant increase of intraneuronal lipofuscin in correlation with age.

Age pigments from the *human* brain as well as ceroid particles from *dogs* with *canine* ceroid lipofuscinosis contained metal ions trapped in a ligand structure (Vistnes et al. 1983). Both Fe^{3+} and Cu^{2+} complexes have been found. The iron complex was of the non-heme high-spin ferric type, whereas the copper ions were of type 2 (non-blue copper ions). The copper is probably ligated to two nitrogen and two oxygen atoms in a square-planar configuration.

Centrophenoxine (80 mg/kg body weight) administered intramuscularly to senile *guinea pigs* for 30–90 days removed lipofuscin from hypothalamic neurones, and pigment elimination continued for a considerable period after stopping drug administration (Spoerri and Glees 1975). The occurrence of vacuoles within the capillary endothelium of the anterior hypothalamus was often encountered after prolonged (13 weeks) drug administration.

Cytoskeletal changes associated with abnormally phosphorylated τ-protein are hallmarks of a variety of *human* neurodegenerative disorders, including Alzheimer's disease. In *human* autopsy cases (mean age: 85.3 years) Schultz et al. (2000) noted periventricular inclusions mainly in the subependymal tissue surrounding the lateral ventricles or the third ventricle. Subpial predilection sites included the mamillary and tuberal regions, the basal forebrain, and the periamygdaloid cortex. In the baboon (*Papio hamadryas*; mean age: 26 years) neurones, astrocytes, and oligodendrocytes showed abnormally phosphorylated τ-protein predominantly in limbic regions as amygdala and hippocampal CA1 pyramidal cells (Schultz et al. 1999).

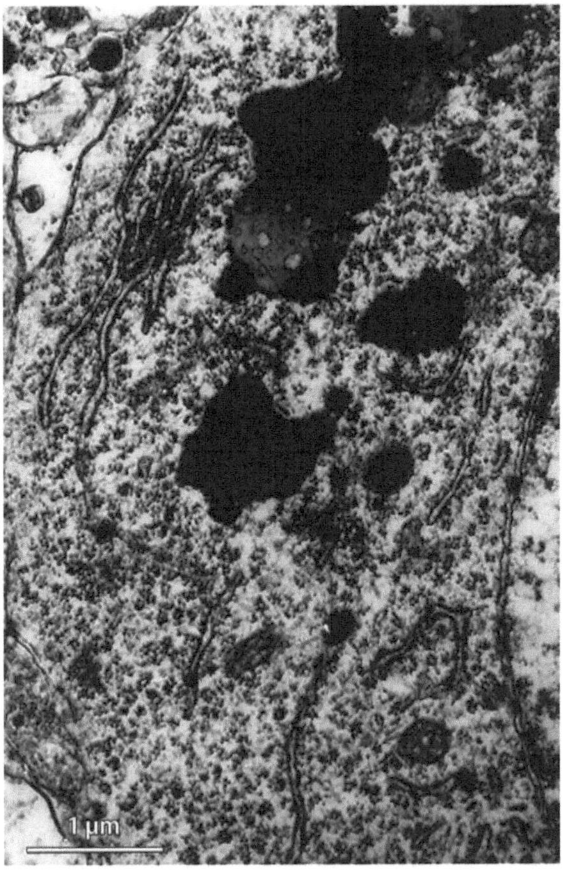

Fig. 332. Lipogenic pigmentation in a hippocampal pyramidal cell (block 2027) from an unmedicated 369 g old Wistar *rat* (No. 8). Under pentobarbital anaesthesia (30 mg/kg), the animal was perfused from the abdominal aorta with 2.5% glutaraldehyde in 0.1 M sodium cacodylate buffer (pH 7.4). Postfixation with 1% osmium tetroxide in sodium cacodylate buffer. Embedded in Epon 812 and sectioned at 50 nm. Lead citrate and uranyl acetate. Film 641-20

In aged rhesus monkeys (*Macaca mulatta*) the expression of ubiquitin, α B-crystallin, and the heat-shock protein 27 was revealed in spheroid bodies predominantly localized in the **globus pallidus** and pars reticulata of the **substantia nigra** (Schultz et al. 2001). The majority of the pallidonigral speroids expressing the stress proteins also accumulated iron, a Fenton catalyst.

In the **cerebellar** Purkinje cells of albino *rats* older than 24 months lipofuscin is easily detected by means of fluorescence microscopy (Heinsen 1979). In 30–38 months old *rats*, quantitatively and qualitatively different types of lipofuscin indicated to the existence of at least three subpopulations of Purkinje cells (Heinsen 1981).

In the cerebellar cortical capillaries of the senile Hannover-Wistar *rat*, processes of pericytes covered less of the capillary surface than in 3-months aged virgin *rats* (H. and Y.L. Heinsen 1983). Arithmetic and harmonic mean thickness of the endothelium and relative volume of mitochondria were reduced in both endothelial cells and pericytes, while luminal diameter of capillaries, harmonic and arithmetic mean thickness of the pericytes and their processes and of the basal laminae between endothelial cells and astrocytes, pericytes and astrocytes, and endothelial cells and pericytes increased.

The effects of ageing in the **vestibular nuclei** could have clinical interest due to the high prevalence of balance control and gait problems in the elderly. ALVAREZ et al. (2000) showed that neuronal loss occurs with ageing in the descending, medial, and lateral vestibular nuclei, but not in the superior.

Brains of patients afflicted with **Alzheimer's disease** show abnormal expression of numerous oxidative stress indicators (PAPPOLLA et al. 1992, SMITH et al. 1994, SAYRE et al. 1997, CALINGASAN et al. 1999) as well as extensive evidence of oxidative damage to proteins (SMITH et al. 1991) and nucleic acids (MECOCCI et al. 1994, LYRAS et al. 1997). Lipoprotein oxidizability measured in cerebrospinal fluid and plasma from 29 patient with Alzheimer's disease was significantly increased in comparison to 29 nondemented controls (SCHIPPLING et al. 2000). The levels of the hydrophilic antioxidant ascorbate were significantly lower in cerebrospinal fluid and plasma from patients with Alzheimer's disease. In plasma, α-carotene was significantly lower in patients with Alzheimer's disease compared to controls while α-tocopherol levels were indistinguishable between patients and controls. CHOI et al. (2002) detected oxidised plasma proteins with new carbonyl groups by reaction with 2,4-dinitrophenylhydrazine, followed by Western blotting with anti-2,4-dinitrophenylhydrazine antibody.

Two types of protein deposits, the amyloid plaques and the neurofibrillary deposits (tangles, neuropil threads) characterize the pathology of Alzheimer's disease. The latter are composed of largely paired helical filaments, which are in turn made up mainly of an insoluble form of the microtubule-associated protein τ (BRION et al. 1985). In the sequence of deposition of β-amyloid, THAL et al. (1999) found the first deposits in the temporal neocortex. The next regions affected were the hippocampal sector CA1 and the entorhinal cortex. Thereafter, the molecular layer of the dentate gyrus and the parvopyramidal layer of the presubiculum became involved in β-amyloidosis. Finally, amyloid plaques occupied the dentate hilus and occasionally even the pre-α layer of the entorhinal cortex.

To explore possible involvement of stress in τ hyperphosphorylation, OKAWA et al. (2003) four brain regions of *mice* subjected to cold water stress.

Human brain tissue obtained from cases of Alzheimer's disease, an endogenous inhibitor (<3500 Da) of antagonist binding to the muscarinic acetylcholine receptor contained free heme, a well-established source of oxidative stress capable of generating free radicals and causing neurotoxicity (VENTERS jr. et al. 1997). While $FeSO_4$, microperoxidase and hemin all inhibited agonist binding to the muscarinic acetylcholine receptor, only hemin shared the inhibitor's requirement for reduced glutathione (GSH). Both the free radical scavengers Trolox and Mn^{2+}, and the metal chelator EDTA, blocked the activity of the endogenous inhibitor. The antioxidants oestrogen, vitamin E and vitamin C all protected the muscarinic acetylcholine receptor from irreversible inhibition by the endogenous inhibitor or hemin.

In screening indole compounds for neuroprotection against amyloid β-protein, potent neuroprotective properties were uncovered for an endogenous related species, indole-3-propionic acid, the capacity of which to scavenge hydroxyl radicals exceeded that of melatonin (CHYAN et al. 1999).

In **Parkinson's disease** oxidative stress may play a principal role in the degeneration of dopaminergic neurones:

- Levodopa has the potential to become a levodopa radical (OGAWA et al. 1993);
- An increase in iron deposition promotes free radical generation (DEXTER et al. 1989);
- High level of lipid peroxidation (DEXTER et al. 1989);
- Decreased activity of complex I in the mitochondrial respiratory chain (SCHAPIRA et al. 1990);
- Glutathione, glutathione peroxidase, and catalase levels are reduced in the brain in Parkinson's disease (AMBANI et al. 1975, KISH et al. 1985).

Accumulation of pigments in the atrophic putamina in striatonigral degeneration, a true supranigral form of parkinsonism, in which the striatal lesion both precedes and exceeds the nigral involvement, was analysed by BORIT et al. (1975). Three kinds of granules were commonly aggregated on groups or even coalesced, and were frequently juxtaposed to a nucleus or surrounded by sheaves of glial filaments. All three types of granules had two common components: (*a*) a global shapes, electron lucent, and (*b*) an irregularly shaped, finely granular and moderately electron-dense component.

17.2 Spinal Cord

In the **cervical anterior horn** of ageing *mice* SEKHON and MAXWELL (1975) found pigment in some nerve cells as early as six weeks after birth. Various types of membrane-bound granules were characterised on the basis of their shape, size and fine structure into primary lysosome-like (L_1) granules (dense bodies), autophagic vacuole-like (L_2) granules and mature (L_3) pigment granules of complex substructure and irregular configuration. L_1, L_2 and L_3 types of granules appear to represent respectively early, intermediate and mature stages in a developmental continuum of lipofuscin pigment granules. Transitional stages suggest that mature L_3 pigment granules evolve by gradual alteration of lysosome-like L_1 and L_2 granules.

17.3 Autonomous Nervous System

In the colonic **myenteric plexus** the number of nNOS-immunoreactive cells and nNOS synthesis were significantly reduced in old *rats* (TAKAHASHI et al. 2000). In contrast, expression of neuronal cytosolic protein, protein gene product 9.5, in colonic tissues was not affected by old age. Basal and veratridin-induced release of L-[^{3}H]citrulline were significantly decreased in colonic tissues from aged *rats*, compared to young *rats*.

Muscarinic cholinergic receptor density investigated by specific binding of quinuclidinyl[^{3}H]benzilate was significantly ($P < 0.01$) decreased in the frontal cortex, the hippocampus, and the striatum of male Wistar *rats* aged 24 months as compared with animals aged 3 months (SCHEUER et al. 1999).

***N*-Methyl-D-aspartate receptor density** investigated by specific binding of tritiated (5*R*, 10*S*)-(+)-5-methyl-10,11-dihydro-5*H*-dibenzo[a,d]-cyclohepten-5,10-imine hydrogenmaleate was significantly ($P < 0.01$) decreased in the frontal cortex, the hippocampus, the striatum, and the cerebellum of male Wistar *rats* aged 24 months as compared with animals aged 3 months (SCHEUER et al. 1999).

17.4 Pineal Gland

In the pineal gland pigmentation begins at puberty (FARINA 1914). In healthy *men* aged 23–50 years who died a violent death, BARGMANN (1943, p.385) very rarely found pigmented pinealocytes. This pigment predominantly may be a wear-and-tear pigment. The number, distribution and composition of the lipid-containing parenchymal cells change with age (QUAY 1973). In youth, they are peripherally distributed in the organ, are few in number and usually do not contain pigment. In adulthood and later life, they are found within lobules throughout most of the organ, and some of them contain yellow to brown pigment granules, some of which are capped with lipid.

A second kind of lipid-containing cell becomes more prominent in adulthood and older age (QUAY 1973). These contain larger lipid droplets and lie primarily within the stromal areas and interlobular septa. They are principally neuroglial and phagocytic in nature, rather than some type of parenchymal cell.

The pineal glands of *rats* 12–28 months old exhibited regressive changes of different intensity with age (BOYA and CALVO 1984). In type I pinealocytes there was a marked increase in dense bodies as well as the occasional appearance of wide cell profiles full of vesicles. Type II pinealocytes showed nuclear infoldings and cytoplasmic deposits of lipofuscin.

Human pineal glands of adult or older age sometimes have iron-rich interlobular pigmented cells.

The claim that melatonin can reverse ageing is based on a study in BALB/c female *mice* 15 months of age where addition of 10 μg melatonin per ml of drinking water during a fixed darkness cycle prolonged their lives from 23.8 to 28.1 months and preserved aspects of their youthful state (PIERPAOLI and REGELSON 1994). Melatonin increased the lag time of formation of oxidised low density lipoprotein only at a concentration of 10 μM (SEEGER et al. 1997). In contrast, 6-hydroxymelatonin, serotonin and *N*-acetylserotonin as well as vitamin E showed inhibitory effects starting at 1 μM. Thus the antioxidative action of melatonin was negligible compared with the effect of its main metabolite, its precursors and of vitamin E.

In male Sprague-Dawley *rats* melatonin treatment for 12 weeks decreased ($P < 0.05$) body weight by 7 % relative to controls, relative intraabdominal adiposity (by 16 %), plasma leptin (by 33 %), and plasma insulin (by 25 %) while increasing ($P < 0.05$) locomotor activity (by 19 %), core body temperature (by 0.5 °C), and morning plasma corticosterone (by 154 %), restoring each of these parameters toward more youthful levels (WOLDEN-HANSON et al. 2000).

17.5 Neurosecretory System

Age-related hypothalamo-pituitary-adrenocortical hyperresponsiveness is evident at multiple levels, prominently including hypothalamic paraventricular nucleus neurones controlling ACTH secretion. These neurones shoe a number of age-related changes in activity that may be ultimately responsible for altered glucocorticoid secretion.

In neurosecretory system of Wistar *rats* aged 15–20 months, PILGRIM (1970) found a strong increase in the number of Herring bodies showing electron dense bodies of various sizes and shape with an amorphous matrix and whirled membranes. DELLMANN and RODRÍGUEZ (1970) distinguished three types of Herring bodies: normal (type I), degenerating (type II), and regenerating (type III). Type II is shown in reoxygenation of *rats* pre-treated with carbocromene and *N*-benzhydryl-*N'*-*p*-hydroxybenzylpiperazine; respectively, and type III in reoxygenation after piracetam pre-treatment (SCHILLER 1998). SLOPER (1955) mentioned a faint sudanophilia of some Herring bodies. With the periodic acid-Schiff technique, which demonstrates unsaturated lipids as well as carbohydrate, there was some reaction in occasional Herring bodies. With one modification of the periodic acid-Schiff technique (SCHIEBLER 1952), Herring bodies were more evident. In some Herring bodies formazan was produced from blue tetrazolium chloride and 2,5-diphenyl-3-4-styrylphenyl tetrazoliumchloride.

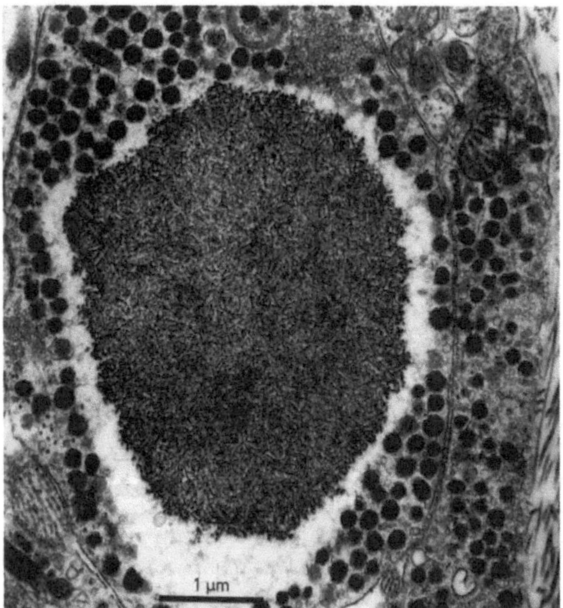

Fig. 333. A type IIa Herring body. Pituitary pars nervosa (block 1992) of a 350 g old Wistar *rat* (No. 6) medicated for 4 months with 400 mg piracetam per kg body weight × day. On October 20, 1976 under pentobarbital anaesthesia (30 mg/kg), the animal was perfused from the abdominal aorta with 2.5% glutaraldehyde in 0.1 M sodium cacodylate buffer (pH 7.4). Postfixation with 1% osmium tetroxide in sodium cacodylate buffer. Embedded in Epon 812 and sectioned at 50 nm. Lead citrate and uranyl acetate. Film 652-21

Vasoactive intestinal peptide (VIP) mRNA selectively decreased in the ageing suprachiasmatic nucleus of the Syrian *hamster* (DUNCAN et al. 2001).

17.6 Pituitary Pars Intermedia

The pituitary pars intermedia of both lean and obese Zucker *rats* as well as of Sprague-Dawley *rats* aged 18 months showed indented nuclear envelops, while those of young animals were generally smooth (SALAND et al. 1990). Lipid droplets and lysosomes, rarely seen in tissue from young animals, were frequently observed in endocrine cells of older *rats*. Most cells had an abundance of secretory granules, suggestive of enhanced storage of peptides in the cytoplasm. Nerve terminals which were present

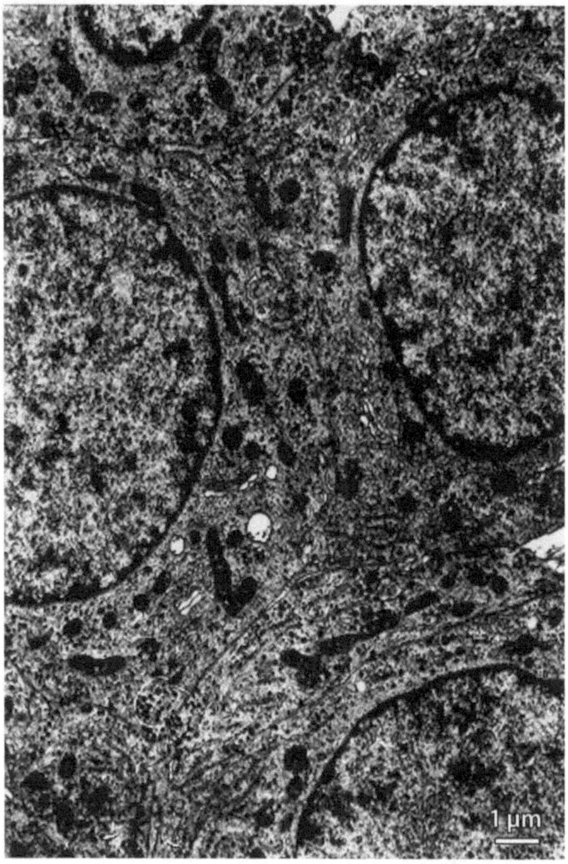

Fig. 334. Pars intermedia of the pituitary gland (block 1993) from a 350 g old Wistar *rat* (No. 6; breeder S. Ivanovas, Kissleg/Allgäu).) medicated for 4 months (June 28 to October 20, 1976) with 400 mg piracetam per kg body weight × day contained in the food (powdered Altromin® R) and drinking water. Under pentobarbital anaesthesia (30 mg/kg), the animal was perfused from the abdominal aorta with 2.5% glutaraldehyde in 0.1 M sodium cacodylate buffer (pH 7.4). Postfixation with 1% osmium tetroxide in sodium cacodylate buffer. Embedded in Epon 812 and sectioned at 50 nm. Lead citrate and uranyl acetate. Film 652-25

among endocrine cells contained myelin figures in some of the old *rats*, and may indicate degenerative changes, while other terminals appeared normal.

17.7 Pituitary Pars Distalis

The pituitary pars distalis of *man* on the light microscopic level with age showed a decrease in acidophils (COOPER 1925, RASMUSSEN 1938), and chromophobes (RASMUSSEN 1938); increased vacuolation in chromophils (PARSON 1935); and an increase in interstitial connective tissue (RASMUSSEN 1934). SEVERINGHAUS (1944) reported cytological evidence of increased activity in chromophils in *women* after menopause, and contrasted this with the inactive appearance of these cells in older *men*.

A combined immunocytochemical, morphometric, and clinicopathologic analysis of growth-hormone-producing and prolactin-producing pituitary cells in 28 subjects ranging in age from 16 to 90 years showed a significant age-related decline in the number and size of growth-hormone-producing cells, which was most marked in the transition from youth to middle age (SUN et al. 1984). There was also a significant age-related decline in the number of pituitary parenchymal cells but not in pituitary weight. Prolactin cells did not show a significant decline in number with age.

Thyrotroph hypertrophy and relative hyperplasia were both present in *human* aged pituitaries (ZEGARELLI-SCHMIDT et al. 1985). No consistent relation with the histological features of the thyroid or adrenal, or with the cause of death, was demonstrable.

In *rats*, according to WOLFE et al. (1938) the percentage of eosinophils decreases, chromophobes increases, and there are degenerative changes in the pituitary cells of older individuals. The majority of the spontaneously occurring pituitary hyperplasias and adenomas in senile *rats* consisted of prolactin-containing cells (EL ETREBY and GÜNZEL 1975). In *rats* of the strain Chbb: THOM (SPF) aged about 2 years, UEBERBERG et al. (1975) differentiated between dark and clear cells. Both of them were mostly free of granules or showed only few granules. The dark cells were characterised by a strongly developed rough endoplasmic reticulum („Nebenkern") and a large Golgi field. The clear cells are degenerative forms.

In the *mouse*'s pituitary, STEIN et al. (1942) found a decrease in both the number and size of the acidophils in old age in both sexes, and an increase in the number of chromophobes in senile females.

In the golden hamster (*Cricetus auratus*), the basophils showed an increase in degranulation and occurrence of pyknotic nuclei (SPAGNOLI and CHARIPPER 1955).

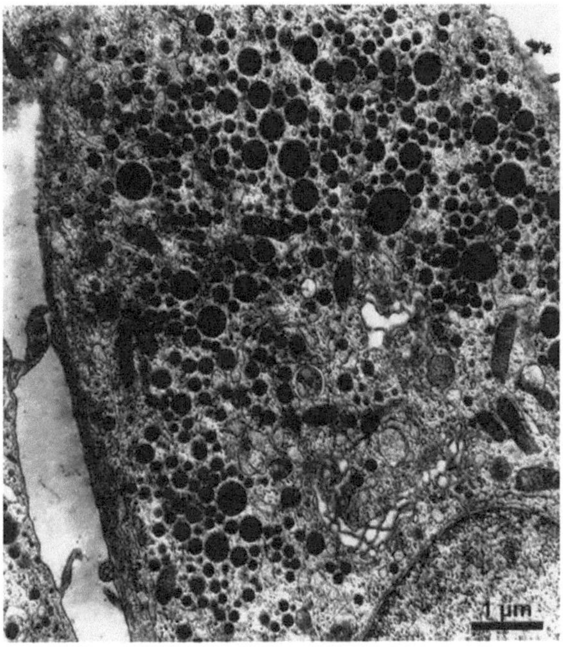

Fig. 335. Pituitary prolactin cell (block 2053) from a unmedicated 331 g old Wistar *rat* (No. 9; breeder S. Ivanovas, Kissleg/Allgäu). Under pentobarbital anaesthesia (30 mg/kg) the animal was perfused from the abdominal aorta with 2.5 % glutaraldehyde in 0.1 M sodium cacodylate buffer (pH 7.4). Postfixation with 1 % osmium tetroxide in sodium cacodylate buffer. Embedded in Epon 812 and sectioned at 50 nm. Lead citrate and uranyl acetate. Film 675/79. – Subplasmalemmal microfilaments and microtubules

The responsiveness to growth hormone-releasing factor was diminishes in ageing male *rats* (CEDA et al. 1986).

Growth hormone and insulin-like growth factor I serum levels decrease with age and appear to contribute to the decline of body functions that is associated with normal ageing (RUDMAN 1985, CORPAS et al. 1993, ROSEN and CONOVER 1997).

Muscarinic receptors were decreased with age in the *rat* adenohypophysis (AVISSAR et al. 1981).

A hypersecretion of the hormone is generally implicated in the pathogenesis of the polypeptide-hormone-derived amyloids (WESTERMARK 1994). The pituitary gland can maintain a prolactin secretory capacity at old age (ROLANDI et al. 1982), and age per se does not seem to alter the rate of secretion of prolactin in *humans* (YAMAJI et al. 1976), nor is the circadian rhythms of prolactin plasma levels modified at old age (TOUITOU et al. 1981).

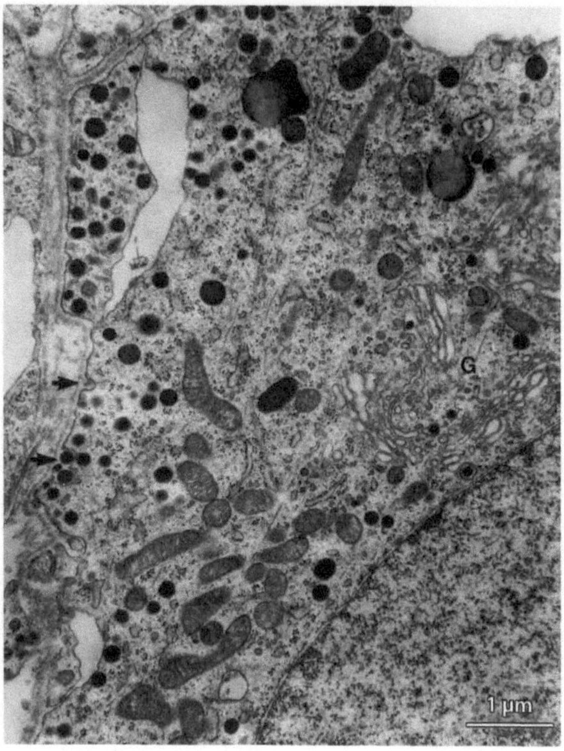

Fig. 336. Prolactin cell (block 2073) from a 335 g old Wistar *rat* (No. 10; breeder S. Ivanovas, Kissleg/Allgäu) medicated for 4 months (June 28 to October 20, 1976) with 400 mg piracetam per kg body weight × day contained in the food (powdered Altromin® R) and drinking water. Under pentobarbital anaesthesia (30 mg/kg) the animal was perfused from the abdominal aorta with 2.5 % glutaraldehyde in 0.1 M sodium cacodylate buffer (pH 7.4). Postfixation with 1 % osmium tetroxide in sodium cacodylate buffer. Embedded in Epon 812 and sectioned at 50 nm. Lead citrate and uranyl acetate. Film 688/79. ù The incretion, which is produced in a large Golgi area (G) as electron dense granules is transferred to the plasmalemma and secreted by emiocytosis (*arrows*)

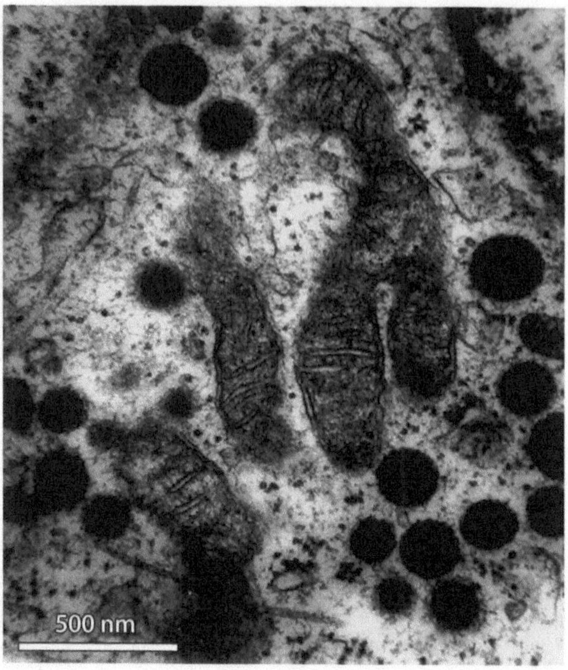

Fig. 337. "Bud-like" outgrowth of mitochondria in a prolactin cell (block 2073) from a 335 g old Wistar *rat* (No. 10; breeder S. Ivanovas, Kissleg/Allgäu) medicated for 4 months (June 28 to October 20, 1976) with 400 mg piracetam per kg body weight × day contained in the food (powdered Altromin® R) and drinking water. Under pentobarbital anaesthesia (30 mg/kg) the animal was perfused from the abdominal aorta with 2.5 % glutaraldehyde in 0.1 M sodium cacodylate buffer (pH 7.4). Postfixation with 1 % osmium tetroxide in sodium cacodylate buffer. Embedded in Epon 812 and sectioned at 50 nm. Lead citrate and uranyl acetate. Film 780/79

17.8 Thyroid

During normal circumstances the weight of the thyroid remains constant in adulthood, with a slow decrease in the senium. Within this time, cell turnover is very rare; only about five divisions per cell during adulthood have been calculated (Coclet et al. 1989). For this reason only a few mitotic cells are found in vivo; they may be due to the compensation of cell death and tissue repair (Rognoni et al. 1987).

In thyroid of ageing domestic *cats* Ives et al. (1975) found an increase in size and number of lipofuscin vesicles located predominantly in the apical areas of the **follicular cells.** In male Sherman *rats* of 14 months or older dense "lysosome-like" granules (0.1-2.5 µm) were more numerous than in 40-45-day-old animals (Youson and Van Heyningen 1968). In Sprague-Dawley *rat* beginning at 56 weeks, some of the thyroid follicles were hyperdistended with colloid, had irregular lumina, and were lined by flattened epithelium (Rao-Rupanagudi et al. 1992). Foci of brown pigmentation were seen in a few rats aged 56 weeks and older. The pigmentation was seen in one or two of the large irregular follicles in the outer part of the thyroid. Sometimes the pigment was accompanied by cellular debris. The pigment has been shown to be iron and periodic acid-Schiff positive. In the cream *hamster* ageing induced an accumulation of lysosomal dense bodies (Nève et al. 1981).

Light microscopy of the thyroid gland of pineal-grafted and control ageing *mice* showed a very remarkable maintenance of a youthful thyroid morphology, as compared to control (Pierpaoli and Regelson 1994).

In the *opossum*, Bensley (1914) and Bargmann (1939) depicted intracytoplasmic proteinaceous crystals in highly activated thyroid follicular cells. Siami and Larras-Regard (1986) found intracytoplasmic crystalline bodies of various sizes in thy-

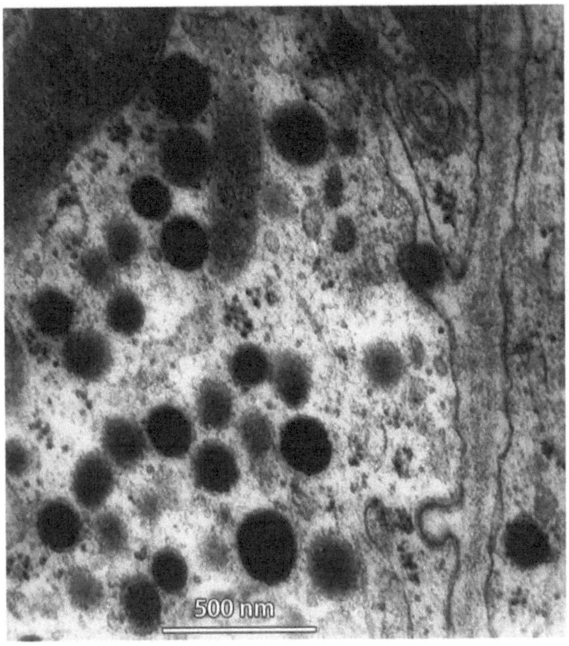

Fig. 338. Exocytosis of specific granules. After exocytosis of matrix material, coated vesicles are formed (bottom right) which later pinch off from the plasmalemma. Prolactin cell (block 2073) from a 335 g old Wistar *rat* (No. 10; breeder S. Ivanovas, Kissleg/Allgäu) medicated for 4 months (June 28 to October 20, 1976) with 400 mg piracetam per kg body weight × day contained in the food (powdered Altromin® R) and drinking water. Under pentobarbital anaesthesia (30 mg/kg) the animal was perfused from the abdominal aorta with 2.5 % glutaraldehyde in 0.1 M sodium cacodylate buffer (pH 7.4). Postfixation with 1 % osmium tetroxide in sodium cacodylate buffer. Embedded in Epon 812 and sectioned at 50 nm. Lead citrate and uranyl acetate. Film 781/79

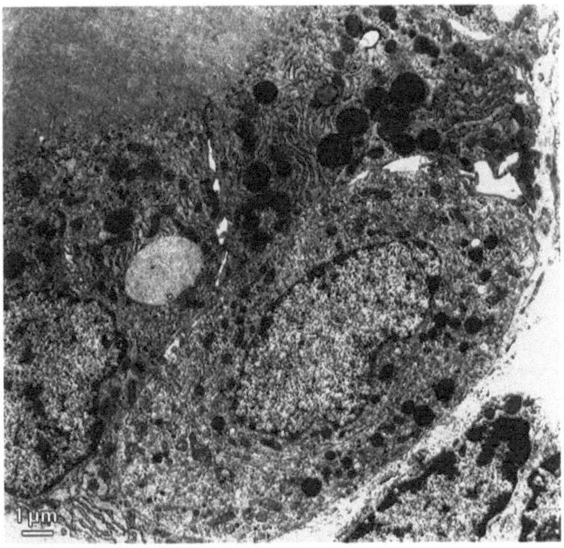

Fig. 339. Follicular cells and a C-cell in the thyroid (block 2184) of an unmedicated 335 g old Wistar *rat* (No. 16; breeder S. Ivanovas, Kissleg/Allgäu) heavily laden with electron dense pigment granules. Under pentobarbital anaesthesia (30 mg/kg), the animal was perfused from the abdominal aorta with 2.5 % glutaraldehyde in 0.1 M sodium cacodylate buffer (pH 7.4). Postfixation with 1 % osmium tetroxide in sodium cacodylate buffer. Embedded in Epon 812 and sectioned at 50 nm. Lead citrate and uranyl acetate. Film 646-24

roid cells of 10-month old *mice* and in younger animals under chronic lithium treatment (3 mg or 0.014 mEq Li). These crystals were frequently surrounded by small microvesicular and dense bodies or enclosed in larger vesicles having a dense content. The crystalline skeleton was a network of protein fibres assembled in a characteristic axis with a periodicity of 8 nm. A deficiency of thyroid cell metabolism related to ageing or lithium gluconate treatment would lead to an accumulation of substances of a crystalline pattern.

Calcitonin immunoreactive cells increase with age (Howieson Gibson et al. 1980). O'Toole et al. (1985) found in *persons* ≥ years 2.97 ± 3.69 calcitonin immunoreactive cells per mm^2 versus 0.99 ± 1.46 cell per mm^2 in patients 16 to 39 years of age. The number of argyrophilic granules – it is not known whether the argyrophilia of the granules is due to their lipoid content or an interaction of the stored amines or peptides with silver salts (Lietz 1971) – decreases with increasing age (Bargmann 1939). The absolute irregularity of the distribution

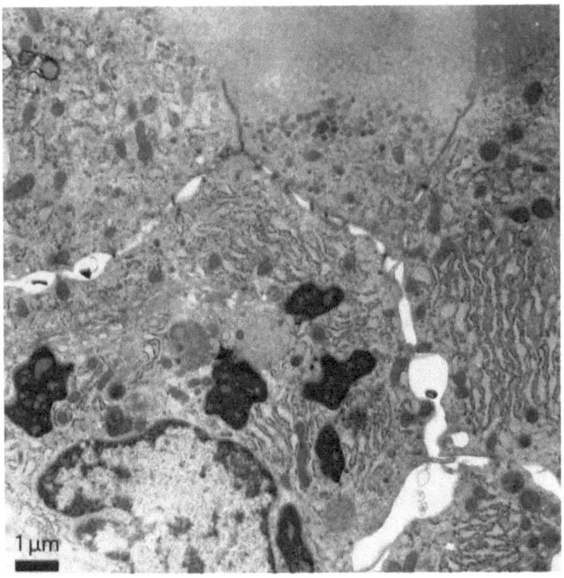

Fig. 340. Lipogenic pigmentation in the follicular cells (block 2144) from a 307 g old Wistar *rat* (No. 14; breeder S. Ivanovas, Kissleg/Allgäu) medicated for 4 months (June 28 to October 20, 1976) with 400 mg piracetam per kg body weight × day contained in the food (powdered Altromin® R) and drinking water. Under pentobarbital anaesthesia (30 mg/kg), the animal was perfused from the abdominal aorta with 2.5 % glutaraldehyde in 0.1 M sodium cacodylate buffer (pH 7.4). Postfixation with 1 % osmium tetroxide in sodium cacodylate buffer. Embedded in Epon 812 and sectioned at 50 nm. Lead citrate and uranyl acetate. Film 645-51

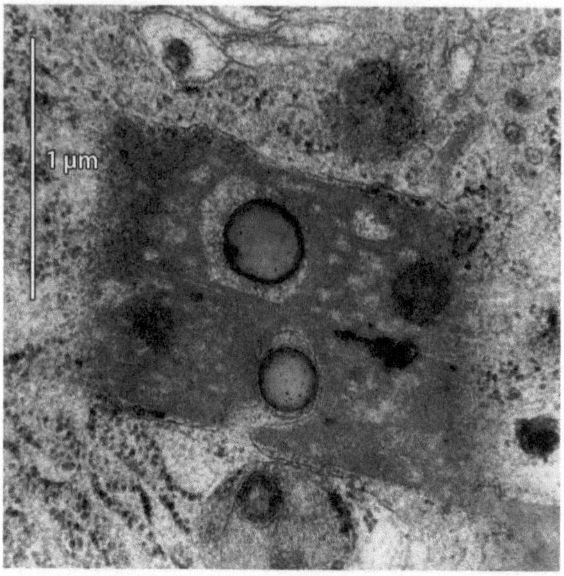

Fig. 341. Crystalloids in a follicular cell (block 1984) from a 350 g old Wistar *rat* (No. 6; breeder S. Ivanovas, Kissleg/Allgäu) medicated for 4 months (June 28 to October 20, 1976) with 400 mg piracetam per kg body weight × day contained in the food (powdered Altromin® R) and drinking water. Under pentobarbital anaesthesia (30 mg/kg), the animal was perfused from the abdominal aorta with 2.5 % glutaraldehyde in 0.1 M sodium cacodylate buffer (pH 7.4). Postfixation with 1 % osmium tetroxide in sodium cacodylate buffer. Embedded in Epon 812 and sectioned at 50 nm. Lead citrate and uranyl acetate. Film 638-30

of the calcitonin cells makes it impossible to confirm an increase or decrease in C-cell number when randomly obtained specimens are used (LIETZ 1971). An attempt by Von Wagenhoff (unpublished) to study the topographic distribution and frequency in human thyroid glands of varying age demonstrated the concentration of C-cells in the dorso-medial areas of both lobed. This is consistent with the findings of ALIAPOULIOS and ROSE (1970) and SOLCIA et al. (1970).

In the Wistar *rat*, the proportion of C-cells to follicular cells was 4.5 % on the day of birth and increased progressively to 10.4 % by 120 days (MARTÍN-LACAVE et al. 1992). The highest density of C-cells was noted in the mid-region of the lobes along a longitudinal axis. From 3 to 24 months of life, 27.5 % of female *rats* showed a normal C-cell pattern, 55.0 % showed C-cell hyperplasia ' and 17.5 % showed C-cell tumours; while 57.5 % of male *rats* showed a normal C-cell pattern, 32.5 % showed C-cell hyperplasia, and 10 % showed C-cell tumours (MARTÍN-LACAVE et al. 1999).

In the Sprague-Dawley *rat*, areas of C-cell hyperplasia appeared with age (RAO-RUPANAGUDI et al. 1992).

Age has a very profound effect on the hypocalcaemic response to calcitonin (MILHAUD et al. 1967). COPP and KUCZERPA (1968) observed that the effect obtained following i.p. injection of 55 mU/100 g to 9-month-old *rats* was only 7 % of that obtained when a similar dose was injected into 5-week-old *rats*.

17.9 Parathyroid

In the *human* parathyroid gland multiple lipid vacuoles could indicate an age-related inactivation (ALTENÄHR 1972). Obviously the lipid bodies correspond to the wear-and-tear pigment described by HAMPERL (1934). In senile *dogs* especially the oxyphil cells and the mitochondria-rich cells of the other type, frequently contained bizarre mitochondria that varied remarkably in shape, size and arrangement (SETOGUTI 1977). The profiles of some mitochondria were strikingly elongated and arranged either in concentric or in parallel fashion. Others were larger in size and oval or elliptical in shape with unusual cristae which frequently formed

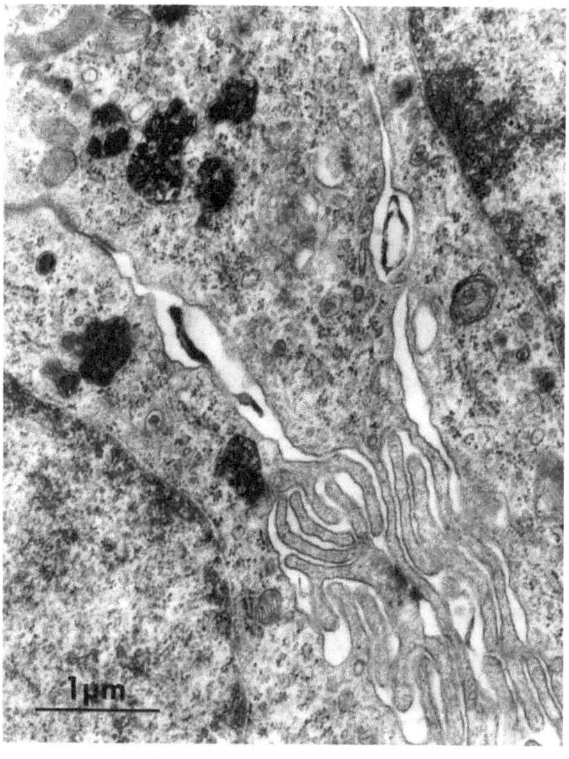

Fig. 342. Lipogenic pigmentation in the parathyroid gland parenchymal cells (block 1911) of an unmedicated old Wistar *rat* (No. 1; breeder S. Ivanovas, Kissleg/Allgäu). Under pentobarbital anaesthesia (30 mg/kg), the animal was perfused from the abdominal aorta with 2.5 % glutaraldehyde in 0.1 M sodium cacodylate buffer (pH 7.4). Postfixation with 1 % osmium tetroxide in sodium cacodylate buffer. Embedded in Epon 812 and sectioned at 50 nm. Lead citrate and uranyl acetate. Film 639-37

a bundle and ran parallel with the long axis to the inner limiting membrane. Still others were cup-shaped, ring-shaped or hook-shaped with one end swollen like a bulb.

17.10 Adrenal Cortex

In 23 healthy aged *subjects* (73.0 ± 8.0 years), the impaired hypothalamo-pituitary-adrenal function became apparent in a most significant decline of the plasma levels of dehydroepiandrosterone and dehydroepiandrosterone-sulphate (MARTÍNEZ-TABOADA et al. 2002). A close correlation between immune changes with ageing and dehydroepiandrosterone response to ACTH stimulation was found. There was an inverse correlation of lymphocyte changes with the plasma levels of steroids, especially dehydroepiandrosterone and its metabolite, dehydroepiandrosterone-sulphate.

In the adrenal cortex of old *rats*, lipogenic pigmentation of parenchymal cells is usually seen in the inner zona fasciculata and zona reticularis. Beginning at the age of about 300 days, macrophages show lipofuscin (VON SEEBACH et al. 1975). In *mice* aged 23 months SETOGUTI et al. (1979) described specific lamellar structures of agranular endoplasmic reticulum in the zona reticularis cells.

In 28-month-old male albino *rats*, the ultrastructure of the adrenocortical cells following 3 h electric stimulation of the ventromedial hypothalamic nuclei revealed a decreased number of lipid droplets in the cytoplasm, primarily in the zona fasciculate, a decrease in the electron opacity of the mitochondrial matrix and a decreased number of intramitochondrial vesicles, as compared with the unstimulated old *rat*, and an increase in the tubule lumen of the endoplasmic reticulum (SHAPOSHNIKOV and BEZRUKOV 1985).

The maximum corticosterone concentration after incubation with ACTH was significantly lower ($P < 0.001$) in the 12-month-old (40 ±7 ng/μg DNA) and 18-month-old (28±3 ng/μg DNA) Sprague-Dawley *rat* compared to that in 2-month-old controls (102±9 ng/μg DNA) (POPPLEWELL et al. 1986).

In the *mouse*, SAMORAJSKI and ORDY (1967) reported that the quantity of pigment increased with age and that many of the cells coalesced to form multinucleated giant cells with unusually large pigment inclusions. Ceroid is rare in untreated *mice* until after 24 months of age, and diethylstilboestrol increases the incidence of the deposition of ceroid in the adrenals of *mice* (FRITH 1983).

In 28-month-old male albino *rats*, the ultrastructure of the adrenomedullary cells following 3 h electric stimulation of the ventromedial hypothalamic nuclei revealed a decrease in the electron opacity of the karyolymphe and swollen nuclei in the chromaffin cells, mitochondria with electron translucent matrix and swollen cristae, and fragmented tubules of the rough endoplasmic reticulum (SHAPOSHNIKOV and BEZRUKOV 1985). The cytoplasm of the cells was oedematous and the number of the secretory granules was insignificantly decreased. Some secretory cells showed numerous mitochondria with dense matrix, vesicle-shaped dyscomplexed cristae and hypertrophied Golgi apparatus. In the norepinephrine cells, there were swollen mitochondria with dyscomplexed cristae, electron lucent matrix and plenty of norepinephrine granules.

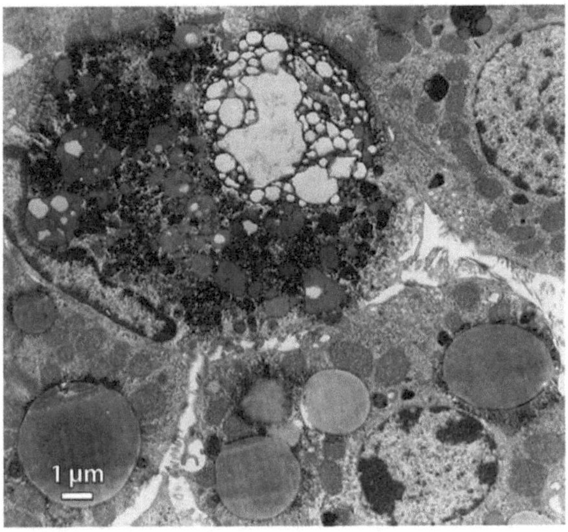

Fig. 343. Lipogenic pigmentation in the adrenal cortex (block 2002) from an unmedicated 335 g old Wistar *rat* (No. 7; breeder S. Ivanovas, Kissleg/Allgäu). Under pentobarbital anaesthesia (30 mg/kg), the animal was perfused from the abdominal aorta with 2.5 % glutaraldehyde in 0.1 M sodium cacodylate buffer (pH 7.4). Postfixation with 1 % osmium tetroxide in sodium cacodylate buffer. Embedded in Epon 812 and sectioned at 50 nm. Lead citrate and uranyl acetate. Film 647-13

17.11 Thymus

Involution of the thymus with age begins early in life and by mid-life the cellular mass, predominantly of the thymic cortex has fallen by 75 to 80 per cent (BOYD 1932). Ageing is associated with a declining number of all thymocytes after the CD3-, CD4-, CD8-, CD25+, CD44- stage of thymocyte development (ASPINALL 1997). The crucial transition from CD3-, CD4-, CD8-, CD25+, CD44- to CD3-, CD4-, CD8-, CD25+, CD44+ thymocytes depends upon the rearrangement of the T cell antigen receptor β chain genes. Thus it is possible that an impaired capacity of thymocytes in old *mice* to arrange their TCR β genes underlies the decline in thymocyte number and thymic involution. A similar age-associated change in thymocyte development has been seen following irradiation of old *mice* (THOMAN 1997).

The involution of the thymic epithelial space is not influenced by puberty-dependent endocrine changes under physiological conditions during ageing (STEINMANN 1986). TOSI et al. (1982) have complemented their study of the relative portion of the cortex during ageing with a non-linear regression analysis. They found that involution of the cortex older than 15 years (puberty) would fit better to a negative exponential function than to a negative linear function. However, the authors reduced their study to a model of postpuberty involution, which does not reflect the involution of the thymic epithelium. According to STEINMANN (1986) the continuous degeneration of the thymic epithelial space starts in the very first years of life and exhibits a constant velocity during the first decade. The velocity of involution decreases progressively. Remnants of thymic epithelial tissue with cortical lymphocyte population are preserved beyond 100 years of age.

By using a sensitive immunoperoxidase technique, serological determinants to terminal deoxynucleotidyl transferase (EC 2.7.7.31)were found in cortical lymphocytes of thymus biopsies of adult and aged persons (21–70 years) of either sex (STEINMANN and MÜLLER-HERMELINK 1984). In contrast to biochemical determinations, specific immunohistological studies showed that terminal deoxynucleotidyl transferase is continuously presnent in very large numbers of cortical thymic lymphocytes during *human* life, despite subtotal physiological involution of the thymus.

Buffalo/Mna *rats* do not show age-related thymic involution, but rather develop thymic hyperplasia with advancing age (HIROKAWA et al. 1990). This thymic growth is expansible and there is no infiltration of the surrounding tissue. Because the enlarging thymus occupies the thoracic cavity, most of the *rats* die of respiratory failure by the age of 24 months. Thymic enlargement is due to primary hyperplasia of cortical epithelial cells and the large number of proliferating lymphocytes. The hyperplasic epithelial cells are bizarre in shape and strongly positive when stained with Th-3 monoclonal antibody, anti-thymosine antibody and anti-EGF antibody, but negative with Th-4 monoclonal antibody. The patterns of distribution of CD-5$^+$, CD-4$^+$ and CD-8$^+$ lymphocytes within the hyperplasic thymus are similar to those seen in young *rats* of other species. The high level of T-cell emigration from the thymus to the periphery appears to persist throughout life, since the percentage of normal splenic T-cells also increase with advancing age and exceed 70 % of the total by 24 months of age.

17.12 Lymphoid Tissues

In addition to thymic involution, there are other important age-associated changes in the structure of the lymphoid tissues. The number of mature lymphocytes and plasma cells in the bone marrow markedly increases while the number of germinal centres in the lymph nodes and spleen is reduced with age (BENNER et al. 1981, GONZALEZ-FERNANDEZ et al. 1994).

17.13 Myocardium

Cardiac function deteriorates with age, and endogenous damage to mitochondrial DNA is believed to be a major contributory factor to ageing. The 7 kb deletion of mtDNA was commonly detected in elderly subjects, and the proportion of deleted mtDNA to normal mtDNA increased with age (HATTORI et al. 1991). A chronic ischaemia-like state is induced in the myocardium, which might contribute to the genesis of ageing heart (presbycardia).

Protein synthesis increased from a low value in 1–3-months old C57Bl/6J *mice* to a maximum in adult animals (8–9 months), followed by a decrease to 42 % of the adult value at 27 months of age (GEARY and FLORINI 1972).

The rate of state 3 respiration of mitochondria isolated from hearts of male *rats* exposed to ischaemia (25 min) was approximate 25 % less than that of controls, independent of age (LUCAS and SZWEDA 1998). Reperfusion (40 min) caused a further decline in the rate of state 3 respiration in hearts isolated from 24- but not 8-month-old *rats*. Furthermore, 4-hydroxy-2-nonenal modification of mitochondrial protein (30 and 44 kDa) occurred only during reperfusion of hearts from 24-month-old *rats*.

Rat myocardium ultrastructurally differed more intensely between 3 and 6 than between 18 and 24 months of age (WELT et al. 2000). Lipid drops and mitochondrial degeneration were more prominent in the older *rats*. Ageing-related alterations were limited to interfibrillar mitochondria, while subsarcolemmal mitochondria remained unaffected (FANNIN et al. 1999). Ageing decreased the rate of oxidative phosphorylation in interfibrillar mitochondria, including when stimulated by electron donors specific for cytochrome oxidase. Cytochrome oxidase enzyme activity was decreased in interfibrillar mitochondria from ageing hearts, while activity in subsarcolemmal mitochondria remained similar to adult controls. Oxidative damage to mitochondrial DNA was inversely related to maximum life span in the heart of *mammals* (Barja and Herrero 2000).

Lipofuscin accumulation in *human* myocardium with age shows a linear relationship (Strehler et al. 1959). In the *dog* the percentage of myocardial volume occupied by lipofuscin pigment is a function of age (Munnell and Getty 1968). In the Japanese monkey, *Macaca fuscata*, cardiac lipofuscin first appears at approximately two years of age (Nakano et al. 1989), in the *rat* at an age of 2–3 months (Feldman and Nararatnam 1981). In the myocardial cells of the left ventricle of ageing C57BL/6J female *mice* between 6 and 27 months of age and also in experimental ageing *mice* subjected to regular endurance-running schedules for two or ten month periods lipofuscin accumulated progressively (Coleman et al. 1982). Both lipid and lipofuscin appeared to be released from the ageing myocardial fibres and to be taken up by macrophages of the surrounding connective tissue. Lipid droplets and lipofuscin were also found in the endothelial cells of the capillaries. No significant qualitative ultrastructural differences were found in lipid or lipofuscin of endurance-trained *mice* as compared with age-matched controls. The 4-hydroxy-2-nonenal modification of mitochondrial protein (~30 and 44 kDa) occurred only during reperfusion of the hearts from senescent (24-month-old) Fisher 344 *rats*, but not from adult (8-month-old) animals (Lucas and Szweda 1998).

Light- and transmission electron microscopic examination of cultured *rat* cardiac myocytes at different ages indicated a progressive time- and oxygen-dependent increase in the quantity and size of organelles containing lipofuscin (Marzabadi et al. 1992). Further support for the involvement of oxidative stress in lipofuscinogenesis was provided by the effects of iron and the iron chelator desferrioxamine on lipofuscin concentration. Augmentation of iron in the culture medium markedly increased the level of lipofuscin accumulation, whereas desferrioxamine had the opposite effect. Both iron and desferrioxamine are endocytosed and will end in the lysosomal vacuome after fusion between endosomes and primary or secondary lysosomes where they influence the intralysosomal milieu and the condition for lipofuscinogenesis. At both 6 and 12 days of culture age, the amount of lipofuscin found in myocytes exposed to 30 μM Fe^{3+} was markedly greater than in the controls. Dense needle-shaped material was shown to be iron when analysed by energy dispersive X-ray analysis (Marzabadi et al. 1988).

Lactate dehydrogenase (EC 1.1.1.27)

$$S\text{-Lactate} + NAD^+ \longrightarrow \text{Pyruvate} + NADH \qquad [125]$$

Succinate dehydrogenase (EC 1.3.99.1) showed an age-dependent activity in the myocardium of the *rat* (Rebel and Stegmann 1973). It decreased with advancing age and increasing weight of the animals.

Superoxide dismutase (EC 1.15.1.1)

$$2\,O_2^{\cdot-} + 2\,H^+ \longrightarrow H_2O_2 + {}^3\Sigma_g^-O_2 \qquad [128]$$

CuZn-superoxide dismutase in 26-month-old *rats* compared with. 8-month-old animals had decreased from $3.89 \pm 0.09 \times 10^2$ units per g heart tissue to $2.31 \pm 1.19 \times 10^2$ units per g heart tissue (Gomi and Matsuo 1995).

Although no free (i.e. non-*N*-acetylated) carnosine (β-alanyl-L-histidine) and anserine (β-alanyl-L-1-methylhistidine) were detected, histidine dipeptide levels significantly ($P < 0.05$) declined 22% in male Fischer *rat* left ventricular cardiac muscle with ageing from 3 to 27 months (Johnson and Hammer 1992). These decreases were consistent with the findings that other antioxidant levels were decreased (Lew and Quintanilha 1991) and the levels of lipid peroxidation were increased (Kihlstrom 1992) as a result of increased production of O_2-derived free radicals (Alessio 1993). In the myocardium, a 23% decrease in the histidine dipeptide level was associated with exercise in normotensive *rats*, and in sedentary rats, the histidine dipeptide level in spontaneously hypertensive *rats* was 19% lower than in normotensive *rats* (Hong and Johnson 1995). In contrast to the pattern observed for the skeletal muscles (quadriceps femuris and longissimus dorsi), myocardial histidine dipeptide levels were not significantly different in the exercised normotensive and spontaneously hypertensive *rats*.

In isolated coronary arterioles of 74- to 82- week-old male Sprague-Dawley *rats*, N^{ω}-nitro-L-arginine methyl ester-sensitive flow-induced dilations were significantly impaired, which could be augmented by superoxide dismutase or 4,5-dihydroxy-1,3-benzene-disulphonic acid (Csiszar et al. 2002). For lucigenin chemiluminescence, superoxide radical generation was significantly greater in arterioles from old compared with those from young *rats*. NADH-driven $O_2^{\cdot-}$ generation was also significantly more productive in vessels from old *rats*. Aged arterioles showed an increased expression of inducible nitric oxide synthase (iNOS), confined to the endothelium. Decreased eNOS mRNA and increased eNOS mRNA expression in old vessels was shown by quantitative reverse transcription- polymerase chain reaction. *In vivo* formation of peroxynitrite was evidenced by Western blotting, and immunohistochemistry showing increased 3-nitrotyrosine content in old vessels.

Endothelin-1-induced contractions were increased in the coronary circulation with age (TSCHUDI and LÜSCHER 1995).

The hypothesis that the ageing-related increase in the risk of heart disease may be causally related to the ageing-related shift in plasma thiol/disulphide redox state is supported by several clinical studies on *N*-acetylcysteine (for review see DRÖGE 2002).

17.14 Lung

The loss of elasticity on the *human* trachea and bronchial tree with advancing age and the growing danger of rupture during intubation in old people was examined by HACKL and KÖNIG (1959). In 25 deceased aged 42 to 97 years they found the ultimate tensile stress decreasing from 775 mm to 425 mm Hg.

In elderly subjects the elastic fibres showed a notable decrease along the alveolar walls while type III collagen increased when compared with that of non-elderly controls (D'ERRICO et al. 1989). No variations of these components were detectable in the alveolar ducts or in the respiratory bronchioli. An increase in the thickness of the alveolar basement membranes were detected in some of the subjects when antibodies against type IV collagen and laminin were used, while antibodies to fibronectin and type V collagen did not reveal any modifications compared with the controls. The modifications revealed in the lungs of the elderly can be related to the alterations of the elastic recoil and pulmonary compliance observed in these subjects.

The elastin content as determined by the hot alkali method did not change significantly with age in the whole lungs of BALB/c and SAMR1 male *mice* (TAKUBO et al. 1999). The total collagen content of the whole lung was significantly higher at 24 months of age, although there were no significant changes with aging neither in the hydroxyproline content per dry lung weight nor in the proportion of type III to type I collagen.

The glycosaminoglycan of the anatomic structures of the lung is determined by gender, smoking and age, more than by acute pathology (GRUNDBOECK-JUSCO et al. 1992).

The bronchial mucous glands increased in elderly people (mean age 79.9 years) as compared to a younger group (mean age 59.4 years) of subjects (HERNANDEZ et al. 1965). This correlation of thickness of the gland layer with age could not be attributed to inhalant exposure since all subjects lived in the same area and did not use tobacco.

The serous cells in the tracheal glands of male Sprague-Dawley *rats* showed marked differences between young (8-17 week-old) and aged (20-30 month-old) subjects (KASUGA 1991).

Lamellar inclusion in type II pneumocytes of *mice* older than 1 year contained a smaller number of lamellae (SIRTORI 1964).

The rates of loss of ^{14}C label from the ribosomes and soluble RNA fractions of lung tissue of two groups of male Wistar *rats*, one aged 12 and the other 24 months, both injected intraperitoneally with [^{14}C]orotic acid (50 μCi per 100 g body weight) did not show any age-associated differences (MENZIES et al. 1972).

Alveolar macrophages freshly harvested from non-sensitised aged *rats* produced less $O_2^{\bullet-}$ than those from young animals (HAYAKAWA et al. 1995). A similar result was obtained in BCG-sensitised *rats*. However, alveolar macrophages from aged *rats* were primed with the *in vitro* treatment with interferon-γ for increased rate of $O_2^{\bullet-}$ production to an equivalent level of that by alveolar macrophages from young animals.

A two-fold increase in reactive oxygen species production by thioglycollate-elicited peritoneal macrophages from senescent female C57BL/6j *mice* was detected by luminol-dependent chemiluminescence after introduction of latex and zymosan phagocytosis (LAVIE and GERSHON 1988).

Superoxide dismutase (EC 1.15.1.1)

$$2\,O_2^{-} + 2\,H^{+} \longrightarrow H_2O_2 + {}^{3}\Sigma_g^{-}O_2 \qquad [128]$$

CuZn-superoxide dismutase in 26-month-old *rats* compared with. 8-month-old animals had decreased from $2.12 \pm 0.26 \times 10^2$ units per g lung tissue to $1.55 \pm 0.82 \times 10^2$ units per g lung tissue (GOMI and MATSUO 1995).

Some parts of the tracheal epithelium of ageing *mice* invaginated into the lamina propria without any noticeable changes in either the epithelial cells or the basal lamina (KAWATA and FUJITA 1983). The lumen of the cyst-like structures usually formed characteristic concentric circles in the central portion and sometimes contained destructed cell debris at the periphery. The concentric circular materials consisted of entangled filaments of 15–20 nm in diameter.

In 28-month-old male Wistar *rats*, vasoactive intestinal polypeptide concentrations were reduced by 60 % in relation to young-adult (3-months-old) animals (GEPPETTI et al. 1988). The density of vasoactive intestinal polypeptide-immunoreactive nerve fibres was remarkably reduced within bronchial smooth muscle and bronchial glands. The number of vasoactive intestinal polypeptide-immunoreactive nerve cell bodies located in intraparenchymal ganglia was decreased in old *rats*.

17.15 Liver

The mitochondria of the liver cells in young and old *mice* of the C57 black strain generally were abundant (ANDREW 1960). In the younger animals, however, the mitochondria generally were more elongated. Study of the large aberrant cells in the liver of senile animals showed relatively little difference in the mitochondrial picture for such cells. There were, however, areas in the senile liver, particularly in regions where there was rather marked sclerosis of the blood vessels, where the mitochondrial picture was quite varied and where some cells were almost completely lacking in the cytoplasmic elements.

In isolated liver mitochondria from 180-days-old male Wistar *rats* the oxidative capacity of flavin adenine dinucleotide (FAD)-linked pathways significantly declined compared with those from younger animals (IOSSA et al. 1998). α-Glycerophosphate dehydrogenase (EC 1.1.1.8) and succinic dehydrogenase (EC 1.3.99.1) activities were significantly decreased in old *rats* compared with younger ones. ANSON et al. (1999) measured several different oxidatively induced base lesions in both mitochondrial and nuclear DNA as a function of age (23–24 months versus 6–7 months). No significant age effects were observed for any lesion. They did not observe elevated levels of oxidatively induced base lesions in mitochondrial DNA. This contrasts with 50-fold differences reported for several lesions between mitochondrial and nuclear DNA from *porcine* liver (ZASTAWNY et al. 1998).

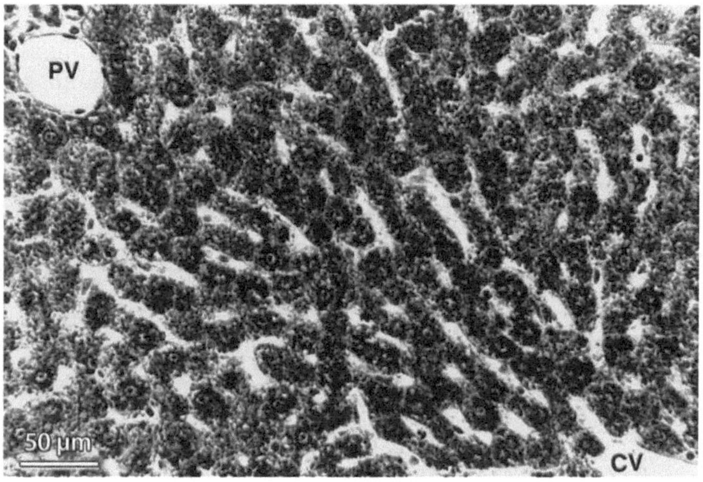

Fig. 344. Uniform distribution of glycogen in the liver lobule (CV = central vein; PV = portal vein branch) an unmedicated 465 g old male Wistar *rat* MR 2000 (No. 2) exsanguinated under ether anaesthesia on March 14, 1977. The tissue was fixed by immersion in Carnoy's fluid (ethanol-chloroform-glacial acetic acid) and embedded in Paraplast. Periodic acid-Schiff reaction. Objective Leitz Pl 40/0.65. Leitz Orthomat® (eyepiece 2×). Film Agfa Pan 25

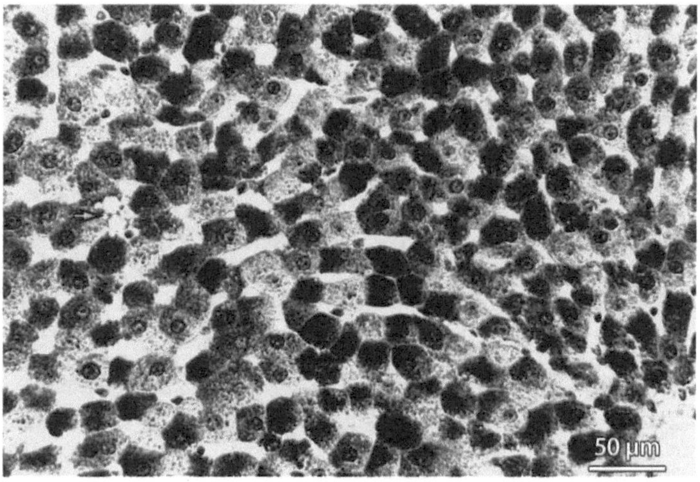

Fig. 345. Random distribution of hepatocytes rich (dark) and poor (light) in glycogen. Sporadic vacuolated cells (*arrow*). Central area of a lobule of an unmedicated 526 g old male Wistar *rat* MR 2000 (No. 9) exsanguinated under ether anaesthesia on March 14, 1977. The tissue was fixed by immersion in Carnoy's fluid (ethanol-chloroform-glacial acetic acid) and embedded in Paraplast. Periodic acid-Schiff reaction. Objective Leitz Pl 40/0.65. Leitz Orthomat® (eyepiece 2×). Film Agfa Pan 25

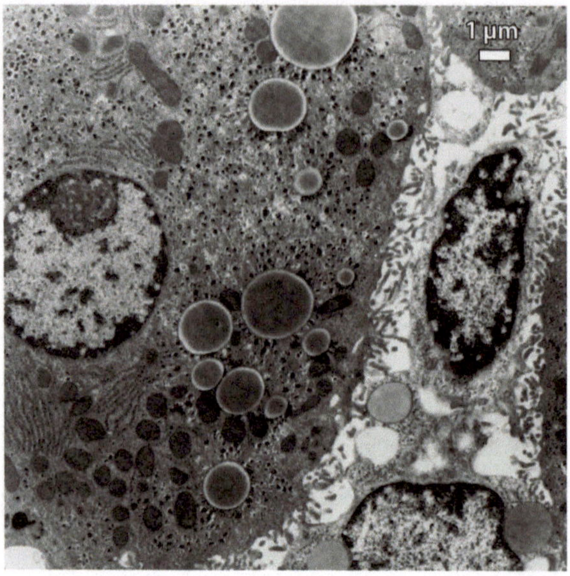

Fig. 346. Liver (block 1952) from a 362 g old Wistar *rat* (No. 4; breeder S. Ivanovas, Kissleg/Allgäu) medicated with 400 mg piracetam per kg body weight × day added to the food (powdered Altromin® R) and drinking water for 4 months (June 28 to October 20, 1976). Under pentobarbital anaesthesia (30 mg/kg), the animal was perfused from the abdominal aorta with 2.5 % glutaraldehyde in 0.1 M sodium cacodylate buffer (pH 7.4). Postfixation with 1 % osmium tetroxide in sodium cacodylate buffer. Embedded in Epon 812 and sectioned at 50 nm. Lead citrate and uranyl acetate. Film 649-30

While foetal *human* livers showed periportal and patchy mid-zone copper-associated protein and lipofuscin granules, together with strong diffuse cytoplasmic and nuclear metallothionein immunostaining, livers of children over 6 months of age showed no copper-associated protein or lipofuscin granules, and minimal or no metallothionein immunostaining (Fuller et al. 1990).

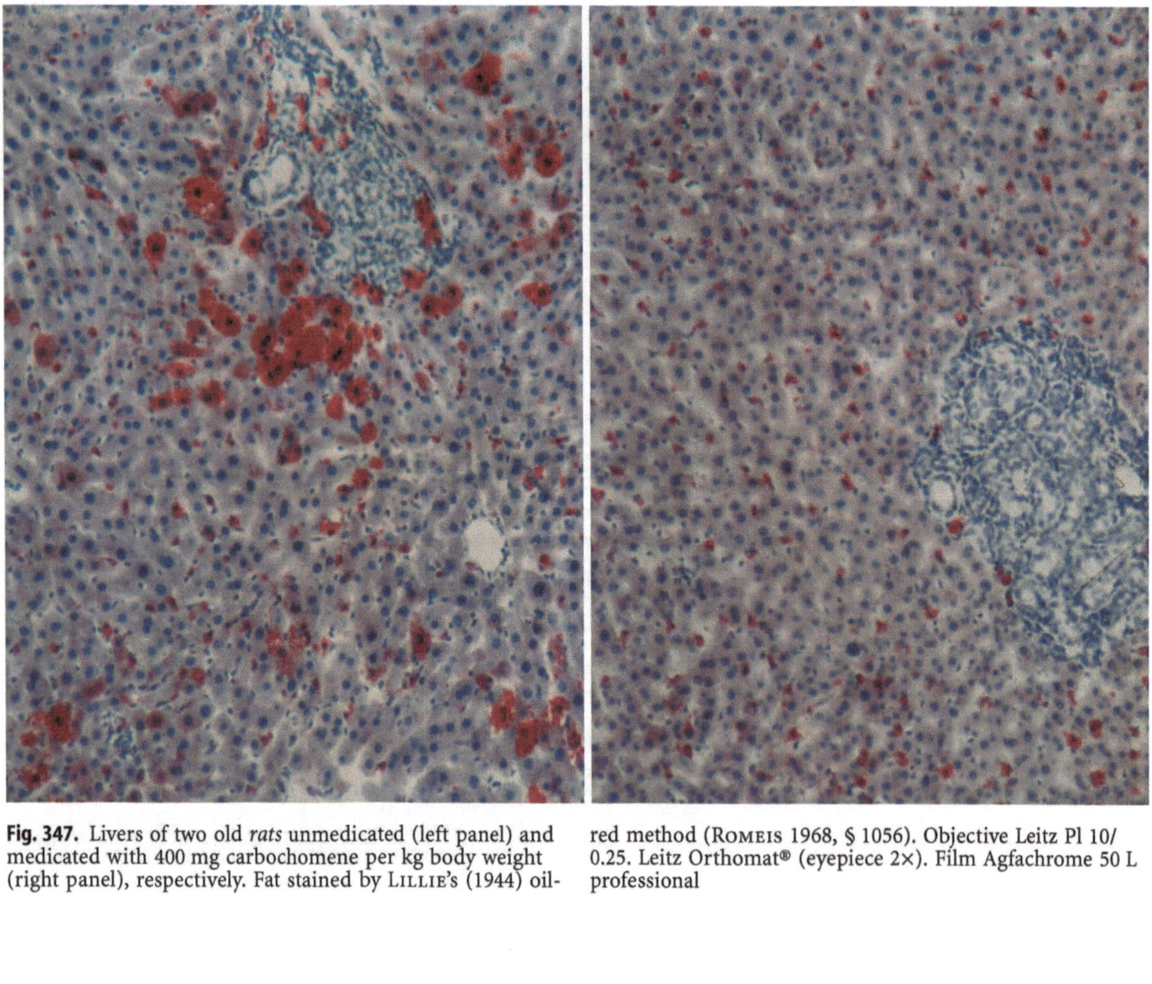

Fig. 347. Livers of two old *rats* unmedicated (left panel) and medicated with 400 mg carbochomene per kg body weight (right panel), respectively. Fat stained by Lillie's (1944) oil-red method (Romeis 1968, § 1056). Objective Leitz Pl 10/0.25. Leitz Orthomat® (eyepiece 2×). Film Agfachrome 50 L professional

The morphometrical analysis of 39 months old male Wistar *rats* revealed a heterogeneous distribution in the liver lobule of the old animals, with a significant elevation of peroxisomal volume density in pericentral over periportal hepatocytes, in contrast to the uniform pattern in young *rats* aged 2 months (Beier et al. 1993). Age-related lobular gradients were also observed by quantitative immunocytochemistry in the peroxisomal concentrations of trifunctional enzyme (central > portal) and, inversely, for catalase (portal > central). The quantitation of core containing peroxisomes exhibited a significant ($P < 0.005$) increase in the core region in livers of old animals both in portal and central regions of the liver lobule. This corroborated the more intense immunostaining of urate oxidase in old animals as shown by immunoblotting, corresponding to an 142 per cent increase.

Chao et al. (2002) investigated peroxisomal β-oxidation activity, a major source of H_2O_2, as well as well as the peroxisomal anti-oxidant enzyme catalase, in male Fischer-344 *rats* of four age groups (4, 10, 50, and 100 week old). In the senescent group, the level of decline in both peroxisomal enzyme activities of 30% was surprisingly similar to the decline observed in the hepatic expression of the retinoid X receptor-α protein.

CuZn-superoxide dismutase (EC 1.15.1.1) in 26-month-old *rats* compared with. 8-month-old animals had decreased from $30.5 \pm 1.21 \times 10^2$ units per g liver tissue to $25.8 \pm 3.64 \times 10^2$ units per g liver tissue (Gomi and Matsuo 1995).

Lysosomal β-acetylglucosaminidase and cathepsin D (EC 3.4.23.5) activities remained constant in 2 years old male albino *rats*, whilst the total activity of β-glucuronidase (EC 3.2.1.31) as well as the bound part of the tree enzymes decreased significantly as compared with the quantities in *rats* aged ten weeks (Platt et al. 1973). After intravenous medication with 100 mg piracetam per kg body weight for 8 days, a significant increase of β-acetylglucosaminidase and cathepsin D activities occurred in the old age group.

Kupffer cells showed an age-related decrease in the capacity to endocytose heat-denatured albumin (Knook et al. 1982). Cathepsin D (EC 3.4.23.5) and aminopeptidase B activities were greatly increased during ageing (Knoop and Sleyster 1978). The activities of acid phosphatase (EC 3.1.3.2) and β-galactosidase (EC 3.2.1.23) remained constant, while arylsulphatase B (EC 3.1.6.1) and β-glucuronidase (EC 3.2.1.31) showed decreased activities in cells from old *rats*.

17.16 Kidney

Kidney tissue is not post-mitotic, but is more limited in its proliferative potential than liver (Goss 1966). Data on two CFN strain male *rats* aged 164 and 798 days given daily intraperitoneal injections of [^{3}H]thymidine showed a higher over-all proportion of labelled nuclei in the old animal than in the young (Falzone et al. 1967).

At birth there is a full complement of glomeruli, and thus of nephrons, numbering 2×10^6 (Dunnill and Halley 1973). After the age of 36 yr the number of glomeruli falls to a value of 1.4×10^6 at 73 yr. The glomerular basement-membrane area was found to be 0.2 m^2 at birth and approximately 1.6 m^2 in adult life. The volume proportion of the cellular component of the glomerular tuft was greatest in the neonate; there was no evidence of an increase in interstitial glomerular tissue in old age. The mean percentage volume of interstitial tissue in the renal cortex was found to be 12.8 ± 5.1; it did not increase significantly with age.

A study by light- and electron-microscopy of 105 *human* cadavers suggested that there is a slow but gradual increase in the thickness of the basement-membranes both of Bowman's capsule and of the convoluted tubules (Darmady et al. 1973).

The histological aspect of the kidney in senile Wistar Institute *rats* is considerably different from that in young and "middle-aged" animals (Andrew and Pruett 1957).

The mean juxtaglomerular granular cell index was 28.87 ± 1.56 in young and 5.7 ± 1.66 in old male Wistar *rats* (Dunihue 1965).

In ischaemic (37 min)/reperfused *rat* kidney ageing increased susceptibility to acute renal failure, an effect that is apparent even during a transition from the adolescent to the mature state (Zager and Alpers 1989).

Hypercholesterolaemia and hyperlipidaemia may contribute to premature renal ageing and renal degeneration by increased generation of reactive oxygen species and oxidatively modified proteins and lipids (Gröne et al. 1997, Scheuer et al. 2000).

Renal and vascular endothelial (eNOS) and inducible (iNOS) nitric oxide synthase isoforms were markedly upregulated in the prehypertensive (4-week-old) and adult hypertensive (12-week-old) spontaneously hypertensive *rats* (Vaziri et al. 1998). Compared to the Wistar-Kyoto control *rats*, untreated spontaneously hypertensive *rats* showed severe hypertension, elevated urinary NO metabolite (NO_x) excretion, marked upregulation of renal and vscular eNOS and iNOS proteins, normal renal function and heart weight at 9 weeks of age (Vaziri et al. 2002). Hypertension control with either AT-1 receptor or calcium channel blockade (felodipine 5 mg/kg × day) mitigated upregulation of NOS isofoms in the young spontaneously hypertensive *rats*. With advanced age (63 weeks), the untreated spontaneously hypertensive *rats* showed increased proteinuria, renal insufficiency, cardiomegaly, reduced urinary NO_x excretion and depressed renal and vascular NOS protein expression as compared to the corresponding Wistar-Kyoto control *rat* group. AT-1 receptor blockade prevented proteinuria, renal insufficiency.

Enzyme excretion related to urinary creatinine (enzyme/creatinine ratio; U/mmol creatinine) significantly decreases with increasing age (JUNG et al. 1990). Sex related differences of some enzyme excretions were found in age groups over 6 years.

Whereas *rats* at 10 weeks of age developed high-level proteinuria upon three repeated injections of $HgCl_2$ (1 mg per kg body weight, i.p.), animals at 18 to 24 months of age did not release urinary protein under these conditions and developed low-level proteinuria with a delayed onset after five repeated injection of $HgCl_2$ (VAN DER MEIDE et al. 1995).

17.17 Blood Vessel Atherogenesis

Accumulating evidence suggests that oxygen stress alters many functions of the endothelium, including modulation of vasomotor tone (CAI and HARRISON 2000). Inactivation of nitric oxide by superoxide and other reactive oxygen species seems to occur in conditions such as hypertension, hypercholesterolaemia, diabetes and cigarette smoking. Loss of •NO associated with these traditional risk factors may in part explain why they dispose to atherosclerosis.

Relatively brief periods (days) of hypercholesterolaemia can exert profound effects on endothelium-dependent functions of the microcirculation, including dilation of arterioles, fluid filtration across capillaries, and regulation of leucocyte recruitment in postcapillary venules. Hypercholesterolaemia appears to convert the normal anti-inflammatory phenotype of the microcirculation to a proinflammatory phenotype. This phenotype change appears to result from a decline in nitric oxide bioavailability that results from a reduction in •NO biosynthesis, inactivation of •NO by $O_2^{\bullet-}$, or both (for review see STOKES at al. 2002).

Many cells that comprise the vasculature generate reactive oxygen (for review see IRANI 2000) and/or nitrogen species. The amount of 8-hydroxy-2'-deoxyguanosine, one of the typical biomarkers of oxidative stress, in DNA isolated from lymphocytes of 43 atherosclerotic *patients* with intermittent claudication (age range from 39 to 78 years) was significantly higher than in 55 healthy persons with an age range from 26 to 87 years (GACKOWSKI et al. 2001).

Vascular **smooth muscle cells** respond to growth factor stimulation with intracellular production of reactive oxygen species. Angiotensin II, a proinflammatory mediator implicated in atherosclerosis, restenosis, and hypertension (ALEXANDER 1995), induces hypertrophic response in smooth muscle cells via the production of both $O_2^{\bullet-}$ and H_2O_2 and activation of p38 mitogen-activated protein kinase (USHIO-FUKAI et al. 1998, ZAFARI et al. 1998). Metallothionein immunoreactivity was seen only within smooth muscle cells, which occurred usually in small clusters and were found mostly near lipid cores and occasionally in the media (GÖBEL et al. 2000). Double immunostaining showed metallothionein-positive smooth muscle cells and matrix metalloproteinase-9 in the same area but not within the same cell. Electron microscopy was done to evaluate the subtype of metallothionein-positive cells and revealed that the majority consisted of synthetic smooth muscle cells. Thus, atherosclerotic plaques in *humans* contain metallothionein known to act as a scavenger for reactive oxygen species.

Cyclophilin A, a member of the immunophilin family, is secreted by vascular smooth muscle cells in response to oxidative stress and mediates extracellular signal-regulated kinase (ERK1/2) activation and vascular smooth muscle cell growth by reactive oxygen species (JIN et al. 2000). *Human* recombinant cyclophilin A can mimic the effects of secreted cyclophilin A to stimulate ERK1/2 and cell growth. The peptidyl-prolyl isomerase activity is required for ERK1/2 activation by cyclophilin A. *In vivo*, cyclophilin A expression and secretion are increased by oxidative stress and vascular injury.

The recruitment of **mononuclear phagocytes** into the arterial wall is one of the earliest events in the pathogenesis of atherosclerosis. Since monocyte chemoattractant protein 1 (MCP-1) plays a pivotal role in the subendothelial recruitment of monocytes, ZEIHER et al. (1995) tested whether •NO modulates the expression of MCP-1 in cultured *human* endothelial cells. Inhibition of basal •NO production by N^G-nitro-L-arginine upregulated endothelial MCP-1 mRNA expression (250 ± 20 %) and protein secretion. Endogenous addition of •NO dose-dependently decreased MCP-1 mRNA expression and secretion. Changes in MCP-1 mRNA expression and secretion were paralleled by corresponding changes in chemotactic activity of cell-conditioned media for monocytes.

The number of adherent and emigrated leucocytes in cremasteric postcapillary venules of wild-type high-cholesterol diet *mice* was significantly higher than that detected in venules of their normal-diet counterparts (STOKES et al. 2001). However, the high-cholesterol diet-induced recruitment of adherent and emigrated leucocytes was not observed in CuZn-superoxide dismutase transgenic *mice*. Whereas hypercholesterolaemic $p47^{phox}$+/– and wild-type mice exhibited similar inflammatory responses, $p47^{phox}$–/–) *mice* did not.

Plaques from *human* coronary and carotid arteries and aorta contained dendritic cells expressing DC-SIGN (dendritic cell-specific ICAM-grabbing non-integrin) (SOILLEUX et al. 2002).

Endothelial cells exposed to shear stress increase their •NO production, which could protect the endothelial cells against different apoptotic stimuli via a cyclic GMP-independent mechanism (DIMMELER and ZEIHER 1997). This indicated that in the normal arterial wall •NO could be protective for different cell types. In atheroscerotic plaques the situation is fundamentally different, since the high output isoform iNOS is expressed (BUTTERY et al. 1996, WILCOX et al. 1997, LUOMA et al. 1998) in an environment with a very high oxidative stress. In this situation, •NO itself or peroxynitrite (BECKMANN et al. 1994) could induce apoptotic cell death an destabilise the athersclerotic plaque. Apoptosis was only found in the advanced athersclerotic plaques in regions that contain numerous foam cells of macrophage origin (KOCKX et al. 1998).

The effects of oral L-arginine on vascular health and disease have been examined both in *human* beings and various *animal* models. In a recent review, PRELI et al. (2002) summarized the results of oral L-arginine supplementation on athersclerotic lesion formation, as well as markers of endothelial function (e.g. macrophage function, platelet aggregation and adhesion, and *in vitro* vascular ring studies).

From salt-sensitive Dahl *rats* made hypertensive using a sodium-rich diet (8 %), in the absence or presence of L-ariginine (1.2 g/l in drinking water), ARTIGUES et al. (2000) incubated isolated, longitudinally opened carotid arteries with 2×10^6 monocytes and counted the adherent cells. Hypertension markedly increased adhesion of monocytes on endothelium, and this was reduced by L-arginine. Hypertension also reduced an index of •NO release at the level of the aorta and the coronary circulation (technique of Langendorff). This impaired release of •NO was partially prevented by L-arginine.

Ascorbate could enhance delivery of •NO to the vascular wall from plasma. Where measured concentrations of •NO in plasma are about 3 nM (STAMLER et al. 1992), •NO can be carried as an S-nitrosothiol on albumin and free cysteine (KEANEY et al. 1993). The plasma concentration of such S-nitrosothiols have been reported to range from 0.45 μM (MARZINZIG et al. 1997) to as high as 7 μM (STAMLER et al. 1992), with 82 % in the form of S-nitrosoalbumin (STAMLER et al. 1992). Treatment of animals with the NOS inhibitor N^G-momomethyl-L-arginine decreased plasma •NO content by 40 %, suggesting that the measured •NO may have derived at least in part from NOS activity (STAMLER et al. 1992).

S-Nitrosothiol + ascorbate
→ reduced thiol + **•NO + ascorbyl radical•** [301]

Donryu *rats* on a **cholesterol-enriched diet** in their aortic endothelial cells showed decreased cNOS, but increased iNOS and endothelin-1 immunoreactivity (ALIEV et al. 2000).

In the presence of physiological concentrations of **homocysteine**, methionine, and folic acid, *human* umbilical vein endothelial cells efficiently convert homocysteine to thiolactone (JAKUBOWSKI et al. 2000). The extent of this conversion is directly proportional to homocysteine concentration and inversely proportional to methionine concentration, suggesting involvement of methionyl-tRNA synthetase. Folic acid inhibited the synthesis of thiolactone by lowering homocysteine and increasing methionine concentrations in endothelial cells. The extent of post-translational protein homocysteinylation increased with increasing homocysteine levels but decreased with increasing folic acid and HDL levels in endothelial cell cultures.

Homozygous apolipoprotein E-knockout *mice* fed **high-fat, low-cholesterol diets** showed significantly decreased atherosclerotic lesions in the aortas (FERRÉ et al. 2001). However, there was an association between those *mice* that were on diets supplemented with palm or coconut oils and a significant increase in hepatic lipid peroxidation. This association was not found in animals fed with olive or sunflower seed oils, the diets with the highest content of vitamin E. The dietary content of vitamin E was significantly correlated ($r = 0.98$; $P < 0.05$) with the hepatic concentration of this compound.

Cholesterol synthesis inhibition leads to cell proliferation inhibition provided the cell has no alternative source of cholesterol, which is easily achieved *in vitro* by incubating the cells in the absence of lipoprotein in the medium (VITOLS et al. 1994). In LDL-receptor deficient cells, the effect of cholesterol synthesis inhibition on cell growth is not prevented by adding LDL to the medium, illustrating the role of LDL receptor in the provision of cholesterol for cell proliferation (CUTHBERT et al. 1986, MARTÍNEZ-BOTAS et al. 1999). Lovastatin blocked HL-60 cell proliferation (MARTÍNEZ-

Botas et al. 2001). At relative low lovastatin concentration (< 10 μM), cells accumulated preferentially in the G_2 phase, an effect which was both prevented and reversed by low-density lipoprotein cholesterol. At higher concentrations (50 μM), the cell cycle was also arrested at G_1 phase.

The increased levels of blood and aortic tissue malondialdehyde and chemiluminescence of polymorphonuclear leucocytes, which were associated with development of atherosclerosis in *rabbits*, suggest a role of reactive oxygen species in the pathogenesis of hypercholesterolemia-induced atherosclerosis (Prasad and Kalra 1993). The protection afforded by vitamin E, which was associated with a decrease in blood and aortic tissue malondialdehyde concentrations in spite of hypercholesterolemia, supports the hypothesis that reactive oxygen species are involved in the development of hypercholesterolemic atherosclerosis.

A high level of **low density lipoprotein** (LDL) is a risk factor for atherosclerosis. It is generally accepted that atherosclerotic lesions are initiated by an enhancement of LDL uptake by monocytes and macrophages (Ross 1993). LDL must be modified prior to uptake; detection of oxidatively modified LDL in atherosclerotic lesions supports a role of this process in vivo (Ylä-Herttuala et al. 1989). Macrophages exposed to oxidised LDL subsequently form foam cells, on the first stage of atherogenesis (Steinberg et al. 1989).

Lipid peroxidation caused a decrease in phospholipid molecule mobility both in the region of polar heads and in the region of acyl chains till the depth of at least 1.7 nm from water–lipid interface (Panasenko et al. 1991). Under relative high levels of oxidation (>6 μmol malondialdehyde/g LDL phospholipid) the polarity of lipid phase increased. The decrease in efficiency of tryptophan fluorescence quenching by nitroxide fragments incorporated into hydrophobic regions at the depth of 2 nm from water–lipid interface indicated that lipid-protein interaction was disturbed as a result of oxidation of LDL lipids.

Fu and Borensztajn (2002) examined whether C-reactive protein-bound aggregates of LDL could be taken up by macrophages in culture. C-reactive protein molecules were aggregated in the presence of calcium and immobilized on the surface of polystyrene microtitre wells. *Human* LDL added to the wells bound to and aggregated on the immobilized C-reactive protein, also in a calcium-dependent manner. On incubation with macrophages, the immobilized C-reactive protein-bound LDL aggregates were readily taken up by the cells, as demonstrated by fluorescence microscopy, by the cellular accumulation of cholesterol and by the overexpression of adipophilin.

Studying the degradation of specifically oxidised cholesteryl esters by P-388D1 *murine* macrophages, Belkner et al. (2000) found that oxidised substrates were hydrolysed preferentially from a 1:1 molar mixture of oxidised and non-oxidised cholesteryl esters. This effect was observed at both neutral and acidic pH. Similar results were obtained with lysates of *human* monocytes and with pure recombinant *human* hormone-sensitive lipase.

The role of the protein moiety in oxidised LDL-induced macrophage activation was shown by Nguyen-Khoa et al. (1999). Compared to native LDL which had no effect, HOCl-oxidised LDL triggered potent lucigenin-amplified chemiluminescence responses in both U937 and THP-1 cells but only when these were fully differentiated into macrophages by 12-*O*-tetradecanoylphorbol-13-acetate. In contrast, Cu-oxidised LDL only triggered a moderate chemiluminescence response of U937 cells and had little effect on THP-1 cells. While delipidation did not affect HOCl-oxidised LDL-induced chemiluminescence response it abolished that induced by Cu-oxidised LDL.

Although various mechanisms have been proposed to explain the ability of Cu(II) to promote LDL modification, the precise reactions involved in initiating the process remain a matter of contention in the literature. In a critical overview of the chemistry of copper-dependent LDL oxidation Burkitt (2001) discusses the key role of α-tocopherol. In addition to its protective, radical-scavenging action, α-tocopherol can also behave as a prooxidant via its reduction of Cu(II) to Cu(I). Generation of Cu(I) greatly facilitates the decomposition of lipid hydroperoxides to chain carrying radicals, but the mechanisms by which the vitamin promotes LDL oxidation in the absence of preformed hydroperoxides remain more speculative.

Otero et al (2002) incubated purified *human* LDL with glucose and LDL oxidation was started by adding $CuCl_2$ to the media. Glucose delayed the vitamin E consumption, but accelerated the formation of conjugated dienes and increased both the formation of thiobarbituric acid-reacting substances and LDL electrophoretic mobility. When LDL was enriched with vitamin E, it showed a delay in the formation of conjugated dienes, even in the presence of glucose.

Minimally oxidized/modified LDL and oxidation products of 1-palmitoyl-2-arachidonyl-*sn*-glycero-3-phosphocholine activated endothelial cells to synthesize monocyte chemotactic protein-1 and interleucin-8 (Lee et al. 2000). Several lines of evidence suggest that this activation is mediated by the lipid-dependent transcription factor peroxysome proliferator-activator receptor α, the most abundant member of the peroxysome proliferator-activator receptor family in *human* aortic endothelial cells.

Although *in vivo* a proportion of foam cells in lesions are demonstrated by immunohistochemical marking to be of smooth muscle cell origin, feeding smooth muscle cells *in vitro* with β-very low-density lipoprotein or chemically modified LDL does not, in contrast to the case with macrophages, lead to foam cell formation (HEINECKE et al. 1991, HOFF et al. 1991, HUFF et al. 1991). The reason for this differences in response relates in the types and properties of the lipoprotein receptors expressed in cultured macrophages and cultured smooth muscle cells. Smooth muscle cells, like fibroblasts (the cell type on which the classical studies were conducted), are typical LDL receptor cells (GOLDSTEIN and BROWN 1977).

Hydrogen abstraction [302], diene conjugation [303] and lipid peroxidation [304]

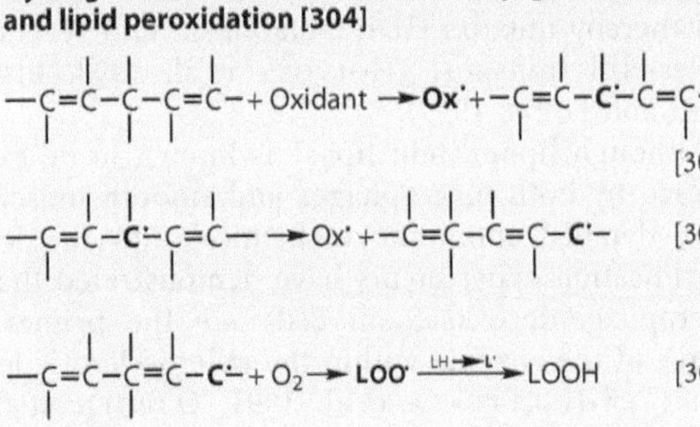

[302] [303] [304]

The potential involvement of oxidised LDL in the pathogenesis of atherosclerotic and fibrotic degeneration of arterial wall was reviewed by BERLINER and HEINECKE (1996), AVIRAM and FUHRMAN (1998), HEINECKE (1998) and STEINBRECHER (1999). LEONARDUZZI et al. (2000) emphasised specific signalling pathways that appear to be modulated by oxidised LDL and by specific lipid oxidation products found in oxidised LDL.

Mouse peritoneal macrophages and *human* monocyte-derived macrophages oxidised cholesteryl linoleate, added to the cultures in the form of an artificial protein, with the production of soluble oxidised lipids, including oxidised sterols, and, in the case of *mouse* peritoneal macrophages, abundant ceroid (CARPENTER et al. 1990, 1991). The oxidation was inhibited by radical scavengers.

Oxidation of LDL is characterised structurally by an increase in electrophoretic mobility due to Schiff-base reactions between lysine residues on apolipoprotein B in LDL and specific aldehydic breakdown products of hydroperoxy fatty acids (STEINBRECHER et al. 1990).

When ICR male *mouse* macrophages were cultured with oxidised *human* LDL, storage of ceroid-like pigments was observed within the cells by light and fluorescent microscopy, and fluorescence spectrometry (SHIMASAKI et al. 1995). The fluorescent products exhibited the characteristics of Schiff base structures, having a fluorescence maximum of 430 nm and an excitation maximum of 355 nm, which has been generally accepted with fluorescent lipid oxidation products.

A cDNA encoding adipophilin was identified in cultured *human* macrophages stimulated with oxidised LDL using mRNA differential display (WANG et al. 1999). Adipophilin is a 50 kDa protein known to be a specific marker for adipocyte cell differentiation and lipid accumulation in a variety of cells. The time-dependent induction of adipophilin mRNA in macrophages was specific to oxidised LDL but not native LDL, and not to various cytokines and serum. In *human* atherosclerotic lesions, adipophilin mRNA expression was localised in the subset of lipid-rich macrophages.

Oxidised LDL is capable of inducing signal transduction leading to activation of protein kinase C (CLAUS et al. 1996) or modulation of transcription factor activities (ARES et al. 1995, SHACKELFORD et al. 1995). Incubation of *human* macrophages with oxidised LDL resulted in an increase in the concentration of ceramide (KINSCHERF et al. 1997), a lipid mediator involved in stress-induced signalling (HANNUN 1996, VERHEIJ et al. 1996).

On exposure to oxidised LDL, tumour necrosis factor-α is released by macrophages (JOVINGE et al. 1996) capable of inducing oxidative stress by the generation of reactive oxygen species from mitochondria (SCHULZE-OSTHOFF et al. 1993, KEANE et al. 1997). Hence, direct of mediated signalling stimulated by oxidised LDL may involve oxidative stress (GOTOH et al. 1993, ROSENFELD 1996).

The specific binding and association of ^{125}I-labelled LDL to *human* monocyte-derived macrophages were not changed under hypoxia compared to normoxia (MATSUMOTO et al. 2000). However, the degradation of ^{125}I-LDL under hypoxia decreased to 60 %. The rate of cholesterol esterification under hypoxia was 2-fold greater on incubation with LDL or 25-hydroxycholesterol. The cellular cholesteryl ester content was also greater under hypoxia on incubation with LDL. Secretion of apolipoprotein E into the medium was not altered under hypoxia, suggesting that apolipoprotein E independent cholesterol efflux may be reduced under hypoxia.

In hypercholesterolemic LDL-receptor-deficient male *mice* NO-containing aspirin (30 mg/kg × day) reduced significantly plasma LDL oxidation compared with aspirin (18 mg/kg × day) and placebo, as shown by the significant reduction of malondialdehyde content ($P < 0.001$) as well as by the prolongation of lag-time ($P < 0.01$) (Napoli et al. 2002). *Mice* treated with NO-aspirin revealed by immunohistochemical analysis of aortic serial sections a significant decrease in the intimal presence of oxidation-specific epitopes of oxidised LDL (E06 monoclonal antibody, $P < 0.01$), and macrophage-derived foam cells (F4/80 monoclonal antibody, $P < 0.01$), compared with placebo or aspirin.

Human monocyte-macrophages that are exposed to 100 μg/ml oxidised LDL undergo **apoptosis**, with maximal effects at 24 h of exposure, as indicated by ultrastructural changes and DNA fragmentation, detected by terminal transferase-mediated DNA nick-end labelling. Apoptosis has been observed in macrophage-rich regions of *human* atherosclerotic lesions (Hegyi et al. 1996). X-ray microanalysis demonstrated that intracellular concentrations of potassium decreased whilst those of sodium increased following 3 h of exposure to 100 μg/ml of oxidised low-density lipoprotein *in vitro* (Skepper et al. 1999).

The possibility that **peroxisome proliferator-activated receptor-γ agonists** might proatherogenic effects has been raised from the findings that peroxisome proliferator-activated receptor-γ is activated by oxidised lipid components derived from LDL and colony stimulating factors in human atheroma monocyte/macrophages. Peroxisome proliferator-activated receptor-γ is also expressed at high levels in the foam cells of athersclerotic lesions. Enhanced peroxisome proliferator-activated receptor-γ expression also induced transcription of a scavenger receptor in monocyte/macrophages so that these cells acquire the ability to bind and internalise oxidised LDL, a potentially pro-atherosclerotic effect (Tontonoz et al. 1998). Chen et al. (1999) have shown that peroxisome proliferator-activated receptor-γ is expressed in *human* vascular endothelial cells and that it can be activated by oxidised LDL and troglitazone; peroxisome proliferator-activated receptor-γ activation is associated with increased expression of ICAM-1 and enhancement of monocyte binding to endothelial cells, suggesting that peroxisome proliferator-activated receptor-γ signalling pathway might contribute to the atherogenicity of oxidised LDL in vascular endothelial cells. This increases ICAM-1 expression, however, contradicts the results of Cominacini et al. (1999) and Pasceri et al. (2000) in which troglitazone has been shown to have an inhibitory effect on ICAM-1 expression in *human* endothelial cells. Peroxisome proliferator-activated receptor-γ activators have been shown to inhibit *human* endothelial cell angiogenesis by inducing apoptosis via caspase-3 mediated process (Bishop-Bailey and Hla 1999). Inappropriate apoptosis in a large vessel may cause structural weakness in an existing athersclerotic plaque and may promote subsequent plaque rupture, which may predispose to embolism or stroke (Newby and Zaltsman 1999).

High-density lipoprotein prevents atherosclerosis by reverting the stimulatory effect of oxidised LDL on monocyte infiltration (Mertens and Holvoet 2001). The HDL-associated enzyme paraoxonase inhibits the oxidation of LDL by hydrolysing lipid peroxides (Mackness et al. 1993, Watson et al. 1995, Aviram et al. 1998). Paraoxonase also renders HDL resistant to oxidation, thereby maintaining the capacity of HDL to induce reverse cholesterol transport. Minimally oxidised LDL inhibits paraoxonase expression (Navab et al. 1997).

Lecithin:cholesterol acyltransferase antioxidant activity prevented the formation of oxidised lipids during lipoprotein oxidation (Vohl et al. 1999). Once minimally oxidised LDL is present, it inhibits plasma lecithin:cholesterol acyltransferase activity and thereby impairs HDL metabolism and reverse cholesterol transport (Holvoet et al. 1998, Bielicki and Forte 1999).

Although **lipoprotein lipase** is known to be expressed by both macrophages and smooth muscle cells, detailed immunocytochemical and in situ hybridisation experiments have demonstrated that macrophage-derived foam cells are the primary source of the enzyme within the atherosclerotic lesion (Ylä-Herttuala et al. 1991, O'Brien et al. 1992). Many effectors have been identified that increase macrophage lipoprotein lipase expression, including hydrogen peroxide (oxidant stress/reactive oxygen species), dexamethasone, glucose, platelet-derived growth factor and macrophage colony-stimulating factor. Renier et al. (1994) showed that lipoprotein lipase induces the expression of the TNF-α gene at the level of both gene transcription and mRNA stability. This induction of TNF-α production by lipoprotein lipase is increased during the differentiation of monocytes into macrophages, occurs via a protein kinase C-dependent pathway and is mediated through the cell surface proteoglycans (Mamputu and Renier 1999). Lipoprotein lipase has also been shown to synergise with IFN-γ in the induction of macrophage nitric oxide synthase mRNA expression at the transcriptional level (Renier and Lambert 1995).

The expression of inducible nitric oxide synthase (iNOS) mRNA and protein was induced in macrophages in the majority of early *human* aortic lesions and in all advanced atherosclerotic lesions (Luoma and Ylä-Herttuala 1999). Epitopes characteristic of oxidised LDL and peroxynitrite-modified protein tended to be colocalised in iNOS-positive lesions.

In vitro, oxidised low-density lipoprotein loaded *murine* resident peritoneal macrophages produced 68–99% less nitrite than non-loaded cells (BOLTON et al. 1994). Failure to detects NOS products from macrophages previously loaded with oxidised low-density lipoprotein appeared from lack of NOS activity, as little active enzyme could be recovered from oxidised low-density lipoprotein loaded cells. However, addition of oxidised low-density lipoprotein to an active cell-free NOS preparation had no direct effect on enzymatic activity. When native LDL was subsequently incubated with these various IFNγ/LPS stimulated cells, cells preloaded with oxidised LDL promoted, on average, a 2-ford greater increase in oxidative modification of the LDL added than either non-loaded or acetylated LDL-loaded cells.

In the coronary arteries of *rats*, the incidence and severity of spontaneous atheromatous lesions increased with age and occurred in over 60 per cent of *rats* more than 500 days old (HUMPHREYS 1957).

In the aorta of normal *rabbits* fed on stock diet, lactic dehydrogenase, NADH-tetrazolium reductase and adenosine triphosphatase are active in smooth-muscle fibres between elastic lamellae (ADAMS et al. 1963). Within 4 weeks on a cholesterol-enriched diet, macrophages infiltrate or proliferate in the thickened intima of the aorta and these cells react strongly with the lactic dehydrogenase, NADH-tetrazolium reductase and adenosine triphosphatase methods.

In vitro, low concentrations of 17β-œstradiol (10 nM) reduced oxidative modification of normal health *men* volunteers' (with total cholesterol <5.5 mM) blood LDL in the presence of either ascorbic acid or tocopherol (HUANG et al. 1999). Introduction of small amounts of esterified 17β-estradiol into lipoproteins by means of incubation of free 17β-œstradiol 17-stearate in plasma did not result in any antioxidant effect (MENG et al. 1999). Using an artificial transfer system (Celite dispersion), larger amounts of 17β-œstradiol esters could be incorporated into lipoproteins. Concentrations ranging between 0.27 and 1.38 molecules/LDL particle for 17β-œstradiol 17-stearate and between 0.36 and 1.93 molecules/LDL particle for 17β-œstradiol 17-oleate resulted in increased Cu^{2+}-induced oxidation resistance of LDL, as indicated by statistically significant lag time prolongations.

The oxidation of LDL *in vitro* by either 10 μM cupric chloride or 5 mM 2,2'-azobis (2-amidinopropane) dihydrochloride was inhibited in a concentration-dependent manner by melatonin (KELLY et al. 1996). After the incubation of LDL with cupric chloride, the associated rise in thiobarbituric acid-reactive substances was reduced when melatonin was also present. Assuming similar relationships *in vivo*, melatonin should be given consideration as an anti-atherogenic factor (REITER et al. 1997).

LAPENNA et al. (1998) found that ticlopidine, a thienopyridine characterised by lipophilic properties, at therapeutically relevant concentrations (2.5–10 μM), but neither aspirin nor salicylate, significantly counteracted copper-driven *human* LDL oxidation. Ticlopidine, at 5 and 10 μM, was also antioxidant on peroxyl radical-induced LDL oxidation; yet it was ineffectual on thiol and ascorbate oxidation mediated by peroxyl radicals themselves, suggesting that drug antioxidant capacity is somehow related to the lipoprotein nature of the oxidizable substrate, but not to radical scavenging. The drug could not indeed react with the stable free radical 1,1-diphenyl-2-picrylhydrazyl, not had apparent metal complexing-inactivating activity.

Glycogen breakdown induced by concentration above 0.2 mM H_2O_2 and *tert*-butylhydroperoxide is due to a five fold increase in the ratio active glycogen phosphorylase *a* to inactive phosphorylase *b* (HEINLE 1982, 1989). Concomitantly, ATP is decreased by about 50% whereas the ATP/ADP ratio is decreased from about 2.3 (as found in fresh arterial tissue) to approximately 1.2. With respect to alterations of contractility, it was found that at concentrations of about 0.3 mM the peroxides were able to induce contraction in relaxed arterial rings (HEINLE 1984). This indicates that cytosolic levels of calcium ions are increased under these conditions. Lower peroxide concentrations did not cause contraction. However, even at concentrations of 10 μM, potentiated contraction enhancement was found when the contractile apparatus was simultaneously activated, e.g., by depolarisation. This finding can be interpreted as a facilitative effect of the hydroperoxides on cytosolic calcium release. The linolenic acid peroxide as synthesised in the reaction with soy bean lipoxidase revealed similar effects on contractility in concentrations up to 20 μM.

The theory that iron may play a significant role in atherogenesis by promoting the formation of free radicals is controversial. The search for epidemiological support for the association between iron status and heart diseases has thus far yielded both positive (BERGE 1994, MAGNUSSON et al. 1994, KIECHL et al. 1997, MEYERS et al. 1997, TUOMAINEN et al. 1997) and negative results (STAMPFER et al. 1993, LIAO et al. 1994, DANESH and APPLEBY 1999). In hypercholesterolaemic *rabbits* (1% cholesterol diet) iron accumulation occurred at the onset of lesion formation (PONRAJ et al. 1999). Weekly bleeding (15 ml/week) decreased the iron uptake into the arterial wall and delayed the onset of atherogenesis.

The Ameroid technique developed by LITVAK (1957) and VINEBERG (1960) was used by SCHILLER

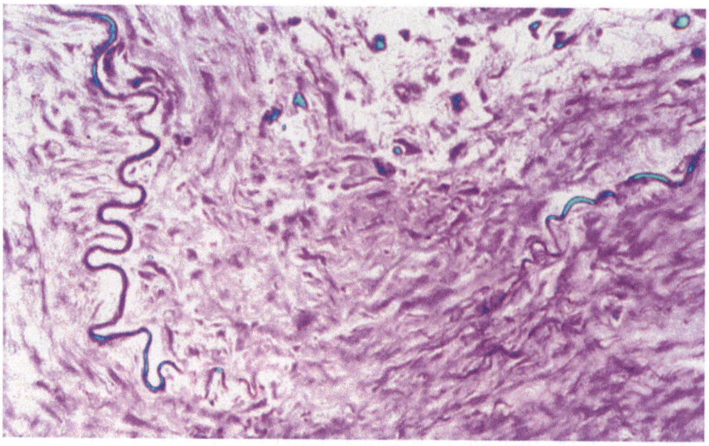

Fig. 348. Disrupted tunica elastica interna of the ramus circumflexus of the left coronary artery of a *dog* (No. 91) sacrificed 180 days after setting Ameroid constrictors (2.5 mm central lumen) and cardioomentopexy. Formalin. Paraffin section, unstained, embedded in glycerol. Interference contrast Fl 50/0.85/ ∞/0.17. Agfachrome 50 L professional

and WERNITSCH (1973) to study the destruction of the artery wall in the place narrowed by the hygroscopic plastic material which increases its volume when taking up tissue fluid. Since the outer dimensions are kept constant by encapsulation into a stainless steel ring the constrictor, when slipped over a coronary artery, narrowed the vessel. Tangential wall stress induced intimal plaques and degenerative changes in the internal elastic membrane as visualised by interference contrast (Fig. 348).

Theaflavin digallate pre-treatment of *mouse* peritoneal resident macrophages or *human* umbilical cord endothelial cells reduced cell-mediated low-density lipoprotein oxidation in a concentration (0–400 μM) and time (0–4 h) dependent manner (YOSHIDA et al. 1999). The inhibitory effect of flavonoids on cell mediated low-density lipoprotein oxidation was in the order of theaflavin digallate > theaflavin ≥ epigallocatechin gallate > epigallocatechin > gallic acid. Theaflavin digallate pretreatment decreased superoxide production of macrophages and chelated iron ions significantly. (–)-Epigallocatechin-3-gallate effectively scavenged $HO^{\bullet}$ with reaction rate of 4.62×10^{11} $M^{-1}s^{-1}$, which is an order of magnitude higher than several well recognised antioxidants, such as ascorbate, glutathione and cysteine (SHI et al. 2000). It also scavenged $O_2^{\bullet-}$ as demonstrated by using xanthine and xanthine oxidase system as a source of $O_2^{\bullet-}$.

The intramuscular injection of the carotenoid compound, crocetin, in rabbits fed an atherosclerosis-producing diet, resulted in greatly reduced severity of the atherosclerosis (GAINER and CHISOLM 1974).

Atherosclerosis → angioplasty intervention → restenosis

Restenosis implicates three pathophysiological mechanisms: thrombosis, recruitment, and proliferation. These complex phenomena may be subdivided into eight main initiating or contributory factors to postangiplasty restenosis:

- Elastic recoil
- Thrombosis (platelet activation, aggregation, and deposition at the vascular injury site)
- Bioactive factor release (growth factors, cytokines)
- Smooth muscle cell migration into the intimal layer
- Neointimal formation
- Reendothelialization
- Matrix formation
- Geometric remodelling

The radical biomediator $^{\bullet}$NO is a natural modulator of several processes contributing to postangiplasty restenosis. At present, $^{\bullet}$NO supplementation represents a unique and potentially powerful approach to help control restenosis, either alone or as a pharmaceutical adjunct to a vascular device (JANERO and EWING 2000).

CHAPTER 18

Possible Strategies to Prevent and Reverse Manifestations of Ageing

Proteins are cellular targets for reactive oxygen species, but, as opposed to nucleic acids where oxidative damage can be repaired by specific enzymatic systems, there are very few enzymatic systems that are able to reverse oxidative damage to proteins (FRIGUET 2002). Only disulphide bridge formation, which occurs upon cysteine oxidation, can be reversed by the thioredoxin/thioredoxin reductase system (HOLMGREN 1989), and methionine sulphoxide (the oxidation product of methionine) can be reversed back to methionine by the peptide methionine sulphoxide reductase enzyme (BROT and WEISSBACH 1983, BROT et al. 1984, MOSKOVITZ et al. 1999). PETROPOULOS et al. (2000) have shown that peptide methionine sulphoxide reductase gene expression is downregulated with age in different *rat* organs such as liver, kidney, and brain.

Oxidation of methionine in peptides

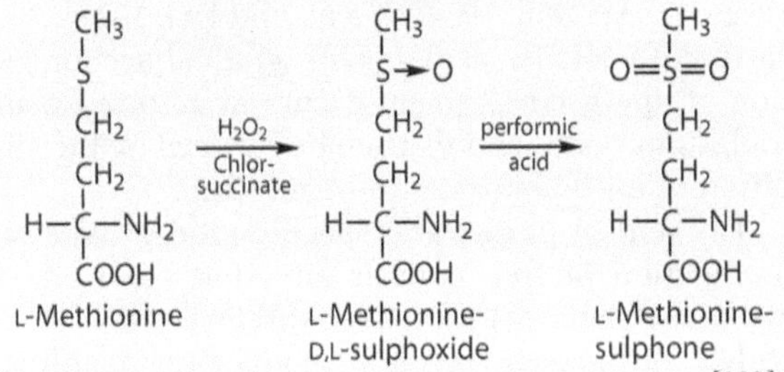

[282]

Antioxidant strategies may be classified in supplementation with antioxidant chemicals, pharmaceuticals, including some antioxidant vitamins and nutrients and one hand, and modifying endogenous antioxidant enzymes at the other. This approach is much more difficult to do. However, at least theoretically, it appears worth trying since some trials of the first category have obvious limitations because of the limited bioavailability of the drugs for kinetic reasons.

The existence of age-related increases in reactive oxygen species production is a well-established fact.

The activity of **catalase** (EC 1.11.1.6), the main enzyme responsible for detoxification against hydrogen peroxide, significantly decreases in prefrontal cortex and hypothalamus of aged *rats* (CIRIOLO et al. 1997). The reductions of the enzyme activity appears to be due to a decreased protein expression rather than impaired function of the native enzyme.

Detoxification of hydrogen peroxide by catalase (EC 1.11.1.6)

$$2\,H_2O_2 \rightarrow 2\,H_2O + {}^3\Sigma g^- O_2 \qquad [127]$$

By means of gene technology ORR and SOHAL (1994) achieved an up-regulation of both Cu,Zn-superoxide dismutase and catalase genes in *Drosophila melanogaster*. These transgenic flies lived for significantly longer times than their wild type counterparts. Studies in *Drosophila* (ARKING et al. 1991) and *Caenorhabditis elegans* (LARSEN 1993) showed that long living mutants were equipped with more efficient antioxidant defence mecha-

Table 54. Statins

Cerivastatin	(3*R*,5*S*,6*E*)-7-[4-(4-fluorophenyl)-2,6-diisopropyl-5-(methoxymethyl)pyrid-3-yl]-3,5-dihydroxy-6-heptenoic acid
Fluvastatin	(±)-(3*R*',5*S*',6*E*)-7-[3-(4-fluorophenyl)-1-isopropyl-2-indolyl-3,5-dihydroxy-6-heptenoic acid
Lovastatin	{1,2,3,7,8,8a-hexahydro-8-[2-(tetrahydro-4-hydroxy-6-oxo-2-pyranyl)ethyl]-3,7-dimethyl-1-naphthyl}-2-methylb 1,1-1,7
Pravastatin	(3*R*,5*R*)-7- {(1*S*,2*S*,6*S*,8*S*,8a*R*)-1,2,6,7,8,8a-hexahydro-6-hydroxy-2-methyl-8-[(*S*)-2-methylbutyryloxy]-1-naphthyl}-3,5-dihydroxyheptanoic acid
Simvastatin	⟨(1-*S*)-1,2,3,7,8α,8aβ-hexahydro-3α,7β-dimethyl-8-{2-[(2*S*,4*S*)-tetrahydro-4-hydroxy-6-oxo-2H-pyran-2yl]-ethyl }-1-naphthyl⟩-2,2-dimethylbutyrate

nisms. However, there is a big gap between these evolutionary lower animals and mammals, especially *humans*. Homozygous transgenic *mice* with a two- to five-fold elevation of Cu,Zn-superoxide dismutase in various tissues showed a slight reduction of life span, whereas hemizygous *mice* with a 1.5- to 3-fold increase in Cu,Zn-superoxide dismutase showed no difference in life span from that of non-transgenic littermate controls (HUANG et al. 2000).

Fluvastatin is a totally synthesised **3-hydroxy-3-methylglutaryl coenzyme A reductase inhibitor** and has clinical anti-hypercholesterolaemic effects (KATHAWALA 1991). The anti-atherogenic properties of fluvastatin may not be limited to its hypocholesterolaemic, but may also be related to its ability to reduced low-density lipoprotein oxidizability (HUSSEIN et al. 1997, BELLOSTA et al. 1998). Antioxidant effects of fluvastatin have also been reported in *humans* (HUSSEIN et al. 1997, LEONHARDT et al. 1997, BELLOSTA et al. 1998) and *rabbits* (BANDOH et al. 1996, MITANI et al. 1996).

Chronic administration of **melatonin** (0.1–10 mg/kg, s.c.) for 30 d significantly reversed the ageing-induced decrease in the *mouse* forbrain total glutathione (RAGHAVENDRA and KULKARNI 2001).

The aromatic amines **phenothiazine, phenoxacine**, and **iminostilbene** (formula [276]) proved to be about two orders of magnitude more effective than common phenolic antioxidants in their protective activity against oxidative nerve cell death (MOOSMANN et al. 2001). This remarkable efficacy could be directly correlated to calculated properties of the compound by means of a novel, quantitative structure-activity relationship model.

Iminostilbene [276]

A specific bile acid, called ursodeoxycholic acid (3α,7β-dihydroxy-5α-cholanoic acid) when orally fed for 3 weeks in *mice* increased liver glutathione *S*-transferase (EC 2.5.1.18) activities (KITANI et al. 1994).

Glutathione *S*-transferase (EC 2.5.1.18)

RX + Glutathione → HX + R–S–G [129]

Gingseng contains phenolic compounds such as maltol, salicylic acid and vanillic acid which are potent radical scavengers, and ginsenosides which were found to enhance antioxidant enzyme activities (CHUNG et al. 1997).

(–)-**Deprenyl**, a monooxygenase B inhibitor, in at least four animal species, such as *rats* (MILGRAM et al. 1990, KITANI et al. 1993), *hamsters* (STOLL et al. 1997), *mice* (ARCHER and HARRISON 1996) and *dogs* (RUEHL et al. 1997) realised a significant prolongation of survivals. The pharmacological modifications of endogenous antioxidant enzymes under this drug are still under discussion (KITANI et al. 1999).

Monoamine oxidase (EC 1.4.3.4)

$RCHNH_2 + H_2O + O_2 \rightarrow RCHO + NH_3 + \mathbf{H_2O_2}$ [270]

Experimental studies indicate that **peroxisome proliferator-activated receptors** may have a preventive role in the pathogenesis of atherosclerosis by regulating cytokine production, adhesion molecule expression on endothelial cells, fibrinolysis, modulation of monocyte-derived macrophages, and proliferation of vascular smooth muscle cells (LOVISCACH and HENRY 1999, LAW et al. 2000). ELANGBAM et al. (2001) reviewed the current understanding of how peroxisome proliferator-activated receptors are involved in modulating inflammation and atherosclerosis, and their possible therapeutic implications.

Several drugs appear to halt or reverse the accumulation of **lipofuscin**; perhaps the most effective is meclofenoxate (2-dimethylaminoethyl-4-chlorophenoxy acetate), which is said to drastically reduce the amount of lipofuscin in the neurones of senile *guinea pigs* and *squirrel monkeys* (NANDY and BOURNE 1966, NANDY 1968, CHEMNITIUS et al. 1970, MEIER and GLEES 1971, HASAN and GLEES 1972, HASAN et al. 1974, NANDY and LAL 1978). According to NANDY et al. (1978), gradual accumulation of lipofuscin pigment occurs in cultured neuroblastoma cells, and this can be reduced by the addition of meclofenoxate to the medium.

The impact of **diet and specific food groups** on modulation of free radicals and thus ageing and age-associated degenerative diseases has been widely recognised. Increased longevity in animal models by caloric restriction is attributed in part to the modulation of free radical production (SOHAL and WEINDRUCH 1996). An inverse association of fruit and vegetable consumption with risk of morbidity and mortality from degenerative diseases such as cardiovascular diseases and cancer lead MEYDANI (1999) to the conclusion that fruits and vegetables containing nutritive and non-nutritive compounds with antioxidant properties may contribute to the overall improvement of the quality and may add a day to our lives.

Ethylnandrol (0.75 mg per day) employed as a lipid-lowering drug in 4 years of treatment caused a reduction in mortality from cerebral infarction and in morbidity in myocardial infarction (TSUSHIMA et al. 1975).

JOSEPH et al. (2002) showed that the ability of COS-7 cells to clear excess Ca^{2+} following oxotremorine stimulation varied as a function of transfected muscarinic acetylcholine receptor subtype, with dopamine (1 mM for 4 h)-treated M1, M2, or M4 cells showing greater decrements in recovery than those transfected with M3 or M5 acetylcholine receptor subtypes. A similar pattern of results in M-1 or M3-transfected dopamine-exposed cells was seen with respect to viability. Viability of the untransfected cells was unaffected by dopamine. Pretreatment with Trolox or PBN (a nitrone trapping agent) did not alter the dopamine effects on cell recovery and viability.

After 4 months of **caloric restriction**, the levels of N^{ε}-(carboxyethyl)lysine, N^{ε}-(carboxymethyl)-lysine, N^{ε}-(malondialdehyde)lysine and glutamic semialdehyde were significantly lower in the *rat* heart mitochondria from caloric restricted animals than in the controls (PAMPLONA et al. 2002).

PART V

Cancer

CHAPTER 19

Carcinogenesis

Cancer is a multi-step process evolving as a result of the accumulation of a number of mutational events. The growing body of evidence implicating genetic instability as a key feature of this evolutionary process and the risk of malignancy associated with chromosomal instability syndromes highlight the importance of understanding the mechanisms that cell use to maintain the integrity of their genomes. To date, chromosomal instability induced by ionising radiation has been the most extensively studies phenotype and it is evident that the expression of inducible instability has a strong dependence on the type of radiation exposure, the cell type irradiated, and the genetic 'predisposition' of the irradiated cell (WRIGHT 1999).

19.1 Reactive Species

19.1.1 Reactive Oxygen Species

Reactive oxygen species induce all forms of DNA damage, including base modifications, base-free (apurinic/apyrimidinic [AP]) sites, strand breakage, and DNA-protein crosslinks, but the specific spectrum of products depends on the reactive species involved (IMLAY and LINN 1988, HALLIWELL and GUTTERIDGE 1989, JOENJE 1989). The majority of lesions induced by reactive oxygen species has been derived from studies using ionizing radiation (TÉOULE and CADET 1978, HUTCHINSON 1985, VON SONTAG 1987, BREIMER 1988, WARD 1988). Homolytic fission of water produces hydroxyl radical which attacks DNA generating a whole series of modified purine and pyrimidine bases.

However, leucocyte-derived reactive oxygen species induce far greater yields of base modifications

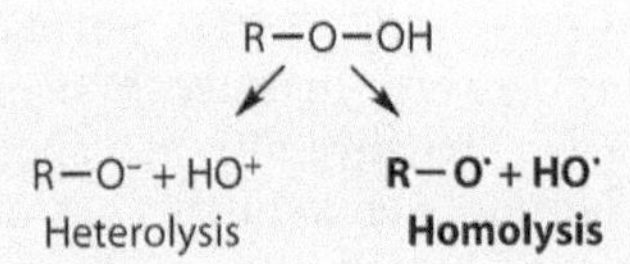

[305]

than does ionizing radiation (FLOYD et al. 1986, FRENKEL et al. 1986). $O_2^{\bullet-}$ and H_2O_2 generated in the xanthine-xanthine-oxidase system are further catalysed to $HO^{\bullet}$ by Fe/EDTA.

A regional difference in the rates of reduction of free radicals by sulphydryl groups may result in the site susceptible to development of *N*-methyl-*N'*-nitro-*N*-nitrosoguanidine-induced gastric cancer in male Wistar *rats* (MIKUNI and TATSUTA 1998).

In a recent review MARNETT (2000) highlights some of the major accomplishments in the study of oxidative DNA damage and its role in carcinogenesis. He also identifies controversies that need to be resolved. Unravelling the contributions to tumorigenesis of DNA damage from endogenous and exogenous sources represents a major challenge for the future.

19.1.2 Reactive Nitrogen Species

Nitric oxide appears to exert a dichotomy of effects within the multistage model of cancer (WINK et al. 1998). Chronic inflammation can lead to the production of chemical intermediates, among them $^{\bullet}NO$, which in turn can mediate damage to DNA. $^{\bullet}NO$ appears to be critical for the tumoricidal activity of the immune system. $^{\bullet}NO$ influences angiogenesis (ZICHE et al. 1994, MONTRUCCHIO et al. 1997, JADESKI et al. 2000) and metastasis. Biopsies of *human* mammary tumours showed that there is greater

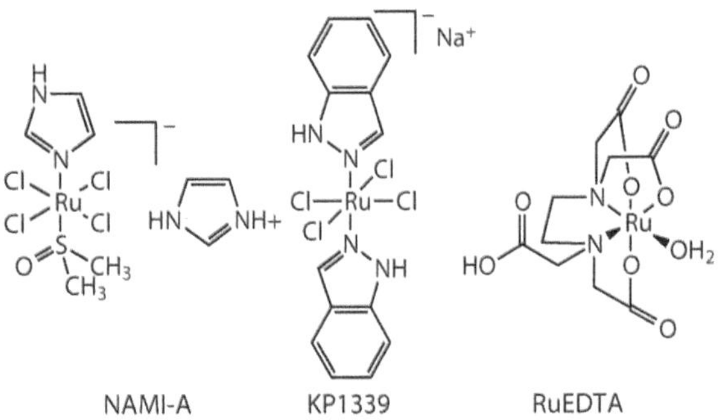

Ruthenium complexes as potential ˙NO scavengers to be used as antiangiogenic/antitumour agent [306]

expression of iNOS in higher tumour grades which tend to be more invasive (JENKINS et al. 1995).

While a cytokine mixture (interleukin-1β/interferon-γ) increased the production of •NO by stomach cancer cells (NCI-N87) in a concentration- and time-dependent manner, pre-treatment with 5-fluorouracil reduced the expression of iNOS and thus inhibited nitric oxide production (JUNG et al. 2002). 5-Fluorouracil stabilized IκBα and inactivated IκB kinase.

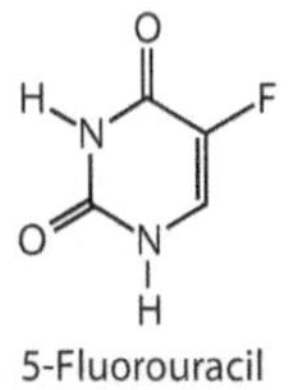

5-Fluorouracil [307]

Viabilities of both B16 melanoma and Lewis lung carcinoma cells were decreased in the presence of *S*-nitroso-*N*-acetyl-DL-penicillamine *in vitro* and the cytotoxicity of *S*-nitroso-*N*-acetyl-DL-penicillamine was reduced dose-dependently by •NO radical scavenger, oxyhemoglobin (HIRANO 1997). Intravenous injection of both cell lines in *mice* exposed to 10–80 ppm NO gas did not reduce the tumour colony formation in the lung. The increase in NO concentration was accompanied by elevation of concomitant nitric dioxide concentration in exposure chambers and exposure to higher concentration of NO appeared to enhance tumour colony formation in the lung.

In cocultures of macrophages and lymphoma cells, •NO generated from macrophages was shown to inhibit cellular respiration in the target cells (HIBBS et al. 1987, STUEHR and NATHAN 1989). •NO derived from macrophages (KLOSTERGAARD et al. 1991, LEU et al. 1991, JIANG et al. 1992), Kupffer cells (CURLEY et al. 1993, KUROSE et al. 1993, FUKUMURA et al. 1996), natural killer cells (CIPONE et al. 1994, XIAO et al. 1995), and endothelial cells (LI et al. 1991) participates in tumoricidal activity against many types of tumours.

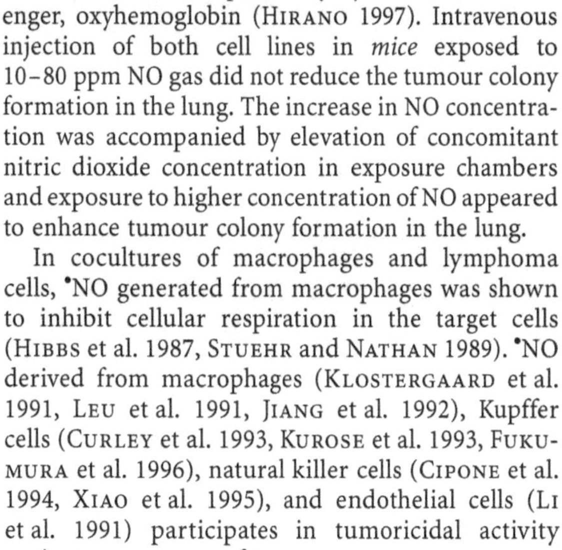

During the normal catalytic turnover of ribonucleotide reductase a tyrosyl radical is formed. Its reaction with •NO has been proposed as a factor in the cytostatic properties of •NO, due to the suppression of DNA synthesis through the salvage pathway. [285]

Several molecular targets, such as aconitase and ribonucleotide reductase, have been implicated in the cytostasis/cytotoxicity mediated by •NO. The reaction between •NO and the tyrosyl radical species formed in ribonucleotide reductase (LEPOIVRE et al. 1994)

Nitric oxide donors appear to reduce the viability of several tumour lines (PETIT et al. 1996) perhaps by deleting intracellular stores of glutathione making the cell susceptible to other toxic mechanisms (WINK et al. 1994).

Peroxynitrite was proposed to be responsible for the strand breakage induced by catechol-estrogens and nitric oxide (YOSHIE and OHSHIMA 1998). However, the DNA stand breakage caused by peroxynitrite was strongly (>70%) inhibited by desferrioxamine (YOSHIE and OHSHIMA 1997), although this compound had no effect upon the strand breakages induced by •NO and carechol-estrogens (YOSHIE and OHSHIMA 1998). These results suggested, that in addition to peroxynitrite, the reaction between catechol-estrogens and •NO might also yield other types of compounds, which could cause DNA damage directly. There could be a reaction between semiquinone radical and •NO, resulting in the formation of a semiquinone-NO adduct(s), which could induce strand breakage by direct reaction with DNA or by NO_x generated from it. Catalase also inhibited strand breakage (YOSHIE and OHSHIMA 1998), suggesting that H_2O_2 may be involved in the DNA damage. NORONHA-DUTRA et al. (1993) reported that the reaction between H_2O_2 and •NO produced an single oxygen-like substance(s). Strand scission in DNA was induced by H_2O_2 and S-nitrosothiols (PARK and KIM 1994). ZINGARELLI et al. (1996) characterised the cytotoxic effect of endogenous •NO and peroxynitrite in J774.2 macrophages immunostimulated with endotoxin which generated superoxide (within 1 h) and •NO (after 8 h). •NO production paralleled an increase in peroxynitrite formation and DNA strand breakage, and a decrease in intracellular NAD^+ content and mitochondrial respiration. A similar pattern of free radical formation and cytotoxicity was observed in peritoneal macrophages from endotoxaemic *rats* (formation of •NO, superoxide, peroxynitrite, and DNA strand breaks).

8-Nitroxanthine is produced is produced as the major nitration product in reactions of 2'-deoxyguanosine or *calf* thymus DNA with nitryl chloride produced by mixing nitrite with hypochlorous acid, and 8-nitroguanidine was a minor product in these reactions (CHEN et al. 2001). Formation of 8-nitroxanthine was also detected by xanthine reaction with various reactive nitrogen species, including nitryl chloride, peroxynitrite, nitronium tetrafluoroborate, and heated nitric and nitrous acids.

19.1.3 Simultaneous Generation of Nitric Oxide and Superoxide

Incubation of *calf* thymus DNA with 3-morpholino-sydnonimine (SIN-1), which simultaneously generates nitric oxide and superoxide (FEELISCH et al. 1989), induced a significant increase of 8-hydroxydeoxyguanosine (INOUE and KAWANISHI 1995). Peroxynitrite also increased 8-hydroxydeoxyguanosine in *calf* thymus DNA. Addition of free hydroxyl radical scavengers inhibited the increase in 8-hydroxydeoxyguanosine by SIN-1 or peroxynitrite. Incubation of ^{32}P-labelled DNA fragment with SIN-1 or peroxynitrite caused DNA cleavage at every nucleotide with a little dominance at guanine residues.

Desamination of guanosine, cytosine, and adenine is mediated in vivo primarily by the nitrosative chemistry of $\mathbf{N_2O_3}$ (WINK et al. 1991, NGUYEN et al. 1992). Nitrosation of an exocyclic amine group has been proposed to lead to the formation of a primary nitrosamine, followed by rapid deamination which culminates in the formation of an hydroxyl group.

$$NH_2\text{-}R + N_2O_3 \longrightarrow R\text{-}NHNO + NO_2^- \quad [88]$$

$$R\text{-}NHNO \longrightarrow RNNOH \longrightarrow R\text{-}OH + N_2. \quad [89]$$

This chemistry would lead to the conversion of cytosine to uracil, guanine to xanthine, methylcytosine to thymine, and adenine to hypoxanthine. Single stranded DNA is far more susceptible to nitrosative chemistry than double stranded DNA (MERCHANT et al. 1996), which suggests that deamination should occur more prevalently during replication and transcription of DNA.

19.1.4 DNA Sugar Radicals

Reactions of DNA sugar radicals lead to formation of altered sugars, which are either released from the DNA chain or bound to DNA with both phosphate linkages still being intact. Altered sugars attached to a broken DNA chain by one phosphate linkage are also formed (for review see VON SONNTAG 1987). For example, the oxidation of the C4'-centred radical leads to 2,5-dideoxypentos-4-ulose, 2,3-dideoxypentos-4-ulose, and 2-deoxypentos-4-ulose, which were isolated by DIZDAROGLU et al. (1975) after γ-irradiation of deoxygenated N_2O-saturated aqueous solutions of DNA (500 mg/l).

19.2 Carcinogenicity of Particulate Air Pollutants

19.2.1 Silica

Meta-analyses of world-wide epidemiological studies show that a critical appraisal of the methods that there is a more than double risk of developing lung

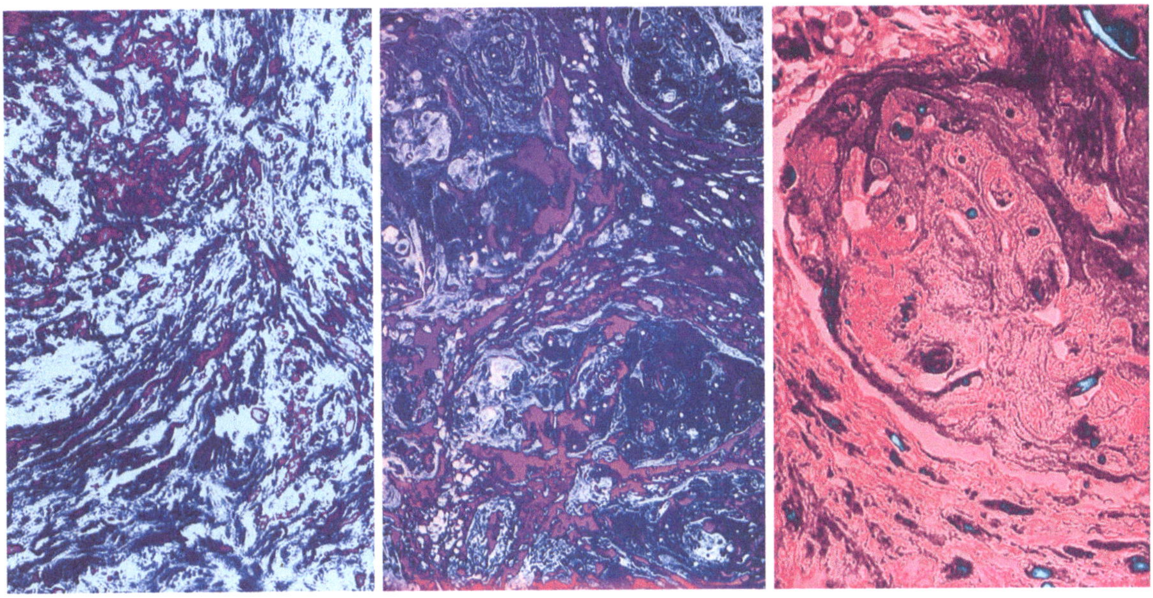

Fig. 349. Autopsy specimen (Duisburg A. 64/61) of a magnocellular pulmonary carcinoma of the right lower lobe with metastases to the hilar lymph nodes. Micronodular silicotuberculosis with marked emphysema. Patient (born in 1902) worked from July 1, 1922 to November 1, 1959 as a trimmer and hewer in a coalmine of the Ruhr district, then for two years on the surface. Fixed in formalin, paraffin section, embedded in glycerol. Interference contrast, achromat 20×/0.35 and fluorite 50×/0.85/0.17, respectively

cancer for persons with silicosis (Woitowitz 1999). Whether there is an increased risk of lung cancer for silica-exposed workers without silicosis has not been clarified (Latza et al. 2000). Lung cancer frequently occurred with simple pneumoconiosis (categories 1 to 3) compared with complicated pneumoconiosis with massive fibrosis (category 4), a difference that was highly significant (60 [33 %] of 183 versus 45 [12 %] of 375, $P < 0.001$) (Katabami et al. 2000). Diffuse interstitial fibrosis also appeared significantly more often with simple pneumoconiosis than with complicated pneumoconiosis (29 [16 %] of 183 versus 26 [7 %] of 373, $P = 0.002$). Neither concurrent lung cancers nor diffuse interstitial fibrosis was associated with occupation (i.e. coal miners, metal miners, or others), duration of dust exposure (years of employment) or other clinical parameters of pneumoconiosis. However, patients with diffuse interstitial fibrosis had a significantly greater tobacco consumption (in terms of pack-years) than had pneumoconiotic patients without diffuse interstitial fibrosis (40.0 ± 26.3 versus 32.0 ± 22.5, $P = 0.02$).

Evaluation of the lung cancer risk in coal miners is complex. On the one hand, the evidence of the carcinogenicity of coal dust is inadequate (for formation of paramagnetic centres during crushing see Lebedev et al. 1978; for the nature and concentration of free radicals in relation to rank, thermal history, and particle size see Petrakis and Grandy 1978), on the other hand, coal miners with exposure to dust with a high quartz concentration can develop silicosis. This so-called stone workers' silicosis in the Ruhr coalmines is of special importance because those workers who have worked above and below the seams "Mausegatt" and "Finefrau" show the same changes as are observed in workers who have worked in quarries in the Ruhr (Zorn and Worth 1952). Since the inflammatory response is generally considered to be a crucial event (Bowden 1987) in dust-induced toxicity, these events invite further testing beyond the animal model. Upregulation of ornithine decarboxylase (EC 4.1.1.17) (Marsh and Mossman 1991) and *c-fos*/*c-jun* proto-oncogenes (Heintz et al. 1993), need further *human* studies.

Silica (Berkeley Min-U-Sil: 5 µm particle size; 50 mg/ml × kg body weight) instilled intratracheally, after 1, 3, and 5 days, respectively, caused a significant increase in 8-hydroxy-2'-deoxyguanosine in the lungs of specific pathogen-free male Wistar *rats* as determined by the modified ECD-equipped high-performance liquid chromatography method of Floyd and Kasai (Yamano et al. 1995). Quartz dust (2.5 mg/*rat*) instilled intratracheally into the lungs of Wistar *rats* induced the formation of 8-oxoguanine 7, 21, and 90 days after the exposures as determined by immunocytochemistry (Nehls et al. 1997). Marked differences between 5 samples of α-quartz, cristobalite and tridymite suspended in 10 mM phosphate buffer (pH 7.4) were found for their levels of oxygen consumption and $HO^{\bullet}$ generation (Daniel et al. 1995). Incubated for 5 days at 37 °C with herring sperm DNA, all samples increased formation of thymine glycol, with wide variations in activity among samples normalised for equal surface area (F600 > cristobalite > Min-U-Sil 5 > Min-U-Sil 5 pre-treated with HF > tridymite > DQ 12 > Chinese standard α-quartz. When normalised for equal surface area, the samples produced different levels of DNA strand breakage. Addition of H_2O_2 strongly accelerated DNA damage – more for cristobalite than for the α-quartz samples.

Primary peripheral lung tumourlets composed predominantly of alveolar type II cells have been induced in Wistar *rats* exposed to α-quartz by inhalation (10 mg/m^3, 56 weeks, 5 days/week, 7 h/d) or intratracheal instillation (Friemann et al. 1995).

The inflammatory mechanism of particle-induced genotoxicity leading to lung tumours in *rats* is supported by a number of observations from both *in vitro* and *in vivo* studies. Driscoll (1996) showed that inflammatory cells of the lungs of *rats* exposed to α-quartz were mutagenic to *rat* alveolar epithelial cells, and that the mutagenic effect could be attenuated by the addition of antioxidants. Polymorphonuclear leucocytes appeared to the most mutagenic inflammatory cells.

19.2.2 Asbestos

Analytical transmission electron microscopy (ATEM) of lung tissue at autopsy was used to evaluate past exposure to fibrous dusts as to asbestos and man-made mineral fibres (McDonald et al. 1990, Arhelger et al. 1992). Roggli et al. (2002) examined 312 cases of mesothelioma for which fibre burden analyses of lung parenchyma had been performed by means of scanning electron microscopy do determine the content of tremolite, non-commercial amphiboles, talc and chrysotile. Tremolite was identified in 166 of 312 cases (53 %) and was increased above background levels in 81 cases (26 %). Fibrous talc was identified in 193 cases (62 %) and correlated strongly with the tremolite content ($P < 0.0001$). Talc levels explained less of the tremolite deviance for cases with an increase tremolite level than for cases a normal range tremolite level (22 versus 42 %). In 14 cases (4.5 %) non-commercial amphibole fibres (tremolite, acti-

Table 55. Experimental induction of mesotheliomas

Dust	Model System	Parameter of Toxicity	Result	Reference	Remarks
	Mesothelioma cell lines	Surface thermodynamic properties	The contact angle of the epitheliomatous mesothelioma was significantly lower than that obtained at normal mesothelium or sarcomatous mesothelioma cells.	Green et al. (1990)	
Chrysotile, crocidolite, amosite, erionite, glass fibre, glass powder	Mesothelial and other cell lines	Chromosomal changes	Interphase cytogenetic studies are suitable in the early detection of mesothelioma	Knuutila et al. (1991)	
MMMF, asbestos, erionite	Mesothelioma cell lines, cultured *human* primary mesothelial cells	Chromosomal damage		Linnainmaa et al. (1991)	
Chrysotile, crocidolite, amosite, anthophyllite, glass fibes code 100 and 110, erionite, potassium octatitanate	Syrian *hamster* embryo cells, C3H10T1/2 cells, Balb 3T3 cells, and *rat* pleural mesothelial cells	Cell transformation	Transformation systems are well suited for detailed studies on the effects of fibres in the cells	Mikalsen (1991)	Review article with 80 references
Erionite, crocidolite, amosite, anthophyllite, chrysotile, JM code 100 glass fibres, glass wool	Incubation with H_2O_2 (5 min, 37 °C, pH 7.0)	Salicylate adduct formation	Erionite, JM code 100 and glass wool were the most effective initiators of $HO^\bullet$ formation, followed, in order, by crocidolite, amosite and chrysotile.	Maples and Johnson (1992)	$HO^\bullet$ formation figures are compared to human and experimental tumour induction.
UICC standard amphibole asbestos amosite and crocidolite, UICC serpentine asbestos chrysotile A (Rhodesian), Oregon erionite, glass wool, rock wool	*Human* primary mesothelial cells, Met-5A cells	Survival, anaphase aberrations	On a comparison by weight, amosite, crocidolite and chrysotile showed similar toxic effects: 2–5 μg/cm² of the asbestos fibres caused 50 % of the cells to die, but erionite was less toxic (10–20 μg/cm² were needed for the same effect). When the doses were converted to the number of fibres/cm² of culture area, amosite was shown to be about 10-times more cytotoxic than crocidolite and chrysotile. Of the MMMF thin glass wool was the most cytotoxic (50 % cell death for 10–20 μg/cm²), followed (in descending order of cytotoxicity) by thin rock wool, coarse glass wool, milled rock wool, milled glass wool and coarse rock wool. All 3 asbestos types studied induced anaphase aberrations at high (near toxic) doses. A statistically significant increase in the number of aberrant anaphases was observed in cultures treated with crocidolite of chrysotile at 5 μg/cm². The increase was caused by lagging chromatids, chromosomes of chromosome fragments.	Pelin et al. (1992)	
Saffil fibres (as manufactured and after 'ageing' at temperatures above 1200 °C for more than 1000 h), aluminosilicate (ceramic) fibre A and B; UICC standard chrysotile A (as positive control)	Intrapleural injection in Wistar-derived *rats*	Incidence of neoplasia	Malignant mesothelioma in 10 rats dosed with asbestos and 3 dosed with aluminosilicate fibre B, but no mesothelioma in any rat dosed with Saffil fibres or aluminosilicate fibre A.	Pigott and Ishmael (1992)	

Table 55. Continued

Dust	Model System	Parameter of Toxicity	Result	Reference	Remarks
Chrysotile and crocidolite asbestos, erionite, man made fibres	*Rat* pleural mesothelial cells	Cytotoxicity assessed by the 3-(4,5-dimethylthiazol-2-yl)-2,5-diphenyltetrazolium bromide assay	The tumorigenic potency of fibres may be related to the fibre dimensions, to their surface properties on in vivo biopersistence.	RENIER et al. (1992)	
Crocidolite, chrysotile	*Rat* pleural mesothelial cells; *hamster* tracheal epithelial cells	c-fos and c-jun mRNA	In contrast to phorbol 12-myristate 13-acetate, which induced rapid and transient increases in c-fos and c-jun mRNA, asbestos caused 2- to 5-fold increases in c-fos and c-jun mRNA that persisted for at least 24 h in mesothelial cells. These inductions were dose-dependent and were most pronounced with crocidolite. Induction of c-jun gene expression by asbestos occurred in tracheal epithelial cells but was not accompanied by a corresonding induction of c-fos gene expression. In both cell types, asbestos induced increases in protein factors specifically binding to DNA sites that mediate gene expression by the AP-1 family of transcription factors.	HEINTZ et al. (1993)	
MMVF 10 and 11, ceramic fibre RCF 1, chrysotile	30 mg/m^3 for MMMF, 10 mg/m^3 for asbestos; *rats*	Wagner Pathology Grading Scale, tumour findings	Respirable fibrous glass → no significant hazard	HESTERBERG et al. (1993)	
Crocidolite	Calf thymus DNA, intraperitoneal injection in female Fischer 344 *rats*	OH^8-2'-Deoxyguanosine formation; incidence of mesotheliomas	Under simple incubation with 2 mg calf thymus DNA 5 mg crocidolite or deferrized crocidolite for 3 h at 37 °C yielded 14.6 and 30.2 OH^8-2'-deoxyguanosine/10^5 2'-deoxyguanosine, respectively. In incubation systems supplemented with 0.5 mM H_2O_2 plus 0.1–1.0 mM Fe_2O_3, 0.5 mM Fe_2O_3 + 0.5 mM ascorbate, 0.5 mM Fe_2O_3 + 1mM EDTA, or 0.5 mM $FeSO_4$ + 0.5 mM ascorbate, deferrized crocidolite induced higher levels of OH^8-2'-deoxyguanosine. Within one year, intraperitoneal injection of 5 mg crocidolite dust in 1 ml saline induced mesotheliomas in 5 of 10 female Fischer 344 *rats*, while deferrized crocidolite did so in 4 of 10, but deferrized crocidolite supplemented by 35 weekly i.p. injections of 2 mg Fe_2O_3 succeeded in 10 of 10 *rats*.	ADACHI et al. (1994)	
Asbestos (Rhodesian chrysotile and crocidolite)	*Rat* pleural mesothelial cells in vitro	3-4-(5-Dimethylthiazol-2-yl)-2,5-diphenyltetrazolium (MTT) bromide colorimetric assay in the pesence or absence of catalase and superoxide dismutase; [^{3}H]thymidine incorporation by non-S-phase cells by means of liquid scintillation counting; proportion of S-phase cells in hydroxyurea-blocked cultures determined by autoradiography	MTT reduction by asbestos-treated rat pleural mesothelial cells fell in a concentration-dependent manner. Both 100 U/ml catalase and 250 U/ml SOD reduced the cytotoxicity of asbestos. H_2O_2 was strongly cytotoxic to rat pleural mesothelial cells and 0.1 mM H_2O_2 produced a complete loss of cell viability. The cytotoxicity of H_2O_2 was completely inhibited by the additon of catalase. [^{3}H]Methylthymidine incorporation was reduced to 40±2 % and 32±3 % of control values in exponentially growing cells treated with chrysotile and crocidolite respectively, at the 2 $\mu g/cm^2$ dose, and to 75±4 % and 91±3 % in confluent mesothelial cells treated with the same dose ($P < 0.01$). Autoradiography of cultures confirmed that asbestos at the 4 $\mu g/cm^2$ dose exposure did not increase the proportion of S-phase cells. In the presence of hydroxyurea the incorporation was enhanced in a concentration-dependent manner. Active catalase prevented the enhancement, but inactivated enzyme also suppressed crocidolite-induced enhancement of unsheduled DNA synthesis at 2 and 4 $\mu g/cm^2$.	DONG et al. (1994)	

Table 55. Continued

Dust	Model System	Parameter of Toxicity	Result	Reference	Remarks
Amosite	Cultured *human* mesothelial cells (MET 5A), hypoxanthine-xanthine oxidase system	DNA single strand breaks, extracellular release of nucleotides and their catabolites, LDH release	Superoxide radical and H_2O_2 exposures resulted in the depletion of adenine nucleotides, accumulation of the products of nucleotide catabolism, induction of DNA single strand breaks.	Kinnula et al. (1994)	Amosite did not induce acute oxidant-type injury to mesothelial cells in vitro.
Synthetic vitreous fibres, refractory ceramic fibre	Inhalation (nose only) in *rats* and *hamsters*	Lung weights, Wagner scores, pulmonary neoplasms	Only refractory ceramic fibre produced a dose-related (3, 16, or 30 mg/m^3) increase in primary lung neoplasma (*rats* only) and mesotheliomas (*rats* and *hamsters*).	McConnell (1994)	
Amosite (short and long)	Lavaged male Wistar *rat* alveolar macrophages	Production of superoxide anion (cytochrome c reduction)	Both long and short fibre samples of amosite without opsonization were ineffective in stimulating isolated rat alveolar macrophages to release $O_2^{\bullet-}$ in vitro. After opsonisation with immunoglobulin, a dramatic enhancement of release of $O_2^{\bullet-}$ was seen with long fibres, but not short. Long fibres bound threefold more immunoglobulin than the short fibres.	Hill et al. (1995)	Increased binding of opsonin to the surface of long fibres may be an important modifying factor in the pathogenicity process.
Amosite	Cultured *human* mesothelial (MET 5A) and bronchial epithelial (BEAS 2B) cells, *human* venous blood polymorphonuclear leucocytes	Luminol-enhanced chemiluminescence, cellular adenine nucleotide depletion, extracellular release of nucleotides and LDH	Amosite-activated (and to a lesser degree nonactivated) PNMs released substantial amounts of reactive oxygen metabolites, whereas the chemiluminescence of amosite-exposed mesothelial cells and epithelial cells did not differ from the background. Amosite treatment (48 h) of the target cells did not change intracellular adenine nucleotides (ATP, ADP, AMP) or or nucleotide catabolites (hypoxanthine, xanthine, uric acid). When target cells were exposed to nonactivated PMNs, significant adenine nucleotide depletion and nucleotide catabolite accumulation was observed in mesothelial cells only.	Kinnula et al. (1995)	Importance of inflammatory cell-derived free radicals in amosite-induced mesothelial cell injury
Amosite, crocidolite, chrysotile A, thin glass wool and thin rook wool fibres; milled glass wool and milled rook wool, titanium dioxide	*Human* primary mesothelial cells, Met-5A cells, rat liver epithelial cells	Multinucleus assay, fibre incorporation study	All 4 fibre types caused statistically significant increases in the amount of binucleated cells in human primary mesothelial cells and MeT-5A cells (in the dose range 0.5–5.0 μg/cm^2). Chrysotile and crocidolite were more effective (1.3-3.0-fold increases) than thin glass wool and thin rock wool fibres (1.3-2.2-fold increases). However, when the fibre dose was expressed as the number of fibres per culture area, the asbestos and MMVF appeared equally effective in human mesothelial cells. In rat liver epithelial cells, chrysotile was the most potent inducer of binucleation (2.9–5.0-fold increases), but the response of the RLE cells to crocidolite, thin glass wool, and thin rook wool fibres was similar to the response of the human mesothelial cells. No statistically significant increases in the number of bi- or multinucleate cells were observed in human primary mesothelial cells or RLE cells exposed to the non-fibrous dusts. In MeT-5A cells exposed to 5 μg/cm^2 of milled glass wool or milled rock wool as well as in cultures exposed to 2 and 5 μg/cm^2 TiO_2, significant increases were, however, observed.	Pelin et al. (1995)	Rodent cells respond differently to mineral fibres than human cells.

Table 55. Continued

Dust	Model System	Parameter of Toxicity	Result	Reference	Remarks
Amphibole (crocidolite and amosite), serpentine (chrysotile), wollastonite, riebeckite, glass beads	*Human* and *rabbit* mesothelial cells in vitro	Oligonucleosomal DNA fragmentation, loss of membrane phospholipid asymmetry, and nuclear condensation	Asbestos fibres, not control particles, induced apoptosis in mesothelial cells by all assays. Induction of apoptosis was dose-dependent for all types of asbestos, with crocidolite (5 µg/cm^2) inducing 15.0±1.1 % apoptosis versus control particles <4 %. Apoptosis induced by asbestos, but not by actinomycin D, was inhibited by extracellular catalase, superoxide dismutase in the presence of catalase, hypoxia (8 % oxygen), deferoxamine, 3-aminobenzamide [an inhibitor of poly(ADP-ribosyl) polymerase], and cytochalasin B. Only catalase and cytochalasin B decreased fibre uptake.	Broaddus et al. (1996)	Escape from asbestos-induced apoptosis could allow the abnormal survival of mesothelial cells with asbestos-induced mutations
Crocidolite asbestos	Primary *murine* mesothelial cell line D9 (2×10^5/cm^2)	Immunocytochemistry for p53 protein expression	After exposure to increasing doses (0–22.5 µg/cm^2) of crocidolite asbestos fibres for 48 h, up to 18 % of late passage cells had micronuclei compared to 4 % of early cytokinesis-arrested passage cells.	Cistulli et al. (1996)	It is hypothesized that loss of the G1 cell cycle checkpoint contributes to genetic instability in *murine* mesothelial cells.
Crocidolite, glass fibres B-01-09	Intraperitoneal injection in female *rats*.	Histology and ultrastructure, monoclonal antibodies against the cytokeratins 5, 6, 8, 17, and 19, α_1-muscle actin, desmin, vimentin, rodent-specific 97 kDa macrophage antigen	Tumours, mostly malignant fibro-histiocytomas or pleomorphic subtypes were found in 15 of 24 rats 11–20 months after injection of crocidolite. 2 of 14 animal died or sacrificed 8–20 months after injection of soluble glass fibres suffered from metastasizing carcinomas of the internal or external genitals.	Friemann et al. (1996)	
Long crocidolite	*Human* mesothelial cell line (MET5A)	EGF-receptor as demonstrated by epifluorescence and confocal microscopy, using a monoclonal antibody directed against the EGF-R external domain	Whereas cells in contact with short asbestos fibres showed no aggregation of EGF-R nor increased fluorescence at sites of fibre contact, cells surrounding long (>50 µm) showed intense and diffuse staining for EGF-R protein.	Pache et al. (1996)	
Amosite	Cultured transformed *human* pleural mesothelial cells (MET 5A)	Northern blot, MnSOD and total SOD, LDH release, high-energy nucleotides (ATP, ADP, AMP), hypoxanthine, xanthine, uric acid	TNF and amosite + TNF caused significant MnSOD mRNA upregulation. Similarly MnSOD specific activity was increased by TNF (290 % increase) and the amosite + TNF combination (313 5 increase) but not by amosite alone. In cell injury experiments amosite and amosite + TNF exposured caused significant cell membrane injury when assessed by lactate dehydrognase release, which was 31 % and 57 % higher than in the unexposed cells. However, only the amosie + TNF combination caused significant depletion of cellular high-energy nucleotide when expressed as percentage of [^{14}C]adenine labeling in cellular high-energy nucleotides. The nucleotide levels were 91.5±2.0 % in the unexposed cells, 89.9±3.9 % in the amosite-exposed cells, 90.1±2.2 % in TNF-exposed cells, and 78.8±9.4 % in amosite±TNF-exposed cells. Amosite + TNF-exposed cells were also most sensitive to menadione (20 µmol/l, 2 h), a compound which generates $O_2^{\bullet-}$ intracellularly.	Pietarinen-Runtti et al. (1996)	*Human* mesothelial cell inflammatoy cytokines but not asbestos fibres per se can cause MnSOD induction.

Table 55. Continued

Dust	Model System	Parameter of Toxicity	Result	Reference	Remarks
Crocidolite, tremolite, 11 vitrous fibre dusts, 5 insulation wools, granular SiC	Intraperitoneal injection in Wistar *rats*.	Tumours in the abdominal cavity	2 mesotheliomas were found in a total 395 *rats* treated with saline or granular SiC (250 and 1000 mg). Dose-dependently, 11 fibre dusts induced mesotheliomas at rates up to 97 %. UICC-like crocidolite ranked at the top, the glass fibres type B-01 with its low biopersistance and type B-09 with a low diameter at the end of the carcinogenicity scale; like tremolite, rock wool MMVF-21 took a front place.	Roller et al. (1996)	
Crocidolite (UICC reference sample)	*Rat* lung mesothelial cells	^{3}H-Thymidine labelled cells	Mesothelial cells showed increased DNA synthesis at 1 week.	Adamson et al. (1997)	
Asbestos (anthophyllite, tremolite-actinolite, amosite, crocidolite, chrysotile), nonasbestos fibres	Analytical electron microscopy of 50 lung tissue samples from Matsubase, where pleural plaques are endemic	Number of asbestos bodies and fibres/ 5 g wet lung tissue; frequency of pleural plaques; size parameter of fibres by mineral type	Anthophyllite (mean length 25.1 μm, mean diameter 0.84 μm) might be responsible for the increased prevalence of pleural plaques in Matsubase. The aspect ratio of anthophyllite (mean = 38.7) was lower than that of amosite (mean = 81.8), which, as reported by Murai and Kitagawa (1992), was found predominantly in cases of pleural mesothelioma.	Murai et al. (1997)	Differences in fibre size may be related to the strength of the carcinogenicity to the pleura.
Amosite, SiC, refractory ceramic fibres 1 and 4, MMVF10, Code 100/475 glass fibre	Incubation of A549 lung epithelial cells with 5 mM H_2O_2 for 30 min and 1 h	Induction of nuclear translocation of NF-ϰB	SiC fibres most potent of the pathogenic fibres, MMVF10 most potent of the non-pathogenic fibres, causing significant nuclear translocation of NF-ϰB	Brown et al. (1999)	Discrimination between pathogenic and non-pathogenic fibres in terms of a key pro-inflammatory event in epithelial cells
Crocidolite, riebeckite	*Rat* pleural mesothelial cells	Epidermal growth factor receptor	Asbestos fibres, but not riebeckite, abolished binding of EGF to EGF receptor. Tyrphostin AG-1478 significantly ameliorated asbestos-induced increases in mRNA levels of *c-fos* but not of *c-jun*.	Zanella et al. (1999)	
Chrysotile (NIEHS and UICC), crocidolite (NIEHS and UICC)	*Rabbit* pleural mesothelial cells	Coating of fibres, phagocytosis and dichloro-fluorescein assay	Vitronectin adsorption to fibres increases fibre phagocytosis and intracellular oxidation. Apoptosis blocked by integrin-ligand blockade with arginine-glycine-aspartic acid peptides	Wu et al. (2000)	

Table 56. Carcinogenicity of fibres: animal experiments versus human pathology (from Friedberg and Schiller 1988)

Amosite	Harrington et al. (1971), Selikoff et al. (1972)
Anthophyllite	Meurman et al. (1974)
Chrysotile	McDonald and Fry (1982)
Crocidolite	McDonald and McDonald (1978)
Tremolite	Churg et al. (1984)
Borosilicate glass	
Aluminium silicate glass	
Mineral wool	Olsen and Jensen (1984)
Aluminium oxide	
Potassium titanate	
Silicon carbide	
Sodium aluminium carbonate	
Wollastonite	Huuskonen et al. (1985)
Attapulgite	

Note: For materials without a reference carcinogenicity in *man* is unknown

nolite and/or anthophyllite) were the only fibre types found above background.

Kidney cancer in persons occupationally exposed to asbestos is unlikely due to the dust exposure; however, high asbestos exposure might entail a small increase in risk (Sali and Boffetta 2000).

19.2.3 Automobile Exhaust Particulates

Howe et al. (1983) found an elevated risk of lung cancer for those pensioners of the Canadian National Railway Company employed in occupations involving exposure to diesel fumes and coal dust, with highly significant dose-response relationships observed. In bus garage workers, the lung cancer risk increased with increasing cumulative exposure to diesel exhaust, but not with cumulative asbestos exposure (Gustavsson et al. 1990).

In the lungs of Osborne Mendel *rats*, only the hydrophobic part (about 75 %-wt) of diesel exhaust condensate which contained polycyclic aromatic compounds resulted in 5 malignant tumours in a group of 35 animals (Grimmer et al. 1987). Polycyclic aromatic compounds consisting of four and more rings (0.8 %-wt) were found to be the most potent subfraction and provoked when proportionally dosed 6 carcinomas in a group of 35 *rats*.

In Chinese *hamster* lung cells V79, an extract of particulate matter from gasoline engine exhaust (0.62–4.95 μg/ml) was found to induce aneuploidy and polyploidy in a dose-dependent manner (Hadnagy and Seemayer 1986).

Glutathione (8 mM), cysteine (8 mM), and 2-mercaptoethanol (4 mM) decreased the cytotoxicity of diesel extract in cultured Chinese *hamster* ovary (CHO) cells (Li 1981).

19.2.4 Beryllium

Starch gel zymograms of lactate dehydrogenase (EC 1.1.1.27) in the pulmonary tissue extract of *rats* exposed daily to the inhalation of $BeSO_4$ aerosol showed well-defined differences in comparison to paired controls (Reeves 1967). Significant increase of both the muscle type and heart-type isozymes was observed during the immediate precancerous phase (8th–10th month of exposure), followed by return to normal or subnormal levels at the time of appearance of the fully grown pulmonary tumours (12th–13th month).

19.2.5 Styrene

A significant increase in 8-hydroxy-2'-deoxyguanosine in the blood of 17 workers exposed to styrene compared to 67 non-exposed healthy volunteers provided a good indication that styrene exposure can result in generation of hydroxyl radicals and oxidative DNA damage (Marczynski et al. 1997).

Styrene is metabolised to styrene oxide, mainly by CYP2B6 followed by CYP1A2, CYP2E1 and CYP2C8 (Nakajima et al.1993). After single i.p. injection of styrene (range 0–4.35 mmol per kg body weight), Pauwels et al. (1996) determined the N-7-guanine adduct of styrene 7,8-oxide in various *mice* tissues. The adducts were most abundant in the lungs (32.8–2056.3 fmol/mg DNA), ca. 30 % more than in liver (11.8–1541.0 fmol/mg DNA) and spleen (58.4–1188.2 fmol/mg DNA). The higher adduct level in lungs could be a result of high CYP-activity converting styrene to styrene 7,8-oxide and for lack of epoxide hydroxylase in that organ. The 1-adenine adduct of styrene 7,8-oxide has been detected in *mice* lungs after inhalation of styrene (Koskinen et al., in press). DNA adducts of styrene 7,8-oxide at the O^6-position of 2'-deoxyguanosine have been identified by the ^{32}P-postlabeling analysis (Otteneder et al. 1999). The two regioisomeric adducts, O^6-(2'-hydroxyl-1-phenylethyl)-2'-deoxyguanosine 3'-phosphate (α-isomer) and O^6-(2'-hydroxyl-2-phenylethyl)-2'-deoxyguanosine 3'-phosphate (β-isomer), were synthesised and used for optimising and quantifying the various analytical steps. The adducts were stable at pH 7 and 10, but not at pH 4. The phosphorylation efficiency with polynucleotide kinase was 5 and 15 % for the α- and β-isomers, respectively.

Phenotyping of the *CYP2E1* alleles in workers exposed to styrene: While the only styrene-exposed worker from a fibreglass-reinforced plastics factory heterozygous for *CYP2E1*1D* allele presented the highest value of chlorzoxazone (500 mg p.o.) metabolic ratio, a trend to lower chlorzoxazone metabolic ratio values for individuals possessing at least one mutant of *CYP2E1*6* allele compared with homozygous wild type was observed by Haufroid et al. (2002).

Vodička and Hemminki (1988) reacted radioactive styrene with double- and single-stranded DNA and characterized the binding products by HPLC after neutral hydrolysis and enzyme digestion of DNA. More products were formed in single-stranded DNA as compared with double-stranded DNA. In single-stranded DNA at least 95 % of the adducts were guanine *N*-7-,N^2- and O^6-alkylation products; they formed in proportions 54:33:12. In double-stranded DNA the respective proportion was 74:23:3.7, indicating a selective suppression of alkylation at atoms N^2 and, particularly, O^6 that take part in hydrogen bonding in double-stranded DNA. The α- and β-isomers of 7-alkylguanine were found in a similar proportion in single- and double-stranded DNA, indicating no steric hindrance.

19.3 Oxidation Reactions of DNA

In a review Cadet (1997) illustrated the complexity of the oxidation reactions of DNA. The association of separation of oxidised bases by high-pressure liquid chromatography and the assessment of oxidative base damage to isolated and cellular DNA by tandem mass spectrometers is more specific and sensitive than by single quadrupolar detectors (Cadet et al. 2002). The increase in specificity comes from the monitoring of a characteristic fragmentation of the molecule to be measured. When tandem spectrometry is used, the first quadrupole filters the major ion, usually the pseudomolecular ion of the molecule, formed during ionisation. Thereafter, in the collision cell (which is usually a quadrupole) fragmentation of the ions is obtained in the presence of a low-pressure inert gas (usually nitrogen). Subsequently, the third quadrupole discriminates the daughter ions, which are then quantified using an electron multiplier detector. The HPLC-MS/MS assay has been first applied to the measurement of 8-oxo-7,8-dihydroxy-2'-deoxyguanosine (Serrano et al. 1996, Ravanat et al. 1998). Application of the method has been extended to other modified DNA bases and nucleosides, including 8-oxo-7,8-dihydroxy-2'-deoxyadenosine (Podmore et al. 2000, Weimann et al. 2001), Fapy-Gua, 5-(hydroxymethyl)-2'-deoxyuridine, 5-formyl-2'-deoxyuridine, 5-hydroxy-2'-deoxyuridine, the *cis* and *trans* diastereoisomers of 5,6-dihydroxy-5,6-dihydrothymidine (Frelon et al. 2000), and also tandem DNA lesions (Bourdat et al. 2000).

Imidazole ring opened guanine is the most frequent product from HO• treatment (Aruoma et al. 1989) followed by **8-hydroxyguanine,**

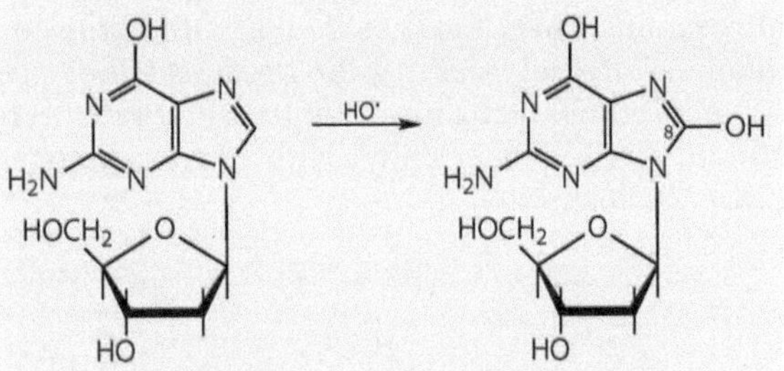

2-Deoxyguanosine 8-Hydroxy-2'-deoxyguanosine

Formation of 8-hydroxy-2'-deoxyguanosine [308]

followed by **imidazole ring opened adenine** and **cytosine glycol.**

The levels of 8-hydroxy-2'-deoxyguanosine in DNA from uterine myoma tissues were significantly ($P < 0.01$) elevated compared with those of tumour-free uterine tissue samples (Foksinski et al. 2000). In tumour-free uterine tissues, the levels of this modified base were higher in the group of premenopausal *women* when compared with postmenopausal ones ($P < 0.05$).

8-Oxo-2'-deoxyguanosine was found to be the main $^1O_2$ oxidation product of DNA upon exposure to either a chemical source of singlet oxygen or excited photosensitizers (Müller et al. 1990, Schneider et al. 1990, Devasagayam et al. 1991).

It is well established that singlet oxygen is able to oxidise DNA with a much higher specificity than hydroxyl radical. The main stable oxidation products of the reaction of $^1O_2$ with 2-deoxyribose guanine were identified as the 4R* and 4S* diastereoisomers of 4-hydroxy-8-oxo-4,8-dihydro-2'-deoxyguanosine on the basis of extensive NMR and mass spectrometry measurements (Ravanat et al. 1992, Ravanat and Cadet 1995). Similar oxidation products were generated by the type II photooxidation reaction of 2'-deoxyguanosylyl -(3'-5')-thymidine (Buchko et al. 1992).

Formation of Singlet Oxygen

$$HO_2^{\cdot} + O_2^{\cdot-} + H^+ \rightarrow {}^1O_2 + H_2O_2 \quad [309]$$

$$O_2^{\cdot-} + H_2O_2 \rightarrow {}^1O_2 + HO^- + HO^{\cdot} \quad [310]$$

$$O_2^{\cdot-} + HO^{\cdot} \rightarrow {}^1O_2 + HO^- \quad [311]$$

Thymine glycol was formed in the presence of H_2O_2, Cu^{2+} and ascorbic acid (Aruoma et al. 1991).

Thymidine glycol has been shown to be an efficient blocking lesion in several replication studies involving DNA polymerases (Ide et al. 1985, Rouet and Essigmann 1985, Clark and Beardsley 1986, 1987, Basu and Essigmann 1988, Basu et al. 1989, Evans et al. 1993). O^4-Ethyldeoxythymidine, but not O^6-ethyldeoxyguanosine, accumulates in hepatocyte DNA of *rats* exposed continuously to diethylnitrosamine (Swenberg et al. 1984). Sinigrin and indole-3-carbinol inhibited the hepatocarcinogenesis induced by diethylnitrosamine when they were administered concurrently with the carcinogen (Tanaka et al. 1990).

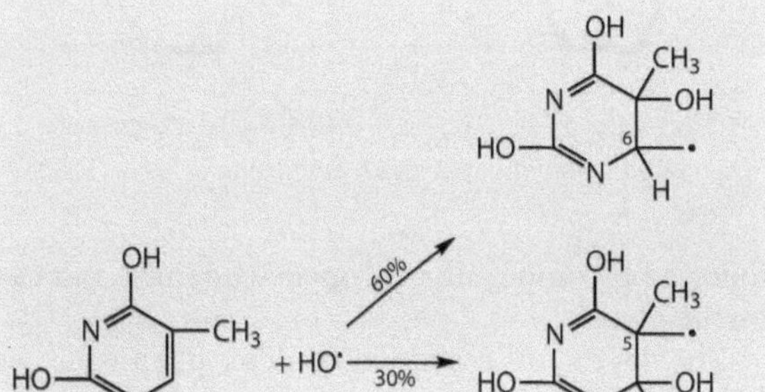

Thymine adducts [312]

The major product resulting from hydroxyl radical attack on DNA **cytosine** is cytosine glycol, which is highly unstable and readily deaminates to form uracil glycol or dehydrates to form 5-hydroxycytosine (Douki et al. 1996). Uracil glycol can also dehydrate to form 5-hydroxyuracil. These latter three products are the major free radical-damaged cytosine products. Uracil glycol, 5-hydroxycytosine, and 5-hydroxyuracil are readily bypassed by DNA polymerases (Purmal et al. 1994, 1998). The uracil derivatives, uracil glycol and 5-hydroxyuracil always pair with adenine, and thus are potentially potent premutagenic lesions (Purmal et al. 1994, 1998). The repair of cytosine lesions was recently reviewed by Wallace (2002).

The main reaction of hydroxyl radical with **2'-deoxycytidine** is the addition across the 5,6-ethylenic bond. The reducing 5-hydroxy-5,6-dihydro-2'-deoxycytidyl-6-yl was found to be preferentially formed, as inferred from the results of pulsed radiolysis experiments using the redox titration method (Hazra and Steenken 1983). On the other hand, the addition of HO$^•$ in position 6 occurred as a minor process (10 %). In a subsequent step molecular oxygen reacts at diffusion controlled rates with 5-hydroxy-5,6-dihydro-2'-deoxycytidyl-6-yl and 6-hydroxy-5,6-dihydro-2'-deoxycytidyl-6-yl (Decarroz 1987), yielding the corresponding peroxyl radicals.

To identify the cytosine modification(s) with the highest mutagenic potential, Feig et al. (1994) treated 2'-deoxycytidine-5'-triphosphate with H_2O_2 in the presence of $FeSO_4$ and ascorbic acid, separated the reaction products by high-performance liquid chromatography and incorporated them into a gene that allows analysis of the formation of mutations in *E. coli*. **5-Hydroxycytosine** was identified as one of the permutagenic modifications that give rise to G:C → A:T transitions. Under cell-free conditions, in addition to guanine both adenine and cytosine were found to be incorporated opposite 5-hydroxycytosine, depending on the sequence context (Purmal et al. 1994). 5-Hydroxycytosine was not excised by endonuclease III from γ-irradiated DNA (Dizdaroglu et al. 1993), while Hatahet et al. (1994) observed that 5-hydroxycytosine in a defined sequence context was recognised by both endonuclease III and Fpg protein.

The first example of the HO$^•$-mediated formation of a **vicinal oxidative base lesion** within a DNA fragment was provided by conduction a model study on 2'-deoxyguanosyl-(3',5')-thymidine (dGpT). X-irradiation of dGpT in aerated aqueous solution was found to give rise to a dinucleoside monophosphate bearing clustered base damage as a primary radiation-induced decomposition product (Box et al. 1993). The guanine moiety has been converted to 9-oxo-7,8-dihydroguanine, whereas the thymine base has been transformed into a formylamine product.

The formation of 7,8-dihydro-8-oxoguanine by electron transfer of hydrogen abstraction from DNA also occurs with excited carbonyl species such as triplet acetone (Epe et al. 1993), which are potential secondary products during lipid oxidation (Sies 1986). The excitation energy of these species is high enough to allow energy transfer to thymine residues as well, and therefore the damage profile consists of pyridine photodimers at levels greater than those of 7,8-dihydro-8-oxoguanine.

Triplet acetone as an excited carbonyl species

$$\underset{\text{Isobutanal}}{(H_3C)_2{-}CH{-}CHO} + O_2 \xrightarrow{\text{Horseradish peroxidase}} \underset{\text{Formic acid}}{HCOOH} + \underset{\text{Triplet acetone}}{CH_3{-}CO^*{-}CH_3} \quad [313]$$

$$\underset{\text{Triplet acetone}}{H_3C{-}CO^*{-}CH_3} \rightarrow H_3C{-}CO{-}CH_3 + hn \quad [314]$$

To maintain the genetic integrity these oxidised bases are **repaired** in *Escherichia coli* **by** several **DNA**

glycosylases, including the Fpg protein. *In vitro*, the Fpg protein excises a broad spectrum of modified purines, in particular, 2,6-diamino-4-hydroxy-5*N*-methylformamidopyrimidine and 7,8-dihydroxy-8-oxoguanine residues (Boiteux et al. 1992, Tchou et al. 1993). Saparbaev et al. (2002) examined, by targeted mutagenesis, the role of two highly conserved amino acid residues, proline 3 and lysine 57, on the catalytic activities of the Fpg protein toward a ring-fragmentation product of thymine (αRT) and 6,6-dihydrothymidine (dHT).

Dodson and Lloyd (2002) discussed the mechanisms by which various DNA glycosylases initiate the base excision repair pathways. Fundamental distinctions are made between "simple glycosylases", that do not form DNA single-strand breaks, and "glycosylases/abasic site lyases", that do form single-strand breaks. Several groupings of base excision repair substrate sites are defined and some interactions between these groupings and glycosylate mechanisms discussed. Two characteristics are proposed to be common among all base excision repair glycosylases: a nucleotide flipping step that serves to expose the scissile glycosyl bond to catalysis, and a glycosylase transition state characterized by substantial tetrahedral character at the base glycosyl atom.

The influence of endogenously and exogenously generated nitric oxide on the generation and repair of oxidative DNA damage in cultured B6 *mouse* fibroblasts was recently analysed by Phoa and Epe (2002). Increased oxidative damage was only observed after exposure to high (toxic) concentrations of exogenous NO generated by decomposition of dipropylene-triamine-NONOate. Under these conditions, the spectrum of DNA modifications was similar to that induced by 3-morpholinosydnonimine (SIN-1), which generates peroxynitrite. The repair rate of additional oxidative DNA base modifications induced by photosensitization was not affected by the endogenous NO generation in iNOS-transfected cells. However, it was completely blocked after pre-treatment with dipropylene-triamine-NONOate at concentrations that did not cause oxidative DNA damage by themselves. In contrast, the repair of DNA single-strand breaks, sites of base loss (AP sites) and UVB-induced pyrimidine, photodimers, was not affected. The endogenous generation of •NO in the iNOS-transfected fibroblasts was associated with a protection from DNA single-strand break formation and micronuclei induction by H_2O_2.

Treatment of DNA (50 μM) with **hypochlorous acid** (100 μM) caused DNA base chlorination as demonstrated by a loss in fluorescence of ethidium-DNA complexes when ethidium bromide (50 μM) was subsequently added as a probe for the functions of DNA (Shishido et al. 2000). HOCl reacted more rapidly with secondary than with primary amines of pyrimidine nucleotides to form chloramines. H_2O_2, ascorbate, and glutathione completely inhibited the decrease in fluorescence.

Reactions of **HOCl** with NO_2^- results in the formation of nitryl chloride (NO_2Cl), a potent oxidising, nitrating and chlorinating species. Exposure of DNA to NO^- alone (up to 250 μM) at pH 7.4 did not induce oxidative DNA damage (Whiteman et al. 1999). However, incubation of DNA with NO_2^- in the presence of HOCl led to increases in thymine glycol, 5-hydroxyhydantoin, 8-hydroxyadenine and 5-chlorouracil to levels higher than those achieved by HOCl alone. No significant increases in 8-hydroxyguanine, xanthine, hypoxanthine, 2-hydroxyadenine. FAPy guanine, FAPy adenine and 8-chloroadenine were observed. HOCl-induced depletion of FAPy guanine and 8-hydroxyguanine was reduced in the presence of NO_2^-.

The photoreduction of benzophenone triplets in micellar solution leads to the generation of isolated radical pairs, the behaviour of which resembles that of biradicals (Scaiano et al. 1982). Radical-pair decay is controlled by intersystem crossing and by radical exit from the micelle.

$$(C_6H_5)_2{-}C{=}O^* + RH \rightarrow (C_6H_5)_2{-}C^{\cdot}OH + R^{\cdot} \quad [315]$$

Assessment of the *in vivo* genotoxicity of 2-hydroxy 4-methoxybenzophenone did not cause any significant increase in chromosomal aberrations (Robison et al. 1994).

Detoxification of hydrogen peroxide by catalase (EC 1.11.1.6)

$$2\,H_2O_2 \rightarrow 2\,H_2O + O_2 \quad [127]$$

Promutagenic etheno (ε) adducts in DNA are generated through reaction of DNA bases with lipid peroxidation products from endogenous sources or from exposure to several xenobiotics. The first evidence that lipid peroxidation products could form ε-adducts with nucleic acid bases was obtained by Sodum and Chung (1988). Measurement of etheno and other exocyclic DNA adducts offers a useful tool in molecular epidemiological studies to elucidate the role of oxidative stress and dietary fat intake on endogenous DNA damage as well as the protective effect of antioxidants (Nair et al. 1999). Serum linoleic acid and oleic acid concentrations were higher in healthy female volunteers taking >15 g than in women taking <5 g linoleic acid per day, whereas the linoleic acid/oleic acid ratios were similar (Hagenlocher et al. 2001). The mean 1,N^6-ethenodeoxyadenine N^2,3-ethenodeoxycytidine levels did not significantly differ in the two groups of test subjects. Correlation analyses revealed a significant inverse correlation for 1,N^6-ethenodeoxyadenine in white blood cell DNA and vegetable or vitamin E consumption. εDNA adduct levels are therefore not determined by linoleic acid intake alone, but might depend on the ratio of ω̄-6 polyunsaturated fatty acid or other fatty acids and of antioxidants consumed in the diet.

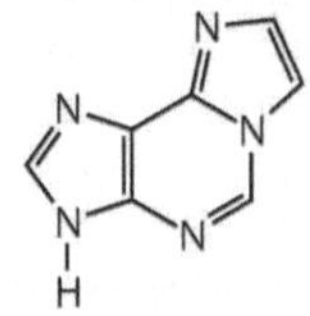

1,N^6-Ethenoadenine [316]

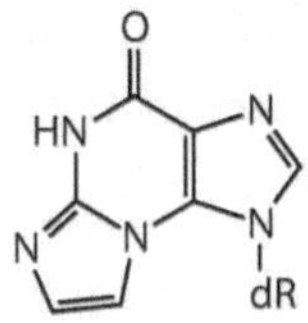

N^2,3-Ethenodeoxycytidine [317]

19.4 Fenton Reactions and Other Metal-Mediated Oxidation Reactions

19.4.1 Copper

Copper ions are significantly more reactive in causing DNA damage than iron ions (Aruoma 1994).

Copper-catalysed Fenton-type reactions induced etheno-DNA adducts in the liver of Long Evans cinnamon (LEC) *rats*, a Long Evans strain with hereditary abnormal copper metabolism, which develop spontaneous hepatitis (Sasaki et al. 1985) and later hepatocellular carcinoma (Nair et al. 1996). These adducts were formed from lipid peroxidation products (El Ghissassi et al. 1995).

One-electron transfer catalysed by copper

$$O_2 + Cu^+ \rightarrow \mathbf{O_2^{\cdot-}} + Cu^{2+} \quad [181]$$

$$\mathbf{O_2^{\cdot-}} + Cu^+ + 2\,H^+ \rightarrow H_2O_2 + Cu^{2+} \quad [182]$$

$$H_2O_2 + Cu^+ \rightarrow \mathbf{HO^{\cdot}} + HO^- + Co^{2+} \quad [183]$$

1,N^6-Ethenodeoxyadenosine (∈dA) and 3,N^4-ethenodeoxycytidine (∈dC) are highly miscoding exocyclic DNA adducts that are also formed in the liver and lung of *rats* exposed to 2000 ppm vinylchloride for 10 subsequent days (7-h exposure on days 1–9 and 24-h exposure on day 10) and sacrificed immediately thereafter (Eberle et al. 1989) and also implicated in vinyl chloride-induced liver angiosarcomas in *humans* (Basu et al. 1993, Bartsch et al. 1994). Drouin et al. (1996) induced base modifications by Cu(II)/ascorbate/H_2O_2 distinguishing base damage from frank strand break. Modified base production predicted by computer simulation at an initial Cu(II) concentration of 50–70 µM was experimentally validated. The copper ion binding sites on DNA were saturated at 50 µM bound copper ion, or when ≈ 40 % of the DNA phosphates were occupied. The k_{app} indicated that DNA base damage occurred slowly in relation to the rate of DNA-Cu(I) oxidation (Stoewe and Prutz 1987). Cupric nitrilotriacetate exposure of HL-60 cells significantly enhanced reactive oxygen species and 8-hydroxydeoxyguanosine formation in the cells (Ma et al. 1998).

The presence of CuZn-superoxide dismutase, Mn-superoxide dismutase or Mn(II) enhanced the frequency of DNA damage induced by H_2O_2 and Cu(II), and altered the site-specificity of the latter: H_2O_2 induced Cu(II)-dependent DNA damage with high frequency at the 5'-guanine of poly G sequences; when superoxide dismutases were added, the frequency of cleavages at thymine and cytosine residues increased (Midorikawa and Kawanishi 2001).

Lipid peroxyl radical formation catalysed by copper

$$LOOH + Cu^{2+} \rightarrow Cu^+ + \mathbf{LOO^{\cdot}} + H^+ \quad [318]$$

LOOH is a lipid hydroperoxide and LOO• is a lipid peroxyl radical. Because LOO• is a long-lived radical, existing tens of seconds on the average, it has an excellent likelihood of undergoing the following reaction:

$$\mathbf{LOO^{\cdot}} + LH \rightarrow LOOH + \mathbf{L^{\cdot}}, \quad [319]$$

to initiate new foci of chain oxidation at distant sites:

$$\mathbf{L^{\cdot}} + O_2 \rightarrow \mathbf{LOO^{\cdot}} \quad [320]$$

Cu^{2+} can promote oxidation of low density lipoprotein by markedly different mechanisms (Ziouzenkowa et al. 1998). Type A oxidations, observed at relatively high Cu^{2+} concentrations of 10–100 Cu^{2+}/LDL, represented the conventional kinetics of LDL oxidation with an inhibition period (= lag-time) followed by a propagation phase. In contrast, type C oxidations proceeded after a negligible short lag time followed by a distinct propagation phase. The rate of this propagation increased rapidly to 0.5 mol diene/mol LDL and then slowed down in the presence of α-, γ-tocopherols and carotenoids, which were consumed faster than tocopherols. The increase in diene absorption was due to the formation of both hydroxides and hydroperoxides suggesting a high initial decomposition of hydroperoxides. At submicromolar concentrations of about 0.1 to 0.5 µM, type C and type A oxidation can be combined resulting in 4 consecutive oxidation phases, i.e. 1st inhibition and 1st propagation (belonging to type C), followed by 2nd inhibition and 2nd propagation (belonging to type A). Increasing copper concentrations lowered the 1st propagation and shortened the 2nd inhibition periods until they melted into one apparent kinetic phase. Decreasing [Cu^{2+}] increased the 1st propagation and 2nd inhibition but lowered the 2nd propagation phase until it completely disappeared. A threshold copper concentration, denoted as Cu_{lim}, can be calculated as a kinetic constant based on the Cu^{2+}-dependence for the rate of 2nd propagation. Below Cu_{lim}, LDL oxidation proceeds only via type C kinetics. The Cu^{2+}-dependence of the oxidation kinetics suggested that LDL contains two different Cu^{2+} binding sites. Cu^{2+} at the low-affinity binding sites, with half-saturation at 5–50 Cu^{2+}/LDL, initiates and accelerates the 2nd propagation by decomposing lipid hydroperoxides. Cu^{2+} bound to the high-affinity binding sites, with half-saturation at 0.3–2.0 Cu^{2+}/LDL, is responsible for the 1st propagation.

α-Tocopherol in Cu(II) reduction

Cu(II) + α-tocopherol-OH → Cu(I) + α-tocopheryl-O˙ + H^+ [321]

α-tocopheryl-O˙ + LH → α-tocopherol-OH + L˙ [322]

The rates of both reactions, [321] and [322], decrease over time as shown by Abuja et al. (1997). The ratio of $v_{[320]}/v_{[319]} \cong 2.5$ is more or less constant during the lag phase, although the absolute values decrease. Perugini et al. (1998) suggested that different mechanisms of Cu(II) reduction, namely α-tocopherol-dependent and independent (likely lipid peroxide-dependent), are progressively recruited during copper-promoted low-density lipoprotein oxidation. The late phase of Cu(II) reduction was strictly related to the availability of copper but was largely independent from α-tocopherol. Neither the amount of Cu(I) generated nor the rate of generation were saturated at concentrations of copper up to 100 μM. Comparable results were obtained by adding bathocuproine, a Cu(I)-specific chelator, at different time points to the low-density lipoprotein-copper mixture, in order to measure at the same time-points both the true rate of Cu(II) reduction and the generation of thiobarbituric acid reactivity during the dynamic process of low-density lipoprotein oxidation.

Thiobarbituric acid reactivity was elevated in serum of breast cancer patients (Punnonen et al. 1994). In the cancerous tissue, catalase activity was lower than in the reference tissue, while the activities of superoxide dismutase, glutathione peroxidase and the hexose monophosphate shunt were elevated.

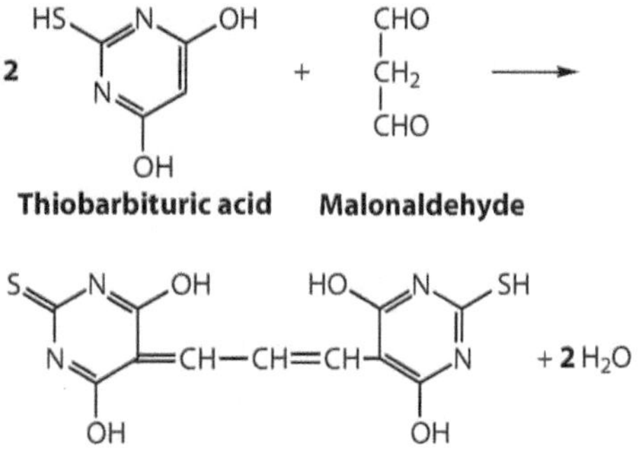

Thiobarbituric acid test [25]

Quercetin (3,3',4',5,7-pentahydroxyflavone), a *rat* intestinal and bladder carcinogen present in bracken fern (*Pteridium aquilinum*), reduced oxygen to superoxide, which in the presence of Cu(II) formed HO• (Fazal et al. 1990). Strand scission of DNA was shown to occur under conditions in which Cu(II), quercetin and either hydrogen peroxide or oxygen were present and superoxide was not a necessary intermediate. Structurally related flavonoids, rutin, galangin, apigenin and fisetin, were ineffective or less effective than quercetin in causing *calf* thymus DNA breakage (Rahman et al. 1989). For the breakage reaction, Cu(II) could be replaced by Fe(III) but not by other ions tested [Fe(II), Co(II), Ni(II), Mn(II) and Ca(II)]. Quercetin and Cu(II) were shown to form a charge transfer complex that decayed in oxygen-dependent reaction(s) and this decay was accelerated by *calf* thymus DNA (Rahman et al. 1990). Yamamoto et al. (1999) confirmed that conjugated quercetin metabolites have an inhibitory effect on copper ion-induced lipid peroxidation in *human* LDL. Quercetin 7-*O*-β-glucopyranoside and rhamnetin (3,3',4',5-tetrahydroxy-7-methoxyflavone) exerted strong inhibition and their effect continued even after complete consumption, similarly to quercetin aglycone. The effect of quercetin 3-*O*-β-glucopyranoside did not continue after its complete consumption, indicating that the antioxidant mechanism of quercetin conjugates lacking a free hydroxyl group at the 3-position is different from that of the other quercetin conjugates. The result that 4'-*O*-β-glucopyranoside and isorhamnetin (3,4',5,7-tetrahydroxy-3'-methoxyflavone) showed little inhibition implies that introduction of a conjugate group to the position of the dihydroxyl group in the B ring markedly decreases the inhibitory effect. The results of azo radical-induced lipid peroxidation of LDL and the measurement of free radical scavenging capacity using stable free radical, 1,1-diphenyl-2-picrylhydrazyl, demonstrated that the *o*-dihydroxyl structure of the B ring is required to exert maximum free radical scavenging activity.

The oxidation of quercetin by horseradish peroxidase/H_2O_2 was studied by Awad et al. (2000). In the absence of reduced glutathione (GSH) at least 20 different products were formed. In the presence of GSH, however, these products were no longer observed and formation of two new products was detected. 6-Gluathionylquercetin and 8-gluathionylquercetin, representing glutathione adducts originating from glutathione conjugation at the A ring instead of at the B ring of quercetin. Glutathione addition at positions 6 and 8 of the A ring can best be explained by taking into consideration of a further oxidation of the quercetin semiquinone, initially formed by the horseradish peroxidase-mediated one-electron oxidation, to give the *o*-quinone, followed by the isomerization of the *o*-quinone to its *p*-quinone methide isomer.

Analyses of quercetin and some of its metabolites (isorhamnetin, tamarixetin, kaempferol) in plasma samples of *pigs* after a single intravenous dose (0.4 mg per kg body weight) and one week later an oral dose of 50 mg quercetin per kg body weight indicated, that the conjugation of orally administered quercetin with glucuronic acid and sulphuric acid appears to occur preferentially in the intestinal wall (Ader et al. 2000).

Kaempferol and morin had antioxidative activity equal to myricetin, quercetin and fisetin in the presence of Cu ions, but were much less effective for Fe, V, or Cd ions (Sugihara et al. 1999).

Alloxan (2,4,5,6 [1H.3H]-pyrimidinetetrone) increased Cu(II)-dependent formation of 8-oxo-2'-deoxyguanosine in the presence of NADH (Murata et al. 1998). Alloxan induced DNA cleavage frequently at thymine and cytosine residues in the presence of NADH and Cu(II). Catalase and bathocuproine almost completely inhibited DNA damage, suggesting the involvement of H_2O_2 and Cu(I). Alloxan induced Cu(II)-dependent production of 8-oxo-2'-deoxyguanosine in *calf* thymus DNA in the presence of NADH. UV-visible and electron spin resonance (ESR) spectroscopic studies showed that superoxide anion radical ($O_2^{\bullet -}$) and alloxan radical were generated by the reduction of alloxan by NADH, and also by autoxidation of dialuric acid, the reduced form of alloxan.

A complex between the chelating agents **1,10-phenanthroline** and copper(II) ions is able to induce the degradation of DNA in the presence of a reducing agent (Sigman et al. 1979, Downey et al. 1980, Graham et al. 1980, Que et al. 1980, Marshall et al. 1981, Reich et al. 1981). The reducing agents employed *in vitro* to facilitate DNA degradation by the Cu-phenanthroline complex have included ascorbate (Que et al. 1980), thiols (Sigman et al. 1979, Downey et al. 1980, Graham et al. 1980, Marshall et al. 1981), NADH (Reich et al. 1981, Gutteridge and Halliwell 1982) and systems generating the superoxide radical $O_2^{\bullet -}$ (Que et al. 1980, Reich et al. 1981, Gutteridge and Halliwell 1982).

The following pathway has been proposed for the generation of hydroxyl radicals by a 1,10-phenanthroline-Cu(II) complex:

$$O_2 + e^- \xrightarrow{\text{xanthine oxidase}} \mathbf{O_2^{\bullet -}} \quad [323]$$

$$2\,\mathbf{O_2^{\bullet -}} + 2\,H^+ \rightarrow H_2O_2 + O_2 \quad [324]$$

$$2\,\mathbf{O_2^{\bullet -}} + \text{OP–Cu(II)} \rightarrow \text{OP–Cu(I)} + O_2 \quad [325]$$

$$\text{OP–Cu(I)} + H_2O_2 \rightarrow \text{OP–Cu(II)} + \mathbf{HO^{\bullet}} + OH^- \quad [326]$$

$$\mathbf{HO^{\bullet}} + \text{DNA} \rightarrow \text{strand scission} \quad [327]$$

In this scheme, hydroxyl radical is generated by the interaction of H_2O_2 with OP-Cu(I) (equation [326]). Superoxide anion, which is required for the production of both H_2O_2 and OP-Cu(I), is produced by the reduction of O_2 by xanthine oxidase via the oxidation of hypoxanthine to xanthine (equation [323]). The superoxide anion thus generated not only produces the required H_2O_2 by spontaneous dismutation (equation [324]), but also it reduces OP–Cu(II) to OP–Cu(I) (equation [325]). Furthermore, the reduction of H_2O_2 by OP-Cu(I) also leads to the regeneration of OP–Cu(II) (equation [326]). Thus, OP–Cu(II) acts catalytically in the generation of $HO^{\bullet}$, being alternately reduced to OP–Cu(I) by $O_2^{\bullet -}$ and deoxidised to OP–Cu(II) by H_2O_2. Superoxide dismutase, which catalyses the dismutation of $O_2^{\bullet -}$ to H_2O_2 would therefore inhibit DNA degradation by diverting all of the superoxide anion to H_2O_2 and preventing the reduction of OP–Cu(II) to OP–Cu(I).

Three representative **isothiocyanates** have abilities to cause site-specific DNA damage in the presence of Cu(II) (Murata et al. 2000). These isothiocyanates induced 8-oxo-7,8-dihydro-2'-deoxyguanosine formation. The extent of DNA damage was dependent on the yield of $O_2^{\bullet -}$, of which generation is associated with formation of the SH group. Relevantly, reduced glutathione, which has an SH group, was found to cause DNA damage in the presence of Cu(II) (John et al. 1993, Milne et al. 1993, Prütz 1994, Oikawa and Kawanishi 1996).

Copper-thiosemicarbazide complexes interact with both the bases and the phosphate groups of native *calf* thymus DNA (Pillai et al. 1977).

Resveratrol catalyses the reduction of Cu(II) to Cu(I), which is accompanied by the formation of

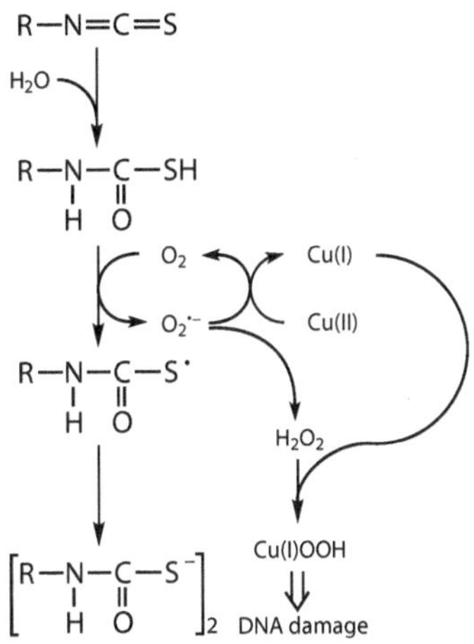

Copper(II)-mediated DNA damage induced by allyl isothiocyanate (MURATA et al. 2000) [328]

'oxidised product(s)' of resveratrol, which in turn also appear the reduction of Cu(II) (AHMAD et al. 2000). Strand scission by resveratrol was found to be biologically active as assayed by bacteriophage inactivation.

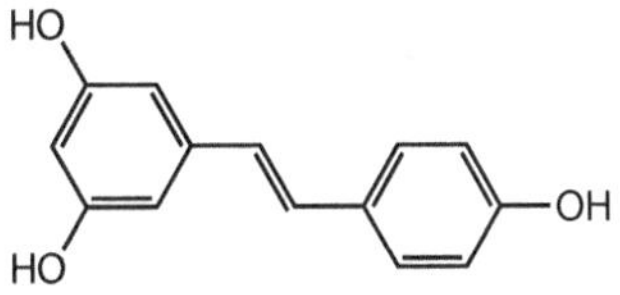

trans-resveratrol [295]

Nitrosamines themselves are not carcinogenic and require metabolic activation by α-oxidation to form α-hydroxynitrosamines (PREUSSMANN and WIESSLER 1987). These enzymatic reactions are mediated by a cytochrome P_{450}-dependent family of enzymes with overlapping substrate specificities (JAKOBY et al. 1982). α-Hydroxylated nitrosamines derived from carcinogenic nitrosamines are stable enough to be conjugated to form β-glucuronides (WIESSLER et al. 1984).

Dimethylnitrosamine appeared more potent than diethylnitrosamine in the induction of lung tumours in different strains of *mice* (TAKAYAMA and OOTA 1965). In the Syrian golden *hamster*, diethylnitrosamine induced tracheal papillomas (WAHNSCHAFFE et al. 1987). Enlarged lysosomes sized up to 10 μm in diameter occurred in a particular type of cells situated at the periphery of tumour lobules. The differentiated phenotype of these cells was secretory in nature.

Hepatocarcinogenic *N*-nitrosodimethylamine known to enhance microsomal lipid peroxidation (JOSE and SLATER 1973) produced oxidative damage in male Wistar *rats*, demonstrated by increased lipid peroxidation *in vivo*, by increased lucigenin-dependent chemiluminescence (maximal effect at 15 μM) and H_2O_2 (maximal effect at 25 μM) release of isolated hepatocytes, and by increased ethane production by microsomes incubated with *N*-nitrosodimethylamine (maximal effect at 6 μM) (AHOTUPA et al. 1987). Simultaneously with the increased lipid peroxidation, *N*-nitrosodimethylamine treatment slightly decreased antioxidant enzyme activities and GSH content in the liver.

N-Nitrosodiethylamine is metabolised by *human* CYP2A6 (CHOLERTON et al. 1992, GONZALEZ and GELBOIN 1994). This minor form of P450 in liver representing up to 1 % of total P450 content. It exhibits a large degree of individual variability in expression (MILES et al. 1990, YAMANO et al. 1990, MAURICE et al. 1991, YUN et al. 1991). CYP2A6 has been found in *rodents'* lung (KIMURA et al. 1989) and *rabbit* nasal epithelia (DING and COON 1988).

Preneoplastic liver nodules induced in *rats* by the Cayama-Farber procedure (*N*-nitrosodiethylamine + 2-acetylaminofluorene + CCl_4) were resistant to the oxidative stress induction caused by redox-enzyme modulation treatment (DENDA et al. 1993). Despite toxic effects in surrounding hepatocytes, no progression pressure was exerted.

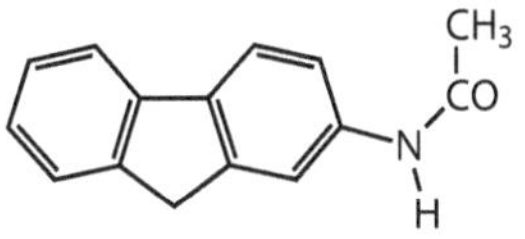

2-Acetylaminofluorene [298]

N-Ethyl-*N*-nitrosourea was decomposed faster with increasing concentrations of Cu^{2+} (2.5–40 μmol/l) at pH 6 and 37 °C in a concentration-dependent manner (PREUSSMANN et al. 1975). Acceleration by Ni^{2+} was markedly weaker. In the presence of Cu^{2+} the half life of *N*-methyl-*N'*-nitro-*N*-nitrosoguanidin was only 39 min instead of 35 h. There was a linear relationship between the rate of degradation and the concentration of Cu^{2+} or Ni^{2+} ions. The metal ions enter the reaction in stoichiometric amounts. The elevation of the rate of decomposition completely agreed with the findings of ZELLER and IVANKOVICH (1972) and IVANKOVICH et al. (1972) who observed an enhancement of the acute toxicity and the carcinogenicity of *N*-ethyl-*N*-nitrosourea administered simultaneously with $CuSO_4$ or $NiSO_4$.

Serum copper levels in cancer cases in comparison with control populations are elevated (FISHER 1979). Serum copper levels have been found to be higher in patients with late stage disease than among those with less severe disease, to be reduced following cancer treatment, and to increase prior to relapse (FISHER 1979). However, evidence from other research suggests that dietary intake of copper and serum copper levels may affect subsequent risk of cancer. In laboratory animals, increased intake of copper has been found to reduce the occurrence of cancer (Committee on diet, nutrition and cancer 1982), but population correlation studies have found the incidence of cancer to vary positively with mean serum copper levels in blood donors (SCHRAUZER 1978). Case-control studies have found increased risk of a subsequent diagnosis of cancer among individuals with elevated serum copper levels (HAINES et al. 1982, KOK et al. 1988) and among those with low serum copper levels (KOK et al. 1988). The finding of COATES et al. (1989) suggested that the presence of cancer may increase serum copper levels several years prior to its diagnosis.

19.4.2 Chromium

Chromate(VI) reacts with H_2O_2 to produce tetraperoxochromate (reaction [121]) which decomposes to chromate(VI) and $HO^•$ and $^1O_2$, which cause DNA damage (KAWANISHI et al. 1986).

$$2\,Cr^{VI}O_4^{2-} + 9\,H_2O_2 + 2\,OH^- \rightarrow 2\,Cr^{V}(O_2)_4^{3-} + 10\,H_2O + O_2 \quad [138]$$

$$2\,Cr^{V}(O_2)_4^{3-} \rightarrow 2\,Cr^{VI}O_4^{2-} + 2\,\mathbf{O_2^{\cdot-}} + 2\,\mathbf{^1O_2} \quad [139]$$

$$2\,H_2O_2 + 2\,\mathbf{O_2^{\cdot-}} \rightarrow 2\,HO^• + 2\,OH^- + 2\,O_2 \quad [140]$$

Equation [138] contains both the reduction of Cr(VI) by hydrogen peroxide and the ligand replacement of O^{2-} with O_2^{2-}. With respect to equation [139] KAWANISHI et al. (1986) presented some evidence for the formation of singlet oxygen and also considered on the basis of the effect of superoxide dismutase that superoxide is generated before the formation of $HO^•$.

Carcinogenic Cr(VI) has been reported to induce DNA lesion *in vivo* and in culture (DE FLORA and WETTERHAHN 1989). KAWANISHI et al. (1994) investigated reactivties of Cr compounds with DNA by the DNA sequencing technique using ^{32}P 5'-end-labelled DNA fragments. Cleavage of piperidine-labile sites in DNA fragments treated with 2.5 mM sodium chromate(VI) in the presence of 25 mM H_2O_2 occurred at every base residue but the cleavage at the guanine positions was more dominant than at the other three bases. AIYAR et al. (1990) reported the enhancing effect of glutathione (GSH) on Cr(VI)-induced formation of $HO^•$. SHI et al. (1992) showed that $HO^•$ generated by Cr(VI)/flavoenzyme/NAD(P)H enzymatic system reacts with 2'-deoxyguanosine to form 8-hydroxy-2'-deoxyguanosine. *In vitro* DNA binding experiments showed that, in the presence of ascorbate (the major intracellular reductant of Cr^{6+}), K_2CrO_4 induces both interstrand cross-links and strand breaks (FLORES and PÉREZ 1999). CASADEVALL et al. (1999) found that mixtures of Cr(VI) and GSH or ascorbate were able to oxidise 2-deoxyribose to yield malondialdehyde, which was detected by reaction with thiobarbituric acid (formula [25]). The characteristic pink chromogen, which forms upon reaction with thiobarbituric acid ' was also observed with *calf* thymus DNA as the substrate. In both experimental systems the addition of catalase prevented the formation deoxyribose breakdown products. Hydroxyl radical did not seem to be important for the generation of DNA damage as the characteristic modified DNA bases could not be detected using gas chromatography – mass spectrometry.

Treatment of cultured A549 *human* lung epithelial cells with sodium dichromate (0–100 μM, 16 h) resulted in a concentration-dependent decrease in the levels of 8-oxo guanine-DNA glycosylase 1 mRNA as measured by both reverse transcription-PCR and ribonuclease protection assay (HODGES and CHIPMAN 2002). Treatment of cells with the pro-oxidant H_2O_2 (0–200 μM, 16 h) had no detectable effect on the levels of 8-oxo guanine-DNA glycosylase 1 mRNA or protein expression suggesting that the effect of sodium dichromate is not mediated by H_2O_2.

Through its antioxidant properties, (-)epigallocatechin-3-gallate exhibited a protective effect against DNA damage induced by Cr(VI) (SHI et al. 2000). (–)Epigallocatechin-3-gallate also inhibited activation of nuclear transcription factor NF-κB induced by Cr(IV).

Cr uptake and the formation of Cr(V), Cr-DNA adducts and 8-hydroxy-2'-deoxyguanosine formation in the liver and kidney of Osteogenic Disorder Shionogi *rats* that lack the ability to synthesise ascorbate were measured by YUANN et al. (1999). Despite a 10-fold difference in tissue ascorbate levels the Cr(V) signal intensity, Cr uptake and total Cr-DNA binding were nor affected in either organ. Treatment of Osteogenic Disorder Shionogi *rats* with Cr(VI) (10 mg/kg), had no substantial effect on the levels of ascorbate and glutathione in these tissues. The levels of Cr(V) and Cr-DNA binding were ~2-fold higher in the liver than in the kidney, although the levels of total Cr uptake were similar in both tissues. Cr uptake levels were significantly in the liver and kidney of Osteogenic Disorder Shionogi *rats* treated with high levels of ascorbate and a high dose of Cr(VI) (40 mg/kg) suggesting a detoxifying played by plasma ascorbate. Similarly, modulation of glutathione levels by *N*-acetyl-L-cysteine, L-buthionine-S,*R*-sulfoximine or phorone in these animals by up to 2-fold had little or no consistent effect on Cr uptake, Cr-DNA binding, Cr(V) levels or 8-hydroxy-2'-deoxyguanosine formation in either organ.

A model high-valent Cr(V) complex, *N,N'*-ethylenebis(salicylideneanimato) oxochromium(V), Cr(V)-Salen, was used to probe the mechanism of interaction between this oxidation state of chromium and DNA. SUGDEN et al. (2001) found this interaction to be specific towards the oxidation of guanine in unmodified single- and double-stranded oligonucleotides as measured by an increased level of DNA strand cleavage at these sites following piperidine treatment. Replacement of a single guanine residue in DNA with a more readily oxidised 7,8-dihydro-8-oxoguanine base allowed for site-specific oxidation at this modified site within the DNA strand by the Cr(V)-Salen complex.

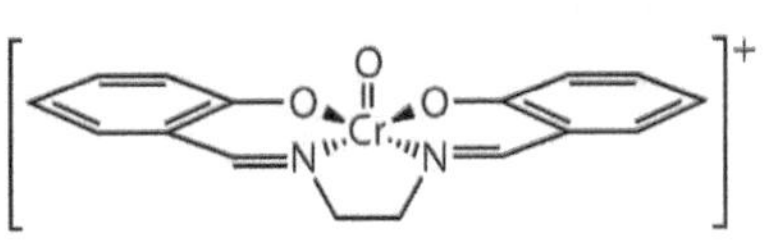

N,N'-ethylenebis(salicylideneanimato)oxochromium(V) [327]

The interactions of reactive species generated in Cr(VI)/catechol(amine) mixtures with plasmid DNA have been investigated to model a potential route to Cr(VI)-induced genotoxicity (PATTISON et al. 2001). Reduction of Cr(VI) by 3,4-dihydroxyphenylalanine (DOPA), dopamine, or adrenaline produces species that cause extensive DNA damage, but the products of similar reactions with catechol or 4-*tert*-butylcatechol do not damage DNA. The Cr(VI)/catechol(amine) reactions have been studies at low added H_2O_2 concentrations, which lead to enhanced DNA cleavage with DOPA and induce DNA cleavage with catechol. The Cr(V) and organic intermediates generated by the reactions of Cr(VI) with DOPA or catechol in the presence of H_2O_2 were characterised by EPR spectroscopy. The detected signals were assigned to Cr(V)-catechol, Cr(V)-peroxo, and mixed Cr(V)-catechol-peroxo complexes. Oxygen consumption during the reactions of Cr(VI) with DOPA, dopamine, catechol, and 4-*tert*-butylcatechol was studied, and H_2O_2 productions was quantified. Reactions of Cr(VI) with DOPA and dopamine, but not catechol and 4-*tert*-butylcatechol, consume considerable amounts of dissolved O_2, and give extensive H_2O_2 production. Extents of oxygen consumption and H_2O_2 production during the reaction of Cr(VI) with enzymatically generated DOPA and *N*-acetyl-DOPA (from the reaction of Tyr and *N*-acetyl-Tyr with tyrosinase, respectively) were correlated with the DNA cleaving abilities of the products of these reactions. The reaction of Cr(VI) with enzymatically generated DOPA produced significant amounts of H_2O_2 and caused significant DNA damage, but the *N*-acetyl-DOPA did not. The extent of *in vitro* DNA damage is reduced considerably by treatment of the Cr(VI)/catechol(amine) mixtures with catalase (EC 1.11.1.6), which shows that the DNA damage is H_2O_2-dependent and that the major reactive intermediates are likely to be Cr(V)-peroxo and mixed Cr(V)-catechol-peroxo complexes, rather than Cr(V)-catechol intermediates.

Detoxification of hydrogen peroxide by catalase (EC 1.11.1.6)

$$2\,H_2O_2 \rightarrow 2\,H_2O + O_2 \qquad [127]$$

Trivalent chromium is generally inactive (FLESSEL 1979, LANGÌRD 1980, LEONARD and LAUWERYS 1980, BIANCHI et al. 1983, LEVIS and BIANCHI 1983). The ability of Cr(III) to interact with DNA and produce different alterations of the DNA structure has been studied mainly by a physicochemical approach, by using techniques such as the alkaline sucrose sedimentation and alkaline elution (WHITING et al. 1979, BRAMBILLA et al. 1980, DOUGLAS et al. 1980, TSAPAKOS et al. 1981). A Cr(III)-DNA complex transfected in bacteria was shown to be mutagenic (SNOW et al. 1989). The mechanisms of Cr(III) genotoxicity may involve a disturbance of DNA processing enzymes in DNA synthesis (TKESHELASHVILI et al. 1980), in transcription (OKADA et al. 1984, WETTERHAHN et al. 1989) or in DNA repair.

In order to understand the role of co-ordinated ligands in controlling the biotoxicity of Cr(III), YAMINI SHRIVASTAVA and NAIR (2000) investigated three types of Cr(III) complexes viz. trans-diaquo [1,2 bis (salicyledeneamino) ethane chromium (III) perchlorate, $[Cr(salen)(H_2O)_2](ClO_4)$; tris (ethylenediamine) chromium (III) chloride, $[Cr(en)_3]Cl_3$, and monosodium ethylene diamine tetraacetato monoaquo chromate (III), $[Cr(EDTA)(H_2O)]Na$ with *bovine* serum albumin. Spectroscopic and equilibrium dialysis studies showed that the two cationic complexes $Cr(salen)(H_2O)_2^+$ and $Cr(en)_3^{3+}$ bind to the protein with a protein-metal ratio of 1:8 and 1:4. The anionic complex $Cr(EDTA)(H_2O)^-$ binds to the protein with a protein-metal ratio of 1:2. The binding constant K_b as estimated from the fluorescence quenching studies has been found to be $7.6 \pm 0.4 \times 10^3$ M^{-1}, $3.1 \pm 0.2 \times 10^2$ M^{-1}, and $1.8 \pm 0.2 \times 10^2$ M^{-1} for $Cr(salen)(H_2O)_2^+$, $Cr(en)_3^{3+}$, and $Cr(EDTA)(H_2O)^-$ respectively indicating that the thermodynamic stability of protein-chromium complex is $Cr(salen)(H_2O)_2^+ > Cr(en)_3^{3+} \approx Cr(EDTA)(H_2O)^-$.The complexes $Cr(salen)(H_2O)_2^+$ and $Cr(EDTA)(H_2O)^-$ in the presence of H_2O_2 have been found to induce protein degradation, whereas $Cr(en)_3^{3+}$ did not induce any protein damage. In the presence of H_2O_2, Qi et al. (2000) observed that Cr(III)-induced formation of 8-hydroxydeoxyguanosine in isolated DNA was dose and time dependent. Melatonin, ascorbate, and vitamin E (Trolox), all of which are free radical scavengers, markedly inhibited the formation of 8-hydroxydeoxyguanosine in a concentration-dependent manner. The concentration that reduced DNA damage by 50 % was 0.51, 30.4, and 36.2 µM for melatonin, ascorbate, and Trolox, respectively. These findings are consistent with the conclusion that the carcinogenic mechanism of Cr(III) is possible due to Cr(III)-mediated Fenton-type reactions and that the highly protective effects of melatonin against Cr(III) relate, at least in part, to its direct hydroxyl radical scavenging ability.

To examine the hypothesis that preconception carcinogenesis involves an increase in the rate of occurrence of neoplasms with spontaneous incidence, Yu et al. (1999) exposed male NIH Swiss *mice* 2 weeks before mating to 1 mmol chromium-(III) chloride per kg body weight. Phaeochromocytomas occurred in both male and female offspring with none in the controls. There was also an increase in incidence of male reproductive gland tumours and of renal non-neoplastic lesions.

19.4.3 Cobalt

Cobalt(II) whose industrial medical importance is in the production of sintered carbides (Schiller 1958, 1961) caused extensive site-specific damage (G>T~C>A) in the presence of H_2O_2. ESR experiments that probably $^1O_2$ and/or a cobalt-oxygen complex were involved in the DNA damage (Yamamoto et al. 1989). Cobalt metal, a mixture of cobalt with tungsten carbide and cobalt chloride were able to induced DNA damage in isolated *human* lymphocytes from three donors, in a dose- and time-dependent way (De Boeck et al. 1998). The DNA-damaging potential of the cobalt-tungsten carbide mixture was higher than that of cobalt metal and cobalt chloride, which had comparable responses. No significant increase of DNA migration was observed when the DNA of cells treated with cobalt metal, cobalt-tungsten carbide or tungsten carbide were incubated with the oxidative lesion-specific enzyme formamidopyrimidine DNA glycosylase. Cobalt metal was able to inhibit the repair of methylmethanesulphonate-induced DNA damage.

Co(II)-induced chemiluminescence increased with increasing concentration of H_2O_2 (Kawanishi et al. 1994). The intensity was enhanced about threefold in D_2O in which the lifetime of singlet oxygen is at least 10 times that in H_2O.

One-electron transfer catalysed by cobalt

$$O_2 + Co^{2+} \rightleftarrows \mathbf{Co^{3+}-O_2^{\cdot -}} \rightarrow \mathbf{O_2^{\cdot -}} + Co^{3+} \quad [145]$$

$$\mathbf{O_2^{\cdot -}} + Co^{2+} + 2\,H^+ \rightarrow H_2O_2 + Co^{3+} \quad [146]$$

$$H_2O_2 + Co^{2+} \rightarrow \mathbf{HO^{\cdot}} + HO^- + Co^{3+} \quad [147]$$

19.4.4 Nickel(II)

Nickel compounds have been shown to have seriously toxic and carcinogenic (Table 3) effects on *humans*. Ni(II) induced strong DNA damage in the presence of H_2O_2 even without piperidine treatment (Kawanishi et al. 1989). Piperidine-labile sites were induced frequently at cytosine, thymidine and guanine residues, and rarely at adenine residue. Diethylene triamine *N,N,N',N',N'*-pentaacetic acid inhibited the DNA damage. In experiment with singlet oxygen scavengers, sodium azide and dGMP inhibited the DNA damage completely, whereas neither 1,2-diazabicyclo[2.2.2]octane nor dimethylfuran inhibited it. Among hydroxyl radical scavengers, dimethylsulfoxide and sodium formate inhibited the DNA damage considerably, whereas ethanol and mannitol did not. Methionine and methional inhibited the DNA damage completely.

The intensity of Ni(II)-mediated DNA damage induced by dithiotreitol was stronger than that by other model endogenous SH compounds, 1,4-dithio-L-threitol and dithioerythritol (Oikawa et al. 2002). DNA damage induced by Ni(II) plus dithiotreitol was observed only when the DNA was treated with piperidine, suggesting that Ni(II) plus dithiotreitol caused only base damage. Formamidopyrimidine-DNA glycosylase, which is known to recognize 8-oxo-7,8-dihydro-2'-deoxyguanisine as well as Fapy residues, treatment induced cleavage sites, mainly guanine residues, particularly at 5'-GG-3', 5'-GGG-3', and 5'-GGGG-3' sequences, in DNA incubated with Ni(II) in the presence of dithiotreitol. Superoxide dismutase and catalase inhibited the DNA damage, suggesting that DNA damage involved superoxide anion and hydrogen peroxide. Sodium azide, a potent and relatively specific scavenger of $^1O_2$, inhibited DNA damage by Ni(II) in the presence of dithiotreitol, whereas the sequence specificity of DNA damage was different from that obtained by $^1O_2$ generating agent. The formation of 8-oxo-7,8-dihydro-2'-deoxyguanisine in *calf* thymus DNA by Ni(II) was observed with the physiological thiols, dihydrolipoic acid and mercaptopyruvate, as well as with dithiotreitol.

There is evidence that Mg is a competitive antagonist of the toxicological effects of Ni. LITTLEFIELD et al. (1991) used a factorial design to examine the interactive influence of Mg and Ni on the deglycosylation and hydroxylation of 2'-deoxyguanosine under a range of pH conditions in which ascorbate and H_2O_2 were added. Formation of guanine (deglycosylation) and 8-hydroxy-2'-deoxyguanosine (hydroxylation) appeared in large amounts in samples in which both ascorbate and H_2O_2 were present. The largest amount of guanine appeared where both Ni and Mg were present. When Mg alone was present, the amounts of guanine were intermediate between these two. Slightly less 8-hydroxy-2'-deoxyguanosine was formed where only Mg was present. The reaction mixtures were more sensitive to the pH than to the respective presence of absence of metals.

One-electron transfer catalysed by nickel

$$O_2 + Ni^{2+} \leftrightarrow \mathbf{Ni^{3+} - O_2^{\bullet-}} \rightarrow \mathbf{O_2^{\bullet-}} + Ni^{3+} \quad [149]$$

$$\mathbf{O_2^{\bullet-}} + Ni^{2+} + 2\,H^+ \rightarrow H_2O_2 + Ni^{3+} \quad [150]$$

$$H_2O_2 + Ni^{2+} \rightarrow \mathbf{HO^\bullet} + HO^- + Ni^{3+} \quad [151]$$

Foetal *human* cortex explants continuously exposed to 5 μg $NiSO_4$/ml after 70–100 days showed foci of phenotypically altered cells (TVEITO et al. 1989). Chromosome changes in the treated cells included ploidy (3n) and abnormalities of chromosomes 1, 7, 9, 11, 13, 14 and 20, increased numbers of chromosomes 17; and loss of normal chromosomes 20 and 22.

Exposure of cultured Chinese *hamster* ovary cells to several nickel compounds, i.e., NiS, Ni_3S_2, NiO (black and green), and $NiCl_2$ increased oxidation of 2'7-dichlorofluorescin to the fluorescent 2',7-dichlorofluorescein, suggesting that nickel compounds increased the concentration of oxidants in CHO cells (HUANG et al. 1994). This fluorescence could be attenuated by addition of exogenous catalase (EC 1.11.1.6) to the extracellular media, indicating that H_2O_2 is one of the oxidants formed in this system. Fluorimetric measurements of chromogens following thiobarbituric acid reaction showed that nickel compounds also induce lipid peroxidation with a decreasing potency NiS, Ni_3S_2 > black NiO > green NiO > $NiCl_2$. In contrast to insoluble nickel, highly water-soluble $NiCl_2$ did not induce lipid peroxidation in CHO cells.

Intratracheal instillation of 0.5 or 1 mg Ni_3S_2, NiO (black), or NiO (green) to Wistar *rats* increased 8-hydroxy-2'-deoxyguanosine in the lungs significantly (KAWANISHI et al. 2001). $NiSO_4$ induced a smaller but significant increase in 8-hydroxy-2'-deoxyguanosine. $^\bullet$NO generation in RAW 264.7

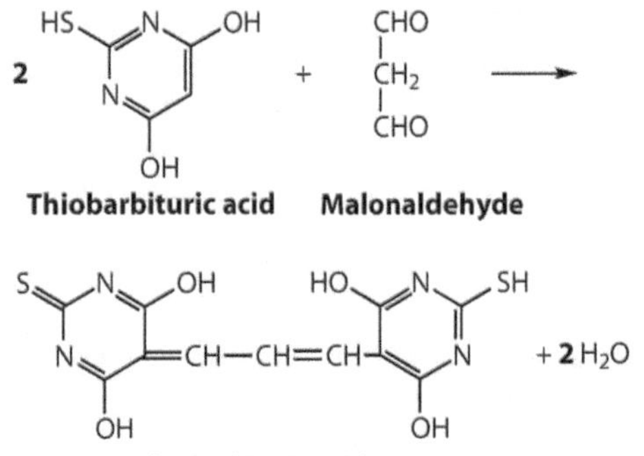

Thiobarbituric acid test [25]

macrophages stimulated with lipopolysaccharide (0.1 µg/ml) was enhanced by all nickel particles.

A single intraperitoneal injection of Ni(II) salt, 95 µM/kg, in male F344 *rats* appeared to be associated with an increased concentration of 8-hydroxy-2'-deoxyguanosine in DNA extracted from kidneys 16–48 h after injection (KASPRZAK et al. 1990).

Ni(II) binding to CH_3CO-Thr-Glu-Ser-His-His-Lys-NH_2, a blocked hexapeptide modelling a part of the C-terminal sequence of the major variant of histone H2A (residues 120–125), revealed the formation of a pseudo-octahedral NiHL complex in weakly acidic and neutral solutions (BAL et al. 1998). Ni(II) is bound to the peptide through imidazole nitrogens on both of its histidine residues and the carboxylate of the side chain of glutamic acid. At higher pH, a series of square planar complexes are formed. The process is accompanied by hydrolytic degradation of the peptide. At pH 7.4, the peptide hydrolyses in a Ni(II)-assisted fashion, yielding the square-planar Ni(II) complexes of Ser-His-His-Lys-NH_2 as the sole product detected by CD, matrix-assisted laser desorption ionisation time-of-flight mass spectrometry, and HPLC. Quantitative analysis of complex stabilities indicated that the Thr-Glu-Ser-His-His-Lys is a very likely binding site for carcinogenic Ni(II) ions in the cell nucleus.

Nickel carbonyl inhalation (80 ppm) for a period of 30 min or chronic, multiple exposures of male Wistar *rats* to $Ni(CO)_4$ (4 ppm) for 30 min 3 times weekly for the remainder of their lives induced squamous cell carcinoma, adenocarcinoma or anaplastic carcinoma of the lung (SUNDERMAN and DONNELLY 1965).

Nickel carbonyl inhibited DNA-dependent RNA polymerase activity in *rat* hepatic nuclei (SUNDERMAN and ESFAHANI 1968).

SCHWERDTLE et al. (2002) investigated the effects of particulate black NiO and soluble $NiCl_2$ on the induction and removal of stable DNA adducts formed by benzo[*a*]pyrene measured by a highly sensitive high performance chromatography/fluorescence assay. With respect to adduct formation,

NiO but not $NiCl_2$ reduced the generation of DNA lesions by ~30 per cent. Regarding the repair, in the absence of nickel compounds, most lesions were removed within 24 h; nevertheless, between 20 and 35 per cent of induced adducts remained even 48 h after treatment. $NiCl_2$ and NiO reduced the removal of adducts in a dos-dependent manner. Thus, 100 μM $NiCl_2$ led to ~80 per cent residual repair capacity; after 500 μM the repair was reduced to about 36 per cent. Even at the completely non-cytotoxic concentration of 0.5 μg/cm² black NiO, lesion removal was reduced to ~35 per cent of control and to 14 per cent at 2.0 μg/cm². Both nickel compounds increased the benzo[*a*]pyrene-7,8-diol 9,10 epoxide-induced cytotoxicity.

19.4.5 Iron

Iron Overload

In black males who died of oesophageal carcinoma, hepatic iron concentrations were significantly higher than those of males in the same age groups who died of other causes (MacPhail et al. 1979).

In healthy subjects, Proteggente et al. (2000) observed no compelling evidence for a pro-oxidant effect of ascorbate supplementation, in the presence or absence of iron (14 mg pro die for 6 weeks), on DNA base damage.

Bhasin et al. (2002) have used various peroxides and hydroperoxides as stage-I and -II tumour promoters and have studied the effect of iron overload on the two stages of tumour promotion in Swiss albino *mice* by injecting iron-dextran. The order in which iron overload was effective in increasing tumour promotion by stage-I tumour promoters was

H_2O_2 > COOH > benzoyl peroxide >
12-*O*-tetradecanoyl phorbol-13-acetate

and the order in which iron overload was effective in increasing tumour promotion by stage-II tumour promoters was

COOH > mezerein > benzoyl peroxide.

The high effectiveness of iron overload in increasing tumour promotion when H_2O_2 was used as a stage-I tumour promoter correlated with its ability to produce most reactive hydroxyl radicals.

Ferric Nitrilotriacetate

Normally the entry of iron(III) into cells and its intracellular distribution is carefully controlled. However, if iron(III) is complexed by nitrilotriacetate (NTA), it circumvents the controlled transferrin path, and may even penetrate into the cell nucleus where it catalyses detrimental oxidative reactions with DNA. The Fe(III) NTA-complex has been shown to cause cancer in *rats* (Ebina et al. 1986) and DNA breaks and increased frequency of sister chromatid exchanges in Chinese *hamster* V79 cells (Hartwig et al. 1993). Ferric nitrilotriacetate catalyses the decompensation of H_2O_2 to produce $HO^•$, which subsequently causes DNA base alterations and backbone breakages (Kawanishi et al. 1994) and formation of 8-hydroxy-2'-deoxyguanosine in isolated chromatin (Dizdaroglu et al. 1991) or in *rat* kidney DNA (Umemura et al. 1990). Base oxidation is GTCA (Inoue and Kawanishi 1987). Kawanishi et al. (1994) summarised the negative activities of other Fe(III)-chelates of aminopolycarboxylic acids (HEDTA, EGTA, EDTA, CDTA, and DTPA) for H_2O_2-dependent DNA damage and $HO^•$ formation from H_2O_2 interpreted by structural considerations. Ferric nitrilotriacetate is supposed to approach the groove of the DNA double helix readily, whereas Fe(III)-HEDTA may not. Since $HO^•$ is short-lived, it damages DNA only when produced in the vicinity of the DNA.

In primary *rat* hepatocyte cultures 100 μM ferric nitrilotriacetate induced five oxidation products of cellular DNA derived from both purines and pyrimidines (Abalea et al. 1999). Addition of increasing concentrations of myricetin (25–50–100 μM) simultaneously with iron prevented both lipid peroxidation and accumulation of oxidation products in DNA. Moreover, as an activation of DNA repair pathways, myricetin stimulated the release of DNA oxidation bases into culture media, especially of purine-derived oxidation products. The removal of highly mutagenic oxidation products from DNA of hepatocytes might correspond to an activation of DNA excision-repair enzymes by myricetin. This was verified by RNA blot analysis of DNA polymerase β gene expression, which was induced by myricetin in a dose-dependent manner.

Iqbal et al. (1999) showed that the toxicity of ferric nitrilotriacetate can be correlated with the tissue accumulation of 4-hydroxy-2-nonenal-modified protein adducts. The toxic manifestations of ferric nitrilotriacetate gradually increased with the increasing age of Wistar *rats*. A dose of ferric nitrilotriacetate which produced almost 100 % mortality in aged rats caused 70 % mortality in adults, 30 % in pups, 20 % in litters, and less than 10 % in neonates. The age-dependent increase in its toxicity was also evident from the data of renal microsomal lipid peroxidation and H_2O_2 generation. The magnitude of ornithine decarboxylase (EC 4.1.1.17) induction and [^{3}H]thymidine incorporation was much higher in aged and adult *rats* in comparison to other groups of animals after ferric nitrilotriacetate treatment.

Evidence was provided that $HO^{\bullet}$ is the reactive oxygen species of the reaction of $[Fe^{II}EDTA]^{-2}$ and H_2O_2 in the absence (GILBERT et al. 1988) and in the presence of ascorbate (POGOZELSKI et al. 1995). The generation of the main oxidising species of the Udenfried reaction requires oxygen instead of H_2O_2 (ITO et al. 1993). The main oxidation product of the reaction of thymine with the latter reagent was identified as *N*-formyl-*N*'-pyruvylurea. This agrees with results on the $HO^{\bullet}$-mediated decomposition of thymine (TÉOULE and CADET 1978). On the other hand, iron(II)-bleomycin is expected to cleave DNA (HECHT 1986) by a mechanism involving an iron-oxo-complex (STUBBE and KOZARICH 1987).

Compounds such as EDTA had long been used in horticulture to assist the growth of fruit trees and iron-fastidious plants, such as rhododendrons, on alkaline soil. When RACKER and KRIMSKY (1947) ascribed the toxic action of the encephalomyelitis virus to it having the ability to transport iron across the blood-brain barrier which is normally impervious to the metal, WILLSON (1982) held a decompartmentalisation and ill-placement of iron responsible for free radical damage causing its carcinogenicity.

In male ddY *mice*, ferric nitrilotriacetate in the precancerous stages induced hyperplasia of acidophilic non-ciliated cells (Clara cells) and alveolar duct cells (KIMOTO et al. 2001). Pulmonary adenomas arose as adenomas developing from areas with papillary hyperplasia of bronchiolar epithelial cells, and the papillary adenomas exhibited vascular stalks lined with iso- or bathyprismatic epithelial cells. In other types or arising adenomas, these consisted of alveolar cells lining the alveolar walls. Malignant progression from adenomas to adenocarcinomas was observed in lungs of *mice* receiving ferric nitrilotriacetate only. In *mice* orally medicated with either Brazilian propolis of artepillin C {3-[4-hydroxy-3,5-bis(3-methyl-2-butenyl)phenyl]-2-propionic acid} adenomas did not progress to carcinomas.

19.4.6 Vanadium

Vanadium(IV) caused molecular oxygen-dependent 2'-deoxyguanosine hydroxylation and DNA strand breaks (SHI et al. 1996). ESR spin trapping measurements demonstrated that the reaction of vanadium(IV) with H_2O_2 generated $HO^{\bullet}$ radicals, which were inhibited by the metal ion chelators, diethylenetriaminepentaacetic acid (DTPA) and deferoxamine. UV-visible measurements indicated that 2'-deoxyguanosine, vanadium(IV) and deferoxamine are able to form a complex, thereby, facilitating site-specific 8-OH-2'-deoxyguanosine formation.

$$V^{IV} + O_2 \rightarrow V^{V} + O_2^{-} \quad [328]$$

$$2\,O_2^{-} + 2\,H^{+} \rightarrow H_2O_2 + O_2 \quad [329]$$

$$V^{IV} + H_2O_2 \rightarrow V^{V} + OH^{-} + HO \quad [330]$$

19.5 Tumour Induction with Chemicals

19.5.1 Arsenics

Oral administration to *mice* of dimethylarsinic acid, a major metabolite of inorganic arsenics, induced lung-specific DNA damage (YAMANAKA and OKADA 1994). The lung-specific strand breaks were not caused by dimethylarsinic acid itself, but by dimethylarsine, a further metabolite of dimethylarsinic acid. An *in vitro* experiment indicated, that DNA single-strand breaks by dimethylarsine were suppressed by the presence of superoxide dismutase and catalase, suggesting that the strand breaks were induced via the production of free radical species including active oxygens. Dimethylarsenic peroxide radical $(CH_3)_2AsOO^{\bullet}$]and superoxide anion radical ($O_2^{\bullet -}$) produced from the reaction between molecular oxygen and dimethylarsine were detected by electron-spin resonance analysis using a spin-trapping agent and the cytochrome *c* method, respectively.

Exposure of three colon cancer cell lines, SW480, DLD-1, and COLO201, to arsenic trioxide in the medium induced a marked concentration-dependent suppression of cell growth (NAKAGAWA et al. 2002). The intracellular content of reduced glutathione (GSH) in these cell lines tended to be inversely correlated with the sensitivity of the cells to arsenic trioxide. The production of reactive oxygen intermediates increased with time after treatment with arsenic trioxide.

19.5.2 Nitrosamines

The formation of tobacco-specific nitrosamines from nicotine and related tobacco alkaloids under mild conditions suggested that they should be present in tobacco and tobacco smoke. Nicotine-derived nirosamino ketone (NNK) and *N*'-nitrosonornicotine (NNN) have the properties of organ specificity in that the former induces lung tumours in *rats* and the latter nasal tumours, independent of the route of administration, these compounds are also effective local carcinogens, which is only known for a few other nitrosamines (PREUSSMANN and STEWART 1984). Following a single dose of NNK, levels of DNA methylation were highest in nasal mucosa followed by liver and lung (HECHT et al. 1986). The rats of removal of O^6-methylguanine from lung and nasal mucosa were slower than from liver. During treatment with 100 mg/kg NNK for 12 days, O^6-methylguanine accumulated and persisted in lung, whereas it was repaired in hepatocytes (BELINSKY et al. 1986). The accumulation and persistence of O^6-methylguanine in lung correlated with inhibition by NNK treatment of the repair enzyme, O^6-methylguanine-DNA methyltransferase in lung (BELINSKY et al. 1986). In contrast, O^4-methylthymidine repair in lung and liver was not inhibited by NNK treatment (BELINSKY et al. 1986). A striking nonlinear relationship of dose of O^6-methylguanine levels in lung was observed; efficiency of methylation was greater at lower doses than at higher doses of NNK during treatment over a 4–12 day period (BELINSKY et al. 1987). Methylation efficiency was particularly high in Clara cells (BELINSKY et al. 1987). These results may be due to the presence in Clara cells of a cytochrome P-450 isozyme with high affinity to the metabolism of NNK by α-hydroxylation (BELINSKY et al. 1987).

19.5.3 *N*-Methyl-1-naphthylcarbamate

The insecticide carbaryl (*N*-methyl-1-naphthylcarbamate), although yielded in relative low amounts after nitrosation under conditions simulating those of the *human* stomach (DERACHE et al. 1982), presents a potential risk of mutagenicity (ELESPURU et al. 1974) and carcinogenicity (LIJINSKY and TAYLOR 1976). The *in vivo* treatment of female Sprague-Dawley *rats* by *N*-nitrosocarbaryl produced a reduction in lipoperoxidative degradation induced *in vitro* by NADPH with regard to the formation of malonaldehyde and conjugated dienes (BERAUD et al. 1989). Carbaryl, its precursor did not affect lipid peroxidation under the same *in vivo* conditions. Moreover, following administration of the two compounds, the activities of NADPH-cytochrome *c* reductase as well as NADPH-neotetrazolium reductase were significantly decreased by *N*-nitrosocarbaryl but not influenced by carbaryl. *N*-Nitrosocarbaryl proved to have a potent inhibitor concentration effect on NADPH-dependent chemiluminescence response *in vitro*; carbaryl was virtually ineffective on this parameter. No significant difference appeared in the affinity of *N*-nitrosocarbaryl and carbaryl for the microsomal phospholipids.

19.5.4 Vinyl Chloride

Exposure to vinyl chloride monomer increased the risk of malignant tumour slightly but not significantly (LAPLANCHE et al. 1992). Three cases of angiosarcoma of the liver occurred in the exposed group of 1099 subjects. Eight cases of lung cancer occurred among exposed subjects and six among 1099 non-exposed control subjects.

In *human* liver angiosarcomas associated with occupational exposure to vinyl chloride, MARION et al. (1991) found that 5 out of 6 tumours contained a Ki-*ras* gene activated by a GC→AT transition at the second base of codon 13. *Ras* gene mutations in vinyl chloride-induced liver tumours are carcinogen-specific but vary with cell type and species (BOIVIN-ANGÈLE et al. 2000).

In the *human* spleen, vinyl chloride-induced lesions were primarily but not a consequence of portal hypertension (HEUSERMANN and STUTTE 1977, STUTTE and HEUSERMANN 1977). Fibre-associated reticulum cells of the red pulp and fibroblastic reticulum cells in white pulp were stimulated to produce excessive amounts of the extracellular elements of connective tissue, especially collagen fibrils.

γ-Glutamyl transpeptidase (EC 2.3.2.2) was found elevated in a strong correlation with the duration of exposure to vinyl chloride (LILIS et al. 1975). Alkaline phosphatase (EC 3.1.3.1) was found to be elevated with significant frequency.

After inhalation of 50 ppm [^{14}C]vinyl chloride monomer for 65 min, with a calculated average of 0.49 mg vinyl chloride per kg body weight, 58 % of the radioactivity was excreted by *rats* in the urine, 2.7 % in the faeces and 9.8 % as expired $^{14}CO_2$ within 15 h (HEFNER et al. 1975). By 70 h, 67 % of the radioactivity had been excreted in the urine, 3.8 % in the faeces and 14 % as expired CO_2. Only 0.02 % was expired as unchanged vinyl chloride. After 75 h, 1.6 % of the radioactivity remained in the liver, 3.6 % in the skin and 0.2 % in the kidneys, while 7.6 % in the remaining carcass.

After exposure of *rats* to [^{14}C]-vinyl chloride gas, vinyl chloride-derived radioactivity incorporated into proteins could mainly be detected in liver, lung, kidney, and spleen (KAPPUS et al. 1976). If *rat* liver microsomes were incubated under atmospheric air containing [^{14}C]-vinyl chloride gas, max. 0.5 nmol of vinyl chloride metabolites was covalently bound to microsomal protein. Also RNA as well as SH-containing proteins, when added to the microsomal incubation, bound vinyl chloride metabolites. All these binding reactions performed by microsomes depended on enzymatically active microsomes, oxygen, NADPH and partial pressure of vinyl chloride in the atmosphere, and could be inhibited by 1-naphthyl-4(5)-imidazole and carbon monoxide. Addition of glutathione plus cytoplasmic fractions decreased covalent binding of vinyl chloride metabolites to microsomal proteins, whereas the total metabolism of vinyl chloride during incubation was enhanced.

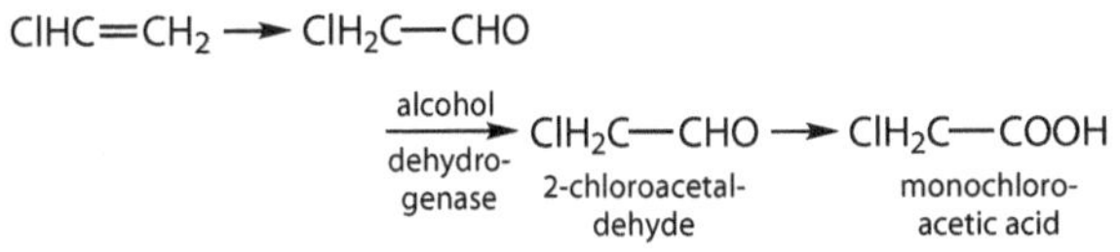

A possible biotransformation of < 100 ppm vinyl chloride as proposed by HEFNER et al. (1975) [331]

Evidence had been obtained *in vitro* for the existence of an alternative pathway for vinyl chloride monomer biotransformation involving microsomal mixed-function oxidase:

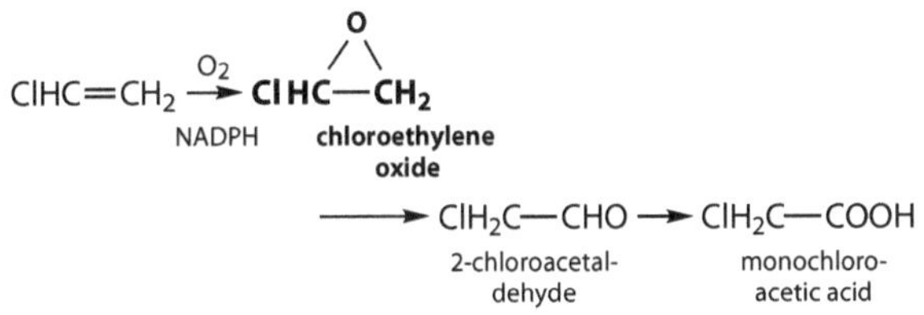

Chloroethylene oxide as the reactive toxic intermediate in the biotransformation of vinyl chloride [296]

ZIEF and SCHRAMM (1964) have shown that chloroethylene oxide spontaneously rearranges to chloroacetaldehyde.

Pre-treatment of *rats* with phenobarbitone, which is known to increase the P_{450} content of microsomal enzymes, increased the mutagenic response to vinyl chloride monomer *in vitro* (BARTSCH et al. 1975). In an aqueous solution at pH 7.4 and 37 °C the epoxy compound had a half-life of 1.6 min, and its rate of hydrolysis followed a first order kinetic. Chloroethylene oxide, but not 2-chloroacetaldehyde, showed a strong alkylating activity as determined by its reaction with 4-(*p*-nitrobenzyl)pyridine (MALAVEILLE et al. 1975). 2-Chloroacetaldehyde is a chemically reactive and toxic compound (LAWRENCE et al. 1972), and is covalently bound to cellular nucleophiles. It reacts at pH 3.5–4.5 and 37 °C with adenosine or cytidine to give fluorescent products, which have been characterised as 3-β-D-ribofuranosyl-imidazo-(2,1-*i*)purine or 5,5-dihydro-5-dihydro-5-oxo-5-β-D-ribofuranosyl-imidazo-(1,2-*c*)pyrimidine (BARRIO et al. 1972).

11-Day-old Sprague-Dawley *rats* exposed to 2000 ppm vinyl chloride for 10 subsequent days (7-h exposure on days 1–9 and 24-h exposure on day 10) in their lungs showed 1,N^6-etheno-2'-deoxyadenosine and 3,N^4-etheno-2'-deoxyadenosine (EBERLE et al. 1989). The biological effects of vinyl chloride are mediated through the conversion by microsomal cytochrome P_{450}-dependent monooxygenases (BARBIN et al. 1975) into the reactive metabolite chloroethylene oxide which can rearrange non-enzymatically to chloroacetaldehyde.

Aliphatic epoxides can alkylate different reactive sites in DNA. In *mice* exposd to vinyl chloride, a preference for the *N*-7 position in guanine has been described for chloroethylene oxide, its reactive metabolite (OSTERMAN-GOLKAR et al. 1977). PETER and CSANÁDY (1990) calculated the reactivity of aliphatic epoxides towards *N*-3 and *N*-7 in both purine bases by the **M**inor **N**eglect of **D**iatomic **O**verlap method (DEWAR and THIEL 1977, BUDA and SYGULA 1987) and correlated them with their **H**ighest **O**ccupied **M**olecular **O**rbital coefficients. They found a greater chemical susceptibility of the *N*-3 position in both purines to aliphatic epoxides when compared with the *N*-7 position. On this background the preference of aliphatic epoxides to the *N*-7 position of guanine and the steric orientation of the *N*-3 (minor groove) and the *N*-7 (major groove) positions of both purines in α-helical DNA.

19.5.5 Acrylamide

Acrylamide is a monomer of great industrial importance. Due residual monomer after polymer formation, possibilities for exposure exist for much larger groups of people from contaminated air, water and foodstuffs.

Acrylamide is metabolised to an 1,2-epoxide by cytochrome P_{450}-dependent (CYP) monooxygenases. Due to the large ring strain associated with the three-membered ring epoxides are reactive molecules.

Cytochrome P_{450} with a molecular weight of 47,000 decreased in liver microsomes of *mice* injected intraperitoneally with 4.5 and 45 mg acrylamide per kg body weight × day for four days (NILSEN et al. 1978).

AGEENKO et al. (1974) studies the radical copolymerisation of ^{14}C-acrylamide and of the incorporation of ^{3}H-thymidine in a *rat* fibroblast culture in the presence of an adenovirus infection.

For 2×10^{-4} M acrylamide solutions in pure water, the fractional yield of $CH_2(Mu)\dot{C}HCONH_2$ radicals is observed by muon-level-crossing-resonance to be 0.2 – consistent with the muonium yield by muon spin rotation (Venkateswaran et al. 1989). The Mu radical yield increased with acrylamide concentration until at ~0.2 M it equals 0.38, which, along with the fraction of muons in a diamagnetic environment, account for the entire muon polarization.

19.5.6 Acrylonitrile

The carcinogenic risk from acrylonitrile has been discussed in Cah Notes Docum Inst Nata Sécur Prév Accid Trav 95:19 (1979). It is a mutagen in the Ames test. The mutagenicity of acrylonitrile is mediated by *rat* liver post-motochondrial (S_9) fractions, and is particularly discernible with strains TA1530, TA1535, and TA1550 strains, which are sensitive to base-substitution mutagens (De Meester et al. 1978). In the *rat*, important urinary metabolites were *N*-acetyl-*S*-(2-cyanoethyl)cysteine, 4-acetyl-3-carboxy-5-cyanotetrahydro-1,4[2*H*]-thiazine, and thiocyanate (Langvardt et al. 1980). The two last-named substances were formed undoubtedly via epoxidation.

In *rats* exposed to acrylonitrile (30 and 300 ppm in drinking water for 21 days), 8-oxodeoxyguanosine levels were two fold grater than in the controls (Whysner et al. 1998). Measures of glutathione levels, glutathione peroxidase and catalase were not significantly changed, but cyst(e)ine was somewhat increased. No changes were found in brain cytochrome oxidase activity, which indicated a lack of metabolic hypoxia. Also, no effects on thiobarbituric acid-reactive substances were found, indicating a lack of lipid peroxidation. In male Sprague-Dawley *rats* exposed to 0 or 100 ppm acrylonitrile in drinking water for 94 days, levels of brain nuclear DNA 8-oxodeoxyguanosine were significantly increased compared with controls.

Using a high performance liquid chromatograph equipped with an electrochemical detector, Murata et al. (2001) revealed that acrylonitrile enhanced the formation of 8-oxo-7,8-dihydro-2'-deoxoguanosine induced by H_2O_2 and Cu(II) whereas acrylonitrile itself did not cause DNA damage. The enhancing effect of acrylonitrile was much more efficient in the double-stranded DNA than that in the single-stranded DNA. Experiments with ^{32}P revealed that the addition of acrylonitrile enhanced the site-specific DNA damage at guanines, particularly at 5'-site of the <u>G</u>G and <u>G</u>GG sequences while H_2O_2/Cu(II) induced piperidine-labile sites at thymine, cytosine, and guanine residues. An electron spin resonance spectroscopy using α-(4-pyridyl-1-oxide)-*N*-*tert*-butylnitrone showed that a nitrogen-centred radical was generated from acrylonitrile in the presence of H_2O_2 and Cu(II).

19.5.7 Trichloroethylene

Trichloroethylene is a versatile chemical compound used extensively in vapour degreasing of metals, and to a more limited degree, as a solvent for dry-cleaning and for adhesives. Although the major rate of elimination of trichloroethylene, regardless of the method of exposure, in the *rat* is by exhalation through the pulmonary system (Daniel 1963), in *man* the nephrotoxic and genotoxic *N*-acetyl-*S*-dichlorovinyl-L-cysteine as an urinary metabolite was found after occupational exposure to 1,1,2-trichloroethylene (Birner et al. 1993).

1,1-Dichloroethylene, a structural analogue of trichloroethylene, in *mice* has been shown to cause Clara cell damage associated with in situ formation of the reactive epoxide (Forkert 1999). Epoxide levels were reduced by pre-treatment with diallyl sulphone (100 mg/kg per os), an inhibitor of CYP2E1. Elfarra et al. (1998) found species- and sex-related differences in metabolism of trichloroethylene to yield chloral and trichloromethanol in *mouse*, *rat*, and *human* liver microsomes. In renal proximal tubule cells there is plenty of smooth endoplasmic reticulum (Bargmann 1978, p. 135). Rovin et al. (1990) found that epithelial cells from the proximal tubules, the cortical collecting duct, and the papillary collecting duct from *rabbit* generated reactive oxygen species in the absence of chemical stimulation. Opsonized zymosan and heat-aggregated IgG enhanced this basal formation of $O_2^{\bullet -}$ and H_2O_2 in a time- and concentration-dependent manner. Superoxide radical generation by renal microsomes of oestrogen-treated *hamsters* was elevated compared to values from untreated controls (Roy and Liehr 1989).

Exposure to trichloroethylene and perchloroethylene produced a dose-dependent and more pronounced accumulation of H_2O_2 in p53-WT NCI-H460 than p53-null NCI-H1299 *human* lung adenocarcinoma cells having constitutively different levels of glutathione (Chen et al. 2002). The accumulation of H_2O_2 was accompanied by severe cellular damage, as indicated by the significant increase of lipid peroxidation and apoptosis in p53-WT H460 cells, but not p53-null H1299 cells. Cotreatment of p53-WT H460 cells with free radical scavengers, such as D-mannitol, uric acid, and sodium selenite, significantly attenuated the trichloroethylene- or perchloroethylene-induced lipid peroxidation. In contrast, depletion of GSH in p53-null H1299 cells enhanced trichloroethylene- or perchloroethylene-induced lipid peroxidation. The levels of p53 and Bax proteins were elevated, while Bcl-2 protein was downregulated in trichloroethylene- or perchloroethylene-treated p53-WT H460 cells. Activity of caspase 3, the apoptotic executioner, was also significantly enhanced in trichloroethylene- or perchloroethylene-treated cells.

The risk of both renal cell cancer and renal pelvic cancer was increased by working in the dry cleaning industry of New South Wales (McCredie and Stewart 1993). Carcinogen involvement in renal cell carcinoma was supported by reports of an increased frequency of renal cell carcinoma among workers exposed to trichloroethylene (Brüning et al. 1997). Tumour DNA was analysed for mutations of the von Hippel-Lindau gene on 3p25.5 by PCR and sequencing analyses (Schraml et al. 1999). The histological analysis of 12 patients revealed 9 clear cell, 2 papillary renal cell carcinomas and 1 oncocytoma.

19.5.8 Pentachlorophenol Metabolites

Pentachlorophenol, a widely used biocide, has been shown to be cancergenic in laboratory *rats* (Chhabra et al. 1999) and *mice* (McConnell et al. 1991) in 2 years chronic bioassays. The chromosomal damage induced by pentachlorophenol may derive primarily from its quinoid metabolites, tetrachlorohydroquinone (Jansson and Jansson 1992) and tetrachloro-1,4-benzoquinone. The ^{32}P-post-labelling assay revealed four major and several minor adducts (3.5 adducts per 10^5 total nucleotides) in *calf* thymus DNA treated with 5 mM tetrachloro-1,4-benzoquinone (Lin et al. 2001).

Autoxidation and/or enzyme-mediated oxidation of pentachlorophenol catechol and hydroquinone to the corresponding semiquinones and quinones followed by subsequent reduction of quinones initiate redox cycling cascades and generate reactive oxygen species, which are believed to be responsible for pentachlorophenol clastogenicity (Naito et al. 1994, Dahlhaus et al. 1996, Wang et al. 1997).

19.5.9 Polychlorinated Biphenyls

Mono- and dichlorinated biphenyls can be metabolised to dihydroxy compounds and further oxidised to reactive metabolites which form adducts with nitrogen and sulphur nucleophiles including DNA (Amaro et al. 1996, Oakley et al. 1996). The former studies also demonstrated that during the metabolism of polychlorinated biphenyls superoxide may be produced. The incubation of 3,4-dichloro-2',5'-dihydroxybiphenyl (100 μM) with calf thymus DNA (300 μg/ml) in the presence of the breast tissue and milk associated enzyme, lactoperoxidase (EC 1.11.1.7), and H_2O_2 resulted in a significant increase in free radical-induced DNA damage as compared to vehicle-treated DNA (Oakley et al. 1996). Substituting $CuCl_2$ (100 μM) for lactoperoxidase/H_2O_2, however, resulted in a substantial increase in 8-oxodeoxyguanosine content. $FeCl_3$ was ineffective, suggesting that $CuCl_2$ but not $FeCl_3$ mediates oxidation of polychlorinated biphenyl dihydroxy metabolites, resulting in oxidative DNA damage. The addition of catalase (100 U/ml) and sodium azide (0.1 M) reduced the effect of $CuCl_2$, while superoxide dismutase (600 U/ml) moderately stimulated and glutathione (100 μM) substantially stimulated 8-oxodeoxyguanosine formation.

19.5.10 4-Aminobiphenyl

The *N*-hydroxy metabolite of 4-aminobiphenyl, a highly efficient rubber antioxidant, was found to cause Cu(II)-mediated DNA damage, especially at thymine residues (Murata et al. 2001). Addition of the endogenous reductant NADH dramatically enhanced this process, while catalase (EC 1.11.1.6) and bathocuproine, a Cu(I)-specific chelator, reduced the amount of DNA damage, suggesting the involvement of H_2O_2 and Cu(I). Increased amounts of 8-hydroxy-2'-deoxyguanosine were found in HL-60 cells compare to the H_2O_2-resistant clone HP100 following *N*-hydroxy-4-aminobiphenyl treatment.

19.5.11 Hydrazines

The hydrazines represent an important class of xenobiotic agents encountered in the environment, in industrial settings, and in medical therapeutics. When incubated in haemolysate, they caused a time- and concentration-dependent strand scission of DNA as monitored using $\phi \times 174$ RF DNA (Runge-Morris et al. 1994). The rank order for hydrazine-mediated damage was phenylhydrazine > phenelzine > hydrazine > hydralazine > methylhydrazine. The free radical spin trap agent dimethylpyrrolidin-*N*-oxide effectively inhibited phenylhydrazine-mediated DNA damage, while the free radical scavenger *N*-acetylcysteine also showed a protective effect against phenylhydrazine-, phenelzine-, hydralazine-, hydrazine-, and methylhydrazine-mediated DNA strand scission. Potassium ferricyanide-mediated methaemoglobin formation and imidazole, a ligand for the heme moiety of haemoglobin, both inhibited phenylhydrazine-stimulated DNA damage in haemolysate demonstrating the importance of oxyhaemoglobin in the process.

When tested by inhalation, hydrazine produced benign and malignant tumours in *rats*, benign nasal polyps, a few colon tumours and thyroid adenomas in *hamsters*, and a slight increase in the incidence of lung adenomas in *mice* (IARC 1987).

Further reference: LATENDRESSE et al. (1995)

19.6 Tumour Induction with Natural Products

19.6.1 Betel

Bursts of reactive oxygen species generation were found in the development of oral cancer in chewers of areca nuts (STICH and ANDERS 1989). Genotoxic compounds appeared within minutes of a chewing session, and disappeared when the chewing mixtures were removed from the mouth. Exfoliative mucosal cells can be sampled using non-invasive procedures from different sites within the oral cavity, and screened for the presence of micronuclei (STICH 1987). The response of the mucosa to the administration of chemopreventive agents with antioxidant capacities can be followed over prolonged time periods (STICH et al. 1984, 1988a, b).

Reactive oxygen species formed from polyphenolic betel quid ingredients and lime at alkaline pH (NAIR et al. 1992) have been implicates as the agents responsible for DNA and tissue damage inducing oral cancer in Southeast Asia and the South Pacific islands. Aqueous extracts of areca nut and catechu were capable of generation $O_2^{\bullet -}$ and H_2O_2 at pH >9.5 (NAIR et al. 1987). The formation of $O_2^{\bullet -}$ was enhanced by Fe^{2+}, Fe^{3*} and Cu^{2+} but inhibited by Mn^{2+}. Tobacco extract failed to generate reactive oxygen species under similar conditions. Saliva was found to inhibit both $O_2^{\bullet -}$ and H_2O_2 formation from betel quid ingredients. Upon incubation of DNA at alkaline pH with areca nut extract and Fe^{3+} or catechu, 8-hydroxydeoxyguanosine was formed as quantified by high performance liquid chromatography/electrochemical detection. NAIR et al.(1995) clearly demonstrated that the $HO^{\bullet}$ radical is formed in the *human* oral cavity during betel quid chewing and is probably implicated in the genetic damage that has been observed in oral epithelial cells of chewers. Fe^{2+} and Mg^{2+} levels in the lime samples were too low to modify the formation of reactive oxygen species, but H_2O_2 formation was almost entirely inhibited by addition of Mg^{2+} to the reaction mixture *in vitro* (NAIR et al. 1990).

Multiple molecular alterations of fragile histidine triad gene were found in betel-associated oral carcinoma (CHANG et al. 2002). Analyses of the coding exons (exons 5–9) identified a deletion of one base in intron 4 in one tumour and a deletion of exon 7 in two tumours. Using bisulphite genomic sequencing, 28 % of the informative subjects exhibited promoter methylation. An aberrant fragile histidine triad transcript spanning from exon 3 to exon 10, which was verified by RT-PCR analysis, was identified in 36 % of the oral squamous cell carcinoma subjects 50 % in the oral pre-invasive lesions, and 5 % of the non-cancerous match tissue. An abnormal immunohistochemical level of fragile histidine triad was detected in 41 % of oral squamous cell carcinoma subjects. A statistically significant association was found between aberrant transcription of the fragile histidine triad gene and an abnormal level of fragile histidine triad immunoreactivity.

19.6.2 Opium

Studies on risk factors for oesophageal cancer in the Caspian littoral of Iran carried out by IARC confirmed that the factors associated with the very high morbidity in this area included nutritional deficiencies and the chewing of the opium residues (dross or sukteh) from pipes or inhaled as opium smoke (GHADIRIAN et al. 1985).

19.6.3 Tobacco

Tobacco smoking causes a major fraction of urinary bladder cancers in *man*, and the relative risk is two to three times higher for those smoking black (air-cured) than for those smoking blond (flue-cured) tobacco. BARTSCH et al. (1993) examined the hypothesis that the aromatic amines were primarily responsible for inducing bladder cancer and found that the higher risk of smokers of black tobacco correlated with the presence in the smoke of black tobacco of two to five times higher concentrations of carcinogenic aromatic amines, notably 4-amionbiphenyl.

High concentrations of reactive oxygen species and nitrogen oxides (NO_x) produced from cigarette smoke (PRYOR et al. 1983, NAKAYAMA and KODAMA 1984, CHURCH and PRYOR 1985, PRYOR 1987) may also be involved in the carcinogenicity (ZHU et al. 1998). Molecular toxicological investigations have demonstrated that cigarette smoke can elevate the expression of heme oxygenase (MÜLLER and GEBEL 1994). CHANG et al. (2001) found increased expression of iNOS in terminal bronchiolar lesions of *rats* exposed to cigarette smoke.

Both of the mean level of 8-nitroguanine levels in peripheral lymphocytes and serum nitrite of light-smoking, moderate smoking, heavy-smoking and cancer heavy-smoking groups were higher than that of non-smoking health controls (HSIEH et al. 2002). In male Wistar *rats* that were cigarette smoke-exposed for 30 min each twice a day on 6 days per week for 1 month, one month later a dose-dependent increase in 8-nitroguanine in the lungs and peripheral lymphocyte DNA was observed.

19.6.4 Aflatoxin B_1

Metabolic conversion of aflatoxin B_1 to the putative reactive electrophilic aflatoxin B_1-2,3-epoxide by a cytochrome P_{450}-linked monooxygenase system is a prerequisite for exerting its carcinogenic and mutagenic potential (SWENSON et al. 1977, FAHMY et al. 1978).

Aflatoxin B_1 administered to *rats* acted on liver parenchymal cells as a substrate for drug-metabolising enzymes (NOVI 1977). An increased rate of formation of metabolically activated aflatoxin B_1 may result in the appearance in the cytoplasm of free radicals that interact electron affinity with nearby intracellular structures bearing the opposite charge, such as ribosome particles. The binding may protect the ribosomes from nuclease attack, resulting in storage of aflatoxin B_1 within the hepatocytes. Increased production of unfavourable metabolites may lead to cell degeneration whereby the carcinogen is released in the liver parenchyma. This endogenous mechanism of storage and release may be responsible for the persistence of the cellular changes induced by aflatoxin B_1 after a withdrawal of an exogenous supply of aflatoxin B_1. Pregnenolone-16α-carbonitrile treatment of male *rats* resulted in a six-fold increase in the 9-hydroxylation of aflatoxin B_1 to aflatoxin Q_1 (AFQ_1); female *rats* showed a 16-fold increase in the formation of AFQ_1 (HALVORSON et al. 1988). The age-dependent decline in constitutive cytochrome-450p levels in female but not in male *rats* resulted in a sex difference in the formation of AFQ_1 liver microsomes from untreated rats (male:female 3:1). The formation AFQ_1 was stimulated up to 4.5-fold when liver microsomes from triacetyloleandomycin-treated *rats* were treated with potassium cyanide, which dissociates the complex between cytochrome P-450p and triacetyloleandomycin. Treatment of male *rats* with the cytochrome P-450p induced, dexamethasone, increased (7-fold) the 9-hydroxylation of AFB_1 to AFQ_1 by liver microsomes, and also enhanced (2-fold) the microsomal activation of AFB1 to metabolites that were mutagenic to *Salmonella typhimurium* TA98 and TA100.

Renal epithelial neoplasias, histologically similar to *human* kidney adenocarcinoma, were found in over one-half of *rats* ingesting 1.0 ppm aflatoxin B_1 and about one-quarter of *rats* given 0.5 or 0.25 ppm for 147 days (EPSTEIN et al. 1969).

19.6.5 Methylazoxymethanol

Cycasin, methylazoxymethanol-β-D-glycopyranoside occurs in roots, leaves and seeds of tropical and subtropical palms processed to starch flour.

Production of duodenal and upper jejunal adenocarcinomas initiated by methyl-azoxymethanol (formula [330]), the aglycone of cycasin, was enhanced when H_2O_2 was given in the drinking water (HIROTA and YOKOYAMA 1981). This effect, which is larger in *mice* with low levels of catalase (ITO et al. 1984), has been ascribed to the promoting action of oxygen radicals, probably $HO^•$ generated from H_2O_2.

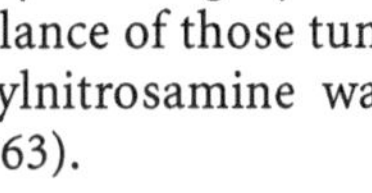

Methylazoxymethanol [332]

Hepatocellular carcinomas and reticuloendothelial neoplasms of the liver and adenomas and undifferentiated tumours of the kidney were induced in *rats* by feeding cycad material and the close resemblance of those tumours to those induced by dimethylnitrosamine was emphasised (LAQUEUR et al. 1963).

19.6.6 Azoxymethane

Colon carcinomas induced by azoxymethane (15 mg/kg body weight once a week for 2 weeks and then maintained with a standard diet, AIN-76A) in male F344 *rats* were found to have an increased expression of iNOS and eNOS proteins as compared to normal colonic mucosa (TAKAHASHI et al. 1997). In particular, the pronounced staining of iNOS protein localised to the luminal surface of carcinoma epithelial cells was not detectable in normal colon epithelium. The neovasculature in tumour tissues also demonstrated intense eNOS immunoreactivity in endothelial cells.

19.7 Peroxisome Proliferators

Peroxisome proliferators caused a dramatic increase in the incidence of liver tumours in *mice* and *rats*. Two major factors, an enhanced cell proliferation and an increased peroxisomal production of H_2O_2 have been implicated (KARAM et al. 1997, CATTLEY et al. 1998). **Nafenopin** (2-methyl-2-[*p*-(1,2,3,4-tetrahydro-1-naphthyl)phenoxy]propionic acid), a peroxisome proliferator, was shown to inhibit liver cell apoptosis in *rat* hepatocyte primary cultures, which effect could also promote carcinogenesis (BAYLY et al. 1994, ROBERTS et al. 1998). Although the responses initiated by **bezafibrate** were quantitatively similar in two rat strains, they differed in their magnitude in a dose-dependent manner, with the Lewis strain exhibiting a more pronounced response than the Sprague-Dawley rats (PILL et al. 1992). A comparative study of wild-type and peroxisome proliferator-activated receptor α KO *mice* fed with [4-chloro-6(2,3-xylidino)2-pyrimidinylthio]acetic acid (**Wy-14643**) suggest that increased cyclin-dependent kinase-1, cyclin dependent kinase-4, cyclin D1, and c-*myc* gene expression might be directly or indirectly proliferator-activated receptor α dependent (PETERS et al. 1998). There are marked species differences in response to peroxisome proliferators, with *mouse* and *rat* being very prone to peroxisome proliferation, while other species, especially *humans*, are unresponsive (HOLDEN and TUGWOOD 1999). In *dogs* treated with **gemfibrozil** (300 mg per kg per day), the number of peroxisomes per hepatocyte was significantly ($P < 0.01$) increased in females only (GRAY and DE LA IGLESIA 1984). The number of peroxisomes in young *rhesus monkeys* did not change after treatment, and the peroxisome volume was decreased in males and increased in females. Aged *monkeys* had increased number of peroxisomes per hepatocyte with increased volume fraction.

Peroxisome-proliferating carcinogens have not been found to be DNA-reactive (KORNBRUST et al. 1984, VON DÄNIKEN et al. 1984) and thus may be carcinogens of the type designated as epigenetic (WEISBURGER and WILLIAMS 1981). This is consistent with the proposal of REDDY et al. (1980) that peroxisome proliferators represent a distinct type of chemical carcinogen that alters cells by a sustained intracellular production of reactive oxygen species as a result of hydrogen peroxide generated by the increased numbers of peroxisomes.

Catalase-positive organelles were found to be more numerous in normal than in colonic neoplastic cells (CABLÉ et al. 1992). The specific activities of catalase, fatty-acyl CoA oxidase and enoyl-CoA hydratase/3 hydroxyacyl-CoA dehydrogenase (the so-called peroxisomal bifunctional enzyme of the β-oxidation system) were found to be diminished in carcinoma cells compared with control tissue.

Peroxisome proliferator-activated receptor γ decreased the growth of certain cancer cells. EIBL et al. (2001) found that six different human pancreatic cancer cell lines (AsPC-1, BxPC-3, Capan-2, HPAF-II, MIA PaCa-2, and PANC-1) expressed PPAR-γ mRNA and synthesised the protein. The endogenous and exogenous PPAR-γ ligands 15-deoxy-$\Delta^{12,14}$-prostaglandin J_2 and ciglitazone decreased cell number, cell viability, and increased floating/attached ratio, in a time and dose-dependent fashion.

19.8 Tumour Promoters

Tumour promoters, the most widely studied of which are the phorbol esters (HECKER 1975), have been demonstrated to stimulate superoxide anion radical production (GOLDSTEIN et al. 1979). 12-*O*-Tetradecanoylphorbol-13-acetate treatment of proliferating *murine* epidermal keratinocytes (MEK) cultured in low Ca^{2+} medium resulted in (i) an initial suppression of proliferation, (ii) the accelerated detachment and differentiation of detached MEKs and (iii) a suppression of catalase induction in the detached population (REINERS et al. 1990). Induction of MEK differentiation by raising the medium Ca^{2+} concentration resulted in a rapid inhibition of cell division and 200 % increases in per cell catalase activities. Addition of 12-*O*-tetradecanoylphorbol-13-acetate immediately prior to Ca^{2+} shift completely suppressed the Ca^{2+}-dependent increases in activity. However, the addition of 12-*O*-tetradecanoylphorbol-13-acetate 48 h after the induction of differentiation by Ca^{2+} shift had no effects on the elevated, pre-existing catalase activities. Per cell catalase activities varies *in vivo* with the stage of MEK differentiation.

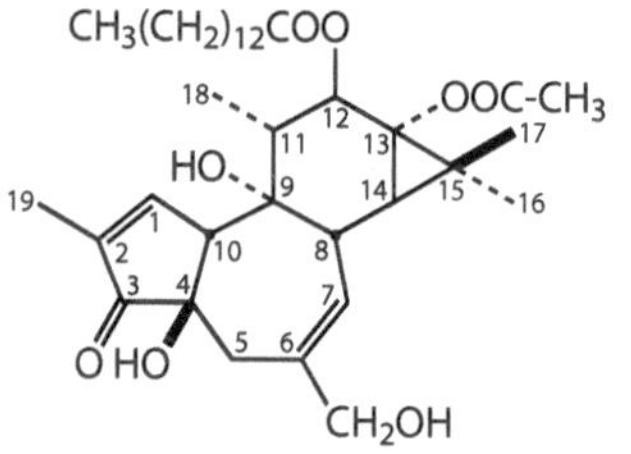

12-*O*-Tetradecanoylphorbol-13-acetate [49]

In polymorphonuclear leucocytes, the tumour promoters, 12-*O*-tetradecanoylphorbol-13-acetate, mezerein, and the alkylated indoleteleocidin caused superoxide radical formation (TROLL et al. 1982). On the other hand, 4-methyl-12-*O*-tetradecanoylphorbol-13-acetate and phorbol-12,13-diacetate that are inactive as tumour promoters do not cause $O_2^{\bullet-}$ formation. Protease inhibitors (TROLL et al.

1982), retinoids (Troll et al. 1982), and dexamethasone, which block tumour promotion by 12-*O*-tetradecanoylphorbol-13-acetate, in skin, also block $O_2^{\delta-}$ formation in polymorphonuclear leucocytes (Witz et al. 1980, Goldstein et al. 1981).

12-*O*-Tetradecanoylphorbol-13-acetate significantly induced extracellular signal-regulated kinase (ERK) 5 activation in Clara-like bronchiolar pulmonary adenocarcinoma (NCI-H441) cells (Reddy et al. 2002). Overexpression of dn-ERK5 strongly suppressed both basal and 12-*O*-tetradecanoylphorbol-13-acetate-inducible transcription of small proline-rich protein 1B, whereas wild type ERK5 upregulated it.

Two lines of *murine* (C57BL/5 *mouse*) 3-methylcholanthrene-induced sarcoma cells isolated from the same parent tumour responded differentially to stimulation with phorbol esters (Batchev et al. 1986). 12-*O*-Tetradecanoylphorbol-13-acetate stimulated the production of prostaglandin E_2 in both cell lines and the rations of the amounts produced in the stimulated versus control cells were very similar. As was seen with the release of radioactivity from [^{3}H]arachidonic acid, however, the amount of prostaglandin E_2 produced in the 1.2 cells was much greater under both control and stimulated conditions than in the low-malignant variant 1.2/anti-B^r cells. 12-*O*-Tetradecanoylphorbol-13-acetate also stimulated the production of leukotriene C_4 in the 1.2 cells but there was no increased production detected in the 1.2/anti-B^r cells.

Gerhäuser et al. (1995) studied the potential of plant extracts to inhibit phorbol ester-induced ornithine decarboxylase (EC 4.1.1.17) activity in cell culture. Ornithine decarboxylase is a key enzyme in the biosynthesis of polyamines and is highly inducible by growth-promoting stimuli including growth factors, steroid hormones, cAMP-elevating agents and tumour promoters. Four active rotenoids, deguelin, tephrosin, (–)-13α-hydroxy tephrosin, and (–)-13α-hydroxydeguelin were obtained from the African legume *Mundulea sericea*.

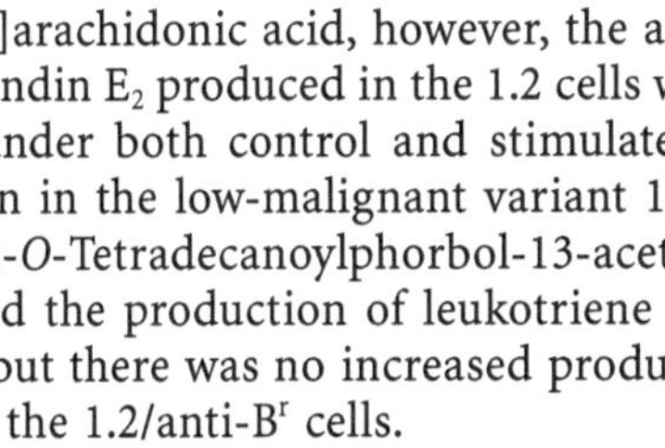

Deguelin [333]

2,3,7,8-Tetrachlorodibenzo-*p*-dioxin (TCDD) is one of the most potent toxins and tumour promoters known to *man*. TCCD acted as a potent tumour promoter but poor initiator in HRS/J *mice*, eliciting the same incidence and multiplicity of skin papillomas as the archetypal tumour promoter, 12-*O*-tetradecanoylphorbol-13-acetate. (Poland et al. 1982). Pitot et al. (1980) and Goldworthy and Pitot (1985) have shown that TCDD promotes liver tumours in *rats* but does not initiate tumour formation. Iron is essential for TCDD-induced lipid peroxidation (Al-Bayati and Stohs 1987) and DNA damage (Wahba et al. 1988, 1989), and the need for iron or other cations as copper can explain the evidence implication $O_2^{\bullet-}$, H_2O_2, and $HO^\bullet$ in TCDD-induced oxidative tissue damage. Pre-treatment of *rats* with antioxidants such as butylated hydroxyanisole for 3 days prior to giving a lethal dose of TCDD and followed by the daily application of butylated hydroxyanisole saved the animals (Hassan et al. 1985). Daily administration of ascorbic acid did not protect against toxicity of TCDD, but enhanced lipid peroxidation (Hassan et al. 1987). Pre-treatment of rats with vitamins A and E provided limited protection against the toxicity of TCDD (Hassan et al. 1985).

Knowledge of the toxic effects of TCDD in *humans* has resulted primarily from clinical and epidemiological studies of workers exposed during the manufacture of chlorinated phenols or a result of explosions and industrial accidents. It is known that TCDD causes chloracne (Taylor 1979). An angiosarcoma of the bony pelvis and proximal right femur with involvement of soft tissue, porphyria tarda and a probable chloracne were described in a worker exposed to waste oil contaminated with TCDD (McConnell et al. 1993).

19.9 Epoxide Formation

Besides carcinogens and promoters there are several other factors the impact of which on the induction of lung cancer is known or may be anticipated. Heinrich et al. (1985) using an adenoma test compared clean air groups of *mice* with groups exposed to total diesel engine exhaust, both treated with 10 μg dibenz[*a*,*h*]anthracene. They obtained a statistically significant lower tumour incidence rate and lower average number of adenomas per lung in the exhaust group. The ratio of biotransformation of dibenz[*a*,*h*]anthracene to its *syn*-diol-epoxide and its **highly mutagenic *anti*-diol-epoxide** depends on the cytochrome P_{450} form induced.

8-Methylbenz[*a*]anthracene is metabolised to an 8,9-diol, presumably via an epoxide (Yang et al. 1979). This finding demonstrates that the presence of a methyl group does not block epoxidation at the methyl substituted double bond. Tierney et al. (1978) have investigated the dihydrodiol formation from a variety of polycyclic hydrocarbons both by liver microsomal fractions and by chemical oxidation.

Epoxides of aromatic rings (arene oxides) rearrange to phenols and are substrates for glutathione-S-epoxide transferase to give glutathione conjugates, the precursors of mercapturic acids (BOYLAND and WILLIAMS 1965).

19.10 Inhibitory Effect of Melatonin

In industrialised countries the incidence of breast cancer has steadily increased during the 20th century. It has been suggested that one cause is exposure to light for longer periods than that afforded by the natural daily light-dark cycle. This so-called melatonin-hypothesis proposes that the suppression of melatonin secretion at night by artificial light increases breast cancer risk by increasing exposure to oestrogen (STEVENS et al. 1992, STEVENS and Davis 1996). A retrospective survey found that severely blind women had about half the incidence of breast cancer of women with normal vision (HAHN 1991). VERKASALO et al. (1999) compared the breast cancer incidence in women with various degrees of visual impairment to that in women with normal vision. Women with less severe visual impairment had intermediate incidence of breast cancer.

Mammary pathological growth as influenced by melatonin was reviewed by Cos and SÁNCHEZ-BARCELÓ (2000). Melatonin downregulates some of the pituitary and gonadal hormones which control mammary gland development and are also responsible for the growth of hormone-dependent mammary tumours. Furthermore, melatonin could act directly on tumour cells, thereby influencing their proliferative rate (BLASK et al. 1986, HILL and BLASK 1988, Cos et al. 1998). From *in vivo* studies on animal models of tumorigenesis, the general conclusion is that experimental manipulations activating the pineal gland or administration of melatonin enlarge the latency or reduce the incidence of growth rate of chemically induced mammary tumours, while pinealectomy has the opposite effects. The direct action of melatonin on mammary tumours has been suggested because of its ability, at physiological doses (1 nM), the *in vitro* proliferation and invasiveness of MCF-7 *human* breast cancer cells (Cos and BLASK 1994, MOLIS et al. 1995, Cos and SÁNCHEZ-BARCELÓ 1997).

The inhibitory effect of melatonin on intestinal carcinogenesis induced by 1,2-dimethylhydrazine in female LIO *rats* was demonstrated by ANISIMOV et al. (1997, 2000). The incidence of carcinomas in the ascending colon was significantly reduced ($P < 0.01$). The multiplicity of total colon tumours per *rat*, as well as the mean number of tumours, ascending and descending tumours per *rat*, was also decreased under the influence of melatonin. Melatonin slightly decreased the depth of tumour invasion and increased number of highly differentiated colon carcinomas induced by 1,2-dimethylhydrazine.

A possible role of •NO on the inhibition by melatonin of *human* breast cancer growth has also been suggested by BLASK and WILSON (1994). Based on the antiproliferative effects of •NO on A375 melanoma cells (MARAGOS et al. 1993), they incubated MCF-7 cells for 5 days with 1 nM melatonin either in the presence or not of N^{ω}-monomethyl-L-arginine. In the presence of N^{ω}-monomethyl-L-arginine, melatonin lacked its antiproliferative activity.

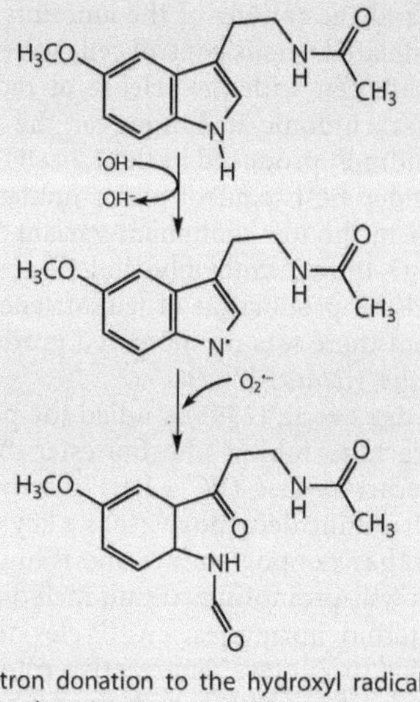

Upon electron donation to the hydroxyl radical (HO•) to form OH^-, melatonin gives rise to the **melatoninyl cation radical**, which is able with the superoxide anion radical ($O_2^{\bullet-}$) to *N*-acetyl-*N*-formyl-5-methoxykynuramide [61]

On pulse-irradiating an N_2O-saturated aqueous solution of 5×10^{-2} mol dm^{-3} KBr, 2×10^{-3} mol dm^{-3} guanosine and $1-5 \times 10^{-5}$ mol dm^{-3} melatonin (pH 7) repair of guanosine was observed, presumably via electron transfer from melatonin (MAHAL et al. 1999).

Melatonin pre-treatment (10 mg/kg body weight every 6 h for 24 h) partially prevented the increase in 8-hydroxy-2'-deoxyguanosine in the renal genomic DNA, which was significantly increased by more than 100 % after $KBrO_3$ treatment (80 mg/kg intraperitoneally) of adult Wistar *rats* (CADENAS and BARJA 1999).

Guanosine• + Melatonin-H → Guanosine-H + Melatonin• [334]

Rat hepatoma 7288CTC uptake of linoleic acid and release of 13-hydroxyoctadecadienoic acid, an important mitogenic signalling molecule within this tumour, were highest during the light phase and lowest during the mid-dark phase, when plasma melatonin levels were lowest and highest, respectively (BLASK et al. 1999). Pinealectomy eliminated this rhythm of tumour linoleic acid uptake and 13-hydroxyoctadecadienoic acid production, indicating that it was driven by the circadian melatonin rhythm.

The tumour growth rate in transplantable tissue-isolated hepatomas (7288 CTC) in adult male Buffalo *rats* treated with melatonin injections was 51 % lower than that in the vehicle-treated animals (BLASK et al. 1997). The latency of onset of tumour appearance was delayed by over one week in the melatonin-treated group compared with the control group. The degree of tumour inhibition observed in this study was virtually identical to that seen by KARASEK et al. (1990) using a transplantable *hamster* hepatoma (Kirkman-Robbins) implanted into male Syrian *hamsters* receiving daily, late-afternoon injections of melatonin (25 μg/animal) for 10 days.

Ehrlich ascites carcinoma cells implanted intraperitoneally in female *mice* were reduced in their viability and volume by oral supplementation of melatonin at 50 mg/kg body weight (EL-MISSIRY and EL-AZIZ 2000). Flow cytometric studies showed that melatonin not only delayed the progression of cells from G0/G1 phase to S-phase of the cell cycle but also reduced DNA synthesis during cell cycle. In addition, the aneuploidy status was depressed in melatonin treated *mice*.

In contrast to melatonin administration, constant light exposure induced a substantially more rapid appearance and growth of hepatoma 7288CTC (BLASK et al. 1997). The markedly enhanced uptake of linoleic acid and its biotransformation to 13-hydroxyoctadecadienoic acid by lipoxygenase under constant light conditions strongly suggests that this is a mechanism by which constant light stimulated the growth of hepatoma 7288CTC. Perfusion studies demonstrated that physiological relevant levels of melatonin markedly inhibited the uptake of linoleic acid and its conversion to 13-hydroxyoctadecadienoic acid.

Table 57. Protective action of melatonin in experimental carcinogenesis

Potassium bromate ($KBrO_3$)	CADENAS and BARJA (1999)
Benzpyrene	BERGMANN and ENGEL (1950)
7,12-Dimethyl-benz(*a*)anthracene	TAMARKIN et al. (1981), COS et al. (1989)
1,2-Dimethylhydrazine	ANISIMOV et al. (1997)
N-Nitroso-*N*-methylurea	BLASK et al. (1991, 1992), MUSATOV et al. (1999)
Safrole	TAN et al. (1993, 1994)

19.11 Indole-3-carbinol

Indole-3-carbinol is a dietary modulator of carcinogenesis that can reduce the level of carcinogen binding to DNA. Indole-3-carbinol-derived products are potent inducers of certain cytochrome P450(CYP)-dependent activities. Pre-treatment of male Sprague Dawley *rats* with 500 μmol indole-3-carbinol/kg per os after 20 h increased ethoxyresorufin *O*-deethylase (associated with CYP1A1 isozyme) activity in lungs 245-fold, compared with the activity in unmedicated controls (PARK and BJELDANES 1992). Intraperitoneal administration of indole-3-carbinol reduced benzo[*a*]pyrene (0.2 μmol/*rat*) binding to DNA in the liver but not in the lung indicating that inhibitors or scavengers produced following intraperitoneal administration of indole-3-carbinol are either not functional or are not present in the lung. Potent inhibition of 2-amino-1-methyl-6-phenylimidazo[4,5-*b*]pyridine (50 mg/kg body weight by oral gavage)-induced aberrant colonic crypt foci occurred in male F344 *rats* following initiation, postinitiation and continuous exposure to 0.1 % indole-3-carbinol in the diet (GUO et al. 1995).

19.12 Antioxidants

Detoxication (phase 2) enzymes, such as glutathione *S*-transferase, NAD(P)H:(quinone acceptor) oxidoreductase, and UPD-glucuronosyltransferase, are induced in animal cells exposed to a variety of electrophilic compounds and phenolic antioxidants (PRESTERA et al. 1993). Induction protects against the toxic and neoplastic effects of carcinogens and is mediated by activation of upstream electrophilic responsive/antioxidant responsive elements. PRESTERA and TALALAY (1995) analysed the mechanism of activation of these enhancers by transient gene expression of growth hormone reporter constructs containing a 41-bp region derived from the mouse glutathione *S*-transferase Ya gene 5'-upstream region that contains the electrophilic responsive/antioxidant responsive element and of constructs in which this element was replaced with either one or two consensus 12-*O*-tetradecanoylphorbol-13-acetate-responsive elements.

Antioxidant Enzyme Expression Induced by Dust Particles

Exposure of *rats* to asbestos resulted in significant increases in steady-state mRNA levels of manganese-containing superoxide dismutase at 3 and 9 days and of glutathione peroxidase at 6 and 9 days (JANSSEN et al. 1992). An increase in steady-state mRNA levels of copper, zinc-containing superoxide dismutase was observed at 6 days. Exposure to asbestos also resulted in overall increased enzyme activities of catalase, glutathione peroxidase and total dismutase in lung. The profiles of antioxidant enzymes were dissimilar during the development of experimental asbestosis or silicosis (α-cristobalite). MnSOD-mRNA expression in *human* BEAS 2 B cells exposed to $\leq 5\ \mu g/cm^2$ crocidolite fibres containing 14 % Fe^{2+} and 13 % Fe^{3+} significantly increased, decreasing at concentrations of 10, 25 and 50 $\mu g/cm^2$ (GILLISSEN et al. 1996).

The inhibitory effect of antioxidants is an indirect evidence supporting a role of free radicals and reactive oxygen species in carcinogenesis. **Vitamin C** (BLOCK 1992) and **vitamin E** (BOSTICK et al. 1993), which have been inversely associated with cancer incidence for a variety of sites in *humans*, have also been shown to inhibit tumour promotion by 12-*O*-tetradecanoylphorbol-13-acetate (PERCHELLET et al. 1985, SMART et al. 1987) and nonphorbol ester-type tumour promoters (IMAMOTO et al. 1990, BATTALORA et al. 1993, OGAWA et al. 1995). Vitamin E suppressed the level of proliferating cell nuclear antigens as a marker of cell proliferation in the lungs of *mice* treated with urethane (YANO et al. 1997).

The phenolic antioxidants, butylated hydroxyanisole and butylated hydroxytoluene, have been shown to inhibit tumour promotion by both, 12-*O*-tetradecanoylphorbol-13-acetate and benzoyl peroxide (SLAGA et al. 1983). The anticarcinogenic effects of naturally occurring flavones such as genistein may also be due to their antioxidant properties (WEI et al. 1993). *In vitro*, genistein inhibited angiogenesis (FOTSIS et al. 1993). Mannitol, a scavenger of hydroxyl radicals, or antioxidants as ascorbic acid plus α-tocopherol, abolished the *in vitro* promoting effects of 12-*O*-tetradecanoylphorbol-13-acetate and 2,3,7,8-tetrachlorodibenzo-*p*-dioxin (WÖLFLE and MARQUARDT 1996).

While H_2O_2 induced DNA damage in normal *human* colonocytes in a dose-dependent manner which was statistically significant at concentrations over 10 μM, pre-incubation of the cells with physiological concentrations of **butyrate** (6.25 and 12.5 mM) reduced H_2O_2 (15 μM) induced damage by 33 and 51 % (ROSIGNOLI et al. 2001). Treatment of cells with a mixture of 25 mM acetate + 10.4 mM propionate + 6.25 mM butyrate did not induce DNA damage, while a mixture of 50 mM acetate + 20.8 mM propionate + 12.5 mM butyrate was weakly genotoxic only towards normal colonocytes. However, both mixtures were able to reduce the H_2O_2-induced DNA damage by about 50 %.

Phytic acid (*myo*-inositol hexaphosphoric acid), which is present in plants, particularly in cereals, nuts, oil seeds, legumes, pollen and spores, is one of the most promising cancer chemopreventive agents. Inositol 1,2,3-trisphosphate and inositol 1,2,3,6-tetrakisphosphate have been used as antioxidants (PHILLIPPY and GRAF 1997). Phytic acid inhibited the formation of 8-oxo-7,8-dihydro-2'-deoxyguanosine in cultured HL-60 cells treated with an H_2O_2-generating system, although it did not scavenge H_2O_2 (MIDORIKAWA et al. 2001). Site-specific DNA damage by H_2O_2 and Cu(II) at GG and GGG sequences was inhibited by phytic acid, but not by *myo*-inositol. Phytic acid alone did not cause DNA damage and thus, it should not act as a prooxidant.

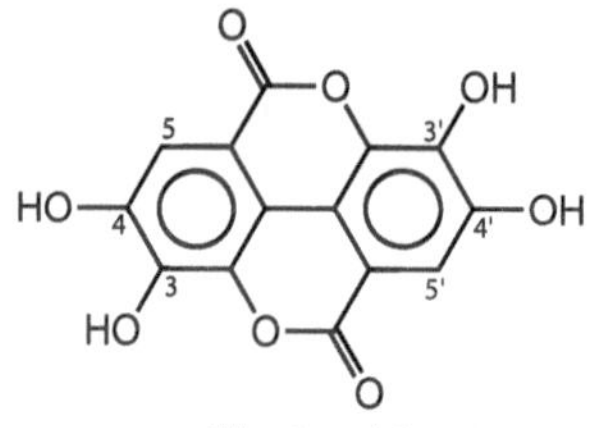

Ellagic acid [51]

Ellagic acid is a naturally occurring dietary anticarcinogen that has been shown to reduce the incidence of a variety of carcinogen-induced tumours (LESCA 1983, MUKHTAR et al. 1984, 1986, CHANG et al. 1985, MANDAL and STONER 1990, PEPIN et al. 1990) through a number of different mechanisms. Ellagic acid inhibits polycyclic aromatic hydrocarbon-induced tumorigenesis by inhibition of the CYP1A1-dependent activation of benzo[*a*]pyrene (MUKHTAR et al. 1984, SHUGART and KAO 1984, DAS et al. 1985, DIXIT et al. 1985, BARCH et al. 1994). Ellagic acid reduces the formation of benzo[*a*]pyrene adducts (MUKHTAR et al. 1984, SHUGART and KAO 1984, DAS et al. 1985, DIXIT et al. 1985) by detoxifying the activated benzo[*a*]pyrene diolepoxide through two additional mechanisms: it induces expression of the phase II detoxification enzyme glutathione *S*-transferase Ya (BARCH et al. 1995), an enzyme known to detoxify the activated benzo[*a*]pyrene diolepoxide (JERNSTROM et al. 1985, PUCHALSKI and FAHL 1990); and its directly binds to and detoxifies the diolepoxide of benzo[*a*]pyrene (SAYER et al. 1982). Ellagic acid has also been shown to induce the expression of the phase II detoxification enzyme NAD(P)H:quinone reductase (BARCH and RUNDHAUGEN 1994). Ellagic acid binds to DNA and inhibits the formation of O^6-methylguanine by methylating carcinogens (DIXIT and GOLD 1986, BARCH and FOX 1988).

Ellagic acid at low concentrations was found to increase antioxidant capacity of *human* hepatocellular carcinoma cells against $ROO^{\bullet}$ and $HO^{\bullet}$ radicals in the oxygen radical absorbance capacity assay (GAMAL-ELDEEN et al. 2001). Ellagic acid significantly increased total intracellular thiol levels and moderately increased the GSH/GSSG ratio. This increase in total thiols was not only due to GSH levels but mainly due to thiol-containing proteins, which might include cysteine-rich proteins like metallothioneins or enzymes like catalase (EC 1.11.1.6) and thioredoxin reductase.

Both the 3-hydroxyl and 4-hydroxyl groups are required for ellagic acid to directly detoxify the diolepoxide of benzo[*a*]pyrene, while only the 4-hydroxyl groups are necessary for ellagic acid to inhibit CYP1A1-dependent benzo[*a*]pyrene hydroxylase activity (BARCH et al. 1996). Induction of glutathione *S*-transferase Ya and NAD(P)H:quinone reductase requires the lactone groups of ellagic acid, but the hydroxyl groups are not required for the induction of these phase II enzymes. The lactone groups, but not the hydroxyl groups, are required for the analogy to reduce the carcinogen-induced formation of O^6-methylguanine.

4-Nitrophenol hydroxylase transforming 4-nitrophenol to 4-nitrocatechol was stronger inhibited by ellagic acid (maximum inhibition near 50 μM) added to dimethylsulphoxide (WILSON et al. 1992).

Xanthohumol was able to scavenge a variety of physiological relevant radicals including peroxyl, hydroxyl, and superoxide anion radials more effectively than the known antioxidant Trolox. It was found to inhibit both the constitutive form of cyclooxygenase-1 and, more importantly, the inducible cyclooxygenase-2, which is linked to carcinogenesis.

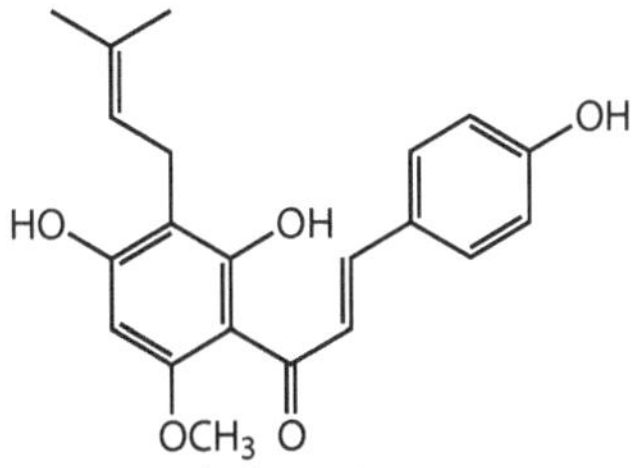

Xanthohumol [335]

In cultured RAW 264.7 *murine* macrophages, xanthohumol was shown to decrease lipopolysaccharide-mediated iNOS induction (GERHÄUSER et al. 2001).

The **lignans** (+)-1-acetoxypinoresinol and (+)-pinoresinol are major components of the phenolic fraction of olive oils (OWEN et al. 2000). These lignans, which are potent antioxidants, are absent in seed oils and virtually absent in refined virgin oils but are present in concentrations of up to 100 mg/kg (mean ± SE, 41.53 ± 3.93 mg/kg; range, 0.65–99.97 mg/kg) in extra virgin oils. As with the simple phenols and secoiridoids, there is considerable interoil variation in lignan concentrations. Foods containing high amounts of lignin precursors have been found to be protective against breast (HIRANO et al. 1990, MARTIN-MORENO et al. 1994, TRICHOPOULOU et al. 1995, LA VECCHIA et al. 1998), colon (BRAGA et al. 1998), and prostate cancers.

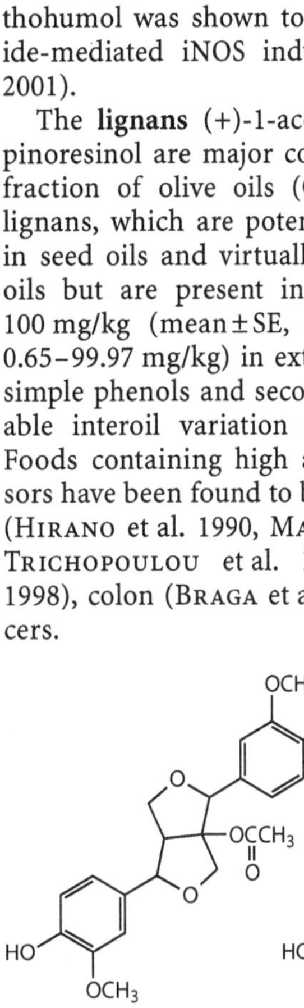

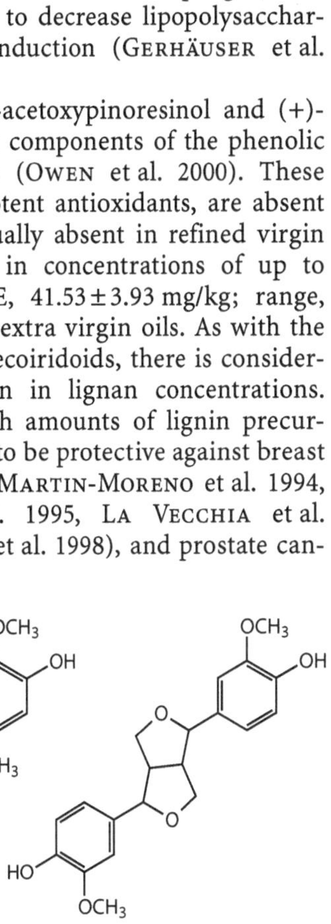

(+)-1-Acetoxypinoresinol [336] (+)-Pinoresinol [337]

19.13 Nitroxides

Nitroxides suppressed tumorigenesis (METODIEWA et al. 1997). The medication of Tempicol-2 [4-hydroxy-4-(2-picolyl-2,2,6,6-tetramethylpiperidine-1-oxyl], a stable free radical, to *rats* bearing 3 day-old Yoshida sarcoma (promotion phase) induced both growth inhibition and apoptotic cell death, comparable to the effects of Tempace and Rutoxyl [rutin/4-acetamide-1-hydroxyl-2,2,6,6-tetramethylpiperidinium] under the same experimental conditions (METODIEWA et al. 1998).

19.14 Selenium

The anticancerogenic effects of selenium have been reviewed by COMBS and GRAY (1998) and GANTHER (1999). GRIFFIN (1979) suggested that the anticancerogenicity of Se may involve cellular Se-dependent glutathione peroxidase, the only selenoenzyme characterised at the time that functions in the metabolic protection from oxidative stress by removing DNA-damaging hydrogen peroxide and lipid peroxides. However, the anticarcinogenic responses observed to supplemental Se by animals fed Se-adequate diets cannot involve the known selenoenzymes, as, for example, animal show maximal glutathione peroxidase activities in most tissues at dietary levels of approximately 0.2 mg/kg. The findings that antitumorigenic amounts of Se (e.g. 1.5 mg/kg) reduced tissue lipid organisation potential only slightly (LANE and MEDINA 1985) or not at all (HORVATH and IP 1983), suggest that those effects are independent of the function of the glutathione peroxidases in cellular antioxidant systems. Therefore, it is possible, that antitumorigenic effects of high levels of Se involve mechanisms unrelated to the activities of the selenoenzymes. Selenium supplemented either before initiation or during initiation and selection/promotion phases of *N*-nitrosodiethylamine-induced hepotocarcinogenesis was found to be effective in altering hepatic lipid peroxidation and antioxidant enzyme activities to a statistically significant level measured either in the hepatoma or in the surrounding liver tissues (THIRUNAVUKKARASU and SAKTHISEKARAN 2001).

The Se-dependent GSH peroxidase is a tetramer of 84 kDa with very high activity towards both H_2O_2 and organic hydroperoxides ($k_{ROOH} \sim 10^8\ M^{-1}s^{-1}$). It contains one residue of selenocysteine per mole at each of the active sites which, according to the crystal structure, is a depression near the molecular surface and readily accessible to substrate. This is consistent with the high reaction rate observed for this enzyme (LADENSTEIN and EPP 1984).

The distribution of Se-dependent and Se-independent GSH transferases varies with the species and the tissue. Despite its greater efficiency, there are several species in which the contribution of the total GSH peroxidase activity is small. Thus, in the liver of the *rat*, *mouse*, *hamster*, *pig*, *chicken*, *sheep*, *human*, and *guinea pig* the percentage of total activity due to the GSH transferases was 35 %, 42 %, 43 %, 67 %, 70 %, 77 % 84 %, and 100 %, respectively (LAWRENCE and BURK 1978).

The selective uptake of [^{75}Se]selenomethionin (3 μCie/kg, i.v.) by primary bronchogenic carcinoma, but not by secondary carcinomas and non-malignant pulmonary disease (CRITCHLEY et al. 1974) showed the biological and metabolic properties of the natural amino acid methionine and when injected intravenously, its rapid incorporation into newly synthesized protein (AWWAD et al. 1967).

The doxorubicin-resistant U-1285dox and GLC_4/ADR sublines of *human* small cell lung carcinoma proved to be 3- and 4-fold, respectively, more sensitive to the cytotoxicity of selenite than the drug-sensitive U-1285 and GLC_4 sublines, whereas no difference was observed between the HL-60 line and its doxorubicin-resistant P-glycoprotein expressing variant (BJÖRKHEM-BERGMAN et al. 2002). The presence of selenite did not significantly affect the expression of the multi-drug resistant proteins (MRP1, LRP and topisomerase IIα).

In *human* hepatoma ($HepG_2$) cells methylseleninic acid ($CH_3SeCOOH$) was shown to deplete intracellular GSH rapidly, preceding the typical apoptotic changes such as DNA fragmentation as measured by the TUNEL assay (SHEN et al. 2002). When the intracellular GSH concentration was enhanced using *N*-acetylcysteine and decreased using buthionine sulphoximine, *N*-acetylcysteine markedly augmented methylseleninic acid-induced apoptosis. Different from the effect of sodium selenite, there was no measurable superoxide anion radical level in methylseleninic acid-treated cells.

19.15 13-Methyltetradecanoic Acid

A saturated branched-chain fatty acid, 13-methyltetradecanoic acid, at a ID_{50} dosage ranging from 10 to 25 µg/ml, induced cell death in *human* cancer cell lines K-562 (leukaemia), MCF7 (mammary adenocarcinoma), DU 145 (prostate carcinoma), NCI-SNU-1 (gastric carcinoma), SNU-423 (liver carcinoma), NCI-H1688 (lung small cell carcinoma), BxPC-3 (pancreatic adenocarcinoma), and HCT 116 (colon carcinoma) (YANG et al. 2000). 13-Methyltetradecanoic acid caused tumour cell death through rapid induction of apoptosis, which could be detected 2 h after treatment. In the nude *mouse*, 13-methyltetradecanoic acid effectively inhibited the growth of orthotoptic tumour implants of the prostate carcinoma cell line DU 145 and hepatocarcinoma LCI-D35.

CHAPTER 20

Tumour Proliferation and Metastasis

20.1 Macrophage-Mediated Tumour Cell Proliferation and Migration

Large numbers of monocytes extravasate from the blood into *human* tumours, where they differentiate into macrophages. In both breast (KELLY et al. 1988, LEEK et al. 1996) and prostate (BURTON et al. 2000) carcinomas, these cells accumulate in areas of low oxygen tension (hypoxia), where they respond to hypoxia with the up-regulation of one or more hypoxia-inducible factors (HIFs). In tumour cell lines, hypoxia-regulated gene expression has been shown to involve stabilization, nuclear accumulation, and DNA binding of the transcription factors hypoxia-inducible factor 1 (WENGER 2000) and -2 (WIESENER 1998). GRIFFITHS et al. (2000) detected HIF-1α mRNA in hypoxic monocyte-derived macrophages, but not HIF-1α protein. BURKE et al. (2002) showed that primary *human* monocyte-derived macrophages accumulate HIF-1α then exposed *in vitro* to the severe hypoxia present in many forms of *human* tumours.

In Balb/c *mice* bearing a methylcholanthrene-induced fibrosarcoma the percentage of tumour-associated macrophages, which were detected on the basis of Fc receptor expression, remained constant in the growing neoplasm, at approximately 23% of the total cell population (VALDEZ et al 1990).

The presence of C3b/iC3b on Lewis lung carcinoma cells enhanced the formation of conjugates with macrophages (LIPARI et al. 1991). In spite of increased contact, macrophages from tumour bearing *mice* were not cytotoxic. Only preactivated macrophages, by in vivo treatment with *Corynebacterium parvum*, were shown to be cytotoxic; this function was potentiated when the target cells were promised with C3b/iC3b.

In vitro, natural cytolytic activity of unstimulated macrophages was generally unable to restrict final mKSA-TU5 tumour cell growth, since it is not coupled with cytostatic capacity (SOLDATESCHI et al. 1984). In contrast, exposure of macrophages *in vitro* to either macrophage activating factor or IFN-β, besides augmenting macrophage cytolytic capacity, induced a very significant cytostatic activity and thus restricted the survival of tumour cells.

In a medium containing 10 μg endotoxin/ml, *rat* macrophage-mediated cytolysis of DHD K12/TS dimethylhydrazine-induced BD IX *rat* colon carcinoma cells ranged from -7 to 36% (JEANNIN et al. 1985). In all the experiments, 1 mM dipalmitoyl phosphatidylcholine small unilamellar liposomes significantly induced or enhanced cytolysis, ranging from 30 – 90%. Liposomes and endotoxins had a synergistic effect on the macrophage cytolytic activity. This effect was dose-dependent on liposome concentration, ranging from 0.25 – 1 mM or 2 mM. Liposomes decreased the endotoxin concentration threshold necessary to induce cytolysis. They did not modify the kinetics of macrophage activation. Liposomes did not modify the binding of tumour cells to macrophages. The optimum synergistic effect was obtained when liposomes were present during the first 18 h of the mixed culture of macrophages and target cells, before adding of endotoxin for the next 18 h. When cholesterol was added to dipalmitoyl phosphatidylcholine (M/M), liposomes did not enhance but rather inhibited macrophage activation by endotoxins.

While at first inflammatory reactions within the stroma of malignant tumours have been considered as a sign of immunological response to the tumour, there is now ample evidence that inflammation is able to promote tumour progression. It is now clear that tumour cells themselves direct the assembly of tumour stroma, using signalling pathways normally confined to inflammatory cells. By producing inflammatory mediators, they are able to recruit inflammatory cells to the tumour site, induce myofibroblasts and the production of extracellular matrix components (HAUPTMANN 2000).

Evidence for macrophage-mediated tumour cell proliferation and migration was achieved in coculture of macrophages and multicellular tumour spheroids (HAUPTMANN et al. 1993). The production of reactive oxygen species by *human* inflammatory macrophages was reduced by 50 per cent, that of resident macrophages by 50%, while colorectal HRT-18 tumour cells did not produce any measurable reactive oxygen species (Hauptmann 2000). Macrophages and HT-29 adenocarcinoma cells in coculture formed nitric oxide (Siegert et al. 2000). The multifaceted roles of •NO in cancer were emphasised by Wink et al. (1998), •NO production by stimulated endothelium and its cytotoxic effect on cancer cells reviewed by Orr et al. (2000).

In a recent review, Bingle et al. (2002) presented evidence for the number and/or distribution of tumour-associated macrophages being linked to prognosis in different types of *human* malignancy.

20.2 Tumour Cell Invasion

The ability of tumour cells to invade into surrounding normal tissues and spread to form secondary lesions at distant host sites is the most devasting aspect of cancer. Invasion appears to the most characteristic step of metastasis, a multistep phenomenon including the release of tumour cells from the primary site, their circulation and lodging at the target organ where they proliferate and form secondary tumour foci (Fig. 350).

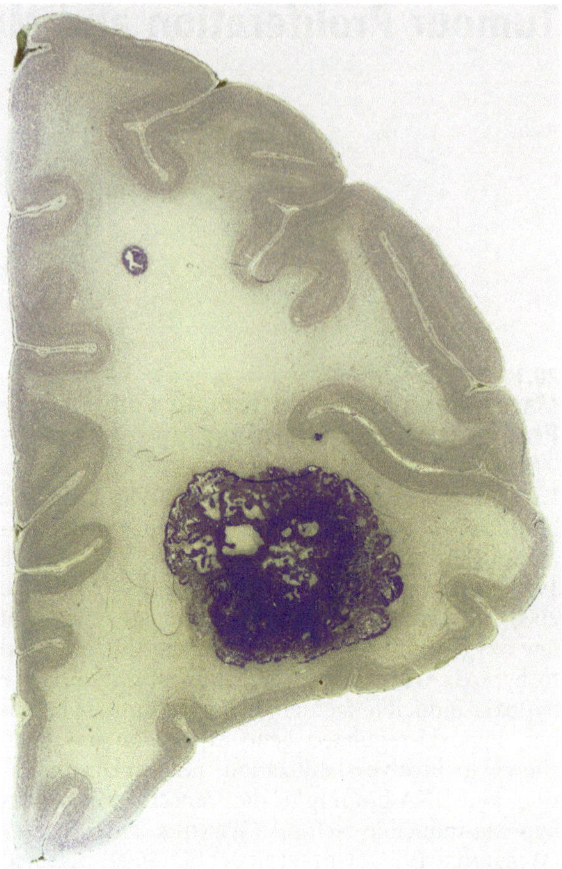

Fig. 350. An extensive and two smaller cerebral metastases of an undifferentiated solid carcinoma with numerous atypical mitoses. Primary tumour unknown. Brain Research Institute at the Department of Neurology, University of Leipzig (49/43). Haematoxylin and eosin

When radiolabelled cancer cells were injected into the tail vein of *mice* during periods of pulmonary endothelial damage induced by bleomycin (120 mg/kg i.v.) or by exposure to 90% O_2 for 2 – 4 days, there was a 3 – 36 fold increase in the number of these cells located in the lung after 24 h (Orr et al. 1988). Subsequently more metastatic tumours formed in the animals with injured lungs. In *rats* injected intravenously with cobra venom factor, the enhanced localisation was prevented by pretreatment of the animals with catalase or with antineutrophil antibodies. Stimulation of *rat* cancer cells by the chemotactic peptide *N*-formyl-methionyl-leucyl-phenylalanine was followed by chemiluminescence, amplified in the presence of luminol. Evidence for the generation of reactive oxygen species by these cells includes inhibition of the response in the absence of oxygen or in the presence of superoxide dismutase, catalase, and mannitol, and dose-dependent reduction of cytochrome *c*.

In order to determine whether angiogenesis is a prognostic marker in lung cancer, Meert et al. (2002) performed a systematic review of the literature to assess the prognostic value on survival of microvessel count in patients with lung cancer. Microvessel count, reflecting the angiogenesis, appeared to be a poor prognostic factor for survival in surgically treated non-small cell lung cancer but standardisation of angiogenesis assessment by the microvessel count is necessary.

20.2.1 Angiogenesis Enhanced by Hypoxia

Mammalian target of rapamycin signalling plays a key role in hypoxia-triggered smooth muscle and endothelial proliferation and angiogenesis *in vitro* (Humar et al. 2002). Hypoxia (3% O_2 and 5% CO_2, balanced with N_2) significantly increased DNA synthesis and proliferative responses to platelet-derived growth factor and fibroblast growth factor in *rat* and *human* muscle and endothelial cells. In an *in vitro* 3-dimensional model of angiogenesis, hypoxia increased platelet-derived growth factor- and fibroblast growth factor-stimulated sprout formation from *rat* and *mouse* aortas. Hypoxia did not modulate platelet-derived growth factor receptor mRNA, protein, or phosphorylation. PI3K activity was essential was essential for cell proliferation under normoxic and hypoxic conditions. Activities of PI3K-downstream target PKB under hypoxia and normoxia were comparable. However, mammalian target of rapamycin inhibition by rapamycin specifically abrogated hypoxia-mediated amplification of proliferation and angiogenesis, but was without effect on proliferation under normoxia. Accordingly, hypoxia-mediated amplification of proliferation was further augmented in mammalian target of rapamycin-overexpressing endothelial cells.

FK228 (a natural cyclic depsipeptide also known as FR901228 or NSC 630176), a specific histone deacetylase inhibitor, inhibited the induction and activity of the hypoxia-inducible factor-1 (HIF-1) in response to hypoxia (LEE et al. 2003). Moreover, FK228 significantly suppressed the induction of vascular endothelial growth factor under hypoxia, suggesting that FK228 contributes to the inhibition of tumour angiogenesis. In Lewis lung carcinoma model, FK228 also blocked angiogenesis induced by hypoxia.

20.2.2 Invasive Pericytes in Tumour Angiogenesis Assay

The involvement of pericytes in physiological or tumour angiogenesis is a matter of debate. PAPOUTSI et al. (2000) studied the expression of pericyte, smooth muscle cell and matrix markers in experimental tumours of the mammary ductal adenoma MDA-MB231 cell line grown on *chick* or *quail* chorioallantoic membrane. Pericyte-like cells may be attracted by MDA-MB231 cells during tumour angiogenesis but failed to interact properly with endothelial cells in the tumour environment (LAUER et al. 2000).

20.2.3 Inhibitors of Angiogenesis

There is a very large body of evidence that the growth of primary and metastatic tumours is angiogenesis-dependent (PLUDA 1997). The role of angiogenesis was recently reviewed by MATTERN (2001). The absence of angiogenesis factors is associated with tumour dormancy, related to increased apoptosis of tumour cells (HOLMGREN et al. 1995). Tumour cells do not only stimulate vascular endothelial cell growth, but are also involved in the production of some of the most potent inhibitors known today (O'REILLY et al. 1994, 1997). A 38 kDa peptide from a subclone of Lewis lung carcinoma cells (MAZURE et al. 1996), called angiostatin, was shown to be a specific and potent inhibitor of endothelial cell proliferation *in vitro*. A 20 kDa peptide (endostatin) isolated from a *murine* haemangioendothelioma was shown to be an even more potent inhibitor of endothelial proliferation, as well as an inhibitor of angiogenesis and a growth suppressor of various different types of tumours in *mice* (O'REILLY et al. 1997). When the treatment of *mice* carrying Lewis lung carcinomas, fibrosarcomas or melanomas with recombinant endostatin to induce tumour regression was ended after regression and the tumours were allowed to regrow before treatment was resumed, it was shown that this could be repeated for up to six cycles, demonstrating that it is not likely that this type of inhibitor of angiogenesis will develop therapy resistance.

In lung non-small cell carcinomas endothelial cells were immunostained by FVIIIRA, CD31, $\alpha 2$, $\alpha 3$, $\alpha 5$, $\alpha 6$, and ICAM-1; 73% of vessels were distributed in the stroma and the remaining 27% in the parenchyma ($P < 0.01$) (VITOLO et al. 2001). Immunohistology on serial sections demonstrated that basal membranes of all stromal and parenchymal vessels were positive for all extracellular matrix proteins except laminin $\alpha 2$ chain (merosin M chain); 22.8% of stromal vessels displayed basal membranes positive for laminin $\alpha 2$ chain, whereas parenchymal vessels were constantly negative ($P < 0.01$). Moreover, all vessels of normal tissues adjacent to the tumour were negative for laminin $\alpha 2$ chain.

Endostatin decreased VEGF-induced formation of endothelial tubes and microvessels sprouting from aortic rings and blocked their network (ERGÜN et al. 2001). After cessation of treatment, the survival time of endostatin plus VEGF-treated tubes was approximately doubled in comparison to VEGF alone. Endostatin antibody blocked VEGF-induced endothelial tube formation and disrupted existing tubes. The expression of collagen XVIII mRNA was increased in tumour blood vessels. Immunohistochemically, endostatin was localised in newly formed blood vessels but not in quiescent blood vessels. Endostatin immunostaining was localised between endothelium and basement membrane and in inter-endothelial junctions of new, but not of quiescent blood vessels.

Since integrin $\alpha_v\beta_3$ is not expressed in quiescent vessels (BROOKS et al. 1994), blocking this key molecule might be a good method to inhibit tumour-induced angiogenesis without damaging pre-existing vessels. KAWAGUCHI et al. (2001) synthesized a new cyclic Arg-Gly-Asp pentapeptide cyclo (-RGDf=V-) and its analogues, cyclo(-RGDfV-) and cyclo(-RGDf-MeV-) (see formula [341]). All three compounds inhibited the adhesion and growth of *human* umbilical endothelial cells in a dose-dependent manner *in vitro*. However, *in vivo*, cyclo(-RGDf=V-) significantly decreased the intratumoral microvessel density in DLD-1 (*human* colon cancer cell) inoculated *mice*, while cyclo (-RGDf-MeV-) had little effect.

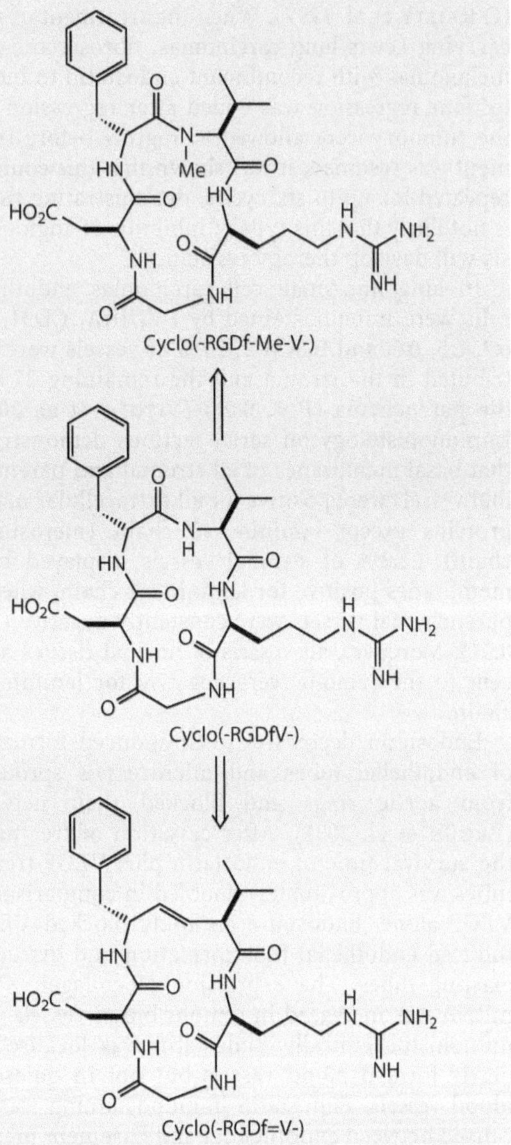

Chemical structure of synthesized cyclo(RGDf=V-) and its analogues [338]

Anti-angiogenic therapies are being developed not only for cancer, but for other pathologies characterised by a neovascular component, including ophthalmic, rheumatoid, paediatric, and AIDS conditions. More than 300 endogenous, natural or synthetic inhibitors of angiogenesis have been described and 31 agents have entered clinical trials at over 140 oncology centres in North America and Europe. The best known at present are interferon-α, TNP-470 (AGM-1470; Masiero et al. 1997), thalidomide (Masiero et al. 1997), CM101 (vascular targeting agent; Quinn et al 1995, DeVore et al. 1997, Harris 1997), Bay 12-9566 (matrix metalloproteinase inhibitor; Ehrlichmann et al. 1998), and Marimastat (metalloproteinase inhibitor).

Arginine deiminase showed anti-angiogenic activity in *human* umbilical vein endothelial cell cultures (Beloussow et al. 2002). The cells in untreated control wells differentiated in morphology, elongating to form networks of capillary-like tubes and loss of area covered by the endothelial cell monolayer, whereas the treated wells had decreased tube formation resulting in sporadic and incomplete networks.

Copper ions required for tumour angiogenesis may be chelated in two tridentate, hydrazone-copper complexes, pyridine-2-carboxaldehyde-2'-pyridylhydrazonato-Cu(II) and salicylaldehydebenzoyl-hydrazonato-Cu(II), which inhibited the growth of an implanted methylcholanthrene-induced fibrosarcoma (MCA 1511) in Balb/C *mice* (Pickart et al. 1983).

20.2.4 Tumour Cell Adhesion to Postcapillary Venules

Adhesion of circulating tumour cells to microvascular endothelium plays an important role in tumour metastasis to distant organs.

Although venules exposed to lipopolysaccharide for 4 h demonstrated an increased adhesivity for RPMI 1846 *hamster* melanoma cells, tumour cell adhesion to lipopolysaccharide-treated was not altered (Kong et al. 1996). Isolated venules exposed to 1 mM DETA/NO $\{H_2NCH_2CH_2N[N(O)NO]^- CH_2CH_2NH_3^+\}$, an NO donor, for 30 min prior to tumour cell perfusion prevented the increment in adhesion induced by lipopolysaccharide and attenuated tumour cell adhesion to naive postcapillary venules.

20.2.5 Lymphatic Vessels of Tumours

Recent studies indicate, that some tumours contain lymphatic vessels, as well as channels that consist of cancer cells and their extracellular matrix. Immunohistochemical analysis of VEGFR3, LEVE-1 and podoplanin expression, as well as that of other lymphatic proteins, has shown the presence of lymphatic vessels in tissues that were previously thought to lack them – both experimental and *human* (Jackson et al. 2001).

Experimental and clinical data strongly indicate that the number, and perhaps the size, of lymphatics in a tumour or around it is an important determinant in the ability of a tumour to metastasise.

CHAPTER 21

Antitumour Drug Therapy

The overexpression of P_{450}s in tumours has potential therapeutic implications. Several P_{450} enzymes are involved in the activation and/or deactivation of a range of anti-cancer drugs in current clinical practice (Kivistö et al. 1995). The tumour-specific expression of CYP1B1 provides a therapeutic target for the development of anti-cancer drugs specifically activated by this P_{450}, or it could be used as a tumour-specific antigen for the development of immunotherapy. Another therapeutic approach to exploiting the presence of specific P_{450} enzymes in tumours is to take advantage of the relative hypoxia of the tumour microenvironment. The drug AQ4N is a topoisomerase II inhibitor which is selectively activated by CYP3A in hypoxic conditions to its active moiety (Raleigh et al. 1998). In normo-oxic conditions, this activation does not occur.

To investigate the importance of NAD(P)H:quinone oxidoreductase 1 (or DT-diaphorase; NQO1) in the bioactivation of antitumour quinones, Winski et al. (2001) established a series of stably transfected cell lines derived from BE *human* colon adenocarcinoma cells. BE cells have no NQO1 activity due to a genetic polymorphism. The new cell lines, BE-NQ, stably express wild-type NQO1. BE-NQ7 cells expressed the highest level of NQO1 and were more susceptible to known antitumour quinones and newer clinical candidates. Inhibition of NQO1 by pre-treatment with an irreversible inhibitor, ES936 {5-methoxy-1,2-dimethyl-3-[(4-nitrophenoxy)methyl]indole-4,7-dione}, protected BE-NQ7 cells from toxicity by streptonigrin, ES921 [5-(aziridin-1-yl)-3-(hydroxymethyl)-1,2-methylindole-4,7-dione], and RH1 [2,5-diaziridinyl-3-(hydroxymethyl)-6-methyl-1,4-benzoquinone]. Cytotoxicity was abrogated by inhibition of NQO1 with ES936 pre-treatment. Using a comet assay to evaluate DNA cross-linking, BE-NQ7 cells demonstrated significantly higher DNA cross-links than did BE cells in response to RH1 treatment. DNA cross-linking in BE-NQ7 cells was observed at very low concentrations of RH1 (5 nM), confirming that NQO1 activates RH1 to a potent cross-linking species.

21.1 Free Radicals-Forming Drugs

Of the numerous anticancer drugs, eight have been reported to form free radicals, namely doxorubicin (= adriamycin = NSC-123127), daunorubicin (= daunomycin = NSC-83142), mitoxantrone (= NSC-301379), bleomycin (= NSC-125066), neocarzinostatin (= NSC-157365), mitomycin C (= NSC-26980), actinomycin D (= NSC-3053), and procarbazine (= NSC-77213).

$$MeNHNHCH_2{-}C_6H_4{-}C(=O)NHPr^i{-}HCl$$

Procarbazine (= NSC-77213) [339]

All three isoforms of nitric oxide synthase (EC 1.14.13.39) exhibited high levels of activity towards tirapazamine, doxorubicin, and menadione (Garner et al. 1999).

21.1.1 Anthracyclines

Doxorubicin known for its cardiotoxicity due to the formation of reactive oxygen species (Bachur et al. 1977, Doroshow 1983, Mimnaugh et al. 1983, Rajagopalan et al. 1988, Thayer 1990, Brehm et al. 1995, Coudray et al. 1996) was demonstrated by fluorescence microscopy in the nuclei of the neurohypophysis and the pituitary stalk of *mice* 15 s to 1 h after intravenous injection (Bigotte et al. 1982, Bigotte de Almeida 1983). In the ependymal zone of the eminentia mediana formed by the tanycyte ependyma almost all the cell nuclei were labelled. The palisade zone also contained many fluorescent nuclei. In animals with survival times shorter than 1 min the neuropil was diffusely stained. Toseland et al. (1996) studied doxorubicin in *chick* embryonic myocardial myocyte reaggregate cultures for up to 24 h. At 5 μM there were nuclear, cytoplasmic and mitochondrial changes: Numerous small focal deposits of heterochromatin were scattered throughout the nucleus. In the cytoplasm there were clear areas containing numerous short unorganised microtubular structures. The mitochondria of the myocytes and, to a lesser extent, of the capsular fibroblastic cells showed an increase in electron density and in many cases disorientation and loss of cristae. When compared with other contractile protein and muscle-specific mRNAs, α cardiac actin mRNA abundance in the *rat* heart was selectively decreased by doxorubicin, daunorubicin and epirubicin (PAPOIAN and LEWIS 1992). Ultrastructural examination of myocardium showed contractile alterations, including a marked loss and disarray of the thin and thick myofilaments, as well as fragmented remnants of the Z-bands. The sarcomeric myofibrillar architecture was disorganised and separated with swollen intermyofibrillar spaces as a result of interstitial oedema. The arrangement and orientation of mitochondria between the myofibrils appeared disorganised, whereas the mitochondrial cristae in the anthracycline-treated myocytes appeared normal and dense as those in the controls.

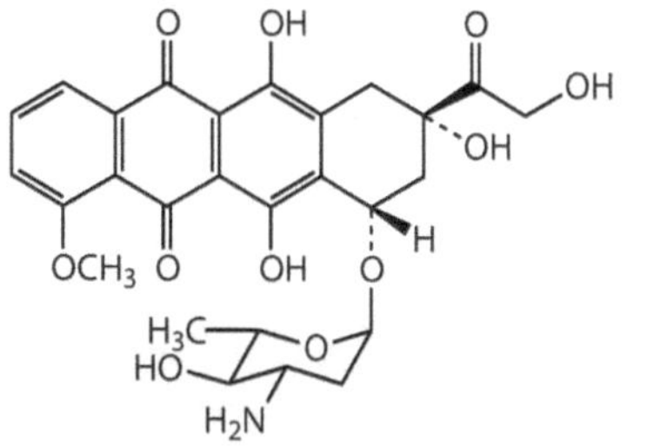

Epirubicin [340]

Table 58. Cytotoxic mechanisms of anthracyclines (MÜLLER et al. 1998)

Inhibition of enzymes: topoisomerase II
Intercalation into DNA
Chelation of iron and generation of reactive oxygen species
Induction of apoptosis

Doxorubicin treatment induced disruption of inner mitochondrial membrane potential $\Delta\psi_m$ that precedes the nuclear signs of apoptosis (DECAUDIN et al. 1997). Both loss of $\Delta\psi_m$ and apoptosis were prevented by Bcl-2 overexpression (ANTOKU et al. 1997, DECAUDIN et al. 1997), suggesting an important role for mitochondria in doxorubicin-induced apoptosis. Western blot analysis of Jurkat cell extracts indicated that caspases 2,3,4,6,7,8,9, and 10 were activated by doxorubicin (GAMEN et al. 2000). Doxorubicin cytotoxicity was blocked by the protein synthesis inhibitor cycloheximide.

The expression of mRNA encoding sarco(endo)-plasmic reticulum Ca^{2+}-ATPase 2, a major Ca^{2+} transport protein in sarcoplasmic reticulum, is markedly decreased in doxorubicin-treated *rabbit* hearts (ARAI et al. 1998). Using cultured *rat* neonatal cardiac myocytes, ARAI et al. (2000) found that the antioxidant *N*-acetylcysteine blocked the doxorubicin-induced decrease in sarco(endo)-plasmic reticulum Ca^{2+}-ATPase 2 mRNA levels, as well as the doxorubicin-induced increase in H_2O_2 concentration; thus H_2O_2 is an intracellular mediator of doxorubicin activity. Using a luciferase reporter assay, ARAI et al. (l.c.) found that the sequence from −284 to −72 bp in the 5' flanking region of sarco(endo)plasmic reticulum Ca^{2+}-ATPase 2 gene has a doxorubicin-responsive element.

Treatment of *rat* hepatocytes isolated by *in vitro* collagenase perfusion with doxorubicin increased catalase but decreased Mn superoxide dismutase mRNA expression (RÖHRDANZ et al. 2000). Doxorubicin almost completely inhibited RNA synthesis and induced lipid peroxidation as measured by the accumulation of malondialdehyde in the medium.

In *rabbit* artery rings denuded of the endothelium and cultured with 0.3 μM doxorubicin for 7 days, the contractions induced by noradrenaline, but not those induced by endothelin-1 or high K^+, were strongly inhibited (MURATA et al. 2001). This reaction was followed by a decrease in the induction of the α_{1A}-adrenoceptor without any change in the mRNA level. Inhibition of noradrenaline-induced contractions by doxorubicin was attenuated by superoxide dismutase (EC 1.15.1.1), and α_{1A}-adrenoceptor protein expression recovered.

Superoxide dismutase (EC 1.15.1.1)

$$2\,O_2^{\cdot-} + 2\,H^+ \longrightarrow H_2O_2 + {}^3\Sigma_g^- O_2 \qquad [128]$$

In spontaneously hypertensive *rats*, the nephropathy and the intestinal toxicity produced by 1 mg doxorubicin per kg × week were more severe than those resulting from 0.25 mg (therapeutically equivalent dose) or 0.5 mg mitoxantrone per kg × week (HERMAN et al. 1997). Apoptosis of cardiac myocytes was not induced by either drug, but involved cardiac dendritic cells in spontaneously hypertensive *rats* given doxorubicin. Apoptosis in renal tubular epithelium was comparable in spontaneously hypertensive *rats* given doxorubicin and the higher dose of mitoxantrone. Doxorubicin induced more frequent apoptosis in intestinal epithelium than did the higher dose of mitoxantrone.

Doxorubicin and daunorubicin formed two well-defined species with Fe(III), which can be formulated as $Fe(HAd)_3$ and $Fe(HDr)_3$, respectively (BERALDO et al. 1985). In these formulas, HAd and HDr stand for adriamycin and daunorubicin in which the 1,4-dihydroxy-anthraquinone moiety is

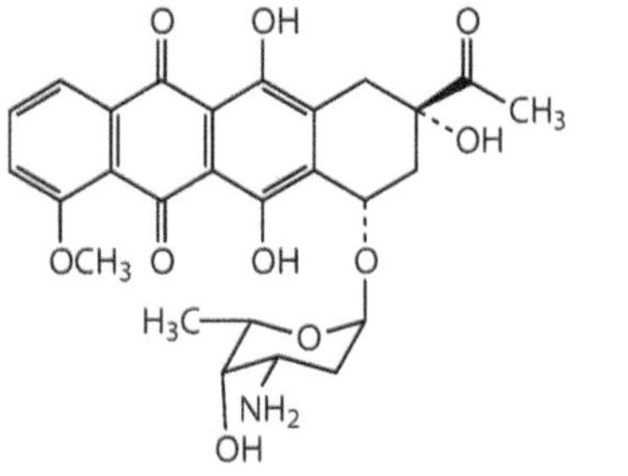

Daunorubicin [341]

half-deprotonated. Both complexes are six-membered chelates. The stability constant is $\beta = (2.5 \pm 0.5) \times 10^{28}$ for both complexes. Interaction with DNA showed that, despite strong coordination to Fe(III), anthracyclines are able to intercalate between DNA base pairs, releasing the metal. These complexes displayed antitumour activity against P 388 leukaemia that compares with that of the free drug. $Fe(HAd)_3$, unlike adriamycin, does nor catalyse the flow of electrons from NADH to molecular oxygen through NADH dehydrogenase. The triferric adriamycin compound so called „quelamycin“ is in fact a mixture of $Fe(HAd)_3$ and polymeric ferric hydroxide.

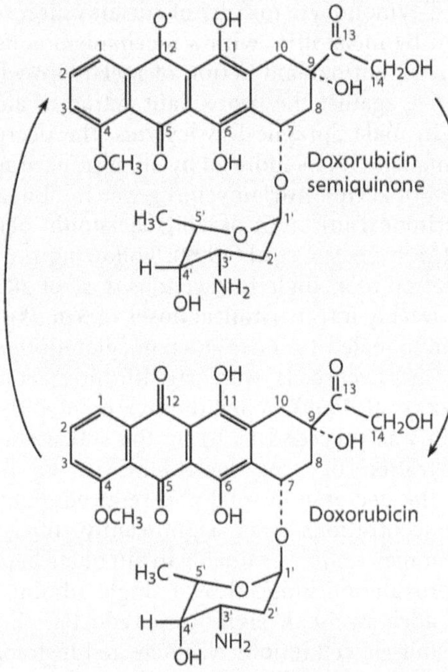

Doxorubicin and its semiquinone radical [342]

Addition of cupric sulphate to neutral solutions of doxorubicin resulted in spectrophotometric, fluorometric, and chromatographic changes indicative of a direct chemical interaction (Wallace 1986). Associated with these changes was a copper-dependent consumption of dissolved oxygen and a superoxide dismutase-sensitive reduction of ferricytochrome *c*, suggesting the liberation of superoxide free radicals ($O_2^{\bullet-}$). Oxygen free radical formation by the drug-copper complex was further implicated by the stimulation of lipid peroxidation, which was completely inhibited by adding ethylenediaminetetraacetic acid. Inhibition by superoxide dismutase (EC 1.15.1.1), catalase (EC 1.11.1.6), and dimethyl urea implicates the involvement of assorted oxygen free radicals in doxorubicin-copper stimulated lipid peroxidation.

The level of formaldehyde above background in MCF-7 breast cancer cell lysates was a function of anthracycline drug concentration (0.5 - 50 μM), treatment time 3 - 24 h), cell density (03. $\times 10^6$ to 7 $\times 10^6$ cells per ml), and cell viability (0 - 100%) (Kato et al. 2000). Higher levels of formaldehyde were observed in lysates of MCF-7 cells treated at higher drug levels, unless the treatment resulted in low cell viability.

Xanthine dehydrogenase (EC 1.1.1.204), the enzymatic precursor of xanthine oxidase (EC 1.1.3.22) reacts with doxorubicin via a two-electron reduction (Yee and Pritsos 1997). This reduction is different from the modified and more extensively studied form xanthine oxidase, which reacts with doxorubicin via a one-electron reduction. Under hypoxic conditions, the formation of large quantities of 7-deoxydoxorubicin aglycone, a deactivation product of doxorubicin metabolism, may serve to moderate the antineoplastic activity of doxorubicin. Under aerobic conditions, however, xanthine dehydrogenase activation led to a greater rate of formation of oxygen radicals than xanthine oxidase thereby possibly potentiating the cytotoxicity of doxorubicin to aerobic tumour cells.

Time resolved near-infrared singlet oxygen luminescence measurements on doxorubicin, adriamycin and 5-iminodaunomycin in ^2H_2O solution and daunomycin in benzene solution at room temperature showed oxygen quenching by the water-soluble anthracyclines and a second-order rate constant of approximately 10^8 $M^{-1} \cdot s^{-1}$ (Andreoni et al. 1989). Electron spin resonance experiments demonstrated that daunomycin photoexcited at $\lambda < 365$ nm gives rise to singlet oxygen as shown by its reaction with 2,2,6,6-tetramethyl-4-piperidone to give the corresponding nitroxyl radical. Pedersen et al (1990) studied the generation of the daunomycin semiquinone in intact erythrocytes under CO atmosphere by ESR spectroscopy. The undialyzed haemolysates and the spin broadening agent chromium oxalate quenched the ESR signal, suggesting external location of the ESR-detectable radicals and their slow diffusion inside. A constant outflow of $O_2^{\bullet-}$ was detected by monitoring the approach to the steady state of the ESR signal of Cu,Zn superoxide dismutase externally added to red blood cells plus daunomycin in air.

Nitric oxide synthase (EC 1.14.13.39) incubated with both doxorubicin and menadione augmented the rate of NADPH oxidation (GARNER et al. 1999). Kinetic parameters by Lineweaver-Burk analysis showed that NOS III had the highest affinity for both doxorubicin and menadione with K_ms of 40.6 and 22.7 μM, respectively. NOS II had the lowest affinity for both compounds with K_ms of 210.5 and 44.1 μM, respectively.

While induction of metallothionein synthesis has been shown to protect organs from the toxic side effects of several anticancer drugs (IMURA et al. 1992), tumours with elevated metallothionein levels showed resistance to these drugs (BASU and LAZO 1990). Resistance to anticancer drugs may be modulated by inhibition of metallothionein synthesis (SATOH et al. 1994). Injection of propargylglycine, an inhibitor of cystathionase, decreased metallothionein induction by 200 μmol $ZnSO_4$/kg in *human* bladder tumour inoculated into ICR nude *mice* and diminished its resistance to adriamycin (20 μmol/kg), cisplatin (40 μmol/kg) and melphalan (60 μmol/kg). At 0.1 μM metallothionein inhibited the DNA cleavage induced by the copper-1,10-phenanthroline complex by about 50% (YANG et al. 2000). At 2.5 μM the cleavage activity was completely inhibited.

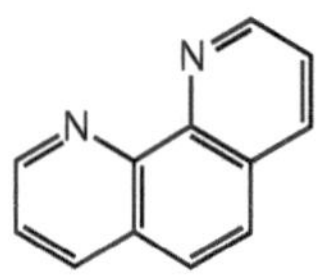

1,10-Phenanthroline [137]

KONOREV et al. (2002) evaluated the pro- and antiapoptotic potential of different metalloporphyrins containing iron, cobalt, zinc, and manganese in adult *rat* cardiomyocytes exposed to doxorubicin. They used electron spin resonance/spin trapping and cytochrome *c* reduction to assess the scavenging of superoxide anion by metalloporphyrins. Superoxide anion was effectively scavenged by 5,10,15,20-tetrakis(benzoic acid)porphyrin iron(III) and 5,10,15,20-tetrakis(benzoic acid)porphyrin manganese(III), but not by 5,10,15,20-tetrakis(benzoic acid)porphyrin cobalt(III) and 5,10,15,20-tetrakis(benzoic acid)porphyrin zinc(II). 5,10,15,20-Tetrakis(benzoic acid)porphyrin iron(III) efficiently scavenged H_2O_2. Both 5,10,15,20-tetrakis(benzoic acid)porphyrin cobalt(III) and 5,10,15,20-tetrakis(benzoic acid)porphyrin iron(III) inhibited doxorubicin-induced cardiomyocyte apoptosis.

An adriamycin-resistant MCF-7 *human* breast tumour cell line showed a developed tolerance to superoxide, most likely because of a twofold increase in superoxide dismutase activity, and a decreased susceptibility to hydrogen peroxide, most likely because of 12-fold augmented selenium-dependent glutathione peroxidase activity (MIMNAUGH et al. 1989). α-Tocopherol in the membrane and cytosolic fractions was 2.8 and 3.0 fold higher, respectively, in Adriamycin-resistant compared with Adriamycin-sensitive cells (WELLS et al. 1995). Supplementation of MCF-7 cells with L-ascorbic acid 2-phosphate (2 and 10 mM) had no effect on Adriamycin-sensitive cell viability after 5 days incubation with up to 0.33 μM Adriamycin. In contrast, supplementation of ADR^R MCF-7 cells with L-ascorbic acid 2-phosphate resulted in enhanced resistance up to 3.4 μM Adriamycin over a 5-day incubation.

Melatonin decreased bone marrow and lymphatic toxicities of adriamycin in CBA *mice* bearing TLX5 lymphoma (RAPOZZI et al. 1998). At the same time, melatonin did not decrease the antitumour action of adriamycin, and significantly reduced the acute host toxicity of this drug, suggesting that an enhancement in the dose intensity of adriamycin can be achieved by its combined administration with exogenous melatonin. Bone marrow and splenic T-lymphocyte toxicity of adriamycin was attenuated by melatonin, with a mechanism consistent with an antioxidant action of melatonin which is effective against the prooxidant action of adriamycin. In male Sprague-Dawley *rats*, the decrease in plasma zinc levels induced by six intraperitoneal injection of 2.5 mg Adriamycin/kg was inhibited by intraperitoneal injection of 4 mg melatonin per kg × day (MORISHIMA et al. 1999). Following the administration of a single intravenous dose of 28 mg/kg or 3 weekly intraperitoneal doses of 5 mg/kg adriamycin to male CBA *mice* at 8 p.m., glutathione levels in the liver cells were significantly reduced (RAPOZZI et al. 1999). When the treatment with adriamycin was preceded (1 h) by the subcutaneous administration of 2 mg melatonin per kg body weight, the decrease in total and reduced glutathione concentrations was significantly obviated. However, melatonin was unable to attenuate hepatic lipid peroxidation induced by a single administration of adriamycin. A significant reduction in *rat* cardiac muscle cell lesions was detected histologically either by the Billingham scale (BILLINGHAM 1991) or by the mean total score technique (HERMAN and FERRANS 1993, PODESTA et al. 1994) during subchronic intoxication with either daunorubicin or doxorubicin when melatonin (10 mg/kg body weight) was given (DZIĘGIEL et al. 2002). Biochemical assays revealed significant decreases in malonyldialdehyde and 4-hydroxyalkenals levels following application of melatonin during acute doxorubicin ($P < 0.05$) or subchronic daunorubicin ($P < 0.01$) intoxication.

L-Carnitine attenuated doxorubicin-induced lipid peroxidation in male Wistar *rats* (Luo et al. 1999). *In vitro*, L-propionylcarnitine showed a dose-dependent superoxide anion-scavenging activity, inhibited peroxidation of linoleic acid, and protected pBR322 DNA from cleavage induced by H_2O_2 UV-photolysis (Vanella et al. 2000).

Vitamin A inhibited doxorubicin-induced membrane lipid peroxidation in *rat* tissue *in vivo* (Ciaccio et al. 1993). Brain and heart membrane preparations from *rats* receiving vitamin A, assayed *in vitro* in the presence of a Fe^{3+} ascorbate induction system, showed a delay in the beginning of the lipid peroxidation and generated lesser amounts of thiobarbituric acid reactive substances, with respect in membranes from control *rats*.

Thymoquinone (10 mg kg^{-1} day^{-1}, p.o.) with drinking water starting 5 days before a single injection of doxorubicin (15 mg kg^{-1}, i.p.) and continuing over the experimental period alleviated doxorubicin-induced cardiotoxicity in male albino *rats* (Nagi and Mansour 2000). Thymoquinone proved to be a potent $O_2^{\bullet-}$ scavenger which was as effective as superoxide dismutase. In addition, thymoquinone had an inhibitory effect on lipid peroxidation induced be Fe^{3+}/ascorbate using *rat* heart homogenate.

(±)-1,2-Bis(3,5-dioxopiperazinyl-1-yl)propane (ICRF-187) protected against doxorubicin cardiotoxicity in BALB/c *mice* acutely (24-h exposure) or chronically (13-week exposure) conclusively demonstrated by electron microscopy (Alderton et al. 1990).

Only 4 synthetic flavonoids proved as potent protectors against doxorubicin-induced cardiotoxicity tested in the isolated *mouse* left atrium (Van Acker et al. 2001). Because of the positive inotropic effect of *N*-[3-(3',4'-dihydroxyflavon-7-yl)oxypropyl]-*N*,*N*,*N*-trimethylammonium chloride and 3-hydroxyethoxy-7,3',4'-trihydroxyflavone on the atrium, 3',4'-dihydroxy-3-glucosylflavone and *N*-[3-(7,3',4'-trihydroxyflavon-3-yl)oxypropyl]-*N*,*N*,*N*-trimethylammonium chloride were selected to be evaluated as cardioprotective agents *in vivo*.

The extract of *Ginkgo biloba* leaves (EGb 761) inhibited both doxorubicin-induced lipid peroxidation of rat brain homogenates and doxorubicin-induced lethality in *mice* (Chatterjee and Gabard 1982). From the leaves of this eucaryote Duke and Salin (1985) isolated an iron-containing, cyanide-insensitive superoxide dismutase (EC 1.15.1.1) exhibiting a sensitivity to hydrogen peroxide. A very faint cuprozinc superoxide dismutase activity band was noted in green leaves, which contain fewer mitochondria and more chloroplasts, from which this isoenzyme of superoxide dismutase is absent.

Using a cardiac microdialysis technique, Kawada et al. (2000) measured dialysate norepinephrine and epinephrine concentrations as induces of myocardial interstitial norepinephrine and epinephrine levels, respectively, in *rabbits* with chronic Adriamycin treatment (4 mg per kg body weight × week for 6 weeks) and in control *rabbits*. Exocytotic release was evoked by the local administration of KCl (100 mM) through the dialysis probe. Basal levels of norepinephrine and epinephrine did not differ between the Adriamycin and control groups. The exocytotic release was suppressed in the Adriamycin compared with the control group.

Prenylamine (200 mg per day) prevented adriamycin-induced cardiotoxicity (Milei et al. 1985, 1987).

Carvedilol increased doxorubicin cytotoxicity by inhibiting P-glycoprotein activity (Jonsson et al. 1999). Verapamil (10 μmol/l), and even more markedly carvedilol (10 μmol/l) increased cellular uptake of P-glycoprotein-transported calcein of a P-glycoprotein-expressing breast cancer cell line (Hs578T-Dox). In the subline (Hs578T) not expressing P-glycoprotein, no effects of carvedilol or verapamil on calcein uptake were seen. Carvedilol and verapamil (10 μmol/l) reduced the LD_{50} of the Hs578T-Dox subline from 200 mg/l to approx. 10 mg/l Dox, whereas the LD_{50} of the Hs578T subline was only marginally affected. Carvedilol (10 μmol/l) reduced P-glycoprotein activity approximately twice as effectively as verapamil at an equimolar concentration. Carvedilol did not affect pyrogallol cytotoxicity and pyrogallol was without effect on calcein accumulation of the Hs578T-Dox cell line, indicating the lack of antioxidative properties affecting P-glycoprotein activity and associated toxicity of the drug.

The newer anthracyclines, both **epirubicin** and **idarubicin**, profoundly inhibited *human* polymorphonuclear superoxide generation ($P < 0.02$ and ($P < 0.01$), respectively (Cairo et al. 1990). Epirubicin treatment of FM3A *murine* breast tumour cells resulted in autophagocytosis of secretory granules (Bilir et al. 2001). Combining medroxyprogesterone with epirubicin enhanced the formation of apoptotic blebs and chromatin fragmentation. When epirubicin was combined with tamoxifen, peculiar nuclear structures were formed.

Idarubicin at concentrations from the range of 0.001 to 10 μM induced DNA damage in normal *human* lymphocytes, measured as the increase in percentage of DNA in the tail of the comet assay in a

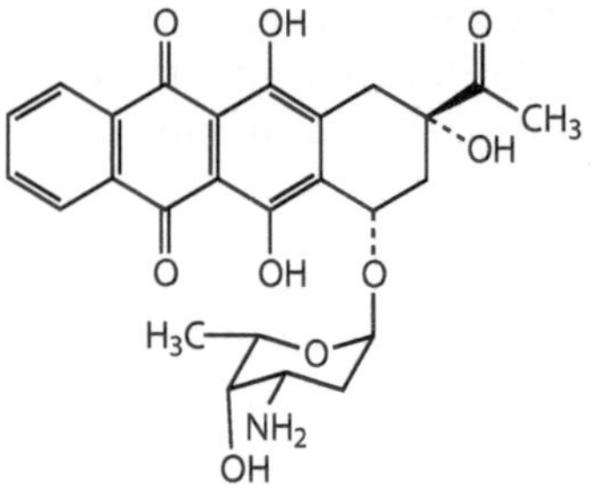

Idarubicin [343]

dose-dependent manner (BŁASIAK et al. 2002). Treated cells were able to recover within a 120-min incubation. Recognised cell protector, amifostine at 14 mM decreased the mean % tail DNA of the cells exposed to idarubicin at all tested concentrations of the drug. So did ascorbic acid at 10 μM, but α-tocopherol at 50 μM increased the % tail DNA. Pretreatment of lymphocytes with nitrone spin traps, *N-tert*-butyl-α-phenylnitrone and α-(4-pyridil-1-oxide)-*N-tert*-butylnitrone decreased the extent of DNA damageevoked by idarubicin.

21.1.2 Further Intercalating Agents

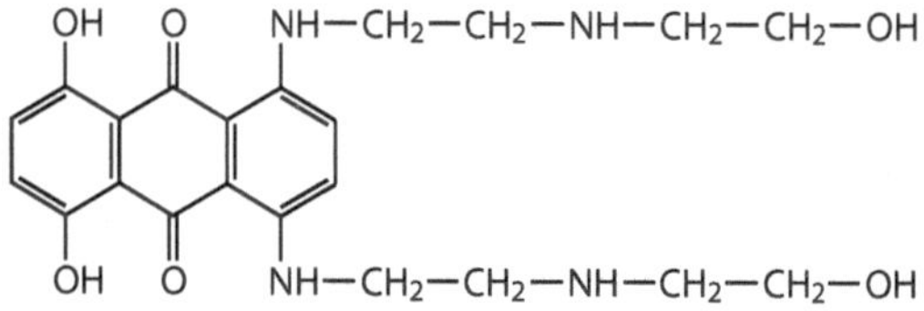

Mitoxantrone (dihydroxyanthracenedione, NSC-301379) [344]

Mitoxantrone exerts cardiotoxic effects that are similar to those of anthracyclines (UNVERFERTH et al. 1983, POWIS 1991, BENJAMIN 1995). HERMAN et al. (1997) showed that mitoxantrone and iron(III) form a strong 2:1 complex, in which the drug may be acting as a tridentate ligand. This complex, like the iron(III)-doxorubicin complex, may be capable of redox cycling and produce reactive oxygen intermediates that damage tissue.

21.1.3 Bleomycin

The biochemical basis for bleomycin cytotoxicity and pulmonary toxicity may involve the generation of reactive oxygen species by a complex of bleomycin with iron. This complex has been shown to catalyse the formation of $O_2^{\bullet -}$ and $HO^{\bullet}$ capable of causing DNA strand excision (SAUSVILLE et al. 1978) and lipid peroxidation (KAMEDA et al. 1979).

Interference with the formation of the bleomycin-iron complex may be the most effective way of controlling bleomycin pulmonary toxicity, a fibrosis model described in *rats* and *mice* (KISSLER 1983). Daily injections of desferrioxamine for a period of weeks after *hamsters* were exposed to bleomycin led to a slight reduction in lung collagen and late phase fibrosis (CHANDLER et al. 1988). In the *rat*, however, desferrioxamine did not inhibit bleomycin-induced lung damage (WARD et al. 1978). In the C57BL/6 *mouse* strain, which is more sensitive than most other strains to the pneumotixic effect of bleomycin, ICRF-187 undergoing intracellular hydrolysis to an open ring derivative, ADR-925, a chelator that removes iron from the iron-bleomycin complex (HASINOFF, unpublished, cited from HERMAN and FERRANS 1995) and more rapidly from the iron-doxorubicin complex (HASINOFF 1989).

BELPERIO et al. (2002) demonstrated that interleukin-13 and a novel CC chemokine, C10 were elevated in the pathogenesis of bleomycin-induced pulmonary fibrosis. Neutralization of IL-13, but not of IL-4, attenuated bleomycin-induced pulmonary fibrosis and levels of C10, suggesting that IL-13 has an important role in the development of pulmonary fibrosis. IL-13 is a potent inducer of C10 *in vivo*, and neutralization attenuated bleomycin-induced pulmonary fibrosis and intrapulmonary macrophage numbers.

Macrolide (clarithromycin, roxithromycin) pretreatment of *mice* inhibited an acute lung injury, which significantly ameliorated the bleomycin-induced increases in the total cell and neutrophil counts in bronchoalveolar lavage fluids and the wet lung weight (KAWASHIMA et al. 2002).

The pulmonary toxic changes produced by antineoplastic agents show a structure of morphological and clinical manifestations, including noncardiogenic pulmonary oedema and pneumonitis/fibrosis. Hypersensitivity reactions may be associated with eosinophilic infiltration and usually lead to little residual damage.

Morphometric methods showed that low temperature (4 °C for 7 days) inhibited bleomycin lung toxicity in the *rat* (BEREND 1983).

The catalytic antioxidant porphyrin manganese (III) tetrakis (4-benzoic acid) porphyrin (5 mg/kg i.p. twice daily) attenuated the severe fibrotic response of the *mouse* lung induced by 3.2 U/kg bleomycin given intratracheally (OURY et al. 2001).

The matrix proteinase inhibitor batimastat (30 mg/kg) significantly reduced bleomycin-induced lung fibrosis, as shown in the *mouse* lung by histopathological examination and by a decrease in hydroxyproline levels (CORBEL et al. 2001). Batimastat also prevented the increase in bronchoalveolar lavage macrophage and lymphocyte numbers, whereas it did not show any effect on the increased expression of active transforming growth factor-β in bronchoalveolar lavage fluid. Batimastat treatment was effective in reducing matrix metalloproteinase-2 and -9 activities as well as the tissue inhibitor of metalloproteinase-1 level in bronchoalveolar lavage fluid.

Older patients appear to be more sensitive to the pulmonary toxic effects of bleomycin (Cooper et al. 1986). This information has been derived from multiple epidemiological studies and appears to be independent of a cumulative dose effect (Ginsberg and Comis 1982).

Acute endothelial injury by bleomycin has been shown to enhance the localisation and metastasis of circulating tumour cells. Adamson et al. (1986) C57bl/6 *mice* with a single intravenous dose of bleomycin (120 mg/kg). After 5 days, severe endothelial injury was demonstrated by morphology and by increased levels of protein in lung lavage fluid. When [^{131}I]-iododeoxyuridine labelled syngeneic fibrosarcoma cells were injected intravenously at this time, a 9-fold increase in their localisation was detected 24 h later in bleomycin-treated lungs compared with saline controls. By electron microscopy tumour cells were observed at sites of denuded vascular basement membrane. There was also a significant increase in the number of gross metastases which developed subsequently and in the percentage of lung occupied by tumour in the bleomycin group. Animals examined 10 days after bleomycin showed less endothelial damage and a smaller increase in tumour cell localisation and metastases. At 21 days, when endothelial structure and alveolar protein levels has returned to normal, and at 6 weeks, when there was a focal fibrosis, no increase in tumour cell localisation or metastases was found.

Bleomycin may participate in a radiorecall phenomenon in which previous thoracic irradiation is a risk factor for development of pulmonary toxicity associated with the drug. Samuels et al. (1976) studied 101 patients with advanced testicular carcinoma undergoing high-dose bleomycin therapy for generalised metastases. Twelve of these patients had previously received thoracic irradiation, all within a year of bleomycin therapy, and 5 of these 12 developed physiologic, radiographic, and histologic signs of pulmonary toxicity. Only 4 of 89 patients who had not received previous irradiation developed similar signs of toxicity.

It cannot be re-emphasised too frequently that toxicity need not be an inevitable consequence of cancer chemotherapy. Great advances have been made in reducing toxicity of cancer chemotherapy without compromising therapeutic efficacy.

In *mice*, 20 mg doxorubicin/kg strongly depressed the enzymatic activities of complex I-III and complex IV of the heart mitochondrial respiratory chain. Pre-treatment with 70 mg AD 20 per kg 15 min before the injection of doxorubicin almost completely restored these activities (Praet et al. 1988). A20 is a *N*-acyldehydroalanine compound capable of stabilising the doxorubicin semiquinone radical as a consequence of its capto-dative properties.

Two radical scavenging plant extract preparations (*Ginkgo biloba*, *Crataegus oxyacantha*) showed significant depression of lucigenin-enhanced chemiluminescence due to scavenging of $O_2^{\bullet -}$ (Brehm et al. 1995). Their combination with doxorubicin, however, resulted in the same antitumour (sarcoma 180 cells *in vitro*) activity as observed with doxorubicin alone.

The potential protective effect of the iron chelator, ICRF-187 [Ib(+)-1,2-bis (3,5-dioxopiperazinyl-1-yl) propane], against bleomycin-induced pulmonary toxicity was assessed in *mice* by Herman et al. (1989, 1990).

21.1.4 Further Antibiotics

Mitomycin C, a clinically widely used bioreductive, alkylating, anticancer drug, contains carbinolamine and aziridine moieties, which can be transformed into electrophilic centres capable of alkylating DNA, the presumed target of this drug *in vivo* (Keyes et al. 1985). The reduction of the quinone moiety of mitomycin C has been shown to proceed through a one-electron reduced semiquinone radical intermediate. Under aerobic conditions this semiquinone radical is able to undergo redox-cycling, whereas under anaerobic conditions the one-electron reduced mitomycin C can lead to DNA alkylation (Keyes et al. 1985, Tomasz et al. 1987). Vromans et al. (1990) studied the role of P-450 and NADPH-cytochrome P-450 reductase of rat liver microsomes in the one-electron reduction of mitomycin C.

Mitomycin C markedly increased *mouse* heart microsomal lipid peroxidation, but had little effect when liver microsomes were used (Mimnaugh et al. 1983).

Tempol may reduce the cytotoxicity of mitomycin C. Under hypoxic conditions the one-electron -reduced product of mitomycin did not react with O_2 to produce cytotoxic $O_2^{\bullet -}$, bur reduced tempol to hydroxylamine (Krishna and Samuni 1993). Tempol also can protect against neocarzinostatin-induced DNA damage (De Graff et al. 1992).

7-Hydroxystaurosporine (UCN-01) selectively enhanced mitomycin C cytotoxicity in p53 defective cells which is mediated through S and/or G2 checkpoint abrogation (Sugiyama et al. 2000).

Actinomycin D (= NSC-3053; 1 μg/ml) in foetal *baboon* lung explants exposed to hyperoxia (95%

O_2), inhibited peroxiredoxin I mRNA induction (Das et al. 2001). Peroxiredoxin is an important antioxidant defence enzyme that reduces hydrogen peroxide to molecular oxygen by using reducing equivalents from thioredoxin (Chae et al. 1994).

In an intracisternal dose of 15 μg per kg *rat*, actinomycin D induced cerebellar lesions as degeneration of Purkinje cells, changes in granule cells and a moderate glial oedema (Rohkamm 1977). Synapses revealed only slight damage.

Federicamycin A (= NSC-305264) spontaneously forms an oxidised free radical with electron transfer to O_2 (Hilton et al. 1986). The observed hyperfine structure of this radical is consistent with one-electron oxidation of the quininoid group. After federicamycin A is exposed to O_2, an electron paramagnetic resonance (EPR) signal is observed with axial symmetry with temperature and power saturation behaviour suggestive of $O_2^{\bullet-}$. Spin-trapping EPR studies demonstrated that the drug reduces O_2 to $O_2^{\bullet-}$. and H_2O_2 to $HO^{\bullet}$.

Neocarzinostatin (= NSC-157365) is a protein antibiotic from *Streptomyces carzinostaticus* variant F-41 (Ishida et al. 1965). Reduction of the neocarzinostatin chromophore by thiol generates a free radical species that may be a peroxyl radical or a "crypto" hydroxyl radical (Edo et al. 1980, Sheridan and Gupta 1981). There is, however, no evidence that $O_2^{\bullet-}$ or $HO^{\bullet}$ is responsible for DNA damage by neocarzinostatin (reviewed by Chin and Goldberg 1986). This occurs because the oxygen radical scavengers superoxide dismutase (EC 1.15.1.1), catalase (EC 1.11.1.6), and hydroxyl radical scavengers do not prevent the DNA damage by neocarzinostatin. Despite the inability to identify the critical free radical species it is generally thought that a free radical mechanism is responsible for the antitumour activity of neocarzinostatin (Goldberg 1987). Thiol containing compounds, glutathione, sodium thioglycolate, L-cysteine, and *N*-(2-mercaptopropionyl)-glycine will block the activity of neocarzinostatin *in vivo* (Ito et al. 1985).

In Chinese *hamster* V79 cells, Tempol (4-hydroxy-2,2,6,6-tetramethylpiperidinyloxy) increased survival from 9% to 80% at 60 ng/ml neocarzinostatin and reduced mutation induction by a factor of approximately 3 (DeGraff et al. 1992).

21.1.5 Cisplatin

Platinum co-ordination complexes share the common formula PtA_2X_2 with only the *cis*-isomers displaying an anti-tumour activity. The active X ligands are monodentate anions of intermediate leaving ability, whereas the amine A ligands influence the solubility of the complex. The stereospecificity of the anti-tumour effect is an important property of the Pt complexes. For ligand sphere of clinically studied platinum compounds see Jakupec et al. (2003).

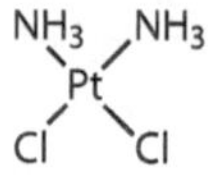

Cis-diamminedichloride platinum, cisplatin [345]

Cis-diamminedichloride platinum, cisplatin, shows a square-planar configuration with a central Pt atom surrounded by Cl atoms and NH_3 in the *cis* position in the horizontal plane. There is evidence that the complex becomes aquated intracellularly to form positively charged Pt species that are likely to interact with nucleophilic sites. The purine and pyrimidine bases of DNA possess such receptor sites and thus provide the locus for effective intracellular binding. The biologically important lesions are most likely to be interstrand and/or intrastrand cross-links (Zwelling and Kohn 1979).

The clinical usefulness of cisplatin is often limited by the inherent cisplatin resistance of certain tumours, and also by the emergence of acquired resistance during treatment. Elevated levels of glutathione and increased activities of glutathione *S*-transferase and glutathione reductase increased resistance to cisplatin and analogues (De Graeff et al. 1988). The amounts of mRNA for glutathione *S*-transferase π were significantly lower in 3 *human* small cell lung cancer cell lines than in 3 non-small lung cancer cell lines (Nakagawa et al. 1988). The sensitivities of the 3 small cell lung cancer cell lines to cisplatin and carboplatin were much higher than those of the 3 non-small lung cancer cell lines. In classic-type small cell lung cancer (SCLO), variant-type SCLO and non-small lung cancer cell (NSLC), the respective mean glutathione *S*-transferase π values were 0.83 ± 0.88, 3.27 ± 2.85 and 2.40 ± 0.76 μg/mg protein (Hida et al 1993). Cell lines not subjected to prior therapy showed a good correlation between glutathione *S*-transferase π levels and chemosensitivity to cisplatin. In cisplatin-resistant *human* leukaemia cells 'GS-X pump' was functionally overexpressed (Ishikawa 1994). Glutathione depletion of K562/B6 and K562/C9 cells by treatment with D,L-buthionine-(*S*,*R*)-sulphoximine, a γ-glutamyl cysteine synthetase inhibitor, restored the susceptibility of cisplatin-resistant chronic myelogenous leukaemia cell lines to natural killer cell-mediated cell death (Dedoussis and Andrikopoulos 2001).

In a cisplatin resistant subline (O-342/DPP) of an intraperitoneally growing transplantable *rat* ovary tumour (O-342), intracellular glutathione (GSH) was approximately doubled (ZELLER et al. 1991). Glutathione reductase (EC 1.6.4.2) activity was higher, although no difference was found for glutathione transferase (EC 2.5.1.18).

KHYNRIAM and PRASAD (2002) studied the effect of cisplatin on five glutathione-related enzymes in liver, kidney, and Dalton lymphoma cells of tumour-bearing *mice*. In liver, the activities of glutathione S-transferase, glutathione peroxidase, catalase, and superoxide dismutase decreased approximately 30-40%, 60-67%, 35-50%, and 70-80%, respectively, while glutathione reductase increased about 36-45% after cisplatin treatment. In kidney, catalase activity decreased by 47-82% at all time points (24-96 h) of cisplatin treatment, while glutathione S-transferase activity decreased significantly (24%) mainly at 72 h of treatment. An increase in glutathione reductase(1.5-2.5 times), glutathione peroxidase (significant at 24 h, 47%) and superoxide dismutase (15-60%) was noted in kidney after the treatment. In Dalton lymphoma cells, the activities of glutathione S-transferase, glutathione peroxidase, and catalase decreased very distinctly (2-5, 2-5 and 5-11 times, respectively) at all time points, but glutathione reductase decreased significantly only at 72 h of cisplatin treatment. Cisplatin treatment caused a decrease in glutathione level in Dalton lymphoma cells (14-20%) and kidney (18-28%) but no change in liver.

Chinese *hamster* V79 lung fibroblast cells pretreated with bolus •NO or •NO delivered from NONOate •NO donors were markedly sensitised to subsequent cisplatin treatment, whereas S-nitrosothiol •NO donors exerted little effect (WINK et al. 1997). The enhancement in cisplatin cytotoxicity from pre-treatment with DEA/NO and PAPA/NO persisted for ~180 and 240 min, respectively; thereafter cytotoxicity returned to a level consistent with cisplatin treatment alone.

The thioredoxin system, composed of nicotinamide adenine dinucleotide phosphate (reduced form), thioredoxin and thioredoxin reductase, may be involved in the cellular sensitivity to *cis*-diamminedichloroplatinum (SASADA et al. 1999). HeLa cells cultured with cisplatin showed a time- and dose-dependent reduction of intracellular thioredoxin reductase activity, which was well correlated with the decrease in cell viability after exposure to cisplatin. In a cell-free system, cisplatin was found to directly inactivate the reduced form of purified *human* thioredoxin reductase. The cisplatin-resistant variants of HeLa cells, established by continuous exposure to cisplatin, exhibited an increased expression and activity of thioredoxin reductase as well as thioredoxin compared with the parental cell. Sodium selenate, an inhibitor of thioredoxin reductase, was found to increase the susceptibility to cisplatin in the cisplatin-resistant cells. The HeLa cells transfected with an antisense thioredoxin reductase RNA expression vector to reduce the intracellular enzyme activity displayed an enhanced sensitivity to cisplatin.

The calcium channel blocker nifedipine enhanced the antitumour action of cisplatin against *murine* tumours which are inherently cisplatin-sensitive (B16a amelanotic melanoma) or inherently cisplatin-resistant (Lewis lung carcinoma; ONODA et al. 1988). In long-term studies, ONODA et al. (1990) reported that combination therapy with nifedipine (10 mg/kg) and cisplatin (4 mg/kg) resulted in a significantly enhanced survival of *mice* injected (10^5 cells) subcutaneously to form primary tumours. Nifedipine alone or cisplatin alone had no significant effect on tumour weight or the number of metastases. Treatment with nifedipine and cisplatin significantly reduced primary tumour ($P < 0.01$) and the incidence of pulmonary metastases ($P < 0.01$) when compared to the cisplatin or nifedipine treated groups or the controls.

Protein kinase C influenced cellular sensitivity to cisplatin. Activators of protein kinase C, such as phorbol 12,13-dibutyrate, enhanced the sensitivity of *human* small cell lung cancer H69 cells to cisplatin by 2-fold but had no effect on the sensitivity of cisplatin-resistant H69 cells to cisplatin (BUSU et al. 1996). The maximum sensitisation was achieved with 10 nM phorbol 12,13-dibutyrate and blocked by down-regulation of protein kinase C with higher concentrations of phorbol 12,13-dibutyrate (1 μM) or bryostatin 1 (0.1 μM). H69 cells expressed conventional protein kinase Cα and -β, novel protein kinase Cδ, atypical protein kinase Cζ and -ι, and novel/atypical protein kinase Cμ. A decrease in conventional protein kinase Cα and -β and an increase in novel protein kinase Cδ were associated with the cisplatin-resistant phenotype.

6-Bromo-6-deoxy-L-ascorbic acid (480 mg per kg body weight) was capable of lowering the toxicity of 10 mg cisplatin per kg *mouse* (ŠVERKO et al. 1999). In *mice* treated with higher doses of cisplatin (15 mg per kg body weight) 6-bromo-6-deoxy-L-ascorbic acid acerbated the toxic effects of cisplatin.

Renal toxicity of high dose *cis*-platinum diammine dichloride was ameliorated by an intravenous infusion of 12.5 g mannitol (50 ml of 25% mannitol) immediately prior to the Pt compound (HAYES et al. 1977). As to survival of Sprague-Dawley *rats* receiving three-weekly intraperitoneal injections of 5 mg *cis*-platinum per kg, catalase (1 mg/kg) appeared to offer some protection (MCGINNESS et al. 1982). Targeting of hexamethylenediamine-conjugated superoxide dismutase to renal proximal tubule cells markedly inhibited the renal injury induced by cisplatin and indicates that $O_2^{\bullet -}$ and/or its hazardous metabolite(s) in and around proximal tubule cells underlie the pathogenesis of the nephrotoxicity of this anticancer agent (NISHIKAWA et al. 2001). α-Tocopherol (50 mg per kg × day) and glutathione (10 mg GSH per kg × day) had no effect on the significant decrease in the activity of lipid peroxidation protecting enzymes induced in the *rat* kidney after cisplatin treatment (SADZUKA et al. 1992).

Lobaplatin [D-19466; 1,2-diamino-methyl-cyclobutane-platinum(II) lactate], showed no nephrotoxic side effects as determined by the measurement of blood urea (VOEGELI et al. 1990). It was cytotoxic *in vitro* for tumour cells in concentrations comparable to or lower than cytotoxic concentrations of cisplatin. It had excellent anticancer activity *in vivo* against a number of *murine* experimental tumours, including a cisplatin-resistant P388 line.

Emesis induced by cisplatin (20 mg/kg i.p.) in the house musk shrew, *Suncus murinus*, was prevented by intraperitoneal injection of *N*-(2-mercaptopropionyl)glycine, a radical scavenging agent, with ID50 value of 130 mg/kg (TORII et al. 1993). Pretreatment with ferric chloride ($FeCl_3$, 16 – 64 mg/kg) increased the number of 20 mg/kg cisplatin-induced vomiting episodes (MATSUKI et al. 1993). Deferoxamine (256 mg/kg) significantly reduced the number of vomiting episodes and prolonged the latency.

Using a short-term ATP bioluminescence assay BOIKE et al. (1990) studied the ability of two methylxanthines (caffeine and pentoxifylline) and an inhibitor of ADP-ribosyltransferase (3-aminobenzamide) to enhance cisplatin cytotoxicity in gynaecological cancer cell lines. Pentoxifylline enhanced cisplatin cytotoxicity in both the *human* cervical cancer cell line ME-180 and the *human* ovarian cancer cell line CAOV-3. Increasing concentrations of pentoxifylline, while relatively nontoxic to both cell lines, produced a concentration-dependent enhancement of cisplatin cytotoxicity.

Allopurinol (50 mg/kg subcutaneously for 5 days) potentiated cisplatin-induced nephrotoxicity in *rats* (ERDINÇ et al. 2000).

Since free platinum intercalates or intracalates in DNA, cisplatin can be cytotoxic towards lung cancer cells, but it can also enhance carcinogenicity in experimental animals (LEOPOLD et al. 1981). (-)-Epigallocatechin gallate can prevent cisplatin-induced lung tumorigenesis in A/J *mice* (MIMOTO et al. 2000).

In polynuclear platinum(II) complexes of spermidine, which showed to have significant cytotoxic and antiproliferative properties on the HeLa cell cancer cell line, the chemical environment of the metal centres of the drug, as well as the coordination pattern of the ligand, were found to be strongly determinant of their cytotoxic ability (MARQUES et al. 2002).

21.1.6
Folic Acid Antagonists

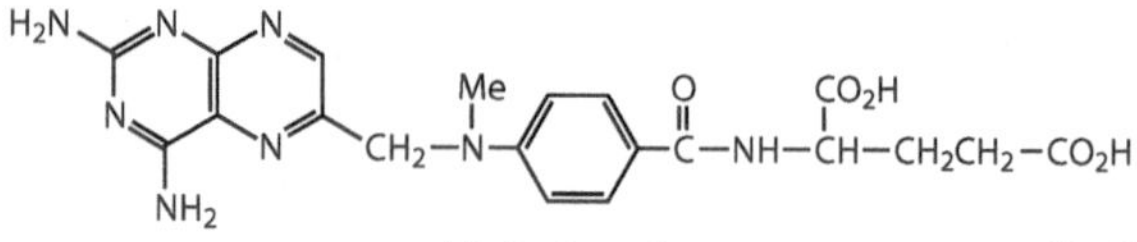

Methotrexate [346]

Methotrexate, a structural analogue to folic acid, inhibited the activities of glutathione reductase (EC 1.6.4.2) and γ-glutamylcystein synthetase (BABIAK et al. 1998) and thus reduced the effectiveness of the antioxidant defence system. Methotrexate induced diffuse interstitial pulmonary fibrosis (KAPLAN and WAITE 1978, BEDROSSIAN et al. 1979). Acute pleuritis, sometimes associated with pleural effusions, is an infrequent complication of high-dose methotrexate therapy (URBAN et al. 1983). Acute respiratory failure due to non-cardiogenic pulmonary oedema has occurred after intrathecal administration of methotrexate (HAMOUS et al. 1983). In the Wistar *rat*, the crypt and villus epithelium associated with Peyer's patches was largely spared from methotrexate (30 mg/kg i.v.)-induced damage, compared with the non-patch epithelium (RENES et al. 2002). Before and after methotrexate treatment, the number of bromo-deoxyuridine-positive cells was higher in Peyer's patch-associated crypts than in non-Peyer's patch-associated crypts. BrdU incorporation was diminished in non-Peyer's patch-associated crypts, while in Peyer's patch-associated crypts incorporation was hardly affected.

21.1.7
Oracin

Oracin, 6-[2-(2-hydroxyethyl)aminoethyl]-5,11-dioxo-5,6-dihydro11*H*-indeno[1,2-*c*]isoquinoline, is a potent antineoplastic agent for oral use. Its biotransformation by *rat* liver microsomes to 3-hydroxyoracin was stimulated by 3-methylcholanthrene-induced cytochrome P4501A, and decreased by the specific P4501A inhibitor α-naphthoflavone (SZOTÁKOVÁ et al. 1999).

21.1.8
Thiosemicarbazones

The class of α-(*N*)-heterocyclic carboxaldehyde thiosemicarbazones includes a large number of compounds that possess significant antineoplastic activity when tested against transplanted animal tumours. Significant regression (38 to 72 per cent) of lymphosarcomatous tumour masses occurred in *dogs* treated with intravenous doses of 2 to 6 mg of the sodium salt of 1-formylisoquinoline thiosemicarbazone per kg for a maximum of 5 consecutive days (CREASEY et al. 1972). Of the label from the 3'-[^{14}C] compound, 20% was excreted in the urine and 2.8% in the faeces in 16 h. During an 8-h period, approximately 2% of the radioactivity was present in the respiratory CO_2 when side-chain labelled (3'-[^{14}C]) but nor ring labelled (1-[^{14}C]) agent was administered. In *dogs*, the half-life of radioactivity in the blood from labelled 1-formylisoquinoline thiosemicarbazone was about 4 h. Between 28 and 46% of the label was excreted in the urine during a 48-h period. Significant amounts of unchanged drug could not be detected in the urine. Desulfuration was extensive, affecting 75% of the total metabolites in urine collected more than 11 h after administration of the thiosemicarbazone. Between 11 and 16% of the side chain was cleaved from the molecule both by apparent azoreductase action leading to urea and thiourea and by hydrolysis to yield semicarbazide and thiosemicarbazide.

Incubating Walker 256 carcinosarcoma (W-256) cells and its nitrogen-mustard resistant variant (W-256-NMR) with 3-ethoxy-2-oxobutyraldehyde bis(-thiosemicarbazone) and several related thiosemicarbazones VAN GIESSEN et al. (1973) showed that cupric ions were required for the antitumour activity of 3-ethoxy-2-oxobutyraldehyde bis(thiosemicarbazone), 3-ethoxy-2-oxobutyraldehyde bis-(N^4-methylthiosemicarbazone) and 3-ethoxy-2-oxobutyraldehyde bis-(N^4,N^4-dimethylthiosemicarbazone). WINKELMANN et al. (1974) reported on the formation constants of some redox and substitution reactions for a series of bis(thiosemicarbazonato) copper(II) complexes. The variation in *in vitro* tumour cell cytotoxicity elicited by the chelates corresponds well with the distribution of the complexes in these correlations, thereby providing an empirical chemical rationale for these differences in biological activity. The linear free energy relationships suggest the operation of a steric factor in the kinetics of the reactions described.

21.1.9
Tirapazamine

Tirapazamine (3-amino-1,2,4-benzotriazine 1,4-dioxide; SR4233; WIN59075; tirazone) is a clinically promising anticancer agent that selectively kills the oxygen-poor (hypoxic) cells found in tumours (BROWN 1993, BROWN and SIIM 1996, BROWN and WANG 1998). In experiments using *human* and *rodent* tumour cell lines, tirapazamine is 15–200 times more toxic to hypoxic cells than it is to normally oxygenated cells. When activated by one-electron enzymatic reduction, tirapazamine induces radical-mediated oxidative DNA strand cleavage (HWANG et al. 1999).

21.1.10
Paclitaxel

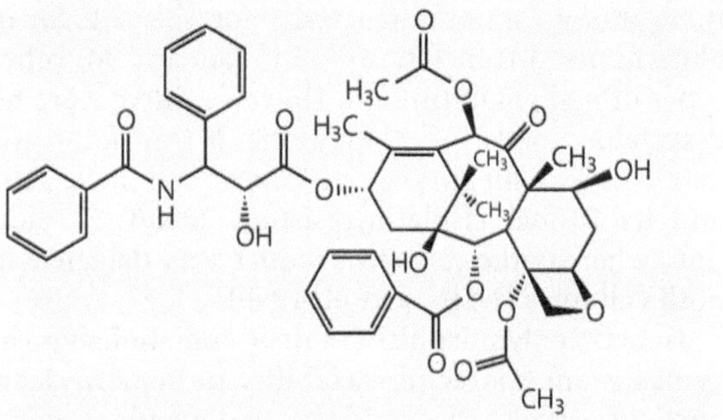

Paclitaxel (Taxol®), $C_{47}H_{51}NO_{14}$, mol. mass 853.9 [347]

EVTODIENKO et al. (1996) suggested that the effect of **paclitaxel** (Taxol®) on the mitochondrial permeability transition pore was mediated through interaction with the cytoskeleton. Mitochondrial permeability transition can lead to cell death, and can be induced by high intracellular organic phosphate, Ca^{2+}, reactive oxygen species, or proapototic proteins such as Bax, whereas it is inhibited by cyclosporine A, bongkrekic acid, Bcl-2, or Bcl-x_L (GREEN and REED 1998). In Percoll-gradient purified *rat* liver mitochondria, Taxol® induced large amplitude swelling in a concentration-dependent manner in the μM range (VARBIRO et al. 2001). Opening of the permeability pore was also confirmed by the access of mitochondrial matrix enzymes for membrane impermeable substrates in Taxol®-treated mitochondria. Taxol® induced the dissipation of mitochondrial membrane potential (ΔΨ) determined by Rhodamine 123 release and induced the release of cytochrome *c* from the intermembrane space. All these effects were inhibited by 2.5 μM cyclosporine A. Taxol significantly increased the formation of reactive oxygen species in both the aqueous and the lipid phases as determined by dihydrorhodamine 123 and resorufin derivative. Cytochrome oxidase inhibitor CN⁻, azide, and •NO abrogated the Taxol-induced mitochondrial reactive oxygen formation while inhibitors of the other respiratory complexes and cyclosporine A had no effect.

21.2 Anticancer Drug Resistance

Anticancer drug resistance may be due to the expression of cytochromes P_{450} (GUENGERICH 1988), epoxide hydroxylase, and glutathione *S*-transferase (HARRISON 1993, 1995).

The **P_{450} subfamilies** CYP1A, CYP2C, and CYP3A were present in 63, 25, and 61 per cent of prostate cancer (MURRAY et al. 1995). Epoxide hydroxylase was identified in 96 per cent of tumours, glutathione *S*-transferases-α and μ were expressed in 29 and 41 per cent of tumours, respectively, while there was no immunoreactivity for the π form of glutathione *S*-transferase – in contrast to other types of malignant tumour. However, there were no detectable levels of glutathione *S*-transferases-α and μ in two *human* ovarian cancer cell lines, 2008 and its 10-fold cisplatin-resistant 2008/C13* subline, whereas the π gene product was detected in both cell lines (NEHMÉ et al. 1995).

Genetic polymorphism of drug-metabolising enzymes is one source of **variability in hepatic clearance of drugs**. At least two independent cytochrome P_{450} enzymes, characterised by debrisoquin hydroxylation (MAHGOUB et al. 1976, EICHELBAUM et al. 1979, PRICE-EVANS 1986) and mephenytoin hydroxylation (KUPFER et al. 1979, WEDLUND et al. 1984, KALOW 1986), are subject to common genetic polymorphism. Teniposide competitively inhibited the 4-hydroxylation of (*S*)-mephenytoin, with a K_i of 12 μM (K_m of the reaction = 65 μM) (RELLING et al. 1989). Etoposide and flavone acetic acid were weaker inhibitors of this reaction. The only agent to inhibit bufuralol hydroxylation was vinblastine, which did so with a K_i of 90 μM (K_m of the enzyme for the substrate =12 μM). In HL-60 *human* promyelocytic cells, etoposide (0.15 mM) stimulated 1,25-dihydroxyvitamin D_3 (10 nM) differentiation activity, hormone binding and hormone receptor expression (TORRES et al. 2000). Etoposide loaded into erythrocytes was mainly localized in the cytoplasmic compartment (LOTERO et al. 2003). Membrane modification of etoposide-loaded erythrocytes with band 3 crosslinkers produced an increased incorporation of the drug into macrophages mainly by phagocytosis process. The toxic effect of etoposide conveyed in this carrier erythrocytes determined as DNA fragmentation in macrophages was higher than that shown by free etoposide added at the same concentration in the culture medium to macrophages.

Expression of *rat* liver glutathione *S*-transferase GSTA5 in cell lines provided increased resistance to the alkylating agents chloramubil and melphalan and toxic aldehydes (KAZI and ELLIS 2002).

Catechol formation by *O*-demethylation from teniposide and etoposide is primarily mediated by CYP3A4 in *human* liver (RELLING et al. 1994). Several substrates for CYP3A4 (e.g. midazolam, erythromycin and cyclosporin) were identified as strong inhibitors of catechol formation from both etoposide and teniposide. The catechols of epipodophyllotoxins are cytotoxic (VAN MAANEN et al. 1987). Cyclosporin has a marked, concentration-dependent effect on the pharmacokinetics of etoposide (LUM et al. 1992). A 10 % decrease in systemic clearance of etoposide was observed during doxorubicin and cyclophosphamide coadministration in patients with breast cancer, which could result from drug interactions affecting renal and/or metabolic elimination of etoposide (BUSSE et al. 2002). Another substrate for CYP3A4, nifedipine, did not interfere with the pharmacokinetics of etoposide (PHILIP et al. 1992).

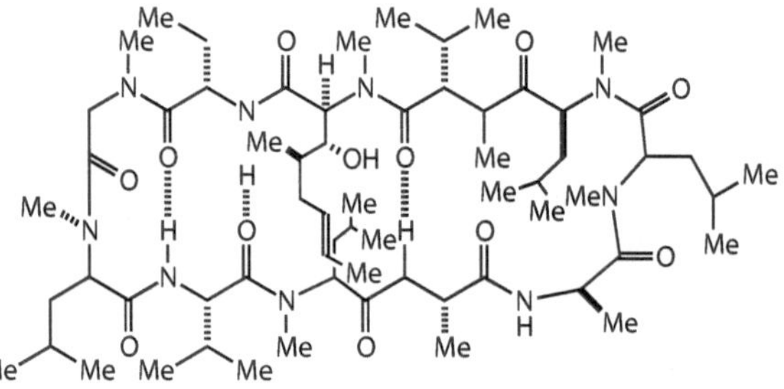

Cyclosporin A [348]

Metallothionein expression in certain tumour cells has been associated with resistance to anticancer drugs. Increased resistance to chlorambucil occurred in cultured cells with a high concentration of cytoplasmic metallothionein (ENDRESEN et al. 1983). Metallothionein inhibited hydroxyl radical-generated DNA degradation (ABEL and RUITER 1989). SATOH et al. (1994) showed that pretreatment of tumour-bearing ICR nude *mice* with zinc salts increased metallothionein content, both in normal and tumour tissues, with a marked reduction in the antitumour activity of cisplatin, Adriamycin®, and melphalan. Metallothionein null cells have increased sensitivity to anticancer drugs (KONDO et al. 1995).

Chloroethylnitrosourea resistance is correlated with its level of activity of the DNA repair protein O^6-alkylguanine-DNA alkyltransferase (CITRON et al. 1991, PEGG et al. 1993, BELANICH et al. 1996).

This repair protein removes alkylation lesions (e.g. methyl, ethyl, chloroethyl adducts) from O^6 position of guanine in DNA by transferring them to an internal cysteine residue at position 145 of its peptide chain thereby inactivating itself in a stoichiometric „suicide reaction" (YAROSH 1985, PEGG 1990, PEGG and BYERS 1992, MITRA and KAINA 1993, PEGG et al. 1993). Increased removal of chloroethylnitrosourea-induced O^6-guanine chloroethyl monoadducts before interstrand cross links are formed results in cell survival and accordingly in drug resistance. In Swiss nu/nu *mice* inoculated with ovarian tumour O-342 locoregional (i.p.) application of an O^6-alkylguanine transferase inhibitor (O^6-benzylguanine) plus 2,3-bis(2-chloroethyl)-1-nitrosourea was superior to 2,3-bis(2-chloroethyl)-1-nitrosourea monotherapy (SEIFERT and ZELLER 1996-1998).

Exposure of cells to nitric oxide resulted in a 4 – 5-fold increase in expression of the DNA-dependent protein-kinase catalytic subunit (XU et al. 2000), one of the key enzymes involved in repairing double-stranded DNA breaks. This NO-mediated increase in enzymatically active DNA-PK not only protected cells from the toxic effects of •NO, but also provided crossprotection against clinically important DNA-damaging agents, such as X-ray radiation, doxorubicin, bleomycin and cisplatin.

In an *in vitro* model, tumour metastases (B16BL6 cells) and invasion (*human* fibrosarcoma HT-1080) were inhibited by repeated addition of 300 μM **2-O-phosphorylated ascorbate** (NAGAO et al. 2000). Intracellular vitamin C increased and both hydroxyl and ascorbyl radicals decreased as quantified by electron spin resonance spectroscopy.

Ehrlich ascites tumour growth was similar in *mice* fed a **L-glutamine**-enriched diet (where 30% of the total dietary nitrogen was from L-glutamine) or a nutritionally complete elemental diet (CARRETERO et al. 2000). As compared with non-tumour-bearing *mice*, tumour growth caused a decrease of blood L-glutamine levels in *mice* fed a nutritionally complete elemental diet but not in those fed a L-glutamine-enriched diet. Tumour cells in *mice* fed a L-glutamine-enriched diet showed higher glutaminase and lower glutamine synthetase activities than did cells isolated from *mice* fed a nutritionally complete elemental diet. Cytosolic glutamate concentration was 2-fold higher in tumour cells from *mice* fed a L-glutamine-enriched diet (~4 mM) than in those fed a nutritionally complete elemental diet. This increase in glutamate content inhibited GSH uptake by tumour mitochondria and led to a selective depletion of mitochondrial GSH content (not found in mitochondria of normal cells such as lymphocytes or hepatocytes) to ~57% of the level found in tumour mitochondria of *mice* fed a nutritionally complete elemental diet. In tumour cells of *mice* fed a L-glutamine-enriched diet, 6-diazo-5-norleucine- or L-glutamate-γ-hydrazine-induced inhibition of glutaminase activity decreased cytosolic glutamate content and restored GSH uptake by mitochondria to the rate found in Ehrlich ascites tumour cells of *mice* fed a nutritionally complete elemental diet. The partial loss of mitochondrial GSH elicited by L-glutamine did not affect generation of reactive oxygen intermediates or mitochondrial functions (e.g., intracellular peroxide levels, $O_2^{\bullet-}$ generation, mitochondrial membrane potential, mitochondrial size, adenosine triphosphate and adenosine diphosphate contents, and oxygen consumption were found similar in tumour cells isolated from *mice* fed a nutritionally complete elemental diet or a L-glutamine-enriched diet); however, mitochondrial production of reactive oxygen intermediates upon TNF-α stimulation was increased.

Tumour necrosis factor (TNF) is selectively cytotoxic for some tumour cells *in vivo* and *in vitro*. FAST et al. (1992) determined whether TNF-mediated cytotoxicity for TNF-sensitive tumour targets was related to TNF-stimulated production of •NO by the tumour cell itself. They found that a cell line that was sensitive to TNF-mediated cytotoxicity produced •NO in response to TNF as measured by the accumulation of nitrite in the supernatants of TNF-stimulated cells. Production of •NO in response to TNF was inhibited by N^G-monomethyl-L-arginine. The kinetics of •NO production in response to TNF indicated that most of the •NO was produced during the first 24 h and peaked after 48 h of culture and that TNF-stimulated •NO production was dose-dependent. TNF-resistant cell lines produced less •NO than a TNF-sensitive cell line, and the amount of nitrite produced correlated with the relative sensitivity of each cell line to TNF-mediated cytotoxicity. Recombinant interferon-γ augmented the amount of •NO produced in response to TNF by both sensitive and resistant cells and correspondingly enhanced the susceptibility of resistant cells to TNF cytotoxicity.

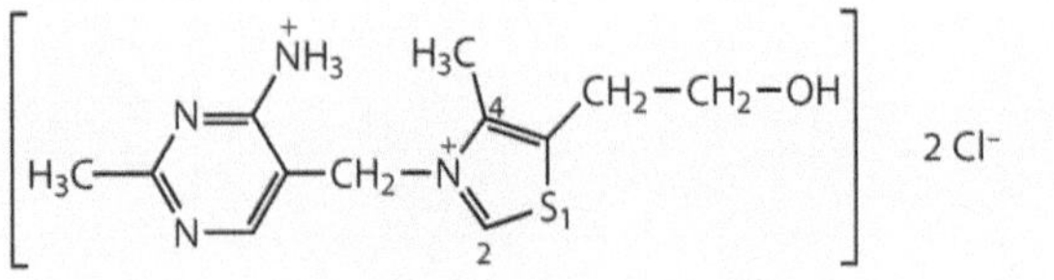

Thiamine [349]

Clinical and experimental data demonstrated increased **thiamine** utilisation of *human* tumours and its interference with experimental chemotherapy (BOROS et al. 1998). Analysis of RNA ribose indicated that glucose carbons contribute to over 90 per cent of ribose synthesis in cultured cervix and pancreatic carcinoma cells and that ribose in synthesised primarily through the thiamine dependent nonoxidative transketolase pathway (> 70 %). The chemically modified co-factor oxythiamine inhibited tumour cell proliferation *in vitro* and *in vivo* by 40 % and 90.7 % in two distinct tumour models, respectively (Boros et al. 1997).

Epilogue

Industrial diseases have been recognized since antiquity. They grew in social significance when free citizens started to succumb to them during the times of PARACELSUS and AGRICOLA and when they were found in all trades during the times of RAMAZZINI. Today, they have gained both economic and environmental importance.

Environmental sciences began when pathologists such as MORGAGNI and VAN DIEMERBROECK found dust in the lungs, when animal experimentalists introduced dusts into the body (ARNOLD), and when cells that engulfed particles, as demonstrated by ZENKER, were cultivated *in vitro* (LAUCHE). Molecular biology and pathology emphasised the importance of the surface of the particles inhaled, and biophysics pointed to their involvement in radical reactions. Surface structure, solubility, radical scavenger and antioxidant activities are – in addition to a genetic disposition (PARRISIUS) to disease – the keys to the problem.

My own research work started 50 years ago, when GÜNTHER WORTH and I were asked by HEINRICH KOST, director general of the German Coal Mining Board, to help the Ruhr miners avoid coalworker's pneumoconiosis by means of clinical and basic research, respectively. Research on the following topics has led to the current state of the art: the effect of hormones on the experimental fibrosis induced by peritoneal application of dusts, the relation of vitamins and hormones including sympathicomimetic drugs applied by metered aerosols, hypoxic (5% O_2) and reoxygenation damage in relation to free radical pathology, carcinogenicity of dusts in relation to hydroxyl radical formation catalysed by iron as found in the pyrite of mineral coal and copper contained in Mansfield copper schist, and last but not least, the problem of asbestos minerals and their iron content, and the continuous oxygen burst induced by incomplete phagocytosis of any fibrous dusts, man-made mineral fibres included. Immunological factors in coalworkers' pneumoconiosis and pure silicosis, and rheumatoid pneumoconioses (Caplan's syndrome) have been discussed since 1949, when FLETCHER defined the term "complicated pneumoconiosis" as a combination of simple pneumoconiosis and progressive massive fibrosis. Although PORCHER et al. (1994) did not find tumour necrosis factor release to be an exposure marker, the observed relation between radiological abnormalities and TNF release is of special interest. TNF quantification might be a useful tool for the clinical examination of persons at risk of progressive massive fibrosis. The problem of malignancy arose when technical dust suppression reduced the dust burden of the lung, and tuberculosis surveillance defeated silicotuberculosis. Miners and ex-miners no longer died before they developed cancer. In this way, pulmonary fibrosis and cor pulmonale were superseded by malignomas. As a first step in chemoprevention, MOONEY (1996) assessed the combined effect of genetic factors and nutritional status on DNA damage in a population of healthy smokers. Subjects with the rare CYP1A1 exon 7 polymorphism had significantly higher levels of DNA adducts than those without. Plasma retinol, α-tocopherol and zeaxanthin were inversely correlated with DNA damage, especially in subjects lacking the "protective" GSTM1 gene. The accumulated evidence was supportive of a protective role for antioxidants in reducing DNA damage. Just as MOONEY's intent was to measure DNA adducts and oxidative DNA damage in peripheral blood cells to determine whether supplementation with antioxidants/vitamins reduces smoking-related DNA damage in all subjects, all men and women exposed to industrial and environmental pollutants of **any** kind should be subject to screening.

Subject Index

A

B

C

D

E

F

G

I

J

K

L

M

N

O

P

Q

R

S

T

U

V